Renewable Energy Sources and Climate Change Mitigation

Special Report of the Intergovernmental Panel on Climate Change

Climate change is one of the great challenges of the 21st century. Its most severe impacts may still be avoided if efforts are made to transform current energy systems. Renewable energy sources have a large potential to displace emissions of greenhouse gases from the combustion of fossil fuels and thereby to mitigate climate change. If implemented properly, renewable energy sources can contribute to social and economic development, to energy access, to a secure and sustainable energy supply, and to a reduction of negative impacts of energy provision on the environment and human health.

This Special Report on Renewable Energy Sources and Climate Change Mitigation (SRREN) impartially assesses the scientific literature on the potential role of renewable energy in the mitigation of climate change for policymakers, the private sector, academic researchers and civil society. It covers six renewable energy sources – bioenergy, direct solar energy, geothermal energy, hydropower, ocean energy and wind energy – as well as their integration into present and future energy systems. It considers the environmental and social consequences associated with the deployment of these technologies, and presents strategies to overcome technical as well as non-technical obstacles to their application and diffusion. The authors also compare the levelized cost of energy from renewable energy sources to recent non-renewable energy costs.

The Intergovernmental Panel on Climate Change (IPCC) is the leading international body for the assessment of climate change. It was established by the United Nations Environment Programme (UNEP) and the World Meteorological Organization (WMO) to provide the world with a clear scientific view on the current state of knowledge on climate change and its potential environmental and socio-economic impacts.

Renewable Energy Sources and Climate Change Mitigation

Special Report of the Intergovernmental Panel on Climate Change

Edited by

Ottmar Edenhofer
Co-Chair Working Group III
Potsdam Institute for Climate
Impact Research (PIK)

Ramón Pichs Madruga
Co-Chair Working Group III
Centro de Investigaciones de la
Economía Mundial (CIEM)

Youba Sokona
Co-Chair Working Group III
African Climate Policy Centre,
United Nations Economic
Commission for Africa (UNECA)

Kristin Seyboth   Patrick Matschoss   Susanne Kadner   Timm Zwickel

Patrick Eickemeier   Gerrit Hansen   Steffen Schlömer   Christoph von Stechow

Technical Support Unit Working Group III
Potsdam Institute for Climate Impact Research (PIK)

CAMBRIDGE UNIVERSITY PRESS
Cambridge, New York, Melbourne, Madrid, Cape Town,
Singapore, São Paulo, Delhi, Tokyo, Mexico City

Cambridge University Press
32 Avenue of the Americas, New York, NY 10013-2473, USA

www.cambridge.org
Information on this title: www.cambridge.org/9781107607101

© Intergovernmental Panel on Climate Change 2012

This publication is in copyright. Subject to statutory exception
and to the provisions of relevant collective licensing agreements,
no reproduction of any part may take place without the written
permission of Cambridge University Press.

First published 2012

Printed in the United States of America

A catalog record for this publication is available from the British Library.

ISBN 978-1-107-02340-6 Hardback
ISBN 978-1-107-60710-1 Paperback

Cambridge University Press has no responsibility for the persistence or accuracy of URLs
for external or third-party Internet Web sites referred to in this publication and does not
guarantee that any content on such Web sites is, or will remain, accurate or appropriate.

Contents

Section I
- Foreword viii
- Preface ix

Section II
- Summary for Policymakers 3
- Technical Summary 27

Section III
- Chapter 1 Renewable Energy and Climate Change 161
- Chapter 2 Bioenergy 209
- Chapter 3 Direct Solar Energy 333
- Chapter 4 Geothermal Energy 401
- Chapter 5 Hydropower 437
- Chapter 6 Ocean Energy 497
- Chapter 7 Wind Energy 535
- Chapter 8 Integration of Renewable Energy into Present and Future Energy Systems 609
- Chapter 9 Renewable Energy in the Context of Sustainable Development 707
- Chapter 10 Mitigation Potential and Costs 791
- Chapter 11 Policy, Financing and Implementation 865

Section IV
- Annex I   Glossary, Acronyms, Chemical Symbols and Prefixes 953
- Annex II   Methodology 973
- Annex III Recent Renewable Energy Cost and Performance Parameters 1001
- Annex IV Contributors to the IPCC Special Report 1023
- Annex V   Reviewers of the IPCC Special Report 1033
- Annex VI Permissions to Publish 1051
- Index 1059

I
Foreword and Preface

Foreword

The IPCC Special Report on Renewable Energy Sources and Climate Change Mitigation (SRREN) provides a comprehensive review concerning these sources and technologies, the relevant costs and benefits, and their potential role in a portfolio of mitigation options.

For the first time, an inclusive account of costs and greenhouse gas emissions across various technologies and scenarios confirms the key role of renewable sources, irrespective of any tangible climate change mitigation agreement.

As an intergovernmental body established in 1988 by the World Meteorological Organization (WMO) and the United Nations Environment Programme (UNEP), the IPCC has successfully provided policymakers over the ensuing period with the most authoritative and objective scientific and technical assessments, which, while clearly policy relevant, never claimed to be policy prescriptive. Moreover, this Special Report should be considered especially significant at a time when Governments are pondering the role of renewable energy resources in the context of their respective climate change mitigation efforts.

The SRREN was made possible thanks to the commitment and dedication of hundreds of experts from various regions and disciplines. We would like to express our deep gratitude to Prof. Ottmar Edenhofer, Dr. Ramon Pichs-Madruga, and Dr. Youba Sokona, for their untiring leadership throughout the SRREN development process, as well as to all Coordinating Lead Authors, Lead Authors, Contributing Authors, Review Editors and Reviewers, and to the staff of the Working Group III Technical Support Unit.

We greatly value Germany's generous support and dedication to the SRREN, as evidenced in particular by its hosting of the Working Group III Technical Support Unit. Moreover, we wish to express our appreciation to the United Arab Emirates, for hosting the plenary session which approved the report; as well as to Brazil, Norway, the United Kingdom and Mexico, which hosted the successive Lead Authors meetings; to all sponsors which contributed to the IPCC work through their financial and logistical support; and finally to the IPCC Chairman, Dr. R. K. Pachauri, for his leadership throughout the SRREN development process.

M. Jarraud
Secretary General
World Meteorological Organization

A. Steiner
Executive Director
United Nations Environment Programme

Preface

The Special Report on Renewable Energy Sources and Climate Change Mitigation (SRREN) of the IPCC Working Group III provides an assessment and thorough analysis of renewable energy technologies and their current and potential role in the mitigation of greenhouse gas emissions. The results presented here are based on an extensive assessment of scientific literature, including specifics of individual studies, but also an aggregate across studies analyzed for broader conclusions. The report combines information on technology specific studies with results of large-scale integrated models, and provides policy-relevant (but not policy-prescriptive) information to decision makers on the characteristics and technical potentials of different resources; the historical development of the technologies; the challenges of their integration and social and environmental impacts of their use; as well as a comparison in levelized cost of energy for commercially available renewable technologies with recent non-renewable energy costs. Further, the role of renewable energy sources in pursuing GHG concentration stabilization levels discussed in this report and the presentation and analysis of the policies available to assist the development and deployment of renewable energy technologies in climate change mitigation and/or other goals answer important questions detailed in the original scoping of the report.

The process

This report has been prepared in accordance with the rules and procedures established by the IPCC and used for previous assessment reports. After a scoping meeting in Lübeck, Germany from the 20th to the 25th of January, 2008, the outline of the report was approved at the 28th IPCC Plenary held in Budapest, Hungary on the 9th and 10th of April, 2008. Soon afterward, an author team of 122 Lead Authors (33 from developing countries, 4 from EIT countries, and 85 from industrialized countries), 25 Review Editors and 132 contributing authors was formed.

The IPCC review procedure was followed, in which drafts produced by the authors were subject to two reviews. 24,766 comments from more than 350 expert reviewers and governments and international organizations were processed. Review Editors for each chapter have ensured that all substantive government and expert review comments received appropriate consideration.

The Summary for Policy Makers was approved line-by-line and the Final Draft of the report was accepted at the 11th Session of the Third Working Group held in Abu Dhabi, United Arab Emirates from the 5th to the 8th of May, 2011. The Special Report was accepted in its entirety at the 33rd IPCC Plenary Session held also in Abu Dhabi from the 10th to the 13th of May, 2011.

Structure of the Special Report

The SRREN consists of three categories of chapters: one introductory chapter; six technology specific chapters (Chapters 2-7); and four chapters that cover integrative issues across technologies (Chapters 8-11).

Chapter 1 is the introductory chapter designed to place renewable energy technologies within the broader framework of climate change mitigation options and identify characteristics common to renewable energy technologies.

Each of the technology chapters (2-7) provides information on the available resource potential, the state of technological and market development and the environmental and social impacts for each renewable energy source including bioenergy, direct solar energy, geothermal energy, hydropower, ocean energy and wind energy. In addition, prospects for future technological innovation and cost reductions are discussed, and the chapters end with a discussion on possible future deployment.

Chapter 8 is the first of the integrative chapters and discusses how renewable energy technologies are currently integrated into energy distribution systems, and how they may be integrated in the future. Development pathways for the strategic use of renewable technologies in the transport, buildings, industry and agricultural sectors are also discussed.

Renewable energy in the context of sustainable development is covered in Chapter 9. This includes the social, environmental and economic impacts of renewable energy sources, including the potential for improved energy access and a secure supply of energy. Specific barriers for renewable energy technologies are also covered.

In a review of over 160 scenarios, Chapter 10 investigates how renewable energy technologies may contribute to varying greenhouse gas emission reduction scenarios, ranging from business-as-usual scenarios to those reflecting ambitious GHG concentration stabilization levels. Four scenarios are analyzed in depth and the costs of extensive deployment of renewable energy technologies are also discussed.

The last chapter of the report, Chapter 11, describes the current trends in renewable energy support policies, as well as trends in financing and investment in renewable energy technologies. It reviews current experiences with RE policies, including effectiveness and efficiency measures, and discusses the influence of an enabling environment on the success of policies.

While the authors of the report included the most recent literature available at the time of publication, readers should be aware that topics covered in this Special Report may be subject to further rapid development. This includes state of development of some renewable energy technologies, as well as the state of knowledge of integration challenges, mitigation costs, co-benefits, environmental and social impacts, policy approaches and financing options. The boundaries and names shown and the designations used on any geographic maps in this report do not imply official endorsement or acceptance by the United Nations. In the geographic maps developed for the SRREN, the dotted line in Jammu and Kashmir represents approximately the Line of Control agreed upon by India and Pakistan. The final status of Jammu and Kashmir has not yet been agreed upon by the parties.

Acknowledgements

Production of this Special Report was a major enterprise, in which many people from around the world were involved, with a wide variety of contributions. We wish to thank the generous contributions by the governments and institutions involved, which enabled the authors, Review Editors and Government and Expert Reviewers to participate in this process.

We are especially grateful for the contribution and support of the German Government, in particular the Bundesministerium für Bildung und Forschung (BMBF), in funding the Working Group III Technical Support Unit (TSU). Coordinating this funding, Gregor Laumann and Christiane Textor of the Deutsches Zentrum für Luft- und Raumfahrt (DLR) were always ready to dedicate time and energy to the needs of the team. We would also like to express our gratitude to the Bundesministerium für Umwelt, Naturschutz und Reaktorsicherheit (BMU). In addition, the Potsdam Institute for Climate Impact Research (PIK) kindly hosted and housed the TSU offices.

We would very much like to thank the governments of Brazil, Norway, the United Kingdom and Mexico, who, in collaboration with local institutions, hosted the crucial lead author meetings in São José dos Campos (January 2009), Oslo (September 2009), Oxford (March 2010) and Mexico City (September 2010). In addition, we would like to thank the government of the United States and the Institute for Sustainability, with the Founder Society Technologies for Carbon Management Project for hosting the SRREN Expert Review meeting in Washington D.C.(February 2010). Finally, we express our appreciation to PIK for welcoming the SRREN Coordinating Lead Authors on their campus for a concluding meeting (January 2011).

This Special Report is only possible thanks to the expertise, hard work and commitment to excellence shown throughout by our Coordinating Lead Authors and Lead Authors, with important assistance by many Contributing Authors. We would also like to express our appreciation to the Government and Expert Reviewers, acknowledging the time and energy invested to provide constructive and useful comments to the various drafts. Our Review Editors were also critical in the SRREN process, supporting the author team with processing the comments and assuring an objective discussion of relevant issues.

It is a pleasure to acknowledge the tireless work of the staff of the Working Group III Technical Support Unit, Patrick Matschoss, Susanne Kadner, Kristin Seyboth, Timm Zwickel, Patrick Eickemeier, Gerrit Hansen, Steffen Schloemer, Christoph von Stechow, Benjamin Kriemann, Annegret Kuhnigk, Anna Adler and Nina Schuetz, who were assisted by Marilyn Anderson, Lelani Arris, Andrew Ayres, Marlen Goerner, Daniel Mahringer and Ashley Renders. Brigitte Knopf, in her role as Senior Advisor to the TSU, consistently provided valuable input and direction. Graphics support by Kay Schröder and his team at Daily-Interactive.com Digitale Kommunikation is gratefully appreciated, as is the layout work by Valarie Morris and her team at Arroyo Writing, LLC.

The Working Group III Bureau – consisting of Antonina Ivanova Boncheva (Mexico), Carlo Carraro (Italy), Suzana Kahn Ribeiro (Brazil), Jim Skea (UK), Francis Yamba (Zambia), and Taha Zatari (Saudi Arabia) and prior to his elevation to IPCC Vice Chair, Ismail A.R. Elgizouli (Sudan) – provided continuous and constructive support to the Working Group III Co-Chairs throughout the SRREN process.

We would like to thank the Renate Christ, Secretary of the IPCC, and the Secretariat staff Gaetano Leone, Mary Jean Burer, Sophie Schlingemann, Judith Ewa, Jesbin Baidya, Joelle Fernandez, Annie Courtin, Laura Biagioni, Amy Smith Aasdam, and Rockaya Aidara, who provided logistical support for government liaison and travel of experts from developing and transitional economy countries.

Our special acknowledgement to Dr. Rajendra Pachauri, Chairman of the IPCC, for his contribution and support during the preparation of this IPCC Special Report.

Ottmar Edenhofer
IPCC WG III Co-Chair

Ramon Pichs-Madruga
IPCC WG III Co-Chair

Youba Sokona
IPCC WG III Co-Chair

Patrick Matshoss
IPCC WG III TSU Head

Kristin Seyboth
IPCC WG III Senior Scientist
SRREN Manager

Preface

This report is dedicated to

Wolfram Krewitt, Germany
Coordinating Lead Author in Chapter 8

Wolfram Krewitt passed away October 8th, 2009. He worked at the Deutsches Zentrum für Luft- und Raumfahrt (DLR) in Stuttgart, Germany.

Raymond Wright, Jamaica
Lead Author in Chapter 10

Raymond Wright passed away July 7th, 2011. He worked at the Petroleum Corporation of Jamaica (PCJ) in Kingston, Jamaica.

Wolfram Krewitt made a significant contribution to this Special Report and his vision for Chapter 8 (Integration of Renewable Energy into Present and Future Energy Systems) remains embedded in the text for which he is acknowledged. Raymond Wright was a critical member of the Chapter 10 (Mitigation Potential and Costs) author team who consistently offered precise insights to the Special Report, ensuring balance and credibility. Both authors were talented, apt and dedicated members of the IPCC author team - their passing represents a deep loss for the international scientific communities working in climate and energy issues. Wolfram Krewitt and Raymond Wright are dearly remembered by their fellow authors.

II Summaries

Summary for Policymakers

Coordinating Lead Authors:
Ottmar Edenhofer (Germany), Ramon Pichs-Madruga (Cuba),
Youba Sokona (Ethiopia/Mali), Kristin Seyboth (Germany/USA)

Lead Authors:
Dan Arvizu (USA), Thomas Bruckner (Germany), John Christensen (Denmark),
Helena Chum (USA/Brazil) Jean-Michel Devernay (France), Andre Faaij (The Netherlands),
Manfred Fischedick (Germany), Barry Goldstein (Australia), Gerrit Hansen (Germany),
John Huckerby (New Zealand), Arnulf Jäger-Waldau (Italy/Germany), Susanne Kadner (Germany),
Daniel Kammen (USA), Volker Krey (Austria/Germany), Arun Kumar (India),
Anthony Lewis (Ireland), Oswaldo Lucon (Brazil), Patrick Matschoss (Germany),
Lourdes Maurice (USA), Catherine Mitchell (United Kingdom), William Moomaw (USA),
José Moreira (Brazil), Alain Nadai (France), Lars J. Nilsson (Sweden), John Nyboer (Canada),
Atiq Rahman (Bangladesh), Jayant Sathaye (USA), Janet Sawin (USA), Roberto Schaeffer (Brazil),
Tormod Schei (Norway), Steffen Schlömer (Germany), Ralph Sims (New Zealand),
Christoph von Stechow (Germany), Aviel Verbruggen (Belgium), Kevin Urama (Kenya/Nigeria),
Ryan Wiser (USA), Francis Yamba (Zambia), Timm Zwickel (Germany)

Special Advisor:
Jeffrey Logan (USA)

This chapter should be cited as:
IPCC, 2011: Summary for Policymakers. In: IPCC Special Report on Renewable Energy Sources and Climate Change Mitigation [O. Edenhofer, R. Pichs-Madruga, Y. Sokona, K. Seyboth, P. Matschoss, S. Kadner, T. Zwickel, P. Eickemeier, G. Hansen, S. Schlömer, C. von Stechow (eds)], Cambridge University Press, Cambridge, United Kingdom and New York, NY, USA.

Table of Contents

1. Introduction 6

2. Renewable energy and climate change 7

3. Renewable energy technologies and markets 7

4. Integration into present and future energy systems 15

5. Renewable energy and sustainable development 18

6. Mitigation potentials and costs 20

7. Policy, implementation and financing 24

8. Advancing knowledge about renewable energy 26

1. Introduction

The Working Group III Special Report on Renewable Energy Sources and Climate Change Mitigation (SRREN) presents an assessment of the literature on the scientific, technological, environmental, economic and social aspects of the contribution of six renewable energy (RE) sources to the mitigation of climate change. It is intended to provide policy relevant information to governments, intergovernmental processes and other interested parties. This Summary for Policymakers provides an overview of the SRREN, summarizing the essential findings.

The SRREN consists of 11 chapters. Chapter 1 sets the context for RE and climate change; Chapters 2 through 7 provide information on six RE technologies, and Chapters 8 through 11 address integrative issues (see Figure SPM.1).

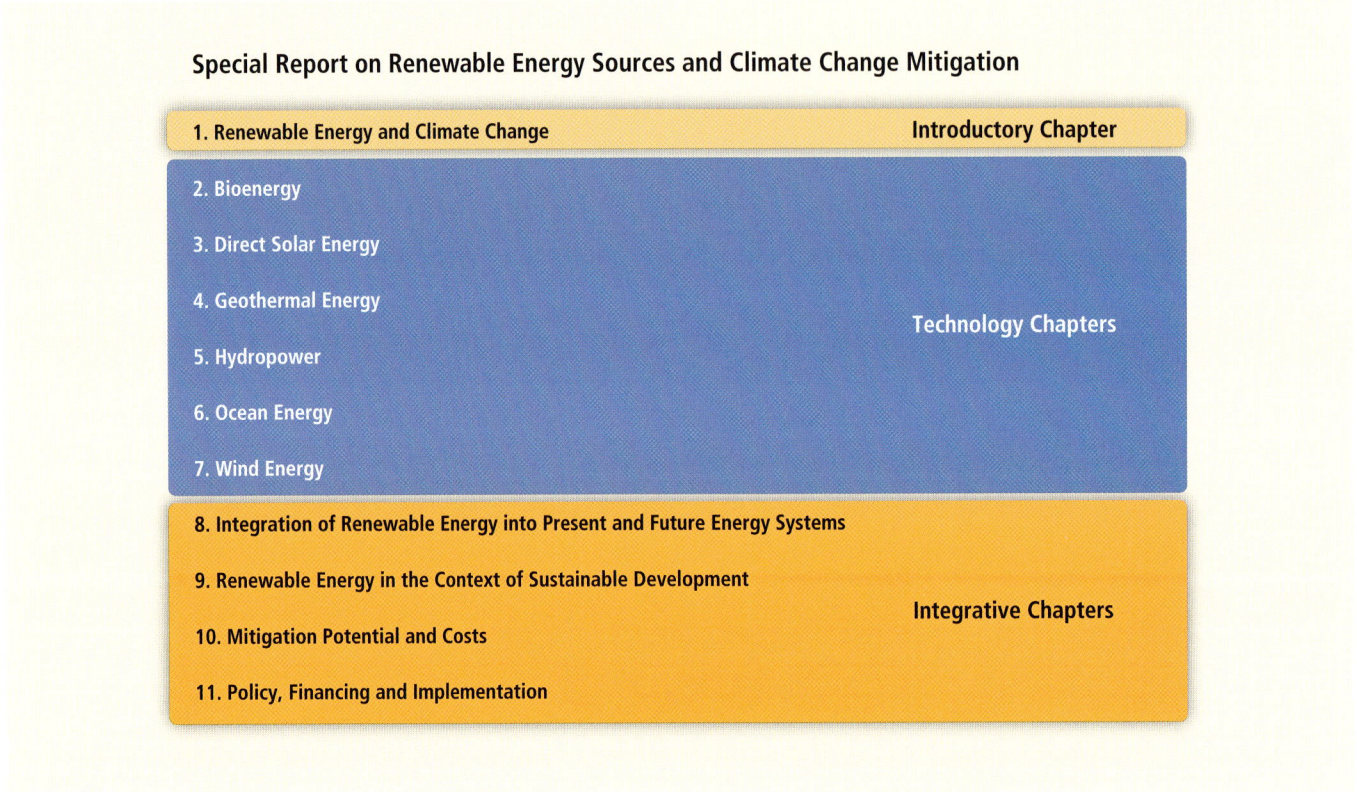

Figure SPM.1 | Structure of the SRREN. [Figure 1.1, 1.1.2]

References to chapters and sections are indicated with corresponding chapter and section numbers in square brackets. An explanation of terms, acronyms and chemical symbols used in this SPM can be found in the glossary of the SRREN (Annex I). Conventions and methodologies for determining costs, primary energy and other topics of analysis can be found in Annex II and Annex III. This report communicates uncertainty where relevant.[1]

---

[1] This report communicates uncertainty, for example, by showing the results of sensitivity analyses and by quantitatively presenting ranges in cost numbers as well as ranges in the scenario results. This report does not apply formal IPCC uncertainty terminology because at the time of the approval of this report, IPCC uncertainty guidance was in the process of being revised.

2. Renewable energy and climate change

Demand for energy and associated services, to meet social and economic development and improve human welfare and health, is increasing. All societies require energy services to meet basic human needs (e.g., lighting, cooking, space comfort, mobility and communication) and to serve productive processes. [1.1.1, 9.3.2] Since approximately 1850, global use of fossil fuels (coal, oil and gas) has increased to dominate energy supply, leading to a rapid growth in carbon dioxide (CO_2) emissions. [Figure 1.6]

Greenhouse gas (GHG) emissions resulting from the provision of energy services have contributed significantly to the historic increase in atmospheric GHG concentrations. The IPCC Fourth Assessment Report (AR4) concluded that "Most of the observed increase in global average temperature since the mid-20th century is very likely[2] due to the observed increase in anthropogenic greenhouse gas concentrations."

Recent data confirm that consumption of fossil fuels accounts for the majority of global anthropogenic GHG emissions.[3] Emissions continue to grow and CO_2 concentrations had increased to over 390 ppm, or 39% above preindustrial levels, by the end of 2010. [1.1.1, 1.1.3]

There are multiple options for lowering GHG emissions from the energy system while still satisfying the global demand for energy services. [1.1.3, 10.1] Some of these possible options, such as energy conservation and efficiency, fossil fuel switching, RE, nuclear and carbon capture and storage (CCS) were assessed in the AR4. A comprehensive evaluation of any portfolio of mitigation options would involve an evaluation of their respective mitigation potential as well as their contribution to sustainable development and all associated risks and costs. [1.1.6] This report will concentrate on the role that the deployment of RE technologies can play within such a portfolio of mitigation options.

As well as having a large potential to mitigate climate change, RE can provide wider benefits. RE may, if implemented properly, contribute to social and economic development, energy access, a secure energy supply, and reducing negative impacts on the environment and health. [9.2, 9.3]

Under most conditions, increasing the share of RE in the energy mix will require policies to stimulate changes in the energy system. Deployment of RE technologies has increased rapidly in recent years, and their share is projected to increase substantially under most ambitious mitigation scenarios [1.1.5, 10.2]. Additional policies would be required to attract the necessary increases in investment in technologies and infrastructure. [11.4.3, 11.5, 11.6.1, 11.7.5]

3. Renewable energy technologies and markets

RE comprises a heterogeneous class of technologies (Box SPM.1). Various types of RE can supply electricity, thermal energy and mechanical energy, as well as produce fuels that are able to satisfy multiple energy service needs [1.2]. Some RE technologies can be deployed at the point of use (decentralized) in rural and urban environments, whereas others are primarily deployed within large (centralized) energy networks [1.2, 8.2, 8.3, 9.3.2]. Though a growing number of RE technologies are technically mature and are being deployed at significant scale, others are in an earlier phase of technical maturity and commercial deployment or fill specialized niche markets [1.2]. The energy output of

---

[2] According to the formal uncertainty language used in the AR4, the term 'very likely' refers to a >90% assessed probability of occurrence.

[3] The contributions of individual anthropogenic GHGs to total emissions in 2004, reported in AR4, expressed as CO_2eq were: CO_2 from fossil fuels (56.6%), CO_2 from deforestation, decay of biomass etc. (17.3%), CO_2 from other (2.8%), methane (14.3%), nitrous oxide (7.9%) and fluorinated gases (1.1%) [Figure 1.1b, AR4, WG III, Chapter 1. For further information on sectoral emissions, including forestry, see also Figure 1.3b and associated footnotes.]

RE technologies can be (i) variable and—to some degree—unpredictable over differing time scales (from minutes to years), (ii) variable but predictable, (iii) constant, or (iv) controllable. [8.2, 8.3]

Box SPM.1 | Renewable energy sources and technologies considered in this report.

Bioenergy can be produced from a variety of biomass feedstocks, including forest, agricultural and livestock residues; short-rotation forest plantations; energy crops; the organic component of municipal solid waste; and other organic waste streams. Through a variety of processes, these feedstocks can be directly used to produce electricity or heat, or can be used to create gaseous, liquid, or solid fuels. The range of bioenergy technologies is broad and the technical maturity varies substantially. Some examples of commercially available technologies include small- and large-scale boilers, domestic pellet-based heating systems, and ethanol production from sugar and starch. Advanced biomass integrated gasification combined-cycle power plants and lignocellulose-based transport fuels are examples of technologies that are at a pre-commercial stage, while liquid biofuel production from algae and some other biological conversion approaches are at the research and development (R&D) phase. Bioenergy technologies have applications in centralized and decentralized settings, with the traditional use of biomass in developing countries being the most widespread current application.[4] Bioenergy typically offers constant or controllable output. Bioenergy projects usually depend on local and regional fuel supply availability, but recent developments show that solid biomass and liquid biofuels are increasingly traded internationally. [1.2, 2.1, 2.3, 2.6, 8.2, 8.3]

Direct solar energy technologies harness the energy of solar irradiance to produce electricity using photovoltaics (PV) and concentrating solar power (CSP), to produce thermal energy (heating or cooling, either through passive or active means), to meet direct lighting needs and, potentially, to produce fuels that might be used for transport and other purposes. The technology maturity of solar applications ranges from R&D (e.g., fuels produced from solar energy), to relatively mature (e.g., CSP), to mature (e.g., passive and active solar heating, and wafer-based silicon PV). Many but not all of the technologies are modular in nature, allowing their use in both centralized and decentralized energy systems. Solar energy is variable and, to some degree, unpredictable, though the temporal profile of solar energy output in some circumstances correlates relatively well with energy demands. Thermal energy storage offers the option to improve output control for some technologies such as CSP and direct solar heating. [1.2, 3.1, 3.3, 3.5, 3.7, 8.2, 8.3]

Geothermal energy utilizes the accessible thermal energy from the Earth's interior. Heat is extracted from geothermal reservoirs using wells or other means. Reservoirs that are naturally sufficiently hot and permeable are called hydrothermal reservoirs, whereas reservoirs that are sufficiently hot but that are improved with hydraulic stimulation are called enhanced geothermal systems (EGS). Once at the surface, fluids of various temperatures can be used to generate electricity or can be used more directly for applications that require thermal energy, including district heating or the use of lower-temperature heat from shallow wells for geothermal heat pumps used in heating or cooling applications. Hydrothermal power plants and thermal applications of geothermal energy are mature technologies, whereas EGS projects are in the demonstration and pilot phase while also undergoing R&D. When used to generate electricity, geothermal power plants typically offer constant output. [1.2, 4.1, 4.3, 8.2, 8.3]

Hydropower harnesses the energy of water moving from higher to lower elevations, primarily to generate electricity. Hydropower projects encompass dam projects with reservoirs, run-of-river and in-stream projects and cover a continuum in project scale. This variety gives hydropower the ability to meet large centralized urban needs as well as decentralized rural needs. Hydropower technologies are mature. Hydropower projects exploit a resource that varies temporally. However, the controllable output provided by hydropower facilities that have reservoirs can be used to meet peak electricity demands and help to balance electricity systems that have large amounts of variable RE generation. The operation of hydropower reservoirs often reflects their multiple uses, for example, drinking water, irrigation, flood and drought control, and navigation, as well as energy supply. [1.2, 5.1, 5.3, 5.5, 5.10, 8.2]

---

4   Traditional biomass is defined by the International Energy Agency (IEA) as biomass consumption in the residential sector in developing countries and refers to the often unsustainable use of wood, charcoal, agricultural residues, and animal dung for cooking and heating. All other biomass use is defined as modern [Annex I].

Ocean energy derives from the potential, kinetic, thermal and chemical energy of seawater, which can be transformed to provide electricity, thermal energy, or potable water. A wide range of technologies are possible, such as barrages for tidal range, submarine turbines for tidal and ocean currents, heat exchangers for ocean thermal energy conversion, and a variety of devices to harness the energy of waves and salinity gradients. Ocean technologies, with the exception of tidal barrages, are at the demonstration and pilot project phases and many require additional R&D. Some of the technologies have variable energy output profiles with differing levels of predictability (e.g., wave, tidal range and current), while others may be capable of near-constant or even controllable operation (e.g., ocean thermal and salinity gradient). [1.2, 6.1, 6.2, 6.3, 6.4, 6.6, 8.2]

Wind energy harnesses the kinetic energy of moving air. The primary application of relevance to climate change mitigation is to produce electricity from large wind turbines located on land (onshore) or in sea- or freshwater (offshore). Onshore wind energy technologies are already being manufactured and deployed on a large scale. Offshore wind energy technologies have greater potential for continued technical advancement. Wind electricity is both variable and, to some degree, unpredictable, but experience and detailed studies from many regions have shown that the integration of wind energy generally poses no insurmountable technical barriers. [1.2, 7.1, 7.3, 7.5, 7.7, 8.2]

On a global basis, it is estimated that RE accounted for 12.9% of the total 492 Exajoules (EJ)[5] of primary energy supply in 2008 (Box SPM.2 and Figure SPM.2). The largest RE contributor was biomass (10.2%), with the majority (roughly 60%) being traditional biomass used in cooking and heating applications in developing countries but with rapidly increasing use of modern biomass as well.[6] Hydropower represented 2.3%, whereas other RE sources accounted for 0.4%. [1.1.5] In 2008, RE contributed approximately 19% of global electricity supply (16% hydropower, 3% other RE) and biofuels contributed 2% of global road transport fuel supply. Traditional biomass (17%), modern biomass (8%), solar thermal and geothermal energy (2%) together fuelled 27% of the total global demand for heat. The contribution of RE to primary energy supply varies substantially by country and region. [1.1.5, 1.3.1, 8.1]

Deployment of RE has been increasing rapidly in recent years (Figure SPM.3). Various types of government policies, the declining cost of many RE technologies, changes in the prices of fossil fuels, an increase of energy demand and other factors have encouraged the continuing increase in the use of RE. [1.1.5, 9.3, 10.5, 11.2, 11.3] Despite global financial challenges, RE capacity continued to grow rapidly in 2009 compared to the cumulative installed capacity from the previous year, including wind power (32% increase, 38 Gigawatts (GW) added), hydropower (3%, 31 GW added), grid-connected photovoltaics (53%, 7.5 GW added), geothermal power (4%, 0.4 GW added), and solar hot water/heating (21%, 31 GW_{th} added). Biofuels accounted for 2% of global road transport fuel demand in 2008 and nearly 3% in 2009. The annual production of ethanol increased to 1.6 EJ (76 billion litres) by the end of 2009 and biodiesel to 0.6 EJ (17 billion litres). [1.1.5, 2.4, 3.4, 4.4, 5.4, 7.4]

Of the approximate 300 GW of new electricity generating capacity added globally over the two-year period from 2008 to 2009, 140 GW came from RE additions. Collectively, developing countries host 53% of global RE electricity generation capacity [1.1.5]. At the end of 2009, the use of RE in hot water/heating markets included modern biomass (270 GW_{th}), solar (180 GW_{th}), and geothermal (60 GW_{th}). The use of decentralized RE (excluding traditional biomass) in meeting rural energy needs at the household or village level has also increased, including hydropower stations, various modern biomass options, PV, wind or hybrid systems that combine multiple technologies. [1.1.5, 2.4, 3.4, 4.4, 5.4]

---

5   1 Exajoule = 10^{18} joules = 23.88 million tonnes of oil equivalent (Mtoe).

6   In addition to this 60% share of traditional biomass, there is biomass use estimated to amount to 20 to 40% not reported in official primary energy databases, such as dung, unaccounted production of charcoal, illegal logging, fuelwood gathering, and agricultural residue use. [2.1, 2.5]

Box SPM.2 | Accounting for primary energy in the SRREN.

There is no single, unambiguous accounting method for calculating primary energy from non-combustible energy sources such as non-combustible RE sources and nuclear energy. The SRREN adopts the 'direct equivalent' method for accounting for primary energy supply. In this method, fossil fuels and bioenergy are accounted for based on their heating value while non-combustible energy sources, including nuclear energy and all non-combustible RE, are accounted for based on the secondary energy that they produce. This may lead to an understatement of the contribution of non-combustible RE and nuclear compared to bioenergy and fossil fuels by a factor of roughly 1.2 up to 3. The selection of the accounting method also impacts the relative shares of different individual energy sources. Comparisons in the data and figures presented in the SRREN between fossil fuels and bioenergy on the one hand, and non-combustible RE and nuclear energy on the other, reflect this accounting method. [1.1.9, Annex II.4]

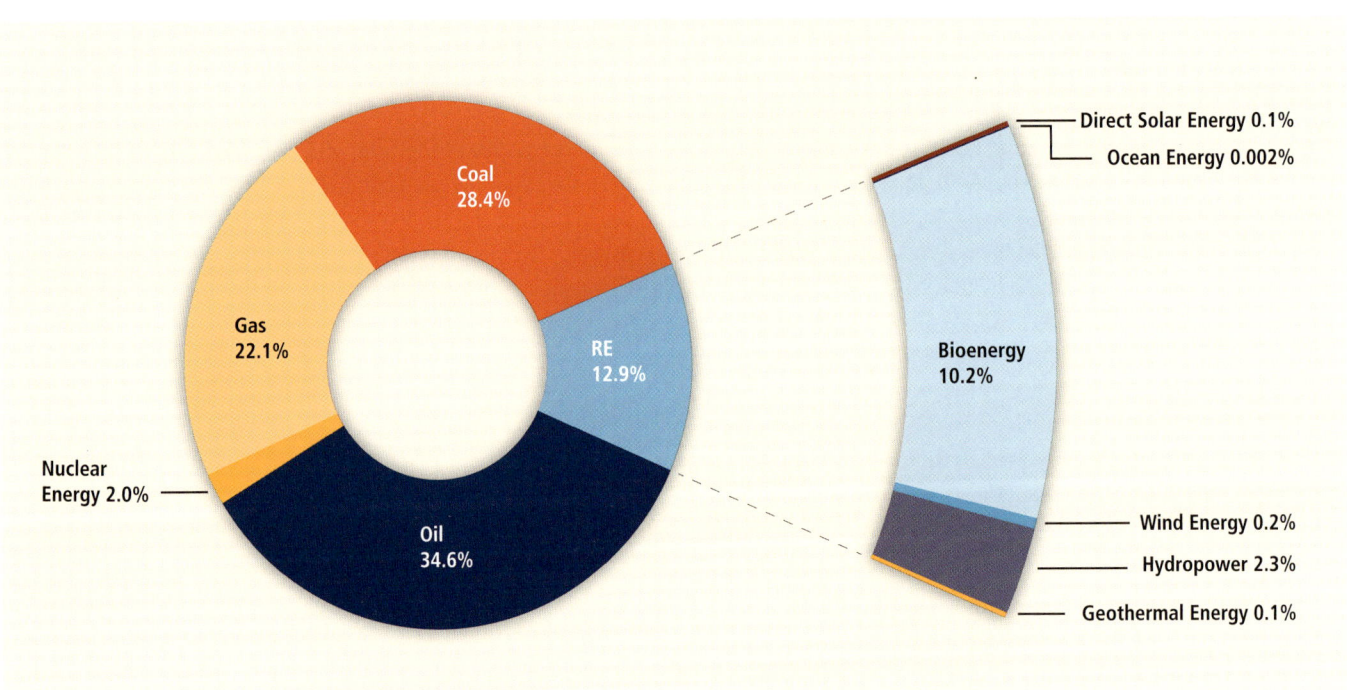

Figure SPM.2 | Shares of energy sources in total global primary energy supply in 2008 (492 EJ). Modern biomass contributes 38% of the total biomass share. [Figure 1.10, 1.1.5]

Note: Underlying data for figure have been converted to the 'direct equivalent' method of accounting for primary energy supply. [Box SPM.2, 1.1.9, Annex II.4]

The global technical potential[7] of RE sources will not limit continued growth in the use of RE. A wide range of estimates is provided in the literature, but studies have consistently found that the total global technical potential for RE is substantially higher than global energy demand (Figure SPM.4) [1.2.2, 10.3, Annex II]. The technical potential for solar energy is the highest among the RE sources, but substantial technical potential exists for all six RE sources. Even in regions with relatively low levels of technical potential for any individual RE source, there are typically significant opportunities for increased deployment compared to current levels. [1.2.2, 2.2, 2.8, 3.2, 4.2, 5.2, 6.2, 6.4, 7.2, 8.2, 8.3, 10.3] In the longer term and at higher deployment levels, however, technical potentials indicate a limit to the

---

[7] Definitions of technical potential often vary by study. 'Technical potential' is used in the SRREN as the amount of RE output obtainable by full implementation of demonstrated technologies or practices. No explicit reference to costs, barriers or policies is made. Technical potentials reported in the literature and assessed in the SRREN, however, may have taken into account practical constraints and when explicitly stated they are generally indicated in the underlying report. [Annex I]

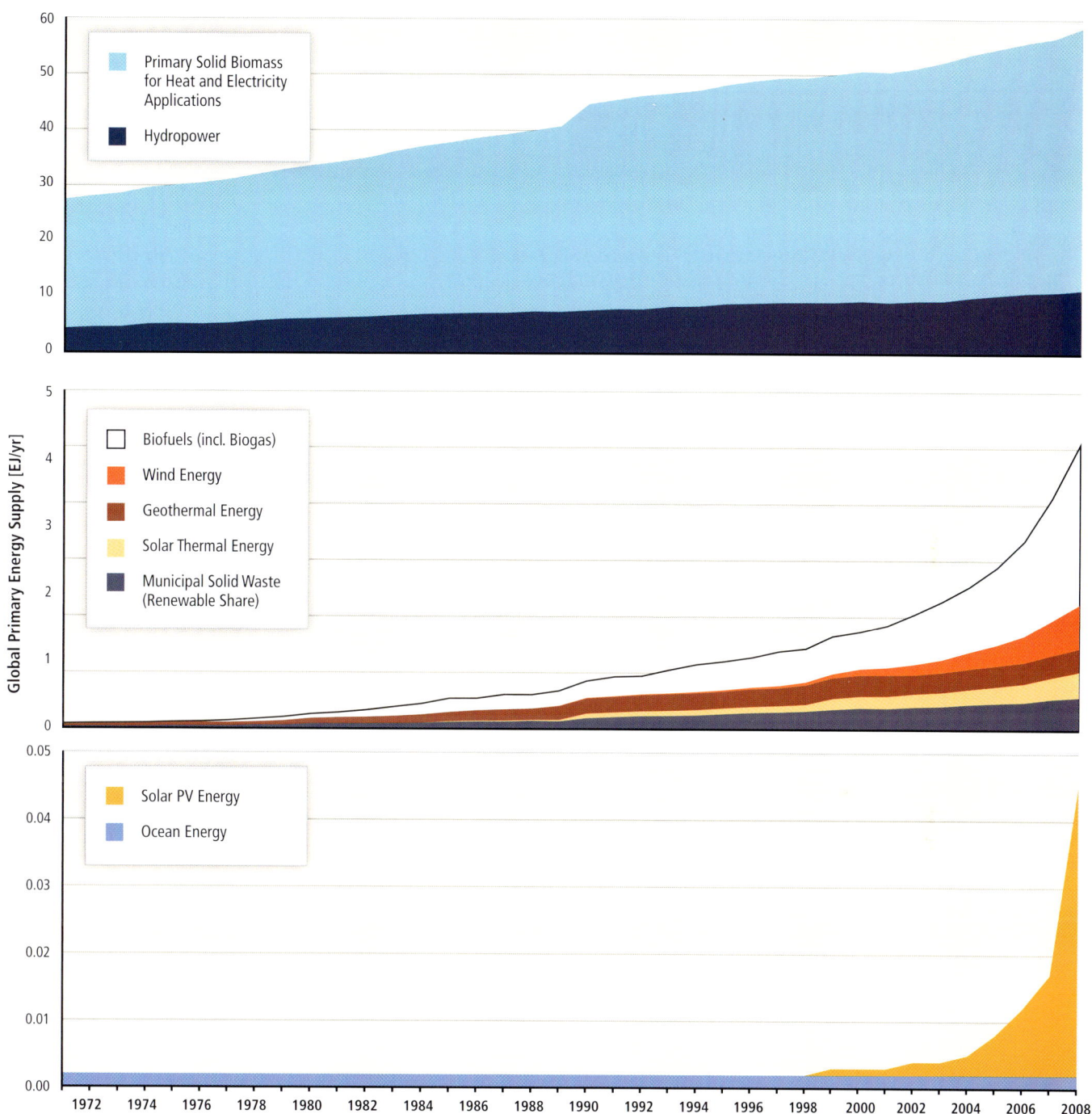

Figure SPM.3 | Historical development of global primary energy supply from renewable energy from 1971 to 2008. [Figure 1.12, 1.1.5]

Notes: Technologies are referenced to separate vertical units for display purposes only. Underlying data for figure has been converted to the 'direct equivalent' method of accounting for primary energy supply [Box SPM.2, 1.1.9, Annex II.4], except that the energy content of biofuels is reported in secondary energy terms (the primary biomass used to produce the biofuel would be higher due to conversion losses. [2.3, 2.4])

contribution of some individual RE technologies. Factors such as sustainability concerns [9.3], public acceptance [9.5], system integration and infrastructure constraints [8.2], or economic factors [10.3] may also limit deployment of RE technologies.

Climate change will have impacts on the size and geographic distribution of the technical potential for RE sources, but research into the magnitude of these possible effects is nascent. Because RE sources are, in many cases, dependent on the climate, global climate change will affect the RE resource base, though the precise nature and magnitude of these impacts is uncertain. The future technical potential for bioenergy could be influenced by climate change through impacts on biomass production such as altered soil conditions, precipitation, crop productivity and other factors. The overall impact of a global mean temperature change of less than 2°C on the technical potential of bioenergy is expected to be relatively small on a global basis. However, considerable regional differences could be expected and uncertainties are larger and more difficult to assess compared to other RE options due to the large number of feedback mechanisms involved. [2.2, 2.6] For solar energy, though climate change is expected to influence the distribution and variability of cloud cover, the impact of these changes on overall technical potential is expected to be small [3.2]. For hydropower the overall impacts on the global technical potential is expected to be slightly positive. However, results also indicate the possibility of substantial variations across regions and even within countries. [5.2] Research to date suggests that climate change is not expected to greatly impact the global technical potential for wind energy development but changes in the regional distribution of the wind energy resource may be expected [7.2]. Climate change is not anticipated to have significant impacts on the size or geographic distribution of geothermal or ocean energy resources. [4.2, 6.2]

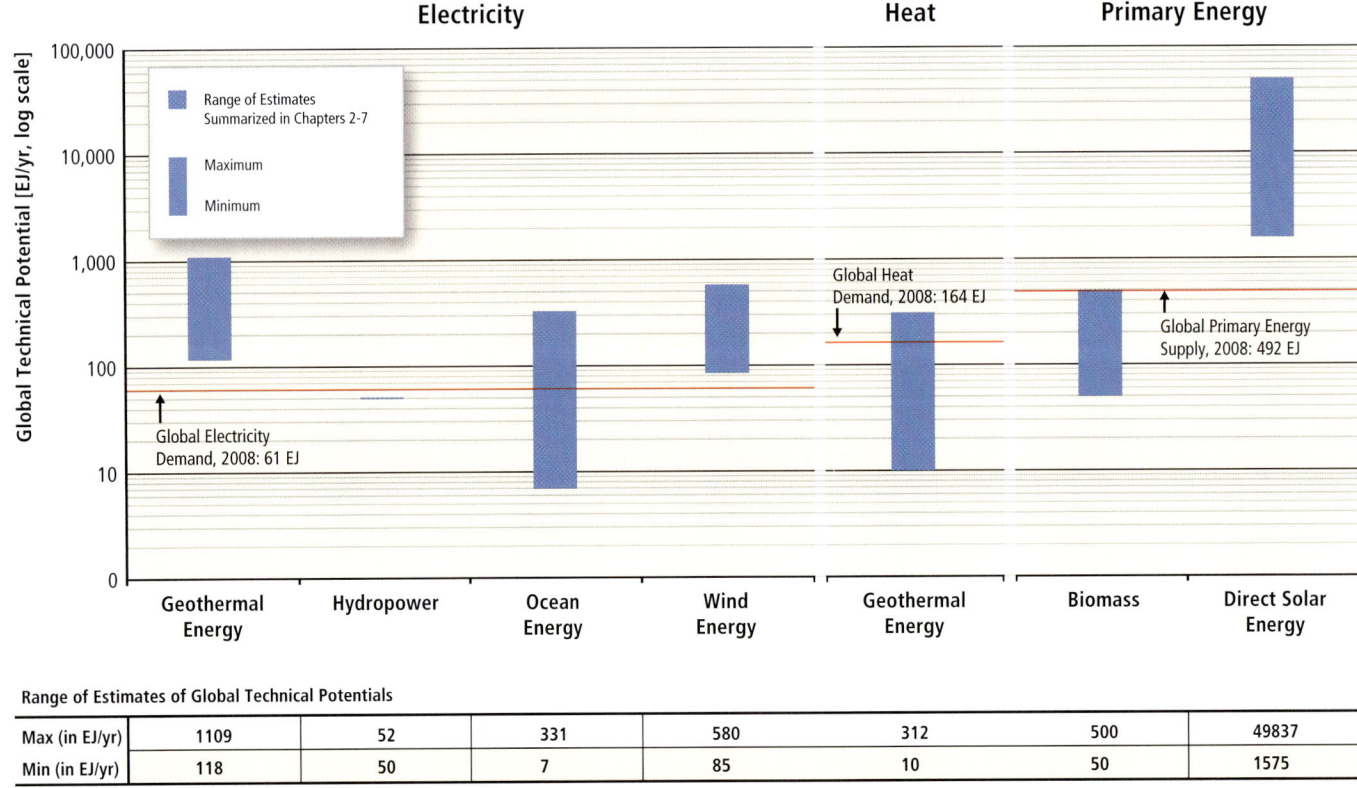

Figure SPM.4 | Ranges of global technical potentials of RE sources derived from studies presented in Chapters 2 through 7. Biomass and solar are shown as primary energy due to their multiple uses; note that the figure is presented in logarithmic scale due to the wide range of assessed data. [Figure 1.17, 1.2.3]

Notes: Technical potentials reported here represent total worldwide potentials for annual RE supply and do not deduct any potential that is already being utilized. Note that RE electricity sources could also be used for heating applications, whereas biomass and solar resources are reported only in primary energy terms but could be used to meet various energy service needs. Ranges are based on various methods and apply to different future years; consequently, the resulting ranges are not strictly comparable across technologies. For the data behind Figure SPM.4 and additional notes that apply, see Chapter 1 Annex, Table A.1.1 (as well as the underlying chapters).

The levelized cost of energy[8] for many RE technologies is currently higher than existing energy prices, though in various settings RE is already economically competitive. Ranges of recent levelized costs of energy for selected commercially available RE technologies are wide, depending on a number of factors including, but not limited to, technology characteristics, regional variations in cost and performance, and differing discount rates (Figure SPM.5). [1.3.2, 2.3, 2.7, 3.8, 4.8, 5.8, 6.7, 7.8, 10.5, Annex III] Some RE technologies are broadly competitive with existing market energy prices. Many of the other RE technologies can provide competitive energy services in certain circumstances, for example, in regions with favourable resource conditions or that lack the infrastructure for other low-cost energy supplies. In most regions of the world, policy measures are still required to ensure rapid deployment of many RE sources. [2.3, 2.7, 3.8, 4.7, 5.8, 6.7, 7.8, 10.5]

Monetizing the external costs of energy supply would improve the relative competitiveness of RE. The same applies if market prices increase due to other reasons (Figure SPM.5). [10.6] The levelized cost of energy for a technology is not the sole determinant of its value or economic competitiveness. The attractiveness of a specific energy supply option depends also on broader economic as well as environmental and social aspects, and the contribution that the technology provides to meeting specific energy services (e.g., peak electricity demands) or imposes in the form of ancillary costs on the energy system (e.g., the costs of integration). [8.2, 9.3, 10.6]

The cost of most RE technologies has declined and additional expected technical advances would result in further cost reductions. Significant advances in RE technologies and associated long-term cost reductions have been demonstrated over the last decades, though periods of rising prices have sometimes been experienced (due to, for example, increasing demand for RE in excess of available supply) (Figure SPM.6). The contribution of different drivers (e.g., R&D, economies of scale, deployment-oriented learning, and increased market competition among RE suppliers) is not always understood in detail. [2.7, 3.8, 7.8, 10.5] Further cost reductions are expected, resulting in greater potential deployment and consequent climate change mitigation. Examples of important areas of potential technological advancement include: new and improved feedstock production and supply systems, biofuels produced via new processes (also called next-generation or advanced biofuels, e.g., lignocellulosic) and advanced biorefining [2.6]; advanced PV and CSP technologies and manufacturing processes [3.7]; enhanced geothermal systems (EGS) [4.6]; multiple emerging ocean technologies [6.6]; and foundation and turbine designs for offshore wind energy [7.7]. Further cost reductions for hydropower are expected to be less significant than some of the other RE technologies, but R&D opportunities exist to make hydropower projects technically feasible in a wider range of locations and to improve the technical performance of new and existing projects. [5.3, 5.7, 5.8]

A variety of technology-specific challenges (in addition to cost) may need to be addressed to enable RE to significantly upscale its contribution to reducing GHG emissions. For the increased and sustainable use of bioenergy, proper design, implementation and monitoring of sustainability frameworks can minimize negative impacts and maximize benefits with regard to social, economic and environmental issues [SPM.5, 2.2, 2.5, 2.8]. For solar energy, regulatory and institutional barriers can impede deployment, as can integration and transmission issues [3.9]. For geothermal energy, an important challenge would be to prove that enhanced geothermal systems (EGS) can be deployed economically, sustainably and widely [4.5, 4.6, 4.7, 4.8]. New hydropower projects can have ecological and social impacts that are very site specific, and increased deployment may require improved sustainability assessment tools, and regional and multi-party collaborations to address energy and water needs [5.6, 5.9, 5.10]. The deployment of ocean energy could benefit from testing centres for demonstration projects, and from dedicated policies and regulations that encourage early deployment [6.4]. For wind energy, technical and institutional solutions to transmission constraints and operational integration concerns may be especially important, as might public acceptance issues relating primarily to landscape impacts. [7.5, 7.6, 7.9]

---

8   The levelized cost of energy represents the cost of an energy generating system over its lifetime; it is calculated as the per-unit price at which energy must be generated from a specific source over its lifetime to break even. It usually includes all private costs that accrue upstream in the value chain, but does not include the downstream cost of delivery to the final customer; the cost of integration, or external environmental or other costs. Subsidies and tax credits are also not included.

Notes: Medium values are shown for the following subcategories, sorted in the order as they appear in the respective ranges (from left to right):

Electricity	Heat	Transport Fuels
Biomass: 1. Cofiring 2. Small scale combined heat and power, CHP (Gasification internal combustion engine) 3. Direct dedicated stoker & CHP 4. Small scale CHP (steam turbine) 5. Small scale CHP (organic Rankine cycle) **Solar Electricity:** 1. Concentrating solar power 2. Utility-scale PV (1-axis and fixed tilt) 3. Commercial rooftop PV 4. Residential rooftop PV **Geothermal Electricity:** 1. Condensing flash plant 2. Binary cycle plant **Hydropower:** 1. All types **Ocean Electricity:** 1. Tidal barrage **Wind Electricity:** 1. Onshore 2. Offshore	**Biomass Heat:** 1. Municipal solid waste based CHP 2. Anaerobic digestion based CHP 3. Steam turbine CHP 4. Domestic pellet heating system **Solar Thermal Heat:** 1. Domestic hot water systems in China 2. Water and space heating **Geothermal Heat:** 1. Greenhouses 2. Uncovered aquaculture ponds 3. District heating 4. Geothermal heat pumps 5. Geothermal building heating	**Biofuels:** 1. Corn ethanol 2. Soy biodiesel 3. Wheat ethanol 4. Sugarcane ethanol 5. Palm oil biodiesel

The lower range of the levelized cost of energy for each RE technology is based on a combination of the most favourable input-values, whereas the upper range is based on a combination of the least favourable input values. Reference ranges in the figure background for non-renewable electricity options are indicative of the levelized cost of centralized non-renewable electricity generation. Reference ranges for heat are indicative of recent costs for oil and gas based heat supply options. Reference ranges for transport fuels are based on recent crude oil spot prices of USD 40 to 130/barrel and corresponding diesel and gasoline costs, excluding taxes.

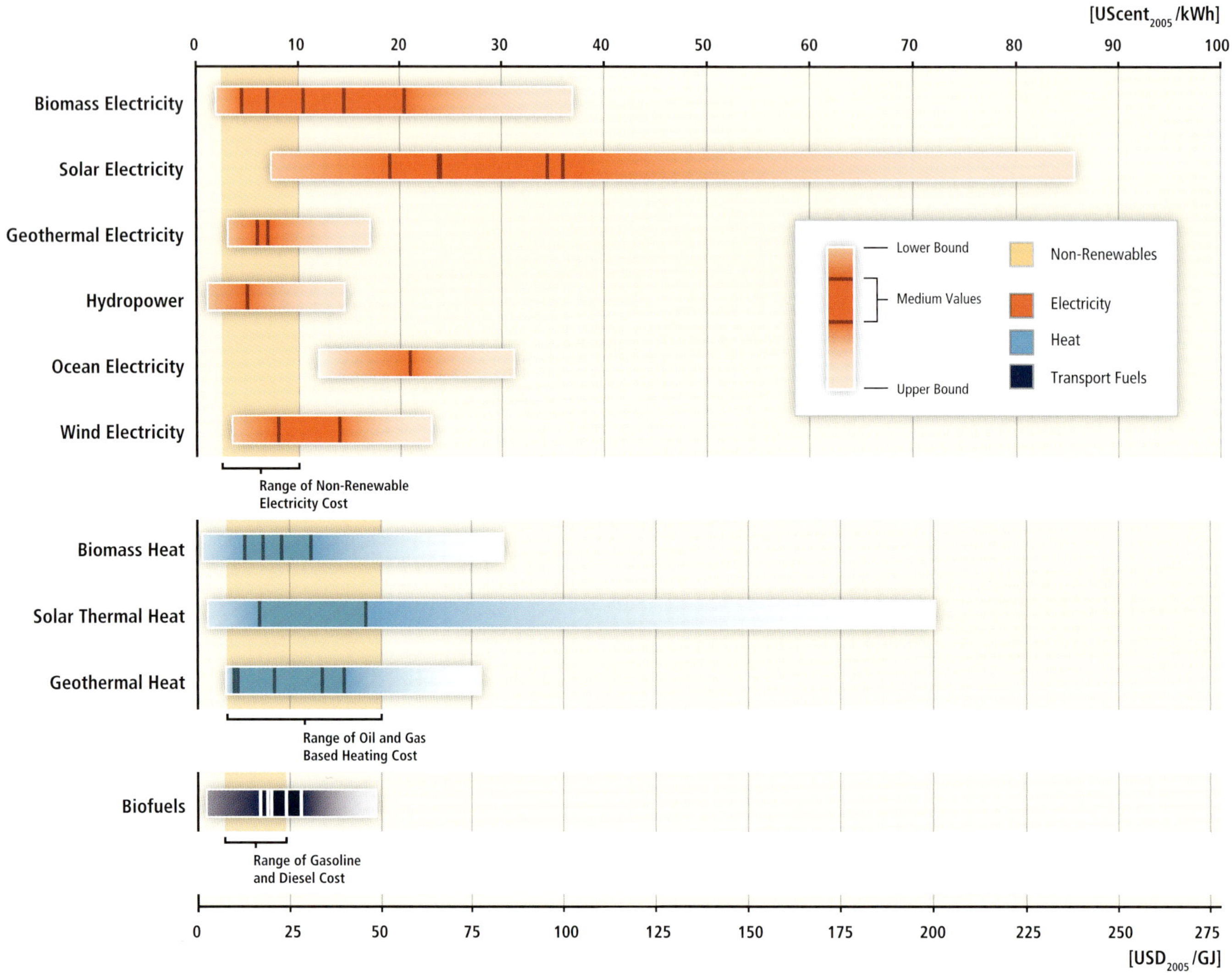

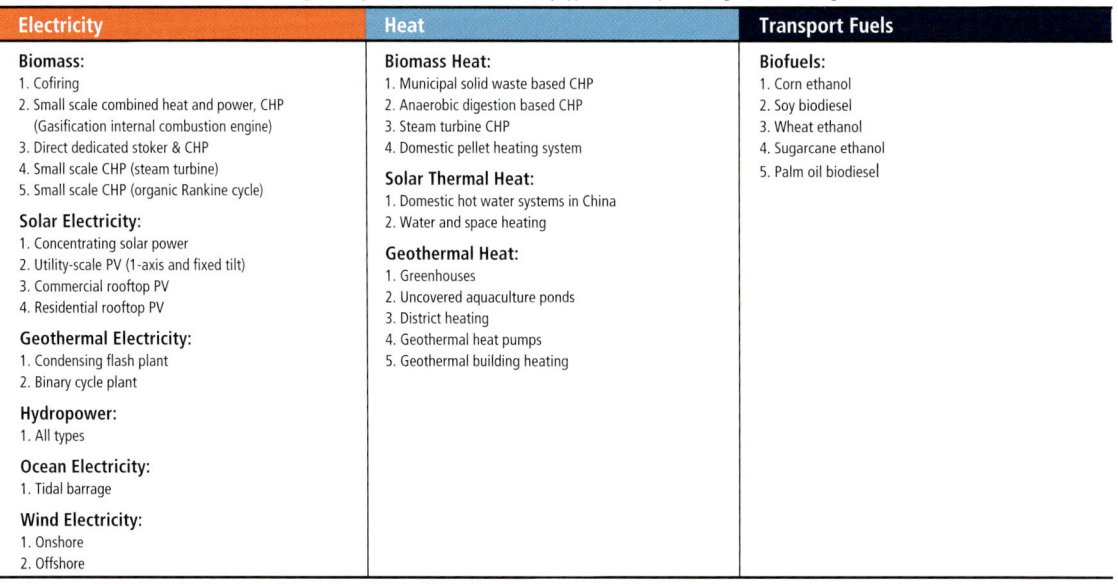

Figure SPM.5 | Range in recent levelized cost of energy for selected commercially available RE technologies in comparison to recent non-renewable energy costs. Technology subcategories and discount rates were aggregated for this figure. For related figures with less or no such aggregation, see [1.3.2, 10.5, Annex III].

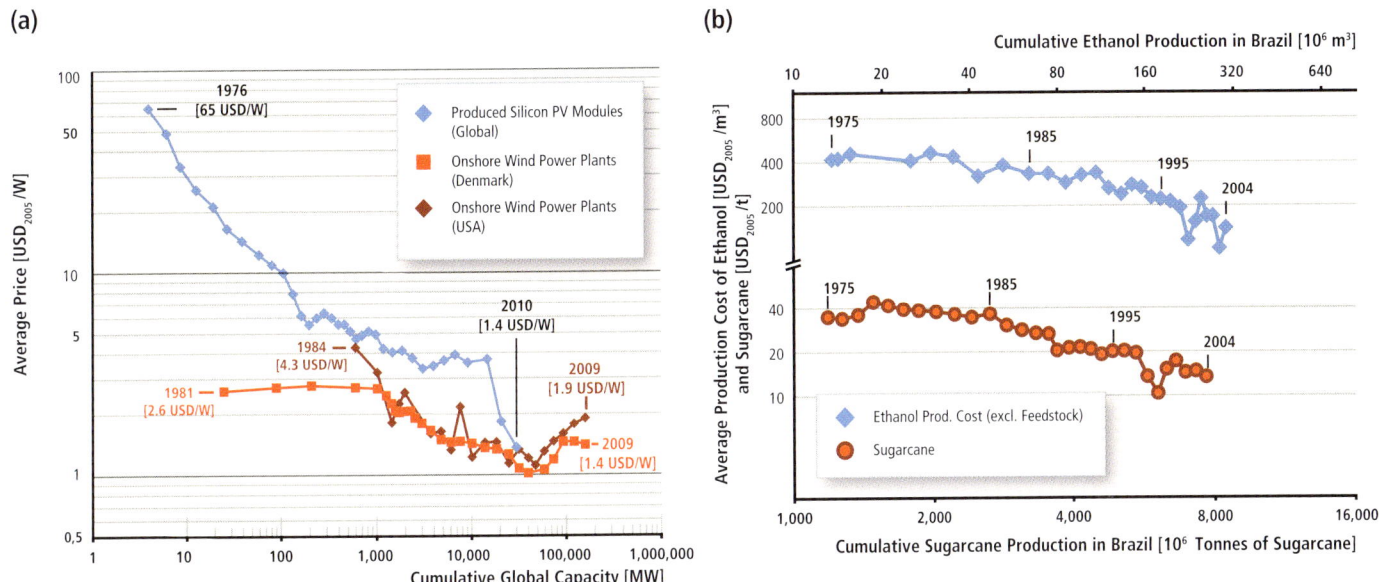

Figure SPM.6 | Selected experience curves in logarithmic scale for (a) the price of silicon PV modules and onshore wind power plants per unit of capacity; and (b) the cost of sugarcane-based ethanol production [data from Figure 3.17, 3.8.3, Figure 7.20, 7.8.2, Figure 2.21, 2.7.2].

Notes: Depending on the setting, cost reductions may occur at various geographic scales. The country-level examples provided here derive from the published literature. No global dataset of wind power plant prices or costs is readily available. Reductions in the cost or price of a technology per unit of capacity understate reductions in the levelized cost of energy of that technology when performance improvements occur. [7.8.4, 10.5]

4. Integration into present and future energy systems

Various RE resources are already being successfully integrated into energy supply systems [8.2] and into end-use sectors [8.3] (Figure SPM.7).

The characteristics of different RE sources can influence the scale of the integration challenge. Some RE resources are widely distributed geographically. Others, such as large-scale hydropower, can be more centralized but have integration options constrained by geographic location. Some RE resources are variable with limited predictability. Some have lower physical energy densities and different technical specifications from fossil fuels. Such characteristics can constrain ease of integration and invoke additional system costs particularly when reaching higher shares of RE. [8.2]

Integrating RE into most existing energy supply systems and end-use sectors at an accelerated rate— leading to higher shares of RE—is technologically feasible, though will result in a number of additional challenges. Increased shares of RE are expected within an overall portfolio of low GHG emission technologies [10.3, Tables 10.4-10.6]. Whether for electricity, heating, cooling, gaseous fuels or liquid fuels, including integration directly into end-use sectors, the RE integration challenges are contextual and site specific and include the adjustment of existing energy supply systems. [8.2, 8.3]

The costs and challenges of integrating increasing shares of RE into an existing energy supply system depend on the current share of RE, the availability and characteristics of RE resources, the system characteristics, and how the system evolves and develops in the future.

- RE can be integrated into all types of *electricity* systems, from large inter-connected continental-scale grids [8.2.1] down to small stand-alone systems and individual buildings [8.2.5]. Relevant system characteristics include the generation mix and its flexibility, network infrastructure, energy market designs and institutional rules, demand location, demand profiles, and control and communication capability. Wind, solar PV energy and CSP without

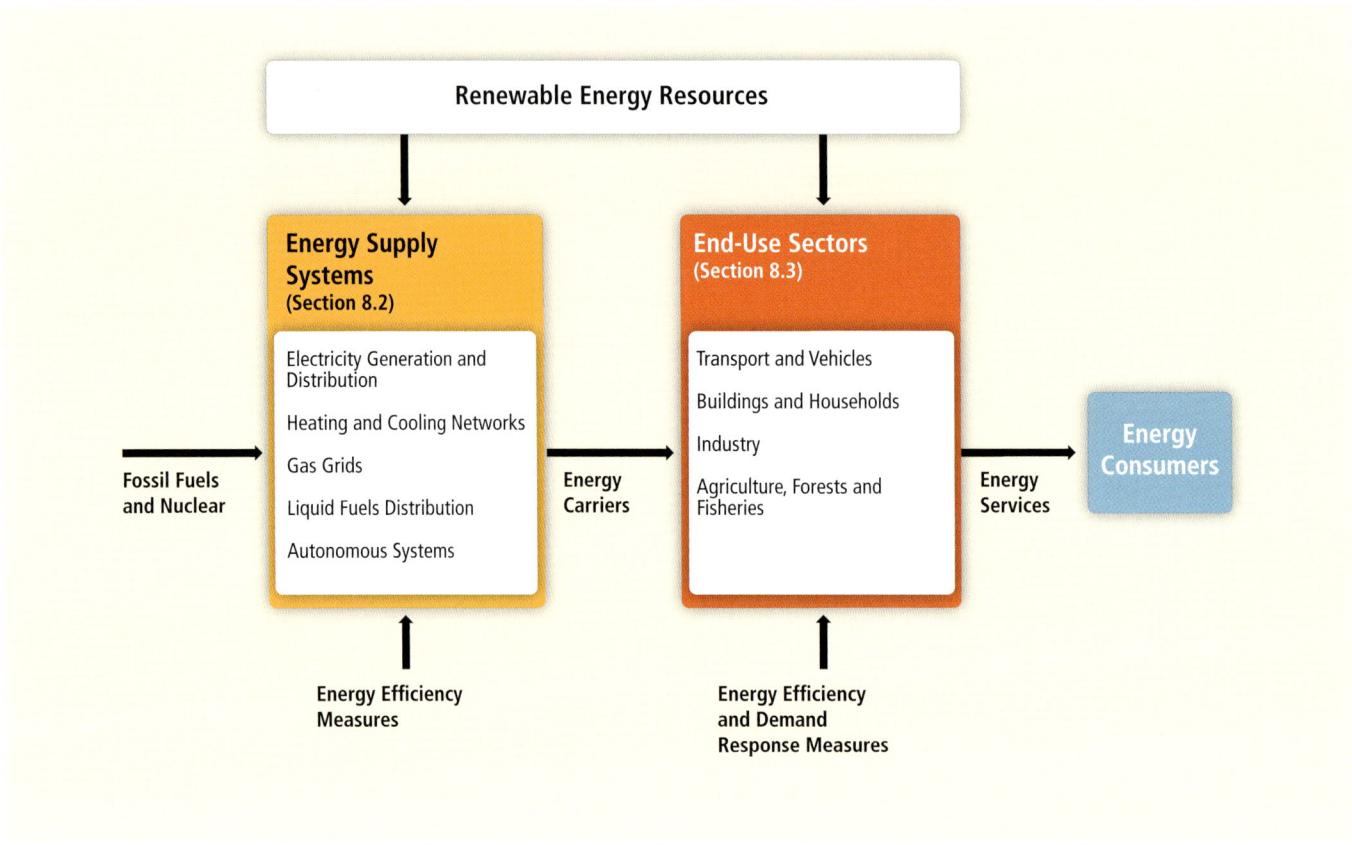

Figure SPM.7 | Pathways for RE integration to provide energy services, either into energy supply systems or on-site for use by the end-use sectors. [Figure 8.1, 8.1]

storage can be more difficult to integrate than dispatchable[9] hydropower, bioenergy, CSP with storage and geothermal energy.

As the penetration of variable RE sources increases, maintaining system reliability may become more challenging and costly. Having a portfolio of complementary RE technologies is one solution to reduce the risks and costs of RE integration. Other solutions include the development of complementary flexible generation and the more flexible operation of existing schemes; improved short-term forecasting, system operation and planning tools; electricity demand that can respond in relation to supply availability; energy storage technologies (including storage-based hydropower); and modified institutional arrangements. Electricity network transmission (including interconnections between systems) and/or distribution infrastructure may need to be strengthened and extended, partly because of the geographical distribution and fixed remote locations of many RE resources. [8.2.1]

- *District heating systems* can use low-temperature thermal RE inputs such as solar and geothermal heat, or biomass, including sources with few competing uses such as refuse-derived fuels. *District cooling* can make use of cold natural waterways. Thermal storage capability and flexible cogeneration can overcome supply and demand variability challenges as well as provide demand response for electricity systems. [8.2.2]

---

[9] Electricity plants that can schedule power generation as and when required are classed as dispatchable [8.2.1.1, Annex I]. Variable RE technologies are partially dispatchable (i.e., only when the RE resource is available). CSP plants are classified as dispatchable when heat is stored for use at night or during periods of low sunshine.

- In *gas distribution grids*, injecting biomethane, or in the future, RE-derived hydrogen and synthetic natural gas, can be achieved for a range of applications but successful integration requires that appropriate gas quality standards are met and pipelines upgraded where necessary. [8.2.3]

- *Liquid fuel systems* can integrate biofuels for transport applications or for cooking and heating applications. Pure (100%) biofuels, or more usually those blended with petroleum-based fuels, usually need to meet technical standards consistent with vehicle engine fuel specifications. [8.2.4, 8.3.1]

There are multiple pathways for increasing the shares of RE across all end-use sectors. The ease of integration varies depending on region, characteristics specific to the sector and the technology.

- For *transport*, liquid and gaseous biofuels are already and are expected to continue to be integrated into the fuel supply systems of a growing number of countries. Integration options may include decentralized on-site or centralized production of RE hydrogen for fuel cell vehicles and RE electricity for rail and electric vehicles [8.2.1, 8.2.3] depending on infrastructure and vehicle technology developments. [8.3.1] Future demand for electric vehicles could also enhance flexible electricity generation systems. [8.2.1, 8.3.1]

- In the *building* sector, RE technologies can be integrated into both new and existing structures to produce electricity, heating and cooling. Supply of surplus energy may be possible, particularly for energy efficient building designs. [8.3.2] In developing countries, the integration of RE supply systems is feasible for even modest dwellings. [8.3.2, 9.3.2]

- Agriculture as well as food and fibre process *industries* often use biomass to meet direct heat and power demands on-site. They can also be net exporters of surplus fuels, heat, and electricity to adjacent supply systems. [8.3.3, 8.3.4] Increasing the integration of RE for use by industries is an option in several sub-sectors, for example through electro-thermal technologies or, in the longer term, by using RE hydrogen. [8.3.3]

The costs associated with RE integration, whether for electricity, heating, cooling, gaseous or liquid fuels, are contextual, site-specific and generally difficult to determine. They may include additional costs for network infrastructure investment, system operation and losses, and other adjustments to the existing energy supply systems as needed. The available literature on integration costs is sparse and estimates are often lacking or vary widely.

In order to accommodate high RE shares, energy systems will need to evolve and be adapted. [8.2, 8.3] Long-term integration efforts could include investment in enabling infrastructure; modification of institutional and governance frameworks; attention to social aspects, markets and planning; and capacity building in anticipation of RE growth. [8.2, 8.3] Furthermore, integration of less mature technologies, including biofuels produced through new processes (also called advanced biofuels or next-generation biofuels), fuels generated from solar energy, solar cooling, ocean energy technologies, fuel cells and electric vehicles, will require continuing investments in research, development and demonstration (RD&D), capacity building and other supporting measures. [2.6, 3.7, 11.5, 11.6, 11.7]

RE could shape future energy supply and end-use systems, in particular for electricity, which is expected to attain higher shares of RE earlier than either the heat or transport fuel sectors at the global level [10.3]. Parallel developments in electric vehicles [8.3.1], increased heating and cooling using electricity (including heat pumps) [8.2.2, 8.3.2, 8.3.3], flexible demand response services (including the use of smart meters) [8.2.1], energy storage and other technologies could be associated with this trend.

As infrastructure and energy systems develop, in spite of the complexities, there are few, if any, fundamental technological limits to integrating a portfolio of RE technologies to meet a majority share of total

energy demand in locations where suitable RE resources exist or can be supplied. However, the actual rate of integration and the resulting shares of RE will be influenced by factors such as costs, policies, environmental issues and social aspects. [8.2, 8.3, 9.3, 9.4, 10.2, 10.5]

5. Renewable energy and sustainable development

Historically, economic development has been strongly correlated with increasing energy use and growth of GHG emissions, and RE can help decouple that correlation, contributing to sustainable development (SD). Though the exact contribution of RE to SD has to be evaluated in a country-specific context, RE offers the opportunity to contribute to social and economic development, energy access, secure energy supply, climate change mitigation, and the reduction of negative environmental and health impacts. [9.2] Providing access to modern energy services would support the achievement of the Millennium Development Goals. [9.2.2, 9.3.2]

- **RE can contribute to social and economic development.** Under favorable conditions, cost savings in comparison to non-RE use exist, in particular in remote and in poor rural areas lacking centralized energy access. [9.3.1, 9.3.2.] Costs associated with energy imports can often be reduced through the deployment of domestic RE technologies that are already competitive. [9.3.3] RE can have a positive impact on job creation although the studies available differ with respect to the magnitude of net employment. [9.3.1]

- **RE can help accelerate access to energy, particularly for the 1.4 billion people without access to electricity and the additional 1.3 billion using traditional biomass.** Basic levels of access to modern energy services can provide significant benefits to a community or household. In many developing countries, decentralized grids based on RE and the inclusion of RE in centralized energy grids have expanded and improved energy access. In addition, non-electrical RE technologies also offer opportunities for modernization of energy services, for example, using solar energy for water heating and crop drying, biofuels for transportation, biogas and modern biomass for heating, cooling, cooking and lighting, and wind for water pumping. [9.3.2, 8.1] The number of people without access to modern energy services is expected to remain unchanged unless relevant domestic policies are implemented, which may be supported or complemented by international assistance as appropriate. [9.3.2, 9.4.2]

- **RE options can contribute to a more secure energy supply, although specific challenges for integration must be considered.** RE deployment might reduce vulnerability to supply disruption and market volatility if competition is increased and energy sources are diversified. [9.3.3, 9.4.3] Scenario studies indicate that concerns regarding secure energy supply could continue in the future without technological improvements within the transport sector. [2.8, 9.4.1.1, 9.4.3.1, 10.3] The variable output profiles of some RE technologies often necessitate technical and institutional measures appropriate to local conditions to assure energy supply reliability. [8.2, 9.3.3]

- **In addition to reduced GHG emissions, RE technologies can provide other important environmental benefits.** Maximizing these benefits depends on the specific technology, management, and site characteristics associated with each RE project.

    - **Lifecycle assessments (LCA) for electricity generation indicate that GHG emissions from RE technologies are, in general, significantly lower than those associated with fossil fuel options, and in a range of conditions, less than fossil fuels employing CCS.** The median values for all RE range from 4 to 46 g CO_2eq/kWh while those for fossil fuels range from 469 to 1,001 g CO_2eq/kWh (excluding land use change emissions) (Figure SPM.8).

    - **Most current bioenergy systems, including liquid biofuels, result in GHG emission reductions, and most biofuels produced through new processes (also called advanced biofuels or next-generation biofuels) could provide higher GHG mitigation.** The GHG balance may be affected by land use changes and corresponding emissions and removals. Bioenergy can lead to avoided GHG emissions from residues and wastes in landfill disposals and co-products; the combination of bioenergy with CCS may provide for further reductions (see Figure SPM.8). The GHG implications related to land management and land use changes in carbon stocks have considerable uncertainties. [2.2, 2.5, 9.3.4.1]

- **The sustainability of bioenergy, in particular in terms of lifecycle GHG emissions, is influenced by land and biomass resource management practices.** Changes in land and forest use or management that, according to a considerable number of studies, could be brought about *directly* or *indirectly* by biomass production for use as fuels, power or heat, can decrease or increase terrestrial carbon stocks. The same studies also

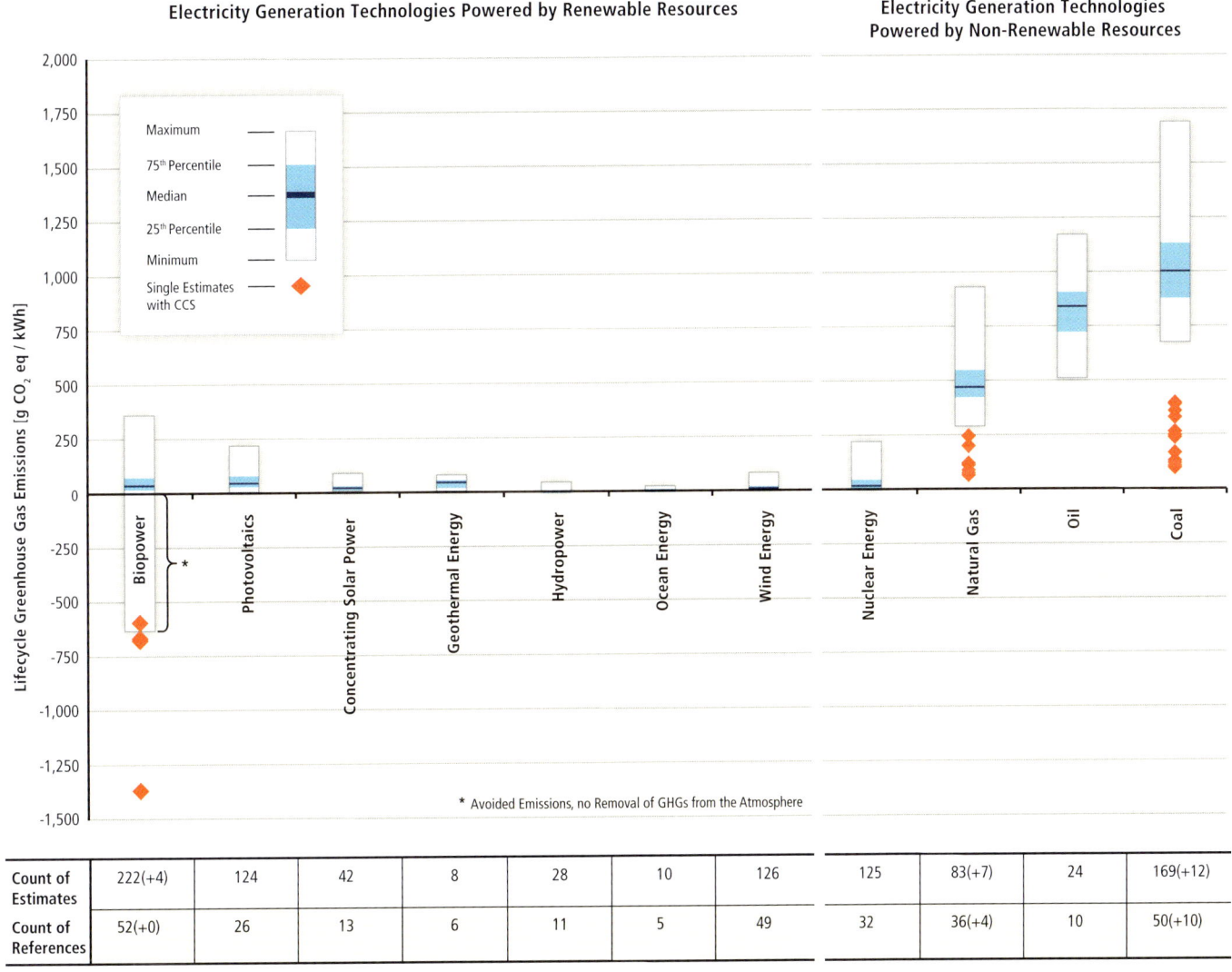

Figure SPM.8 | Estimates of lifecycle GHG emissions (g CO_2eq/kWh) for broad categories of electricity generation technologies, plus some technologies integrated with CCS. Land use-related net changes in carbon stocks (mainly applicable to biopower and hydropower from reservoirs) and land management impacts are excluded; negative estimates[10] for biopower are based on assumptions about avoided emissions from residues and wastes in landfill disposals and co-products. References and methods for the review are reported in Annex II. The number of estimates is greater than the number of references because many studies considered multiple scenarios. Numbers reported in parentheses pertain to additional references and estimates that evaluated technologies with CCS. Distributional information relates to estimates currently available in LCA literature, not necessarily to underlying theoretical or practical extrema, or the true central tendency when considering all deployment conditions. [Figure 9.8, 9.3.4.1]

---

10 'Negative estimates' within the terminology of lifecycle assessments presented in the SRREN refer to avoided emissions. Unlike the case of bioenergy combined with CCS, avoided emissions do not remove GHGs from the atmosphere.

show that indirect changes in terrestrial carbon stocks have considerable uncertainties, are not directly observable, are complex to model and are difficult to attribute to a single cause. Proper governance of land use, zoning, and choice of biomass production systems are key considerations for policy makers. [2.4.5, 2.5.1, 9.3.4, 9.4.4] Policies are in place that aim to ensure that the benefits from bioenergy, such as rural development, overall improvement of agricultural management and the contribution to climate change mitigation, are realized; their effectiveness has not been assessed. [2.2, 2.5, 2.8]

- **RE technologies, in particular non-combustion based options, can offer benefits with respect to air pollution and related health concerns.** [9.3.4.3, 9.4.4.1] Improving traditional biomass use can significantly reduce local and indoor air pollution (alongside GHG emissions, deforestation and forest degradation) and lower associated health impacts, particularly for women and children in developing countries. [2.5.4, 9.3.4.4]

- **Water availability could influence choice of RE technology.** Conventional water-cooled thermal power plants may be especially vulnerable to conditions of water scarcity and climate change. In areas where water scarcity is already a concern, non-thermal RE technologies or thermal RE technologies using dry cooling can provide energy services without additional stress on water resources. Hydropower and some bioenergy systems are dependent on water availability, and can either increase competition or mitigate water scarcity. Many impacts can be mitigated by siting considerations and integrated planning. [2.5.5.1, 5.10, 9.3.4.4]

- **Site-specific conditions will determine the degree to which RE technologies impact biodiversity.** RE-specific impacts on biodiversity may be positive or negative. [2.5, 3.6, 4.5, 5.6, 6.5, , 9.3.4.6]

- **RE technologies have low fatality rates.** Accident risks of RE technologies are not negligible, but their often decentralized structure strongly limits the potential for disastrous consequences in terms of fatalities. However, dams associated with some hydropower projects may create a specific risk depending on site-specific factors. [9.3.4.7]

6. Mitigation potentials and costs

A significant increase in the deployment of RE by 2030, 2050 and beyond is indicated in the majority of the 164 scenarios reviewed in this Special Report.[11] In 2008, total RE production was roughly 64 EJ/yr (12.9% of total primary energy supply) with more than 30 EJ/yr of this being traditional biomass. More than 50% of the scenarios project levels of RE deployment in 2050 of more than 173 EJ/yr reaching up to over 400 EJ/yr in some cases (Figure SPM.9). Given that traditional biomass use decreases in most scenarios, a corresponding increase in the production level of RE (excluding traditional biomass) anywhere from roughly three-fold to more than ten-fold is projected. The global primary energy supply share of RE differs substantially among the scenarios. More than half of the scenarios show a contribution from RE in excess of a 17% share of primary energy supply in 2030 rising to more than 27% in 2050. The scenarios with the highest RE shares reach approximately 43% in 2030 and 77% in 2050. [10.2, 10.3]

RE can be expected to expand even under baseline scenarios. Most baseline scenarios show RE deployments significantly above the 2008 level of 64 EJ/yr and up to 120 EJ/yr by 2030. By 2050, many baseline scenarios reach RE deployment levels of more than 100 EJ/yr and in some cases up to about 250 EJ/yr (Figure SPM.9). These baseline deployment levels result from a range of assumptions, including, for example, continued demand growth for energy services throughout the century, the ability of RE to contribute to increased energy access and the limited long-term

---

[11] For this purpose a review of 164 global scenarios from 16 different large-scale integrated models was conducted. Although the set of scenarios allows for a meaningful assessment of uncertainty, the reviewed 164 scenarios do not represent a fully random sample suitable for rigorous statistical analysis and do not represent always the full RE portfolio (e.g., so far ocean energy is only considered in a few scenarios) [10.2.2]. For more specific analysis, a subset of 4 illustrative scenarios from the set of 164 was used. They represent a span from a baseline scenario without specific mitigation targets to three scenarios representing different CO_2 stabilization levels. [10.3]

availability of fossil resources. Other assumptions (e.g., improved costs and performance of RE technologies) render RE technologies increasingly economically competitive in many applications even in the absence of climate policy. [10.2]

RE deployment significantly increases in scenarios with low GHG stabilization concentrations. Low GHG stabilization scenarios lead on average to higher RE deployment compared to the baseline. However, for any given long-term GHG concentration goal, the scenarios exhibit a wide range of RE deployment levels (Figure SPM.9). In scenarios that stabilize the atmospheric CO_2 concentrations at a level of less than 440 ppm, the median RE deployment level in 2050 is 248 EJ/yr (139 in 2030), with the highest levels reaching 428 EJ/yr by 2050 (252 in 2030). [10.2]

Many combinations of low-carbon energy supply options and energy efficiency improvements can contribute to given low GHG concentration levels, with RE becoming the dominant low-carbon energy supply option by 2050 in the majority of scenarios. This wide range of results originates in assumptions about factors such as developments in RE technologies (including bioenergy with CCS) and their associated resource bases and costs; the comparative attractiveness of other mitigation options (e.g., end-use energy efficiency, nuclear energy, fossil energy with CCS); patterns of consumption and production; fundamental drivers of energy services demand (including future population and economic growth); the ability to integrate variable RE sources into power grids; fossil fuel resources; specific policy approaches to mitigation; and emissions trajectories towards long-term concentration levels. [10.2]

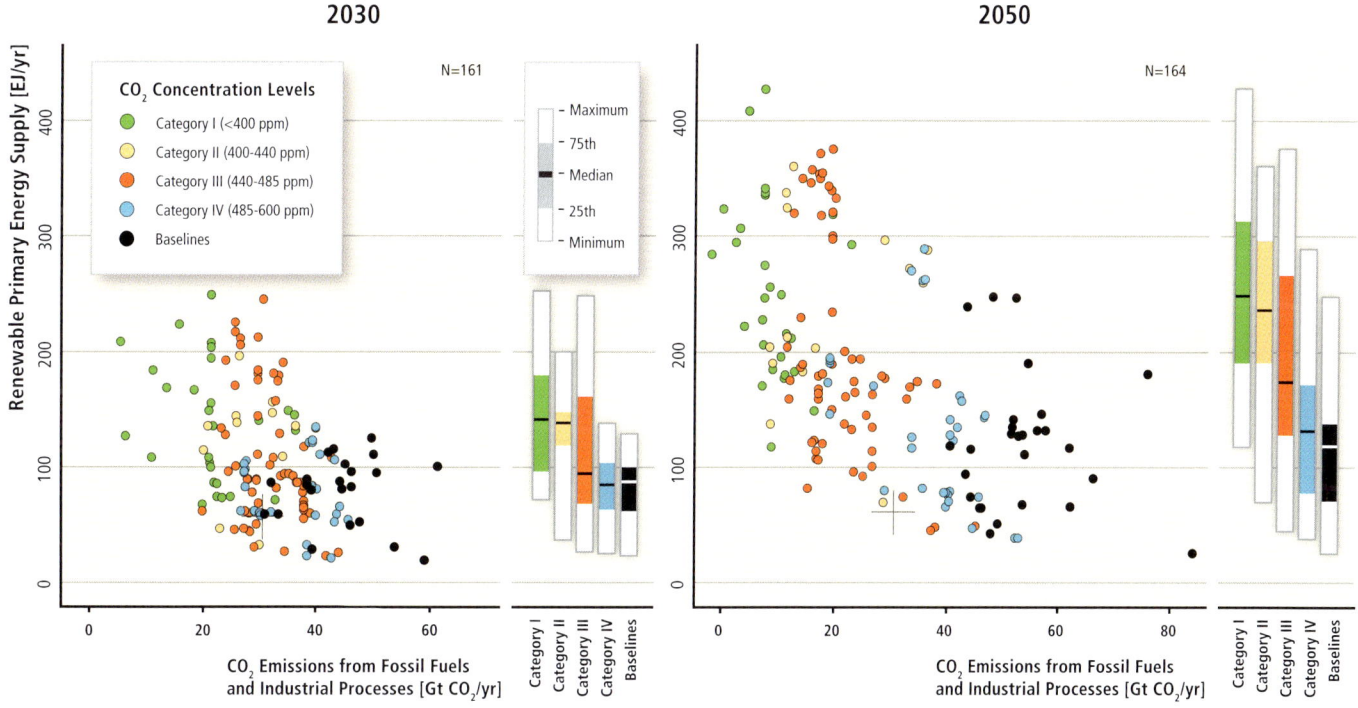

Figure SPM.9 | Global RE primary energy supply (direct equivalent) from 164 long-term scenarios versus fossil and industrial CO_2 emissions in 2030 and 2050. Colour coding is based on categories of atmospheric CO_2 concentration stabilization levels that are defined consistently with those in the AR4. The panels to the right of the scatterplots show the deployment levels of RE in each of the atmospheric CO_2 concentration categories. The thick black line corresponds to the median, the coloured box corresponds to the inter-quartile range (25th to 75th percentile) and the ends of the white surrounding bars correspond to the total range across all reviewed scenarios. The grey crossed lines show the relationship in 2007. [Figure 10.2, 10.2.2.2]

Notes: For data reporting reasons only 161 scenarios are included in the 2030 results shown here, as opposed to the full set of 164 scenarios. RE deployment levels below those of today are a result of model output and differences in the reporting of traditional biomass. For details on the use of the 'direct equivalent' method of accounting for primary energy supply and the implied care needed in the interpretation of scenario results, see Box SPM.2. Note that categories V and above are not included and category IV is extended to 600 ppm from 570 ppm, because all stabilization scenarios lie below 600 ppm CO_2 in 2100 and because the lowest baseline scenarios reach concentration levels of slightly more than 600 ppm by 2100.

The scenario review in this Special Report indicates that RE has a large potential to mitigate GHG emissions. Four illustrative scenarios span a range of global cumulative CO_2 savings between 2010 and 2050, from about 220 to 560 Gt CO_2 compared to about 1,530 Gt cumulative fossil and industrial CO_2 emissions in the IEA World Energy Outlook 2009 Reference Scenario during the same period. The precise attribution of mitigation potentials to RE depends on the role scenarios attribute to specific mitigation technologies, on complex system behaviours and, in particular, on the energy sources that RE displaces. Therefore, attribution of precise mitigation potentials to RE should be viewed with appropriate caution. [10.2, 10.3, 10.4]

Scenarios generally indicate that growth in RE will be widespread around the world. Although the precise distribution of RE deployment among regions varies substantially across scenarios, the scenarios are largely consistent in indicating widespread growth in RE deployment around the globe. In addition, the total RE deployment is higher over the long term in the group of non-Annex I countries[12] than in the group of Annex I countries in most scenarios (Figure SPM.10). [10.2, 10.3]

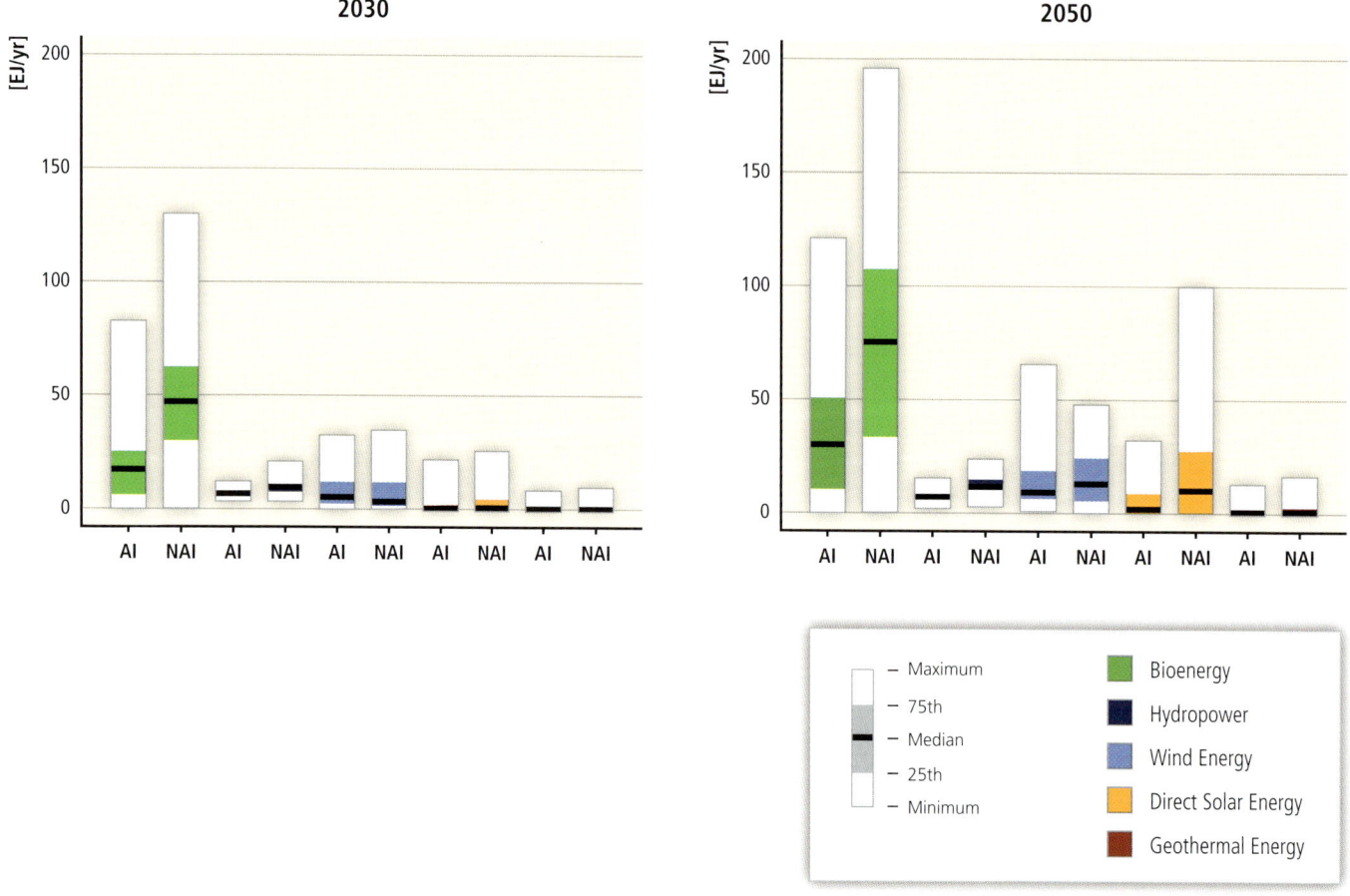

Figure SPM.10 | Global RE primary energy supply (direct equivalent) by source in the group of Annex I (AI) and the group of Non-Annex I (NAI) countries in 164 long-term scenarios by 2030 and 2050. The thick black line corresponds to the median, the coloured box corresponds to the inter-quartile range (25th to 75th percentile) and the ends of the white surrounding bars correspond to the total range across all reviewed scenarios. [Figure 10.8, 10.2.2.5]

Notes: For details on the use of the 'direct equivalent' method of accounting for primary energy supply and the implied care needed in the interpretation of scenario results, see Box SPM.2. More specifically, the ranges of secondary energy provided from bioenergy, wind energy and direct solar energy can be considered of comparable magnitude in their higher penetration scenarios in 2050. Ocean energy is not presented here as only very few scenarios consider this RE technology.

---

12 The terms 'Annex I' and 'non-Annex I' are categories of countries that derive from the United Nations Framework Convention on Climate Change (UNFCCC).

Scenarios do not indicate an obvious single dominant RE technology at a global level; in addition, the global overall technical potentials do not constrain the future contribution of RE. Although the contribution of RE technologies varies across scenarios, modern biomass, wind and direct solar commonly make up the largest contributions of RE technologies to the energy system by 2050 (Figure SPM.11). All scenarios assessed confirm that technical potentials will not be the limiting factors for the expansion of RE at a global scale. Despite significant technological and regional differences, in the four illustrative scenarios less than 2.5% of the global available technical RE potential is used. [10.2, 10.3]

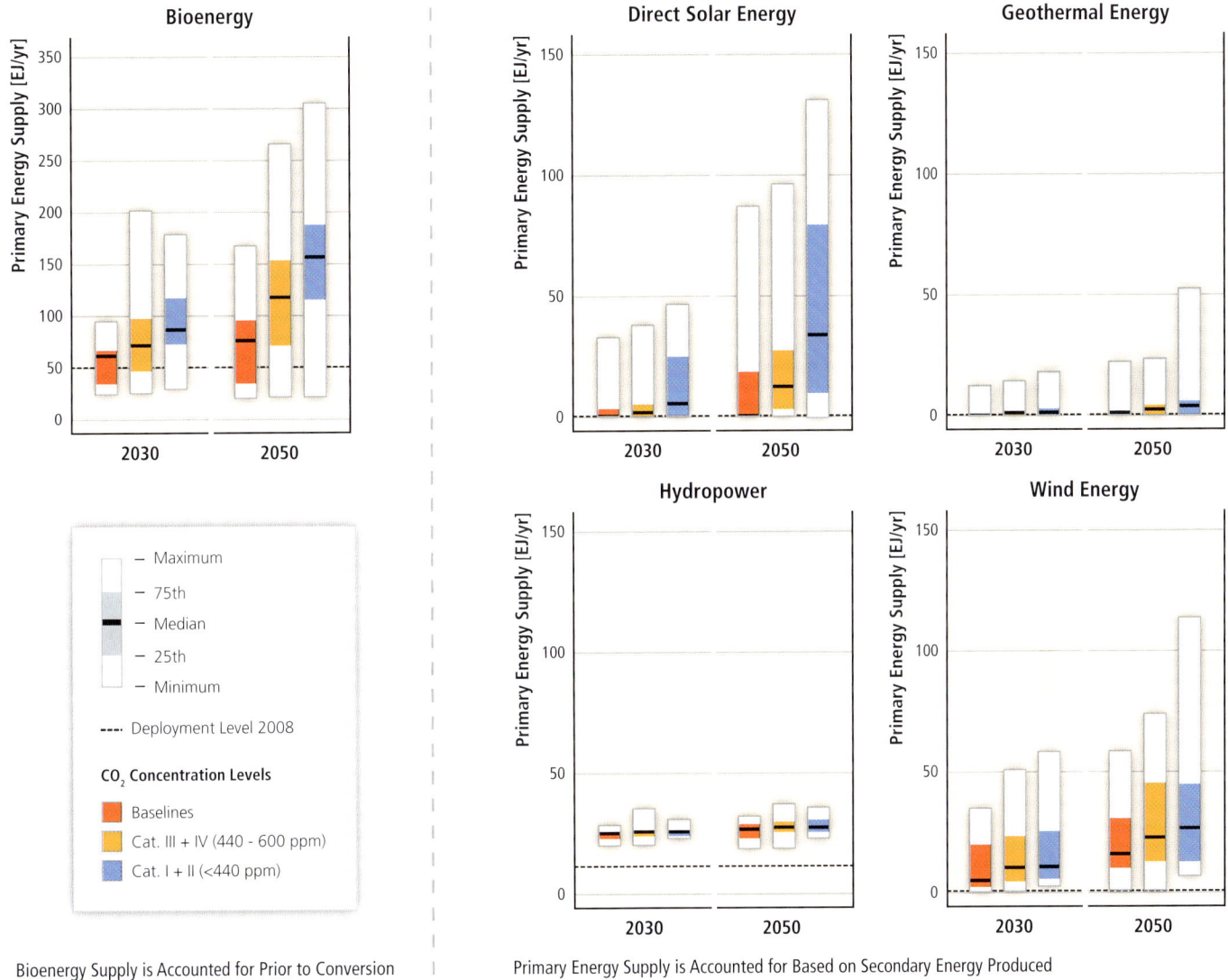

Figure SPM.11 | Global primary energy supply (direct equivalent) of bioenergy, wind, direct solar, hydro, and geothermal energy in 164 long-term scenarios in 2030 and 2050, and grouped by different categories of atmospheric CO_2 concentration level that are defined consistently with those in the AR4. The thick black line corresponds to the median, the coloured box corresponds to the inter-quartile range (25th to 75th percentile) and the ends of the white surrounding bars correspond to the total range across all reviewed scenarios. [Excerpt from Figure 10.9, 10.2.2.5]

Notes: For details on the use of the 'direct equivalent' method of accounting for primary energy supply and the implied care needed in the interpretation of scenario results, see Box SPM.2. More specifically, the ranges of secondary energy provided from bioenergy, wind energy and direct solar energy can be considered of comparable magnitude in their higher penetration scenarios in 2050. Ocean energy is not presented here as only very few scenarios consider this RE technology. Note that categories V and above are not included and category IV is extended to 600 ppm from 570 ppm, because all stabilization scenarios lie below 600 ppm CO_2 in 2100 and because the lowest baselines scenarios reach concentration levels of slightly more than 600 ppm by 2100.

Individual studies indicate that if RE deployment is limited, mitigation costs increase and low GHG concentration stabilizations may not be achieved. A number of studies have pursued scenario sensitivities that assume constraints on the deployment of individual mitigation options, including RE as well as nuclear and fossil energy with CCS. There is little agreement on the precise magnitude of the cost increase. [10.2]

A transition to a low-GHG economy with higher shares of RE would imply increasing investments in technologies and infrastructure. The four illustrative scenarios analyzed in detail in the SRREN estimate global cumulative RE investments (in the power generation sector only) ranging from USD_{2005} 1,360 to 5,100 billion for the decade 2011 to 2020, and from USD_{2005} 1,490 to 7,180 billion for the decade 2021 to 2030. The lower values refer to the IEA World Energy Outlook 2009 Reference Scenario and the higher ones to a scenario that seeks to stabilize atmospheric CO_2 (only) concentration at 450 ppm. The annual averages of these investment needs are all smaller than 1% of the world's gross domestic product (GDP). Beyond differences in the design of the models used to investigate these scenarios, the range can be explained mainly by differences in GHG concentrations assessed and constraints imposed on the set of admissible mitigation technologies. Increasing the installed capacity of RE power plants will reduce the amount of fossil and nuclear fuels that otherwise would be needed in order to meet a given electricity demand. In addition to investment, operation and maintenance (O&M) and (where applicable) feedstock costs related to RE power plants, any assessment of the overall economic burden that is associated with their application will have to consider avoided fuel and substituted investment costs as well. Even without taking the avoided costs into account, the lower range of the RE power investments discussed above is lower than the respective investments reported for 2009. The higher values of the annual averages of the RE power sector investment approximately correspond to a five-fold increase in the current global investments in this field. [10.5, 11.2.2]

7. Policy, implementation and financing

An increasing number and variety of RE policies—motivated by many factors—have driven escalated growth of RE technologies in recent years. [1.4, 11.2, 11.5, 11.6] Government policies play a crucial role in accelerating the deployment of RE technologies. Energy access and social and economic development have been the primary drivers in most developing countries whereas secure energy supply and environmental concerns have been most important in developed countries [9.3, 11.3]. The focus of policies is broadening from a concentration primarily on RE electricity to include RE heating and cooling and transportation. [11.2, 11.5]

RE-specific policies for research, development, demonstration and deployment help to level the playing field for RE. Policies include regulations such as feed-in-tariffs, quotas, priority grid access, building mandates, biofuel blending requirements, and bioenergy sustainability criteria. [2.4.5.2, 2.ES, TS.2.8.1] Other policy categories are fiscal incentives such as tax policies and direct government payments such as rebates and grants; and public finance mechanisms such as loans and guarantees. Wider policies aimed at reducing GHG emissions such as carbon pricing mechanisms may also support RE.

Policies can be sector specific, can be implemented at the local, state/provincial, national and in some cases regional level, and can be complemented by bilateral, regional and international cooperation. [11.5]
Policies have promoted an increase in RE capacity installations by helping to overcome various barriers. [1.4, 11.1, 11.4, 11.5, 11.6] Barriers to RE deployment include:

- Institutional and policy barriers related to existing industry, infrastructure and regulation of the energy system;

- Market failures, including non-internalized environmental and health costs, where applicable;

- Lack of general information and access to data relevant to the deployment of RE, and lack of technical and knowledge capacity; and

- Barriers related to societal and personal values and affecting the perception and acceptance of RE technologies. [1.4, 9.5.1, 9.5.2.1]

Public R&D investments in RE technologies are most effective when complemented by other policy instruments, particularly deployment policies that simultaneously enhance demand for new technologies. Together, R&D and deployment policies create a positive feedback cycle, inducing private sector investment. Enacting deployment policies early in the development of a given technology can accelerate learning by inducing private R&D, which in turn further reduces costs and provides additional incentives for using the technology. [11.5.2]

Some policies have been shown to be effective and efficient in rapidly increasing RE deployment. However, there is no one-size-fits-all policy. Experience shows that different policies or combinations of policies can be more effective and efficient depending on factors such as the level of technological maturity, affordable capital, ease of integration into the existing system and the local and national RE resource base. [11.5]

- Several studies have concluded that some feed in tariffs have been effective and efficient at promoting RE electricity, mainly due to the combination of long-term fixed price or premium payments, network connections, and guaranteed purchase of all RE electricity generated. Quota policies can be effective and efficient if designed to reduce risk; for example, with long-term contracts. [11.5.4]

- An increasing number of governments are adopting fiscal incentives for RE heating and cooling. Obligations to use RE heat are gaining attention for their potential to encourage growth independent of public financial support. [11.5.5]

- In the transportation sector, RE fuel mandates or blending requirements are key drivers in the development of most modern biofuel industries. Other policies include direct government payments or tax reductions. Policies have influenced the development of an international biofuel trade. [11.5.6]

The flexibility to adjust as technologies, markets and other factors evolve is important. The details of design and implementation are critical in determining the effectiveness and efficiency of a policy. [11.5]. Policy frameworks that are transparent and sustained can reduce investment risks and facilitate deployment of RE and the evolution of low-cost applications. [11.5, 11.6]

'Enabling' policies support RE development and deployment. A favourable, or enabling, environment for RE can be created by addressing the possible interactions of a given policy with other RE policies as well as with energy and non-energy policies (e.g., those targeting agriculture, transportation, water management and urban planning); by easing the ability of RE developers to obtain finance and to successfully site a project; by removing barriers for access to networks and markets for RE installations and output; by increasing education and awareness through dedicated communication and dialogue initiatives; and by enabling technology transfer. In turn, the existence of an 'enabling' environment can increase the efficiency and effectiveness of policies to promote RE. [9.5.1.1, 11.6]

Two separate market failures create the rationale for the additional support of innovative RE technologies that have high potential for technological development, even if an emission market (or GHG pricing policy in general) exists. The first market failure refers to the external cost of GHG emissions. The second market failure is in the field of innovation: if firms underestimate the future benefits of investments into learning RE technologies or if they cannot appropriate these benefits, they will invest less than is optimal from a macroeconomic perspective. In addition to GHG pricing policies, RE-specific policies may be appropriate from an economic point of view if the related opportunities for technological development are to be addressed (or if other goals beyond climate mitigation are pursued). Potentially adverse consequences such as lock-in, carbon leakage and rebound effects should be taken into account in the design of a portfolio of policies. [11.1.1, 11.5.7.3]

The literature indicates that long-term objectives for RE and flexibility to learn from experience would be critical to achieve cost-effective and high penetrations of RE. This would require systematic development of policy frameworks that reduce risks and enable attractive returns that provide stability over a time frame relevant to the investment. An appropriate and reliable mix of policy instruments, including energy efficiency policies, is even more important where energy infrastructure is still developing and energy demand is expected to increase in the future. [11.5, 11.6, 11.7]

8. Advancing knowledge about renewable energy

Enhanced scientific and engineering knowledge should lead to performance improvements and cost reductions in RE technologies. Additional knowledge related to RE and its role in GHG emissions reductions remains to be gained in a number of broad areas including: [for details, see Table 1.1]

- Future cost and timing of RE deployment;

- Realizable technical potential for RE at all geographical scales;

- Technical and institutional challenges and costs of integrating diverse RE technologies into energy systems and markets;

- Comprehensive assessments of socioeconomic and environmental aspects of RE and other energy technologies;

- Opportunities for meeting the needs of developing countries with sustainable RE services; and

- Policy, institutional and financial mechanisms to enable cost-effective deployment of RE in a wide variety of contexts.

Knowledge about RE and its climate change mitigation potential continues to advance. The existing scientific knowledge is significant and can facilitate the decision-making process. [1.1.8]

Technical Summary

Lead Authors:

Dan Arvizu (USA), Thomas Bruckner (Germany), Helena Chum (USA/Brazil), Ottmar Edenhofer (Germany), Segen Estefen (Brazil) Andre Faaij (The Netherlands), Manfred Fischedick (Germany), Gerrit Hansen (Germany), Gerardo Hiriart (Mexico), Olav Hohmeyer (Germany), K. G. Terry Hollands (Canada), John Huckerby (New Zealand), Susanne Kadner (Germany), Ånund Killingtveit (Norway), Arun Kumar (India), Anthony Lewis (Ireland), Oswaldo Lucon (Brazil), Patrick Matschoss (Germany), Lourdes Maurice (USA), Monirul Mirza (Canada/Bangladesh), Catherine Mitchell (United Kingdom), William Moomaw (USA), José Moreira (Brazil), Lars J. Nilsson (Sweden), John Nyboer (Canada), Ramon Pichs-Madruga (Cuba), Jayant Sathaye (USA), Janet L. Sawin (USA), Roberto Schaeffer (Brazil), Tormod A. Schei (Norway), Steffen Schlömer (Germany), Kristin Seyboth (Germany/USA), Ralph Sims (New Zealand), Graham Sinden (United Kingdom/Australia), Youba Sokona (Ethiopia/Mali), Christoph von Stechow (Germany), Jan Steckel (Germany), Aviel Verbruggen (Belgium), Ryan Wiser (USA), Francis Yamba (Zambia), Timm Zwickel (Germany)

Review Editors:

Leonidas O. Girardin (Argentina), Mattia Romani (United Kingdom/Italy)

Special Advisor:

Jeffrey Logan (USA)

This Technical Summary should be cited as:

Arvizu, D., T. Bruckner, H. Chum, O. Edenhofer, S. Estefen, A. Faaij, M. Fischedick, G. Hansen, G. Hiriart, O. Hohmeyer, K. G. T. Hollands, J. Huckerby, S. Kadner, Å. Killingtveit, A. Kumar, A. Lewis, O. Lucon, P. Matschoss, L. Maurice, M. Mirza, C. Mitchell, W. Moomaw, J. Moreira, L. J. Nilsson, J. Nyboer, R. Pichs-Madruga, J. Sathaye, J. Sawin, R. Schaeffer, T. Schei, S. Schlömer, K. Seyboth, R. Sims, G. Sinden, Y. Sokona, C. von Stechow, J. Steckel, A. Verbruggen, R. Wiser, F. Yamba, T. Zwickel, 2011: Technical Summary. In IPCC Special Report on Renewable Energy Sources and Climate Change Mitigation [O. Edenhofer, R. Pichs-Madruga, Y. Sokona, K. Seyboth, P. Matschoss, S. Kadner, T. Zwickel, P. Eickemeier, G. Hansen, S. Schlömer, C. von Stechow (eds)], Cambridge University Press, Cambridge, United Kingdom and New York, NY, USA.

Table of Contents

1.	**Overview of Climate Change and Renewable Energy**	33
1.1	Background	33
1.2	Summary of renewable energy resources and potential	38
1.3	Meeting energy service needs and current status	40
1.4	Opportunities, barriers, and issues	40
1.5	Role of policy, research and development, deployment and implementation strategies	44
2.	**Bioenergy**	46
2.1	Introduction to biomass and bioenergy	46
2.2	Bioenergy resource potential	46
2.3	Bioenergy technology and applications	48
2.4	Global and regional status of markets and industry deployment	48
2.5	Environmental and social impacts	50
2.6	Prospects for technology improvement and integration	53
2.7	Current costs and trends	53
2.8	Potential deployment levels	55
3.	**Direct Solar**	60
3.1	Introduction	60
3.2	Resource potential	60
3.3	Technology and applications	60
3.4	Global and regional status of market and industry deployment	63
3.5	Integration into the broader energy system	65
3.6	Environmental and social impacts	65

3.7	Prospects for technology improvements and innovation	66
3.8	Cost trends	68
3.9	Potential deployment	71

4. Geothermal Energy 71

4.1	Introduction	71
4.2	Resource potential	72
4.3	Technology and applications	73
4.4	Global and regional status of market and industry development	74
4.5	Environmental and social impacts	74
4.6	Prospects for technology improvement, innovation and integration	77
4.7	Cost trends	77
4.8	Potential deployment	78

5. Hydropower 80

5.1	Introduction	80
5.2	Resource potential	80
5.3	Technology and applications	80
5.4	Global and regional status of market and industry development	82
5.5	Integration into broader energy systems	82
5.6	Environmental and social impacts	83
5.7	Prospects for technology improvement and innovation	84
5.8	Cost trends	84

5.9	Potential deployment	86
5.10	Integration into water management systems	87

6. Ocean Energy ... 87

6.1	Introduction	87
6.2	Resource potential	87
6.3	Technology and applications	89
6.4	Global and regional status of the markets and industry development	90
6.5	Environmental and social impacts	92
6.6	Prospects for technology improvement, innovation and integration	93
6.7	Cost trends	93
6.8	Potential deployment	94

7. Wind Energy ... 95

7.1	Introduction	95
7.2	Resource potential	95
7.3	Technology and applications	96
7.4	Global and regional status of market and industry development	97
7.5	Near-term grid integration issues	98
7.6	Environmental and social impacts	99
7.7	Prospects for technology improvement and innovation	100
7.8	Cost trends	101
7.9	Potential deployment	103

8.	**Integration of Renewable Energy into Present and Future Energy Systems**	103
8.1	Introduction	103
8.2	Integration of renewable energy into electrical power systems	107
8.3	Integration of renewable energy into heating and cooling networks	110
8.4	Integration of renewable energy into gas grids	111
8.5	Integration of renewable energy into liquid fuels	112
8.6	Integration of renewable energy into autonomous systems	113
8.7	End-use sectors: Strategic elements for transition pathways	113
9.	**Renewable Energy in the Context of Sustainable Development**	119
9.1	Introduction	119
9.2	Interactions between sustainable development and renewable energy	119
9.3	Social, environmental and economic impacts: Global and regional assessment	120
9.4	Implication of sustainable development pathways for renewable energy	125
9.5	Barriers and opportunities for renewable energy in the context of sustainable development	129
9.6	Synthesis, knowledge gaps and future research needs	130
10.	**Mitigation Potential and Costs**	130
10.1	Introduction	130
10.2	Synthesis of mitigation scenarios for different renewable energy strategies	131
10.3	Assessment of representative mitigation scenarios for different renewable energy strategies	133
10.4	Regional cost curves for mitigation with renewable energy sources	135
10.5	Cost of commercialization and deployment	137
10.6	Social and environmental costs and benefits	144

11.	**Policy, Financing and Implementation**	146
11.1	Introduction	146
11.2	Current trends: Policies, financing and investment	148
11.3	Key drivers, opportunities and benefits	148
11.4	Barriers to renewable energy policymaking, implementation and financing	148
11.5	Experience with and assessment of policy options	150
11.6	Enabling environment and regional issues	155
11.7	A structural shift	158

1. Overview of Climate Change and Renewable Energy

1.1 Background

All societies require energy services to meet basic human needs (e.g., lighting, cooking, space comfort, mobility, communication) and to serve productive processes. For development to be sustainable, delivery of energy services needs to be secure and have low environmental impacts. Sustainable social and economic development requires assured and affordable access to the energy resources necessary to provide essential and sustainable energy services. This may mean the application of different strategies at different stages of economic development. To be environmentally benign, energy services must be provided with low environmental impacts and low greenhouse gas (GHG) emissions. However, the IPCC Fourth Assessment Report (AR4) reported that fossil fuels provided 85%[1] of the total primary energy in 2004, which is the same value as in 2008. Furthermore, the combustion of fossil fuels accounted for 56.6% of all anthropogenic GHG emissions (CO_2eq)[2] in 2004. [1.1.1, 9.2.1, 9.3.2, 9.6, 11.3]

Renewable energy (RE) sources play a role in providing energy services in a sustainable manner and, in particular, in mitigating climate change. This Special Report on *Renewable Energy Sources and Climate Change Mitigation* explores the current contribution and potential of RE sources to provide energy services for a sustainable social and economic development path. It includes assessments of available RE resources and technologies, costs and co-benefits, barriers to up-scaling and integration requirements, future scenarios and policy options. In particular, it provides information for policymakers, the private sector and civil society on:

- Identification of RE resources and available technologies and impacts of climate change on these resources [Chapters 2–7];
- Technology and market status, future developments and projected rates of deployment [Chapters 2–7,10];
- Options and constraints for integration into the energy supply system and other markets, including energy storage, modes of transmission, integration into existing systems and other options [Chapter 8];
- Linkages among RE growth, opportunities and sustainable development [Chapter 9];
- Impacts on secure energy supply [Chapter 9];
- Economic and environmental costs, benefits, risks and impacts of deployment [Chapters 9, 10];
- Mitigation potential of RE resources [Chapter 10];
- Scenarios that demonstrate how accelerated deployment might be achieved in a sustainable manner [Chapter 10];
- Capacity building, technology transfer and financing [Chapter 11]; and
- Policy options, outcomes and conditions for effectiveness [Chapter 11].

The report consists of 11 chapters. Chapter 1 sets the scene on RE and climate change; Chapters 2 through 7 provide information on six RE technologies while Chapters 8 through 11 deal with integrative issues (see Figure TS.1.1). The report communicates uncertainty where relevant.[3] This Technical Summary (TS) provides an overview of the report, summarizing the essential findings.

While the TS generally follows the structure of the full report, references to the various applicable chapters and sections are indicated with corresponding chapter and section numbers in square brackets. An explanation of terms, acronyms and chemical symbols used in the TS can be found in Annex I. Conventions and methodologies for determining costs, primary energy and other topics of analysis can be found in Annex II. Information on levelized costs of RE can be found in Annex III.

GHG emissions associated with the provision of energy services is a major cause of climate change. The AR4 concluded that "Most of the observed increase in global average temperature since the mid-20th century is very likely due to the observed increase in anthropogenic GHG (greenhouse gas) concentrations." Concentrations have continued to grow since the AR4 to over 390 ppm CO_2 or 39% above pre-industrial levels by the end of 2010. Since approximately 1850, global use of fossil fuels (coal, oil and gas) has increased to dominate energy supply, leading to a rapid growth in carbon dioxide (CO_2) emissions [Figure 1.6]. The amount of carbon in fossil fuel reserves and resources not yet burned [Figure 1.7] has the potential to add quantities of CO_2 to the atmosphere—if burned over coming centuries—that would exceed the range of any scenario considered in the AR4 [Figure 1.5] or in Chapter 10 of this report. [1.1.3, 1.1.4]

Despite substantial associated decarbonization, the overwhelming majority of the non-intervention emission projections exhibit considerably higher emissions in 2100 compared with those in 2000, implying rising GHG concentrations and, in turn, an increase in global mean temperatures. To avoid such adverse impacts of climate change on water resources, ecosystems, food security, human health and coastal settlements with potentially irreversible abrupt changes in the climate system,

---

1 The number from AR4 is 80% and has been converted from the physical content method for energy accounting to the direct equivalent method as the latter method is used in this report. Please refer to Section 1.1.9 and Annex II (Section A.II.4) for methodological details.

2 The contributions from other sources and/or gases are: CO_2 from deforestation, decay of biomass etc. (17.3%), CO_2 from other (2.8%), CH_4 (14.3%), N_2O (7.9%) and fluorinated gases (1.1%).

3 This report communicates uncertainty, for example, by showing the results of sensitivity analyses and by quantitatively presenting ranges in cost numbers as well as ranges in the scenario results. This report does not apply formal IPCC uncertainty terminology because at the time of the approval of this report, IPCC uncertainty guidance was in the process of being revised.

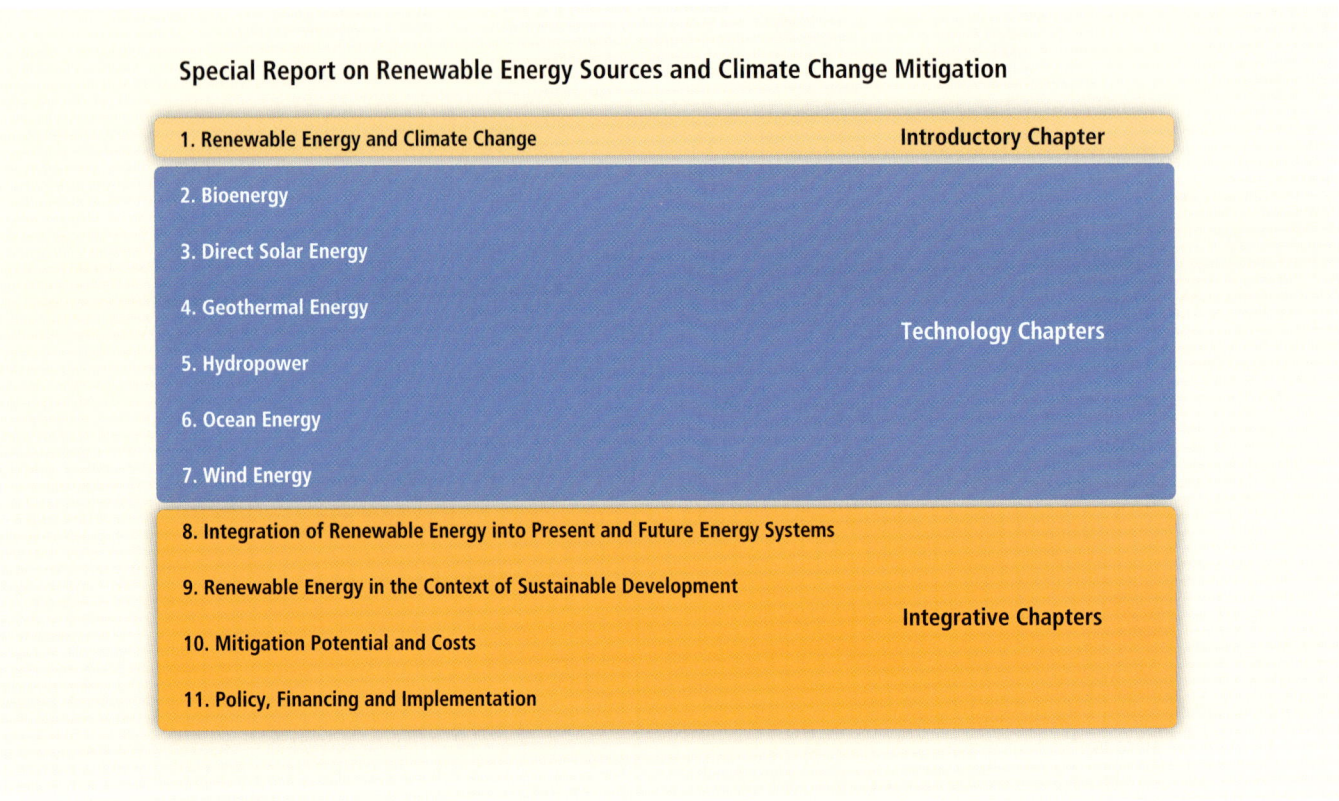

Figure TS.1.1 | Structure of the report. [Figure 1.1]

the Cancun Agreements call for limiting global average temperature rises to no more than 2°C above pre-industrial values, and agreed to consider limiting this rise to 1.5°C. In order to be confident of achieving an equilibrium temperature increase of only 2°C to 2.4°C, atmospheric GHG concentrations would need to be stabilized in the range of 445 to 490 ppm CO_2eq in the atmosphere. This in turn implies that global emissions of CO_2 will need to decrease by 50 to 85% below 2000 levels by 2050 and begin to decrease (instead of continuing their current increase) no later than 2015. [1.1.3]

To develop strategies for reducing CO_2 emissions, the Kaya identity can be used to decompose energy-related CO_2 emissions into four factors: 1) population, 2) gross domestic product (GDP) per capita, 3) energy intensity (i.e., total primary energy supply (TPES) per GDP) and 4) carbon intensity (i.e., CO_2 emissions per TPES). [1.1.4]

CO_2 emissions = Population x (GDP/population) x (TPES/GDP) x (CO_2/TPES)

The annual change in these four components is illustrated in Figure TS.1.2. [1.1.4]

While GDP per capita and population growth had the largest effect on emissions growth in earlier decades, decreasing energy intensity significantly slowed emissions growth in the period from 1971 to 2008. In the past, carbon intensity fell because of improvements in energy efficiency and switching from coal to natural gas and the expansion of nuclear energy in the 1970s and 1980s that was particularly driven by Annex I countries.[4] In recent years (2000 to 2007), increases in carbon intensity have been driven mainly by the expansion of coal use in both developed and developing countries, although coal and petroleum use have fallen slightly since 2007. In 2008 this trend was broken due to the financial crisis. Since the early 2000s, the energy supply has become more carbon intensive, thereby amplifying the increase resulting from growth in GDP per capita. [1.1.4]

On a global basis, it is estimated that RE accounted for 12.9% of the 492 EJ of total primary energy supply in 2008. The largest RE contributor was biomass (10.2%), with the majority (roughly 60%) of the biomass fuel used in traditional cooking and heating applications in developing countries but with rapidly increasing use of modern biomass as well.[5] Hydropower represented 2.3%, whereas other RE sources accounted for 0.4%. (Figure TS.1.3). In 2008, RE contributed approximately 19% of global electricity supply (16% hydropower, 3% other RE). [1.1.5]

Deployment of RE has been increasing rapidly in recent years. Under most conditions, increasing the share of RE in the energy mix will require policies to stimulate changes in the energy system. Government policy, the declining cost of many RE technologies, changes in the prices of fossil

---

[4]   See Glossary (Annex I) for a definition of Annex I countries.

[5]   Not accounted for here or in official databases is the estimated 20 to 40% of additional traditional biomass used in informal sectors. [2.1]

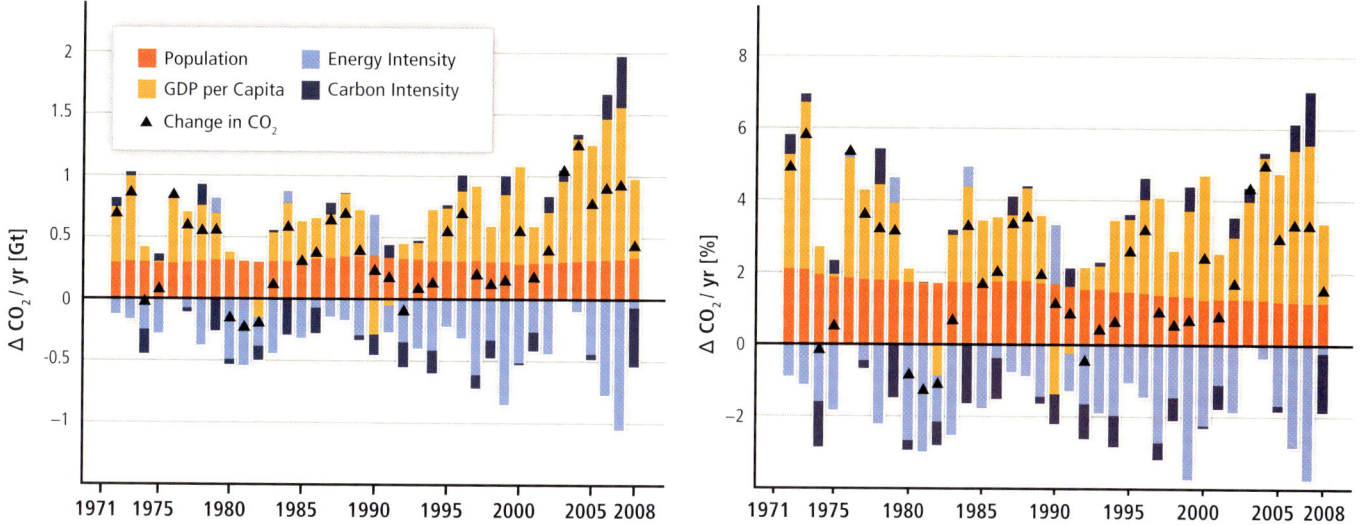

Figure TS.1.2 | Decomposition of (left) annual absolute change and (right) annual growth rate in global energy-related CO_2 emissions by the factors in the Kaya identity; population (red), GDP per capita (orange), energy intensity (light blue) and carbon intensity (dark blue) from 1971 to 2008. The colours show the changes that would occur due to each factor alone, holding the respective other factors constant. Total annual changes are indicated by a black triangle. [Figure 1.8]

fuels and other factors have supported the continuing increase in the use of RE. While the RE share is still relatively small, its growth has accelerated in recent years as shown in Figure TS.1.4. In 2009, despite global financial challenges, RE capacity continued to grow rapidly, including wind power (32%, 38 GW added), hydropower (3%, 31 GW added), grid-connected photovoltaics (53%, 7.5 GW added), geothermal power (4%, 0.4 GW added), and solar hot water/heating (21%, 31 GW_{th} added). Biofuels accounted for 2% of global road transport fuel demand in 2008 and nearly 3% in 2009. The annual production of ethanol increased to 1.6 EJ (76 billion litres) by the end of 2009 and biodiesel production increased to 0.6 EJ (17 billion litres). Of the approximate 300 GW of new electricity generating capacity added globally from 2008 to 2009, about 140 GW came from RE additions. Collectively, developing countries host 53% of global RE electricity generation capacity (including all sizes of hydropower), with China adding more RE power capacity than any other country in 2009. The USA and Brazil accounted for 54 and 35% of global bioethanol production in 2009, respectively, while China led in the use of solar hot water. At the end of 2009, the use of RE in hot water/heating

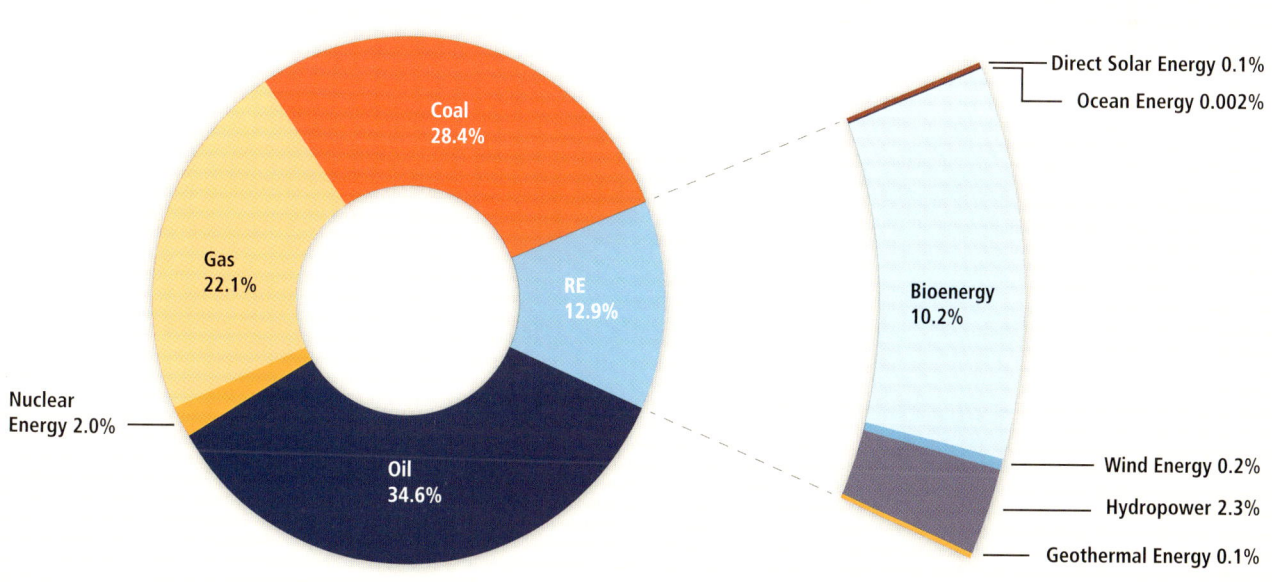

Figure TS.1.3 | Shares of energy sources in total global total primary energy supply in 2008 (492 EJ). Modern biomass contributes 38% of the total biomass share. [Figure 1.10]

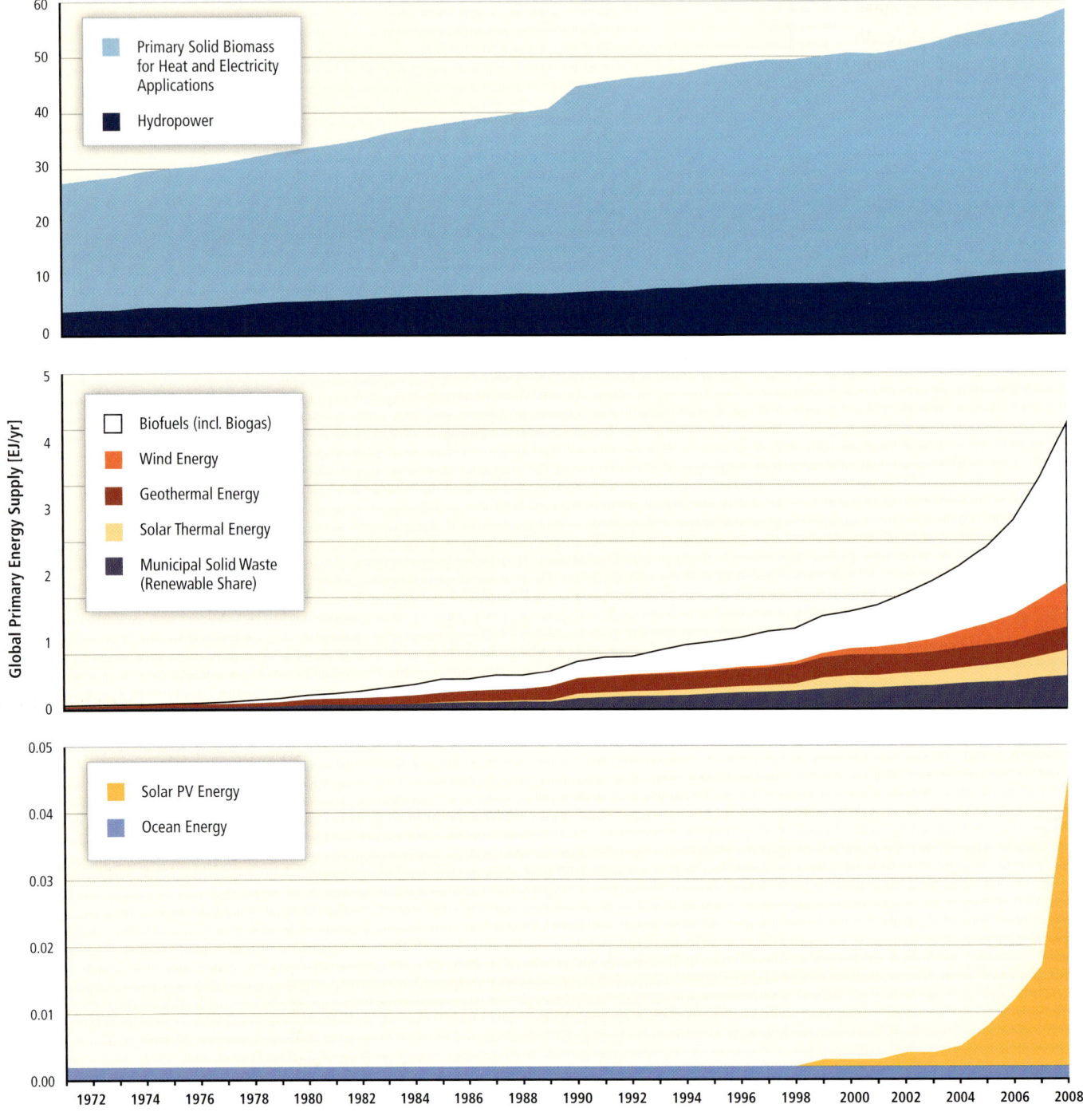

Figure TS.1.4 | Historical development of global primary energy supply from renewable energy from 1971 to 2008. [Figure 1.12]

Note: Technologies are referenced to separate vertical units for display purposes only. Underlying data for the figure has been converted to the 'direct equivalent' method of accounting for primary energy supply [1.1.9, Annex II.4], except that the energy content of biofuels is reported in secondary energy terms (the primary biomass used to produce the biofuel would be higher due to conversion losses [2.3, 2.4]).

markets included modern biomass (270 GW$_{th}$), solar energy (180 GW$_{th}$), and geothermal energy (60 GW$_{th}$). The use of RE (excluding traditional biomass) in meeting rural energy needs has also increased, including small-scale hydropower stations, various modern biomass options, and household or village photovoltaic (PV), wind or hybrid systems that combine multiple technologies. [1.1.5]

There are multiple means for lowering GHG emissions from the energy system while still providing desired energy services. The AR4 identified a number of ways to lower heat-trapping emissions from energy sources while still providing energy services: [1.1.6]

- Improve supply side efficiency of energy conversion, transmission and distribution, including combined heat and power.

- Improve demand side efficiency in the respective sectors and applications (e.g., buildings, industrial and agricultural processes, transportation, heating, cooling and lighting).

- Shift from high-GHG energy carriers such as coal and oil to lower-GHG energy carriers such as natural gas, nuclear fuels and RE sources.

- Utilize CO_2 capture and storage (CCS) to prevent post-combustion or industrial process CO_2 from entering the atmosphere. CCS has the potential for removing CO_2 from the atmosphere when biomass is processed, for example, through combustion or fermentation.

- Change behaviour to better manage energy use or to use fewer carbon- and energy-intensive goods and services.

The future share of RE applications will heavily depend on climate change mitigation goals, the level of requested energy services and resulting energy needs as well as their relative merit within the portfolio of zero- or low-carbon technologies (Figure TS.1.5). A comprehensive evaluation of any portfolio of mitigation options would involve an evaluation of their respective mitigation potential as well as all associated risks, costs and their contribution to sustainable development. [1.1.6]

Setting a climate protection goal in terms of the admissible change in global mean temperature broadly defines a corresponding GHG concentration limit with an associated CO_2 budget and subsequent time-dependent emission trajectory, which then defines the admissible amount of freely emitting fossil fuels. The complementary contribution of zero- or low-carbon energies to the primary energy supply is influenced by the 'scale' of the requested energy services. [1.1.6]

As many low-cost options to improve overall energy efficiency are already part of the non-intervention scenarios, the *additional* opportunities to decrease energy intensity in order to mitigate climate change are limited. In order to achieve ambitious climate protection goals, energy efficiency improvements alone do not suffice, requiring additional zero- or low-carbon technologies. The contribution RE will provide within the portfolio of these low-carbon technologies heavily depends on the economic competition between these technologies, a comparison of the relative environmental burden (beyond climate change) associated with them, as well as security and societal aspects (Figure TS.1.5). [1.1.6]

The body of scientific knowledge on RE and on the possible contribution of RE towards meeting GHG mitigation goals, as compiled and assessed in this report, is substantial. Nonetheless, due in part to the site-specific nature of RE, the diversity of RE technologies, the multiple end-use energy service needs that those technologies might serve, the range of markets and regulations governing integration, and the complexity of energy system transitions, knowledge about RE and its climate mitigation potential continues to advance. Additional knowledge remains to be gained in a number of broad areas related to RE and its possible role in GHG emissions reductions: [1.1.8]

- Future cost and timing of RE deployment;
- Realizable technical potential for RE at all geographical scales;
- Technical and institutional challenges and costs of integrating diverse RE technologies into energy systems and markets;
- Comprehensive assessment of socioeconomic and environmental aspects of RE and other energy technologies;
- Opportunities for meeting the needs of developing countries with sustainable RE services; and
- Policy, institutional and financial mechanisms to enable cost-effective deployment of RE in a wide variety of contexts.

Though much is already known in each of these areas, as compiled in this report, additional research and experience would further reduce uncertainties and thus facilitate decision making related to the use of RE in the mitigation of climate change. [1.1.6]

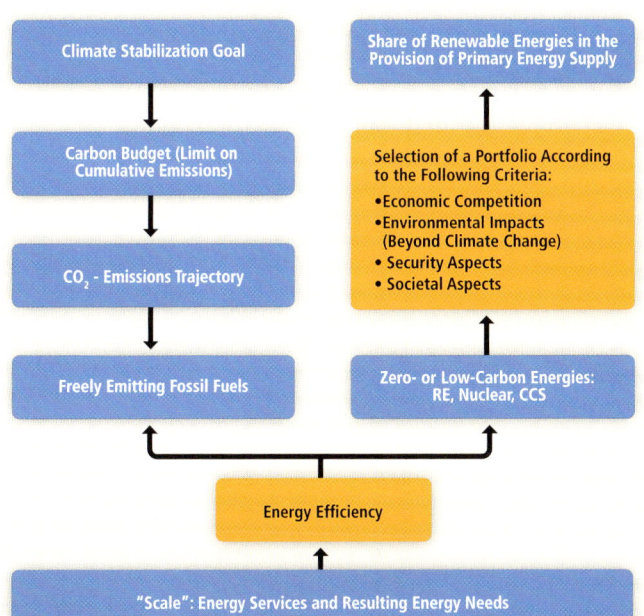

Figure TS.1.5 | The role of renewable energies within the portfolio of zero- or low-carbon mitigation options (qualitative description). [Figure 1.14]

1.2 Summary of renewable energy resources and potential

RE is any form of energy from solar, geophysical or biological sources that is replenished by natural processes at a rate that equals or exceeds its rate of use. RE is obtained from the continuing or repetitive flows of energy occurring in the natural environment and includes resources such as biomass, solar energy, geothermal heat, hydropower, tide and waves, ocean thermal energy and wind energy. However, it is possible to utilize biomass at a greater rate than it can grow or to draw heat from a geothermal field at a faster rate than heat flows can replenish it. On the other hand, the rate of utilization of direct solar energy has no bearing on the rate at which it reaches the Earth. Fossil fuels (coal, oil, natural gas) do not fall under this definition, as they are not replenished within a time frame that is short relative to their rate of utilization. [1.2.1]

There is a multi-step process whereby primary energy is converted into an energy carrier, and then into an energy service. RE technologies are diverse and can serve the full range of energy service needs. Various types of RE can supply electricity, thermal energy and mechanical energy, as well as produce fuels that are able to satisfy multiple energy service needs. Figure TS.1.6 illustrates the multi-step conversion processes. [1.2.1]

Since it is energy services and not energy that people need, the process should be driven in an efficient manner that requires less primary energy consumption with low-carbon technologies that minimize CO_2 emissions. Thermal conversion processes to produce electricity (including biomass and geothermal) suffer losses of approximately 40 to 90%, and losses of around 80% occur when supplying the mechanical energy needed for transport based on internal combustion engines. These conversion losses raise the share of primary energy from fossil fuels, and the primary energy required from fossil fuels to produce electricity and mechanical energy from heat. Direct energy conversions from solar PV, hydro, ocean and wind energy to electricity do not suffer thermodynamic power cycle (heat to work) losses although they do experience other conversion inefficiencies in extracting energy from natural energy flows that may also be relatively large and irreducible (chapters 2-7). [1.2.1]

Some RE technologies can be deployed at the point of use (decentralized) in rural and urban environments, whereas others are primarily employed within large (centralized) energy networks. Though many

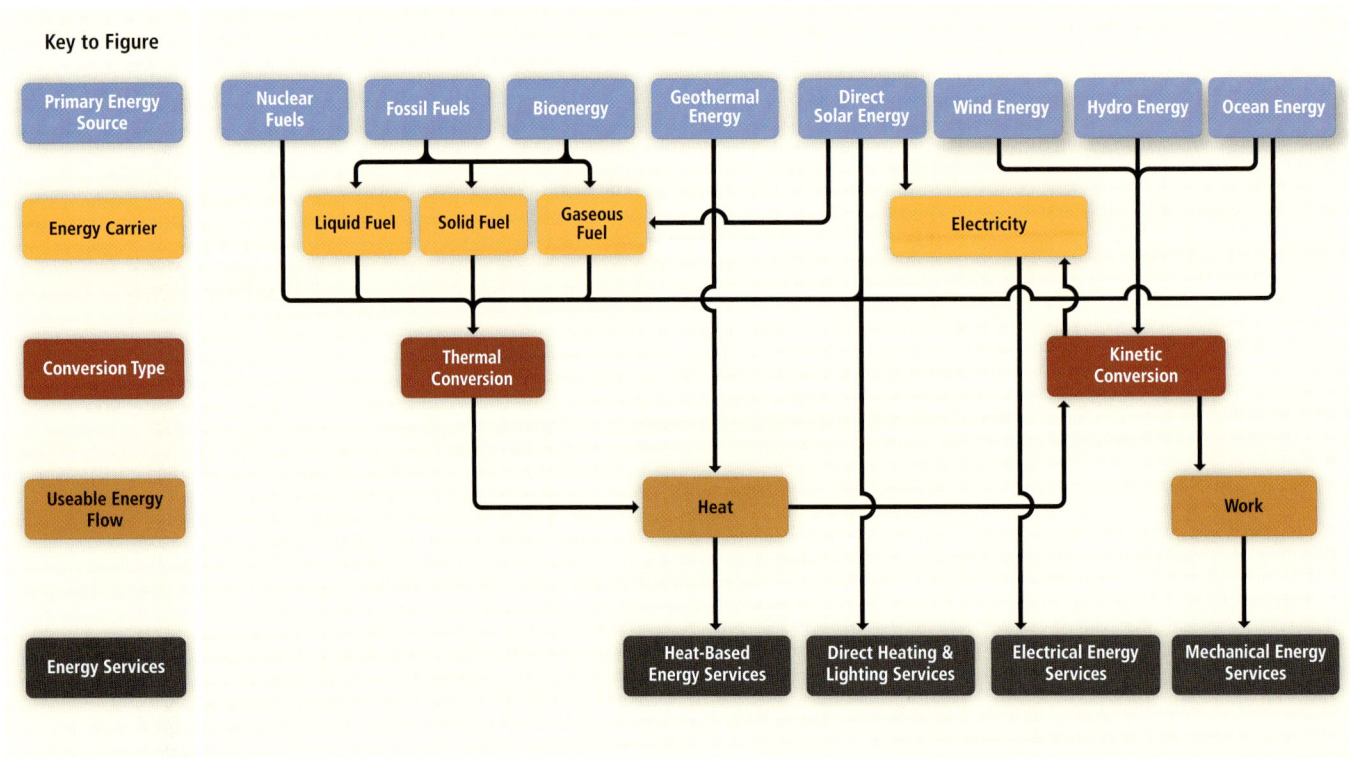

Figure TS.1.6 | Illustrative paths of energy from source to service. All connected lines indicate possible energy pathways. The energy services delivered to the users can be provided with differing amounts of end-use energy. This in turn can be provided with more or less primary energy from different sources, and with differing emissions of CO_2 and other environmental impacts. [Figure 1.16]

RE technologies are technically mature and are being deployed at significant scale, others are in an earlier phase of technical maturity and commercial deployment. [1.2.1]

The theoretical potential for RE exceeds current and projected global energy demand by far, but the challenge is to capture and utilize a sizable share of that potential to provide the desired energy services in a cost-effective and environmentally sound manner. [1.2.2]

The global technical potential of RE sources will also not limit continued market growth. A wide range of estimates are provided in the literature but studies have consistently found that the total global technical potential for RE is substantially higher than both current and projected future global energy demand. The technical potential for solar energy is the highest among the RE sources, but substantial technical potential exists for all forms of RE. The absolute size of the global technical potential for RE as a whole is unlikely to constrain RE deployment. [1.2.3]

Figure TS.1.7 shows that the technical potential[6] exceeds by a considerable margin the global electricity and heat demand, as well as the global primary energy supply, in 2008. While the figure provides a perspective for the reader to understand the relative sizes of the RE resources in the context of current energy demand and supply, note that the technical potentials are highly uncertain. Table A.1.1 in the Annex to Chapter 1 includes more detailed notes and explanations. [1.2.3]

RE can be integrated into all types of electricity systems from large, interconnected continental-scale grids down to small autonomous buildings. Whether for electricity, heating, cooling, gaseous fuels or liquid fuels, RE integration is contextual, site specific and complex. Partially dispatchable wind and solar energy can be more difficult to integrate than fully dispatchable hydropower, bioenergy and geothermal energy. As the penetration of partially dispatchable RE electricity increases, maintaining system reliability becomes more challenging and costly. A portfolio of solutions to minimize the risks and costs of RE integration can include the development of complementary flexible generation, strengthening and extending network infrastructure and interconnections, electricity demand that can respond in relation to supply availability, energy storage technologies (including reservoir-based hydropower), and modified institutional arrangements

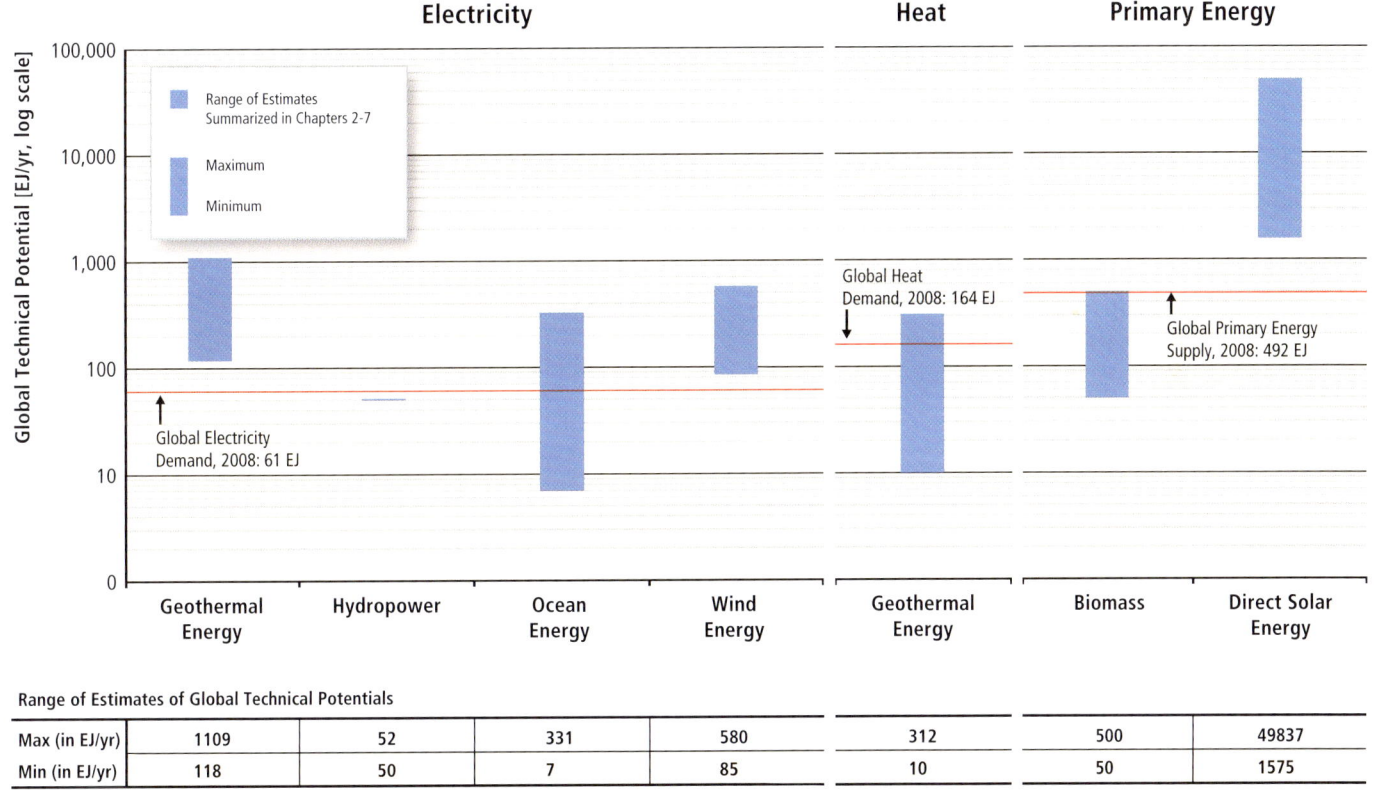

Range of Estimates of Global Technical Potentials							
	Geothermal Energy	Hydropower	Ocean Energy	Wind Energy	Geothermal Energy	Biomass	Direct Solar Energy
Max (in EJ/yr)	1109	52	331	580	312	500	49837
Min (in EJ/yr)	118	50	7	85	10	50	1575

Figure TS.1.7 | Ranges of global technical potentials of RE sources derived from studies presented in Chapters 2 through 7. Biomass and solar are shown as primary energy due to their multiple uses. Note that the figure is presented in logarithmic scale due to the wide range of assessed data. [Figure 1.17]

Notes: Technical potentials reported here represent total worldwide potentials for annual RE supply and do not deduct any potential that is already being utilized. Note that RE electricity sources could also be used for heating applications, whereas biomass and solar resources are reported only in primary energy terms but could be used to meet various energy service needs. Ranges are based on various methods and apply to different future years; consequently, the resulting ranges are not strictly comparable across technologies. For the data behind the figure and additional notes that apply, see Table A.1.1 (as well as the underlying chapters).

---

6   See Annex I for a complete definition of technical potential.

including regulatory and market mechanisms. As the penetration level of RE increases, there is need for a mixture of inexpensive and effective communications systems and technologies, as well as smart meters. [1.2.4]

Energy services are the tasks performed using energy. A specific energy service can be provided in many ways and may therefore be characterized by high or low energy efficiency, implying the release of relatively smaller or larger amounts of CO_2 (under a given energy mix). Reducing energy needs at the energy services delivery stage through energy efficiency is an important means of reducing primary energy demand. This is particularly important for RE sources since they usually have lower power densities than fossil or nuclear fuels. Efficiency measures are often the lowest-cost option to reducing end-use energy demand. This report provides some specific definitions for different dimensions of efficiency. [1.2.5]

Energy savings resulting from efficiency measures are not always fully realized in practice. There may be a rebound effect in which some fraction of the measure is offset because the lower total cost of energy (due to less energy use) to perform a specific energy service may lead to utilization of more energy services. It is estimated that the rebound effect is probably limited by saturation effects to between 10 and 30% for home heating and vehicle use in Organisation for Economic Co-operation and Development (OECD) countries, and is very small for more efficient appliances and water heating. An efficiency measure that is successful in lowering economy-wide energy demand, however, lowers the price of energy as well, leading in turn to a decrease in economy-wide energy costs and additional cost savings (lower energy prices and less energy use). It is expected that the rebound effect may be greater in developing countries and among poor consumers. For climate change, the main concern with any rebound effect is its influence on CO_2 emissions. [1.2.5]

Carbon leakage may also reduce the effectiveness of carbon reduction policies. If carbon reduction policies are not applied uniformly across sectors and political jurisdictions, then it may be possible for carbon emitting activities to move to a sector or country without such policies. Recent research suggests, however, that estimates of carbon leakage are too high. [1.2.5]

1.3 Meeting energy service needs and current status

Global renewable energy flows from primary energy through carriers to end uses and losses in 2008 are shown in Figure TS.1.8. [1.3.1]

Globally in 2008, around 56% of RE was used to supply heat in private households and in the public and services sector. Essentially, this refers to wood and charcoal, widely used in developing countries for cooking. On the other hand, only a small amount of RE is used in the transport sector. Electricity production accounts for 24% of the end-use consumption. Biofuels contributed 2% of global road transport fuel supply in 2008, and traditional biomass (17%), modern biomass (8%), solar thermal and geothermal energy (2%) together fuelled 27% of the total global demand for heat in 2008. [1.3.1]

While the resource is obviously large and could theoretically supply all energy needs long into the future, the levelized cost of energy for many RE technologies is currently higher than existing energy prices, though in various settings RE is already economically competitive. Ranges of recent levelized costs of energy for selected commercially available RE technologies are wide, depending on a number of factors, including, but not limited to, technology characteristics and size, regional variations in cost and performance and differing discount rates (Figure TS.1.9). [1.3.2, 2.3, 2.7, 3.8, 4.8, 5.8, 6.7, 7.8, 10.5, Annex III]

The cost of most RE technologies has declined and additional expected technical advances would result in further cost reductions. Such cost reductions as well as monetizing the external cost of energy supply would improve the relative competitiveness of RE. The same applies if market prices increase due to other reasons. [1.3.2, 2.6, 2.7, 3.7, 3.8, 4.6, 4.7, 5.3, 5.7, 5.8, 6.6, 6.7, 7.7, 7.8, 10.5]

The contribution of RE to primary energy supply varies substantially by country and region. The geographic distribution of RE manufacturing, use and export is now being diversified from the developed world to other developing regions, notably Asia including China. In terms of installed renewable power capacity, China now leads the world followed by the USA, Germany, Spain and India. RE is more evenly distributed than fossil fuels and there are countries or regions rich in specific RE resources. [1.3.3]

1.4 Opportunities, barriers, and issues

The major global energy challenges are securing energy supply to meet growing demand, providing everybody with access to energy services and curbing energy's contribution to climate change. For developing countries, especially the poorest, energy is needed to stimulate production, income generation and social development, and to reduce the serious health problems caused by the use of fuel wood, charcoal, dung and agricultural waste. For industrialized countries, the primary reasons to encourage RE include emission reductions to mitigate climate change, secure energy supply concerns and employment creation. RE can open opportunities for addressing these multiple environmental, social and economic development dimensions, including adaptation to climate change. [1.4, 1.4.1]

Some form of renewable resource is available everywhere in the world, for example, solar radiation, wind, falling water, waves, tides and stored ocean heat or heat from the Earth. Furthermore, technologies exist that can harness these forms of energy. While the opportunities [1.4.1] seem great, there are barriers [1.4.2] and issues [1.4.3] that slow the introduction of RE into modern economies. [1.4]

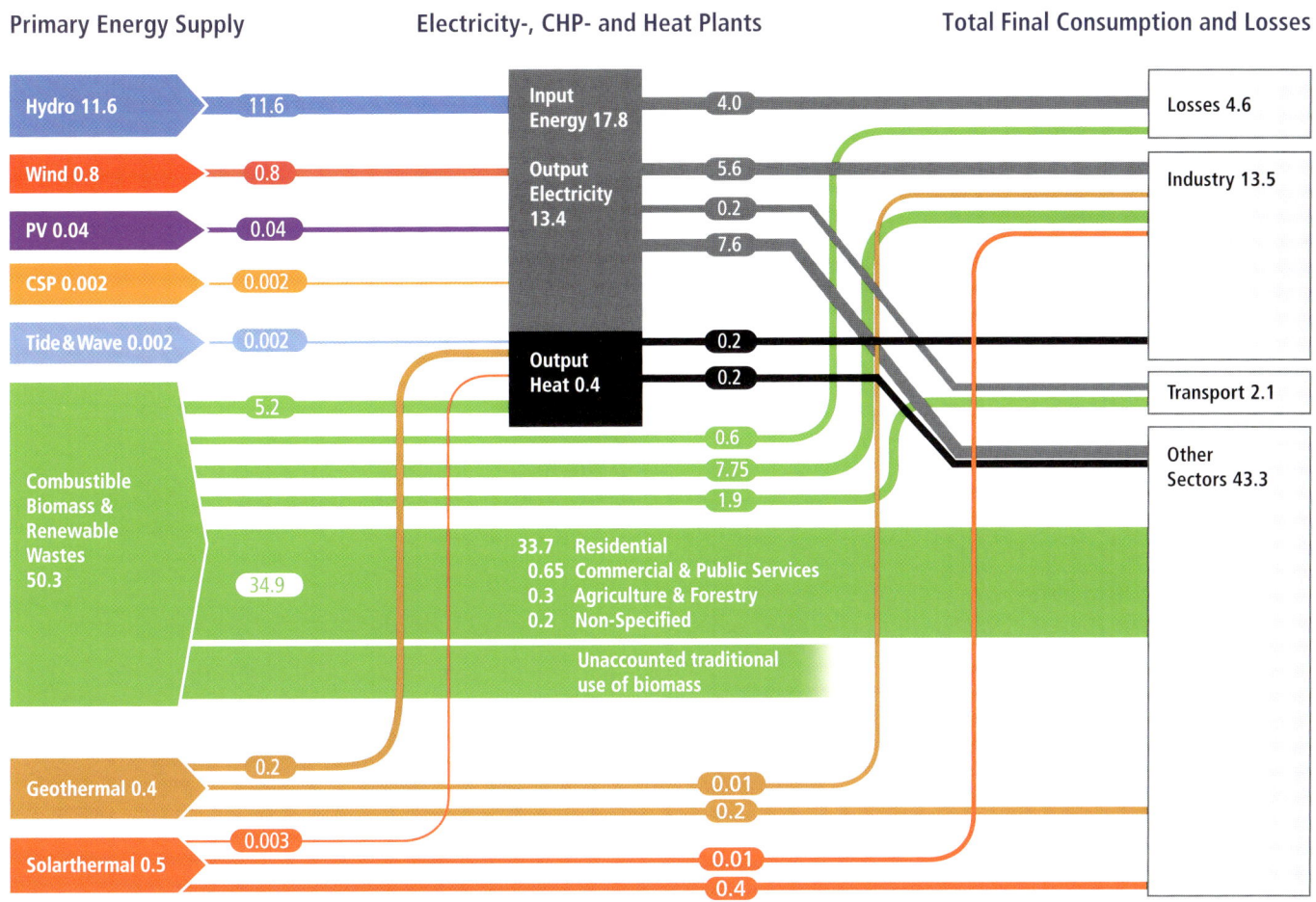

Figure TS.1.8 | Global energy flows (EJ in 2008) from primary RE through carriers to end-uses and losses (based on International Energy Agency (IEA) data). 'Other sectors' include agriculture, commercial and residential buildings, public services and non-specified other sectors. 'Transport sector' includes road transport, international aviation and international marine bunkers. [Figure 1.18]

Opportunities can be defined as circumstances for action with the attribute of a chance character. In the policy context that could be the anticipation of additional benefits that may go along with the deployment of RE but that are not intentionally targeted. These include four major opportunity areas: social and economic development; energy access; energy security; and climate change mitigation and the reduction of environmental and health impacts. [1.4.1, 9.2–9.4]

Globally, per capita incomes as well as broader indicators such as the Human Development Index (HDI) are positively correlated with per capita energy use, and economic growth can be identified as the most relevant factor behind increasing energy consumption in the last decades. Economic development has been associated with a shift from direct combustion of fuels to higher quality electricity. [1.4.1, 9.3.1]

Particularly for developing countries, the link between social and economic development and the need for modern energy services is evident. Access to clean and reliable energy constitutes an important prerequisite for fundamental determinants of human development, contributing, inter alia, to economic activity, income generation, poverty alleviation, health, education and gender equality. Due to their decentralized nature, RE technologies can play an important role in fostering rural development. The creation of (new) employment opportunities is seen as a positive long-term effect of RE in both developed and developing countries. [1.4.1, 9.3.1.4, 11.3.4]

Access to modern energy services can be enhanced by RE. In 2008, 1.4 billion people around the world lacked electricity, some 85% of them in rural areas, and the number of people relying on the traditional use of biomass for cooking is estimated to be 2.7 billion. In particular, reliance on RE in rural applications, use of locally produced bioenergy to produce electricity, and access to clean cooking facilities will contribute to attainment of universal access to modern energy services. The transition to modern energy access is referred to as moving up the energy ladder and implies a progression from traditional to more modern devices/fuels that are more environmentally benign and have fewer negative health impacts. This transition is influenced by income level. [1.4.1, 9.3.2]

Energy security concerns that may be characterized as availability and distribution of resources, as well as variability and reliability of energy supply, may also be enhanced by the deployment of RE. As RE technologies help to diversify the portfolio of energy sources and to reduce the economy's

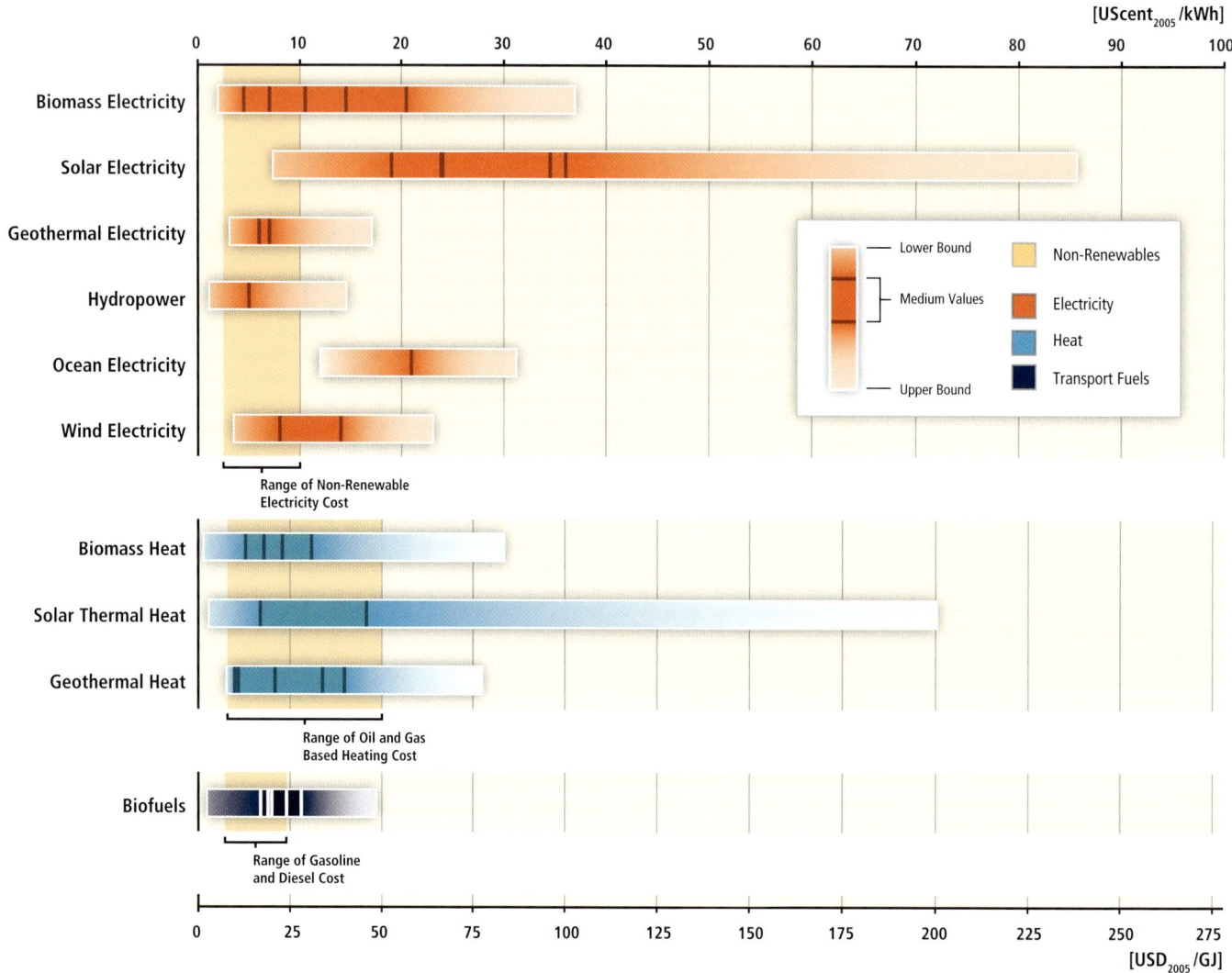

Figure TS.1.9 | (Preceding page) Range in recent levelized cost of energy for selected commercially available RE technologies in comparison to recent non-renewable energy costs. Technology subcategories and discount rates were aggregated for this figure. For related figures with less or no such aggregation, see [1.3.2, 10.5, Annex III]. Additional information concerning the cost of non-renewable energy supply options is given in [10.5]. [Figure 10.28]

vulnerability to price volatility and redirect foreign exchange flows away from energy imports, they reduce social inequities in energy supply. Current energy supplies are dominated by fossil fuels (petroleum and natural gas) whose prices have been volatile with significant implications for social, economic and environmental sustainability in the past decades, especially for developing countries and countries with high shares of imported fuels. [1.4.1, 9.2.2, 9.3.3, 9.4.3]

Climate change mitigation is one of the key driving forces behind a growing demand for RE technologies. In addition to reducing GHG emissions, RE technologies can also offer benefits with respect to air pollution and health compared to fossil fuels. However, to evaluate the overall burden from the energy system on the environment and society, and to identify potential trade-offs and synergies, environmental impacts apart from GHG emissions and categories have to be taken into account as well. The resource may also be affected by climate change. Lifecycle assessments facilitate a quantitative comparison of 'cradle to grave' emissions across different energy technologies. Figure TS.1.10 illustrates the lifecycle structure for CO_2 emission analysis, and qualitatively indicates the relative GHG implications for RE, nuclear power and fossil fuels. [1.4.1, 9.2.2, 9.3.4, 11.3.1]

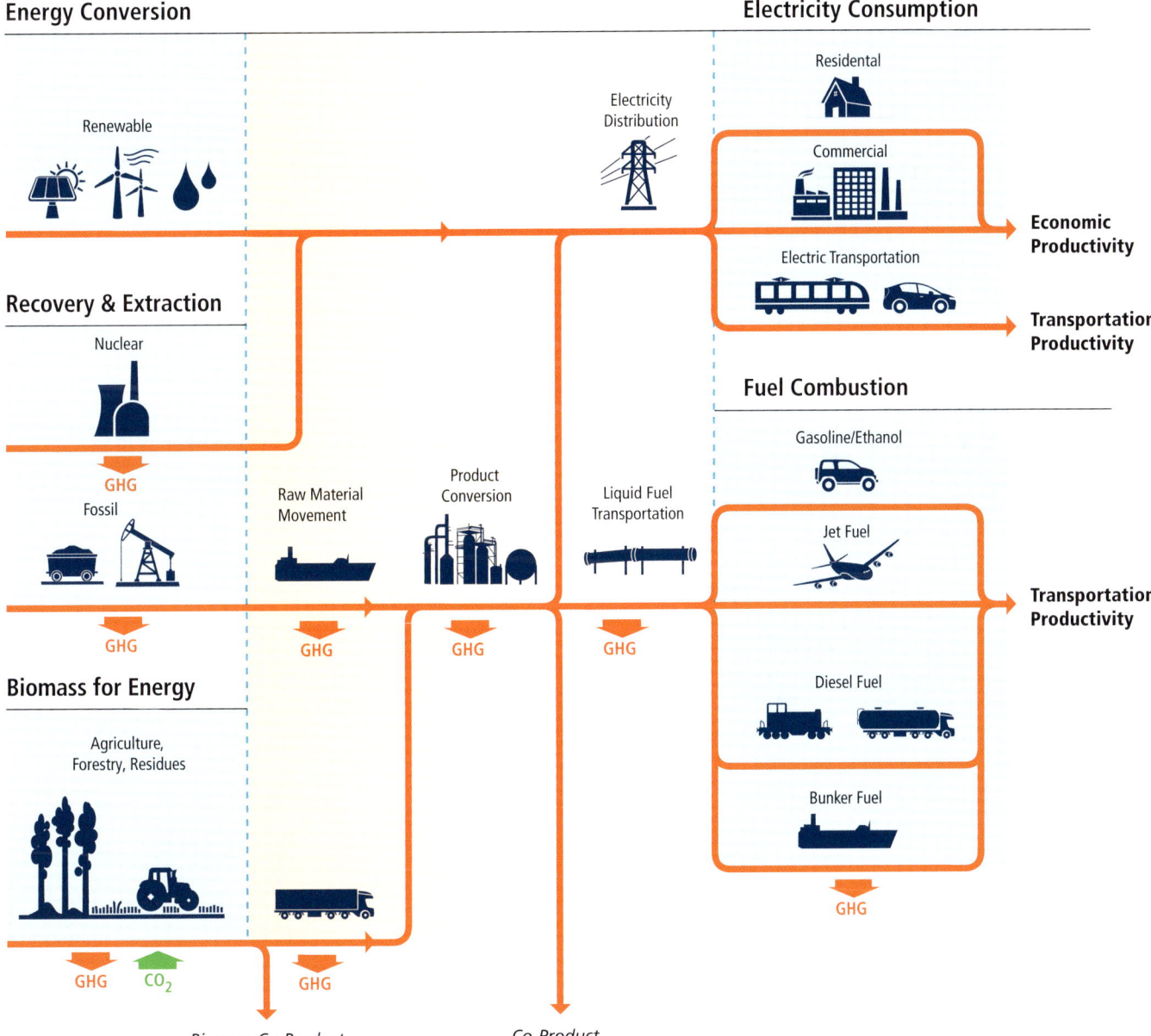

Figure TS.1.10 | Illustrative system for energy production and use illustrating the role of RE along with other production options. A systemic approach is needed to conduct lifecycle assessments. [Figure 1.22]

Traditional biomass use results in health impacts from the high concentrations of particulate matter and carbon monoxide, among other pollutants. In this context, non-combustion-based RE power generation technologies have the potential to significantly reduce local and regional air pollution and lower associated health impacts compared to fossil-based power generation. Improving traditional biomass use can reduce negative sustainable development (SD) impacts, including local and indoor air pollution, GHG emissions, deforestation and forest degradation. [1.4.1, 2.5.4, 9.3.4, 9.3.4, 9.4.2]

Impacts on water resources from energy systems strongly depend on technology choice and local conditions. Electricity production with wind and solar PV, for example, requires very little water compared to thermal conversion technologies, and has no impacts on water or air quality. Limited water availability for cooling thermal power plants decreases their efficiency, which can affect plants operating on coal, biomass, gas, nuclear and concentrating solar power. There have been significant power reductions from nuclear and coal plants during drought conditions in the USA and France in recent years. Surface-mined coal in particular produces major alterations of land; coal mines can create acid mine drainage and the storage of coal ash can contaminate surface and ground waters. Oil production and transportation have led to significant land and water spills. Most renewable technologies produce lower conventional air and water pollutants than fossil fuels, but may require large amounts of land as, for example, reservoir-based hydropower, wind and biofuels. Since a degree of climate change is now inevitable, adaptation to climate change is also an essential component of sustainable development. [1.4.1, 9.3.4]

Barriers are defined in AR4 as "any obstacle to reaching a goal, adaptation or mitigation potential that can be overcome or attenuated by a policy programme or measure". The various barriers to RE use can be categorized as market failures and economic barriers, information and awareness barriers, socio-cultural barriers and institutional and policy barriers. Policies and financing mechanisms to overcome those barriers are extensively assessed in Chapter 11. When a barrier is particularly pertinent to a specific technology, it is examined in the appropriate 'technology' chapters of this report [Chapters 2–7]. A summary of barriers and potential policy instruments to overcome these barriers is shown in Table 1.5 of Chapter 1. Market failures are often due to external effects. These arise from a human activity, when agents responsible for the activity do not take full account of the activity's impact on others. Another market failure is rent appropriation by monopolistic entities. In the case of RE deployment, these market failures may appear as underinvestment in invention and innovation in RE technologies, un-priced environmental impacts and risks of energy use as well as the occurrence of monopoly (one seller) or monopsony (one buyer) powers in energy markets. Other economic barriers include up-front investment cost and financial risks, the latter sometimes due to immaturity of the technology. [1.4.2, 1.5, 11.4]

Informational and awareness barriers include deficient data about natural resources, often due to site-specificity (e.g., local wind regimes), lack of skilled human resources (capacity) especially in rural areas of developing countries as well as the lack of public and institutional awareness. Socio-cultural barriers are intrinsically linked to societal and personal values and norms that affect the perception and acceptance of RE and may be slow to change. Institutional and policy barriers include existing industry, infrastructure and energy market regulation. Despite liberalization of energy markets in several countries in the 1990s, current industry structures are still highly concentrated and regulations governing energy businesses in many countries are still designed around monopoly or near-monopoly providers. Technical regulations and standards have evolved under the assumption that energy systems are large and centralized, and of high power density and/or high voltage. Intellectual property rights, tariffs in international trade and lack of allocation of government financial support may constitute further barriers. [1.4.2]

Issues are not readily amenable to policies and programmes. An issue is that the resource may be too small to be useful at a particular location or for a particular purpose. Some renewable resources such as wind and solar energy are variable and may not always be available for dispatch when needed. Furthermore, the energy density of many renewable sources is relatively low, so that their power levels may be insufficient on their own for some purposes such as very large-scale industrial facilities. [1.4.3]

1.5 Role of policy, research and development, deployment and implementation strategies

An increasing number and variety of RE policies—motivated by a variety of factors—have driven escalated growth in RE technologies in recent years. For policymakers wishing to support the development and deployment of RE technologies for climate change mitigation goals, it is critical to consider the potential of RE to reduce emissions from a lifecycle perspective, as addressed in each technology chapter of this report. Various policies have been designed to address every stage of the development chain involving research and development (R&D), testing, deployment, commercialization, market preparation, market penetration, maintenance and monitoring, as well as integration into the existing system. [1.4.1, 1.4.2, 9.3.4, 11.1.1, 11.2, 11.4, 11.5]

Two key market failures are typically addressed: 1) the external cost of GHG emissions are not priced at an appropriate level; and 2) deployment of low-carbon technologies such as RE create benefits to society beyond those captured by the innovator, leading to under-investment in such efforts. [1.4, 1.5, 11.1, 11.4]

Policy- and decision-makers approach the market in a variety of ways. No globally-agreed list of RE policy options or groupings exists. For the purpose of simplification, R&D and deployment policies have been organized within the following categories in this report: [1.5.1, 11.5]

- **Fiscal incentive:** actors (individuals, households, companies) are granted a reduction of their contribution to the public treasury via income or other taxes;

- **Public finance:** public support for which a financial return is expected (loans, equity) or financial liability is incurred (guarantee); and

- **Regulation:** rule to guide or control conduct of those to whom it applies.

R&D, innovation, diffusion and deployment of new low-carbon technologies create benefits to society beyond those captured by the innovator, resulting in under-investment in such efforts. Thus, government R&D can play an important role in advancing RE technologies. Public R&D investments are most effective when complemented by other policy instruments, particularly RE deployment policies that simultaneously enhance demand for new RE technologies. [1.5.1, 11.5.2]

Some policy elements have been shown to be more effective and efficient in rapidly increasing RE deployment, but there is no one-size-fits-all policy. Experience shows that different policies or combinations of policies can be more effective and efficient depending on factors such as the level of technological maturity, affordable capital, ease of integration into the existing system and the local and national RE resource base:

- Several studies have concluded that some feed-in tariffs have been effective and efficient at promoting RE electricity, mainly due to the combination of long-term fixed price or premium payments, network connections, and guaranteed purchase of all RE electricity generated. Quota policies can be effective and efficient if designed to reduce risk; for example, with long-term contracts.

- An increasing number of governments are adopting fiscal incentives for RE heating and cooling. Obligations to use RE heat are gaining attention for their potential to encourage growth independent of public financial support.

- In the transportation sector, RE fuel mandates or blending requirements are key drivers in the development of most modern biofuel industries. Other policies include direct government payments or tax reductions. Policies have influenced the development of an international biofuel and pellet trade.

One important challenge will be finding a way for RE and carbon-pricing policies to interact such that they take advantage of synergies rather than tradeoffs. In the long-term, support for technological learning in RE can help reduce costs of mitigation, and putting a price on carbon can increase the competitiveness of RE. [1.5.1, 11.1, 11.4, 11.5.7]

RE technologies can play a greater role if they are implemented in conjunction with 'enabling' policies. A favourable, or 'enabling', environment for RE can be created by addressing the possible interactions of a given policy with other RE policies as well as with other non-RE policies and the existence of an 'enabling' environment can increase the efficiency and effectiveness of policies to promote RE. Since all forms of RE capture and production involve spatial considerations, policies need to consider land use, employment, transportation, agricultural, water, food security and trade concerns, existing infrastructure and other sectoral specifics. Government policies that complement each other are more likely to be successful. [1.5.2, 11.6]

Advancing RE technologies in the electric power sector, for example, will require policies to address their integration into transmission and distribution systems both technically [Chapter 8] and institutionally [Chapter 11]. The grid must be able to handle both traditional, often more central, supply as well as modern RE supply, which is often variable and distributed. [1.5.2, 11.6.5]

In the transport sector, infrastructure needs for biofuels, recharging hydrogen, battery or hybrid electric vehicles that are 'fuelled' by the electric grid or from off-grid renewable electrical production need to be addressed.

If decision makers intend to increase the share of RE and, at the same time, to meet ambitious climate mitigation targets, then long-standing commitments and flexibility to learn from experience will be critical. To achieve international GHG concentration stabilization levels that incorporate high shares of RE, a structural shift in today's energy systems will be required over the next few decades. The available time span is restricted to a few decades and RE must develop and integrate into a system constructed in the context of an existing energy structure that is very different from what might be required under higher-penetration RE futures. [1.5.3, 11.7]

A structural shift towards a world energy system that is mainly based on RE might begin with a prominent role for energy efficiency in combination with RE. Additional policies are required that extend beyond R&D to support technology deployment; the creation of an enabling environment that includes education and awareness raising; and the systematic development of integrative policies with broader sectors, including agriculture, transportation, water management and urban planning. The appropriate and reliable mix of instruments is even more important where energy infrastructure is not yet developed and energy demand is expected to increase significantly in the future. [1.2.5, 1.5.3, 11.7, 11.6, 11.7]

2. Bioenergy

2.1 Introduction to biomass and bioenergy

Bioenergy is embedded in complex ways in global biomass systems for food, fodder and fibre production and for forest products as well as in wastes and residue management. Perhaps most importantly, bioenergy plays an intimate and critical role in the daily livelihoods of billions of people in developing countries. Figure TS.2.1 shows the types of biomass used for bioenergy in developing and developed countries. Expanding bioenergy production significantly will require sophisticated land and water use management; global feedstock productivity increases for food, fodder, fibre, forest products and energy; substantial conversion technology improvements; and a refined understanding of the complex social, energy and environmental interactions associated with bioenergy production and use.

In 2008, biomass provided about 10% (50.3 EJ/yr) of the global primary energy supply (see Table TS.2.1). Major biomass uses fall into two broad categories:

- Low-efficiency traditional biomass[7] such as wood, straws, dung and other manures are used for cooking, lighting and space heating, generally by the poorer populations in developing countries. This biomass is mostly combusted, creating serious negative impacts on health and living conditions. Increasingly, charcoal is becoming secondary energy carrier in rural areas with opportunities to create productive chains. As an indicator of the magnitude of traditional biomass use, Figure TS.2.1(b) illustrates that the global primary energy supply from traditional biomass parallels the world's industrial wood production. [2.5.4, 2.3, 2.3.2.2, 2.4.2, 2.5.7]

- High-efficiency modern bioenergy uses more convenient solids, liquids and gases as secondary energy carriers to generate heat, electricity, combined heat and power (CHP), and transport fuels for various sectors. Liquid biofuels include ethanol and biodiesel for global road transport and some industrial uses. Biomass derived gases, primarily methane, from anaerobic digestion of agricultural residues and municipal solid waste (MSW) treatment are used to generate electricity, heat or both. The most important contribution to these energy services is based on solids, such as chips, pellets, recovered wood previously used and others. Heating includes space and hot water heating such as in district heating systems. The estimated total primary biomass supply for modern bioenergy is 11.3 EJ/yr and the secondary energy delivered to end-use consumers is roughly 6.6 EJ/yr. [2.3.2, 2.4, 2.4.6, 2.6.2]

Additionally, the industry sector, such as the pulp and paper, forestry, and food industries, consumes approximately 7.7 EJ of biomass annually, primarily as a source for industrial process steam. [2.7.2, 8.3.4]

2.2 Bioenergy resource potential

The inherent complexity of biomass resources makes the assessment of their combined technical potential controversial and difficult to characterize. Estimates in the literature range from zero technical potential (no biomass available for energy production) to a maximum theoretical potential of

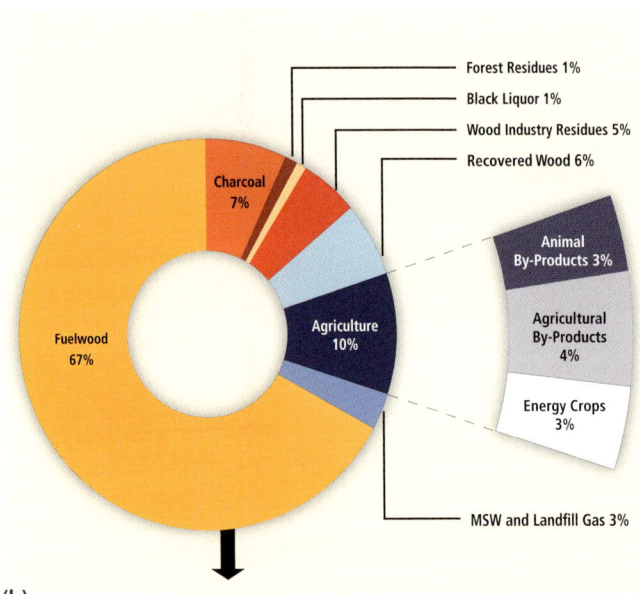

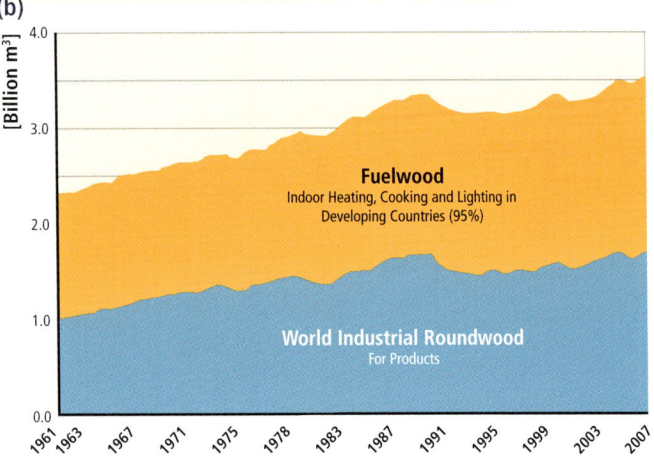

Figure TS.2.1 | (a) Shares of global primary biomass sources for energy; and (b) fuelwood used in developing countries parallels world industrial roundwood[1] production levels. [Figure 2.1]

Note: 1. Roundwood products are saw logs and veneer logs for the forest products industry and wood chips that are used for making pulpwood used in paper, newsprint and Kraft paper. In 2009, reflecting the downturn in the economy, there was a decline to 3.25 (total) and 1.25 (industrial) billion m³.

---

[7] Traditional biomass is defined as biomass consumption in the residential sector in developing countries and refers to the often unsustainable use of wood, charcoal, agricultural residues and animal dung for cooking and heating. All other biomass use is defined as modern biomass; this report further differentiates between highly efficient modern bioenergy and industrial bioenergy applications with varying degrees of efficiency. [Annex I] The renewability and sustainability of biomass use is primarily discussed in Sections 2.5.4 and 2.5.5, respectively (see also Section 1.2.1 and Annex I).

Table TS.2.1 | Examples of traditional and select modern biomass energy flows in 2008; see Table 2.1 for notes on specific flows and accounting challenges. [Table 2.1]

Type	Approximate Primary Energy (EJ/yr)	Approximate Average Efficiency (%)	Approximate Secondary Energy (EJ/yr)
Traditional Biomass			
Accounted for in IEA energy balance statistics	30.7	10–20	3–6
Estimated for informal sectors (e.g., charcoal) [2.1]	6–12		0.6–2.4
Total Traditional Biomass	37–43		3.6–8.4
Modern Bioenergy			
Electricity and CHP from biomass, MSW, and biogas	4.0	32	1.3
Heat in residential, public/commercial buildings from solid biomass and biogas	4.2	80	3.4
Road Transport Fuels (ethanol and biodiesel)	3.1	60	1.9
Total Modern Bioenergy	11.3	58	6.6

about 1,500 EJ from global modelling efforts. Figure TS.2.2 presents a summary of technical potentials found in major studies, including data from the scenario analysis of Chapter 10. To put biomass technical potential for energy in perspective, global biomass used for energy currently amounts to approximately 50 EJ/yr and all harvested biomass used for food, fodder and fibre, when expressed in a caloric equivalent, contains about 219 EJ/yr (2000 data); nearly the entire current global biomass harvest would be required to achieve a 150 EJ/yr deployment level of bioenergy by 2050. [2.2.1]

An assessment of technical potential based on an analysis of the literature available in 2007 and additional modelling studies arrived at the conclusion that the upper bound of the technical potential in 2050 could amount to about 500 EJ, shown in the stacked bar of Figure TS.2.2. The study assumes policy frameworks that secure good governance of land use and major improvements in agricultural management and takes into account water limitations, biodiversity protection, soil degradation and competition with food. Residues originating from forestry, agriculture and organic wastes (including the organic fraction of MSW, dung, process residues, etc.) are estimated to amount to 40 to 170 EJ/yr, with a mean estimate of around 100 EJ/yr. This part of the technical potential is relatively certain, but competing applications may push net availability for energy applications to the lower end of the range. Surplus forestry products other than from forestry residues have an additional technical potential

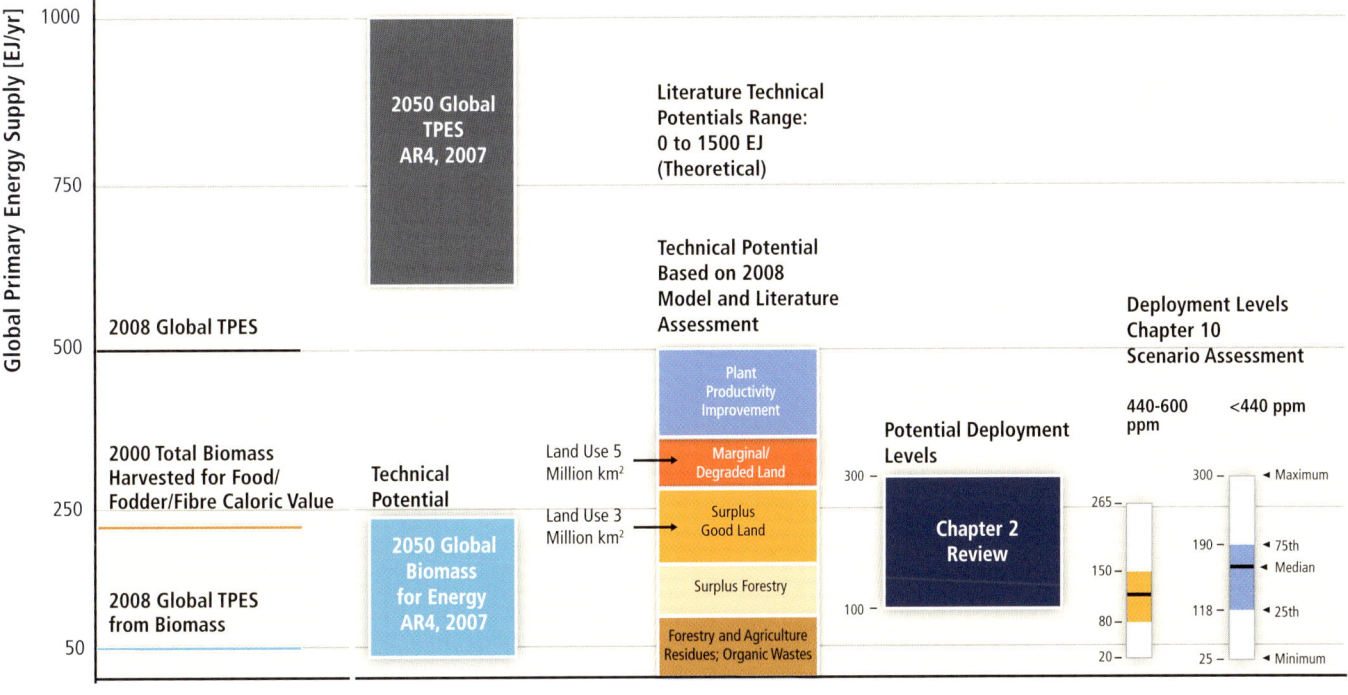

Figure TS.2.2 | A summary of major 2050 projections of global terrestrial biomass technical potential for energy and possible deployment levels compared to 2008 global total primary energy and biomass supply as well as the equivalent energy of world total biomass harvest. [Figure 2.25]

of about 60 to 100 EJ/yr. A lower estimate for energy crop production on possible surplus, good quality agricultural and pasture lands is 120 EJ/yr. The potential contribution of water-scarce, marginal and degraded lands could amount to up to an additional 70 EJ/yr. This would comprise a large area where water scarcity imposes limitations and soil degradation is more severe. Assuming strong learning in agricultural technology for improvements in agricultural and livestock management would add 140 EJ/yr. The three categories added together lead to a technical potential from this analysis of up to about 500 EJ/yr (Figure TS 2.2).

Developing this technical potential would require major policy efforts, therefore, actual deployment would likely be lower and the biomass resource base will be largely constrained to a share of the biomass residues and organic wastes, some cultivation of bioenergy crops on marginal and degraded lands, and some regions where biomass is a cheaper energy supply option compared to the main reference options (e.g., sugarcane-based ethanol production). [2.2.2, 2.2.5, 2.8.3]

The expert review conclusions based on available scientific literature are: [2.2.2–2.2.4]

- Important factors include (1) population and economic/technology development, food, fodder and fibre demand (including diets), and developments in agriculture and forestry; (2) climate change impacts on future land use including its adaptation capability; and (3) the extent of land degradation, water scarcity and biodiversity and nature conservation requirements.

- Residue flows in agriculture and forestry and unused (or extensively used thus becoming marginal/degraded) agricultural land are important sources for expansion of biomass production for energy, both in the near- and longer term. Biodiversity-induced limitations and the need to ensure maintenance of healthy ecosystems and avoidance of soil degradation set limits on residue extraction in agriculture and forestry.

- The cultivation of suitable plants (e.g., perennial crops or woody species) can allow for higher technical potentials by making it possible to produce bioenergy on lands less suited for conventional food crops—also when considering that the cultivation of conventional crops on such lands can lead to soil carbon emissions.

- Multi-functional land use systems with bioenergy production integrated into agriculture and forestry systems could contribute to biodiversity conservation and help restore/maintain soil productivity and healthy ecosystems.

- Regions experiencing water scarcity may have limited production. The possibility that conversion of lands to biomass plantations reduces downstream water availability needs to be considered. The use of suitable drought-tolerant energy crops can help adaptation in water-scarce situations. Assessments of biomass resource potentials need to more carefully consider constraints and opportunities in relation to water availability and competing uses.

Following the restrictions outlined above, the expert review concludes that potential deployment levels of biomass for energy by 2050 could be in the range of 100 to 300 EJ. However, there are large uncertainties in this potential, such as market and policy conditions, and there is strong dependence on the rate of improvements in the agricultural sector for food, fodder and fibre production and forest products. One example from the literature suggests that bioenergy can expand from around 100 EJ/yr in 2020 to 130 EJ/yr in 2030, and could reach 184 EJ/yr in 2050. [2.2.1, 2.2.2, 2.2.5]

To reach the upper range of the expert review deployment level of 300 EJ/yr (shown in Figure TS.2.2) would require major policy efforts, especially targeting improvements and efficiency increases in the agricultural sector and good governance, such as zoning, of land use.

2.3 Bioenergy technology and applications

Commercial bioenergy technology applications include heat production—with scales ranging from home cooking with stoves to large district heating systems; power generation from biomass via combustion, CHP, or co-firing of biomass and fossil fuels; and first-generation liquid biofuels from oil crops (biodiesel) and sugar and starch crops (ethanol) as shown in the solid lines of Figure TS.2.3. The figure also illustrates developing feedstocks (e.g., aquatic biomass), conversion routes and products.[8] [2.3, 2.6, 2.7, 2.8]

Section 2.3 addresses key issues related to biomass production and the logistics of supplying feedstocks to the users (individuals for traditional and modern biomass, firms that use and produce secondary energy products or, increasingly, an informal sector of production and distribution of charcoal). The conversion technologies that transform biomass to convenient secondary energy carriers use thermochemical, chemical or biochemical processes, and are summarized in Sections 2.3.1–2.3.3 and 2.6.1–2.6.3. Chapter 8 addresses energy product integration with the existing and evolving energy systems. [2.3.1–2.3.3, 2.6.1–2.6.3]

2.4 Global and regional status of markets and industry deployment

A review of biomass markets and policy shows that bioenergy has seen rapid developments in recent years such as the use of modern biomass for liquid and gaseous energy carriers (an increase of 37% from 2006 to 2009). Projections from the IEA, among others, count on biomass delivering a substantial increase in the share of RE, driven in some cases by national targets. International trade in biomass and biofuels has

---

8 Biofuels produced via new processes are also called advanced or next-generation biofuels, e.g. lignocellulosic.

also become much more important over recent years, with roughly 6% (reaching levels of up to 9% in 2008) of biofuels (ethanol and biodiesel only) traded internationally and one-third of all pellet production for energy use in 2009. The latter facilitated both increased utilization of biomass in regions where supplies were constrained as well as mobilized resources from areas lacking demand. Nevertheless, many barriers remain in developing effective commodity trading of biomass and biofuels that, at the same time, meets sustainability criteria. [2.4.1, 2.4.4]

In many countries, the policy context for bioenergy and, in particular, biofuels, has changed rapidly and dramatically in recent years. The debate surrounding biomass in the food versus fuel competition, and growing concerns about other conflicts, have resulted in a strong push for the development and implementation of sustainability criteria and frameworks as well as changes in target levels and schedules for bioenergy and biofuels. Furthermore, support for advanced biorefinery and next-generation biofuel[9] options is driving bioenergy to be more sustainable. [2.4.5]

Persistent and stable policy support has been a key factor in building biomass production capacity and markets, requiring infrastructure and conversion capacity that gets more competitive over time. These conditions have led to the success of the Brazilian programme to the point that ethanol production costs are now lower than those for gasoline. Sugarcane fibre bagasse generates heat and electricity, with an energy portfolio mix that is substantially based on RE and that minimizes foreign oil imports. Sweden and Finland also have shown significant growth in renewable electricity and in management of integrated resources, which steadily resulted in innovations such as industrial symbiosis of collocated industries. The USA has been able to quickly ramp up production with alignment of national and sub-national policies for power in the 1980s to 1990s and for biofuels in the 1990s to the present, as

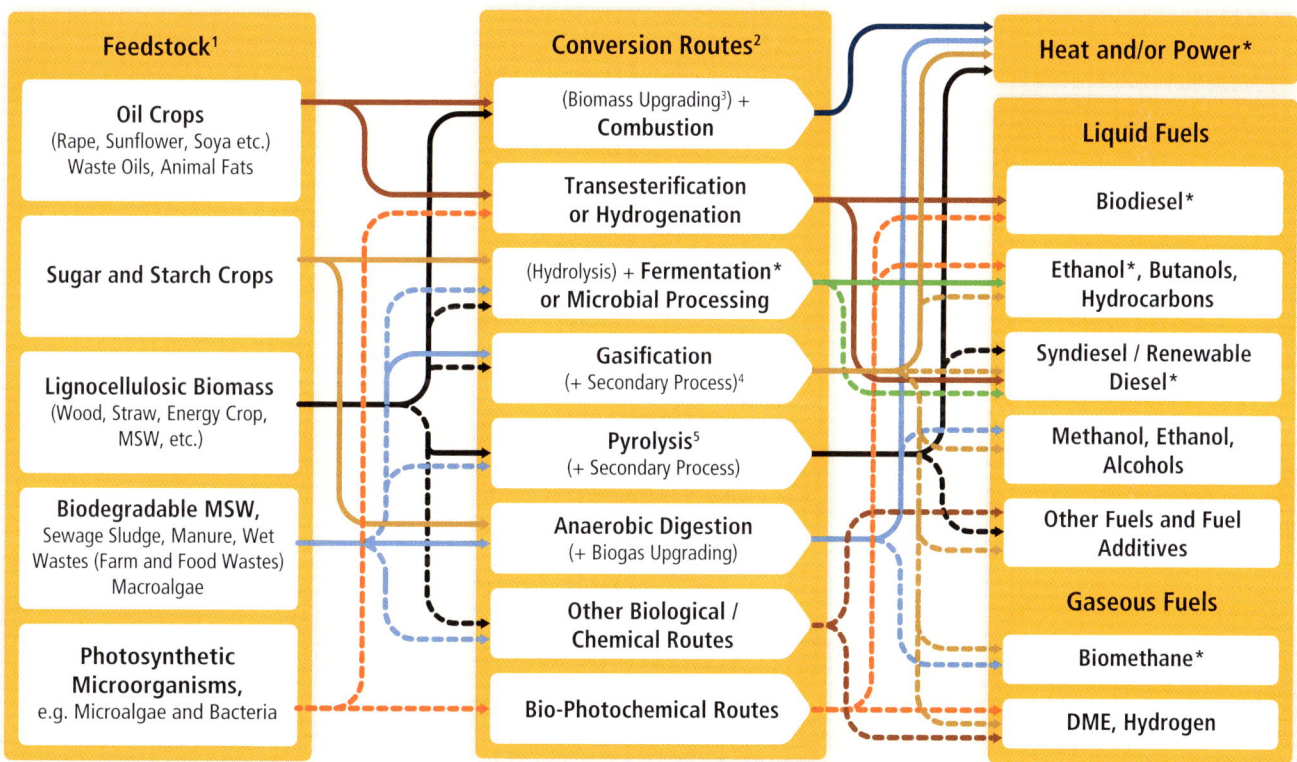

Figure TS.2.3 | Schematic view of the variety of commercial (solid lines) and developing bioenergy routes (dotted lines) from biomass feedstocks through thermochemical, chemical, biochemical and biological conversion routes to heat, power, CHP and liquid or gaseous fuels. Commercial products are marked with an asterisk. [Figure 2.2, 2.1.1]

Notes: 1. Parts of each feedstock could be used in other routes. 2. Each route can also make coproducts. 3. Biomass upgrading includes densification processes (such as pelletization, pyrolysis, torrefaction, etc.). 4. Anaerobic digestion processes to various gases which can be upgraded to biomethane, essentially methane, the major component of natural gas. 5. Could be other thermal processing routes such as hydrothermal, liquefaction, etc. Other chemical routes include aqueous phase reforming. DME=dimethyl ether.

---

[9] Biofuels produced by new processes (e.g. from lignocellulosic biomass) are also called advanced biofuels.

petroleum prices and instability in key producing countries increased and to foster rural development and a secure energy supply. [2.4.5]

Countries differ in their priorities, approaches, technology choices and support schemes for further developing bioenergy. Market and policy complexities emerge when countries seek to balance specific priorities in agriculture and land use, energy policy and security, rural development and environmental protection while considering their unique stage of development, geographic access to resources, and availability and costs of resources. [2.4.5, 2.4.7]

One overall trend is that as policies surrounding bioenergy and biofuels become more holistic, sustainability becomes a stronger criterion at the starting point. This is true for the EU, the USA and China, but also for many developing countries such as Mozambique and Tanzania. This is a positive development, but by no means settled. The registered 70 initiatives worldwide by 2009 to develop and implement sustainability frameworks and certification systems for bioenergy and biofuels, as well as agriculture and forestry, can lead to a fragmentation of efforts. The need for harmonization and international and multilateral collaboration and dialogue are widely stressed. [2.4.6, 2.4.7]

2.5 Environmental and social impacts

Bioenergy production has complex interactions with other social and environmental systems. Concerns—ranging from health and poverty to biodiversity and water scarcity and quality—vary depending upon many factors including local conditions, technology and feedstock choices, sustainability criteria design, and the design and implementation of specific projects. Perhaps most important is the overall management and governance of land use when biomass is produced for energy purposes on top of meeting food and other demands from agricultural, livestock and fibre production. [2.5]

Direct land use change (dLUC) occurs when bioenergy feedstock production modifies an existing land use, resulting in a change in above- and below-ground carbon stocks. Indirect LUC (iLUC) occurs when a change in production level of an agricultural product (i.e., a reduction in food or feed production induced by agricultural land conversion to produce a bioenergy feedstock) leads to a market-mediated shift in land management activities (i.e., dLUC) outside the region of primary production expansion. iLUC is not directly observable and is complex to model and difficult to attribute to a single cause as multiple actors, industry, countries, policies and markets dynamically interact. [2.5.3, 9.3.4.1]

In cases where increases in land use due to biomass production for bioenergy are accompanied by improvements in agricultural management (e.g., intensification of perennial crop and livestock production in degraded lands), undesirable (i)LUC effects can be avoided. If left unmanaged, conflicts can emerge. The overall performance of bioenergy production systems is therefore interlinked with management of land

and water resources use. Trade-offs between those dimensions exist and need to be managed through appropriate strategies and decision making (Figure TS.2.4). [2.5.8]

Most bioenergy systems can contribute to climate change mitigation if they replace traditional fossil fuel use and if the bioenergy production emissions are kept low. High nitrous oxide emissions from feedstock production and use of fossil fuels (especially coal) in the biomass conversion process can strongly impact the GHG savings. Options to lower GHG emissions include best practices in fertilizer management, process integration to minimize losses, utilization of surplus heat, and use of biomass or other low-carbon energy sources as process fuel. However, the displacement efficiency (GHG emissions relative to carbon in biomass) can be low when additional biomass feedstock is used for process energy in the conversion process - unless the displaced energy is generated from coal. If the biomass feedstock can produce both liquid fuel and electricity, the displacement efficiency can be high. [2.5.1–2.5.3]

There are different methods to evaluate the GHG emissions of key first- and second-generation biofuel options. Well-managed bioenergy projects can reduce GHG emissions significantly compared to fossil alternatives, especially for lignocellulosic biomass used in power generation and heat, and when that feedstock is commercially available. Advantages can be achieved by making appropriate use of agricultural residues and organic wastes, principally animal residues. Most current biofuel production systems have significant reductions in GHG emissions relative to the fossil fuels displaced, if no iLUC effects are considered. Figure TS.2.5 shows a snapshot of the ranges of lifecycle GHG emissions associated with various energy generation technologies from modern biomass compared to the respective fossil reference systems commonly used in these sectors. Commercial chains such as biomass direct power, anaerobic digestion biogas to power, and very efficient modern heating technologies are shown on the right side and provide significant GHG savings compared to the fossil fuels. More details of the GHG meta-analysis study comparing multiple biomass electricity generating technologies are available in Figure 2.11, which shows that the majority of lifecycle GHG emission estimates cluster between about 16 and 74 g CO_2eq/kWh.

The transport sector is addressed for today's and tomorrow's technologies. For light-duty vehicle applications, sugarcane today and lignocellulosic feedstocks in the medium term can provide significant emissions savings relative to gasoline. In the case of diesel, the range of GHG emissions depends on the feedstock carbon footprint. Biogas-derived biomethane also offers emission reductions (compared to natural gas) in the transport sector. [2.5.2, 9.3.4.1]

When land high in carbon (notably forests and especially drained peat soil forests) is converted to bioenergy production, upfront emissions may cause a time lag of decades to centuries before net emission savings are achieved. In contrast, the establishment of bioenergy plantations on marginal and degraded soils can lead to assimilation of CO_2 into soils

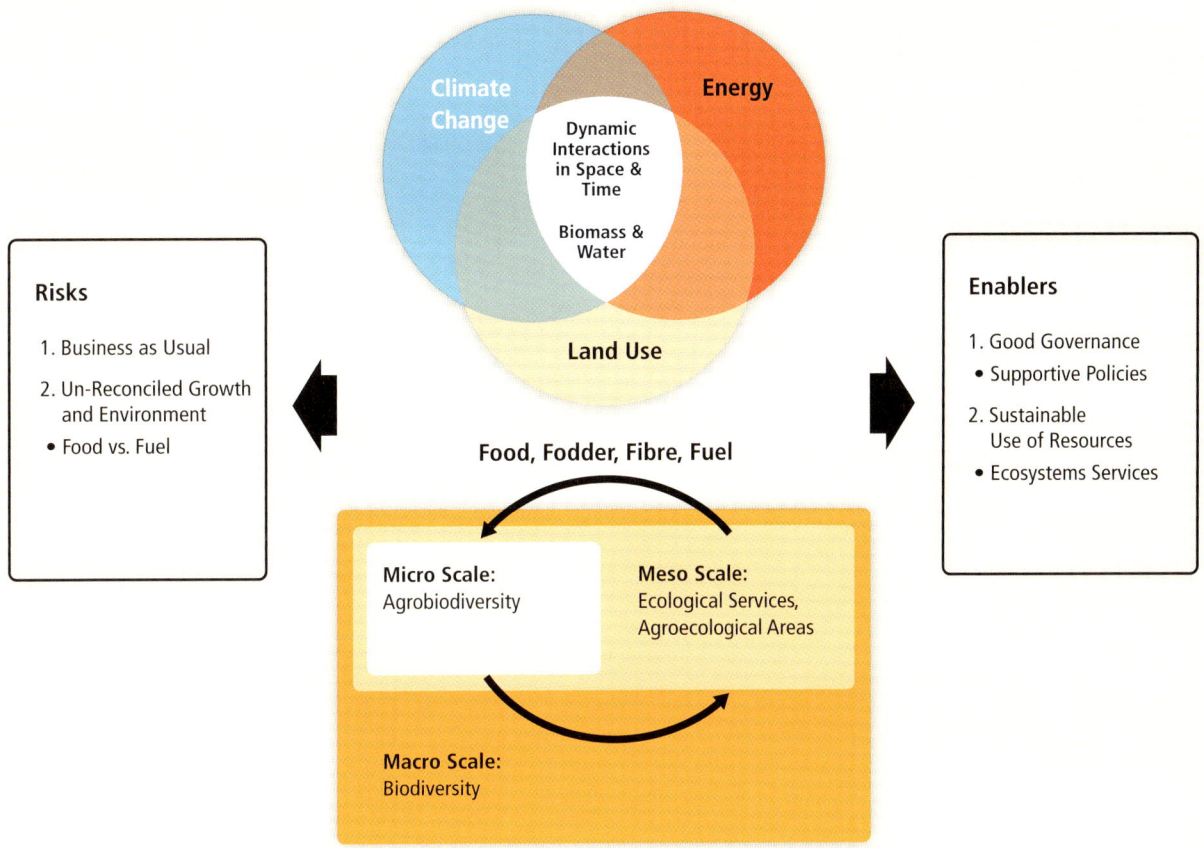

Figure TS.2.4 | The complex dynamic interactions among society, energy and the environment associated with bioenergy. Approaches of uncoordinated production of food and fuel that emerge in poor governance of land use are examples of business as usual practices. [Figure 2.15]

and aboveground biomass and when harvested for energy production it will replace fossil fuel use. Appropriate governance of land use (e.g., proper zoning) and choice of biomass production systems are crucial to achieve good performance. The use of post-consumer organic waste and by-products from the agricultural and forest industries does not cause LUC if these biomass sources were not utilized for alternative purposes. [2.5.3]

Lignocellulosic feedstocks for bioenergy can decrease the pressure on prime cropland. Stimulating increased productivity in all forms of land use reduces the LUC pressure. [2.2.4.2, 2.5.2]

The assessment of available iLUC literature indicates that initial models were lacking in geographic resolution leading to higher proportions of assignments of land use to deforestation. While a 2008 study claimed an iLUC factor of 0.8 (losing 0.8 ha of forest land for each hectare of land used for bioenergy) later (2010) studies that coupled macro-economic to biophysical models reported a reduction to 0.15 to 0.3. Major factors are the rate of improvement in agricultural and livestock management and the rate of deployment of bioenergy production. The results from increased model sophistication and improved data on the actual dynamics of land distribution in the major biofuel producing countries are leading to lower overall LUC impacts, but still with wide uncertainties. All studies acknowledge that land use management at large is a key. Research to improve LUC assessment methods and increase the availability and quality of information on current land use, bioenergy-derived products and other potential LUC drivers can facilitate evaluation and provide tools to mitigate the risk of bioenergy-induced LUC. [2.5.3, 9.3.4.1]

Air pollution effects of bioenergy depend on both the bioenergy technology (including pollution control technologies) and the displaced energy technology. Improved biomass cookstoves for traditional biomass use can provide large and cost-effective mitigation of GHG emissions with substantial co-benefits for the 2.7 billion people that rely on traditional biomass for cooking and heating in terms of health and quality of life. [2.5.4, 2.5.5]

Without proper management, increased biomass production could come with increased competition for water in critical areas, which is highly undesirable. Water is a critical issue that needs to be better analyzed at a regional level to understand the full impact of changes in vegetation and land use management. Recent studies indicate that considerable improvements can be made in water use efficiency in conventional

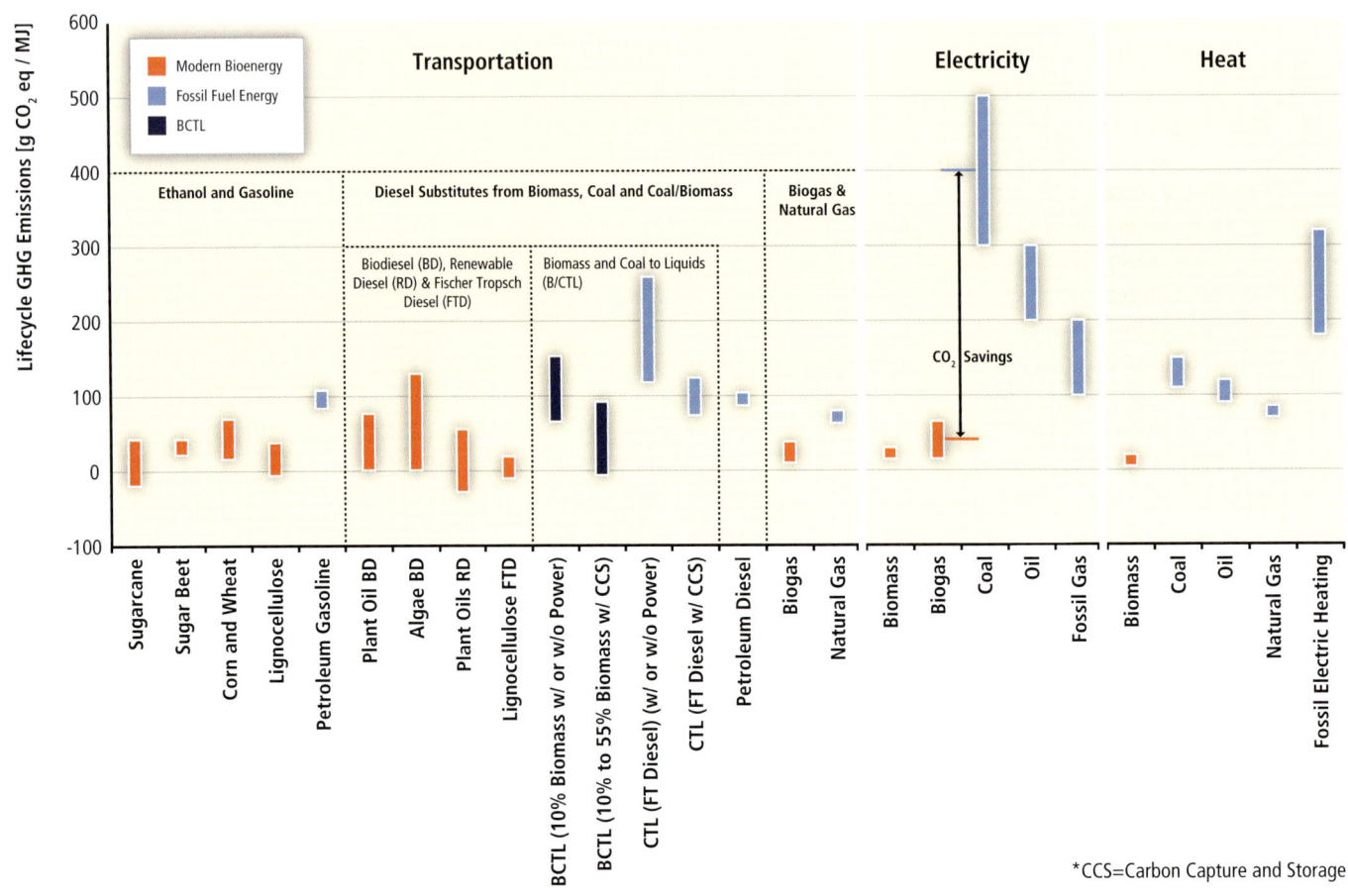

Figure TS.2.5 | Ranges of GHG emissions per unit energy output (MJ) from major modern bioenergy chains compared to current and selected advanced fossil fuel energy systems (land use-related net changes in carbon stocks and land management impacts are excluded). Commercial and developing (e.g., algae biofuels, Fischer-Tropsch) systems for biomass and fossil technologies are illustrated. When CCS technologies are developed, capture and sequestration of biomass carbon emissions can compensate fossil fuel-based energy production emissions. [Figure 2.10]

agriculture, bioenergy crops and, depending on location and climate, perennial cropping systems by improving water retention and lowering direct evaporation from soils. [2.5.5, 2.5.5.1]

Similar remarks can be made with respect to biodiversity, although more scientific uncertainty exists due to ongoing debates on methods of biodiversity impact assessment. Clearly, development of large-scale monocultures at the expense of natural areas is detrimental for biodiversity, as highlighted in the 2007 Convention on Biological Diversity. However, integrating different perennial grasses and woody crops into agricultural landscapes can also increase soil carbon and productivity, reduce shallow landslides and local 'flash floods', provide ecological corridors, reduce wind and water erosion and reduce sediment and nutrients transported into river systems. Forest biomass harvesting can improve conditions for replanting, improve productivity and growth of the remaining stand and reduce wildfire risk. [2.5.5.3]

Social impacts associated with large expansions in bioenergy production are very complex and difficult to quantify. The demand for biofuels represents one driver of demand growth in the agricultural and forestry sectors and therefore contributes to global food price increases. Even considering the benefit of increased prices to poor farmers, higher food prices adversely affect poverty levels, food security, and malnourishment of children. On the other hand, biofuels can also provide opportunities for developing countries to make progress in rural development and agricultural growth, especially when this growth is economically sustainable. In addition, expenditures on imported fossil fuels can be reduced. However, whether such benefits end up with rural farmers depends largely on the way production chains are organized and how land use is governed. [2.5.7.4–2.5.7.6, 9.3.4]

The development of sustainability frameworks and standards can reduce potential negative impacts associated with bioenergy production and lead to higher efficiency than today's systems. Bioenergy can contribute to climate change mitigation, a secure and diverse energy supply, and economic development in developed and developing countries alike, but the effects of bioenergy on environmental sustainability may be positive or negative depending upon local conditions, how criteria are defined, and how projects are designed and implemented, among many other factors. [2.4.5.2, 2.8.3, 2.5.8, 2.2.5, 9.3.4]

2.6 Prospects for technology improvement and integration

Further improvements in biomass feedstock production and conversion technologies are quite possible and necessary if bioenergy is to contribute to global energy supply to the degree reflected in the high end of deployment levels shown in Figure TS.2.2. Increasing land productivity, whether for food or energy purposes, is a crucial prerequisite for realizing large-scale future deployment of biomass for energy since it would make more land available for growing biomass and reduce the associated demand for land. In addition, multi-functional land and water use systems could develop with bioenergy and biorefineries integrated into agricultural and forestry systems, contributing to biodiversity conservation and helping to restore/maintain soil productivity and healthy ecosystems. [2.6.1]

Lignocellulosic feedstocks offer significant promise because they 1) do not compete directly with food production, 2) can be bred specifically for energy purposes, enabling higher production per unit land area and a large market for energy products, 3) can be harvested as residues from crop production and other systems that increase land use efficiency, and 4) allow the integration of waste management operations with a variety of other industries offering prospects for industrial symbiosis at the local level. Literature on and investment trends in conversion technologies indicate that the industry is poised to increase product diversification, as did the petroleum industry, with increased interest in the high energy density fuels for air transport, an application for which other non-carbon fuels have not been identified. [2.6.4]

A new generation of aquatic feedstocks that produce algal lipids for diesel, jet fuels, or higher value products from CO_2 and water with sunlight can provide strategies for lower land use impacts, as algae can grow in brackish waters, lands inappropriate for cultivation, and industrial waste water. Algal organisms can operate in the dark and metabolize sugars for fuels and chemicals. Many microbes could become microscopic factories to produce specific products, fuels and materials that decrease society's dependence on fossil energy sources. [2.6.1.2, 2.7.3]

Although significant technical progress has been made, the more complex processing required by solid lignocellulosic biomass and the integration of a number of new steps takes time and support to bring development through the 'Valley of Death' in demonstration plants, first-of-a-kind plants and early commercialization. Projected costs of biofuels from a wide range of sources and process variables are very sensitive to feedstock cost and range from USD_{2005} 10 to 30/GJ. The US National Academies project a 40% reduction in operating costs for biochemical routes by 2035 to USD_{2005} 12 to 15/GJ. [2.6.3, 2.6.4]

Biomass gasification currently provides about 1.4 GW_{th} in industrial applications, thermal applications and co-firing. Small-scale systems ranging from cooking stoves and anaerobic digestion systems to small gasifiers have been improving in efficiency over time. Many stakeholders have had a special interest in integrated gasification combined-cycle (IGCC) power plants that use bioenergy as a feedstock. These plants are projected to be more efficient than traditional steam turbine systems but have not yet reached full commercialization. However, they also have the potential to be integrated into CCS systems more effectively. In addition to providing power, syngas from gasification plants can be used to produce a wide range of fuels (methanol, ethanol, butanols and syndiesel) or can be used in a combined power and fuels approach. Technical and engineering challenges have so far prevented more rapid deployment of this technology option. Biomass to liquids conversion uses commercial technology developed for fossil fuels. Figure TS.2.5 illustrates projected emissions from coal to liquid fuels and the offsetting emissions that biomass could offer all the way to removal of GHG from the atmosphere when coupled with CCS technologies. Gaseous products (hydrogen, methane, synthetic natural gas) have lower estimated production costs and are in an early commercialization phase. [2.6.3, 2.6.4]

Pyrolysis and hydrothermal oils are low-cost transportable oils, used in heat or CHP applications and could become a feedstock for upgrading either in stand-alone facilities or coupled to a petrochemical refinery. [2.3.4, 2.6.3, 2.6.4, 2.7.1]

The production of biogas from a variety of waste streams and its upgrading to biomethane is already penetrating small markets for multiple applications, including transport in small networks in Sweden and for heat and power in Nordic and European countries. A key factor is the combination of waste streams, including agriculture residues. Improved upgrading and reducing costs is also needed. [2.6.3, 2.6.4]

Many bioenergy/biofuels routes enable CCS with significant opportunities for emissions reductions and sequestration. As CCS technologies are further developed and verified, coupling fermentation with concentrated CO_2 streams or IGCC offers opportunities to achieve carbon-neutral fuels, and in some cases negative net emissions. Achieving this goal will be facilitated by well-designed systems that span biomass selection, feedstock supply system, conversion to a secondary energy carrier and integration of this carrier into the existing and future energy systems. [2.6.3, 2.6.4, 9.3.4]

2.7 Current costs and trends

Biomass production, supply logistics, and conversion processes contribute to the cost of final products. [2.3, 2.6, 2.7]

The economics and yields of feedstocks vary widely across world regions and feedstock types with costs ranging from USD_{2005} 0.9 to 16/GJ (data from 2005 to 2007). Feedstock production for bioenergy competes with the forestry and food sectors, but integrated production systems such as agro-forestry or mixed cropping may provide synergies along with additional environmental services. Handling and transport of biomass from production sites to conversion plants may contribute 20 to up to 50% of the total costs of bioenergy production. Factors such as scale increase and technological innovations increase competition and contribute to a decrease in economic and energy costs of supply chains by more than 50%. Densification via pelletization or briquetting is required for transportation distances over 50 km. [2.3.2, 2.6.2]

Several important bioenergy systems today, most notably sugarcane-based ethanol and heat and power generation from residues and waste biomass, can be deployed competitively. [Tables 2.6, 2.7]

Based on a standardized methodology outlined in Annex II, and the cost and performance data summarized in Annex III, the estimated production costs for commercial bioenergy systems at various scales and with some consideration of geographical regions are summarized in Figure TS.2.6. Values include production, supply logistics and conversion costs. [1.3.2, 2.7.2, 10.5.1, Annex II, Annex III]

Costs vary by world regions, feedstock types, feedstock supply costs, the scale of bioenergy production, and production time during the year, which is often seasonal. Examples of estimated commercial bioenergy levelized[10] cost ranges are roughly USD_{2005} 2 to 48/GJ for liquid and gaseous biofuels; roughly 3.5 to 25 US $cents_{2005}$/kWh (USD_{2005} 10 to 50/GJ) for electricity or CHP systems larger than about 2 MW (with feedstock costs of USD_{2005} 3/GJ feed and a heat value of USD_{2005} 5/GJ for steam or USD_{2005} 12/GJ for hot water); and roughly USD_{2005} 2 to 77/GJ for domestic or district heating systems with feedstock costs in the range of USD_{2005} 0 to 20/GJ (solid waste to wood pellets). These calculations refer to 2005 to 2008 data and are in expressed USD_{2005} at a 7% discount rate. The cost ranges for biofuels in Figure TS.2.6 cover the Americas, India, China and European countries. For heating systems, the costs are primarily European and the electricity and CHP costs come from primarily large user countries. [2.3.1–2.3.3, 2.7.2, Annex III]

In the medium term, the performance of existing bioenergy technologies can still be improved considerably, while new technologies offer the prospect of more efficient and competitive deployment of biomass for energy (and materials). Bioenergy systems, namely for ethanol and biopower production, show technological learning and related cost reductions with learning rates comparable to those of other RE technologies. This applies to cropping systems (following progress in agricultural management for sugarcane and maize), supply systems and logistics (as observed in Nordic countries and international logistics) and in conversion (ethanol production, power generation and biogas) as shown in Table TS.2.2.

Although not all bioenergy options discussed in Chapter 2 have been investigated in detail with respect to technological learning, several important bioenergy systems have reduced their cost and improved environmental performance. However, they usually still require government subsidies provided for economic development (e.g., poverty reduction and a secure energy supply) and other country-specific reasons. For traditional biomass, charcoal made from biomass is a major fuel in developing countries, and should benefit from the adoption of higher-efficiency kilns. [2.3, 2.6.1, 2.6.2, 2.6.3, 2.7.2, 10.4, 10.5]

The competitive production of bio-electricity (through methane or biofuels) depends on the integration with the end-use systems, performance of alternatives such as wind and solar energy, developing CCS technologies coupled with coal conversion, and nuclear energy. The implications of successful deployment of CCS in combination with biomass conversion could result in removal of GHGs from the atmosphere and attractive mitigation cost levels but have so far received limited attention. [2.6.3.3, 8.2.1, 8.2.3, 8.2.4, 8.3, 9.3.4]

Table TS.2.3 illustrates that costs for some key bioenergy technology are expected to decline over the near- to mid-term. With respect to lignocellulosic biofuels, recent analyses have indicated that the improvement potential is large enough for competition with oil at prices of USD_{2005} 60 to 80/barrel (USD_{2005} 0.38 to 0.44/litre). Currently available scenario analyses indicate that if shorter-term R&D and market support is strong, technological progress could allow for their commercialization around 2020 (depending on oil and carbon prices). Some scenarios also indicate that this would mean a major shift in the deployment of biomass for energy, since competitive production would decouple deployment from policy targets (mandates) and demand for biomass would move away from food crops to biomass residues, forest biomass and perennial cropping systems. The implications of such a (rapid) shift are so far poorly studied. [2.8.4, 2.4.3, 2.4.5]

Lignocellulosic ethanol development and demonstration continues in several countries. A key development step is the pretreatment to overcome the recalcitrance of the cell wall of woody, herbaceous or agricultural residues to make carbohydrate polymers accessible to hydrolysis (e.g., by enzymes) and fermentation of sugars to ethanol (or butanol) and lignin for process heat or electricity. Alternatively, multiple steps can be combined and bio-processed with multiple organisms simultaneously. A review of progress in the enzymatic area suggests that a 40% reduction in cost could be expected by 2030 from process improvements, which would bring down the estimated cost of production from USD_{2005} 18 to 22/GJ (pilot data) to USD 12 to 15/GJ, a competitive range. [2.6.3]

Biomass pyrolysis routes and hydrothermal concepts are also developing in conjunction with the oil industry and have demonstrated technically that upgrading of oils to blendstocks of gasoline or diesel and even jet fuel quality products is possible. [2.6.3]

Photosynthetic organisms such as algae biologically produce (using CO_2, water and sunlight) a variety of carbohydrates and lipids that can be used directly or for biofuels. These developments have significant long-term potential because algae photosynthetic efficiency is much higher

---

10 As in the electricity production in CHP systems in which calculations assumed a value for the co-produced heat, for biofuels systems, there are cases in which two co-products are obtained; for instance, sugarcane to sugar, ethanol, and electricity. Sugar co-product revenue could be about $US\$_{2005}$ 2.6/GJ and displace the ethanol cost by that amount.

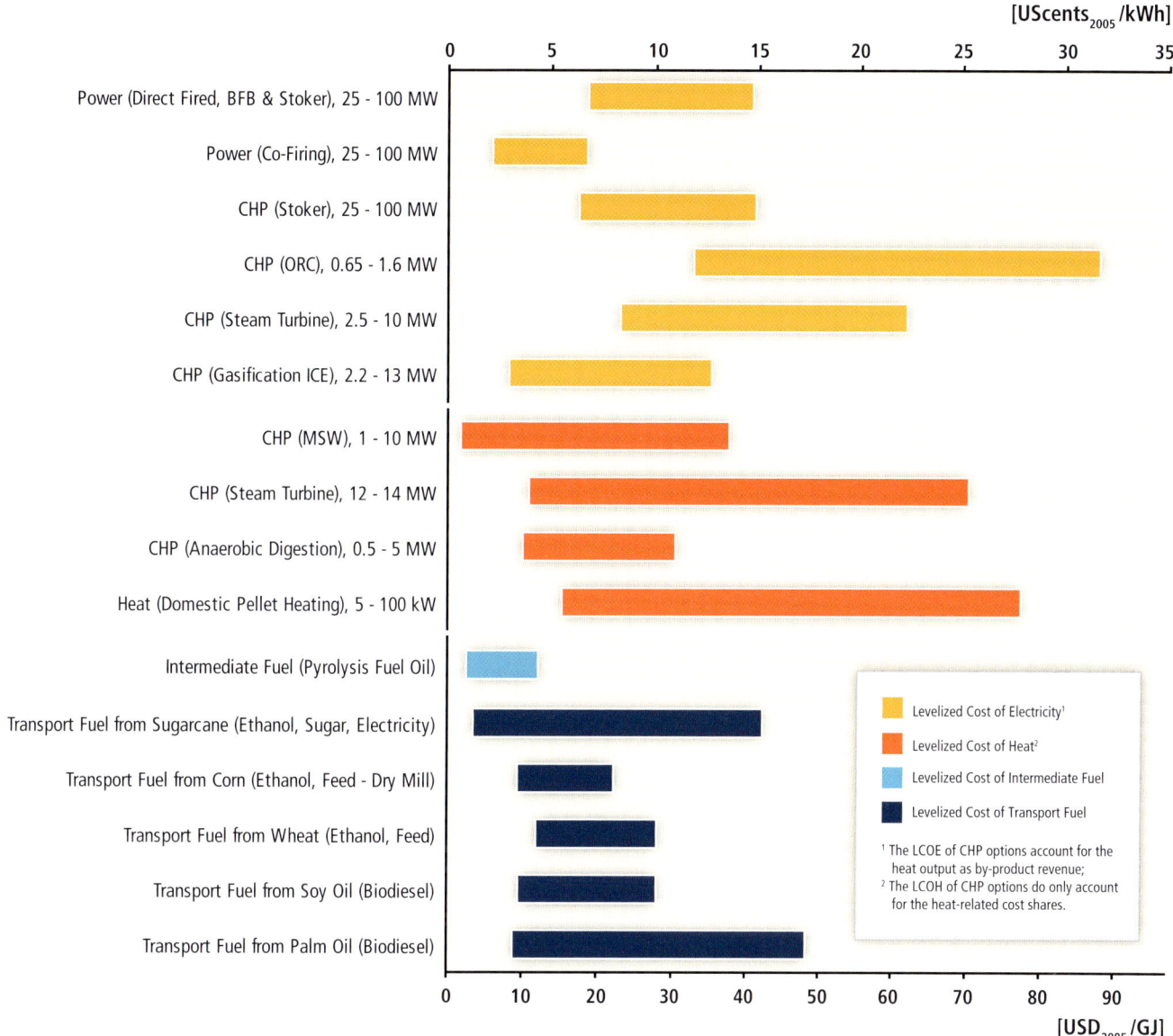

Figure TS.2.6 | Typical recent levelized cost of energy services from commercially available bioenergy systems at a 7% discount rate, calculated over a year of feedstock costs, which differ between technologies. These costs do not include interest, taxes, depreciation and amortization. [Figure 2.18] Levelized costs of electricity (LCOE), heat (LCOH), fuels (LCOF), intermediate fuel (LCOIF), BFB: Bubbling Fluidized Bed, ORC: Organic Rankine Cycle and ICE: Internal Combustion Engine. For biofuels, the range of LCOF represents production in a wide range of countries whereas LCOE and LCOH are given only for major user markets of the technologies for which data were available. Calculations are based on High Heating Value.

than that of oil crops. Potential bioenergy supplies from plants are very uncertain, but because their development can utilize brackish waters and heavily saline soils, their use is a strategy for low LUC impacts. [2.6.2, 3.3.5, 3.7.6]

Data availability is limited with respect to production of biomaterials, while cost estimates for chemicals from biomass are rare in peer-reviewed literature and future projections and learning rates even more so. This condition is linked, in part, to the fact that successful bio-based products are entering the market place either as partial components of otherwise fossil-derived products or as fully new synthetic polymers such as polylactides based on lactic acid derived from sugar fermentation. In addition to producing biomaterials to replace fossil fuels, analyses indicate that cascaded use of biomaterials and subsequent use of waste material for energy can offer more effective and larger mitigation impacts per hectare or tonne of biomass used. [2.6.3.5]

2.8 Potential deployment levels

Between 1990 and 2008, bioenergy use increased at an average annual growth rate of 1.5% for solid biomass, while the more modern biomass use for secondary carriers such as liquid and gaseous forms increased at 12.1 and 15.4% respectively. As a result, the share of biofuels in global road transport was 2% in 2008. The production of ethanol and biodiesel increased by 10 and 9%, respectively, in 2009, to 90 billion litres, such that biofuels contributed nearly 3% of global road transport in 2009, as oil demand decreased for the first time since 1980. Government

Table TS.2.2 | Experience curves for major components of bioenergy systems and final energy carriers expressed as reduction (%) in cost (or price) per doubling of cumulative production, the Learning Rate (LR); N: number of doublings of cumulative production; R2 is the correlation coefficient of the statistical data; O&M: Operations and Maintenance. [Table 2.17]

Learning system	LR (%)	Time frame	Region	N	R^2
Feedstock production					
Sugarcane (tonnes sugarcane)	32±1	1975–2005	Brazil	2.9	0.81
Corn (tonnes corn)	45±1.6	1975–2005	USA	1.6	0.87
Logistic chains					
Forest wood chips (Sweden)	15–12	1975–2003	Sweden/Finland	9	0.87–0.93
Investment and O&M costs					
CHP plants	19-25	1983–2002	Sweden	2.3	0.17–0.18
Biogas plants	12	1984–1998		6	0.69
Ethanol production from sugarcane	19±0.5	1975–2003	Brazil	4.6	0.80
Ethanol production from corn (only O&M costs)	13±0.15	1983–2005	USA	6.4	0.88
Final energy carriers					
Ethanol from sugarcane	7	1970–1985	Brazil		N/A
	29	1985–2002		~6.1	N/A
Ethanol from sugarcane	20±0.5	1975–2003	Brazil	4.6	0.84
Ethanol from corn	18±0.2	1983–2005	USA	6.4	0.96
Electricity from biomass CHP	9-8	1990–2002	Sweden	~9	0.85–0.88
Electricity from biomass	15	Unknown	OECD	N/A	N/A
Biogas	0–15	1984–2001	Denmark	~10	0.97

Table TS.2.3 | Projected production cost ranges for developing technologies. [Table 2.18]

Selected Bioenergy Technologies	Energy Sector (Electricity, Thermal, Transport)[6]	2020-2030 Projected Production Costs (USD$_{2005}$/GJ)
Integrated gasification combined cycle [1]	Electricity and/or transport	12.8–19.1 (4.6–6.9 cents/kWh)
Oil plant-based renewable diesel and jet fuel	Transport and electricity	15–30
Lignocellulose sugar-based biofuels[2]	Transport	6–30
Lignocellulose syngas-based biofuels[3]		12–25
Lignocellulose pyrolysis-based biofuels[4]		14–24 (fuel blend components)
Gaseous biofuels[5]	Thermal and transport	6–12
Aquatic plant-derived fuels, chemicals	Transport	30–140

Notes: 1. Feed cost USD$_{2005}$ 3.1/GJ, IGCC (future) 30 to 300 MW, 20-yr life, 10% discount rate. 2. Ethanol, butanols, microbial hydrocarbons and microbial hydrocarbons from sugar or starch crops or lignocellulose sugars. 3. Syndiesel, methanol and gasoline, etc.; syngas fermentation routes to ethanol. 4. Biomass pyrolysis and catalytic upgrading to gasoline and diesel blend components or to jet fuels. 5. Synfuel to synthetic natural gas, methane, dimethyl ether, hydrogen from biomass thermochemical and anaerobic digestion (larger scale). 6. Several applications can be coupled with CCS when these technologies, including CCS, are mature and thus could remove GHG from the atmosphere.

policies in various countries led to a five-fold increase in global biofuels production from 2000 to 2008. Biomass and renewable waste power generation was 259 TWh (0.93 EJ) in 2007 and 267 TWh (0.96 EJ) in 2008 representing 1% of the world's electricity and a doubling since 1990 (from 131 TWh (0.47 EJ)). [2.4]

The expected continued deployment of biomass for energy in the 2020 to 2050 time frame varies considerably between studies. A key message from the review of available insights is that large-scale biomass deployment strongly depends on sustainable development of the resource base, governance of land use, development of infrastructure and cost reduction of key technologies, for example, efficient and complete use of primary biomass for energy from the most promising first-generation feedstocks and new-generation lignocellulosic biomass. [2.4.3, 2.8]

The scenario results summarized in Figure TS.2.7 derive from a diversity of modelling teams and a wide range of assumptions including energy demand growth, cost and availability of competing low-carbon technologies, and cost and availability of RE technologies. Traditional biomass use is projected to decline in most scenarios while the use of liquid biofuels, biogas and electricity and hydrogen produced from biomass tends to increase. Results for biomass deployment for energy under these scenarios for 2020, 2030 and 2050 are presented for three GHG stabilization ranges based on the AR4: Categories III and IV (440-600 ppm CO_2), Categories I and II (<440 ppm CO_2) and Baselines (>600 ppm CO_2) all by 2100. [10.1–10.3]

Global biomass deployment for energy is projected to increase with more ambitious GHG concentration stabilization levels indicating its long-term role in reducing global GHG emissions. Median levels are 75

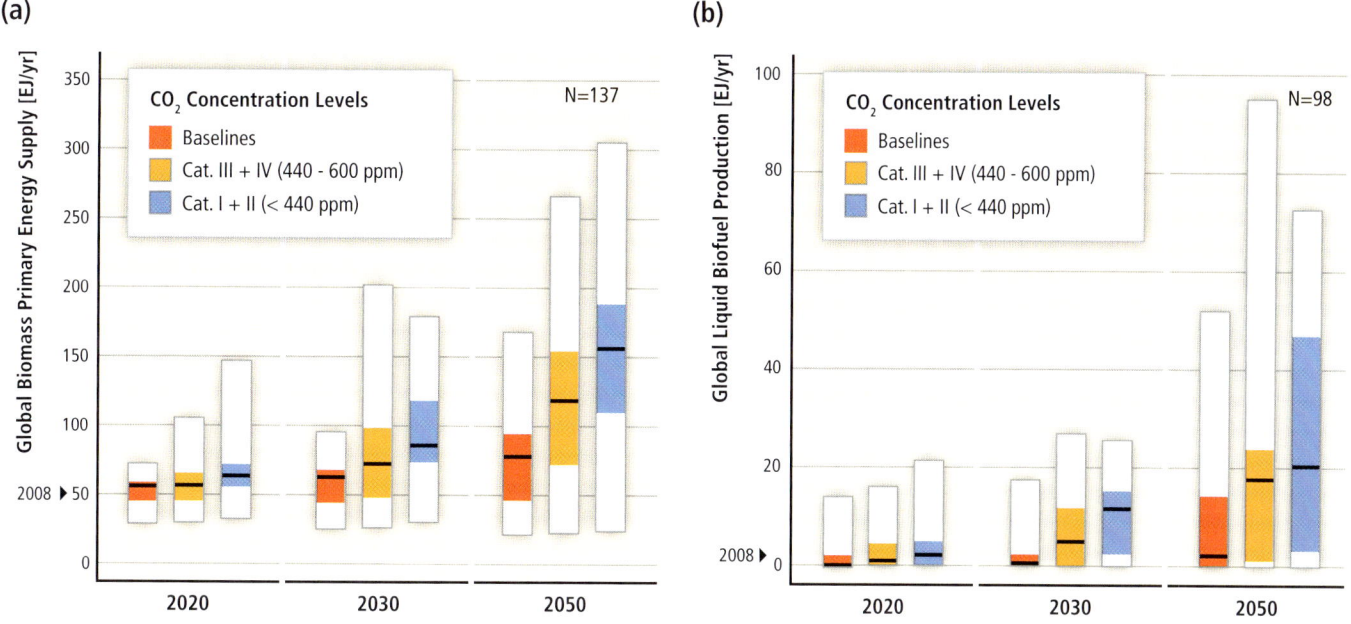

Figure TS.2.7 | (a) The global primary energy supply from biomass in long-term scenarios for electricity, heat and biofuels, all accounted for as primary energy; and (b) global biofuels production in long-term scenarios reported in secondary energy terms. For comparison, the historical levels in 2008 are indicated in the small black arrows on the left axis. [Figure 2.23]

to 85 EJ and 120 to 155 EJ for the two mitigation scenarios in 2030 and 2050, respectively, almost two and three times the 2008 deployment level of 50 EJ. These deployment levels are similar to the expert review mid-range levels for 2050. Global biofuels production shown in Figure TS.2.7(b) for 2020 and 2030 are at fairly low levels, but most models lack a detailed description of different conversion pathways and related learning potential. [2.7.3] For the <440 ppm mitigation scenario, biofuels production reaches six (2030) and ten (2050) times the 2008 actual value of 2 EJ. [2.2.5, 2.8.2, 2.5.8, 2.8.3]

The sector-level penetration of bioenergy is best explained using a single model with detailed transport sector representation such as the 2010 IEA World Energy Outlook (WEO) that also models both traditional and modern biomass applications and takes into account anticipated industrial and government investments and goals. This model projects very significant increases in modern bioenergy and a decrease in traditional biomass use. These projections are in qualitative agreement with the results from Chapter 10. In 2030, for the WEO 450-ppm mitigation scenario, the IEA projects that 11% of global transport fuels will be provided by biofuels with second-generation biofuels contributing 60% of the projected 12 EJ and half of this amount is projected to be supplied owing to continuation of current policies. Biomass and renewable wastes would supply 5% of the world's electricity generation or 1,380 TWh/yr (5 EJ/yr) of which 555 TWh/yr (2 EJ/yr) are a result of the stringent climate mitigation strategy. Biomass industrial heating applications for process steam and space and hot water heating for buildings (3.3 EJ in 2008) would each double in absolute terms from 2008 levels. However, the total heating demand is projected to decrease because of assumed traditional biomass decline. Heating is seen as a key area for continued modern bioenergy growth. Biofuels are projected to mitigate 17% of road and 3% of air transport emissions by 2030. [2.8.3]

2.8.1 Conclusions regarding deployment: Key messages about bioenergy

The long-term scenarios reviewed in Chapter 10 show increases in bioenergy supply with increasingly ambitious GHG concentration stabilization levels, indicating that bioenergy could play a significant long-term role in reducing global GHG emissions. [2.8.3]

Bioenergy is currently the largest RE source and is likely to remain one of the largest RE sources for the first half of this century. There is considerable growth potential, but it requires active development. [2.8.3]

- Assessments in the recent literature show that the technical potential of biomass for energy may be as large as 500 EJ/yr by 2050. However, large uncertainty exists about important factors such as market and policy conditions that affect this potential. [2.8.3]

- The expert assessment in Chapter 2 suggests potential deployment levels by 2050 in the range of 100 to 300 EJ/yr. Realizing this potential represents a major challenge but would make a substantial contribution to the world's primary energy demand in 2050—roughly equal to the equivalent heat content of today's worldwide biomass extraction in agriculture and forestry. [2.8.3]

- Bioenergy has significant potential to mitigate GHGs if resources are sustainably developed and efficient technologies are applied.

Certain current systems and key future options, including perennial crops, forest products and biomass residues and wastes, and advanced conversion technologies, can deliver significant GHG mitigation performance—an 80 to 90% reduction compared to the fossil energy baseline. However, land conversion and forest management that lead to a large loss of carbon stocks and iLUC effects can lessen, and in some cases more than neutralize, the net positive GHG mitigation impacts. [2.8.3]

- In order to achieve the high potential deployment levels of biomas for energy, increases in competing food and fibre demand must be moderate, land must be properly managed and agricultural and forestry yields must increase substantially. Expansion of bioenergy in the absence of monitoring and good governance of land use carries the risk of significant conflicts with respect to food supplies, water resources and biodiversity, as well as a risk of low GHG benefits. Conversely, implementation that follows effective sustainability frameworks could mitigate such conflicts and allow realization of positive outcomes, for example, in rural development, land amelioration and climate change mitigation, including opportunities to combine adaptation measures. [2.8.3]

- The impacts and performance of biomass production and use are region- and site-specific. Therefore, as part of good governance of land use and rural development, bioenergy policies need to consider regional conditions and priorities along with the agricultural (crops and livestock) and forestry sectors. Biomass resource potentials are influenced by and interact with climate change impacts but the specific impacts are still poorly understood; there will be strong regional differences in this respect. Bioenergy and new (perennial) cropping systems also offer opportunities to combine adaptation measures (e.g., soil protection, water retention and modernization of agriculture) with production of biomass resources. [2.8.3]

- Several important bioenergy options (i.e., sugarcane ethanol production in Brazil, select waste-to-energy systems, efficient biomass cookstoves, biomass-based CHP) are competitive today and can provide important synergies with longer-term options. Lignocellulosic biofuels to replace gasoline, diesel and jet fuels, advanced bio-electricity options, and biorefinery concepts can offer competitive deployment of bioenergy for the 2020 to 2030 timeframe. Combining biomass conversion with CCS raises the possibility of achieving GHG removal from the atmosphere in the long term—a necessity for substantial GHG emission reductions. Advanced biomaterials are promising as well for economics of bioenergy production and mitigation, though the potential is less well understood as is the potential role of aquatic biomass (algae), which is highly uncertain. [2.8.3]

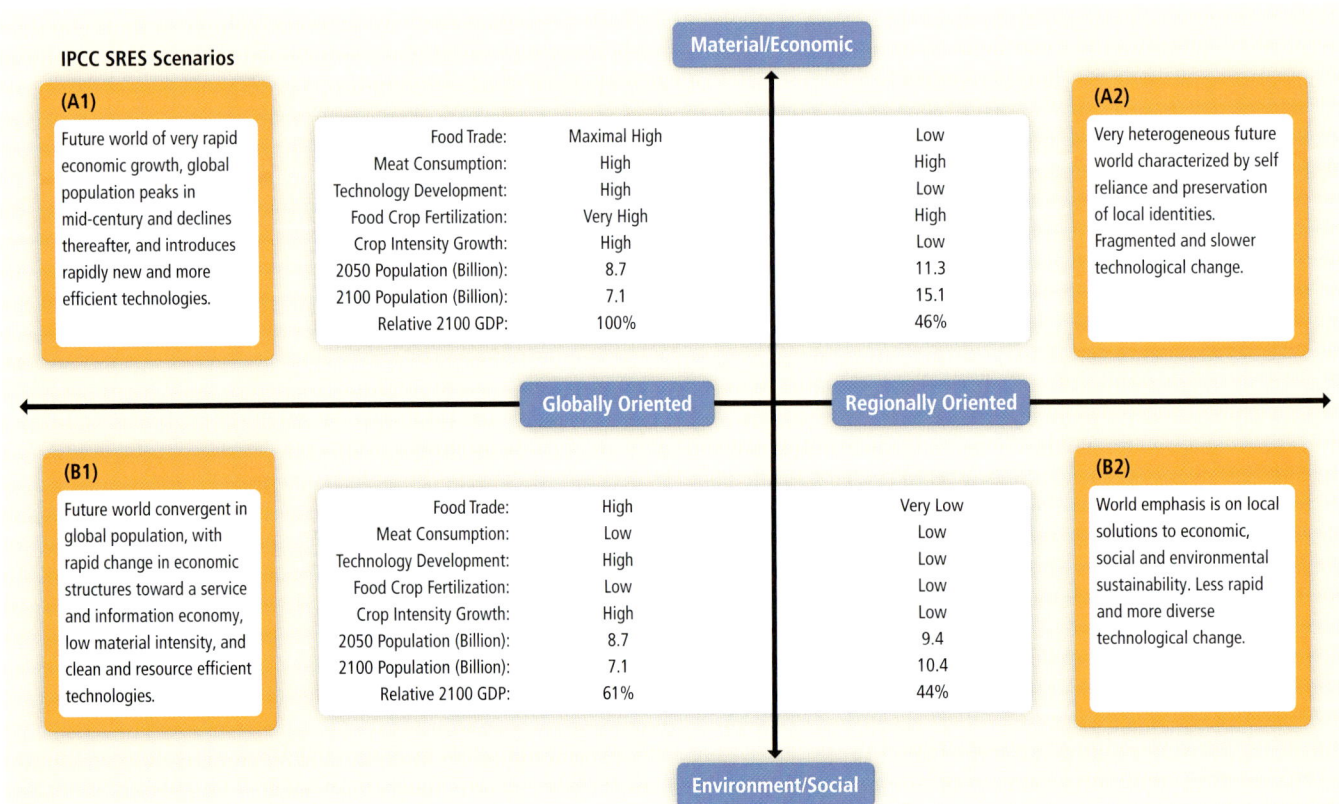

Figure TS.2.8 | Storylines for the key SRES scenario variables used to model biomass and bioenergy, the basis for the 2050 sketches adapted to this report and used to derive the stacked bar showing the biomass technical potential in Figure TS.2.2. [Figure 2.26]

- Rapidly changing policy contexts, recent market-based activities, the increasing support for advanced biorefineries and lignocellulosic biofuel options, and in particular the development of sustainability criteria and frameworks, all have the potential to drive bioenergy systems and their deployment in sustainable directions. Achieving this goal will require sustained investments that reduce costs of key technologies, improved biomass production and supply infrastructure, and implementation strategies that can gain public and political acceptance. [2.8.3]

In conclusion and for illustrating the interrelations between scenario variables (see Figure TS.2.8), key preconditions under which bioenergy production capacity is developed and what the resulting impacts may be, Figure TS.2.8 presents four different sketches for biomass deployment for energy at a global scale by 2050. The 100 to 300 EJ range that follows from the resource potential review delineates the lower and upper limit for deployment. The assumed storylines roughly follow the IPCC Special Report on Emissions Scenarios (SRES) definitions, applied to bioenergy and summarized in Figure TS.2.9 and which were also used

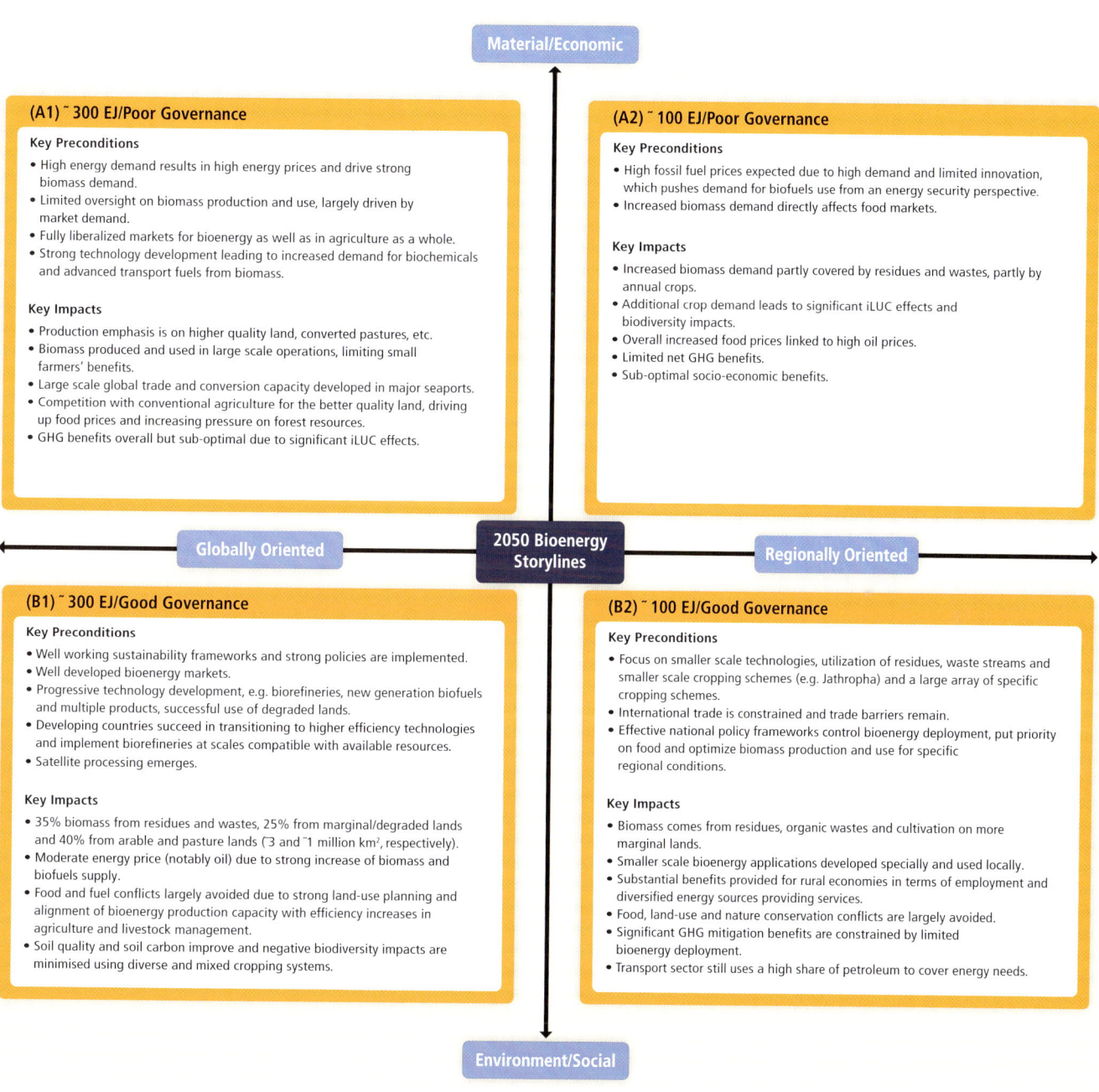

Figure TS.2.9 | Possible futures for 2050 biomass deployment for energy: Four illustrative contrasting sketches describing key preconditions and impacts following world conditions typical of the IPCC SRES storylines summarized in Figure TS.2.8. [Figure 2.27]

to derive the technical potential shown on the stacked bar of Figure TS.2.2. [2.8.3]

Biomass and its multiple energy products can be developed alongside food, fodder, fibre and forest products in both sustainable and unsustainable ways. As viewed through IPCC scenario storylines and sketches, high and low penetration levels can be reached with and without taking into account sustainable development and climate change mitigation pathways. Insights into bioenergy technology developments and integrated systems can be gleaned from these storylines. [2.8.3]

3. Direct Solar

3.1 Introduction

Direct solar energy technologies are diverse in nature. Responding to the various ways that humans use energy—such as heating, electricity, and fuels—they constitute a family of technologies. This summary focuses on four major types: 1) solar thermal, which includes both active and passive heating of buildings, domestic and commercial solar water heating, swimming pool heating and process heat for industry; 2) photovoltaic (PV) electricity generation via direct conversion of sunlight to electricity by photovoltaic cells; 3) concentrating solar power (CSP) electricity generation by optical concentration of solar energy to obtain high-temperature fluids or materials to drive heat engines and electrical generators; and 4) solar fuels production methods, which use solar energy to produce useful fuels. [3.1]

The term 'direct' solar energy refers to the energy base for those RE technologies that draw on the Sun's energy directly. Certain renewable technologies, such as wind and ocean thermal, use solar energy after it has been absorbed on the Earth and converted to other forms. (In the remainder of this section, the adjective 'direct' applied to solar energy will often be deleted as being understood.) [3.1]

3.2 Resource potential

Solar energy constitutes the thermal radiation emitted by the Sun's outer layer. Just outside Earth's atmosphere, this radiation, called solar irradiance, has a magnitude that averages 1,367 W/m² for a surface perpendicular to the Sun's rays. At ground level (generally specified as sea level with the sun directly overhead), this irradiance is attenuated by the atmosphere to about 1,000 W/m² in clear sky conditions within a few hours of noon—a condition called 'full sun'. Outside the atmosphere, the Sun's energy is carried in electromagnetic waves with wavelengths ranging from about 0.25 to 3 μm. Part of the solar irradiance is contributed by rays arriving directly from the sun without being scattered in the atmosphere. This 'beam' irradiance, which is capable of being concentrated by mirrors and lenses, is most available in low cloud-cover areas. The remaining irradiance is called the diffuse irradiance. The sum of the beam and diffuse irradiance is called global solar irradiation. [3.2]

The theoretical solar energy potential, which indicates the amount of irradiance at the Earth's surface (land and ocean) that is theoretically available for energy purposes, has been estimated at 3.9×10^6 EJ/yr. This number, clearly intended for illustrative purposes only, would require the full use of all available land and sea area at 100% conversion efficiency. A more useful metric is the technical potential; this requires assessing the fraction of land that is of practical use for conversion devices using a more realistic conversion efficiency. Estimates for solar energy's technical potential range from 1,575 to 49,837 EJ/yr, that is, roughly 3 to 100 times the world's primary energy consumption in 2008. [3.2, 3.2.2]

3.3 Technology and applications

Figure TS.3.1 illustrates the types of passive and active solar technologies currently in use to capture the Sun's energy to provide both residential energy services and direct electricity. In this summary, only technologies for active heating and electricity are treated in depth. [3.3.1–3.3.4]

Solar thermal: The key component in active solar thermal systems is the solar collector. A flat-plate solar collector consists of a blackened plate with attached conduits, through which passes a fluid to be heated. Flat-plate collectors may be classified as follows: unglazed, which are suitable for delivering heat at temperatures a few degrees above ambient temperature; glazed, which have a sheet of glass or other transparent material placed parallel to the plate and spaced a few centimetres above it, making it suitable for delivering heat at temperatures of about 30°C to 60°C; or evacuated, which are similar to glazed, but the space between the plate and the glass cover is evacuated, making this type of collector suitable for delivering heat at temperatures of about 50°C to 120°C. To withstand the vacuum, the plates of an evacuated collector are usually put inside glass tubes, which constitute both the collector's glazing and its container. In the evacuated type, a special black coating called a 'selective surface' is put on the plate to help prevent re-emission of the absorbed heat; such coatings are often used on the non-evacuated glazed type as well. Typical efficiencies of solar collectors used in their proper temperature range extend from about 40 to 70% at full sun. [3.3.2.1]

Flat-plate collectors are commonly used to heat water for domestic and commercial use, but they can also be used in active solar heating to provide comfort heat for buildings. Solar cooling can be obtained by using solar collectors to provide heat to drive an absorption refrigeration cycle. Other applications for solar-derived heat are industrial process heat, agricultural applications such as drying of crops, and for cooking. Water tanks are the most commonly used items to store heat during

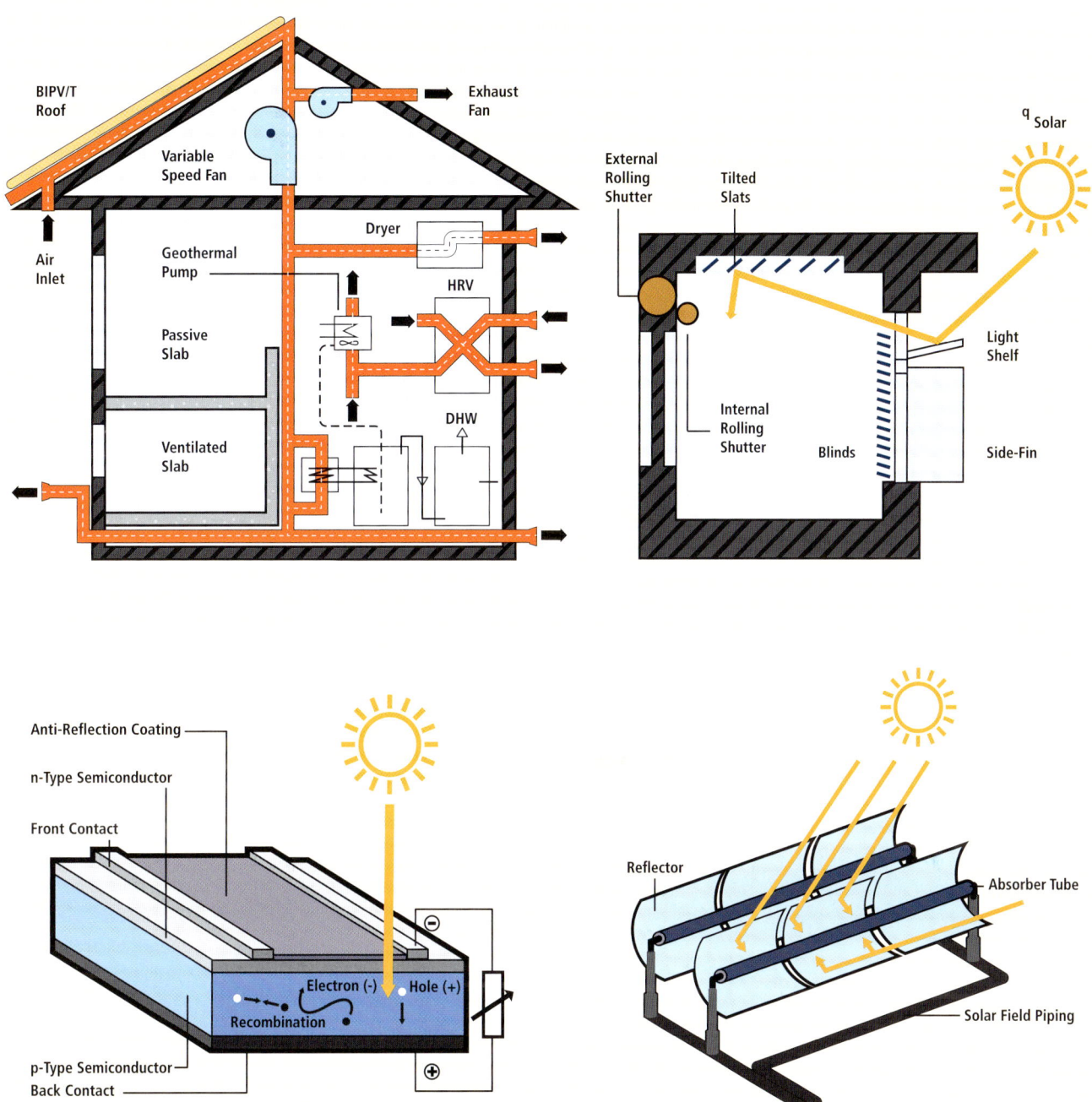

Figure TS.3.1 | Selected examples of (top) solar thermal, both passive and active integrated into a building; (bottom left) a photovoltaic device schematic for direct solar to electricity conversion; and (bottom right) one common type of concentrating solar power technology, a trough collector. [Derived from Figures 3.2, 3.5, 3.7]

the day/night period or short periods of cloudy weather. Supplemented by other energy sources, these systems typically provide 40 to 80% of the demand for heat energy of the target application. [3.3.2.2–3.3.2.4]

For passive solar heating, the building itself—particularly its windows—acts as the solar collector, and natural methods are used to distribute and store the heat. The basic elements of passive heating architecture are high-efficiency equatorial-facing windows and large internal thermal mass. The building must also be well insulated and incorporate methods such as shading devices to prevent it from overheating. Another feature of passive solar is 'daylighting', which incorporates special strategies to maximize the use of natural (solar) lighting in the building. Studies have shown that with current technology, using these strategies in new buildings in northern Europe or North America can reduce the building heating demands by as much as 40%. For existing, rather than new, buildings retrofitted with passive heating concepts, reductions of as much as 20% are achievable. [3.3.1]

Photovoltaic electricity generation: A detailed description of how PV conversion works is available in many textbooks. In the simplest terms, a thin sheet of semiconductor material such as silicon is placed in the Sun. The sheet, known as a cell, consists of two distinct layers formed by introducing impurities into the silicon resulting in an n-type layer and a p-type layer that form a junction at the interface. Solar photons striking the cell generate electron-hole pairs that are separated spatially by an internal electric field at the junction. This creates negative charges on one side of the interface and positive charges are on the other side. This resulting charge separation creates a voltage. When the two sides of the illuminated cell are connected to a load, current flows from one side of the device via the load to the other side of the cell generating electricity. [3.3.3]

Various PV technologies have been developed in parallel. Commercially available PV technologies include wafer-based crystalline silicon PV, as well as the thin-film technologies of copper indium/gallium disulfide/(di) selenide (CIGS), cadmium telluride (CdTe), thin-film silicon (amorphous and microcrystalline silicon), and dye-sensitized solar cells. In addition, there are commercially available concentrating PV concepts, in which very high efficiency cells (such as gallium arsenide (GaAs)-based materials) are placed at the focus of concentrating mirrors or other collectors such as Fresnel lenses. Mono- and multi- crystalline (sometimes called "polycrystalline") silicon wafer PV (including ribbon technologies) are the dominant technologies on the PV market, with a 2009 market share of about 80%. Peak efficiencies achieved by various cell types include more than 40% for GaAs-based concentrator cells, about 25% for monocrystalline, 20% for multicrystalline and CIGS, 17% for CdTe, and about 10% for amorphous silicon. Typically, groups of cells are mounted side by side under a transparent sheet (usually glass) and connected in series to form a 'module' with dimensions of up to 1 m by 1 m. In considering efficiencies, it is important to distinguish between cell efficiencies (quoted above) and module efficiencies; the latter are typically 50 to 80% of the former. Manufacturers continue to improve performance and reduce costs with automation, faster cell processing, and low-cost, high-throughput manufacturing. The performance of modules is typically guaranteed by manufacturers for 20 to 30 years. [3.3.3.1, 3.3.3.2]

The application of PV for useful power involves more than just the cells and modules; the PV system, for example, will often include an inverter to convert the DC power from the cells to AC power to be compatible with common networks and devices. For off-grid applications, the system may include storage devices such as batteries. Work is ongoing to make these devices more reliable, reduce their cost, and extend their lifetime to be comparable with that of the modules. [3.3.3.4]

PV power systems are classified as two major types: off-grid and grid-connected. Grid-connected systems are themselves classified into two types: distributed and centralized. The distributed system is made up of a large number of small local power plants, some of which supply the electricity mainly to an on-site customer, and the remaining electricity feeds the grid. The centralized system, on the other hand, works as one large power plant. Off-grid systems are typically dedicated to a single or small group of customers and generally require an electrical storage element or back-up power. These systems have significant potential in non-electrified areas. [3.3.3.5]

Concentrating solar power electricity generation: CSP technologies produce electricity by concentrating the Sun's rays to heat a medium that is then used (either directly or indirectly) in a heat engine process (e.g., a steam turbine) to drive an electrical generator. CSP uses only the beam component of solar irradiation, and so its maximum benefit tends to be restricted to a limited geographical range. The concentrator brings the solar rays to a point (point focus) when used in central-receiver or dish systems and to a line (line focus) when used in trough or linear Fresnel systems. (These same systems can also be used to drive thermo-chemical processes for fuel production, as described below.) In trough concentrators, long rows of parabolic reflectors that track the movement of the Sun concentrate the solar irradiation on the order of 70 to 100 times onto a heat-collection element (HCE) mounted along the reflector's focal line. The HCE comprises a blackened inner pipe (with a selective surface) and a glass outer tube, with an evacuated space between the two. In current commercial designs, a heat transfer oil is circulated through the steel pipe where it is heated (to nearly 400°C), but systems using other heat transfer materials such as circulating molten salt or direct steam are currently being demonstrated. [3.3.4]

The second kind of line-focus system, the linear Fresnel system, uses long parallel mirror strips as the concentrator, again with a fixed linear receiver. One of the two point-focus systems, the central-receiver (also called the 'power tower'), uses an array of mirrors (heliostats) on the ground, each tracking the Sun on two axes so as to focus the Sun's rays at a point on top of a tall tower. The focal point is directed onto a receiver, which comprises either a fixed inverted cavity and/or tubes in which the heat transfer fluid circulates. It can reach higher temperatures (up to 1,000°C) than the line-focus types, which allows the heat engine to convert (at least theoretically) more of the collected heat to power. In the second type of point-focus system, the dish concentrator, a single paraboloidal reflector (as opposed to an array of reflectors) tracking the sun on two axes is used for concentration. The dish focuses the solar rays onto a receiver that is not fixed, but moves with the dish, being only about one dish diameter away. Temperatures on the receiver engine can reach as high as 900°C. In one popular realization of this concept, a Stirling engine driving an electrical generator is mounted at the focus. Stirling dish units are relatively small, typically producing 10 to 25 kW, but they can be aggregated in field configuration to realize a larger central station-like power output. [3.3.4]

The four different types of CSP plants have relative advantages and disadvantages. [3.3.4] All four have been built and demonstrated. An important advantage of CSP technologies (except for dishes) is the ability to store thermal energy after it has been collected at the receiver and before going to the heat engine. Storage media considered include molten salt, pressurized air or steam accumulators (for short-term storage only), solid ceramic particles, high-temperature, phase-change materials, graphite, and high-temperature concrete. Commercial CSP plants are being built with thermal storage capacities reaching 15 hours, allowing CSP to offer dispatchable power. [3.3.4]

Solar fuel production: Solar fuel technologies convert solar energy into chemical fuels such as hydrogen, synthetic gas and liquids such as methanol and diesel. The three basic routes to solar fuels, which can work alone or in combination, are: (1) electrochemical; (2) photochemical/photo-biological; and (3) thermo-chemical. In the first route, hydrogen is produced by an electrolysis process driven by solar-derived electrical power that has been generated by a PV or CSP system. Electrolysis of water is an old and well-understood technology, typically achieving 70% conversion efficiency from electricity to hydrogen. In the second route, solar photons are used to drive photochemical or photo-biological reactions, the products of which are fuels: that is, they mimic what plants and organisms do. Alternatively, semiconductor material can be used as a solar light-absorbing anode in photoelectrochemical cells, which also generate hydrogen by water decomposition. In the third route, high-temperature solar-derived heat (such as that obtained at the receiver of a central-receiver CSP plant) is used to drive an endothermic chemical reaction that produces fuel. Here, the reactants can include combinations of water, CO_2, coal, biomass and natural gas. The products, which constitute the solar fuels, can be any (or combinations) of the following: hydrogen, syngas, methanol, dimethyl ether and synthesis oil. When a fossil fuel is used as the reactant, overall calorific values of the products will exceed those of the reactants, so that less fossil fuel needs to be burned for the same energy release. Solar fuel can also be synthesized from solar hydrogen and CO_2 to produce hydrocarbons compatible with existing energy infrastructures. [3.3.5]

3.4 Global and regional status of market and industry deployment

3.4.1 Installed capacity and generated energy

Solar thermal: Active solar heating and cooling technologies for residential and commercial buildings represent a mature market. This market, which is distributed to various degrees in most countries of the world, grew by 34.9% from 2007 to 2009 and continues to grow at a rate of about 16% per year. At the end of 2009, the global installed capacity of thermal power from these devices was estimated to be 180 GW_{th}. The global market for sales of active solar thermal systems reached an estimated 29.1 GW_{th} in 2008 and 31 GW_{th} in 2009. Glazed collectors comprise the majority of the world market. China accounted for 79% of the installation of glazed collectors in 2008, and the EU accounted for about 14.5%. In the USA and Canada, swimming pool heating is still the dominant application, with an installed capacity of 12.9 GW_{th} of unglazed plastic collectors. Notably in 2008, China led the world in installed capacity of flat-plate and evacuated-tube collectors with 88.7 GW_{th}. Europe had 20.9 GW_{th} and Japan 4.4 GW_{th}. In Europe, the market size more than tripled between 2002 and 2008. Despite these gains, solar thermal still accounts for only a relatively small portion of the demand for hot water in Europe. For example, in Germany, with the largest market, about 5% of one- and two-family homes are using solar thermal energy. One measure of the market penetration is the per capita annual usage of solar energy. The lead country in this regard is Cyprus, where the figure is 527 kW_{th} per 1,000 people. Note that there is no available information on passive solar regarding the status of its market and its deployment by industry. Consequently, the preceding numbers refer only to active solar. [3.4.1]

Photovoltaic electricity generation: In 2009, about 7.5 GW of PV systems were installed. That brought the cumulative installed PV capacity worldwide in 2009 to about 22 GW—a capacity able to generate up to 26 TWh (93,600 TJ) per year. More than 90% of this capacity is installed in three leading markets: the EU with 73% of the total, Japan with 12% and the USA with 8%. Roughly 95% of the PV installed capacity in the OECD countries is grid connected, the remainder being off-grid. Growth in the top eight PV markets through 2009 is illustrated in Figure TS.3.2. Spain and Germany have seen, by far, the largest amounts of solar installed in recent years. [3.4.1]

Concentrating solar power: CSP has reached a cumulative installed capacity of about 0.7 GW, with another 1.5 GW under construction. The capacity factors for a number of these CSP plants are expected to range from 25 to 75%; these can be higher than for PV because CSP plants contain the opportunity to add thermal storage where there is a commensurate need to overbuild the collector field to charge the thermal storage. The lower end of the capacity factor range is for no thermal storage and the upper end is for up to 15 hours of thermal storage. [3.8.4] The earliest commercial CSP plants were the Solar Electric Generating Systems in California

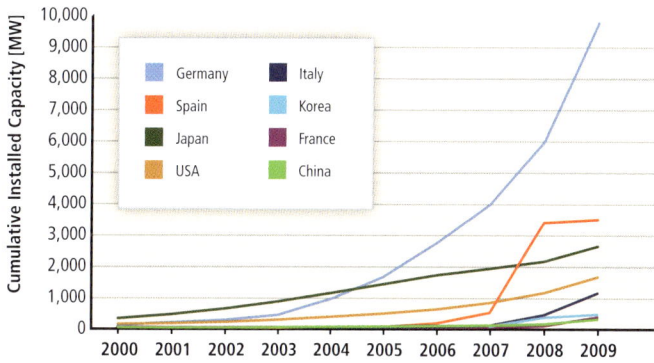

Figure TS.3.2 | Installed PV capacity for the years 2000 to 2009 in eight markets. [Figure 3.9]

capable of producing 354 MW of power; installed between 1985 and 1991, they are still operating today. The period from 1991 to the early 2000s was slow for CSP, but since about 2004, there has been strong growth in planned generation. The bulk of the current operating CSP generation consists of trough technology, but central-receiver technology comprises a growing share, and there is strong proposed commercial activity in dish-Stirling. In early 2010, most of the planned global capacity was in the USA and Spain, but recently other countries announced commercial plans. Figure TS.3.3 shows the current and planned deployment of CSP capacity through the year 2015. [3.3.4, 3.4.1]

Solar fuel production: Currently, solar fuel production is in the pilot-plant phase. Pilot plants in the power range of 300 to 500 kW have been built for the carbo-thermic reduction of zinc oxide, steam methane reforming, and steam gasification of petcoke. A 250-kW steam-reforming reactor is operating in Australia. [3.3.4, 3.4.1]

3.4.2 Industry capacity and supply chain

Solar thermal: In 2008, manufacturers produced approximately 41.5 million m² of solar collectors, a scale large enough to adapt to mass production, even though production is spread among a large number of companies around the world. Indeed, large-scale industrial production levels have been attained in most parts of the industry. In the manufacturing process, a number of readily available materials—including copper, aluminium, stainless steel, and thermal insulation—are being applied and combined through different joining technologies to produce the absorber plate. This box is topped by the cover glass, which is almost always low-iron glass, now readily available. Most production is in China, where it is aimed at internal consumption. Evacuated collectors, suitable for mass production techniques, are starting to dominate that market. Other important production sites are in Europe, Turkey, Brazil and India. Much of the export market comprises total solar water heating systems rather than solar collectors per se. The largest exporters of solar water heating systems are Australia, Greece, the USA and France. Australian exports constitute about 50% of its production. [3.4.2]

For passive solar heating, part of the industry capacity and supply chain lies in people: namely, the engineers and architects who must systematically collaborate to produce a passively heated building. Close collaboration between the two disciplines has often been lacking in the past, but the dissemination of systematic design methodologies issued by different countries has improved the design capabilities. Windows and glazing are an important part of passively heated buildings, and the availability of a new generation of high-efficiency (low-emissivity, argon-filled) windows is having a major impact on solar energy's contribution to heating requirements in the buildings sector. These windows now constitute the bulk of new windows being installed in most northern-latitude countries. There do not appear to be any issues of industrial capacity or supply chains hindering the adoption of better windows. Another feature of passive design is adding internal mass to the building's structure. Concrete and bricks, the most commonly used storage materials, are readily available; phase-change materials (e.g., paraffin), considered to be the storage materials of the future, are not expected to have supply-chain issues. [3.4.2]

Photovoltaic electricity generation: The compound annual growth rate in PV manufacturing production from 2003 to 2009 exceeded 50%. In 2009, solar cell production reached about 11.5 GW per year (rated at peak capacity) split among several economies: China had about 51% of world production (including 14% from the Chinese province of Taiwan); Europe about 18%; Japan about 14%; and the USA about 5%. Worldwide, more than 300 factories produce solar cells and modules. In 2009, silicon-based solar cells and modules represented about 80% of the worldwide market. The remaining 20% mostly comprised cadmium telluride, amorphous silicon, and copper indium gallium diselenide. The total market is expected to increase significantly during the next few years, with thin-film module production gaining market share. Manufacturers are moving towards original design of manufacturing units and are also moving components of module production closer to the final market. Between 2004 and early 2008, the demand for crystalline silicon (or polysilicon) outstripped supply, which led to a price hike. With the new price, ample supplies have become available; the PV market is now driving its own supply of polysilicon. [3.4.2]

Concentrating solar power: In the past several years, the CSP industry has experienced a resurgence from a stagnant period to more than 2 GW being either commissioned or under construction. More than 10 different companies are now active in building or preparing for commercial-scale plants. They range from start-up companies to large organizations, including utilities, with international construction

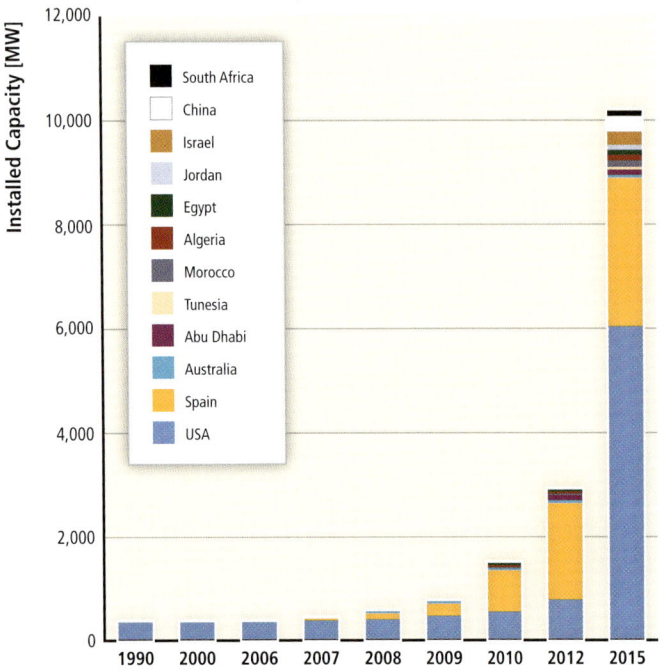

Figure TS.3.3 | Installed and planned concentrated solar power plants by country. [Figure 3.10]

management expertise. None of the supply chains for construction of plants are limited by the availability of raw material. Expanded capacity can be introduced with a lead time of about 18 months. [3.4.2]

Solar fuel production: Solar fuel technology is still at an emerging stage, and there is no supply chain in place at present for commercial applications. Solar fuels will comprise much of the same solar-field technology as is being deployed for other high-temperature CSP systems, in addition to downstream technologies similar to those in the petrochemical industry. [3.4.2]

3.4.3 Impact of policies

Direct solar energy technologies face a range of potential barriers to achieving wide-scale deployment. Solar technologies differ in levels of maturity, and although some applications are already competitive in localized markets, they generally face one common barrier: the need to reduce costs. Utility-scale CSP and PV systems face different barriers than distributed PV and solar heating and cooling technologies. Important barriers include: siting, permitting, and financing challenges to develop land with favourable solar resources for utility-scale projects; lack of access to transmission lines for large projects far from electric load centres; complex access laws, permitting procedures, and fees for smaller-scale projects; lack of consistent interconnection standards and time-varying utility rate structures that capture the value of distributed generated electricity; inconsistent standards and certifications and enforcement of these issues; and lack of regulatory structures that capture environmental and risk-mitigation benefits across technologies. Through appropriate policy designs, governments have shown that they can support solar technologies by funding R&D and by providing incentives to overcome economic barriers. Price-driven incentive frameworks, for example, were popularized after FIT policies boosted levels of PV deployment in Germany and Spain. Quota-driven frameworks such as renewable portfolio standards and government bidding are common in the USA and China, respectively. In addition to these regulatory frameworks, fiscal policies and financing mechanisms (e.g., tax credits, soft loans and grants) are often employed to support the manufacturing of solar goods and to increase consumer demand. Most successful solar policies are tailored to the barriers imposed by specific applications, and the most successful policies are those that send clear, long-term and consistent signals to the market. [3.4.3]

3.5 Integration into the broader energy system

Solar technologies have a number of attributes that allow their advantageous integration into a broader energy system. In this section, only the integration features unique to solar technologies are summarized. These include low-capacity energy demand, district heating and other thermal loads, PV generation characteristics and smoothing effects, and CSP generation characteristics and grid stabilization. [3.5.1–3.5.4]

For applications that have low power consumption, such as lighting or solar-derived hot water, solar technologies sometimes have a comparative advantage relative to non-renewable fuel technologies. In addition, solar technologies allow small decentralized applications as well as larger centralized ones. In some regions of the world, integration of solar energy into district heating and other thermal loads has proven to be an effective strategy, especially because highly insulated buildings can be heated effectively with relatively low-temperature energy carriers. In some locations, a district cooling and heating system can provide additional advantages compared to decentralized cooling, including cost advantages for economies of scale, diversity of cooling demand of different buildings, reducing noise and structural load, and equipment space savings. Also, by combining biomass and low-temperature solar thermal energy, system capacity factor and emissions profiles can be improved. [3.5.1, 3.5.2]

For PV power generation at a specific location, electricity varies systematically during a day and a year, but also randomly according to weather conditions. This variation can, in some instances, have a large impact on voltage and power flow in the local transmission and distribution system from the early penetration stage, and the supply-demand balance in total power system operation in the high-penetration stage. This effect can potentially constrain PV system integration. However, modelling and system simulations suggest that numerous PV systems in a broad area should have less-random and slower variations, which are sometimes referred to as the 'smoothing effect'. Studies are underway to evaluate and quantify actual smoothing effects at a large scale (1,000 sites at distances from 2 to 200 km) and at time scales of 1 minute or less. [3.5.3]

In a CSP plant, even without storage, the inherent thermal mass in the collector system and spinning mass in the turbine tend to significantly reduce the impact of rapid solar transients on electrical output, and thus, lead to a reduced impact on the grid. By including integrated thermal storage systems, capacity factors typical of base-load operation could be achieved in the future. In addition, integrating CSP plants with fossil fuel generators, especially with gas-fired integrated solar combined-cycle systems (with storage), can offer better fuel efficiency and extended operating hours and ultimately be more cost effective than operating separate CSP and/or combined-cycle plants. [3.5.4]

3.6 Environmental and social impacts

3.6.1 Environmental impacts

Apart from its benefits in GHG reduction, the use of solar energy can reduce the release of pollutants—such as particulates and noxious gases—from the older fossil fuel plants that it replaces. Solar thermal and PV technologies do not generate any type of solid, liquid or gaseous by-products when producing electricity. The family of solar energy technologies may create other types of air, water, land and ecosystem impacts, depending on how they are managed. The PV industry uses some toxic, explosive gases as well as corrosive liquids in its production lines. The presence and amount of those materials depend strongly on the cell type. However, the intrinsic needs of the productive process of the PV industry force the use of quite rigorous control methods that minimize the emission of potentially hazardous elements during module production. For other solar energy technologies, air and water pollution impacts are generally expected to be relatively minor. Furthermore, some solar technologies in certain regions may require water usage for cleaning to maintain performance. [3.6.1]

Lifecycle assessment estimates of the GHGs associated with various types of PV modules and CSP technologies are provided in Figure TS.3.4. The majority of estimates for PV modules cluster between 30 and 80 g of CO_2eq/kWh. Lifecycle GHG emissions for CSP-generated electricity have recently been estimated to range from about 14 to 32 g of CO_2eq/kWh. These emission levels are about an order of magnitude lower than those of natural gas-fired power plants. [3.6.1, 9.3.4]

Land use is another form of environmental impact. For roof-mounted solar thermal and PV systems, this is not an issue, but it can be an issue for central-station PV as well as for CSP. Environmentally sensitive lands may pose a special challenge for CSP permitting. One difference for CSP vis-à-vis PV is that it needs a method to cool the working fluid, and such cooling often involves the use of scarce water. Using local air as the coolant (dry cooling) is a viable option, but this can decrease plant efficiency by 2 to 10%. [3.6.1]

3.6.2 Social impacts

The positive benefits of solar energy in the developing world provide arguments for its expanded use. About 1.4 billion people do not have access to electricity. Solar home systems and local PV-powered community grids can provide electricity to many areas for which connection to a main grid is cost prohibitive. The impact of electricity and solar energy technologies on the local population is shown through a long list of important benefits: the replacement of indoor-polluting kerosene lamps and inefficient cook stoves; increased indoor reading; reduced time gathering firewood for cooking (allowing the women and children who normally gather it to focus on other priorities); street lighting for security; improved health by providing refrigeration for vaccines and food products; and, finally, communications devices (e.g., televisions, radios). All of these provide a myriad of benefits that improve the lives of people. [3.6.2]

Job creation is an important social consideration associated with solar energy technology. Analysis indicates that solar PV has the highest job-generating potential among the family of solar technologies. Approximately 0.87 job-years per GWh are created through solar PV, followed by CSP with 0.23 job-years per GWh. When properly put forward, these job-related arguments can help accelerate social acceptance and increase public willingness to tolerate the perceived disadvantages of solar energy, such as visual impacts. [3.6.2]

3.7 Prospects for technology improvements and innovation

Solar thermal: If integrated at the earliest stages of planning, buildings of the future could have solar panels – including PV, thermal collector, and combined PV-thermal (hybrids) – making up almost all viewed components of the roof and façades. Such buildings could be established not just through the personal desires of individual builders/owners, but also as a result of public policy mandates, at least in some areas. For example, the vision of the European Solar Thermal Technology Platform is to establish the 'Active Solar Building' as a standard for new buildings by 2030, where an Active Solar Building, on average, covers all of its energy demand for water heating and space conditioning. [3.7.2]

In highlighting the advances in passive solar, two climates can be distinguished between: those that are dominated by the demand for heating and those dominated by the demand for cooling. For the former, a wider-scale adoption of the following items can be foreseen: evacuated (as opposed to sealed) glazing, dynamic exterior night-time insulation, and translucent glazing systems that can automatically change solar/visible transmittance and that also offer improved insulation values. For the latter, there is the expectation for an increased use of cool roofs (i.e., light-coloured roofs that reflect solar energy); heat-dissipation techniques such as use of the ground and water as heat sinks; methods that improve the microclimate around the buildings; and solar control devices that allow penetration of the lighting, but not the thermal, component of solar energy. For both climates, improved thermal storage is expected to be embedded in building materials. Also anticipated are improved methods for distributing the absorbed solar heat around the building and/or to the outside air, perhaps using active methods such as fans. Finally, improved design tools are expected to facilitate these various improved methods. [3.7.1]

Photovoltaic electricity generation: Although now a relatively mature technology, PV is still experiencing rapid improvements in performance and cost, and a continuation of this steady progress is expected. The efforts required are being taken up in a framework of intergovernmental cooperation, complete with roadmaps. For the different PV technologies, four broad technological categories, each requiring specific R&D approaches, have been identified: 1) cell efficiency, stability, and lifetime; 2) module productivity and manufacturing; 3) environmental sustainability; and 4) applicability, all of which include standardization and harmonization. Looking to the future, PV technologies can by categorized in three major classes: current; emerging, which represent medium risk with a mid-term (10 to 20 year) time line; and the high-risk technologies aimed at 2030 and beyond, which have extraordinary potential but require technical breakthroughs. Examples of emerging cells are multiple-junction, polycrystalline thin films and crystalline silicon in the sub-100-μm thickness range. Examples of high-risk cells are organic solar cells, biomimetic devices and quantum dot designs that have the potential to substantially increase the maximum efficiency. Finally, there is important work to be done on the balance of systems (BOS), which comprises inverters, storage, charge controllers, system structures and the energy network. [3.7.3]

	All Values	Mono-Crystalline Silicon (m-Si)	Poly-Crystalline Silicon (p-Si)	Amorphous Silicon (a-Si)	Cadmium Telluride (CdTe)	Nano-Crystalline Dye Sensitized (DSC)	Concentrator	Ribbon Silicon	Cadmium Selenide Quantum Dot (QDPV)
Estimates:	124	30	56	12	13	4	6	2*	1
References:	26	9	15	3	3	1	2	2	1

*same value

	All Values	Trough	Tower	Stirling	Fresnel
Estimates:	42	20	14	4	4
References:	13	7	5	3	1

Figure TS.3.4 | GHG emissions from the life cycles of (top) PV modules and (bottom) CSP technologies. See Annex II for details of literature search and citations of literature contributing to the estimates displayed. [Figures 3.14, 3.15]

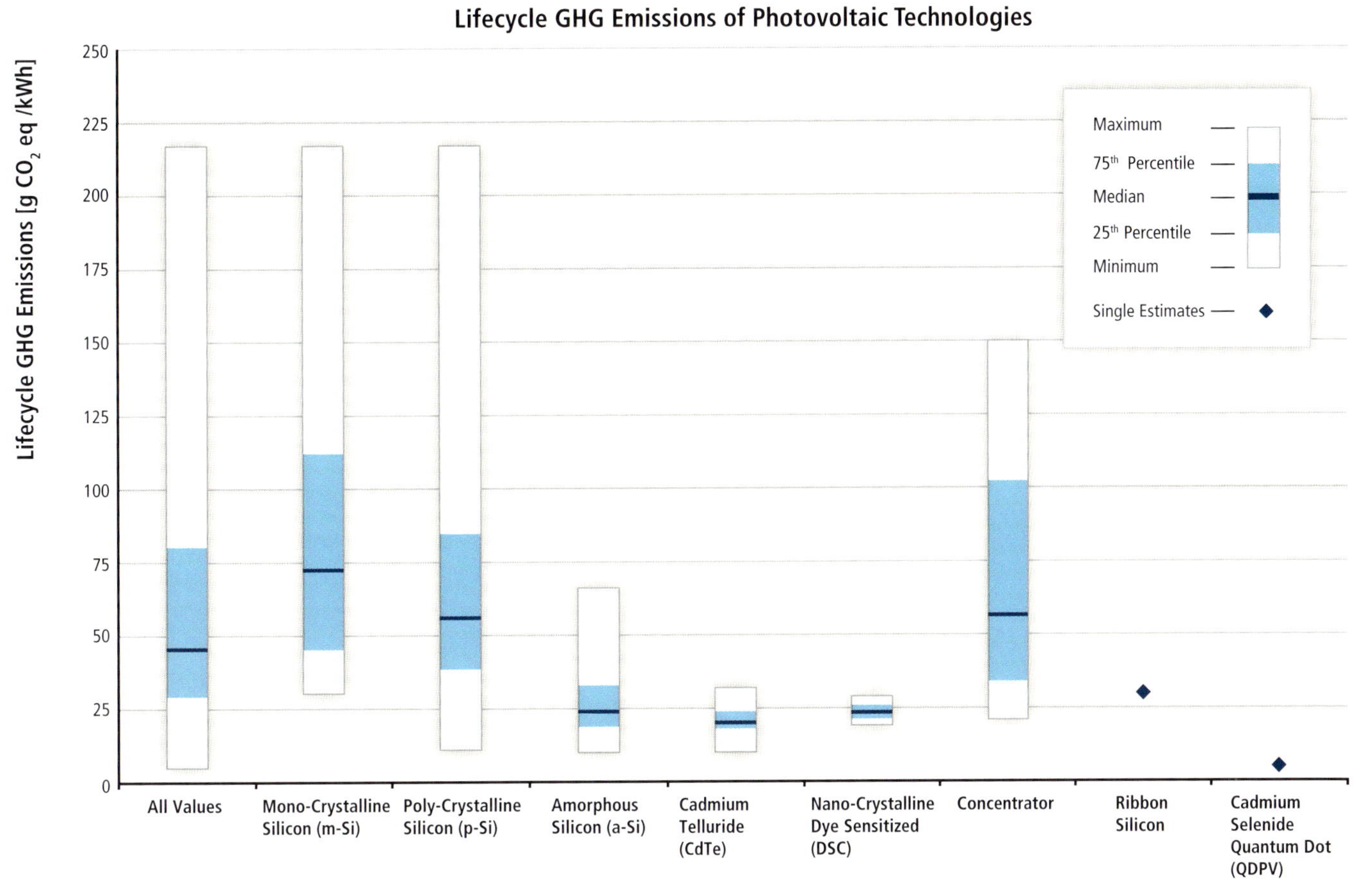

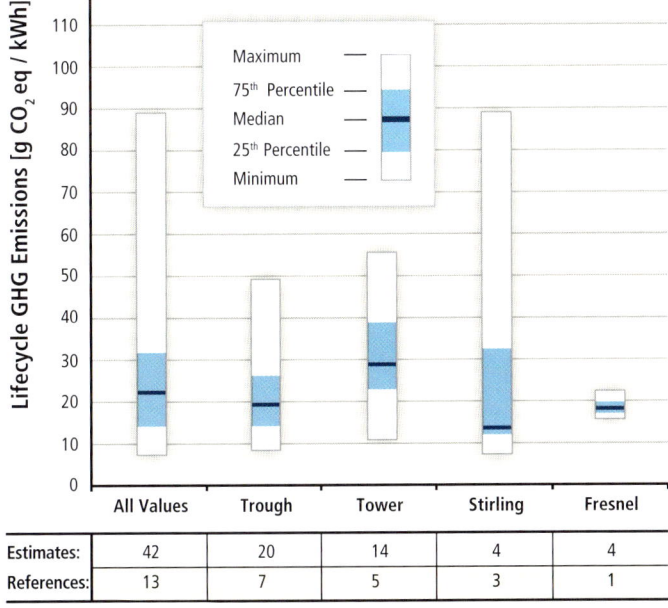

CSP electricity generation: Although CSP is now a proven technology at the utility scale, technology advances are still taking place. As plants are built, both mass production and economies of scale are leading to cost reductions. There is scope for continuing improvement in solar-to-electricity efficiency, partly through higher collector temperatures. To increase temperature and efficiency, alternatives to the use of oil as the heat-transfer fluid—such as water (boiling in the receiver) or molten salts—are being developed, permitting higher operating temperatures. For central-receiver systems, the overall efficiencies can be higher because the operating temperatures are higher, and further improvements are expected to achieve peak efficiencies (solar to electricity) almost twice those of existing systems, up to 35%. Trough technology will benefit from continuing advances in solar-selective surfaces, and central receivers and dishes will benefit from improved receiver/absorber designs that afford high levels of solar irradiance at the focus. Capital cost reduction is expected to come from the benefits of mass production, economies of scale and learning from previous experience. [3.7.4]

Solar fuel production: Solar electrolysis using PV or CSP is available for niche applications, but it remains costly. Many paths are being pursued to develop a technology that will reduce the cost of solar fuels. These include solid-oxide electrolysis cells, the photoelectrochemical cell (which combines all the steps in solar electrolysis into a single unit), advanced thermo-chemical processes, and photochemical and photobiological processes—sometimes in combinations that integrate artificial photosynthesis in man-made biomimetic systems and photobiological hydrogen production in living organisms. [3.7.5]

Other potential future applications: Other methods under investigation for producing electricity using solar thermal technologies without an intermediate thermodynamic cycle include thermoelectric, thermionic, magnetohydrodynamic and alkali-metal methods. Space solar power, in which solar power collected in space is beamed via microwaves to receiving antennae on the ground, has also been proposed. [3.7.6]

3.8 Cost trends

Although the cost of solar energy varies widely by technology, application, location and other factors, costs have been reduced significantly during the past 30 years, and technical advances and supportive public policies continue to offer the potential for additional cost reductions. The degree of continued innovation will have a significant bearing on the level of solar deployment. [3.7.2–3.7.5, 3.8.2–3.8.5]

Solar thermal: The economics of solar heating applications depend on appropriate design of the system with regard to energy service needs, which often involves the use of auxiliary energy sources. In some regions, for example, in southern parts of China, solar water heating (SWH) systems are cost competitive with traditional options. SWH systems are generally more competitive in sunny regions, but this picture changes for space heating based on its usually higher overall heating load. In colder regions capital costs can be spread over a longer heating season, and solar thermal can then become more competitive. [3.8.2]

The investment costs for solar thermal heating systems vary widely depending on the complexity of the technology used as well as the market conditions in the country of operation. The costs for an installed system vary from as low as USD_{2005} 83/m² for SWH systems in China to more than USD_{2005} 1,200/m² for certain space-heating systems. The levelized cost of heat (LCOH) mirrors the wide variation in investment cost, and depends on an even larger number of variables, including the particular type of system, investment cost of the system, available solar irradiance in a particular location, conversion efficiency of the system, operating costs, utilization strategy of the system and the applied discount rate. Based on a standardized methodology outlined in Annex II and the cost and performance data summarized in Annex III, the LCOH for solar thermal systems over a large set and range of input parameters has been calculated to vary widely from USD_{2005} 9 to 200/GJ, but can be estimated for more specific settings with parametric analysis. Figure TS.3.5 shows the LCOH over a somewhat narrower set and range of input parameters. More specifically, the figure shows that for SWH systems with costs in the range of USD_{2005} 1,100 to 1,200/kW$_{th}$ and conversion efficiencies of roughly 40%, LCOH is expected to range from slightly more than USD_{2005} 30/GJ to slightly less than USD_{2005} 50/GJ in regions comparable to Central and Southern European locations and up to almost USD_{2005} 90/GJ for regions with less solar irradiation. Not surprisingly, LCOH estimates are highly sensitive to all of the parameters shown in Figure TS.3.5, including investment costs and capacity factors. [3.8.2, Annex II, Annex III]

Over the last decade, for each 50% increase in installed capacity of solar water heaters, investment costs have fallen 20% in Europe. According to the IEA, further cost reductions in OECD countries will come from the use of cheaper materials, improved manufacturing processes, mass production, and the direct integration into buildings of collectors as multi-functional building components and modular, easy-to-install systems. Delivered energy costs in OECD countries are anticipated by the IEA to eventually decline by around 70 to 75%. [3.8.2]

PV electricity generation: PV prices have decreased by more than a factor of 10 during the last 30 years; however, the current levelized cost of electricity (LCOE) from solar PV is generally still higher than wholesale market prices for electricity. In some applications, PV systems are already competitive with other local alternatives (e.g., for electricity supply in certain rural areas in developing countries). [3.8.3, 8.2.5, 9.3.2]

The LCOE of PV highly depends on the cost of individual system components, with the highest cost share stemming from the PV module. The LCOE also includes BOS components, cost of labour for installation, operation and maintenance (O&M) cost, location and capacity factor, and the applied discount rate. [3.8.3]

The price for PV modules dropped from USD_{2005} 22/W in 1980 to less than USD_{2005} 1.50/W in 2010. The corresponding historical learning rate ranges from 11 to 26%, with a median learning rate of 20%. The price in USD/W for an entire system, including the module, BOS, and installation costs, has also decreased steadily, reaching numbers as low as USD_{2005} 2.72/W for some thin-film technologies by 2009. [3.8.3]

The LCOE for PV depends not only on the initial investment; it also takes into account operation costs and the lifetime of the system components, local solar irradiation levels and system performance. Based on the standardized methodology outlined in Annex II and the cost and performance data summarized in Annex III, the recent LCOE for different types of PV systems has been calculated. It shows a wide variation from as low as USD_{2005} 0.074/kWh to as high as USD_{2005} 0.92/kWh, depending on a large set and range of input parameters. Narrowing the range of parameter variations, the LCOE in 2009 for utility-scale PV electricity generation in regions of high solar irradiance in Europe and the USA were in the range of about USD_{2005} 0.15/kWh to USD_{2005} 0.4/kWh at a

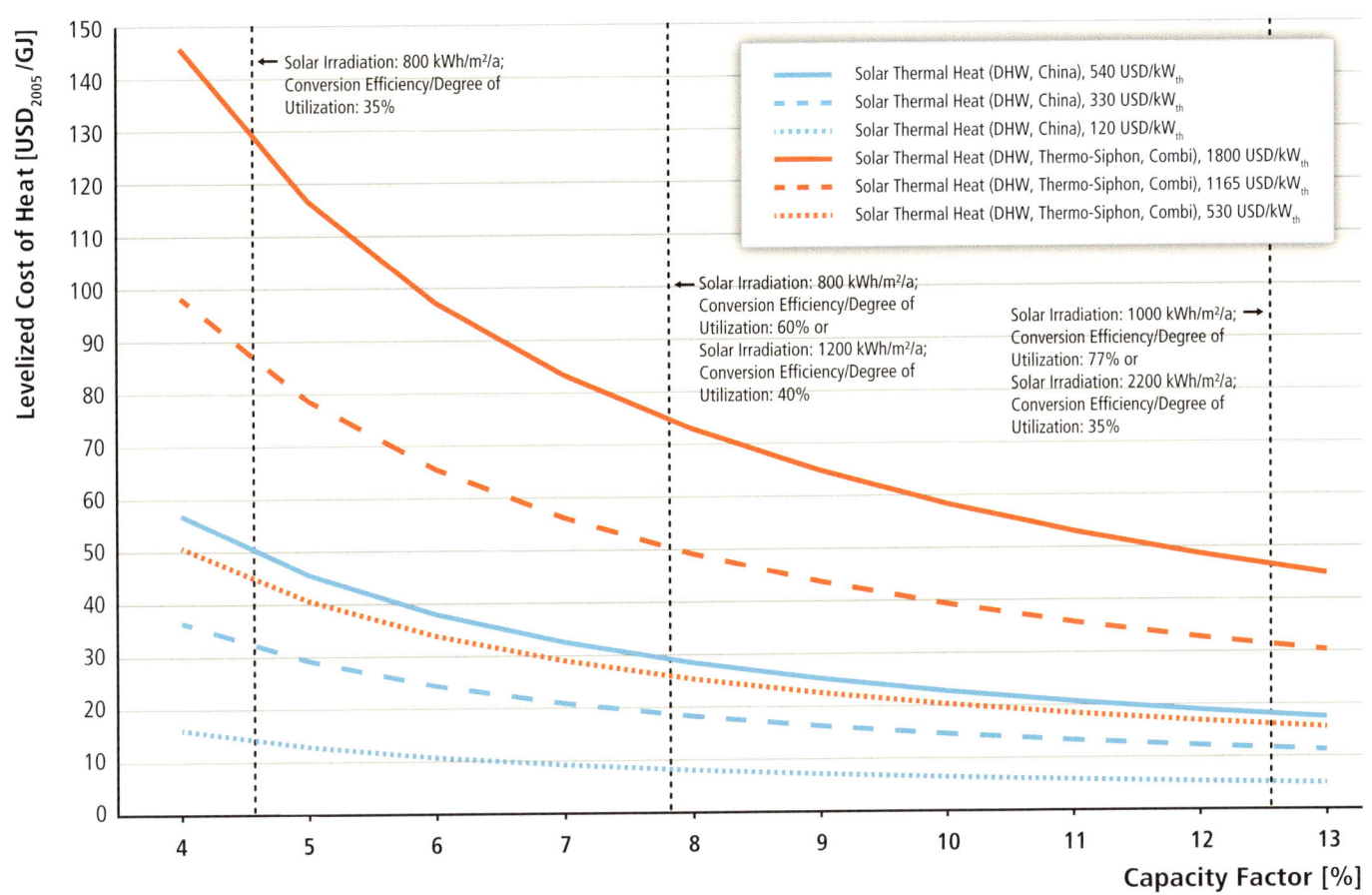

Figure TS.3.5 | Sensitivity of levelized cost of heat with respect to investment cost as a function of capacity factor. (Discount rate assumed to be 7%, annual operation and maintenance cost USD$_{2005}$ 5.6 and 14/kW, and lifetimes set at 12.5 and 20 years for domestic hot water (DHW) systems in China and various types of systems in OECD countries, respectively.) [Figure 3.16]

7% discount rate, but may be lower or higher depending on the available resource and on other framework conditions. Figure TS.3.6 shows a wide variation of LCOE for PV depending on the type of system, investment cost, discount rates and capacity factors. [1.3.2, 3.8.3, 10.5.1, Annex II, Annex III]

Costs of electricity generation or LCOE are projected by the IEA to reach the following in 2020: US cent$_{2005}$ 14.5/kWh to US cent$_{2005}$ 28.6/kWh for the residential sector and US cent$_{2005}$ 9.5/kWh to US cent$_{2005}$ 19/kWh for the utility sector under favourable conditions of 2,000 kWh/kW (equivalent to a 22.8% capacity factor) and less favourable conditions of 1,000 kWh/kW (equivalent to a 11.4% capacity factor), respectively. The goal of the US Department of Energy is even more ambitious, with an LCOE goal of US cent$_{2005}$ 5/kWh to US cent$_{2005}$ 10/kWh, depending on the end user, by 2015. [3.8.3]

CSP electricity generation: CSP electricity systems are a complex technology operating in a complex resource and financial environment; so many factors affect the LCOE. The publicized investment costs of CSP plants are often confused when compared to other renewable sources, because varying levels of integrated thermal storage increase the investment, but also improve the annual output and capacity factor of the plant. For large, state-of-the-art trough plants, current investment costs are estimated to be USD$_{2005}$ 3.82/W (without storage) to USD$_{2005}$ 7.65/W (with storage) depending on labour and land costs, technologies, the amount and distribution of beam irradiance and, above all, the amount of storage and the size of the solar field. Performance data for modern CSP plants are limited, particularly for plants equipped with thermal storage, because new plants only became operational from 2007 onward. Capacity factors for early plants without storage were up to 28%. For modern plants without storage, capacity factors of roughly 20 to 30% are envisioned; for plants with thermal storage, capacity factors of 30 to 75% may be achieved. Based on the standardized methodology outlined in Annex II and the cost and performance data summarized in Annex III, the LCOE for a solar trough plant with six hours of thermal storage in 2009 over a large set and range of input parameters has been calculated to range from slightly more than US cent$_{2005}$ 10/kWh to about US cent$_{2005}$ 30/kWh. Restricting the range of discount rates to 10% results in a somewhat narrower range of about US cent$_{2005}$ 20/kWh to US cent$_{2005}$ 30/kWh, which is roughly in line with the range of US cent$_{2005}$ 18 to US cent$_{2005}$ 27/kWh available in the literature. Particular cost

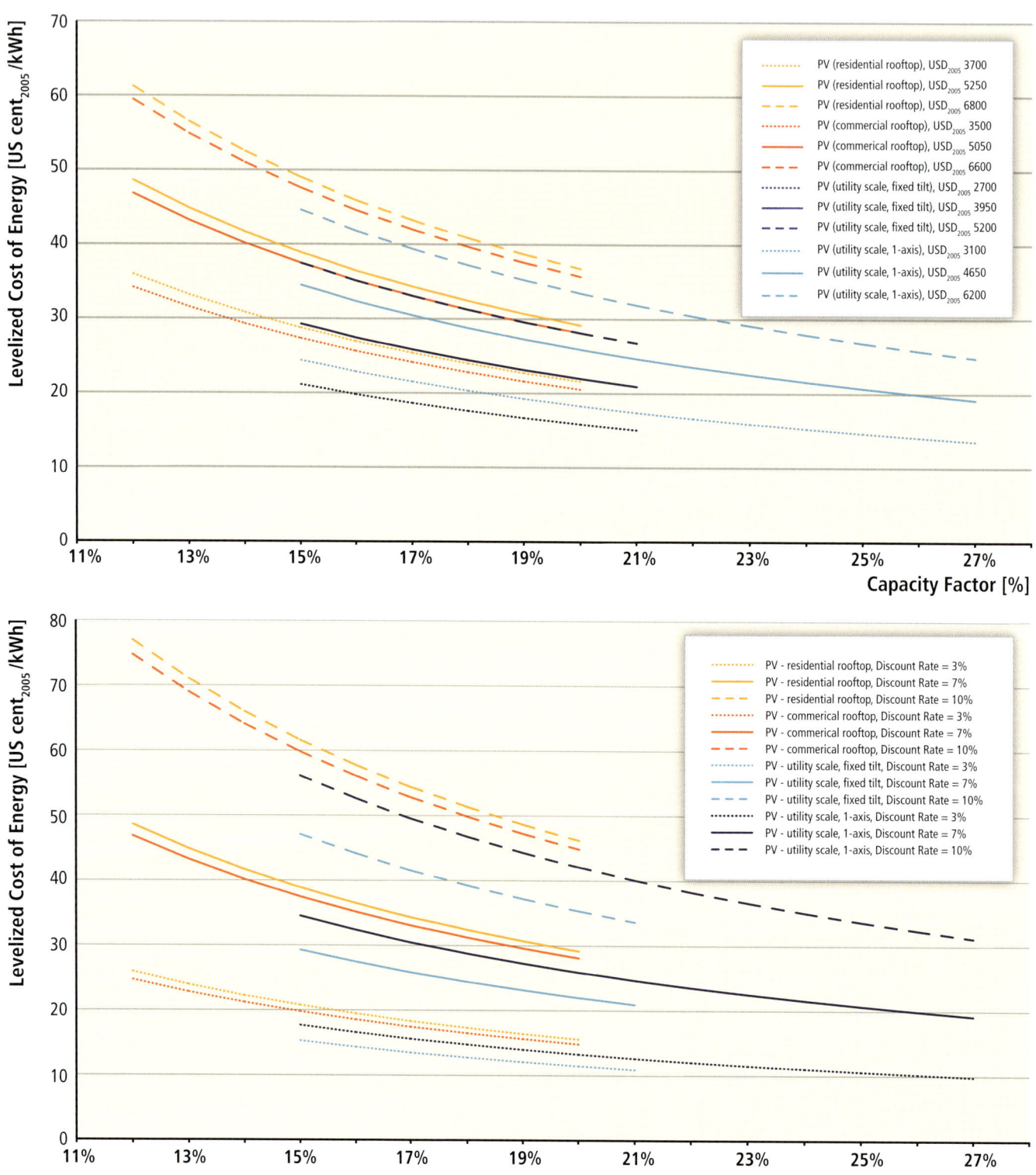

Figure TS.3.6 | Levelized cost of PV electricity generation, 2008–2009: (top) as a function of capacity factor and investment cost*,***; and (bottom) as a function of capacity factor and discount rate**,***. [Figure 3.19]

Notes: * Discount rate assumed to equal 7%. ** Investment cost for residential rooftop systems assumed at USD 5,500 US/kW, for commercial rooftop systems at USD 5,150, for utility-scale fixed tilt projects at USD 3,650/kW and for utility-scale one-axis projects at USD 4,050/kW. ***Annual O&M cost assumed at USD 41 to 64/kW, lifetime at 25 years.

and performance parameters, including the applied discount rate and capacity factor, affect the specific LCOE estimate, although the LCOE of different system configurations for otherwise identical conditions are expected to differ only marginally. [3.8.4]

The learning ratio for CSP, excluding the power block, has been estimated at 10 ± 5%. Specific LCOE goals for the USA are US cent$_{2005}$ 6/kWh to US cent$_{2005}$ 8/kWh with 6 hours storage by 2015 and US cent$_{2005}$ 50/kWh to US cent$_{2005}$ 60/kWh with 12 to 17 hours of storage by 2020. The EU is pursuing similar goals. [3.8.4]

3.9 Potential deployment

3.9.1 Near-term (2020) forecasts

Table TS.3.1 summarizes findings from the available studies on potential deployment up to 2020, as taken from the literature. Sources for the tabulated data are the following: European Renewable Energy Council (EREC) – Greenpeace (Energy [r]evolution, reference and advanced scenarios); and IEA (CSP and PV Technology Roadmaps). With regard to the solar thermal entries, note that passive solar contributions are not included in these data; although this technology reduces the demand for energy, it is not part of the supply chain considered in energy statistics. [3.9]

3.9.2 Long-term deployment in the context of carbon mitigation

Figure TS.3.7 presents the results of more than 150 long-term modelling scenarios described in Chapter 10. The potential deployment scenarios vary widely—from direct solar energy playing a marginal role in 2050 to it becoming one of the major sources of energy supply. Although direct solar energy today provides only a very small fraction of the world energy supply, it remains undisputed that this energy source has one of the largest potential futures.

Reducing cost is a key issue in making direct solar energy more commercially relevant and in position to claim a larger share of the worldwide energy market. This can only be achieved if solar technologies' costs are reduced as they move along their learning curves, which depend primarily on market volumes. In addition, continuous R&D efforts are required to ensure that the slopes of the learning curves do not flatten too early. The true costs of deploying solar energy are still unknown because the main deployment scenarios that exist today consider only a single technology. These scenarios do not take into account the co-benefits of a renewable/sustainable energy supply via a range of different RE sources and energy efficiency measures.

Potential deployment depends on the actual resources and availability of the respective technology. However, to a large extent, the regulatory and legal framework in place can foster or hinder the uptake of direct solar energy applications. Minimum building standards with respect to building orientation and insulation can reduce the energy demand of buildings significantly and can increase the share of RE supply without increasing the overall demand. Transparent, streamlined administrative procedures to install and connect solar power sources to existing grid infrastructures can further lower the cost related to direct solar energy.

4. Geothermal Energy

4.1 Introduction

Geothermal resources consist of thermal energy from the Earth's interior stored in both rock and trapped steam or liquid water, and are used to generate electric energy in a thermal power plant or in other domestic and agro-industrial applications requiring heat as well as in CHP applications. Climate change has no significant impacts on the effectiveness of geothermal energy. [4.1]

Geothermal energy is a renewable resource as the tapped heat from an active reservoir is continuously restored by natural heat production, conduction and convection from surrounding hotter regions, and the extracted geothermal fluids are replenished by natural recharge and by reinjection of the cooled fluids. [4.1]

Table TS.3.1 | Evolution of cumulative solar capacities. [Table 3.7]

		Low-Temperature Solar Heat (GW$_{th}$)			Solar PV Electricity (GW)			CSP Electricity (GW)		
	Year	2009	2015	2020	2009	2015	2020	2009	2015	2020
Name of Scenario	Current cumulative installed capacity	180			22			0.7		
	EREC – Greenpeace (reference scenario)		180	230		44	80		5	12
	EREC – Greenpeace ([r]evolution scenario)		715	1,875		98	335		25	105
	EREC – Greenpeace (advanced scenario)		780	2,210		108	439		30	225
	IEA Roadmaps		N/A			95[1]	210		N/A	148

Note: 1. Extrapolated from average 2010 to 2020 growth rate.

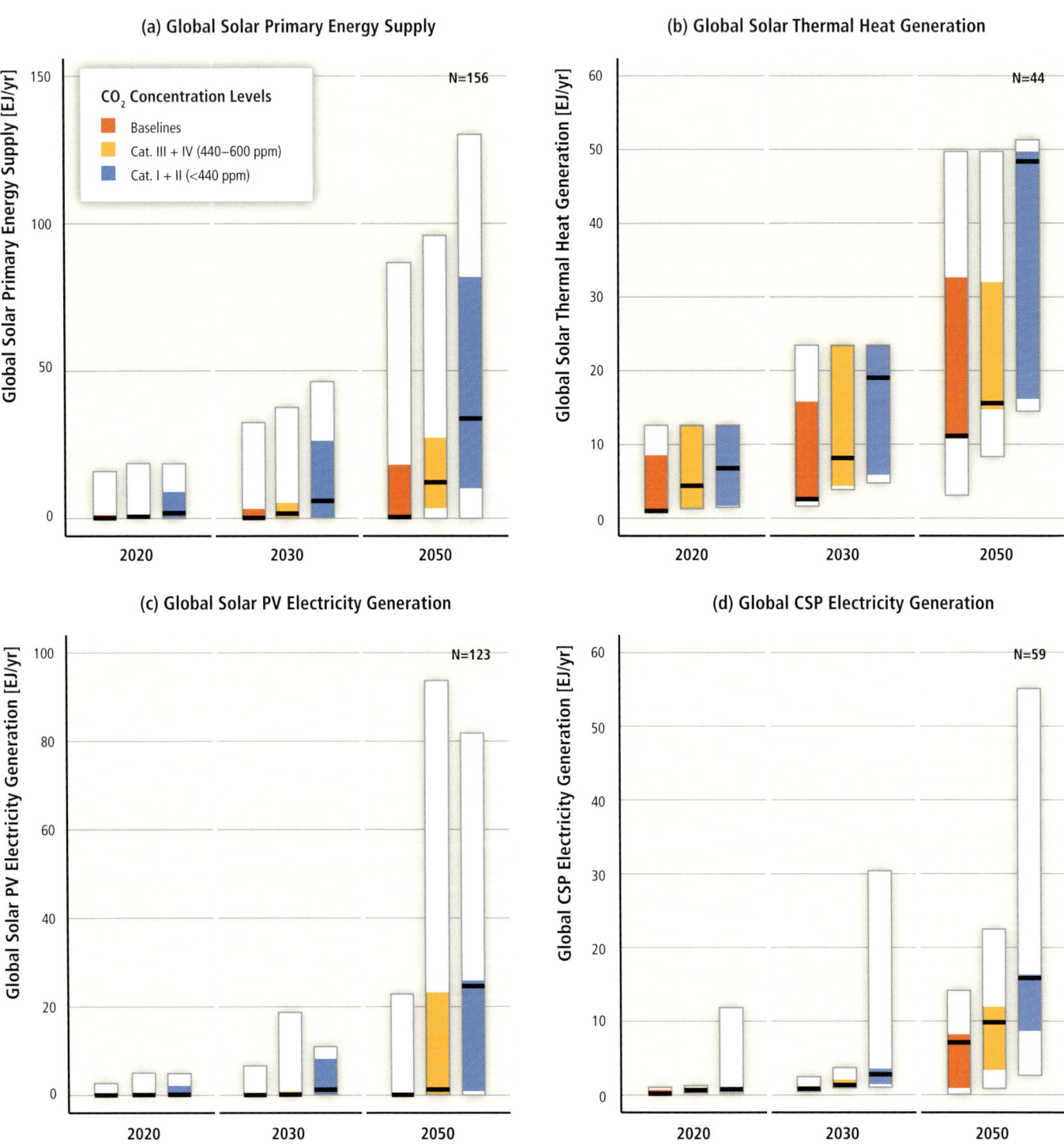

Figure TS.3.7 | Global solar supply and generation in long-term scenarios (median, 25th to 75th percentile range, and full range of scenario results; colour coding is based on categories of atmospheric CO_2 concentration level in 2100; the specific number of scenarios underlying the figure is indicated in the upper right-hand corner). (a) Global solar primary energy supply; (b) global solar thermal heat generation; (c) global solar PV electricity generation; and (d) global CSP electricity generation. [Figure 3.22]

4.2 Resource potential

The accessible stored heat from hot dry rocks in the Earth is estimated to range from 110 to 403 × 10^6 EJ down to 10 km depth, 56 to 140 × 10^6 EJ down to 5 km depth, and around 34 × 10^6 EJ down to 3 km depth. Using previous estimates for hydrothermal resources and calculations for enhanced (or engineered) geothermal systems derived from stored heat estimates at depth, geothermal technical potentials for electric generation range from 118 to 146 EJ/yr (at 3 km depth) to 318 to 1,109 EJ/yr (at 10 km depth), and for direct uses range from 10 to 312 EJ/yr (Figure TS.4.1). [4.2.1]

Technical potentials are presented on a regional basis in Table TS.4.1. The regional breakdown is based on the methodology applied by the Electric Power Research Institute to estimate theoretical geothermal

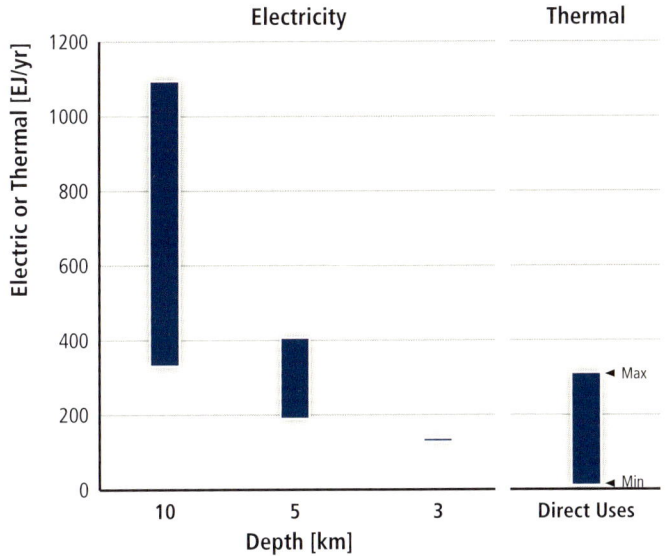

Figure TS.4.1 | Geothermal technical potentials for electricity and direct uses (heat). Direct uses usually do not require development to depths greater than about three km. [Figure 4.2]

potentials for each country, and then countries are grouped regionally. Thus, the present disaggregation of global technical potential is based on factors accounting for regional variations in the average geothermal gradient and the presence of either a diffuse geothermal anomaly or a high-temperature region associated with volcanism or plate boundaries. The separation into electric and thermal (direct uses) potentials is somewhat arbitrary in that most higher-temperature resources could be used for either, or both, in CHP applications depending on local market conditions. [4.2.2]

The heat extracted to achieve the technical potentials can be fully or partially replenished over the long term by the continental terrestrial heat flow of 315 EJ/yr at an average flux of 65 mW/m². [4.2.1]

4.3 Technology and applications

Geothermal energy is currently extracted using wells and other means that produce hot fluids from: (a) hydrothermal reservoirs with naturally high permeability, or (b) Enhanced or engineered geothermal systems (EGS) with artificial fluid pathways (Figure TS.4.2). Technology for electricity generation from hydrothermal reservoirs is mature and reliable, and has been operating for about 100 years. Technologies for direct heating using geothermal heat pumps (GHPs) for district heating and for other applications are also mature. Technologies for EGS are in the demonstration stage. [4.3]

Electric power from geothermal energy is especially suitable for supplying base-load power, but also can be dispatched and used to meet peak demand. Hence, geothermal electric power can complement variable electricity generation. [4.3]

Since geothermal resources are underground, exploration methods (including geological, geochemical and geophysical surveys) have been developed to locate and assess them. The objectives of geothermal exploration are to identify and rank prospective geothermal reservoirs prior to drilling. Today, geothermal wells are drilled over a range of depths up to 5 km using conventional rotary drilling methods similar to those for accessing oil and gas reservoirs. Advanced drilling technologies allow for high-temperature operation and provide directional capability. [4.3.1]

The basic types of geothermal power plants in use today are steam condensing turbines and binary cycle units. Condensing plants can be of the flash or dry-steam type (the latter do not require brine separation, resulting in simpler and cheaper plants) and are more common than binary units. They are installed in intermediate- and high-temperature resources (≥150°C) with capacities often between 20 and 110 MW$_e$.

Table TS.4.1 | Geothermal technical potentials on continents for the IEA regions. [Table 4.3]

REGION[1]	Electric technical potential (EJ/yr) at depths to:						Technical potentials (EJ/yr) for direct uses	
	3 km		5 km		10 km			
	Lower	Upper	Lower	Upper	Lower	Upper	Lower	Upper
OECD North America	25.6	31.8	38.0	91.9	69.3	241.9	2.1	68.1
Latin America	15.5	19.3	23.0	55.7	42.0	146.5	1.3	41.3
OECD Europe	6.0	7.5	8.9	21.6	16.3	56.8	0.5	16.0
Africa	16.8	20.8	24.8	60.0	45.3	158.0	1.4	44.5
Transition Economies	19.5	24.3	29.0	70.0	52.8	184.4	1.6	51.9
Middle East	3.7	4.6	5.5	13.4	10.1	35.2	0.3	9.9
Developing Asia	22.9	28.5	34.2	82.4	62.1	216.9	1.8	61.0
OECD Pacific	7.3	9.1	10.8	26.2	19.7	68.9	0.6	19.4
Total	**117.5**	**145.9**	**174.3**	**421.0**	**317.5**	**1,108.6**	**9.5**	**312.2**

Note: 1. For regional definitions and country groupings see Annex II.

In binary cycle plants, the geothermal fluid passes through a heat exchanger heating another working fluid with a low boiling point, which vaporizes and drives a turbine. They allow for use of lower-temperature hydrothermal reservoirs and of EGS reservoirs (generally from 70°C to 170°C), and are often constructed as linked modular units of a few MW_e in capacity. Combined or hybrid plants comprise two or more of the above basic types to improve versatility, increase overall thermal efficiency, improve load-following capability, and efficiently cover a wide resource temperature range. Finally, cogeneration plants, or CHP plants, produce both electricity and hot water for direct use. [4.3.3]

EGS reservoirs require stimulation of subsurface regions where temperatures are high enough for effective utilization. A reservoir consisting of a fracture network is created or enhanced to provide well-connected fluid pathways between injection and production wells. Heat is extracted by circulating water through the reservoir in a closed loop and can be used for power generation and for industrial or residential heating (see Figure TS.4.2). [4.3.4]

Direct use provides heating and cooling for buildings including district heating, fish ponds, greenhouses, bathing, wellness and swimming pools, water purification/desalination and industrial and process heat for agricultural products and mineral drying. Although it can be debated whether GHPs are a 'true' application of geothermal energy, they can be utilized almost anywhere in the world for heating and cooling, and take advantage of the relatively constant ground or groundwater temperature in the range of 4°C to 30°C. [4.3.5]

4.4 Global and regional status of market and industry development

For nearly a century, geothermal resources have been used to generate electricity. In 2009, the global geothermal electric market had a wide range of participants with 10.7 GW_e of installed capacity. Over 67 TWh_e (0.24 EJ) of electricity were generated in 2008 in 24 countries (Figure TS.4.3), and provided more than 10% of total electricity demand in 6 of them. There were also 50.6 GW_{th} of direct geothermal applications operating in 78 countries, which generated 121.7 TWh_{th} (0.44 EJ) of heat in 2008. GHPs contributed 70% (35.2 GW_{th}) of this installed capacity for direct use. [4.4.1, 4.4.3]

The global average annual growth rate of installed geothermal electric capacity over the last five years (2005-2010) was 3.7%, and over the last 40 years (1970-2010), 7.0%. For geothermal direct uses rates were 12.7% (2005-2010), and 11% between 1975 and 2010. [4.4.1]

EGS is still in the demonstration phase, with one small plant in operation in France and one pilot project in Germany. In Australia considerable investment has been made in EGS exploration and development in recent years, and the USA has recently increased support for EGS research, development and demonstration as part of a revived national geothermal programme. [4.4.2]

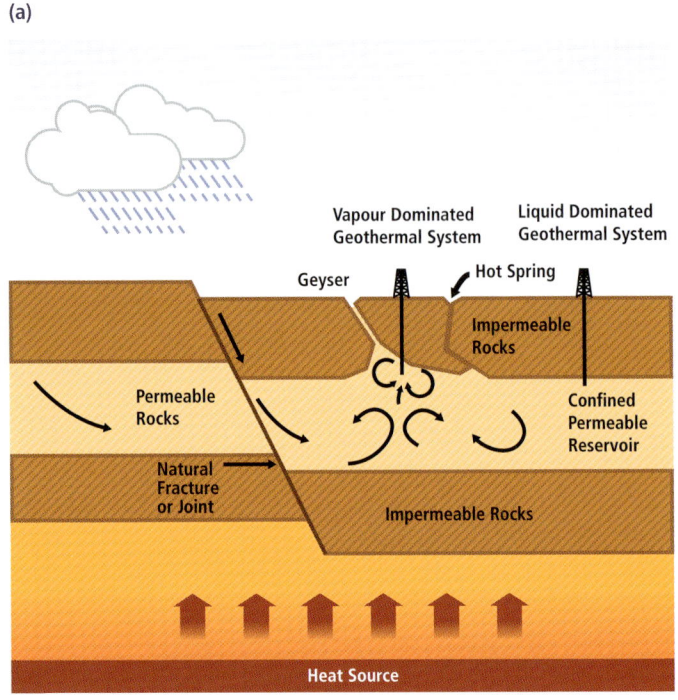

Figure TS.4.2a | Scheme showing convective (hydrothermal) resources. [Figure 4.1a]

In 2009, the main types (and relative percentages) of direct geothermal applications in annual energy use were: space heating of buildings (63%), bathing and balneology (25%), horticulture (greenhouses and soil heating) (5%), industrial process heat and agricultural drying (3%), aquaculture (fish farming) (3%) and snow melting (1%). [4.4.3]

For geothermal to reach its full capacity in climate change mitigation it is necessary to overcome technical and non-technical barriers. Policy measures specific to geothermal technology can help overcome these barriers. [4.4.4]

4.5 Environmental and social impacts

Environmental and social impacts related to geothermal energy do exist, and are typically site- and technology-specific. Usually, these impacts are manageable, and the negative environmental impacts are minor. The main GHG emission from geothermal operations is CO_2, although it is not created through combustion, but emitted from naturally occurring sources. A field survey of geothermal power plants operating in 2001 found a wide spread in the direct CO_2 emission rates, with values ranging from 4 to 740 g/kWh_e depending on technology design and composition of the geothermal fluid in the underground reservoir. Direct CO_2 emissions for direct use applications are negligible, while EGS power plants are likely to be designed as liquid-phase closed-loop circulation systems, with zero direct emissions. Lifecycle assessments anticipate that CO_2-equivalent emissions are less than 50 g/kWh_e for geothermal power plants; less than 80 g/kWh_e for projected EGS; and

(b)

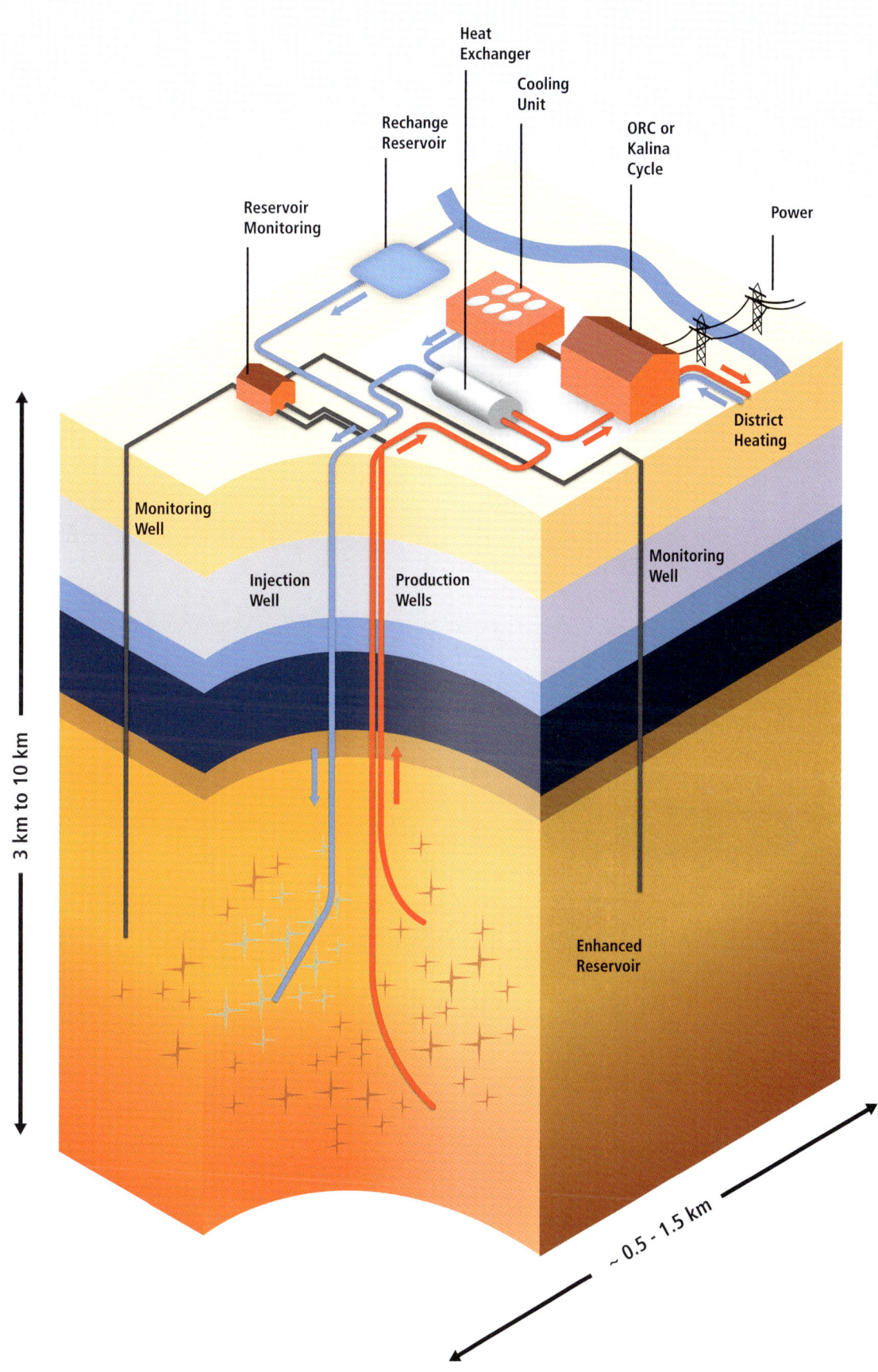

Figure TS.4.2b | Scheme showing conductive (EGS) resources. [Figure 4.1b]

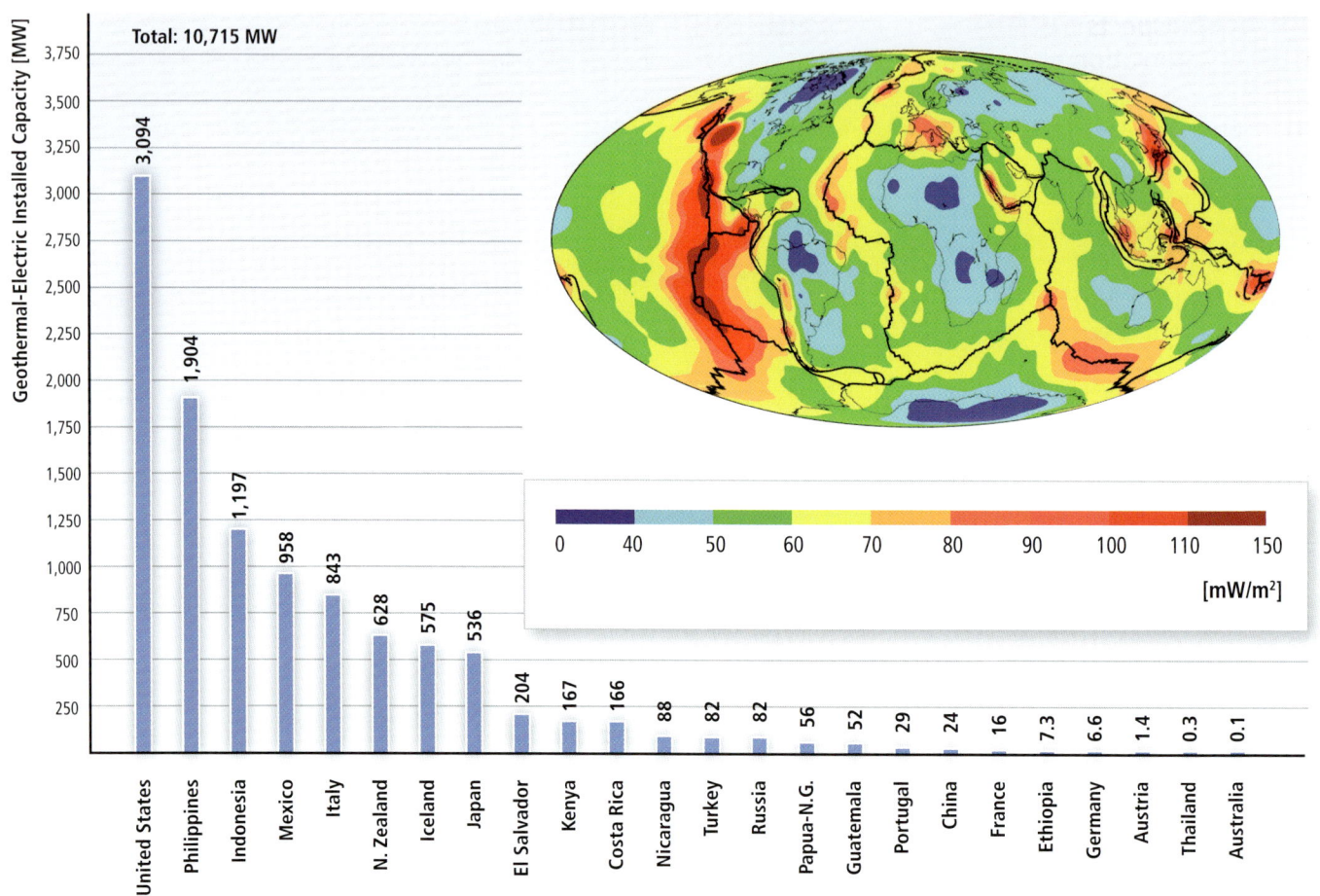

Figure TS.4.3 | Geothermal electric installed capacity by country in 2009. Figure shows worldwide average heat flow in mW/m² and tectonic plate boundaries. [Figure 4.5]

between 14 and 202 g/kWh$_{th}$ for district heating systems and GHPs. [4.5, 4.5.1, 4.5.2]

Environmental impacts associated with geothermal projects involve consideration of a range of local air, land and water use impacts during both construction and operational phases that are common to most energy projects as well as specific to geothermal energy. Geothermal systems involve natural phenomena, and typically discharge gases mixed with steam from surface features, and minerals dissolved in water from hot springs. Some gases may be dangerous, but are typically either treated or monitored during production. In the past, surface disposal of separated water was more common, but today happens only in exceptional circumstances. Geothermal brine is usually injected back into the reservoir to support reservoir pressures and to avoid adverse environmental effects. Surface disposal, if significantly in excess of natural hot-spring flow rates, and if not strongly diluted, can have adverse effects on the ecology of rivers, lakes or marine environments. [4.5.3.1]

Local hazards arising from natural phenomena, such as micro-earthquakes, hydrothermal steam eruptions and ground subsidence may be influenced by the operation of geothermal fields. During 100 years of development, no buildings or structures within a geothermal operation or local community have been significantly damaged by shallow earthquakes originating from either geothermal production or injection activities. Some EGS demonstration projects, particularly in populated areas of Europe, have raised social opposition. The process of high-pressure injection of cold water into hot rock generates small seismic events. Induced seismic events have not been large enough to lead to human injury or significant property damage, but proper management of this issue will be an important step to facilitating significant expansion of future EGS projects. [4.5.3.2]

Land use requirements range from 160 to 290 m²/GWh$_e$/yr excluding wells, and up to 900 m²/GWh/yr including wells. Specific geothermal impacts on land use include effects on outstanding natural features such as springs, geysers and fumaroles. Land use issues in many settings (e.g., Japan, the USA and New Zealand) can be a serious impediment to further expansion of geothermal development. [4.5.3.3]

Geothermal resources may also have significant environmental advantages compared to the energy use they otherwise offset. [4.5.1]

4.6 Prospects for technology improvement, innovation and integration

Geothermal resources can be integrated into all types of electrical power supply systems, from large, interconnected continental transmission grids to onsite use in small, isolated villages or autonomous buildings. Since geothermal energy typically provides base-load electric generation, integration of new power plants into existing power systems does not present a major challenge. For geothermal direct uses, no integration problems have been observed, and for heating and cooling, geothermal energy (including GHPs) is already widespread at the domestic, community and district scales. Section 8 of this summary addresses integration issues in greater depth. [4.6]

Several prospects for technology improvement and innovation can reduce the cost of producing geothermal energy and lead to higher energy recovery, longer field and plant lifetimes, and better reliability. Advanced geophysical surveys, injection optimization, scaling/corrosion inhibition, and better reservoir simulation modelling will help reduce the resource risks by better matching installed capacity to sustainable generation capacity. [4.6]

In exploration, R&D is required to locate hidden geothermal systems (e.g., with no surface manifestations) and for EGS prospects. Refinement and wider usage of rapid reconnaissance geothermal tools such as satellite- and airborne-based hyper-spectral, thermal infrared, high-resolution panchromatic and radar sensors could make exploration efforts more effective. [4.6.1]

Special research in drilling and well construction technology is needed to improve the rate of penetration when drilling hard rock and to develop advanced slim-hole technologies, with the general objectives of reducing the cost and increasing the useful life of geothermal production facilities. [4.6.1]

The efficiency of the different system components of geothermal power plants and direct uses can still be improved, and it is important to develop conversion systems that more efficiently utilize the energy in the produced geothermal fluid. Another possibility is the use of suitable oil and gas wells potentially capable of supplying geothermal energy for power generation. [4.6.2]

EGS projects are currently at a demonstration and experimental stage. EGS require innovative methods to hydraulically stimulate reservoir connectivity between injection and production wells to attain sustained, commercial production rates while reducing the risk of seismic hazard, and to improve numerical simulators and assessment methods to enable reliable predictions of chemical interaction between geo-fluids and geothermal reservoirs rocks. The possibility of using CO_2 as a working fluid in geothermal reservoirs, particularly in EGS, is also under investigation since it could provide a means for enhancing the effect of geothermal energy deployment, lowering CO_2 emissions beyond just generating electricity with a carbon-free renewable resource. [4.6.3]

Currently there are no technologies in use to tap submarine geothermal resources, but in theory electrical energy could be produced directly from a hydrothermal vent. [4.6.4]

4.7 Cost trends

Geothermal projects typically have high upfront investment costs, due to the need to drill wells and construct power plants, and relatively low operational costs. Though costs vary by project, the LCOE of power plants using hydrothermal resources are often competitive in today's electricity markets; the same is true for direct uses of geothermal heat. EGS plants remain in the demonstration phase, but estimates of EGS costs are higher than those for hydrothermal reservoirs. [4.7]

The investment costs of a typical geothermal electric project are: (a) exploration and resource confirmation (10 to 15% of the total); (b) drilling of production and injection wells (20 to 35% of the total); (c) surface facilities and infrastructure (10 to 20% of the total); and (d) power plant (40 to 81% of the total). Current investment costs vary worldwide between USD$_{2005}$ 1,800 and 5,200/kW$_e$. [4.7.1]

Geothermal electric O&M costs, including make-up wells (i.e., new wells to replace failed wells and restore lost production or injection capacity), have been calculated to be USD$_{2005}$ 152 to 187/kW$_e$/yr, but in some countries can be significantly lower (e.g., USD$_{2005}$ 83 to 117/kW$_e$/yr in New Zealand). [4.7.2]

Power plant longevity and capacity factor are also important economic parameters. The worldwide capacity factor average in 2008 for existing geothermal power plants was 74.5%, with newer installations above 90%. [4.7.3]

Based on a standardized methodology outlined in Annex II and the cost and performance data summarized in Annex III, the LCOE for hydrothermal geothermal projects over a large set and range of input parameters has been calculated to range from US cents$_{2005}$ 3.1/kWh to US cents$_{2005}$ 17/kWh, depending on the particular type of technology and project-specific conditions. Using a narrower set and range of parameters, Figure TS.4.4 shows that, at a 7% discount rate, recently installed green-field hydrothermal projects operating at the global average capacity factor of 74.5% (and under other conditions specified in [4.7.4]) have LCOE in the range from US cents$_{2005}$ 4.9/kWh to US cents$_{2005}$ 7.2/kWh for condensing flash plants and, for binary cycle plants, from US cents$_{2005}$ 5.3/kWh to US cents$_{2005}$ 9.2/kWh. The LCOE is shown to vary substantially with capacity factor, investment cost and discount rate. No LCOE data exist for EGS, but some projections have been made using different models for several cases with diverse temperatures and depths, for example, US cents$_{2005}$ 10/kWh to US cents$_{2005}$ 17.5/kWh for relatively high-grade EGS resources. [1.3.2, 4.7.4, 10.5.1, Annex II, Annex III]

Estimates of possible cost reductions from design changes and technical advances rely solely on expert knowledge of the geothermal process

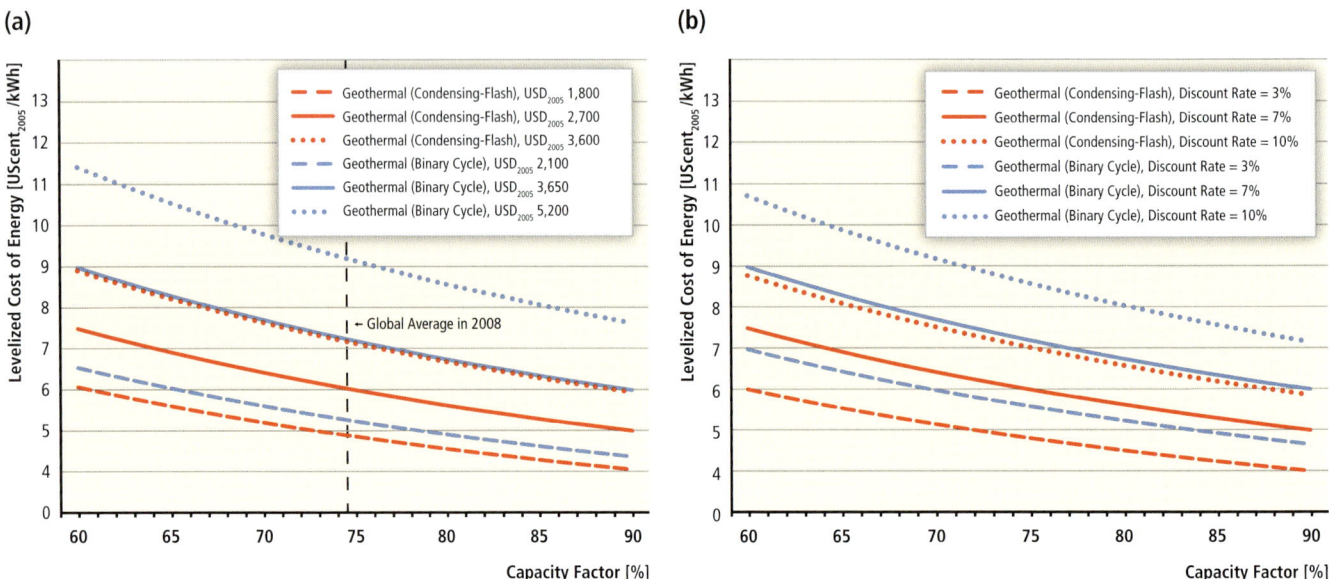

Figure TS.4.4 | Levelized cost of geothermal power, 2008: a) as a function of capacity factor and cost*,***; and b) as a function of capacity factor and discount rate**,***. [Figure 4.8]

Notes: * Discount rate assumed to equal 7%. ** Investment cost for condensing flash plants assumed at USD 2,700/kW and for binary-cycle plants at USD 3,650/kW. ***Annual O&M cost assumed to be USD 170/kW and lifetime 27.5 years.

value chain, as published learning curve studies are limited. Engineering improvements in design and stimulation of geothermal reservoirs, and improvements in materials, operation and maintenance are expected to have the greatest impact on LCOE in the near term, for example, leading to higher capacity factors and a lower contribution of drilling cost to overall investment costs. For green-field projects in 2020, the worldwide average projected LCOE is expected to range from US cents$_{2005}$ 4.5/kWh to US cents$_{2005}$ 6.6/kWh for condensing flash plants and from US cents$_{2005}$ 4.9/kWh to US cents$_{2005}$ 8.6/kWh for binary cycle plants ranges, given an average worldwide capacity factor of 80%, a 27.5-year lifetime and a discount rate of 7%. Therefore, a global average LCOE reduction of about 7% is expected for geothermal flash and binary plants by 2020. Future costs of EGS are expected to decline to lower levels as well. [4.7.5]

The LCOH for direct-use projects has a wide range, depending upon specific use, temperature and flow rate required, associated O&M and labour costs, and output of the produced product. In addition, costs for new construction are usually less than costs for retrofitting older structures. The cost figures given in Table TS.4.2 are based on a climate typical of the northern half of the USA or Europe. Heating loads would be higher for more northerly climates such as Iceland, Scandinavia and Russia. Most figures are based on cost in the USA, but would be similar in developed countries and lower in developing countries. [4.7.6]

Industrial applications are more difficult to quantify, as they vary widely depending upon the energy requirements and the product to be produced. These plants normally require higher temperatures and often compete with power plant use; however, they do have a high load factor of 0.40 to 0.70, which improves the economics. Industrial applications vary from large food, timber and mineral drying plants (USA and New Zealand) to pulp and paper plants (New Zealand). [4.7.6]

4.8 Potential deployment

Geothermal energy can contribute to near- and long-term carbon emissions reduction. In 2008, global geothermal energy use represented only about 0.1% of the global primary energy supply. However, by 2050, geothermal could meet roughly 3% of the global electricity demand and 5% of the global demand for heating and cooling. [4.8]

Taking into account the geothermal electric projects under construction or planned in the world, installed geothermal capacity is expected to reach 18.5 GW$_e$ by 2015. Practically all the new power plants expected to be on line by 2015 will be flash-condensing and binary utilizing hydrothermal resources, with a small contribution from EGS projects. Geothermal direct uses (heat applications including GHP) are expected to grow at the same historic annual rate (11% between 1975 and 2010) to reach 85.2 GW$_{th}$. By 2015, total electric generation could reach 121.6 TWh/yr (0.44 EJ/yr) while direct generation of heat could reach 224 TWh$_{th}$/yr (0.8 EJ/yr), with the regional breakdown presented in Table TS.4.3. [4.8.1]

The long-term potential deployment of geothermal energy based on a comprehensive assessment of numerous model-based scenarios is mentioned in Section 10 of this summary and spans a broad range. The scenario medians for three GHG concentration stabilization ranges, based

Table TS.4.2 | Investment costs and calculated levelized cost of heat (LCOH) for several direct geothermal applications. [Table 4.8]

Heat application	Investment cost (USD$_{2005}$/kW$_{th}$)	LCOH (USD$_{2005}$/GJ) at discount rates of:		
		3%	7%	10%
Space heating (buildings)	1,600–3,940	20–50	24–65	28–77
Space heating (districts)	570–1,570	12–24	14–31	15–38
Greenhouses	500–1,000	7.7–13	8.6–14	9.3–16
Uncovered aquaculture ponds	50–100	8.5–11	8.6–12	8.6–12
GHP (residential and commercial)	940–3,750	14–42	17–56	19–68

Table TS.4.3 | Regional current and forecast installed capacity for geothermal power and direct uses (heat) and forecast generation of electricity and heat by 2015. [Table 4.9]

REGION[1]	Current capacity (2010)		Forecast capacity (2015)		Forecast generation (2015)	
	Direct (GW$_{th}$)	Electric (GW$_e$)	Direct (GW$_{th}$)	Electric (GW$_e$)	Direct (TW$_{th}$)	Electric (TWh$_e$)
OECD North America	13.9	4.1	27.5	6.5	72.3	43.1
Latin America	0.8	0.5	1.1	1.1	2.9	7.2
OECD Europe	20.4	1.6	32.8	2.1	86.1	13.9
Africa	0.1	0.2	2.2	0.6	5.8	3.8
Transition Economies	1.1	0.1	1.6	0.2	4.3	1.3
Middle East	2.4	0	2.8	0	7.3	0
Developing Asia	9.2	3.2	14.0	6.1	36.7	40.4
OECD Pacific	2.8	1.2	3.3	1.8	8.7	11.9
TOTAL	50.6	10.7	85.2	18.5	224.0	121.6

Notes: 1. For regional definitions and country groupings see Annex II. Estimated average annual growth rate for 2010 to 2015 is 11.5% for power and 11% for direct uses. Average worldwide capacity factors of 75% (for electric) and 30% (for direct use) were assumed by 2015.

Table TS.4.4 | Potential geothermal deployments for electricity and direct uses in 2020 through 2050. [Table 4.10]

Year	Use	Capacity[1] (GW)	Generation (TWh/yr)	Generation (EJ/yr)	Total (EJ/yr)
2020	Electricity	25.9	181.8	0.65	2.01
	Direct	143.6	377.5	1.36	
2030	Electricity	51.0	380.0	1.37	5.23
	Direct	407.8	1,071.7	3.86	
2050	Electricity	150.0	1,182.8	4.26	11.83
	Direct	800.0	2,102.3	7.57	

Notes: 1. Installed capacities for 2020 and 2030 are extrapolated from 2015 estimates using a 7% annual growth rate for electricity and 11% for direct uses, and for 2050 are the middle value between projections cited in Chapter 4. Generation was estimated with average worldwide capacity factors of 80% (2020), 85% (2030) and 90% (2050) for electricity and of 30% for direct uses.

on the AR4 baselines (>600 ppm CO$_2$), 440 to 600 ppm (Categories III and IV) and <440 ppm (Categories I and II), range from 0.39 to 0.71 EJ/yr for 2020, 0.22 to 1.28 EJ/yr for 2030 and 1.16 to 3.85 EJ/yr for 2050.

Carbon policy is likely to be one of the main driving factors for future geothermal development, and under the most favourable GHG concentration stabilization policy (<440 ppm), geothermal deployment by 2020, 2030 and 2050 could be significantly higher than the median values noted above. By projecting the historic average annual growth rates of geothermal power plants (7%) and direct uses (11%) from the estimates for 2015, the installed geothermal capacity in 2020 and 2030 for electricity and direct uses could be as shown in Table TS.4.4. By 2050, the geothermal-electric capacity would be as high as 150 GW$_e$ (with half of that comprised of EGS plants), and up to an additional 800 GW$_{th}$ of direct-use plants (Table TS.4.4). [4.8.2]

Even the highest estimates for the long-term contribution of geothermal energy to the global primary energy supply (52.5 EJ/yr by 2050) are within the technical potential ranges (118 to 1,109 EJ/yr for electricity and 10 to 312 EJ/yr for direct uses) and even within the upper range of hydrothermal resources (28.4 to 56.8 EJ/yr). Thus, technical potential is not likely to be a barrier to reaching more ambitious levels of geothermal deployment (electricity and direct uses), at least on a global basis. [4.8.2]

Evidence suggests that geothermal supply could meet the upper range of projections derived from a review of about 120 energy and GHG-reduction scenarios. With its natural thermal storage capacity, geothermal is especially suitable for supplying base-load power. Considering its technical potential and possible deployment, geothermal energy could meet roughly 3% of global electricity demand by 2050, and also has the potential to provide roughly 5% of the global demand for heating and cooling by 2050. [4.8.3]

5. Hydropower

5.1 Introduction

Hydropower is a renewable energy source where power is derived from the energy of water moving from higher to lower elevations. It is a proven, mature, predictable and cost-competitive technology. The mechanical power of falling water is an old tool used for various services from the time of the Greeks more than 2,000 years ago. The world's first hydroelectric station of 12.5 kW was commissioned on 30 September 1882 on Fox River at the Vulcan Street Plant in Appleton, Wisconsin, USA. Though the primary role of hydropower in global energy supply today is in providing centralized electricity generation, hydropower plants also operate in isolation and supply independent systems, often in rural and remote areas of the world. [5.1]

5.2 Resource potential

The annual global technical potential for hydropower generation is 14,576 TWh (52.47 EJ) with a corresponding estimated total capacity potential of 3,721 GW—four times the currently installed global hydropower capacity (Figure TS.5.1). Undeveloped capacity ranges from about 47% in Europe to 92% in Africa, indicating large and well-distributed opportunities for hydropower development worldwide (see Table TS.5.1). Asia and Latin America have the largest technical potentials and the largest undeveloped resources. Africa has highest portion of total potential that is still undeveloped. [5.2.1]

It is noteworthy that the total installed capacities of hydropower in North America, Latin America, Europe and Asia are of the same order of magnitude and, in Africa and Australasia/Oceania, an order of magnitude less; Africa due to underdevelopment and Australasia/Oceania because of size, climate and topography. The global average capacity factor for hydropower plants is 44%. Capacity factor can be indicative of how hydropower is employed in the energy mix (e.g., peaking versus base-load generation) or water availability, or can be an opportunity for increased generation through equipment upgrades and operational optimization. [5.2.1]

The resource potential for hydropower could change due to climate change. Based on a limited number of studies to date, the climate change impacts on existing global hydropower systems is expected to be slightly positive, even though individual countries and regions could have significant positive or negative changes in precipitation and runoff. Annual power production capacity in 2050 could increase by 2.7 TWh (9.72 PJ) in Asia under the SRES A1B scenario, and decrease by 0.8 TWh (2.88 PJ) in Europe. In other regions, changes are found to be even smaller. Globally, the changes caused by climate change in the existing hydropower production system are estimated to be less than 0.1%, although additional research is needed to lower the uncertainty of these projections. [5.2.2]

5.3 Technology and applications

Hydropower projects are usually designed to suit particular needs and specific site conditions, and are classified by project type, head (i.e., the vertical height of water above the turbine) or purpose (single- or multi-purpose). Size categories (installed capacity) are based on national definitions and differ worldwide due to varying policies. There is no immediate, direct link between installed capacity as a classification criterion and general properties common to all hydropower plants (HPPs) above or below that MW limit. All in all, classification according to size, while both common and administratively simple, is—to a degree—arbitrary: general concepts like 'small' or 'large' hydropower are not technically or scientifically rigorous indicators of impacts, economics or characteristics. It may be more useful to evaluate a hydropower project on its sustainability or economic performance thus setting out more realistic indicators. The cumulative relative environmental and social impacts of large versus small hydropower development remain unclear and context dependent. [5.3.1]

Hydropower plants come in three main project types: run-of-river (RoR), storage and pumped storage. RoR HPPs have small intake basins with no storage capacity. Power production therefore follows the hydrological cycle of the watershed. For RoR HPPs the generation varies as water availability changes and thus they may be operated as variable in small streams or as base-load power plants in large rivers. Large-scale RoR HPPs may have some limited ability to regulate water flow, and if they operate in cascades in unison with storage hydropower in upstream reaches, they may contribute to the overall regulating and balancing ability of a fleet of HPPs. A fourth category, in-stream (hydrokinetic) technology, is less mature and functions like RoR without any regulation. [5.3.2]

Hydropower projects with a reservoir (storage hydropower) deliver a broad range of energy services such as base load, peak, and energy storage, and act as a regulator for other sources. In addition they often deliver services that go beyond the energy sector, including flood control, water supply, navigation, tourism and irrigation. Pumped storage plants store water as a source for electricity generation. By reversing the

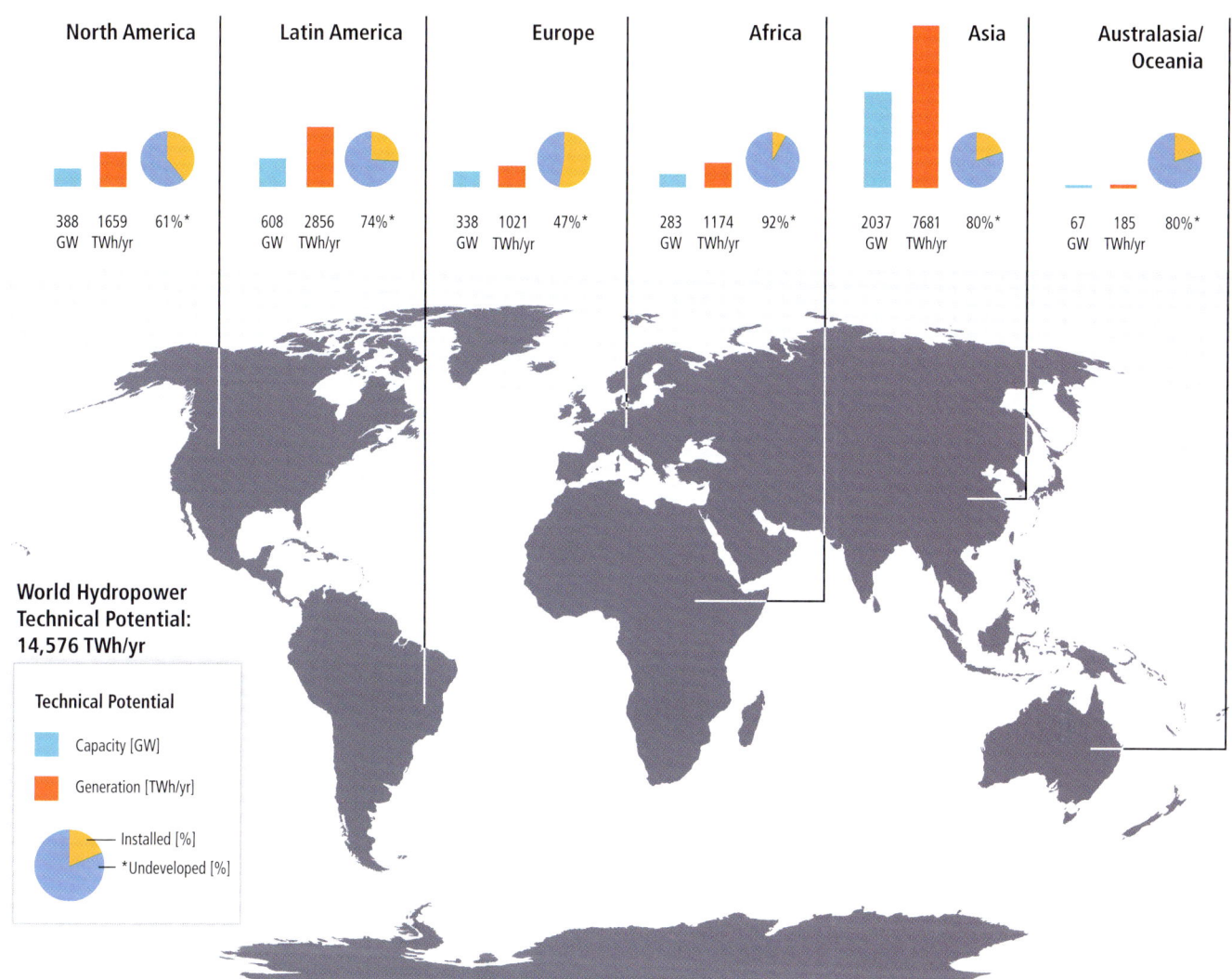

Figure TS.5.1 | Regional hydropower technical potential in terms of annual generation and installed capacity and the percentage of undeveloped technical potential in 2009. [Figure 5.2]

Table TS.5.1 | Regional hydro power technical potential in terms of annual generation and installed capacity (GW); and current generation, installed capacity, average capacity factors and resulting undeveloped potential as of 2009. [Table 5.1]

World region	Technical potential, annual generation TWh/yr (EJ/yr)	Technical potential, installed capacity (GW)	2009 Total generation TWh/yr (EJ/yr)	2009 Installed capacity (GW)	Undeveloped potential (%)	Average regional capacity factor (%)
North America	1,659 (5.971)	388	628 (2.261)	153	61	47
Latin America	2,856 (10.283)	608	732 (2.635)	156	74	54
Europe	1,021 (3.675)	338	542 (1.951)	179	47	35
Africa	1,174 (4.226)	283	98 (0.351)	23	92	47
Asia	7,681 (27.651)	2,037	1,514 (5.451)	402	80	43
Australasia/Oceania	185 (0.666)	67	37 (0.134)	13	80	32
World	14,576 (52.470)	3,721	3,551 (12.783)	926	75	44

flow of water, electrical energy can be produced on demand, with a very fast response time. Pumped storage is the largest-capacity form of grid energy storage now available. [5.3.2.2–5.3.2.3]

Sediment transport and reservoir sedimentation are problems that need to be understood as they have a number of negative effects on HPP performance: depletion of reservoir storage capacity over time; an increase in downstream degradation; increased flood risk upstream of reservoirs; generation losses due to reductions in turbine efficiency; increased frequency of repair and maintenance; and reductions in turbine lifetime and in regularity of power generation. The sedimentation problem may ultimately be controlled through land use policies and the protection of vegetation coverage. Hydropower has the best conversion efficiency of all known energy sources (about 90% efficiency, water to wire) and a very high energy payback ratio. [5.3.3]

Normally the life of a hydroelectric power plant is 40 to 80 years. Electrical and mechanical components and control equipment wear out early compared to civil structures, typically in 30 to 40 years, after which they require renovation. Upgrading/up-rating of HPPs calls for a systematic approach as there are a number of factors (hydraulic, mechanical, electrical and economic) that play a vital role in deciding the course of action. From a techno-economic viewpoint, up-rating should be considered along with renovation and modernization measures. Hydropower generating equipment with improved performance can be retrofitted, often to accommodate market demands for more flexible, peaking modes of operation. Most of the 926 GW of hydropower equipment in operation today (2010) will need to be modernized by 2030 to 2040. Refurbishment of existing hydropower plants often results in enhanced hydropower capacity, both where turbine capacity is being renovated/up-rated or where existing civil infrastructure (like barrages, weirs, dams, canal tunnels, etc.) is being reworked to add new hydropower facilities. [5.3.4]

5.4 Global and regional status of market and industry development

Hydropower is a mature, predictable and price-competitive technology. It currently provides approximately 16% of the world's total electricity production and 86% of all electricity from renewable sources. While hydropower contributes to some level of power generation in 159 countries, 5 countries make up more than half of the world's hydropower production: China, Canada, Brazil, the USA and Russia. The importance of hydroelectricity in the electricity matrix of these countries differs widely, however. While Brazil and Canada are heavily dependent on hydropower to produce 84% and 59% of total generation, respectively, Russia and China produce only 19% and 16% of their total electricity from hydropower, respectively. Despite the significant growth of hydroelectric production around the globe, the percentage share of hydroelectricity has dropped during the last three decades (1973 to 2008) from 21 to 16%, because electricity load and other generation sources have grown more rapidly than has hydropower. [5.4.1]

Carbon credits benefit hydropower projects by helping to secure financing and to reduce risks. Financing is the most decisive step in the entire project development process. Hydropower projects are one of the largest contributors to the flexible mechanisms of the Kyoto Protocol and therefore to existing carbon credit markets. Out of the 2,062 projects registered by the Clean Development Mechanism (CDM) Executive Board by 1 March 2010, 562 are hydropower projects. With 27% of the total number of projects, hydropower is the CDM's leading deployed RE source. China, India, Brazil and Mexico represent roughly 75% of the hosted projects. [5.4.3.1]

Many economical hydropower projects are financially challenged. High up-front costs are a deterrent for investment. Also, hydropower tends to have lengthy lead times for planning, permitting and construction. In the evaluation of lifecycle costs, hydropower often has a very high performance, with annual O&M costs being a fraction of the capital investment. As hydropower and its industry are old and mature, it is expected that the hydropower industry will be able to meet the demand that will be created by the predicted deployment rate in the years to come. For example, in 2008 the hydropower industry managed to install more than 41 GW of new capacity worldwide. [5.4.3.2]

The development of more appropriate financing models is a major challenge for the hydropower sector, as is finding the optimum roles for the public and private sectors. The main challenges for hydropower relate to creating private-sector confidence and reducing risk, especially prior to project permitting. Green markets and trading in emissions reductions will undoubtedly provide incentives. Also, in developing regions, such as Africa, interconnection between countries and the formation of power pools is building investor confidence in these emerging markets. [5.4.3.2]

The concepts of classifying HPPs as 'small' or 'large', as defined by installed capacity (MW), can act as a barrier to the development of hydropower. For example, these classifications can impact the financing of new hydropower plants, determining how hydropower is treated in climate change and energy policies. Different incentives are used for small-scale hydropower (FITs, green certificates and bonuses) depending on the country, but no incentives are available for large-scale HPPs. The EU Linking Directive sets a limit for carbon credits issued from HPPs to 20 MW. The same limit is found in the UK Renewables Obligation, a green certificate market-based mechanism. Likewise, in several countries FITs do not apply to hydropower above a certain size limit (e.g., France 12 MW, Germany 5 MW, India 5 and 25 MW). [5.4.3.4]

The UNFCCC CDM Executive Board has decided that storage hydropower projects will have to follow the power density indicator (PDI: installed capacity/reservoir area in W/m^2) to be eligible for CDM credits. The PDI rule seems to presently exclude storage hydropower from qualifying for CDM (or Joint Implementation) credits and may lead to suboptimal development of hydropower resources as the non-storage RoR option will be favoured.

5.5 Integration into broader energy systems

Hydropower's large capacity range, its flexibility, storage capability (when coupled with a reservoir), and ability to operate in a stand-alone mode or in grids of all sizes enables it to deliver a broad range of services. [5.5]

Hydropower can be delivered through the national and regional electric grid, mini-grids and also in isolated mode. Realization has been growing in developing countries that small-scale hydropower schemes have an important role to play in the socioeconomic development of remote rural, especially hilly, areas as those can provide power for industrial, agricultural and domestic uses. In China, small-scale HPPs have been one of the most successful examples of rural electrification, where over 45,000 small HPPs totalling over 55,000 MW of capacity and producing 160 TWh (576 PJ) of generation annually benefit over 300 million people. [5.5.2]

With a very large reservoir relative to the size of the hydropower plant (or very consistent river flows), HPPs can generate power at a near-constant level throughout the year (i.e., operate as a base-load plant). Alternatively, in the case that the hydropower capacity far exceeds the amount of reservoir storage, the hydropower plant is sometimes referred to as energy-limited. An energy-limited hydro plant would exhaust its 'fuel supply' by consistently operating at its rated capacity throughout the year. In this case, the use of reservoir storage allows hydropower generation to occur at times that are most valuable from the perspective of the power system rather than at times dictated solely by river flows. Since electrical demand varies during the day and night, during the week and seasonally, storage hydropower generation can be timed to coincide with times where the power system needs are the greatest. In part, these times will occur during periods of peak electrical demand. Operating hydropower plants in a way to generate power during times of high demand is referred to as peaking operation (in contrast to base-load). Even with storage, however, hydropower generation will still be limited by the size of the storage, the rated electrical capacity of the hydropower plant, and downstream flow constraints for irrigation, recreation or environmental uses of the river flows. Hydropower peaking may, if the outlet is directed to a river, lead to rapid fluctuations in river flow, water-covered area, depth and velocity. In turn this may, depending on local conditions, lead to negative impacts in the river unless properly managed. [5.5.3]

In addition to hydropower supporting fossil and nuclear generation technologies, it can also help reduce the challenges with integrating variable renewable resources. In Denmark, for example, the high level of variable wind energy (>20% of the annual energy demand) is managed in part through strong interconnections (1 GW) to Norway, which has substantial storage hydropower. More interconnectors to Europe may further support increasing the share of wind power in Denmark and Germany. Increasing variable generation will also increase the amount of balancing services, including regulation and load following, required by the power system. In regions with new and existing hydropower facilities, providing these services from hydropower may avoid the need to rely on increased part-load and cycling of conventional thermal plants to provide these services. [5.5.4]

Though hydro has the potential to offer significant power system services in addition to energy and capacity, interconnecting and reliably utilizing HPPs may also require changes to power systems. The interconnection of hydropower to the power system requires adequate transmission capacity from HPPs to demand centres. Adding new HPPs has in the past required network investments to extend the transmission network. Without adequate transmission capacity, HPP operation can be constrained such that the services offered by the plant are less than what it could offer in an unconstrained system. [5.5.5]

5.6 Environmental and social impacts

Like all energy and water management options, hydropower projects have negative and positive environmental and social impacts. On the environmental side, hydropower may have a significant environmental footprint at local and regional levels but offers advantages at the macro-ecological level. With respect to social impacts, hydropower projects may entail the relocation of communities living within or nearby the reservoir or the construction sites, compensation for downstream communities, public health issues, and others. A properly designed hydropower project may, however, be a driving force for socioeconomic development, though a critical question remains about how these benefits are shared. [5.6]

All hydroelectric structures affect a river's ecology, mainly by inducing a change into its hydrologic characteristics and by disrupting the ecological continuity of sediment transport and fish migration through the building of dams, dikes and weirs. However, the extent to which a river's physical, chemical, biological and ecosystem characteristics are modified depends largely on the type of HPP. Whereas RoR hydropower projects do not alter a river's flow regime, the creation of a reservoir for storage hydropower entails a major environmental change by transforming a fast-running river ecosystem into a still-standing artificial lake. [5.6.1.1–5.6.1.6]

Similar to a hydropower project's ecological effects, the extent of its social impacts on the local and regional communities, land use, economy, health and safety or heritage varies according to project type and site-specific conditions. While RoR projects generally introduce little social change, the creation of a reservoir in a densely populated area can entail significant challenges related to resettlement and impacts on the livelihoods of the downstream populations. Restoration and improvement of living standards of affected communities is a long-term and challenging task that has been managed with variable success in the past. Whether HPPs can contribute to fostering socioeconomic development depends largely on how the generated services and revenues are shared and distributed among different stakeholders. HPPs can also have positive impacts on the living conditions of local communities and the regional economy, not only by generating electricity but also by facilitating through the creation of freshwater storage schemes multiple other water-dependent activities, such as irrigation, navigation, tourism, fisheries or sufficient water supply to municipalities and industries while protecting against floods and droughts. [5.6.1.7–5.6.1.11]

The assessment and management of environmental and social impacts associated with, especially, larger HPPs represent a key challenge for hydropower development. Emphasizing transparency and an open, participatory decision-making process, the stakeholder consultation approach is driving both present-day and future hydropower projects towards increasingly more environmentally friendly and sustainable solutions. In many countries, a national legal and regulatory framework has been put in place to determine how hydropower projects shall be developed and operated, while numerous multilateral financing agencies have developed their own guidelines and requirements to assess the economic, social and environmental performance of hydropower projects. [5.6.2]

One of hydropower's main environmental advantages is that it creates no atmospheric pollutants or waste associated with fuel combustion. However, all freshwater systems, whether they are natural or man-made, emit GHGs (e.g., CO_2, methane) due to decomposing organic material. Lifecycle assessments (LCAs) carried out on hydropower projects have so far demonstrated the difficulty of generalizing estimates of lifecycle GHG emissions for hydropower projects in all climatic conditions, pre-impoundment land cover types, ages, hydropower technologies, and other project-specific circumstances. The multipurpose nature of most hydropower projects makes allocation of total impacts to the several purposes challenging. Many LCAs to date allocate all impacts of hydropower projects to the electricity generation function, which in some cases may overstate the emissions for which they are 'responsible'. LCAs (Figure TS.5.2) that evaluate GHG emissions of HPPs during construction, operation and maintenance, and dismantling, show that the majority of lifecycle GHG emission estimates for hydropower cluster between about 4 and 14 g CO_2eq/kWh, but under certain scenarios there is potential to emit much larger quantities of GHGs, as shown by the outliers. [5.6.3.1]

While some natural water bodies and freshwater reservoirs may even absorb more GHGs than they emit, there is a definite need to properly assess the net change in GHG emissions induced by the creation of such reservoirs. All LCAs included in these assessments evaluated only gross GHG emissions from reservoirs. Whether reservoirs are net emitters of GHGs, considering emissions that would have occurred without the reservoir, is an area of active research. When considering net anthropogenic emissions as the difference in the overall carbon cycle between the situations with and without the reservoir, there is currently no consensus on whether reservoirs are net emitters or net sinks. Presently two international processes are investigating this issue: the UN Educational, Scientific and Cultural Organization/International Hydrological Programme research project and the IEA Hydropower Agreement Annex XII. [5.6.3.2]

5.7 Prospects for technology improvement and innovation

Though hydropower is a proven and well-advanced technology, there is still room for further improvement, for example, by optimizing operations, mitigating or reducing environmental impacts, adapting to new social and environmental requirements and implementing more robust and cost-effective technological solutions. Large hydropower turbines are now close to the theoretical limit for efficiency, with up to 96% efficiency when operated at the best efficiency point, but this is not always possible and continued research is needed to make more efficient operation possible over a broader range of flows. Older turbines can have lower efficiency by design or reduced efficiency due to corrosion and cavitation. There is therefore the potential to increase energy output by retrofitting with new higher efficiency equipment and usually also with increased capacity. Most of the existing electrical and mechanical equipment in operation today will need to be modernized during the next three decades, allowing for improved efficiency and higher power and energy output. Typically, generating equipment can be upgraded or replaced with more technologically advanced electro-mechanical equipment two or three times during the lifetime of the project, making more effective use of the same flow of water. [5.7]

There is much ongoing technology innovation and material research aiming to extend the operational range in terms of head and discharge, and also to improve environmental performance, reliability and reduce costs. Some of the promising technologies under development are variable-speed and matrix technologies, fish-friendly turbines, hydrokinetic turbines, abrasive-resistant turbines, and new tunnelling and dam technologies. New technologies aiming at utilizing low (<15 m) or very low (<5 m) head may open up many sites for hydropower that have not been within reach of conventional technology. As most of the data available on hydropower potential are based on field work produced several decades ago, when low-head hydropower was not a high priority, existing data on low-head hydropower potential may not be complete. Finally, there is a significant potential for improving operation of HPPs by utilizing new methods for optimizing plant operation. [5.7.1–5.7.8]

5.8 Cost trends

Hydropower is often economically competitive with current market energy prices, though the cost of developing, deploying and operating new hydropower projects will vary from project to project. Hydropower projects often require a high initial investment, but have the advantage of very low O&M costs and a long lifespan. [5.8]

Investment costs for hydropower include costs of planning; licensing; plant construction; impact reductions for fish and wildlife, recreational, historical and archaeological sites; and water quality monitoring. Overall, there are two major cost groups: the civil construction costs, which normally are the greatest costs of the hydropower project; and electromechanical equipment costs. The civil construction costs follow the price trends in the country where the project is going to be developed. In the case of countries with economies in transition, the costs are likely to be relatively low due to the use of local labour and local materials. The costs of electromechanical equipment follow the tendency of prices at a global level. [5.8.1]

Based on a standardized methodology outlined in Annex II and the cost and performance data summarized in Annex III, the LCOE for hydropower projects over a large set and range of input parameters has been

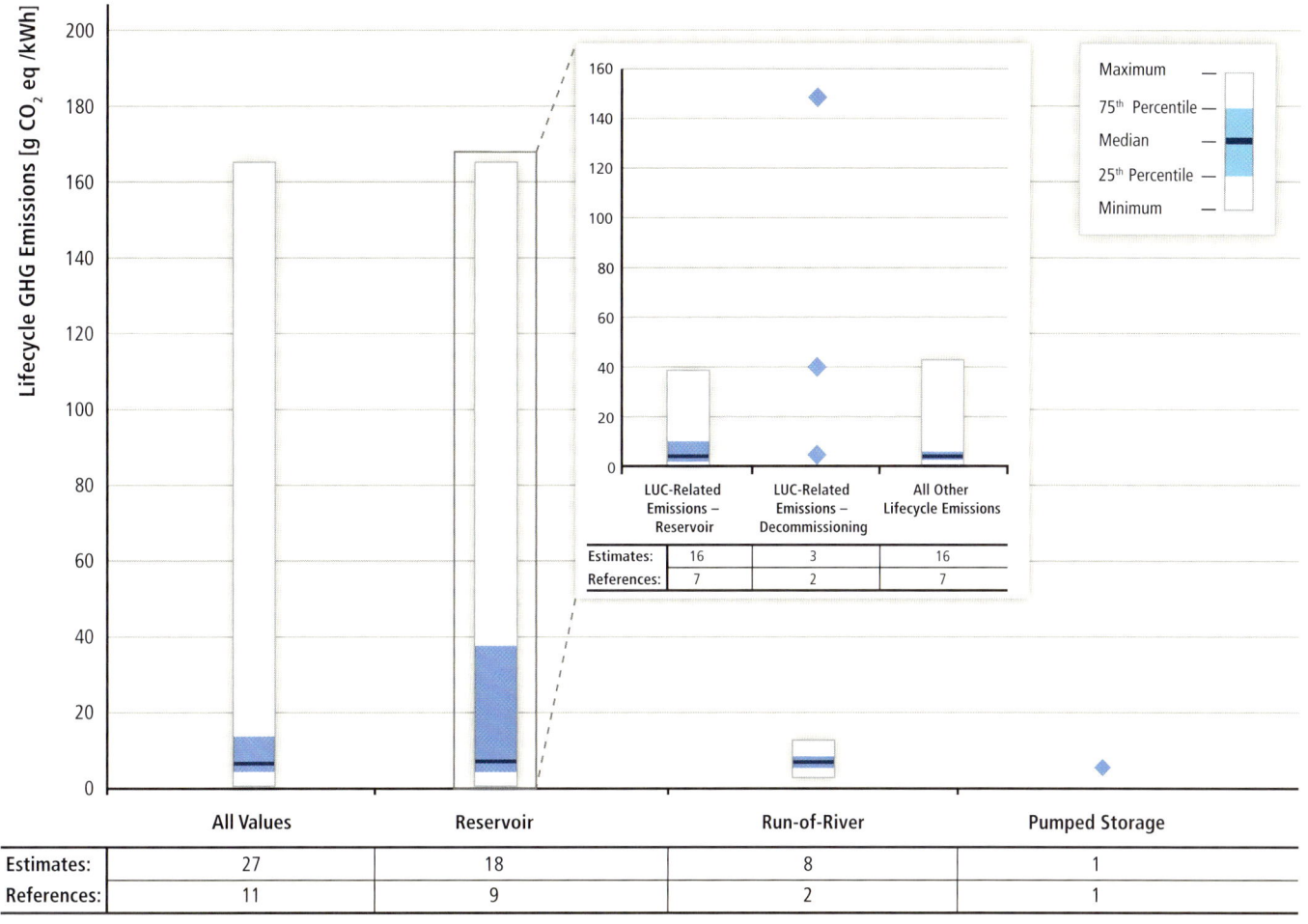

Figure TS.5.2 | Life-cycle GHG emissions of hydropower technologies (unmodified literature values, after quality screen). See Annex I for details of literature search and citations of literature contributing to the estimates displayed. Surface emissions from reservoirs are referred to as gross GHG emissions. [Figure 5.15]

calculated to range from as low as US cent$_{2005}$ 1.1/kWh to US cent$_{2005}$ 15/kWh, depending on site-specific parameters for investment costs of each project and on assumptions regarding the discount rate, capacity factor, lifetime and O&M costs. [1.3.2, 5.8, 10.5.1, Annex II, Annex III]

Figure TS.5.3 presents the LCOE for hydropower projects over a somewhat different and more typical set and range of parameters consistent with the majority of hydropower projects, and does so as a function of capacity factor while applying different investment costs and discount rates.

Capacity factors will be determined by hydrological conditions, installed capacity and plant design, and the way the plant is operated. For power plant designs intended for maximum energy production (base-load) and/or with some regulation, capacity factors will often be from 30 to 60%, with average capacity factors for different world regions shown in the graph. For peaking-type power plants, the capacity factor can be even lower, whereas capacity factors for RoR systems vary across a wide range (20 to 95%) depending on the geographical and climatological conditions, technology, and operational characteristics. For an average capacity factor of 44% and investment costs between USD$_{2005}$ 1,000/kW and USD$_{2005}$ 3,000/kW, the LCOE ranges from US cent$_{2005}$ 2.5/kWh to US cent$_{2005}$ 7.5/kWh.

Most of the projects developed in the near-term future (up to 2020) are expected to have investment costs and LCOE in this range, though projects with both lower and higher costs are possible. Under good conditions, the LCOE of hydropower can be in the range of US cent$_{2005}$ 3/kWh to US cent$_{2005}$ 5/kWh. [5.8.3, 8.2.1.2, Annex III]

There is relatively little information on historical trends in hydropower costs in the literature. One reason for this—besides the fact that project costs are highly site-specific—may be the complex cost structure for hydropower plants, where some components may have decreasing cost trends (e.g., tunnelling costs), while others may have increasing cost trends (e.g., social and environmental mitigation costs). [5.8.4]

One complicating factor when considering the cost of hydropower is that, for multipurpose reservoirs, there is a need to share or allocate the cost of serving other water uses like irrigation, flood control, navigation, roads, drinking water supply, fish, and recreation. There are

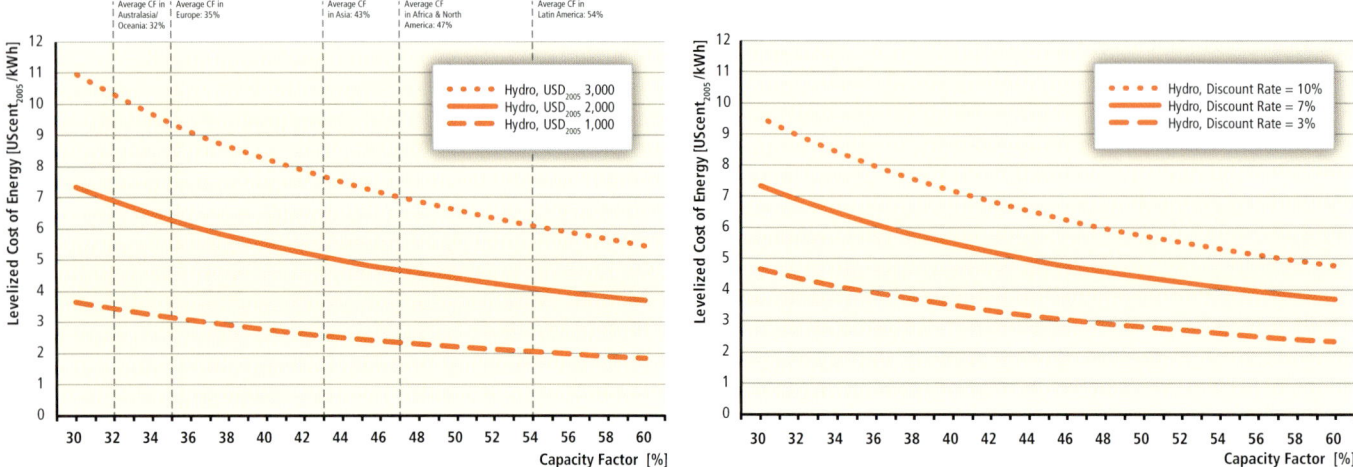

Figure TS.5.3 | Recent and near-term estimated levelized cost of hydropower (a) as a function of capacity factor and investment cost*, ***; and (b) as a function of capacity factor and discount rate**,***. [Figure 5.20]

Notes: * Discount rate is assumed to equal 7%. ** Investment cost is assumed to be USD 2,000/kW. *** Annual O&M cost is assumed at 2.5%/yr of investment cost and plant lifetime as 60 years.

different methods of allocating the cost to individual purposes, each of which has advantages and drawbacks. The basic rules are that the allocated cost to any purpose does not exceed that benefit of that purpose and each purpose will be carried out at its separable cost. Separable cost for any purpose is obtained by subtracting the cost of a multipurpose project without that purpose from the total cost of the project with the purpose included. Merging economic elements (energy and water selling prices) with social benefits (supplying water to farmers in case of lack of water) and the value of the environment (to preserve a minimum environmental flow) is becoming a tool for consideration of cost sharing for multipurpose reservoirs. [5.8.5]

5.9 Potential deployment

Hydropower offers a significant potential for near- and long-term carbon emissions reduction. On a global basis, the hydropower resource is unlikely to constrain further development in the near to medium term, though environmental and social concerns may limit deployment opportunities if not carefully managed. [5.9]

So far, only 25% of the hydropower potential has been developed across the world (that is, 3,551 TWh out of 14,575 TWh) (12.78 EJ out of 52.47 EJ). The different long-term prospective scenarios propose a continuous increase for the next decades. The increase in hydropower capacity over the last 10 years is expected by several studies to continue in the near to medium term: from 926 GW in 2009 to between 1,047 and 1,119 GW by 2015; an annual addition ranging from 14 to 25 GW. [5.9, 5.9.1]

The reference-case projections presented in Chapter 10 (based on 164 analyzed longer-term scenarios) show hydropower's role in the global energy supply covering a broad range, with a median of roughly 13 EJ (3,600 TWh) in 2020, 16 EJ (4,450 TWh) in 2030 and 19 EJ (5,300 TWh) in 2050. 12.78 EJ was reached already in 2009 and thus the average estimate of 13 EJ for 2020 has probably been exceeded today. Also, some scenario results provide lower values than the current installed capacity for 2020, 2030 and 2050, which is counterintuitive given, for example, hydropower's long lifetimes, its significant market potential and other important services. These results could maybe be explained by model/scenario weaknesses (see discussions in Section 10.2.1.2 of this report). Growth of hydropower is therefore projected to occur even in the absence of GHG mitigation policies, even with hydropower's median contribution to global electricity supply dropping from about 16% today to less than 10% by 2050. As GHG mitigation policies are assumed to become more stringent in the alternative scenarios, the contribution of hydropower grows: by 2030, hydropower's median contribution equals roughly 16.5 EJ (4,600 TWh) in the 440 to 600 and <440 ppm CO_2 stabilization ranges (compared to the median of 15 EJ in the baseline cases), increasing to about 19 EJ by 2050 (compared to the median of 18 EJ in the baseline cases). [5.9.2]

Regional projections of hydropower generation in 2035 show a 98% increase in the Asia Pacific region compared to 2008 levels and a 104% increase in Africa. Brazil is the main driving force behind the projected 46% increase in hydropower generation in the South and Central America region over the same time period. North America and Europe/Eurasia expect more modest increases of 13 and 27%, respectively, over the period. [5.9.2]

Overall, evidence suggests that relatively high levels of deployment in the next 20 years are feasible. Even if hydropower's share in global electricity supply decreases by 2050, hydropower would remain an attractive RE source within the context of global carbon mitigation scenarios. Furthermore, increased development of storage hydropower may enable investment into water management infrastructure, which is needed in response to growing problems related to water resources. [5.9.3]

5.10 Integration into water management systems

Water, energy and climate change are inextricably linked. Water availability is crucial for many energy technologies, including hydropower, while energy is needed to secure water supply for agriculture, industries and households, in particular in water-scarce areas in developing countries. This close relationship has led to the understanding that the water-energy nexus must be addressed in a holistic way, in particular with regard to climate change and sustainable development. Providing energy and water for sustainable development may require improved regional and global water governance. As it is often associated with the creation of water storage facilities, hydropower is at the crossroads of these issues and can play an important role in enhancing both energy and water security. [5.10]

Today, about 700 million people live in countries experiencing water stress or scarcity. By 2035, it is projected that three billion people will be living in conditions of severe water stress. Many countries with limited water availability depend on shared water resources, increasing the risk of conflict over these scarce resources. Therefore, adaptation to climate change impacts will become very important in water management. [5.10.1]

In a context where multipurpose hydropower can be a tool to mitigate both climate change and water scarcity, these projects may have an enabling role beyond the electricity sector as a financing instrument for reservoirs, helping to secure freshwater availability. However, multiple uses may increase the potential for conflicts and reduce energy production during times of low water levels. As major watersheds are shared by several nations, regional and international cooperation is crucial. Both intergovernmental agreements and initiatives by international institutions are actively supporting these important processes. [5.10.2, 5.10.3]

6. Ocean Energy

6.1 Introduction

Ocean energy offers the potential for long-term carbon emissions reduction but is unlikely to make a significant short-term contribution before 2020 due to its nascent stage of development. The theoretical potential of 7,400 EJ/yr contained in the world's oceans easily exceeds present human energy requirements. Government policies are contributing to accelerate the deployment of ocean energy technologies, heightening expectations that rapid progress may be possible. The six main classes of ocean energy technology offer a diversity of potential development pathways, and most offer potentially low environmental impacts as currently understood. There are encouraging signs that the investment cost of ocean energy technologies and the levelized cost of electricity generated will decline from their present non-competitive levels as R&D and demonstrations proceed, and as deployment occurs. Whether these cost reductions are sufficient to enable broad-scale deployment of ocean energy is the most critical uncertainty in assessing the future role of ocean energy in mitigating climate change. [6 ES, 6.1]

6.2 Resource potential

Ocean energy can be defined as energy derived from technologies that utilize seawater as their motive power or harness the water's chemical or heat potential. The RE resource in the ocean comes from six distinct sources, each with different origins and each requiring different technologies for conversion. These sources are:

Wave energy derived from the transfer of the kinetic energy of the wind to the upper surface of the ocean. The total theoretical wave energy resource is 32,000 TWh/yr (115 EJ/yr), but the technical potential is likely to be substantially less and will depend on development of wave energy technologies. [6.2.1]

Tidal range (tidal rise and fall) derived from gravitational forces of the Earth-Moon-Sun system. The world's theoretical tidal power potential is in the range of 1 to 3 TW, located in relatively shallow waters. Again, technical potential is likely to be significantly less than theoretical potential. [6.2.2]

Tidal currents derived from water flow that results from the filling and emptying of coastal regions associated with tides. Current regional estimates of tidal current technical potential include 48 TWh/yr (0.17 EJ) for Europe and 30 TWh/yr (0.11 EJ/yr) for China. Commercially attractive sites have also been identified in the Republic of Korea, Canada, Japan, the Philippines, New Zealand and South America. [6.2.3]

Ocean currents derived from wind-driven and thermohaline ocean circulation. The best-characterized system of ocean currents is the Gulf Stream in North America, where the Florida Current has a technical potential for 25 GW of electricity capacity. Other regions with potentially promising ocean circulation include the Agulhas/Mozambique Currents off South Africa, the Kuroshio Current off East Asia and the East Australian Current. [6.2.4]

Ocean thermal energy conversion (OTEC) derived from temperature differences arising from solar energy stored as heat in upper ocean layers and colder seawater, generally below 1,000 m. Although the energy density of OTEC is relatively low, the overall resource potential is much larger than for other forms of ocean energy. One 2007 study estimates that about 44,000 TWh/yr (159 EJ/yr) of steady-state power may be possible. [6.2.5]

Salinity gradients (osmotic power) derived from salinity differences between fresh and ocean water at river mouths. The theoretical potential of salinity gradients is estimated at 1,650 TWh/yr (6 EJ/yr). [6.2.6]

Figure TS.6.1 provides examples of how selected ocean energy resources are distributed across the globe. Some ocean energy resources, such as ocean currents or power from salinity gradients, are globally distributed. Ocean thermal energy is principally located in the Tropics around the equatorial latitudes (latitudes 0° to 35°), whilst the highest annual wave power occurs between latitudes of 30° to 60°. Wave power in the southern hemisphere undergoes smaller seasonal variation than in the northern hemisphere. Ocean currents, ocean thermal energy, salinity gradients and, to some extent, wave energy are consistent enough to generate base-load power. Given the early state of the available literature and the substantial uncertainty in ocean energy's technical potential, the estimates for technical ocean energy potential vary widely. [6.2.1–6.2.6]

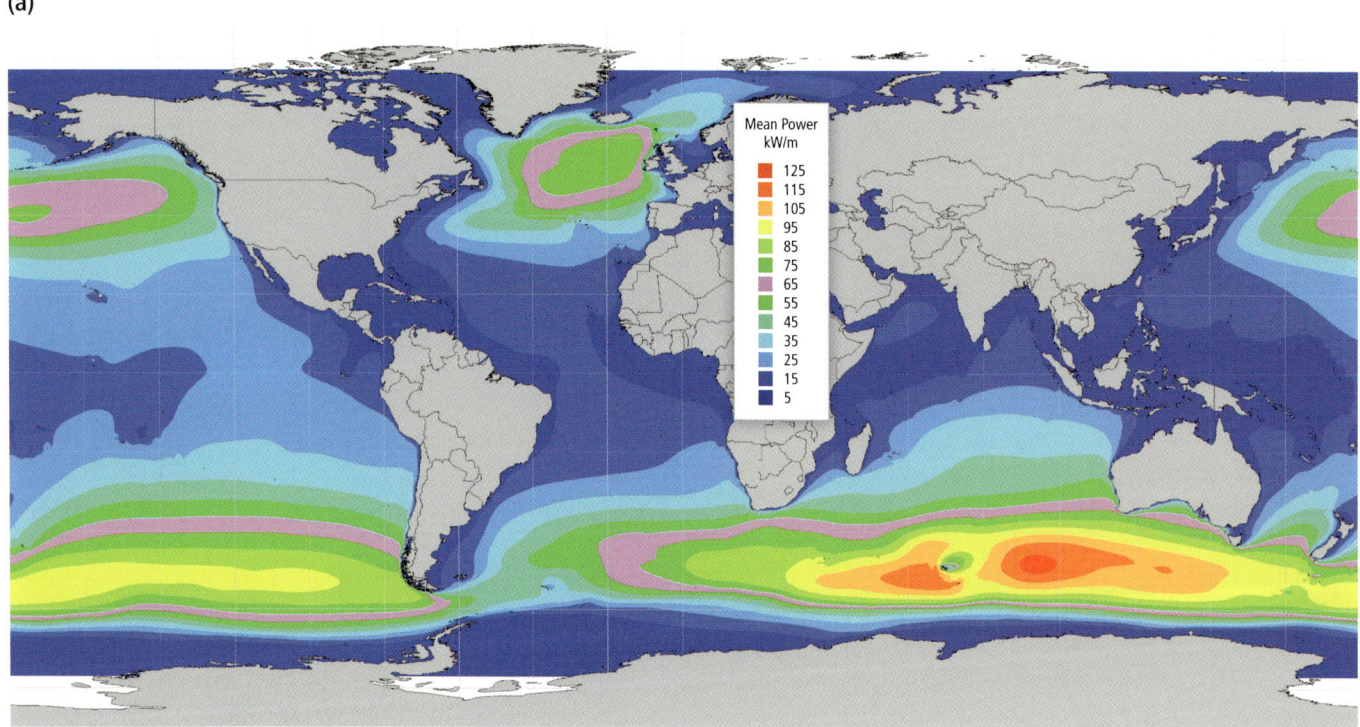

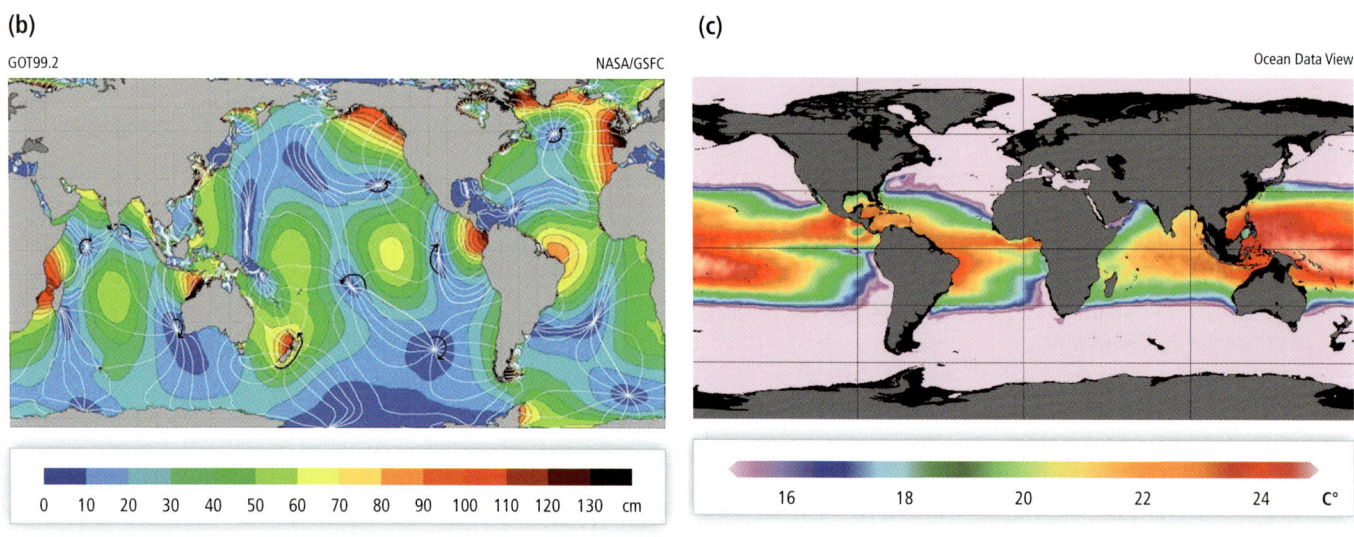

Figure TS.6.1a-c | Global distribution of various ocean energy resources: (a) Wave power; (b) Tidal range, (c) Ocean thermal energy. [Figures 6.1, 6.2, 6.4]

(d)

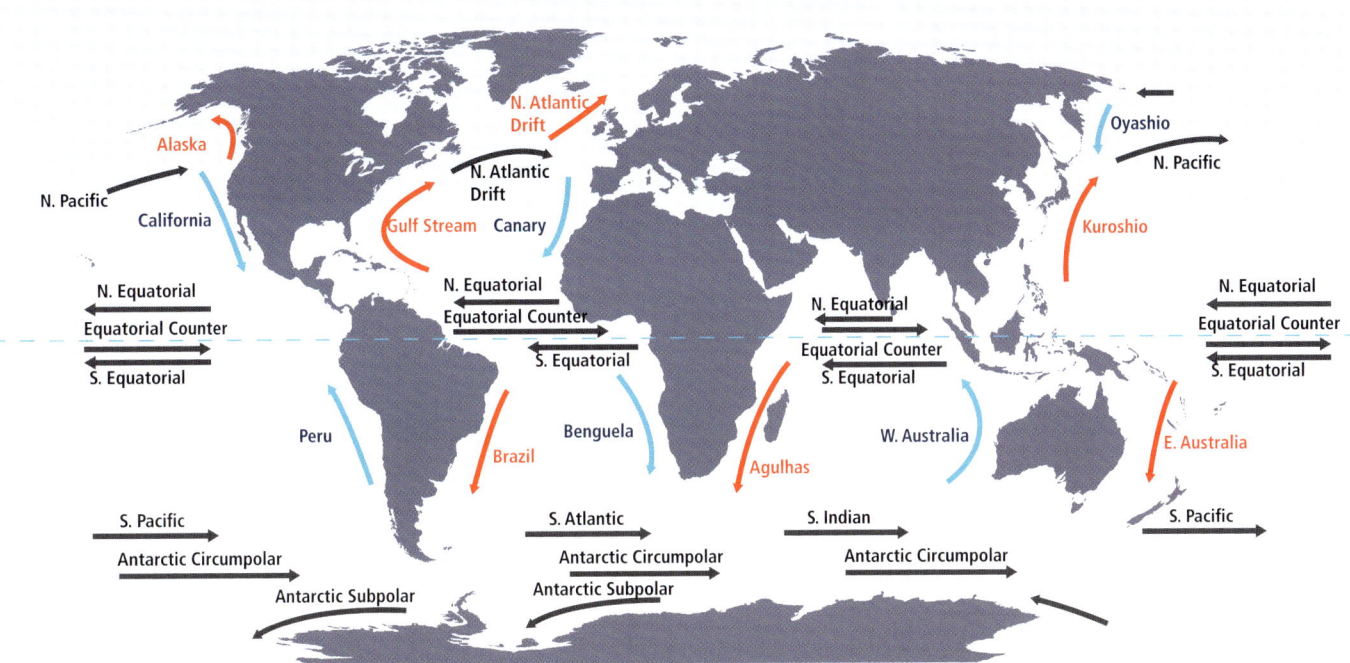

Figure TS.6.1d | Global distribution of various ocean energy resources: (d) Ocean currents. [Figure 6.3]

6.3 Technology and applications

The current development status of ocean energy technologies ranges from the conceptual and pure R&D stages to the prototype and demonstration stage, and only tidal range technology can be considered mature. Presently there are many technology options for each ocean energy source and, with the exception of tidal range barrages, technology convergence has not yet occurred. Over the past four decades, other marine industries (primarily offshore oil and gas) have made significant advances in the fields of materials, construction, corrosion, submarine cables and communications. Ocean energy is expected to directly benefit from these advances. [6.3.1]

Many wave energy technologies representing a range of operating principles have been conceived, and in many cases demonstrated, to convert energy from waves into a usable form of energy. Major variables include the method of wave interaction with respective motions (heaving, surging, pitching) as well as water depth (deep, intermediate, shallow) and distance from shore (shoreline, near-shore, offshore). Wave energy technologies can be classified into three groups: oscillating water columns (OWC: shore-based, floating), oscillating bodies (surface buoyant, submerged), and overtopping devices (shore-based, floating). [6.2.3] Principles of operation are presented in Figure TS.6.2.

Tidal range energy can be harnessed by the adaptation of river-based hydroelectric dams to estuarine situations, where a barrage encloses an estuary. The barrage may generate electricity on both the ebb and flood tides and some future barrages may have multiple basins to enable almost continuous generation. The most recent technical concepts are stand-alone offshore 'tidal lagoons'. [6.3.3]

Technologies to harness power from tidal and ocean currents are also under development, but tidal energy turbines are more advanced. Some of the tidal/ocean current energy technologies are similar to mature wind turbine generators but submarine turbines must also account for reversing flow, cavitation at blade tips and harsh underwater marine conditions. Tidal currents tend to be bidirectional, varying with the tidal cycle, and relatively fast-flowing, compared with ocean currents, which are usually unidirectional and slow-moving but continuous. Converters are classified by their principle of operation into axial flow turbines, cross flow turbines and reciprocating devices as presented in Figure TS.6.3. [6.3.4]

Ocean thermal energy conversion (OTEC) plants use the temperature differences between warm seawater from the ocean surface and cool seawater from depth (1,000 m is often used as a reference level) to produce electricity. Open-cycle OTEC systems use seawater directly as the circulating fluid, whilst closed-cycle systems use heat exchangers and a secondary working fluid (most commonly ammonia) to drive a turbine. Hybrid systems use both open- and closed-cycle operation. Although there have been trials of OTEC technologies, problems have been encountered with maintenance of vacuums, heat exchanger biofouling and corrosion issues. Current research is focused on overcoming these problems. [6.3.5]

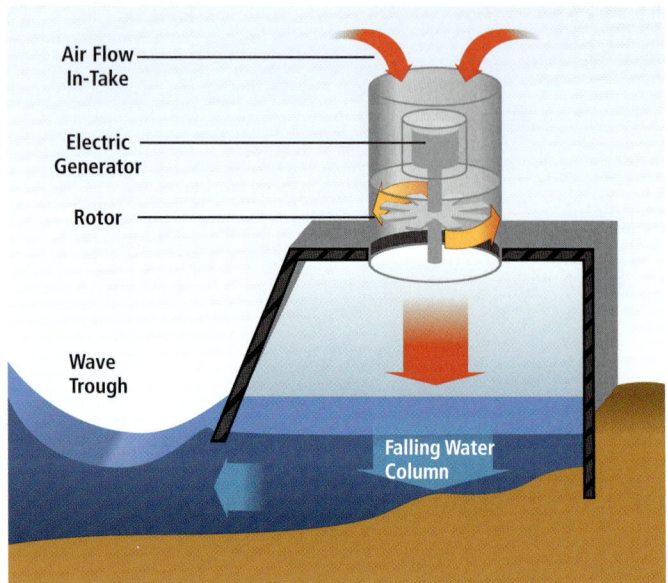
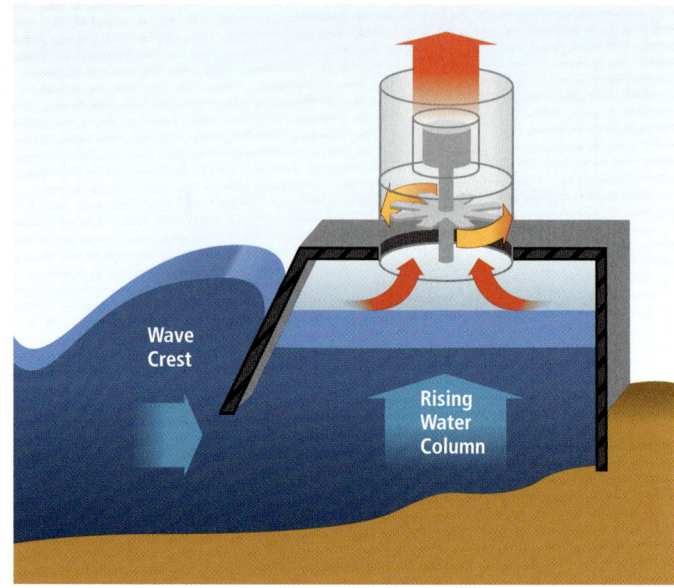

Figure TS.6.2a/b | Type of wave energy converter and its operation: oscillating water column device. [Figure 6.6] (design by the National Renewable Energy Laboratory (NREL))

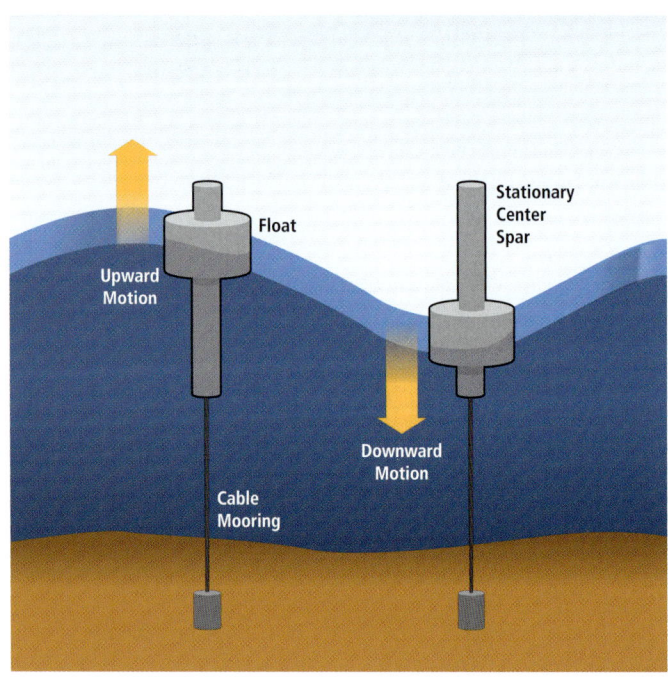
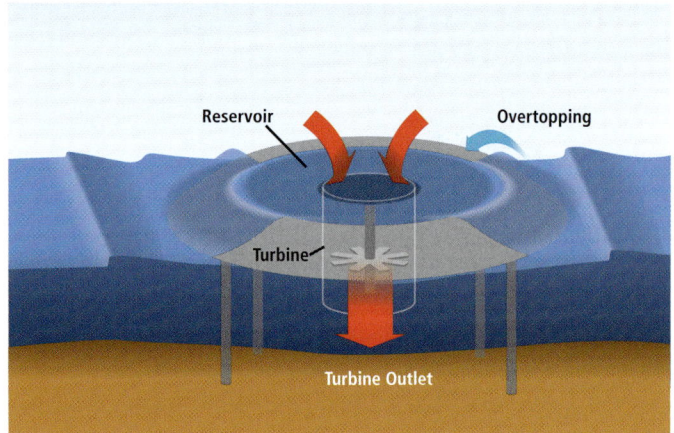

Figure TS.6.2c/d | Wave energy converters and their operation: (left) oscillating body device; and (right) overtopping device. [Figure 6.6] (design by the National Renewable Energy Laboratory (NREL))

The salinity gradient between freshwater from rivers and seawater can be utilized as a source of power with at least two concepts under development. The reversed electro dialysis (RED) process is a concept in which the difference in chemical potential between the two solutions is the driving force (Figure TS.6.4). The pressure-retarded osmosis, or osmotic power process, utilizes the concept of naturally occurring osmosis, a hydraulic pressure potential, caused by the tendency of freshwater to mix with seawater due to the difference in salt concentration (Figure TS.6.5). [6.3.6]

6.4 Global and regional status of the markets and industry development

R&D projects on wave and tidal current energy technologies have proliferated over the past two decades, with some now reaching the full-scale pre-commercial prototype stage. Presently, the only full-size and operational ocean energy technology available is the tidal barrage, of which the best example is the 240 MW La Rance Barrage in north-western France, completed in 1966. The 254 MW Sihwa Barrage (South Korea) is due to become operational in 2011. Technologies to develop other ocean energy sources including OTEC, salinity gradients and ocean currents are still at the conceptual, R&D or early prototype stages. Currently, more than 100 different ocean energy technologies are under development in over 30 countries. [6.4.1]

The principal investors in ocean energy R&D and deployments are national, federal and state governments, followed by major energy utilities and investment companies. National and regional governments are

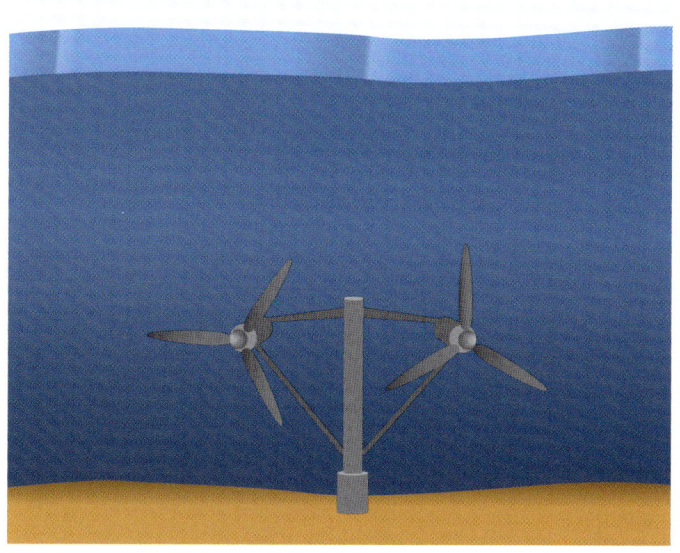
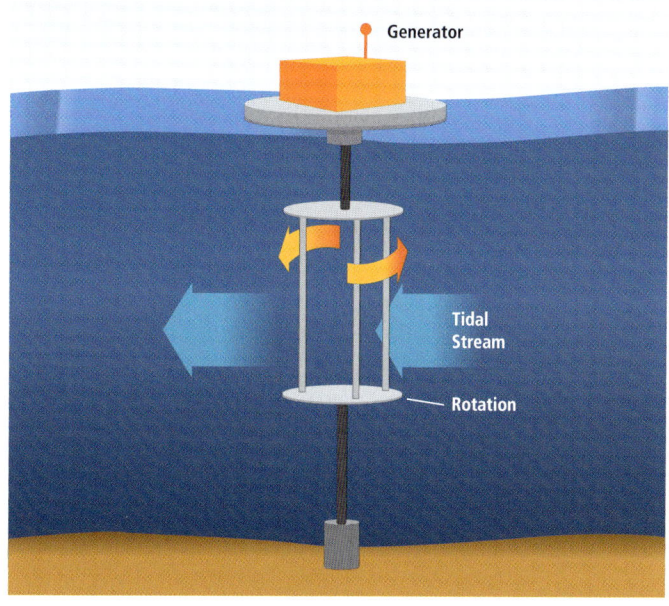
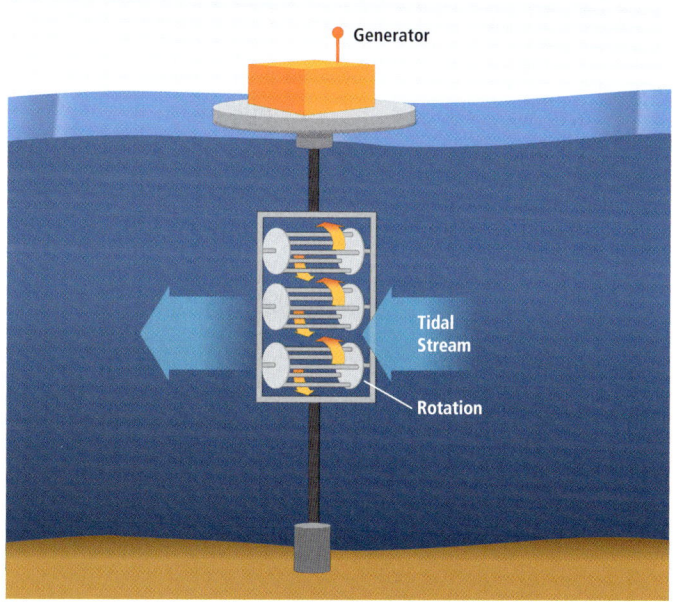

Figure TS.6.3 | Tidal current energy converters and their operation: (Top left) twin turbine horizontal axis device; (Bottom left) cross-flow device; and (Top right) vertical axis device. [Figure 6.8]

particularly supportive of ocean energy through a range of financial, regulatory and legislative initiatives to support developments. [6.4.7]

Industrial involvement in ocean energy is at a very early stage and there is no manufacturing industry for these technologies at present. The growth of interest may lead to the transfer of capacity, skills and capabilities from related industries, combined with new specific innovative aspects. One interesting feature of ocean energy is the development of a number of national marine energy testing centres and these are becoming foci for device testing, certification and advanced R&D. [6.4.1.2]

The status of industry development can be assessed by the current and recent deployments of ocean energy systems.

Wave energy: A number of shore-based wave energy prototypes are operating around the world. Two OWC devices have been operational in Portugal and Scotland for approximately a decade, while two other offshore OWC devices have been tested at prototype scale in Australia and Ireland. Another OWC was operational off the southern coast of India between 1990 and 2005. A number of companies in Australia, Brazil, Denmark, Finland, Ireland, Norway, Portugal, Spain, Sweden, New Zealand, the UK and the USA have been testing pilot scale or pre-commercial prototypes at sea, with the largest being 750 kW. [6.4.2]

Tidal range: The La Rance 240 MW plant in France has been operational since 1966. Other smaller projects have been commissioned since then in China, Canada and Russia. The Sihwa barrage 254 MW plant in Korea will be commissioned during 2011, and several other large projects are under consideration. [6.4.3]

Tidal and ocean currents: There are probably more than 50 tidal current devices at the proof-of-concept or prototype development stage, but large-scale deployment costs are yet to be demonstrated. The most advanced example is the SeaGen tidal turbine, which was installed near Northern Ireland and has delivered electricity into the electricity grid for more than one year. An Irish company has tested its open-ring turbine in Scotland, and more recently in Canada. Two companies have demonstrated horizontal-axis turbines at full scale in Norway and Scotland, whilst another has demonstrated a vertical-axis turbine in Italy. Lastly,

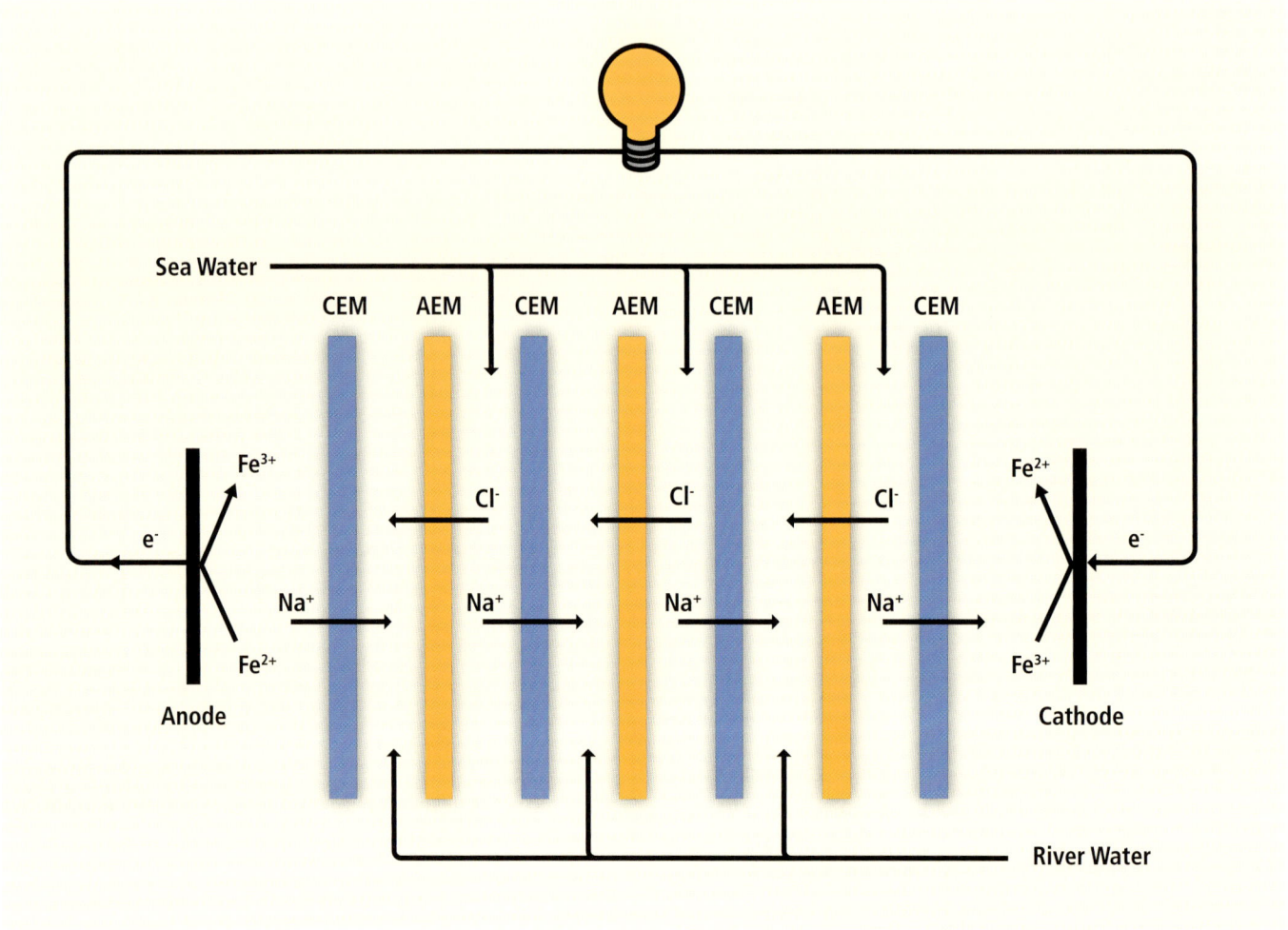

Figure TS.6.4 | Reversed electro dialysis (RED) system. [Figure 6.9]

Notes: CEM = cation exchange membrane; AEM = anion exchange membrane, Na = sodium, Cl = Chlorine, Fe = iron.

a reciprocating device was demonstrated in the UK in 2009. No pilot or demonstration plants have been deployed for ocean currents to date, although much larger scales are envisioned if technologies are able to capture the slower-velocity currents. [6.4.4]

OTEC: Japan, India, the USA and several other countries have tested pilot OTEC projects. Many have experienced engineering challenges related to pumping, vacuum retention and piping. Larger-scale OTEC developments could have significant markets in tropical maritime nations, including the Pacific Islands, Caribbean Islands, and Central American and African nations if the technology develops to the point of being a cost-effective energy supply option. [6.4.5]

Salinity gradients: Research into osmotic power is being pursued in Norway, with a prototype in operation since 2009 as part of a drive to deliver a commercial osmotic power plant. At the same time, the RED technology has been proposed for retrofitting the 75-year-old Afsluitdijk dike in The Netherlands. [6.4.6]

6.5 Environmental and social impacts

Ocean energy does not directly emit CO_2 during operation; however, GHG emissions may arise from different aspects of the lifecycle of ocean energy systems, including raw material extraction, component manufacturing, construction, maintenance and decommissioning. A comprehensive review of lifecycle assessment studies published since 1980 suggests that lifecycle GHG emissions from wave and tidal energy systems are less than 23 g CO_2eq/kWh, with a median estimate of lifecycle GHG emissions of around 8 g CO_2eq/kWh for wave energy. Insufficient studies are available to estimate lifecycle emissions from the other classes of ocean energy technology. Regardless, in comparison to fossil energy generation technologies, the lifecycle GHG emissions from ocean energy devices appear low. [6.5.1]

The local social and environmental impacts of ocean energy projects are being evaluated as actual deployments multiply, but can be estimated based on the experience of other maritime and offshore

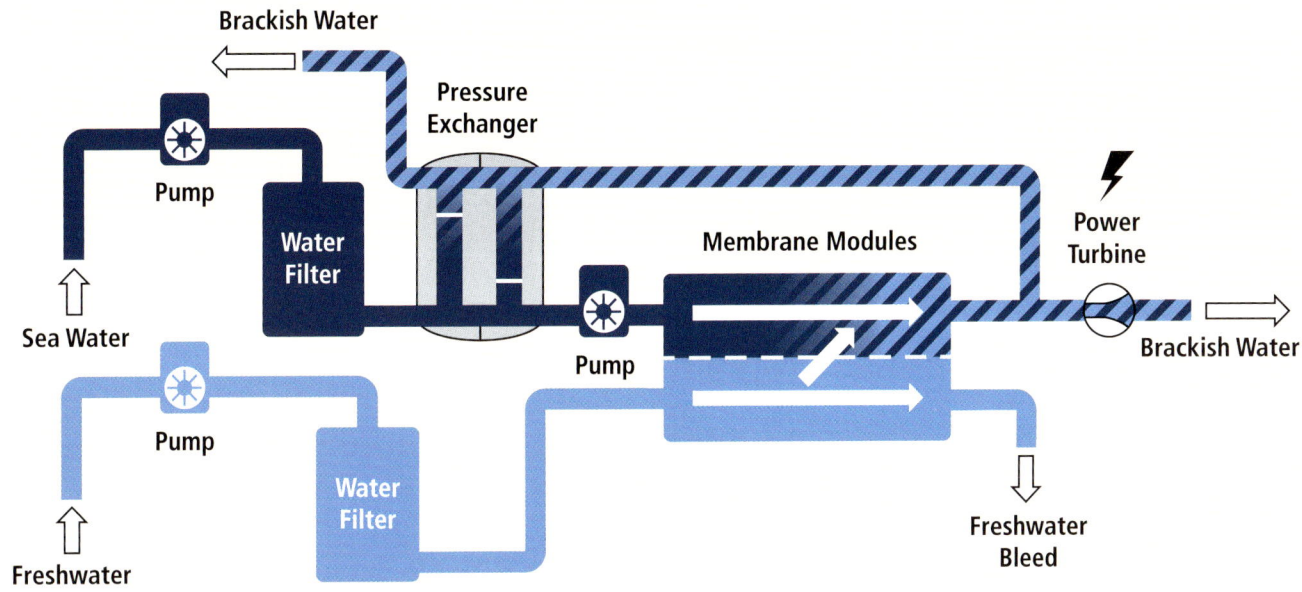

Figure TS.6.5 | Pressure-retarded osmosis (PRO) process. [Figure 6.10]

industries. Environmental risks from ocean energy technologies appear to be relatively low, but the early stage of ocean energy deployment creates uncertainty about the degree to which social and environmental concerns might eventually constrain development. [6 ES]

Each ocean power technology has its own specific set of environmental and social impacts. Possible positive effects from ocean energy may include avoidance of adverse effects on marine life by virtue of reducing other human activities in the area around the ocean devices, and the strengthening of energy supply and regional economic growth, employment and tourism. Negative effects may include a reduction in visual amenity and loss of access to space for competing users, noise during construction, noise and vibration during operation, electromagnetic fields, disruption to biota and habitats, water quality changes and possible pollution, for instance from chemical or oil leaks, and other limited specific impacts on local ecosystems. [6.5.2]

6.6 Prospects for technology improvement, innovation and integration

As emerging technologies, ocean energy devices have the potential for significant technological advances. Not only will device-specific R&D and deployment be important to achieving these advances, but technology improvements and innovation in ocean energy converters are also likely to be influenced by developments in related fields. [6.6]

Integration of ocean energy into wider energy networks will need to recognize the widely varying generation characteristics arising from the different resources. For example, electricity generation from tidal stream resources shows very high variability over one to four hours, yet extremely limited variability over monthly or longer time horizons. [6.6]

6.7 Cost trends

Commercial markets are not yet driving marine energy technology development. Government-supported R&D and national policy incentives are the key motivations. Because none of the ocean energy technologies but tidal barrages are mature (experience with other technologies is only now becoming available for validation of demonstration/prototype devices), it is difficult to accurately assess the economic viability of most ocean energy technologies. [6.7.1]

Table TS.6.1 shows the best available data for some of the primary cost factors that affect the levelized cost of electricity by each of the ocean energy sub-types. In most cases, these cost and performance parameters are based on sparse information due to the lack of peer-reviewed reference data and actual operating experience, and in many cases therefore reflect estimated cost and performance assumptions based on engineering knowledge. Present-day investment costs were found in a few instances but are based on a small sample of projects and studies, which may not be representative of the entire industry. [6.7.1]

Based on a standardized methodology outlined in Annex II and the cost and performance data summarized in Annex III, the LCOE for tidal barrages (which is currently the only commercially available ocean energy technology) over a large set and range of input parameters has been

Table TS.6.1 | Summary of core available cost and performance parameters for all ocean energy technology sub-types. [Table 6.3]

Ocean Energy Technology	Investment Costs (USD$_{2005}$/kW)	Annual O&M Costs (USD$_{2005}$/kW)	Capacity Factor (CF) (%)	Design Life (years)
Wave	6,200–16,100	180	25–40	20
Tidal Range	4,500–5,000	100	22.5–28.5	40
Tidal Current	5,400–14,300	140	26–40	20
Ocean Current	N/A	N/A	N/A	20
Ocean Thermal	4,200–12,300[1]	N/A	N/A	20
Salinity Gradient	N/A	N/A	N/A	20

Note: 1. Cost figures for ocean thermal energy have not been converted to 2005 USD.

calculated to range from US cent$_{2005}$ 12/kWh to US cent$_{2005}$ 32/kWh. This range should, however, only be considered as indicative given the present state of deployment experience. [1.3.2, 6.7.1, 6.7.3, 10.5.1, Annex II, Annex III]

Because of the early stage of technology development, estimates of future costs for ocean energy should be considered speculative. Nonetheless, the cost of ocean energy is expected to decline over time as R&D, demonstrations, and deployments proceed. [6.7.1–6.7.5]

6.8 Potential deployment

Until about 2008, ocean energy was not considered in any of the major global energy scenario modelling activities and therefore its potential impact on future world energy supplies and climate change mitigation is just now beginning to be investigated. As such, the results of the published scenarios literature as they relate to ocean energy are sparse and preliminary, reflecting a wide range of possible outcomes. Specifically, scenarios for ocean energy deployment are considered in only three major sources here: Energy [R]evolution (E[R]) 2010, IEA World Energy Outlook (WEO) 2009 and Energy Technology Perspectives (ETP) 2010. Multiple scenarios were considered in the E[R] and the ETP reports and a single reference scenario was documented in the WEO report. Each scenario is summarized in Table TS.6.2.

This preliminary presentation of scenarios that describe alternative levels of ocean energy deployment is among the first attempts to review the potential role of ocean energy in the medium- to long-term scenarios literature with the intention of establishing the potential contribution of ocean energy to future energy supplies and climate change mitigation. As shown by the limited number of existing scenarios, ocean energy has the potential to help mitigate long-term climate change by offsetting GHG emissions with projected deployments resulting in energy delivery of up to 1,943 TWh/yr (~7 EJ/yr) by 2050. Other scenarios have been developed that indicate deployment as low as 25 TWh/yr (0.9 EJ/yr) from ocean energy. The wide range in results is based in part on uncertainty about the degree to which climate change mitigation will drive energy

Table TS.6.2 | Main characteristics of medium- to long-term scenarios from major published studies that include ocean energy. [Table 6.5]

| Scenario | Deployment TWh/yr (PJ/yr) | | | | GW | Notes |
	2010	2020	2030	2050	2050	
Energy [R]evolution - Reference	N/A	3 (10.8)	11 (36.6)	25 (90)	N/A	No policy changes
Energy [R]evolution	N/A	53 (191)	128 (461)	678 (2,440)	303	Assumes 50% carbon reduction
Energy [R]evolution – Advanced	N/A	119 (428)	420 (1,512)	1,943 (6,994)	748	Assumes 80% carbon reduction
WEO 2009	N/A	3 (10.8)	13 (46.8)	N/A	N/A	Basis for E[R] reference case
ETP BLUE map 2050	N/A	N/A	N/A	133 (479)	N/A	Power sector is virtually decarbonized
ETP BLUE map no CCS 2050	N/A	N/A	N/A	274 (986)	N/A	BLUE Map Variant – Carbon capture and storage is found to not be possible
ETP BLUE map hi NUC 2050	N/A	N/A	N/A	99 (356)	N/A	BLUE Map Variant – Nuclear share is increased to 2,000 GW
ETP BLUE Map hi REN 2050	N/A	N/A	N/A	552 (1,987)	N/A	BLUE Map Variant – Renewable share is increased to 75%
ETP BLUE map 3%	N/A	N/A	N/A	401 (1,444)	N/A	BLUE Map Variant – Discount rates are set to 3% for energy generation projects.

sector transformation, but for ocean energy, is also based on inherent uncertainty as to when and if various ocean energy technologies become commercially available at attractive costs. To better understand the possible role of ocean energy in climate change mitigation, not only will continued technical advances be necessary, but the scenarios modelling process will need to increasingly incorporate the range of potential ocean energy technology sub-types, with better data for resource potential, present and future investment costs, O&M costs, and anticipated capacity factors. Improving the availability of the data at global and regional scales will be an important ingredient to improving coverage of ocean energy in the scenarios literature. [6.8.4]

7. Wind Energy

7.1 Introduction

Wind energy has been used for millennia in a wide range of applications. The use of wind energy to generate electricity on a commercial scale, however, became viable only in the 1970s as a result of technical advances and government support. A number of different wind energy technologies are available across a range of applications, but the primary use of wind energy of relevance to climate change mitigation is to generate electricity from larger, grid-connected wind turbines, deployed either on land ('onshore') or in sea- or freshwater ('offshore').[11] [7.1]

Wind energy offers significant potential for near-term (2020) and long-term (2050) GHG emissions reductions. The wind power capacity installed by the end of 2009 was capable of meeting roughly 1.8% of worldwide electricity demand, and that contribution could grow to in excess of 20% by 2050 if ambitious efforts are made to reduce GHG emissions and to address other impediments to increased wind energy deployment. Onshore wind energy is already being deployed at a rapid pace in many countries, and no insurmountable technical barriers exist that preclude increased levels of wind energy penetration into electricity supply systems. Moreover, though average wind speeds vary considerably by location, ample technical potential exists in most regions of the world to enable significant wind energy deployment. In some areas with good wind resources, the cost of wind energy is already competitive with current energy market prices, even without considering relative environmental impacts. Nonetheless, in most regions of the world, policy measures are still required to ensure rapid deployment. Continued advancements in on- and offshore wind energy technology are expected, however, further reducing the cost of wind energy and improving wind energy's GHG emissions reduction potential. [7.9]

---
[11] Smaller wind turbines, higher-altitude wind electricity, and the use of wind energy in mechanical and propulsion applications are only briefly discussed in Chapter 7.

7.2 Resource potential

The global technical potential for wind energy is not fixed, but is instead related to the status of the technology and assumptions made regarding other constraints to wind energy development. Nonetheless, a growing number of global wind resource assessments have demonstrated that the world's technical potential exceeds current global electricity production. [7.2]

No standardized approach has been developed to estimate the global technical potential of wind energy: the diversity in data, methods, assumptions, and even definitions for technical potential complicate comparisons. The AR4 identified the technical potential for onshore wind energy as 180 EJ/yr (50,000 TWh/yr). Other estimates of the global technical potential for wind energy that consider relatively more development constraints range from a low of 70 EJ/yr (19,400 TWh/yr) (onshore only) to a high of 450 EJ/yr (125,000 TWh/yr) (on- and near-shore). This range corresponds to roughly one to six times global electricity production in 2008, and may understate the technical potential due to several of the studies relying on outdated assumptions, the exclusion or only partial inclusion of offshore wind energy in some of the studies, and methodological and computing limitations. Estimates of the technical potential for offshore wind energy alone range from 15 EJ/yr to 130 EJ/yr (4,000 to 37,000 TWh/yr) when only considering relatively shallower and near-shore applications; greater technical potential is available if also considering deeper-water applications that might rely on floating wind turbine designs. [7.2.1]

Regardless of whether existing estimates under- or overstate the technical potential for wind energy, and although further advances in wind resource assessment methods are needed, it is evident that the technical potential of the resource itself is unlikely to be a limiting factor for global wind energy deployment. Instead, economic constraints associated with the cost of wind energy, institutional constraints and costs associated with transmission access and operational integration, and issues associated with social acceptance and environmental impacts are likely to restrict growth well before any absolute limit to the global technical potential is encountered. [7.2.1]

In addition, ample technical potential exists in most regions of the world to enable significant wind energy deployment. The wind resource is not evenly distributed across the globe nor uniformly located near population centres, however, and wind energy will therefore not contribute equally in meeting the needs of every country. The technical potentials for onshore wind energy in OECD North America and Eastern Europe/Eurasia are found to be particularly sizable, whereas some areas of non-OECD Asia and OECD Europe appear to have more limited onshore technical potential. Figure TS.7.1, a global wind resource map, also shows limited technical potential in certain areas of Latin America and Africa, though other portions of those continents have significant

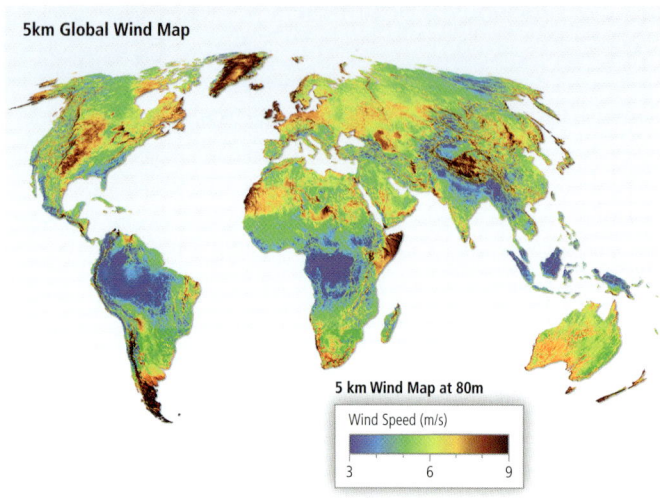

Figure TS.7.1 | Example global wind resource map with 5 km x 5 km resolution. [Figure 7.1]

covering northern Europe suggests that multi-year annual mean wind power densities will likely remain within ±50% of current values. Fewer studies have been conducted for other regions of the world. Though research in this field is nascent and additional study is warranted, research to date suggests that global climate change may alter the geographic distribution of the wind resource, but that those effects are unlikely to be of a magnitude to greatly impact the global potential for wind energy deployment. [7.2.3]

7.3 Technology and applications

Modern, commercial grid-connected wind turbines have evolved from small, simple machines to large, highly sophisticated devices. Scientific and engineering expertise and advances, as well as improved computational tools, design standards, manufacturing methods and O&M procedures, have all supported these technology developments. [7.3]

Generating electricity from the wind requires that the kinetic energy of moving air be converted to electrical energy, and the engineering challenge for the wind energy industry is to design cost-effective wind turbines and power plants to perform this conversion. Though a variety of turbine configurations have been investigated, commercially available turbines are primarily horizontal-axis machines with three blades positioned upwind of the tower. In order to reduce the levelized cost of wind energy, typical wind turbine sizes have grown significantly (Figure TS.7.2), with the largest fraction of onshore wind turbines installed globally in 2009 having a rated capacity of 1.5 to 2.5 MW. As of 2010, onshore wind turbines typically stand on 50- to 100-m towers, with rotors that are often 50 to 100 m in diameter; commercial machines

technical potential. Recent, detailed regional assessments have generally found the size of the wind resource to be greater than estimated in previous assessments. [7.2.2]

Global climate change may alter the geographic distribution and/or the inter- and intra-annual variability of the wind resource, and/or the quality of the wind resource, and/or the prevalence of extreme weather events that may impact wind turbine design and operation. Research to date suggests that it is unlikely that multi-year annual mean wind speeds will change by more than a maximum of ±25% over most of Europe and North America during the present century, while research

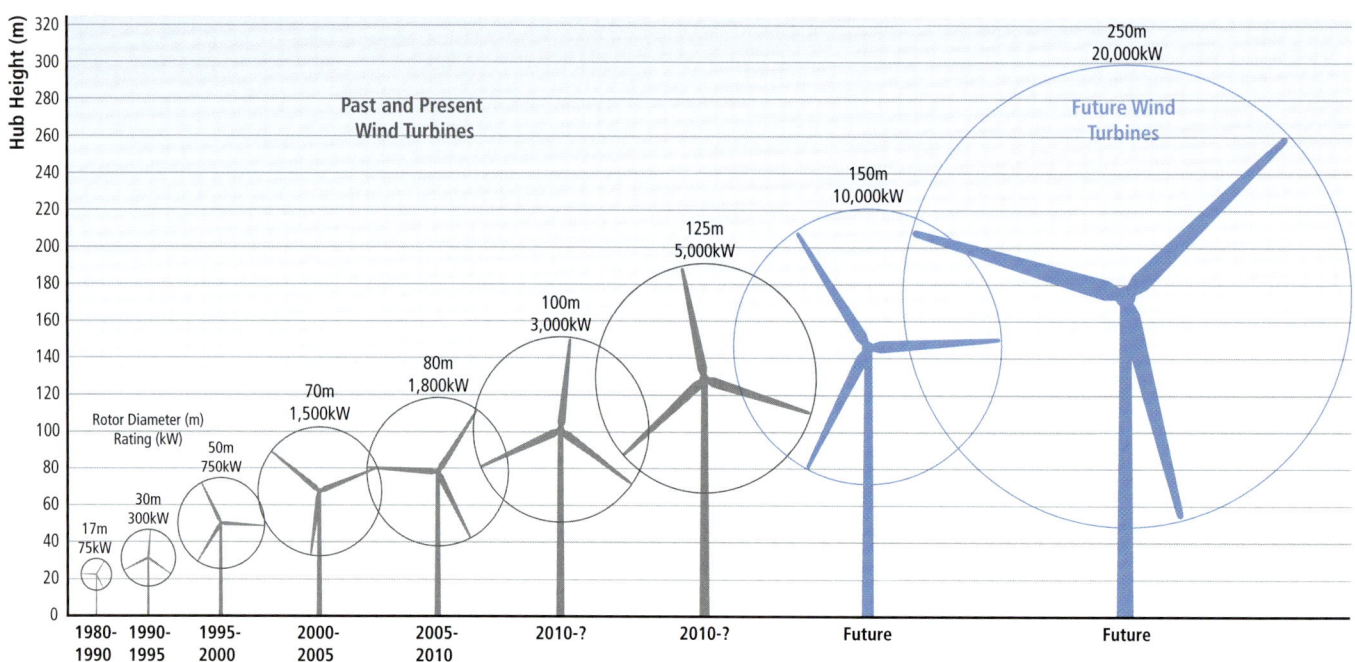

Figure TS.7.2 | Growth in size of typical commercial wind turbines. [Figure 7.6]

with rotor diameters and tower heights in excess of 125 m are operating, and even larger machines are under development. Onshore wind energy technology is already being commercially manufactured and deployed at a large scale. [7.3.1]

Offshore wind energy technology is less mature than onshore, with higher investment costs. Lower power plant availabilities and higher O&M costs have also been common both because of the comparatively less mature state of the technology and because of the inherently greater logistical challenges of maintaining and servicing offshore turbines. Nonetheless, considerable interest in offshore wind energy exists in the EU and, increasingly, in other regions. The primary motivation to develop offshore wind energy is to provide access to additional wind resources in areas where onshore wind energy development is constrained by limited technical potential and/or by planning and siting conflicts with other land uses. Other motivations include the higher-quality wind resources located at sea; the ability to use even larger wind turbines and the potential to thereby gain additional economies of scale; the ability to build larger power plants than onshore, gaining plant-level economies of scale; and a potential reduction in the need for new, long-distance, land-based transmission infrastructure to access distant onshore wind energy. To date, offshore wind turbine technology has been very similar to onshore designs, with some modifications and with special foundations. As experience is gained, water depths are expected to increase and more exposed locations with higher winds will be utilized. Wind energy technology specifically tailored for offshore applications will become more prevalent as the offshore market expands, and it is expected that larger turbines in the 5 to 10 MW range may come to dominate this segment. [7.3.1.3]

Alongside the evolution of wind turbine design, improved design and testing methods have been codified in International Electrotechnical Commission standards. Certification agencies rely on accredited design and testing bodies to provide traceable documentation demonstrating conformity with the standards in order to certify that turbines, components or entire wind power plants meet common guidelines relating to safety, reliability, performance and testing. [7.3.2]

From an electric system reliability perspective, an important part of the wind turbine is the electrical conversion system. For modern turbines, variable-speed machines now dominate the market, allowing for the provision of real and reactive power as well as some fault ride-through capability, but no intrinsic inertial response (i.e., turbines do not increase or decrease power output in synchronism with system power imbalances); wind turbine manufacturers have recognized this latter limitation and are pursuing a variety of solutions. [7.3.3]

7.4 Global and regional status of market and industry development

The wind energy market has expanded substantially, demonstrating the commercial and economic viability of the technology and industry. Wind energy expansion has been concentrated in a limited number of regions, however, and further expansion, especially in regions with little wind energy deployment to date and in offshore locations, is likely to require additional policy measures. [7.4]

Wind energy has quickly established itself as part of the mainstream electricity industry. From a cumulative capacity of 14 GW at the end of 1999, global installed capacity increased twelve-fold in 10 years to reach almost 160 GW by the end of 2009. The majority of the capacity has been installed onshore, with offshore installations primarily in Europe and totalling a cumulative 2.1 GW. The countries with the highest installed capacity by the end of 2009 were the USA (35 GW), China (26 GW), Germany (26 GW), Spain (19 GW) and India (11 GW). The total investment cost of new wind power plants installed in 2009 was USD_{2005} 57 billion, while worldwide direct employment in the sector in 2009 has been estimated at approximately 500,000. [7.4.1, 7.4.2]

In both Europe and the USA, wind energy represents a major new source of electric capacity additions. In 2009, roughly 39% of all capacity additions in the USA and the EU came from wind energy; in China, 16% of the net capacity additions in 2009 came from wind energy. On a global basis, from 2000 through 2009, roughly 11% of all newly installed net electric capacity additions came from new wind power plants; in 2009 alone, that figure was probably more than 20%. As a result, a number of countries are beginning to achieve relatively high levels of annual wind electricity penetration in their respective electric systems. By the end of 2009, wind power capacity was capable of supplying electricity equal to roughly 20% of Denmark's annual electricity demand, 14% of Portugal's, 14% of Spain's, 11% of Ireland's and 8% of Germany's. [7.4.2]

Despite these trends, wind energy remains a relatively small fraction of worldwide electricity supply. The total wind power capacity installed by the end of 2009 would, in an average year, meet roughly 1.8% of worldwide electricity demand. Additionally, though the trend over time has been for the wind energy industry to become less reliant on European markets, with significant recent expansion in the USA and China, the market remains concentrated regionally: Latin America, Africa and the Middle East, and the Pacific regions have installed relatively little wind power capacity despite significant technical potential for wind energy in each region (Figure TS.7.3). [7.4.1, 7.4.2]

The deployment of wind energy must overcome a number of challenges, including: the relative cost of wind energy compared to energy market prices, at least if environmental impacts are not internalized and monetized; concerns about the impact of wind energy's variability; challenges of building new transmission; cumbersome and slow planning, siting and permitting procedures; the technical advancement needs and higher cost of offshore wind energy technology; and lack of institutional and technical knowledge in regions that have not yet experienced substantial wind energy deployment. As a result, growth is affected by a wide range of government policies. [7.4.4]

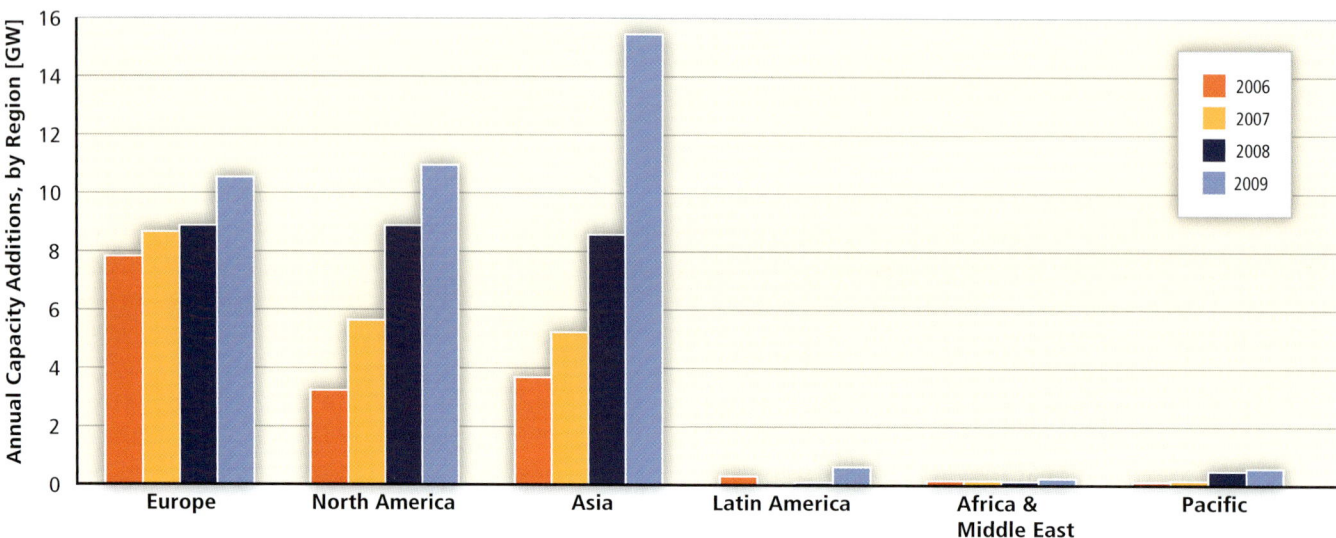

Figure TS.7.3 | Annual wind power capacity additions by region. [Figure 7.10]

Note: Regions shown in the figure are defined by the study.

7.5 Near-term grid integration issues

As wind energy deployment has increased, so have concerns about the integration of that energy into electric systems. The nature and magnitude of the integration challenge will depend on the characteristics of the existing electric system and the level of wind electricity penetration. Moreover, as discussed in Chapter 8, integration challenges are not unique to wind energy. Nevertheless, analysis and operating experience primarily from certain OECD countries suggests that, at low to medium levels of wind electricity penetration (defined here as up to 20% of total annual average electrical energy demand)[12], the integration of wind energy generally poses no insurmountable technical barriers and is economically manageable. At the same time, even at low to medium levels of wind electricity penetration, certain (and sometimes system-specific) technical and/or institutional challenges must be addressed. Concerns about (and the costs of) wind energy integration will grow with wind energy deployment, and even higher levels of penetration may depend on or benefit from the availability of additional technological and institutional options to increase flexibility and maintain a balance between supply and demand, as discussed further in Chapter 8 (Section 8.2). [7.5]

Wind energy has characteristics that present integration challenges, and that must be considered in electric system planning and operation to ensure the reliable and economical operation of the electric power system. These include: the localized nature of the wind resource with possible implications for new transmission for both on- and offshore wind energy; the variability of wind power output over multiple time scales; and the lower levels of predictability of wind power output than are common for many other types of power plants. The aggregate variability and uncertainty of wind power output depends, in part, on the degree of correlation between the output of different geographically dispersed wind power plants: generally, the outputs of wind power plants that are farther apart are less correlated with each other, and variability over shorter time periods (minutes) is less correlated than variability over longer time periods (multiple hours). Forecasts of wind power output are also more accurate over shorter time periods, and when multiple plants are considered together. [7.5.2]

Detailed system planning for new generation and transmission infrastructure is used to ensure that the electric system can be operated reliably and economically in the future. To do so, planners need computer-based simulation models that accurately characterize wind energy. Additionally, as wind power capacity has increased, so has the need for wind power plants to become more active participants in maintaining the operability and power quality of the electric system, and technical standards for grid connection have been implemented to help prevent wind power plants from adversely affecting the electric system during normal operation and contingencies. Transmission adequacy evaluations, meanwhile, must account for the location dependence of the wind resource, and consider any trade-offs between the costs of expanding the transmission system to access higher-quality wind resources in comparison to the costs of accessing lower-quality wind resources that require less transmission investment. Even at low to medium levels of wind electricity penetration, the addition of large quantities of on- or offshore wind energy in areas with higher-quality wind resources may require significant new additions or upgrades to the transmission system. Depending on the legal and regulatory framework in any particular region, the institutional challenges of transmission expansion can be substantial. Finally, planners need to account for wind

---
[12] This level of penetration was chosen to loosely separate the integration needs for wind energy in the relatively near term from the broader, longer- term, and non-wind-specific discussion of power system changes provided in Chapter 8.

power output variability in assessing the contribution of wind energy to generation adequacy and therefore the long-term reliability of the electric system. Though methods and objectives vary from region to region, the contribution of wind energy to generation adequacy usually depends on the correlation of wind power output with the periods of time when there is a higher risk of a supply shortage, typically periods of high electricity demand. The marginal contribution of wind energy to generation adequacy typically declines as wind electricity penetration increases, but aggregating wind power plants over larger areas may slow this decline if adequate transmission capacity is available. The relatively low average contribution of wind energy to generation adequacy (compared to fossil units) suggests that electric systems with large amounts of wind energy will also tend to have significantly more total nameplate generation capacity to meet the same peak electricity demand than will electric systems without large amounts of wind energy. Some of this generation capacity will operate infrequently, however, and the mix of other generation will therefore tend (on economic grounds) to increasingly shift towards flexible 'peaking' and 'intermediate' resources and away from 'base-load' resources. [7.5.2]

The unique characteristics of wind energy also have important implications for electric system operations. Because wind energy is generated with a very low marginal operating cost, it is typically used to meet demand when it is available; other generators are then dispatched to meet demand minus any available wind energy (i.e., 'net demand'). As wind electricity penetration grows, the variability of wind energy results in an overall increase in the magnitude of changes in net demand, and also a decrease in the minimum net demand. As a result of these trends, wholesale electricity prices will tend to decline when wind power output is high and transmission interconnector capacity to other energy markets is constrained, and other generating units will be called upon to operate in a more flexible manner than required without wind energy. At low to medium levels of wind electricity penetration, the increase in minute-to-minute variability is expected to be relatively small. The more significant operational challenges relate to the need to manage changes in wind power output over one to six hours. Incorporating wind energy forecasts into electric system operations can reduce the need for flexibility from other generators, but even with high-quality forecasts, system operators will need a broad range of strategies to actively maintain the supply/demand balance, including the use of flexible power generation technologies, wind energy output curtailment, and increased coordination and interconnection between electric systems. Mass-market demand response, bulk energy storage technologies, large-scale deployment of electric vehicles and their associated contributions to system flexibility through controlled battery charging, diverting excess wind energy to fuel production or local heating, and geographic diversification of wind power plant siting will also become increasingly beneficial as wind electricity penetration rises. Despite the challenges, actual operating experience in different parts of the world demonstrates that electric systems can operate reliably with increased contributions of wind energy; in four countries (Denmark, Portugal, Spain, Ireland), wind energy in 2010 was already able to supply from 10 to roughly 20% of annual electricity demand. Experience is limited, in particular with regard to system faults at high instantaneous penetration levels, however, and as more wind energy is deployed in diverse regions and electric systems, additional knowledge about wind energy integration will be gained. [7.5.3]

In addition to actual operating experience, a number of high-quality studies of the increased transmission and generation resources required to accommodate wind energy have been completed, primarily covering OECD countries. These studies employ a wide variety of methodologies and have diverse objectives, but the results demonstrate that the cost of integrating up to 20% wind energy into electric systems is, in most cases, modest but not insignificant. Specifically, at low to medium levels of wind electricity penetration, the available literature (again, primarily from a subset of OECD countries) suggests that the additional costs of managing electric system variability and uncertainty, ensuring generation adequacy, and adding new transmission to accommodate wind energy will be system specific but generally in the range of US cent$_{2005}$ 0.7/kWh to US cent$_{2005}$ 3/kWh. The technical challenges and costs of integration are found to increase with wind electricity penetration. [7.5.4]

7.6 Environmental and social impacts

Wind energy has significant potential to reduce (and is already reducing) GHG emissions. Moreover, attempts to measure the relative impacts of various electricity supply technologies suggest that wind energy generally has a comparatively small environmental footprint. [9.3.4, 10.6] As with other industrial activities, however, wind energy has the potential to produce some detrimental impacts on the environment and on human activities and well being, and many local and national governments have established planning and siting requirements to reduce those impacts. As wind energy deployment increases and as larger wind power plants are considered, existing concerns may become more acute and new concerns may arise. [7.6]

Although the major environmental benefits of wind energy result from displacing electricity generated from fossil fuel-based power plants, estimating those benefits is somewhat complicated by the operational characteristics of the electric system and the investment decisions that are made about new power plants. In the short run, increased wind energy will typically displace the operations of existing fossil fuel-fired plants. In the longer term, however, new generating plants may be needed, and the presence of wind energy can influence what types of power plants are built. The impacts arising from the manufacture, transport, installation, operation and decommissioning of wind turbines should also be considered, but a comprehensive review of available studies demonstrates that the energy used and GHG emissions produced during these steps are small compared to the energy generated and emissions avoided over the lifetime of wind power plants. The GHG emissions intensity of wind energy is estimated to range from 8 to 20 g CO_2/kWh in most instances, whereas energy payback times are between 3.4 and 8.5 months. In addition, managing the variability of wind power output has not been found to significantly degrade the GHG emissions benefits of wind energy. [7.6.1]

Other studies have considered the local ecological impacts of wind energy development. The construction and operation of both on- and offshore wind power plants impacts wildlife through bird and bat collisions and through habitat and ecosystem modifications, with the nature and magnitude of those impacts being site- and species-specific. For offshore wind energy, implications for benthic resources, fisheries and marine life more generally must be considered. Research is also underway on the potential impact of wind power plants on the local climate. Bird and bat fatalities through collisions with wind turbines are among the most publicized environmental concerns. Though much remains unknown about the nature and population-level implications of these impacts, avian fatality rates have been reported at between 0.95 and 11.67 per MW per year. Raptor fatalities, though much lower in absolute number, have raised special concerns in some cases, and as offshore wind energy has increased, concerns have also been raised about seabirds. Bat fatalities have not been researched as extensively, but fatality rates ranging from 0.2 to 53.3 per MW per year have been reported; the impact of wind power plants on bat populations is of particular contemporary concern. The magnitude and population-level consequences of bird and bat collision fatalities can also be viewed in the context of other fatalities caused by human activities. The number of bird fatalities at existing wind power plants appears to be orders of magnitude lower than other anthropogenic causes of bird deaths, it has been suggested that onshore wind power plants are not currently causing meaningful declines in bird population levels, and other energy supply options also impact birds and bats through collisions, habitat modifications and contributions to global climate change. Improved methods to assess species-specific population-level impacts and their possible mitigation are needed, as are robust comparisons between the impacts of wind energy and of other electricity supply options. [7.6.2]

Wind power plants can also impact habitats and ecosystems through avoidance of or displacement from an area, habitat destruction and reduced reproduction. Additionally, the impacts of wind power plants on marine life have moved into focus as offshore development has increased. The impacts of offshore wind energy on marine life vary between the installation, operation and decommissioning phases, depend greatly on site-specific conditions, and may be negative or positive. Potential negative impacts include underwater sounds and vibrations, electromagnetic fields, physical disruption and the establishment of invasive species. The physical structures may, however, create new breeding grounds or shelters and act as artificial reefs or fish aggregation devices. Additional research is warranted on these impacts and their long-term and population-level consequences, but they do not appear to be disproportionately large compared to onshore wind energy. [7.6.2]

Surveys have consistently found wind energy to be widely accepted by the general public. Translating this support into increased deployment, however, often requires the support of local host communities and/or decision makers. To that end, in addition to ecological concerns, a number of concerns are often raised about the impacts of wind power plants on local communities. Perhaps most importantly, modern wind energy technology involves large structures, so wind turbines are unavoidably visible in the landscape. Other impacts of concern include land and marine usage (including possible radar interference), proximal impacts such as noise and flicker, and property value impacts. Regardless of the type and degree of social and environmental concerns, addressing them is an essential part of any successful wind power planning and plant siting process, and engaging local residents is often an integral aspect of that process. Though some of the concerns can be readily mitigated, others—such as visual impacts—are more difficult to address. Efforts to better understand the nature and magnitude of the remaining impacts, together with efforts to minimize and mitigate those impacts, will need to be pursued in concert with increasing wind energy deployment. In practice, planning and siting regulations vary dramatically by jurisdiction, and planning and siting processes have been obstacles to wind energy development in some countries and contexts. [7.6.3]

7.7 Prospects for technology improvement and innovation

Over the past three decades, innovation in wind turbine design has led to significant cost reductions. Public and private R&D programmes have played a major role in these technical advances, leading to system- and component-level technology improvements, as well as improvements in resource assessment, technical standards, electric system integration, wind energy forecasting and other areas. From 1974 to 2006, government R&D budgets for wind energy in IEA countries totalled USD$_{2005}$ 3.8 billion, representing 1% of total energy R&D expenditure. In 2008, OECD research funding for wind energy totalled USD$_{2005}$ 180 million. [7.7, 7.7.1]

Though onshore wind energy technology is already commercially manufactured and deployed at a large scale, continued incremental advances are expected to yield improved turbine design procedures, more efficient materials usage, increased reliability and energy capture, reduced O&M costs and longer component lifetimes. In addition, as offshore wind energy gains more attention, new technology challenges arise and more radical technology innovations are possible. Wind power plants and turbines are complex systems that require integrated design approaches to optimize cost and performance. At the plant level, considerations include the selection of a wind turbine for a given wind resource regime; wind turbine siting, spacing and installation procedures; O&M methodologies; and electric system integration. Studies have identified a number of areas where technology advances could result in changes in the investment cost, annual energy production, reliability, O&M cost and electric system integration of wind energy. [7.3.1, 7.7.1, 7.7.2]

At the component level, a range of opportunities are being pursued, including: advanced tower concepts that reduce the need for large cranes and minimize materials demands; advanced rotors and blades through better designs, coupled with better materials and advanced manufacturing methods; reduced energy losses and improved availability through advanced turbine control and condition monitoring; advanced drive trains, generators and power electronics; and manufacturing learning improvements. [7.7.3]

In addition, there are several areas of possible advancement that are more specific to offshore wind energy, including O&M procedures, installation and assembly schemes, support structure design, and the development of larger turbines, possibly including new turbine concepts. Foundation structure innovation, in particular, offers the potential to access deeper waters, thereby increasing the technical potential of wind energy. Offshore turbines have historically been installed primarily in relatively shallow water, up to 30 m deep, on a mono-pile structure that is essentially an extension of the tower, but gravity-based structures have become more common. These approaches, as well as other concepts that are more appropriate for deeper waters, including floating platforms, are depicted in Figure TS.7.4. Additionally, offshore turbine size is not restricted in the same way as onshore wind turbines, and the relatively higher cost of offshore foundations provides motivation for larger turbines. [7.7.3]

Wind turbines are designed to withstand a wide range of challenging conditions with minimal attention. Significant effort is therefore needed to enhance fundamental understanding of the operating environment in which turbines operate in order to facilitate a new generation of reliable, safe, cost-effective wind turbines, and to further optimize wind power plant siting and design. Research in the areas of aeroelastics, unsteady aerodynamics, aeroacoustics, advanced control systems, and atmospheric science, for example, is anticipated to lead to improved design tools, and thereby increase the reliability of the technology and encourage further design innovation. Fundamental research of this nature will help improve wind turbine design, wind power plant performance estimates, wind resource assessments, short-term wind energy forecasting, and estimates of the impact of large-scale wind energy deployment on the local climate, as well as the impact of potential climate change effects on wind resources. [7.7.4]

7.8 Cost trends

Though the cost of wind energy has declined significantly since the 1980s, policy measures are currently required to ensure rapid deployment in most regions of the world. In some areas with good wind resources, however, the cost of wind energy is competitive with current energy market prices, even without considering relative environmental impacts. Moreover, continued technology advancements are expected, supporting further cost reduction. [7.8]

The levelized cost of energy from on- and offshore wind power plants is affected by five primary factors: annual energy production; investment costs; O&M costs; financing costs; and the assumed economic life of

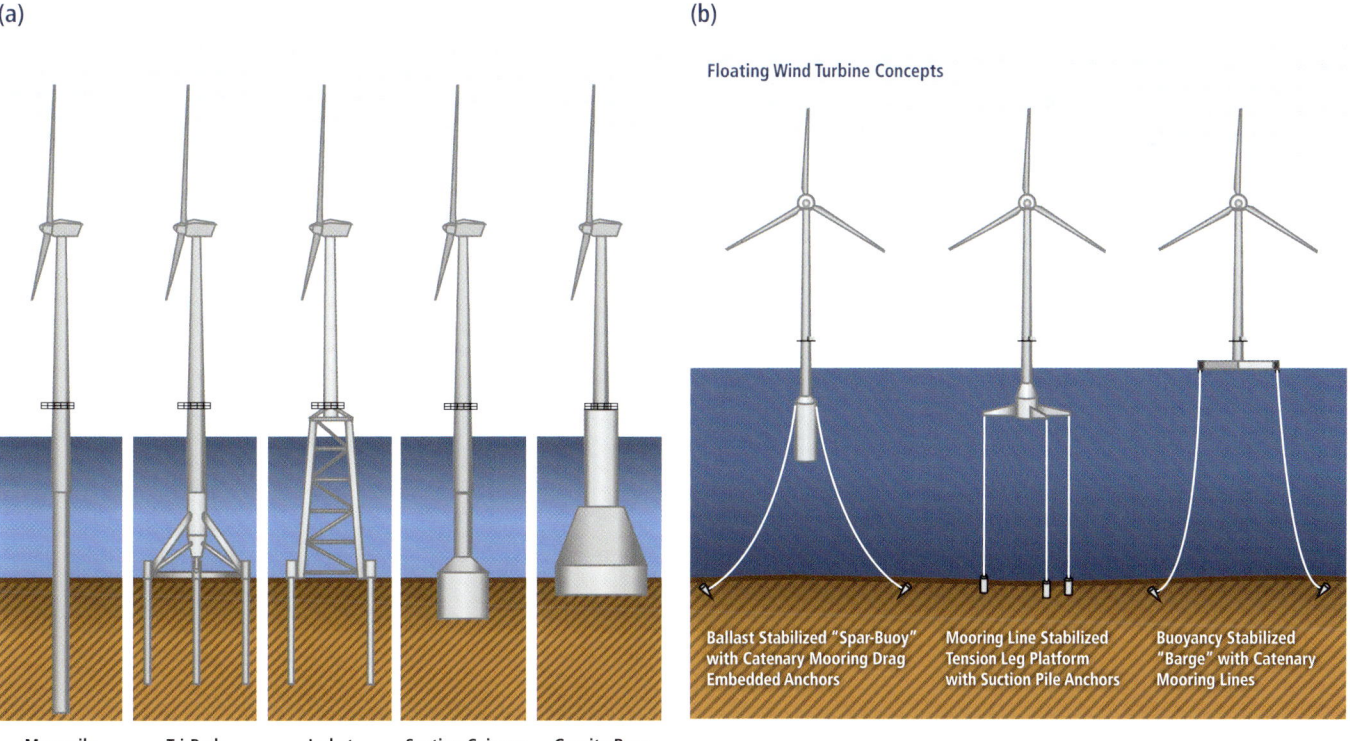

Figure TS.7.4 | Offshore wind turbine foundation designs: (a) near-term concepts and (b) floating offshore turbine concepts. [Figure 7.19]

the power plant.[13] From the 1980s to roughly 2004, the investment cost of onshore wind power plants dropped. From 2004 to 2009, however, investment costs increased, the primary drivers of which were: escalation in the cost of labour and materials inputs; increasing profit margins among turbine manufacturers and their suppliers; the relative strength of the Euro currency; and the increased size of turbine rotors and hub heights. In 2009, the average investment cost for onshore wind power plants installed worldwide was approximately USD$_{2005}$ 1,750/kW, with many plants falling in the range of USD$_{2005}$ 1,400 to 2,100/kW; investment costs in China in 2008 and 2009 were around USD$_{2005}$ 1,000 to 1,350/kW. There is far less experience with offshore wind power plants, and the investment costs of offshore plants are highly site-specific. Nonetheless, the investment costs of offshore plants have historically been 50 to more than 100% higher than for onshore plants; O&M costs are also greater for offshore plants. Offshore costs have also been influenced by some of the same factors that caused rising onshore costs from 2004 through 2009, as well as by several unique factors. The most recently installed or announced offshore plants have investment costs that are reported to range from roughly USD$_{2005}$ 3,200/kW to USD$_{2005}$ 5,000/kW. Notwithstanding the increased water depth of offshore plants over time, the majority of the operating plants have been built in relatively shallow water. The performance of wind power plants is highly site-specific, and is primarily governed by the characteristics of the local wind regime, but is also impacted by wind turbine design optimization, performance and availability, and by the effectiveness of O&M procedures. Performance therefore varies by location, but has also generally improved with time. Offshore wind power plants are often exposed to better wind resources. [7.8.1–7.8.3]

Based on a standardized methodology outlined in Annex II and the cost and performance data summarized in Annex III, the LCOE for on- and offshore wind power plants over a large set and range of input parameters has been calculated to range from US cent$_{2005}$ 3.5/kWh to US cent$_{2005}$ 17/kWh and from US cent$_{2005}$ 7.5/kWh to US cent$_{2005}$ 23/kWh, respectively. [1.3.2, 10.5.1, Annex II, Annex III]

Figure TS.7.5 presents the LCOE of on- and offshore wind energy over a somewhat different set and range of parameters, and shows that the LCOE varies substantially depending on assumed investment costs, energy production and discount rates. For onshore wind energy, estimates are provided for plants built in 2009; for offshore wind energy, estimates are provided for plants built from 2008 to 2009 as well as those plants that were planned for completion in the early 2010s. The LCOE for onshore wind energy in good to excellent wind resource regimes are estimated to average approximately US cent$_{2005}$ 5/kWh to US cent$_{2005}$ 10/kWh, and can reach more than US cent$_{2005}$ 15/kWh in lower-resource areas. Though

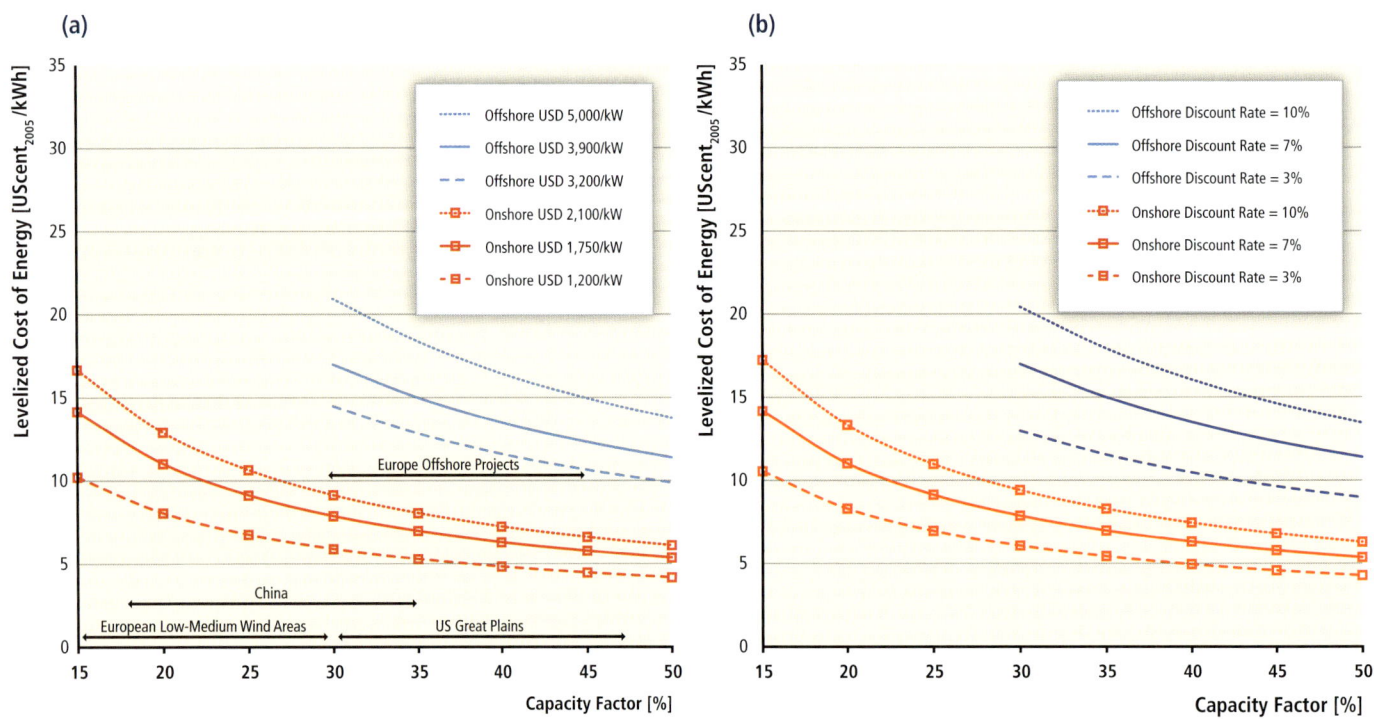

Figure TS.7.5 | Estimated levelized cost of on- and offshore wind energy, 2009: (a) as a function of capacity factor and investment cost* and (b) as a function of capacity factor and discount rate**. [Figure 7.23]

Notes: * Discount rate assumed to equal 7%. ** Onshore investment cost assumed at USD$_{2005}$ 1,750/kW, and offshore at USD$_{2005}$ 3,900/kW.

---

13 The economic competitiveness of wind energy in comparison to other energy sources, which necessarily must also include other factors such as subsidies and environmental externalities, is not covered in this section.

the offshore cost estimates are more uncertain, typical LCOE are estimated to range from US cent$_{2005}$ 10/kWh to more than US cent$_{2005}$ 20/kWh for recently built or planned plants located in relatively shallow water. Where the exploitable onshore wind resource is limited, offshore plants can sometimes compete with onshore plants. [7.8.3, Annex II, Annex III]

A number of studies have developed forecasted cost trajectories for on- and offshore wind energy based on differing combinations of learning curve estimates, engineering models and/or expert judgement. Among these studies, the starting year of the forecasts, the methodological approaches and the assumed wind energy deployment levels vary. Nonetheless, a review of this literature supports the idea that continued R&D, testing and experience could yield reductions in the levelized cost of onshore wind energy of 10 to 30% by 2020. Offshore wind energy is anticipated to experience somewhat deeper cost reductions of 10 to 40% by 2020, though some studies have identified scenarios in which market factors lead to cost increases in the near to medium term. [7.8.4]

7.9 Potential deployment

Given the commercial maturity and cost of onshore wind energy technology, increased utilization of wind energy offers the potential for significant near-term GHG emission reductions: this potential is not conditioned on technology breakthroughs, and no insurmountable technical barriers exist that preclude increased levels of wind energy penetration into electricity supply systems. As a result, in the near to medium term, the rapid increase in wind power capacity from 2000 to 2009 is expected by many studies to continue. [7.9, 7.9.1]

Moreover, a number of studies have assessed the longer-term potential of wind energy, often in the context of GHG concentration stabilization scenarios. [10.2, 10.3] Based on a review of this literature (including 164 different long-term scenarios), and as summarized in Figure TS.7.6, wind energy could play a significant long-term role in reducing global GHG emissions. By 2050, the median contribution of wind energy among the scenarios with GHG concentration stabilization ranges of 440 to 600 ppm CO_2 and <440 ppm CO_2 is 23 to 27 EJ/yr (6,500 to 7,600 TWh/yr), increasing to 45 to 47 EJ/yr at the 75th percentile of scenarios (12,400 to 12,900 TWh/yr), and to more than 100 EJ/yr in the highest study (31,500 TWh). Achieving this contribution would require wind energy to deliver around 13 to 14% of global electricity supply in the median scenario result by 2050, increasing to 21 to 25% at the 75th percentile of the reviewed scenarios. [7.9.2]

Achieving the higher end of this range of global wind energy utilization would likely require not only economic support policies of adequate size and predictability, but also an expansion of wind energy utilization regionally, increased reliance on offshore wind energy in some regions, technical and institutional solutions to transmission constraints and operational integration concerns, and proactive efforts to mitigate and manage social and environmental concerns. Additional R&D is expected to lead to incremental cost reductions for onshore wind energy, and enhanced R&D expenditures may be especially important for offshore wind energy technology. Finally, for those markets with good wind resource potential but that are new to wind energy deployment, both knowledge and technology transfer may help facilitate early wind power plant installations. [7.9.2]

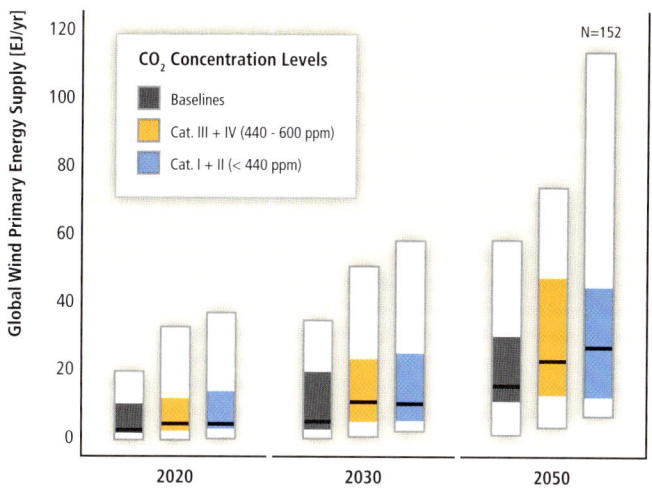

Figure TS.7.6 | Global primary energy supply of wind energy in long-term scenarios (median, 25th to 75th percentile range, and full range of scenario results; colour coding is based on categories of atmospheric CO_2 concentration level in 2100; the specific number of scenarios underlying the figure is indicated in the right upper corner). [Figure 7.24]

8. Integration of Renewable Energy into Present and Future Energy Systems

8.1 Introduction

In many countries, energy supply systems have evolved over decades, enabling the efficient and cost-effective distribution of electricity, gas, heat and transport energy carriers to provide useful energy services to end users. The transition to a low-carbon future that employs high shares of RE may require considerable investment in new RE technologies and infrastructure, including more flexible electricity grids, expansion of district heating and cooling schemes, distribution systems for RE-derived gases and liquid fuels, energy storage systems, novel methods of transport, and innovative distributed energy and control systems in buildings. Enhanced RE integration can lead to the provision of the full range of energy services for large and small communities in both developed and developing countries. Regardless of the energy supply system presently in place, whether in energy-rich or energy-poor communities, over the long term, and through measured system planning and integration, there are few, if any, technical limits to increasing the shares of RE at the national, regional and local scales as well as for individual buildings, although other barriers may need to be overcome. [8.1, 8.2]

Energy supply systems are continuously evolving, with the aim of increasing conversion technology efficiencies, reducing losses and lowering the costs of providing energy services to end users. To provide a greater share of RE heating, cooling, transport fuels and electricity may require modification of current policies, markets and existing energy supply systems over time so that they can accommodate higher rates of deployment leading to greater supplies of RE. [8.1]

All countries have access to some RE resources and in many parts of the world these are abundant. The characteristics of many of these resources distinguish them from fossil fuels and nuclear systems. Some resources, such as solar and ocean energy, are widely distributed, whereas others, such as large-scale hydropower, are constrained by geographic location and hence integration options are more centralized. Some RE resources are variable and have limited predictability. Others have lower energy densities and their technical specifications differ from solid, liquid and gaseous fossil fuels. Such RE resource characteristics can constrain the ease of integration and invoke additional system costs, particularly when reaching higher shares of RE. [8.1, 8.2]

Following the structural outline of Chapter 8, RE resources can be used through integration into energy supply networks delivering energy to consumers using energy carriers with varying shares of RE embedded or by direct integration into the transport, buildings, industry and agriculture end-use sectors (Figure TS.8.1). [8.2, 8.3]

The general and specific requirements for enhanced integration of RE into energy supply systems are reasonably well understood. However, since integration issues tend to be site-specific, analyses of typical additional costs for RE integration options are limited and future research is required for use in scenario modelling. For example, it is not clear how the possible trend towards more decentralized energy supply systems might affect the future costs for developing further centralized heat and power supplies and the possible avoidance of constructing new infrastructure. [8.2]

Centralized energy systems, based mainly on fossil fuels, have evolved to provide reasonably cost-effective energy services to end users using

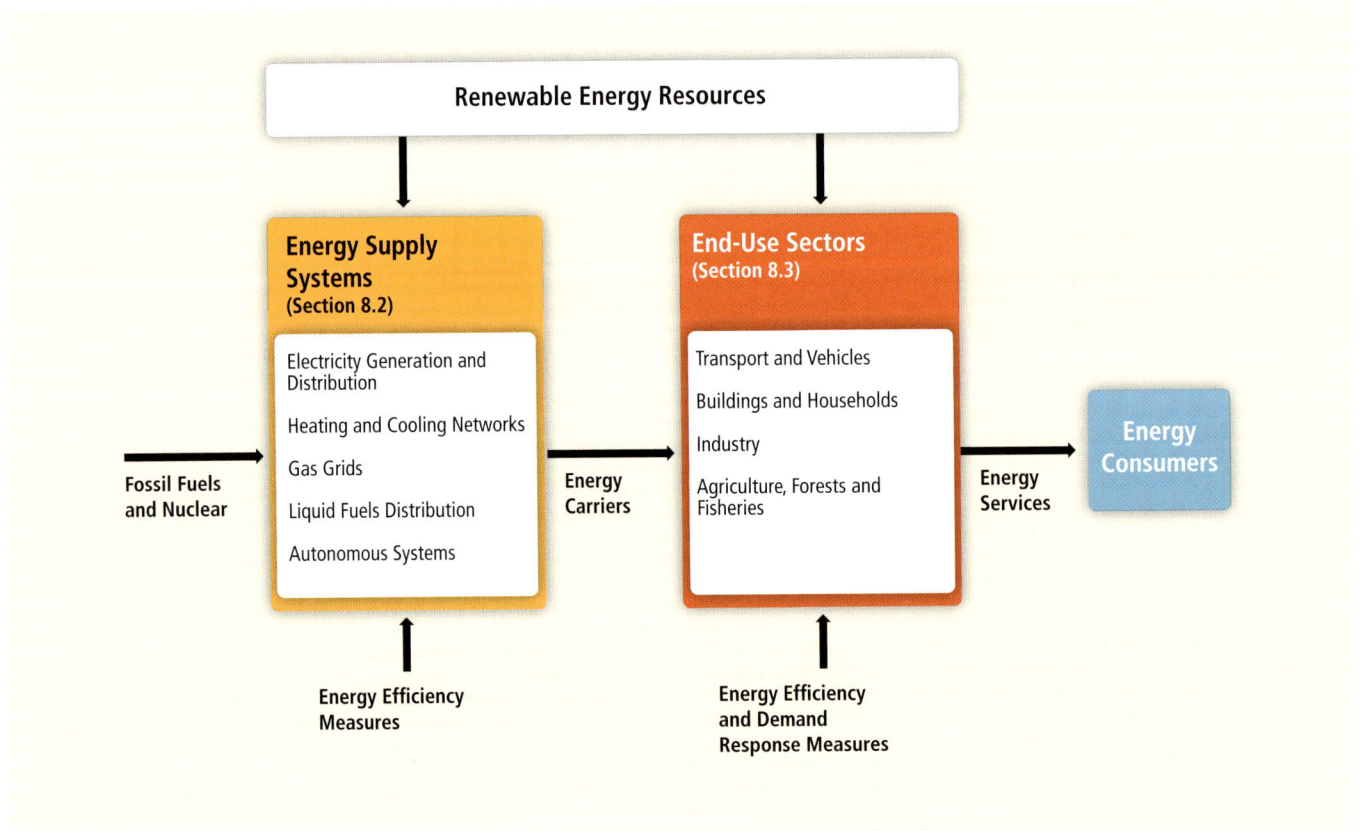

Figure TS.8.1 | Pathways for RE integration to provide energy services, either into energy supply systems or on-site for use by the end-use sectors. [Figure 8.1]

a range of energy carriers including solid, liquid and gaseous fuels, electricity, and heat. Increasing the deployment of RE technologies requires their integration into these existing systems by overcoming the associated technical, economic, environmental and social barriers. The advent of decentralized energy systems could open up new deployment opportunities. [8.1, 8.2]

In some regions, RE electricity systems could become the dominant future energy supply, especially if heating and transport demands are also to be met by electricity. This could be driven by parallel developments in electric vehicles, increased heating and cooling using electricity (including heat pumps), flexible demand response services (including the use of smart meters), and other innovative technologies. [8.1, 8.2.1.2, 8.2.2, 8.3.1–8.3.3]

The various energy systems differ markedly between countries and regions around the world and each is complex. As a result, a range of approaches are needed to encourage RE integration, whether centralized or decentralized. Prior to making any significant change in an energy supply system that involves increasing the integration of RE, a careful assessment of the RE resource availability; the suitability of existing technologies; institutional, economic and social constraints; the potential risks; and the need for related capacity building and skills development should be undertaken. [8.1, 8.2]

The majority of scenarios that stabilize atmospheric GHG concentrations around 450 ppm CO_2eq show that RE will exceed a 50% share of low-carbon primary energy by 2050. This transition can be illustrated by many scenarios, the single example of increasing market shares shown in Figure TS.8.2 being based on the IEA's World Energy Outlook 2010 '450 Policy Scenario'. To achieve such increased shares of primary and consumer energy from RE by 2035 would require the annual average incremental growth in primary RE to more than treble from today's level to around 4.0 EJ/yr. [8.1, 10.2, 10.2.2.4]

In order to gain greater RE deployment in each of the transport, building, industry and agriculture sectors, strategic elements need to be better understood, as do the social issues. Transition pathways for increasing the shares of each RE technology through integration depend on the specific sector, technology and region. Facilitating a smoother integration with energy supply systems and providing multiple benefits for energy end users should be the ultimate aims. [8.2, 8.3]

Several mature RE technologies have already been successfully integrated into a wide range of energy supply systems, mostly at relatively low shares but with some examples (including small- and large-scale hydropower, wind power, geothermal heat and power, first-generation biofuels and solar water heating systems) exceeding 30%. This was due mainly to their improved cost-competitiveness, an increase in support policies and growing public support due to the threats of an insecure energy supply and climate change. Exceptional examples are large-scale hydropower in Norway and hydro and geothermal power in Iceland approaching 100% of RE electricity, as has also been achieved by several small islands and towns. [8.2.1.3, 8.2.5.5, 11.2, 11.5]

Other less mature technologies require continuing investment in research, development, and demonstration (RD&D), infrastructure, capacity building and other supporting measures over the longer term. Such technologies include advanced biofuels, fuel cells, solar fuels, distributed power generation control systems, electric vehicles, solar absorption cooling and enhanced geothermal systems. [11.5, 11.6]

The current status of RE use varies for each end-use sector. There are also major regional variations in future pathways to enhance further integration by removal of barriers. For example, in the building sector, integrating RE technologies is vastly different for commercial high-rise buildings and apartments in mega-cities than for integration into small, modest village dwellings in developing countries that currently have limited access to energy services. [8.3.2]

Most energy supply systems can accommodate a greater share of RE than at present, particularly if the RE share is at relatively low levels (usually assumed to be below a 20% share of electricity, heat, pipeline gas blend or biofuel blend). To accommodate higher RE shares in the future, most energy supply systems will need to evolve and be adapted. In all cases, the maximum practical RE share will depend on the technologies involved, the RE resources available and the type and age of the present energy system. Further integration and increased rates of deployment can be encouraged by local, national and regional initiatives. The overall aim of Chapter 8 is to present the current knowledge on opportunities and challenges relating to RE integration for governments wishing to develop a coherent framework in preparation for future higher levels of RE penetration. Existing power supply systems, natural gas grids, heating/cooling schemes, petroleum-based transport fuel supply distribution networks and vehicles can all be adapted to accommodate greater supplies of RE than at present. RE technologies range from mature to those at the early concept demonstration stage. New technologies could enable increased RE uptake and their integration will depend upon improved cost-effectiveness, social acceptance, reliability and political support at national and local government levels in order to gain greater market shares. [8.1.2, 11.5]

Taking a holistic approach to the whole energy system may be a prerequisite to ensure efficient and flexible RE integration. This would include achieving mutual support between the different energy sectors, an intelligent forecasting and control strategy and coherent long-term planning. Together, these would enable the provision of electricity, heating, cooling and mobility to be more closely inter-linked. The optimum combination of technologies and social mechanisms to enable RE integration to reach high shares varies with the limitations of specific site conditions, characteristics of the available RE resources, and local energy demands. Exactly how present energy supply and demand systems can be adapted and developed to accommodate higher shares of RE, and the additional costs involved for their integration, depend on the specific circumstances, so

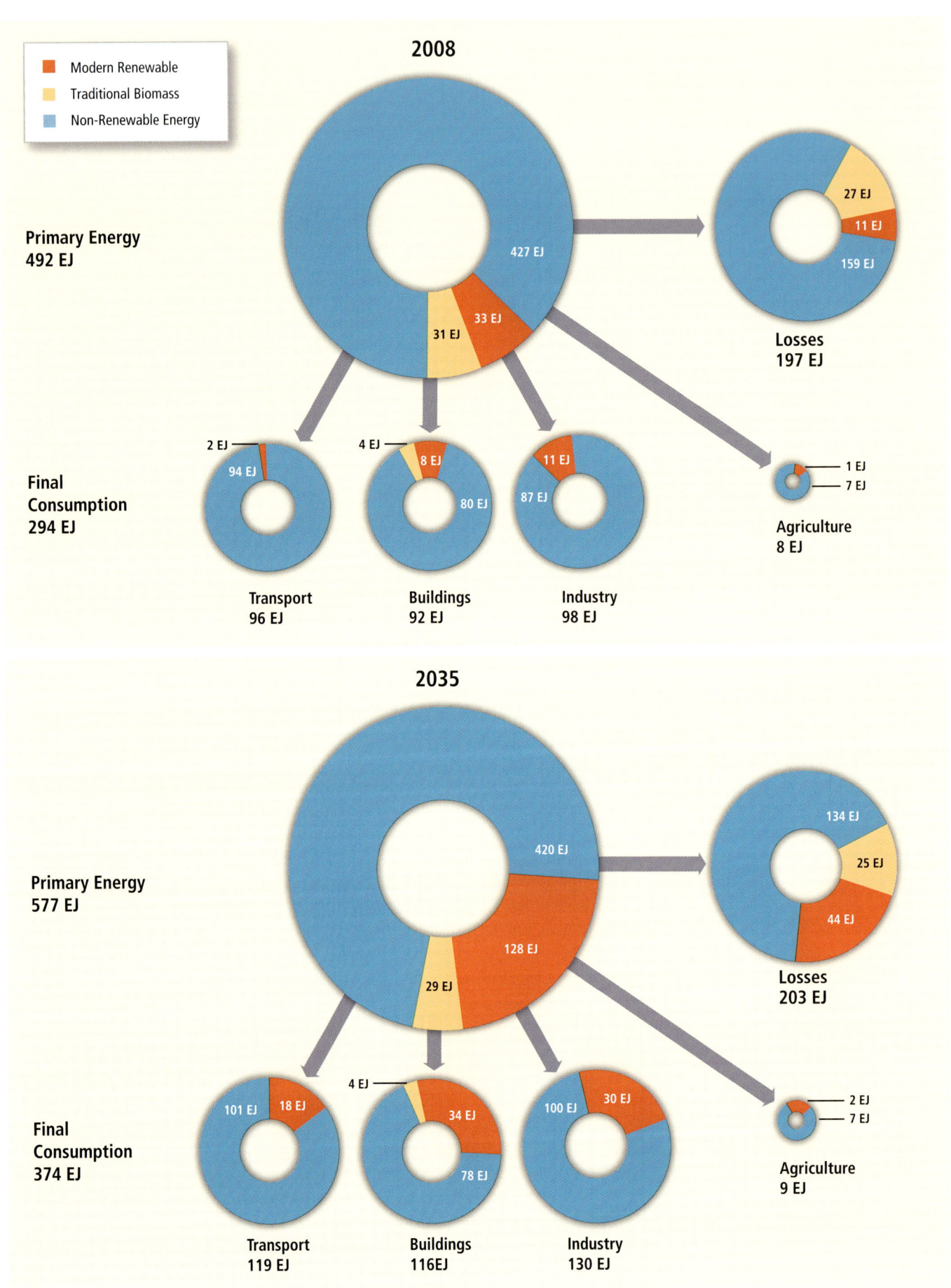

Figure TS.8.2 | (Preceding page) RE shares (red) of primary and final consumption energy in the transport, buildings (including traditional biomass), industry and agriculture sectors in 2008 and an indication of the projected increased RE shares needed by 2035 in order to be consistent with a 450 ppm CO_2eq stabilization level. [Figure 8.2]

Notes: Area of circles are approximately to scale. Energy system losses occur during the conversion, refining and distribution of primary energy sources to produce energy services for final consumption. 'Non-renewable' energy (blue) includes coal, oil, natural gas (with and without CCS by 2035) and nuclear power. This scenario example is based on data taken from the IEA World Energy Outlook 2010 but converted to direct equivalents. [Annex II.4] Energy efficiency improvements above the baseline are included in the 2035 projection. RE in the buildings sector includes traditional solid biomass fuels (yellow) for cooking and heating for 2.7 billion people in developing countries [2.2] along with some coal. By 2035, some traditional biomass has been partly replaced by modern bioenergy conversion systems. Excluding traditional biomass, the overall RE system efficiency (when converting from primary to consumer energy) remains around 66%.

further studies will be required. This is particularly the case for the electricity sector due to the wide variety of existing power generation systems and scales that vary with country and region. [8.2.1, 8.2.2, 8.3]

8.2 Integration of renewable energy into electrical power systems

Electrical power systems have been evolving since the end of the 19th century. Today, electrical power systems vary in scale and technological sophistication from the synchronized Eastern Interconnection in North America to small individual diesel-powered autonomous systems, with some systems, as in China, undergoing rapid expansion and transformation. Within these differences, however, electrical power systems are operated and planned with a common purpose of providing a reliable and cost-effective supply of electricity. Looking forward, electric power systems are expected to continue to expand in importance given that they supply modern energy, enable the transport of energy over long distances, and provide a potential pathway for delivering low-carbon energy. [8.2.1]

Electric power systems have several important characteristics that affect the challenges of integrating RE. The majority of electric power systems operate using alternating current (AC) whereby the majority of generation is synchronized and operated at a frequency of approximately either 50 or 60 Hz, depending on the region. The demand for electricity varies throughout the day, week and season, depending on the needs of electricity users. The aggregate variation in demand is matched by variation in schedules and dispatch instructions for generation in order to continuously maintain a balance between supply and demand. Generators and other power system assets are used to provide active power control to maintain the system frequency and reactive power control to maintain voltage within specified limits. Minute-to-minute variations in supply and demand are managed with automatic control of generation through services called regulation and load following, while changes over longer time scales of hours to days are managed by dispatching and scheduling generation (including turning generation on or off, which is also known as unit commitment). This continuous balancing is required irrespective of the mechanism used to achieve it. Some regions choose organized electricity markets in order to determine which generation units should be committed and/or how they should be dispatched. Even autonomous systems must employ methods to maintain a balance between generation and demand (via controllable generators, controllable loads, or storage resources like batteries). [8.2.1.1]

In addition to maintaining a balance between supply and demand, electric power systems must also transfer electricity between generation and demand through transmission and distribution networks with limited capacity. Ensuring availability of adequate generation and network capacity requires planning over multiple years. Planning electrical power systems incorporates the knowledge that individual components of the system, including generation and network components, will periodically fail (a contingency). A target degree of reliability can be met, however, by building adequate resources. One important metric used to determine the contribution of generation—fossil-fuel based or renewable—to meeting demand with a target level of reliability is called the capacity credit. [8.2.1.1]

Based on the features of electrical power systems, several RE characteristics are important for integrating RE into power systems. In particular, variability and predictability (or uncertainty) of RE is relevant for scheduling and dispatch in the electrical power system, the location of RE resources is a relevant indicator for impact on needs for electrical networks, and capacity factor, capacity credit and power plant characteristics are indicators relevant for comparison, for example, with thermal generation. [8.2.1.2]

Some RE electricity resources (particularly ocean, solar PV, wind) are variable and only partially dispatchable: generation from these resources can be reduced if needed, but maximum generation depends on availability of the RE resource (e.g., tidal currents, sun or wind). The capacity credit can be low if the generation is not well correlated with times of high demand. In addition, the variability and partial predictability of some RE increases the burden on dispatchable generation or other resources to ensure balance between supply and demand given deviations in RE. In many cases variability and partial predictability are somewhat mitigated by geographic diversity—changes and forecast errors will not always occur at the same time in the same direction. A general challenge for most RE, however, is that renewable resources are location specific, therefore concentrated renewably generated electricity may need to be transported over considerable distances and require network expansion. Dispatchable renewable sources (including hydropower, bioenergy, geothermal energy, and CSP with thermal storage) can in many cases offer extra flexibility for the system to integrate other renewable sources and often have a higher capacity credit. [8.2.1.2]

A very brief summary of the particular characteristics for a selection of the technologies is given in Table TS.8.1. [8.2.1.3]

Table TS.8.1 | Summary of integration characteristics for a selection of RE technologies. [Table 8.1]

Technology		Plant size range (MW)	Variability: Characteristic time scales for power system operation (Time scale)	Dispatchability (See legend)	Geographical diversity potential (See legend)	Predictability (See legend)	Capacity factor range %	Capacity credit range %	Active power, frequency control (See legend)	Voltage, reactive power control (See legend)
Bioenergy		0.1–100	Seasons (depending on biomass availability)	+++	+	++	50–90	Similar to thermal and CHP	++	++
Direct solar energy	PV	0.004–100 modular	Minutes to years	+	++	+	12–27	<25–75	+	+
	CSP with thermal storage[1]	50–250	Hours to years	++	+[2]	++	35–42	90	++	++
Geothermal energy		2–100	Years	+++	N/A	++	60–90	Similar to thermal	++	++
Hydro power	Run of river	0.1–1,500	Hours to years	++	+	++	20–95	0–90	++	++
	Reservoir	1–20,000	Days to years	+++	+	++	30–60	Similar to thermal	++	++
Ocean Energy	Tidal range	0.1–300	Hours to days	+	+	++	22.5–28.5	<10%	++	++
	Tidal current	1–200	Hours to days	+	+	++	19–60	10–20	+	++
	Wave	1–200	Minutes to years	+	++	+	22–31	16	+	+
Wind energy		5–300	Minutes to years	+	++	+	20–40 onshore, 30–45 offshore	5–40	+	++

Notes: 1. Assuming a CSP system with six hours of thermal storage in US Southwest. 2. In areas with direct-normal irradiance (DNI) >2,000 kWh/m²/yr (7,200 MJ/m²/yr).

Plant size: range of typical rated plant capacity.

Characteristic time scales: time scales where variability significant for power system integration occurs.

Dispatchability: degree of plant dispatchability: + low partial dispatchability, ++ partial dispatchability, +++ dispatchable.

Geographical diversity potential: degree to which siting of the technology may mitigate variability and improve predictability, without substantial need for additional network: +moderate potential, ++ high diversity potential.

Predictability: Accuracy to which plant output power can be predicted at relevant time scales to assist power system operation: + moderate prediction accuracy (typical <10% Root Mean Square (RMS) error of rated power day ahead), ++ high prediction accuracy.

Active power and frequency control: technology possibilities enabling plant to participate in active power control and frequency response during normal situations (steady state, dynamic) and during network fault situations (for example active power support during fault ride-through): + good possibilities, ++ full control possibilities.

Voltage and reactive power control: technology possibilities enabling plant to participate in voltage and reactive power control during normal situations (steady state, dynamic) and during network fault situations (for example reactive power support during fault ride-through): + good possibilities, ++ full control possibilities.

There is already significant experience with operating electrical power systems with a large share of renewable sources, in particular hydropower and geothermal power. Hydropower storage and strong interconnections help manage fluctuations in river flows. Balancing costs for variable generation are incurred when there are differences between the scheduled generation (according to forecasts) and the actual production. Variability and uncertainty increase balancing requirements. Overall, balancing is expected to become more difficult to achieve as partially dispatchable RE penetrations increase. Studies show clearly that combining different variable renewable sources, and resources from larger geographical areas, will be beneficial in smoothing the variability and decreasing overall uncertainty for the power systems. [8.2.1.3]

The key issue is the importance of *network infrastructure*, both to deliver power from the generation plant to the consumer as well as to enable larger regions to be balanced. Strengthening connections within an electrical power system and introducing additional interconnections to other systems can directly mitigate the impact of variable and uncertain RE sources. Network expansion is required for most RE, although the level is dependent on the resource and location relative to existing network infrastructure. Amongst other challenges will be expanding network infrastructure within the context of public opposition to overhead network infrastructure. In general, major changes will be required in the generation plant mix, the electrical power systems' infrastructure and operational procedures to make the transition to increased renewable generation while maintaining cost and environmental effectiveness. These changes will require major investments far enough in advance to maintain a reliable and secure electricity supply. [8.2.1.3]

In addition to improving network infrastructure, several other important integration options have been identified through operating experience or studies:

Increased generation flexibility: An increasing penetration of variable renewable sources implies a greater need to manage variability and uncertainty. Greater flexibility is required from the generation mix. Generation provides most of a power system's existing flexibility to cope with variability and uncertainty through ramping up or down and cycling as needed. Greater need for flexibility can imply either investment in new flexible generation or improvements to existing power plants to enable them to operate in a more flexible manner. [8.2.1.3]

Demand side measures: Although demand side measures have historically been implemented only to reduce average demand or demand during peak load periods, demand side measures may potentially contribute to meeting needs resulting from increased variable renewable generation. The development of advanced communications technology, with smart electricity meters linked to control centres, offers the potential to access much greater levels of flexibility from demand. Electricity users can be provided with incentives to modify and/or reduce their consumption by pricing electricity differently at different times, in particular with higher prices during higher demand periods. This reduction in demand during high demand periods can mitigate the impact of the low capacity credit of some types of variable generation. Furthermore, demand that can quickly be curtailed without notice during any time of the year can provide reserves rather than requiring generation resources to provide this reserve. Demand that can be scheduled to be met at anytime of the day or that responds to real-time electricity prices can participate in intra-day balancing thereby mitigating operational challenges that are expected to become increasingly difficult with variable generation. [8.2.1.3]

Electrical energy storage: By storing electrical energy when renewable output is high and the demand low, and generating when renewable output is low and the demand high, the curtailment of RE can be reduced, and the base-load units on the system will operate more efficiently. Storage can also reduce transmission congestion and may reduce the need for, or delay, transmission upgrades. Technologies such as batteries or flywheels that store smaller amounts of energy (minutes to hours) can in theory be used to provide power in the intra-hour timeframe to regulate the balance between supply and demand. [8.2.1.3]

Improved operational/market and planning methods: To help cope with the variability and uncertainty associated with variable generation sources, forecasts of their output can be combined with improved operational methods to determine both the required reserve to maintain the demand-generation balance, and also optimal generation scheduling. Making scheduling decisions closer to real time (i.e., shorter gate closure time in markets) and more frequently allows newer, more accurate information to be used in dispatching generating units. Moving to larger balancing areas, or shared balancing between areas, is also desirable with large amounts of variable generation, due to the aggregation benefits of multiple, dispersed renewable sources. [8.2.1.3]

In summary, RE can be integrated into all types of electrical power systems from large interconnected continental-scale systems to small autonomous systems. System characteristics including the network infrastructure, demand pattern and its geographic location, generation mix, control and communication capability combined with the location, geographical footprint, variability and predictability of the renewable resources determine the scale of the integration challenge. As the amounts of RE resources increase, additional electricity network infrastructure (transmission and/or distribution) will generally have to be constructed. Variable renewable sources, such as wind, can be more difficult to integrate than dispatchable renewable sources, such as bioenergy, and with increasing levels maintaining reliability becomes more challenging and costly. These challenges and costs can be minimized by deploying a portfolio of options including electrical network interconnection, the development of complementary flexible generation, larger balancing areas, sub-hourly markets, demand that can respond in relation to supply availability, storage technologies, and better forecasting, system operating and planning tools.

8.3 Integration of renewable energy into heating and cooling networks

A district heating (DH) or district cooling (DC) network allows multiple energy sources (Figure TS.8.3) to be connected to many energy consumers by pumping the energy carriers (hot or cold water and sometimes steam) through insulated underground pipelines. Centralized heat production can facilitate the use of low-cost and/or low-grade RE heat from geothermal or solar thermal sources or combustion of biomass (including refuse-derived fuels and waste by-products that are often unsuitable for use by individual heating systems). Waste heat from CHP generation and industrial processes can also be used. This flexibility produces competition among various heat sources, fuels and technologies. Centralized heat production can also facilitate the application of cost-effective measures that reduce local air pollution compared with having a multitude of small individual boilers. Being flexible in the sources of heat or cold utilized, district heating and cooling systems allow for the continuing uptake of several types of RE so that a gradual or rapid substitution of competing fossil fuels is usually feasible. [8.2.2]

Occupiers of buildings and industries connected to a network can benefit from a professionally managed central system, hence avoiding the need to operate and maintain individual heating/cooling equipment.

Several high-latitude countries already have a district heating market penetration of 30 to 50%, with Iceland reaching 96% using its geothermal resources. World annual delivery of district heat has been estimated to be around 11 EJ though heat data are uncertain. [8.2.2.1]

DH schemes can provide electricity through CHP system designs and can also provide demand response options that can facilitate increased integration of RE, including by using RE electricity for heat pumps and electric boilers. Thermal storage systems can bridge the heat supply/demand gap resulting from variable, discontinuous or non-synchronized heating systems. For short-term storage (hours and days), the thermal capacity of the distribution network itself can be used. Thermal storage systems with storage periods up to several months at temperatures up to hundreds of degrees Celsius use a variety of materials and corresponding storage mechanisms that can have capacities up to several TJ. Combined production of heat, cold and electricity (tri-generation), as well as the possibility for diurnal and seasonal storage of heat and cold, mean that high overall system efficiency can be obtained and higher shares of RE achieved through increased integration. [8.2.2.2, 8.2.2.3]

Many commercial geothermal and biomass heat and CHP plants have been successfully integrated into DH systems without government support. Several large-scale solar thermal systems with collector areas

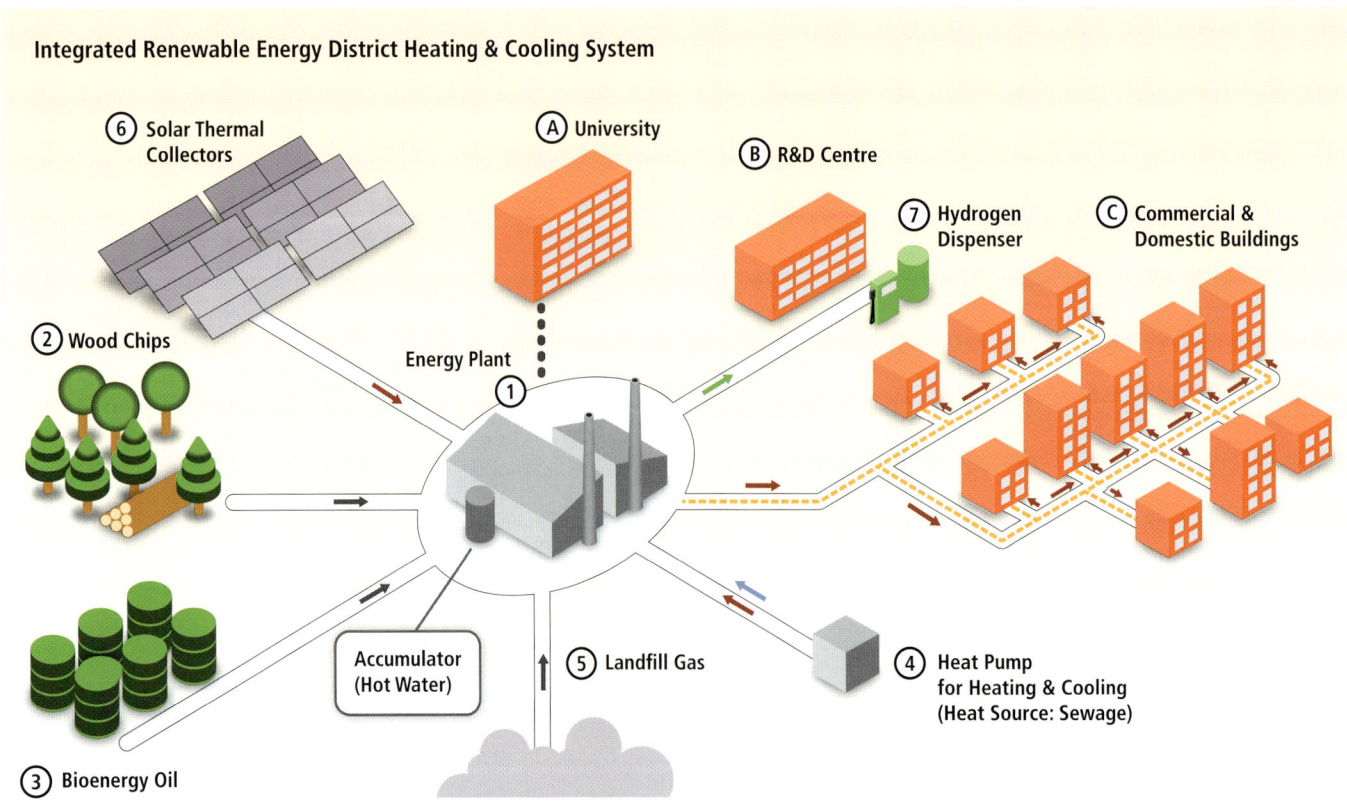

Figure TS.8.3 | An integrated RE-based energy plant in Lillestrøm, Norway, supplying the University, R&D Centre and a range of commercial and domestic buildings using a district heating and cooling system incorporating a range of RE heat sources, thermal storage and a hydrogen production and distribution system. (Total investment around USD$_{2005}$ 25 million and due for completion in 2011.) 1) Central energy system with 1,200 m³ accumulator hot water storage tank; (2) 20 MW$_{th}$ wood burner system (with flue gas heat recovery); (3) 40 MW$_{th}$ bio-oil burner; (4) 4.5 MW$_{th}$ heat pump; (5) 1.5 MW$_{th}$ landfill gas burner and a 5 km pipeline; (6) 10,000 m² solar thermal collector system; and (7) RE-based hydrogen production (using water electrolysis and sorption-enhanced steam methane reforming of landfill gas) and vehicle dispensing system. [Figure 8.3]

of around 10,000 m² (Figure TS.8.3) have also been built in Denmark, Norway and elsewhere. The best mix of hot and cold sources, and heat transfer and storage technologies, depends strongly on local conditions, including user demand patterns. As a result, the heat energy supply mix varies widely between different systems. [3.5.3, 8.2.2]

Establishing or expanding a DH scheme involves high up-front capital costs for the piping network. Distribution costs alone can represent roughly half of the total cost but are subject to large variations depending on the heat demand density and the local conditions for building the insulated piping network. Increasing urbanization facilitates DH since network capital costs are lower for green-field sites and distribution losses per unit of heat delivered are lower in areas with higher heat demand densities. Heat distribution losses typically range from 5 to 30% but the extent to which high losses are considered a problem depends on the source and cost of the heat. [8.2.2.1, 8.2.2.3]

Expanding the use of deep geothermal and biomass CHP plants in DH systems can facilitate a higher share of RE sources, but to be economically viable this usually requires the overall system to have a large heat load. Some governments therefore support investments in DH networks as well as provide additional incentives for using RE in the system. [8.2.2.4]

Modern building designs and uses have tended to reduce their demand for additional heating whereas the global demand for cooling has tended to increase. The cooling demand to provide comfort has increased in some low-latitude regions where countries have become wealthier and in some higher latitudes where summers have become warmer. Cooling load reductions can be achieved by the use of passive cooling building design options or active RE solutions including solar absorption chillers. As for DH, the rate of uptake of energy efficiency to reduce cooling demand, deployment of new technologies, and the structure of the market, will determine the viability of developing a DC scheme. Modern DC systems, ranging from 5 to 300 MW$_{th}$, have been operating successfully for many years using natural aquifers, waterways, the sea or deep lakes as the sources of cold, classed as a form of RE. [8.2.2.4]

DH and DC schemes have typically been developed in situations where strong planning powers have existed, such as centrally planned economies, US university campuses, Western European countries with multi-utilities, and urban areas controlled by local municipalities.

8.4 Integration of renewable energy into gas grids

Over the past 50 years, large natural gas networks have been developed in several parts of the world. And more recently there has been increasing interest to 'green' them by integrating RE-based gases. Gaseous fuels from RE sources originate largely from biomass and can be produced either by anaerobic digestion to produce biogas (mainly methane and CO_2) or thermo-chemically to give synthesis (or producer) gas (mainly hydrogen and carbon monoxide). Biomethane, synthesis gas and, in the longer term, RE-based hydrogen can be injected into existing gas pipelines for distribution at the national, regional or local level. Differences in existing infrastructure, gas quality, and production and consumption levels can make planning difficult for increasing the RE share of gases by integration into an existing grid. [8.2.3, 8.2.3.1]

Biogas production is growing rapidly and several large gas companies are now making plans to upgrade large quantities for injection at the required quality into national or regional transmission gas pipelines. Most of the biomethane currently produced around the world is already distributed in local gas pipeline systems primarily dedicated for heating purposes. This can be a cheaper option per unit of energy delivered (Figure TS.8.4) than when transported by trucks (usually to filling stations for supplying gas-powered vehicles) depending on distance and the annual volume to be transported. [8.2.3.4]

Gas utilization can be highly efficient when combusted for heat; used to generate electricity by fuelling gas engines, gas boilers or gas turbines; or used in vehicles either compressed or converted to a range of liquid fuels using various processes. For example, biogas or landfill gas can be combusted onsite to produce heat and/or electricity; cleaned and upgraded to natural gas quality biomethane for injection into gas grids; or, after compressing or liquefying, distributed to vehicle filling stations for use in dedicated or dual gas-fuelled vehicles. [8.2.3.2–8.2.3.4]

Technical challenges relate to gas source, composition and quality. Only biogas and syngas of a specified quality can be injected into existing gas

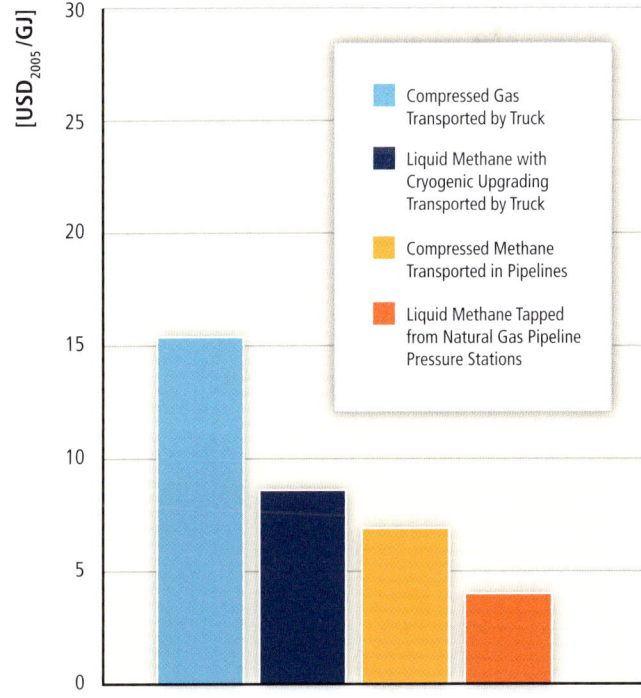

Figure TS.8.4 | Relative costs for distributing and dispensing biomethane (either compressed or liquefied) at the medium scale by truck or pipeline in Europe. [Figure 8.9]

grids so clean-up is a critical step to remove water, CO_2 (thereby increasing the heating value) and additional by-products from the gas stream. The cost of upgrading varies according to the scale of the facility and the process, which can consume around 3 to 6% of the energy content of the gas. RE gas systems are likely to require significant storage capacity to account for variability and seasonality of supply. The size and shape of storage facilities and the required quality of the gas will depend on the primary energy source of production and its end use. [8.2.3]

Hydrogen gas can be produced from RE sources by several routes including biomass gasification, the reformation of biomethane, or electrolysis of water. The potential RE resource base for hydrogen is therefore greater than for biogas or syngas. Future production of hydrogen from variable RE resources, such as wind or solar power by electrolysis, will depend significantly on the interaction with existing electricity systems and the degree of surplus capacity. In the short term, blending of hydrogen with natural gas (up to 20% by volume) and transporting it long distances in existing gas grids could be an option. In the longer term, the construction of pipelines for carrying pure hydrogen is possible, constructed from special steels to avoid embrittlement. The rate-limiting factors for deploying hydrogen are likely to be the capital and time involved in building a new hydrogen infrastructure and any additional cost for storage in order to accommodate variable RE sources. [8.2.3.2, 8.2.3.4]

In order to blend a RE gas into a gas grid, the gas source needs to be located near to the existing system to avoid high costs of additional pipeline construction. In the case of remote plant locations due to resource availability, it may be better to use the gas onsite where feasible to avoid the need for transmission and upgrading. [8.2.3.5]

8.5 Integration of renewable energy into liquid fuels

Most of the projected demand for liquid biofuels is for transport purposes, though industrial demand could emerge for bio-lubricants and bio-chemicals such as methanol. In addition, large amounts of traditional solid biomass could eventually be replaced by more convenient, safer and healthier liquid fuels such as RE-derived dimethyl ether (DME) or ethanol gels. [8.2.4]

Producing bioethanol and biodiesel fuels from various crops, usually used for food, is well understood (Figure TS.8.5). The biofuels produced can take advantage of existing infrastructure components already used for petroleum-based fuels including storage, blending, distribution and dispensing. However, sharing petroleum-product infrastructure (storage tanks, pipelines, trucks) with ethanol or blends can lead to problems from water absorption and equipment corrosion, so may require investment in specialized pipeline materials or linings. Decentralized biomass production, seasonality and remote agricultural locations away from existing oil refineries or fuel distribution centres, can impact the supply chain logistics and storage of biofuels. Technologies continue to evolve to produce biofuels from non-food feedstocks and biofuels that are more compatible with existing petroleum fuels and infrastructure. Quality control procedures need to be implemented to ensure that such biofuels meet all applicable product specifications. [8.2.4.1, 8.2.4.3, 8.2.4.4]

The use of blended fuels produced by replacing a portion (typically 5 to 25% but can be up to 100% substitution) of gasoline with ethanol,

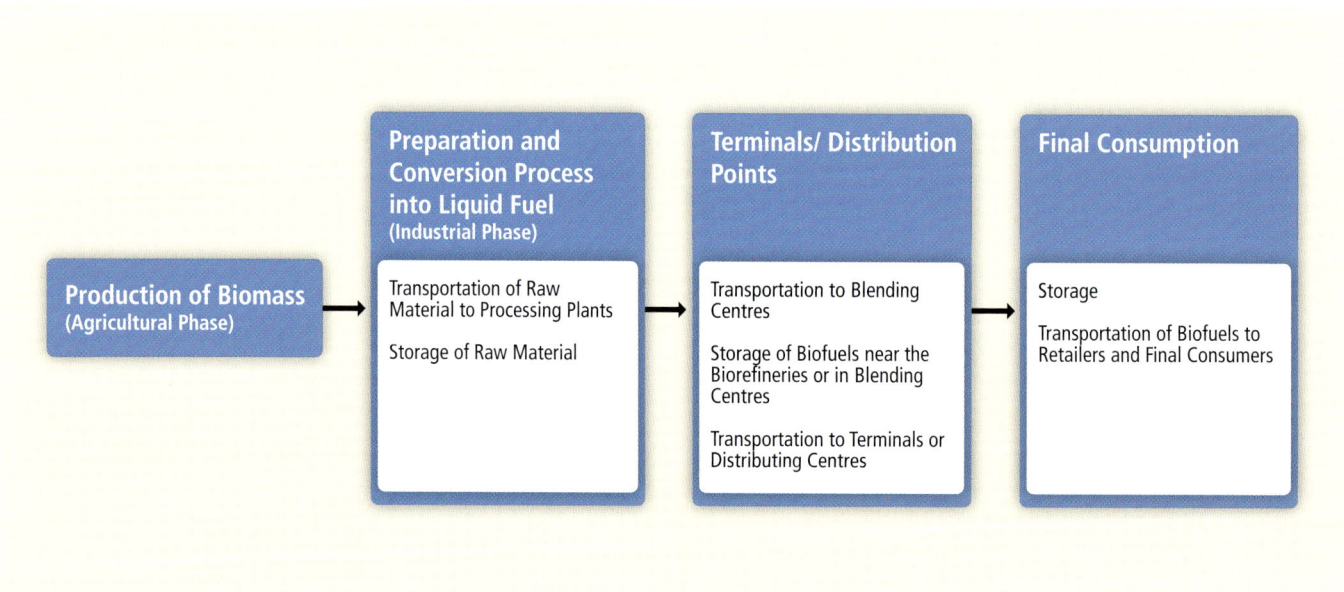

Figure TS.8.5 | The production, blending and distribution system for a range of liquid biofuels is similar regardless of the biomass feedstock. [Figure 8.11]

or diesel with biodiesel, requires investment in infrastructure including additional tanks and pumps at vehicle service stations. Although the cost of biofuel delivery is a small fraction of the overall cost, the logistics and capital requirements for widespread integration and expansion could present major hurdles if not well planned. Since ethanol has only around two-thirds the energy density (by volume) of gasoline, larger storage systems, more rail cars or vessels, and larger capacity pipelines are needed to store and transport the same amount of energy. This increases the fuel storage and delivery costs. Although pipelines would, in theory, be the most economical method of delivery, and pipeline shipments of ethanol have been successfully achieved, a number of technical and logistical challenges remain. Typically, current volumes of ethanol produced in an agricultural region to meet local demand, or for export, are usually too low to justify the related investment costs and operational challenges of constructing a dedicated pipeline. [8.2.4.3]

8.6 Integration of renewable energy into autonomous systems

Autonomous energy supply systems are typically small scale and are often located in off-grid remote areas, on small islands, or in individual buildings where the provision of commercial energy is not readily available through grids and networks. Several types of autonomous systems exist and can make use of either single energy carriers, for example, electricity, heat, or liquid, gaseous or solid fuels, or a combination of carriers. [8.2.5, 8.2.5.1]

In principle, RE integration issues for autonomous systems are similar to centralized systems, for example, for supply/demand balancing of electricity supply systems, selection of heating and cooling options, production of RE gases and liquid biofuel production for local use. However, unlike larger centralized supply systems, smaller autonomous systems often have fewer RE supply options that are readily available at a local scale. Additionally, some of the technical and institutional options for managing integration within larger networks become more difficult or even implausible for smaller autonomous systems, such as RE supply forecasting, probabilistic unit commitment procedures, stringent fuel quality standards, and the smoothing effects of geographical and technical diversity. [8.2.1–8.2.5]

RE integration solutions typically become more restricted as supply systems become smaller. Therefore greater reliance must be placed on those solutions that are readily available. Focusing on variable RE resources, because of restricted options for interconnection and operating and planning procedures, autonomous systems will naturally have a tendency to focus on energy storage options, various types of demand response, and highly flexible fossil fuel generation to help match supply and demand. RE supply options that better match local load profiles, or that are dispatchable, may be chosen over other lower-cost options that do not have as strong a match with load patterns or are variable. Managing RE integration within autonomous systems will, all else being equal, be more costly than in larger integrated networks because of the restricted set of options, but in most instances, such as on islands or in remote rural areas, there is no choice for the energy users. One implication is that autonomous electricity system users and designers can face difficult trade-offs between a desire for reliable and continuous supply and minimizing overall supply costs. [8.2.5]

The integration of RE conversion technologies, balancing options and end-use technologies in an autonomous energy system depend on the site-specific availability of RE resources and the local energy demand. These can vary with local climate and lifestyles. The balance between cost and reliability is critical when designing and deploying autonomous power systems, particularly for rural areas of developing economies because the additional cost of providing continuous and reliable supply may become higher for smaller autonomous systems. [8.2.5.2]

8.7 End-use sectors: Strategic elements for transition pathways

RE technology developments have continued to evolve, resulting in increased deployment in the transport, building, industry, and agriculture, forestry and fishery sectors. In order to achieve greater RE deployment in all sectors, both technical and non-technical issues should be addressed. Regional variations exist for each sector due to the current status of RE uptake, the wide range of energy system types, the related infrastructure currently in place, the different possible pathways to enhance increased RE integration, the transition issues yet to be overcome, and the future trends affected by variations in national and local ambitions and cultures. [8.3, 8.3.1]

8.7.1 Transport

Recent trends and projections show strong growth in transport demand, including the rapidly increasing number of vehicles worldwide. Meeting this demand, whilst achieving a low-carbon, secure energy supply, will require strong policy initiatives, rapid technological change, monetary incentives and/or the willingness of customers to pay additional costs. [8.3.1]

In 2008, the combustion of fossil fuels for transport consumed around 19% of global primary energy use, equivalent to 30% of total consumer energy and producing around 22% of GHG emissions, plus a significant share of local air-polluting emissions. Light duty vehicles (LDVs) accounted for over half of transport fuel consumption worldwide, with heavy duty vehicles (HDVs) accounting for 24%, aviation 11%, shipping 10% and rail 3%. Demand for mobility is growing rapidly with the number of motorized vehicles projected to triple by 2050 and with a similar growth in air travel. Maintaining a secure supply of energy is therefore a serious concern for the transport sector with about 94% of transport fuels presently coming from oil products that, for most countries, are imported. [8.3.1]

There are a number of possible fuel/vehicle pathways from the conversion of the primary energy source to an energy carrier (or fuel) through to the end use, whether in advanced internal combustion engine vehicles (ICEVs), electric battery vehicles (EVs), hybrid electric vehicles (HEVs), plug-in hybrid electric vehicles (PHEVs) or hydrogen fuel cell vehicles (HFCVs) (Figure TS.8.6). [8.3.1.2]

Improving the efficiency of the transport sector, and decarbonizing it, have been identified as being critically important to achieving long-term, deep reductions in global GHG emissions. The approaches to reducing transport-related emissions include a reduction in travel demand, increased vehicle efficiency, shifting to more efficient modes of transport, and replacing petroleum-based fuels with alternative low- or near-zero-carbon fuels (including biofuels, electricity or hydrogen produced from low-carbon primary energy sources). Scenario studies strongly suggest that a combination of technologies will be needed to accomplish 50 to 80% reductions (compared to current rates) in GHG emissions by 2050 whilst meeting the growing transport energy demand (Figure TS.8.7). [8.3.1.1]

The current use of RE for transport is only a few percent of the total energy demand, mainly through electric rail and the blending of liquid biofuels with petroleum products. Millions of LDVs capable of running on high-biofuel blends are already in the world fleet and biofuel technology is commercially mature, as is the use of compressed biomethane in vehicles suitable for running on compressed natural gas. [8.2.3]

However, making a transition to new fuels and engine types is a complex process involving technology development, cost, infrastructure, consumer acceptance, and environmental and resource impacts. Transition issues vary for biofuels, hydrogen, and electric vehicles (Table TS.8.2) with no one option seen to be a clear 'winner' and all needing several decades to be deployed at a large scale. Biofuels are well proven, contributing around 2% of road transport fuels in 2008, but there are issues of sustainability. [2.5] Many hydrogen fuel cell vehicles have been demonstrated, but these are unlikely to be commercialized until at least 2015 to 2020 due to the barriers of fuel cell durability, cost, onboard hydrogen storage issues and hydrogen infrastructure availability. For EVs and PHEVs, the cost and relatively short life of present

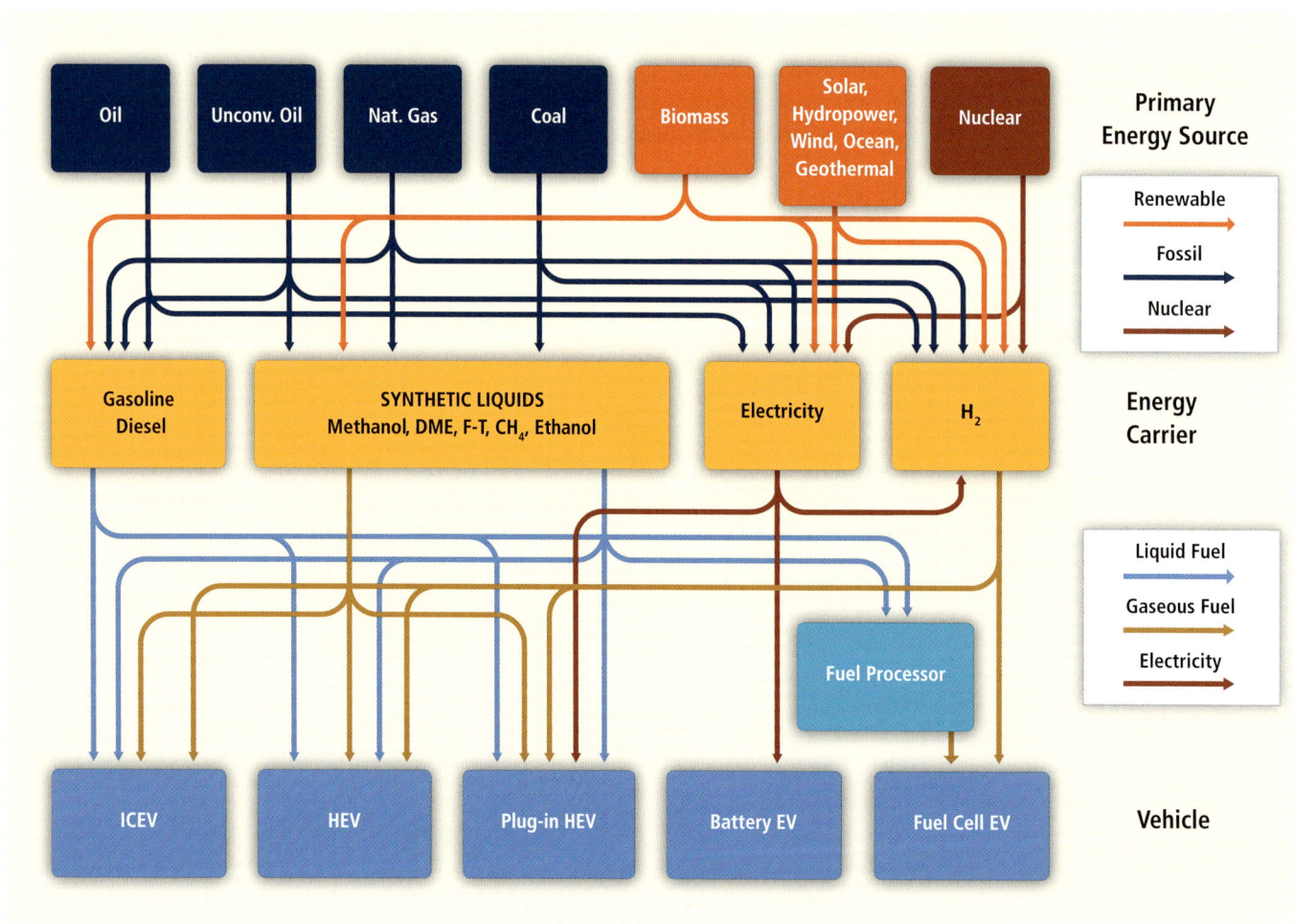

Figure TS.8.6 | A range of possible light duty vehicle fuel pathways, from primary energy sources (top), through energy carriers, to end-use vehicle drive train options (bottom) (with RE resources highlighted in green). [Figure 8.13]

Notes: F-T= Fischer-Tropsch process; DME = dimethyl ether; ICE = internal combustion engine; HEV = hybrid electric vehicle; EV = electric vehicle; 'unconventional oil' refers to oil sands, oil shale and other heavy crudes.

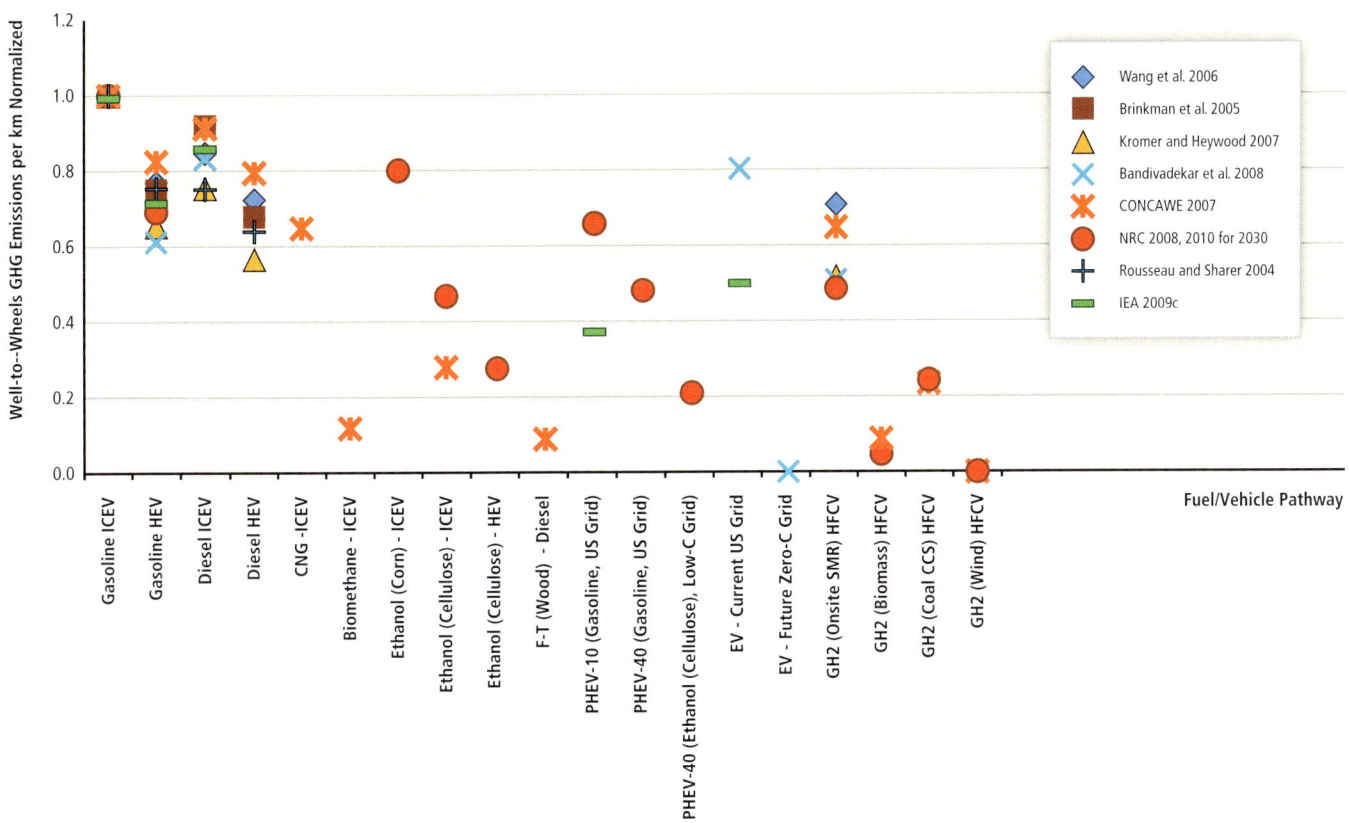

Figure TS.8.7 | Well-to-wheels (WTW) GHG emission reductions per kilometre travelled, with ranges shown taken from selected studies of alternative light duty fuel/vehicle pathways, normalized to the GHG emissions of a gasoline, internal combustion engine, light-duty vehicle. [Figure 8.17]

Notes: To allow for easier comparison among studies, WTW GHG emissions per km were normalized to emissions from a gasoline ICEV (such that 'Gasoline ICEV' = 1) taken from each study and ranging from 170 to 394 g CO_2/km. For all hydrogen pathways, hydrogen is stored onboard the vehicle as a compressed gas (GH2). CNG = compressed natural gas; SMR = steam methane reformer.

battery technologies, the limited vehicle range between recharging, and the time for recharging, can be barriers to consumer acceptance. EV and PHEV designs are undergoing rapid development, spurred by recent policy initiatives worldwide, and several companies have announced plans to commercialize them. One strategy could be to introduce PHEVs initially while developing and scaling up battery technologies. For hydrogen and electric vehicles, it may take several decades to implement a practical transport system by developing the necessary infrastructure at the large scale.

An advantage of *biofuels* is their relative compatibility with the existing liquid fuel infrastructure. They can be blended with petroleum products and most ICE vehicles can be run on blends, some even on up to 100% biofuel. They are similar to gasoline or diesel in terms of vehicle performance[14] and refuelling times, though some have limits on the concentrations that can be blended and they typically cannot be easily distributed using existing fuel pipelines without modifications. The sustainability of the available biomass resource is a serious issue for some biofuels. [2.5, 8.2.4, 8.3.1.2]

Hydrogen has the potential to tap vast new energy resources to provide transport with zero or near-zero emissions. The technology for hydrogen from biomass gasification is being developed, and could become competitive beyond 2025. Hydrogen derived from RE sources by electrolysis has cost barriers rather than issues of technical feasibility or resource availability. Initially RE and other low-carbon technologies will likely be used to generate electricity, a development that could help enable near-zero-carbon hydrogen to be co-produced with electricity or heat in future energy complexes. Hydrogen is not yet widely distributed compared to electricity, natural gas, gasoline, diesel or biofuels but could be preferred in the future for large HDVs that have a long range and need relatively fast refuelling times. Bringing hydrogen to large numbers of vehicles would require building a new refuelling infrastructure that could take several decades to construct. The first steps to provide hydrogen to test fleets and demonstrate refuelling technologies in mini-networks have begun in several countries. [2.6.3.2, 8.3.1, 8.3.1.2]

For RE *electricity* to supply high numbers of EVs and PHEVs in future markets, several innovations must occur such as development of batteries and low-cost electricity supply available for recharging when the EVs need it. If using night-time, off-peak recharging, new capacity is less likely to be needed and in some locations there may be a good temporal match with

---

14 Performance in this instance excludes energy content. The energy content of biofuels is generally lower than their equivalent petroleum product.

Table TS.8.2 | Transition issues for the use of biofuels, hydrogen and electricity as transport fuels for light duty vehicles. [Summarized from 8.3.1]

Technology Status	Biofuels	Hydrogen	Electricity
Existing and potential primary resources	Sugar, starch, oil crops; cellulosic crops; forest, agricultural and solid wastes; algae and other biological oils.	Fossil fuels; nuclear; all RE. Potential RE resource base is large but inefficiencies and costs of converting to H_2 can be an issue.	Fossil fuels, nuclear, all RE. Potential RE resource base is large.
Fuel production	First generation: ethanol from sugar and starch crops, biomethane, biodiesel. Advanced second-generation biofuels, e.g., from cellulosic biomass, bio-wastes, bio-oils, and algae after at least 2015.	Fossil H_2 commercial for large-scale industrial applications, but not competitive as transport fuel. Renewable H_2 generally more costly.	Commercial power readily available. RE electricity can be more costly, but preferred for transport due to low GHG emissions on a lifecycle basis.
Vehicles	Millions of flexi-fuel vehicles exist that use high shares of ethanol. Conventional ICEVs limited to low concentration blends of ethanol (<25%). Some commercial agricultural tractors and machinery can run on 100% biodiesel.	Demonstration HFCVs. Commercial HFCVs not until 2015 to 2020.	Demonstration PHEVs, Commercial PHEVs not until 2012 to 2015. Limited current use of EVs. Commercial EVs not until 2015 to 2020.
Costs[1] compared with gasoline ICE vehicles			
Incremental vehicle price compared to future gasoline ICEV (USD$_{2005}$)	Similar price.	HFCV experience (by 2035) price increment >USD 5,300	Experience (by 2035) price increment: PHEVs >USD 5,900; EVs >USD 14,000
Fuel cost (USD$_{2005}$/km)	Fuel cost per km varies with biofuel type and level of agricultural subsidy. Biofuel can compete if price per unit of energy equates to gasoline/diesel price per unit of energy. Ethanol in Brazil competes without subsidies.	Target fuel cost at USD 3 to 4/kg for mature H_2 infrastructure—may prove optimistic. When used in HFCVs, competes with gasoline in HCEVs at USD 0.40 to 0.53/l. Assumes HFCV has twice fuel economy of gasoline ICEV. RE-derived H_2 around 1.5 to 3 times more expensive than other from sources.	Electricity cost per km, when the power is purchased at USD 0.10 to 0.30/kWh, competes with gasoline when purchased at USD 0.3 to 0.9/l (assuming the EV has fuel economy 3 times that of the gasoline ICEV).
Compatibility with existing infrastructure	Partly compatible with existing petroleum distribution system. Separate distribution and storage infrastructure may be needed for ethanol.	New H_2 infrastructure needed, as well as renewable H_2 production sources. Infrastructure deployment must be coordinated with vehicle market growth.	Widespread electric infrastructure in place. Need to add in-home and public recharger costs, RE generation sources, and upgrading of transmission and distribution (especially for fast chargers).
Consumer acceptance	Depends upon comparative fuel costs. Alcohol vehicles can have shorter range than gasoline. Potential cost impact on food crops. Land use and water issues can be factors.	Depends upon comparative vehicle and fuel costs. Public perception of safety. Poor public refuelling station availability in early markets.	High initial vehicle cost. High electricity cost of charging on-peak. Limited range unless PHEV. Modest to long recharging time, but home recharging possible. Significantly degraded performance in extreme cold winters or hot summers. Poor public refuelling station availability in early markets
GHG emissions	Depends on feedstock, pathway and land use issue[2]. Low for fuels from biomass residues including sugarcane. Near-term can be high for corn ethanol. Advanced second-generation biofuels likely to be lower.	Depends on H_2 production mix. Compared to future hybrid gasoline ICEVs, WTW GHG emissions for HFCVs using H_2 from natural gas can be slightly more or less depending on assumptions. WTW GHG emissions can approach zero for RE or nuclear pathways.	Depends on grid mix. Using coal-dominated grid mix, EVs and PHEVs have WTW GHG emissions similar or higher than gasoline HEV. With larger fraction of RE and low-carbon electricity, WTW emissions are lower.
Petroleum consumption	Low for blends	Very low	Very low
Environmental and sustainability issues			
Air pollution	Similar to gasoline. Additional issues for ethanol due to permeation of volatile organic compounds through fuel tank seals. Aldehyde emissions.	Zero emission vehicle	Zero emission vehicle.
Water use	More than gasoline depending on feedstock and crop irrigation needs.	Potentially low but depends on pathway as electrolysis and steam reformation depend on water.	Potentially very low but depends on pathway used for power generation.
Land use	Might compete with food and fibre production on cropland.	Depends on pathway.	Depends on pathway.
Materials use		Platinum in fuel cells. Neodymium and other rare earths in electric motors. Material recycling.	Lithium in batteries. Neodymium and other rare earths in electric motors. Material recycling.

Notes: 1. Costs quoted do not always include payback of incremental first vehicle costs. 2. Indirect land use-related GHG emissions linked to biofuels is not included.

wind or hydropower resources. Grid flexibility and/or energy storage may also be needed to balance vehicle recharging electricity demand with RE source availability. [8.2.1]

Other than LDVs, it is possible to introduce RE options and lower GHG emissions in the other transport sectors: HDVs, aviation, maritime and rail. The use of biofuels is key for increasing the share of RE in these sub-sectors but current designs of ICEs would probably need to be modified to operate on high-biofuel blends (above 80%). Aviation has perhaps less potential for fuel switching than the other sub-sectors due to safety needs and to minimize fuel weight and volume. However, various airlines and aircraft manufacturers have flown demonstration test flights using various biofuel blends, but significantly more processing is needed than for road fuels to ensure that stringent aviation fuel specifications are met, particularly at cold temperatures. For rail transport, as around 90% of the industry is powered by diesel fuel, greater electrification and the increased use of biodiesel are the two primary options for introducing RE. [8.3.1.5]

Given all these uncertainties and cost reduction challenges, it is important to maintain a portfolio approach over a long time line that includes behavioural changes (for example to reduce annual vehicle kilometres travelled or kilometres flown), more energy efficient vehicles, and a variety of low-carbon fuels. [8.3.1.5]

8.7.2 Buildings and households

The building sector provides shelter and a variety of energy services to support the livelihoods and well-being of people living in both developed and developing countries. In 2008, it accounted for approximately 120 EJ (about 37%) of total global final energy use (including between 30 and 45 EJ of primary energy from traditional biomass used for cooking and heating). The high share of total building energy demand for heating and cooling is usually met by fossil fuels (oil burners, gas heaters) and electricity (fans and air-conditioners). In many regions, these can be replaced economically by district heating and cooling (DHC) schemes or by the direct use of RE systems in buildings, such as modern biomass pellets and enclosed stoves, heat pumps (including ground source), solar thermal water and space heating, and solar sorption cooling systems. [2.2, 8.2.2, 8.3.2]

RE electricity generation technologies integrated into buildings (such as solar PV panels) provide the potential for buildings to become energy suppliers rather than energy consumers. Integration of RE into existing urban environments, combined with energy efficient appliances and 'green building' designs, are key to further deployment. For both household and commercial building sub-sectors, energy vectors and energy service delivery systems vary depending on the local characteristics and RE resources of a region, its wealth, and the average age of the current buildings and infrastructure impacting stock turnover. [8.3.2]

The features and conditions of energy demands in an existing or new building, and the prospects for RE integration, differ with location and between one building design and another. In both urban and rural settlements in developed countries, most buildings are connected to electricity, water and sewage distribution schemes. With a low building stock turnover rate of only around 1% per year in developed countries, future retrofitting of existing buildings will need to play a significant role in RE integration as well as energy efficiency improvements. Examples include installation of solar water heaters and ground source heat pumps and development or extensions of DHC systems that, being flexible on sources of heat or cold, allow for a transition to a greater share of RE over time. These can involve relatively high up-front investment costs and long payback periods, but these can possibly be offset by amended planning consents and regulations so they become more enabling, improved energy efficient designs, and the provision of economic incentives and financial arrangements. [8.2.2, 8.3.2.1]

Grid electricity supply is available in most urban areas of developing countries, although often the supply system has limited capacity and is unreliable. Increased integration of RE technologies using local RE resources could help ensure a secure energy supply and also improve energy access. In urban and rural settlements in developing countries, energy consumption patterns often include the unsustainable use of biomass and charcoal. The challenge is to reverse the increasing traditional biomass consumption patterns by providing improved access to modern energy carriers and services and increasing the share of RE through integration measures. The distributed nature of solar and other RE resources is beneficial for their integration into new and existing buildings however modest they might be, including dwellings in rural areas not connected to energy supply grids. [8.2.2.2, 8.2.5]

8.7.3 Industry

Manufacturing industries account for about 30% of global final energy use, although the share differs markedly between countries. The sector is highly diverse, but around 85% of industrial energy use is by the more energy-intensive 'heavy' industries including iron and steel, non-ferrous metals, chemicals and fertilizers, petroleum refining, mineral mining, and pulp and paper. [8.3.3.1]

There are no severe technical limits to increasing the direct and indirect use of RE in industry in the future. However, integration in the short term may be limited by factors such as land and space constraints or demands for high reliability and continuous operation. In addition to the integration of higher shares of RE, key measures to reduce industrial energy demands and/or GHG emissions include energy efficiency, recycling of materials, CCS for CO_2-emitting industries such as cement manufacturing, and the substitution of fossil fuel feedstocks. In addition, industry can provide demand-response facilities that are likely to achieve greater prominence in future electricity systems that have a higher penetration of variable RE sources. [8.3.3.1]

The main opportunities for RE integration in industry include:

- Direct use of biomass-derived fuels and process residues for onsite production, and use of biofuels, heat and CHP; [2.4.3]
- Indirect use through increased use of RE-based electricity, including electro-thermal processes; [8.3.3]
- Indirect use through other purchased RE-based energy carriers including heat, liquid fuels, biogas, and, possibly to a greater degree in the future, hydrogen; [8.2.2–8.2.4]
- Direct use of solar thermal energy for process heat and steam demands although few examples exist to date; [3.3.2] and
- Direct use of geothermal resources for process heat and steam demands. [4.3.5]

Industry is not only a potential user of RE but also a potential supplier of bioenergy as a co-product. The current direct use of RE in industry is dominated by biomass produced in the pulp and paper, sugar and ethanol industries as process by-products and used for cogenerated heat and electricity, mainly onsite for the process but also sold off-site. Biomass is also an important fuel for many small and medium enterprises such as brick making, notably as charcoal in developing countries. [8.3.3.1]

Possible pathways for increased use of RE in energy-intensive industries vary between the different industrial sub-sectors. Biomass, for example, is technically able to replace fossil fuels in boilers, kilns and furnaces or to replace petrochemicals with bio-based chemicals and materials. However, due to the scale of many industrial operations, access to sufficient volumes of local biomass may be a constraint. Use of solar technologies can be constrained in some locations with low annual sunshine hours. The direct supply of hydropower to aluminium smelters is not unusual but, for many energy-intensive processes, the main option is indirect integration of RE through switching to RE electricity from the grid, or, in the future, to hydrogen. The broad range of options for producing low-carbon electricity, and its versatility of use, implies that electro-thermal processes could become more important in the future for replacing fossil fuels in a range of industrial processes. [8.3.3.2]

Less energy-intensive 'light' industries, including food processing, textiles, light manufacturing of appliances and electronics, automotive assembly plants, and saw-milling, although numerous, account for a smaller share of total energy use than do the heavy industries. Much of the energy demand by these 'light' industries reflects the energy use in commercial buildings for lighting, space heating, cooling, ventilation and office equipment. In general, light industries are more flexible and offer more readily accessible opportunities for the integration of RE than do energy-intensive industries. [8.3.3.3]

RE integration for process heat is practical at temperatures below around 400°C using the combustion of biomass (including charcoal) as well as solar thermal or direct geothermal energy. To meet process heat demand above 400°C, RE resources, with the exception of high-temperature solar, are less suitable (Figure TS.8.8). [8.3.3.3]

The potentials and costs for increasing the use of RE in industry are poorly understood due to the complexity and diversity of industry and the various geographical and local climatic conditions. Near-term opportunities for achieving higher RE shares could result from the increased utilization of process residues, CHP in biomass-based industries, and substitution of fossil fuels used for heating. Solar thermal technologies are promising with further development of collectors, thermal storage, back-up systems, process adaptation and integration under evaluation. RE integration using electricity generated from RE sources for electro-technologies may have the largest impact both in the near and long term. [8.3.3.2, 8.3.3.3]

Use of RE in industry has had difficulty in competing in the past in many regions due to relatively low fossil fuel prices together with low, or

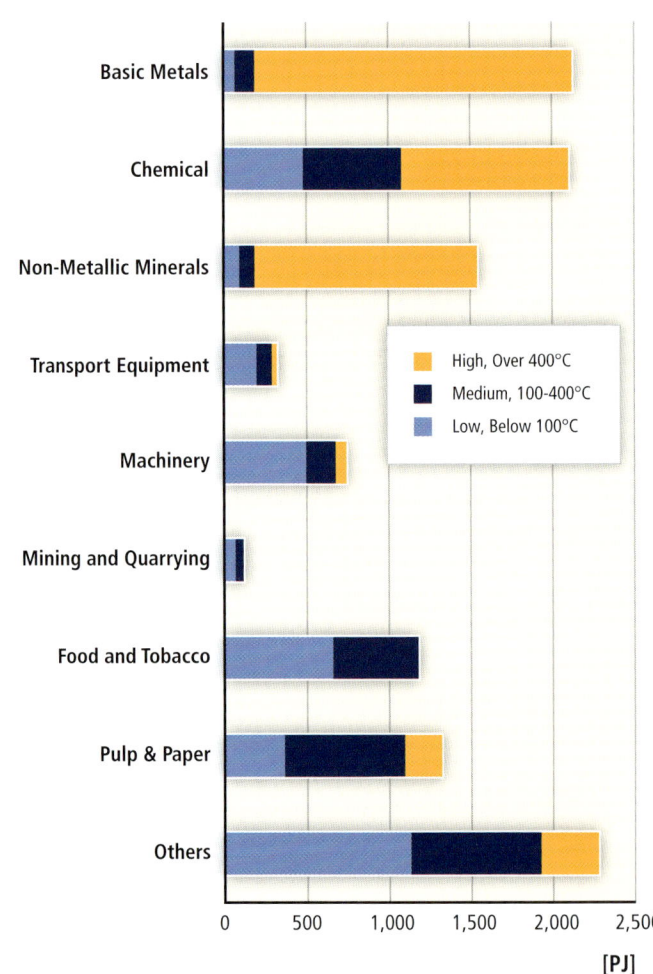

Figure TS.8.8 | Industrial heat demands for various temperature quality ranges by the heavy industrial and light manufacturing sub-sectors, based on an assessment within 32 European countries. [Figure 8.23]

non-existent, energy and carbon taxes. RE support policies in different countries tend to focus more on the transport and building sectors than on industry and consequently the potential for RE integration is relatively uncertain. Where support policies have been applied, successful RE deployment has resulted. [8.3.3.3]

8.7.4 Agriculture, forestry and fishing

Agriculture is a relatively low energy-consuming sector, utilizing only around 3% of total global consumer energy. The sector includes large corporate-owned farms and forests as well as subsistence farmers and fisher-folk in developing countries. The relatively high indirect energy use for the manufacture of fertilizers and machinery is included in the industry sector. Pumping water for irrigation usually accounts for the highest on-farm energy demand, along with diesel use for machinery and electricity for milking, refrigeration and fixed equipment. [8.3.4.1]

In many regions, land under cultivation could simultaneously be used for RE production. Multi-use of land for agriculture and energy purposes is becoming common, such as wind turbines constructed on grazing land; biogas plants used for treating animal manure with the nutrients recycled to the land; waterways used for small- and micro-hydropower systems; crop residues collected and combusted for heat and power; and energy crops grown and managed specifically to provide a biomass feedstock for liquid biofuels, heat and power generation (with co-products possibly used for feed and fibre). [2.6, 8.3.4.2, 8.3.4.3]

Since RE resources including wind, solar, crop residues and animal wastes are often abundant in rural areas, their capture and integration can enable the landowner or farm manager to utilize them locally for the farming operations. They can also earn additional revenue when energy carriers such as RE electricity or biogas are exported off the farm. [8.3.4]

Despite barriers to greater RE technology deployment including high capital costs, lack of available financing and remoteness from energy demand, it is likely that RE will be used to a greater degree by the global agricultural sector in the future to meet energy demands for primary production and post-harvest operations at both large and small scales. [8.3.4.1–8.3.4.2]

Integration strategies that could increase the deployment of RE in the primary sector will partly depend upon the local and regional RE resources, on-farm energy demand patterns, project financing opportunities and existing energy markets. [8.3.4.3]

9. Renewable Energy in the Context of Sustainable Development

9.1 Introduction

Sustainable development (SD) addresses concerns about relationships between human society and nature. Traditionally, SD has been framed in the three-pillar model—Economy, Ecology, and Society—allowing a schematic categorization of development goals, with the three pillars being interdependent and mutually reinforcing. Within another conceptual framework, SD can be oriented along a continuum between the two paradigms of weak sustainability and strong sustainability. The two paradigms differ in assumptions about the substitutability of natural and human-made capital. RE can contribute to the development goals of the three-pillar model and can be assessed in terms of both weak and strong SD, since RE utilization is defined as sustaining natural capital as long as the resource use does not reduce the potential for future harvest. [9.1]

9.2 Interactions between sustainable development and renewable energy

The relationship between RE and SD can be viewed as a hierarchy of goals and constraints that involve both global and regional or local considerations. Though the exact contribution of RE to SD has to be evaluated in a country-specific context, RE offers the opportunity to contribute to a number of important SD goals: (1) social and economic development; (2) energy access; (3) energy security; and (4) climate change mitigation and the reduction of environmental and health impacts. The mitigation of dangerous anthropogenic climate change is seen as one strong driving force behind the increased use of RE worldwide. [9.2, 9.2.1]

These goals can be linked to both the three-pillar model and the weak and strong SD paradigms. SD concepts provide useful frameworks for policymakers to assess the contribution of RE to SD and to formulate appropriate economic, social and environmental measures. [9.2.1]

The use of indicators can assist countries in monitoring progress made in energy subsystems consistent with sustainability principles, although there are many different ways to classify indicators of SD. The assessments carried out for the report and Chapter 9 are based on different methodological tools, including bottom-up indicators derived from attributional lifecycle assessments (LCA) or energy statistics, dynamic integrated modelling approaches, and qualitative analyses. [9.2.2]

Conventional economic growth metrics (GDP) as well as the conceptually broader Human Development Index (HDI) are analyzed to evaluate the contribution of RE to social and economic development. Potential employment opportunities, which serve as a motivation for some countries to support RE deployment, as well as critical financing questions for developing countries are also addressed. [9.2.2]

Access to modern energy services, whether from renewable or non-renewable sources, is closely correlated with measures of development, particularly for those countries at earlier development stages. Providing access to modern energy for the poorest members of society is crucial for the achievement of any single of the eight Millennium Development Goals. Concrete indicators used include per capita final energy consumption related to income, as well as breakdowns of electricity access (divided into rural and urban areas), and numbers for those parts of the population using coal or traditional biomass for cooking. [9.2.2]

Despite the lack of a commonly accepted definition, the term 'energy security' can best be understood as robustness against (sudden) disruptions of energy supply. Two broad themes can be identified that are relevant to energy security, whether for current systems or for the planning of future RE systems: availability and distribution of resources; and variability and reliability of energy supply. The indicators used to provide information about the energy security criterion of SD are the magnitude of reserves, the reserves-to-production ratio, the share of imports in total primary energy consumption, the share of energy imports in total imports, as well as the share of variable and unpredictable RE sources. [9.2.2]

To evaluate the overall burden from the energy system on the environment, and to identify potential trade-offs, a range of impacts and categories have to be taken into account. These include mass emissions to air (in particular GHGs) and water, and usage of water, energy and land per unit of energy generated and these must be evaluated across technologies. While recognizing that LCAs do not give the only possible answer as to the sustainability of a given technology, they are a particularly useful methodology for determining total system impacts of a given technology, which can serve as a basis for comparison. [9.2.2]

Scenario analyses provide insights into what extent integrated models take account of the four SD goals in different RE deployment pathways. Pathways are primarily understood as scenario results that attempt to address the complex interrelations among the different energy technologies at a global scale. Therefore, Chapter 9 mainly refers to global scenarios derived from integrated models that are also at the core of the analysis in Chapter 10. [9.2.2]

9.3 Social, environmental and economic impacts: Global and regional assessment

Countries at different levels of development have different incentives to advance RE. For developing countries, the most likely reasons to adopt RE technologies are providing access to energy, creating employment opportunities in the formal (i.e., legally regulated and taxable) economy, and reducing the costs of energy imports (or, in the case of fossil energy exporters, prolonging the lifetime of their natural resource base). For industrialized countries, the primary reasons to encourage RE include reducing carbon emissions to mitigate climate change, enhancing energy security, and actively promoting structural change in the economy, such that job losses in declining manufacturing sectors are softened by new employment opportunities related to RE. [9.3]

9.3.1 Social and economic development

Globally, per capita incomes are positively correlated with per capita energy use and economic growth can be identified as the most relevant factor behind increasing energy consumption in the last decades. However, there is no agreement on the direction of the causal relationship between energy use and increased macroeconomic output. [9.3.1.1]

As economic activity expands and diversifies, demands for more sophisticated and flexible energy sources arise: from a sectoral perspective, countries at an early stage of development consume the largest part of total primary energy in the residential (and to a lesser extent agricultural) sector; in emerging economies the manufacturing sector dominates, while in fully industrialized countries services and transport account for steadily increasing shares (see Figure TS.9.1). [9.3.1.1]

Despite the close correlation between GDP and energy use, a wide variety of energy use patterns across countries prevails: some have achieved high levels of per capita incomes with relatively low energy consumption. Others remain rather poor despite elevated levels of energy use, in particular countries abundantly endowed with fossil fuel resources, in which energy is often heavily subsidized. One hypothesis suggests that economic growth can largely be decoupled from energy use by steady declines in energy intensity. Further, it is often asserted that developing economies and economies in transition can 'leapfrog', that is, limit their energy use by adopting modern, highly efficient energy technologies. [9.3.1.1, Box 9.5]

Access to clean and reliable energy constitutes an important prerequisite for fundamental determinants of human development, such as health, education, gender equality and environmental safety. Using the HDI as a proxy indicator of development, countries that have achieved high HDI levels in general consume relatively large amounts of energy per capita and no country has achieved a high or even a medium HDI without significant access to non-traditional energy supplies. A certain minimum amount of energy is required to guarantee an acceptable standard of living (e.g., 42 GJ per capita), after which raising energy consumption yields only marginal improvements in the quality of life. [9.3.1.2]

Estimates of current net employment effects of RE differ due to disagreements regarding the use of the appropriate methodology. Still, there seems to be agreement about the positive long-term effects of RE

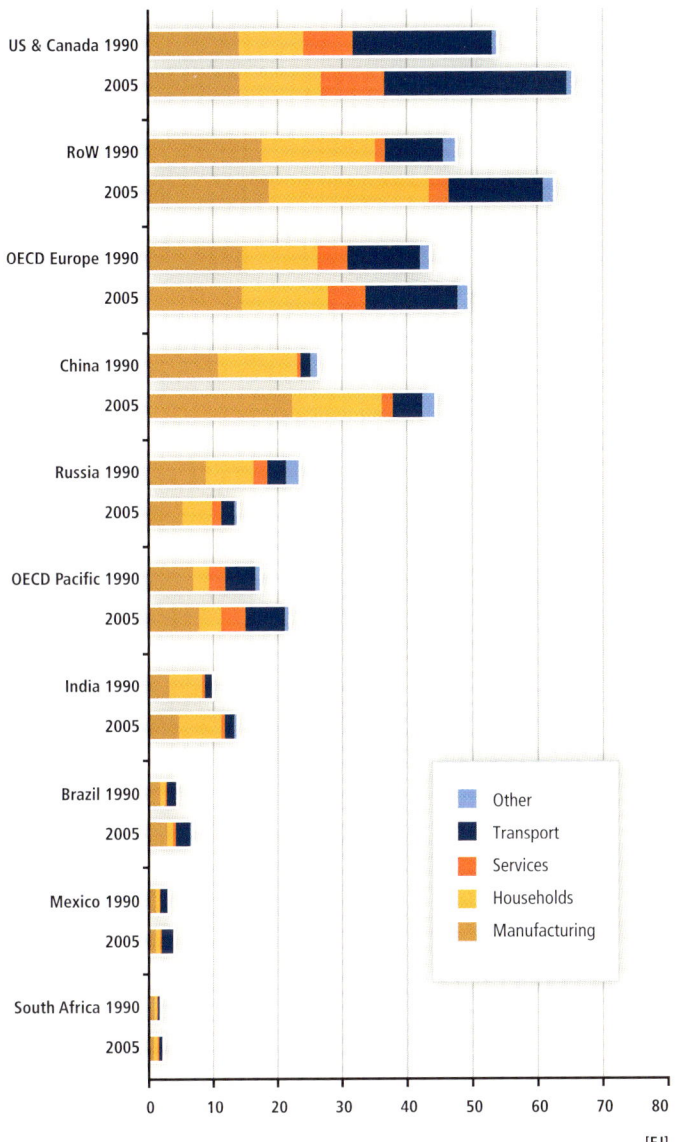

Figure TS.9.1 | Energy use (EJ) by economic sector. Note that the underlying data are calculated using the IEA physical content method, not the direct equivalent method.[1]

Notes: RoW = Rest of World. [Figure 9. 2] 1. Historical energy data have only been available for energy use by economic sector. For a conversion of the data using the direct equivalent method, the different energy carriers used by each economic sector would need to be known.

as an important contribution to job creation, which has been stressed in many national green-growth strategies. [9.3.1.3]

In general, the purely economic costs of RE exceed those of fossil fuel-based energy production in most instances. Especially for developing countries, the associated costs are a major factor determining the desirability of RE to meet increasing energy demand, and concerns have been voiced that increased energy prices might endanger industrializing countries' development prospects. Overall, cost considerations cannot be discussed independently of the burden-sharing regime adopted, that is, without specifying who assumes the costs for the benefits brought about from reduced GHG emissions, which can be characterized as a global public good. [9.3.1.4]

9.3.2   Energy access

Significant parts of the global population today have no or limited access to modern and clean energy services. From a sustainable development perspective, sustainable energy expansion needs to increase the availability of energy services to groups that currently have no or limited access to them: the poor (measured by wealth, income or more integrative indicators), those in rural areas and those without connections to the grid. [9.3.2]

Acknowledging the existing constraints regarding data availability and quality, 2009 estimates of the number of people without access to electricity are around 1.4 billion. The number of people relying on traditional biomass for cooking is around 2.7 billion, which causes significant health problems (notably indoor air pollution) and other social burdens (e.g., time spent gathering fuel) in the developing world. Given the strong correlation between household income and use of low quality fuels (Figure TS.9.2), a major challenge is to reverse the pattern of inefficient biomass consumption by changing the present, often unsustainable, use to more sustainable and efficient alternatives. [9.3.2]

By defining energy access as 'access to clean, reliable and affordable energy services for cooking and heating, lighting, communications and productive uses', the incremental process of climbing the steps of the energy ladder is illustrated; even basic levels of access to modern energy services can provide substantial benefits to a community or household. [9.3.2]

In developing countries, decentralized grids based on RE have expanded and improved energy access; they are generally more competitive in rural areas with significant distances to the national grid and the low levels of rural electrification offer significant opportunities for RE-based mini-grid systems. In addition, non-electrical RE technologies offer opportunities for direct modernization of energy services, for example, using solar energy for water heating and crop drying, biofuels for transportation, biogas and modern biomass for heating, cooling, cooking and lighting, and wind for water pumping. While the specific role of RE in providing energy access in a more sustainable manner than other energy sources is not well understood, some of these technologies allow local communities to widen their energy choices; they stimulate economies, provide incentives for local entrepreneurial efforts and meet basic needs and services related to lighting and cooking, thus providing ancillary health and education benefits. [9.3.2]

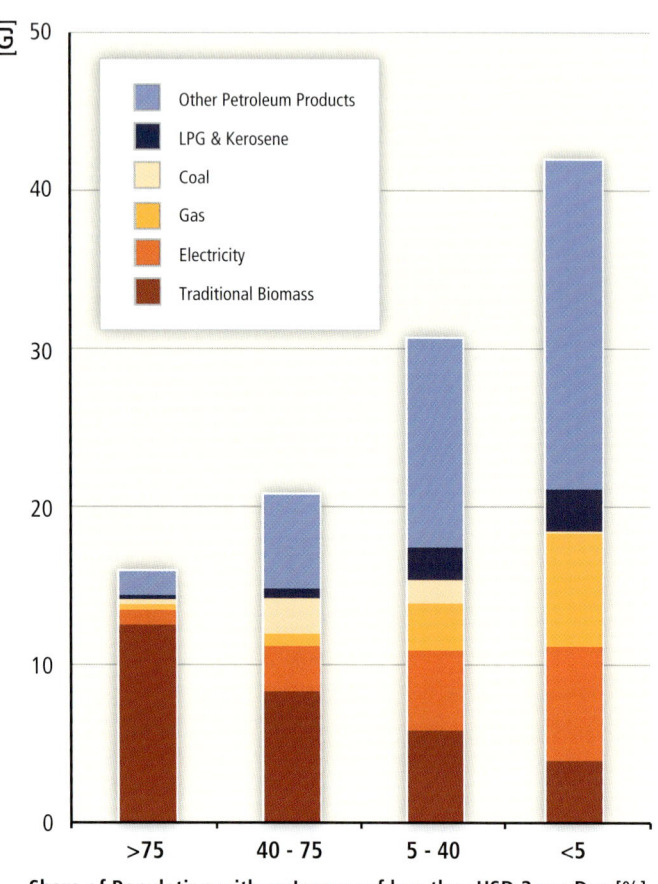

Figure TS.9.2 | The relationship between per capita final energy consumption and income in developing countries. Data refer to the most recent year available during the period 2000 to 2008. [Figure 9.5]

Note: LPG = liquid petroleum gas.

9.3.3 Energy security

The use of RE permits substitution away from increasingly scarce fossil fuel supplies; current estimates of the ratio of proven reserves to current production show that globally oil and natural gas would be exhausted in about four and six decades, respectively. [9.3.3.1]

As many renewable sources are localized and not internationally tradable, increasing their share in a country's energy portfolio diminishes the dependence on imports of fossil fuels, whose spatial distribution of reserves, production and exports is very uneven and highly concentrated in a few regions (Figure TS.9.3). As long as RE markets are not characterized by such geographically concentrated supply, this helps to diversify the portfolio of energy sources and to reduce the economy's vulnerability to price volatility. For oil-importing developing countries, increased uptake of RE technologies could be an avenue to redirect foreign exchange flows away from energy imports towards imports of goods that cannot be produced locally, such as high-tech capital goods. For example, Kenya and Senegal spend more than half of their export earnings for importing energy, while India spends over 45%. [9.3.3.1]

However, import dependencies can also occur in relation to the technologies needed for implementation of RE, with the secure access to required scarce inorganic mineral raw materials at reasonable prices constituting an upcoming challenge for all industries. [9.3.3.1]

The variable output profiles of some RE technologies often necessitate technical and institutional measures appropriate to local conditions to assure a constant and reliable energy supply. Reliable energy access is a particular challenge in developing countries and indicators for the reliability of infrastructure services show that in sub-Saharan Africa, almost 50% of firms maintain their own generation equipment. Many developing countries therefore specifically link energy access and security issues by broadening the definition of energy security to include stability and reliability of local supply. [9.3.3.2]

9.3.4 Climate change mitigation and reduction of environmental and health impacts

Sustainable development must ensure environmental quality and prevent undue environmental harm. No large-scale technology deployment comes without environmental trade-offs and a large body of literature is available that assesses various environmental impacts of the broad range of energy technologies (RE, fossil and nuclear) from a bottom-up perspective. [9.3.4]

Impacts on the climate through GHG emissions are generally well covered, and LCAs [Box 9.2] facilitate a quantitative comparison of 'cradle to grave' emissions across technologies. While a significant number of studies report on air pollutant emissions and operational water use, evidence is scarce for lifecycle emissions to water, land use, and health impacts other than those linked to air pollution. The assessment concentrates on those sectors which are best covered by the literature, such as electricity generation and transport fuels for GHG emissions. Heating and household energy are discussed only briefly, in particular with regards to air pollution and health. Impacts on biodiversity and ecosystems are mostly site-specific, difficult to quantify and are presented in a more qualitative manner. To account for burdens associated with accidents as opposed to normal operation, an overview of risks associated with energy technologies is provided. [9.3.4]

LCAs for electricity generation indicate that *GHG emissions from RE technologies* are, in general, considerably lower than those associated with fossil fuel options, and in a range of conditions, less than fossil fuels employing CCS. The maximum estimate for CSP, geothermal, hydropower, ocean and wind energy is less than or equal to 100 g CO_2eq/kWh, and median values for all RE range from 4 to 46 g CO_2eq/kWh. The upper quartile of the distribution of estimates for PV and biopower extend two to three times above the maximum for other RE technologies. However, GHG balances of bioenergy production have more uncertainties: excluding LUC, biopower could reduce GHG emissions compared to fossil fuelled systems and can lead to avoided GHG emissions from residues and wastes in landfill disposals and co-products; the combination of

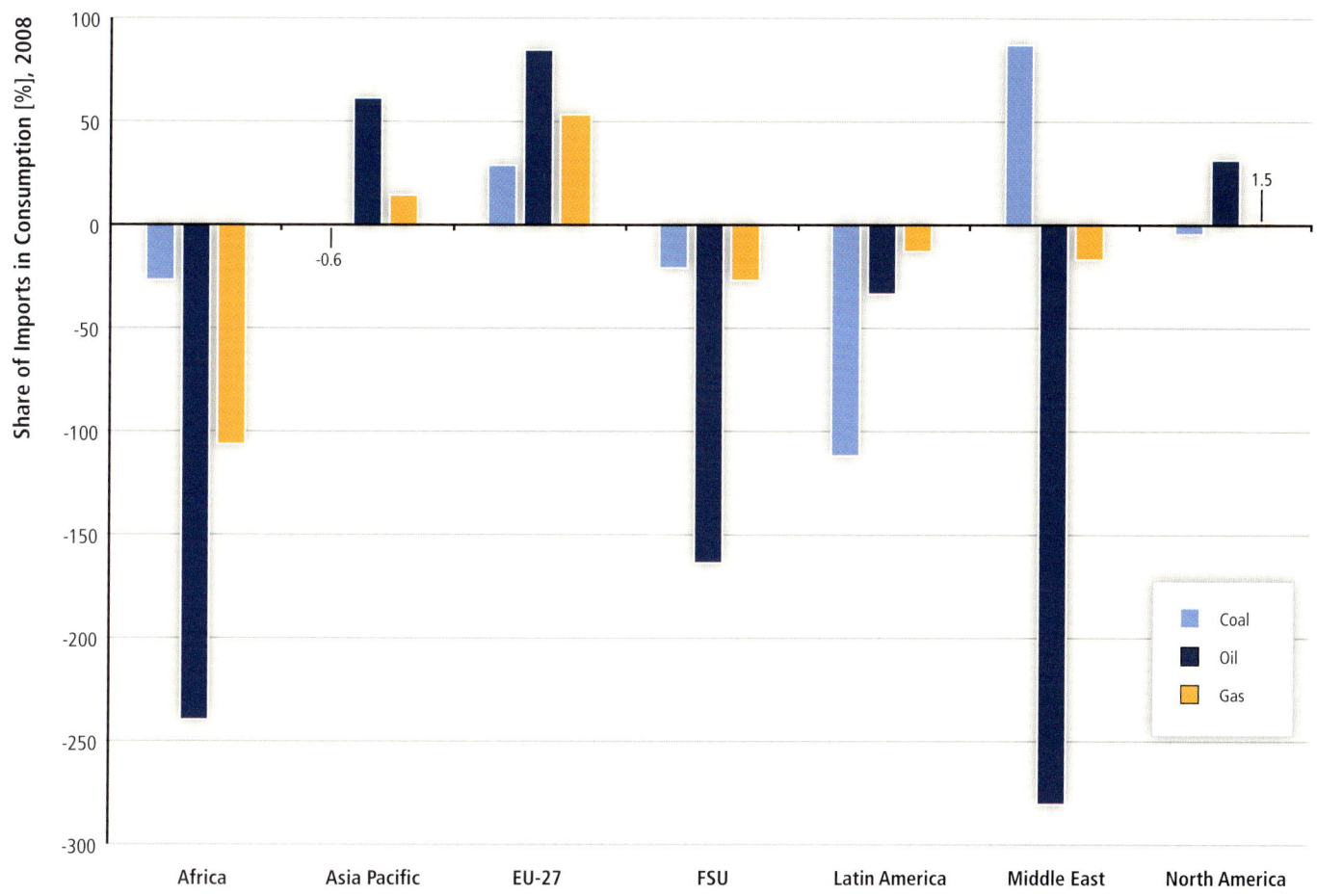

Figure TS.9.3 | Energy imports as the share of total primary energy consumption (%) for coal (hard coal and lignite), crude oil and natural gas for selected world regions in 2008. Negative values denote net exporters of energy carriers. [Figure 9.6]

bioenergy with CCS may provide for further reductions (Figure TS.9.4). [9.3.4.1]

Accounting for differences in the quality of power produced, potential impacts to grid operation related to the addition of variable generation sources, and for direct or indirect LUC could reduce the GHG emissions benefit from switching to renewable electricity generation, but is not likely to negate the benefit. [9.3.4.1]

Measures such as the energy payback time, describing the energetic efficiency of technologies or fuels, have been declining rapidly for some RE technologies over recent years (e.g., wind and PV) due to technological advances and economies of scale. Fossil and nuclear power technologies are characterized by the continuous energy requirements for fuel extraction and processing, which might become increasingly important as qualities of conventional fuel supply decline and shares of unconventional fuels rise. [9.3.4.1]

For the assessment of *GHG emissions from transportation fuels*, selected petroleum fuels, first-generation biofuels (i.e., sugar- and starch-based ethanol, oilseed-based biodiesel and renewable diesel), and selected next-generation biofuels derived from lignocellulosic biomass (i.e.,

ethanol and Fischer-Tropsch diesel) are compared on a well-to-wheel basis. In this comparison, GHG emissions from LUC (direct and indirect) and other indirect effects (e.g., petroleum consumption rebound) have been excluded, but are separately considered below. Substituting biofuels for petroleum-based fuels has the potential to reduce lifecycle GHG emissions directly associated with the fuel supply chain. While first-generation biofuels result in relatively modest GHG mitigation potential (-19 to 77 g CO_2eq/MJ for first-generation biofuels versus 85 to 109 g CO_2eq/MJ for petroleum fuels), most next-generation biofuels (with lifecycle GHG emissions between -10 and 38 g CO_2eq/MJ) could provide greater climate benefits. Estimates of lifecycle GHG emissions are variable and uncertain for both biofuels and petroleum fuels, primarily due to assumptions about biophysical parameters, methodological issues and where and how the feedstocks are produced. [9.3.4.1]

Lifecycle *GHG emissions from LUC* are difficult to quantify, with land and biomass resource management practices strongly influencing any GHG emission reduction benefits and as such the sustainability of bioenergy. Changes to land use or management, brought about directly or indirectly by biomass production for use as fuels, power or heat, can lead to changes in terrestrial carbon stocks. Depending on the converted land's prior condition, this can either cause significant upfront emissions, requiring a time

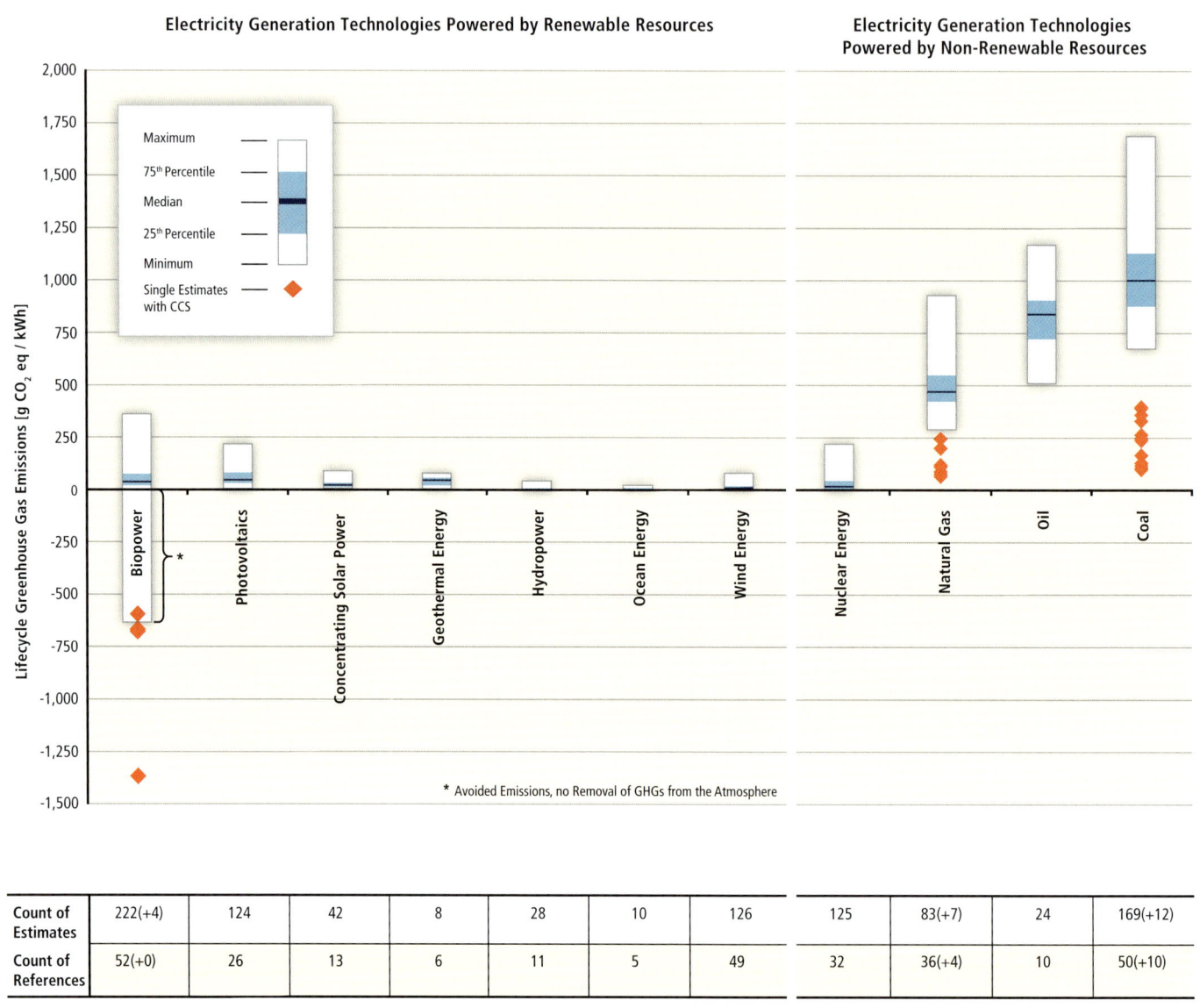

Figure TS.9.4 | Estimates of lifecycle GHG emissions (g CO_2eq/kWh) for broad categories of electricity generation technologies, plus some technologies integrated with CCS. Land-use related net changes in carbon stocks (mainly applicable to biopower and hydropower from reservoirs) and land management impacts are excluded; negative estimates[1] for biopower are based on assumptions about avoided emissions from residues and wastes in landfill disposals and co-products. References and methods for the review are reported in Annex II. The number of estimates is greater than the number of references because many studies considered multiple scenarios. Numbers reported in parentheses pertain to additional references and estimates that evaluated technologies with CCS. Distributional information relates to estimates currently available in LCA literature, not necessarily to underlying theoretical or practical extrema, or the true central tendency when considering all deployment conditions. [Figure 9.8]

Note: 1. 'Negative estimates' within the terminology of lifecycle assessments presented in this report refer to avoided emissions. Unlike the case of bioenergy combined with CCS, avoided emissions do not remove GHGs from the atmosphere.

lag of decades to centuries before net savings are achieved, or improve the net uptake of carbon into soils and aboveground biomass. Assessments of the net GHG effects of bioenergy are made difficult by challenges in observation, measurement, and attribution of indirect LUC, which depends on the environmental, economic, social and policy context and is neither directly observable nor easily attributable to a single cause. Illustrative estimates of direct and indirect LUC-related GHG emissions induced by several first-generation biofuel pathways provide central tendencies (based on different reporting methods) for a 30-year timeframe: for ethanol (EU wheat, US maize, Brazilian sugarcane) 5 to 82 g CO_2eq/MJ and for diesel (soy and rapeseed) 35 to 63 g CO_2eq/MJ. [9.3.4.1]

Impacts from *local and regional air pollution* constitute another important assessment category, with air pollutants (including particulate matter (PM), nitrous oxides (NO_x), sulphur dioxide (SO_2) and non-methane volatile organic compounds (NMVOC)) having effects at the global [Box 9.4], regional and local scale. Compared to fossil-based power generation, non-combustion-based RE power generation technologies have the potential to significantly reduce regional and local air pollution and associated health impacts (see this section below). For transportation fuels, however, the effect of switching to biofuels on tailpipe emissions is not yet clear. [9.3.4.2]

Local air pollutant emissions from fossil fuels and biomass combustion constitute the most important energy related impacts on *human health*. Ambient air pollution, as well as exposure to indoor air pollution from the combustion of coal and traditional biomass, has major health impacts and is recognized as one of the most important causes of morbidity and mortality worldwide, particularly for women and children in developing countries. In 2000, for example, comparative quantifications of health risks showed that more than 1.6 million deaths and over 38.5 million of disability-adjusted life-years (DALYs) were attributable to indoor smoke from solid fuels. Besides a fuel switch, mitigation options include improved cookstoves, ventilation and building design and behavioural changes. [9.3.4.3]

Impacts on *water* relate to operational and upstream water consumption of energy technologies and to water quality. These impacts are site specific and need to be considered with respect to local resources and needs. RE technologies like hydropower and some bioenergy systems, for example, are dependent on water availability and can either increase competition or mitigate water scarcity. In water-scarce areas, non-thermal RE technologies (e.g., wind and PV) can provide clean electricity without putting additional stress on water resources. Conventionally cooled thermal RE technologies (e.g., CSP, geothermal, biopower) can use more water during operation than non-RE technologies, yet dry cooling configurations can reduce this impact (Figure TS.9.5). Water use in upstream processes can be high for some energy technologies, particularly for fuel extraction and biomass feedstock production; including the latter, the current water footprint for electricity generation from biomass can be up to several hundred times greater than operational water consumption requirements for thermal power plants. Feedstock production, mining operations and fuel processing can also affect water quality. [9.3.4.4]

Most energy technologies have substantial *land requirements* when the whole supply chain is included. While the literature on lifecycle estimates for land use by energy technologies is scarce, the available evidence suggests that lifecycle land use by fossil energy chains can be comparable to or higher than land use by RE sources. For most RE sources, land use requirements are largest during the operational stage. An exception is the land intensity of bioenergy from dedicated feedstocks, which is significantly higher than for any other energy technology and shows substantial variations in energy yields per hectare for different feedstocks and climatic zones. A number of RE technologies (wind, wave and ocean) occupy large areas, but allow secondary uses such as farming, fishing and recreational activities. [9.3.4.5] Connected to land use are (site-specific) impacts on *ecosystems and biodiversity*. Occurring through various pathways, the most evident ones are through large-scale direct physical alteration of habitats and, more indirectly, habitat deterioration. [9.3.4.6]

The comparative assessment of *accident risks* is a pivotal aspect in a comprehensive evaluation of energy security aspects and sustainability performance associated with current and future energy systems. Risks of various energy technologies to society and the environment occur not only during the actual energy generation, but at all stages of energy chains. Accident risks of RE technologies are not negligible, but the technologies' often decentralized structure strongly limits the potential for disastrous consequences in terms of fatalities. While RE technologies overall exhibit low fatality rates, dams associated with some hydropower projects may create a specific risk depending on site-specific factors. [9.3.4.7]

9.4 Implication of sustainable development pathways for renewable energy

Following the more static analysis of the impacts of current and emerging RE systems on the four SD goals, the SD implications of possible future RE deployment pathways are assessed in a more dynamic manner and thus incorporate the intertemporal component of SD. Since the interaction of future RE and SD pathways cannot be anticipated by relying on a partial analysis of individual energy technologies, the discussion is based on results from the scenario literature that typically treats the portfolio of technological alternatives in the framework of a global or regional energy system. [9.4]

The vast majority of models used to generate the scenarios reviewed (see Chapter 10, Section 10.2) capture the interactions between different options for supplying, transforming and using energy. The models range from regional, energy-economic models to integrated assessment models (IAMs) and are here referred to as integrated models. Historically, these models have focused much more on the technological and macroeconomic aspects of energy transitions, and in the process have produced largely aggregated measures of technological penetration or energy generated by particular sources of supply. The value of these models in generating long-term scenarios and their potential to help understand the interrelation between SD and RE rests on their ability to consider interactions across a broad set of human activities over different regional and time scales. Integrated models continually undergo developments, some of which will be crucial for the representation of sustainability concerns in the future, for example, increasing their temporal and spatial resolution, allowing for a better representation of the distribution of wealth across the population and incorporating greater detail in human and physical Earth system characterization. [9.4]

The assessment focuses on what model-based analyses currently have to say with respect to SD pathways and the role of RE and evaluates how model-based analyses can be improved to provide a better understanding of sustainability issues in the future. [9.4]

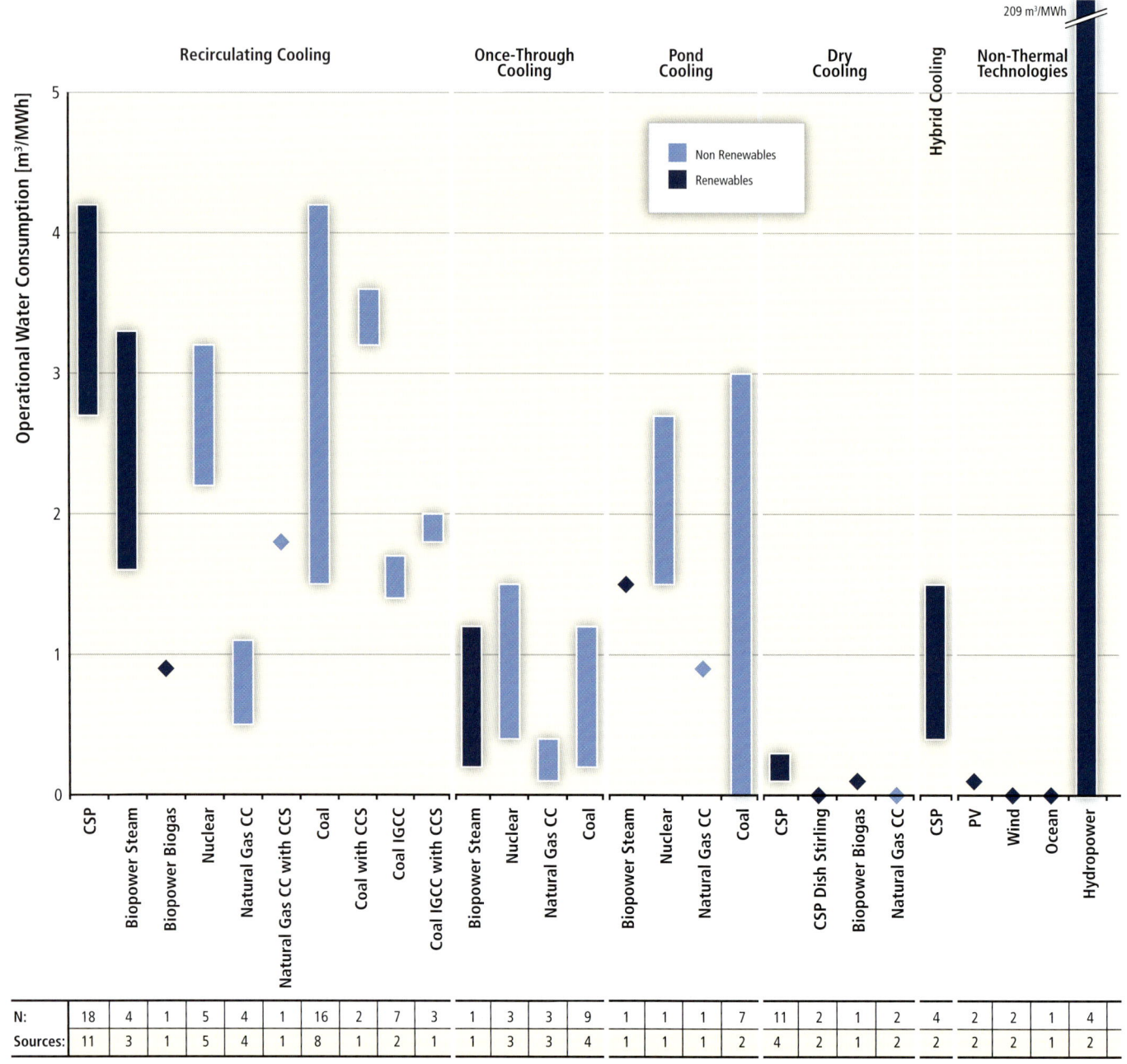

Figure TS.9.5 | Ranges of rates of operational water consumption by thermal and non-thermal electricity-generating technologies based on a review of available literature (m³/MWh). Bars represent absolute ranges from available literature, diamonds single estimates; n represents the number of estimates reported in the sources. Methods and references used in this literature review are reported in Annex II. Note that upper values for hydropower result from a few studies measuring gross evaporation values, and may not be representative (see Box 5.2). [Figure 9.14]

Notes: CSP: concentrated solar power; CCS: carbon capture and storage; IGCC: integrated gasification combined cycle; CC: combined cycle; PV: photovoltaic.

9.4.1 Social and economic development

Integrated models usually have a strong macro-perspective and do not consider advanced welfare measures. [9.2.2, 9.3.1] Instead, they focus on economic growth, which in itself is an insufficient measure of sustainability, but can be used as an indicative welfare measure in the context of different stabilization pathways. Mitigation scenarios usually include a tentative strong sustainability constraint by putting an upper limit on future GHG emissions. This results in welfare losses (usually measured as GDP or consumption foregone) based on assumptions about the availability and costs of mitigation technologies. Limiting the availability of technological alternatives for constraining GHGs further increases welfare losses. Studies that specifically assess the implications of constraining RE for different GHG concentration stabilization levels show that the wide availability of all RE technologies is essential in order to reach low stabilization levels and that the full availability of low-carbon technologies, including RE, is crucial for keeping mitigation costs at relatively low levels, even for less strict stabilization levels. [9.4.1]

With respect to regional effects, scenario analyses show that developing countries are likely to see most of the expansion in RE production. With the challenge to overcome high LCOEs of RE technologies still to be met, these results hint at the potential of developing countries to leapfrog the emission-intensive developing paths that developed countries have taken so far. Regional mitigation opportunities will, however, vary, depending on many factors including technology availability, but also population and economic growth. Costs will also depend on the allocation of tradable emission permits, both initially and over time, under a global climate mitigation regime. [9.4.1]

In general, scenario analyses point to the same links between RE, mitigation and economic growth in developed and developing countries, only the forces are generally larger in non-Annex I countries than in Annex I countries due to more rapid assumed economic growth and the consequently increasing mitigation burden over time. However, the modelling structures used to generate long-term global scenarios generally assume perfectly functioning economic markets and institutional infrastructures across all regions of the globe. They also discount the special circumstances that prevail in all countries, particularly in developing countries where these assumptions are particularly tenuous. These sorts of differences and the influence they might have on social and economic development among countries should be an area of active future research. [9.4.1]

9.4.2 Energy access

Integrated models thus far have often been based on developed country information and experience and assumed energy systems in other parts of the world and at different stages of development to behave likewise. Usually, models do not capture important and determinative dynamics in developing countries, such as fuel choices, behavioural heterogeneity and informal economies. This impedes an assessment of the interaction between RE and the future availability of energy services for different populations, including basic household level tasks, transportation, and energy for commerce, manufacturing and agriculture. However, some models have started to integrate factors such as potential supply shortages, informal economies and diverse income groups, and to increase the distributional resolution. [9.4.2]

Available scenario analyses are still characterized by large uncertainties. For India, results suggested that income distribution in a society is as important for increasing energy access as income growth. Also, increasing energy access is not necessarily beneficial for all aspects of SD, as a shift to modern energy away from, for example, traditional biomass could simply be a shift to fossil fuels. In general, available scenario analyses highlight the role of policies and finance for increased energy access, even though forced shifts to RE that would provide access to modern energy services could negatively affect household budgets. [9.4.2]

Further improvements in the distribution resolution and structural rigidity (inability of many models to capture social phenomena and structural changes that underlie peoples' utilization of energy technologies) are particularly challenging. An explicit representation of the energy consequences for the poorest, women, specific ethnic groups within countries, or those in specific geographical areas, tends to be outside the range of current global model output. In order to provide a more comprehensive view of the possible range of energy access options, future energy models should aim for a more explicit representation of relevant determinants (such as traditional fuels, modes of electrification, and income distribution) and link these to representations of alternative development pathways. [9.4.2]

9.4.3 Energy security

RE can influence energy security by mitigating concerns with respect to both availability and distribution of resources, as well as to the variability of energy sources. [9.2.2, 9.3.1] To the extent that RE deployment in mitigation scenarios reduces the overall risk of disruption by diversifying the energy portfolio, the energy system is less susceptible to (sudden) energy supply disruption. In scenarios, this role of RE will vary with the energy form. Solar, wind and ocean energy, which are closely associated with electricity production, have the potential to replace concentrated and increasingly scarce fossil fuels in the buildings and the industry sector. With appropriate carbon mitigation policies in place, electricity generation can be relatively easily decarbonized. In contrast, the demand for liquid fuels in the transport sector remains inelastic if no technological breakthrough can be achieved. While bioenergy could play an important role, this will depend on the availability of CCS that could divert its use to power generation with CCS—resulting in negative net carbon emissions for the system and smoothing the overall mitigation efforts significantly. [9.4.1, 9.4.3]

Against this background, energy security concerns raised in the past that related to oil supply disruptions are likely to remain relevant in the future. For developing countries the issue will become even more important, as their share in global total oil consumption increases in all assessed scenarios (Figure TS.9.6b). As long as technological alternatives for oil, for example, biofuels and/or the electrification of the transportation sector, do not play a dominant role in scenario analyses,

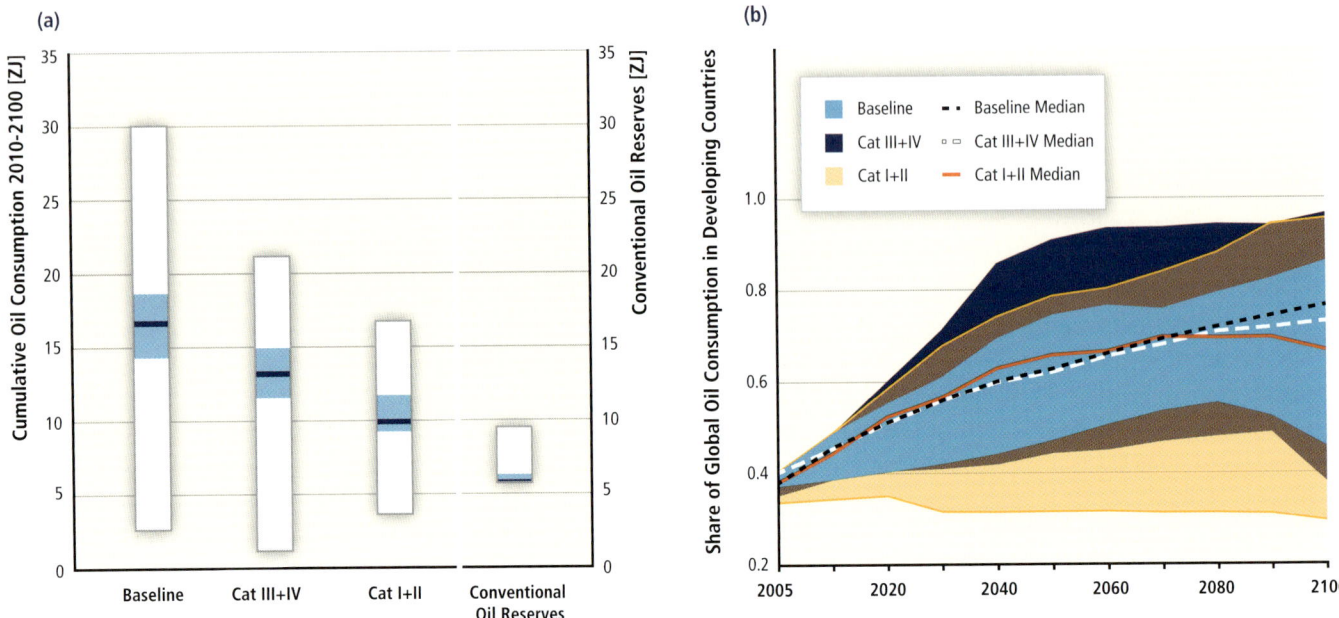

Figure TS.9.6 | (a) Conventional oil reserves compared to projected cumulative oil consumption (ZJ) from 2010 to 2100 in scenarios assessed in Chapter 10 for different scenario categories: baseline scenarios, Category III and IV scenarios and low stabilization (Category I+II) scenarios. The thick dark blue line corresponds to the median, the light blue bar corresponds to the inter-quartile range (25th to 75th percentile) and the white surrounding bar corresponds to the total range across all reviewed scenarios. The last column shows the range of proven recoverable conventional oil reserves (light blue bar) and estimated additional reserves (white surrounding bar). (b) Range of share of global oil consumed in non-Annex I countries for different scenario categories over time, based on scenarios assessed in Chapter 10. [Figure 9.18]

most mitigation scenarios do not see dramatic differences between the baseline and policy scenarios with respect to cumulative oil consumption (Figure TS.9.6a). [9.4.3]

An increased market for bioenergy could raise additional energy security concerns in the future if it was characterized by a small number of sellers and thus showed parallels to today's oil market. In such an environment, the risk that food prices could be linked to volatile bioenergy markets would have to be mitigated to impede severe impacts on SD as high and volatile food prices would clearly hurt the poor. [9.4.3]

The introduction of variable RE technologies also adds new concerns, such as vulnerability to extreme natural events or international price fluctuations, which are not yet satisfactorily addressed by large integrated models. Additional efforts to increase system reliability are likely to add costs and involve balancing needs (such as holding stocks of energy), the development of complementary flexible generation, strengthening network infrastructure and interconnections, energy storage technologies and modified institutional arrangements including regulatory and market mechanisms [7.5, 8.2.1, 9.4.3]

Energy security considerations today usually focus on the most prominent energy security issues in recent memory. However, energy security aspects of the future might go well beyond these issues, for example, in relation to critical material inputs for RE technologies. These broader concerns as well as options for addressing them, for example, recycling, are largely absent from future scenarios of mitigation and RE. [9.4.3]

9.4.4 Climate change mitigation and environmental and health impacts in scenarios of the future

Replacing fossil fuels with RE or other low-carbon technologies can significantly contribute to the reduction of NO_x and SO_2 emissions. Several models have included explicit representation of factors, such as sulphate pollution, that are linked to environmental or health impacts. Some scenario results show that climate policy can help drive improvements in local air pollution (i.e., PM), but air pollution reduction policies alone do not necessarily drive reductions in GHG emissions. Another implication of some potential energy trajectories is the possible diversion of land to support biofuel production. Scenario results have pointed at the possibility that, if not accompanied by other policy measures, climate policy could drive widespread deforestation, with land use being shifted to bioenergy crops with possibly adverse SD implications, including GHG emissions. [9.4.4]

Unfortunately, existing scenario literature does not explicitly treat the many non-emissions related elements of sustainable energy development, such as water use, the impacts of energy choices on household-level services, or indoor air quality. This can be partly explained by models being designed to look at fairly large world regions without income or geographic distributional detail. For a broad assessment of environmental impacts at the regional and local level, models would need to look at smaller scales of geographical impacts, which is currently a matter of ongoing research. Finally, many models do not explicitly allow for incorporation of LCA results of the technological alternatives. What these impacts are, whether and how to compare them across categories, and whether they might be incorporated into future scenarios would constitute useful areas for future research. [9.4.4]

9.5 Barriers and opportunities for renewable energy in the context of sustainable development

Pursuing a renewable energy deployment strategy in the context of SD implies that most environmental, social and economic effects are taken explicitly into account. Integrated planning, policy and implementation processes can support this by anticipating and overcoming potential barriers to and exploiting opportunities of RE deployment. [9.5]

Barriers that are particularly pertinent in a sustainable development context and that may either impede RE deployment or result in trade-offs with SD criteria relate to socio-cultural, information and awareness, market-related and economic barriers. [9.5.1]

Socio-cultural barriers or concerns have different origins and are intrinsically linked to societal and personal values and norms. Such values and norms affect the perception and acceptance of RE technologies and the potential impacts of their deployment by individuals, groups and societies. From a sustainable development perspective, barriers may arise from inadequate attention to such socio-cultural concerns, which include barriers related to behaviour; natural habitats and natural and human heritage sites, including impacts on biodiversity and ecosystems; landscape aesthetics; and water/land use and water/land use rights, as well as their availability for competing uses. [9.5.1.1]

Public awareness and acceptance is an important element in the need to rapidly and significantly scale up RE deployment to help meet climate change mitigation goals. Large-scale implementation can only be undertaken successfully with the understanding and support of the public. This may require dedicated communication efforts related to the achievements and the opportunities associated with wider-scale applications. At the same time, however, public participation in planning decisions as well as fairness and equity considerations in the distribution of the benefits and costs of RE deployment play an equally important role and cannot be side-stepped. [9.5.1.1]

In developing countries, limited technical and business skills and the absence of technical support systems are particularly apparent in the energy sector, where awareness of and information dissemination regarding available and appropriate RE options among potential consumers is a key determinant of uptake and market creation. This gap in awareness is often perceived as the single most important factor affecting the deployment of RE and development of small and medium enterprises that contribute to economic growth. Also, there is a need to focus on the capacity of private actors to develop, implement and deploy RE technologies, which includes increasing technical and business capability at the micro or firm level. [9.5.1.2]

Attitudes towards RE in addition to rationality are driven by emotions and psychological issues. To be successful, RE deployment and information and awareness efforts and strategies need to take this explicitly into account. [9.5.1.2]

To assess the economics of RE in the context of SD, social costs and benefits need to be explicitly considered. RE should be assessed against quantifiable criteria targeted at cost effectiveness, regional appropriateness, and environmental and distributional consequences. Grid size and technologies are key determinants of the *economic viability* of RE and of the competitiveness of RE compared to non-renewable energy. Appropriate RE technologies that are economically viable are often found to be available for expanding rural off-grid energy access, in particular smaller off-grid and mini-grid applications. [9.5.1.3]

In cases where deployment of RE is viable from an economic perspective, other economic and financial barriers may affect its deployment. High upfront costs of investments, including high installation and grid connection costs, are examples of frequently identified barriers to RE deployment. In developing countries, policy and entrepreneurial support systems are needed along with RE deployment to stimulate economic growth and SD and catalyze rural and peri-urban cash economies. Lack of adequate resource potential data directly affects uncertainty regarding resource availability, which may translate into higher risk premiums for investors and project developers. The internalization of environmental and social externalities frequently results in changes in the ranking of various energy sources and technologies, with important lessons for SD objectives and strategies. [9.5.1.3]

Strategies for SD at international, national and local levels as well as in private and nongovernmental spheres of society can help overcome barriers and create opportunities for RE deployment by integrating RE and SD policies and practices. [9.5.2]

Integrating RE policy into national and local SD strategies (explicitly recognized at the 2002 World Summit on Sustainable Development) provides a framework for countries to select effective SD and RE strategies and to align those with international policy measures. To that end, national strategies should include the removal of existing financial mechanisms that work against SD. For example, the removal of fossil fuel subsidies may have the potential to open up opportunities for more extensive use or even market entry of RE, but any subsidy reform towards the use of RE technologies needs to address the specific needs of the poor and demands a case-specific analysis. [9.5.2.1]

The CDM established under the Kyoto Protocol is a practical example of a mechanism for SD that internalizes environmental and social externalities. However, there are no international standards for sustainability assessments (including comparable SD indicators) to counter weaknesses in the existing system regarding sustainability approval. As input to the negotiations for a post-2012 climate regime, many suggestions have been made about how to reform the CDM to better achieve new and improved mechanisms for SD. [9.5.2.1]

Opportunities for RE to play a role in national strategies for SD can be approached by integrating SD and RE goals into development policies and by development of sectoral strategies for RE that contribute to goals for green growth and low-carbon and sustainable development including leapfrogging. [9.5.2.1]

At the local level, SD initiatives by cities, local governments, and private and nongovernmental organizations can be drivers of change and contribute to overcome local resistance to RE installations. [9.5.2.2]

9.6 Synthesis, knowledge gaps and future research needs

RE can contribute to SD and the four goals assessed to varying degrees. While benefits with respect to reduced environmental and health impacts may appear more clear-cut, the exact contribution to, for example, social and economic development is more ambiguous. Also, countries may prioritize the four SD goals according to their level of development. To some extent, however, these SD goals are also strongly interlinked. Climate change mitigation constitutes in itself a necessary prerequisite for successful social and economic development in many developing countries. [9.6.6]

Following this logic, climate change mitigation can be assessed under the strong SD paradigm, if mitigation goals are imposed as constraints on future development pathways. If climate change mitigation is balanced against economic growth or other socioeconomic criteria, the problem is framed within the paradigm of weak SD allowing for trade-offs between these goals and using cost-benefit type analyses to provide guidance in their prioritization. [9.6.6]

However, the existence of uncertainty and ignorance as inherent components of any development pathway, as well as the existence of associated and possibly 'unacceptably high' opportunity costs, will make continued adjustments crucial. In the future, integrated models may be in a favourable position to better link the weak and strong SD paradigms for decision-making processes. Within well-defined guardrails, integrated models could explore scenarios for different mitigation pathways, taking account of the remaining SD goals by including important and relevant bottom-up indicators. According to model type, these alternative development pathways might be optimized for socially beneficial outcomes. Equally, however, the incorporation of GHG emission-related LCA data will be crucial for a clear definition of appropriate GHG concentration stabilization levels in the first place. [9.6.6]

In order to improve the knowledge regarding the interrelations between SD and RE and to find answers to the question of effective, economically efficient and socially acceptable transformations of the energy system, it is necessary to develop a closer integration of insights from social, natural and economic sciences (e.g., through risk analysis approaches), reflecting the different dimensions of sustainability (especially intertemporal, spatial, and intergenerational). So far, the knowledge base is often limited to very narrow views from specific branches of research, which do not fully account for the complexity of the issue. [9.7]

10. Mitigation Potential and Costs

10.1 Introduction

Future GHG emission estimates are highly dependent on the evolution of many variables, including, among others, economic growth, population growth, energy demand, energy resources and the future costs and performance of energy supply and end-use technologies. Mitigation and other non-mitigation policy structures in the future will also influence deployment of mitigation technologies and therefore GHG emissions and the ability to meet climate goals. Not only must all these different forces be considered simultaneously when exploring the role of RE in climate mitigation [see Figure 1.14], it is not possible to know today with any certainty how these different key forces might evolve decades into the future. [10.1]

Questions about the role that RE sources are likely to play in the future, and how they might contribute to GHG mitigation pathways, need to be explored within this broader context. Chapter 10 provides such an exploration through the review of 164 existing medium- to long-term scenarios from large-scale, integrated models. The comprehensive review explores the range of global RE deployment levels emerging in recent published scenarios and identifies many of the key forces that drive the variation among scenarios (note that the chapter relies exclusively on existing published scenarios and does not create any new scenarios). It does so both at the scale of RE as a whole and also in the context of individual RE technologies. The review highlights the importance of interactions and competition with other technologies as well as the evolution of energy demand more generally. [10.2]

This large-scale review is complemented with a more detailed discussion of future RE deployment, using 4 of the 164 scenarios as illustrative examples. The chosen scenarios span a range of different future expectations about RE characteristics, are based on different methodologies and cover different GHG concentration stabilization levels. This approach provides a next level of detail for exploring the role of RE in climate change mitigation, distinguishing between different applications (electricity generation, heating and cooling, transport) and regions. [10.3]

As the resulting role of RE is significantly determined by cost factors, a more general discussion about cost curves and cost aspects is then provided. This discussion starts with an assessment of the strengths and shortcomings of supply curves for RE and GHG mitigation, and then reviews the existing literature on regional RE supply curves, as well as abatement cost curves, as they pertain to mitigation using RE sources. [10.4]

Costs of RE commercialization and deployment are then addressed. The chapter reviews present RE technology costs, as well as expectations about how these costs might evolve into the future. To allow an assessment of future market volumes and investment needs, based on the results of the four illustrative scenarios investments in RE are discussed in particular with respect to what might be required if ambitious climate protection goals are to be achieved. [10.5]

Standard economic measures do not cover the full set of costs. Therefore, social and environmental costs and benefits of increased deployment of RE in relation to climate change mitigation and SD are synthesized and discussed. [10.6]

10.2 Synthesis of mitigation scenarios for different renewable energy strategies

An increasing number of integrated scenario analyses that are able to provide relevant insights into the potential contribution of RE to future energy supplies and climate change mitigation has become available. To provide a broad context for understanding the role of RE in mitigation and the influence of RE on the costs of mitigation, 164 recent medium- to long-term scenarios from 16 global energy-economic and integrated assessment models were reviewed. The scenarios were collected through an open call. The scenarios cover a large range of CO_2 concentrations (350 to 1,050 ppm atmospheric CO_2 concentration by 2100), representing both mitigation and baseline scenarios. [10.2.2.1]

Although these scenarios represent some of the most recent and sophisticated thinking regarding climate mitigation and the role of RE in climate mitigation in the medium- to long-term, they, as with any analysis looking decades into the future, must be interpreted carefully. All of the scenarios were developed using quantitative modelling, but there is enormous variation in the detail and structure of the models used to construct the scenarios. In addition, the scenarios do not represent a random sample of possible scenarios that could be used for formal uncertainty analysis. Some modelling groups provided more scenarios than others. In scenario ensemble analyses based on collecting scenarios from different studies, such as the review here, there is an inevitable tension between the fact that the scenarios are not truly a random sample and the sense that the variation in the scenarios does still provide real and often clear insights into our knowledge about the future, or lack thereof. [10.2.1.2, 10.2.2.1]

A fundamental question relating to the role of RE in climate mitigation is how closely RE deployment levels are correlated with long-term atmospheric CO_2 concentration or related climate goals. The scenarios indicate that although there is a strong correlation between fossil and industrial CO_2 emissions pathways and long-term CO_2 concentration goals across the scenarios, the relationship between RE deployment and CO_2 concentration goals is far less robust (Figure TS.10.1). RE deployment generally increases with the stringency of the CO_2 concentration goal, but there is enormous variation among RE deployment levels for any given CO_2 concentration goal. For example, in scenarios that stabilize the atmospheric CO_2 concentration at a level of less than 440 ppm (Categories I and II), the median RE deployment levels are 139 EJ/yr in 2030 and 248 EJ/yr in 2050, with the highest levels reaching 252 EJ/yr in 2030 and up to 428 EJ/yr in 2050. These levels are considerably higher than the corresponding RE deployment levels in baseline scenarios, although it has to be acknowledged that the range of RE deployment in each of the CO_2 stabilization categories is wide. [10.2.2.2]

At the same time, it is also important to note that despite the variation, the absolute magnitudes of RE deployment are dramatically higher than those of today in the vast majority of the scenarios. In 2008, global renewable primary energy supply in direct equivalent stood at roughly 64 EJ/yr. The majority of this, about 30 EJ/yr, was traditional biomass. In contrast, by 2030, many scenarios indicate a doubling of RE deployment or more compared to today, and this is accompanied in most scenarios by a reduction in traditional biomass, implying substantial growth in non-traditional RE sources. By 2050, RE deployment levels in most scenarios are higher than 100 EJ/yr (median at 173 EJ/yr), reach 200 EJ/yr in many of the scenarios and more than 400 EJ/yr in some cases. Given that traditional biomass use decreases in most scenarios, the scenarios represent an increase in RE production (excluding traditional biomass) of anywhere from roughly three- to more than ten-fold. More than half of the scenarios show a contribution of RE in excess of a 17% share of primary energy supply in 2030, rising to more than 27% in 2050. The scenarios with the highest RE shares reach approximately 43% in 2030 and 77% in 2050. Deployments after 2050 are even larger. This is an extraordinary expansion in energy production from RE. [10.2.2.2]

Indeed, RE deployment is quite large in many of the baseline scenarios with no assumed GHG concentration stabilization level. By 2030, RE deployment levels of up to about 120 EJ/yr are projected, with many baseline scenarios reaching more than 100 EJ/yr in 2050 and in some cases up to 250 EJ/yr. These large RE baseline deployments result from a range of underlying scenario assumptions, for example, the assumption that energy consumption will continue to grow substantially throughout the century, assumptions about the ability of RE to contribute to increased energy access, assumptions about the availability of fossil resources, and other assumptions (e.g., improved costs and performance of RE technologies) that would render RE technologies economically increasingly competitive in many applications even absent climate policy. [10.2.2.2]

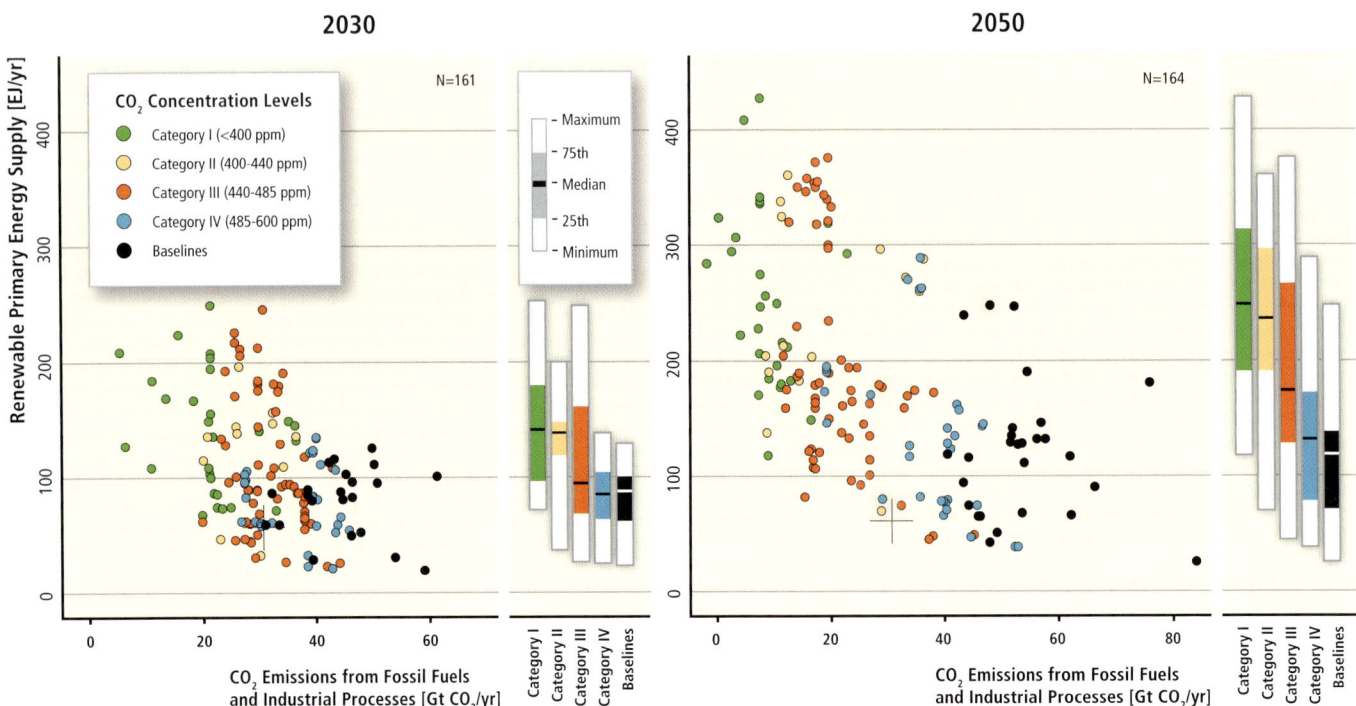

Figure TS.10.1 | Global RE primary energy supply (direct equivalent) from 164 long-term scenarios as a function of fossil and industrial CO_2 emissions in 2030 and 2050. Colour coding is based on categories of atmospheric CO_2 concentration level in 2100. The panels to the right of the scatterplots show the deployment levels of RE in each of the atmospheric CO_2 concentration categories. The thick black line corresponds to the median, the coloured box corresponds to the inter-quartile range (25th to 75th percentile) and the ends of the white surrounding bars correspond to the total range across all reviewed scenarios. The blue crossed-lines show the relationship in 2007. Pearson's correlation coefficients for the two data sets are -0.40 (2030) and -0.55 (2050). For data reporting reasons, only 161 scenarios are included in the 2030 results shown here, as opposed to the full set of 164 scenarios. RE deployment levels below those of today are a result both of model output as well as differences in the reporting of traditional biomass. [Figure 10.2]

The uncertainty in RE's role in climate mitigation results from uncertainty regarding a number of important forces that influence the deployment of RE. Two important factors are energy demand growth and the competition with other options to reduce CO_2 emissions (primarily nuclear energy and fossil energy with CCS). Meeting long-term climate goals requires a reduction in the CO_2 emissions from energy and other anthropogenic sources. For any given climate goal, this reduction is relatively well defined; there is a tight relationship between fossil and industrial CO_2 emissions and the deployment of freely emitting fossil energy across the scenarios (Figure TS.10.2). The demand for low-carbon energy (including RE, nuclear energy and fossil energy with CCS) is simply the difference between total primary energy demand and the production of freely-emitting fossil energy; that is, whatever energy cannot be supplied by freely-emitting fossil energy because of climate constraints must be supplied either by low-carbon energy or by measures that reduce energy consumption. However, scenarios indicate enormous uncertainty about energy demand growth, particularly many decades into the future. This variation is generally much larger than the effect of mitigation on energy consumption. Hence, there is substantial variability in low-carbon energy for any given CO_2 concentration goal due to variability in energy demand (Figure TS.10.2). [10.2.2.3]

The competition between RE, nuclear energy, and fossil energy with CCS then adds another layer of variability in the relationship between RE deployment and the CO_2 concentration goal. The cost, performance and availability of the competing supply side options—nuclear energy and fossil energy with CCS—is also uncertain. If the option to deploy these other supply-side mitigation technologies is constrained—because of cost and performance, but also potentially due to environmental, social or national security barriers—then, all things being equal, RE deployment levels will be higher (Figure TS.10.3). [10.2.2.4]

There is also great variation in the deployment characteristics of individual RE technologies. The absolute scales of deployments vary considerably among technologies and also deployment magnitudes are characterized by greater variation for some technologies relative to others (Figures TS.10.4 and TS.10.5). Further, the time scale of deployment varies across different RE sources, in large part representing differences in deployment levels today and (often) associated assumptions about relative technological maturity. [10.2.2.5]

The scenarios generally indicate that RE deployment is larger in non-Annex I countries over time than in the Annex I countries. Virtually all scenarios include the assumption that economic and energy demand growth will be larger at some point in the future in the non-Annex I countries than in the Annex I countries. The result is that the non-Annex I countries account for an increasingly large proportion of CO_2 emissions in baseline, or no-policy, cases and must therefore make larger emissions reductions over time (Figure TS.10.4). [10.2.2.5]

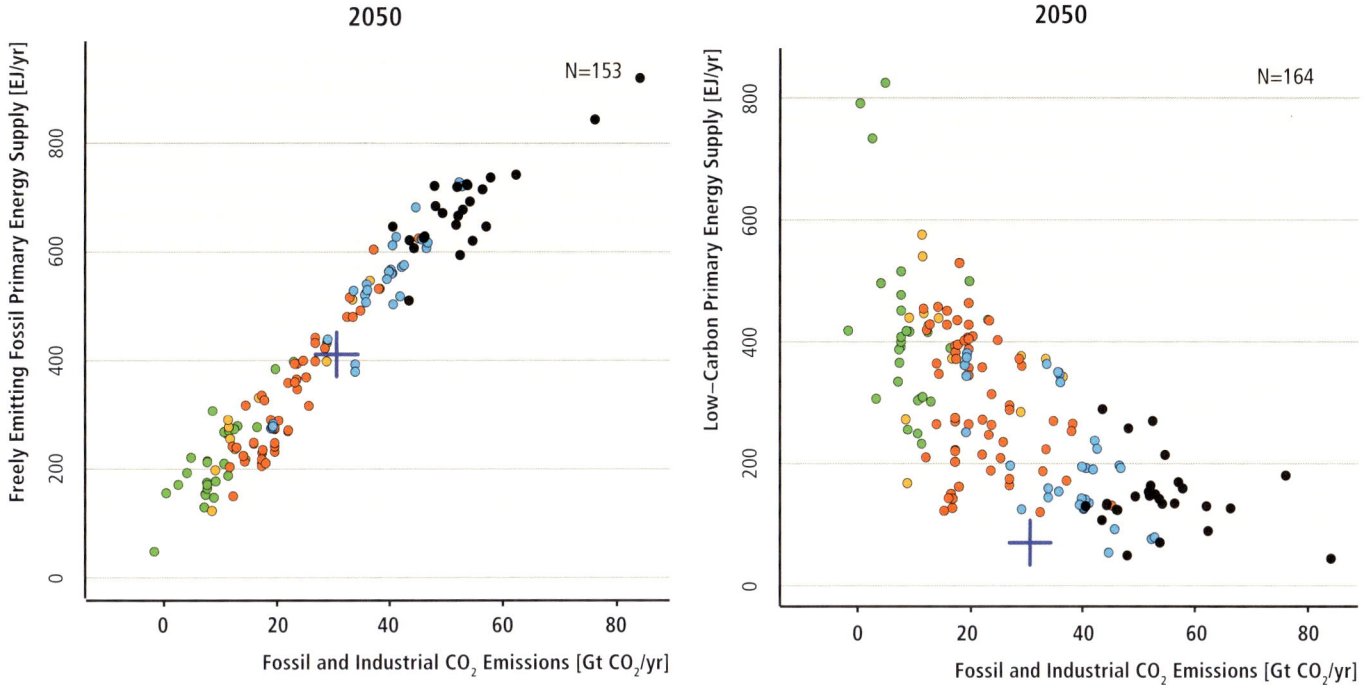

Figure TS.10.2 | Global freely emitting fossil fuel (left panel; direct equivalent) and low-carbon primary energy supply (right panel; direct equivalent) in 164 long-term scenarios in 2050 as a function of fossil and industrial CO_2 emissions. Low-carbon energy refers to energy from RE, fossil energy with CCS, and nuclear energy. Colour coding is based on categories of atmospheric CO_2 concentration level in 2100. The blue crossed lines show the relationship in 2007. Pearson's correlation coefficients for the two data sets are 0.97 (freely emitting fossil) and -0.68 (low-carbon energy). For data reporting reasons, only 153 scenarios and 161 scenarios are included in the freely-emitting fossil and low-carbon primary energy results shown here, respectively, as opposed to the full set of 164 scenarios. [Figure 10.4, right panel, Figure 10.5, right panel]

Another fundamental question regarding RE and mitigation is the relationship between RE and mitigation costs. A number of studies have pursued scenario sensitivities that assume constraints on the deployment of individual mitigation options, including RE as well as nuclear energy and fossil energy with CCS (Figures TS.10.6 and TS.10.7). These studies indicate that mitigation costs are higher when options, including RE, are not available. Indeed, the cost penalty for limits on RE is often at least of the same order of magnitude as the cost penalty for limits on nuclear energy and fossil energy with CCS. The studies also indicate that more aggressive concentration goals may not be possible when RE options, or other low-carbon options, are not available. At the same time, when taking into account the wide range of assumptions across the full range of scenarios explored in this assessment, the scenarios demonstrate no meaningful link between measures of cost (e.g., carbon prices) and absolute RE deployment levels. This variation is a reflection of the fact that large-scale integrated models used to generate scenarios are characterized by a wide range of carbon prices and mitigation costs based on both parameter assumptions and model structure. To summarize, while there is an agreement in the literature that mitigation costs will increase if the deployment of RE technologies is constrained and that more ambitious concentration stabilization levels may not be reachable, there is little agreement on the precise magnitude of the cost increase. [10.2.2.6]

10.3 Assessment of representative mitigation scenarios for different renewable energy strategies

An in-depth analysis of 4 selected illustrative scenarios from the larger set of 164 scenarios allowed a more detailed look at the possible contribution of specific RE technologies in different regions and sectors. The IEA's World Energy Outlook (IEA WEO 2009) was selected as an example of a baseline scenario, while the other scenarios set clear GHG concentration stabilization levels. The chosen mitigation scenarios are ReMIND-RECIPE from the Potsdam Institute, MiniCAM EMF 22 from the Energy Modelling Forum Study 22 and the Energy [R] evolution scenario from the German Aerospace Centre, Greenpeace International and EREC (ER 2010). The scenarios work as illustrative examples, but they are not representative in a strict sense. However they represent four different future paths based on different methodologies and a wide range of underlying assumptions. Particularly, they stand for different RE deployment paths reaching from a typical

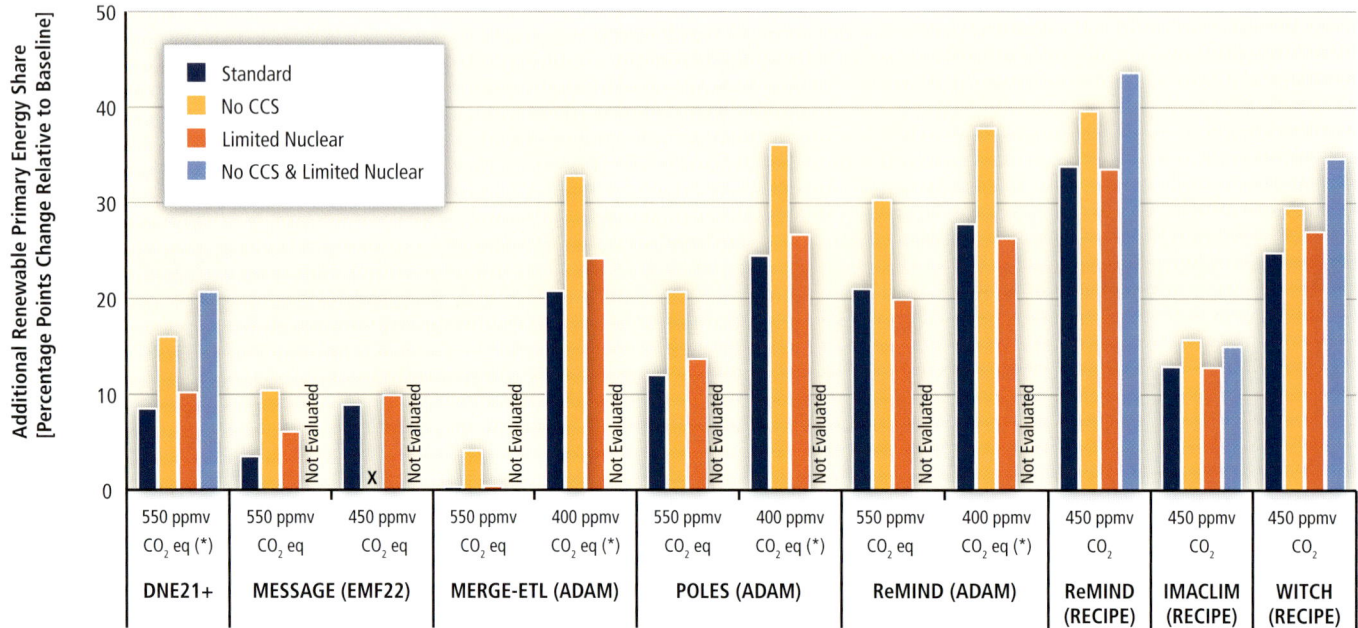

Figure TS.10.3 | Increase in global renewable primary energy share (direct equivalent) in 2050 in selected constrained technology scenarios compared to the respective baseline scenarios. The 'X' indicates that the respective concentration level for the scenario was not achieved. The definition of 'lim Nuclear' and 'no CCS' cases varies across models. The DNE21+, MERGE-ETL and POLES scenarios represent nuclear phase-outs at different speeds; the MESSAGE scenarios limit the deployment to 2010; and the ReMIND, IMACLIM and WITCH scenarios limit nuclear energy to the contribution in the respective baseline scenarios, which can still imply a significant expansion compared to current deployment levels. The REMIND (ADAM) 400 ppmv no CCS scenario refers to a scenario in which cumulative CO_2 storage is constrained to 120 Gt CO_2. The MERGE-ETL 400 ppmv no CCS case allows cumulative CO_2 storage of about 720 Gt CO_2. The POLES 400 ppmv CO_2eq no CCS scenario was infeasible and therefore the respective concentration level of the scenario shown here was relaxed by approximately 50 ppm CO_2. The DNE21+ scenario is approximated at 550 ppmv CO_2eq based on the emissions pathway through 2050. [Figure 10.6]

baseline perspective to a scenario that follows an optimistic application path for RE assuming that amongst others driven by specific policies the current high dynamic (increase rates) in the sector can be maintained. [10.3.1]

Figure TS.10.8 provides an overview of the resulting primary energy production by source for the four selected scenarios for 2020, 2030 and 2050 and compares the numbers with the range of the global primary energy supply. Using the direct equivalent methodology as done here, in 2050 bioenergy has the highest market share in all selected scenarios, followed by solar energy. The total RE share in the primary energy mix by 2050 has a substantial variation across all four scenarios. With 15% by 2050—more or less about today's level (12.9% in 2008)—the IEA WEO 2009 projects the lowest primary RE share, while the ER 2010 with 77% marks the upper level. The MiniCam EMF 22 expects that 31% and ReMIND-RECIPE that 48% of the world's primary energy demand will be provided by RE in 2050. The wide ranges of RE shares are a function of different assumptions for technology cost and performance data, availability of other mitigation technologies (e.g., CCS, nuclear power), infrastructure or integration constraints, non-economic barriers (e.g., sustainability aspects), specific policies and future energy demand projections. [10.3.1.4]

In addition, although deployment of the different technologies significantly increases over time, the resulting contribution of RE in the scenarios for most technologies in the different regions of the world is much lower than their corresponding technical potentials (Figure TS.10.9). The overall total global RE deployment by 2050 in all analyzed scenarios represents less than 3% of the available technical RE potential. On a regional level, the maximum deployment share out of the overall technical potential for RE in 2050 was found for China, with a total of 18% (ER 2010), followed by OECD Europe with 15% (ER 2010) and India with 13% (MiniCam EMF 22). Two regions have deployment rates of around 6% of the regional available technical RE potential by 2050: 7% in Developing Asia (MiniCam EMF 22) and 6% in OECD North America (ER 2010). The remaining five regions use less than 5% of the available technical potential for RE. [10.3.2.1]

Based on the resulting RE deployment for the selected four illustrative scenarios, the corresponding GHG mitigation potential has been calculated. For each sector, emission factors have been specified, addressing the kind of electricity generation or heat supply that RE displaces. As the substituted energy form depends on the overall system behaviour, this cannot be done exactly without conducting new and consistent scenario analysis or complex power plant dispatching analysis. Therefore, the calculation is necessarily based on simplified assumptions and can only be seen as indicative. Generally, attribution of precise mitigation potentials to RE should be viewed with caution. [10.3.3]

Very often RE applications are supposed to fully substitute for the existing mix of fossil fuel use, but in reality that may not be true as RE can compete, for instance, with nuclear energy or within the RE portfolio itself. To cover the uncertainties even partly for the specification of the emission factor, three different cases have been distinguished

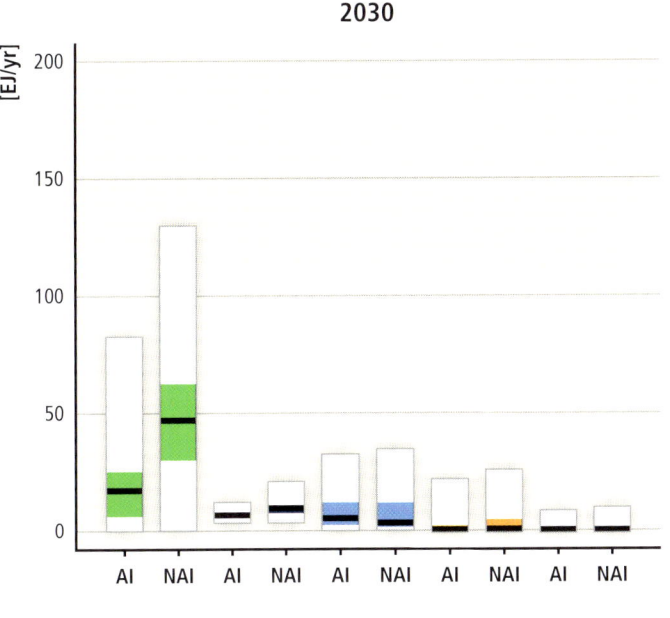
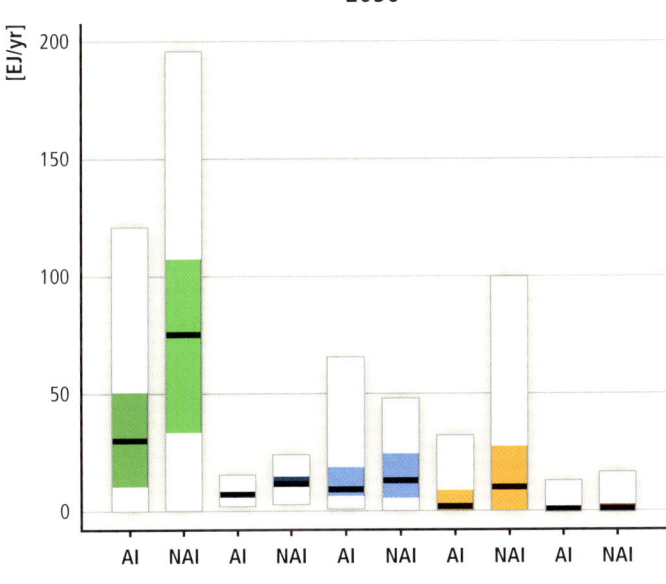
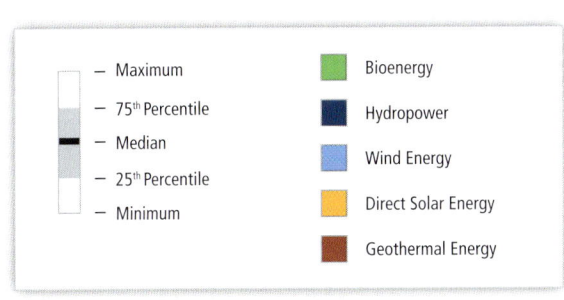

Figure TS.10.4 | Global RE primary energy supply (direct equivalent) by source in Annex I (AI) and Non-Annex I (NAI) countries in 164 long-term scenarios by 2030 and 2050. The thick black line corresponds to the median, the coloured box corresponds to the inter-quartile range (25th to 75th percentile) and the ends of the white surrounding bars correspond to the total range across all reviewed scenarios. Depending on the source, the number of scenarios underlying these figures varies between 122 and 164. Although instructive for interpreting the information, it is important to note that the 164 scenarios are not explicitly a random sample meant for formal statistical analysis. (One reason that bioenergy supply appears larger than supplies from other sources is that the direct equivalent method is used to represent primary energy in this figure. Bioenergy is accounted for prior to conversion to fuels such as ethanol or electricity. The other technologies produce primarily (but not entirely) electricity, and they are accounted for based on the electricity produced. If primary equivalents were used, based on the substitution method, rather than direct equivalents, then energy production from non-biomass RE would be of the order of three times larger than shown here.) Ocean energy is not presented here as only very few scenarios consider this RE technology. [Figure 10.8]

(upper case: specific average CO_2 emissions of the fossil generation mix under the baseline scenario; medium case: specific average CO_2 emissions of the overall generation mix under the baseline scenario; and lower case: specific average CO_2 emissions of the generation mix of the particular analyzed scenario). Biofuels and other RE options for transport are excluded from the calculation due to limited data availability.

Additionally, to reflect the embedded GHG emissions from bioenergy used for direct heating, only half of the theoretical CO_2 savings have been considered in the calculation. Given the high uncertainties and variability of embedded GHG emissions, this is necessarily once more a simplified assumption. [10.3.3]

Figure TS.10.10 shows cumulative CO_2 reduction potentials from RE sources up to 2020, 2030 and 2050 resulting from the four scenarios reviewed here in detail. The analyzed scenarios outline a cumulative reduction potential (2010 to 2050) in the medium-case approach of between 244 Gt CO_2 (IEA WEO 2009) under the baseline conditions, 297 Gt CO_2 (MiniCam EMF 22), 482 Gt CO_2 (ER 2010) and 490 Gt CO_2 (ReMIND-RECIPE scenario). The full range across all calculated cases and scenarios is cumulative CO_2 savings of 218 Gt CO_2 (IEA WEO 2009) to 561 Gt CO_2 (ReMIND-RECIPE) compared to about 1,530 Gt CO_2 cumulative fossil and industrial CO_2 emissions in the WEO 2009 Reference scenario during the same period. However, these numbers exclude CO_2 savings for RE use in the transport sector (including biofuels and electric vehicles). The overall CO_2 mitigation potential can therefore be higher. [10.3.3]

10.4 Regional cost curves for mitigation with renewable energy sources

The concept of supply curves of carbon abatement, energy, or conserved energy all rest on the same foundation. They are curves consisting typically of discrete steps, each step relating the marginal cost of the abatement measure/energy generation technology or measure to conserve energy to its potential; these steps are ranked according to their cost. Graphically, the steps start at the lowest cost on the left with the next highest cost added to the right and so on, making an upward sloping left-to-right marginal cost curve. As a result, a curve is obtained that can be interpreted similarly to the concept of supply curves in traditional economics. [10.4.2.1]

The concept of energy conservation supply curves is often used, but it has common and specific limitations. The most often cited limitations in

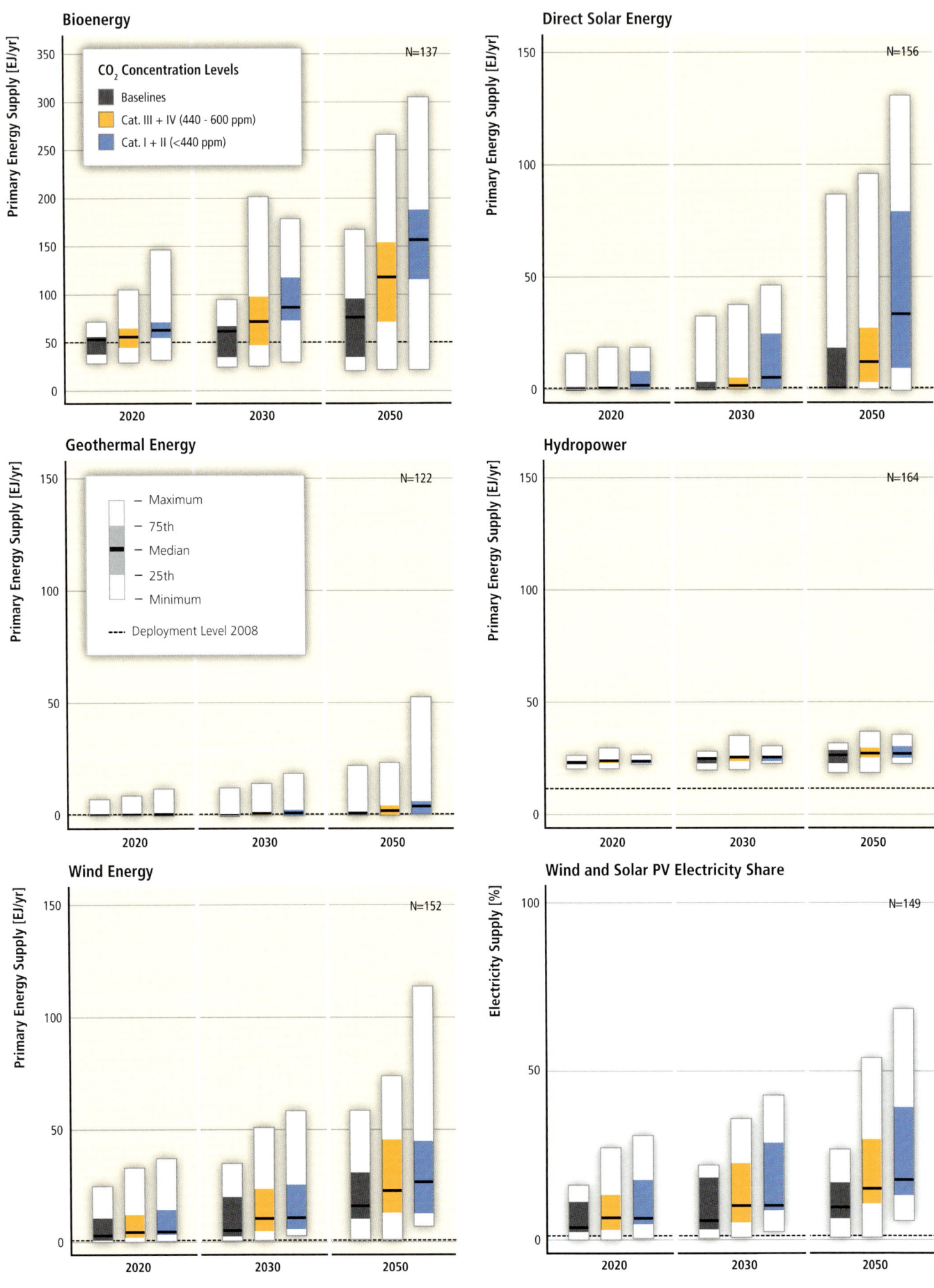

Figure TS.10.5 | (Preceding page) Global primary energy supply (direct equivalent) of biomass, wind, solar, hydro, and geothermal energy in 164 long-term scenarios in 2020, 2030 and 2050, and grouped by different categories of atmospheric CO_2 concentration level in 2100. The thick black line corresponds to the median, the coloured box corresponds to the inter-quartile range (25th to 75th percentile) and the ends of the white surrounding bars correspond to the total range across all reviewed scenarios. [Figure 10.9]

Notes: For data reporting reasons, the number of scenarios included in each of the panels shown here varies considerably. The number of scenarios underlying the individual panels, as opposed to the full set of 164 scenarios, is indicated in the right upper corner of each panel. One reason that bioenergy supply appears larger than supplies from other sources is that the direct equivalent method is used to represent primary energy in this figure. Bioenergy is accounted for prior to conversion to fuels such as biofuels, electricity and heat. The other technologies produce primarily (but not entirely) electricity and heat, and they are accounted for based on this secondary energy produced. If primary equivalents based on the substitution method were used rather than direct equivalent accounting, then energy production from non-biomass RE would be of the order of two to three times larger than shown here. Ocean energy is not presented here as scenarios so far seldom consider this RE technology. Finally, categories V and above are not included and Category IV is extended to 600 ppm from 570 ppm, because all stabilization scenarios lie below 600 ppm CO_2 in 2100, and because the lowest baselines scenarios reach concentration levels of slightly more than 600 ppm by 2100.

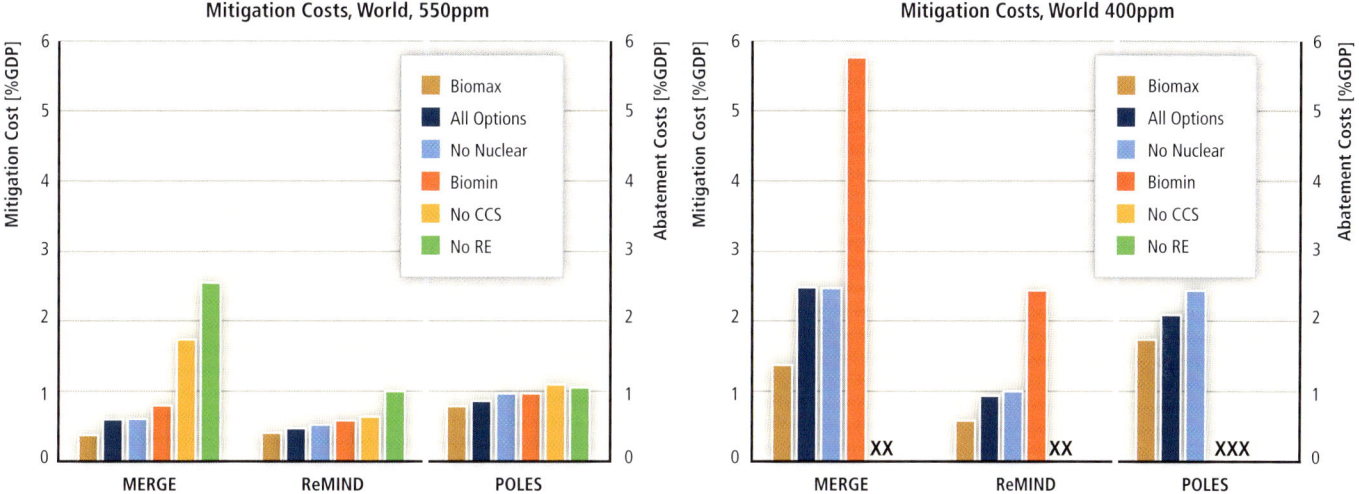

Figure TS.10.6 | Global mitigation costs (measured in terms of consumption loss) from the ADAM project under varying assumptions regarding technology availability for long-term stabilization levels of 550 and 400 ppmv CO_2eq. 'All options' refers to the standard technology portfolio assumptions in the different models, while 'biomax' and 'biomin' assume double and half the standard biomass potential of 200 EJ respectively. 'noccs' excludes CCS from the mitigation portfolio and 'nonuke' and 'norenew' constrain the deployment levels of nuclear and RE to the baseline level, which still potentially means a considerable expansion compared to today. The 'X' in the right panel indicates non-attainability of the 400 ppmv CO_2eq level in the case of limited technology options. [Figure 10.11]

this context are: controversy among scientists about potentials at negative costs; simplification of reality as actors also base their decisions on other criteria than those reflected in the curves; economic and technological uncertainty inherent to predicting the future, including energy price developments and discount rates; further uncertainty due to strong aggregation; high sensitivity relative to baseline assumptions and the entire future generation and transmission portfolio; consideration of individual measures separately, ignoring interdependencies between measures applied together or in different order; and, for carbon abatement curves, high sensitivity to (uncertain) emission factor assumptions. [10.4.2.1]

Having these criticisms in mind, it is also worth noting that it is very difficult to compare data and findings from RE abatement cost and supply curves, as very few studies have used a comprehensive and consistent approach that details their methodologies. Many of the regional and country studies provide less than 10% abatement of the baseline CO_2 emissions over the medium term at abatement costs under approximately USD_{2005} 100/t CO_2. The resulting low-cost abatement potentials are quite low compared to the reported mitigation potentials of many of the scenarios reviewed here. [10.4.3.2]

10.5 Cost of commercialization and deployment

Some RE technologies are broadly competitive with current market energy prices. Many of the other RE technologies can provide competitive energy services in certain circumstances, for example, in regions with favourable resource conditions or that lack the infrastructure for other low-cost energy supplies. In most regions of the world, however, policy measures are still required to ensure rapid deployment of many RE sources. [2.7, 3.8, 4.6, 5.8, 6.7, 7.8, 10.5.1, Figure TS.1.9]

Figures TS.10.11 and TS.10.12 provide additional data on levelized costs of energy (LCOE), also called levelized unit costs or levelized generation costs, for selected renewable power technologies and for renewable heating technologies, respectively. Figure TS.10.13 shows the levelized

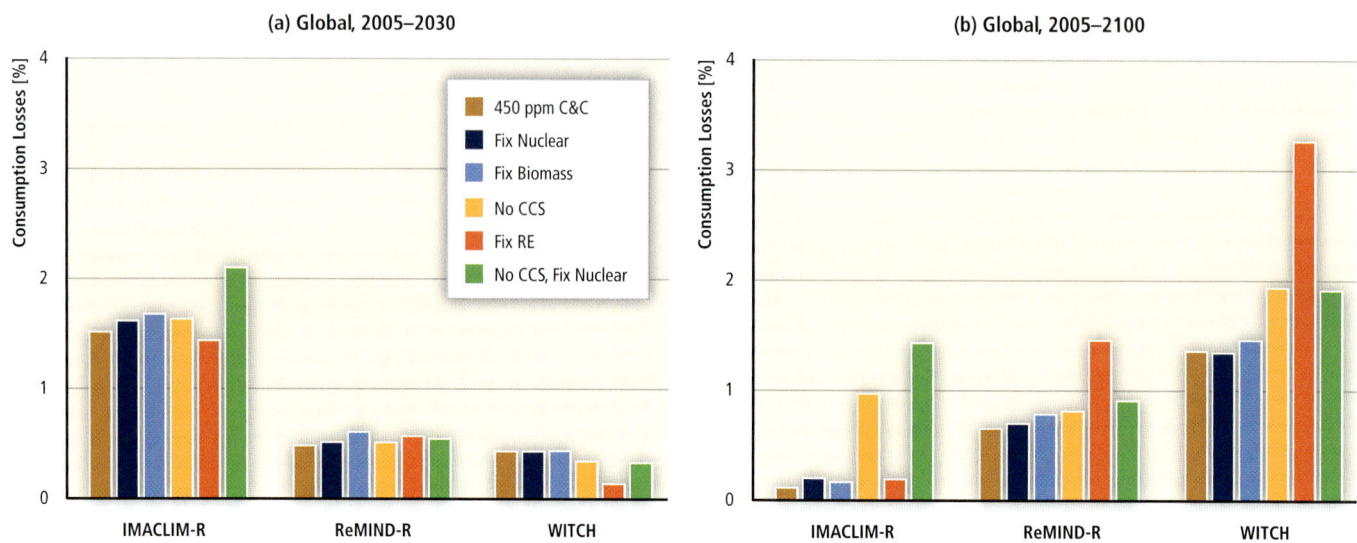

Figure TS.10.7 | Mitigation costs from the RECIPE project under varying assumptions regarding technology availability for a long-term stabilization level of 450 ppmv CO_2. Option values of technologies in terms of consumption losses for scenarios in which the option indicated is foregone (CCS) or limited to baseline levels (all other technologies) for the periods a) 2005 to 2030 and b) 2005 to 2100. Option values are calculated as differences in consumption losses for a scenario in which the use of certain technologies is limited with respect to the baseline scenario. Note that for WITCH, the generic backstop technology was assumed to be unavailable in the 'fix RE' scenario. [Figure 10.12]

cost of transport fuels (LCOF). LCOEs capture the full costs (i.e., investment costs, O&M costs, fuel costs and decommissioning costs) of an energy conversion installation and allocate these costs over the energy output during its lifetime, although not taking into account subsidies or policy incentives. As some RE technologies (e.g., PV, CSP and wind energy) are characterized by high shares of investment costs relative to variable costs, the applied discount rate has a prominent influence on the LCOE of these technologies (see Figures TS.10.11, TS.10.12 and TS.10.13). [10.5.1] The LCOEs are based on literature reviews and represent the most current cost data available. The respective ranges are rather broad as the levelized cost of identical technologies can vary across the globe depending on the RE resource base and local costs of investment, financing and O&M. Comparison between different technologies should not be based solely on the cost data provided in Figures TS 1.9,

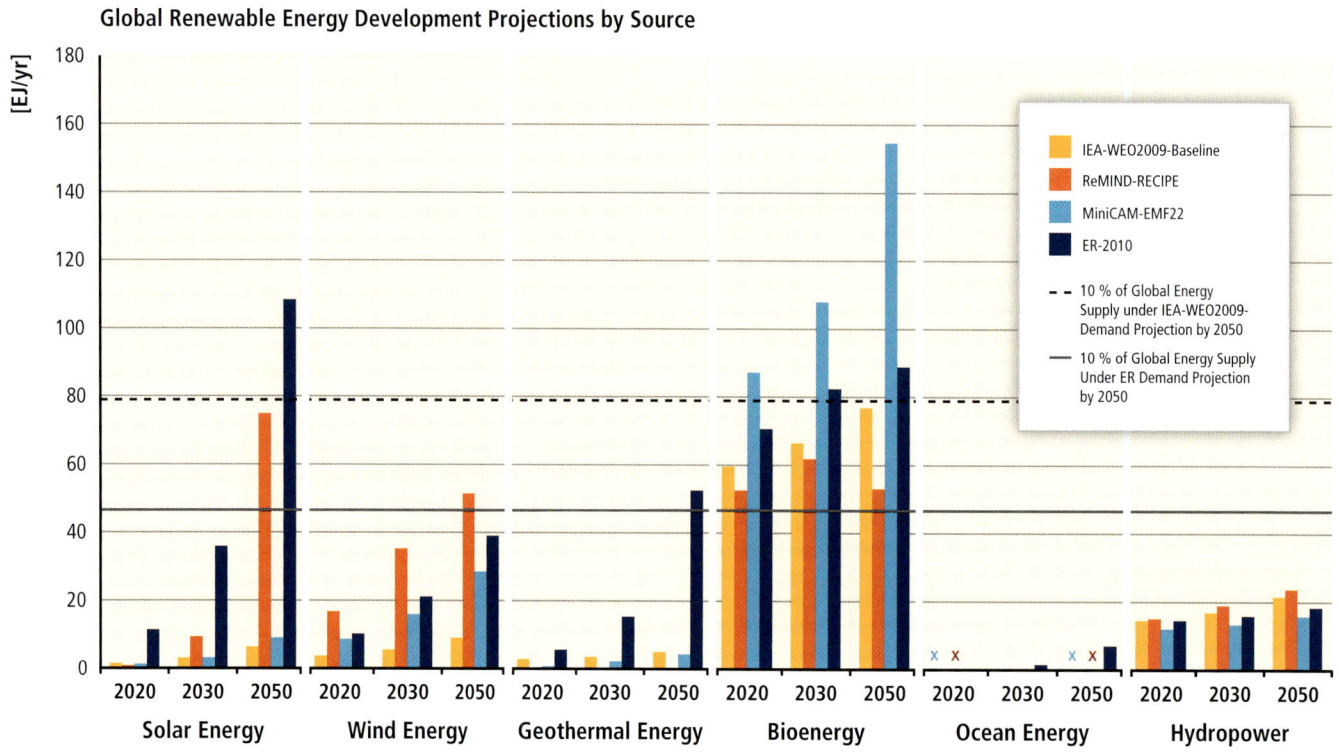

Figure TS.10.8 | Global RE development projections by source and global primary RE shares by source for a set of four illustrative scenarios. [Figure 10.14]

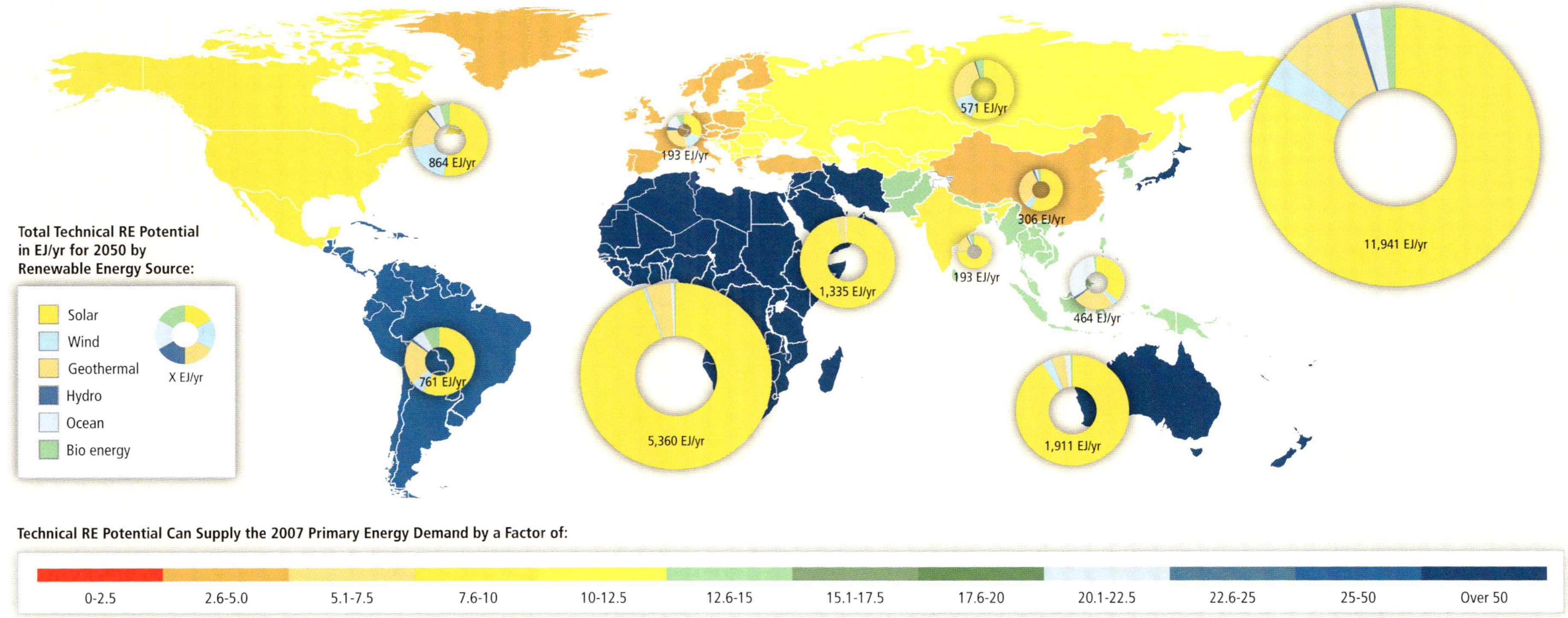

Summaries

Technical Summary

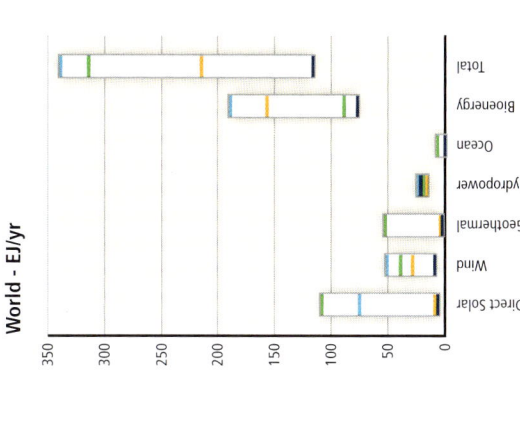

Total Technical RE Potential in EJ/yr for 2050 by Renewable Energy Source:
- Solar
- Wind
- Geothermal
- Hydro
- Ocean
- Bio energy

Technical RE Potential Can Supply the 2007 Primary Energy Demand by a Factor of:

0-2.5 | 2.6-5.0 | 5.1-7.5 | 7.6-10 | 10-12.5 | 12.6-15 | 15.1-17.5 | 17.6-20 | 20.1-22.5 | 22.6-25 | 25-50 | Over 50

Regional values (EJ/yr): 11,941; 464; 1,911; 306; 571; 193; 1,335; 5,360; 193; 864; 761

Range graphs: Level of RE Deployment in 2050 by Scenario and Renewable Energy, in EJ/yr:
- IEA-WEO2009-Baseline
- ReMIND-RECIPE
- MiniCAM-EMF22
- ER-2010
- Range

RE potential analysis: Technical RE potentials reported here represent total worldwide and regional potentials based on a review of studies published before 2009 by Krewitt et al. (2009). They do not deduct any potential that is already being utilized for energy production. Due to methodological differences and accounting methods among studies, strict comparability of these estimates across technologies and regions, as well as to primary energy demand, is not possible. Technical RE potential analyses published after 2009 show higher results in some cases but are not included in this figure. However, some RE technologies may compete for land which could lower the overall RE potential.

Scenario data: IEA WEO 2009 Reference Scenario (International Energy Agency (IEA), 2009; Teske et al., 2010), ReMIND-RECIPE 450ppm Stabilization Scenario (Luderer et al., 2009), MiniCAM EMF22 1st-best 2.6 W/2 Overshoot Scenario (Calvin et al., 2009), Advanced Energy [R]evolution 2010 (Teske et al., 2010)

139

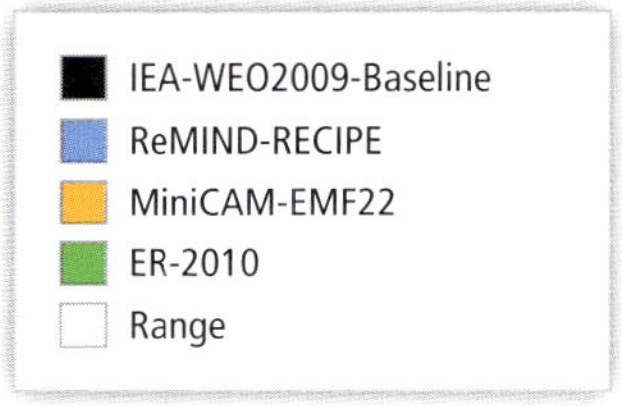

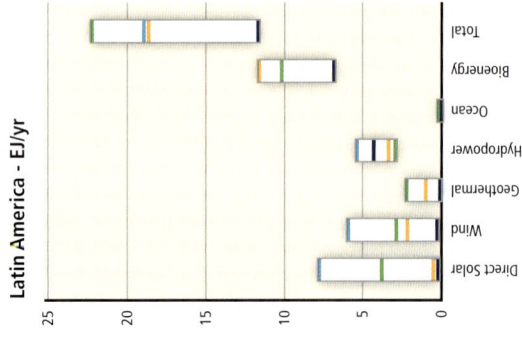
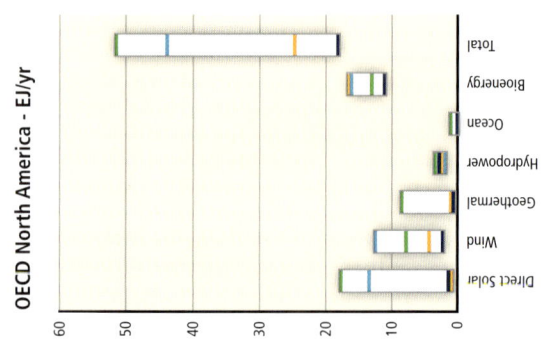
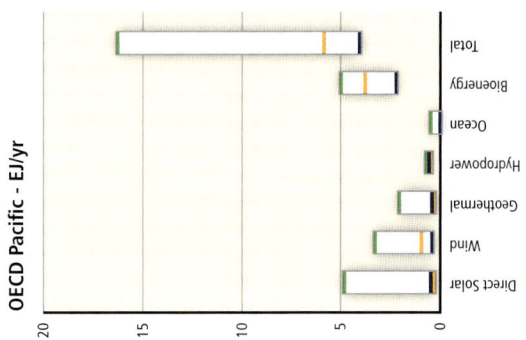
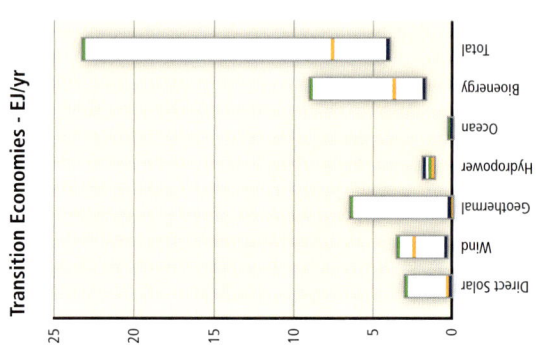
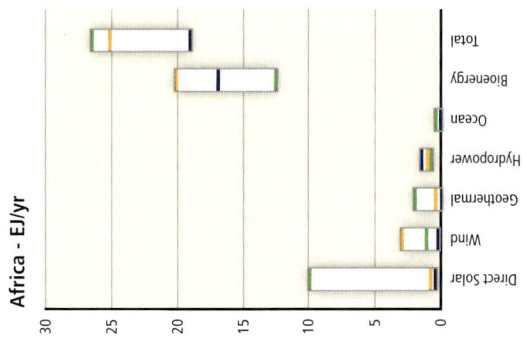
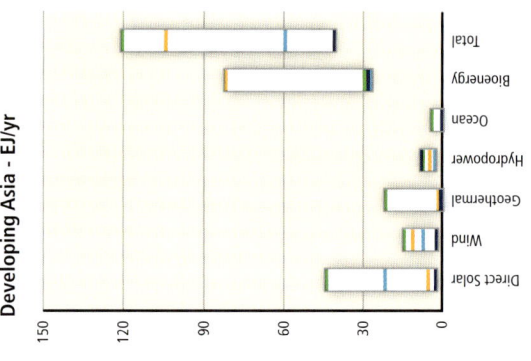
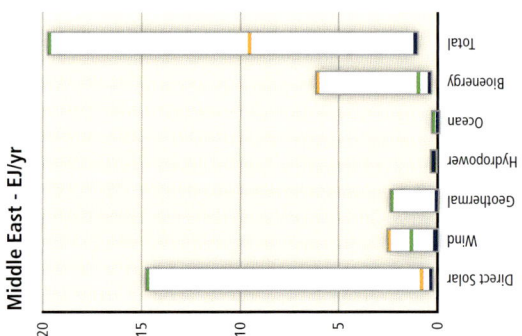
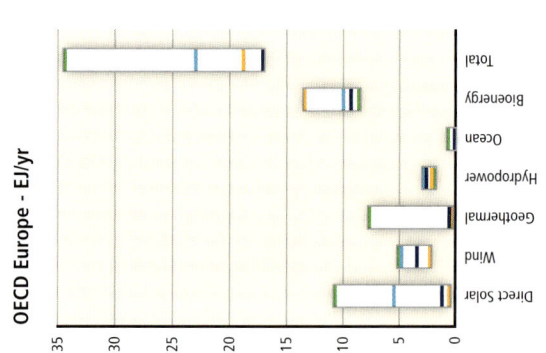

Figure TS.10.9 | (Preceding pages) Regional breakdown of RE deployment in 2050 for an illustrative set of four scenarios and comparison of the potential deployment to the corresponding technical potential for different technologies. The selected four illustrative scenarios are a part of the comprehensive survey of 164 scenarios. They represent a span from a reference scenario (IEA WEO 2009) without specific GHG concentration stabilization levels to three scenarios representing different CO_2 concentration categories, one of them (REMind-RECIPE) Category III (440 to 485 ppm) and two of them (MiniCam EMF 22 and ER 2010 Category I (<400 ppm). Of the latter, MiniCam EMF 22 includes nuclear energy and CCS as mitigation options and allows overshoot to get to the concentration level, while ER 2010 follows an optimistic application path for RE. Transition economies are countries that changed from a former centrally planned economy to a free market system. [Figure 10.19]

TS 10.11, TS.10.12 and TS.10.13; instead site, project and/or investor-specific conditions should be taken into account. The technology chapters [2.7, 3.8, 4.7, 5.8, 6.7, 7.8] provide useful sensitivities in this respect. [10.5.1]

The cost ranges provided here do not reflect costs of integration (Chapter 8), external costs or benefits (Chapter 9) or costs of policies (Chapter 11). Given suitable conditions, the lower ends of the ranges indicate that some RE technologies already can compete with traditional forms at current energy market prices in many regions of the world. [10.5.1]

The supply cost curves presented [10.4.4, Figures 10.23, 10.25, 10.26, and 10.27] provide additional information about the available resource base (given as a function of the LCOE associated with harvesting it). The supply cost curves discussed [10.3.2.1, Figures 10.15–10.17], in contrast, illustrate the amount of RE that is harnessed (once again as a function of the associated LCOE) in different regions once specific trajectories for the expansion of RE are followed. In addition, it must be emphasized that most of the supply cost curves refer to future points in time (e.g., 2030 or 2050), whereas the LCOE given in the cost sections of the technology chapters as well as those shown in Figures TS.10.11, TS.10.12, and TS.10.13 (and in Annex III) refer to current costs. [10.5.1]

Significant advances in RE technologies and associated cost reductions have been demonstrated over the last decades, though the contribution and mutual interaction of different drivers (e.g., learning by searching, learning by doing, learning by using, learning by interacting, upsizing of technologies, and economies of scale) is not always understood in detail. [2.7, 3.8, 7.8, 10.5.2]

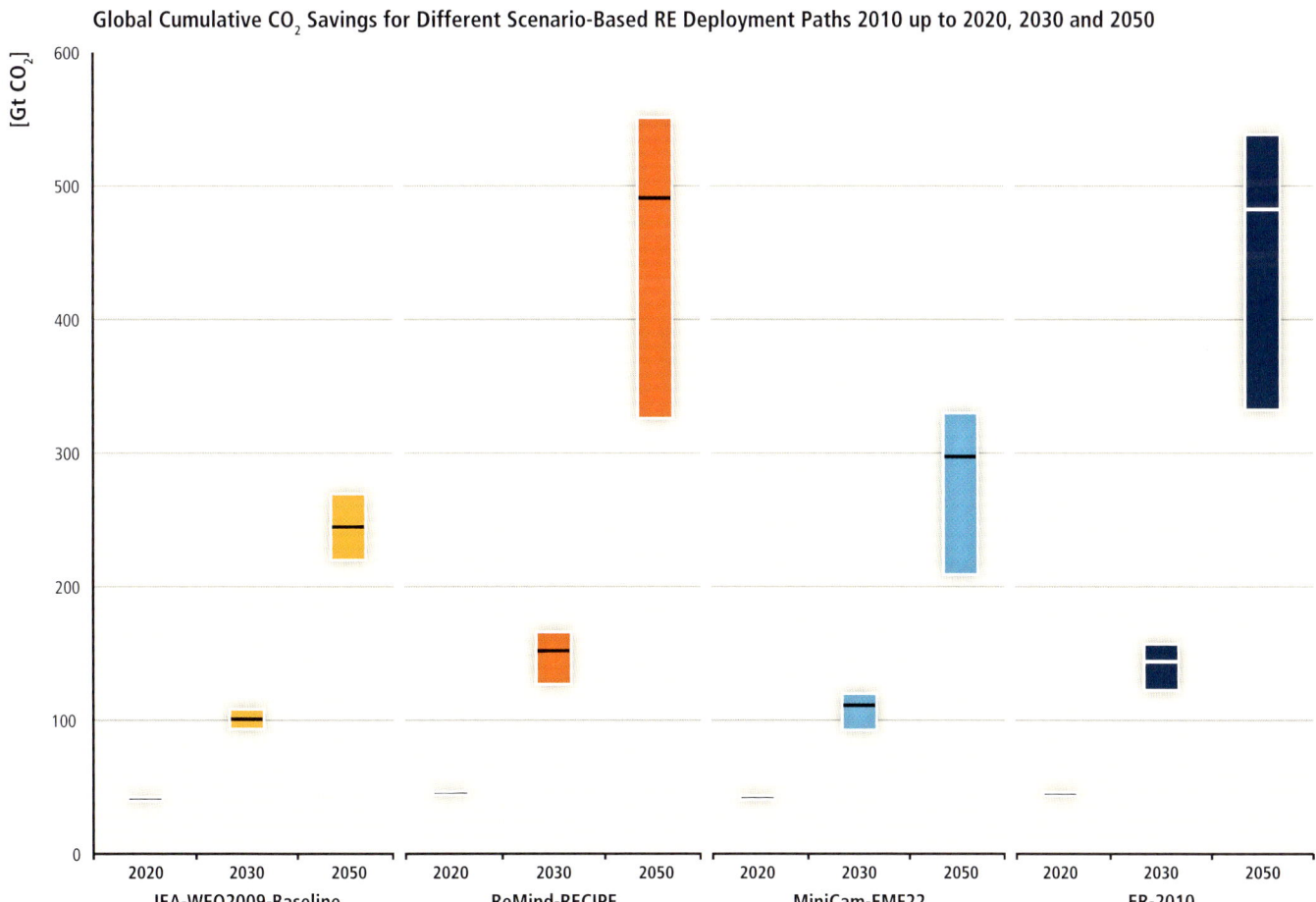

Figure TS.10.10 | Global cumulative CO_2 savings between 2010 and 2050 for four illustrative scenarios. The presented ranges mark the high uncertainties regarding the substituted conventional energy source. While the upper limit assumes a full substitution of high-carbon fossil fuels, the lower limit considers specific CO_2 emissions of the analyzed scenario itself. The line in the middle was calculated assuming that RE displaces the specific energy mix of a reference scenario. [Figure 10.22]

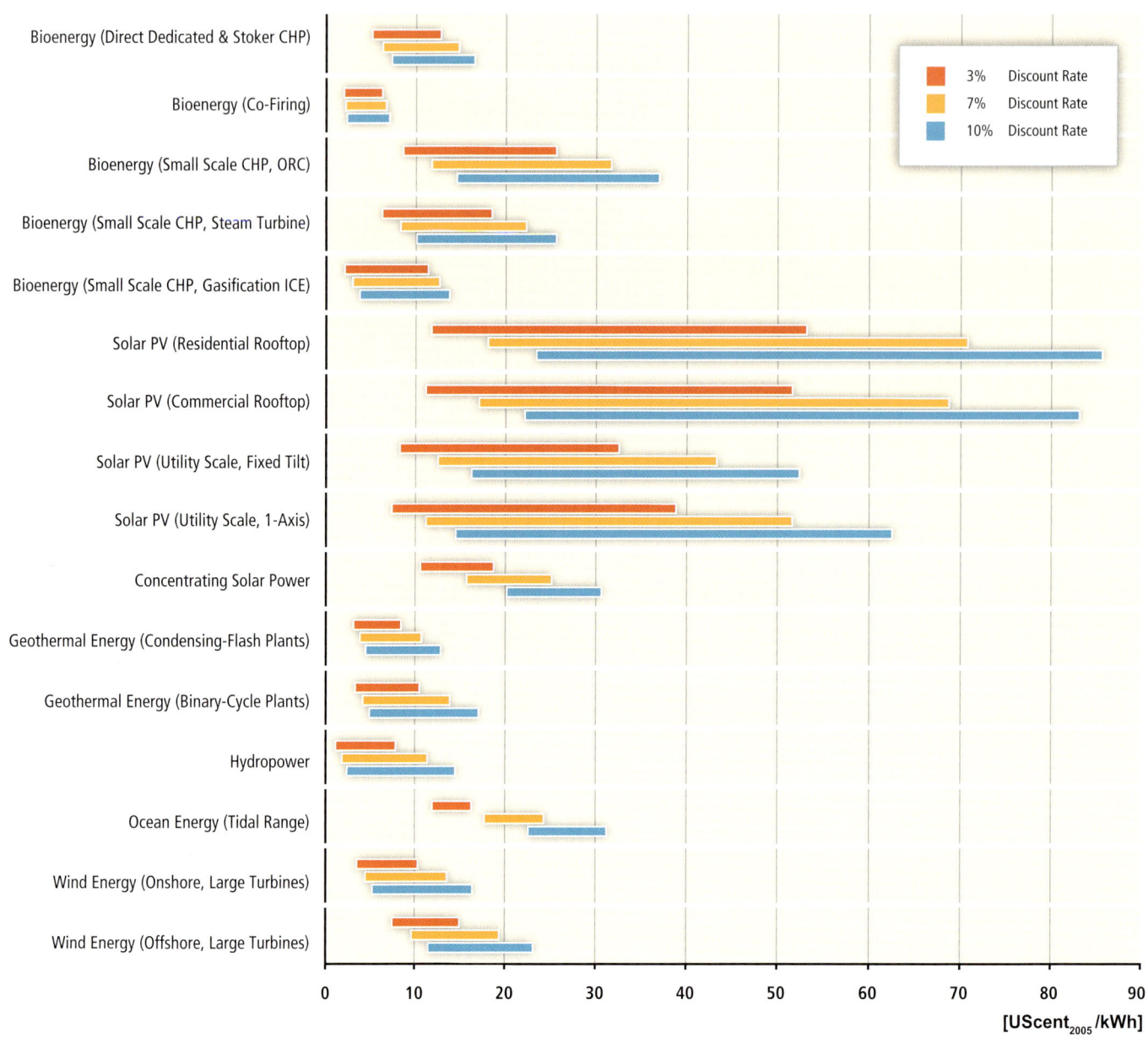

Figure TS.10.11 | Levelized cost of electricity for commercially available RE technologies at 3, 7 and 10% discount rates. The levelized cost of electricity estimates for all technologies are based on input data summarized in Annex III and the methodology outlined in Annex II. The lower bound of the levelized cost range is based on the low ends of the ranges of investment, operations and maintenance (O&M), and (if applicable) feedstock cost and the high ends of the ranges of capacity factors and lifetimes as well as (if applicable) the high ends of the ranges of conversion efficiencies and by-product revenue. The higher bound of the levelized cost range is accordingly based on the high end of the ranges of investment, O&M and (if applicable) feedstock costs and the low end of the ranges of capacity factors and lifetimes as well as (if applicable) the low ends of the ranges of conversion efficiencies and by-product revenue. Note that conversion efficiencies, by-product revenue and lifetimes were in some cases set to standard or average values. For data and supplementary information see Annex III. (CHP: combined heat and power; ORC: organic Rankine cycle, ICE: internal combustion engine.) [Figure 10.29]

From an empirical point of view, the resulting cost decrease can be described by experience (or 'learning') curves. For a doubling of the (cumulative) installed capacity, many technologies showed a more or less constant percentage decrease in the specific investment costs (or in the levelized costs or unit price, depending on the selected cost indicator). The numerical value describing this improvement is called the learning rate (LR). A summary of observed learning rates is provided in Table TS.10.1. [10.5.2]

Any efforts to assess future costs by extrapolating historic experience curves must take into account the uncertainty of learning rates as well as caveats and knowledge gaps discussed. [10.5.6, 7.8.4.1] As a supplementary approach, expert elicitations could be used to gather additional information about future cost reduction potentials, which might be contrasted with the assessments gained by using learning rates. Furthermore, engineering model analyses to identify technology improvement potentials could also provide additional information for developing cost projections. [2.6, 3.7, 4.6, 6.6, 7.7, 10.5.2]

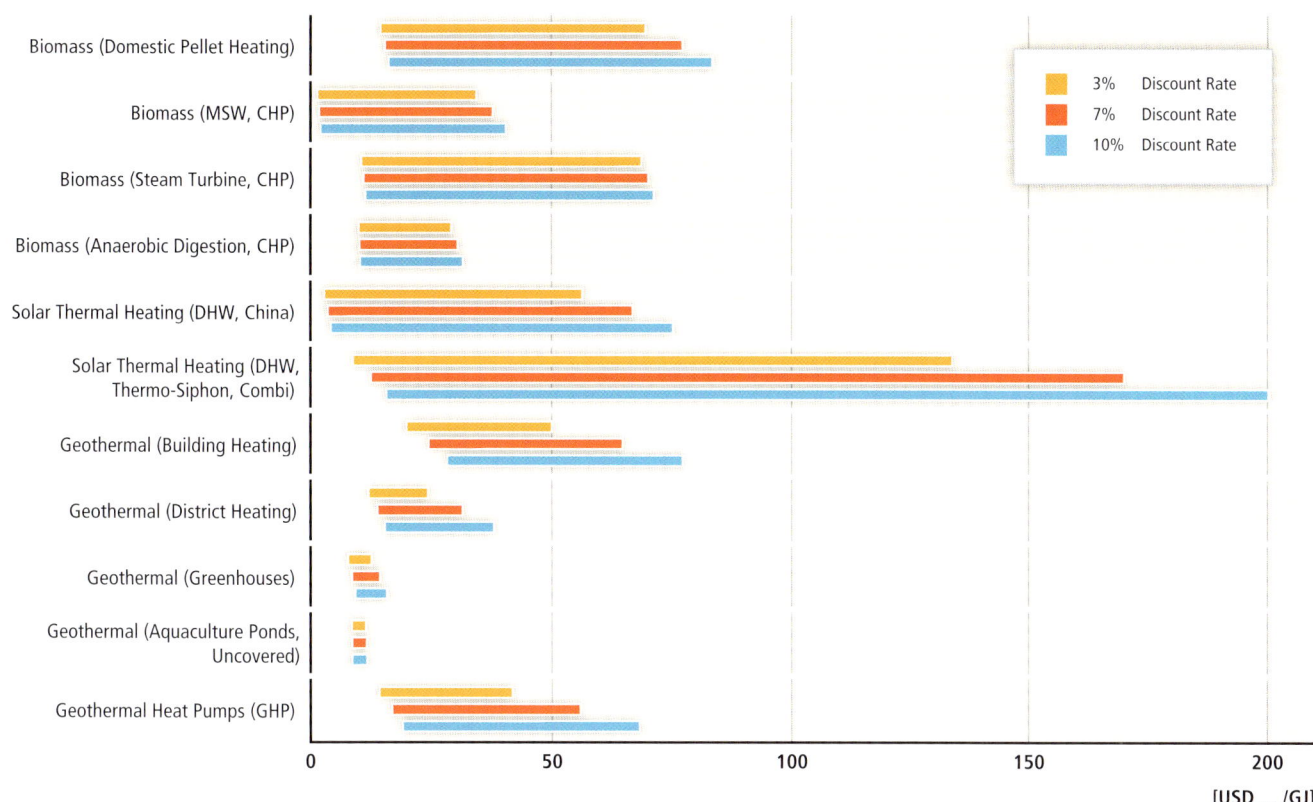

Figure TS.10.12 | Levelized cost of heat (LCOH) for commercially available RE technologies at 3, 7 and 10% discount rates. The LCOH estimates for all technologies are based on input data summarized in Annex III and the methodology outlined in Annex II. The lower bound of the levelized cost range is based on the low ends of the ranges of investment, operations and maintenance (O&M), and (if applicable) feedstock cost and the high ends of the ranges of capacity factors and lifetimes as well as (if applicable) the high ends of the ranges of conversion efficiencies and by-product revenue. The higher bound of the levelized cost range is accordingly based on the high end of the ranges of investment, O&M and (if applicable) feedstock costs and the low end of the ranges of capacity factors and lifetimes as well as (if applicable) the low ends of the ranges of conversion efficiencies and by-product revenue. Note that capacity factors and lifetimes were in some cases set to standard or average values. For data and supplementary information see Annex III. (MSW: municipal solid waste; DHW: domestic hot water.) [Figure 10.30]

Important potential technological advances and associated cost reductions, for instance, are expected in (but are not limited to) the following application fields: next-generation biofuels and biorefineries; advanced PV and CSP technologies and manufacturing processes; enhanced geothermal systems; multiple emerging ocean technologies; and foundation and turbine designs for offshore wind energy. Further cost reductions for hydropower are likely to be less significant than some of the other RE technologies, but R&D opportunities exist to make hydropower projects technically feasible in a wider range of natural conditions and to improve the technical performance of new and existing projects. [2.6, 3.7, 4.6, 5.3, 5.7, 5.8, 6.6, 7.7]

An answer to the question whether or not upfront investments in a specific innovative technology are justified cannot be given as long as the technology is treated in isolation. In a first attempt to clarify this issue and, especially, to investigate the mutual competition of prospective climate protection technologies, integrated assessment modellers have started to model technological learning in an endogenous way. The results obtained from these modelling comparison exercises indicate that—in the context of stringent climate goals—upfront investments in learning technologies can be justified in many cases. [10.5.3.]

However, as the different scenarios considered in Figure TS.10.14 and other studies clearly show, considerable uncertainty surrounds the exact volume and timing of these investments. [10.5.4]

The four illustrative scenarios that were analyzed in detail in Section 10.3 span a range of cumulative global decadal investments (in the power generation sector) ranging from USD$_{2005}$ 1,360 to 5,100 billion (for the decade 2011 to 2020) and from USD$_{2005}$ 1,490 to 7,180 billion (for the decade 2021 to 2030). These numbers allow the assessment of future market volumes and resulting investment opportunities. The lower values refer to the IEA World Energy Outlook 2009 Reference Scenario and the higher ones to a scenario that seeks to stabilize atmospheric CO$_2$ (only) concentration at 450 ppm. The average annual investments in the reference scenario are slightly lower than the respective investments reported for 2009. Between 2011 and 2020, the higher values of the annual averages of the RE power generation sector investment approximately correspond to a three-fold increase in the current global investments in this field. For the next decade (2021 to 2030), a five-fold increase is projected. Even the upper level of the annual investments is smaller than 1% of the world's GDP. Additionally, increasing the installed capacity of

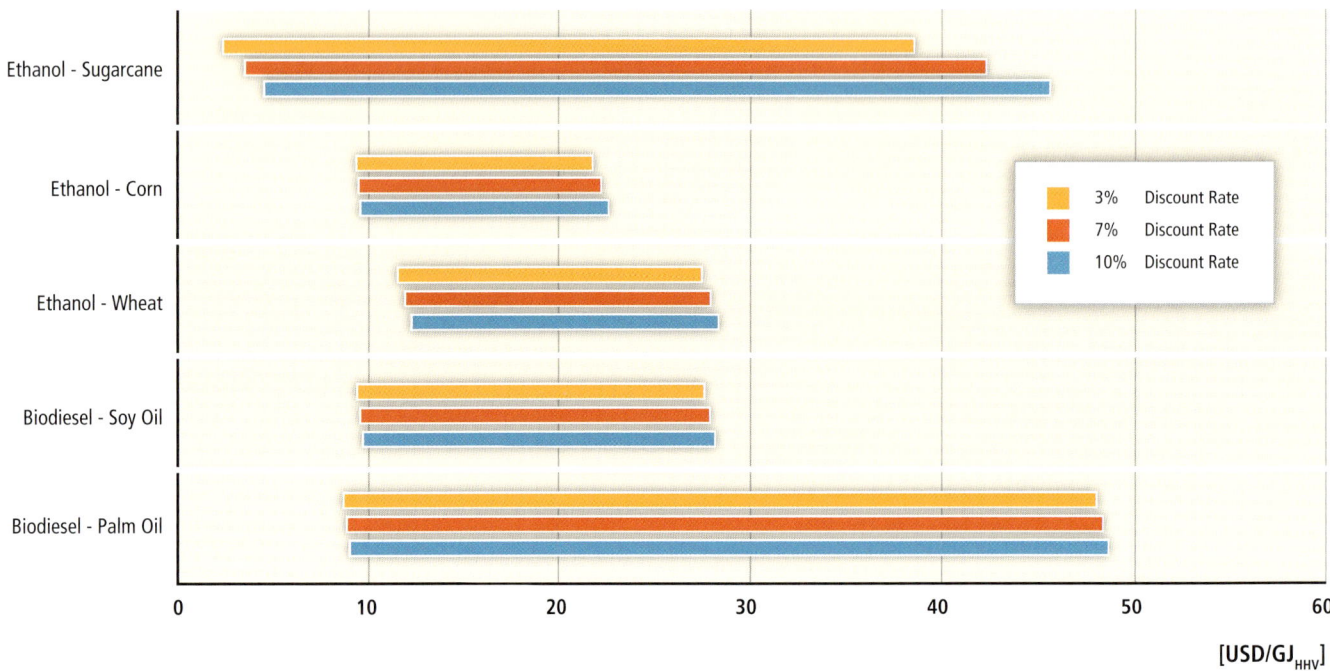

Figure TS.10.13 | Levelized cost of fuels (LCOF) for commercially available biomass conversion technologies at 3, 7 and 10% discount rates. LCOF estimates for all technologies are based on input data summarized in Annex III and the methodology outlined in Annex II. The lower bound of the levelized cost range is based on the low ends of the ranges of investment, O&M and feedstock cost. The higher bound of the levelized cost range is accordingly based on the high end of the ranges of investment, O&M and feedstock costs. Note that conversion efficiencies, by-product revenue, capacity factors and lifetimes were set to average values. For data and supplementary information see Annex III. (HHV: higher heating value.) [Figure 10.31]

RE power plants will reduce the amount of fossil and nuclear fuels that otherwise would be needed in order to meet a given electricity demand. [10.5.4]

10.6 Social and environmental costs and benefits

Energy extraction, conversion and use cause significant environmental impacts and external costs. Although replacing fossil fuel-based energy with RE often can reduce GHG emissions and also to some extent other environmental impacts and external costs, RE technologies can also have environmental impacts and external costs themselves, depending on the energy source and technology. These impacts and costs should be considered if a comprehensive cost assessment is required. [10.6.2]

Figure TS.10.15 shows the large uncertainty ranges of two dominant external cost components, namely climate- and health-related external costs. Small-scale biomass fired CHP plants cause relatively high external costs due to health effects via particulate emissions. Offshore wind energy seems to cause the smallest external cost. External cost estimates for nuclear power are not reported here because the character and assessment of external costs and risk from release of radionuclides due to low-probability accidents or due to leakages from waste repositories in a distant future are very different, for example, from climate change and air pollution, which are practically unavoidable. Those external impacts related to nuclear power can be, however, considered by discussion and judgment in the society. Accident risks in terms of fatalities due to various energy production chains (e.g., coal, oil, gas and hydro) are generally higher in non-OECD countries than in OECD countries. [10.6.3, 9.3.4.7]

As only external costs of individual technologies are shown in Figure TS.10.15, benefits can be derived when assuming that one technology replaces another one. RE sources and the technologies using them for electricity generation have mostly lower external costs per produced electricity than fossil fuel-based technologies. However, case-specific considerations are needed as there can also be exceptions. [10.6.3]

There are, however, considerable uncertainties in the assessment and valuation of external impacts of energy sources. The assessment of physical, biological and health damages includes considerable uncertainty and the estimates are based typically on calculational models, the results of which are often difficult to validate. The damages or changes seldom have market values that could be used in cost estimation, thus indirect information or other approaches must be used for damage valuation. Further, many of the damages will take place far in the future or in societies very different from those benefiting from the use of the considered energy production, which complicates the

Table TS.10.1 | Observed learning rates for various energy supply technologies. Note that values cited by older publications are less reliable as these refer to shorter time periods. [Table 10.10]

Technology	Source	Country / region	Period	Learning rate (%)	Performance measure
Onshore wind					
	Neij, 1997	Denmark	1982-1995	4	Price of wind turbine (USD/kW)
	Mackay and Probert, 1998	USA	1981-1996	14	Price of wind turbine (USD/kW)
	Neij, 1999	Denmark	1982-1997	8	Price of wind turbine (USD/kW)
	Durstewitz, 1999	Germany	1990-1998	8	Price of wind turbine (USD/kW)
	IEA, 2000	USA	1985-1994	32	Electricity production cost (USD/kWh)
	IEA, 2000	EU	1980-1995	18	Electricity production cost (USD/kWh)
	Kouvaritakis et al., 2000	OECD	1981-1995	17	Price of wind turbine (USD/kW)
	Neij, 2003	Denmark	1982-1997	8	Price of wind turbine (USD/kW)
	Junginger et al., 2005a	Spain	1990-2001	15	Turnkey investment costs (EUR/kW)
	Junginger et al., 2005a	UK	1992-2001	19	Turnkey investment costs (EUR/kW)
	Söderholm and Sundqvist, 2007	Germany, UK, Denmark	1986-2000	5	Turnkey investment costs (EUR/kW)
	Neij, 2008	Denmark	1981-2000	17	Electricity production cost (USD/kWh)
	Kahouli-Brahmi, 2009	Global	1979-1997	17	Investment costs (USD/kW)
	Nemet, 2009	Global	1981-2004	11	Investment costs (USD/kW)
	Wiser and Bolinger, 2010	Global	1982-2009	9	Investment costs (USD/kW)
Offshore wind					
	Isles, 2006	8 EU countries	1991-2006	3	Investment cost of wind farms (USD/kW)
Photovoltaics (PV)					
	Harmon, 2000	Global	1968-1998	20	Price PV module (USD/Wpeak)
	IEA, 2000	EU	1976-1996	21	Price PV module (USD/Wpeak)
	Williams, 2002	Global	1976-2002	20	Price PV module (USD/Wpeak)
	ECN, 2004	EU	1976-2001	20-23	Price PV module (USD/Wpeak)
	ECN, 2004	Germany	1992-2001	22	Price of balance of system costs
	van Sark et al., 2007	Global	1976-2006	21	Price PV module (USD/Wpeak)
	Kruck and Eltrop, 2007	Germany	1977-2005	13	Price PV module (EUR/Wpeak)
	Kruck and Eltrop, 2007	Germany	1999-2005	26	Price of balance of system costs
	Nemet, 2009	Global	1976-2006	15-21	Price PV module (USD/Wpeak)
Concentrating Solar Power (CSP)					
	Enermodal, 1999	USA	1984-1998	8-15	Plant investment cost (USD/kW)
Biomass					
	IEA, 2000	EU	1980-1995	15	Electricity production cost (USD/kWh)
	Goldemberg et al., 2004	Brazil	1985-2002	29	Prices for ethanol fuel (USD/m^3)
	Junginger et al., 2005b	Sweden, Finland	1975-2003	15	Forest wood chip prices (EUR/GJ)
	Junginger et al., 2006	Denmark	1984-1991	15	Biogas production costs (EUR/Nm^3)
	Junginger et al., 2006	Sweden	1990-2002	8-9	Biomass CHP power (EUR/kWh)
	Junginger et al., 2006	Denmark	1984-2001	0-15	Biogas production costs (EUR/Nm^3)
	Junginger et al., 2006	Denmark	1984-1998	12	Biogas plants (€/m^3 biogas/day)
	Van den Wall Bake et al., 2009	Brazil	1975-2003	19	Ethanol from sugarcane (USD/m^3)
	Goldemberg et al., 2004	Brazil	1980-1985	7	Ethanol from sugarcane (USD/m^3)
	Goldemberg et al., 2004	Brazil	1985-2002	29	Ethanol from sugarcane (USD/m^3)
	Van den Wall Bake et al., 2009	Brazil	1975-2003	20	Ethanol from sugarcane (USD/m^3)
	Hettinga et al., 2009	USA	1983-2005	18	Ethanol from corn (USD/m^3)
	Hettinga et al., 2009	USA	1975-2005	45	Corn production costs (USD/t corn)
	Van den Wall Bake et al., 2009	Brazil	1975-2003	32	Sugarcane production costs (USD/t)

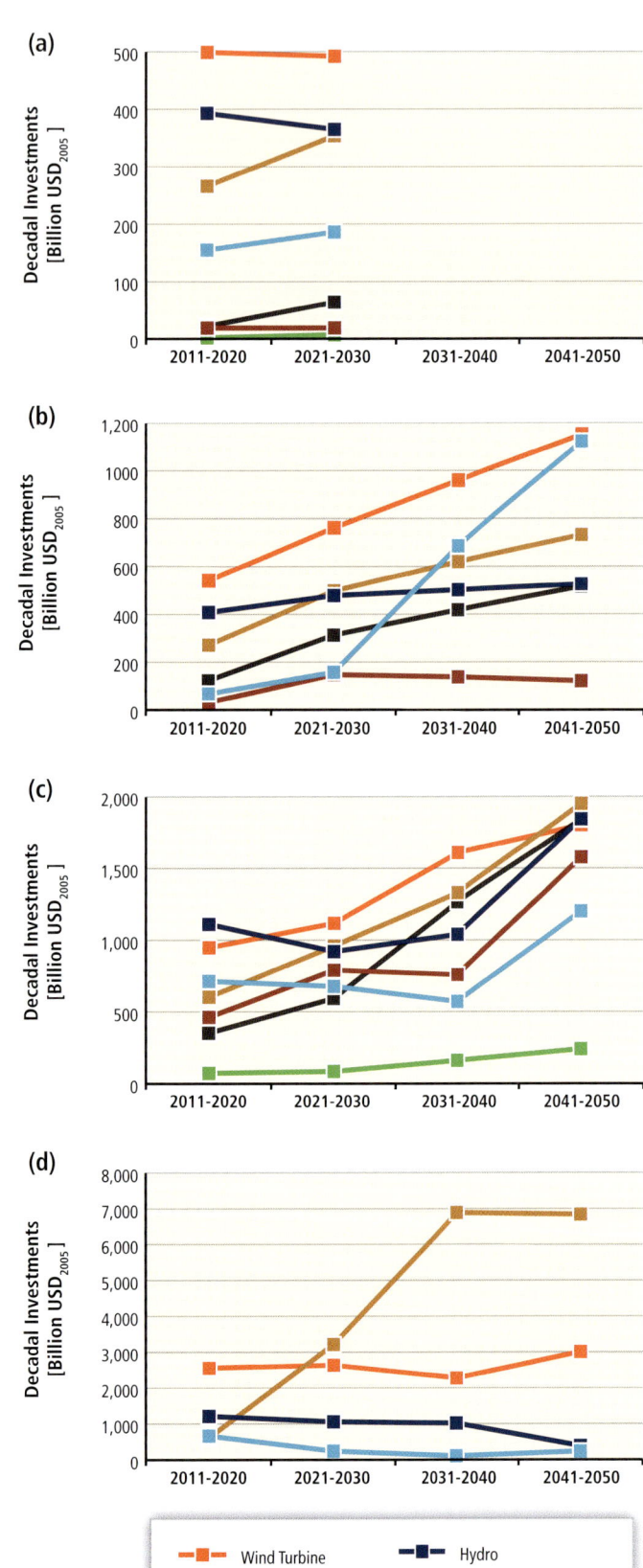

Figure TS.10.14 | Illustrative global *decadal* investments (in billion USD$_{2005}$) needed in order to achieve ambitious climate protection goals: (b) MiniCAM-EMF22 (first-best 2.6 W/m² overshoot scenario, nuclear and carbon capture technologies are permitted); (c) ER-2010 (450 ppm CO$_2$eq, nuclear and carbon capture technologies are not permitted); and (d) ReMIND-RECIPE (450 ppm CO$_2$, nuclear power plants and carbon capture technologies are permitted). Compared to the other scenarios, the PV share is high in (d) as concentrating solar power has not been considered. For comparison, (a) shows the IEA-WEO2009-Baseline (baseline scenario without climate protection). Sources: (a) IEA (2009); (b) Calvin et al. (2009); (c) Teske et al. (2010); and (d) Luderer et al. (2009).

considerations. These factors contribute to the uncertainty of external costs. [10.6.5]

However, the knowledge about external costs and benefits due to RE sources can provide some guidance for society to select best alternatives and to steer the energy system towards overall efficiency and high welfare gains. [10.6.5]

11. Policy, Financing and Implementation

11.1 Introduction

RE capacity is increasing rapidly around the world, but a number of barriers continue to hold back further advances. Therefore, if RE is to contribute substantially to the mitigation of climate change, and to do so quickly, various forms of economic support policies as well as policies to create an enabling environment are likely to be required. [11.1]

RE policies have promoted an increase in RE shares by helping to overcome various barriers that impede technology development and deployment of RE. RE policies might be enacted at all levels of government—from local to state/provincial to national to international—and range from basic R&D for technology development through to support for installed RE systems or the electricity, heat or fuels they produce. In some countries, regulatory agencies and public utilities may be given responsibility for, or on their own initiative, design and implement support mechanisms for RE. Nongovernmental actors, such as international agencies and development banks, also have important roles to play. [1.4, 11.1, 11.4, 11.5]

RE may be measured by additional qualifiers such as time and reliability of delivery (availability) and other metrics related to RE's integration into networks. There is also much that governments and other actors can do to create an environment conducive for RE deployment. [11.1, 11.6]

11.1.1 The rationale of renewable energy-specific policies in addition to climate change policies

Renewable energies can provide a host of benefits to society. Some RE technologies are broadly competitive with current market energy prices.

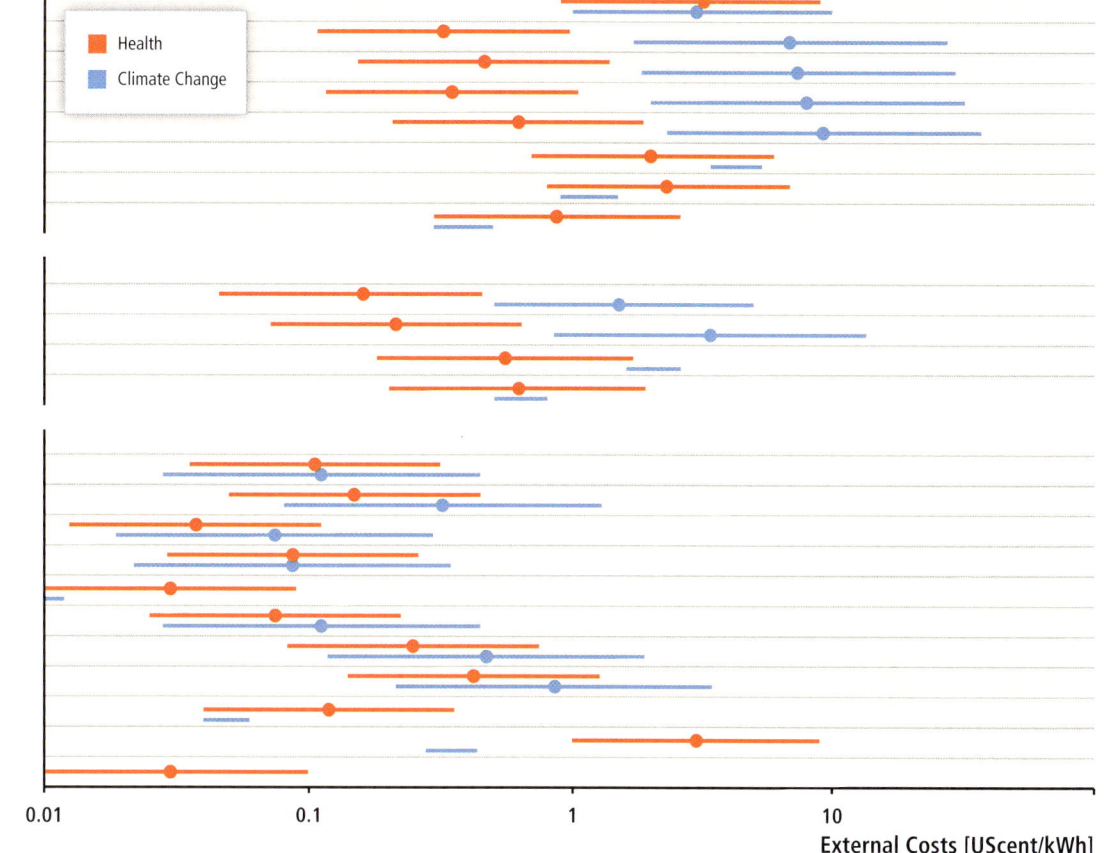

Figure TS.10.15 | Illustration of external costs due to the lifecycle of electricity production based on RE and fossil energy. Note the logarithmic scale of the figure. The black lines indicate the range of the external cost due to climate change and the red lines indicate the range of the external costs due to air pollutant health effects. External costs due to climate change mainly dominate in fossil energy if not equipped with CCS. Comb.C: Combined Cycle; Postcom: Post-Combustion; η: efficiency factor. The results are based on four studies having different assumptions (A–D). The uncertainty for the external costs of health impacts is assumed to be a factor of three. [Figure 10.36]

Of the other RE technologies that are not yet broadly competitive, many can provide competitive energy services in certain circumstances. In most regions of the world, however, policy measures are still required to facilitate an increasing deployment of RE. [11.1, 10.5]

Climate policies (carbon taxes, emissions trading or regulatory policies) decrease the relative costs of low-carbon technologies compared to carbon-intensive technologies. It is questionable, however, whether climate policies (e.g., carbon pricing) alone are capable of promoting RE at sufficient levels to meet the broader environmental, economic and social objectives related to RE. [11.1.1]

Two separate market failures create the rationale for the additional support of innovative RE technologies that have high potential for technological development, even if an emission market (or GHG pricing policy in general) exists. The first market failure refers to the external cost of GHG emissions. The second market failure is in the field of innovation: if firms underestimate the future benefits of investments into learning RE technologies or if they cannot appropriate these benefits, they will invest less than is optimal from a macroeconomic perspective. In addition to GHG pricing policies, RE-specific policies may be appropriate from an economic point of view if the related opportunities for technological development are to be addressed (or if the goals beyond climate change mitigation are pursued). Potentially adverse consequences such as lock-in, carbon leakage and rebound effects should be taken into account in the design of a portfolio of policies. [11.1.1, 11.5.7.3]

11.1.2 Policy timing and strength

The timing, strength and level of coordination of R&D versus deployment policies have implications for the efficiency and effectiveness of the policies, and for the total cost to society in three main ways: 1) whether a country promotes RE immediately or waits until costs have declined further; 2) once a country has decided to support RE, the timing, strength and coordination of when R&D policies give way to deployment policies; and 3) the cost and benefit of accelerated versus slower 'market demand' policy implementation. With regard to the first, in order to achieve full competitiveness with fossil fuel technologies, significant upfront investments in RE will be required until the break-even point is achieved. When those investments should be made depends on the goal. If the international community aims to stabilize global temperature increases at 2°C, then investments in low-carbon technologies must start almost immediately.

11.2 Current trends: Policies, financing and investment

An increasing number and variety of RE policies have driven substantial growth in RE technologies in recent years. Until the early 1990s, few countries had enacted policies to promote RE. Since then, and particularly since the early- to mid-2000s, policies have begun to emerge in a growing number of countries at the municipal, state/provincial and national levels, as well as internationally (see Figure TS.11.1). [1.4, 11.1, 11.2.1, 11.4, 11.5]

Initially, most policies adopted were in developed countries, but an increasing number of developing countries have enacted policy frameworks at various levels of government to promote RE since the late 1990s and early 2000s. Of those countries with RE electricity policies by early 2010, approximately half were developing countries from every region of the world. [11.2.1]

Most countries with RE policies have more than one type of mechanism in place, and many existing policies and targets have been strengthened over time. Beyond national policies, the number of international policies and partnerships is increasing. Several hundred city and local governments around the world have also established goals or enacted renewable promotion policies and other mechanisms to spur local RE deployment. [11.2.1]

The focus of RE policies is shifting from a concentration almost entirely on electricity to include the heating/cooling and transportation sectors. These trends are matched by increasing success in the development of a range of RE technologies and their manufacture and implementation (see Chapters 2 through 7), as well as by a rapid increase in annual investment in RE and a diversification of financing institutions, particularly since 2004/2005. [11.2.2]

In response to the increasingly supportive policy environment, the overall RE sector globally has seen a significant rise in the level of investment since 2004-2005. Financing occurs over what is known as the 'continuum' or stages of technology development. The five segments of the continuum are: 1) R&D; 2) technology development and commercialization; 3) equipment manufacture and sales; 4) project construction; and 5) the refinancing and sale of companies, largely through mergers and acquisitions. Financing has been increasing over time in each of these stages, providing indications of the RE sector's current and expected growth, as follows: [11.2.2]

- Trends in (1) R&D funding and (2) technology investment are indicators of the long- to mid-term expectations for the sector—investments are being made that will begin to pay off in several years' time, once the technology is fully commercialized. [11.2.2.2, 11.2.2.3]

- Trends in (3) manufacturing and sales investment are an indicator of near-term expectations for the sector—essentially, that the growth in market demand will continue. [11.2.2.4]

- Trends in (4) construction investment are an indicator of current sector activity, including the extent to which internalizing costs associated with GHGs can result in new financial flows to RE projects. [11.2.2.5]

- Trends in (5) industry mergers and acquisitions can reflect the overall maturity of the sector, and increasing refinancing activity over time indicates that larger, more conventional investors are entering the sector, buying up successful early investments from first movers. [11.2.2.6]

11.3 Key drivers, opportunities and benefits

Renewable energy can provide a host of benefits to society. In addition to the reduction of CO_2 emissions, governments have enacted RE policies to meet any number of objectives, including the creation of local environmental and health benefits; facilitation of energy access, particularly for rural areas; advancement of energy security goals by diversifying the portfolio of energy technologies and resources; and improving social and economic development through potential employment opportunities and economic growth. [11.3.1–11.3.4]

The relative importance of the drivers for RE differ from country to country, and may vary over time. Energy access has been described as the primary driver in developing countries whereas energy security and environmental concerns have been most important in developed countries. [11.3]

11.4 Barriers to renewable energy policymaking, implementation and financing

RE policies have promoted an increase in RE shares by helping to overcome various barriers that impede technology development and deployment of RE. Barriers specific to RE policymaking, to implementation and to financing (e.g., market failures) may further impede deployment of RE. [1.4, 11.4]

Barriers to making and enacting policy include a lack of information and awareness about RE resources, technologies and policy options; lack of understanding about best policy design or how to undertake energy transitions; difficulties associated with quantifying and internalizing external costs and benefits; and lock-in to existing technologies and policies. [11.4.1]

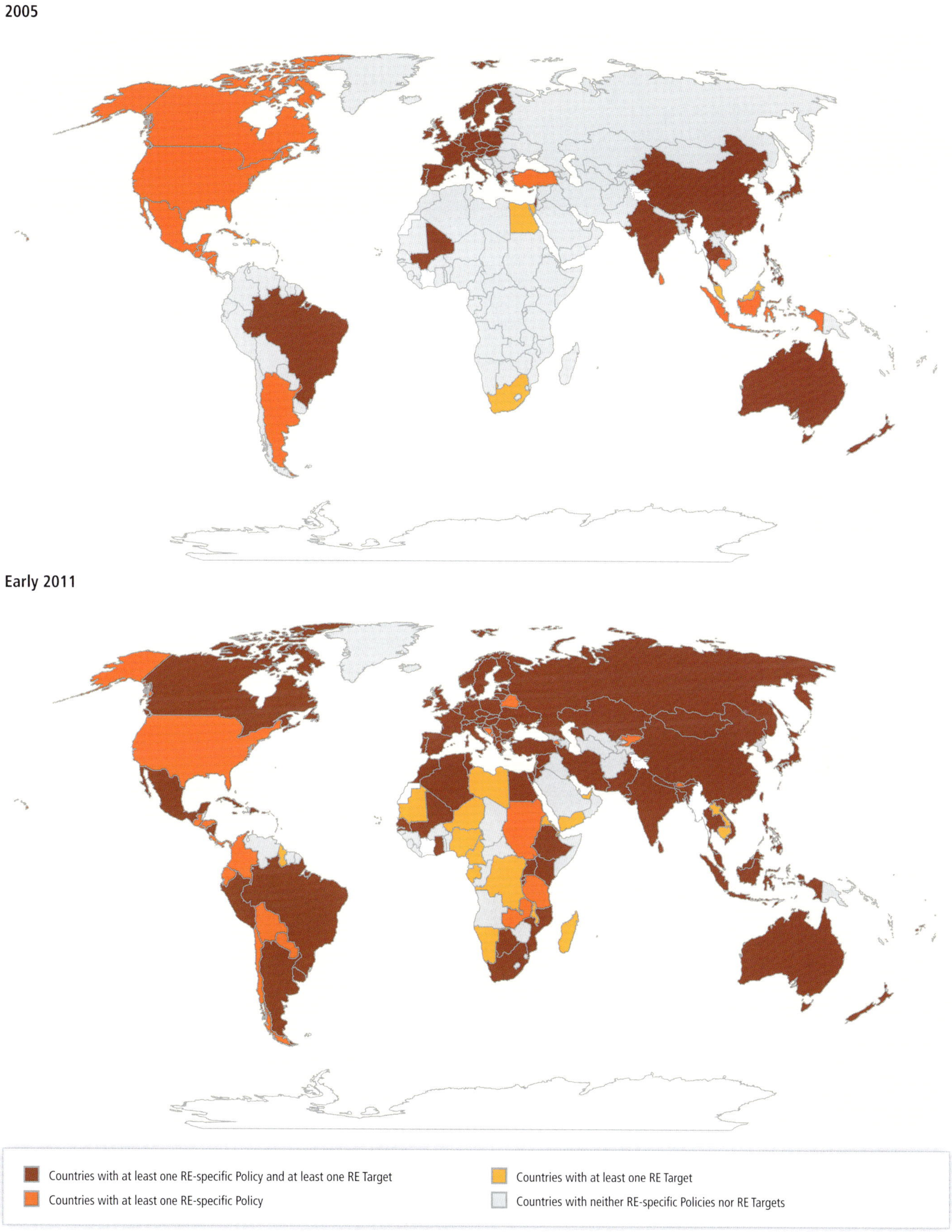

Summaries   Technical Summary

2005

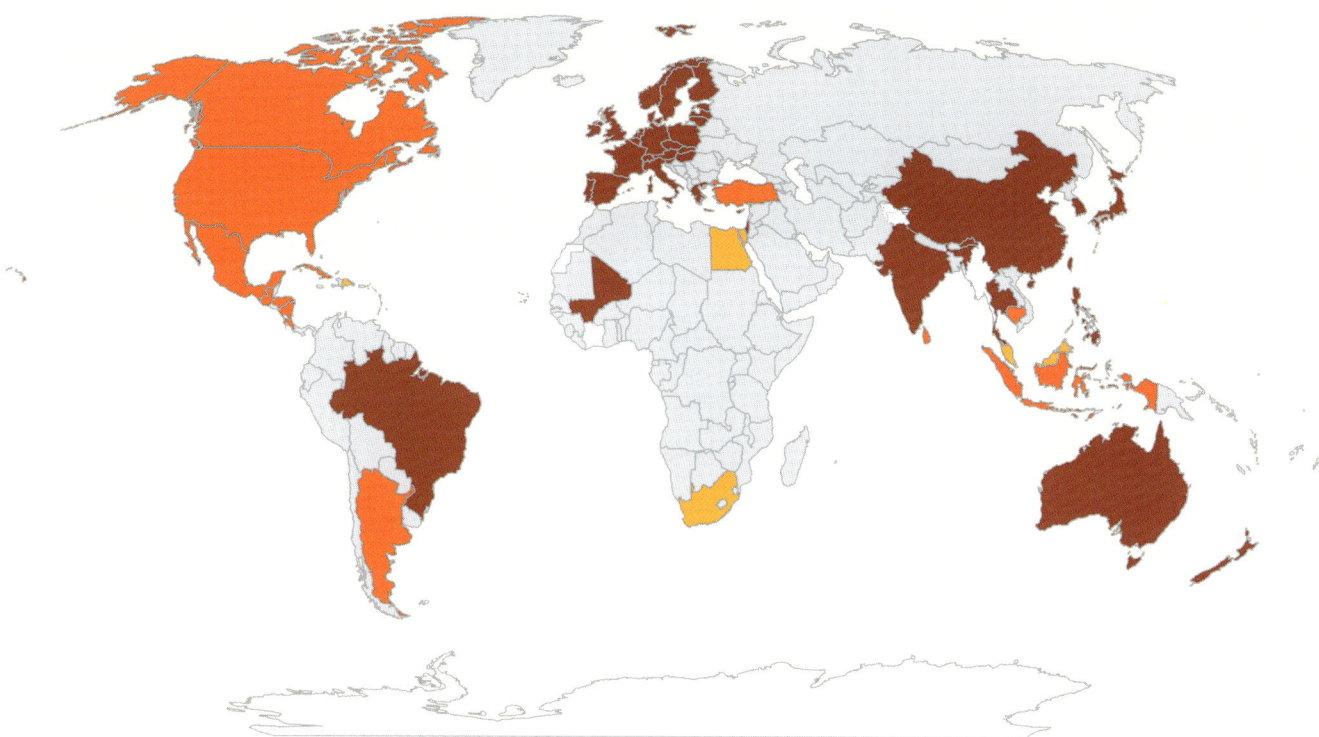

Early 2011

■ Countries with at least one RE-specific Policy and at least one RE Target   ■ Countries with at least one RE Target
■ Countries with at least one RE-specific Policy   ■ Countries with neither RE-specific Policies nor RE Targets

Figure TS.11.1 | Countries with at least one RE target and/or at least one RE-specific policy, in mid-2005 and in early 2011. This figure includes only national-level targets and policies (not municipal or state/provincial) and is not necessarily all-inclusive. [Figure 11.1]

Barriers related to policy implementation include conflicts with existing regulations; lack of skilled workers; and/or lack of institutional capacity to implement RE policies. [11.4.2]

Barriers to financing include a lack of awareness among financiers and lack of timely and appropriate information; issues related to financial structure and project scale; issues related to limited track records; and, in some countries, institutional weakness, including imperfect capital markets and insufficient access to affordable financing, all of which increase perceived risk and thus increase costs and/or make it more difficult to obtain RE project financing. Most importantly, many RE technologies are not economically competitive with current energy market prices, making them financially unprofitable for investors absent various forms of policy support, and thereby restricting investment capital. [11.4.3]

11.5 Experience with and assessment of policy options

Many policy options are available to support RE technologies, from their infant stages to demonstration and pre-commercialization, and through to maturity and wide-scale deployment. These include government R&D policies (supply-push) for advancing RE technologies, and deployment policies (demand-pull) that aim to create a market for RE technologies. Policies could be categorized in a variety of ways and no globally-agreed list of RE policy options or groupings exists. For the purpose of simplification, R&D and deployment policies have been organized within the following categories [11.5]:

- **Fiscal incentive:** actors (individuals, households, companies) are allowed a reduction of their contribution to the public treasury via income or other taxes or are provided payments from the public treasury in the form of rebates or grants.

- **Public finance:** public support for which a financial return is expected (loans, equity) or financial liability is incurred (guarantee); and

- **Regulation:** rule to guide or control conduct of those to whom it applies.

Although targets are a central component of policies, policies in place may not need specific targets to be successful. Further, targets without policies to deliver them are unlikely to be met. [11.5]

The success of policy instruments is determined by how well they are able to achieve various objectives or criteria, including:

- **Effectiveness:** extent to which intended objectives are met;

- **Efficiency:** ratio of outcomes to inputs, or RE targets realized for economic resources spent;

- **Equity:** the incidence and distributional consequences of a policy; and

- **Institutional feasibility:** the extent to which a policy instrument is likely to be viewed as legitimate, gain acceptance, and be adopted and implemented, including the ability to implement a policy once it has been designed and adopted. [11.5.1]

Most literature focuses on effectiveness and efficiency of policies. Elements of specific policy options make them more or less apt to achieve the various criteria, and how these policies are designed and implemented can also determine how well they meet these criteria. The selection of policies and details of their design ultimately will depend on the goals and priorities of policymakers. [11.5.1]

11.5.1 Research and development policies for renewable energy

R&D, innovation, diffusion and deployment of new low-carbon technologies create benefits to society beyond those captured by the innovator, resulting in under-investment in such efforts. Thus, government R&D can play an important role in advancing RE technologies. Not all countries can afford to support R&D with public funds, but in the majority of countries where some level of support is possible, public R&D for RE enhances the performance of nascent technologies so that they can meet the demands of initial adopters. Public R&D also improves existing technologies that already function in commercial environments. [11.5.2]

Government R&D policies include fiscal incentives, such as academic R&D funding, grants, prizes, tax credits, and use of public research centres; as well as public finance, such as soft or convertible loans, public equity stakes, and public venture capital funds. Investments falling under the rubric of R&D span a wide variety of activities along the technology development lifecycle, from RE resource mapping to improvements in commercial RE technologies. [11.5.2]

The success of R&D policies depends on a number of factors, some of which can be clearly determined, and others which are debated in the literature. Successful outcomes from R&D programmes are not solely related to the total amount of funding allocated, but are also related to the consistency of funding from year to year. On-off operations in R&D are detrimental to technical learning, and learning and cost reductions depend on continuity, commitment and organization of effort, and where and how funds are directed, as much as they rely on the scale of effort. In the literature, there is some debate as to the most successful approach to R&D policy in terms of timing: bricolage (progress via research aiming at incremental improvements) versus breakthrough (radical technological advances) with arguments favouring either option or a combination of both. Experience has shown that it is important that subsidies for R&D (and beyond) are designed to have an 'exit-strategy' whereby the subsidies are progressively phased out as the technology commercializes, leaving a functioning and sustainable sector in place. [11.5.2.3]

One of the most robust findings, from both the theoretical literature and technology case studies, is that R&D investments are most effective when complemented by other policy instruments—particularly, but not limited to, policies that simultaneously enhance demand for new RE technologies. Relatively early deployment policies in a technology's development accelerate learning, whether learning through R&D or learning through utilization (as a result of manufacture) and cost reduction. Together, R&D and deployment policies create a positive feedback cycle, inducing private sector investment in R&D (See Figure TS.11.2). [11.5.2.4]

11.5.2 Policies for deployment

Policy mechanisms enacted specifically to promote deployment of RE are varied and can apply to all energy sectors. They include fiscal incentives (grants, energy production payments, rebates, tax credits, reductions and exemptions, variable or accelerated depreciation); public finance (equity investment, guarantees, loans, public procurement); and regulations (quotas, tendering/bidding, FITs, green labelling and green energy purchasing, net metering, priority or guaranteed access, priority dispatch). While regulations and their impacts vary quite significantly from one end-use sector to another, fiscal incentives and public finance apply generally to all sectors. [11.5.3.1]

Fiscal incentives can reduce the costs and risks of investing in RE by lowering the upfront investment costs associated with installation, reducing the cost of production, or increasing the payment received for RE generated. Fiscal incentives also compensate for the various market failures that leave RE at a competitive disadvantage compared to fossil fuels and nuclear energy, and help to reduce the financial burden of investing in RE. [11.5.3.1]

Fiscal incentives tend to be most effective when combined with other types of policies. Incentives that subsidize production are generally preferable to investment subsidies because they promote the desired outcome—energy generation. However, policies must be tailored to particular technologies and stages of maturation, and investment subsidies can be helpful when a technology is still relatively expensive or when the technology is applied at a small scale (e.g., small

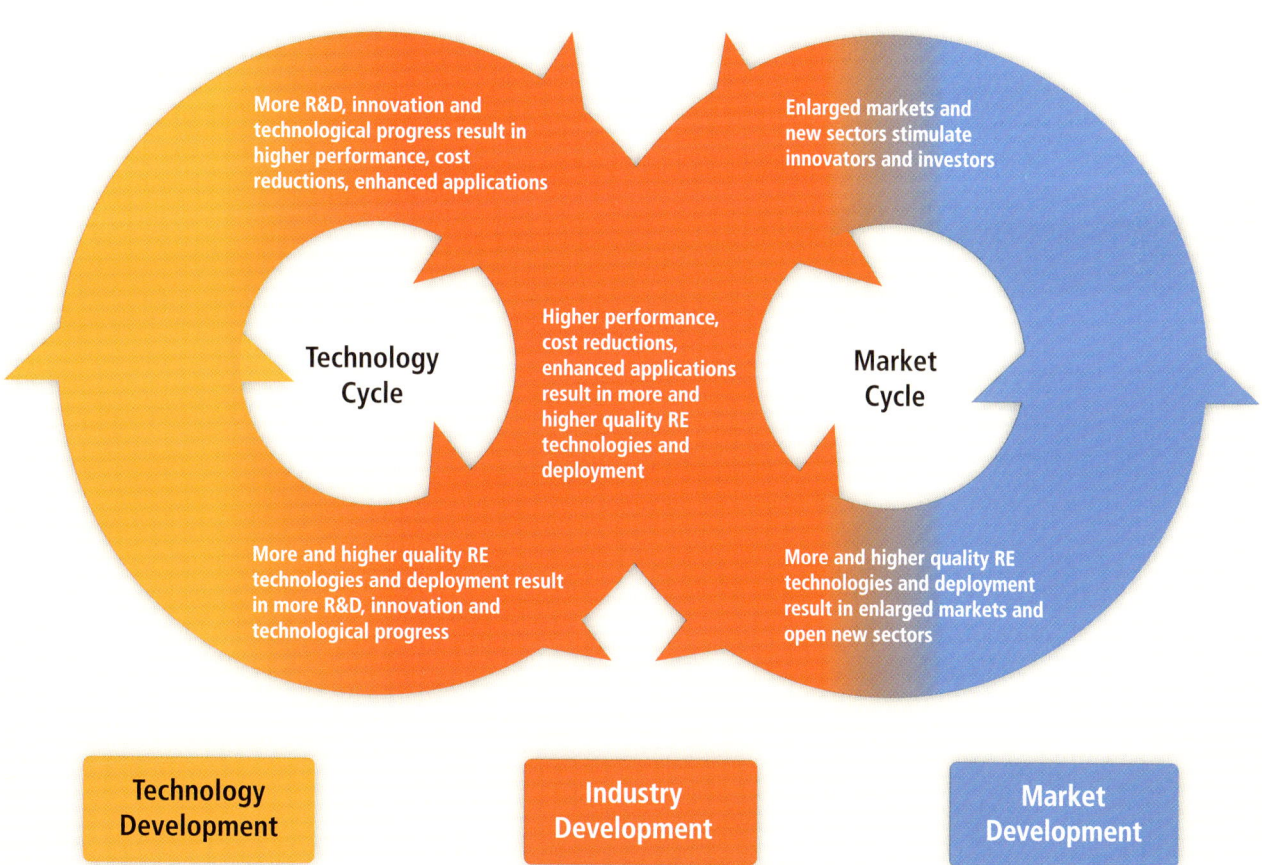

Figure TS.11.2 | The mutually-reinforcing cycles of technology development and market deployment drive down technology costs. [Figure 11.5]

rooftop solar systems), particularly if they are paired with technology standards and certification to ensure minimum quality of systems and installation. Experience with wind energy policies suggests that production payments and rebates may be preferable to tax credits because the benefits of payments and rebates are equal for people of all income levels and thus promote broader investment and use. Also, because they are generally provided at or near the time of purchase or production, they result in more even growth over time (rather than the tendency to invest in most capacity toward the end of a tax period). Tax-based incentives have historically tended to be used to promote only the most mature and cheapest available technologies. Generally, tax credits work best in countries where there are numerous profitable, tax-paying private sector firms that are in a position to take advantage of them. [11.5.3.1]

Public finance mechanisms have a twofold objective: to directly mobilize or leverage commercial investment into RE projects, and to indirectly create scaled-up and commercially sustainable markets for these technologies. In addition to the more traditional public finance policies such as soft loans and guarantees, a number of innovative mechanisms are emerging at various levels of government, including the municipal level. These include financing of RE projects through long-term loans to property owners that allow repayment to be matched with energy savings (for example, Property Assessed Clean Energy in California), and the 'recycling' of government funds for multiple purposes (e.g., using public funds saved through energy efficiency improvements for RE projects). [11.5.3.2]

Public procurement of RE technologies and energy supplies is a frequently cited but not often utilized mechanism to stimulate the market for RE. Governments can support RE development by making commitments to purchase RE for their own facilities or encouraging clean energy options for consumers. The potential of this mechanism is significant: in many nations, governments are the largest consumer of energy, and their energy purchases represent the largest components of public expenditures. [11.5.3.2]

Regulatory policies include quantity- and price-driven policies such as quotas and FITs; quality aspects and incentives; and access instruments such as net metering. Quantity-driven policies set the quantity to be achieved and allow the market to determine the price, whereas price-driven policies set the price and allow the market to determine quantity. Quantity-driven policies can be used in all three end-use sectors in the form of obligations or mandates. Quality incentives include green energy purchasing and green labelling programmes (occasionally mandated by governments, but not always), which provide information to consumers about the quality of energy products to enable consumers to make voluntary decisions and drive demand for RE. [11.5.3.3]

Policies for deployment: Electricity

To date, far more policies have been enacted to promote RE for electricity generation than for heating and cooling or transport. These include fiscal incentives and public finance to promote investment in and generation of RE electricity, as well as a variety of electricity-specific regulatory policies. Although governments use a variety of policy types to promote RE electricity, the most common policies in use are FITs and quotas or Renewable Portfolio Standards (RPS). [11.5.4]

There is a wealth of literature assessing quantity-based (quotas, RPS; and tendering/bidding policies) and price-based (fixed-price and premium-price FITs) policies, primarily quotas and FITs, and with a focus on effectiveness and efficiency criteria. A number of historical studies, including those carried out for the European Commission, have concluded that 'well-designed' and 'well–implemented' FITs have to date been the most efficient (defined as comparison of total support received and generation cost) and effective (ability to deliver an increase in the share of RE electricity consumed) support policies for promoting RE electricity. [11.5.4]

One main reason for the success of well-implemented FITs is that they usually guarantee high investment security due to the combination of long-term fixed-price payments, network connection, and guaranteed grid access for all generation. Well-designed FITs have encouraged both technological and geographic diversity, and have been found to be more suitable for promoting projects of varying sizes. The success of FIT policies depends on the details. The most effective and efficient policies have included most or all of the following elements [11.5.4.3]:

- Utility purchase obligation;
- Priority access and dispatch;
- Tariffs based on cost of generation and differentiated by technology type and project size, with carefully calculated starting values;
- Regular long-term design evaluations and short-term payment level adjustments, with incremental adjustments built into law in order to reflect changes in technologies and the marketplace, to encourage innovation and technological change, and to control costs;
- Tariffs for all potential generators, including utilities;
- Tariffs guaranteed for a long enough time period to ensure an adequate rate of return;
- Integration of costs into the rate base and shared equally across country or region;
- Clear connection standards and procedures to allocate costs for transmission and distribution;
- Streamlined administrative and application processes; and
- Attention to preferred exempted groups, for example, major users on competitiveness grounds or low-income and other vulnerable customers.

Experiences in several countries demonstrate that the effectiveness of quota schemes can be high and compliance levels achieved if RE certificates are delivered under well-designed policies with long-term contracts that mute (if not eliminate) price volatility and reduce risk. However, they have been found to benefit the most mature, least-cost technologies. This effect can be addressed in the design of the policy if different RE options are distinguished or are paired with other incentives. The most effective and efficient quantity-based mechanisms have included most if not all of the following elements, particularly those that help to minimize risk [11.5.4.3]:

- Application to large segment of the market (quota only);
- Clearly defined eligibility rules including eligible resources and actors (applies to quotas and tendering/bidding);
- Well-balanced supply-demand conditions with a clear focus on new capacities—quotas should exceed existing supply but be achievable at reasonable cost (quota only);
- Long-term contracts/specific purchase obligations and end dates, and no time gaps between one quota and the next (quota only);
- Adequate penalties for non-compliance, and adequate enforcement (applies to quotas and tendering/bidding);
- Long-term targets, of at least 10 years (quota only);
- Technology-specific bands or carve-outs to provide differentiated support (applies to quotas and tendering/bidding); and
- Minimum payments to enable adequate return and financing (applies to quotas and tendering/bidding).

Net metering enables small producers to 'sell' into the grid, at the retail rate, any renewable electricity that they generate in excess of their total demand in real time as long as that excess generation is compensated for by excess customer load at other times during the designated netting period. It is considered a low-cost, easily administered tool for motivating customers to invest in small-scale, distributed power and to feed it into the grid, while also benefiting providers by improving load factors if RE electricity is produced during peak demand periods. On its own, however, it is generally insufficient to stimulate significant growth of less competitive technologies like PV at least where generation costs are higher than retail prices. [11.5.4]

Policies for deployment: Heating and cooling

An increasing number of governments are adopting incentives and mandates to advance RE heating and cooling (H/C) technologies. Support for RE H/C presents policymakers with a unique challenge due to the often distributed nature of heat generation. Heating and cooling services can be provided via small- to medium-scale installations that service a single dwelling, or can be used in large-scale applications to provide district heating and cooling. Policy instruments for both RE heating (RE-H) and cooling (RE-C) need to specifically address the more heterogeneous characteristics of resources, including their wide range in scale, varying ability to deliver different levels of temperature, widely distributed demand, relationship to heat load, variability of use, and the absence of a central delivery or trading mechanism. [11.5.5]

The number of policies to support RE sources of heating and cooling has increased in recent years, resulting in increasing generation of RE H/C. However, a majority of support mechanisms have been focused on RE-H. Policies in place to promote RE-H include fiscal incentives such as rebates and grants, tax reductions and tax credits; public finance policies like loans; regulations such as use obligations; and educational efforts. [11.5.5.1–11.5.5.3, 11.6]

To date, fiscal incentives have been the prevalent policy in use, with grants being the most commonly applied. Tax credits available after the installation of a RE-H system (i.e., ex-post) may be logistically advantageous over, for example, grants requiring pre-approval before installation, though there is limited experience with this option. Regulatory mechanisms like use obligations and quotas have attracted increased interest for their potential to encourage growth of RE-H independent of public budgets, though there has been little experience with these policies to date. [11.5.5]

Similar to RE electricity and RE transport, RE H/C policies will be better suited to particular circumstances/locations if, in their design, consideration is given to the state of maturity of the particular technology, of the existing markets and of the existing supply chains. Production incentives are considered be more effective for larger H/C systems, such as district heating grids, than they are for smaller, distributed onsite H/C generation installations for which there are few cost-effective metering or monitoring procedures. [11.5.5]

Though there are some examples of policies supporting RE-C technologies, in general policy aiming to drive deployment of RE-C solely is considerably less well-developed than that for RE-H. Many of the mechanisms described in the above paragraphs could also be applied to RE-C, generally with similar advantages and disadvantages. The lack of experience with deployment policies for RE-C is probably linked to the early levels of technological development of many RE-C technologies. R&D support as well as policy support to develop the early market and supply chains may be of particular importance for increasing the deployment of RE-C technologies in the near future. [11.5.5.4]

Policies for deployment: Transportation

A range of policies has been implemented to support the deployment of RE for transport, though the vast majority of these policies and related experiences have been specific to biofuels. Biofuel support policies aim to promote domestic consumption via fiscal incentives (e.g., tax exemptions for biofuel at the pump) or regulations (e.g., blending mandates), or to promote domestic production via public finance (e.g., loans) for production facilities, via feedstock support or tax incentives (e.g., excise tax exemptions). Most commonly, governments enact a combination of policies. [11.5.6]

Tax incentives are commonly used to support biofuels because they change their cost-competitiveness relative to fossil fuels. They can be installed along the whole biofuel value chain, but are most commonly provided to either biofuel producers (e.g., excise tax exemptions/credits) and/or to end consumers (e.g., tax reductions for biofuels at the pump). [11.5.6]

However, several European and other G8+5 countries have begun gradually shifting from the use of tax breaks for biofuels to blending mandates. It is difficult to assess the level of support under biofuel mandates because prices implied by these obligations are generally not public (in contrast to the electricity sector, for example). While mandates are key drivers in the development and growth of most modern biofuels industries, they are found to be less appropriate for the promotion of specific types of biofuel because fuel suppliers tend to blend low-cost biofuels. By nature, mandates need to be carefully designed and accompanied by further requirements in order to reach a broader level of distributional equity and to minimize potential negative social and environmental impacts. Those countries with the highest share of biofuels in transport fuel consumption have had hybrid systems that combine mandates (including penalties) with fiscal incentives (tax exemptions foremost). [11.5.6]

Synthesis
Some policy elements have been shown to be more effective and efficient in rapidly increasing RE deployment and enabling governments and society to achieve specific targets. The details of policy design and implementation can be as important in determining effectiveness and efficiency as the specific policies that are used. Key policy elements include [11.5.7]:

- Adequate value derived from subsidies, FITs, etc. to cover cost such that investors are able to recover their investment at a rate of return that matches their risk.

- Guaranteed access to networks and markets or at a minimum clearly defined exceptions to that guaranteed access.

- Long-term contracts to reduce risk thereby reducing financing costs.

- Provisions that account for diversity of technologies and applications. RE technologies are at varying levels of maturity and with different characteristics, often facing very different barriers. Multiple RE sources and technologies may be needed to mitigate climate change, and some that are currently less mature and/or more costly than others could play a significant role in the future in meeting energy needs and reducing GHG emissions.

- Incentives that decline predictably over time as technologies and/or markets advance.

- Policy that is transparent and easily accessible so that actors can understand the policy and how it works, as well as what is required to enter the market and/or to be in compliance. Also includes longer-term transparency of policy goals, such as medium- and long-term policy targets.

- Inclusive, meaning that the potential for participation is as broad as possible on both the supply side (traditional producers, distributors of technologies or energy supplies, whether electricity, heat or fuel), and the demand side (businesses, households, etc.), which can 'self-generate' with distributed RE, enabling broader participation that unleashes more capital for investment, helps to build broader public support for RE, and creates greater competition.

- Attention to preferred exempted groups, for example, major users on competitiveness grounds or low-income and vulnerable customers on equity and distributional grounds.

It is also important to recognize that there is no one-size-fits-all policy, and policymakers can benefit from the ability to learn from experience and adjust programmes as necessary. Policies need to respond to local political, economic, social, ecological, cultural and financial needs and conditions, as well as factors such as the level of technological maturity, availability of affordable capital, and the local and national RE resource base. In addition, a mix of policies is generally needed to address the various barriers to RE. Policy frameworks that are transparent and sustained—from predictability of a specific policy, to pricing of carbon and other externalities, to long-term targets for RE—have been found to be crucial for reducing investment risks and facilitating deployment of RE and the evolution of low-cost applications. [11.5.7]

Macroeconomic impacts of renewable energy policies
Payment for supply-push type RE support tends to come from public budgets (multinational, national, local), whereas the cost of demand-pull mechanisms often lands on the end users. For example, if a renewable electricity policy is added to a countries' electricity sector, this additional cost is often borne by electricity consumers, although exemptions or re-allocations can reduce costs for industrial or vulnerable customers where necessary. Either way, there are costs to be paid. If the goal is to transform the energy sector over the next several decades, then it is important to minimize costs over this entire period; it is also important to include all costs and benefits to society in that calculation. [11.5.7.2]

Conducting an integrated analysis of costs and benefits of RE is extremely demanding because so many elements are involved in determining net impacts. Effects fall into three categories: direct and indirect costs of the system as well as benefits of RE expansion; distributional effects (in which economic actors or groups enjoy benefits or suffer burdens as a result of RE support); and macroeconomic aspects such as impacts on GDP or employment. For example, RE policies provide opportunities for potential economic growth and job creation, but measuring net effects is complex and uncertain because the additional costs of RE support create distributional and budget effects on the economy. Few studies have examined such impacts on national or regional economies; however, those that have been carried out have generally found net positive economic impacts. [11.3.4, 11.5.7.2]

Interactions and potential unintended consequences of renewable energy and climate policies
Due to overlapping drivers and rationales for RE deployment and overlapping jurisdictions (local, national, international) substantial interplay

may occur among policies at times with unintended consequences. Therefore, a clear understanding of the interplay among policies and the cumulative effects of multiple policies is crucial. [11.3, 11.5.7, 11.6.2]

If not applied globally and comprehensively, both carbon pricing and RE policies create risks of 'carbon leakage', where RE policies in one jurisdiction or sector reduce the demand for fossil fuel energy in that jurisdiction or sector, which *ceteris paribus* reduces fossil fuel prices globally and hence increases demand for fossil energy in other jurisdictions or sectors. Even if implemented globally, suboptimal carbon prices and RE policies could potentially lead to higher carbon emissions. For example, if fossil fuel resource owners fear more supportive RE deployment policies in the long term, they could increase resource extraction as long as RE support is moderate. Similarly, the prospect of future carbon price increases may encourage owners of oil and gas wells to extract resources more rapidly, while carbon taxes are lower, undermining policymakers' objectives for both the climate and the spread of RE technology. The conditions of such a 'green paradox' are rather specific: carbon pricing would have to begin at low levels and increase rapidly. Simultaneously, subsidized RE would have to remain more expensive than fossil fuel-based technologies. However, if carbon prices and RE subsidies begin at high levels from the beginning, such green paradoxes become unlikely. [11.5.7]

The cumulative effect of combining policies that set fixed carbon prices, like carbon taxes, with RE subsidies is largely additive: in other words, extending a carbon tax with RE subsidies decreases emissions and increases the deployment of RE. However, the effect on the energy system of combining endogenous-price policies, like emissions trading and/or RE quota obligations, is usually not as straightforward. Adding RE policies on top of an emissions trading scheme usually reduces carbon prices which, in turn, makes carbon-intensive (e.g., coal-based) technologies more attractive compared to other non-RE abatement options such as natural gas, nuclear energy and/or energy efficiency improvements. In such cases, although overall emissions remain fixed by the cap, RE policies reduce the costs of compliance and/or improve social welfare only if RE technologies experience specific externalities and market barriers to a greater extent than other energy technologies. [11.5.7]

Finally, RE policies alone (i.e., without carbon pricing) are not necessarily an efficient instrument to reduce carbon emissions because they do not provide enough incentives to use all available least-cost mitigation options, including non-RE low-carbon technologies and energy efficiency improvements. [11.5.7]

11.6 Enabling environment and regional issues

RE technologies can play a greater role in climate change mitigation if they are implemented in conjunction with broader 'enabling' policies that can facilitate change in the energy system. An 'enabling' environment encompasses different institutions, actors (e.g., the finance community, business community, civil society, government), infrastructures (e.g., networks and markets), and political outcomes (e.g., international agreements/cooperation, climate change strategies) (see Table TS.11.1). [11.6]

A favourable or 'enabling' environment for RE can be created by encouraging innovation in the energy system; addressing the possible interactions of a given policy with other RE policies as well as with other non-RE policies; easing the ability of RE developers to obtain finance and to successfully site a project; removing barriers for access to networks and markets for RE installations and output; enabling technology transfer and capacity building; and by increasing education and awareness raising at the institutional level and within communities. In turn, the existence of an 'enabling' environment can increase the efficiency and effectiveness of policies to promote RE. [11.6.1–11.6.8]

A widely accepted conclusion in innovation literature is that established socio-technical systems tend to narrow the diversity of innovations because the prevailing technologies develop a fitting institutional environment. This may give rise to strong path dependencies and exclude (or lock out) rivalling and potentially better-performing alternatives. For these reasons, socio-technical system change takes time, and it involves change that is systemic rather than linear. RE technologies are being integrated into an energy system that, in much of the world, was constructed to accommodate the existing energy supply mix. As a result, infrastructure favours the currently dominant fuels, and existing lobbies and interests all need to be taken into account. Due to the intricacies of technological change, it is important that all levels of government (from local through to international) encourage RE development through policies, and that nongovernmental actors also be involved in policy formulation and implementation. [11.6.1]

Government policies that complement each other are more likely to be successful, and the design of individual RE policies will also affect the success of their coordination with other policies. Attempting to actively promote the complementarities of policies across multiple sectors—from energy to agriculture to water policy, etc.—while also considering the independent objectives of each, is not an easy task and may create win-win and/or win-lose situations, with possible trade-offs. This implies a need for strong central coordination to eliminate contradictions and conflicts among sectoral policies and to simultaneously coordinate action at more than one level of governance. [11.6.2]

A broader enabling environment includes a financial sector that can offer access to financing on terms that reflect the specific risk/reward profile of a RE technology or project. The cost of financing and access to it depends on the broader financial market conditions prevalent at the time of investment, and on the specific risks of a project, technology, and actors involved. Beyond RE-specific policies, broader conditions can

Table TS.11.1 | Factors and participants contributing to a successful RE governance regime. [Table 11.4]

Dimensions of an Enabling Environment >> Factors and actors contributing to the success of RE policy	Section 11.6.2 Integrating Policies (national/supranational policies)	Section 11.6.3 Reducing Financial and Investment Risk	Section 11.6.4 Planning and Permitting at the local level	Section 11.6.5 Providing infrastructures networks and markets for RE technology	Section 11.6.6 Technology Transfer and Capacity Building	Section 11.6.7 Learning from actors beyond government
Institutions	Integrating RE policies with other policies at the design level reduces potential for conflict among government policies	Development of financing institutions and agencies can aid cooperation between countries, provide soft loans or international carbon finance (CDM). Long-term commitment can reduce the perception of risk	Planning and permitting processes enable RE policy to be integrated with non-RE policies at the local level	Policymakers and regulators can enact incentives and rules for networks and markets, such as security standards and access rules	Reliability of RE technologies can be ensured through certification Institutional agreements enable technology transfer	Openness to learning from other actors can complement design of policies and enhance their effectiveness by working within existing social conditions
Civil society (individuals, households, NGOs, unions …)	Municipalities or cities can play a decisive role in integrating state policies at the local level	Community investment can share and reduce investment risk Public-private partnerships in investment and project development can contribute to reducing risks associated with policy instruments Appropriate international institutions can enable an equitable distribution of funds	Participation of civil society in local planning and permitting processes might allow for selection of the most socially relevant RE projects	Civil society can become part of supply networks through co-production of energy and new decentralized models.	Local actors and NGOs can be involved in technology transfer through new business models bringing together multi-national companies / NGOs / Small and Medium Enterprises	Civil society participation in open policy processes can generate new knowledge and induce institutional change Municipalities or cities may develop solutions to make RE technology development possible at the local level People (individually or collectively) have a potential for advancing energy-related behaviours when policy signals and contextual constraints are coherent
Finance and business communities		Public private partnerships in investment and project development can contribute to reducing risks associated with policy instruments	RE project developers can offer know-how and professional networks in : i) aligning project development with planning and permitting requirements ; ii) adapting planning and permitting processes to local needs and conditions Businesses can be active in lobbying for coherent and integrated policies	Clarity of network and market rules improves investor confidence	Financing institutions and agencies can partner with national governments, provide soft loans or international carbon finance (CDM).	Multi-national companies can involve local NGOs or SMEs as partners in new technology development (new business models) Development of corporations and international institutions reduces risk of investment
Infrastructures	Policy integration with network and market rules can enable development of infrastructure suitable for a low-carbon economy	Clarity of network and market rules reduces risk of investment and improves investor confidence		Clear and transparent network and market rules are more likely to lead to infrastructures complementary to a low-carbon future		City and community level frameworks for the development of long-term infrastructure and networks can sustain the involvement of local actors in policy development

Continued next Page →

Dimensions of an Enabling Environment >> Factors and actors contributing to the success of RE policy	Section 11.6.2 Integrating Policies (national/ supranational policies)	Section 11.6.3 Reducing Financial and Investment Risk	Section 11.6.4 Planning and Permitting at the local level	Section 11.6.5 Providing infrastructures networks and markets for RE technology	Section 11.6.6 Technology Transfer and Capacity Building	Section 11.6.7 Learning from actors beyond government
Politics (international agreements / cooperation, climate change strategy, technology transfer...)	Supra-national guidelines (e.g., EU on "streamlining", ocean planning, impact study) may contribute to integrating RE policy with other policies	Long-term political commitment to RE policy reduces investors risk in RE projects	Supra- national guidelines may contribute to evolving planning and permitting processes	Development cooperation helps sustain infrastructure development and allows easier access to low-carbon technologies	CDMs, Intellectual property rights (IPR) and patent agreements can contribute to technology transfer	Appropriate input from non-government institutions stimulates more agreements that are socially connected UNFCCC process mechanisms such as Expert Group on Technology Transfer (EGTT), the Global Environment Facility (GEF), and the Clean Development Mechanism (CDM) and Joint Implementation (JM) may provide guidelines to facilitate the involvement of non-state actors in RE policy development

include political and currency risks, and energy-related issues such as competition for investment from other parts of the energy sector, and the state of energy sector regulations or reform. [11.6.3]

The successful deployment of RE technologies to date has depended on a combination of favourable planning procedures at both national and local levels. Universal procedural fixes, such as 'streamlining' of permitting applications, are unlikely to resolve conflicts among stakeholders at the level of project deployment because they would ignore place- and scale-specific conditions. A planning framework to facilitate the implementation of RE might include the following elements: aligning stakeholder expectations and interests; learning about the importance of context for RE deployment; adopting benefit-sharing mechanisms; building collaborative networks; and implementing mechanisms for articulating conflict for negotiation. [11.6.4]

After a RE project receives planning permission, investment to build it is only forthcoming once its economic connection to a network is agreed; when it has a contract for the 'off-take' of its production into the network; and when its sale of energy, usually via a market, is assured. The ability, ease and cost of fulfilling these requirements is central to the feasibility of a RE project. Moreover, the methods by which RE is integrated into the energy system will have an effect on the total system cost of RE integration and the cost of different scenario pathways. In order to ensure the timely expansion and reinforcement of infrastructure for and connection of RE projects, economic regulators may need to allow 'anticipatory' or 'proactive' network investment and/or allow projects to connect in advance of full infrastructure reinforcement. [11.6.5, 8.2.1.3]

For many countries, a major challenge involves gaining access to RE technologies. Most low-carbon technologies, including RE technologies, are developed and concentrated in a few countries. It has been argued that many developing nations are unlikely to 'leapfrog' pollution-intensive stages of industrial development without access to clean technologies that have been developed in more advanced economies. However, technologies such as RE technologies typically do not flow across borders unless environmental policies in the recipient country provide incentives for their adoption. Further, technology transfer should not replace but rather should complement domestic efforts at capacity building. In order to have the capacity to adapt, install, maintain, repair and improve on RE technologies in communities without ready access to RE, investment in technology transfer must be complemented by investment in community-based extension services that provide expertise, advice and training regarding installation, technology adaptation, repair and maintenance. [11.6.6]

In addition to technology transfer, institutional learning plays an important role in advancing deployment of RE. Institutional learning is conducive to institutional change, which provides space for institutions to improve the choice and design of RE policies. It also encourages a stronger institutional capacity at the deeper, often more local, level where numerous decisions are made on siting and investments in RE projects. Institutional learning can occur if policymakers can draw on nongovernmental actors, including private actors (companies, etc.) and civil society for collaborative approaches in policymaking. Information and education are often emphasized as key policy tools for influencing energy-related behaviours. However, the effectiveness of education- and information-based policies is limited by contextual factors, which cautions against an over-reliance on information- and education-based policies alone. Changes in energy-related behaviours are the outcome of a process in which personal norms or attitudes interact with prices, policy signals, and the RE technologies themselves, as well as the social context in which individuals find themselves. These contextual factors point to the importance of collective action as a more effective, albeit more complex medium for change than individual action. This supports coordinated, systemic policies that go beyond narrow 'attitude-behaviour-change' policies if policymakers wish to involve individuals in the RE transition. [11.6.7, 11.6.8]

11.7 A structural shift

If decision makers intend to increase the share of RE and, at the same time, meet ambitious climate mitigation targets, then long-standing commitments and flexibility to learn from experience will be critical. To achieve GHG concentration stabilization levels with high shares of RE, a structural shift in today's energy systems will be required over the next few decades. Such a transition to low-carbon energy differs from previous energy transitions (e.g., from wood to coal, or coal to oil) because the available time span is restricted to a few decades, and because RE must develop and integrate into a system constructed in the context of an existing energy structure that is very different from what might be required under higher penetration RE futures. [11.7]

A structural shift towards a world energy system that is mainly based on renewable energy might begin with a prominent role for energy efficiency in combination with RE. This requires, however, a reasonable carbon pricing policy in the form of a tax or emission trading scheme that avoids carbon leakage and rebound effects. Additional policies are required that extend beyond R&D to support technology deployment; the creation of an enabling environment that includes education and awareness raising; and the systematic development of integrative policies with broader sectors, including agriculture, transportation, water management and urban planning. [11.6, 11.7] The policy frameworks that induce the most RE investment are those designed to reduce risks and enable attractive returns, and to provide stability over a time frame relevant to the investment. [11.5] The appropriate and reliable mix of instruments is even more important where energy infrastructure is not yet developed and energy demand is expected to increase significantly in the future. [11.7]

III

Chapters 1 to 11

1. Renewable Energy and Climate Change

Coordinating Lead Authors:

William Moomaw (USA), Francis Yamba (Zambia)

Lead Authors:

Masayuki Kamimoto (Japan), Lourdes Maurice (USA), John Nyboer (Canada), Kevin Urama (Kenya/Nigeria), Tony Weir (Fiji/Australia)

Contributing Authors:

Thomas Bruckner (Germany), Arnulf Jäger-Waldau (Italy/Germany), Volker Krey (Austria/Germany), Ralph Sims (New Zealand), Jan Steckel (Germany), Michael Sterner (Germany), Russell Stratton (USA), Aviel Verbruggen (Belgium), Ryan Wiser (USA)

Review Editors:

Jiahua Pan (China) and Jean-Pascal van Ypersele (Belgium)

This chapter should be cited as:

Moomaw, W., F. Yamba, M. Kamimoto, L. Maurice, J. Nyboer, K. Urama, T. Weir, 2011: Introduction. In IPCC Special Report on Renewable Energy Sources and Climate Change Mitigation [O. Edenhofer, R. Pichs-Madruga, Y. Sokona, K. Seyboth, P. Matschoss, S. Kadner, T. Zwickel, P. Eickemeier, G. Hansen, S. Schlömer, C.von Stechow (eds)], Cambridge University Press, Cambridge, United Kingdom and New York, NY, USA.

Table of Contents

Executive Summary		164
1.1	**Background**	167
1.1.1	Introduction	167
1.1.2	The Special Report on Renewable Energy Sources and Climate Change Mitigation	167
1.1.3	Climate change	168
1.1.4	Drivers of carbon dioxide emissions	169
1.1.5	Renewable energy as an option to mitigate climate change	172
1.1.6	Options for mitigation	174
1.1.7	Trends in international policy on renewable energy	177
1.1.8	Advancing knowledge about renewable energy	178
1.1.9	Metrics and definitions	178
1.2	**Summary of renewable energy resources**	178
1.2.1	Definition, conversion and application of renewable energy	178
1.2.2	Theoretical potential of renewable energy	181
1.2.3	Technical potential of renewable energy technologies	181
1.2.4	Special features of renewable energy with regard to integration	184
1.2.5	Energy efficiency and renewable energy	185
1.3	**Meeting energy service needs and current status**	187
1.3.1	Current renewable energy flows	187
1.3.2	Current cost of renewable energy	187
1.3.3	Regional aspects of renewable energy	190

1.4	**Opportunities, barriers and issues**	190
1.4.1	**Opportunities**	191
1.4.1.1	Social and economic development	191
1.4.1.2	Energy access	191
1.4.1.3	Energy security	191
1.4.1.4	Climate change mitigation and reduction of environmental and health impacts	192
1.4.2	**Barriers**	192
1.4.2.1	Market failures and economic barriers	193
1.4.2.2	Informational and awareness barriers	194
1.4.2.3	Socio-cultural barriers	195
1.4.2.4	Institutional and policy barriers	196
1.4.3	**Issues**	196
1.5	**Role of policy, research and development, deployment, scaling up and implementation strategies**	197
1.5.1	**Policy options: trends, experience and assessment**	197
1.5.2	**Enabling environment**	199
1.5.2.1	Complementing renewable energy policies and	199
	non-renewable energy policies	199
1.5.2.2	Providing infrastructure, networks and markets for renewable energy	199
1.5.3	**A structural shift**	199

References 200

Appendix to Chapter 1 206

Executive Summary

All societies require energy services to meet basic human needs (e.g., lighting, cooking, space comfort, mobility, communication) and to serve productive processes. For development to be sustainable, delivery of energy services needs to be secure and have low environmental impacts. Sustainable social and economic development requires assured and affordable access to the energy resources necessary to provide essential and sustainable energy services. This may mean the application of different strategies at different stages of economic development. To be environmentally benign, energy services must be provided with low environmental impacts and low greenhouse gas (GHG) emissions. However, 85% of current primary energy driving global economies comes from the combustion of fossil fuels and consumption of fossil fuels accounts for 56.6% of all anthropogenic GHG emissions.

Renewable energy sources play a role in providing energy services in a sustainable manner and, in particular, in mitigating climate change. This Special Report on Renewable Energy Sources and Climate Change Mitigation explores the current contribution and potential of renewable energy (RE) sources to provide energy services for a sustainable social and economic development path. It includes assessments of available RE resources and technologies, costs and co-benefits, barriers to up-scaling and integration requirements, future scenarios and policy options.

GHG emissions associated with the provision of energy services are a major cause of climate change. The IPCC Fourth Assessment Report (AR4) concluded that "Most of the observed increase in global average temperature since the mid-20th century is *very likely* due to the observed increase in anthropogenic greenhouse gas concentrations." Concentrations of CO_2 have continued to grow and by the end of 2010 had reached 390 ppm CO_2 or 39% above pre-industrial levels.

The long-term baseline scenarios reviewed for the AR4 show that the expected decrease in the energy intensity will not be able to compensate for the effects of the projected increase in the global gross domestic product. As a result, most of the scenarios exhibit a strong increase in primary energy supply throughout this century. In the absence of any climate policy, the overwhelming majority of the baseline scenarios exhibit considerably higher emissions in 2100 compared to 2000, implying rising CO_2 concentrations and, in turn, enhanced global warming. Depending on the underlying socioeconomic scenarios and taking into account additional uncertainties, global mean temperature is expected to rise and to approach a level between 1.1°C and 6.4°C over the 1980 to 1999 average by the end of this century.

To avoid adverse impacts of such climate change on water resources, ecosystems, food security, human health and coastal settlements with potentially irreversible abrupt changes in the climate system, the Cancun Agreements call for limiting global average temperature rises to no more than 2°C above pre-industrial values, and agreed to consider limiting this rise to 1.5°C. In order to be confident of achieving an equilibrium temperature increase of only 2°C to 2.4°C, GHG concentrations would need to be stabilized in the range of 445 to 490 ppm CO_2eq in the atmosphere.

There are multiple means for lowering GHG emissions from the energy system, while still providing desired energy services. RE technologies are diverse and can serve the full range of energy service needs. Various types of RE can supply electricity, thermal energy and mechanical energy, as well as produce fuels that are able to satisfy multiple energy service needs. RE is any form of energy from solar, geophysical or biological sources that is replenished by natural processes at a rate that equals or exceeds its rate of use. Unlike fossil fuels, most forms of RE produce little or no CO_2 emissions.

The contribution RE will provide within the portfolio of low carbon technologies heavily depends on the economic competition between these technologies, their relative environmental burden (beyond climate change), as well as on security and societal aspects. A comprehensive evaluation of any portfolio of mitigation options would involve an evaluation of their respective mitigation potential as well as all associated risks, costs and their contribution to sustainable development. Even without a push for climate change mitigation, scenarios that are examined in this report find that the increasing demand for energy services is expected to drive RE to levels exceeding today's energy usage.

On a global basis, it is estimated that RE accounted for 12.9% of the total 492 EJ of primary energy supply in 2008. The largest RE contributor was biomass (10.2%), with the majority (roughly 60%) of the biomass fuel used in traditional cooking and heating applications in developing countries but with rapidly increasing use of modern biomass as well.[1] Hydropower represented 2.3%, whereas other RE sources accounted for 0.4%. In 2008, RE contributed approximately 19% of global electricity supply (16% hydropower, 3% other RE), biofuels contributed 2% of global road transport fuel supply, and traditional biomass (17%), modern biomass (8%), solar thermal and geothermal energy (2%) together fuelled 27% of the total global demand for heat. The contribution of RE to primary energy supply varies substantially by country and region. Scenarios of future low greenhouse gas futures consider RE and RE in combination with nuclear, and coal and natural gas with carbon capture and storage.

While the RE share of global energy consumption is still relatively small, deployment of RE has been increasing rapidly in recent years. Of the approximately 300 GW of new electricity generating capacity added globally over the two-year period from 2008 to 2009, 140 GW came from RE additions. Collectively, developing countries hosted 53% of global RE power generation capacity in 2009. Under most conditions, increasing the share of RE in the energy mix will require policies to stimulate changes in the energy system. Government policy, the declining cost of many RE technologies, changes in the prices of fossil fuels and other factors have supported the continuing increase in the use of RE. These developments suggest the possibility that RE could play a much more prominent role in both developed and developing countries over the coming decades.

Some RE technologies can be deployed at the point of use (decentralized) in rural and urban environments, whereas others are primarily employed within large (centralized) energy networks. Though many RE technologies are technically mature and are being deployed at significant scale, others are in an earlier phase of technical maturity and commercial deployment.

The theoretical potential for RE greatly exceeds all the energy that is used by all economies on Earth. The global technical potential of RE sources will also not limit continued market growth. A wide range of estimates are provided in the literature but studies have consistently found that the total global technical potential for RE is substantially higher than both current and projected future global energy demand. The technical potential for solar energy is the highest among the RE sources, but substantial technical potential exists for all forms of RE. The absolute size of the global technical potential for RE as a whole is unlikely to constrain RE deployment.

Some RE, including wind and solar power, are variable and may not always be available for dispatch when needed. The energy density of some RE is also relatively lower, so that reducing the delivered energy needed to supply end-use energy services is especially important for RE even though benefiting all forms of energy.

The levelized cost of energy for many RE technologies is currently higher than existing energy prices, though in various settings RE is already economically competitive. Ranges of recent levelized costs of energy for selected commercially available RE technologies are wide, depending on a number of factors including, but not limited to, technology characteristics, regional variations in cost and performance and differing discount rates.

RE may provide a number of opportunities and can not only address climate change mitigation but may also address sustainable and equitable economic development, energy access, secure energy supply and local environmental and health impacts. Market failures, up-front costs, financial risk, lack of data as well as capacities and public and institutional awareness, perceived social norms and value structures, present infrastructure and current

---

1   Not accounted for here or in official databases is the estimated 20 to 40% of additional traditional biomass used in informal sectors (Section 2.1).

energy market regulation, inappropriate intellectual property laws, trade regulations, lack of amenable policies and programs, lower power of RE and land use conflicts are amongst existing barriers and issues to expanding the use of RE.

Some governments have successfully introduced a variety of RE policies, motivated by a variety of factors, to address these various components of RE integration into the energy system. These policies have driven escalated growth in RE technologies in recent years. These policies can be categorized as fiscal incentives, public finance and regulation. They typically address two market failures: 1) the external cost of GHG emissions are not priced at an appropriate level; and 2) RE creates benefits to society beyond those captured by the innovator, leading to under-investment in such efforts. Several studies have concluded that some feed-in tariffs have been effective and efficient at promoting RE electricity. Quota policies can be effective and efficient if designed to reduce risk. An increasing number of governments are adopting fiscal incentives for RE heating and cooling. In the transportation sector, RE fuel mandates or blending requirements are key drivers in the development of most modern biofuel industries. Policies have influenced the development of an international biofuel trade. One important challenge will be finding a way for RE and carbon-pricing policies to interact such that they take advantage of synergies rather than trade-offs. RE technologies can play a greater role if they are implemented in conjunction with 'enabling' policies.

1.1 Background

1.1.1 Introduction

All societies require energy services to meet basic human needs (e.g., lighting, cooking, space comfort, mobility, communication) and to serve productive processes. The quality of energy is important to the development process (Cleveland et al., 1984; Brookes, 2000; Kaufmann, 2004). For development to be sustainable, delivery of energy services needs to be secure and have low environmental impacts. Sustainable social and economic development requires assured and affordable access to the energy resources necessary to provide essential and sustainable energy services. This may mean the application of different strategies at different stages of economic development. To be environmentally benign, energy services must be provided with low environmental impacts, including GHG emissions.

The IPCC Fourth Assessment Report (AR4) reported that fossil fuels provided 85% of the total primary energy in 2004 (Sims et al., 2007),[2] which is the same value as in 2008 (IEA 2010a; Table A.II.1). Furthermore, the combustion of fossil fuels accounted for 56.6% of all anthropogenic GHG emissions (CO_2eq) in 2004 (Rogner et al., 2007).[3] To maintain both a sustainable economy that is capable of providing essential goods and services to the citizens of both developed and developing countries, and to maintain a supportive global climate system, requires a major shift in how energy is produced and utilized (Nfah et al., 2007; Kankam and Boon, 2009). However, renewable energy technologies, which release much lower amounts of CO_2 than fossil fuels are growing. Chapter 10 examines more than 100 scenarios in order to explore the potential for RE to contribute to the development of a low-carbon future.

1.1.2 The Special Report on Renewable Energy Sources and Climate Change Mitigation

Renewable energy (RE) sources play a role in providing energy services in a sustainable manner and, in particular, in mitigating climate change. This Special Report on *Renewable Energy Sources and Climate Change Mitigation* explores the current contribution and potential of RE sources to provide energy services for a sustainable social and economic development path. It includes assessments of available RE resources and technologies, costs and co-benefits, barriers to up-scaling and integration requirements, future scenarios and policy options. It consists of 11 chapters (Figure 1.1). Chapter 1 provides an overview of RE and climate change; Chapters 2 through 7 provide information on six types of RE technologies (biomass, solar, geothermal, hydro, ocean and wind)

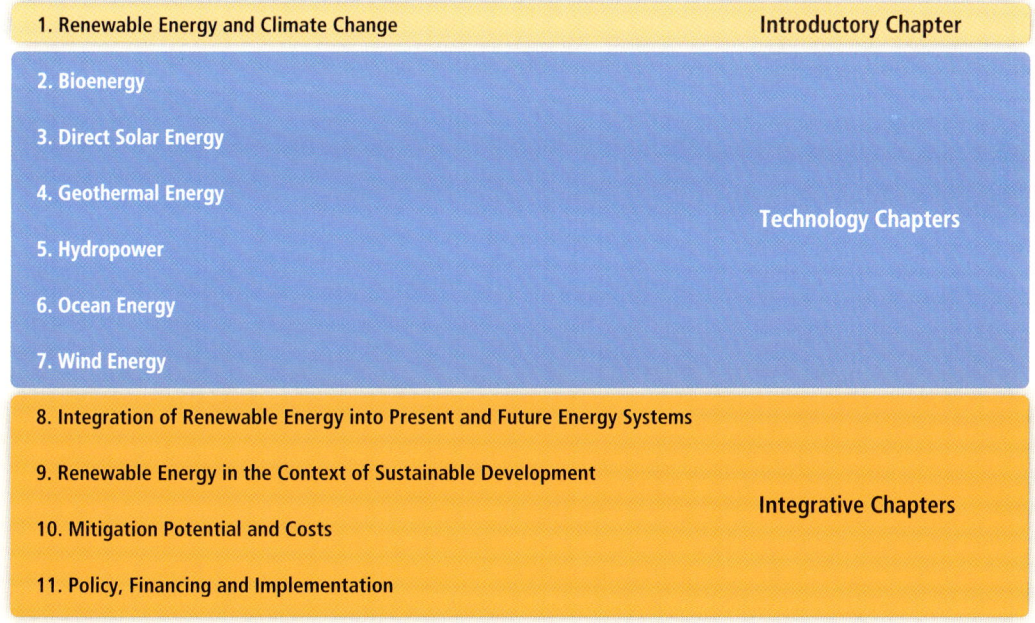

Figure 1.1 | Structure of the report.

---

[2] The number from the AR4 is 80% and has been converted from the physical content method for energy accounting to the direct equivalent method, as the latter method is used in this report. Please refer to Section 1.1.9 and Annex II (Section A.II.4) for methodological details.

[3] The contributions from other sources and/or gases (see Figure 1.1b in Rogner et al., 2007) are: CO_2 from deforestation, decay of biomass etc. (17.3%), CO_2 from other (2.8%), CH_4 (14.3%), N_2O (7.9%) and fluorinated gases (1.1%). For further information on sectoral emissions, including from forestry, see also Figure 1.3b in Rogner et al. (2007) and associated footnotes.

while Chapters 8 through 11 deal with integrative issues (integration of RE into present and future energy systems; RE in the context of sustainable development; mitigation potential and costs; and policy, financing and implementation). The report communicates uncertainty where relevant.[4] It provides the following information on the potential for renewable energy sources to meet GHG reduction goals:

- Identification of RE resources and available technologies and impacts of climate change on these resources (Chapters 2 through 7);
- Technology and market status, future developments and projected rates of deployment (Chapters 2 through 7 and 10);
- Options and constraints for integration into the energy supply system and other markets, including energy storage, modes of transmission, integration into existing systems and other options (Chapter 8);
- Linkages among RE growth, opportunities and sustainable develoment (Chapter 9);
- Impacts on secure energy supply (Chapter 9);
- Economic and environmental costs, benefits, risks and impacts of deployment (Chapters 9 and 10);
- Mitigation potential of RE sources (Chapter 10);
- Scenarios that demonstrate how accelerated deployment might be achieved in a sustainable manner (Chapter 10);
- Capacity building, technology transfer and financing (Chapter 11); and
- Policy options, outcomes and conditions for effectiveness (Chapter 11).

1.1.3 Climate change

GHG emissions associated with the provision of energy services are a major cause of climate change. The AR4 concluded that "Most of the observed increase in global average temperature since the mid-20th century is *very likely* due to the observed increase in anthropogenic greenhouse gas concentrations." (IPCC, 2007a). Concentrations of CO_2 have continued to grow since the AR4 to about 390 ppm CO_2 or 39% above pre-industrial levels by the end of 2010 (IPCC, 2007b; NOAA, 2010). The global average temperature has increased by 0.76°C (0.57°C to 0.95°C) between 1850 to 1899 and 2001 to 2005, and the warming trend has increased significantly over the last 50 years (IPCC, 2007b). While this report focuses on the energy sector, forest clearing and burning and land use change, and the release of non-CO_2 gases from industry, commerce and agriculture also contribute to global warming (IPCC, 2007b).

An extensive review of long-term scenarios (Fisher et al., 2007) revealed that economic growth is expected to lead to a significant increase in gross domestic product (GDP) during the 21st century (see Figure 1.2 left panel), associated with a corresponding increase in the demand for energy services. Historically, humankind has been able to reduce the primary energy input required to produce one GDP unit (the so-called primary energy intensity) and is expected to do so further in the future (see Figure 1.2 right panel).

Within the considered scenarios, the increase in energy efficiency is more than compensated for by the anticipated economic growth. In the

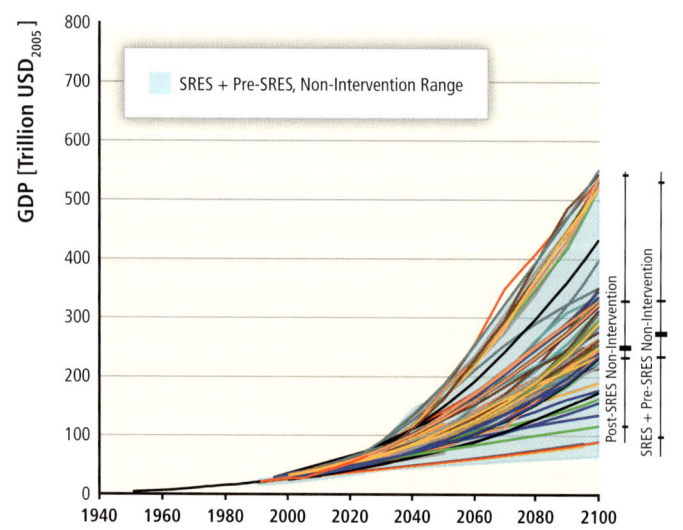
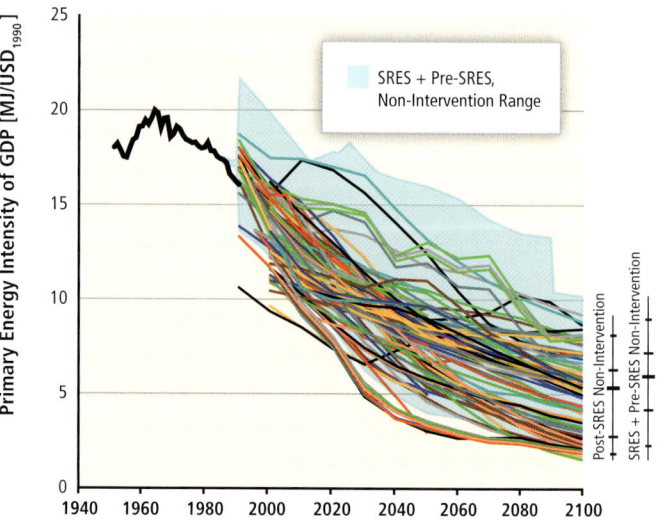

Figure 1.2 | Left panel: Comparison of GDP projections in post-SRES (Special Report on Emission Scenarios) emissions scenarios with those used in previous scenarios. The median of the new scenarios is about 7% below the median of the pre-SRES and SRES scenario literature. The two vertical bars on the right extend from the minimum to maximum of the distribution of scenarios by 2100. Right panel: Development of primary energy intensity of GDP: historical development and projections from SRES and pre-SRES scenarios compared to post-SRES scenarios. Adapted from Fisher et al., 2007, pp. 180 and 184.

---

4   This report communicates uncertainty, for example, by showing the results of sensitivity analyses and by quantitatively presenting ranges in cost numbers as well as ranges in the scenario results. This report does not apply formal IPCC uncertainty terminology because at the time of the approval of this report, IPCC uncertainty guidance was in the process of being revised.

business-as-usual case, the demand for global primary energy therefore is projected to increase substantially during the 21st century (see Figure 1.3 left panel).

Similarly to the behaviour of primary energy intensity, carbon intensity (the amount of CO_2 emissions per unit of primary energy) is—with few exceptions—expected to decrease as well (see Figure 1.3 right panel). Despite the substantial associated decarbonization, the overwhelming majority of the non-intervention emission projections exhibit considerably higher emissions in 2100 compared with those in 2000 (see the shaded area in Figure 1.4 left panel). Because emission rates substantially exceed natural removal rates, concentrations will continue to increase, which will raise global mean temperature. Figure 1.4 right panel shows the respective changes for representative emission scenarios (so-called SRES (Special Report on Emissions Scenarios) scenarios; see IPCC (2000a)) taken from the set of emissions scenarios shown in Figure 1.4 left panel.

In the absence of additional climate policies, the IPCC (2007a; see Figure 1.4) projected that global average temperature will rise over this century by between 1.1°C and 6.4°C over the 1980 to 1999 average, depending on socioeconomic scenarios (IPCC, 2000a). This range of uncertainty arises from uncertainty about the amount of GHGs that will be emitted in the future, and from uncertainty about the climate sensitivity. In addition to an investigation of potentially irreversible abrupt changes in the climate system, the IPCC assessed the adverse impacts of such climate change (and the associated sea level rise and ocean acidification) on water supply, ecosystems, food security, human health and coastal settlements (IPCC, 2007c).

The Cancun Agreements (2010) call for limiting global average temperature rise to no more than 2°C above pre-industrial values, and agreed to consider a goal of 1.5°C. The analysis shown in Figure 1.5 concludes that in order to be confident of achieving an equilibrium temperature increase of only 2°C to 2.4°C, atmospheric GHG concentrations would need to be in the range of 445 to 490 ppm CO_2eq. This in turn implies that global emissions of CO_2 will need to decrease by 50 to 85% below 2000 levels by 2050 and begin to decrease (instead of continuing their current increase) no later than 2015 (IPCC, 2007a). Note that there is a considerable range of probable temperature outcomes at this concentration range. Additional scenario analysis and mitigation costs under various GHG concentration stabilization levels are analyzed in Chapter 10. This report does not analyze the economic cost of damages from climate change.

1.1.4 Drivers of carbon dioxide emissions

Since about 1850, global use of fossil fuels (coal, oil and gas) has increased to dominate energy supply, both replacing many traditional uses of bioenergy and providing new services. The rapid rise in fossil fuel combustion (including gas flaring) has produced a corresponding rapid growth in CO_2 emissions (Figure 1.6).

The amount of carbon in fossil fuel reserves and resources (unconventional oil and gas resources as well as abundant coal) not yet burned has the potential to add quantities of CO_2 to the atmosphere—if burned over coming centuries—that would exceed the range of any of the scenarios considered in Figure 1.5 or in Chapter 10 (Moomaw et al., 2001; Knopf et al., 2010). Figure 1.7 summarizes current estimates of fossil fuel resources and reserves in terms of carbon content, and compares them with the amount already released to the atmosphere as CO_2. Reserves refer to what is extractable with today's technologies at current energy

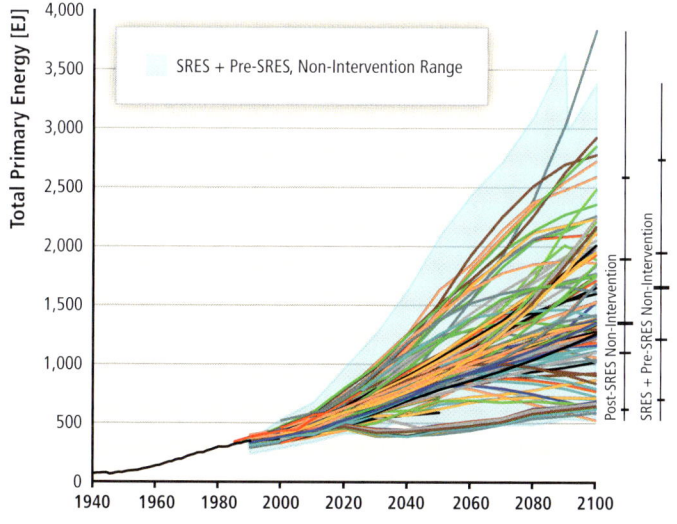
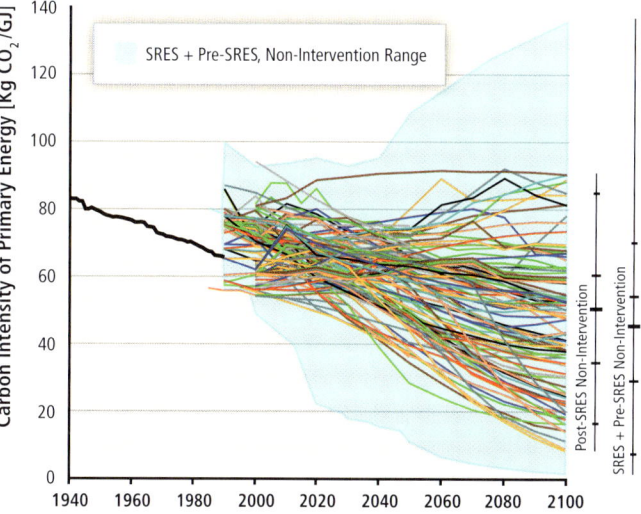

Figure 1.3 | Left panel: Projected increase in primary energy supply. Comparison of 153 SRES and pre-SRES baseline energy scenarios in the literature compared with the 133 more recent, post-SRES scenarios. The ranges are comparable, with small changes in the lower and upper boundaries. Right panel: Expected carbon intensity changes. Historical development and projections from SRES and pre-SRES scenarios compared to post-SRES scenarios. Adapted from Fisher et al., 2007, pp. 183 and 184.

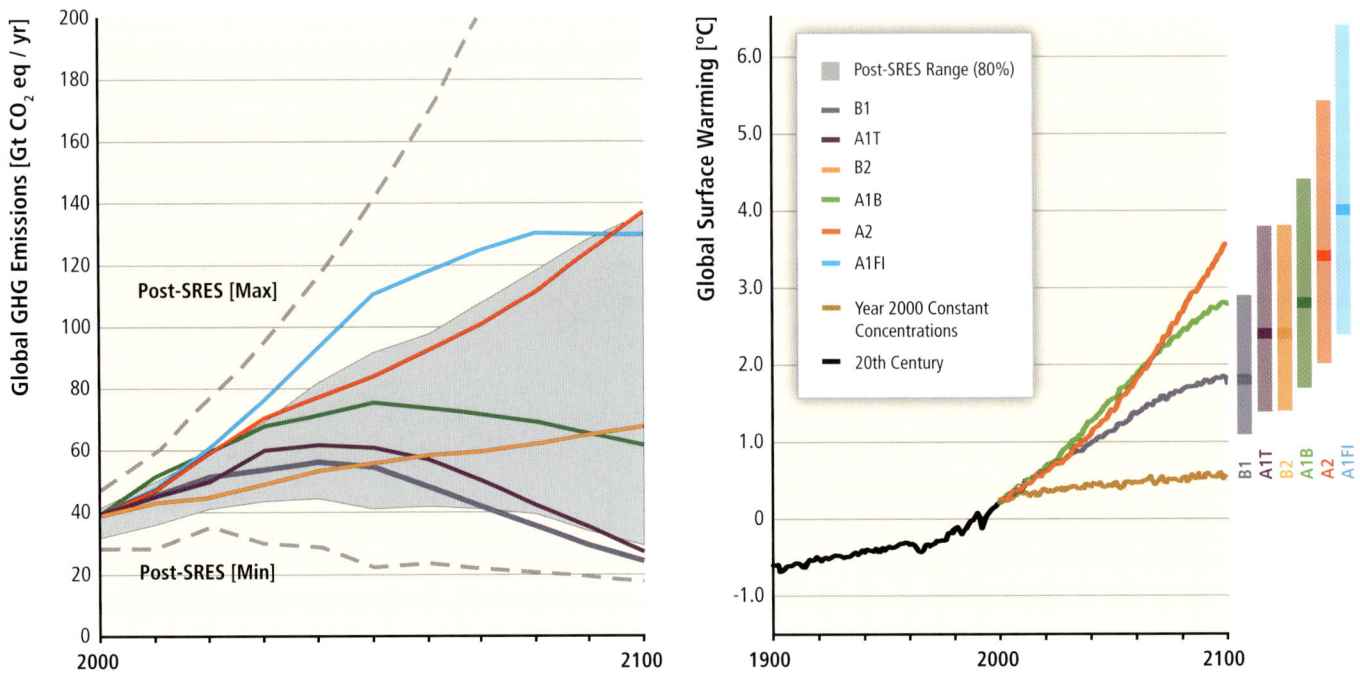

Figure 1.4 | Left panel: Global GHG emissions (Gt CO$_2$eq) in the absence of climate policies: six illustrative SRES marker scenarios (coloured lines) and the 80th percentile range of recent scenarios published since SRES (post-SRES) (grey shaded area). Dashed lines show the full range of post-SRES scenarios. The emissions include CO$_2$, methane (CH$_4$), nitrous oxide (N$_2$O) and fluorinated gases. Right panel: Solid lines are multi-model global averages of projected surface warming for SRES scenarios A2, A1B and B1, shown as continuations of the 20th-century simulations. These projections also take into account emissions of short-lived GHGs and aerosols. The brown line is not a scenario, but is for atmosphere-ocean general circulation model simulations where atmospheric concentrations are held constant at year 2000 values. The bars at the right of the figure indicate the best estimate (solid line within each bar) and the likely range assessed for the six SRES marker scenarios for 2090 to 2099. All temperatures are relative to the period 1980 to 1999 (IPCC, 2007a, Figure SPM 5, page 7).

prices. Resources represent the total amount estimated to be available without regard to the technical or economic feasibility of extracting it (IEA, 2005).

In developing strategies for reducing CO$_2$ emissions it is useful to consider the Kaya identity that analyzes energy-related CO$_2$ emissions as a function of four factors: 1) Population; 2) GDP per capita; 3) energy intensity (i.e., total primary energy supply (TPES) per GDP); and 4) carbon intensity (i.e., CO$_2$ emissions per TPES) (Ehrlich and Holdren, 1971; Kaya, 1990).

The Kaya identity is then:

CO_2 emissions = Population x (GDP/population) x (TPES/GDP) x (CO$_2$/TPES)

This is sometimes referred to as:

CO_2 emissions = (Population x Affluence x Energy intensity x Carbon intensity)

Renewable energy supply sources are effective in lowering CO$_2$ emissions because they have low carbon intensity with emissions per unit of energy output typically 1 to 10% that of fossil fuels (see Figure 1.13 and Chapter 10). Further reductions can also be achieved by lowering the energy intensity required to provide energy services. The role of these two strategies and their interaction is discussed in more detail in Section 1.2.6.

The absolute (a) and percentage (b) annual changes in global CO$_2$ emissions are shown in terms of the Kaya factors in Figure 1.8 (Edenhofer et al., 2010).

While GDP per capita and population growth had the largest effect on emissions growth in earlier decades, decreasing energy intensity significantly slowed emissions growth in the period from 1971 to 2008. In the past, carbon intensity fell because of improvements in energy efficiency and switching from coal to natural gas and the expansion of nuclear energy in the 1970s and 1980s that was particularly driven by Annex I countries.[5] In recent years (2000 to 2007), increases in carbon intensity have mainly been driven by the expansion of coal use by both developed and developing countries, although coal and petroleum use have fallen slightly since 2007. In 2008 this trend was broken due to the financial crisis. Since the early 2000s, the energy supply has become more carbon intensive, thereby amplifying the increase resulting from growth in GDP per capita (Edenhofer et al., 2010).

---

5 See Glossary (Annex I) for a definition of Annex I countries.

Chapter 1

Renewable Energy and Climate Change

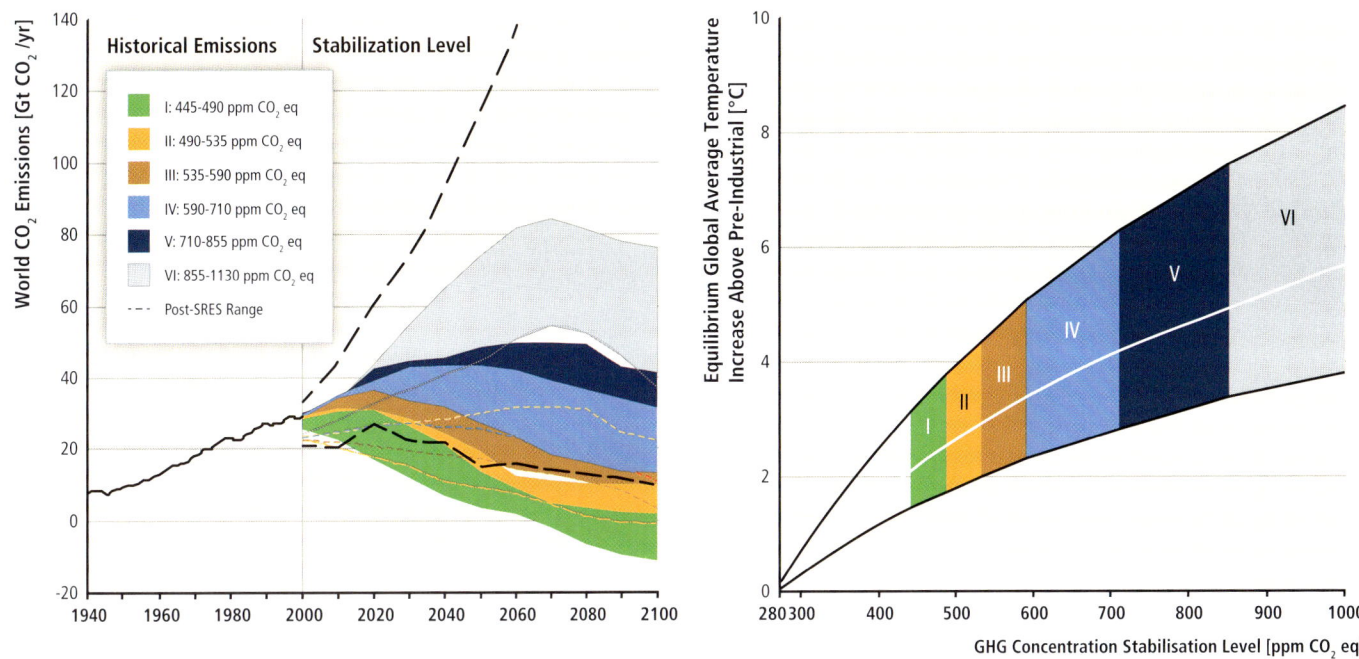

Figure 1.5 | Global CO_2 emissions from 1940 to 2000 and emissions ranges for categories of stabilization scenarios from 2000 to 2100 (left panel); and the corresponding relationship between the stabilization target and the likely equilibrium global average temperature increase above pre-industrial (right panel). Coloured shadings show stabilization scenarios grouped according to different targets (stabilization categories I to VI). The right panel shows ranges of global average temperature change above pre-industrial, using (i) 'best estimate' climate sensitivity of 3°C (line in middle of shaded area); (ii) upper bound of likely range of climate sensitivity of 4.5°C (line at top of shaded area); and (iii) lower bound of likely range of climate sensitivity of 2°C (line at bottom of shaded area) (IPCC, 2007a, Figure SPM-11, page 21).

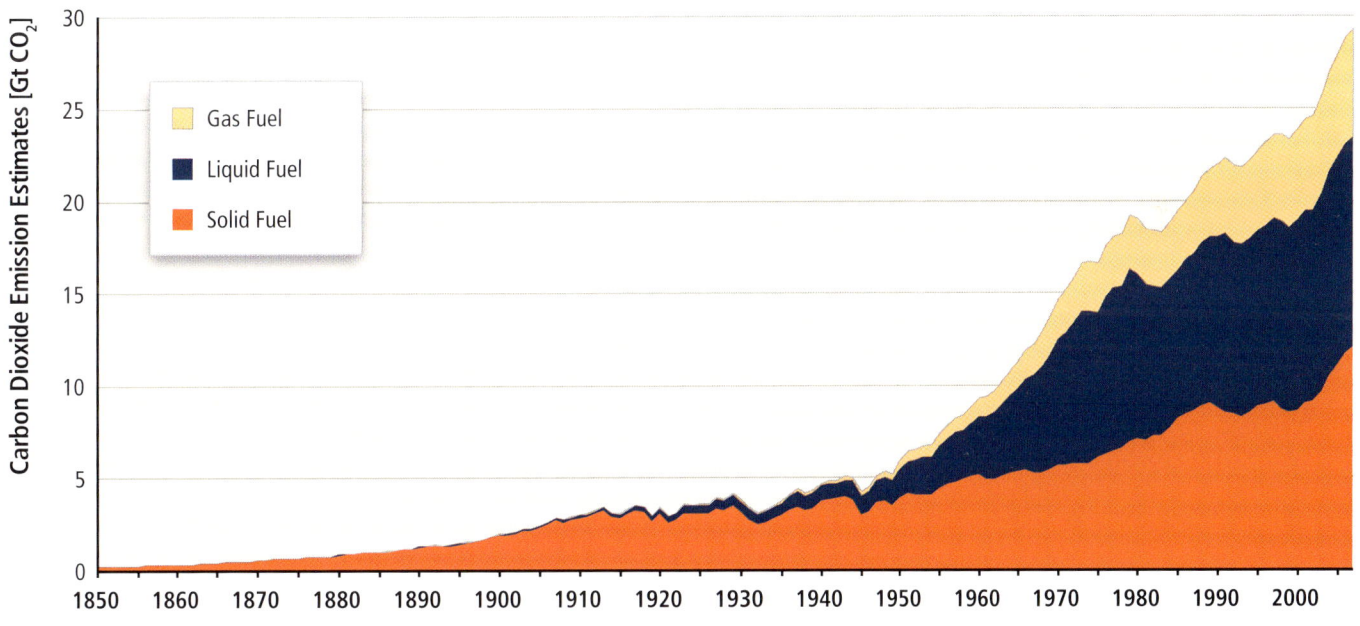

Figure 1.6 | Global CO_2 emissions from fossil fuel burning, 1850 to 2007. Gas fuel includes flaring of natural gas. All emission estimates are expressed in Gt CO_2. Data Source: (Boden and Marland, 2010).

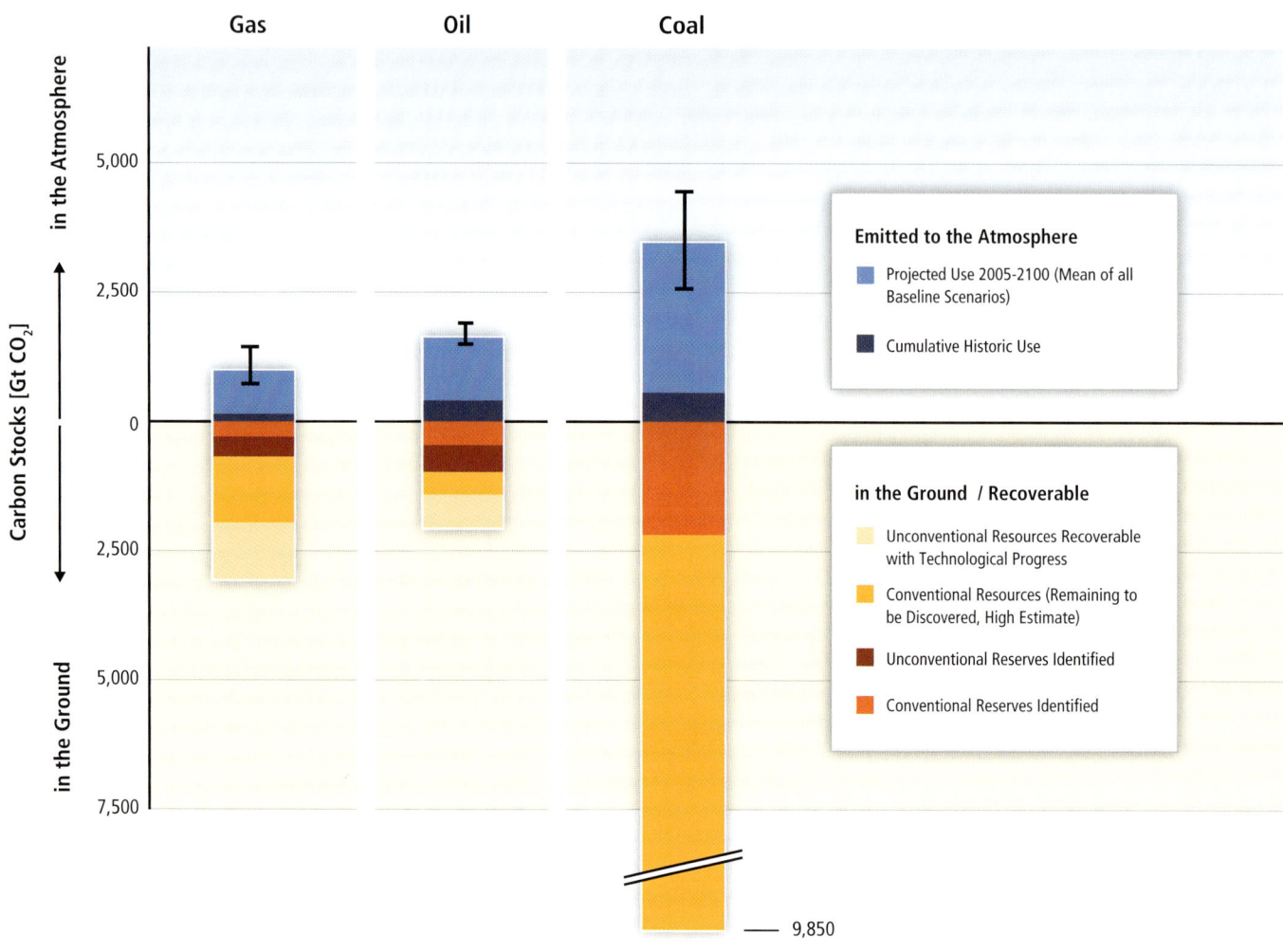

Figure 1.7 | CO_2 released to the atmosphere (above zero) and stocks of recoverable carbon from fossil fuels in the ground (below zero, converted to CO_2). Estimates of carbon stocks in the ground are taken from IPCC (2000a, Table 3-5). Estimates of carbon stocks remaining are provided by BGR (2009), cumulative historic carbon consumption (1750 to 2004) is from Boden et al. (2009) and estimated future consumption (2005 to 2100) from the mean of the baseline scenarios of the energy-economic and integrated assessment models considered in the analysis of Chapter 10 (Table 10.1). Only those scenarios where the full data set until 2100 was available were considered (i.e., 24 scenarios from 12 models). The light blue stacked bar shows the mean and the black error bars show the standard deviation of the baseline projections. Fossil energy stocks were converted to CO_2 emissions by using emission factors from IPCC (2006). Adapted from Knopf et al. (2010).

Historically, developed countries have contributed the most to cumulative global CO_2 emissions, and still have the highest total historical emissions and largest emissions per capita (World Bank, 2009). Recently, developing country annual emissions have risen to more than half of the total, and China surpassed the USA in annual emissions in 2007 (IEA, 2010f). Figure 1.9 examines the annual change in absolute emissions by country and country groups between 1971 and 2008 (Edenhofer et al., 2010).

1.1.5 Renewable energy as an option to mitigate climate change

On a global basis, it is estimated that RE accounted for 12.9% of the total 492 EJ of primary energy supply in 2008 (IEA, 2010a). The largest RE contributor was biomass (10.2%), with the majority (roughly 60%) of the biomass fuel used in traditional cooking and heating applications in developing countries but with rapidly increasing use of modern biomass as well.[6] Hydropower represented 2.3%, whereas other RE sources accounted for 0.4% (Figure 1.10).

RE's contribution to electricity generation is summarized in Figure 1.11. In 2008, RE contributed approximately 19% of global electricity supply (16% hydropower, 3% other RE). Global electricity production in 2008 was 20,181 TWh (or 72.65 EJ) (IEA, 2010a).

Deployment of RE has been increasing rapidly in recent years. Under most conditions, increasing the share of RE in the energy mix will require policies to stimulate changes in the energy system. Government policy, the declining cost of many RE technologies, changes in the prices of fossil fuels and other factors have supported the continuing increase

---

[6] In addition, biomass use estimated to amount to 20 to 40% is not reported in official databases, such as dung, unaccounted production of charcoal, illegal logging, fuelwood gathering, and agricultural residue use (Section 2.1).

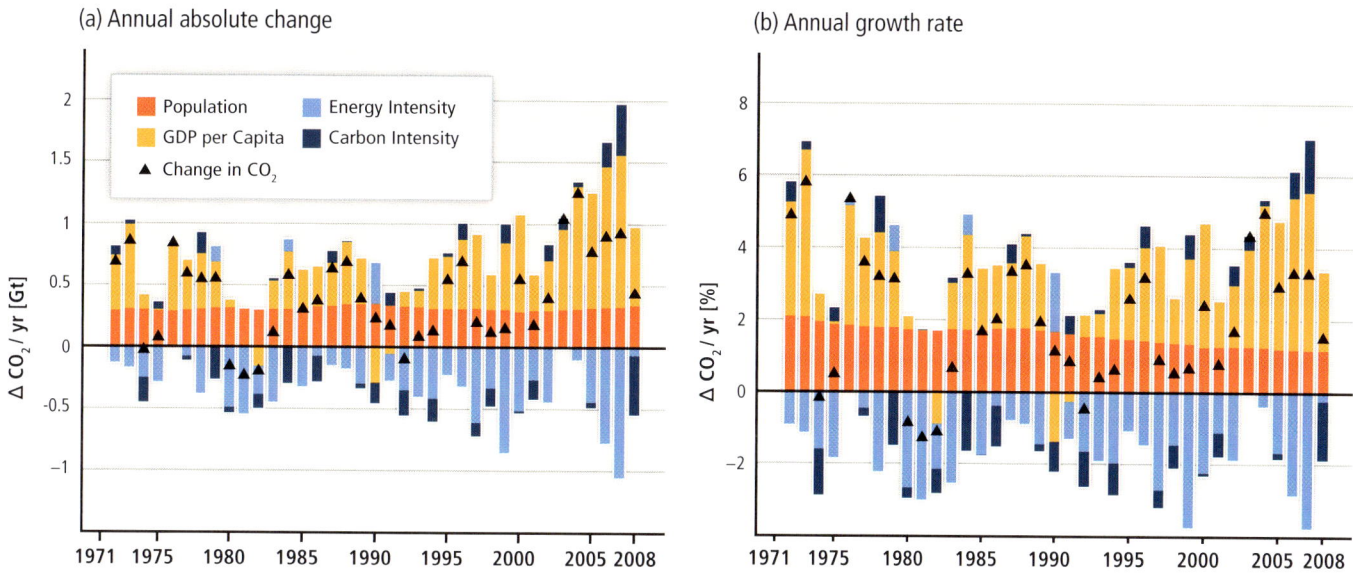

Figure 1.8 | Decomposition of (a) annual absolute change and (b) annual growth rate in global energy-related CO_2 emissions by the factors in the Kaya identity; population (red), GDP per capita (orange), energy intensity (light blue) and carbon intensity (dark blue) from 1971 to 2008. The colours show the changes that would occur due to each factor alone, holding the respective other factors constant. Total annual changes are indicated by a black triangle. Data source: IEA (2010a).

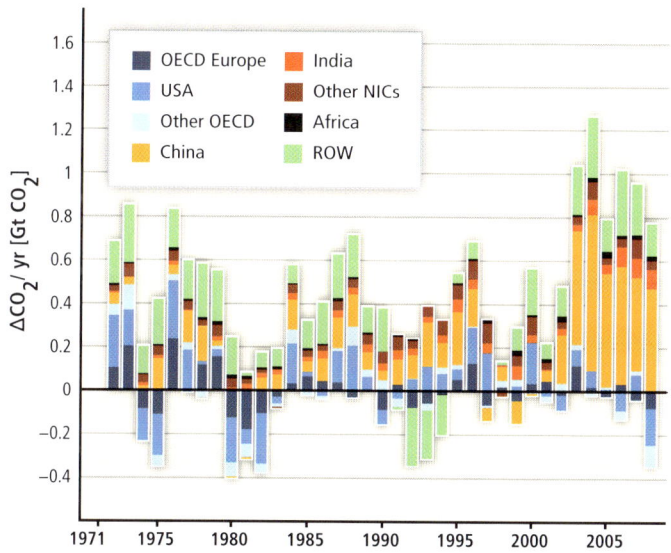

Figure 1.9 | Influence of selected countries and country groups on global changes in CO_2 emissions from 1971 to 2008. ROW: rest of world. Data source: IEA (2010a).

Note: "OECD" is the Organisation for Economic Co-operation and Development; "Other Newly Industrializing Countries (NICs)" include Brazil, Indonesia, the Republic of Korea, Mexico and South Africa; "Other OECD" does not include the Republic of Korea and Mexico; and "Africa" does not include South Africa.

in the use of RE (see Section 1.5.1 and Chapter 11). While RE is still relatively small, its growth has accelerated in recent years, as shown in Figure 1.12. In 2009, despite global financial challenges, RE capacity continued to grow rapidly, including wind power (32%, 38 GW added), hydropower (3%, 31 GW added), grid-connected photovoltaics (53%, 7.5 GW added), geothermal power (4%, 0.4 GW), and solar hot water/heating (21%, 31 GW_{th}) (REN21, 2010). Biofuels accounted for 2% of global road transport fuel demand in 2008 and nearly 3% in 2009 (IEA, 2010c). The annual production of ethanol increased to 1.6 EJ (76 billion litres) by the end of 2009 and biodiesel production increased to 0.6 EJ (17 billion litres). Of the approximate 300 GW of new electricity generating capacity added globally over the two-year period from 2008 to 2009, 140 GW came from RE additions. Collectively, by the end of 2009 developing countries hosted 53% of global RE power generation capacity (including all sizes of hydropower), with China adding more capacity than any other country in 2009. The USA and Brazil accounted for 54 and 35% of global bioethanol production in 2009, respectively, while China led in the use of solar hot water. At the end of 2009, the use of RE in hot water/heating markets included modern biomass (270 GW_{th}), solar (180 GW_{th}) and geothermal (60 GW_{th}). The use of RE (excluding traditional biomass) in meeting rural energy needs is also increasing, including small hydropower stations, various modern bioenergy options, and household or village PV, wind or hybrid systems that combine multiple technologies (REN21, 2010).

UNEP found that in 2008, despite a decline in overall energy investments, global investment in RE power generation rose by 5% to USD 140 billion (USD_{2005} 127 billion), which exceeded the 110 billion (USD_{2005} 100 billion) invested in fossil fuel generation capacity (UNEP, 2009).

These developments suggest the possibility that RE could play a much more prominent role in both developed and developing countries over the coming decades (Demirbas, 2009). New policies, especially in the USA, China and the EU, are supporting this effort (Chapter 11).

Estimates of the lifecycle CO_2 intensity for electric power-producing renewable energy technologies relative to fossil fuels and nuclear power are shown in Figure 1.13 and are discussed in more detail in Chapter 9. Renewable energy and nuclear technologies produce one to two orders of

Renewable Energy and Climate Change

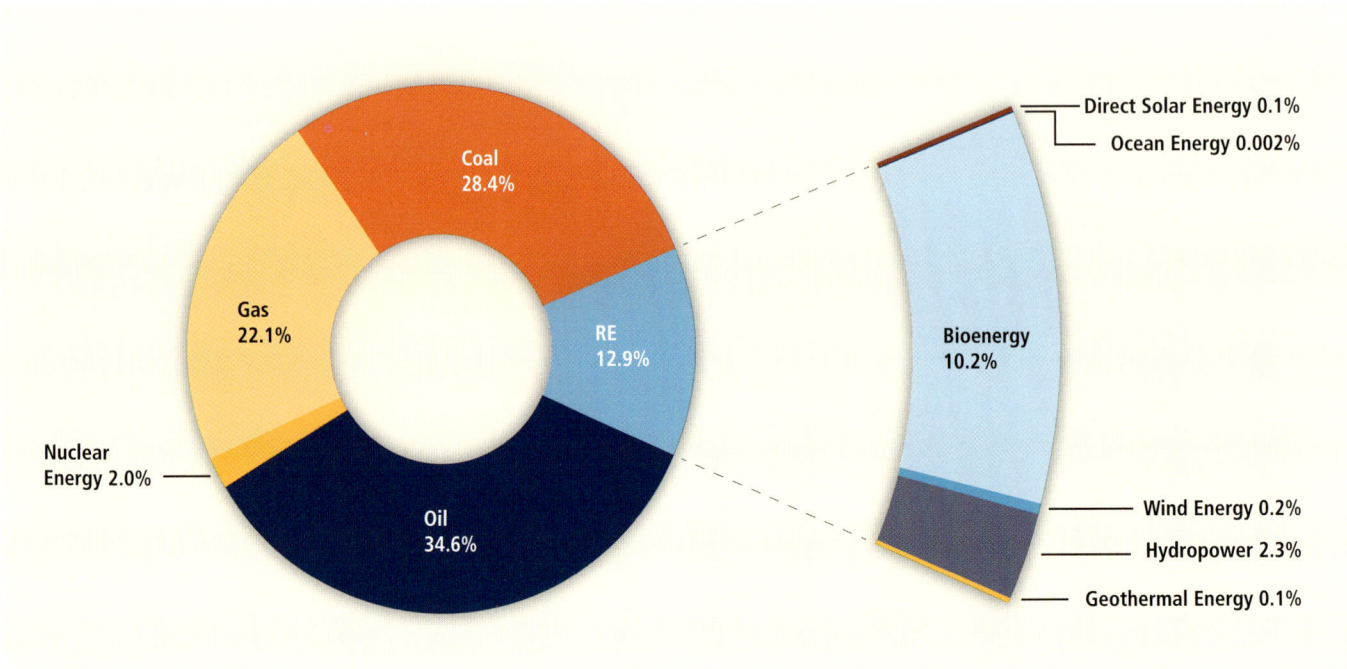

Figure 1.10 | Shares of energy sources in total global primary energy supply in 2008 (492 EJ). Modern biomass contributes 38% to the total biomass share. Data source: IEA (2010a).

Notes: Underlying data for figure have been converted to the direct equivalent method of accounting for primary energy supply (Annex II.4).

magnitude lower CO_2 emissions than fossil fuels in grams of CO_2 per kWh of electricity produced (Weisser, 2007; Sovacool, 2008; Jacobson, 2009).

Most RE technologies have low specific emissions of CO_2 into the atmosphere relative to fossil fuels, which makes them useful tools for addressing climate change (see Figure 1.13). For a RE resource to be sustainable, it must be inexhaustible and not damage the delivery of environmental goods and services including the climate system. For example, to be sustainable, biofuel production should not increase net CO_2 emissions, should not adversely affect food security, or require excessive use of water and chemicals or threaten biodiversity. To be sustainable, energy must also be economically affordable over the long term; it must meet societal needs and be compatible with social norms now and in the future. Indeed, as use of RE technologies accelerates, a balance will have to be struck among the several dimensions of sustainable development. It is important to assess the entire lifecycle of each energy source to ensure that all of the dimensions of sustainability are met (Sections 1.4.1.4 and 9.3.4).

1.1.6 Options for mitigation

There are multiple means for lowering GHG emissions from the energy system while still providing energy services (Pacala and Socolow, 2004; IPCC, 2007d). Energy services are the tasks to be performed using energy. Many options and combinations are possible for reducing emissions. In order to assess the potential contribution of RE to mitigating global climate change, competing mitigation options therefore must be considered as well (Chapter 10).

Chapter 4 of AR4 (Sims et al., 2007) identified a number of ways to lower heat-trapping emissions from energy sources while still providing energy services. They include:

- Improve supply side efficiency of energy conversion, transmission and distribution including combined heat and power.
- Improve demand side efficiency in the respective sectors and applications (e.g., buildings, industrial and agricultural processes, transportation, heating, cooling, lighting) (see also von Weizsäcker et al., 2009).
- Shift from high GHG energy carriers such as coal and oil to lower GHG energy carriers such as natural gas, nuclear fuels and RE sources (Chapters 2 through 7).
- Utilize carbon capture and storage (CCS) to prevent post-combustion or industrial process CO_2 from entering the atmosphere. CCS has the potential for removing CO_2 from the atmosphere when biomass is burned (see also IPCC, 2005).
- Change behaviour to better manage energy use or to use fewer carbon- and energy-intensive goods and services (see also Dietz et al., 2009).

Two additional means of reducing GHGs include enhancing the capacity of forests, soils and grassland sinks to absorb CO_2 from the atmosphere (IPCC, 2000b), and reducing the release of black carbon aerosols and particulates

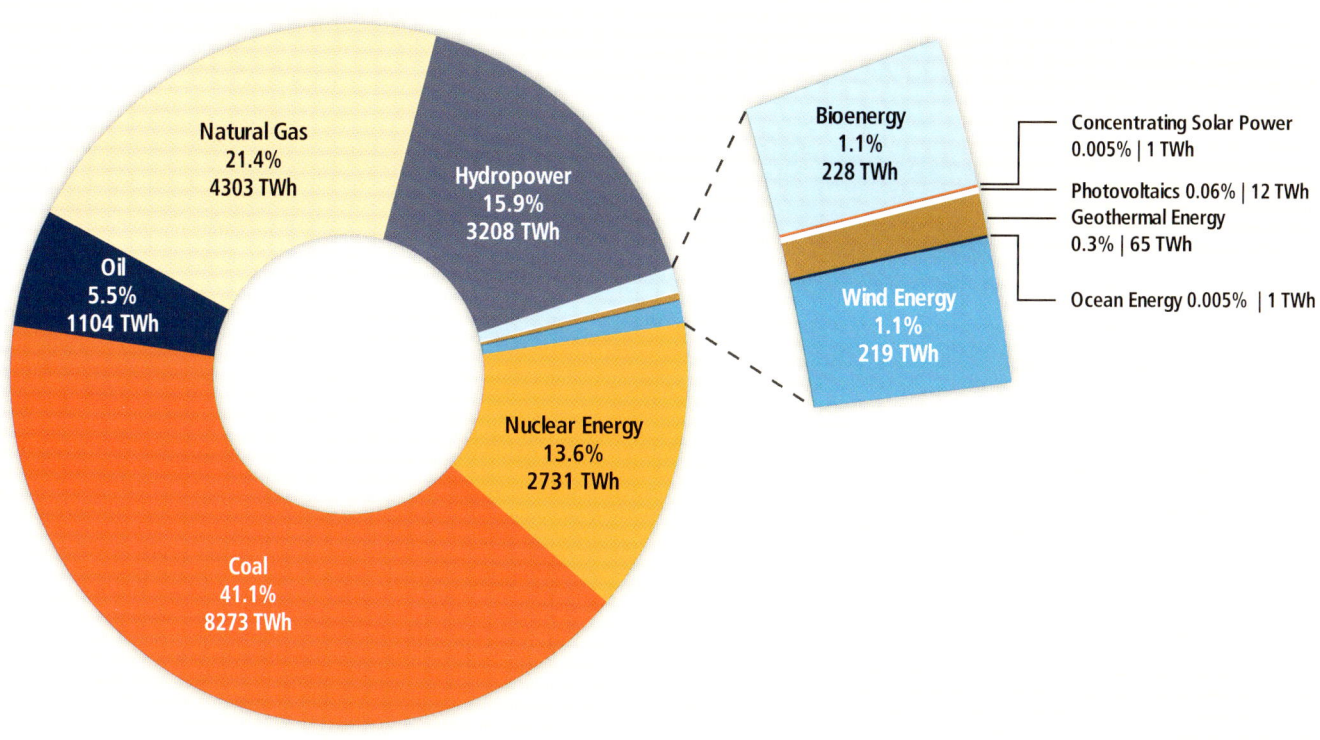

Figure 1.11 | Share of primary energy sources in world electricity generation in 2008. Data for renewable energy sources from IEA (2010a); for fossil and nuclear from IEA (2010d).

from diesel engines, biomass fuels and from the burning of agricultural fields (Bond and Sun, 2005). Additional reductions in non-CO_2 heat-trapping GHGs (CH_4, N_2O, hydrofluorocarbons, sulphur hexafluoride) can also reduce global warming (Moomaw et al., 2001, their Appendix; Sims et al., 2007).

Geoengineering solutions have been proposed to address other aspects of climate change, including altering the heat balance of the Earth by increasing surface albedo (reflectivity), or by reflecting incoming solar radiation with high-altitude mirrors or with atmospheric aerosols. Enhanced CO_2 absorption from the atmosphere through ocean fertilization with iron has also been proposed and tested (Robock et al., 2009; Royal Society, 2009).

There are multiple combinations of these means that can reduce the extent of global warming. A comprehensive evaluation of any portfolio of mitigation options would involve an evaluation of their respective mitigation potential as well as all associated risks, costs and their contribution to sustainable development. This report focuses on substitution of fossil fuels with low-carbon RE to reduce GHGs, and examines the competition between RE and other options to address global climate change (see Figure 1.14).

Setting a climate protection goal in terms of the admissible change in global mean temperature broadly defines (depending on the assumed climate sensitivity) a corresponding atmospheric CO_2 concentration limit and an associated carbon budget over the long term (see Figure 1.5, right panel) (Meinshausen et al., 2009). This budget, in turn, can be broadly translated into a time-dependent emission trajectory that serves as an upper bound or (if the remaining time flexibility is taken into account) in an associated corridor of admissible emissions (Figure 1.5, left panel). Subtracting any expected CO_2 emissions from land use change and land cover change constrains the admissible CO_2 emissions that could be realized by freely emitting carbon fuels (i.e., coal, oil, and gas burned without applying carbon capture technologies).

The corresponding fossil fuel supply is part of the total primary energy supply (see Figure 1.14). The remainder of the TPES is provided by zero- or low-carbon energy technologies, such as RE, nuclear or the combustion of fossil fuels combined with CCS (Clarke et al., 2009).

Whereas the admissible amount of freely emitting fossil fuels is mainly fixed by the climate protection goal, the complementary contribution of zero- or low-carbon energies to the primary energy supply is influenced by the 'scale' of the requested energy services and the overall efficiency with which these services can be provided.

As Figure 1.2 right panel clearly shows, the energy intensity is already expected to decrease significantly in the non-intervention scenarios. Technical improvements and structural changes are expected to result in considerably lower emissions than otherwise would be projected. As many low-cost options to improve the overall energy efficiency are

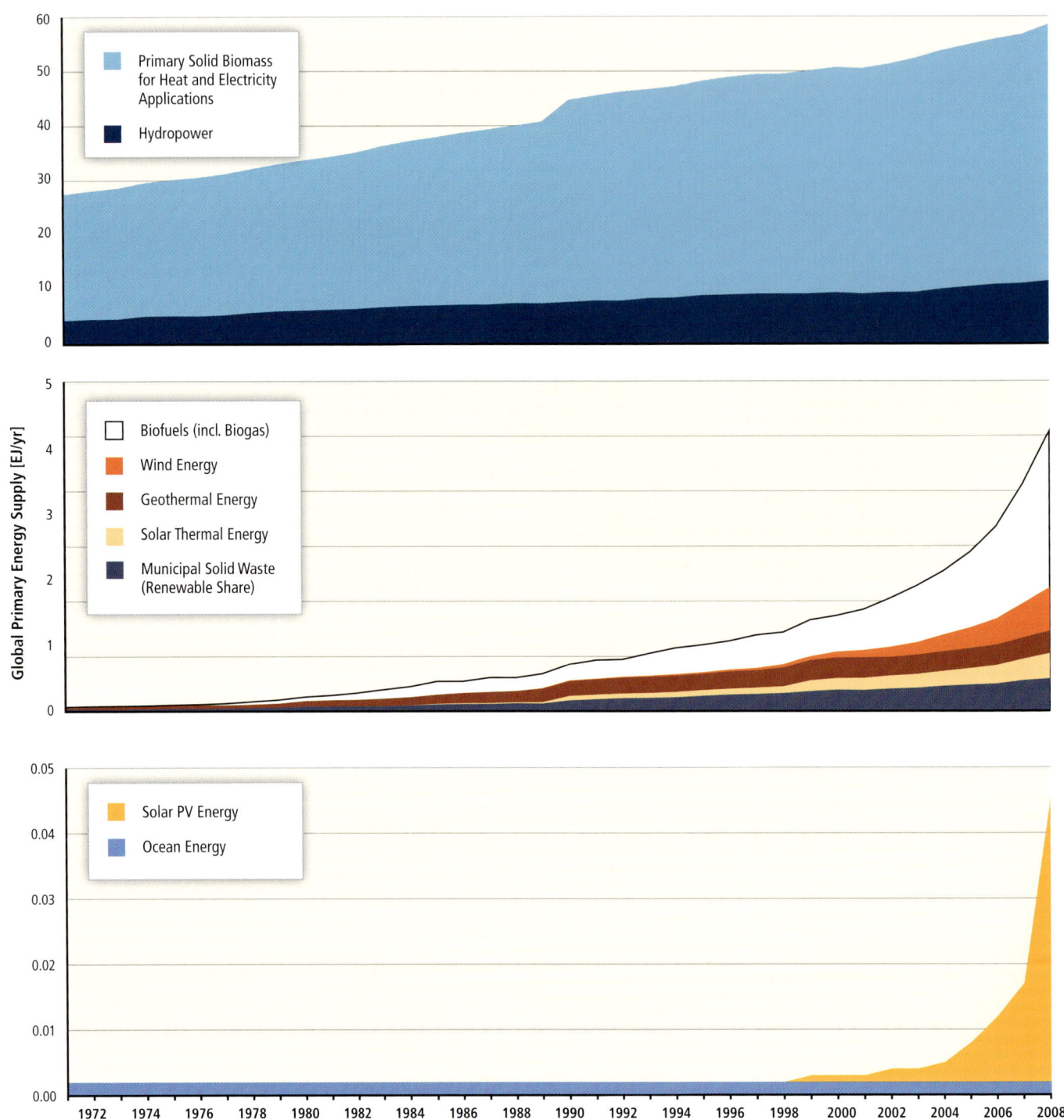

Figure 1.12 | Historical development of global primary energy supply from renewable energy from 1971 to 2008. Data Source: IEA (2010a).

Note: Technologies are referenced to separate vertical units for display purposes only. Underlying data for figure have been converted to the 'direct equivalent' method of accounting for primary energy supply (Section 1.1.9 and Annex II.4), except that the energy content of biofuels is reported in secondary energy terms (the primary biomass used to produce the biofuel would be higher due to conversion losses (Sections 2.3 and 2.4)).

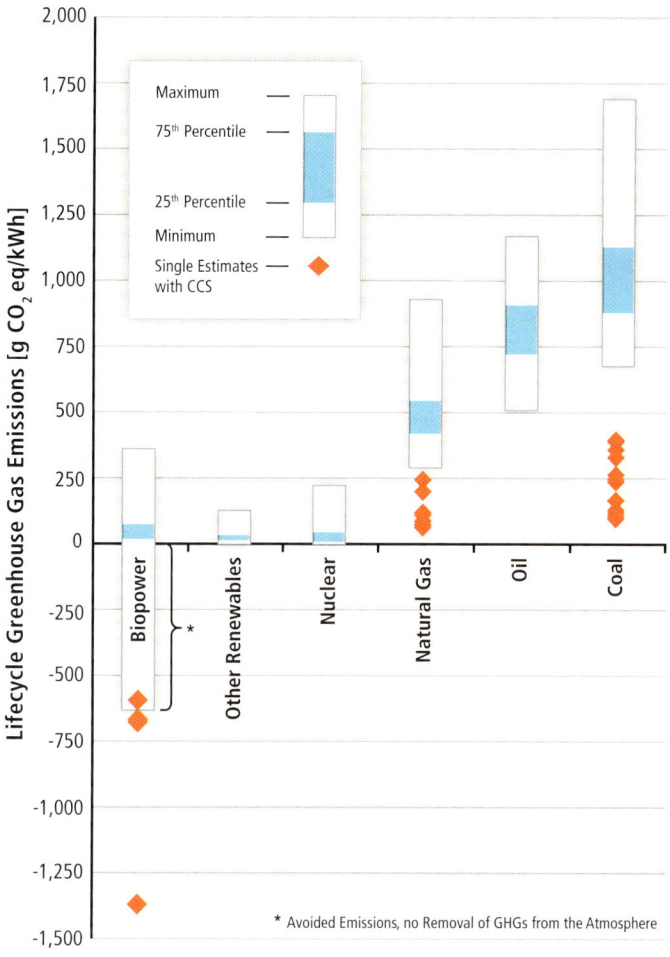

Figure 1.13 | Lifecycle GHG emissions of renewable energy, nuclear energy and fossil fuels (Chapter 9, Figure 9.8).

technologies, the ability of RE technologies to overcome initial cost barriers, preferences, environmental considerations and other barriers.

1.1.7 Trends in international policy on renewable energy

The international community's discussions of RE began with the fuel crises of the 1970s, when many countries began exploring alternative energy sources. Since then, RE has featured prominently in the United Nations agenda on environment and development through various initiatives and actions (WIREC, 2008; Hirschl, 2009).

The 1981 UN Conference on New and Renewable Sources of Energy adopted the Nairobi Programme of Action. The 1992 UN Conference on Environment and Development, and Action Plan for implementing sustainable development through sustainable energy and protection of the atmosphere was reinforced by the 2002 World Summit on Sustainable Development where several RE Partnerships were signed. 'Energy for Sustainable Development' highlighted the importance of RE at the 2001 UN Commission on Sustainable Development (CSD, 2001). Major RE meetings were held in Bonn in 2004, Beijing in 2005 and in Washington, DC, in 2008.

The International Energy Agency (IEA) has provided a forum for discussing energy issues among OECD countries, and provides annual reports on all forms of energy including RE. The IEA also prepares scenarios of alternative futures utilizing differing combinations of primary energy

already part of the non-intervention scenarios (Fisher et al., 2007), the *additional* opportunities to decrease energy intensity in order to mitigate climate change are limited (Bruckner et al., 2010). In order to achieve ambitious climate protection goals, for example, stabilization below the aforementioned 2°C global mean temperature change, energy efficiency improvements alone do not suffice. In addition, low-carbon technologies become imperative.

Chapter 10 includes a comprehensive analysis of over 100 scenarios of energy supply and demand to assess the costs and benefits of RE options to reduce GHG emissions and thereby mitigate climate change. The contribution RE will provide within the portfolio of these low-carbon technologies heavily depends on the economic competition between these technologies (Chapter 10), a comparison of the relative environmental burdens (beyond climate change) associated with them, as well as secure energy supply and societal aspects (Figure 1.14). However, even without a push for climate change mitigation, scenarios that are examined in this report find that the increasing demand for energy services is expected to drive RE to levels exceeding today's energy usage. There are large uncertainties in projections, including economic and population growth, development and deployment of higher efficiency

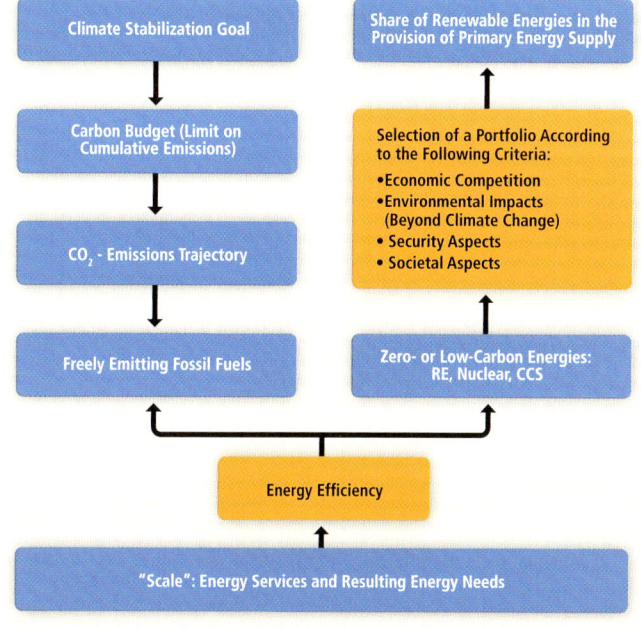

Figure 1.14 | The role of renewable energies within the portfolio of zero- or low-carbon mitigation options (qualitative description).

sources, energy efficiency and CO_2 emissions. REN 21, a nongovernmental organization, compiles recent data on RE resources based upon industrial and governmental reports. A new international organization, the International Renewable Energy Agency (IRENA), was also established in 2009 and has 149 signatories and 57 member countries[7].

1.1.8 Advancing knowledge about renewable energy

The body of scientific knowledge on RE and on the possible contribution of RE towards meeting GHG mitigation goals, as compiled and assessed in this report, is substantial. Nonetheless, due in part to the site-specific nature of RE, the diversity of RE technologies, the multiple end-use energy service needs that those technologies might serve, the range of markets and regulations governing integration, and the complexity of energy system transitions, knowledge about RE and its climate mitigation potential continues to advance. Additional knowledge remains to be gained in a number of broad areas related to RE and its possible role in GHG emissions reductions.

Though much is already known in each of these areas, as compiled in this report, additional research and experience would further reduce uncertainties and thus facilitate decision making related to the use of RE in the mitigation of climate change.

Though not comprehensive, a broad and selective listing of areas of anticipated present and future knowledge advancement is provided in Table 1.1.

1.1.9 Metrics and definitions

A glossary of terms is provided in Annex I. Conventions, conversion factors and methodologies are described in Annex II. A cost table for RE technologies is provided in Annex III.

To have a common comparison for all low-carbon sources, primary energy is measured according to the direct equivalent method rather than the physical content method favoured by IEA. The two methods treat all combustion technologies the same, but the direct equivalent method only counts the electric or thermal energy that is produced as primary energy for nuclear power or geothermal power, while the physical content method counts the total heat that is released. See Box 1.1 and Annex II where the differences between these methods are described in further detail.

---
7 See www.irena.org/

1.2 Summary of renewable energy resources

1.2.1 Definition, conversion and application of renewable energy

Renewable energy is any form of energy from solar, geophysical or biological sources that is replenished by natural processes at a rate that equals or exceeds its rate of use. RE is obtained from the continuing or repetitive flows of energy occurring in the natural environment and includes resources such as biomass, solar energy, geothermal heat, hydropower, tide and waves and ocean thermal energy, and wind energy. However, it is possible to utilize biomass at a greater rate than it can grow, or to draw heat from a geothermal field at a faster rate than heat flows can replenish it. On the other hand, the rate of utilization of direct solar energy has no bearing on the rate at which it reaches the Earth. Fossil fuels (coal, oil, natural gas) do not fall under this definition, as they are not replenished within a time frame that is short relative to their rate of utilization.

There is a multi-step process whereby primary energy is converted into an energy carrier (heat, electricity or mechanical work), and then into an energy service. RE technologies are diverse and can serve the full range of energy service needs. Various types of RE can supply electricity, thermal energy and mechanical energy, as well as produce fuels that are able to satisfy multiple energy service needs (Figure 1.16).

Since it is energy services and not energy that people need, the goal is to meet those needs in an efficient manner that requires less primary energy consumption with low-carbon technologies that minimize CO_2 emissions (Haas et al., 2008). Thermal conversion processes to produce electricity (including from biomass and geothermal) suffer losses of approximately 40 to 90%, and losses of around 80% occur when supplying the mechanical energy needed for transport based on internal combustion engines. These conversion losses raise the share of primary energy from fossil fuels, and the primary energy required from fossil fuels to produce electricity and mechanical energy from heat (Jacobson, 2009; LLNL, 2009; Sterner, 2009). Direct energy conversions from solar PV, hydro, ocean, and wind energy to electricity do not suffer thermodynamic power cycle (heat to work) losses although they do experience other conversion inefficiencies in extracting energy from natural energy flows that may also be relatively large and irreducible (Chapters 2 through 7). To better compare low-carbon sources that produce electricity over time, this report has adopted the *direct equivalent method* in which primary energy of *all non-combustible sources* is defined as one unit of secondary energy, for example, electricity,

Table 1.1 | Select areas of possible future knowledge advancement

Future cost and timing of RE deployment	• Cost of emerging and non-electricity RE technologies, in diverse regional contexts • Future cost reduction given uncertainty in research and development (R&D)-driven advances and deployment-oriented learning • Cost of competing conventional and low-carbon energy technologies • Ability to analyze variable and location-dependent RE technologies in large-scale energy models, including the contribution of RE towards sustainable development and energy access • Further assessments of RE deployment potentials at global, regional and local scales • Analysis of technology-specific mitigation potential through comparative scenario exercises considering uncertainties • Impacts of policies, barriers and enabling environments on deployment volume and timing
Realizable technical potential for RE at all geographic scales	• Regional/local RE resource assessments • Improved resource assessments for emerging technologies and non-electricity RE technologies • Future impacts of climate change on RE technical potential • Competition for RE resources, such as biomass, between RE technologies and other human activities and needs • Location of RE resources relative to the location of energy demand (i.e., population centres)
Technical and institutional challenges and costs of integrating diverse RE technologies into energy systems and markets	• Comparative assessment of the short- and long-term technical/institutional solutions and costs of integrating high penetrations of RE • Specific technical/institutional challenges of integrating variable RE into electricity markets that differ from those of the OECD, for RE resources other than wind, and the challenges and costs of cycling coal and nuclear plants • Benefits and costs of combining multiple RE sources for the purpose of integration into energy markets • Institutional and technical barriers to integrating RE into heating and transport networks • Impacts of possible future changes in energy systems (including more or less centralization or decentralization, degree of demand response, and the level of integration of the electricity sector with the presently distinct heating and transport sectors) on integration challenges and cost
Comprehensive assessment of socioeconomic and environmental aspects of RE and other energy technologies	• Net lifecycle carbon emissions of certain RE technologies (e.g., some forms of bioenergy, hydropower) • Assessment of local and regional impacts on ecosystems and the environment • Assessment of local and regional impacts on human activities and well-being • Balancing widely varying positive and negative impacts over different geographic and temporal scales • Policies to effectively minimize and manage negative impacts, and realize positive benefits • Understanding and methods to address public acceptance concerns of local communities
Opportunities for meeting the needs of developing countries with sustainable RE services	• Impacts of RE deployment on multiple indicators of sustainable development • Regional/local RE resource assessments in developing countries • Advantages and limitations of improving energy access with decentralized forms of RE • Local human resource needs to ensure effective use of RE technologies • Financing mechanisms and investment tools to ensure affordability • Effective capacity building, as well as technology and knowledge transfer
Policy, institutional and financial mechanisms to enable cost-effective deployment of RE in a wide variety of contexts	• The combination of policies that are most efficient and effective for deploying different RE technologies in different countries. • How to address equity concerns while encouraging significant increases in RE investment. • How to design a policy such that potential co-benefits of RE deployment are maximized, for example security, equity and environmental benefits • Optimizing the balance of design and of timing of RE-specific versus carbon-pricing policies to take best advantage of the synergies between these two policy types. • Finding the most effective way to overcome the inherent advantage of current energy technologies including regulations and standards that lock-out RE technologies and what needs to change in order to allow RE to penetrate the energy system

Box 1.1 | Implications of different primary energy accounting conventions for energy and emission scenarios.

Primary energy for combustible energy sources is defined as the heat released when it is burned in air. As discussed in Annex II (A.II.4) and Table 1.A.1, there is no single, unambiguous accounting method for calculating primary energy from non-combustible energy sources such as nuclear energy and all RE sources with the exception of bioenergy. The *direct equivalent method* is used throughout this report. The direct equivalent method treats all non-combustible energy sources in an identical way by counting one unit of secondary energy provided from non-combustible sources as one unit of primary energy, that is, 1 kWh of electricity or heat is accounted for as 1 kWh = 3.6 MJ of primary energy. Depending on the type of secondary energy produced, this may lead to an understatement of the contribution of non-combustible RE and nuclear compared to bioenergy and fossil fuels by a factor of roughly 1.2 up to 3 (using indicative fossil fuel to electricity and heat conversion efficiencies of 38 and 85%, respectively). The implications of adopting the direct equivalent method in contrast to the other two most prominent methods—the physical energy content method and the substitution method—are illustrated in Figure 1.15 and Table 1.2 based on a selected climate stabilization scenario. The scenario is from Loulou et al. (2009) and is referred to as 1B3.7MAX in that publication. CO_2-equivalent concentrations of the Kyoto gases reach 550 ppm by 2100.

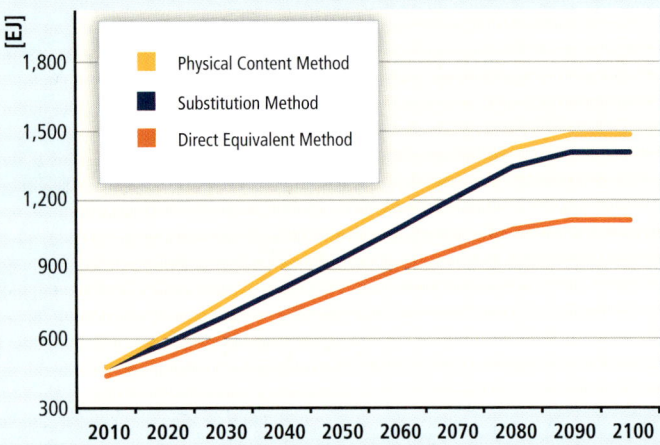

Differences from applying the three accounting methods to current energy consumption remain limited. However, substantial differences arise when applying the methods to long-term scenarios when RE reaches higher shares. For the selected scenario, the accounting gap between methods grows substantially over time, reaching about 370 EJ by 2100. There are significant differences in the accounting for individual non-combustible sources by 2050, and even the share of total renewable primary energy supply varies between 24 and 37% across the three methods. The biggest absolute gap for a single source is geothermal energy, with about 200 EJ difference between the direct equivalent and the physical energy content method. The gaps for hydro and nuclear energy remain considerable. For more details on the different approaches, see Annex II.

Figure 1.15 | Comparison of global total primary energy supply between 2010 and 2100 using different primary energy accounting methods based on a 550 ppm CO_2eq stabilization scenario.

Table 1.2 | Comparison of global total primary energy supply in 2050 using different primary energy accounting methods based on a 550 ppm CO_2eq stabilization scenario.

	Physical content method		Direct equivalent method		Substitution method	
	EJ	%	EJ	%	EJ	%
Fossil fuels	586.56	55.24	581.56	72.47	581.56	61.71
Nuclear	81.10	7.70	26.76	3.34	70.43	7.47
RE	390.08	37.05	194.15	24.19	290.37	30.81
Bioenergy	119.99	11.40	119.99	14.95	119.99	12.73
Solar	23.54	2.24	22.04	2.75	35.32	3.75
Geothermal	217.31	20.64	22.88	2.85	58.12	6.17
Hydro	23.79	2.26	23.79	2.96	62.61	6.64
Ocean	0.00	0.00	0.00	0.00	0.00	0.00
Wind	5.45	0.52	5.45	0.68	14.33	1.52
Total	1,052.75	100.00	802.47	100.00	942.36	100.00

instead of wind kinetic energy, geothermal heat, uranium fuel or solar radiation (Macknick, 2009; Nakicenovic et al., 1998). Hence any losses between the original sources and electricity are not counted in the amount of primary energy from these non-combustible sources (Annex II, A.II.4). Hence, primary energy requirements to produce a unit of electricity or other work from these sources are generally lower than for fossil fuels or biomass combustion processes.

Some RE technologies can be deployed at the point of use (decentralized) in rural and urban environments, whereas others are primarily employed within large (centralized) energy networks. Though many RE technologies are technically mature and are being deployed at significant scale, others are in an earlier phase of technical maturity and commercial deployment. The overview of RE technologies and applications in Table 1.3 provides an abbreviated list of the major renewable primary energy sources and technologies, the status of their development and the typical or primary distribution method (centralized network/grid required or decentralized, local standalone supply). The list is not considered to be comprehensive, for example, domestic animals and obtaining energy from plant biomass provide an important energy service in transportation and agriculture in many cultures but are not considered in this report. The table is constructed from the information and findings in the respective technology chapters.

1.2.2 Theoretical potential of renewable energy

The theoretical potential of RE is much greater than all of the energy that is used by all the economies on Earth. The challenge is to capture it and utilize it to provide desired energy services in a cost-effective manner. Estimated annual fluxes of RE and a comparison with fossil fuel reserves and 2008 annual consumption of 492 EJ are provided in Table 1.4.

1.2.3 Technical potential of renewable energy technologies

Technical potential is defined as the amount of RE output obtainable by full implementation of demonstrated and likely to develop technologies or practices.[8] The literature related to the technical potential of the different RE types assessed in this report varies considerably (Chapters 2 through 7 contain details and references). Among other things, this variation is due to methodological differences among studies, variant definitions of technical potential and variation due to differences between authors about how technologies and resource capture techniques may change over time. The global technical potential of RE sources will not limit continued market growth. A wide range of estimates is

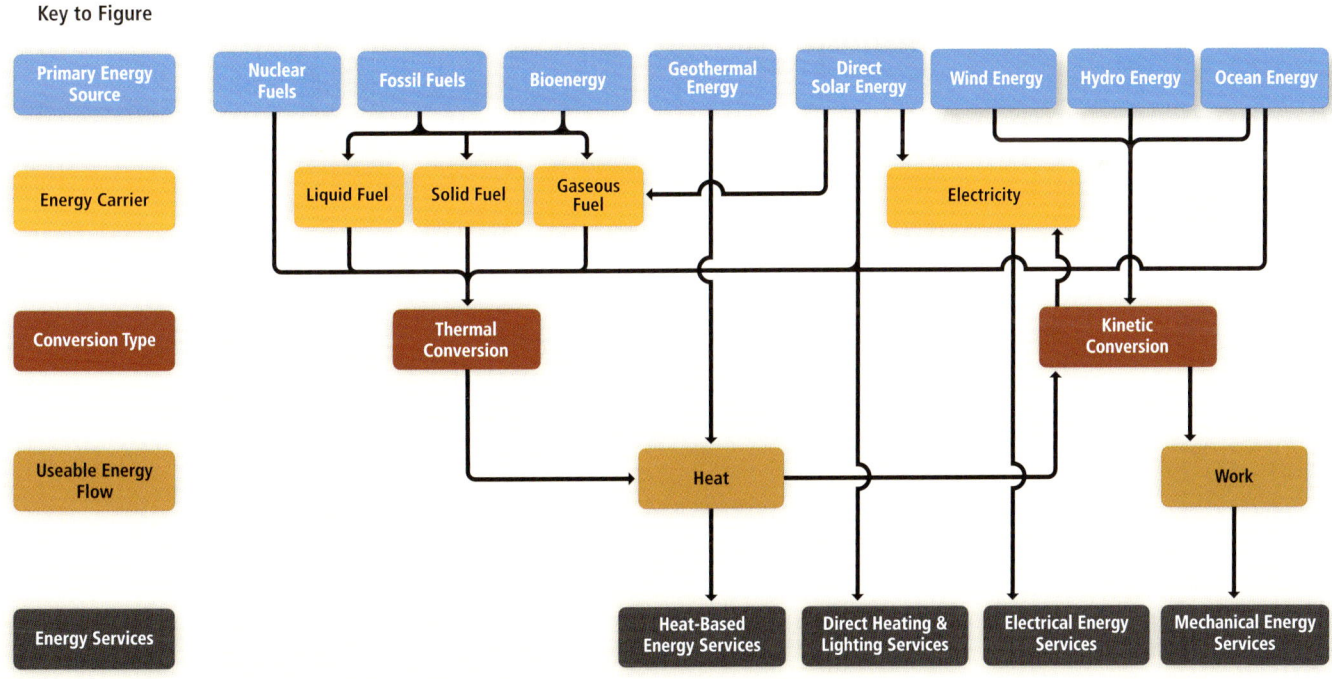

Figure 1.16 | Illustrative paths of energy from source to service. All connected lines indicate possible energy pathways. The energy services delivered to the users can be provided with differing amounts of end-use energy. This in turn can be provided with more or less primary energy from different sources, and with differing emissions of CO_2 and other environmental impacts.

---

8   The Glossary (Annex I) provides a more comprehensive definition of this term and of economic and market potential.

Table 1.3 | Overview of renewable energy technologies and applications (Chapters 2 through 7)

Renewable Energy Source	Select Renewable Energy Technologies	Primary Energy Sector (Electricity, Thermal, Mechanical, Transport)[1]	Technology Maturity[2]				Primary Distribution Method[3]	
			R & D	Demo & Pilot Project	Early-Stage Com'l	Later-Stage Com'l	Centralized	Decentralized
Bioenergy[4]	Traditional Use of Fuelwood/Charcoal	Thermal				•		•
	Cookstoves (Primitive and Advanced)	Thermal				•		•
	Domestic Heating Systems (pelletbased)	Thermal				•		•
	Small- and Large-Scale Boilers	Thermal				•	•	•
	Anaerobic Digestion for Biogas Production	Electricity/Thermal/Transport				•	•	•
	Combined Heat and Power (CHP)	Electricity/Thermal				•	•	•
	Co-firing in Fossil Fuel Power Plant	Electricity				•	•	
	Combustion-based Power Plant	Electricity				•	•	•
	Gasification-based Power Plant	Electricity			•		•	•
	Sugar- and Starch-Based Crop Ethanol	Transport				•	•	
	Plant- and Seed Oil-Based Biodiesel	Transport				•	•	
	Lignocellulose Sugar-Based Biofuels	Transport		•			•	
	Lignocellulose Syngas-Based Biofuels	Transport			•		•	
	Pyrolysis-Based Biofuels	Transport		•			•	
	Aquatic Plant-Derived Fuels	Transport	•				•	
	Gaseous Biofuels	Thermal				•		
Direct Solar	Photovoltaic (PV)	Electricity				•	•	•
	Concentrating PV (CPV)	Electricity			•		•	•
	Concentrating Solar Thermal Power (CSP)	Electricity			•		•	•
	Low Temperature Solar Thermal	Thermal				•		•
	Solar Cooling	Thermal		•				•
	Passive Solar Architecture	Thermal				•		•
	Solar Cooking	Thermal				•		•
	Solar Fuels	Transport	•				•	
Geothermal	Hydrothermal, Condensing Flash	Electricity				•	•	
	Hydrothermal, Binary Cycle	Electricity				•	•	
	Engineered Geothermal Systems (EGS)	Electricity		•			•	
	Submarine Geothermal	Electricity	•				•	
	Direct Use Applications	Thermal				•	•	•
	Geothermal Heat Pumps (GHP)	Thermal				•		•
Hydropower	Run-of-River	Electricity/Mechanical				•	•	•
	Reservoirs	Electricity				•	•	•
	Pumped Storage	Electricity				•	•	
	Hydrokinetic Turbines	Electricity/Mechanical		•			•	•
Ocean Energy	Wave	Electricity		?			?	
	Tidal Range	Electricity				?	?	
	Tidal Currents	Electricity		?			?	
	Ocean Currents	Electricity	?				?	
	Ocean Thermal Energy Conversion	Electricity/Thermal		?			?	
	Salinity Gradients	Electricity		?			?	

Continued next Page →

Renewable Energy Source	Select Renewable Energy Technologies	Primary Energy Sector (Electricity, Thermal, Mechanical, Transport)[1]	Technology Maturity[2]				Primary Distribution Method[3]	
			R & D	Demo & Pilot Project	Early-Stage Com'l	Later-Stage Com'l	Centralized	Decentralized
Wind Energy	Onshore, Large Turbines	Electricity				•	•	
	Offshore, Large Turbines	Electricity			•		•	
	Distributed, Small Turbines	Electricity				•		•
	Turbines for Water Pumping / Other Mechanical	Mechanical				•		•
	Wind Kites	Transport		•				•
	Higher-Altitude Wind Generators	Electricity	•				•	

Notes: 1. Primary energy sector as used here is intended to refer to the primary current or expected use(s) of the RE technology. In practice, RE-generated fuels may be used to meet a variety of energy service needs (not only transportation); electricity can be used to meet thermal and transportation needs; etc. 2. The highest level of maturity within each technology category is identified in the table; less mature technologies exist within some technology categories. 3. Centralized refers to energy supply that is distributed to end users through a network; decentralized refers to energy supply that is created onsite. Categorization is based on the 'primary' distribution method, recognizing that virtually all technologies can, in some circumstances, be used in both a centralized and decentralized fashion. 4. Bioenergy technologies can also be combined with CCS, though CCS technology is at an earlier stage of maturity.

Table 1.4 | Renewable energy theoretical potential expressed as annual energy fluxes of EJ/yr compared to 2008 global primary energy supply.

Renewable source	Annual Flux (EJ/yr)	Ratio (Annual energy flux/ 2008 primary energy supply)	Total reserve
Bioenergy	1,548[d]	3.1	—
Solar Energy	3,900,000[a]	7,900	—
Geothermal Energy	1,400[c]	2.8	—
Hydropower	147[a]	0.30	—
Ocean Energy	7,400[a]	15	—
Wind Energy	6,000[a]	12	—
Annual Primary energy source	**Annual Use 2008 (EJ/yr)**	**Lifetime of Proven Reserve (years)**	**Total Reserve (EJ)**
Total Fossil	418[b]	112	46,700
Total Uranium	10[b]	100–350	1,000–3,500
Total RE	64[b]	—	—
Primary Energy Supply	492 (2008)[b]	—	—

Sources: a. Rogner et al. (2000); b. IEA (2010c) converted to direct equivalent method (Annex II; IEA, 2010d); c. Pollack et al. (1993); d. Smeets et al. (2007).

provided in the literature but studies have consistently found that the total global technical potential for RE is substantially higher than both current and projected future global energy demand. Figure 1.17 summarizes the ranges of technical potential for the different RE technologies based on the respective chapter discussions. These ranges are compared to a comprehensive literature review by Krewitt et al. (2009) in Table 1.A.1 including more detailed notes and explanations in the Appendix to this chapter.[9] The technical potential for solar energy is the highest among the RE sources, but substantial technical potential exists for all forms of RE. According to the definition of technical potential in the Glossary (Annex I), many of the studies summarized in Table 1.A.1 to some extent take into account broader economic and socio-political considerations. For example, for some technologies, land suitability or other sustainability factors are included, which result in lower technical potential estimates. However, the absolute size of the global technical potential for RE as a whole is unlikely to constrain RE deployment.

Taking into account the uncertainty of the technical potential estimates, Figure 1.17 and Table 1.A.1 provide a perspective for the reader to understand the relative technical potential of the RE resources in the context of current global electricity and heat demand as well as of global primary energy supply. Aspects related to technology evolution, sustainability, resource availability, land use and other factors that relate to this technical potential are explored in the various

---

9 The definition of technical potential in Loulou et al. (2009) is similar but not identical to the definition here in that it is bounded by local/geographical availability and technological limitations associated with conversion efficiencies and the capture and transfer of the energy. See footnotes to Table 1.A.1.

Renewable Energy and Climate Change

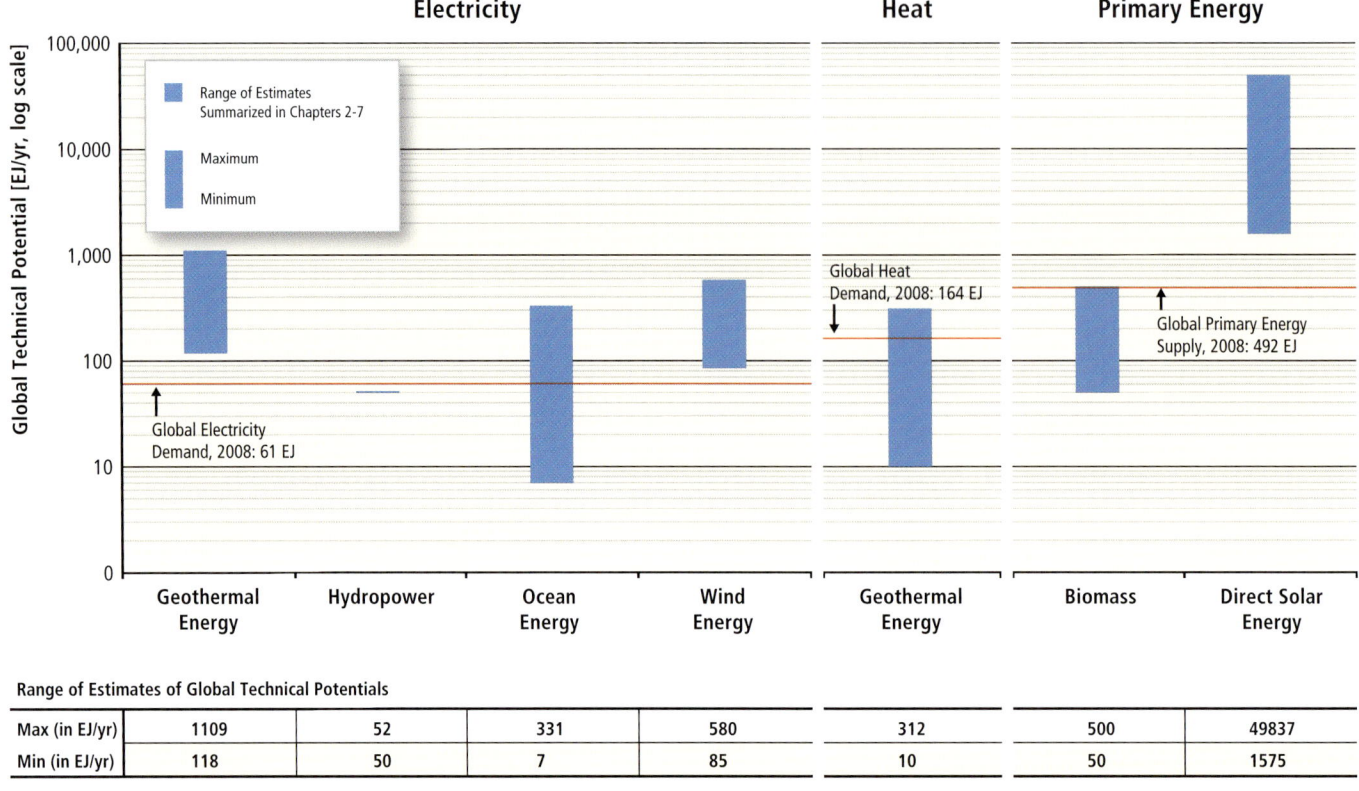

Figure 1.17 | Ranges of global technical potentials of RE sources derived from studies presented in Chapters 2 through 7. Biomass and solar are shown as primary energy due to their multiple uses. Note that the figure is presented in logarithmic scale due to the wide range of assessed data.

Notes: Technical potentials reported here represent total worldwide potentials for annual RE supply and do not deduct any potential that is already being utilized. Note that RE electricity sources could also be used for heating applications, whereas biomass and solar resources are reported only in primary energy terms but could be used to meet various energy service needs. Ranges are based on various methods and apply to different future years; consequently, the resulting ranges are not strictly comparable across technologies. For the data behind the figure and additional notes that apply, see Table 1.A.1 (as well as the underlying chapters).

chapters. The regional distribution of technical potential is addressed in Chapter 10.

Note also that the various types of energy cannot necessarily be added together to estimate a total, because each type was estimated independently of the others (e.g., the assessment did not take into account land use allocation; for example, PV and concentrating solar power cannot occupy the same space even though a particular site is suitable for either of them).

In addition to the theoretical and technical potential discussions, this report also considers the economic potential of RE sources that takes into account all social costs and assumes perfect information (covered in Section 10.6) and the market potential of RE sources that depends upon existing and expected real-world market conditions (covered in Section 10.3) shaped by policies, availability of capital and other factors, each of which is discussed in AR4 and defined in Annex I.

1.2.4 Special features of renewable energy with regard to integration

The costs and challenges of integrating increasing shares of RE into an existing energy supply system depend on the system characteristics, the current share of RE, the RE resources available and how the system evolves and develops in the future. Whether for electricity, heating, cooling, gaseous fuels or liquid fuels, RE integration is contextual, site specific and complex. The characteristics of RE specific to integration in existing energy networks are discussed in detail in Chapter 8.

RE can be integrated into all types of electricity systems from large, interconnected continental-scale grids (Section 8.2.1) down to small autonomous buildings (Sections 1.3.1, 8.2.5). System characteristics are important, including the generation mix, network infrastructure, energy market designs and institutional rules, demand location, demand profiles, and control and communication capability. Combined with the location, distribution, variability and predictability of the RE resources, these characteristics determine the scale of the integration challenge. Partially dispatchable wind and solar energy can be more difficult to integrate than fully dispatchable hydropower, bioenergy and geothermal energy. Partly because of the geographical distribution and fixed remote locations of many RE resources, as the penetration level of RE increases, there is need for a mixture of inexpensive and effective communications systems and technologies, as well as smart meters (Section 8.2.1).

As the penetration of partially dispatchable RE electricity increases, maintaining system reliability becomes more challenging and costly. A portfolio of solutions to minimize the risks and costs of RE integration can include the development of complementary flexible generation, strengthening and extending network infrastructure and interconnections, electricity demand that can respond in relation to supply availability, energy storage technologies (including reservoir hydropower), and modified institutional arrangements including regulatory and market mechanisms (Section 8.2.1).

Integration of RE into district heating and cooling networks (Section 8.2.2), gas distribution grids (Section 8.2.3) and liquid fuel systems (Section 8.2.4) has different system requirements and challenges than those of electrical power systems. Storage is an option for heating and cooling networks that incorporate variable RE sources. For RE integration into gas distribution grids, it is important that appropriate gas quality standards are met. Various RE technologies can also be utilized directly in all end-use sectors (such as first-generation biofuels, building-integrated solar water heaters and wind power) (Section 8.3).

The full utilization of variable renewable sources such as wind and solar power can be enhanced by energy storage. Storing energy as heat is commonly practised today, and multiple means of storing electricity have been developed. Pumped water storage is a well-developed technology that can utilize existing dams to provide electricity when variable sources are not providing. Other technologies include flywheel storage of kinetic energy, compressed air storage and batteries. Battery and other storage technologies are discussed in Chapter 8. If electric vehicles become a major fraction of the fleet, it is possible to utilize their batteries in a vehicle-to-grid system for managing the variability of RE supply (Moomaw, 1991; Kempton and Tomic, 2005; Hawken et al., 2010).

1.2.5 Energy efficiency and renewable energy

Energy services are the tasks to be performed using energy. A specific energy service can be provided in many ways. Lighting, for example, may be provided by daylight, candles or oil lamps or by a multitude of different electric lamps. The efficiency of the multiple conversions of energy from primary source to final output may be high or low, and may involve the release of large or small amounts of CO_2 (under a given energy mix). Hence there are many options as to how to supply any particular service.

In this report, some specific definitions for different dimensions of efficiency are utilized.

Energy efficiency is the ratio of useful energy or other useful physical outputs obtained from a system, conversion process, transmission or storage activity to its energy input (measured as kWh/kWh, tonnes/kWh or any other physical measure of useful output like tonne-km transported, etc.). Energy efficiency can be understood as the reciprocal of energy intensity. Hence the fraction of solar, wind or fossil fuel energy that can be converted to electricity is the conversion efficiency. There are fundamental limitations on the efficiency of conversions of heat to work in an automobile engine or a steam or gas turbine, and the attained conversion efficiency is always significantly below these limits. Current supercritical coal-fired steam turbines seldom exceed a 45% conversion of heat to electric work (Bugge et al., 2006), but a combined-cycle steam and gas turbine operating at higher temperatures has achieved 60% efficiencies (Pilavachi, 2000; Najjar et al., 2004).

Energy intensity is the ratio of energy use to output. If output is expressed in physical terms (e.g., tonnes of steel output), energy intensity is the reciprocal of energy productivity or energy efficiency. Alternatively (and often more commonly), output is measured in terms of populations (i.e., per capita) or monetary units such as contribution to gross domestic product (GDP) or total value of shipments or similar terms. At the national level, energy intensity is the ratio of total domestic primary (or final) energy use to GDP. Energy intensity can be decomposed as a sum of intensities of particular activities weighted by the activities' shares of GDP. At an aggregate macro level, energy intensity stated in terms of energy per unit of GDP or in energy per capita is often used for a sector such as transportation, industry or buildings, or to refer to an entire economy.

Energy savings arise from decreasing energy intensity by changing the activities that demand energy inputs. For example, turning off lights when not needed, walking instead of taking vehicular transportation, changing the controls for heating or air conditioning to avoid excessive heating or cooling or eliminating a particular appliance and performing a task in a less energy intensive manner are all examples of energy savings (Dietz et al., 2009). Energy savings can be realized by technical, organizational, institutional and structural changes and by changed behaviour.

Studies suggest that energy savings resulting from efficiency measures are not always fully realized in practice. There may be a rebound effect in which some fraction of the measure is offset because the lower total cost of energy to perform a specific energy service may lead to utilization of more energy services. Rebound effects can be distinguished at the micro and macro level. At the micro level, a successful energy efficiency measure may be expected to lead to lower energy costs for the entity subject to the measure because it uses less energy. However, the full energy saving may not occur because a more efficient vehicle reduces the cost of operation per kilometre, so the user may drive more kilometres. Or a better-insulated home may not achieve the full saving because it is now possible to achieve greater comfort by using some of the saved energy. The analysis of this effect is filled with many methodological difficulties (Guerra and Sancho, 2010), but it is estimated that the rebound effect is probably limited by saturation effects to between 10 and 30% for home heating and vehicle use in OECD countries, and is very small for more efficient appliances and water heating (Sorrell et al., 2009). An efficiency measure that is successful in lowering economy-wide energy demand, however, lowers the price of energy as well. This leads to a decrease in economy-wide energy costs leading to additional cost savings for the entities that are subject to the efficiency measure (lower energy price and less energy use) as well as cost savings for the rest of the economy that may not be subject to the measure but benefits from the lower energy price. Studies that examine changes in energy intensity in OECD countries find that at the macro level, there is a reduction that appears related to energy efficiency gains, and any rebound effect is small (Schipper and Grubb, 2000). One analysis suggests that when all effects of lower energy prices are taken into account, there are offsetting factors that can outweigh a positive rebound effect (Turner, 2009). It is expected that the rebound effect may be greater in developing countries and among poor consumers (Orasch and Wirl, 1997). These analyses of the rebound effect do not examine whether an energy user might spend his economic savings on something other than the energy use whose efficiency was just improved (i.e., on other activities that involve either higher or lower energy intensity than the saved energy service), nor do there appear to be studies of corporate efficiency, where the savings might pass through to the bottom profit line. For climate change, the main concern with any rebound effect is its influence on CO_2 emissions, which can be addressed effectively with a price on carbon (Chapter 11).

The role of energy efficiency in combination with RE is somewhat more complex and less studied. It is necessary to examine the total cost of end-use efficiency measures plus RE technology, and then determine whether there is rebound effect for a specific case.

Furthermore, carbon leakage may also reduce the effectiveness of carbon reduction policies. Carbon leakage is defined as the increase in CO_2 emissions outside of the countries taking domestic mitigation action divided by the reduction in the emissions of these countries. If carbon reduction policies are not applied uniformly across sectors and political jurisdictions, then it is possible for carbon-emitting activities that are controlled in one place to move to another sector or country where such activities are not restricted (Kallbekken, 2007; IEA, 2008a). Recent research suggests, however, that estimates of carbon leakage are too high (Paltsev, 2001; Barker et al., 2007; Di Maria and van der Werf, 2008).

Reducing energy needed at the energy services delivery stage is an important means of reducing the primary energy required for all energy supply fuels and technologies. Because RE sources usually have a lower power density than fossil or nuclear fuels, energy savings at the end-use stage are often required to utilize a RE technology for a specific energy service (Twidell and Weir, 2005). For example, it may not be possible to fuel all vehicles on the planet with biofuels at their current low engine efficiencies, but if vehicle fuel efficiency were greater, a larger fraction of vehicles could be run on biofuels. Similarly, by lowering demand, the size and cost of a distributed solar system may become competitive (Rezaie et al., 2011). The importance of end-use efficiency in buildings in order for renewable technology to be a viable option has been documented (Frankl et al., 1998). Furthermore, electricity distribution and management is simplified and system balancing costs are lower if the energy demands are smaller (see Chapter 8). Energy efficiency at the end-use stage thus facilitates the use of RE.

Often the lowest cost option is to reduce end-use energy demand through efficiency measures, which include both new technologies and more efficient practices (Hamada et al., 2001; Venema and Rehman, 2007; Ambrose, 2009; Harvey, 2009). Examples can be found in efficient appliances for lighting, as well as heating and cooling in the building sector. For example, compact fluorescent or light-emitting diode lamps use much less electricity to produce a lumen of light than does a traditional incandescent lamp (Mehta et al., 2008). Properly sized variable-speed electric motors and improved efficiency compressors for refrigerators, air conditioners and heat pumps can lower primary energy use in many applications (Ionel, 1986; Sims et al., 2007; von Weizsäcker et al., 2009). Efficient houses and small commercial buildings such as the Passivhaus design from Germany are so air tight and well insulated that they require only about one-tenth the energy of more conventional dwellings (Passivhaus, 2010). Energy efficient design of high-rise buildings in tropical countries could reduce emissions from cooling at a substantial cost savings (Ossen et al., 2005; Ambrose, 2009).

Examples from the transportation sector include utilizing engineering improvements in traditional internal combustion engines to reduce fuel consumption rather than enhancing acceleration and performance (Ahman and Nilsson, 2008). Significant efficiency gains and substantial CO_2 emission reductions have also been achieved through the use of hybrid electric systems, battery electric systems and fuel cells (see Section 8.3.1). Biofuels become more economically feasible for aircraft as engine efficiency improves (Lee, 2010). Examples that raise energy efficiency in the power supply and industrial sectors include combined heat and power systems (Casten, 2008; Roberts, 2008), and recovery of otherwise wasted thermal or mechanical energy (Bailey and Worrell, 2005; Brown et al., 2005) thereby avoiding burning additional fuel for commercial and industrial heat. These latter examples are also applicable to enhancing the overall delivery of energy from RE such as capturing and utilizing the heat from PV or biomass electricity systems, which is done frequently in the forest products industry.

1.3 Meeting energy service needs and current status

1.3.1 Current renewable energy flows

Global renewable energy flows from primary energy through carriers to end uses and losses in 2008 (IEA, 2010a) are shown in Figure 1.18. 'RE' here includes combustible biomass, forest and crop residues and renewable municipal waste as well as the other types of RE considered in this report: direct solar (PV and solar thermal) energy, geothermal energy, hydropower, and ocean and wind energy.

'Other sectors' include agriculture, commercial and residential buildings, public services and non-specified other sectors. The 'transport sector' includes international aviation and international marine bunkers. Data for the renewable electricity and heat flows to the end-use sectors are not available. Considering that most of the renewable electricity is grid-connected, they are estimated on the assumption that their allocations to industries, transport and other sectors are proportional to those of the total electricity and heat, which are available from the IEA (IEA, 2010a).

At the global level, on average, RE supplies increased by 1.8% per annum between 1990 and 2007 (IEA, 2009b), nearly matching the growth rate in total primary energy consumption (1.9%).

Globally in 2008, around 56% of RE was used to supply heat in private households and in the public and services sector. Essentially, this refers to wood and charcoal, widely used in developing countries for cooking. On the other hand, only a small amount of RE is used in the transport sector. Electricity production accounts for 24% of the end-use consumption (IEA, 2010a). Biofuels contributed 2% of global road transport fuel supply in 2008, and traditional biomass (17%), modern biomass (8%), solar thermal and geothermal energy (2%) together fuelled 27% of the total global demand for heat in 2008 (IEA, 2010c).

1.3.2 Current cost of renewable energy

While the resource is obviously large and could theoretically supply all energy needs long into the future, the levelized cost of energy (LCOE) for many RE technologies is currently higher than existing energy prices,

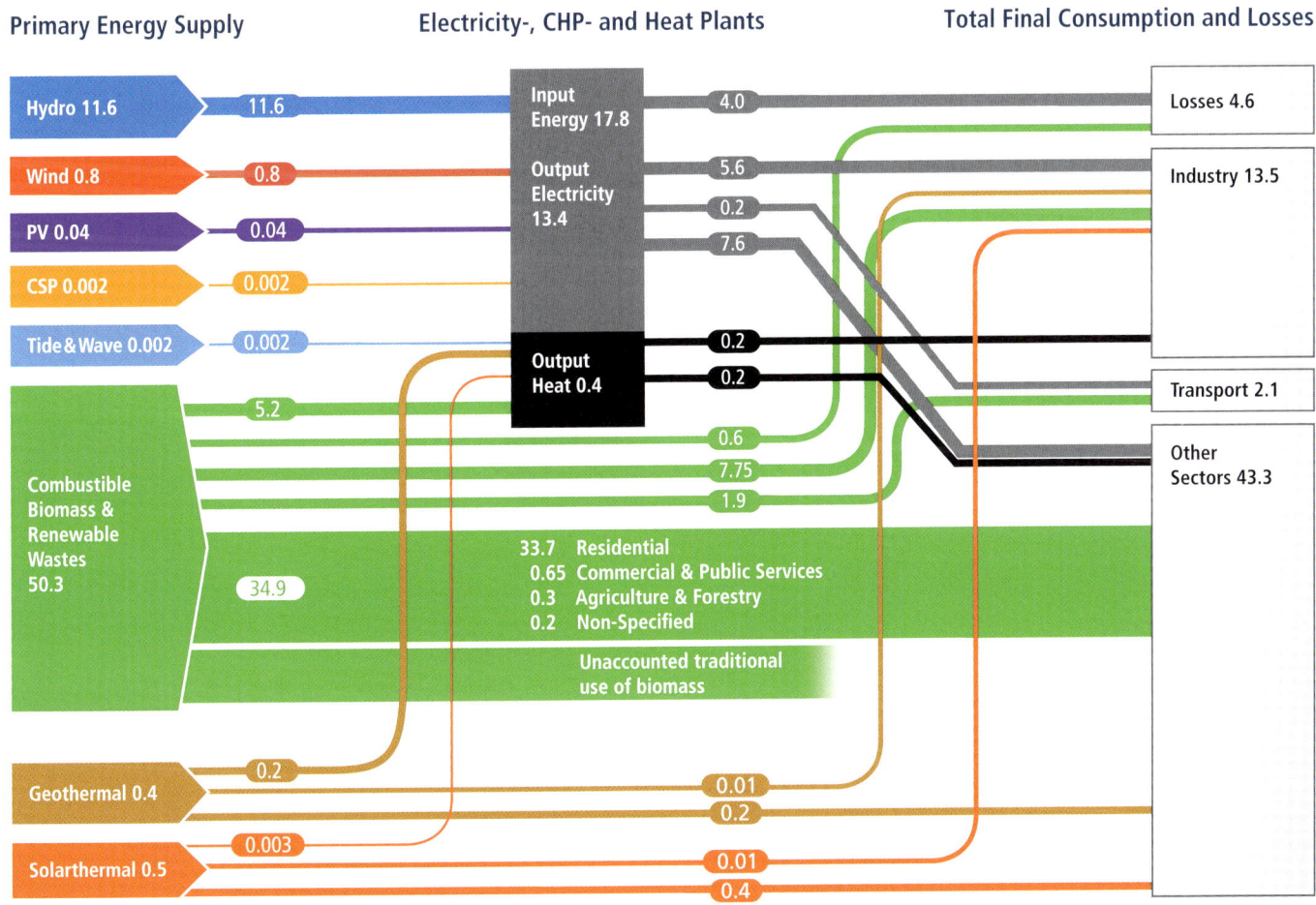

Figure 1.18 | Global energy flows (EJ) in 2008 from primary RE through carriers to end uses and losses. Data Source: (IEA, 2010a).

though in various settings RE is already economically competitive. Even though the LCOE of a particular energy technology is not the sole determinant of its value or economic competitiveness, ranges of recent LCOE are provided in this report as one of several benchmark values.[10] Figures 1.19, 1.20 and 1.21 provide a comparison of LCOE ranges associated with selected RE technologies that are currently commercially available to provide electricity, heat and transportation fuels, respectively. The ranges of recent LCOE for some of these RE technologies are wide and depend, inter alia, on technology characteristics, regional variations in cost and performance, and differing discount rates.

These cost ranges in these figures are broad and do not resolve the significant uncertainties surrounding the costs, if looked at from a very general perspective. Hence, as with the technical potential described above, the data are meant to provide context only (as opposed to precise comparison).

The levelized costs of identical technologies can vary across the globe, depending on services rendered, RE quality and local costs of investment, financing, operation and maintenance. The breadth of the ranges can be narrowed if region-, country-, project- and/or investor-specific conditions are taken into account. Chapters 2 through 7 provide some detail on the sensitivity of LCOE to such framework conditions; Section 10.5 shows the effect of the choice of the discount rate on levelized costs; and Annex III provides the full set of data and additional sensitivity analysis.

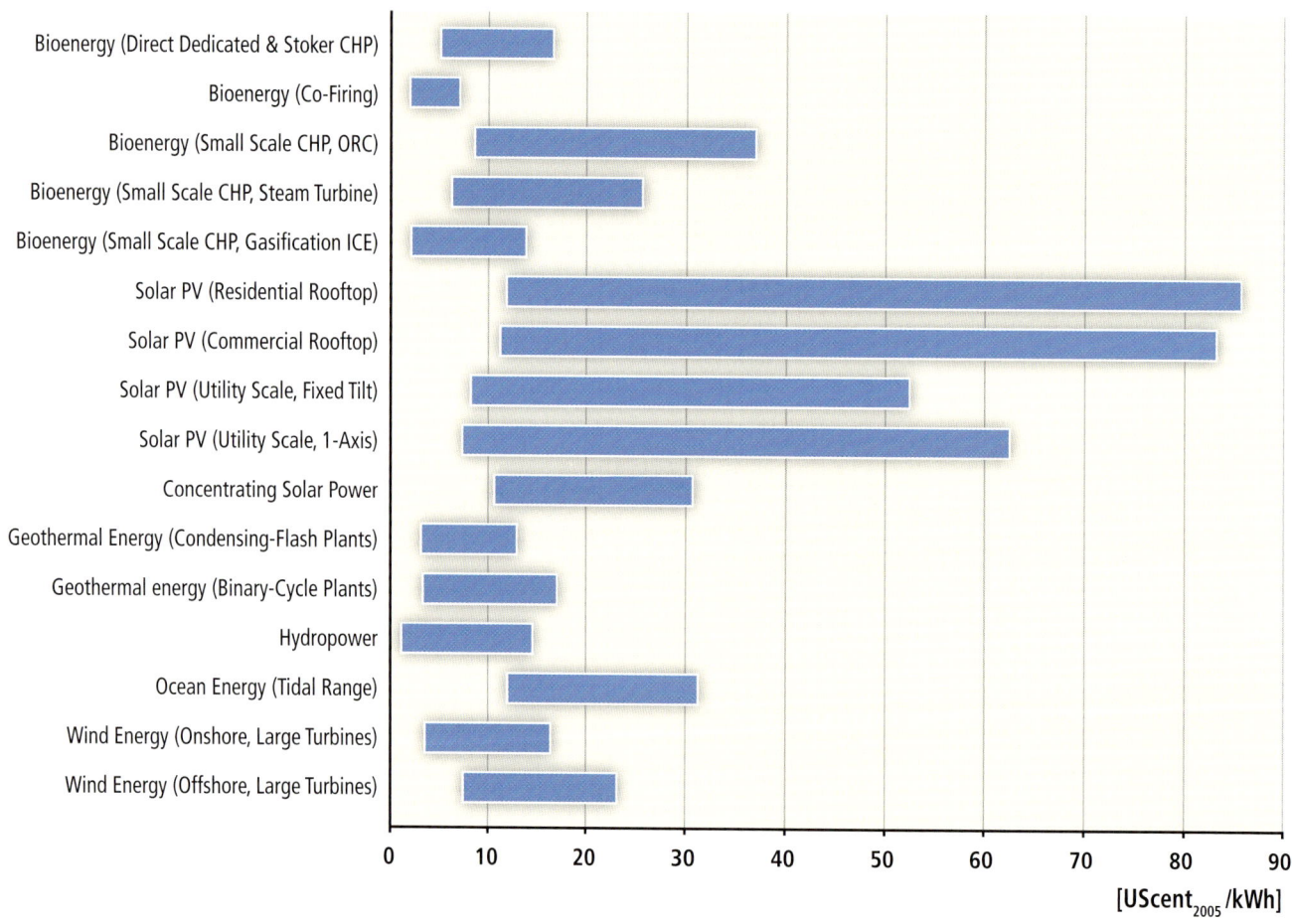

Figure 1.19 | Levelized cost of electricity (LCOE) for commercially available RE technologies covering a range of different discount rates. The LCOE estimates for all technologies are based on input data summarized in Annex III and the methodology outlined in Annex II. The lower bound of the levelized cost range is based on a 3% discount rate applied to the low ends of the ranges of investment, operations and maintenance (O&M), and (if applicable) feedstock cost and the high ends of the ranges of capacity factors and lifetimes as well as (if applicable) the high ends of the ranges of conversion efficiencies and by-product revenue. The higher bound of the levelized cost range is accordingly based on a 10% discount rate applied to the high end of the ranges of investment, O&M and (if applicable) feedstock costs and the low end of the ranges of capacity factors and lifetimes as well as (if applicable) the low ends of the ranges of conversion efficiencies and by-product revenue. Note that conversion efficiencies, by-product revenue and lifetimes were in some cases set to standard or average values. For data and supplementary information see Annex III.

---

10 Cost and performance data were gathered by the authors of Chapters 2 through 7 from a variety of sources in the available literature. They are based on the most recent information available in the literature. Details can be found in the respective chapters and are summarized in a data table in Annex III. All costs were assessed using standard discounting analysis at 3, 7 and 10% as described in the Annex II. A number of default assumptions about costs and performance parameters were made to define the levelized cost if data were unavailable and are also laid out in Annex III.

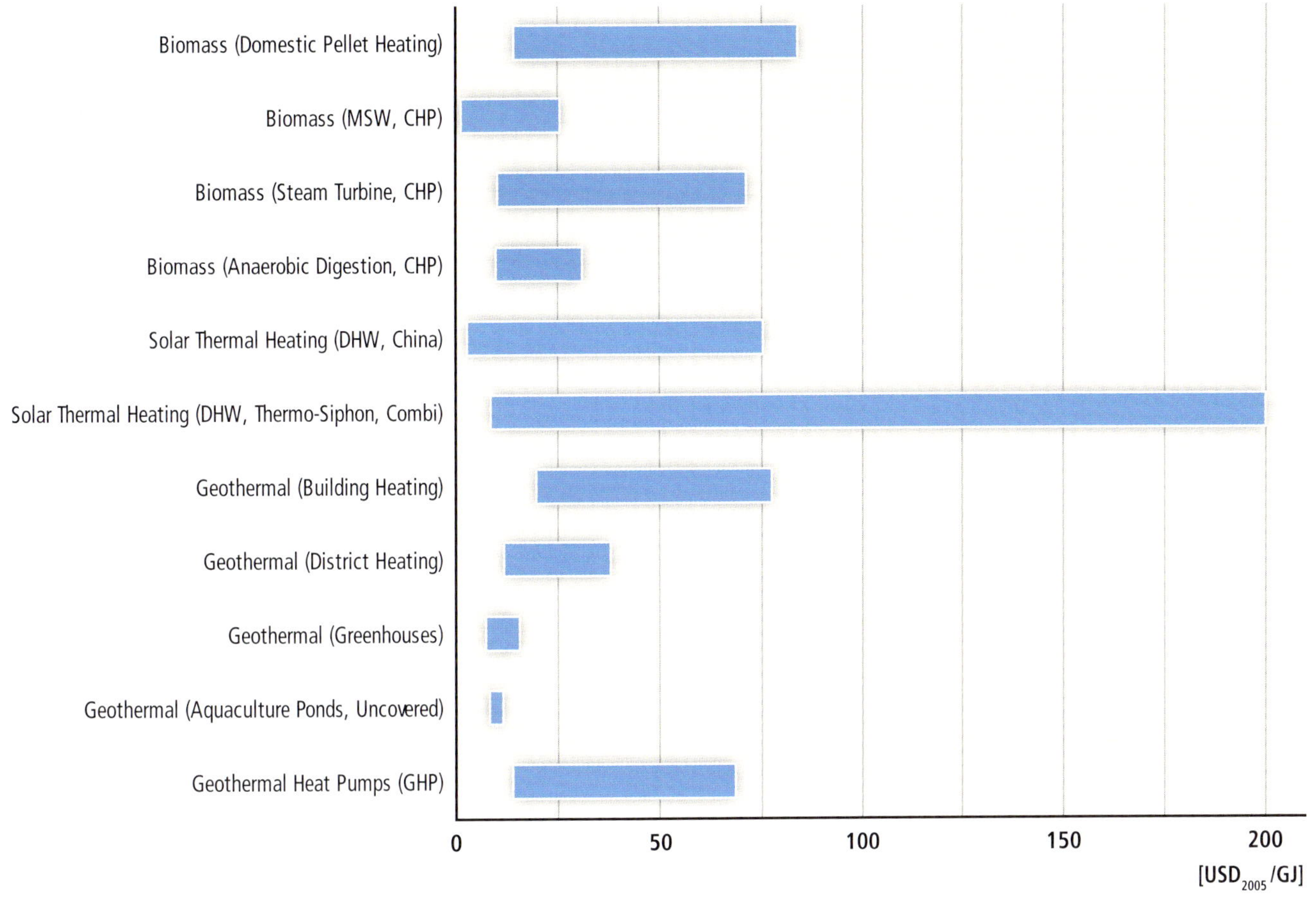

Figure 1.20 | Levelized cost of heat (LCOH) for commercially available RE technologies covering a range of different discount rates. The LCOH estimates for all technologies are based on input data summarized in Annex III and the methodology outlined in Annex II. The lower bound of the levelized cost range is based on a 3% discount rate applied to the low ends of the ranges of investment, operations and maintenance (O&M), and (if applicable) feedstock cost and the high ends of the ranges of capacity factors and lifetimes as well as (if applicable) the high ends of the ranges of conversion efficiencies and by-product revenue. The higher bound of the levelized cost range is accordingly based on a 10% discount rate applied to the high end of the ranges of investment, O&M and (if applicable) feedstock costs and the low ends of the ranges of capacity factors and lifetimes as well as (if applicable) the low ends of the ranges of conversion efficiencies and by-product revenue. Note that capacity factors and lifetimes were in some cases set to standard or average values. For data and supplementary information see Annex III.

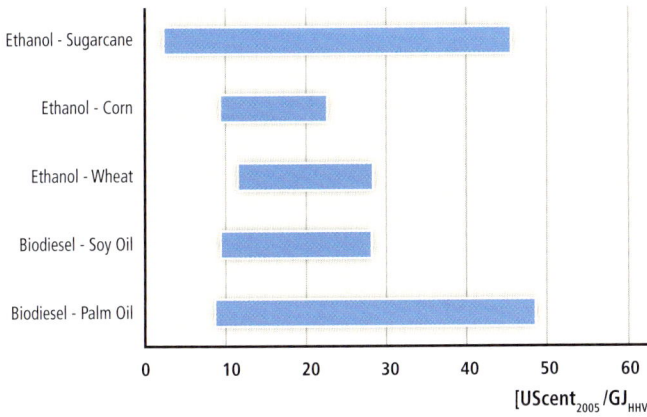

Figure 1.21 | Levelized Cost of Fuels (LCOF) for commercially available biomass conversion technologies covering a range of different discount rates. LCOF estimates for all technologies are based on input data summarized in Annex III and the methodology outlined in Annex II. The lower bound of the levelized cost range is based on a 3% discount rate applied to the low ends of the ranges of investment, operations and maintenance (O&M) and feedstock cost. The higher bound of the levelized cost range is accordingly based on a 10% discount rate applied to the high end of the ranges of investment, O&M and feedstock costs. Note that conversion efficiencies, by-product revenue, capacity factors and lifetimes were set to average values. HHV stands for 'higher heating value'. For data and supplementary information see Annex III.

Given favourable conditions, however, the lower ends of the ranges indicate that some RE technologies are broadly competitive at existing energy prices (see also Section 10.5). Monetizing the external costs of energy supply would improve the relative competitiveness of RE. The same applies if market prices increase due to other reasons (see Section 10.6). That said, these graphs provide no indication of the technical potential that can be utilized. Section 10.4 provides more information in this regard, for example, in discussing the concept of energy supply curves.

Furthermore, the levelized cost for a technology is not the sole determinant of its value or economic competitiveness. The attractiveness of a specific energy supply option depends also on broader economic as well as environmental and social aspects and the contribution that the technology makes to meeting specific energy services (e.g., peak electricity demands) or imposes in the form of ancillary costs on the energy system (e.g., the costs of integration). Chapters 8 to 11 offer important complementary perspectives on such cost issues covering, for example, the cost of integration, external costs and benefits, economy-wide costs and costs of policies.

The costs of most RE technologies have declined and additional expected technical advances would result in further cost reductions. Significant advances in RE technologies and associated long-term cost reductions have been demonstrated over the last decades, though periods of rising prices have sometimes been experienced (due to, for example, increasing demand for RE in excess of available supply) (see Section 10.5). The contribution of different drivers (e.g., R&D, economies of scale, deployment-oriented learning and increased market competition among RE suppliers) is not always understood in detail (see Sections 2.7, 3.8, 7.8, and 10.5).

Historical and potential future cost drivers are discussed in most of the technology chapters (Chapters 2 through 7) as well as in Chapter 10, including in some cases an assessment of historical learning rates and the future prospects for cost reductions under specific framework conditions. Further cost reductions are expected, resulting in greater potential deployment and consequent climate change mitigation. Examples of important areas of potential technological advancement include: new and improved feedstock production and supply systems; biofuels produced via new processes (also called next-generation or advanced biofuels, e.g., lignocellulosic) and advanced biorefining (Section 2.6); advanced PV and CSP technologies and manufacturing processes (Section 3.7); enhanced geothermal systems (EGS) (Section 4.6); multiple emerging ocean technologies (Section 6.6); and foundation and turbine designs for offshore wind energy (Section 7.7). Further cost reductions for hydropower are expected to be less significant than some of the other RE technologies, but R&D opportunities exist to make hydropower projects technically feasible in a wider range of locations and to improve the technical performance of new and existing projects (Sections 5.3, 5.7, and 5.8).

1.3.3 Regional aspects of renewable energy

The contribution of RE to primary energy supply varies substantially by country and region. Geographic distribution of RE manufacturing, use and export is now being diversified from the developed world to other developing regions, notably Asia including China (UNStats, 2010). In China, growing energy needs for solar cooking and hot water production have promoted RE development. China is now the leading producer, user and exporter of solar thermal panels for hot water production, and has been rapidly expanding its production of solar PV, most of which is exported, and has recently become the leading global producer. In terms of capacity, in 2008, China was the largest investor in thermal water heating and third in bioethanol production (REN21, 2009). China has been doubling its wind turbine installations every year since 2006, and was second in the world in installed capacity in 2009. India has also become a major producer of wind turbines and now is among the top five countries in terms of installation. In terms of installed renewable power capacity, China now leads the world followed by the USA, Germany, Spain and India (REN21, 2009, 2010).

As noted earlier, RE is more evenly distributed than fossil fuels. There are countries or regions rich in specific RE resources. Twenty-four countries utilize geothermal heat to produce electricity. The share of geothermal energy in national electricity production is above 15% in El Salvador, Kenya, the Philippines and Iceland (Bromley et al., 2010). More than 60% of primary energy is supplied by hydropower and geothermal energy in Iceland (IEA, 2010a). In some years, depending on the level of precipitation, Norway produces more hydroelectricity than it needs and exports its surplus to the rest of Europe. Brazil, New Zealand and Canada also have a high share of hydroelectricity in total electricity: 80, 65 and 60%, respectively (IEA, 2010c). Brazil relies heavily on and is the second-largest producer of bioethanol, which it produces from sugarcane (EIA, 2010; IEA, 2010e).

As regards biomass as a share of regional primary energy consumption, Africa is particularly high, with a share of 48.0%, followed by India at 26.5%, non-OECD Asia excluding China and India at 23.5%, and China at 10% (IEA, 2010a). Heat pump systems that extract stored solar energy from the air, ground or water have penetrated the market in developed countries, sometimes in combination with renewable technologies such as PV and wind. Heat pump technology is discussed in Chapter 4.

Sun-belt areas such as deserts and the Mediterranean littoral are abundant in direct normal radiation (cloudless skies) and suitable for concentrated solar thermal power plants. Export of solar- and wind-generated electricity from the countries rich in these resources could become important in the future (Desertec, 2010).

1.4 Opportunities, barriers and issues

The major global energy challenges are securing energy supply to meet growing demand, providing everybody with access to energy services and curbing energy's contribution to climate change. For developing countries, especially the poorest, energy is needed to stimulate production, income generation and social development, and to reduce the serious health problems caused by the use of fuel wood, charcoal, dung and agricultural waste. For industrialized countries, the primary reasons to encourage RE include emission reductions to mitigate climate change, secure energy supply concerns and employment creation. RE can open opportunities for addressing these multiple environmental, social and economic development dimensions, including adaptation to climate change, which is described in Section 1.4.1.

Some form of renewable resource is available everywhere in the world—for example, solar radiation, wind, falling water, waves, tides and stored ocean heat, heat from the earth or biomass—furthermore, technologies that can harness these forms of energy are available and are improving rapidly (Asif and Muneer, 2007). While the opportunities seem great and are discussed in Section 1.4.1, there are barriers (Section 1.4.2) and issues (Section 1.4.3) that slow the introduction of RE into modern economies.

1.4.1 Opportunities

Opportunities can be defined as circumstances for action with the attribute of a chance character. In the policy context, that could be the anticipation of additional benefits that may go along with the deployment of RE (and laid out below) but that are not intentionally targeted. There are four major opportunity areas that RE is well suited to address, and these are briefly described here and in more detail in Section 9.2.2. The four areas are social and economic development, energy access, energy security, and climate change mitigation and the reduction of environmental and health impacts.

1.4.1.1 Social and economic development

Globally, per capita incomes as well as broader indicators such as the Human Development Index are positively correlated with per capita energy use, and economic growth can be identified as the most relevant factor behind increasing energy consumption in the last decades. As economic activity expands and diversifies, demands for more sophisticated and flexible energy sources arise. Economic development has therefore been associated with a shift from direct combustion of fuels to higher quality electricity (Kaufmann, 2004; see Section 9.3.1).

Particularly for developing countries, the link between social and economic development and the need for modern energy services is evident. Access to clean and reliable energy constitutes an important prerequisite for fundamental determinants of human development, contributing, inter alia, to economic activity, income generation, poverty alleviation, health, education and gender equality (Kaygusuz, 2007; UNDP, 2007). Because of their decentralized nature, RE technologies can play an important role in fostering rural development (see Section 1.4.1.2).

The creation of (new) employment opportunities is seen as a positive long-term effect of RE both in developed and developing countries and was stressed in many national green-growth strategies. Also, policymakers have supported the development of domestic markets for RE as a means to gain competitive advantage in supplying international markets (see Sections 9.3.1.4 and 11.3.4).

1.4.1.2 Energy access

In 2009, more than 1.4 billion people globally lacked access to electricity, 85% of them in rural areas, and the number of people relying on traditional biomass for cooking was estimated to be around 2.7 billion (IEA, 2010c). By 2015, almost 1.2 billion more people will need access to electricity and 1.9 billion more people will need access to modern fuels to meet the Millennium Development Goal of halving the proportion of people living in poverty (UNDP/WHO, 2009).

The transition to modern energy access is referred to as moving up the energy ladder and implies a progression from traditional to more modern devices/fuels that are more environmentally benign and have fewer negative health impacts. Various initiatives, some of them based on RE, particularly in the developing countries, aim at improving universal access to modern energy services through increased access to electricity and cleaner cooking facilities (REN 21, 2009; see Sections 9.3.2 and 11.3.2). In particular, reliance on RE in rural applications, use of locally produced bioenergy to produce electricity, and access to clean cooking facilities will contribute to attainment of universal access to modern energy services (IEA, 2010d).

For electricity, small and standalone configurations of RE technologies such as PV (Chapter 3), hydropower (Chapter 5), and bioenergy (Chapter 2) can often meet energy needs of rural communities more cheaply than fossil fuel alternatives such as diesel generators. For example, PV is attractive as a source of electric power to provide basic services, such as lighting and clean drinking water. For greater local demand, small-scale hydropower or biomass combustion and gasification technologies may offer better solutions (IEA, 2010d). For bioenergy, the progression implies moving from the use of, for example, firewood, cow dung and agricultural residues to, for example, liquid propane gas stoves, RE-based advanced biomass cookstoves or biogas systems (Clancy et al., 2007; UNDP, 2005; IEA, 2010d; see Sections 2.4.2 and 9.3.2).

1.4.1.3 Energy security

At a general level, energy security can best be understood as robustness against (sudden) disruptions of energy supply. More specifically, availability and distribution of resources, as well as variability and reliability of energy supply can be identified as the two main themes.

Current energy supplies are dominated by fossil fuels (petroleum and natural gas) whose price volatility can have significant impacts, in particular for oil-importing developing countries (ESMAP, 2007). National security concerns about the geopolitical availability of fuels have also been a major driver for a number of countries to consider RE. For example, in the USA, the military has led the effort to expand and diversify fuel supplies for aviation and cites improved energy supply security as the major driving force for sustainable alternative fuels (Hileman et al., 2009; Secretary of the Air Force, 2009; USDOD, 2010).

Local RE options can contribute to energy security goals by means of diversifying energy supplies and diminishing dependence on limited suppliers, although RE-specific challenges to integration must be considered. In addition, the increased uptake of RE technologies could be an avenue to redirect foreign exchange flows away from energy imports towards imports of goods that cannot be produced locally, such as high-tech capital goods. This may be particularly important for oil-importing developing countries with high import shares (Sections 9.3.3, 9.4.3 and 11.3.3).

1.4.1.4 Climate change mitigation and reduction of environmental and health impacts

Climate change mitigation is one of the key driving forces behind a growing demand for RE technologies (see Section 11.3.1). In addition to reducing GHG emissions, RE technologies can also offer benefits with respect to air pollution and health compared to fossil fuels (see Section 9.3.4). Despite these important advantages of RE, no large-scale technology deployment comes without trade-offs, such as, for example, induced land use change. This mandates an assessment of the overall burden from the energy system on the environment and society, taking account of the broad range of impact categories with the aim of identifying possible trade-offs and potential synergies.

Lifecycle assessments facilitate a quantitative comparison of 'cradle to grave' emissions across different energy technologies (see Section 9.3.4.1). Figure 1.22 illustrates the lifecycle structure for CO_2 emission analysis, and qualitatively indicates the relative GHG implications for RE, nuclear power and fossil fuels. Alongside the commonly known CO_2 production pathways from fossil fuel combustion, natural gas production (and transportation) and coal mines are a source of methane, a potent greenhouse gas, and uncontrolled coal mine fires release significant amounts of CO_2 to the atmosphere.

Traditional biomass use results in health impacts from the high concentrations of particulate matter and carbon monoxide, among other pollutants. Long-term exposure to biomass smoke increases the risk of a child developing an acute respiratory infection and is a major cause of morbidity and mortality in developing countries (WEC/FAO, 1999).

In this context, non-combustion-based RE power generation technologies have the potential to significantly reduce local and regional air pollution and lower associated health impacts compared to fossil-based power generation. Improving traditional biomass use can reduce negative impacts on sustainable development, including local and indoor air pollution, GHG emissions, deforestation and forest degradation (see Sections 2.5.4, 9.3.4.2, 9.3.4.3 and 9.4.2).

Impacts on water resources from energy systems strongly depend on technology choice and local conditions. Electricity production with wind and solar PV, for example, requires very little water compared to thermal conversion technologies, and has no impacts on water quality. Limited water availability for cooling thermal power plants decreases their efficiency, which can affect plants operating on coal, biomass, gas, nuclear and concentrating solar power (see Section 9.3.4.4). There have been significant power reductions from nuclear and coal plants during drought conditions in the USA and France in recent years.

Surface-mined coal in particular produces major alterations of land; coal mines can create acid mine drainage and the storage of coal ash can contaminate surface and ground waters. Oil production and transportation have lead to significant land and water spills. Most renewable technologies produce lower conventional air and water pollutants than fossil fuels, but may require large amounts of land as, for example, reservoir hydropower (which can also release methane from submerged vegetation), wind energy and biofuels (see Section 9.3.4.5).

Since a degree of climate change is now inevitable, adaptation to climate change is an essential component of sustainable development (IPCC, 2007e). Adaptation can be either anticipatory or reactive to an altered climate. Some RE technologies may assist in adapting to change, and are usually anticipatory in nature. AR4 includes a chapter on the linkage between climate mitigation (reducing emissions of GHGs) and climate adaptation including the potential to assist adaptation to climate change (Klein et al., 2007a, b).

- Active and passive solar cooling of buildings helps counter the direct impacts on humans of rising mean temperatures (Chapter 3);

- Dams (used for hydropower) may also be important in managing the impacts of droughts and floods, which are projected to increase with climate change. Indeed, this is one of reasons for building such dams in the first place (Section 5.10; see also World Commission on Dams (WCD, 2000);

- Solar PV and wind require no water for their operation, and hence may become increasingly important as droughts and high river temperatures limit the power output of thermal power plants (Section 9.3.4);

- Water pumps in rural areas remote from the power grid can utilize PV (Chapter 3) or wind (Chapter 7) for raising agricultural productivity during climate-induced increases in dry seasons and droughts; and

- Tree planting and forest preservation along coasts and riverbanks is a key strategy for lessening the coastal erosion impacts of climate change. With suitable choice of species and silvicultural practices, these plantings can also yield a sustainable source of biomass for energy, for example, by coppicing (Section 2.5).

1.4.2 Barriers

A barrier was defined in the AR4 as 'any obstacle to reaching a goal, adaptation or mitigation potential that can be overcome or attenuated by a policy, programme or measure'(IPCC, 2007d; Verbruggen et al., 2010). For example, the technology as currently available may not suit the desired scale of application. This barrier could be attenuated in principle by a program of technology development (R&D).

This section describes some of the main barriers and issues to using RE for climate change mitigation, adaptation and sustainable development. As throughout this introductory chapter, the examples are illustrative and not comprehensive. Section 1.5 (briefly) and Section 11.4 (in more

Chapter 1 — Renewable Energy and Climate Change

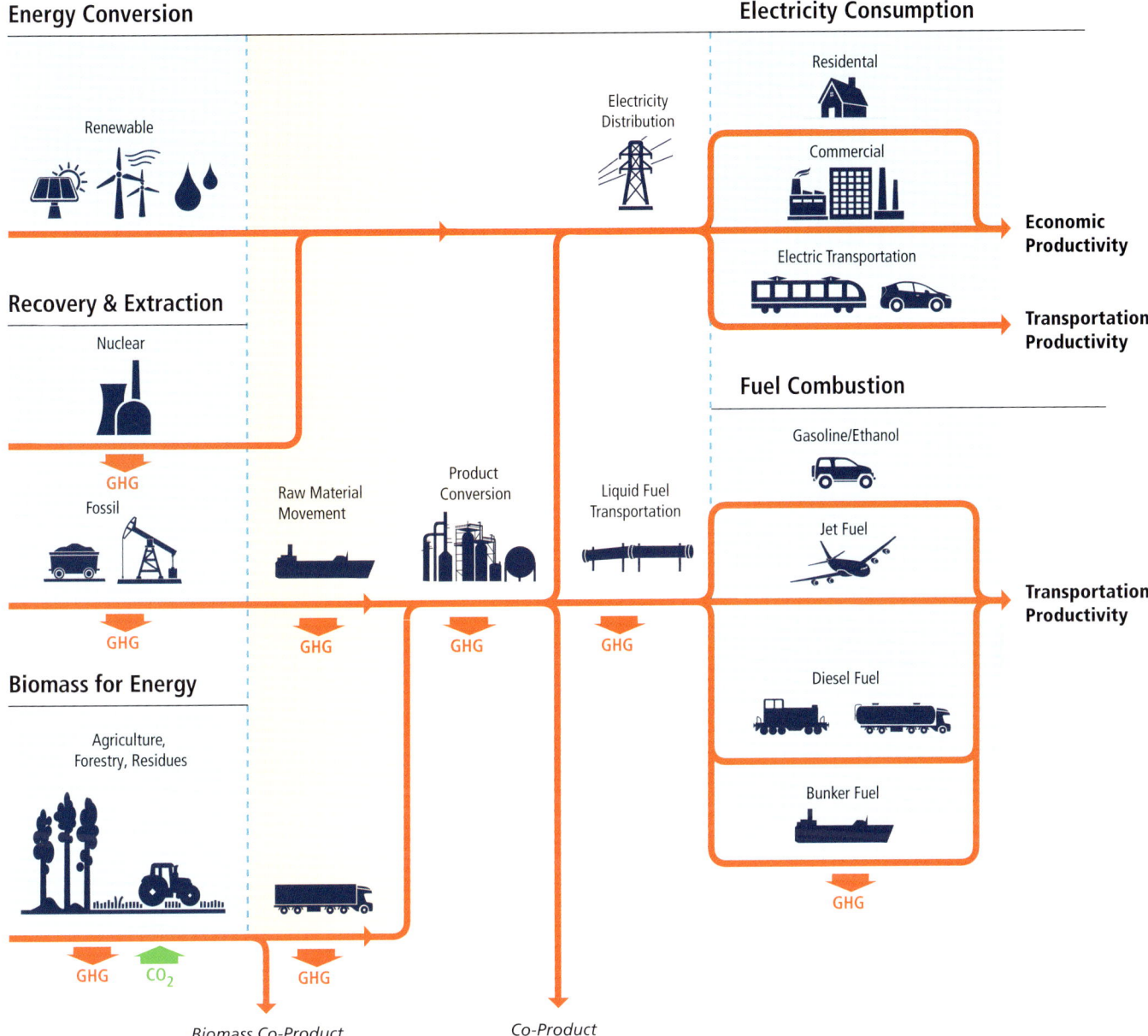

Figure 1.22 | Illustrative system for energy production and use illustrating the role of RE along with other production options. A systemic approach is needed to conduct lifecycle systems analysis.

detail) look at policies and financing mechanisms that may overcome them. When a barrier is particularly pertinent to a specific technology, it is examined in the appropriate technology chapter (i.e., Chapters 2 to 7).

The various barriers are categorized as 1) market failures and economic barriers, 2) information and awareness barriers, 3) socio-cultural barriers and 4) institutional and policy barriers (see Table 1.5). This categorization is somewhat arbitrary since, in many cases, barriers extend across several categories. More importantly, for a particular project or set of circumstances it will usually be difficult to single out one particular barrier. They are interrelated and need to be dealt with in a comprehensive manner.

1.4.2.1 Market failures and economic barriers

Market Failures

In economics a distinction is often made between *market failures* and *barriers*. With reference to the theoretical ideal market conditions (Debreu, 1959; Becker, 1971), all real-life markets fail to some degree (Bator, 1958; Meade, 1971; Williamson, 1985), evidenced by losses in welfare. Market failures (imperfections) are often due to externalities or external effects. These arise from a human activity when agents responsible for the activity do not take full account of the activity's impact on others. Externalities may be negative (external costs) or positive (external benefits). External benefits lead to an undersupply of beneficial

Table 1.5 | A categorization of barriers to RE deployment

Section	Type of barrier	Some potential policy instruments (see Chapter 11)
1.4.2.1	Market failures and economic barriers	Carbon taxes, emission trading schemes, public support for R&D, economic climate that supports investment, microfinance
1.4.2.2	Information and awareness barriers	Energy standards, information campaigns, technical training
1.4.2.3	Socio-cultural barriers	Improved processes for land use planning
1.4.2.4	Institutional and policy barriers	Enabling environment for innovation, revised technical regulations, international support for technology transfer (e.g., under the UNFCCC), liberalization of energy industries

activities (e.g., public goods) from a societal point of view because the producer is not fully rewarded. External costs lead to a too-high demand for harmful activities because the consumer does not bear the full (societal) cost. Another market failure is rent appropriation by monopolistic entities. In the case of RE deployment, these may appear as:

- Underinvestment in invention and innovation in RE technologies because initiators cannot benefit from exclusive property rights for their efforts (Margolis and Kammen, 1999; Foxon and Pearson, 2008).

- Un-priced environmental impacts and risks of energy use when economic agents have no obligation to internalize the full costs of their actions (Beck, 1995; Baumol and Oates, 1998). The release of GHG emissions and the resulting climate change is a clear example (Stern, 2007; Halsnaes et al., 2008), but the impacts and risks of some RE projects and of other low-carbon technologies (nuclear, CCS) may not always be fully priced either.

- The occurrence of monopoly (one seller) or monopsony (one buyer) powers in energy markets limits competition among suppliers or demanders and reduces opportunities for free market entry and exit (see Section 1.4.2.4). Monopoly and oligopoly power may be due to deliberate concentration, control and collusion. Regulated interconnected network industries (e.g., electric, gas and heat transmission grids) within a given area are natural monopolies because network services are least-cost when provided by a single operator (Baumol et al., 1982, p.135).

Characterizing these imperfections as market failures, with high likelihoods of welfare losses and of the impotence of market forces in clearing the imperfections, provides strong economic arguments for public policy intervention to repair the failures (Coase, 1960; Bromley, 1986). On top of imperfections classified as market failures, various factors affect the behaviour of market agents, and are categorized here as other types of barriers.

Up-front Investment Cost
The initial investment cost of a unit of RE capacity may be higher than for a non-RE energy system. Because the cost of such systems is largely up-front, it would be unaffordable to most potential customers, especially in developing countries, unless a financial mechanism is established to allow them to pay for the RE energy service month by month as they do for kerosene. Even if the initial equipment is donated by an overseas agency, such a financial mechanism is still needed to pay for the technical support, spare parts and eventual replacement of the system. Failure to have these institutional factors properly set up has been a major inhibitor to the use of RE in the Pacific Islands, where small-scale PV systems would appear to be a natural fit to the scattered tropical island communities (Johnston and Vos, 2005; Chaurey and Kandpal, 2010).

Financial risk
All power projects carry financial risk because of uncertainty in future electricity prices, regardless of its source, making it difficult for a private or public investor to anticipate future financial returns on investment. Moreover, the financial viability of an RE system strongly depends on the availability of capital and its cost (interest rates) because the initial capital cost comprises most of the economic cost of an RE system. While the predictability of such costs is a relative advantage of RE systems, many RE technologies are still in their early development phase, so that the risks related to the first commercial projects are high. The private capital market requires higher returns for such risky investments than for established technologies, raising the cost of RE projects (Gross et al., 2010; Bazilian and Roques, 2008).

An example of financial risk from an RE system outside the power sector is the development of biofuels for aviation. In 2009, neither the potential bio-jet fuel refiners nor the airlines fully understood how to structure a transaction that was credit worthy and as a result might get financed if there were interested financial institutions. (Slade et al., 2009)

1.4.2.2 Informational and awareness barriers

Deficient data about natural resources
RE is widely distributed but is site-specific in a way that fossil fuel systems are not. For example, the output of a wind turbine depends strongly on the wind regime at that place, unlike the output of a diesel generator. While broad-scale data on wind is reasonably well available from meteorological records, it takes little account of local topography, which may mean that the output of a particular turbine could be 10 to 50 % higher on top of a local hill than in the valley a few hundred metres away (Petersen et al., 1998). To obtain such site-specific data requires onsite measurement for at least a year and/or detailed modelling. Similar data

deficiencies apply to most RE resources, but can be attenuated by specific programs to better measure those resources (Hammer et al., 2003).

Skilled human resources (capacity)

To develop RE resources takes skills in mechanical, chemical and electrical engineering, business management and social science, as with other energy sources. But the required skill set differs in detail for different technologies and people require specific training. Developing the skills to operate and maintain the RE 'hardware' is exceedingly important for a successful RE project (Martinot, 1998). Where these barriers are overcome as in Bangladesh, significant installations of RE systems in developing countries has occurred (Barua et al., 2001; Ashden Awards for Sustainable Energy, 2008; Mondal et al., 2010). It is also important that the user of RE technology understand the specific operational aspects and availability of the RE source. One case where this is important is in the rural areas of developing countries. Technical support for dispersed RE, such as PV systems in the rural areas of developing countries, requires many people with basic technical skill rather than a few with high technical skill as tends to be the case with conventional energy systems. Training such people and ensuring that they have ready access to spare parts requires establishment of new infrastructure.

More generally, in some developing countries, the lack of an ancillary industry for RE (such as specialized consulting, engineering and procurement, maintenance, etc.) implies higher costs for project development and is an additional barrier to deployment.

Public and institutional awareness

The oil (and gas) price peaks of 1973, 1980, 1991 and 2008 made consumers, governments and industry in both industrialized and developing counties search for alternative sources of energy. While these price surges caused some shift to coal for power production, they also generated actions to adopt more RE, especially solar, wind and biomass (Rout et al., 2008; van Ruijven and van Vuuren, 2009; Chapter 7). There is, however, limited awareness of the technical and financial issues of implementing a sustained transition to alternative primary energy sources—especially RE (Henriques and Sadorsky, 2008). The economic and transactional costs of shifting away from vulnerable and volatile fossil fuels like oil are overestimated, and there is always a shift back to these fuels once price shocks abate. The reluctance to make a shift away from a known energy source is very high because of institutional, economic and social lock-in (Unruh and Carillo-Hermosilla, 2006). One means of motivation might be a realization that the economic welfare cost of high oil prices exceeds that of effective climate polices (Viguier and Vielle, 2007).

1.4.2.3 Socio-cultural barriers

Socio-cultural barriers or concerns have different origins and are intrinsically linked to societal and personal values and norms. Such values and norms affect the perception and acceptance of RE technologies and the potential impacts of their deployment by individuals, groups and societies. Barriers may arise from inadequate attention to such socio-cultural concerns and may relate to impacts on behaviour, natural habitats and natural and human heritage sites, including impacts on biodiversity and ecosystems, landscape aesthetics, and water/land use and water/land use rights as well as their availability for competing uses (see Section 9.5.1.1).

Farmers on whose land wind farms are built rarely object; in fact they usually see turbines as a welcome extra source of income either as owners (Denmark) or as leasers of their land (USA), as they can continue to carry on agricultural and grazing activities beneath the turbines. Other forms of RE, however, preclude multiple uses of the land (Kotzebue et al., 2010). Dams for hydropower compete for recreational or scenic use of rivers (Hynes and Hanley, 2006), and the reservoirs may remove land from use for agriculture, forests or urban development. Large-scale solar or wind may conflict with other values (Simon, 2009) and may conflict with other social values of land such as nature preserves or scenic vistas (Groothuis et al., 2008; Valentine, 2010). Specific projects may also have negative implications for poor populations (Mariita, 2002). Land use can be just as contentious in some developing countries. In Papua New Guinea, for example, villagers may insist on being paid for the use of their land, for example, for a mini-hydro system of which they are the sole beneficiaries (Johnston and Vos, 2005).

Hence, social acceptance is an important element in the need to rapidly and significantly scale up RE deployment to help meet climate change mitigation goals, as large-scale implementation can only be successfully undertaken with the understanding and support of the public. Social acceptance of RE is generally increasing; having domestic solar energy PV or domestic hot water systems on one's roof has become a mark of the owner's environmental commitment (Bruce et al., 2009). However, wind farms still have to battle local opposition before they can be established (Pasqualetti et al., 2002; Klick and Smith, 2010; Webler and Tuler, 2010) and there is opposition to aboveground transmission lines from larger-scale renewable generation facilities (as well as from conventional power sources) (Furby et al., 1988; Hirst and Kirby, 2001; Gerlach, 2004; Vajjhala and Fischbeck, 2007; Puga and Lesser, 2009).

To overcome such barriers may require dedicated communication efforts related to such subjective and psychological aspects as well as the more objective opportunities associated with wider-scale applications of RE technologies. At the same time, public participation in planning decisions as well as fairness and equity considerations in the distribution of the benefits and costs of RE deployment play an equally important role and cannot be side-stepped (see Section 9.5.2). See Chapters 7 and 11 for more discussion of how such local planning issues impact the uptake of RE. Chapter 11 also includes a wider discussion of the enabling social and institutional environment required for the transition to RE systems. Opposition to unwanted projects can be influenced by policies but social acceptance may be slow to change.

1.4.2.4 Institutional and policy barriers

Existing industry, infrastructure and energy market regulation
Apart from constituting a market failure (see above), monopoly power can be perceived as an institutional barrier if not addressed adequately by energy market regulation.

The energy industry in most countries is based on a small number of companies (sometimes only one in a particular segment such as electricity or gas supply) operating a highly centralized infrastructure. These systems evolved as vertically integrated monopolies that may become committed to large conventional central power facilities supported by policies to ensure they deliver affordable and reliable electricity or gas. They are sometimes unreceptive to distributed smaller supply technologies (World Bank, 2006).

Therefore, regulations governing energy businesses in many countries are still designed around monopoly or near-monopoly providers and technical regulations and standards have evolved under the assumption that energy systems are large and centralized and of high power density and/or high voltage, and may therefore be unnecessarily restrictive for RE systems. In the process of historical development, most of the rules governing sea lanes and coastal areas were written long before offshore wind power and ocean energy systems were being developed and do not consider the possibility of multiple uses that include such systems (See Chapter 7).

Liberalization of energy markets occurred in several countries in the 1990s and more extensively in Europe in the past decade. Some of these changes in regulations allow independent power producers to operate, although in the USA many smaller proposed RE projects were often excluded due to the scales required by regulation (Markard and Truffer, 2006). In many countries, current regulations remain that protect the dominant centralized production, transmission and distribution system and make the introduction of alternative technologies, including RE, difficult. An examination and modification of existing laws and regulations is a first step in the introduction of RE technologies, especially for integrating them into the electric power system (Casten, 2008).

In addition to regulations that address the power generation sector, local building codes sometimes prevent the installation of rooftop solar panels or the introduction of wind turbines for aesthetic or historical preservation reasons (Bronin, 2009; Kooles, 2009).

Intellectual property rights
Intellectual property rights play a complex and conflicting role. Technological development of RE has been rapid in recent years, particularly in PV and wind power (Lior, 2010; see Chapters 8 and 11). Much of the basic technology is in the public domain, which can lead to underinvestment in the industry. Patents protect many of these new developments thereby promoting more private investment in R&D (Beck, 1995; Baumol and Oates, 1998). Countering this benefit are concerns that have been raised that patents may unduly restrict low-cost access to these new technologies by developing countries, as has happened with many new pharmaceuticals (Barton, 2007; Ockwell et al., 2010; Chapters 3 and 7). There are certainly circumstances where developing country companies need patent protection for their products as well.

Tariffs in international trade
Tariff barriers (import levies) and non-trade barriers imposed by some countries significantly reduce trade in some RE technologies. Discussions about lowering or eliminating tariffs on environmental goods and services including RE technologies have been part of the Doha round of trade negotiations since 2001. Many developing countries argue that reducing these tariffs would primarily benefit developed countries economically, and no resolution has been achieved so far. Developed countries have levied tariffs on imported biofuels, much of which originates in developing countries, thereby discouraging their wider use (Elobeid and Tokgoz, 2008; see Section 2.4.6.2).

Allocation of government financial support
Since the 1940s, governments in industrialized countries have spent considerable amounts of public money on energy-related research, development, and demonstration. By far the greatest proportion of this has been on nuclear energy systems (IEA, 2008b; see also Section 10.5). However, following the financial crisis of 2008 and 2009, some governments used part of their 'stimulus packages' to encourage RE or energy efficiency (Section 9.3.1.3). Tax write-offs for private spending have been similarly biased towards non-RE sources (e.g., in favour of oil exploration or new coal-burning systems), notwithstanding some recent tax incentives for RE (GAO, 2007; Lior, 2010). The policy rationale for government support for developing new energy systems is discussed in Section 1.5 and Chapter 11.

1.4.3 Issues

Issues are not readily amenable to policies and programs.

An issue is that the resource may be too small to be useful at a particular location or for a particular purpose. For example, the wind speed may be too low or too variable to produce reliable power, the topography may be either too flat or there may be insufficient flow to sustain low-head hydro or run-of-river systems for hydropower, or the demands of industry may be too large to be supplied by a local renewable source (Painuly, 2001).

Some renewable resources such as wind and solar are variable and may not always be available for dispatch when needed (Chapter 8). Furthermore, the energy density of many renewable sources is relatively low, so that their power levels may be insufficient on their own for some purposes such as very large-scale industrial facilities. Extensive planting for biomass production or building of large-area reservoirs can lead to displacement of forests with associated negative effects, such as the direct and indirect release of CO_2 and/or methane and soil loss (Melillo et al., 2009; Chapter 2 and Section 5.6.1).

1.5 Role of policy, research and development, deployment, scaling up and implementation strategies

An increasing number and variety of RE policies—motivated by a variety of factors—have driven escalated growth in RE technologies in recent years (Section 11.2). In addition to the reduction of CO_2 emissions, governments have enacted RE policies to meet a number of objectives, including the creation of local environmental and health benefits; facilitation of energy access, particularly for rural areas; advancement of energy security goals by diversifying the portfolio of energy technologies and resources; and improving social and economic development through potential employment opportunities. In general, energy access has been the primary driver in developing countries whereas energy security and environmental concerns have been most important in developed countries (Chapter 9 and Section 11.3).

For policymakers wishing to support the development and deployment of RE technologies for climate change mitigation goals, it is critical to consider the potential of RE to reduce emissions from a lifecycle perspective, an issue that each technology chapter addresses. For example, while the use of biofuels can offset GHG emissions from fossil fuels, direct and indirect land use changes must be also be evaluated in order to determine net benefits.[11] In some cases, this may even result in increased GHG emissions, potentially overwhelming the gains from CO_2 absorption (Fargione et al., 2008; Scharlemann and Laurance, 2008; Searchinger et al., 2008; Krewitt et al., 2009; Melillo et al., 2009). A full discussion of this effect can be found in Sections 2.5.3 and 9.3.4.

Various policies have been designed to address every stage of the development chain, involving R&D, testing, deployment, commercialization, market preparation, market penetration, maintenance and monitoring, as well as integration into the existing system. These policies are designed and implemented to overcome the barriers and markets failures discussed above (Sections 1.4.2, 11.1.1, 11.4 and 11.5).

Two key market failures are typically addressed: 1) the external costs of GHG emissions are not priced at an appropriate level; and 2) deployment of low-carbon technologies such as RE creates benefits to society beyond those captured by the innovator, leading to under-investment in such efforts (Sections 11.1 and 11.4). Implementing RE policies (i.e., those promoting exclusively RE) in addition to climate change mitigation policies (i.e., encouraging low-carbon technologies in general) can be justified if a) the negative consequences of innovation market failures should be mitigated and/or b) other goals beyond climate protections are to be addressed.

1.5.1 Policy options: trends, experience and assessment

The focus of RE policies is shifting from a concentration almost entirely on electricity to include the heating/cooling and transportation sectors. These trends are matched by increasing success in the development of a range of RE technologies and their manufacture and implementation (see Chapters 2 through 7), as well as by a rapid increase in annual investment in RE and a diversification of financing institutions, particularly since 2004/2005 (Section 11.2.2).

Policy and decision makers approach the market in a variety of ways: level the playing field in terms of taxes and subsidies; create a regulatory environment for effective utilization of the resource; internalize externalities of all options or modify or establish prices through taxes and subsidies; create command and control regulations; provide government support for R&D; provide for government procurement priorities; or establish market oriented regulations, all of which shape the markets for new technologies. Some of these options, such as price, modify relative consumer preferences, provide a demand pull and enhance utilization for a particular technology. Others, such as government-supported R&D, attempt to create new products through supply push (Freeman and Soete, 2000; Sawin, 2001; Moore, 2002). No globally-agreed list of RE policy options or groupings exists. For the purpose of simplification, R&D and deployment policies have been organized within the following categories in this report (Section 11.5):

- **Fiscal incentives:** actors (individuals, households, companies) are granted a reduction of their contribution to the public treasury via income or other taxes;

- **Public finance:** public support for which a financial return is expected (loans, equity) or financial liability is incurred (guarantee); and

- **Regulation:** rule to guide or control conduct of those to whom it applies.

Research and development, innovation, diffusion and deployment of new low-carbon technologies create benefits to society beyond those captured by the innovator, resulting in under-investment in such efforts. Thus, government R&D can play an important role in advancing RE technologies. Not all countries can afford to support R&D with public funds, but in the majority of countries where some level of support is possible, public R&D for RE enhances the performance of nascent technologies so that they can meet the demands of initial adopters. Public R&D also improves existing technologies that already function in commercial environments. A full discussion of R&D policy options can be found in Section 11.5.2.

---

11 Note that such land use changes are not restricted to biomass based RE. For example, wind generation and hydro developments as well as surface mining for coal and storage of combustion ash also incur land use impacts.

Public R&D investments are most effective when complemented by other policy instruments, particularly RE deployment policies that simultaneously enhance demand for new RE technologies. Together R&D and deployment policies create a positive feedback cycle, inducing private sector investment in R&D. Relatively early deployment policies in a technology's development accelerate learning through private R&D and/or through utilization and cost reduction (Section 11.5.2). The failure of many worthy technologies to move from R&D to commercialization has been coined the 'valley of death' for new products (Markham, 2002; Murphy and Edwards, 2003; IEA, 2009b; Section 11.5). Attempts to move renewable technology into mainstream markets following the oil price shocks failed in most developed countries (Roulleau and Loyd, 2008). Many of the technologies were not sufficiently developed or had not reached cost competitiveness and, once the price of oil came back down, interest in implementing these technologies faded. Solar hot water heaters were a technology that was ready for the market and with tax incentives many such systems were installed. But once the tax advantage was withdrawn, the market largely collapsed (Dixit and Pindyck, 1994).

Some policy elements have been shown to be more effective and efficient in rapidly increasing RE deployment, but there is no one-size-fits-all policy, and the mix of policies and their design and implementation vary regionally and depend on prevailing conditions. Experience shows that different policies or combinations of policies can be more effective and efficient depending on factors such as the level of technological maturity, availability of affordable capital, and the local and national RE resource base. Key policy elements include adequate value to cover costs and account for social benefits, inclusiveness and ease of administration. Further, the details of policy design and implementation—including flexibility to adjust as technologies, markets and other factors evolve—can be as important in determining effectiveness and efficiency as the specific policies that are used (Section 11.5). Transparent, sustained, consistent signals—from predictability of a specific policy, to pricing of carbon and other externalities, to long-term targets for RE—have been found to be crucial for reducing the risk of investment sufficiently to enable appropriate rates of deployment and the evolution of low-cost applications (Sections 11.2, 11.4 and 11.5).

For deployment policies with a focus on RE electricity, there is a wealth of literature assessing quantity-based (quotas, renewable portfolio standards that define the degree to which electricity generated must be from renewable sources, and tendering/bidding policies) and price-based (fixed-price and premium-price feed-in tariffs (FIT)) policies, primarily quotas and FITs, and with a focus on effectiveness and efficiency criteria. Several studies have concluded that some FITs have been effective and efficient at promoting RE electricity, mainly due to the combination of long-term fixed price or premium payments, network connections, and guaranteed purchase of all RE electricity generated. A number of studies have concluded that 'well-designed' and 'well-implemented' FITs have to date been the most efficient (defined as comparison of total support received and generation cost) and effective (ability to deliver an increase in the share of RE electricity consumed) support policies for promoting RE electricity (Ragwitz et al., 2005; Stern, 2007; de Jager and Rathmann, 2008; Section 11.5.4). Quota policies have been moderately successful in some cases. They can be effective and efficient if designed to reduce risk; for example, with long-term contracts.

An increasing number of governments are adopting fiscal incentives for RE heating and cooling. To date, fiscal incentives have been the prevalent policy in use to support RE heating and cooling, with grants the most commonly applied incentive. Obligations to use RE heat are gaining attention for their potential to encourage growth independent of public financial support (Section 11.5.5).

A range of policies has been implemented to support the deployment of RE for transport, though the vast majority of these policies and related experiences have been specific to biofuels. RE fuel mandates or blending requirements are key drivers in the development of most modern biofuel industries. Other policies include direct government payments or tax reductions. Those countries with the highest share of biofuels in transport fuel consumption have had hybrid systems that combine mandates (including penalties) with fiscal incentives (foremost tax exemptions). Policies have influenced the development of an international biofuel trade (Section 11.5.6).

There is now considerable experience with several types of policies designed to increase the use of renewable technology. Denmark became a world leader in the manufacture and deployment of large-scale wind turbines by setting long-term contracts for renewably generated electricity production (REN21, 2009). Germany and Spain (among others) have used a similar demand-pull mechanism through FITs that assured producers of RE electricity sufficiently high rates for a long and certain time period. Germany is the world's leading installer of solar PV, and in 2008 had the largest installed capacity of wind turbines (REN21, 2009). The USA has relied mostly on government subsidies for RE technologies and this supply-push approach has been less successful than demand pull (Lewis and Wiser, 2007; Butler and Neuhoff, 2008). China has encouraged renewable technology for water heating, solar PV and wind turbines by investing in these technologies directly. China is already the leading producer of solar hot water systems for both export and domestic use, and is now the largest producer of PV technology (REN 21, 2009).

One important challenge will be finding a way for RE and carbon-pricing policies to interact such that they take advantage of synergies rather than tradeoffs (Section 11.5.7). Impacts can be positive or negative, depending on policy choice, design and the level of implementation (local, regional, national or global). Negative effects would include the risk of carbon leakage and rebound effects, which need to be taken into account when designing policies. In the long term, enhancing knowledge for the implementers and regulators of RE supply technologies and processes can help reduce costs of mitigation, and putting a price on carbon can increase the competitiveness of RE (Sections 11.1.1 and 11.5.7).

1.5.2 Enabling environment

RE technologies can play a greater role if they are implemented in conjunction with 'enabling' policies. A favourable, or 'enabling', environment for RE can be created by addressing the possible interactions of a given policy with other RE policies as well as with other non-RE policies; by understanding the ability of RE developers to obtain finance and planning permission to build and site a project; by removing barriers for access to networks and markets for RE installations and output; by increasing education and awareness raising; and by enabling technology transfer. In turn, existence of an 'enabling' environment can increase the efficiency and effectiveness of policies to promote RE (Section 11.6).

1.5.2.1 Complementing renewable energy policies and non-renewable energy policies

Since all forms of RE capture and production involve spatial considerations, policies need to consider land use, employment, transportation, agricultural, water, food security, trade concerns, existing infrastructure and other sector-specific issues. Government policies that complement each other are more likely to be successful, and the design of individual RE policies will also affect the success of their coordination with other policies. Attempting to actively promote the complementarities of policies across multiple sectors—from energy to agriculture to water policy, etc.—while also considering the independent objectives of each, is not an easy task and may create win-lose and/or win-win situations, with possible trade-offs.

1.5.2.2 Providing infrastructure, networks and markets for renewable energy

Advancing RE in the electric power sector, for example, will require policies to address its integration into transmission and distribution systems both technically (Chapter 8) and institutionally (Chapter 11). The grid must be able to handle both traditional, often more central, supply as well as modern RE supply, which is often variable and distributed (Quezada et al., 2006; Cossent et al., 2009) and the governance of the system may need to be adjusted to ease or harmonize access; current regulations and laws, designed to assure the reliability of the current centralized grid, may prevent the wide-scale introduction of renewable electric generating technology.

In the transport sector, issues exist related to the necessary infrastructure for biofuels, recharging hydrogen, battery or hybrid electric vehicles that are 'fuelled' by the electric grid or from off-grid renewable electrical production (Tomic and Kempton, 2007; Sections 1.4.2.4 and 11.6.5).

Brazil has been especially effective in establishing a rural agricultural development program around sugarcane. Bioethanol produced from sugarcane in Brazil is currently responsible for about 40% of the spark ignition travel and it has been demonstrated for use in diesel buses and even in a crop duster aircraft. The bagasse, which is otherwise wasted, is gasified and used to operate gas turbines for electricity production while the 'waste' heat is used in the sugar to bioethanol refining process (Pousa et al., 2007; Searchinger et al., 2008).

1.5.3 A structural shift

If decision makers intend to increase the share of RE and, at the same time, to meet ambitious climate mitigation targets, then long-standing commitments and flexibility to learn from experience will be critical. Some analyses conclude that large, low-carbon facilities such as nuclear power, or large coal (and natural gas) plants with CCS can be scaled up rapidly enough to meet CO_2 reduction goals if they are available (MIT, 2003, 2007, 2009). Alternatively, the expansion of natural gas-fired turbines during the past few decades in North America and Europe, and the rapid growth in wind and solar technologies for electric power generation (see Figure 1.12) demonstrate that modularity and more widely distributed smaller-scale units can also scale rapidly to meet large-scale energy demands. The technological and economic potential for each of these approaches and their costs have important implications for the scale and role of RE in addressing climate change (Pilavachi, 2002; MIT, 2003, 2007, 2009; Onovwiona and Ugursal, 2006). To achieve GHG concentration stabilization levels that incorporate high shares of RE, a structural shift in today's energy systems will be required over the next few decades. Such a transition to low-carbon energy differs from previous ones (e.g., from wood to coal, or coal to oil) because the available time span is restricted to a few decades, and because RE must develop and integrate into a system constructed in the context of an existing energy structure that is very different from what might be required under higher penetration RE futures (Section 11.7 and Chapter 10).

A structural shift towards a world energy system that is mainly based on renewable energy might begin with a prominent role for energy efficiency in combination with RE; policies that extend beyond R&D to support technology deployment; the creation of an enabling environment that includes education and awareness raising; and the systematic development of integrative policies with broader sectors, including agriculture, transportation, water management and urban planning (Sections 11.6 and 11.7). The appropriate and reliable mix of instruments is even more important where energy infrastructure is not yet developed and energy demand is expected to increase significantly in the future (Section 11.7).

References

Ahman, M., and L.J. Nilsson (2008). Path dependency and the future of advanced vehicles and biofuels. *Utilities Policy*, **16**(2), pp. 80-89.

Ambrose, M. (2009). Energy-efficient planning and design. In: *Technology, Design, and Process Innovation in the Built Environment*. P. Newton, K.D. Hampson, and R. Drogemuller (eds.), Taylor and Francis, New York, pp. 238-249.

Archer, C.L. and Jacobson M.Z (2005). Evaluation of global wind power. *Journal of Geophysical Research*, **110**, D12110.

Ashden Awards for Sustainable Energy, 2008. *Rapidly growing solar installer also provides clean cooking*. Ashden Awards for Sustainable Energy, London, UK. Available at: www.ashdenawards.org/winners/grameen08.

Asif, M., and T. Muneer (2007). Energy supply, its demand and security issues for developed and emerging economies. *Renewable and Sustainable Energy Reviews*, **11**(7), pp. 1388-1413.

Bailey, O., and E. Worrell (2005). *Clean Energy Technologies: A Preliminary Inventory for the Potential for Electricity Generation*. Ernest Orlando Lawrence Berkeley National Laboratory, Berkeley, CA, USA.

Barker, T., I. Bashmakov, A. Alharthi, M. Amann, L. Cifuentes, J. Drexhage, M. Duan, O. Edenhofer, B. Flannery, M. Grubb, M. Hoogwijk, F.I. Ibitoye, C.J. Jepma, W.A. Pizer, and K. Yamaji (2007). Mitigation from a cross-sectoral perspective. In: *Climate Change 2007: Mitigation of Climate Change. Contribution of Working Group III to the Fourth Assessment Report of the Intergovernmental Panel on Climate Change*. B. Metz, O.R. Davidson, P.R. Bosch, R. Dave, and L.A. Meyer (eds), Cambridge University Press.

Barton, J.H. (2007). *Intellectual Property and Access to Clean Energy Technologies in Developing Countries: An Analysis of Solar Photovoltaic, Biofuel, and Wind Technologies*. Issue Paper No. 2, International Center for Trade and Sustainable Development, Geneva, Switzerland.

Barua, D.C., T.P. Urmee, S. Kumar, and S.C. Bhattachary (2001). A photovoltaic solar home system dissemination model. *Progress in Photovoltaics: Research and Applications*, **9**(4), pp. 313-322.

Bator, F.M. (1958). The anatomy of market failure. *The Quarterly Journal of Economics*, **72**(3), pp. 351-379.

Baumol, W.J., and W.E. Oates (1998). *The Theory of Environmental Policy*. Cambridge University Press.

Baumol, W.J., J.C. Panzar, and T.D. Willig (1982). *Contestable Markets and the Theory of Industry Structure*. Harcourt Brace Jovanovich, New York, NY, USA.

Bazilian, M., and F. Roques (2008). *Analytical Methods for Energy Diversity And Security: Mean-Variance Optimization for Electric Utilities Planning: A Tribute to the Work of Dr. Shimon Awerbuch*. 1st ed. Elsevier, Boston, MA, USA.

Beck, U. (1995). *Ecological Politics in an Age of Risk*. Blackwell Publishers, Inc., Malden, MA, USA.

Becker, G.S. (1971). *Economic Theory*. Alfred A. Knopf, New York, NY, USA.

BGR (2009). *Energierohstoffe 2009. Reserven, Ressourcen, Verfügbarkeit*. Bundesamt für Geowissenschaften und Rohstoffe, Hannover, Germany.

Bodart, M., and D. Herde (2001). Global energy savings in office buildings by the use of daylighting. *Energy and Buildings*, **34**, pp. 421-429.

Boden, T. and G. Marland (2010). *Global CO_2 Emissions from Fossil-Fuel Burning, Cement Manufacture, and Gas Flaring: 1751-2007*. Carbon Dioxide Information Analysis Center, Oak Ridge National Laboratory, Oak Ridge, TN, USA. Available at: cdiac.ornl.gov/ftp/ndp030/global.1751_2007.ems.

Boden, T.A., G. Marland, and R.J. Andres (2009). *Global, Regional, and National Fossil-Fuel CO_2 Emissions*. Carbon Dioxide Information Analysis Center, Oak Ridge National Laboratory, US Department of Energy, Oak Ridge, TN, USA. Available at: cdiac.ornl.gov/trends/emis/overview_2007.html.

Bond, T., and H. Sun (2005). Can reducing black carbon emissions counteract global warming? *Environmental Science and Technology*, **39**(16), pp. 5921-5926.

Bromley, C.J., M.A. Mongillo, B. Goldstein, G. Hiriart, R. Bertani, E. Huenges, H. Muraoka, A. Ragnarsson, J. Tester, and V. Zui, (2010). Contribution of geothermal energy to climate change mitigation: the IPCC renewable energy report. In: *Proceedings World Geothermal Congress 2010, Bali, Indonesia, 25-30 April 2010*. Available at: www.geothermal-energy.org/pdf/IGAstandard/WGC/2010/0225.pdf.

Bromley, D.W. (ed.) (1986). *Natural Resource Economics: Policy Problems and Contemporary Analysis*. Kluwer Nijhoff, Hingham, MA, USA.

Bronin, S.C. (2009). Solar rights. *Boston University Law Review*, **89**, pp. 1217.

Brookes, L.G. (2000). Energy efficiency fallacies revisited. *Energy Policy*, **28**(6-7), pp. 355-366.

Brown, D., F. Marechal, and J. Paris (2005). A dual representation for targeting process retrofit, application to a pulp and paper process. *Applied Thermal Engineering*, **25**(7), pp. 1067-1082.

Bruce, A., M.E. Watt, and R. Passey (2009). Who buys PV systems? A survey of New South Wales residential PV rebate recipients. In: *47th Annual Conference of the Australia and New Zealand Solar Energy Society*, Townsville, Australia, 29 September – 2 October 2009.

Bruckner, T., O. Edenhofer, H. Held, M. Haller, M. Lüken, N. Bauer, and N. Nakicenovic (2010). Robust options for decarbonisation. In: *Global Sustainability – A Nobel Cause*. H.J. Schellnhuber, M. Molina, N. Stern, V. Huber, and S. Kadner (eds.), Cambridge University Press, pp. 189-204.

Bugge, J., S. Kjaer, and R. Blum (2006). High-efficiency coal-fired power plants development and perspectives. *Energy*, **31**(10-11), pp. 1437-1445.

Butler, L., and K. Neuhoff (2008). Comparison of feed-in tariff, quota, and auction mechanisms to support wind power development. *Renewable Energy*, **33**(8), pp. 1854-1867.

Cancun Agreements (2010). Cancun Agreements. *United Nations Framework Convention on Climate Change (UNFCCC)*, Cancun, Mexico. Available at: http://unfccc.int/documentation/decisions/items/3597.php?such=j&volltext=%22cancun%20agreements%22#beg

Casten, T.R. (2008). Recycling energy to reduce costs and mitigate climate change. In: *Sudden and Disruptive Climate Change: Exploring the Real Risks and How We Can Avoid Them*. M.C. MacCracken, F. Moore, and J.C. Topping, Jr. (eds.), Earthscan, London, UK.

Chasapis, D., I. Papamechael, A. Aidnis, and R. Blanchard (2008). Monitoring and operational results of a hybrid solar-biomass heating system. *Renewable Energy*, **33**, pp. 1759-1767.

Chaurey, A., and T. Kandpal (2010). Assessment and evaluation of PV based decentralized rural electrification: An overview. *Renewable and Sustainable Energy Reviews*, **14**(8), pp. 2266-2278.

Clancy, J.S., F.U. Malik, I. Shakya, and G. Kelkar (2007). Appropriate gender-analysis tools for unpacking the gender-energy-poverty nexus. *Gender & Development*, **15**(2), pp. 241-257.

Clarke, L., J. Edmonds, V. Krey, R. Richels, S. Rose, and M. Tavoni (2009). International climate policy architectures: Overview of the EMF 22 International Scenarios. *Energy Economics*, **31**(Supplement 2), pp. 64-81.

Cleveland, C.J., R. Costanza, C.A.S. Hall, and R.K. Kaufmann (1984). Energy and the US economy: A biophysical perspective. *Science*, **225**, pp. 890-897.

Coase, R.H. (1960). The problem of social cost. *Journal of Law and Economics*, **3**, pp. 1-44.

Cossent, R., T. Gomez, and P. Frias (2009). Towards a future with large penetration of distributed generation: Is the current regulation of electricity distribution ready? Regulatory recommendations under a European perspective. *Energy Policy*, **37**, pp. 1145-1155.

CSD (2001). *Commission on Sustainable Development. Report of the Ninth Sessions, 5 May 2000 and 16-27 April 2001*. Official Records 2001, Supplement No. 9, Economic and Social Council, United Nations, New York, NY, USA. Available at: www.un.org/esa/sustdev/csd/ecn172001-19e.htm.

de Jager, D., and M. Rathmann (2008). *Policy Instrument Design to Reduce Financing Costs in Renewable Energy Technology Projects*. ECOFYS, Utrecht, The Netherlands, 142 pp.

Debreu, G. (1959). *Theory of Value: An Axiomatic Analysis of Economic Equilibrium*. Yale University Press, New Haven, CT, USA.

Demirbas, A. (2009). Global renewable energy projections. *Energy Sources*, **4**, pp. 212-224.

Desertec (2010). *Desertec: An Overview of the Concept*. Desertec Foundation, Hamburg, Germany.

de Vries, B., M.M. Hoogwijk, and D. van Vuuren (2007). Renewable energy sources: Their global potential for the first half of the 21st century at a global level: An integrated approach. *Energy Policy*, **35**(4), pp. 2590-2610.

Di Maria, C., and E. van der Werf (2008). Carbon leakage revisited: unilateral climate policy with directed technical change. *Environmental and Resource Economics*, **39**(2), pp. 55-74.

Dietz, T., G.T. Gardner, J. Gilligan, P.C. Stern, and M.P. Vandenburgh (2009). Household actions can provide a behavioral way to rapidly reduce US carbon emissions. *Proceedings of the National Academy of Sciences*, **106**(44), pp. 18452-18456.

Dixit, A.K., and R.S. Pindyck (1994). *Investment Under Uncertainty*. Princeton University Press, Princeton, NJ, USA.

Dornburg, V., D. van Vuuren, G. van de Ven, H. Langeveld, M. Meeusen, M. Banse, M. van Oorschot, J. Ros, G.J. van den Born, H. Aiking, M. Londo, H. Mozaffarian, P. Verweij, E. Lysen, and A. Faaij (2010). Bioenergy revisited: Key factors in global potentials of bioenergy. *Energy & Environmental Science*, **3**, pp. 258-267.

Edenhofer, O., B. Knopf, G. Luderer, J. Steckel, and T. Bruckner (2010). *More Heat than Light? On the Economics of Decarbonization*. Routledge, London, UK.

Ehrlich, P.R., and J.P. Holdren (1971). Impact of population growth. *Science*, **171**(3977), pp. 1212-1217.

EIA (2010). *Independent Statistics and Analysis, International Data*. United States Energy Information Administration, Washington, DC, USA.

Elobeid, A. and S. Tokgoz (2008). Removing distortions in the U.S. ethanol market: What does it imply for the United States and Brazil? *American Journal of Agricultural Economics*, **90**(4), pp. 918-932.

EPRI (1978). *Geothermal Energy Prospects for the Next 50 Years*. ER-611-SR, Special Report for the World Energy Conference, Electric Power Research Institute, Palo Alto, CA, USA.

ESMAP (2007). *Technical and Economic Assessment of Off-grid, Mini-grid and Grid Electrification Technologies*. Energy Sector Management Assistance Program, The World Bank Group, Washington, DC, USA.

Fargione, J., J. Hill, D. Tilman, S. Polasky, and P. Hawthorne (2008). Land clearing and the biofuel carbon debt. *Science*, **319**(5867), pp. 1235-1238.

Fellows, A. (2000) *The Potential of Wind Energy to Reduce Carbon Dioxide Emissions*. Garrad Hassan and Partners Ltd., Glasgow, Scotland.

Fisher, B., N. Nakicenovic, K. Alfsen, J. Corfee Morlot, F. de la Chesnaye, J.-Ch. Hourcade, K. Jiang, M. Kainuma, E. La Rovere, A. Matysek, A. Rana, K. Riahi, R. Richels, S. Rose, D. van Vuuren, and R. Warren (2007). Issues related to mitigation in the long term context. In: *Climate Change 2007: Mitigation of Climate Change. Contribution of Working Group III to the Fourth Assessment Report of the Intergovernmental Panel on Climate Change*. B. Metz, O.R. Davidson, P.R. Bosch, R. Dave, and L.A. Meyer (eds.), Cambridge University Press, pp. 169-250.

Foxon, T., and P. Pearson (2008). Overcoming barriers to innovation and diffusion of cleaner technologies: some features of a sustainable innovation policy regime. *Journal of Cleaner Production*, **16**(1, Supplement 1), pp. S148-S161.

Frankl, P., A. Masini, M. Gambrale, and D. Toccaceli (1998). Simplified life-cycle analysis of PV in buildings: Present situation and future trends. *Progress in Photovoltaics: Research and Applications*, **6**, pp. 137-146.

Freeman, C., and L. Soete (2000). *The Economics of Industrial Innovation*. MIT Press, Cambridge, MA, USA.

Furby, L., P. Slovic, B. Fischoff, and G. Gregory (1988). Public perceptions of electric powerlines. *Journal of Environmental Psychology*, **8**(1), pp. 19-43.

GAO (2007). *Federal Electricity Subsidies: Information on Research Funding, Tax Expenditures and Other Activities that Support Electricity Production*. Report to Congressional Requesters, Government Accountability Office, Washington, DC, USA.

Gerlach, L.P. (2004). Public reaction to electricity transmission lines. In: *Encyclopedia of Energy*. Elsevier, New York, pp. 145-167.

Groothuis, P.A., J.D. Groothuis, and J. C. Whitehead (2008). Green vs. green: Measuring the compensation required to site electrical generation windmills in a viewshed. *Energy Policy*, **36**(4), pp. 1545-1550.

Gross, R., W. Blyth, and P. Heptonstall (2010). Risks, revenues and investment in electricity generation: Why policy needs to look beyond costs. *Energy Economics*, **32**(4), pp. 796-804.

Guerra, A.-I., and F. Sancho (2010). Rethinking economy-wide rebound measures: An unbiased proposal. *Energy Policy*, **38**(11), pp. 6684-6694.

Haas, R., N. Nakicenovic, A. Ajanovic, T. Faber, L. Kranzl, A. Mueller, and G. Resch (2008). Towards sustainability of energy systems: A primer on how to apply the concept of energy services to identify necessary trends and policies. *Energy Policy*, **36**(11), pp. 4012-4021.

Halsnaes, K., P.R. Shukla, and A. Garg (2008). Sustainable development and climate change: lessons from country studies. *Climate Policy*, **8**(2), pp. 202-219.

Hamada, Y., M. Nakamura, K. Ochifuji, K. Nagano, and S. Yokoyama (2001). Field performance of a Japanese low energy home relying on renewable energy. *Energy and Buildings*, **33**, pp. 805-814.

Hammer, A., D. Heinemann, C. Hoyer, R. Kuhlemann, E. Lorenz, R. Muller, and H. Beyer (2003). Solar energy assessment using remote sensing technologies. *Remote Sensing of Environment*, **86**, pp. 423-432.

Harvey, L. (2009). Reducing energy use in the buildings sector: measures, costs, and examples. *Energy Efficiency*, **2**(2), pp. 139-163.

Hawken, P., A.B. Lovins, and L.H. Lovins (2010). *The Next Industrial Revolution*. 2nd ed., Earthscan, London, UK and Sterling, VA, USA.

Henriques, I., and P. Sadorsky (2008). Oil prices and the stock prices of alternative energy companies. *Energy Economics*, **30**(3), pp. 998-1010.

Hileman, J.I., D.S. Ortiz, J.T. Bartis, H.M. Wong, P.E. Donohoo, M.A. Weiss, and I.A. Waitz (2009). *Near-Term Feasibility of Alternative Jet Fuels.* RAND Corporation and Massachusetts Institute of Technology, Santa Monica, CA, USA.

Hirschl, B. (2009). International renewable energy policy–between marginalization and initial approaches. *Energy Policy*, **37**(11), pp. 4407-4416.

Hirst, E., and B. Kirby (2001). Key transmission planning issues. *The Electricity Journal*, **8**, pp. 59-70.

Hofman, Y., D. de Jager, E. Molenbroek, F. Schilig, and M. Voogt, (2002) *The potential of solar electricity to reduce CO_2 emissions.* Ecofys, Utrecht, The Netherlands, 106 pp.

Hoogwijk, M. (2004). *On the Global and Regional Potential of Renewable Energy Sources.* Utrecht University, Department of Science, Technology and Society, Utrecht, The Netherlands.

Hynes, S., and N. Hanley (2006). Preservation versus development on Irish rivers: whitewater kayaking and hydro-power in Ireland. *Land Use Policy*, **23**(2), pp. 170-180.

IEA (2005). *Resources to Reserves, Oil and Gas Technologies for the Energy Markets of the Future.* International Energy Agency, Paris, France.

IEA (2008a). *Issues Behind Competitiveness and Carbon Leakage, Focus on Heavy Industry.* International Energy Agency, Paris, France.

IEA (2008b). *World Energy Outlook 2008.* International Energy Agency, Paris, France, 578 pp.

IEA (2009a). *Statistics and Balances.* International Energy Agency, Paris, France.

IEA (2009b). *World Energy Outlook 2009.* International Energy Agency, Paris, France, 696 pp.

IEA (2010a). *Energy Balances of Non-OECD Countries.* International Energy Agency, Paris, France.

IEA (2010b). *Key World Energy Statistics 2010.* International Energy Agency, Paris, France.

IEA (2010c). *World Energy Outlook 2010.* International Energy Agency, Paris, France, 736 pp.

IEA (2010d). *World Energy Outlook 2010. Energy Poverty: How to make modern energy access universal.*, International Energy Agency, Paris, France.

IEA (2010e). *Midterm Oil and Gas Markets 2010.* International Energy Agency, Paris, France.

IEA (2010f). *CO_2 Emissions from Fuel Combustion Highlights 2010.* International Energy Agency, Paris, France.

Ionel, I.I. (1986). *Pumps and Pumping.* Elsevier, New York, NY, USA.

IPCC (2000a). *Special Report on Emission Scenarios.* N. Nakicenovic and R. Swart (eds.), Cambridge University Press, 570 pp.

IPCC (2000b). *Land Use, Land Use Change and Forestry.* R.T. Watson, I.R. Noble, B. Bolin, N.H. Ravindranath, D.J. Verardo and D.J. Dokken (eds.), Cambridge University Press, 375 pp.

IPCC (2005). *Special Report on Carbon Dioxide Capture and Storage.* B. Metz, O. Davidson, H. de Coninck, M. Loos, and L. Meyer (eds.), Cambridge University Press, 431 pp.

IPCC (2006). *2006 IPCC Guidelines for National Greenhouse Gas Inventories.* Institute for Global Environmental Strategies (IGES) for the IPCC, Hayama, Kanagawa, Japan.

IPCC (2007a). *Climate Change 2007: Synthesis Report. Contribution of Working Groups I, II and III to the Fourth Assessment Report of the Intergovernmental Panel on Climate Change.* Core Writing Team, R.K. Pachauri, and A. Reisinger (eds.), Cambridge University Press, 104 pp.

IPCC (2007b). Summary for policymakers. In: *Climate Change 2007: The Physical Science Basis. Contribution of Working Group I to the Fourth Assessment Report of the Intergovernmental Panel on Climate Change.* S. Solomon, D. Qin, M. Manning, Z. Chen, M. Marquis, K.B. Averyt, M. Tignor, and H.L. Miller (eds.), Cambridge University Press, pp. 1-18.

IPCC (2007c). Summary for Policymakers. In: *Climate Change 2007: Impacts, Adaptation and Vulnerability. Contribution of Working Group II to the Fourth Assessment Report of the Intergovernmental Panel on Climate Change.* M.L. Parry, O.F. Canziani, J.P. Palutikof, P.J. van der Linden, and C.E. Hanson (eds.), Cambridge University Press, pp. 7-22.

IPCC (2007d). *Climate Change 2007: Mitigation of Climate Change. Contribution of Working Group III to the Fourth Assessment Report of the Intergovernmental Panel on Climate Change.* B. Metz, O.R. Davidson, P.R. Bosch, R. Dave, and L.A. Meyer (eds.), Cambridge University Press, 851 pp.

Jacobson, M. (2009). Review of solutions to global warming, air pollution, and energy security. *Energy and Environmental Science*, **2**, pp. 148-173.

Johnston, P., and J. Vos (2005). *Pacific Regional Energy Assessment 2004: A Regional Overview Report.* Secretariat of the Pacific Regional Environmental Programme (SPREP), Apia, Samoa.

Kallbekken, S. (2007). Why the CDM will reduce carbon leakage. *Climate Policy*, **7**, pp. 197-211.

Kankam, S., and E. Boon (2009). Energy delivery and utilization for rural development: Lessons from Northern Ghana. *Energy for Sustainable Development*, **13**(3), pp. 212-218.

Kaufmann, R.K. (2004). The mechanisms of autonomous energy efficiency increases: a cointegration analysis of the US energy/GDP ratio. *The Energy Journal*, **25**, pp. 63-86.

Kaya, Y. (1990). Impact of carbon dioxide emission control on GNP growth: Interpretation of proposed scenarios. In: *IPCC Energy and Industry Subgroup, Response Strategies Workshop*, Paris, France.

Kaygusuz, K. (2007). Energy for sustainable development: Key issues and challenges. *Energy Sources*, **2**, pp. 73-83.

Kempton, W., and J. Tomic (2005). Vehicle to grid implementation: from stabilizing the grid to supporting large-scale renewable energy. *Journal of Power Sources*, **144**(1), pp. 280-294.

Klein, R.J.T., S.E.H. Eriksen, L.O. Naess, A. Hammill, T.M. Tanner, C. Robledo, and K.L. O'Brien (2007a). Portfolio screening to support the mainstreaming of adaptation to climate change into development assistance. *Climatic Change*, **84**(1), pp. 23-44.

Klein, R.J.T., S. Huq, F. Denton, T.E. Downing, R.G. Richels, J.B. Robinson, and F.L. Toth (2007b). Inter-relationships between adaptation and mitigation. In: *Climate Change 2007: Impacts, Adaptation and Vulnerability. Contribution of Working Group II to the Fourth Assessment Report of the Intergovernmental Panel on Climate Change.* M.L. Parry, O.F. Canziani, J.P. Palutikof, P.J. van der Linden, and C.E. Hanson (eds.), Cambridge University Press, pp. 745-777.

Klick, H., and E.R.A.N. Smith (2010). Public understanding of and support for wind power in the United States. *Renewable Energy*, **35**(7), pp. 1585-1591.

Knopf, B., O. Edenhofer, C. Flaschland, M. Kok, H. Lotze-Campen, G. Luderer, A. Popp, and D. van Vuuren (2010). Managing the low carbon transition: from model results to policies. *Energy Journal*, **31**(Special Issue), pp. 223-245.

Kooles, K. (2009). Adapting historic district guidelines for solar and other green technologies. *Forum Journal*, **24**, pp. 24-29.

Kotzebue, J., H. Bressers, and C. Yousif (2010). Spatial misfits in a multi-level renewable energy policy implementation process on the Small Island State of Malta. *Energy Policy*, **38**(10), pp. 5967-5976.

Krewitt, W., K. Nienhaus, C. Kleßmann, C. Capone, E. Stricker, W. Graus, M. Hoogwijk, N. Supersberger, U. von Winterfeld, and S. Samadi (2009). *Role and Potential of Renewable Energy and Energy Efficiency for Global Energy Supply.* Climate Change 18/2009, ISSN 1862-4359, Federal Environment Agency, Dessau-Roßlau, Germany, 336 pp.

Lee, J. (2010). Can we accelerate the improvement of energy efficiency in aircraft systems? *Energy Conversion and Management*, **51**(1), pp. 189-196.

Leutz, R., T. Ackermann, A. Suzuki, A. Akisawa, and T. Kashiwagi (2001). Technical offshore wind energy potentials around the globe. In: *Proceedings of the European Wind Energy Conference and Exhibition*, Copenhagen, Denmark, 2-6 July 2001.

Lewis, J., and R. Wiser (2007). Fostering a renewable energy technology industry: an international comparison of wind industry policy support mechanisms. *Energy Policy*, **35**(3), pp. 1844-1857.

Lior, N. (2010). Sustainable energy development: The present (2009) situation and possible paths to the future. *Energy*, **32**(8), pp. 1478-1483.

LLNL (2009). *Estimated World Energy Use in 2006.* Lawrence Livermore National Laboratory, Livermore, CA, USA.

Loulou, R., M. Labriet, and A. Kanudia (2009). Deterministic and stochastic analysis of alternative climate targets under differentiated cooperation regimes. *Energy Economics*, **31**(Supplement 2), pp. S131-S143.

Lund, H. (2007). Renewable energy strategies for sustainable development. *Energy*, **32**, pp. 912-919.

Lund, J. (2003). Direct-use of geothermal energy in the USA. *Applied Energy*, **74**, pp. 33-42.

Macknick, J. (2009). *Energy and Carbon Dioxide Emission Data Uncertainties.* IR-09-032, International Institute for Applied Systems Analysis, Laxenburg, Austria.

Margolis, R.M., and D.M. Kammen (1999). Underinvestment: the energy technology and R&D policy challenge. *Science*, **285**(5428), pp. 690-692.

Mariita, N. (2002). The impact of large-scale renewable energy development on the poor: environmental and socio-economic impact of a geothermal power plant on a poor rural community in Kenya. *Energy Policy*, **30**(11-12), pp. 1119-1128.

Markard, J., and B. Truffer (2006). Innovation processes in large technical systems: Market liberalization as a driver for radical change? *Research Policy*, **35**(5), pp. 609-625.

Markham, S.K. (2002). Moving technologies from lab to market. *Research-Technology Management*, **45**(6), pp. 31.

Martinot, E. (1998). Energy efficiency and renewable energy in Russia: Transaction barriers, market intermediation and capacity building. *Energy Policy*, **26**, pp. 905-915.

Meade, J.E. (1971). *The Controlled Economy.* State University of New York Press, Albany, NY.

Mehta, R., D. Deshpande, K. Kulkarni, S. Sharma, and D. Divan (2008). LEDs: A competitive solution for general lighting applications. In: *Energy 2030 Conference: ENERGY 2008 IEEE*, Atlanta, GA, USA, 17-18 November 2008, doi:10.1109/ENERGY.2008.4781063.

Meinshausen, M., N. Meinshausen, W. Hare, S.C.B. Raper, K. Frieler, R. Knutti, D.J. Frame, and M.R. Allen (2009). Greenhouse-gas emission targets for limiting global warming to 2° C. *Nature*, **458**, pp. 1158-1163, doi:10.1038/nature08017.

Melillo, J.M., J.M. Reilly, D.W. Kicklighter, A.C. Gurgel, T.W. Cronin, S. Palstev, B.S. Felzer, X. Wang, A.P. Sokolov, and C.A. Schlosser (2009). Indirect emissions from biofuels: how important? *Science*, **326**, pp. 1397-1399.

MIT (2003). *The Future of Nuclear Power: An Interdisciplinary Study.* Massachusetts Institute of Technology, Cambridge, MA, USA.

MIT (2007). *The Future of Coal: An Interdisciplinary MIT Study.* Massachusetts Institute of Technology, Cambridge, MA, USA.

MIT (2009). *Update of the MIT 2003 Future of Nuclear Power: An Interdisciplinary Study.* Massachusetts Institute of Technology, Cambridge, MA, USA.

Moomaw, W. (1991). Photovoltaics and materials science: helping to meet the environmental imperatives of clean air and climate change. *Journal of Crystal Growth*, **109**, pp. 1-11.

Moomaw, W.R., J.R. Moreira, K. Blok, D. Greene, K. Gregory, T. Jaszay, T. Kashiwagi, M. Levine, M. MacFarland, N.S. Prasad, L. Price, H. Rogner, R. Sims, F. Zhou, E. Alsema, H. Audus, R.K. Bose, G.M. Jannuzzi, A. Kollmuss, L. Changsheng, E. Mills, K. Minato, S. Plotkin, A. Shafer, A.C. Walter, R. Ybema, J. de Beer, D. Victor, R. Pichs-Madruga, and H. Ishitani (2001). Technological and economic potential of greenhouse gas emissions reduction. In: *Climate Change 2001: Mitigation of Climate Change. Contribution of Working Group III to the Third Assessment Report of the Intergovernmental Panel on Climate Change.* Cambridge University Press, pp. 167-299.

Mondal, Md. A.H., L.M. Kamp, and N. I. Pachova (2010). Drivers, barriers, and strategies for implementation of renewable energy technologies in rural areas in Bangladesh-An innovative system analysis. *Energy Policy*, **38**, pp. 4626-4634.

Moore, G. (2002). *Crossing the Chasm: Marketing and Selling Products to Mainstream Customers.* Harper, New York, NY.

Murphy, L.M., and P.L. Edwards (2003). *Bridging the Valley of Death: Transitioning from Public to Private Sector Financing.* National Renewable Energy Laboratory, Golden, CO, USA.

Najjar, Y.S.H., A.S. Alghamdi, and M.H. Al-Beirutty (2004). Comparative performance of combined gas turbine systems under three different blade cooling schemes. *Applied Thermal Engineering*, **24**, pp. 1919-1934.

Nakicenovic, N., A. Grubler, and A. McDonald (eds.) (1998). *Global Energy Perspectives.* Cambridge University Press.

Nfah, E., J. Ngundam, and R. Tchinda (2007). Modelling of solar/diesel/battery hybrid power systems for far-north Cameroon. *Renewable Energy*, **32**(5), pp. 832-844.

NOAA (2010). *Trends in Carbon Dioxide.* National Oceanic and Atmospheric Administration Earth Systems Research Laboratory, Washington, D.C, USA.

Ockwell, D., R. Haum, A. Mallett, and J. Watson (2010). Intellectual property rights and low carbon technology transfer: conflicting discourses of diffusion and development. *Global Environmental Change*, **20**(4), pp. 729-738.

Onovwiona, H., and V. Ugursal (2006). Residential cogeneration systems: review of the current technology. *Renewable and Sustainable Energy Reviews*, **11**, pp. 482-496.

Orasch, W., and F. Wirl (1997). Technological efficiency and the demand for energy (road transport). *Energy Policy*, **25**(14-15), pp. 1129-1136.

Ossen, D., M. Hamdan Ahmad, and N.H. Madros (2005). Optimum overhang geometry for building energy saving in tropical climates. *Journal of Asian Architecture and Building Engineering*, **4**(2), pp. 563-570.

Ozgener, O., and A. Hapbasili (2004). A review on the energy and exergy analysis of solar assisted heat pump systems. *Renewable and Sustainable Energy Reviews*, **11**, pp. 482-496.

Pacala, S., and R. Socolow (2004). Stabilization wedges: Solving the climate problem for the next 50 years with current technologies. *Science*, **305**(5686), pp. 968-972.

Painuly, J. (2001). Barriers to renewable energy penetration: a framework for analysis. *Renewable Energy*, **24**(1), pp. 83-89.

Paltsev, S. (2001). The Kyoto Protocol: Regional and sectoral contributions to the carbon leakage. *Energy Journal*, **22**(4), pp. 53-79.

Pasqualetti, M.J., P. Gipe, and R.W. Righter (2002). *Wind power in view : energy landscapes in a crowded world.* Academic Press, San Diego, xi, 234 pp.

Passivhaus (2010). *What is a Passive House?* Passivhaus Institute, Darmstadt, Germany. Available at: www.passiv.de/07_eng/index_e.html.

Petersen, E.L., N.G. Mortensen, L. Landberg, J.R. Højstrup, and H.P. Frank (1998). Wind power meteorology. Part II: siting and models. *Wind Energy*, **1**(2), pp. 55-72.

Pilavachi, P.A. (2000). Power generation with gas turbine systems and combined heat and power. *Applied Thermal Engineering*, **20**, pp. 1421-1429.

Pilavachi, P.A. (2002). Mini- and micro-gas turbines for combined heat and power. *Applied Thermal Engineering*, **22**(18), pp. 2003-2014.

Pollack, H.N., S.J. Hurter, and J.R. Johnson (1993). Heat flow from the Earth's interior: Analysis of the global data set. *Reviews of Geophysics*, **31**(3), 267–280, doi:10.1029/93RG01249.

Pousa, G.P.A.G., A.L.F. Santos, and P.A.Z. Suarez (2007). History and policy of biodiesel in Brazil. *Energy Policy*, **35**(11), pp. 5393-5398.

Puga, J., and J. Lesser (2009). Public policy and private interests: why transmission planning and cost-allocation methods continue to stifle renewable energy policy goals. *The Electricity Journal*, **22**(10), pp. 7-19.

Quezada, V., J. Abbad, and T. San Roman (2006). Assessment of energy distribution losses for increasing penetration of distributed generation. *IEEE Transactions on Power Systems*, **21**(2), pp. 533-540.

Ragwitz, M., A. Held, G. Resch, T. Faber, C. Huber, and R. Haas, 2005. *Final Report: Monitoring and Evaluation of Policy Instruments to Support Renewable Electricity in EU Member States.* Fraunhofer Institute Systems and Innovation Research and Energy Economics Group, Karlsruhe, Germany and Vienna, Austria.

REN21 (2009). *Renewables Global Status Report: 2009 Update.* Renewable Energy Policy Network for the 21st Century Secretariat, Paris, France.

REN21 (2010). *Renewables 2010: Global Status Report.* Renewable Energy Policy Network for the 21st Century Secretariat, Paris, France, 80 pp.

Rezaie, B., E. Esmailzadeh, and I. Dincer (2011). Renewable energy options for buildings: case studies. *Energy and Buildings*, **43**(1), pp. 56-65.

Richter, A. (2007). *United States Geothermal Energy Market Report.* GlitnirBank, Reykjavik, Iceland.

Roberts, S. (2008). Infrastructure and challenges for the built environment. *Energy Policy*, **36**(12), pp. 4563-4567.

Robock, A., A. Marquardt, B. Kravitz, and G. Stenchikov (2009). Benefits, risks, and costs of stratospheric geoengineering. *Geophysical Research Letters*, **36**, L19703.

Rogner, H.-H., F. Barthel, M. Cabrera, A. Faaij, M. Giroux, D. Hall, V. Kagramanian, S. Kononov, T. Lefevre, R. Moreira, R. Nötstaller, P. Odell, and M. Taylor (2000). Energy resources. In: *World Energy Assessment. Energy and the Challenge of Sustainability.* United Nations Development Programme, United Nations Department of Economic and Social Affairs, World Energy Council, New York, USA, pp. 30-37.

Rogner, H.-H., D. Zhou, R. Bradley. P. Crabbé, O. Edenhofer, B.Hare (Australia), L. Kuijpers, and M. Yamaguchi (2007). Introduction. In: *Climate Change 2007: Mitigation of Climate Change. Contribution of Working Group III to the Fourth Assessment Report of the Intergovernmental Panel on Climate Change.* B. Metz, O.R. Davidson, P.R. Bosch, R. Dave, and L.A. Meyer (eds.), Cambridge University Press, pp. 95-116.

Roulleau, T., and C. Loyd (2008). International policy issues regarding solar water heating, with a focus on New Zealand. *Energy Policy*, **36**(6), pp. 1843-1857.

Rout, U., K. Akomoto, F. Sano, J. Oda, T. Homma, and T. Tomada (2008). Impact assessment of the increase in fossil fuel prices on the global energy system, with and without CO_2 concentration stabilization. *Energy Policy*, **36**(9), pp. 3477-3484.

Rowley, J.C. (1982). Worldwide geothermal resources. In: *Handbook of Geothermal Energy.* L.M. Edwards, G.V. Chilingar, H.H. Rieke III, and W.H. Fertl (eds.), Gulf Publishing, Houston, TX, USA, pp. 44-176.

Royal Society (2009). *Geoengineering the Climate: Science, Governance, and Uncertainty.* The Royal Society, London, UK.

Rubin, E., C. Chen, and A. Rao (2007). Cost and performance of fossil fuel power plants with CO_2 capture and storage. *Energy Policy*, **35**, pp. 4444-4454.

Sanner, B., C. Karytsas, D. Mendrinos, and L. Rybach (2003). Current status of ground source heat pumps and underground thermal energy storage in Europe. *Geothermics*, **32**, pp. 579-588.

Sawin, J. (2001). *The Role of Government in the Development and Diffusion of Renewable Energy Technologies.* UMI Proquest, Ann Arbor, MI, USA, 618 pp.

Scharlemann, J.P.W., and W.F. Laurance (2008). How green are biofuels? *Science*, **319**(5859), pp. 43-44.

Schipper, L., and M. Grubb (2000). On the rebound? Feedback between energy intensity and energy use in IEA countries. *Energy Policy*, **28**, pp. 367-388.

Searchinger, T., R. Heimlich, R.A. Houghton, F. Dong, A. Elobeid, J. Fabiosa, S. Tokgoz, D. Hayes, and T.-H. Yu (2008). Use of U.S. croplands for biofuels increases greenhouse gases through emissions from land-use change. *Science*, **319**, pp. 1238-1240.

Secretary of the Air Force (2009). *Air Force Energy Policy Memorandum.* Secretary of the Air Force, Department of Defense, Washington, DC, USA.

Simon, C. (2009). Cultural constraints on wind and solar energy in the U.S. context. *Comparative Technology Transfer and Society*, **7**(3), pp. 251-269.

Sims, R.E.H., R.N. Schock, A. Adegbululgbe, J. Fenhann, I. Konstantinaviciute, W. Moomaw, H.B. Nimir, B. Schlamadinger, J. Torres-Martínez, C. Turner, Y. Uchiyama, S.J.V. Vuori, N. Wamukonya, and X. Zhang (2007). Energy supply. In: *Climate Change 2007: Mitigation of Climate Change. Contribution of Working Group III to the Fourth Assessment Report of the Intergovernmental Panel on Climate Change.* B. Metz, O.R. Davidson, P.R. Bosch, R. Dave, and L.A. Meyer (eds.), Cambridge University Press, pp. 251-322.

Slade, R., C. Panoutsou, and A. Bauen (2009). Reconciling bio-energy policy and delivery in the UK: Will UK policy initiatives lead to increased deployment? *Biomass and Bioenergy*, **33**(4), pp. 679-688.

Smeets, E.M.W., A.P.C. Faaij, I.M. Lewandowski, and W.C. Turkenburg (2007). A bottom-up assessment and review of global bio-energy potentials to 2050. *Progress in Energy and Combustion Science*, **33**, pp. 56-106.

Sorrell, S., J. Dimitropoulos, and M. Sommerville (2009). Empirical estimates of the direct rebound effect: a review. *Energy Policy*, **37**, pp. 1356-1371.

Sovacool, B. (2008). Valuing the emissions from nuclear power: a critical survey. *Energy Policy*, **26**, pp. 2940-2953.

Stern, N. (2007). *The Economics of Climate Change.* Cambridge University Press, 712 pp. Available at: webarchive.nationalarchives.gov.uk/+/http://www.hm-treasury.gov.uk/sternreview_index.htm.

Sterner, M. (2009). *Bioenergy and Renewable Power Methane in Integrated 100% Renewable Energy Systems: Limiting Global Warming by Transforming Energy Systems.* University of Kassel, Kassel, Germany.

Tester, J.W., E.M. Drake, M.W. Golay, M.J. Driscoll, and W.A. Peters (2005). *Sustainable Energy – Choosing Among Options.* MIT Press, Cambridge, MA, USA, 850 pp.

Tester, J.W., B.J. Anderson, A.S. Batchelor, D.D. Blackwell, R. DiPippo, and E.M. Drake (2006). *The Future of Geothermal Energy: Impact of Enhanced Geothermal Systems on the United States in the 21st Century.* Prepared by the Massachusetts Institute of Technology, under Idaho National Laboratory Subcontract No. 63 00019 for the U.S. Department of Energy, Assistant Secretary for Energy Efficiency and Renewable Energy, Office of Geothermal Technologies, Washington, DC, USA, 358 pp. Available at: geothermal.inl.gov.

Tomic, J., and W. Kempton (2007). Using fleets of electric-drive vehicles for grid support. *Journal of Power Sources*, **168**(2), pp. 459-468.

Trieb, F. (2005). *Concentrating Solar Power for the Mediterranean Region. Final Report.* German Aerospace Centre (DLR), Stuttgart, Germany, 285 pp.

Trieb, F., M. O'Sullivan, T. Pregger, C. Schillings, and W. Krewitt (2009). *Characterisation of Solar Electricity Import Corridors from MENA to Europe - Potential, Infrastructure and Cost.* German Aerospace Centre (DLR), Stuttgart, Germany, 172 pp.

Turner, K. (2009). Negative rebound and disinvestment effects in response to an improvement in energy efficiency in the UK economy. *Energy Economics*, **31**(5), pp. 648-666.

Twidell, J., and A.D. Weir (2005). *Renewable Energy Resources.* 2nd Ed. Taylor & Francis, London, UK and New York, NY, USA.

UNEP (2009). *Global Trends in Sustainable Energy Investment 2009: Analysis of Trends and Issues in the Financing of Renewable Energy and Energy Efficiency.* United Nations Environment Programme, Paris, France.

UNDP (2005). *Energy Challenge for Achieving the Millennium Development Goals 2010.* United Nations Development Programme, New York, NY, USA.

UNDP (2007). *Human Development Report 2007*, United Nations Development Programme, New York, NY, USA.

UNDP/WHO (2009). *The Energy Access Situation in Developing Countries: A Review Focusing on the Least Developed Countries and sub-Saharan Africa.* United Nations Development Programme, New York, NY, USA.

Unruh, G.C., and J. Carillo-Hermosilla (2006). Globalizing carbon lock-in. *Energy Policy*, **34**, pp. 1185-1197.

UNStats (2010). *Energy Balances and Electricity Profiles – Concepts and Definitions.* United Nations Statistics Division, New York, NY, USA.

USDOD (2010). *Quadrennial Defense Review Report.* US Department of Defense, Washington, DC, USA.

Vajjhala, S., and P. Fischbeck (2007). Quantifying siting difficulty: A case study of US transmission line siting. *Energy Policy*, **35**(1), pp. 650-671.

Valentine, S. (2010). A step toward understanding wind power development policy barriers in advanced economies. *Renewable and Sustainable Energy Reviews*, **14**(9), pp. 2796-2807.

van Ruijven, B., and D. van Vuuren (2009). Oil and natural gas prices and greenhouse gas emission mitigation. *Energy Policy*, **37**(11), pp. 4797-4808.

Venema, H., and I. Rehman (2007). Decentralized renewable energy and the climate change mitigation-adaptation nexus. *Mitigation and Adaptation Strategies for Global Change*, **12**, pp. 875-900.

Verbruggen, A., M. Fischedick, W. Moomaw, T. Weir, A. Nadaï, L.J. Nilsson, J. Nyboer, and J. Sathaye (2010). Renewable energy costs, potentials, barriers: Conceptual issues. *Energy Policy*, **38**(2), pp. 850-861.

Viguier, L., and M. Vielle (2007). On the climate change effects of high oil prices. *Energy Policy*, **35**(2), pp. 844-849.

von Weizsäcker, E., K. Hargroves, M.H. Smith, C. Desha, and P. Stasinopoulos (2009). *Factor Five: Transforming the Global Economy through 80% Improvements in Resource Productivity: A Report to the Club of Rome.* Earthscan/The Natural Edge Project, London, UK, 400 pp.

WCD (2000). *Dams and Development: A New Framework for Decision-Making: The Report of the World Commission on Dams.* World Commission on Dams, Earthscan, London, UK. Available at: www.dams.org/index.php?option=com_content&view=article&id=49&Itemid=29.

Webler, T., and S.P. Tuler (2010). Getting the engineering right is not always enough: Researching the human dimensions of the new energy technologies. *Energy Policy*, **38**(6), pp. 2690-2691.

WEC (1994). *New Renewable Energy Resources: A Guide to the Future.* World Energy Council, Kogan Page, London, UK.

WEC/FAO (1999). *The Challenge of Rural Energy Poverty in Developing Countries.* World Energy Council and Food and Agricultural Organization of the United Nations, London, UK.

Weisser, D. (2007). A guide to life-cycle greenhouse gas (GHG) emissions from electric supply technologies. *Energy*, **32**, pp. 1543-1559.

Williamson, O.E. (1985). *The Economic Institutions of Capitalism.* The Free Press, New York, NY, USA.

WIREC (2008). WIREC 2008: The power of independence. In: *Washington International Renewable Energy Conference*, Washington, DC, 4-6 March 2008. Available at: www.iisd.ca/ymb/wirec2008/html/ymbvol95num8e.html.

World Bank (2006). *Reforming Power Markets in Developing Countries: What Have We Learned?* World Bank, Washington, DC, USA.

World Bank (2009). *World Development Indicators – CO_2 emissions (kt).* World Bank, Washington, DC, USA. Available at: data.worldbank.org/indicator/EN.ATM.CO2E.KT/countries/1W?display=graph.

Appendix to Chapter 1

Table 1.A.1 | Global technical potential of RE sources (compared to global primary energy supply in 2008 of 492 EJ).[1]

		Technical Potential (EJ/yr)					Notes and Sources for Range of Estimates and Notes on Krewitt et al. (2009) estimates
		Krewitt et al. (2009)[2]			Range of Estimates Summarized in Chapters 2-7[3]		
		2020	2030	2050	Low	High	
Electric Power (EJ/yr)	Solar PV[4]	1,126	1,351	1,689	1,338	14,778	Chapter 3 – Hofman et al. (2002); Hoogwijk (2004); de Vries et al. (2007). The methodology used by Krewitt et al. (2009) differs between PV and CSP; details are described in Chapter 3.
	Solar CSP[4]	5,156	6,187	8,043	248	10,791	Chapter 3 – Hofman et al. (2002); Trieb (2005); Trieb et al. (2009). The methodology used by Krewitt et al. (2009) differs between PV and CSP; details are described in Chapter 3.
	Geothermal[5]	4,5	18	45	118	1,109	Hydrothermal and EGS: Chapter 4 – EPRI (1978); Rowley (1982); Stefansson (2005); Tester et al. (2005, 2006).
	Hydropower	48	49	50	50	52	Chapter 5 – Krewitt et al. (2009); International Journal of Hydro & Dams (2010).
	Ocean[6]	66	166	331	7	331	Chapter 6 – Sims et al. (2007); Krewitt et al. (2009); technical potential estimates may not include all ocean energy technologies; Sims et al. (2007) estimate is referred to as 'exploitable estimated available energy resource'.
	Wind On-Shore	362	369	379	70	450	Chapter 7 – low estimate from WEC (1994), high estimate from Archer and Jacobson (2005) and includes 'near-shore', more recent estimates tend towards higher end of range.
	Wind Off-Shore[7]	26	36	57	15	130	Chapter 7 – low estimate from Fellows (2000), high estimate from Leutz et al. (2001), only considering relatively shallow water and near-shore applications; greater technical potential exists if one considers deeper water applications (Lu et al., 2009; Capps and Zender, 2010).
Heat (EJ/yr)	Solar	113	117	123	N/A	N/A	Technical potential is mainly limited by the demand for heat. Krewitt et al. (2009) base estimates on available rooftop area and only solar water heating; technical potential considering non-rooftop applications and process heat would far exceed these estimates.
	Geothermal	104	312	1,040	10	312	Hydrothermal: Chapter 4 – Stefansson (2005). Although the estimates from Krewitt et al. (2009) are also based on Stefansson (2005), Krewitt et al. (2009) assume a higher capacity factor than Chapter 4.
Primary Energy (EJ/yr)	Solar[8]	N/A	N/A	N/A	1,575	49,837	Total solar energy technical potential: Chapter 3 – Rogner et al. (2000)
	Biomass Energy Crops[9]	43	61	96	small	120	Dedicated biomass production on surplus agriculture and pasture lands: Chapter 2 – Dornburg et al. (2010).
					small	140	Further intensification of agriculture: Chapter 2 – Dornburg et al. (2010).
					small	70	Dedicated biomass production on marginal/degraded lands: Chapter 2 – Dornburg et al. (2010).
					small	100	More intensive forest management: Chapter 2 – Dornburg et al. (2010).
	Biomass Residues[9]	59	68	88	40	100	Agriculture and forestry residues, other organic wastes, dung etc.: Chapter 2 – Dornburg et al. (2010).
	Biomass Total[9]	**102**	**129**	**184**	**50**[10]	**500**[11]	Rounded figures based on Chapter 2 expert review of technical potential assessments.

Notes:

1 Technical potentials reported here represent total worldwide potentials for annual RE supply and do not deduct any potential that is already being utilized for energy production. In 2008, total primary energy supply from RE sources on a direct equivalent basis equalled: bioenergy (50.33 EJ); hydropower (11.55 EJ); wind (0.79 EJ); solar (0.50 EJ); geothermal (0.41 EJ); and ocean (0.002 EJ). According to the definition of technical potential in the Glossary (see Annex I), many of the studies summarized here take into some account broader economic and socio-political considerations. For example, for some technologies, land suitability or other sustainability factors are included, which result in lower technical potential estimates.

2 Technical potential estimates for 2020, 2030 and 2050 are based on a review of studies in Krewitt et al. (2009). Due to differences in methodologies and accounting methods between studies, comparison of these estimates across technologies and regions, as well as to primary energy demand, should be exercised with caution. Data presented in Chapters 2 through 7 may disagree with these figures due to differing methodologies. Krewitt et al. (2009), as well as many of the other studies reported in the table, assume that technical potential increases over time due, in part, to technological advancements.

Continued next Page →

3   Range of estimates derives from studies presented in Chapters 2 through 7 (occasionally including some of the studies reported in the Krewitt et al. (2009) review). As a result, ranges do not always encompass the figures presented in Krewitt et al. (2009). Ranges are based on various methods and apply to different future years; consequently, as with Krewitt et al. (2009), the resulting ranges are not strictly comparable across technologies.

4   Estimates for PV and CSP in Krewitt et al. (2009) are based on different data and methodologies, which tend to significantly understate the technical potential for PV relative to CSP. In part as a result, a range for total solar energy technical potential is provided in the primary energy category based on Rogner et al. (2000). Note that this technical potential for total solar primary energy is not the sum of the three listed technologies (PV, CSP and solar heat) due to different studies used. Also note that the technical potentials for PV, CSP and solar heat listed in the table are not strictly additive due to possible competition for land among specific solar technologies.

5   Estimates for geothermal electricity in Krewitt et al. (2009) appear to largely consider only hydrothermal resources. The range of estimates presented in Chapter 4 derives from EPRI (1978), Rowley (1982), Stefansson (2005), and Tester et al. (2005, 2006) and includes both hydrothermal and EGS potential.

6   The absolute range of technical potential for ocean energy is highly uncertain, because few technical potential estimates have been conducted due to the fact that the technologies are still largely in the R&D phase and have not been commercially deployed at scale.

7   Estimates for offshore wind energy in Krewitt et al. (2009) and the range of estimates provided in the literature as presented in the table are both based on relatively shallow water and near-shore applications. Greater technical potential for offshore wind energy is found when considering deeper-water applications that might rely on floating wind turbine designs.

8   The technical potential for total solar primary energy is not the sum of the three listed technologies (PV, CSP and solar heat) due to different studies used; also note that possible competition for land among specific solar technologies makes it inappropriate to add the technical potential estimates for PV, CSP and solar heat to derive a total solar technical potential. The estimates of the total solar energy technical potential provided in the table do not differentiate between the different solar conversion technologies, but just take into account average conversion efficiency, available land area and meteorological conditions. At certain geographical locations all listed solar technologies could be used and users will decide what service they need from which technology.

9   Primary energy from biomass (in direct equivalent terms) could be used to meet electricity, thermal or transportation needs, all with a conversion loss from primary energy ranging from roughly 20 to 80%. As a result, comparisons of the technical potential for biomass in primary energy terms to the technical potentials of other RE sources in delivering secondary energy supply (i.e., electric power and heat) should be made with care.

10 The conditions under the low technical potential estimate could emerge when agricultural productivity increases stall worldwide combined with high food demand and no surplus land for energy crops being available. It is also assumed that marginal and degraded lands are not utilized and a large fraction of biomass residue flows is assumed to be used as feedstock in other sectors rather than for bioenergy. However, low-grade residues, dung and municipal waste will in such a situation likely still remain available for bioenergy.

11 The higher end of the biomass potential is conditional and assumes proper land management and substantial increases in agricultural yields and intensified forestry management. Achieving such a potential will be sustainable only if monitoring and good governance of land use is effective, and sustainability frameworks are in place.

2 Bioenergy

Coordinating Lead Authors:
Helena Chum (USA/Brazil), Andre Faaij (The Netherlands), José Moreira (Brazil)

Lead Authors:
Göran Berndes (Sweden), Parveen Dhamija (India), Hongmin Dong (China), Benoît Gabrielle (France), Alison Goss Eng (USA), Wolfgang Lucht (Germany), Maxwell Mapako (South Africa/Zimbabwe), Omar Masera Cerutti (Mexico), Terry McIntyre (Canada), Tomoaki Minowa (Japan), Kim Pingoud (Finland)

Contributing Authors:
Richard Bain (USA), Ranyee Chiang (USA), David Dawe (Thailand, USA), Garvin Heath (USA), Martin Junginger (The Netherlands), Martin Patel (The Netherlands), Joyce Yang (USA), Ethan Warner (USA)

Review Editors:
David Paré (Canada) and Suzana Kahn Ribeiro (Brazil)

This chapter should be cited as:

Chum, H., A. Faaij, J. Moreira, G. Berndes, P. Dhamija, H. Dong, B. Gabrielle, A. Goss Eng, W. Lucht, M. Mapako, O. Masera Cerutti, T. McIntyre, T. Minowa, K. Pingoud, 2011: Bioenergy. In IPCC Special Report on Renewable Energy Sources and Climate Change Mitigation [O. Edenhofer, R. Pichs-Madruga, Y. Sokona, K. Seyboth, P. Matschoss, S. Kadner, T. Zwickel, P. Eickemeier, G. Hansen, S. Schlömer, C. von Stechow (eds)], Cambridge University Press, Cambridge, United Kingdom and New York, NY, USA.

Table of Contents

Executive Summary 214

2.1 Introduction 216

2.1.1 Current pattern of biomass and bioenergy use and trends 216

2.1.2 Previous Intergovernmental Panel on Climate Change assessments 219

2.2 Resource potential 220

2.2.1 Introduction 220
2.2.1.1 Methodology assessment 220
2.2.1.2 Total aboveground net primary production of biomass 222
2.2.1.3 Human appropriation of terrestrial net primary production 222

2.2.2 Global and regional technical potential 223
2.2.2.1 Literature assessment 223
2.2.2.2 The contribution from residues, dung, processing by-products and waste 223
2.2.2.3 The contribution from unutilized forest growth 223
2.2.2.4 The contribution from biomass plantations 224

2.2.3 Economic considerations in biomass resource assessments 227

2.2.4 Factors influencing biomass resource potentials 228
2.2.4.1 Residue supply in agriculture and forestry 229
2.2.4.2 Dedicated biomass production in agriculture and forestry 229
2.2.4.3 Use of marginal lands 231
2.2.4.4 Biodiversity protection 231

2.2.5 Possible impact of climate change on resource potential 232

2.2.6 Synthesis 232

2.3 Technologies and applications 233

2.3.1 Feedstocks 233
2.3.1.1 Feedstock production and harvest 233
2.3.1.2 Synergies with the agriculture, food and forest sectors 235

2.3.2 Logistics and supply chains for energy carriers from modern biomass 236
2.3.2.1 Solid biomass supplies and market development for utilization 236
2.3.2.2 Solid biomass and charcoal supplies in developing countries 237
2.3.2.3 Wood pellet logistics and supplies 237

2.3.3	Conversion technologies to electricity, heat, and liquid and gaseous fuels	238
2.3.3.1	Development stages of conversion technologies	238
2.3.3.2	Thermochemical processes	238
2.3.3.3	Chemical processes	240
2.3.3.4	Biochemical processes	240
2.3.4	Bioenergy systems and chains: Existing state-of-the-art systems	240
2.3.4.1	Bioenergy chains for power, combined heat and power, and heat	241
2.3.4.2	Bioenergy chains for liquid transport fuels	241
2.3.5	Synthesis	244
2.4	**Global and regional status of market and industry development**	**246**
2.4.1	Current bioenergy production and outlook	246
2.4.2	Traditional biomass, improved technologies and practices, and barriers	248
2.4.2.1	Improved biomass cook stoves	249
2.4.2.2	Biogas systems	250
2.4.3	Modern biomass: Large-scale systems, improved technologies and practices, and barriers	250
2.4.4	Global trade in biomass and bioenergy	251
2.4.5	Overview of support policies for biomass and bioenergy	253
2.4.5.1	Intergovernmental platforms for exchange on bioenergy policies and standardization	254
2.4.5.2	Sustainability frameworks and standards	254
2.4.6	Main opportunities and barriers for the market penetration and international trade of bioenergy	255
2.4.6.1	Opportunities	255
2.4.6.2	Barriers	255
2.4.7	Synthesis	257
2.5	**Environmental and social impacts**	**257**
2.5.1	Environmental effects	258
2.5.2	Modern bioenergy: Climate change excluding land use change effects	259
2.5.3	Modern bioenergy: Climate change including land use change effects	263
2.5.4	Traditional biomass: Climate change effects	268

2.5.5	**Environmental impacts other than greenhouse gas emissions**	268
2.5.5.1	Impacts on air quality and water resources	268
2.5.5.2	Biodiversity and habitat loss	269
2.5.5.3	Impacts on soil resources	269
2.5.6	**Environmental health and safety implications**	270
2.5.6.1	Feedstock issues	270
2.5.6.2	Biofuels production issues	270
2.5.7	**Socioeconomic aspects**	271
2.5.7.1	Socioeconomic impact studies and sustainability criteria for bioenergy systems	271
2.5.7.2	Socioeconomic impacts of small-scale systems	271
2.5.7.3	Socioeconomic aspects of large-scale bioenergy systems	272
2.5.7.4	Risks to food security	273
2.5.7.5	Impacts on rural and social development	274
2.5.7.6	Trade-offs between social and environmental aspects	274
2.5.8	**Synthesis**	274
2.6	**Prospects for technology improvement and innovation**	**276**
2.6.1	**Improvements in feedstocks**	276
2.6.1.1	Yield gains	276
2.6.1.2	Aquatic biomass	277
2.6.2	**Improvements in biomass logistics and supply chains**	278
2.6.3	**Improvements in conversion technologies for secondary energy carriers from modern biomass**	280
2.6.3.1	Liquid fuels	281
2.6.3.2	Gaseous fuels	285
2.6.3.3	Biomass with carbon capture and storage: long-term removal of greenhouse gases from the atmosphere	286
2.6.3.4	Biorefineries	286
2.6.3.5	Bio-based products	286
2.6.4	**Synthesis**	287
2.7	**Cost trends**	**288**
2.7.1	**Determining factors**	288
2.7.1.1	Recent levelized costs of electricity, heat and fuels for selected commercial systems	288

2.7.2	Technological learning in bioenergy systems	292
2.7.3	**Future scenarios of cost reduction potentials**	293
2.7.3.1	Future cost trends of commercial bioenergy systems	293
2.7.3.2	Future cost trends for pre-commercial bioenergy systems	295
2.7.4	**Synthesis**	295

2.8 Potential Deployment ... 296

2.8.1	**Current deployment of bioenergy**	296
2.8.2	**Near-term forecasts**	297
2.8.3	**Long-term deployment in the context of carbon mitigation**	297
2.8.4	**Conditions and policies: Synthesis of resource potentials, technology and economics, and environmental and social impacts of bioenergy**	300
2.8.4.1	Resource potentials	300
2.8.4.2	Bioenergy technologies, supply chains and economics	302
2.8.4.3	Social and environmental impacts	304
2.8.5	**Conclusions regarding deployment: Key messages about bioenergy**	306

References ... 309

Executive Summary

Bioenergy has a significant greenhouse gas (GHG) mitigation potential, provided that the resources are developed sustainably and that efficient bioenergy systems are used. Certain current systems and key future options including perennial cropping systems, use of biomass residues and wastes and advanced conversion systems are able to deliver 80 to 90% emission reductions compared to the fossil energy baseline. However, land use conversion and forest management that lead to a loss of carbon stocks (direct) in addition to indirect land use change (d+iLUC) effects can lessen, and in some cases more than neutralize, the net positive GHG mitigation impacts. Impacts of climate change through temperature increases, rainfall pattern changes and increased frequency of extreme events will influence and interact with biomass resource potential. This interaction is still poorly understood, but it is likely to exhibit strong regional differences. Climate change impacts on biomass feedstock production exist but if global temperature rise is limited to less than 2°C compared with the pre-industrial record, it may pose few constraints. Combining adaptation measures with biomass resource production can offer more sustainable opportunities for bioenergy and perennial cropping systems.

Biomass is a primary source of food, fodder and fibre and as a renewable energy (RE) source provided about 10.2% (50.3 EJ) of global total primary energy supply (TPES) in 2008. Traditional use of wood, straws, charcoal, dung and other manures for cooking, space heating and lighting by generally poorer populations in developing countries accounts for about 30.7 EJ, and another 20 to 40% occurs in unaccounted informal sectors including charcoal production and distribution. TPES from biomass for electricity, heat, combined heat and power (CHP), and transport fuels was 11.3 EJ in 2008 compared to 9.6 EJ in 2005 and the share of modern bioenergy was 22% compared to 20.6%.

From the expert review of available scientific literature, potential deployment levels of biomass for energy by 2050 could be in the range of 100 to 300 EJ. However, there are large uncertainties in this potential such as market and policy conditions, and it strongly depends on the rate of improvement in the production of food and fodder as well as wood and pulp products.

The upper bound of the technical potential of biomass for energy may be as large as 500 EJ/yr by 2050. Reaching a substantial fraction of the technical potential will require sophisticated land and water management, large worldwide plant productivity increases, land optimization and other measures. Realizing this potential will be a major challenge, but it could make a substantial contribution to the world's primary energy supply in 2050. For comparison, the equivalent heat content of the total biomass harvested worldwide for food, fodder and fibre is about 219 EJ/yr today.

A scenario review conducted in Chapter 10 indicates that the contribution of bioenergy in GHG stabilization scenarios of different stringency can be expected to be significantly higher than today. By 2050, in the median case bioenergy contributes 120 to 155 EJ/yr to global primary energy supply, or 150 to 190 EJ/yr for the 75th percentile case, and even up to 265 to 300 EJ/yr in the highest deployment scenarios. This deployment range is roughly in line with the IPCC Special Report on Emission Scenarios (SRES) regionally oriented A2 and B2 and globally oriented A1 and B1 conditions and storylines. Success in implementing sustainability and policy frameworks that ensure good governance of land use and improvements in forestry, agricultural and livestock management could lead to both high (B1) and low (B2) potentials. However, biomass supplies may remain limited to approximately 100 EJ/yr in 2050 if such policy frameworks and enforcing mechanisms are not introduced and if there is strong competition for biomaterials from other (innovative future) sectors. In that environment, further biomass expansion could lead to significant regional conflicts for food supplies, water resources and biodiversity, and could even result in additional GHG emissions, especially due to iLUC and loss of carbon stocks. In another deployment scenario, biomass resources may be constrained to use of residues and organic waste, energy crops cultivated on marginal/degraded and poorly utilized lands, and to supplies in endowed world regions where bioenergy is a cheaper energy option compared to market alternatives (e.g., sugarcane ethanol production in Brazil).

Bioenergy has complex societal and environmental interactions, including climate change feedback, biomass production and land use. The impact of bioenergy on social and environmental issues (e.g., health, poverty, biodiversity) may be positive or negative depending on local conditions and the design and implementation of specific projects. The policy context for bioenergy, and particularly biofuels, has changed rapidly and dramatically in recent years. The food versus fuel debate and growing concerns about other conflicts are driving a strong push for the development and implementation of sustainability criteria and frameworks. Many conflicts can be reduced if not avoided by encouraging synergisms in the management of natural resource, agricultural and livestock sectors as part of good governance of land use that increases rural development and contributes to poverty alleviation and a secure energy supply.

Costs vary by world regions, feedstock types, feedstock supply costs for conversion processes, the scale of bioenergy production and production time during the year. Examples of estimated commercial bioenergy levelized cost ranges are roughly USD_{2005} 2 to 48/GJ for liquid and gaseous biofuels; roughly US $cents_{2005}$ 3.5 to 25/kWh (USD_{2005} 10 to 50/GJ) for electricity or CHP systems larger than about 2 MW (with feedstock costs of USD_{2005} 3/GJ_{feed} and a heat value of USD_{2005} 5/GJ for steam or USD_{2005} 12/GJ for hot water); and roughly USD_{2005} 2 to 77/GJ for domestic or district heating systems with feedstock costs in the range of USD_{2005} 0 to 20/GJ (solid waste to wood pellets). These calculations refer to 2005 to 2008 data and are expressed in USD_{2005} at a 7% discount rate.

Recent analyses of lignocellulosic biofuels indicate potential improvements that enable them to compete at oil prices of USD_{2005} 60 to 70/barrel (USD_{2005} 0.38 to 0.44/litre) assuming no revenue from carbon dioxide (CO_2) mitigation. Scenario analyses indicate that strong short-term research and development (R&D) and market support could allow for commercialization around 2020 depending on oil and carbon pricing. In addition to ethanol and biodiesel, a range of hydrocarbons and chemicals/materials similar to those currently derived from oil could provide biofuels for not only vehicles but also for the aviation and maritime sectors. Biomass is the only renewable resource that can currently provide high energy density liquid fuels. A wider variety of bio-based products can also be produced at biorefineries to enhance the economics of the overall conversion process. Short-term options (some of them already competitive) that can deliver long-term synergies include co-firing, CHP, heat generation and sugarcane-based ethanol and bioelectricity co-production. Development of working bioenergy markets and facilitation of international bioenergy trade can help achieve these synergies.

Further improvements in power generation technologies, supply systems of biomass and production of perennial cropping systems can bring bioenergy costs down. There is clear evidence that technological learning and related cost reductions occur in many biomass technologies with learning rates comparable to other RE technologies. This is true for cropping systems where improvements in agricultural management of annual crops, supply systems and logistics, conversion technologies to produce energy carriers such as heat, electricity and ethanol from sugarcane or maize, and biogas have demonstrated significant cost reductions.

Combining biomass conversion with developing carbon capture and storage (CCS) could lead to long-term substantial removal of GHGs from the atmosphere (also referred to as negative emissions). Advanced biomaterials are promising as well from both an economic and a GHG mitigation perspective, though the relative magnitude of their mitigation potential is not well understood. The potential role of aquatic biomass (algae) is highly uncertain but could reduce land use conflict. More experience, research, development and demonstration (RD&D), and detailed analyses of these options are needed.

Multiple drivers for bioenergy systems and their deployment in sustainable directions are emerging.
Examples include rapidly changing policy contexts, recent market-based activities, the increasing support for advanced biorefinery and lignocellulosic biofuel options and, in particular, development of sustainability criteria and frameworks. Sustained cost reductions of key technologies in biomass production and conversion, supply infrastructure development, and integrated systems research can lead to the implementation of strategies that facilitate sustainable land and water use and gain public and political acceptance.

2.1 Introduction

Bioenergy is embedded in complex ways in global biomass systems for food, fodder and fibre production and for forest products; in wastes and residue management; and in the everyday living of the developing countries' poor. Bioenergy includes different sets of technologies for applications in various sectors.

2.1.1 Current pattern of biomass and bioenergy use and trends

Biomass provided about 10.2% (50.3 EJ/yr) of the annual global primary energy supply in 2008, from a wide variety of biomass sources feeding numerous sectors of society (see Table 2.1; IEA, 2010a). The biomass feedstocks used for energy are shown in Figure 2.1 (top), and more than 80% are derived from wood (trees, branches, residues) and shrubs. The remaining bioenergy feedstocks came from the agricultural sector (energy crops, residues and by-products) and from various commercial and post-consumer waste and by-product streams (biomass product recycling and processing or the organic biogenic fraction of municipal solid waste[1] (MSW)).

Biomass is used (see Table 2.1) with varying degrees of energy efficiency in various sectors:

- Low-efficiency *traditional biomass*[2] such as wood, straws, dung and other manures are used for cooking, lighting and space heating, generally by the poorer populations in developing countries. This biomass is mostly combusted, creating serious negative impacts on health and living conditions. Increasingly, charcoal is becoming a secondary energy carrier in rural areas. As an indicator of the magnitude of traditional biomass use, Figure 2.1 (bottom) illustrates that the global primary energy supply from traditional biomass parallels the world's industrial roundwood production.

In the International Energy Agency's (IEA) World Energy Statistics (IEA, 2010a) and World Energy Outlook (WEO: IEA, 2010b) TPES from traditional biomass amounts to 30.7 EJ/yr based on national databases that tend to systematically underestimate fuelwood consumption. Although international forestry and energy data (FAO, 2005) are the main reference sources for policy analyses, they are

Table 2.1 | Examples of traditional and select modern biomass energy flows in 2008 according to the IEA (2010 a,b) and supplemented by Masera et al., 2005, 2006; Drigo et al., 2007, 2009.

Type	Approximate Primary Energy (EJ/yr)	Approximate Average Efficiency (%)	Approximate Secondary Energy (EJ/yr)
Traditional Biomass			
Accounted for in IEA energy statistics	30.7	10–20	3–6
Estimated for informal sectors (e.g., charcoal)	6–12		0.6–2.4
Total Traditional Biomass	37–43		3.6–8.4
Modern Bioenergy			
Electricity and CHP from biomass, MSW, and biogas	4.0	32	1.3
Heat in residential, public/commercial buildings from solid biomass and biogas	4.2	80	3.4
Road transport fuels (ethanol and biodiesel)	3.1	60	1.9
Total Modern Bioenergy	11.3	58	6.6

Notes: According to the IEA (2010a,b), the 2008 TPES from biomass of 50.3 EJ was composed primarily of solid biomass (46.9 EJ); biogenic MSW used for heat and CHP (0.58 EJ); and biogas (secondary energy) for electricity and CHP (0.41 EJ) and heating (0.33 EJ). The contribution of ethanol, biodiesel, and other biofuels (e.g., ethers) used in the transport sector amounted to 1.9 EJ in secondary energy terms. Examples of specific flows: output electricity from biomass was 0.82 EJ (biomass power plants including pulp and paper industry surplus, biogas and MSW) and output heating from CHP was 0.44 EJ. Modern residential heat consumption was calculated by subtracting the IEA estimate of traditional use of biomass (30.7 EJ) from the total residential heat consumption (33.7 EJ).

Some table numbers were taken directly from the IEA global energy statistics, such as secondary biofuels at 1.9 EJ (whereas the derived primary energy input is based on the assumed efficiency of 60% which could be lower) as well as output electricity and heat at 1.3 EJ for all feedstocks. Primary input for MSW and biogas (secondary) and the corresponding output were available and efficiencies are calculated. Solid biomass primary input was calculated from the average efficiency for MSW. Not included in the numbers above are solid biomass (3.4 EJ) used to make charcoal (1.15 EJ) for heating (0.88 EJ, traditional mostly) and industry, such as the iron/steel industry (0.22 EJ), mostly in Brazil. Heat for making charcoal is included in Figure 1.18 in the 5.2 EJ from biomass for electricity, CHP, and heat plants. Not included in Table 2.1 is the industry sector that consumed 7.7 EJ, but the electricity sold by the pulp and paper industry is included.

---

[1] MSW is used throughout the chapter with the same meaning as the term municipal wastes as defined by EUROSTAT.

[2] Traditional biomass is defined as biomass consumption in the residential sector in developing countries and refers to the often unsustainable use of wood, charcoal, agricultural residues and animal dung for cooking and heating (IEA, 2010b and Annex I). All other biomass use is defined as modern biomass; this report further differentiates between highly efficient modern bioenergy and industrial bioenergy applications with varying degrees of efficiency (Annex I). The renewability and sustainability of biomass use is primarily discussed in Sections 2.5.4 and 2.5.5, respectively (see also Section 1.2.1 and Annex I).

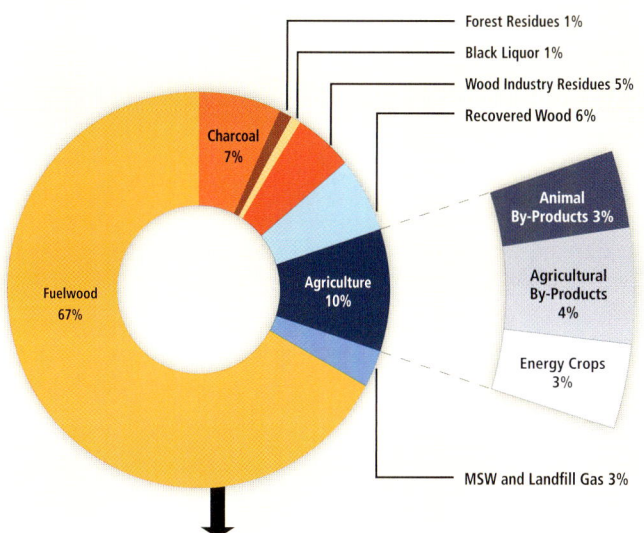

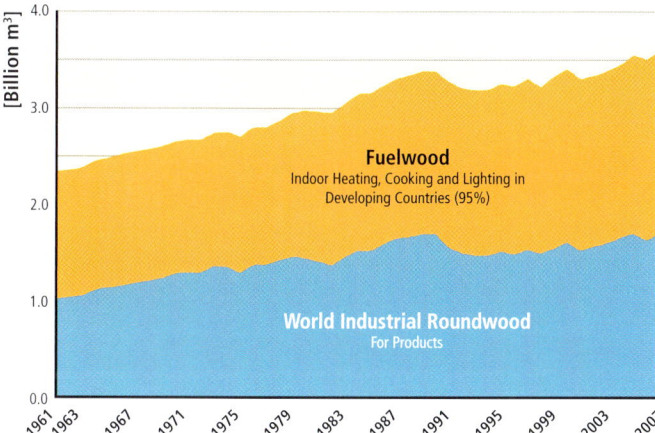

Figure 2.1 | Top: Shares of global primary biomass sources for energy (IPCC, 2007a,d; IEA Bioenergy, 2009); Bottom: Fuelwood used in developing countries parallels world industrial roundwood[1] production levels (UNECE/FAO Timber Database, 2011).

Note: 1. Roundwood products are saw logs and veneer logs for the forest products industry and wood chips that are used for making pulpwood used in paper, newsprint and Kraft paper. In 2009, reflecting the downturn in the economy, there was a decline to 3.25 (total) and 1.25 (industrial) billion m³; the data can be retrieved from a presentation on Global Forest Resources and Market Developments: timber.unece.org/fileadmin/DAM/other/GlobalResMkts300311.pdf.

often in contradiction when it comes to estimates of biomass consumption for energy, because production and trade of these solid biomass fuels are largely informal.[3] A supplement of 20 to 40% to the global TPES of biomass in Table 2.1 is based on detailed, multi-scale, spatially explicit analyses performed in more than 20 countries (e.g., Masera et al., 2005, 2006; Drigo et al., 2007, 2009). Traditional biomass is discussed in later sections on feedstock logistics and supply (Section 2.3.2.2), improved technologies, practices and barriers (Sections 2.4.2.1 and 2.4.2.2), climate change effects (Section 2.5.4) and socioeconomic aspects (Section 2.5.7).

---

[3] See the Glossary in Annex I for a definition of informal sector/economy.

- High-efficiency *modern bioenergy* uses more convenient solids, liquids and gases as secondary energy carriers to generate heat, electricity, combined heat and power (CHP) and transport fuels for various sectors (Figure 2.2). Many entities in the process industry, municipalities, districts and cooperatives generate these energy products, in some cases for their own use, but also for sale to national and international markets in the increasingly global trade. Liquid biofuels, such as ethanol and biodiesel, are used for global road transport and some industrial uses. Biomass-derived gases, primarily methane from anaerobic digestion of agricultural residues and waste treatment streams, are used to generate electricity, heat or CHP for multiple sectors. The most important contribution to these energy services is, however, based on solids, such as chips, pellets, recovered wood previously used etc. Heating includes space and hot water heating such as in district heating systems. The estimated TPES from modern bioenergy is 11.3 EJ/yr and the secondary energy delivered to end-use consumers is roughly 6.6 EJ/yr (IEA, 2010a,b). Modern bioenergy feedstocks such as short-rotation trees (poplars or willows) and herbaceous plants (*Miscanthus* or switchgrass) are discussed in Sections 2.3.1 and 2.6.1. The discussion of modern bioenergy includes biomass logistics and supply chains (Sections 2.3.2 and 2.6.2); conversion of biomass into secondary carriers or energy through existing (Section 2.3.3) or developing (Section 2.6.3) technologies; integration into bioenergy systems and supply chains (Section 2.3.4); and market and industry development (Section 2.4).

- High energy efficiency biomass conversion is found typically in the *industry* sector (with a total consumption of ~7.7 EJ/yr) associated with the pulp and paper industry, forest products, food and chemicals. Examples are fibre products (e.g., paper), energy, wood products, and charcoal for steel manufacture. Industrial heating is primarily steam generation for industrial processes, often in conjunction with power generation. The industry sector's final consumption of biomass is not shown in Table 2.1 since it cannot be unambiguously assigned. Also see Section 8.3.4, which addresses the biomass industry sector.

Global bioenergy use has steadily grown worldwide in absolute terms in the last 40 years, with large differences among countries. In 2006, China led all countries and used 9 EJ of biomass for energy, followed by India (6 EJ), the USA (2.3 EJ) and Brazil (2 EJ) (GBEP, 2008). Bioenergy provides a relatively small but growing share of TPES (1 to 4 % in 2006) in the largest industrialized countries (grouped as the G8 countries: the USA, Canada, Germany, France, Japan, Italy, the UK and Russia). The use of solid biomass for electricity production is particularly important in pulp and paper plants and in sugar mills. Bioenergy's share in total energy consumption is generally increasing in the G8 countries through the use of modern biomass forms (e.g., co-combustion or co-firing for electricity generation, space heating with pellets) especially in Germany, Italy and the UK (see Figure 2.8; GBEP, 2008).

By contrast, in 2006, bioenergy provided 5 to 27% of TPES in the largest developing countries (China, India, Mexico, Brazil and South Africa),

Bioenergy

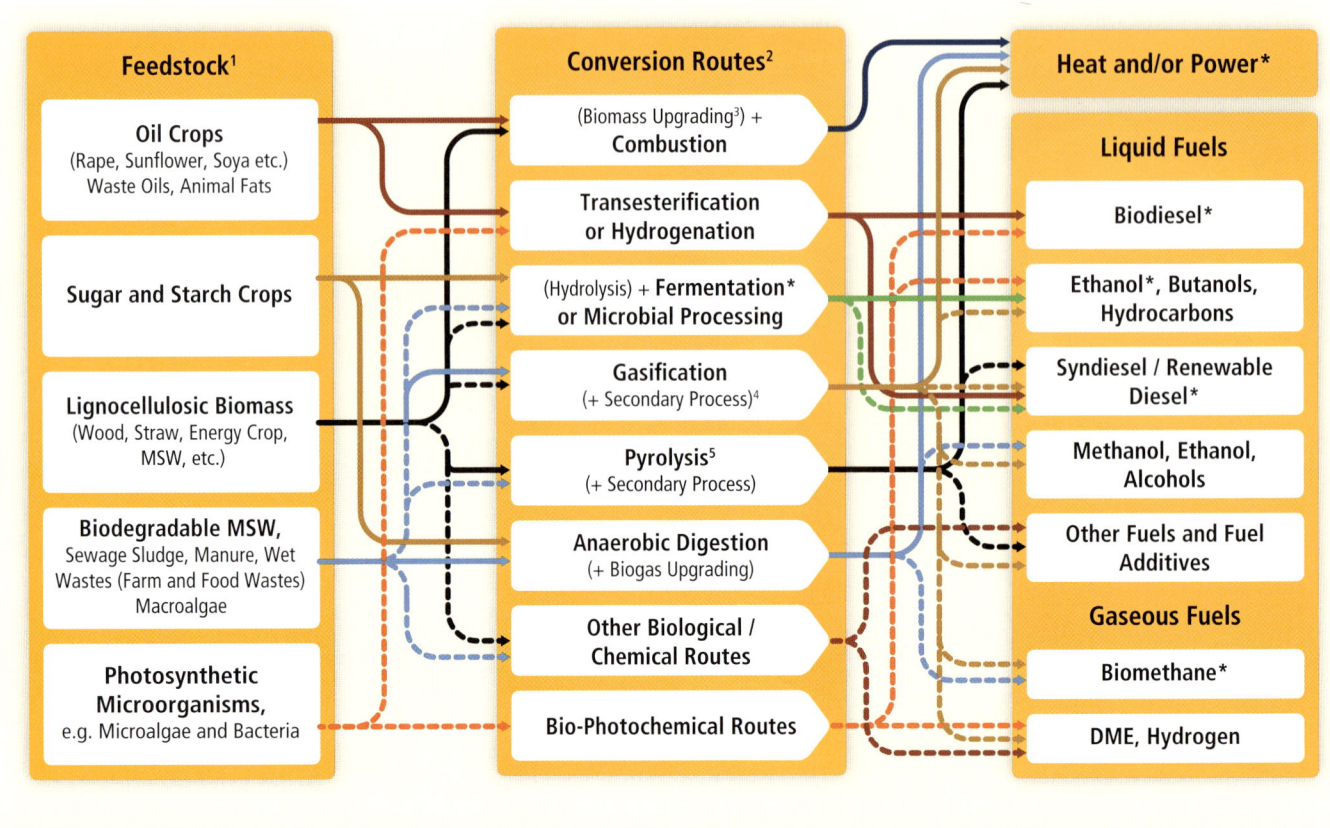

Figure 2.2 | Schematic view of the variety of commercial (solid lines, see Figure 2.6) and developing bioenergy routes (dotted lines) from biomass feedstocks through thermochemical, chemical, biochemical and biological conversion routes to heat, power, CHP and liquid or gaseous fuels (modified from IEA Bioenergy, 2009). Commercial products are marked with an asterisk.

Notes: 1. Parts of each feedstock, for example, crop residues, could also be used in other routes. 2. Each route also gives coproducts. 3. Biomass upgrading includes any one of the densification processes (pelletization, pyrolysis, torrefaction, etc.). 4. Anaerobic digestion processes release methane and CO_2 and removal of CO_2 provides essentially methane, the major component of natural gas; the upgraded gas is called biomethane. 5. Could be other thermal processing routes such as hydrothermal, liquefaction, etc. DME=dimethyl ether.

mainly through the use of traditional forms, and more than 80% of TPES in the poorest countries. The bioenergy share in India, China and Mexico is decreasing, mostly as traditional biomass is substituted by kerosene and liquefied petroleum gas within large cities. However, consumption in absolute terms continues to grow. This trend is also true for most African countries, where demand has been driven by a steady increase in wood fuels, particularly in the use of charcoal in booming urban areas (GBEP, 2008).

Turning from the technological perspectives of bioenergy to environmental and social aspects, the literature assessments in this chapter reveal positive and negative aspects of bioenergy. Sustainably produced and managed, bioenergy can provide a substantial contribution to climate change mitigation through increasing carbon stocks in the biosphere (e.g., in degraded lands), reducing carbon emissions from unsustainable forest use and replacing fossil fuel-based systems in the generation of heat, power and modern fuels. Additionally, bioenergy may provide opportunities for regional economic development (see Sections 9.3.1 and 2.5.4). Advanced bioenergy systems and end-use technologies can also substantially reduce the emissions of black carbon and other short-lived GHGs such as methane and carbon monoxide (CO), which are related to the burning of biomass in traditional open fires and kilns. If improperly designed or implemented, the large-scale expansion of bioenergy systems is likely to have negative consequences for climate and sustainability, for example, by inducing d+iLUC that can alter surface albedo and release carbon from soils and vegetation, reducing biodiversity or negatively impacting local populations in terms of land tenure or reduced food security, among other effects.

The literature on the resource potential of biomass is covered in Section 2.2, which discusses a variety of global modelling studies and the factors that influence the assessments. Section 2.2 also presents examples of resource assessments from countries and specific regions, which provide cost dimensions for these resources. The overall technology portfolio is shown in Figure 2.2 and includes commercial and developing energy carriers from modern biomass. The commercially available energy products and (conversion) technologies are discussed in Section 2.3. These are based on sugar crops (perennial sugarcane and beets), starch crops (maize, wheat, cassava etc.), and oil crops (soy, rapeseed) as feedstocks, and they expand food and fodder processing to bioenergy production. Current bioenergy production is also coupled with forest products industry residues and the pulping industry that has traditionally self generated heat and power; with dry and wet municipal wastes; with sewage sludge; and with a variety of organic wet wastes from various sectors. These wastes and residues, if left untreated, can have a major impact on climate through methane emission releases. The bioenergy market is described in Section 2.4 for traditional and modern forms, as are evolving international trade and sustainability frameworks for bioenergy. The advanced technologies for production of feedstocks and conversion to energy products are discussed in Section 2.6.

In Section 2.5, the environmental and social impacts of biomass use are addressed with emphasis on the climate change effects of bioenergy. Because of the complexity of GHG impacts and of the bioenergy chains, impacts are analyzed without and with LUC separately. These impacts span micro-, meso- and macro- scales and depend on the land cover conversion and water availability, among other factors, in specific regions. Direct land use impacts occur locally by changes in crop use or the dedication of a crop to bioenergy. The iLUC results from a market-mediated shift in land management activities (i.e., dLUC) outside the region of primary production expansion. Both are addressed in Section 2.5. The social impacts of modern and traditional biomass use are presented and related to key issues such as the impact of bioenergy on food production and sustainable development in Section 2.5.7 (also refer to Sections 9.3 and 9.4).

To reach high levels of bioenergy production and minimize environmental and social impacts, it is necessary to develop a variety of lignocellulosic biomass sources and a portfolio of conversion routes for power, heat and gaseous and liquid fuels that satisfy existing and future energy needs (Figure 2.2). With these prospects for technology improvement, innovation and integration, key conversion intermediates derived from biomass such as sugars, syngas, pyrolysis oils (or oils derived from other thermal treatments), biogas and vegetable oils (lipids) can be upgraded in conversion facilities that are capable of making a variety of products including biofuels, power and process heat, alongside other products as discussed in Section 2.6. In Section 2.7, the costs of existing commercial technologies and their trends are discussed, highlighting that over the past 25 years technological learning occurred in a variety of bioenergy systems in specific countries. Finally, Section 2.8 addresses the potential deployment of biomass for energy. It also compares biomass resource assessments from Section 2.2, informed by environmental and social impacts discussions, with the levels of deployment indicated by the scenario literature review described in Chapter 10. The role of biomass and its multiple energy products alongside food, fodder, fibre and forest products is viewed through IPCC scenario storylines (IPCC, 2000a,d) to reach significant penetration levels with and without taking into account sustainable development and climate change mitigation pathways. High and low penetration levels can be reached with (and without) climate change mitigation and sustainable development strategies. Many insights into bioenergy technology developments and integrated systems can be gleaned from these sketches, and they will be useful in further developing bioenergy sustainably with climate mitigation.

2.1.2 Previous Intergovernmental Panel on Climate Change assessments

Bioenergy has not been examined in detail in previous IPCC reports. In the most recent Fourth Assessment Report (AR4), the analysis of GHG mitigation from bioenergy was scattered among seven chapters, making it difficult to obtain an integrated and cohesive picture of the resource and mitigation potential, challenges and opportunities. The main conclusions from the AR4 report (IPCC, 2007b,d) are as follows:

- **Biomass energy demand.** Primary biomass requirements for the production of transportation fuels were largely based on the WEO (IEA, 2006) global projections, with a relatively wide range of about 14 to 40 EJ/yr of primary biomass, or 8 to 25 EJ/yr of biofuels in 2030. However, higher demand estimates of 45 to 85 EJ/yr for primary biomass in 2030 (roughly 30 to 50 EJ/yr of biofuel) were also included. For comparison, the scenario review in Chapter 10 shows biofuel production ranges of 0 to 14 EJ/yr in 2030 and 2 to 50 EJ/yr in 2050 with median values of 5 to 12 EJ/yr and 18 to 20 EJ/yr in the two GHG mitigation scenario categories analyzed. The demand for biomass-generated heat and power was stated to be strongly influenced by the availability and introduction of competing technologies such as CCS, nuclear power, wind energy, solar heating and others. The projected biomass demand in 2030 would be around 28 to 43 EJ according to the data used in the AR4. These estimates focus on electricity generation. Heat was not explicitly modelled or estimated in the WEO (IEA, 2006), on which the AR4 was based, therefore underestimating the total demand for biomass.

  Potential future demand for biomass in industry (especially new uses such as biochemicals, but also expansion of charcoal use for steel production) and the built environment (heating as well as increased use of biomass as a building material) was also highlighted as important, but no quantitative projections were included in the potential demand for biomass at the medium and longer term.

- **Biomass resource potential (supply).** According to the AR4, the largest contribution to technical potential could come from energy crops on arable land, assuming that efficiency improvements in agriculture are fast enough to outpace food demand so as to avoid increased pressure on forests and nature areas. A range of 20 to 400 EJ/yr is presented for 2050, with a best estimate of 250 EJ/yr. Using degraded lands for biomass production (e.g., in reforestation schemes: 8 to 110 EJ/yr) can contribute significantly. Although such low-yielding biomass production generally results in more expensive biomass supplies, competition with food production is almost absent and various co-benefits, such as regeneration of soils (and carbon storage), improved water retention and protection from (further) erosion may also offset part of the establishment costs. A current example of such biomass production schemes is the establishment of *Jatropha* crops (oilseeds) on marginal lands.

The technical potential in residues from forestry is estimated at 12 to 74 EJ/yr, that from agriculture at 15 to 70 EJ/yr and that from waste at 13 EJ/yr. These biomass resource categories are largely available before 2030, but also partly uncertain. The uncertainty comes from possible competing uses (e.g., increased use of biomaterials such as fibreboard production from forest residues and use of agricultural residues for fodder and fertilizer) and differing assumptions about sustainability criteria deployed with respect to forest management and agricultural intensity. The technical potential for biogas fuel from waste, landfill gas and digester gas is much smaller.

- **Carbon mitigation potential.** The mitigation potential for electricity generation from biomass reaches 1,220 Mt CO_2eq for the year 2030, a substantial fraction of it at costs lower than USD_{2005} 19.5/t CO_2. From a top-down assessment, the economic mitigation potential of biomass energy supplied from agriculture is estimated to range from 70 to 1,260 Mt CO_2eq/yr at costs of up to USD_{2005} 19.5/t CO_2eq, and from 560 to 2,320 Mt CO_2eq/yr at costs of up to USD_{2005} 48.5/t CO_2eq. The overall mitigation from biomass energy coming from the forest sector is estimated to reach 400 Mt CO_2/yr up to 2030.

2.2 Resource potential

2.2.1 Introduction

Bioenergy production interacts with food, fodder and fibre production as well as with conventional forest products in complex ways. Bioenergy demand constitutes a benefit to conventional plant production in agriculture and forestry by offering new markets for biomass flows that earlier were considered to be waste products; it can also provide opportunities for cultivating new types of crops and integrating bioenergy production with food and forestry production to improve overall resource management. However, biomass for energy production can intensify competition for land, water and other production factors, and can result in overexploitation and degradation of resources. For example, too-intensive biomass extraction from the land can lead to soil degradation, and water diversion to energy plantations can impact downstream and regional ecological functions and economic services.

As a consequence, the magnitude of the biomass resource potential depends on the priority given to bioenergy products versus other products obtained from the land—notably food, fodder, fibre and conventional forest products such as sawn wood and paper—and on how much total biomass can be mobilized in agriculture and forestry. This in turn depends on natural conditions (climate, soils, topography), on agronomic and forestry practices, and on how societies understand and prioritize nature conservation and soil/water/biodiversity protection and on how production systems are shaped to reflect these priorities (Figure 2.3).

This section focuses on long-term biomass resource potential and how it has been estimated based on considerations of the Earth's biophysical resources (ultimately net primary production: NPP) and restrictions on their energetic use arising from competing requirements, including non-extractive requirements such as soil quality maintenance/improvement and biodiversity protection. Additionally, approaches to assessing biomass resource potentials—and results from selected studies—are presented with an account of the main determining factors. These factors are treated explicitly, including the constraints on their utilization. The section ends by summarizing conclusions about biomass resource assessments, including uncertainties.

2.2.1.1 Methodology assessment

Studies quantifying biomass resource potential have assessed the resource base in a variety of ways. They differ in the extent to which the influence of natural conditions (and how these can change in the future) are considered as well as in the extent to which the types and details of important additional factors are taken into account, such as socioeconomic considerations, the character and development of agriculture and forestry, and factors connected to nature conservation and soil/water/biodiversity preservation (Berndes et al., 2003). Different types of resource potentials are assessed but the following are commonly referred to (see Glossary in Annex I):

- **Theoretical potential** refers to the biomass supply as limited only by biophysical conditions (see discussion below in this same sub-section);

- **Technical potential** considers the limitations of the biomass production practices assumed to be employed and also takes into account concurrent demand for food, fodder, fibre, forest products and area requirements for human infrastructure. Restrictions connected to nature conservation and soil/water/biodiversity preservation can also be considered. In such cases, the term *sustainable potential* is sometimes used (see Section 2.2.2); and

- **Market potential** refers to the part of the technical potential that can be produced given a specified requirement for the level of economic profit in production. This depends not only on the cost of production but also on the price of the biomass feedstock, which is determined by a range of factors such as the characteristics of biomass conversion technologies, the price of competing energy technologies and the prevailing policy regime (see Section 2.2.3).

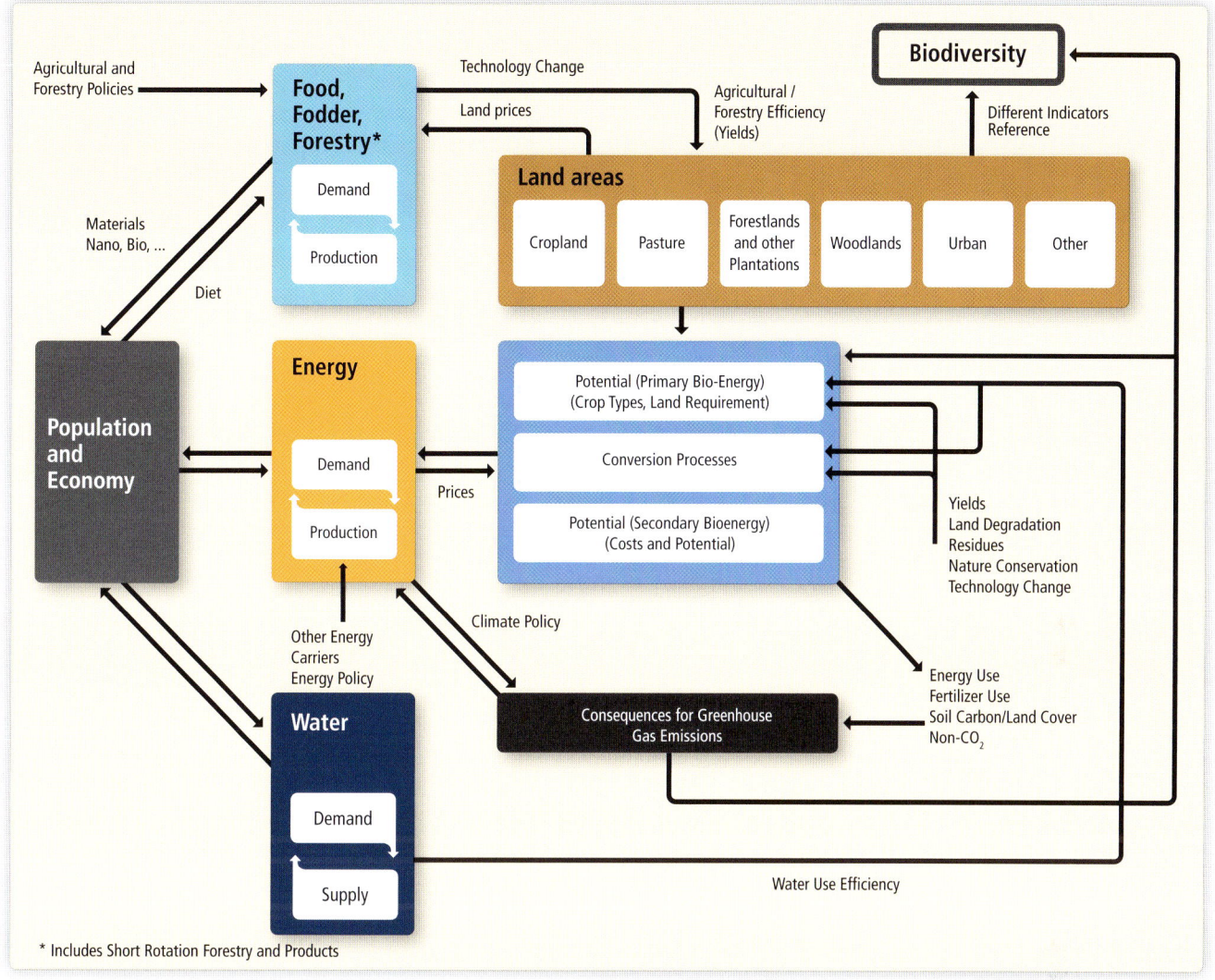

Figure 2.3 | Overview of key relationships relevant to assessment of biomass resource potentials (modified from Dornburg et al., 2010). Indirect land use and social issues are not displayed. Reproduced with permission from the Royal Society of Chemistry.

Three principal categories are—more or less comprehensively—considered in assessments of biomass resource potentials (see also Section 2.3.1.1):

- Primary residues from conventional food and fibre production in agriculture and forestry, such as cereal straw and logging residues;

- Secondary and tertiary residues in the form of organic food/forest industry by-products and retail/post consumer waste; and

- Plants produced for energy supply, including conventional food/fodder/industrial crops, surplus roundwood forestry products, and new agricultural, forestry or aquatic plants.

Given that resource potential assessments quantify the availability of residue flows in the food and forest sectors, the definition of how these sectors develop is central for the outcome. As discussed below, consideration of various environmental and socioeconomic factors as a rule reduces the assessed resource potential to lower levels.

Most assessments of the biomass resource potential considered in this section are variants of technical/market potentials employing a 'food/fibre first principle', applied with the objective of quantifying biomass resource potentials under the condition that global requirements for food and conventional forest products such as sawn wood and paper are met with priority (see, e.g., WBGU, 2009; Smeets and Faaij, 2007).

Studies that start out from such principles should not be understood as providing guarantees that a certain level of biomass can be supplied for energy purposes without competing with food or fibre production. They quantify how much bioenergy could be produced in a certain future year based on using resources not required for meeting food and fibre demands, given a specified development in the world or in a region. But they do not analyze how bioenergy expansion towards such a future level of production would—or should—interact with food and fibre production.

Studies using integrated energy/industry/land use cover models (see, e.g., Leemans et al., 1996; Strengers et al., 2004; Johansson and Azar, 2007; van Vuuren et al., 2007; Fischer et al., 2009; Lotze-Campben, 2009; Melillo et al., 2009; Wise et al., 2009; Figure 2.4) can provide insights into how an expanding bioenergy sector interacts with other sectors in society including land use and the management of biospheric carbon stocks. Studies focused on sectors can contain more detailed information on interactions with other biomass uses. Restricted scope (only selected biofuel/land uses and/or regions covered) or lack of sufficiently detailed empirical data can limit the confidence in results—especially in prospective studies. This is further discussed in Sections 2.5 and 2.8.

By considering the upper level of productivity of biomass plantations on land while assuming theoretical potentials also for worldwide agriculture and fully taking into account conservation of a viable biosphere, global modelling studies by Smeets et al. (2007) derived a maximum global potential of biomass for energy of 1,548 EJ/yr.[4] In this chapter, this figure is considered to be an estimate of theoretical potential.

2.2.1.2   Total aboveground net primary production of biomass

A first qualitative understanding of biomass technical potentials can be gained from considering the total annual aboveground net primary production (NPP: the net amount of carbon assimilated in a time period by vegetation) on the Earth's terrestrial surface. This is estimated to be about 35 Gt carbon, or 1,260 EJ/yr assuming an average carbon content of 50% and 18 GJ/t average heating value (Haberl et al., 2007), which can be compared to the current world primary energy supply of about 500 EJ/yr (IEA, 2010a). This comparison shows that total terrestrial aboveground NPP is larger, but by no more than a factor of around three, than what is required to meet society's energy demand. Establishing bioenergy as a major source of future primary energy requires that a significant part of global terrestrial NPP takes place within production systems that provide bioenergy feedstocks (removing their NPP from the trophic chains of ecosystems). In addition, total terrestrial NPP may have to be increased through fertilizer, irrigation and other inputs on lands managed for food, fodder, fibre, forest products and bioenergy.

2.2.1.3   Human appropriation of terrestrial net primary production

A comparison with biomass production in agriculture and forestry can give a perspective on the potential bioenergy supply in relation to what is presently harvested. Today's global industrial roundwood production corresponds to 15 to 20 EJ/yr, and the global harvest of major crops (cereals, oil crops, sugar crops, roots, tubers and pulses) corresponds to about 60 EJ/yr (FAOSTAT, 2011). One immediate conclusion from this comparison is that biomass extraction by agriculture and forestry will have to increase substantially in order to provide feedstocks for a bioenergy sector large enough to make a significant contribution to the future energy supply.

Studies estimating the overall human appropriation of terrestrial NPP across all human uses of biomass (HANPP, taking into account all NPP gained or lost due to human activities, including harvesting and backflows) suggest that societies already appropriate a substantial share of the world's aboveground terrestrial NPP. This provides a context for prospective future biomass extraction for bioenergy. Estimates of HANPP vary depending on its definition as well as the models and data used for the calculations. A spatially explicit calculation by Haberl et al. (2007) estimated that in the year 2000, aboveground HANPP amounted to nearly 29% of the modelled global aboveground NPP. Total human biomass harvest alone was estimated to amount to about 20% (including utilized residues and grazing), with all harvested biomass used by humans containing an energy of 219 EJ/yr (Krausmann et al., 2008).

Other HANPP estimates range from a similar level down to about half of this level (D. Wright, 1990; Imhoff et al., 2004). The HANPP concept cannot directly be used to define a certain level of biomass use that would be 'safe' or 'sustainable' because the impacts of human land use depend on how agriculture and forestry systems are shaped (Bai et al., 2008). However, it can be used as a measure of the human domination of the biosphere and provide a reference for assessing the comparative magnitude of prospective additional biomass resource potentials.

Besides biophysical factors, socioeconomic conditions also influence the biomass resource potential by defining how—and how much—biomass can be produced without causing socioeconomic impacts that might be considered unacceptable. Socioeconomic restrictions vary around the world, change as society develops and depend on how societies prioritize bioenergy in relation to other socioeconomic objectives (see also Sections 2.5 and 2.8).

---

4   Smeets et al. (2007) model a scenario with a fully landless animal production system with globally high feed conversion efficiency and a 4.6-fold increase in global agricultural productivity by 2050 due to technological progress and deployment that is considerably faster than has historically ever been achieved (a 1.9-fold increase for Europe and a 7.7-fold increase in sub-Saharan Africa). In that case, 72% of current agricultural area could be used for bioenergy production in 2050 and supply a theoretical potential of 1,548 EJ/yr, which is of the same magnitude as the total energy content of the world's natural aboveground net primary production on land.

2.2.2 Global and regional technical potential

2.2.2.1 Literature assessment

In an assessment of technical potential based on an analysis of the literature available in 2007 and additional modelling, Dornburg et al. (2008, 2010) arrived at the conclusion that the upper bound of the technical potential in 2050 can amount to about 500 EJ. The study assumes policy frameworks that secure good governance of land use and major improvements in agricultural management and takes into account water limitations, biodiversity protection, soil degradation and competition with food. Residues originating from forestry, agriculture and organic wastes (including the organic fraction of MSW, dung, process residues etc.) are estimated to amount to 40 to 170 EJ/yr, with a mean estimate of around 100 EJ/yr. This part of the technical potential is relatively certain, but competing applications may push net availability for energy applications to the lower end of the range. Surplus forestry other than from forestry residues has an additional technical potential of 60 to 100 EJ/yr.

The findings of the Dornburg et al. (2008, 2010) reviews for biomass produced via cropping systems is that a lower estimate for energy crop production on possible surplus, good quality agricultural and pasture lands is 120 EJ/yr. The potential contribution of water-scarce, marginal and degraded lands could amount up to an additional 70 EJ/yr. This would comprise a large area where water scarcity provides limitations and soil degradation is more severe. Assuming strong learning in agricultural technology for improvements in agricultural and livestock management would add 140 EJ/yr. The three categories added together lead to a technical potential from this analysis of up to about 500 EJ/yr (Dornburg et al., 2008, 2010). For example, Hoogwijk et al. (2005, 2009) estimate that the biomass technical potential could expand from 290 to 320 EJ/yr in 2020 to 330 to 400 EJ/yr in 2030. Developing the technical potential would require major policy efforts; therefore, actual deployment is likely to be lower and the biomass resource base will be largely constrained to a share of the biomass residues and organic wastes, some cultivation of bioenergy crops on marginal and degraded lands, and some regions where biomass is a cheaper energy supply option compared to the main reference options (e.g., sugarcane-based ethanol production), amounting to a minimum of about 50 EJ/yr (Dornburg et al., 2008, 2010).

Table 2.2 shows ranges in the assessed global technical potential for the year 2050 explicitly for various biomass categories. The wide ranges shown are due to differences in the studies' approaches to considering important factors, which are in themselves uncertain: population, economic and technology development assumed or computed can vary and evolve at different regional paces; biodiversity, nature conservation and other environmental requirements are difficult to assess and depend on numerous factors and social preferences; and the magnitude and pattern of climate change and land use can strongly influence the biophysical capacity of the environment. Furthermore, technical potentials cannot be determined precisely while uncertainties remain regarding societal preferences with respect to trade-offs in environmental impacts and the implications of increased intensification in food and fibre production, and regarding potential synergies between different forms of land use.

Although assessments employing improved data and modelling capacity have not succeeded in providing narrow distinct estimates of the technical potential of biomass, they do indicate the most influential factors that affect this technical potential. This is further discussed below, where approaches used in the assessments are treated in more detail.

2.2.2.2 The contribution from residues, dung, processing by-products and waste

As can be seen in Table 2.2, biomass resource assessments indicate that retail/post-consumer waste, dung and primary residues/processing by-products in the agriculture and forestry sectors have prospects for providing a substantial share of the total global biomass supply in the longer term. Yet, the sizes of these biomass resources are ultimately determined by the demand for conventional agriculture and forestry products and the sustainability of the land resources.

Assessments of the potential contribution from these sources to the future biomass supply combine data on future production of agriculture and forestry products obtained from food/forest sector scenarios, the possibility of use of degraded lands, and the residue factors that account for the amount of residues generated per unit of primary product produced. For example, harvest residue generation in agricultural crops cultivation is estimated based on harvest index data, that is, the ratio of harvested product to total aboveground biomass (e.g., Wirsenius, 2003; Lal, 2005; Krausmann et al., 2008; Hakala et al., 2009). The generation of logging residues in forestry, and of additional biomass flows such as thinning wood and process by-products, is estimated using similar methods (see Ericsson and Nilsson, 2006; Smeets and Faaij, 2007).

The shares of the biomass flows that are available for energy (i.e., recoverability fractions) are then estimated based on consideration of other extractive uses and requirements (e.g., soil conservation, animal feeding or bedding in agriculture, and fibre board production in the forest sector).

2.2.2.3 The contribution from unutilized forest growth

In addition to the residue flows that are linked to industrial roundwood production and processing into conventional forest products, forest growth currently not harvested is considered in some studies. This biomass resource is quantified based on estimates of the biomass increment in parts of forests that are assessed as being available for wood supply. This increment is compared with the estimated level of forest biomass extraction for conventional industrial roundwood production—and sometimes for traditional biomass, notably heating and cooking—to obtain the unutilized forest growth. Smeets and Faaij (2007) provide illustrative quantifications showing how this technical potential of biomass can vary from being a major source of bioenergy to being practically zero as a consequence of competing demand and economic and ecological considerations. A comparison with the present industrial roundwood production of about 15 to 20 EJ/yr shows that a drastic increase in forest biomass output is required to reach the higher-end technical potential assessed for the forest biomass category in Table 2.2. A special case that can play a role is forest growth that becomes available after extensive tree mortality from insect outbreaks or fires (Dymond et al., 2010).

2.2.2.4 The contribution from biomass plantations

Table 2.2 indicates that substantial supplies from biomass plantations are required for reaching the high end of the technical potential range. Land availability (and its suitability) for dedicated biomass plantations and the biomass yields that can be obtained on the available lands are two critical determinants of the technical potential. Given that surplus agricultural land is commonly identified as the major land resource for the plantations, food sector development is critical. Methods for determining land availability and suitability should consider requirements for maintaining the economic, ecological and social value of ecosystems. There are different approaches for considering such requirements, as described for a selection of studies below.

Most earlier assessments of biomass resource potentials used rather simplistic approaches to estimating the technical potential of biomass plantations (Berndes et al., 2003), but the continuous development of modelling tools that combine databases containing biophysical information (soil, topography, climate) with analytical representations of relevant crops and agronomic systems and the use of economic and full biogeochemical vegetation models has resulted in improvements over time (see, e.g., van Vuuren et al., 2007; Fischer et al., 2008; Lotze-Campen et al., 2009; Melillo et al., 2009; WBGU, 2009; Wise et al., 2009;

Table 2.2 | Global technical potential overview for a number of categories of land-based biomass supply for energy production (primary energy numbers have been rounded). The total assessed technical potential can be lower than the present biomass use of about 50 EJ/yr in the case of high future food and fibre demand in combination with slow productivity development in land use, leading to strong declines in biomass availability for energetic purposes.

Biomass category	Comment	2050 Technical potential (EJ/yr)
Category 1. **Residues from agriculture**	By-products associated with food/fodder production and processing, both primary (e.g., cereal straw from harvesting) and secondary (e.g., rice husks from rice milling) residues.	15 – 70
Category 2. **Dedicated biomass production on surplus agricultural land**	Includes both conventional agriculture crops and dedicated bioenergy plants including oil crops, lignocellulosic grasses, short-rotation coppice and tree plantations. Only land not required for food, fodder or other agricultural commodities production is assumed to be available for bioenergy. However, surplus agriculture land (or abandoned land) need not imply that its development is such that less total land is needed for agriculture: the lands may become excluded from agriculture use in modelling runs due to land degradation processes or climate change (see also 'marginal lands' below). Large technical potential requires global development towards high-yielding agricultural production and low demand for grazing land. Zero technical potential reflects that studies report that food sector development can be such that no surplus agricultural land will be available.	0 – 700
Category 3. **Dedicated biomass production on marginal lands**	Refers to biomass production on deforested or otherwise degraded or marginal land that is judged unsuitable for conventional agriculture but suitable for some bioenergy schemes (e.g., via reforestation). There is no globally established definition of degraded/marginal land and not all studies make a distinction between such land and other land judged as suitable for bioenergy. Adding categories 2 and 3 can therefore lead to double counting if numbers come from different studies. High technical potential numbers for categories 2 and 3 assume biomass production on an area exceeding the present global cropland area (ca. 1.5 billion ha or 15 million km^2). Zero technical potential reflects low potential for this category due to land requirements for, for example, extensive grazing management and/or subsistence agriculture or poor economic performance if using the marginal lands for bioenergy.	0 – 110
Category 4. **Forest biomass**	Forest sector by-products including both primary residues from silvicultural thinning and logging, and secondary residues such as sawdust and bark from wood processing. Dead wood from natural disturbances, such as fires and insect outbreaks, represents a second category. Biomass growth in natural/semi-natural forests that is not required for industrial roundwood production to meet projected biomaterials demand (e.g., sawn wood, paper and board) represents a third category. By-products provide up to about 20 EJ/yr implying that high forest biomass technical potentials correspond to a much larger forest biomass extraction for energy than what is presently achieved in industrial wood production. Zero technical potential indicates that studies report that demand from sectors other than the energy sector can become larger than the estimated forest supply capacity.	0 – 110
Category 5. **Dung**	Animal manure. Population development, diets and character of animal production systems are critical determinants.	5 – 50
Category 6. **Organic wastes**	Biomass associated with materials use, for example, organic waste from households and restaurants and discarded wood products including paper, construction and demolition wood; availability depends on competing uses and implementation of collection systems.	5 – >50
Total		<50 – >1000

Notes: Based on Fischer and Schrattenholzer (2001); Hoogwijk et al. (2003, 2005, 2009); Smeets and Faaij (2007); Dornburg et al. (2008, 2010); Field et al. (2008); Hakala et al. (2009); IEA Bioenergy (2009); Metzger and Huttermann (2009); van Vuuren et al. (2009); Haberl et al. (2010); Wirsenius et al. (2010); Beringer et al. (2011).

Beringer et al., 2011). Important conclusions are: a) the effects of LUC associated with bioenergy expansion can considerably influence the climate benefit of bioenergy (see Section 2.5) and b) biofuel yields from crops have frequently been overestimated by neglecting spatial variations in productivity (Johnston et al., 2009).

Figure 2.4—representing one example (Fischer et al., 2009)—shows the modelled global land suitability for selected first-generation biofuel feedstocks and for lignocellulosic plants (see caption to Figure 2.4 for information about plants included). By overlaying spatial data on global land cover derived from the best available remote sensing data combined

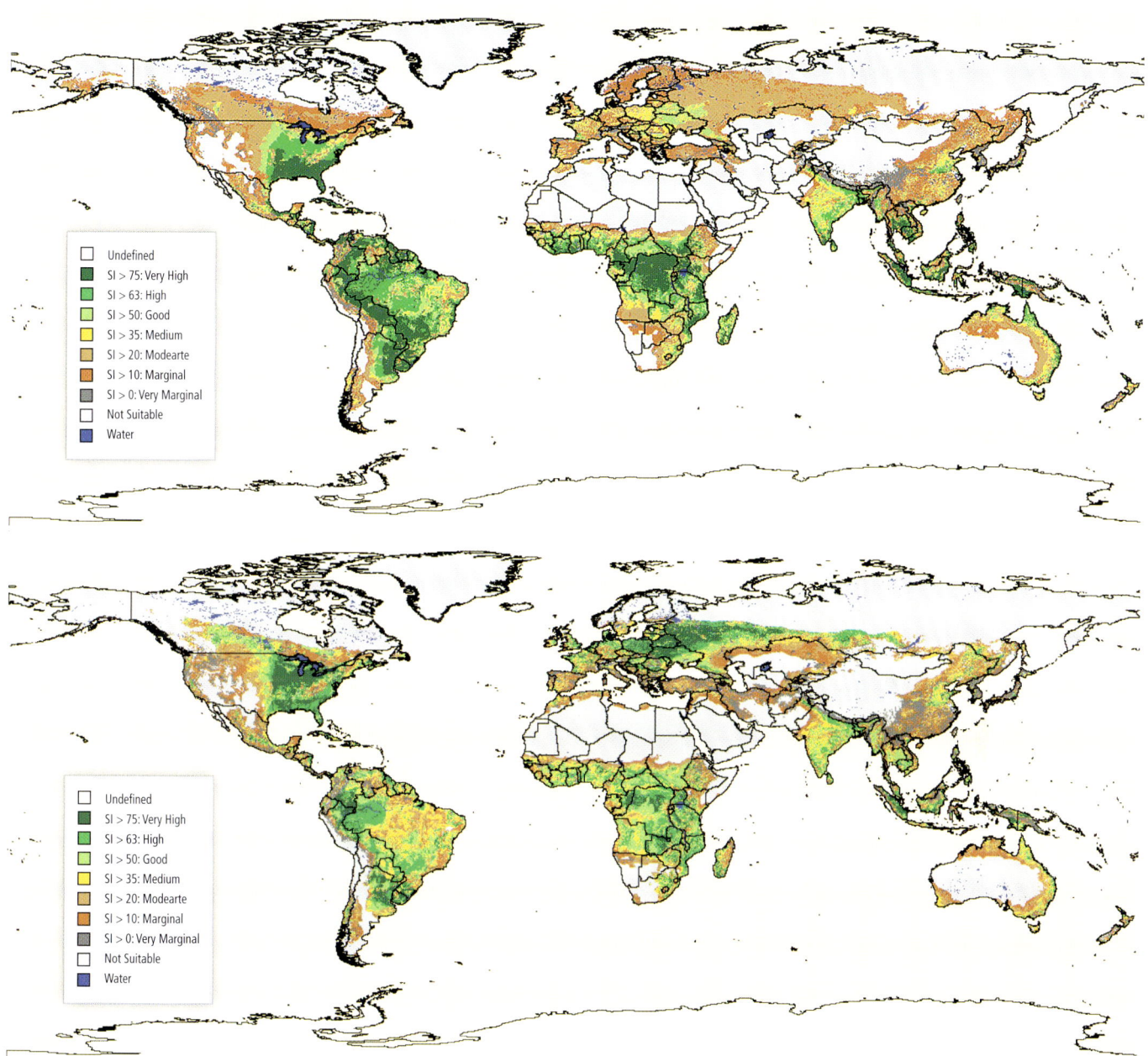

Figure 2.4 | Global land suitability for bioenergy plantations. The upper map shows suitability for herbaceous and woody lignocellulosic plants (*Miscanthus*, switchgrass, reed canary grass, poplar, willow, eucalyptus) and the lower map shows suitability for first-generation biofuel feedstocks (sugarcane, maize, cassava, rapeseed, soybean, palm oil, *Jatropha*). The suitability index (SI)[1] describes the spatial suitability of each pixel and reflects the match between crop requirements and prevailing climate, soil and terrain conditions. The map shows suitability under rain-fed cultivation and advanced management systems that assume availability of sufficient nutrients, adequate pest control and mechanization, and other practices. Results for irrigated conditions or low-input management systems would result in different pictures (Fischer et al., 2009; reproduced with permission from the International Institute for Applied Systems Analysis (IIASA)).

Note: 1. SI: suitability index. The SI used reflects the spatial suitability of each pixel and is calculated as $SI = VS*0.9+S*0.7+MS*0.5+mS*0.3$, where VS, S, MS and mS correspond to yield levels at 80–100%, 60–80%, 40–60% and 20–40% of the modelled maximum, respectively (Fischer et al., 2009).

with statistical information and data on protected areas, it is possible to quantify suitable lands for different land cover types. A suitability index has been used in order to represent both yield potentials[5] and suitability (see caption to Figure 2.4). For instance, almost 700 Mha (7,000 km^2), or about 20%, of currently unprotected grasslands and woodlands are assessed as suitable for soybean while less than 50 Mha (500 km^2) are assessed as suitable for oil palm (note that these land suitability numbers cannot be added because areas overlap). Considering unprotected forest land, an area roughly 10 times larger (almost 500 Mha or 5,000 km^2) is suitable for oil palm cultivation (Fischer et al., 2009, their Annex 5 and 6). However, converting large areas of forests into biomass plantations would negatively impact biodiversity and might—depending on the carbon density of converted forests—also lead to large initial CO_2 emissions that can drastically reduce the annual accumulated climate benefit of substituting fossil fuels with the bioenergy derived from such plantations. Converting grass- and woodlands with high soil carbon content to intensively cultivated annual crops can similarly lead to large CO_2 emissions, while if degraded and C-depleted pastures are cultivated with herbaceous and woody lignocellulosic plants soil carbon may instead accumulate, enhancing the climate benefit. This is further discussed in Section 2.5.

under a 'food and environment first' paradigm excluding forests and land currently used for food and fodder production. The latter includes estimates of unprotected grassland and woodland required today for ruminant livestock feeding. Calculations are based on FAOSTAT data on fodder utilization of crops, and national livestock numbers, estimated fodder energy requirements of the national herds and derived fodder gaps filled by grassland and pastures. Grassland and woodland with very low productivity or steep sloping conditions were considered unsuitable for lignocellulosic feedstock production. The results, shown in Table 2.3, represent one example of estimates of regional technical potentials of biomass resulting from a specific set of assumptions with respect to nature protection requirements, biofuel feedstock crop choice and agronomic practice determining attainable yield levels and livestock production systems determining grazing requirements. Furthermore, the results represent current agriculture practice and productivity, population, diets, climate etc. Quantifications of the technical potential of the future biomass resource need to consider how such parameters change over time.

A similar analysis (WBGU, 2009; Beringer et al., 2011) reserved current and near-future agricultural land for food and fibre production and also

Table 2.3 | Example of the technical potential of rain-fed lignocellulosic plants on unprotected grassland and woodland (i.e., forests excluded) where land requirements for food production, including grazing, have been considered at 2000 levels. Calculated based on Fischer et al. (2009); reproduced with permission from the International Institute for Applied Systems Analysis (IIASA).

Region	Total grass- and woodland area (Mha) [million km^2]	Protected areas (Mha) [million km^2]	Unproductive or very low productive areas (Mha) [million km^2]	Bioenergy area also excluding grazing land (Mha) [million km^2]	Technical potential (average yield,[1] GJ/ha/yr) [GJ/km^2/yr]	Technical Potential[2] (total, EJ/yr)
North America	659 [6.59]	103 [1.03]	391 [3.91]	111 [1.11]	165 [16,500]	19
Europe and Russia	902 [9.02]	76 [0.76]	618 [6.18]	122 [1.22]	140 [14,000]	17
Pacific OECD	515 [5.15]	7 [0.07]	332 [3.32]	97 [0.97]	175 [17,500]	17
Africa	1,086 [10.68]	146 [1.46]	386 [3.86]	275 [2.75]	250 [2,500]	69
South and East Asia	556 [5.56]	92 [0.92]	335 [3.35]	14 [0.14]	285 [28,500]	4
Latin America	765 [7.65]	54 [0.54]	211 [2.11]	160 [1.6]	280 [28,000]	45
Middle East and North Africa	107 [1.07]	2 [0.02]	93 [0.93]	1 [0.01]	125 [12,500]	0.2
World	4,605 [46.05]	481 [4.81]	2,371 [23.71]	780 [7.80]	220 [22,000]	171

Notes: 1. Calculated based on average yields of rain-fed lignocellulosic feedstocks on grass- and woodland area given in Fischer et al. (2009, p.174) and assuming an energy content of 18 GJ/t dry matter (rounded numbers). 2. If livestock grazing area can be freed up by intensification of agricultural practices and pasture use, these areas could be used for additional bioenergy production. The technical potential in this case could increase from 171 up to 288 EJ/yr.

Technical potentials of biomass plantations can thus be calculated based on assessed land availability and corresponding yield levels. Based on the results as shown in Figure 2.4, Fischer et al. (2009) estimated regional land balances of unprotected grassland and woodland potentially available for rain-fed lignocellulosic biofuel feedstock production

excluded unmanaged land from bioenergy production if its conversion to biomass plantations would lead to large net CO_2 emissions to the atmosphere, or if the land was degraded, a wetland, environmentally protected or rich in biodiversity. If dedicated biomass plantations were established in the available lands, an estimated 26 to 116 EJ/yr could be produced (52 to 174 EJ with irrigation). The spatial variation of technical potential was computed from biogeochemical principles, that is, photosynthesis, transpiration, soil quality and climate. Haberl et

---
[5] Yield potential is the yield obtained when an adapted cultivar (cultivated variety of a plant) is grown with the minimal possible stress that can be achieved with best management practices, a functional definition by Cassman (1999).

al. (2010) considered the land available after meeting prospective future food, fodder and nature conservation targets, also taking into account spatial variation in projected future productivity of bioenergy plantations, and arrived at a technical potential in 2050 in the range of 160 to 270 EJ/yr. Of the 210 EJ/yr average technical potential, 81 EJ/yr are provided by dedicated plantations, 27 EJ/yr by residues in forestry and 100 EJ/yr by crop residues, manure and organic wastes, emphasizing the importance of process optimization and cascading biomass use.

Water constraints are highlighted in the literature for agriculture (UN-Water, 2007) and for bioenergy (Berndes, 2002; Molden, 2007; De Fraiture et al., 2008; Sections 9.3.4.4 and 2.5.5.1). In a number of regions the technical potential can decrease to lower levels than what is assessed based on approaches that do not involve explicit geo-hydrological modelling (Rost et al., 2009). Such modelling can lead to improved quality bioenergy potential assessments. Planting of trees and other perennial vegetation can decrease erosive water run-off and replenish groundwater but may lead to substantial reductions in downstream water availability (Calder et al., 2004; Farley et al., 2005).

Illustrative of this, Zomer et al. (2006) report that large areas deemed suitable for afforestation within the Clean Development Mechanism (CDM) would exhibit evapotranspiration increases and/or decreases in runoff if they become forested, that is, a decrease in water potentially available offsite for other uses. This would be particularly evident in drier areas, the semi-arid tropics, and in conversion from grasslands and subsistence agriculture. Similarly, based on a global analysis of 504 annual catchment observations, Jackson et al. (2005) report that afforestation dramatically decreased stream flow within a few years of planting. Across all plantation ages in the database, afforestation of grasslands, shrublands or croplands decreased stream flow by, on average, 38%. Average losses for 10- to 20-year-old plantations were even greater, reaching 52% of stream flow.

Studies by Hoogwijk et al. (2003), Wolf et al. (2003), Smeets et al. (2007) and van Minnen et al., (2008) also illustrate the importance of biomass plantations for reaching a higher global technical potential, and how different determining parameters greatly influence the technical potential. For instance, in a scenario with rapid population growth and slow technology progress, where agriculture productivity does not increase from its present level and little biomass is traded, Smeets et al. (2007) found that no land would be available for bioenergy plantations. In a contrasting scenario where all critical parameters were instead set to be very favourable, up to 3.5 billion hectares (35 million km^2) of former agricultural land—mainly pastures and with large areas in Latin America and sub-Saharan Africa—were assessed as not required for food in 2050. A substantial part of this area was assessed as technically suitable for bioenergy plantations.

2.2.3 Economic considerations in biomass resource assessments

Some studies exclude areas where attainable yields are below a certain minimum level. Other studies exclude biomass resources judged as being too expensive to mobilize, given a certain biomass price level. These assessments address biomass resource availability and cost for given levels of production so that an owner of a facility for secondary energy production from modern biomass could assess a location and the size of a facility for a cost-effective business with a guaranteed supply of biomass throughout the year. Costs models are based on combining land availability, yield levels and production costs to obtain plant- and region-specific cost-supply curves (Walsh, 2008). These are based on projections or scenarios for the development of cost factors, including opportunity cost of land, and can be produced for different contexts and scales—including feasibility studies of supplying individual bioenergy plants and estimating the future global cost-supply curve. Studies using this approach at different scales include Dornburg et al. (2007), Hoogwijk et al. (2009), de Wit et al. (2010) and van Vuuren et al. (2009). P. Gallagher et al. (2003) exemplify the production of cost-supply curves for the case of crop harvest residues and Gerasimov and Karjalainen (2009) for the case of forest wood.

The biomass production costs can be combined with technological and economic data for related logistic systems and conversion technologies to derive market potentials at the level of secondary energy carriers such as bioelectricity and biofuels for transport (e.g., Gan, 2007; Hoogwijk et al., 2009; van Dam et al., 2009c). Using biomass cost and availability data as exogenously defined input parameters in scenario-based energy system modelling can provide information about levels of implementation in relation to a specific energy system context and possible climate and energy policy targets. Cost trends are discussed further in Section 2.7.

Figure 2.5(a) shows projections of European market potential estimated based on food sector scenarios for 2030, considering also nature protection requirements and infrastructure development (Fischer et al., 2010). Estimated production cost supply curves shown in Figure 2.5(b) were subsequently produced including biomass plantations and forest/agriculture residues (de Wit and Faaij, 2010). The key factor determining the size of the market potential was the development of agricultural land productivity, including animal production.

Figure 2.5(c) data for the USA are based on recent assessments of lignocellulosic feedstock supply cost curves conducted at county-level resolution (Walsh, 2008; Perlack et al., 2005; US DOE, 2011). Figure 2.5(d) illustrates the delivered price of biomass to the conversion facility under the baseline conditions for various production levels of lignocellulosic feedstocks.[6] Total market potential for crop-based ethanol and

---

6 For instance, at a biomass feedstock price of USD$_{2005}$ 3/GJ delivered to the conversion facility, the three types of feedstocks shown in Figure 2.5(d) would provide 5.5 EJ. At higher prices there is more feedstock up to a point, for example, 1.5 EJ for the forest residues in the figure.

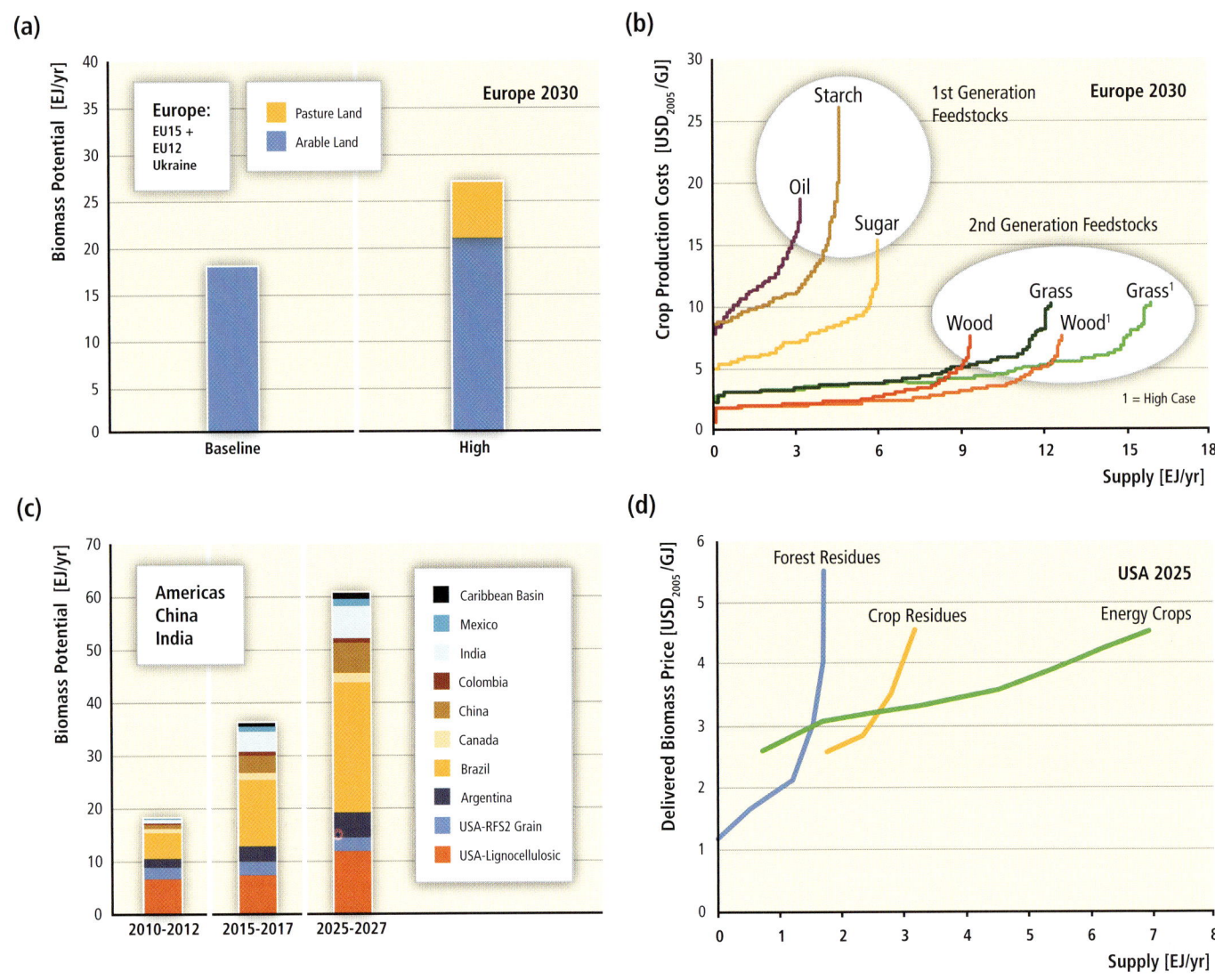

Figure 2.5 | Examples of preliminary market potentials based on feedstock cost supply curves shown in (b) for European countries and (d) for the USA. The feedstock cost supply curves for these assessments are from recent studies conducted at levels of: (a) region; (c) country based on state/province except for the USA, which is performed at a county level. In (c) the US data are for the baseline case and the other countries' cases are for a high-growth scenario (a total of 45 EJ/yr, which would decrease to 25 EJ/yr in the base case and to around 8 EJ/yr in the low case) by 2025. See text for further information. Sources: (a) Fischer et al. (2010); (b) de Wit and Faaij (2010); (c) Kline et al. (2007); Walsh (2008); EPA (2010); (d) Walsh (2008), US DOE (2011).

biodiesel are from EPA (2010) projections. In addition, Figure 2.5(c) includes preliminary estimates of high-growth scenarios of market potentials for the Americas, China and India based on historic production trends and average production costs at the state/province level (Kline et al., 2007), considering multiple crops, residues and perennial biomass crops. Market potentials were estimated based on arable land availability for bioenergy plants and some degree of environmental protection and infrastructure. High-growth market potentials are shown for years 2012, 2017 and 2027 (Kline et al., 2007). The largest supplier, Brazil, is using AgroEcological Zoning (EMBRAPA, 2010) to limit expansion to unrestricted areas with appropriate soil and climate, with no or low irrigation requirements, and low slopes for mechanized harvesting.

Similar zoning is available for oil palm.[7] These steps are recommended by several of the organizations developing sustainability criteria (van Dam et al., 2010, and see Section 2.4.5).

2.2.4 Factors influencing biomass resource potentials

As described briefly above, many studies that quantify the biomass resource potential consider a range of factors that reduce it to lower levels than if they are not included. These factors are also connected to impacts arising from the exploitation of biomass resources, which are further discussed in Section 2.5. The most important factors are

---

7 DECRETO Nº 7172, DE 07 DE MAIO DE 2010, Brazil.

discussed below in relation to how they influence the future biomass resource potential.

2.2.4.1 Residue supply in agriculture and forestry

Soil conservation and biodiversity requirements influence technical potentials for both agriculture and forestry residues. In forestry, the combination of residue harvest and nutrient (including wood ash) input can avoid nutrient depletion and acidification and can in some areas improve environmental conditions due to reduced nutrient leaching from forests (Börjesson, 2000; Eisenbies et al., 2009). Even so, organic matter at different stages of decay plays an important ecological role in conserving soil quality as well as for biodiversity in soils and above ground (Grove and Hanula, 2006). Thresholds for desirable amounts of dead wood in forest stands are difficult to set and the most demanding species require amounts of dead wood that are difficult to reach in managed forests (Ranius and Fahrig, 2006). Dymond et al. (2010) report that estimates from studies taking into account the need for on-site sustainability can be several times lower than those that do not. Large differences were also reported by Gronowska et al. (2009). Titus et al. (2009) report wide ranges (0 to 100%) in allowed residue recovery rates for large-scale logging residue inventories and propose a 50% retention proportion as an appropriate level, noting that besides soil sustainability additional aspects (e.g., biodiversity and water quality) need to be considered.

Development of technologies for stump harvesting after felling increases the availability of residues during logging (Näslund-Eriksson and Gustavsson, 2008). Stump harvesting can also reduce the cost of site preparation for replanting (Saarinen, 2006). It can reduce damage from insects and spreading of root rot fungus, but can also lead to negative effects including reduced forest soil carbon and nutrient stocks, increased soil erosion and soil compaction (Zabowski et al., 2008; Walmsley and Godbold, 2010).

In agriculture, overexploitation of harvest residues is one important cause of soil degradation in many places in the world (Wilhelm et al., 2004; Ball et al., 2005; Blanco-Canqui et al., 2006; Lal, 2008). Fertilizer inputs can compensate for nutrient removals connected to harvest and residue extraction, but maintenance or improvement of soil fertility, structural stability and water-holding capacity requires recirculation of organic matter to the soil (Lal and Pimentel, 2007; Wilhelm et al., 2007; Blanco-Canqui and Lal, 2009). Residue recirculation leading to nutrient replenishment and carbon storage in soils and dead biomass not only contributes positively to climate change mitigation by withdrawing carbon from the atmosphere but also by reducing soil degradation and improving soil productivity. This leads to higher yields and consequently less need to convert land to croplands for meeting future food/fibre/bioenergy demand (i.e., fewer GHG emissions arising from vegetation removal and ploughing of soils). Residue removal can, all other things being equal, be increased when total biomass production per hectare becomes higher and if 'waste' from processing of crop residues that is rich in refractory compounds such as lignin is returned to the field (J. Johnson et al., 2004; Reijnders, 2008; Lal, 2008).

Principles, criteria and indicators are developed to ensure ecological sustainability (e.g., van Dam et al., 2010; Lattimore et al., 2009; Section 2.4.3) but these cannot easily be used to derive sustainable residue extraction rates. Large uncertainties are also linked to the possible future development of several factors determining residue generation rates. Population growth, economic development and dietary changes influence the demand for products from agriculture and forestry, and materials management strategies (including recycling and cascading use of material) influence how this demand translates into demand for basic food commodities and industrial roundwood.

Furthermore, changes in food and forestry sectors influence the residue/waste generation per unit of product output up or down: crop breeding leads to improved harvest index, reducing residue generation rates; implementation of no-till/conservation agriculture requires that harvest residues are left on the fields to maintain soil cover and increase organic matter in soils (Lal, 2004); shifts in livestock production to more confined and intensive systems can increase recoverability of dung but reduce overall dung production at a given level of livestock product output; and increased occurrence of silvicultural treatments such as early thinning to improve stand growth will lead to increased availability of small roundwood suitable for energy uses.

Consequently, the longer-term technical potential connected to residue/waste flows will continue to be uncertain even if more comprehensive assessment approaches are used. It should be noted that it does not necessarily follow that more comprehensive assessments of determining factors will lead to a lower technical potential of residues; earlier studies may have used conservative residue recovery rates as a precaution in the face of uncertainties (S. Kim and Dale, 2004). However, modelling studies indicate that the cost of soil productivity loss may restrict residue removal intensity to much lower levels than the quantity of biomass physically available in forestry (Gan and Smith, 2010).

2.2.4.2 Dedicated biomass production in agriculture and forestry

Studies indicate significant potential for intensifying conventional long-rotation forestry to increase forest growth and total biomass output—for instance, by fertilizing selected stands and using shorter rotations (Nohrstedt, 2001; Saarsalmi and Mälkönen, 2001)—especially in regions of the world with large forest areas that currently practice extensive forest management. Yet, the prospects for intensifying conventional long-rotation forestry to increase forest growth are not thoroughly investigated in the assessed studies of biomass resource potentials. Instead, the major source of increased forest biomass output is assumed to be fast-growing tree plantations. Besides tree plantations, short-rotation coppicing plants such as willow and perennial grasses such as switchgrass and *Miscanthus* are considered candidate bioenergy plants to become established on these lands.

It is commonly assumed that biomass plantations are established on surplus agricultural land. Intensification in agriculture is therefore a key aspect in essentially all of the assessed studies because it influences both land availability for biomass plantations (indirectly by determining the land requirements in the food sector) and the biomass yield levels obtained. High assessed technical potentials for energy plantations rely on high-yielding agricultural systems and international bioenergy trade leading to the result that biomass plantations are established globally where the production conditions are most favourable. Increasing yields from existing agricultural land is also proposed as a key component for agricultural development (Ausubel, 2000; Fischer et al., 2002; Tilman et al., 2002; Cassman et al., 2003; Evans, 2003; Balmford et al., 2005; Green et al., 2005; D. Lee et al., 2006; Bruinsma, 2009). Studies also point to the importance of diets and the food sector's biomass use efficiency in determining land requirements (both cropland and grazing land) for food (Gerbens-Leenes and Nonhebel, 2002; Smil, 2002; Carlsson-Kanyama and Shanahan, 2003; de Boer et al., 2006; Elferink and Nonhebel, 2007; Stehfest et al., 2009; Wirsenius et al., 2010).

Studies of agricultural development (e.g., Koning, 2008; Alexandratos, 2009; IAASTD, 2009) show lower expected yield growth than studies of the biomass resource potential that report very high technical potentials for biomass plantations (Johnston et al., 2009). Some observations indicate that it can be a challenge to maintain yield growth in several main producer countries and that much cropland and grazing land undergoes degradation and productivity loss as a consequence of improper land use (Cassman, 1999; Pingali and Heisey, 1999; Fischer et al., 2002). The possible consequences of climate change for crop yields are not firmly established but indicate net global negative impact, where damages will be concentrated in developing countries that will lose agriculture production potential while developed countries might gain (Fischer et al., 2002; Cline, 2007; Easterling et al., 2007; Schneider et al., 2007; Lobell et al., 2008; Fischer et al., 2009). Water scarcity can limit both intensification possibilities and the prospects for expansion of bioenergy plantations (Berndes, 2008a,b; de Fraiture et al., 2008; de Fraiture and Berndes, 2009; Rost et al., 2009; van Vuuren et al., 2009) but can be partially alleviated through on-site water management (Rost et al., 2009). Biomass resource potential studies that use biophysical data sets and modelling are able to consider water limitations on land productivity. However, assumptions about productivity growth in land use may implicitly presume irrigation development that could lead to problems in regional water availability, use and distribution among users. Empirical data are needed for use in hydrological process models to better understand and predict the hydrological effects of various land use options at the landscape level (Malmer et al., 2010). Water and land use-related aspects are further discussed in Section 2.5.

Conversely, some observations indicate that rates of gain obtained from breeding have increased in recent years after previous stagnation and that yields might increase faster again as newer hybrids are adopted more widely (Edgerton, 2009). Theoretical limits also appear to leave scope for further increasing the genetic yield potential (Fischer et al., 2009). It should be noted that studies finding high technical potential for bioenergy plantations point primarily to tropical developing countries as major contributors. These countries still have substantial yield gaps to exploit and large opportunities for productivity growth—not the least in livestock production (Fischer et al., 2002; Edgerton, 2009; Wirsenius et al., 2010). There is also a large yield growth potential for dedicated bioenergy plants that have not been subject to the same breeding efforts as the major food crops. Selection and development of suitable plant species and genotypes for given locations to match specific soil types, climate and conversion technologies are possible, but are at an early stage of understanding for some energy plants (Bush and Leach, 2007; Chapple et al., 2007; Lawrence and Walbot, 2007; Carpita and McCann, 2008; Karp and Shield, 2008). Traditional plant breeding, selection and hybridization techniques are slow, particularly for woody plants but also for grasses, but new biotechnological routes to produce both genetically modified (GM) and non-GM plants are possible (Brunner et al., 2007). GM energy plant species may be more acceptable to the public than GM food crops, but there are concerns about the potential environmental impacts of such plants, including gene flow from non-native to native plant relatives (Chapotin and Wolt, 2007; Firbank, 2008; Warwick et al., 2009; see Section 2.5.6.1).

There can be limitations on and negative aspects of further intensification aiming at farm yield increases, for example, high crop yields depending on large inputs of nutrients, fresh water and pesticides can contribute to negative ecosystem effects, such as changes in species composition in the surrounding ecosystems, groundwater contamination and eutrophication with harmful algal blooms, oxygen depletion and anoxic 'dead' zones in oceans (Donner and Kucharik, 2008; Simpson et al., 2009; Sections 2.5.5.1 and 2.6.1.2). However, intensification is not necessarily equivalent to an industrialization of agriculture, as agricultural productivity can be increased in many regions and systems with conventional or organic farming methods (Badgley et al., 2007). The potential to increase the currently low productivity of rain-fed agriculture exists in large parts of the world through improved soil and water conservation (Lal, 2003; Rockström et al., 2007, 2010), fertilizer use and crop selection (Cassman, 1999; Keys and McConnell, 2005). Available best practices[8] are not at present applied in many world regions (Godfray et al., 2010), due to a lack of dissemination, capacity building, availability of resources and access to capital and markets, with distinct regional differences (Neumann et al., 2010).

---

8 For example, mulching, low tillage, contour ploughing, bounds, terraces, rainwater harvesting and supplementary irrigation, drought adapted crops, crop rotation and fallow time reduction.

Conservation agriculture and mixed production systems (double-cropping, crop with livestock and/or crop with forestry) hold potential to sustainably increase land productivity and water use efficiency as well as carbon sequestration and to improve food security and efficiency in the use of limited resources such as phosphorous (Kumar, 2006; Heggenstaller et al., 2008; Herrero et al., 2010). Integration can also be based on integrating feedstock production with conversion—typically producing animal feed that can replace cultivated feed such as soy and corn (Dale et al., 2009, 2010) and also reduce grazing requirements (Sparovek et al., 2007).

Investment in agricultural research, development and deployment could produce a considerable increase in land and water productivity (Rost et al., 2009; Herrero et al., 2010; Sulser et al., 2010) as well as improve robustness of plant varieties (Reynolds and Borlaug, 2006; Ahrens et al., 2010). Multi-functional systems (IAASTD, 2009) providing multiple ecosystem services (Berndes et al., 2004, 2008a,b; Folke et al., 2004, 2009) represent alternative options for the production of bioenergy on agricultural lands that could contribute to development of farming systems and landscape structures that are beneficial for the conservation of biodiversity (Vandermeer and Perfecto, 2006).

2.2.4.3 Use of marginal lands

Biomass resource potential studies also point to marginal/degraded lands—where productive capacity has declined temporarily or permanently—as lands that can be used for biomass production. Advances in plant breeding and genetic modification of plants not only raise the genetic yield potential but also may adapt plants to more challenging environmental conditions (Fischer et al., 2009). Improved drought tolerance can improve average yields in drier areas and in rain-fed systems in general by reducing the effects of sporadic drought (Nelson et al., 2007; Castiglioni et al., 2008) and can also reduce water requirements in irrigated systems. Thus, besides reducing land requirements for meeting food and materials demand by increasing yields, plant breeding and genetic modification could make lands initially considered unsuitable available for rain-fed or irrigated production.

Some studies show a significant technical potential of marginal/degraded land, but it is uncertain how much of this technical potential can be realized. The main challenges in relation to the use of marginal/degraded land for bioenergy include (1) the large efforts and long time periods required for the reclamation and maintenance of more degraded land; (2) the low productivity levels of these soils; and (3) ensuring that the needs of local populations that use degraded lands for their subsistence are carefully addressed. Studies point to the benefits of local stakeholder participation in appraising and selecting appropriate measures (Schwilch et al., 2009) and suggest that land degradation control could benefit from addressing aspects of biodiversity and climate change and that this could pave the way for funding via international financing mechanisms and major donors (Knowler, 2004; Gisladottir and Stocking, 2005). In this context, the production of properly selected plant species for bioenergy can be an opportunity, where additional benefits involve carbon sequestration in soils and aboveground biomass and improved soil quality over time.

2.2.4.4 Biodiversity protection

Considerations regarding biodiversity can limit residue extraction as well as intensification and expansion of agricultural land area. WBGU (2009) shows that the way biodiversity is considered can have a larger impact on technical potential than either irrigation or climate change. The common way of considering biodiversity requirements as a constraint is by including requirements for land reservation for biodiversity protection. Biomass resource potential assessments commonly exclude nature conservation areas from being available for biomass production, but the focus is as a rule on forest ecosystems and takes the present level of protection as a basis. Other natural ecosystems also require protection—not least grassland ecosystems—and the present status of nature protection for biodiversity may not be sufficient for given targets. While many highly productive lands have low natural biodiversity, the opposite is true for some marginal lands and, consequently, the largest impacts on biodiversity could occur with widespread use of marginal lands.

Some studies indirectly consider biodiversity constraints on productivity by assuming a certain expansion of alternative agriculture production (to promote biodiversity) that yields less than conventional agriculture and therefore requires more land for food production (EEA, 2007; Fischer et al., 2009). However, for multi-cropping systems a general assumption of lower yields from alternative cropping systems is not consistent. Biodiversity loss may also occur indirectly, such as when productive land use displaced by energy crops is re-established by converting natural ecosystems into croplands or pastures elsewhere. Integrated energy system and land use/vegetation cover modelling have better prospects for analyzing these risks.

Bioenergy plantations can play a role in promoting biodiversity, particularly when multiple species are planted and mosaic landscapes are established in uniform agricultural landscapes and in some currently poor or degraded areas (Hartley, 2002). Agro-forestry systems combining biomass and food production can support biodiversity conservation in human-dominated landscapes (Bhagwat et al., 2008). Biomass resource potential assessments, however, as a rule assume yield levels corresponding to those achieved in monoculture plantations and therefore provide little insight into how much biomass could be produced if a significant part of the biomass plantation were shaped to contribute to biodiversity preservation.

2.2.5 Possible impact of climate change on resource potential

Technical potentials are influenced by climate change. The magnitude and spatial pattern of climate change remain uncertain[9] despite high scientific confidence that global warming and an intensification of the hydrological cycle will be a consequence of increased GHG concentrations in the atmosphere (IPCC, 2007c). Furthermore, the effect of unhistorical new changes in temperature, irradiation and soil moisture on the growth of agricultural plants is frequently uncertain (Lobell and Burke, 2008), as is the adaptive response of farmers. As a consequence, the overall magnitude and pattern of climate change effects on agricultural production, including bioenergy plantations, remain uncertain. While positive effects on plant growth may occur, detrimental impacts on productivity cannot at present be precluded for many important regions.

Uncertainty also remains about the concurrent ecophysiological effect of elevated atmospheric CO_2 concentration on plant productivity—the CO_2 fertilization effect. Under elevated CO_2 supply, the growth of plants with C_3 photosynthesis is increased unless it is hampered by increased water stress or nutrient depletion (Oliver et al., 2009). The long-term magnitude of the carbon fertilization effect is disputed, with increases in annual NPP of around 25% possible and observed in some field experiments for a doubling of atmospheric CO_2 concentration (the effect levels off at higher CO_2 concentrations), while some expect smaller gains due to co-limitations and eventual adaptations (Ainsworth and Long, 2005; Körner et al., 2007). The magnitude of the effect under agricultural management and breeding conditions may be different and is not well known.

Under climate warming, the increased requirement for transpiration water by vegetation is partially countered by increased water use efficiency (increased stomatal closure) under elevated atmospheric CO_2 concentrations, with variable regional patterns (Gerten et al., 2005). Changes in precipitation patterns and magnitude can increase or decrease plant production depending on the direction of change. Generally, some semi-arid marginal lands are projected to be more productive due to increased water use efficiency under CO_2 fertilization (Lioubimtseva and Adams, 2004). As crop production is projected to mostly decline with warming of more than 2°C (Easterling et al., 2007), particularly in the tropics, biomass for energy production could be similarly affected. Overall, the effects of climate change on biomass technical potential are found to be smaller than the effects of management, breeding and area planted (WBGU, 2009), but in any particular region they can be strong. Which regions will be most affected remains uncertain, but tropical regions are most likely to see the strongest negative impact.

2.2.6 Synthesis

As discussed, narrowing down the technical potential of the biomass resource to precise numbers is not possible. A number of studies show that between less than 50 and several hundred EJ per year can be provided for energy in the future, the latter strongly conditional on favourable developments. From an assessment of the findings, it can be concluded that:

- The size of the future technical potential is dependent on a number of factors that are inherently uncertain and will continue to make long-term technical potentials unclear. Important factors are population and economic/technology development and how these translate into fibre, fodder and food demand (especially share and type of animal food products in diets) and development in agriculture and forestry.

- Additional important factors include (1) climate change impacts on future land use including its adaptation capability; (2) considerations set by biodiversity and nature conservation requirements; and (3) consequences of land degradation and water scarcity.

- Studies point to residue flows in agriculture and forestry and unused (or extensively used) agricultural land as an important basis for expansion of biomass production for energy, both in the near term and in the longer term. Consideration of biodiversity and the need to ensure maintenance of healthy ecosystems and avoid soil degradation set bounds on residue extraction in agriculture and forestry (further discussed in Section 2.5.5).

- Grasslands and marginal/degraded lands are considered to have potential for supporting substantial bioenergy production, but biodiversity considerations and water shortages may limit this potential. The possibility that conversion of such lands to biomass plantations reduces downstream water availability needs to be considered.

- The cultivation of suitable plants can allow for higher technical potentials by making it possible to produce bioenergy on lands less suited for conventional food crops—also when considering that the cultivation of conventional crops on such lands can lead to soil carbon emissions (further discussed in Section 2.5.2).

- Landscape approaches integrating bioenergy production into agriculture and forestry systems to produce multi-functional land use systems could contribute to the development of farming systems and landscape structures that are beneficial for the conservation of biodiversity and help restore/maintain soil productivity and healthy ecosystems.

---

[9] Uncertainties arise because future GHG emission trajectories cannot be known (and are therefore studied using a variety of scenarios), the computed sensitivities of climate models to GHG forcing vary (i.e., the amount of warming that follows from a given emission scenario), and the spatial pattern and seasonality of changes in precipitation vary greatly between models, particularly for some tropical and subtropical regions (Li et al., 2006).

- Water constraints may limit production in regions experiencing water scarcity. But the use of suitable energy crops that are drought tolerant can also help adaptation in water-scarce situations. Assessments of biomass resource potentials need to more carefully consider constraints and opportunities in relation to water availability and competing uses.

Based on this expert review of the available scientific literature, deployment levels of biomass for energy could reach a range of 100 to 300 EJ/yr around 2050 (see Section 2.8.4.1 for more detail). This can be compared with the present biomass use for energy of about 50 EJ/yr. While recent assessments employing improved data and modelling capacity have not succeeded in providing narrow, distinct estimates of the biomass resource potential, they have advanced the understanding of how influential various factors are on the resource potential and that both positive and negative effects may follow from increased biomass use for energy. One important conclusion is that the effects of LUC associated with bioenergy expansion can considerably influence the climate benefit of bioenergy (Section 2.5.5). The insights from the resource assessments can improve the prospects for bioenergy by pointing out the areas where development is most crucial and where research is needed. A summary is given in Section 2.8.4.3.

2.3 Technologies and applications

This section reviews commercial technologies for biomass feedstock production, pretreatment of solid biomass and logistics of supply chains bringing feedstocks to direct users. The users can be individuals (e.g., fuelwood for cooking or heating) or firms (e.g., industrial users or processors). Pretreated and converted energy carriers are more convenient and can be used in more applications than the original biomass and are modern solid (e.g., pellets), liquid (e.g., ethanol) and gaseous (e.g., methane) fuels from which electricity and/or heat or mobility services are produced (see Figure 2.2). The integration of modern biomass with existing and evolving electricity, natural gas, heating (residential and district, commercial and public services), industrial, agriculture/forestry, and fossil liquid fuels systems is discussed thoroughly in Chapter 8.

This section is organized along the supply chain of bioenergy and thus discusses feedstock production and the synergies with related sectors before turning to pretreatment, logistics and supply chains of solid biomass. The section then explains different state-of-the-art conversion technologies for energy carriers from modern biomass before discussing the costs, directly available from relevant literature, of these broader bioenergy systems and supply chains. Section 2.6 provides prospects for technology improvement, innovation and integration before Section 2.7 addresses relevant cost information in terms of levelized cost of production for many world regions.

2.3.1 Feedstocks

2.3.1.1 Feedstock production and harvest

The performance characteristics of major biomass production systems, dedicated plants or primary residues across the world regions are summarized in Table 2.4. The management of energy plants includes the provision of seeds or seedlings, stand establishment and harvest, soil tillage, irrigation, and fertilizer and pesticide inputs. The latter depend on crop requirements, target yields and local pedo-climatic conditions, and may vary across world regions for similar species (Table 2.4). Strategies such as integrated pest management or organic farming may alleviate the need for synthetic inputs for a given output of biomass (Pimentel et al., 2005).

Wood for energy is obtained as fuelwood or as residue. While fuelwood is derived from the logging of natural or planted forests or trees and shrubs grown in agriculture fields, residues are derived from wood waste and by-products. While natural forests are not managed for production per se, problems arise if fuelwood extraction exceeds the regeneration capacity of the forests, which is the case in many parts of the world. The management of planted forests involves silvicultural techniques similar to those used in cropping systems and includes stand establishment and tree felling (Nabuurs et al., 2007).

Biomass may be harvested several times per year (for forage-type feedstocks such as hay or alfalfa), once per year (for annual species such as wheat or perennial grasses), or every 2 to 50 years or more (for short-rotation coppice and conventional forestry, respectively). Sugarcane is harvested annually but planted every 4 to 7 years and grown in ratoons; it is considered a perennial grass. Harvested biomass is typically transported to a collection point on the farm or at the edge of the road before being transported to the bioenergy unit or to an intermediate storage facility. It may be preconditioned and densified to facilitate storage, transport and handling (see Section 2.3.2).

The species listed in Table 2.4 have different possible energy end uses and require diverse conversion technologies (see Figure 2.6). Starch and oil crops are grown and harvested annually as feedstocks for what are called first-generation liquid biofuels (ethanol and biodiesel, see Section 2.3.3). Only a fraction of the total aboveground biomass is used for biofuels, with the rest being processed for animal feed or lignocellulosic residues. Sugarcane plants are feedstocks for the production of sugar and ethanol and, increasingly, sugarcane bagasse and straw, which serve as sources of process heat and extra power in many sugar- and ethanol-producing countries (Macedo et al., 2008; Dantas et al., 2009; Seabra et al., 2010) resulting in favourable environmental footprints for these biorefinery products. Lignocellulosic plants such as perennial grasses or short-rotation coppice may be entirely converted to energy, and feature two to five times higher

Table 2.4 | Typical characteristics of the production technologies for dedicated species and their primary residues. Yields are expressed as GJ of energy content in biomass prior to conversion to energy, or of the ethanol end product for sugar and starch crops. Costs refer to private production costs or market price when costs were unavailable (data from 2005 to 2009). Key to management inputs: +: low; ++: moderate; +++: high requirements.

Feedstock type	Region	Yield	Management			Co-products	Costs	Refs.
		GJ/ha/yr [TJ/km^2/yr]	Fertilizer use[1]	Water needs	Pesticides		Examples (2005-2009) USD/GJ	
OIL CROPS		As oil						
Oilseed rape	Europe	60–70 [6.7–7.0]	+++	+	+++	Rape cake, straw	7.2–16.0	1,2,3,22
Soybean	North America	16–19 [1.6–1.9]	++	+	+++	Soy cake, straw	11.7	3,12
	Brazil	18–21 [1.8–2.1]	++	+	+++		N/A	
Palm oil	Asia	135–200 [13.5–20.0]	++	+	+++	Fruit bunches, press fibres	N/A	
	Brazil	169 [16.9]	++	+	+++		12.6[2]	3
Jatropha	World	17–88 [1.7–8.8]	+/++	+	+	Seed cake (toxic), wood, shells	3.2	3,4,5,10,11
STARCH CROPS		As ethanol						
Wheat	Europe	54–58 [5.4–5.8]	+++	++	+++	Straw, DDGS[3]	5.2	3
Maize	North America	72–79 [7.2–7.9]	+++	+++	+++	Corn stover, DDGS	10.9	3
Cassava	World	43 [4.3]	++	+	++	DDGS	3.3–4	3
SUGAR CROPS		As ethanol						
Sugarcane	Brazil	116–149 [11.6–14.9]	++	+	+++	Bagasse, straw	1.0–2.0[2]	3,17
	India	95–112 [9.5–11.2]					N/A	3
Sugar beet	Europe	116–158 [11.6–15.8]	++	++	+++	Molasses, pulp	5.2–9.6	3,13,22
Sorghum (sweet)	China	105–160 [10.5–16.0]	+++	+	++	Bagasse	4.4	2,21
LIGNOCELLULOSIC CROPS		As ethanol						
Miscanthus	Europe	190–280 [19.0–28.0]	+/++	++	+		4.8–16	6,8
Switchgrass	Europe	120–225 [12.0–22.5]	++	+	+		2.4–3.2	10,14
	North America	103–150 [10.3–15.0]	++	+	+		4.4	
Short rotation (SR)	Southern Europe	90–225 [9.0–22.5]	+	++	+	Tree bark	2.9–4	10,14
Eucalyptus	South America	150–415 [15.0–41.5]	+/++	+	+		2.7	16,19
SR Willow	Europe	140 [14.0]					4.4	2,7
Fuelwood (chopped)	Europe	110 [11.0]				Forest residues	3.4–13.6	15
Fuelwood (renewable, native forest)	Central America	80–150 [8.0–15.0]					1.8–2.0	23
PRIMARY RESIDUES								
Wheat straw	Europe	60 [6.0]	+			Not Applicable	1.9	2
	USA	7–75 [0.7–7.5]					N/A	14, 20
Sugarcane straw	Brazil	90–126 [9.0–12.6]	+				N/A	17
Corn stover	North America	15–155 [1.5–15.5]	+				N/A	9,14
	India	22–30 [2.2–3.0]	+				0.9	18
Sorghum stover	World	85 [8.5]	+				N/A	9
Forest residues	Europe	2–15 [0.2–1.5]					1–7.7	15

Notes: 1. Nitrogen, phosphorus, and potassium; 2. Market price; 3. DDGS: Dried Distillers Grain with Solubles. These are illustrative cost figures or market prices from the literature. See Annex II for ranges of costs for specific commercial feedstocks over a year period.

References: 1: EEA (2006); 2: Edwards et al. (2007); 3: Bessou et al. (2010); 4: Jongschaap et al. (2007); 5: Openshaw (2000); 6: Clifton-Brown et al. (2004); 7: Ericsson et al. (2009); 8: Fagernäs et al. (2006); 9: Lal (2005); 10: WWI, (2006); 11: Maes et al. (2009); 12: Gerbens-Leenes et al. (2009);13: Berndes (2008a,b); 14: Perlack et al. (2005); 15: Asikainen et al. (2008); 16: Scolforo (2008); 17: Folha (2005); 18: Guille (2007); 19: Diaz-Balteiro and Rodriguez (2006); 20: Lal (2005); 21: Grassi et al. (2006); 22: Faaij (2006); 23: T. Johnson et al. (2009). See Bessou et al. (2010) for specific biofuel volumes per hectare for various countries; see also IEA Renewable Energy Division (2010) for additional country information.

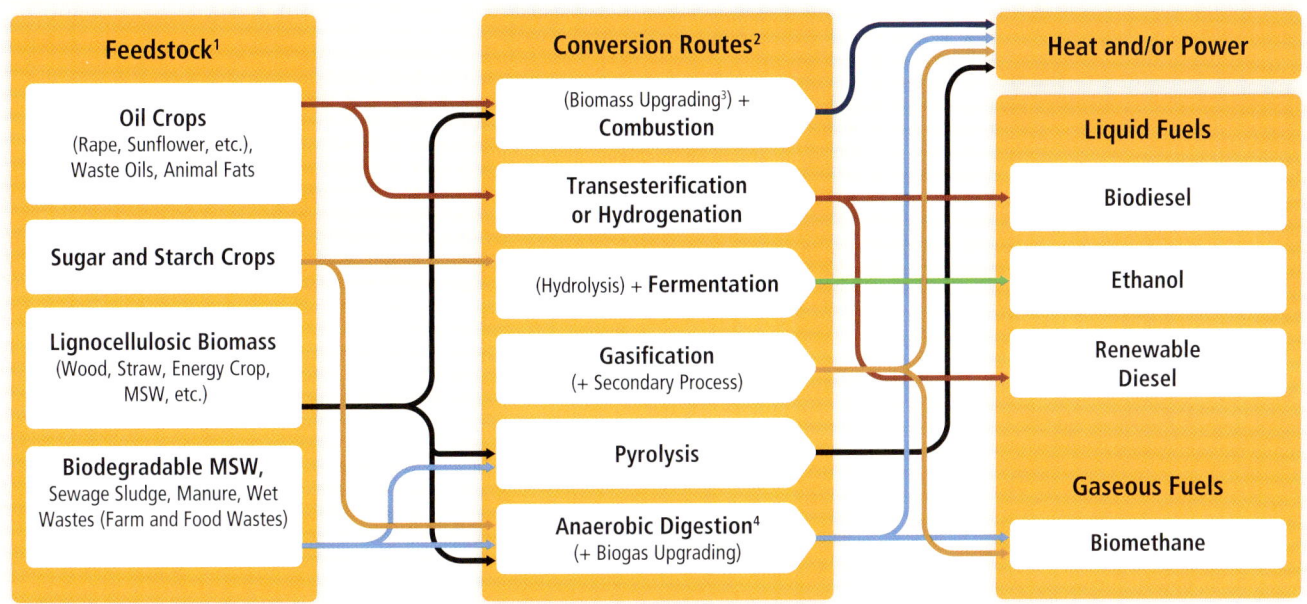

Figure 2.6 | Schematic view of commercial bioenergy routes (modified from IEA, Bioenergy, 2009).

Notes: 1. Parts of each feedstock, for example, crop residues, could also be used in other routes. 2. Each route also gives co-products. 3. Biomass upgrading includes any one of the densification processes (pelletization, pyrolysis, etc.). 4. Anaerobic digestion processes release methane and CO_2 and removal of CO_2 provides essentially methane, the main component of natural gas; the upgraded gas is called biomethane.

yields per hectare than most of the other feedstock types, while requiring far fewer synthetic inputs when managed carefully (Hill, 2007). However, their impact on soil organic matter after the removal of stands is not well understood (Wilhelm et al., 2007; Anderson-Teixeira et al., 2009). Research is underway to assess site-specific removal levels as a function of time and strategies to mitigate weather impacts on residue removal (e.g., Karlen, 2010; Zhang et al., 2010). With technologies that are currently commercial, lignocellulosic feedstocks are only providing heat and power whereas the harvest products of oil, sugar and starch crops are being converted readily to liquid biofuels and in some cases together with heat and power.

Production and harvest costs for dedicated plants vary widely according to the prices of inputs, machinery, labour and land-related costs (Ericsson et al., 2009; Table 2.4). If energy plantations are to compete with land dedicated to food production, the opportunity cost of land (the price that a farmer needs to receive in order to switch from the known annual crop cultivation to an energy crop) could be quite significant and may escalate proportionally with the demand for energy feedstocks (Bureau et al., 2010). Cost-supply curves scaling from farm to the regional level are needed to account for possible large-scale deployment scenario effects (see examples in Figures 2.5(b) and 2.5(d) for feedstock supplies in Europe (cost) and the USA (delivered price), respectively, as a function of feedstock production level, with the unit price per GJ growing several-fold as the total demand for biomass increases).

The cost of forest products depends heavily on harvesting and other logistical practices. In particular labour costs, machinery and the distance from the logging site to the conversion plant are important (Asikainen et al., 2008). This favours local, non-centralized markets especially in developing countries where forests are the dominant fuel source for households (Bravo et al., 2010).

2.3.1.2 Synergies with the agriculture, food and forest sectors

As emphasized in Section 2.2.1, bioenergy feedstock production competes with other uses for resources, chiefly land, with possible negative effects on biodiversity, water availability, soil quality and climate (see Sections 2.2.4 and 2.5). However, synergistic effects may also emerge through the design of integrated production systems, which also provide additional environmental services. Intercropping and mixed cropping are options to maximize the output of biomass per unit area farmed (WWI, 2006). Mixed cropping systems result in increased yields compared to single crops, and may provide both food/fodder and energy feedstocks from the same field (Jensen, 1996; Tilman et al., 2006b). Double-cropping systems have the potential to generate additional feedstocks for bioenergy and livestock utilization and potentially higher yields of biofuel from two crops in the same area in a year (Heggenstaller et al., 2008).

Agro-forestry systems make it possible to use land for food, fodder, timber and energy purposes with mutual benefits for the associated species (R. Bradley et al., 2008). The associated land equivalent ratios may reach up to 1.5, meaning a 50% saving in land area when combining trees with arable crops compared to monocultures (Dupraz and Liagre, 2008) and therefore an equal reduction in indirect LUC effects (see Section

2.5.3). Another option is growing an understory food crop and coppicing the lignocellulosic species to produce residual biomass for energy, similarly to short-rotation coppice (Dupraz and Liagre, 2008). Perennial plants create positive externalities such as erosion control, improved fertilizer use efficiency and reduction in nitrate leaching relative to annual plants (see Section 2.2.4.2). Lastly, the revenues generated from growing bioenergy feedstocks may provide access to technologies or inputs enhancing the yields of food crops, drive additional investments in the agricultural sector and contribute to productivity gains (De La Torre Ugarte and Hellwinckel, 2010), provided feedstock benefits are distributed to local communities (Practical Action Consulting, 2009).

2.3.2 Logistics and supply chains for energy carriers from modern biomass

Because biomass is mostly available in low-density form, it demands more storage space, transport and handling than fossil equivalents, with consequent cost implications. Biomass often needs to be processed (pretreated) to improve handling. For most bioenergy systems and chains, handling and transport of biomass from the source location to the conversion plant is an important contributor to the overall costs of energy production. Crop harvesting, storage, transport, pretreatment and delivery can amount to 20 to 50% of the total costs of energy production (J. Allen et al., 1998).

Use of a single agricultural biomass feedstock for year-round energy generation requires relatively large storage because biomass is only available for a short time following harvest in many places. In addition to such seasonal variations in biomass availability, other characteristics complicate the biomass supply chain and should be taken into account. These include multiple feedstocks with their own complex supply chains, and storage challenges such as space constraints, fire hazards, moisture control and health risks from fungi and spores (Junginger et al., 2001; Rentizelas et al., 2009).

2.3.2.1 Solid biomass supplies and market development for utilization

Over time, several stages may be observed in biomass utilization and market developments in biomass supplies. Different countries seem to follow these stages over time, but clearly differ in their respective stages of development (Faaij, 2006; Sims et al., 2010).

1. Waste treatment (e.g., MSW and use of process residues (paper industry, food industry) onsite at production facilities) is generally the starting phase of a developing bioenergy system. Resources are available and often have a disposal cost (could have a negative value) making utilization profitable and simultaneously solving waste management problems. Large- and small-scale developments are evolving along with integrated resource management.

2. Local utilization of resources from forest management and agriculture. Such resources are more expensive to collect and transport, but usually still economically attractive. Infrastructure development is needed.

3. Biomass market development at regional scale; larger-scale conversion units with increasing fuel flexibility are deployed; increasing average transport distances further improves economies of scale. Increasing costs of biomass supplies make more energy-efficient conversion facilities necessary as well as feasible. Policy support measures such as feed-in tariffs (FITs) are usually needed to develop into this stage.

4. Development of national markets with increasing numbers of suppliers and buyers; creation of a marketplace; increasingly complex logistics. Availability often increases due to improved supply systems and access to markets. Price levels may therefore decrease (see, e.g., Junginger et al., 2005).

5. Increasing scale of markets and transport distances, including cross-border transport of biofuels; international trade in biomass resources (and energy carriers derived from biomass). Biomass is increasingly becoming a globally traded energy commodity (see, e.g., Junginger et al., 2008). Bio-ethanol trade has come closest to that situation (see, e.g., Walter et al., 2008).

6. Growing role for dedicated fuel supply systems (biomass production largely or only for energy purposes). So far, most energy crops are grown because of agricultural interests and support (subsidies for farmers, use of set-aside subsidies), which concentrate on oil crops (such as rapeseed) and surplus food crops (cereals and sugar beets).

Countries that have gained substantial commercial experience with biomass supplies and biomass markets are generally able to obtain substantial cost reductions in biomass supply chains over time. In Finland and Sweden, delivery costs decreased from USD$_{2005}$ 12 to 5/GJ from 1975 to 2003, due to factors such as scale increases, technological innovations or increased competition (Junginger et al., 2005). Similar trends are observed in the corn ethanol industry in the USA and the sugarcane ethanol industry in Brazil (see Table 2.17).

Analyses of regional and international biomass supply chains show that road transport of untreated and bulky biomass becomes uncompetitive and energy-inefficient when crossing distances of 50 to 150 km (Dornburg and Faaij, 2001; McKeough et al., 2005). When long-distance transport is required, early pretreatment and densification in the supply chain (see Sections 2.3.2.3 and 2.6.2) pays off to minimize transport costs. Taking into account energy use and related GHG emissions, well-organized logistic chains can require less than 10% of the initial energy content of the biomass (Hamelinck et al., 2005b; Damen and Faaij, 2006), but this requires substantial scale in transport, efficient pretreatment and minimization of road transport of untreated biomass.

Such organization is observed in the rapidly developing international wood pellet markets (see Sections 2.3.2.3 and 2.4.4). Furthermore, (long distance) transport costs of liquid fuels such as ethanol and vegetable oils contribute only a minor fraction of overall costs and energy use of bioenergy chains (Hamelinck et al., 2005b).

2.3.2.2 Solid biomass and charcoal supplies in developing countries

The majority of poorest households in the developing world depend on solid biomass fuels such as charcoal for cooking, and millions of small industries (such as brick and pottery kilns) generate process heat from these fuels (FAO, 2010a; IEA, 2010b; see Section 1.4.1.2). Despite this pivotal role of biomass, the sector remains largely unregulated, poorly understood, and the supply chains are predominantly in the hands of the informal sector (Sepp, 2008).

When fuelwood is marketed, trees are usually felled and cut into large pieces and transported to local storage facilities where they are collected by merchants and delivered to wholesale and retail facilities, mainly in rural areas. Some of the wood is converted to charcoal in kilns, packed into large bags and transported by hand, animal-drawn carts and small trucks to roadside sites where it is collected by trucks and sent to urban wholesale and retail sites. Thus charcoal making is an enterprise for rural populations to supply urban markets. Crop residues and dung are normally used by animal owners as a seasonal supplement to fuelwood (FAO 2010a).

Shredded biomass residues may be densified by briquetting or pelletizing, typically in screw or piston presses that compress and extrude the biomass (FAO, 1985). Briquettes and pellets can be good substitutes for coal, lignite and fuelwood because they are renewable and have consistent quality and size, better thermal efficiency, and higher density than loose biomass.

There are briquetting plants in operation in India and Thailand, using a range of secondary residues and with different capacities, but none as yet in other Asian countries. There have been numerous, mostly development agency-funded, briquetting projects in Africa, and most have failed technically and/or commercially. The reasons for failure include deployment of new test units that were not proven technically, selection of very expensive machines that did not make economic sense given the location, low local capacity to fabricate components and provide maintenance, and lack of markets for the briquettes due to uncompetitive cost and low acceptance (Erikson and Prior, 1990).

Wood pellets are made of wood waste such as sawdust and grinding dust. Pelletization machines are based on fodder-making technology and produce somewhat lighter and smaller pellets of biomass compared to briquetting. Wood pellets are easy to handle and burn because their shape and characteristics are uniform, transportation efficiency is high and energy density is high. Wood pellets are used as fuel in many countries for cooking and heating applications (Peksa-Blanchard et al., 2007).

Chips are mainly produced from plantations' waste wood and wood residues (branches and presently even spruce stumps) as a by-product of conventional forestry. They require less processing and are cheaper than pellets. Depending on end use, chips may be produced onsite, or the wood may be transported to the chipper. Chips are commonly used in automated heating systems, and can be used directly in coal-fired power stations or for CHP production (Fagernäs et al., 2006).

Charcoal is obtained by heating woody biomass to high temperatures in the absence of oxygen, and has a twice higher calorific value than the original feedstock. It burns without smoke and has a low bulk density, which reduces transport costs. In rural areas in many African countries, charcoal is produced in traditional kilns with efficiencies as low as 10% (Adam, 2009), and typically sold to urban households while rural households use fuelwood. Hardwoods are the most suitable raw material for charcoal, because softwoods incur possibly high losses during handling/transport. Charcoal from granular materials like coffee shells, sawdust and straw is in powder form and needs to be briquetted with or without a binder. Charcoal is also used in large-scale industries, particularly in Brazil from high-yielding eucalyptus plantations (Scolforo, 2008), and in many cases, in conjunction with sustainably produced wood, and also increasingly as a co-firing feedstock in oil-based electric power plants. The projected costs for charcoal production from Brazilian eucalyptus plantations are USD_{2005} 5.7 to 9.8/GJ (Fallot et al., 2009) using industrial carbonizing process.

Charcoal in Africa is predominantly produced in inefficient traditional kilns in the informal sector, often illegally. Current production, packaging and transport of charcoal are characterized by low efficiencies and poor handling, leading to losses. Introducing change to this industry requires that it be recognized and legalized, where it is found to be sustainable and not contradictory to environmental protection goals. Once legalized, it would be possible to regulate it and introduce standards addressing fuel quality, packaging and production kiln standards and better enforcement of which tree species should be used to produce charcoal (Kituyi, 2004).

2.3.2.3 Wood pellet logistics and supplies

Wood pellets are one of the most successful bioenergy-based commodities traded internationally. Wood pellets offer several advantages over other solid biomass fuels: they generally have a low moisture content and a relatively high heating value (about 17 GJ/t), which allow long-distance transport by ship without affecting the energy balance (Junginger et al., 2008). Local transport is carried out by trucks, which sets a feasible upper limit for transportation of 50 km for raw biomass (150 km for pellets) and together with the necessary storage usually represents more than 50% of the final cost. Bulk delivery of pellets is very similar to delivery of home heating oil and is carried out by the lorry driver blowing pellets into the storage space, while a suction pump takes away any dust. Storage solutions include underground tanks, container units, silos or storage within the boiler room. Design of more efficient pellet storage, charging and combustion systems for domestic users is ongoing (Peksa-Blanchard et al., 2007). International trade by ships to ports that are properly equipped for handling pellets is a major logistical barrier.[10] Freight costs are another barrier very sensitive to international trade demand. For instance, in 2004, the average price of pellets at a mill in Canada was USD$_{2005}$ 3.4/GJ; shipped to the Netherlands, USD$_{2005}$ 4.1/GJ (Free on Board); and delivered to the Rotterdam harbour, USD$_{2005}$ 7.5/GJ (Junginger et al., 2008; see also Sikkema et al., 2011).

2.3.3 Conversion technologies to electricity, heat, and liquid and gaseous fuels

Commercial bioenergy routes are shown in Figure 2.6 and start with feedstocks such as forest- or agriculture-based crops or industrial, commercial or municipal waste streams and by-products. These routes deliver electricity or heat from biomass directly or as CHP, biogas and liquid biofuels, including ethanol from sugarcane or corn and biodiesel from oilseed crops. Current biomass-based commercial processes produce a limited range of liquid fuels compared to the variety of petroleum-based fuels and products.

Figure 2.2 presented a complex set of developing technological options based on second- (lignocellulosic herbaceous or woody species) and higher- (aquatic plants) generation feedstocks and a variety of second- (or higher-) generation conversion processes.[11] It also included the commercial (Figure 2.6) first-generation (oil, sugar and starch crops) and solid biomass feedstocks and conversion processes (fermentation, transesterification, combustion, gasification, pyrolysis and anaerobic digestion). Second-generation feedstocks and conversion processes can produce higher-efficiency electricity and heat, as well as a wider range of liquid hydrocarbon fuels, alcohols (including some with higher energy density), ethers, chemical products and polymers (biobased materials) in the developing biorefineries that are discussed in more detail in Section 2.6.3.4. Initial R&D on producing hydrocarbon fuels is starting with sugar and starch crops and covers the range of gasoline, diesel and jet fuel with an increasing focus on chemicals. Both improved first-generation crops (e.g., perennial sugarcane-derived) and second-generation plants suited to specific geographic regions have the potential to provide a variety of energy products, along with high-volume chemicals and materials traditionally derived from the petrochemical industry, maximizing the outputs of end products per unit of feedstock.

---

[10] In most countries with export potential, ports are not yet equipped with storage and modern handling equipment or are poorly managed, which implies high shipping costs.

[11] Biofuels produced via new processes are also called advanced or next-genereation biofuels, e.g. from lignocellulosic biomass.

2.3.3.1 Development stages of conversion technologies

The development stages of selected thermochemical, biochemical and chemical routes from solid lignocellulosic biomass, wet waste streams, sugars from sugarcane or starch crops, and vegetable oils are shown in Table 2.5 for the production of heat, power and fuels. For instance, while biomass combustion coupled with electricity generators such as turbines using steam cycles is a commercial system for electricity production (or CHP), coupling with the Stirling engine is still developing, and the Organic Rankine Cycle (ORC) is just starting commercial penetration (van Loo and Koppejan, 2002). Generally, solid wood or waste biomass is processed by thermochemical routes, and wet feedstocks and sugar or starch crops are processed biochemically or chemically and, in the case of the vegetable oils, after a mechanical pressing step (Bauen et al., 2009a). The development stages are roughly divided into R&D, demonstration, early commercial and full commercial products and processes. Precise allocation to these different stages is difficult and somewhat arbitrary, because many developments are taking place in industry and are not often documented in the peer-reviewed literature (Regalbuto, 2009; Bacovsky et al., 2010a,b). Usually, those processes that are deployable throughout the world are fully commercial technologies because their technical risk is small and financing can be obtained (Kirkels and Verbong, 2011).

Synergies between biomass industries and waste management are already established and additional synergies are evolving with the petroleum refining, chemicals, natural gas and coal industries (King et al., 2010; Kirkels and Verbong, 2011). Many bioenergy systems that are moving towards commercialization still have a high technical risk. Section 2.6.3 will describe these additional advancing conversion processes in more detail.

2.3.3.2 Thermochemical processes

Biomass combustion is a process where carbon and hydrogen in the fuel react with excess oxygen to form CO_2 and water and release heat. Direct burning of biomass is popular in rural areas for cooking. Wood and charcoal are also used as a fuel in the industry. Combustion processes are well understood and a wide range of existing commercial technologies are tailored to the characteristics of the biomass and the scale of their applications. Biomass can also be co-combusted with coal in coal-fired plants (van Loo and Koppejan, 2002; Faaij, 2006; Egsgaard et al., 2009).

Pyrolysis is the thermal decomposition of biomass occurring in the absence of oxygen (anaerobic environment) that produces a solid (charcoal), a liquid (pyrolysis oil or bio-oil) and a gas product. The relative amounts of the three co-products depend on the operating temperature and the residence time used in the process. High heating rates of the biomass feedstocks at moderate temperatures (450°C to 550°C) result in oxygenated oils as the major products (70 to 80%), with the remainder split between a biochar and gases. Slow pyrolysis (also known

Table 2.5 | Examples of stages of development of bioenergy: thermochemical (orange), biochemical (blue), and chemical routes (red) for heat, power, and liquid and gaseous fuels from solid lignocellulosic and wet waste biomass streams, sugars from sugarcane or starch crops, and vegetable oils (IEA Bioenergy, 2009; Alper and Stephanopoulos, 2009; Regalbuto, 2009).

Type of Plant	Type of Product	Stage of Development of Process for Product(s) or System(s)			
		Basic and Applied R&D	**Demonstration**	**Early Commercial**	**Commercial**
Low Moisture Lignocellulosic	Densified Biomass	Torrefaction	Hydrothermal Oil (Hy Oil)	Pyrolysis Oil (Py Oil)	Pelletization
	Charcoal	Pyrolysis (Biochar)			Carbonization
	Heat			Small Scale Gasification	Combustion Stoves
		Combustion		Py/Hy Oil	Home/District/Industrial
	Power or CHP	Combustion Coupled with → Stirling Engine		ORC[1]	Steam Cycles
		Co-Combustion or Co-Firing with Coal → Indirect		Parallel	Direct
		Gasification (G) or Integrated Gasification (IG) → IG-Fuel Cell IG-Gas Turbine			
		IG-Combined Cycle		G and Steam Cycle	
Wet Waste	Heat or Power or Fuel	Anaerobic Digestion to Biogas → 2-Stage			Landfills (1-Stage)
				Reforming to Hydrogen (H$_2$)	Small Manure Digesters
		Microbial Fuel Cell		Biogas Upgrading to Methane	
		Hydrothermal Processing to Oils or Gaseous Fuels			
Sugar or Starch Crops	Fuels	Sugar Fermentation	Butanol		Ethanol
		Microbial Processing[2] → H$_2$	Gasoline/Diesel/Jet Fuel	Biobutanol/Butanols[3]	
Oils Vegetable or Waste	Fuels	Extraction and Esterification			Biodiesel
		Extraction and Hydrogenation		Renewable Diesel	
		Extraction and Refining	Jet Fuel		

Notes: 1. ORC: Organic Rankine Cycle; 2. genetically engineered yeasts or bacteria to make, for instance, isobutanol (or hydrocarbons) developed either with tools of synthetic biology or through metabolic engineering. 3. Several four-carbon alcohols are possible and isobutanol is a key chemical building block for gasoline, diesel, kerosene and jet fuel and other products.

as carbonization) is practiced throughout the world, for example, in traditional stoves in developing countries, in barbecues in Western countries, and in the Brazilian steel industry (Bridgwater et al., 2003; Laird et al., 2009).

Biomass Gasification occurs when a partial oxidation of biomass happens upon heating. This produces a combustible gas mixture (called producer gas or fuel gas) rich in CO and hydrogen (H$_2$) that has an energy content of 5 to 20 MJ/Nm^3 (depending on the type of biomass and whether gasification is conducted with air, oxygen or through indirect heating). This energy content is roughly 10 to 45% of the heating value of natural gas. Fuel gas can then be upgraded to a higher-quality gas mixture called biomass synthesis gas or syngas (Faaij, 2006). A gas turbine, a boiler or a steam turbine are options to employ unconverted gas fractions for electricity co-production. Coupled with electricity generators, syngas can be used as a fuel in place of diesel in suitably designed or adapted internal combustion engines. Most commonly available gasifiers use wood or woody biomass and specially designed gasifiers can convert non-woody biomass materials (Yokoyama and Matsumura, 2008). Biomass gasifier stoves are also being used in many rural industries for heating and drying, for instance, in India and China (Yokoyama and Matsumura, 2008; Mukunda et al., 2010). Compared to combustion, gasification is more efficient, providing better controlled heating, higher efficiencies in power production and the possibility for co-producing chemicals and fuels (Kirkels and Verbong, 2011).

2.3.3.3 Chemical processes

Transesterification is the process through which alcohols (often methanol) react in the presence of a catalyst (acid or base) with triglycerides contained in vegetable oils or animal fats to form an alkyl ester of fatty acids and a glycerine by-product. Vegetable oil is extracted from the seeds, usually with mechanical crushing or chemical solvents prior to transesterification. The fatty acid alkyl esters are typically referred to as 'biodiesel' and can be blended with petroleum-based diesel fuel. The protein-rich residue, also known as cake, is typically sold as animal feed or fertilizer, but may also be used to synthesize higher-value chemicals (WWI, 2006; Bauen et al., 2009a; Demirbas, 2009; Balat, 2011).

The **hydrogenation** of vegetable oil, animal fats or recycled oils in the presence of a catalyst yields a renewable diesel fuel—hydrocarbons that can be blended in any proportion with petroleum-based diesel and propane as products. This process involves reacting vegetable oil or animal fats with H_2 (typically sourced from an oil refinery) in the presence of a catalyst (Bauen et al., 2009a). Although at an earlier stage of development and deployment than transesterification, hydrogenation of vegetable oils and animal fats can still be considered a first-generation route as it is demonstrated at a commercial scale.[12] Hydrogenated biofuels have a high cetane number, low sulphur content and high viscosity (Knothe, 2010).

2.3.3.4 Biochemical processes

Biochemical processes use a variety of microorganisms to perform reactions under milder conditions and typically with greater specificity compared to thermochemical processes. These reactions can be part of the organisms' metabolic functions or they can be modified for a specific product through metabolic engineering (Alper and Stephanopoulos, 2009). For instance, *fermentation* is the process by which microorganisms such as yeasts metabolize sugars under low or no oxygen to produce ethanol. Among bacteria, the most commonly employed is *Escherichia (E.) coli*, often used to perform industrial synthesis of biochemical products, including ethanol, lactic acid and others. *Saccharomyces cerevisiae* is the most common yeast used for industrial ethanol production from sugars. The major raw feedstocks for biochemical conversion today are sugarcane, sweet sorghum, sugar beet and starch crops (such as corn, wheat or cassava) and the major commercial product from this process is ethanol, which is predominantly used as a gasoline substitute in light-duty transport.

Anaerobic digestion (AD) involves the breakdown of organic matter in agricultural feedstocks such as animal dung, human excreta, leafy plant materials, urban solid and liquid wastes, or food processing waste streams by a consortium of microorganisms in the absence of oxygen to produce biogas, a mixture of methane (50 to 70%) and CO_2. In this process, the organic fraction of the waste is segregated and fed into a closed container (biogas digester). In the digester, the segregated biomass undergoes biodegradation in the presence of methanogenic bacteria under anaerobic conditions, producing methane-rich biogas and effluent. The biogas can be used either for cooking and heating or for generating motive power or power through dual-fuel or gas engines, low-pressure gas turbines, or steam turbines. The biogas can also be upgraded through enrichment to a higher heat content biomethane (85 to 90% methane) gas and injected in the natural gas grid (Bauen et al., 2009a; Petersson and Wellinger, 2009). The residue from AD, after stabilization, can be used as an organic soil amendment or a fertilizer. The residue can be sold as manure depending upon the composition of the input waste.

Many developing countries, for example India and China, are making use of AD technology extensively in rural areas. Many German and Swedish companies are market leaders in large biogas plant technologies (Faaij, 2006; Petersson and Wellinger, 2009). In Sweden, multiple wastes and manures (co-digestion) are also used and the biogas is upgraded to biomethane, a higher methane content gas, which can be distributed via natural gas pipelines and can also be used directly in vehicles.[13]

2.3.4 Bioenergy systems and chains: Existing state-of-the-art systems

Literature examples of relevant commercial bioenergy systems operating in various countries today by type of energy product(s), feedstock, major process, current and estimated future (2020 to 2030) efficiency, and estimated current and future (2020) production costs are presented in Tables 2.6 and 2.7. Current markets and potential are reviewed in Section 2.4.

Production costs presented in Tables 2.6 and 2.7 are taken directly from the available literature with no attempt to harmonize the literature data because the underlying techno-economic parameters are not always sufficiently transparent to assess the specific conditions under which

---

12 Many companies throughout the world have patents, demonstration plants, and have tested this technology at a commercial scale for diesel, including Neste Oil's commercial facility in Singapore (Bauen et al., 2009a; Bacovsky et al., 2010b).

13 See, for instance, the Linköping example at www.iea-biogas.net/_download/linkoping_final.pdf (IEA Bioenergy Task 37 success story).

comparable production costs can be achieved, except in cases analyzing multiple products. Section 2.7 presents complementary information on the levelized costs of various bioenergy systems and discusses specific cost determinants based on the methods specified in Annex II and the assumptions summarized in Annex II (note that only a few of the underlying assumptions included in Tables 2.6 and 2.7 were used as inputs to the data presented in Annex III).

2.3.4.1 Bioenergy chains for power, combined heat and power, and heat

Liquid biofuels from biomass have higher production costs than solid biomass (at USD_{2005} ~2 to 5/GJ) used for heat and power. Unprocessed solid biomass is less costly than pre-processed types (via densification, e.g., delivered wood pellets at USD_{2005} 10 to 20/GJ), but entails higher logistic costs and is a reason why both types of solid biomass markets developed (Sections 2.3.2.2 and 2.3.2.3). Because of economies of scale, some of the specific technologies that have proven successful at a large scale (such as combustion for electricity generation) cannot be directly applied to small-scale applications in a cost-effective fashion, making it necessary to identify suitable alternative technologies, usually adapting existing technologies used with carbonaceous fuels. This is the case for ORC technologies, which are entering the commercial stage, and Stirling engine technologies, which are still in developmental phase, or moving from combustion to gasification, coupled to an engine (IEA, 2008a).

An intermediate liquid fuel from pyrolysis is part of evolving heating and power in co-firing applications because it is a transportable fuel (see Table 2.6) and is under investigation for stationary power and for upgrading to transport fuel (see Sections 2.3.3.2 and 2.6.3.1). Pyrolysis oils are a commercial source of low-volume specialty chemicals (see Bridgwater et al., 2003, 2007).

Many bioenergy chains employ cogeneration in their systems where the heat generated as a by-product of power generation is used as steam to meet process heating requirements, with an overall efficiency of 60% or even higher (over 90%) in some cases (IEA, 2008a; Williams et al., 2009). Technologies available for high-temperature/high-pressure steam generation using bagasse as a fuel, for example, make it possible for sugar mills to operate at higher levels of energy efficiency and generate more electricity than what they require. Sugarcane bagasse and now increasingly sugarcane field residues from cane mechanical harvesting are used for process heat and power (Maués, 2007; Macedo et al., 2008; Dantas et al., 2009; Seabra et al., 2010) to such an extent that in 2009, 5% of Brazil's electricity was provided by bagasse cogeneration (EPE, 2010). Similarly, black liquor, an organic pulping product containing pulping chemicals, is produced in the paper and pulp industry and is being burnt efficiently in boilers to produce energy that is then used as process heat (Faaij, 2006). Cogeneration-based district heating in Nordic and European countries is also very popular.

A significant number of electricity generation routes are available, including co-combustion (co-firing) with non-biomass fuels, which is a relatively efficient use of solid biomass compared to direct combustion. Due to economies of scale, small-scale plants usually provide heat and electricity at a higher production cost than do larger systems, although that varies somewhat with location. Heat and power systems are available in a variety of sizes and with high efficiency. Biomass gasification currently provides an annual supply of about 1.4 GW_{th} in industrial applications, CHP and co-firing (Kirkels and Verbong, 2011). Small-scale systems ranging from cooking stoves and anaerobic digestion systems to small gasifiers have been improving in efficiency over time. Several European countries are developing digestion systems using a mixture of solid biomass, municipal waste and manures, producing either electricity or high-quality methane. At the smallest scales, the primary use of biomass is for lighting, heating and cooking (see Table 2.6).

Many region-specific factors determine the production costs of bioenergy carriers, including land and labour costs, biomass distribution density, and seasonal variation. Also, other markets and applications partly determine the value of biomass. For many bioenergy systems, biomass supply costs represent a considerable proportion of total production costs. The scale of biofuel conversion technologies, local legislation and environmental standards can also differ considerably from country to country. Even the operation of conversion systems (e.g., load factor) varies, depending on, for example, climatic conditions (e.g., winter district heating) or crop harvesting cycles (e.g., sugarcane harvest cycles and climate impact). The result is a wide range of production costs that varies not only by technology and resource type, but also by numerous regional and local factors (see examples of such ranges in Section 2.7 and Annex III).

2.3.4.2 Bioenergy chains for liquid transport fuels

Bioenergy chains for liquid transportation fuels are similarly diverse and are described below under three subsections: (1) integrated ethanol, power, and sugar from sugarcane; (2) ethanol and fodder products; and (3) biodiesel. Also covered here are 2008 to 2009 biofuels production costs by feedstock and region. Though liquid biofuels are mainly used in the transport sector, in many developing and in some developed countries they are also used to generate electricity or peak power.

Integrated ethanol, power and sugar from sugarcane
Ethanol from sugarcane is primarily made from pressed juices and molasses or from by-products of sugar mills. The fermentation takes place in single-batch, fed-batch or continuous processes, the latter becoming widespread and being more efficient because yeasts can be recycled. The ethanol content in the fermented liquor is 7 to 10% in Brazil (BNDES/CGEE, 2008), and is subsequently distilled to increase purity to about 93%. To be blended with gasoline in most applications,

Table 2.6 | Current and projected estimated production costs and efficiencies of bioenergy chains at various scales in world regions for power, heat, and biomethane from wastes directly taken from available literature data.

Feedstock/ Country/ Region	Major Process	Efficiency, Application and Production Costs; Eff. = bioenergy/biomass energy Component costs in USD$_{2005}$/GJ	Estimated Production Costs USD$_{2005}$/GJ US cents$_{2005}$/kWh	Potential Advances USD$_{2005}$/GJ US cents$_{2005}$/kWh
Wood log, residues, chips/ Ag. Wastes/ Worldwide	Co-combustion with coal	5 to 100 MW$_e$, Eff. ~30 to 40%.[1,2] >50 power plants operated or carried on experimental operation using wood logs/ residues, of which 16 are operational and using coal. More than 20 pulverized coal plants in operation.[3] Wood chips (straw) used in at least 5 (10) operating power plants in co-firing with coal.[3]	8.1 – 15 2.9 – 5.3 Inv. Cost (USD/kW): 100 – 1,300[1]	Reduce fuel cost by improved pretreatment, characterization and measurement methods.[4] Torrefied biomass is a solid uniform product with low moisture and high energy content and more suitable for co-firing in pulverized coal plants.[3] Cost reduction and corrosion-resistant materials for coal plant needed.[5]
Wood log, residues, chips/ Ag. Wastes/ Worldwide	Direct combustion	10 to 100 MW$_e$, Eff. ~20 to 40%.[1,2] Well deployed in Scandinavia and North America; various advanced concepts give high efficiency, low costs and high flexibility.[2] Major variable is biomass supply costs.[2]	20 – 25 7.2 – 9.2 Inv. Cost (USD/kW): 1,600 – 2,500[1]	U.S. 2020 cost projections:[6] 6.3 – 7.8 Stoker fired boilers: 7.5 – 8.1
MSW/ Worldwide	Direct combustion (gasification and co-combustion with coal)	50 to 400 MW$_e$, Eff. ~22%, due to low-temperature steam to avoid corrosion.[7,8] Commercially deployed incineration has higher capital costs and lower (average) efficiency.[2] Four coal-based plants co-fire MSW.[3]	9.1 – 26 3.3 – 9.4[7]	New CHP plant designs using MSW are expected to reach 28 to 30% electrical efficiency, and above 85 to 90% overall efficiency in CHP.[8]
Wood/ Ag. Wastes/ Worldwide	Small scale/gas engine gasification	5 to 10 MW$_e$, Eff. ~15 to 30%.[1,2] First-generation concepts prove capital intensive.[2]	29 – 38 10 – 14 Inv. Cost (USD/kW): 2,500 – 5,600[1]	Increased efficiency of the gasification and performance of the integrated system. Decrease tars and emissions.[1]
Wood pellets/ EU	Direct coal co-firing or co-gasification	12.5 to 300 MW$_e$.[9] Used in 2 operating power plants in co-firing with coal.[3] Costs highly dependent on shipment size and distances.[9]	14 – 36 5.0 – 13[9,10]	See PELLETS@LAS Pellet Handbook and www.pelletsatlas.info.
Pyrolysis oil /EU	Coal co-combustion/ gasification	12.5 to 1,200 MW$_e$.[9] Costs highly dependent on shipment size and distances.[9]	19 – 42 7.0 – 15[9,10]	Develop direct conventional oil refinery integrated and/or upgrading processes allowing for direct use in diesel blends.[1]
Fuelwood/ Mostly in developing countries	Combustion for heat	0.005 to 0.05 MW$_{th}$, Eff. ~10 to 20%.[2] Traditional devices are inefficient and generate indoor pollution. Improved cook stoves are available that reduce fuel use (up to 60%) and cut 70% of indoor pollution. Residential use (cooking) application.[2]	Inv. Cost (USD/kW): 100[2]	New stoves with 35 to 50% efficiency also reduce indoor air pollution more than 90%.[2] See Section 2.5.7.2.
		1 to 5 MW$_{th}$, Eff. ~70 to 90% for modern furnaces.[2] Existing industries have highly polluting low-efficiency kilns.[11]	Inv. Cost (USD/kW): 300 – 800[2]	More widespread use of improved kilns to cut consumption by 50 to 60% and reduce pollution.[11]
Organic Waste/MSW/ Worldwide	Landfill with methane recovery	Eff. ~10 to 15% (electricity).[2] Widely applied for electricity and part of waste treatment policies of many countries.[2]	Biogas: 1.3 – 1.7[12]	Continued efficiency increases are expected.
Organic Waste/MSW/ Manures/ Sweden/ EU in expansion	Anaerobic co-digestion, gas clean up, compression, and distribution	Widely applied for homogeneous wet organic waste streams and waste water.[2] To a lesser extent used for heterogeneous wet wastes such as organic domestic wastes.[2]	Fuel: 2.4 – 6.6[13] Elec.: 48 – 59[1] 17 – 21[1]	Improvements in biomass pretreatment, the biogas cleansing processes, the thermophilic process, and biological digestion (already at R&D stage).[1,17]
		Costs do not include credits for sale of fertilizer by-product.[14]	Fuel: 15 – 16 Inv. Cost (USD/kW): 13,000[14]	In commercial use in Sweden, other EU countries. State of California study shows potential for the augmentation of natural gas distribution.[14]
Manures/ Worldwide	Household digestion	Cooking, heating and electricity applications. By-product liquid fertilizer credit possible.	1 to 2 years payback time	Large reductions in costs by using geomembranes. Improved designs and reduction in digestion times.[15]

Continued next Page →

Feedstock/ Country/ Region	Major Process	Efficiency, Application and Production Costs; Eff. = bioenergy/biomass energy Component costs in USD$_{2005}$/GJ	Estimated Production Costs USD$_{2005}$/GJ US cents$_{2005}$/kWh	Potential Advances USD$_{2005}$/GJ US cents$_{2005}$/kWh
Manures/Finland	Farms	Biogas from farms 0.018 to 0.050 MW$_e$.[16]	Elec.: 77 – 110 Inv. Cost (USD/kW): 14000 – 23000[16]	Improved designs and reduction in digestion times. Improvements in the understanding of anaerobic digestion, metagenomics of complex consortia of microorganisms.[12]
Manures/Food residues	Farms/Food Industry	Biogas from farm animal residues and food processing residues at 0.15 to 0.29 MW$_e$.[16]	Elec.: 70 – 89 Inv. Cost (USD/kW): 12000 – 15000[16]	

Abbreviations: Inv. = Investment; Elec. = Electricity. References: 1. Bauen et al. (2009a); 2. IEA Bioenergy (2007); 3. Cremers (2009) (see IEA co-firing database at www.ieabcc.nl/database/cofiring.php); 4. Econ Poyry (2008); 5. Egsgaard et al. (2009); 6. NRC (2009b); 7. Koukouzas et al. (2008); 8. IEA (2008a); 9. Hamelinck (2004); 10. Uslu et al. (2008); 11. REN21 (2007); 12. Cirne et al. (2007); 13. Sustainable Transport Solutions (2006); 14. Krich et al. (2005); 15. Müller, (2007); 16. Kuuva and Ruska (2009); 17. Petersson and Wellinger, 2009.

ethanol should be anhydrous and the mixture has to be further dehydrated to reach a grade of 99.8 to 99.9% (WWI, 2006).

Ethanol and fodder products

The dominant dry mill (or dry grind) process (88% of US production) for ethanol fuel manufactured from corn starts with hammer milling the whole grain into a coarse flour, which is cooked into a slurry, then hydrolyzed with alpha amylase enzymes to form dextrins, next hydrolyzed by gluco-amylases to form glucose that is finally fermented by yeasts (the last two processes can be combined). The byproduct is distillers' grains with solubles, an animal feed (McAloon et al., 2000; Rendleman and Shapouri, 2007) that can be sold wet to feedlots near the biorefinery or be dried for stabilization and sold. The most common source of process heat is natural gas. From the early 1980s to 2005, the energy intensity of average dry mill plants in North America has been reduced by 14% for every cumulative doubling of production (learning rate, see Table 2.17; Hettinga et al., 2007, 2009). Since then, 10 cumulative doublings (see also Section 2.7.2) have occurred and the industry continues to improve its energy performance with, for instance, CHP ((S&T)^2 Consultants, 2009). The impacts of this and other process improvements have been estimated to continue such that, by 2022, the projected production cost is USD$_{2005}$ 16/GJ, reduced from USD$_{2005}$ 17.5/GJ in 2009 (EPA, 2010). Table 2.7 presents examples of process improvements from membrane separation for ethanol to enzymes operating at lower temperature, etc. A similar process to corn dry milling is wheat-to-ethanol processing, starting with a malting step, and either enzyme or acid hydrolysis leading to sugars for fermentation.

Biodiesel

Biodiesel is produced from oil seed crops like rapeseed or soybeans, or from trees such as oil seed palms. It is also produced from a variety of greases and wastes from cooking oils or animal fats. This wide range of feedstocks, from low-cost wastes to more expensive vegetable oils, produces biodiesel fuels with more variable properties that follow those of the starting oil seed plant. Fuel standards' harmonization is still under development as are a variety of non-edible oil seed plants (Knothe, 2010; Balat, 2011). Examples of producing regions are shown in Figure 2.7.

Snapshot of 2008 to 2009 biofuels costs from multiple feedstocks and world regions

A snapshot of ranges of biofuels production costs for 2008 to 2009 (primarily 2009) is shown in Figure 2.7 for various world regions based on a variety of feedstocks including wastes and processing streams from the manufacture of sugar (molasses). The snapshot is based on various literature sources such as the recent comparison of costs for Asian Pacific Economic Countries (Milbrandt and Overend, 2008, updated),[14] and data from Table 2.7.[15] For production volumes of these countries see Figure 2.9. For ethanol production, feedstock costs represent about 60 to 80% of the total production cost while, for biodiesel from oil seeds, the proportion is higher (80 to 90%) (data from 2008 to 2009). Latin and Central American sugarcane ethanol is found to have had the lowest production costs over this period, followed by Asian, Pacific and North American starch crops, then by European Union (EU) sugar beet and finally EU grains. Molasses production costs are lower in India and Pacific countries than in Other Asia countries. For biodiesel production, Latin America has the lowest costs, followed by Other Asia countries palm oil, Other Asia rapeseed and soybean, and then North American soybean and EU rapeseed. Biodiesel production costs are generally somewhat higher than for ethanol, but can reach those of ethanol for countries with higher-productivity plants or a lower cost base such as Indonesia/Malaysia and Argentina.

There is significant room for feedstock improvement, mainly its productivity (see also Section 2.6.1), and also for its conversion to products based on the projected increases in efficiency shown in Table 2.7. In an analysis of US biofuel production, the US Environmental Protection Agency (EPA) projected costs based on the Forest and Agricultural Sector Optimization Model (FASOM) and found significant room for improvement (see

---

14 The study addressed biofuels production, feedstock availability, economics, refuelling infrastructure, use of alternative fuel vehicles, trade, and policies.

15 The ranges of production costs shown here include a variety of waste streams and feedstocks with a broader geographic distribution than those summarized in Section 2.7 and detailed in Annex III. Data in Annex III cover broad ranges of a few feedstocks varying their costs, investment capital, co-products, and financial assumptions. From these transparent techno-economic data, it is possible for the reader to change assumptions and recalculate approximate production costs in specific regions.

Table 2.7 | Current and projected estimated production costs and efficiencies of commercial biofuels in various countries directly taken from available literature data. Also provided is the range of direct reductions of GHG emissions from these routes compared to the fossil fuel replaced (see Section 2.5 for detailed GHG emissions discussion). Parts A and B address ethanol and biodiesel fuels, respectively.

A: Ethanol

Feedstock/ Process	Country/ Region	Efficiency, Application and Production Costs; Eff. = bioenergy/ biomass energy Component costs in USD$_{2005}$/GJ	Estimated Production Costs USD$_{2005}$/GJ	Direct GHG Reduction (%) from Fossil Reference (FR)	Potential Advances in Cost Reductions and Efficiency USD$_{2005}$/GJ
Sugarcane pressed, juice fermented to ethanol, bagasse to process heat and power, and increasingly sale of electricity.	Brazil	Eff. ~38%,[1] ~41% (ethanol only);[2] 170 million l/yr, FC: 11.1; CC*: 3.7 w/o CR.[2]	14.8 w/o CR.[2]	79 to 86% (w/o and w/ CPC); FR: gasoline.[4]	9 – 10.[1] Eff. ~50%.[5] Mechanized harvest and efficient use of sugarcane straw and leaves.[6] Biorefineries with multiple products.[5] Improved yeasts.
	Australia	Eff. ~38%, ~41% (ethanol only), FC: 24.8; CC*: 7 w/o CR.[3]	31.8 w/o CR.[3]		
Corn grain dry milling process for ethanol, fodder (DGS) for animal feed		Eff. ~62%;[2,8] 89% of production.[5] 30% co-product feed DGS sold wet.[5,8] 250 million l/yr plant, FC: 14.1[2] – 29.4[11]; CC*: 6 and CR: 3.8 – 4.4.[2]	20–21 w/ CR[2,15,19] 17.5[5] 31 w/ CR.[11]	35 to 56% for various CPC methods; FR: gasoline 35% (system expansion); Process Heat: NG.[12,13]	Eff. ~64%.[11] Industry Eff. ~65 to 68%. Estimated production cost:16.[5,8] US projected low temperature starch enzyme hydrolysis/fermentation, corn dry fractionation, biodiesel from oil in 90% of mills, membrane ethanol separation, and CHP.[5]
	France	170 million l/yr, FC: 29.3; CC*: 10.5 and CR: 5.[11]	34.8 w/ CR.[11]	60%[9,14]	
Wheat similar to corn to ethanol, fodder (DGS)	EU (UK)	Eff. ~53 to 59%.[11,16] 250 million l/yr plant, FC: 36.2; CC*: 10.5 and CR: 6.[11]	40.7 w/ CR.[11]	40%, DGS to energy.[17] 2 to 80% w/ DGS to energy -8 to 70% w/ DGS to feed.[18]	2020 Eff. ~64%.[11]
	Australia (from waste)	30 million l/yr plant, FC: 14.4; CC*: 8.6 and CR: 0.2.[3]	22.8 w/ CR.[3]	55% wheat starch NG, 27% wheat-coal, 59% wheat w/ straw firing.[3]	
Sugar beet crushing, ferment sugar to ethanol and residue	EU (UK)	Eff. ~12%.[1,16,19] 250 million l/yr plant, FC: 21.6; CC*: 11 and CR: 8.2.[11]	24.4 w/ CR.[11]	28 to 66%, alternate co-product use.[17,18]	2020 Eff. ~15%.[1]
Cassava mashing, cooking, fermentation to ethanol	Thailand/ China	Thailand's process with 38 million l, and feed productivity 20 to 21 t/ha.[16,20,21] China ethanol plant operating at partial capacity.[22]	Thailand: 26[23]	Thailand: 45%.[24] China: 20% with anaerobic digestion energy.[25]	
Molasses by-product of sugar production	Thailand/ Australia	About 3% of molasses could be used for ethanol in Thailand. FC: 10.9 and 10; CC*: 10.1 and CR: 5.7.[23]	Thailand: 21[23] Australia: 16[3]	27 to 59% depending on co-product credit method (Australia).[26,27]	

Continued next Page →

Table 2.7; EPA, 2010). The IEA has similarly estimated cost reductions for Organisation for Economic Co-operation and Development (OECD) countries' rapeseed biodiesel by 2030 (IEA Bioenergy, 2007). Further discussions of historical and future cost expectations are provided in Section 2.7.

2.3.5 Synthesis

The key currently commercial technologies are heat production (ranging from home cooking to district heating), power generation from biomass via combustion, CHP, co-firing of biomass and fossil fuels, and first-generation liquid biofuels from oil crops (biodiesel) and sugar and starch crops (ethanol). Several bioenergy systems have been deployed competitively, most notably sugarcane ethanol and heat and power generation from wastes and residues. Other biofuels have also undergone cost and environmental impact reductions and reached significant scales but still require government subsidies.

Modern bioenergy systems involve a wide range of feedstock types, residues from agriculture and forestry, various streams of organic waste, and dedicated crops or perennial systems. Existing bioenergy systems rely mostly on wood, residues and waste for heat and power production, and agricultural crops for liquid biofuels. The economics and yields of feedstocks vary widely across world regions and feedstock types. Energy yields per unit area range from 16 to 200 GJ/ha (1.6 to 20 TJ/km^2) for

B: Biodiesel

Feedstock/ Process	Country	Efficiency, Application and Production Costs; Eff. = bioenergy/biomass energy Component costs in USD$_{2005}$/GJ	Estimated Production Costs USD$_{2005}$/GJ	Direct GHG Reduction (%) from Fossil Reference (FR)	Potential Advances in Cost Reductions and Efficiency USD$_{2005}$/GJ
Rape seed	Germany	Eff. ~29%; for the total system it is assumed that surpluses of straw are used for power production.[27]	31 – 50.[1]	31 to 70%, alternate co-product use.[9,17,28]	25 – 37 for OECD.[1] New methods using bio-catalysts; Supercritical alcohol processing. Heterogeneous catalysts or bio-catalysts. New uses for glycerine. Improved feedstock productivity.[30]
	France	55 GJ/ha/yr (EU), 220 million l/yr plant, FC: 40.5; CC*: 2.7 and CR: 1.7.[11]	41.5 w/ CR.[11]		
	UK	220 million l/yr plant, FC: 35.6; CC*: 4.2 and CR: 11.3.[11]	28.5 w/ CR.[11]		
Oil palm	Indonesia Malaysia Asian countries[20]	163 GJ/ha/yr. 220 million l/yr plant, FC: 25.1; CC*: 2.7 and CR: 1.7.[11]	26.1 w/ CR.[11]	35 to 66%, alternate co-product use.[31] (tropical fallow land, residue to power, good management).[28]	
Vegetable oils	109 countries	Costs neglect some countries with high production costs. FC: 0.6 – 21; CC*: 2.3 – 3.7 and CR: 0 – 6.2.[3,11,29]	4.2 – 17.9.[3,11,31]	N/A	US projected 2020 waste oil ester cost 14.[5] About 50 billion l projected from 119 countries.[29]

Abbreviations: *Conversion costs (CC) include investment costs and operating expenses; CR = Co-product Revenue; CPC = coproduct credit; FC = feedstock cost; FR = fossil reference; N/A = not available.

References: 1. IEA Bioenergy (2007a); 2. Tao and Aden (2009); 3. Beer and Grant 2007; 4. Macedo et al. (2008); 5. EPA (2010); 6. Seabra et al. (2010); 7. UK DfT (2003); 8. Rendleman and Shapouri (2007); 9. Bessou et al. (2010); 10. Wang et al. (2011); 11. Bauen et al. (2009a); 12. Wang et al. (2010); 13. Plevin (2009); 14. Ecobilan (2002); 15. Bain (2007); 16. Fulton et al. (2004); 17. Edwards et al. (2008); 18. Edwards et al. (2007); 19. Hamelinck (2004); 20. Koizumi and Ohga (2008); 21. Milbrandt and Overend (2008); 22. GAIN (2009a); for China); 23. GAIN (2009c; for Thailand); 24. Nguyen and Gheewala et al. (2008); 25. Leng et al. (2008); 26. Beer et al. (2001); 27. Beer et al. (2000); 28. Reinhardt et al. (2006); 29. Johnston and Holloway (2007); 30. Bhojvaid (2007); 31. Wicke et al. (2008).

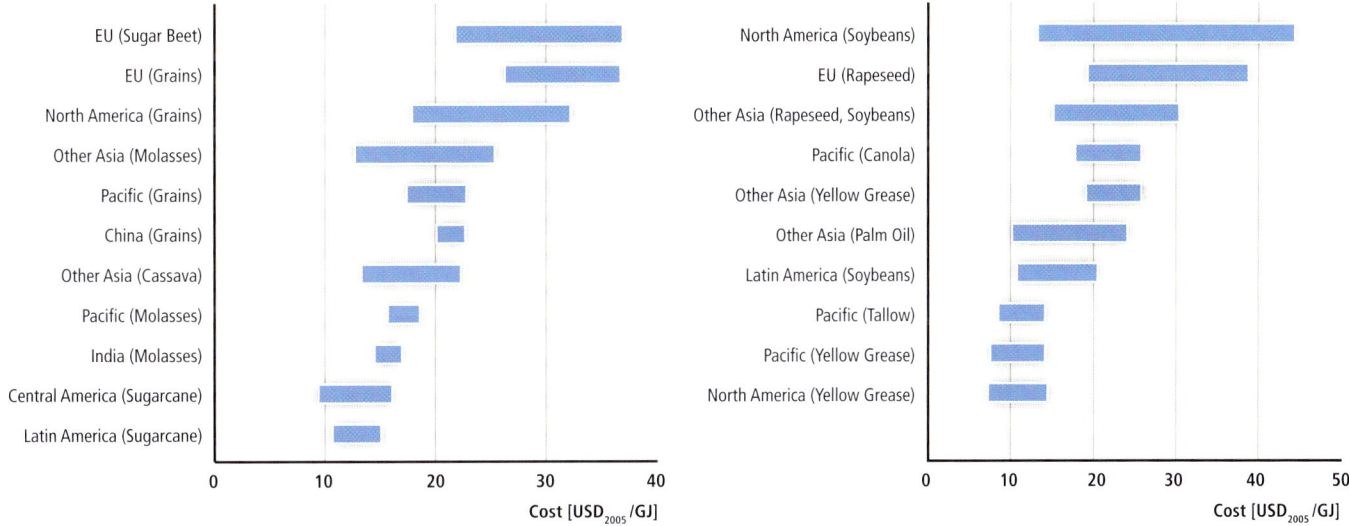

Figure 2.7 | Snapshots of regional ranges of current (2008-2009) estimated production costs for ethanol and biodiesel from various biomass feedstocks and wastes based on Milbrandt and Overend (2008) and Table 2.7.

Notes: The upper value of the range of soybean diesel in North America is due to the single point estimate of Bauen et al. (2009a). Other estimates are in the USD$_{2005}$ 12 to 32/GJ range.

biofuel feedstocks, from 80 to 415 GJ/ha (8 to 41.5 TJ/km²) for lignocellulosic feedstocks, and from 2 to 155 GJ/ha (0.2 to 15.5 TJ/km²) for residues, while costs range from USD$_{2005}$ 0.9 to 16/GJ/ha (USD$_{2005}$ 0.09 to 1.6/TJ/km²). Feedstock production competes with the forestry and food sectors, but the design of integrated production systems such as agro-forestry or mixed cropping may provide synergies along with additional environmental services.

Handling and transport of biomass from production sites to conversion plants may contribute 20 to 50% of the total costs of bioenergy production. Factors such as scale increases, technological innovation and increased competition have contributed to decrease the economic and energy costs of supply chains by more than 50%. Densification via pelletization or briquetting is required for transport distances over 50 km. International costs of delivering densified feedstocks are sensitive to trade and are in the USD$_{2005}$ 10 to 20/GJ range for pellet fuels, and competitive with other market fuels in several regions, thus explaining why such markets are increasing. Charcoal made from biomass is a major fuel in developing countries, and should benefit from the adoption of higher-efficiency kilns and densification technologies.

A significant number of electricity generation routes are available and co-combustion (co-firing) is a relatively efficient way to use solid biomass compared to direct combustion. Small-scale plants usually provide heat and electricity at a higher production cost than larger systems, although this varies somewhat with location. Heat and power systems are available in a variety of sizes and efficiencies. Biomass gasification currently provides about 1.4 GW$_{th}$ of industrial applications, CHP and co-firing. Small-scale systems ranging from cooking stoves and anaerobic digestion systems to small gasifiers have been improving in efficiency over time. Several European countries are developing digestion systems using a mixture of solid biomass, municipal waste and manures, producing either electricity or high-quality methane from upgrading. Many applications, including transport systems, are developing and have the potential to further increase their effectiveness. Technologies at small scales, primarily stoves for heating, continue to improve but diffusion is slow.

Sugarcane-, sugar beet-, and cereal grain-derived ethanol production reached a high level of energy efficiency in major producing countries such as Brazil, the USA, and the EU. The ethanol industry in Center South Brazil significantly increased its cogeneration efficiency and supplied 5% of the country's electricity in 2009. Development of ethanol from waste streams from sugar processing is occurring in India, Pacific and other Asian countries that produce relatively low-cost ethanol but with limited production volumes. Biodiesel production from waste fats and greases has a lower feedstock cost than from rapeseed and soybean but waste fat and grease volumes are limited.

Biofuel production economics is of key importance for future expansion of the biofuels industry. The future development of sustainable biofuels also depends on a balanced scorecard that includes economic, environmental, and social metrics (see Section 2.5). Resolution of technical, economic, social, environmental and regulatory issues remains critical to further development of biofuels. The development of a global market and industry is described in the next section.

2.4 Global and regional status of market and industry development

2.4.1 Current bioenergy production and outlook[16]

Biomass provides about 10% (50.3 EJ in 2008) of the annual global primary energy supply. As presented in Table 2.1, about 60% (IEA accounted) to 70% (including unaccounted informal sector) of this biomass is used in rural areas and relates to charcoal, wood, agricultural residues and manure used for cooking, lighting and space heating, generally by the poorer part of the population in developing countries. Modern bioenergy use (for power generation and CHP, heat or transport fuels) accounted for a primary biomass supply of 11.3 EJ (IEA, 2010a,b; see Table 2.1) in 2008, up from 9.6 EJ[17] in 2004 (IPCC, 2007d), and a rough estimate of 8 EJ in 2000 (IEA Bioenergy, 2007).

The use of solid biomass for energy increased at an average annual growth rate of 1.5%, but secondary energy carriers from modern biomass such as liquid and gaseous fuels increased at 12.1 and 15.4% average annual growth rates, respectively, from 1990 to 2008 (IEA, 2010a). As a result, biofuels' share of global road transport fuel use was 2% in 2008. In 2009, the production of ethanol and biodiesel increased by 10 and 9%, respectively, to 90 billion litres; biofuels provided nearly 3% of global road transport fuel use in 2009, as oil demand decreased for the first time since 1980 (IEA, 2010b). Government policies in various countries led to a five-fold increase in global biofuels production from 2000 to 2008. Biomass and renewable waste power generation was 259 TWh (0.93 EJ) in 2007 and 267 TWh (0.96 EJ) in 2008, representing 1% of the world's electricity, which doubled since 1990 (from 131 TWh or 0.47 EJ). Industrial biomass heating accounts for 8 EJ while space and water heating for building applications account for 3.4 EJ (IEA, 2010b; see Table 2.1).

Most of the increase in the use of biofuels in 2007 and 2008 occurred in the OECD, mainly in North America and Europe. Excess capacity was installed in expectation of increased demand with mandates and subsidies in many countries; however, feedstock and oil price increases and the worsening overall economic conditions during and after the credit crunch made many of these facilities unprofitable. As a result, some are underutilized, more so in biodiesel than in ethanol production. Some plants are not in operation and some businesses failed. Asia Pacific and Latin American markets are growing, primarily

---

16 This sub-section is largely based on the WEO 2009 (IEA, 2009b) and 2010 (IEA, 2010b) and the Global Biofuels Center assessments, web-based biofuels news, reports, trade, and market information (Hart Energy Publishing, LP, www.globalbiofuelscenter.com/).

17 The 9.6 EJ is an estimated equivalent primary biomass energy deducting the non-biogenic MSW that was included in the AR4 study (IPCC, 2007d), or about 0.4 EJ of plastics (estimated based on subsequent IEA 2005 data).

in developing countries due to economic development. Despite this anticipated short-term downturn, world use of biofuels for road transport is projected to recover in the next few years (IEA, 2010b).

The WEO (IEA, 2010b) projections for 2020 to 2035 are summarized in Table 2.8 (in terms of global TPES from biomass); Table 2.9 (in terms of global biofuel demand, i.e., secondary energy); and Table 2.10 (in terms of global electricity generation)—all of them comparing a baseline case (Current Policies) and a mitigation scenario reaching an atmospheric CO_2 concentration of 450 ppm by 2100.

The overall TPES from biomass in the 450 ppm CO_2 stabilization scenario increases to 83 (95) EJ/yr in 2030 (2035) adding 14 (12) EJ to the Reference (Current Policies) scenario (see Table 2.8).

and many of the technologies needed are at the demonstration to early commercialization stages of development in 2011 (see Tables 2.5 and 2.15; IEA Renewable Energy Division, 2010).

Global biomass and renewable waste electricity generation is also projected to increase in both scenarios, reaching 5.6% of global electricity generation by 2035 in the 450-ppm scenario as shown in Table 2.10. The climate change driver nearly doubles the anticipated penetration levels of biopower compared to the projected levels owing to continuation of current policies.

In the WEO (IEA, 2010b), biomass industrial heating applications for process steam and space and hot water heating for buildings would each double in absolute terms from 2008 levels by 2035, offsetting

Table 2.8 | IEA WEO scenarios: global TPES from biomass projections (EJ/yr) for 2020 to 2035 (IEA, 2010b).

Year	2007	2008	2020		2030		2035	
Scenario	Actual	Actual	Baseline	450 ppm	Baseline	450 ppm	Baseline	450 ppm
EJ/yr	48	50	60	63	66	83	70	95
Delta, EJ		2		3		17		25

Table 2.9 | IEA WEO scenarios: global biofuels demand projections (EJ/yr) for 2020 to 2035 reported in secondary energy terms of the delivered product according to IEA data (IEA, 2010b).

Year	2008	2009	2020		2030		2035	
Scenario	Actual	Actual	Baseline	450 ppm	Baseline	450 ppm	Baseline	450 ppm
EJ/yr	1.9	2.1	4.5	5.1	5.9	11.8	6.8	16.2
% Global road transport	2	3	4.4	7	4.4	11 (and air)	5	14 (and air)
% Advanced biofuels			Deployment			60		66

Table 2.10 | IEA WEO scenarios: primary biomass and renewable waste electricity generation projections for 2030 (IEA, 2009, 2010b) and 2035 (IEA, 2010b).

Year	2008	2030		2035	
Scenario	Actual	Baseline, Reference case	450 ppm Scenario	Current Policies	450 ppm Scenario
TWh/yr (EJ/yr)	267 (0.96)	825 (3.0)	1380 (5.0)	1052 (3.8)	1890 (6.8)
% Global electricity	0.96	2.4	4.5	2.7	5.6
TWh/yr (EJ/yr)		840 (3.0)	1450 (5.2)		
% Global electricity		2.4	4.8		

The use of liquid and gaseous energy carriers from modern biomass is growing, in particular biofuels, with a 37% increase from 2006 to 2009 (IEA, 2010c). Regions that currently have strong policy support for biofuels are projected to take the largest share of the eight-fold increase in the market for biofuels that occurs from 2008 to 2035. This is led by the USA (where one-third of the increase occurs), followed by Brazil, the EU and China. To highlight the scale, 7 EJ of advanced biofuels (second generation) is greater than, for example, India's 2007 oil consumption,

some of the expected decrease in the major component of the heating category, traditional biomass, as the total heating demand is projected to decrease in 2035. Industrial and building heating is seen as an area for continued biomass growth. In fact, biomass is very efficiently used in CHP plants, supplying a district heating network. Biomass combustion to produce electricity and heat in CHP plants is an efficient and mature technology and is already competitive with fossil fuels in certain locations (IEA, 2008a).

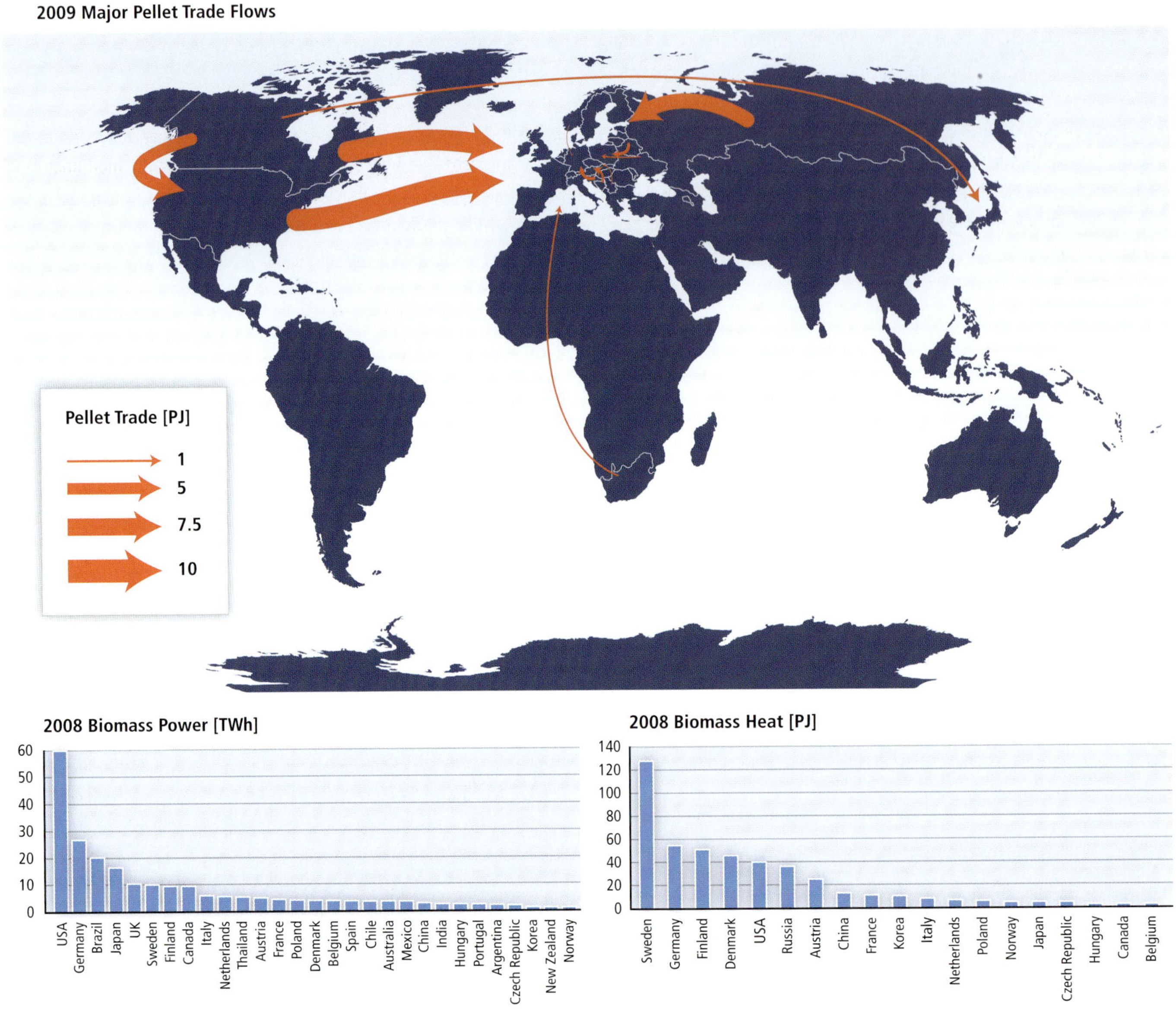

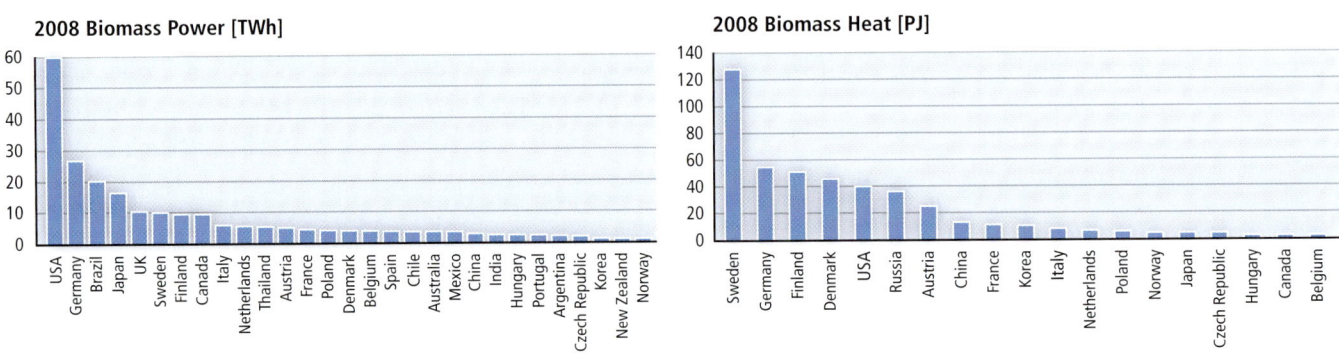

Figure 2.8 | Examples of biomass electricity generation and heating for select countries in 2008 and of the 2009 global trade in wood pellets. Sources: bar chart data from IEA (2010c); trade flow data reproduced from Sikkema et al. (2011) with permission from the Society of Chemical Industry and John Wiley & Sons, Ltd.

The use of solid biomass for electricity production is important, especially from pulp and paper plants and sugar mills. Bioenergy's share of total energy consumption is increasing in the G8 countries (e.g., co-combustion for electricity generation, building heating with pellets), especially in Germany, Italy and the UK (IEA, 2009b). The electricity generation and biomass heating are shown in Figure 2.8. Worldwide biomass heating statistics are uncertain (Sims, 2007) for developed countries. In Europe, biomass heating applications in the building sector are cost competitive and are shown in Figure 2.8. For developing countries, the statistics are less developed, as tools to collect data from informal sectors are lacking (see Table 2.1).

2.4.2 Traditional biomass, improved technologies and practices, and barriers

Biomass is an important traditional fuel in developing countries, where on average it accounts for 22% of the energy mix;[18] in the poorest countries it accounts for more than 80% (see IEA, 2010c). Traditional sources of biomass include mostly wood fuels but also agriculture residues and dung, and they contribute essentially to domestic heating and cooking. The number of people dependent on biomass for cooking is estimated at

---

18 Average contribution to the energy mix from renewable and waste combustibles was 48, 20, 24, 27, and 10% for Africa, Latin America, India, Non-OECD Asia, and China, respectively, while only 4% for the OECD countries in 2008 (IEA, 2010c).

2.7 billion (for 2008) and is projected to increase to 2.8 billion by 2030 (IEA, 2010b). Many thousand biomass-based small industries—such as brick making, food, charcoal, bakeries and others—provide employment and income to people. Most of these technologies are resource intensive, highly polluting and exhibit low efficiencies (see Tables 2.1 and 2.6; FAO, 2010b). However, there is currently a significant and growing market for improved technologies. Also, several programmes at the global, national and local levels are in place to disseminate more efficient technology options.

2.4.2.1 Improved biomass cook stoves

Most developing countries have initiated some type of improved cook stove (ICS) programme since the 1980s. The World Bank Energy Sector Management Assistance Program (World Bank, 2010) reviewed in depth the international experience on improved stoves and summarized significant lessons learned for developing countries and, in particular, for Bangladesh, the objective of the study. For Eastern African countries, see Karekezi and Turyareeba (1995). Many programmes are in operation, sponsored by development agencies, governments, nongovernmental organizations (NGOs) and the private sector. By the end of 2009, 173 million energy saving stoves were in use in China. Other countries were not very successful in disseminating ICS. Over the past 10 years, a whole new generation of advanced biomass stoves and dissemination approaches have been developed, and the field is now bursting with innovations (World Bank, 2010).

A variety of technologies are used, including direct combustion, small-scale gasification, small-scale anaerobic digestion, direct use of a liquid fuel (ethanol) or combinations of technologies.[19] As a result, combustion efficiency has been greatly improved relative to the alternative open fires. The cost ranges from less than USD 10 for the simpler models to more than USD 100 or more for more sophisticated models and USD 100 to 300 for institutional stoves (e.g., schools, hospitals, and barracks) according to 2007 to 2009 cost range data. Fuel savings are 30 to 60%, measured in field conditions, to more than 90%, measured in pilot testing of the most advanced models (Berrueta et al., 2008; World Bank, 2010). There are also significant reductions in GHG emissions and indoor air pollutants (Section 2.5.4).

By 2008 an estimated 820 million people (around 30% of the 2.7 billion that rely on traditional biomass for cooking, see Section 1.4.1.2) in the world were using some type of improved cook stove for cooking (Legros et al., 2009), and more than 160 stove programmes are in place worldwide, with recently launched large-scale national programmes in India, Mexico and Peru, as well as large donor-based programmes in Africa. The UN Foundation-led Global Alliance for Clean Cookstoves started in 2010 to promote the dissemination and adoption of 100 million advanced cook stoves by 2020.[20]

Two main lines of technology development have been followed. Mass-scale approaches—some of which use state-of-the-art manufacturing facilities—rely on centralized production of stoves or critical components, with distribution channels that can even include different countries. As a result, there are companies that produce more than 100,000 stoves per year (Bairiganjan et al., 2010). A second approach relies more on strengthening regional capabilities, giving more emphasis to local employment creation; sometimes the stoves are built onsite rather than sold on markets, such as the Patsari Stove in Mexico and Groupe Energies Renouvelables, Environnement et Solidarités (GERES) in Cambodia (Bairiganjan et al., 2010). Improved stove designs to appeal to consumers, market segmentation and microfinance mechanisms have also been developed (Hilman et al., 2007).

Incentives and barriers

Cookstove programmes have been successful in countries where proper assessment was made of the local needs in terms of technology, cooking devices, user needs and institutional setting. Financial incentives have helped with the dissemination, while an enabling institutional environment by governments—such as in China—has also helped promote new technologies. Finally, accurate monitoring and evaluation has been critical for successful stove adoption and use (Bairiganjan et al., 2010; Venkataraman et al., 2010). Other drivers for increased adoption of ICS have included: (1) cooking environments where users feel smoke is a health problem and annoyance; (2) a short consumer payback (few months); (3) donor or government support extended over at least five years; and (4) financial support to build local institutions and develop local expertise. Government assistance has been more effective in technical advice and quality control. Carbon offset projects are increasingly providing new financing for these activities, either through the Voluntary Market (Gold Standard) or, increasingly, through the CDM. Successful programmes with low-cost but efficient ICS report that local poor residents purchased cookstoves without support of programmes because of fuel savings (World Bank, 2010).

Several barriers need to be overcome for a rapid diffusion of ICS. There are needs for (1) substantial increases in R&D;[21] (2) more field testing and stove customization for users' needs; and (3) strict product specifications and testing and certification programmes. Finally, it is important to better understand the patterns of stove adoption given the multiple devices and fuels as well as mechanisms to foster their long-term use.

---

19 These ICS technologies include improvements in the combustion chamber (such as the Rocket 'elbow'), insulation materials, heat transfer ratios, stove geometry and air flow (Still et al., 2003). The most reliable of these use small electric blowers to stabilize the combustion, but there are also designs using natural air flow (World Bank, 2010).

20 See www.cleancookstoves.org.

21 Particularly for new insulating materials as well as robust designs that endure several years of rough use, and small-scale gasification.

2.4.2.2 Biogas systems

Convenient cooking and lighting are also provided by biogas production using household-scale biodigesters.[22] Biodigesters have the distinct co-benefits of enhancing the fertilizer value of the dung in addition to reducing the pathogen risks of human waste. Early stage results have been mixed because of quality control and management problems, which have resulted in a large number of failures. Smaller-scale biogas experience in Africa has been often disappointing at the household level as the capital cost, maintenance and management support required have been higher than expected. The experience gained, new technology developments (such as the use of geo-membranes), better understanding of the resources available to users, such as dung, and better market segmentation are improving the success of new programmes (Kishore et al., 2004).[23]

Incentives and barriers

Key factors for project success include a proper understanding of users' needs and resources.[24] For example, the role of NGOs, networks and associations in transfer, capacity building, extension and adoption of biogas plants in rural India was found to be very important (Myles, 2001). Financial mechanisms, including microfinance schemes and carbon offset projects under the CDM, are also important in the implementation of household biogas programmes. Barriers to increased biogas adoption include lack of proper technical standards; insufficient financial mechanisms to achieve desired profits relative to the digesters' investment, installation and equipment costs; and relatively high costs of technologies and of labour (e.g., geological investigations into proper site installations). Other related barriers include poor reliability and performance of the designs and construction, and limited application of knowledge gained from the operation of existing plants to the design of new plants.

Many other small-scale bioenergy applications are emerging, including systems aimed at transport and productive uses of energy and electricity. The market penetration is still limited, but many of these systems show important benefits in terms of livelihood, new income, revenues and efficiency (Practical Action Consulting, 2009).

2.4.3 Modern biomass: Large-scale systems, improved technologies and practices, and barriers

The deployment of large-scale bioenergy systems faces a wide range of barriers. Economic barriers appear most prominent for currently commercial technologies constrained by feedstock availability and by meeting sustainability requirements (Fagernäs et al., 2006; Mayfield et al., 2007), while technical barriers predominate for developing technologies such as second-generation biofuels (Cheng and Timilsina, 2010). Non-technical barriers are related to deployment policies (fiscal incentives, regulations and public finance), market creation, supply chain, infrastructure development, community engagement, collaboration and education (Mayfield et al., 2007; Adams et al., 2011). No single barrier appears to be most critical, but the interactions among different individual barriers seem to impede rapid bioenergy expansion. The relative importance of the barriers hinges on the particular value chain and context considered. In particular, national regulations, such as price-driven FITs for bioelectricity and quantity-driven blending level mandates for biofuels, play a major role in the emergence of large-scale projects, alongside public finance through government loans or guarantee programmes (Table 2.11; Section 11.5.3; Chum and Overend, 2003; Fagernäs et al., 2006). The priorities also depend on the stakeholder groups involved in the value chain and differ from feedstock producers to fuel producers and through to end users (Adams et al., 2011). Scale also matters, because barriers perceived by national governments differ from those perceived by stakeholders and communities in the vicinity of bioenergy projects.[25]

Technical and non-technical barriers may be overcome by appropriate policy frameworks, economic instruments such as government support tied to private investment support for first-of-a kind commercial plants to decrease investment risk,[26] sustained RD&D efforts, and catalysis of coordinated multiple private sector activities[27] (IATA, 2009; Regalbuto, 2009; Sims et al., 2010). In 2009, global public RD&D efforts were USD 0.6 billion and 0.2 billion for biofuels and biomass to energy, respectively, and biofuels public funding increased by 88% from 2008. Corporate RD&D efforts were USD 0.2 billion each for the two areas (UNEP/SEFI/Bloomberg, 2010). Venture capital and private equity investing was

---

22 By the end of 2009, there were 35 million household biodigesters in China and in India (Gerber, 2008; REN21, 2009, 2010). There is also significant experience with commercial biogas use in Nepal. Müller (2007) reviewed existing biogas technologies and case studies with contributions from China, Thailand, India, South Africa, Kenya, Rwanda, and Ghana.

23 For example, the high first cost (which can run up to USD 300 for some systems, including the digestion chamber unit) of traditional systems is being reduced considerably by new designs that reduce the digestion time, increase the specific methane yield and use alternate or multiple feedstocks (such as leafy material and food wastes), substantially reducing the size and cost of the digestion unit (Lehtomäki et al., 2007).

24 The Hedon Household Network provides references to the experience in the field at www.hedon.info. One example is www.hedon.info/docs/20060531_Report_(final)_on_Biogas_Experts_Network_Meeting_Hanoi.pdf.

25 For instance, the impacts of bioenergy development on landscapes are a barrier to adoption of new bioenergy conversion plants by some farmers as local acceptance decreases with increased local traffic to supply biomass (van der Horst and Evans, 2010). Some governments are more sensitive to increased efficiencies in GHG abatement and competitiveness of bioenergy with other energy sources, which often means increased scale (Adams et al., 2011) unless technologies succeed in increasing their throughput to accommodate smaller-scale applications without as large of a cost penalty (see Section 2.6.2).

26 See, for instance, the US Department of Energy's integrated biorefinery projects, including first-of-a-kind commercial plants, www1.eere.energy.gov/biomass/integrated_biorefineries.html; see also the IEA Bioenergy Task 39 interactive site with pilot, demonstration and commercial biofuels plants: biofuels.abc-energy.at/demoplants/projects/mapindex.

27 See, for instance, the European Industrial Bioenergy Initiative, a multi-industry partnership across the bioenergy value chains, www.biofuelstp.eu/eibi.html.

Table 2.11 | Key policy instruments in selected countries where E = electricity, H = heat, T = transport, Eth = ethanol and BD = biodiesel (modified after GBEP, 2008; updated with data from the REN21 global interactive map (see note 4 to Figure 2.9); reproduced with permission from GBEP).

Country	Binding Targets/Mandates[1]	Voluntary Targets[1]	Direct Incentives[2]	Grants	Feed-in Tariffs	Compulsory grid connection	Sustainability Criteria	Tariffs
Brazil	E, T		T					removed
China		E, T[4]	T	E, T	E, H	E, H		n/a
India	T, (E[3])	T(BD)	E	E, H, T	E			n/a
Mexico	(E[3])	(T)	(E)			(E)		Eth
South Africa	T, E	E, (T)	(E), T					n/a
Canada	E, T, H	E[4], T[4]	T	E, H, T				Eth
France		E[3], H[3], T	E, H, T		E			as EU below
Germany	E[3], T		H	H	E	E	(E, H, T)	as EU below
Italy	E[3]	E[3], T	T	E, H, T	E	E		as EU below
Japan		E, H, T				E		Eth, B-D
Russia		(E, H, T)	(T)					n/a
UK	E[3], T[3]	E[3], T	E, H, T	E, H, T	E		T	as EU below
USA	T, T[4], E[4]	E[4]	E, H, T	E, T	E			Eth
EU	E[3], T	E[3], H[3], T	T	E, H, T		E	(T)	Eth, B-D

Notes: 1. blending or market penetration; 2. fiscal incentives: tax reductions; public finance: loan support/guarantees; 3. target applies to all RE sources; 4. target is set at a sub-national level.

estimated at USD 1.1 billion and 0.4 billion for biofuels and biomass to energy, respectively (UNEP/SEFI/Bloomberg, 2010). A significant fraction of the venture capital investment was in the USA (Curtis, 2010). There was significant first-generation biofuels industry consolidation in the USA and in Brazil. Major global oil company investments occurred in both countries and in the EU (IATA, 2009; Curtis, 2010; IEA, 2010b; UNEP/SEFI/Bloomberg, 2010).

Addressing knowledge gaps in the sustainability of bioenergy systems, as discussed in Section 2.5, is reported as crucial to enable public and private decision making and increase public acceptance. Those gaps are mostly related to feedstock production and the associated impacts on land use, biodiversity, water, and food prices (WWI, 2006; Adams et al., 2011). Other suggested R&D avenues include more sustainable feedstocks and conversion technologies (WWI, 2006), increased conversion efficiency (Cheng and Timilsina, 2010) and overall chain optimization (Fagernäs et al., 2006).

Integrating bioenergy production with other industries/sectors (such as forest, food/fodder, power, or chemical industries) should improve competitiveness and utilize raw materials more efficiently (Fagernäs et al., 2006). For instance, industrial symbiosis evolved over 50 years in the city of Kalundborg, Denmark, as a community of businesses located together on a common property voluntarily entered into several bilateral contracts to enhance environmental, economic and social performance in managing environmental and resource issues by sharing resources in close cooperation with government authorities (Grann, 1997).[28] The Kalundborg experience increased the viability of the businesses involved over the years and developed a community thinking systems approach that could be applied to many other industrial settings (Jacobsen, 2006).

2.4.4 Global trade in biomass and bioenergy

Global trade in biomass feedstocks (e.g., wood chips, raw vegetable oils, agricultural residues) and especially of energy carriers from modern

---

28 The latest addition is a wheat straw-to-ethanol demonstration plant to the complex of a coal power plant, an oil refinery, biotechnology companies, district heating, fish aquaculture, landfill plant with gas collection, fertilizer production, gypsum (plaster), soil remediation and water treatment facilities, and others. Waste products (e.g., heat, gas and sulphur, ash, hot water, yeasts, fertilizers, waste slurries, solid wastes) from one company become a resource for use by one or more companies, and a nearby town, in a well-functioning industrial ecosystem. (See, for instance, www.kalundborg.dk/Erhvervsliv/The_Green_Industrial_Municipality/Cluster_Biofuels_Denmark_(CBD).aspx and www.inbicon.com/Biomass Refinery/Pages/Inbicon_Biomass_Refinery_at_Kalundborg.aspx.)

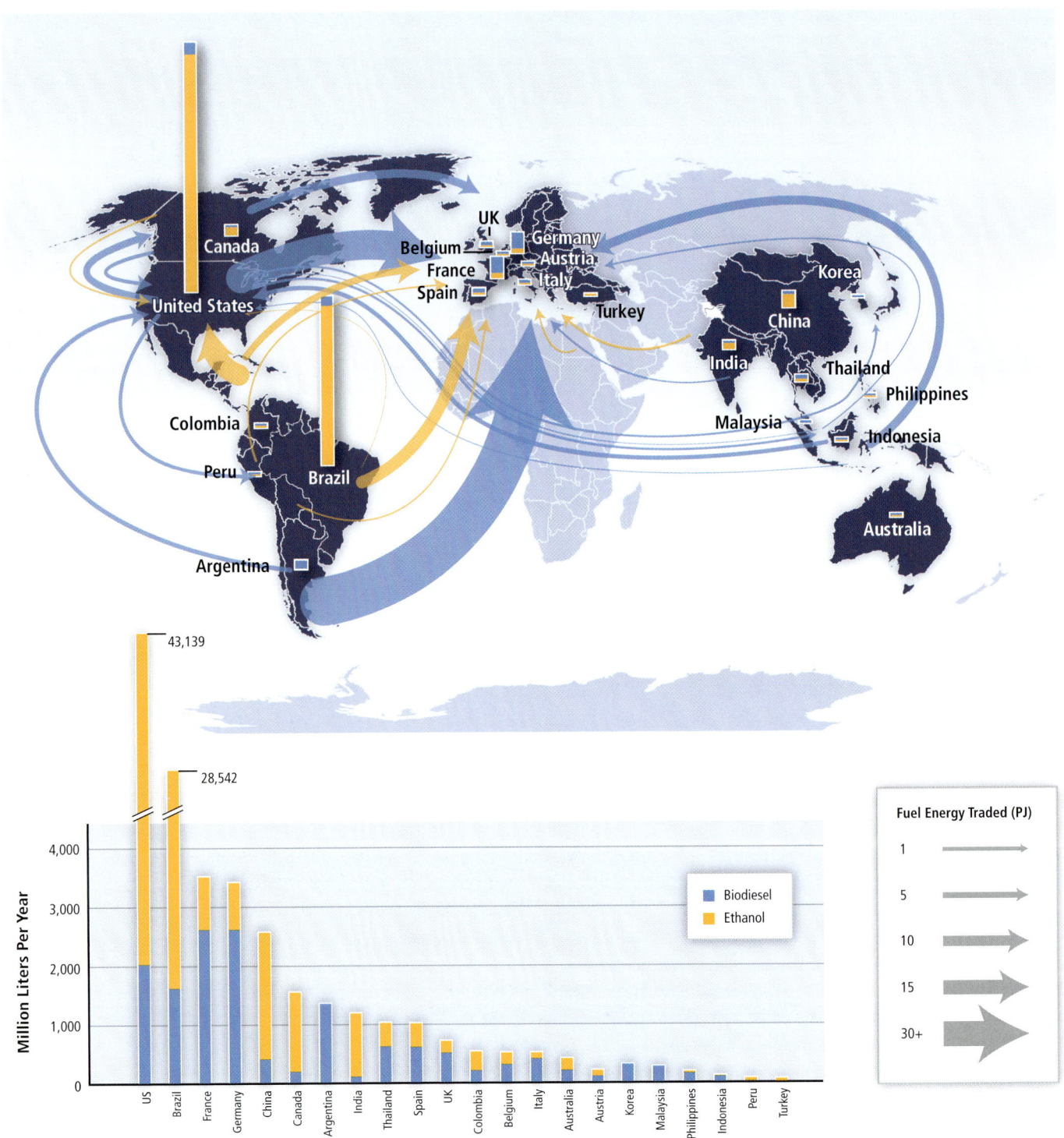

Figure 2.9 | Global biofuels production and main international trade, 2009. Biofuel volume sources: GAIN (2009a,b,[1] 2010a-j[2]); EIA (2010a); EurObserv'ER (2010); RFA (2010);[3] REN21 (2010).[4] Trade flows: Lamers et al. (2010).[5] The total intra-EU biodiesel and ethanol trade corresponds to 78 and 116 PJ, respectively (Lamers et al., 2011).

Notes: 1. Data for China and Indonesia. 2. Data for Argentina, Australia, Brazil, Canada, India, Korea, Malaysia, Peru, The Philippines, Thailand and Turkey. 3. www.ethanolrfa.org/pages/statistics. 4. See www.ren21.net/REN21Activities/ for updated information on biofuels volumes and targets for the various countries and other policy information and interactive tools (www.map.ren21.net). 5. For trade flows used in Figure 2.9 see www.chem.uu.nl/nws; for detailed data see Lamers et al. (2011).

bioenergy (e.g., ethanol, biodiesel, wood pellets) is growing rapidly. While practically no liquid biofuels or wood pellets were traded in 2000, the world net trade of liquid biofuels amounted to 120 to 130 PJ in 2009 (Figure 2.9), compared to about 75 PJ for wood pellets (Figure 2.8). Larger quantities of these products are expected to be traded internationally in the future, with Latin America and sub-Saharan Africa as potential net exporters and North America, Europe and Asia expected as net importers (Heinimö and Junginger, 2009). Trade can therefore become an important component of the sustained growth of the bioenergy sector. Figure 2.9 shows 2009 biofuels production in many countries along with the net global trade streams of bioethanol and biodiesel (see also Table 2.9). In 2008, around 9% of global biofuel production was traded internationally (Junginger et al., 2010). Production and trade of these three commodities are discussed in more detail below.

Global fuel *ethanol production* grew from around 0.375 EJ in 2000 to more than 1.6 EJ in 2009 (Lamers et al., 2011). The USA and Brazil, the two leading ethanol producers and consumers, accounted for about 85% of the world's production. In the EU, total consumption of ethanol for transport in 2009 was 94 PJ (3.6 Mt), with the largest users being France, Germany, Sweden and Spain (Lamers et al., 2011; EurObserv'ER, 2010). Data related to fuel *bioethanol trade* are imprecise on account of the various potential end uses of ethanol (i.e., fuel, industrial and beverage use) and also because of the lack of proper codes for biofuels in global trade statistics. As an estimate, a net amount of 40 to 51 PJ of fuel ethanol was traded in 2009 (Lamers et al., 2011).

World *biodiesel production* started below 20 PJ in 2000 and reached about 565 PJ in 2009 (Lamers et al., 2011). The EU produced 334 PJ (roughly two-thirds of the global production), with Germany, France, Spain and Italy being the top EU producers (EurObserv'ER, 2010). EU27 biodiesel production rates levelled off towards 2008 (FAPRI, 2009).[29] The intra-European biodiesel market has become more competitive, and the 2009 overcapacity has already led to the closure of (smaller, less vertically integrated, less efficient, remote, etc.) biodiesel plants in Germany, Austria and the UK. As shown in Figure 2.9, other main biodiesel producers include the USA, Argentina and Brazil. Biodiesel consumption in the EU amounted to about 403 PJ (8.5 Mt) (EurObserv'ER, 2010), with Germany and France consuming almost half of this amount. Net international *biodiesel trade* was below 1 PJ before 2005 but grew very fast from this small base to more than 80 PJ in 2009, as shown in Figure 2.9 (Lamers et al., 2011).

Production, consumption and trade of *wood pellets* have grown strongly within the last decade and are comparable to ethanol and biodiesel in terms of global trade volumes. As a rough estimate, in 2009, more than 13 Mt (230 PJ) of *wood pellets* were produced primarily in 30 European countries, the USA and Canada (Figure 2.8). Consumption was high in many EU countries and the USA. The largest EU consumers were Sweden (1.8 Mt or 32 PJ), Denmark, the Netherlands, Belgium, Germany and Italy (roughly 1 Mt or 18 PJ each). Main *wood pellet trade* routes lead from Canada and the USA to Europe (especially Sweden, the Netherlands and Belgium) and to the USA. In 2009, other minor trade flows were also reported, for example, from Australia, Argentina and South Africa to the EU. Canadian producers also started to export small quantities to Japan. Total imports of wood pellets by European countries in 2009 were estimated to be about 3.9 Mt (69 PJ), of which about half can be assumed to be intra-EU trade (Sikkema et al., 2010, 2011).

2.4.5 Overview of support policies for biomass and bioenergy[30]

Typical examples of support policies are shown in Table 2.11. For instance, *liquid biofuels* policies include the (former) Brazilian Proálcool programme, regulations in the form of mandates in many EU countries and the USA fiscal incentives such as tax exemptions, production tax credits and accelerated depreciation (WWI, 2007). The majority of successful policies for *heat* from biomass in recent decades have focused on more centralized applications for heat or CHP in district heating and industry (Bauen et al., 2009a). For these sectors, a combination of direct support schemes with indirect incentives has been successful in several countries, such as Sweden (Junginger, 2007). Both quota systems and FITs have been implemented in support of bioenergy *electricity* generation, though FITs have gradually become the more popular incentive. The effectiveness and efficiency of FITs and quota systems for promoting RE generation (including for bioenergy) has been thoroughly debated. A full discussion of these instruments can be found in Section 11.5.3. Next to FITs or quotas, almost all countries that have successfully stimulated bioenergy development have applied additional public finance relating to investment support and soft loans along with fiscal measures (GBEP, 2008). Additionally, grid access for renewable power is an important issue that needs to be addressed. Priority grid access for renewable sources is applied in most countries where bioenergy technologies have been successfully deployed (Sawin, 2004).

Support policies (see Table 2.11) have strongly contributed in past decades to the growth of bioenergy for electricity, heat and transport fuels. However, several reports also point out the costs and risks associated with support policies for biofuels. According to the WEO (IEA, 2010b), the annual global government support for biofuels in 2009, 2008 and 2007 was USD_{2009} 20 billion, 17.5 billion and 14 billion, respectively, with corresponding EU spending of USD_{2009} 7.9 billion, 8.0 billion and 6.3 billion and corresponding US spending of USD_{2009} 8.1 billion, 6.6 billion and 4.9 billion. The US spending was driven by energy security and fossil fuel import reduction goals. Concerns about food prices, GHG emissions and environmental impacts have also led to many countries rethinking biofuels blending targets. For example, Germany revised its blending target for 2009 downward from 6.25 to 5.25%.[31] Addressing these concerns led also to the incorporation of environmental and social

---

29 While most EU Member States (MS) increased their production volumes, the German biodiesel market shrunk both in supply and demand due to a change in the policy framework phasing out tax exemptions for neat biodiesel at the pump. At the same time biodiesel export to other EU MS became less and less feasible for German (and other) producers due to increasing shares of competitively priced biodiesel imports, mainly from the USA in the period from 2006 to 2008 and also from Argentina in the years 2008 and 2009 (Lamers et al., 2011).

30 Non-technology-specific policy issues are covered in Chapter 11 of this report.

31 Bundesministerium für Umwelt, Naturschutz und Reaktorsicherheit decision published on 22.10.2008 and available at www.bmu.de/pressearchiv/16_legislaturperiode/pm/42433.php.

sustainability criteria for biofuels in the EU Renewable Energy Directive. Although seemingly effective in supporting domestic farmers, the effectiveness of biofuel policies in reaching the climate change and secure energy supply objectives is coming under increasing scrutiny. It has been argued that these policies have been costly and have tended to introduce new distortions to already severely distorted and protected agricultural markets—at both domestic and global levels. This has not tended to favour an efficient international production pattern for biofuels and their feedstocks (FAO, 2008a; Bringezu et al., 2009). An overall biomass strategy would have to consider all types of use of food and non-food biomass (Bringezu et al., 2009).

The main drivers behind government support for the sector have been concerns over climate change and energy security as well as the desire to support the agricultural sector through increased demand for agricultural products (FAO, 2008a). According to the REN21 global interactive map (see note 4 to Figure 2.9) a total of 69 countries had one or several biomass support policies in place in 2009 (REN21, 2010; Section 11.2).

2.4.5.1 Intergovernmental platforms for exchange on bioenergy policies and standardization

Several multi-stakeholder initiatives exist in which policymakers can find advice, support and the possibility of exchanging experiences on policymaking for bioenergy. Examples of such international organizations and forums supporting the further development of sustainability criteria and methodological frameworks for assessing GHG mitigation benefits of bioenergy include the Global Bioenergy Partnership (GBEP from the G8+5),[32] the IEA Bioenergy Agreement,[33] the International Bioenergy Platform at the Food and Agriculture Organization (FAO),[34] the OECD Roundtable on Sustainable Development,[35] and standardization organizations such as the European Committee for Standardization[36] and the International Organization for Standardization[37] (ISO) that are actively working toward the development of sustainability standards.

2.4.5.2 Sustainability frameworks and standards

Governments are stressing the importance of ensuring sufficient climate change mitigation and avoiding unacceptable negative effects of bioenergy as they implement regulating instruments. For example, the Renewable Energy Directive (European Commission, 2009) provides mandatory sustainability requirements for liquid transport fuels.[38] Also, in the USA, the Renewable Fuel Standard—included in the 2007 Energy Independence and Security Act (EISA, 2007)—mandates minimum GHG reductions from renewable fuels, discourages use of food and fodder crops as feedstocks, permits use of cultivated land and estimates (indirect) LUC effects to set thresholds of GHG emission reductions for categories of fuels (EPA, 2010; see also Section 2.5). The California Low Carbon Fuel Standard set an absolute carbon intensity reduction standard and periodic evaluation of new information, for instance, on indirect land use impacts.[39] Other examples are the UK Renewable Transport Fuel Obligation, the German Biofuel Sustainability Ordinance, and the Cramer Report (The Netherlands). With the exception of Belgium, no mandatory sustainability criteria for solid biomass (e.g., wood pellets) have been implemented—the European Commission will review this at the end of 2011 (European Commission, 2010).

The development of impact assessment frameworks and sustainability criteria involves significant challenges in relation to methodology, process development and harmonization. As of a 2010 review, nearly 70 ongoing certification initiatives exist to safeguard the sustainability of agriculture and forestry products, including those used as feedstock for the production of bioenergy (van Dam et al., 2010). Within the EU, a number of initiatives started or have already set up certification schemes in order to guarantee a more sustainable cultivation of energy crops and production of energy carriers from modern biomass (e.g., ISCC[40]; REDCert[41] 2010 in Germany; or the NTA8080/8081 (NEN[42]) in the Netherlands). Many initiatives focus on the sustainability of liquid biofuels including primarily environmental principles, although some of them, such as the Council for Sustainable Biomass Production and the Better Sugarcane Initiative, the Roundtable for Sustainable Biofuels (RSB) and the Roundtable for Responsible Soy, include explicit socioeconomic impacts of bioenergy production. Principles such as those from the RSB have already led to a Biofuels Sustainability Scorecard used by the Inter-American Development Bank for the development of projects.

---

32 The GBEP provides a forum to inform policy development frameworks, promote sustainable biomass and bioenergy development, facilitate investments in bioenergy, promote project development and implementation, and foster R&D and commercial bioenergy activities. Membership includes individual countries, multilateral organizations, and associations.

33 The IEA Bioenergy Agreement provides an umbrella organization and structure for a collective effort in the field of bioenergy including non-OECD countries interested in the topics from RD&D to policies. It brings together policy and decision makers and national experts from research, government and industry across the member countries.

34 See ftp.fao.org/docrep/fao/009/A0469E/A0469E00.pdf.

35 See www.oecd.org/dataoecd/14/3/46063741.pdf.

36 See www.cen.eu/cen/Sectors/TechnicalCommitteesWorkshops/CENTechnicalCommittees/Pages/default.aspx TC335 for solid biofuels standards, TC19 for liquid biofuels, and TC 383 for sustainability criteria for biofuels.

37 See www.iso.org/iso/standards_development/technical_committees/list_of_iso_technical_committees.htm TC 248 for sustainability criteria for biofuels, TC 238 for solid biofuels, TC255 for biogas, and TC 28/SC 7 for liquid biofuels.

38 These requirements are: specific GHG emission reductions must be achieved, and the biofuels in question must not be produced from raw materials being derived from land of high value in terms of biological diversity or high carbon stocks.

39 The California Air Resources Board requires 10% absolute emissions reductions from fossil energy sources by 2020 and considers direct lifecycle emissions of the biofuels and also indirect LUC as required by legislation (CARB, 2009).

40 International Sustainability and Carbon Certification, Koeln, Germany, www.iscc-system.org/index_eng.html

41 REDcert Certification System, www.redcert.org

42 NTA 8080 - Sustainabley Produced Biomass. Dutch Normalization Institute (NEN), Delft, The Netherlands, www.sustainable-biomass.org/publicaties/3950

The proliferation of standards that has taken place over the past four years, and continues, shows that certification has the potential to influence local impacts related to the environmental and social effects of direct bioenergy production. Many of the bodies involved conclude that for an efficient certification system there is a need for further harmonization, availability of reliable data, and linking indicators at micro, meso and macro levels (see Figure 2.15). Considering the multiple spatial scales, certification should be combined with additional measurements and tools at regional, national and international levels.

The role of bioenergy production in iLUC is still uncertain; current initiatives have rarely captured impacts from iLUC in their standards, and the time scale becomes another important variable in assessing such changes (see Section 2.5.3). Addressing unwanted LUC requires overall sustainable agricultural production and good governance first of all, regardless of the end use of the product or of the feedstocks.

2.4.6 Main opportunities and barriers for the market penetration and international trade of bioenergy

2.4.6.1 Opportunities[43]

The prospects for biofuels for road transport depend on developments in competing low-carbon and oil-reducing technologies for road transport (e.g., electric vehicles). Biofuels may in the longer term be increasingly used within the aviation industry, for which high energy density carbon fuels are necessary (see Section 2.6.3), and also in marine shipping.

The development of international markets for bioenergy has become an essential driver to develop available biomass resources and market potential, which are currently underutilized in many world regions. This is true for both (available) residues as well as possibilities for dedicated biomass production (through energy crops or multifunctional systems such as agro-forestry). Export of biomass-derived commodities for the world's energy market can provide a stable and reliable income for rural communities in many (developing) countries, thus creating an important incentive and market access.[44]

Also on the demand side, large biomass users that rely on a stable supply of biomass can benefit from international bioenergy trade, as this enables (often very large) investments in infrastructure and conversion capacity.[45]

Introduction of incentives based on political decisions is a driving force and has triggered an expansion of bioenergy trade. For example, wood pellet imports in the Netherlands and Belgium have been driven respectively through a feed-in premium system and a Green Certificate system. However, the success of policies has varied, due partly to the nature of the design and implementation of the given policy but also to the fact that the institutions related to the incentives are different. For a full discussion of influencing factors outside of policies (e.g., institutions, network access), see Section 11.6.

Another driver is the utilization of established logistics for existing commodities. Taking again the example of wood pellet co-firing in large power plants, the existing infrastructure at ports and storage facilities used to supply coal and other dry bulk goods can (partially, and after adaptations) also be used for wood pellets, making cost-efficient transport and handling possible. Another form of integrated supply chain is bark, sawdust and other residues from imported roundwood, which is common in, for example, Northern Europe. Finally, the concept of regional biomass processing centres has been proposed to deal with supply side challenges and also to help address social sustainability concerns (Carolan et al., 2007).

2.4.6.2 Barriers

Major risks and barriers to deployment are found all along the bioenergy value chain and concern all final energy products (bioheat, biopower, and biofuel for transport).[46] On the supply side, there are challenges related to securing quantity, quality and price of biomass feedstock, irrespective of the origin of the feedstock (energy crops, wastes or residues). There are also technology challenges related to the varied physical properties and chemical composition of the biomass feedstock and challenges associated with the poor economics of current power and biofuel technologies at small scales. On the demand side, the main challenges are the stability and supportiveness of policy frameworks and investors' confidence in the sector and its technologies, in particular to overcome financing challenges associated with demonstrating the reliable operation of new technologies at commercial scale.[47] In the power and heat sectors, competition with other RE sources may also be an issue. Public acceptance and public perception are also critical factors in gaining support for energy crop production and bioenergy facilities.

Specifically for the bioenergy trade, Junginger et al. (2010) identified a number of (potential) barriers:

Tariffs. As of January 2007, import tariffs apply in many countries, especially for ethanol and biodiesel. Tariffs (expressed in local currency and year) are applied on bioethanol imports by both the EU (€ 0.192 per litre) and the USA (USD 0.1427 per litre and an additional 2.5%

---

43 This sub-section is largely based on Junginger et al. (2008).

44 Exports of ethanol from Brazil and wood pellets from Canada are examples where export opportunities (at least partially) were drivers to further develop the supply side.

45 Utilities in the Netherlands and Belgium import large amounts of wood pellets to co-fire with coal, as domestic biomass resources are very limited and of varying quality.

46 Most of the remainder of this paragraph is based on Bauen et al. (2009a).

47 Some governments have jointly financed first-of-a-kind commercial technological development with the private sector in the past five years, but the financial crisis is making it difficult to complete the private financing needed to continue to obtain government financing.

ad valorem subsidy). In general, the most-favoured nation tariffs range from roughly 6 to 50% on an ad valorem equivalent basis in the OECD, and up to 186% in the case of India (Steenblik, 2007). Biodiesel used to be subject to lower import tariffs than bioethanol, ranging from 0% in Switzerland to 6.5% in the EU and the USA (Steenblik, 2007). However, in July 2009, the European Commission confirmed a five-year temporary imposition of anti-dumping and anti-subsidy rights on American biodiesel imports, with fees standing between € 213 and 409 per tonne (local currency and year) (EurObserv'ER, 2010). These trade tariffs were a reaction to the so-called 'splash-and-dash' practice, in which biodiesel blended with a 'splash' of fossil diesel was eligible for a USD 1 per gallon subsidy (equivalent to USD 300/t) in 2008-2009; see Lamers et al. (2011) for detailed information on the various tariffs, trade regimes, and policies worldwide.

Technical standards describe in detail the physical and chemical properties of fuels. Regulations pertaining to the technical characteristics of liquid transport fuels (including biofuels) exist in all countries. These have been established in large part to ensure the safety of the fuels and to protect consumers from buying fuels that could damage their vehicles' engines. Regulations include maximum percentages of biofuels that can be blended with petroleum fuels and regulations pertaining to the technical characteristics of the biofuels themselves. In the case of biodiesel, the latter may depend on the vegetable oils used for the production, and thus regulations might be used to favour biodiesel from domestic feedstocks over biodiesel from imported feedstocks. Technical barriers for the bioethanol trade also exist. For example, the different demands for maximum water content have negative impacts on trade. However, in practice, most market actors have indicated that they see technical standards as an opportunity enabling international trade rather than as a barrier (Junginger et al., 2010).

Sustainability criteria and biomass and biofuels certification have been developed in increasing numbers in recent years as voluntary or mandatory systems (see Section 2.4.5.2); such criteria, so far, do not apply to conventional fossil fuels. Three major concerns in relation to the international bioenergy trade are:

1. Criteria, especially those related to environmental and social issues, could be too stringent or inappropriate to local environmental and technological conditions in producing developing countries (van Dam et al., 2010). The fear of many developing countries is that if the selected criteria are too strict or are based on the prevailing conditions in the countries setting up the certification schemes, only producers from those countries may be able to meet the criteria, and thus these criteria may act as trade barriers. As the criteria are extremely diverse, ranging from purely commercial aims to rainforest protection, there is a danger that a compromise could result in overly detailed rules that lead to compliance difficulties, or, on the other hand, in standards so general that they become meaningless. Implementing binding requirements is also limited by World Trade Organization rules.

2. With current developments by the European Commission, different European governments, several private sector initiatives, and initiatives of round tables and NGOs, there is a risk that in the short term a multitude of different and partially incompatible systems will arise, creating trade barriers (van Dam et al., 2010). If they are not developed globally or with clear rules for mutual recognition, such a multitude of systems could potentially become a major barrier for international bioenergy trade instead of promoting the use of sustainable biofuels production. A lack of transparency in the development of some methodologies, for example, in the EU legislation, is an issue. Also, the eventual existence of different demands for proving compliance with the criteria for locally produced biomass sources and imported ones is a potential barrier. Finally, lack of international systems may cause market distortions.

Production of 'uncertified' biofuel feedstocks will continue and enter other markets in countries with lower standards or for non-biofuel applications that may not have the same standards. The existence of a 'two-tier' system would result in failure to achieve the safeguards envisaged (particularly for LUC and socioeconomic impacts).

3. Finally, note that to ensure that biomass commodities are being produced in a sustainable manner, some chain of custody (CoC) method must be used to track biomass and biofuels from production to end use. Generally, the three types of CoC methods are segregation (also known as track-and-trace), book-and-claim and mass-balance. While this is not necessarily a major barrier, it may cause additional cost and administrative burdens.

Logistics are a pivotal part of the system and essential to set up biomass fuel supply chains for large-scale biomass systems. Various studies have shown that long-distance international transport by ship is feasible in terms of energy use and transportation costs (e.g., Sikkema et al., 2010, 2011), but availability of suitable vessels and meteorological conditions (e.g., winter in Scandinavia and Russia) need to be considered. One logistical barrier is a general lack of technically mature technologies to densify biomass at low cost to facilitate transport, although technologies are being developed (Sections 2.3.2 and 2.6.2).

Sanitary and phytosanitary (SPS) measures may be faced by feedstocks for liquid biofuels or technical regulations applied at borders. SPS measures mainly affect feedstocks that, because of their biological origin, can carry pests or pathogens. One of the most common SPS measures is a limit on pesticide residues. Meeting pesticide residue limits is usually not difficult but on occasion has led to the rejection of imported shipments of crop products, especially from developing countries (Steenblik, 2007).

2.4.7 Synthesis

The review of developments in biomass use, markets and policy shows that bioenergy has seen rapid developments over the past years. The use of modern biomass for liquid and gaseous energy carriers is growing, in particular biofuels (with a 37% increase from 2006 to 2009). Projections from the IEA, among others, but also many national targets, count on biomass delivering a substantial increase in the share of RE. International trade in biomass and biofuels has also become much more important over recent years, with roughly 6% (reaching levels of up to 9% in 2008) of biofuels (ethanol and biodiesel only), and one-third of all pellet production for energy use, traded internationally in 2009. Pellets have proven to be an important facilitating factor in both increasing utilization of biomass in regions where supplies are constrained as well as mobilizing resources from areas where demand is lacking. Nevertheless, many barriers remain to developing well-working commodity trading of biomass and biofuels that at the same time meets sustainability criteria.

The policy context for bioenergy, and in particular biofuels, in many countries has changed rapidly and dramatically in recent years. The debate on food versus fuel competition and the growing concerns about other conflicts have resulted in a strong push for the development and implementation of sustainability criteria and frameworks as well as changes in temporization of targets for bioenergy and biofuels. Furthermore, the support for advanced biorefinery and second-generation biofuel options is driving bioenergy in more sustainable directions.

Persistent policy and stable policy support has been a key factor in building biomass production capacity and working markets, required infrastructure and conversion capacity that gets more competitive over time. These conditions have led to the success of the Brazilian programme to the point that ethanol production costs are lower than those of gasoline. Brazil achieved an energy portfolio mix that is substantially renewable and that minimized foreign oil imports. Sweden, Finland, and Denmark also have shown significant growth in renewable electricity and in management of integrated resources, which steadily resulted in innovations such as industrial symbiosis of collocated industries. The USA has been able to quickly ramp up production with the alignment of national and sub-national policies for power in the 1980s and for biofuels in the 1990s to present, as petroleum prices and instability in key producing countries increased; however, as oil prices decreased, policy support and bioenergy production decreased for biopower and is increasing again with environmental policies and sub-national targets.

Countries differ in their priorities, approaches, technology choices and support schemes for further development of bioenergy. Although this means increased complexity of the bioenergy market, this also reflects the many aspects that affect bioenergy deployment—agriculture and land use, energy policy and security, rural development and environmental policies. Priorities, stage of development and geographic access to the resources, and their availability and costs differ widely from country to country.

As policies surrounding bioenergy and biofuels become more holistic, using sustainability demands as a starting point is becoming an overall trend. This is true for the EU, the USA and China, but also for many developing countries such as Mozambique and Tanzania. This is a positive development but is by no means settled (see also Section 2.5). The 70 initiatives registered worldwide by 2009 to develop and implement sustainability frameworks and certification systems for bioenergy and biofuels, as well as agriculture and forestry, can lead to a fragmentation of efforts (van Dam et al., 2010). The needs for harmonization and for international and multilateral collaboration and dialogue are widely stressed at present.

2.5 Environmental and social impacts[48]

Recent studies have highlighted both positive and negative environmental and socioeconomic effects of bioenergy and the associated agriculture and forestry LUC (IPCC, 2000b; Millennium Ecosystem Assessment, 2005). Like conventional agriculture and forestry systems, bioenergy can exacerbate soil and vegetation degradation associated with overexploitation of forests, too intensive crop and forest residue removal, and water overuse (Koh and Ghazoul, 2008; Robertson et al., 2008). Diversion of crops or land into bioenergy production can influence food commodity prices and food security (Headey and Fan, 2008). With proper operational management, the positive effects can include enhanced biodiversity (C. Baum et al., 2009; Schulz et al., 2009), soil carbon increases and improved soil productivity (Tilman et al., 2006a; S. Baum et al., 2009), reduced shallow landslides and local flash floods, reduced wind and water erosion and reduced sediment volume and nutrients transported into river systems (Börjesson and Berndes, 2006). For forests, bioenergy can improve growth and productivity, improve site conditions for replanting and reduce wildfire risk (Dymond et al., 2010). However, forest residue harvesting can have negative impacts such as the loss of coarse woody debris that provides essential habitat for forest species.

Biofuels derived from purpose-grown agricultural feedstocks are water intensive (see Section 9.3.4.4 for comparisons of renewable and non-renewable power sources; Berndes, 2002; King and Weber, 2008; Chiu et al., 2009; Dominguez-Faus et al., 2009; Gerbens-Leenes et al., 2009; Wu et al., 2009; Fingerman et al., 2010). Their influence on water resources and the wider hydrologic cycle depends on where, when and how the biofuel feedstock is produced. Among different bioenergy supply chains, across the spectrum of feedstocks, cultivation systems and conversion technologies, water demand varies greatly (Wu et al., 2009; Fingerman et al., 2010, De La Torre Ugarte, et al., 2010). While biofuel made from irrigated crops requires extraction of large volumes of water from lakes, rivers and aquifers, use of agricultural or forestry residues as bioenergy feedstocks does not generally require much additional land or water. Rain-fed feedstock production does not require water extraction from

---

48 A comprehensive assessment of social and environmental impacts of all RE sources covered in this report can be found in Chapter 9.

water bodies, but it can still reduce downstream water availability by redirecting precipitation from runoff and groundwater recharge to crop evapotranspiration. Using water for bioenergy has very different social and ecological consequences depending upon the state of the resource base from which that water was drawn.

Few universal conclusions about the socioeconomic and environmental implications of bioenergy can currently be drawn, given the multitude of rapidly evolving bioenergy sources, the complexities of physical, chemical and biological conversion processes, the multiple energy products, and the variability in environmental conditions. Thus, the positive and negative effects of bioenergy are a function of the socioeconomic and institutional context, the types of lands and feedstocks used, the scale of bioenergy programmes and production practices, the conversion processes, and the rate of implementation (e.g., Kartha et al., 2006; Firbank, 2008; E. Gallagher, 2008; OECD-FAO, 2008; Royal Society, 2008; UNEP, 2008b; Howarth et al., 2009; Pacca and Moreira, 2009; Purdon et al., 2009; Rowe et al., 2008).

Bioenergy system impact assessments (IAs) must be compared to the IAs of replaced systems.[49] The methodologies and underlying assumptions for assessing environmental (Sections 2.5.1 through 2.5.6) and socioeconomic (Section 2.5.7) effects (see Table 2.12 for examples of these impacts) differ greatly and therefore the conclusions reached by these studies are inconsistent (H. Kim et al., 2009). One particular challenge for socioeconomic IAs is that their boundaries are difficult

2.5.1 Environmental effects

Studies of environmental effects, including those focused on energy balances and GHG emission balances, usually employ methodologies in line with the principles, framework, requirements and guidelines in the ISO 14040:2006 and 14044:2006 standards for Life Cycle Assessment (LCA) discussed in Section 9.3.4.1. An earlier specific method for assessing GHG balances of biomass and bioenergy systems was developed by Schlamadinger et al. (1997).

Key issues for bioenergy LCAs are system definition including spatial and dynamic boundaries, functional units, reference system, and the selection of methods for considering energy and material flows across system boundaries (Soimakallio et al., 2009a; Cherubini and Strømman, 2010). As part of cascading cycles, many processes create multiple products; for example, biomass is used to produce biomaterials while co-products and the biomaterial itself are used for energy after their useful life (Dornburg and Faaij, 2005). Such cascading results in significant data and methodological challenges because environmental effects can be distributed over several decades and in different geographical locations (Cherubini et al., 2009b).

Most of the assumptions and data used in LCA studies of existing bioenergy systems are related to first-generation biofuels and to conditions and practices in Europe or the USA, although studies are becoming

Table 2.12 | Environmental and socioeconomic impacts of bioenergy: example areas of concern with selected impact categories (synthesized from the literature review by van Dam et al., 2010).

Example areas of concern	Examples of impact categories
Global, regional, off-site environmental effects	GHGs; albedo; acidification; eutrophication; water availability and quality; regional air quality
Local/onsite environmental effects	Soil quality; local air quality; water availability and quality; biodiversity and habitat loss
Technology	Hazards; emissions; congestion; safety; genetically modified organisms/plants
Human rights and working conditions	Freedom of association; access to social security; job creation and average wages; freedom from discrimination; no child labour and minimum age of workers; freedom of labour (no forced labour); rights of indigenous people; acknowledgment of gender issues
Health and safety	Impacts on workers and users; safety conditions at work
Food security	Replacement of staple crops; safeguarding local food security
Land and property rights	Acknowledgment of customary and legal rights of land owners; proof of ownership; compensation systems available; agreements by consent
Participation and well-being of local communities	Cultural and religious values; contribution to local economy and activities; compensation for use of traditional knowledge; support to local education; local procurement of services and inputs; special measures to target vulnerable groups

to quantify and are a complex composite of numerous interrelated factors, many of which are poorly understood or unknown. Social processes have feedbacks that are difficult to clearly define with an acceptable level of confidence. Environmental IAs include many quantifiable impact categories but still lack data and are uncertain in many areas. The outcome of an environmental IA depends on methodological choices, which are not yet standardized or uniformly applied throughout the world.

available for Brazil, China and other countries (see examples in Tables 2.7, 2.13, and 2.15). Ongoing development of biomass production and conversion technologies makes many of these studies of commercial technologies outdated.[50] LCA studies of prospective bioenergy options involve projections of technology performance and have relatively greater uncertainties (see, e.g., Figure 9.9). The way that uncertainties

---

49 A 'rebound effect' could be included, usually fossil fuels, but also other primary energy sources (Barker et al., 2009).

50 For instance, using a 2006 reference that analyzed an industrial system in 2002 will not represent the industry in 2010 because learning occurred in commercial technologies that exhibited a significant accumulation of production volume such as in the USA and in Brazil; an example of wide-spread adoption of a different technology in this industry is the USA where dry milling has become the major route to ethanol production (see Sections 2.3.4 and 2.7.2).

and parameter sensitivities are handled across the supply chain to fuel production significantly impacts the results (Sections 2.5.2 through 2.5.6). Studies combining several LCA models and/or Monte Carlo analysis provide bioenergy system uncertainties and levels of confidence for some bioenergy options (e.g., Soimakallio et al., 2009b; Hsu et al., 2010; Spatari and MacLean, 2010).

Most bioenergy system LCAs are designated as attributional to the defined process system boundaries. Consequential LCAs analyze bioenergy systems beyond these boundaries, in the context of the economic interactions, chains of cause and effect in bioenergy production and use, and effects of policies or other initiatives that increase bioenergy production and use. Consequential LCAs can investigate systemic responses to bioenergy expansion (e.g., how the food system changes if increasing volumes of cereals are used as biofuel feedstock or how petroleum markets respond if increased biofuels production results in reduced petroleum demand—see Section 2.5.3 and Figure 2.13). The outcome of *any* measure to reduce a certain use can be affected by a rebound effect—in the case of bioenergy, if increased production of solid, liquid and gaseous biofuels leads to lower demand for fossil fuels, this in turn could lead to lower fossil fuel prices and increased fossil fuel demand (Rajagopal et al., 2011; Stoft, 2010).[51] Similarly, when considering co-products, LCAs should ideally model displacement of alternative products as a dynamic result of market interactions. Consequential LCAs therefore require auxiliary tools such as economic equilibrium models.

2.5.2 Modern bioenergy: Climate change excluding land use change effects

The ranges of GHG emissions for bioenergy systems and their fossil alternatives per unit energy output are shown in Figure 2.10 for several uses (transport, power, heat) calculated based on LCA methodologies (land use-related net changes in carbon stocks and land management impacts

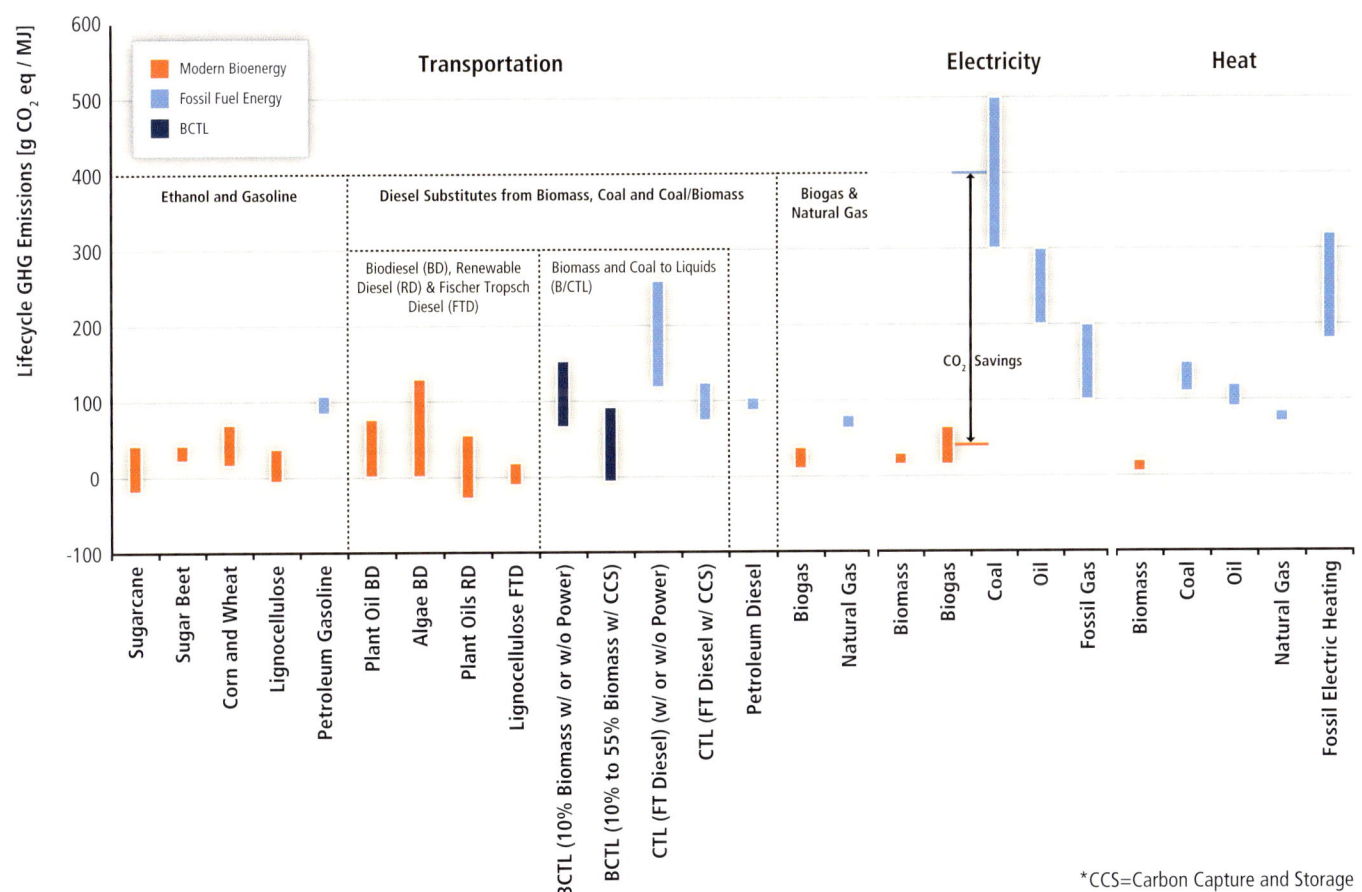

Figure 2.10 | Ranges of GHG emissions per unit energy output (MJ) from major modern bioenergy chains compared to conventional and selected advanced fossil fuel energy systems (land use-related net changes in carbon stocks and land management impacts are excluded). Commercial and developing (e.g., algae biofuels, Fischer-Tropsch) systems for biomass and fossil technologies are illustrated.

Data sources: Wu et al. (2005); Fleming et al. (2006); Hill et al. (2006, 2009); Beer and Grant (2007); Wang et al. (2007, 2010); Edwards et al. (2008); Kreutz et al. (2008); Macedo and Seabra (2008); Macedo et al. (2008); NETL (2008, 2009a,b); CARB (2009); Cherubini et al. (2009a); Huo et al. (2009); Kalnes et al. (2009); van Vliet et al. (2009); EPA (2010); Hoefnagels et al. (2010); Kaliyan et al. (2010); Larson et al. (2010); 25th to 75th percentile of all values from Figure 2.11.

---

51 The same rebound effect applies to other RE technologies displacing incumbent fossil technologies.

are excluded). Meta-analyses to quantify the influence of bioenergy systems on climate are complicated because of the multitude of existing and rapidly evolving bioenergy sources, the complexities of physical, chemical and biological conversion processes, and feedstock diversity and variability in site-specific environmental conditions—together with differences between studies in method interpretation, assumptions and data. Due to this, review studies report varying estimates of GHG emissions and a wide range of results have been reported for the same bioenergy options, even when temporal and spatial considerations are constant (see, e.g., S. Kim and Dale, 2002; Fava, 2005; Farrell et al., 2006; Fleming et al., 2006; Larson, 2006; von Blottnitz and Curran, 2007; Rowe et al., 2008; Börjesson, 2009; Cherubini et al., 2009a; Menichetti and Otto, 2009; Soimakallio et al., 2009b; Hoefnagels et al., 2010; Wang et al., 2010, 2011).

For electricity generated by various technologies, GHG emissions per kWh generated are detailed in Figure 2.11, based on published estimates from lifecycle GHG emissions (land use-related net changes in carbon stocks and land management impacts are excluded) of an extensive review of biopower LCAs.[52] Figure 2.11 shows that the majority of lifecycle GHG emission estimates cluster between about 16 and 74 g CO_2eq/kWh (4.4 and 21 g CO_2eq/MJ), with one estimate reaching

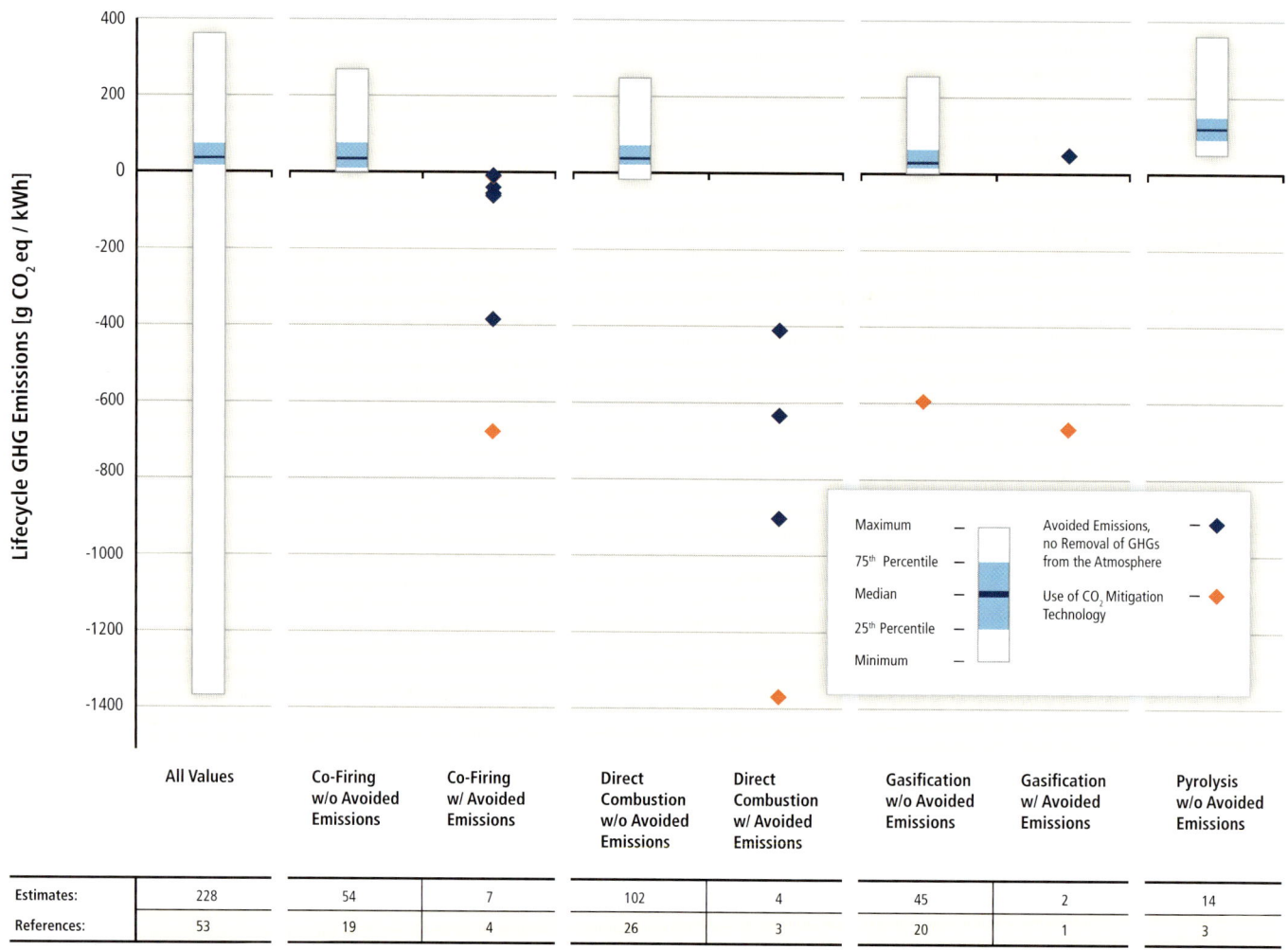

Figure 2.11 | Lifecycle GHG emissions of biopower technologies per unit of electricity generation, including supply chain emissions (land use-related net changes in carbon stocks and land management impacts are excluded). Co-firing is shown for the biomass portion only (without GHG emissions and electricity output associated with coal). Included in the avoided GHG emissions category are only estimates in which the use of the feedstock itself (e.g. residues and wastes) leads to avoided emissions, for example, in the form of avoided methane emissions from landfills (most common in the literature).[1] Estimates that include avoided emissions from the production of co-products are not included in the avoided GHG emissions category. Individual data points were used instead of box plots for estimates with avoided emissions because of high variability. Red diamonds indicate that a carbon mitigation technology (CCS or carbonate formation by absorption) was considered. Along the bottom of the figure and aligned with each column are the number of estimates and the number of references (CCS estimates in parentheses) producing the distributions.

Note: 1. 'Negative estimates' within the terminology of lifecycle assessments presented in this report refer to avoided emissions. Unlike the case of bioenergy combined with CCS, avoided emissions do not remove GHGs from the atmosphere. Due to the inclusion of a non-CCS carbon sequestration technology and non-landfilling related reference cases of avoided emissions credits, estimates displayed here vary slightly from the aggregated values in Figure 9.8.

[52] See Annex II for the complete list of references providing estimates for this figure and description of the literature review method.

360 g CO_2eq/kWh (100 g CO_2eq/MJ).[53] Again, variability is caused by differences in study methods, agricultural practice, technology performance and maturity of development (see Section 2.3.3). While the range and central tendency of each evaluated technology are similar to each other, the figure shows that depending on business-as-usual assumptions, avoided GHG emissions (here, mostly methane from landfills) from non-harvest wastes and residues can more than outweigh the GHG emissions associated with the biomass supply chains. Technologies with high conversion efficiency reach lower GHG emissions per kWh generated than less efficient technologies do. Though not displayed here, CHP and other integrated systems with many products could also be an effective way to minimize GHG emissions per unit of primary energy (e.g., in terms of primary energy), though the way co-products are considered in the quantification and allocation of GHG emissions can lead to different results. In the end, the economic value of outputs plays a decisive role, but climate policies that influence the cost of GHG emissions may alter the balance of products.

LCA aspects found to be especially important for GHG results are: (1) assumptions regarding GHG emissions from biomass production where LUC emissions (see Section 2.5.3) and nitrous oxide (N_2O) emissions are especially important; (2) methods used for considering co-products; (3) assumptions about conversion process design, process integration and the type of process fuel used in the conversion of biomass to solid or fluid fuels; (4) the performance of end-use technology, that is, vehicle technology or power/heat plant performance; and (5) the reference system.

N_2O emissions can have an important impact on the overall GHG balance of biofuels (Smeets et al., 2009; Soimakallio et al., 2009b). N_2O emissions vary considerably with environmental and management conditions, including soil water content, temperature, texture, carbon availability, and, most importantly, nitrogen fertilizer input (Bouwman et al., 2002; Stehfest and Bouwman, 2006). Emission factors are used to quantify N_2O emissions as a function of nitrogen fertilizer input. Crutzen et al. (2007) proposed that N_2O emissions from fresh anthropogenic nitrogen are considerably higher than results based on the IPCC's recommended tier 1 method and that N_2O emissions from biofuels consequently have been underestimated by a factor of two to three. IPCC tier 1 and Crutzen et al. (2007) estimates use different accounting approaches. About one-third of agricultural N_2O emissions are due to newly-fixed nitrogen fertilizer (A. Mosier et al., 1998) and two-thirds occur as nitrogen is recycled internally in animal production or by using plant residues as fertilizers. Recent modelling efforts by Davidson (2009) support the conclusion that emission factors based on Crutzen et al. (2007) overestimate the emissions. Using N_2O emissions factors from Crutzen et al. (2007) makes a specific bioenergy plantation responsible for all N_2O emissions taking place subsequently, even for the part of the applied nitrogen that is recirculated into other agriculture systems and substituted for other nitrogen input. See Bessou et al. (2010) for an overview of reactive nitrogen emissions impacts on LCAs.

Process fuel choice is critical and the use of coal especially can drastically reduce the climate benefit of bioenergy. Process integration and the use of biomass fuels or surplus heat from nearby energy/industrial plants can lower net GHG emissions from the biomass conversion process. For example, Wang et al. (2007) showed that GHG emissions for US corn ethanol can vary significantly—from a 3% increase if coal is the process fuel to a 52% reduction if wood chips are used or if improved dry milling processes are used (Wang et al., 2011). Similarly, the low fossil GHG emissions reported for Swedish cereal ethanol plants are explained by their use of biomass-based process energy (Börjesson, 2009). Sugarcane ethanol plants that use the fibrous by-product bagasse as process fuel can provide their own heat, steam and electricity and export surplus electricity to the grid (Macedo et al., 2008). Further improvements are possible as mechanical harvesting becomes established practice, because harvest residues can also be used for energy (Seabra et al., 2010).

However, the marginal benefit of using surplus heat or biomass for the conversion process depends on local economic circumstances and on alternative uses for the surplus heat and biomass (e.g., it could displace coal-based heat or power generation elsewhere). GHG reductions per unit weight of total biomass could be small when biomass is used both as a feedstock and as a process fuel for conversion to biofuels. This underscores the importance of using several indicators in bioenergy option evaluations (see also Section 9.3.4).

Practical uses of indicators to design and establish projects
As shown above, climate change effects can be evaluated based on indicators such as g CO_2eq per MJ (Figure 2.10) or per kWh (Figure 2.11), for which the reference system matters greatly (cf. bioenergy GHG emissions with those from coal and natural gas). Other indicators include mileage per hectare or per unit weight of biomass or per vehicle-km (see Section 8.3.1.3).[54] Limiting resources may define the extent to which land management and biomass-derived fuels can contribute to climate change mitigation, making the following indicators relevant in different contexts (Schlamadinger et al., 2005).

The *displacement factor* indicator describes the reduction in GHG emissions from the displaced energy system per unit of biomass used (e.g., tonne of carbon equivalent per tonne of carbon contained in the biomass that generated the reduction). This indicator does not discourage fossil inputs in the bioenergy chain if these inputs increase the displacement efficiency but it does not consider costs.

The indicator *relative GHG savings* describes the percentage emissions reduction with respect to the fossil alternative for a specific biomass

---

53 Note that the distributions in Figure 2.11 do not represent an assessment of likelihood; the figure simply reports the distribution of currently published literature estimates that passed screens for quality and relevance.

54 For example, the higher land use efficiency of electric vehicles using bioelectricity compared to ethanol cars reported by Campbell et al. (2009) is partly due to the assumed availability of advanced future drive trains for the bioelectricity option but not for the ethanol option.

use.[55] GHG savings favour biomass options with low GHG emissions. However, this indicator alone cannot distinguish between different biomass uses, such as transport fuel, heat, electricity or CHP, to determine which use reduces emissions more. It ignores the amount of biomass, land or money required, and it can be distorted as each use can have different reference systems.

The indicator *GHG savings per ha (or m² or km²) of land* favours biomass yield and conversion efficiency but ignores costs.[56] Intensified land use that increases the associated GHG emissions (e.g., due to higher fertilizer input) can still improve the indicator value if the amount of biomass produced increases sufficiently.

The indicator *GHG savings per monetary unit input* tends to favour the lowest cost, commercially available bioenergy options. Prioritization based on monetary indicators can lock in current technologies and delay (or preclude) future, more cost-effective or GHG reduction-efficient bioenergy options because their near-term costs are higher.

The usefulness of two indicators for considering local and regional bioenergy options is shown in Table 2.13. In the Finnish study, the use of logging residues in modern CHP plants receives a high ranking in relative GHG savings whether the displaced fossil source is coal or natural gas. However, the displacement factor indicator is only high when coal is displaced and is medium for natural gas displacement. The biodiesel from annual crops option receives the lowest ranking (<1) for both indicators, while the Fischer-Tropsch diesel, with or without electricity from wood residues, receives different rankings depending on indicator and plant configuration but is in all cases higher than crop-derived biodiesel. The standalone plant is the best option from the perspective of relative GHG savings. But if the displacement factor is used the integrated plant is preferable. From the plant owner's perspective, local monetary indicators enable assessment of additional costs of the integrated plant, the relative prices for biomass versus electricity, relative prices for fossil diesel versus CO_2 emissions, as well as existing policy support (and its duration). The differences between the two indicators highlight the need to consider the biomass system when planning bioenergy projects at specific locations. For example, in cases where the displacement factor is less than 1, using biomass to displace fossil fuels would increase net emissions (with respect to the global carbon sink baseline) at least within the next decades. The use of such biomass resources could be sustainable; but is not climate or emissions neutral during that period. Additional fossil carbon reductions may then be needed to achieve low GHG concentration stabilization levels.

For North American corn ethanol, technology improvements from 1995 to 2005 are reflected in both indicators. Implementation of improvements in plant efficiency with existing cogeneration systems brings

Table 2.13 | Two indicators of GHG performance facilitate ranking of new technologies using forest residues and comparison with current agricultural biofuel. Two indicators show improvement of technology performance with time for commercial ethanol systems and project the impact of technology improvements. Ranking: High >70; Low <30.

		Fossil energy reference	Displacement factor[1]	Relative GHG savings[2] (%)
Finnish modern CHP plant (from logging residues)		Coal	78	86[e]
		Natural gas	30	86[e]
Finnish Fischer-Tropsch diesel[3] as a standalone plant or integrated with a pulp and paper mill plant; with/without electricity	Standalone plant	Fossil diesel	39[a]	78[f]
	Integrated plant, minimize biomass		50[b]	55[g]
	Integrated plant, minimize electricity		50[c]	78[h]
Finnish biodiesel (rapeseed oil)		Fossil diesel	-9[d]	-15[i]
North American ethanol (corn) powered by natural gas (NG) dry mill 1995 2005 2015 with CHP[3] 2015 with CHP and CCS[3]		Fossil gasoline	18 24 31 51	26 39 55 72
Brazilian ethanol (sugarcane) 2005–2006 (average 44 mills) 2020 CHP[3] (mechanical harvest) 2020 CHP and CCS[3]		Fossil gasoline/ electricity marginal NG	29 36 51	79 120 160

Notes: 1. Tonne of carbon equivalent displaced per tonne of biomass carbon in the feedstock. 2. With respect to the fossil alternative and excluding LUC. 3. Projected performance
Uncertainty ranges: For displacement factors a. 35–46; b. 21–61; c. 45–57; d. -107–7. For relative GHG savings e. 60–94; f. 67–90; g. 31–86; h. 69–89; i. -150–5
References: Finland, Soimakallio et al. (2009b); North America, (S&T)2 Consultants (2009); and Brazil, Möllersten et al. (2003) and Macedo et al. (2008).

---

55 Relative GHG savings are used, for instance, in the EU Directive on Renewable Energy (European Commission, 2009).

56 See Bessou et al. (2010) for examples of LCA emissions as a function of area needed for a variety of feedstocks and biofuels in specific countries.

both indicators to medium range but improves the GHG reduction more than the displacement factor indicator. Application of developing CCS is projected to improve both indicators significantly and bring the GHG reduction indicator to high. In all Brazilian sugarcane ethanol cases, the GHG reduction indicator is high while the displacement factor is low to medium, which is expected because marginal natural gas, not coal, is the displaced fossil fuel and this is a site characteristic (EPE, 2010). The land use indicator differentiates the corn and sugarcane ethanol systems as producing 3,500 and 7,500 litres/ha, respectively. By 2020, biomass productivity increases and also CHP are projected to increase the land use indicator for corn and sugarcane ethanol systems to 4,500 and 12,000 litres/ha, respectively (Möllersten et al., 2003; Macedo et al., 2008; (S&T)[2] Consultants, 2009). See also Wang et al. (2011) for more recent data confirming these trends.

2.5.3 Modern bioenergy: Climate change including land use change effects

Bioenergy is different from the other RE technologies in that it is a part of the terrestrial carbon cycle. The CO_2 emitted due to bioenergy use was earlier sequestered from the atmosphere and will be sequestered again if the bioenergy system is managed sustainably, although emissions and sequestration are not necessarily in temporal balance with each other (e.g., due to long rotation periods of forest stands). In addition to changes in atmospheric carbon, bioenergy use may cause changes in terrestrial carbon stocks. The significance of land use and LUC (e.g., Leemans et al., 1996) and forest rotation (Marland and Schlamadinger, 1997) was demonstrated in the 1990s when dLUC effects were also considered in LCA studies (e.g., Reinhardt, 1991; DeLuchi, 1993). DeLuchi (1993) also called for consideration of indirect effects and iLUC. These effects were first considered about 10 years later (Jungk and Reinhardt, 2000), but most LCA studies have not considered iLUC. LUC can affect GHG emissions in a number of ways, including when biomass is burned in the field during land clearing; when the land management practice changes so that the carbon stocks in soils and vegetation change and/or non-CO_2 emissions (N_2O, ammonium (NH_4^+)) change; and when LUC results in changes in rates of carbon sequestration, that is, CO_2 assimilation by the land increases or decreases relative to the case in which LUC is absent.

Schlamadinger et al. (2001) proposed that bioenergy can have direct/indirect, positive/negative effects on biospheric carbon stocks and that crediting under the CDM could stimulate development of systems that function as a positive carbon sink. Recently, negative effects have been re-emphasized, and studies have estimated LUC emissions associated with, primarily, biofuels for transport. Other bioenergy systems and impact categories (e.g., biodiversity, eutrophication; see Section 2.2.4) have received less attention (see Section 9.3.4). There has been little connection with earlier research in the area of land use, LUC and forestry that partly addressed similar concerns, for example, direct environmental and socioeconomic impacts and leakage (Watson, 2000b).

The quantification of the net GHG effects of dLUC occurring on the site used for bioenergy feedstock production requires definition of reference land use and carbon stock data for relevant land types. Carbon stock data can be uncertain but still appear to allow quantification of dLUC emissions with sufficient confidence for guiding policy (see, e.g., Gibbs et al., 2008).

The quantification of the GHG effects of iLUC is more uncertain. Existing methods for studying iLUC effects employ either (1) a deterministic approach where global LUC is allocated to specific biofuels/feedstocks grown on specified land types (Fritsche et al., 2010); or (2) economic equilibrium models integrating biophysical information and/or biophysical models (Edwards et al., 2010; EPA, 2010; Hertel et al., 2010a,b; Plevin et al., 2010). In the second approach, the amount (and approximate location) of additional land required to produce a specified amount of bioenergy is typically projected. This land is then distributed over land cover categories in line with historic LUC patterns, and iLUC emissions are calculated in the same way as dLUC emissions are. There are inherent uncertainties in this approach because models are calibrated against historic data and are best suited for studying existing production systems and land use regimes. Difficult aspects to model include innovation and paradigm shifts in land use including the presently little-used biomass and mixed production systems described in Sections 2.3 and 2.6. There are also studies that compare scenarios with and without increases in bioenergy to derive LUC associated with the bioenergy expansion (e.g., Fischer et al., 2009). Despite the uncertainties, important conclusions can be drawn from these studies.

Production and use of bioenergy influences climate change through:

- Emissions from the bioenergy chain including non-CO_2 GHG and fossil CO_2 emissions from auxiliary energy use in the biofuel chain.

- GHG emissions related to changes in biospheric carbon stocks often caused by associated LUC.

- Other non-GHG related climatic forcers including particulate and black carbon emissions from small-scale bioenergy use (Ramanathan and Carmichael, 2008), aerosol emissions associated with forests (Carslaw et al., 2010) and changes in surface albedo. Reduction in albedo due to the introduction of perennial green vegetative cover can counteract the climate change mitigation benefit of bioenergy in regions with seasonal snow cover or a seasonal dry period (e.g., savannas). Conversely, albedo increases associated with the conversion of forests to energy crops (e.g., annual crops and grasses) may reduce the net climate change effect from the deforestation (Schwaiger and Bird, 2010).

- Effects due to the bioenergy use, such as price effects on petroleum that impact consumption levels. The net effect is the difference between the influence of the bioenergy system and of the energy system (often fossil-based) that is displaced. Current fossil energy chains and evolving non-conventional sources have land use impacts (Gorissen et al., 2010; Liska and Perrin, 2010; Yeh et al., 2010), but LUC has a tighter link to bioenergy because of its close association with agriculture and forestry.

- Other factors include the extent and timing of the reversion of cultivated land when the use for bioenergy production ends and how future climate change impacts relative to present impacts are treated (DeLucchi, 2010).

Mitigation efforts over the next two to three decades will influence prospects for achieving lower stabilization levels (van Vuuren et al., 2007; den Elzen et al., 2010). For instance, the dynamics of terrestrial carbon stocks in LUC and long-rotation forestry lead to GHG mitigation trade-offs between biomass extraction for energy use and the alternative to leave the biomass as a carbon store that could further sequester more carbon over time (Marland and Schlamadinger, 1997; Marland et al., 2007; Righelato and Spracklen, 2007). Observations indicate that old forests can be net carbon sinks (Luyssaert et al., 2008; Lewis et al., 2009) but fires, insect outbreaks and other natural disturbances can quickly convert a forest from a net sink to an emitter (Kurz et al., 2008a,b; Lindner et al., 2010).

Short- and long-term indicators

Indicators such as *carbon debt* (Fargione et al., 2008) and *ecosystem carbon payback time* (Gibbs et al., 2008) focus on upfront LUC emissions arising from the conversion of land to bioenergy production. The balance between short- and long-term emissions and the climate benefits of bioenergy projects are reflected in indicators that describe the dynamic effect of GHG emissions (see also Section 9.3.4), for example, *cumulative warming impacts* or *global warming potential* (Kirschbaum, 2003, 2006; Dornburg and Marland, 2008; Fearnside, 2008). These indicators have been used, to a limited extent, to describe bioenergy dynamic climate effects (Kendall et al., 2009; Kirkinen et al., 2009; Levasseur et al., 2010; O'Hare et al., 2009).

Figure 2.12 shows dLUC effects on GHG balances for liquid biofuels using the ecosystem carbon payback time indicator. The left diagram shows payback times with current yields and conversion efficiencies and the right diagram shows the effect of higher yields (set to equal the top 10% of area-weighted yields). The payback times in Figure 2.12 neglect the GHG emissions associated with production and distribution of the transport fuels. Because these emissions currently tend to be higher for biofuels than for gasoline and diesel, the payback times are underestimated. The payback times in Figure 2.12 are calculated assuming constant GHG savings from the gasoline/diesel displacement. Higher GHG savings, that is, reducing the payback times, would be achieved if the biofuels conversion efficiency improved, if more carbon intensive transport fuels were replaced, or if the produced biomass displaced carbon-intensive fossil options for heat/power (Figure 2.10). Further biomass yield increases would reduce payback times but may require higher agronomic inputs that lead to increased GHG emissions,

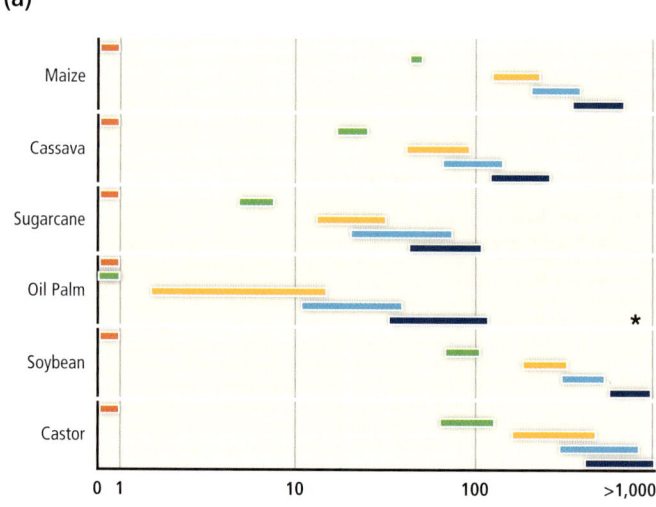

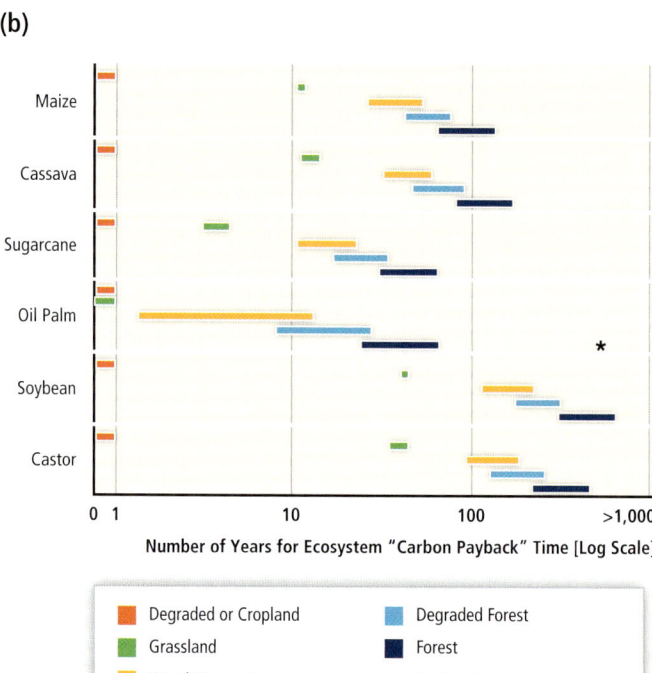

Figure 2.12 | The ecosystem carbon payback time for potential biofuel crop expansion pathways across the tropics comparing the year 2000 agricultural system shown in (a) with a future higher yield scenario (b) which was set to equal the top 10% of area-weighted yields. The asterisk represents oil palm crops grown in peatlands with payback times greater than 900 years in the year 2000 compared to 600 years for a 10% increase in crop productivity. Based on Gibbs et al. (2008) and reproduced with permission from IOP Publishing Ltd.

notably N_2O. The payback times would increase if the feedstock production resulted in land degradation over time, impacting yield levels or requiring increased input to maintain yield levels.

As shown, all biofuel options have significant payback times when dense forests are converted into bioenergy plantations. The starred points represent very long payback times for oil palm establishment on tropical peat swamp forests because drainage leads to peat oxidation and causes CO_2 emissions that occur over several decades and that can be several times higher than the displaced emissions of fossil diesel (Hooijer et al., 2006; Edwards et al., 2008, 2010). Under natural conditions, these tropical peat swamp forests have negligible CO_2 emissions and small methane emissions (Jauhiainen et al., 2008). Payback times are practically zero when degraded land or cropland is used, and they are relatively low for the most productive systems when grasslands and woody savannas are used (not considering the iLUC that can arise if these lands were originally used, for example, for grazing).

Targeting unused marginal and degraded lands for bioenergy production can thus mitigate dLUC emissions. For some options (e.g., perennial grasses, woody plants, mechanically harvested sugarcane), net gains of soil and aboveground carbon can be obtained (Tilman et al., 2006b; Liebig et al., 2008; Robertson et al., 2008; Anderson-Teixeira et al., 2009; Dondini et al., 2009; Hillier et al., 2009; Galdos et al., 2010). In this context, land application of biochar produced via pyrolysis could be an option to sequester carbon in a more stable form and improve the structure and fertility of soils (Laird et al., 2009; Woolf et al., 2010).

Bioenergy does not always result in LUC. Bioenergy feedstocks can be produced in combination with food and fibre, avoiding land use displacement and improving the productive use of land (Section 2.2). These possibilities may be available for bioenergy options that can use lignocellulosic biomass but also for some other options that use waste oil and oil seeds such as *Jatropha* (Section 2.3). The use of post-consumer organic waste and by-products from the agricultural and forest industries does not cause LUC if these biomass sources are wastes, that is, they were not utilized for alternative purposes. On the other hand, if not utilized for bioenergy, some biomass sources (e.g., harvest residues left in the forest) would retain organic carbon for a longer time than if used for energy. Such delayed GHG emissions can be considered a benefit in relation to near-term GHG mitigation, and this is an especially relevant factor in longer-term accounting for regions where biomass degradation is slow (e.g., boreal forests). However, as noted above, natural disturbances can convert forests from net sinks to net sources of GHGs, and dead wood left in forests can be lost in fires. In forest lands susceptible to periodic fires, good silviculture practices can lead to less frequent, lower intensity fires that accelerate forest growth rates and soil carbon storage. Using biomass removed in such practices for bioenergy can provide GHG and particulate emission reductions.

For different world regions, Edwards et al. (2010) describe the comparison of six equilibrium models to quantify LUC associated with a standard biofuel shock defined as a marginal increase in demand for first-generation ethanol or biodiesel from a base year.[57] All models showed significant LUC (dLUC and iLUC were not considered separable) with variations between models in terms of the extent of LUC and its distribution over regions and crops. A follow-on study by Hiederer et al. (2010) compared the ranges of LUC emissions shown in Figure 2.13 for common biofuel crops as a function of the 'biofuel shock' (0.2 to 1.5 EJ) for select studies. Figure 2.13 also shows the 2010 EPA model results with a relatively high resolution of land use distribution[58] for Brazil resulting in mid-range LUC emissions for sugarcane ethanol (5 to 10 g CO_2eq/MJ), similar to the European study (Al-Riffai et al., 2010) estimate of 12 g CO_2eq/MJ. The Brazilian study with measured LUC dynamics for common crops and native vegetation between 2005 and 2008 by Nassar et al. (2010) obtained 8 g CO_2eq/MJ for iLUC and dLUC, with the latter being nearly zero. Fischer et al. (2010) obtained 28 g CO_2eq/MJ using a deterministic methodology and assuming a high risk of deforestation. Model results from Figure 2.13 show all other crops as having higher LUC values than sugarcane ethanol. In the US maize ethanol case, Plevin et al. (2010) report a plausible range of 25 to 150 g CO_2eq/MJ based on uncertainty analysis of various model parameters and assumptions.

The utility of these models to study scenarios is illustrated with an analysis of the relative contributions of changes in yield and land area to increased crop output along with assumptions about trade-critical factors in model-based LUC estimates (D. Keeney and Hertel, 2009). Subsequent model improvements incorporate crop yields, by-product markets interactions, and trade and policy assumptions, and analyze past and project future usage with existing (2010) EU and US policies, finding LUC in other countries such as Latin America and Oceania to be primarily at the expense of pastureland followed by commercial forests (Hertel et al., 2010a,b).

Lywood et al. (2009b) report that the extent to which output change comes from increased crop yield or land area changes varies between crops and regions. They estimate that yield growth contributed 80 and 60% of the incremental output growth for EU cereals and US maize, respectively, between 1961 and 2007. Conversely, area expansion

---

57 Biofuel shock (Hertel et al., 2010a,b) is introduced in general equilibrium models by changing some economic parameters (e.g., subsidies to ethanol production) to reach predetermined volume levels (i.e., sum of government mandates for a certain year). The comparison of new and previously determined equilibrium enables estimates of land area changes impacted directly to meet mandates and those indirectly involved to compensate for that agricultural production no longer available, its co-products and its impact throughout the global economic chain. These studies have high uncertainties. Partial equilibrium models were also included in Edwards et al. (2010).

58 Based on the Nassar et al. (2009) Brazilian Land Use Model, which shows a lower share of LUC due to deforestation. More recently, Nassar et al. (2010) obtained elasticities for models from direct data (statistical and satellite-based) of land use substitution over time. The matrix elasticity results for major crops in various regions provide a deterministic estimate for the d+iLUC of sugarcane ethanol of about 8 g CO_2eq/MJ. Higher substitution coefficients are found for soy into native vegetation.

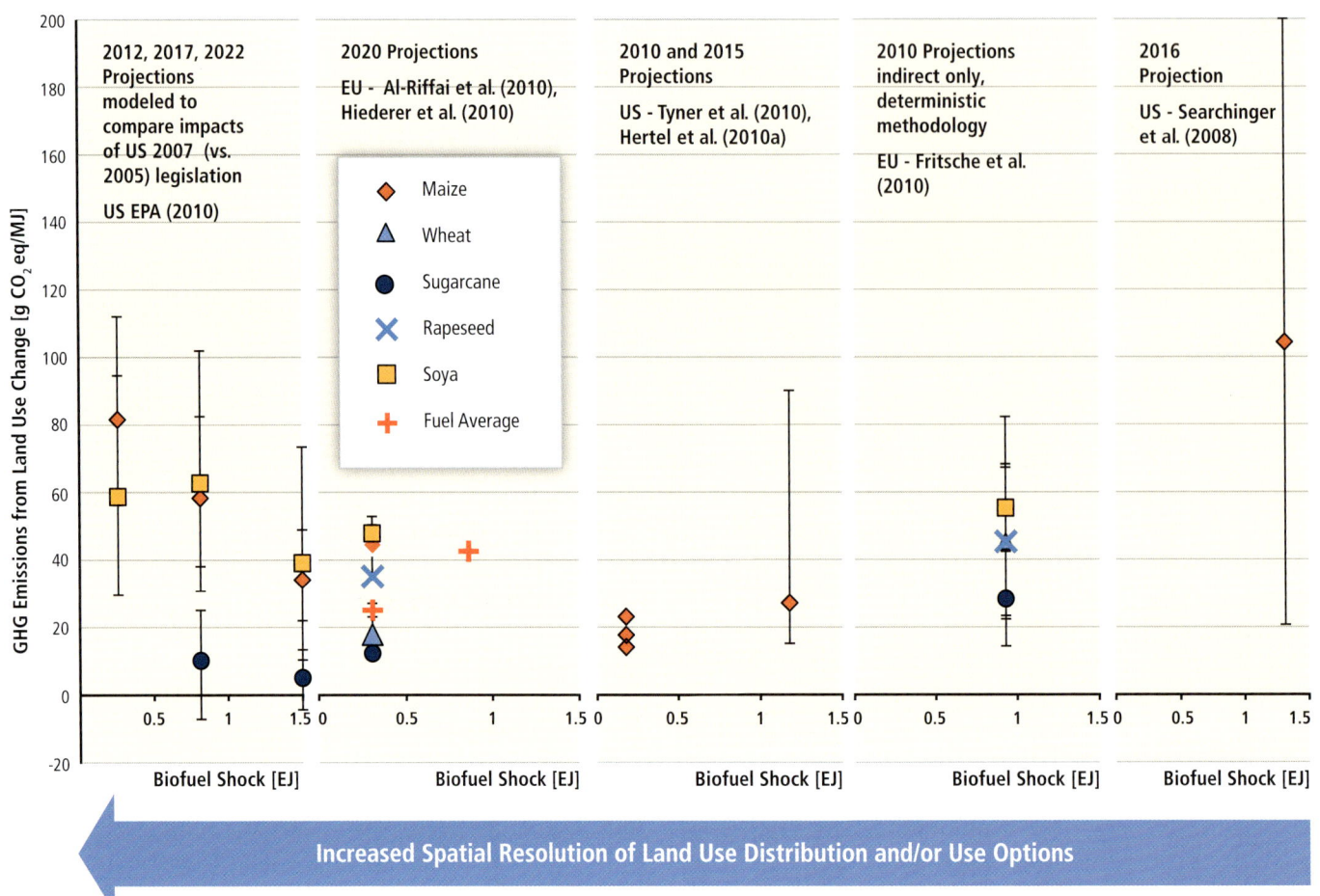

Figure 2.13 | Select model-based estimates of LUC emissions for major biofuel crops given a certain level of demand, a biofuel shock, expressed in EJ (30-year accounting framework). Mid-range values of multiple studies (g CO_2eq/MJ): 14 to 82 for US maize ethanol with high-resolution models and 100 for earlier models; 5 to 28 for sugarcane ethanol; 18 to 45 for European wheat ethanol; 40 to 63 for soy biodiesel (uncertain); and 35 to 45 for rapeseed biodiesel. Points for Tyner et al. (2010) and Hertel et al. (2010a) represent model improvements with the lowest value including feedstock yield and population increases (baseline 2006). Fritsche et al. (2010) value ranges derive from a deterministic methodology representing risk values of 25 and 75% of the theoretical worst case of LUC scenarios, such as high deforestation, to calculate iLUC.

contributed to more than 60% of output growth for EU rapeseed, Brazilian sugarcane, South American soy, and Southeast Asia oil palm. Studies report price-yield relationships; there is a weak basis for deriving these relationships (D. Keeney and Hertel, 2008) although rising oil prices and fuel tax exemptions show strong correlations for the USA and EU, respectively. Edwards et al. (2010) state that the marginal area requirement per additional unit output of a particular biofuel should increase due to decreasing productivity of additional land converted to biofuel feedstock production (also reflected in, e.g., R. Keeney and Hertel, 2005; Tabeau et al., 2006). Lywood et al. (2009b), however, state that in the case of EU cereals and US corn, there is no evidence that average yields decline as more land is used. The assumed or modelled displacement effect of process co-products used as feed can also have a strong influence on LUC values.

For European biofuels, if soy meal and cereals for feed are displaced, the net land area required to produce biofuel from EU cereal, rapeseed and sugar beet is much lower than the gross land requirement (e.g., only 6% for ethanol from feed wheat in northwestern Europe (Lywood et al., 2009a). Lywood et al. (2008) obtained large improvements in net GHG savings for European cereal ethanol and rapeseed biodiesel based on co-products displacing imported soy as animal feed, which reduces deforestation and other LUC for soy cultivation in Brazil. Conversely, increased corn cultivation at the cost of soy cultivation, in response to increasing ethanol demand in the USA, has been reported to increase soy cultivation in other countries such as Brazil (Laurance, 2007). Trade assumptions are critical and differ in the various models. In addition, marginal displacement effects of co-products may have a saturation level (McCoy, 2006; Edwards et al., 2010), although new uses may be developed, for example, to produce more biofuels (Yazdani and Gonzalez, 2007).

Bioenergy options that use lignocellulosic feedstocks are projected to have lower LUC values than those of first-generation biofuels (see, e.g., EPA, 2010; Hoefnagels et al., 2010; see Figure 9.9). As noted above, some of these feedstock sources can be used without causing LUC. Lower LUC values might be expected because of high biomass productivity, multiple products (e.g., animal feed) or avoided competition for prime cropland by using more marginal lands (Sections 2.2 and 2.3). The lower productivity of marginal lands, however, results in higher land requirements per given biomass output and presents particular challenges as discussed in Section 2.2. Also, as many lignocellulosic plants are grown under longer rotations, they should be less responsive to price increases because the average yield over a plantation lifetime can only be influenced through agronomic means (notably increased fertilizer input) and by variety selection at the time of replanting. Thus, output growth in response to increasing demand is more readily obtained by area expansion.

Depending on the atmospheric lifetime of specific GHGs, the trade-off between emitting more now and less in the future is not one-to-one in general. But the relationship for CO_2 is practically one-to-one, so that one additional (less) tonne CO_2 emitted today requires a future reduction (allows a future increase) by one tonne. This relationship is due to the close to irreversible climate effect of CO_2 emissions (Matthews and Caldeira, 2008; M. Allen et al., 2009; Matthews et al., 2009; Solomon et al., 2009).

Integrated energy-industry-land use/cover models can give insights into how an expanding bioenergy sector interacts with other sectors in society, influencing longer-term energy sector development, land use, management of biospheric carbon stocks, and global cumulative GHG emissions. In an example of early studies, Leemans et al. (1996) implemented in the IMAGE model (Integrated Model to Assess the Global Environment) the LESS (low CO_2-emitting energy supply system) scenario, which was developed for the IPCC Second Assessment Report (IPCC, 1996). This study showed that the required land use expansion to provide biomass feedstock can cause significant food-bioenergy competition and influence deforestation rates with significant consequences for environmental issues such as biodiversity, and that the outcome is sensitive to regional emissions and feedback in the carbon cycle. More recently, using linked economic and terrestrial biogeochemistry models, Melillo et al. (2009) found a similar level of cumulative CO_2 emissions associated with LUC from an expanded global cellulosic biofuels programme over the 21st century. The study concluded that iLUC was a larger source of carbon loss than dLUC; fertilizer N_2O emissions were a substantial source of global warming; and forest protection and best practices for nitrogen fertilizer use could dramatically reduce emissions associated with biofuels production.

Wise et al. (2009) also stressed the importance of limiting terrestrial carbon emissions and showed how the design of mitigation regimes can strongly influence the nature of bioenergy development and associated environmental consequences, including the net GHG savings from bioenergy. Including both fossil and LUC emissions in a carbon tax regime, instead of taxing only fossil emissions, was found to lower the cost of meeting environmental goals. However, this tax regime was also found to induce rising food crop and livestock prices and expansion of unmanaged ecosystems and forests. Improved crop productivity was proposed as a potentially important means for GHG emissions reduction, with the caution that non-CO_2 emissions (not modelled) need to be considered.

Biospheric carbon pricing as a sufficient mechanism to protect forests was proposed by Wise et al. (2009) and supported by Venter et al. (2009) and others. Persson and Azar (2010) acknowledge that pricing LUC carbon emissions could potentially make many of the current proximate causes of deforestation unprofitable (e.g., extensive cattle ranching, small-scale slash-and-burn agriculture and fuelwood use) but they question whether it will suffice to make deforestation for bioenergy production unprofitable because these bioenergy systems are highly productive according to the Wise et al. (2009) assumptions of generic feedstock productivity and biofuel conversion efficiency. A higher carbon price will increase not only the cost of forest clearing but also the revenues from certain bioenergy production systems. The upfront cost of land conversion may also be reduced if the bioenergy industry partners with the timber and pulp industries that seek access to timber revenues from clear felling forests as the first step in plantation development (Fitzherbert et al., 2008).

Three tentative conclusions are:

1. Additional, and stronger, protection measures may be needed to meet the objective of tropical forest preservation. A strict focus on the climate benefits of ecosystem preservation may put undue pressure on valuable ecosystems that have a relatively low carbon density. While this may have a small impact in terms of climate change mitigation, it may negatively impact other parts of the ecosystem, for example, biodiversity and water tables.

2. From a strict climate and cost efficiency perspective, in some places a certain level of upfront LUC emissions may be acceptable in converting forest to highly productive bioenergy plantations due to the climate benefits of subsequent continued biofuel production and fossil fuel displacement. The balance between bioenergy expansion benefits and LUC impacts on biodiversity, water and soil conservation is delicate. Climate change mitigation is just one of many rationales for ecosystem protection.

3. iLUC effects strongly (up to fully) depend on the rate of improvement in agricultural and livestock management and the rate of deployment of bioenergy production. Subsequently, implementation of bioenergy production and energy cropping schemes that follow effective sustainability frameworks and start from simultaneous improvements in agricultural management could mitigate conflicts and allow realization of positive outcomes, for example, in rural development, land amelioration and climate change mitigation including opportunities to combine adaptation measures.

2.5.4 Traditional biomass: Climate change effects

Traditional open fires and simple low-efficiency stoves have low combustion efficiency, producing large amounts of incomplete combustion products (CO, methane, particle matter, non-methane volatile organic compounds, and others) that have negative consequences for climate change and local air pollution (Smith et al., 2000; see also Box 9.4 in Section 9.3.4.2). When biomass is harvested renewably—for example, from standing trees or agricultural residues—CO_2 already emitted to the atmosphere is sequestered as biomass re-grows. Because the products of incomplete combustion also include important short-lived greenhouse pollutants and black carbon, even sustainable harvesting does not make such fuel cycles GHG neutral. Worldwide, it is estimated that household fuel combustion causes approximately 30% of the warming due to black carbon and CO emissions from human sources, about 15% of ozone-forming chemicals, and a few percent of methane and CO_2 emissions (Wilkinson et al., 2009).

Improved cookstoves (ICS) and other advanced biomass systems for cooking are cost-effective for achieving large benefits in energy use reduction and climate change mitigation. Fuel savings of 30 to 60% are reported (Berrueta et al., 2008; Jetter and Kariher, 2009). The savings in GHG emissions associated with these efficient stoves are difficult to derive because of the wide range of fuel types, stove designs, cooking practices and environmental conditions across the world. However, advanced biomass systems, such as small-scale gasifier stoves and biogas stoves, have had design improvements that increase combustion efficiency and dramatically reduce the production of short-lived GHGs by up to 90% relative to traditional stoves. Some of these new stoves even reach performance levels similar to liquid propane gas (Jetter and Kariher, 2009). Patsari improved stoves in rural Mexico save between 3 and 9 t CO_2eq/stove/yr relative to open fires, with renewable or non-renewable harvesting of biomass, respectively (M. Johnson et al., 2009).

Venkataraman et al. (2010) estimate that the dissemination of 160 million advanced ICS in India may result in the mitigation of 80 Mt CO_2eq/yr, or more than 4% of India's total estimated GHG emissions, plus a 30% reduction in India's human-caused black carbon emissions. Worldwide, with GHG mitigation per unit at 1 to 4 t CO_2eq/stove/yr compared to traditional open fires, the global mitigation potential of advanced ICS was estimated to be between 0.6 and 2.4 Gt CO_2eq/yr. This estimate does not consider the additional potential reduction in black carbon emissions. Actual figures depend on the renewability of the biomass fuel production, stove and fuel characteristics, and the actual adoption and sustained used of improved cookstoves. Reduction in fuelwood and charcoal use due to the adoption of advanced ICS may help reduce pressure on forest and agricultural areas and improve aboveground biomass stocks and soil and biodiversity conservation (Ravindranath et al., 2006; García-Frapolli et al., 2010).

2.5.5 Environmental impacts other than greenhouse gas emissions

2.5.5.1 Impacts on air quality and water resources

Air pollutant emissions from bioenergy production depend on technology, fuel properties, process conditions and installed emission reduction technologies. Compared to coal and oil stationary applications, sulphur dioxide (SO_2) and nitrous oxide (NO_x) emissions from bioenergy applications are mostly lower (see also Section 9.3.4.2). When biofuel replaces gasoline and diesel in the transport sector, SO_2 emissions are reduced, but changes in NO_x emissions depend on the substitution pattern and technology. The effects of replacing gasoline with ethanol and biodiesel also depend on engine features. Biodiesel can have higher NO_x emissions than petroleum diesel in traditional direct-injected diesel engines that are not equipped with NO_x control catalysts (e.g., Verhaeven et al., 2005; Yanowitz and McCormick, 2009).

Bioenergy production can have both positive and negative effects on water resources (see also Section 9.3.4.4). Bioenergy production generally consumes more water than gasoline production (Wu et al., 2009; Fingerman et al., 2010). However, this relationship and the water impacts of bioenergy production are highly dependent on location, the specific feedstock, production methods and the supply chain element.

Feedstock cultivation can lead to leaching and emission of nutrients that increase eutrophication of aquatic ecosystems (Millennium Ecosystem Assessment, 2005; SCBD, 2006; Spranger et al., 2008). Pesticide emissions to water bodies may also negatively impact aquatic life. Given that several types of energy crops are perennials grown in arable fields being used temporarily as a pasture for grazing animals or woody crops grown in multi-year rotations, the increasing bioenergy demand may drive land use towards systems with substantially higher water productivity. On the other hand, shifting demand to alternative—mainly lignocellulosic—bioenergy can decrease water competition. Perennial herbaceous crops and short-rotation woody crops generally require fewer agronomic inputs and have reduced impacts compared to annual crops, although large-scale production can require high levels of nutrient input (see Sections 2.2.4.2 and 2.3.1). Water impacts can also be mitigated by integrating lignocellulosic feedstocks in agricultural landscapes as vegetation filters to capture nutrients in passing water (Börjesson and Berndes, 2006). A prolonged growing season may redirect unproductive soil evaporation and runoff to plant transpiration (Berndes, 2008a,b). Crops that provide a continuous cover over the year can also conserve soil outside the growing season of annual crops by diminishing the erosion from precipitation and runoff (Berndes, 2008a,b). A number of bioenergy crops can be grown on a wide spectrum of land types that are not suitable for conventional food or feed crops. These marginal lands, pastures and grasslands could become available for feedstock production under sustainable management practices (if adverse downstream water impacts can be mitigated).

The subsequent processing of the feedstock into biofuels and electricity can increase chemical and thermal pollution loads from effluents and generate waste to aquatic systems (Martinelli and Filoso 2007, Simpson et al., 2008). These environmental impacts can be reduced if suitable equipment is installed (Wilkie et al., 2000; BNDES/CGEE, 2008).

Water demand for bioenergy can be reduced substantially through process changes and recycling (D. Keeney and Muller, 2006; BNDES/CGEE, 2008). Currently, most water is lost to the atmosphere through evapotranspiration during the production of cultivated feedstock (Berndes, 2002). Feedstock processing into fuels and electricity requires much less water (Aden et al., 2002; Berndes, 2002; D. Keeney and Muller, 2006; Phillips et al., 2007; NRC, 2008; Wang et al., 2010), but water needs to be extracted from lakes, rivers and other water bodies.

2.5.5.2 Biodiversity and habitat loss

Habitat loss is one of the major drivers of biodiversity decline globally and is projected to be the major driver of biodiversity loss and decline over the next 50 years (Sala et al., 2000; UNEP, 2008b; see Sections 9.3.4.5 and 9.3.4.6). Increased biomass output for bioenergy can directly impact wild biodiversity through conversion of natural ecosystems into bioenergy plantations or through changed forest management. Habitat and biodiversity loss may also occur indirectly, such as when productive land use displaced by energy crops is re-established by converting natural ecosystems into croplands or pastures elsewhere. Because biomass feedstocks can generally be produced most efficiently in tropical regions, there are strong economic incentives to replace tropical natural ecosystems—many of which host high biodiversity values (Doornbosch and Steenblik, 2008). However, forest clearing is mostly influenced by local social, economic, technological, biophysical, political and demographic forces (Kline and Dale, 2008).

Increasing demand for oilseed has put pressure on areas designated for conservation in some OECD member countries (Steenblik, 2007). Similarly, the rising demand for palm oil has contributed to extensive deforestation in parts of Southeast Asia (UNEP, 2008a). The palm oil plantations support significantly fewer species than the forest they replaced (Fitzherbert et al., 2008).

To the extent that bioenergy systems are based on conventional food and feed crops, biodiversity impacts from pesticide and nutrient loading can be expected from bioenergy expansion. Bioenergy production can also impact agricultural biodiversity when large-scale monocultures, based on a narrow pool of genetic material, reduce the use of traditional varieties.

Depending on a variety of factors, bioenergy expansion can also lead to positive outcomes for biodiversity. Using bioenergy to replace fossil fuels can reduce climate change, which is expected to be a major driver of habitat loss. Establishment of perennial herbaceous plants or short-rotation woody crops in agricultural landscapes has been found to improve biodiversity (Lindenmayer and Nix, 1993; Semere and Slater, 2007; Royal Society, 2008). Bioenergy plantations that are cultivated as vegetation filters can improve biodiversity by reducing the nutrient load and eutrophication in water bodies (Foley et al., 2005; Börjesson and Berndes, 2006) and providing a varied landscape.

Bioenergy plantations can be located in the agricultural landscape to provide ecological corridors through which plants and animals can move between spatially separated natural and semi-natural ecosystems. Thus, bioenergy plantations can reduce the barrier effect of agricultural lands (Firbank, 2008). However, bioenergy plantations can contribute to habitat fragmentation, as has occurred with some oil palm plantations (Danielsen et al. 2009; Fitzherbert, 2008).

Properly located biomass plantations can also protect biodiversity by reducing the pressure on nearby natural forests. A study from Orissa, India, showed that introducing village biomass plantations increased biomass consumption (as a consequence of increased availability) while decreasing pressure on the surrounding natural forests (Köhlin and Ostwald, 2001; Francis et al., 2005).

When crops are grown on degraded or abandoned land, such as previously deforested areas or degraded crop- and grasslands, the production of feedstocks for biofuels could have positive impacts on biodiversity by restoring or conserving soils, habitats and ecosystem functions (Firbank, 2008). For instance, several experiments with selected trees and intensive management on severely degraded Indian wastelands (such as alkaline, sodic or salt-affected lands) showed increases in soil carbon, nitrogen and available phosphorous within eight years (Garg, 1998).

2.5.5.3 Impacts on soil resources

The considerable soil impacts of increased biofuel production include soil carbon oxidation, changed rates of soil erosion, and nutrient leaching. However, these effects are heavily dependent on agronomic techniques and the feedstock under consideration (UNEP, 2008a). Land preparation required for feedstock production, as well as nutrient demand, varies widely across feedstocks. For instance, wheat, rapeseed and corn require significant tillage compared to oil palm, sugarcane and switchgrass (FAO, 2008a; UNEP, 2008a). In sugarcane production, soil quality benefits greatly from recycled nutrients from sugar mill and distillery wastes (IEA, 2006).

Using agricultural residues without proper management can lead to detrimental impacts on soil organic matter through increased erosion. However, this impact depends heavily on management, yield, soil type and location. In some areas, the impact of residue removal may be minimal.

Certain cultivation practices, including conservation tillage and crop rotations, can mitigate adverse impacts and in some cases improve environmental benefits of biofuel production. For example, *Jatropha* can stabilize soils and store moisture while it grows (Dufey, 2006). Other potential benefits of planting feedstocks on degraded or marginal lands include reduced nutrient leaching, increased soil productivity and increased carbon content (Berndes, 2002). If lignocellulosic energy crop plantations, which require low-intensity management and few fossil energy inputs relative to current biofuel systems, are established on abandoned agricultural or degraded land, soil carbon and soil quality could increase over time. This beneficial effect would be especially significant with perennial species.

2.5.6 Environmental health and safety implications

2.5.6.1 Feedstock issues

Currently, many crops used in fuel ethanol manufacturing are also traditional feed sources (e.g., maize, soy, canola and wheat). However, considerable efforts are focused on new crops that either enhance fuel ethanol production (e.g., high-starch corn) or that are not traditional food or feed crops (e.g., switchgrass). If the resultant distillers' grains from these new crops are used as livestock feed or could inadvertently end up in livestock feeds, pre-market assessment of their acceptability in feed prior to their use in fuel ethanol production will be necessary (Hemakanthi and Heller, 2010).

Concerns about cross-pollination, hybridization, pest resistance and disruption of ecosystem functions (FAO, 2004; FAO, 2008; IAASTD, 2009) have limited the use of genetically engineered (GE) crops in some regions. Transgene movement leading to weediness or invasiveness of the crop itself or of its wild or weedy relatives is a major reason (Warwick et al., 2009). Clarity, predictability and established risk assessment processes are literature recommendations to decrease GE crop use concerns (Warwick et al., 2009).[59] The first assessment (NRC, 2010) of the impact of GE crops in use in the USA since 1996 found that benefits to the farmer included increased worker safety from pesticide handling; indicated that water quality improves with GE crops; and acknowledged that more work needs to be done, particularly to install infrastructure to measure water quality impacts, develop weed management practices, and address the needs of farmers whose markets depend on the absence of GE traits.

Several grasses and woody species that are candidates for biofuel production have traits commonly found in invasive species (Howard and Ziller, 2008). These traits include rapid growth, high water-use efficiency and long canopy duration (Clifton-Brown et al., 2000). There are fears that if these crops are introduced, they could become invasive, displace indigenous species and decrease biodiversity. For example, *Jatropha curcas* is considered weedy in several countries, including India and many South American states (Low and Booth, 2007). Warnings have been raised about *Miscanthus* and switchgrass (*Panicum virgatum*). *Sorghum halepense* (Johnson grass), *Arundo donax* (giant reed) and *Phalaris arundinacea* (reed canary grass) are known to be invasive in the USA. A number of protocols have evolved that allow for a systematic assessment and evaluation of the inherent risk associated with species introduction (McWhorter, 1971; Randall, 1996; Molofsky et al., 1999; Dudley, 2000; Forman, 2003; Raghu et al., 2006). DiTomaso et al. (2010) address policies to keep these agro-ecosystems in check while developing desirable biofuels crops, such as preventive actions prior to and during cultivation of biofuel plants.

2.5.6.2 Biofuels production issues

Globally, most biofuels are produced with conventional production technologies (see Section 2.3) that have been used in many industries for many years (Gunderson, 2008; Abbasi and Abbasi, 2010). Hazards associated with most of these technologies are well characterized, and it is possible to limit risks to very low levels by applying existing knowledge and standards (see, e.g., Astbury, 2008; Hollebone and Yang, 2009; Marlair et al., 2009; Williams et al., 2009) and their typology is under development (Rivière and Marlair, 2009, 2010).

The literature highlights environmental health and safety areas for further evaluation as new technologies (see Section 2.6) are developed (e.g., Madsen et al., 2004; Madsen, 2006; Vinnerås et al., 2006; Narayanan et al., 2007; Gunderson, 2008; McLeod et al., 2008; Hill et al., 2009; Martens and Böhm, 2009; Moral et al., 2009; Perry, 2009; Sumner and Layde, 2009). Key areas include:

- Health risk to workers using engineered microorganisms or their metabolites.

- Potential ecosystem effects from the release of engineered microorganisms.

- Impact to workers, biofuel consumers or the environment from pesticides and mycotoxins that accumulate in processing intermediates, residues or products (e.g., spent grains, spent oil seeds).

- Risks to workers from infectious agents that can contaminate feedstocks in production facilities.

- Exposure to toxic substances, particularly for workers at biomass thermochemical processing facilities that use routes not currently practised by the fossil fuels industry.

- Fugitive air emissions and site runoff impacts on public health, air quality, water quality and ecosystems.

---

59 Other concerns include: reduction in crop diversity, increases in herbicide use, herbicide resistance (increased weediness), loss of farmer's sovereignty over seed, ethical concerns over transgenes origin, lack of access to intellectual property rights held by the private sector, and loss of markets owing to moratoriums on genetically modifed organisms (GMOs) (IAASTD, 2009).

- Exposure to toxic substances, particularly if production facilities become as commonplace as landfill sites or natural gas-fired electricity generating stations.

- Cumulative environmental impacts from the siting of multiple biofuel/bioenergy production facilities in the same air- and/or watershed.

2.5.7 Socioeconomic aspects

The large-scale and global development of bioenergy will be associated with a complex set of socioeconomic issues and trade-offs, ranging from local issues (e.g., income and employment generation, improved health conditions, agrarian structure, land tenure, land use competition and strengthening of regional economies) to national issues (e.g., food security, a secure energy supply and balance of trade). Participation of local stakeholders, in particular small farmers and poor households, is essential to ensure socioeconomic benefits from bioenergy projects.

2.5.7.1 Socioeconomic impact studies and sustainability criteria for bioenergy systems

The complex nature of bioenergy, with many conversion routes and the multifaceted potential socioeconomic impacts, makes the overall impact analysis difficult to conduct. Also, many impacts are not easily quantifiable in monetary or numerical terms. To overcome these problems, semi-quantitative methods based on stakeholder involvement have been used to assess social criteria such as societal product benefit and social dialogue[60] (von Geibler et al., 2006).

Regarding economic impacts, the most commonly reported variables are private production costs over the value chain, assuming a fixed set of prices for basic commodities (e.g., for fossil fuels and fertilizers). The bioenergy costs are usually compared to alternatives already on the market (fossil-based) to judge the potential competitiveness. Bioenergy systems are mostly analyzed at a micro-economic level, although interactions with other sectors cannot be ignored because of the competition for land and other resources. Opportunity costs may be calculated from food commodity prices and gross margins to account for food-bioenergy interactions. Social impact indicators include consequences for local employment, although this impact is difficult to assess because of possible offsets between fossil and bioenergy chains. Impacts at a macro-economic level include the social costs incurred because of fiscal measures (e.g., tax exemptions) to support bioenergy chains (DeLucchi, 2005). Fossil energy's negative externalities also need to be assessed (Bickel and Friedrich, 2005).

Several sustainability frameworks and certification systems have been proposed to better document and integrate the socioeconomic impacts of bioenergy systems, particularly at the project level (Bauen et al., 2009b; WBGU, 2009; van Dam et al., 2010; see also Section 2.4). Specifically, criteria and indicators related to the development of liquid biofuels have been proposed for these issues: human rights, including gender issues; working and wage conditions, including health and safety issues; local food security; rural and social development, with special regard to poverty reduction; and land rights (Table 2.12). So far, while rural and local development are included, specific economic criteria for the cost-effectiveness of the projects, level of subsidies and other financial aspects have not been included in the sustainability frameworks. Most of the frameworks are still under development. The progress of certification systems was reviewed by van Dam et al. (2008, 2010). The FAO's Bioenergy and Food Security Criteria and Indicators project has compiled bioenergy sustainability initiatives (see also Sections 2.4.5.1 and 2.4.5.2).

2.5.7.2 Socioeconomic impacts of small-scale systems

The inefficient use of biomass in traditional devices such as open fires has significant socioeconomic impacts including drudgery for getting the fuel, the cost of satisfying cooking needs, and significant health impacts from the very high levels of indoor air pollution, especially for women and children (Masera and Navia, 1997; Pimentel et al., 2001; Biran et al., 2004; Bruce et al., 2006; Romieu et al., 2009). Indoor air pollutants include respirable particles, CO, oxides of nitrogen and sulphur, benzene, formaldehyde, 1, 3-butadiene, and polyaromatic compounds such as benzo(a)pyrene (Smith et al., 2000). Wood smoke exposure can increase respiratory symptoms and problems (Thorn et al., 2001; Mishra et al., 2004; Schei et al., 2004; Boman et al., 2006). Exposures of household members have been measured to be many times higher than World Health Organization guidelines and national standards (Smith et al., 2000; Bruce et al., 2006) (see also Sections 9.3.4.3 and 9.4.4). More than 200 studies over the past two decades have assessed levels of indoor air pollutants in households using solid fuels. The burden from related diseases was estimated at 1.6 million excess deaths per year, including 900,000 children under five, and a loss of 38.6 million DALY (Disability Adjusted Life Year) per year (Smith and Haigler, 2008). This burden is similar in magnitude to the burden of disease from malaria and tuberculosis (Ezzati et al., 2002).

Properly designed and implemented ICS projects, based on the new generation of biomass stoves, have led to significant health improvements (von Schirnding et al., 2001; Ezzati et al., 2004). ICS health benefits include a 70 to 90% reduction in indoor air pollution, a 50% reduction in human exposure, and reductions in respiratory and other illnesses (Armendáriz et al., 2008; Romieu et al., 2009). Substantial health benefits can accrue even with modest reductions in exposure to indoor air pollutants. For example, in Guatemala, a 50% reduction in exposure has been shown to produce a 40% improvement in childhood pneumonia cases. In India, the health benefits from the dissemination of advanced ICS have been estimated to be potentially equivalent to eliminating nearly half the entire cancer burden in 2020. These health benefits include 240,000 averted premature deaths from acute lower

---
60 Multi Criteria Analysis methods have been applied in the bioenergy field during the past 15 years (Buchholz et al., 2009).

respiratory infections in children younger than five years and more than 1.8 million averted premature adult deaths from ischemic heart disease and chronic obstructive pulmonary disease (Bruce et al., 2006; Wilkinson et al., 2009).

Figure 2.14 shows the cost effectiveness of treatment options for the eight major risk factors that account for 40% of the global disease burden (Glass, 2006). ICS are among the most cost-effective options in terms of the cost per avoided DALY. Overall, ICS and other small-scale biomass systems represent a very cost-effective intervention with benefits to cost ratios of 5.6:1, 20:1 and 13:1 found in Malawi, Uganda and Mexico, respectively (Frapolli et al., 2010).

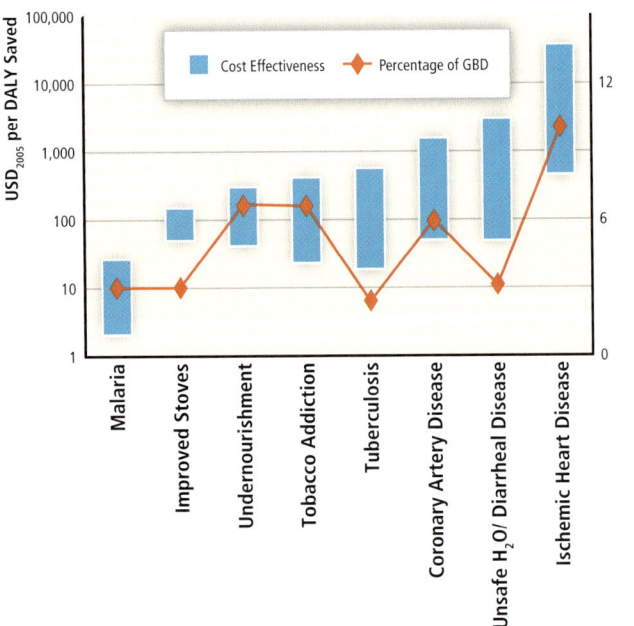

Figure 2.14 | Cost effectiveness of interventions expressed in dollars per disability adjusted life year (DALY) saved (Glass, 2006) on the left scale (logarithmic scale), and contributions to the global burden of disease (GBD) from eight major risk factors and diseases (in %, right scale). The figure shows that the dissemination of improved biomass stoves—depicted here as an intervention to reduce the health effects of indoor air pollution due to fuelwood use—compares well with the cost of interventions aimed at combating major health problems and diseases such as undernourishment, tuberculosis, heart diseases and others (Bailis et al., 2009 with permission from Elsevier B.V.).

Increased use of ICS frees up time for women to engage in income-generating activities. Reduced fuel collection times and savings in cooking time can also translate into increased time for education of rural children, especially girls (Karekezi and Majoro, 2002). ICS use fosters improvements in local living conditions, kitchens and homes, and quality of life (Masera et al., 2000). The manufacture and dissemination of ICS also represents an important source of income and employment for thousands of local small businesses around the world (Masera et al., 2005). Similar impacts were found for small-scale biogas plants, which have the added benefits of providing lighting for individual households and villages and increasing the quality of life. More efficient technologies than currently employed in small-scale industries (such as improved brick and charcoal kilns) are available that increase work productivity, quality of products and overall working conditions (FAO, 2006, 2010b).

2.5.7.3 Socioeconomic aspects of large-scale bioenergy systems

Large-scale bioenergy systems have sparked heated controversies around food security, income generation, rural development and land tenure. The controversy makes clear that there may be both advantages and disadvantages to the further development of large-scale bioenergy systems, depending on their characteristics, local conditions and the mode of implementation.

Impacts on job and income generation

Increased demand for agricultural and forestry waste materials (i.e., residues) can supplement farmers' and foresters' incomes, particularly if the wastes were previously burned or landfilled. Bioenergy can also generate jobs; in general, bioenergy generates more jobs per unit of energy delivered than other energy sources, largely due to feedstock production, especially in developing countries and rural areas (FAO, 2010b).

Wage income is a key contribution to the livelihoods of many poor rural dwellers (Ivanic and Martin, 2008). The benefits from bioenergy jobs depend on the relative labour intensity of the feedstock crop compared to the crop that was previously grown on the same land. For example, cultivation of perennial energy crops requires less labour than cereal crop cultivation, and this displacement effect should be taken into account (Thornley et al., 2009). While increased employment is an important potential benefit, highly labour-intensive operations might also reduce competitiveness (depending on the relative prices of labour and capital) (see Section 9.3.1.3).

The number of jobs created is very location-specific and varies considerably with plant size, the degree of feedstock production mechanization (Berndes and Hansson, 2007) and the contribution of imports to meeting demand (Nusser et al., 2007; Wydra, 2009). Estimates of the employment creation potential of bioenergy options differ substantially, but liquid biofuels based on traditional agricultural crops seem to provide the most employment, especially when the biofuel conversion plants are small (Berndes and Hansson, 2007). Even within liquid biofuel options, the use of different crops introduces wide differences. For ethanol, the number of direct and indirect jobs generated ranges from 45 (corn) to 2,200 (sugarcane) jobs/PJ of ethanol. For biodiesel, the number of direct and indirect jobs generated ranges from 100 (soybean) to 2,000 (oil palm) jobs/PJ of biodiesel (Dias de Moraes, 2007; Clayton et al., 2010). For electricity production, mid-scale power plants in developing countries using a low-mechanized system (25 MW) are estimated to generate approximately 400 jobs/plant or 250 jobs/PJ, of which 94% are in the production and harvesting of feedstocks. For instance, in a detailed UK study, 1.27 jobs/GWh were calculated for power generation from a 25 MW_e plant using dedicated crops (woody or *Miscanthus*). During the complete lifecycle, 4,000 to 6,000 person-year jobs are created, representing on a yearly basis 200 jobs/PJ (15, 73, and 12% at the electricity plant, feedstock production and delivery, and induced, respectively) (Thornley et al., 2008).

In Europe, if the EU25 scenario is followed, Berndes and Hansson (2007) estimate that biomass production for energy can create employment at a magnitude that is significant relative to total agricultural employment (up to 15% in selected countries) but small compared to the total industrial employment in a country. The latest analysis also shows some trade-offs—for instance, agricultural options for liquid biofuels create more employment, but forest-based options for electricity and heat production produce more climate benefits. In Brazil, the biofuel sector accounted for about one million jobs in rural areas in 2001, mostly for unskilled labour related to manual harvesting after field burning of sugarcane (Moreira, 2006). Indeed, mechanization, already ongoing in about 50% of the Center South production (responsible for 90% of the country's harvest), reduces demand for unskilled labour for manual harvest but produces an environmental benefit. Meanwhile, worker productivity continues to grow and part of the workforce is retrained for the skilled higher-paying jobs required for mechanized operations (Oliveira, 2009).

2.5.7.4 Risks to food security

Unless the feedstocks are grown on abandoned land or use residues that previously had no economic value, liquid biofuel production creates additional demand for food and agricultural commodities that places additional pressure on natural resources such as land and water and thus raises food commodity prices (Chakravorty et al., 2009; B. Wright, 2009). Lignocellulosic biofuels, because they can be grown more easily on land that is not suitable for food production, can reduce but not eliminate competition (Chakravorty et al., 2009). To the extent that domestic food markets are linked to international food markets, even countries that do not produce bioenergy may be affected by the higher prices.

Commodity prices are determined by a complex set of factors, of which biofuels is only one, and projections of future prices are highly uncertain. Nevertheless, several studies have examined the contribution of increased biofuels production to the surge in food prices that occurred in the mid-2000s. These studies use different analytical methods and report their results in different ways (for a comprehensive review of these studies, see DEFRA, 2009). For example, the OECD-FAO Agricultural Outlook (OECD-FAO, 2008) model found that if biofuel production were frozen at 2007 levels, coarse grains prices would be 12% lower and vegetable oil prices 15% lower in 2017 compared with a situation where biofuels production continues to increase as expected. Rosegrant et al. (2008) estimated that world maize prices would be 26% higher under a scenario of continued biofuel expansion according to the existing national development plans and more than 70% higher under a drastic biofuel expansion scenario where biofuel demand is double that under the first scenario (these scenarios are relative to a baseline of modest biofuel development where biofuel production remains constant at 2010 levels in most countries). IFPRI (2008) estimated that 30% of the weighted average increase in world cereal prices was attributable to biofuels between 2000 and 2007. Elobeid and Hart (2007) compared two modelled scenarios, with and without biofuel utilization barriers, and found that removing utilization barriers doubled the projected increases in corn and food basket prices. These studies generally agree that increased biofuels production played some role in increased food prices, but there is no consensus about the size of this contribution (FAO, 2008a; Mitchell, 2008; DEFRA, 2009; Baffes and Haniotis, 2010). Other factors include the weak US dollar, increased energy costs, increased agricultural production costs, speculation on commodities, and adverse weather conditions (Headey and Fan, 2008; Mitchell, 2008; DEFRA, 2009; Baffes and Haniotis, 2010). The eventual impact of biofuels on prices will depend, among other factors, on the specific technology used, the strength of government mandates for biofuel use, the design of trade policies that favour inefficient methods of biofuel production, and oil prices.

The impact of higher prices on the welfare of the poor depends on whether the poor are net sellers of food (benefit from higher prices) or net buyers of food (harmed by higher prices). On balance, the evidence indicates that higher prices will adversely affect poverty and food security in developing countries, even after taking into account the benefits of higher prices for farmers (Ivanic and Martin, 2008; Zezza et al., 2008). A major FAO study on the socioeconomic impacts of the expansion of liquid biofuels (FAO, 2008a) indicates that poor urban consumers and poor net food buyers in rural areas are particularly at risk. Rosegrant et al. (2008) estimated that the number of malnourished children would double under the two scenarios mentioned above.

A significant increase in the cultivation of crops for bioenergy indicates a close coupling of the markets for energy and food (Schmidhuber, 2008), and an analysis by the World Bank (2009) confirmed a strong association between food and energy prices when oil prices are above USD_{2005} 45 per barrel. Thus, if energy prices increase, there may be spillovers into food markets that increase food insecurity.

Meeting the food demands of the world's growing population will require a 70% increase in global food production by 2050 (Bruinsma, 2009). At the same time, FAO (2008b) estimates that the increase in arable land between 2005 and 2050 will be just 5% (Alexandratos et al., 2009). This limited increase indicates that economically exploitable arable land is scarce. Because biomass production is land-intensive, there could be significant competition between food and fuel for the use of agricultural land (Chakravorty et al., 2009). Increased biofuels production could also reduce water availability for food production, as more water is diverted to production of biofuel feedstocks (Chakravorty et al., 2009; Hoekstra et al., 2010).

2.5.7.5 Impacts on rural and social development

Growing demand for biofuels and the resulting rise in agricultural commodity prices can present an opportunity for promoting agricultural growth and rural development in developing countries (Schmidhuber, 2008). The development potential critically depends on whether the bioenergy market is economically sustainable without government subsidies. If long-term subsidies are required, fewer government funds will be available for the wide range of other public goods that are essential for economic and social development, such as agricultural research, rural roads, and education. Even short-term subsidies need to be considered very carefully, as once subsidies are implemented they can be difficult to remove. Latin American experience shows that governments that use agricultural budgets for investment in public goods experience faster growth and alleviate poverty and environmental degradation more rapidly than those that apply them for subsidies (López and Galinato, 2007).

Bioenergy may reduce dependence on fossil fuel imports and increase energy supply security. In many cases these benefits are not likely to be large, although the contribution could be substantial for countries with large amounts of arable land per person (FAO, 2008a). Recent analyses of the use of indigenous resources implies that much of the expenditure on energy is retained locally and recirculated within the local or regional economy, but there are trade-offs to consider. For example, the increased use of biomass for electricity production and the corresponding increase in demand for some types of biomass (e.g., pellets) could cause a temporary lack of biomass supply during periods of high demand. Households are particularly vulnerable to this market distortion.

The biofuels production technologies and institutions will also be an important determinant of rural development outcomes. In some instances, private investors will look to establish biofuel plantations to ensure security of supply. If plantations are established on non-productive land without harming the environment, there should be benefits to the economy. It is essential not to overlook the uses of land that are important to the poor. Governments may need to establish clear criteria for determining whether land is marginal or productive, and these criteria must protect vulnerable communities and female farmers who may have less secure land rights (FAO, 2008a). Research in Mozambique shows that, compared with a more capital-intensive plantation approach, an out-grower approach to producing biofuels helps to reduce poverty due to the greater use of unskilled labour and accrual of land rents to smallholders (Arndt et al., 2010).

Increased investment in rural areas will be crucial for making biofuels a positive development force. If governments rely exclusively on short-term farm-level supply side economic response, the negative effects of higher food prices will predominate. If higher prices motivate greater public and private investment in agriculture (e.g., rural roads and education, R&D), there is tremendous potential for sparking medium- and long-term rural development (De La Torre Ugarte and Hellwinckel, 2010). As one example, proposed biofuel investments in Mozambique could increase annual economic growth by 0.6% and reduce the incidence of poverty by about 6% over a 12-year period between 2003 and 2015 (Arndt et al., 2010).

2.5.7.6 Trade-offs between social and environmental aspects

Some important trade-offs between environmental and social criteria exist and need to be considered in future bioenergy developments. In the case of sugarcane, the environmental sustainability criteria promoted by certification frameworks (such as the Roundtable for Sustainable Biofuels) favour mechanical harvesting due to the avoided emissions from sugarcane field burning required in manual systems. Several other organizations are concerned about the large number of workers that will be displaced by these new systems. Also, the mechanized model tends to favour further concentration of land ownership, potentially excluding small- and medium-scale farmers and reducing employment opportunities for rural workers (Huertas et al., 2010).

Strategies for addressing such concerns can include providing support for small- and medium-size stakeholders that lack the capacity to meet the certification system requirements and/or developing alternative income possibilities for the seasonal workers that presently earn a substantial part of their annual income by cutting sugarcane (Huertas et al., 2010). Retraining workers from manual to skilled labour, such as truck driving, is already taking place in Center South Brazil (Oliveira, 2009).

2.5.8 Synthesis

As a component of the much larger agriculture and forestry systems of the world, traditional and modern biomass affects social and environmental issues ranging from health and poverty to biodiversity and water quality. Land and water resources need to be properly managed in concert with each specific region's economic development situation and suitable types of bioenergy. Bioenergy has the opportunity to contribute positively to climate change mitigation, secure energy supply and diversity goals, and economic development in developed and developing countries alike. However, the effects of bioenergy on environmental sustainability may also be negative depending upon local conditions, how criteria are defined, and how actual projects are designed and implemented, among many other factors.

- Climate change and biomass production can be influenced by interactions and feedbacks among land and water use, energy and climate at scales that range from micro through macro (see Figure 2.15). Social and environmental trade-offs may be present but can be minimized to a large extent with appropriate project design and implementation.

- Although crops grown as biofuels feedstocks currently use less than 1% of the world's agricultural land, the expansion of large-scale bioenergy systems raises several important socioeconomic issues including food security, income generation, rural development, land tenure and water scarcity in specific regions.

- Estimates of LUC effects require value judgments about the temporal scale of analysis, the land use under the assumed 'no action' scenario, the expected uses in the longer term, and the allocation of impacts among different uses over time. Regardless, a system that ensures consistent and accurate inventory of and reporting on carbon stocks is considered an important first step towards LUC carbon accounting.

- Emissions of pollutants, like SO_2 and NO_x, are generally lower for bioenergy than for coal, gasoline and diesel, though the NO_x results for biodiesel are more variable. Thus, bioenergy can reduce negative impacts on air quality. Bioenergy impacts on water resources can be positive or negative, depending on the particular feedstock, supply chain element and processing methodologies. Bioenergy systems similar to conventional food and feed crop systems can contribute to loss of habitat and biodiversity, but bioenergy plantations can be designed to provide filters for nutrient loss, to function as ecological corridors, to reduce pressure on natural forests and to restore degraded or abandoned land. Genetically engineered and potentially invasive bioenergy crops have raised concerns. More research and protocols are needed to monitor and evaluate the introduction of new or modified species.

- Advanced ICS for traditional biomass use can provide large and cost-effective mitigation of GHG emissions (GHG mitigation potential of 0.6 to 2.4 Gt CO_2eq/yr) with substantial co-benefits in health and living conditions, particularly for the poorest 2.7 billion people in the world. Efficient technologies for cooking are cost-effective and comparable to major health interventions such as those for tobacco addiction, undernourishment or tuberculosis.

- Biofuel production has contributed to increases in food prices, but additional factors affect food prices, including weather conditions, changes in food demand and increasing energy costs. Even considering the benefit of increased prices to poor farmers, increased food prices have adversely affected poverty, food security and malnourishment of children. On the other hand, biofuels can also

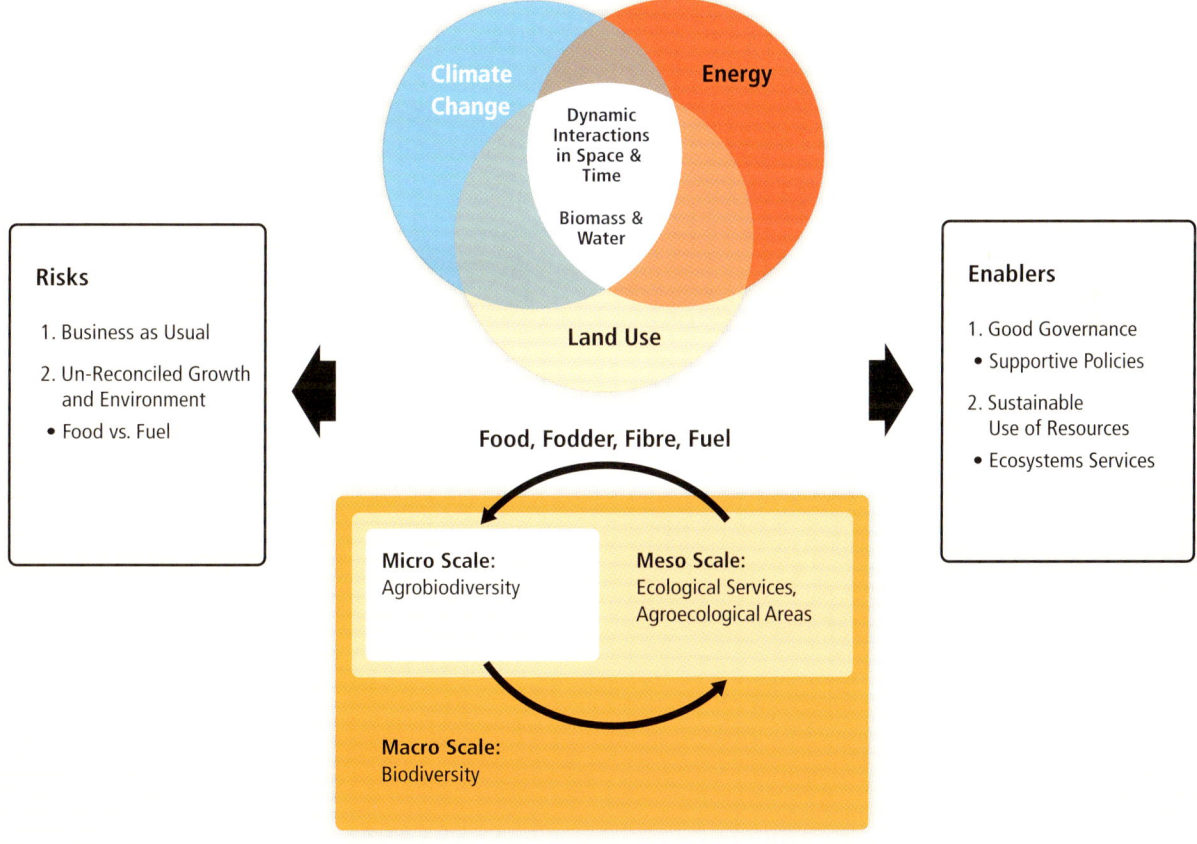

Figure 2.15 | Bioenergy's complex, dynamic interactions among society, energy and the environment include climate change feedbacks, biomass production and land use with direct and indirect impacts at various spatial and temporal scales on all resource uses for food, fodder, fibre and energy (Dale et al., 2011). Biomass resources need to be produced in sustainable ways as their impacts can be felt from micro to macro scales (van Dam et al., 2010). Risks are maintenance of business-as-usual approaches with uncoordinated production of food and fuel. Opportunities are many and include good governance and sustainability frameworks that generate effective policies that also lead to sustainable ecosystem services.

provide opportunities for developing countries to make progress in rural development and agricultural growth, especially when this growth is economically sustainable. Proper design, implementation, monitoring and adherence to sustainability frameworks may help minimize negative socioeconomic impacts and maximize benefits, particularly for local people.

- These social and environmental impacts should be compared with those of the energy systems they replace. Many lifecycle assessments that characterize the amount of RE provided relative to fossil energy used in biofuel production and compare that with the reference system show GHG emission savings for biofuels. These studies can be expanded to use multiple indicators and more comprehensively analyze the whole chain from feedstock to final energy use.

2.6 Prospects for technology improvement and innovation[61]

This section provides a literature overview of the sets of developing technologies, their performance characteristics and projections of cost performance for biomass feedstocks, logistics and supply chains, and conversion routes to a variety of biofuels alone or in combination with heat and power or with other bio-based products. Advanced power routes are also discussed. As illustrated in Figure 2.2 and Table 2.5, many such advanced biomass energy chains are commercial or in development at various stages ranging from small-scale R&D through near commercialization for each component of the chain, including some examples of integrated systems. Linkages are made with the various applications, with the suppliers of feedstocks, which can be residues from urban or rural areas, and with the existing and developing biomass conversion industry to products. The integration of biomass energy and related products into the electricity, natural gas, heating (residential and district, commercial and public services), industrial and fossil liquid fuels systems for transport is discussed more thoroughly in Chapter 8. The structure of this section parallels that of Section 2.3, following the bioenergy supply chain from feedstocks (Section 2.6.1) to logistics (Section 2.6.2) to end products (e.g., various advanced secondary energy carriers in gaseous or liquid states) made by various conversion technologies (Section 2.6.3).

2.6.1 Improvements in feedstocks

2.6.1.1 Yield gains

Increasing land productivity, whether for food or energy purposes, is a crucial prerequisite for realizing large-scale future deployment of biomass for energy because it would make more land available for growing biomass and reduce the associated demand for land. Much of

---
61 Section 10.5 offers a complementary perspective on drivers and trends of technological progress across RE technologies.

the increase in agricultural productivity over the past 50 years came about through plant breeding and improved agricultural management practices including irrigation, fertilizer and pesticide use. The adoption of these techniques in the developing world is most advanced in Asia, where productivity grew strongly during the past 50 years, and also in Brazil, with sugarcane. Considerable potential exists for extending the same kind of gains to other regions, particularly sub-Saharan Africa, Latin America, Eastern Europe and Central Asia, where adoption of these techniques has been slower (Evenson and Gollin, 2003; FAO, 2008a). A recent long-term forecast by the FAO expects global agricultural production to rise by 1.5% per year for the next three decades, still significantly faster than projected population growth (World Bank, 2009). For the major food staple crops, maximum attainable yields may increase by more than 30% by switching from rain-fed to irrigated and optimal rainwater use production (Rost et al., 2009), while moving from intermediate- to high-input technology may result in 50% increases in tropical regions and 40% increases in subtropical and temperate regions. The yield increase when moving from low- to intermediate-input levels can reach 100% for wheat, 50% for rice and 60% for maize (Table 2.14), due to better pest control and adequate nutrient supply. However, important environmental trade-offs may be involved with agricultural intensification, and avenues for more sustainable management practices may need exploration and adoption (IAASTD, 2009).

Biotechnologies or conventional plant breeding could improve biomass production by focusing on traits relevant to energy production such as biomass per hectare, increased oil or fermentable sugar yields, or other characteristics that facilitate their conversion to energy end-products (e.g., Sannigrahi et al., 2010). Also, considerable genetic improvement is still possible for drought-tolerant plants (Nelson et al., 2007; Castiglioni et al., 2008; FAO, 2008b).

The projected increases in productivity reflect present knowledge and technology (Duvick and Cassman, 1999; Fischer and Schrattenholzer, 2001) and vary across the regions of the world (FAO, 2008a). In developed countries where cropping systems are already highly input-intensive, productivity increases will be more limited. Also, projections do not always account for the strong environmental limitations in many regions, such as water or temperature (Nelson et al., 2007; Castiglioni et al., 2008; FAO, 2008b).

Doubling the current yields of perennial grasses appears achievable through genetic manipulation such as marker-assisted breeding (Turhollow, 1994; Eaton et al., 2008; Tobias et al., 2008; Okada et al., 2010). Shifts to sustainable farming practices and large improvements in crop and residue yield could increase the outputs of residues from arable crops (Paustian et al., 2006).

Future feedstock production cost projections are scant because of their connections with food markets (which are, as all commodities, volatile and uncertain) and because many candidate feedstock types are still in the R&D phase. Cost figures for growing these feedstock species in commercial farms are not well understood yet but will likely reduce over time

Table 2.14 | Prospects for yield improvements by 2030 relative to 2007 to 2009 data from Table 2.4.

Feedstock type	Regions	Yield trend (%/yr)	Potential yield increase by 2030 (%)	Improvement routes	Ref.
DEDICATED CROPS					
Wheat	Temperate	0.7	20-50	New energy-oriented varieties	1,10
	Subtropics		30-100	Higher input rates, irrigation	
Maize	N America	0.7	20-35	New varieties, GMOs, higher plantation density, reduced tillage	
	Subtropics		20-60		
	Tropics		50	Higher input rates, irrigation	
Soybean	USA	0.7	15-35	Breeding	2,3,10
	Brazil	1.0	20-60		
Oil palm	World	1.0	30	Breeding, mechanization	3
Sugarcane	Brazil	1.5	20-40	Breeding, GMOs, irrigation inputs	2,3,8,10
SR Willow	Temperate	—	50	Breeding, GMOs	3
SR Poplar	Temperate	—	45		
Miscanthus	World	—	100	Breeding for minimal input, improved management	
Switchgrass	Temperate	—	100	Genetic manipulation	
Planted forest	Europe	1.3	20	Species choice, breeding, fertilization, shorter rotations, increased rooting depth	4,9
	Canada		20		11
PRIMARY RESIDUES					
Cereal straw	World	—	15	Improved collection equipment, breeding for higher residue-to-grain ratios (soybean)	5,6
Soybean straw	N America	—	50		
Forest residues	Europe	1.0	25	Ash recycling, cutting increases, increased roundwood, productivity	4,7

Abbreviations: SR = short rotation; GMO = genetically modified organism.

References: 1. Fischer and Schrattenholzer (2001); 2. Bauen et al. (2009a); 3. WWI (2006); 4. Nabuurs et al. (2002); 5. Paustian et al. (2006); 6. Perlack et al. (2005); 7. EEA (2007); 8. Matsuoka et al. (2009); 9. Loustau et al. (2005); 10. Jaggard et al. (2010). 11. APEC (2003).

as farmers descend the learning curves, as past experience has shown in Brazil (van den Wall Bake et al., 2009).

Under temperate conditions, the expenses for the farm- or forest-gate supply of lignocellulosic biomass from perennial grasses or short-rotation coppice are expected to fall to less than USD$_{2005}$ 2.5/GJ by 2020 (WWI, 2006) from a USD$_{2005}$ 3 to 16/GJ range today (Table 2.6, without land rental cost). However, these are marginal costs, which do not account for the competition for land with other sectors and markets that would increase unit costs as the demand for biomass increases. This is reflected in supply curves (see Section 2.2 and Figure 2.5(b)). Recent studies in Northern Europe that include such land-related costs thus report somewhat higher projections, in a USD$_{2005}$ 2 to 7.5/GJ range for herbaceous grasses and USD$_{2005}$ 1.5 to 6/GJ range for woody biomass (Ericsson et al., 2009; de Wit and Faaij, 2010). For perennial species, the transaction costs required to secure a supply of energy feedstock from farmers may increase the production costs by 15% (Ericsson et al., 2009). Delivered prices for herbaceous crops are shown in Figure 2.5(d) for the USA and about 8 EJ could be delivered at USD$_{2005}$ 5/GJ to the conversion facility.

In recent decades, forest productivity has increased more than 1% per year in temperate and boreal regions due to higher CO_2 concentrations and nitrogen deposition or fertilization rates (Table 2.14). This trend is projected to continue until 2030 when productivity might plateau due to increased stand ages and increased respiration rates in response to warmer temperatures (Nabuurs et al., 2002). However, yield trends vary across climatic zones at a finer scale. Water limitations in Mediterranean/semi-arid environments lead to zero or even negative variations in biomass yield increments by 2030 (Loustau et al., 2005). This may be counteracted by adaptive measures such as choosing species more tolerant to water stress or using appropriate thinning regimes (Loustau et al., 2005). Where water is non-limiting, productivity may be maximized by more intensive silvicultural practices, including shorter rotations, optimum row spacing, fertilization and improved breeding stock (Loustau et al., 2005; Feng et al., 2006). Increased roundwood extraction would also generate extra logging residues and carbon sequestration in forest soils as a co-benefit, outweighing several-fold the GHG emissions generated by management practices (Markewitz, 2006).

2.6.1.2 Aquatic biomass

Aquatic phototrophic organisms dominate the world's oceans, producing 350 to 500 billion tonnes of biomass annually and include 'algae', both microalgae (such as *Chlorella* and *Spirulina*) and macroalgae (i.e., seaweeds) and cyanobacteria (also called 'blue-green algae') (Garrison, 2008). Oleaginous microalgae such as *Schizochytrium* and *Nannochloropsis* can accumulate neutral lipids, analogous to seed oil triacylglycerides, at greater than 50% of their dry cell weight (Chisti, 2007). Weyer et al. (2009) reported yields of 40 x 10^3 to 50 x 10^3 litres/ha/yr (0.04 to 0.05 litres/m^2/yr) in unrefined algal oil from biomass grown in the Equator region and containing 50% oil. Assuming a neutral lipid yield ranging from 30 to 50%, algae productivity can be several-fold higher than palm oil productivity at 4.7 x 10^3 litres/ha/yr (0.0047 litres/m^2/yr). Photosynthetic cyanobacteria used to produce nutraceuticals at commercial scales (J. Lee, 1997; Colla et al., 2007) could also directly produce fuels such as H_2 (Hu et al., 2008; Sections 3.3.5 and 3.7.5).

Macroalgae do not accumulate lipids like microalgae do. Instead, they synthesize polysaccharides from which various fuels could be made (see Figure 2.6). Uncultivated macroalgae can have polysaccharide yields higher than those of terrestrial plants (per unit area) (Zemke-White and Ohno, 1999; Ross et al., 2009) and can live in marine environments. Halophiles, another group of phototrophic organisms, live in environments with high salt concentration.

Microalgae can photoproduce chemicals, fuels or materials in non-agricultural land such as brackish waters and highly saline soils. Hundreds of microalgae species, out of hundreds of thousands of species, have been tested or used for industrial purposes. Understanding the genetic potential, lipid productivity, growth rates and control, and use of genetic engineering allows broader use of land and decreases the LUC impacts of biofuels production (Hu et al., 2008). Microalgae can be cultivated in open ponds and closed photobioreactors (PBRs) (Sheehan et al., 1998a; van Iersel et al., 2009) but scale-up can involve logistical challenges, can require high cost to produce the biomass, and requires water consumption minimization (Borowitzka et al., 1999; Molina Grima et al., 2003). Production costs using low- to high-productivity scenarios currently range approximately from USD$_{2005}$ 30 to 80/GJ for open ponds and from USD$_{2005}$ 50 to 140/GJ for PBR (EPA, 2010).

Macroalgae are typically grown in offshore cultivation systems (Ross et al., 2009; van Iersel et al., 2009) that require shallow waters for light penetration (Towle and Pearse, 1973). The impact of biofuel production on competing uses (fisheries, leisure) and on marine ecosystems needs assessment. Using aquatic biomass harvested from algal blooms may provide multiple benefits (Wilkie and Evans, 2010).

The bioenergy potential from aquatic plants is usually excluded from resource potential determinations because of insufficient data available for such an assessment. However, the potential may be substantial compared to conventional energy crops, considering the high yield potential of cultivated microalgae production (up to 150 dry t/ha/yr, 0.015 t/m^2/yr) (Kheshgi et al., 2000; Smeets et al., 2007). With the large number of diverse algal species in the world, upper range productivity potentials of up to several hundred EJ for microalgae and up to several thousand EJ for macroalgae (Sheehan et al., 1998a; van Iersel et al., 2009) have been reported. Figure 2.10 shows very approximate ranges for GHG reductions relative to the fossil fuel replaced. Comparable or increased emission reductions relative to crop biodiesel could be achieved with successful RD&D and commercialization (EPA, 2010).

Some key conclusions from current efforts (US DOE, 2009; IEA Bioenergy, 2010; Darzins et al., 2010) are the following: (1) Microalgae can offer productivity levels above those possible with terrestrial plants. (2) There are currently several significant barriers to widespread deployment and many information gaps and opportunities for improvement and breakthroughs. (3) Various systems suited to different types of algal organisms, climatic conditions, and products are still being considered. (4) Basic information related to genomics, industrial design and performance is still needed. (5) Cost estimates for algal biofuels production vary widely, but the best estimates are promising at this early stage of technology development. (6) The cost of processing algae solely for fuel production is still too high. Producing a range of products for the food, fodder and fuel markets offers opportunities for economical operation of algal biorefineries. (7) Lifecycle assessments are needed to guide future developments of sustainable fuel production systems.

2.6.2 Improvements in biomass logistics and supply chains

Optimization of supply chains includes achieving economies of scale in transport, in pretreatment and in conversion technologies. Relevant factors include spatial distribution and seasonal supply patterns of the biomass resources, transportation, storage, handling and pretreatment costs, and economies of scale benefiting from large centralized plants (Dornburg and Faaij, 2001; Nagatomi et al., 2008). Smart utilization of a combination of biomass resources over time can help conversion plants gain economies of scale through year-round supplies of biomass and thus efficiently utilize the investment cost (Junginger et al., 2001; McKeough et al., 2005; Nishi et al., 2005; Ileleji et al., 2010; Kang et al., 2010) and technology transfer (Asikainen et al., 2010).

Over time the lower-cost biomass residue resources are increasingly depleted and more expensive (e.g., cultivated) biomass needs to cover the growing demand for bioenergy. Part of this growing demand may be met by learning and optimization, but, for example, future heat generation from pellets in the UK may be more costly (2020) than it is today due to a shift from local to imported feedstocks (E4tech, 2010). Similar effects are found in scenarios for large-scale deployment of biofuels in Europe (Londo et al., 2010).

Learning and optimization in the past one to two decades in Europe (Scandinavia and the Baltic in particular), North America, Brazil and also in various developing countries have shown steady progress in market development and cost reduction of biomass supplies (Section 2.7.2; Junginger et al., 2006). Well-working international biomass markets and substantial investments in logistics capacity are key prerequisites to achieve this (see also Section 2.4).

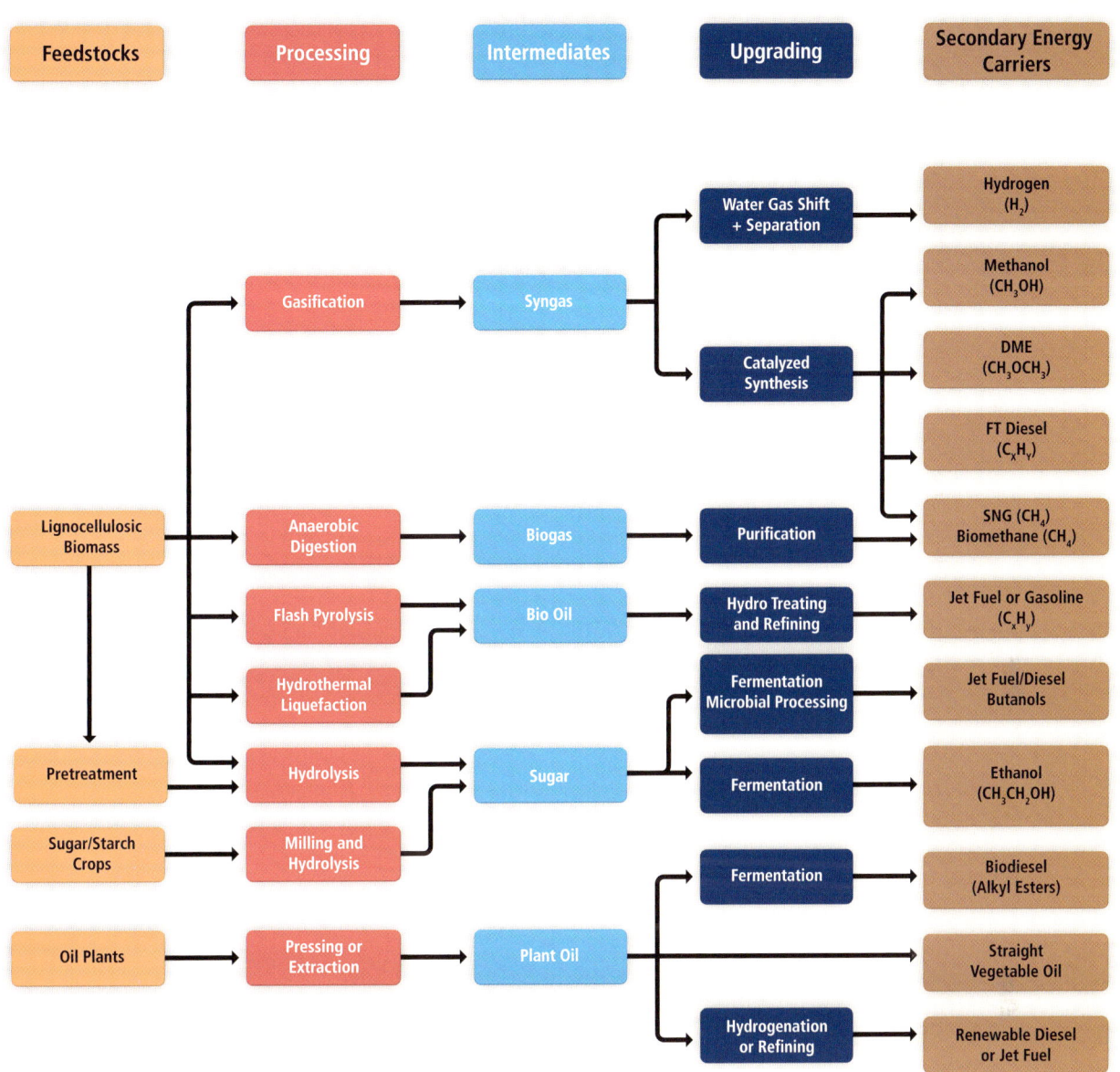

Figure 2.16 | Overview of lignocellulosic biomass, sugar/starch crops and oil plants (feedstocks) and the processing routes to key intermediates, which can be upgraded through various routes to secondary energy carriers, such as liquid and gaseous biofuels. Fuel product examples are (1) oxygenated biofuels to blend with current gasoline and diesel fuels or to use in pure form, such as ethanol, butanols, methanol, liquid ethers, biodiesel, and gaseous DME (dimethyl ether); (2) hydrocarbon biofuels such as Fischer Tropsch (FT) liquids, renewable diesel and some microbial fuels (which are compatible with the current infrastructure of liquid fuels because their chemical composition is similar to that of gasoline, diesel, and jet fuels (see Table 2.15.C)), or the simplest hydrocarbon methane for natural gas replacement (SNG) from gasification or biomethane from anaerobic digestion; and (3) H_2 for future transportation (adapted from Hamelinck and Faaij, 2006 and reproduced with permission from Elsevier B.V.).

Notes: Microbial fuels include hydrocarbons derived from isoprene, the component of natural rubber; a variety of non-fermentative alcohols with three to six carbon atoms including butanols (four carbons); and fatty acids which can be processed as plant oils to hydrocarbons (Rude and Schirmer, 2009).[1] For sugar and starch crops the sugar box indicates six-carbon sugars, while for lignocellulosic biomass this box is more complex and has mixtures of six- and five-carbon sugars, with proportions dependent on the feedstock type. Hardwoods and agricultural residues contain xylan and other polymers of five-carbon sugars in addition to cellulose that yield glucose, a six-carbon sugar.

1. Not shown are the aquatic plants (see Section 2.6.1.2) that can utilize the same types of processing shown for their vegetable oil and carbohydrate fractions.

Pretreatment technologies

Torrefied wood is manufactured by heating wood in a process similar to charcoal production. At temperatures up to 160ºC, wood loses water, but it keeps its physical and mechanical properties and typically maintains 70% of its initial weight and 90% of the original energy content (D. Bradley et al., 2009). Torrefied wood only absorbs 1 to 6% moisture (Uslu et al., 2008). Torrefaction can produce uniform quality feedstock, which eliminates inefficient and expensive methods designed to handle feedstock variations and thus makes conversion more efficient (Badger, 2000) and more predictable.

Pyrolysis processes convert solid biomass to liquid bio-oil, a complex mixture of oxidized hydrocarbons. Although this liquid product is toxic and needs stabilization for longer-term storage, bio-oil is relatively easy to transport. Pyrolysis oil production is more expensive and less efficient per unit of energy delivered compared to torrefaction of wood pellets. Section 2.3.4 discusses the cost data for multiple countries based on Bain (2007); McKeough et al. (2005) arrive at similar figures of USD_{2005} 6.2 to 7.0/GJ. The process allows for separation of a solid fraction (biochar) that contains the bulk of the nutrients of the biomass. With proper handling, such biochars could be used to improve soil quality and productivity, recycle nutrients and possibly store carbon in the soil for long periods of time (Laird, 2008; Laird et al., 2009; Woolf et al., 2010).

2.6.3 Improvements in conversion technologies for secondary energy carriers from modern biomass

Different conversion technologies (or combinations) including mechanical, thermochemical, biochemical and chemical steps, as shown in Figure 2.2, are needed to transform the variety of potential feedstocks into a broader range of secondary energy carriers. In addition to electricity and heat as products, a variety of liquid and gaseous fuels or products can be made from biomass as illustrated in Figure 2.16, where key chemical intermediates that could make identical, similar or new products as energy carriers, chemicals and materials are highlighted (see Section 2.6.3.4 for further detail):

- **Sugars**, mixtures of five- and six-carbon sugars from lignocellulosic materials, are converted primarily through biochemical or chemical processes into liquid or gaseous fuels and a variety of chemical products.

- **Syngas** from thermochemical gasification processes, which can be converted in integrated gasification combined cycle (IGCC) systems to electricity, through a variety of thermal/catalytic processes to gaseous or liquid fuels, or through biological processes at low temperature to H_2 or polymers.

- **Oils** from pyrolysis or hydrothermal treatment, which can be upgraded into a variety of fuels and chemicals.

- **Lipids** from plant oils, seeds or microalgae, which can be converted into a wide variety of fuels, such as diesel or jet fuels, and chemicals.

- **Biogas** is a mixture of methane and CO_2 released from anaerobic degradation of organic materials with a lower heat content than its upgraded form, mostly methane, called biomethane. If upgraded, it can be added to natural gas grids or used for transport.

Table 2.15 contains process efficiency and projected improvements along with cost information expressed in USD_{2005}/GJ for several bioenergy systems and chains, in various stages of development, from various studies from multiple sources. Part A details processes for alcohols; Part B summarizes microalgal fuels; Part C details hydrocarbon fuels; and Part D includes gaseous fuels and electricity from IGCC. Financial assumptions are provided at the end of the table; some groups of references use the same assumptions but not all. First-of-a-kind plants are more expensive as there are technical uncertainties in the chemical, biochemical, thermochemical or mechanical component steps in a route, as shown by Kazi et al. (2010) and Swanson et al. (2010) compared to Bauen et al. (2009a) or Foust et al. (2009). Such combination of steps is often significantly more complex than a similar petroleum industry process because of the characteristics of solid biomass. Scaling up is conducted after initial bench-scale experimentation and encouraging initial techno-economic evaluation. As experience in operating the process and correcting design or operating parameters is gained, cost evaluations are conducted and the plant is operated until costs decrease at a slower pace. At this point, the technical and economic risks of the plant have decreased and the production costs have reached so-called nth plant status. The uncertainties in these studies are variable and higher for the least-developed concepts (Bauen et al., 2009a).

An overview of advanced pilot, demonstration and commercial-scale bioenergy projects in 33 countries is provided by Bacovsky et al. (2010a,b), including the site at Kalundborg, Denmark, where a wheat straw ethanol is made in the pilot plant and sold to a gasoline distributor in 2010.[62] The number of actual projects moving to pilot and demonstration scale is probably larger. The reference contains descriptions of most of the development projects listed in Table 2.15. See also the IEA (Renewable Energy Division, 2010) report on global sustainable second-generation technologies and future perspectives in the context of the transport sector and the recently published technology roadmap for biofuels (IEA, 2011).

This section focuses on bioenergy products to avoid repetition of technology descriptions provided in Section 2.3—for instance, a thermochemical technology such as gasification can produce multiple fuels and electricity. Similarly, a variety of end products can be made from sugars.

An initial meta-analysis of advanced conversion routes (Hamelinck and Faaij, 2006) for methanol, H_2, Fischer-Tropsch liquids and biochemical ethanol produced from lignocellulosic biomass under comparable financial assumptions suggests that these systems compare favourably with starch-based biofuels and offer more competitive fuel prices and opportunities in the longer term because of their inherently lower feedstock costs and because of the variety of sources of lignocellulosic biomass, including agricultural residues from cereal crop production, and forest residues. The feedstock cost range used in this meta-analysis is in line with costs highlighted in Section 2.6.1.1 and the low range of the supply curves shown in Figure 2.5. In the EU study, Northern Europe projected production costs are in the USD_{2005} 2 to 7.5/GJ range for herbaceous grasses and USD_{2005} 1.5 to 6/GJ for woody biomass (land-related costs included). For perennial species, transaction costs may need to increase by 15% to secure a supply of energy feedstock from farmers. This additional cost (e.g., transport to the conversion plant and payment to secure the feedstock) is already built into the prices of the US supply

---

[62] An interactive website with this information is maintained by the IEA Bioenergy Task 39: biofuels.abc-energy.at/demoplants.

Table 2.15 | Summary of developing technologies costs projected for 2030 biofuel production and their 2010 industrial development level. Using today's performance for a pioneer plant built in the near term increases costs, and the majority of the references assumes that technology learning will occur upon development, referred to as nth plant costs. Costs expressed in USD$_{2005}$.

A: Fuels – Alcohols by Biochemical and Gasification Processes

Process	Feedstock	Efficiency and process economics. Eff. = Energy product/biomass energy Component costs in USD$_{2005}$/GJ	% GHG reduction from fossil reference	Potential technical advances and challenges	Production cost by 2030 (USD$_{2005}$/GJ)	Industrial development (see Bacovsky et al., 2010a,b)
Consolidated bioprocessing (CBP)	Lignocellulosic	Eff. ~49% for wood and 42% for straw (ethanol) + 5% power.[19]	Scenarios analyzed[30]	Lignin engineering cellulose access.[7] Develop CBP organisms.[44]	15.5[19] future	
Separate hydrolysis/ co-fermentation		Eff. ~39% (ethanol) + 10% power.[1]		Efficient 5-carbon sugar conversion.[2,3] R&D investment.[5] Advanced enzyme.[6]	25[1]–27[19] 28–35[48]	
Simultaneous saccharification/ co- fermentation	Barley straw	Steam explosion, enzyme hydrolysis, ethanol fermentation.[9] High solids 15%.	N/A	System integration, high solids, decrease toxicity for fermentation.	30[9] (Finland) from pilot data	
Simultaneous saccharification and fermentation	Corn stover	Dilute acid hydrolysis, 260 million L/yr; FC: 6.6, CC*: 10.1, CR: 1.1 for ethanol.[24]	83–88 Depending on co-product credit method[25]	Pretreatment, process integration, enzyme costs.[24]	15.5 (US) nth plant, future[24]	Demonstration and pilots. Reduce enzyme and pretreatment costs. Several pilots in many countries. First commercial plants. Lignin residues co-firing.[32]
	Lignocellulosic Various Eff. 35% ethanol + 4% power.[1]	Generic; 90 million L/yr; FC:14; CC*:14. At 360 million L/yr; FC:14; CC*:10; CR:0.5.[45]		Meta-analysis conditions.[45]	28 (2015)[45] 23.5 (2022)[45]	
		Eff. kg/L ethanol (poplar, *Miscanthus*, switchgrass, corn stover, wheat: 3.7, 3.2, 2.6, 2.6, 2.4). Plant sizes 1,500 to 1,000 t/day. FC 50% of total.[10]		Process integration—capital costs per installed litre of product USD 0.9 to 1.3 for plants of 150 to 380 million litres/yr (2020 estimates). Project a 25% operating cost reduction by 2025 and a 40% operating cost reduction by 2035.[10]	18–22[10] (2020) breakeven USD 100/barrel; + CCS USD 95/ barrel; USD 50/t CO$_2$	
	Bagasse	Standalone plant[35] 370 L/t dry (ethanol) + 0.56 kWh/L ethanol (elec.).	86 Advanced CHP: 120% (replace NG peak power).[36]	Mechanical harvest improvements sugarcane residues (occurring).[35,36]	6[35]–15[35] w/o and w FC	
Gasification/catalytic synthesis ethanol	Lignocellulosic	170 million L per year plant (varies in size).[18] By-product propanol/butanols.	90[38]	Improvements in catalyst development and syngas cleaning.	12[49]–15[18] 14.5[24]	RD&D, pilot.
Fermentation; product compatible with gasoline infrastructure to butanols, in particular biobutanol	Sugar/starch	Development of an integrated biobutanol production and removal systems using the solvent-producing bacteria *Clostridia* improved by genetic engineering.[29] Initial acetone, butanol, and ethanol (ABE) fermentation is costly.	5–31 Depending on co-product credit method.[29]	For high selectivity to biobutanol: (1) mutated strain of *Clostridium beijernekei* BA101, or protein engineering in *E. coli* to increase selectivity/lower cost to biobutanol.[15,16] (2) dual fermentation to butyric acid and reduction to butanols.	29.6 for ABE;[18] 25.2 for mutated *Clostridia*[17] or 21.6 for dual process[17]	Large and small venture companies in different routes, including yeast host. Hydrocarbon precursor.
Gasification to butanols	Lignocellulosic	Catalytic process for synthesis of predominantly butanols.	N/A	Estimated production costs include return on capital.[17]	13[17]	N/A
Gasification/synthesis to methanol for fuel and/or power	Lignocellulosic	Eff. 55% fuel only[19] Eff. 48% fuel and 12% power.[19]	90[27]	Methanol (and dimethyl ether) production possible in various configurations that co-produce power.	12–18 (fuel)[19] 7.1–9.5 (fuel and power)[19]	Pilots, demos, and first commercial.

Continued next Page →

curves based on county-level data; the projected price of delivery to the conversion facility for forest and related residues is USD$_{2005}$ 1 to 3/GJ up to about 1.5 EJ, and for woody and herbaceous plants and sorghum delivered to the conversion facility the projected price is USD$_{2005}$ 2 to 4/GJ up to about 5 EJ (or more at higher price).

2.6.3.1 Liquid fuels

Alcohols. Estimated production costs for various fuel processes are assembled in Part A of Table 2.15, and they range from USD$_{2005}$ 13 to 30/GJ.

B: Fuels – Algae

Process	Feedstock	Efficiency and process economics Eff. = Energy product/biomass energy Component costs in USD$_{2005}$/GJ	% GHG reduction from fossil reference	Potential technical advances and challenges	Production cost by 2030 (USD$_{2005}$/GJ)	Industrial development
Lipid production, extraction, and conversion of microalgae neutral lipids to biodiesel or renewable diesel. Remainder of algal mass digested or used in other process	Microalgae lipids; see Section 2.6.1.2	Assuming biomass production capacity of 10,000 t/yr, cost of production per kg is USD 0.47 and 0.60 for photobioreactors (PBR) and raceways, respectively.[23]	28–76 Scenarios for open pond and bioreactor[34]	Assuming[34] biomass contains 30% oil by weight, cost of biomass for providing a litre of oil would be USD 1 to 3 and USD 1.5 to 5 for algae of low productivity = 2.5 g/m²/day or high productivity = 10 g/m²/day in open ponds or photobiological reactors.	Preliminary Results 95 or more[23] 30–80[34] for open ponds 50–140[34] for PBR going from low to high productivity	Active R&D by companies small and large including pilots pursuing jet and diesel fuel substitutes.

C: Fuels – Hydrocarbons by Gasification, Pyrolysis, Hydrogenation and Isomerization of Vegetable Oils and Wastes

Process	Feedstock	Efficiency and process economics Eff. = Energy product/biomass energy Component costs in USD$_{2005}$/GJ	% GHG reduction from fossil reference	Potential technical advances and challenges	Production cost by 2030 (USD$_{2005}$/GJ)	Industrial development
Gasification to syndiesel followed by FT (Fischer-Tropsch) process. Known as biomass to liquids. With and without CCS. Process makes hydrocarbons fuels (number of carbon atoms) for gasoline (5–10); kerosene (jet fuel) (10–15); diesel (15–20); fuel oil (20–30)	Lignocellulosic	Eff. = 0.42 fuel only; 0.45 fuel + power.[19]	91[27] (EU)	CCS for CO_2 from processing.	14–20 (fuel only) 8–11 (fuel/power)[19] 15.2-18.6[43]	One first commercial plant (wood) under way. Many worldwide demonstration and pilot processes under way.
		80 million L/yr; FC:12, CC*17 (2015); 280 million L/yr; FC:12, CC*8 (2022).[45]		Meta-analysis conditions.[45]	20–29.5[45]	
		Eff. = 0.52 w/o CCS and 0.5 w/ CCS + 35 and 24 MW$_e$. 4000 t/day switchgrass. Plant cost ~ USD 650 Mi.[10]	90[26] (US)	Gas clean-up costs and scale/volume. Breakeven with barrel of crude oil of USD 122 (USD 113 with CCS and USD 50/t CO_2).[10]	25[10] (w/o CCS US) 30[10] (w/ CCS US) see[38] for cost breakdown (2020)	
		Eff. = 0.52 + 22 MW$_e$. Capital USD 500 million; wide range of densified feeds imported into EU for processing.[39]	Detailed Well-to-Wheel EU[39] US[14] scenarios	Breakeven with barrel of crude oil of USD 75. Mixture of 50% biomass and coal is climate neutral.	16–22.5[39]	
		Coal and biomass co-gasification.	See Fig. 2.10	Switchgrass and mixed prairie grasses.	29[38]	
Hydrogenation to renewable diesel	Plant oils, animal fat, waste	Technology well known. Cost of feedstock is the barrier.	63–130[26] Depending on the co-product treatment method	Feedstock costs drive this process. Process is standard in petrochemical operations.	17–18[34]	One large and few small commercial (see, e.g., footnote 68 in the main text); many demos.
Biomass pyrolysis[4] and catalytic upgrading to diesel/jet fuel; vegetable oils processed directly into a refinery[33]	Biomass/wastes, plant oils, animal fat, waste oils	Developing pyrolysis[8,13] process (also from hydrothermal processing)[46] to a blendstock for a refinery,[33] for direct coupled firing in a boiler (e.g., with coal)[32] or a final product.		Catalyst development, process yield improvements with biomass.	14–24[47] for pyrolysis oils to refinery blendstocks	Demos and fuel product tests in USA, Brazil, EU. Test flights using biojet fuels from plant oils conducted.[33]

Continued next Page →

While some methanol, butanols and other alcohol production processes from biomass exist in various stages of technical development, the most predominant alcohol production pathways have ethanol as their finished product. Lignocellulosic ethanol technologies have many possible process chains (e.g., Sánchez and Cardona, 2008; Sims et al., 2010). Those with the highest sugar yields and with low environmental impact were considered more promising (Wooley et al., 1999) and involve chemical/biochemical, mechanical/chemical/biochemical, and biological/chemical/biochemical processing steps. Most of these chains involve a pretreatment step to overcome the recalcitrance of the plant cell wall, with separate and partial hydrolysis of the cellulose and hemicelluloses fibres to release the complex streams of five- and six-carbon sugars for fermentation. Simultaneous saccharification and fermentation (SSF), simultaneous saccharification and co-fermentation (SSCF) and consolidated bioprocessing

D: Gaseous Fuels, Power and Heat from Gasification

Process	Feedstock	Efficiency and process economics Eff. = product energy/biomass energy Component costs in USD$_{2005}$/GJ	% GHG reduction from fossil reference	Potential technical advances and challenges	Production cost by 2030 (USD$_{2005}$/GJ)	Industrial development
Gasification/syngas processing of H$_2$ to fuel and power	Lignocellulosic	Eff. 60% (fuel only). Needs 0.19 GJ of elect. per GJ H$_2$ for liquid estimated at USD 11–14/GJ (long term), wood USD 2.4/GJ, USD 568/kW$_{th}$ capital.[19]	88[30]	Co-production H$_2$ and power (55% fuel efficiency, 5% power) in the longer term.[19] USD 426/kW$_{th}$ capital.[19]	4–5[19] (longer) 6[20]–12[12] 5.5–7.7[41]	R&D stage.
Gasification/methanation to methane for fuel, heat and/or power	Lignocellulosic	Eff. ~60% (or higher for dry feed).[42] Combined fuel and power production possible.	98[27]	RD&D on gas clean up and methanation catalysts. For wet feedstocks wet gasification developing.[46]	10.6–11.5[42] wood USD 2.8/GJ	RD&D stage.
Anaerobic digestion, upgrading of gas, liquefaction	Organic wastes, sludges	Eff. ~20 to 30%; includes mixtures of animal and agriculture residues.		Improve technology robustness with new metagenomic tools, reduce costs.	15–16[21]	
Integrated gasification combined cycle for CHP	Lignocellulosic	District heating; power-to-heat ratio 0.8 to 1.2; power production efficiency 40 to 45%; total efficiency 85 to 90%. Investment USD 1,200/kW$_{th}$. Wood residues in Finland.[22]	96[31]	Gas cleaning, increased efficiency cycles, cost reductions.	8–11[11]	Demos at 5 to 10 MW projected cost at USD 29–38/GJ or US cents 10–13.5/kWh.[45]
				IGCC at 30 to 300 MW[45] with a capital cost of USD 1,150 to 2,300/kW$_e$, at 10% discount rate, 20 year plant life, and USD 3/GJ. Meta-analysis conditions.	13–19[45] or US cents 4.5–6.9/kWh	

Notes: Abbreviations: *Conversion costs (CC) include investment costs and operating expenses; CR = Co-product Revenue; FC = feedstock cost; CC = conversion cost. All CC, CR, FC costs are given in USD$_{2005}$/GJ.

System Boundaries: Many references use a 10% discount rate, 20-yr plant life referred to as meta-analysis conditions. 17. Production costs include return on capital; 24.10% IRR (Internal Rate of Return), 39% tax rate, 20-yr plant life, Double-declining-balance depreciation method, 100% equity, nth plant, for the biochemical pathway costs are FC: 6, CC*: 10.6, CR: 1.1 and for thermochemical pathway costs are FC: 6.7, CC*: 10, CR: 2.5; 30.12% IRR, 39% tax rate, 25-yr plant life, Modified Accelerated Cost Recovery System depreciation method (MACRS dep.), 65/35 equity/debt, 7% debt interest, nth plant, FC: 8.2, CC*: 16.9, CR: 2.6; 37. Pioneer (first-of-a-kind) plant example: 10% IRR, 39% tax rate, 20-yr plant life, MACRS dep., 100% equity, FC: 12.2–20.7, CC*: 27.3–38, CR: 0–6; 38. 7% discount rate, 39% tax rate, 20-yr plant life, MACRS dep., 45/55 equity/debt, 4.4% debt interest, nth plant, FC w/ CCS: 16, FC w/o CCS: 8.8, CC* w/ CCS: 14.7, CC* w/o CSS: 15.7, CR w/ CCS: 2, CR w/o CCS: 2.1; 39.10% discount rate, 10-yr plant life; 40. Pioneer plant example: 10% IRR, 39% tax rate, 20-yr plant life, MACRS dep, 100% equity, FC: 9.5, CC*: 24.5, CR: 1.1; 41.10% IRR, 15-yr plant life.

References: 1. Hamelinck et al. (2005a); 2. Jeffries (2006); 3. Jeffries et al. (2007); 4. Balat et al. (2009) and see IEA Bioenergy Pyrolysis Task (www.pyne.co.uk); 5. Sims et al. (2008); 6. Himmel et al. (2010); 7. Sannigrahi et al. (2010); 8. Bain (2007); 9. von Weyman (2007); 10. NRC (2009a); 11. IEA Bioenergy (2007); 12. Kinchin and Bain (2009); 13. McKeough et al 2005; 14. Wu et al. (2005); 15. Ezeji et al. (2007a); 16. Ezeji et al. (2007b); 17. Cascone (2008); 18. Tao and Aden (2009); 19. Hamelinck and Faaij (2006); 20. Hoogwijk (2004); 21. Sustainable Transport Solutions (2006); 22. Helynen et al. (2002); 23. Chisti (2007); 24. Foust et al. (2009); 25. Wang et al. (2010); 26. Kalnes et al. (2009); 27. Edwards et al. (2008); 28. Huo et al. (2009); 29. Wu et al. (2008); 30. Laser et al. (2009); 31. Daugherty (2001); 32. Cremers (2009) (see IEA co-firing database at www.ieabcc.nl/database/cofiring.php); 33. IATA (2009); 34. EPA (2010); 35. Seabra et al. (2010); 36. Macedo et al. (2008); 37. Kazi et al. (2010); 38. Larson et al. (2009); 39. van Vliet et al. (2009); 40. Swanson et al. (2010); 41. Hamelinck and Faaij (2002); 42. Mozaffarian et al. (2004); 43. Hamelinck et al. (2004); 44. van Zyl et al. (2007); 45. Bauen et al. (2009a); 46. Elliott (2008); 47. Holmgren et al. (2008); 48. Dutta et al. (2010); 49. Phillips et al. (2007).

(CBP), which combines all of the hydrolysis, fermentation and enzyme production steps into one, were defined as short-, medium- and longer-term approaches, respectively. For CBP, efficiencies and yields are expected to increase and costs to decrease by 35 and 66% relative to SSF and SSCF, respectively (Hamelinck et al., 2005a, and see Table 2.15).

Pretreatment is one of the key technical barriers causing high costs, and a multitude of possible options exist. So far, no 'best' technology has been identified (da Costa Sousa et al., 2009; Sims et al., 2010). Pretreatment overcomes the recalcitrance of the cell wall of woody, herbaceous or agricultural residues and makes carbohydrate polymers accessible to hydrolysis (e.g., by enzymes) and in some cases liberates a portion of the sugars for fermentation to ethanol (or butanols) and the lignin for process heat or electricity. Alternatively, multiple steps (including pretreatment) can be combined with other downstream conversion steps and material can be bioprocessed with multiple organisms simultaneously. To evaluate pretreatment options,[63] the use of common

---

63 The areas of biomass pretreatment and low-cost ethanol emerged as essential in 2009 with fourteen core papers establishing a biology/biochemistry/biomass chemical analysis concentration area (sciencewatch.com/dr/tt/2009/09-octtt-BIO/). Included were coordinated pretreatment research in multiple US and Canadian institutions, investigating common samples and analytical methodology and conducting periodic joint evaluation of technical and economic performance of these processes.

feedstocks and common analytical methodology (Wyman et al., 2005) is needed to differentiate between the performance of the many chains and combinations. For corn stover, among the evaluated options of ammonia fibre expansion (AFEX), dilute acid and hot water pretreatments, dilute acid pretreatment had the lowest cost and the hot water process cost was the highest by 25%. This ranking, however, does not hold for other feedstocks (Elander et al., 2009). On-site enzyme preparation increased the cost of the dilute acid pretreatment by 4.5% (Kazi et al., 2010). Apart from pretreatment, enzymes are another key variable cost and are the focus of major global efforts in RD&D and cost reduction (e.g., Himmel et al., 2010; Sims et al., 2010). Finally, all of the key individual conversion steps (e.g., pretreatment, enzymatic hydrolysis and fermentation) are highly interdependent. Therefore, process integration is another very important focus area, as many steps are either not yet optimized or have not been optimized in a fully integrated process.

The US National Academies analyzed liquid transport fuels from biomass (NRC, 2009a), and their cost analysis found the breakeven point for cellulosic ethanol with crude oil to be USD_{2005} 100/barrel (USD_{2005} 0.64/litre) in 2020, which translates to USD_{2005} 18 to 22/GJ. This projection is similar[64] to the USD_{2005} 23.5/GJ projected by Bauen et al. (2009a) for 2022. The National Research Council (NRC, 2009a) projects that by 2035, process improvements could reduce the plant-related costs by up to 40%, or to within USD_{2005} 12 to 15/GJ, in line with estimates for nth plant costs of USD_{2005} 15.5/GJ (Foust et al., 2009). Further cost reductions in some of the processing pathways may come from converting bagasse to ethanol, as the feedstock is already at the conversion facility, and the bagasse has the potential to produce an additional 30 to 40% yield of ethanol per unit land area in Brazil (Seabra et al., 2010). A similar strategy is currently being employed in the USA, where the coupling of crop residue collection and collocation of the second-generation (residue) and first-generation (corn) ethanol facilities are being pursued by two of the first commercial cellulosic ethanol plant developments by the U.S. Department of Energy.[65]

Several strains of microorganisms have been selected or genetically modified to increase the enzyme production efficiency (FAO, 2008b) for SSF (Himmel et al., 2010), for SSCF (e.g., Dutta et al., 2010) and for CPB (van Zyl et al., 2007; Himmel et al., 2010). Many of the current commercially available enzymes are produced in closed fermenters from genetically modified (GM) microorganisms. The final enzyme product does not contain GM microorganisms (Royal Society, 2008), which facilitates acceptance of the routes (FAO, 2008b).

Microbial fuels. Industrial microorganisms[66] with imported genes to accelerate bioprocessing functions (Rude and Schirmer, 2009) can make hydrocarbon fuels, higher alcohols, lipids and chemicals from sugars. Researchers in synthetic biology have imported pathways, and more recently used artificial biology to design alternative biological paths into microorganisms, which may lead to increased efficiency of fuels and chemicals production (Keasling and Chou, 2008; S. Lee et al., 2008). Another route is to alter microorganisms' existing functions with metabolic engineering tools. Detailed production costs are not available in the literature but Regalbuto (2009) and E4tech (2009) summarize some data.[67] Additionally, some microalgae can metabolize sugars in the absence of light (heterotrophically) to make lipids (similar to plant oils) that are easily converted downstream to biodiesel and/or renewable diesel or jet fuel. With additional genetic engineering, the microorganisms can excrete lipids, leading to a decrease in production costs. Microbial biofuels and chemicals are under active development (Alper and Stephanopoulos, 2009; Rude and Schirmer, 2009).

Gasification-derived products (see Table 2.15.A and B)
Gasification of biomass to syngas (CO and H_2) followed by catalytic upgrading to either ethanol or butanols has estimated production costs (USD_{2005} 12 to 20/GJ) comparable to the biochemical chains discussed above. The lowest-cost liquid fuel is methanol (produced in combination with power) at USD_{2005} 7 to 10/GJ (USD_{2005} 12 to 18/GJ for fuel only). Further reduction in production costs of fuels derived from gasification will depend on significant development of IGCC (currently at the 5 to 10 MW_e demonstration phase) to obtain practical experience and reduce technical risks. Costs are projected to be USD_{2005} 13 to 19/GJ (US $cents_{2005}$ 4.6 to 6.9/kWh) for 30 to 300 MW_e plants (see Table 2.15; Bauen et al., 2009a). Although process reliability is still an issue for some designs, niche markets have begun to develop (Kirkels and Verbong, 2011).

Even though the cost bases are not entirely comparable, the recent estimates for Fischer-Tropsch (FT) syndiesel from Bauen et al. (2009a), van Vliet et al. (2009), the NRC (2009a) and Larson et al. (2009) are (in USD_{2005}/GJ), respectively: 20 to 29.5, 16 to 22, 25 to 30, and 28 (coal and biomass). The breakeven point would occur around USD_{2005} 80 to 120/barrel (USD_{2005} 0.51 to 0.74/litre). High efficiency gains are expected, especially in the case of polygeneration with FT fuels (Hamelinck and Faaij, 2006; Laser et al., 2009; Williams et al., 2009).

Process intensification is the combination of multiple unit operations conducted in a chemical plant into one thus reducing its footprint and

---

64 See Table 2.15 for financial assumptions that are not identical; Bauen et al. (2009a) and Foust et al. (2009) are close.

65 Impact Assessment of first-of-a-kind commercial ethanol from corn stover and cobs collocated with grain ethanol facilities is provided by the Integrated Bioenergy Projects. U.S. DOE Golden Field Office web site: www.eere.energy.gov/golden/Reading_Room.aspx; www.eere.energy.gov/golden/PDFs/ReadingRoom/NEPA/Final_Range_Fuels_EA_10122007.pdf; www.eere.energy.gov/golden/PDFs/ReadingRoom/NEPA/POET_Project_LIBERTY_Final_EA.pdf; and www.biorefineryprojecteis-abengoa.com/Home_Page.html.

66 E.g., *Escherichia coli* and *Saccharomyces cerevisiae* have well-established genetic tools and industrial use.

67 Rude and Schimer (2009) report stoichiometric data, for example, per tonne of glucose the number of litres is 297 of farnesene (for diesel), and 384 of microbial biocrude oil (for jet fuel) compared with 648 of ethanol (for gasoline). Metabolic mass yields are 25 and 30% for farnesene and biocrude, respectively, compared to 51% for ethanol. The routes grow the intermediate cell mass that then starts producing biofuels or intermediates—these steps are usually aerobic and require air and agitation that reduce the overall energy efficiency.

capital costs and enabling plants to operate more cost effectively at smaller scale. Therefore chemical/thermal processing that previously could only be conducted at very large scale could now be downsized to match the supply of biomass cost effectively. Efficient heat and mass transfer in micro-channel reactors has been explored to compact reactors by 1-2 orders of magnitude in water-gas-shift, steam reforming and FT processes for conventional natural gas or coal gasification streams (Nehlsen et al., 2007) and significantly reduce capital costs (Schouten et al., 2002; Sharma, 2002; Tonkovich et al., 2004). Such intensification could lead to distributed biomass to liquids (BTL) production, as capital requirements would be significantly reduced (as they would be for coal to liquids (CTL) or gas to liquids (GTL) (Shah, 2007). Methanol/DME synthesis could be intensified as well. Additionally, combined biomass/coal gasification options could capture some of the economies of scale while taking advantage of biomass' favourable CO_2 mitigation potential.

Other intermediates: vegetable or pyrolysis/ hydrothermal processing oils

For **diesel substitution**, hydrogenation technologies are already commercially producing direct hydrocarbon diesel substitutes from hydrogenation of vegetable oils to renewable diesel in 2011.[68] Costs depend on the vegetable oil prices and subsidies (see Table 2.15.C and Section 2.3.4). Lignocellulosic residues from vegetable oil production could provide the energy for standalone hydrogenation. The downstream processing of the lipids/plant oils to finished fuels is often conducted in conjunction with a petroleum refinery, in which case jet fuel and other products can be made.

Fast **pyrolysis** processes or **hydrothermal liquefaction** processing of biomass make low-cost intermediate oil products (Bain, 2007; Barth and Kleinert, 2008; Section 2.7.1). Holmgren et al. (2008) estimated production costs for lignocellulose pyrolysis upgrading to a blendstock (component that can be blended with gasoline at a refinery) as USD_{2005} 14 to 24/GJ, from bench scale data.

Under mild conditions of **aqueous phase reforming** and in the presence of multifunctional supported metal catalysts, biomass-derived sugars and other oxygenated organics can be combined and chemically rearranged (with retention of carbon and hydrogenation) to make hydrocarbon fuels. These processes can also make hydrogen at moderate temperature and pressure (Cortright et al., 2002; Huber et al., 2004, 2005, 2006; Davda et al., 2005; Gurbuz et al., 2010). These developments have reached the pilot and demonstration phase (Regalbuto, 2009).

From carbon dioxide, water and light energy with photosynthetic algae (Table 2.15.B)

Microalgal lipids (microalgal oil) are at an early stage of R&D and currently have significant feedstock production and processing costs,

ranging from USD_{2005} 30 to 140/GJ (EPA, 2010). Exploring the biodiversity of microbial organisms for their chemical composition and their innate microbial pathways can lead to use of highly saline lands, brackish waters or industrial waste waters, avoiding competition with land for food crops but the potential of microalgae is highly uncertain.

Prospects. In the near to medium term, the biofuel industry, encompassing first- and second-generation technologies that meet agreed-upon environmental and economic sustainability and policy goals, will grow at a steady rate. It is expected that the transition to an integrated first- and second-generation biofuel landscape will likely require another decade or two (Sims et al., 2008, 2010; NRC, 2009a; Darzins et al., 2010).

2.6.3.2 Gaseous fuels

Part D of Table 2.15 compares estimated production costs for the production of gaseous fuels from lignocellulosic biomass and various waste streams:

Anaerobic digestion. Production of methane from a variety of waste streams, alone or combined with agricultural residues, is being used throughout the world at various levels of performance. The estimated production costs depend strongly on the application: USD_{2005} 1 to 2/GJ for landfill gas, USD_{2005} 15 to 20/GJ for natural gas or transport applications, USD_{2005} 50 to 60/GJ for on-farm digesters/small engines and USD_{2005} 100 to 120/GJ for distributed electricity generation (see Tables 2.6 and 2.15). The reliability, predictability and cost of individual technologies and assembled systems could be decreased using advanced metagenomics tools[69] and microbial morphology and population structure (Cirne et al., 2007). Also, control and automation technologies and improved gas clean-up and upgrading and quality standards are needed to permit injection into natural gas lines, which could result in more widespread application. Avoided methane emissions provide a significant climate benefit with simultaneous generation of energy and other products.

Synthesis gas-derived methane (a substitute for natural gas), methanol-dimethyl ether (DME), and H_2 are gaseous products from biomass gasification that are projected to be produced in the USD_{2005} 5 to 18/GJ range. After suitable gas cleaning and tar removal, the syngas is converted in a catalytic synthesis reactor into other products by designing catalysts and types of reactors used (e.g., nickel/magnesium catalysts will lead to SNG, while copper/zinc oxide will preferentially make methanol and DME). Processes developed for use with multiple feedstocks in various proportions can decrease investment risks by ensuring continuous feedstock availability throughout the year and decreasing vulnerability to weather and climate. Methanol synthesis from natural gas (and coal) is practised commercially, and synthesis from biomass is being developed at demonstration and first commercial plants. H_2 production has the lowest potential costs, but more developed infrastructure

---

68 Renewable Diesel is currently produced by Neste Oil in Singapore from Malaysian palm oil and then shipped to Germany (see biofuelsdigest.com/bdigest/2011/03/11/neste-oil-opens-giant-renewable-diesel-plant-in-singapore/). The development of the process took about 10 years from proof of principle as described in www.climatechange.ca.gov/events/2006-06-27+28_symposium/presentations/CalHodge_handout_NESTE_OIL.PDF (nesteoil.com/).

69 See, for instance, www.jgi.doe.gov/sequencing/why/99203.html.

is needed for transportation applications (Kirkels and Verbong, 2011). DME is another product from gasification and upgrading (jointly produced with methanol). It can be made from wood residues and black liquor and is being pursued as a transportation fuel. Sweden considered scenarios for multiple bioenergy products, including a substantial replacement of diesel fuel and gasoline with DME and methanol (Gustavsson et al., 2007).

Microbial fuel cells using organic matter as a source of energy are being developed for direct generation of electricity. Electricity is generated through what may be called a microbiologically mediated oxidation reaction, which implies that overall conversion efficiencies are potentially higher for microbial fuel cells compared to other biofuel processes (Rabaey and Verstraete, 2005). Microbial fuel cells could be applied for the treatment of liquid waste streams and initial pilot winery wastewater treatment is described by Cusick et al. (2011).

2.6.3.3 Biomass with carbon capture and storage: long-term removal of greenhouse gases from the atmosphere

Bioenergy technologies coupled with CCS (Obersteiner et al., 2001; Möllersten et al., 2003; Yamashita and Barreto, 2004; IPCC, 2005; Rhodes and Keith, 2008; Pacca and Moreira, 2009) could substantially increase the role of biomass-based GHG mitigation if the geological technologies of CCS can be developed, demonstrated and verified to maintain the stored CO_2 over time. These technologies may become a cost-effective indirect mitigation, for instance, through offsets of emission sources that are expensive to mitigate directly (IPCC, 2005; Rhodes and Keith, 2008; Azar et al., 2010; Edenhofer et al., 2010; van Vuuren et al., 2010).

Corn ethanol manufacturers in the USA supply CO_2 for carbonated beverages, flash freezing meat and to enhance oil recovery in depleted fields, but due to the low commercial value of CO_2 markets and requirements for regional proximity, the majority of the ethanol plants vent it into the air. CO_2 capture from sugar fermentation to ethanol is thus possible (Möllersten et al., 2003) and may now be used for carbon sequestration. Demonstrations of these technologies are proceeding.[70] The impact of this technology was projected to reduce the lifecycle GHG emissions of a natural gas-fired ethanol plant from 39 to 70% relative to the fossil fuel ethanol replaced, while the energy balance is degraded by only 3.5% (see Table 2.13 for performance in different functional units) ((S&T)² Consultants, 2009).

Similarly, van Vliet et al. (2009) estimated that a net neutral climate change impact could be achieved by combining 50% BTL and 50% coal FTL fuels with CCS, if biomass gasification and CCS can be made to work at an industrial scale and the feedstock is obtained in a climate-neutral manner (see Figure 2.10). Perhaps additional removal could be achieved by using crops that increase soil carbon content (e.g., on degraded lands) as indicated by Larson et al. (2009).

2.6.3.4 Biorefineries

The concept of biorefining is analogous to petroleum refining in that a wide array of products including liquid fuels, chemicals and other products (Kamm et al., 2006) can be produced. Even today's first generation biorefineries are making a variety of products (see Table 2.7), many of which are associated with food and fodder production. For example, sugarcane ethanol biorefineries produce multiple energy products (EPE, 2008, 2010). Sustainable lignocellulosic biorefineries can also enhance the integration of energy and material flows (e.g., Cherubini and Strohman 2010). These biorefineries optimize the use of biomass and resources in general (including water and nutrients) while mitigating GHG emissions (Ragauskas et al., 2006). The World Economic Forum (King et al., 2010) projects that biorefinery revenue potentials with existing policies along the entire value chain could be significant and could reach about USD_{2005} 295 billion by 2020.[71]

2.6.3.5 Bio-based products

Bio-based products are defined as non-food products derived from biomass. The term is typically used for new non-food products and materials such as bio-based plastics, lubricants, surfactants, solvents and chemical building blocks. Plastics represent 73% of the total petrochemical product mix, followed by synthetic fibres, solvents, detergents and synthetic rubber (2007 data; Gielen et al., 2008). Bio-based products can therefore be expected to play a pivotal role in these product categories, in particular plastics and fibres.

The four principal ways of producing polymers and other organic chemicals from biomass are: (1) direct use of several naturally occurring polymers, usually modified with some thermal treatment, chemical transformation or blending; (2) thermochemical conversion (e.g., pyrolysis or gasification) followed by synthesis and further processing; (3) fermentation (for most bulk products) or enzymatic conversion (mainly for specialty and fine chemicals) of biomass-derived sugars or other intermediates; and (4) bioproduction of polymers or precursors in genetically modified field crops such as potatoes or *Miscanthus*.

Worldwide production of recently emerging bio-based plastics is expected to grow from less than 0.4 Mt in 2007 to 3.45 Mt in 2020 (Shen et al., 2009). Cost-effective bio-based products with properties superior to those in conventional materials, not just renewability, are

---

70 See sequestration.org/report.htm and www.netl.doe.gov/technologies/carbon_seq/database/index.html. In the USA, through the Midwest Geological Sequestration Consortium, a coal-fired wet-milled ethanol plant is planning over three years to inject 1 Mt of CO_2 into the Mount Simon sandstone saline formation in central Illinois at a depth of about 2 km in a verification phase test project including monitoring, verification and accounting, which is in the characterization phase (June 2010).

71 Approximate values (USD_{2005} billion by 2020) of business potential for the various parts of the value chain were estimated as: agricultural inputs (15), biomass production (89), biomass trading (30), biorefining inputs (10), biorefining fuels (80), biorefining chemicals and products (6), and biomass power and heat (65).

projected to penetrate the markets (King et al., 2010). For synthetic organic materials production, scenario studies indicate that at a productivity of 0.15 ha/t, an area of 75 million hectares globally could supply the equivalent of 15 to 30 EJ of value-added products (Patel et al., 2006).

Given the early stage of development, the GHG abatement costs differ substantially. The current abatement costs for polylactic acid are estimated at USD$_{2005}$ 100 to 200/t of abated CO_2. Today's abatement costs for bio-based polyethylene, if produced from sugarcane-based ethanol, may be of the order of USD$_{2005}$ 100/t CO_2 or lower. For all processes, technological progress in chemical and biochemical conversion and the combined production of bioenergy is likely to reduce abatement costs by USD$_{2005}$ 50 to 100/t CO_2 in the medium term (Patel et al., 2006).

2.6.4 Synthesis

Lignocellulosic feedstocks offer significant promise because they (1) do not compete directly with food production; (2) can be bred specifically for energy purposes (or energy-specific products), enabling higher production per unit land area, and have a very large market for the products; (3) can be harvested as residues from crop production and other systems that increase land use efficiency; and (4) allow the integration of waste management operations with a variety of other industries offering prospects for industrial symbiosis at the local level.

Drivers and challenges for converting biomass to fuels, power, heat and multiple products are economic growth and development, environmental awareness, social needs, and energy and climate security. The estimated revenue potential along the entire value chain could be of the order of USD$_{2005}$ 295 billion in 2020 with current policies (King et al., 2010).

Residues from crop harvests and from planted forests are projected to increase on average by about 20% by 2030 to 2050 in comparison to 2007 to 2009. Production costs of bioenergy from perennial grasses or short rotation coppice are expected to fall to under USD$_{2005}$ 2.5/GJ by 2020 (WWI, 2006), from a range of USD$_{2005}$ 3 to 16/GJ today. Supply curves projecting the costs and quantities available at specific sites are needed, and they should also consider competing uses as shown in examples in Figure 2.5. For example, EU and US lignocellulosic supply curves show more than 20 EJ at reasonable delivered costs by 2025 to 2030.

A new generation of aquatic feedstocks that use sunlight to produce algal lipids for diesel, jet fuels or higher-value products from CO_2 and water can provide strategies for lowering land use impacts because they enable use of lands with brackish waters or industrial waste water. Today's estimated production costs are very uncertain and range from USD$_{2005}$ 30 to 140/GJ in open ponds and engineered reactors.

Many microbes could become microscopic factories to produce specific products, fuels or materials that decrease society's dependence on fossil energy sources.

Although significant technical progress has been made, the more complex processing required by lignocellulosic biomass and the integration of a number of new steps take time and support to bring development through the 'Valley of Death' in demonstration plants, first-of-a kind plants and early commercialization. Projected costs from a wide range of sources and process variables are very sensitive to feedstock cost and range from USD$_{2005}$ 10 to 30/GJ. The US National Academies project a 40% reduction in operating costs for biochemical routes by 2035.

Cost projections for pilot integrated gasification combined cycle plants in many countries are USD$_{2005}$ 13 to 19/GJ (US cents$_{2005}$ 4.6 to 6.9/kWh at USD$_{2005}$ 3/GJ feedstock cost). In addition to providing power, syngas can be used to produce a wide range of fuels or can be used in a combined power and fuels approach. Estimated projected costs are in the range of USD$_{2005}$ 12 to 25/GJ for methanol, ethanol, butanols and syndiesel. Biomass to liquids technology uses a commercial process already developed for fossil fuel feedstocks. Gaseous products (H_2, methane, SNG) have lower estimated production costs (USD$_{2005}$ 6 to 12/GJ) and are in an early commercialization phase.

The production of biogas from a variety of waste streams and its upgrading to biomethane is already penetrating small markets for multiple applications, including transport in Sweden and heat and power in Nordic and European countries. A key factor is the combination of waste streams with agriculture residues. Improved upgrading and further cost reductions are still needed.

Pyrolysis oil/hydrothermal oils are low-cost transportable oils (see Sections 2.3.4 and 2.7.2) that could become a feedstock for upgrading either in standalone facilities or coupled to a petrochemical refinery. Pyrolysis oils have low estimated production costs of about USD$_{2005}$ 7/GJ and provide options for electricity, heat and chemicals production. Pyrolysis-oil stabilization and subsequent upgrading still require cost reductions and are active areas of research.

Many bioenergy/biofuels routes enable CCS with significant opportunities for removal of GHGs from the atmosphere. As CCS technologies are further developed and verified, coupling concentrated CO_2 streams from fermentation or IGCC for electricity or biomass and coal to liquids through Fischer-Tropsch processes with CCS offer opportunities to achieve carbon-neutral fuels, and in some cases carbon-negative fuels, within the next 35 years. Achieving this goal will be facilitated by well-designed systems that span biomass selection, feedstock supply systems, conversion technologies to secondary energy carriers, and integration of these carriers into the existing energy systems of today and tomorrow.

2.7 Cost trends[72]

2.7.1 Determining factors

Determining the production costs of energy (or materials) from biomass is complex because of the regional variability in the costs of feedstock production and supply and the wide variety of deployed and possible biomass conversion technology combinations. Key factors that affect the costs of bioenergy production are:

- For crop production: the cost of land and labour, crop yields, prices of various inputs (such as fertilizer), water supply and the management system (e.g., mechanized versus manual harvesting) (Sections 2.3.1 and 2.6.1; see Wiskerke et al., 2010 for a local specific example).

- For delivering biomass to a conversion facility: spatial distribution of biomass resources, transport distance, mode of transport and the deployment (and timing) of pretreatment technologies in the chain. Supply chains range from onsite use (e.g., fuelwood or use of bagasse in the sugar industry, or biomass residues in other conversion facilities) all the way to international supply chains with shipped pellets or liquid fuels such as ethanol (Sections 2.3.2 and 2.6.2); see Dornburg and Faaij (2001) on regional transport for power; Hamelinck et al. (2005b) on international supply chains.

- For final conversion to energy carriers (or biomaterials): the scale of conversion, financing mechanisms, load factors, production and value of co-products and ultimate conversion costs (in the production facility). These key factors vary between technologies and locations. The type of energy carrier used in the conversion process influences the climate mitigation potential (Wang et al., 2011).

The analyses of Hoogwijk et al. (2009) provide a global and long-term outlook for potential biomass production costs (focused on perennial cropping systems) of different IPCC SRES scenarios (IPCC, 2000) discussed in Sections 2.8.4 and 2.8.5 (see Table 2.16 and Figure 2.17). Land rental/lease costs, although a smaller cost factor in most world regions, are dependent on intensity of land use in the underlying scenarios. Capital costs vary due to different levels of mechanization. Based on these analyses, a sizeable part (100 to 300 EJ) of the long-range technical potentials based on perennial cropping systems could cost around USD_{2005} 2.3/GJ. The cost range depends on the assumed scenario conditions, and is shown in Figure 10.23 (Hoogwijk et al., 2009; see also cost supply curves and potentials shown in Figure 2.5 for near-term production). More details on costs of both annual and perennial energy crop production are described in Sections 2.3.1 and 2.6.1.

Biomass supplies are, as with any commodity, subject to complex pricing mechanisms. Biomass supplies are strongly affected by fossil fuel prices (OECD-FAO, 2008; Schmidhuber, 2008; Tyner and Taheripour, 2008) and by agricultural commodity and forest product markets. In an ideal situation, demand and supply will balance and price levels will provide a good measure of actual production and supply costs (see also Section 2.5.3 for discussions on LUC). At present, market dynamics determine the costs of the most important biofuel feedstocks, such as corn, rapeseed, palm oil and sugarcane. For wood pellets, another important internationally traded feedstock for modern bioenergy production, prices have been strongly influenced by oil prices, because wood pellets partly replace heating oil, and by supportive measures to stimulate green electricity production, such as FITs for co-firing (Section 2.4; Junginger et al., 2008). In addition, prices of solid and liquid biofuels are determined by national settings, and specific policies and the market value of biomass residues for which there may be alternative applications is often determined by price mechanisms of other markets influenced by national policies (see Junginger et al., 2001 for a specific example for Thailand).

2.7.1.1 Recent levelized costs of electricity, heat and fuels for selected commercial systems

The factors discussed above make it clear that it is difficult to generate generic cost information for bioenergy that is valid worldwide. Nonetheless, this section provides estimates for the recent levelized cost of electricity (LCOE), heat (LCOH) and fuels (LCOF) typical of selected commercial bioenergy systems, some of which are described in more technological detail in Section 2.3.4.[73] The methodology for calculating levelized cost is described in Annex II. Data and assumptions used to produce these figures are provided in Annex III, with those assumptions derived in part from the literature summarized earlier.

The results of the LCOE, LCOH and LCOF calculations for a selected set of commercially available bioenergy options, and based on recent costs, are summarized in Figure 2.18 and discussed below.

To calculate the LCOE for electricity generation, a standardized range of feedstock cost of USD_{2005} 1.25 to 5/GJ was assumed (based on High Heating Value, HHV). To calculate the LCOE of CHP plants where both electricity and heat are produced, the heat was counted as a co-product with revenue that depended on the assumed quality and application of the heat. For large-scale CHP plants, where steam is generated for process heat, the co-product revenue was set at USD_{2005} 5/GJ. For small-scale CHP plants, on the other hand, the revenue was effectively set according to the cost of hot water, or USD_{2005} 13/GJ (applicable, e.g., in Nordic countries and Europe).

The LCOH for heating systems illustrated in the light blue bars of Figure 2.18 is less certain due to a more limited set of available literature. For

---

72 Discussion of costs in this section is largely limited to the perspective of private investors producing secondary energy carriers. Chapters 1 and 8 to 11 offer complementary perspectives on cost issues covering e.g. costs of integration, external costs and benefits, economy-wide costs and costs of policies.

73 The levelized cost of energy represents the cost of an energy generating system over its lifetime; it is calculated as the per-unit price at which energy must be generated from a specific source over its lifetime to break even. It usually includes all private costs that accrue upstream in the value chain, but does not include the downstream cost of delivery to the final customer the cost of integration or external environmental or other costs. Subsidies and tax credits are also not included.

Table 2.16 | Estimated regional technical potential of energy crops for 2050 (in EJ) on abandoned agricultural land and rest of land at various cut-off costs (in USD$_{2005}$/GJ biomass harvested, including local transport) for the two extreme SRES land use scenarios A1 and A2 (Hoogwijk et al., 2009; reproduced with permission from Elsevier B.V.).

Region	A1: high crop growth intensity and maximum international trade in 2050			A2: low crop growth intensity and minimum trade and low technology development in 2050		
cut-off cost	<1.15 USD/GJ	<2.3 USD/GJ	<4.6 USD/GJ	<1.15 USD/GJ	<2.3 USD/GJ	<4.6 USD/GJ
Canada	0	11.4	14.3	0.0	7.9	9.4
USA	0	17.8	34.0	0.0	6.9	18.7
C America	0	7.0	13.0	0.0	2.0	2.9
S America	0	11.7	73.5	0.0	5.3	14.8
N Africa	0	0.9	2.0	0.0	0.7	1.3
W Africa	6.6	26.4	28.5	7.9	14.6	15.5
E Africa	8.1	23.8	24.4	3.6	6.2	6.4
S Africa	0	12.5	16.6	0.1	0.3	0.7
W Europe	0	3.0	11.5	0.0	5.6	12.5
E Europe	0	6.8	8.9	0.0	6.2	6.3
Former USSR	0	78.6	84.9	0.8	41.9	46.6
Middle East	0	0.1	3.0	0.0	0.0	1.3
South Asia	0.1	12.1	15.3	0.6	8.2	9.8
East Asia	0	16.3	63.6	0.0	0.0	5.8
SE Asia	0	8.8	9.7	0.0	6.9	7.0
Oceania	0.7	33.4	35.2	1.6	16.6	18.0
Japan	0	0.0	0.1	0.0	0.0	0.0
Global	15.5	271	438	14.6	129	177

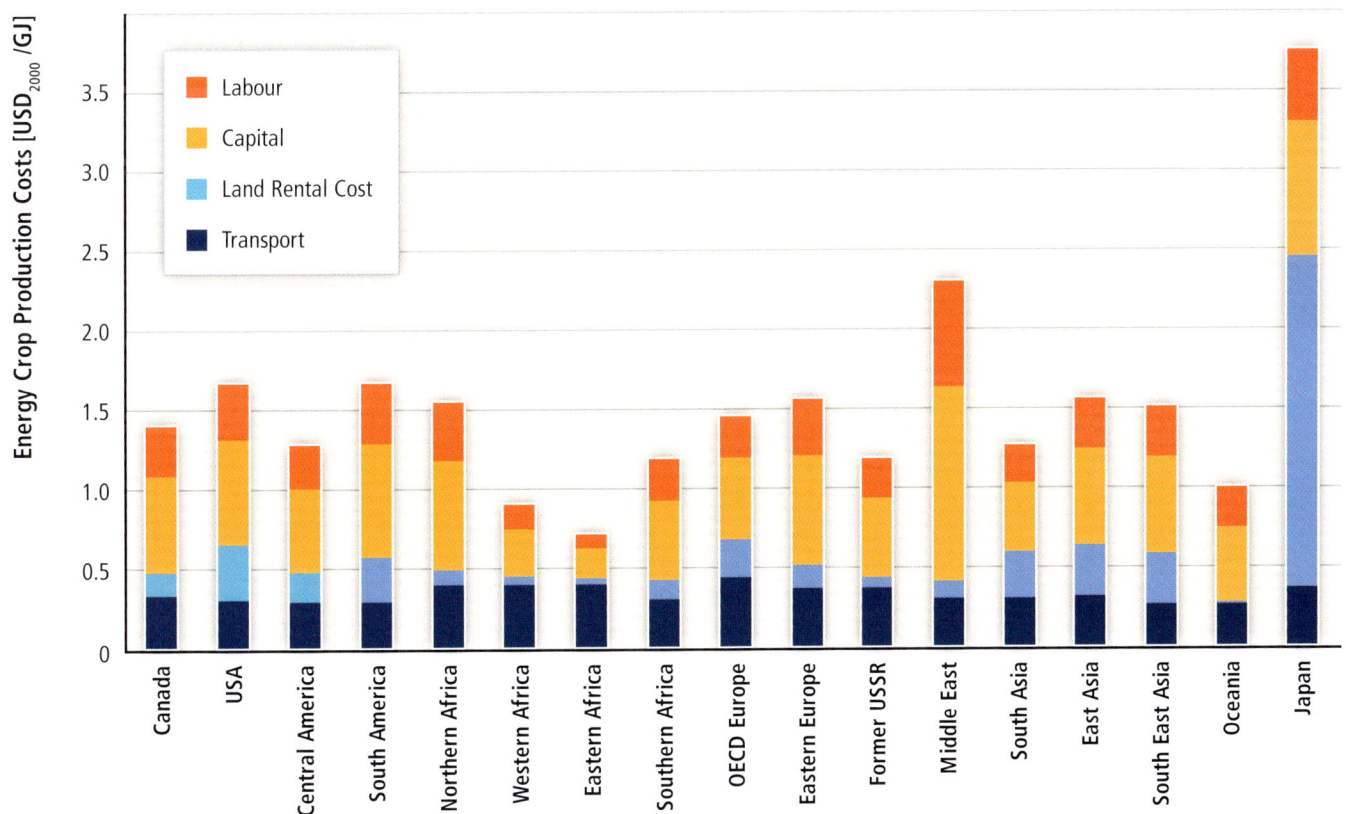

Figure 2.17 | Cost breakdown for energy crop production costs in the grid cells with the lowest production costs within each region for the SRES A1 scenario (IPCC, 2000) in 2050 (in USD$_{2000}$ instead of USD$_{2005}$)(Hoogwijk et al., 2009; reproduced with permission from Elsevier B.V.).

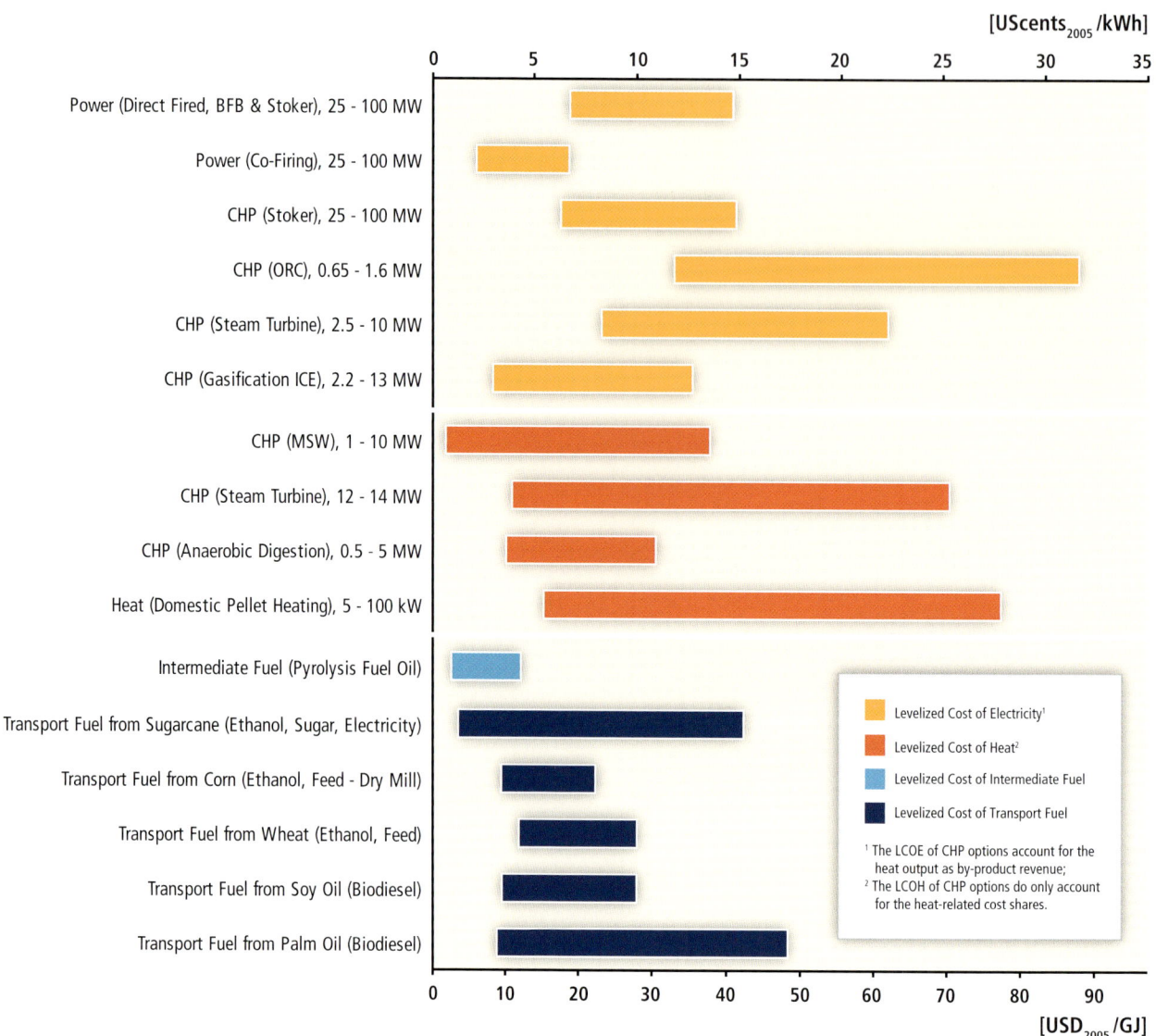

Figure 2.18 | Typical recent levelized cost of energy service from commercially available bioenergy systems at 7% discount rate. Feedstock cost ranges differ between technologies. For levelized cost at other discount rates (3 and 10%) see Annex III and Section 10.5. For biofuels, the range of LCOF represents production in a wide range of countries whereas LCOE and LCOH are given only for major user markets of the technologies for which data were available. The underlying cost and performance assumptions used in the calculations are summarized in Annex III. Calculations are based on HHV.

Abbreviations: BFB: Bubbling fluidized bed; ORC: Organic Rankine cycle; ICE: Internal combustion engine.

heating applications, investment cost assumptions came principally from literature from European and Nordic countries, which are major users of these applications (see Figure 2.8). Feedstock cost ranges came from the same literature and therefore may not be representative of other world regions: feedstock costs were assumed to be USD_{2005} 0 to 3.0/GJ for MSW and low-cost residues, USD_{2005} 2.5 to 3.7/GJ for anaerobic digestion, USD_{2005} 3.7 to 6.2/GJ for steam turbine and USD_{2005} 10 to 20/GJ for pellets. The LCOH figures presented here are therefore most representative of European systems.

LCOF estimates were derived from a techno-economic evaluation of the production of biofuels in multiple countries (Bain, 2007).[74] Underlying feedstock cost assumptions represent the maximum and minimum recent feedstock cost in the respective regions, and are provided in Annex III. All routes for biofuel production take into account sometimes multiple co-product revenues, which were subtracted from expenditures to calculate the LCOF. In the case of ethanol from sugarcane, for example,

[74] The study was done in conjunction with a preliminary economic characterization of feedstock supply curves for the Americas, China and India (Kline et al., 2007) described in Section 2.2.3. The biomass market potential associated with these calculations (Alfstad, 2008) is shown in Figure 2.5(c) (45 EJ, 25 EJ and 8 EJ respectively for the high-growth, baseline and low-growth cases for these countries).

the revenue from sugar was set at USD$_{2005}$ 4.3/GJ$_{feed}$, though this value varies with sugar market prices and can go up to about USD$_{2005}$ 5.6/GJ$_{feed}$. For the LCOF calculations, however, average by-product revenues were assumed. Along with ethanol and sugar (and potentially other biomaterials in the future), the third co-product is electricity, revenues for which were also assumed to be deducted in calculating the LCOF. A similar approach was used for other biofuel pathways (see Annex III). This single example, however, illustrates the complexity of biofuel production cost assessments.

Finally, the levelized cost of pyrolysis oil as an intermediate fuel, a densified energy carrier, was also assessed, because pyrolysis oils are already used for heating and CHP applications and are also being investigated for stationary power and transport applications (see Sections 2.3.3.2, 2.6.2 and 2.6.3.1).

Figure 2.18 presents a broad range of values, driven by variations not only in feedstock costs but also investment costs, efficiencies, plant lifetimes and other factors. Feedstock costs, however, not only vary substantially by region but also represent a sizable fraction of the total levelized cost of many bioenergy applications. The effect of different feedstock cost levels on the LCOE of the electricity generation technologies considered here is shown more clearly in Figure 2.19, where variations are also shown for investment costs and capacity factors.[75] Similar effects are shown for the levelized cost of biofuels (LCOF) in Figure 2.20. (Though a figure is not shown for heating systems, a similar relationship would

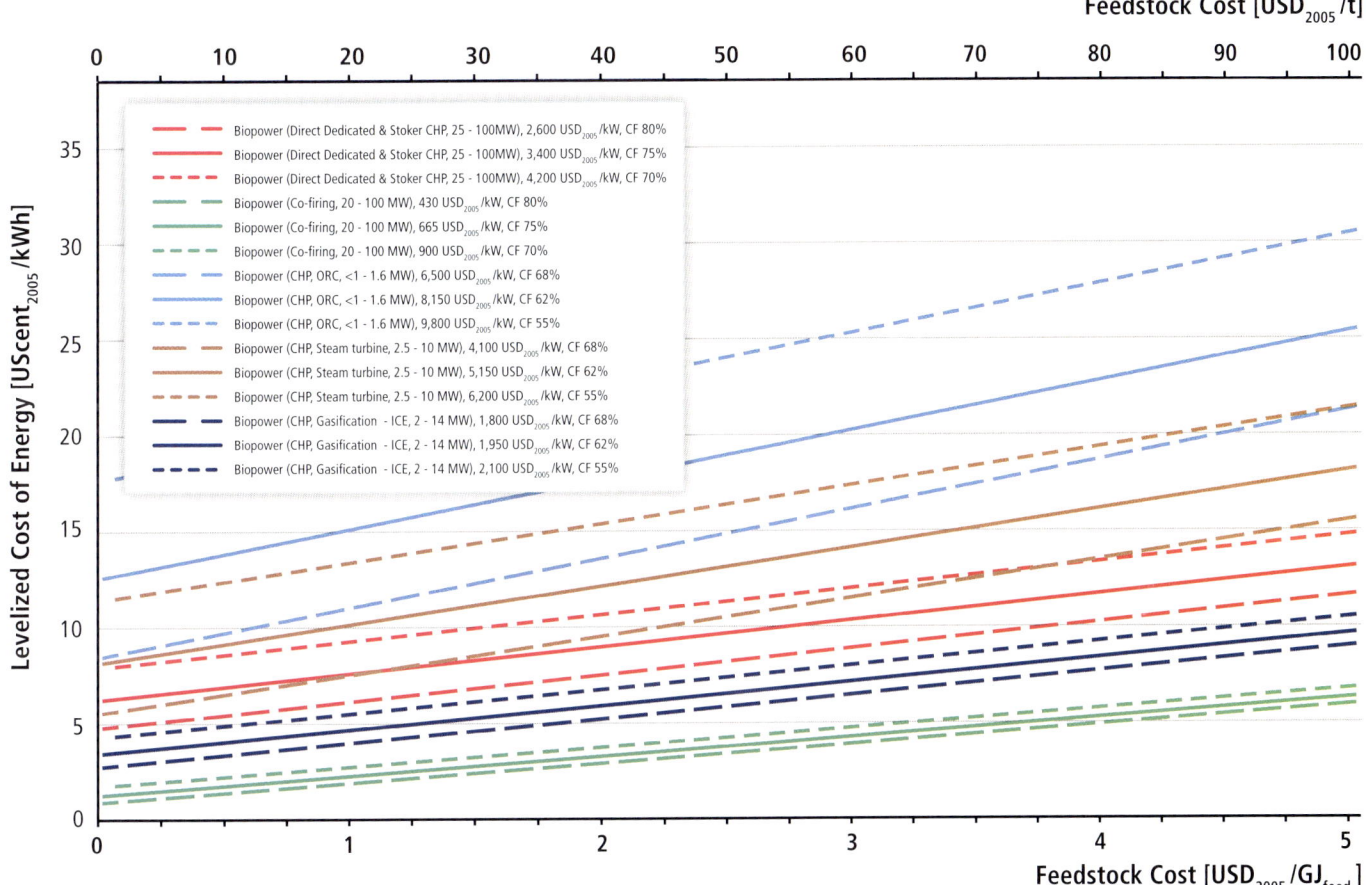

Figure 2.19 | Sensitivity of LCOE with respect to feedstock cost for a variety of investment costs and plant capacity factors (CF). LCOE is based on a 7% discount rate, the mid-value of the operations and maintenance (O&M) cost range, and the mid-value of the lifetime range (see Annex III). Calculations are based on HHV.

References: DeMeo and Galdo (1997); Bain et al. (2003); EIA (2009); Obernberger and Thek (2004); Sims (2007); McGowin (2008); Obernberger et al. (2008); EIA (2010b); Rauch (2010); Skjoldborg (2010); Bain (2011); OANDA (2011).

[75] Note that large-scale power only and CHP technologies have been aggregated in Figure 2.18, while they are shown separately in Figure 2.19.

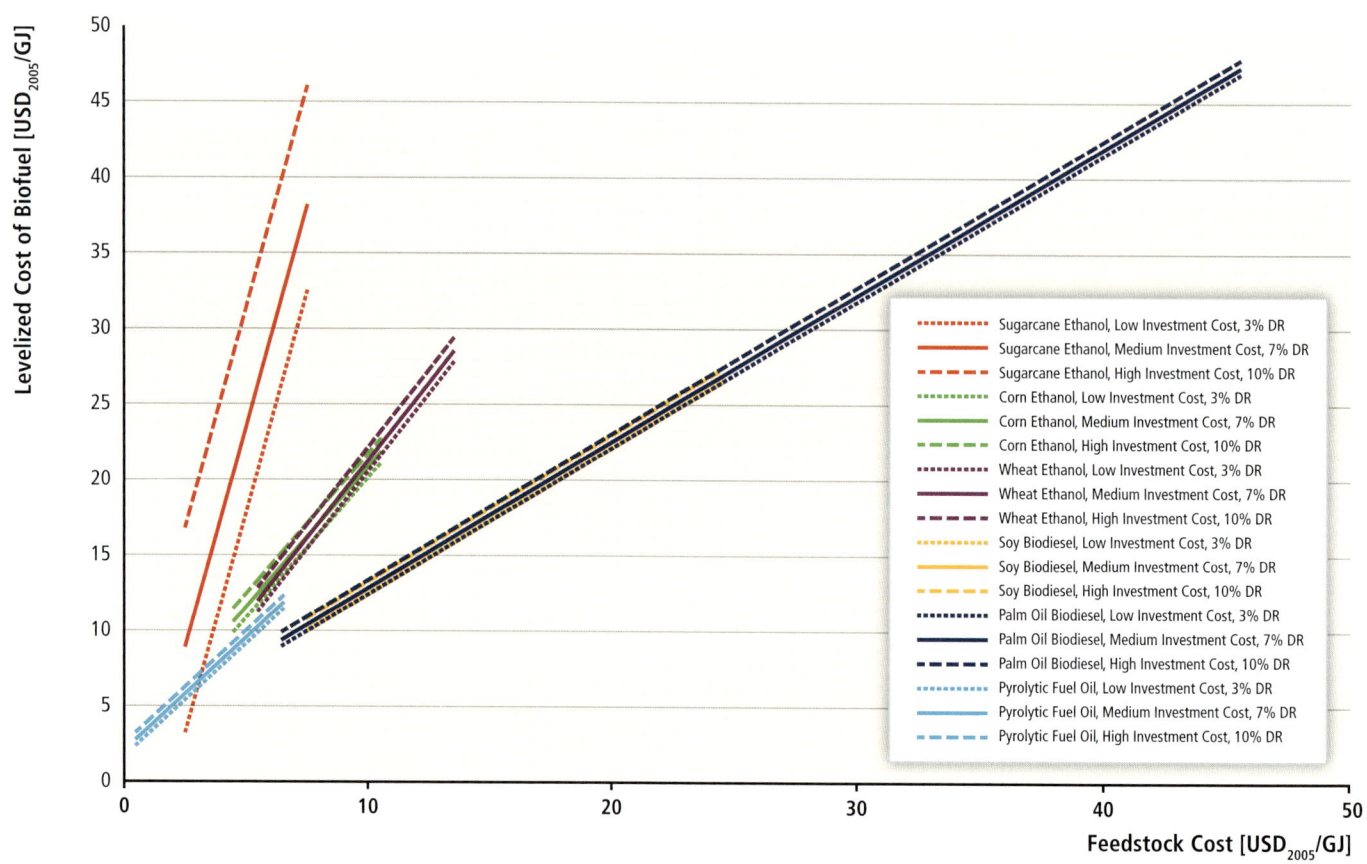

Figure 2.20 | Sensitivity of LCOF with respect to feedstock cost for different discount rates and the mid-values of other cost components from multiple countries (see Annex III). Calculations are based on HHV.

References: Delta-T Corporation (1997); Sheehan et al. (1998b); McAloon et al. (2000); Rosillo-Calle et al. (2000); McDonald and Schrattenholzer (2001); Ibsen et al. (2005); Jechura (2005); Bohlmann (2006); CBOT (2006); Haas et al. (2006); Oliverio (2006); Oliverio and Ribeiro (2006); Ringer et al. (2006); Shapouri and Salassi (2006); USDA (2006); Bain (2007); Kline et al. (2007); USDA (2007); Alfstad (2008); RFA (2011); University of Illinois (2011).

exist.) References used to generate the cost data are assembled in notes to the figures.

2.7.2 Technological learning in bioenergy systems

Cost trends and technological learning in bioenergy systems are not as well described as those for solar or wind energy technologies. Recent literature, however, gives more detailed insights into the learning curves of various bioenergy systems. Table 2.17 and Figure 2.21 summarize a number of analyses that have quantified learning, expressed by learning rates (LR) and learning (or experience) curves, for three commercial biomass systems:

1. Sugarcane-based ethanol production (van den Wall Bake et al., 2009);
2. Corn-based ethanol production (Hettinga et al., 2009);
3. Wood fuel chips and CHP in Scandinavia (Junginger et al., 2005 and a number of other sources).

The LR is the rate of a unit cost decline associated with each doubling of cumulative production (see Section 10.2.5 for a more detailed discussion). For example, a LR of 20% implies that after one doubling of cumulative production, unit costs decreased by 20% of the original costs. The definition of the 'unit' depends on the study variable.

Learning curve studies have accuracy limitations (Junginger et al., 2008; see also Section 10.5.3). Yet, there are a number of general factors that drive cost reductions that can be identified: For biomass feedstocks for ethanol production such as sugar crops (sugarcane) and starch crops (corn), increasing crop yields have been the driving force behind cost reductions.

- For sugarcane, cost reductions have come from R&D efforts to develop varieties with increased sucrose content and thus ethanol yield, increasing the number of harvests from the crop ratoon (from shoots) before replanting the field, increasingly efficient manual harvesting and the use of larger trucks for transportation. More recently, mechanical harvesting of sugarcane is replacing manual harvest, increasing the amount of residues for electricity production (van den Wall Bake et al., 2009; Seabra et al., 2010).

- For the production of corn, the highest cost decline occurred in costs for capital, land and fertilizer until 2005. Additional drivers behind cost reductions were increased plant sizes through cooperatives that

Table 2.17 | Experience curves for major components of bioenergy systems and final energy carriers expressed as reduction (%) in cost (or price) per doubling of cumulative production.

Learning system	LR (%)	Time frame	Region	N	R^2
Feedstock production					
Sugarcane (tonnes sugarcane)[1]	32±1	1975–2005	Brazil	2.9	0.81
Corn (tonnes corn)[2]	45±1.5	1975–2005	USA	1.6	0.87
Logistic chains					
Forest wood chips (Sweden)[3]	12–15	1975–2003	Sweden/Finland	9	0.87–0.93
Investment and O&M costs					
CHP plants[3]	19–25	1983–2002	Sweden	2.3	0.17–0.18
Biogas plants[4]	12	1984–1998		6	0.69
Ethanol production from sugarcane[1]	19±0.5	1975–2003	Brazil	4.6	0.80
Ethanol production from corn (only O&M costs)[2]	13±0.15	1983–2005	USA	6.4	0.88
Final energy carriers					
Ethanol from sugarcane[5]	7 / 29	1970–1985 / 1985–2002	Brazil	~6.1	n.a.
Ethanol from sugarcane[1]	20±0.5	1975–2003	Brazil	4.6	0.84
Ethanol from corn[2]	18±0.2	1983–2005	USA	7.2	0.96
Electricity from biomass CHP[4]	8–9	1990–2002	Sweden	~9	0.85–0.88
Electricity from biomass[6]	15	Unknown	OECD	n.a.	n.a.
Biogas[4]	0–15	1984–2001	Denmark	~10	0.97

Notes: Abbreviations: LR: Learning Rate, N: Number of doublings of cumulative production, R^2: Correlation coefficient of the statistical data.
References: 1. van den Wall Bake et al. (2009); 2. Hettinga et al. (2009); 3. Junginger et al. (2005); 4. Junginger et al. (2006); 5. Goldemberg et al. (2004); 6. IEA (2000).

enabled higher production volumes, efficient feedstock collection, decreased investment risk through government loans and the introduction of improved efficiency natural gas-fired ethanol plants, which are responsible for nearly 90% of ethanol production in the USA (Hettinga et al., 2009). Higher yields were achieved from corn hybrids genetically modified to have higher pest resistance and increased adoption of no-till practices that improved water quality (NRC, 2010). While it is difficult to quantify the effects of these factors, it seems clear that R&D efforts (realizing better plant varieties), technology improvements and learning by doing (e.g., more efficient harvesting) played important roles.

For ethanol production, industrial costs from both sugarcane and corn mainly decreased because of increasing scales of the ethanol plants.

- Cost breakdowns of the sugarcane production process showed reductions of around 60% within all sub processes from 1975 to 2005. Ethanol production costs (excluding feedstock costs) declined by a factor of three between 1975 and 2005 (in real terms, i.e., corrected for inflation). Investment and operation and maintenance costs declined mainly due to economies of scale. Other fixed costs, such as administrative costs and taxes, did not fall dramatically, but cost reductions can be ascribed to automated administration systems. Decreased costs can be primarily ascribed to increased scales and load factors (van den Wall Bake et al., 2009).

- For ethanol from corn, the conversion costs (without costs for corn) declined by 45% from USD_{2005} 240/m³ in the early 1980s to USD_{2005} 130/m³ in 2005. Costs for energy, labour and enzymes contributed in particular to the overall decline in costs. Additional drivers behind these reductions are higher ethanol yields, the introduction of automation and control technologies that require less energy and labour and the up-scaling of average dry grind plants (Hettinga et al., 2009).

2.7.3 Future scenarios of cost reduction potentials

2.7.3.1 Future cost trends of commercial bioenergy systems

For the production of ethanol from sugarcane and corn, future production cost scenarios based on direct experience curve analysis were found in the literature:

For Brazilian sugarcane ethanol (van den Wall Bake et al., 2009), total production costs in 2005 were approximately USD_{2005} 340/m³ (USD_{2005} 16/GJ). Based on the experience curves for the cost components shown in Figure 2.21 (feedstock and ethanol without feedstock costs), total ethanol production costs in 2020 are estimated between USD_{2005} 200 and 260/m³ (USD_{2005} 9.2 to 12.2/GJ). These costs compare well with those in Table 2.7 for Brazil with a current production cost estimate of USD_{2005} 14.8/GJ and projected 2020 cost of USD_{2005} 9 to 10/GJ. Ethanol production costs without feedstocks are in a range of USD_{2005} 139 to 183/m³ (USD_{2005} 6.5 to 8.6/GJ) in 2005 and could reach about USD_{2005} 113/m³ (USD_{2005} 6.6/GJ) by 2020, assuming a constant 82 m³ hydrous ethanol per t of sugarcane.

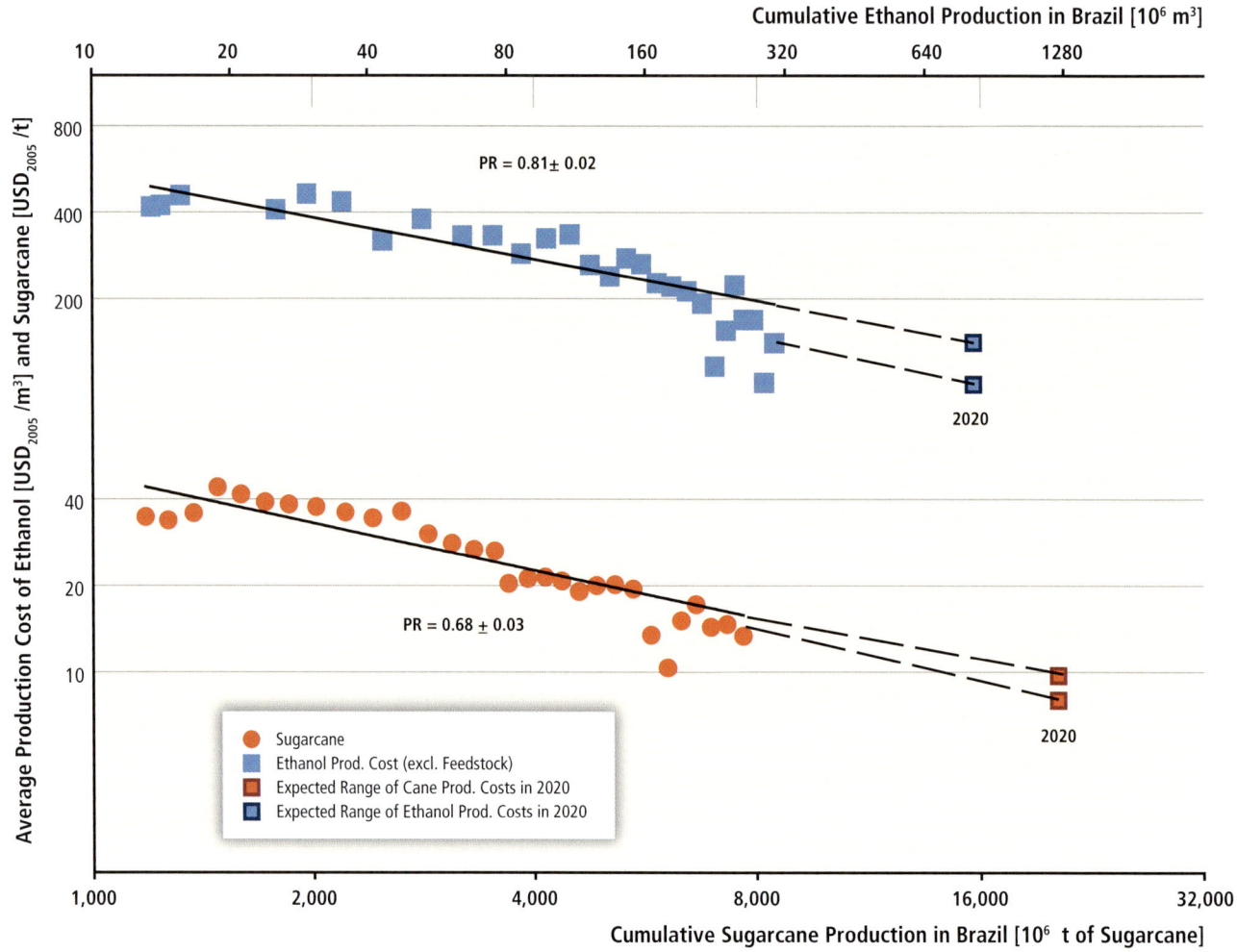

Figure 2.21 | Brazilian sugarcane and ethanol production cost learning curves for between 1975 and 2005 and extrapolated to 2020 (in USD_{2005}). Progress ratio (PR=1-LR) is obtained by best fit to data (van den Wall Bake et al., 2009; reproduced with permission from Elsevier B.V.).

For US ethanol from corn (Hettinga et al., 2009), costs of corn production and ethanol processing are estimated respectively as USD_{2005} 75/t and USD_{2005} 60 to 77/m³ by 2020. Overall ethanol production costs could decline from a current level of USD_{2005} 310/m³ to USD_{2005} 248/m³ (USD_{2005} 14.7 to 11.7/GJ) by 2020. This estimate excludes the investment costs and the effect of future corn prices. The EPA (2010) Regulatory Impact Analysis of the Renewable Fuel Standard 2 modelled the current corn ethanol industry in detail and projected a decrease in total production cost from USD_{2005} 17.5 to 16/GJ by 2022 by taking into account both feedstock and process improvements listed in Table 2.7 and the anticipated co-product revenue.

Confirming the trend and supporting the projections to 2020, Table 2.13 illustrates key indicators for environmental performance of a North American corn dry-grind natural gas-fired mill and the Brazilian sugarcane benchmark of 44 mills in terms of GHG emissions per carbon content of the biomass feedstock (displacement factor), emissions reductions relative to the reference fossil fuel in the production region (GHG savings), and a land use efficiency (volume of production per unit area) indicator. The commercial North American system's performance improved with time; for instance, using the relative GHG savings, which were 26% in 1995 and 39% in 2005, and the projected efficiency improvements through application of commercial CHP systems alone or in combination with CCS, would lead to 55 and 72% emissions savings by 2015, respectively. Similarly, the Brazilian sugarcane ethanol/electricity/sugar mill would go from 79 to 120 and 160% in relative GHG savings for the 2005-2006 baseline and the CHP and CCS scenarios, respectively.

In the Renewable Fuels for Europe project that focused on deployment of biofuels in Europe (de Wit et al., 2010; Londo et al., 2010), specific attention was paid to the effects of learning for lignocellulosic biofuels technologies on projections of future costs. The analyses showed two key points:

- Lignocellulosic biofuels have considerable potential for improvement in the areas of crop production, supply systems and the conversion technology. For conversion in particular, economies of scale are a very important element of the future cost reduction potential as specific capital costs can be reduced (partly due to improved conversion efficiency). Biomass resources may become somewhat more expensive due to a reduced share of (less costly) residues over time. It was estimated that lignocellulosic biofuel production cost could compete with gasoline and diesel from oil at USD_{2005} 60 to 70/barrel by 2030 (USD_{2005} 0.38 to 0.44/litre) (Hamelinck and Faaij, 2006).

- The penetration of lignocellulosic biofuel options depends considerably on the rate of learning. This rate is in turn dependent on increased market penetration (which allows for producing with larger production facilities), which makes the LR partly dependent on market support or mandates in earlier phases of market penetration.

The IEA Energy Technology Perspectives report (IEA, 2008a) and the WEO (IEA, 2009b) project a rapid increase in production of lignocellulosic biofuels, especially between 2020 and 2030, accounting for all incremental biomass increases after 2020. The biofuels analysis projects an almost complete phase-out of cereal- and corn-based ethanol production and edible oilseed-based biodiesel after 2030. The potential cost reductions from current demonstration projects to future commercial-scale facilities for production of specific lignocellulosic biofuels are shown in Figure 2.22. Such potential cost reductions are also quantified in Hamelinck and Faaij (2006) and van Vliet et al. (2009).

2.7.3.2 Future cost trends for pre-commercial bioenergy systems

A number of bioenergy systems are evolving, as shown in Figure 2.2 and discussed in Section 2.6. The key intermediates that enable generation of bioenergy from modern biomass include syngas, sugars, vegetable oils/lipids, thermochemical oils derived from biomass (pyrolysis or other thermal treatments), and biogas. These intermediates can produce higher efficiency electricity and heat, a wider range of liquid hydrocarbon fuels, alcohols (including some with higher energy density), ethers, and chemical products and polymers (bio-based materials) in the developing biorefineries that are discussed in Section 2.6. Initial R&D on producing hydrocarbon fuels is starting with sugar and starch crops and covering the range of gasoline, diesel and higher-energy content transport fuels such as jet fuels and chemicals. Both improved first-generation crops, perennial sugarcane-derived, in particular, and second-generation plants have the potential to provide a variety of energy products suited to specific geographic regions, and high-volume chemicals and materials traditionally derived from the petrochemical industry, maximizing the outputs of end products per unit of feedstock.

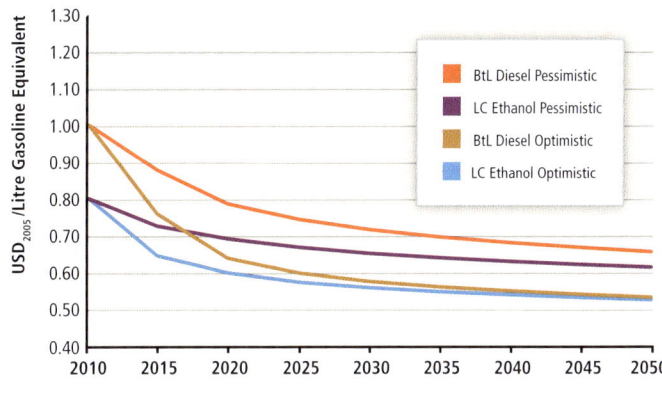

Figure 2.22 | Cost projections for lignocellulosic ethanol and BTL diesel (*Energy Technology Perspectives 2008*, © OECD/IEA, Figure 9.11, p. 335 in IEA (2008a); for additional future cost considerations see also Sims et al. (2008), IEA Renewable Energy Division (2010) and IEA (2011)).

Table 2.18 presents projected ranges of production costs for developing technologies such as integrated gasification combined cycle for the production of higher efficiency electricity and gasification-(syngas) derived fuels, including diesel, jet fuel, and H_2, methane, dimethyl ether and other oxygenated fuels through catalytic upgrading of the syngas. The sugar intermediates, lignocellulosic for instance, can be converted through biochemical routes to a variety of fuels with the properties of petroleum-based fuels. Similarly, pyrolysis oil-based hydrocarbon fuels are under development. Oilseed crop and tree seed oil development could also expand the range of fuel products with properties of petroleum fuels because they are readily upgraded to hydrocarbons. Finally, algae for biomass production are photosynthetic, using CO_2, water, and sunlight to biologically produce a variety of carbohydrates, lipids, plastics, chemicals or fuels like H_2, along with oxygen. In addition, heterotrophic microbes, such as certain algae are engineered to metabolize sugars and excrete lipids in the dark. Microorganisms or their consortia can consolidate various processing steps; genetically engineered yeasts or bacteria can make specific fuel products, including hydrocarbons and lipids, developed either with tools from synthetic biology or through metabolic engineering (see also IEA, 2011).

2.7.4 Synthesis

Despite the complexities of determining the economic performance and regional specificities of bioenergy systems, several key conclusions can be drawn from available experiences and literature:

- Several important bioenergy systems today can be deployed competitively, most notably sugarcane-based ethanol and heat and power generation from residues and waste.

- Although not all bioenergy options discussed in this chapter have been investigated in detail with respect to technological learning, several important bioenergy systems have reduced their cost and improved environmental performance over time. These systems still

Table 2.18 | Projected production cost ranges estimated for developing technologies (see Section 2.6.3).

Selected Bioenergy Technologies	Energy Sector (Electricity, Thermal, Transport)*	2020-2030 Projected Production Costs (USD$_{2005}$/GJ)
IGCC[1]	Electricity and/or transport	12.8–19.1 (4.6–6.9 cents/kWh)
Oil plant-based renewable diesel and jet fuel	Transport and electricity	15–30
Lignocellulose sugar-based biofuels[2]	Transport	6–30
Lignocellulose syngas-based biofuels[3]	Transport	12–25
Lignocellulose pyrolysis-based biofuels[4]	Transport	14–24 (fuel blend components)
Gaseous biofuels[5]	Thermal and transport	6–12
Aquatic plant-derived fuels, chemicals	Transport	30–140

Notes: 1. Feed cost USD$_{2005}$ 3.1/GJ, IGCC (future) 30 to 300 MW, 20-yr life, 10% discount rate; 2. ethanol, butanols, microbial hydrocarbons from sugar or starch crops or lignocellulose sugars; 3. syndiesel, methanol and gasoline, etc.; syngas fermentation routes to ethanol; 4. biomass pyrolysis (or other thermal treatment) and catalytic upgrading to gasoline and diesel fuel blend components or to jet fuels; 5. synfuel to SNG, methane, dimethyl ether, or H$_2$ from biomass thermochemical and anaerobic digestion (larger scale).
*Several applications could be coupled with CCS when these technologies, including CCS, are mature and thus could remove GHGs from the atmosphere.

require government subsidies that are put in place for economic development, poverty reduction, a secure and diverse energy supply, and other reasons.

- There is clear evidence that further improvements in power generation technologies, production of perennial cropping systems and development of supply systems can bring the costs of power (and heat) generation from biomass down to attractive cost levels in many regions. With the deployment of carbon taxes of up to USD$_{2005}$ 50/t, biomass can, in many cases, also be competitive with coal-based power generation. Nevertheless, the competitive production of bio-electricity depends also on the performance of alternatives such as wind and solar energy, CCS coupled with coal, and nuclear energy (see Section 10.2.2.4 and Chapter 8).

- Bioenergy systems for ethanol and biopower production show technological learning and related cost reductions with LRs comparable to those of other RE technologies. This applies to cropping systems (following progress in agricultural management of annual crops), supply systems and logistics (as clearly observed in Scandinavia, as well as international logistics) and in conversion (ethanol production, power generation and biogas).

- With respect to lignocellulosic biofuels, recent analyses have indicated that the improvement potential is large enough to make them competitive with oil prices of USD$_{2005}$ 60 to 70/barrel (USD 0.38 to 0.44/litre). Currently available scenario analyses indicate that if shorter-term R&D and market support are strong, technological progress could allow for commercialization around 2020 (depending on oil price developments and level of carbon pricing). Some scenarios also indicate that this would mean a major shift in the deployment of biomass for energy, because competitive production would decouple deployment from policy targets (mandates) and demand for biomass would move away from food crops to biomass residues, forest biomass and perennial cropping systems. The implications of such a (rapid) shift have not been studied.

- Data about the production of biomaterials and cost estimates for chemicals from biomass are rare in peer-reviewed literature. Future projections and LRs are even rarer, because successful bio-based products are just now entering the market place. Two examples are as partial components of otherwise fossil-derived products (e.g., poly(1,3)-propylene terephthalates based on 1,2-propanediol derived from sugar fermentation) or as fully new synthetic polymers such as polylactides based on lactic acid derived from sugar fermentation. This is also the case for biomass conversion coupled with CCS (see Section 2.6.3.3) concepts, which are not developed at present and for which cost trends are not available in literature. CO$_2$ from ethanol fermentation is commercially sold to carbonate beverages, flash freeze meats or enhance oil recovery, and demonstrations of CCS are ongoing (see Section 2.6.3.3). Nevertheless, recent scenario analyses indicate that advanced biomaterials (and cascaded use of biomass) as well as other biomass conversion coupled to CCS may become attractive medium-term mitigation options. It is therefore important to gain experience so that more detailed analyses on those options can be conducted in the future.

2.8 Potential Deployment[76]

2.8.1 Current deployment of bioenergy

Modern biomass use (for electricity and CHP for the power sector; modern residential, commercial, and public buildings heating; or transport fuels) already provides a significant contribution of about 11.3 EJ (see Table 2.1; IEA, 2010a,b) out of the 2008 TPES from biomass of 50.3 EJ. Between 60 and 70% of the total biomass supply is used in rural areas and relates to charcoal, wood, agricultural residues and manure used for cooking, lighting and space heating, generally by the poorer part of the population in developing countries. From 1990 to 2008, the

---
[76] Complementary perspectives on potential deployment based on a comprehensive assessment of numerous model-based scenarios of the energy system are presented in Sections 10.2 and 10.3 of this report.

average annual growth rate of solid biomass use for bioenergy was 1.5%, while the average annual growth rate of modern liquid and gaseous biofuels use was 12.1 and 15.4%, respectively, during the same period (IEA, 2010c). As a result, biofuels' share of global road transport fuels was about 2% in 2008; and nearly 3% of global road transport fuels in 2009, as oil demand decreased for the first time since 1980 (IEA, 2010b). Government policies in various countries fostered the five-fold increase in global biofuels production from 2000 to 2008. Biomass and renewable waste power generation was 259 TWh (0.93 EJ) in 2007 and 267 TWh (0.96 EJ) in 2008, representing 1% of the world's electricity and a doubling since 1990 (from 131 TWh, 0.47 EJ) (Section 2.4.1). Modern bioenergy heating applications, including space and hot water heating systems such as for district heating, account for 3.4 EJ (see Table 2.1 and Section 2.4.1).

International trade in biomass and biofuels has also become much more important over the recent years, with roughly 6% (reaching levels of up to 9% in 2008) of biofuels (ethanol and biodiesel only) traded internationally and one-third of pellet production dedicated to energy use in 2009 (Figures 2.8 and 2.9; Junginger et al., 2010; Lamers et al., 2010; Sikkema et al., 2011). The latter has proven to be an important facilitating factor in both increased utilization of biomass in regions where supplies are constrained and mobilizing resources from areas where demand is lacking.

The policy context for bioenergy and particularly biofuels has changed rapidly and dramatically since the mid-2000s in many countries. The food versus fuel debate and growing concerns about other conflicts created a strong push for the development and implementation of sustainability criteria and frameworks and changes in temporization of targets for bioenergy and biofuels. Furthermore, the support for advanced biorefinery and second-generation biofuel options drives bioenergy in more sustainable directions.

Nations like Brazil, Sweden, Finland and the USA have shown that persistent and stable policy support is a key factor in building biomass production capacity and working markets, required competitive infrastructure and conversion capacity (see also Section 2.4) and results in considerable economic activity.

2.8.2   Near-term forecasts

Countries differ in their priorities, approaches, technology choices and support schemes for bioenergy development. Although on the one hand complex for the market, this is also a reflection of the many aspects that affect bioenergy deployment: agriculture and land use; forestry and industry development; energy policy and security; rural development; and environmental policies. Priorities, the stage of technology development, and access to, availability of and cost of resources differ widely from country to country and in different settings.

The near-term forecasts reflect that the policies already in place, as shown in Table 2.11, are driving current forecasts. For instance, the WEO (IEA, 2010b) projects that the bioenergy industry will continue the growth observed in the past five years and reach about 60 EJ by 2020 in the Current Policies scenario (which replaces the former Reference scenario), with slightly higher levels of up to 63 EJ in the more ambitious New Policies and 450-ppm CO_2 scenarios (Section 2.4.1). Considering the 2008 starting point at 50 EJ/yr, this represents a 10 to 13 EJ increase in bioenergy consumption over 10 years. Much of the increase happens in the transport sector, with biofuel consumption starting from 2.1 EJ in 2009 and increasing to 4.5 to 5.1 EJ in 2020 in the three presented scenarios. Most of this growth is therefore already expected due to existing policies, and additional growth relying on new policies is expected to only foster an additional 10% increase. The global primary biomass supply (efficiency of about 65% for first-generation biofuels) needed to deliver this amount of biofuels ranges between 7.4 and 8.4 EJ. The increase at the global level goes along with further regional diversification of biofuels adoption. While the currently dominant biofuels markets in Brazil, the USA and the EU are projected to roughly double consumption by 2020, many other regions with very little or no biofuels consumption currently are expected to adopt biofuel policies, resulting in significant growth, most notably in Asia. Electricity generation increases by 85% from 265 TWh/yr (0.96 EJ/yr) in 2008 to 493 TWh/yr (1.8 EJ/yr) in the Current Policies scenario, again with relatively modest additional growth (20%) in the more ambitious policy scenarios (up to 594 TWh/yr or 2.1 EJ/yr) (Table 2.10).

2.8.3   Long-term deployment in the context of carbon mitigation

The AR4 (IPCC, 2007d) demand projections for primary biomass for production of transportation fuel were largely based on WEO (IEA, 2006) global projections, with a relatively wide range of about 14 to 40 EJ of primary biomass, or 8 to 25 EJ of biofuels in 2030. However, higher estimates were also included, in the range of 45 to 85 EJ of demand for primary biomass for electricity generation in 2030 (equivalent to roughly 30 to 50 EJ of biofuel). Demand for biomass for heat and power was stated to be strongly influenced by (availability and introduction of) competing technologies such as CCS, nuclear power and non-biomass RE. The demand in 2030 for biomass was estimated in the AR4 to be around 28 to 43 EJ. These estimates focus on electricity generation. Heat was not explicitly modelled or estimated in the WEO (on which the AR4 was based); therefore it underestimates total demand for biomass. Also, potential future demand for biomass in industry (especially new uses such as biochemicals, but also expansion of charcoal use for iron and steel production) and the built environment (heating as well as increased use of biomass as building material) was highlighted as important, but no quantitative projections were included in potential demand for biomass at the medium or longer term.

A summary of the literature on the possible future contribution of RE supplies in meeting global energy needs under a range of GHG stabilization scenarios is provided in Chapter 10. Focussing specifically on bioenergy, Figure 2.23 presents modelling results for global primary energy supply from biomass (a) and global biofuels production in secondary energy terms (b). Between about 100 and 140 different long-term scenarios underlie Figure 2.23 (Section 10.2). These scenario results derive from a diversity of modelling teams and cover a wide range of assumptions about—among other variables—energy demand growth, the cost and availability of competing low-carbon technologies and the cost and availability of RE technologies (including bioenergy). A description of the literature from which the scenarios have been taken (Section 10.2.2) and how changes in some of these variables impact RE deployment outcomes are displayed in Figure 10.9.

in most scenarios, which means that modern use of biomass as liquid biofuels, biogas, and electricity and H_2 produced from biomass tends to increase even more strongly than suggested by the above primary energy numbers. This trend is also illustrated by the example of liquid biofuels production shown in the right panel of Figure 2.23(b). With increasingly ambitious GHG concentration stabilization levels, bioenergy supply increases, indicating that bioenergy could play a significant long-term role in reducing global GHG emissions. The median levels of biomass deployment for energy in the most stringent mitigation categories I and II (<440 ppm atmospheric CO_2 concentration by 2100) increase significantly compared to the baseline levels to 63, 85 and 155 EJ/yr by 2020, 2030 and 2050, respectively.

Despite these robust trends, there is by no means an agreement about

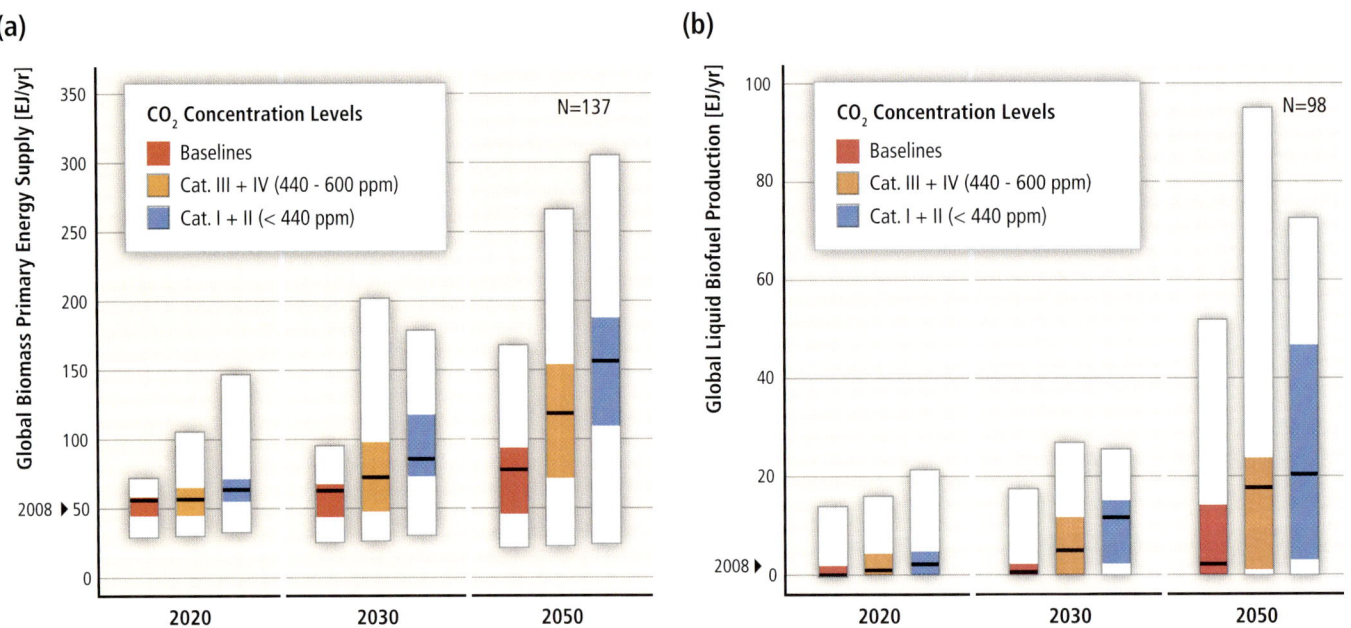

Figure 2.23 | (a) The global primary energy supply from biomass in long-term scenarios; (b) global biofuels production in long-term scenarios reported in secondary energy terms of the delivered product (median, 25th to 75th percentile range and full range of scenario results; colour coding is based on categories of atmospheric CO_2 concentration levels in 2100; the number of scenarios underlying the figure is indicated in the right upper corner) (adapted from Krey and Clarke, 2011). For comparison, the historic levels in 2008 are indicated by the small black arrows on the left axis.

In Figure 2.23, the results for biomass deployment for energy under these scenarios for 2020, 2030 and 2050 are presented for three GHG stabilization ranges based on the AR4: Categories I and II (<440 ppm CO_2), Categories III and IV (440-600 ppm CO_2) and Baselines (>600 ppm CO_2) all by 2100. Results are presented for the median scenario, the 25th to 75th percentile range among the scenarios, and the minimum and maximum scenario results. Figure 2.23(a) shows a clear increase in global primary energy supply from biomass over time in the baseline scenarios, that is, absent climate policies, reaching about 55, 62 and 77 EJ/yr in the median cases by 2020, 2030 and 2050, respectively. At the same time, traditional use of solid biomass is projected to decline

the precise future role of bioenergy across the scenarios, leading to fairly wide deployment ranges in the different GHG stabilization categories. For 2030, primary biomass supply estimates for energy vary (rounded) between 30 and 200 EJ for the full range of results obtained. The 25th to 75th percentiles cover a range of 45 to 120 EJ, with a comparatively narrower range of 44 to 67 EJ/yr in the baselines and much wider ranges of 47 to 98 EJ/yr in the 440 to 600 ppm stabilization category and 73 to 120 EJ/yr in the <440 ppm category. By 2050, the contribution of biomass to primary energy supply in the two GHG stabilization categories ranges from 70 to 120 EJ/yr at the 25th percentile to about 150 to 190 EJ/yr at the 75th percentile, and to about 265-300 EJ/yr in the highest ranges. It should be noted that the net GHG mitigation impact of

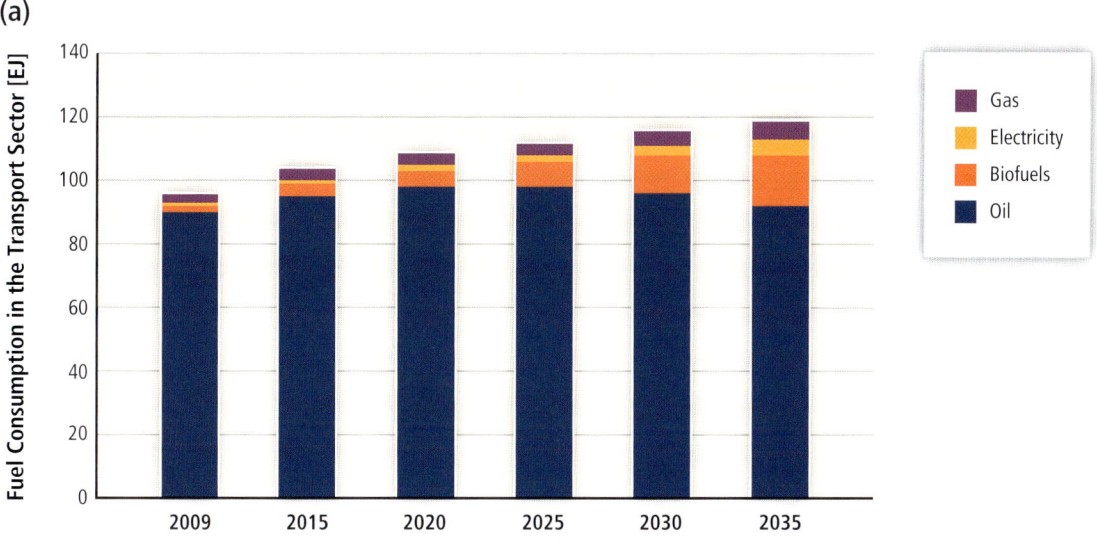

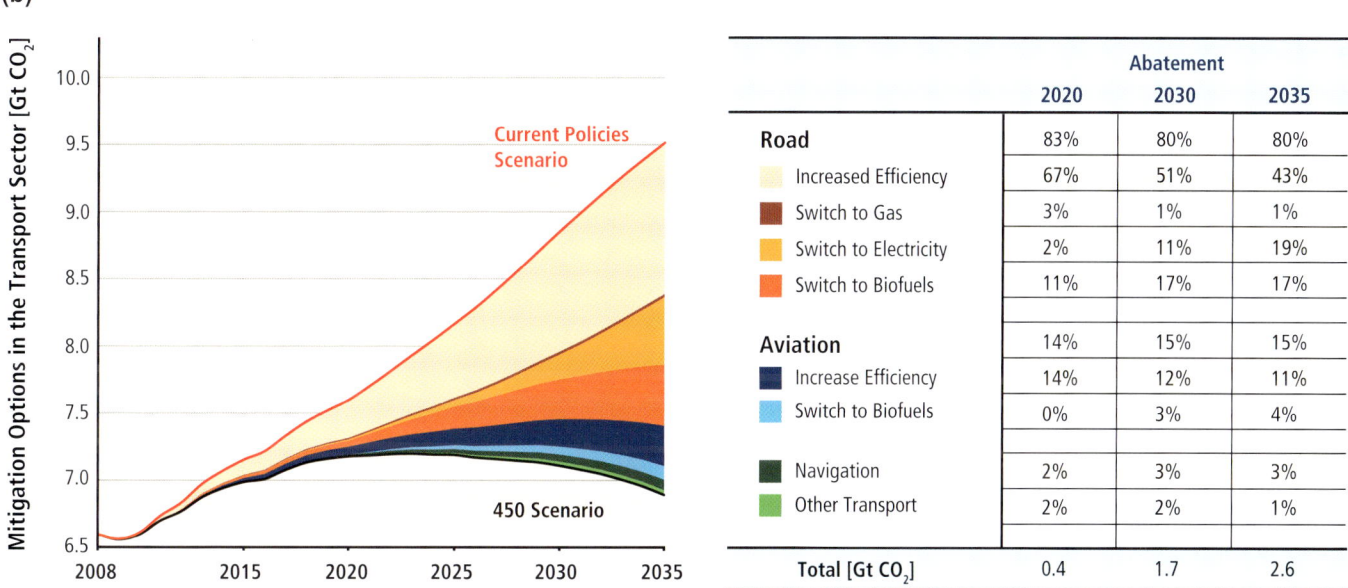

Figure 2.24 | (a) Evolution of fuel consumption in the transport sector including biofuels (*World Energy Outlook 2010*, © OECD/IEA, figure 14.12, page 429 in IEA (2010b)) and (b) shares of carbon mitigation by various technologies including biofuels for road and aviation transport from current policies baseline (upper red line) to the 450 ppm bottom curve of the mitigation scenario. (*World Energy Outlook 2010*, © OECD/IEA, figure 14.14, page 432 in IEA (2010b)).

bioenergy deployment is not straightforward because different options result in different GHG savings, and savings depend on how land use is managed, which is a central reason for the wide ranges in the stabilization scenarios.

The sector-level penetration of bioenergy is best explained using a model with detailed transport sector representation such as the WEO (IEA, 2010b) that is also modelling both traditional and modern biomass applications, and includes second-generation biofuels evolution. Additionally, the WEO model takes into account anticipated industrial and government investments and goals. It projects very significant increases in modern bioenergy and a decrease in traditional biomass use, in qualitative agreement with the results from Chapter 10. By 2030, for the 450-ppm mitigation scenario, the model projects that 11% of global transport fuels will be provided by biofuels with second-generation biofuels contributing 60% of the projected 12 EJ, and half of this production is projected to be supplied owing to continuation of current policies (see Table 2.9). Biomass and renewable wastes would supply 5% of the world's electricity generation, or 1,380 TWh/yr (5 EJ/yr) of which 555 TWh/yr (2 EJ/yr) result from the 450 ppm strategy by 2030 (see Table 2.10). Biomass industrial heating applications for process steam and space and hot water heating for buildings would each double in absolute terms from 2008 levels. However, the total heating demand is projected to decrease because of assumed traditional biomass decline. Heating is seen as a key area for continued modern bioenergy growth.

The evolution of biofuels in the transport sector is shown in Figure 2.24a. Biofuels penetration is projected to be significant in both in global road transport and in air transport. Second-generation technologies are projected to provide 66% of the biofuels by 2035 and 14% of world transport energy demand in the 450-ppm scenario (see Figure 2.24a and Table 2.9). Figure 2.24b shows the projected GHG emissions mitigation of biofuels relative to projected road and air transport applications from the current policies to the 450 ppm scenario. For instance, by 2030, 17% of road transport emissions and 3% of air transport emissions could be mitigated by biofuels in the 450-ppm stabilization scenario. A biofuels technology roadmap was recently developed (IEA, 2011).

The potential demand of biomass for materials is not explicitly addressed by many of the scenarios, but it could become significant and add up to several dozens of EJ (Section 2.6.3.5; Hoogwijk et al., 2003).

The expected deployment of biomass for energy in the 2020 to 2050 time frame differs considerably between studies, also due to varying detail in bioenergy system representation in the relevant models. A key message from the review of available insights is that large-scale biomass deployment strongly depends on sustainable development of the resource base, governance of land use, development of infrastructure and cost reduction of key technologies, for example, efficient and complete use of primary biomass for energy from the most promising first-generation feedstocks and second-generation lignocellulosic biomass. The results discussed above are consistent with the *Energy Technology Perspectives* report (IEA, 2008a), which projects a rapid penetration of second-generation biofuels after 2010 and an almost complete phase-out of cereal- and corn-based ethanol production and oilseed-based biodiesel after 2030.[77]

2.8.4 Conditions and policies: Synthesis of resource potentials, technology and economics, and environmental and social impacts of bioenergy

2.8.4.1 Resource potentials

The inherent complexity of biomass resources makes the assessment of their combined technical potential controversial and difficult to characterize. Literature studies range from zero (no biomass potential available as energy) to around 1,500 EJ, the theoretical potential for terrestrial biomass based on modelling studies exploring the widest potential ranges of favourable conditions (Smeets et al., 2007).

Figure 2.25 presents a summary of technical potential found in major studies, including potential deployment data from the scenario analysis of Chapter 10 compared to global TPES (projections). To put technical potential in perspective, because global biomass used for energy currently amounts to approximately 50 EJ/yr, and all harvested biomass used for food, fodder, fibre and forest products, when expressed in equivalent heat content, equals 219 EJ/yr (2000 data, Krausmann et al., 2008), the entire current global biomass harvest would be required to achieve a 200 EJ/yr deployment level of bioenergy by 2050 (Section 2.2.1).

From a detailed assessment, the upper-bound technical potential of biomass was about 500 EJ with a minimum of about 50 EJ in the case that even residues had significant competition with other uses. The assessment of each contributing category performed by Dornburg et al. (2008, 2010) was based on literature up to 2007 (stacked bar of Figure 2.25) and is roughly in line with the conditions sketched in the IPCC SRES A1 and B1 storylines (IPCC, 2000), assuming sustainability and policy frameworks to secure good governance of land use and major improvements in agricultural management (summarized in Figure 2.26). The resources used are:

- Residues originating from forestry, agriculture and organic wastes (including the organic fraction of MSW, dung, process residues etc.) were estimated at around 100 EJ/yr. This part of the technical potential of biomass supply is relatively certain, but competing applications may push net availability for energy applications to the lower end of the range.

- Surplus forestry other than from forestry residues had an additional technical potential of about 60 to 100 EJ/yr.

- Biomass produced via cropping systems had a lower range estimate for energy crop production on possible surplus good quality agricultural and pasture lands of 120 EJ/yr. The potential contribution of water-scarce, marginal and degraded lands could amount to an additional 70 EJ/yr, corresponding to a large area where water scarcity provides limitations and soil degradation is more severe. Assuming strong learning in agricultural technology leading to improvements in agricultural and livestock management would add 140 EJ/yr.

Adding these categories together leads to a technical potential of up to about 500 EJ in 2050, with temporal data on the development of biomass potential ramping from 290 to 320 EJ/yr in 2020 to 330 to 400 EJ/yr in 2030 (Hoogwijk et al., 2005, 2009; Dornburg et al., 2008, 2010).

From the expert review of available scientific literature in this chapter, *potential deployment levels of biomass for energy by 2050 could be in the range of 100 to 300 EJ* (Sections 2.2.1, 2.2.2, and 2.2.5).

Values in this range are described in van Vuuren et al. (2009), which focused on an intermediate development scenario within the SRES scenario family. The lower estimates of Smeets et al. (2007) and Hoogwijk et al. (2005, 2009) are in line with those figures, and further confirmation for such a range is given by Beringer et al. (2011), who report a 26 to 116 EJ range for energy crops alone in 2050 without irrigation (and 52 to 174 EJ with irrigation), and Haberl et al. (2010), who report 160 to 270 EJ/yr in 2050 across all biomass categories. Krewitt et al. (2009), following Seidenberger et al. (2008), also estimated the technical potential to be 184 EJ/yr in 2050 using strong sustainability

---

77 Contrast these projections with the 2007 and 2008 WEO studies (IEA, 2007b, 2008b), where second-generation biofuels were excluded from the scenario analysis and thus biofuels at large played a marginal role in the 2030 projections.

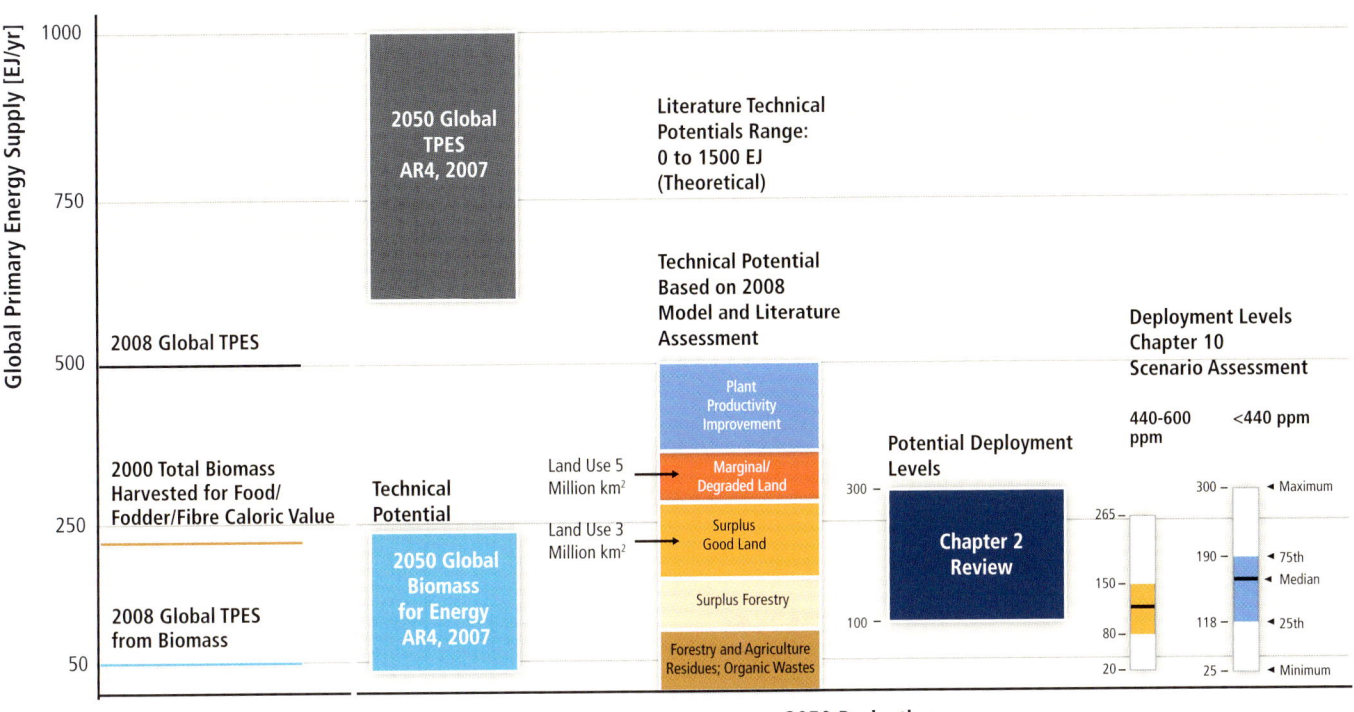

Figure 2.25 | On the left-hand side, the lines represent the 2008 global primary energy supply from biomass, the primary energy supply, and the equivalent energy of the world's total harvest for food, fodder and fibre in 2000. A summary of major global 2050 projections of primary energy supply from biomass is shown from left to right: (1) The global AR4 (IPCC, 2007d) estimates for primary energy supply and technical potential for primary biomass for energy; (2) the theoretical primary biomass potential for energy and the upper bound of biomass technical potential based on integrated global assessment studies using five resource categories indicated on the stacked bar chart and limitations and criteria with respect to biodiversity protection, water limitations, and soil degradation, assuming policy frameworks that secure good governance of land use (Dornburg et al., 2010, reproduced with permission from the Royal Society of Chemistry); (3) from the expert review of available scientific literature, potential deployment levels of terrestrial biomass for energy by 2050 could be in the range of 100 to 300 EJ; and (4) deployment levels of biomass for energy from long-term scenarios assessed in Chapter 10 in two cases of climate mitigation levels (CO_2 concentrations by 2100 of 440 to 600 ppm (orange) or <440 ppm (blue) bars or lines, see Figure 2.23(a)). Biomass deployment levels for energy from model studies described in (4) are consistent with the expert review of potential biomass deployment levels for energy depicted in (3). The most likely range is 80 to 190 EJ/yr with upper levels in the range of 265 to 300 EJ/yr.

criteria and including 88 EJ/yr from residues. They project a ramping-up to this potential from around 100 EJ/yr in 2020 and 130 EJ/yr in 2030.

The expert review conclusions based on available scientific literature (Sections 2.2.2 through 2.2.5) are:

- Important uncertainties include:

  - Population and economic/technology development; food, fodder and fibre demand (including diets); and development in agriculture and forestry;

  - Climate change impacts on future land use including its adaptation capability (IPCC, 2007a; Lobell et al., 2008; Fischer et al., 2009); and

  - Extent of land degradation, water scarcity, and biodiversity and nature conservation requirements (Molden, 2007; Bai et al., 2008; Berndes, 2008a,b; WBGU, 2009; Dornburg et al., 2010; Beringer et al., 2011).

- Residue flows in agriculture and forestry and unused (or extensively used thus becoming marginal/degraded) agricultural land are important sources for expansion of biomass production for energy, both in the near and longer term. Biodiversity-induced limitations and the need to ensure maintenance of healthy ecosystems and avoid soil degradation set limits on residue extraction in agriculture and forestry (Lal, 2008; Blanco-Canqui and Lal, 2009; WBGU, 2009).

- The cultivation of suitable (especially perennial) crops and woody species can lead to higher technical potential. These crops can produce bioenergy on lands less suited for the cultivation of conventional food crops that would also lead to larger soil carbon emissions than perennial crops and woody species. Multifunctional land use systems with bioenergy production integrated into agriculture and forestry systems could contribute to biodiversity conservation and help restore/maintain soil productivity and healthy ecosystems (Hoogwijk et al., 2005; Berndes et al., 2008; Folke et al., 2009; IAASTD, 2009; Malézieux et al., 2009; Dornburg et al., 2010).

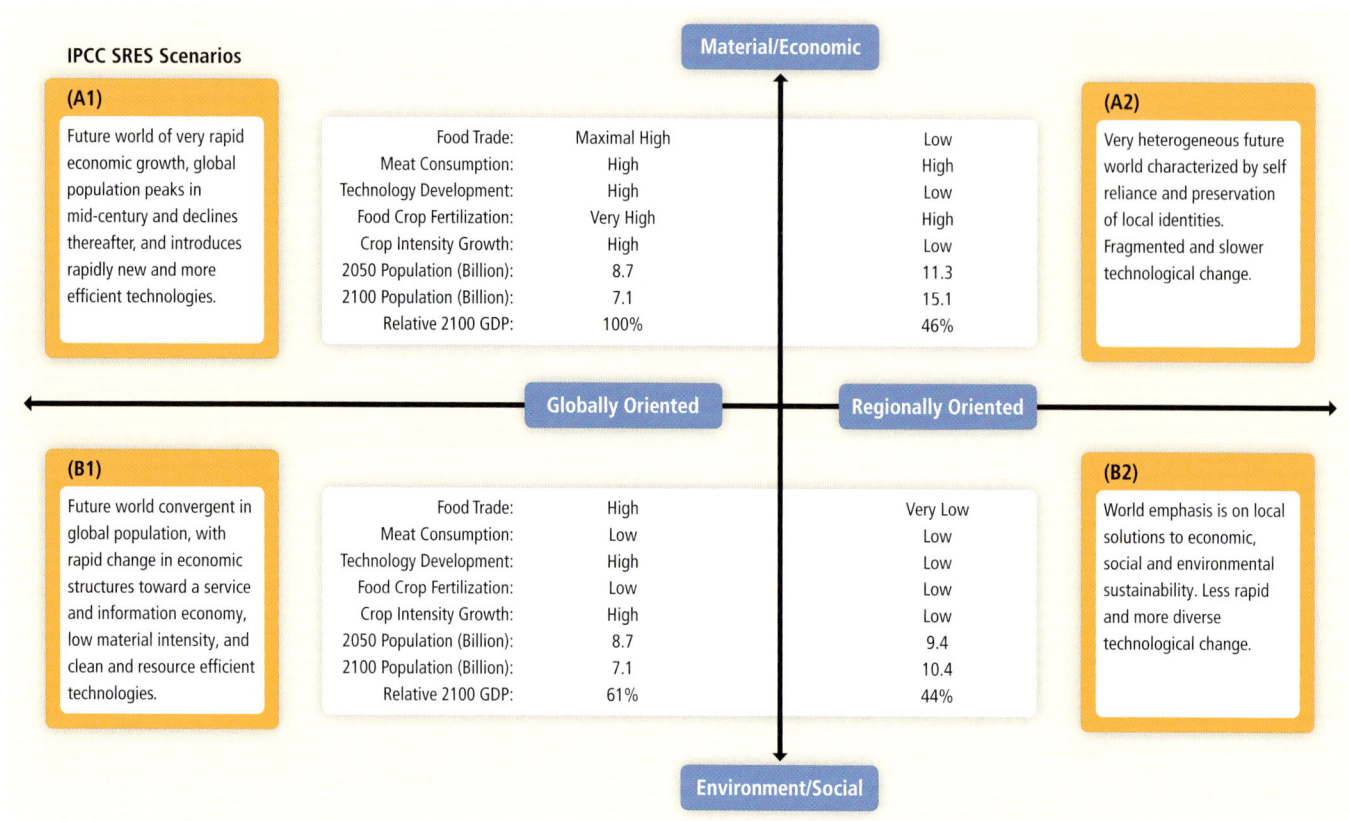

Figure 2.26 | Storylines for the key scenario variables of the IPCC SRES (IPCC, 2000) used to model biomass and bioenergy by Hoogwijk et al. (2005, reproduced with permission from Elsevier B.V.), the basis for the 2050 sketches adapted for this report and used to derive the stacked bar showing the upper bound of the biomass technical potential for energy in Figure 2.25.

- Regions experiencing water scarcity may have limited production. The possibility that conversion of lands to biomass plantations reduces downstream water availability needs to be considered. The use of suitable energy crops that are drought tolerant can help adaptation in water-scarce situations. Assessments of biomass resource potentials need to more carefully consider constraints and opportunities in relation to water availability and competing uses (Jackson et al., 2005; Zomer et al., 2006; Berndes et al., 2008; de Fraiture and Berndes, 2009).

To reach the *upper range of the deployment level* of 300 EJ/yr shown in Figure 2.25 would require major policy efforts, especially targeting improvements and efficiency increases in the agricultural sector and good governance, such as zoning, of land use.

Review scenario studies (as included in Dornburg et al., 2008) that calculate the amount of biomass used if energy demands are supplied cost-efficiently for different carbon tax regimes estimate that in 2050, between about 50 and 250 EJ/yr of biomass are used (cf. Figure 2.25). This is roughly in line with the scenarios reviewed in Chapter 10 (see Figure 2.23, which shows that the maximum demand is 300 EJ and the median value is about 155 EJ; note that the high end is only reached under the stringent mitigation scenarios of Categories I+II (<440 ppm CO_2) only).

2.8.4.2 Bioenergy technologies, supply chains and economics

A wide array of technologies and bioenergy systems exist to produce heat, electricity and fuels for transport, at commercial or development stages. Furthermore, biomass conversion to energy can be integrated with the production of biomaterials and biochemicals in cascading schemes that maximize the outputs of end products per unit input feedstock and land used.

The key currently commercial technologies are heat production at scales ranging from home cooking to district heating; power generation from biomass via combustion, CHP, or co-firing of biomass and fossil fuels; and first-generation liquid biofuels from oil crops (biodiesel) and sugar and starch crops (ethanol).

Modern biomass systems involve a wide range of feedstock types, including dedicated crops or trees, residues from agriculture and forestry, and various organic waste streams. Existing bioenergy systems rely mostly on wood, residues and waste for heat and power production and agricultural crops for liquid biofuels. The economics and yields of feedstocks vary widely across world regions and feedstock types. Energy yields per unit area range from 16 to 200 GJ/ha (1.6 to 20.0 TJ/km^2) for crops and oil seeds (biofuel feedstocks), from 80 to 415 GJ/ha (8.0 to 41.5 TJ/km^2) for lignocellulosic biomass, and from 2 to 155 GJ/ha

(0.2 to 15.5 TJ/km^2) for residues, while costs range from USD$_{2005}$ 0.9 to 16/GJ (data from 2005 to 2007). Feedstock production competes with the forestry and food sectors, but integrated production systems such as agro-forestry or mixed cropping may provide synergies along with additional environmental services.

Handling and transport of biomass from production sites to conversion plants may contribute 20 to up to 50% of the total costs of biomass production. Factors such as scale increase, technological innovations and increased competition contributed to decrease the economic and energy costs of supply chains by more than 50%. Densification via pelletization or briquetting is required for transportation distances over 50 km. Charcoal made from biomass is a major fuel in developing countries, and it should benefit from the adoption of higher-efficiency kilns.

Different end-use applications require that biomass be processed through a variety of conversion steps depending on the physical nature and the chemical composition of feedstocks. Costs vary by world regions, feedstock types, feedstock supply costs for conversion processes, the scale of bioenergy production, and production time during the year. Examples of estimated commercial bioenergy levelized cost ranges are roughly USD 2 to 48/GJ for liquid and gaseous biofuels; roughly US cents$_{2005}$ 3.5 to 25/kWh (USD$_{2005}$ 10 to 50/GJ) for electricity or CHP systems larger than about 2 MW (with feedstock costs of USD$_{2005}$ 3/GJ based on high heating value and a heat value of USD$_{2005}$ 5/GJ (steam) or USD$_{2005}$ 12/GJ (hot water)); and roughly USD$_{2005}$ 2 to 77/GJ for domestic or district heating systems with feedstock costs in the range of USD$_{2005}$ 0 to 20/GJ (solid waste to wood pellets). These calculations refer to 2005 to 2008 data and are expressed in USD$_{2005}$ at a 7% discount rate. Several bioenergy systems have deployed competitively, most notably sugarcane ethanol and heat and power generation from wastes and residues. Other biofuels have also undergone cost and environmental impact reductions but still require government subsidies.

In the medium term, the performance of existing bioenergy technologies can still be improved considerably, while new technologies offer the prospect of more efficient and competitive deployment of biomass for energy (as well as materials). Bioenergy systems, namely for ethanol and biopower production, show rates of technological learning and related cost reductions with learning comparable to those of other RE technologies. This applies to cropping systems (following progress in agricultural management when annual crops are concerned), to supply systems and logistics (as clearly observed in Scandinavia, as well as international logistics) and in conversion (e.g., ethanol production, power generation and biogas). Although not all bioenergy options discussed in this chapter have been investigated in detail with respect to technological learning, several important bioenergy systems have reduced their cost and improved environmental performance (Sections 2.3.4.2 and 2.7.2; Table 2.13). However, they usually still require government subsidies provided for economic development, poverty reduction and a secure energy supply or other country-specific reasons.

There is clear evidence that further improvements in power generation technologies (e.g., via biomass IGCC technology), supply systems for biomass, and production of perennial cropping systems can bring the costs of power (and heat or fuels) generation from biomass down to attractive cost levels in many regions. Nevertheless, the competitive production of bio-electricity (through methane or biofuels) depends on the integration with the end-use systems (Sections 8.2 and 8.3), performance of alternatives such as wind and solar energy, developing CCS technologies coupled with coal conversion, and nuclear energy (Sections 10.2.2.4, 10.2.2.6, 9.3, and 9.4). The implications of successful deployment of CCS in combination with biomass conversion could result in removal of GHG from the atmosphere and attractive mitigation cost levels but have so far received limited attention (Section 2.6.3.3).

With respect to lignocellulosic biofuels, recent analyses have indicated that the improvement potential is large enough for competition with oil at oil prices of USD$_{2005}$ 60 to 80/barrel (USD$_{2005}$ 0.38 to 0.44/litre). Currently available scenario analyses indicate that if shorter-term R&D and market support is strong, technological progress could allow for their commercialization around 2020 (depending on oil and carbon prices). Some scenarios also indicate that this would mean a major shift in the deployment of biomass for energy, because competitive production would decouple deployment from policy targets (mandates), and demand for biomass would move away from food crops to biomass residues, forest biomass and perennial cropping systems. The implications of such a (rapid) shift are so far poorly studied.

Integrated biomass gasification is a major avenue for the development of a variety of biofuels, with equivalent properties to gasoline, diesel and jet fuel (see Table 2.15.C for composition of hydrocarbon fuels). An option highlighted as promising in the literature is fuel product generation passing syngas through the catalytic reactor only once with the unreacted gas going to the power generation system instead of being recycled through the catalytic reactor. Other hybrid biochemical and thermochemical concepts have also been contemplated (Laser et al., 2009). Biomass pyrolysis routes and hydrothermal concepts are also developing in conjunction with the oil industry and have demonstrated that upgrading of oils to blendstocks of gasoline or diesel or even jet fuel quality products is technically possible (IATA, 2009).

Lignocellulosic ethanol development and demonstration continues in several countries. A key development step is pretreatment to overcome the recalcitrance of the cell wall of woody, herbaceous or agricultural residues to release the simple sugar components of biomass polymers and lignin. A review of the progress in this area suggests that a 40% reduction in cost could be expected by 2025 from process improvements, which would bring down the estimated cost of pilot plant production from USD$_{2005}$ 18 to 22/GJ to USD$_{2005}$ 12 to 15/GJ (Hamelinck et al., 2005a; Foust et al., 2009; NRC, 2009a) and into a competitive range.

Photosynthetic organisms, such as algae, use CO_2, water, and sunlight to biologically produce a variety of carbohydrates and lipids, chemicals, fuels like H_2, other molecules and oxygen with high photosynthetic efficiency and possibly high potentials (Sections 2.6.1, 3.3.5 and 3.7.6). Estimates of potential bioenergy supply from aquatic plants are very uncertain because of the lack of sufficient data for their assessment (Kheshgi et al., 2000; Smeets et al., 2009). Nevertheless these species need to be explored further because their development can utilize brackish waters and heavily saline soils and thus represent a strategy for low LUC impacts (Chisti, 2007; Weyer et al., 2009). The prospects of algae-based fuels and chemicals are at this stage uncertain, with wide ranges for potential production costs reported in the literature.

Data availability is limited with respect to production of biomaterials; cost estimates for chemicals from biomass are rare in the peer-reviewed literature, and future projections and LRs are even rarer. This condition is linked, in part, to the fact that successful bio-based products are entering the market place either as partial components of otherwise fossil-derived products or as fully new synthetic polymers, such as polylactides based on lactic acid derived from sugar fermentation. Analyses indicate that, in addition to producing biomaterials to replace fossil fuels, cascaded use of biomaterials and subsequent use of waste material for energy can offer more effective and larger mitigation impacts per hectare or tonne of biomass used (e.g., Dornburg and Faaij, 2005).

The benefits of biomass gasification and CCS alone or with coal are significant (see Figures 2.10 and 2.11). Similarly, capturing CO_2 from fermentation processes offers a significant option in many regions of the world, and coupling with CCS may become an attractive medium-term mitigation option. However, such concepts are not deployed at present and cost trends are not available in the literature, making investments in biomass (or coal) gasification technologies risky. Also, geologic sequestration reliability and the uncertainty of the regulatory environment pose further barriers. More detailed analysis is desired in this field.

2.8.4.3 Social and environmental impacts

The effects of bioenergy on social and environmental issues—ranging from health and poverty to biodiversity and water quality—may be positive or negative depending upon local conditions, the specific feedstock production system and technology paths chosen, how criteria and the alternative scenarios are defined, and how actual projects are designed and implemented, among other variables (Sections 9.2 through 9.5). Perhaps most important is the overall management and governance of land use when biomass is produced for energy on top of meeting food and other demands from agricultural production (as well as livestock). In cases where increases in land use due to biomass production are balanced out by improvements in agricultural management, undesirable iLUC effects can be avoided, while if unmanaged, conflicts may emerge. The overall performance of bioenergy production systems is therefore interlinked with management of land use and water resources. Trade-offs between those dimensions exist and need to be resolved through appropriate strategies and decision making. Such strategies are currently emerging due to many efforts targeting the deployment of sustainability frameworks and certification systems for bioenergy production (see also Section 2.4.5), setting standards for GHG performance (including LUC effects), addressing environmental issues and taking into consideration a number of social aspects.

Most bioenergy systems can contribute to climate change mitigation if they replace fossil-based energy that was causing high GHG emissions and if the bioenergy production emissions—including those arising due to LUC or temporal imbalance of terrestrial carbon stocks—are kept low (examples given in Sections 2.3 and 2.6). High N_2O emissions from feedstock production and the use of high carbon intensity fossil fuels in the biomass conversion process can strongly impact the GHG savings. Best fertilizer management practices, process integration minimizing losses, surplus heat utilization, and biomass use as a process fuel can reduce GHG emissions. But in cold climates the displacement efficiency (see Section 2.5.3) can become low when biomass is used both as feedstock and as fuel in the conversion process.

Given the lack of studies on how biomass resources may be distributed over various demand sectors, no detailed allocation of the different biomass supplies for various applications is suggested here. Furthermore, the net avoidance costs per tonne of CO_2 for biomass usage depend on various factors, including the biomass resource and supply (logistics) costs, conversion costs (which in turn depend on availability of improved or advanced technologies) and fossil fuel prices, most notably of oil.

A GHG performance evaluation of key biofuel production systems deployed today and possible second-generation biofuels using different calculation methods is available (Sections 2.5.2, 2.5.3 and 9.3.4; Hoefnagels et al., 2010). Recent insights converge by concluding that well-managed bioenergy production and utilization chains can deliver high GHG mitigation percentages (80 to 90%) compared to their fossil counterparts, especially for lignocellulosic biomass used for power generation and heat and, when the technology would be commercially available, for lignocellulosic biofuels. The use of most residues and organic wastes, principally animal residues, for energy result in such good performance. Also, most current biofuel production systems have positive GHG balances, and for some of them this situation persists even when significant iLUC effects are incorporated (see below).

LUC can strongly affect those scores, and when conversion of land with large carbon stocks takes place for the purpose of biofuel production, emission benefits can shift to negative levels in the near term. This is most extreme for palm oil-based biodiesel production, where extreme carbon emissions are obtained if peatlands are drained and converted to oil palm (Wicke et al., 2008). The GHG mitigation effect of biomass use for energy (and materials) therefore strongly depends on location (in particular avoidance of converting carbon-rich lands to carbon-poor cropping systems), feedstock choice and avoiding iLUC (see below). In contrast, using perennial cropping systems can store large amounts of carbon and enhance sequestration on marginal and degraded soils, and biofuel production can replace fossil fuel use. Governance of land use, proper zoning and choice of biomass production systems are therefore key factors to achieve good performance.

The assessment of available iLUC literature (Figures 2.13, 9.10, and 9.11) indicated that initial models were lacking in geographic resolution, leading to higher proportions than necessary of land use assigned to deforestation, as the models did not have other kinds of lands (e.g., pastures in Brazil) for use. While the early paper of Searchinger et al. (2008) claimed an iLUC factor of 0.8 (losing 0.8 ha of forest land for each hectare of land used for bioenergy), later (2010) studies that coupled macro-economic to biophysical models tuned that down to 0.15 to 0.3 (see, e.g., Al-Riffai et al., 2010). Models used to estimate iLUC effects vary in their estimates of land displacement. Partial and general equilibrium models have different assumptions and reflect different time frames, and thus they incorporate more or less adjustment. More detailed evaluations (e.g., Al-Riffai et al., 2010; Lapola et al., 2010; see Section 2.5.3) do estimate significant iLUC impacts but also suggest that any iLUC effect strongly (up to fully) depends on the rate of improvement in agricultural and livestock management and the rate of deployment of bioenergy production. This balance in development is also the basis for the recent European biomass resource potential analysis, for which expected gradual productivity increments in agriculture are the basis for possible land availability (as reported in Fischer et al. (2010) and de Wit and Faaij (2010); see Figure 2.5(a)) minimizing competition with food (or nature) as a starting point. Increased model sophistication to adapt to the complex type of analysis required and improved data on the actual dynamics of land distribution in the major biofuel-producing countries are now producing results that show lower overall LUC impacts (Figure 9.11) and acknowledge that land use management at large is key (Berndes et al., 2010).

Bioenergy projects can result in gains or losses in associated biospheric stocks and in both direct and indirect LUC, the latter being inherently difficult to quantify. Even so, it can be concluded that LUC can affect GHG balances in several ways, with beneficial or detrimental outcomes for bioenergy's contribution to climate change mitigation, depending on conditions and context. When land high in carbon (notably forests and especially peat soil forests) is converted to bioenergy, upfront emissions may cause a time lag of decades to centuries before net emission savings are achieved. But the establishment of bioenergy plantations can also lead to assimilation of CO_2 into soils and aboveground biomass in the short term. Increased utilization of forest biomass can reduce forest carbon stocks. The longer-term net effect on forest carbon stocks can be positive or negative depending on natural conditions (including disturbances such as insect outbreaks and fires) and forest management practices. The use of post-consumer organic waste and by-products from the agricultural and forest industries does not cause LUC if these biomass sources were not utilized for alternative purposes. Bioenergy feedstocks can be produced in combination with food and fibre, avoiding land use displacement and improving the productive use of land. Lignocellulosic feedstocks for bioenergy can decrease the pressure on prime cropping land. Stimulation of increased productivity in all forms of land use reduces the LUC pressure.

Air pollution effects of bioenergy depend on both the bioenergy technology (including pollution control technologies) and the displaced energy technology (e.g., inefficient coal versus modern natural gas combustion) (Figure 9.12). Improved biomass cookstoves for traditional biomass use can provide large and cost-effective mitigation of GHG emissions with substantial co-benefits in terms of health and living conditions, particularly for the 2.7 billion people in the world that rely on traditional biomass for cooking and heating (Sections 2.5.4, 9.3.4, 9.3.4.2 and 9.3.4.3). Efficient technologies for cooking are even cost-effective compared to other major interventions in health, such as those addressing tobacco, undernourishment or tuberculosis (Figures 2.14 and 9.13).

Other key environmental impacts cover water use, biodiversity and other emissions (Sections 2.5.5 and 9.3.4). Just as for GHG impacts, proper management determines emission levels to water, air and soil. Development of standards or criteria (and continuous improvement processes) will push bioenergy production to lower emissions and higher efficiency than today's systems.

Water is a critical issue that needs to be better analyzed at a regional level to understand the full impact of changes in vegetation and land use management. Recent studies (Berndes, 2002; Dornburg et al., 2008; Rost et al., 2009; Wu et al., 2009) indicate that considerable improvements can be made in water use efficiency in conventional agriculture, bioenergy crops and, depending on location and climate, perennial cropping systems, by improving water retention and lowering direct evaporation from soils (Figure 9.14). Nevertheless, without proper management, increased biomass production could come with increased competition for water in critical areas, which is highly undesirable (Fingerman et al., 2010).

Similar remarks can be made with respect to biodiversity, although more scientific uncertainty exists due to ongoing debates about methods of biodiversity impacts assessment. Clearly, development of large-scale monocultures at the expense of natural areas is detrimental for biodiversity (for example, highlighted in UNEP. 2008b). However, as discussed in Section 2.5, bioenergy can also lead to positive effects by integrating different perennial grasses and woody crops into agricultural landscapes, which could also increase soil carbon and productivity, reduce shallow landslides and local 'flash floods', reduce wind and water erosion, and reduce sediment and nutrients transported into river systems. Forest residue harvesting improves forest site conditions for replanting, and thinning generally improves productivity and growth of the remaining stand. Removal of biomass from overly-dense stands can reduce wildfire risk.

The impact assessments for all these areas deserve considerably more research, data collection and proper monitoring, as exemplified by ongoing activities of governments (see footnote 64) and roundtables[78] for pilot studies.

---

78 See Roundtable on Sustainable Biofuels pilot studies at www2.epfl.ch/energycenter-jahia4/page65660.html.

Social impacts from a large expansion of bioenergy are very complex and difficult to quantify. Crops grown as biofuel feedstock currently use less than 1% of the world's agricultural land, but demand for biofuels has represented one driver of demand growth and therefore contributed to global food price increases. Increased demand for food and feed, increases in oil prices, speculation on international food markets, and incidental poor harvests due to extreme weather events are examples of events that have likely also had an impact on global food prices. Even considering the benefit of increased prices to poor farmers, increased food prices adversely affect the level of poverty, food security, and malnourishment of children. On the other hand, biofuels can also provide opportunities for developing countries to make progress in rural development and agricultural growth, especially when this growth is economically sustainable.

In general, bioenergy options have a much larger positive impact on job creation in rural areas than other energy sources, for example, 50 to 2,200 jobs/PJ (Section 2.5.7.3). Also when the intensification of conventional agriculture frees up land that could be used for bioenergy, the total job impact and added value generated in rural regions increases when bioenergy production increases. Effective pasture/agriculture land use management could increase the rain-fed production potential significantly (see Table 2.3; Wicke et al., 2009). For many developing countries, the potential of bioenergy to generate employment, economic activity in rural areas, and fuel supply security are key drivers. In addition, expenditures on fossil fuel (imports) can be (strongly) reduced. However, whether such benefits end up with rural farmers depends largely on the way production chains are organized and how land use is governed.

The bioenergy options that are developed, the way they are developed, and under what conditions will have a profound influence on whether impacts will largely be positive or negative (Argentina scenarios; van Dam et al., 2009a,b). The development of standards or criteria (and continuous improvement processes) can push bioenergy production to lower or positive impacts and higher efficiency than today's systems. Bioenergy has the opportunity to contribute to climate change mitigation, a secure and diverse energy supply, and economic development in developed and developing countries alike, but the effects of bioenergy on environmental sustainability may be positive or negative depending upon local conditions, how criteria are defined, and how actual projects are designed and implemented, among many other factors.

2.8.5 Conclusions regarding deployment: Key messages about bioenergy

Bioenergy is currently the largest RE source and is likely to remain one of the largest RE sources for the first half of this century. There is considerable growth potential, but it requires active development.

- Assessments in the recent literature show that the technical potential of biomass for energy may be as large as 500 EJ/yr by 2050. However, large uncertainty exists about important factors such as market and policy conditions that affect this potential.

- The expert assessment in this chapter suggests potential deployment levels by 2050 in the range of 100 to 300 EJ/yr. Realizing this potential represents a major challenge but would make a substantial contribution to the world's primary energy demand in 2050—roughly equal to the equivalent heat content of today's worldwide biomass extraction in agriculture and forestry.

- Bioenergy has significant potential to mitigate GHGs if resources are sustainably developed and efficient technologies are applied. Certain current systems and key future options including perennial crops, forest products and biomass residues and wastes, and advanced conversion technologies, can deliver significant GHG mitigation performance—an 80 to 90% reduction compared to the fossil energy baseline. However, land conversion and forest management that lead to a large loss of carbon stocks and iLUC effects can lessen, and in some cases more than neutralize, the net positive GHG mitigation impacts.

- In order to achieve the high potential deployment levels of biomass for energy, increases in competing food and fibre demand must be moderate, land must be properly managed and agricultural and forestry yields must increase substantially. Expansion of bioenergy in the absence of monitoring and good governance of land use carries the risk of significant conflicts with respect to food supplies, water resources and biodiversity, as well as a risk of low GHG benefits. Conversely, implementation that follows effective sustainability frameworks could mitigate such conflicts and allow realization of positive outcomes, for example, in rural development, land amelioration and climate change mitigation, including opportunities to combine adaptation measures.

- The impacts and performance of biomass production and use are region- and site-specific. Therefore, as part of good governance of land use and rural development, bioenergy policies need to consider regional conditions and priorities along with the agricultural (crops and livestock) and forestry sectors. Biomass resource potentials are influenced by and interact with climate change impacts but the specific impacts are still poorly understood; there will be strong regional differences in this respect. Bioenergy and new (perennial) cropping systems also offer opportunities to combine adaptation measures (e.g., soil protection, water retention and modernization of agriculture) with production of biomass resources.

- Several important bioenergy options (i.e., sugarcane ethanol production in Brazil, select waste-to-energy systems, efficient biomass cookstoves, biomass-based CHP) are competitive today and can provide important synergies with longer-term options. Lignocellulosic biofuels replacing gasoline, diesel and jet fuels, advanced bioelectricity options and biorefinery concepts can offer competitive deployment of bioenergy for the 2020 to 2030 timeframe. Combining biomass conversion with CCS raises the possibility of achieving GHG removal from the atmosphere in the long term—a necessity for substantial GHG emission reductions. Advanced biomaterials are promising as well for the economics of bioenergy production and mitigation, though the potential is less well understood as is the potential role of aquatic biomass (algae), which is highly uncertain.

- Rapidly changing policy contexts, recent market-based activities, the increasing support for advanced biorefineries and lignocellulosic biofuel options, and in particular the development of sustainability criteria and frameworks, all have the potential to drive bioenergy systems and their deployment in sustainable directions. Achieving this goal will require sustained investments that reduce costs of key technologies, improved biomass production and supply infrastructure, and implementation strategies that can gain public and political acceptance.

In conclusion and for illustrating the interrelations between scenario variables (see Figure 2.26), key preconditions under which bioenergy production capacity is developed and what the resulting impacts may be, Figure 2.27 presents four different sketches for biomass deployment for energy on a global scale by 2050. The 100 to 300 EJ range that follows from the resource potential review delineates the lower and upper limit for deployment. The assumed storylines roughly follow the IPCC SRES definitions, applied to bioenergy and summarized in Figure 2.26 (Hoogwijk et al., 2005), that were also used to derive the technical potential shown on the stacked bar of Figure 2.25 (Dornburg et al., 2008, 2010).

Biomass and its multiple energy products can be developed alongside food, fodder, fibre and forest products in both sustainable and unsustainable ways. As viewed through the IPCC scenario storylines and sketches, high and low penetration levels can be reached with and without taking into account sustainable development and climate change mitigation pathways. Insights into bioenergy technology developments and integrated systems can be gleaned from these sketches.

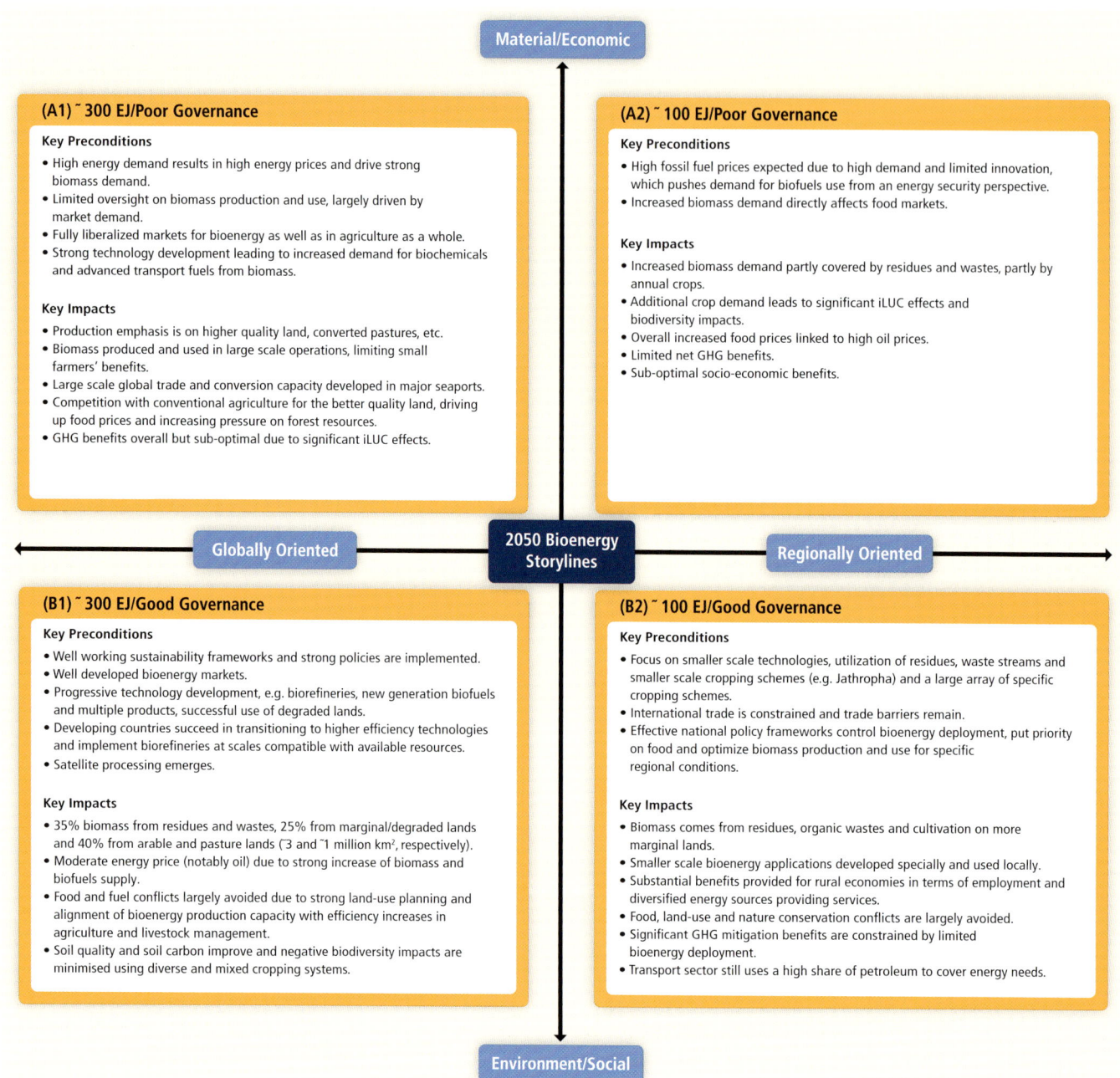

Figure 2.27 | Possible futures for 2050 biomass deployment for energy: Four illustrative contrasting sketches describing key preconditions and impacts following world conditions typical of the IPCC SRES storylines (IPCC, 2000) summarized in Figure 2.26.

References

Abbasi, T., and S.A. Abbasi (2010). Biomass energy and the environmental impacts associated with its production and utilization. *Renewable and Sustainable Energy Reviews*, **14**(3), pp. 919-937.

Adam, J.C. (2009). Improved and more environmentally friendly charcoal production system using a low-cost retort-kiln (Eco-charcoal). *Renewable Energy*, **34**(8), pp. 1923-1925.

Adams, P.W., G.P. Hammond, M.C. McManus, and W.G. Mezzullo (2011). Barriers to and drivers for UK bioenergy development. *Renewable and Sustainable Energy Reviews*, **15**(2), pp. 1217-1227.

Aden, A., M. Ruth, K. Ibsen, J. Jechura, K. Neeves, J. Sheehan, B. Wallace, L. Montague, A. Slayton, and J. Lukas (2002). *Lignocellulosic Biomass to Ethanol Process Design and Economics Utilizing Co-Current Dilute Acid Prehydrolysis and Enzymatic Hydrolysis for Corn Stover*. NREL/TP-510-32438, National Renewable Energy Laboratory, Golden, Colorado, USA, 154 pp.

Ahrens, T.D., D.B. Lobell, J.I. Ortiz-Monasterio, Y. Li, and P.A. Matson (2010). Narrowing the agronomic yield gap with improved nitrogen use efficiency: a modeling approach. *Ecological Applications*, **20**(1), pp. 91-100.

Ainsworth, E.A., and S.P. Long (2005). What have we learned from 15 years of free-air CO_2 enrichment (FACE)? A meta-analytic review of the responses of photosynthesis, canopy properties and plant production to rising CO_2. *New Phytologist*, **165**(2), pp. 351-372.

Al-Riffai, P., B. Dimaranan, and L. Laborde (2010). *Global Trade and Environmental Impact Study of the EU Biofuels Mandate*. Project Report - Specific Contract No SI2.537.787 implementing Framework Contract No TRADE/07/A2, International Food Policy Research Institute, Washington, DC, USA, 123 pp.

Alexandratos, N. (2009). World food and agriculture to 2030/50: highlights and views from mid-2009. In: *Proceedings of the Expert Meeting on How to Feed the World in 2050*, Rome, Italy, 24-26 June 2009. Economic and Social Development Department, Food and Agriculture Organization of the United Nations, Rome, Italy, pp. 78. Available at: www.fao.org/docrep/012/ak542e/ak542e00.htm.

Alfstad, T. (2008). *World Biofuels Study: Scenario Analysis of Global Biofuels Markets*. BNL-80238-2008, Brookhaven National Laboratory, New York, NY, USA, 67 pp.

Allen, J., M. Browne, A. Hunter, J. Boyd, and H. Palmer (1998). Logistics management and costs of biomass fuel supply. *International Journal of Physical Distribution & Logistics Management*, **28**(6), pp. 463-477.

Allen, M.R., D.J. Frame, C. Huntingford, C.D. Jones, J.A. Lowe, M. Meinshausen, and N. Meinshausen (2009). Warming caused by cumulative carbon emissions towards the trillionth tonne. *Nature*, **458**(7242), pp. 1163-1166.

Alper, H., and G. Stephanopoulos (2009). Engineering for biofuels: exploiting innate microbial capacity or importing biosynthetic potential? *Nature Reviews Microbiology*, **7**(10), pp. 715-723.

Anderson-Teixeira, K.J., S.C. Davis, M.D. Masters, and E.H. Delucia (2009). Changes in soil organic carbon under biofuel crops. *Global Change Biology Bioenergy*, **1**(1), pp. 75-96.

APEC (2003). *The New Brunswick Forest Industry: The Potential Economic Impact of Proposals to Increase the Wood Supply*. Atlantic Provinces Economic Council, Halifax, Canada.

Armendáriz, C.A., R.D. Edwards, M. Johnson, M. Zuk, L. Rojas, R.D. Jiménez, H. Riojas-Rodriguez, and O. Masera (2008). Reduction in personal exposures to particulate matter and carbon monoxide as a result of the installation of a Patsari improved cook stove in Michoacan Mexico. *Indoor Air*, **18**(2), pp. 93-105.

Arndt, C., R. Benfica, F. Tarp, and R. Uaiene (2010). Biofuels, poverty, and growth: A computable general equilibrium analysis of Mozambique. *Environment and Development Economics*, **15**(1), pp. 81-105.

Asikainen, A., H. Liiri, S. Peltola, T. Karjalainen, and J. Laitila (2008). *Forest Energy Potential in Europe (EU 27)*. Working Papers of the Finnish Forest Research Institute, No. 69, Finnish Forest Research Institute, Helsinki, Finland, 33 pp.

Asikainen, A., P. Anttila, J. Heinimö, T. Smith, I. Stupak, and W. Ferreira Quirino (2010). Forest and bioenergy production. In: *Forest and Society – Responding to Global Drivers of Change*. G. Mery, P. Katila, G. Galloway, R.I. Alfaro, M. Kanninen, M. Lobovikov, and J. Varjo (eds.). International Union of Forest Research Organizations, Vienna, Austria, pp. 183-200.

Astbury, G. (2008). A review of the properties and hazards of some alternative fuels. *Process Safety and Environmental Protection*, **86**(6), pp. 397-414.

Ausubel, J.H. (2000). The great reversal: nature's chance to restore land and sea. *Technology in Society*, **22**, pp. 289-301.

Azar, C., K. Lindgren, M. Obersteiner, K. Riahi, D. van Vuuren, K. den Elzen, K. Möllersten, and E. Larson (2010). The feasibility of low CO_2 concentration targets and the role of bio-energy with carbon capture and storage (BECCS). *Climatic Change*, **100**(1), pp. 195-202.

Bacovsky, D., M. Dallos, and M. Worgetter (2010a). *Status of 2nd Generation Biofuels Demonstration Facilities in June 2010*. T39-P1b, IEA Bioenergy Task 39, 126 pp. Available at: www.ascension-publishing.com/BIZ/IEATask39-0610.pdf.

Bacovsky, D., W. Mabee, and M. Wörgetter (2010b). How close are second-generation biofuels? *Biofuels, Bioproducts and Biorefining*, **4**(3), pp. 249-252.

Badger, P. (2000). New process for torrefied wood manufacturing. *Bioenergy Update*, **2**(4), pp. 1-4. Available at: www.bioenergyupdate.com/magazine/security/NL0400/gbinl0400.pdf.

Badgley, C., J. Moghtader, E. Quintero, E. Zakem, M.J. Chappell, K. Avilés-Vázquez, A. Samulon, and I. Perfecto (2007). Organic agriculture and the global food supply. *Renewable Agriculture and Food Systems*, **22**(02), pp. 86-86.

Baffes, J., and T. Haniotis (2010). *Placing the 2006/08 Commodity Price Boom into Perspective*. Policy Research Working Paper 5371, The World Bank Group, Washington, DC, USA, 42 pp.

Bai, Z.G., D.L. Dent, L. Olsson, and M.E. Schaepman (2008). Proxy global assessment of land degradation. *Soil Use and Management*, **24**(3), pp. 223-234.

Bailis, R., A. Cowan, V. Berrueta, and O. Masera (2009). Arresting the killer in the kitchen: The promises and pitfalls of commercializing improved cookstoves. *World Development*, **37**(10), pp. 1694-1705.

Bain, R.L. (2007). *World Biofuels Assessment, Worldwide Biomass Potential: Technology Characterizations*. NREL/MP-510-42467, National Renewable Energy Laboratory, Golden, CO, USA, 140 pp.

Bain, R.L. (2011). *Biopower Technologies in Renewable Electricity Alternative Futures*. National Renewable Energy Laboratory, Golden, CO, USA, in press.

Bain, R.L., W.P. Amos, M. Downing, and R.L. Perlack (2003). *Biopower Technical Assessment: State of the Industry and the Technology*. TP-510-33123, National Renewable Energy Laboratory, Golden, CO, USA, 277 pp.

Bairiganjan, S., R. Cheung, E.A. Delio, D. Fuente, S. Lall, and S. Singh (2010). *Power to the People: Investing in Clean Energy for the Base of the Pyramid in India*. Centre for Development Finance, c/o Institute for Financial Management and Research, World Resources Institute, Washington, DC, USA, 74 pp.

Balat, M. (2011). Potential alternatives to edible oils for biodiesel production - A review of current work. *Energy Conversion and Management*, **52**(2), pp. 1479-1492.

Ball, B.C., I. Bingham, R.M. Rees, C.A. Watson, and A. Litterick (2005). The role of crop rotations in determining soil structure and crop growth conditions. *Canadian Journal of Soil Science*, **85**, pp. 557-577.

Balmford, A., R.E. Green, and J.P.W. Scharlemann (2005). Sparing land for nature: exploring the potential impact of changes in agricultural yield on the area needed for crop production. *Global Change Biology*, **11**(10), pp. 1594-1605.

Barker, W.T., C.L. Lura, and J.L. Richardson (2009). Soil organic matter and native plant production on the Missouri Coteau. *North Dakota Farm Research*, **44**(2), pp. 14-21.

Barth, T., and M. Kleinert (2008). Motor fuels from biomass pyrolysis. *Chemical Engineering & Technology*, **31**(5), pp. 773-781.

Bauen, A., G. Berndes, M. Junginger, M. Londo, F. Vuille, R. Ball, T. Bole, C. Chudziak, A. Faaij, and H. Mozaffarian (2009a). *Bioenergy: A Sustainable and Reliable Energy Source: A Review of Status and Prospects*. IEA Bioenergy: ExCo:2009:06, 108 pp.

Bauen, A., F. Vuille, P. Watson, and K. Vad (2009b). *The RSB GHG accounting scheme. Feasibility of a meta-methodology and way forward*. E4tech version 4.1, Roundtable on Sustainable Biofuels., Lausanne, Switzerland, 91 pp. Available at: rsb.epfl.ch/files/content/sites/rsb2/files/Biofuels/Documents%20and%20Resources/09-10-08_E4Tech%20Report%20GHG%20Accounting_V4%201_08October09.pdf.

Baum, C., P. Leinweber, M. Weih, N. Lamersdorf, and I. Dimitriou (2009). Effects of short rotation coppice with willows and poplar on soil ecology. *Landbauforschung vTI Agriculture and Forestry Research*, **59**(3), pp. 183-186.

Baum, S., M. Weih, G. Busch, F. Kroiher, and A. Bolte (2009). The impact of Short Rotation Coppice plantations on phytodiversity. *Landbauforschung vTI Agriculture and Forestry Research*, **59**(3), pp. 163-170.

Beer, T., and T. Grant (2007). Life-cycle analysis of emissions from fuel ethanol and blends in Australian heavy and light vehicles. *Journal of Cleaner Production*, **15**(8-9), pp. 833-837.

Beer, T., T. Grant, R. Brown, J. Edwards, P. Nelson, H. Watson, and D. Williams (2000). *Life Cycle Emissions Analysis of Alternative Fuels for Heavy Vehicles*. Atmospheric Research Report C/0411.1.1/F2, Australian Commonwealth Scientific and Research Organization, Aspendale, Australia, 148 pp.

Beer, T., T. Grant, G. Morgan, J. Lapszewicz, P. Anyon, J. Edwards, P. Nelson, H. Watson, and D. Williams (2001). *Comparison of Transport Fuels*. EV45A/2/F3C, Australian Commonwealth Scientific and Research Organization, Aspendale, Australia, 485 pp.

Beringer, T.I.M., W. Lucht, and S. Schaphoff (2011). Bioenergy production potential of global biomass plantations under environmental and agricultural constraints. *Global Change Biology Bioenergy*, doi:10.1111/j.1757-1707.2010.01088.x.

Berndes, G. (2002). Bioenergy and water--the implications of large-scale bioenergy production for water use and supply. *Global Environmental Change*, **12**(4), pp. 253-271.

Berndes, G. (2008a). Future biomass energy supply: The consumptive water use perspective. *International Journal of Water Resources Development*, **24**(2), pp. 235-245.

Berndes, G. (2008b). *Water Demand for Global Bioenergy Production: Trends, Risks and Opportunities*. Report commissioned by the German Advisory Council on Global Change. Wissenschaftlicher Beirat der Bundesregierung Globale Umweltveränderungen, Berlin, Germany, 46 pp.

Berndes, G., and J. Hansson (2007). Bioenergy expansion in the EU: Cost-effective climate change mitigation, employment creation and reduced dependency on imported fuels. *Energy Policy*, **35**(12), pp. 5965-5979.

Berndes, G., M. Hoogwijk, and R. van den Broek (2003). The contribution of biomass in the future global energy supply: a review of 17 studies. *Biomass and Bioenergy*, **25**(1), pp. 1-28.

Berndes, G., F. Fredrikson, and P. Börjesson (2004). Cadmium accumulation and Salix-based phytoextraction on arable land in Sweden. *Agriculture, Ecosystems & Environment*, **103**(1), pp. 207-223.

Berndes, G., P. Börjesson, M. Ostwald, and M. Palm (2008). Multifunctional biomass production systems – an overview with presentation of specific applications in India and Sweden. *Biofuels, Bioproducts and Biorefining*, **2**(1), pp. 16-25.

Berndes, G., N. Bird, and A. Cowie (2010). *Bioenergy, Land Use Change and Climate Change Mitigation*. IEA Bioenergy: ExCo:2010:03, International Energy Agency, Whakarewarewa, Rotorua, New Zealand, 20 pp. Available at: www.ieabioenergy.com/LibItem.aspx?id=6770.

Berrueta, V.M., R.D. Edwards, and O.R. Masera (2008). Energy performance of wood-burning cookstoves in Michoacan, Mexico. *Renewable Energy*, **33**(5), pp. 859-870.

Bessou, C., F. Ferchaud, B. Gabrielle, and B. Mary (2010). Biofuels, greenhouse gases and climate change. A review. *Agronomy for Sustainable Development*, doi:10.1051/agro/2009039,132 pp.

Bhagwat, S.A., K.J. Willis, H.J.B. Birks, and R.J. Whittaker (2008). Agroforestry: a refuge for tropical biodiversity? *Trends in Ecology & Evolution*, **23**(5), pp. 261-267.

Bhojvaid, P.P. (2007). Recent trends in biodiesel production. In: *Biofuels: Towards a Greener and Secure Energy Future*. P.P. Bhojvaid (ed.), The Energy and Resources Institute (TERI) Press, New Delhi, India, pp. 119-136.

Bickel, P., and R. Friedrich (eds.) (2005). *ExternE - Externalities of Energy: Methodology 2005 Update*. European Commission, Brussels, Belgium, 270 pp.

Biran, A., J. Abbot, and R. Mace (2004). Families and firewood: A comparative analysis of the costs and benefits of children in firewood collection and use in two rural communities in sub-Saharan Africa. *Human Ecology*, **32**(1), pp. 1-25.

Blanco-Canqui, H., and R. Lal (2009). Corn stover removal for expanded uses reduces soil fertility and structural stability. *Soil Science Society of America Journal*, **73**(2), pp. 418-426.

Blanco-Canqui, H., R. Lal, W.M. Post, R.C. Izaurralde, and L.B. Owens (2006). Rapid changes in soil carbon and structural properties due to stover removal from no-till corn plots. *Soil Science*, **171**(6), pp. 468-482.

BNDES/CGEE (2008). *Sugarcane-Based Bioethanol: Energy for Sustainable Development*. Brazilian Development Bank and Center for Strategic Studies and Management Science, Technology and Innovation, Rio de Janeiro, Brazil, 304 pp.

Bohlmann, G.M., and M.A. Cesar (2006). The Brazilian opportunity for biorefineries. *Industrial Biotechnology*, **2**(2), pp. 127-132.

Boman, C., B. Forsberg, and T. Sandström (2006). Shedding new light on wood smoke: a risk factor for respiratory health. *European Respiratory Journal*, **27**(3), pp. 446-447.

Bon, E.P.S., and M.A. Ferrara (2007). Bioethanol production via enzymatic hydrolysis of cellulosic biomass. In: *FAO Symposium on the Role of Agricultural Biotechnologies for Production of Bioenergy in Developing Countries*, 12 October 2007, Food and Agriculture Organization, Rome, Italy, 11 pp. Available at: http://www.fao.org/biotech/docs/bon.pdf.

Börjesson, P. (2000). Economic valuation of the environmental impact of logging residue recovery and nutrient compensation. *Biomass and Bioenergy*, **19**(3), pp. 137-152.

Börjesson, P. (2009). Good or bad bioethanol from a greenhouse gas perspective – What determines this? *Applied Energy*, **86**(5), pp. 589-594.

Börjesson, P., and G. Berndes (2006). The prospects for willow plantations for wastewater treatment in Sweden. *Biomass and Bioenergy*, **30**(5), pp. 428-438.

Borowitzka, M.A., R. Osinga, J. Tramper, J.G. Burgess, and R.H. Wijffels (1999). Commercial production of microalgae: ponds, tanks, and fermenters. *Progress in Industrial Microbiology*, **35**, pp. 313-321.

Bouwman, A.F., L.J.M. Boumans, and N.H. Batjes (2002). Emissions of N_2O and NO from fertilized fields: Summary of available measurement data. *Global Biogeochemical Cycles*, **16**(4), 13 pp.

Bradley, D., F. Diesenreiter, M. Wild, and E. Tromborg (2009). *World Biofuel Maritime Shipping Study.* IEA Bioenergy Task 40, IEA, Paris, France, 38 pp.

Bradley, R.L., A. Olivier, N. Thevathasan, and J. Whalen (2008). Environmental and economic benefits of tree-based intercropping systems. *Policy Options*, February, pp. 46-49. Available at: www.irpp.org/po/archive/feb08/bradley.pdf.

Bravo, R., O. Masera, T. Chalico, D. Pandey, E. Riegelhaupt, and A. Uhlig (2010). *Case-studies from Brazil, India and Mexico.* Food and Agriculture Organization, Rome, Italy, 80 pp.

Bridgwater, A., S. Czernik, J. Diebold, D. Meier, A. Oasmaa, C. Peacocke, J. Piskorz, and D. Radlein (2003). *Fast Pyrolysis of Biomass: A Handbook.* CPL Press, Berks, UK, 180 pp.

Bridgwater, A. Ed. (2007). *Success & Visions for Bioenergy: Thermal processing of biomass for bioenergy, biofuels and bioproducts.* CPL Scientific Publishing Services Ltd., CD-ROM, ISBN 9781872691282.

Bringezu, S., H. Schuetz, M. O Brien, L. Kauppi, R. Howarth, and J. McNeely (2009). *Towards Sustainable Production and Use of Resources: Assessing Biofuels.* United Nations Environment Programme, Paris, France, 102 pp.

Bruce, N., E. Rehfuess, S. Mehta, G. Hutton, and K.R. Smith (2006). Indoor air pollution. In: *Disease Control Priorities in Developing Countries.* 2nd ed. World Bank Group, Washington, DC, USA, pp. 793-815.

Bruinsma, J. (2009). The resource outlook to 2050: by how much do land, water and crop yields need to increase by 2050? In: *Expert Meeting on How to Feed the World in 2050*, 24-26 June 2009, Economic and Social Development Department, Food and Agriculture Organization of the United Nations, Rome, Italy, 33 pp.

Brunner, A.M., S.P. DiFazio, and A.T. Groover (2007). Forest genomics grows up and branches out. *New Phytologist*, **174**(4), pp. 710-713.

Buchholz, T., E. Rametsteiner, T. Volk, and V. Luzadis (2009). Multi Criteria Analysis for bioenergy systems assessments. *Energy Policy*, **37**(2), pp. 484-495.

Bureau, J.-C., H. Guyomard, F. Jacquet, and D. Tréguer (2010). European biofuel policy: How far will public support go? In: *Handbook of Bioenergy Economics and Policy.* M. Khanna, J. Scheffran, and D. Zilberman (eds.), Springer, Heidelberg, Germany, pp. 401-423.

Bush, D.R., and J.E. Leach (2007). Translational genomics for bioenergy production: There's room for more than one model. *The Plant Cell*, **19**(10), pp. 2971-2973.

Calder, I., J. Amezaga, J. Bosch, B. Aylward, and L. Fuller (2004). Forest and water policies: the need to reconcile public and science conceptions. *Geologica acta*, **2**(2), pp. 157-166.

Campbell, J.E., D.B. Lobell, and C.B. Field (2009). Greater transportation energy and GHG offsets from bioelectricity than ethanol. *Science*, **324**(5930), pp. 1055-1057.

CARB (2009). *Proposed Regulation to Implement the Low Carbon Fuel Standard.* California Environmental Protection Agency, Air Resources Board, Stationary Source Division, Sacramento, California, 374 pp.

Carlsson-Kanyama, A., and H. Shanahan (2003). Food and life cycle energy inputs: consequences of diet and ways to increase efficiency. *Ecological Economics*, **44**(2-3), pp. 293-307.

Carolan, J., S. Joshi, and B.E. Dale (2007). Technical and financial feasibility analysis of distributed bioprocessing using regional biomass pre-processing centers. *Journal of Agricultural & Food Industrial Organization*, **5**(2) Article 10, 29 pp. Available at: www.bepress.com/jafio/vol5/iss2/art10.

Carpita, N.C., and M.C. McCann (2008). Maize and sorghum: genetic resources for bioenergy grasses. *Trends in Plant Science*, **13**(8), pp. 415-420.

Carslaw, K.S., O. Boucher, D.V. Spracklen, G.W. Mann, J.G.L. Rae, S. Woodward, and M. Kulmala (2010). A review of natural aerosol interactions and feedbacks within the Earth system. *Atmospheric Chemistry and Physics*, **10**(4), pp. 1701-1737.

Cascone, R. (2008). Biobutanol: a replacement for bioethanol? In: *Chemical Engineering Progress Special Edition Biofuels*, **104**, pp. S4–S9.

Cassman, K.G. (1999). Ecological intensification of cereal production systems: Yield potential, soil quality, and precision agriculture. *Proceedings of the National Academy of Sciences of the United States of America*, **96**(11), pp. 5952-5959.

Cassman, K.G., A. Dobermann, D.T. Walters, and H. Yang (2003). Meeting cereal demand while protecting natural resources and improving environmental quality. *Annual Review of Environment and Resources*, **28**(1), pp. 315-358.

Castiglioni, P., D. Warner, R.J. Bensen, D.C. Anstrom, J. Harrison, M. Stoecker, M. Abad, G. Kumar, S. Salvador, R. D'Ordine, S. Navarro, S. Back, M. Fernandes, J. Targolli, S. Dasgupta, C. Bonin, M.H. Luethy, and J.E. Heard (2008). Bacterial RNA chaperones confer abiotic stress tolerance in plants and improved grain yield in maize under water-limited conditions. *Plant Physiology*, **147**(2), pp. 446-455.

CBOT (2006). *CBOT® Soybean Crush Reference Guide.* Board of Trade of the City of Chicago, Chicago, IL, USA.

Chakravorty, U., M. Hubert, and L. Nostbakken (2009). Fuel versus food. *Annual Review of Resource Economics*, **1**(1), pp. 645-663.

Chapotin, S.M., and J. Wolt (2007). Genetically modified crops for the bioeconomy: meeting public and regulatory expectations. *Transgenic Research*, **16**(6), pp. 675-688.

Chapple, C., M. Ladisch, and R. Meilan (2007). Loosening lignin's grip on biofuel production. *Nature Biotechnology*, **25**(7), pp. 746-748.

Cheng, J.J., and G.R. Timilsina (2010). *Advanced Biofuel Technologies: Status and Barriers.* Policy Research Working Paper 5411, The World Bank, Washington, DC, USA, 47 pp.

Cherubini, F., and A.H. Strømman (2010). Life cycle assessment of bioenergy systems: State of the art and future challenges. *Bioresource Technology*, **102**(2), pp. 437-451.

Cherubini, F., N.D. Bird, A. Cowie, G. Jungmeier, B. Schlamadinger, and S. Woess-Gallasch (2009a). Energy- and greenhouse gas-based LCA of biofuel and bioenergy systems: Key issues, ranges and recommendations. *Resources, Conservation and Recycling*, **53**(8), pp. 434-447.

Cherubini, F., G. Jungmeier, M. Mandl, M. Wellisch, C. Philips, and H. Joergensen (2009b). *Biorefineries: Adding Value to the Sustainable Utilisation of Biomass.* IEA Bioenergy Publication T42:2009:01, IEA Bioenergy Task 42 on Biorefineries, 16 pp.

Chisti, Y. (2007). Biodiesel from microalgae. *Biotechnology Advances*, **25**(3), pp. 294-306.

Chiu, Y.-W., B. Walseth, and S. Suh (2009). Water embodied in bioethanol in the United States. *Environmental Science & Technology*, **43**(8), pp. 2688-2692.

Chum, H.L., and R.P. Overend (2003). Biomass and bioenergy in the United States. In: *Advances in Solar Energy: An Annual Review of Research and Development.* Vol. 15. D.Y. Goswami (ed.), American Solar Energy Society (ASES), Boulder, CO, USA, pp. 83-148.

Cirne, D.G., A. Lehtomäki, L. Björnsson, and L.L. Blackall (2007). Hydrolysis and microbial community analyses in two-stage anaerobic digestion of energy crops. *Journal of Applied Microbiology*, **103**(3), pp. 516-527.

Clayton, R., G. McDougall, M. Perry, D. Doyle, J. Doyle, and D. O'Connor (2010). *A Study of Employment Opportunities from Biofuel Production in APEC Economies.* APEC#210-RE-01.9, APEC Energy Working Group, Asia-Pacific Economic Cooperation (APEC), Singapore, Singapore, 82 pp.

Clifton-Brown, J.C., and I. Lewandowski (2000). Water use efficiency and biomass partitioning of three different Miscanthus genotypes with limited and unlimited water supply. *Annals of Botany*, **86**(1), pp. 191-200.

Clifton-Brown, J., P.F. Stampfl, and M.B. Jones (2004). Miscanthus biomass production for energy in Europe and its potential contribution to decreasing fossil fuel carbon emissions. *Global Change Biology*, **10**(4), pp. 509-518.

Cline, W.R. (2007). *Global Warming and Agriculture: Impact Estimates by Country.* Peterson Institute for International Economics, Washington, DC, USA, 201 pp.

Colla, L.M., C. Oliveira Reinehr, C. Reichert, and J.A.V. Costa (2007). Production of biomass and nutraceutical compounds by Spirulina platensis under different temperature and nitrogen regimes. *Bioresource Technology*, **98**(7), pp. 1489-1493.

Cortright, R.D., R.R. Davda, and J.A. Dumesic (2002). Hydrogen from catalytic reforming of biomass-derived hydrocarbons in liquid water. *Nature*, **418**(6901), pp. 964-967.

Cremers, M.F.G. (2009). *Deliverable 4, Technical Status of Biomass Co-firing.* 50831165-Consulting 09-1654, IEA Bioenergy Task 32, Arnhem, The Netherlands, 43 pp.

Crutzen, P., A. Moiser, K. Smith, and W. Winiwarter (2007). N_2O release from agro-biofuel production negates global warming reduction by replacing fossil fuels. *Atmospheric Chemistry and Physics*, **7**, pp. 11191-11205.

Curtis, B. (2010). *2008-2009 Review U.S. Biofuels Industry: Mind the Gap.* Concentric Energies & Resource Group, Inc., Los Gatos, CA, USA, 52 pp.

Cusick, R.D., B. Bryan, D.S. Parker, M.D. Merrill, M. Mehanna, P.D. Kiely, G. Liu, and B.E. Logan (2011). Performance of a pilot-scale continuous flow microbial electrolysis cell fed winery wastewater. *Applied Microbiology and Biotechnology*, **89**, pp. 2053-2063.

da Costa Sousa, L., S.P.S. Chundawat, V. Balan, and B.E. Dale (2009). 'Cradle-to-grave' assessment of existing lignocellulose pretreatment technologies. *Current Opinion in Biotechnology*, **20**(3), pp. 339-347.

Dale, B.E., M.S. Allen, M. Laser, and L.R. Lynd (2009). Protein feeds coproduction in biomass conversion to fuels and chemicals. *Biofuels, Bioproducts and Biorefining*, **3**(2), pp. 219-230.

Dale, B.E., B.D. Bals, S. Kim, and P. Eranki (2010). Biofuels done right: Land efficient animal feeds enable large environmental and energy benefits. *Environmental Science & Technology*, **44**(22), pp. 8385-8389.

Dale, V.H., R.A. Efroymson, and K.L. Kline (2011). The land use-climate change-energy nexus. *Landscape Ecology*, in press.

Damen, K., and A. Faaij (2006). A greenhouse gas balance of two existing international biomass import chains. *Mitigation and Adaptation Strategies for Global Change*, **11**(5), pp. 1023-1050.

Danielsen, F., H. Beukema, N.D. Burgess, F. Parish, C.A. Brühl, P.F. Donald, D. Murdiyarso, B.E.N. Phalan, L. Reijnders, M. Struebig, and E.B. Fitzherbert (2009). Biofuel plantations on forested lands: Double jeopardy for biodiversity and climate (Plantaciones de biocombustible en terrenos boscosos: Doble peligro para la biodiversidad y el clima). *Conservation Biology*, **23**(2), pp. 348-358.

Dantas, D.N., F.F. Mauad, and A.R. Ometto (2009). Potential for generation of thermal and electrical energy from biomass of sugarcane: A exergetic analysis. In: *11th International Conference on Advanced Materials*, Rio de Janeiro, Brazil, 20-25 September 2009, p. 131.

Darzins, A., P. Pienkos, and L. Edye (2010). *Current Status and Potential for Algal Biofuels Production.* Report T39-02, IEA Bioenergy Task 39, 128 pp.

Daugherty, E. (2001). *Biomass Energy Systems Efficiency Analyzed through a LCA Study.* Masters Thesis, Lund University, Gothenburg, Sweden, 45 pp.

Davda, R.R., J.W. Shabaker, G.W. Huber, R.D. Cortright, and J.A. Dumesic (2005). A review of catalytic issues and process conditions for renewable hydrogen and alkanes by aqueous-phase reforming of oxygenated hydrocarbons over supported metal catalysts. *Applied Catalysis B: Environmental*, **56**(1-2), pp. 171-186.

Davidson, E.A. (2009). The contribution of manure and fertilizer nitrogen to atmospheric nitrous oxide since 1860. *Nature Geoscience*, **2**(9), pp. 659-662.

de Boer, J., M. Helms, and H. Aiking (2006). Protein consumption and sustainability: Diet diversity in EU-15. *Ecological Economics*, **59**(3), pp. 267-274.

de Fraiture, C., and G. Berndes (2009). Biofuels and water. In: *Biofuels: Environmental Consequences and Interactions with Changing Land Use.* R.W. Howarth and S. Bringezu (eds.), Proceedings of the SCOPE - Scientific Committee on Problems of the Environment International Biofuels Project Rapid Assessment, Gummersbach, Germany, 22-25 September 2009, pp. 139-152.

de Fraiture, C., M. Giordano, and Y. Liao (2008). Biofuels and implications for agricultural water uses: blue impacts of green energy. *Water Policy*, **10**(S1), pp. 67-81.

De La Torre Ugarte, D.G., and C.C. Hellwinckel (2010). The problem is the solution: The role of biofuels in the transition to a regenerative agriculture. In: *Plant Biotechnology for Sustainable Production of Energy and Co-products.* P.N. Mascia, J. Scheffran, and J.M. Widholm (eds.), Springer-Verlag, Berlin and Heidelberg, Germany, 475 pp.

De La Torre Ugarte, D.G., L. He, K.L. Jensen, and B.C. English (2010). Expanded ethanol production: Implications for agriculture, water demand, and water quality. *Biomass and Bioenergy*, **34**(11), pp. 1586-1596.

de Wit, M., and A. Faaij (2010). European biomass resource potential and costs. *Biomass and Bioenergy*, **34**(2), pp. 188-202.

de Wit, M., M. Junginger, S. Lensink, M. Londo, and A. Faaij (2010). Competition between biofuels: Modeling technological learning and cost reductions over time. *Biomass and Bioenergy*, **34**(2), pp. 203-217.

DEFRA (2009). *The 2007/08 Agricultural Price Spikes: Causes and Policy Implications.* Department for Environment Food and Rural Affairs (DEFRA), HM Government, London, UK, 123 pp.

Delta-T Corporation (1997). *Proprietary information.* Williamsburg, VA, USA.

DeLucchi, M.A. (2005). *A Multi-Country Analysis of Lifecycle Emissions from Transportation Fuels and Motor Vehicles.* UCD-ITS-RR-05-10, Institute of Transportation Studies, University of California at Davis, Davis, CA, USA, 205 pp.

DeLucchi, M.A. (2010). *A Conceptual Framework for Estimating Bioenergy-Related Land-Use Change and Its Impacts over Time.* Biomass and Bioenergy, 1-24, doi:10.1016/j.biombioe.2010.11.028 (2010).

DeLuchi, M.A. (1993). Greenhouse-gas emissions from the use of new fuels for transportation and electricity. *Transportation Research Part A*, **27A**(3), pp. 187-191.

DeMeo, E.A., and J.F. Galdo (1997). *Renewable Energy Technology Characterizations.* TR-109496, U.S. Department of Energy and Electric Power Research Institute, Washington, DC, USA, 283 pp.

Demirbas, A. (2009). Progress and recent trends in biodiesel fuels. *Energy Conversion and Management*, **50**, pp. 14-34.

den Elzen, M., D. van Vuuren, and J. van Vliet (2010). Postponing emission reductions from 2020 to 2030 increases climate risks and long-term costs. *Climatic Change*, **99**(1), pp. 313-320.

Dias de Moraes, M.A.F. (2007). Indicadores do mercado de trabalho do sistema agroindustrial da cana-de-açúcar do Brasil no período 1992-2005. *Estudos Econ. (São Paulo)*, **37**(4), pp. 875-902.

Diaz-Balteiro, L., and L.C.E. Rodriguez (2006). Optimal rotations on Eucalyptus plantations including carbon sequestration – A comparison of results in Brazil and Spain. *Forest Ecology and Management*, **229**(1-3), pp. 247-258.

DiTomaso, J.M., J.K. Reaser, C.P. Dionigi, O.C. Doering, E. Chilton, J.D. Schardt, and J.N. Barney (2010). Biofuel vs. bioinvasion: seeding policy priorities. *Environmental Science & Technology*, **44**(18), pp. 6906-6910.

Dominguez-Faus, R., S.E. Powers, J.G. Burken, and P.J. Alvarez (2009). The water footprint of biofuels: A drink or drive issue? *Environmental Science & Technology*, **43**(9), pp. 3005-3010.

Dondini, M., A. Hastings, G. Saiz, M.B. Jones, and P. Smith (2009). The potential of Miscanthus to sequester carbon in soils: comparing field measurements in Carlow, Ireland to model predictions. *Global Change Biology Bioenergy*, **1**(6), pp. 413-425.

Donner, S.D., and C.J. Kucharik (2008). Corn-based ethanol production compromises goal of reducing nitrogen export by the Mississippi River. *Proceedings of the National Academy of Sciences*, **105**(11), pp. 4513-4518.

Doornbosch, V., and R. Steenblik (2008). Biofuels: Is the cure worse than the disease? In: *Round Table on Sustainable Development*, 11-12 September 2007, Organisation for Economic Co-operation and Development, Paris, France.

Dornburg, V., and A.P.C. Faaij (2001). Efficiency and economy of wood-fired biomass energy systems in relation to scale regarding heat and power generation using combustion and gasification technologies. *Biomass and Bioenergy*, **21**(2), pp. 91-108.

Dornburg, V., and A.P.C. Faaij (2005). Cost and CO_2-emission reduction of biomass cascading: methodological aspects and case study of SRF poplar. *Climatic Change*, **71**(3), pp. 373-408.

Dornburg, V., and G. Marland (2008). Temporary storage of carbon in the biosphere does have value for climate change mitigation: a response to the paper by Miko Kirschbaum. *Mitigation and Adaptation Strategies for Global Change*, **13**(3), pp. 211-217.

Dornburg, V., J. Vandam, and A. Faaij (2007). Estimating GHG emission mitigation supply curves of large-scale biomass use on a country level. *Biomass and Bioenergy*, **31**(1), pp. 46-65.

Dornburg, V., A. Faaij, P. Verweij, H. Langeveld, G. van de Ven, F. Wester, H. van Keulen, K. van Diepen, M. Meeusen, M. Banse, J. Ros, D. van Vuuren, G.J. van den Born, M. van Oorschot, F. Smout, J. van Vliet, H. Aiking, M. Londo, H. Mozaffarian, K. Smekens, E. Lysen, and S. van Egmond (2008). *Assessment of Global Biomass Potentials and their Links to Food, Water, Biodiversity, Energy Demand and Economy.* WAB 500102 012, The Netherlands Environmental Assessment Agency, Bilthoven, The Netherlands, 108 pp.

Dornburg, V., D. van Vuuren, G. van de Ven, H. Langeveld, M. Meeusen, M. Banse, M. van Oorschot, J. Ros, G.J. van den Born, H. Aiking, M. Londo, H. Mozaffarian, P. Verweij, E. Lysen, and A. Faaij (2010). Bioenergy revisited: Key factors in global potentials of bioenergy. *Energy & Environmental Science*, **3**, pp. 258-267.

Drigo, R., G. Chirici, B. Lasserre, and M. Marchetti (2007). Analisi su base geografica della domanda e dell'offerta di combustibili legnosi in Italia. (Geographical analysis of demand and supply of woody fuel in Italy). *L'italia Forestale e Montana*, **LXII**(5/6), pp. 303-324.

Drigo, R., A. Anschau, N. Flores Marcos, and S. Carballo (2009). *Análisis del balance de energia derivada de biomasa en Argentina – WISDOM Argentina.* Project TCP/ARG/3103 of FAO Dendroenergy, FAO Forestry Department, Rome, Italy, 102 pp.

Dudley, T.L. (2000). Arundo donax. In: *Invasive Plants of California's Wildlands.* C. Bossard and J. Randall (eds.), University of California Press, Berkeley, CA, USA, pp. 53-58.

Dufey, A. (2006). *Biofuels Production, Trade and Sustainable Development: Emerging Issues.* International Institute for Environment and Development, London, UK, 62 pp.

Dupraz, C., and F. Liagre (2008). *Agroforestry: Trees and Crops.* La France Agricole, Paris, France.

Dutta, A., N. Dowe, K.N. Ibsen, D.J. Schell, and A. Aden (2010). An economic comparison of different fermentation configurations to convert corn stover to ethanol using Z. mobilis and Saccharomyces. *Biotechnology Progress*, **26**(1), pp. 64-72.

Duvick, D.N., and K.G. Cassman (1999). Post-green revolution trends in yield potential of temperate maize in the north-central United States. *Crop Science*, **39**(6), pp. 1622-1630.

Dymond, C.C., B.D. Titus, G. Stinson, and W.A. Kurz (2010). Future quantities and spatial distribution of harvesting residue and dead wood from natural disturbances in Canada. *Forest Ecology and Management*, **260**(2), pp. 181-192.

E4tech (2009). *Review of the Potential for Biofuels in Aviation.* Committee on Climate Change (CCC) of the U.K. Government, London, UK, 117 pp.

E4tech (2010). *Biomass Prices in the Heat and Electricity Sectors in the UK.* Report prepared for the UK Department of Energy and Climate Change. URN 10D/546, E4tech, London, UK, 33 pp.

Easterling, W.E., J. Morton, J.F. Soussana, J. Schmidhuber, F.N. Tubiello, P.K. Aggarwal, P. Batima, K.M. Brander, L. Erda, S.M. Howden, and A. Kirilenko (2007). Food, fibre and forest products. In: *Climate Change 2007: Impacts, Adaptation and Vulnerability: Contribution of Working Group II to the Fourth Assessment Report of the Intergovernmental Panel on Climate Change.* M.L. Parry, O.F. Canziani, J.P. Palutikof, P.J. van der Linden, and C.E. Hanson (eds.), Cambridge University Press, pp. 273-313.

Eaton, J., N. McGoff, K. Byrne, P. Leahy, and G. Kiely (2008). Land cover change and soil organic carbon stocks in the Republic of Ireland 1851–2000. *Climatic Change*, **91**(3), pp. 317-334.

Ecobilan (2002). *Energy and Greenhouse Gas Balances of Biofuels' Production Chains in France.* Ecobilan, Neuilly-sur-Seine Cedex, France, 9 pp.

Econ Poyry (2008). *Current Bioenergy Application and Conversion Technologies in the Nordic Countries.* 2008-052, Nordic Energy Research, Copenhagen, Denmark, 37 pp.

Edenhofer, O., B. Knopf, M. Leimbach, and N. Bauer (2010). The economics of low stabilization. *The Energy Journal*, **31**(Special Issue 1), pp. 57-90.

Edgerton, M.D. (2009). Increasing crop productivity to meet global needs for feed, food, and fuel. *Plant Physiology*, **149**(1), pp. 7-13.

Edwards, R., D. Mulligan, and L. Marelli (2010). *Indirect Land Use Change from Increased Biofuel Demand. Comparison of Models and Results for Marginal Biofuels Production from Different Feedstocks.* EUR 24485 EN-2010, European Commission, Joint Research Centre, Institute for Energy, Ispra, Italy, 150 pp.

Edwards, R., J.F. Larivé, V. Mahieu, and P. Rouveirolles (2007). *Well-To-Wheels Analysis of Future Automotive Fuels and Power Trains in the European Context.* Version 2c, Joint Research Center, Brussels, Belgium, 88 pp.

Edwards, R., J.F. Larivé, V. Mahieu, and P. Rouveirolles (2008). *Well-To-Wheels analysis of future automotive fuels and power trains in the European context.* Version 3, Joint Research Center, Brussels, Belgium, 43 pp.

EEA (2006). *How Much Bioenergy can Europe Produce without Harming the Environment?* European Environment Agency (EEA), Copenhagen, Denmark, 72 pp.

EEA (2007). *Estimating the Environmentally Compatible Bio-energy Potential from Agriculture.* European Environment Agency (EEA), Copenhagen, Denmark.

Egsgaard, H.U., P. Hansen, J. Arendt, P. Glarborg, and C. Nielsen (2009). Combustion and gasification technologies. In: *Risø Energy Report 2.* Risø, Roskilde, Denmark, pp. 35-39.

EIA (2009). *2006 Energy Consumption by Manufacturers – Data Table 7.2.* U.S. Energy Information Administration, U.S. Department of Energy, Washington, DC, USA.

EIA (2010a). *Annual Energy Review*, DOE/EIA-0384(2009) updated August 2011, Energy Information Administration, U.S. Department of Energy, Washington, DC, USA.

EIA (2010b). *Updated Capital Costs for Electricity Generation Plants.* U.S. Energy Information Administration, U.S. Department of Energy, Washington, DC, USA.

EISA, 2007. *Energy Independence and Security Act*, United States Congress, Public Law 110-140. U.S. Government Printing Office, Washington, DC, USA.

Eisenbies, M., E. Vance, W. Aust, and J. Seiler (2009). Intensive utilization of harvest residues in southern pine plantations: Quantities available and implications for nutrient budgets and sustainable site productivity. *BioEnergy Research*, **2**(3), pp. 90-98.

Elander, R., B. Dale, M. Holtzapple, M. Ladisch, Y. Lee, C. Mitchinson, J. Saddler, and C. Wyman (2009). Summary of findings from the Biomass Refining Consortium for Applied Fundamentals and Innovation (CAFI): corn stover pretreatment. *Cellulose*, **16**(4), pp. 649-659.

Elferink, E.V., and S. Nonhebel (2007). Variations in land requirements for meat production. *Journal of Cleaner Production*, **15**, pp. 1778-1786.

Elliott, D.C. (2008). Catalytic hydrothermal gasification of biomass. *Biofuels, Bioproducts, Biorefining*, **2**, pp. 254-265.

Elobeid, A., and C. Hart (2007). Ethanol expansion in the food versus fuel debate: How will developing countries fare? *Journal of Agricultural & Food Industrial Organization*, **5**(2), pp. 1-21.

EMBRAPA (2010). *Brazilian Sugarcane Agroecological Zoning (in English), Zoneamento Agroecológico da Cana de Açúcar (in Portuguese).* Empresa Brasileira de Pesquisa Agropecuária (Brazilian Agricultural Research Corporation, EMBRAPA), Brasilia, Brazil, 58 pp.

EPA (2010). *Renewable Fuel Standard Program (RFS2) Regulatory Impact Analysis.* EPA-420-R-10-006, Environmental Protection Agency, Washington, DC, USA, 1120 pp.

EPE (2008). *Plano Decenal De Expansao De Energia 2008 – 2017.* Ministry of Mines and Energy Secretariat of Planning and Energy Development, Brasilia, Brazil, 354 pp.

EPE (2010). *2010 Balanco Energetico Nacional - Ano Base 2009.* CDU 620.9:553.04(81), Ministry of Mines and Energy and Energy Planning Enterprise (EPE), Brasilia, Brazil, 271 pp.

EPRI (2008). *Technical Performance Indicators for Biomass Energy.* Electric Power Research Institute, Palo Alto, CA, USA, 86 pp.

Ericsson, K., and L.J. Nilsson (2006). Assessment of the potential biomass supply in Europe using a resource-focused approach. *Biomass and Bioenergy*, **30**(1), pp. 1-15.

Ericsson, K., H. Rosenqvist, and L.J. Nilsson (2009). Energy crop production costs in the EU. *Biomass and Bioenergy*, **33**(11), pp. 1577-1586.

Erikson, S., and M. Prior (1990). *The Briquetting of Agricultural Wastes for Fuel.* FAO paper No. 11, Food and Agriculture Organization, Rome, Italy.

EurObserv'ER (2010). Biofuels barometer/Baromètre Biocarburant Observ'ER. *Systèmes Solaires le journal des énergies renouvelables*, Paris, France, pp. 72-96.

European Commission (2009). *Directive 2009/28/EC of the European Parliament and of the Council of 23 April 2009 on the promotion of the use of energy from renewable sources.* European Commission, Brussels, Belgium.

European Commission (2010). *Report from the commission to the council and the European parliament on sustainability requirements for the use of solid and gaseous biomass sources in electricity, heating and cooling.* SEC(2010) 65, European Commission, Brussels, Belgium, 20 pp.

Evans, L.T. (2003). Agricultural intensification and sustainability. *Outlook on Agriculture*, **32**(2), pp. 83-89.

Evenson, R.E., and D. Gollin (2003). Assessing the impact of the Green Revolution, 1960 to 2000. *Science*, **300**(5620), pp. 758-762.

Ezeji, T., N. Quereshi, and H.P. Blaschek (2007a). Butanol production from agricultural residues: impact of degradation products on Clostridium beijerinckii growth and butanol fermentation. *Biotechnology and Bioengineering*, **97**(6), pp. 1460-1467.

Ezeji, T.C., N. Qureshi, and H.P. Blaschek (2007b). Bioproduction of butanol from biomass: from genes to bioreactors. *Current Opinion in Biotechnology*, **18**(3), pp. 220-227.

Ezzati, M., A. Lopez, S. Vander Hoorn, A. Rodgers, and C.L.J. Murray (2002). Comparative Risk Assessment Collaborative Group. Selected major risk factors and global regional burden of disease. *The Lancet*, **360**(9343), pp. 1347-1360.

Ezzati, M., R. Bailis, D.M. Kammen, T. Holloway, L. Price, L.A. Cifuentes, B. Barnes, A. Chaurey, and K.N. Dhanapala (2004). Energy management and global health. *Annual Review of Environment and Resources*, **29**(1), pp. 383-419.

Faaij, A. (2006). Modern biomass conversion technologies. *Mitigation and Adaptation Strategies for Global Change*, **11**(2), pp. 335-367.

Fagernäs, L., A. Johansson, C. Wilén, K. Sipilä, T. Mäkinen, S. Helynen, E. Daugherty, H. den Uil, J. Vehlow, T. Kåberger, and M. Rogulska (2006). *Bioenergy in Europe: Opportunities and Barriers*. VTT Res. Notes 2532, VTT, Espoo, Finland, 122 pp.

Fallot, A., L. Saint-Andre, G. Le Maire, J.-P. Laclau, Y. Nouvellon, C. Marsden, J.-P. Bouillet, T. Silva, M.-G. Piketty, and O. Hamel (2009). Biomass sustainability, availability and productivity. *Revue de Métallurgie Paris*, **106**(10), pp. 410-418.

FAO (1985). Industrial charcoal making technologies. In: *Industrial Charcoal Making*. FAO Forestry Paper 63, FAO Forestry Department, Food and Agriculture Organization, Rome, Italy. Available at: www.fao.org/docrep/X5555E/x5555e02.htm#1.1%20what%20are%20industrial%20charcoal%20making%20methods.

FAO (2004). *The State of Food and Agriculture 2003-2004 – Agricultural Biotechnology: Meeting the Needs of the Poor?* 0081-4539, Food and Agriculture Organization, Rome, Italy, 196 pp.

FAO (2005). *World Forest Assessment*. Food and Agriculture Organization, Rome, Italy, 166 pp.

FAO (2006). *Energy and Gender in Rural Sustainable Development*. Food and Agriculture Organization, Rome, Italy, 46 pp.

FAO (2008a). *The State of Food and Agriculture 2008 – Biofuels: Prospects, Risks, and Opportunities*. Food and Agriculture Organization, Rome, Italy, 138 pp.

FAO (2008b). The role of agricultural biotechnologies for production of bioenergy in developing countries. In: *Background Document to Conference 15 of the FAO Biotechnology Forum*, 10 November to 14 December 2008, Food and Agriculture Organization, Rome, Italy.

FAO (2010a). *What Woodfuels can do to Mitigate Climate Change*. 0258-6150, Food and Agricultural Organization (FAO), Rome, Italy, 98 pp.

FAO (2010b). *Criteria and Indicators for Sustainable Biofuels*. FAO Forestry Paper 160, Food and Agricultural Organization (FAO), Rome, Italy, 102 pp.

FAOSTAT (2011). *FAOSTAT*. Food and Agriculture Organization, Rome, Italy. Available at: faostat.fao.org/default.aspx.

FAPRI (2009). *FAPRI 2009: U.S. and World Agricultural Outlook*. Food and Agricultural Policy Research Institute, Ames, Iowa, 411 pp.

Fargione, J., J. Hill, D. Tilman, S. Polasky, and P. Hawthorne (2008). Land clearing and the biofuel carbon debt. *Science*, **319**(5867), pp. 1235-1238.

Farley, K.A., E.G. Jobbagy, and R.B. Jackson (2005). Effects of afforestation on water yield: a global synthesis with implications for policy. *Global Change Biology*, **11**(10), pp. 1565-1576.

Farrell, A.E., R.J. Plevin, B.T. Turner, A.D. Jones, M. O'Hare, and D.M. Kammen (2006). Ethanol can contribute to energy and environmental goals. *Science*, **311**(5760), pp. 506-508.

Fearnside, P.M. (2008). The roles and movements of actors in the deforestation of Brazilian Amazonia. *Ecology and Society*, **13**(23), pp. 23-55.

Feng, Z., K.J. Stadt, and V.J. Lieffers (2006). Linking juvenile white spruce density, dispersion, stocking, and mortality to future yield. *Canadian Journal of Forest Research*, **36**, pp. 3173-3182.

Field, C.B., J.E. Campbell, and D.B. Lobell (2008). Biomass energy: the scale of the potential resource. *Trends in Ecology & Evolution*, **23**(2), pp. 65-72.

Fingerman, K.R., M.H. Torn, M.S. O'Hare, and D.M. Kammen (2010). Accounting for the water impacts of ethanol production. *Environmental Research Letters*, **5**(1), 014020 (7pp.).

Firbank, L. (2008). Assessing the ecological impacts of bioenergy projects. *BioEnergy Research*, **1**(1), pp. 12-19.

Fischer, G., and L. Schrattenholzer (2001). Global bioenergy potentials through 2050. *Biomass and Bioenergy*, **20**(3), pp. 151-159.

Fischer, G., H. van Velthuizen, M. Shah, and F. Nachtergaele (2002). *Global Agro-ecological Assessment for Agriculture in the 21st Century: Methodology and Results*. RR-02-02, International Institute for Applied Systems Analysis, Laxenburg, Austria, 156 pp.

Fischer, G., F. Nachtergaele, S. Prieler, F. Teixeira, H. van Velthuizen, L. Verelst, and D. Wiberg (2008). *Global Agro-ecological Zones Assessment for Agriculture (GAEZ 2008)*. Version 2, International Institute for Applied Systems Analysis and Food and Agriculture Organization, Laxenburg, Austria and Rome, Italy, 43 pp.

Fischer, G., E. Hizsnyik, S. Prieler, M. Shah, and H. van Velthuizen (2009). *Biofuels and Food Security*. The OPEC Fund for International Development (OFID) and International Institute of Applied Systems Analysis (IIASA), Vienna, Austria, 228 pp.

Fischer, G., S. Prieler, H. van Velthuizen, G. Berndes, A. Faaij, M. Londo, and M. de Wit (2010). Biofuel production potentials in Europe: Sustainable use of cultivated land and pastures, Part II: Land use scenarios. *Biomass and Bioenergy*, **34**(2), pp. 173-187.

Fitzherbert, E.B., M.J. Struebig, A. Morel, F. Danielsen, C.A. Brühl, P.F. Donald, and B. Phalan (2008). How will oil palm expansion affect biodiversity? *Trends in Ecology & Evolution*, **23**(10), pp. 538-545.

Fleming, J.S., S. Habibi, and H.L. MacLean (2006). Investigating the sustainability of lignocellulose-derived fuels for light-duty vehicles. *Transportation Research Part D: Transport and Environment*, **11**(2), pp. 146-159.

Foley, J.A., R. DeFries, G.P. Asner, B. Carol, G. Bonan, S.R. Carpenter, F.S. Chapin, M.T. Coe, G.C. Daily, H.K. Gibbs, J.H. Helkowski, T. Holloway, E.A. Howard, C.J. Kucharik, C. Monfreda, J.A. Patz, I.C. Prentice, N. Ramankutty, and P.K. Snyder (2005). Global consequences of land use. *Science*, **309**(5734), pp. 570-574.

Folha (2005). Bagaço da cana será usado para fabricação de papel. *Folha da Região - Araçatuba*. São Paulo, Brazil.

Folke, C., S. Carpenter, B. Walker, M. Scheffer, T. Elmqvist, L. Gunderson, and C.S. Holling (2004). Regime shifts, resilence, and biodiversity in ecosystem management. *Annual Review of Ecology, Evolution, and Systematics*, **35**(1), pp. 557-581.

Folke, C., F.S. Chapin, and P. Olsson (2009). Transformations in ecosystem stewardship. In: *Principles of Ecosystem Stewardship: Resilience-Based Natural Resource Management in a Changing World*. F.S. Chapin III, G.P. Kofinas, and C. Folke (eds.), Springer Verlag, New York, NY, USA, pp. 103-128.

Forman, J. (2003). The introduction of American plant species into Europe: issues and consequences. In: *Plant Invasions: Ecological Threats and Management Solutions*. L. Child, J. H. Brock, G. Brundu, K. Prach, P. Pysek, P. M. Wade, and M. Williamson (eds.), Backhuys Publishers, Leiden, Netherlands, pp. 17-39.

Foust, T.D., A. Aden, A. Dutta, and S. Phillips (2009). Economic and environmental comparison of a biochemical and a thermochemical lignocellulosic ethanol conversion process. *Cellulose*, **16**, pp. 547-565.

Francis, G., R. Edinger, and K. Becker (2005). A concept for simultaneous wasteland reclamation, fuel production, and socio-economic development in degraded areas in India: Need, potential and perspectives of Jatropha plantations. *Natural Resources Forum*, **29**(1), pp. 12-24.

Fritsche, U., K. Hennenberg, and K. Hünecke (2010). *The "iLUC factor" as a Means to Hedge Risks of GHG Emissions from Indirect Land Use Change.* Öko Institute, Darmstadt, Germany, 64 pp.

Fulton, L., T. Howes, and J. Hardy (2004). *Biofuels for Transport - An International Perspective.* Organization for Economic Cooperation and Development and International Energy Agency, Paris, France, 210 pp.

GAIN (2009a). *Biofuel's Impact on Food Crops (China).* CH9059, Global Agriculture Information Network, United States Department of Agriculture, Washington, DC, USA, 14 pp.

GAIN (2009b). *Biofuel's Impact on Food Crops (Indonesia).* ID9017, Global Agriculture Information Network, United States Department of Agriculture, Washington, DC, 7 pp.

GAIN (2009c). *Biofuel's Impact on Food Crops (Thailand).* TH9047, Global Agriculture Information Network, United States Department of Agriculture, Washington, DC, USA, 11 pp.

GAIN (2010a). *Biofuel's Impact on Food Crops (Argentina).* Global Agriculture Information Network, United States Department of Agriculture, Washington, DC, USA, 15 pp.

GAIN (2010b). *Biofuel's Impact on Food Crops (Brazil).* BR10006, Global Agriculture Information Network, United States Department of Agriculture, Washington, DC, USA, 52 pp.

GAIN (2010c). *Biofuel's Impact on Food Crops (Canada).* CA0023, Global Agriculture Information Network, United States Department of Agriculture, Washington, DC, USA, 34 pp.

GAIN (2010d). *Biofuel's Impact on Food Crops (India).* IN1058, Global Agriculture Information Network, United States Department of Agriculture, Washington, DC, USA, 11 pp.

GAIN (2010e). *Biofuel's Impact on Food Crops (Korea).* KS1001, Global Agriculture Information Network, United States Department of Agriculture, Washington, DC, USA, 6 pp.

GAIN (2010f). *Biofuel's Impact on Food Crops (Malaysia).* MY0008, Global Agriculture Information Network, United States Department of Agriculture, Washington, DC, USA, 9 pp.

GAIN (2010g). *Biofuel's Impact on Food Crops (Peru).* Global Agriculture Information Network, United States Department of Agriculture, Washington, DC, USA, 7 pp.

GAIN (2010h). *Biofuel's Impact on Food Crops (Philippines).* Global Agriculture Information Network, United States Department of Agriculture, Washington, DC, USA, 21 pp.

GAIN (2010i). *Biofuel's Impact on Food Crops (Thailand).* TH0098, Global Agriculture Information Network, United States Department of Agriculture, Washington, DC, USA, 15 pp.

GAIN (2010j). *Biofuel's Impact on Food Crops (Turkey).* Global Agriculture Information Network, United States Department of Agriculture, Washington, DC, USA, 6 pp.

Galdos, M.V., C.C. Cerri, R. Lal, M. Bernoux, B. Feigl, and C.E.P. Cerri (2010). Net greenhouse gas fluxes in Brazilian ethanol production systems. *Global Change Biology Bioenergy*, **2**(1), pp. 37-44.

Gallagher, E. (2008). *The Gallagher Review of the Indirect Effects of Biofuels Production.* Renewable Fuels Agency, London, UK, 92 pp.

Gallagher, P., M. Dikeman, J. Fritz, E. Wailes, W. Gauther, and H. Shapouri (2003). *Biomass from Crop Residues: Cost and Supply Estimates.* Agricultural Economic Report No. 819, Economic Research Service, United States Department of Agriculture, Washington, DC, USA, 30 pp.

Gan, J. (2007). Supply of biomass, bioenergy, and carbon mitigation: Method and application. *Energy Policy*, **35**(12), pp. 6003-6009.

Gan, J., and C. Smith (2010). Integrating biomass and carbon values with soil productivity loss in determining forest residue removals. *Biofuels*, **1**(4), pp. 539-546.

García-Frapolli, E., A. Schilmann, V.M. Berrueta, H. Riojas-Rodríguez, R.D. Edwards, M. Johnson, A. Guevara-Sanginés, C. Armendariz, and O. Masera (2010). Beyond fuelwood savings: Valuing the economic benefits of introducing improved biomass cookstoves in the Purépecha region of Mexico. *Ecological Economics*, **69**(12), pp. 2598-2605.

Garg, V.K. (1998). Interaction of tree crops with a sodic soil environment: potential for rehabilitation of degraded environments. *Land Degradation & Development*, **9**(1), pp. 81-93.

Garrison, T. (2008). *Essentials of Oceanography.* 5th ed., Brooks/Cole Cengage Learning, Belmont, CA, USA, 464 pp.

GBEP (2008). *A Review of the Current State of Bioenergy Development in G8+5 Countries.* Global Bioenergy Partnership (GBEP), Food and Agriculture Organization of the United Nations, Rome, Italy, 278 pp.

Gerasimov, Y., and T. Karjalainen (2009). *Estimation of supply and delivery cost of energy wood from Northwest Russia.* Working Papers of the Finnish Forest Research Institute Number 123, Finnish Forest Research Institute, Vantaa, Finland, 21 pp.

Gerbens-Leenes, P., and S. Nonhebel (2002). Consumption patterns and their effects on land required for food. *Ecological Economics*, **42**(1-2), pp. 185-199.

Gerbens-Leenes, W., A.Y. Hoekstra, and T.H. van der Meer (2009). The water footprint of bioenergy. *Proceedings of the National Academy of Sciences*, **106**(25), pp. 10219-10223.

Gerber, N. (2008). *Bioenergy and Rural Development in Developing Countries: A Review of Existing Studies.* Discussion Papers On Development Policy No. 122, Center for Development Research (ZEF), Bonn, Germany, 58 pp.

Gerten, D., W. Lucht, S. Schaphoff, W. Cramer, T. Hickler, and W. Wagner (2005). Hydrologic resilience of the terrestrial biosphere. *Geophysical Research Letters*, **32**(21), L21408.

Gibbs, H.K., M. Johnston, J.A. Foley, T. Holloway, C. Monfreda, N. Ramankutty, and D. Zaks (2008). Carbon payback times for crop-based biofuel expansion in the tropics: the effects of changing yield and technology. *Environmental Research Letters*, **3**(3), 034001 (10 pp.).

Gielen, D., J. Newman, and M.K. Patel (2008). Reducing industrial energy use and CO_2 emissions: The role of materials science. In: *MRS Bulletin Harnessing Materials for Energy*, **33**, pp. 471-477.

Gisladottir, G., and M. Stocking (2005). Land degradation control and its global environmental benefits. *Land Degradation & Development*, **16**(2), pp. 99-112.

Glass, R. (2006). Disease control priorities in developing countries. *The New England Journal of Medicine*, **355**(10), pp. 1074-1075.

Godfray, H.C.J., C. Toulmin, J.R. Beddington, I.R. Crute, L. Haddad, D. Lawrence, J.F. Muir, J. Pretty, S. Robinson, and S.M. Thomas (2010). Food security: The challenge of feeding 9 billion people. *Science*, **327**(5967), pp. 812-818.

Goldemberg, J., S.T. Coelho, P.M. Nastari, and L. O (2004). Ethanol learning curve - the Brazilian experience. *Biomass and Bioenergy*, **26**, pp. 301-304.

Gorissen, L., V. Buytaert, D. Cuypers, T. Dauwe, and L. Pelkmans (2010). Why the debate about land use change should not only focus on biofuels. *Environmental Science & Technology*, **44**(11), pp. 4046-4049.

Grann, H. (1997). The industrial symbiosis at Kalundborg, Denmark. In: *The Industrial Green Game. Implications for Environmental Design and Management.* D.J. Richards (ed.), National Academy Press, Washington, DC, USA, pp. 117-123.

Grassi, G., A. Nardi, and S. Vivarelli (2006). Low cost production of bioethanol from sweet sorghum. In: *Proceedings of the 14th European Biomass Conference*, Paris, France, 17-21 October 2005, pp. 91-95.

Green, R.E., S.J. Cornell, J.P.W. Scharlemann, and A. Balmford (2005). Farming and the fate of wild nature. *Science*, **307**(5709), pp. 550-555.

Gronowska, M., S. Joshi, and H.L. MacLean (2009). A review of U.S. and Canadian biomass supply studies. *BioResources*, **4**(1), pp. 341-369.

Grove, S. and J. Hanula (eds) (2006). *Insect Biodiversity and Dead Wood: Proceedings of the 22nd International Congress of Entomology.* Report SRS–93, United States Department of Agriculture, Asheville, NC, USA, 109 pp.

Guille, T. (2007). *Evaluation of the Potential Uses of Agricultural Residues for Energy Purposes.* Master's Thesis, Montpellier SupAgro, Paris, France.

Gunderson, P. (2008). Biofuels and North American agriculture - Implications for the health and safety of North American producers. *Journal of Agromedicine*, **13**(4), pp. 219-224.

Gurbuz, E.I., E.L. Kunkes, and J.A. Dumesic (2010). Dual-bed catalyst system for C–C coupling of biomass-derived oxygenated hydrocarbons to fuel-grade compounds. *Green Chemistry*, **12**(2), pp. 223-227.

Gustavsson, L., J. Holmberg, V. Dornburg, R. Sathre, T. Eggers, K. Mahapatra, and G. Marland (2007). Using biomass for climate change mitigation and oil use reduction. *Energy Policy*, **35**(11), pp. 5671-5691.

Haas, M.J., A.J. McAloon, W.C. Yee, and T.A. Foglia (2006). A process model to estimate biodiesel production costs. *Bioresource Technology*, **97**(4), pp. 671-678.

Haberl, H., K.H. Erb, F. Krausmann, V. Gaube, A. Bondeau, C. Plutzar, S. Gingrich, W. Lucht, and M. Fischer-Kowalski (2007). Quantifying and mapping the human appropriation of net primary production in earth's terrestrial ecosystems. *Proceedings of the National Academy of Sciences*, **104**(31), pp. 12942-12947.

Haberl, H., T. Beringer, S.C. Bhattacharya, K.-H. Erb, and M. Hoogwijk (2010). The global technical potential of bio-energy in 2050 considering sustainability constraints. *Current Opinion in Environmental Sustainability*, **2**(5-6), pp. 394-403.

Hakala, K., M. Kontturi, and K. Pahkala (2009). Field biomass as global energy source. *Agricultural and Food Science*, **18**, pp. 347-365.

Hamelinck, C.N. (2004). *Outlook for Advanced Biofuels.* Master's Thesis, University of Utrecht, Utrecht, The Netherlands, 232 pp.

Hamelinck, C.N., and A.P.C. Faaij (2002). Future prospects for production of methanol and hydrogen from biomass. *Journal of Power Sources*, **111**(1), pp. 1-22.

Hamelinck, C.N., and A.P.C. Faaij (2006). Outlook for advanced biofuels. *Energy Policy*, **34**(17), pp. 3268-3283.

Hamelinck, C.N., A.P.C. Faaij, H. den Uil, and H. Boerrigter (2004). Production of FT transportation fuels from biomass; technical options, process analysis and optimisation, and development potential. *Energy*, **29**(11), pp. 1743-1771.

Hamelinck, C.N., G.v. Hooijdonk, and A.P.C. Faaij (2005a). Ethanol from lignocellulosic biomass: techno-economic performance in short-, middle- and long-term. *Biomass and Bioenergy*, **28**(4), pp. 384-410.

Hamelinck, C.N., R.A.A. Suurs, and A.P.C. Faaij (2005b). International bioenergy transport costs and energy balance. *Biomass and Bioenergy*, **29**(2), pp. 114-134.

Hartley, M.J. (2002). Rationale and methods for conserving biodiversity in plantation forests. *Forest Ecology and Management*, **155**(1-3), pp. 81-95.

Headey, D., and S. Fan (2008). Anatomy of a crisis: the causes and consequences of surging food prices. *Agricultural Economics*, **39**, pp. 375-391.

Heggenstaller, A.H., R.P. Anex, M. Liebman, D.N. Sundberg, and L.R. Gibson (2008). Productivity and nutrient dynamics in bioenergy double-cropping systems. *Agronomy Journal*, **100**(6), pp. 1740-1748.

Heinimö, J., and M. Junginger (2009). Production and trading of biomass for energy – An overview of the global status. *Biomass and Bioenergy*, **33**(9), pp. 1310-1320.

Helynen, S., M. Flyktman, T. Mäkinen, K. Sipilä, and P. Vesterinen (2002). *The possibilities of bioenergy in reducing greenhouse gases.* VTT Tiedotteita 2145, VTT Technical Research Centre of Finland, Espoo, Finland, 110 pp.

Hemakanthi, D.A., and D.N. Heller (2010). Multiclass, multiresidue method for the detection of antibiotic residues in distillers grains by liquid chromatography and ion trap tandem mass spectrometry. *Journal of Chromatography A*, **1217**(18), pp. 3076-3084.

Herrero, M., J. van de Steeg, J. Lynam, P.P. Rao, S. Macmillan, B. Gerard, J. McDermott, C. Sere, M. Rosegrant, P.K. Thornton, A.M. Notenbaert, S. Wood, S. Msangi, H.A. Freeman, D. Bossio, J. Dixon, and M. Peters (2010). Smart investments in sustainable food production: Revisiting mixed crop-livestock systems. *Science*, **327**(5967), pp. 822-825.

Hertel, T.W., A.A. Golub, A.D. Jones, M. O'Hare, R.J. Plevin, and D.M. Kammen (2010a). Effects of US maize ethanol on global land use and greenhouse gas emissions: Estimating market mediated responses. *BioScience*, **60**(3), pp. 223-231.

Hertel, T.W., W.E. Tyner, and D.K. Birur (2010b). Global impacts of biofuels. *Energy Journal*, **31**(1), pp. 35-100.

Hettinga, W.G., H.M. Junginger, M. Hoogwijk, A. McAloon, and T. Hickler (2007). Technological learning in U.S. ethanol production. In: *15th European Biomass Conference and Exhibition*, Berlin, Germany, 7-11 May 2007.

Hettinga, W.G., H.M. Junginger, S.C. Dekker, M. Hoogwijk, A.J. McAloon, and K.B. Hicks (2009). Understanding the reductions in US corn ethanol production costs: An experience curve approach. *Energy Policy*, **37**(1), pp. 190-203.

Hiederer, R., F. Ramos, C. Capitani, R. Koeble, V. Blujdea, O. Gomez, D. Mulligan, and L. Marelli (2010). *Biofuels: A New Methodology to Estimate GHG Emissions from Global Land Use Change, A methodology involving spatial allocation of agricultural land demand and estimation of CO_2 and N_2O emissions.* EUR 24483 EN - 2010, Joint Research Center, European Commission, Ispa, Italy, 168 pp.

Hill, J. (2007). Environmental costs and benefits of transportation biofuel production from food- and lignocellulose-based energy crops. A review. *Agronomy for Sustainable Development*, **27**(1), pp. 1-12.

Hill, J., E. Nelson, D. Tilman, S. Polasky, and D. Tiffany (2006). Environmental, economic, and energetic costs and benefits of biodiesel and ethanol biofuels. *Proceedings of the National Academy of Sciences*, **103**(30), pp. 11206-11210.

Hill, J., S. Polasky, E. Nelson, D. Tilman, H. Huo, L. Ludwig, J. Neumann, H. Zheng, and D. Bonta (2009). Climate change and health costs of air emissions from biofuels and gasoline. *Proceedings of the National Academy of Sciences*, **106**(6), pp. 2077-2082.

Hillier, J., C. Whittaker, G. Dailey, M. Aylott, E. Casella, G.M. Richter, A. Riche, R. Murphy, G. Taylor, and P. Smith (2009). Greenhouse gas emissions from four bioenergy crops in England and Wales: Integrating spatial estimates of yield and soil carbon balance in life cycle analyses. *Global Change Biology Bioenergy*, **1**(4), pp. 267-281.

Hilman, H.J., E. Gidwani, P. Morris, S. Sagar, and S. Chowdhary (2007). *Using Microfinance to Expand Access to Energy Services: The Emerging Experiences in Asia of Self-Employed Women's Association Bank (SEWA), Sarvodaya Economic Enterprise Development Services (SEEDS), Nirdhan Utthan Bank Limited (NUBL), and AMRET.* The SEEP Network, Washington, DC, USA, 124 pp.

Himmel, M.E., Q. Xu, Y. Luo, S.-Y. Ding, R. Lamed, and E.A. Bayer (2010). Microbial enzyme systems for biomass conversion: emerging paradigms. *Biofuels*, **1**(2), pp. 323-341.

Hoefnagels, R., E. Smeets, and A. Faaij (2010). Greenhouse gas footprints of different biofuel production systems. *Renewable and Sustainable Energy Reviews*, **14**(7), pp. 1661-1694.

Hoekstra, A.Y., P.W. Gerbens-Leenes, and T.H. Van der Meer (2010). The water footprint of bio-energy. In: *Climate Change and Water: International Perspectives on Mitigation and Adaptation.* C. Howe, J.B. Smith, and J. Henderson (eds.), American Water Works Association, IWA Publishing, London, UK, pp. 81-95.

Hollebone, B.P., and Z. Yang (2009). Biofuels in the environment: A review of behaviours, fates, effects and possible remediation techniques. In: *Proceedings of the 32nd Arctic and Marine Oil Spill Program (AMOP) Technical Seminar on Environmental Contamination and Response*, Emergencies Science Division, Environment Canada, Ottawa, Canada, pp. 127-139.

Holmgren, J., R. Marinangeli, P. Nair, D. Elliott, and R. Bain (2008). Consider upgrading pyrolysis oils into renewable fuels. *Hydrocarbon Processing*, **87**(9), pp. 95-113.

Hoogwijk, M. (2004). *On the Global and Regional Potential of Renewable Energy Sources*. PhD Thesis, Utrecht University, Utrecht, The Netherlands, 257 pp.

Hoogwijk, M., A. Faaij, R. van den Broek, G. Berndes, D. Gielen, and W. Turkenburg (2003). Exploration of the ranges of the global potential of biomass for energy. *Biomass and Bioenergy*, **25**(2), pp. 119-133.

Hoogwijk, M., A. Faaij, B. Eickhout, B. de Vries, and W. Turkenburg (2005). Potential of biomass energy out to 2100, for four IPCC SRES land-use scenarios. *Biomass and Bioenergy*, **29**(4), pp. 225-257.

Hoogwijk, M., A. Faaij, B. de Vries, and W. Turkenburg (2009). Exploration of regional and global cost-supply curves of biomass energy from short-rotation crops at abandoned cropland and rest land under four IPCC SRES land-use scenarios. *Biomass and Bioenergy*, **33**(1), pp. 26-43.

Hooijer, A., M. Silvius, H. Wösten, and S. Page (2006). *PEAT- CO_2 Assessment of CO_2 Emissions from Drained Peatlands in SE Asia.* Delft Hydraulics report Q3943, WL Delft Hydraulics, Delft, The Netherlands, 41 pp.

Howard, D., and S. Ziller (2008). Alien alert – plants for biofuel may be invasive. *Bioenergy Business*, July/August, pp. 14-16.

Howarth, R.W., S. Bringezu, L.A. Martinelli, R. Santoro, D. Messem, and O.E. Sala (2009). Introduction: biofuels and the environment in the 21st century. In: *Biofuels: Environmental Consequences and Interactions with Changing Land Use.* R.W. Howarth and S. Bringezu (eds.), Proceedings of the SCOPE - Scientific Committee on Problems of the Environment International Biofuels Project Rapid Assessment, Gummersbach, Germany, 22-25 September 2009, pp. 15-36.

Hsu, D.D., D. Inman, G.A. Heath, E.J. Wolfrum, M.K. Mann, and A. Aden (2010). Life cycle environmental impacts of selected U.S. ethanol production and use pathways in 2022. *Environmental Science & Technology*, **44**, pp. 5289-5297.

Hu, Q., M. Sommerfeld, E. Jarvis, M. Ghirardi, M. Posewitz, M. Seibert, and A. Darzins (2008). Microalgal triacylglycerols as feedstocks for biofuel production: perspectives and advances. *The Plant Journal*, **54**, pp. 621-639.

Huber, G.W., R.D. Cortright, and J.A. Dumesic (2004). Renewable alkanes by aqueous-phase reforming of biomass-derived oxygenates. *Angewandte Chemie*, **116**(12), pp. 1575-1577.

Huber, G.W., J.N. Chheda, C.J. Barrett, and J.A. Dumesic (2005). Production of liquid alkanes by aqueous-phase processing of biomass-derived carbohydrates. *Science*, **308**(5727), pp. 1446-1450.

Huber, G.W., S. Iborra, and A. Corma (2006). Synthesis of transportation fuels from biomass: chemistry, catalysts, and engineering. *Chemical Reviews*, **106**(9), pp. 4044-4098.

Huertas, D.A., G. Berndes, M. Holmen, and G. Sparovek (2010). Sustainability certification of bioethanol. How is it perceived by Brazilian stakeholders? *Biomass and Bioenergy*, **4**(4), pp. 369-384.

Huo, H., M. Wang, C. Bloyd, and V. Putsche (2009). Life-cycle assessment of energy use and greenhouse gas emissions of soybean-derived biodiesel and renewable fuels. *Environmental Science & Technology*, **43**(3), pp. 750-756.

IAASTD (2009). *Agriculture at a Crossroads.* International Assessment of Agricultural Knowledge, Science and Technology for Development, Washington, DC, USA, 606 pp.

IATA (2009). *Report on Alternative Fuels.* International Air Transport Association, Montreal, Canada, 92 pp.

Ibsen, K., R. Wallace, S. Jones, and T. Werpy (2005). *Evaluating Progressive Technology Scenarios in the Development of the Advanced Dry Mill Biorefinery.* FY05-630, National Renewable Energy Laboratory, Golden, CO, USA, 35 pp.

IEA (2000). *Experience Curves for Energy Technology Policy.* International Energy Agency, Paris, France, 133 pp.

IEA (2006). *World Energy Outlook 2006.* International Energy Agency, Paris, France, 601 pp.

IEA (2007a). *World Energy Outlook 2007.* International Energy Agency, Paris, France, 600 pp.

IEA (2007b). *Biomass for Power Generation and CHP.* Energy Technology Essentials ETE03, International Energy Agency, Paris, France, 4pp.

IEA (2008a). *Energy Technology Perspectives 2008. Scenarios and Strategies to 2050.* International Energy Agency, Paris, France, 646 pp.

IEA (2008b). *World Energy Outlook 2008.* International Energy Agency, Paris, France, 578 pp.

IEA (2009). *World Energy Outlook 2009.* International Energy Agency, Paris, France, 696 pp.

IEA (2010a). *World Energy Statistics 2010.* International Energy Agency, Paris, France.[1]

IEA (2010b). *World Energy Outlook 2010.* International Energy Agency, Paris, France, 736 pp.

IEA (2010c). *Renewables Information 2010 with 2009 Data.* International Energy Agency, Paris, France, 428 pp (ISBN 978-92-64-08416-2).

IEA (2011). *Technology Roadmap: Biofuels for Transport.* International Energy Agency, Renewable Energy Division, Paris, France, 52 pp.

IEA Bioenergy (2007). *Potential Contribution of Bioenergy to the World's Future Energy Demand.* IEA Bioenergy: ExCo: 2007:02, 12 pp.

---

1 Combination of four IEA publications: IEA energy balances and statistics databases for most of the OECD countries for the years 1960 to 2008, with supply estimates for 2009, and energy statistics and balances for more than 100 non-OECD countries for the years 1971 to 2008.

IEA Bioenergy (2009). *Bioenergy: A Sustainable and Reliable Energy Source. Main Report.* IEA Bioenergy: ExCo:2009:06, 108 pp.

IEA Bioenergy (2010). *Algae – The Future for Bioenergy?* Summary and conclusions from the IEA Bioenergy ExCo64 Workshop. IEA Bioenergy: ExCo: 2010:02, 16 pp.

IEA Renewable Energy Division (2010). *Sustainable Production of Second-Generation Biofuels: Potential and Perspectives in Major Economies and Developing Countries.* International Energy Agency, Paris, France, 217 pp.

IFPRI (2008). *High Food Prices: The What, Who, and How of Proposed Policy Actions.* International Food Policy Research Institute, Washington, DC, USA, 12 pp.

Ileleji, K.E., S. Sokhansanj, and J.S. Cundiff (2010). Farm-gate to plant-gate delivery of lignocellulosic feedstocks from plant biomass for biofuel production. In: *Biofuels from Agricultural Wastes and Byproducts.* H.P. Blaschek, T. Ezeji, and J. Scheffran (eds.), Wiley-Blackwell, Oxford, UK, pp. 117-159.

Imhoff, M.L., L. Bounoua, T. Ricketts, C. Loucks, R. Harriss, and W.T. Lawrence (2004). Global patterns in human consumption of net primary production. *Nature*, **429**(6994), pp. 870-873.

IPCC (1996). *Climate Change 1995: Impacts, Adaptation, and Mitigation: Scientific-Technical Analyses.* R.T. Watson, M.C. Zinyowera, and R.H. Moss (eds.), Cambridge University Press, 878 pp.

IPCC (2000a). *Special Report on Emissions Scenarios.* N. Nakicenovic and R. Swart (eds.), Cambridge University Press, 570 pp.

IPCC (2000b). *Land Use, Land Use Change and Forestry.* R.T. Watson, I.R. Noble, B. Bolin, N.H. Ravindranath, D.J. Verardo and D.J. Dokken (eds.), Cambridge University Press, 375 pp.

IPCC (2005). *Special Report on Carbon Dioxide Capture and Storage.* B. Metz, O. Davidson, H. de Coninck, M. Loos, and L. Meyer (eds.), Cambridge University Press, 431 pp.

IPCC (2007a). *Climate Change 2007: Impacts, Adaptation and Vulnerability. Contribution of Working Group II to the Fourth Assessment Report of the Intergovernmental Panel on Climate Change.* M.L. Parry, O.F. Canziani, J.P. Palutikof, P.J. van der Linden, and C.E. Hanson (eds.),Cambridge University Press, 979 pp.

IPCC (2007b). *Climate Change 2007: Synthesis Report. Contributions of Working Groups I, II and II to the Fourth Assessment Report of the Intergovernmental Panel on Climate Change.* Core Writing Team, R.K. Pachauri, and A. Reisinger (eds.), Cambridge University Press, 104 pp.

IPCC (2007c). *Climate Change 2007: The Physical Science Basis. Contribution of Working Group I to the Fourth Assessment Report of the Intergovernmental Panel on Climate Change.* S. Solomon, D. Qin, M. Manning, Z. Chen, M. Marquis, K.B. Averyt, M. Tignor, and H.L. Miller (eds.), Cambridge University Press, 996 pp.

IPCC (2007d). *Climate Change 2007: Mitigation of Climate Change. Contribution of Working Group III to the Fourth Assessment Report of the Intergovernmental Panel on Climate Change.* B. Metz, O.R. Davidson, P.R. Bosch, R. Dave, and L.A. Meyer (eds.), Cambridge University Press, 851 pp.

IUCN (2009). *Guidelines on Biofuels and Invasive Species.* IUCN, Gland, Switzerland, 20 pp. (ISBN: 978-2-8317-1222-2).

Ivanic, M., and W. Martin (2008). Implications of higher global food prices for poverty in low-income countries. *Agricultural Economics*, **39**, pp. 405-416.

Jackson, R.B., B.C. Murray, E.G. Jobbagy, R. Avissar, S.B. Roy, D.J. Barrett, C.W. Cook, K.A. Farley, D.C. le Maitre, and B.A. McCarl (2005). Trading water for carbon with biological carbon sequestration. *Science*, **310**(5756), pp. 1944-1947.

Jacobsen, N.B. (2006). Industrial symbiosis in Kalundborg, Denmark: A quantitative assessment of economic and environmental aspects. *Journal of Industrial Ecology*, **10**(1-2), pp. 239-255.

Jaggard, K.W., A. Qi, and E.S. Ober (2010). Possible changes to arable crop yields by 2050. *Philosophical Transactions of the Royal Society B: Biological Sciences*, **365**(1554), pp. 2835-2851.

Jauhiainen, J., S. Limin, H. Silvennoinen, and H. Vasander (2008). Carbon dioxide and methane fluxes in drained tropical peat before and after hydrological restoration. *Ecology*, **89**(12), pp. 3503-3514.

Jechura, J. (2005). *Dry Mill Cost-By-Area: ASPEN Case Summary.* National Renewable Energy Laboratory, Golden, CO, USA, 2 pp.

Jeffries, T.W. (2006). Engineering yeasts for xylose metabolism. *Current Opinion in Biotechnology*, **17**(3), pp. 320-326.

Jeffries, T.W., I.V. Grigoriev, J. Grimwood, J.M. Laplaza, A. Aerts, A. Salamov, J. Schmutz, E. Lindquist, P. Dehal, H. Shapiro, Y.-S. Jin, V. Passoth, and P.M. Richardson (2007). Genome sequence of the lignocellulose-bioconverting and xylose-fermenting yeast Pichia stipitis. *Nature Biotechnology*, **25**(3), pp. 319-326.

Jensen, E.S. (1996). Grain yield, symbiotic N_2 fixation and interspecific competition for inorganic N in pea-barley intercrops. *Plant and Soil*, **182**(1), pp. 25-38.

Jetter, J.J., and P. Kariher (2009). Solid-fuel household cook stoves: Characterization of performance and emissions. *Biomass and Bioenergy*, **33**(2), pp. 294-305.

Johansson, D., and C. Azar (2007). A scenario based analysis of land competition between food and bioenergy production in the US. *Climatic Change*, **82**(3), pp. 267-291.

Johnson, J.M.F., D. Reicosky, B. Sharratt, M. Lindstrom, W. Voorhees, and L. Carpenter-Boggs (2004). Characterization of soil amended with the by-product of corn stover fermentation. *Soil Science Society of America Journal*, **68**(1), pp. 139-147.

Johnson, M., R. Edwards, A. Ghilardi, V. Berrueta, D. Gillen, C.A. Frenk, and O. Masera (2009). Quantification of carbon savings from improved biomass cookstove projects. *Environmental Science & Technology*, **43**(7), pp. 2456-2462.

Johnson, T.M., C. Alatorre, Z. Romo, and F. Liu (2009). *Low-Carbon Development for Mexico.* The International Bank for Reconstruction and Development, The World Bank, Washington, DC, USA.

Johnston, M., and T. Holloway (2007). A global comparison of national biodiesel production potentials. *Environmental Science & Technology*, **41**(23), pp. 7967-7973.

Johnston, M., J.A. Foley, T. Holloway, C. Kucharik, and C. Monfreda (2009). Resetting global expectations from agricultural biofuels. *Environmental Research Letters*, **4**(1), 014004.

Jongschaap, R.E.E., W.J. Corré, P.S. Bindraban, and W.A. Brandenburg (2007). *Claims and Facts on Jatropha curcas L. Global Jatropha curcas Evaluation, Breeding And Propagation Programme.* Report 158, Plant Research International BV, Wageningen, The Netherlands and Stichting Het Groene Woudt, Laren, The Netherlands, 66 pp.

Junginger, M. (2007). *Lessons from (European) Bioenergy Policies; Results of a Literature Review for IEA Bioenergy Task 40.* Utrecht University, Utrecht, Netherlands, 14 pp. Available at: www.bioenergytrade.org/downloads/junginerlessonsfromeuropeanbioenergypolicies.pdf.

Junginger, M., A. Faaij, R. van den Broek, A. Koopmans, and W. Hulscher (2001). Fuel supply strategies for large-scale bio-energy projects in developing countries. Electricity generation from agricultural and forest residues in Northeastern Thailand. *Biomass and Bioenergy*, **21**(4), pp. 259-275.

Junginger, M., A. Faaij, R. Björheden, and W.C. Turkenburg (2005). Technological learning and cost reductions in wood fuel supply chains in Sweden. *Biomass and Bioenergy*, **29**(6), pp. 399-418.

Junginger, M., E. de Visser, K. Hjort-Gregersen, J. Koornneef, R. Raven, A. Faaij, and W. Turkenburg (2006). Technological learning in bioenergy systems. *Energy Policy*, **34**(18), pp. 4024-4041.

Junginger, M., T. Bolkesjø, D. Bradley, P. Dolzan, A. Faaij, J. Heinimö, B. Hektor, Ø. Leistad, E. Ling, M. Perry, E. Piacente, F. Rosillo-Calle, Y. Ryckmans, P.-P. Schouwenberg, B. Solberg, E. Trømborg, A.d.S. Walter, and M.d. Wit (2008). Developments in international bioenergy trade. *Biomass and Bioenergy*, **32**(8), pp. 717-729.

Junginger, M., A. Faaij, J. van Dam, S. Zarrilli, A. Mohammed, and D. Marchal (2010). *Opportunities and Barriers for International Bioenergy Trade and Strategies to Overcome Them.* IEA Bioenergy Task 40, 17 pp. Available at: www.bioenergytrade.org/downloads/t40opportunitiesandbarriersforbioenergytradefi.pdf.

Jungk, N., and G. Reinhardt (2000). *Landwirtschaftliche Referenzsysteme in Ökologischen Bilanzierungen: eine Basis Analyse.* Institute für Energie- und Umweltforschung, Heidelberg, Germany.

Kaliyan, N., R.V. Morey, and D.G. Tiffany (2010). Reducing life cycle greenhouse gas emissions of corn ethanol by integrating biomass to produce heat and power at ethanol plants. *Biomass and Bioenergy*, **35**(3), pp. 1103-1113.

Kalnes, T.N., K.P. Koers, T. Marker, and D.R. Shonnard (2009). A technoeconomic and environmental life cycle comparison of green diesel to biodiesel and syndiesel. *Environmental Progress & Sustainable Energy*, **28**(1), pp. 111-120.

Kamm, B., P.R. Gruber, and M. Kamm (eds.) (2006). *Biorefineries - Industrial Processes and Products: Status Quo and Future Directions.* Wiley-VCH, Weinheim, UK, 949 pp.

Kang, S., M. Khanna, J. Scheffran, D. Zilberman, H. Önal, Y. Ouyang, and Ü.D. Tursun (2010). Optimizing the biofuels infrastructure: Transportation networks and biorefinery locations in Illinois. In: *Handbook of Bioenergy Economics and Policy.* M. Khanna, J. Scheffran, and D. Zilberman (eds.), Springer New York, New York, NY, USA, pp. 151-173.

Karekezi, S., and L. Majoro (2002). Improving modern energy services for Africa's urban poor. *Energy Policy*, **30**(11-12), pp. 1015-1028.

Karekezi, S., and P. Turyareeba (1995). Woodstove dissemination in Eastern Africa - a review. *Energy for Sustainable Development*, **1**(6), pp. 12-19.

Karlen, D.L. (2010). Corn stover feedstock trials to support predictive modeling. *Global Change Biology Bioenergy*, **2**(5), pp. 235-247.

Karp, A., and I. Shield (2008). Bioenergy from plants and the sustainable yield challenge. *New Phytologist*, **179**(1), pp. 15-32.

Kartha, S., P. Hazel, and R.K. Pachauri (2006). Environmental effects of bioenergy. In: *Bioenergy and Agriculture: Promises and Challenges.* P.B.R. Hazell and R.K. Pachauri (eds.), International Food Policy Research Institute, Washington, D.C., USA, 2 pp.

Kazi, F.K., J.A. Fortman, R.P. Anex, D.D. Hsu, A. Aden, A. Dutta, and G. Kothandaraman (2010). Techno-economic comparison of process technologies for biochemical ethanol production from corn stover. *Fuel*, **89**(Supplement 1), pp. S20-S28.

Keasling, J.D., and H. Chou (2008). Metabolic engineering delivers next-generation biofuels. *Nature Biotechnology*, **26**(3), pp. 298-299.

Keeney, D., and T.W. Hertel (2008). *Yield Response to Prices: Implications for Policy Modeling.* Working paper 08/13, Department of Agricultural Economics, Purdue University, West Lafayette, IN, USA, 37 pp.

Keeney, D., and T.W. Hertel (2009). The indirect land use impacts of United States biofuel policies: The importance of acreage, yield, and bilateral trade responses. *American Journal of Agricultural Economics*, **91**, pp. 895-909.

Keeney, D., and M. Muller (2006). *Water Use by Ethanol Plants: Potential Challenges.* Institute for Agriculture and Trade Policy, Minneapolis, MN, USA, 7 pp.

Keeney, R., and T. Hertel (2005). *GTAP-AGR : A Framework for Assessing the Implications of Multilateral Changes in Agricultural Policies.* GTAP Technical Paper No.24, Center for Global Trade Analysis, Department of Agricultural Economics, Purdue University, West Lafayette, IN, USA, 61 pp.

Kendall, A., B. Chang, and B. Sharpe (2009). Accounting for time-dependent effects in biofuel life cycle greenhouse gas emissions calculations. *Environmental Science & Technology*, **43**(18), pp. 7142-7147.

Keys, E., and W. McConnell (2005). Global change and the intensification of agriculture in the tropics. *Global Environmental Change*, **15**(4), pp. 320-337.

Kheshgi, H.S., R.C. Prince, and G. Marland (2000). The potential of biomass fuels in the context of global climate change: Focus on transportation fuels. *Annual Review of Energy and the Environment*, **25**(1), pp. 199-244.

Kim, H., S. Kim, and B.E. Dale (2009). Biofuels, land use change, and greenhouse gas emissions: Some unexplored variables. *Environmental Science & Technology*, **43**(3), pp. 961-967.

Kim, S., and B. Dale (2002). Allocation procedure in ethanol production system from corn grain i. system expansion. *The International Journal of Life Cycle Assessment*, **7**(4), pp. 237-243.

Kim, S., and B.E. Dale (2004). Global potential bioethanol production from wasted crops and crop residues. *Biomass and Bioenergy*, **26**(4), pp. 361-375.

Kinchin, C. and R.L. Bain (2009). *Hydrogen Production from Biomass via Indirect Gasification: The Impact of NREL Process Development Unit Gasifier Correlations.* NREL Report No. TP-510-44868. National Renewable Energy Laboratory, Golden, CO, USA, 27 pp.

King, C.W., and M.E. Webber (2008). Water intensity of transportation. *Environmental Science & Technology*, **42**(21), pp. 7866-7872.

King, D., O.R. Inderwildi, and A. Williams (2010). *The Future of Industrial Biorefineries.* 210610, World Economic Forum White Paper, Cologn/Geneva, Switzerland, 40 pp.

Kirkels, A., and G. Verbong (2011). Biomass gasification: Still promising? A 30-year global overview. *Renewable and Sustainable Energy Reviews*, **15**(1), pp. 471-481.

Kirkinen, J., A. Sahay, and I. Savolainen (2009). Greenhouse impact of fossil, forest residues and Jatropha diesel: a static and dynamic assessment. *Progress in Industrial Ecology, An International Journal*, **6**, pp. 185-206.

Kirschbaum, M.U.F. (2003). To sink or burn? A discussion of the potential contributions of forests to greenhouse gas balances through storing carbon or providing biofuels. *Biomass and Bioenergy*, **24**(4-5), pp. 297-310.

Kirschbaum, M.U.F. (2006). Temporary carbon sequestration cannot prevent climate change. *Mitigation and Adaptation Strategies for Global Change*, **11**(5), pp. 1151-1164.

Kishore, V.V.N., P.M. Bhandari, and P. Gupta (2004). Biomass energy technologies for rural infrastructure and village power – opportunities and challenges in the context of global climate change concerns. *Energy Policy*, **32**(6), pp. 801-810.

Kituyi, E. (2004). Towards sustainable production and use of charcoal in Kenya: exploring the potential in life cycle management approach. *Journal of Cleaner Production*, **12**(8-10), pp. 1047-1057.

Kline, K.L., and V.H. Dale (2008). Biofuels: Effects on land and fire. *Science*, **321**(5886), pp. 199-201.

Kline, K.L., G. Oladosu, A. Wolfe, R.D. Perlack, and M. McMahon (2007). *Biofuel Feedstock Assessment for Selected Countries.* ORNL/TM-2007/224, Oak Ridge National Laboratory, Oak Ridge, TN, USA, 243 pp.

Knothe, G. (2010). Biodiesel and renewable diesel: a comparison. *Progress in Energy and Combustion Science*, **36**(3), pp. 364-373.

Knowler, D.J. (2004). The economics of soil productivity: local, national and global perspectives. *Land Degradation & Development*, **15**(6), pp. 543-561.

Koh, L.P., and J. Ghazoul (2008). Biofuels, biodiversity, and people: Understanding the conflicts and finding opportunities. *Biological Conservation*, **141**(10), pp. 2450-2460.

Köhlin, G., and M. Ostwald (2001). Impact of plantations on forest use and forest status in Orissa, India. *AMBIO*, **30**(1), pp. 37-42.

Koizumi, T., and K. Ohga (2008). Biofuels policies in Asian countries: Impact of the expanded biofuels programs on world agricultural markets. *Journal of Agricultural & Food Industrial Organization*, **5**(2), Article 8, 22 pp.

Koning, N. (2008). Long-term global availability of food: continued abundance or new scarcity? *NJAS - Wageningen Journal of Life Sciences*, **55**(3), pp. 229-292.

Körner, C., J.A. Morgan, and R.J. Norby (2007). CO_2 fertilization: When, where and how much? In: *Terrestrial ecosystems in a changing world.* Springer, Berlin, Germany, pp. 9-22.

Koukouzas, N., A. Katsiadakis, E. Karlopoulos, and E. Kakaras (2008). Co-gasification of solid waste and lignite - A case study for Western Macedonia. *Waste Management*, **28**(7), pp. 1263-1275.

Krausmann, F., K.-H. Erb, S. Gingrich, C. Lauk, and H. Haberl (2008). Global patterns of socioeconomic biomass flows in the year 2000: A comprehensive assessment of supply, consumption and constraints. *Ecological Economics*, **65**(3), pp. 471-487.

Kreutz, T.G., E.D. Larson, G. Liu, and R.H. Williams (2008). Fischer-Tropsch fuels from coal and biomass. In: *Proceedings of the 25th International Pittsburgh Coal Conference.* Princeton Environmental Institute, Pittsburgh, PA, 29 September – 2 October 2008.

Krewitt, W., K. Nienhaus, C. Kleßmann, C. Capone, E. Stricker, W. Graus, M. Hoogwijk, N. Supersberger, U. von Winterfeld, and S. Samadi (2009). *Role and Potential of Renewable Energy and Energy Efficiency for Global Energy Supply.* ISSN 1862-4359, Federal Environment Agency, Dessau-Roßlau, Germany, 336 pp.

Krey, V., and L. Clarke (2011). Role of renewable energy in climate change mitigation: A synthesis of recent scenarios. *Climate Policy*, in press.

Krich, K., D. Augenstein, J.P. Batmale, J. Benemann, B. Rutledge, and D. Salour (2005). *Biomethane from Dairy: A Sourcebook for the Production and Use of Renewable Natural Gas in California.* Report prepared for Western United Dairymen with funding from USDA Rural Development, Washington, DC, USA, 282 pp. Available at: www.suscon.org/cowpower/biomethaneSourcebook/Full_Report.pdf.

Kumar, B.M. (2006). Agroforestry: the new old paradigm for Asian food security. *Journal of Tropical Agriculture*, **44**(1-2), pp. 1-14.

Kumar, L.N.V., and S. Maithel (2007). Alternative feedstock for Bio-ethanol production in India. In: *Biofuels: Towards a Greener and Secure Energy Future.* P.P. Bhojvaid (ed.), The Energy Resources Institute Press (TERI Press), New Delhi, India, pp. 89-104.

Kurz, W.A., G. Stinson, and G. Rampley (2008a). Could increased boreal forest ecosystem productivity offset carbon losses from increased disturbances? *Philosophical Transactions of the Royal Society B: Biological Sciences*, **363**(1501), pp. 2259-2268.

Kurz, W.A., C.C. Dymond, G. Stinson, G.J. Rampley, E.T. Neilson, A.L. Carroll, T. Ebata, and L. Safranyik (2008b). Mountain pine beetle and forest carbon feedback to climate change. *Nature*, **452**(7190), pp. 987-990.

Kuuva, K., and L. Ruska (2009). *Final Report of Feed-in Tariff Task Force.* 59/2009, Ministry of Employment and the Economy of Finland, Helsinki, Finland, 101 pp.

Laird, D.A. (2008). The charcoal vision: A win-win-win scenario for simultaneously producing bioenergy, permanently sequestering carbon, while improving soil and water quality. *Agronomy Journal*, **100**(1), pp. 178-181.

Laird, D.A., R.C. Brown, J.E. Amonette, and J. Lehmann (2009). Review of the pyrolysis platform for coproducing bio-oil and biochar. *Biofuels, Bioproducts and Biorefining*, **3**(5), pp. 547-562.

Lal, R. (2003). Offsetting global CO_2 emissions by restoration of degraded soils and intensification of world agriculture and forestry. *Land Degradation & Development*, **14**(3), pp. 309-322.

Lal, R. (2004). Soil carbon sequestration impacts on global climate change and food security. *Science*, **304**(5677), pp. 1623-1627.

Lal, R. (2005). World crop residues production and implications of its use as a biofuel. *Environment International*, **31**(4), pp. 575-584.

Lal, R. (2008). Crop residues as soil amendments and feedstock for bioethanol production. *Waste Management*, **28**(4), pp. 747-758.

Lal, R., and D. Pimentel (2007). Bio-fuels from crop residues. *Soil & Tillage Research*, **93**(2), pp. 237-238.

Lamers, P., C. Hamelinck, M. Junginger, and A. Faaij (2011). International Bioenergy, Trade - A Review of Past Developments in the Liquid Biofuels Market. *Renewable and Sustainable Energy Reviews*, **15**(6), pp. 2655-2676.

Lapola, D.M., R. Schaldach, J. Alcamo, A. Bondeau, J. Koch, C. Koelking, and J.A. Priess (2010). Indirect land-use changes can overcome carbon savings from biofuels in Brazil. *Proceedings of the National Academy of Sciences*, **107**(8), pp. 3388-3393.

Larson, E. (2006). A review of life-cycle analysis studies on liquid biofuel systems for the transport sector. *Energy for Sustainable Development*, **10**(2), pp. 109-126.

Larson, E.D., G. Fiorese, G. Liu, R.H. Williams, T.G. Kreutz, and S. Consonni (2009). Co-production of synfuels and electricity from coal + biomass with zero net carbon emissions: An Illinois case study. *Energy Procedia*, **1**(1), pp. 4371-4378.

Larson, E.D., G. Fiorese, G. Liu, R.H. Williams, T.G. Kreutz, and S. Consonni (2010). Co-production of decarbonized synfuels and electricity from coal + biomass with CO_2 capture and storage: an Illinois case study. *Energy & Environmental Science*, **3**(1), pp. 28-42.

Laser, M., E. Larson, B. Dale, M. Wang, N. Greene, and L.R. Lynd (2009). Comparative analysis of efficiency, environmental impact, and process economics for mature biomass refining scenarios. *Biofuels, Bioproducts and Biorefining*, **3**(2), pp. 247-270.

Lattimore, B., C.T. Smith, B.D. Titus, I. Stupak, and G. Egnell (2009). Environmental factors in woodfuel production: Opportunities, risks, and criteria and indicators for sustainable practices. *Biomass and Bioenergy*, **33**(10), pp. 1321-1342.

Laurance, W.F. (2007). Switch to corn promotes Amazon deforestation. *Science*, **318**(5857), pp. 1721-1724.

Lawrence, C.J., and V. Walbot (2007). Translational genomics for bioenergy production from fuelstock grasses: Maize as the model species. *Plant Cell*, **19**(7), pp. 2091-2094.

Lee, D.R., C.B. Barrett, and J.G. McPeak (2006). Policy, technology, and management strategies for achieving sustainable agricultural intensification. *Agricultural Economics*, **34**(2), pp. 123-127.

Lee, J. (1997). Biological conversion of lignocellulosic biomass to ethanol. *Journal of Biotechnology*, **56**(1), pp. 1-24.

Lee, S.K., H. Chou, T.S. Ham, T.S. Lee, and J.D. Keasling (2008). Metabolic engineering of microorganisms for biofuels production: from bugs to synthetic biology to fuels. *Current Opinion in Biotechnology*, **19**(6), pp. 556-563.

Leemans, R., A. van Amstel, C. Battjes, E. Kreileman, and S. Toet, (1996). The land cover and carbon cycle consequences of large-scale utilizations of biomass as an energy source. *Global Environmental Change*, **6**(4), pp. 335-357.

Legros, G., I. Havet, N. Bruce, and S. Bonjour (2009). *The Energy Access Situation in Developing Countries: A Review Focusing on the Least Developed Countries and Sub-Saharan Africa.* United Nations Development Programme and the World Health Organization, New York, NY, USA, 142 pp.

Lehtomäki, A., S. Huttunen, and J.A. Rintala (2007). Laboratory investigations on co-digestion of energy crops and crop residues with cow manure for methane production: Effect of crop to manure ratio. *Resources, Conservation and Recycling*, **51**(3), pp. 591-609.

Leng, R., C. Wang, C. Zhang, D. Dai, and G. Pu (2008). Life cycle inventory and energy analysis of cassava-based fuel ethanol in China. *Journal of Cleaner Production*, **16**(3), pp. 374-384.

Levasseur, A., P. Lesage, M. Margni, L. Deschênes, and R. Samson (2010). Considering time in LCA: Dynamic LCA and its application to global warming impact assessments. *Environmental Science & Technology*, **44**(8), pp. 3169-3174.

Lewis, S.L., G. Lopez-Gonzalez, B. Sonke, K. Affum-Baffoe, T.R. Baker, L.O. Ojo, O.L. Phillips, J.M. Reitsma, L. White, J.A. Comiskey, M.-N.D. K, C.E.N. Ewango, T.R. Feldpausch, A.C. Hamilton, M. Gloor, T. Hart, A. Hladik, J. Lloyd, J.C. Lovett, J.-R. Makana, Y. Malhi, F.M. Mbago, H.J. Ndangalasi, J. Peacock, K.S.H. Peh, D. Sheil, T. Sunderland, M.D. Swaine, J. Taplin, D. Taylor, S.C. Thomas, R. Votere, and H. Woll (2009). Increasing carbon storage in intact African tropical forests. *Nature*, **457**(7232), pp. 1003-1006.

Li, W., R. Fu, and R.E. Dickinson (2006). Rainfall and its seasonality over the Amazon in the 21st century as assessed by the coupled models for the IPCC AR4. *Journal of Geophysical Research*, **111**(D2), D02111.

Liebig, M., M. Schmer, K. Vogel, and R. Mitchell (2008). Soil carbon storage by switchgrass grown for bioenergy. *BioEnergy Research*, **1**(3), pp. 215-222.

Lindenmayer, D.B., and H.A. Nix (1993). Ecological principles for the design of wildlife corridors. *Conservation Biology*, **7**(3), pp. 627-631.

Lindner, M., M. Maroschek, S. Netherer, A. Kremer, A. Barbati, J. Garcia-Gonzalo, R. Seidl, S. Delzon, P. Corona, M. Kolström, M.J. Lexer, and M. Marchetti (2010). Climate change impacts, adaptive capacity, and vulnerability of European forest ecosystems. *Forest Ecology and Management*, **259**(4), pp. 698-709.

Lioubimtseva, E., and J.M. Adams (2004). Possible implications of increased carbon dioxide levels and climate change for desert ecosystems. *Environmental Management*, **33**(Supplement 1), pp. S388-S404.

Liska, A.J., and R.K. Perrin (2010). Securing foreign oil: A case for including military operations in the climate change impact of fuels. *Environment: Science and Policy for Sustainable Development*, **52**(4), pp. 9-12.

Lobell, D.B., and M.B. Burke (2008). Why are agricultural impacts of climate change so uncertain? The importance of temperature relative to precipitation. *Environmental Research Letters*, **3**(3), 034007.

Lobell, D.B., M.B. Burke, C. Tebaldi, M.D. Mastrandrea, W.P. Falcon, and R.L. Naylor (2008). Prioritizing climate change adaptation needs for food security in 2030. *Science*, **319**(5863), pp. 607-610.

Londo, M., A. Faaij, M. Junginger, G. Berndes, S. Lensink, A. Wakker, G. Fischer, S. Prieler, H. van Velthuizen, and M. de Wit (2010). The REFUEL EU road map for biofuels in transport: Application of the project's tools to some short-term policy issues. *Biomass and Bioenergy*, **34**(2), pp. 244-250.

López, R., and G.I. Galinato (2007). Should governments stop subsidies to private goods? Evidence from rural Latin America. *Journal of Public Economics*, **91**(5-6), pp. 1071-1094.

Lotze-Campen, H., A. Popp, T. Beringer, C. Müller, A. Bondeau, S. Rost, and W. Lucht (2009). Scenarios of global bioenergy production: The trade-offs between agricultural expansion, intensification and trade. *Ecological Modelling*, **221**(18), pp. 2188-2196.

Loustau, D., A. Bosc, A. Colin, J. Ogee, H. Davi, C. Francois, E. Dufrene, M. Deque, E. Cloppet, D. Arrouays, C. Le Bas, N. Saby, G. Pignard, N. Hamza, A. Granier, N. Breda, P. Ciais, N. Viovy, and F. Delage (2005). Modeling climate change effects on the potential production of French plains forests at the sub-regional level. *Tree Physiology*, **25**(7), pp. 813-823.

Low, T., and C. Booth (2007). *The Weedy Truth about Biofuels.* Invasive Species Council, Melbourne, Fairfield, Australia, 46 pp.

Luyssaert, S., E.D. Schulze, A. Börner, A. Knohl, D. Hessenmöller, B.E. Law, P. Ciais, and J. Grace (2008). Old-growth forests as global carbon sinks. *Nature*, **455**(7210), pp. 213-215.

Lywood, W. (2008). *Indirect Effects of Biofuels.* Renewable Fuels Agency, East Sussex, UK.

Lywood, W., J. Pinkney, and S.A.M. Cockerill (2009a). Impact of protein concentrate coproducts on net land requirement for European biofuel production. *Global Change Biology Bioenergy*, **1**(5), pp. 346-359.

Lywood, W., J. Pinkney, and S.A.M. Cockerill (2009b). The relative contributions of changes in yield and land area to increasing crop output. *Global Change Biology Bioenergy*, **1**(5), pp. 360-369.

Macedo, I.C., and J.E.A. Seabra (2008). Mitigation of GHG emissions using sugarcane bioethanol. In: *Sugarcane Ethanol: Contributions to Climate Change Mitigation and the Environment.* P. Zuubier and J. van de Vooren (eds.), Wageningen Academic Publishers, Wageningen, The Netherlands, pp. 95-112.

Macedo, I.C., J.E.A. Seabra, and J.E.A.R. Silva (2008). Green house gases emissions in the production and use of ethanol from sugarcane in Brazil: The 2005/2006 averages and a prediction for 2020. *Biomass and Bioenergy*, **32**(7), pp. 582-595.

Madsen, A.M. (2006). Exposure to airborne microbial components in autumn and spring during work at Danish biofuel plants. *Annals of Occupational Hygiene*, **50**(8), pp. 821-831.

Madsen, A.M., L. Martensson, T. Schneider, and L. Larsson (2004). Microbial dustiness and particle release of different biofuels. *Annals of Occupational Hygiene*, **48**(4), pp. 327-338.

Maes, W., W. Achten, and B. Muys (2009). Use of inadequate data and methodological errors lead to a dramatic overestimation of the water footprint of *Jatropha curcas*. *Nature Precedings*, hdl:10101/npre.2009.3410.1, 3 pp.

Malézieux, E., Y. Crozat, C. Dupraz, M. Laurans, D. Makowski, H. Ozier-Lafontaine, B. Rapidel, S. de Tourdonnet, and M. Valantin-Morison (2009). Mixing plant species in cropping systems: concepts, tools and models. A review. *Agronomy for Sustainable Development*, **29**(1), pp. 43-62.

Malmer, A., D. Murdiyarso, L.A.S. Bruijnzeel, and U. Lstedt (2010). Carbon sequestration in tropical forests and water: a critical look at the basis for commonly used generalizations. *Global Change Biology*, **16**(2), pp. 599-604.

Markewitz, D. (2006). Fossil fuel carbon emissions from silviculture: Impacts on net carbon sequestration in forests. *Forest Ecology and Management*, **236**(2-3), pp. 153-161.

Marlair, G., P. Rotureau, S. Breulet, and S. Brohez (2009). Booming development of biofuels for transport: Is fire safety of concern? *Fire and Materials*, **33**(1), pp. 1-19.

Marland, G., and B. Schlamadinger (1997). Forests for carbon sequestration or fossil fuel substitution? A sensitivity analysis. *Biomass and Bioenergy*, **13**(6), pp. 389-397.

Marland, G., M. Obersteiner, and B. Schlamadinger (2007). The carbon benefits of fuels and forests. *Science*, **318**(5853), pp. 1066.

Martens, W., and R. Böhm (2009). Overview of the ability of different treatment methods for liquid and solid manure to inactivate pathogens. *Bioresource Technology*, **100**(22), pp. 5374-5378.

Martinelli, L.A., and S. Filoso (2007). Polluting effects of Brazil's sugar-ethanol industry. *Nature*, **445**(7126), pp. 364-364.

Masera, O.R., and J. Navia (1997). Fuel switching or multiple cooking fuels: Understanding interfuel substitution patterns in rural Mexican households. *Biomass and Bioenergy*, **12**(5), pp. 347-361.

Masera, O.R., B.D. Saatkamp, and D.M. Kammen (2000). From linear fuel switching to multiple cooking strategies: A critique and alternative to the energy ladder model. *World Development*, **28**(12), pp. 2083-2103.

Masera, O.R., R. Diaz, and V. Berrueta (2005). From cookstoves to cooking systems: the integrated program on sustainable household energy use in Mexico. *Energy for Sustainable Development*, **9**(1), pp. 25-36.

Masera, O., A. Ghilardi, R. Drigo, and M. Angel Trossero (2006). WISDOM: A GIS-based supply demand mapping tool for woodfuel management. *Biomass and Bioenergy*, **30**(7), pp. 618-637.

Matsuoka, S., J. Ferro, and P. Arruda (2009). The Brazilian experience of sugarcane ethanol industry. *In Vitro Cellular & Developmental Biology - Plant*, **45**(3), pp. 372-381.

Matthews, H.D., and K. Caldeira (2008). Stabilizing climate requires near-zero emissions. *Geophysical Research Letters*, **35**(4), L04705.

Matthews, H.D., N.P. Gillett, P.A. Stott, and K. Zickfeld (2009). The proportionality of global warming to cumulative carbon emissions. *Nature*, **459**(7248), pp. 829-832.

Maués, J.A. (2007). Maximizacao da geracao eletrica a partir do bagaco e palha em usinas de acucar e alcool. *Revista Engenharia*, **583**, pp. 88-98.

Mayfield, C.A., C.D. Foster, C.T. Smith, J. Gan, and S. Fox (2007). Opportunities, barriers, and strategies for forest bioenergy and bio-based product development in the Southern United States. *Biomass & Bioenergy*, **31**, pp. 631-637.

McAloon, A., F. Taylor, W. Lee, K. Ibsen, and R. Wooley (2000). *Determining the Cost of Producing Ethanol from Corn Starch and Lignocellulosic Feedstocks*. NREL/TP-580-28893, National Renewable Energy Laboratory, Golden, CO, USA, 43 pp.

McCoy, M. (2006). Glycerin surplus. *Chemical & Engineering News*, **83**(4), pp. 21-22.

McDonald, A., and L. Schrattenholzer (2001). Learning rates for energy technologies. *Energy Policy*, **29**(4), pp. 255-261.

McGowin, C. (2008). *Renewable Energy Technical Assessment Guide*. TAG-RE:2007, Electric Power Research Institute, Palo Alto, CA, USA.

McKeough, P., and E. Kurkela (2008). *Process Evaluations and Design Studies in the UCG Project 2004-2007*. VTT Research Notes 2434, VTT Technical Research Centre of Finland, Espoo, Finland, 49 pp.

McKeough, P., Y. Solantausta, H. Kyllönen, A. Faaij, C. Hamelinck, M. Wagener, D. Beckman and B. Kjellström (2005). *Technoeconomic Analysis of Biotrade Chains: Upgraded Biofuels from Russia and Canada to the Netherlands*. VTT Research Notes 2312, VTT Technical Research Centre of Finland, Espoo, Finland, 65 pp.

McLeod, J.E., J.E. Nunez, and S.S. Rivera (2008). A discussion about how to model biofuel plants for the risk optimization. In: *Proceedings of the World Congress on Engineering 2008, Vol. II*, London, UK, 2-4 July 2008, pp. 1214-1219.

McWhorter, C. (1971). Introduction and spread of johnsongrass in the United States. *Weed Science*, **19**(5), pp. 496-500.

Melillo, J.M., J.M. Reilly, D.W. Kicklighter, A.C. Gurgel, T.W. Cronin, S. Paltsev, B.S. Felzer, X. Wang, A.P. Sokolov, and C.A. Schlosser (2009). Indirect emissions from biofuels: How important? *Science*, **326**(5958), pp. 1397-1399.

Menichetti, E., and M. Otto (2009). Energy balance and greenhouse gas emissions of biofuels from a product life-cycle perspective. In: *Biofuels: Environmental Consequences and Interactions with Changing Land Use*. R.W. Howarth and S. Bringezu (eds.), Proceedings of the SCOPE - Scientific Committee on Problems of the Environment International Biofuels Project Rapid Assessment, Gummersbach, Germany, 22-25 September 2009, pp. 81-109.

Metzger, J.O., and A. Hüttermann (2009). Sustainable global energy supply based on lignocellulosic biomass from afforestation of degraded areas. *Naturwissenschaften*, **96**(2), pp. 279-288.

Milbrandt, A., and R.P. Overend (2008). *Future of Liquid Biofuels for APEC Economies*. NREL/TP-6A2-43709, report prepared for Asia-Pacific Economic Cooperation by the National Renewable Energy Laboratory, Golden, CO, USA, 103 pp. Available at: www.biofuels.apec.org/pdfs/ewg_2008_liquid_biofuels.pdf.

Millennium Ecosystem Assessment (2005). *Ecosystems and Human Well-being*. Island Press, Washington, DC, USA, 36 pp.

Mishra, V., X. Dai, K. Smith, and L. Mike (2004). Maternal exposure to biomass smoke and reduced birth weight in Zimbabwe. *Annals of Epidemiology*, **14**(10), pp. 740-747.

Mitchell, D. (2008). *A Note on Rising Food Prices*. Policy Research Working Paper 4682, The World Bank Group, Washington, DC, USA, 22 pp.

Molden, D. (ed.) (2007). *Water for Food, Water for Life: A Comprehensive Assessment of Water Management in Agriculture*. Earthscan, London, UK.

Molina Grima, E., E.H. Belarbi, F.G. Acien Fernandez, A. Robles Medina, and Y. Chisti (2003). Recovery of microalgal biomass and metabolites: process options and economics. *Biotechnology Advances*, **20**(7-8), pp. 491-515.

Möllersten, K., J. Yan, and J. R. Moreira (2003). Potential market niches for biomass energy with CO_2 capture and storage – Opportunities for energy supply with negative CO_2 emissions. *Biomass and Bioenergy*, **25**(3), pp. 273-285.

Molofsky, J., S.L. Morrison, and C.J. Goodnight (1999). Genetic and environmental controls on the establishment of the invasive grass, Phalaris arundinacea. *Biological Invasions*, **1**(2), pp. 181-188.

Moral, R., C. Paredes, M.A. Bustamante, F. Marhuenda-Egea, and M.P. Bernal (2009). Utilisation of manure composts by high-value crops: Safety and environmental challenges. *Bioresource Technology*, **100**(22), pp. 5454-5460.

Moreira, J.R. (2006). *Bioenergy and Agriculture, Promises and Challenges: Brazil's Experience with Bioenergy.* International Food Policy Research Institute, Washington, DC, USA, 2 pp.

Mosier, A., C. Kroeze, C. Nevison, O. Oenema, S. Seitzinger, and O. van Cleemput (1998). Closing the global N_2O budget: nitrous oxide emissions through the agricultural nitrogen cycle. *Nutrient Cycling in Agroecosystems*, **52**(2-3), pp. 1385-1314.

Mosier, N., C. Wyman, B. Dale, R. Elander, Y.Y. Lee, M. Holtzapple, and M. Ladisch (2005). Features of promising technologies for pretreatment of lignocellulosic biomass. *Bioresource Technology*, **96**(6), pp. 673-686.

Mozaffarian, H., R.W.R. Zwart, H. Boerrigeter, and E.P. Deurwaarder (2004). Biomass and waste-related SNG production technologies; technical, economic and ecological feasibility. In: *2nd World Conference and Technology Exhibition on Biomass for Energy, Industry and Climate Protection*, Rome, Italy, 10-14 May 2004. Available at: www.biosng.com/fileadmin/biosng/user/documents/reports/rx04024.pdf.

Mukunda, H.S., S. Dasappa, P.J. Paul, N.K.S. Rajan, M. Yagnaraman, D.R. Kumar, and M. Deogaonkar (2010). Gasifier stoves - science, technology and field outreach. *Current Science (Bangalore)*, **98**(5), pp. 627-638.

Müller, C. (2007). *Anaerobic Digestion of Biodegradable Solid Waste in Low- and Middle-Income Countries - Overview Over Existing Technologies and Relevant Case Studies.* Eawag Aquatic Research, Dübendorf, Switzerland, 63 pp.

Myles, R. (2001). Implementation of Household Biogas Plant by NGOs in India-Practical Experience in Implementation Household Biogas Technology, Lessons Learned, Key Issues and Future Approach for Sustainable Village Development. In: *VODO International Conference on Globalisation and Sustainable Development*, Antwerp, Brussels, 19-21 Nov 2001, 19 pp. Available at: www.inseda.org/Additional%20material/Lessons%20learnt%20NGOs%20Biogas%20programme.pdf.

Nabuurs, G.-J., A. Pussinen, T. Karjalainen, M. Erhard, and K. Kramer (2002). Stemwood volume increment changes in European forests due to climate change—a simulation study with the EFISCEN model. *Global Change Biology*, **8**(4), pp. 304-316.

Nabuurs, G.J., O. Masera, K. Andrasko, P. Benitez-Ponce, R. Boer, M. Dutschke, E. Elsiddig, J. Ford-Robertson, P. Frumhoff, T. Karjalainen, O. Krankina, W.A. Kurz, M. Matsumoto, W. Oyhantcabal, N.H. Ravindranath, M.Z. Sanz Sanchez, and X. Zhang (2007). Forestry. In: *Climate Change 2007: Mitigation of Climate Change. Contribution of Working Group III to the Fourth Assessment Report of the Intergovernmental Panel on Climate Change.* B. Metz, O.R. Davidson, P.R. Bosch, R. Dave, and L.A. Meyer (eds.), Cambridge University Press, pp. 541-584.

Nagatomi, Y., Y. Hiromi, Y. Kenji, I. Hiroshi, and Y. Koichi (2008). A system analysis of energy utilization and competing technology using oil palm residue in Malaysia. *Journal of Japan Society of Energy and Resources*, **29**(5), pp. 1-7.

Narayanan, D., Y. Zhang, and M.S. Mannan (2007). Engineering for Sustainable Development (ESD) in bio-diesel production. *Process Safety and Environmental Protection*, **85**(5), pp. 349-359.

Näslund-Eriksson, L., and L. Gustavsson (2008). Biofuels from stumps and small roundwood – Costs and CO_2 benefits. *Biomass and Bioenergy*, **32**(10), pp. 897-902.

Nassar, A., L. Harfurch, M.M.R. Moreira, L.C. Bachion, L.B. Antoniazzi, and G. Sparovek (2009). *Impacts on Land Use and GHG Emissions from a Shock on Brazilian Sugarcane Ethanol Exports to the United States using the Brazilian Land Use Model (BLUM).* EPA HQ OAR 2005 0161, Institute for International Negotiations, Geneva, Switzerland, 32 pp.

Nassar, A., L.B. Antoniazzi, M.R. Moreira, L. Chiodi, and L. Harfurch (2010). *An Allocation Methodology to Assess GHG Emissions Associated with Land Use Change.* Institute for International Negotiations, Geneva, Switzerland, 31 pp.

NRC (2008). *Water Implications of Biofuels Production in the United States.* National Research Council, The National Academies Press, Washington, DC, USA, 88 pp.

NRC (2009a). *Liquid Transportation Fuels from Coal and Biomass Technological Status, Costs, and Environmental Impacts.* National Research Council, National Academies Press, Washington, DC, USA, 50 pp.

NRC (2009b). *America's Energy Future: Electricity from Renewable Resources: Status, Prospects, and Impediments.* National Research Council, The National Academies Press, Washington, DC, 367 pp.

NRC (2010). *Impact of Genetically Engineered Crops on Farm Sustainability in the United States, 2010.* Committee on the Impact of Biotechnology on Farm-Level Economics and Sustainability; National Research Council, Washington, DC, USA, 35 pp.

Nehlsen, J., M. Mukherjee, and R.V. Porcelli (2007). Apply an integrated approach to catalytic process design. *Chemical Engineering Progress*, **103**(2), pp. 31-41.

Nelson, D.E., P.P. Repetti, T.R. Adams, R.A. Creelman, J. Wu, D.C. Warner, D.C. Anstrom, R.J. Bensen, P.P. Castiglioni, M.G. Donnarummo, B.S. Hinchey, R.W. Kumimoto, D.R. Maszle, R.D. Canales, K.A. Krolikowski, S.B. Dotson, N. Gutterson, O.J. Ratcliffe, and J.E. Heard (2007). Plant nuclear factor Y (NF-Y) B subunits confer drought tolerance and lead to improved corn yields on water-limited acres. *Proceedings of the National Academy of Sciences*, **104**(42), pp. 16450-16455.

NETL (2008). *Development of Baseline Data and Analysis of Life Cycle Greenhouse Gas Emissions of Petroleum-Based Fuels.* DOE/NETL-2009/1362, National Energy Technology Laboratory, Pittsburgh, PA, USA.

NETL (2009a). *An Evaluation of the Extraction, Transport and Refining of Imported Crude Oils and the Impact on Life Cycle Greenhouse Gas Emissions.* DOE/NETL-2009/1362, National Energy Technology Laboratory, Pittsburgh, PA, USA.

NETL (2009b). *Affordable, Low Carbon Diesel Fuel from Domestic Coal and Biomass.* DOE/NETL-2009/1349, National Energy Technology Laboratory, Pittsburgh, PA, USA.

Neumann, K., P.H. Verburg, E. Stehfest, and C. Müller (2010). The yield gap of global grain production: A spatial analysis. *Agricultural Systems*, **103**(5), pp. 316-326.

Nguyen, T.L.T., and S.H. Gheewala (2008). Life cycle assessment of fuel ethanol from Cassava in Thailand. *International Journal of Life Cycle Assessment*, **13**(2), pp. 147-154.

Nishi, T., M. Konishi, and S. Hasebe (2005). An autonomous decentralized supply chain planning system for multi-stage production processes. *Journal of Intelligent Manufacturing*, **16**(3), pp. 259-275.

Nohrstedt, H.O. (2001). Response of coniferous forest ecosystems on mineral soils to nutrient additions: A review of Swedish experiences. *Scandinavian Journal of Forest Research*, **16**(6), pp. 555-573.

Nusser, M., B. Hüsing, and S. Wydra (2007). *Potenzialanalyse der Industriellen, Weißen Biotechnologie, Karlsruhe.* Fraunhofer-Institut für System- und Innovationsforschung, Karlsruhe, Germany, 426 pp.

O'Hare, M., R.J. Plevin, J.I. Martin, A.D. Jones, A. Kendall, and E. Hopson (2009). Proper accounting for time increases crop-based biofuels' greenhouse gas deficit versus petroleum. *Environmental Research Letters*, **4**(2), 024001 (7 pp.).

OANDA (2011). *Historical Exchange Rates.* Oanda Corporation, New York, NY, USA. Available at: www.oanda.com/currency/historical-rates/.

Obernberger, I., and G. Thek (2004). *Techno-economic evaluation of selected decentralised CHP applications based on biomass combustion in IEA partner countries.* BIOS Bioenergiesysteme GmbH, Graz, Austria, 87 pp.

Obernberger, I., G. Thek, and D. Reiter (2008). *Economic evaluation of decentralised CHP applications based on biomass combustion and biomass gasification.* BIOS Bioenergiesysteme GmbH, Graz, Austria, 19 pp.

Obersteiner, M., K. Mollersten, J. Moreira, S. Nilsson, P. Read, K. Riahi, B. Schlamadinger, Y. Yamagata, J. Yan, J.P. Ypserle, C. Azar, and P. Kauppi (2001). Managing climate risk. *Science*, **294**(5543), pp. 786-787.

OECD-FAO (2008). *Agricultural Outlook 2008-2017.* Organisation for Economic Cooperation and Development and Food and Agriculture Organization, Paris, France, 73 pp.

Okada, M., C. Lanzatella, M.C. Saha, J. Bouton, R. Wu, and C.M. Tobias (2010). Complete switchgrass genetic maps reveal subgenome collinearity, preferential pairing and multilocus interactions. *Genetics*, **185**(3), pp. 745-760.

Oliveira, F.C.R. (2009). *Ocupação, emprego e remuneração na cana-de-açúcar e em outras atividades agropecuárias no Brasil, de 1992 a 2007.* Master's Thesis, Universidade de São Paulo, São Paulo, Brazil, 168 pp.

Oliver, R.J., J.W. Finch, and G. Taylor (2009). Second generation bioenergy crops and climate change: a review of the effects of elevated atmospheric CO_2 and drought on water use and the implications for yield. *Global Change Biology Bioenergy*, **1**(2), pp. 97-114.

Oliverio, J.L. (2006). Technological evolution of the Brazilian sugar and alcohol sector: Dedini's contribution. *International Sugar Journal*, **108**(1287), pp. 120-129.

Oliverio, J.L., and J.E. Ribeiro (2006). Cogeneration in Brazilian sugar and bioethanol mills: Past, present and challenges. *International Sugar Journal*, **108**(191), pp. 391-401.

Openshaw, K. (2000). A review of Jatropha curcas: an oil plant of unfulfilled promise. *Biomass and Bioenergy*, **19**(1), pp. 1-15.

Pacca, S., and J.R. Moreira (2009). Historical carbon budget of the Brazilian ethanol program. *Energy Policy*, **37**(11), pp. 4863-4873.

Patel, M.K., M. Crank, V. Dornburg, B. Hermann, L. Roes, B. Hüsing, L. van Overbeek, F. Terragni, and E. Recchia (2006). *Medium and Long-Term Opportunities and Risks of the Biotechnological Production of Bulk Chemicals from Renewable Resources.* Copernicus Institute for Sustainable Development and Innovation, Utrecht, The Netherlands, 420 pp.

Paustian, K., J.M. Antle, J. Sheehan, and A.P. Eldor (2006). *Agriculture's Role in Greenhouse Gas Mitigation.* Pew Center on Global Climate Change, Arlington, VA, USA.

Peksa-Blanchard, M., P. Dolzan, A. Grassi, J. Heinimö, M. Junginger, T. Ranta, and A. Walter (2007). *Global Wood Pellets Markets and Industry: Policy Drivers, Market Status and Raw Material Potential.* IEA Bioenergy Task 40 Publication, 120 pp. Available at: www.canbio.ca/documents/publications/ieatask40pelletandrawmaterialstudynov2007final.pdf.

Perlack, R.D., L.L. Wright, A.F. Turhollow, R.L. Graham, B.J. Stokes, and D.C. Erbach (2005). *Biomass as Feedstock for a Bioenergy and Bioproducts Industry: The Technical Feasibility of a Billion-Ton Annual Supply.* ORNL/TM-2005/66, Oak Ridge National Laboratory, Oak Ridge, TN, USA, 78 pp.

Perry, J.A. (2009). Catastrophic incident prevention and proactive risk management in the new biofuels industry. *Environmental Progress & Sustainable Energy*, **28**(1), pp. 72-82.

Persson, U.M., and C. Azar (2010). Preserving the world's tropical forests – A price on carbon may not do. *Environmental Science & Technology*, **44**(1), pp. 210-215.

Petersson, A. and A. Wellinger (2009). *Biogas Upgrading Technologies – Developments and Innovations.* IEA Bioenergy Task 37 publication, Malmo, Sweden, 20 pp. Available at: www.biogasmax.com/media/iea_2biogas_upgrading_tech__025919000_1434_30032010.pdf.

Phillips, S., A. Aden, J. Jechura, D. Dayton, and T. Eggeman (2007). *Thermochemical Ethanol via Indirect Gasification and Mixed Alcohol Synthesis of Lignocellulosic Biomass.* NREL/TP-510-41168, National Renewable Energy Laboratory, Golden, CO, USA, 132 pp.

Pimentel, D., S. McNair, J. Janecka, J. Wightman, C. Simmonds, C. O'Connell, E. Wong, L. Russel, J. Zern, T. Aquino, and T. Tsomondo (2001). Economic and environmental threats of alien plant, animal, and microbe invasions. *Agriculture, Ecosystems & Environment*, **84**(1), pp. 1-20.

Pimentel, D., P. Hepperly, J. Hanson, D. Douds, and R. Seidel (2005). Environmental, energetic, and economic comparisons of organic and conventional farming systems. *BioScience*, **55**(7), pp. 573-582.

Pingali, P.L., and P.W. Heisey (1999). *Cereal Crop Productivity in Developing Counties: Past Trends and Future Prospects.* Working Paper 99-03, International Maize and Wheat Improvement Center, Texcoco, Mexico, 34 pp.

Plevin, R.J. (2009). Modeling corn ethanol and climate. *Journal of Industrial Ecology*, **13**(4), pp. 495-507.

Plevin, R.J., O.H. Michael, A.D. Jones, M.S. Torn, and H.K. Gibbs (2010). Greenhouse gas emissions from biofuels indirect land use change are uncertain but may be much greater than previously estimated. *Environmental Science & Technology*, **44**(21), pp. 8015-8021.

Practical Action Consulting (2009). *Small Scale Bioenergy can Benefit Poor: Brief Description and Preliminary Lessons on Livelihood Impacts from Case Studies in Asia, Latin America and Africa.* Food and Agriculture Organization and Policy Innovation Systems for Clean Energy Security, Rome, Italy, 142 pp.

Purdon, M., S. Bailey-Stamler, and R. Samson (2009). Better bioenergy: Rather than picking bioenergy "winners," effective policy should let a lifecycle analysis decide. *Alternatives Journal*, **35**(2), pp. 23-29.

Rabaey, K., and W. Verstraete (2005). Microbial fuel cells: novel biotechnology for energy generation. *Trends in Biotechnology*, **23**(6), pp. 291-298.

Ragauskas, A.J., C.K. Williams, B.H. Davison, G. Britovsek, J. Cairney, C.A. Eckert, W.J. Frederick, Jr., J.P. Hallett, D.J. Leak, C.L. Liotta, J.R. Mielenz, R. Murphy, R. Templer, and T. Tschaplinski (2006). The path forward for biofuels and biomaterials. *Science*, **311**(5760), pp. 484-489.

Raghu, S., R.C. Anderson, C.C. Daehler, A.S. Davis, R.N. Wiedenmann, D. Simberloff, and R.N. Mack (2006). Adding biofuels to the invasive species fire? *Science*, **313**(5794), pp. 1742.

Rajagopal, D., G. Hochman, and D. Zilberman (2011). Indirect fuel use change (IFUC) and the lifecycle environmental impact of biofuel policies. *Energy Policy*, **39**(1), pp. 228-233.

Ramanathan, V., and G. Carmichael (2008). Global and regional climate changes due to black carbon. *Nature Geoscience*, **1**(4), pp. 221-227.

Randall, J.M. (1996). Plant Invaders: How Non-native Species Invade & Degrade Natural Areas Invasive Plants. In: *Weeds of the Global Garden.* J. Marinelli and J.M. Randall (eds.), Brooklyn Botanic Garden, Brooklyn, NY, pp. 7-12.

Ranius, T., and L. Fahrig (2006). Targets for maintenance of dead wood for biodiversity conservation based on extinction thresholds. *Scandinavian Journal of Forest Research*, **21**(3), pp. 201-208.

Rauch, R. (2010). Indirect Gasification. In: *IEA Bioenergy Joint Tasks 32 &33 Workshop, State-of-the-Art Technologies for Small Biomass Co-generation*, International Energy Agency, Copenhagen, Denmark, 7 Oct 2010. Available at: www.ieabcc.nl/meetings/task32_Copenhagen/09%20TU%20Vienna.pdf.

Ravindranath, N.H., P. Balachandra, S. Dasappa, and K. Usha Rao (2006). Bioenergy technologies for carbon abatement. *Biomass and Bioenergy*, **30**(10), pp. 826-837.

Regalbuto, J.R. (2009). Cellulosic biofuels – got gasoline? *Science*, **325**(5942), pp. 822-824.

Reijnders, L. (2008). Ethanol production from crop residues and soil organic carbon. *Resources, Conservation and Recycling*, **52**(4), pp. 653-658.

Reinhardt, G. (1991). *Biofuels: Energy and GHG balances: Methodology and Case Study Rape Seed Biodiesel.* Institut für Energie- und Umweltforschung Heidelberg, Heidelberg, Germany.

Reinhardt, G., S. Gartner, A. Patyk, and N. Rettenmaier (2006). *Ökobilanzen zu BTL: Eine ökologische Einschätzung.* 2207104, ifeu Institut für Energie- und Umweltforschung Heidelberg gGmbH, Heidelberg, Germany, 108 pp.

REN21 (2007). *Global Status Report.* Renewable Energy Policy Network for the 21st Century Secretariat, Paris, France, 54 pp.

REN21 (2009). *Renewables Global Status Report: 2009 Update.* Renewable Energy Policy Network for the 21st Century Secretariat, Paris, France, 42 pp.

Rendleman, C.M., and H. Shapouri (2007). *New Technologies in Ethanol Production.* Report Number 842, United States Department of Agriculture, Washington, DC, USA.

RFA (2010). *U.S. Cellulosic.* Renewable Fuels Association, Washington, DC, USA. Available at: www.ethanolrfa.org/pages/cellulosic-ethanol

RFA (2011). *Biorefinery Plant Locations.* Renewable Fuels Association, Washington, DC, USA. Available at: www.ethanolrfa.org/bio-refinery-locations/.

Rentizelas, A.A., A.J. Tolis, and I.P. Tatsiopoulos (2009). Logistics issues of biomass: The storage problem and the multi-biomass supply chain. *Renewable and Sustainable Energy Reviews*, **13**(4), pp. 887-894.

Reynolds, M.P., and N.E. Borlaug (2006). Applying innovations and new technologies for international collaborative wheat improvement. *The Journal of Agricultural Science*, **144**(02), pp. 95-95.

Rhodes, J.S., and D.W. Keith (2008). Biomass with capture: negative emissions within social and environmental constraints: an editorial comment. *Climatic Change*, **87**(3-4), pp. 321-328.

Righelato, R., and D.V. Spracklen (2007). Carbon mitigation by biofuels or by saving and restoring forests? *Science*, **317**(5840), pp. 902.

Ringer, M., V. Putsche, and J. Scahill (2006). *Large-Scale Pyrolysis Oil Production: A Technology Assessment and Economic Analysis.* TP-510-37779, National Renewable Energy Laboratory, Golden, CO, USA, 93 pp.

Riviére, C., and G. Marlair (2009). BIOSAFUEL®, a pre-diagnosis tool of risks pertaining to biofuels chains. *Journal of Loss Prevention in the Process Industries*, **22**(2), pp. 228-236.

Riviére, C., and G. Marlair (2010). The use of multiple correspondence analysis and hierarchical clustering to identify incident typologies pertaining to the biofuel industry. *Biofuels, Bioproducts and Biorefining*, **4**(1), pp. 53-65.

Robertson, G.P., V.H. Dale, O.C. Doering, S.P. Hamburg, J.M. Melillo, M.M. Wander, W.J. Parton, P.R. Adler, J.N. Barney, R.M. Cruse, C.S. Duke, P.M. Fearnside, R.F. Follett, H.K. Gibbs, J. Goldemberg, D.J. Mladenoff, D. Ojima, M.W. Palmer, A. Sharpley, L. Wallace, K.C. Weathers, J.A. Wiens, and W.W. Wilhelm (2008). Sustainable biofuels redux. *Science*, **322**(5898), pp. 49-50.

Rockström, J., M. Lannerstad, and M. Falkenmark (2007). Assessing the water challenge of a new green revolution in developing countries. *Proceedings of the National Academy of Sciences*, **104**(15), pp. 6253-6260.

Rockström, J., L. Karlberg, S.P. Wani, J. Barron, N. Hatibu, T. Oweis, A. Bruggeman, J. Farahani, and Z. Qiang (2010). Managing water in rainfed agriculture – The need for a paradigm shift. *Agricultural Water Management*, **97**(4), pp. 543-550.

Romieu, I., H. Riojas-Rodriguez, A.T. Marron-Mares, A. Schilmann, R. Perez-Padilla, and O. Masera (2009). Improved biomass stove intervention in rural Mexico: Impact on the respiratory health of women. *American Journal of Respiratory and Critical Care Medicine*, **180**(7), pp. 649-656.

Rosegrant, M.W., T. Zhu, S. Msangi, and T. Sulser (2008). Biofuels: Long-run implications for food security and the environment. *Review of Agricultural Economics*, **30**(3), pp. 495-505.

Rosillo-Calle, F., S.V. Bajay, and H. Rothman (eds.) (2000). *Industrial Uses of Biomass Energy: The Example of Brazil.* Taylor & Francis, London, UK.

Ross, A.B., K. Anastasakis, M. Kubacki, and J.M. Jones (2009). Investigation of the pyrolysis behaviour of brown algae before and after pre-treatment using PY-GC/MS and TGA. *Journal of Analytical and Applied Pyrolysis*, **85**(1-2), pp. 3-10.

Rost, S., D. Gerten, H. Hoff, W. Lucht, M. Falkenmark, and J. Rockstrom (2009). Global potential to increase crop production through water management in rainfed agriculture. *Environmental Research Letters*, **4**(4), 044002 (9 pp.).

Rowe, R., J. Whitaker, J. Chapman, D. Howard, and G. Talor (2008). *Life-Cycle Assessment in Bioenergy Sector: Developing a Systematic Review Working Paper.* UKERC/WP/FSE/2008/002, UK Energy Research Centre, London, UK, 20 pp.

Royal Society (2008). *Sustainable Biofuels: Prospects and Challenges.* Policy document 01/08, The Royal Society, London, UK, 90 pp.

Rude, M.A., and A. Schirmer (2009). New microbial fuels: a biotech perspective. *Current Opinion in Microbiology*, **12**(3), pp. 274-281.

(S&T)[2] Consultants (2009). *An Examination of the Potential for Improving Carbon/Energy Balance of Bioethanol.* T39-TR1, (S&T)[2] Consultants Inc., Delta, Canada, 72 pp.

Saarinen, V.-M. (2006). The effects of slash and stump removal on productivity and quality of forest regeneration operations – preliminary results. *Biomass and Bioenergy*, **30**(4), pp. 349-356.

Saarsalmi, A., and E. Mälkönen (2001). Forest fertilization research in Finland: A literature review. *Scandinavian Journal of Forest Research*, **16**(6), pp. 514-535.

Sala, O.E., A. Kinzig, R. Leemans, D.M. Lodge, H.A. Mooney, M. Oesterheld, N.L. Poff, M.T. Sykes, B.H. Walker, M. Walker, D.H. Wall, F.S. Chapin, J.J. Armesto, E. Berlow, J. Bloomfield, R. Dirzo, E. Huber-Sanwald, L.F. Huenneke, and R.B. Jackson (2000). Global biodiversity scenarios for the year 2100. *Science*, **287**(5459), pp. 1770-1774.

Sánchez, Ó.J., and C.A. Cardona (2008). Trends in biotechnological production of fuel ethanol from different feedstocks. *Bioresource Technology*, **99**(13), pp. 5270-5295.

Sannigrahi, P., A.J. Ragauskas, and G.A. Tuskan (2010). Poplar as a feedstock for biofuels: A review of compositional characteristics. *Biofuels, Bioproducts and Biorefining*, **4**(2), pp. 209-226.

Sawin, J.L. (2004). *National Policy Instruments: Policy Lessons for the Advancement and Diffusion of Renewable Energy Technologies around the World.* World Watch Institute, Washington, DC, USA.

SCBD (2006). *Global Biodiversity Outlook 2.* Secretariat of the Convention on Biological Diversity, Montreal, Canada, 92 pp. (ISBN-92-9225-040-X).

Schei, M.A., J.O. Hessen, K.R. Smith, N. Bruce, J. McCracken, and V. Lopez (2004). Childhood asthma and indoor woodsmoke from cooking in Guatemala. *Journal of Exposure Analysis and Environmental Epidemiology*, **14**, pp. S110-S117.

Schlamadinger, B., F. Bohlin, L. Gustavsson, G. Jungmeier, G. Marlandii, K. Pingoudt, and I. Savolaine (1997). Towards a standard methodology for greenhouse gas balances of bioenergy systems in comparison with fossil energy systems. *Biomass and Bioenergy*, **13**(6), pp. 359-375.

Schlamadinger, B., M. Grubb, C. Azar, A. Bauen, and G. Berndes (2001). Carbon sinks and the CDM: could a bioenergy linkage offer a constructive compromise? *Climate Policy*, **1**, pp. 411-417.

Schlamadinger, B., B. Bosquet, C. Streck, I. Noble, M. Dutschke, and N. Bird (2005). Can the EU emission trading scheme support CDM forestry. *Climate Policy*, **5**, pp. 199-208.

Schmidhuber, J. (2008). Impact of an increased biomass use on agricultural markets, prices and food security: A longer-term perspective. In: *Energy Security in Europe: Proceedings from the Conference 'Energy Security in Europe'*, Lund, Sweden, 24-25 September 2007, Lund University, Lund, Sweden, pp. 133-170. Available at: www.cfe.lu.se/upload/LUPDF/CentrumforEuropaforskning/Confpap2.pdf.

Schneider, U.E., B.A. McCarl, and E. Schmid (2007). Agricultural sector analysis on greenhouse gas mitigation in US agriculture and forestry. *Agricultural Systems*, **94**(2), pp. 128-140

Schouten, J.C., E.V. Rebrov, and M.H.J.M. de Croon (2002). Miniaturization of heterogeneous catalytic reactors: prospects for new developments in catalysis and process engineering. *CHIMIA International Journal for Chemistry*, **56**(11), pp. 627-635.

Schulz, U., O. Brauner, and H. Gruss (2009). Animal diversity on short-rotation coppices – a review. *Landbauforschung vTI Agriculture and Forestry Research 3*, **59**(3), pp. 171-182.

Schwaiger, H.P., and D.N. Bird (2010). Integration of albedo effects caused by land use change into the climate balance: Should we still account in greenhouse gas units? *Forest Ecology and Management*, **260**(3), pp. 278-286.

Schwilch, G., F. Bachmann, and H. Liniger (2009). Appraising and selecting conservation measures to mitigate desertification and land degradation based on stakeholder participation and global best practices. *Land Degradation & Development*, **20**(3), pp. 308-326.

Scolforo, J.R. (2008). *Mundo Eucalipto – Os Fatos E Mitos De Sua Cultura.* Editora Mar de Ideias, Rio de Janeiro, Brazil, 72 pp.

Seabra, J.E.A., L. Tao, H.L. Chum, and I.C. Macedo (2010). A techno-economic evaluation of the effects of centralized cellulosic ethanol and co-products refinery options with sugarcane mill clustering. *Biomass and Bioenergy*, **34**(8), pp. 1065-1078.

Searchinger, T., R. Heimlich, R.A. Houghton, F. Dong, A. Elobeid, J. Fabiosa, S. Tokgoz, D. Hayes, and T.H. Yu (2008). Use of U.S. croplands for biofuels increases greenhouse gases through emissions from land-use change. *Science*, **319**, pp. 1238-1240.

Seidenberger, T., D. Thran, R. Offermann, U. Seyfert, M. Buchhorn, and J. Zeddies (2008). *Global Biomass Potentials.* Report prepared for Greenpeace International by the German Biomass Research Center, Leipzig, Germany, 137 pp.

Semere, T., and F. Slater (2007). Ground flora, small mammal and bird species diversity in miscanthus (*Miscanthus×giganteus*) and reed canary-grass (*Phalaris arundinacea*) fields. *Biomass and Bioenergy*, **31**(1), pp. 20-29.

Sepp, S. (2008). *Analysis of Charcoal Value Chains – General Considerations.* GTZ Household Energy Programme, Eschborn, Germany, 11 pp.

Shah, S. (2007). Modular mini-plants: A new paradigm. *Chemical Engineering Progress*, **103**(3), pp. 36-41.

Shapouri, H., and M. Salassi (2006). *The Economic Feasibility of Ethanol Production in the United States.* United States Department of Agriculture, Washington, DC, USA, 69 pp.

Sharma, M.M. (2002). Strategies of conducting reactions on a small scale. Selectivity engineering and process intensification. *Pure and Applied Chemistry*, **74**(12), pp. 2265-2269.

Sheehan, J., T. Dunahay, J. Benemann, G. Roessler, and C. Weissman (1998a). *A Look Back at the U.S. Department of Energy's Aquatic Species Program: Biodiesel from Algae.* NREL/TP-580-24190, National Renewable Energy Laboratory, Golden, CO, USA, 295 pp.

Sheehan, J., V. Camobreco, J. Duffield, M. Graboski, and H. Shapouri (1998b). *Life Cycle Inventory of Biodiesel and Petroleum Diesel for Use in an Urban Bus.* NREL/SR-580-24089, National Renewable Energy Laboratory, Golden, CO, USA, 314 pp.

Sheehan, J., A. Aden, K. Paustian, K. Killian, J. Brenner, M. Walsh, and R. Nelson (2003). Energy and environmental aspects of using corn stover for fuel ethanol. *Journal of Industrial Ecology*, **7**(3-4), pp. 117-146.

Shen, L., J. Haufe, and M.K. Patel (2009). *Product Overview and Market Projection of Emerging Bio-based Plastics: PRO-BIP 2009.* Copernicus Institute for Sustainable Development and Innovation for the European Polysaccharide Network of Excellence (EPNOE) and European Bioplastics, Utrecht University, Utrecht, The Netherlands, 78 pp.

Sikkema, R., M. Junginger, W. Pichler, S. Hayes, and A.P.C. Faaij (2010). The international logistics of wood pellets for heating and power production in Europe: Costs, energy-input and greenhouse gas balances of pellet consumption in Italy, Sweden and the Netherlands. *Biofuels, Bioproducts and Biorefining*, **4**(2), pp. 132-153.

Sikkema, R., M. Steiner, M. Junginger, W. Hiegl, M.T. Hansen, and A. Faaij (2011). The European wood pellet markets: current status and prospects for 2020. *Biofuels, Bioproducts and Biorefining*, doi:10.1002/bbb.277.

Simpson, T.W., A.N. Sharpley, R.W. Howarth, H.W. Paerl, and K.R. Mankin (2008). The new gold rush: Fueling ethanol production while protecting water quality. *Journal of Environmental Quality*, **37**(2), pp. 318-324.

Simpson, T.W., L.A. Martinelli, A.N. Sharpley, R.W. Howarth, and S. Bringezu (2009). Impact of ethanol production on nutrient cycles and water quality: the United States and Brazil as case studies. In: *Biofuels: Environmental Consequences and Interactions with Changing Land Use*. R.W. Howarth and S. Bringezu (eds.), Proceedings of the SCOPE - Scientific Committee on Problems of the Environment International Biofuels Project Rapid Assessment, Gummersbach, Germany, 22-25 September 2009,, pp. 153-167.

Sims, R. (ed.) (2007). *Renewables for Heating and Cooling. An Untapped Potential*. Joint report for the Renewable Energy Technology Deployment Implementing Agreement and the Renewable Energy Working Party published by the International Energy Agency, Paris, France, 210 pp.

Sims, R., M. Taylor, J. Saddler, W. Mabee, and J. Riese (2008). *From 1st-to 2nd Generation Biofuel Technologies. An overview of current industry and R&D activities.* IEA Bioenergy, Paris, France, 124 pp.

Sims, R.E.H., W. Mabee, J.N. Saddler, and M. Taylor (2010). An overview of second generation biofuel technologies. *Bioresource Technology*, **101**(6), pp. 1570-1580.

Skjoldborg, B. (2010). Optimization of I/S Skive District Heating Plant. In: IEA Joint Task 32 & 33 Workshop, *State-of-the-Art Technologies for Small Biomass Co-generation*, International Energy Agency, Copenhagen, Denmark, 7 Oct 2010. Available at: www.ieabcc.nl/meetings/task32_Copenhagen/11%20Skive.pdf.

Smeets, E.M.W., and A.P.C. Faaij (2007). Bioenergy potentials from forestry in 2050. *Climatic Change*, **81**(3-4), pp. 353-390.

Smeets, E.M.W., A. Faaij, I. Lewandowski, and W. Turkenburg (2007). A bottom-up assessment and review of global bio-energy potentials to 2050. *Progress in Energy and Combustion Science*, **33**(1), pp. 56-106.

Smeets, E.M.W., L.F. Bouwman, E. Stehfest, D.P. van Vuuren, and A. Posthuma (2009). Contribution of N_2O to the greenhouse gas balance of first-generation biofuels. *Global Change Biology*, **15**(1), pp. 1-23.

Smil, V. (2002). Worldwide transformation of diets, burdens of meat production and opportunities for novel food proteins. *Enzyme and Microbial Technology*, **30**(3), pp. 305-311.

Smith, K.R., and E. Haigler (2008). Co-benefits of climate mitigation and health protection in energy systems: Scoping methods. *Annual Review of Public Health*, **29**(1), pp. 11-25.

Smith, K.R., R. Uma, V.V.N. Kishore, J. Zhang, V. Joshi, and M.A.K. Khalil (2000). Greenhouse implications of household stoves: An analysis for India. *Annual Review of Energy and the Environment*, **25**(1), pp. 741-763.

Soimakallio, S., R. Antikainen, and R. Thun (2009a). *Assessing the Sustainability of Liquid Biofuels from Evolving Technologies*. VTT Research Notes 2482, VTT Technical Research Centre of Finland, Espoo, Finland, 268 pp.

Soimakallio, S., T. Mäkinen, T. Ekholm, K. Pahkala, H. Mikkola, and T. Paappanen (2009b). Greenhouse gas balances of transportation biofuels, electricity and heat generation in Finland – Dealing with the uncertainties. *Energy Policy*, **37**(1), pp. 80-90.

Solomon, S., G.K. Plattner, R. Knutti, and P. Friedlingstein (2009). Irreversible climate change due to carbon dioxide emissions. *Proceedings of the National Academy of Sciences*, **106**(6), pp. 1704-1709.

Sparovek, G., G. Berndes, A. Egeskog, F.L.M. de Freitas, S. Gustafsson, and J. Hansson (2007). Sugarcane ethanol production in Brazil: an expansion model sensitive to socioeconomic and environmental concerns. *Biofuels, Bioproducts and Biorefining*, **1**(4), pp. 270-282.

Spatari, S., and H.L. MacLean (2010). Characterizing model uncertainties in the life cycle of lignocellulose-based ethanol fuels. *Environmental Science & Technology*, **44**(22), pp. 8773-8780.

Spranger, T., J.P. Hettelingh, J. Slootweg, and M. Posch (2008). Modelling and mapping long-term risks due to reactive nitrogen effects: An overview of LRTAP convention activities. *Environmental Pollution*, **154**(3), pp. 482-487.

Steenblik, R. (2007). *Subsidies: The Distorted Economics of Biofuels.* Discussion Paper No. 2007-3, International Transport Forum, Organisation for Economic Co-operation and Development, Geneva, Switzerland, 66 pp.

Stehfest, E., and L. Bouwman (2006). N_2O and NO emission from agricultural fields and soils under natural vegetation: summarizing available measurement data and modeling of global annual emissions. *Nutrient Cycling in Agroecosystems*, **74**(3), pp. 207-228.

Stehfest, E., L. Bouwman, D.P. Vuuren, M.G.J. Elzen, B. Eickhout, and P. Kabat (2009). Climate benefits of changing diet. *Climatic Change*, **95**(1-2), pp. 83-102.

Still, D., M. Pinnell, D. Ogle, and B. van Appel (2003). Insulative ceramics for improved cooking stoves. *Boiling Point*, **49**, pp. 7-10.

Stoft, S. (2010). *Renewable Fuel and the Global Rebound Effect.* Research Paper No. 10-06, Global Energy policy Center, Berkeley, CA, USA, 19 pp.

Strengers, B., R. Leemans, B. Eickhout, B. Vries, and L. Bouwman (2004). The land-use projections and resulting emissions in the IPCC SRES scenarios as simulated by the IMAGE 2.2 model. *GeoJournal*, **61**(4), pp. 381-393.

Sulser, T.B., C. Ringler, T. Zhu, S. Msangi, E. Bryan, and M.W. Rosegrant (2010). Green and blue water accounting in the Ganges and Nile basins: Implications for food and agricultural policy. *Journal of Hydrology*, **384**(3-4), pp. 276-291.

Sumner, S.A., and P.M. Layde (2009). Expansion of renewable energy industries and implications for occupational health. *Journal of the American Medical Association*, **302**(7), pp. 787-789.

Sustainable Transport Solutions (2006). *Biogas as a Road Transport Fuel: An Assessment of the Potential Role of Biogas as a Renewable Transport Fuel.* National Society for Clean Air and Environmental Protection, London, UK, 66 pp.

Swanson, R.M., A. Platon, J.A. Satrio, and R.C. Brown (2010). Techno-economic analysis of biomass-to-liquids production based on gasification. *Fuel*, **89**(Supplement 1), pp. S11-S19.

Tabeau, A., B. Eickhout, and H. van Meijl (2006). Endogenous agricultural land supply: estimation and implementation in the GTAP model. In: *Ninth Annual Conference on Global Economic Analysis*, Addis Ababa, Ethiopia, 15-17 June 2006. Available at: https://www.gtap.agecon.purdue.edu/resources/res_display.asp?RecordID=2007.

Tao, L., and A. Aden (2009). The economics of current and future biofuels. *In Vitro Cellular & Developmental Biology - Plant*, **45**(3), pp. 199-217.

Thorn, J., J. Brisman, and K. Toren (2001). Adult-onset asthma is associated with self-reported mold or environmental tobacco smoke exposures in the home. *Allergy*, **56**(4), pp. 287-292.

Thornley, P., J. Rogers, and Y. Huang (2008). Quantification of employment from biomass power plants. *Renewable Energy*, **33**(8), pp. 1922-1927.

Thornley, P., P. Upham, Y. Huang, S. Rezvani, J. Brammer, and J. Rogers (2009). Integrated assessment of bioelectricity technology options. *Energy Policy*, **37**(3), pp. 890-903.

Tilman, D., K.G. Cassman, P.A. Matson, R. Naylor, and S. Polasky (2002). Agricultural sustainability and intensive production practices. *Nature*, **418**(6898), pp. 671-677.

Tilman, D., J. Hill, and C. Lehman (2006a). Carbon-negative biofuels from low-input high-diversity grassland biomass. *Science*, **314**(5805), pp. 1598-1600.

Tilman, D., P.B. Reich, and J.M.H. Knops (2006b). Biodiversity and ecosystem stability in a decade-long grassland experiment. *Nature*, **441**(7093), pp. 629-632.

Titus, B.D., D.G. Maynard, C.C. Dymond, G. Stinson, and W.A. Kurz (2009). Wood energy: Protect local ecosystems. *Science*, **324**(5933), pp. 1389-1390.

Tobias, C.M., G. Sarath, P. Twigg, E. Lindquist, J. Pangilinan, B.W. Penning, K. Barry, M.C. McCann, N.C. Carpita, and G.R. Lazo (2008). Comparative genomics in switchgrass using 61,585 high-quality expressed sequence tags. *The Plant Genome*, **1**(2), pp. 111-124.

Tonkovich, A.Y., S. Perry, Y. Wang, D. Qiu, T. LaPlante, and W.A. Rogers (2004). Microchannel process technology for compact methane steam reforming. *Chemical Engineering Science*, **59**(22-23), pp. 4819-4824.

Towle, D.W., and J.S. Pearse (1973). Production of the giant kelp, Macrocystis, estimated by in situ incorporation of ^{14}C in polyethylene bags. *American Society of Limnology and Oceanography*, **18**(1), pp. 155-159.

Turhollow, A. (1994). The economics of energy crop production. *Biomass and Bioenergy*, **6**(3), pp. 229-241.

Tyner, W.E., and F. Taheripour (2008). Policy options for integrated energy and agricultural markets. *Applied Economic Perspectives and Policy*, **30**(3), pp. 387-396.

Tyner, W., F. Taheripour, Q. Zhuang, D. Birur, and U. Baldos (2010). *Land Use Changes and Consequent CO_2 Emissions due to U.S. Corn Ethanol Production: A Comprehensive Analysis.* GTAP Resource 3288, Department of Agricultural Economics, Purdue University, West Lafayette, IN, USA, 90 pp.

UN-Water (2007). *Coping with Water Scarcity: Challenge of the Twenty-First Century.* Prepared for World Water Day 2007, United Nations, New York, NY, USA, 29 pp.

UNECE/FAO TIMBER Database (2011). Market Presentations at http: timber.unece. org/. United Nations Economic Commission for Europe and Food and Agricultural Organization.

UNEP (2008a). *UNEP Year Book 2008: an Overview of Our Changing Environment.* United Nations Environment Programme, Nairobi, Kenya.

UNEP (2008b). *The Potential Impacts of Biofuels on Biodiversity, Notes by the Executive Secretary.* UNEP/CBD/COP/9/26, United Nations Environment Programme, Bonn, Germany, 16 pp.

UNEP/SEFI/Bloomberg (2010). *Global Trends in Sustainable Energy Investment 2010.* DTI/1186/PA, Sustainable Energy Finance Initiative, United Nations Environment Programme and Bloomberg New Energy Finance, Geneva, Switzerland, 61 pp.

UK DfT (2003). *International Resource Costs of Biodiesel and Bioethanol.* AEAT/ENV/ ED50273/R1, United Kingdom Department for Transport, Oxfordshire, UK, 51 pp.

University of Illinois (2011). *farmdoc: Historical Corn Prices.* University of Illinois, Urbana, Illinois, USA. Available at: www.farmdoc.illinois.edu/manage/pricehistory/price_history.htm.

US DOE (2009). *National Algal Biofuels Technology Roadmap.* U.S. Department of Energy Biomass Program, Washington, DC, USA, 214 pp.

US DOE (2011). *Biomass as Feedstock for a Bioenergy and Bioproducts Industry: An Update to the Billion-Ton Annual Supply.* R.D. Perlack and B.J. Stokes (Leads), ORNL/TM-2010/xx. U.S. Department of Energy, Oak Ridge National Laboratory, Oak Ridge, TN, USA, in press.

USDA (2006). *Oil Crops Yearbook/OCS-2006/Mar21.* Economic Research Service, US Department of Agriculture (USDA), Washington, DC, USA.

USDA (2007). *Wheat Data: Yearbook Tables.* Economic Research Service, US Department of Agriculture (USDA), Washington, DC, USA.

Uslu, A., A.P.C. Faaij, and P.C.A. Bergman (2008). Pre-treatment technologies, and their effect on international bioenergy supply chain logistics. Techno-economic evaluation of torrefaction, fast pyrolysis and pelletisation. *Energy*, **33**(8), pp. 1206-1223.

van Dam, J., M. Junginger, A. Faaij, I. Jürgens, G. Best, and U. Fritsche (2008). Overview of recent developments in sustainable biomass certification. *Biomass and Bioenergy*, **32**(8), pp. 749-780.

van Dam, J., A.P.C. Faaij, J. Hilbert, H. Petruzzi, and W.C. Turkenburg (2009a). Large-scale bioenergy production from soybeans and switchgrass in Argentina: Part A: Potential and economic feasibility for national and international markets. *Renewable and Sustainable Energy Reviews*, **13**(8), pp. 1710-1733.

van Dam, J., A.P.C. Faaij, J. Hilbert, H. Petruzzi, and W.C. Turkenburg (2009b). Large-scale bioenergy production from soybeans and switchgrass in Argentina: Part B. Environmental and socio-economic impacts on a regional level. *Renewable and Sustainable Energy Reviews*, **13**(8), pp. 1679-1709.

van Dam, J., A.P.C. Faaij, I. Lewandowski, and B. Van Zeebroeck (2009c). Options of biofuel trade from Central and Eastern to Western European countries. *Biomass and Bioenergy*, **33**(4), pp. 728-744.

van Dam, J., M. Junginger, and A.P.C. Faaij (2010). From the global efforts on certification of bioenergy towards an integrated approach based on sustainable land use planning. *Renewable and Sustainable Energy Reviews*, **14**(9), pp. 2445-2472.

van den Wall Bake, J.D. (2006). *Cane as Key in Brazilian Ethanol Industry.* Master's Thesis, Copernicus Institute, Utrecht University, Utrecht, The Netherlands, 83 pp.

van den Wall Bake, J.D., M. Junginger, A. Faaij, T. Poot, and A. Walter (2009). Explaining the experience curve: Cost reductions of Brazilian ethanol from sugarcane. *Biomass and Bioenergy*, **33**(4), pp. 644-658.

van der Horst, D., and J. Evans (2010). Carbon claims and energy landscapes: Exploring the political ecology of biomass. *Landscape Research*, **35**(2), pp. 173-193.

van Iersel, S., L. Gamba, A. Rossi, S. Alberici, B. Dehue, J. van de Staaij, and A. Flammini (2009). *Algae-Based Biofuels: A Review of Challenges and Opportunities for Developing Countries.* Food and Agriculture Organization, Rome, Italy, 59 pp.

van Loo, S., and J. Koppejan (eds.) (2002). *Handbook of Biomass Combustion and Cofiring.* 1st ed. Twente University Press, Enschede, The Netherlands, 442 pp.

van Minnen, J.G., B.J. Strengers, B. Eickhout, R.J. Swart, and R. Leemans (2008). Quantifying the effectiveness of climate change mitigation through forest plantations and carbon sequestration with an integrated land-use model. *Carbon Balance and Management*, **3**(3), 20 pp.

van Vliet, O.P.R., A.P.C. Faaij, and W.C. Turkenburg (2009). Fischer-Tropsch diesel production in a well-to-wheel perspective: A carbon, energy flow and cost analysis. *Energy Conversion and Management*, **50**(4), pp. 855-876.

van Vuuren, D., M. den Elzen, P. Lucas, B. Eickhout, B. Strengers, B. van Ruijven, S. Wonink, and R. van Houdt (2007). Stabilizing greenhouse gas concentrations at low levels: an assessment of reduction strategies and costs. *Climatic Change*, **81**(2), pp. 119-159.

van Vuuren, D.P., J. van Vliet, and E. Stehfest (2009). Future bio-energy potential under various natural constraints. *Energy Policy*, **37**(11), pp. 4220-4230.

van Vuuren, D.P., E. Bellevrat, A. Kitous, and M. Isaac (2010). Bio-energy use and low stabilization scenarios. *The Energy Journal*, **31**(Special Issue), pp. 192-222.

van Zyl, W., L. Lynd, R. den Haan, and J. McBride (2007). Consolidated bioprocessing for bioethanol production using Saccharomyces cerevisiae. *Advanced Biochemical Engineering Biotechnology*, **108**, pp. 205-235.

Vandermeer, J., and I. Perfecto (2006). Response to comments on "A Keystone Mutualism Drives Pattern in a Power Function". *Science*, **313**(5794), pp. 1739.

Venkataraman, C., A.D. Sagar, G. Habib, N. Lam, and K.R. Smith (2010). The Indian National Initiative for Advanced Biomass Cookstoves: The benefits of clean combustion. *Energy for Sustainable Development*, **14**(2), pp. 63-72.

Venter, O., E. Meijaard, H. Possingham, R. Dennis, D. Sheil, S. Wich, L. Hovani, and K. Wilson (2009). Carbon payments as a safeguard for threatened tropical mammals. *Conservation Letters*, **2**(3), pp. 123-129.

Verhaeven, E., L. Pelkmans, L. Govaerts, R. Lamers, and F. Theunissen (2005). Results of demonstration and evaluation projects of biodiesel from rapeseed and used frying oil on light and heavy duty vehicles. In: *2005 SAE Brasil Fuels & Lubricants Meeting*. SAE International, Rio De Janeiro, Brasil, May 2005, 7 pp., doi:10.4271/2005-01-2201.

Vinnerås, B., C. Schönning, and A. Nordin (2006). Identification of the microbiological community in biogas systems and evaluation of microbial risks from gas usage. *Science of the Total Environment*, **367**(2-3), pp. 606-615.

von Blottnitz, H., and M.A. Curran (2007). A review of assessments conducted on bio-ethanol as a transportation fuel from a net energy, greenhouse gas, and environmental life cycle perspective. *Journal of Cleaner Production*, **15**(7), pp. 607-619.

von Geibler, J., C. Liedtke, H. Wallbaum, and S. Schaller (2006). Accounting for the social dimension of sustainability: experiences from the biotechnology industry. *Business Strategy and the Environment*, **15**(5), pp. 334-346.

von Schirnding, Y., N. Bruce, K. Smith, G. Ballard-Treemer, M. Ezzati, and K. Lvovsky (2001). *Addressing the Impact of Household Energy and Indoor Air Pollution on the Health of the Poor - Implications for Policy Action and Intervention Measures*. WHO/HDE/HID/02.9, Commission on Macroeconomics and Health, World Health Organization, Geneva, Switzerland, 52 pp.

von Weyman, N. (2007). *Bioetanolia maatalouden selluloosavirroista (Bioethanol from agricultural lignocellulosic residues)*. VTT Tiedotteita 2412, VTT Technical Research Centre of Finland, Espoo, Finland, 48 pp.

Walmsley, J.D., and D.L. Godbold (2010). Stump harvesting for bioenergy - A review of the environmental impacts. *Forestry*, **83**(1), pp. 17-38.

Walsh, M.E. (2008). *U.S. Cellulosic Biomass Feedstock Supplies and Distribution*. Ag Econ Search - Research in Agricultural and Applied Economics, University of Minnesota, St. Paul, Minnesota, 47 pp. Available at: ageconsearch.umn.edu/bitstream/7625/2/U.S.%20Biomass%20Supplies.pdf.

Walter, A., F. Rosillo-Calle, P. Dolzan, E. Piacente, and K. Borges da Cunha (2008). Perspectives on fuel ethanol consumption and trade. *Biomass and Bioenergy*, **32**(8), pp. 730-748.

Wang, M., C. Saricks, and D. Santini (1999). *Effects of fuel ethanol use on fuel-cycle energy and greenhouse gas emissions*. ANL/ESD-38, Argonne Energy Systems Division, Argonne National Laboratory, Argonne, IL, USA, 39 pp.

Wang, M., M. Wu, and H. Huo (2007). Life-cycle energy and greenhouse gas emission impacts of different corn ethanol plant types. *Environmental Research Letters*, **2**(2), 024001.

Wang, M., H. Huo, and S. Arora (2010). Methods of dealing with co-products of biofuels in life-cycle 3 analysis and consequent results within the U.S. context. *Energy Policy*, in press, doi:10.1016/j.enpol.2010.03.052.

Wang, M.Q., J. Han, Z. Haq, W.E. Tyner, M. Wu, and A. Elgowainy (2011). Energy and greenhouse gas emission effects of corn and cellulosic ethanol with technology improvements and land use changes. *Biomass and Bioenergy*, **35**(5), pp. 1885-1896.

Warwick, S.I., H.J. Beckie, and L.M. Hall (2009). Gene flow, invasiveness, and ecological impact of genetically modified crops. *Annals of the New York Academy of Sciences*, **1168**(1), pp. 72-99.

WBGU (2009). *World in Transition – Future Bioenergy and Sustainable Land Use*. German Advisory Council on Global Change (WBGU), Berlin, Germany, 393 pp. (ISBN 978-1-84407-841-7).

Weyer, K., D. Bush, A. Darzins, and B. Willson (2009). Theoretical maximum algal oil production. *BioEnergy Research*, **3**(2), pp. 204-213.

Wicke, B., V. Dornburg, M. Junginger, and A. Faaij (2008). Different palm oil production systems for energy purposes and their greenhouse gas implications. *Biomass and Bioenergy*, **32**(12), pp. 1322-1337.

Wicke, B., E. Smeets, A. Tabeau, J. Hilbert, and A. Faaij (2009). Macroeconomic impacts of bioenergy production on surplus agricultural land – A case study of Argentina. *Renewable and Sustainable Energy Reviews*, **13**(9), pp. 2463-2473.

Wilhelm, W.W., J.M.F. Johnson, J.L. Hatfield, W.B. Voorhees, and D.R. Linden (2004). Crop and soil productivity response to corn residue removal: A literature review. *Agronomy Journal*, **96**(1), pp. 1-17.

Wilhelm, W.W., J.M.E. Johnson, D.L. Karlen, and D.T. Lightle (2007). Corn stover to sustain soil organic carbon further constrains biomass supply. *Agronomy Journal*, **99**, pp. 1665-1667.

Wilkie, A.C., and J.M. Evans (2010). Aquatic plants: an opportunity feedstock in the age of bioenergy. *Biofuels*, **1**(2), pp. 311-321.

Wilkie, A.C., K.J. Riedesel, and J.M. Owens (2000). Stillage characterization and anaerobic treatment of ethanol stillage from conventional and cellulosic feedstocks. *Biomass and Bioenergy*, **19**(2), pp. 63-102.

Wilkinson, P., K.R. Smith, M. Davies, H. Adair, B.G. Armstrong, M. Barrett, N. Bruce, A. Haines, I. Hamilton, T. Oreszczyn, I. Ridley, C. Tonne, and Z. Chalabi (2009). Public health benefits of strategies to reduce greenhouse-gas emissions: household energy. *The Lancet*, **374**(9705), pp. 1917-1929.

Williams, R.H., E.D. Larson, G. Liu, and T.G. Kreutz (2009). Fischer-Tropsch fuels from coal and biomass: Strategic advantages of once-through ("polygeneration") configurations. *Energy Procedia*, **1**(1), pp. 4379-4386.

Wirsenius, S. (2003). Efficiencies and biomass appropriation of food commodities on global and regional levels. *Agricultural Systems*, **77**(3), pp. 219-255.

Wirsenius, S., C. Azar, and G. Berndes (2010). How much land is needed for global food production under scenarios of dietary changes and livestock productivity increases in 2030? *Agricultural Systems*, **103**(9), pp. 621-638.

Wise, M., K. Calvin, A. Thomson, L. Clarke, B. Bond-Lamberty, R. Sands, S.J. Smith, A. Janetos, and J. Edmonds (2009). Implications of limiting CO_2 concentrations for land use and energy. *Science*, **324**(5931), pp. 1183-1186.

Wiskerke, W.T., V. Dornburg, C.D.K. Rubanza, R.E. Malimbwi, and A.P.C. Faaij (2010). Cost/benefit analysis of biomass energy supply options for rural smallholders in the semi-arid eastern part of Shinyanga Region in Tanzania. *Renewable and Sustainable Energy Reviews*, **14**(1), pp. 148-165.

Wolf, J., P.S. Bindraban, J.C. Luijten, and L.M. Vleeshouwers (2003). Exploratory study on the land area required for global food supply and the potential global production of bioenergy. *Agricultural Systems*, **76**, pp. 841-861.

Wooley, R., M. Ruth, D. Glassner, and J. Sheehan (1999). Process design and costing of bioethanol technology: A tool for determining the status and direction of research and development. *Biotechnology Progress*, **15**(5), pp. 794-803.

Woolf, D., J.E. Amonette, F.A. Street-Perrott, J. Lehmann, and S. Joseph (2010). Sustainable biochar to mitigate global climate change. *Nature Communications*, **1**(56), pp. 1-9.

World Bank (2009). *The World Bank - Global Economic Prospects - Commodities at the Crossroad.* The World Bank, The International Bank for Reconstruction and Development, Washington, DC, USA, 196 pp. (see page 73).

World Bank (2010). *Improved Cookstoves and Better Health in Bangladesh: Lessons from Household Energy and Sanitation Programs.* The World Bank, The International Bank for Reconstruction and Development, Washington, DC, USA, 136 pp.

Wright, B. (2009). *International Gain Reserves and Other Instruments to Address Volatility in Grain Market.* Policy Research Working Paper 5028, World Bank Group, Washington, DC, USA, 61 pp.

Wright, D.H. (1990). Human impacts on energy flow through natural ecosystems, and implications for species endangerment. *AMBIO: A Journal of the Human Environment*, **19**(4), pp. 189-194.

Wu, M., Y. Wu, and M. Wang (2005). *Mobility Chains Analysis of Technologies for Passenger Cars and Light Duty Vehicles Fueled with Biofuels: Application of the Greet Model to Project the Role of Biomass in America's Energy Future (RBAEF) project.* ANL/ESD/07-11, Argonne National Laboratory, Argonne, IL, USA, 84 pp.

Wu, M., M. Wang, J. Liu, and H. Huo (2008). Assessment of potential life-cycle energy and greenhouse gas emission effects from using corn-based butanol as a transportation fuel. *Biotechnology Progress*, **24**(6), pp. 1204-1214.

Wu, M., M. Mintz, M. Wang, and S. Arora (2009). Water consumption in the production of ethanol and petroleum gasoline. *Environmental Management*, **44**(5), pp. 981-997.

WWI (2006). *Biofuels for Transportation – Global Potential and Implications for Sustainable Agriculture and Energy in the 21st Century.* World Watch Institute, Washington, DC, USA, 417 pp.

Wydra, S. (2009). *Production and Employment Impacts of New Technologies – Analysis for Biotechnology.* Discussion Paper 08-2009, Forschungszentrum Innovation und Dienstleistung, Universität Hohenheim, Stuttgart, Germany, 31 pp.

Wyman, C.E., B.E. Dale, R.T. Elander, M. Holtzapple, M.R. Ladisch, and Y.Y. Lee (2005). Coordinated development of leading biomass pretreatment technologies. *Bioresource Technology*, **96**(18), pp. 1959-1966.

Yamashita, K., and L. Barreto (2004). *Biomass Gasification for the Co-production of Fischer-Tropsch Liquids and Electricity.* Interim Report IR-04-047, International Institute for Applied Systems Analysis (IIASA), Laxenburg, Austria, 50 pp.

Yanowitz, J., and R.L. McCormick (2009). Effect of biodiesel blends on North American heavy-duty diesel engine emissions. *European Journal of Lipid Science and Technology*, **111**(8), pp. 763-772.

Yazdani, S.S., and R. Gonzalez (2007). Anaerobic fermentation of glycerol: a path to economic viability for the biofuels industry. *Current Opinion in Biotechnology*, **18**(3), pp. 213-219.

Yeh, S., S.M. Jordaan, A.R. Brandt, M.R. Turetsky, S. Spatari, and D.W. Keith (2010). Land use greenhouse gas emissions from conventional oil production and oil sands. *Environmental Science & Technology*, **44**(22), pp. 8766-8772.

Yokoyama, S., and Y. Matsumura (eds.) (2008). *The Asian Biomass Handbook: A Guide for Biomass Production and Utilization.* The Japan Institute of Energy, Tokyo, Japan, 326 pp.

Zabowski, D., D. Chambreau, N. Rotramel, and W.G. Thies (2008). Long-term effects of stump removal to control root rot on forest soil bulk density, soil carbon and nitrogen content. *Forest Ecology and Management*, **255**(3-4), pp. 720-727.

Zemke-White, L., and M. Ohno (1999). World seaweed utilisation: An end-of-century summary. *Journal of Applied Phycology*, **11**(4), pp. 369-376.

Zezza, A., B. Davis, C. Azzarri, K. Covarrubias, L. Tasciotti, and G. Anriquez (2008). *The Impact of Rising Food Prices on the Poor.* ESA Working Paper No. 08-07, Agricultural Development Economics Division, Food and Agriculture Organization, Rome, Italy, 37 pp.

Zhang, X., R.C. Izaurralde, D. Manowitz, T.O. West, W.M. Post, A.M. Thomson, V.P. Bandaru, J. Nichols, and J.R. Williams (2010). An integrative modeling framework to evaluate the productivity and sustainability of biofuel crop production systems. *Global Change Biology Bioenergy*, **2**(5), pp. 258-277.

Zomer, R.J., A. Trabucco, O. van Straaten, and D.A. Bossio (2006). *Carbon, Land and Water: A Global Analysis of the Hydrologic Dimensions of Climate Change Mitigation through Afforestation/Reforestation.* IWMI Research Report 101, International Water Management Institute, Colombo, Sri Lanka, 48 pp.

3 Direct Solar Energy

Coordinating Lead Authors:
Dan Arvizu (USA) and Palani Balaya (Singapore/India)

Lead Authors:
Luisa F. Cabeza (Spain), K.G. Terry Hollands (Canada), Arnulf Jäger-Waldau (Italy/Germany),
Michio Kondo (Japan), Charles Konseibo (Burkina Faso), Valentin Meleshko (Russia),
Wesley Stein (Australia), Yutaka Tamaura (Japan), Honghua Xu (China),
Roberto Zilles (Brazil)

Contributing Authors:
Armin Aberle (Singapore/Germany), Andreas Athienitis (Canada), Shannon Cowlin (USA),
Don Gwinner (USA), Garvin Heath (USA), Thomas Huld (Italy/Denmark), Ted James (USA),
Lawrence Kazmerski (USA), Margaret Mann (USA), Koji Matsubara (Japan),
Anton Meier (Switzerland), Arun Mujumdar (Singapore), Takashi Oozeki (Japan),
Oumar Sanogo (Burkina Faso), Matheos Santamouris (Greece), Michael Sterner (Germany),
Paul Weyers (Netherlands)

Review Editors:
Eduardo Calvo (Peru) and Jürgen Schmid (Germany)

This chapter should be cited as:

Arvizu, D., P. Balaya, L. Cabeza, T. Hollands, A. Jäger-Waldau, M. Kondo, C. Konseibo, V. Meleshko, W. Stein, Y. Tamaura, H. Xu, R. Zilles, 2011: Direct Solar Energy. In IPCC Special Report on Renewable Energy Sources and Climate Change Mitigation [O. Edenhofer, R. Pichs-Madruga, Y. Sokona, K. Seyboth, P. Matschoss, S. Kadner, T. Zwickel, P. Eickemeier, G. Hansen, S. Schlömer, C. von Stechow (eds)], Cambridge University Press, Cambridge, United Kingdom and New York, NY, USA.

Table of Contents

Executive Summary 337

3.1 Introduction 340

3.2 Resource potential 341
3.2.1 Global technical potential 341
3.2.2 Regional technical potential 342
3.2.3 Sources of solar irradiance data 342
3.2.4 Possible impact of climate change on resource potential 343

3.3 Technology and applications 343
3.3.1 Passive solar and daylighting technologies 344
3.3.2 Active solar heating and cooling 346
3.3.2.1 Solar heating 346
3.3.2.2 Solar cooling 349
3.3.2.3 Thermal storage 349
3.3.2.4 Active solar heating and cooling applications 350
3.3.3 Photovoltaic electricity generation 351
3.3.3.1 Existing photovoltaic technologies 351
3.3.3.2 Emerging photovoltaic technologies 352
3.3.3.3 Novel photovoltaic technologies 353
3.3.3.4 Photovoltaic systems 353
3.3.3.5 Photovoltaic applications 353
3.3.4 Concentrating solar power electricity generation 355
3.3.5 Solar fuel production 358

3.4 Global and regional status of market and industry development 359
3.4.1 Installed capacity and generated energy 359
3.4.2 Industry capacity and supply chain 362
3.4.3 Impact of policies 366

3.5	Integration into the broader energy system	367
3.5.1	Low-capacity electricity demand	367
3.5.2	District heating and other thermal loads	367
3.5.3	Photovoltaic generation characteristics and the smoothing effect	368
3.5.4	Concentrating solar power generation characteristics and grid stabilization	369
3.6	Environmental and social impacts	369
3.6.1	Environmental impacts	369
3.6.2	Social impacts	372
3.7	Prospects for technology improvements and innovation	373
3.7.1	Passive solar and daylighting technologies	373
3.7.2	Active solar heating and cooling	374
3.7.3	Photovoltaic electricity generation	375
3.7.4	Concentrating solar power electricity generation	377
3.7.5	Solar fuel production	377
3.7.6	Other potential future applications	378
3.8	Cost trends	378
3.8.1	Passive solar and daylighting technologies	378
3.8.2	Active solar heating and cooling	379
3.8.3	Photovoltaic electricity generation	380
3.8.4	Concentrating solar power electricity generation	382
3.8.5	Solar fuel production	385

3.9	**Potential deployment**	386
3.9.1	Near-term forecasts	386
3.9.2	Long-term deployment in the context of carbon mitigation	386
3.9.3	Conclusions regarding deployment	390

References 391

Executive Summary

Solar energy is abundant and offers significant potential for near-term (2020) and long-term (2050) climate change mitigation. There are a wide variety of solar technologies of varying maturities that can, in most regions of the world, contribute to a suite of energy services. Even though solar energy generation still only represents a small fraction of total energy consumption, markets for solar technologies are growing rapidly. Much of the desirability of solar technology is its inherently smaller environmental burden and the opportunity it offers for positive social impacts. The cost of solar technologies has been reduced significantly over the past 30 years and technical advances and supportive public policies continue to offer the potential for additional cost reductions. Potential deployment scenarios range widely—from a marginal role of direct solar energy in 2050 to one of the major sources of energy supply. The actual deployment achieved will depend on the degree of continued innovation, cost reductions and supportive public policies.

Solar energy is the most abundant of all energy resources. Indeed, the rate at which solar energy is intercepted by the Earth is about 10,000 times greater than the rate at which humankind consumes energy. Although not all countries are equally endowed with solar energy, a significant contribution to the energy mix from direct solar energy is possible for almost every country. Currently, there is no evidence indicating a substantial impact of climate change on regional solar resources.

Solar energy conversion consists of a large family of different technologies capable of meeting a variety of energy service needs. Solar technologies can deliver heat, cooling, natural lighting, electricity, and fuels for a host of applications. Conversion of solar energy to *heat* (i.e., thermal conversion) is comparatively straightforward, because any material object placed in the sun will absorb thermal energy. However, maximizing that absorbed energy and stopping it from escaping to the surroundings can take specialized techniques and devices such as evacuated spaces, optical coatings and mirrors. Which technique is used depends on the application and temperature at which the heat is to be delivered. This can range from 25°C (e.g., for swimming pool heating) to 1,000°C (e.g., for dish/Stirling concentrating solar power), and even up to 3,000°C in solar furnaces.

Passive solar heating is a technique for maintaining comfortable conditions in buildings by exploiting the solar irradiance incident on the buildings through the use of glazing (windows, sun spaces, conservatories) and other transparent materials and managing heat gain and loss in the structure without the dominant use of pumps or fans. Solar *cooling* for buildings can also be achieved, for example, by using solar-derived heat to drive thermodynamic refrigeration absorption or adsorption cycles. Solar energy for lighting actually requires no conversion since solar lighting occurs naturally in buildings through windows. However, maximizing the effect requires specialized engineering and architectural design.

Generation of *electricity* can be achieved in two ways. In the first, solar energy is converted directly into electricity in a device called a photovoltaic (PV) cell. In the second, solar thermal energy is used in a concentrating solar power (CSP) plant to produce high-temperature heat, which is then converted to electricity via a heat engine and generator. Both approaches are currently in use. Furthermore, solar driven systems can deliver process heat and cooling, and other solar technologies are being developed that will deliver energy carriers such as hydrogen or hydrocarbon fuels—known as *solar fuels*.

The various solar technologies have differing maturities, and their applicability depends on local conditions and government policies to support their adoption. Some technologies are already competitive with market prices in certain locations, and in general, the overall viability of solar technologies is improving. Solar thermal can be used for a wide variety of applications, such as for domestic hot water, comfort heating of buildings, and industrial process heat. This is significant, as many countries spend up to one-third of their annual energy usage for heat. Service hot water heating for domestic and commercial buildings is now a mature technology growing at a rate of about 16% per year and employed in most countries of the world. The world installed capacity of solar thermal systems at the end of 2009 has been estimated to be 180 GW_{th}.

Passive solar and daylighting are conserving energy in buildings at a highly significant rate, but the actual amount is difficult to quantify. Well-designed passive solar systems decrease the need for additional comfort heating requirements by about 15% for existing buildings and about 40% for new buildings.

The generation of electricity using PV panels is also a worldwide phenomenon. Assisted by supportive pricing policies, the compound annual growth rate for PV production from 2003 to 2009 was more than 50%—making it one of the fastest-growing energy technologies in percentage terms. As of the end of 2009, the installed capacity for PV power production was about 22 GW. Estimates for 2010 give a consensus value of about 13 GW of newly added capacity. Most of those installations are roof-mounted and grid-connected. The production of electricity from CSP installations has seen a large increase in planned capacity in the last few years, with several countries beginning to experience significant new installations.

Integration of solar energy into broader energy systems involves both challenges and opportunities. Energy provided by PV panels and solar domestic water heaters can be especially valuable because the energy production often occurs at times of peak loads on the grid, as in cases where there is a large summer daytime load associated with air conditioning. PV and solar domestic water heaters also fit well with the needs of many countries because they are modular, quick to install, and can sometimes delay the need for costly construction or expansion of the transmission grid. At the same time, solar energy typically has a variable production profile with some degree of unpredictability that must be managed, and central-station solar electricity plants may require new transmission infrastructure. Because CSP can be readily coupled with thermal storage, the production profile can be controlled to limit production variability and enable dispatch capability.

Solar technologies offer opportunities for positive social impacts, and their environmental burden is small. Solar technologies have low lifecycle greenhouse gas emissions, and quantification of external costs has yielded favourable values compared to fossil fuel-based energy. Potential areas of concern include recycling and use of toxic materials in manufacturing for PV, water usage for CSP, and energy payback and land requirements for both. An important social benefit of solar technologies is their potential to improve the health and livelihood opportunities for many of the world's poorest populations—addressing some of the gap in availability of modern energy services for the roughly 1.4 billion people who do not have access to electricity and the 2.7 billion people who rely on traditional biomass for home cooking and heating needs. On the downside, some solar projects have faced public concerns regarding land requirements for centralized CSP and PV plants, perceptions regarding visual impacts, and for CSP, cooling water requirements. Land use impacts can be minimized by selecting areas with low population density and low environmental sensitivity. Similarly, water usage for CSP could be significantly reduced by using dry cooling approaches. Studies to date suggest that none of these issues presents a barrier against the widespread use of solar technologies.

Over the last 30 years, solar technologies have seen very substantial cost reductions. The current levelized costs of energy (electricity and heat) from solar technologies vary widely depending on the upfront technology cost, available solar irradiation as well as the applied discount rates. The levelized costs for solar thermal energy at a 7% discount rate range between less than USD_{2005} 10 and slightly more than USD_{2005} 20/GJ for solar hot water generation with a high degree of utilization in China to more than USD_{2005} 130/GJ for space heating applications in Organisation for Economic Co-operation and Development (OECD) countries with relative low irradiation levels of 800 kWh/m^2/yr. Electricity generation costs for utility-scale PV in regions of high solar irradiance in Europe and the USA are in the range of approximately 15 to 40 US $cents_{2005}$/kWh at a 7% discount rate, but may be lower or higher depending on the available resource and on other framework conditions. Current cost data are limited for CSP and are highly dependent on other system factors such as storage. In 2009, the levelized costs of energy for large solar troughs with six hours of thermal storage ranged from below 20 to approximately 30 US $cents_{2005}$/kWh. Technological improvements and cost reductions are expected, but the learning curves and subsequent cost reductions of solar technologies depend on production volume, research and development (R&D), and other factors such as access to capital, and not on the mere passage of time. Private capital is flowing into all the technologies, but government support and stable political conditions can lessen the risk of private investment and help ensure faster deployment.

Potential deployment scenarios for solar energy range widely—from a marginal role of direct solar energy in 2050 to one of the major sources of global energy supply. Although it is true that direct solar energy provides only a very small fraction of global energy supply today, it has the largest technical potential of all energy sources. In concert with technical improvements and resulting cost reductions, it could see dramatically expanded use in the decades to come. Achieving continued cost reductions is the central challenge that will influence the future deployment of solar energy. Moreover, as with some other forms of renewable energy, issues of variable production profiles and energy market integration as well as the possible need for new transmission infrastructure will influence the magnitude, type and cost of solar energy deployment. Finally, the regulatory and legal framework in place can also foster or hinder the uptake of direct solar energy applications.

3.1 Introduction

The aim of this chapter is to provide a synopsis of the state-of-the-art and possible future scenarios of the full realization of direct solar energy's potential for mitigating climate change. It establishes the resource base, describes the many and varied technologies, appraises current market development, outlines some methods for integrating solar into other energy systems, addresses its environmental and social impacts, and finally, evaluates the prospects for future deployment.

Some of the solar energy absorbed by the Earth appears later in the form of wind, wave, ocean thermal, hydropower and excess biomass energies. The scope of this chapter, however, does not include these other indirect forms. Rather, it deals with the *direct* use of solar energy.

Various books have been written on the history of solar technology (e.g., Butti and Perlin, 1980). This history began when early civilizations discovered that buildings with openings facing the Sun were warmer and brighter, even in cold weather. During the late 1800s, solar collectors for heating water and other fluids were invented and put into practical use for domestic water heating and solar industrial applications, for example, large-scale solar desalination. Later, mirrors were used (e.g., by Augustin Mouchot in 1875) to boost the available fluid temperature, so that heat engines driven by the Sun could develop motive power, and thence, electrical power. Also, the late 1800s brought the discovery of a device for converting sunlight directly into electricity. Called the photovoltaic (PV) cell, this device bypassed the need for a heat engine. The modern silicon solar cell, attributed to Russell Ohl working at American Telephone and Telegraph's (AT&T) Bell Labs, was discovered around 1940.

The modern age of solar research began in the 1950s with the establishment of the International Solar Energy Society (ISES) and increased research and development (R&D) efforts in many industries. For example, advances in the solar hot water heater by companies such as Miromit in Israel and the efforts of Harry Tabor at the National Physical Laboratory in Jerusalem helped to make solar energy the standard method for providing hot water for homes in Israel by the early 1960s. At about the same time, national and international networks of solar irradiance measurements were beginning to be established. With the oil crisis of the 1970s, most countries in the world developed programs for solar energy R&D, and this involved efforts in industry, government labs and universities. These policy support efforts, which have, for the most part, continued up to the present, have borne fruit: now one of the fastest-growing renewable energy (RE) technologies, solar energy is poised to play a much larger role on the world energy stage.

Solar energy is an abundant energy resource. Indeed, in just one hour, the solar energy intercepted by the Earth exceeds the world's energy consumption for the entire year. Solar energy's potential to mitigate climate change is equally impressive. Except for the modest amount of carbon dioxide (CO_2) emissions produced in the manufacture of conversion devices (see Section 3.6.1) the direct use of solar energy produces very little greenhouse gases, and it has the potential to displace large quantities of non-renewable fuels (Tsilingiridis et al., 2004).

Solar energy conversion is manifest in a family of technologies having a broad range of energy service applications: lighting, comfort heating, hot water for buildings and industry, high-temperature solar heat for electric power and industry, photovoltaic conversion for electrical power, and production of solar fuels, for example, hydrogen or synthesis gas (syngas). This chapter will further detail all of these technologies.

Several solar technologies, such as domestic hot water heating and pool heating, are already competitive and used in locales where they offer the least-cost option. And in jurisdictions where governments have taken steps to actively support solar energy, very large solar electricity (both PV and CSP) installations, approaching 100 MW of power, have been realized, in addition to large numbers of rooftop PV installations. Other applications, such as solar fuels, require additional R&D before achieving significant levels of adoption.

In pursuing any of the solar technologies, there is the need to deal with the variability and the cyclic nature of the Sun. One option is to store excess collected energy until it is needed. This is particularly effective for handling the lack of sunshine at night. For example, a 0.1-m thick slab of concrete in the floor of a home will store much of the solar energy absorbed during the day and release it to the room at night. When totalled over a long period of time such as one year, or over a large geographical area such as a continent, solar energy can offer greater service. The use of both these concepts of time and space, together with energy storage, has enabled designers to produce more effective solar systems. But much more work is needed to capture the full value of solar energy's contribution.

Because of its inherent variability, solar energy is most useful when integrated with another energy source, to be used when solar energy is not available. In the past, that source has generally been a non-renewable one. But there is great potential for integrating direct solar energy with other RE technologies.

The rest of this chapter will include the following topics. Section 3.2 summarizes research that characterizes this solar resource and discusses the global and regional technical potential for direct solar energy as well as the possible impacts of climate change on this resource. Section 3.3 describes the five different technologies and their applications: passive solar heating and lighting for buildings (Section 3.3.1), active solar heating and cooling for buildings and industry (Section 3.3.2), PV electricity generation (Section 3.3.3), CSP electricity generation (Section 3.3.4), and solar fuel production (Section 3.3.5). Section 3.4 reviews the current status of market development, including installed capacity and energy currently being generated (Section 3.4.1), and the industry capacity and supply chain (Section 3.4.2). Following this are sections on the integration of solar technologies into other energy systems (Section 3.5), the environmental and social impacts (Section 3.6), and the prospects for future technology innovations (Section 3.7). The two final sections cover cost trends (Section 3.8) and the policies needed to achieve the goals for deployment (Section 3.9). Many of the sections, such as Section 3.3, are segmented into subsections, one for each of the five solar technologies.

3.2 Resource potential

The solar resource is virtually inexhaustible, and it is available and able to be used in all countries and regions of the world. But to plan and design appropriate energy conversion systems, solar energy technologists must know how much irradiation will fall on their collectors.

Iqbal (1984), among others, has described the character of solar irradiance, which is the electromagnetic radiation emitted by the Sun. Outside the Earth's atmosphere, the solar irradiance on a surface perpendicular to the Sun's rays at the mean Earth-Sun distance is practically constant throughout the year. Its value is now accepted to be 1,367 W/m² (Bailey et al., 1997). With a clear sky on Earth, this figure becomes roughly 1,000 W/m² at the Earth's surface. These rays are actually electromagnetic waves—travelling fluctuations in electric and magnetic fields. With the Sun's surface temperature being close to 5800 Kelvin, solar irradiance is spread over wavelengths ranging from 0.25 to 3 μm. About 40% of solar irradiance is visible light, while another 10% is ultraviolet radiation, and 50% is infrared radiation. However, at the Earth's surface, evaluation of the solar irradiance is more difficult because of its interaction with the atmosphere, which contains clouds, aerosols, water vapour and trace gases that vary both geographically and temporally. Atmospheric conditions typically reduce the solar irradiance by roughly 35% on clear, dry days and by about 90% on days with thick clouds, leading to lower average solar irradiance. On average, solar irradiance on the ground is 198 W/m² (Solomon et al., 2007), based on ground surface area (Le Treut et al., 2007).

The solar irradiance reaching the Earth's surface (Figure 3.1) is divided into two primary components: beam solar irradiance on a horizontal surface, which comes directly from the Sun's disk, and diffuse irradiance, which comes from the whole of the sky except the Sun's disk. The term 'global solar irradiance' refers to the sum of the beam and the diffuse components.

There are several ways to assess the global resource potential of solar energy. The *theoretical* potential, which indicates the amount of irradiance at the Earth's surface (land and ocean) that is theoretically available for energy purposes, has been estimated at 3.9×10^6 EJ/yr (Rogner et al., 2000; their Table 5.18). *Technical potential* is the amount of solar irradiance output obtainable by full deployment of demonstrated and likely-to-develop technologies or practices (see Annex I, Glossary).

3.2.1 Global technical potential

The amount of solar energy that could be put to human use depends significantly on local factors such as land availability and meteorological conditions and demands for energy services. The technical potential varies over the different regions of the Earth, as do the assessment methodologies. As described in a comparative literature study (Krewitt et al., 2009) for the German Environment Agency, the solar electricity technical potential of PV and CSP depends on the available solar irradiance, land use exclusion factors and the future development of technology improvements. Note that this study used different assumptions for the land use factors for PV and CSP. For PV, it assumed that 98% of the technical potential comes from centralized PV power plants and that the suitable land area in the world for PV deployment averages 1.67% of total land area. For CSP, all land areas with high direct-normal irradiance (DNI)—a minimum DNI of 2,000 kWh/m²/yr (7,200 MJ/m²/yr)—were defined as suitable, and just 20% of that land was excluded for other uses. The

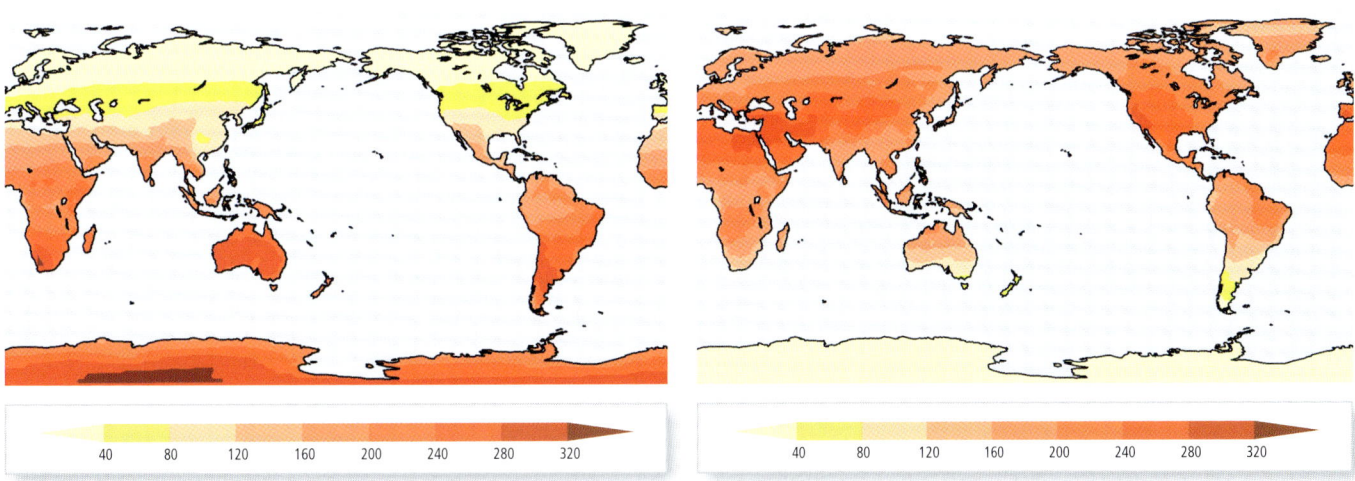

Figure 3.1 | The global solar irradiance (W/m²) at the Earth's surface obtained from satellite imaging radiometers and averaged over the period 1983 to 2006. Left panel: December, January, February. Right panel: June, July, August (ISCCP Data Products, 2006).

resulting technical potentials for 2050 are 1,689 EJ/yr for PV and 8,043 EJ/yr for CSP.

Analyzing the PV studies (Hofman et al., 2002; Hoogwijk, 2004; de Vries et al., 2007) and the CSP studies (Hofman et al., 2002; Trieb, 2005; Trieb et al., 2009a) assessed by Krewitt et al. (2009), the technical potential varies significantly between these studies, ranging from 1,338 to 14,778 EJ/yr for PV and 248 and 10,791 EJ/yr for CSP. The main difference between the studies arises from the allocated land area availabilities and, to some extent, on differences in the power conversion efficiency used.

The technical potential of solar energy for heating purposes is vast and difficult to assess. The deployment potential is mainly limited by the demand for heat. Because of this, the technical potential is not assessed in the literature except for REN21 (Hoogwijk and Graus, 2008) to which Krewitt et al. (2009) refer. In order to provide a reference, REN21 has made a rough assessment of the technical potential of solar water heating by taking the assumed available rooftop area for solar PV applications from Hoogwijk (2004) and the irradiation for each of the regions. Therefore, the range given by REN21 is a lower bound only.

3.2.2 Regional technical potential

Table 3.1 shows the minimum and maximum estimated range for total solar energy technical potential for different regions, not differentiating the ways in which solar irradiance might be converted to secondary energy forms. For the minimum estimates, minimum annual clear-sky irradiance, sky clearance and available land used for installation of solar collectors are assumed. For the maximum estimates, maximum annual clear-sky irradiance and sky clearance are adopted with an assumption of maximum available land used. As Table 3.1 also indicates, the worldwide solar energy technical potential is considerably larger than the current primary energy consumption.

3.2.3 Sources of solar irradiance data

The calculation and optimization of the energy output and economical feasibility of solar energy systems such as buildings and power plants requires detailed solar irradiance data measured at the site of the solar installation. Therefore, it is essential to know the overall global solar energy available, as well as the relative magnitude of its two primary components: direct-beam irradiation and diffuse irradiation from the sky including clouds. Additionally, sometimes it is necessary to account for irradiation received by reflection from the ground and other surfaces. The details on how solar irradiance is measured and calculated can be found in the *Guide to Meteorological Instruments and Methods of Observation* (WMO, 2008). Also important are the patterns of seasonal availability, variability of irradiation, and daytime temperature onsite. Due to significant interannual variability of regional climate conditions in different parts of the world, such measurements must be generated over several years for many applications to provide sufficient statistical validity.

In regions with a high density of well-maintained ground measurements of solar irradiance, sophisticated gridding of these measurements can be expected to provide accurate information about the local solar irradiance. However, many parts of the world have inadequate ground-based sites (e.g., central Asia, northern Africa, Mexico, Brazil, central South America). In these regions, satellite-based irradiance measurements are

Table 3.1 | Annual total technical potential of solar energy for various regions of the world, not differentiated by conversion technology (Rogner et al., 2000; their Table 5.19).

REGIONS	Range of Estimates	
	Minimum, EJ	Maximum, EJ
North America	181	7,410
Latin America and Caribbean	113	3,385
Western Europe	25	914
Central and Eastern Europe	4	154
Former Soviet Union	199	8,655
Middle East and North Africa	412	11,060
Sub-Saharan Africa	372	9,528
Pacific Asia	41	994
South Asia	39	1,339
Centrally planned Asia	116	4,135
Pacific OECD	73	2,263
TOTAL	1,575	49,837
Ratio of technical potential to primary energy supply in 2008 (492 EJ)	3.2	101

Note: Basic assumptions used in assessing minimum and maximum technical potentials of solar energy are given in Rogner et al. (2000):
- Annual minimum clear-sky irradiance relates to horizontal collector plane, and annual maximum clear-sky irradiance relates to two-axis-tracking collector plane; see Table 2.2 in WEC (1994).
- Maximum and minimum annual sky clearance assumed for the relevant latitudes; see Table 2.2 in WEC (1994).

the primary source of information, but their accuracy is inherently lower than that of a well-maintained and calibrated ground measurement. Therefore, satellite radiation products require validation with accurate ground-based measurements (e.g., the Baseline Surface Radiation Network). Presently, the solar irradiance at the Earth's surface is estimated with an accuracy of about 15 W/m^2 on a regional scale (ISCCP Data Products, 2006). The Satellite Application Facility on Climate Monitoring project, under the leadership of the German Meteorological Service and in partnership with the Finnish, Belgian, Dutch, Swedish and Swiss National Meteorological Services, has developed methodologies for irradiance data from satellite measurements.

Various international and national institutions provide information on the solar resource, including the World Radiation Data Centre (Russia), the National Renewable Energy Laboratory (USA), the National Aeronautics and Space Administration (NASA, USA), the Brasilian Spatial Institute (Brazil), the German Aerospace Center (Germany), the Bureau of Meteorology Research Centre (Australia), and the Centro de Investigaciones Energéticas, Medioambientales y Tecnológicas (Spain), National Meteorological Services, and certain commercial companies. Table 3.2 gives references to some international and national projects that are collecting, processing and archiving information on solar irradiance resources at the Earth's surface and subsequently distributing it in easily accessible formats with understandable quality metrics.

3.2.4 Possible impact of climate change on resource potential

Climate change due to an increase of greenhouse gases (GHGs) in the atmosphere may influence atmospheric water vapour content, cloud cover, rainfall and turbidity, and this can impact the resource potential of solar energy in different regions of the globe. Changes in major climate variables, including cloud cover and solar irradiance at the Earth's surface, have been evaluated using climate models and considering anthropogenic forcing for the 21st century (Meehl et al., 2007; Meleshko et al., 2008). These studies found that the pattern of variation of monthly mean global solar irradiance does not exceed 1% over some regions of the globe, and it varies from model to model. Currently, there is no other evidence indicating a substantial impact of global warming on regional solar resources. Although some research on global dimming and global brightening indicates a probable impact on irradiance, no current evidence is available. Uncertainty in pattern changes seems to be rather large, even for large-scale areas of the Earth.

3.3 Technology and applications

This section discusses technical issues for a range of solar technologies, organized under the following categories: passive solar and daylighting,

Table 3.2 | International and national projects that collect, process and archive information on solar irradiance resources at the Earth's surface.

Available Data Sets	Responsible Institution/Agency
Ground-based solar irradiance from 1,280 sites for 1964 to 2009 provided by national meteorological services around the world.	World Radiation Data Centre, Saint Petersburg, Russian Federation (wrdc.mgo.rssi.ru)
National Solar Radiation Database that includes 1,454 ground locations for 1991 to 2005. The satellite-modelled solar data for 1998 to 2005 provided on 10-km grid. The hourly values of solar data can be used to determine solar resources for collectors.	National Renewable Energy Laboratory, USA (www.nrel.gov)
European Solar Radiation Database that includes measured solar radiation complemented with other meteorological data necessary for solar engineering. Satellite images from METEOSAT help in improving accuracy in spatial interpolation. Test Reference Years were also included.	Supported by Commission of the European Communities, National Weather Services and scientific institutions of the European countries
The Solar Radiation Atlas of Africa contains information on surface radiation over Europe, Asia Minor and Africa. Data covering 1985 to 1986 were derived from measurements by METEOSAT 2.	Supported by the Commission of the European Communities
The solar data set for Africa based on images from METEOSAT processed with the Heliosat-2 method covers the period 1985 to 2004 and is supplemented with ground-based solar irradiance.	Ecole des Mines de Paris, France
Typical Meteorological Year (Test Reference Year) data sets of hourly values of solar radiation and meteorological parameters derived from individual weather observations in long-term (up to 30 years) data sets to establish a typical year of hourly data. Used by designers of heating and cooling systems and large-scale solar thermal power plants.	National Renewable Energy Laboratory, USA. National Climatic Data Center, National Oceanic and Atmospheric Administration, USA. (www.ncdc.noaa.gov)
The solar radiation data for solar energy applications. IEA/SHC Task36 provides a wide range of users with information on solar radiation resources at Earth's surface in easily accessible formats with understandable quality metrics. The task focuses on development, validation and access to solar resource information derived from surface- and satellite-based platforms.	International Energy Agency (IEA) Solar Heating and Cooling Programme (SHC). (swera.unep.net)
Solar and Wind Energy Resource Assessment (SWERA) project aimed at developing information tools to simulate RE development. SWERA provides easy access to high-quality RE resource information and data for users. Covered major areas of 13 developing countries in Latin America, the Caribbean, Africa and Asia. SWERA produced a range of solar data sets and maps at better spatial scales of resolution than previously available using satellite- and ground-based observations.	Global Environment Facility-sponsored project. United Nations Environment Programme (swera.unep.net)

active heating and cooling, PV electricity generation, CSP electricity generation and solar fuel production. Each section also describes applications of these technologies.

3.3.1 Passive solar and daylighting technologies

Passive solar energy technologies absorb solar energy, store and distribute it in a natural manner (e.g., natural ventilation), without using mechanical elements (e.g., fans) (Hernandez Gonzalvez, 1996). The term 'passive solar building' is a qualitative term describing a building that makes significant use of solar gain to reduce heating energy consumption based on the natural energy flows of radiation, conduction and convection. The term 'passive building' is often employed to emphasize use of passive energy flows in both heating and cooling, including redistribution of absorbed direct solar gains and night cooling (Athienitis and Santamouris, 2002).

Daylighting technologies are primarily passive, including windows, skylights and shading and reflecting devices. A worldwide trend, particularly in technologically advanced regions, is for an increased mix of passive and active systems, such as a forced-air system that redistributes passive solar gains in a solar house or automatically controlled shades that optimize daylight utilization in an office building (Tzempelikos et al., 2010).

The basic elements of passive solar design are windows, conservatories and other glazed spaces (for solar gain and daylighting), thermal mass, protection elements, and reflectors (Ralegaonkar and Gupta, 2010). With the combination of these basic elements, different systems are obtained: direct-gain systems (e.g., the use of windows in combination with walls able to store energy, solar chimneys, and wind catchers), indirect-gain systems (e.g., Trombe walls), mixed-gain systems (a combination of direct-gain and indirect-gain systems, such as conservatories, sunspaces and greenhouses), and isolated-gain systems. Passive technologies are integrated with the building and may include the following components:

- Windows with high solar transmittance and a high thermal resistance facing towards the Equator as nearly as possible can be employed to maximize the amount of direct solar gains into the living space while reducing heat losses through the windows in the heating season and heat gains in the cooling season. Skylights are also often used for daylighting in office buildings and in solaria/sunspaces.

- Building-integrated thermal storage, commonly referred to as thermal mass, may be sensible thermal storage using concrete or brick materials, or latent thermal storage using phase-change materials (Mehling and Cabeza, 2008). The most common type of thermal storage is the direct-gain system in which thermal mass is adequately distributed in the living space, absorbing the direct solar gains. Storage is particularly important because it performs two essential functions: storing much of the absorbed direct solar energy for slow release, and maintaining satisfactory thermal comfort conditions by limiting the maximum rise in operative (effective) room temperature (ASHRAE, 2009). Alternatively, a collector-storage wall, known as a Trombe wall, may be used, in which the thermal mass is placed directly next to the glazing, with possible air circulation between the cavity of the wall system and the room. However, this system has not gained much acceptance because it limits views to the outdoor environment through the fenestration. Hybrid thermal storage with active charging and passive heat release can also be employed in part of a solar building while direct-gain mass is also used (see, e.g., the EcoTerra demonstration house (Figure 3.2, left panel), which uses solar-heated air from a building-integrated photovoltaic/thermal system to heat a ventilated concrete slab). Isolated thermal storage passively coupled to a fenestration system or solarium/sunspace is another option in passive design.

- Well-insulated opaque envelope appropriate for the climatic conditions can be used to reduce heat transfer to and from the outdoor environment. In most climates, this energy efficiency aspect must be integrated with the passive design. A solar technology that may be used with opaque envelopes is transparent insulation (Hollands et al., 2001) combined with thermal mass to store solar gains in a wall, turning it into an energy-positive element.

- Daylighting technologies and advanced solar control systems, such as automatically controlled shading (internal, external) and fixed shading devices, are particularly suited for daylighting applications in the workplace (Figure 3.2, right panel). These technologies include electrochromic and thermochromic coatings and newer technologies such as transparent photovoltaics, which, in addition to a passive daylight transmission function, also generate electricity. Daylighting is a combination of energy conservation and passive solar design. It aims to make the most of the natural daylight that is available. Traditional techniques include: shallow-plan design, allowing daylight to penetrate all rooms and corridors; light wells in the centre of buildings; roof lights; tall windows, which allow light to penetrate deep inside rooms; task lighting directly over the workplace, rather than lighting the whole building interior; and deep windows that reveal and light room surfaces to cut the risk of glare (Everett, 1996).

- Solariums, also called sunspaces, are a particular case of the direct-gain passive solar system, but with most surfaces transparent, that is, made up of fenestration. Solariums are becoming increasingly attractive both as a retrofit option for existing houses and as an integral part of new buildings (Athienitis and Santamouris, 2002). The major driving force for this growth is the development of new advanced energy-efficient glazing.

Some basic rules for optimizing the use of passive solar heating in buildings are the following: buildings should be well insulated to reduce overall heat losses; they should have a responsive, efficient heating system; they should face towards the Equator, that is, the glazing should

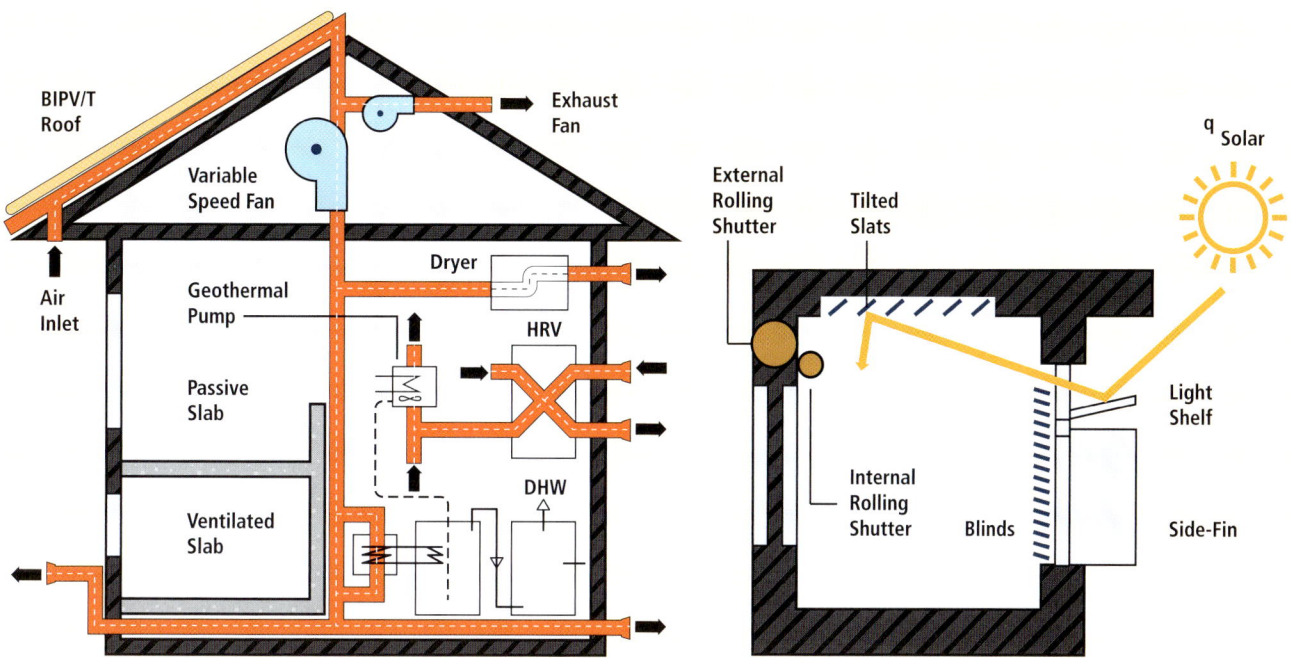

Figure 3.2 | Left: Schematic of thermal mass placement and passive-active systems in a house; solar-heated air from building-integrated photovoltaic/thermal (BIPV/T) roof heats ventilated slab or domestic hot water (DHW) through heat exchanger; HRV is heat recovery ventilator. Right: Schematic of several daylighting concepts designed to redistribute daylight into the office interior space (Athienitis, 2008).

be concentrated on the equatorial side, as should the main living rooms, with rooms such as bathrooms on the opposite side; they should avoid shading by other buildings to benefit from the essential mid-winter sun; and they should be 'thermally massive' to avoid overheating in the summer and on certain sunny days in winter (Everett, 1996).

Clearly, passive technologies cannot be separated from the building itself. Thus, when estimating the contribution of passive solar gains, the following must be distinguished: 1) buildings specifically designed to harness direct solar gains using passive systems, defined here as solar buildings, and 2) buildings that harness solar gains through near-equatorial facing windows; this orientation is more by chance than by design. Few reliable statistics are available on the adoption of passive design in residential buildings. Furthermore, the contribution of passive solar gains is missing in existing national statistics. Passive solar is reducing the demand and is not part of the supply chain, which is what is considered by the energy statistics.

The passive solar design process itself is in a period of rapid change, driven by the new technologies becoming affordable, such as the recently available highly efficient fenestration at the same prices as ordinary glazing. For example, in Canada, double-glazed low-emissivity argon-filled windows are presently the main glazing technology used; but until a few years ago, this glazing was about 20 to 40% more expensive than regular double glazing. These windows are now being used in retrofits of existing homes as well. Many homes also add a solarium during retrofit. The new glazing technologies and solar control systems allow the design of a larger window area than in the recent past.

In most climates, unless effective solar gain control is employed, there may be a need to cool the space during the summer. However, the need for mechanical cooling may often be eliminated by designing for passive cooling. Passive cooling techniques are based on the use of heat and solar protection techniques, heat storage in thermal mass and heat dissipation techniques. The specific contribution of passive solar and energy conservation techniques depends strongly on the climate (UNEP, 2007). Solar-gain control is particularly important during the 'shoulder' seasons when some heating may be required. In adopting larger window areas—enabled by their high thermal resistance—active solar-gain control becomes important in solar buildings for both thermal and visual considerations.

The potential of passive solar cooling in reducing CO_2 emissions has been shown recently (Cabeza et al., 2010; Castell et al., 2010). Experimental work demonstrates that adequate insulation can reduce by up to 50% the cooling energy demand of a building during the hot season. Moreover, including phase-change materials in the already-insulated building envelope can reduce the cooling energy demand in such buildings further by up to 15%—about 1 to 1.5 kg/yr/m^2 of CO_2 emissions would be saved in these buildings due to reducing the energy consumption compared to the insulated building without phase-change material.

Passive solar system applications are mainly of the direct-gain type, but they can be further subdivided into the following main application categories: multi-story residential buildings and two-story detached or semi-detached solar homes (see Figure 3.2, left panel), designed to have a large equatorial-facing façade to provide the potential for a large solar capture area (Athienitis, 2008). Perimeter zones and their fenestration systems in office buildings are designed primarily based on daylighting performance. In this application, the emphasis is usually on reducing cooling loads, but passive heat gains may be desirable as well during the heating season (see Figure 3.2, right panel, for a schematic of shading devices).

In addition, residential or commercial buildings may be designed to use natural or hybrid ventilation systems and techniques for cooling or fresh air supply, in conjunction with designs for using daylight throughout the year and direct solar gains during the heating season. These buildings may profit from low summer night temperatures by using night hybrid ventilation techniques that utilize both mechanical and natural ventilation processes (Santamouris and Asimakopoulos, 1996; Voss et al., 2007).

In 2010, passive technologies played a prominent role in the design of net-zero-energy solar homes—homes that produce as much electrical and thermal energy as they consume in an average year. These houses are primarily demonstration projects in several countries currently collaborating in the International Energy Agency (IEA) Task 40 of the Solar Heating and Cooling (SHC) Programme (IEA, 2009b)—Energy Conservation in Buildings and Community Systems Annex 52—which focuses on net-zero-energy solar buildings. Passive technologies are essential in developing affordable net-zero-energy homes. Passive solar gains in homes based on the Passive House Standard are expected to reduce the heating load by about 40%. By extension, systematic passive solar design of highly insulated buildings at a community scale, with optimal orientation and form of housing, should easily result in a similar energy saving of 40%. In Europe, according to the Energy Performance of Buildings Directive recast, Directive 2010/31/EC (The European Parliament and the Council of the European Union, 2010), all new buildings must be nearly zero-energy buildings by 31 December 2020, while EU member states should set intermediate targets for 2015. New buildings occupied and owned by public authorities have to be nearly zero-energy buildings after 31 December 2018. The nearly zero or very low amount of energy required should to a very significant level be covered by RE sources, including onsite energy production using combined heat and power generation or district heating and cooling, to satisfy most of their demand. Measures should also be taken to stimulate building refurbishments into nearly zero-energy buildings.

Low-energy buildings are known under different names. A survey carried out by Concerted Action Energy Performance of Buildings (EPBD) identified 17 different terms to describe such buildings across Europe, including: low-energy house, high-performance house, passive house ('Passivhaus'), zero-carbon house, zero-energy house, energy-savings house, energy-positive house and 3-litre house. Concepts that take into account more parameters than energy demand again use special terms such as eco-building or green building.

Another IEA Annex—Energy Conservation through Energy Storage Implementing Agreement (ECES IA) Annex 23—was initiated in November 2009 (IEA ECES, 2004). The general objective of the Annex is to ensure that energy storage techniques are properly applied in ultra-low-energy buildings and communities. The proper application of energy storage is expected to increase the likelihood of sustainable building technologies.

Another passive solar application is natural drying. Grains and many other agricultural products have to be dried before being stored so that insects and fungi do not render them unusable. Examples include wheat, rice, coffee, copra (coconut flesh), certain fruits and timber (Twidell and Weir, 2006). Solar energy dryers vary mainly as to the use of the solar heat and the arrangement of their major components. Solar dryers constructed from wood, metal and glass sheets have been evaluated extensively and used quite widely to dry a full range of tropical crops (Imre, 2007).

3.3.2 Active solar heating and cooling

Active solar heating and cooling technologies use the Sun and mechanical elements to provide either heating or cooling; various technologies are discussed here, as well as thermal storage.

3.3.2.1 Solar heating

In a solar heating system, the solar collector transforms solar irradiance into heat and uses a carrier fluid (e.g., water, air) to transfer that heat to a well-insulated storage tank, where it can be used when needed. The two most important factors in choosing the correct type of collector are the following: 1) the service to be provided by the solar collector, and 2) the related desired range of temperature of the heat-carrier fluid. An uncovered absorber, also known as an unglazed collector, is likely to be limited to low-temperature heat production (Duffie and Beckman, 2006).

A solar collector can incorporate many different materials and be manufactured using a variety of techniques. Its design is influenced by the system in which it will operate and by the climatic conditions of the installation location.

Flat-plate collectors are the most widely used solar thermal collectors for residential solar water- and space-heating systems. They are also used in air-heating systems. A typical flat-plate collector consists of an absorber, a header and riser tube arrangement or a single serpentine tube, a transparent cover, a frame and insulation (Figure 3.3a). For low-temperature applications, such as the heating of swimming pools, only a single plate is used as an absorber (Figure 3.3b). Flat-plate collectors demonstrate a good price/performance ratio, as well as a broad range of mounting possibilities (e.g., on the roof, in the roof itself, or unattached).

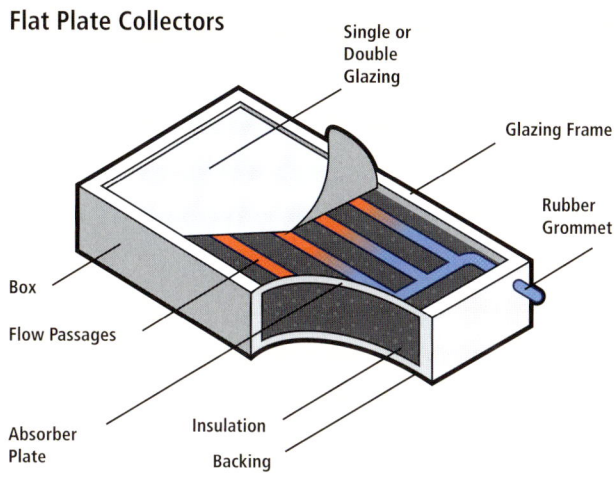

Figure 3.3a | Schematic diagram of thermal solar collectors: Glazed flat-plate.

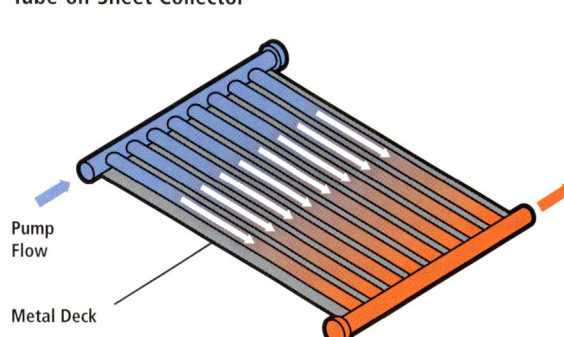

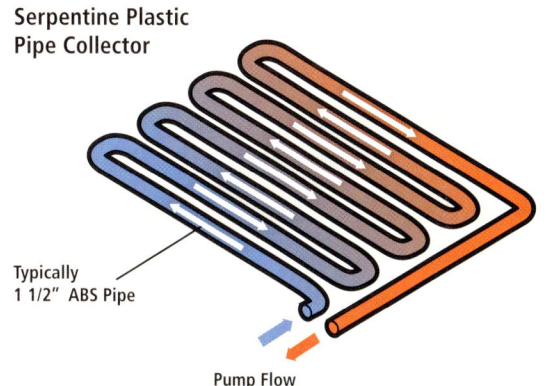

Figure 3.3b | Schematic diagram of thermal solar collectors: Unglazed tube-on-sheet and serpentine plastic pipe.

Evacuated-tube collectors are usually made of parallel rows of transparent glass tubes, in which the absorbers are enclosed, connected to a header pipe (Figure 3.3c). To reduce heat loss within the frame by convection, the air is pumped out of the collector tubes to generate a vacuum. This makes it possible to achieve high temperatures, useful

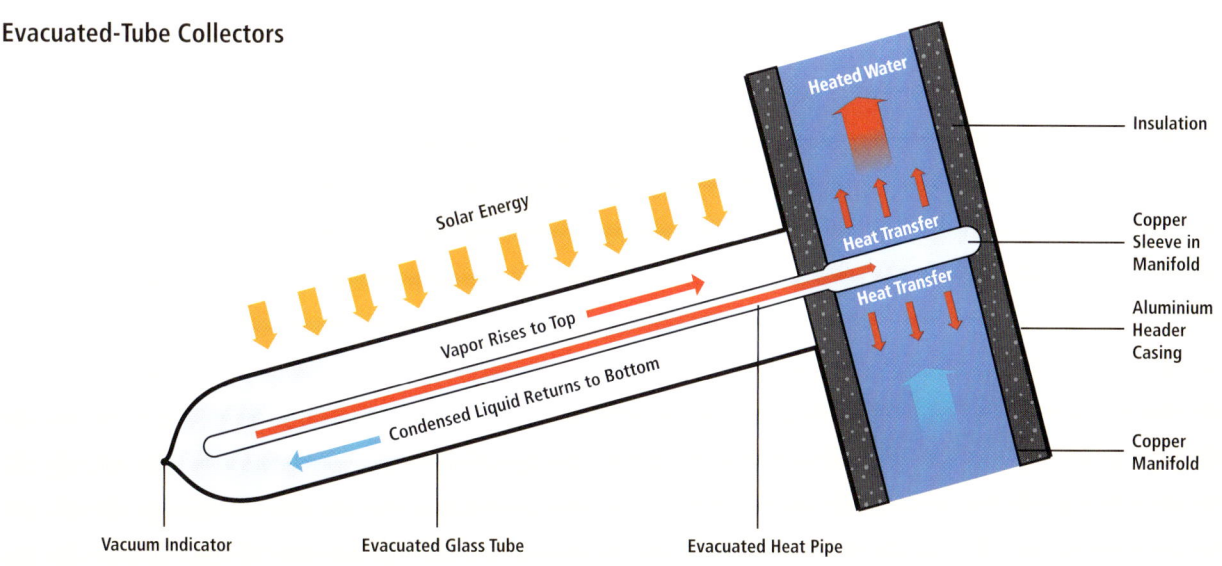

Figure 3.3c | Schematic diagram of thermal solar collectors: Evacuated-tube collectors.

for cooling (see below) or industrial applications. Most vacuum tube collectors use heat pipes for their core instead of passing liquid directly through them. Evacuated heat-pipe tubes are composed of multiple evacuated glass tubes, each containing an absorber plate fused to a heat pipe. The heat from the hot end of the heat pipes is transferred to the transfer fluid of a domestic hot water or hydronic space-heating system.

Solar water-heating systems used to produce hot water can be classified as passive or active solar water heaters (Duffie and Beckman, 2006). Also of interest are active solar cooling systems, which transform the hot water produced by solar energy into cold water.

Passive solar water heaters are of two types (Figure 3.4). Integral collector-storage (ICS) or 'batch' systems include black tanks or tubes in an insulated glazed box. Cold water is preheated as it passes through the solar collector, with the heated water flowing to a standard backup water heater. The heated water is stored inside the collector itself. In thermosyphon (TS) systems, a separate storage tank is directly above the collector. In direct (open-loop) TS systems, the heated water rises from the collector to the tank and cool water from the tank sinks back into the collector. In indirect (closed-loop) TS systems (Figure 3.4, left), heated fluid (usually a glycol-water mixture) rises from the collector to an outer tank that surrounds the water storage tank and acts as a heat exchanger (double-wall heat exchangers) for separation from potable water. In climates where freezing temperatures are unlikely, many collectors include an integrated storage tank at the top of the collector. This design has many cost and user-friendly advantages compared to a system that uses a separate standalone heat-exchanger tank. It is also appropriate in households with significant daytime and evening hot water needs; but it does not work well in households with predominantly morning draws because sometimes the tanks can lose most of the collected energy overnight.

Active solar water heaters rely on electric pumps and controllers to circulate the carrier fluid through the collectors. Three types of active solar water-heating systems are available. Direct circulation systems use pumps to circulate pressurized potable water directly through the collectors. These systems are appropriate in areas that do not freeze for long periods and do not have hard or acidic water. Antifreeze indirect-circulation systems pump heat-transfer fluid, which is usually a glycol-water mixture, through collectors. Heat exchangers transfer the heat from the fluid to the water for use (Figure 3.4, right). Drainback indirect-circulation systems use pumps to circulate water through the collectors. The water in the collector and the piping system drains into a reservoir tank when the pumps stop, eliminating the risk of freezing in cold climates. This system should be carefully designed and installed to ensure that the piping always slopes downward to the reservoir tank. Also, stratification should be carefully considered in the design of the water tank (Hadorn, 2005).

A *solar combisystem* provides both solar space heating and cooling as well as hot water from a common array of solar thermal collectors, usually backed up by an auxiliary non-solar heat source (Weiss, 2003). Solar combisystems may range in size from those installed in individual properties to those serving several in a block heating scheme. A large number of different types of solar combisystems are produced. The systems on the market in a particular country may be more restricted, however, because different systems have tended to evolve in different countries.

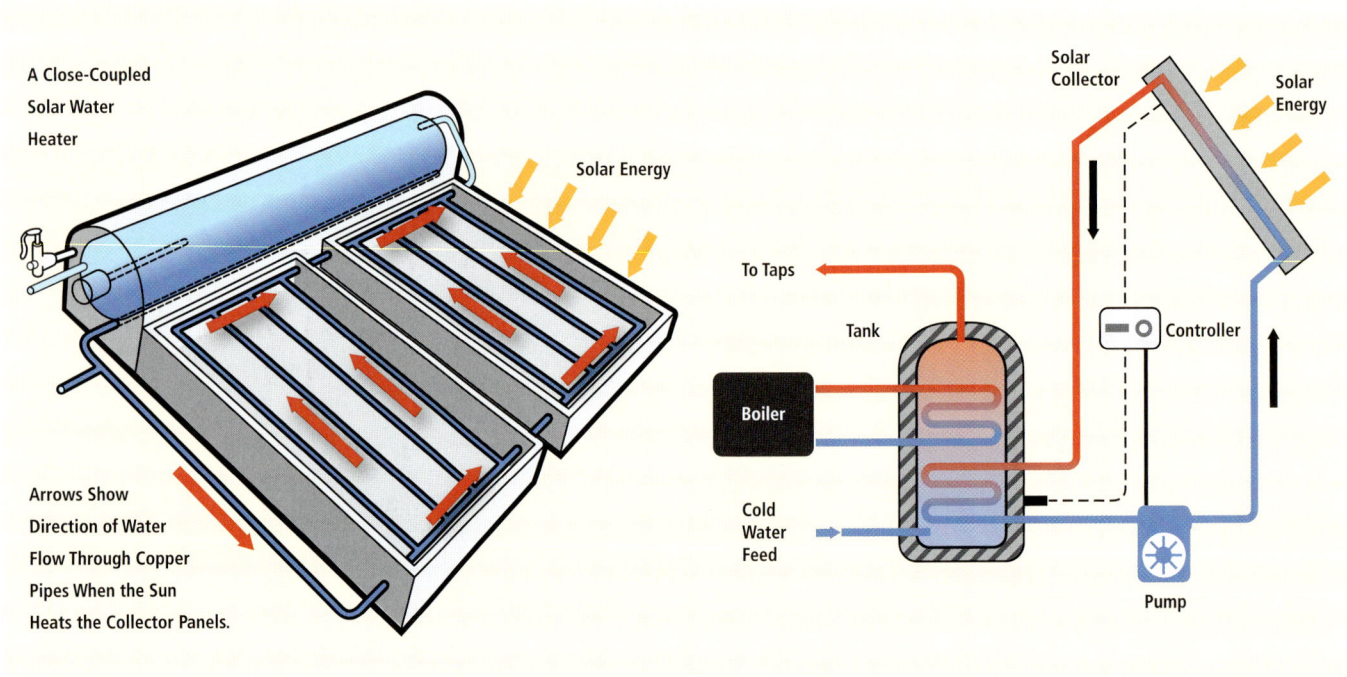

Figure 3.4 | Generic schematics of thermal solar systems. Left: Passive (thermosyphon). Right: Active system.

Depending on the size of the combisystem installed, the annual space heating contribution can range from 10 to 60% or more in ultra-low energy Passivhaus-type buildings, and even up to 100% where a large seasonal thermal store or concentrating solar thermal heat is used.

3.3.2.2 Solar cooling

Solar cooling can be broadly categorized into solar electric refrigeration, solar thermal refrigeration, and solar thermal air-conditioning. In the first category, the solar electric compression refrigeration uses PV panels to power a conventional refrigeration machine (Fong et al., 2010). In the second category, the refrigeration effect can be produced through solar thermal gain; solar mechanical compression refrigeration, solar absorption refrigeration, and solar adsorption refrigeration are the three common options. In the third category, the conditioned air can be directly provided through the solar thermal gain by means of desiccant cooling. Both solid and liquid sorbents are available, such as silica gel and lithium chloride, respectively.

Solar electrical air-conditioning, powered by PV panels, is of minor interest from a systems perspective, unless there is an off-grid application (Henning, 2007). This is because in industrialized countries, which have a well-developed electricity grid, the maximum use of photovoltaics is achieved by feeding the produced electricity into the public grid.

Solar thermal air-conditioning consists of solar heat powering an absorption chiller and it can be used in buildings (Henning, 2007). Deploying such a technology depends heavily on the industrial deployment of low-cost small-power absorption chillers. This technology is being studied within the IEA Task 25 on solar-assisted air-conditioning of buildings, SHC program and IEA Task 38 on solar air-conditioning and refrigeration, SHC program.

Closed heat-driven cooling systems using these cycles have been known for many years and are usually used for large capacities of 100 kW and greater. The physical principle used in most systems is based on the sorption phenomenon. Two technologies are established to produce thermally driven low- and medium-temperature refrigeration: absorption and adsorption.

Open cooling cycle (or desiccant cooling) systems are mainly of interest for the air conditioning of buildings. They can use solid or liquid sorption. The central component of any open solar-assisted cooling system is the dehumidification unit. In most systems using solid sorption, this unit is a desiccant wheel. Various sorption materials can be used, such as silica gel or lithium chloride. All other system components are found in standard air-conditioning applications with an air-handling unit and include the heat recovery units, heat exchangers and humidifiers. Liquid sorption techniques have been demonstrated successfully.

3.3.2.3 Thermal storage

Thermal storage within thermal solar systems is a key component to ensure reliability and efficiency. Four main types of thermal energy storage technologies can be distinguished: sensible, latent, sorption and thermochemical heat storage (Hadorn, 2005; Paksoy, 2007; Mehling and Cabeza, 2008; Dincer and Rosen, 2010).

Sensible heat storage systems use the heat capacity of a material. The vast majority of systems on the market use water for heat storage. Water heat storage covers a broad range of capacities, from several hundred litres to tens of thousands of cubic metres.

Latent heat storage systems store thermal energy during the phase change, either melting or evaporation, of a material. Depending on the temperature range, this type of storage is more compact than heat storage in water. Melting processes have energy densities of the order of 100 kWh/m^3 (360 MJ/m^3), compared to 25 kWh/m^3 (90 MJ/m^3) for sensible heat storage. Most of the current latent heat storage technologies for low temperatures store heat in building structures to improve thermal performance, or in cold storage systems. For medium-temperature storage, the storage materials are nitrate salts. Pilot storage units in the 100-kW range currently operate using solar-produced steam.

Sorption heat storage systems store heat in materials using water vapour taken up by a sorption material. The material can either be a solid (adsorption) or a liquid (absorption). These technologies are still largely in the development phase, but some are on the market. In principle, sorption heat storage densities can be more than four times higher than sensible heat storage in water.

Thermochemical heat storage systems store heat in an endothermic chemical reaction. Some chemicals store heat 20 times more densely than water (at a $\Delta T \approx 100°C$); but more typically, the storage densities are 8 to 10 times higher. Few thermochemical storage systems have been demonstrated. The materials currently being studied are the salts that can exist in anhydrous and hydrated form. Thermochemical systems can compactly store low- and medium-temperature heat. Thermal storage is discussed with specific reference to higher-temperature CSP in Section 3.3.4.

Underground thermal energy storage is used for seasonal storage and includes the various technologies described below. The most frequently used storage technology that makes use of the underground is *aquifer thermal energy storage*. This technology uses a natural underground layer (e.g., sand, sandstone or chalk) as a storage medium for the temporary storage of heat or cold. The transfer of thermal energy is realized by extracting groundwater from the layer and by re-injecting it at the modified temperature level at a separate location nearby. Most applications are for the storage of winter cold to be used for the cooling of large office buildings and industrial processes. Aquifer cold storage is gaining interest because savings on electricity bills for chillers are about 75%, and in many cases, the payback time for additional investments is shorter than five years. A major condition for the application of this technology is the availability of a suitable geologic formation.

3.3.2.4 Active solar heating and cooling applications

For active solar heating and cooling applications, the amount of hot water produced depends on the type and size of the system, amount of sun available at the site, seasonal hot-water demand pattern, and installation characteristics of the system (Norton, 2001).

Solar heating for industrial processes is at a very early stage of development in 2010 (POSHIP, 2001). Worldwide, less than 100 operating solar thermal systems for process heat are reported, with a total capacity of about 24 MW_{th} (34,000 m² collector area). Most systems are at an experimental stage and relatively small scale. However, significant potential exists for market and technological developments, because 28% of the overall energy demand in the EU27 countries originates in the industrial sector, and much of this demand is for heat below 250°C. Education and knowledge dissemination are needed to deploy this technology.

In the short term, solar heating for industrial processes will mainly be used for low-temperature processes, ranging from 20°C to 100°C. With technological development, an increasing number of medium-temperature applications—up to 250°C—will become feasible within the market. According to Werner (2006), about 30% of the total industrial heat demand is required at temperatures below 100°C, which could theoretically be met with solar heating using current technologies. About 57% of this demand is required at temperatures below 400°C, which could largely be supplied by solar in the foreseeable future.

In several specific industry sectors—such as food, wine and beverages, transport equipment, machinery, textiles, and pulp and paper—the share of heat demand at low and medium temperatures (below 250°C) is around 60% (POSHIP, 2001). Tapping into this low- and medium-temperature heat demand with solar heat could provide a significant opportunity for solar contribution to industrial energy requirements. A substantial opportunity for solar thermal systems also exists in chemical industries and in washing processes.

Among the industrial processes, desalination and water treatment (e.g., sterilization) are particularly promising applications for solar thermal energy, because these processes require large amounts of medium-temperature heat and are often necessary in areas with high solar irradiance and high energy costs.

Some process heat applications can be met with temperatures delivered by 'ordinary' low-temperature collectors, namely, from 30°C to 80°C. However, the bulk of the demand for industrial process heat requires temperatures from 80°C to 250°C.

Process heat collectors are another potential application for solar thermal heat collectors. Typically, these systems require a large capacity (hence, large collector areas), low costs, and high reliability and quality. Although low- and high-temperature collectors are offered in a dynamically growing market, process heat collectors are at a very early stage of development and no products are available on an industrial scale. In addition to 'concentrating' collectors, improved flat collectors with double and triple glazing are currently being developed, which could meet needs for process heat in the range of up to 120°C. Concentrating-type solar collectors are described in Section 3.3.4.

Solar refrigeration is used, for example, to cool stored vaccines. The need for such systems is greatest in peripheral health centres in rural communities in the developing world, where no electrical grid is available.

Solar cooling is a specific area of application for solar thermal technology. High-efficiency flat plates, evacuated tubes or parabolic troughs can be used to drive absorption cycles to provide cooling. For a greater coefficient of performance (COP), collectors with low concentration levels can provide the temperatures (up to around 250°C) needed for double-effect absorption cycles. There is a natural match between solar energy and the need for cooling.

A number of closed heat-driven cooling systems have been built, using solar thermal energy as the main source of heat. These systems often have large cooling capacities of up to several hundred kW. Since the early 2000s, a number of systems have been developed in the small-capacity range, below 100 kW, and, in particular, below 20 kW and down to 4.5 kW. These small systems are single-effect machines of different types, used mainly for residential buildings and small commercial applications.

Although open-cooling cycles are generally used for air conditioning in buildings, closed heat-driven cooling cycles can be used for both air conditioning and industrial refrigeration.

Other solar applications are listed below. The production of potable water using solar energy has been readily adopted in remote or isolated regions (Narayan et al., 2010). Solar stills are widely used in some parts of the world (e.g., Puerto Rico) to supply water to households of up to 10 people (Khanna et al., 2008). In appropriate isolation conditions, solar detoxification can be an effective low-cost treatment for low-contaminant waste (Gumy et al., 2006). Multiple-effect humidification (MEH) desalination units indirectly use heat from highly efficient solar thermal collectors to induce evaporation and condensation inside a thermally isolated, steam-tight container. These MEH systems are now beginning to appear in the market. Also see the report on water desalination by CSP (DLR, 2007) and the discussion of SolarPACES Task VI (SolarPACES, 2009b).

In solar drying, solar energy is used either as the sole source of the required heat or as a supplemental source, and the air flow can be generated by either forced or free (natural) convection (Fudholi et al., 2010). Solar cooking is one of the most widely used solar applications in developing countries (Lahkar and Samdarshi, 2010) though might still be considered an early stage commercial product due to limited overall deployment in comparison to other cooking methods. A solar cooker uses sunlight as its energy source, so no fuel is needed and operating costs are zero. Also, a reliable solar cooker can be constructed easily and quickly from common materials.

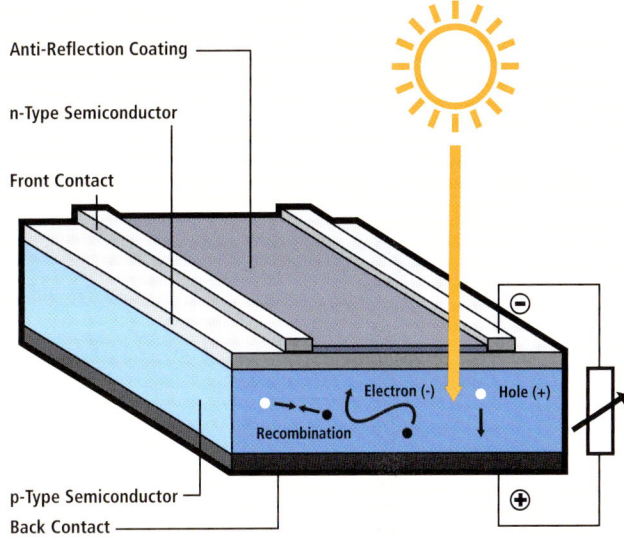

Figure 3.5 | Generic schematic cross-section illustrating the operation of an illuminated solar cell.

3.3.3 Photovoltaic electricity generation

Photovoltaic (PV) solar technologies generate electricity by exploiting the photovoltaic effect. Light shining on a semiconductor such as silicon (Si) generates electron-hole pairs that are separated spatially by an internal electric field created by introducing special impurities into the semiconductor on either side of an interface known as a p-n junction. This creates negative charges on one side of the interface and positive charges are on the other side (Figure 3.5). This resulting charge separation creates a voltage. When the two sides of the illuminated cell are connected to a load, current flows from one side of the device via the load to the other side of the cell. The conversion efficiency of a solar cell is defined as a ratio of output power from the solar cell with unit area (W/cm^2) to the incident solar irradiance. The maximum potential efficiency of a solar cell depends on the absorber material properties and device design. One technique for increasing solar cell efficiency is with a multijunction approach that stacks specially selected absorber materials that can collect more of the solar spectrum since each different material can collect solar photons of different wavelengths.

PV cells consist of organic or inorganic matter. Inorganic cells are based on silicon or non-silicon materials; they are classified as wafer-based cells or thin-film cells. Wafer-based silicon is divided into two different types: monocrystalline and multicrystalline (sometimes called 'polycrystalline').

3.3.3.1 Existing photovoltaic technologies

Existing PV technologies include wafer-based crystalline silicon (c-Si) cells, as well as thin-film cells based on copper indium/gallium disulfide/diselenide (CuInGaSe$_2$; CIGS), cadmium telluride (CdTe), and thin-film silicon (amorphous and microcrystalline silicon). Mono- and multicrystalline silicon wafer PV (including ribbon technologies) are the dominant technologies on the PV market, with a 2009 market share of about 80%; thin-film PV (primarily CdTe and thin-film Si) has the remaining 20% share. Organic PV (OPV) consists of organic absorber materials and is an emerging class of solar cells.

Wafer-based silicon technology includes solar cells made of monocrystalline or multicrystalline wafers with a current thickness of around 200 μm, while the thickness is decreasing down to 150 μm. Single-junction wafer-based c-Si cells have been independently verified to have record energy conversion efficiencies of 25.0% for monocrystalline silicon cells and 20.3% for multicrystalline cells (Green et al., 2010b) under standard test conditions (i.e., irradiance of 1,000 W/m^2, air-mass 1.5, 25°C). The theoretical Shockley-Queisser limit of a single-junction cell with an energy bandgap of crystalline silicon is 31% energy conversion efficiency (Shockley and Queisser, 1961).

Several variations of wafer-based c-Si PV for higher efficiency have been developed, for example, heterojunction solar cells and interdigitated back-contact (IBC) solar cells. Heterojunction solar cells consist of a crystalline silicon wafer base sandwiched by very thin (~5 nm) amorphous silicon layers for passivation and emitter. The highest-efficiency heterojunction solar cell is 23.0% for a 100.4-cm^2 cell (Taguchi et al., 2009). Another advantage is a lower temperature coefficient. The efficiency of conventional c-Si solar cells declines with elevating ambient temperature at a rate of -0.45%/°C, while the heterojunction cells show a lower rate of -0.25%/°C (Taguchi et al., 2009). An IBC solar cell, where both the base and emitter are contacted at the back of the cell, has the advantage of no shading of the front of the cell by a top electrode. The highest efficiency of such a back-contact silicon wafer cell is 24.2% for 155.1 cm^2 (Bunea et al., 2010). Commercial module efficiencies for wafer-based silicon PV range from 12 to 14% for multi-crystalline Si and from 14 to 20% for monocrystalline Si.

Commercial thin-film PV technologies include a range of absorber material systems: amorphous silicon (a-Si), amorphous silicon-germanium, microcrystalline silicon, CdTe and CIGS. These thin-film cells have an absorber layer thickness of a few μm or less and are deposited on glass, metal or plastic substrates with areas of up to 5.7 m^2 (Stein et al., 2009).

The a-Si solar cell, introduced in 1976 (Carlson and Wronski, 1976) with initial efficiencies of 1 to 2%, has been the first commercially successful thin-film PV technology. Because a-Si has a higher light absorption coefficient than c-Si, the thickness of an a-Si cell can be less than 1 μm—that is, more than 100 times thinner than a c-Si cell. Developing higher efficiencies for a-Si cells has been limited by inherent material quality and by light-induced degradation identified as the Staebler-Wronski effect (Staebler and Wronski, 1977). However, research efforts have successfully lowered the impact of the Staebler-Wronski effect to around 10% or less by controlling the microstructure of the film. The highest stabilized efficiency—the efficiency after the light-induced degradation—is reported as 10.1% (Benagli et al., 2009).

Higher efficiency has been achieved by using multijunction technologies with alloy materials, e.g., germanium and carbon or with microcrystalline silicon, to form semiconductors with lower or higher bandgaps, respectively, to cover a wider range of the solar spectrum (Yang and Guha, 1992; Yamamoto et al., 1994; Meier et al., 1997). Stabilized efficiencies of 12 to 13% have been measured for various laboratory devices (Green et al., 2010b).

CdTe solar cells using a heterojunction with cadmium sulphide (CdS) have a suitable energy bandgap of 1.45 electron-volt (eV) (0.232 aJ) with a high coefficient of light absorption. The best efficiency of this cell is 16.7% (Green et al., 2010b) and the best commercially available modules have an efficiency of about 10 to 11%.

The toxicity of metallic cadmium and the relative scarcity of tellurium are issues commonly associated with this technology. Although several assessments of the risk (Fthenakis and Kim, 2009; Zayed and Philippe, 2009) and scarcity (Green et al., 2009; Wadia et al., 2009) are available, no consensus exists on these issues. It has been reported that this potential hazard can be mitigated by using a glass-sandwiched module design and by recycling the entire module and any industrial waste (Sinha et al., 2008).

The CIGS material family is the basis of the highest-efficiency thin-film solar cells to date. The copper indium diselenide (CuInSe$_2$)/CdS solar cell was invented in the early 1970s at AT&T Bell Labs (Wagner et al., 1974). Incorporating Ga and/or S to produce CuInGa(Se,S)$_2$ results in the benefit of a widened bandgap depending on the composition (Dimmler and Schock, 1996). CIGS-based solar cells have been validated at an efficiency of 20.1% (Green et al., 2010b). Due to higher efficiencies and lower manufacturing energy consumptions, CIGS cells are currently in the industrialization phase, with best commercial module efficiencies of up to 13.1% (Kushiya, 2009) for CuInGaSe$_2$ and 8.6% for CuInS$_2$ (Meeder et al., 2007). Although it is acknowledged that the scarcity of In might be an issue, Wadia et al. (2009) found that the current known economic indium reserves would allow the installation of more than 10 TW of CIGS-based PV systems.

High-efficiency solar cells based on a multijunction technology using III-V semiconductors (i.e., based on elements from the III and V columns of the periodic chart), for example, gallium arsenide (GaAs) and gallium indium phosphide (GaInP), can have superior efficiencies. These cells were originally developed for space use and are already commercialized. An economically feasible terrestrial application is the use of these cells in concentrating PV (CPV) systems, where concentrating optics are used to focus sunlight onto high efficiency solar cells (Bosi and Pelosi, 2007). The most commonly used cell is a triple-junction device based on GaInP/GaAs/germanium (Ge), with a record efficiency of 41.6% for a lattice-matched cell (Green et al., 2010b) and 41.1% for a metamorphic or lattice-mismatched device (Bett et al., 2009). Sub-module efficiencies have reached 36.1% (Green et al., 2010b). Another advantage of the concentrator system is that cell efficiencies increase under higher irradiance (Bosi and Pelosi, 2007), and the cell area can be decreased in proportion to the concentration level. Concentrator applications, however, require direct-normal irradiation, and are thus suited for specific climate conditions with low cloud coverage.

3.3.3.2 Emerging photovoltaic technologies

Emerging PV technologies are still under development and in laboratory or (pre-) pilot stage, but could become commercially viable within the next decade. They are based on very low-cost materials and/or processes and include technologies such as dye-sensitized solar cells, organic solar cells and low-cost (printed) versions of existing inorganic thin-film technologies.

Electricity generation by dye-sensitized solar cells (DSSCs) is based on light absorption in dye molecules (the 'sensitizers') attached to the very large surface area of a nanoporous oxide semiconductor electrode (usually titanium dioxide), followed by injection of excited electrons from the dye into the oxide. The dye/oxide interface thus serves as the separator of negative and positive charges, like the p-n junction in other devices. The negatively charged electrons are then transported through the semiconductor electrode and reach the counter electrode through the load, thus generating electricity. The injected electrons from the dye molecules are replenished by electrons supplied through a liquid electrolyte that penetrates the pores of the semiconductor electrode, providing the electrical path from the counter electrode (Graetzel, 2001). State-of-the-art DSSCs have achieved a top conversion efficiency of 10.4% (Chiba et al., 2005). Despite the gradual improvements since its discovery in 1991 (O'Regan and Graetzel, 1991), long-term stability against ultraviolet light irradiation, electrolyte leakage and high ambient temperatures continue to be key issues in commercializing these PV cells.

Organic PV (OPV) cells use stacked solid organic semiconductors, either polymers or small organic molecules. A typical structure of a small-molecule OPV cell consists of a stack of p-type and n-type organic semiconductors forming a planar heterojunction. The short-lived nature of the tightly bound electron-hole pairs (excitons) formed upon light absorption limits the thickness of the semiconductor layers that can be used—and therefore, the efficiency of such devices. Note that excitons need to move to the interface where positive and negative charges can be separated before they recombine. If the travel distance is short, the 'active' thickness of material is small and not all light can be absorbed within that thickness.

The efficiency achieved with single-junction OPV cells is about 5% (Li et al., 2005), although predictions indicate about twice that value or higher can be achieved (Forrest, 2005; Koster et al., 2006). To decouple exciton transport distances from optical thickness (light absorption), so-called bulk-heterojunction devices have been developed. In these devices, the absorption layer is made of a nanoscale mixture of p- and n-type materials to allow excitons to reach the interface within their lifetime, while also enabling a sufficient macroscopic layer thickness. This bulk-heterojunction structure plays a key role in improving the efficiency, to a record value of 7.9% in 2009 (Green et al., 2010a). The developments in cost and processing (Brabec, 2004; Krebs, 2005) of materials have caused OPV research to advance further. Also, the main development challenge is to achieve a sufficiently high stability in combination with a reasonable efficiency.

3.3.3.3 Novel photovoltaic technologies

Novel technologies are potentially disruptive (high-risk, high-potential) approaches based on new materials, devices and conversion concepts. Generally, their practically achievable conversion efficiencies and cost structure are still unclear. Examples of these approaches include intermediate-band semiconductors, hot-carrier devices, spectrum converters, plasmonic solar cells, and various applications of quantum dots (Section 3.7.3). The emerging technologies described in the previous section primarily aim at very low cost, while achieving a sufficiently high efficiency and stability. However, most of the novel technologies aim at reaching very high efficiencies by making better use of the entire solar spectrum from infrared to ultraviolet.

3.3.3.4 Photovoltaic systems

A photovoltaic system is composed of the PV module, as well as the balance of system (BOS) components, which include an inverter, storage devices, charge controller, system structure, and the energy network. The system must be reliable, cost effective, attractive and match with the electric grid in the future (US Photovoltaic Industry Roadmap Steering Committee, 2001; Navigant Consulting Inc., 2006; EU PV European Photovoltaic Technology Platform, 2007; Kroposki et al., 2008; NEDO, 2009).

At the component level, BOS components for grid-connected applications are not yet sufficiently developed to match the lifetime of PV modules. Additionally, BOS component and installation costs need to be reduced. Moreover, devices for storing large amounts of electricity (over 1 MWh or 3,600 MJ) will be adapted to large PV systems in the new energy network. As new module technologies emerge in the future, some of the ideas relating to BOS may need to be revised. Furthermore, the quality of the system needs to be assured and adequately maintained according to defined standards, guidelines and procedures. To ensure system quality, assessing performance is important, including on-line analysis (e.g., early fault detection) and off-line analysis of PV systems. The knowledge gathered can help to validate software for predicting the energy yield of future module and system technology designs.

To increasingly penetrate the energy network, PV systems must use technology that is compatible with the electric grid and energy supply and demand. System designs and operation technologies must also be developed in response to demand patterns by developing technology to forecast the power generation volume and to optimize the storage function. Moreover, inverters must improve the quality of grid electricity by controlling reactive power or filtering harmonics with communication in a new energy network that uses a mixture of inexpensive and effective communications systems and technologies, as well as smart meters (see Section 8.2.1).

3.3.3.5 Photovoltaic applications

Photovoltaic applications include PV power systems classified into two major types: those not connected to the traditional power grid (i.e., off-grid applications) and those that are connected (i.e., grid-connected applications). In addition, there is a much smaller, but stable, market segment for consumer applications.

Off-grid PV systems have a significant opportunity for economic application in the un-electrified areas of developing countries. Figure 3.6 shows the ratio of various off-grid and grid-connected systems in the Photovoltaic Power Systems (PVPS) Programme countries. Of the total capacity installed in these countries during 2009, only about 1.2% was installed in off-grid systems that now make up 4.2% of the cumulative installed PV capacity of the IEA PVPS countries (IEA, 2010e).

Off-grid centralized PV mini-grid systems have become a reliable alternative for village electrification over the last few years. In a PV mini-grid system, energy allocation is possible. For a village located in an isolated area and with houses not separated by too great a distance, the power may flow in the mini-grid without considerable losses. Centralized systems for local power supply have different technical advantages concerning electrical performance, reduction of storage needs, availability

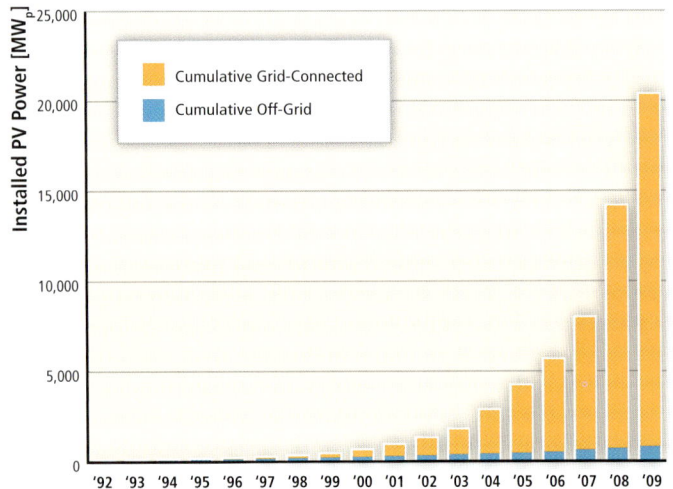

Figure 3.6 | Historical trends in cumulative installed PV power of off-grid and grid-connected systems in the OECD countries (IEA, 2010e). Vertical axis is in peak megawatts.

of energy, and dynamic behaviour. Centralized PV mini-grid systems could be the least-cost options for a given level of service, and they may have a diesel generator set as an optional balancing system or operate as a hybrid PV-wind-diesel system. These kinds of systems are relevant for reducing and avoiding diesel generator use in remote areas (Munoz et al., 2007; Sreeraj et al., 2010).

Grid-connected PV systems use an inverter to convert electricity from direct current (DC)—as produced by the PV array—to alternating current (AC), and then supply the generated electricity to the electricity network. Compared to an off-grid installation, system costs are lower because energy storage is not generally required, since the grid is used as a buffer. The annual output yield ranges from 300 to 2,000 kWh/kW (Clavadetscher and Nordmann, 2007; Gaiddon and Jedliczka, 2007; Kurokawa et al., 2007; Photovoltaic Geographic Information System, 2008) for several installation conditions in the world. The average annual performance ratio—the ratio between average AC system efficiency and standard DC module efficiency—ranges from 0.7 to 0.8 (Clavadetscher and Nordmann, 2007) and gradually increases further to about 0.9 for specific technologies and applications.

Grid-connected PV systems are classified into two types of applications: distributed and centralized. Grid-connected *distributed* PV systems are installed to provide power to a grid-connected customer or directly to the electricity network. Such systems may be: 1) on or integrated into the customer's premises, often on the demand side of the electricity meter; 2) on public and commercial buildings; or 3) simply in the built environment such as on motorway sound barriers. Typical sizes are 1 to 4 kW for residential systems, and 10 kW to several MW for rooftops on public and industrial buildings.

These systems have a number of advantages: distribution losses in the electricity network are reduced because the system is installed at the point of use; extra land is not required for the PV system, and costs for mounting the systems can be reduced if the system is mounted on an existing structure; and the PV array itself can be used as a cladding or roofing material, as in building-integrated PV (Eiffert, 2002; Ecofys Netherlands BV, 2007; Elzinga, 2008).

An often-cited disadvantage is the greater sensitivity to grid interconnection issues, such as overvoltage and unintended islanding (Kobayashi and Takasaki, 2006; Cobben et al., 2008; Ropp et al., 2008). However, much progress has been made to mitigate these effects, and today, by Institute of Electrical and Electronics Engineers (IEEE) and Underwriter Laboratories standards (IEEE 1547 (2008), UL 1741), all inverters must have the function of the anti-islanding effect.

Grid-connected *centralized* PV systems perform the functions of centralized power stations. The power supplied by such a system is not associated with a particular electricity customer, and the system is not located to specifically perform functions on the electricity network other than the supply of bulk power. Typically, centralized systems are mounted on the ground, and they are larger than 1 MW.

The economical advantage of these systems is the optimization of installation and operating cost by bulk buying and the cost effectiveness of the PV components and balance of systems at a large scale. In addition, the reliability of centralized PV systems can be greater than distributed PV systems because they can have maintenance systems with monitoring equipment, which can be a smaller part of the total system cost.

Multi-functional PV, daylighting and solar thermal components involving PV or solar thermal that have already been introduced into the built environment include the following: shading systems made from PV and/or solar thermal collectors; hybrid PV/thermal (PV/T) systems that generate electricity and heat from the same 'panel/collector' area; semi-transparent PV windows that generate electricity and transmit daylight from the same surface; façade collectors; PV roofs; thermal energy roof systems; and solar thermal roof-ridge collectors. Currently, fundamental and applied R&D activities are also underway related to developing other products, such as transparent solar thermal window collectors, as well as façade elements that consist of vacuum-insulation panels, PV panels, heat pump, and a heat-recovery system connected to localized ventilation.

Solar energy can be integrated within the building envelope and with energy conservation methods and smart-building operating strategies. Much work over the last decade or so has gone into this integration, culminating in the 'net-zero' energy building.

Much of the early emphasis was on integrating PV systems with thermal and daylighting systems. Bazilian et al. (2001) and Tripanagnostopoulos (2007) listed methods for doing this and reviewed case studies where the methods had been applied. For example, PV cells can be laid on the absorber plate of a flat-plate solar collector. About 6 to 20% of the solar energy absorbed on the cells is converted to electricity; the remaining roughly 80% is available as low-temperature heat to be transferred to the fluid being heated. The resulting unit produces both heat and electricity and requires only slightly more than half the area used if the two conversion devices had been mounted side by side and worked independently. PV cells have also been developed to be applied to windows to allow daylighting and passive solar gain. Reviews of recent work in this area are provided by Chow (2010) and Arif Hasan and Sumathy (2010).

Considerable work has also been done on architecturally integrating the solar components into the building. Any new solar building should be very well insulated, well sealed, and have highly efficient windows and heat recovery systems. Probst and Roecker (2007), surveying the opinions of more than 170 architects and engineers who examined numerous existing solar buildings, concluded the following: 1) best integration is achieved when the solar component is integrated as a construction element, and 2) appearance—including collector colour, orientation and jointing—must sometimes take precedence over performance in the overall design. In describing 16 case studies of building-integrated photovoltaics, Eiffert and Kiss (2000) identified two main products available on the architectural market: façade systems and roof systems. Façade systems include curtain wall products, spandrel panels and glazings; roofing products include tiles, shingles, standing-seam products and skylights. These can be integrated as components or constitute the entire structure (as in the case of a bus shelter).

The idea of the net-zero-energy solar building has sparked recent interest. Such buildings send as much excess PV-generated electrical energy to the grid as the energy they draw over the year. An IEA Task is considering how to achieve this goal (IEA NZEB, 2009). Recent examples for the Canadian climate are provided by Athienitis (2008). Starting from a building that meets the highest levels of conservation, these homes use hybrid air-heating/PV panels on the roof; the heated air is used for space heating or as a source for a heat pump. Solar water-heating collectors are included, as is fenestration permitting a large passive gain through equatorial-facing windows. A key feature is a ground-source heat pump, which provides a small amount of residual heating in the winter and cooling in the summer.

Smart solar-building control strategies may be used to manage the collection, storage and distribution of locally produced solar electricity and heat to reduce and shift peak electricity demand from the grid. An example of a smart solar-building design is given by Candanedo and Athienitis (2010), where predictive control based on weather forecasts one day ahead and real-time prediction of building response are used to optimize energy performance while reducing peak electricity demand.

3.3.4 Concentrating solar power electricity generation

Concentrating solar power (CSP) technologies produce electricity by concentrating direct-beam solar irradiance to heat a liquid, solid or gas that is then used in a downstream process for electricity generation. The majority of the world's electricity today—whether generated by coal, gas, nuclear, oil or biomass—comes from creating a hot fluid. CSP simply provides an alternative heat source. Therefore, an attraction of this technology is that it builds on much of the current know-how on power generation in the world today. And it will benefit not only from ongoing advances in solar concentrator technology, but also as improvements continue to be made in steam and gas turbine cycles.

Any concentrating solar system depends on direct-beam irradiation as opposed to global horizontal irradiation as for flat-plate systems. Thus, sites must be chosen accordingly, and the best sites for CSP are in near-equatorial cloud-free regions such as the North African desert. The average capacity factor of a solar plant will depend on the quality of the solar resource.

Some of the key advantages of CSP include the following: 1) it can be installed in a range of capacities to suit varying applications and conditions, from tens of kW (dish/Stirling systems) to multiple MWs (tower and trough systems); 2) it can integrate thermal storage for peaking loads (less than one hour) and intermediate loads (three to six hours); 3) it has modular and scalable components; and 4) it does not require exotic materials. This section discusses various types of CSP systems and thermal storage for these systems.

Large-scale CSP plants most commonly concentrate sunlight by reflection, as opposed to refraction with lenses. Concentration is either to a line (linear focus) as in trough or linear Fresnel systems or to a point (point focus) as in central-receiver or dish systems. The major features of each type of CSP system are illustrated in Figure 3.7 and are described below.

In trough concentrators, long rows of parabolic reflectors concentrate the solar irradiance by the order of 70 to 100 times onto a heat collection element (HCE) mounted along the reflector's focal line. The troughs track the Sun around one axis, with the axis typically being oriented north-south. The HCE comprises a steel inner pipe (coated with a solar-selective surface) and a glass outer tube, with an evacuated space in between. Heat-transfer oil is circulated through the steel pipe and heated to about 390°C. The hot oil from numerous rows of troughs is passed through a heat exchanger to generate steam for a conventional steam turbine generator (Rankine cycle). Land requirements are of the order of 2 km^2 for a 100-MW$_e$ plant, depending on the collector technology and assuming no storage. Alternative heat transfer fluids to the synthetic oil commonly used in trough receivers, such as steam and molten salt, are being developed to enable higher temperatures and overall efficiencies, as well as integrated thermal storage in the case of molten salt.

Linear Fresnel reflectors use long lines of flat or nearly flat mirrors, which allow the moving parts to be mounted closer to the ground, thus reducing structural costs. (In contrast, large trough reflectors presently use thermal bending to achieve the curve required in the glass surface.) The receiver is a fixed inverted cavity that can have a simpler construction than evacuated tubes and be more flexible in sizing. The attraction of

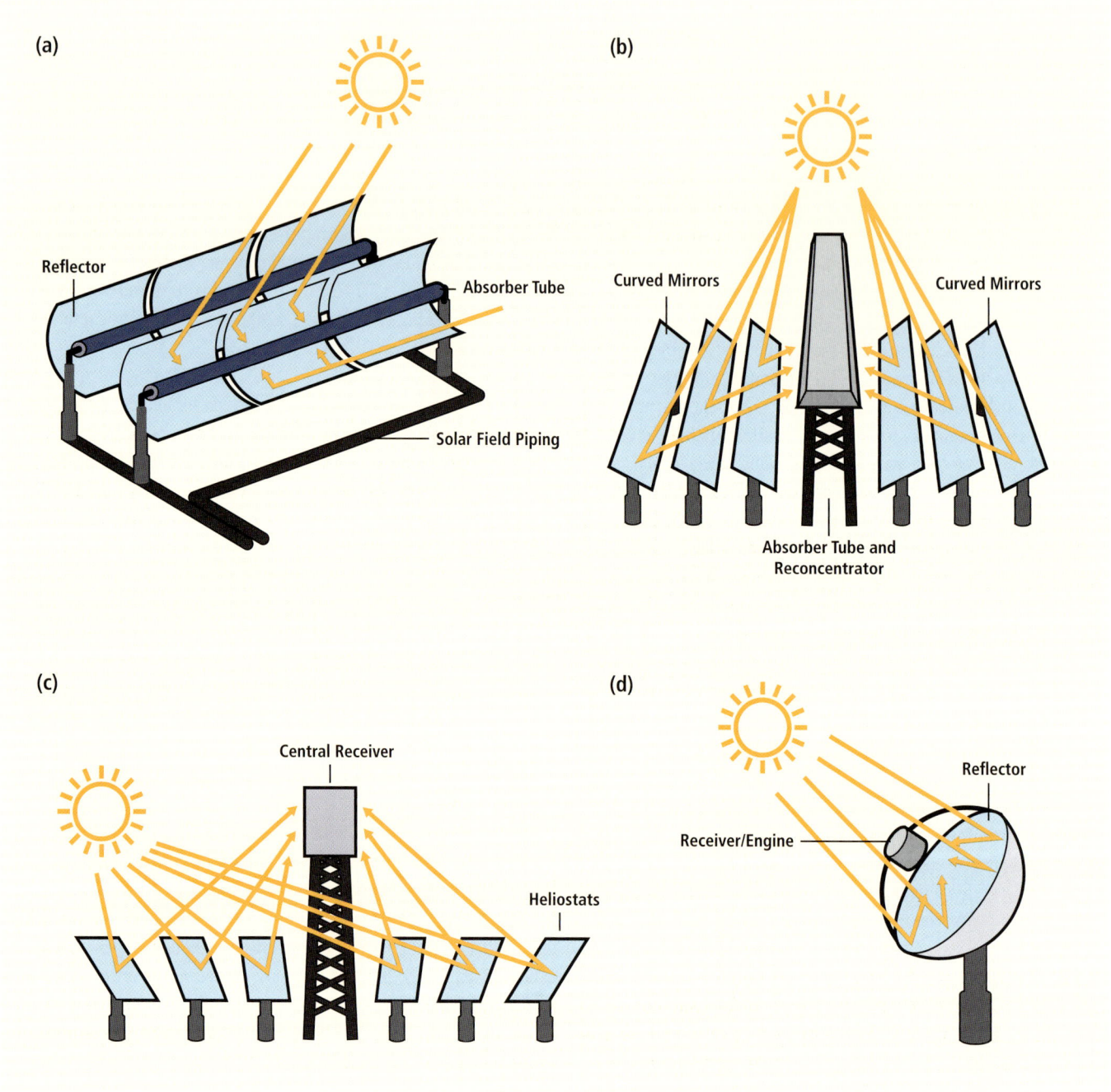

Figure 3.7 | Schematic diagrams showing the underlying principles of four basic CSP configurations: (a) parabolic trough, (b) linear Fresnel reflector, (c) central receiver/power tower, and (d) dish systems (Richter et al., 2009).

linear Fresnel reflectors is that the installed costs on a per square metre basis can be lower than for trough systems. However, the annual optical performance is less than that for a trough.

Central receivers (or power towers), which are one type of point-focus collector, are able to generate much higher temperatures than troughs and linear Fresnel reflectors, although requiring two-axis tracking as the Sun moves through solar azimuth and solar elevation. This higher temperature is a benefit because higher-temperature thermodynamic cycles used for generating electricity are more efficient. This technology uses an array of mirrors (heliostats), with each mirror tracking the Sun and reflecting the light onto a fixed receiver atop a tower. Temperatures of more than 1,000°C can be reached. Central receivers can easily generate the maximum temperatures of advanced steam turbines, can use high-temperature molten salt as the heat transfer fluid, and can be used to power gas turbine (Brayton) cycles.

Dish systems include an ideal optical reflector and therefore are suitable for applications requiring high temperatures. Dish reflectors are paraboloid and concentrate the solar irradiation onto a receiver mounted at the focal point, with the receiver moving with the dish. Dishes have been used to power Stirling engines at 900°C, and also for steam generation. There is now significant operational experience with dish/Stirling engine systems, and commercial rollout is planned. In 2010, the capacity of each Stirling engine is small—on the order of 10 to 25 $kW_{electric}$. The largest solar dishes have a 485-m^2 aperture and are in research facilities or demonstration plants.

In *thermal storage*, the heat from the solar field is stored prior to reaching the turbine. Thermal storage takes the form of sensible or latent heat storage (Gil et al., 2010; Medrano et al., 2010). The solar field needs to be oversized so that enough heat can be supplied to both operate the turbine during the day and, in parallel, charge the thermal storage. The term 'solar multiple' refers to the total solar field area installed divided by the solar field area needed to operate the turbine at design point without storage. Thermal storage for CSP systems needs to be at a temperature higher than that needed for the working fluid of the turbine. As such, system temperatures are generally between 400°C and 600°C, with the lower end for troughs and the higher end for towers. Allowable temperatures are also dictated by the limits of the media available. Examples of storage media include molten salt (presently comprising separate hot and cold tanks), steam accumulators (for short-term storage only), solid ceramic particles, high-temperature phase-change materials, graphite, and high-temperature concrete. The heat can then be drawn from the storage to generate steam for a turbine, as and when needed. Another type of storage associated with high-temperature CSP is thermochemical storage, where solar energy is stored chemically. This is discussed more fully in Sections 3.3.5 and 3.7.5.

Thermal energy storage integrated into a system is an important attribute of CSP. Until recently, this has been primarily for operational purposes, providing 30 minutes to 1 hour of full-load storage. This eases the impact of thermal transients such as clouds on the plant, assists start-up and shut-down, and provides benefits to the grid. Trough plants are now designed for 6 to 7.5 hours of storage, which is enough to allow operation well into the evening when peak demand can occur and tariffs are high. Trough plants in Spain are now operating with molten-salt storage. In the USA, Abengoa Solar's 280-MW Solana trough project, planned to be operational by 2013, intends to integrate six hours of thermal storage. Towers, with their higher temperatures, can charge and store molten salt more efficiently. Gemasolar, a 17-MW_e solar tower project under construction in Spain, is designed for 15 hours of storage, giving a 75% annual capacity factor (Arce et al., 2011).

Thermal storage is a means of providing dispatchability. Hybridization with non-renewable fuels is another way in which CSP can be designed to be dispatchable. Although the back-up fuel itself may not be renewable (unless it is biomass-derived), it provides significant operational benefits for the turbine and improves solar yield.

CSP applications range from small distributed systems of tens of kW to large centralized power stations of hundreds of MW.

Stirling and Brayton cycle generation in CSP can be installed in a wide range from small distributed systems to clusters forming medium- to large-capacity power stations. The dish/Stirling technology has been under development for many years, with advances in dish structures, high-temperature receivers, use of hydrogen as the circulating working fluid, as well as some experiments with liquid metals and improvements in Stirling engines—all bringing the technology closer to commercial deployment. Although the individual unit size may only be of the order of tens of kW_e, power stations having a large capacity of up to 800 MW_e have been proposed by aggregating many modules. Because each dish represents a stand-alone electricity generator, from the perspective of distributed generation there is great flexibility in the capacity and rate at which units are installed. However, the dish technology is less likely to integrate thermal storage.

An alternative to the Stirling engine is the Brayton cycle, as used by gas turbines. The attraction of these engines for CSP is that they are already in significant production, being used for distributed generation fired with landfill gas or natural gas. In the solarized version, the air is instead heated by concentrated solar irradiance from a tower or dish reflector. It is also possible to integrate with a biogas or natural gas combustor to back up the solar. Several developments are currently underway based on solar tower and micro-turbine combinations.

Centralized CSP benefits from the economies of scale offered by large-scale plants. Based on conventional steam and gas turbine cycles, much of the technological know-how of large power station design and practice is already in place. However, although larger capacity has significant cost benefits, it has also tended to be an inhibitor until recently because of the much larger investment commitment required from investors. In addition, larger power stations require strong infrastructural support, and new or augmented transmission capacity may be needed.

The earliest commercial CSP plants were the 354 MW of Solar Electric Generating Stations in California—deployed between 1985 and 1991—that continue to operate commercially today. As a result of the positive experiences and lessons learned from these early plants, the trough systems tend to be the technology most often applied today as the CSP industry grows. In Spain, regulations to date have mandated that the largest capacity unit that can be installed is 50 MW_e to help stimulate industry competition. In the USA, this limitation does not exist, and proposals are in place for much larger plants—280 MW_e in the case of troughs and 400-MW_e plants (made up of four modules) based on towers. There are presently two operational solar towers of 10 and 20 MW_e, and all tower developers plan to increase capacity in line with technology development, regulations and investment capital. Multiple dishes have also been proposed as a source of aggregated heat, rather than distributed-generation Stirling or Brayton units.

CSP or PV electricity can also be used to power reverse-osmosis plants for desalination. Dedicated CSP desalination cycles based on pressure and temperature are also being developed for desalination (see Section 3.3.2).

3.3.5 Solar fuel production

Solar fuel technologies convert solar energy into chemical fuels, which can be a desirable method of storing and transporting solar energy. They can be used in a much wider variety of higher-efficiency applications than just electricity generation cycles. Solar fuels can be processed into liquid transportation fuels or used directly to generate electricity in fuel cells; they can be employed as fuels for high-efficiency gas-turbine cycles or internal combustion engines; and they can serve for upgrading fossil fuels, CO_2 synthesis, or for producing industrial or domestic heat. The challenge is to produce large amounts of chemical fuels directly from sunlight in cost-effective ways and to minimize adverse effects on the environment (Steinfeld and Meier, 2004).

Solar fuels that can be produced include synthesis gas (syngas, i.e., mixed gases of carbon monoxide and hydrogen), pure hydrogen (H_2) gas, dimethyl ether (DME) and liquids such as methanol and diesel. The high energy density of H_2 (on a mass basis) and clean conversion give it attractive properties as a future fuel and it is also used as a feedstock for many industrial processes. H_2 has a higher energy density than batteries, although batteries have a higher round-trip efficiency. However, its very low energy density on a volumetric basis poses economic challenges associated with its storage and transport. It will require significant new distribution infrastructure and either new designs of internal combustion engine or a move to fuel cells. Additionally, the synthesis of hydrogen with CO_2 can produce hydrocarbon fuels that are compatible with existing infrastructures. DME gas is similar to liquefied petroleum gas (LPG) and easily stored. Methanol is liquid and can replace gasoline without significant changes to the engine or the fuel distribution infrastructure. Methanol and DME can be used for fuel cells after reforming, and DME can also be used in place of LPG. Fischer-Tropsch processes can produce hydrocarbon fuels and electricity (see Sections 2.6 and 8.2.4).

There are three basic routes, alone or in combination, for producing storable and transportable fuels from solar energy: 1) the electrochemical route uses solar electricity from PV or CSP systems followed by an electrolytic process; 2) the photochemical/photobiological route makes direct use of solar photon energy for photochemical and photobiological processes; and 3) the thermochemical route uses solar heat at moderate and/or high temperatures followed by an endothermic thermochemical process (Steinfeld and Meier, 2004). Note that the electrochemical and thermochemical routes apply to any RE technology, not exclusively to solar technologies.

Figure 3.8 illustrates possible pathways to produce H_2 or syngas from water and/or fossil fuels using concentrated solar energy as the source of high-temperature process heat. Feedstocks include *inorganic* compounds such as water and CO_2, and *organic* sources such as coal, biomass and natural gas (NG). See Chapter 2 for parallels with biomass-derived syngas.

Electrolysis of water can use solar electricity generated by PV or CSP technology in a conventional (alkaline) electrolyzer, considered a benchmark for producing solar hydrogen. With current technologies, the overall solar-to-hydrogen energy conversion efficiency ranges between 10 and 14%, assuming electrolyzers working at 70% efficiency and solar electricity being produced at 15% (PV) and 20% (CSP) annual efficiency. The electricity demand for electrolysis can be significantly reduced if the electrolysis of water proceeds at higher temperatures (800° to 1,000°C) via solid-oxide electrolyzer cells (Jensen et al., 2007). In this case, concentrated solar energy can be applied to provide both the high-temperature process heat and the electricity needed for the high-temperature electrolysis.

Thermolysis and thermochemical cycles are a long-term sustainable and carbon-neutral approach for hydrogen production from water. This route involves energy-consuming (endothermic) reactions that make use of concentrated solar irradiance as the energy source for high-temperature process heat (Abanades et al., 2006). Solar thermolysis requires temperatures above 2,200°C and raises difficult challenges for reactor materials and gas separation. Water-splitting thermochemical cycles allow operation at lower temperature, but require several chemical reaction steps and also raise challenges because of inefficiencies associated with heat transfer and product separation at each step.

Decarbonization of fossil fuels is a near- to mid-term transition pathway to solar hydrogen that encompasses the carbothermal reduction of metal oxides (Epstein et al., 2008) and the decarbonization of fossil fuels via solar cracking (Spath and Amos, 2003; Rodat et al., 2009), reforming (Möller et al., 2006) and gasification (Z'Graggen and Steinfeld, 2008; Piatkowski et al., 2009). These routes are being pursued by European, Australian and US academic and industrial research consortia. Their technical feasibility has been demonstrated in concentrating solar chemical pilot plants at the power level of 100 to 500 kW$_{th}$. Solar hybrid fuel can be produced by supplying concentrated solar thermal energy to the endothermic processes of methane and biomass reforming—that is, solar heat is used for process energy only, and fossil fuels are still a required input. Some countries having vast solar and natural gas resources, but a relatively small domestic energy market (e.g., the Middle East and Australia) are in a position to produce and export solar energy in the form of liquid fuels.

Solar fuel synthesis from solar hydrogen and CO_2 produces hydrocarbons that are compatible with existing energy infrastructures such as the natural gas network or existing fuel supply structures. The renewable methane process combines solar hydrogen with CO_2 from the

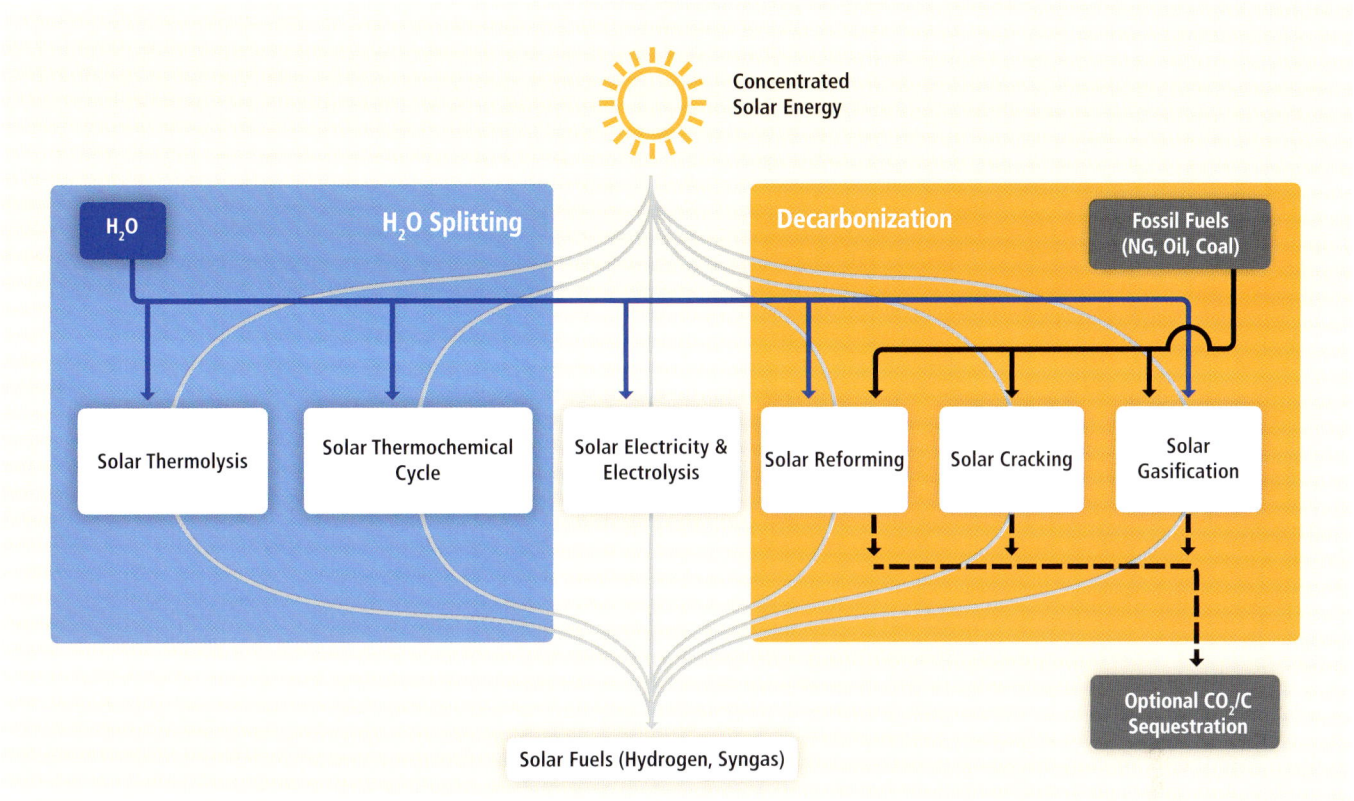

Figure 3.8 | Thermochemical routes for solar fuels production, indicating the chemical source of H_2: water (H_2O) for solar thermolysis and solar thermochemical cycles to produce H_2 only; fossil or biomass fuels as feedstock for solar cracking to produce H_2 and carbon (C); or a combination of fossil/biomass fuels and H_2O/CO_2 for solar reforming and gasification to produce syngas, H_2 and carbon monoxide (CO). For the solar decarbonization processes, sequestration of the CO_2/C may be considered (from Steinfeld and Meier, 2004; Steinfeld, 2005).

atmosphere or other sources in a synthesis reactor with a nickel catalyst. In this way, a substitute for natural gas is produced that can be stored, transported and used in gas power plants, heating systems and gas vehicles (Sterner, 2009).

Solar methane can be produced using water, air, solar energy and a source of CO_2. Possible CO_2 sources are biomass, industry processes or the atmosphere. CO_2 is regarded as the carrier for hydrogen in this energy system. By separating CO_2 from the combustion process of solar methane, CO_2 can be recycled in the energy system or stored permanently. Thus, carbon sink energy systems powered by RE can be created (Sterner, 2009). The first pilot plants at the kW scale with atmospheric CO_2 absorption have been set up in Germany, proving the technical feasibility. Scaling up to the utility MW scale is planned in the next few years (Specht et al., 2010).

In an alternative conversion step, liquid fuels such as Fischer-Tropsch diesel, DME, methanol or solar kerosene (jet fuel) can be produced from solar energy and CO_2/water (H_2O) for long-distance transportation. The main advantages of these solar fuels are the same range as fossil fuels (compared to the generally reduced range of electric vehicles), less competition for land use, and higher per-hectare yields compared to biofuels. Solar energy can be harvested via natural photosynthesis in biofuels with an efficiency of 0.5%, via PV power and solar fuel conversion (technical photosynthesis) with an efficiency of 10% (Sterner, 2009) and via solar-driven thermochemical dissociation of CO_2 and H_2O using metal oxide redox reactions, yielding a syngas mixture of carbon monoxide (CO) and H_2, with a solar-to-fuel efficiency approaching 20% (Chueh et al., 2010). This approach would provide a solution to the issues and controversy surrounding existing biofuels, although the cost of this technology is a possible constraint.

3.4 Global and regional status of market and industry development

This section looks at the five key solar technologies, first focusing on installed capacity and generated energy, then on industry capacity and supply chains, and finally on the impact of policies specific to these technologies.

3.4.1 Installed capacity and generated energy

This subsection discusses the installed capacity and generated energy within the five technology areas of passive solar, active solar heating and cooling, PV electricity generation, CSP electricity generation, and solar fuel production.

For *passive solar technologies*, no estimates are available at this time for the installed capacity of passive solar or the energy generated or saved through this technology.

For *active solar heating*, the total installed capacity worldwide was about 149 GW$_{th}$ in 2008 and 180 GW$_{th}$ in 2009 (Weiss and Mauthner, 2010; REN21, 2010).

In 2008, new capacity of 29.1 GW$_{th}$, corresponding to 41.5 million m^2 of solar collectors, was installed worldwide (Weiss and Mauthner, 2010). In 2008, China accounted for about 79% of the installations of glazed collectors, followed by the EU with 14.5%.

The overall new installations grew by 34.9% compared to 2007. The growth rate in 2006/2007 was 18.8%. The main reasons for this growth were the high growth rates of glazed water collectors in China, Europe and the USA.

In 2008, the global market had high growth rates for evacuated-tube collectors and flat-plate collectors, compared to 2007. The market for unglazed air collectors also increased significantly, mainly due to the installation of 23.9 MW$_{th}$ of new systems in Canada.

Compared to 2007, the 2008 installation rates for new unglazed, glazed flat-plate, and evacuated-tube collectors were significantly up in Jordan, Cyprus, Canada, Ireland, Germany, Slovenia, Macedonia (FYROM), Tunisia, Poland, Belgium and South Africa.

New installations in China, the world's largest market, again increased significantly in 2008 compared to 2007, reaching 21.7 GW$_{th}$. After a market decline in Japan in 2007, the growth rate was once again positive in 2008.

Market decreases compared to 2007 were reported for Israel, the Slovak Republic and the Chinese province of Taiwan.

The main markets for unglazed water collectors are still found in the USA (0.8 GW$_{th}$), Australia (0.4 GW$_{th}$), and Brazil (0.08 GW$_{th}$). Notable markets are also in Austria, Canada, Mexico, The Netherlands, South Africa, Spain, Sweden and Switzerland, with values between 0.07 and 0.01 GW$_{th}$ of new installed unglazed water collectors in 2008.

Comparison of markets in different countries is difficult due to the wide range of designs used for different climates and different demand requirements. In Scandinavia and Germany, a solar heating system will typically be a combined water-heating and space-heating system, known as a solar combisystem, with a collector area of 10 to 20 m^2. In Japan, the number of solar domestic water-heating systems is large, but most installations are simple integral preheating systems. The market in Israel is large due to a favourable climate, as well as regulations mandating installation of solar water heaters. The largest market is in China, where there is widespread adoption of advanced evacuated-tube solar collectors. In terms of per capita use, Cyprus is the leading country in the world, with an installed capacity of 527 kW$_{th}$ per 1,000 inhabitants.

The type of application of solar thermal energy varies greatly in different countries (Weiss and Mauthner, 2010). In China (88.7 GW$_{th}$), Europe (20.9 GW$_{th}$) and Japan (4.4 GW$_{th}$), flat-plate and evacuated-tube collectors mainly prepare hot water and provide space heating. However, in the USA and Canada, swimming pool heating is still the dominant application, with an installed capacity of 12.9 GW$_{th}$ of unglazed plastic collectors.

The biggest reported solar thermal system for industrial process heat was installed in China in 2007. The 9 MW$_{th}$ plant produces heat for a textile company. About 150 large-scale plants (>500 m^2; 350 kW$_{th}$)[1] with a total capacity of 160 MW$_{th}$ are in operation in Europe. The largest plants for solar-assisted district heating are located in Denmark (13 MW$_{th}$) and Sweden (7 MW$_{th}$).

In Europe, the market size more than tripled between 2002 and 2008. However, even in the leading European solar thermal markets of Austria, Greece, and Germany, only a minor portion of residential homes use solar thermal. For example, in Germany, only about 5% of one- and two-family homes are using solar thermal energy.

The European market has the largest variety of different solar thermal applications, including systems for hot-water preparation, plants for space heating of single- and multi-family houses and hotels, large-scale plants for district heating, and a growing number of systems for air-conditioning, cooling and industrial applications.

Advanced applications such as solar cooling and air conditioning (Henning, 2004, 2007), industrial applications (POSHIP, 2001) and desalination/water treatment are in the early stages of development. Only a few hundred first-generation systems are in operation.

For *PV electricity generation*, newly installed capacity in 2009 was about 7.5 GW, with shipments to first point in the market at 7.9 GW (Jäger-Waldau, 2010a; Mints, 2010). This addition brought the cumulative installed PV capacity worldwide to about 22 GW—a capacity able to generate up to 26 TWh (93,600 TJ) per year. More than 90% of this capacity is installed in three leading markets: the EU27 with 16 GW (73%), Japan with 2.6 GW (12%), and the USA with 1.7 GW (8%) (Jäger-Waldau, 2010b). These markets are dominated by grid-connected PV systems, and growth within PV markets has been stimulated by various government programmes around the world. Examples of such programmes include feed-in tariffs in Germany and Spain, and various mechanisms in the USA, such as buy-down incentives, investment tax credits, performance-based incentives and RE quota systems. For 2010,

---

1   To enable comparison, the IEA's Solar Heating and Cooling Programme, together with the European Solar Thermal Industry Federation and other major solar thermal trade associations, publish statistics in kW$_{th}$ (kilowatt thermal) and use a factor of 0.7 kW$_{th}$/m^2 to convert square metres of collector area into installed thermal capacity (kW$_{th}$).

the market is estimated between 9 and 24 GW of additional installed PV systems, with a consensus value in the 13 GW range (Jäger-Waldau, 2010a).

Figure 3.9 illustrates the cumulative installed capacity for the top eight PV markets through 2009, including Germany (9,800 MW), Spain (3,500 MW), Japan (2,630 MW), the USA (1,650 MW), Italy (1,140 MW), Korea (460 MW), France (370 MW) and the People's Republic of China (300 MW). By far, Spain and Germany have seen the largest amounts of growth in installed PV capacity in recent years, with Spain seeing a huge surge in 2008 and Germany having experienced steady growth over the last five years.

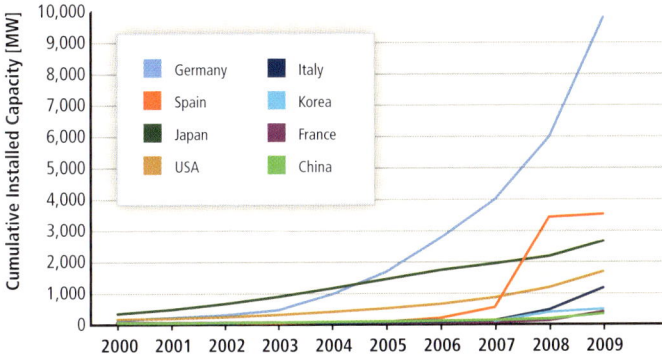

Figure 3.9 | Installed PV capacity in eight markets. Data sources: EurObserv'ER (2009); IEA (2009c); REN21 (2009); and Jäger-Waldau (2010b).

Concentrating photovoltaics (CPV) is an emerging market with about 17 MW of cumulative installed capacity at the end of 2008. The two main tracks are high-concentration PV (>300 times or 300 suns) and low- to medium-concentration PV with a concentration factor of 2 to about 300 (2 to ~300 suns). To maximize the benefits of CPV, the technology requires high direct-beam irradiance, and these areas have a limited geographical range—the 'Sun Belt' of the Earth. The market share of CPV is still small, but an increasing number of companies are focusing on CPV. In 2008, about 10 MW of CPV were installed, and market estimates for 2009 are in the 20 to 30 MW range; for 2010, about 100 MW are expected.

Regarding CSP *electricity generation*, at the beginning of 2009, more than 700 MW_e of grid-connected CSP plants were installed worldwide, with another 1,500 MW_e under construction (Torres et al., 2010). The majority of installed plants use parabolic trough technology. Central-receiver technology comprises a growing share of plants under construction and those announced. The bulk of the operating capacity is installed in Spain and the south-western United States.

In 2007, after a hiatus of more than 15 years, the first major CSP plants came on line with Nevada Solar One (64 MW_e, USA) and PS10 (11 MW_e, Spain). In Spain, successive Royal Decrees have been in place since 2004 and have stimulated the CSP industry in that country. Royal Decree 661/2007 has been a major driving force for CSP plant construction and expansion plans. As of November 2009, 2,340 MW_e of CSP projects had been preregistered for the tariff provisions of the Royal Decree. In the USA, more than 4,500 MW_e of CSP are currently under power purchase agreement contracts. The different contracts specify when the projects must start delivering electricity between 2010 and 2015 (Bloem et al., 2010). More than 10,000 MW_e of new CSP plants have been proposed in the USA. More than 50 CSP electricity projects are currently in the planning phase, mainly in North Africa, Spain and the USA. In Australia, the federal government has called for 1,000 MW_e of new solar plants, covering both CSP and PV, under the Solar Flagships programme. Figure 3.10 shows the current and planned deployment to add more CSP capacity in the near future.

Hybrid solar/fossil plants have received increasing attention in recent years, and several integrated solar combined-cycle (ISCC) projects have been either commissioned or are under construction in the Mediterranean region and the USA. The first plant in Morocco (Ain Beni Mathar: 470 MW total, 22 MW solar) began operating in June 2010, and two additional plants in Algeria (Hassi R'Mel: 150 MW total, 30 MW solar) and Egypt (Al Kuraymat: 140 MW total, 20 MW solar) are under construction. In Italy, another example of an ISCC project is Archimede; however, the plant's 31,000-m² parabolic trough solar field will be the first to use molten salt as the heat transfer fluid (SolarPACES, 2009a).

Solar fuel production technologies are in an earlier stage of development. The high-temperature solar reactor technology is typically being developed at a laboratory scale of 1 to 10 kW_{th} solar power input.

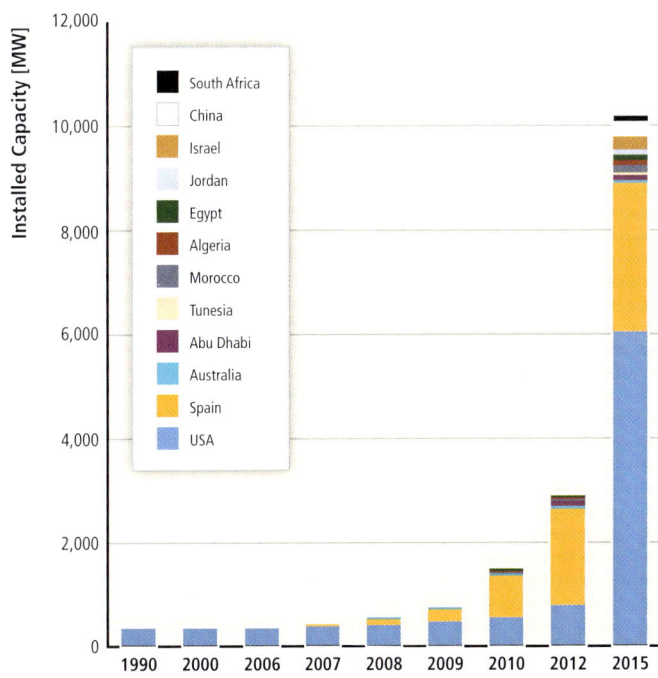

Figure 3.10 | Installed and planned concentrated solar power plants by country (Bloem et al., 2010).

Scaling up thermochemical processes for hydrogen production to the 100-kW$_{th}$ power level is reported for a medium-temperature mixed iron oxide cycle (800°C to 1,200°C) (Roeb et al., 2006, 2009) and for the high-temperature zinc oxide (ZnO) dissociation reaction at above 1,700°C (Schunk et al., 2008, 2009). Pilot plants in the power range of 300 to 500 kW$_{th}$ have been built for the carbothermic reduction of ZnO (Epstein et al., 2008), the steam reforming of methane (Möller et al., 2006), and the steam gasification of petcoke (Z'Graggen and Steinfeld, 2008). Solar-to-gas has been demonstrated at a 30-kW scale to drive a commercial natural gas vehicle, applying a nickel catalyst (Specht et al., 2010). Demonstration at the MW scale should be warranted before erecting commercial solar chemical plants for fuels production, which are expected to be available only after 2020 (Pregger et al., 2009).

Direct conversion of solar energy to fuel is not yet widely demonstrated or commercialized. But two options appear commercially feasible in the near to medium term: 1) the solar hybrid fuel production system (including solar methane reforming and solar biomass reforming), and 2) solar PV or CSP electrolysis.

Australia's Commonwealth Scientific and Industrial Research Organisation is running a 250-kW$_{th}$ reactor and plans to build a MW-scale demonstration plant using solar steam-reforming technology, with an eventual move to CO_2 reforming for higher performance and less water usage. With such a system, liquid solar fuels can be produced in sunbelts such as Australia and solar energy shipped on a commercial basis to Asia and beyond.

Oxygen gas produced by solar (PV or CSP) electrolysis can be used for coal gasification and partial oxidation of natural gas. With the combined process of solar electrolysis and partial oxidation of coal or methane, theoretically 10 to 15% of solar energy is incorporated into the methanol or DME. Also, the production cost of the solar hybrid fuel can be lower than the solar hydrogen produced by the solar electrolysis process only.

3.4.2 Industry capacity and supply chain

This subsection discusses the industry capacity and supply chain within the five technology areas of passive solar, active solar heating and cooling, PV electricity generation, CSP electricity generation and solar fuel production.

In passive solar technologies, people make up part of the industry capacity and the supply chain: namely, the engineers and architects who collaborate to produce passively heated buildings. Close collaboration between the two disciplines has often been missing in the past, but the dissemination of systematic design methodologies issued by different countries has improved the design capabilities (Athienitis and Santamouris, 2002).

The integration of passive solar systems with the active heating/cooling air-conditioning systems both in the design and operation stages of the building is essential to achieve good comfort conditions while saving energy. However, this is often overlooked because of inadequate collaboration for integrating building design between architects and engineers. Thus, the architect often designs the building envelope based solely on qualitative passive solar design principles, and the engineer often designs the heating-ventilation-air-conditioning system based on extreme design conditions without factoring in the benefits due to solar gains and natural cooling. The result may be an oversized system and inappropriate controls incompatible with the passive system and that can cause overheating and discomfort (Athienitis and Santamouris, 2002). Collaboration between the disciplines involved in building design is now improving with the adoption of computer tools for integrated analysis and design.

The design of high-mass buildings with significant near-equatorial-facing window areas is common in some areas of the world such as Southern Europe. However, a systematic approach to designing such buildings is still not widely employed. This is changing with the introduction of the passive house standard in Germany and other countries (PHPP, 2004), the deployment of the European Directives, and new national laws such as China's standard based on the German one.

Glazing and window technologies have made substantial progress in the last 20 years (Hollands et al., 2001). New-generation windows result in low energy losses, high daylight efficiency, solar shading, and noise reduction. New technologies such as transparent PV and electrochromic and thermochromic windows provide many possibilities for designing solar houses and offices with abundant daylight. The change from regular double-glazed to double-glazed low-emissivity argon windows is presently occurring in Canada and is accelerated by the rapid drop in prices of these windows.

The primary materials for low-temperature thermal storage in passive solar systems are concrete, bricks and water. A review of thermal storage materials is given by Hadorn (2008) under IEA SHC Task 32, focusing on a comparison of the different technologies. Phase-change material (PCM) thermal storage (Mehling and Cabeza, 2008) is particularly promising in the design, control and load management of solar buildings because it reduces the need for structural reinforcement required for heavier traditional sensible storage in concrete-type construction. Recent developments facilitating integration include microencapsulated PCM that can be mixed with plaster and applied to interior surfaces (Schossig et al., 2005). PCM in microencapsulated polymers is now on the market and can be added to plaster, gypsum or concrete to enhance the thermal capacity of a room. For renovation, this provides a good alternative to new heavy walls, which would require additional structural support (Hadorn, 2008).

In spite of the advances in PCM, concrete has certain advantages for thermal storage when a massive building design approach is used, as in many of the Mediterranean countries. In this approach, the concrete also serves as the structure of the building and is thus likely more cost effective than thermal storage without this added function.

For active solar heating and cooling, a number of different collector technologies and system approaches have been developed due to different applications—including domestic hot water, heating, preheating and combined systems—and varying climatic conditions.

In some parts of the production process, such as selective coatings, large-scale industrial production levels have been attained. A number of different materials, including copper, aluminium and stainless steel, are applied and combined with different welding technologies to achieve a highly efficient heat-exchange process in the collector. The materials used for the cover glass are structured or flat, low-iron glass. The first antireflection coatings are coming onto the market on an industrial scale, leading to efficiency improvements of about 5%.
In general, vacuum-tube collectors are well-suited for higher-temperature applications. The production of vacuum-tube collectors is currently dominated by the Chinese Dewar tubes, where a metallic heat exchanger is integrated to connect them with the conventional hot-water systems. In addition, some standard vacuum-tube collectors, with metallic heat absorbers, are on the market.

The largest exporters of solar water-heating systems are Australia, Greece and the USA. The majority of exports from Greece are to Cyprus and the near-Mediterranean area. France also sends a substantial number of systems to its overseas territories. The majority of US exports are to the Caribbean region. Australian companies export about 50% of production (mainly thermosyphon systems with external horizontal tanks) to most of the areas of the world that do not have hard-freeze conditions.

PV electricity generation is discussed under the areas of overall solar cell production, thin-film module production and polysilicon production. The development characteristic of the PV sector is much different than the traditional power sector, more closely resembling the semiconductor market, with annual growth rates between 40 to 50% and a high learning rate. Therefore, scientific and peer-reviewed papers can be several years behind the actual market developments due to the nature of statistical time delays and data consolidation. The only way to keep track of such a dynamic market is to use commercial market data. Global PV cell production[2] reached more than 11.5 GW in 2009.

---

2 Solar cell production capacities mean the following: for wafer-silicon-based solar cells, only the cells; for thin films, the complete integrated module. Only those companies that actually produce the active circuit (solar cell) are counted; companies that purchase these circuits and then make modules are not counted.

Figure 3.11 plots the increase in production from 2000 through 2009, showing regional contributions (Jäger-Waldau, 2010a). The compound annual growth rate in production from 2003 to 2009 was more than 50%.

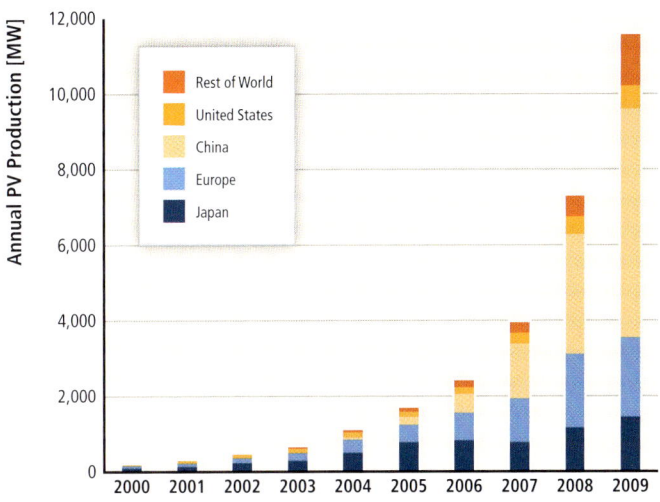

Figure 3.11 | Worldwide PV production from 2000 to 2009 (Jäger-Waldau, 2010b).

The announced production capacities—based on a survey of more than 300 companies worldwide—increased despite very difficult economic conditions in 2009 (Figure 3.12) (Jäger-Waldau, 2010b). Only published announcements from the respective companies, not third-party information, were used. April 2010 was the cut-off date for the information included. This method has the drawback that not all companies announce their capacity increases in advance; also, in times of financial tightening, announcements of scale-backs in expansion plans are often delayed to prevent upsetting financial markets. Therefore, the capacity figures provide a trend, but do not represent final numbers.

In 2008 and 2009, Chinese production capacity increased overproportionally. In actual production, China surpassed all other countries,

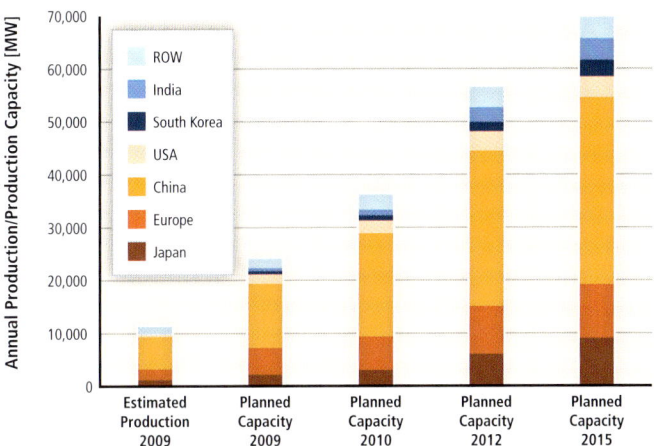

Figure 3.12 | Worldwide annual PV production in 2009 compared to the announced production capacities (Jäger-Waldau, 2010a).

estimated in 2009 at between 5.4 and 6.1 GW (including 1.5 to 1.7 GW production in the Chinese province of Taiwan), Europe had 2.0 to 2.2 GW, and was followed by Japan, with 1.5 to 1.7 GW (Jäger-Waldau, 2010b). In terms of production, First Solar (USA/Germany/France/Malaysia) was number one (1,082 MW), followed by Suntech (China) estimated at 750 MW and Sharp (Japan) estimated at 580 MW.

If all these ambitious plans can be realized by 2015, then China will have about 51% (including 16% in the Chinese province of Taiwan) of the worldwide production capacity of 70 GW, followed by Europe (15%) and Japan (13%).

Worldwide, more than 300 companies produce solar cells. In 2009, *silicon-based solar cells and modules* represented about 80% of the worldwide market (Figure 3.13). In addition to a massive increase in production capacities, the current development predicts that thin-film-based solar cells will increase their market share to over 30% by 2012.

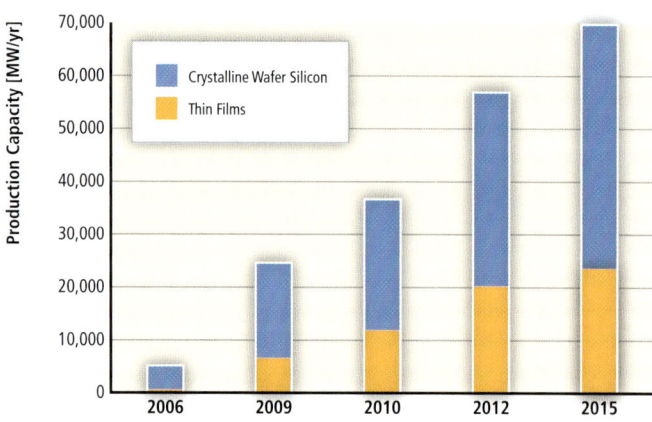

Figure 3.13 | Actual (2006) and announced (2009 to 2015) production capacities of thin-film and crystalline silicon-based solar modules (Jäger-Waldau, 2010b).

In 2005, production of *thin-film PV modules* grew to more than 100 MW per year. Since then, the compound annual growth rate of thin-film PV module production was higher than that of the industry—thus increasing the market share of thin-film products from 6% in 2005 to about 20% in 2009. Most of this thin-film share comes from the largest PV company.

More than 150 companies are involved in the thin-film solar cell production process, ranging from R&D activities to major manufacturing plants. The first 100-MW thin-film factories became operational in 2007, and the announcements of new production capacities accelerated again in 2008. If all expansion plans are realized in time, thin-film production capacity could be 20.0 GW, or 35% of the total 56.7 GW in 2012, and 23.5 GW, or 34% of a total of 70 GW in 2015 (Jäger-Waldau, 2009,

2010b). The first thin-film factories with GW production capacity are already under construction for various thin-film technologies.

The rapid growth of the PV industry since 2000 led to the situation between 2004 and early 2008 where the demand for polysilicon outstripped the supply from the semiconductor industry. This led to a silicon shortage, which resulted in silicon spot-market prices as high as USD_{2005} 450/kg (USD_{2005}, assumed 2008 base) in 2008 compared to USD_{2005} 25.5/kg in 2003 and consequently higher prices for PV modules. This extreme price hike triggered the massive capacity expansion, not only of established companies, but of many new entrants as well.

The six companies that reported shipment figures delivered together about 43,900 tonnes of polysilicon in 2008, as reported by Semiconductor Equipment and Materials International (SEMI, 2009a). In 2008, these companies had a production capacity of 48,200 tonnes of polysilicon (Service, 2009). However, all polysilicon producers, including new entrants with current and alternative technologies, had a production capacity of more than 90,000 tonnes of polysilicon in 2008. Considering that not all new capacity actually produced polysilicon at nameplate capacity in 2008, it was estimated that 62,000 tonnes of polysilicon could be produced. Subtracting the needs of the semiconductor industry and adding recycling and excess production, the available amount of silicon for the PV industry was estimated at 46,000 tonnes of polysilicon. With an average material need of 8.7 g/W_p (p = peak), this would have been sufficient for the production of 5.3 GW of crystalline silicon PV cells.

The drive to reduce costs and secure key markets has led to the emergence of two interesting trends. One is the move to large original design manufacturing units, similar to the developments in the semiconductor industry. A second is that an increasing number of solar manufacturers move part of their module production close to the final market to demonstrate the local job creation potential and ensure the current policy support. This may also be a move to manufacture in low-cost or subsidized markets.

The regional distribution of polysilicon production capacities is as follows: China 20,000 tonnes, Europe 17,500 tonnes, Japan 12,000 tonnes, and USA 37,000 tonnes (Service, 2009).

In 2009, solar-grade silicon production of about 88,000 tonnes was reported, sufficient for about 11 GW of PV assuming an average materials need of 8 g/W_p (Displaybank, 2010). China produced about 18,000 tonnes or 20% of world demand, fulfilling about half of its domestic demand (Baoshan, 2010).

Projections of silicon production capacities for solar applications in 2012 span a range between 140,000 tonnes from established polysilicon producers, up to 250,000 tonnes including new producers (e.g., Bernreuther and Haugwitz, 2010; Ruhl et al., 2010). The possible solar cell production will also depend on the material use per W_p. Material consumption could decrease from the current 8 g/W_p to 7 g/W_p or even 6 g/W_p (which could increase delivered PV capacity from 31 to 36 to 42 GW, respectively), but this may not be achieved by all manufacturers.

Forecasts of the future costs of vital materials have a high-profile history, and there is ongoing public debate about possible material shortages and competition regarding some (semi-)metals (e.g., In and Te) used in thin-film cell production. In a recent study, Wadia et al. (2009) explored material limits for PV expansion by examining the dual constraints of material supply and least cost per watt for the most promising semiconductors as active photo-generating materials. Contrary to the commonly assumed scarcity of indium and tellurium, the study concluded that the currently known economic reserves of these materials would allow about 10 TW of CdTe or $CuInS_2$ solar cells to be installed.

In CSP electricity generation, the solar collector field is readily scalable, and the power block is based on adapted knowledge from the existing power industry such as steam and gas turbines. The collectors themselves benefit from a range of existing skill sets such as mechanical, structural and control engineers, and metallurgists. Often, the materials or components used in the collectors are already mass-produced, such as glass mirrors.

By the end of 2010, strong competition had emerged and an increasing number of companies had developed industry-level capability to supply materials such as high-reflectivity glass mirrors and manufactured components. Nonetheless, the large evacuated tubes designed specifically for use in trough/oil systems for power generation remain a specialized component, and only two companies (Schott and Solel) have been capable of supplying large orders of tubes, with a third company (Archimedes) now emerging. The trough concentrator itself comprises know-how in both structures and thermally sagged glass mirrors. Although more companies are now offering new trough designs and considering alternatives to conventional rear-silvered glass (e.g., polymer-based reflective films), the essential technology of concentration remains unchanged. Direct steam generation in troughs is under demonstration, as is direct heating of molten salt, but these designs are not yet commercially available. As a result of its successful operational history, the trough/oil technology comprised most of the CSP installed capacity in 2010.

Linear Fresnel and central-receiver systems comprise a high level of know-how, but the essential technology is such that there is the potential for a greater variety of new industry participants. Although only a couple of companies have historically been involved with central receivers, new players have entered the market over the last few years. There are also technology developers and projects at the demonstration level (China, USA, Israel, Australia, Spain). Central-receiver developers are aiming for higher temperatures, and, in some cases, alternative heat transfer fluids such as molten salts. The accepted standard to date has been to use large heliostats, but many of the new entrants are pursuing much smaller heliostats to gain potential cost reductions through high-volume mass production. The companies now interested in heliostat development range from optics companies to the automotive industry looking to diversify. High-temperature steam receivers will benefit from existing knowledge in the boiler industry. Similarly, with linear Fresnel, a range of new developments are occurring, although not yet as developed as the central-receiver technology.

Dish technology is much more specialized, and most effort presently has been towards developing the dish/Stirling concept as a commercial product. Again, the technology can be developed as specialized components through specific industry know-how such as the Stirling engine mass-produced through the automotive industry.

Within less than 10 years prior to 2010, the CSP industry has gone from negligible activity to over 2,400 MW_e either commissioned or under construction. A list of new CSP plants and their characteristics can be found at the IEA SolarPACES web site.[3] More than ten different companies are now active in building or preparing for commercial-scale plants, compared to perhaps only two or three who were in a position to build a commercial-scale plant three years ago. These companies range from large organizations with international construction and project management expertise who have acquired rights to specific technologies, to start-ups based on their own technology developed in-house. In addition, major independent power producers and energy utilities are playing a role in the CSP market.

The supply chain does not tend to be limited by raw materials, because the majority of required materials are bulk commodities such as glass, steel/aluminium, and concrete. The sudden new demand for the specific solar salt mixture material for molten-salt storage is claimed to have impacted supply. At present, evacuated tubes for trough plants can be produced at a sufficient rate to service several hundred MW per year. However, expanded capacity can be introduced readily through new factories with an 18-month lead time.

Solar fuel technology is still at an emerging stage—thus, there is no supply chain in place at present for commercial applications. However, solar fuels will comprise much of the same solar-field technology being deployed for other high-temperature CSP systems, with solar fuels requiring a different receiver/reactor at the focus and different downstream processing and control. Much of the downstream technology, such as Fischer-Tropsch liquid fuel plants, would come from existing expertise in the petrochemical industry. The scale of solar fuel demonstration plants is being ramped up to build confidence for industry, which will eventually expand operations.

---

3   See: www.solarpaces.org.

Hydrogen has been touted as a future transportation fuel due to its versatility, pollutant-free end use and storage capability. The key is a sustainable, CO_2-free source of hydrogen such as solar, cost-effective storage and appropriate distribution infrastructure. The production of solar hydrogen, in and of itself, does not produce a hydrogen economy because many factors are needed in the chain. The suggested path to solar hydrogen is to begin with solar enhancement of existing steam reforming processes, with a second generation involving solar electricity and advanced electrolysis, and a third generation using thermolysis or advanced thermochemical cycles, with many researchers aiming for the production of fuels from concentrated solar energy, water, and CO_2. In terms of making a transition, solar hydrogen can be mixed with natural gas and transported together in existing pipelines and distribution networks to customers, thus enhancing the solar portion of the global energy mix.

Steam reforming of natural gas for hydrogen production is a conventional industrial-scale process that produces most of the world's hydrogen today, with the heat for the process derived from burning a significant proportion of the fossil fuel feedstock. Using concentrated solar power, instead, as the source of the heat embodies solar energy in the fuel. The solar steam-reforming of natural gas and other hydrocarbons, and the solar steam-gasification of coal and other carbonaceous materials yields a high-quality syngas, which is the building block for a wide variety of synthetic fuels including Fischer-Tropsch-type chemicals, hydrogen, ammonia and methanol (Steinfeld and Meier, 2004).

The solar cracking route refers to the thermal decomposition of natural gas and other hydrocarbons. Besides H_2 and carbon, other compounds may also be formed, depending on the reaction kinetics and on the presence of impurities in the raw materials. The thermal decomposition yields a carbon-rich condensed phase and a hydrogen-rich gas phase. The carbonaceous solid product can either be sequestered without CO_2 release or used as material commodity (carbon black) under less severe CO_2 restraints. It can also be applied as reducing agent in metallurgical processes. The hydrogen-rich gas mixture can be further processed to high-purity hydrogen that is not contaminated with oxides of carbon; thus, it can be used in proton-exchange-membrane fuel cells without inhibiting platinum electrodes. From the perspective of carbon sequestration, it is easier to separate, handle, transport and store solid carbon than gaseous CO_2. Further, thermal cracking removes and separates carbon in a single step. The major drawback of thermal cracking is the energy loss associated with the sequestration of carbon. Thus, solar cracking may be the preferred option for natural gas and other hydrocarbons with a high H_2/C ratio (Steinfeld and Meier, 2004).

3.4.3   Impact of policies[4]

Direct solar energy technologies support a broad range of applications, and their deployment is confronted by many of the barriers outlined in Chapter 1. Solar technologies differ in levels of maturity, and although some applications are already competitive in localized markets, they generally face one common barrier: the need to achieve cost reductions (see Section 3.8). Utility-scale CSP and PV systems face different barriers than distributed PV and solar heating and cooling technologies. Important barriers include: 1) siting, permitting and financing challenges to develop land with favourable solar resources for utility-scale projects; 2) lack of access to transmission lines for large projects far from electric load centres; 3) complex access laws, permitting procedures and fees for smaller-scale projects; 4) lack of consistent interconnection standards and time-varying utility rate structures that capture the value of distributed generated electricity; 5) inconsistent standards and certifications and enforcement of these issues; and 6) lack of regulatory structures that capture environmental and risk mitigation benefits across technologies (Denholm et al., 2009).

Through appropriate policy designs (see Chapter 11), governments have shown that they can support solar technologies by funding R&D and by providing incentives to overcome economic barriers. Price-driven instruments (see Section 11.5.2), for example, were popularized after feed-in tariff (FIT) policies boosted levels of PV deployment in Germany and Spain. In 2009, various forms of FIT policies were implemented in more than 50 countries (REN21, 2010) and some designs offer premiums for building-integrated PV. Quota-driven frameworks such as renewable portfolio standards (RPS) and government bidding are common in the USA and China, respectively (IEA, 2009a). Traditional RPS frameworks are designed to be technology-neutral, and this puts at a disadvantage many solar applications that are more costly than alternatives such as wind power. In response, features of RPS frameworks (set-asides and credits) increasingly are including solar-specific policies, and such programs have led to increasing levels of solar installations (Wiser et al., 2010). In addition to these regulatory frameworks, fiscal policies and financing mechanisms (e.g., tax credits, soft loans and grants) are often employed to support the manufacturing of solar goods and to increase consumer demand (Rickerson et al., 2009). The challenge for solar projects to secure financing is a critical barrier, especially for developing technologies in market structures dominated by short-term transactions and planning.

Most successful solar policies are tailored to the barriers posed by specific applications. Across technologies, there is a need to offset relatively high upfront investment costs (Denholm et al., 2009). Yet, in the case of utility-scale CSP and PV projects, substantial and long-term investments are required at levels that exceed solar applications in distributed markets. Solar heating and cooling technologies are included in many policies, yet the characteristics of their applications differ from electricity-generating technologies. Policies based on energy yield rather than collector surface area are generally preferred for various types of solar thermal collectors (IEA, 2007). See Section 1.5 for further discussion.

Similar to other renewable sources, there is ongoing discussion about the merits of existing solar policies to spur innovation and accelerate deployment using cost-effective measures. Generally—and as discussed

---

4   Non-technology-specific policy issues are covered in Chapter 11 of this report.

3.5 Integration into the broader energy system[5]

This section discusses how direct solar energy technologies are part of the broader energy framework, focusing specifically on the following: low-capacity energy demand; district heating and other thermal loads; PV generation characteristics and the smoothing effect; and CSP generation characteristics and grid stabilization. Chapter 8 addresses the broader technical and institutional options for managing the unique characteristics, production variability, limited predictability and locational dependence of some RE technologies, including solar, as well as existing experience with and studies associated with the costs of that integration.

3.5.1 Low-capacity electricity demand

There can be comparative advantages for using solar energy rather than non-renewable fuels in many developing countries. Within a country, the advantages can be higher in un-electrified rural areas compared to urban areas. Indeed, solar energy has the advantage, due to being modular, of being able to provide small and decentralized supplies, as well as large centralized ones. For more on integrated buildings and households, see Section 8.3.2.

In a wide range of countries, particularly those that are not oil producers, solar energy and other forms of RE can be the most appropriate energy source. If electricity demand exceeds supply, the lack of electricity can prevent development of many economic sectors. Even in countries with high solar energy sustainable development potential, RE is often only considered to satisfy high-power requirements such as the industrial sector. However, large-scale technologies such as CSP are often not available to them due, for example, to resource conditions or suitable land area availability. In such cases, it is reasonable to keep the electricity generated near the source to provide high amounts of power to cover industrial needs. Applications that have low power consumption, such as lighting in rural areas, can primarily be satisfied using onsite PV—even if the business plan for electrification of the area indicates that a grid connection would be more profitable. Furthermore, the criteria to determine the most suitable technological option for electrifying a rural area should include benefits such as local economic development, exploiting natural resources, creating jobs, reducing the country's dependence on imports, and protecting the environment.

---
[5] Non-technology-specific issues related to integration of RE sources in current and future energy systems are covered in Chapter 8 of this report.

3.5.2 District heating and other thermal loads

Highly insulated buildings can be heated easily with relatively low-temperature district-heating systems, where solar energy is ideal, or quite small quantities of renewable-generated electricity (Boyle, 1996). A district cooling and heating system (DCS) can provide both cooling and heating for blocks of buildings. Since the district heating system already makes the outdoor pipe network available, a district cooling system becomes a viable solution to the cooling demand of buildings. There are already many DCS installations in the USA, Europe, Japan and other Asian countries because this system has many advantages compared to a decentralized cooling system. For example, it takes full advantage of economy of scale and diversity of cooling demand of different buildings, reduces noise and structure load, and saves considerable equipment area. It also allows greater flexibility in designing the building by removing the cooling tower on the roof and chiller plant in the building or on the roof, and it can provide more reliable and flexible services through a specialized professional team in cold-climate areas (Shu et al., 2010). For more on RE integration in district heating and cooling networks, see Section 8.2.2.3.

In China, Greece, Cyprus and Israel, solar water heaters make a significant contribution to supplying residential energy demand. In addition, solar water heating is widely used for pool heating in Australia and the USA. In countries where electricity is a major resource for water heating (e.g., Australia, Canada and the USA), the impact of numerous solar domestic water heaters on the operation of the power grid depends on the utility's load management strategy. For a utility that uses centralized load switching to manage electric water heater load, the impact is limited to fuel savings. Without load switching, the installation of many solar water heaters may have the additional benefit of reducing peak demand on the grid. For a utility that has a summer peak, the time of maximum solar water heater output corresponds with peak electrical demand, and there is a capacity benefit from load displacement of electric water heaters. Large-scale deployment of solar water heating can benefit both the customer and the utility. Another benefit to utilities is emissions reduction, because solar water heating can displace the marginal and polluting generating plant used to produce peak-load power.

Combining biomass and low-temperature solar thermal energy could provide zero emissions and high capacity factors to areas with less frequent direct-beam solar irradiance. In the short term, local tradeoffs exist for areas that have high biomass availability due to increased cloud cover and rainfall. However, solar technology is more land-efficient for energy production and greatly reduces the need for biomass growing area and biomass transport cost. Some optimum ratio of CSP and biomass supply is likely to exist at each site. Research is being conducted on tower and dish systems to develop technologies—such as solar-driven gasification of biomass—that optimally combine both these renewable resources. In the longer term, greater interconnectedness across different climate regimes may provide more stability of supply as a total grid system; this situation could reduce the need for occasional fuel supply for each individual CSP system.

3.5.3 Photovoltaic generation characteristics and the smoothing effect

At a specific location, the generation of electricity by a PV system varies systematically during a day and a year, but also randomly according to weather conditions. The variation of PV generation can, in some instances, have a large impact on voltage and power flow of the local transmission/distribution system from the early penetration stage, and on supply-demand balance in a total power system operation in the high-penetration stage (see also Section 8.2.1 for a further discussion of solar electricity characteristics, and the implications of those characteristics for electricity market planning, operations, and infrastructure).

Various studies have been published on the impact of supply-demand balance for a power system with a critical constraint of PV systems integration (Lee and Yamayee, 1981; Chalmers et al., 1985; Chowdhury and Rahman, 1988; Jewell and Unruh, 1990; Bouzguenda and Rahman, 1993; Asano et al., 1996). These studies generally conclude that the economic value of PV systems is significantly reduced at increasing levels of system penetration due to the high variability of PV. Today's base-load generation has a limited ramp rate—the rate at which a generator can change its output—which limits the feasible penetration of PV systems. However, these studies generally lack high-time-resolution PV system output data from multiple sites. The total electricity generation of numerous PV systems in a broad area should have less random and fast variation—because the generation output variations of numerous PV systems have low correlation and cancel each other in a 'smoothing effect'. The critical impact on supply-demand balance of power comes from the total generation of the PV systems within a power system (Piwko et al., 2007, 2010; Ogimoto et al., 2010).

Some approaches for analyzing the smoothing effect use modelling and measured data from around the world. Cloud models have been developed to estimate the smoothing effect of geographic diversity by considering regions ranging in size from 10 to 100,000 km^2 (Jewell and Ramakumar, 1987) and down to 0.2 km^2 (Kern and Russell, 1988). Using measured data, Kitamura (1999) proposed a set of specifications for describing fluctuations, considering three parameters: magnitude, duration of a transition between clear and cloudy, and speed of the transition, defined as the ratio of magnitude and duration; he evaluated the smoothing effect in a small area (0.1 km by 0.1 km). A similar approach, 'ramp analysis', was proposed by Beyer et al. (1991) and Scheffler (2002).

In a statistical approach, Otani et al. (1997) characterized irradiance data by the fluctuation factor using a high-pass filtered time series of solar irradiance. Woyte et al. (2001, 2007) analyzed the fluctuations of the instantaneous clearness index by means of a wavelet transform. To demonstrate the smoothing effect, Otani et al. (1998) demonstrated that the variability of sub-hourly irradiance even within a small area of 4 km by 4 km can be reduced due to geographic diversity. They analyzed the non-correlational irradiation/generation characteristics of several PV systems/sites that are dispersed spatially.

Wiemken et al. (2001) used data from actual PV systems in Germany to demonstrate that five-minute ramps in normalized PV power output at one site may exceed ±50%, but that five-minute ramps in the normalized PV power output from 100 PV systems spread throughout the country never exceed ±5%. Ramachandran et al. (2004) analyzed the reduction in power output fluctuation for spatially dispersed PV systems and for different time periods, and they proposed a cluster model to represent very large numbers of small, geographically dispersed PV systems. Results from Curtright and Apt (2008) based on three PV systems in Arizona indicate that 10-minute step changes in output can exceed 60% of PV capacity at individual sites, but that the maximum of the aggregate of three sites is reduced. Kawasaki et al. (2006) similarly analyzed the smoothing effect within a small (4 km by 4 km) network of irradiance sensors and concluded that the smoothing effect is most effective during times when the irradiance variability is most severe—particularly days characterized as partly cloudy.

Murata et al. (2009) developed and validated a method for estimating the variability of power output from PV plants dispersed over a wide area that is very similar to the methods used for wind by Ilex Energy Consulting Ltd et al. (2004) and Holttinen (2005). Mills and Wiser (2010) measured one-minute solar insolation for 23 sites in the USA and characterized the variability of PV with different degrees of geographic diversity, comparing the variability of PV to the variability of similarly sited wind. They determined that the relative aggregate variability of PV plants sited in a dense ten by ten array with 20-km spacing is six times less than the variability of a single site for variability on time scales of less than 15 minutes. They also found that for PV and wind plants similarly sited in a five by five grid with 50-km spacing, the variability of PV is only slightly more than the variability of wind on time scales of 5 to 15 minutes.

Oozeki et al. (2010) quantitatively evaluated the smoothing effect in a load-dispatch control area in Japan to determine the importance of data accumulation and analysis. The study also proposed a methodology to calculate the total PV output from a limited number of measurement data using Voronoi Tessellation. Marcos et al. (2010) analyzed one-second data collected throughout a year from six PV systems in Spain, ranging from 1 to 9.5 MW$_p$, totalling 18 MW. These studies concluded that over shorter and longer time scales, the level of variability is nearly identical because the aggregate fluctuation of PV systems spread over the large area depends on the correlation of the fluctuation between PV systems. The correlation of fluctuation, in turn, is a function both of the time scale and distance between PV systems. Variability is less correlated for PV systems that are further apart and for variability over shorter time scales.

Currently, however, not enough data on generation characteristics exist to evaluate the smoothing effect. Data collection from a sufficiently large number of sites (more than 1,000 sites and at distances of 2 to 200 km), periods and time resolution (one minute or less) had just begun in mid-2010 in several areas in the world. The smoothed generation characteristics of PV penetration considering area and multiple sites will be analyzed precisely after collecting reliable measurement data with sufficient time resolution and time synchronization. The results will contribute to the economic and reliable integration of PV into the energy system.

3.5.4 Concentrating solar power generation characteristics and grid stabilization

In a CSP plant, even without integrated storage, the inherent thermal mass in the collector system and spinning mass in the turbine tend to significantly reduce the impact of rapid solar transients on electrical output, and thus, lead to less impact on the grid (also see Section 8.2.1). By including integrated thermal storage systems, base-load capacity factors can be achieved (IEA, 2010b). This and the ability to dispatch power on demand during peak periods are key characteristics that have motivated regulators in the Mediterranean region, starting with Spain, to support large-scale deployment of this technology with tailored FITs. CSP is suitable for large-scale 10- to 300-MW$_e$ plants replacing non-renewable thermal power capacity. With thermal storage or onsite thermal backup (e.g., fossil or biogas), CSP plants can also produce power at night or when irradiation is low. CSP plants can reliably deliver firm, scheduled power while the grid remains stable.

CSP plants may also be integrated with fossil fuel-fired plants such as displacing coal in a coal-fired power station or contributing to gas-fired integrated solar combined-cycle (ISCC) systems. In ISCC power plants, a solar parabolic trough field is integrated in a modern gas and steam power plant; the waste heat boiler is modified and the steam turbine is oversized to provide additional steam from a solar steam generator. Better fuel efficiency and extended operating hours make combined solar/fossil power generation much more cost-effective than separate CSP and combined-cycle plants. However, without including thermal storage, solar steam could only be supplied for some 2,000 of the 6,000 to 8,000 combined-cycle operating hours of a plant in a year. Furthermore, because the solar steam is only feeding the combined-cycle turbine—which supplies only one-third of its power—the maximum solar share obtainable is under 10%. Nonetheless, this concept is of special interest for oil- and gas-producing sunbelt countries, where solar power technologies can be introduced to their fossil-based power market (SolarPACES, 2008).

3.6 Environmental and social impacts[6]

This section first discusses the environmental impacts of direct solar technologies, and then describes potential social impacts. However, an overall issue identified at the start is the small number of peer-reviewed studies on impacts, indicating the need for much more work in this area.

---
6 A comprehensive assessment of social and environmental impacts of all RE sources covered in this report can be found in Chapter 9.

3.6.1 Environmental impacts

No consensus exists on the premium, if any, that society should pay for cleaner energy. However, in recent years, there has been progress in analyzing environmental damage costs, thanks to several major projects to evaluate the externalities of energy in the USA and Europe (Gordon, 2001; Bickel and Friedrich, 2005; NEEDS, 2009; NRC, 2010). Solar energy has been considered desirable because it poses a much smaller environmental burden than non-renewable sources of energy. This argument has almost always been justified by qualitative appeals, although this is changing.

Results for damage costs per kilogram of pollutant and per kWh were presented by the International Solar Energy Society in Gordon (2001). The results of studies such as NEEDS (2009), summarized in Table 3.3 for PV and in Table 3.4 for CSP, confirm that RE is usually comparatively beneficial, though impacts still exist. In comparison to the figures presented for PV and CSP here, the external costs associated with fossil generation options, as summarized in Chapter 10.6, are considerably higher, especially for coal-fired generation.

Considering passive solar technology, higher insulation levels provide many benefits, in addition to reducing heating loads and associated costs (Harvey, 2006). The small rate of heat loss associated with high levels of insulation, combined with large internal thermal mass, creates a more comfortable dwelling because temperatures are more uniform. This can indirectly lead to higher efficiency in the equipment supplying the heat. It also permits alternative heating systems that would not

Table 3.3 | Quantifiable external costs for photovoltaic, tilted-roof, single-crystalline silicon, retrofit, average European conditions; in US$_{2005}$ cents/kWh (NEEDS, 2009).

	2005	2025	2050
Health Impacts	0.17	0.14	0.10
Biodiversity	0.01	0.01	0.01
Crop Yield Losses	0.00	0.00	0.00
Material Damage	0.00	0.00	0.00
Land Use	N/A	0.01	0.01
Total	0.18	0.17	0.12

Table 3.4 | Quantifiable external costs for concentrating solar power; in US$_{2005}$ cents/kWh (NEEDS, 2009).

	2005	2025	2050
Health Impacts	0.65	0.10	0.06
Biodiversity	0.03	0.00	0.00
Crop Yield Losses	0.00	0.00	0.00
Material Damage	0.01	0.00	0.00
Land Use	N/A	N/A	N/A
Total	0.69	0.10	0.06

otherwise be viable, but which are superior to conventional heating systems in many respects. Better-insulated houses eliminate moisture problems associated, for example, with thermal bridges and damp basements. Increased roof insulation also increases the attenuation of outside sounds such as from aircraft.

For active solar heating and cooling, the environmental impact of solar water-heating schemes in the UK would be very small according to Boyle (1996). For example, in the UK, the materials used are those of everyday building and plumbing. Solar collectors are installed to be almost indistinguishable visually from normal roof lights. In Mediterranean countries, the use of free-standing thermosyphon systems on flat roofs can be visually intrusive. However, the collector is not the problem, but rather, the storage tank above it. A study of the lifecycle environmental impact of a thermosyphon domestic solar hot water system in comparison with electrical and gas water heating shows that these systems have improved LCA indices over electrical heaters, but the net gain is reduced by a factor of four when the primary energy source is natural gas instead of electricity (Tsilingiridis et al., 2004).

With regard to complete solar domestic hot water systems, the energy payback time requires accounting for any difference in the size of the hot water storage tank compared to the non-solar system and the energy used to manufacture the tank (Harvey, 2006). It is reported that the energy payback time for a solar/gas system in southern Australia is 2 to 2.5 years, despite the embodied energy being 12 times that of a tankless system. For an integrated thermosyphon flat-plate solar collector and storage device operating in Palermo (Italy), a payback time of 1.3 to 4.0 years is reported (Harvey, 2006).

PV systems do not generate any type of solid, liquid or gaseous by-products when producing electricity. Also, they do not emit noise or use non-renewable resources during operation. However, two topics are often considered: 1) the emission of pollutants and the use of energy during the full lifecycle of PV manufacturing, installation, operation and maintenance (O&M) and disposal; and 2) the possibility of recycling the PV module materials when the systems are decommissioned.

Starting with the latter concern, the PV industry uses some toxic, explosive gases, GHGs, as well as corrosive liquids, in its production lines. The presence and amount of those materials depend strongly on the cell type (see Section 3.3.3). However, the intrinsic needs of the production process of the PV industry force the use of quite rigorous control methods that minimize the emission of potentially hazardous elements during module production.

Recycling the material in PV modules is already economically viable, mainly for concentrated and large-scale applications. Projections are that between 80 and 96% of the glass, ethylene vinyl acetate, and metals (Te, selenium and lead) will be recycled. Other metals, such as Cd, Te, tin, nickel, aluminium and Cu, should be saved or they can be recycled by other methods. For discussions of Cd, for example, see Sinha et al. (2008), Zayed and Philippe (2009) and Wadia et al. (2009).

It is noted that, in certain locations, periodic cleaning of the PV panels may be necessary to maintain performance, resulting in non-negligible water requirements.

With respect to lifecycle GHG emissions, Figure 3.14 shows the result of a comprehensive literature review of PV-related lifecycle assessment (LCA) studies published since 1980 conducted by the National Renewable Energy Laboratory. The majority of lifecycle GHG emission estimates cluster between about 30 and 80 g CO_2eq/kWh, with potentially important outliers at greater values (Figure 3.14). Note that the distributions shown in Figure 3.14 do not represent an assessment of likelihood; the figure simply reports the distribution of currently published literature estimates passing screens for quality and relevance. Refer to Annex II for a description of literature search methods and complete reference list, and Section 9.3.4.1 for further details on interpretation of LCA data. Variability in estimates stems from differences in study context (e.g., solar resource, technological vintage), technological performance (e.g., efficiency, silicon thickness) and methods (e.g., LCA system boundaries). Efforts to harmonize the methods and assumptions of these studies are recommended such that more robust estimates of central tendency and variability can be realized, as well as a better understanding of the upper-quartile estimates. Further LCA studies are also needed to increase the number of estimates for some technologies (e.g., CdTe).

As for the energy payback of PV (see also Box 9.3), Perpinan et al. (2009) report paybacks of 2.0 and 2.5 years for microcrystalline silicon and monocrystalline silicon PV, respectively, taking into account use in locations with moderate solar irradiation levels of around 1,700 kWh/m^2/yr (6,120 MJ/m^2/yr). Fthenakis and Kim (2010) show payback times of grid-connected PV systems that range from 2 to 5 years for locations with global irradiation ranges from 1,900 to 1,400 kWh/m^2/yr (6,840 MJ/m^2/yr).

For CSP plants, the environmental consequences vary depending on the technology. In general, GHG emissions and other pollutants are reduced without incurring additional environmental risks. Each square metre of CSP concentrator surface is enough to avoid the annual production of 0.25 to 0.4 t of CO_2. The energy payback time of CSP systems can be as low as five months, which compares very favourably with their lifespan of about 25 to 30 years (see Box 9.3 for further discussion). Most CSP solar field materials can be recycled and reused in new plants (SolarPACES, 2008).

Land consumption and impacts on local flora and wildlife during the build-up of the heliostat field and other facilities are the main environmental issues for CSP systems (Pregger et al., 2009). Other impacts are associated with the construction of the steel-intensive infrastructure for solar energy collection due to mineral and fossil resource consumption,

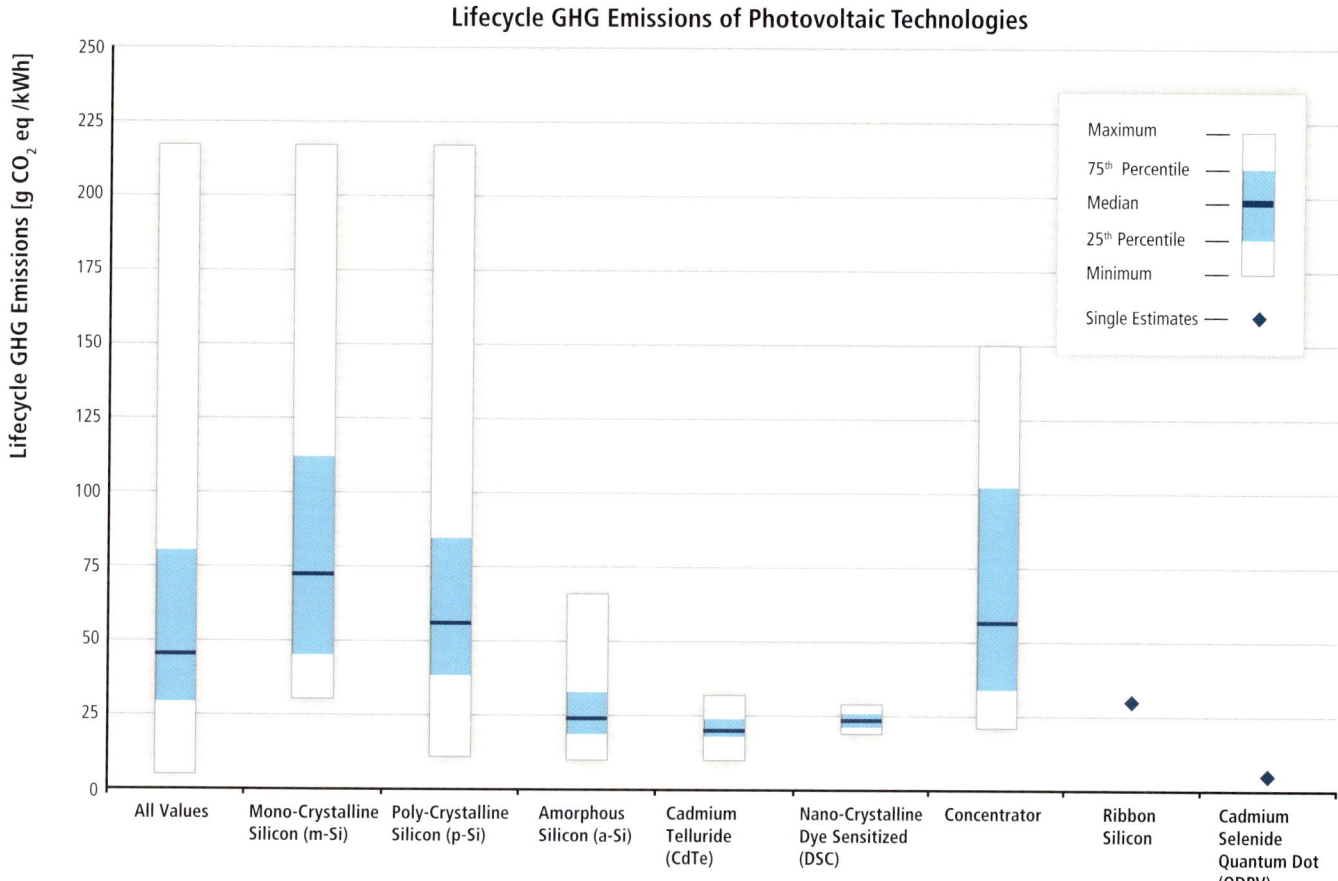

Figure 3.14 | Lifecycle GHG emissions of PV technologies (unmodified literature values, after quality screen). See Annex II for details of the literature search and citations of literature contributing to the estimates displayed.

as well as discharge of pollutants related to today's steel production technology (Felder and Meier, 2008).

The cost of land generally represents a very minor cost proportion of the whole plant. A 100-MW CSP plant with a solar multiple of one (see Section 3.3.4) would require 2 km² of land. However, the land does need to be relatively flat (particularly for linear trough and Fresnel systems), ideally near transmission lines and roads for construction traffic, and not on environmentally sensitive land. Although the mirror area itself is typically only about 25 to 35% of the land area occupied, the site of a solar plant will usually be arid. Thus, it is generally not suitable for other agricultural pursuits, but may still have protected or sensitive species. For this kind of system, sunny deserts close to electricity infrastructure are ideal. As CSP plant capacity is increased, however, the economics of longer electricity transmission distances improves. So, more distant siting might be expected with according increases in transmission infrastructure needs. Attractive sites exist in many regions of the world, including southern Europe, northern and southern African countries, the Middle East, Central Asian countries, China (Tibet, Xinjan),

India (Rajasthan and Gujarat states), Australia, Chile, Peru, Mexico and south-western USA.

In the near term, water availability may be important to minimize the cost of Rankine cycle-based CSP systems. Water is also needed for steam-cycle make-up and mirror cleaning, although these two uses represent only a few percent of that needed if wet cooling is used. However, there will be otherwise highly favourable sites where water is not available for cooling. In these instances, water use can be substantially reduced if dry or hybrid cooling is used, although at an additional cost. The additional cost of electricity from a dry-cooled plant is 2 to 10% (US DOE, 2009), although it depends on many factors such as ambient conditions and technology, for example, tower plants operating at higher temperatures require less cooling per MWh than troughs. Tower and dish Brayton and Stirling systems are being developed for their ability to operate efficiently without cooling water.

In a manner similar to that for PV, NREL conducted an analogous search for CSP lifecycle assessments. Figure 3.15 displays distributions

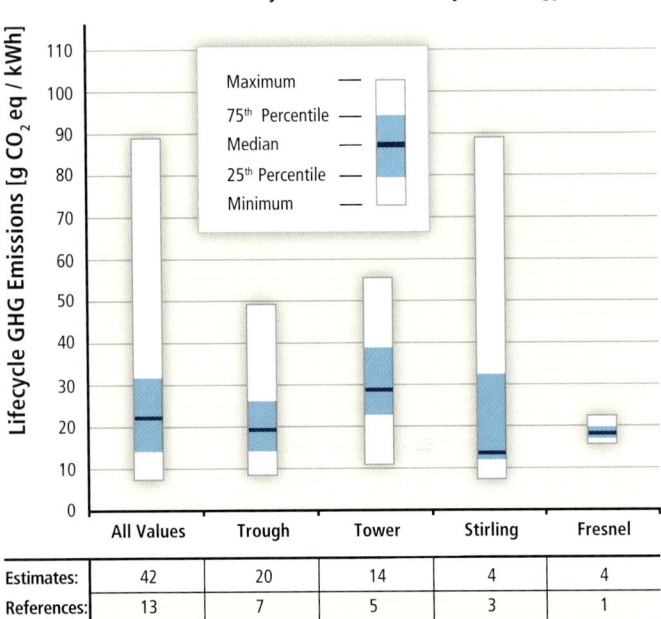

Figure 3.15 | Lifecycle GHG emissions of CSP technologies (unmodified literature values, after quality screen). See Annex II for details of literature search and citations of literature contributing to the estimates displayed.

of as-published estimates of lifecycle GHG emissions. The majority of estimates fall between 14 and 32 g CO_2eq/kWh for trough, tower, Stirling and Fresnel systems, and no great difference between technologies emerges from the available literature. Less literature is available to evaluate CSP systems than for some PV designs; however, the current state of knowledge of lifecycle GHG emissions for these technologies appears fairly consistent, although augmentation with additional LCAs is recommended.

In *solar fuel production*, solar thermal processes use concentrated solar irradiance as the main or sole source of high-temperature process heat. Such a plant consists of a central-receiver system comprising a heliostat field focusing direct solar irradiance on a receiver mounted on a tower. The receiver comprises a chemical reactor or a heat-exchanging device. Direct CO_2 emissions released by the thermochemical processes are negligible or significantly lower than from current processes (Pregger et al., 2009). All other possible effects are comparable to the conventional processes or can be prevented by safety measures and equipment that are common practice in the chemical industry.

3.6.2 Social impacts

Solar energy has the potential to meet rising energy demands and decrease GHG emissions, but solar technologies have faced resistance due to public concerns among some groups. The land area requirements for centralized CSP and PV plants raise concerns about visual impacts, which can be minimized during the siting phase by choosing locations in areas with low population density, although this will usually be the case for suitable solar sites anyway. Visual concerns also exist for distributed solar systems in built-up areas, which may find greater resistance for applications on historical or cultural buildings versus modern construction. By avoiding conservation areas and incorporating solar technologies into building design, these conflicts can be minimized. Noise impacts may be of concern in the construction phase, but impacts can be mitigated in the site-selection phase and by adopting good work practices (Tsoutsos et al., 2005). Community engagement throughout the planning process of renewable projects can also significantly increase public acceptance of projects (Zoellner et al., 2008).

Increased deployment of consumer-purchased systems still faces barriers with respect to costs, subsidy structures that may be confusing, and misunderstandings about reliability and maintenance requirements (Faiers and Neame, 2006). Effective marketing of solar technologies—including publicizing impacts relative to traditional power generation facilities, environmental benefits and contribution to a secure energy supply—have helped to accelerate social acceptance and increase willingness to pay (Batley et al., 2001). Government spending on solar technologies through fiscal incentives and R&D could garner increased public support through increased quantification and dissemination of the economic impacts associated with those programs. A recent study comparing job impacts across energy technologies showed that solar PV had the greatest job-generating potential at an average of 0.87 job-years per GWh, whereas CSP yielded an average of 0.23 job-years per GWh, both of which exceeded estimated job creation for fossil technologies (Wei et al., 2010). Section 9.3.1 discusses qualifications and limitations of assessing the job market impact of RE.

Solar technologies can also improve the health and livelihood opportunities for many of the world's poorest populations. Solar technologies have the potential to address some of the gap in availability of modern energy services for the roughly 1.4 billion people who do not have access to electricity and the more than 2.7 billion people who rely on traditional biomass for home cooking and heating needs (IEA, 2010d; see Section 9.3.2).

Solar home systems and PV-powered community grids can provide economically favourable electricity to many areas for which connection to a main grid is impractical, such as in remote, mountainous and delta regions. Electric lights are the most frequently owned and operated household appliance in electrified households, and access to electric lighting is widely accepted as the principal benefit of electrification programs (Barnes, 1988). Electric lighting may replace light supplied by kerosene lanterns, which are generally associated with poor-quality light and high household fuel expenditures, and which pose fire and poisoning risks. The improved quality of light allows for increased reading by household members, study by children, and home-based enterprise activities after dark, resulting in increased education and income opportunities for the household. Higher-quality light can also be provided through solar lanterns, which can afford the same benefits achieved through solar home system-generated lighting. Solar lantern models can be stand-alone or can require central-station charging, and programs of manufacture, distribution and maintenance can provide micro-enterprise opportunities. Use of solar lighting can represent a significant cost savings to households over the lifetime of the technology compared to kerosene, and it can reduce the 190 Mt of estimated annual CO_2 emissions attributed to fuel-based lighting (Mills, 2005). Solar-powered street lights and lights for community buildings can increase security and safety and provide night-time gathering locations for classes or community meetings. PV systems have been effectively deployed in disaster situations to provide safety, care and comfort to victims in the USA and Caribbean and could be similarly deployed worldwide for crisis relief (Young, 1996).

Solar home systems can also power televisions, radios and cellular telephones, resulting in increased access to news, information and distance education opportunities. A study of Bangladesh's Rural Electrification Program revealed that in electrified households all members are more knowledgeable about public health issues, women have greater knowledge of family planning and gender equality issues, the income and gender discrepancies in adult literacy rates are lower, and immunization guidelines for children are adhered to more regularly when compared with non-electrified households (Barkat et al., 2002). Electrified households may also buy appliances such as fans, irons, grinders, washing machines and refrigerators to increase comfort and reduce the drudgery associated with domestic tasks (ESMAP, 2004).

Indoor smoke from solid fuels is responsible for more than 1.6 million deaths annually and 3.6% of the global burden of disease. This mortality rate is similar in scale to the 1.7 million annual deaths associated with unsafe sanitation and more than twice the estimated 0.8 million yearly deaths from exposure to urban air pollution (Ezzati et al., 2002; see Sections 9.3.2 and 9.3.4.3). In areas where solar cookers can satisfactorily produce meals, these cookers can reduce unhealthy exposure to high levels of particulate matter from traditional use of solid fuels for cooking and heating and the associated morbidity and mortality from respiratory and other diseases. Decreased consumption of firewood will correspondingly reduce the time women spend collecting firewood. Studies in India and Africa have collected data showing that this time can total 2 to 15 hours per week, and this is increasing in areas of diminishing fuelwood supply (Brouwer et al., 1997; ESMAP, 2004). Risks to women collecting fuel include injury, snake bites, landmines and sexual violence (Manuel, 2003; Patrick, 2007); when children are enlisted to help with this activity, they may do so at the expense of educational opportunities (Nankhuni and Findeis, 2004). Well-being may be acutely at risk in refugee situations, as are strains on the natural resource systems where fuel is collected (Lynch, 2002). Solar cookers do not generally fulfil all household cooking needs due to technology requirements or their inability to cook some traditional foods; however, even partial use of solar cookers can realize fuelwood savings and reductions in exposure to indoor air pollution (Wentzel and Pouris, 2007).

Solar technologies also have the potential to combat other prevalent causes of morbidity and mortality in poor, rural areas. Solar desalination and water purification technologies can help combat the high prevalence of diarrhoeal disease brought about by lack of access to potable water supplies. PV systems for health clinics can provide refrigeration for vaccines and lights for performing medical procedures and seeing patients at all hours. Improved working conditions for rural health-care workers can also lead to decreased attrition of talented staff to urban centres.

Solar technologies can improve the economic opportunities and working conditions for poor rural populations. Solar dryers can be used to preserve foods and herbs for consumption year round and produce export-quality products for income generation. Solar water pumping can minimize the need for carrying water long distances to irrigate crops, which can be particularly important and impactful in the dry seasons and in drought years. Burdens and risks from water collection parallel those of fuel collection, and decreased time spent on this activity can also increase the health and well-being of women, who are largely responsible for these tasks.

3.7 Prospects for technology improvements and innovation[7]

This section considers technical innovations that are possible in the future for a range of solar technologies, under the following headings: passive solar and daylighting technologies; active solar heat and cooling; PV electricity generation; CSP electricity generation; solar fuel production; and other possible applications.

3.7.1 Passive solar and daylighting technologies

Passive solar technologies, particularly the direct-gain system, are intrinsically highly efficient because no energy is needed to move collected energy to storage and then to a load. The collection, storage and use are all integrated. Through technological advances such as low-emissivity coatings and the use of gases such as argon in glazings, near-equatorial-facing windows have reached a high level of performance at increasingly affordable cost. Nevertheless, in heating-dominated climates, further advances are possible, such as the following: 1) reduced thermal conductance by using dynamic exterior night insulation (night shutters); 2) use of evacuated glazing units; and 3) translucent glazing systems, which may include materials that change solar/visible transmittance with temperature (including a

---

7 Section 10.5 offers a complementary perspective on drivers and trends of technological progress across RE technologies.

possible phase change) while providing increased thermal resistance in the opaque state.

Increasingly larger window areas become possible and affordable with the drop in prices of highly efficient double-glazed and triple-glazed low-emissivity argon-filled windows (see Sections 3.4.1 and 3.4.2). These increased window areas make systematic solar gain control essential in mild and moderate climatic conditions, but also in continental areas that tend to be cold in winter and hot in summer. Solar gain control techniques may increasingly rely on active systems such as automatically controlled blinds/shades or electrochromic, thermochromic and gasochromic coatings to admit the solar gains when they are desirable or keep them out when overheating in the living space is detected or anticipated. Solar gain control, thermal storage design and heating/cooling system control are three strongly linked aspects of passive solar design and control.

Advances in thermal storage integrated in the interior of direct-gain zones are still possible, such as phase-change materials integrated in gypsum board, bricks, or tiles and concrete. The target is to maximize energy storage per unit volume/mass of material so that such materials can be integrated in lightweight wood-framed homes common in cold-climate areas. The challenge for such materials is to ensure that they continue to store and release heat effectively after 10,000 cycles or more while meeting other performance requirements such as fire resistance. Phase-change materials may also be used systematically in plasters to reduce high indoor temperatures in summer.

Considering cooling-load reduction in solar buildings, advances are possible in areas such as the following: 1) cool-roof technologies involving materials with high solar reflectivity and emissivity; 2) more systematic use of heat-dissipation techniques such as using the ground and water as a heat sink; 3) advanced pavements and outdoor structures to improve the microclimate around the buildings and decrease urban ambient temperatures; and 4) advanced solar control devices allowing penetration of daylight, but not thermal energy.

In any solar building, there are normally some direct-gain zones that receive high solar gains and other zones behind that are generally colder in winter. Therefore, it is beneficial to circulate air between the direct-gain zones and back zones in a solar home, even when heating is not required. With forced-air systems commonly used in North America, this is increasingly possible and the system fan may be run at a low flow rate when heating is not required, thus helping to redistribute absorbed direct solar gains to the whole house (Athienitis, 2008).

During the summer period, hybrid ventilation systems and techniques may be used to provide fresh air and reduce indoor temperatures (Heiselberg, 2002). Various types of hybrid ventilation systems have been designed, tested and applied in many types of buildings. Performance tests have found that although natural ventilation cannot maintain appropriate summer comfort conditions, the use of a hybrid system is the best choice—using at least 20% less energy than any purely mechanical system.

Finally, design tools are expected to be developed that will facilitate the simultaneous consideration of passive design, daylighting, active solar gain control, heating, ventilation and air-conditioning (HVAC) system control, and hybrid ventilation at different stages of the design of a solar building. Indeed, systematically adopting these technologies and their optimal integration is essential to move towards the goal of cost-effective solar buildings with net-zero annual energy consumption (IEA, 2009b). Optimal integration of passive with active technologies requires smart buildings with optimized energy generation and use (Candanedo and Athienitis, 2010). A smart solar house would rely on predictions of the weather to optimally control solar gains and their storage, ensure good thermal comfort, and optimize its interaction with the electricity grid, applying a mixture of inexpensive and effective communications systems and technologies (see Section 8.2.1).

3.7.2 Active solar heating and cooling

Improved designs for solar heating and cooling systems are expected to address longer lifetimes, lower installed costs and increased temperatures. The following are some design options: 1) the use of plastics in residential solar water-heating systems; 2) powering air-conditioning systems using solar energy systems, especially focusing on compound parabolic concentrating collectors; 3) the use of flat-plate collectors for residential and commercial hot water; and 4) concentrating and evacuated-tube collectors for industrial-grade hot water and thermally activated cooling (see Section 3.3.4).

Heat storage represents a key technological challenge, because the wide deployment of active solar buildings, covering 100% of their demand for heating (and cooling, if any) with solar energy, largely depends on developing cost-effective and practical solutions for seasonal heat storage (Hadorn, 2005; Dincer and Rosen, 2010). The European Solar Thermal Technology Platform vision assumes that by 2030, heat storage systems will be available that allow for seasonal heat storage with an energy density eight times higher than water (ESTTP, 2006).

In the future, active solar systems—such as thermal collectors, PV panels, and PV-thermal systems—will be the obvious components of roof and façades, and will be integrated into the construction process at the earliest stages of building planning. The walls will function as a component of the active heating and cooling systems, supporting thermal energy storage by applying advanced materials (e.g., phase-change materials). One central control system will lead to optimal regulation of the whole HVAC system, maximizing the use of solar energy within the comfort parameters set by users. Heat- and cold-storage systems will play an increasingly important role in reaching maximum solar thermal contributions to cover the thermal requirements in buildings.

Solar-assisted air-conditioning technology is still in an early stage of development (Henning, 2007). However, increased efforts in technological development will help to increase the competitiveness of this technology in the future. The major trends are as follows:

- Research in providing thermally driven cooling equipment in the low cooling power range (less than 20 kW);

- Developing single-effect cycles with increased COP values at low driving temperatures;

- Studying new approaches to enhance heat transfer in compartments containing sorption material to improve the power density and thermal performance of adsorption chillers;

- Developing new schemes and new working fluids for steam jet cycles and promising candidates for closed cycles to produce chilled water; and

- Research activities on cooled open sorption cycles for solid and liquid sorbents.

3.7.3 Photovoltaic electricity generation

This subsection discusses photovoltaic technology improvements and innovation within the areas of solar PV cells and the entire PV system. Photovoltaic modules are the basic building blocks of flat-plate PV systems. Further technological efforts will likely lead to reduced costs, enhanced performance and improved environmental profiles. It is useful to distinguish between technology categories that require specific R&D approaches.

Funding of PV R&D over the past four decades has supported innovation and gains in PV cell quality, efficiencies and price. In 2008, public budgets for R&D programs in the IEA Photovoltaic Power Systems Programme countries collectively reached about USD_{2005} 390 million (assumed 2008 base), a 30% increase compared to 2007, but stagnated in 2009 (IEA, 2009c, 2010e).

For wafer-based crystalline silicon, existing thin-film technologies, and emerging and novel technologies (including 'boosters' to the first two categories), the following paragraphs list R&D topics that have highest priority. Further details can be found in the various PV roadmaps, for example, the Strategic Research Agenda for Photovoltaic Solar Energy Technology (US Photovoltaic Industry Roadmap Steering Committee, 2001; European Commission, 2007; NEDO, 2009).

- **Efficiency, energy yield, stability and lifetime.** Research often aims at optimizing rather than maximizing these parameters, which means that additional costs and gains are critically compared. Because research is primarily aimed at reducing the cost of electricity generation, it is important not to focus only on initial costs (USD/W_p), but also on lifecycle gains, that is, actual energy yield (kWh/W_p or kJ/W_p over the economic or technical lifetime).

- **High-productivity manufacturing, including in-process monitoring and control.** Throughput and yield are important parameters in low-cost manufacturing and essential to achieve the cost targets. In-process monitoring and control are crucial tools to increase product quality and yield. Focused effort is needed to bring PV manufacturing to maturity.

- **Environmental sustainability.** The energy and materials requirements in manufacturing, as well as the possibilities for recycling, are important parameters in the overall environmental quality of the product. Further shortening of the energy payback time, design for recycling and, ideally, avoiding the use of materials that are not abundant on Earth are the most important issues to be addressed.

- **Applicability.** As discussed in more detail in the paragraphs on BOS and systems, standardization and harmonization are important to bring down the investment costs of PV. Some related aspects are addressed on a module level. In addition, improved ease of installation is partially related to module features. Finally, aesthetic quality of modules (and systems) is an important aspect for large-scale use in the built environment.

Advanced technologies include those that have passed some proof-of-concept phase or can be considered as 10- to 20-year development options for the PV approaches discussed in Section 3.3.3 (Green, 2001, 2003; Nelson, 2003). These emerging PV concepts are medium to high risk and are based on extremely low-cost materials and processes with high performance. Examples are four- to six-junction concentrators (Marti and Luque, 2004; Dimroth et al., 2005), multiple-junction polycrystalline thin films (Coutts et al., 2003), crystalline silicon in the sub-100-µm-thick regime (Brendel, 2003), multiple-junction organic PV (Yakimov and Forrest, 2002; Sun and Sariciftci, 2005) and hybrid solar cells (Günes and Sariciftci, 2008).

Even further out on the timeline are concepts that offer exceptional performance and/or very low cost but are yet to be demonstrated beyond some preliminary stages. These technologies are truly high risk, but have extraordinary technical potential involving new materials, new device architectures and even new conversion concepts (Green, 2001, 2003; Nelson, 2003). They go beyond the normal Shockley-Queisser limits (Shockley and Queisser, 1961) and may include biomimetic devices (Bar-Cohen, 2006), quantum dots (Conibeer et al., 2010), multiple-exciton generation (Schaller and Klimov, 2004; Ellingson et al., 2005) and plasmonic solar cells (Catchpole and Polman, 2008).

PV concentrator systems are considered a separate category, because the R&D issues are fundamentally different compared to flat-plate technologies. As mentioned in Section 3.3.3, CPV offers a variety of technical solutions that are provided at the system level. Research issues can be divided into the following activities: 1) concentrator solar cell manufacturing; 2) optical system; 3) module assembly and fabrication method of concentrator modules and systems; and 4) system aspects, such as tracking, inverter and installation issues.

However, it should be clearly stated once more: CPV is a system approach. The whole system is optimized only if all the interconnections between the components are considered. A corollary is that an optimized component is not necessarily the best choice for the optimal CPV system. Thus, strong interactions are required among the various research groups.

A photovoltaic system is composed of the PV module, as well as the *balance-of-system components and system*, which can include an inverter, storage, charge controller, system structure and the energy network. Users meet PV technology at the system level, and their interest is in a reliable, cost-effective and attractive solution to their energy supply needs. This research agenda concentrates on topics that will achieve one or more of the following: 1) reduce costs at the component and/or system level; 2) increase the overall performance of the system, including increased and harmonized component lifetimes, reduced performance losses and maintenance of performance levels throughout system life; and 3) improve the functionality of and services provided by the system, thus adding value to the electricity produced (US Photovoltaic Industry Roadmap Steering Committee, 2001; Navigant Consulting Inc., 2006; EU PV European Photovoltaic Technology Platform, 2007; Kroposki et al., 2008; NEDO, 2009).

At the component level, a major objective of BOS development is to extend the lifetime of BOS components for grid-connected applications to that of the modules, typically 20 to 30 years.

For off-grid systems, component lifetime should be increased to around 10 years, and components for these systems need to be designed so that they require little or no maintenance. Storage devices are necessary for off-grid PV systems and will require innovative approaches to the short-term storage of small amounts of electricity (1 to 10 kWh, or 3,600 to 36,000 kJ), and for providing a single streamlined product (such as integrating the storage component into the module) that is easy to use in off-grid and remote applications.

For on-grid systems, high penetration of distributed PV may raise concerns about potential impacts on the stability and operation of the grid, and these concerns may create barriers to future expansion (see also Section 8.2.1). An often-cited disadvantage is the greater sensitivity to grid interconnection issues such as overvoltage and unintended islanding in the low- or middle-voltage network (Kobayashi and Takasaki, 2006; Cobben et al., 2008; Ropp et al., 2008). Moreover, imbalance between demand and supply is often discussed with respect to the variation of PV system output (Braun et al., 2008; NEDO, 2009; Piwko et al., 2010). PV system designs and operation technologies can address these issues to a degree through technical solutions and through more accurate solar energy forecasting. Moreover, PV inverters can help to improve the quality of grid electricity by controlling reactive power or filtering harmonics with communication in a new energy network that applies a mixture of inexpensive and effective communications systems and technologies, including smart meters (see Section 8.2.1).

As new module technologies emerge in the future, some ideas relating to BOS, such as micro-converters, may need to be revised. Furthermore, the quality of the system needs to be assured and adequately maintained according to defined standards, guidelines and procedures. To assure system quality, assessing performance is important, including on-line analysis (e.g., early fault detection) and off-line analysis of PV systems. The gathered knowledge can help to validate software for predicting the energy yield of future module and system technology designs.

Furthermore, very-large-scale PV systems with capacities ranging from several MW to GW are beginning to be planned for deployment (Komoto et al., 2009). In the long term, these systems may play an important role in the worldwide energy network (DESERTEC Foundation, 2007), but may demand new transmission infrastructure and new technical and institutional solutions for electricity system interconnection and operational management.

Standards, quality assurance, and safety and environmental aspects are other important issues. National and especially local authorities and utilities require that PV systems meet agreed-upon standards (such as building standards, including fire and electrical safety requirements). In a number of cases, the development of the PV market is being hindered by either: 1) existing standards, 2) differences in local standards (e.g., inverter requirements/settings) or 3) the lack of standards (e.g., PV modules/PV elements not being certified as a building element because of the lack of an appropriate standard). Standards and/or guidelines are required for the whole value chain. In many cases, developing new and adapted standards and guidelines implies that dedicated R&D is required.

Quality assurance is an important tool that assures the effective functioning of individual components in a PV system, as well as the PV system as a whole. Standards and guidelines are an important basis for quality assurance. In-line production control procedures and guidelines must also be developed. At the system level, monitoring techniques must be developed for early fault detection.

Recycling is an important building block to ensure a sustainable PV industry. Through 2010, most attention has focused on recycling crystalline silicon and CdTe solar modules. Methods for recycling other thin-film modules and BOS components (where no recycling procedures exist) must be addressed in the future. LCA studies are an important tool for evaluating the environmental profile of the various RE sources. Reliable LCA data are required to assure the position of PV with respect to other sources. From these data, properties such as the CO_2 emission per kWh or kJ of electricity produced and the energy payback time can be calculated. In addition, the results of LCA analyses can be used in the design phase of new processes and equipment for cell and module production lines.

3.7.4 Concentrating solar power electricity generation

CSP is a proven technology at the utility scale. The longevity of components has been established over two decades, O&M aspects are understood, and there is enough operational experience to have enabled O&M cost-reduction studies not only to recommend, but also to test, those improvements. In addition, field experience has been fed back to industry and research institutes and has led to improved components and more advanced processes. Importantly, there is now substantial experience that allows researchers and developers to better understand the limits of performance, the likely potential for cost reduction, or both. Studies (Sargent and Lundy LLC Consulting Group, 2003) have concluded that cost reductions will come from technology improvement, economies of scale and mass production. Other innovations related to power cycles and collectors are discussed below.

CSP is a technology driven largely by thermodynamics. Thus, the *thermal energy conversion cycle* plays a critical role in determining overall performance and cost. In general, thermodynamic cycles with higher temperatures will perform more efficiently. Of course, the solar collectors that provide the higher-temperature thermal energy to the process must be able to perform efficiently at these higher temperatures, and today, considerable R&D attention is on increasing the operating temperature of CSP systems. Although CSP works with turbine cycles used by the fossil-fuel industry, there are opportunities to refine turbines such that they can better accommodate the duties associated with thermal cycling invoked by solar inputs.

Considerable development is taking place to optimize the linkage between solar collectors and higher-temperature thermodynamic cycles. The most commonly used power block to date is the steam turbine (Rankine cycle). The steam turbine is most efficient and most cost effective in large capacities. Present trough plants using oil as the heat transfer fluid limit steam turbine temperatures to 370°C and turbine cycle efficiencies to around 37%, leading to design-point solar-to-electric efficiencies of the order of 18% and annual average efficiency of 14%. To increase efficiency, alternatives to the use of oil as the heat transfer fluid—such as producing steam directly in the receiver or using molten salts—are being developed for troughs.

These fluids and others are already preferred for central receivers. Central receivers and dishes are capable of reaching the upper temperature limits of these fluids (around 600°C for present molten salts) for advanced steam turbine cycles, whether subcritical or supercritical, and they can also provide the temperatures needed for higher-efficiency cycles such as gas turbines (Brayton cycle) and Stirling engines. Such high-temperature cycles have the capacity to boost design-point solar-to-electricity efficiency to 35% and annual average efficiency to 25%. The penalty for dry cooling is also reduced, and at higher temperatures thermal storage is more efficient.

The *collector* is the single largest area for potential cost reduction in CSP plants. For CSP collectors, the objective is to lower their cost while achieving the higher optical efficiency necessary for powering higher-temperature cycles. Trough technology will benefit from continuing advances in solar-selective surfaces, and central receivers and dishes will benefit from improved receiver/absorber design that allows collection of very high solar fluxes. Linear Fresnel is attractive in part because the inverted-cavity design can reduce some of the issues associated with the heat collection elements of troughs, although with reduced annual optical performance.

Improved overall efficiency yields a corresponding decrease in the area of mirrors needed in the field, and thus, lower collector cost and lower O&M cost. Investment cost reduction is expected to come primarily from the benefits of mass production of key components that are specific to the solar industry, and from economies of scale as the fixed price associated with manufacturing tooling and installation is spread over larger and larger capacities. In addition, the benefits of 'learning by doing' cannot be overestimated. A more detailed assessment of future technology improvements that would benefit CSP can be found in ECOSTAR (2005), a European project report edited by the German Aerospace Center.

3.7.5 Solar fuel production

The ability to store solar energy in the form of a fuel may be desirable not only for the transportation industry, but also for high-efficiency electricity generation using today's combined cycles, improved combined cycles using advances in gas turbines, and fuel cells. In addition, solar fuels offer a form of storage for solar electricity generation.

Future solar fuel processes will benefit from the continuing development of high-temperature solar collectors, but also from other fields of science such as electrochemistry and biochemistry. Many researchers consider hydrogen to offer the most attraction for the future, although intermediate and transitional approaches are also being developed. Hydrogen is considered in this section, with other solar fuels having been covered in previous sections.

Future technology innovation for solar electrolysis is the photoelectrochemical (PEC) cell, which converts solar irradiance into chemical energy such as H_2. A PEC cell is fabricated using an electrode that absorbs the solar light, two catalytic films, and a membrane separating H_2 and oxygen (O_2). Semiconductor material can be used as a solar light-absorbing anode in PEC cells (Bolton, 1996; Park and Holt, 2010).

Promising *thermochemical* processes for future 'clean' hydrogen mass production encompass the hybrid-sulphur cycle and metal oxide-based cycles. The hybrid-sulphur cycle is a two-step water-splitting process using an electrochemical, instead of thermochemical, reaction for one of the two steps. In this process, sulphur dioxide depolarizes the anode of the electrolyzer, which results in a significant decrease in the reversible cell potential—and, therefore, the electric power requirement for the electrochemical reaction step. A number of solar reactors applicable to solar thermochemical metal oxide-based cycles have been developed, including a 100-kW$_{th}$ monolithic dual-chamber solar reactor for a mixed-iron-oxide cycle, demonstrated within the European R&D project *HYDROSOL-2* (Roeb et al., 2009); a rotary solar reactor for the ZnO/Zn process being scaled up to 100 kW$_{th}$ (Schunk et al., 2009); the Tokyo Tech rotary-type solar reactor (Kaneko et al., 2007); and the Counter-Rotating-Ring Receiver/Reactor/Recuperator, a device using recuperation of sensible heat to efficiently produce H$_2$ in a two-step thermochemical process (Miller et al., 2008).

High temperatures demanded by the thermodynamics of the thermochemical processes pose considerable material challenges and also increase re-radiation losses from the reactor, thereby lowering the absorption efficiency (Steinfeld and Meier, 2004). The overall energy conversion efficiency is improved by reducing thermal losses at high temperatures through improved mirror optics and cavity-receiver design, and by recovering part of the sensible heat from the thermochemical processes.

High-temperature thermochemical processes require thermally and chemically stable reactor-wall materials that can withstand the extreme operating conditions of the various solar fuel production processes. For many lower-temperature processes (e.g., sulphur-based thermochemical cycles), the major issue is corrosion. For very high-temperature metal-oxide cycles, the challenge is the thermal shock resistance of the ceramic wall materials. Near-term solutions include surface modification of thermally compatible refractory materials such as graphite and silicon carbide. Longer-term solutions include modifications of bulk materials. Novel reactor designs may prevent wall reactions.

A key aspect is integrating the chemical process into the solar concentrating system. The concentrating optics—consisting of heliostats and secondary concentrators (compound parabolic concentrator)—need to be further developed and specifically optimized to obtain high solar-flux intensities and high temperatures in solar chemical reactors for producing fuels.

Photochemical and photobiological processes are other strong candidates for solar fuel conversion. Innovative technologies are being developed for producing biofuels from modified photosynthetic microorganisms and photocatalytic cells for fuel production. Both approaches have the potential to provide fuels with solar energy conversion efficiencies far greater than those based on field crops (Turner et al., 2008). Solar-driven fuel production requires biomimetic nanotechnology, where scientists must develop a series of fundamental and technologically advanced multi-electron redox catalysts coupled to photochemical elements. Hydrogen production by these methods at scale has vast technical potential and promising avenues are being vigorously pursued.

A combination of all three forms is found in the *synthesis of biogas*, a mixture of methane and CO$_2$, with solar-derived hydrogen. Solar hydrogen is added by electrochemical water-splitting. Bio-CO$_2$ reacts with hydrogen in a thermochemical process to generate hydrocarbons such as synthetic natural gas or liquid solar fuels (Sterner, 2009). These approaches are still nascent, but could become viable in the future as energy market prices increase and solar power generation costs continue to decrease.

3.7.6 Other potential future applications

There are also methods for producing electricity from solar thermal energy without the need for an intermediate thermodynamic cycle. This direct solar thermal power generation includes such concepts as thermoelectric, thermionic, magnetohydrodynamic and alkali-metal methods. The thermoelectric concept is the most investigated to date, and all have the attraction that the absence of a heat engine should mean a quieter and theoretically more efficient method of producing electricity, with suitability for distributed generation. Specialized applications include military and space power.

Space-based solar power (SSP) is the concept of collecting vast quantities of solar power in space using large satellites in Earth orbit, then sending that power to receiving antennae (rectennae) on Earth via microwave power beaming. The concept was first introduced in 1968 by Peter Glaser. NASA and the US Department of Energy (US DOE) studied SSP extensively in the 1970s as a possible solution to the energy crisis of that time. Scientists studied system concepts for satellites large enough to send GW of power to Earth and concluded that the concept seemed technically feasible and environmentally safe, but the state of enabling technologies was insufficient to make SSP economically competitive. Since the 1970s, however, great advances have been made in these technologies, such as high-efficiency PV cells, highly efficient solid-state microwave power electronics, and lower-cost space launch vehicles (Mankins, 1997, 2002, 2009; Kaya et al., 2001; Hoffert et al., 2002). Still, significant breakthroughs will be required to achieve cost-competitive terrestrial base-load power (NAS, 2004).

3.8 Cost trends[8]

3.8.1 Passive solar and daylighting technologies

High-performance building envelopes entail greater upfront construction costs, but lower energy-related costs during the lifetime of the building (Harvey, 2006). The total investment cost of the building may or may not be higher, depending on the extent to which heating and cooling systems can be downsized, simplified or eliminated altogether as a result of the high-performance envelope. Any additional investment cost will be compensated for, to some extent, by reduced energy costs over the lifetime of the building.

---

8   Discussion of costs in this section is largely limited to the perspective of private investors. Chapters 1 and 8 to 11 offer complementary perspectives on cost issues covering, for example, costs of integration, external costs and benefits, economy-wide costs and costs of policies.

The reduction in the cost of furnaces or boilers due to substantially better thermal envelopes is normally only a small fraction of the additional cost of the better thermal envelope. However, potentially larger cost savings can occur through downsizing or eliminating other components of the heating system, such as ducts to deliver warm air or radiators (Harvey, 2006). High-performance windows eliminate the need for perimeter heating. A very high-performance envelope can reduce the heating load to that which can be met by ventilation airflow alone. High-performance envelopes also lead to a reduction in peak cooling requirements, and hence, in cooling equipment sizing costs, and they permit use of a variety of passive and low-energy cooling techniques.

If a fully integrated design takes advantage of all opportunities facilitated by a high-performance envelope, savings in the cost of mechanical systems may offset all or much of the additional cost of the high-performance envelope.

In considering daylighting, the economic benefit for most commercial buildings is enhanced when sunlight is plentiful because daylighting reduces electricity demand for artificial lighting. This is also when the daily peak in electricity demand tends to occur (Harvey, 2006). Several authors report measurements and simulations with annual electricity savings from 50 to 80%, depending on the hours and the location. Daylighting can lead to reduced cooling loads if solar heat gain is managed and an integrated thermal-daylighting design of the building is followed (Tzempelikos et al., 2010). This means that replacing artificial light with just the amount of natural light needed reduces internal heating. Savings in lighting plus cooling energy use of 22 to 86%, respectively, have been reported (Duffie and Beckman, 2006).

Daylighting and passive solar features in buildings can have significant financial benefits not easily addressed in standard lifecycle and payback analysis. They generally add value to the building, and in the case of office buildings, can contribute to enhanced productivity (Nicol et al., 2006).

3.8.2 Active solar heating and cooling

Solar drying of crops and timber is common worldwide, either by using natural processes or by concentrating the heat in specially designed storage buildings. However, market data are not available.

Advanced applications—such as solar cooling and air conditioning, industrial applications and desalination/water treatment—are in the early stages of development, with only a few hundred first-generation systems in operation. Considerable cost reductions are expected if R&D efforts are increased over the next few years.

Solar water heating is characterized by a higher first cost investment and low operation and maintenance (O&M) costs. Some solar heating applications require an auxiliary energy source, and then annual loads are met by a combination of different energy sources. Solar thermal hot water systems are generally more competitive in sunny regions but this picture changes for space heating due to its usually higher overall heating load. In colder regions, capital costs can be spread over a longer heating season and solar thermal can then become more competitive (IEA, 2007).

The investment costs for solar water heating depend on the complexity of the technology used as well as the market conditions in the country of operation (IEA, 2007; Chang et al., 2009; Han et al., 2010). The costs for an installed solar hot-water system vary from as low as USD_{2005} 83/m^2 to more than USD_{2005} 1,200/m^2, which is equivalent to the USD_{2005} 120 to 1,800/kW$_p$[9] used in Annex III and the resulting levelized cost of heat (LCOH) calculations presented here as well as in Chapters 1 and 10. For the costs of the delivered heat, there is an additional geographic variable related to the available solar irradiation and the number of heating degree days (Mills and Schleich, 2009).

Based on the data and assumptions provided in Annex III, and the methods specified in Annex II, the plot in Figure 3.16 shows the sensitivity of the LCOH with respect to investment cost as a function of capacity factor.

Research to decrease the cost of solar water-heating systems is mainly oriented towards developing the next generation of low-cost, polymer-based systems for mild climates. The focus includes testing the durability of materials. The work to date includes unpressurized polymer integral collector-storage systems that use a load-side immersed heat exchanger and direct thermosyphon systems.

Over the last decade, for each 50% increase in the installed capacity of solar water heaters, investment costs have fallen by around 20% in Europe (ESTTP, 2008). According to the IEA (2010a), cost reductions in OECD countries will come from the use of cheaper materials, improved manufacturing processes, mass production, and the direct integration into buildings of collectors as multi-functional building components and modular, easy to install systems. Delivered energy costs are anticipated by the IEA to eventually decline by around 70 to 75%. One measure suggested by the IEA to realize those cost reductions are more research, development and demonstration (RD&D) investments. Priority areas for attention include new flat-plate collectors that can be more easily integrated into building façades and roofs, especially as multi-functional building components.

Energy costs should fall with ongoing decreases in the costs of individual system components and with better optimization and design. For example, Furbo et al. (2005) show that better design of solar domestic hot-water storage tanks when combined with an auxiliary energy source can improve the utilization of solar energy by 5 to 35%, thereby permitting a smaller collector area for the same solar yield.

---

9   1 m^2 of collector area is converted into 0.7 kW$_{th}$ of installed capacity (see Section 3.4.1).

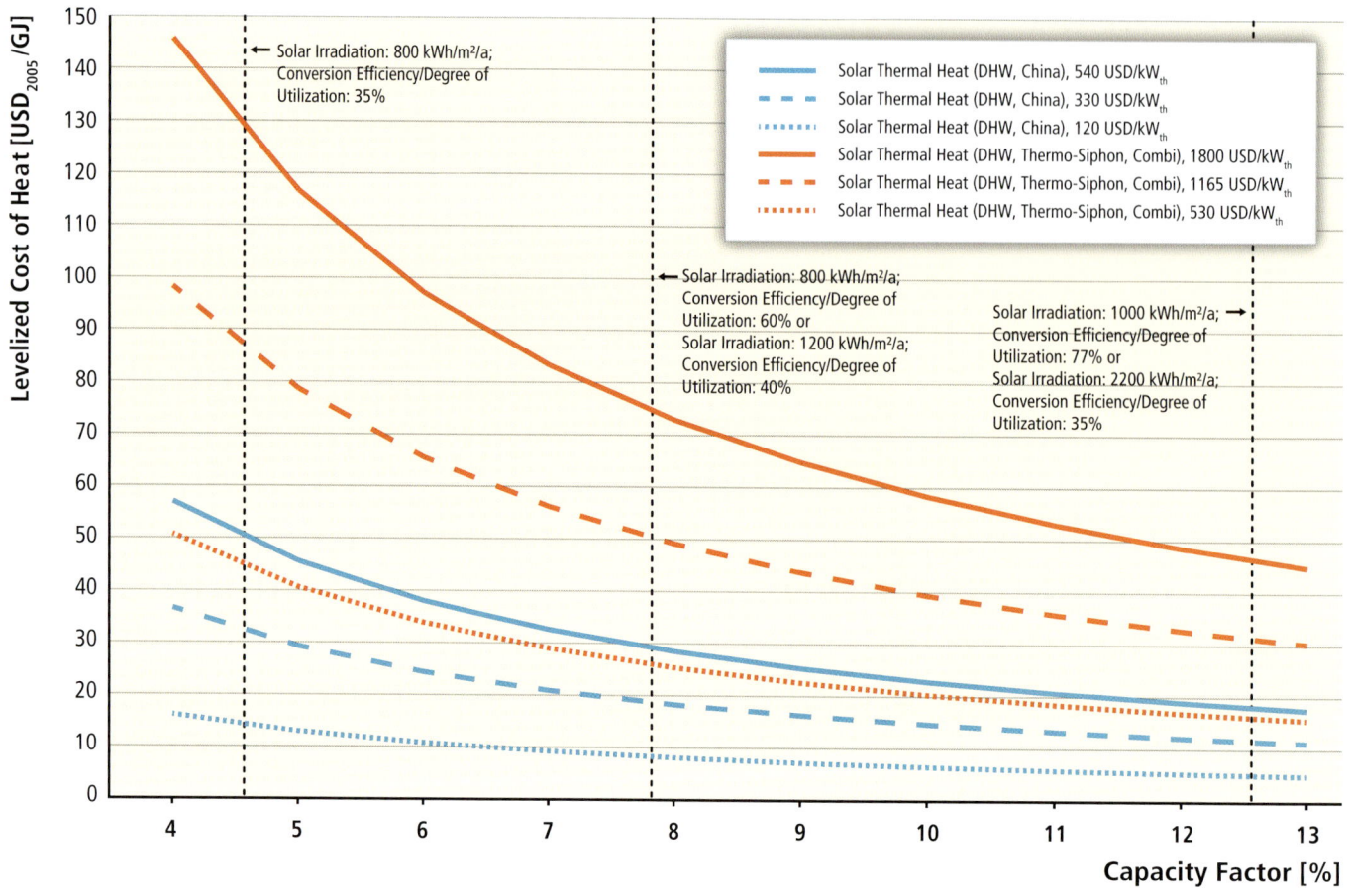

Figure 3.16 | Sensitivity of LCOH with respect to investment cost as a function of capacity factor (Source: Annex III).

3.8.3 Photovoltaic electricity generation

PV prices have decreased by more than a factor of 10 over the last 30 years; however, the current levelized cost of electricity (LCOE) from solar PV is generally still higher than wholesale market prices for electricity.[10] The competitiveness in other markets depends on a variety of local conditions.

The LCOE of PV systems is generally highly dependent on the cost of individual system components as well as on location and other factors affecting the overall system performance. The largest component of the investment cost of PV systems is the cost of the PV module. Other cost factors that affect the LCOE include—but are not limited to—BOS components, labour cost of installation and O&M costs. Due to the dynamic development of the cost of PV systems, this section focuses on cost trends rather than current cost. Nonetheless, recent costs are presented in the discussion of individual cost factors and resulting LCOE below.

Average global PV module factory prices dropped from about USD_{2005} 22/W in 1980 to less than USD_{2005} 1.5/W in 2010 (Bloomberg, 2010).

Most studies about learning curve experience in photovoltaics focus on PV modules because they represent the single-largest cost item of a PV system (Yang, 2010). The PV module historical learning experience ranges between 11 and 26% (Maycock, 2002; Parente et al., 2002; Neij, 2008; IEA, 2010c) with a median progress ratio of 80%, and consequently, a median historical learning rate (price experience factor) of 20%, which means that the price was reduced by 20% for each doubling of cumulative sales (Hoffmann, 2009; Hoffmann et al., 2009). Figure 3.17 depicts the price developments for crystalline silicon modules over the last 35 years. The huge growth of demand after 2003 led to an increase in prices due to the supply-constrained market, which then changed into a demand-driven market leading to a significant price reduction due to module overcapacities in the market (Jäger-Waldau, 2010a).

The second-largest technical-related costs are the BOS components, and therein, the single largest item is the inverter. While the overall BOS experience curve was between 78 and 81%, or a 19 to 22% learning rate, quite similar to the module rates, learning rates for inverters were just in the range of 10% (Schaeffer et al., 2004). A similar trend was found in the USA for cost reduction for labour costs attributed to installed PV systems (Hoff et al., 2010).

The average investment cost of PV systems, that, the sum of the costs of the PV module, BOS components and labour cost of installation, has also

---

10 LCOE is not the sole determinant of its value or economic competitiveness (relative environmental and social impacts must be considered, as well as the contribution that the technology provides to meeting specific energy services, for example, peak electricity demands, or integration costs).

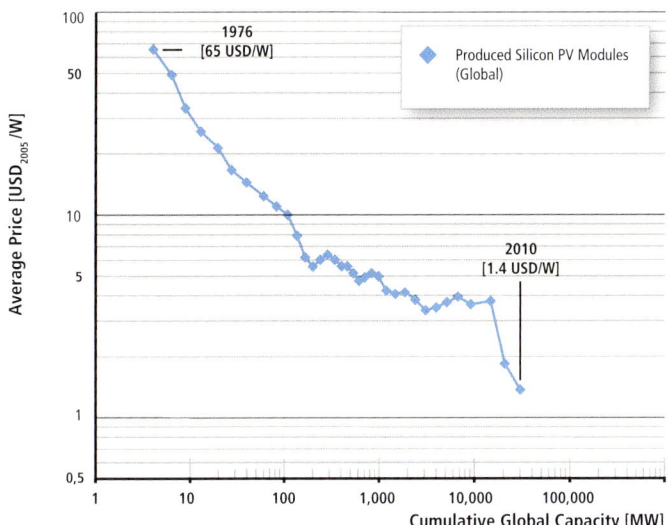

Figure 3.17 | Solar price experience or learning curve for silicon PV modules. Data displayed follow the supply and demand fluctuations. Data source: Maycock (1976-2003); Bloomberg (2010).

decreased significantly over the past couple of decades and is projected to continue decreasing rapidly as PV technology and markets mature. However, the system price decrease[11] varies significantly from region to region and depends strongly on the implemented support schemes and maturity of markets (Wiser et al., 2009). Figure 3.18 shows the system price developments in Europe, Japan, and the USA.

The capacity-weighted average investment costs of PV systems installed in the USA declined from USD_{2005} 9.7/W in 1998 to USD_{2005} 6.8/W in 2008. This decline was attributed primarily to a drop in non-module (BOS) costs. Figure 3.18 also shows that PV system prices continued to decrease considerably since the second half of 2008. This decrease is considered to be due to huge increases in production capacity and production overcapacities and, as a result, increased competition between PV companies (LBBW, 2009; Barbose et al., 2010; Mints, 2011). More generally, Figure 3.18 shows that the gap between PV system prices or investment cost between and within different world regions narrowed until 2005. In the period from 2006 to 2008, however, the cost spread widened at least temporarily. The first-quarter 2010 average PV system price in Germany dropped to € 2,864/kW_p (USD_{2005} 3,315/kW_p) for systems below 100 kW_p (Bundesverband Solarwirtschaft e.V., 2010). In 2009, thin-film projects at utility scale were realized at costs as low as USD_{2005} 2.72/W_p (Bloomberg, 2010).

O&M costs of PV electricity generation systems are low and are found to be in a range between 0.5 and 1.5% annually of the initial investment costs (Breyer et al., 2009; IEA, 2010c).

---

11   System prices determine the investment cost for independent project developers. Since, prices can contain profit mark-ups, the investment cost may be higher for independent project developers than for vertically integrated companies that are engaged in the production of PV systems or components thereof.

The main parameter that influences the capacity factor of a PV system is the actual annual solar irradiation at a given location given in $kWh/m^2/yr$. Capacity factors for PV installations are found to be between 11 and 24% (Sharma, 2011), which is in line with earlier findings of the IEA Implementing Agreement PVPS (IEA, 2007), which found that most of the residential PV systems had capacity factors in the range of 11 to 19%. Utility-scale systems currently under construction or in the planning phase are projected to have 20 to 30% capacity factors (Sharma, 2011).

Based on recent data representative of the global range of investment cost around 2008 as discussed above, assumptions provided in Annex III of this report, and the methods specified in Annex II, the following two plots show the sensitivity of the LCOE of various types of PV systems with respect to investment cost (Figure 3.19a) and discount rates (Figure 3.19b) as a function of the capacity factor.

Note that 1-axis tracking for utility-scale PV systems range from 15-20% increase in investment cost over fixed utility-scale PV systems. Modeling studies for c-Si indicate 16% increase for 1-axis tracking over fixed utility-scale PV systems (Goodrich et al., 2011). In 2008 and 2009, commercial rooftop PV systems of 20 to 500 kW were reported to be roughly 5% lower in investment cost than residential rooftop PV systems of 4 to 10 kW (NREL, 2011).

These figures highlight that the LCOE of individual projects depends strongly on the particular combination of investment costs, discount rates and capacity factors as well as on the type of project (residential, commercial, utility-scale).

Several studies have published LCOEs for PV electricity generation based on different assumptions and methodologies. Based on investment cost for thin-film projects of USD_{2005} 2.72/W_p in 2009 and further assumptions, Bloomberg (2010) finds LCOEs in the range of 14.5 and 36.3 US $cent_{2005}$/kWh. Breyer et al. (2009) find LCOEs in the range of 19.2 to 22.6 US $cent_{2005}$/kWh in regions of high solar irradiance (>1,800 $kWh/m^2/yr$) in Europe and the USA in 2009. All of these ranges can be considered to be reasonably achievable according to the LCOE ranges shown in Figure 3.19 and included in Annex III.

Assuming the PV market will continue to grow at more than 35% per year, the cost is expected to drop more than 50% to about 7.3 US $cent_{2005}$/kWh by 2020 (Breyer et al., 2009). Table 3.5 shows the 2010 IEA PV roadmap projections, which are somewhat less ambitious, but still show significant reductions (IEA, 2010c). The underlying deployment scenario assumes 3,155 GW of cumulative installed PV capacity by 2050.

The goal of the US DOE Solar Program's Technology Plan is to make PV-generated electricity cost-competitive with market prices in the USA by 2015. Their ambitious energy cost targets for various market sectors are 8 to 10 US $cents_{2005}$/kWh for residential, 6 to 8 US $cents_{2005}$/kWh for commercial

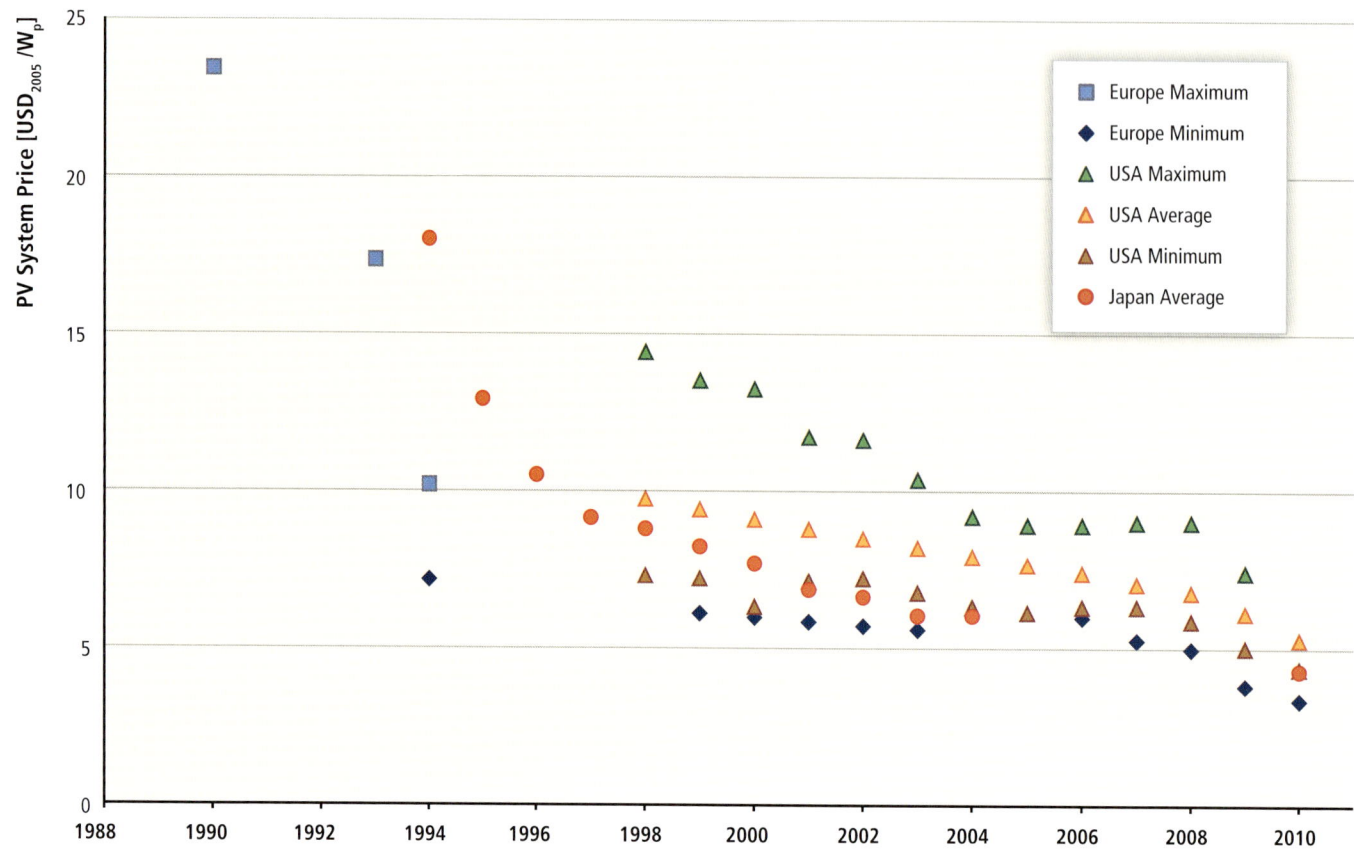

Figure 3.18 | Installed cost of PV systems smaller than 100 kW$_p$ in Europe, Japan and the USA. Data sources: Urbschat et al. (2002); Jäger-Waldau (2005); Wiser et al. (2009); Bundesverband Solarwirtschaft e.V. (2010); SEIA (2010a,b).

and 5 to 7 US cents$_{2005}$/kWh for utilities (US DOE, 2008). All of these cost targets are just below what seems to be possible to achieve for projects of similar type realized around 2008 even under very optimistic conditions (see Figure 3.19 as well as Annex III). Given continued cost reductions in the near term, these cost targets appear to be well within reach for projects that can be realized under favourable conditions. Relatively more progress will be required, however, to allow achieving such costs on a broader scale.

3.8.4 Concentrating solar power electricity generation

Concentrating solar power electricity systems are a complex technology operating in a complex resource and financial environment, so many factors affect the LCOE (Gordon, 2001). A study for the World Bank (World Bank Global Environment Facility Program, 2006) suggested four phases of cost reduction for CSP technology and forecast that cost competitiveness with non-renewable fuel could be reached by 2025. Figure 3.20 shows that cost reductions for CSP technologies are expected to come from plant economies of scale, reducing costs of components through material improvements and mass production, and implementing higher-efficiency processes and technologies.

The total investment for the nine plants comprising the Solar Electric Generating Station (SEGS) in California was USD$_{2005}$ 1.18 billion, and construction and associated costs for the Nevada Solar One plant amounted to 245 million (USD$_{2005}$, assumed 2007 base).

The publicized investment costs of CSP plants are often confused when compared with other renewable sources, because varying levels of integrated thermal storage increase the investment, but also improve the annual output and capacity factor of the plant.

The two main parameters that influence the solar capacity factor of a CSP plant are the solar irradiation and the amount of storage or the availability of a gas-fired boiler as an auxiliary heater, for example, the SEGS plants in California (Fernández-García et al., 2010). In case of solar-only CSP plants, the capacity factor is directly related to the available solar irradiation. With storage, the capacity factor could in theory be increased to 100%; however, this is not an economic option and trough plants are now designed for 6 to 7.5 hours of storage and a capacity factor of 36 to 41% (see Section 3.3.4). Tower plants, with their higher temperatures, can charge and store molten salt more efficiently, and projects designed for up to

Direct Solar Energy

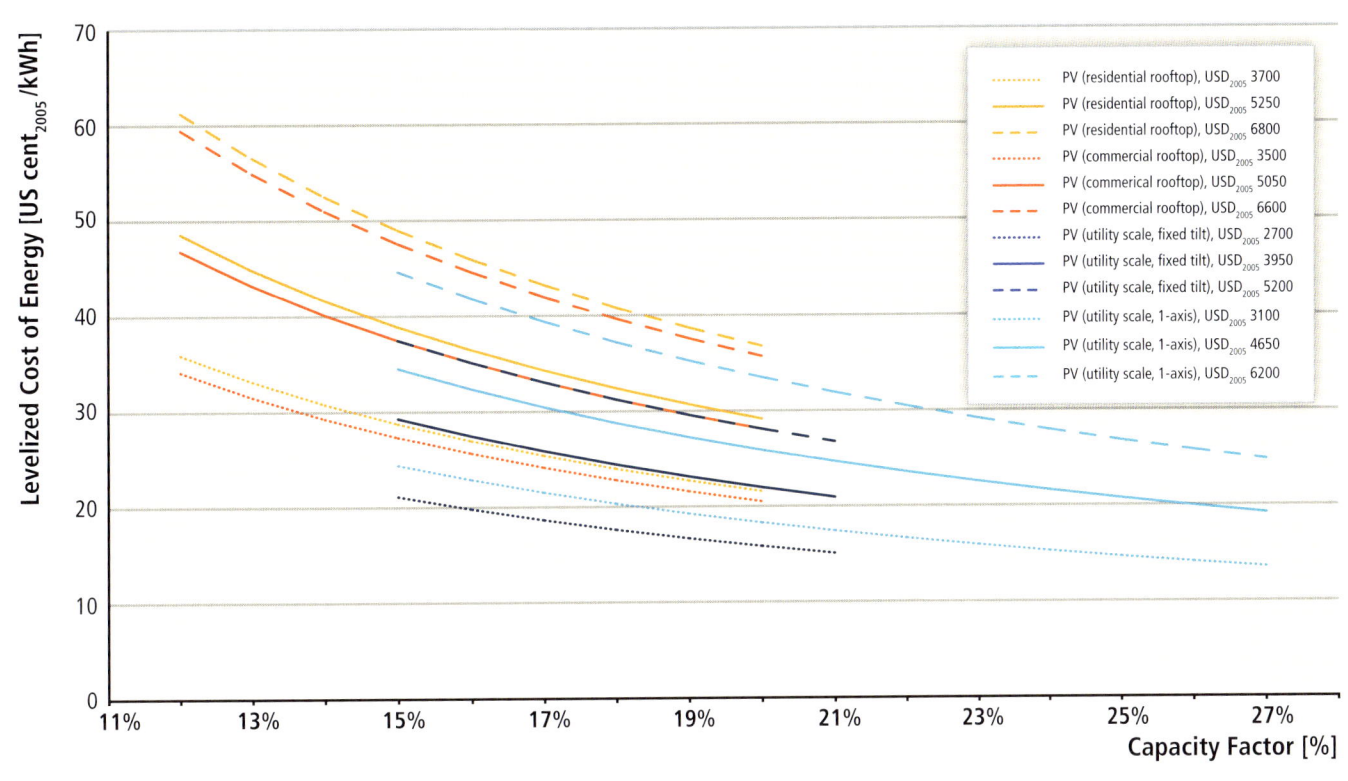

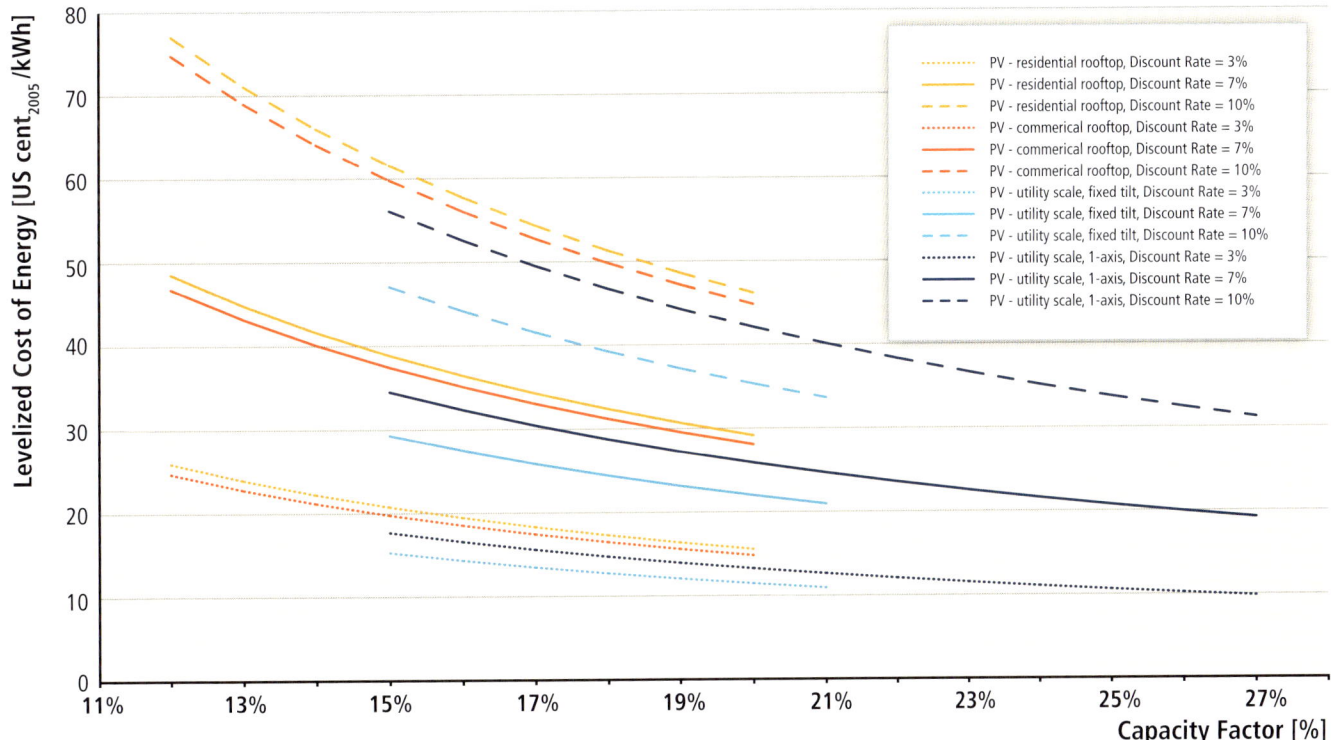

Figure 3.19 | Levelized cost of PV electricity generation, 2009. Upper panel: Cost of PV electricity generation as a function of capacity factor and investment cost[1,3]. Lower panel: Cost of PV electricity generation as a function of capacity factor and discount rate[2,3]. Source: (Annex III).

Notes: 1. Discount rate assumed to equal 7%. 2. Investment cost for residential rooftop systems assumed at USD$_{2005}$ 5,250/kW, for commercial rooftop systems at USD$_{2005}$ 5,050/kW, for utility-scale fixed tilt projects at USD$_{2005}$ 3,950/kW and for utility-scale one-axis projects at USD$_{2005}$ 4,650/kW. 3. Annual O&M cost assumed at USD$_{2005}$ 41 to 64/kW, lifetime at 25 years.

Table 3.5 | IEA price forecasts for 2020 and 2050. The ranges are given for 2,000 kWh/kW_p and 1,000 kWh/kW_p (IEA, 2010c).

	2020 (US cents$_{2005}$)		2050 (US cents$_{2005}$)	
Energy yields (kWh/kW_p)	2000	1000	2000	1000
Equivalent Capacity Factor	22.8%	11.4%	22.8%	11.4%
Residential PV	14.5	28.6	5.9	12.2
Utility-scale PV	9.5	19.0	4.1	8.2

15 hours of storage, giving a 75% annual capacity factor, are under construction.

Because, other than the SEGS plants, new CSP plants only became operational from 2007 onwards, few actual performance data are available. For the SEGS plants, capacity factors of between 12.5 and 28% are reported (Sharma, 2011). The predicted yearly average capacity factor of a number of European CSP plants in operation or close to completion of construction is given as 22 to 29% without thermal storage and 27 to 75% with thermal storage (Arce et al., 2011). These numbers are well in line with the capacity figures given in the IEA CSP Roadmap (IEA, 2010b) and the US Solar Vision Study (US DOE, 2011). However, the limited available performance data for the thermal storage state should be noted.

For large, state-of-the-art trough plants, current investment costs are reported as USD$_{2005}$ 3.82/W (without storage) to USD$_{2005}$ 7.65/W (with storage) depending on labour and land costs, technologies, the amount and distribution of direct-normal irradiance and, above all, the amount of storage and the size of the solar field (IEA, 2010b). Storage increases the investment costs due to the storage itself, as well as the additional collector area needed to charge the storage. But it also improves the ability to dispatch electricity at times of peak tariffs in the market or when balancing power is needed. Thus, a strategic approach to storage can improve a project's internal rate of return.

The IEA (2010b) estimates LCOEs for large solar troughs in 2009 to range from USD$_{2005}$ 0.18 to 0.27/kWh for systems with different amounts of thermal storage and for different levels of solar irradiation. This is broadly in line with the range of LCOEs derived for a system with six hours of storage at a 10% discount rate (as applied by the IEA), although the full range of values derived for different discount rates is broader (see Annex III). Based on the data and assumptions provided in Annex III of this report, and the methods specified in Annex II, the following two

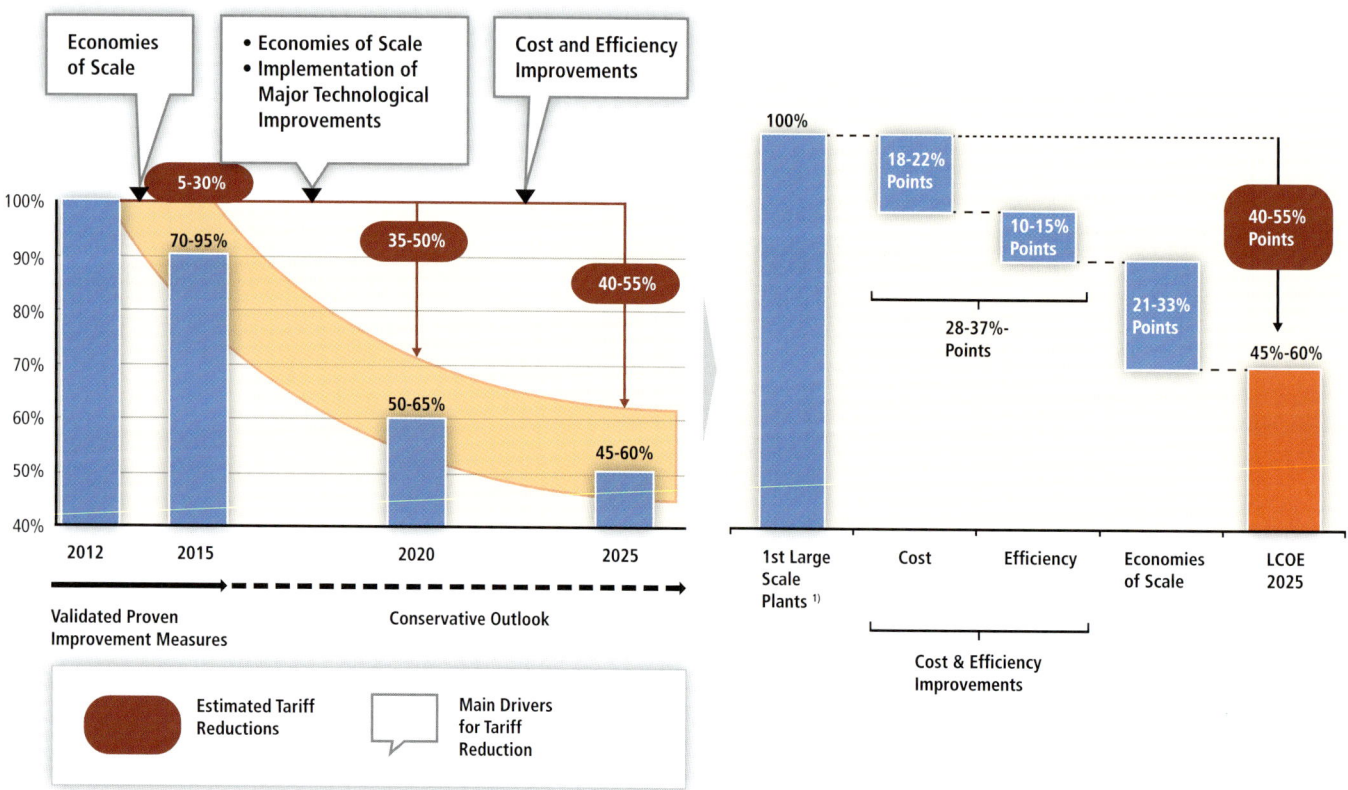

Figure 3.20 | Expected cost decline for CSP plants from 2012 to 2025. The cost number includes the cost of the plant plus financing (A.T. Kearney, 2010). As reduction ranges for cost, efficiency and economies of scale in the right panel overlap, their total contribution in 2025 amounts to less than their overall total.

Note: General. Tariffs equal the minimum required tariff, and are compared to 2012 tariffs. 1. Referring to 2010 to 2013 according to planned commercialization date of each technology (reference plant).

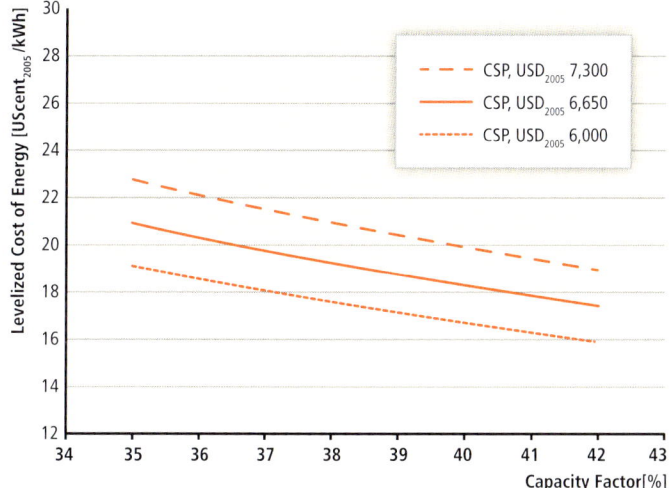

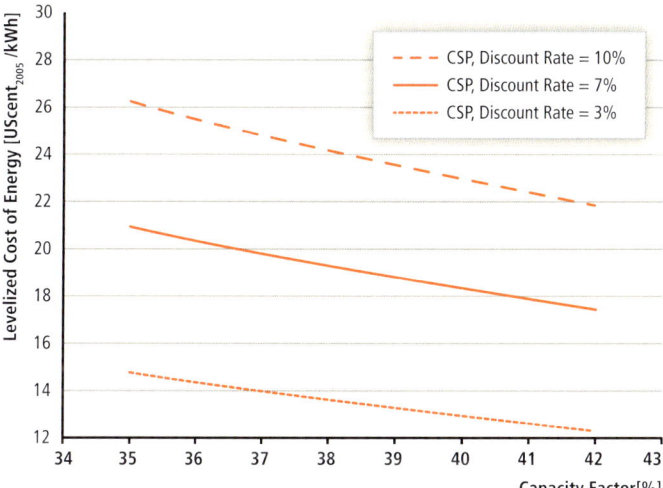

Figure 3.21 | Levelized cost of CSP electricity generation, 2009. Upper panel: Cost of CSP electricity generation as a function of capacity factor and investment cost[1,3]. Lower panel: Cost of CSP electricity generation as a function of capacity factor and discount rate[2,3]. Source: Annex III.

Notes: 1. Discount rate assumed to equal 7%. 2. Investment cost for CSP plant with six hours of thermal storage assumed at USD_{2005} 6,650/kW. 3. Annual O&M cost assumed at USD_{2005} 71/kW, lifetime at 25 years.

plots show the sensitivity of the LCOE of CSP plants with six hours of thermal storage with respect to investment cost (Figure 3.21, upper) and discount rates (Figure 3.21, lower) as a function of capacity factor.

The learning ratio for CSP, excluding the power block, is given as 10 ±5% by Neij (2008; IEA, 2010b). Other studies provide learning rates according to CSP components: Trieb et al. (2009b) give 10% for the solar field, 8% for storage, and 2% for the power block, whereas NEEDS (2009) and Viebahn et al. (2010) state 12% for the solar field, 12% for storage, and 5% for the power block.

Cost reductions for trough plants of the order of 30 to 40% within the next decade are considered achievable. Central-receiver technology is less commercially mature than troughs and thus presents slightly higher investment costs than troughs at the present time; however, cost reductions of 40 to 75% are predicted for central-receiver technology (IEA, 2010b).

The US DOE (2011) states its CSP goals for the USA in terms of USD/kWh, rather than USD/W, because the Solar Energy Technologies Program is designed to affect the LCOE and includes significant storage. The specific CSP goals are the following: 9 to 11 US $cents_{2005}$/kWh by 2010; 6 to 8 US $cents_{2005}$/kWh (with 6 hours of thermal storage) by 2015; and 5 to 6 US $cents_{2005}$/kWh (with 12 to 17 hours of thermal storage) by 2020 (USD_{2005}, assumed 2009 base). The EU is pursuing similar goals through a comprehensive RD&D program.

3.8.5 Solar fuel production

Direct conversion of solar energy to fuel is not yet widely demonstrated or commercialized. Thermochemical cycles along with electrolysis of water are the most promising processes for 'clean' hydrogen production in the future. In a comparison study, both the hybrid-sulphur cycle and a metal-oxide-based cycle were operated by solar tower technology for multi-stage water splitting (Graf et al., 2008). The electricity required for the alkaline electrolysis was produced by a parabolic trough power plant. For each process, the investment, operating and hydrogen production costs were calculated on a 50-MW_{th} scale. The study points out the market potential of sustainable hydrogen production using solar energy and thermochemical cycles compared to commercial electrolysis. A sensitivity analysis was done for three different cost scenarios: conservative, standard and optimistic (Table 3.6).

As a result, variation of the chosen parameters has the least impact on the hydrogen production costs of the hybrid-sulphur process, ranging from USD_{2005} 4.4 to 6.4/kg (Graf et al., 2008). The main cost factor for electrolysis is the electricity: just the variation of electricity costs leads to hydrogen costs of between USD_{2005} 2.4 to 7.7/kg. The highest range of hydrogen costs is obtained with the metal oxide-based process: USD_{2005} 4.0 to 14.5/kg. The redox system has the largest impact on the costs for the metal oxide-based cycle. The high electrical energy demand for nitrogen recycling influences the result significantly.

A substitute natural gas can be produced by the combination of solar hydrogen and CO_2 in a thermochemical synthesis at cost ranges from 12 to 14 US $cents_{2005}$/kWh_{th} with renewable power costs of 2 to 6 US $cents_{2005}$/kWh_e (Sterner, 2009). These costs depend highly on the operation mode of the plant and can be reduced by improving efficiency and reducing electricity costs.

The weakness of current economic assessments is primarily related to the uncertainties in the viable efficiencies and investment costs of the various solar components due to their early stage of development and their economy of scale as well as the limited amount of available literature data.

Table 3.6 | Overview of parameters for sensitivity (Graf et al., 2008).

	Cost scenario		
	Conservative	Standard	Optimistic
Heliostat costs (USD$_{2005}$/m^2)	159	136	114
Lifetime (years)	20	25	30
Redox system costs (USD$_{2005}$/kg)	1,700	170	17
Electricity costs (USD$_{2005}$/kWh$_e$)	0.14	0.11	0.05
Electrolyzer (decrease in %)	0	-10	-20
Chemical application (decrease in %)	0	-10	-20
Recycling of nitrogen (decrease in %)	0	-20	-40

3.9 Potential deployment[12]

Forecasts for the future deployment of direct solar energy may be underestimated, because direct solar energy covers a wide range of technologies and applications, not all of which are adequately captured in the energy scenarios literature. Nonetheless, this section presents near-term (2020) and long-term (2030 to 2050) forecasts for solar energy deployment. It then comments on the prospects and barriers to solar energy deployment in the longer-term scenarios, and the role of the deployment of solar energy in reaching different GHG concentration stabilization levels. This discussion is based on energy-market forecasts and carbon and energy scenarios published in recent literature.

3.9.1 Near-term forecasts

In 2010, the main market drivers are the various national support programs for solar-powered electricity systems or low-temperature solar heat installations. These programs either support the installation of the systems or the generated electricity. The market support for the different solar technologies varies significantly between the technologies, and also varies regionally for the same technology. This leads to very different thresholds and barriers for becoming competitive with existing technologies. Regardless, the future deployment of solar technologies depends strongly on public support to develop markets, which can then drive down costs due to learning. It is important to remember that learning-related cost reductions depend, in part at least, on actual production and deployment volumes, not just on the passage of time, though other factors such as R&D also act to drive costs down (see Section 10.5).

Table 3.7 presents the results of a selection of scenarios for the growth in solar deployment capacities in the near term, until 2020. It should be highlighted that passive solar gains are not included in these statistics, because this technology reduces demand and is therefore not part of the supply chain considered in energy statistics. The same PV technology can be applied for stand-alone, mini-grid, or hybrid systems in remote areas without grid connection, as well as for distributed and centralized grid-connected systems. The deployment of CSP technology is limited by regional availability of good-quality direct-normal irradiance of 2,000 kWh/m^2 (7,200 MJ/m^2) or more in the Earth's sunbelt. As shown in Table 3.7, solar capacity is expected to expand even in reference or baseline scenarios, but that growth is anticipated to accelerate dramatically in alternative scenarios that seek a more dramatic transformation of the global energy sector towards lower carbon emissions.

Photovoltaic market projections at the end of 2009 for the short term until 2013 indicate a steady increase, with annual growth rates ranging between 10 and more than 50% (UBS, 2009; EPIA, 2010; Fawer and Magyar, 2010). Several countries are discussing and proposing ambitious targets for the accelerated deployment of solar technologies. If fully implemented, the following policies could drive global markets in the period up to 2020:

- The National Development and Reform Commission (NDRC) expects non-fossil energy to supply 15% of China's total energy demand by 2020. Specifically for installed solar capacity, the NDRC's 2007 'Medium and Long-Term Development Plan for Renewable Energy in China' set a target of 1,800 MW by 2020. However, these goals have been discussed as being too low, and the possibility of reaching 20 GW or more seems more likely.

- The 2009 European Directive on the Promotion of Renewable Energy set a target of 20% RE in 2020 (The European Parliament and the Council of the European Union, 2010), and the Strategic Energy Technology plan is calling for electricity from PV in Europe of up to 12% in 2020 (European Commission, 2007).

- The 2009 Indian Solar Plan ('India Solar Mission') calls for a goal of 20 GW of solar power in 2022: 12 GW are to come specifically from ground-mounted PV and CSP plants; 3 GW from rooftop PV systems; another 3 GW from off-grid PV arrays in villages; and 2 GW from other PV projects, such as on telecommunications towers (Ministry of New and Renewable Energy, 2009).

- Relating to US cumulative installed capacity by 2030, the USDOE-sponsored Solar Vision Study (US DOE, 2011) is exploring the following two scenarios: a 10% solar target of 180 GW PV (120 GW central, 60 GW distributed); and a 20% solar target of 300 GW PV (200 GW central, 100 GW distributed).

3.9.2 Long-term deployment in the context of carbon mitigation

The IPCC Fourth Assessment Report estimated the available (technical) solar energy resource as 1,600 EJ/yr for PV and 50 EJ/yr for CSP; however, this estimate was given as very uncertain, with sources reporting values orders of magnitude higher (Sims et al., 2007). On the other hand, the projected deployment of direct solar in the IPCC Fourth Assessment Report gives an economic potential contribution of

---

12 Complementary perspectives on potential deployment based on a comprehensive assessment of numerous model-based scenarios of the energy system are presented in Sections 10.2 and 10.3.

Table 3.7 | Evolution of cumulative solar capacities based on different scenarios reported in EREC-Greenpeace (Teske et al., 2010) and IEA Roadmaps (IEA, 2010b,c).

Cumulative installed capacity	Low-Temperature Solar Heat (GW$_{th}$)			Solar PV Electricity (GW)			CSP Electricity (GW)		
	2009	2015	2020	2009	2015	2020	2009	2015	2020
Current value	180			22			0.7		
EREC – Greenpeace (reference scenario)		180	230		44	80		5	12
EREC – Greenpeace ([r]evolution scenario)		715	1,875		98	335		25	105
EREC – Greenpeace (advanced scenario)		780	2,210		108	439		30	225
IEA Roadmaps		N/A			95[1]	210		N/A	148

Note: 1. Extrapolated from average 2010 to 2020 growth rate.

direct solar to the world electricity supply by 2030 of 633 TWh (2.3 EJ/yr) (Sims et al., 2007).

Chapter 10 provides a summary of the literature on the possible future contribution of RE supplies in meeting global energy needs under a range of GHG concentration stabilization scenarios. Focusing specifically on solar energy, Figure 3.22(a) presents modelling results for the global supply of solar energy. Figure 3.22(b) shows solar thermal heat generation, and Figures 3.22(c) and (d) present solar PV and CSP electricity generation respectively, all at the global scale. Depending on the quantity shown, between 44 and about 156 different long-term scenarios underlie these figures derived from a diversity of modelling teams and spanning a wide range of assumptions about—among other variables—energy demand growth, cost and availability of competing low-carbon technologies, and cost and availability of RE technologies (including solar energy). Chapter 10 discusses how changes in some of these variables impact RE deployment outcomes, with Section 10.2.2 describing the literature from which the scenarios have been taken. Figures 3.22(a) to 3.22(d) present the solar energy deployment results under these scenarios for 2020, 2030 and 2050 for three GHG concentration stabilization ranges, based on the IPCC's Fourth Assessment Report: >600 ppm CO_2 (Baselines), 440 to 600 ppm (Categories III and IV) and <440 ppm (Categories I and II), all by 2100. Results are presented for the median scenario, the 25th to 75th percentile range among the scenarios, and the minimum and maximum scenario results.[13]

In the baseline scenarios, that is, without any climate policies assumed, the median deployment levels for solar energy remain very low, in the range of today's solar primary energy supply of below 1 EJ/yr, until 2050. It is worthwhile noting that the much smaller set of scenarios that reports solar thermal heat generation (44 compared to the full set of 156 that report solar primary energy) shows substantially higher median deployment levels of solar thermal heat of up to about 12 EJ/yr by 2050 even in the baseline cases. In contrast, electricity generation from solar PV and CSP is projected to stay at very low levels.

The picture changes with increasingly low GHG concentration stabilization levels that exhibit significantly higher median contributions from solar energy than the baseline scenarios. By 2030 and 2050, the median deployment levels of solar energy reach 1.6 and 12.2 EJ/yr, respectively, in the intermediate stabilization categories III and IV that result in atmospheric CO_2 concentrations of 440-600 ppm by 2100. In the most ambitious stabilization scenario category, where CO_2 concentrations remain below 440 ppm by 2100, the median contribution of solar energy to primary energy supply reaches 5.9 and 39 EJ/yr by 2030 and 2050, respectively.

The scenario results suggest a strong dependence of the deployment of solar energy on the climate stabilization level, with significant growth expected in the median cases until 2030 and in particular until 2050 in the most ambitious climate stabilization scenarios. Breaking down the development by individual technology, it appears that solar PV deployment is most dependent on climate policies to reach significant deployment levels while CSP and even more so solar thermal heat deployment show a lower dependence on climate policies. However, this interpretation should be applied with care, because CSP electricity and solar thermal heat generation were reported by significantly fewer scenarios than solar PV electricity generation.

The ranges of solar energy deployment at the global level are extremely large, also compared to other RE sources (see Section 10.2.2.5), indicating

---

13 In scenario ensemble analyses such as the review underlying the figures, there is a constant tension between the fact that the scenarios are not truly a random sample and the sense that the variation in the scenarios does still provide real and often clear insights into collective knowledge or lack of knowledge about the future (see Section 10.2.1.2 for a more detailed discussion).

Direct Solar Energy

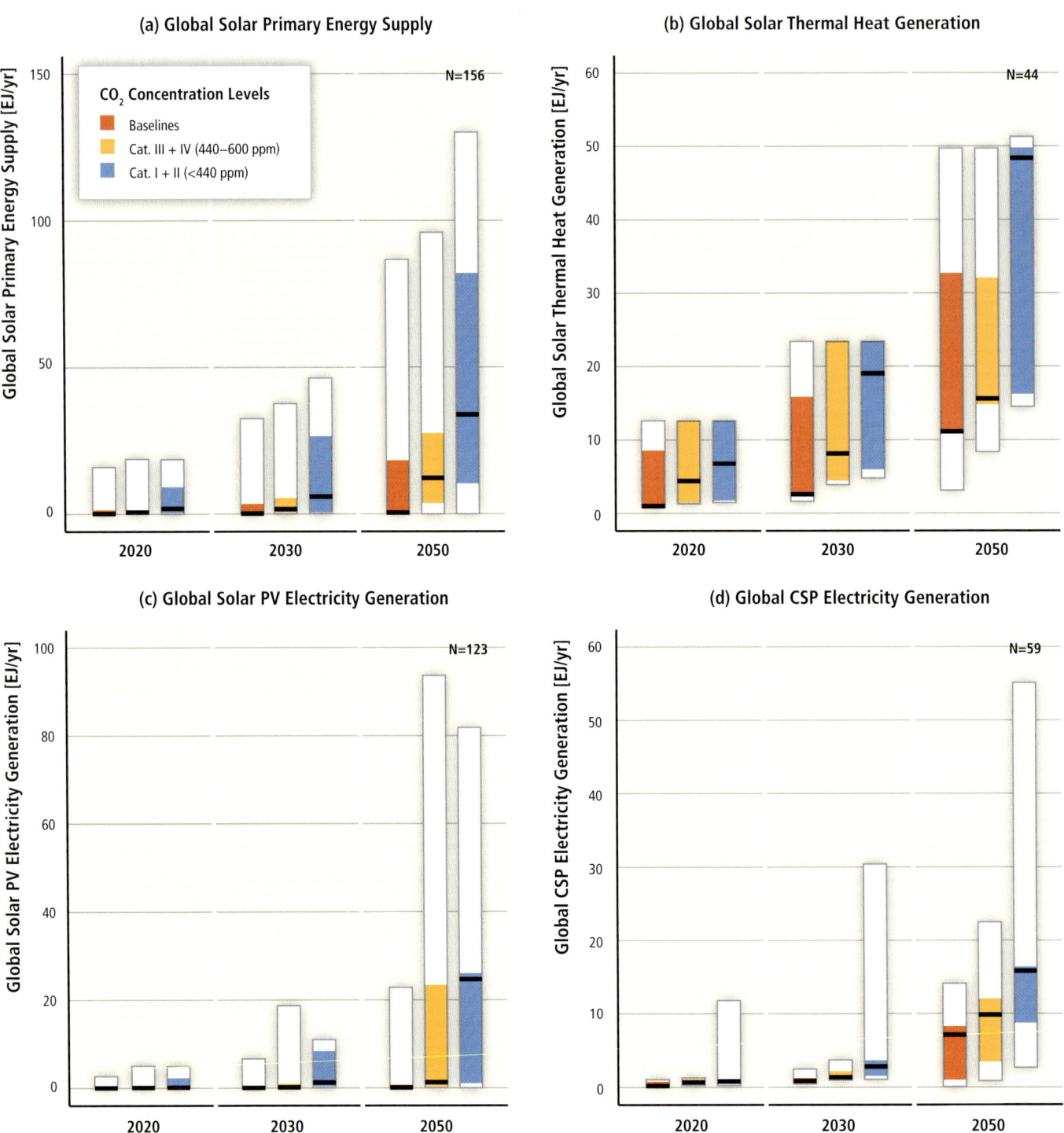

Figure 3.22 | Global solar energy supply and generation in long-term scenarios (median, 25th to 75th percentile range, and full range of scenario results; colour coding is based on categories of atmospheric CO_2 concentration level in 2100; the specific number of scenarios underlying the figure is indicated in the right upper corner): (a) Global solar primary energy supply; (b) global solar thermal heat generation; (c) global PV electricity generation; and (d) Global CSP electricity generation (adapted from Krey and Clarke, 2011; see also Chapter 10).

a very wide range of assumptions about the future development of solar technologies in the reviewed scenarios. In the majority of baseline scenarios the solar deployment remains low until 2030, with the 75th percentile reaching some 3 EJ/yr and only very few scenarios showing significantly higher levels. By 2050, this relatively narrow deployment range in the baselines disappears; the 75th percentile shows roughly a 30-fold increase compared to the median baseline case, reaching about 15 EJ/yr and even much higher levels in the uppermost quartile. A combination of increasing relative prices of fossil fuels with more optimistic assumptions about cost declines for solar technologies is likely to be responsible for the higher baseline deployment levels.

In the most ambitious climate stabilization scenarios, the 75th percentiles of the solar primary energy supply by 2030 reach up to 26 EJ/yr, a five-fold increase compared to the median of the same category and the highest estimates even reach up to 50 EJ/yr. For 2050 the equivalent numbers are 82 EJ/yr (75th percentile) and 130 EJ/yr (maximum level), which can be attributed to a large extent to solar PV electricity generation, which reaches deployment levels of more than 80 EJ/yr by 2050, but CSP electricity and solar thermal heat also contribute significantly under these very high solar deployment levels. The share of solar PV in global electricity generation in the most extreme scenarios reaches up to about 12% by 2030 and up to one-third by 2050, but in the vast majority of scenarios remains in the single digit percentage range.

To achieve the higher levels of deployment envisioned by some of these scenarios, policies to reduce GHG emissions and/or increase RE supplies are likely to be necessary, and those policies would need to be of adequate economic attractiveness *and* predictability to motivate substantial private investment (see Chapter 11). A variety of other possible challenges to rapid solar energy growth also deserve discussion, as do factors that can contribute to it.

Resource potential. The solar resource is virtually inexhaustible, and it is available and able to be used in most countries and regions of the world. The worldwide technical potential of solar energy is considerably larger than the current primary energy consumption (IEA, 2008), and will not serve as a primary barrier to even the most ambitious deployment paths included in the scenarios literature summarized above.

Regional deployment. Industry-driven scenarios with regional visions for up to 100% of RE supply by 2050 have been developed in various parts of the world, often with substantial levels of solar energy deployment.

The Semiconductor Equipment and Materials International Association developed PV roadmaps for China and India that go far beyond the targets of the national governments (SEMI, 2009b,c). These targets are about 20 GW by 2020 and 100 GW by 2050 for electricity generation in China and 20 GW and 200 GW in India (both PV and CSP) (Ministry of New and Renewable Energy, 2009; Zhang et al., 2010).

In Europe, the European Renewable Energy Council developed a 100% Renewable Energy vision based on the inputs of the various European industrial associations (Zervos et al., 2010). Assumptions for 2020 about final electricity, heating and cooling, as well as transport demand are based on the European Commission's New Energy Policy (NEP) scenario with both a moderate and high price environment as outlined in the Second Strategic Energy Review (European Commission, 2008). The scenarios for 2030 and 2050 assume a massive improvement in energy efficiency to realize the 100% RE goals. For Europe, this scenario assumes that solar can contribute about 557 TWh (2,005 PJ) and 1415 TWh (5,094 PJ) heating and cooling in 2030 and 2050, respectively. For electricity generation, about 556 TWh (2,002 PJ) from PV and 141 TWh (508 PJ) from CSP are anticipated for 2030 and 1,347 TWh (4,849 PJ) and 385 TWh (1,386 PJ) for 2050, respectively.

In Japan, the New Energy Development Organisation, the Ministry for Economy, Trade and Industry, the Photovoltaic Power Generation Technology Research Association and the Japan Photovoltaic Energy Association drafted the 'PV Roadmap Towards 2030' in 2004 (Kurokawa and Aratani, 2004). In 2009, the roadmap was revised: the target year was extended from 2030 to 2050, and a goal was set to cover between 5 and 10% of domestic primary energy demand with PV power generation in 2050. The targets for electricity from PV systems range between 35 TWh (126 PJ) for the reference scenario and 89 TWh (320 PJ) for the advanced scenario in 2050 (Komiyama et al., 2009).

In the USA, the industry associations—the Solar Electric Power Association and the Solar Energy Industry Association—are working together with the USDOE and other stakeholders to develop scenarios for electricity from solar resources (PV and CSP) of 10 and 20% in 2030. The results of the Solar Vision Study (USDOE, 2011) are expected in 2011.

Achieving the higher *global* scenario results for solar energy would clearly require substantial solar deployment in every region of the world. The *regional* scenarios presented here suggest that regional deployment paths may exist to support such a global result. Nonetheless, enabling this growth in regions new to solar energy may present cost and institutional challenges that would require active management; institutional and technical knowledge transfer from those regions that are already witnessing substantial solar energy activity may be required.

Supply chain issues. Passive solar energy markets and industries have largely developed locally to this point because the building market itself is local. Enabling high-penetration solar energy futures may require a globalization of at least knowledge on passive solar technologies to enable broader market penetration. Low-temperature solar thermal is implemented all over the world within local markets, with local suppliers, but a global market is starting to be developed. The PV industry is already global in scope, with a global supply chain, while CSP is starting to develop a global supply chain—in 2010, the CSP market was driven by Spain and the USA, but other countries such as Germany and India are also helping to expand the market. In general, supply chain and materials constraints may impact the speed and scope of solar energy deployment in certain regions and at certain times, but such factors are unlikely to restrict the ability of solar energy technologies to meet the higher penetrations envisioned by the more aggressive scenarios presented earlier. In fact, the modular nature of many of the solar technologies, both in manufacturing and use, as well as the diverse applications for solar energy suggest that supply chain issues are unlikely to constrain growth.

Technology and economics. The technical maturity and economic competitiveness of solar technologies vary. Passive solar consists of well-established technologies, though with room for improvement;

however, the awareness of the building sector is not always available. The economics are understood, but they depend on local solar resources and local support and building regulations. Low-temperature solar thermal is also a well-established technology, with economics that depend on the solar resource, the applications, and the cost of competing technologies—some regions may need support programs to create markets and enable growth, whereas in other regions solar thermal is already competitive.

PV is already an established technology, but substantial further technological advances are possible with the prospect for continued cost reduction. To this point, however, the deployment of PV technology has strongly depended on local support programs in most markets. Similarly, CSP technology has substantial room for additional improvement, but CSP costs have to this point exceeded market energy prices.

Continued cost reductions are therefore likely to be needed if solar energy is to meet the higher global scenario results presented earlier. Support programs to encourage solar deployment and R&D may both play an important role in seeking to achieve the necessary reductions.

Integration and transmission. Integration and transmission are not a central concern for passive solar applications. Integration issues in low-temperature solar, on the other hand, are especially important for larger systems where integration into local district heating systems is needed, and where the temporal variability of solar output needs to be matched with other supply sources to meet customer demands (see Chapter 8). Due to the availability of the resource only during the day and the short-time-period variability associated with passing clouds, proactive technical and institutional solutions to operational integration concerns will need to be implemented to enable large-scale PV penetration; CSP, if implemented with thermal storage, would not impose similar requirements. Moreover, high-penetration PV and CSP scenarios that involve larger-scale developments are likely to require additional transmission infrastructure in order to access the highest-quality solar sites. Section 8.2.1 identifies a variety of the technical and institutional challenges associated with increased deployment of variable generation sources, and also highlights the variety of solutions for managing those challenges. Though Chapter 8 finds no insurmountable technical barriers to increased variable renewable energy supply, as solar deployment increases, transmission expansion and operational integration costs are also expected to rise, potentially constraining growth on economic terms. Proactively managing these challenges is likely to be central to achieving the high-penetration solar energy scenarios described earlier.

Social and environmental concerns. Direct solar energy appears to have relatively few social and environmental concerns. Rather, the main benefit of passive solar is in reducing the energy demand of buildings. Similarly, low-temperature solar thermal applications are comparatively benign from an environmental perspective. One concern for some PV technologies is that the PV industry uses some toxic materials and corrosive liquids in its production lines. The presence and amount of those materials depend strongly on the cell type, however, and rigorous control methods are used to minimize the risk of accidental releases. Recycling of PV materials may also become more common as deployment continues. Water availability and consumption is the main environmental concern for CSP, though dry cooling technology can substantially reduce water usage. Finally, especially for central-station PV and CSP installations, the ecological, social and visual impacts associated with plant infrastructure may be of concern. Efforts to better understand the nature and magnitude of these impacts, together with efforts to minimize and mitigate them, may need to be pursued in concert with increasing solar energy deployment.

3.9.3 Conclusions regarding deployment

Potential deployment scenarios range widely—from a marginal role of direct solar energy in 2050 to one of the major sources of energy supply. Although direct solar energy provides only a very small fraction of global energy supply in 2011, it has the largest technical potential of all energy sources and, in concert with technical improvements and resulting cost reductions, could see dramatically expanded use in the decades to come.

Achieving continued cost reductions is the central challenge that will influence the future deployment of solar energy. Reducing cost, meanwhile, can only be achieved if the solar technologies decrease their costs along their learning curves, which depends in part on the level of solar energy deployment. In addition, continuous R&D efforts are required to ensure that the slopes of the learning curves do not flatten before solar is widely cost competitive with other energy sources.

The true costs of and potential for deploying solar energy are still unknown because the main deployment scenarios that exist today often consider only a single solar technology: PV. In addition, scenarios often do not account for the co-benefits of a renewable/sustainable energy supply (but see Section 9.4 for some research in this area). At the same time, as with some other forms of RE, issues of variable production profiles and energy market integration as well as the possible need for new transmission infrastructure will influence the magnitude, type and cost of solar energy deployment.

Finally, the regulatory and legal framework in place can also foster or hinder the uptake of direct solar energy applications. For example, minimum building standards with respect to building orientation and insulation can reduce the energy demand of buildings significantly, increasing the share of RE supply without increasing the overall demand, while building and technical standards can also support or hinder the installation of rooftop solar systems. Transparent, streamlined administrative procedures to site, permit, install and connect solar power sources can further support the deployment of direct solar energy.

References

A.T. Kearney (2010). *Solar Thermal Electricity 2025--Clean Electricity On Demand: Attractive STE Cost Stabilize Energy Production*. A.T. Kearney GmbH, Duesseldorf, Germany, 52 pp.

Abanades, S., P. Charvin, G. Flamant, and P. Neveu (2006). Screening of water-splitting thermochemical cycles potentially attractive for hydrogen production by concentrated solar energy. *Energy*, **31**(14), pp. 2805-2822.

Arce, P., M. Medrano, A. Gil, E. Oró, and L.F. Cabeza (2011). Overview of thermal energy storage (TES) potential energy savings and climate change mitigation in Spain and Europe. *Applied Energy*. **88**(8), pp. 2764-2774.

Arif Hasan, M., and K. Sumathy (2010). Photovoltaic thermal module concepts and their performance analysis: A review. *Renewable and Sustainable Energy Reviews*, **14**, pp. 1845-1859.

Asano, H., K. Yajima, and Y. Kaya (1996). Influence of photovoltaic power generation on required capacity for load frequency control. *IEEE Transactions on Energy Conversion*, **11**(1), pp. 188-193.

ASHRAE (2009). *ASHRAE Handbook – Fundamentals*. American Society of Heating, Refrigerating, and Air-Conditioning Engineers, Atlanta, GA, USA.

Athienitis, A.K. (2008). Design of advanced solar homes aimed at net-zero annual energy consumption in Canada. In: *ISES-AP – 3rd International Solar Energy Society Conference – Asia Pacific Region. Incorporating the 46th ANZSES Conference*, Sydney, Australia, 25-28 Nov. 2008, pp. 14.

Athienitis, A.K., and M. Santamouris (2002). *Thermal Analysis and Design of Passive Solar Buildings*. James & James, London, UK, 288 pp.

Bailey, S., D. Brinker, H. Curtis, P. Jenkins, and D. Scheiman (1997). *NASA Technical Memorandum 113155 – Solar Cell Calibration and Measurement Techniques*. IECEC-97534, NASA, Cleveland, OH, USA.

Baoshan, L. (2010). Research on the progress of silicon materials in China. In: *6th China SoG Silicon and PV Conference (CSPV) 2010*, Shanghai, PR China, 16-18 March 2010.

Bar-Cohen, Y. (ed.) (2006). *Biomimetics: Biologically Inspired Technologies*. Taylor & Francis, Boca Raton, FL, USA.

Barbose, G., N. Darghouth, and R. Wiser (2010). *The Installed Cost of Photovoltaics in the U.S. from 1998-2009*. Lawrence Berkeley National Laboratory, Berkeley, CA, USA.

Barkat, A., S.H. Khan, M. Rahman, S. Zaman, A. Poddar, S. Halim, N.N. Ratna, M. Majid, A.K.M. Maksud, A. Karim, and S. Islam (2002). Economic and social impact evaluation study of the rural electrification program in Bangladesh. In: *Electric Cooperatives in Bangladesh*. National Rural Electric Cooperative Association (NRECA) International, Dhaka, Bangladesh, pp. 41.

Barnes, D.F. (1988). *Electric Power for Rural Growth: How Electricity Affects Rural Life in Developing Countries*. Westview Press, Boulder, CO, USA, 236 pp.

Batley, S.L., D. Colbourne, P.D. Fleming, and P. Urwin (2001). Citizen versus consumer: challenges in the UK green power market. *Energy Policy*, **29**, pp. 479-487.

Bazilian, M.D., F. Leenders, B.G.C. Van der Ree, and D. Prasad (2001). Photovoltaic cogeneration in the built environment. *Solar Energy*, **71**(1), pp. 57-69.

Benagli, S., D. Borrello, E. Vallat-Sauvain, J. Meier, U. Kroll, J. Hoetzel, J. Bailat, J. Steinhauser, M. Marmelo, and L. Castens (2009). High-efficiency amorphous silicon devices on LPCVD-ZnO-TCO prepared in industrial KaiTM-M R&D reactor. In: *Proceedings of the 24th European Photovoltaic Solar Energy Conference*, Hamburg, Germany, 21-25 September 2009, pp. 2293-2298.

Bernreuther, J., and F. Haugwitz (2010). *The Who's Who of Silicon Production*. Bernreuther Consulting, Wuerzburg, Germany.

Bett, A.W., F. Dimroth, W. Gulter, R. Hoheisel, E. Oliva, S.P. Philips, J. Schone, G. Siefer, M. Steiner, A. Wekkeli, E. Welser, M. Meusel, W. Kostler, and G. Strobl (2009). Highest efficiency multi-junction solar cell for terrestrial and space applications. In: *Proceedings of the 24th European Photovoltaic Solar Energy Conference*, Hamburg, Germany, 21-25 September 2009, pp. 1-6.

Beyer, H.G., B. Decker, J. Luther, and R. Steinberger-Willms (1991). Spatial and temporal characteristics of short term fluctuations in solar radiation for PV-plant applications. In: *Proceedings of the 10th E.C. Photovoltaic Solar Energy Conference*, Lisbon, Portugal, 8-12 Apr 1991, pp. 453–456.

Bickel, P., and R. Friedrich (eds.) (2005). *ExternE--Externalities of Energy: Methodology 2005 Update*. Office for Official Publications of the European Communities, Luxembourg, 270 pp.

Bloem, H., F. Monforti-Ferrario, M. Szabo, and A. Jäger-Waldau (2010). *Renewable Energy Snapshots 2010*. EUR 24440 EN, European Commission, Joint Research Centre, Institute for Energy, Ispra, Italy, 52 pp.

Bloomberg (2010). *Bloomberg New Energy Finance – Renewable Energy Data*. Subscriber info at: bnef.com/bnef/markets/renewable-energy/solar/.

Bolton, J.R. (1996). Solar photoproduction of hydrogen: A review. *Solar Energy*, **57**(1), pp. 37-50.

Bosi, M., and C. Pelosi (2007). The potential of III-V semiconductors as terrestrial photovoltaic devices. *Progress in Photovoltaics: Research and Applications*, **15**(1), pp. 51-68.

Bouzguenda, M., and S. Rahman (1993). Value analysis of intermittent generation sources from the system operations perspective. *IEEE Transactions on Energy Conversion*, **8**(3), pp. 484-490.

Boyle, G. (1996). *Renewable Energy: Power for a Sustainable Future*. Oxford University Press in association with the Open University, Oxford, UK, 479 pp.

Brabec, C.J. (2004). Organic photovoltaics: technology and market. *Solar Energy Materials and Solar Cells*, **83**(2-3), pp. 273-292.

Braun, M., Y.-M. Saint-Drenan, T. Glotzbach, T. Degner, and S. Bofinger (2008). Value of PV energy in Germany – Benefit from the substitution of conventional power plants and local power generation. In: *Proceedings of the 23rd European Photovoltaic Solar Energy Conference*, Valencia, Spain, 1-5 September 2008, pp. 3645-3652.

Brendel, R. (2003). *Thin Film Crystalline Silicon Solar Cells*. Wiley-VCH, Weinheim, Germany.

Breyer, C., A. Gerlach, J. Mueller, H. Behacker, and A. Milner (2009). Grid-parity analysis for EU and US regions and market segments – Dynamics of grid-parity and dependence on solar irradiance, local electricity prices and PV progress ratio. In: *Proceedings of the 24th European Photovoltaic Solar Energy Conference*, Hamburg, Germany, 21-25 September 2009, pp. 4492-4500 (ISBN: 3-936338-25-6).

Brouwer, I.D., J.C. Hoorweg, and M.J. van Liere (1997). When households run out of fuel: responses of rural households to decreasing fuelwood availability, Ntcheu District, Malawi. *World Development*, **25**(2), pp. 255-266.

Bundesverband Solarwirtschaft e.V. (2010). *Statistische Zahlen der deutschen Solarstrombranche (photovoltaik)*. Bundesverband Solarwirtschaft e.V. (BSW Solar), Berlin, Germany, 4 pp.

Bunea, M.M., K. Johnston, C.M. Bonner, P. Cousins, D.D. Smith, D.H. Rose, W.P. Mulligan, and R.M. Swanson (2010). Simulation and characterization of high efficiency back contact solar cells for low concentration photovoltaics. In: *Proceedings of the 35th Institute of Electrical and Electronics Engineers (IEEE) Photovoltaic Specialists Conference*, Honolulu, Hawaii, 20-25 June 2010, pp. 823-826 (ISSN: 0160-8371).

Butti, K., and J. Perlin (1980). *A Golden Thread: 2500 Years of Solar Architecture and Technology*. Cheshire Books, Frodsham, UK, 289 pp.

Cabeza, L., A. Castell, M. Medrano, I. Martorell, G. Pérez, and I. Fernández (2010). Experimental study on the performance of insulation materials in Mediterranean construction. *Energy and Buildings*, **42**(5), pp. 630-636.

Candanedo, J., and A.K. Athienitis (2010). A simulation study of anticipatory control strategies in a net zero energy solar-optimized house. *ASHRAE Transactions*, **116**(1), pp. 246-260.

Carlson, D.E., and C.R. Wronski (1976). Amorphous silicon solar cell. *Applied Physics Letters*, **28**(11), pp. 671-673.

Castell, A., I. Martorell, M. Medrano, G. Peréz, and L.F. Cabeza (2010). Experimental study using PCM in brick constructive solutions for passive cooling. *Energy and Buildings*, **42**(4), pp. 534-540.

Catchpole, K.R., and A. Polman (2008). Plasmonic solar cells. *Optics Express*, **16**(26), pp. 21793-21800.

Chalmers, S., M. Hitt, J. Underhill, P. Anderson, P. Vogt, and R. Ingersoll (1985). The effect of photovoltaic power generation on utility operation. *IEEE Transactions on Power Apparatus and Systems*, **104**(3), pp. 524-530.

Chang, K.C., W.M. Lin, T.S. Lee, and K.M. Chung (2009). Local market of solar water heaters in Taiwan: Review and perspectives. *Renewable and Sustainable Energy Reviews*, **13**(9), pp. 2605-2612.

Chiba, Y., A. Islam, K. Kakutani, R. Komiya, N. Koide, and L. Han (2005). High efficiency dye sensitized solar cells. In: *Technical Digest, 15th International Photovoltaic Science and Engineering Conference*, Shanghai, China, 10-15 October 2005, pp. 665-666.

Chow, T.T. (2010). A review on photovoltaic/thermal hybrid solar technology. *Applied Energy*, **87**, pp. 365-379.

Chowdhury, B., and S. Rahman (1988). Is central station photovoltaic power dispatchable? *IEEE Transactions on Energy Conversion*, **3**(4), pp. 747-754.

Chueh, W.C., C. Falter, M. Abbott, D. Scipio, P. Furler, S.M. Haile, and A. Steinfeld (2010). High-flux solar-driven thermochemical dissociation of CO_2 and H_2O using nonstoichiometric ceria. *Science*, **330**, pp. 1797-1801.

Clavadetscher, L., and T. Nordmann (2007). *Cost and Performance Trends in Grid-Connected Photovoltaic System and Case Studies*. IEA Photovoltaic Power Systems Program (PVPS). Task 2, Report IEA-PVPS T2-06:2007, IEA PVPS Pool Switzerland for the IEA Photovoltaic Power Systems Program, 54 pp. Available at: www.iea-pvps-task2.org/public/download/T2_Cost_and_Performance.pdf.

Cobben, S., B. Gaiddon, and H. Laukamp (2008). *Impact of Photovoltaic Generation on Power Quality in Urban areas with High PV Population: Results from Monitoring Campaigns*. PVupscale, WP4 - Deliverable 4.3, Contract EIE/05/171/SI2.420208, Intelligent Energy Europe, PVupscale project, 53 pp. Available at: www.pvupscale.org/IMG/pdf/WP4_D4-3_public_v1c.pdf

Conibeer, G., M.A. Green, D. König, I. Perez-Wurfl, S. Huang, X. Hao, D. Di, L. Shi, S. Shrestha, B. Puthen-Veetil, Y. So, B. Zhang, and Z. Wan (2010). Silicon quantum dot based solar cells: addressing the issues of doping, voltage and current transport. *Progress in Photovoltaics: Research and Applications*, doi:10.1002/pip.1045.

Coutts, T.J., J.S. Ward, D.L. Young, K.A. Emery, T.A. Gessert, and R. Noufi (2003). Critical issues in the design of polycrystalline, thin-film tandem solar cells. *Progress in Photovoltaics: Research and Applications*, **11**(6), pp. 359-375.

Curtright, A., and J. Apt (2008). The character of power output from utility-scale photovoltaic systems. *Progress in Photovoltaics: Research and Applications*, **16**(3), pp. 241-247.

de Vries, B., D. van Vuuren, and M.M. Hoogwijk (2007). Renewable energy sources: Their global potential for the first half of the 21st century at a global level: An integrated approach. *Energy Policy*, **35**(4), pp. 2590-2610.

Denholm, P., E. Drury, R. Margolis, and M. Mehos (2009). *Solar Energy: The Largest Energy Resource, Generating Electricity in a Carbon Constrained World*. Elsevier, Paris, France.

DESERTEC Foundation (2007). *Clean Power from Deserts - The DESERTEC Concept for Energy, Water and Climate Security*. 4th ed. WhiteBook, Protext Verlag, Bonn, Germany.

Dimmler, B., and H.W. Schock (1996). Scaling-up of CIS technology for thin-film solar modules. *Progress in Photovoltaics: Research and Applications*, **4**(6), pp. 425-433.

Dimroth, F., C. Baur, A.W. Bett, M. Meusel, and G. Strobl (2005). 3-6 junction photovoltaic cells for space and terrestrial concentrator. In: *31st Institute of Electrical and Electronics Engineers (IEEE) Photovoltaic Specialists Conference*, Lake Buena Vista, Florida, 3-7 January 2005, pp. 525-529.

Dincer, I., and M. Rosen (2010). *Thermal Energy Storage. Systems and Applications*. 2nd ed. Wiley-Blackwell, Bogner Regis, UK.

Displaybank (2010). *Displaybank Briefing*. January 28, 2010. Displaybank Consulting and Research, San Jose, CA, USA. Available for subscribers at: www.displaybank.com/eng/info/sread.php?id=5724.

DLR (2007). *AQUA-CSP. Concentrating Solar Power for Seawater Desalination. Final Report*. Institute of Technical Thermodynamics, Section Systems Analysis and Technology Assessment, German Aerospace Center (DLR), Stuttgart, Germany, 279 pp.

Duffie, J.A., and W.A. Beckman (2006). *Solar Engineering of Thermal Processes*. 3rd ed. Wiley, New York, NY, USA, 928 pp.

Ecofys Netherlands BV (2007). *Current State-of-the-Art and Best Practices of BIPV. D6.1.1. European Commission Sixth Framework Programme*. Contract No. 019718, Ecofys Netherlands BV, Utrecht, The Netherlands.

ECOSTAR (2005). *European Concentrated Solar Thermal Road-Mapping*. German Aerospace Centre (DLR), Stuttgart, Germany.

Eiffert, P. (2002). *Guidelines for Economic Evaluation of Building Integrated Photovoltaic Power Systems.* IEA Photovoltaic Power Systems Program (PVPS), Task 7, NREL/TP-550-31977, National Renewable Energy Laboratory, Golden, CO, USA, 52 pp. Available at: www.nrel.gov/docs/fy03osti/31977.pdf.

Eiffert, P., and G.K. Kiss (2000). *Building-Integrated Photovoltaic Designs for Commercial and Institutional Structures: A Source Book for Architects.* National Renewable Energy Laboratory, Golden, CO, USA, 90 pp.

Ellingson, R.J., M.C. Beard, J.C. Johnson, P. Yu, O.I. Micic, A.J. Nozik, A. Shabaev, and A.L. Efros (2005). Highly efficient multiple exciton generation in colloidal PbSe and PbS quantum dots. *Nano Letters*, **5**(5), pp. 865-871.

Elzinga, D. (2008). *Urban BIPV in the New Residential Construction Industry.* IEA Photovoltaic Power Systems Program (PVPS), Task 10, Activity 3.1, Report IEA-PVPS-T10-03:2008, Natural Resources Canada for the International Energy Agency Photovoltaic Power Systems Program, Ottawa, Canada, 58 pp. Available at: www.iea-pvps-task10.org/IMG/pdf/IEA-PVPS_T10-03-2006__Urban_BIPV_in_the_New_Residential_Construction_Industry.pdf.

EPIA (2010). *Global Market Outlook for Photovoltaics until 2014.* European Photovoltaic Industry Association (EPIA), Brussels, Belgium, 28 pp.

Epstein, M., G. Olalde, S. Santen, A. Steinfeld, and C. Wieckert (2008). Towards the industrial solar carbothermic production of zinc. *Journal of Solar Energy Engineering*, **130**(1), 014505.

ESMAP (2004). *The Impact of Energy on Women's Lives in Rural India.* Energy Sector Management Assistance Programme (ESMAP) Paper No. ESM276, Joint UNDP/World Bank ESMAP, Washington, DC, 96 pp.

ESTTP (2006). *Solar Thermal Vision 2030. Vision of the Usage and Status of Solar Thermal Energy Technology in Europe and the Corresponding Research Topics to Make the Vision a Reality.* European Solar Thermal Technology Platform (ESTTP) Secretariat, Brussels, Belgium, 123 pp.

ESTTP (2008). *Solar Heating and Cooling for a Sustainable Energy Future in Europe – Strategic Research Agenda.* European Solar Thermal Technology Platform (ESTTP) Secretariat, Brussels, Belgium, 123 pp.

EU PV European Photovoltaic Technology Platform (2007). *A Strategic Research Agenda for Photovoltaic Solar Energy Technology.* European Communities, Sixth European Framework Programme for Research and Technological Development, Luxembourg, 76 pp.

EurObserv'ER (2009). *Photovoltaic Energy Barometer.* EurObserv'ER, Observatoire des Energies Renouvelables, Paris, France.

European Commission (2007). *A European Strategic Energy Technology Plan (SET-Plan) - Technology Map.* European Commission, Brussels, Belgium.

European Commission (2008). *Communication from the Commission to the European Parliament, the Council, the European Economic and Social Committee and the Committee of the Regions. Second Strategic Energy Review. An EU Energy Security and Solidarity Action Plan.* European Commission, Brussels, Belgium, 781 pp.

Everett, B. (1996). Solar Thermal Energy. In: *Renewable Energy: Power for a Sustainable Future.* G. Boyle (ed.), Oxford University Press, Oxford, UK, pp. 41-88.

Ezzati, M., A.D. Lopez, A. Rodgers, S. Vander Hoorn, and C.J.L. Murray (2002). Selected major risk factors and global and regional burden of disease. *Lancet*, **360**(9343), pp. 1347-1360.

Faiers, A., and C. Neame (2006). Consumer attitudes towards domestic solar power systems. *Energy Policy*, **34**, pp. 1797-1806.

Fawer, M., and B. Magyar (2010). *Solar Industry – Entering New Dimensions: Comparison of Technologies, Markets and Industries. Solar Energy 2010.* Sustainable Investment Division, Bank Sarasin, Basel, Switzerland, 56 pp. Available at: www.esocialsciences.com/data/articles/Document110122010400.3811609.pdf.

Felder, R., and A. Meier (2008). Well-to-wheel analysis of solar hydrogen production and utilization for passenger car transportation. *Journal of Solar Energy Engineering*, **130**(1), 011017.

Fernández-García, A., E. Zarza, L. Valenzuela, and M. Pérez (2010). Parabolic-trough solar collectors and their applications. *Renewable and Sustainable Energy Reviews*, **14**, pp. 1695-1721.

Fong, K.F., T.T. Chow, C.K. Lee, Z. Lin, and L.S. Chan (2010). Comparative study of different solar cooling systems for buildings in subtropical city. *Solar Energy*, **84**, pp. 227-244.

Forrest, S.R. (2005). The limits to organic photovoltaic cell efficiency. *Materials Research Society (MRS) Bulletin*, **30**(1), pp. 28-32.

Fthenakis, V., and H.C. Kim (2009). Land use and electricity generation: A life-cycle analysis. *Renew. Renewable and Sustainable Energy Reviews*, **13**(6-7), pp. 1465-1474.

Fthenakis, V.M., and H.C. Kim (2010). Photovoltaics: Life-cycle analyses. *Solar Energy*, doi:10.1016/j.solener.2009.10.002.

Fudholi, A., K. Sopian, M.H. Ruslan, M.A. Alghoul, and M.Y. Sulaiman (2010). Review of solar dryers for agricultural and marine products. *Renewable and Sustainable Energy Reviews*, **14**(1), pp. 1-30.

Furbo, S., E. Andersen, S. Knudsen, N.K. Vejen, and L.J. Shah (2005). Smart solar tanks for small solar domestic hot water systems. *Solar Energy*, **78**(2), pp. 269-279.

Gaiddon, B., and M. Jedliczka (2007). *Compared Assessment of Selected Environmental Indicators of Photovoltaic Electricity in OECD Cities.* IEA Photovoltaic Power Systems Program (PVPS), Task 10, Activity 4.4, Report IEA-PVPS-T10-01:2006, French Agency for Environment and Energy Management for the International Energy Agency, Hespu, Villeurbanne, France, 44 pp. Available at: www2.epia.org/documents/NL_0606_002.pdf.

Gil, A., M. Medrano, I. Martorell, A. Lazaro, P. Dolado, B. Zalba, and L.F. Cabeza (2010). State of the art on high temperature thermal energy storage for power generation. Part 1-Concepts, materials and modellization. *Renewable and Sustainable Energy Reviews*, **14**(1), pp. 31-55.

Goodrich, A. C., M. Woodhouse, and T. James (2011). Solar PV manufacturing cost model group: Installed solar PV system prices. *Presentation to SEGIS_ADEPT Power Electronic in Photovoltaic Systems Workshop*, Arlington, Virginia, 8 February 2001. NREL/PR-6A20-50955. Available at: arpa-e.energy.gov/LinkClick.aspx?fileticket=2WF9d-ukumA%3D&tabid=408.

Gordon, J.M. (ed.) (2001). *Solar Energy: The State of the Art. ISES Position Papers.* James & James London, UK, 706 pp.

Graf, D., N. Monnerie, M. Roeb, M. Schmitz, and C. Sattler (2008). Economic comparison of solar hydrogen generation by means of thermochemical cycles and electrolysis. *International Journal of Hydrogen Energy*, **33**(17), pp. 4511-4519.

Graetzel, M. (2001). Photoelectrochemical cells. *Nature*, **414**(6861), pp. 338-344.

Green, M.A. (2001). Third generation photovoltaics: Ultra-high conversion efficiency at low cost. *Progress in Photovoltaics: Research and Applications*, **9**(2), pp. 123-135.

Green, M.A. (2003). *Third Generation Photovoltaics: Advanced Solar Energy Conversion.* Springer, Berlin, Germany.

Green, M.A., K. Emery, Y. Hishikawa, and W. Warta (2009). Solar cell efficiency tables (version 34). *Progress in Photovoltaics: Research and Applications*, **17**(5), pp. 320-326.

Green, M.A., K. Emery, Y. Hishikawa, and W. Warta (2010a). Solar cell efficiency tables (version 35). *Progress in Photovoltaics: Research and Applications*, **18**(2), pp. 144-150.

Green, M.A., K. Emery, Y. Hishikawa, and W. Warta (2010b). Solar efficiency tables (version 36). *Progress in Photovoltaics: Research and Applications*, **18**(5), pp. 346-352.

Gumy, D., A.G. Rincon, R. Hajdu, and C. Pulgarin (2006). Solar photocatalysis for detoxification and disinfection of water: Different types of suspended and fixed TiO_2 catalysts study. *Solar Energy*, **80**, pp. 1376-1381.

Günes, S., and N.S. Sariciftci (2008). Hybrid solar cells. *Inorganica Chimica Acta*, **361**, pp. 581-588.

Hadorn, J.C. (ed.) (2005). *Thermal Energy Storage for Solar and Low Energy Buildings: State of the Art.* IEA Solar Heating and Cooling Task 32, International Energy Agency, Solar Heating and Cooling Programme, Servei de Publicacions Universidad Lleida, Lleida, Spain, 170 pp.

Hadorn, J.C. (2008). Thermal energy storage – Overview of technologies and status for solar heat. In: *EuroSun 2008, 1st International Conference on Solar Heating, Cooling and Buildings*, International Solar Energy Society, Lisbon, Portugal, 7-10 October 2008, pp.1-8.

Han, J., A.P.J. Mol, and Y. Lu (2010). Solar water heaters in China: A new day dawning. *Energy Policy*, **38**(1), pp. 383-391.

Harvey, L.D.D. (2006). *A Handbook on Low-Energy Buildings and District-Energy Systems: Fundamentals, Techniques and Examples.* Earthscan, Sterling, VA, USA, 701 pp.

Heiselberg, P. (ed.) (2002). *Principles of Hybrid Ventilation.* IEA Energy Conservation in Buildings and Community Systems Programme, Annex 35, Hybrid Ventilation and New and Retrofitted Office Buildings, IEA Energy Conservation in Buildings and Community Systems Programme, Birmingham, UK, 73 pp.

Henning, H.-M. (2004). *Solar-Assisted Air-Conditioning in Buildings: A Handbook for Planners.* Springer, New York, NY, USA, 136 pp.

Henning, H.-M. (2007). Solar assisted air conditioning of buildings – an overview. *Applied Thermal Engineering*, **27**, pp. 1734-1749.

Hernandez Gonzalvez, C. (1996). *Manual de Energía Solar Térmica.* Instituto para la Diversificación de la Energía (IDAE), Madrid, Spain, 123 pp.

Hoff, T.E., B.J. Pasquier, and J.M. Peterson (2010). Market transformation benefits of a PV incentive program. In: *SOLAR 2010 Conference Proceedings*, American Solar Energy Society, Phoenix, Arizona, 17-22 May 2010.

Hoffert, M.I., K. Caldeira, G. Benford, D.R. Criswell, C. Green, H. Herzog, A.K. Jain, H.S. Kheshgi, K.S. Lackner, J.S. Lewis, H.D. Lightfoot, W. Manheimer, J.C. Mankins, M.E. Mauel, L.J. Perkins, M.E. Schlesinger, T. Volk, and T.M.L. Wigley (2002). Advanced technology paths to global climate stability: Energy for a greenhouse planet. *Science*, **298**, pp. 981-987.

Hoffmann, W. (2009). The role of PV solar electricity to power the 21st century's global prime energy demand. *IOP Conference Series: Earth and Environmental Sciences*, **8**, 012007.

Hoffmann, W., S. Wieder, and T. Pellkofer (2009). Differentiated price experience curves as evaluation tool for judging the further development of crystalline silicon and thin film PV solar electricity products. In: *Proceedings of the 24th European Photovoltaic Solar Energy Conference*, Hamburg, Germany, 21-25 September 2009, pp. 4387-4394 (ISBN: 3-936338-25-6).

Hofman, Y., D. de Jager, E. Molenbroek, F. Schilig, and M. Voogt (2002). *The Potential of Solar Electricity to reduce CO_2 Emissions.* Ecofys, Utrecht, The Netherlands, 106 pp.

Hollands, K.G.T., J.L. Wright, and C.G. Granqvist (2001). Glazing and coatings. In: *Solar Energy: The State of the Art. ISES Position Papers.* J.M. Gordon (ed.), James & James, London, UK, pp. 24-107.

Holttinen, H. (2005). Hourly wind power variations in the Nordic countries. *Wind Energy*, **8**(2), pp. 173-195.

Hoogwijk, M. (2004). *On the Global and Regional Potential of Renewable Energy Sources.* Department of Science, Technology and Society, Utrecht University, Utrecht, the Netherlands.

Hoogwijk, M., and W. Graus (2008). *Global Potential of Renewable Energy Sources: A Literature Assessment.* Ecofys, Utrecht, The Netherlands, 45 pp.

IEA (2007). *Renewables for Heating and Cooling – Untapped Potential.* International Energy Agency, Paris, 205 pp.

IEA (2008). *World Energy Outlook 2008.* International Energy Agency, Paris, France, 578 pp.

IEA (2009a). *Global Renewable Energy Policies and Measures Database.* International Energy Agency, Paris, France. Available at: www.iea.org/textbase/pm/?mode=re.

IEA (2009b). *Towards Net Zero Energy Solar Buildings.* IEA Energy Conservation in Buildings and Community Systems (ECBCS) Programme, Solar Heating and Cooling Programme, SHC Task 40, ECBCS Annex 52, International Energy Agency, Paris, France. Available at: www.iea-shc.org/task40/.

IEA (2009c). *Trends in Photovoltaic Applications: Survey Report of Selected IEA Countries between 1992 and 2008.* IEA Photovoltaic Power Systems Program (PVPS), Task 1, Report IEA-PVPS T1-18:2009, International Energy Agency, Paris, France, 44 pp. Available at: www.iea-pvps.org/index.php?id=32.

IEA (2010a). *Energy Technology Perspectives 2010. Scenarios & Strategies to 2050.* International Energy Agency, Paris, France, 708 pp.

IEA (2010b). *Technology Roadmap, Concentrating Solar Power.* International Energy Agency, Paris, France, 48 pp.

IEA (2010c). *Technology Roadmap, Solar Photovoltaic Energy.* International Energy Agency, Paris, France, 48 pp.

IEA (2010d). *World Energy Outlook 2010.* International Energy Agency, Paris, France, 736 pp.

IEA (2010e). *Trends in Photovoltaic Applications: Survey Report of Selected IEA Countries between 1992 and 2009.* IEA Photovoltaic Power Systems Program (PVPS), Task 1, Report IEA-PVPS T1-10-2010, International Energy Agency, Paris, France, 44 pp. Available at: www.iea-pvps.org/index.php?id=32.

IEA ECES (2004). *Energy Conservation through Energy Storage (ECES) Implementing Agreement*, International Energy Agency, Paris, France, 24 pp. Available at: www.iea-eces.org/files/iaeces_2004.pdf.

IEA NZEB (2009). *IEA Joint Project: Towards Net Zero Energy Solar Buildings (NZEBs).* SHC Task 40, ECBCS Annex 52, Revised 25 February 2009, IEA Solar Heating and Cooling Programme. Available at: www.iea-shc.org/publications/downloads/task40-Net_Zero_Energy_Solar_Buildings.pdf.

Weiss, W., and F. Mauthner (2010). *Solar Heat Worldwide – Markets and Contribution to the Energy Supply 2008.* AEE - Institute for Sustainable Technologies, Gleisdorf, Austria for the International Energy Agency Solar Heating and Cooling Programme, 52 pp. Available at: www.iea-shc.org/publications/downloads/Solar_Heat_Worldwide-2010.pdf.

Ilex Energy Consulting Ltd., Electricity Research Centre, Electric Power and Energy Systems Research Group, and Manchester Centre for Electrical Energy (2004). *Operating Reserve Requirements as Wind Power Penetration Increases in the Irish Electricity System.* Report No. 04-RERDD-011-R-01, Sustainable Energy Ireland, Dublin, Ireland. Available at: www.seai.ie/uploadedfiles/InfoCentre/IlexWindReserrev2FSFinal.pdf.

Imre, L. (2007). Solar drying. In: *Handbook of Industrial Drying.* 3rd ed. A.S. Mujumdar (ed.), Taylor & Francis, Philadelphia, PA, USA, pp. 307-361.

Iqbal, M. (1984). *An Introduction to Solar Radiation.* Academic Press, New York, 390 pp.

ISCCP Data Products (2006). International Satellite Cloud Climatology Project (ISCCP). Available at: isccp.giss.nasa.gov/projects/flux.html.

Jäger-Waldau, A. (2005). *Photovoltaics Status Report 2005: Research, Solar Cell Production and Market Implementation of Photovoltaics.* Euro-Report EUR 21836 EN, European Commission, Joint Research Centre, Renewable Energies Unit, Luxembourg.

Jäger-Waldau, A. (2009). *Photovoltaics Status Report 2009: Research, Solar Cell Production and Market Implementation of Photovoltaics.* Euro-Report EUR 24027 EN. Office for Official Publications of the European Union, Luxembourg.

Jäger-Waldau, A. (2010a). *Photovoltaics Status Report 2010: Research, Solar Cell Production and Market Implementation of Photovoltaics.* Office for Official Publications of the European Union, Luxembourg.

Jäger-Waldau, A. (2010b). Status and perspectives of thin film photovoltaics. In: *Thin Film Solar Cells: Current Status and Future Trends.* A. Bosio and A. Romeo (eds.), Nova Publishers, New York, NY, USA, pp. 1-24.

Jensen, S.H., P.H. Larsen, and M. Mogensen (2007). Hydrogen and synthetic fuel production from renewable energy sources. *International Journal of Hydrogen Energy*, **32**(15), pp. 3253-3257.

Jewell, W., and R. Ramakumar (1987). The effect of moving clouds on electric utilities with dispersed photovoltaic generation. *IEEE Transactions on Energy Conversion*, **2**(4), pp. 570-576.

Jewell, W., and T. Unruh (1990). Limits on cloud-induced fluctuation in photovoltaic generation. *IEEE Transaction on Energy Conversion*, **5**(1), pp. 8-14.

Kaneko, H., T. Miura, A. Fuse, H. Ishihara, S. Taku, H. Fukuzumi, Y. Naganuma, and Y. Tamaura (2007). Rotary-type solar reactor for solar hydrogen production with two-step water splitting process. *Energy Fuels*, **21**(4), pp. 2287-2293.

Kawasaki, N., T. Oozeki, K. Otani, and K. Kurokawa (2006). An evaluation method of the fluctuation characteristics of photovoltaic systems by using frequency analysis. *Solar Energy Materials and Solar Cells*, **90**(18-19), pp. 3356-3363.

Kaya, N., J.C. Mankins, B. Erb, D. Vassaux, G. Pignolet, D. Kassing, and P. Collins (2001). Report of workshop on clean and inexhaustible space solar power at Unispace III Conference. *Acta Astronautica*, **49**(11), pp. 627-630.

Kern, E.J., and M. Russell (1988). Spatial and temporal irradiance variations over large array fields. In: *Conference Record of the 20th IEEE Photovoltaic Specialists Conference*, Las Vegas, NV, 26-30 September 2008, 2, pp. 1043-1050.

Khanna, R.K., R.S. Rathore, and C. Sharma (2008). Solar still an appropriate technology for potable water need of remote villages of desert state of India - Rajasthan. *Desalination*, **220**, pp. 645-653.

Kitamura, A. (1999). *Demonstration Test Results for Grid Interconnected Photovoltaic Power Systems.* Report IEA-PVPS T5-02:1999, International Energy Agency, Paris, France.

Kobayashi, H., and M. Takasaki (2006). Demonstration study of autonomous demand area power system. In: *Transmission and Distribution Conference and Exhibition 2005/2006*, Institute of Electrical and Electronics Engineers (IEEE) Power Engineering Society, Dallas, TX, 21-24 May 2006, pp. 548-555.

Komiyama, R., C. Marnay, M. Stadler, J. Lai, S. Borgeson, B. Coffey, and I. Lima Azevedo (2009). *Japan's Long-term Energy Demand and Supply Scenario to 2050 – Estimation for the Potential of Massive CO_2 Mitigation.* The Institute for Energy Economics, Tokyo, Japan.

Komoto, K., M. Ito, P. van der Vleuten, D. Faiman, and K. Kurokawa (eds.) (2009). *Energy from the Desert: Very Large Scale Photovoltaic Systems: Socio-economic, Financial, Technical and Environmental Aspects.* Earthscan Publishers, London, UK.

Koster, L.J.A., V.D. Mihailetchi, and P.W.M. Blom (2006). Ultimate efficiency of polymer/fullerene bulk heterojunction solar cells. *Applied Physics Letters*, **88**(9), 093511, doi:10.1063/1.2181635.

Krebs, F.C. (2005). Alternative PV: Large scale organic photovoltaics. *REfocus*, **6**(3), pp. 38-39.

Krewitt, W., K. Nienhaus, C. Kleßmann, C. Capone, E. Stricker, W. Graus, M. Hoogwijk, N. Supersberger, U. von Winterfeld, and S. Samadi (2009). *Role and Potential of Renewable Energy and Energy Efficiency for Global Energy Supply.* Climate Change 18/2009, ISSN 1862-4359, Federal Environment Agency, Dessau-Roßlau, Germany, 336 pp.

Krey, V., and L. Clarke (2011). Role of renewable energy in climate change mitigation: a synthesis of recent scenarios. *Climate Policy*, in press.

Kroposki, B., R. Margolis, G. Kuswa, J. Torres, W. Bower, T. Key, and D. Ton (2008). *Renewable Systems Interconnection.* National Renewable Energy Laboratory, Golden, CO, USA, 23 pp.

Kurokawa, K., and F. Aratani (2004). Perceived technical issues accompanying large PV development and Japanese "PV2030". In: *19th European Photovoltaic Solar Energy Conference and Exhibition*, Paris, France, 7-11 June 2004.

Kurokawa, K., K. Komoto, P. van der Vleuten, and D. Faiman (eds.) (2007). *Energy from the Desert: Practical Proposals for Very Large Scale Photovoltaic Systems.* Earthscan, London, UK.

Kushiya, K. (2009). Key near-term R&D issues for continuous improvement in CIS-based thin-film PV modules. *Solar Energy Materials & Solar Cells*, **93**, pp. 1037-1041.

Lahkar, P.J., and S.K. Samdarshi (2010). A review of the thermal performance parameters of box type solar cookers and identification of their correlations. *Renewable and Sustainable Energy Reviews*, **14**, pp. 1615-1621.

LBBW (2009). *Branchenanalyse Photovoltaik.* Landesbank Baden-Württemberg (LBBW), Stuttgart, Germany.

Le Treut, H., R. Somerville, U. Cubasch, Y. Ding, C. Mauritzen, A. Mokssit, T. Peterson, and M. Prather (2007). Historical overview of climate change science. In: *Climate Change 2007: The Physical Science Basis. Working Contribution of Working Group I to the Fourth Assessment Report of the Intergovernmental Panel on Climate Change*. S. Solomon, D. Qin, M. Manning, Z. Chen, M. Marquis, K.B. Averyt, M. Tignor, and H.L. Miller (eds.). Cambridge University Press, pp. 93-127.

Lee, S., and Z. Yamayee (1981). Load-following and spinning-reserve penalties for intermittent generation. *IEEE Transactions on Power Apparatus and Systems*, **100**(3), pp. 1203-1211.

Li, G., V. Shrotriya, J.S. Huang, Y. Yao, T. Moriarty, K. Emery, and Y. Yang (2005). High-efficiency solution processable polymer photovoltaic cells by self-organization of polymer blends. *Nature Materials*, **4**(11), pp. 864-868.

Lynch, M. (2002). Reducing environmental damage caused by the collection of cooking fuel by refugees. *Refuge*, **21**(1), pp. 18-27.

Mankins, J.C. (1997). A fresh look at space solar power: New architectures, concepts, and technologies. *Acta Astronautica*, **41**(4-10), pp. 347-359.

Mankins, J.C. (2002). A technical overview of the "Suntower" solar power satellite concept. *Acta Astronautica*, **50**(6), pp. 369-377.

Mankins, J.C. (2009). New directions for space solar power. *Acta Astronautica*, **65**, pp. 146-156.

Manuel, J. (2003). The quest for fire: Hazards of a daily struggle. *Environmental Health Perspectives*, **111**(1), pp. A28-A33.

Marcos, J., L. Marroyo, E. Lorenzo, D. Alvira, and E. Izco (2010). Power output fluctuations in large scale PV plants: One year observations with one second resolution and a derived analytic model. *Progress in Photovoltaics: Research and Applications*, doi:10.1002/pip.1016.

Marti, A., and A. Luque (eds.) (2004). *Next Generation Photovoltaics: High Efficiency through Full Spectrum Utilization*. Institute of Physics Publishing, Bristol, UK and Philadelphia, PA, USA, 332 pp.

Maycock, P.D. (1976-2003). *PV News*. PV Energy Systems 1982 (vol 1) through 2003 (vol 22). PV Energy Systems, Williamsburg, VA, USA.

Maycock, P.D. (2002). *The World Photovoltaic Market – Report* (January). PV Energy Systems, Williamsburg, VA, USA.

Medrano, M., A. Gil, I. Martorell, X. Potau, and L.F. Cabeza (2010). State of the art on high-temperature thermal energy storage for power generation. Part 2-Case studies. *Renewable and Sustainable Energy Reviews*, **14**(1), pp. 56-72.

Meeder, A., A. Neisser, U. Rühle, and N. Mayer (2007). Manufacturing the first MW of large-area CuInS$_2$-based solar modules – Recent experiences and progress. In: *Proceedings of the 22nd European Photovoltaic Solar Energy Conference*, Milan, Italy, 3-7 September 2007, pp. 2115.

Meehl, G.A., T.F. Stocker, W.D. Collins, P. Friedlingstein, A.T. Gaye, J.M. Gregory, A. Kitoh, R. Knutti, J.M. Murphy, A. Noda, S.C.B. Raper, I.G. Watterson, A.J. Weaver, and Z.-C. Zhao (2007). Global climate projections. In: *Climate Change 2007: The Physical Science Basis. Contribution of Working Group I to the Fourth Assessment Report of the Intergovernmental Panel on Climate Change*, 2007. S. Solomon, D. Qin, M. Manning, Z. Chen, M. Marquis, K.B. Averyt, M. Tignor, and H.L. Miller (eds.), Cambridge University Press, pp. 747-846.

Mehling, H., and L.F. Cabeza (2008). *Heat and Cold Storage with PCM : An Up to Date Introduction into Basics and Applications*. Springer, Berlin, Germany and London, UK, 308 pp.

Meier, J., S. Dubail, R. Platz, P. Torres, U. Kroll, J.A. Selvan, N. Pellaton Vaucher, C. Hof, D. Fischer, H. Keppner, R. Flückiger, A. Shah, S. Shklover, and K.-D. Ufert (1997). Towards high-efficiency thin-film silicon solar cells with the "micromorph" concept. *Solar Energy Materials and Solar Cells*, **49**(1-4), pp. 35-44.

Meleshko, V.P., V.M. Kattsov, B.A. Govorkova, P.V. Sporyshev, I.M. Skolnik, and B.E. Sneerov (2008). Climate of Russia in the 21st century. Part 3. Future climate change calculated with an ensemble of coupled atmosphere-ocean general circulation CMIP3 models. *Meteorology and Hydrology*, **33**(9), pp. 541-552. Original Russian text published in Meteorologiya i Gidrologiya, no. 9, 2008.

Miller, J.E., M.D. Allendorf, R.B. Diver, L.R. Evans, N.P. Siegel, and J.N. Stuecker (2008). Metal oxide composites and structures for ultra-high temperature solar thermochemical cycles. *Journal of Materials Science*, **43**(14), pp. 4714-4728.

Mills, A., and R. Wiser (2010). *Implications of Wide-Area Geographic Diversity for Short-Term Variability of Solar Power*. DE-AC02-05CH11231, LBNL-3884E, Ernest Orlando Lawrence Berkeley National Laboratory, Berkeley, CA, USA.

Mills, B.F., and J. Schleich (2009). Profits or preferences? Assessing the adoption of residential solar thermal technologies. *Energy Policy*, **37**(10), pp. 4145-4154.

Mills, E. (2005). The specter of fuel-based lighting. *Science*, **308**(5726), pp. 1263-1264.

Ministry of New and Renewable Energy (2009). *Jawaharlal Nehru National Solar Mission Towards Building SOLAR INDIA*. Ministry of New and Renewable Energy, New Delhi, India, 15 pp.

Mints, P. (2010). The PV industry's black swan. *Photovoltaics World*, 18 March 2010.

Mints, P. (2011). PV sector market forecast – Thin-film in the era of cheap crystalline PV. *Renewable Energy World Magazine*. 11 February 2011.

Möller, S., D. Kaucic, and C. Sattler (2006). Hydrogen production by solar reforming of natural gas: A comparison study of two possible process configurations. *Journal of Solar Energy Engineering*, **128**(1), pp. 16-23.

Munoz, J., L. Narvarte, and E. Lorenzo (2007). Experience with PV-diesel hybrid village power systems in southern Morocco. *Progress in Photovoltaics: Research and Applications*, **15**, pp. 529-539.

Murata, A., H. Yamaguchi, and K. Otani (2009). A method of estimating the output fluctuation of many photovoltaic power generation systems dispersed in a wide area. *Electrical Engineering in Japan*, **166**(4), pp. 9-19.

Nankhuni, F.J., and J.L. Findeis (2004). Natural resource-collection work and children's schooling in Malawi. *Agricultural Economics*, **31**(2-3), pp. 123-134.

Narayan, G.P., M.H. Sharqawy, E.K. Summers, J.H. Lienhard, S.M. Zubair, and M.A. Antar (2010). The potential of solar-driven humidification-dehumidification desalination for small-scale decentralized water production. *Renewable and Sustainable Energy Reviews*, **14**, pp. 1187-1201.

NAS (2004). *Laying the Foundation for Space Solar Power – An Assessment of NASA's Space Solar Power Investment Strategy*. National Academy of Sciences (NAS), Washington, DC, USA.

NRC (2010). *Hidden Costs of Energy: Unpriced Consequences of Energy Production and Use*. National Research Council (NRC), The National Academies Press, Washington, DC, USA, 506 pp.

Navigant Consulting Inc. (2006). *A Review of PV Inverter Technology Cost and Performance Projections*. NREL/SR-620-38771, National Renewable Energy Laboratory, Golden, CO, USA, 100 pp.

NEDO (2009). *The Roadmap PV2030+*. New Energy and Industrial Technology Organization (NEDO), Kawasaki, Japan.

NEEDS (2009). *New Energy Externalities Development for Sustainability (NEEDS). Final Report and Database*. New Energy Externalities Development for Sustainability, Rome, Italy.

Neij, L. (2008). Cost development of future technologies for power generation – A study based on experience curves and complementary bottom-up assessments. *Energy Policy*, **36**, pp. 2200-2211.

Nelson, J. (2003). Over the limit: Strategies for high efficiency. In: *The Physics of Solar Cells*. Imperial College Press, London, England, pp. 289-323.

Nicol, F., M. Wilson, and C. Chiancarella (2006). Using field measurements of desktop illuminance in European offices to investigate its dependence on outdoor conditions and its effect on occupant satisfaction, and the use of lights and blinds. *Energy and Buildings*, **38**(7), pp. 802-813.

Norton, B. (2001). Solar process heat: Distillation, drying, agricultural and industrial uses. In: *Solar Energy: The State of the Art. ISES Position Papers*. J.M. Gordon (ed.), James & James, London, UK, pp. 477-496.

NREL (2011). *The Open PV Project* (online database). National Renewable Energy Laboratory (NREL), Golden, CO, USA. Available at: openpv.nrel.gov.

O'Regan, B., and M. Graetzel (1991). A low-cost, high-efficiency solar-cell based on dye-sensitized colloidal TiO_2 films. *Nature*, **353**(6346), pp. 737-740.

Ogimoto, K., T. Oozeki, and Y. Ueda (2010). Long-range power demand and supply planning analysis including photovoltaic generation penetration. *The Institute of Electrical Engineers of Japan (IEEJ) Transactions on Power and Energy, Tokyo*, **130-B**(6), pp. 575-583.

Oozeki, T., T. Takashima, K. Otani, Y. Hishikawa, G. Koshimizu, Y. Uchida, and K. Ogimoto (2010). Statistical analysis of the smoothing effect for photovoltaic systems in a large area. *The Institute of Energy Economics, Japan (IEEJ) Transactions on Power and Energy, Tokyo*, **130-B**(5), pp. 491-500.

Otani, K., J. Minowa, and K. Kurokawa (1997). Study on areal solar irradiance for analyzing areally-totalized PV systems. *Solar Energy Materials and Solar Cells*, **47**(1-4), pp. 281-288.

Otani, K., A. Murata, K. Sakuta, J. Minowa, and K. Kurokawa (1998). Statistical smoothing of power delivered to utilities by distributed PV systems. In: *2nd World Conference and Exhibition on Photovoltaic Solar Energy Conversion. Proceedings of the International Conference*, Vienna, Austria, 6-10 July 1998.

Paksoy, H. (2007). *Thermal Energy Storage for Sustainable Energy Consumption: Fundamentals, Case Studies and Design*. Springer, London, UK and Berlin, Germany.

Parente, V., J. Goldemberg, and R. Zilles (2002). Comments on experience curves for PV modules. *Progress in Photovoltaics: Research and Applications*, **10**, pp. 571-574.

Park, H.G., and J.K. Holt (2010). Recent advances in nanoelectrode architecture for photochemical hydrogen production. *Energy & Environmental Science*, **3**(8), pp. 1028-1036.

Patrick, E. (2007). Sexual violence and fuelwood collection in Darfur. *Forced Migration Review*, **27**, pp. 40-41.

Perpinan, O., E. Lorenzo, M.A. Castro, and R. Eyras (2009). Energy payback time of grid connected PV systems: Comparison between tracking and fixed systems. *Progress in Photovoltaics: Research and Applications*, **17**(2), pp. 137-147.

Photovoltaic Geographic Information System (2008). *Solar Radiation and Photovoltaic Electricity Potential Country and Regional Maps for Europe (Africa)*. Institute for Energy, Renewable Energy Unit, European Commission, Joint Research Centre, Ispra, Italy.

PHPP (2004). *PassivHaus Planning Package (PHPP). Technical Information PHI-2004/1(E) - Specifications for Quality Approved Passive Houses*. PassiveHause Institute, Darmstadt, Germany.

Piatkowski, N., C. Wieckert, and A. Steinfeld (2009). Experimental investigation of a packed-bed solar reactor for the steam-gasification of carbonaceous feedstocks. *Fuel Processing Technology*, **90**(3), pp. 360-366.

Piwko, R.J., X. Bai, K. Clark, G.A. Jordan, and N.W. Miller (2007). *Intermittency Analysis Project: Appendix B: Impact of Intermittent Generation on Operation of California Power Grid*. California Energy Commission, PIER Research Development & Demonstration Program, Sacramento, CA, USA.

Piwko, R., K. Clark, L. Freeman, G. Jordan, and N. Miller (2010). *Western Wind and Solar Integration Study*. National Renewable Energy Laboratory, Golden, CO, USA.

POSHIP (2001). *Calor Solar Para Provesos Industriales: Proyecto POSHIP (Potential of Solar Heat for Industrial Processes)*. Instituto Para Lad Diversificacion y Ahorro de Energia (IDAE), Madrid, Spain.

Pregger, T., D. Graf, W. Krewitt, C. Sattler, M. Roeb, and S. Moeller (2009). Prospects of solar thermal hydrogen production processes. *International Journal of Hydrogen Energy*, **34**(10), pp. 4256-4267.

Probst, M.M., and C. Roecker (2007). Towards an improved architectural quality of building integrated solar thermal systems (BIST). *Solar Energy*, **81**(9), pp. 1104-1116.

Ralegaonkar, R.V., and R. Gupta (2010). Review of intelligent building construction: A passive solar architecture approach. *Renewable and Sustainable Energy Reviews*, **14**, pp. 2238-2242.

Ramachandran, J., N.M. Pearsall, and G.A. Putrus (2004). Reduction in solar radiation fluctuation by spatial smoothing effect. In: *19th European Photovoltaic Solar Energy Conference. Proceedings of the International Conference*, WIP-Renewable Energies, Paris, France, 7-11 July 2004, pp. 2900-2903.

REN21 (2009). *Renewables Global Status Report. 2009 Update*. Renewable Energy Policy Network for the 21st Century Secretariat, Paris, France, 32 pp.

REN21 (2010). *Renewables 2010 Global Status Report*. Renewable Energy Policy Network for the 21st Century Secretariat, Paris, France, 80 pp.

Richter, C., S. Teske, and R. Short (2009). *Concentrating Solar Power: Global Outlook 2009 – Why Renewable Energy is Hot*. Greenpeace International, SolarPACES (Solar Power and Chemical Energy Storage), and ESTELA (European Solar Thermal Electricity Association), 88 pp. Available at: www.greenpeace.org/raw/content/international/press/reports/concentrating-solar-power-2009.pdf.

Rickerson, W., T. Halfpenny, and S. Cohan (2009). The emergence of renewable heating and cooling policy in the United States. *Policy and Society*, **27**(4), pp. 365-377.

Rodat, S., S. Abanades, and G. Flamant (2009). High-temperature solar methane dissociation in a multitubular cavity-type reactor in the temperature range 1823-2073 K. *Energy & Fuels*, **23**, pp. 2666-2674.

Roeb, M., C. Sattler, R. Kluser, N. Monnerie, L. de Oliveira, A.G. Konstandopoulos, C. Agrafiotis, V.T. Zaspalis, L. Nalbandian, A. Steele, and P. Stobbe (2006). Solar hydrogen production by a two-step cycle based on mixed iron oxides. *Journal of Solar Energy Engineering*, **128**(2), pp. 125-133.

Roeb, M., M. Neises, J.P. Sack, P. Rietbrock, N. Monnerie, J. Dersch, M. Schmitz, and C. Sattler (2009). Operational strategy of a two-step thermochemical process for solar hydrogen production. *International Journal of Hydrogen Energy*, **34**(10), pp. 4537-4545.

Rogner, H.-H., F. Barthel, M. Cabrera, A. Faaij, M. Giroux, D. Hall, V. Kagramanian, S. Kononov, T. Lefevre, R. Moreira, R. Nötstaller, P. Odell, and M. Taylor (2000). Energy resources. In: *World Energy Assessment. Energy and the Challenge of Sustainability.* United Nations Development Programme, United Nations Department of Economic and Social Affairs, World Energy Council, New York, USA, 508 pp.

Ropp, M., J. Newmiller, C. Whitaker, and B. Norris (2008). Review of potential problems and utility concerns arising from high penetration levels of photovoltaics in distribution systems. In: *Proceedings of the 33rd Institute of Electrical and Electronics Engineers (IEEE) Photovoltaic Specialists Conference*, IEEE, San Diego, CA, 11-16 May, 2008, pp. 518-523.

Ruhl, V., F. Luetter, C. Schmidt, J. Wackerbauer, U. Triebswetter (2008). *Standortgutachten Photovoltaik in Deutschland.* EuPD Research and IFO Institut fur Wirtschaftforschung, Universitat Munchen, Bonn, Germany and München, Germany, 28 pp.

Santamouris, M., and D. Asimakopoulos (eds.) (1996). *Passive Cooling of Buildings.* James & James, London, 472 pp.

Sargent and Lundy LLC Consulting Group (2003). *Assessment of Parabolic Trough and Power Tower Solar Technology Cost and Performance Forecasts.* National Renewable Energy Laboratory, Golden, CO, USA, 344 pp.

Schaeffer, G.J., A.J. Seebregts, L.W.M. Beurskens, H.H.C. Moor, E.A. Alsema, W. Sark, M. Durstewicz, M. Perrin, P. Boulanger, H. Laukamp, and C. Zuccaro (2004). *Learning from the Sun: Analysis of the Use of Experience Curves for Energy Policy Purposes – The Case of Photovoltaic Power. Final Report of the Photex Project.* DEGO: ECN-C-04-035, Energy Research Centre of the Netherlands, Petten, The Netherlands.

Schaller, R., and V. Klimov (2004). High efficiency carrier multiplication in PbSe nanocrystals: Implications for solar energy conversion. *Physical Review Letters*, **92**(18), 186601.

Scheffler, J. (2002). *Bestimmung der maximal zulässigen Netzanschlussleistung photovoltaischer Energiewandlungsanlagen in Wohnsiedlungsgebieten.* Fakultät Elektrotechnik und Informatik, Technische Universität Chemnitz, Chemnitz, Germany.

Schossig, P., H.-M. Henning, S. Gschwander, and T. Haussmann (2005). Micro-encapsulated phase-change materials integrated into construction materials. *Solar Energy Materials & Solar Cells*, **89**, pp. 297-306.

Schunk, L.O., P. Haeberling, S. Wepf, D. Wuillemin, A. Meier, and A. Steinfeld (2008). A receiver-reactor for the solar thermal dissociation of zinc oxide. *Journal of Solar Energy Engineering*, **130**(2), 021009.

Schunk, L.O., W. Lipinski, and A. Steinfeld (2009). Heat transfer model of a solar receiver-reactor for the thermal dissociation of ZnO - Experimental validation at 10 kW and scale-up to 1 MW. *Chemical Engineering Journal*, **150**(2-3), pp. 502-508.

SEIA (2010a). *US Solar Industry – Year in Review 2009.* US Solar Energy Industries Association (SEIA), Washington, DC, USA.

SEIA (2010b). *US Solar Market Insight, 2nd Quarter 2010 Executive Summary.* US Solar Energy Industries Association (SEIA), Washington, DC, USA.

SEMI (2009a). *Polysilicon shipments reach 43,901 Mt in 2008*. Press Release, 10 March 2009, Semiconductor Equipment and Materials International (SEMI), San Jose, CA, USA. Available at: www.semi.org/en/Press/CTR_028736.

SEMI (2009b). *China's Solar Future (SEMI China White Paper).* Semiconductor Equipment and Materials International (SEMI), San Jose, CA, USA.

SEMI (2009c). *The Solar PV Landscape in India – An Industry Perspective (SEMI India White Paper).* Semiconductor Equipment and Materials International (SEMI), San Jose, CA, USA.

Service, R.F. (2009). Sunlight in your tank. *Science*, **326**(5959), pp. 1471-1475.

Sharma, A. (2011). A comprehensive study of solar power in India and World. *Renewable and Sustainable Energy Reviews*, **15**, pp. 1767–1776.

Shockley, W., and H.J. Queisser (1961). Detailed balance limit of efficiency of p-n junction solar cells. *Journal of Applied Physics*, **32**(3), pp. 510-519.

Shu, H., L. Duanmu, C. Zhang, and Y. Zhu (2010). Study on the decision-making of district cooling and heating systems by means of value engineering. *Renewable Energy*, **35**, pp. 1929-1939.

Sims, R.E.H., R.N. Schock, A. Adegbululgbe, J. Fenhann, I. Konstantinaviciute, W. Moomaw, H.B. Nimir, B. Schlamadinger, J. Torres-Martínez, C. Turner, Y. Uchiyama, S.J.V. Vuori, N. Wamukonya, and X. Zhang (2007). Energy supply. In: *Climate Change 2007: Mitigation of Climate Change. Contribution of Working Group III to the Fourth Assessment Report of the Intergovernmental Panel on Climate Change*. B. Metz, O.R. Davidson, P.R. Bosch, R. Dave and L.A. Meyer (eds.), Cambridge University Press, pp. 251-322.

Sinha, P., C.J. Kriegner, W.A. Schew, S.W. Kaczmar, M. Traister, and D.J. Wilson (2008). Regulatory policy governing cadmium-telluride photovoltaics: A case study contrasting life-cycle management with the precautionary principle. *Energy Policy*, **36**, pp. 381-387.

SolarPACES (2008). *SolarPACES Annual Report 2007.* International Energy Agency, Paris, France, 204 pp.

SolarPACES (2009a). *SolarPACES Annual Report 2008. Task I: Solar Thermal Electric Systems.* International Energy Agency, Paris, France, pp. 3.1-3.14.

SolarPACES (2009b). *SolarPACES Annual Report 2008. Task VI: Solar Energy & Water Processes and Applications.* International Energy Agency, Paris, France, pp. 8.1-12.

Solomon, S., D. Qin, M. Manning, R.B. Alley, T. Berntsen, N.L. Bindoff, Z. Chen, A. Chidthaisong, J.M. Gregory, G.C. Hegerl, M. Heimann, B. Hewitson, B.J. Hoskins, F. Joos, J. Jouzel, V. Kattsov, U. Lohmann, T. Matsuno, M. Molina, N. Nicholls, J. Overpeck, G. Raga, V. Ramaswamy, J. Ren, M. Rusticucci, R. Somerville, T.F. Stocker, P. Whetton, R.A. Wood, and D. Wratt (2007). Technical summary. In: *Climate Change 2007: The Physical Science Basis. Contribution of Working Group I to the Fourth Assessment Report of the Intergovernmental Panel on Climate Change, 2007.* S. Solomon, D. Qin, M. Manning, Z. Chen, M. Marquis, K.B. Averyt, M. Tignor, and H.L. Miller (eds.), Cambridge University Press, pp.19-91.

Spath, P.L., and W.A. Amos (2003). Using a concentrating solar reactor to produce hydrogen and carbon black via thermal decomposition of natural gas: Feasibility and economics. *Journal of Solar Energy Engineering*, **125**(2), pp. 159-164.

Specht, M., F. Baumgart, B. Feigl, V. Frick, B. Stuermer, U. Zuberbuehler, M. Sterner, and G. Waldstein (2010). Speicherung von Bioenergie und erneuerbarem Strom im Erdgasnetz (Storage of bioenergy and renewable power in the natural gas network). In: *FVEE Annual Meeting 2009. Forschen für globale Märkte erneuerbarer Energien*, Berlin, Germany.

Sreeraj, E.S., K. Chatterjee, and S. Bandyopadhyay (2010). Design of isolated renewable hybrid power systems. *Solar Energy*, **84**(7), pp. 1124-1136.

Staebler, D.L., and C.R. Wronski (1977). Reversible conductivity changes in discharge-produced amorphous Si. *Applied Physics Letters*, **31**(4), pp. 292-294.

Stein, W., B. Antonioli, S. Bönisch, M. Hausotte, A. Hofmann, R. Joziak, H. Krautz, P. Minton, B. Mone, J.-C. Müller, J. Pantförder, A Schuster, G. Springer, J. Springer, and S. Klein (2009). Status of thin-film Si high-efficiency tandem junction module fabrication on ultra-large substrates of 2.20 x 2060 m^2 at Sunfilm. In: *Proceedings of the 24th EU PV Solar Energy Conference, 6th European PV Industry Forum*, Hamburg, Germany, 23 September 2009.

Steinfeld, A. (2005). Solar thermochemical production of hydrogen - a review. *Solar Energy*, **78**(5), pp. 603-615.

Steinfeld, A., and A. Meier (2004). Solar Fuels and Materials. In: *Encyclopedia of Energy*. Vol. 5. Elsevier, Amsterdam, The Netherlands, pp. 623-637.

Sterner, M. (2009). *Bioenergy and Renewable Power Methane in Integrated 100% Renewable Energy Systems. Limiting Global Warming by Transforming Energy Systems*. Dissertation, Kassel University, Kassel, Germany.

Sun, S.-S., and N.S. Sariciftci (eds.) (2005). *Organic Photovoltaics: Mechanisms, Materials, and Devices*. CRC Press, Taylor & Francis, Boca Raton, FL, USA.

Taguchi, M., Y. Tsunomura, H. Inoue, S. Taira, T. Nakashima, T. Baba, H. Sakata, and E. Maruyama (2009). High-efficiency HIT solar cell on thin (< 100 μm) silicon wafer. In: *Proceedings of the 24th European Photovoltaic Solar Energy Conference*, Hamburg, Germany, 21-25 September 2009, pp. 1690.

Teske, S., T. Pregger, S. Simon, T. Naegler, W. Graus, and C. Lins (2010). Energy [R]evolution 2010—a sustainable world energy outlook. *Energy Efficiency*, doi:10.1007/s12053-010-9098-y.

The European Parliament and the Council of the European Union (2010). *Directive 2010/31/EU of the European Parliament and of the Council of 19 May 2010 on the Energy Performance of Buildings*. 2010/31/EU. Official Journal of the European Union, pp. 23.

Torres, J.M.M., N.G. Löpez, and C. Márquez (2010). *The Global Concentrator Solar Power Industry Report 2010-2011*. First Conferences Ltd., London, UK.

Trieb, F. (2005). *Concentrating Solar Power for the Mediterranean Region, Final Report*, German Aerospace Centre (DLR), Stuttgart, 285 pp.

Trieb, F., M. O'Sullivan, T. Pregger, C. Schillings, and W. Krewitt (2009a). *Characterisation of Solar Electricity Import Corridors from MENA to Europe - Potential, Infrastructure and Cost*. German Aerospace Centre (DLR), Stuttgart, Germany, 172 pp.

Trieb, F., C. Schillings, M. O'Sullivan, T. Pregger, and C. Hoyer-Klick (2009b). Global potential of concentrating solar power. In: *SolarPACES Conference*, Berlin, Germany, 15-18 September 2009.

Tripanagnostopoulos, Y. (2007). Aspects and improvements of hybrid photovoltaic/thermal solar energy systems. *Solar Energy*, **81**(9), pp. 1117-1131.

Tsilingiridis, G., G. Martinopoulos, and N. Kyriakis (2004). Life cycle environmental impact of a thermosyphonic domestic solar hot water system in comparison with electrical and gas water heating. *Renewable Energy*, **29**, pp. 1277-1288.

Tsoutsos, T., N. Fratezeskaki, and V. Gekas (2005). Environmental impacts from the solar energy technologies. *Energy Policy*, **33**, pp. 289-296.

Turner, J., G. Sverdrup, M.K. Mann, P.-C. Maness, B. Kroposki, M. Ghirardi, R.J. Evans, and D. Blake (2008). Renewable hydrogen production. *International Journal of Energy Research*, **32**, pp. 379-407.

Twidell, J. and A.D. Weir (2006). *Renewable Energy Resources*. Taylor & Francis, Oxon, UK.

Tzempelikos, A., A.K. Athienitis, and A. Nazos (2010). Integrated design of perimeter zones with glass facades. *ASHRAE Transactions*, **116**(1), pp. 461-478.

UBS (2009). *UBS Wealth Management Research (5 March 2009). Solar Energy.* Union Bank of Switzerland, 37 pp. Available at: http://www.cleantechsandiego.org/reports/UBS_Solar_Energy_Report_3-5-09.pdf.

UNEP (2007). *Buildings and Climate Change – Status, Challenges and Opportunities.* United Nations Environment Programme, Nairobi, Kenya.

Urbschat, C., F. Barban, B. Baumgartner, M. Beste, M. Herr, A. Schmid-Kieninger, F. Rossani, G. Stry-Hipp, and M. Welke (2002). *Sunrise 2002 – The Solar Thermal and Photovoltaic Markets in Europe. Market Survey.* eclareon GmbH, Berlin, Germany.

US Photovoltaic Industry Roadmap Steering Committee (2001). *Solar-Electric Power: The U.S. Photovoltaic Industry Roadmap.* Sandia National Laboratories, Albuquerque, NM, USA, 36 pp.

US DOE (2008). *Solar Energy Technologies Programme, Multi Year Program Plan 2008-2012.* U.S. Department of Energy (US DOE), Washington, DC, USA.

US DOE (2009). *Concentrating Solar Power Commercial Application Study: Reducing Water Consumption of Concentrating Solar Power Electricity Generation. Report to Congress.* U.S. Department of Energy (US DOE), Washington, DC, USA, 35 pp.

US DOE (2011). *Solar Vision Study (draft).* U.S. Department of Energy (US DOE), Washington DC, USA, final publication currently postponed.

Viebahn, P., Y. Lechon, and F. Trieb (2010). The potential role of concentrated solar power (CSP) in Africa and Europe: A dynamic assessment of technology development, cost development and life cycle inventories until 2050. *Energy Policy*, doi:10.1016/j.enpol.210.09.026.

Voss, K., S. Herkel, J. Pfafferott, G. Löhnert, and A. Wagner (2007). Energy efficient office buildings with passive cooling – Results and experiences from a research and demonstration programme. *Solar Energy*, **81**, pp. 424-434.

Wadia, C., A.P. Alivisatos, and D.M. Kammen (2009). Materials availability expands the opportunity for large-scale photovoltaic deployment. *Environmental Science & Technology*, **43**(6), pp. 2072-2077.

Wagner, S., J.L. Shay, P. Migliorato, and H.M. Kasper (1974). CuInSe$_2$/CdS heterojunction photovoltaic detectors. *Applied Physics Letters*, **25**, pp. 434.

WEC (1994). *New Energy Resources.* World Energy Council (WEC), London, UK.

Wei, M., S. Patadia, and D. Kammen (2010). Putting renewables to work: How many jobs can the clean energy industry generate in the US? *Energy Policy*, **38**, pp. 919-931.

Weiss, W. (2003). *Solar Heating Systems for Houses. A Design Handbook for Solar Combisystems.* Earthscan, London, UK.

Weiss, W., and F. Mauthner (2010). *Solar Heat Worldwide – Markets and Contribution to the Energy Supply 2008.* AEE - Institute for Sustainable Technologies, Gleisdorf, Austria for the International Energy Agency Solar Heating and Cooling Programme, 52 pp. Available at: www.iea-shc.org/publications/downloads/Solar_Heat_Worldwide-2010.pdf.

Wentzel, M., and A. Pouris (2007). The development impact of solar cookers: A review of solar cooking impact research in South Africa. *Energy Policy*, **35**(3), pp. 1909-1919.

Werner, S. (2006). Ecoheatcool. WP1: The European Heat Market 2003, WP4: District Heating Possibilities, WP2: The European Cooling Market. In: *Nordic Energy Perspectives Conference*, Helsinki, Finland, 24 January 2006. Available at: www.nordicenergyperspectives.org/doc24jan06.asp.

Wiemken, E., H.G. Beyer, W. Heydenreich, and K. Kiefer (2001). Power characteristics of PV ensembles: Experiences from the combined power production of 100 grid connected PV systems distributed over the area of Germany. *Solar Energy*, **70**(6), pp. 513-518.

Wiser, R., G. Barbose, and C. Peterman (2009). *Tracking the Sun: The Installed Cost of Photovoltaics in the U.S. from 1998-2007.* Lawrence Berkeley National Laboratory, Berkeley, CA, USA, 42 pp.

Wiser, R., G. Barbose, and E. Holt (2010). *Supporting Solar Power in Renewable Portfolio Standards: Experience from the United States.* LBNL- 3984E, Lawrence Berkeley National Laboratory, Berkeley, CA, USA, 38 pp.

WMO (2008). *Guide to Meteorological Instruments and Methods of Observation.* WMO-No. 8, World Meteorological Organization (WMO), Geneva, Switzerland, 681 pp.

World Bank Global Environment Facility Program (2006). *Assessment of the World Bank/GEF Strategy for the Market Development of Concentrating Solar Thermal Power.* The International Bank for Reconstruction and Development, The World Bank, Washington, DC, USA, 149 pp.

Woyte, A., R. Belmans, and J. Nijs (2001). Power flow fluctuations in distribution grids with high PV penetration. In: *Proceedings of the 17th EC PV Solar Energy Conference*, Munich, Germany, 22-26 October 2001, pp. 2414-2417.

Woyte, A., R. Belmans, and J. Nijs (2007). Fluctuations in instantaneous clearness index: Analysis and statistics. *Solar Energy*, **81**(2), pp. 195-206.

Yakimov, A., and S.R. Forrest (2002). High photovoltaics multiple-heterojunction organic solar cells incorporating interfacial metallic nanoclusters. *Applied Physics Letters*, **80**(9), pp. 1667-1669.

Yamamoto, K., A. Nakashima, T. Suzuki, M. Yoshimi, H. Nishio, and M. Izumina (1994). Thin-film polycrystalline Si solar cell on glass substrate fabricated by a novel low temperature process. *Japanese Journal of Applied Physics*, **33**, pp. L1751-L1754.

Yang, C.J. (2010). Reconsidering solar grid parity. *Energy Policy*, **38**, pp. 3270-3273.

Yang, J., and S. Guha (1992). Double-junction amorphous silicon-based solar cells with 11-percent stable efficiency. *Applied Physics Letters*, **61**(24), pp. 2917-2919.

Young, W.R. (1996). *History of Applying Photovoltaics to Disaster Relief.* Florida Solar Energy Center, Cocoa, FL, USA, 16 pp.

Z'Graggen, A., and A. Steinfeld (2008). Hydrogen production by steam-gasification of carbonaceous materials using concentrated solar energy – V. Reactor modeling, optimization, and scale-up. *International Journal of Hydrogen Energy*, **33**(20), pp. 5484-5492.

Zayed, J., and S. Philippe (2009). Acute oral and inhalation toxicities in rats with cadmium telluride. *International Journal of Toxicology*, **28**(4), pp. 259-265.

Zervos, A., C. Lins, and J. Muth (2010). *RE-thinking 2050 – A 100% Renewable Energy Vision for the European Union.* European Renewable Energy Council, Brussels, Belgium.

Zhang, X., W. Ruoshui, H. Molin, and E. Martinot (2010). A study of the role played by renewable energies in China's sustainable energy supply. *Energy*, **35**(11), pp.4392-4399, doi:10.1016/j.energy.2009.05.030.

Zoellner, J., P. Schweizer-Ries, and C. Wemheuer (2008). Public acceptance of renewable energies: Results from case studies in Germany. *Energy Policy*, **26**, pp. 4136-4141.

4. Geothermal Energy

Coordinating Lead Authors:
Barry Goldstein (Australia) and Gerardo Hiriart (Mexico)

Lead Authors:
Ruggero Bertani (Italy), Christopher Bromley (New Zealand),
Luis Gutiérrez-Negrín (Mexico), Ernst Huenges (Germany), Hirofumi Muraoka (Japan),
Arni Ragnarsson (Iceland), Jefferson Tester (USA), Vladimir Zui (Republic of Belarus)

Contributing Authors:
David Blackwell (USA), Trevor Demayo (USA/Canada), Garvin Heath (USA),
Arthur Lee (USA), John W. Lund (USA), Mike Mongillo (New Zealand),
David Newell (Indonesia/USA), Subir Sanyal (USA), Kenneth H. Williamson (USA),
Doone Wyborne (Australia)

Review Editors:
Meseret Teklemariam Zemedkun (Ethiopia) and David Wratt (New Zealand)

This chapter should be cited as:
Goldstein, B., G. Hiriart, R. Bertani, C. Bromley, L. Gutiérrez-Negrín, E. Huenges, H. Muraoka, A. Ragnarsson, J. Tester, V. Zui, 2011: Geothermal Energy. In IPCC Special Report on Renewable Energy Sources and Climate Change Mitigation [O. Edenhofer, R. Pichs-Madruga, Y. Sokona, K. Seyboth, P. Matschoss, S. Kadner, T. Zwickel, P. Eickemeier, G. Hansen, S. Schlömer, C. von Stechow (eds)], Cambridge University Press, Cambridge, United Kingdom and New York, NY, USA.

Table of Contents

Executive Summary 404

4.1 Introduction 406

4.2 Resource Potential 408

4.2.1 Global technical potential 408

4.2.2 Regional technical potential 410

4.2.3 Possible impact of climate change on resource potential 410

4.3 Technology and applications 410

4.3.1 Exploration and drilling 411

4.3.2 Reservoir engineering 411

4.3.3 Power plants 412

4.3.4 Enhanced Geothermal Systems (EGS) 412

4.3.5 Direct use 412

4.4 Global and regional status of market and industry development 414

4.4.1 Status of geothermal electricity from conventional geothermal resources 415

4.4.2 Status of EGS 416

4.4.3 Status of direct uses of geothermal resources 416

4.4.4 Impact of policies 417

4.5 Environmental and social impacts 418

4.5.1 Direct greenhouse gas emissions 418

4.5.2 Lifecycle assessment 418

4.5.3	Local environmental impacts	419
4.5.3.1	Other gas and liquid emissions during operation	419
4.5.3.2	Potential hazards of seismicity and other phenomena	420
4.5.3.3	Land use	420
4.5.4	Local social impacts	420
4.6	**Prospects for technology improvement, innovation and integration**	**421**
4.6.1	Improvements in exploration, drilling and assessment technologies	421
4.6.2	Efficient production of geothermal power, heat and/or cooling	422
4.6.3	Technological and process challenges in enhanced geothermal systems	422
4.6.4	Technology of submarine geothermal generation	423
4.7	**Cost trends**	**423**
4.7.1	Investment costs of geothermal-electric projects and factors that affect them	424
4.7.2	Geothermal-electric operation and maintenance costs	425
4.7.3	Geothermal-electric performance parameters	425
4.7.4	Levelized costs of geothermal electricity	425
4.7.5	Prospects for future cost trends	426
4.7.6	Costs of direct uses and geothermal heat pumps	427
4.8	**Potential deployment**	**428**
4.8.1	Near-term forecasts	428
4.8.2	Long-term deployment in the context of carbon mitigation	429
4.8.3	Conclusions regarding deployment	432
References		**433**

Executive Summary

Geothermal energy has the potential to provide long-term, secure base-load energy and greenhouse gas (GHG) emissions reductions. Accessible geothermal energy from the Earth's interior supplies heat for direct use and to generate electric energy. Climate change is not expected to have any major impacts on the effectiveness of geothermal energy utilization, but the widespread deployment of geothermal energy could play a meaningful role in mitigating climate change. In electricity applications, the commercialization and use of engineered (or enhanced) geothermal systems (EGS) may play a central role in establishing the size of the contribution of geothermal energy to long-term GHG emissions reductions.

The natural replenishment of heat from earth processes and modern reservoir management techniques enable the sustainable use of geothermal energy as a low-emission, renewable resource. With appropriate resource management, the tapped heat from an active reservoir is continuously restored by natural heat production, conduction and convection from surrounding hotter regions, and the extracted geothermal fluids are replenished by natural recharge and by injection of the depleted (cooled) fluids.

Global geothermal technical potential is comparable to global primary energy supply in 2008. For electricity generation, the technical potential of geothermal energy is estimated to be between 118 EJ/yr (to 3 km depth) and 1,109 EJ/yr (to 10 km depth). For direct thermal uses, the technical potential is estimated to range from 10 to 312 EJ/yr. The heat extracted to achieve these technical potentials can be fully or partially replenished over the long term by the continental terrestrial heat flow of 315 EJ/yr at an average flux of 65 mW/m^2. Thus, technical potential is not likely to be a barrier to geothermal deployment (electricity and direct uses) on a global basis. Whether or not the geothermal technical potential will be a limiting factor on a regional basis depends on the availability of EGS technology.

There are different geothermal technologies with distinct levels of maturity. Geothermal energy is currently extracted using wells or other means that produce hot fluids from: a) hydrothermal reservoirs with naturally high permeability; and b) EGS-type reservoirs with artificial fluid pathways. The technology for electricity generation from hydrothermal reservoirs is mature and reliable, and has been operating for more than 100 years. Technologies for direct heating using geothermal heat pumps (GHP) for district heating and for other applications are also mature. Technologies for EGS are in the demonstration stage. Direct use provides heating and cooling for buildings including district heating, fish ponds, greenhouses, bathing, wellness and swimming pools, water purification/desalination and industrial and process heat for agricultural products and mineral drying.

Geothermal resources have been commercially used for more than a century. Geothermal energy is currently used for base load electric generation in 24 countries, with an estimated 67.2 TWh/yr (0.24 EJ/yr) of supply provided in 2008 at a global average capacity factor of 74.5%; newer geothermal installations often achieve capacity factors above 90%. Geothermal energy serves more than 10% of the electricity demand in 6 countries and is used directly for heating and cooling in 78 countries, generating 121.7 TWh/yr (0.44 EJ/yr) of thermal energy in 2008, with GHP applications having the widest market penetration. Another source estimates global geothermal energy supply at 0.41 EJ/yr in 2008.

Environmental and social impacts from geothermal use are site and technology specific and largely manageable. Overall, geothermal technologies are environmentally advantageous because there is no combustion process emitting carbon dioxide (CO_2), with the only direct emissions coming from the underground fluids in the reservoir. Historically, direct CO_2 emissions have been high in some instances with the full range spanning from close to 0 to 740 g CO_2/kWh$_e$ depending on technology design and composition of the geothermal fluid in the underground reservoir. Direct CO_2 emissions for direct use applications are negligible and EGS power plants are likely to be designed with zero direct emissions. Life cycle assessment (LCA) studies estimate that full lifecycle CO_2-equivalent emissions for geothermal energy technologies are less than 50 g CO_2eq/kWh$_e$ for flash steam geothermal power plants, less than 80 g CO_2eq/kWh$_e$ for projected EGS power plants, and between 14 and 202 g CO_2eq/kWh$_{th}$ for district heating systems and GHP. Local hazards arising from natural phenomena, such as micro-earthquakes, may be influenced by the operation of geothermal fields. Induced seismic events have not been large enough to lead to human injury or relevant property damage, but proper management of this issue will be an important step to facilitating significant expansion of future EGS projects.

Several prospects exist for technology improvement and innovation in geothermal systems. Technical advancements can reduce the cost of producing geothermal energy and lead to higher energy recovery, longer field and plant lifetimes, and better reliability. In exploration, research and development (R&D) is required for hidden geothermal systems (i.e., with no surface manifestations such as hot springs and fumaroles) and for EGS prospects. Special research in drilling and well construction technology is needed to reduce the cost and increase the useful life of geothermal production facilities. EGS require innovative methods to attain sustained, commercial production rates while reducing the risk of seismic hazard. Integration of new power plants into existing power systems does not present a major challenge, but in some cases can require extending the transmission network.

Geothermal-electric projects have relatively high upfront investment costs but often have relatively low levelized costs of electricity (LCOE). Investment costs typically vary between USD_{2005} 1,800 and 5,200 per kW, but geothermal plants have low recurring 'fuel costs'. The LCOE of power plants using hydrothermal resources are often competitive in today's electricity markets, with a typical range from US $cents_{2005}$ 4.9 to 9.2 per kWh considering only the range in investment costs provided above and medium values for other input parameters; the range in LCOE across a broader array of input parameters is US $cents_{2005}$ 3.1 to 17 per kWh. These costs are expected to decrease by about 7% by 2020. There are no actual LCOE data for EGS power plants, as EGS plants remain in the demonstration phase, but estimates of EGS costs are higher than those for hydrothermal reservoirs. The cost of geothermal energy from EGS plants is also expected to decrease by 2020 and beyond, assuming improvements in drilling technologies and success in developing well-stimulation technology.

Current levelized costs of heat (LCOH) from direct uses of geothermal heat are generally competitive with market energy prices. Investment costs range from USD_{2005} 50 per kW_{th} (for uncovered pond heating) to USD_{2005} 3,940 per kW_{th} (for building heating). Low LCOHs for these technologies are possible because the inherent losses in heat-to-electricity conversion are avoided when geothermal energy is used for thermal applications.

Future geothermal deployment could meet more than 3% of global electricity demand and about 5% of the global demand for heat by 2050. Evidence suggests that geothermal supply could meet the upper range of projections derived from a review of about 120 energy and GHG reduction scenarios summarized in Chapter 10. With its natural thermal storage capacity, geothermal energy is especially suitable for supplying base-load power. By 2015, geothermal deployment is roughly estimated to generate 122 TWh_e/yr (0.44 EJ/yr) for electricity and 224 TWh_{th}/yr (0.8 EJ/yr) for heat applications. In the long term (by 2050), deployment projections based on extrapolations of long-term historical growth trends suggest that geothermal could produce 1,180 TWh_e/yr (~4.3 EJ/yr) for electricity and 2,100 TWh_{th}/yr (7.6 EJ/yr) for heat, with a few countries obtaining most of their primary energy needs (heating, cooling and electricity) from geothermal energy. Scenario analysis suggests that carbon policy is likely to be one of the main driving factors for future geothermal development, and under the most favourable climate policy scenario (<440 ppm atmospheric CO_2 concentration level in 2100) considered in the energy and GHG scenarios reviewed for this report, geothermal deployment could be even higher in the near and long term.

High-grade geothermal resources have restricted geographic distribution—both cost and technology barriers exist for the use of low-grade geothermal resources and EGS. High-grade geothermal resources are already economically competitive with market energy prices in many locations. However, public and private support for research along with favourable deployment policies (drilling subsidies, targeted grants for pre-competitive research and demonstration to reduce exploration risk and the cost of EGS development) may be needed to support the development of lower-grade hydrothermal resources as well as the demonstration and further commercialization of EGS and other geothermal resources. The effectiveness of these efforts may play a central role in establishing the magnitude of geothermal energy's contributions to long-term GHG emissions reductions.

4.1 Introduction

Geothermal resources consist of thermal energy from the Earth's interior stored in both rock and trapped steam or liquid water. As presented in this chapter, climate change has no major impacts on the effectiveness of geothermal energy utilization, but its widespread deployment could play a significant role in mitigating climate change by reducing greenhouse gas (GHG) emissions as an alternative for capacity addition and/or replacement of existing base load fossil fuel-fired power and heating plants.

Geothermal systems as they are currently exploited occur in a number of geological environments where the temperatures and depths of the reservoirs vary accordingly. Many high-temperature (>180°C) hydrothermal systems are associated with recent volcanic activity and are found near plate tectonic boundaries (subduction, rifting, spreading or transform faulting), or at crustal and mantle hot spot anomalies. Intermediate- (100 to 180°C) and low-temperature (<100°C) systems are also found in continental settings, where above-normal heat production through radioactive isotope decay increases terrestrial heat flow or where aquifers are charged by water heated through circulation along deeply penetrating fault zones. Under appropriate conditions, high-, intermediate- and low-temperature geothermal fields can be utilized for both power generation and the direct use of heat (Tester et al., 2005).

Geothermal resources can be classified as convective (hydrothermal) systems, conductive systems and deep aquifers. Hydrothermal systems include liquid- and vapour-dominated types. Conductive systems include hot rock and magma over a wide range of temperatures (Mock et al., 1997) (Figure 4.1). Deep aquifers contain circulating fluids in porous media or fracture zones at depths typically greater than 3 km, but lack a localized magmatic heat source. They are further subdivided into systems at hydrostatic pressure and systems at pressure higher than hydrostatic (geo-pressured). Enhanced or engineered geothermal system (EGS) technologies enable the utilization of low permeability and low porosity conductive (hot dry rock) and low productivity convective and aquifer systems by creating fluid connectivity through hydraulic stimulation and advanced well configurations. In general, the main types of geothermal systems are hydrothermal and EGS.

Resource utilization technologies for geothermal energy can be grouped under types for electrical power generation, for direct use of the heat, or for combined heat and power in cogeneration applications. Geothermal heat pump (GHP) technologies are a subset of direct use. Currently, the only commercially exploited geothermal systems for power generation and direct use are hydrothermal (of continental subtype). Table 4.1 summarizes the resources and utilization technologies.

Hydrothermal, convective systems are typically found in areas of magmatic intrusions, where temperatures above 1,000°C can occur at less than 10 km depth. Magma typically emits mineralized liquids and gases, which then mix with deeply circulating groundwater. Such systems can last hundreds of thousands of years, and the gradually cooling magmatic heat sources can be replenished periodically with fresh intrusions from a deeper magma chamber. Heat energy is also transferred by conduction, but convection is the most important process in magmatic systems.

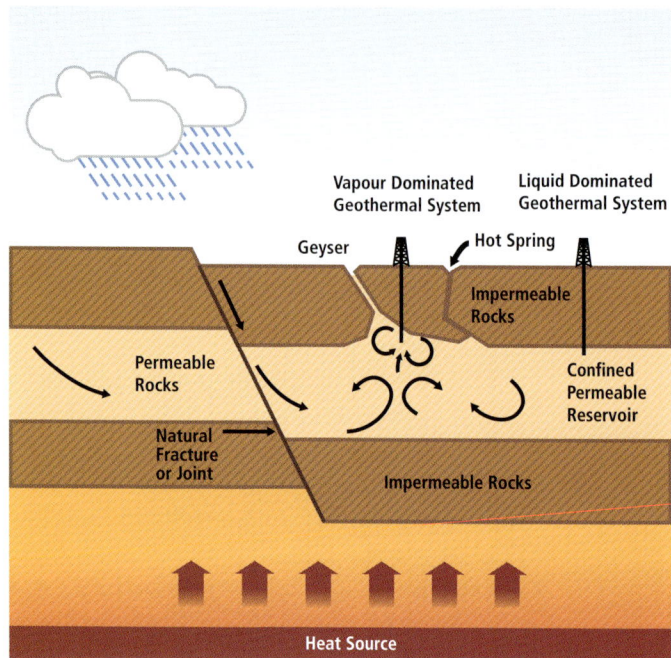

Figure 4.1a | Scheme showing convective (hydrothermal) resources. Adapted from Mock et al. (1997) and from US DOE publications.

Subsurface temperatures increase with depth and if hot rocks within drillable depth can be stimulated to improve permeability, using hydraulic fracturing, chemical or thermal stimulation methods, they form a potential EGS resource that can be used for power generation and direct heat applications. EGS resources include hot dry rock (HDR), hot fractured rock (HFR) and hot wet rock (HWR), among other terms. They occur in all geothermal environments, but are likely to be economic in geological settings where the thermal gradient is high enough to permit exploitation at depths of less than 5 km. In the future, given average geothermal gradients of 25 to 30°C/km, EGS resources at relatively high temperature (≥180°C) may be exploitable in broad areas at depths as shallow as 7 km, which is well within the range of existing drilling technology (~10 km depth). Geothermal resources of different types may occur at different depths below the same surface location. For example, fractured and water-saturated hot-rock EGS resources lie below deep-aquifer resources in the Australian Cooper Basin (Goldstein et al., 2009).

Direct use of geothermal energy has been practised at least since the Middle Palaeolithic when hot springs were used for ritual or routine bathing (Cataldi, 1999), and industrial utilization began in Italy by exploiting boric acid from the geothermal zone of Larderello, where in 1904 the first kilowatts of geothermal electric energy were generated and in 1913 the first 250-kW$_e$ commercial geothermal power unit was installed (Burgassi, 1999). Larderello is still active today.

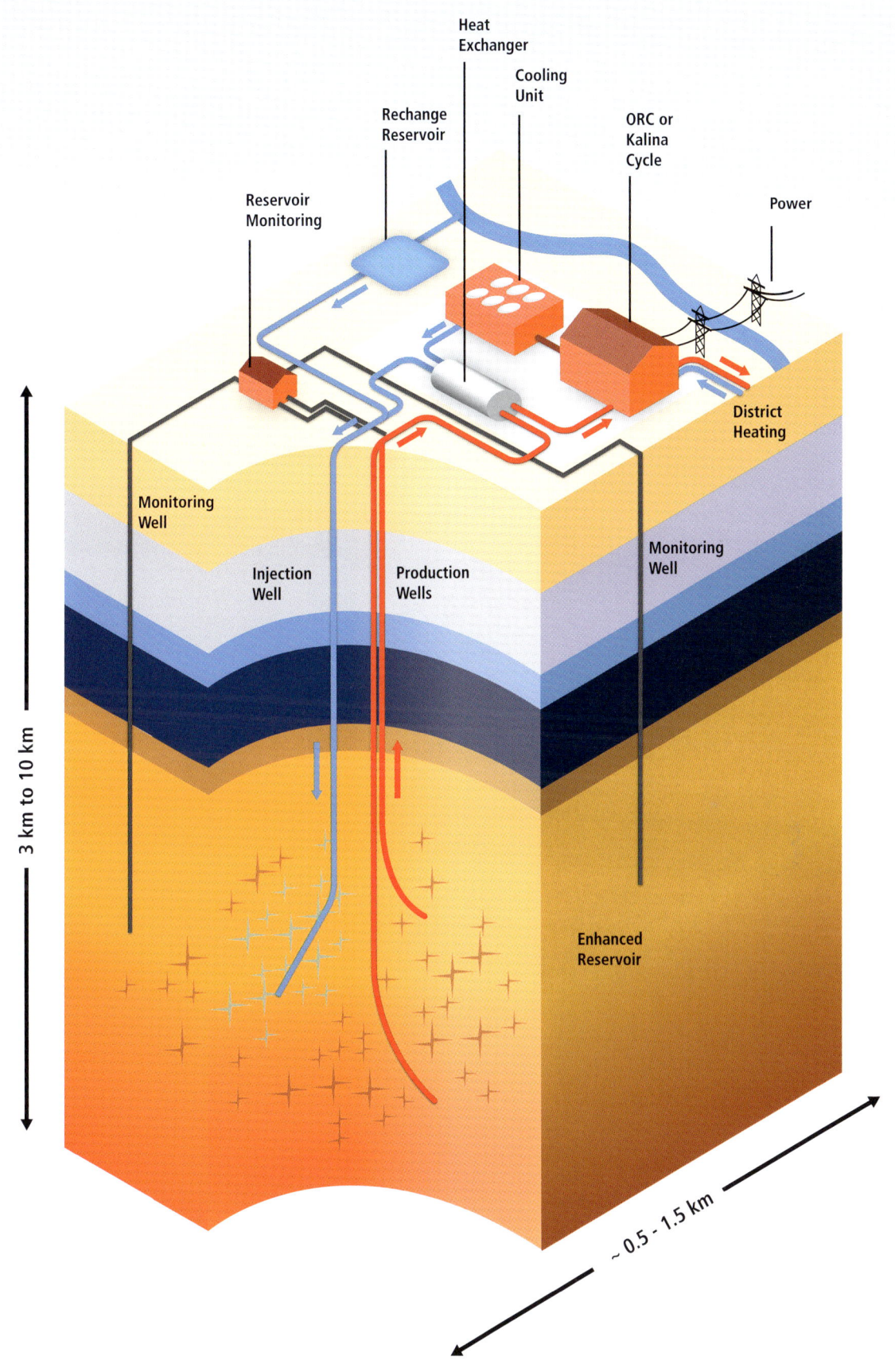

Figure 4.1b | Scheme showing conductive (EGS) resources. Adapted from Mock et al. (1997) and from US DOE publications.

Table 4.1 | Types of geothermal resources, temperatures and uses.

Type	In-situ fluids	Subtype	Temperature Range	Utilization	
				Current	Future
Convective systems (hydrothermal)	Yes	Continental	H, I & L	Power, direct use	
		Submarine	H	None	Power
Conductive systems	No	Shallow (<400 m)	L	Direct use (GHP)	
		Hot rock (EGS)	H, I	Prototypes	Power, direct use
		Magma bodies	H	None	Power, direct use
Deep aquifer systems	Yes	Hydrostatic aquifers	H, I & L	Direct use	Power, direct use
		Geo-pressured		Direct use	Power, direct use

Note: Temperature range: H: High (>180°C), I: Intermediate (100-180°C), L: Low (ambient to 100°C). EGS: Enhanced (or engineered) geothermal systems. GHP: Geothermal heat pumps.

Geothermal energy is classified as a renewable resource (see Chapter 1) because the tapped heat from an active reservoir is continuously restored by natural heat production, conduction and convection from surrounding hotter regions, and the extracted geothermal fluids are replenished by natural recharge and by injection of the depleted (cooled) fluids. Geothermal fields are typically operated at production rates that cause local declines in pressure and/or in temperature within the reservoir over the economic lifetime of the installed facilities. These cooler and lower-pressure zones are subsequently recharged from surrounding regions when extraction ceases.

There are many examples where for economical reasons high extraction rates from hydrothermal reservoirs have resulted in local fluid depletion that exceeded the rate of its recharge, but detailed modelling studies (Pritchett, 1998; Mégel and Rybach, 2000; O'Sullivan and Mannington, 2005) have shown that resource exploitation can be economically feasible in practical situations, and still be renewable on a time scale of the order of 100 years or less, when non-productive recovery periods are considered. Models predict that replenishment will occur in hydrothermal systems on time scales of the same order as the lifetime of the geothermal production cycle where the extraction rate is designed to be sustainable over a 20 to 30 year period (Axelsson et al., 2005, 2010).

This chapter includes a brief discussion of the theoretical potential of geothermal resources, the global and regional technical potential, and the possible impacts of climate change on the resource (Section 4.2), the current technology and applications (Section 4.3) and the expected technological developments (Section 4.6), the present market status (Section 4.4) and its probable future evolution (Section 4.8), environmental and social impacts (Section 4.5) and cost trends (Section 4.7) in using geothermal energy to contribute to reduced GHG emissions.

4.2 Resource Potential

The total thermal energy contained in the Earth is of the order of 12.6×10^{12} EJ and that of the crust of the order of 5.4×10^9 EJ to depths of up to 50 km (Dickson and Fanelli, 2003). The main sources of this energy are due to the heat flow from the Earth's core and mantle, and that generated by the continuous decay of radioactive isotopes in the crust itself. Heat is transferred from the interior towards the surface, mostly by conduction, at an average of 65 mW/m² on continents and 101 mW/m² through the ocean floor. The result is a global terrestrial heat flow rate of around 1,400 EJ/yr. Continents cover ~30% of the Earth's surface and their terrestrial heat flow has been estimated at 315 EJ/yr (Stefansson, 2005).

Stored thermal energy down to 3 km depth on continents was estimated to be 42.67×10^6 EJ by EPRI (1978), consisting of 34.14×10^6 EJ (80%) from hot dry rocks (or EGS resources) and 8.53×10^6 EJ (20%) from hydrothermal resources. Within 10 km depth, Rowley (1982) estimated the continental stored heat to be 403×10^6 EJ with no distinction between hot dry rock and hydrothermal resources, and Tester et al. (2005) estimated it to be 110.4×10^6 EJ from hot dry rocks and only 0.14×10^6 EJ from hydrothermal resources. A linear interpolation between the EPRI (1978) values for 3 km depth and the values from Rowley (1982) results in 139.5×10^6 EJ down to 5 km depth, while linear interpolation between the EPRI (1978) values and those from Tester et al. (2005) only for EGS resources results in 55.9×10^6 EJ down to 5 km depth (see second column of Table 4.2). Based on these estimates, the theoretical potential is clearly not a limiting factor for global geothermal deployment.

In practice geothermal plants can only utilize a portion of the stored thermal energy due to limitations in drilling technology and rock permeability. Commercial utilization to date has concentrated on areas in which geological conditions create convective hydrothermal reservoirs where drilling to depths up to 4 km can access fluids at temperatures of 180°C to more than 350°C.

4.2.1 Global technical potential

Regarding geothermal technical potentials,[1] one recent and comprehensive estimate for conventional hydrothermal resources in the world was presented by Stefansson (2005). For electric generation, he calculated the global geothermal technical potential for identified hydrothermal

---

[1] Definition of technical potential is included in the Glossary (Annex I).

Table 4.2 | Global continental stored heat and EGS technical potentials for electricity.

Depth range (km)	Technically accessible stored heat from EGS		Estimated technical potential (electric) for EGS (EJ/yr)
	(10^6 EJ)	Source	
0–10	403	Rowley, 1982	1051.8
0–10	110.4	Tester et al., 2005	288.1
0–5	139.5	Interpolation between values from Rowley (1982) and EPRI (1978)	364.2
0–5	55.9	Interpolation between values from Tester et al. (2005) and EPRI (1978)	145.9
0–3	34.1	EPRI, 1978	89.1

resources as 200 GW_e (equivalent to 5.7 EJ/yr with a capacity factor (CF)[2] of 90%), with a lower limit of 50 GW_e (1.4 EJ/yr). He assumed that unidentified, hidden resources are 5 to 10 times more abundant than the identified ones and then estimated the upper limit for the worldwide geothermal technical potential as between 1,000 and 2,000 GW_e (28.4 and 56.8 EJ/yr at 90% CF), with a mean value of 1,500 GW_e (~42.6 EJ/yr). Mainly based on those numbers, Krewitt et al. (2009) estimated geothermal technical potential for 2050 at 45 EJ/yr, largely considering only hydrothermal resources.

No similar recent calculation of global technical potential for conductive (EGS) geothermal resources has been published, although the study by EPRI (1978) included some estimates as did others (Armstead and Tester, 1987). Estimating the technical potential of EGS is complicated due to the lack of commercial experience to date. EGS field demonstrations must achieve sufficient reservoir productivity and lifetime to prove both the viability of stimulation methods and the scalability of the technology. Once these features have been demonstrated at several locations, it will be possible to develop better assessments of technical potential, and it is possible that EGS will become a leading geothermal option for electricity and direct use globally because of its widespread availability and lower exploration risk relative to hydrothermal systems.

More recently, Tester et al. (2006; see their Table 1.1) estimated the accessible conductive resources in the USA (excluding Alaska, Hawaii and Yellowstone National Park) and calculated that the stored heat at depths less than 10 km is 13.4 x 10^6 EJ (in conduction-dominated EGS of crystalline basement and sedimentary rock formations). Assuming that 2% of the heat is recoverable and that average temperatures drop 10°C below initial conditions during exploitation, and taking into account all losses in the conversion of recoverable heat into electricity over a lifespan of 30 years, electrical generating capacity from EGS in the USA was estimated at 1,249 GW_e, corresponding to 35.4 EJ/yr of electricity at a CF of 90% (Tester et al., 2006; see their Table 3.3). Based on the same assumptions for the USA,[3] estimates for the global technical potential of EGS-based energy supply can be derived from estimates of the heat stored in the Earth's crust that is both accessible and recoverable (see Table 4.2, fourth column).

Therefore, the global technical potential of geothermal resources for electricity generation can be estimated as the sum of the upper (56.8 EJ/yr) and lower (28.4 EJ/yr) of Stefansson's estimate for hydrothermal resources (identified and hidden) and the EGS technical potentials of Table 4.2 (fourth column), obtaining a lower value of 117.5 EJ/yr (down to 3 km depth) to a maximum of 1,108.6 EJ/yr down to 10 km depth (Figure 4.2). It is important to note that the heat extracted to achieve these technical potentials can be fully or partially replenished over the long term by the continental terrestrial heat flow of 315 EJ/yr (Stefansson, 2005) at an average flux of 65 mW/m^2. Although hydrothermal resources are only a negligible fraction of total theoretical potential given in Tester et al. (2005), their contribution to technical potential might be considerably higher than implied by the conversion from theoretical potential data to technical potential data. This is the rationale for considering the Rowley (1982) estimate for EGS technical potential only and adding the estimate for hydrothermal technical potential from Stefansson (2005).

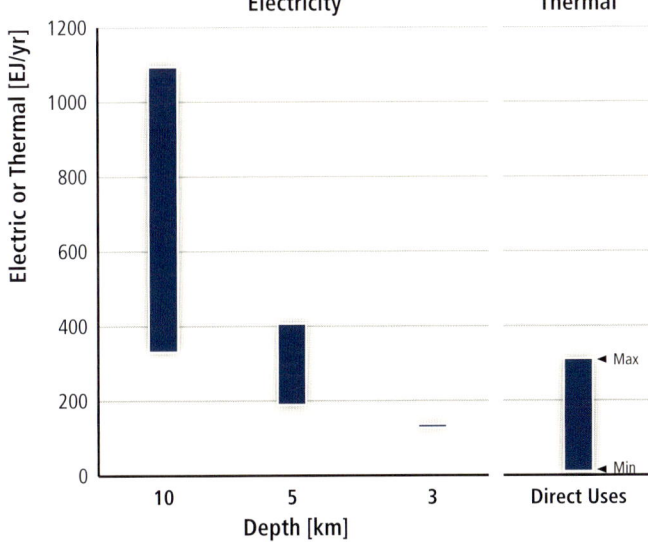

Figure 4.2 | Geothermal technical potentials for electricity and direct uses (heat). Direct uses do not require development to depths greater than approximately 3 km (Prepared with data from Tables 4.2 and 4.3).

---

2   Capacity factor (CF) definition is included in the Glossary (Annex I).

3   1 x 10^6 EJ stored heat equals approximately 2.61 EJ/yr of technical potential for electricity at a 90% CF for 30 years.

For hydrothermal submarine vents, an estimate of >100 GW$_e$ (>2.8 EJ/yr) offshore technical potential has been made (Hiriart et al., 2010). This is based on the 3,900 km of ocean ridges confirmed as having hydrothermal vents,[4] with the assumption that only 1% could be developed for electricity production using a recovery factor of 4%. This assumption is based on capturing part of the heat from the flowing submarine vent without any drilling, but considering offshore drilling, a technical potential of 1,000 GW$_e$ (28.4 EJ/yr) from hydrothermal vents may be possible. However, the technical potential of these resources is still highly uncertain, and is therefore not included in Figure 4.2.

For geothermal direct uses, Stefansson (2005) estimated 4,400 GW$_{th}$ from hydrothermal systems as the world geothermal technical potential from resources <130°C, with a minimum of 1,000 GW$_{th}$ and a maximum, considering hidden resources, of 22,000 to 44,000 GW$_{th}$. Taking a worldwide average CF for direct uses of 30%, the geothermal technical potential for heat can be estimated to be 41.6 EJ/yr with a lower value of 9.5 EJ/yr and an upper value of 312.2 EJ/yr (equivalent to 33,000 GW$_{th}$ of installed capacity) (Figure 4.2). Krewitt et al. (2009) used the same values estimated by Stefansson (2005) in GW$_{th}$, but a CF of 100% was assumed when converted into EJ/r, leading to an average upper limit of 33,000 GW$_{th}$, or 1,040 EJ/yr.

In comparison, the IPCC Fourth Assessment Report (AR4) estimated an available energy resource for geothermal (including potential reserves) of 5,000 EJ/yr (Sims et al., 2007; see their Table 4.2). This amount cannot be properly considered as technical potential and looks overestimated compared with the geothermal technical potentials presented in Figure 4.2. It is important to note, however, that technical potentials tend to increase as technology progresses and overcomes some of the technical constraints of accessing theoretically available resources.

4.2.2 Regional technical potential

The assessed geothermal technical potentials included in Table 4.2 and Figure 4.2 are presented on a regional basis in Table 4.3. The regional breakdown in Table 4.3 is based on the methodology applied by EPRI (1978) to estimate theoretical geothermal potentials for each country, and then countries were grouped into the IEA regions. Thus, the present disaggregation of the global technical potentials is based on factors accounting for regional variations in the average geothermal gradient and the presence of either a diffuse geothermal anomaly or a high-temperature region, associated with volcanism or plate boundaries as estimated by EPRI (1978). Applying these factors to the global technical potentials listed in Table 4.2 gives the values stated in Table 4.3. The separation into electric and thermal (direct uses) technical potentials is somewhat arbitrary in that most higher-temperature resources could be used for either or both in combined heat and power applications depending on local market conditions and the distance between geothermal facilities and the consuming centres. Technical potentials for direct uses include only identified and hidden hydrothermal systems as estimated by Stefansson (2005), and are presented independently from depth since direct uses of geothermal energy usually do not require developments over 3 km in depth.

4.2.3 Possible impact of climate change on resource potential

Geothermal resources are not dependent on climate conditions and climate change is not expected to have a significant impact on the geothermal resource potential. The operation of geothermal heat pumps will not be affected significantly by a gradual change in ambient temperature associated with climate change, but in some power plants it may affect the ability to reject heat efficiently and perhaps adversely impact power generation (Hiriart, 2007). On a local basis, the effect of climate change on rainfall distribution may have a long-term effect on the recharge to specific groundwater aquifers, which in turn may affect discharges from some hot springs, and could have an effect on water levels in shallow geothermally heated aquifers. Also, the availability of cooling water from surface water supplies could be affected by changes in rainfall patterns, and this may require air-cooled power plant condensers (Saadat et al., 2010). However, each of these effects, if they occur, can be remedied by adjustments to the technology, generally for an incremental cost. Regarding future EGS projects, water management may impact the development of EGS particularly in water-deficient regions, where availability is an issue.

4.3 Technology and applications

For the last 100 years, geothermal energy has provided safe, reliable, environmentally benign energy used in a sustainable manner to generate electric power and provide direct heating services from hydrothermal-type resources, using mature technologies. Geothermal typically provides base-load generation, but it has also been used for meeting peak demand. Today's technologies for using hydrothermal resources have demonstrated high average CFs (up to 90% in newer plants, see DiPippo (2008)) in electric generation with low GHG emissions. However, technologies for EGS-type geothermal resources are still in demonstration (see Section 4.3.4).

Geothermal energy is currently extracted using wells or other means that produce hot fluids from: (a) hydrothermal reservoirs with naturally high permeability; or (b) EGS-type reservoirs with artificial fluid pathways. Production wells discharge hot water and/or steam. In high-temperature hydrothermal reservoirs, as pressure drops a fraction of the liquid water component 'flashes' to steam. Separated steam is piped to a turbine to generate electricity and the remaining hot water may be flashed again at lower pressures (and temperatures) to obtain more steam. The

---

4 Some discharge thermal energy of up to 60 MW$_{th}$ (Lupton, 1995) but there are other submarine vents, such as the one known as 'Rainbow', with an estimated output of 1 to 5 GW$_{th}$ (German et al., 1996).

Table 4.3 | Geothermal technical potentials on continents for the International Energy Agency (IEA) regions (prepared with data from EPRI (1978) and global technical potentials described in section 4.2.1).

REGION*	Electric technical potential in EJ/yr at depths to:						Technical potentials (EJ/yr) for direct uses	
	3 km		5 km		10 km			
	Lower	Upper	Lower	Upper	Lower	Upper	Lower	Upper
OECD North America	25.6	31.8	38.0	91.9	69.3	241.9	2.1	68.1
Latin America	15.5	19.3	23.0	55.7	42.0	146.5	1.3	41.3
OECD Europe	6.0	7.5	8.9	21.6	16.3	56.8	0.5	16.0
Africa	16.8	20.8	24.8	60.0	45.3	158.0	1.4	44.5
Transition Economies	19.5	24.3	29.0	70.0	52.8	184.4	1.6	51.9
Middle East	3.7	4.6	5.5	13.4	10.1	35.2	0.3	9.9
Developing Asia	22.9	28.5	34.2	82.4	62.1	216.9	1.8	61.0
OECD Pacific	7.3	9.1	10.8	26.2	19.7	68.9	0.6	19.4
Total	117.5	145.9	174.3	421.0	317.5	1108.6	9.5	312.2

Note: *For regional definitions and country groupings see Annex II.

remaining brine is sent back to the reservoir through injection wells or first cascaded to a direct-use system before injecting. A few reservoirs, such as The Geysers in the USA, Larderello in Italy, Matsukawa in Japan, and some Indonesian fields, produce vapour as 'dry' steam (i.e., pure steam, with no liquid water) that can be sent directly to the turbine. In these cases, control of steam flow to meet power demand fluctuations is easier than in the case of two-phase production, where continuous up-flow in the well bore is required to avoid gravity collapse of the liquid phase. Hot water produced from intermediate-temperature hydrothermal or EGS reservoirs is commonly utilized by extracting heat through a heat exchanger for generating power in a binary cycle, or in direct use applications. Recovered fluids are also injected back into the reservoir (Armstead and Tester, 1987; Dickson and Fanelli, 2003; DiPippo, 2008).

Key technologies for exploration and drilling, reservoir management and stimulation, and energy recovery and conversion are described below.

4.3.1 Exploration and drilling

Since geothermal resources are underground, exploration methods (including geological, geochemical and geophysical surveys) have been developed to locate and assess them. The objectives of geothermal exploration are to identify and rank prospective geothermal reservoirs prior to drilling, and to provide methods of characterizing reservoirs (including the properties of the fluids) that enable estimates of geothermal reservoir performance and lifetime. Exploration of a prospective geothermal reservoir involves estimating its location, lateral extent and depth with geophysical methods and then drilling exploration wells to test its properties, minimizing the risk. All these exploration methods can be improved (see Section 4.6.1).

Today, geothermal wells are drilled over a range of depths down to 5 km using methods similar to those used for oil and gas. Advances in drilling technology have enabled high-temperature operation and provide directional drilling capability. Typically, wells are deviated from vertical to about 30 to 50° inclination from a 'kick-off point' at depths between 200 and 2,000 m. Several wells can be drilled from the same pad, heading in different directions to access larger resource volumes, targeting permeable structures and minimizing the surface impact. Current geothermal drilling methods are presented in more detail in Chapter 6 of Tester et al. (2006). For other geothermal applications such as GHP and direct uses, smaller and more flexible rigs have been developed to overcome accessibility limitations.

4.3.2 Reservoir engineering

Reservoir engineering efforts are focused on two main goals: (a) to determine the volume of geothermal resource and the optimal plant size based on a number of conditions such as sustainable use of the available resource; and (b) to ensure safe and efficient operation during the lifetime of the project. The modern method of estimating reserves and sizing power plants is to apply reservoir simulation technology. First a conceptual model is built, using available data, and is then translated into a numerical representation, and calibrated to the unexploited, initial thermodynamic state of the reservoir (Grant et al., 1982). Future behaviour is forecast under selected load conditions using a heat and mass transfer algorithm (e.g., TOUGH2)[5], and the optimum plant size is selected.

Injection management is an important aspect of geothermal development, where the use of isotopic and chemical tracers is common. Cooling of production zones by injected water that has had insufficient contact with hot reservoir rock can result in production declines. In some circumstances, placement of wells could also aim to enhance deep hot recharge through production pressure drawdown, while suppressing shallow inflows of peripheral cool water through injection pressure increases.

---

5   More information is available on the TOUGH2 website: esd.lbl.gov/TOUGH2/.

Given sufficient, accurate calibration with field data, geothermal reservoir evolution can be adequately modelled and proactively managed. Field operators monitor the chemical and thermodynamic properties of geothermal fluids, and map their flow and movement in the reservoir. This information, combined with other geophysical data, is fed back to recalibrate models for better predictions of future production (Grant et al., 1982).

4.3.3 Power plants

The basic types of geothermal power plants in use today are steam condensing turbines and binary cycle units. Steam condensing turbines[6] can be used in flash or dry-steam plants operating at sites with intermediate- and high-temperature resources (≥150°C). The power plant generally consists of pipelines, water-steam separators, vaporizers, de-misters, heat exchangers, turbine generators, cooling systems, and a step-up transformer for transmission into the electrical grid (see Figure 4.3, top). The power unit size usually ranges from 20 to 110 MW_e (DiPippo, 2008), and may utilize a multiple flash system, flashing the fluid in a series of vessels at successively lower pressures, to maximize the extraction of energy from the geothermal fluid. The only difference between a flash plant and a dry-steam plant is that the latter does not require brine separation, resulting in a simpler and cheaper design.

Binary-cycle plants, typically organic Rankine cycle (ORC) units, are commonly installed to extract heat from low- and intermediate-temperature geothermal fluids (generally from 70 to 170°C), from hydrothermal- and EGS-type reservoirs. Binary plants (Figure 4.3, bottom) are more complex than condensing ones since the geothermal fluid (water, steam or both) passes through a heat exchanger heating another working fluid. This working fluid, such as isopentane or isobutene with a low boiling point, vaporizes, drives a turbine, and then is air cooled or condensed with water. Binary plants are often constructed as linked modular units of a few MW_e in capacity.

There are also combined or hybrid plants, which comprise two or more of the above basic types, such as using a binary plant as a bottoming cycle with a flash steam plant, to improve versatility, increase overall thermal efficiency, improve load-following capability, and efficiently cover a wide resource temperature range.

Cogeneration plants, or combined or cascaded heat and power plants (CHP), produce both electricity and hot water for direct use. Relatively small industries and communities of a few thousand people provide sufficient markets for CHP applications. Iceland has three geothermal cogeneration plants with a combined capacity of 580 MW_{th} in operation (Hjartarson and Einarsson, 2010). At the Oregon Institute of Technology, a CHP plant provides most of the electricity needs and all the heat demand (Lund and Boyd, 2009).

4.3.4 Enhanced Geothermal Systems (EGS)

EGS require stimulation of subsurface regions where temperatures are high enough for effective utilization. A reservoir consisting of a fracture network is created or enhanced to provide well-connected fluid pathways between injection and production wells (see Figure 4.1). Heat is extracted by circulating water through the reservoir in a closed loop and can be used for power generation with binary-cycle plants and for industrial or residential heating (Armstead and Tester, 1987; Tester et al., 2006).

Knowledge of temperature at drillable depth is a prerequisite for site selection for any EGS development. The thermo-mechanical signature of the lithosphere and crust are equally important as they provide critical constraints affecting the crustal stress field, heat flow and temperature gradients. Recently developed analogue and numerical models provide insights useful for geothermal exploration and production, including improved understanding of fundamental mechanisms for predicting crustal stress and basin and basement heat flow (Cloetingh et al., 2010).

EGS projects are currently at a demonstration and experimental stage in a number of countries. The key challenge for EGS is to stimulate and maintain multiple reservoirs with sufficient volumes to sustain long-term production at acceptable rates, and flow impedances, while managing water losses and risk from induced seismicity (Tester et al., 2006).

4.3.5 Direct use

Direct use provides heating and cooling for buildings[7] including district heating, fish ponds, greenhouses, bathing, wellness and swimming pools, water purification/desalination, and industrial and process heat for agricultural products and mineral extraction and drying.

For space heating, two basic types of systems are used: open or closed loop. Open loop (single pipe) systems utilize directly the geothermal water extracted from a well to circulate through radiators (Figure 4.4, top). Closed loop (double pipe) systems use heat exchangers to transfer heat from the geothermal water to a closed loop that circulates heated freshwater through the radiators (Figure 4.4, bottom). This system is commonly used because of the chemical composition of the geothermal water. In both cases the spent geothermal water is disposed of into injection wells and a conventional backup boiler may be provided to meet peak demand.

---

6 A condensing turbine will expand steam to below atmospheric pressure to maximize power production. Vacuum conditions are usually maintained by a direct contact condenser. Back-pressure turbines, much less common and less efficient than condensing turbines, let steam down to atmospheric pressure and avoid the need for condensers and cooling towers.

7 Space and water heating are significant parts of the energy budget in large parts of the world. In Europe, 30% of energy use is for space and water heating alone, representing 75% of total building energy use (Lund et al., 2010a).

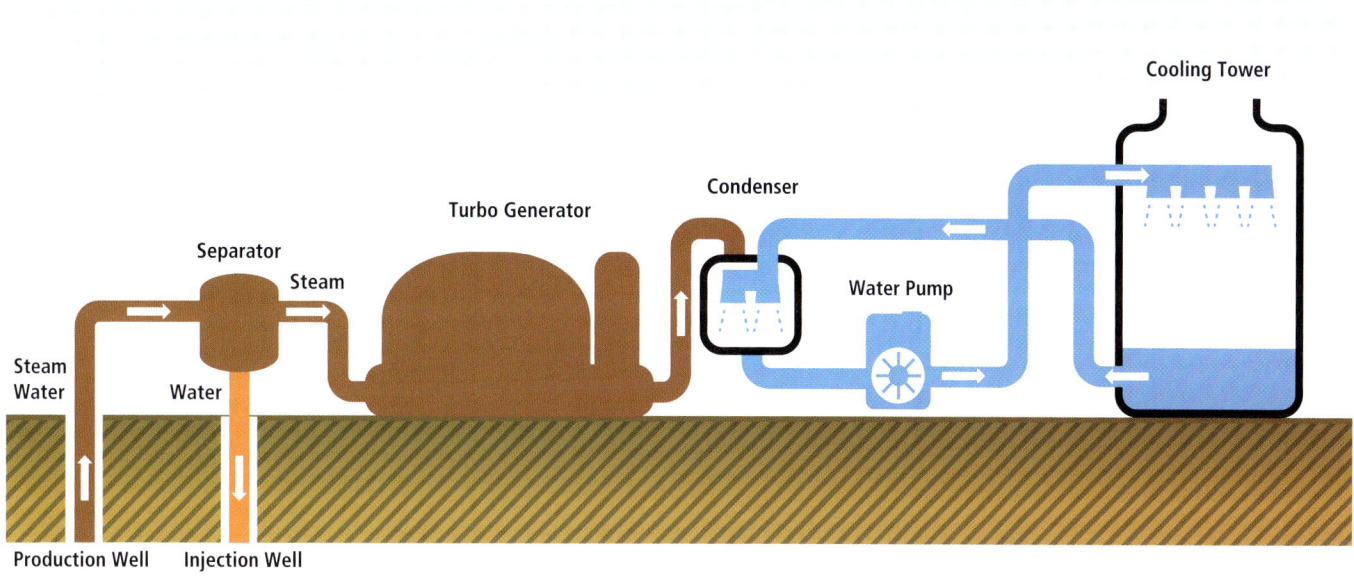
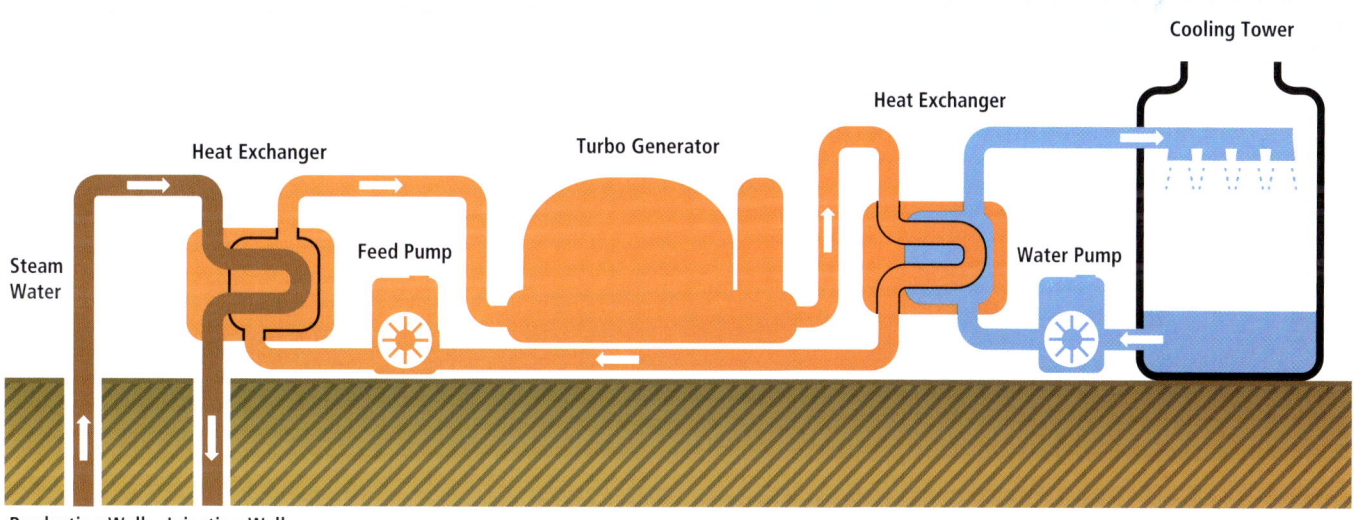

Figure 4.3 | Schematic diagram of a geothermal condensing steam power plant (top) and a binary-cycle power plant (bottom) (adapted from Dickson and Fanelli (2003)).

Transmission pipelines consist mostly of steel insulated by rock wool (surface pipes) or polyurethane (subsurface). However, several small villages and farming communities have successfully used plastic pipes (polybutylene) with polyurethane insulation, as transmission pipes. The temperature drop is insignificant in large-diameter pipes with a high flow rate, as observed in Iceland where geothermal water is transported up to 63 km from the geothermal fields to towns.

Although it is debatable whether geothermal heat pumps, also called ground source heat pumps (GHP), are a 'true' application of geothermal energy or whether they are partially using stored solar energy, in this chapter they are treated as a form of direct geothermal use. GHP technology is based on the relatively constant ground or groundwater temperature ranging from 4°C to 30°C to provide space heating, cooling and domestic hot water for all types of buildings. Extracting energy during heating periods cools the ground locally. This effect can be minimized by dimensioning the number and depth of probes in order to avoid harmful impacts on the ground. These impacts are also reduced by storing heat underground during cooling periods in the summer months.

There are two main types of GHP systems: closed loop and open loop. In ground-coupled systems a closed loop of plastic pipe is placed into

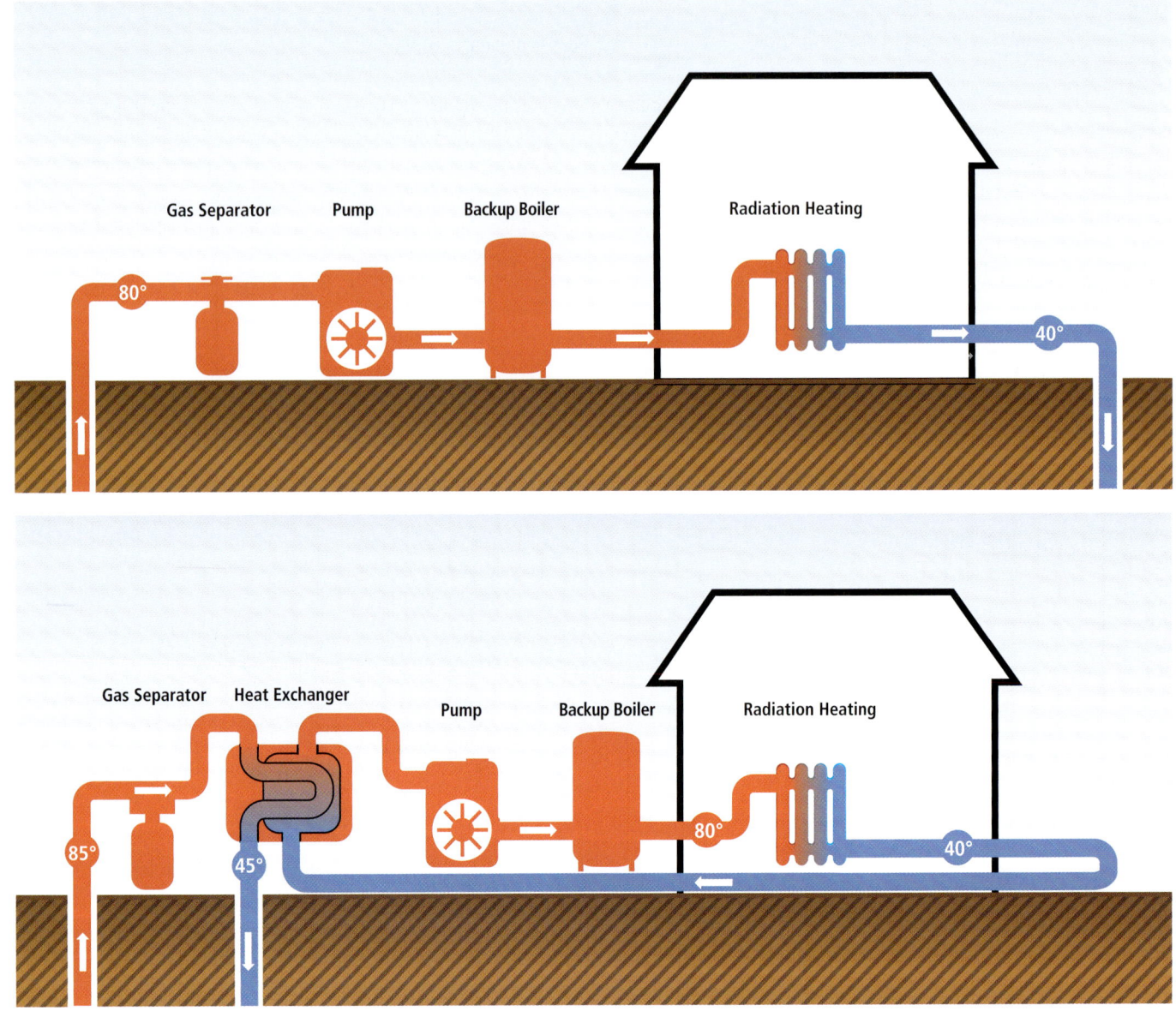

Figure 4.4 | Two main types of district heating systems: top, open loop (single pipe system), bottom, closed loop (double pipe system) (adapted from Dickson and Fanelli, (2003)).

the ground, either horizontally at 1 to 2 m depth or vertically in a borehole down to 50 to 250 m depth. A water-antifreeze solution is circulated through the pipe. Heat is collected from the ground in the winter and rejected to the ground in the summer. An open loop system uses groundwater or lake water directly as a heat source in a heat exchanger and then discharges it into another well or into the same water reservoir (Lund et al., 2003).

Heat pumps operate similarly to vapour compression refrigeration units with heat rejected in the condenser used for heating or extracted in the evaporator used for cooling. GHP efficiency is described by a coefficient of performance (COP) that scales the heating or cooling output to the electrical energy input, and typically lies between 3 and 4 (Lund et al., 2003; Rybach, 2005). The seasonal performance factor (SPF) provides a metric of the overall annual efficiency. It is the ratio of useful heat to the consumed driving energy (both in kWh/yr), and it is slightly lower than the COP.

4.4 Global and regional status of market and industry development

Electricity has been generated commercially by geothermal steam since 1913. Currently, the geothermal industry has a wide range of participants, including major energy companies, private and public utilities, equipment manufacturers and suppliers, field developers and drilling companies. The geothermal-electric market appears to be accelerating compared to previous years, as indicated by the increase in installed and planned capacity (Bertani, 2010; Holm et al., 2010).

4.4.1 Status of geothermal electricity from conventional geothermal resources

In 2009, electricity was being produced from conventional (hydrothermal) geothermal resources in 24 countries with an installed capacity of 10.7 GW_e (Figure 4.5), with an annual increase of 405 MW (3.9%) over the previous year (Bertani, 2010, see his Table X). The worldwide use of geothermal energy for power generation was 67.2 TWh/yr (0.24 EJ/yr)[8] in 2008 (Bertani, 2010) with a worldwide CF of 74.5% (see also Table 4.7). Many developing countries are among the top 15 in geothermal electricity production.

Conventional geothermal resources currently used to produce electricity are either high-temperature systems (>180°C), using steam power cycles (either flash or dry steam driving condensing turbines), or low to intermediate temperature (<180°C) using binary-cycle power plants.

Around 11% of the installed capacity in the world in 2009 was composed of binary plants (Bertani, 2010).

In 2009, the world's top geothermal producer was the USA with almost 29% of the global installed capacity (3,094 MW_e; Figure 4.5). The US geothermal industry is currently expanding due to state Renewable Portfolio Standards (RPS) and various federal subsidies and tax incentives (Holm et al., 2010). US geothermal activity is concentrated in a few western states, and only a fraction of the geothermal technical potential has been developed so far.

Outside of the USA, about 29% of the global installed geothermal capacity in 2009 was located in the Philippines and Indonesia. Mexico, Italy, Japan, Iceland and New Zealand together account for one-third of the global installed geothermal capacity. Although some of these markets have seen relatively limited growth over the past few years, others

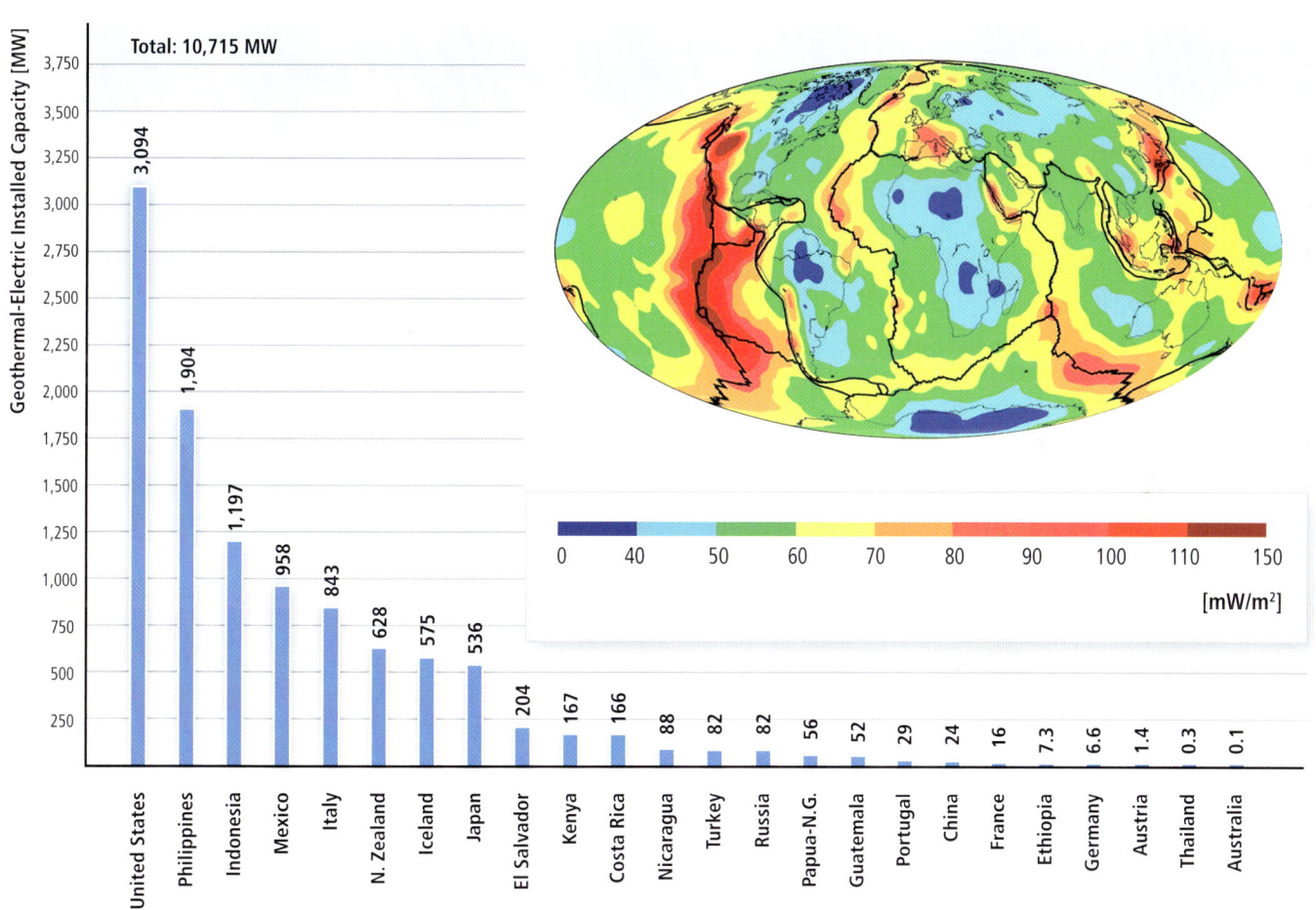

Figure 4.5 | Geothermal-electric installed capacity by country in 2009. Inset figure shows worldwide average heat flow in mW/m² and tectonic plates boundaries (figure from Hamza et al. (2008), used with kind permission from Springer Science+Business Media B.V.; data from Bertani (2010)).

---

8   Based on IEA data presented in Chapter 1, electricity production from geothermal energy in 2008 equaled 65 TWh/yr.

such as Iceland and New Zealand doubled the installed capacity from 2005 to 2009 (IEA-GIA, 2009). Moreover, attention is turning to new markets such as Chile, Germany and Australia.

The majority of existing geothermal assets are operated by state-owned utilities or independent power producers. Currently, more than 30 companies globally have an ownership stake in at least one geothermal field. Altogether, the top 20 owners of geothermal capacity control approximately 90% of the installed global market (Bertani, 2010).

At the end of 2008, geothermal electricity contributed only about 0.3% of the total worldwide electric generation. However, 6 of the 24 countries shown in Figure 4.5 (El Salvador, Kenya, Philippines, Iceland, Costa Rica and New Zealand) obtained more than 10% of their national electricity production from high-temperature geothermal resources (Bromley et al., 2010).

Worldwide evolution of geothermal power and geothermal direct uses during the last 40 years is presented in Table 4.4, including the annual average rate of growth over each period. The average annual growth of geothermal-electric installed capacity over the last 40 years is 7%, and for geothermal direct uses (heat applications) is 11% over the last 35 years.

Table 4.4 | Average annual growth rate in geothermal power capacity and direct uses (including GHP) in the last 40 years (prepared with data from Lund et al., 2005, 2010a; Fridleifsson and Ragnarsson, 2007; Gawell and Greenberg, 2007; Bertani, 2010).

Year	Electric capacity		Direct uses capacity	
	MW_e	%	MW_{th}	%
1970	720	—	N/A	—
1975	1,180	10.4	1,300	—
1980	2,110	12.3	1,950	8.5
1985	4,764	17.7	7,072	29.4
1990	5,834	4.1	8,064	2.7
1995	6,833	3.2	8,664	1.4
2000	7,972	3.1	15,200	11.9
2005	8,933	2.3	27,825	12.9
2010*	10,715	3.7	50,583	12.7
Total annual average:		7.0		11.0

Notes:
%: Average annual growth in percent over the period.
N/A: Reliable data not available.
*End of 2009.

4.4.2 Status of EGS

While there are no commercial-scale operating EGS plants, a number of demonstrations are active in Europe, the USA and Australia. In the latter, by 2009, 50 companies held about 400 geothermal exploration licences to develop EGS (AL-AGEA, 2009) with investments of USD_{2005} 260 million and government grants of USD_{2005} 146 million (Goldstein et al., 2009). In France, the EU project 'EGS Pilot Plant' at Soultz-sous-Forêts started in 1987 and has recently commissioned the first power plant (1.5 MW_e) to utilize the enhanced fracture permeability at 200°C. In Landau, Germany, a 2.5 to 2.9 MW_e EGS plant went into operation in late 2007 (Hettkamp et al., 2010). Deep sedimentary aquifers are being tapped at the geothermal test site in Groß Schönebeck, Germany, using two research wells (Huenges et al., 2009). These demonstration prototypes have provided data on the performance of the EGS concepts subject to real field conditions. Nonetheless, sustained multiyear commitments to field-scale demonstrations in different geologic settings are still needed to reduce technical and economic risks.

The USA has recently increased support for EGS research, development and demonstration as part of a revived national geothermal program. Currently the main short-term goals for the US program are to demonstrate commercial viability of EGS and upscale to several tens of megawatts (Holm et al., 2010). A US commitment to multiyear EGS demonstrations covering a range of resource grades is less certain.

The availability of water, other lower-cost renewable resources, transmission and distribution infrastructure, and most importantly project financing, will play major roles in regional growth trends of EGS projects (Tester et al., 2006).

4.4.3 Status of direct uses of geothermal resources

The world installed capacity of direct-use geothermal energy in 2009 was estimated at 50.6 GW_{th} (Table 4.4), with a total thermal energy usage of about 121.7 TWh_{th}/yr (0.44 EJ/yr) in 2008, distributed in 78 countries, with an annual average CF of 27.5% (Lund et al., 2010a). Another source (REN21, 2010) estimates geothermal direct use at 60 GW_{th} as of the end of 2009.

Direct heat supply temperatures are typically close to actual process temperatures in district heating systems that range from approximately 60°C to 120°C. In 2009 the main types (and relative percentages) of direct applications in annual energy use were: space heating of buildings[9] (63%), bathing and balneology (25%), horticulture (greenhouses and soil heating) (5%), industrial process heat and agricultural drying (3%), aquaculture (fish farming) (3%) and snow melting (1%) (Lund et al., 2010a).

When the resource temperature is too low for other direct uses, it is possible to use GHP. GHP contributed 70% (35.2 GW_{th}) of the worldwide installed geothermal heating capacity in 2009, and has been the fastest growing form of all geothermal direct use since 1995 (Rybach, 2005; Lund et al., 2010a).

---

[9] China is the world's largest user of geothermal heat for space heating (Lund et al., 2010a).

Bathing, swimming and balneology are globally widespread. In addition to the thermal energy, the chemicals dissolved in the geothermal fluid are used for treating various skin and health diseases. Greenhouses heated by geothermal energy and heating soil in outdoor agricultural fields have been developed in several countries. A variety of industrial processes utilize heat applications, including drying of forest products, food and minerals industries as in the USA, Iceland and New Zealand. Other applications are process heating, evaporation, distillation, sterilization, washing, and CO_2 and salt extraction. Aquaculture using geothermal heat allows better control of pond temperatures, with tilapia, salmon and trout the most common fish raised. Low-temperature geothermal water is used in some colder climate countries for snow melting or de-icing. City streets, sidewalks and parking lots are equipped with buried piping systems carrying hot geothermal water (Lund et al., 2005, 2010a).

Geothermal direct uses have experienced a significant global increase in the last 15 years (Table 4.4) after a period of stagnation (1985 to 1995), mainly due to the increasing costs of fossil fuels for heating and cooling and the need to replace them with renewable sources. The technical potential of direct-use applications for heating and cooling buildings is still largely unrealized (Lund et al., 2010a).

4.4.4 Impact of policies[10]

For geothermal to reach its full capacity in climate change mitigation it is necessary to address the following technical and non technical barriers (Wonstolen, 1980; Mock et al., 1997; Imolauer et al., 2010).

Technical barriers. Distributions of potential geothermal resources vary from being almost site-independent (for GHP technologies and EGS) to being much more site-specific (for hydrothermal sources). The distance between electricity markets or centres of heat demand and geothermal resources, as well as the availability of transmission capacity, can be a significant factor in the economics of power generation and direct use.

Non-technical barriers.
- Information and awareness barriers. Lack of clarity in understanding geothermal energy is often a barrier, which could be overcome by dissemination of information on reliable and efficient geothermal technologies to enhance governmental and public knowledge. On the other hand, for deep geothermal drilling and reservoir management, skilled companies and well-trained personnel are currently concentrated in a few countries. For GHP installation and district heating, there is also a correlation between local availability and awareness of service companies and technology uptake. This constraint could be overcome by an improved global infrastructure of services and education programs (geothermal engineering programs) for an expanding workforce to replace retiring staff.

- Market failures and economic barriers, due to un-priced or under-priced environmental impacts of energy use, and poor availability of capital risk insurance.

- Institutional barriers due in many countries to the lack of specific laws governing geothermal resources, which are commonly considered as mining or water resources.

Policies set to drive uptake of geothermal energy work better if local demand and risk factors are taken into account (Rybach, 2010). For example, small domestic heat customers can be satisfied using GHP technologies, which require relatively small budgets. For other countries, district heating systems and industrial heat applications are more efficient and provide greater mitigation of CO_2 emissions, but these markets typically require larger-scale investments and a different policy framework.

Policies that support improved applied research and development would benefit all geothermal technologies, but especially emerging technologies such as EGS. Specific incentives for geothermal development can include fiscal incentives, public finance and regulation policies such as targeted grants for pre-competitive research and demonstration, subsidies, guarantees, tax write-offs to cover the commercial upfront exploration costs, including the higher-risk initial drilling costs, feed-in tariffs and additional measures like portfolio standards (Rybach, 2010). Feed-in tariffs (FITs, see Section 11.5.4.3) with defined geothermal pricing have been very successful in attracting commercial investment in some European countries such as Germany, Switzerland, Belgium, Austria, Spain and Greece, among others (Rybach, 2010). Direct subsidies for new building heating, refurbishment of existing buildings with GHP, and for district heating systems may be also applicable.

Experience has shown that the relative success of geothermal development in particular countries is closely linked to their government's policies, regulations, incentives and initiatives. Successful policies have taken into account the benefits of geothermal energy, such as its independence from weather conditions and its suitability for base-load power. Another important policy consideration is the opportunity to support the price of geothermal kWh (both power and direct heating and cooling) through the United Nations' Clean Development Mechanism (CDM) program. A recent example is the Darajat III geothermal power plant, developed by a private company in Indonesia in 2007, and registered with the CDM. The plant currently generates about 650,000 carbon credits (or certified emission reductions, CER) per year, thus reducing the lifecycle cost of geothermal energy by about 2 to 4% (Newell and Mingst, 2009).

---
10 Non-technology-specific policy issues are covered in Chapter 11 of this report.

4.5 Environmental and social impacts[11]

In general, negative environmental impacts associated with geothermal energy utilization are minor. Hot fluid production can emit varying quantities of GHGs, which are usually small. These originate from naturally sourced CO_2 fluxes that would eventually be released into the atmosphere through natural surface venting. The exploitation of geothermal energy does not ultimately create any additional CO_2 from the subsurface, since there is no combustion process, though the rate of natural emissions can be altered by geothermal production depending on the plant configuration.

Water is not a limiting factor for geothermal power generation, since geothermal fluids are usually brines (i.e., not competing with other uses). Flash power plants do not consume potable water for cooling and yield condensed water that can, with proper treatment, be used for agricultural and industrial purposes. Binary power plants can minimize their water use with air cooling.

Potential adverse effects from disposal of geothermal fluids and gases, induced seismicity and ground subsidence can be minimized by sound practices. Good practice can also optimize water and land use, improve long-term sustainability of production and protect natural thermal features that are valued by the community. The following sections address these issues in more detail.

4.5.1 Direct greenhouse gas emissions

The main GHG emitted by geothermal operations is CO_2. Geothermal fluids contain minerals leached from the reservoir rock and variable quantities of gas, mainly CO_2 and a smaller amount of hydrogen sulphide. The gas composition and quantity depend on the geological conditions encountered in the different fields. Depending on technology, most of the mineral content of the fluid and some of the gases are re-injected back into the reservoir. The gases are often extracted from a steam turbine condenser or two-phase heat exchanger and released through a cooling tower. CO_2, on average, constitutes 90% of these non-condensable gases (Bertani and Thain, 2002). A field survey of geothermal power plants operating in 2001 found a wide spread in the direct CO_2 emission rates. The average weighted by generation was 122 g CO_2/kWh, with values ranging from 4 to 740 g CO_2/kWh (Bertani and Thain, 2002). In closed-loop binary-cycle power plants, where the extracted geothermal fluid is passed through a heat exchanger and then completely injected, the operational CO_2 emission is near zero.

In direct heating applications, emissions of CO_2 are also typically negligible (Fridleifsson et al., 2008). For instance, in Reykjavik, Iceland, the CO_2 content of thermal groundwater used for district heating (0.05 mg CO_2/kWh$_{th}$) is lower than that of the cold groundwater. In China (Beijing, Tianjin and Xianyang) it is less than 1 g CO_2/kWh$_{th}$. In places such as Iceland, co-produced CO_2, when sufficiently pure, may also be used in greenhouses to improve plant growth, or extracted for use in carbonated beverages. In the case of Iceland, the replacement of fossil fuel with geothermal heating has avoided the emission of approximately 2 Mt of CO_2 annually and significantly reduced air pollution (Fridleifsson et al., 2008). Other examples of the environmental benefits of geothermal direct use are at Galanta in Slovakia (Fridleifsson et al., 2008), the Pannonian Basin in Hungary (Arpasi, 2005), and the Paris Basin (Laplaige et al., 2005).

EGS power plants are likely to be designed as liquid-phase closed-loop circulation systems, with zero direct emissions, although, if gas separation occurs within the circulation loop, some gas extraction and emission is likely. If the current trend towards more use of lower-temperature resources and binary plants continues, there will be a reduction in average emissions.

4.5.2 Lifecycle assessment

Life-cycle assessment (LCA) analyzes the whole lifecycle of a product 'from cradle to grave'. For geothermal power plants, all GHG emissions directly and indirectly related to the construction, operation and decommissioning of the plant are considered in LCA.

Figure 4.6 shows the result of a comprehensive literature review of geothermal electricity generation LCA studies published since 1980, which were screened for quality and completeness (see Annex II for details on methodology). All estimates of lifecycle GHG emissions are less than 50

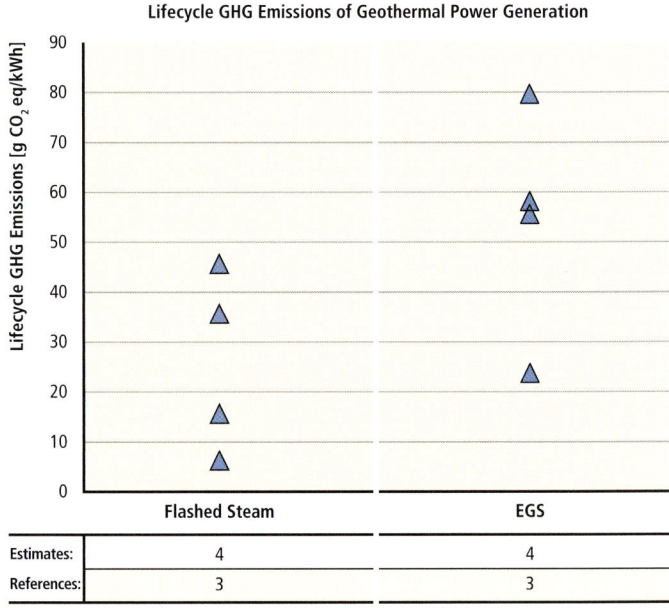

Figure 4.6 | Estimates of lifecycle GHG emissions from geothermal power generation (flashed steam and EGS technologies). Unmodified literature values, after quality screen. (See Annex II and Section 9.3.4.1 for details of literature search and citations.)

---
11 A comprehensive assessment of social and environmental impacts of all RE sources covered in this report can be found in Chapter 9.

g CO_2eq/kWh for flash steam plants and less than 80 g CO_2eq/kWh for projected EGS plants.

The Bertani and Thain (2002) estimates are higher than these for several reasons. First, Bertani and Thain collected information from a very large fraction of global geothermal facilities (85% of world geothermal capacity in 2001), whereas qualifying LCA studies were few. Some open-loop facilities with high dissolved CO_2 concentrations can emit CO_2 at very high rates, though this is relevant for a minority of installed capacity only. For closed-loop geothermal systems with more common dissolved CO_2 concentrations, most lifecycle GHG emissions are embodied in plant materials and emitted during construction. These were the cases examined in the qualifying LCA literature displayed in Figure 4.6. Despite few available studies, it is tentatively observed that systems using flashed or dry geothermal steam appear to have lower GHG emissions than do systems combining EGS reservoir development with binary power conversion systems, though this difference is small relative to, for instance, coal-fired electricity generation GHG emissions (see Section 9.3.4.1). A key factor contributing to higher reported emissions for EGS/binary systems versus steam-driven geothermal systems is higher energy and materials requirements for EGS' well-field development. Additional LCA studies to increase the number of estimates for all geothermal energy technologies are needed.

Frick et al. (2010) compared LCA environmental indicators to those of European and German reference power mixes, the latter being composed of lignite coal (26%), nuclear power (26%), hard coal (24%), natural gas (12%), hydropower (4%), wind power (4%), crude oil (1%) and other fuels (3%), and observed that geothermal GHG emissions fall in a range between 8 and 12% of these reference mixes. At sites with above-average geological conditions, low-end GHG emissions from closed loop geothermal power systems can be less than 1% of corresponding emissions for coal technologies.

For lifecycle GHG emissions of geothermal energy, Kaltschmitt (2000) published figures of 14.3 to 57.6 g CO_2eq/kWh$_{th}$ for low-temperature district heating systems, and 180 to 202 g CO_2eq/kWh$_{th}$ for GHP, although the latter values depend significantly on the mix of electricity sources that power them.

The LCA of intermediate- to low-temperature geothermal developments is dominated by larger initial material and energy inputs during the construction of the wells, power plant and pipelines. For hybrid electricity/district heating applications, greater direct use of the heat generally provides greater environmental benefits.

In conclusion, the LCA assessments show that geothermal is similar to other RE and nuclear energy in total lifecycle GHG emissions (see 9.3.4.1), and it has significant environmental advantages relative to a reference electricity mix dominated by fossil fuel sources.

4.5.3 Local environmental impacts

Environmental impact assessments for geothermal developments involve consideration of a range of local land and water use impacts during both construction and operation phases that are common to most energy projects (e.g., noise, vibration, dust, visual impacts, surface and ground water impacts, ecosystems, biodiversity) as well as specific geothermal impacts (e.g., effects on outstanding natural features such as springs, geysers and fumaroles).

4.5.3.1 Other gas and liquid emissions during operation

Geothermal systems involve natural phenomena, and typically discharge gases mixed with steam from surface features, and minerals dissolved in water from hot springs. Apart from CO_2, geothermal fluids can, depending on the site, contain a variety of other minor gases, such as hydrogen sulphide (H_2S), hydrogen (H_2), methane (CH_4), ammonia (NH_3) and nitrogen (N_2). Mercury, arsenic, radon and boron may be present. The amounts depend on the geological, hydrological and thermodynamic conditions of the geothermal field, and the type of fluid collection/ injection system and power plant utilized.

Of the minor gases, H_2S is toxic, but rarely of sufficient concentration to be harmful after venting to the atmosphere and dispersal. Removal of H_2S released from geothermal power plants is practised in parts of the USA and Italy. Elsewhere, H_2S monitoring is a standard practice to provide assurance that concentrations after venting and atmospheric dispersal are not harmful. CH_4, which has warming potential, is present in small concentrations (typically a few percent of the CO_2 concentration).

Most hazardous chemicals in geothermal fluids are in aqueous phase. If present, boron and arsenic are likely to be harmful to ecosystems if released at the surface. In the past, surface disposal of separated water has occurred at a few fields. Today, this happens only in exceptional circumstances, and geothermal brine is usually injected back into the reservoir to support reservoir pressures, as well as avoid adverse environmental effects. Surface disposal, if significantly in excess of natural hot spring flow rates, and if not strongly diluted, can have adverse effects on the ecology of rivers, lakes or marine environments. Shallow groundwater aquifers of potable quality are protected from contamination by injected fluids by using cemented casings, and impermeable linings provide protection from temporary fluid disposal ponds.

Such practices are typically mandated by environmental regulations. Geochemical monitoring is commonly undertaken by the field operators to investigate, and if necessary mitigate, such adverse effects (Bromley et al., 2006).

4.5.3.2 Potential hazards of seismicity and other phenomena

Local hazards arising from natural phenomena, such as micro-earthquakes, hydrothermal steam eruptions and ground subsidence may be influenced by the operation of a geothermal field (see also Section 9.3.4.7). As with other (non-geothermal) deep drilling projects, pressure or temperature changes induced by stimulation, production or injection of fluids can lead to geo-mechanical stress changes and these can affect the subsequent rate of occurrence of these phenomena (Majer et al., 2008). A geological risk assessment may help to avoid or mitigate these hazards.

Routine seismic monitoring is used as a diagnostic tool and management and protocols have been prepared to measure, monitor and manage systems proactively, as well as to inform the public of any hazards (Majer et al., 2008). In the future, discrete-element models would be able to predict the spatial location of energy releases due to injection and withdrawal of underground fluids. During 100 years of development, although turbines have been tripped offline for short periods, no buildings or structures within a geothermal operation or local community have been significantly damaged by shallow earthquakes originating from geothermal production or injection activities.

With respect to induced seismicity, ground vibrations or noise have been a social issue associated with some EGS demonstration projects, particularly in populated areas of Europe. The process of high-pressure injection of cold water into hot rock generates small seismic events. Induced seismic events have not been large enough to lead to human injury or significant property damage, but proper management of this issue will be an important step to facilitating significant expansion of future EGS projects. Collaborative research initiated by the IEA-GIA (Bromley and Mongillo, 2008), the USA and Australia (International Partnership for Geothermal Technology: IPGT)[12] and in Europe (GEISER)[13], is aimed at better understanding and mitigating induced seismicity hazards, and providing risk management protocols.

Hydrothermal steam eruptions have been triggered at a few locations by shallow geothermal pressure changes (both increases and decreases). These risks can be mitigated by prudent field design and operation.

Land subsidence has been an issue at a few high-temperature geothermal fields where pressure decline has affected some highly compressible formations causing them to compact anomalously and form local subsidence 'bowls'. Management by targeted injection to maintain pressures at crucial depths and locations can minimize subsidence effects. Some minor subsidence may also be related to thermal contraction and minor tumescence (inflation) can overlie areas of injection and rising pressure.

4.5.3.3 Land use

Good examples exist of unobtrusive, scenically landscaped developments (e.g., Matsukawa, Japan), and integrated tourism/energy developments (e.g., Wairakei, New Zealand and Blue Lagoon, Iceland). Nonetheless, land use issues still seriously constrain new development options in some countries (e.g., Indonesia, Japan, the USA and New Zealand) where new projects are often located within or adjacent to national parks or tourist areas. Spa resort owners are very sensitive to the possibility of depleted hot water resources. Potential pressure and temperature interference between adjacent geothermal developers or users can be another issue that affects all types of heat and fluid extraction, including heat pumps and EGS power projects (Bromley et al., 2006). Good planning should take this into account by applying predictive simulation models when allocating permits for energy extraction.

Table 4.5 presents the typical operational footprint for conventional geothermal power plants, taking into account surface installations (drilling pads, roads, pipelines, fluid separators and power-stations). Due to directional drilling techniques, and appropriate design of pipeline corridors, the land area above geothermal resources that is not covered by surface installations can still be used for other purposes such as farming, horticulture and forestry, as occurs, for example, at Mokai and Rotokawa in New Zealand (Koorey and Fernando, 2010), and a national park at Olkaria, Kenya.

Table 4.5 | Land requirements for typical geothermal power generation options expressed in terms of square meter per generation capacity and per annual energy output.

Type of power plant	Land Use	
	m^2/MW_e	$m^2/GWh/yr$
110-MW_e geothermal flash plants (excluding wells)	1,260	160
56-MW_e geothermal flash plant (including wells, pipes, etc.)	7,460	900
49-MW_e geothermal FC-RC plant (excluding wells)	2,290	290
20-MW_e geothermal binary plant (excluding wells)	1,415	170

Notes: FC: Flash cycle. RC: Rankine cycle (data from Tester et al. (2006) taken from DiPippo (1991); the CFs originally used to calculate land use vary between 90 and 95% depending on the plant type).

4.5.4 Local social impacts

The successful realization of geothermal projects often depends on the level of acceptance by local people. Prevention or minimization of detrimental impacts on the environment, and on land occupiers, as well as

---
12 A description of the project IPGT is available at: internationalgeothermal.org/IPGT.html.

13 A description of the GEISER project is available at: www.gfz-potsdam.de.

the creation of benefits for local communities, is indispensable to obtain social acceptance. Public education and awareness of the probability and severity of detrimental impacts are also important. The necessary prerequisites to secure agreement of local people are: (a) prevention of adverse effects on people's health; (b) minimization of environmental impacts; and (c) creation of direct and ongoing benefits for the resident communities (Rybach, 2010). Geothermal development creates local job opportunities during the exploration, drilling and construction period (typically four years minimum for a greenfield project). It also creates permanent and full-time jobs when the power plant starts to operate (Kagel, 2006) since the geothermal field from which the fluids are extracted must be operated locally. This can alleviate rural poverty in developing countries, particularly in Asia, Central and South America, and Africa, where geothermal resources are often located in remote mountainous areas. Some geothermal companies and government agencies have approached social issues by improving local security, building roads, schools, medical facilities and other community assets, which may be funded by contributions from profits obtained from operating the power plant (De Jesus, 2005).

Multiple land use arrangements that promote employment by integrating subsurface geothermal energy extraction with labour-intensive agricultural activities are also useful. In many developing countries, geothermal energy is also an appropriate energy source for small-scale distributed generation, helping accelerate development through access to energy in remote areas. This has occurred, for example, in Maguarichi, Mexico (Sánchez-Velasco et al., 2003).

4.6 Prospects for technology improvement, innovation and integration[14]

Geothermal resources can be integrated into all types of electrical power supply systems, from large, interconnected continental transmission grids to onsite use in small, isolated villages or autonomous buildings. They can be utilized in a variety of sustainable power generating modes, including continuous low power rates, long-term (decades long) cycles of high power rates separated by recovery periods and long-term, uninterrupted high power rates sustained with effective fluid reinjection (Bromley et al., 2006). Since geothermal typically provides base-load electric generation, integration of new power plants into existing power systems does not present a major challenge. Indeed, in some configurations, geothermal energy can provide valuable flexibility, such as the ability to increase or decrease production or start up/shut down as required. In some cases, however, the location dependence of geothermal resources requires new transmission infrastructure investments in order to deliver geothermal electricity to load centres.

For geothermal direct uses, no integration problems have been observed. For heating and cooling, geothermal (including GHP) is already widespread at the domestic, community and district scales. District heating networks usually offer flexibility with regard to the primary energy source and can therefore use low-temperature geothermal resources or cascaded geothermal heat (Lund et al., 2010b).

For technology improvement and innovation, several prospects can reduce the cost of producing geothermal energy and lead to higher energy recovery, longer field lifetimes, and better reliability. With time, better technical solutions are expected to improve power plant performance and reduce maintenance down time. The main technological challenges and prospects are described below.

4.6.1 Improvements in exploration, drilling and assessment technologies

In exploration, R&D is required to locate hidden geothermal systems (i.e., with no surface manifestations such as hot springs and fumaroles) and for EGS prospects. Refinement and wider usage of rapid reconnaissance geothermal tools such as satellite-based hyper-spectral, thermal infrared, high-resolution panchromatic and radar sensors could make exploration efforts more effective. Once a regional focus area has been selected, availability of improved cost-effective reconnaissance survey tools to detect as many geothermal indicators as possible is critical in providing rapid coverage of the geological environment being explored at an appropriate resolution.

Special research is needed to improve the rate of penetration when drilling hard rock and to develop advanced slim-hole technologies, and also in large-diameter drilling through ductile, creeping or swelling formations. Drilling must minimize formation damage that occurs as a result of a complex interaction of the drilling fluid (chemical, filtrate and particulate) with the reservoir fluid and formation. The objectives of new-generation geothermal drilling and well construction technologies are to reduce the cost and increase the useful life of geothermal production facilities through an integrated effort (see Table 4.6).

Improvements and innovations in deep drilling are expected as a result of the international Iceland Deep Drilling Project. The aim of this project is to penetrate into supercritical geothermal fluids, which can be a potential source of high-grade geothermal energy. The concept behind it is to flow supercritical fluid to the surface in such a way that it changes directly to superheated (>450°C) hot steam at sub-critical pressures. This would provide up to ten-fold energy output of approximately 50 MW$_e$ as compared to average high enthalpy geothermal wells (Fridleifsson et al., 2010).

All tasks related to the engineering of the reservoir require a more sophisticated modelling of the reservoir processes and interactions to be

---

[14] Chapter 10.5 offers a complementary perspective on drivers of and trends in technological progress across RE technologies. Chapter 8 deals with other integration issues more widely.

Table 4.6 | Priorities for advanced geothermal research (HTHF: high temperature and high flow rate).

Complementary research & share knowledge	Education / training
Standard geothermal resource & reserve definitions	Improved HTHF hard rock drill equipment
Predictive reservoir performance modelling	Improved HTHF multiple zone isolation
Predictive stress field characterization	Reliable HTHF slim-hole submersible pumps
Mitigate induced seismicity / subsidence	Improve resilience of casings to HTHF corrosion
Condensers for high ambient surface temperatures	Optimum HTHF fracture stimulation methods
Use of CO_2 as a circulating fluid for heat exchangers	HTHF logging tools and monitoring sensors
Improve power plant design	HTHF flow survey tools
Technologies & methods to minimize water use	HTHF fluid flow tracers
Predict heat flow and reservoirs ahead of the bit	Mitigation of formation damage, scale and corrosion

able to predict reservoir behaviour with time, to recommend management strategies for prolonged field operation and to minimize potential environmental impacts.

4.6.2 Efficient production of geothermal power, heat and/or cooling

Equipment needed to provide heating/cooling and/or electricity from geothermal wells is already available on the market. However, the efficiency of the different system components can still be improved, and it is even more important to develop conversion systems that more efficiently utilize energy in the produced geothermal fluid at competitive costs. It is basically inevitable that more efficient plants (and components) will have higher investment costs, but the objective would be to ensure that the increased performance justifies these costs. Combined heat and power (CHP) or cogeneration applications provide a means for significantly improving utilization efficiency and economics of geothermal projects, but one of the largest technical barriers is the inability in some cases to fully utilize the thermal energy produced (Bloomquist et al., 2001).

New and cost-effective materials for pipes, casing liners, pumps, heat exchangers and other components for geothermal plants is considered a prerequisite for reaching higher efficiencies.

Another possibility for an efficient type of geothermal energy production is the use of suitable oil fields. There are three types of oil and gas wells potentially capable of supplying geothermal energy for power generation: medium- to high-temperature (>120°C or so) producing wells with a sufficient water cut; abandoned wells due to a high water cut; and geo-pressured brine with dissolved gas. All of these types have been assessed and could be developed depending on the energy market evolution (Sanyal and Butler, 2010). The primary benefit from such a possibility is that the drilling is already in place and can greatly reduce the first costs associated with geothermal project development. However, these savings may be somewhat offset by the need to handle (separate and clean up) multi-phase co-produced fluids, consisting of water, hydrocarbons and other gases.

The potential development of valuable by-products may improve the economics of geothermal development, such as recovery of the condensate for industrial applications after an appropriate treatment, and in some cases recovery of valuable minerals from geothermal brines (such as lithium, zinc, high grade silica and in some cases, gold).

4.6.3 Technological and process challenges in enhanced geothermal systems

EGS require innovative methods, some of which are also applicable to power plants and direct-use projects based on hydrothermal resources. Among these are (Tester et al., 2006):

- Improvement and innovation in well drilling, casing, completion and production technologies for the exploration, appraisal and development of deep geothermal reservoirs (as generalized in Table 4.6).

- Improvement of methods to hydraulically stimulate reservoir connectivity between injection and production wells to attain sustained, commercial production rates. Reservoir stimulation procedures need to be refined to significantly enhance the productivity, while reducing the risk of seismic hazard. Imaging fluid pathways induced by hydraulic stimulation treatments through innovative technology would facilitate this. Technology development to create functional EGS reservoirs independent of local subsurface conditions will be essential.

- Development/adaptation of data management systems for interdisciplinary exploration, development and production of geothermal reservoirs, and associated teaching tools to foster competence and capacity amongst the people who will work in the geothermal sector.

- Improvement of numerical simulators for production history matching and predicting coupled thermal-hydraulic-mechanical-chemical processes during development and exploitation of reservoirs. In order to accurately simulate EGS reservoirs, computer codes must fully couple flow, chemistry, poro-elasticity and temperature. Development of suitable fully coupled reservoir simulators, including nonlinear deformability of fractures, is a necessity. Modern laboratory facilities capable of testing rock specimens under simulated down-hole conditions of pressure and temperature are also needed.

- Improvement in assessment methods to enable reliable predictions of chemical interactions between geo-fluids and geothermal reservoir rocks, geothermal plants and equipment, enabling optimized, well, plant and field lifetimes.

- Performance improvement of thermodynamic conversion cycles for a more efficient utilization of the thermal heat sources in district heating and power generation applications.

Conforming research priorities for EGS and magmatic resources as determined in Australia (DRET, 2008), the USA, the EU ((ENGINE, 2008), the Joint Programme on Geothermal Energy of the European Energy Research Alliance)[15] and the already-mentioned IPGT (see footnote in Section 4.5.3.2) are summarized in Table 4.6. Successful deployment of the associated services and equipment is also relevant to many conventional geothermal projects.

The required technology development would clearly reflect assessment of environmental impacts including land use and induced micro-seismicity hazards or subsidence risks (see Section 4.5).

The possibility of using CO_2 as a working fluid in geothermal reservoirs, particularly in EGS, has been under investigation. Recent modelling studies show that CO_2 would achieve heat extraction at higher rates than aqueous fluids, and that in fractured reservoirs CO_2 arrival at production wells would occur a few weeks after starting CO_2 injection. A two-phase water-CO_2 mixture could be produced for a few years followed by production of a single phase of supercritical CO_2 (Pruess and Spycher, 2010). In addition, it could provide a means for enhancing the effect of geothermal energy deployment for lowering CO_2 emissions beyond just generating electricity with a carbon-free renewable resource: a 5 to 10% loss rate of CO_2 from the system ('sequestered'), which is equivalent to the water loss rate observed at the Fenton Hill test in the USA, leads to 'sequestration' of 3 MW of coal burning per 1 MW of EGS electricity (Pruess, 2006). As of 2010, much remains to be done before such an approach is technically proven.

4.6.4 Technology of submarine geothermal generation

Currently no technologies are in use to tap submarine geothermal resources. However, in theory, electric energy could be produced directly from a hydrothermal vent using an encapsulated plant, like a submarine, containing an organic Rankine cycle (ORC) binary plant, as described by Hiriart and Espíndola (2005). The operation would be similar to other binary-cycle power plants using evaporator and condenser heat exchangers, with internal efficiency of the order of 80%. The overall efficiency for a submarine vent at 250°C of 4% (electrical power generated/thermal power) is a reasonable estimate for such an installation (Hiriart et al., 2010). Critical challenges for these resources include the distance from shore, water depth, grid connection costs, the current cable technology that limits ocean depths, and the potential impact on unique marine life around hydrothermal vents.

4.7 Cost trends[16]

Geothermal projects typically have high upfront investment costs due to the need to drill wells and construct power plants and relatively low operational costs. Operational costs vary depending on plant capacity, make-up and/or injection well requirements, and the chemical composition of the geothermal fluids. Without fuel costs, operating costs for geothermal plants are predictable in comparison to combustion-based power plants that are subject to market fluctuations in fuel prices. This section describes the fundamental factors affecting the levelized cost of electricity (LCOE) from geothermal power plants: upfront investment costs; financing costs (debt interest and equity rates); taxes; operation and maintenance (O&M) costs; decommissioning costs; capacity factor and the economic lifetime of the investment. This section also includes some historic and probable future trends, and presents investment and levelized costs of heat (LCOH) for direct uses of geothermal energy in addition to electric production.

Cost estimates for geothermal installations may vary widely (up to 20 to 25% not including subsidies and incentives) between countries (e.g., between Indonesia, the USA and Japan). EGS projects are expected to be more capital intensive than high-grade hydrothermal projects. Because there are no commercial EGS plants in operation, estimated costs are subject to higher uncertainties.

---

15 The Joint Programme on Geothermal Energy (JPGE) is described at: www.eera-set.eu/index.php?index=36.

16 Discussion of costs in this section is largely limited to the perspective of private investors. Chapters 1 and 8 to 11 offer complementary perspectives on cost issues covering, for example, costs of integration, external costs and benefits, economy-wide costs and costs of policies. All values are expressed in USD_{2005}.

4.7.1 Investment costs of geothermal-electric projects and factors that affect them

Investment costs of a geothermal-electric project are composed of the following components: (a) exploration and resource confirmation; (b) drilling of production and injection wells; (c) surface facilities and infrastructure; and (d) the power plant. Component costs and factors influencing them are usually independent from each other, and each component is described in the text that follows, including its impact on total investment costs.

The first component (a) includes lease acquisition, permitting, prospecting (geology and geophysics) and drilling of exploration and test wells. Drilling of exploration wells in greenfield areas is reported to have a success rate of typically about 50 to 60%, and the first exploration well of 25% (Hance, 2005), although other sources (GTP, 2008) reduce the percentage success to 20 to 25%. Confirmation costs are affected by well parameters (mainly depth and diameter), rock properties, well productivity, rig availability, time delays in permitting or leasing land, and interest rates. This first component represents between 10 and 15% of the total investment cost (Bromley et al., 2010) but for expansion projects may be as low as 1 to 3%.

Drilling of production and injection wells (component b) has a success rate of 60 to 90% (Hance, 2005; GTP, 2008). Factors influencing the cost include well productivity (permeability and temperature), well depths, rig availability, vertical or directional design, special circulation fluids, special drilling bits, number of wells and financial conditions in a drilling contract (Hance, 2005; Tester et al., 2006). This component (b) represents 20 to 35% of the total investment (Bromley et al., 2010).

The surface facilities and infrastructure component (c) includes facilities for gathering steam and processing brine: separators, pumps, pipelines and roads. Vapour-dominated fields have lower facility costs since brine handling is not required. Factors affecting this component are reservoir fluid chemistry, commodity prices (steel, cement), topography, accessibility, slope stability, average well productivity and distribution (pipeline diameter and length), and fluid parameters (pressure, temperature, chemistry) (Hance, 2005). Surface facilities and infrastructure costs represent 10 to 20% of the investment (Bromley et al., 2010) although in some cases these costs could be <10%, depending upon plant size and location.

Power plant components (d) include the turbines, generator, condenser, electric substation, grid hook-up, steam scrubbers and pollution abatement systems. Power plant design and construction costs depend upon type (flash, dry steam, binary, or hybrid), location, size (a larger unit and plant size is cheaper per unit of production (Dickson and Fanelli, 2003; Entingh and Mines, 2006), fluid enthalpy (resource temperature) and chemistry, type of cooling cycle used (water or air cooling) and cooling water availability if using water. This component varies between 40 and 81% of the investment (Hance, 2005; Bromley et al., 2010).

Some historic and current investment costs for typical geothermal-electric projects are shown in Figure 4.7. For condensing flash power plants, the current (2009) worldwide range is estimated to be USD_{2005} 1,780 to 3,560/kW_e, and for binary cycle plants USD_{2005} 2,130 to 5,200/kW_e (Bromley et al., 2010).

One additional factor affecting the investment cost of a geothermal-electric project is the type of project: field expansion projects may cost 10 to 15% less than a greenfield project, since investments have already been made in infrastructure and exploration and valuable resource information has been learned from drilling and producing start-up wells (Stefansson, 2002; Hance, 2005).

Most geothermal projects are financed with two different kinds of capital with different rates of return: equity and debt interest. Equity rates can be up to 20% while debt interest rates are lower (6 to 8%). The capital structure of geothermal-electric projects is commonly composed of 55 to 70% debt and 30 to 45% equity, but in the USA, debt lenders usually require 25% of the resource capacity to be proven before lending money. Thus, the early phases of the project often have to be financed by equity due to the higher risk of failure in these phases (Hance, 2005). Real and perceived risks play major roles in setting equity rates and in determining the availability of debt interest financing.

From the 1980s until about 2003-2004, investment costs remained flat or even decreased (Kagel, 2006; Mansure and Blankenship, 2008). Since then project costs have increased (Figure 4.7) due to increases in the cost of engineering, commodities such as steel and cement, and particularly drilling rig rates. This cost trend was not unique to geothermal and was mirrored across most other power sectors.

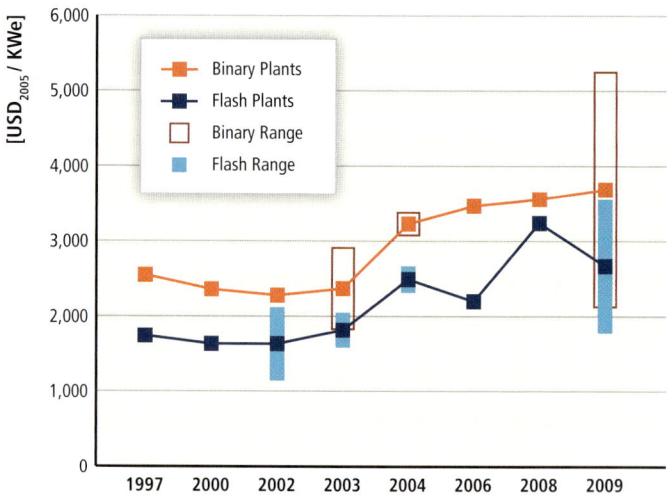

Figure 4.7 | Historic and current investment costs for typical turnkey (installed) geothermal-electric projects (rounded values taken from Kutscher, 2000; Owens, 2002; Stefansson, 2002; Hance, 2005; GTP, 2008; Cross and Freeman, 2009; Bromley et al., 2010; Hjartarson and Einarsson, 2010).

4.7.2 Geothermal-electric operation and maintenance costs

O&M costs consist of fixed and variable costs directly related to the electricity production phase. O&M per annum costs include field operation (labour and equipment), well operation and work-over and facility maintenance. For geothermal plants, an additional factor is the cost of make-up wells, that is, new wells to replace failed wells and restore lost production or injection capacity. Costs of these wells are typically lower than those for the original wells, and their success rate is higher.

Each geothermal power plant has specific O&M costs that depend on the quality and design of the plant, the characteristics of the resource, environmental regulations and the efficiency of the operator. The major factor affecting these costs is the extent of work-over and make-up well requirements, which can vary widely from field to field and typically increase with time (Hance, 2005). For the USA, O&M costs including make-up wells have been calculated to be between US cents$_{2005}$ 1.9 and 2.3/kWh (Lovekin, 2000; Owens, 2002), and Hance (2005) proposed an average cost of US cents$_{2005}$ 2.5/kWh. In terms of installed capacity, current O&M costs range between USD$_{2005}$ 152 and 187/kW per year, depending of the size of the power plant. In New Zealand, O&M costs range from US cents$_{2005}$ 1.0 to 1.4/kWh for 20 to 50 MW$_e$ plant capacity (Barnett and Quinlivan, 2009), which are equivalent to USD$_{2005}$ 83 to 117/kW per year.

4.7.3 Geothermal-electric performance parameters

One important performance parameter is the economic lifetime of the power plant. Twenty-five to thirty years is the common planned lifetime of geothermal power plants worldwide, although some of them have been in operation for more than 30 years, such as Units 1 and 2 in Cerro Prieto, Mexico (since 1973; Gutiérrez-Negrín et al., 2010), Eagle Rock and Cobb Creek in The Geysers, USA (since 1975 and 1979, respectively), and Mak-Ban A and Tiwi A, the Philippines (since 1979) (Bertani, 2010). This payback period allows for refurbishment or replacement of aging surface plants at the end of the plant lifetime, but is not equivalent to the economic lifetime of the geothermal reservoir, which is typically longer, for example, Larderello, The Geysers, Wairakei, Olkaria and Cerro Prieto, among others. In some reservoirs, however, the possibility of resource degradation over time is one of several factors that affect the economics of continuing plant operation.

Another performance parameter is the capacity factor (CF). The evolution of the worldwide average CF of geothermal power plants since 1995 is provided in Table 4.7, calculated from the installed capacity and the average annual generation as reported in different country updates gathered by Bertani (2010). For 2008, the installed capacity worldwide was 10,310 MW$_e$ (10,715 MW$_e$ as of the end of 2009, reduced by the 405 MW$_e$ added in 2009, according to Table X in Bertani (2010)), with an average CF of 74.5%. This worldwide average varies significantly by country and field. For instance, the annual average gross CF in 2008 for

Table 4.7 | World installed capacity, electricity production and capacity factor of geothermal power plants from 1995 to 2009 (adapted from data from Bertani (2010).

Year	Installed Capacity (GW$_e$)	Electricity Production (GWh/yr)	Capacity Factor (%)
1995	6.8	38,035	63.5
2000	8.0	49,261	70.5
2005	8.9	55,709	71.2
2008-2009[1]	10.7	67,246	74.5

Note: 1. Installed capacity as of December 2009, and electricity production as of December 2008. Installed capacity in 2008 was 10.3 GW$_e$ and was used to estimate the capacity factor of 74.5% shown here.

Mexico was 84% (data from Gutiérrez-Negrín et al., 2010), while for the USA it was 62% (Lund et al., 2010b) and in Indonesia it was 78% (Darma et al., 2010; data from their Table 1).

The geothermal CF worldwide average increased significantly between 1995 and 2000, with a lower increase in the last decade. This lower increase can be partially explained by the degradation in resource productivity (temperature, flow, enthalpy or combination of these) in geothermal fields operated for decades, although make-up drilling can offset this effect. The complementary explanation is that in the last decade some operating geothermal turbines have exceeded their economic lifetime, and thus require longer periods of shut-down for maintenance or replacement. For instance, out of the 48 geothermal-electric power units of >55 MW$_e$ operating in the world in 2009, 13 (27%) had been in operation for 27 years or more (Bertani, 2010, Table IX). Moreover, 15 new power plants, with a combined capacity of 456 MW$_e$, started to operate during 2008, but their generation contributed for only part of the year (Bertani, 2010, Table X). Typical CFs for new geothermal power plants are over 90% (Hance, 2005; DiPippo, 2008; Bertani, 2010).

4.7.4 Levelized costs of geothermal electricity

The current LCOE for geothermal installations (including investment cost for exploration, drilling and power plant and O&M costs) are shown in Figure 4.8.

The LCOE is presented as a function of CF, investment cost and discount rates (3, 7 and 10%), assuming a 27.5-year lifetime and using the values for worldwide investment and O&M costs shown in Figure 4.7 for 2009 and as presented in Section 4.7.2 (Bromley et al., 2010). As can be expected, the main conclusions from the figure are that the LCOE is proportional to investment cost and discount rate, and inversely proportional to CF, assuming the same average O&M costs. When lower O&M costs can be achieved, as is currently the case in New Zealand (Barnett and Quinlivan, 2009), the resulting LCOE would be proportionally lower. For greenfield projects, the LCOE for condensing flash plants currently ranges from US cents$_{2005}$ 4.9 to 7.2/kWh and, for binary-cycle plants, the LCOE ranges from US cents$_{2005}$ 5.3 to 9.2/kWh, at a CF of 74.5%, a 27.5-year economic design lifetime, and a discount rate of 7% and using the

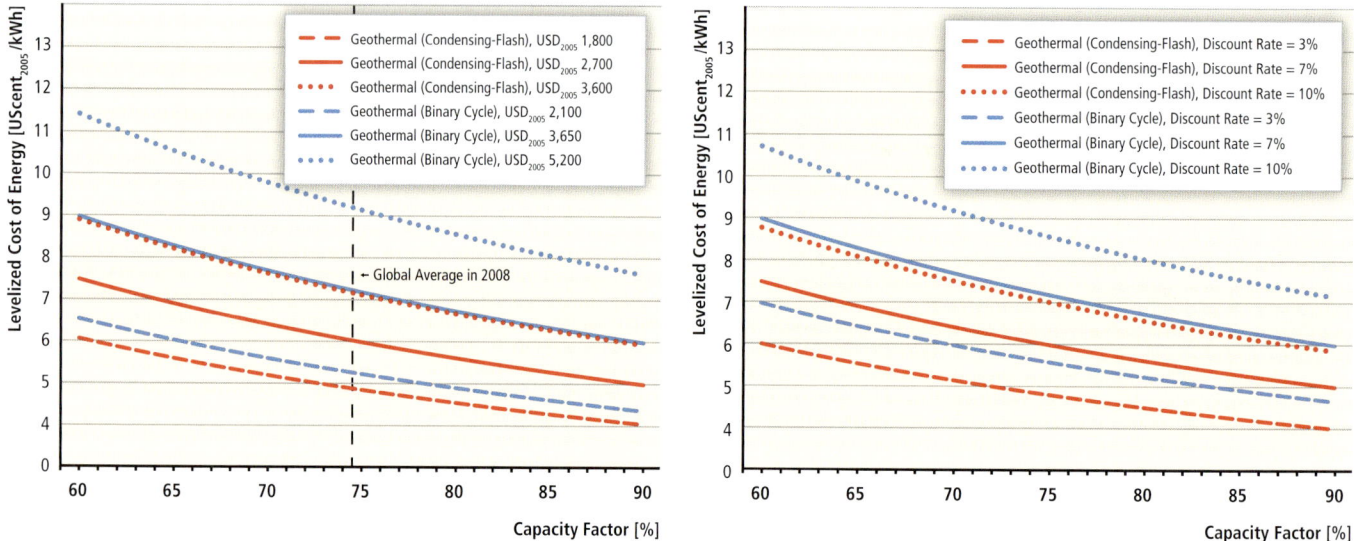

Figure 4.8 | Current LCOE for geothermal power generation as a function of (left panel) capacity factor and investment cost (discount rate at 7%, mid-value of the O&M cost range, and mid-value of the lifetime range), and (right panel) capacity factor and discount rate (mid-value of the investment cost range, mid-value of the O&M cost range, and mid-value of the lifetime range) (see also Annex III).

lowest and highest investment cost, respectively. Achieving a 90% lifetime average CF in new power plants can lead to a roughly 17% lower LCOE (Figure 4.8). The complete range of LCOE estimates, considering variations in plant lifetime, O&M costs, investment costs, discount rates and CFs, can vary from US cents$_{2005}$ 3.1 to 13/kWh for condensing flash plants and from US cents$_{2005}$ 3.3 to 17/kWh for binary plants (see also Annex III and Chapters 1 and 10).

No actual LCOE data exist for EGS, but some projections have been made using different models for several cases with diverse temperatures and depths (Table 9.5 in Tester et al., 2006). These projections do not include projected cost reductions due to future learning and technology improvements, and all estimates for EGS carry higher uncertainties than for conventional hydrothermal resources. The obtained LCOE values for the Massachusetts Institute of Technology EGS model range from US cents$_{2005}$ 10 to 17.5/kWh for relatively high-grade EGS resources (250°C to 330°C, 5-km depth wells) assuming a base case present-day productivity of 20 kg/s per well. Another model for a hypothetical EGS project in Europe considers two wells at 4 km depth, 125°C to 165°C reservoir temperature, 33 to 69 kg/s flow rate and a binary power unit of 1.6 MW$_e$ running with an annual capacity factor of 86%, and obtains LCOE values of US cents$_{2005}$ 30 to 37/kWh (Huenges and Frick, 2010).[17]

4.7.5 Prospects for future cost trends

The prospects for technical improvements outlined in Section 4.6 indicate that there is potential for cost reductions in the near and longer term for both conventional geothermal technology and EGS. Additionally, the future costs for geothermal electricity are likely to vary widely because

future deployment will include an increasing percentage of unconventional development types, such as EGS, as mentioned in Section 4.8.

The following estimates are based on possible cost reductions from design changes and technical advancements, relying solely on expert knowledge of the geothermal process value chain. Published learning curve studies for geothermal are limited, so the other major approach to forecasting future costs, extrapolating from historical learning rates, is not pursued here. See Section 10.5 for a more complete discussion of learning curves, including their advantages and limitations.

Foreseeable technological advances were presented in Section 4.6. Those potentially having the greatest impact on LCOEs in the near term are: (a) engineering improvements in design and stimulation of geothermal reservoirs; and (b) improvements in materials, operation and maintenance mentioned in Section 4.6.3 as well as some from Section 4.6.1. These changes will increase energy extraction rates and lead to a better plant performance, and less frequent and shorter maintenance periods, all of which will result in better CFs. With time, more efficient plants (with CFs of 90 and 95%) are expected to replace the older ones still in operation, increasing the average CF to between 80 and 95% (Fridleifsson et al., 2008). Accordingly, the worldwide average CF for 2020 is projected to be 80%, and could be 85% in 2030 and as high as 90% in 2050.

Important improvements in drilling techniques described in Section 4.6.2 are expected to reduce drilling costs. Drilling cost reductions due to increasing experience are also based on historic learning curves for deep oil and gas drilling (Tester et al., 2006). Since drilling costs represent at least between 20 and 35% of total investment cost (Section 4.7.1), and also impact the O&M cost due to the cost of make-up wells, a lower LCOE can be expected as drilling cost decreases. Additionally, an increased success rate for exploration, development and make-up

---

17 Further assumptions, for example, about O&M costs, lifetime, CFs and the discount rate may be available from the references.

wells is also foreseeable. Nevertheless, these reductions are unlikely to be achieved in the near term, and were not included in projections for LCOE reductions by 2020. Other improvements in exploration, surface installations, materials and power plants mentioned in Sections 4.6.2 and 4.6.3 are likely, and should lead to reduced costs.

Based on those premises, future potential LCOEs were calculated for 2020. For greenfield projects the worldwide average projected LCOE for condensing flash plants with a distribution of investment costs ranges from US cents$_{2005}$ 4.5 to 6.6/kWh and for binary-cycle plants ranges from US cents$_{2005}$ 4.9 to 8.6/kWh, at a CF of 80%, 27.5-year lifetime and discount rate of 7%. Therefore, a global average LCOE reduction of about 7% is expected for geothermal flash and binary plants by 2020.

For projected future costs for EGS, a sensitivity analysis of model variables carried out in Australia obtained near-term LCOE estimates of between AU$ 92 and AU$ 110 per MWh, equivalent to US cents$_{2005}$ 6.3 and 7.5/kWh, which are slightly higher than comparable estimates from Credit Suisse (Cooper et al., 2010). Another model (Sanyal et al., 2007) suggested that the LCOE for EGS will decline with increasing stimulated volume and replication of EGS units, with increasing the maximum practicable pumping rate from a well, and with the reduced rate of cooling of the produced fluid (LCOE increases approximately US cents$_{2005}$ 0.45/kWh per additional degree Celsius of cooling per year), which in turn can be achieved by improving the effectiveness of stimulation by closely spaced fractures (Sanyal, 2010). Tester et al. (2006) suggested that a four-fold improvement in productivity to 80 kg/s per well by 2030 would be possible and that the projected LCOE values would range from US cents$_{2005}$ 3.6 to 5.2/kWh for high-grade EGS resources, and for low-grade geologic settings (180°C to 220°C, 5- to 7-km depth wells) LCOE would also become more economically viable at about US cents$_{2005}$ 5.9 to 9.2/kWh.[18]

---

[18] Further assumptions, for example, about future O&M costs, lifetime, CFs and the discount rate may be available from the references.

4.7.6 Costs of direct uses and geothermal heat pumps

Direct-use project costs have a wide range, depending upon specific use, temperature and flow rate required, associated O&M and labour costs, and output of the produced product. In addition, costs for new construction are usually less than costs for retrofitting older structures. The cost figures given in Table 4.8 are based on a climate typical of the northern half of the USA or Europe. Heating loads would be higher for more northerly climates such as Iceland, Scandinavia and Russia. Most figures are based on cost in the USA (in USD$_{2005}$), but would be similar in developed countries and lower in developing countries (Lund and Boyd, 2009).

Some assumptions for the levelized cost of heat (LCOH) estimates presented in Table 4.8 are mentioned in Annex III. For building heating, assumptions included a load factor of 25 to 30%, investment cost of USD$_{2005}$ 1,600 to 3,900/kW$_{th}$ and a lifetime of 20 years, and for district heating, the same load factor, USD$_{2005}$ 600 to 1,600/kW$_{th}$ and a lifetime of 25 years. Thermal load density (heating load per unit of land area) is critical to the feasibility of district heating because it is one of the major determinants of the distribution network capital and operating costs. Thus, downtown high-rise buildings are better candidates than a single family residential area (Bloomquist et al., 2001). Generally, a thermal load density of about 1.2 × 10^9 J/hr/ha (120,000 J/hr/m^2) is recommended.

The LCOH calculation for greenhouses assumed a load factor of 0.50, and 0.60 for uncovered aquaculture ponds and tanks, with a lifespan of 20 years. Covered ponds and tanks have higher investment costs than uncovered ones, but lower heating requirements.

GHP project costs vary between residential installations and commercial/institutional installations. Heating and/or cooling large buildings lowers the investment cost and LCOH. In addition, the type of installation, closed loop (horizontal or vertical) or open loop using groundwater,

Table 4.8 | Investment costs and calculated levelized cost of heat (LCOH) for several geothermal direct applications (investment costs are rounded and taken from Lund, 1995; Balcer, 2000; Radeckas and Lukosevicius, 2000; Reif, 2008; Lund and Boyd, 2009).

Heat application	Investment cost USD$_{2005}$/kW$_{th}$	LCOH in USD$_{2005}$/GJ at discount rates of		
		3%	7%	10%
Space heating (buildings)	1,600–3,940	20–50	24–65	28–77
Space heating (districts)	570–1,570	12–24	14–31	15–38
Greenhouses	500–1,000	7.7–13	8.6–14	9.3–16
Uncovered aquaculture ponds	50–100	8.5–11	8.6–12	8.6–12
GHP (residential and commercial)	940–3,750	14–42	17–56	19–68

has a large influence on the installed cost (Lund and Boyd, 2009). The LCOH reported in Table 4.8 assumed 25 to 30% as the load factor and 20 years as the operational lifetime. It is worth taking into account that actual LCOH are influenced by electricity market prices, as operation of GHPs requires auxiliary power input. In the USA, recent trends in lower natural gas prices have resulted in poor GHP project economics compared to alternative options for heat supply, and drilling costs continue to be the largest barrier to GHP deployment.

Industrial applications are more difficult to quantify, as they vary widely depending upon the energy requirements and the product to be produced. These plants normally require higher temperatures and often compete with power plant use; however, they do have a high load factor of 0.40 to 0.70, which improves the economics. Industrial applications vary from large food, timber and mineral drying plants (USA and New Zealand) to pulp and paper plants (New Zealand).

4.8 Potential deployment[19]

Geothermal energy can contribute to near- and long-term carbon emissions reductions. In 2008, the worldwide geothermal-electric generation was 67.2 TWh$_e$ (Sections 4.4.1 and 4.7.3) and the heat generation from geothermal direct uses was 121.7 TWh$_{th}$ (Section 4.4.3). These amounts of energy are equivalent to 0.24 EJ/yr and 0.44 EJ/yr, respectively, for a total of 0.68 EJ/yr (direct equivalent method). The IEA (2010) reports only 0.41 EJ/yr (direct equivalent method) as the total primary energy supply from geothermal resources in 2008 (see Chapter 1); the reason for this difference is unclear. Regardless, geothermal resources provided only about 0.1% of the worldwide primary energy supply in 2008. By 2050, however, geothermal could meet roughly 3% of global electricity demand and 5% of the global demand for heating and cooling, as shown in Section 4.8.2.

This section starts by presenting near-term (2015) global and regional deployments expected for geothermal energy (electricity and heat) based on current geothermal-electric projects under construction or planned, observed historic growth rates, as well as the forecast generation of electricity and heat. Subsequently, this section presents the middle- and long-term (2020, 2030, 2050) global and regional deployments, compared to the IPCC AR4 estimate, displays results from scenarios reviewed in Chapter 10 of this report, and discusses their feasibility in terms of technical potential, regional conditions, supply chain aspects, technological-economic conditions, integration-transmission issues, and environmental and social concerns. Finally, the section presents a short conclusion regarding potential deployment.

---

19 Complementary perspectives on potential deployment based on a comprehensive assessment of numerous model-based scenarios of the energy system are presented in Chapter 10 and Sections 10.2 and 10.3 of this report.

4.8.1 Near-term forecasts

Reliable sources for near-term geothermal power deployment forecasts are the country updates recently presented at the *World Geothermal Congress 2010*. This congress is held every five years, and experts on geothermal development in several countries are asked to prepare and present a paper on the national status and perspectives. According to projections included in those papers, which are based on the capacity of geothermal-electric projects stated as under construction or planned, the geothermal-electric installed capacity in the world is expected to reach 18.5 GW$_e$ by 2015 (Bertani, 2010). This represents an annual average growth of 11.5% between 2010 and 2015, based on the present conditions and expectations of geothermal markets. This annual growth rate is larger than the historic rates observed between 1970 and 2010 (7%, Table 4.4), and reflects increased activity in several countries, as mentioned in Section 4.4.

Assuming the countries' projections of geothermal-electric deployment are fulfilled in the next five years, which is uncertain, the regional deployments by 2015 are shown in Table 4.9. Note that each region has its own growth rate but the average global rate is 11.5%. Practically all the new power plants expected to be on line by 2015 will be conventional (flash and binary) utilizing hydrothermal resources, with a small contribution from EGS projects. The worldwide development of EGS is forecasted to be slow in the near term and then accelerate, as expected technological improvements lower risks and costs (see Section 4.6).

The country updates did not include projections for geothermal direct uses (heat applications, including GHP). Projecting the historic annual growth rate in the period 1975 to 2010 (Table 4.4) for the following five years results in a global projection of 85.2 GW$_{th}$ of geothermal direct uses by 2015. The expected deployments and thermal generation by region are also presented in Table 4.9. By 2015, total electric generation could reach 121.6 TWh/yr (0.44 EJ/yr) while direct generation of heat, including GHP, could attain 224 TWh$_{th}$/yr (0.8 EJ/yr).

On a regional basis, the forecast deployment for harnessing identified and hidden hydrothermal resources varies significantly in the near term. In Europe, Africa and Central Asia, large deployment is expected in both electric and direct uses of geothermal, while in India and the Middle East, only a growing deployment in direct uses is projected with no electric uses projected over this time frame.

The existing installed capacity in North America (USA and Mexico) of 4 GW$_e$, mostly from mature developments, is expected to increase almost 60% by 2015, mainly in the USA (from 3,094 to 5,400 MW$_e$, according to Lund et al. (2010b) and Bertani (2010). In Central America, the future geothermal-electric deployment has been estimated at 4 GW$_e$ (Lippmann, 2002), of which 12% has been harnessed so far (~0.5 GW$_e$). South American countries, particularly along the

Table 4.9 | Regional current and forecast installed capacity for geothermal power and direct uses (heat, including GHP) and forecast generation of electricity and heat by 2015.

REGION*	Current capacity (2010)		Forecast capacity (2015)		Forecast generation (2015)	
	Direct (GW_{th})	Electric (GW_e)	Direct (GW_{th})	Electric (GW_e)	Direct (TW_{th}/yr)	Electric (TWh_e/yr)
OECD North America	13.9	4.1	27.5	6.5	72.3	43.1
Latin America	0.8	0.5	1.1	1.1	2.9	7.2
OECD Europe	20.4	1.6	32.8	2.1	86.1	13.9
Africa	0.1	0.2	2.2	0.6	5.8	3.8
Transition Economies	1.1	0.08	1.6	0.2	4.3	1.3
Middle East	2.4	0	2.8	0	7.3	0
Developing Asia	9.2	3.2	14.0	6.1	36.7	40.4
OECD Pacific	2.8	1.2	3.3	1.8	8.7	11.9
TOTAL	50.6	10.7	85.2	18.5	224.0	121.6

Notes: * For regional definitions and country groupings see Annex II.

Current and forecast data for electricity taken from Bertani (2010), and for direct uses from Lund et al. (2010a), both as of December 2009. Estimated average annual growth rate in 2010 to 2015 is 11.5% for power and 11% for direct uses. Average worldwide capacity factors of 75% (for electric) and 30% (for direct use) were assumed by 2015.

Andes mountain chain, also have significant untapped—and under-explored—hydrothermal resources (Bertani, 2010).

For island nations with mature histories of geothermal development, such as New Zealand, Iceland, the Philippines and Japan, identified geothermal resources could allow for a future expansion potential of two to five times existing installed capacity, although constraints such as limited grid capacity, existing or planned generation (from other renewable energy sources) and environmental factors (such as national park status of some resource areas) may limit the hydrothermal geothermal deployment. Indonesia is thought to be one of the world's richest countries in geothermal resources and, along with other volcanic islands in the Pacific Ocean (Papua-New Guinea, Solomon, Fiji, etc.) and the Atlantic Ocean (Azores, Caribbean, etc.) has significant potential for growth from known hydrothermal resources, but is market-constrained in growth potential.

Remote parts of Russia (Kamchatka) and China (Tibet) contain identified high-temperature hydrothermal resources, the use of which could be significantly expanded given the right incentives and grid access to load centres. Parts of other South-East Asian nations and India contain numerous hot springs, inferring the possibility of potential, as yet unexplored, hydrothermal resources.

Additionally, small-scale distributed geothermal developments could be an important base-load power source for isolated population centres in close proximity to geothermal resources, particularly in areas of Indonesia, the Philippines and Central and South America.

4.8.2 Long-term deployment in the context of carbon mitigation

The IPCC Fourth Assessment Report (AR4) estimated a potential contribution of geothermal to world electricity supply by 2030 of 633 TWh/yr (2.28 EJ/yr), equivalent to about 2% of the total (Sims et al., 2007). Other forecasts for the same year range from 173 TWh/yr (0.62 EJ/yr) (IEA, 2009) to 1,275 TWh/yr (4.59 EJ/yr) (Teske et al., 2010).

A summary of the literature on the possible future contribution of RE supplies in meeting global energy needs under a range of GHG concentration stabilization scenarios is provided in Chapter 10. Focusing specifically on geothermal energy, Figure 4.9 (left) presents modelling results for the global supply of geothermal energy in EJ/yr. About 120 different long-term scenarios underlie Figure 4.9 that derive from a diversity of modelling teams, and span a wide range of assumptions for—among other variables—energy demand growth, the cost and availability of competing low-carbon technologies, and the cost and availability of RE technologies (including geothermal energy).

Chapter 10 discusses how changes to some of these variables impact RE deployment outcomes, with Section 10.2.2 providing a description of the literature from which the scenarios have been taken. In Figure 4.9 (left) the geothermal energy deployment results under these scenarios for 2020, 2030 and 2050 are presented for three GHG concentration stabilization ranges, based on the AR4: Baselines (>600 ppm CO_2), Categories III and IV (440 to 600 ppm) and Categories I and II (<440 ppm), all by 2100. Results are presented for the median scenario, the 25th to 75th percentile range among the scenarios, and the minimum and maximum scenario results. Primary energy is provided as direct equivalent, that is, each unit of heat or electricity is accounted for as one unit at the primary energy level.[20]

The long-term projections presented in Figure 4.9 (left) span a broad range. The 25th to 75th percentile ranges of all three scenarios are 0.07

---

[20] In scenario ensemble analyses such as the review underlying Figure 4.9, there is a constant tension between the fact that the scenarios are not truly a random sample and the sense that the variation in the scenarios does still provide real and often clear insights into collective knowledge or lack of knowledge about the future (see Section 10.2.1.2 for a more detailed discussion).

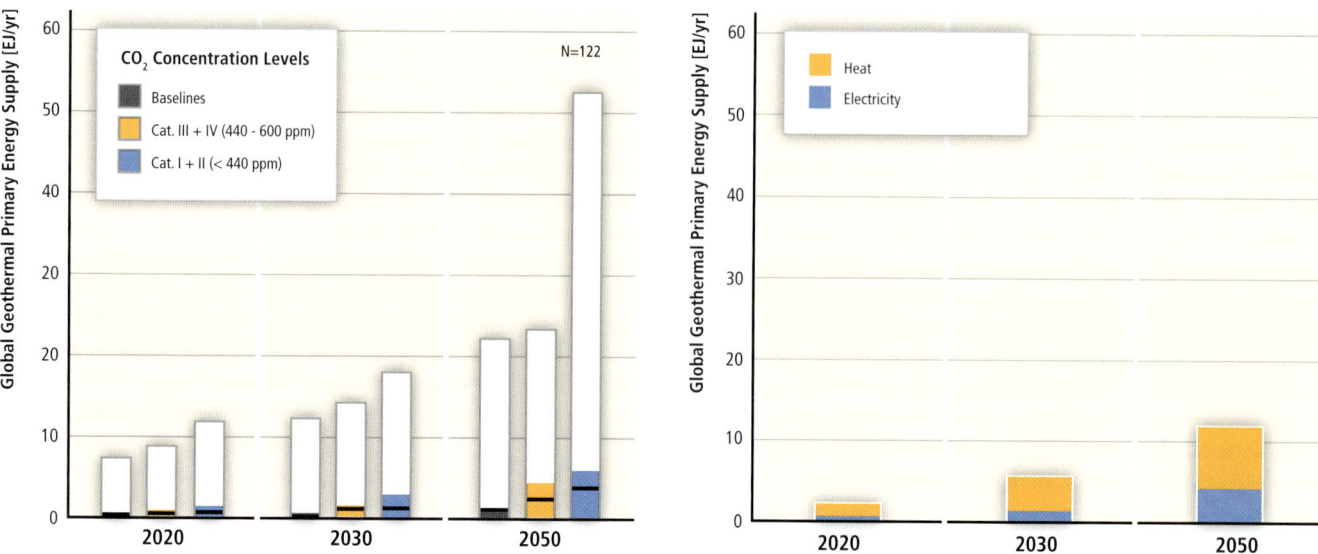

Figure 4.9 | Global primary energy supply of geothermal energy. Left panel: In long-term scenarios (median, 25th to 75th percentile range, and full range of scenario results; colour coding is based on categories of atmospheric CO_2 concentration level in 2100; the specific number of scenarios underlying the figure is indicated in the right upper corner) (adapted from Krey and Clarke, 2011; see also Chapter 10). Right panel: Estimated in Section 4.8.2 as potential geothermal deployments for electricity and heat applications.

Table 4.10 | Potential geothermal deployments for electricity and direct uses in 2020 through 2050.

Year	Use	Capacity[1] (GW)	Generation (TWh/yr)	Generation (EJ/yr)	Total (EJ/yr)
2020	Electricity	25.9	181.8	0.65	2.01
	Direct	143.6	377.5	1.36	
2030	Electricity	51.0	380.0	1.37	5.23
	Direct	407.8	1,071.7	3.86	
2050	Electricity	150.0	1,182.8	4.26	11.83
	Direct	800.0	2,102.3	7.57	

Note: 1. Installed capacities for 2020 and 2030 are extrapolated from 2015 estimates at 7% annual growth rate for electricity and 11% for direct uses, and for 2050 are the middle value between projections from Bertani (2010) and Goldstein et al. (2011). Generation was estimated with an average worldwide CF of 80% (2020), 85% (2030) and 90% (2050) for electricity and of 30% for direct uses.

to 1.38 EJ/yr by 2020, 0.10 to 2.85 EJ/yr by 2030 and 0.11 to 5.94 EJ/yr by 2050. The scenario medians range from 0.39 to 0.71 EJ/yr for 2020, 0.22 to 1.28 EJ/yr for 2030 and 1.16 to 3.85 EJ/yr for 2050. The medians for 2030 are lower than the IPCC AR4 estimate of 2.28 EJ/yr, which is for electric generation only, although the latter lies in the 25th to 75th percentile range of the most ambitious GHG concentration stabilization scenarios presented in Figure 4.9 (left). Figure 4.9 (left) shows that geothermal deployment is sensitive to the GHG concentration level, with greater deployment correlated with lower GHG concentration stabilization levels.

Based on geothermal technical potentials and market activity discussed in Sections 4.2 and 4.4, and on the expected geothermal deployment by 2015, the projected medians for geothermal energy supply and the 75th percentile amounts of all the modelled scenarios are technically reachable for 2020, 2030 and 2050.

As indicated above, climate policy is likely to be one of the main driving factors of future geothermal development, and under the most favourable policy of CO_2 emissions (<440 ppm) geothermal deployment by 2020, 2030 and 2050 could be higher than the 75th percentile estimates of Figure 4.9, as a simple extrapolation exercise shows. By projecting the historic average annual growth rates of geothermal power plants (7%) and direct uses (11%) from the estimates for 2015 (Table 4.9), the geothermal deployment in 2020 and 2030 would reach the figures shown in Table 4.10 (see also Figure 4.9, right).

By 2050 the projected installed capacity of geothermal power plants would be between 140 GW_e (Bertani, 2010) and 160 GW_e (Goldstein et al., 2011), with one-half of them being of EGS type, while the potential installed capacity for direct uses could reach 800 GW_{th} (Bertani, 2010). Potential deployment and generation for 2050 are also shown in Table 4.10 and Figure 4.9 (right).

The total contribution (thermal and electric) of geothermal energy would be 2 EJ/yr by 2020, 5.2 EJ/yr by 2030 and 11.8 EJ/yr by 2050 (Table 4.10), where each unit of heat or electricity is accounted for as one unit at the primary energy level. These estimates practically double the estimates for the 75th percentile of Figure 4.9, because many of the approximately 120 reviewed scenarios have not included the potential for EGS development in the long term.

Future geothermal deployment may not follow its historic growth rate between 2015 and 2030. In fact, it could be higher (e.g., Krewitt et al., (2009) adopted an annual growth rate of 10.4% for electric deployment between 2005 and 2030), or lower. Yet the results from this extrapolation exercise indicate that future geothermal deployment may reach levels in the 75 to 100% range of Figure 4.9 rather than in the 25 to 75% range.

Note that for 2030, the extrapolated geothermal electric generation of 380 TWh/yr (1.37 EJ/yr) is lower than the IPCC AR4 estimate (633 TWh/yr or 2.28 EJ/yr).

Teske et al. (2010) estimate the electricity demand to be 25,851 to 27,248 TWh/yr by 2020, 30,133 to 34,307 TWh/yr in 2030 and 37,993 to 46,542 TWh/yr in 2050. The geothermal share would be around 0.7% of global electric demand by 2020, 1.1 to 1.3% by 2030 and 2.5 to 3.1% by 2050.

Teske et al. (2010) project the global demand for heating and cooling by 2020 to be 156.8 EJ/yr, 162.4 EJ/yr in 2030 and 161.7 EJ/yr in 2050. Geothermal would then supply about 0.9% of the total demand by 2020, 2.4% by 2030 and 4.7% by 2050.

The high levels of deployment shown in Figure 4.9 could not be achieved without economic incentive policies to reduce GHG emissions and increase RE. Policy support for research and development (subsidies, guarantees and tax write-offs for initial deep drilling) would assist in the demonstration and commercialization of some geothermal technologies such as EGS and other non-conventional geothermal resource development. Feed-in tariffs with confirmed geothermal prices, and direct subsidies for district and building heating would also help to accelerate deployment. The deployment of geothermal energy can also be fostered with drilling subsidies, targeted grants for pre-competitive research and demonstration to reduce exploration risk and the cost of EGS development. In addition, the following issues are worth noting.

Resource potential: Even the highest estimates for the long-term contribution of geothermal energy to the global primary energy supply (52.5 EJ/yr by 2050, Figure 4.9, left) are well within the technical potentials described in Section 4.2 (118 to 1,109 EJ/yr for electricity and 10 to 312 EJ/yr for heat, see Figure 4.2) and even within the upper range of hydrothermal resources (28.4 to 56.8 EJ/yr). Thus, technical potential is not likely to be a barrier in reaching more ambitious levels of geothermal deployment (electricity and direct uses), at least on a global basis.

Regional deployment: Future deployment of geothermal power plants and direct uses are not the same for every region. Availability of financing, water, transmission and distribution infrastructure and other factors will play major roles in regional deployment rates, as will local geothermal resource conditions. For instance, in the USA, Australia and Europe, EGS concepts are already being field tested and deployed, providing advantages for accelerated deployment in those regions as risks and uncertainties are reduced. In other rapidly developing regions in Asia, Africa and South America, as well as in remote and island settings where distributed power supplies are needed, factors that would affect deployment include market power prices, population density, market distance, electricity and heating and cooling demand.

Supply chain issues: No mid- or long-term constraints to materials supply, labour availability or manufacturing capacity are foreseen from a global perspective.

Technology and economics: GHP, district heating, hydrothermal and EGS methods are available, with different degrees of maturity. GHP systems have the widest market penetration, and an increased deployment can be supported by improving the coefficient of performance and installation efficiency. The direct use of thermal fluids from deep aquifers, and heat extraction using EGS, can be increased by further technical advances in accessing and fracturing geothermal reservoirs. Combined heat and power applications may also be particularly attractive for EGS and low-temperature hydrothermal resource deployment. To achieve a more efficient and sustainable geothermal energy supply, subsurface exploration risks need to be reduced and reservoir management needs to be improved by optimizing injection strategies and avoiding excessive depletion. Improvement in energy utilization efficiency from cascaded use of geothermal heat is an effective deployment strategy when markets permit. Evaluation of geothermal plants performance, including heat and power EGS installations, needs to take into account heat quality of the fluid by considering the useful energy that can be converted to electric power. These technological improvements will influence the economics of geothermal energy.

Integration and transmission: The site-specific geographic location of conventional hydrothermal resources results in transmission constraints for future deployment. However, no integration problems have been observed once transmission issues are solved, due to the base-load characteristic of geothermal electricity. In the long term, fewer transmission constraints are foreseen since EGS developments are less geography-dependent, even though EGS' resource grades can vary substantially on a regional basis.

Social and environmental concerns: Concerns expressed about geothermal energy development include the possibility of induced local seismicity for EGS, water usage by geothermal power plants in arid regions, land subsidence in some circumstances, concerns about water and soil contamination and potential impacts of facilities on scenic quality and use of natural areas and features (such as geysers) that might otherwise be used for tourism. Sustainable practices will help protect natural thermal features valued by the community, optimize water and land use and minimize adverse effects from disposal of geothermal fluids and gases, induced seismicity and ground subsidence.

4.8.3 Conclusions regarding deployment

Overall, the geothermal-electric market appears to be accelerating compared to previous years, as indicated by the increase in installed and planned power capacity. The gradual introduction of new technology improvements, including EGS, is expected to boost the deployment, which could reach 140 to 160 GW_e by 2050 if certain conditions are met. Some new technologies are entering the field demonstration phase to evaluate commercial viability (e.g., EGS), or the early investigation stage to test practicality (e.g., utilization of supercritical temperature and submarine hydrothermal vents). Power generation with binary plants permits the possibility of producing electricity in countries that have no high-temperature resources, though overall costs are higher than for high-temperature resources.

Direct use of geothermal energy for heating and cooling is competitive in certain areas, using accessible, hydrothermal resources. A moderate increase can be expected in the future development of such resources for direct use, but a sustained compound annual growth is expected with the deployment of GHP. Direct use in lower-grade regions for heating and/or cooling in most parts of the world could reach 800 GW_{th} by 2050 (Section 4.8.2). Cogeneration and hybridization with other thermal sources may provide additional opportunities.

Evidence suggests that geothermal supply could meet the upper range of projections derived from a review of about 120 energy and GHG-reduction scenarios. With its natural thermal storage capacity, geothermal is especially suitable for supplying base-load power. Considering its technical potential and possible deployment, geothermal energy could meet roughly 3% of global electricity demand by 2050, and also has the potential to provide roughly 5% of the global demand for heating and cooling by 2050.

References

AL-AGEA (2009). *Geothermal Energy in Australia.* Prepared by Activated Logic Pty Ltd. for the Australian Geothermal Energy Association (AGEA), Unley, South Australia, 53 pp. Available at: www.agea.org.au.

Armstead, H.C.H., and J.W. Tester (1987). *Heat Mining.* E&FN Spon Ltd, London, UK and New York, NY, USA, 478 pp (ISBN 0-419-12230-3).

Arpasi, M. (2005). Geothermal update of Hungary 2000-2004. In: *Proceedings World Geothermal Congress 2005*, Antalya, Turkey, 24-29 April 2005 (ISBN 9759833204). Available at: www.geothermal-energy.org/pdf/IGAstandard/WGC/2005/0127.pdf.

Axelsson, G., V. Stefánsson, G. Björnsson, and J. Liu (2005). Sustainable management of geothermal resources and utilisation for 100 – 300 years. In: *Proceedings World Geothermal Congress 2005*, Antalya, Turkey, 24-29 April 2005 (ISBN 9759833204). Available at: www.geothermal-energy.org/pdf/IGAstandard/WGC/2005/0507.pdf.

Axelsson, G.V., C.J. Bromley, M.A. Mongillo, and L. Rybach (2010). Sustainability task of the International Energy Agency's Geothermal Implementing Agreement. In: *Proceedings World Geothermal Congress 2010*, Bali, Indonesia, 25-30 April 2010. Available at: www.geothermal-energy.org/pdf/IGAstandard/WGC/2010/0512.pdf.

Balcer, M. (2000). Infrastruktura techniczna zakladu geotermalnego w Mszczonowie (in Polish). In: *Symposium on the Role of Geothermal Energy in the Sustainable Development of the Mazovian and Lodz Regions (Rola energii geotermalnej w zrównowazonym rozwoju regionów Mazowieckiego i Lodzkiego)*, Mineral and Energy Economy Research Institute, Polish Academy of Sciences, Cracow, Poland, 4-6 October 2000, pp. 107-114 (ISBN 83-87854-62-X).

Barnett, P., and P. Quinlivan (2009). *Assessment of Current Costs of Geothermal Power Generation in New Zealand (2007 Basis).* Report by SKM for New Zealand Geothermal Association, Wellington, New Zealand. Available at: www.nzgeothermal.org.nz\industry_papers.html.

Bertani, R. (2010). Geothermal power generation in the world - 2005–2010 update report. In: *Proceedings World Geothermal Congress 2010*, Bali, Indonesia, 25-30 April 2010. Available at: www.geothermal-energy.org/pdf/IGAstandard/WGC/2010/0008.pdf.

Bertani, R., and I. Thain (2002). Geothermal power generating plant CO_2 emission survey. *International Geothermal Association (IGA) News*, **49**, pp. 1-3 (ISSN: 0160-7782). Available at: www.geothermal-energy.org/308,iga_newsletter.html.

Bloomquist, R.G., J. Nimmons, and M. Spurr (2001). *Combined Heat and Power – Legal, Institutional, Regulatory.* No. WSUCEEP 01-013, Washington State University, Cooperative Extension, Energy Program, Olympia, Washington, 122 pp.

Bromley, C.J., and M.A. Mongillo (2008). Geothermal energy from fractured reservoirs: dealing with induced seismicity. *IEA OPEN Energy Technology Bulletin Feb. 2008*, **48**, 7 pp. Available at: www.iea.org/impagr/cip/pdf/Issue48Geothermal.pdf.

Bromley, C.J., L. Rybach, M.A. Mongillo, and I. Matsunaga (2006). Geothermal resources – utilization strategies to promote beneficial environmental effects and to optimize sustainability. In: *Proceedings RE 2006*, Chiba, Japan, 9-13 October 2006, pp. 1637-1640. In: *Science and Technology in Japan*, **25**(100), 2007.

Bromley, C.J., M.A. Mongillo, B. Goldstein, G. Hiriart, R. Bertani, E. Huenges, H. Muraoka, A. Ragnarsson, J. Tester, and V. Zui (2010). Contribution of geothermal energy to climate change mitigation: the IPCC renewable energy report. In: *Proceedings World Geothermal Congress 2010*, Bali, Indonesia, 25-30 April 2010. Available at: www.geothermal-energy.org/pdf/IGAstandard/WGC/2010/0225.pdf.

Burgassi, P.D. (1999). Historical outline of geothermal technology in the Larderello region to the middle of the 20th century. In: *Stories from a Heated Earth.* R. Cataldi, S. Hodgson and J.W. Lund (eds.), Geothermal Resources Council and International Geothermal Association, Sacramento, CA, USA, pp. 195-219 (ISBN: 0934412197).

Cataldi, R. (1999). The year zero of geothermics. In: *Stories from a Heated Earth.* R. Cataldi, S. Hodgson and J.W. Lund (eds.), Geothermal Resources Council and International Geothermal Association, Sacramento, CA, USA, pp. 7-17 (ISBN: 0934412197).

Cloetingh, S., J.D.v. Wees, P.A. Ziegler, L. Lenkey, F. Beekman, M. Tesauro, A. Förster, B. Norden, M. Kaban, N. Hardebol, D. Bonté, A. Genter, L. Guillou-Frottier, M.T. Voorde, D. Sokoutis, E. Willingshofer, T. Cornu, and G. Worum (2010). Lithosphere tectonics and thermo-mechanical properties: An integrated modelling approach for Enhanced Geothermal Systems exploration in Europe. *Earth-Science Reviews*, **102**(3-4), pp. 159-206.

Cooper, G.T., G.R. Beardsmore, B.S. Waining, N. Pollington, and J.P. Driscoll (2010). The relative cost of Engineered Geothermal Systems exploration and development in Australia. In: *Proceedings World Geothermal Congress 2010*, Bali, Indonesia, 25-29 April, 2010. Available at: www.geothermal-energy.org/pdf/IGAstandard/WGC/2010/3115.pdf.

Cross, J., and J. Freeman (2009). *2008 Geothermal Technologies Market Report.* Geothermal Technologies Program of the US Department of Energy, Washington, DC, USA, 46 pp. Available at: www1.eere.energy.gov/geothermal/pdfs/2008_market_report.pdf.

Darma, S., S. Harsoprayitno, B. Setiawan, Hadyanto, R. Sukhyar, A.W. Soedibjo, N. Ganefianto, and J. Stimac (2010). Geothermal energy update: Geothermal energy development and utilization in Indonesia. In: *Proceedings World Geothermal Congress 2010*, Bali, Indonesia, 25-29 April, 2010. Available at: www.geothermal-energy.org/pdf/IGAstandard/WGC/2010/0128.pdf.

De Jesus, A.C. (2005). Social issues raised and measures adopted in Philippine geothermal projects. In: *Proceedings World Geothermal Congress 2005*, Antalya, Turkey, 24-29 April, 2005 (ISBN 9759833204). Available at: www.geothermal-energy.org/pdf/IGAstandard/WGC/2005/0219.pdf.

Dickson, M.H., and M. Fanelli (2003). *Geothermal energy: Utilization and technology.* Renewable Energy Series, United Nations Educational, Scientific and Cultural Organization, Paris, France, 205 pp. (ISBN: 92-3-103915-6).

DiPippo, R. (1991). Geothermal energy: Electricity generation and environmental impact. *Energy Policy*, **19**, pp. 798-807.

DiPippo, R. (2008). *Geothermal Power Plants: Principles, Applications, Case Studies and Environmental Impact.* Elsevier, London, UK, 493 pp. (ISBN: 9780750686204).

DRET, 2008: *Australian Geothermal Industry Development Framework.* Department of Resources, Energy and Tourism, Commonwealth of Australia, Canberra, Australia. (ISBN 978-1-921516-11-5 [paperback], ISBN 978-1-921516-15-3 [pdf]). Available at: www.ret.gov.au/energy/clean_energy_technologies/energy_technology_framework_and_roadmaps/hydrogen_technology_roadmap/Documents/GEOTHERMAL%20FRAMEWORK.pdf.

ENGINE (2008). Propositions for the definition of research areas on Enhanced Geothermal Systems. *ENGINE Newsletter No. 11 – June 2008*, Enhanced Geothermal Innovative Network for Europe, Orléans, France, pp. 4-7. Available at: engine.brgm.fr/bulletins/ENGINE_Newsletter11_062008.pdf.

Entingh, D.J., and G. Mines (2006). A framework for evaluating research to improve U.S. geothermal power systems. *Transactions of the Geothermal Resources Council*, **30**, pp. 741-746.

EPRI (1978). *Geothermal Energy Prospects for the Next 50 Years – ER-611-SR, Special Report for the World Energy Conference.* Electric Power Research Institute, Palo Alto, CA, USA.

Frick, S., G. Schröder, and M. Kaltschmitt (2010). Life cycle analysis of geothermal binary power plants using enhanced low temperature reservoirs. *Energy*, **35**(5), pp. 2281-2294.

Fridleifsson, I.B., and A. Ragnarsson (2007). Geothermal energy. In: *2007 Survey of Energy Resources.* World Energy Council, London, UK, pp. 427-437 (ISBN: 0946121 26 5). Available at: www.worldenergy.org/documents/ser2007_final_online_version_1.pdf.

Fridleifsson, I.B., R. Bertani, E. Huenges, J.W. Lund, A. Ragnarsson, and L. Rybach (2008). The possible role and contribution of geothermal energy to the mitigation of climate change. In: *IPCC Scoping Meeting on Renewable Energy Sources*, Luebeck, Germany, 21-25 January 2008, pp. 36. Available at: www.ipcc.ch/pdf/supporting-material/proc-renewables-lubeck.pdf.

Fridleifsson, G.O., B. Pálsson, B. Stefánsson, A. Albertsson, E. Gunnlaugsson, J. Ketilsson, R. Lamarche, and P.E. Andersen (2010). Iceland Deep Drilling Project. The first IDDP drill hole drilled and completed in 2009. In: *Proceedings World Geothermal Congress 2010*, Bali, Indonesia, 25-29 April, 2010. Available at: www.geothermal-energy.org/pdf/IGAstandard/WGC/2010/3902.pdf.

Gawell, K., and G. Greenberg. (2007). *2007 Interim Report. Update on World Geothermal Development.* Geothermal Energy Association, Washington, DC, USA. Available at: www.geo-energy.org/reports/GEA%20World%20Update%202007.pdf.

German, C.R., G.P. Klinkhammer, and M.D. Rudnicki (1996). The Rainbow Hydrothermal Plume, 36°15'N, MAR. *Geophysical Research Letters*, **23**(21), pp. 2979-2982.

Goldstein, B.A., A.J. Hill, A. Long, A.R. Budd, B. Ayling, and M. Malavazos (2009). Hot rocks down under – Evolution of a new energy industry. *Transactions of the Geothermal Resources Council*, **33**, pp. 185-198.

Goldstein, B.A., G. Hiriart, J.W. Tester, R. Bertani, C.J. Bromley, L.C. Gutiérrez-Negrín, E. Huenges, A. Ragnarsson, M.A. Mongillo, H. Muraoka, and V.I. Zui (2011). Great expectations for geothermal energy to 2100. In: *Proceedings of the Thirty-Sixth Workshop of Geothermal Reservoir Engineering*, Stanford University, Stanford, CA, 31 January – 2 February 2011, SGP-TR-191, pp. 5-12.

Grant, M.A., I.G. Donaldson, and P.F. Bixley (1982). *Geothermal Reservoir Engineering.* Academic Press, New York, NY, USA.

GTP (2008). *Geothermal Tomorrow 2008.* DOE-GO-102008-2633, Geothermal Technologies Program of the US Department of Energy, Washington, DC, USA, 36 pp.

Gutiérrez-Negrín, L.C.A., R. Maya-González, and J.L. Quijano-León (2010). Current status of geothermics in Mexico. In: *Proceedings World Geothermal Congress 2010*, Bali, Indonesia, 25-29 April 2010. Available at: www.geothermal-energy.org/pdf/IGAstandard/WGC/2010/0101.pdf.

Hamza, V.M., R.R. Cardoso, and C.F.P. Neto (2008). Spherical harmonic analysis of Earth's conductive heat flow. *International Journal of Earth Sciences*, **97**(2), pp. 205-226.

Hance, C.N. (2005). *Factors Affecting Costs of Geothermal Power Development.* Geothermal Energy Association, for the U.S. Department of Energy, Washington, DC, USA, 64 pp. Available at: www.geo-energy.org/reports/Factors%20Affecting%20Cost%20of%20Geothermal%20Power%20Development%20-%20August%202005.pdf.

Hettkamp, T., J. Baumgärtner, D. Teza, P. Hauffe, and B. Rogulic (2010). *Erfahrungen aus dem Geothermieprojekt in Landau (Experience from the geothermal project in Landau).* Institute for Geotechnik at TU Bergakademie, Freiberg, Publisher: H. Konietzky, S., pp. 43-54 (ISSN 1611-1605).

Hiriart, G. (2007). Impacto del cambio climático sobre la generación eléctrica con fuente de energía geotérmica (Impact of climate change on power generation from geothermal energy source). In: *Escenarios de energías renovables en México bajo cambio climático*. A. Tejeda-Martínez, C. Gay-García, G. Cuevas-Guillaumín and C.O. Rivera-Blanco (eds.), pp. 83-105. Available at: www.ine.gob.mx/descargas/cclimatico/e2007q.pdf.

Hiriart, G., and S. Espíndola (2005). Aprovechamiento de las ventilas hidrotermales para generar electricidad. In: *Memorias de la VIII Conferencia Anual de la AMEE*, Colegio de México, 28-29 November 2005, pp. 153-159. Available at: http://www.iie.org.mx/economia-energetica/MemoriasVIIIConfAnualAMEE.pdf.

Hiriart, G., R.M. Prol-Ledesma, S. Alcocer, and S. Espíndola (2010). Submarine geothermics: Hydrothermal vents and electricity generation. In: *Proceedings World Geothermal Congress 2010*, Bali, Indonesia, 25-29 April, 2010. Available at: http://www.geothermal-energy.org/pdf/IGAstandard/WGC/2010/3704.pdf.

Hjartarson, A., and J.G. Einarsson (2010). *Geothermal resources and properties of HS Orka, Reyjanes Peninsula, Iceland.* Independent Technical Report prepared by Mannvit Engineering for Magma Energy Corporation, Vancouver, Canada, 151 pp. Available upon request at: www.mannvit.com.

Holm, A., L. Blodgett, D. Jennejohn, and K. Gawell (2010). *Geothermal Energy: International Market Update.* The Geothermal Energy Association, Washington, DC, USA, 77 pp. Available at: www.geo-energy.org/pdf/reports/GEA_International_Market_Report_Final_May_2010.pdf.

Huenges, E., and S. Frick (2010). Costs of CO_2 mitigation by deployment of Enhanced Geothermal Systems plants. In: *Proceedings World Geothermal Congress 2010*, Bali, Indonesia, 25-30 April 2010. Available at: www.geothermal-energy.org/pdf/IGAstandard/WGC/2010/0238.pdf.

Huenges, E., K. Erbas, I. Moeck, G. Blöcher, W. Brandt, T. Schulte, A. Saadat, G. Kwiatek, and G. Zimmermann (2009). The EGS project Groß Schönebeck – Current status of the large scale research project in Germany. *Transactions of the Geothermal Resources Council*, **39**, pp. 403-408.

IEA (2009). *World Energy Outlook 2009.* International Energy Agency, Paris, France, 696 pp.

IEA (2010). *Key World Energy Statistics 2010.* International Energy Agency, Paris, France.

IEA-GIA (2009). *IEA Geothermal Energy 12th Annual Report 2008.* International Energy Agency – Geothermal Implementing Agreement, Paris, France, 257 pp.

Imolauer, K., B. Richter, and A. Berger (2010). Non-technical barriers of geothermal projects. In: *Proceedings World Geothermal Congress*, Bali, Indonesia, 25-30 April 2010. Available at: www.geothermal-energy.org/pdf/IGAstandard/WGC/2010/0314.pdf.

Kagel, A. (2006). *A Handbook on the Externalities, Employment, and Economics of Geothermal Energy.* Geothermal Energy Association, Washington, DC, USA, 65 pp. Available at: www.geo-energy.org/reports/Socioeconomics%20Guide.pdf.

Kaltschmitt, M. (2000). Environmental effects of heat provision from geothermal energy in comparison to other resources of energy. In: *Proceedings World Geothermal Congress 2000*, Kyushu-Tohoku, Japan, 28 May - 10 June 2000 (ISBN: 0473068117). Available at: www.geothermal-energy.org/pdf/IGAstandard/WGC/2000/R0908.PDF.

Koorey, K.J., and A.D. Fernando (2010). Concurrent land use in geothermal steamfield developments. In: *Proceedings World Geothermal Congress 2010*, Bali, Indonesia, 25-29 April 2010. Available at: www.geothermal-energy.org/pdf/IGAstandard/WGC/2010/0207.pdf.

Krewitt, W., K. Nienhaus, C. Kleßmann, C. Capone, E. Stricker, W. Graus, M. Hoogwijk, N. Supersberger, U. von Winterfeld, and S. Samadi (2009). *Role and Potential of Renewable Energy and Energy Efficiency for Global Energy Supply.* Climate Change 18/2009, ISSN 1862-4359, Federal Environment Agency, Dessau-Roßlau, Germany, 336 pp.

Krey, V., and L. Clarke (2011). Role of renewable energy in climate change mitigation: a synthesis of recent scenarios. *Climate Policy*, in press.

Kutscher, C. (2000). *The Status and Future of Geothermal Electric Power.* NREL/CP-550-28204, National Renewable Energy Laboratory, Golden, CO, USA, 9 pp. Available at: www.nrel.gov/docs/fy00osti/28204.pdf.

Laplaige, P., J. Lemale, S. Decottegnie, A. Desplan, O. Goyeneche, and G. Delobelle (2005). Geothermal resources in France – current situation and prospects. In: *Proceedings World Geothermal Congress 2005*, Antalya, Turkey, 24-29 April 2005 (ISBN 9759833204).

Lippmann, M.J. (2002). Geothermal and the electricity market in Central America. *Transactions of the Geothermal Resources Council*, **26**, pp. 37-42.

Lovekin, J. (2000). The economics of sustainable geothermal development. In: *Proceedings World Geothermal Congress 2000*, Kyushu-Tohoku, Japan, 28 May – 10 June 2000 (ISBN: 0473068117). Available at: www.geothermal-energy.org/pdf/IGAstandard/WGC/2000/R0123.PDF.

Lund, J.W. (1995). Onion dehydration. *Transactions of the Geothermal Resources Council*, **19**, pp. 69-74.

Lund, J.W., and T.L. Boyd (2009). Geothermal utilization on the Oregon Institute of Technology campus, Klamath Falls, Oregon. In: *Proceedings of the 34th Workshop on Geothermal Reservoir Engineering*, Stanford University, Stanford, CA, USA, 9-11 February, 2009 (ISBN: 9781615673186).

Lund, J.W., B. Sanner, L. Rybach, R. Curtis, and G. Hellström (2003). Ground-source heat pumps – A world overview. *Renewable Energy World*, **6**(14), pp. 218-227.

Lund, J.W., D.H. Freeston, and T.L. Boyd (2005). Direct application of geothermal energy: 2005 worldwide review. *Geothermics*, **24**, pp. 691-727.

Lund, J.W., D.H. Freeston, and T.L. Boyd (2010a). Direct utilization of geothermal energy 2010 worldwide review. In: *Proceedings World Geothermal Congress 2010*, Bali, Indonesia, 25-30 April 2010. Available at: www.geothermal-energy.org/pdf/IGAstandard/WGC/2010/0007.pdf.

Lund, J.W., K. Gawell, T.L. Boyd, and D. Jennejohn (2010b). The United States of America country update 2010. In: *Proceedings World Geothermal Congress 2010*, Bali, Indonesia, 25-30 April 2010. Available at: www.geothermal-energy.org/pdf/IGAstandard/WGC/2010/0102.pdf.

Lupton, J. (1995). Hydrothermal plumes: Near and far field. In: *Seafloor Hydrothermal Systems: Physical, Chemical, Biological, and Geological Interactions*, S. Humphris, R. Zierenberg, L. Mullineaux and R. Thomson (eds.), Geophysical Monograph 91, American Geophysical Union, Washington, DC, USA, pp. 317-346 (ISBN 0875900488).

Majer, E., E. Bayer, and R. Baria (2008). *Protocol for induced seismicity associated with enhanced geothermal systems.* International Energy Agency – Geothermal Implementing Agreement (incorporating comments by C. Bromley, W. Cumming., A. Jelacic and L. Rybach), Paris, France. Available at: www.iea-gia.org/documents/ProtocolforInducedSeismicityEGS-GIADoc25Feb09.pdf.

Mansure, A.J., and D.A. Blankenship (2008). Geothermal well cost analyses. *Transactions of the Geothermal Resources Council*, **32**, pp. 43-48.

Mégel, T., and L. Rybach (2000). Production capacity and sustainability of geothermal doublets. In: *Proceedings World Geothermal Congress 2000*, Kyushu-Tohoku, Japan, 28 May – 10 June 2000 (ISBN: 0473068117). Available at: www.geothermal-energy.org/pdf/IGAstandard/WGC/2000/R0102.PDF.

Mock, J.E., J.W. Tester, and P.M. Wright (1997). Geothermal energy from the Earth: Its potential impact as an environmentally sustainable resource. *Annual Review of Energy and the Environment*, **22**, pp. 305-356 (ISBN: 978-0-8243-2322-6).

Newell, D., and A. Mingst (2009). Power from the Earth. *Trading Carbon*, **2**(10), p. 24.

O'Sullivan, M., and W. Mannington (2005). Renewability of the Wairakei-Tauhara geothermal resource. In: *Proceedings World Geothermal Congress 2005*, Antalya, Turkey, 24-29 April 2005 (ISBN 9759833204). Available at: http://www.geothermal-energy.org/pdf/IGAstandard/WGC/2005/0508.pdf.

Owens, B. (2002). *An Economic Valuation of a Geothermal Production Tax Credit.* NREL/TP-620-31969, National Renewable Energy Laboratory, Golden, CO, USA, 24 pp. Available at: www.nrel.gov/docs/fy02osti/31969.pdf.

Pritchett, R. (1998). Modeling post-abandonment electrical capacity recovery for a two-phase geothermal reservoir. *Transactions of the Geothermal Resources Council*, **22**, pp. 521-528.

Pruess, K. (2006). Enhanced geothermal systems (EGS) using CO_2 as a working fluid - A novel approach for generating renewable energy with simultaneous sequestration of carbon. *Geothermics*, **35**, pp. 351-367.

Pruess, K., and N. Spycher (2010). Enhanced Geothermal Systems (EGS) with CO_2 as heat transmission fluid – A scheme for combining recovery or renewable energy with geologic storage of CO_2. In: *Proceedings World Geothermal Congress 2010*, Bali, Indonesia, 25-30 April 2010. Available at: www.geothermal-energy.org/pdf/IGAstandard/WGC/2010/3107.pdf.

Radeckas, B., and V. Lukosevicius (2000). Klaipeda Geothermal demonstration project. In: *Proceedings World Geothermal Congress 2000*, Kyushu-Tohoku, Japan, 28 May – 10 June 2000, pp. 3547-3550 (ISBN: 0473068117). Available at: www.geothermal-energy.org/pdf/IGAstandard/WGC/2000/R0237.PDF.

Reif, T. (2008). Profitability analysis and risk management of geothermal projects. *Geo-Heat Center Quarterly Bulletin*, **28**(4), pp. 1-4. Available at: geoheat.oit.edu/bulletin/bull28-4/bull28-4-all.pdf.

REN21 (2010). *Renewables 2010: Global Status Report.* Renewable Energy Policy Network for the 21st Century (REN21) Secretariat, Paris, France. Available at: www.ren21.net/Portals/97/documents/GSR/REN21_GSR_2010_full_revised%20Sept2010.pdf.

Rowley, J.C. (1982). Worldwide geothermal resources. In: *Handbook of Geothermal Energy.* Gulf Publishing, Houston, TX, USA, pp. 44-176 (ISBN 0-87201-322-7).

Rybach, L. (2005). The advance of geothermal heat pumps world-wide. *International Energy Agency (IEA) Heat Pump Centre Newsletter*, **23**, pp. 13-18.

Rybach, L. (2010). Legal and regulatory environment favourable for geothermal development investors. In: *Proceedings World Geothermal Congress 2010*, Bali, Indonesia, 25-30 April 2010. Available at: www.geothermal-energy.org/pdf/IGAstandard/WGC/2010/0303.pdf.

Saadat, A., S. Frick, S. Kranz, and S. Regenspurg (2010). Energy use of EGS reservoirs. In: *Geothermal Energy Systems: Exploration, Development and Utilization.* E. Huenges and P. Ledru (eds.), Wiley-VCH, Berlin, Germany, pp. 303-372 (ISBN: 978-3527408313).

Sánchez-Velasco, R., M. López-Díaz, H. Mendoza, and R. Tello-Hinojosa (2003). Magic at Maguarichic. *Geothermal Resources Council Bulletin* (March-April 2003), pp. 67-70.

Sanyal, S.K. (2010). On minimizing the levelized cost of electric power from Enhanced Geothermal Systems. In: *Proceedings World Geothermal Congress 2010*, Bali, Indonesia, 25-30 April 2010. Available at: www.geothermal-energy.org/pdf/IGAstandard/WGC/2010/3154.pdf.

Sanyal, S.K., and S.J. Butler (2010). Geothermal power capacity from petroleum wells – Some case histories and assessment. In: *Proceedings World Geothermal Congress 2010*, Bali, Indonesia, 25-30 April 2010. Available at: www.geothermal-energy.org/pdf/IGAstandard/WGC/2010/3713.pdf.

Sanyal, S.K., J.W. Morrow, S.J. Butler, and A. Robertson-Tait (2007). Is EGS commercially feasible? *Transactions of the Geothermal Resources Council*, **31**, pp. 313-322.

Sims, R.E.H., R.N. Schock, A. Adegbululgbe, J. Fenhann, I. Konstantinaviciute, W. Moomaw, H.B. Nimir, B. Schlamadinger, J. Torres-Martínez, C. Turner, Y. Uchiyama, S.J.V. Vuori, N. Wamukonya, and X. Zhang (2007). Energy supply. In: *Climate Change 2007: Mitigation of Climate Change. Contribution of Working Group III to the Fourth Assessment Report of the Intergovernmental Panel on Climate Change*, B. Metz, O.R. Davidson, P.R. Bosch, R. Dave and L.A. Meyer (eds.), Cambridge University Press, pp. 251-322.

Stefansson, V. (2002). Investment cost for geothermal power plants. *Geothermics*, **31**, pp. 263-272.

Stefansson, V. (2005). World geothermal assessment. In: *Proceedings World Geothermal Congress 2005*, Antalya, Turkey, 24-29 April 2005 (ISBN: 9759833204). Available at: www.geothermal-energy.org/pdf/IGAstandard/WGC/2005/0001.pdf.

Teske, S., T. Pregger, S. Simon, T. Naegler, W. Graus, and C. Lins (2010). Energy [R]evolution 2010—a sustainable world energy outlook. *Energy Efficiency*, doi:10.1007/s12053-010-9098-y.

Tester, J.W., E.M. Drake, M.W. Golay, M.J. Driscoll, and W.A. Peters (2005). *Sustainable Energy – Choosing Among Options.* MIT Press, Cambridge, Massachusetts, USA, 850 pp (ISBN 0-262-20153-4).

Tester, J.W., B.J. Anderson, A.S. Batchelor, D.D. Blackwell, R. DiPippo, E.M. Drake, J. Garnish, B. Livesay, M.C. Moore, K. Nichols, S. Petty, M.N, Toksöks, and R.W. Veatch Jr. (2006). *The Future of Geothermal Energy: Impact of Enhanced Geothermal Systems on the United States in the 21st Century.* Prepared by the Massachusetts Institute of Technology, under Idaho National Laboratory Subcontract No. 63 00019 for the U.S. Department of Energy, Assistant Secretary for Energy Efficiency and Renewable Energy, Office of Geothermal Technologies, Washington, DC, USA, 358 pp (ISBN-10: 0486477711, ISBN-13: 978-0486477718). Available at: geothermal.inel.gov/publications/future_of_geothermal_energy.pdf.

Wonstolen, K. (1980). Geothermal legislative policy concerns. In: *Proceedings Geothermal Symposium: Potential, Legal Issues, Economics, Financing*, Bloomquist, R.G. and K. Wonstolen (eds.), Seattle, WA, 2 June 1980. Available at: www.osti.gov/bridge/product.biblio.jsp?query_id=0&page=0&osti_id=6238884.

5 Hydropower

Coordinating Lead Authors:
Arun Kumar (India) and Tormod Schei (Norway)

Lead Authors:
Alfred Ahenkorah (Ghana), Rodolfo Caceres Rodriguez (El Salvador),
Jean-Michel Devernay (France), Marcos Freitas (Brazil), Douglas Hall (USA),
Ånund Killingtveit (Norway), Zhiyu Liu (China)

Contributing Authors:
Emmanuel Branche (France), John Burkhardt (USA), Stephan Descloux (France),
Garvin Heath (USA), Karin Seelos (Norway)

Review Editors:
Cristobal Diaz Morejon (Cuba) and Thelma Krug (Brazil)

This chapter should be cited as:
Kumar, A., T. Schei, A. Ahenkorah, R. Caceres Rodriguez, J.-M. Devernay, M. Freitas, D. Hall, Å. Killingtveit, Z. Liu, 2011: Hydropower. In IPCC Special Report on Renewable Energy Sources and Climate Change Mitigation [O. Edenhofer, R. Pichs-Madruga, Y. Sokona, K. Seyboth, P. Matschoss, S. Kadner, T. Zwickel, P. Eickemeier, G. Hansen, S. Schlömer, C. von Stechow (eds)], Cambridge University Press, Cambridge, United Kingdom and New York, NY, USA.

Table of Contents

Executive Summary		441
5.1	**Introduction**	443
5.1.1	Source of energy	443
5.1.2	History of hydropower development	443
5.2	**Resource potential**	444
5.2.1	Global Technical Potential	444
5.2.2	Possible impact of climate change on resource potential	447
5.2.2.1	Projected changes in precipitation and runoff	447
5.2.2.2	Projected impacts on hydropower generation	447
5.3	**Technology and applications**	449
5.3.1	Classification by head and size	450
5.3.2	Classification by facility type	451
5.3.2.1	Run-of-River	451
5.3.2.2	Storage Hydropower	451
5.3.2.3	Pumped storage	451
5.3.2.4	In-stream technology using existing facilities	452
5.3.3	Status and current trends in technology development	452
5.3.3.1	Efficiency	452
5.3.3.2	Tunnelling capacity	453
5.3.3.3	Technical challenges related to sedimentation management	454
5.3.4	Renovation, modernization and upgrading	454
5.4	**Global and regional status of market and industry development**	455
5.4.1	Existing generation	455
5.4.2	The hydropower industry	456
5.4.3	Impact of policies	457
5.4.3.1	International carbon markets	457
5.4.3.2	Project financing	457

5.4.3.3	Administrative and licensing process	457
5.4.3.4	Classification by size	458

5.5 Integration into broader energy systems 458

5.5.1	Grid-independent applications	458
5.5.2	Rural electrification	458
5.5.3	Power system services provided by hydropower	459
5.5.4	Hydropower support of other generation including renewable energy	460
5.5.5	Reliability and interconnection needs for hydropower	460

5.6 Environmental and social impacts 461

5.6.1	**Typical impacts and possible mitigation measures**	462
5.6.1.1	Hydrological regimes	463
5.6.1.2	Reservoir creation	464
5.6.1.3	Water quality	464
5.6.1.4	Sedimentation	465
5.6.1.5	Biological diversity	465
5.6.1.6	Barriers for fish migration and navigation	466
5.6.1.7	Involuntary population displacement	467
5.6.1.8	Affected people and vulnerable groups	467
5.6.1.9	Public health	467
5.6.1.10	Cultural heritage	468
5.6.1.11	Sharing development benefits	468
5.6.2	**Guidelines and regulations**	468
5.6.3	**Lifecycle assessment of environmental impacts**	470
5.6.3.1	Current lifecycle estimates of greenhouse gas emissions	471
5.6.3.2	Quantification of gross and net emissions from reservoirs	472

5.7 Prospects for technology improvement and innovation 474

5.7.1	Variable-speed technology	475
5.7.2	Matrix technology	475
5.7.3	Fish-friendly turbines	475

5.7.4	Hydrokinetic turbines	475
5.7.5	New materials	476
5.7.6	Tunnelling technology	476
5.7.7	Dam technology	476
5.7.8	Optimization of operation	476

5.8 Cost trends 477

5.8.1	Investment cost of hydropower projects and factors that affect it	477
5.8.2	Other costs occurring during the lifetime of hydropower projects	480
5.8.3	Performance parameters affecting the levelized cost of hydropower	481
5.8.4	Past and future cost trends for hydropower projects	482
5.8.5	Cost allocation for other purposes	483

5.9 Potential deployment 484

5.9.1	Near-term forecasts	484
5.9.2	Long-term deployment in the context of carbon mitigation	485
5.9.3	Conclusions regarding deployment	487

5.10 Integration into water management systems 488

5.10.1	The need for climate-driven water management	488
5.10.2	Multipurpose use of reservoirs and regulated rivers	488
5.10.3	Regional cooperation and sustainable watershed management	489

References 491

Executive Summary

Hydropower offers significant potential for carbon emissions reductions. The installed capacity of hydropower by the end of 2008 contributed 16% of worldwide electricity supply, and hydropower remains the largest source of renewable energy in the electricity sector. On a global basis, the technical potential for hydropower is unlikely to constrain further deployment in the near to medium term. Hydropower is technically mature, is often economically competitive with current market energy prices and is already being deployed at a rapid pace. Situated at the crossroads of two major issues for development, water and energy, hydro reservoirs can often deliver services beyond electricity supply. The significant increase in hydropower capacity over the last 10 years is anticipated in many scenarios to continue in the near term (2020) and medium term (2030), with various environmental and social concerns representing perhaps the largest challenges to continued deployment if not carefully managed.

Hydropower is a renewable energy source where power is derived from the energy of water moving from higher to lower elevations. It is a proven, mature, predictable and typically price-competitive technology. Hydropower has among the best conversion efficiencies of all known energy sources (about 90% efficiency, water to wire). It requires relatively high initial investment, but has a long lifespan with very low operation and maintenance costs. The levelized cost of electricity for hydropower projects spans a wide range but, under good conditions, can be as low as 3 to 5 US cents$_{2005}$ per kWh. A broad range of hydropower systems, classified by project type, system, head or purpose, can be designed to suit particular needs and site-specific conditions. The major hydropower project types are: run-of-river, storage- (reservoir) based, pumped storage and in-stream technologies. There is no worldwide consensus on classification by project size (installed capacity, MW) due to varying development policies in different countries. Classification according to size, while both common and administratively simple, is—to a degree—arbitrary: concepts like 'small' or 'large hydro' are not technically or scientifically rigorous indicators of impacts, economics or characteristics. Hydropower projects cover a continuum in scale and it may ultimately be more useful to evaluate hydropower projects based on their sustainability or economic performance, thus setting out more realistic indicators.

The total worldwide technical potential for hydropower generation is 14,576 TWh/yr (52.47 EJ/yr) with a corresponding installed capacity of 3,721 GW, roughly four times the current installed capacity. Worldwide total installed hydropower capacity in 2009 was 926 GW, producing annual generation of 3,551 TWh/y (12.8 EJ/y), and representing a global average capacity factor of 44%. Of the total technical potential for hydropower, undeveloped capacity ranges from about 47% in Europe and North America to 92% in Africa, which indicates large opportunities for continued hydropower development worldwide, with the largest growth potential in Africa, Asia and Latin America. Additionally, possible renovation, modernization and upgrading of old power stations are often less costly than developing a new power plant, have relatively smaller environment and social impacts, and require less time for implementation. Significant potential also exists to rework existing infrastructure that currently lacks generating units (e.g., existing barrages, weirs, dams, canal fall structures, water supply schemes) by adding new hydropower facilities. Only 25% of the existing 45,000 large dams are used for hydropower, while the other 75% are used exclusively for other purposes (e.g., irrigation, flood control, navigation and urban water supply schemes). Climate change is expected to increase overall average precipitation and runoff, but regional patterns will vary: the impacts on hydropower generation are likely to be small on a global basis, but significant regional changes in river flow volumes and timing may pose challenges for planning.

In the past, hydropower has acted as a catalyst for economic and social development by providing both energy and water management services, and it can continue to do so in the future. Hydro storage capacity can mitigate freshwater scarcity by providing security during lean flows and drought for drinking water supply, irrigation, flood control and navigation services. Multipurpose hydropower projects may have an enabling role beyond the electricity sector as a financing instrument for reservoirs that help to secure freshwater availability. According to the World Bank, large hydropower projects can have important multiplier effects, creating an additional USD$_{2005}$ 0.4 to 1.0 of indirect benefits for every dollar of value generated. Hydropower can serve both in large, centralized and small, isolated grids, and small-scale hydropower is an option for rural electrification.

Environmental and social issues will continue to affect hydropower deployment opportunities. The local social and environmental impacts of hydropower projects vary depending on the project's type, size and local conditions and are often controversial. Some of the more prominent impacts include changes in flow regimes and water quality, barriers to fish migration, loss of biological diversity, and population displacement. Impoundments and reservoirs stand out as the source of the most severe concerns but can also provide multiple beneficial services beyond energy supply. While lifecycle assessments indicate very low carbon emissions, there is currently no consensus on the issue of land use change-related net emissions from reservoirs. Experience gained during past decades in combination with continually advancing sustainability guidelines and criteria, innovative planning based on stakeholder consultations and scientific know-how can support high sustainability performance in future projects. Transboundary water management, including the management of hydropower projects, establishes an arena for international cooperation that may contribute to promoting sustainable economic growth and water security.

Technological innovation and material research can further improve environmental performance and reduce operational costs. Though hydropower technologies are mature, ongoing research into variable-speed generation technology, efficient tunnelling techniques, integrated river basin management, hydrokinetics, silt erosion resistive materials and environmental issues (e.g., fish-friendly turbines) may ensure continuous improvement of future projects.

Hydropower can provide important services to electric power systems. Storage hydropower plants can often be operated flexibly, and therefore are valuable to electric power systems. Specifically, with its rapid response load-following and balancing capabilities, peaking capacity and power quality attributes, hydropower can play an important role in ensuring reliable electricity service. In an integrated system, reservoir and pumped storage hydropower can be used to reduce the frequency of start-ups and shutdowns of thermal plants; to maintain a balance between supply and demand under changing demand or supply patterns and thereby reduce the load-following burden of thermal plants; and to increase the amount of time that thermal units are operated at their maximum thermal efficiency, thereby reducing carbon emissions. In addition, storage and pumped storage hydropower can help reduce the challenges of integrating variable renewable resources such as wind, solar photovoltaics, and wave power.

Hydropower offers significant potential for carbon emissions reductions. Baseline projections of the global supply of hydropower rise from 12.8 EJ in 2009 to 13 EJ in 2020, 15 EJ in 2030 and 18 EJ in 2050 in the median case. Steady growth in the supply of hydropower is therefore projected to occur even in the absence of greenhouse gas (GHG) mitigation policies, though demand growth is anticipated to be even higher, resulting in a shrinking percentage share of hydropower in global electricity supply. Evidence suggests that relatively high levels of deployment over the next 20 years are feasible, and hydropower should remain an attractive renewable energy source within the context of global GHG mitigation scenarios. That hydropower can provide energy and water management services and also help to manage variable renewable energy supply may further support its continued deployment, but environmental and social impacts will need to be carefully managed.

5.1 Introduction

This chapter describes hydropower technology. It starts with a brief historical overview of how the technology has evolved (Section 5.1), a discussion of resource potential and how it may be affected by climate change (Section 5.2), and a description of the technology (Section 5.3) and its social and environmental impacts (Section 5.6). Also included is a summary of the present global and regional status of the hydropower industry (Section 5.4) and the role of hydropower in the broader energy system (Section 5.5), as well as a summary of the prospects for technology improvement (Section 5.7), cost trends (Section 5.8), and potential deployment in both the near term (2020) and long term (2050) (Section 5.9). The chapter also covers the integration of hydropower into broader water management solutions (Section 5.10). In this chapter, the focus is largely on the generation and storage of electrical energy from water; the use of hydropower in meeting mechanical energy demands is covered only peripherally.

5.1.1 Source of energy

Hydropower is generated from water moving in the hydrological cycle, which is driven by solar radiation. Incoming solar radiation is absorbed at the land or sea surface, heating the surface and creating evaporation where water is available. A large percentage—close to 50% of all the solar radiation reaching the Earth's surface—is used to evaporate water and drive the hydrological cycle. The potential energy embedded in this cycle is therefore huge, but only a very limited amount may be technically developed. Evaporated water moves into the atmosphere and increases the water vapour content in the air. Global, regional and local wind systems, generated and maintained by spatial and temporal variations in the solar energy input, move the air and its vapour content over the surface of the Earth, up to thousands of kilometres from the origin of evaporation. Finally, the vapour condenses and falls as precipitation, about 78% on oceans and 22% on land. This creates a net transport of water from the oceans to the land surface of the Earth, and an equally large flow of water back to the oceans as river and groundwater runoff. It is the flow of water in rivers that can be used to generate hydropower, or more precisely, the energy of water moving from higher to lower elevations on its way back to the ocean, driven by the force of gravity.

5.1.2 History of hydropower development

Prior to the widespread availability of commercial electric power, hydropower was used for irrigation and operation of various machines, such as watermills, textile machines and sawmills. By using water for power generation, people have worked with nature to achieve a better lifestyle. The mechanical power of falling water is an old resource used for services and productive uses. It was used by the Greeks to turn water wheels for grinding wheat into flour more than 2,000 years ago. In the 1700s, mechanical hydropower was used extensively for milling and pumping. During the 1700s and 1800s, water turbine development continued. The first hydroelectric power plant was installed in Cragside, Rothbury, England in 1870. Industrial use of hydropower started in 1880 in Grand Rapids, Michigan, when a dynamo driven by a water turbine was used to provide theatre and storefront lighting. In 1881, a brush dynamo connected to a turbine in a flour mill provided street lighting at Niagara Falls, New York. The breakthrough came when the electric generator was coupled to the turbine and thus the world's first hydroelectric station (of 12.5 kW capacity) was commissioned on 30 September 1882 on Fox River at the Vulcan Street Plant, Appleton, Wisconsin, USA, lighting two paper mills and a residence.[1]

Early hydropower plants were much more reliable and efficient than the fossil fuel-fired plants of the day (Baird, 2006). This resulted in a proliferation of small- to medium-sized hydropower stations distributed wherever there was an adequate supply of moving water and a need for electricity. As electricity demand grew, the number and size of fossil fuel, nuclear and hydropower plants increased. In parallel, concerns arose around environmental and social impacts (Thaulow et al., 2010).

Hydropower plants (HPP) today span a very large range of scales, from a few watts to several GW. The largest projects, Itaipu in Brazil with 14,000 MW[2] and Three Gorges in China with 22,400 MW,[3] both produce between 80 to 100 TWh/yr (288 to 360 PJ/yr). Hydropower projects are always site-specific and thus designed according to the river system they inhabit. Historical regional hydropower generation from 1965 to 2009 is shown in Figure 5.1.

The great variety in the size of hydropower plants gives the technology the ability to meet both large centralized urban energy needs as well as decentralized rural needs. Though the primary role of hydropower in the global energy supply today is in providing electricity generation as part of centralized energy networks, hydropower plants also operate in isolation and supply independent systems, often in rural and remote areas of the world. Hydro energy can also be used to meet mechanical energy needs, or to provide space heating and cooling. More recently hydroelectricity has also been investigated for use in the electrolysis process for hydrogen fuel production, provided there is abundance of hydropower in a region and a local goal to use hydrogen as fuel for transport (Andreassen et al., 2002; Yumurtacia and Bilgen, 2004; Silva et al., 2005)

Hydropower plants do not consume the water that drives the turbines. The water, after power generation, is available for various other essential uses. In fact, a significant proportion of hydropower projects are designed for multiple purposes (see Section 5.10.2). In these instances, the dams help to prevent or mitigate floods and droughts, provide the possibility to irrigate agriculture, supply water for domestic, municipal and industrial use, and can improve conditions for navigation, fishing, tourism or leisure

---

1 United States Bureau of Reclamation: www.usbr.gov/power/edu/history.html.

2 Itaipu Binacional hydroelectric power plant (www.itaipu.gov.br).

3 China Three Gorges Project Corporation Annual Report 2009 (www.ctgpc.com).

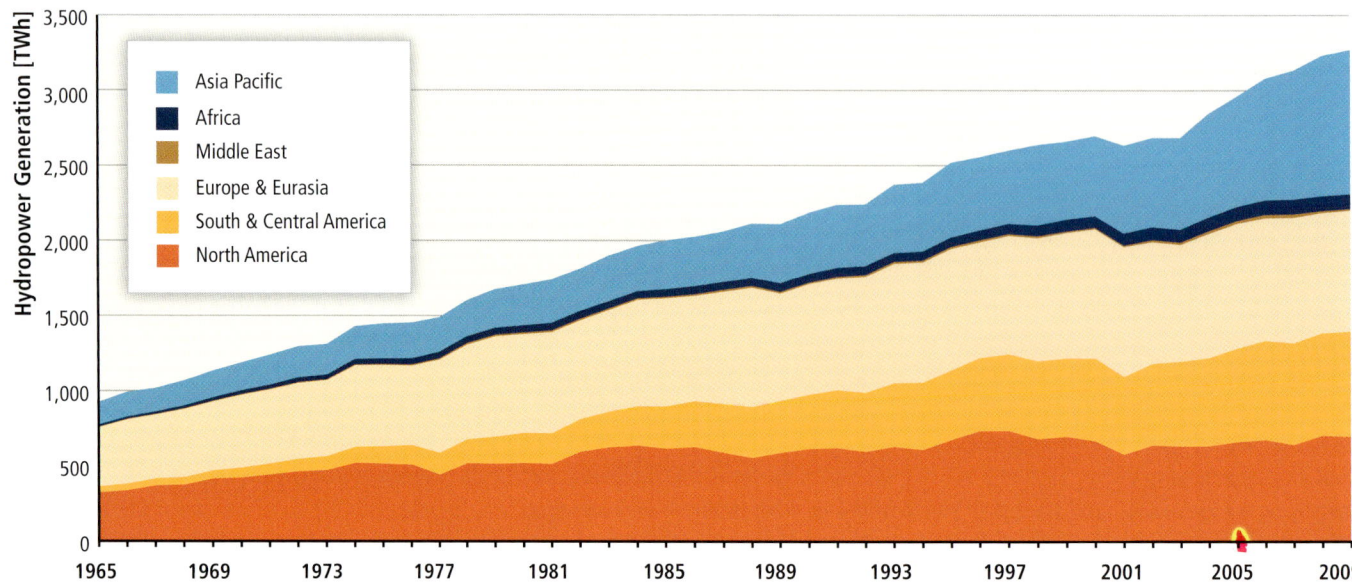

Figure 5.1 | Hydropower generation (TWh) by region (BP, 2010).

activities. One aspect often overlooked when addressing hydropower and the multiple uses of water is that the power plant, as a generator of revenue, in some cases can help pay for the facilities required to develop other water uses that might not generate sufficient direct revenues to finance their construction.

5.2 Resource potential

Hydropower resource potential can be derived from total available flow multiplied by head and a conversion factor. Since most precipitation usually falls in mountainous areas, where elevation differences (head) are the largest, the largest potential for hydropower development is in mountainous regions, or in rivers coming from such regions. The total annual runoff has been estimated as 47,000 km^3, out of which 28,000 km^3 is surface runoff, yielding a theoretical potential for hydropower generation of 41,784 TWh/yr (147 EJ/yr) (Rogner et al., 2004). This value of theoretical potential is similar to a more recent estimate of 39,894 TWh/yr (144 EJ/yr) (IJHD, 2010) (see Chapter 1).

Section 5.2.1 discusses the global technical potential, considering that gross theoretical potential is of no practical value and what is economically feasible is variable depending on energy supply and pricing, which can vary with time and by location.

5.2.1 Global Technical Potential

The International Journal on Hydropower & Dams *2010 World Atlas & Industry Guide* (IJHD, 2010) provides the most comprehensive inventory of current hydropower installed capacity and annual generation, and hydropower resource potential. The Atlas provides three measures of hydropower resource potential, all in terms of annual generation (TW/yr): gross theoretical, technically feasible,[4] and economically feasible. The total worldwide technical potential for hydropower is estimated at 14,576 TWh/yr (52.47 EJ/yr) (IJHD, 2010), over four times the current worldwide annual generation.[5]

This technical potential corresponds to a derived estimate of installed capacity of 3,721 GW.[6] Technical potentials in terms of annual generation and estimated capacity for the six world regions[7] are shown in Figure 5.2. Pie charts included in the figure provide a comparison of current annual generation to technical potential for each region and the percentage of undeveloped potential compared to total technical potential. These charts illustrate that the percentages of undeveloped potential range from 47% in Europe and North America to 92% in Africa, indicating large opportunities for hydropower development worldwide.

There are several notable features of the data in Figure 5.2. North America and Europe, which have been developing their hydropower resources for more than a century, still have sufficient technical potential to double their hydropower generation, belying the perception that the hydropower resources in these highly developed parts of the world are

---

4 Equivalent to the technical potential definition provided in Annex I (Glossary).

5 Chapter 1 presents current and future technical potential estimates for all RE sources as assessed by Krewitt et al. (2009), based on a review of several studies. There, hydropower technical potential by 2050 is estimated to be 50 EJ/y. However, this chapter will exclusively rely on IJHD (2010) for technical potential estimates.

6 Derived value of potential installed nameplate capacity based on regional generation potentials and average capacity factors shown in Figure 5.3.

7 The Latin America region includes Central and South America, consistent with the IEA world regions. This differs from the regions in IJHD (2010), which includes Central America as part of North America. Data from the reference have been re-aggregated to conform to regions used in this document.

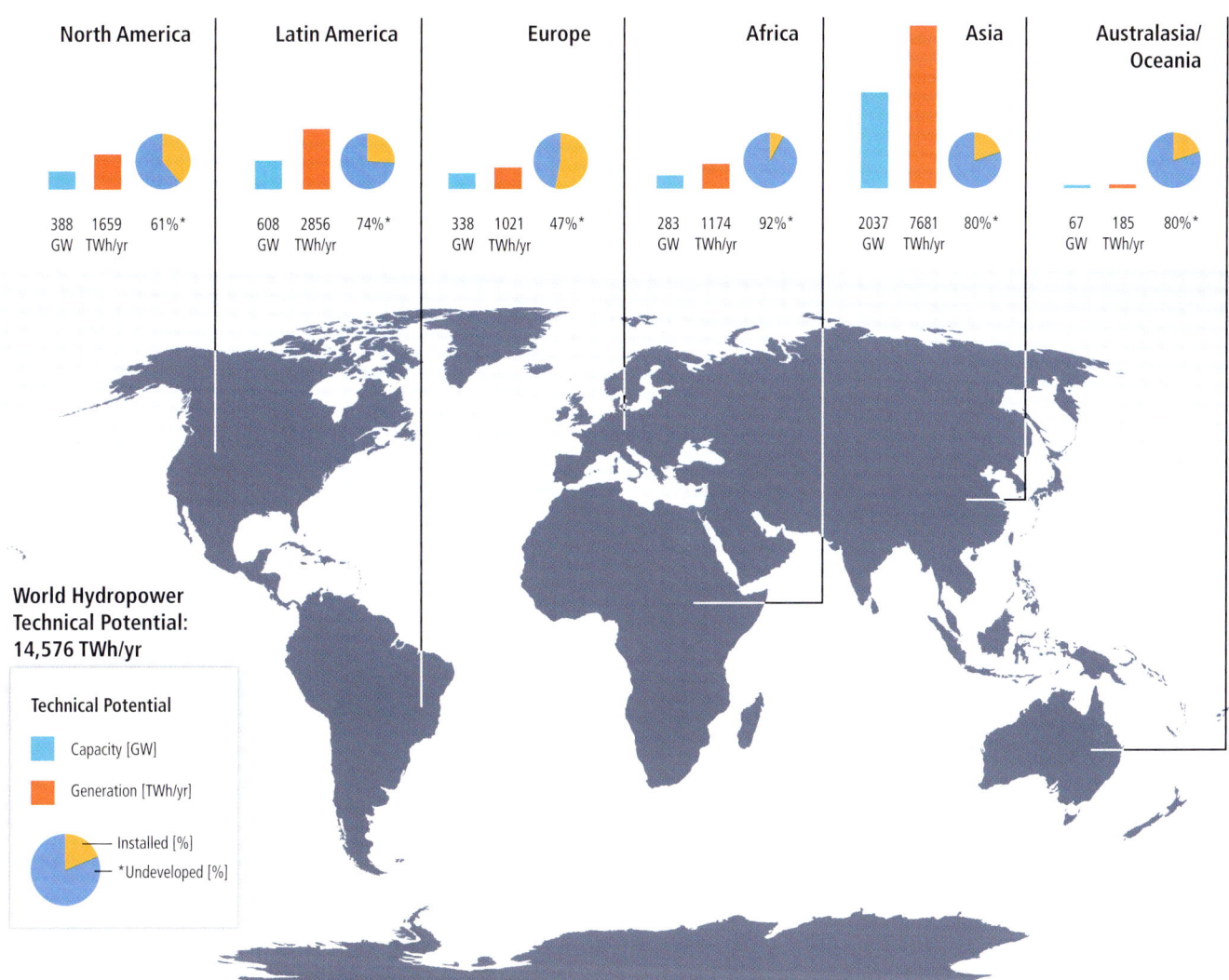

Figure 5.2 | Regional hydropower technical potential in terms of annual generation and installed capacity, and percentage of undeveloped technical potential in 2009. Source: IJHD (2010).

exhausted. However, how much of this untapped technical potential is economically feasible is subject to time-dependent economic conditions. Actual development will also be impacted by sustainability concerns and related policies. Notably, Asia and Latin America have comparatively large technical potentials and, along with Australasia/Oceania, the fraction of total technical potential that is undeveloped is quite high in these regions. Africa has a large technical potential and could develop 11 times its current level of hydroelectric generation in the region. An overview of regional technical potentials for hydropower is given in Table 5.1.

Understanding and appreciation of hydropower technical potential can also be obtained by considering the current (2009) total regional hydropower installed capacity and annual generation shown in Figure 5.3. The reported worldwide total installed hydropower capacity is 926 GW producing a total annual generation of 3,551 TWh/yr (12.8 EJ/yr) in 2009. Figure 5.3 also includes regional average capacity factors calculated using current regional total installed capacity and annual generation (capacity factor = generation/(installed capacity x 8,760 hrs)).

It is interesting to note that North America, Latin America, Europe and Asia have the same order of magnitude of total installed capacity while Africa and Australasia/Oceania have an order of magnitude less—Africa due in part to the lack of available investment capital and Australasia/Oceania in part because of size, climate and topography. The average capacity factors are in the range of 32 to 55%. Capacity factor can be indicative of how hydropower is employed in the energy mix (e.g., peaking versus base-load generation), water availability, or an opportunity for increased generation through equipment upgrades and operation optimization. Generation increases that have been achieved by equipment upgrades and operation optimization have generally not been assessed in detail, but are briefly discussed in Sections 5.3.4 and 5.8.

The regional technical potentials presented above are for conventional hydropower corresponding to sites on natural waterways where there is significant topographic elevation change to create useable hydraulic head. Hydrokinetic technologies that do not require hydraulic head but rather extract energy in-stream from the current of a waterway are being developed. These technologies increase the potential for energy

Table 5.1 | Regional hydropower technical potential in terms of annual generation and installed capacity (GW); and current generation, installed capacity, average capacity factors in percent and resulting undeveloped potential as of 2009. Source: IJHD (2010).

World region	Technical potential, annual generation TWh/yr (EJ/yr)	Technical potential, installed capacity (GW)	2009 Total generation TWh/yr (EJ/yr)	2009 Installed capacity (GW)	Un-developed potential (%)	Average regional capacity factor (%)
North America	1,659 (5.971)	388	628 (2.261)	153	61	47
Latin America	2,856 (10.283)	608	732 (2.635)	156	74	54
Europe	1,021 (3.675)	338	542 (1.951)	179	47	35
Africa	1,174 (4.226)	283	98 (0.351)	23	92	47
Asia	7,681 (27.651)	2,037	1,514 (5.451)	402	80	43
Australasia/Oceania	185 (0.666)	67	37 (0.134)	13	80	32
World	14,576 (52.470)	3,721	3,551 (12.783)	926	75	44

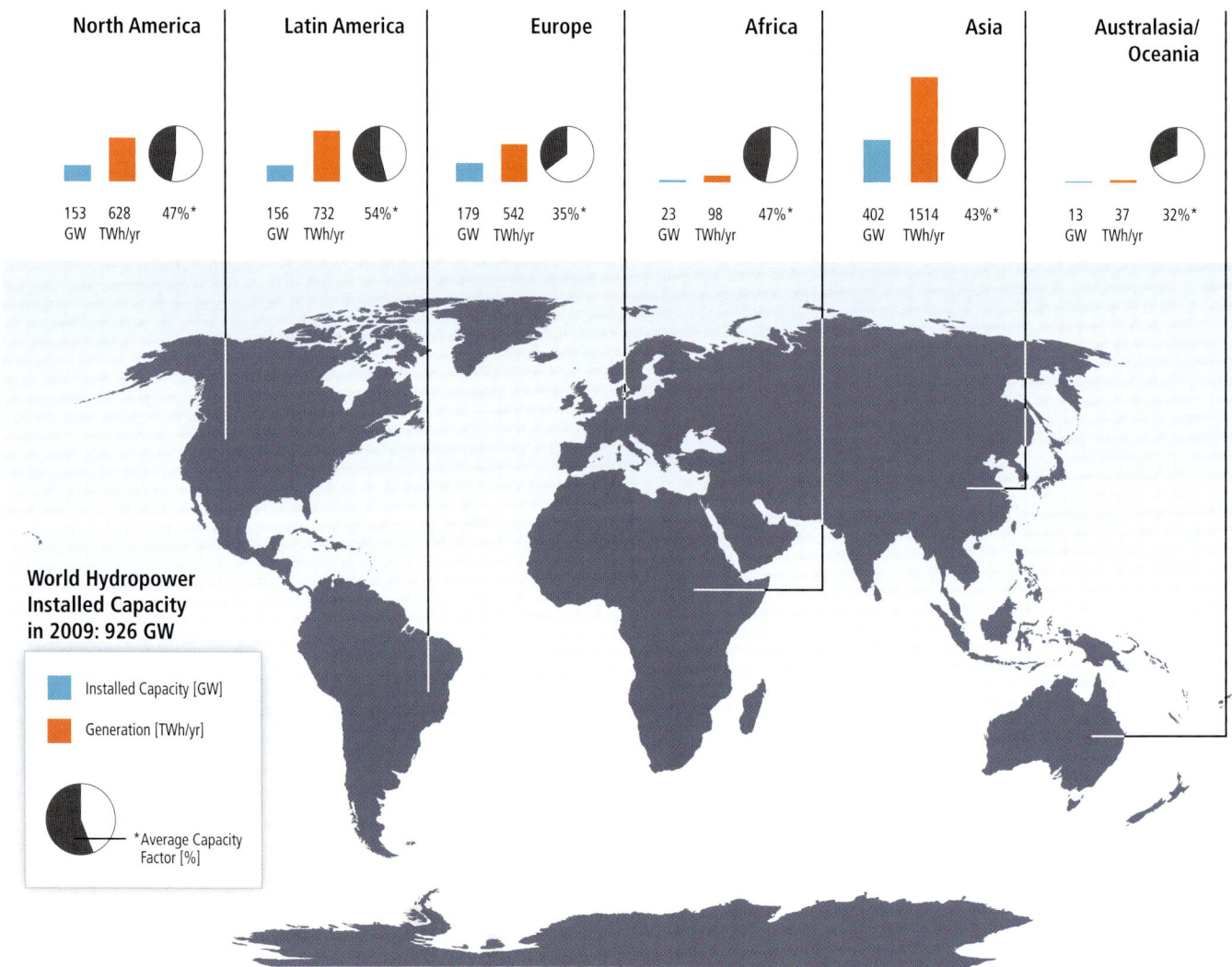

Figure 5.3 | Total regional installed hydropower capacity and annual generation in 2009, and average regional capacity factors (derived as stated above). Source: IJHD (2010).

production at sites where conventional hydropower technology cannot operate. Non-traditional sources of hydropower are also not counted in the regional technical potentials presented above. Examples are constructed waterways such as water supply and treatment systems, aqueducts, canals, effluent streams and spillways. Applicable conventional and hydrokinetic technologies can produce energy using these resources.

While the total technical potentials of in-stream and constructed waterway resources have not been assessed, they may prove to be significant given their large extent.

5.2.2 Possible impact of climate change on resource potential

The resource potential for hydropower is currently based on historical data for the present climatic conditions. With a changing climate, this resource potential could change due to:

- Changes in river flow (runoff) related to changes in local climate, particularly in precipitation and temperature in the catchment area. This may lead to changes in runoff volume, variability of flow and seasonality of the flow (e.g., by changing from spring/summer high flow to more winter flow), directly affecting the resource potential for hydropower generation.

- Changes in extreme events (floods and droughts) may increase the cost and risk for the hydropower projects.

- Changes in sediment loads due to changing hydrology and extreme events. More sediment could increase turbine abrasions and decrease efficiency. Increased sediment load could also fill up reservoirs faster and decrease the live storage, reducing the degree of regulation and decreasing storage services.

The work of IPCC Working Group II (reported in IPCC, 2007b) includes a discussion of the impact of climate change on water resources. Later, a technical paper on water was prepared based on the material included in the previous IPCC reports as well as other sources (Bates et al., 2008). The information presented in this section is mostly based on these two sources, with a few additions from more recent papers and reports, as presented, for example, in a recent review by Hamududu et al. (2010).

5.2.2.1 Projected changes in precipitation and runoff

A wide range of possible future climatic projections have been presented, with corresponding variability in projection of precipitation and runoff (IPCC, 2007c; Bates et al., 2008). Climate projections using multi-model ensembles show increases in globally averaged mean water vapour, evaporation and precipitation over the 21st century. At high latitudes and in part of the tropics, nearly all models project an increase in precipitation, while in some subtropical and lower mid-latitude regions, precipitation is projected to decrease. Between these areas of robust increase or decrease, even the sign of projected precipitation change is inconsistent across the current generation of models (Bates et al., 2008).

Changes in river flow due to climate change will primarily depend on changes in volume and timing of precipitation, evaporation and snowmelt. A large number of studies of the effect on river flow have been published and were summarized in IPCC (2007b). Most of these studies use a catchment hydrological model driven by climate scenarios based on climate model simulations. Before data can be used in the catchment hydrological models, it is necessary to downscale data, a process where output from the global climate model is converted to corresponding climatic data in the catchments. Such downscaling can be both temporal and spatial, and it is currently a high priority research area to find the best methods for downscaling.

A few global-scale studies have used runoff simulated directly by climate models (Egré and Milewski, 2002; IPCC, 2007b). The results of these studies show increasing runoff in high latitudes and the wet tropics and decreasing runoff in mid-latitudes and some parts of the dry tropics. Figure 5.4 illustrates projected changes in runoff by the end of the century, based on the IPCC A1B scenario[8] (Bates et al., 2008).

Uncertainties in projected changes in the hydrological systems arise from internal variability in the climatic system, uncertainty about future greenhouse gas and aerosol emissions, the translations of these emissions into climate change by global climate models, and hydrological model uncertainty. Projections become less consistent between models as the spatial scale decreases. The uncertainty of climate model projections for freshwater assessments is often taken into account by using multi-model ensembles (Bates et al., 2008). The multi-model ensemble approach is, however, not a guarantee of reducing uncertainty in mathematical models.

Global estimates as shown in Figure 5.4 represent results at a large scale, and cannot be applied to shorter temporal and smaller spatial scales. In areas where rainfall and runoff are very low (e.g., desert areas), small changes in runoff can lead to large percentage changes. In some regions, the sign of projected changes in runoff differs from recently observed trends. Moreover, in some areas with projected increases in runoff, different seasonal effects are expected, such as increased wet season runoff and decreased dry season runoff. Studies using results from fewer climate models can be considerably different from the results presented here (Bates et al., 2008).

5.2.2.2 Projected impacts on hydropower generation

Though the average global or continent-wide impacts of climate change on hydropower resource potential might be expected to be relatively small, more significant regional and local effects are possible. Hydropower resource potential depends on topography and the volume, variability and seasonal distribution of runoff. Not only are these regionally and locally determined, but an increase in climate variability,

---

8 Four scenario families or 'storylines' (A1, A2, B1 and B2) were developed by the IPCC and reported in the IPCC *Special Report On Emission Scenarios* (SRES) as a basis for projection of future climate change, where each represents different demographic, social, economic, technological and environmental development over the 21st century (IPCC, 2000). Therefore, a wide range of possible future climatic projections have been presented based on the resulting emission scenarios, with corresponding variability in projections of precipitation and runoff (IPCC, 2007b).

Hydropower　　　　　　　　　　　　　　　　　　　　　　　　　　　　　　Chapter 5

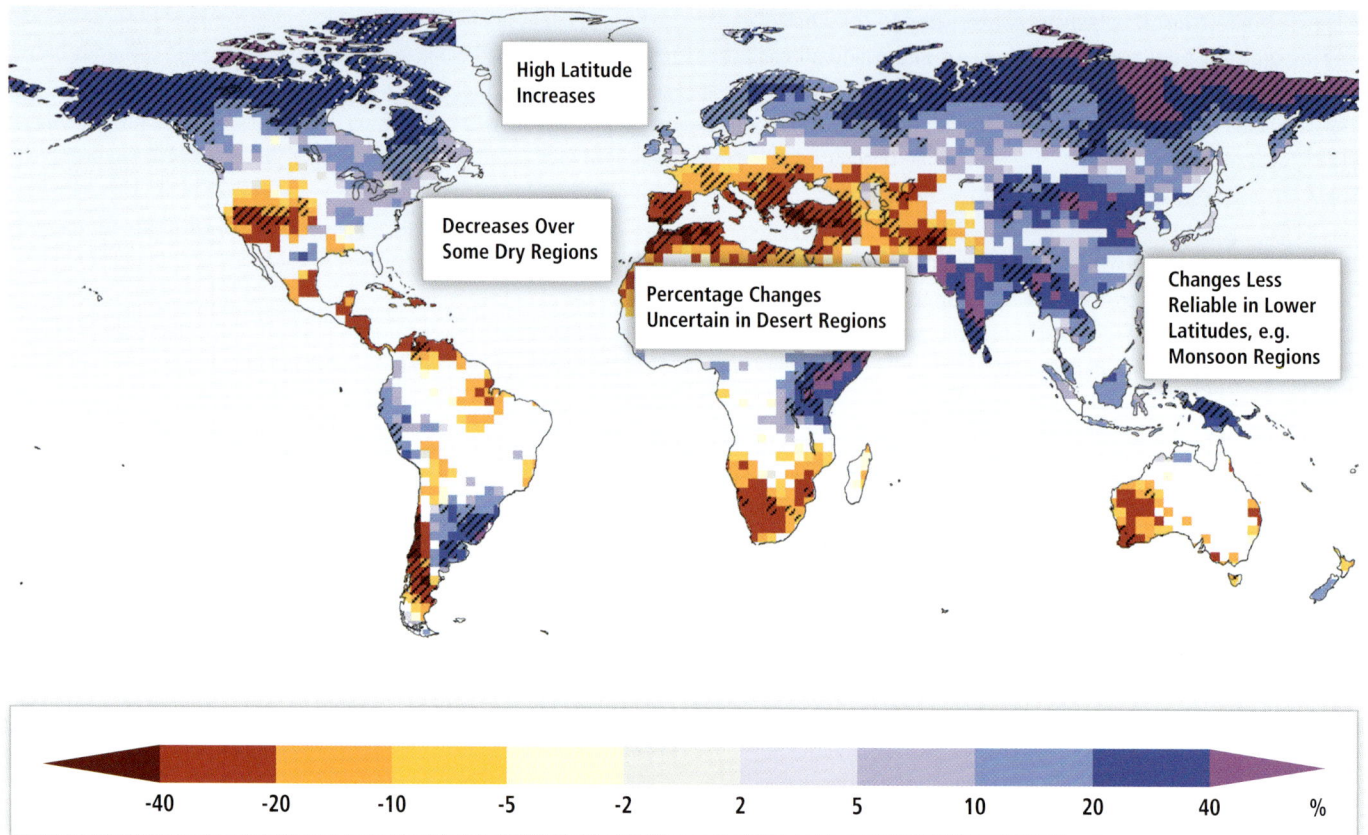

Figure 5.4 | Large-scale changes in annual runoff (water availability, in percent) for the period 2090 to 2099, relative to 1980 to 1999. Values represent the median of 12 climate model projections using the SRES A1B scenario. White areas are where less than 66% of the 12 models agree on the sign of change and hatched areas are where more than 90% of models agree on the sign of change. Source: IPCC (2007a).

even with no change in average runoff, can lead to reduced hydropower production unless more reservoir capacity is built and operations are modified to account for the new hydrology that may result from climate change.

In order to make accurate quantitative predictions of regional effects it is therefore necessary to analyze both changes in average flow and changes in the temporal distribution of flow, using hydrological models to convert time series of climate scenarios into time series of runoff scenarios. In catchments with ice, snow and glaciers it is of particular importance to study the effects of changes in seasonality, because a warming climate will often lead to increasing winter runoff and decreasing runoff in spring and summer. A shift in winter precipitation from snow to rain due to increased air temperature may lead to a temporal shift in peak flow and winter conditions (Stickler and Alfredsen, 2009) in many continental and mountain regions. The spring snowmelt peak would then be brought forward or eliminated entirely, with winter flow increasing. As glaciers retreat due to warming, river flows would be expected to increase in the short term but decline once the glaciers disappear (Bates et al., 2008; Milly et al., 2008).

Summarizing available studies up to 2007, IPCC (2007b) and Bates et al. (2008) found examples of both positive and negative regional effects on hydropower production, mainly following the expected changes in river runoff. Unfortunately, few quantitative estimates of the effects on technical potential for hydropower were found. The regional distribution of studies was also skewed, with most studies done in Europe and North America, and a weak literature base for most developing country regions, in particular for Africa. The summary below is based on findings summarized in Bates et al. (2008) and IPCC (2007b) unless additional sources are given.

In Africa, the electricity supply in a number of states is largely based on hydroelectric power. However, few available studies examine the impacts of climate change on hydropower resource potential in Africa. Observations deducted from general predictions for climate change and runoff point to a reduction in hydropower resource potential with the exception of East Africa (Hamududu et al., 2010).

In major hydropower-generating Asian countries such as China, India, Iran, Tajikistan etc., changes in runoff are found to potentially have a significant effect on the power output. Increased risks of landslides and glacial lake outbursts, and impacts of increased variability, are of particular concern to Himalayan countries (Agrawala et al., 2003). The possibility of accommodating increased intensity of seasonal precipitation by increasing storage capacities may become of particular importance (Iimi, 2007).

In Europe, by the 2070s, hydropower potential for the whole of Europe has been estimated to potentially decline by 6%, translated into a 20 to 50% decrease around the Mediterranean, a 15 to 30% increase in northern and Eastern Europe, and a stable hydropower pattern for western and central Europe (Lehner et al., 2005).

In New Zealand, increased westerly wind speed is very likely to enhance wind generation and spill over precipitation into major South Island watersheds, and to increase winter rain in the Waikato catchment. Warming is virtually certain to increase melting of snow, the ratio of rainfall to snowfall, and to increase river flows in winter and early spring. This is very likely to increase hydroelectric generation during the winter peak demand period, and to reduce demand for storage.

In Latin America, hydropower is the main electrical energy source for most countries, and the region is vulnerable to large-scale and persistent rainfall anomalies due to El Niño and La Niña, as observed in Argentina, Colombia, Brazil, Chile, Peru, Uruguay and Venezuela. A combination of increased energy demand and droughts caused a virtual breakdown of hydroelectricity in most of Brazil in 2001 and contributed to a reduction in gross domestic product (GDP). Glacier retreat is also affecting hydropower generation, as observed in the cities of La Paz and Lima.

In North America, hydropower production is known to be sensitive to total runoff, to its timing, and to reservoir levels. During the 1990s, for example, Great Lakes levels fell as a result of a lengthy drought, and in 1999, hydropower production was down significantly both at Niagara and Sault St. Marie. For a 2°C to 3°C warming in the Columbia River Basin and BC Hydro service areas, the hydroelectric supply under worst-case water conditions for winter peak demand is likely to increase (high confidence). Similarly, Colorado River hydropower yields are likely to decrease significantly, as will Great Lakes hydropower. Northern Québec hydropower production would be likely to benefit from greater precipitation and more open-water conditions, but hydropower plants in southern Québec would be likely to be affected by lower water levels. Consequences of changes in the seasonal distribution of flows and in the timing of ice formation are uncertain.

In a recent study (Hamududu and Killingtveit, 2010), the regional and global changes in hydropower generation for the existing hydropower system were computed, based on a global assessment of changes in river flow by 2050 (Milly et al., 2005, 2008) for the SRES A1B scenario using 12 different climate models. The computation was done at the country or political region (USA, Canada, Brazil, India, China, Australia) level, and summed up to regional and global values (see Table 5.2).

Table 5.2 | Power generation capacity in GW and TWh/yr (2005) and estimated changes (TWh/yr) due to climate change by 2050. Results are based on an analysis using the SRES A1B scenario in 12 different climate models (Milly et al., 2008), UNEP world regions and data for the hydropower system in 2005 (US DOE, 2009) as presented in Hamududu and Killingtveit (2010).

REGION	Power Generation Capacity (2005)		Change by 2050 TWh/yr (PJ/yr)
	GW	TWh/yr (PJ/yr)	
Africa	22	90 (324)	0.0 (0)
Asia	246	996 (3,586)	2.7 (9.7)
Europe	177	517 (1,861)	-0.8 (-2.9)
North America	161	655 (2,358)	0.3 (≈1)
South America	119	661 (2,380)	0.3 (≈1)
Oceania	13	40 (144)	0.0 (0)
TOTAL	737	2931 (10,552)	2.5 (9)

In general the results given in Table 5.2 are consistent with the (mostly qualitative) results given in previous studies (IPCC, 2007b; Bates et al., 2008). For Europe, the computed reduction (-0.2%) has the same sign, but is less than the -6% found by Lehner et al. (2005). One reason could be that Table 5.2 shows changes by 2050 while Lehner et al. (2005) give changes by 2070, so a direct comparison is difficult.

It can be concluded that the overall impacts of climate change on the existing global hydropower generation may be expected to be small, or even slightly positive. However, results also indicated substantial variations in changes in energy production across regions and even within countries (Hamududu and Killingtveit, 2010).

Insofar as a future expansion of the hydropower system will occur incrementally in the same general areas/watersheds as the existing system, these results indicate that climate change impacts globally and averaged across regions may also be small and slightly positive.

Still, uncertainty about future impacts as well as increasing difficulty of future systems operations may pose a challenge that must be addressed in the planning and development of future HPP (Hamududu et al., 2010).

Indirect effects on water availability for energy purposes may occur if water demand for other uses such as irrigation and water supply for households and industry rises due to the climate change. This effect is difficult to quantify, and it is further discussed in Section 5.10.

5.3 Technology and applications

Head and also installed capacity (size) are often presented as criteria for the classification of hydropower plants. The main types of hydropower, however, are run-of-river, reservoir (storage hydro), pumped storage, and in-stream technology. Classification by head and classification by size are discussed in Section 5.3.1. The main types of hydropower are presented in Section 5.3.2. Maturity of the technology, status and current trends in technology development, and trends in renovation and modernization follow in Sections 5.3.3 and 5.3.4 respectively.

5.3.1 Classification by head and size

A classification by head refers to the difference between the upstream and the downstream water levels. Head determines the water pressure on the turbines that together with discharge are the most important parameters for deciding the type of hydraulic turbine to be used. Generally, for high heads, Pelton turbines are used, whereas Francis turbines are used to exploit medium heads. For low heads, Kaplan and Bulb turbines are applied. The classification of what 'high head' and 'low head are varies widely from country to country, and no generally accepted scales are found.

Classification according to size has led to concepts such as 'small hydro' and 'large hydro', based on installed capacity measured in MW as the defining criterion. Small-scale hydropower plants (SHP) are more likely to be run-of-river facilities than are larger hydropower plants, but reservoir (storage) hydropower stations of all sizes will utilize the same basic components and technologies. Compared to large-scale hydropower, however, it typically takes less time and effort to construct and integrate small hydropower schemes into local environments (Egré and Milewski, 2002). For this reason, the deployment of SHPs is increasing in many parts of the world, especially in remote areas where other energy sources are not viable or are not economically attractive.

Nevertheless, there is no worldwide consensus on definitions regarding size categories (Egré and Milewski, 2002). Various countries or groups of countries define 'small hydro' differently. Some examples are given in Table 5.3. From this it can be inferred that what presently is named 'large hydro' spans a very wide range of HPPs. IJHD (2010) lists several more examples of national definitions based on installed capacity.

This broad spectrum in definitions of size categories for hydropower may be motivated in some cases by national licensing rules (e.g., Norway[9]) to determine which authority is responsible for the process or in other cases by the need to define eligibility for specific support schemes (e.g., US Renewable Portfolio Standards). It clearly illustrates that different countries have different legal definitions of size categories that match their local energy and resource management needs.

Regardless, there is no immediate, direct link between installed capacity as a classification criterion and general properties common to all HPPs above or below that MW limit. Hydropower comes in manifold project types and is a highly site-specific technology, where each project is a tailor-made outcome for a particular location within a given river basin to meet specific needs for energy and water management services. While run-of-river facilities may tend to be smaller in size, for example, large numbers of small-scale storage hydropower stations are also in operation worldwide. Similarly, while larger facilities will tend to have lower costs on a USD/kW basis due to economies of scale, that tendency will only hold on average. Moreover, one large-scale hydropower project of 2,000 MW located in a remote area of one river basin might have fewer negative impacts than the cumulative impacts of 400 5-MW hydropower projects in many river basins (Egré and Milewski, 2002). For that reason, even the cumulative relative environmental and social impacts of large versus small hydropower development remain unclear, and context dependent.

All in all, classification according to size, while both common and administratively simple, is—to a degree—arbitrary: general concepts like 'small' or 'large hydro' are not technically or scientifically rigorous indicators of impacts, economics or characteristics (IEA, 2000c). Hydropower projects cover a continuum in scale, and it may be more useful to evaluate a hydropower project on its sustainability or economic performance (see Section 5.6 for a discussion of sustainability), thus setting out more realistic indicators.

Table 5.3 | Small-scale hydropower by installed capacity (MW) as defined by various countries

Country	Small-scale hydro as defined by installed capacity (MW)	Reference Declaration
Brazil	≤30	Brazil Government Law No. 9648, of May 27, 1998
Canada	<50	Natural Resources Canada, 2009: canmetenergy-canmetenergie.nrcan-rncan.gc.ca/eng/renewables/small_hydropower.html
China	≤50	Jinghe (2005); Wang (2010)
EU Linking Directive	≤20	EU Linking directive, Directive 2004/101/EC, article 11a, (6)
India	≤25	Ministry of New and Renewable Energy, 2010: www.mnre.gov.in/
Norway	≤10	Norwegian Ministry of Petroleum and Energy. Facts 2008. Energy and Water Resources in Norway; p.27
Sweden	≤1.5	European Small Hydro Association, 2010: www.esha.be/index.php?id=13
USA	5–100	US National Hydropower Association. 2010 Report of State Renewable Portfolio Standard Programs (US RPS)

---

9 Norwegian Water Resources and Energy Directorate, Water resource act and regulations, 2001.

5.3.2 Classification by facility type

Hydropower plants are often classified in three main categories according to operation and type of flow. Run-of-river (RoR), storage (reservoir) and pumped storage HPPs all vary from the very small to the very large scale, depending on the hydrology and topography of the watershed. In addition, there is a fourth category called in-stream technology, which is a young and less-developed technology.

5.3.2.1 Run-of-River

A RoR HPP draws the energy for electricity production mainly from the available flow of the river. Such a hydropower plant may include some short-term storage (hourly, daily), allowing for some adaptations to the demand profile, but the generation profile will to varying degrees be dictated by local river flow conditions. As a result, generation depends on precipitation and runoff and may have substantial daily, monthly or seasonal variations. When even short-term storage is not included, RoR HPPs will have generation profiles that are even more variable, especially when situated in small rivers or streams that experience widely varying flows.

In a RoR HPP, a portion of the river water might be diverted to a channel or pipeline (penstock) to convey the water to a hydraulic turbine, which is connected to an electricity generator (see Figure 5.5). RoR projects may form cascades along a river valley, often with a reservoir-type HPP in the upper reaches of the valley that allows both to benefit from the cumulative capacity of the various power stations. Installation of RoR HPPs is relatively inexpensive and such facilities have, in general, lower environmental impacts than similar-sized storage hydropower plants.

5.3.2.2 Storage Hydropower

Hydropower projects with a reservoir are also called storage hydropower since they store water for later consumption. The reservoir reduces the dependence on the variability of inflow. The generating stations are located at the dam toe or further downstream, connected to the reservoir through tunnels or pipelines. (Figure 5.6). The type and design of reservoirs are decided by the landscape and in many parts of the world are inundated river valleys where the reservoir is an artificial lake. In geographies with mountain plateaus, high-altitude lakes make up another kind of reservoir that often will retain many of the properties

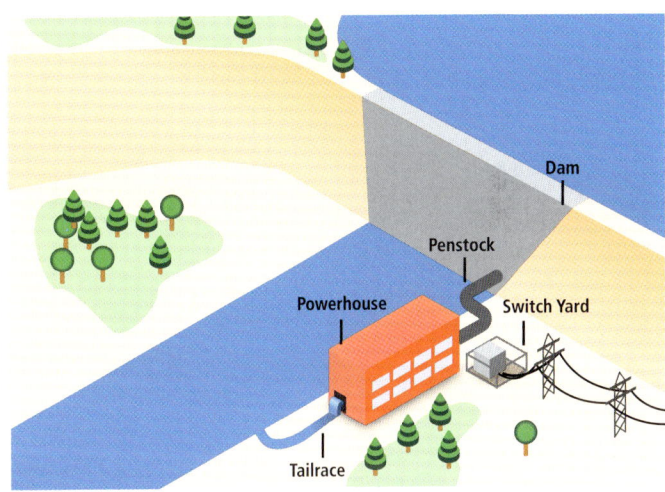

Figure 5.6 | Typical hydropower plant with reservoir.

of the original lake. In these types of settings, the generating station is often connected to the lake serving as reservoir via tunnels coming up beneath the lake (lake tapping). For example, in Scandinavia, natural high-altitude lakes are the basis for high pressure systems where the heads may reach over 1,000 m. One power plant may have tunnels coming from several reservoirs and may also, where opportunities exist, be connected to neighbouring watersheds or rivers. The design of the HPP and type of reservoir that can be built is very much dependent on opportunities offered by the topography.

5.3.2.3 Pumped storage

Pumped storage plants are not energy sources, but are instead storage devices. In such a system, water is pumped from a lower reservoir into an upper reservoir (Figure 5.7), usually during off-peak hours, while flow is reversed to generate electricity during the daily peak load period or at

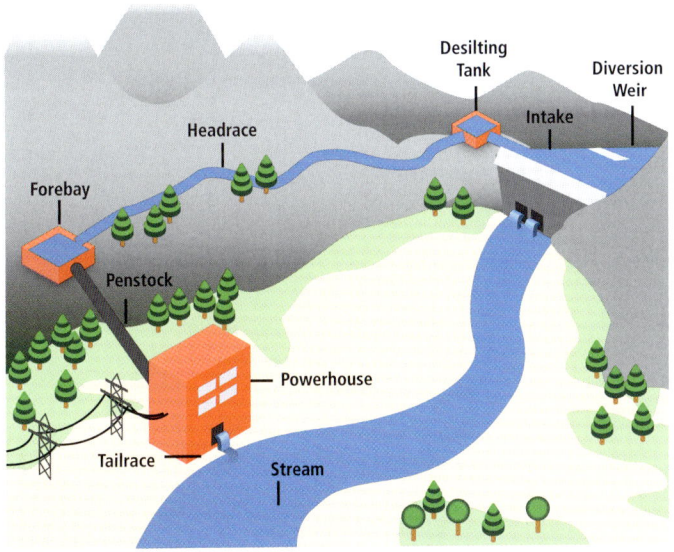

Figure 5.5 | Run-of-river hydropower plant.

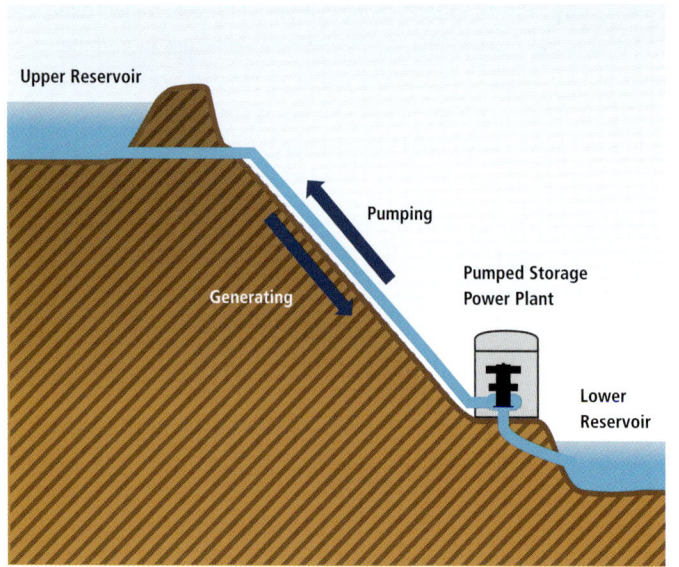

Figure 5.7 | Typical pumped storage project.

5.3.3 Status and current trends in technology development

Hydropower is a proven and well-advanced technology based on more than a century of experience—with many examples of hydropower plants built in the 19th century still in operation today. Hydropower today is an extremely flexible power technology with among the best conversion efficiencies of all energy sources (~90%, water to wire) due to its direct transformation of hydraulic energy to electricity (IEA, 2004). Still, there is room for further improvements, for example, by improving operation, reducing environmental impacts, adapting to new social and environmental requirements and by developing more robust and cost-effective technological solutions. The status and current trends are presented below, and options and prospects for future technology innovations are discussed in Section 5.7.

other times of need. Although the losses of the pumping process make such a plant a net energy consumer overall, the plant is able to provide large-scale energy storage system benefits. In fact, pumped storage is the largest-capacity form of grid energy storage now readily available worldwide (see Section 5.5.5).

5.3.2.4 In-stream technology using existing facilities

To optimize existing facilities like weirs, barrages, canals or falls, small turbines or hydrokinetic turbines can be installed for electricity generation. These basically function like a run-of-river scheme, as shown in Figure 5.8. Hydrokinetic devices being developed to capture energy from tides and currents may also be deployed inland in both free-flowing rivers and in engineered waterways (see Section 5.7.4).

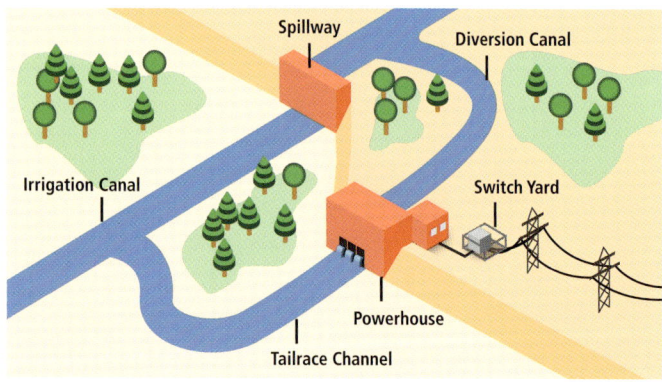

Figure 5.8 | Typical in-stream hydropower plant using existing facilities.

5.3.3.1 Efficiency

The potential for energy production in a hydropower plant is determined by the following parameters, which are dependent on the hydrology, topography and design of the power plant:

- The amount of water available;

- Water loss due to flood spill, bypass requirements or leakage;

- The difference in head between upstream intake and downstream outlet;

- Hydraulic losses in water transport due to friction and velocity change; and

- The efficiency in energy conversion of electromechanical equipment.

The total amount of water available at the intake will usually not be possible to utilize in the turbines because some of the water will be lost or will not be withdrawn. This loss occurs because of water spill during high flows when inflow exceeds the turbine capacity, because of bypass releases for environmental flows, and because of leakage.

In the hydropower plant the potential (gravitational) energy in water is transformed into kinetic energy and then mechanical energy in the turbine and further to electrical energy in the generator. The energy transformation process in modern hydropower plants is highly efficient, usually with well over 90% mechanical efficiency in turbines and over 99% in the generator. The inefficiency is due to hydraulic loss in the water circuit (intake, turbine and tailrace), mechanical loss in the turbo-generator group and electrical loss in the generator. Old turbines can have lower efficiency, and efficiency can also be reduced due to wear and abrasion caused by sediments in the water. The rest of the potential energy is lost as heat in the water and in the generator.

In addition, some energy losses occur in the headrace section where water flows from the intake to the turbines, and in the tailrace section taking water from the turbine back to the river downstream. These losses, called head loss, reduce the head and hence the energy potential for the power plant. These losses can be classified either as friction losses or singular losses. Friction losses depend mainly on water velocity and the roughness in tunnels, pipelines and penstocks.

The total efficiency of a hydropower plant is determined by the sum of these three loss components. Hydraulic losses can be reduced by increasing the turbine capacity or by increasing the reservoir capacity to get better regulation of the flow. Head losses can be reduced by increasing the area of headrace and tailrace, by decreasing the roughness in these and by avoiding too many changes in flow velocity and direction. The efficiency of electromechanical equipment, especially turbines, can be improved by better design and also by selecting a turbine type with an efficiency profile that is best adapted to the duration curve of the inflow. Different turbine types have quite different efficiency profiles when the turbine discharge deviates from the optimal value (see Figure 5.9). Improvements in turbine design by computational fluid dynamics software and other innovations are discussed in Section 5.7.

5.3.3.2 Tunnelling capacity

In hydropower projects, tunnels in hard and soft rock are often used for transporting water from the intake to the turbines (headrace), and from the turbine back to the river, lake or fjord downstream (tailrace). In addition, tunnels are used for a number of other purposes when the power station is placed underground, for example for access, power cables, surge shafts

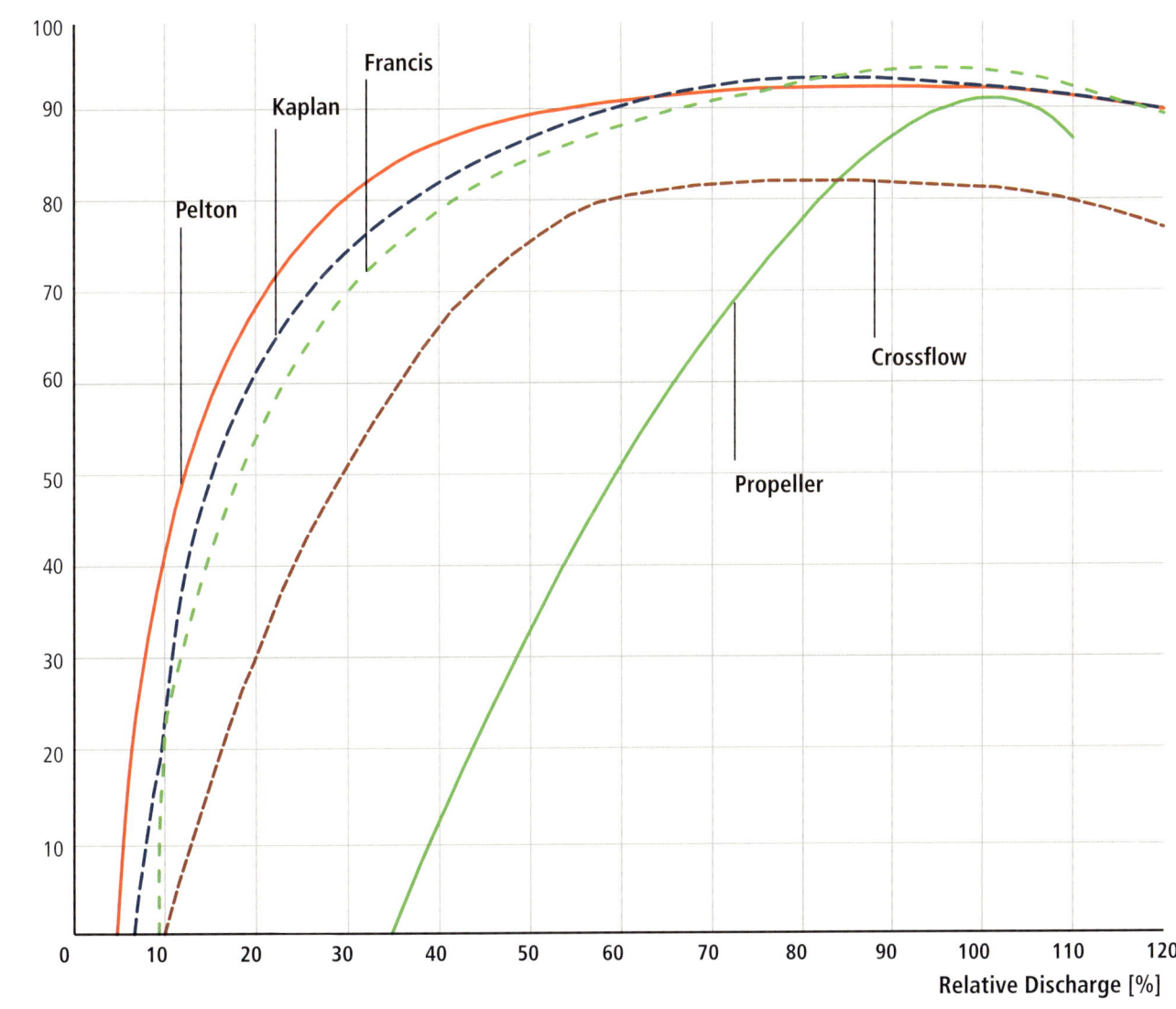

Figure 5.9 | Typical efficiency curves for different types of hydropower turbines (Vinogg and Elstad, 2003).

and ventilation. Tunnels are increasingly favoured for hydropower construction as a replacement for surface structures like canals and penstocks.

Tunnelling technology has improved greatly due to the introduction of increasingly efficient equipment, as illustrated by Figure 5.10 (Zare and Bruland, 2007). Today, the two most important technologies for hydropower tunnelling are the drill and blast method and the use of tunnel-boring machines (TBM).

The drill and blast method is the conventional method for tunnel excavation in hard rock. Thanks to the development in tunnelling technology, excavation costs have been reduced by 25%, or 0.8%/yr, over the past 30 years (see Figure 5.10).

TBMs excavate the entire cross section in one operation without the use of explosives. TBMs carry out several successive operations: drilling, support of the ground traversed and construction of the tunnel. The diameter of tunnels constructed can be from <1 m ('micro tunnelling') up to 15 m. The excavation progress of the tunnel is typically from 30 up to 60 m/day.

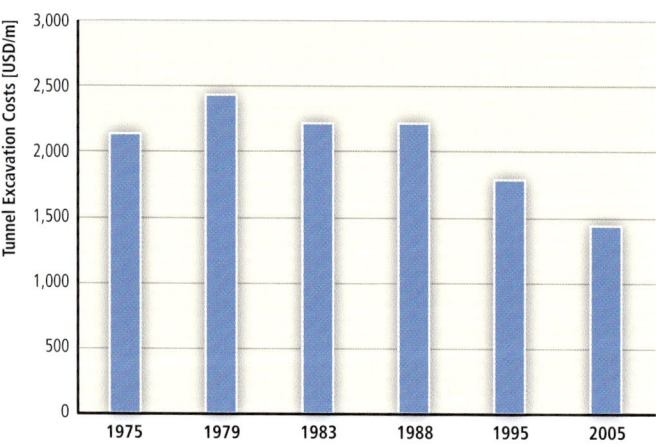

Figure 5.10 | Developments in tunnelling technology: the trend in excavation costs for a 60 m² tunnel, in USD$_{2005}$ per metre (adapted from Zare and Bruland, 2007).

5.3.3.3 Technical challenges related to sedimentation management

Although sedimentation problems are not found in all rivers (see Section 5.6.1.4), operating a hydropower project in a river with a large sediment load comes with serious technical challenges.

Specifically, increased sediment load in the river water induces wear on hydraulic machinery and other structures of the hydropower plant. Deposition of sediments can obstruct intakes, block the flow of water through the system and also impact the turbines. The sediment-induced wear of the hydraulic machinery is more serious when there is no room for storage of sediments.

In addition, for HPPs with reservoirs, their storage capacity can be filled up by sediments, which requires special technical mitigation measures or plant design.

Lysne et al. (2003) reported that the effects of sediment-induced wear of turbines in power plants can be, among others:

- Generation loss due to reduction in turbine efficiency;
- Increase in frequency of repair and maintenance;
- Increase in generation losses due to downtime;
- Reduction in lifetime of the turbine; and
- Reduction in regularity of power generation.

All of these effects are associated with revenue losses and increased maintenance costs. Several promising concepts for sediment control at intakes and mechanical removal of sediment from reservoirs and for settling basins have been developed and practised. A number of authors (Mahmood, 1987; Morris and Fan, 1997; ICOLD, 1999; Palmieri et al., 2003; White, 2005) have reported measures to mitigate the sedimentation problems by better management of land use practices in upstream watersheds to reduce erosion and sediment loading, mechanical removal of sediment from reservoirs and design of hydraulic machineries aiming to resist the effect of sediment passing through them.

5.3.4 Renovation, modernization and upgrading

Renovation, modernization and upgrading (RM&U) of old power stations is often less costly than developing a new power plant, often has relatively smaller environment and social impacts, and requires less time for implementation. Capacity additions through RM&U of old power stations can therefore be attractive. Selective replacement or repair of identified hydro powerhouse components like turbine runners, generator windings, excitation systems, governors, control panels or trash cleaning devices can reduce costs and save time. It can also lead to increased efficiency, peak power and energy availability of the plant (Prabhakar and Pathariya, 2007). RM&U may allow for restoring or improving environmental conditions in already-regulated areas. Several national programmes for RM&U are available. For example, the Research Council of Norway recently initiated a program with the aim to increase power production in existing hydropower plants and at the same time improve environmental conditions.[10] The US Department of Energy has been using a similar approach to new technology development since 1994 when it started the Advanced Hydropower Turbine Systems Program that emphasized simultaneous improvements in energy and environmental performance (Odeh, 1999; Cada, 2001; Sale et al., 2006a).

Normally the life of hydroelectric power plants is 40 to 80 years. Electromechanical equipment may need to be upgraded or replaced after 30 to 40 years, however, while civil structures like dams, tunnels

---

10 Centre for Environmental Design of Renewable Energy: www.cedren.no/.

etc. usually function longer before they requires renovation. The lifespan of properly maintained hydropower plants can exceed 100 years. Using modern control and regulatory equipment leads to increased reliability (Prabhakar and Pathariya, 2007). Upgrading hydropower plants calls for a systematic approach, as a number of hydraulic, mechanical, electrical and economic factors play a vital role in deciding the course of action. For techno-economic reasons, it can also be desirable to consider up-rating (i.e., increasing the size of the hydropower plant) along with RM&U/life extension. Hydropower generating equipment with improved performance can also be retrofitted, often to accommodate market demands for more flexible, peaking modes of operation. Most of the existing worldwide hydropower equipment in operation will need to be modernized to some degree by 2030 (SER, 2007). Refurbished or up-rated hydropower plants also result in incremental increases in hydropower generation due to more efficient turbines and generators.

In addition, existing infrastructure without hydropower plants (like existing barrages, weirs, dams, canal fall structures, water supply schemes) can also be reworked by adding new hydropower facilities. The majority of the world's 45,000 large dams were not built for hydropower purposes, but for irrigation, flood control, navigation and urban water supply schemes (WCD, 2000). Retrofitting these with turbines may represent a substantial potential, because only about 25% of large reservoirs are currently used for hydropower production. For example, from 1997 to 2008 in India, about 500 MW have been developed on existing facilities. A recent study in the USA indicated some 20 GW could be installed by adding hydropower capacity to 2,500 dams that currently have none (UNWWAP, 2006).

5.4 Global and regional status of market and industry development

5.4.1 Existing generation

In 2008, the generation of electricity from hydroelectric plants was 3,288 TWh (11.8 EJ)[11] compared to 1,295 TWh (4.7 EJ) in 1973 (IEA, 2010a), which represented an increase of roughly 25% in this period, and was mainly a result of increased production in China and Latin America, which reached 585 TWh (2.1 EJ) and 674 TWh (2.5 EJ), respectively (Figures 5.11 and 5.12).

Hydropower provides some level of power generation in 159 countries. Five countries make up more than half of the world's hydropower production: China, Canada, Brazil, the USA and Russia. The

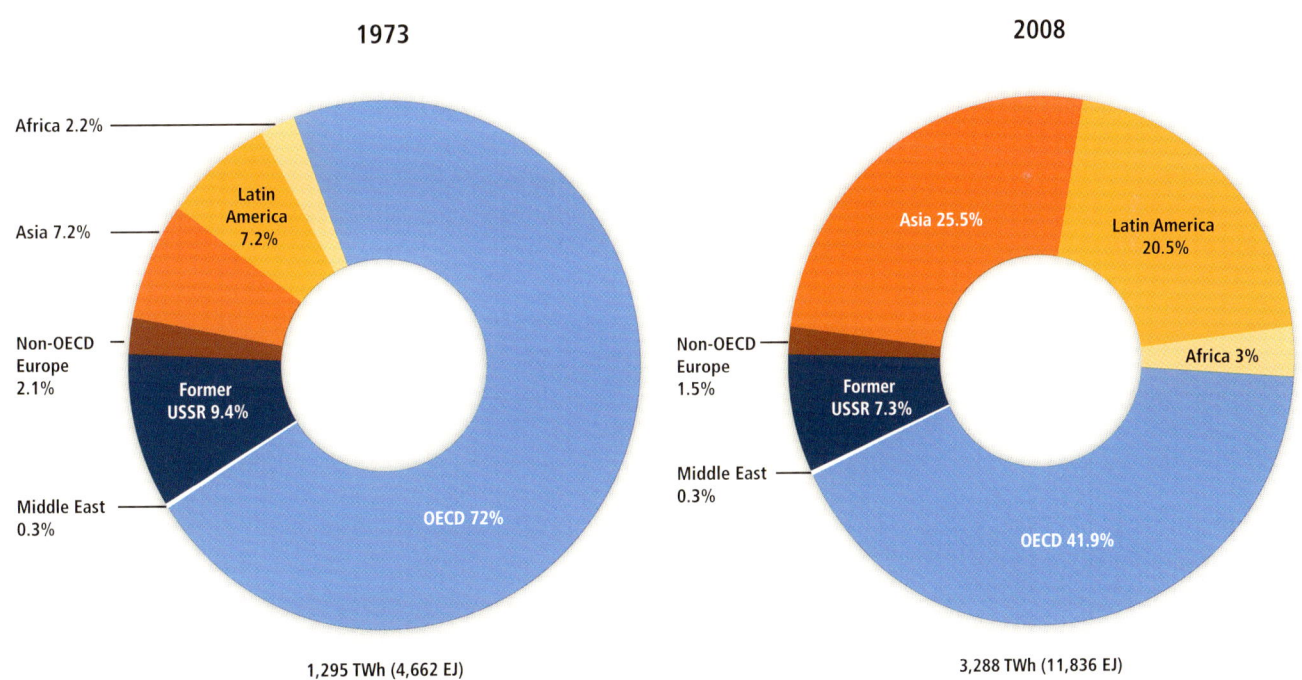

Figure 5.11 | 1973 and 2008 regional shares of hydropower production (IEA, 2010a).

---

11 These figures differ slightly from those presented in Chapter 1.

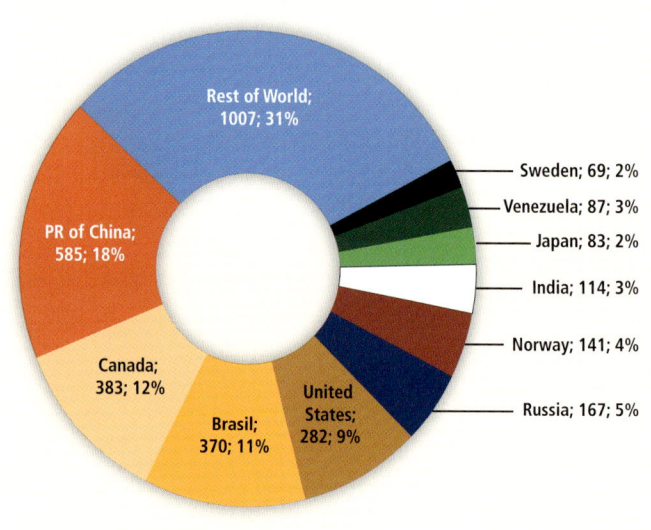

Figure 5.12 | Hydropower generation in 2008 by country, indicating total generation (TWh) and respective global share (IEA, 2010a).

importance of hydroelectricity in the electricity mix of these countries is, however, different (Table 5.4). On the one hand, Brazil and Canada are heavily dependent on this source, with a percentage share of total domestic electricity generation of 83.9% and 59%, respectively, whereas in Russia the share is 19.0% and in China 15.5%.

China, Canada, Brazil and the USA together account for over 46% of the production (TWh/EJ) of hydroelectricity in the world and are also the four largest in terms of installed capacity (GW) (IEA, 2010a). Figure 5.12 shows hydropower generation by country. It is noteworthy that 5 out of the 10 major producers of hydroelectricity are among the world's most industrialized countries: Canada, the USA, Norway, Japan and Sweden. This is no coincidence, given that the possibility of drawing on the hydroelectric resource was important for the introduction and consolidation of the main electro-intensive sectors on which the industrialization process in these countries was based during a considerable part of the 20th century.

Despite the significant growth in hydroelectric production, the percentage share of hydroelectricity on a global basis has dropped during the last three decades (1973 to 2008), from 21 to 16%. This is because electricity demand and the deployment of other energy technologies have increased more rapidly than hydropower generating capacity.

5.4.2 The hydropower industry

In developed markets such as the Europe, the USA, Canada, Norway and Japan, where many hydropower plants were built 30 to 60 years ago, the hydropower industry is focused on re-licensing and renovation as well as on adding new hydropower generation to existing dams. In emerging markets such as China, Brazil, Ethiopia, India, Malaysia, Iran, Laos, Turkey, Venezuela, Ecuador and Vietnam, utilities and private developers are pursuing large-scale new hydropower construction (116 GW of capacity is under construction; IJHD, 2010). Canada is still on the list of the top five hydropower markets for new installations worldwide. Orders for hydropower equipment were lower in 2009 and 2010 compared to the peaks in 2007 and 2008, though the general high level after 2006, when the hydropower market almost doubled, is anticipated to continue for the near future. With increasing policy support of governments for new hydropower (see Sections 5.4.3 and 5.10.3) construction, hydropower industrial activity is expected to be higher in the coming years compared to the average since 2000 (IJHD, 2010). As hydropower and its industry are mature, it is expected that the industry will be able to meet the demand that materializes (see Section 5.9). In 2008, the hydropower industry installed more than 40 GW of new capacity worldwide (IJHD, 2010), with 31 GW added in 2009 (REN21, 2010; see Chapter 1).

Table 5.4 | Major hydroelectricity producer countries with total installed capacity and percentage of hydropower generation in the electricity mix. Source: IJHD (2010).

Country	Installed Capacity (GW)	Country Based on Top 10 Producers	Percent of Hydropower in Total Domestic Electricity Generation (%)
China	200	Norway	99
Brazil	84	Brazil	83.9
USA	78.2	Venezuela	73.4
Canada	74.4	Canada	59.0
Russia	49.5	Sweden	48.8
India	38	Russia	19.0
Norway	29.6	India	17.5
Japan	27.5	China	15.5
France	21	Italy	14.0
Italy	20	France	8.0
Rest of the world	301.6	Rest of the world[1]	14.3
World	**926.1**	**World**	**15.9**

Note: 1. Excluding countries with no hydropower production.

5.4.3 Impact of policies[12]

Hydropower infrastructure development is closely linked to national, regional and global development policies. Beyond its role in contributing to a secure energy supplysecurity and reducing a country's dependence on fossil fuels, hydropower offers opportunities for poverty alleviation and sustainable development. Hydropower also can contribute to regional cooperation, as good practice in managing water resources requires a river basin approach regardless of national borders (see also Section 5.10). In addition, multipurpose hydropower can strengthen a country's ability to adapt to climate change-induced hydrological variability (World Bank, 2009).

The main challenges for hydropower development are linked to a number of associated risks such as poor identification and management of environmental and social impacts, insufficient hydrological data, unexpected adverse geological conditions, lack of comprehensive river basin planning, shortage of financing, scarcity of local skilled human resources and lack of regional collaboration. These challenges can be and are being addressed to varying degrees at the policy level by a number of governments, international financing institutions, professional associations and nongovernmental organizations (NGOs). Examples of policy initiatives dealing with the various challenges can be found in Sections 5.6.2 and 5.10.

Challenges posed by various barriers can be addressed and met by public policies, bearing in mind the need for an appropriate environment for investment, a stable regulatory framework and incentives for research and technological development (Freitas and Soito, 2009; see Chapter 11). A variety of policies have been enacted in individual countries to support certain forms and types of hydropower, as highlighted generally in Chapter 11. More broadly, in addition to country-specific policies, several larger policy issues have been identified as particularly important for the development of hydropower, including carbon markets, financing, administration and licensing procedures, and size-based classification schemes.

5.4.3.1 International carbon markets

As with other carbon reduction technologies, carbon credits can benefit hydropower projects by bringing additional funding and thus helping to reduce project risk and thereby secure financing. Though the Clean Development Mechanism (CDM) is not unique to hydropower, hydropower projects are one of the largest contributors to the CDM and Joint Implementation (JI) mechanisms and therefore to existing carbon credit markets. In part, this is due to the fact that new hydropower development is targeted towards developing countries that are in need of investment capital, and international carbon markets offer one possible route to that capital. Out of the 2,062 projects registered by the CDM Executive Board (EB) by 1 March 2010, 562 were hydropower projects. When considering the predicted volumes of Certified Emission Reductions to be delivered, registered hydropower projects are expected to avoid more than 50 Mt of carbon dioxide (CO_2) emissions per year by 2012. China, India, Brazil and Mexico represent roughly 75% of the hosted projects.

5.4.3.2 Project financing

Hydropower projects can often deliver electricity at comparatively low costs relative to existing market energy prices (see Section 5.8). Nonetheless, many otherwise economically feasible hydropower projects are financially challenging because high upfront costs are often a deterrent to investment. Related to this, hydropower projects tend to have lengthy lead times for planning, permitting and construction, increasing development risk and delaying revenue generation. A key challenge, then, is to create sufficient private sector confidence in hydropower investment, especially prior to project permitting. Deployment policies of the types described in Chapter 11 are being used in some countries to encourage investment. Also, in developing regions such as Africa, interconnection between countries and the formation of larger energy markets is helping to build investor confidence by reducing the risk of a monopsony buyer. Feasibility and impact assessments carried out by the public sector, prior to developer tendering, can also help ensure greater private sector interest in hydropower development (WEC, 2007; Taylor, 2008). Nonetheless, the development of appropriate financing models that consider the uncertainty imposed by long planning and regulatory processes, and finding the optimum roles for the public and private sectors, remain key challenges for hydropower development.

5.4.3.3 Administrative and licensing process

Hydropower is often regarded as a public resource (Sternberg, 2008), emphasized by the operating life of a reservoir that may be more than 100 years. Legal frameworks vary from country to country, however, including practices in the award and structuring of concessions, for instance, regarding concession periods, royalties, water rights etc. Environmental licensing procedures also vary greatly. With growing involvement of the private sector in what was previously managed by public sector, contractual arrangements surrounding hydropower have become increasingly complex. There are now more parties involved and much greater commercial accountability, with a strong awareness of environmental and social indicators and licensing processes. Clearly, the policies and procedures established by governments in granting licenses and concessions will impact hydropower development outcomes.

---
12   Non-technology-specific policy issues are covered in Chapter 11 of this report.

5.4.3.4 Classification by size

Finally, many governments and international bodies have relied upon various distinctions between 'small' and 'large' hydro, as defined by installed capacity (MW), in establishing the eligibility of hydropower plants for certain programs. While it is well known that large-scale HPPs can create conflicts and concerns (WCD, 2000), the environmental and social impacts of a HPP cannot be deduced by size in itself, even if increasing the physical size may increase the overall impacts of a specific HPP (Egré and Milewski, 2002; Sternberg, 2008). Despite their lack of robustness (see Section 5.3.1), these classifications have had significant policy and financing consequences (Egré and Milewski, 2002).

In the UK Renewables Obligation,[13] eligible hydropower plants must be below 20 MW in size. Likewise, in several countries, feed-in tariffs are targeted only towards smaller projects. For example, in France, only projects with an installed capacity not exceeding 12 MW are eligible,[14] and in Germany, a 5 MW maximum capacity has been established.[15] In India, projects below 5 and 25 MW in capacity obtain promotional support that is unavailable to projects of larger sizes. Similar approaches exist in many developed and developing countries around the world, for example, in Indonesia.[16] Because project size is neither a perfect indicator of environmental and social impact nor of the financial need of a project for addition policy support, these categorizations may, at times, impede the development of socially beneficial projects.

Similar concerns have been raised with respect to international and regional climate policy. Though hydropower is recognized as a contributor to reducing GHG emissions and is included in the Kyoto Protocol's flexible mechanisms, those mechanisms differentiate HPPs depending on size and type. The United Nations Framework Convention on Climate Change (UNFCCC) CDM EB, for example, has established that storage hydropower projects are to follow the power density indicator (PDI), W/m² (installed capacity/reservoir area), to be eligible for CDM credits. The PDI indicates tentative GHG emissions from reservoirs. The CDM Executive Board stated (February 2006) that "Hydroelectric power plants with power densities greater than 4 W/m² but less than or equal to 10 W/m² can use the currently approved methodologies, with an emission factor of 90 g CO_2eq/kWh for projects with reservoir emissions", while "less than or equal to 4 W/m² cannot use current methodologies". There is little link, however, between installed capacity, the area of a reservoir and the various biogeochemical processes active in a reservoir. Hypothetically, two identical storage HPPs would, according to the PDI, have the same emissions independent of climate zones or of inundated biomass and carbon fluxes (see Section 5.6.3). As such, the PDI rule may inadvertently impede the development of socially beneficial hydropower projects, while at the same time supporting less beneficial projects. The European Emission Trading Scheme and related trading markets similarly treat small- and large-scale hydropower stations differently.[17]

5.5 Integration into broader energy systems

Hydropower's large capacity range, flexibility, storage capability when coupled with a reservoir, and ability to operate in a stand-alone mode or in grids of all sizes enables hydropower to deliver a broad range of services. Hydropower's various roles in and services to the energy system are discussed below (see also Chapter 8).

5.5.1 Grid-independent applications

Hydropower can be delivered through national and regional interconnected electric grids, through local mini-grids and isolated grids, and can also serve individual customers through captive plants. Water mills in England, Nepal, India and elsewhere, which are used for grinding cereals, for lifting water and for powering machinery, are early testimonies of hydropower being used as captive power in mechanical and electrical form. The tea and coffee plantation industries as well as small islands and developing states have used and still make use of hydropower to meet energy needs in isolated areas.

Captive power plants (CPPs) are defined here as plants set up by any person or group of persons to generate electricity primarily for the person or the group's members (Indian Electricity Act, 2003). CPPs are often found in decentralized isolated systems and are generally built by private interests for their own electricity needs. In deregulated electricity markets that allow open access to the grid, hydropower plants are also sometimes installed for captive purposes by energy-intensive industries such as aluminium smelters, pulp and paper mills, mines and cement factories in order to weather short-term market uncertainties and volatility (Shukla et al., 2004). For governments of emerging economies such as India facing shortages of electricity, CPPs are also a means to cope with unreliable power supply systems and higher industrial tariffs by encouraging decentralized generation and private participation (Shukla et al., 2004).

5.5.2 Rural electrification

According to the International Energy Agency (IEA, 2010c), 1.4 billion people have no access to electricity (see Section 9.3.2). Related to the discussion in Section 5.5.1, small-scale hydropower (SHP) can sometimes be an economically viable supply source in these circumstances, as SHP can provide a decentralized electricity supply in those rural areas

---

[13] The Renewables Obligation Order 2006, No. 1004 (ROO 2006): www.statutelaw.gov.uk.

[14] Décret n°2000-1196, Decree on capacity limits for different categories of systems for the generation of electricity from renewable sources that are eligible for the feed-in tariff: www.legifrance.gouv.fr.

[15] EEG, 2009 - Act on Granting Priority to Renewable Energy and Mineral Sources: bundesrecht.juris.de/eeg_2009/.

[16] Regulation of the Minister of Energy and Mineral Resources, No.31, 2009.

[17] Directive 2004/101/E, C article 11a(6), www.eur-lex.europa.eu.

that have adequate hydropower technical potential (Egré and Milewski, 2002). In fact, SHPs already play an important role in the economic development of some remote rural areas. Small-scale hydropower-based rural electrification in China has been one of the most successful examples, where over 45,000 small hydropower plants totalling 55 GW have been built that are producing 160 TWh (0.58 EJ) annually. Though many of these plants are used in centralized electricity networks, SHPs constitute one-third of China's total hydropower capacity and are providing services to over 300 million people (Liu and Hu, 2010). More generally, SHP is found in isolated grids as well as in off-grid and central-grid settings. As 75% of costs are site-specific, proper site selection is a key challenge. Additionally, in isolated grid systems, natural seasonal flow variations might require that hydropower plants be combined with other generation sources in order to ensure continuous supply during dry periods (World Bank, 2008) and may have excess production during wet seasons; such factors need to be considered in the planning process (Sundqvist and Wårlind, 2006).

In general, SHPs

- Are often but certainly not always RoR schemes;
- Can use existing infrastructure such as dams or irrigation channels;
- Are located close to villages to avoid expensive high-voltage distribution equipment;
- Can use pumps as turbines and motors as generators for a turbine/generator set; and
- Have a high level of local content both in terms of materials and work force during the construction period and local materials for the civil works.

A recent example from western Canada[18] shows that SHP might also be a solution for remote communities in developed countries by replacing fossil-fired diesel generation with hydropower generation.

All in all, the development of SHP for rural areas involves environmental, social, technical and economic considerations. Local management, ownership and community participation, technology transfer and capacity building are basic issues for sustainable SHP plants in such circumstances.

5.5.3 Power system services provided by hydropower

Hydroelectric generation differs from thermal generation in that the quantity of 'fuel' (i.e., water) that is available at any given time is determined by river flows leading to the hydroelectric plant. Run-of-river HPPs lack a reservoir to store large quantities of water, though large RoR HPPs may have some limited ability to regulate river flow. Storage hydropower, on the other hand, can largely decouple the timing of hydropower generation and variable river flows. For large storage reservoirs, the storage may be sufficient to buffer seasonal or multi-seasonal changes in river flows, whereas for smaller reservoirs the storage may buffer river flows on a daily or weekly basis.

With a very large reservoir relative to the size of the hydropower plant (or very consistent river flows), HPPs can generate power at a near-constant level throughout the year (i.e., operate as a base-load plant). Alternatively, in the case that the hydropower capacity far exceeds the amount of reservoir storage, the hydropower plant is sometimes referred to as energy-limited. An energy-limited hydropower plant would exhaust its 'fuel supply' by consistently operating at its rated capacity throughout the year. In this case, the use of reservoir storage allows hydropower generation to occur at times that are most valuable from the perspective of the power system rather than at times dictated solely by river flows. Since electrical demand varies during the day and night, during the week and seasonally, storage hydropower generation can be timed to coincide with times where the power system needs are the greatest. In part, these times will occur during periods of peak electrical demand. Operating hydropower plants in a way that generates power during times of high demand is referred to as peaking operation (in contrast to base-load). Even with storage, however, hydropower generation will still be limited by the size of the storage, the rated electrical capacity of the hydropower plant, and downstream flow constraints for irrigation, recreation or environmental uses of the river flows. Hydropower peaking may, if the outlet is directed to a river, lead to rapid fluctuations in river flow, water-covered area, depth and velocity. In turn this may, depending on local conditions, lead to negative impacts in the river (see Section 5.6.1.5) unless properly managed.

Hydropower generation that consistently occurs during periods with high system demand can offset the need for thermal generation to meet that same demand. The ratio of the amount of demand that can be reliably met by adding hydropower to the nameplate capacity of the hydropower plant is called the capacity credit. Even RoR hydropower that consistently has river flows during periods of high demand can earn a high capacity credit, while adding reservoir storage can increase the capacity credit to levels comparable to thermal power plants (see Section 8.2.1.2).

In addition to providing energy and capacity to meet electrical demand, hydropower generation often has several characteristics that enable it to provide other services to reliably operate power systems. Because hydropower plants utilize gravity instead of combustion to generate electricity, hydropower plants are often less susceptible to the sudden loss of generation than is thermal generation. Hydropower plants also offer operating flexibility in that they can start generating electricity with very short notice and low start-up costs, provide rapid changes in generation, and have a wide range of generation levels over which power can be generated efficiently (i.e., high part-load efficiency) (Haldane and Blackstone, 1955; Altinbilek et al., 2007). The ability to rapidly change

---

18 Natural Resources Canada. 2009. Isolated-grid case study: the Hluey Lake project in British Columbia: www.retscreen.net/ang/case_studies_2900kw_isolated_grid_internal_load_canada.php.

output in response to system needs without suffering large decreases in efficiency makes hydropower plants well suited to providing the balancing services called regulation and load-following. RoR HPPs operated in cascades in unison with storage hydropower in upstream reaches may similarly contribute to the overall regulating and balancing ability of a fleet of HPPs. With the right equipment and operating procedures, hydropower can also provide the ability to restore a power station to operation without relying on the electric power transmission network (i.e., black start capability) (Knight, 2001).

Overall, with its important load-following and balancing capabilities, peaking capacity and power quality attributes, hydropower can play a significant role in ensuring reliable electricity service (US Department of the Interior, 2005).

5.5.4 Hydropower support of other generation including renewable energy

Electricity systems worldwide rely upon widely varying amounts of hydropower today. In this range of hydropower capabilities, electric system operators have developed economic dispatch methodologies that take into account the unique role of hydropower, including coordinating the operation of hydropower plants with other types of generating units. In particular, many thermal power plants (coal, gas or liquid fuel, or nuclear energy) require considerable lead times (often four hours for gas turbines and over eight hours for steam turbines) before they attain an optimum thermal efficiency at which point fuel consumption and emissions per unit output are minimum. In an integrated system, the considerable flexibility provided by storage HPPs can be used to reduce the frequency of start ups and shut downs of thermal plants; to maintain a balance between supply and demand under changing demand or supply patterns and thereby reduce the load-following burden on thermal plants; and to increase the amount of time that thermal units are operated at their maximum thermal efficiency. In some regions, for instance, hydroelectric power plants are used to follow varying peak load demands while nuclear or fossil fuel power plants are operated as base-load units.

Pumped hydropower storage can further increase the support of other resources. In cases with pumped hydropower storage, pumps can use the output from thermal plants during times that they would otherwise operate less efficiently at part load or be shut down (i.e., low load periods). The pumped storage plant then keeps water in reserve for generating power during peak period demands. Pumped storage has much the same ability as storage HPPs to provide balancing and regulation services.

Pumped storage hydropower is usually not a source for energy, however. The hydraulic, mechanical and electrical efficiencies of pumped storage determine the overall cycle efficiency, ranging from 65 to 80% (Egré and Milewski, 2002). If the upstream pumping reservoir is also used as a traditional reservoir the inflow from the watershed may balance out the energy loss caused by pumping. If not, net losses lead to pumped hydropower being a net energy consumer. A traditional storage HPP may also be retrofitted with pump technologies to combine the properties of storage and pump storage HPPs (SRU, 2010). The use and benefit of pumped storage hydropower in the power system will depend on the overall mix of existing generating plants and the architecture of the transmission system. Pumped storage represents about 2.2% of all generation capacity in the USA, 10.2 % in Japan and 18.7 % in Austria (Deane et al., 2010). Various technologies for storing electricity in the grid are compared by Vennemann et al. (2010) in Figure 5.13 for selected large storage sites in different parts of the world.

In addition to hydropower supporting fossil and nuclear generation technologies, hydropower can also help reduce the challenges of integrating variable renewable resources. In Denmark, for example, the high level of variable wind (>20% of the annual energy demand) is managed in part through strong interconnections (1 GW) to Norway, where there is substantial storage hydropower (Nordel, 2008). More interconnectors to Europe may further support increasing the share of wind power in Denmark and Germany (SRU, 2010; see also Section 11.6.5). From a technical viewpoint, Norway alone has a long-term potential to establish pumped storage facilities in the 10 to 25 GW range, enabling energy storage over periods from hours to several weeks in existing reservoirs, and more or less doubling the present installed capacity of 29 GW (IEA-ENARD, 2010).

Increasing variable generation will also increase the amount of balancing services, including regulation and load following, required by the power system (e.g., Holttinen et al., 2009). In regions with new and existing hydropower facilities, providing these services with hydropower may avoid the need to rely on increased part-load and cycling of thermal plants to provide these services. Similarly, in systems with high shares of variable renewable resources that provide substantial amounts of energy but limited capacity, the potential for a high capacity credit of hydropower can be used to meet peak demand rather than requiring peaking thermal plants.

5.5.5 Reliability and interconnection needs for hydropower

Though hydropower has the potential to offer significant power system services in addition to energy and capacity, interconnecting and reliably utilizing hydropower plants may also require changes to power systems. The interconnection of hydropower to the power system requires adequate transmission capacity from hydropower plants to demand centres. Adding new hydropower plants has in the past required network investments to extend the transmission network (see Section 8.2.1.3). Without adequate transmission capacity, hydropower plant operation can be constrained such that the services offered by the hydropower plant are less than what it could offer in an unconstrained system.

Aside from network expansion, changes in the river flow between a dry year and a wet year can be a significant concern for ensuring

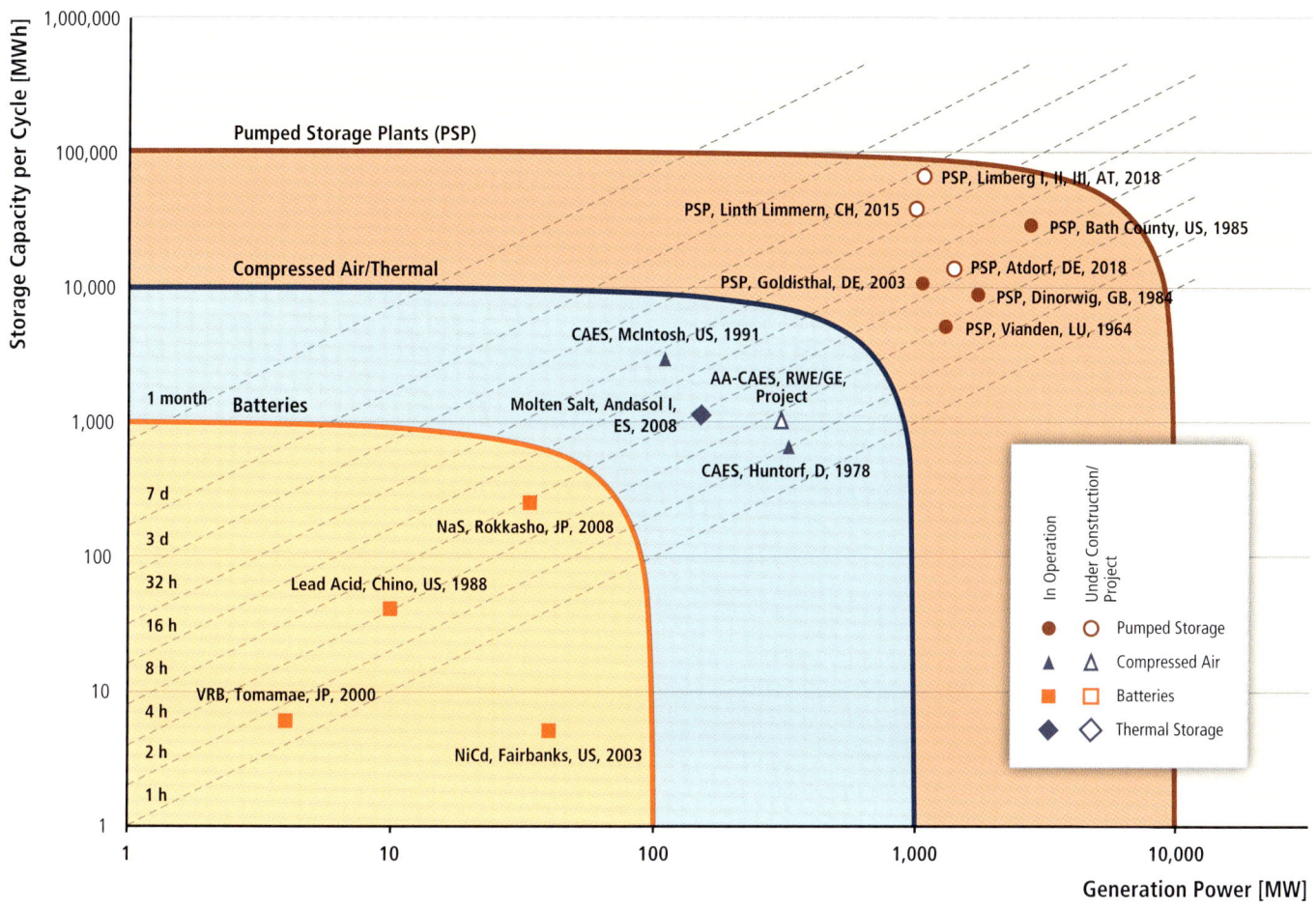

Figure 5.13 | Storage and installed capacity of selected large electricity storage sites (Vennemann et al., 2010).

Note: PSP = Pumped storage plants; CAES = compressed air energy storage, AA-CAES = advanced adiabatic compressed air energy storage; Batteries: NaS = sodium-sulphur, NiCd = nickel cadmium, VRB = vanadium redox battery.

that adequate total annual energy demand can be met. Strong interconnections between diverse hydropower resources or between hydro-dominated and thermal-dominated power systems have been used in existing systems to ensure adequate energy generation (see Section 8.2.1.3). In the future, interconnection to other renewable resources could also ensure adequate energy. Wind and direct solar power, for instance, can be used to reduce demands on hydropower, either by allowing dams to save their water for later release in peak periods or letting storage or pumped storage HPPs consume excess energy produced in off-peak hours.

5.6 Environmental and social impacts[19]

Like all energy and water management options, hydropower projects have negative and positive environmental and social impacts. On the environmental side, hydropower may have a significant environmental footprint at local and regional levels but offers advantages at the macro-ecological level. With respect to social impacts, hydropower projects may entail the relocation of communities living within or nearby the reservoir or the construction sites, compensation for downstream communities, public health issues etc. A properly designed hydropower project may, however, be a driving force for socioeconomic development (see Box 5.1), though a critical question remains about how these benefits are shared.

Because each hydropower plant is uniquely designed to fit the site-specific characteristics of a given geographical site and the surrounding society and environment, the magnitude of environmental and social impacts as well as the extent of their positive and negative effects is highly site dependent. Though the size of a HPP is not, alone, a relevant criterion to predict environmental performance, many impacts are related to the impoundment and existence of a reservoir, and therefore do not apply to all HPP types (see Table 5.5). Section 5.6.1 summarizes

---

[19] A comprehensive assessment of social and environmental impacts of all RE sources covered in this report can be found in Chapter 9.

> **Box 5.1 | Possible multiplier effects of hydropower projects.**
>
> Dam projects generate numerous impacts both on the region where they are located, as well as at an inter-regional, national and even global level (socioeconomic, health, institutional, environmental, ecological and cultural impacts). The World Commission on Dams (WCD) and numerous other studies have discussed the importance and difficulties of evaluating a number of these impacts. One of the issues raised by these studies is the need to extend consideration to indirect benefits and costs of dam projects (Bhatia et al., 2003). According to the WCD's Final Report (WCD, 2000) "a simple accounting for the direct benefits provided by large dams—the provision of irrigation water, electricity, municipal and industrial water supply, and flood control—often fails to capture the full set of social benefits associated with these services. It also misses a set of ancillary benefits and indirect economic (or multiplier) benefits of dam projects". Indirect impacts are called multiplier impacts, and result from both inter-industry linkage impacts (increase in the demand for an increase in outputs of other sectors) and consumption-induced impacts (increase in incomes and wages generated by the direct outputs). Multipliers are summary measures expressed as a ratio of the total effects (direct and indirect) of a project to its direct effects. A multi-country study on multiplier effects of large hydropower projects was performed by the World Bank (2005), which estimates that the multiplier values for large scale hydropower projects vary from 1.4 to 2.0, meaning that for every dollar of value generated by the sectors directly involved in dam-related activities, another 40 to 100 cents could be generated indirectly in the region. Though these multiplier benefits are not unique to hydropower projects, but accompany—to varying degrees—any energy project, they nonetheless represent benefits that might be considered by communities considering hydropower development.

Table 5.5 | Types of hydropower projects, their main services and distinctive environmental and social characteristics (adapted from IEA, 2000d; Egré and Milewski, 2002). The number of subsections within section 5.6.1 that address specific impacts are given in parentheses.

HPP Type	Energy and water management services	Main environmental and social characteristics (corresponding subsection)
All	Renewable electricity generation Increased water management options	Barrier for fish migration and navigation (1,6), and sediment transport (4) Physical modification of riverbed and shorelines (1)
Run-of-river	Limited flexibility and increased variability in electricity generation output profile Water quality (but no water quantity) management	Unchanged river flow when powerhouse in dam toe; when localized further downstream reduced flow between intake and powerhouse (1)
Reservoir (Storage)	Storage capacity for energy and water Flexible electricity generation output Water quantity and quality management; groundwater stabilization; water supply and flood management, see also Section 5.10	Alteration of natural and human environment by impoundment (2), resulting in impacts on ecosystems and biodiversity (1, 5, 6) and communities (7–11) Modification of volume and seasonal patterns of river flow (1), changes in water temperature and quality (3), land use change-related GHG emissions (see Section 5.6.2)
Multipurpose	As for reservoir HPPs; Dependent on water consumption of other uses	As for reservoir HPP; Possible water use conflicts; Driver for regional development (see Box 5.1)
Pumped storage	Storage capacity for energy and water; net consumer of electricity due to pumping No water management options	Impacts confined to a small area; often operated outside the river basin as a separate system that only exchanges the water from a nearby river from time to time

the main environmental and social impacts that can arise from development of the various types of hydropower projects, as well as a number of practicable mitigation measures that can be implemented to minimize negative effects and maximize positive outcomes. More information about existing guidance for sustainable hydropower development is provided in Section 5.6.2. Hydropower creates no direct atmospheric pollutants or waste during operation, and GHG emissions associated with most lifecycle stages are minor. However, methane (CH_4) emissions from reservoirs might be substantial under certain conditions. Thus, there is a need to properly assess the net change in GHG emissions induced by the creation of such reservoirs. The lifecycle GHG emissions of hydropower are discussed in Section 5.6.3, including the scientific status of the carbon balances of reservoirs and other lifecycle aspects.

5.6.1 Typical impacts and possible mitigation measures

Although the type and magnitude of impacts will vary from project to project, it is possible to describe some typical effects, along with the experience that has been gained throughout the past decades in managing and solving problems. Though some impacts are unavoidable, they can be minimized or compensated for, as experience in successful mitigation demonstrates. Information has been systematically gathered on effective assessment and management of impacts related to various types of hydropower (IEA, 2000a; UNEP, 2007). By far the most effective measure is impact avoidance, by weeding out less sustainable alternatives early in the design stage.

All hydroelectric structures affect a river's ecology mainly by inducing a change in its hydrologic characteristics and by disrupting the ecological continuity of sediment transport and fish migration through the building of dams, dikes and weirs. However the extent to which a river's physical, chemical and biological characteristics are modified depends largely on the type of HPP. Whereas run-of-river HPPs do not alter a river's flow regime, the creation of a reservoir for storage hydropower entails a major environmental change by transforming a fast-running fluvial ecosystem into a still-standing lacustrine one. The extent to which a hydropower project has adverse impacts on the riverbed morphology, on water quality and on fauna and flora is highly site-specific and to a certain degree dependent on what resources can be invested into mitigation measures. A more detailed summary of ecological impacts and their possible management measures are discussed in Sections 5.6.1.1 though 5.6.1.6.

Similar to a HPPs environmental effects, the extent of its social impacts on the local and regional communities, land use, the economy, health and safety or heritage varies according to project type and site-specific conditions. While run-of-river projects generally introduce little social change, the creation of a reservoir in a densely populated area can entail significant challenges related to resettlement and impacts on the livelihoods of the downstream populations. Restoration and improvement of living standards of affected communities is a long-term and challenging task that has been managed with variable success in the past (WCD, 2000). Whether HPPs can contribute to fostering socioeconomic development depends largely on how the generated services and revenues are shared and distributed among different stakeholders. As documented by Scudder (2005), HPPs can also have positive impacts on the living conditions of local communities and the regional economy, not only by generating electricity but also by facilitating, through the creation of freshwater storage schemes, multiple other water-dependent activities, such as irrigation, navigation, tourism, fisheries or sufficient water supply to municipalities and industries while protecting against floods and droughts. Yet, inevitably questions arise about the sharing of these revenues among the local affected communities, government, investors and the operator. Key challenges in this domain are the fair treatment of affected communities and especially vulnerable groups like indigenous people, resettlement if necessary, and public health issues, as well as appropriate management of cultural heritage values that will be discussed in more detail in Sections 5.6.1.7 through 5.6.1.11.

All in all, for the sake of sustainability it is important to assess the negative and positive impacts of a hydropower project in the light of a region's needs for energy and water management services. An overview of the main energy and water management services and distinctive environmental characteristics in relation to the different HP project types are presented in the Table 5.5.

According to the results of decade-long IEA research focusing on hydropower and the environment, 11 sensitive issues have been identified that need to be carefully assessed and managed to achieve sustainable hydropower projects. These peer-reviewed reports were produced under the IEA Implementing Agreement on Hydropower Technologies between 1996 and 2006 in collaboration with private agencies, governmental institutions, universities, research institutions and international organizations with relevance to the subject. They are based on more than 200 case studies, involving more than 112 experts from 16 countries, and are considered to be the most comprehensive international information source presently available with regard to managing social and environmental issues related to hydropower. Unless a different reference is mentioned, Sections 5.6.1.1 to 5.6.1.11 are based on the outcomes of these five IEA reports (IEA, 2000a,b,c,d,e).

5.6.1.1 Hydrological regimes

A hydropower project may modify a river's flow regime if the project includes a reservoir. Run-of-river projects change the river's flow pattern marginally, thus creating fewer impacts downstream from the project.

Hydropower plants with reservoirs significantly modify the downstream flow regime (i.e., the magnitude and timing of discharge and hence water levels), and may also alter water temperature over short stretches downstream. Some RoR hydropower projects with river diversions may alter flows along the diversion routes. Physical and biological changes are related to such variations in water level, timing and temperature. Major changes in the flow regime may also cause changes in the river's estuary, where the extent of salt water intrusion depends on the freshwater discharge.

The slope, current velocity and water depth are also important factors influencing sediment-carrying capacity and erosion (Section 5.6.1.4). The construction of a major dam decreases in general the sediment loading to river deltas.

The change in the annual flow pattern may affect significantly natural aquatic and terrestrial habitats in the river and along the shore. The disappearance of heavy natural floods as the result of regulating watercourses alters the natural lifecycle of the floodplains located downstream from the structure. This may affect vegetation species and community structure, which in turn affect the mammalian and avian fauna. On the other hand, frequent (daily or weekly) fluctuations in the water level downstream from a hydropower reservoir and a tailrace area might create problems for both mammals and birds. Sudden water releases could not only drown animals and wash away waterfowl nests, but also represent a public security issue for other water users. The magnitude of these changes can be mitigated by proper power plant operation and discharge management, regulating ponds, information and warning systems as well as access limitations. A thorough flow-management program can prevent loss of habitats and resources. Further possible mitigation measures might be the release of controlled floods in critical periods and building of weirs in order to maintain water levels in rivers with reduced flow or to prevent salt intrusion from the estuary.

5.6.1.2 Reservoir creation

Creating a reservoir entails not only the transformation of a terrestrial ecosystem into an aquatic one, it also makes important modifications to river flow regimes by transforming a relatively fast-flowing water course into a still-standing water body: an artificial lake. For this reason, the most suitable site for a reservoir needs to be thoroughly studied, as the most effective impact avoidance action is to limit the extent of flooding on the basis of technical, economic, social and environmental considerations.

Fluctuations in water levels often lead to erosion of the reservoir shoreline (draw-down zone) and along the downstream riverbanks. Measures to promote vegetation or erosion control following reservoir impoundment include bank restoration, riparian vegetation enhancement, installation of protective structures (e.g., gravel embankments, riprap, gabions) as well as bioengineering for shore protection and enhancement.

The creation of a reservoir causes profound changes in fish habitats. Generally, the transformation of a river into a lake favours species that are adapted to still-standing waters to the detriment of those species requiring faster flowing water (see Section 5.6.1.5).Due to the high phytoplankton productivity of reservoirs, the fish biomass tends to increase overall. However, the impacts of reservoirs on fish species may only be perceived as positive if species are of commercial value or appreciated for sport and subsistence fishing. If water quality proves to be inadequate, measures to enhance the quality of other water bodies for valued species should be considered in cooperation with affected communities. Other options to foster the development of fish communities and fisheries in and beyond the reservoir zone are, for example, to create spawning and rearing habitat; to install fish incubators; to introduce fish farming technologies; to stock fish species of commercial interest that are well adapted to reservoirs as long as this is compatible with the conservation of biodiversity within the reservoir and does not conflict with native species; to develop facilities for fish harvesting, processing and marketing; to build access roads, ramps and landing areas or to cut trees prior to impoundment along navigation corridors and fishing sites; to provide navigation maps and charts; and to recover floating debris.

As reservoirs replace terrestrial habitats, it is also important to protect and/or recreate the types of habitats lost through inundation (WCD, 2000). In general, long-term compensation and enhancement measures have turned out to be beneficial. Further possible mitigation measures might be to protect areas and wetlands that have an equivalent or better ecological value than the land lost; to preserve valuable land bordering the reservoir for ecological purposes and erosion prevention; to conserve flooded emerging forest in some areas for brood-rearing waterfowl; to enhance the habitat of reservoir islands for conservation purposes; to develop or enhance nesting areas for birds and nesting platforms for raptors; to practice selective wood cutting for herbivorous mammals; and to implement wildlife rescue and management plans. Good-practice examples show that some hydropower reservoirs have even been recognized as new, high-value ecosystems by being registered as 'Ramsar' reservoirs in the Ramsar List of Wetlands of International Importance.[20]

5.6.1.3 Water quality

In some densely populated areas with rather poor water quality, RoR hydropower plants are regularly used to improve oxygen levels and filter tonnes of floating waste out of the river, or to reduce high water temperature levels from thermal power generating outlets. However, maintaining the water quality of reservoirs is often a challenge, as reservoirs constitute a focal point for the river basin catchment. In cases where municipal, industrial and agricultural waste waters entering the reservoir are exacerbating water quality problems, it might be relevant that proponents and stakeholders cooperate in the context of an appropriate land and water use plan encompassing the whole catchment area, preventing, for example, excessive usage of fertilizers and pesticides.

Water quality issues related to reservoirs depend on several factors: climate, reservoir morphology and depth, water retention time in the reservoir, water quality of tributaries, quantity and composition of the inundated soil and vegetation, and rapidity of impounding, which affects the quantity of biomass available over time. Also, the operation of the HPP and thus the reservoir can significantly affect water quality, both negatively and positively.

Water quality issues can often be managed by site selection and appropriate design, taking the future reservoir morphology and hydraulic characteristics into consideration. The primary goals are to reduce the submerged area and to minimize water retention in the reservoir. The release of poor-quality water (due to thermal stratification, turbidity and temperature changes both within and downstream of the reservoir) may be reduced by the use of selective or multi-level water intakes. This may also help to reduce oxygen depletion and the volume of anoxic waters. Since the absence of oxygen may contribute to the formation of methane during the first few years after impoundment, especially in warm climates, measures to prevent the formation of anoxic reservoir zones

---

[20] The Ramsar Convention on Wetlands of International Importance is an intergovernmental treaty that provides the framework for national action and international cooperation on the conservation and wise use of wetlands and their resources. The convention was signed in Ramsar, Iran, in 1971 and entered into force in 1975. The Ramsar List of Wetlands of International Importance (2009) and other information is available at http://www.ramsar.org.

will also help mitigate potential methane emissions (see Section 5.6.3 for more details).

Spillways, stilling basins or structures that promote degassing, such as aeration weirs, may help to avoid downstream gas super-saturation. While some specialists recommend pre-impoundment clearing of the reservoir area, this must be carried out carefully because (i) in some cases, significant re-growth may occur prior to impoundment (and will be rapidly degraded once flooded) and (ii) the massive and sudden release of nutrients (in the case of vegetation clearance through burning) may lead to algal blooms and water quality problems. In some situations, filling up and then flushing out the reservoir prior to commercial operation might contribute to water quality improvement. Planning periodic peak flows can increase aquatic weed drift and decrease suitable substrates for weed growth, reducing problems with undesired invasive species. Increased water turbidity can be mitigated by protecting shorelines that are highly sensitive to erosion, or by managing flow regimes in a manner that reduces downstream erosion.

5.6.1.4 Sedimentation

The sediment-carrying capacity of a river depends on its hydrologic characteristics (slope, current velocity, water depth), the nature of the sediments in the riverbed and the material available in the catchment. In general, a river's sediment load is composed of sediments from the riverbed and sediments generated by erosion in the drainage basin. Dams reduce current velocity and the slope of the water body. The result is a decrease in sediment-carrying capacity. Flow reduction contributes to lower sediment transport capacity and increased sediment deposition, which could lead to the raising of riverbed and an increase in flood risk, as, for example, experienced in the lower reaches of the Yellow River (Xu, 2002). The scope of the impact depends on the natural sediment load of the river basin, which varies according to geomorphologic composition of the riverbed, as well as the soil composition and the vegetation coverage of the drainage basin. In areas dominated by rocky granite, such as in Canada and Norway, sedimentation is generally not an issue. Rivers with large sediment loads are found mainly in arid and semi-arid or mountainous regions with fine soil composition. A World Bank study (Mahmood, 1987) estimated that about 0.5 to 1% of the total freshwater storage capacity of existing reservoirs is lost each year due to sedimentation. Similar conditions were also reported by WCD (2000) and ICOLD (2004). Climate change may affect sediment generation, transport processes, sediment flux in a river and sedimentation in reservoirs, due to changes in hydrological processes and, in particular, floods (Zhu et al., 2007).

In countries with extensive sediment control works such as Japan, the riverbed is often lowered in the middle to downstream reaches of rivers, causing serious scoring of bridge piers and disconnection between water use or intake facilities and the lowered river water table (Takeuchi, 2004). Virtually no sediment has been discharged from the Nile River below Aswan High Dam since its construction (completed in 1970), which has resulted in a significant erosion of the riverbed and banks and retreat of its estuary (Takeuchi et al., 1998). The bed of the Nile, downstream of the High Aswan Dam, was reported to be lowered by some 2 to 3 m in the years following completion of the dam, with irrigation intakes left high and dry and bridges undermined (Helland-Hansen et al., 2005).

Besides exposing the machinery and other technical installations to significant wear and tear (see Section 5.3.3.3), sedimentation also has a major impact on reservoirs by depleting not only their storage capacity over time due to sediment deposition, but also by increasing the risk of upstream flooding due to continuous accumulation of sediments in the backwater region (Goodwin et al., 2001; Wang and Hu, 2004).

In order to gain precise knowledge about long-term sediment inflow characteristics and to support proper site selection, the Revised Universal Soil Loss Equation is a method that is widely utilized to estimate soil erosion from a particular land area (Renard et al., 1997). The Geographic Information System (GIS)-based model includes calibration and the use of satellite images to determine vegetation coverage for the entire basin, which determines the erosion potential of the sub-basins as well as the critical areas. If excessive reservoir sedimentation cannot be avoided by proper site selection, appropriate provision of storage volume that is compatible with the required project life has to be planned. If sediment loading occurs, it can be reduced by opening the spillway gates to allow for sediment flushing during flooding or by adding sluices to the main dam. Different sediment-trapping devices or conveyance systems have also been used with success, along with extraction of coarse material from the riverbed and dredging of sediment deposits However, adequate bank protection in the catchment area and the protection of the natural vegetation in the watershed is one of the best ways to minimize erosion and prevent sediment loading.

5.6.1.5 Biological diversity

Although existing literature related to ecological effects of river regulations on wildlife is extensive (Nilsson and Dynesius, 1993; WCD, 2000), the knowledge is mainly restricted to and based on environmental impact assessments. A restricted number of long-term studies have been carried out that enable predictions of species-specific effects of hydropower development on fish, mammals and birds. In general, four types of environmental disturbances are singled out:

- Habitat changes;
- Geological and climatic changes;
- Direct mortality; and
- Increased human use of the area.

Most predictions are, however, very general and only able to focus on the type of change, without quantifying the short- and long-term effects. Thus, it is generally realized that current knowledge cannot provide a basis for precise predictions. The impacts are, however, highly species-, site-, seasonal- and construction-specific.

The most serious causes of ecological effects from hydropower development on wildlife are, in general:

- Permanent loss of habitat and special biotopes through inundation;
- Loss of flooding;
- Fluctuating water levels (and habitat change);
- Introduction and dispersal of exotic species; and
- Obstacles to fish migration.

Fish are among the main organisms of aquatic wildlife to be affected by a HPP. Altered flow regimes, changes in temperature and habitat modifications are known types of negative impacts (Helland-Hansen et al., 2005) impacting fish. Rapidly changing water levels following hydropower peaking operations are another type of impact that may also affect the downstream fish populations. Yet, in some cases, the effects on the river system from various alterations following regulation may also be positive. For instance, L'Abée-Lund et al. (2006) compared 22 Norwegian rivers, both regulated and non-regulated, based on 128 years of catch statistics. For the regulated rivers they observed no significant effect of hydropower development on the annual catch of anadromous salmonids. For two of the regulated rivers the effect was positive. In addition, enhancement measures such as stocking and building fish ladders significantly increased annual catches. A review by Bain (2007) looking at several hydropower peaking cases in North America and Europe indicates clearly that the impacts from HPPs in the operational phase are variable, but may have a positive effect on downstream areas.

On the other hand, peaking may lead to rapid shifts in the water level where the HPP discharges into a river (as opposed to lakes or the ocean). Sudden shutdown of the peaking HPP may lead to a rapid fall in the water table downstream and a possibility for so-called stranding of fish, where especially small species or fry may be locked in pools, between rocks of various sizes, or in the gravel. An example is salmonid fry that may use dewatered areas. Experiments indicate that if the water level, after a shutdown of the HPP, falls at a rate of below 10 to 15 cm/hr, stranding in most cases will not be a problem, depending on local conditions (Saltveit et al., 2001). However, there are individual differences and fish may also be stranded at lower rates (Halleraker et al., 2003), and even survive for several hours in the substrate after dewatering (Saltveit et al., 2001).

A submerged land area loses all terrestrial animals, and many animals will be dispelled or sometimes drown when a new reservoir is filled. This can be partly mitigated through implementation of a wildlife rescue program, although it is generally recognized that these programs may have a limited effect on the wild populations on the long term (WCD, 2000; Ledec and Quintero, 2003). Endangered species attached to specific biotopes require particular attention and dedicated management programs prior to impoundment. Increased aquatic production caused by nutrient leakage from the inundated soil immediately after damming has been observed to affect both invertebrates and vertebrates positively for some time, that is, until the soil nutrients have been washed out. An increase in aquatic birds associated with this damming effect in the reservoir has also been observed.

Whereas many natural habitats are successfully transformed for human purposes, the natural value of certain other areas is such that they must be used with great care or left untouched. The choice can be made to preserve natural environments that are deemed sensitive or exceptional. To maintain biological diversity, the following measures have proven to be effective: establishing protected areas; choosing a reservoir site that minimizes loss of ecosystems; managing invasive species through proper identification, education and eradication; and conducting specific inventories to learn more about the fauna, flora and specific habitats within the studied area.

5.6.1.6   Barriers for fish migration and navigation

Dams may create obstacles for the movement of migratory fish species and for river navigation. They may reduce access to spawning grounds and rearing zones, leading to a decrease in migratory fish populations and fragmentation of non-migratory fish populations. However, natural waterfalls also constitute obstacles to upstream fish migration and river navigation. Dams that are built on such waterfalls therefore do not constitute an additional barrier to passage. Solutions for upstream fish migrations are now widely available: a variety of solutions have been tested for the last 30 years and have shown acceptable to high efficiency. Fish ladders can partly restore the upstream migration, but they must be carefully designed, and well suited to the site and species considered (Larinier and Marmulla, 2004). High-head schemes are usually off limits for fish ladders. Conversely, downstream fish migration remains more difficult to address. Most fish injuries or mortalities during downstream movement are due to their passage through turbines and spillways. In low-head HPPs, improvement in turbine design (for instance 'fish-friendly turbines'), spillway design or overflow design has proven to successfully reduce fish injury or mortality rates, especially for eels, and to a lesser extent salmonids (Amaral et al., 2009). More improvements may be obtained by adequate management of the power plant flow regime or through spillway openings during downstream movement of migratory species. Once the design of the main components (plant, spillway, overflow) has been optimized for fish passage, some avoidance systems may be installed (screens, strobe and laser lights, acoustic cannons, bubbles, electric fields etc.). However, their efficiency is highly site- and species-dependent, especially in large rivers. In some cases, it may be more useful to capture fish in the headrace or upstream and release the individuals downstream. Other common devices include bypass channels, fish elevators with attraction flow or leaders to guide fish to fish ladders and the installation of avoidance systems upstream of the power plant.

To ensure navigation at a dam site, ship locks are the most effective technique available. For small craft, lifts and elevators can be used with success. Navigation locks can also be used as fish ways with some adjustments to the equipment. Sometimes, it is necessary to increase the upstream attraction flow. In some projects, bypass or diversion channels have been dug around the dam.

5.6.1.7 Involuntary population displacement

Although not all hydropower projects require resettlement, involuntary displacement is one of the most sensitive socioeconomic issues surrounding hydropower development (WCD, 2000; Scudder, 2005). It consists of two closely related, yet distinct processes: displacing and resettling people as well as restoring their livelihoods through the rebuilding or 'rehabilitation' of their communities.

When involuntary displacement cannot be avoided, the following measures might contribute to optimize resettlement outcomes:

- Involving affected people in defining resettlement objectives, in identifying reestablishment solutions and in implementing them; rebuilding communities and moving people in groups, while taking special care of indigenous peoples and other vulnerable social groups;

- Publicizing and disseminating project objectives and related information through community outreach programs, to ensure widespread acceptance and success of the resettlement process;

- Improving livelihoods by fostering the adoption of appropriate regulatory frameworks, by building required institutional capacities, by providing necessary income restoration and compensation programs and by ensuring the development and implementation of long-term integrated community development programs;

- Allocating resources and sharing benefits, based upon accurate cost assessments and commensurate financing, with resettlement timetables tied to civil works construction and effective executing organizations that respond to local development needs, opportunities and constraints.

5.6.1.8 Affected people and vulnerable groups

Like in all other large-scale interventions, it is important during the planning of hydropower projects to identify through a proper social impact study who will benefit from the project and especially who will be exposed to negative impacts. Project-affected people are individuals living in the region that is impacted by a hydropower project's preparation, implementation and/or operation. These may be within the catchment, reservoir area, downstream, or in the periphery where project-associated activities occur, and also can include those living outside of the project-affected area who are economically affected by the project.

A massive influx of workers and creation of transportation corridors also have a potential impact on the environment and surrounding communities if not properly controlled and managed. In addition, workers should be in a position once demobilized at least to return to their previous activities, or to have access to other construction sites due to their increased capacities and experience.

Particular attention needs to be paid to groups that might be considered vulnerable with respect to the degree to which they are marginalized or impoverished and their capacity and means to cope with change. Although it is very difficult to mitigate or fully compensate the social impacts of reservoir hydropower projects on indigenous or other culturally vulnerable communities for whom major transformations to their physical environment run contrary to their fundamental beliefs, special attention has to be paid to those groups in order to ensure that their needs are integrated into project design and adequate measures are taken.

Negative impacts can be minimized for such communities if they are willing partners in the development of a hydropower project, rather than perceiving it as a development imposed on them by an outside agency with conflicting values. Such communities require sufficient lead time, appropriate resources and communication tools to assimilate or think through the project's consequences and to define on a consensual basis the conditions in which they would be prepared to proceed with the proposed development. Granting long-term financial support for activities that define local cultural specificities may also be a way to minimize impacts as well as ensure early involvement of concerned communities in project planning in order to reach agreements on proposed developments and economic spin-offs between concerned communities and proponents. Furthermore, granting legal protections so that affected communities retain exclusive rights to the remainder of their traditional lands and to new lands obtained as compensation might be an appropriate mitigation measure as well as to restrict access of non-residents to the territory during the construction period while securing compensation funds for the development of community infrastructure and services such as access to domestic water supply or to restore river crossings and access roads. Also, it is possible to train community members for project-related job opportunities.

5.6.1.9 Public health

In warmer climate zones, the creation of still-standing water bodies such as reservoirs can lead to increases in waterborne diseases like malaria, river blindness, dengue or yellow fever, which need to be taken into account when designing and constructing reservoirs for supply security, which may be one of the most pressing needs in these regions.

In other zones, a temporary increase in mercury may have to be managed in the reservoir, due to the liberation of mercury from the soil through bacteria, which can then enter the food chain in the form of methyl mercury. In some areas, human activities like coal burning (North America) and mining represent a significant contributor.

Moreover, higher incidences of behavioural diseases linked to increased population densities are frequent consequences of large construction sites. Therefore, public health impacts should be considered and addressed from the outset of the project.

Reservoirs that are likely to become the host of waterborne disease vectors require provisions for covering the cost of health care services to improve health conditions in affected communities. In order to manage health effects related to substantial population growth around hydropower reservoirs, options may include controlling the influx of migrant workers or migrant settlers as well as planning the announcement of the project in order to avoid early population migration to an area not prepared to receive them. Moreover, mechanical and/or chemical treatment of shallow reservoir areas could be considered to reduce the proliferation of insects carrying diseases, while planning and implementing disease prevention programs. Additional options include increasing access to good quality medical services in project-affected communities and in areas where population densities are likely to increase as well as establishing detection and epidemiological monitoring programs, establishing public health education programs directed at the populations affected by the project and implementing a health plan for the work force and along the transportation corridor to reduce the risk of transmittable diseases (e.g., sexually transmitted diseases).

5.6.1.10 Cultural heritage

Cultural heritage is the present manifestation of the human past and refers to sites, structures and remains of archaeological, historical, religious, cultural and aesthetic value (World Bank, 1994). Exceptional natural landscapes or physical features of the environment are also an important part of human heritage as landscapes are endowed with a variety of meanings. The creation of a reservoir might lead to the disappearance of valued exceptional landscapes such as spectacular waterfalls and canyons. Long-term landscape modifications can also occur through soil erosion, sedimentation and low water levels in reservoirs as well as through associated infrastructure impacts (e.g., new roads, transmission lines). It is therefore important that appropriate measures be taken to preserve natural beauty in the project area and to protect cultural properties with high historic value.

Possible measures to minimize negative impacts are, for example: ensuring on-site protection; conserving and restoring, relocating and/or re-creating important physical and cultural resources; creating a museum in partnership with local communities to make archaeological findings, documentation and record keeping accessible; including landscape architecture competences into the project design to optimize harmonious integration of the infrastructure into the landscape; using borrow pits and quarries for construction material that will later disappear through impoundment; re-vegetating dumping sites for soil and excavation material with indigenous species; putting transmission lines and power stations underground in areas of exceptional natural beauty; incorporating residual flows to preserve important waterfalls at least during the tourism high season; keeping as much as possible the natural appearance of river landscapes by constructing weirs to adjust the water level using local rocks instead of concrete; and by constructing small islands in impounded areas, which might be of ecological interest for waterfowl and migrating birds.

5.6.1.11 Sharing development benefits

The economic importance of hydropower and irrigation dams for densely populated countries that are affected by scarce water resources for agriculture and industry, limited access to indigenous sources of oil, gas or coal, and frequent shortages of electricity may be substantial. In many cases, however, hydropower projects have resulted both in winners and losers: affected local communities have often born the brunt of project-related economic and social losses, while people outside the project area have benefited from better access to affordable power and improved flood/drought protection. Although the overall economic gains may be substantial, special attention has to be paid to those local and regional communities that have to cope with the negative impacts of a HPP to ensure that they get a faire share of benefits from the project as compensation. This may take many forms including business partnerships, royalties, development funds, equity sharing, job creation and training, jointly managed environmental mitigation and enhancement funds, improvements of roads and other infrastructure, recreational and commercial facilities (e.g., tourism, fisheries), sharing of revenues, payment of local taxes, or granting preferential electricity rates and fees for other water-related services to local companies and project-affected populations.

5.6.2 Guidelines and regulations

The assessment and management of the above impacts represents a key challenge for hydropower development. The issues at stake are complex and have long been the subject of intense controversy (Goldsmith and Hilyard, 1984). Moreover, unsolved socio-political issues, which are often not project related, tend to come to the forefront of the decision-making process in a large-scale infrastructure development (Beauchamp, 1997).

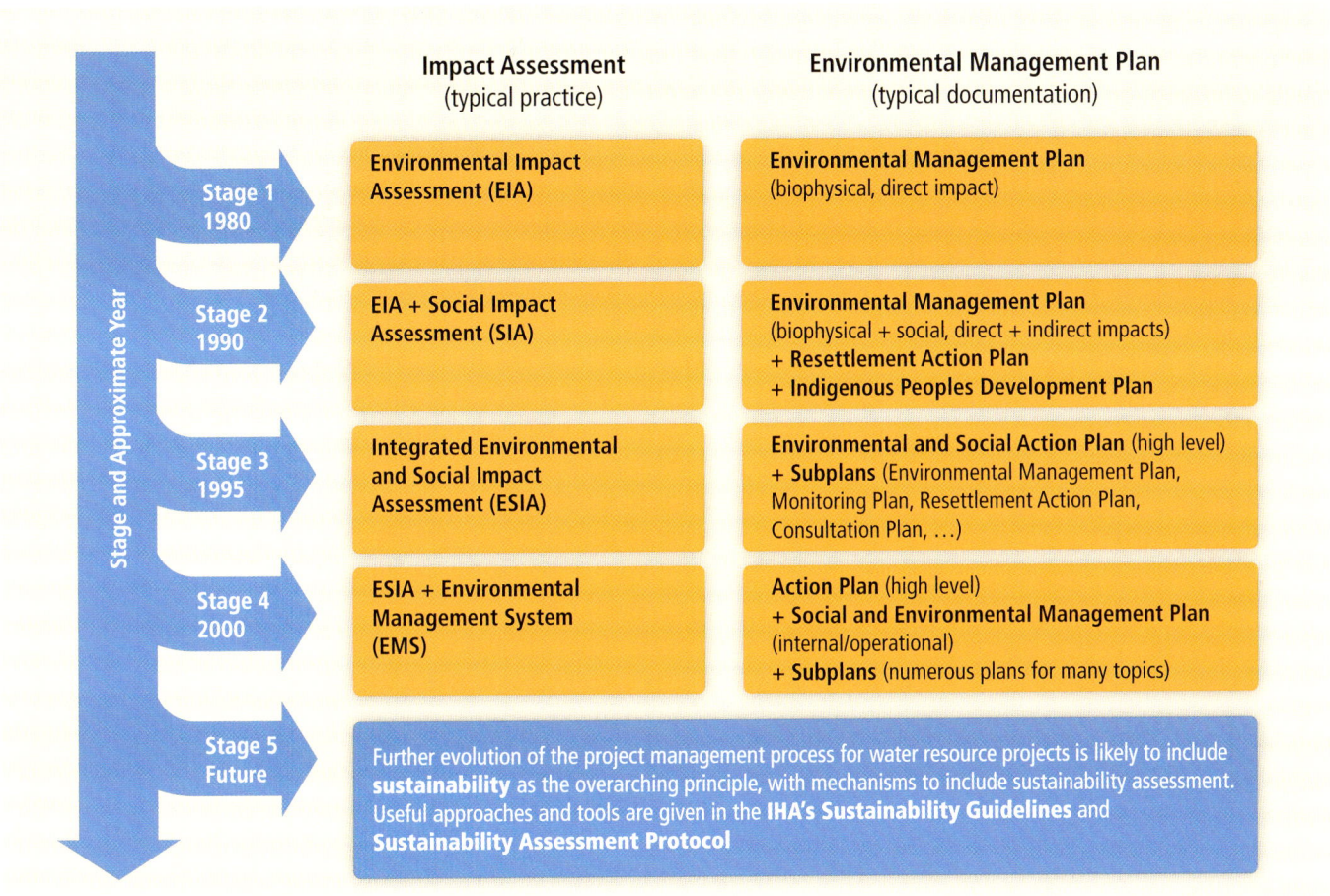

Figure 5.14 | Evolution of environmental and social impact assessment and management (adapted from UNEP, 2007).

Throughout the past decades, project planning has increasingly witnessed a paradigm shift from a technocratic approach to a participative one (Healey, 1992). This shift is also reflected in the evolution of the environmental and social impact assessment and management process that is summarized in Figure 5.14. Today, stakeholder consultation has become an essential tool to improve project outcomes. It is therefore important to identify key stakeholders such as local, national or regional authorities, affected populations, or environmental NGOs, early in the development process in order to ensure positive and constructive consultations, and develop a clear and common understanding of the associated environmental and social impacts, risks and opportunities. Emphasizing transparency and an open, participatory decision-making process, this new approach is driving both present-day and future hydropower projects towards increasingly more environment-friendly and sustainable solutions. At the same time, the concept and scope of environmental and social management associated with hydropower development and operation have changed, moving from a mere impact assessment process to a global management plan encompassing all sustainability aspects.

In particular, the planning of larger hydropower developments mandates guidelines and regulations to ensure that impacts are assessed as objectively as possible and managed in an appropriate manner. In many countries a strong national legal and regulatory framework has been put in place to determine how hydropower projects shall be developed and operated, through a licensing process and follow-up obligations enshrined into the operating permit often also known as concession agreement. Yet, discrepancies between various national regulations as well as controversies have lead to the need to establish international guidelines on how to avoid, minimize or compensate negative impacts while maximizing the positive ones.

Besides the international financing agencies' safeguard policies, one of the first initiatives was launched in 1996 by countries like Canada, Norway, Sweden, Spain and the USA for which hydropower is an important energy resource. Their governments set up, in collaboration with their mainly state-owned hydropower utilities and research institutions, a five-year research program under the auspices of the International Energy Agency (IEA, 2000c) called 'Hydropower and the Environment'. In 1998, the World Commission on Dams (WCD) was established to review the development effectiveness of large dams, to assess alternatives for water and power development, and to develop acceptable criteria, guidelines and standards, where appropriate, for the planning, design, appraisal, construction, operation, monitoring and decommissioning of dams. As a result, 5 core values,[21] 8 strategic priorities[22] and 26 guidelines were suggested (WCD, 2000). While governments, financiers and the industry have widely endorsed the WCD core values and strategic priorities, they consider the guidelines to be only partly applicable to hydropower dams. As a consequence, international financial institutions such as the World Bank, the Asian Development Bank, the African Development Bank and the European Bank for Reconstruction and Development have not endorsed the WCD report as a whole, in particular not its guidelines, but they have kept or developed their own guidelines and criteria (World Bank, 2001). All major export credit agencies have done the same (Knigge et al., 2008). Whereas the WCD's work focused on analyzing the reasons for shortcomings with respect to poorly performing dams, its follow-up initiative, the 'Dams and Development Project' hosted by the UN Environment Programme (UNEP), put an emphasis on gathering good practice into a compendium (UNEP, 2007). With a similar goal, the IEA launched in 2000 a second hydropower-specific five-year research program called 'Hydropower Good Practice' (IEA, 2006) to further document effective management of key environmental and social issues.

Even though each financing agency has developed its own set of quality control criteria to ensure acceptable environmental and social project performance (e.g., World Bank Safeguard, International Finance Corporation's Performance Standards, etc.), there is still no broadly accepted standard to assess the economic, social and environmental performance specifically for hydropower projects. In order to meet this need, the International Hydropower Association (IHA) has produced Sustainability Guidelines (IHA, 2004) and a Hydropower Sustainability Assessment Protocol (IHA, 2006), both of which are based on the broadly shared five core values and seven strategic priorities of the WCD report,

taking the hydropower-specific previous IEA study as starting point. This industry-initiated process may be further improved by a multi-stakeholder review initiative called the Hydropower Sustainability Assessment Forum. This cross-sector working group is comprised of representatives from governments of developed and developing countries, as well as from international financial institutions, NGOs and industry groups.[23] A recommended Final Draft Protocol was published in November 2010 (IHA, 2010) and a continuous improvement process has been put in place for its further application and review.

5.6.3 Lifecycle assessment of environmental impacts

Life cycle assessment (LCA) aims at comparing the full range of environmental impacts assignable to products and services, across their lifecycle, including all processes upstream and downstream of operation or use of the product/service. The following subsection focuses on LCA for GHG emissions, while other metrics are briefly discussed in Box 5.2, and more comprehensively in Section 9.3.4.

The lifecycle of hydropower plants consists of three main stages:

- **Construction:** In this phase, GHGs are emitted from the production and transportation of materials (e.g., concrete, steel etc.) and the use of civil work equipment and materials for construction of the facility (e.g., diesel engines).

- **Operation and maintenance:** GHG emissions can be generated by operation and maintenance activities, for example, building

---

Box 5.2 | Energy payback and lifecycle water use.

The **energy payback** ratio is the ratio of total energy produced during a system's normal lifespan to the energy required to build, maintain and fuel that system. Other metrics that refer to the same basic calculation include the energy returned on energy invested, or the energy ratio (see Annex II). A high energy payback ratio indicates good performance. Lifecycle energy payback ratios for well-performing hydropower plants reach the highest values of all energy technologies, ranging from 170 to 267 for run-of-river, and from 205 to 280 for reservoirs (Gagnon, 2008). However, the range of performances is wider, with literature reporting minimum values of 30 to 50 (Gagnon et al., 2002) or even lower values (Kubiszewski et al., 2010; see also Box 9.2).

Hydropower relies upon water in large quantities, but the majority of this is simply passed through the turbines with negligible losses. As up- and downstream stages require little water, **lifecycle water use** is close to zero for run-of-river hydropower plants (Fthenakis and Kim, 2010). However, consumptive use in the form of evaporation can occur from hydroelectric reservoirs. Global assessments for lifecycle water consumption of reservoirs are not available, and published regional results show high ranges for different climatic and project conditions (Gleick, 1993; LeCornu, 1998; Torcellini et al., 2003; Mielke et al., 2010). Allocation schemes for determining water consumption from various reservoir uses in the case of multipurpose reservoirs can significantly influence reported water consumption values (see also Section 9.3.4.4). Also, research may be needed to determine the net effect of reservoir construction on the evaporation in the specific watershed.

---

21 Equity, efficiency, participatory decision making, sustainability, and accountability.

22 Gaining public acceptance, comprehensive options assessment, addressing existing dams, sustaining rivers and livelihoods, recognizing entitlements and sharing benefits, ensuring compliance, sharing rivers for peace, development and security.

23 For example, the World Bank, the Equator Principles Financial Institutions, the World Wide Fund for Nature, the Nature Conservancy, Transparency International, Oxfam and the IHA.

heating/cooling systems, auxiliary diesel generating units, or onsite staff transportation for maintenance activities. Furthermore, land use change induced by reservoir creation and the associated modification of the terrestrial carbon cycle must be considered, and may lead to net GHG emissions from the reservoir during operation (see Section 5.6.3.1).

- **Dismantling:** Dams can be decommissioned for economic, safety or environmental reasons. Up to now, only a small number of small-size dams have been removed, mainly in the USA. Therefore, emissions related to this stage have rarely been included in LCAs so far.

5.6.3.1 Current lifecycle estimates of greenhouse gas emissions

LCAs carried out on hydropower projects up to now have demonstrated the difficulty of generalizing estimates of lifecycle GHG emissions for hydropower projects across climatic conditions, pre-impoundment land cover types and hydropower technologies. An important issue for hydropower is the multipurpose nature of most reservoir projects, and allocation of total impacts to the several purposes that is then required. Many LCAs to date allocate all impacts to the electricity generation function, which in some cases may overstate the emissions for which they are 'responsible'.

Figure 5.15 displays results of a review of the LCA literature reporting estimates of lifecycle GHG emissions from hydropower technologies published since 1980 (see Annex II for further description of review methods and list of references). The majority of lifecycle GHG emission estimates for hydropower cluster between about 4 and 14 g CO_2eq/kWh, but under certain scenarios there is the potential for much larger quantities of GHG emissions, as shown by the outliers. Note that the distributions shown in Figure 5.15 do not represent an assessment of likelihood; the figure simply reports the distribution of currently published literature estimates passing screens for quality and relevance. As depicted in Figure 5.15, reservoir hydropower has been shown to potentially emit over

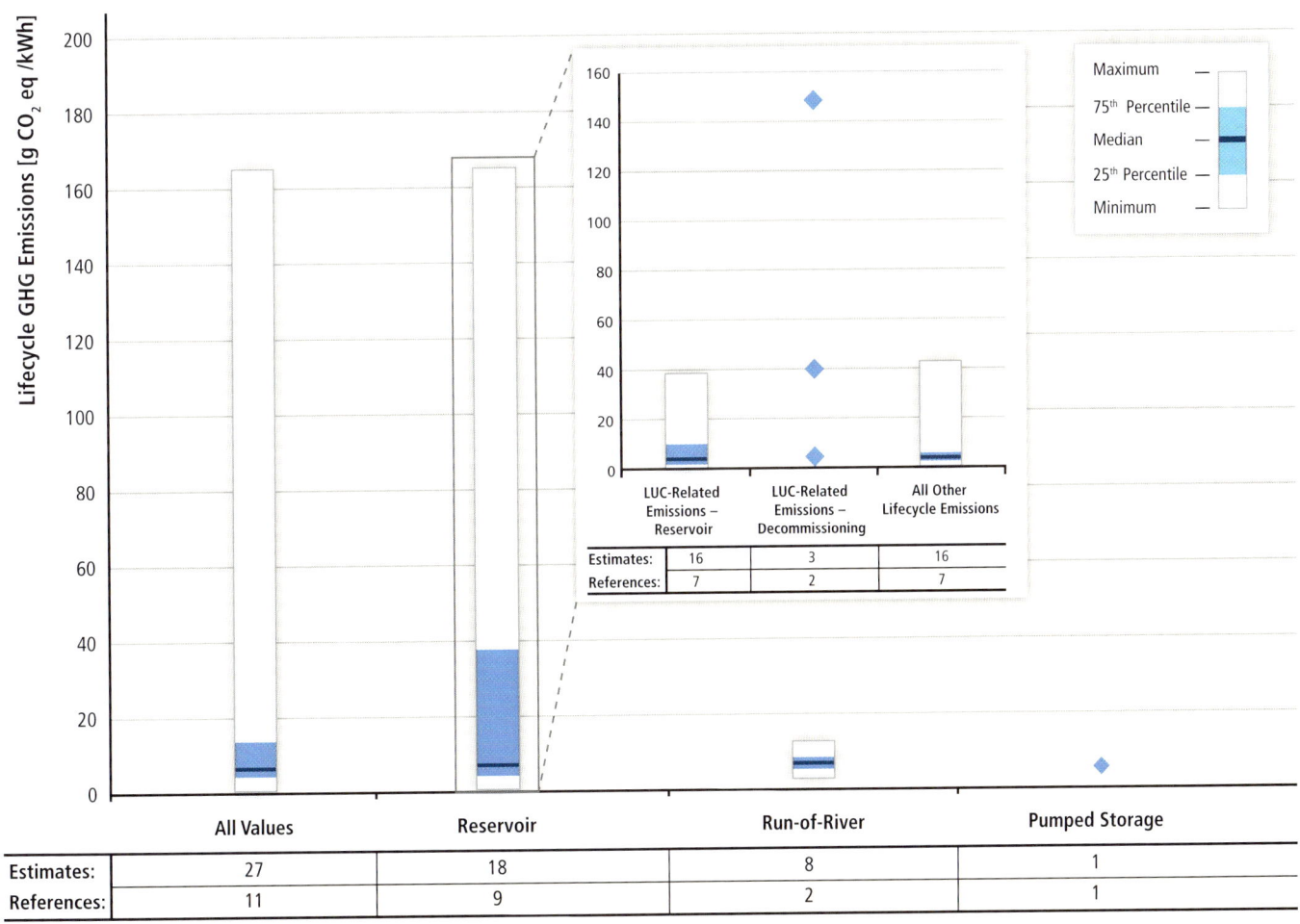

Figure 5.15 | Lifecycle GHG emissions of hydropower technologies (unmodified literature values, after quality screen). See Annex II for details of literature search and citations of literature contributing to the estimates displayed. Emissions from reservoirs are referred to as gross GHG emissions.

150 g CO_2eq/kWh, which is significantly higher than run-of-river or pumped storage, though fewer GHG emission estimates exist for the latter two technologies.

The outliers stem from studies that included assessments of GHG emissions from land use change (LUC) from reservoir hydropower. While the magnitude of potential LUC-related emissions from reservoir hydropower (caused by inundation) is significant, uncertainty in the quantification of these emissions is also high. LUC emissions can be both ongoing, (i.e., methane emitted from the reservoir from soil and vegetation decomposition), and from decommissioning (release of GHGs from large quantities of silt collected over the life of the plant). The LCAs evaluated in this assessment only accounted for gross LUC-related GHG emissions. Characterizing a reservoir as a net emitter of GHGs implies consideration of emissions that would have occurred without the reservoir, which is an area of active research and currently without consensus (see Section 5.6.4.2). LUC-related emissions from decommissioning have only been evaluated in two studies (Horvath, 2005; Pacca, 2007) that provided three estimates (see Figure 5.15). Both reported significantly higher estimates of lifecycle GHG emissions than the other literature owing to this differentiating factor. However, caution should be used in applying these two estimates of the impact of decommissioning broadly to all hydropower systems as they may not be representative of other technologies, sites, or dam sizes.

Variability in estimates stems from differences in study context (e.g., climate, carbon stock of flooded area), technological performance (e.g., turbine efficiency, lifetime, residence time of water) and methods (e.g., LCA system boundaries) (UNESCO/IHA, 2008). For instance, the assumed operating lifetime of a dam can significantly influence the estimate of lifecycle GHG emissions as it amortizes the construction- and dismantling-related emissions over a shorter or longer period. Completion of additional LCA studies is needed to increase the number of estimates and the breadth of their coverage in terms of climatic zones, technology types, dam sizes etc.

5.6.3.2 Quantification of gross and net emissions from reservoirs

With respect to studies that have explored GHG impacts of reservoirs, research and field surveys on GHG balances of freshwater systems involving 14 universities and 24 countries (Tremblay et al., 2005) have led to the following conclusions:

- All freshwater systems, whether they are natural or manmade, emit GHGs due to decomposing organic material. This means that lakes, rivers, estuaries, wetlands, seasonal flooded zones and reservoirs emit GHGs. They also bury some carbon in the sediments (Cole et al., 2007).

- Within a given region that shares similar ecological conditions, reservoirs and natural water systems produce similar levels of CO_2 emissions per unit area. In some cases, natural water bodies and freshwater reservoirs absorb more CO_2 than they emit.

Reservoirs are collection points for material coming from the whole drainage basin area upstream. As part of the natural cycle, organic matter is flushed into these collection points from the surrounding terrestrial ecosystems. In addition, domestic sewage, industrial waste and agricultural pollution may also enter these systems and produce GHG emissions. Therefore, the assessment of man-made net emissions involves a) appropriate estimation of the natural emissions from the terrestrial ecosystem, wetlands, rivers and lakes that were located in the area before impoundment; and b) abstracting the effect of carbon inflow from the terrestrial ecosystem, both natural and related to human activities, on the net GHG emissions before and after impoundment.

The main GHGs produced in freshwater systems are CO_2 and methane (CH_4). Nitrous oxide (N_2O) may be of importance, particularly in reservoirs with large drawdown zones[24] or in tropical areas, but no global estimate of these emissions presently exists. Results from reservoirs in boreal environments indicate a low quantity of N_2O emissions, while a recent study of tropical reservoirs does not give clear evidence of whether tropical reservoirs act as sources of N_2O to the atmosphere (Guerin et al., 2008).

Two pathways of GHG emissions to the atmosphere are usually studied: diffusive fluxes from the surface of the reservoir and bubbling (Figure 5.16). Bubbling refers to the discharge of gaseous substances resulting from carbonation, evaporation or fermentation from a water body (UNESCO/IHA, 2010). In addition, studies at Petit-Saut, Samuel and Balbina have investigated GHG emissions downstream of the dams (degassing just downstream of the dam and diffusive fluxes along the river course downstream of the dam). CH_4 transferred through diffusive fluxes from the bottom to the water surface of the reservoir may undergo oxidation (i.e., be transformed into CO_2) in the water column nearby the oxycline when methanotrophic bacteria are present. Regarding N_2O, Guerin et al. (2008) have identified several possible pathways for N_2O emissions: these could occur via diffusive flux, degassing and possibly through macrophytes, but this last pathway has never been quantified for either boreal or tropical environments.

Still, for the time being, only a limited amount of studies appraising the net emissions from freshwater reservoirs (i.e., excluding unrelated anthropogenic sources and pre-existing natural emissions) is available, whereas gross fluxes have been investigated in boreal (e.g., Rudd et al.,

---

24 The drawdown zone is defined as the area temporarily inundated depending on the reservoir level variation during operation.

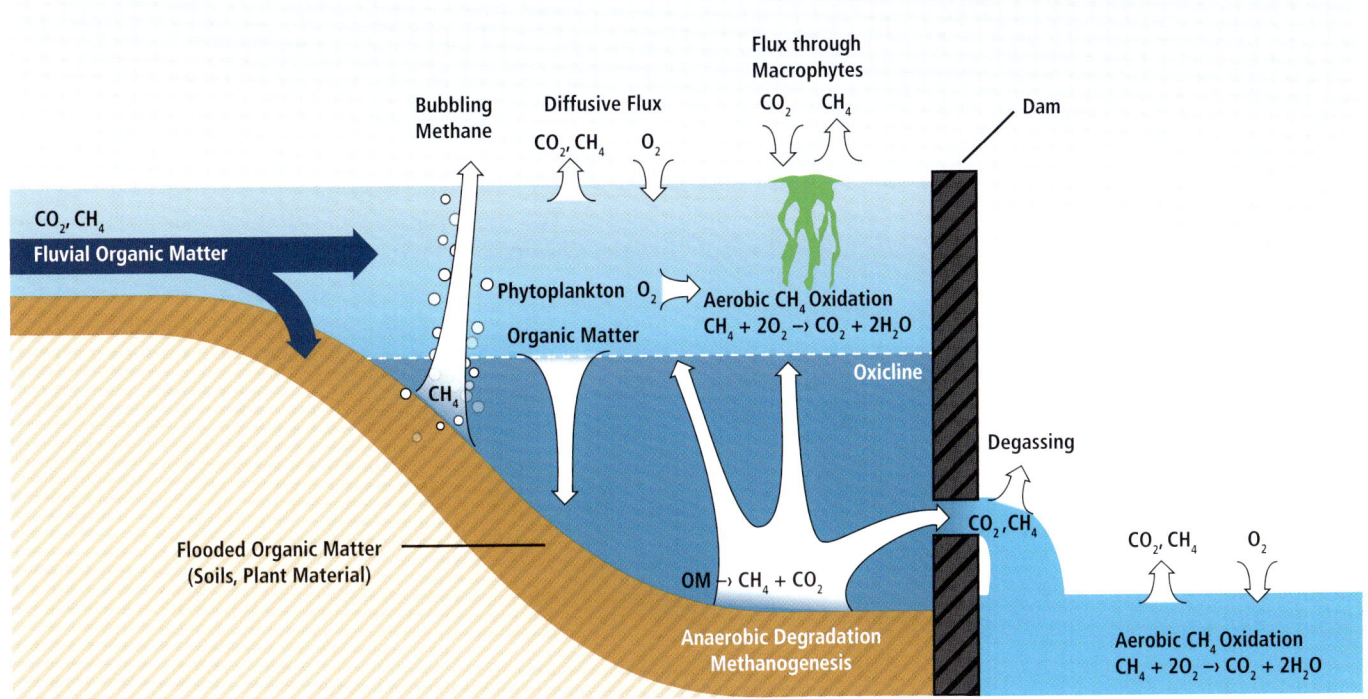

Figure 5.16 | Carbon dioxide and methane pathways in a freshwater reservoir with an anoxic hypolimnion (adapted from Guerin, 2006).

Table 5.6 | Range of gross CO_2 and CH_4 emissions from hydropower freshwater reservoirs; numbers of studied reservoirs are given in parentheses (UNESCO-RED, 2008).

GHG pathway	Boreal and temperate		Tropical	
	CO_2 (mmol/m²/d)	CH_4 (mmol/m²/d)	CO_2 (mmol/m²/d)	CH_4 (mmol/m²/d)
Diffusive fluxes	-23 to 145 (107)	-0.3 to 8 (56)	-19 to 432 (15)	0.3 to 51 (14)
Bubbling	0	0 to 18 (4)	0	0 to 88 (12)
Degassing[1]	~0.2 (2) to 0.1 (2)	n.a.	4 to 23 (1)	4 to 30 (2)
River below the dam	n.a.	n.a.	500 to 2500 (3)	2 to 350 (3)

Note: 1. The degassing (generally in mg/d) is attributed to the surface of the reservoir and is expressed in the same units as the other fluxes (mmol/m²/d).

1993; Tremblay et al., 2005), temperate (Casper et al., 2000; Soumis et al., 2004; Therrien et al., 2005) and tropical/subtropical (e.g., Guerin et al., 2008) regions. Gross emissions measurements are summarized in Table 5.6.

Gross emissions measurements in boreal and temperate regions from Canada, Finland, Iceland, Norway, Sweden and the USA imply that highly variable results can be obtained for CO_2 emissions, so that reservoirs can act as sinks, but also can present significant CO_2 emissions. In some cases, small CH_4 emissions were observed in these studies. Under boreal and temperate conditions, significant CH_4 emissions are expected only for reservoirs with large drawdown zones and high organic and nutrient inflows.

In tropical regions, high temperatures coupled with important demand for oxygen due to the degradation of substantial organic matter (OM)

amounts favour the production of CO_2, the establishment of anoxic conditions, and thus the production of CH_4. In new reservoirs, OM mainly comes from submerged biomass and soil organic carbon with different absolute and relative contents of OM (Galy-Lacaux et al., 1999; Blais et al., 2005; Descloux et al., 2010). Later, OM may also come from primary production or other biological processes within the reservoir.

According to the UN Educational, Scientific and Cultural Organization (UNESCO) and the IHA (UNESCO/IHA, 2008), measurements of gross emissions have been taken in the tropics at four Amazonian locations and 16 additional sites in central and southern Brazil. They have shown, in some cases, significant gross GHG emissions. Measurements are not available from reservoirs in other regions of the tropics or subtropics except for Gatum in Panama, Petit-Saut in French Guyana and Nam Theun 2, Nam Ngum and Nam Leuk in Lao People's Democratic Republic (UNESCO/IHA, 2009). Preliminary studies of Nam Ngum and Nam Leuk indicate that an old reservoir might act as a carbon sink under certain conditions (Harby et al., 2009). This underlines the necessity to also monitor old reservoirs. The age of the reservoir has proven to be an important issue as well as the organic carbon standing stock, water residence time, type of vegetation, season, temperature, oxygen and local primary production, themselves dependent on the geographic area (Fearnside, 2002). According to the IPCC (2006), evidence suggests that CO_2 emissions for approximately the first 10 years after flooding are the results of decay of some of the organic matter on the land prior to flooding, but, beyond this time period, these emissions are sustained by the input of inorganic and organic carbon material transferred into the flooded area from the watershed or by internal processes in the reservoir. In boreal and temperate conditions, GHG emissions have been observed to return to the levels found in neighbouring natural lakes after the two to four years following impoundment (Tremblay et al., 2005). Further measurements could resolve this question for tropical conditions. Comparisons of these results are not easy to achieve, as different methodologies and data (e.g., concerning equipment, procedures, units of measurement) were applied for each study. Few measurements of material transported into or out of the reservoir have been reported, and few studies have measured carbon accumulation in reservoir sediments (UNESCO-RED, 2008).

Since 2008, UNESCO and IHA have been hosting an international research project, with the aim of establishing a robust methodology to accurately estimate the net effect on GHG emissions caused by the creation of a reservoir, and to identify gaps in knowledge. The project published *GHG Measurement Guidelines for Freshwater Reservoirs* in 2010 (UNESCO/IHA, 2010) to enable standardized measurements and calculations worldwide, and aims at delivering a database of results and characteristics of the measurement specification guidance being applied to a representative set of reservoirs worldwide. The final outcome will be building predictive modelling tools to assess the GHG status of unmonitored reservoirs and new reservoir sites, and guidance on mitigation for vulnerable sites. Recently, the IEA has set up a program called IEA Hydropower Agreement Annex XII that will work in parallel with IHA and UNESCO to solve the GHG issue regarding reservoirs.

5.7 Prospects for technology improvement and innovation[25]

Though hydropower is a proven and well-advanced technology, there is still room for further improvement, for example, through optimization of operation, mitigating or reducing environmental impacts, adapting to new social and environmental requirements and more robust and cost-effective technological solutions.

Large hydropower turbines are now close to the theoretical limit for efficiency, with up to 96% efficiency when operated at the best efficiency point, but this is not always possible and continued research is needed to make more efficient operation possible over a broader range of flows. Older turbines can have lower efficiency by design or reduced efficiency due to corrosion and cavitation damage.

Potential therefore exists to increase energy output by retrofitting new equipment with improved efficiency and usually also with increased capacity. Most of the existing hydropower equipment in operation today will need to be modernized during the next three decades, allowing for improved efficiency and higher power and energy output (UNWWAP, 2006) but also for improved environmental solutions by utilizing environmental design principles.

The structural elements of a hydropower project, which tend to take up to 70% of the initial investment cost for large hydropower projects, have a projected life of up to 100 years or more. On the equipment side, some refurbishment can be an attractive option after 30 years. Advances in technology can justify the replacement of key components or even complete generating sets. Typically, generating equipment can be upgraded or replaced with more technologically advanced electro-mechanical equipment two or three times during the life of the project, making more effective use of the same flow of water (UNWWAP, 2006).

The US Department of Energy reported that a 6.3% generation increase could be achieved in the USA from efficiency improvements if plant units fabricated in 1970 or prior years, having a total capacity of 30,965 MW, are replaced. Based on work done for the Tennessee Valley Authority and other hydroelectric plant operators, a generation improvement of 2 to 5.2% has also been estimated for conventional hydropower in the USA (75,000 MW) from installing new equipment and technology, and optimizing water use (Hall et al., 2003). In Norway it has been estimated that an increase in energy output from existing hydropower of 5 to 10% is possible with a combination of improved efficiency in new equipment, increased capacity, reduced head loss and reduced water losses and improved operation.

There is much ongoing research aiming to extend the operational range in terms of head and discharge, and also to improve environmental performance and reliability and reduce costs. Some of the promising technologies under development are described briefly in the following section. Most of the new technologies under development aim at utilizing low (<15 m) or very low (<5 m) head, opening up many sites for hydropower that have not been possible to use with conventional technology. Use of computational fluid dynamics (CFD) is an important tool, making it possible to design turbines with high efficiency over a broad range of discharges. Other techniques like artificial intelligence, neural networks, fuzzy logic and genetic algorithms are increasingly used to improve operation and reduce the cost of maintenance of hydropower equipment.

Most of the data available on hydropower technical potential are based on field work produced several decades ago, when low-head

---
[25] Section 10.5 offers a complementary perspective on drivers and trends of technological progress across RE technologies.

hydropower was not a high priority. Thus, existing data on low-head hydropower technical potential may not be complete. As an example, in Canada, a market potential of 5,000 MW has recently been identified for low-head hydropower (in Canada, low head is defined as below 5 m) alone (Natural Resources Canada, 2009). As another example, in Norway, the environmentally feasible small-scale hydropower (<10 MW) market potential was previously assumed to be 7 TWh (25.2 PJ). A study conducted from 2002 to 2004, however, revealed this market potential to be nearly 25 TWh (90 PJ) at a cost below 6 US cents per kWh, and 32 TWh (115 PJ) at a cost below 9 US cents per kWh (Jensen, 2009).

5.7.1 Variable-speed technology

Usually, hydropower turbines are optimized for an operating point defined by speed, head and discharge. At fixed-speed operation, any head or discharge deviation involves some decrease in efficiency. The application of variable-speed generation in hydroelectric power plants offers a series of advantages, based essentially on the greater flexibility of the turbine operation in situations where the flow or the head deviate substantially from their nominal values. In addition to improved efficiency, the abrasion from silt in the water will also be reduced. Substantial increases in production in comparison to a fixed-speed plant have been found in simulation studies (Terens and Schafer, 1993; Fraile et al., 2006).

5.7.2 Matrix technology

A number of small identical units comprising turbine and generator can be inserted in a frame in the shape of a matrix where the number of (small) units is adapted to the available flow. During operation, it is possible to start and stop any number of units so those in operation can always run under optimal flow conditions. This technology can be installed at existing structures, for example, irrigation dams, low-head weirs, ship locks etc where water is released at low heads (Schneeberger and Schmid, 2004).

5.7.3 Fish-friendly turbines

Fish-friendly turbine technology is an emerging technology that provides a safe approach for fish passing though low-head hydraulic turbines by minimizing the risk of injury or death (Cada, 2001). While conventional hydropower turbine technologies focus solely on electrical power generation, a fish-friendly turbine brings about benefits for both power generation and protection of fish species.[26] Alden Laboratory (USA) predicts that their fish-friendly turbine will have a maximum efficiency of 90.5% with a survival rate for fish of between 94 and 100% (Amaral et al., 2009). One turbine manufacturer predicts approximately 98% fish survival through fish-friendly improvements on their Kaplan turbines.[27]

5.7.4 Hydrokinetic turbines

Generally, projects with a head under 1.5 or 2 m are not viable with traditional technology. New technologies are being developed to take advantage of these small water elevation changes, but they generally rely on the kinetic energy in the stream flow as opposed to the potential energy due to hydraulic head. These technologies are often referred to as kinetic hydropower or hydrokinetic (see Section 6.3 for more details on this technology). Hydrokinetic devices being developed to capture energy from tides and currents may also be deployed inland in both free-flowing rivers and in engineered waterways such as canals, conduits, cooling water discharge pipes or tailraces of existing dams. One type of these systems relies on underwater turbines, either horizontal or vertical. Large turbine blades would be driven by the moving water, just as windmill blades are moved by the wind; these blades would turn the generators and capture the energy of the water flow (Wellinghoff et al., 2008).

'Free flow' or 'hydrokinetic' generation captures energy from moving water without requiring a dam or diversion. While hydrokinetic technology includes generation from ocean tides, currents and waves, it is believed that its most practical application in the near term is likely to be in rivers and streams (see Section 6.3.4). Hydrokinetic turbines have low energy density.

A study from 2007 concluded that the current generating capacity of hydropower of 75,000 MW in the USA (excluding pumped storage) could be nearly doubled, including a contribution from hydrokinetic generation in rivers and constructed waterways of 12,800 MW (EPRI, 2007).

In a 'Policy Statement' issued on 30 November 2007 by the US Federal Energy Regulatory Commission (FERC, 2007) it is stated that:

> *"Estimates suggest that new hydrokinetic technologies, if fully developed, could double the amount of hydropower production in the United States, bringing it from just under 10 percent to close to 20 percent of the national electric energy supply. Given the potential benefits of this new, clean power source, the Commission has taken steps to lower regulatory barriers to its development."*

The potential contributions from very low head projects and hydrokinetic projects are usually not included in existing resource assessments for hydropower (see Section 5.2). The assessments are also usually based on rather old data and lower energy prices than today and future values. It is therefore highly probable that the hydropower resource potential

---

26 See: canmetenergy-canmetenergie.nrcan-rncan.gc.ca/eng/renewables/small_hydropower/fishfriendly_turbine.html.

27 Fish friendliness, Voith Hydro, June 2009, pp 18-21; www.voithhydro.com/media/Hypower_18_18.pdf.

will increase significantly as these new sources are more closely investigated and technology is improved.

5.7.5 New materials

Corrosion, cavitation damages and abrasion are major wearing effects on hydropower equipment. An intensified use of suitable proven materials such as stainless steel and the invention of new materials for coatings limit the wear on equipment and extend lifespan. Improvements in material development have been performed for almost every plant component. Examples include: a) penstocks made of fibreglass; b) better corrosion protection systems for hydro-mechanical equipment; c) better understanding of electrochemical corrosion leading to a suitable material combination; and d) trash rack systems with plastic slide rails.

Water in rivers often contains large amounts of sediments, especially during flood events when soil erosion creates high sediment loads. In reservoirs the sediments may have time to settle, but in run-of-the-river projects most of the sediments may follow the water flow up to the turbines. If the sediments contain hard minerals like quartz, the abrasive erosion of guide vanes, runners and other steel parts may become very high and quickly reduce efficiency or destroy turbines completely within a very short time (Lysne et al., 2003; Gummer, 2009). Erosive wear of hydropower turbine runners is a complex phenomenon, depending on different parameters such as particle size, density and hardness, concentration, velocity of water and base material properties. The efficiency of the turbine decreases with the increase in the erosive wear. The traditional solution to the problem has been to build de-silting chambers to trap the silt and flush it out in bypass outlets, but it is very difficult to trap all particles, especially the fines. New solutions are being developed by coating steel surfaces with a very hard ceramic coating, protecting against erosive wear or delaying the process.

The problem of abrasive particles in hydropower plants is not new, but is becoming more acute with increasing hydropower development in developing countries with sediment-rich rivers. For example, many new projects in India, China and South America are planned in rivers with high sediment concentrations (Gummer, 2009). The problem may also become more important in cases of increased use of hydropower plants in peaking applications.

Modern turbine design using three-dimensional flow simulation provides not only better efficiencies in energy conversion by improved shape of turbine runners and guide/stay vanes, but also leads to a decrease in cavitation damages at high-head power plants and to reduced abrasion effects when dealing with heavy sediment-loaded propulsion water. Other inventions concern, for example, improved self-lubricating bearings with lower damage potential and the use of electrical servo motors instead of hydraulic ones.

5.7.6 Tunnelling technology

Recently, new equipment for very small tunnels (0.7 to 1.3 m diameter) based on oil-drilling technology has been developed and tested in hard rock in Norway, opening up the possibility of directional drilling of 'penstocks' for small hydropower directly from the power station up to intakes, up to 1 km or more from the power station (Jensen, 2009). This could lower cost and reduce the environmental and visual impacts from above-ground penstocks for small hydropower, and open up even more sites for small hydropower.

5.7.7 Dam technology

The International Commission on Large Dams (ICOLD) recently decided to focus on better planning of existing and new (planned) hydropower dams. It is believed that the annual worldwide investment in dams will be about USD 30 billion during the next decade, and the cost can be reduced by 10 to 20% by more cost-effective solutions. ICOLD also wants to promote multipurpose dams and better planning tools for multipurpose water projects (Berga, 2008). Another main issue ICOLD is focusing on is that of small-scale dams between 5 and 15 m high.

The roller-compacted concrete dam is relatively new dam type, originating in Canada in the 1970s. This dam type is built using much drier concrete than in other gravity dams, and it allows a quicker and more economical dam construction (as compared to conventional concrete placing methods). It is assumed that this type of dams will be much more used in the future, lowering the construction cost and thereby also the cost of energy for hydropower projects.

5.7.8 Optimization of operation

Hydropower generation can be increased at a given plant by optimizing a number of different aspects of plant operations, including the settings of individual units, the coordination of multiple unit operations, and release patterns from multiple reservoirs. Based on the experience of federal agencies such as the Tennessee Valley Authority and on strategic planning workshops with the hydropower industry, it is clear that substantial operational improvements can be made in hydropower systems, given new investments in R&D and technology transfer (Sale et al., 2006b). In the future, improved hydrological forecasts combined with optimization models are likely to improve operation and water use, increasing the energy output from existing power plants significantly.

5.8 Cost trends[28]

Hydropower generation is a mature RE technology and can provide electricity as well as a variety of other services at low cost compared to many other power technologies. A variety of prospects for improvement of currently available technology as outlined in the above section exist, but these are unlikely to result in a clear and sustained cost trend due to other counterbalancing factors.

This section describes the fundamental factors affecting the levelized cost of electricity (LCOE) of hydropower plants: a) upfront investment costs; b) operation and maintenance (O&M) costs; c) decommissioning costs; d) the capacity factor; e) the economic lifetime of the investment; and f) the cost of project financing (discount rate).

Discussion of costs in this section is largely limited to the perspective of private investors. Chapters 1, 8, 10 and 11 offer complementary perspectives on cost issues covering, for example, costs of integration, external costs and benefits, economy-wide costs and costs of policies.

Historic and probable future cost trends are presented throughout this section drawing mainly on a number of studies that were published from 2003 up to 2010 by the IEA and other organizations. Box 5.3 contains brief descriptions of each of those studies to provide an overview of the material assessed for this section. The LCOEs provided in the studies themselves are not readily comparable, but have to be considered in conjunction with the underlying cost parameters that affect them. The parameters and resulting study-specific LCOE estimates range are summarized in Table 5.7a for recent conditions and Table 5.7b with a view to future costs.

Later in this section, some of the underlying cost and performance parameters that impact the delivered cost of hydroelectricity are used to estimate recent LCOE figures for hydropower plants across a range of input assumptions. The methodology used in these calculations is described in Annex II, while the input parameters and the resulting range of LCOEs are also listed in Annex III to this report and are reported in Chapters 1 and 10.

It is important to recognize, however, that the LCOE is not the sole determinant of the economic value or profitability of hydropower projects. Hydropower plants designed to meet peak electricity demands, for instance, may have relatively high LCOEs. However, in these instances, not only is the cost per unit of power usually higher, but also average power prices during periods of peak demand and thus revenues per unit of power sold to the market.

Since hydropower projects may provide multiple services in addition to the supply of electric power, the allocation of total cost to individual purposes also matters for the resulting LCOE. Accounting for costs of multipurpose projects is dealt with in Section 5.8.5.

5.8.1 Investment cost of hydropower projects and factors that affect it

Basically, there are two major cost groups for hydropower projects: a) the civil construction costs, which normally are the major costs of the hydropower project, and b) the cost related to electromechanical equipment for energy transformation. Additionally, investment costs include the costs of planning, environmental impact analysis, licensing, fish and wildlife mitigation, recreation mitigation, historical and archaeological mitigation and water quality monitoring and mitigation.

The civil construction costs follow the price trend of the country where the project is going to be developed. In the case of countries with economies in transition, the civil construction costs are usually lower than in developed countries due to the use of local labour and local construction materials.

Civil construction costs are always site specific, mainly due to the inherent characteristics of the topography, geological conditions and the construction design of the project. This could lead to different investment cost and LCOE even for projects of the same capacity.

The costs of electromechanical equipment—in contrast to civil construction cost—follow world market prices for these components. Alvarado-Ancieta (2009) presents the typical cost of electromechanical equipment from various hydropower projects in Figure 5.17.

Figure 5.18 shows the investment cost trend for a large number of investigated projects of different sizes in the USA. The figure is from a study by Hall et al. (2003) that presents typical plant investment costs for new sites.

Figure 5.18 shows that while there is a general tendency of increasing investment cost as the capacity increases, there is also a wide range of cost for projects of the same capacity, given by the spread from the general (blue) trend line. For example, a project of 100 MW in size has an average investment cost of USD_{2002} 200 million (USD_{2002} 2,000/kW) but the range of costs is from less than USD_{2002} 100 million (USD_{2002} 1,000/kW) and up to more than USD_{2002} 400 million (USD_{2002} 4,000/kW). (There could of course also be projects with higher costs, but these have already been excluded from analysis in the selection process).

In hydropower projects where the installed capacity is less than 5 MW, the electromechanical equipment costs tend to dominate. As the capacity increases, the costs are increasingly influenced by the cost of civil structures. The components of the construction project that impact the civil construction costs most are dams, intakes, hydraulic pressure conduits (tunnels and penstocks) and power stations; therefore,

---

28 Chapter 10.5 offers a complementary perspective on drivers and trends of technological progress across RE technologies.

> **Box 5.3 | Brief description of some important hydropower cost studies.**
>
> **Hall et al. (2003)** published a study for the USA where 2,155 sites with a total potential capacity of 43,036 MW were examined and classified according to investment cost. The distribution curve shows investment costs that vary from less than USD 500/kW up to over USD 6000/kW (Figure 5.18). Except for a few projects with very high cost, the distribution curve is nearly linear for up to 95% of the projects. The investment cost of hydropower as defined in the study included the cost of licensing, plant construction, fish and wildlife mitigation, recreation mitigation, historical and archaeological mitigation and water quality monitoring cost.
>
> **VLEEM-2003** (Very Long Term Energy-Environment Model) was an EU-funded project executed by a number of research institutions in France, Germany, Austria and the Netherlands. One of the reports contains detailed information, including cost estimates, for 250 hydropower projects worldwide with a total capacity of 202,000 MW, with the most in-depth focus on Asia and Western Europe (Lako et al., 2003). The projects were planned for commissioning between 2002 and 2020.
>
> **WEA-2004.** The World Energy Assessment (WEA) was first published in 2000 by the United Nations Development Programme (UNDP), the United Nations Department of Economic and Social Affairs (UNDESA) and the World Energy Council (WEC). An update to the original report (UNDP/UNDESA/WEC, 2000) was issued in 2004 (UNDP/UNDESA/WEC, 2004), and data from this version are used here. The report gives cost estimates for both current and future hydropower development. The cost estimates are given both as turnkey investment cost in USD per kW and as energy cost in US cents per kWh. Both cost estimates and capacity factors are given as a range with separate values for small and large hydropower.
>
> **IEA** has published several reports, including *World Energy Outlook 2008* (IEA, 2008a), *Energy Technology Perspectives 2008* (IEA, 2008b) and *Projected Costs of Generating Electricity 2010 Edition* (IEA, 2010b) where cost data can be found both for existing and future hydropower projects.
>
> **EREC/Greenpeace.** The European Renewable Energy Council (EREC) and Greenpeace presented a study in 2008 called *Energy [R] evolution: A Sustainable World Energy Outlook* (Teske et al., 2010). The report presents a global energy scenario with increasing use of renewable energy, in particular wind and solar energy. It contains a detailed analysis up to 2050 and perspectives for beyond, up to 2100. Hydropower is included and future scenarios for cost are given from 2008 up to 2050.
>
> **BMU Lead Study 2008.** *Further development of the strategy to increase the use of renewable energies within the context of the current climate protection goals of Germany and Europe* (BMU, 2008) was commissioned by the German Federal Ministry for the Environment, Nature Conservation and Nuclear Safety (BMU) and published in October 2008. It contains estimated cost for hydropower development up to 2050.
>
> **Krewitt et al. (2009)** reviewed and summarized findings from a number of studies from 2000 through 2008. The main sources of data for future cost estimates were UNDP/UNDESA/WEC (2000), Lako et al. (2003), UNDP/UNDESA/WEC (2004) and IEA (2008).
>
> **REN21.** The global status reports by the Renewable Energy Policy Network for the 21st Century (REN21) are published regularly, with the last update in 2010 (REN21, 2010).
>
> **ECOFYS 2008.** In the background paper *Global Potential of Renewable Energy sources: A Literature Assessment*, provided by Ecofys for REN21, data can be found both for assumed hydropower resource potential and cost of development for undeveloped technical potential (Hoogwijk and Graus, 2008).

these elements have to be optimized carefully during the engineering design stage.

The same overall generating capacity can be achieved with a few large or several smaller generating units. Plants using many small generating units have higher costs per kW than plants using fewer, but larger units. Higher costs per kW installed capacity associated with a higher number of generating units are justified by greater efficiency and flexibility of the hydroelectric plants' integration into the electric grid.

Table 5.7a | Cost ranges for hydropower: Summary of main cost parameters from 10 studies.

Source	Investment cost (IC) (USD$_{2005}$/kW)	O&M cost (% of IC)	Capacity Factor (%)	Lifetime (years)	Discount rate (%)	LCOE (cents/kWh)	Comments
Hall et al. 2003 Ref: Hall et al. (2003)	<500 – 6,200 Median 1,650 90% below 3,250		41 – 61				2,155 Projects in USA 43,000 MW in total Annual Capacity factor (except Rhode Island)
VLEEM-2003 Ref: Lako et al. (2003)	<500 – 4,500 Median 1,000 90% below 1,700		55 – 60				250 Projects for commissioning 2002–2020 Total Capacity 202,000 MW Worldwide but mostly Asia and Europe
WEA 2004 Ref: UNDP/UNDESA/WEC (2004)	1,000 – 3,500 700 – 8,000		35 – 60 20 – 90			2 – 10 2 – 12	Large Hydro Small Hydro (<10 MW) (Not explicitly stated as levelized cost in report)
IEA-WEO 2008 Ref: IEA (2008a)	2,184	2.5	45	40	10	7.1	
IEA-ETP 2008 Ref: IEA (2008b)	1,000 – 5,500 2,500 – 7,000	2.2 – 3			10 10	3 – 12 5.6 – 14	Large Hydro Small Hydro
EREC/Greenpeace Ref: Teske et al. (2010)	2,880 in 2010	4	45	40	10	10.4	
BMU Lead Study 2008 Ref: BMU (2008)	2,440				6	7.3	Study applies to Germany only
Krewitt et al 2009 Ref: Krewitt et al. (2009)	1,000 – 5,500	4	33	30		9,8	Indicative average LCOE year 2000
IEA-2010 Ref: IEA (2010b)	750 – 19,000 in 2010 (1,278 average)		51	80 80		2.3 – 45.9 4.8	Range for 13 projects from 0.3 to 18,000 MW Weighted average for all projects
REN21 Ref: REN21 (2010)						5 – 12 3 – 5 5 – 40	Small Hydro (<10 MW) Large Hydro (>10 MW) Off-Grid (<1 MW)

Table 5.7b | Future cost of hydropower: Summary of main cost parameters from five studies.

Source	Investment cost (IC) (USD$_{2005}$/kW)	O&M cost (% of IC)	Capacity Factor (%)	Lifetime (years)	Discount rate (%)	LCOE (cents/kWh)	Comments
WEA 2004 Ref: UNDP/UNDESA/WEC (2004)						2 – 10	No trend—Future cost same as in 2004 Same for small and large hydro
IEA-WEO 2008 Ref: IEA (2008a)	2,194 in 2030 2,202 in 2050	2.5 2.5	45 45	40 40	10 10	7.1 7.1	
IEA-ETP 2008 Ref: IEA (2008b)	1,000 – 5,400 in 2030 1,000 – 5,100 in 2050 2,500 – 7,000 in 2030 2,000 – 6,000 in 2050	2.2 – 3			10 10 10 10	3 – 11.5 3 – 11 5.2 – 13 4.9 – 12	Large Hydro Large Hydro Small Hydro Small Hydro
EREC/Greenpeace Ref: Teske et al. (2010)	3,200 in 2030 3,420 in 2050	4 4	45 45	40 40	10 10	11.5 12.3	
Krewitt et al 2009 Ref: Krewitt et al. (2009)	1,000 – 5,400 in 2030 1,000 – 5,100 in 2050	4 4	33 33	30 30		10.8 11.9	Indicative average LCOE in 2030 Indicative average LCOE in 2050

Specific investment costs (per installed kW) tend to be reduced for a higher head and higher installed capacity of the project. With higher head, the hydropower project can be set up to use less volume flow, and therefore smaller hydraulic conduits or passages. The size of the equipment is also smaller and related costs are lower.

Results from two of the studies listed in Box 5.3 and Table 5.7a can be used to illustrate the characteristic distribution of investment costs within certain geographic areas. The detailed investment cost surveys provide an assessment of how much of the technical potential can be exploited at or below specific investment costs. Such studies are not readily available in the published literature for many regions. The results of two studies on cumulative investment costs are presented in Figure 5.19. A summary from a study of investment cost typical of the USA by Hall et al. (2003) shows a range of investment costs for 2,155 hydropower projects with a total capacity of 43,000 MW from less than USD$_{2005}$ 500/kW up to more than USD$_{2005}$ 6,000/kW. Twenty-five percent of the assessed technical potential can be developed at an investment cost of up to USD$_{2005}$ 960/kW, an

Hydropower

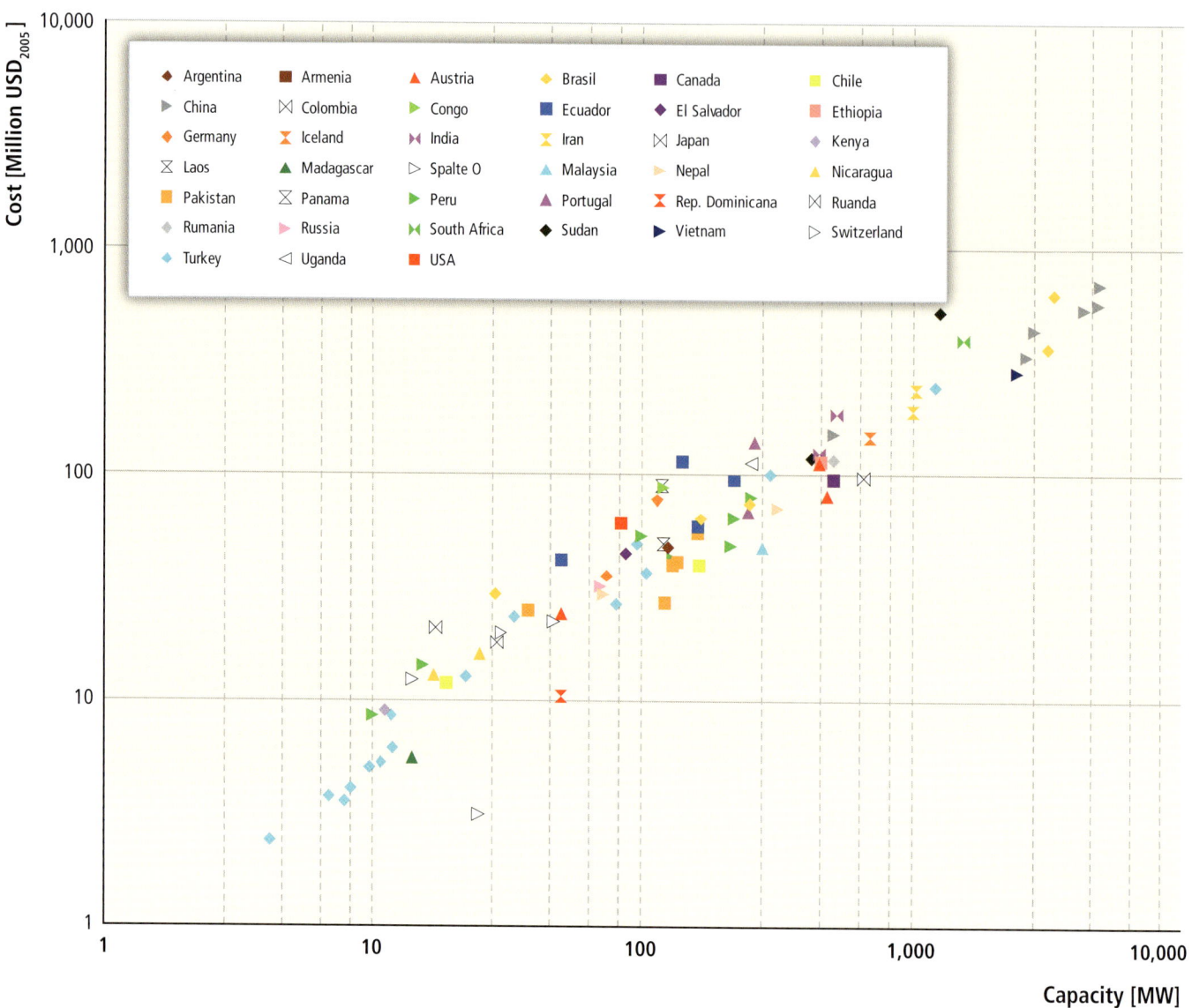

Figure 5.17 | Costs of electrical and mechanical equipment as a function of installed capacity in 81 hydropower plants in America, Asia, Europe and Africa in USD$_{2008}$. Source: Alvarado-Ancieta (2009).

additional 25% at costs between USD$_{2005}$ 960 and 1,650/kW, and another 25% at costs between USD$_{2005}$ 1,650 and 2,700/kW.

A similar summary of cost estimates for 250 projects worldwide with a total capacity of 202,000 MW has been compiled in the VLEEM-2003 study (Lako et al., 2003). Here, the range of investment costs are from USD$_{2005}$ 450/kW up to more than USD$_{2005}$ 4500/kW. Weighted costs (percentiles) are: 25% can be developed at costs up to USD$_{2005}$ 660/kW, 50% (median) at costs up to USD$_{2005}$ 1,090/kW, and 75% at costs up to USD$_{2005}$ 1,260/kW. In general, these and other studies suggest average recent investment cost figures for storage hydropower projects of USD$_{2005}$ 1,000 to 3,000/kW. Small projects in certain areas may sometimes have investment costs that exceed these figures, while lower investment costs are also sometimes feasible. For the purpose of the LCOE calculations that follow, however, a range of USD$_{2005}$ 1,000 to 3,000/kW is considered representative of most hydropower projects.

5.8.2 Other costs occurring during the lifetime of hydropower projects

Operation and maintenance (O&M) costs: Once built and put in operation, hydropower plants usually require very little maintenance and operation costs can be kept low, since hydropower plants do not have recurring fuel costs. O&M costs are usually given as a percentage of investment cost per kW. The EREC/Greenpeace study (Teske et al., 2010) and Krewitt et al. (2009) used 4%, which may be appropriate for small-scale hydropower but is too high for large-scale hydropower plants.

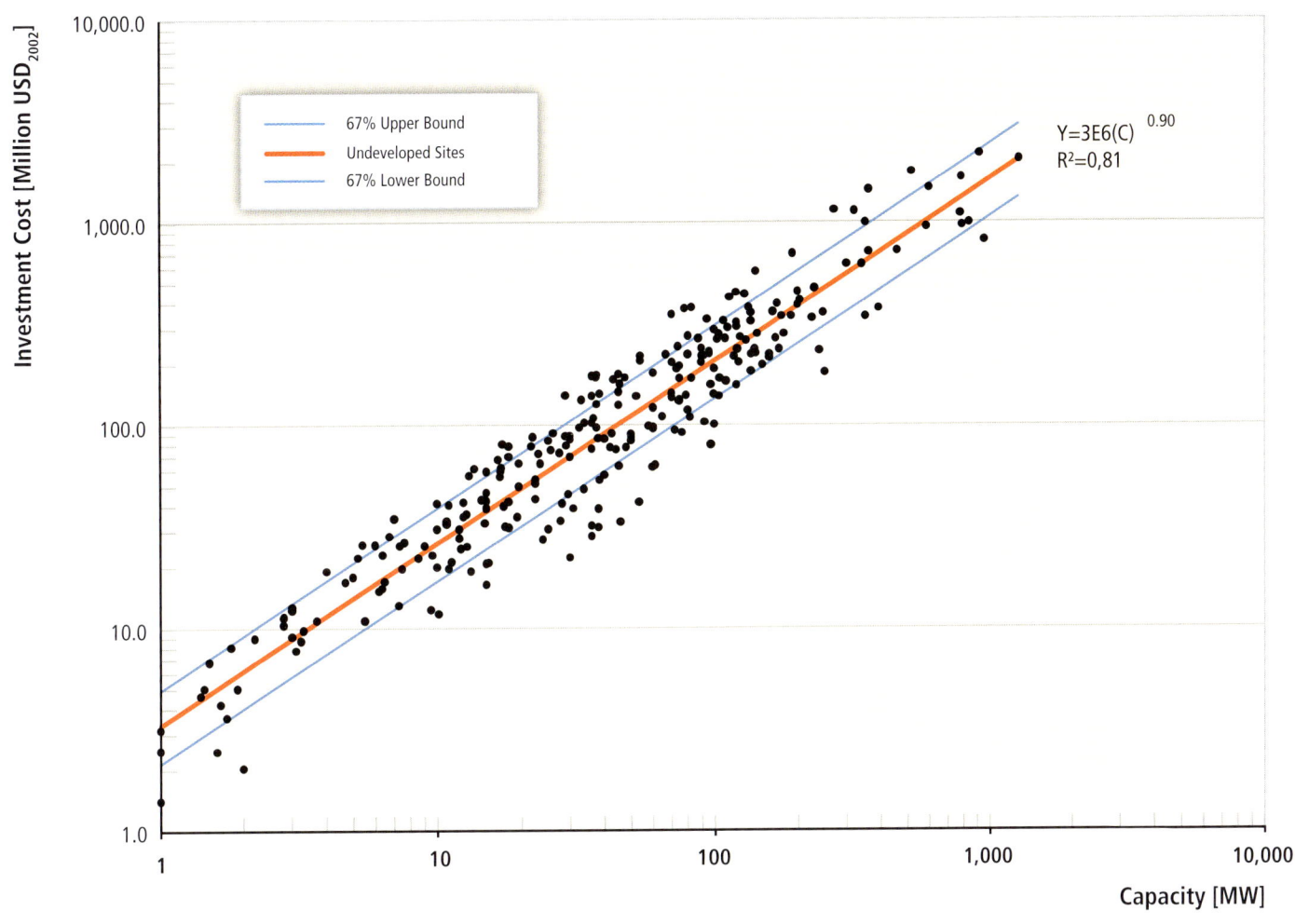

Figure 5.18 | Hydropower plant investment cost as a function of plant capacity for undeveloped sites. Adapted from Hall et al. (2003) (Note: both axes have a logarithmic scale).

The IEA WEO used 2.5% (IEA, 2008a) and 2.2% for large hydropower increasing to 3% for smaller and more expensive projects in IEA-ETP (IEA, 2008b). A typical average O&M cost for hydropower is 2.5%, and this figure is used in the LCOE calculations that follow.

Decommissioning cost: Hydropower plants are rarely decommissioned and it is therefore very difficult to find information about decommissioning costs in the literature. An alternative to decommissioning is project re-licensing and continued operation. A few cases of dam decommissioning are reported in the literature, but these dams are usually not hydropower dams. Due to the long lifetime of hydropower projects (see Section 5.8.3), the decommissioning costs occurring 40 to 80 years into the future are unlikely to contribute significantly to the LCOE. Therefore, decommissioning costs are usually not included in LCOE analyses for hydropower.

5.8.3 Performance parameters affecting the levelized cost of hydropower

Capacity factor: For variable energy sources like solar, wind and waves, the statistical distribution of the energy resource will largely determine the capacity factor. For hydropower, however, the capacity factor is usually designed in the planning and optimization of the project, by considering both the statistical distribution of flow and the market demand characteristics for power. A peaking power plant will be designed to have a low capacity factor, for example 10 to 20%, in order to supply peaking power to the grid only during peak hours. On the other hand, a power plant designed for supplying energy to aluminium plants may be designed to have a capacity factor of 80% or more, in order to supply a nearly constant base load. Reservoirs may be built in order to increase the stability of flow for base-load production, but could also be designed for supplying highly variable (but reliable) flow to a peaking power plant.

A low capacity factor gives low production and higher LCOE. Krewitt et al. (2009) used a low value for hydropower, 2,900 hours or 33%, while, for example, IEA (2010b) used an average of 4,470 hours or 51%. An analysis of energy statistics from the IEA shows that typical capacity factors for existing hydropower systems are in the range from below 40 to nearly 60% (USA 37%, China 42%, India 41%, Russia 43%, Norway 49%, Brazil 56%, Canada 56%). In Figure 5.3, average capacity factors are given for each region, with 32% in Australasia/Oceania, 35% in

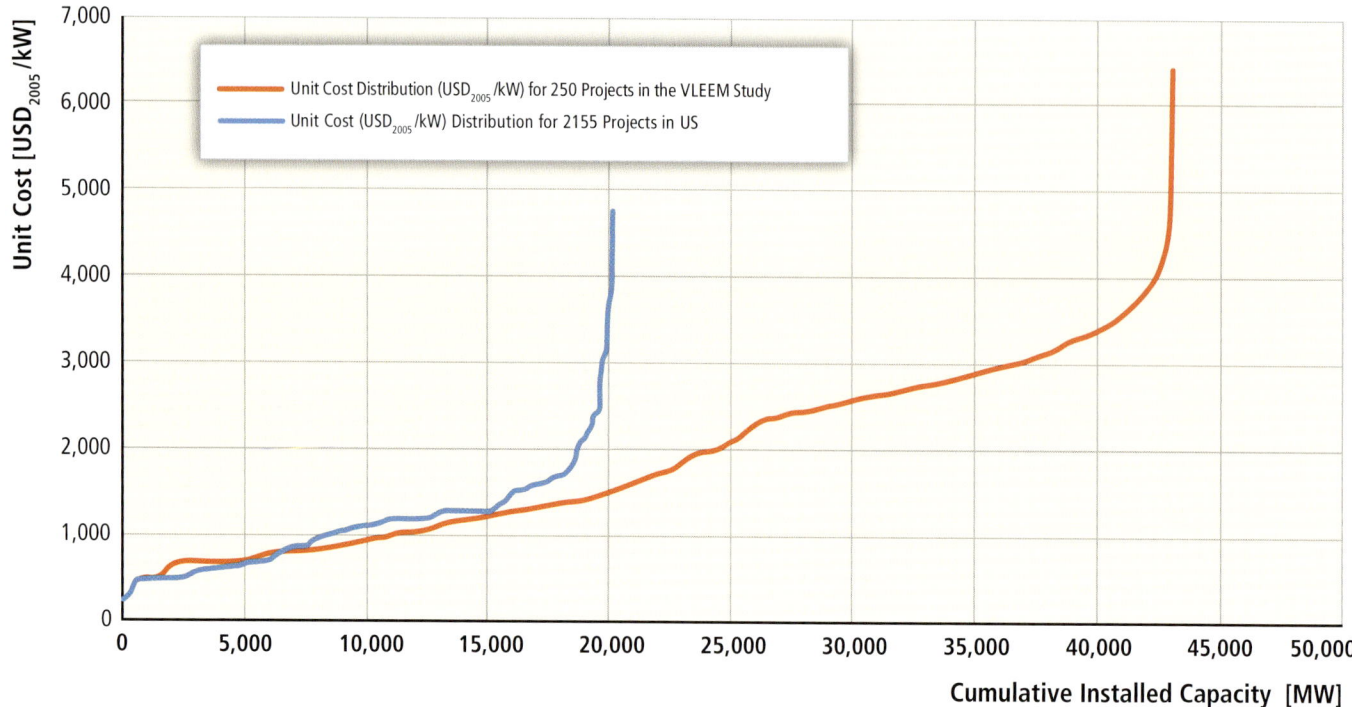

Figure 5.19 | Distribution of investment cost (USD$_{2005}$/kW) for 2,155 hydropower project sites studied in the USA (Hall et al., 2003), and for 250 hydropower project sites worldwide studied in the VLEEM project (Lako et al., 2003). This graph is also called a cumulative capacity curve.

Europe, 43% in Asia, 47% in North America, 47% in Africa and 54% in Latin America. The weighted world average in 2009 was roughly 44%.

Based on the parameters listed in Annex III and methods described in Annex II, Figure 5.20 (upper) illustrates the effect of capacity factors in the range of 30 to 60% on the LCOE of hydropower under three different investment cost scenarios: USD$_{2005}$ 1,000/kW, 2,000/kW and 3,000/kW; other parameter assumptions include a 2.5%/yr O&M cost as a proportion of investment cost, a 60-year economic design lifetime, and a 7% discount rate. Average regional hydropower capacity factors from Figure 5.3 are also shown in the graph.

Lifetime: For hydropower, and in particular large hydropower, the largest cost components are civil structures with very long lifetimes, like dams, tunnels, canals, powerhouses etc. Electrical and mechanical equipment, with much shorter lifetimes, usually contribute less to the cost. It is therefore common to use a longer lifetime for hydropower than for other electricity generation sources. Krewitt et al. (2009) used 30 years, IEA-WEO 2008 (IEA, 2008a) and Teske et al. (2010) used 40 years and the IEA (2010b) used 80 years as the lifetime for hydropower projects. A range of 40 to 80 years is used in the LCOE calculations presented in Annex III as well as in Chapters 1 and 10.

Discount rate:[29] The discount rate is not strictly a performance parameter. Nonetheless, it can have a critical influence on the LCOE depending on the patterns of expenditures and revenues that typically occur over the lifetime of the investment. Private investors usually choose discount rates according to the risk-return characteristics of available investment alternatives. A high discount rate will be beneficial for technologies with low initial investment and high running costs. A low discount rate will generally favour RE sources, as many of these, including hydropower, have relatively high upfront investment cost and low recurring costs. This effect will be even more pronounced for technologies with long lifetimes like hydropower. In some of the studies, it is not stated clearly what discount rate was used to calculate the LCOE. The BMU Lead Study 2008 (BMU, 2008) used 6%. In IEA (2010b) energy costs were computed for both 5 and 10% discount rates. For hydropower, an increase from 5 to 10% gives an increase in the LCOE of nearly 100%. The relationship between the discount rate and resulting LCOE is illustrated in Figure 5.20 (lower) for discount rates of 3, 7 and 10% as used in this report over a range of capacity factors, and using other input assumptions as follows: investment costs of USD$_{2005}$ 2,000/kW, O&M cost of 2.5%/yr of investment cost, and an economic design lifetime of 60 years.

5.8.4 Past and future cost trends for hydropower projects

There is relatively little information on historical trends of hydropower cost in the literature. Such information could be compiled by studying a large number of already-implemented projects, but because hydropower projects are so site-specific it would be difficult to identify trends in project component costs unless a very detailed and time-consuming analysis was completed for a large sample of projects. It is therefore difficult to present historical trends in investment costs and LCOE.

---

[29] For a general discussion of the effect of the choice of the discount rate on LCOE, see Section 10.5.1.

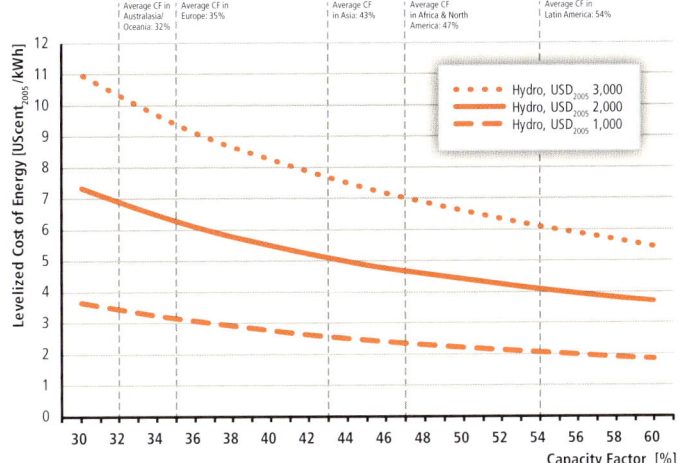

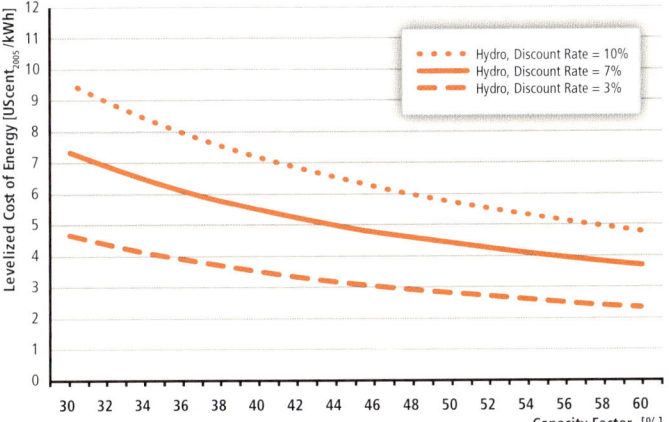

Figure 5.20 | Recent estimated levelized cost of hydropower. Upper panel: Cost of hydropower as a function of capacity factor and investment cost. Lower panel: Cost of hydropower as a function of capacity factor and discount rate. Source: Annex III.

Note: In the upper panel the discount rate is assumed to equal 7%, in the lower panel the investment cost is assumed to be USD 2,000/kW, and in both panels the annual O&M cost is assumed at 2.5%/yr of investment cost and plant lifetime as 60 years.

As a general trend, it can be assumed that projects with low cost will tend to be developed first, and once the best projects have been developed, increasingly costly projects will be developed. (There are, however, many barriers and the selection of the 'cheapest projects first' may not always be possible. Some of these barriers are discussed in Section 5.4.5.) Overall, this general trend could lead to a gradually increasing cost for new projects.

On the other hand, technological innovation and improvements (as discussed in Section 5.7) could lower the cost in the future. Empirical evidence for reductions in the cost of specific components of hydropower systems is provided for tunnelling costs in Figure 5.10. However, evidence for an overall trend with respect to the specific investment cost of hydropower projects or the levelized cost of hydropower cannot be deduced from such information and is very limited. Kahouli-Brahmi (2008) found historical learning rates in the range from 0.5 to 2% for the investment cost of hydropower (for different types of hydropower with varying regional scope and time periods).

In the studies included in Box 5.3 and Table 5.7b, there is no consensus on the future cost trend. Some studies predict a gradually lowering cost (IEA, 2008b; Krewitt et al., 2009), some a gradually increasing cost and one no trend (UNDP/UNDESA/WEC, 2004).

A reason for this may be the complex cost structure of hydropower plants, where some components may have decreasing cost trends (for example tunnelling costs), while other may have increasing cost trends (for example social and environmental mitigation costs). This is discussed, for example, in WEA-2004 (see Box 5.3) where the conclusion is that these factors probably balance each other.

There is significant technical potential for increased hydropower development, as discussed in other sections of this chapter. Since hydropower projects are site-specific, this technical potential necessarily includes projects with widely varying costs, likely ranging from under USD_{2005} 500/kW up to and over USD_{2005} 5,000/kW.

Investment costs based on studies in Table 5.7a (recent) and Table 5.7b (future) are typically in the range from USD_{2005} 1,000 to 3,000/kW, though higher and lower cost possibilities exist, as discussed earlier. Since different studies do not agree on trends in future cost, the present cost range is assumed as typical for the near-term future up to 2020. With investment costs ranging from USD_{2005} 1,000 to 3,000/kW and capacity factor and O&M costs as discussed earlier, typical values for the LCOE of hydropower can be computed for different discount rates (3, 7, 10) and lifetimes (40 and 80 years). The results are shown in Table 5.8, giving an indication of the typical LCOE for hydropower in the near-term future up to 2020. The O&M cost was fixed at 2.5% per year and capacity factor at 45% for the purpose of the results presented in the table.

The LCOE values in Table 5.8 are well within the typical range of cost estimates given in Table 5.7a, (UNDP/UNDESA/WEC, 2004; BMU, 2008; IEA, 2008b; IEA, 2010b; REN21, 2010) but somewhat lower than the values found by Teske et al. (2010) and Krewitt et al. (2009). The results demonstrate that LCOE is very sensitive to investment costs and interest rates, but less sensitive to lifetime, within the lifetime range typical for hydropower (40 to 80 years). Particularly small projects would be expected to have higher investment costs on a dollar per kW basis, and therefore may tend towards the higher end of the range presented in Table 5.8, and may in some instances fall above that range.

5.8.5 Cost allocation for other purposes

Hydropower stations can be installed along with multiple purposes such as irrigation, flood control, navigation, provision of roads,

Table 5.8 | LCOE estimation for parameters typical of current and near-term future hydropower projects in US cents$_{2005}$ (2010 up to 2020).

Investment cost (USD$_{2005}$/kW)	Discount rate (%)	O&M cost (%/yr)	Capacity factor (%)	Lifetime (years)	LCOE (cents/kWh)	Lifetime (years)	LCOE (cents/kWh)
1,000	3	2.5	45	40	1.7	80	1.5
1,000	7	2.5	45	40	2.5	80	2.4
1,000	10	2.5	45	40	3.2	80	3.2
2,000	3	2.5	45	40	3.5	80	2.9
2,000	7	2.5	45	40	5.1	80	4.8
2,000	10	2.5	45	40	6.5	80	6.3
3,000	3	2.5	45	40	5.2	80	4.4
3,000	7	2.5	45	40	7.6	80	7.3
3,000	10	2.5	45	40	9.7	80	9.5

drinking water supply, fish supply and recreation. Many of the purposes cannot be served alone as they have consumptive use of water and may have different priority of use. There are different methods of allocating the cost to individual purposes, each of which has advantages and drawbacks. The basic rules for cost allocation are that the allocated cost to any purpose does not exceed the benefit of that purpose and each purpose will carry its separable cost. Separable cost for any purpose is obtained by subtracting the cost of a multipurpose project without that purpose from the total cost of the project with the purpose included (Dzurik, 2003). Three commonly used cost allocation methods are: the separable cost-remaining benefits method (US Inter-Agency Committee on Water Resources, 1958), the alternative justifiable expenditure method (Petersen, 1984) and the proportionate use-of-facilities method (Hutchens, 1999).

Historically, reservoirs were mostly funded and owned by the public sector, thus project profitability was not the highest consideration or priority in the decision. Today, the liberalization of the electricity market has set new economic standards for the funding and management of dam-based projects. The investment decision is based on an evaluation of viability and profitability over the full lifecycle of the project. The merging of economic elements (energy and water selling prices) with social benefits (flood protection, supplying water to farmers in case of lack of water) and the value of the environment (to preserve a minimum environmental flow) are becoming tools for consideration of cost sharing for multipurpose reservoirs (Skoulikaris, 2008).

Votruba et al. (1988) reported the practice in Czechoslovakia for cost allocation in proportion to benefits and side effects expressed in monetary units. In the case of the Hirakund project in India, the principle of the alternative justifiable expenditure method was followed, with the allocation of the costs of storage capacities between flood control, irrigation and power in the ratio of 38:20:42 (Jain, 2007). The Government of India later adopted the use-of-facilities method for allocation of joint costs of multipurpose river valley projects (Jain, 2007).

5.9 Potential deployment[30]

Hydropower offers significant potential for near- and long-term carbon emissions reductions. The hydropower capacity installed by the end of 2008 delivered roughly 16% of worldwide electricity supply: hydropower is by far the largest current source of RE in the electricity sector (representing 86% of RE electricity in 2008). On a global basis, the hydropower resource is unlikely to constrain further development in the near to medium term (Section 5.2), though environmental and social concerns may limit deployment opportunities if not carefully managed (Section 5.6). Hydropower technology is already being deployed at a rapid pace (see Sections 5.3 and 5.4), therefore offering an immediate option for reducing carbon emissions from the electricity sector. With good conditions, the LCOE can be around 3 to 5 cents/kWh (see Section 5.8). Hydropower is a mature technology and is at the crossroads of two major issues for development: water and energy. This section begins by highlighting near-term forecasts (2015) for hydropower deployment (Section 5.9.1). It then discusses the prospects for and potential barriers to hydropower deployment in the longer term (up to 2050) and the potential role of that deployment in reaching various GHG concentration stabilization levels (Section 5.9.2). Both sections are largely based on energy market forecasts and carbon and energy scenarios literature published in the 2006 to 2010 time period.

5.9.1 Near-term forecasts

The rapid increase in hydropower capacity over the last 10 years is expected by several studies, among them EIA (2010) and IEA (2010c), to continue in the near term (see Table 5.9). Much of the recent global increase in renewable electricity supply has been fuelled by hydropower and wind power. From the 945 GW of hydropower capacity, including pumped storage power plants, installed at the end of 2008, the IEA (2010c) and US Energy Information Administration (EIA, 2010) reference-case forecasts predict growth to 1,119 and 1,047 GW, respectively, by 2015 (e.g., and additional 25 and 30 GW/yr, respectively, by 2015).

---

30 Complementary perspectives on potential deployment based on a comprehensive assessment of numerous model based scenarios of the energy system are presented in Sections 10.2 and 10.3 of this report.

Table 5.9 | Near-term (2015) hydropower energy forecasts.

Study	Hydropower situation				Hydropower forecast for 2015		
	Reference year	Installed capacity (GW)	Electricity generation (TWh/EJ)	Percent of global electricity supply (%)	Installed capacity (GW)	Electricity generation (TWh/EJ)	Percent of global electricity supply (%)
IEA (2010c)	2008	945[1]	3 208/11.6	16	1,119	3,844/13.9	16%
EIA (2010)	2006	776	2 997/10.8	17	1,047	3,887/14	17%

Note: 1. Including pumped storage hydropower plants.

Non-OECD countries, and in particular Asia (China and India) and Latin America, are projected to lead in hydropower additions over this period.

5.9.2 Long-term deployment in the context of carbon mitigation

The IPCC's Fourth Assessment Report (AR4) assumed that hydropower could contribute 17% of global electricity supply by 2030, or 5,382 TWh/yr (~19.4 EJ/yr) (Sims et al., 2007). This figure is not much higher than some commonly cited business-as-usual cases. The IEA's World Energy Outlook 2010 reference scenario, for example, projects 5,232 TWh/yr (18.9 EJ/yr) of hydropower by 2030, or 16% of global electricity supply (IEA, 2010c). The EIA forecasts 4,780 TWh/yr (17.2 EJ/yr) of hydropower in its 2030 reference case projection, or 15% of net electricity production (EIA, 2010).

Beyond the reference scenario, the IEA's World Energy Outlook 2010 presents three additional GHG mitigation scenarios (IEA, 2010c). In the most stringent 450 ppm stabilization scenarios in 2030, installed capacity of new hydropower increases by 689 GW compared to 2008 or 236 GW compared to the Existing Policies scenario in 2030. The report highlights that there is an increase in hydropower supply with increasingly low GHG concentration stabilization levels. Hydropower is estimated to increase annually by roughly 31 GW in the most ambitious mitigation scenario (i.e., 450 ppm) until 2030.

A summary of the literature on the possible future contribution of RE supplies in meeting global energy needs under a range of GHG concentration stabilization scenarios is provided in Chapter 10. Focusing specifically on hydro energy, Figures 5.21 and 5.22 present modelling results on the global supply of hydro energy in EJ/yr and as a percent of global electricity demand, respectively. About 160 different long-term scenarios underlie Figures 5.21 and 5.22. Those scenario results derive from a diversity of modelling teams, and span a wide range of assumptions for—among other variables—energy demand growth, the cost and availability of competing low-carbon technologies and the cost and availability of RE technologies (including hydro energy). Chapter 10 discusses how changes in some of these variables impact RE deployment outcomes, with Section 10.2.2 providing a description of the literature from which the scenarios have been taken. In Figures 5.21 and 5.22, the hydro energy deployment results under these scenarios for 2020, 2030 and 2050 are presented for three GHG concentration stabilization ranges, based on the AR4: Baselines (>600 ppm CO_2), Categories III and IV (440 to 600 ppm CO_2) and Categories I and II (<440 ppm CO_2), all by 2100. Results are presented for the median scenario, the 25th to 75th percentile range among the scenarios, and the minimum and maximum scenario results.[31]

The baseline projections of hydropower's role in global energy supply span a broad range, with medians of roughly 13 EJ in 2020,[32] 15 EJ in 2030 and 18 EJ in 2050 (Figure 5.21). Some growth of hydropower is therefore projected to occur even in the absence of GHG mitigation policies, but with hydropower's median contribution to global electricity supply dropping from about 16% today to less than 10% by 2050. The decreasing share of hydroelectricity despite considerable absolute growth in hydropower supply is a result of expected energy demand growth and continuing electrification. The contribution of hydropower grows to some extent as GHG mitigation policies are assumed to become more stringent: by 2030, hydropower's median contribution equals roughly 16.5 EJ in the 440 to 600 and <440 ppm CO_2 stabilization ranges (compared to the median of 15 EJ in the baseline cases), increasing to about 19 EJ by 2050 (compared to the median of 18 EJ in the baseline cases).

The large diversity of approaches and assumptions used to generate these scenarios results in a wide range of findings. Baseline results for hydropower supply in 2050 range from 14 to 21 EJ at the 25th and 75th percentiles (median 18 EJ), or 7 to 11% (median 9%) of global electricity supply. In the most stringent <440 ppm stabilization scenarios, hydropower supply in 2050 ranges from 16 to 24 EJ at the 25th and 75th percentiles (median 19 EJ), equivalent to 8 to 12% (median 10%) of global electricity supply.

---

31 In scenario ensemble analyses such as the review underlying the figures, there is a constant tension between the fact that the scenarios are not truly a random sample and the sense that the variation in the scenarios does still provide real and often clear insights into collective knowledge or lack of knowledge about the future (see Section 10.2.1.2 for a more detailed discussion).

32 12.78 EJ was reached already in 2009 and thus the average estimates of 13 EJ for 2020 will be exceeded soon, probably already in 2010. Also, some scenario results provide lower values than the current installed capacity for 2020, 2030 and 2050, which is counterintuitive given, for example, hydropower's long lifetimes, its significant market potential and other important services. These results could maybe be explained by model/scenario weaknesses (see discussions in Section 10.2.1.2 of this report).

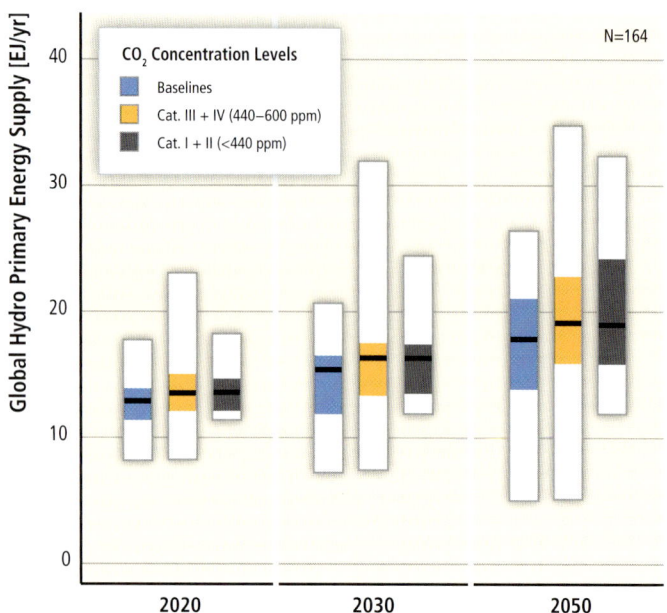

Figure 5.21 | Global primary energy supply from hydro energy in long-term scenarios (median, 25th to 75th percentile range, and full range of scenario results; colour coding is based on categories of atmospheric CO_2 concentration level in 2100; the specific number of scenarios underlying the figure is indicated in the right upper corner) (adapted from Krey and Clarke, 2011; see also Chapter 10).

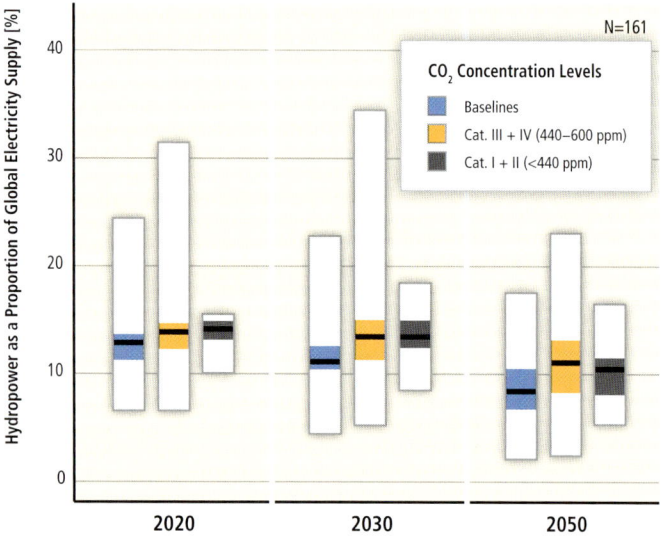

Figure 5.22 | Hydropower electricity share of total global electricity supply in the long-term scenarios (median, 25th to 75th percentile range, and full range of scenario results; colour coding is based on categories of atmospheric CO_2 concentration level in 2100; the specific number of scenarios underlying the figure is indicated in the right upper corner) (adapted from Krey and Clarke, 2011; see also Chapter 10).

Despite this wide range, hydropower has the lowest range compared to other renewable energy sources (see Chapter 10). Moreover, the AR4 estimate for potential hydropower supply of 19.4 EJ by 2030 appears somewhat conservative compared to the more recent scenarios literature presented above, which reaches 24 EJ in 2030 for the IEA's 450 ppm scenario (IEA, 2010c).

Although the literature summarized in Figure 5.21 shows an increase in hydropower supply for scenarios aiming at lower GHG concentration stabilization levels, that impact is smaller than for bioenergy, geothermal, wind and solar energy, where increasingly stringent GHG concentration stabilization ranges lead to more substantial increases in technology deployment (Section 10.2.2.5). One explanation for this result is that hydropower is already mature and economically competitive; as a result, deployment is projected to proceed steadily even in the absence of ambitious efforts to reduce GHG emissions.

The scenarios literature also shows that hydropower could play an important continuing role in reducing global carbon emissions: by 2050, the median contribution of hydropower in the two stabilization categories is around 19 EJ, increasing to 23 EJ at the 75th percentile, and to 35 EJ in the highest scenario. To achieve this contribution requires hydropower to deliver around 11% of global electricity supply in the medium case, or 14% at the 75th percentile. Though this implies a decline in hydropower's contribution to the global electricity supply on a percentage basis, it would still require significant absolute growth in hydropower generation.

Assuming that lower hydropower costs prevail and that growth continues based on the current trend (e.g., the same used in the IEA (2010c) 450 ppm scenario), the hydropower industry forecasts a hydropower market potential of more than 8,700 TWh/yr or 32.2 EJ/yr (IJHD, 2010) to be reached in 2050. The long lifetime of HPPs (in many cases more than 100 years, no/or very few decommissioning cases), along with hydropower's significant market potential, the ability of storage hydropower as a controllable RE source to be used to balance variable RE, and the multipurpose aspects of hydropower, could be taken as support for this view. However, to achieve these levels of deployment, a variety of possible challenges to the growth of hydropower deserve discussion.

Resource Potential: Even the highest estimates for long-term hydropower production are within the global technical potential presented in Section 5.2, suggesting that—on a global basis, at least—technical potential is unlikely to be a limiting factor to hydropower deployment. Moreover, ample market potential exists in most regions of the world to enable significant hydro energy development on an economic basis. In certain countries or regions, however, higher deployment levels will begin to constrain the most economical resource supply, and hydro energy will therefore not contribute equally to meeting the needs of every country (see Section 10.3).

Regional Deployment: Hydropower would need to expand beyond its current status, where most of the resource potential developed so far has been in Europe and North America. The IEA reference case forecast projects the majority (57%) of hydropower deployment by 2035 to come from non-OECD Asia countries (e.g., 33% in China and 13% in India), 16% from non-OECD Latin America (e.g., 7% in Brazil) and only 11% in OECD countries (see Table 5.10). Regional collaboration would be required to combine power systems development with sound

Table 5.10 | Regional distribution of global hydropower generation in 2008 and projection for 2035 in TWh and EJ (percentage of hydropower generation in regional electricity generation, CAAGR: 'compounded average annual growth rate' from 2008 to 2035) for the IEA New Policies Scenario[1] (IEA, 2010c).

Hydropower generation by region		2008			2035			CAAGR 2008–2035 (%)
		TWh/yr	EJ/yr	% of global electricity supply	TWh/yr	EJ/yr	% of global electricity supply	
	World	3,208	11.58	16	5,533	19.97	16	2.0
OECD	OECD total	1,312	4.74	12	1,576	5.69	12	0.7
	North America	678	2.45	13	771	2.78	12	0.5
	USA	257	0.93	6	310	1.12	6	0.7
	OECD Europe	521	1.88	14	653	2.36	15	0.8
	EU	327	1.18	10	402	1.45	10	0.8
	OECD Pacific	114	0.41	6	152	0.55	7	1.1
Non-OECD	Non-OECD Total	1,895	6.84	20	3,958	14.29	18	2.8
	Eastern Europe/Eurasia	284	1.03	17	409	1.48	17	1.4
	Russia	165	0.60	16	251	0.91	18	1.6
	Non-OECD Asia Total	834	3.01	16	2,168	7.83	14	3.6
	China	585	2.11	17	1,348	4.87	14	3.1
	India	114	0.41	14	408	1.47	13	4.8
	Africa	95	0.34	15	274	0.99	23	4.0
	Latin America Total	673	2.43	63	1,054	3.81	59	1.7
	Brazil	370	1.34	80	528	1.91	64	1.3

Note: 1. The 'new policy scenario' reflects conditions set forth by the UNFCCC's Copenhagen accord, and is considered a reference scenario by the IEA.

integrated water resources management, as was observed, for example, in the Nile Basin Initiative and the Greater Mekong Subregion program (see Section 5.10.3).

Supply chain issues: 40 GW of new hydropower capacity was added globally in 2008, which is equivalent to the highest annual long-term IEA forecast scenario in its 450 ppm scenario (IEA, 2010c). As such, though some efforts may be required to ensure an adequate supply of labour and materials in the long term, no fundamental long-term constraints to materials supply, labour availability or manufacturing capacity are envisioned if policy frameworks for hydropower are sufficiently attractive.

Technology and Economics: Hydropower is a mature technology that under many circumstances is already cost-competitive compared to market energy prices. Though additional technical advances are anticipated, they are not central to achieving the lower ranges of GHG concentration stabilization levels described earlier. Hydropower also comes in a broad range of types and size, and can meet both large centralized needs and small decentralized consumption, ensuring that hydropower might be used to meet the electricity needs of many countries and in many different contexts.

Integration and Transmission: Hydropower development occurs in synergy with other RE deployment. Indeed hydropower with reservoirs and/or pumped storage power plants (PSPP) provide a storage capacity that can help transmission system operators to operate their networks in a safe and flexible way by providing balancing generation for variable RE (e.g., wind and solar PV). Hydropower is useful for ancillary services and for balancing unstable transmission networks, as hydropower is the most responsive energy source for meeting peak demand (see Chapter 8). PSPPs and storage hydropower can therefore ensure transmission, and also distribution, security and quality of services.

Social and Environmental Concerns: Social and environmental impacts of hydropower projects vary depending on type, size and local conditions. The most prominent impacts include barriers to fish migration, GHG emissions and water quality degradation in some tropical reservoirs, loss of biological diversity, and population displacement (Section 5.6.1). Impoundments and the existence of reservoirs stand out as the source of the most severe concerns, but can also provides multiple beneficial services beyond energy supply. Efforts to better understand the nature and magnitude of these impacts, together with efforts to mitigate any remaining concerns, will need to be pursued in concert with increasing hydropower deployment. This work has been initiated by the WCD (2000), and has been endorsed and improved by the IHA (2006), providing guidelines and best practice examples.

5.9.3 Conclusions regarding deployment

Overall, evidence suggests that relatively high levels of deployment in the next 20 years are feasible. Even if hydropower's share of the global electricity supply decreases by 2050 (from 16% in 2008 to about 10 to 14% according to different long-term scenarios), hydropower remains an attractive RE source within the context of global carbon mitigation scenarios. Furthermore, increased development of storage hydropower may enable investment into water management infrastructure, which is needed in response to growing problems related to water resources, including climate change adaptation (see Section 5.10).

5.10 Integration into water management systems

Water, energy and climate change are inextricably linked. On the one hand, water availability is crucial for many energy technologies, including hydropower (see Section 9.3.4.4), and on the other hand, energy is needed to secure water supply for agriculture, industries and households, particularly in water-scarce areas in developing countries (Sinha et al., 2006; Mukherji, 2007; Kahrl and Roland-Holst, 2008). This mutual dependence has lead to the understanding that the water-energy nexus must be addressed in a holistic way, especially regarding climate change and sustainable development (Davidson et al., 2003; UNESCO-RED, 2008; WBCSD, 2009). Providing energy and water for sustainable development will require improved regional and global water governance, and since hydroelectric facilities are often associated with the creation of water storage facilities, hydropower is at the crossroads of these issues and can play an important role in enhancing both energy and water security.

Therefore, hydropower development is part of water management systems as much as energy management systems, both of which are increasingly becoming climate driven.

5.10.1 The need for climate-driven water management

As described in Section 5.2.2, climate change will probably lead to changes in the hydrological regime in many countries, including increased variability and more frequent hydrological extremes (floods and droughts). This will introduce additional uncertainty into water resource management. For poor countries that have always faced hydrologic variability and have not yet achieved water security, climate change will make water security even more difficult and costly to achieve. Climate change may also reintroduce water security challenges in countries that for 100 years have enjoyed water security. Today, about 700 million people live in countries experiencing water stress or scarcity. By 2035, it is projected that three billion people will be living in conditions of severe water stress (World Bank, 2011). Many countries with limited water availability depend on shared water resources, increasing the risk of conflict. Therefore, adaptation to climate change impacts on often scarce resources will become very important in water management (World Bank, 2009). Major international financial institutions are aware of the growing need for water storage. For example, the World Bank recognizes the need for better security against climate variability by investing in major hydraulic infrastructure (e.g., dams, canals, dykes and inter-basin transfer schemes). In the Bank's Resource Sector Strategy it is mentioned that developing countries have as little as 1% of the hydraulic infrastructure of developed countries with comparable climatic variability. It was suggested that developing countries construct well-performing hydraulic infrastructures to be used for hydropower generation and water management that also meet environmental and social standards (World Bank, 2004).

Climate change affects the function and operation of existing water infrastructure as well as water management practices. Adverse climate effects on freshwater systems aggravate the impacts of other stresses, such as population growth, changing economic activity, land use change and urbanization. Globally, water demand will grow in the coming decades, primarily due to population growth and increased affluence; regionally, climate change may lead to large changes in irrigation water demand. Current water management practices may be inadequate to reduce the negative impacts of climate change on water supply reliability, flood risk, health, energy and aquatic ecosystems. Improved incorporation of current climate variability into water-related management would make adaptation to future climate change easier.

The need for climate-driven water management positions hydropower systems as key components of future multipurpose water infrastructure projects.

5.10.2 Multipurpose use of reservoirs and regulated rivers

Creating reservoirs is often the only way to adjust the uneven distribution of water in space and time that occurs in the unmanaged environment. Reservoirs add great benefit to hydropower projects, because of the possibility to store water (and energy) during periods of water surplus, and release the water during periods of deficit, making it possible to produce energy according to the demand profile. This is necessary because of large seasonal and year-to-year variability in the inflow. Such hydrological variability is found in most regions in the world, caused by climatic variability in rainfall and/or air temperature. Most reservoirs are built for supplying seasonal storage, but some also have capacity for multi-year regulation, where water from two or more wet years can be stored and released during a later sequence of dry years. The need for water storage also exists for many other types of water use, such as irrigation, water supply and navigation and for flood control. In addition to these primary objectives, reservoirs can provide a number of other uses like recreation and aquaculture. Reservoirs that are created to serve more than one purpose are known as multipurpose reservoirs. Harmonious and economically optimal operation of such multipurpose schemes may involve trade-offs between the various uses, including hydropower generation.

According to the WCD, about 75% of the existing 45,000 large dams in the world were built for the purpose of irrigation, flood control, navigation and urban water supply schemes (WCD, 2000). About 25% of large reservoirs are used for hydropower alone or in combination with other uses, as multipurpose reservoirs (WCD, 2000).

For instance, China is constructing more than 90,000 MW of new hydropower capacity and much of this development is designed for multipurpose utilization of water resources. For the Three Gorges Project (22,400 MW of installed capacity) the primary purpose of the project is flood control (Zhu et al., 2007). In Brazil, it has been recommended that hydropower generation be sustained and expanded, given the uncertainties of the current climate models when predicting future rainfall patterns in the Brazilian and its trans-boundary drainage basins (Freitas, 2009; Freitas and Soito, 2009). On the other hand, significant potential exists for increased hydropower deployment by upgrading existing dams, or using low-head waterways at irrigation dams and conveyance systems (see Sections 5.3.5 and 5.7).

In a context where multipurpose hydropower can be a tool to mitigate both climate change and water scarcity, multipurpose hydropower projects may play an enabling role beyond the electricity sector as a financing instrument for reservoirs, thereby helping to secure freshwater availability. However, multiple uses may increase the potential for conflicts and reduce energy production in times of low water levels. As many watersheds are shared by several nations, regional and international cooperation is crucial to reach consensus on dam and river management.

5.10.3 Regional cooperation and sustainable watershed management

The availability and movement of water may cross political or administrative boundaries. There are 263 trans-boundary river basins and 33 nations have over 95% of their territory within international river basins. While most trans-boundary river basins are shared between two countries, this number is much higher in some river basins. Worldwide, 13 river basins are shared between five to eight countries. Five river basins, namely the Congo, Niger, Nile, Rhine and Zambezi, are shared between 9 to 11 countries. The Danube River flows through the territory of 18 countries, which is the highest number of states for any basin (CWC, 2009). Management of trans-boundary waters poses a difficult and delicate problem, but the vital nature of freshwater also provides a powerful natural incentive for cooperation. Fears have been expressed that conflicts over water might be inevitable as water scarcity increases. International cooperation is required to ensure that the mutual benefits of a shared watercourse are maximized and optimal utilization of the water resources is achieved. This cooperation will be key to facilitate economic development and maintain peaceful relations in the face of water scarcity.

Hamner and Wolf (1998) studied the details of 145 water treaties and found that 124 (86%) are bilateral and the remaining multilateral. Twenty-one (14%) are multilateral; two of the multilateral treaties are unsigned agreements or drafts (Hamner and Wolf, 1998). Most treaties focus on hydropower and water supplies: 57 (39%) treaties discuss hydroelectric generation and 53 (37%) water distribution for consumption. Nine (6%) mention industrial uses, six (4%) navigation, and six (4%) primarily discuss pollution. Thirteen of the 145 (9%) focus on flood control (Hamner and Wolf, 1998). Mountainous nations at the headwaters of the world's rivers are signatories to the bulk of the hydropower agreements. Disputes regarding treaties are resolved through technical commissions, basin commissions or via government officials.

International treaties may be a tool for establishing cooperation in trans-boundary water management. The 1997 UN Convention on the Non-Navigational Uses of International Watercourses (UN IWC, 1997) is the only universal treaty dealing with the use of freshwater resources. Of bilateral treaties, Nepal alone has four with India (the Kosi River agreements, 1954, 1966 and 1978 and the Gandak Power Project, 1959) to exploit the huge power potential in the region. Itaipu Hydropower on the river Parana in Brazil and Paraguay and Victoria Lake hydropower in Uganda, Tanzania and Kenya are other instances of regional cooperation for hydropower development.

The inter-governmental agreements signed between Laos and its neighbouring countries (Thailand, Vietnam, Cambodia) create the necessary institutional framework for the development of major trans-boundary projects such as the 1,088 MW Nam Theun 2 project developed under a public-private partnership model (Viravong, 2008). The support of the World Bank and other international financial institutions has greatly helped in mobilizing private loans and equity. The sales of electricity to Thailand started in March 2010. Over the 25-year concession period, the revenues for the Government of Laos will amount to USD 2 billion, which will be used to serve the country's development objectives through a Poverty Reduction Fund and environmental programmes (Fozzard, 2005).

Several initiatives by international institutions, or intergovernmental agreements, focus on the development of hydropower in a broader context of sustainable development, for example:

- The UN 'Beijing Declaration on Hydropower and Sustainable Development' (UN, 2004) underscores the strategic importance of hydropower for sustainable development, calling on governments and the hydropower industry to disseminate good practices, policies, frameworks and guidelines and build on those to mainstream hydropower development in an economically, socially and environmentally sustainable way, and in a river basin context. The Declaration also calls for tangible action to assist developing countries with financing sustainable hydropower.[33]

---
33 See: www.un.org/esa/sustdev/sdissues/energy/hydropower_sd_beijingdeclaration.pdf.

- The Action Plan elaborated during the *African Ministerial Conference on Hydropower* held in Johannesburg in 2006 aimed, inter alia, at strengthening regional collaboration, fostering the preparation of feasibility studies, strengthening legal and regulatory frameworks and human capacity, promoting synergies between hydropower and other renewable technologies, ensuring proper benefit sharing, and expanding the use of the CDM for financing hydropower projects in Africa (ADB, 2006).

- In 2009, the World Bank Group (WBG) released its *Directions in Hydropower* that outlines the rationale for hydropower sector expansion and describes the WBG portfolio and renewed policy framework for tackling the challenges and risks associated with scaling up hydropower development. WBG's lending to hydropower increased from less than USD 250 million per year during the period 2002 to 2004 to over USD 1 billion in 2008 (World Bank, 2009).

- The Nile basin initiative,[34] comprised of nine African countries (Uganda, Sudan, Egypt, Ethiopia, Zaire, Kenya, Tanzanian, Rwanda and Burundi), aims at developing the Nile River in a cooperative manner, sharing substantial socioeconomic benefits, and promoting regional peace and security in a region that is characterized by water scarcity, poverty, a long history of dispute and insecurity, and rapidly growing populations and demand for water.

- The Greater Mekong sub-region (GMS), comprised of Cambodia, the People's Republic of China, Lao People's Democratic Republic, Myanmar, Thailand and Viet Nam, established a program of sub-regional economic cooperation[35] in 1992 to enhance their economic relations, building on their shared histories and cultures. The program covers nine priority sectors: agriculture, energy, environment, human resource development, investment, telecommunications, tourism, transport infrastructure, and transport and trade facilitation.

- In India, following the announcement of a 50,000 MW hydropower initiative by the Prime Minister in 2003, the Federal Government has taken a number of legislative and policy initiatives, including preparation of a shelf of well-investigated projects and streamlining of statutory clearances and approval, establishment of independent regulatory commissions, provision for long-term financing, increased flexibility in sale of power, etc. India is also cooperating with Bhutan and Nepal for the development of their hydropower resource potential (Ramanathan and Abeygunawardena, 2007).

---

34 See: www.nilebasin.org/.

35 See: www.adb.org/gms/.

References

ADB (2006). African Ministerial Conference on Hydropower and Sustainable Development. In: *ADB FINESSE Africa Newsletter*, African Development Bank, Tunisia, April 2006.

Agrawala, S., V. Raksakulthai, M. van Aalst, P. Larsen, J. Smith, and J. Reynolds (2003). *Development and Climate Change in Nepal: Focus on Water Resources and Hydropower.* COM/ENV/EPOC/DCD/DAC(2003)1/FINAL, 2003, Environment Directorate, Development Co-Operation Directorate of Nepal, Organisation for Economic Co-operation and Development, Paris, France, 64 pp. Available at: www.oecd.org/dataoecd/6/51/19742202.pdf.

Altinbilek, D., R. Abdel-Malek, J.-M. Devernay, R. Gill, S. Leney, T. Moss, H.P. Schiffer, and R.M. Taylor (2007). Hydropower's contribution to energy security. In: *20th World Energy Congress*, Rome, Italy, 11-15 November 2007. Available at: www.worldenergy.org/documents/congresspapers/P000960.pdf.

Alvarado-Ancieta, C.A. (2009). Estimating E&M powerhouse costs. *International Water Power and Dam Construction*, 17 February 2009, pp. 21-25.

Amaral, S., G. Allen, G. Hecker, D. Dixon, and Fisher. R. (2009). Development and application of an advanced fish-friendly hydro turbine. In: *Proceedings of the HYDRO 2009 conference*, Session 2, Lyon, France, 26-28 October 2009, 14 pp.

Andreassen, K., U.H. Buenger, N. Henriksen, A. Oyvann, and O. Ullmann (2002). Norwegian hydro energy in Germany. *International Journal of Hydrogen Energy*, **18**(4), pp. 325-336.

Bain, M.B. (2007). *Hydropower Operations and Environmental Conservation: St. Mary's River, Ontario and Michigan.* Project Report for the International Lake Superior Board of Control, Washington, DC, USA and Ottawa, Canada, 56 pp.

Baird, S. (2006). The future of electricity. *Electricity Today*, **18**(5), pp. 9-11.

Bates, B.C., Z.W. Kundzewicz, S. Wu, and J.P. Palutikof (2008). *Climate Change and Water.* Technical Paper VI, Intergovernmental Panel on Climate Change Secretariat, Geneva, Switzerland, 210 pp.

Beauchamp, A. (1997). *Environnement et consensus social.* Editions l'Essentiel, Montréal, Canada, 141 pp. (ISBN 9782921970051).

Berga, L. (2008). Solving dam engineering challenges. *Hydro Review*, **16**(1), pp. 14-15.

Bhatia, R., M. Scatasta, and R. Cestti (2003). Study on the multiplier effects of dams: Methodology issues and preliminary results. In: *Third World Water Forum*, Kyoto, Japan, 16-23 March 2003.

Blais, A.-M., S. Lorrain, Y. Plourde, and L. Varfalvy (2005). Organic carbon densities of soils and vegetation of tropical, temperate and boreal forests. In: *Greenhouse Gas Emissions - Fluxes and Processes.* A. Tremblay, L. Varfalvy, C. Roehm, and M. Garneau (eds.), Springer-Verlag, Berlin, Germany, pp. 155-185.

BMU (2008). *Lead Study 2008 - Further development of the Strategy to increase the use of renewable energies within the context of the current climate protection goals of Germany and Europe.* German Federal Ministry for the Environment, Nature Conservation and Nuclear Safety (BMU), Berlin, Germany, 118 pp.

BP (2010). *British Petroleum Statistical Review of World Energy 2010.* British Petroleum, London, UK, 50 pp.

Cada, G.F. (2001). The development of advanced hydroelectric turbines to improve fish passage survival. *Fisheries*, **26**(9), pp. 14-23.

Casper, P., S.C.H. Maberly, G. H. Hall, and B.J. Finlay (2000). Fluxes of methane and carbon dioxide from a small productive lake to the atmosphere. *Biogeochemistry*, **49**, pp. 1-19.

Cole, J.J., Y.T. Prairie, N.F. Caraco, W.H. McDowell, L.J. Tranvik, R.R. Striegl, C.M. Duarte, P. Kortelainen, J.A. Downing, J. Middleburg, and J.M. Melack (2007). Plumbing the global carbon cycle: Integrating inland waters into the terrestrial carbon budget. *Ecosystems*, **10**, pp. 171-184

CWC (2009). *World Water Day 2009 - Theme paper on Transboundary waters.* Government of India, Ministry of Water Resources, Central Water Commission, New Delhi, India. Available at: www.cwc.nic.in/main/downloads/Theme%20Paper%20WWD-2009.pdf.

Davidson, O., K. Halsnæs, S. Huq, M. Kok, B. Metz, Y. Sokona, and J. Verhagen (2003). The development and climate nexus: the case of sub-Saharan Africa. *Climate Policy*, **3**(Supplement 1), pp. S97-S113.

Deane, J.P., B.P. Ó Gallachóir, and E.J. McKeogh (2010). Techno-economic review of existing and new pumped hydro energy storage plant. *Renewable and Sustainable Energy Reviews*, **14**(4), pp. 1293-1302.

Descloux, S., V. Chanudet, H. Poilvé, and A. Grégoire (2010). Co-assessment of biomass and soil organic carbon stocks in a future reservoir area located in Southeast Asia. *Environmental Monitoring and Assessment*, **173**(1-4), pp. 723-741.

Dzurik, A.A. (2003). *Water Resources Planning.* 3rd ed. Rowman & Littlefield, Lanham, MD, USA, 393 pp.

Egré, D., and J.C. Milewski (2002). The diversity of hydropower projects. *Energy Policy*, **30**(14), pp. 1225-1230.

EIA (2010). *International Energy Outlook.* U.S. Energy Information Agency, US Department of Energy, Washington, DC, USA, 328 pp.

EPRI (2007). *Assessment of Waterpower Potential and Development Needs.* Electrical Power Research Institute, Palo Alto, CA, USA, 110 pp.

Fearnside, P.M. (2002). Greenhouse gas emissions from hydroelectric reservoir (Brazil's Tucurui dam) and the energy policy implications. *Water, Air, and Soil Pollution*, **133**, pp. 69-96.

FERC (2007). *Assessment of Demand Response and Advanced Metering.* Federal Energy Regulatory Commission, Washington, DC, USA, 92 pp.

Fozzard, A. (2005). *Revenue and Expenditure Management, Nam Theun 2 Hydroelectric Project.* World Bank. Washington, DC, USA, 346 pp.

Fraile, A.J., I.S. Moreno, J.I.P. Diaz, J.R.W. Ayza, and J. Fraile-Mora (2006). Speed Optimization Module of a Hydraulic Francis turbine based on Artificial Neural Networks. Application to the dynamic analysis and control of an adjustable speed hydro plant. In: *International Joint Conference on Neural Networks, 2006*, IJCNN, Vancouver, Canada, 16-21 July 2006, pp. 4104-4110.

Freitas, M.A.V. (2009). Vulnerabilidades da Hidroenergia. *Special Edition Scientific American Brasil*, **32**, pp. 50-55.

Freitas, M.A.V., and J.L. Soito (2009). Vulnerability to climate change and water management: hydropower generation in Brazil. *WIT Transactions on Ecology and the Environment*, **124**, pp. 217-228.

Fthenakis, V., and H.C. Kim (2010). Life-cycle uses of water in U.S. electricity generation. *Renewable and Sustainable Energy Reviews*, **14**, pp. 2039-2048.

Gagnon, L. (2008). Energy payback. *Energy Policy*, **36**, pp. 3317-3322.

Gagnon, L., C. Bélanger, and Y. Uchiyama (2002). Life-Cycle assessment of electricity generation options: The status of research in year 2001. *Energy Policy*, **30**(14), pp. 1267-1278.

Galy-Lacaux, C., R. Delmas, G. Kouadio, S. Richard, and P. Gosse (1999). Long term greenhouse gas emission from a hydroelectric reservoir in tropical forest regions. *Global Biogeochemical Cycles*, **13**, pp. 503-517.

Gleick, P.H. (1993). *Water in Crisis: A Guide to the World's Fresh Water Resources.* Oxford University Press, New York, NY, USA, 437 pp.

Goldsmith, E., and N. Hilyard (1984). *The Social and Environmental Effects of Large Dams. Volume 1: Overview.* Wadebridge Ecological Centre, Worthyvale Manor, Camelford, Cornwall, UK (ISBN: 0 9504111 5 9).

Goodwin, P., M. Falte, and A.D.K. Betts (2001). *Managing for Unforeseen Consequences of Large Dam Operations.* World Commission on Dams, Vlaeberg, Cape Town, South Africa.

Guerin, F. (2006). *Emissions de Gaz a Effet de Serre (CO_2 CH_4) par une Retenue de Barrage Hydroelectrique en Zone Tropicale (Petit-Saut, Guyane Francaise): Experimentation et Modelization.* Thèse de doctorat de l'Université Paul Sabatier (Toulouse III).

Guerin, F., A. Gwenaël, A. Tremblay, and R. Delmas (2008). Nitrous oxide emissions from tropical hydroelectric reservoirs. *Geophysical Research Letters*, **35**, L06404.

Gummer, J.H. (2009). Combating silt erosion in hydraulic turbines. *Hydro Review Worldwide*, **17**(1), pp. 28-34.

Haldane, T.G.N., and P.L. Blackstone (1955). Problems of hydro-electric design in mixed thermal hydro-electric systems. *Proceedings of the IEE - Part A: Power Engineering*, **102**(3), pp. 311-322.

Hall, D.G., Hunt, R.T., Reeves, K.S. and Carroll, G.R., (2003). *Estimation of Economic Parameters of U.S. Hydropower Resources.* INEEL/EXT-03-00662, U.S. Department of Energy Idaho Operations Office, Idaho Falls, ID, USA, 25 pp.

Halleraker, J.H., S.J. Saltveit, A. Harby, J.V. Arnekleiv, H.-P. Fjeldstad, and B. Kohler. (2003). Factors influencing stranding of wild juvenile Brown Trout (*Salmo trutta*) during rapid and frequent flow decreases in an artificial stream. *River Research and Applications*, **19**, pp. 589-603.

Hamner, J.H., and A.T. Wolf (1998). Patterns in international water resource treaties: The Transboundary Freshwater Dispute Database. *Colorado Journal of International, Environmental Law and Policy. 1997 Yearbook. Supplement*, pp. 155-177

Hamududu, B., and Å. Killingtveit (2010). Estimating effects of climate change on global hydropower production. In: *Hydropower'10, 6th International Conference on Hydropower, Hydropower supporting other renewables.* Tromsø, Norway, 1-3 February 2010, 13 pp.

Hamududu, B., E. Jjunju and Å. Killingtveit (2010). Existing studies of hydropower and climate change: A review. In: *Hydropower'10, 6th International Conference on Hydropower, Hydropower supporting other renewables*, Tromsø, Norway, 1-3 February 2010, 18 pp.

Harby, A., O.G. Brakstad, B.H. Hansen, H. Sundt, S. Descloux, V. Chanudet, and F. Becerra (2009). Net GHG emissions from Lao Reservoirs. Poster, In: *IHA World Congress*, EDF/SINTEF, Reykjavik, Iceland, June 2009.

Healey, P. (1992). Planning through debate: The communicative turn in planning theory. *Town Planning Review*, **63**(2), pp. 143-163.

Helland-Hansen, E., T. Holtedahl, and O.A. Lye (2005). *Environmental Effects Update*. NTNU, Department of Hydraulic and Environmental Engineering, Trondheim, Norway, 219 pp.

Holttinen, H., P. Meibom, A. Orths, F.V. Hulle, B. Lange, M. O'Malley, J. Pierik, B. Ummels, J.O. Tande, A. Estanqueiro, M. Matos, E. Gomez, L. Söder, G. Strbac, A. Shakoor, J. Ricardo, J.C. Smith, M. Milligan, and E. Ela (2009). *Design and Operation of Power Systems with Large Amounts of Wind Power*. VTT Technical Research Centre of Finland, Helsinki, Finland, 199 pp.

Hoogwijk, M and W. Graus (2008). *Global Potential of Renewable Energy Sources: A Literature Assessment.* Ecofys, Utrecht, The Netherlands, 45 pp.

Horvath, A. (2005). *Decision-making in Electricity Generation Based on Global Warming Potential and Life-cycle Assessment for Climate Change.* University of California Energy Institute, Berkeley, CA, USA, 12 pp.

Hutchens, A.O. (1999). *Example Allocations of Operating and Maintenance Costs of Interstate Water Control Facilities Employing the Use-of-Facilities Method.* Asian Development Bank, Jakarta, 120 pp. Available at: www.adb.org/Documents/Reports/CAREC/Water-Energy-Nexus/appendix-A-19.PDF.

ICOLD (1999). *Dealing with sedimentation.* Bulletin 115, International Commission on Large Dams, Paris, France, 102 pp.

ICOLD (2004). *Management of Reservoir Water Quality – Introduction and Recommendations.* Bulletin 128, International Commission on Large Dams, Paris, France, 111 pp.

IEA-ENARD (2010). *Harnessing the North Sea Offshore Wind. Grid Policy Workshop.* International Energy Agency, Electricity Networks Analysis, Research & Development (ENARD), Paris, France, 28 April 2010.

IEA (2000a). *Hydropower and the Environment: Effectiveness of Mitigation Measures.* IEA Hydropower Agreement, Annex III – Subtask 6. International Energy Agency, Paris, France. Available at: www.ieahydro.org/reports/IEA_AIII_ST6.pdf.

IEA (2000b). *Hydropower and the Environment: Survey of the Environmental and Social Impacts and the Effectiveness of Mitigation Measures in Hydropower Development.* IEA Hydropower Agreement Technical Report Subtask 1, International Energy Agency, Paris, France, 111 pp. Available at: www.ieahydro.org/reports/IEA%20AIII%20ST1%20Vol%20I.pdf.

IEA (2000c). *Hydropower and the Environment: Present Context and Guidelines for Future Action. Volume I: Summary and Recommendations.* Implementing Agreement for Hydropower Technologies and Programmes, Annex III, International Energy Agency, Paris, France. Available at: www.ieahydro.org/uploads/files/hya3s5v1.pdf.

IEA (2000d). *Hydropower and the Environment: Present Context and Guidelines for Future Action. Volume II: Main Report.* Implementing Agreement for Hydropower Technologies and Programmes, Annex III, International Energy Agency, Paris, France, 172 pp. Available at: www.ieahydro.org/reports/HyA3S5V2.pdf.

IEA (2000e). *Hydropower and the Environment: Present Context and Guidelines for Future Action. Volume III: Appendices.* Implementing Agreement for Hydropower Technologies and Programmes, Annex III, International Energy Agency, Paris, France, 165 pp. Available at: www.ieahydro.org/uploads/files/hya3s5v3.pdf.

IEA (2004). *World Energy Outlook 2004*. International Energy Agency, Paris, France, 570 pp.

IEA (2006). *World Energy Outlook 2006*. International Energy Agency, Paris, France, 601 pp.

IEA (2008a). *World Energy Outlook 2008*. International Energy Agency, Paris, France, 578 pp.

IEA (2008b). *Energy Technology Perspectives 2008. Scenarios and Strategies to 2050*. International Energy Agency, Paris, France, 646 pp.

IEA (2010a). *Key World Energy Statistics*. International Energy Agency, Paris, France, 76 pp.

IEA (2010b). *Projected Costs of Generating Electricity*. International Energy Agency, Paris, France, 218 pp.

IEA (2010c). *World Energy Outlook 2010*. International Energy Agency, Paris, France, 736 pp.

IHA (2004). *Sustainability Assessment Protocol*. International Hydropower Association, London, UK.

IHA (2006). *Sustainability Assessment Protocol*. International Hydropower Association, London, UK, 64 pp.

IHA (2010). *Hydropower Sustainability Assessment Protocol – Background Document*. International Hydropower Association, London, UK.

Iimi, A. (2007). *Estimating Global Climate Change Impacts on Hydropower Projects: Applications in India, Sri Lanka and Vietnam*. World Bank, Washington, DC, USA, 38 pp.

IJHD (2010). *World Atlas & Industry Guide*. International Journal of Hydropower and Dams, Wallington, Surrey, UK, 405 pp.

Indian Electricity Act (2003). *Indian Electricity Act*. Ministry of Power, New Delhi, India. Available at: www.powermin.nic.in.

IPCC (2000). *Special Report on Emissions Scenarios*. N. Nakicenovic and R. Swart (eds.), Cambridge University Press, 570 pp.

IPCC (2006). Agriculture, forestry and other land use. In: *IPCC Guidelines for National Greenhouse Gas Inventories*. Prepared by the National Greenhouse Gas Inventories Programme, H.S. Eggleston, L. Buendia, K. Miwa, T. Ngara, and K. Tanabe (eds), Institute for Global Environmental Strategies (IGES), Japan.

IPCC (2007a): *Climate Change 2007: Synthesis Report. Contributions of Work Groups I, II and II to the Fourth Assessment Report of the Intergovernmental Panel of Climate Change*. Core Writing Team, R.K. Pachauri, and A. Reisinger (eds.), Cambridge University Press, 104 pp.

IPCC (2007b). *Climate Change 2007: Impacts, Adaptation and Vulnerability. Contribution of Working Group II to the Fourth Assessment Report of the Intergovernmental Panel on Climate Change*. M.L. Parry, O.F. Canziani, J.P. Palutikof, P.J. van der Linden, and C.E. Hanson (eds.), Cambridge University Press, 979 pp.

IPCC (2007c). *Climate Change 2007: The Physical Science Basis. Contribution of Working Group I to the Fourth Assessment Report of the Intergovernmental Panel on Climate Change*. S. Solomon, D. Qin, M. Manning, Z. Chen, M. Marquis, K.B. Averyt, M. Tignor, and H.L. Miller (eds.), Cambridge University Press, 996 pp.

Jain, S.K. (2007). Hydrology and water resources of India. *Water Science and Technology Library*, **57**, pp. 388-389.

Jensen, T. (2009). Building small hydro in Norway. *Hydro Review*, **16**(4), pp. 20-27.

Jinghe, L. (2005). Status and prospects of the small-scale hydro technologies in China. In: *Proceedings of China Renewable Energy Development Strategy Workshop*, 28 October 2005, Tsinghua University, Beijing, China, pp. 95.

Kahouli-Brahmi, S. (2008). Technological learning in energy-environment-economy modelling: A survey. *Energy Policy*, **36**(1), pp. 138-162.

Kahrl, F., and D. Roland-Holst (2008). China's water-energy nexus. *Water Policy*, **10**(Supplement 1), pp. 51-65.

Knigge, M., B. Goerlach., A. Hamada, C. Nuffort, and R.A. Kraemer (2008). *The Use of Environmental and Social Criteria in Export Credit Agencies' Practices*. Deutsche Gesellschaft für Technische Zusammenarbeit GmbH (GTZ) and Ecologic, Berlin and Bonn, Germany, 54 pp.

Knight, U.G. (2001). The 'Black Start' situation. Section 7.5 In: *Power Systems in Emergencies: From Contingency Planning to Crisis Management*. John Wiley & Sons, 394 pp.

Krewitt, W., K. Nienhaus, C. Kleßmann, C. Capone, E. Stricker, W. Graus, M. Hoogwijk, N. Supersberger, U. von Winterfeld, and S. Samadi (2009). *Role and Potential of Renewable Energy and Energy Efficiency for Global Energy Supply*. Climate Change 18/2009. ISSN 1862-4359. Federal Environment Agency, Dessau-Roßlau, 336 pp.

Krey, V., and L. Clarke (2011). Role of renewable energy in climate change mitigation: a synthesis of recent scenarios. *Climate Policy*, in press.

Kubiszewski, I., C.J. Cleveland, and P.K. Endres (2010). Meta-analysis of net energy return for wind power systems. *Journal of Renewable Energy*, **35**, pp. 218-225.

L'Abée-Lund, J., T.O. Haugen, and L.A. Vøllestad (2006). Disentangling from macroenvironmental effects: quantifying the effect of human encroachments based on historical river catches of anadromous salmonids. *Canadian Journal of Fisheries and Aquatic Sciences*, **63**, pp. 2318-2369.

Lako, P., H. Eder, M. de Noord, and H. Reisinger (2003). *Hydropower Development with a Focus on Asia and Western Europe. Overview in the Framework of VLEEM 2*. ECN-C--03-027, Verbundplan, Vienna, Austria, 96 pp.

Larinier, M., and G. Marmulla (2004). Fish passes: types, principles and geographical distribution – an overview. In: *Proceedings of the Second International Symposium on the Management of Large Rivers for Fisheries*. R. Welcomme and T. Petr (eds.), RAP Publication 2004/17, FAO Regional Office for Asia and the Pacific (RAP), Bangkok, Thailand. pp. 183-205.

LeCornu, J. (1998). Dams and water management. Report of the Secretary General, International Commission on Large Dams to the *Conférence Internationale Eau et Développement Durable*, Paris, France, 19-21 March 1998.

Ledec, G., and J.D. Quintero (2003). *Good Dams and Bad Dams: Environmental Criteria for Site Selection of Hydroelectric Projects*. World Bank, Washington, DC, USA, 16 pp.

Lehner, B., G. Czisch, and S. Vassolo (2005). The impact of global change on the hydropower potential of Europe: A model-based analysis. *Energy Policy*, **33**(7), pp. 839-855.

Liu, H., and X.-b. Hu (2010). The development & practice of SHP CDM Project in China. *China Water Power & Electrification*, **69**(9), pp. 8-14.

Lysne, D., B. Glover, H. Stole, and E. Tesakar (2003). *Hydraulic Design*. Publication No. 8, Norwegian Institute of Technology, Trondheim, Norway.

Mahmood, K. (1987). *Reservoir Sedimentation: Impact, Extent, and Mitigation*. World Bank, Washington, DC, USA, pp. 113-118.

Mielke, E., L.D. Anadon, and V. Narayanamurti (2010). *Water Consumption of Energy Resource Extraction, Processing and Conversion*. Discussion Paper No. 2010-15, Harvard Kennedy School, Cambridge, MA, USA.

Milly, P.C.D., K.A. Dunne, and A.V. Vecchia (2005). Global pattern of trends in streamflow and water availability in a changing climate. *Nature*, **438**(7066), pp. 347-350.

Milly, P.C.D., J. Betancourt, M. Falkenmark, R.M. Hirsch, Z.W. Kundzewicz, D.P. Lettenmaier, and R.J. Stouffer (2008). Climate change: Stationarity is dead: Whither water management? *Science*, **319**(5863), pp. 573-574.

Morris, G.L., and J. Fan (1997). *Reservoir Sedimentation Handbook: Design and Management of Dams, Reservoirs and Watershed for Sustainable Use*. McGraw-Hill, New York, NY, USA, 805 pp.

Mukherji, A. (2007). The energy-irrigation nexus and its impact on groundwater markets in eastern Indo-Gangetic basin: Evidence from West Bengal, India. *Energy Policy*, **35**(12), pp. 6413-6430.

Natural Resources Canada (2009). *Small Hydropower, Low Head and Very Low Head Hydro Power Generation*. Natural Resources Canada, Ottawa, Canada. Available at: www.canmetenergy-canmetenergie.nrcan-rncan.gc.ca/eng/renewables/small_hydropower.html.

Nilsson, C., and M. Dynesius (1993). Ecological effects of river regulation on mammals and birds: A review. *Regulated Rivers: Research and Management*, **9**(1), pp. 45-53.

Nordel (2008). *Annual Statistics 2008*. Organisation of the Transmission System Operators (TSOs) of Denmark, Finland, Iceland, Norway, and Sweden (www.nordel.org). Now merged into European Network of Transmission System Operators for Electricity, 22 pp. Available at: https://www.entsoe.eu/fileadmin/user_upload/_library/publications/nordic/annualstatistics/Annual%20Statistics%202008.pdf.

Odeh, M. (1999). *A Summary of Environmentally Friendly Turbine Design Concepts*. DOE/ID/13741, US Department of Energy, Idaho Operations Office, Idaho Falls, Idaho, 39 pp.

Pacca, S. (2007). Impacts from decommissioning of hydroelectric dams: a life cycle perspective. *Climatic Change*, **84**(3-4), pp. 281-294.

Palmieri, A., F. Shah, G.W. Annandale, and A. Dinar (2003). *Reservoir Conservation Volume 1: The RESCON Approach. Economic and engineering evaluation of alternative strategies for managing sedimentation in storage reservoirs*. The World Bank, Washington, DC, USA, 102 pp.

Petersen, M.S. (ed.) (1984). *Water Resources Planning and Development*. Prentice Hall, Eaglewood Cliffs, NJ, USA (ISBN 0-13-945908-1).

Prabhakar, B., and G.K. Pathariya (2007). Recent renovation and modernization technologies for existing hydro turbine. In: *International Conference on Small Hydropower – Hydro Sri Lanka*, Kandy, Sri Lanka, 22-24 October 2007, pp 161-168. Available at: www.ahec.org.in/links/International%20conference%20on%20SHP%20Kandy%20Srilanka%20All%20Details%5CPapers%5CTechnical%20Aspects-A%5CA20.pdf.

Ramanathan, K., and P. Abeygunawardena (2007). *Hydropower Development in India, A Sector Assessment*. Asian Development Bank, Mandaluyong City, The Philippines, 84 pp.

REN21 (2010). *Renewables 2010: Global Status Report*. Renewable Energy Policy Network for the 21st Century Secretariat, Paris, France, 80 pp.

Renard, K.G., G.R. Foster, G.A. Weesies, D.K. McCool, and D.C. Yoder (1997). *Predicting Soil Erosion by Water: A Guide to Conservation Planning with the Revised Universal Soil Loss Equation (RUSLE)*. Agriculture Handbook 703, US Department of Agriculture, Washington, DC, USA, 440 pp.

Rogner, H.-H., F. Barthel, M. Cabrera, A. Faaij, M. Giroux, D. Hall, V. Kagramanian, S. Kononov, T. Lefevre, R. Moreira, R. Nötstaller, P. Odell, and M. Taylor (2000). Energy resources. In: *World Energy Assessment. Energy and the Challenge of Sustainability*. United Nations Development Programme, United Nations Department of Economic and Social Affairs, World Energy Council, New York, USA, 508 pp.

Rudd, J.W.M., R. Harris, C.A. Kelly, and R.E. Hecky (1993). Are hydroelectric reservoirs significant sources of greenhouse gases? *Ambio*, **22**, pp. 246-248.

Sale, M.J., G.F. Cada, and D.D. Dauble (2006a). Historical perspective on the U.S. Department of Energy's Hydropower Program. In: *Proceedings of HydroVision 2006*, HCI Publications, Kansas City, MO, USA, pp. 1-12.

Sale, M.J., T.L. Acker, M.S. Bevelhimer, G.F. Cada, T. Carlson, D.D. Dauble, D.G. Hall, B.T. Smith, and F. Sotiropoulos (2006b). *DOE Hydropower Program Biannual Report for FY 2005-2006*. ORNL/TM-2006/97, Oak Ridge National Laboratory, Oak Ridge, TN, USA.

Saltveit, S.J., J.H. Halleraker, J.V. Arnekleiv, and A. Harby (2001). Field experiments on stranding in juvenile Atlantic salmon (*Salmo salar*) and brown trout (*Salmo trutta*) during rapid flow decreases caused by hydropeaking. *Regulated Rivers: Research and Management*, **17**, pp. 609-622.

Schneeberger, M., and H. Schmid (2004). StrafloMatrix™, 2004 – Further refinement to the HYDROMATRIX® technology. In: *Proceedings of Hydro 2004 Conference*, Porto, Portugal, 18-20 October 2004.

Scudder, T. (2005). *The Future of Large Dams: Dealing with Social, Environmental, Institutional and Political Costs*. Earthscan, London, UK, 432 pp.

SER (2007). *Ecology-Based Restoration in a Changing World. 18th Annual Meeting of the Society for Ecological Restoration International (SER)*, San Jose, CA, USA, 5-10 August 2007.

Shukla, P.R., D. Biswas, T. Nag, A. Yajnik, T. Heller, and D.G. Victor (2004). *Captive Power Plants: Case Study of Gujarat India*. Working Paper 22, Stanford University, Stanford, CA, USA, 40 pp. Available at: iis-db.stanford.edu/pubs/20454/wp22_cpp_5mar04.pdf.

Silva, E.P.D., A.J.M. Neto, P.F.P. Ferreira, J.C. Camargo, F.R. Apolinário, and C.S. Pinto (2005). Analysis of hydrogen production from combined photovoltaics, wind energy and secondary hydroelectricity supply in Brazil. *Solar Energy*, **78**, pp. 670-677.

Sims, R.E.H., R.N. Schock, A. Adegbululgbe, J. Fenhann, I. Konstantinaviciute, W. Moomaw, H.B. Nimir, B. Schlamadinger, J. Torres-Martínez, C. Turner, Y. Uchiyama, S.J.V. Vuori, N. Wamukonya, and X. Zhang (2007). Energy supply. In: *Climate Change 2007: Mitigation of Climate Change. Contribution of Working Group III to the Fourth Assessment Report of the Intergovernmental Panel on Climate Change*. B. Metz, O.R. Davidson, P.R. Bosch, R. Dave and L.A. Meyer (eds.), Cambridge University Press, pp. 251-322.

Sinha, S., B. Sharma, and C.A. Scott (2006). Understanding and managing the water–energy nexus: Moving beyond the energy debate. In: *Groundwater Research and Management: Integrating Science into Management Decisions*. B.R. Sharma, K.G. Villholth, and K.D. Sharma (eds.), International Water Management Institute (IWMI), South Asia Office, New Delhi, India, pp. 242-257.

Skoulikaris, C. (2008). *Mathematical Modeling Applied to Integrated Water Resources Management: The Case of Mesta-Nestos Basin*. PhD Thesis, Dissertation No. 398, Hydrologie et Hydrogéologie Quantitatives, GEOSC-Centre de Géosciences, ENSMP, Greece, 306 pp.

Soumis, N., E. Duchemin, R. Canuel, and M. Lucotte (2004). Greenhouse gas emissions from reservoirs of the western United States. *Global Biogeochemical Cycles*, **18**, GB2022.

SRU (2010). 100% erneuerbare Stromversorgung bis 2050: klimaverträglich, sicher, bezahlbar. *Sachverständigenrat für Umweltfragen*, Nr. **15**, May 2010, ISSN 1612 – 2968.

Sternberg, R. (2008). Hydropower: Dimensions of social and environmental coexistence. *Renewable and Sustainable Energy Reviews*, **12**(6), pp. 1588-1621.

Stickler, M., and K.T. Alfredsen (2009). Anchor ice formation in streams: a field study. *Hydrological Processes*, **23**(16), pp. 2307-2315.

Sundqvist, E., and D. Wårlind (2006). *The Importance of Micro Hydropower for Rural Electrification in Lao PDR*. Lund University, Lund, Sweden, 123 pp.

Takeuchi, K. (2004). Importance of sediment research in global water systems. In: *Proceedings from Ninth International Symposium on River Sedimentation*, Yichang, China, 18-21 October 2004, **1**, pp 10-18 (ISBN 7-302-09684-8).

Takeuchi, K., M. Hamlin, Z.W. Kundzewicz, D. Rosbjerg, and S.P. Simonovic (1998). *Sustainable Reservoir Development and Management*. International Association of Hydrological Sciences (IAHS) Publication No. 251, Institute for Hydrology, Wallingford, UK, 187 pp.

Taylor, R. (2008). The possible role and contribution of hydropower to the mitigation of climate change. In: *Proceedings of IPCC Scoping Meeting on Renewable Energy Resources, Working Group III*, 20-25 January 2008, Lubeck, Germany, pp. 81-91.

Terens, L., and R. Schafer (1993). Variable speed in hydropower generation utilizing static frequency converters. In: *Proceedings of the International Conference on Hydropower, Waterpower 93*, Nashville, TN, USA, 10-13 August 1993, pp. 1860-1869.

Teske, S., T. Pregger, S. Simon, T. Naegler, W. Graus, and C. Lins (2010). Energy [R]evolution 2010—a sustainable world energy outlook. *Energy Efficiency*, doi:10.1007/s12053-010-9098-y.

Thaulow, H., A. Tvede, T.S. Pedersen, and K. Seelos (2010). Managing catchments for hydropower generation. In: *Handbook of Catchment Management*. R.C. Ferrier and A. Jenkins (eds.), Blackwell Publishing (ISBN 978-1-4051-7122-9).

Therrien, J., A. Tremblay, and R. Jacques (2005). CO_2 emissions from semi-arid reservoirs and natural aquatic ecosystems. In: *Greenhouse Gas Emissions: Fluxes and Processes. Hydroelectric Reservoirs and Natural Environments*. A. Tremblay, L. Varfalvy, C. Roehm, and M. Garneau (eds.), Environmental Science Series, Springer, New York, NY, USA, pp. 233-250.

Torcellini, P., N. Long, and R. Judkoff (2003). *Consumptive Water Use for U.S. Power Production*. Technical Report-TP-550-33905, National Renewable Energy Laboratory, Golden, CO, USA, 18 pp.

Tremblay, A., L. Varfalvy, C. Roehm, and M. Garneau (2005). *Greenhouse Gas Emissions: Fluxes and Processes. Hydroelectric Reservoirs and Natural Environments*. Springer, New York, NY, USA, 732 pp.

UN (2004). *Beijing Declaration on Hydropower and Sustainable Development*. United Nations. Beijing, China, 29 October 2004. Available at: www.un.org/esa/sustdev/sdissues/energy/hydropower_sd_beijingdeclaration.pdf.

UN IWC (1997). *Convention on the Law of the Non-navigational Uses of International Watercourses Organization*, United Nations, New York, NY, USA, 18 pp. Available at: untreaty.un.org/ilc/texts/instruments/english/conventions/8_3_1997.pdf.

UNDP/UNDESA/WEC (2000). *World Energy Assessment – Energy and the Challenge of Sustainability*. United Nations Development Programme, United Nations Department of Economic and Social Affairs, and World Energy Council, New York, NY, USA.

UNDP/UNDESA/WEC (2004). *World Energy Assessment Overview: 2004 Update*. J. Goldemberg, and T.B. Johansson (eds.), United Nations Development Programme, United Nations Department of Economic and Social Affairs, and World Energy Council, New York, New York, USA, 85 pp.

UNEP (2007). *A Compendium of Relevant Practices for Improved Decision-Making, Planning and Management of Dams and Their Alternatives*. United Nations Environment Programme Dams and Development Project, Nairobi, Kenya, 52 pp.

UNESCO-RED (2008). *Resolving the Water-Energy Nexus – Assessment and Recommendations*. United Nations Educational Scientific and Cultural Organization – International Hydropower Program and Red Ethique, In: *International Symposium on Resolving the Water-Energy Nexus*, Paris, France, 26-28 November 2008.

UNESCO/IHA (2008). *Assessment of the GHG Status of Freshwater Reservoirs - Scoping Paper*. IHA/GHG-WG/3, United Nations Educational, Scientific and Cultural Organization Working Group on Greenhouse Gas Status of Freshwater Reservoirs and the International Hydropower Association, London, UK, 28 pp. Available at: unesdoc.unesco.org/images/0018/001817/181713e.pdf.

UNESCO/IHA (2009). *Measurement Specification Guidance for Evaluating the GHG Status of Man-Made Freshwater Reservoirs, Edition 1*. IHA/GHG-WG/5. United Nations Educational Scientific and Cultural Organization and the International Hydropower Association, London, UK, 55 pp. Available at: http://unesdoc.unesco.org/images/0018/001831/183167e.pdf.

UNESCO/IHA (2010). *GHG Measurement Guidelines for Freshwater Reservoirs*. United Nations Educational, Scientific and Cultural Organization and the International Hydropower Association, London, UK, 138 pp. (ISBN 978-0-9566228-0-8).

UNWWAP (2006). *Water – A Shared Responsibility*. United Nations World Water Development Report 2, World Water Assessment Programme, United Nations Educational, Scientific and Cultural Organization, Paris, France, 584 pp. Available at: www.unesco.org/water/wwap.

US Department of the Interior (2005). *Renewable Energy for America's Future*. US Department of the Interior, Washington, DC, USA. Available at: www.doi.gov/initiatives/renewable_energy.pdf.

US DOE (2009). *Annual Energy Outlook 2009 with Projection to 2030*. US Department of Energy. Washington, DC, USA, 221 pp.

US Inter-Agency Committee on Water Resources (1958). *Proposed Practices for Economic Analysis of River Basin Projects*. May 1950; revised May 1958. US Interagency Committee on Water Resources, Washington, DC, USA, 56 pp.

Vennemann, P., L. Thiel, and H.C. Funke (2010). Pumped storage plants in the future power supply system. *Journal VGB Power Tech*, **90**(1/2), pp. 44-49.

Vinogg, L., and I. Elstad (2003). *Mechanical Equipment*. Norwegian University of Science and Technology, Trondheim, Norway, 130 pp.

Viravong, V. (2008). Lao PDR – powering progress. In: *Regional Consultation on Mekong River Commission's Hydropower Programme*, Vientiane, PR Laos, 25-27 September 2008. Available at: www.mrcmekong.org/download/Presentations/regional-hydro/2.2%20MRC-consultation_LAOPDR.pdf.

Votruba, L., Z. Kos, K. Nachazel, A. Patera, and V. Zeman (1988). *Analysis of Water Resource Systems*. Elsevier (ISBN 9780 449 89444).

Wang, X. (2010). Installed capacity of small hydropower in China reaching 55.12 GW. *Guangxi Electric Power*, **120**(4), pp. 11 (in Chinese).

Wang, Z., and C. Hu (2004). Interaction between fluvial systems and large scale hydro-projects. In: *Proceedings from Ninth International Symposium on River Sedimentation 2009*, Yichang, China, 18-24 October 2004, pp. 20.

WBCSD (2009). *Water, Energy and Climate Change. A contribution from the business community*. World Business Council for Sustainable Development, Geneva, Switzerland and Washington, DC, USA, 20 pp.

WCD (2000). *Dams and Development – A New Framework for Decision-Making*. World Commission on Dams, Earthscan, London, UK, 356 pp.

WEC (2007). Hydropower. In: *2007 Survey of Energy Resources*. World Energy Council (WEC), London, UK, pp 271-313 (ISBN: 0 946121 26 5).

Wellinghoff, H.J., J. Pederson, and D.L. Morenoff (2008). Facilitating hydrokinetic energy development through regulatory innovation. *Energy Law Journal*, **29**(398-419)

White, W.R. (2005). *World Water Storage in Man-Made Reservoirs: Review of Current Knowledge*. Foundation for Water Research, Marlow, UK, 53 pp.

World Bank (1994). *Culture Heritage and Environmental Assessment*. Environmental Assessment Sourcebook Update, No. 8, World Bank, Washington, DC, USA, 8 pp.

World Bank (2001). The World Bank position on the report of the World Commission on Dams. In: *Water Resources Sector Strategy: Strategic Directions for World Bank Engagement*. International Bank for Reconstruction and Development/The World Bank, Washington, DC, USA, pp. 75–78 (report published 2004).

World Bank (2004). *Water Resources Sector Strategy: Strategic Directions for World Bank Engagement*. International Bank for Reconstruction and Development/World Bank, Washington, DC, USA.

World Bank (2005). *Shaping the Future of Water for Agriculture: A Sourcebook for Investment in Agricultural Water Management*. World Bank, Washington, DC, USA.

World Bank (2008). *Designing Sustainable Off-Grid Rural Electrification Projects: Principles and Practices - Operational Guidance for World Bank Group Staff*. World Bank, Washington, DC, USA, 17 pp.

World Bank (2009). *Directions in Hydropower: Scaling Up for Hydropower*. Water Sector Board Practitioner Notes (P-Notes), Sustainable Development Network of the World Bank Group, Washington, DC, USA, 16 pp.

World Bank (2011). *Water Resources Management*. Available at: http://water.worldbank.org/water/topics/water-resources-management.

Xu, J. (2002). River sedimentation and channel adjustment of the lower Yellow River as influenced by low discharges and seasonal channel dry-ups. *Geomorphology*, **43**(1-2), pp. 151-164.

Yumurtacia, Z., and E. Bilgen (2004). Hydrogen production from excess power in small hydroelectric installations. *International Journal of Hydrogen Energy*, **29**, pp. 687-693.

Zare, S., and A. Bruland (2007). Progress of drill and blast tunnelling efficiency with relation to excavation time and cost. In: *33rd ITA World Tunnel Congress*, Prague, Czech Republic, 5-10 May 2007, pp. 805-809.

Zhu, Y.-M., X.X. Lu, and Y. Zhou (2007). Sediment flux sensitivity to climate change: A case study in the Longchuanjiang catchment of the upper Yangtze River, China. *Global and Planetary Change*, **60**(3-4), pp. 429-442.

6

Ocean Energy

Coordinating Lead Authors:
Anthony Lewis (Ireland) and Segen Estefen (Brazil)

Lead Authors:
John Huckerby (New Zealand), Kwang Soo Lee (Republic of Korea), Walter Musial (USA), Teresa Pontes (Portugal), Julio Torres-Martinez (Cuba)

Contributing Authors:
Desikan Bharathan (USA), Howard Hanson (USA), Garvin Heath (USA), Frederic Louis (France), Sandvik Øystein Scråmestø (Norway)

Review Editors:
Amjad Abdulla (Republic of Maldives), José M. Moreno (Spain) and Yage You (China)

This chapter should be cited as:
Lewis, A., S. Estefen, J. Huckerby, W. Musial, T. Pontes, J. Torres-Martinez, 2011: Ocean Energy. In IPCC Special Report on Renewable Energy Sources and Climate Change Mitigation [O. Edenhofer, R. Pichs-Madruga, Y. Sokona, K. Seyboth, P. Matschoss, S. Kadner, T. Zwickel, P. Eickemeier, G. Hansen, S. Schlömer, C. von Stechow (eds)], Cambridge University Press, Cambridge, United Kingdom and New York, NY, USA.

Table of Contents

Executive Summary 501

6.1 Introduction 503

6.2 Resource potential 503
- 6.2.1 Wave energy 503
- 6.2.2 Tidal range 505
- 6.2.3 Tidal currents 506
- 6.2.4 Ocean currents 506
- 6.2.5 Ocean thermal energy conversion 507
- 6.2.6 Salinity gradients 507

6.3 Technology and applications 507
- 6.3.1 Introduction 507
- 6.3.2 Wave energy 507
  - 6.3.2.1 Oscillating water columns 508
  - 6.3.2.2 Oscillating-body systems 508
  - 6.3.2.3 Overtopping devices 509
  - 6.3.2.4 Power take-off systems 509
- 6.3.3 Tidal range 510
- 6.3.4 Tidal and ocean currents 510
- 6.3.5 Ocean thermal energy conversion 511
- 6.3.6 Salinity gradients 512
  - 6.3.6.1 Reversed electro dialysis 512
  - 6.3.6.2 Pressure-retarded osmosis 513

6.4 Global and regional status of market and industry development 513
- 6.4.1 Introduction 513
  - 6.4.1.1 Markets 513
  - 6.4.1.2 Industry development 514

6.4.2	Wave energy	514
6.4.3	Tidal range	515
6.4.4	Tidal and ocean currents	515
6.4.5	Ocean thermal energy conversion	516
6.4.6	Salinity gradients	516
6.4.7	Impact of Policies	516

6.5	**Environmental and Social Impacts**	**517**
6.5.1	Lifecycle greenhouse gas emissions	517
6.5.2	Other environmental and social impacts	518
6.5.2.1	Wave energy	518
6.5.2.2	Tidal range	519
6.5.2.3	Tidal and ocean currents	519
6.5.2.4	Ocean thermal energy conversion	520
6.5.2.5	Salinity gradients	520

6.6	**Prospects for technology improvement, innovation and integration**	**520**
6.6.1	Wave energy	520
6.6.2	Tidal range	521
6.6.3	Tidal and ocean currents	521
6.6.4	Ocean thermal energy conversion	521
6.6.5	Salinity gradients	521

6.7	**Cost trends**	**522**
6.7.1	Introduction	522
6.7.2	Wave and tidal current energy	522
6.7.3	Tidal range	524

6.7.4	Ocean thermal energy conversion	525
6.7.5	Salinity gradients	526

6.8 Potential deployment 526

6.8.1	Deployment scenarios with ocean energy coverage	526
6.8.2	Near-term forecasts	527
6.8.3	Long-term deployment in the context of carbon mitigation	527
6.8.4	Conclusions regarding deployment	528

References 530

Executive Summary

Ocean energy offers the potential for long-term carbon emissions reduction but is unlikely to make a significant short-term contribution before 2020 due to its nascent stage of development. In 2009, additionally installed ocean capacity was less than 10 MW worldwide, yielding a cumulative installed capacity of approximately 300 MW by the end of 2009. All ocean energy technologies, except tidal barrages, are conceptual, undergoing research and development (R&D), or are in the pre-commercial prototype and demonstration stage. The performance of ocean energy technologies is anticipated to improve steadily over time as experience is gained and new technologies are able to access poorer quality resources. Whether these technical advances lead to sufficient associated cost reductions to enable broad-scale deployment of ocean energy is the most critical uncertainty in assessing the future role of ocean energy in mitigating climate change. Though technical potential is not anticipated to be a primary global barrier to ocean energy deployment, resource characteristics will require that local communities in the future select among multiple available ocean technologies to suit local resource conditions.

Though ocean energy resource assessments are at a preliminary phase, the theoretical potential for ocean energy easily exceeds present human energy requirements. Ocean energy is derived from technologies that utilize seawater as their motive power or harness its chemical or heat potential. The renewable energy (RE) resource in the ocean comes from six distinct sources, each with different origins and requiring different technologies for conversion: waves; tidal range; tidal currents; ocean currents; ocean thermal energy conversion (OTEC); and salinity gradients. Ocean energy could be used not only to supply electricity but also for direct potable water production or to meet thermal energy service needs. The theoretical potential for ocean energy technologies has been estimated at 7,400 EJ/yr, well exceeding current and future human energy needs. Relatively few assessments have been conducted on the technical potential of the various ocean energy technologies and such potentials will vary based on future technology developments. One assessment places the global technical potential for 2050 at 331 EJ/yr, dominated by OTEC (300 EJ/yr) and wave energy (20 EJ/yr), whereas on the other end of the spectrum, another assessment lists the 'exploitable estimated available energy resource' at just 7 EJ/yr. Whilst some potential ocean energy resources, such as ocean currents and osmotic power from salinity gradients, are globally distributed, other forms of ocean energy have complementing distributions. Ocean thermal energy is principally distributed in the tropics around the Equator (latitudes 0° to 35°), whilst wave energy principally occurs between latitudes of 30° to 60°. Some ocean energy resources, such as ocean thermal, ocean currents and salinity gradients may be used to generate base-load electricity, whereas others have variable generation profiles that differ in their predictability. Though the available literature is limited, the impact of climate change on the technical potential for ocean energy is anticipated to be modest.

Ocean energy systems are at an early stage of development, but technical advances may progress rapidly given the number of technology demonstrations. With the exception of tidal range energy, which can be harnessed by the adaptation of river-based hydroelectric dams to estuarine situations, most ocean energy technologies have not yet been developed beyond the prototype stage. Although basic concepts have been known for decades, if not centuries, ocean energy technology development really began in the 1970s, only to languish in the post-oil-price crisis period of the 1980s. Research and development on a wide range of ocean energy technologies was rejuvenated at the start of the 2000s and some technologies, specifically wave and tidal current energy, have reached full-scale prototype deployments. Unlike wind turbine generators, there is presently no convergence on a single design configuration for ocean energy converters and, given the range of options for energy extraction, a single device design is unlikely. Worldwide developments of devices are accelerating with a large number of prototype wave and tidal current devices under development.

Government policies are contributing to accelerate the implementation of ocean energy technologies. Some national and regional governments are supporting ocean energy development through a range of initiatives, including R&D and capital grants to device developers; performance incentives for produced electricity; marine infrastructure development; standards, protocols and regulatory interventions for permitting; and space and resource allocation.

Ocean energy has the potential to deliver long-term carbon emissions reductions and appears to have low environmental impacts. Ocean energy technologies do not generate GHGs in operation and have low lifecycle GHG emissions, providing the potential to significantly contribute to emissions reductions. Utility-scale deployments with transmission grid connections can be used to displace carbon-emitting energy supplies, while smaller-scale developments may supply electricity and/or drinking water to remote communities. As shown by a review of a limited number of existing global energy scenarios, ocean energy has the potential to help mitigate long-term climate change by offsetting GHG emissions with projected deployments resulting in energy delivery of up to 1,943 TWh/yr (~7 EJ/yr) by 2050. The local social and environmental impacts of ocean energy projects are being evaluated as actual deployments multiply, but can be estimated based on the experience of other maritime and offshore industries. Environmental risks from ocean energy technologies appear to be relatively low, but the early stage of ocean energy deployment creates uncertainty on the degree to which social and environmental concerns might eventually constrain development.

Successful deployment will lead to cost reductions. Although ocean energy technologies are at an early stage of development, there are encouraging signs that the investment cost of technologies and the levelized cost of electricity generated will decline from their present non-competitive levels as R&D and demonstrations proceed, and as deployment occurs. Whether these cost reductions are sufficient to enable broad-scale deployment of ocean energy is the most critical uncertainty in assessing the future role of ocean energy in mitigating climate change.

6.1 Introduction

This chapter discusses the potential contribution that energy derived from the ocean can make to overall energy supply and hence its potential contribution to climate mitigation. The RE resource in the ocean comes from six distinct sources, each with different origins and requiring different technologies for conversion. These sources are:

- **Waves**, derived from the transfer of the kinetic energy of the wind to the upper surface of the ocean;
- **Tidal Range (tidal rise and fall)**, derived from the gravitational forces of the Earth-Moon-Sun system;
- **Tidal Currents**, water flow resulting from the filling and emptying of coastal regions as a result of the tidal rise and fall;
- **Ocean Currents**, derived from wind-driven and thermohaline ocean circulation;
- **Ocean Thermal Energy Conversion (OTEC)**, derived from temperature differences between solar energy stored as heat in upper ocean layers and colder seawater, generally below 1,000 m; and
- **Salinity Gradients (osmotic power)**, derived from salinity differences between fresh and ocean water at river mouths.

Marine biomass farming—production of biofuels from seaweed and/or algae—is covered in Chapter 2, whereas submarine geothermal energy—high-temperature water issuing from submarine vents at seabed ocean ridges—is covered in Chapter 4.

All ocean energy technologies, except tidal barrages, are conceptual, undergoing R&D, or are in the pre-commercial prototype and demonstration stage. The globally distributed resources and relatively high energy density associated with most ocean energy sources provide ocean energy with the potential to make an important contribution to energy supply and to the mitigation of climate change in the coming decades, if technical challenges can be overcome and costs thereby reduced. Accordingly, a range of initiatives are being employed by some governments to promote and accelerate the development and deployment of ocean energy technologies.

Information on the environmental and social impacts is limited mainly due to the lack of experience in deploying and operating ocean technologies, although adverse environment effects are foreseen to be relatively low. The current and future costs of most ocean energy technologies are also difficult to assess as little fabrication and deployment experience is available for validation of cost assumptions.

This chapter is presented in eight sections covering different aspects of ocean energy. Resource potential from different ocean sources is treated in Section 6.2, with a focus on both theoretical and technical potentials. The present state of development of ocean technologies and applications is considered in Section 6.3. Discussion about markets and industry developments, including government policies, is presented in Section 6.4. Environmental and social impacts are covered in Section 6.5. Finally, prospects for technology improvement, cost trends and potential deployment are considered in Sections 6.6, 6.7 and 6.8, respectively.

6.2 Resource potential

Relatively few assessments have been conducted on the technical potential of the various ocean energy technologies, and such potentials will vary based on future technology developments. As presented in Chapter 1, the theoretical potential for ocean energy technologies has been estimated to be 7,400 EJ/yr (Rogner et al., 2000), whereas Krewitt et al. (2009) report a global technical potential for 2050 of 331 EJ/yr, dominated by OTEC (300 EJ/yr) and wave energy (20 EJ/yr). On the other end of the spectrum, the IPCC Fourth Assessment Report reports what it lists as an 'exploitable estimated available energy resource' of just 7 EJ/yr (Sims et al., 2007). Given the early state of the available literature and the substantial uncertainty in ocean energy's technical potential, this section covers selected estimates of both theoretical and technical potential. Moreover, because of the inherent differences among the various ocean energy sources, resource potential assessments are discussed for each ocean energy source in turn.

Also discussed in this section is the potential impact of climate change on the technical potential for ocean energy. In summary, though the available literature is limited, the impact of climate change is anticipated to be modest. In a number of instances, climate variables simply have little to no influence on the underlying energy sources (e.g., tidal range, tidal current), whereas in other cases the impacts do not seem likely to greatly influence global technical potential estimates (e.g., OTEC, wave, salinity gradient, ocean current).

6.2.1 Wave energy

Ocean wave energy (as distinct from internal waves or tsunamis) is energy that has been transferred from the wind to the ocean. As the wind blows over the ocean, air-sea interaction transfers some of the wind energy to the water, forming waves, which store this energy as potential energy (in the mass of water displaced from the mean sea level) and kinetic energy (in the motion of water particles). The size and period of the resulting waves depend on the amount of transferred energy, which is a function of the wind speed, the length of time the wind blows (order of days) and the length of ocean over which the wind blows (fetch). Waves are very efficient at transferring energy, and can travel long distances over the ocean surface beyond the storm area and are then classed as swells (Barber and Ursell, 1948; Lighthill, 1978). The most energetic waves on earth are generated between 30° and 60° latitudes by extra-tropical storms. Wave energy availability typically varies seasonally and over shorter time periods, with seasonal variation typically being greater in the northern hemisphere. Annual variations in the wave climate are usually estimated by the use of long-term averages in modelling, using global databases with reasonably long histories.

A map of the global offshore average annual wave power distribution (Figure 6.1) shows that the largest power levels occur off the west coasts of the continents in temperate latitudes, where the most energetic winds and greatest fetch areas occur.

decrease of 8% from the total theoretical wave energy potential above (it excludes areas with less than 5 kW/m), but should still be considered an estimate of theoretical potential. The technical potential of wave energy will be substantially below this figure and will depend upon

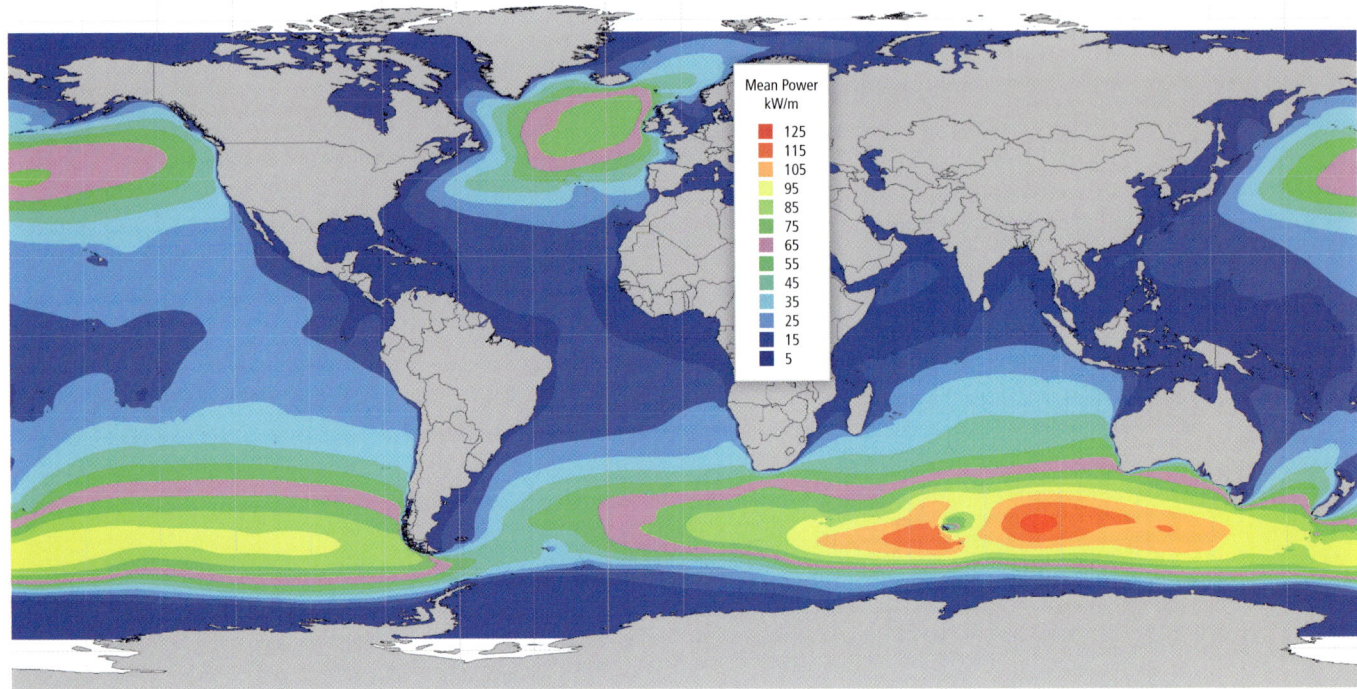

Figure 6.1 | Global offshore annual wave power level distribution (Cornett, 2008).

The total theoretical wave energy potential is estimated to be 32,000 TWh/yr (115 EJ/yr) (Mørk et al., 2010), roughly twice the global electricity supply in 2008 (16,800 TWh/yr or 54 EJ/yr). This figure is unconstrained by geography, technical or economic considerations. The regional distribution of the annual wave energy incident on the coasts of countries or regions has been obtained for areas where theoretical wave power P ≥ 5 kW/m and latitude ≤66.5° (Table 6.1). The theoretical wave energy potential listed in Table 6.1 (29,500 TWh/yr or 10^6 EJ/yr) represents a

technical developments in wave energy devices. Sims et al. (2007) estimate a global technical potential of 500 GW for wave energy, assuming that offshore wave energy devices have an efficiency of 40% and are only installed near coastlines with wave climates of >30 kW/m, whereas Krewitt et al. (2009) report a wave energy potential of 20 EJ/yr.

Potential changes in wind patterns, caused by climate change, are likely to affect the long-term wave climate distribution (Harrison and Wallace,

Table 6.1 | Regional theoretical potential of wave energy (Mørk et al., 2010).

REGION	Wave Energy TWh/yr (EJ/yr)
Western and Northern Europe	2,800 (10.1)
Mediterranean Sea and Atlantic Archipelagos (Azores, Cape Verde, Canaries)	1,300 (4.7)
North America and Greenland	4,000 (14.4)
Central America	1,500 (5.4)
South America	4,600 (16.6)
Africa	3,500 (12.6)
Asia	6,200 (22.3)
Australia, New Zealand and Pacific Islands	5,600 (20.2)
TOTAL	29,500 (106.2)

Note: The results presented in Mørk et al. (2010) regarding the overall theoretical global potential for wave energy are consistent with other studies (Cornett, 2008). No further studies of regional theoretical potential of wave energy are available to validate the data provided in Table 6.1.

2005; MCCIP, 2008), though the impact of those changes is likely to have only a modest impact on the global technical potential for wave energy given the ability to relocate wave energy devices as needed over the course of decades.

A range of devices are used to measure waves:

- Wave-measuring buoys are used in water depths greater than 20 m (see Allender et al., 1989). Seabed-mounted (pressure and acoustic) probes are used in shallower waters. Capacity/resistive probes or down-looking infrared and laser devices can be used when offshore structures are available (e.g., oil or gas platforms).

- Satellite-based measurements have been made regularly since 1991 by altimeters that provide measurements of significant wave height and wave period with accuracies similar to wave buoys (Pontes and Bruck, 2008). The main drawback of satellite data is the long interval between measurements (several days) and the corresponding large distance between adjacent tracks (0.8° to 2.8° along the Equator).

- The results of numerical wind-wave models are now quite accurate, especially for average wave conditions. Such models compute directional spectra over the oceans, taking as input wind fields provided by atmospheric models; they are by far the largest source of wave information.

The different types of wave information are complementary and should be used together for best results. For a review of wave data sources, atlases and databases, see Pontes and Candelária (2009).

6.2.2 Tidal range

Tides are the regular and predictable change in the height of the ocean, driven by gravitational and rotational forces between the Earth, Moon and Sun, combined with centrifugal and inertial forces. Many coastal areas experience roughly two high tides and two low tides per day (called 'semi-diurnal'); in some locations there is only one tide per day (called diurnal). The lunar day of 24 hrs and 50 min means that the timing of subsequent high and low tides advances each day as this constituent is the predominant one. Diurnal and semi-diurnal tides also occur at different times in different locations around the Earth.

During the year, the amplitude of the tides varies depending on the respective positions of the Earth, the Moon and the Sun. Spring tides (maximum tidal range) occur when the Sun, Moon and Earth are aligned (at full moon and at new moon). Neap tides (minimum tidal range) occur when the gravitational forces of the Earth-Moon axis are at 90 degrees to the Earth-Sun axis. The spring-neap tide cycle is driven by the 29.5 day orbit of the Moon around the Earth and is experienced throughout the world at the same time. Longer-period fluctuations in tide height also occur, but are of very low magnitude compared to diurnal, semi-diurnal and spring-neap cycles (Sinden, 2007).

The timing and magnitude of the tide varies depending on global position and also on the shape of the ocean bed, the shoreline geometry and Coriolis acceleration. Within a tidal system there are points where the tidal range is nearly zero, called amphidromic points (Figure 6.2). However, even at these points tidal currents will generally flow with high velocity as the water surface on either side of the amphidromic point is at different levels. This is a result of the Coriolis effect and interference within oceanic basins, seas and bays, creating a tidal wave pattern (called an amphidromic system), which rotates around the amphidromic point. See Pugh (1987) for full details of tidal behaviour.

Tidal periodicities can resonate with the natural oscillatory frequencies of estuaries and bays, resulting in greatly increased tidal range. Consequently, the locations with the largest tidal ranges are at resonant estuaries, such as the Bay of Fundy in Canada (17 m tidal range), the Severn Estuary in the UK (15 m) and Baie du Mont Saint Michel in France (13.5 m) (Kerr, 2007). In other places (e.g., the Mediterranean Sea), the tidal range is less than 1 m (Shaw, 1997; Usachev, 2008).

Tidal range can be forecast with a high level of accuracy, even centuries in advance: while the resultant power is variable, there is no resource risk due to climate change. The world's theoretical tidal power potential (tidal range plus tidal currents) is in the range of 3 TW, with 1 TW located in relatively shallow waters (Charlier and Justus, 1993), though Sims et al. (2007) and Krewitt et al. (2009) note that only a fraction of the theoretical potential is likely to be exploited.

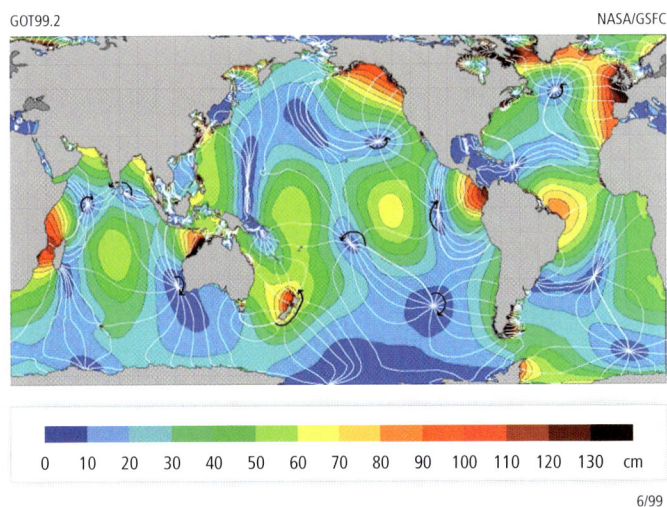

Figure 6.2 | World map of M2 tidal amplitude (NASA, 2006).

Notes: M2 is the largest (semidiurnal) tidal constituent, whose amplitude is about 60% of the total tidal range. The white lines are cotidal lines—where tides are at the same point of rising or falling, spaced at phase intervals of 30° (a bit over 1 hr). The amphidromic points are the dark blue areas where the cotidal lines meet. Tides rotate about these points where little or no tidal rise and fall occurs but where there can be strong tidal currents.

6.2.3 Tidal currents

Tidal currents are the ocean water mass response to tidal range (see Section 6.2.2). Tidal currents are generated by horizontal movements of water, modified by seabed bathymetry, particularly near coasts or other constrictions (e.g., islands). Tidal current flows result from the rise and fall of the tide; although these flows can be slightly influenced by short-term weather fluctuations, their timing and magnitude are highly predictable and largely insensitive to climate change influences.

A number of methods for the assessment of the tidal current energy resource potential have been discussed (Hagerman et al., 2006; Mackay, 2008). In the energy flux method, which is widely used, the potential power of a tidal current is proportional to the cube of the current velocity. Hence, the power density (in W/m^2) of tidal currents increases substantially with small increases in velocity. For near-shore currents such as those occurring in channels between mainland and islands or in estuaries, current velocity varies systematically and predictably in relation to the tide. In the specific case of tidal channels, however, there is a further limitation on the calculation of the overall resource (Garrett and Cummins, 2005, 2008; Karsten et al., 2008; Sutherland et al., 2008).

An atlas of wave energy and tidal current resource potential has been developed for the UK (UK Department of Trade and Industry, 2004). Similar resource estimates have been published for the EU (CEC, 1996; Carbon Trust, 2004), Canada (Cornett, 2006) and China (CEC, 1998).

In Europe, the tidal current energy resource potential is of special interest for the UK, Ireland, Greece, France and Italy. Over 106 promising locations have been identified, mostly in the UK (CEC, 1996). Using present-day state-of-the-art technologies, these sites have been estimated to have a technical potential of 48 TWh/yr (0.17 EJ/yr) (CEC, 1996). China has estimated that around 14 GW of tidal current power is available (Wang and Lu, 2009). Commercially attractive sites have also been identified in the Republic of Korea, Canada, Japan, the Philippines, New Zealand and South America.

6.2.4 Ocean currents

In addition to near-shore tidal currents, significant current flows also exist in the open ocean. These currents flow continuously in the same direction and have low variability. Large-scale circulation of the oceans is concentrated in various regions, notably the western boundary currents associated with wind-driven circulations. Some of these offer sufficient current velocities (~2 m/s) to drive present-day technologies (Leaman et al., 1987). These include the Agulhas/Mozambique Currents off South Africa, the Kuroshio Current off East Asia, the East Australian Current, and the Gulf Stream off eastern North America (Figure 6.3). Other ocean currents may also have potential for development as improvements in turbine systems occur.

The potential for power generation from the Florida Current of the Gulf Stream system was recognized decades ago. The 'MacArthur Workshop' concluded that the Florida Current had a technical potential of 25 GW (Stewart, 1974; Raye, 2001). It has a core region 15 to 30 km off the coast near the surface and flows strongly year-round as part of the North Atlantic Ocean subtropical gyre (Niiler and Richardson, 1973; Johns et al., 1999).

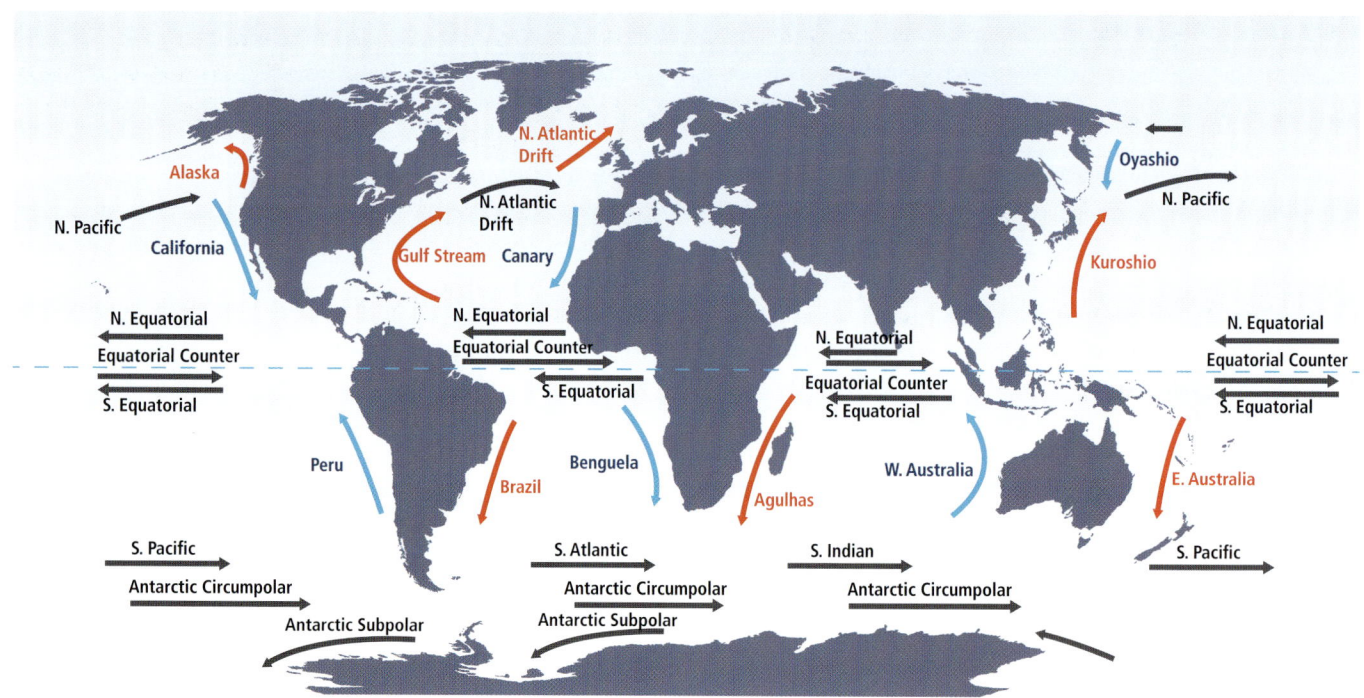

Figure 6.3 | Surface ocean currents, showing warm (red) and cold (blue) systems.

6.2.5 Ocean thermal energy conversion

About 15% of the total solar input to the ocean is retained as thermal energy, with absorption concentrated at the top layers, declining exponentially with depth as the thermal conductivity of sea water is low. Sea surface temperature can exceed 25°C in tropical latitudes, while temperatures 1 km below the surface are between 5°C and 10°C (Charlier and Justus, 1993).

A minimum temperature difference of 20°C is considered necessary to operate an OTEC power plant. Both coasts of Africa and India, the tropical west and south-eastern coasts of the Americas and many Caribbean and Pacific islands have sea surface temperature of 25°C to 30°C, declining to 4°C to 7°C at depths varying from 750 to 1,000 m. The OTEC resource map showing annual average temperature differences between surface waters and the water at 1,000-m depth shows a wide tropical area with a potential greater than 20° C temperature difference (Figure 6.4). A number of Pacific and Caribbean countries could develop OTEC plants close to their shores (UN, 1984). It seems unlikely that climate change would have a meaningful impact on the size of the global technical potential for OTEC.

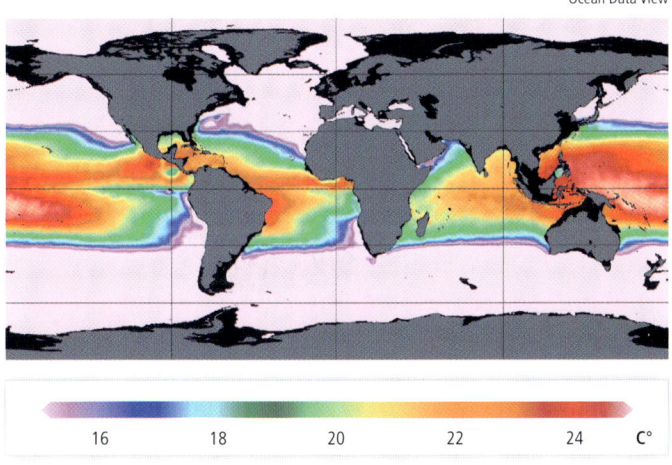

Figure 6.4 | Worldwide average ocean temperature differences (°C) between 20 and 1,000 m water depth (Nihous, 2010).

Among ocean energy sources, OTEC is one of the continuously available renewable resources that could contribute to base-load power supply (there is a slight variation from summer to winter), although compared to wave and tidal current energy, its energy density is very low.

The resource potential for OTEC is considered to be much larger than for other ocean energy forms (World Energy Council, 2000). It also has a widespread distribution between the two tropics. An optimistic estimate of the global theoretical potential is 30,000 to 90,000 TWh/yr (108 to 324 EJ/yr) (Charlier and Justus, 1993). More recently, Nihous (2007) calculated that about 44,000 TWh/yr (159 EJ/yr) of steady-state power may be possible. Up to 88,000 TWh/yr (318 EJ/yr) of power could be generated from OTEC without affecting the ocean's thermal structure (Pelc and Fujita, 2002).

6.2.6 Salinity gradients

The mixing of freshwater and seawater releases energy as heat. Harnessing the chemical potential between the two water sources, across a semi-permeable membrane, can capture this energy as pressure, rather than heat, which can then be converted into useful energy forms.

Since freshwater from rivers discharging into saline seawater is globally distributed, osmotic power could be generated and used in all regions wherever there is a sufficient supply of freshwater. River mouths are most appropriate, because of the potential for large adjacent volumes of freshwater and seawater.

Recently, the technical potential for power generation was calculated as 1,650 TWh/yr (6 EJ/yr) (Scråmestø et al., 2009). Salinity gradients could potentially generate base-load electricity, if cost-effective technologies can be developed.

6.3 Technology and applications

6.3.1 Introduction

The current development status of ocean energy technologies ranges from the conceptual and pure R&D stages to the prototype and demonstration stage, and only tidal range technology can be considered mature. Presently there are many technology options for each ocean energy source and, with the exception of tidal range barrages, technology convergence has not yet occurred. Over the past four decades, other marine industries (primarily offshore oil and gas) have made significant advances in the fields of materials, construction, corrosion, submarine cables and communications. Ocean energy is expected to directly benefit from these advances.

Competitive ocean energy technologies could emerge in the present decade, but only if significant technical progress is achieved. Ocean energy technologies are suitable for the production of both electricity and potable water, whilst OTEC can also be used to provide thermal energy services (e.g., seawater cooling for air conditioners). A general overview is given in Krishna (2009).

6.3.2 Wave energy

Many wave energy technologies representing a range of operating principles have been conceived, and in many cases demonstrated, to convert energy from waves into a usable form of energy. Major variables include the method of wave interaction with respective motions (heaving, surging, pitching) as well as water depth (deep, intermediate, shallow) and distance from shore (shoreline, near-shore, offshore). Efficient operation of floating devices requires large motions, which can be achieved by resonance or by latching, that is, with hold/release of moving parts until potential energy has accumulated.

A generic scheme for characterizing ocean wave energy generation devices consists of primary, secondary and tertiary conversion stages (Khan et al., 2009). The primary interface subsystem represents fluid-mechanical processes and feeds mechanical power to the next stage. The secondary subsystem can incorporate direct drive or include short-term storage, so that power processing can be facilitated before the electrical machine is operated. The tertiary conversion utilizes electromechanical and electrical processes.

Recent reviews have identified more than 50 wave energy devices at various stages of development (Falcão, 2009; Khan and Bhuyan, 2009; US DOE, 2010). The dimensional scale constraints of wave devices have not been fully investigated in practice. The dimension of wave devices in the direction of wave propagation is generally limited to lengths below the scale of the dominant wavelengths that characterize the wave power density spectrum at a particular site. Utility-scale electricity generation from wave energy will require device arrays, rather than larger devices and, as with wind turbine generators, devices are likely to be chosen for specific site conditions.

Several methods have been proposed to classify wave energy systems (e.g., Falcão, 2009; Khan and Bhuyan, 2009; US DOE, 2010). The classification system proposed by Falcão (2009) (Figure 6.5) is based mainly upon the principle of operation. The first column is the genus, the second column is the location and the third column represents the mode of operation as outlined in the subsections below. A small number of prototype devices based upon novel uses of electro-polymers and bulging tubes fall outside of this classification scheme.

6.3.2.1 Oscillating water columns

Oscillating water columns (OWC) are wave energy converters that use wave motion to induce varying pressure levels between the air-filled chamber and the atmosphere (Falcão et al., 2000; Falcão, 2009). High-velocity air exhausts through an air turbine coupled to an electrical generator, which converts the kinetic energy into electricity (Figure 6.6, top left). When the wave recedes, the airflow reverses and fills the chamber, generating another pulse of energy (Figure 6.6, top right). The air turbine rotates in the same direction, regardless of the flow, through either its design or variable-pitch turbine blades. An OWC device can be a fixed structure located above the breaking waves (cliff-mounted or part of a breakwater), it can be bottom mounted near shore or it can be a floating system moored in deeper waters.

6.3.2.2 Oscillating-body systems

Oscillating-body (OB) wave energy conversion devices use the incident wave motion to induce oscillatory motions between two bodies; these motions are then used to drive the power take-off system (Falcão,

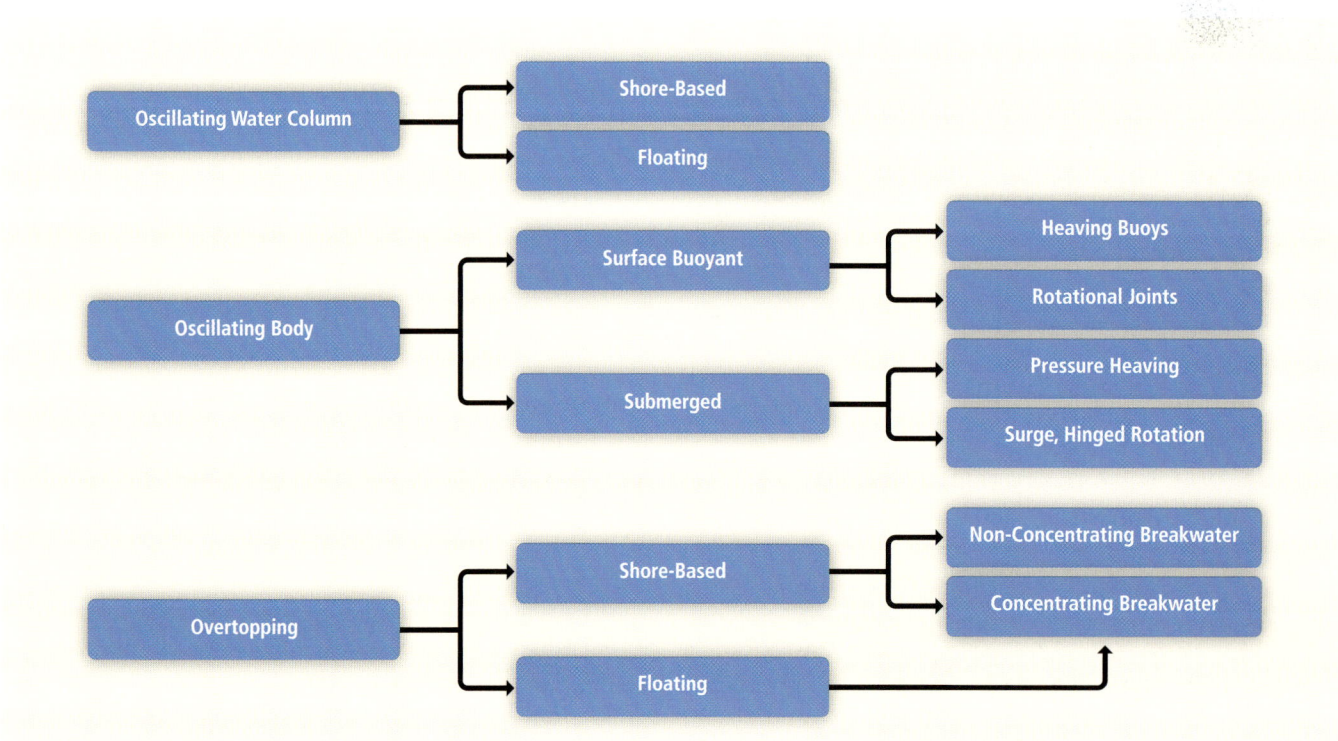

Figure 6.5 | Wave energy technologies: Classification based on principles of operation (Falcão, 2009).

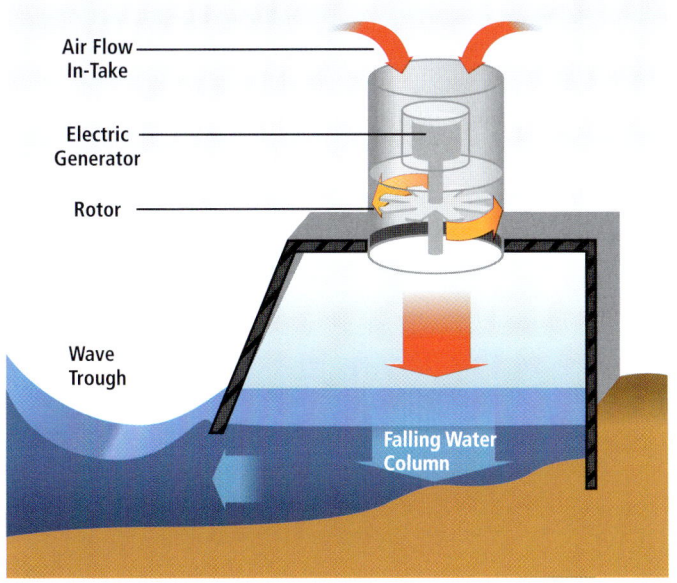
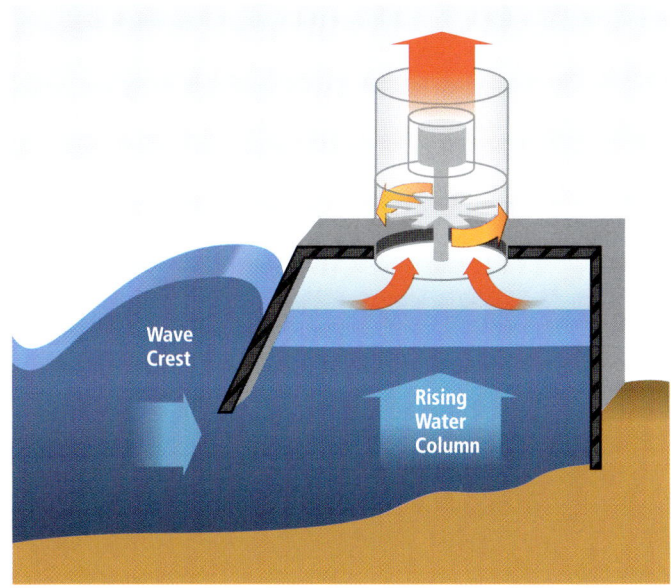

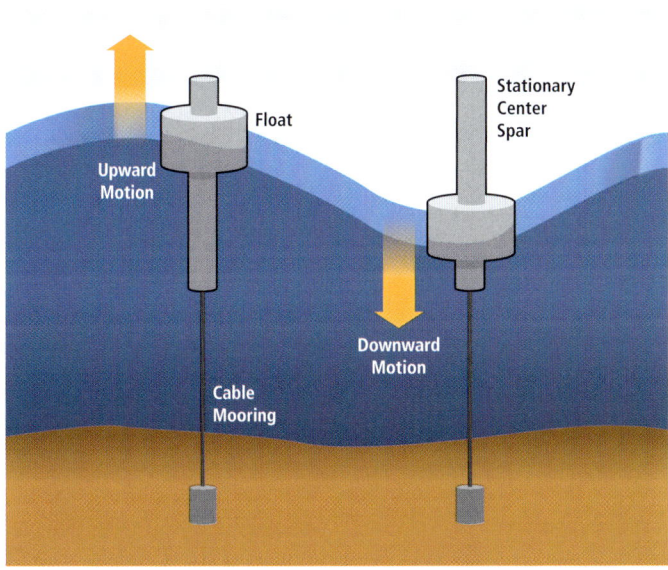
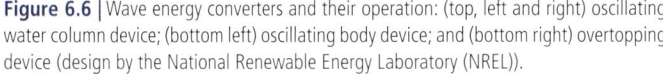

Figure 6.6 | Wave energy converters and their operation: (top, left and right) oscillating water column device; (bottom left) oscillating body device; and (bottom right) overtopping device (design by the National Renewable Energy Laboratory (NREL)).

2009). OBs can be surface devices or, more rarely, fully submerged. Commonly, axi-symmetric surface flotation devices (buoys) use buoyant forces to induce heaving motion relative to a secondary body that can be restrained by a fixed mooring (Figure 6.6, bottom left). Generically, these devices are referred to as 'point absorbers', because they are non-directional. Another variation of floating surface device uses angularly articulating (pitching) buoyant cylinders linked together. The waves induce alternating rotational motions of the joints that are resisted by the power take-off device. Some OB devices are fully submerged and rely on oscillating hydrodynamic pressure to extract the wave energy.

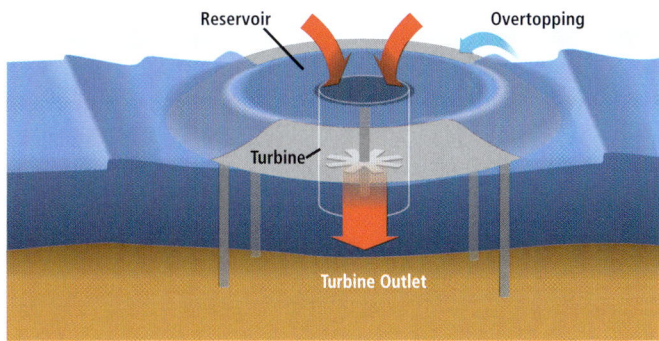

Lastly, there are hinged devices, which sit on the seabed relatively close to shore and harness the horizontal surge energy of incoming waves.

6.3.2.3 Overtopping devices

An overtopping device is a type of wave terminator that converts wave energy into potential energy by collecting surging waves into a water reservoir at a level above the free water surface (Falcão, 2009). The reservoir drains down through a conventional low-head hydraulic turbine. These systems can be offshore floating devices or incorporated into shorelines or man-made breakwaters (Figure 6.6, bottom right).

6.3.2.4 Power take-off systems

Power take-off systems are used to convert the kinetic energy, air flow or water flow generated by the wave energy device into a useful form, usually electricity. There are a large number of different options depending upon the technology adopted and these are fully described in Khan and Bhuyan (2009). Real-time wave oscillations will produce corresponding electrical power oscillations that may degrade power quality from a single device. In practice, some method of short-term energy storage (durations of seconds) may be needed to smooth energy delivery. The cumulative power generated by several devices will be smoother than from a single device, so device arrays are likely to be common. Most oscillating-body devices use resonance to derive optimal energy absorption, which requires that the geometry, mass or size of the structure must be linked to wave frequency. Maximum power can only be extracted by advanced control systems.

6.3.3 Tidal range

The development of tidal range hydropower has usually been based on estuarine developments, where a barrage encloses an estuary, which creates a single reservoir (basin) behind it and incorporates conventional low-head hydro turbines. Alternative barrage configurations have been proposed based on multiple-basin operations. Basins are filled and emptied at different times with turbines located between the basins. Multi-basin schemes may offer more flexible power generation availability over normal schemes, such that it is possible to generate power almost continuously.

The most recent advances focus on offshore basins (single or multiple) located away from estuaries, called 'tidal lagoons', which offer greater flexibility in terms of capacity and output with little or no impact on delicate estuarine environments.

This technology uses commercially available systems and the conversion mechanism most widely used to produce electricity from tidal range is the bulb-turbine (Bosc, 1997). The 240 MW power plant at La Rance in northern France has bulb turbines that can generate in both directions (on the ebb and flood tides) and also offer the possibility of pumping, when the tide is high, in order to increase storage in the basin at low head (Andre, 1976; De Laleu, 2009). The 254 MW Sihwa Barrage in the Republic of Korea, which is nearing completion, will employ ten 25.4 MW bulb turbines in a single flood tide mode (Paik, 2008).

Some favourable sites, such as very gradually sloping coastlines, are well suited to tidal range power plants, such as the Severn Estuary between southwest England and South Wales. Current feasibility studies there include options such as barrages and tidal lagoons. Conventional tidal range power stations will generate electricity for only part of each tide cycle. Consequently, the average capacity factor for tidal power stations has been estimated to vary from 25 to 35% (Charlier, 2003); ETSAP (2010b), meanwhile, reports a capacity factor range of 22.5 to 28.5%.

6.3.4 Tidal and ocean currents

Technologies to extract kinetic energy from tidal, river and ocean currents are under development, with tidal energy converters the most common to date. River current devices are covered in Chapter 5. The principal difference between tidal and river/ocean current turbines is that river and ocean currents flows are unidirectional, whilst tidal currents reverse flow direction between ebb and flood cycles. Consequently, tidal current turbines have been designed to generate in both directions.

Several classification schemes for tidal and ocean current energy systems have been proposed (Khan et al., 2009; US DOE, 2010). Usually they are classified based on the principle of operation, such as axial-flow turbines, cross-flow turbines and reciprocating devices (Bernitsas et al., 2006, see Figure 6.7). Some devices have multiple turbines on a single device (Figure 6.8, top left). Axial-flow turbines (Figure 6.8, top left) operate about a horizontal axis whilst cross-flow turbines may operate about a vertical axis (Figure 6.8, bottom left and right) or a horizontal axis with or without a shroud to accentuate the flow.

Many of the water current energy conversion systems resemble wind turbine generators. However, marine turbine designers must also take into account factors such as reversing flows, cavitation and harsh underwater marine conditions (e.g., salt water corrosion, debris, fouling, etc). Axial flow turbines must be able to respond to reversing flow directions, while cross-flow turbines continue to operate regardless of current flow direction. Axial-flow turbines will either reverse nacelle direction about 180º with each tide or, alternatively, the nacelle will have a fixed position but the rotor blades will accept flow from both directions. Rotor shrouds (also known as cowlings or ducts) enhance hydrodynamic performance by increasing the flow velocity through the rotor and reducing tip losses. To be economically beneficial, the additional energy capture must offset the cost of the shroud over the life of the device.

Reciprocating devices (not illustrated) are generally based on basic fluid flow phenomena such as vortex shedding or passive and active flutter systems (usually hydrofoils), and normal hydrofoils (e.g., tidal sails), which induce mechanical oscillations in a direction transverse to the water flow.

Most of these devices are in the conceptual stage of development, although two prototype oscillating devices have been trialled at open sea locations in the UK (Engineering Business, 2003; TSB, 2010).

The development of the tidal current resource will require multiple machines deployed in a similar fashion to a wind farm, thus the turbine siting is important especially in relation to wake effects (Peyrard et al., 2006).

Capturing the energy of open-ocean current systems is likely to require the same basic technology as for tidal flows but some of the infrastructure involved will differ. For deep-water applications, neutrally buoyant turbine/generator modules with mooring lines and anchor systems may replace fixed bottom support structures. Alternatively, they can be attached to other structures, such as offshore platforms (VanZwieten et al., 2005). These modules will also have hydrodynamic lifting designs to allow optimal and flexible vertical positioning (Venezia and Holt, 1995; Raye, 2001; VanZwieten et al., 2005). In addition, open ocean currents

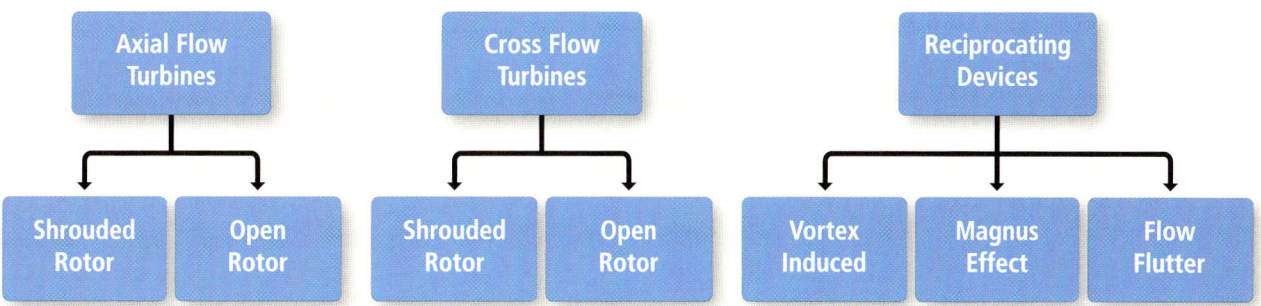

Figure 6.7 | Classification of current tidal and ocean energy technologies (principles of operation).

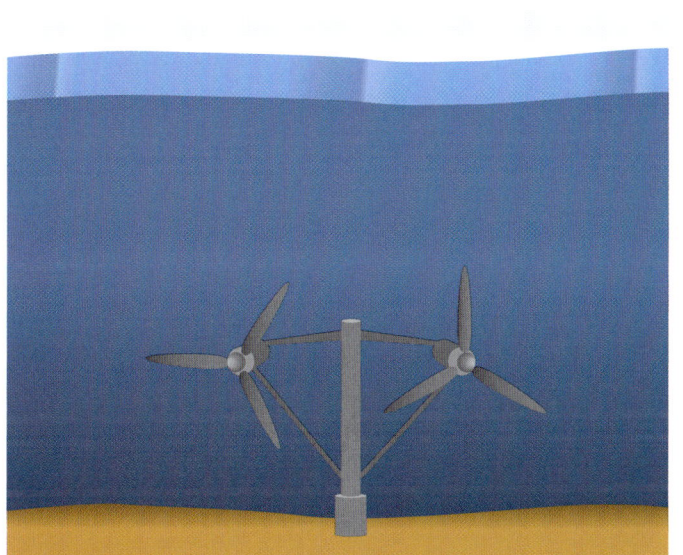

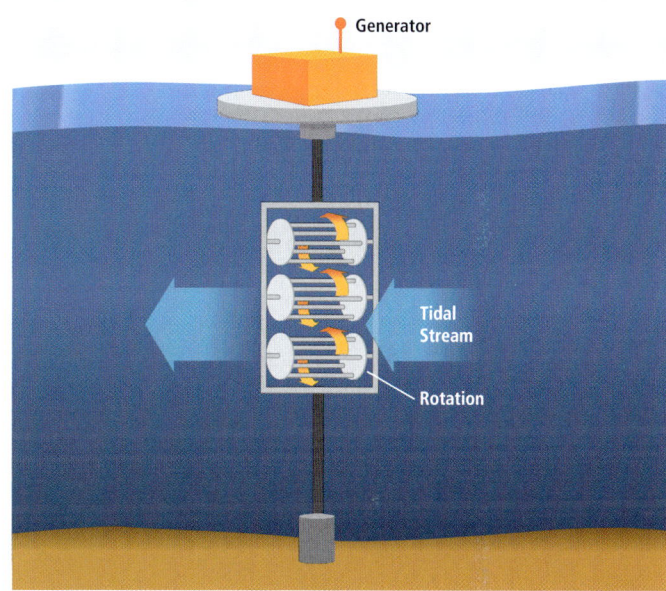

Figure 6.8 | Tidal current energy converters and their operation: twin turbine horizontal axis device (top left); cross-flow device (top right); and vertical axis device (bottom left) (design by NREL).

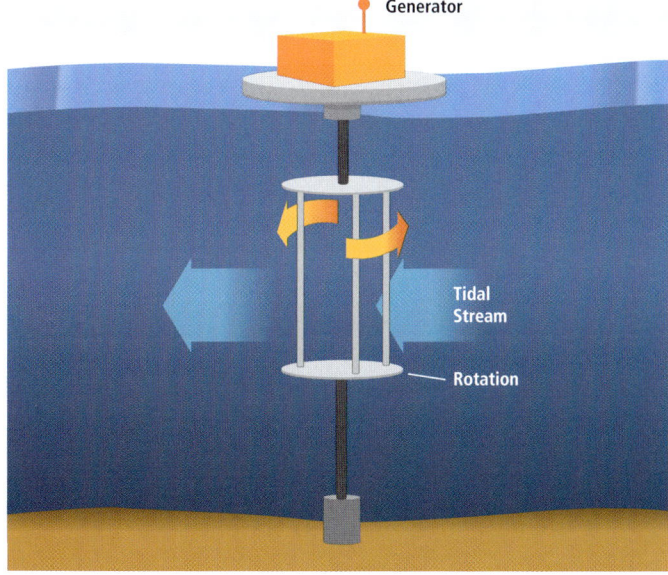

will not impose a size restriction on the rotors due to lack of channel constraints and therefore, ocean current systems may have larger rotors.

6.3.5 Ocean thermal energy conversion

Ocean thermal energy conversion (OTEC) plants have three conversion schemes: open, closed and hybrid (Charlier and Justus, 1993). In the open conversion cycle, about 0.5% of the warm surface seawater is flash-evaporated in a vacuum chamber. This steam is the cycle's working fluid, which passes through a power-generating turbine before being condensed by deep cold seawater. Desalinated water can be obtained as an additional product by employing an appropriate cycle.

Closed conversion cycles offer more efficient thermal performance, with warm seawater from the ocean surface being pumped through heat exchangers to vaporize a secondary working fluid (such as ammonia, propane or chlorofluorocarbon (CFC)) creating a high-pressure vapour to drive a turbine. The vapour is subsequently cooled by seawater to return it to a liquid phase. Closed-cycle turbines may be smaller than open-cycle turbines because the secondary working fluid operates at a higher pressure.

A hybrid conversion cycle combines both open and closed cycles, with steam generated by flash evaporation acting as the heat source for a closed Rankine cycle, using ammonia or another working fluid.

Although there have been trials of OTEC technologies, problems have been encountered with maintenance of vacuums, heat exchanger biofouling and corrosion issues. However, there are a large number of potential by-products, including hydrogen, lithium and other rare elements, which enhance the economic viability of this technology.

Ocean thermal energy can also be used for seawater air conditioning, thereby providing thermal energy services (Nihous, 2009).

6.3.6 Salinity gradients

The mixing of freshwater and seawater, such as where a river flows into a saline ocean, releases energy and causes a very small increase in local water temperature (Scråmestø et al., 2009). Reversed electro dialysis (RED) and pressure-retarded osmosis (PRO) are among the concepts identified for converting this heat into electricity. This form of energy conversion is often called osmotic power and the first 5 kW PRO pilot power plant was commissioned in Norway in 2009.

6.3.6.1 Reversed electro dialysis

The RED process harnesses the difference in chemical potential between two solutions. Concentrated salt solution and freshwater are brought into contact through an alternating series of anion and cation exchange membranes (AEM and CEM) (Figure 6.9). The chemical potential difference generates a voltage across each membrane; the overall potential of the system is the sum of the potential differences over the sum of the membranes. The first prototype to test this concept is being built in the Netherlands (van den Ende and Groeman, 2007).

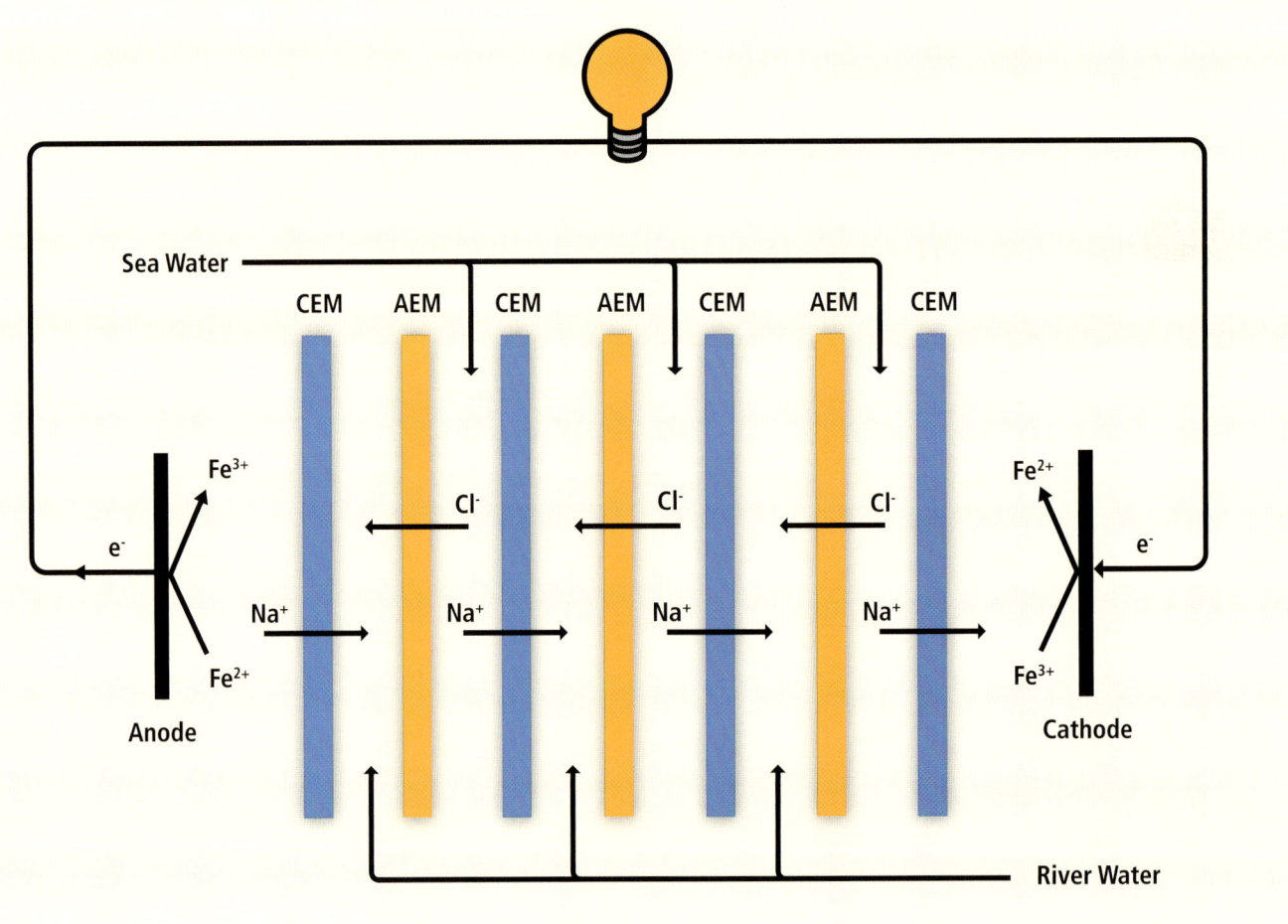

Figure 6.9 | Reversed electro dialysis (RED) system (van den Ende and Groeman, 2007).

Notes: CEM = cation exchange membrane; AEM = anion exchange membrane, Na = sodium, Cl = Chlorine, Fe = iron.

6.3.6.2 Pressure-retarded osmosis

Pressure-retarded osmosis (PRO), also known as osmotic power, is a process where the chemical potential is exploited as pressure (Figure 6.10) and was first proposed in the 1970s (Loeb and Norman, 1975).

The PRO process utilizes naturally occurring osmosis caused by the difference in salt concentration between two liquids (for example, seawater and freshwater). Seawater and freshwater have a strong tendency to mix and this will occur as long as the pressure difference between the liquids is less than the osmotic pressure difference. For seawater and freshwater the osmotic pressure difference will be in the range of 2.4 to 2.6 MPa (24 to 26 bar), depending on seawater salinity.

Before entering the PRO membrane modules, seawater is pressurized to approximately half the osmotic pressure, about 1.2 to 1.3 MPa (12 to 13 bar). In the membrane module, freshwater migrates through the membrane and into pressurized seawater. The resulting brackish water is then split into two streams (Skråmestø et al., 2009). One-third is used for power generation (corresponding to approximately the volume of freshwater passing through the membrane) in a hydropower turbine, whilst the remainder passes through a pressure exchanger in order to pressurize the incoming seawater. The brackish water can be fed back to the river or into the sea, where the two original sources would have eventually mixed.

6.4 Global and regional status of market and industry development

6.4.1 Introduction

Since the 1990s, R&D projects on wave and tidal current energy technologies have proliferated, with some now reaching the full-scale pre-commercial prototype stage. Presently, the only full-size and operational ocean energy technology available is the tidal barrage, of which the best example is the 240 MW La Rance Barrage in northwestern France, completed in 1966 (540 GWh/yr; De Laleu, 2009). The 254 MW Sihwa Barrage (South Korea) is due to become operational in 2011. Technologies to develop the other ocean energy sources—ocean thermal energy conversion (OTEC), salinity gradients and ocean currents—are still at the conceptual, R&D or early prototype stages. Currently, more than 100 different ocean energy technologies are under development in over 30 countries (Khan and Bhuyan, 2009).

6.4.1.1 Markets

Apart from tidal barrages, all ocean energy technologies are conceptual, undergoing R&D or in the pre-commercial prototype stage. Consequently, there is virtually no commercial market for ocean energy technologies at present.

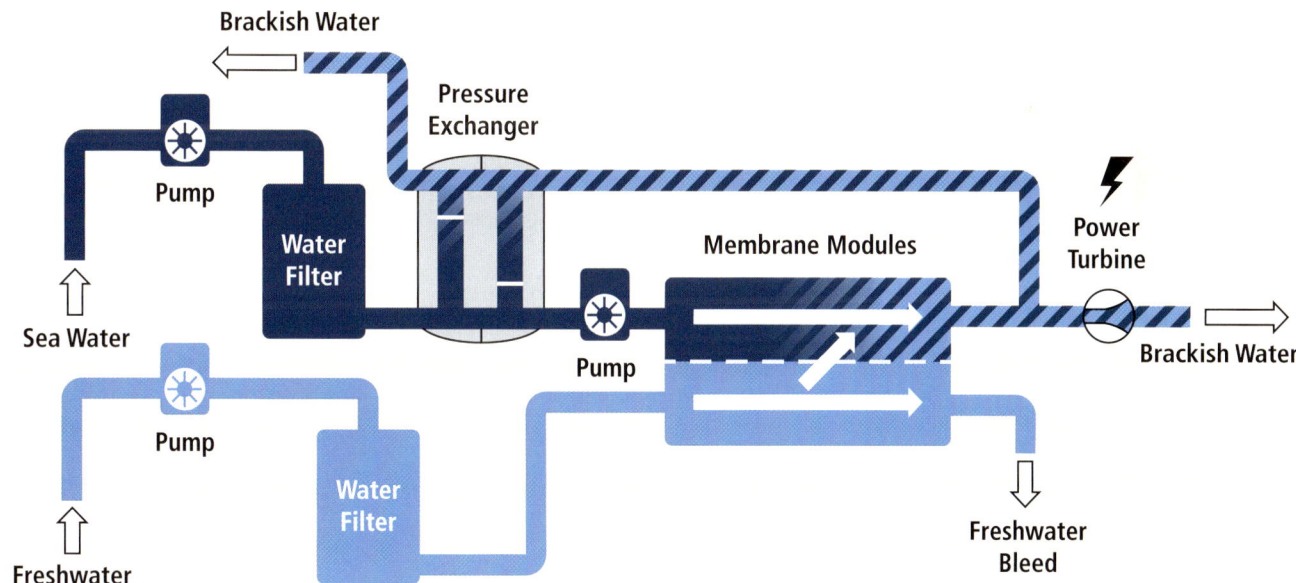

Figure 6.10 | Pressure-retarded osmosis (PRO) process (Skråmestø et al., 2009).

Some governments are using a range of initiatives and incentives to promote and accelerate the implementation of ocean energy technologies. These are described in Section 6.4.7. The north-eastern Atlantic coastal countries lead the development of the market for ocean energy technologies and their produced electricity. Funding mechanisms such as the Clean Development Mechanism (CDM) or Joint Implementation (JI) projects enable governments to secure additional external funding for ocean energy projects in developing nations. The Sihwa barrage project in the Republic of Korea was funded, in part, by CDM finance (UNFCCC, 2005).

Since there are ocean energy technologies being developed that produce pressurized or potable water as well as, or instead of, electricity, they may be able to compete in the market for water.

6.4.1.2 Industry development

As the marine energy industry moves from its present R&D phase, capacity and expertise from existing industries, such as electrical and marine engineering and offshore operations, will be drawn in, encouraging rapid growth of industry supply chains. The industry is presently underpinned by a large number of independent, entrepreneurial companies with limited investment from the finance sector. Large utility investment in device developments has become more commonplace in the last 10 years and some governments have recognized the skills and knowledge transfer benefits from other industries.

An unusual feature of ocean energy is the emergence of an international network of national marine energy testing centres, which includes the European Marine Energy Centre[1] (EMEC) in Scotland—the first of a growing number of testing centres worldwide—where device developers can reduce the costs of testing their prototypes by using existing infrastructure, particularly the offshore cable, power purchase agreements and permits. These centres are accelerating the development of a wide range of wave and tidal current technologies by effectively allowing device developers to share the costs of device prototype testing.

Industry development road maps and supply chain studies have been developed for Scotland, the UK and New Zealand (AWATEA, 2008; Mueller and Jeffrey, 2008; MEG, 2009). The USA (Thresher, 2010) and Ireland (SEAI, 2010) have completed road mapping exercises and Canada has begun road mapping exercises. Similar road maps have been produced for the EU countries (EOEA, 2010) and European marine energy science research (ESF MB, 2010). These countries have begun to assess the market potential for ocean energy as an industry or regional development initiative. Regions supporting industry cluster development, leading to scalable power developments, seek to attract concentrations of industry.

A series of global and regional initiatives now exist for collaborative development of ocean energy markets and industry. These are assisting in the development of international networks, information flow, removal of barriers and efforts to accelerate marine energy uptake. The presently active initiatives include the following:

- The International Energy Agency's Ocean Energy Systems Implementing Agreement.[2] This initiative has members from the developing countries who can see an opportunity for the transfer of knowledge to exploit their local ocean energy resources.

- The Equitable Testing and Evaluation of Marine Energy Extraction Devices (EquiMar). This EU-funded initiative intends to deliver a suite of protocols for the evaluation of wave and tidal stream energy converters.[3]

- The Wave Energy PLanning And Marketing (WavePLAM) project. This European industry initiative addresses non-technical barriers to wave energy.[4]

6.4.2 Wave energy

Wave energy technologies started to be developed after the first oil crisis in 1974. Many different converter types have been, and continue to be, proposed and tested but they are still at the pre-commercial phase. Recently, governments and developers have begun to use Technology Readiness Levels to guide their structured development of marine energy devices (Holmes and Nielsen, 2010). It is usual to test devices at a small scale in laboratory test-tank facilities (1:15 to 1:50 scale) before the first open-sea prototype testing (1:4 to 1:10 scale). Pre-commercial testing may be at half or full scale. Presently only a handful of devices have been built and tested at full scale. Pre-commercial trials of individual modules and small arrays began in recent years and are expected to accelerate through this decade. Given the early stage of development, the costs for wave energy are relatively high, but significant potential for cost reductions exist. Programmes such as the Marine Energy Accelerator programme (Callaghan, 2006) and incentives for pilot markets are intended to accelerate the cost reduction experience to seek to make wave energy technologies commercially competitive in the future.

A coast-attached oscillating water column device has been operational in Portugal since 1999 (Falcão et al., 2000; Aqua-RET, 2008) and a somewhat similar device (Voith Hydro Wavegen's LIMPET device)[5] has been operating almost continuously on the island of Islay in Scotland since 2000. Two offshore oscillating water column devices have been tested at prototype scale in Australia (Energetech/Oceanlinx)[6] since 2006

---

1 See www.emec.org.uk for Centre description.

2 See www.iea-oceans.org for description of activity.

3 See www.equimar.org for description of project outcomes.

4 See www.waveplam.eu for description of project outcomes.

5 See www.wavegen.co.uk for description of technology.

6 See www.oceanlinx.com for description of technology.

(Denniss, 2005) and Ireland (the OE Buoy)[7]. An oscillating water column device was operational off the southern coast of India between 1990 and 2005, when several experiments on the power modules were conducted and wave-powered desalination was demonstrated (Ravindran et al., 1997; Sharmila et al., 2004).

The most maturely developed oscillating-body device is the 750 kW Pelamis Wavepower[8] attenuator device, which has been tested in Scotland and deployed in Portugal. The Portuguese devices were sold as part of a commercial demonstration project. The other near-commercial oscillating-body technology is Ocean Power Technologies' PowerBuoy,[9] a small (40 to 250 kW) vertical axis device, which has been deployed in Hawaii, New Jersey and on the north Spanish coast. Other oscillating-body devices under development include the Irish device, Wavebob,[10] the WET-NZ device[11] and the Brazilian hyperbaric converter (Estefen et al., 2010).

Two Danish overtopping devices have been built at prototype scale and deployed at sea (Wave Dragon[12] and WavePlane[13]). Finally, two surge devices have been tested. Aquamarine Power[14] deployed its first full-scale 'Oyster' unit at EMEC in November 2009, whilst AW Energy (Finland) will deploy its Waveroller[15] surge device off the coast of Portugal.

6.4.3 Tidal range

Presently, only estuary-type tidal power stations are in operation. They rely on a barrage, equipped with generating units, closing the estuary. Though the technology itself is mature, the only utility-scale tidal power station in the world is the 240 MW La Rance power station, which has been in successful operation since 1966. Other smaller projects have been commissioned since then in China, Canada and Russia. The 254 MW Sihwa barrage is expected to be commissioned in 2011 and will then become the largest tidal power station in the world. The Sihwa power station is being retrofitted to an existing 12.7 km sea dyke that was built in 1994. The project will generate electricity whilst also improving flushing in the reservoir basin to improve water quality.

By the end of 2011, the world's installed capacity of tidal range power will still be less than 600 MW, assuming that the Sihwa power plant comes on line. However, numerous projects have been identified, some of them with very large capacities, including in the UK (Severn Estuary), India, Korea and Russia (the White Sea and Sea of Okhotsk). Total installed capacity under consideration is approximately 43.7 GW, or 64.05 TWh/yr (233 PJ/yr) (Kerr, 2007).

6.4.4 Tidal and ocean currents

There are probably more than 50 tidal current devices at the proof-of-concept or prototype development stage, but large-scale deployment costs are yet to be demonstrated. The most advanced example is the SeaGen[16] 1.2 MW capacity tidal turbine, which was installed in Strangford Lough in Northern Ireland and has delivered electricity into the electricity grid for more than one year. An Irish company, Open Hydro,[17] has tested its open-ring turbine at EMEC in Scotland, and more recently in Canada (Bay of Fundy). A number of devices have also been tested in China (Zhang and Sun, 2007).

Two companies have demonstrated horizontal axis turbines at full scale: Hammerfest Strom[18] in Norway and Atlantis Resources Corporation[19] in Scotland, whilst Ponte di Archimede[20] has demonstrated a vertical-axis turbine in the Straits of Messina (Italy). Finally, Pulse Tidal Limited[21] demonstrated a reciprocating device off the Humber Estuary in the UK in 2009.

The resource for tidal current energy is not widespread, with potentially economically viable sites located where tidal current velocities are accelerated around headlands or through channels between islands. Potential sites have been identified in Europe (particularly Scotland, Ireland, the UK and France), China, Korea, Canada, Japan, the Philippines, Australasia and South America. A number of development projects will begin during the present decade: experience and scale-up in these projects is expected to drive down costs.

Open ocean currents, such as the Gulf Stream, are being explored for development. Because they are slower moving and unidirectional, harnessing open ocean currents may require different technologies from those presently being developed for the faster, more restricted tidal stream currents (MMS, 2006). No pilot or demonstration plants have been deployed to date. Given the scale of open ocean currents, which involve much larger water volumes than tidal currents, there is a promise

---

7   See www.oceanenergy.ie/index.html for description of technology.

8   See www.pelamiswave.com for description of technology.

9   See www.oceanpowertechnologies.com for description of technology.

10 See www.wavebob.com for description of technology.

11 See www.wavenergy.co.nz for description of technology.

12 See www.wavedragon.net for description of technology.

13 See www.waveplane.com for description of technology.

14 See www.aquamarinepower.com for description of technology.

15 See www.aw-energy.com for description of technology.

16 See www.marinecurrentturbines.com for description of technology.

17 See www.openhydro.com/home.html for description of technology.

18 See www.hammerfeststrom.com for description of technology.

19 See www.atlantisresourcescorporation.com for description of technology.

20 See www.pontediarchimede.it/language_us for description of technology.

21 See www.pulsetidal.co.uk/our-technology.html for description of technology.

of significant project scale if technologies can be developed to harness the lower-velocity currents.

6.4.5 Ocean thermal energy conversion[22]

Presently only a small number of OTEC test facilities have been trialled globally. A small 'Mini-OTEC' prototype plant was tested in the USA in 1979. Built on a floating barge, the plant used an ammonia-based closed-cycle system with a 28,200 rpm radial inflow turbine. Although the prototype had a rated capacity of 53 kW, pump efficiency problems reduced its output to 18 kW. A second floating OTEC plant (OTEC-1) using the same closed-cycle system but without a turbine was built in 1980. Rated at 1 MW, it was primarily used for testing and demonstration, including studies of issues with the heat exchanger and water pipe, during its four months of operation in 1981.

In 1982 and 1983 in the Republic of Nauru, a 120-kW plant that used a Freon-based closed-cycle system and a cold water pipe to a depth of 580 m was operated for several months. It was connected to the electric grid and generated a peak of 31.5 kW of power.

An open-cycle OTEC plant was built in Hawaii in 1992 that operated between 1993 and 1998, with peak production of 103 kW and 0.4 l/s of desalinated water. Operational issues included seawater out-gassing in the vacuum chamber, problems with the vacuum pump, varying output from the turbogenerator and the connection to the electrical grid.

In 1984, India designed a 1 MW ammonia-based closed-cycle OTEC system. Construction began in 2000 but could not be completed due to difficulties in deployment of the long cold water pipe (Ravindran and Raju, 2002). A 10-day experiment was conducted on the same barge off Tuticorin in 2005, and desalination using ocean thermal gradients was demonstrated in shallow depths.

By the early 2000s, Japan had tested a number of OTEC power plants (Kobayashi et al., 2004). In 2006, the Institute of Ocean Energy at Saga University built a prototype 30-kW hybrid OTEC plant that uses a mixed water/ammonia working fluid and continues to generate electrical power.

Larger-scale OTEC developments could have significant markets in tropical maritime nations, including the Pacific Islands, Caribbean Islands, Central American and African nations, if the technology develops to the point of being a cost-effective energy supply option.

6.4.6 Salinity gradients

Salinity gradient power is still a concept under development (Scråmestø et al., 2009), with two research/demonstration projects under development, using two different technology concepts (Section 6.3.6). The parallel development of related technologies, such as desalination, is expected to benefit the development of osmotic power systems.

Research into osmotic power is being pursued in Norway, with a prototype becoming operational in 2009 (Statkraft, 2009) as part of a drive to deliver a commercial osmotic power plant. At the same time, the RED technology has been proposed for retrofitting to the 75-year-old Afsluitdijk dike in the Netherlands (Willemse, 2007).

6.4.7 Impact of Policies[23]

Presently the north-western European coastal countries lead development of ocean energy technologies, with the North and South American, north-western Pacific and Australasian countries also involved. Ocean energy technologies could offer emission-free electricity generation and potable water production, and a number of governments have introduced policy initiatives to promote and accelerate the uptake of marine energy. Chapter 11 gives more details of policies and initiatives that promote renewable energy technology uptake. Some of these policies and initiatives are applied to ocean energy and fall into six main categories:

1. Capacity or generation targets;
2. Capital grants and financial incentives, including prizes;
3. Market incentives;
4. Industry development;
5. Research and testing facilities and infrastructure; and
6. Permitting/space/resource allocation regimes, standards and protocols.

Generally, the countries that have ocean energy-specific policies in place are also the most advanced with respect to technology developments and deployments, and given the early state of the technology, government support for ocean energy is likely to be critical to the pace at which technologies and projects are developed.

There are a variety of targets both aspirational and legislated. Most ocean energy-specific targets relate to proposed installed capacity, complementing other general targets, such as for proportional increases in other RE generation. Some European countries, such as Portugal and Ireland, have preferred 'market pull' mechanisms, such as feed-in tariffs (i.e., additional payments for produced electricity from specific technologies), whilst the UK and the Scottish Government have utilized enhanced banded Renewable Obligations Certificates schemes, that is, tradable certificates awarded to generators of electricity using ocean energy technologies. The Scottish Government introduced the Saltire Prize in 2008, which is a prize for the first device developer to meet a cumulative electricity generation target of 100 GWh over a continuous two-year period.

---

22 The contents of Section 6.4.5 are primarily derived from Vega (1999) and Khan and Bhuyan (2009) except where stated.

23 Non-technology-specific policy issues are covered in Chapter 11 of this report.

Most countries offer R&D grants for RE technologies but some have ocean energy-specific grant programs. The UK has had the longest, largest and most comprehensive programs, though the US Federal Government has increased investment significantly since 2008. Capital grant programs for device deployments have been implemented by both the UK and New Zealand as 'supply push' mechanisms but both countries have a range of policy instruments in place (Table 6.2). Note that Table 6.2 shows only examples of ocean energy policies existing at the end of 2010.

6.5 Environmental and Social Impacts[24]

6.5.1 Lifecycle greenhouse gas emissions

Ocean energy does not directly emit CO_2 during operation; however, GHG emissions may arise from different aspects of the lifecycle of ocean energy systems, including raw material extraction, component manufacturing, construction, maintenance and decommissioning. A comprehensive review of lifecycle assessment (LCA) studies published

Table 6.2 | Examples of ocean energy-specific policies (modified from Huckerby and McComb, 2008).

Policy Instrument	Country	Example Description
Capacity or Generation Targets		
Aspirational Targets And Forecasts	UK Spain (Basque Government) Canada	3% of UK electricity from ocean energy by 2020 5 MW off Basque coast by 2020 Canada is developing a roadmap for 2050 (Ocean Renewable Energy Group)[1]
Legislated Targets (Total Energy Or Electricity)	Ireland Portugal	Specific targets for marine energy installations 500 MW by 2020 off Ireland 550 MW by 2020 off Portugal
Capital Grants and Financial Incentives		
R&D Programs/Grants	USA China	US Department of Energy Wind & WaterPower Program (capital grants for R&D and market acceleration) High Tech Research & Development Programme (#863)
Prototype Deployment Capital Grants	UK New Zealand China	Marine Renewables Proving Fund Marine Energy Deployment Fund Ocean Energy Major Projects
Project Deployment Capital Grants	UK	Marine Renewables Deployment Fund
Prizes	Scotland	Saltire Prize (GBP 10 million for first ocean energy device to deliver over 100 GWh of electricity over a continuous two-year period)
Market Incentives		
Feed-In Tariffs	Portugal Ireland/Germany	Guaranteed price (in $/kWh or equivalent) for ocean energy-generated electricity
Tradable certificates and Renewables Obligation	UK	Renewable Obligation Scheme - tradable certificates (in $/MWh or equivalent) for ocean energy-generated electricity
Industry Development		
Industry & Regional Development Grants	Scotland, UK and others	Cluster developments
Industry Association Support	Ireland New Zealand	Government financial support for establishment of industry associations
Research and Testing Facilities and Infrastructure		
National Marine Energy Centres	USA	Two centres established (Oregon/Washington for wave/tidal and Hawaii for OTEC/wave)
Marine Energy Testing Centres	Scotland, Canada and others	European Marine Energy Centre[2] and Fundy Ocean Research Centre for Energy, Canada[3]
Offshore Hubs	UK	Wave hub, connection infrastructure for devices
Permitting/Space/Resource Allocation Regimes, Standards And Protocols		
Standards/Protocols	International Electrotechnical Commission	Development of international standards for wave, tidal and ocean currents
Permitting Regimes	UK	Crown Estate competitive tender for Pentland Firth licences
Space/Resource Allocation Regimes	USA	Department of Interior permitting regime in US Outer Continental Shelf

Notes: 1. See www.oreg.ca for description of roadmap. 2. See www.emec.org.uk for description of Centre. 3. See www.fundyforce.ca for description of Centre.

---

24 A comprehensive assessment of social and environmental impacts of all RE sources covered in this report can be found in Chapter 9.

since 1980 suggests that lifecycle GHG emissions from wave and tidal energy systems are less than 23 g CO_2eq/kWh, with a median estimate of lifecycle GHG emissions of around 8 g CO_2eq/kWh for wave energy (Figure 6.11). (Note that the distributions shown in Figure 6.11 do not represent an assessment of likelihood; the figure simply reports the distribution of currently published literature estimates passing screens for quality and relevance. See Annex II for further description of the literature search methods and list of references.)

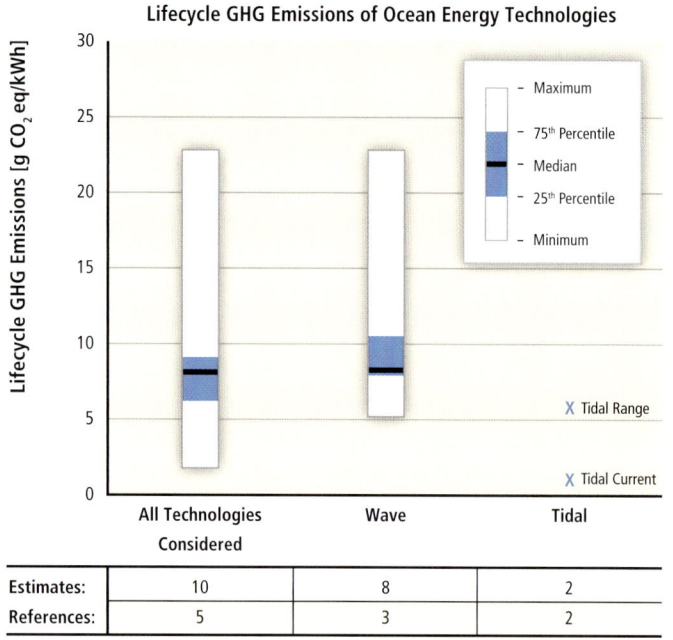

Figure 6.11 | Estimates of life-cycle GHG emissions of wave and tidal range technologies (unmodified literature values, after quality screen). See Annex II for details of literature search and citations of literature contributing to the estimates displayed.

Insufficient studies have been conducted on wave and tidal range devices to determine whether there are any significant differences between them regarding GHG emissions; studies of tidal and ocean current, ocean thermal energy conversion and salinity gradient devices that pass the quality screens are lacking. Further LCA studies to increase the number of estimates for all ocean energy technologies are needed. Regardless, in comparison to fossil energy generation technologies, the lifecycle GHG emissions from ocean energy devices appear low.

6.5.2 Other environmental and social impacts

Ocean energy projects may be long-lived, more than 25 years in general and over 100 years for tidal barrages (Sustainable Development Commission, 2007), so the long-term effects of their development need to be considered. While the transfer of experience from other offshore technologies (such as oil and gas operations and offshore wind energy) may be appropriate, the lack of experience in deploying and operating ocean energy technologies means that there is presently little information regarding their local environmental or social impacts.

In 2001, the British Government concluded that "the adverse environmental impact of wave and tidal energy devices is minimal and far less than that of nearly any other source of energy, but further research is required to establish the effect of real installations" (House of Commons, 2001). At the same time, some European and North American governments are undertaking strategic environmental assessments to plan for the potential environmental effects of ocean energy projects, which would typically include the effects of deployment scale, design, installation, operation and maintenance (O&M) and decommissioning on the physical and biological environment. Any type of large-scale ocean energy development is likely to require extensive social and environmental impact assessments to fully evaluate all development options. A description of potential environmental effects is given by Boehlert and Gill (2010).

Besides climate change mitigation, possible positive effects from ocean energy may include avoidance of adverse effects on marine life by virtue of reducing other human activities in the area around the ocean devices, and the strengthening of energy supply and regional economic growth, employment and tourism. As one example, it has been estimated that Scotland has the possibility to create between 630 and 2,350 jobs in ocean energy by 2020 (AEA Technology & Poyry Energy Consulting, 2006). In another example, ocean energy systems have become tourist attractions in their own right, providing jobs in tourism and services (e.g., La Rance tidal barrage: Lang, 2008; De Laleu, 2009).

Negative effects may include a reduction in visual amenity and loss of access to space for competing users, noise during construction, and other limited specific impacts on local ecosystems. Project-specific effects will vary, depending on the specific qualities of the project, the environment where the project will be located and the communities that live near it. Technology-specific strategies, such as mobile OTEC plants that limit concentrated environmental effects, are one approach to mitigating possible negative impacts. The specific environmental and social impacts of ocean energy technologies will depend in part on the technology in question and so the following sections describe the potential impacts for each energy source in turn.

6.5.2.1 Wave energy

The environmental impacts of wave energy technologies are difficult to assess due to the lack of deployment experience. The potential effects will vary by technology and location, but may include competition for space, noise and vibration, electromagnetic fields, disruption to biota and habitats, water quality changes and possible pollution. Pilot projects and pre-commercial deployments are likely to generate useful data on potential environmental effects and their mitigation.

The visual impacts of wave energy converters are likely to be negligible, since most devices are partially or completely submerged, except where large arrays of devices are located near shore. For the same reason, the potential effects on bird migration routes, feeding and nesting are expected to be negligible.

Deploying wave devices may have effects similar to other existing marine structures, although the extent of some effects may be smaller than for existing uses (see Boehlert et al., 2007). Noise and vibration are likely to be most disruptive during construction and decommissioning, while electromagnetic fields around devices and electrical connection/export cables that connect arrays to the shore may be problematic to sharks, skates and rays (elasmobranchii) that use electromagnetic fields to navigate and locate prey. Chemical leakage due to abrasion (of paints and anti-fouling chemicals) and leaks, for example, oil leaks from hydraulic power take-off systems are potential impacts. All of these effects will require R&D to understand, eliminate or mitigate. Energy capture and thus downstream effects could cause changes in sedimentation (e.g., seabed scouring or sediment accumulation) as well as wave height reductions. Wave energy farms could reduce swell conditions at adjacent beaches and modify wave dynamics along the shoreline. These aspects can be assessed through numerical and tank testing studies.

In addition to electricity generation with low lifecycle GHG emissions, the possible benefits of wave energy include industry stimulation for local shipyards (device construction and/or assembly), transportation, installation and maintenance. In addition, exclusion areas for wave farms may create wildlife refuges, which may be a net benefit to fishery resources (House of Commons, 2001).

6.5.2.2 Tidal range

Estuaries are complex, unique and dynamic natural environments that require very specific and careful attention. The impacts on the natural environment have to be addressed for both the construction phase and for future operations and decommissioning.

Construction impacts will differ depending on the construction techniques employed, with some long-term effects being positive for species diversity and abundance (Retiere and Kirby, 2006). At the La Rance power plant, although the estuary was closed for the construction period, biodiversity comparable to that of neighbouring estuaries was reportedly restored less than 10 years after commissioning (De Laleu, 2009). Other construction methods, such as floating caissons being submerged in place, may further reduce short- and longer-term impacts (Lang, 2008). The environmental impacts during construction of the Sihwa tidal power plant have been very limited, in large part because the barrage into which the plant has been inserted already existed.

Operation of a barrage will affect the amplitude and timing of the tides inside the basin, and modify fish and bird life and habitat, water salinity and sediment movements in the estuary (Bonnot-Courtois, 1993). Some of these impacts can be mitigated through adopting appropriate operational practices: for example, the La Rance barrage maintains two tides a day inside the basin, which has resulted in the restoration of a 'natural' biodiversity in the basin. However, sediments accumulating towards the upstream end of the basin require regular dredging.

Construction and operation of offshore tidal lagoons is less likely to have adverse impacts on delicate near-shore ecosystems; however, it will impact the area covered by the new lagoon.

With respect to social impacts, tidal range projects constructed to date have not required any relocation of nearby inhabitants, and this should continue to be so for future projects. Moreover, the construction phase will generate local employment opportunities and associated benefits for local communities. Following construction, barrages may provide new and shorter road transport routes along the top of the barrage walls, and this also may improve the socioeconomic conditions for local communities.

6.5.2.3 Tidal and ocean currents

Tidal currents
Tidal current technologies are likely to involve large submarine structures, although some devices have surface-piercing structures. Environmental effects may be somewhat limited because devices will be located in already energetic, moving water environments, which have low species diversity and abundances.

While current technologies have moving parts (rotating rotor blades or flapping hydrofoils) that may harm marine life, there is no evidence to date of harm from tidal current devices to marine life, such as whales, dolphins, seals and sharks. This may be due in part to the limited number and duration of device deployments, but it may also be due to slow rotation speeds (relative to escape velocities of the marine fauna) compared with ship propulsion.

Ocean currents
Possible impacts from full-scale commercial deployments of ocean current energy systems can be grouped into four broad categories: the physical environment (the ocean itself); benthic (ocean-bottom) communities; marine life in the water column; and competing uses for marine space (Charlier and Justus, 1993; Van Walsum, 2003).

Physical effects on the ocean are expected to be limited: ocean current energy devices will not be of sufficient scale to alter ocean circulation or net mass transport. For example, the equatorward drift in wind-driven circulation, for which western boundary currents are the poleward return flow, is independent of the basin's dissipative mechanisms (e.g., Stommel, 1966). Systems could, however, alter meander patterns and upper-ocean mixing processes. These effects need to be fully evaluated prior to full site development. Modelling studies of the Florida Current are underway to assess these potential impacts (Chassignet et al., 2007).

Open-ocean energy generation systems are likely to operate below the draught of even the largest surface vessels, so hazards to commercial navigation will be minimal. Submarine naval operations could be impacted, although the stationary nature of the systems will make avoidance relatively simple. Underwater structures may affect fish habitats and behaviour. Because underwater structures are known to become fish aggregating devices (Relini et al., 2000), possible user conflicts, including line entanglement issues, must be considered. Associated alterations to pelagic habitats, particularly for large-scale installations, may become issues as well (Battin, 2004).

6.5.2.4 Ocean thermal energy conversion

Potential changes in the regional properties of seawater due to OTEC pumping operations may be an environmental concern. Large volumes of cold deep water and warm shallow water will be pumped to the heat exchangers and mixed. Mixing will modify the temperature and nutrient characteristics of the waters before discharge into ambient ocean water near the site. For this reason, shipboard (or 'grazing') OTEC projects have been proposed so that the large volumes of discharged water do not have a long-term impact on the discharge site (Nihous and Vega, 1993). Discharging the water at depth may minimize the environmental effects, but no robust evidence is currently available (Marti, 2008).

Under normal operating conditions, OTEC power plants will release few emissions to the atmosphere and will not adversely affect local air quality. Plankton (and perhaps food web) growth could occur as nutrient-rich deepwater effluents are released; this might only occur if sufficient light is also available at the stabilized plume depth (generally deeper than the discharge depth). Marine organisms, mainly plankton will be attracted by marine nutrients in the OTEC plant's discharge pipe, which can cause biofouling and corrosion (Panchal, 2008).

6.5.2.5 Salinity gradients

The mixing of seawater and freshwater is a natural process in estuarine environments (van den Ende and Groeman, 2007), and salinity gradient power plants would replicate this process by mixing freshwater and seawater before returning the brackish water to the ocean. Though normal brackish water is the main waste product, its concentrated discharge may alter the environment and have impacts on animals and plants living in the location.

Major cities and industrial areas are often sited at the mouths of major rivers, so power plants could be constructed on 'brown-field' sites. The plants could also be constructed partly or completely underground to reduce the visual impact on the local environment.

6.6 Prospects for technology improvement, innovation and integration[25]

As emerging technologies, ocean energy devices have the potential for significant technological advances. Not only will device-specific R&D and deployment be important to achieving these advances, but technology improvements and innovations in ocean energy converters are also likely to be influenced by developments in related fields. Rapidly growing deployments of offshore wind power plants, for example, may lead to the possibility of wave or tidal current projects being combined with them to share infrastructure (Stoutenburg et al., 2010). Similarly some breakwater-attached wave energy converters may benefit from synergies with new construction used for other purposes such as the Mutriku plant, Portugal (Torre-Enciso et al., 2009) and in China (Liu et al., 2009).

Integration of ocean energy into wider energy networks will need to recognize the widely varying generation characteristics arising from the different resources. For example, electricity generation from tidal stream resources shows very high variability over one to four hours, yet extremely limited variability over monthly or longer time horizons (Sinden, 2007). By comparison, hour-to-hour variability of wave energy tends to be lower than that of wind power, and many times lower than that of tidal stream power, while retaining significant seasonal and interannual variability (Sinden, 2007). These patterns of resource availability have implications for the large-scale integration of ocean energy into electricity networks (see Chapter 8), and on the requirements for, and utilization of, transmission capacity.

6.6.1 Wave energy

Wave energy technologies are still largely at an early stage of development and all are pre-commercial (Falcão, 2009). Any cost or reliability projections have a high level of uncertainty, because they require assumptions to be made about optimized systems that have not yet been proven at or beyond the prototype level. 'Time in the water' is critical for prototype wave devices, so developers can gain enough operating experience. Demonstrated survivability in extreme conditions will be required to advance technology developments. As has happened with wind turbine generators, wave energy devices are expected to evolve to the scale of the largest practical machine. This will minimize the number of aggregate O&M service visits, reduce installation and decommissioning costs and limit mooring requirements.

Cost reductions may in part arise from maximizing power production by individual wave energy converters, even if deployed in arrays, and from manufacturing and installation experience. This will likely require

---

25 Section 10.5 offers a complementary perspective on drivers of and trends in technological progress across RE technologies. Chapter 8 deals with other integration issues more widely.

efficient capture devices and dependable, efficient conversion systems, together with dedicated manufacturing and installation infrastructure.

6.6.2 Tidal range

Tidal range power projects rely on proven hydropower technologies built and operated in an estuarine environment. There are basically three areas where technology improvements can still be achieved: development of offshore tidal lagoons may allow the implementation of cost-effective projects (Friends of the Earth, 2004); multiple tidal basins may increase the value of projects by reducing the variability and even allowing base-load or dispatchable electricity (Baker, 1991); and turbine efficiency improvements (e.g., Nicholls-Lee et al., 2008), particularly in bi-directional flows (including pumping), may reduce overall costs of electricity delivery.

Technologies may be further improved, for instance, with gears allowing different rotation speeds for the turbine and the generator or with variable frequency generation, allowing better outputs. Power plants may be built onsite within cofferdams or be pre-fabricated in caissons (steel or reinforced concrete) and floated to the site.

6.6.3 Tidal and ocean currents

Like wave energy converters, tidal and ocean current technologies are at an early stage of development. Extensive operational experience with horizontal-axis wind turbines may give axial flow water current turbines a developmental advantage, since the operating principles are similar. Future water current designs are likely to increase swept area (i.e., rotor diameter) to the largest practical machine size to increase generation capacity, minimize the number of aggregate O&M service visits, reduce installation and decommissioning costs and minimize substructure requirements. A key area for R&D is likely to be in the development of deployment and recovery equipment, since periods of slack water in tidal channels can be very brief. The same applies to O&M requirements.

The total tidal and ocean current energy resource could be increased, if commercial threshold current velocities can be reduced. Tidal energy device optimization will follow a path of increasingly large turbines in lower flow regimes (BWEA, 2005). A similar trend is well documented in the wind energy industry in the USA, where wind turbine technology developments targeted less energetic sites, creating a 20-fold increase in the available resource (Wiser and Bolinger, 2010).

As with wave energy, performance and reliability will be top priorities for future tidal and ocean current energy arrays, as commercialization and economic viability will depend on systems that need minimal servicing, thus producing power reliably without costly maintenance. New materials that resist degradation caused by corrosion, cavitation, water absorption and debris impact could reduce operational costs.

6.6.4 Ocean thermal energy conversion

OTEC is also at an early stage of development. The heat exchanger system is one of the key components of closed-cycle ocean thermal energy conversion power plants. Evaporator and condenser units must efficiently convert the working fluid from liquid to gaseous phase and back to liquid phase with low temperature differentials. Thermal conversion efficiency is highly dependent on heat exchangers, which can cause substantial losses in terms of power production and reduce economic viability of systems (Panchal, 2008). Evaporator and condenser units represent 20 to 40% of the total plant cost, so most research efforts are directed towards improving heat exchanger performance. A second key component of an OTEC plant is the large diameter pipe, which carries deep, cold water to the surface (Miller, 2010). Experience obtained in the last decade with large-diameter risers for offshore oil and gas production can be transferred to the cold water pipe design.

A number of options are available for the closed-cycle working fluid, which has to boil at the low temperature of ocean surface water and condense at the lower temperature of deep sea water. Three major candidates are ammonia, propane and a commercial refrigerant R-12/31.

6.6.5 Salinity gradients

The first osmotic power prototype plant became operational in October 2009 at Tofte, near Oslo in south-eastern Norway. The location has sufficient access to seawater and freshwater from a nearby lake (Scråmestø et al., 2009).

The main objective of the prototype is to confirm that the designed system can produce power reliably 24 hours per day. The plant will be used for further testing of technology developed to increase the efficiency. These activities will focus on membrane modules, pressure exchanger equipment and power generation (i.e., the turbine and generator). Further development of control systems, water pretreatment equipment and the water inlets and outlets is needed (Scråmestø et al., 2009).

The developers of the Dutch RED system have identified the Afsluitdijk causeway in the Netherlands, which separates the salty North Sea from the less brackish Lake Ijsselmeer, as the potential site for a 200 MW power plant (Ecofys, 2007). Further R&D will focus on material selection for effectiveness of the membranes and the purification of the water flows.

6.7 Cost trends[26]

6.7.1 Introduction

Commercial markets are not yet driving marine energy technology development. Government-supported R&D and national policy incentives are the key motivation for most technology development and deployment (IEA, 2009). The cost of most ocean energy technologies is difficult to assess, because very little fabrication and deployment experience is available for validation of cost assumptions. Table 6.3 shows the best available data for some of the primary cost factors that affect the levelized cost of electricity (LCOE)[27] delivered by each of the ocean energy subtypes.

In most cases these cost and performance parameters are based on sparse information due to the lack of peer-reviewed reference data and actual operating experience, and in many cases therefore reflect estimated cost and performance assumptions based on engineering knowledge. Present-day investment costs were found in a few instances but are based on a small sample of projects and studies, which may not be representative of the entire industry. However, these parameter sets can be used to assess the overall validity of the levelized cost values published in the non-peer-reviewed literature and—to some extent—the validity and likelihood of the underlying assumptions. This is done by recalculating the LCOE based on a standard methodology outlined in Annex II and the above input data for 3, 7 and 10% discount rates and then comparing the results to previously published data. Focusing on the three ocean energy technologies for which full parameter sets are shown in Table 6.3, Figure 6.12 presents the resulting LCOE values.

Callaghan (2006) calculates LCOEs in the range of US cents$_{2005}$ 21 to 79/kWh for wave energy, which are broadly in line with the values based on the data set in Table 6.3 and shown in Figure 6.12. The EPRI study (Previsic, 2004), assessing one particular project design, is more optimistic. Besides, Callaghan (2006) calculates the LCOE for tidal current technology in the range of US cents$_{2005}$ 16 to 32/kWh. Similar LCOE values for tidal current of US cents$_{2005}$ 1 to 3/kWh are also obtained by the California Energy Commission (2010), but based on investment costs of approximately USD$_{2005}$ 2,000 to 3,000/kW that are envisaged for the year 2018, which are much lower than those estimated by Callaghan (2006) and ETSAP (2010b) for current conditions (see Table 6.3). A consistent set of input data and resulting LCOE are contained in ETSAP (2010b). The medium LCOE values that it found for wave energy, tidal range and tidal current projects are US cents$_{2005}$ 36, 24 and 31/kWh, respectively, for a 10% discount rate. The ETSAP (2010b) values for both wave and tidal current technology are at the low end of the range determined on the basis of the data in Table 6.3 for the 10% discount rate. The calculated LCOE values for tidal range shown in Figure 6.12 are based exclusively on the input data from ETSAP (2010b) and are in line with those reported by ETSAP.

The LCOE presented in Sections 1.3.2 and 10.5 and included in Annex III only include tidal range systems as this was the only ocean technology that had reached commercial maturity.

Future cost estimates come with an even larger degree of uncertainty and should be considered highly speculative. One of the methods, however, that can be used to derive possible future cost is based on the concept of learning. The accumulation of experience from increased deployment of new technologies usually leads to cost reductions. Empirical studies have quantified the link between cumulative deployment and cost reductions yielding so-called learning rates.[28] Applying such learning rates that have been found for technologies broadly similar to ocean energy allows estimation of future cost under certain deployment scenarios. Several estimates of the future costs of ocean energy technologies have been published. The underlying deployment scenarios and detailed cost assumptions, however, remain largely unclear. The following subsections assess some of the published future cost estimates by examining the conditions under which those cost levels can be achieved.

6.7.2 Wave and tidal current energy

Some studies have estimated costs for wave and tidal current energy devices by extrapolating from available prototype cost data (Binnie Black & Veatch, 2001; Previsic, 2004; Callaghan, 2006; Li and Florig, 2006).

Wave and tidal current devices are at approximately the same early stage of development. Investment costs could potentially decline with experience to costs achieved by other RE technologies such as wind energy (Bedard et al., 2006). This can only be demonstrated by extrapolation from a few limited data, since there is limited actual operating experience. Present investment cost estimates were derived from single prototypes, whose costs are likely to be higher than more mature future commercial versions. Some O&M cost data appears in Table 6.3, for both wave and tidal current energy, but it should be acknowledged that this data was extrapolated from a limited amount of operating data.

---

26 Discussion of costs in this section is largely limited to the perspective of private investors. Chapters 1 and 8 to 11 offer complementary perspectives on cost issues covering, for example, costs of integration, external costs and benefits, economy-wide costs and costs of policies.

27 LCOE is a widely used measure that allows for a comparison of the cost of alternative ways of generating electricity. The concept of levelized costs and the methodology used to calculate them is explained in Annex II of the report. However, even from the perspective of a private investor the LCOE is not the sole determinant of the value of a particular project. Risks associated with a particular project and the timing of electricity generation, for instance, are further relevant factors, to name just a few.

28 An overview of the theory and empiricism of learning can, for instance, be found in Section 10.5. Several technology chapters also provide information on technology-specific assessments of learning effects.

Table 6.3 | Summary of core available cost and performance parameters for all ocean energy technology subtypes.

Ocean Energy Technology	Investment costs (USD$_{2005}$/kW) [i]	Annual O&M Costs (USD$_{2005}$/kW)	Capacity Factor (CF)[ii] (%)	Design Life[iii] (years)
Wave	6,200–16,100[iv,v,vi]	180[v,vi]	25–40[v,vi]	20
Tidal Range	4,500–5,000[vi]	100[vi]	22.5–28.5[vi]	40[vii]
Tidal Current	5,400–14,300[iv,vi]	140[vi]	26–40[vi]	20
Ocean Current	N/A	N/A	N/A	20
Ocean Thermal	4,200–12,300[viii]	N/A	N/A	20
Salinity Gradient	N/A	N/A	N/A	20

Notes and References:

i. Cost figures for ocean thermal technologies are in different year-dollars.

ii. Capacity factors are estimated based on technology and resource characteristics, not on actual in-the-field hardware experience.

iii. Design life estimates are based on expert knowledge. A standard assumption is to set the design lifetime of an ocean energy device to 20 years.

iv. Callaghan (2006). Higher ranges of investment cost based on this source.

v. Previsic (2004) published a assessment of future cost based on 213 x 500 kW Pelamis wave energy converters with investment cost of USD$_{2005}$ 2,620/kW, annual O&M cost of USD$_{2005}$ 123/kW and additional retrofit cost after 10 years of USD$_{2005}$ 264/kW. Assumed CF was 38%; the design lifetime 20 years.

vi. ETSAP (2010b). Lower ranges of investment cost for wave, tidal range and tidal current are all based on this source. Note that ETSAP (2010a) estimated that investment cost could be as low as USD$_{2005}$ 5,200/kW for wave and as low as USD$_{2005}$ 4,500/kW for tidal current technology. Later in the same year, however, ETSAP (2010b) adjusted its estimates for both wave and tidal stream technologies up significantly to the lower bounds stated in the table, while the estimated investment cost for tidal barrages remained stable. With respect to CFs, the more recent source (ETSAP, 2010b) is more optimistic. The ranges stated in the table are based on both references.

vii. Tidal barrages resemble hydropower plants, which in general have very long design lives. There are many examples of hydropower plants that have been in operation for more than 100 years, with regular upgrading of electromechanical systems but no major upgrades of the most expensive civil structures (dams, tunnels etc). Tidal barrages are therefore assumed to have a similar economic design lifetime as large hydropower plants that can safely be set to at least 40 years (see Chapter 5).

viii. Cost estimates for ocean thermal technologies are in different-year dollars and cover a range of different technologies and locations. Most are for plants of 100 MW size. Many are highly speculative (see, e.g., Francis, 1985; SERI, 1989; Vega, 2002; Lennard, 2004; Cohen, 2009). The most current costs available for OTEC come from Lockheed-Martin, which estimates investment costs at USD 32,500/kW for a 10 MW pilot plant, which shrink to an estimated USD 10,000/kW for a commercial 100 MW plant (Cooper et al., 2009).

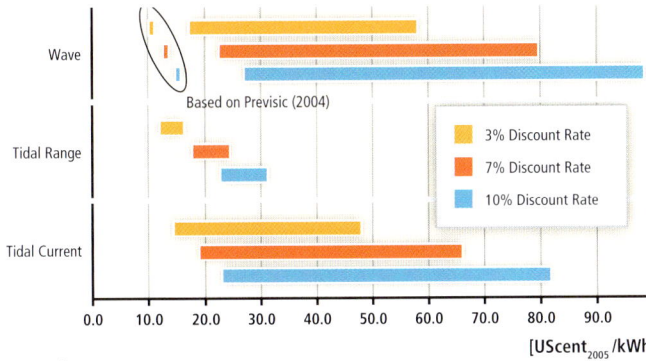

Figure 6.12 | LCOE of wave energy, tidal range and tidal current technology based on primary cost and performance parameters drawn from various studies and listed in Table 6.3.

One of the few studies that provides analysis on future costs was commissioned by the Electric Power Research Institute (EPRI) in the United States to examine theoretical commercial-scale project costs, using Pelamis wave energy converters off the California coast (Previsic, 2004). Overall plant size was assumed to be 213 x 500 kW devices (106.5 MW). The LCOE was calculated based on a 20-year design life and 95% availability. Energy capture technical potential was assumed to take advantage of near-term R&D improvement opportunities not yet realized but which were thought to be achievable at current assumed investment costs. The study concluded that an LCOE of US cents$_{2005}$ 13.4/kWh could be achieved, based upon an investment cost of USD$_{2005}$ 279 million (USD$_{2005}$ 2,620/kW), a discount rate of 7.5%, a capacity factor of 38% and annual O&M costs of USD$_{2005}$ 13.1 million (USD$_{2005}$ 123/kW/yr), with an assumed retrofit cost of USD$_{2005}$ 28.1 million (USD$_{2005}$ 264/kW) after 10 years.

In 2006 the UK Carbon Trust (Callaghan, 2006) published the results of a survey of current costs for prototype and pre-commercial wave and tidal energy converters from which much of the investment cost data was derived. Wave energy converters had investment costs ranging from USD$_{2005}$ 7,700 to 16,100/kW with a midpoint of USD$_{2005}$ 11,875/kW. Similarly, prototype tidal current energy generator costs ranged from USD$_{2005}$ 8,600 to 14,300/kW with a midpoint of USD$_{2005}$ 11,400/kW. Some tidal current device concepts may have even greater investment costs. The same study estimated that energy from early UK wave energy farms would have LCOEs of between US cents$_{2005}$ 21 and 79/kWh, whilst early tidal current farms had estimated LCOEs of between US cents$_{2005}$ 16 and 32/kWh. The Carbon Trust studies did not account for economies of scale, R&D improvements or learning curve effects (Callaghan, 2006).

A recent study undertaken for the California RE Transmission Initiative showed that tidal current generation (deployed in California) would cost US cents$_{2005}$ 1 to 3/kWh (Klein, 2009).

The theoretical analyses for wave energy devices appear to provide plausible benchmarks to demonstrate that near-term wave energy projects might have LCOEs comparable to wind energy in the 1980s. It is less clear how the LCOE levels published by the Callaghan (2006) and Klein (2009) could be achieved, unless the costs were lower or the performance parameters were significantly better than the ranges published. The greatest uncertainties in estimating the LCOE for ocean energy are in the long-term estimation of capacity factor and O&M costs, which require operational data to determine. To achieve economically competitive LCOE estimates, capacity factors near 40%, excellent availability (near 95%) and high efficiency commensurate with mature technology must be assumed for wave energy converters (Previsic, 2004; Buckley, 2005).

Learning curve effects could be an important downward cost driver for LCOE but have a high degree of uncertainty due to lack of industry experience from which to extrapolate. As deployments multiply, costs could be reduced due to learning that is derived from natural production efficiency gains, assimilated experience, economies of scale and R&D innovations. Learning rates for wind power plants over a three-decade span from the early 1980s to 2008 have been estimated at 11%, without including an R&D factor (Wiser and Bolinger, 2009). As a first-order estimate, ocean energy industries (except tidal range, which is already comparatively mature) could follow the same 11% learning curve.[29] Beginning with the midpoints for the investment costs given by Callaghan (2006), such a learning rate implies a decline in investments

costs of nearly three times corresponding to approximately nine capacity doublings from 2010 capacity levels (Figure 6.13).

Investment costs for wave and tidal current energy technologies under this scenario reduce to a range from USD$_{2005}$ 2,600 to 5,400/kW (average: USD$_{2005}$ 4,000/kW), assuming worldwide deployments of 2 to 5 GW by 2020. Note that this level of deployment is likely to be highly dependent on sustained policies of the UK, the USA, Canada and other ocean technology countries.

Figure 6.14 shows projections of the LCOE for wave and tidal current energy in 2020 as a function of capacity factor and investment costs, using the methods summarized in Annex II, and with other assumptions as used earlier in calculating LCOE values.

Figure 6.14 shows the possible impact of the capacity factor on LCOE but is included for illustrative purposes only. These results are based on only a single reference (Callaghan, 2006) and the previous learning curve analysis applied to estimate possible 2020 costs given a deployment rate of 2 to 5 GW. The three curves correspond to the calculated high, middle and low investment cost curves, that is, USD$_{2005}$ 5,600, 4,000 and 2,600/kW, estimated for the year 2020.

Figure 6.14 further shows that, if wave and tidal current devices can be developed to operate with capacity factors in the range of 30 to 40% at the above level of investment cost (USD$_{2005}$ 2,600 to 5,600/kW), they can potentially generate electricity at rates comparable with some of the other renewable technologies. Devices must be reliable and located in a high-quality wave or tidal current resource to achieve such capacity factors. Realization of the necessary investment cost levels may require cost reductions that could potentially be derived from manufacturing economies, new technology designs, knowledge and experience transfer from other industries and design modifications realized through operation and experience.

Although no definitive cost studies are available in the public domain for ocean current technologies, the cost and economics for open-ocean current technologies may have attributes similar to tidal current technologies.

6.7.3 Tidal range

Tidal barrages are considered the most mature of the ocean energy technologies reviewed in this report, since there are a number of examples of sustained plant operation, although very little data on cost was available. Tidal barrage projects usually require a very high capital investment, with relatively long construction periods. Civil construction in the marine environment—with additional infrastructure to protect against the harsh sea conditions—is complex and expensive. Consequently, investment costs associated with tidal range technologies are high when

---

29 The 11% learning rate is based on wind energy market analysis and is only used in making preliminary projections of ocean energy's future cost potential. Actual learning rates are not yet known. Theoretical and empirical literature on learning as a driver of cost reductions is presented in Section 10.5.2

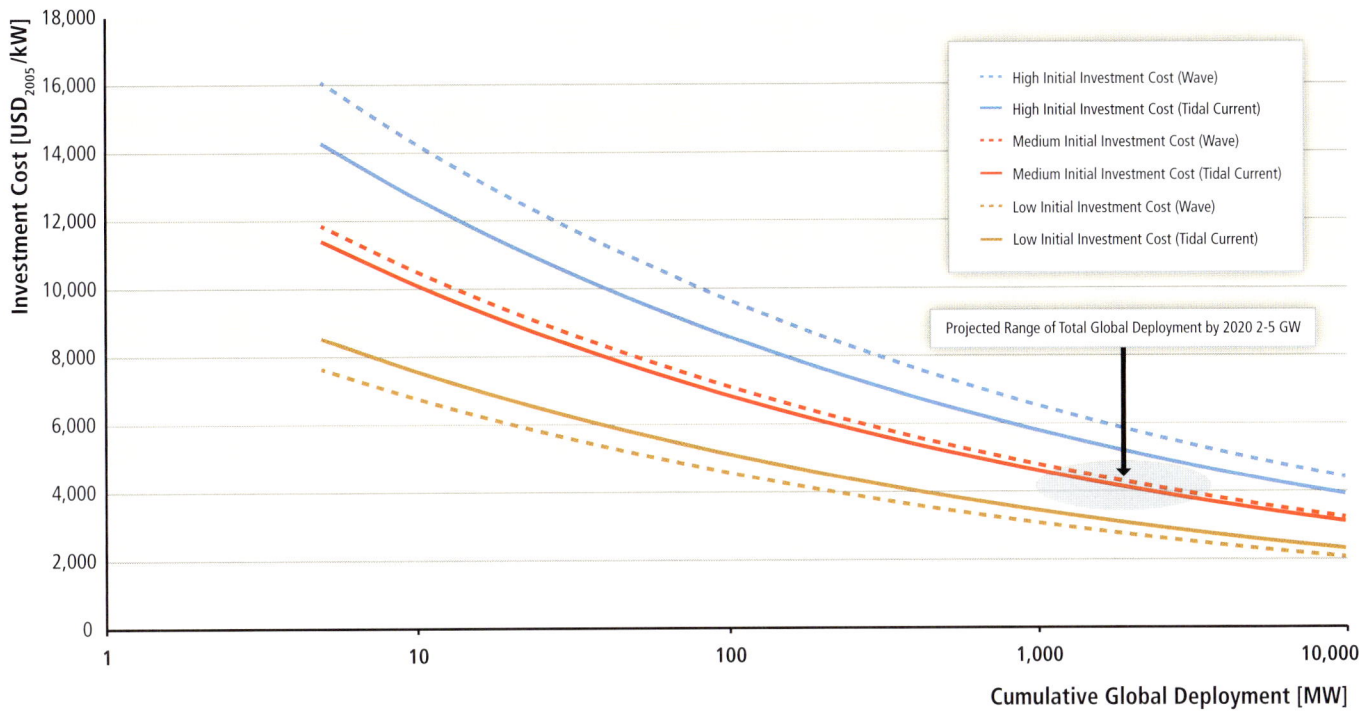

Figure 6.13 | Potential reductions in investment costs for wave and tidal current energy devices based on estimated current cost (Callaghan, 2006) and 11% cost reduction per doubling of cumulative installed capacity (Wiser and Bolinger, 2009).

Note: Initial deployments are assumed to be 5 MW for both subtypes.

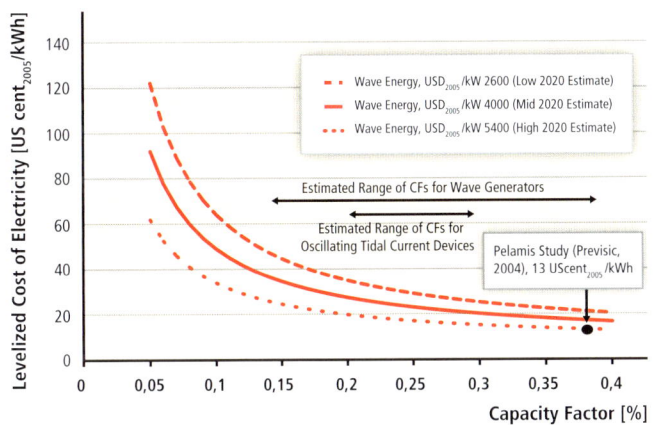

Figure 6.14 | Capacity factor effect on LCOE for estimated 2020 wave and tidal current investment costs. The data point showing the EPRI conceptual design, using Pelamis 500 kW machines at 38% capacity factor, is based on Previsic (2004).

compared to other sources of energy. Innovative techniques, including construction of large civil components onshore and flotation to the site, are expected to allow for substantial reductions in risks and costs. To date, tidal barrage projects have been larger in scale than other ocean energy projects, as the scale reduces the unit cost of generation.

Tidal barrage costs were estimated to be between USD_{2005} 4,500 and 5,000/kW with O&M costs of approximately USD_{2005} 100/kW/yr (ETSAP, 2010b). The design life of a tidal range energy project is expected to exceed 20 years and can be compared to hydroelectric facilities, which can reach economic lives of 40 to 100 years or more.

6.7.4 Ocean thermal energy conversion

There has been no long-term, sustained field experience with OTEC technologies, so it is difficult to predict current costs and future trends. Investment costs for individual projects are high, so technology development has been slow. Published cost estimates are presented in Table 6.4. These cost estimates are presented to provide some insight about what has been documented to date. They do not imply that OTEC technologies have achieved significant maturity. The figures presented have not been converted to 2005 USD, so they appear in different-year dollars and cover a range of different technologies and locations. Many are also highly speculative.

The most current costs available for OTEC come from Lockheed-Martin, which estimates investment costs at USD 32,500/kW for a 10 MW pilot plant, which drop to an estimated USD 10,000/kW for a commercial 100 MW plant (Cooper et al., 2009).

Advances in new materials and construction techniques in other fields in recent years may improve OTEC economics and technical feasibility.

Table 6.4 | Published investment costs and LCOE for OTEC pilot projects and concepts.

Source of Cost Data	Investment Cost (USD/kW)	LCOE (US cents/kWh)	Notes
Vega (2002)	12,300	22	100 MW closed-cycle, 400 km from shore
SERI (1989)	12,200	—	40 MW plant planned at Kahe Point, Oahu
Cohen (2009)	8,000–10,000	16–20	100 MW early commercial plant
Francis (1985)	5,000–11,000	—	—
Lennard (2004)	9,400	18 [11]	10 MW closed-cycle; LCOE in brackets apply if also producing potable water
SERI (1989)	7,200	—	Onshore, open-cycle
Vega (2002)	6,000	10	100 MW closed-cycle, 100 km from shore
Vega (2002)	4,200	7	100 MW closed-cycle, 10 km from shore
Plocek et al. (2009)	8,000	15	Estimate for 75 MW commercial floating plant off Puerto Rico

Note: LCOEs listed in this table are from the published literature. Underlying assumptions are not always known. Neither investment cost nor LCOE have been converted to 2005 USD.

6.7.5 Salinity gradients

Salinity gradient technologies are immature and current costs are not available. Statkraft has estimated that the future LCOE for salinity gradients power may fall in the same range as other more mature renewable technologies, such as wind, based on their current hydropower knowledge, general desalination (reverse osmosis) engineering and a specific membrane technology. Achieving competitive costs will, however, be dependent on the development of reliable, large-scale and low-cost membranes. Statkraft estimates that investment costs will be much higher than other RE technologies, but that capacity factors could be very high, with 8,000 hours of operation annually (Scråmestø et al., 2009).

6.8 Potential deployment[30]

Ocean energy may offer the potential for long-term carbon emissions reduction but is unlikely to make a significant short-term contribution before 2020 due to its nascent stage of development. In 2009, additionally installed ocean capacity was less than 10 MW worldwide (Renewable UK, 2010), yielding a cumulative installed capacity of about 300 MW (REN21, 2010) at present.

6.8.1 Deployment scenarios with ocean energy coverage

Until about 2008, ocean energy was not considered in any of the major energy scenario modelling activities worldwide and therefore its potential impact on future world energy supplies and climate change mitigation is just now beginning to be investigated. As such, the results of the published scenarios literature as it relates to ocean energy are sparse and preliminary, reflecting a wide range of possible outcomes.

Specifically, scenarios for ocean energy deployment are considered in only three major sources here: Energy [R]evolution (E[R]) (Teske et al., 2010), IEA World Energy Outlook (WEO) (IEA, 2009), and IEA Energy Technology Perspectives (ETP) (IEA, 2010). Multiple scenarios were considered in the E[R] and the ETP reports and a single reference scenario was documented in the WEO report. Note that the E[R] Reference scenario is based on the WEO 2009 Reference case and therefore deployment levels until 2030 are very close (Teske et al., 2010). The main characteristics of the considered scenarios, including the deployment levels of ocean energy are summarized in Table 6.5.

The treatment of ocean energy in each of these scenarios reflects a very preliminary state of analysis. In most cases, the inputs have not been fully validated and may not represent the diverse characteristics of the multiple ocean energy resource technologies. In most scenarios, all ocean energy technologies are represented as a single aggregate. This approach is taken out of convenience, and because relevant disaggregated data (e.g., detailed resource assessments with global coverage) are limited (see Chapter 10.2.4 for a more detailed discussion). Many of the technologies are still at an early stage of development and do not have fully established estimates for current and future investment cost, O&M cost, and capacity factors, or even technical potential. Disaggregation into the technology subtypes in future scenario studies may provide further insight into the possible role of ocean energy, but doing so would require a level of data fidelity that does not yet exist for ocean energy technologies.

Regardless of the limitations of the existing scenarios, they do provide a first-order analysis of possible ocean energy technology deployments from which to build a more refined analysis. Specifically, the scenarios indicate a wide range of possible deployments for ocean energy from a conservative baseline case presented by the IEA WEO 2009 to the most aggressive Advanced E[R] scenario, which assumes an 80% CO_2 emissions reduction by 2050.

---

30 Complementary perspectives on potential deployment based on a comprehensive assessment of numerous model-based scenarios of the energy system are presented in Chapter 10 and Sections 10.2 and 10.3 of this report.

Table 6.5 | Main characteristics of medium- to long-term scenarios from major published studies that include ocean energy.

Scenario	Source	Deployment TWh/yr (PJ/yr)				GW	Notes
		2010	2020	2030	2050	2050	
Energy [R]evolution - Reference	(Teske et al., 2010)	N/A	3 (10.8)	11 (36.6)	25 (90)	N/A	No policy changes
Energy [R]evolution	(Teske et al., 2010)	N/A	53 (191)	128 (461)	678 (2,440)	303	Assumes 50% carbon reduction
Energy [R]volution - Advanced	(Teske et al., 2010)	N/A	119 (428)	420 (1,512)	1943 (6,994)	748	Assumes 80% carbon reduction
WEO 2009	(IEA, 2009)	N/A	3 (10.8)	13 (46.8)	N/A	N/A	Basis for E[R] reference case
ETP BLUE map 2050	(IEA, 2010)	N/A	N/A	N/A	133 (479)	N/A	Power sector is virtually decarbonized
ETP BLUE map no CCS 2050	(IEA, 2010)	N/A	N/A	N/A	274 (986)	N/A	BLUE Map Variant – Carbon capture and storage is found to not be possible
ETP BLUE map hi NUC 2050	(IEA, 2010)	N/A	N/A	N/A	99 (356)	N/A	BLUE Map Variant – Nuclear share is increased to 2000-GW
ETP BLUE Map hi REN 2050	(IEA, 2010)	N/A	N/A	N/A	552 (1,987)	N/A	BLUE Map Variant – Renewable share is increased to 75%
ETP BLUE map 3%	(IEA, 2010)	N/A	N/A	N/A	401 (1,444)	N/A	BLUE Map Variant – Discount rates are set to 3% for energy generation projects.

6.8.2 Near-term forecasts

Most near-term ocean energy deployment will likely be policy driven in those countries where government-sponsored research programs and policy incentives have been implemented to promote ocean energy (IEA, 2009). In those cases, near-term forecasts for ocean energy deployment may be related to any country-specific deployment targets that have been established for ocean energy. Some countries have, in fact, proposed non-binding deployment targets and timelines to achieve prescribed ocean energy capacity. The UK government has a target of 2 GW by 2020 (Mueller and Jeffrey, 2008). Canada, the USA, Portugal and Ireland are working on establishing deployment targets for a similar timeframe. Most countries with significant ocean resources have not yet quantified their resource potentials, however, and have not established national deployment goals. And, in those countries that have established ocean energy goals, those goals are rarely obligatory.

Regardless of the drivers for near-term deployment, in general, the near-term forecasts for ocean energy among the scenarios reviewed in this chapter and summarized in Table 6.5 do not envisage a substantial contribution to near-term carbon mitigation. From the scenarios shown in Table 6.5, the near-term (2020) deployment for ocean energy ranges from 3 to 119 TWh/yr (10.8 to 428 PJ/yr), with the highest case being the Advanced E[R] scenario. This wide range reflects the high degree of uncertainty embodied in the scenario assumptions, as well as the different frames of the analysis as the reference case is intended to be a business-as-usual case in which new policies are not enacted, whereas the ambitious Advanced E[R] scenario seeks to dramatically reduce carbon emissions.

6.8.3 Long-term deployment in the context of carbon mitigation

The potential for ocean energy supply to make contributions to the mitigation of climate change is expected to increase to more significant levels in the longer term. By 2050, the deployment scenarios indicated in Table 6.5 range from the Reference E[R] case of only 25 TWh/yr (90 PJ/yr) to the Advanced E[R] case of 1,943 TWh/yr (6,994 PJ/yr). Since ocean energy technologies are presently at an early stage of development, current deployments are very limited. Significant deployments are not forecast until after 2030, though commercial deployments would be expected to continue well beyond the 2050 modelling horizon.

To achieve these higher levels of deployment in the longer term, a variety of possible challenges to the growth of ocean energy deserve discussion.

Resource potential: Resource potential assessments for ocean energy are at a preliminary stage. Nonetheless, even the highest estimates for long-term (2050) ocean energy supply (7 EJ/yr) presented above are well within the theoretical and technical potential for the resource, suggesting that—on a global basis, at least—technical potential is unlikely to be a limiting factor to ocean energy deployment. As presented earlier, OTEC may have the highest technical potential of the available ocean energy options, but even excluding OTEC, the technical potential for ocean energy has been found to exceed 7 EJ/yr. Moreover, though the available literature is limited, the impact of climate change on the technical potential for ocean energy is anticipated to be modest. Regardless, certain regional limitations to resource supply are possible. Wave energy sites are globally dispersed over all coastal boundaries, for example, but the availability of mid-latitude sites (30° to 60°) with lower levels of seasonal variation, adequate incident wave energy, and that are close to load centres may become a barrier in some regions under high penetration scenarios or in populated areas with competing uses. Similarly, limited site availability may prevent widespread deployment of tidal power plants, tidal current energy and ocean current energy beyond certain areas, while OTEC and salinity gradient opportunities are also not equally distributed globally.

Regional deployment: Whether the more ambitious levels of deployment considered in Table 6.5 are feasible will depend, in part, on whether locations of ocean energy resource potential are correlated with areas that demand ocean energy services. Wave and tidal energy technologies are under development in countries bordering the North Atlantic and North Pacific, as well as Australasia, where government-sponsored programmes support R&D and deployments, with pro-active policy incentives to promote early-stage projects. OTEC projects are likely to be developed off the coasts of tropical islands and states. Tidal current, ocean current and salinity gradient projects are most likely to be limited to specific locations where resource quality is strong. These locations are likely to become more numerous and widespread as the efficiencies of these technologies mature. Overall, while technical potential is not anticipated to be a primary global barrier to ocean energy deployment, resource characteristics will require that local communities in the future select among multiple available ocean technologies to suit local resource conditions.

Supply chain issues: Wave, tidal current and some other ocean energy technologies require a sophisticated O&M infrastructure of sufficient scale to be cost effective. Different technologies require different support vessels due to differences in insertion and extraction methods. Until there is a critical mass of deployment for some of the ocean technologies, lack of sufficient infrastructure could be a significant barrier to industry growth. Some benefits may be realized from offshore wind energy development, which may contribute to this infrastructure requirement (in terms of deployment vessels, moorings and export cable access) in advance of significant ocean energy deployment.

Technology and economics: All ocean energy technologies, except tidal barrages, are conceptual, undergoing R&D, or are in the pre-commercial prototype and demonstration stage. The technical performance of ocean energy technologies is anticipated to improve steadily over time as experience is gained and new technologies are able to access poorer quality resources. Technical improvements can reduce capital costs, enhance efficiency, reduce O&M requirements and enhance capacity factors, giving access to sites that are more remote and providing improved methods for harnessing poorer-quality resources. Concurrently with these technical improvements, the LCOE for ocean energy technologies should decline. Whether the technical advances lead to sufficient associated cost reductions to enable broad-scale deployment of ocean energy is the most critical uncertainty in assessing the future role of ocean energy in meeting ambitious long-term deployment targets.

Integration and transmission: The integration of ocean energy into wider energy networks will need to recognize the widely varying generation characteristics arising from the different resources. These patterns of resource availability have implications for the large-scale integration of ocean energy into electricity networks (see Chapter 8), and on the requirements for, and utilization of, transmission capacity, including the need for and value of offshore transmission networks. To effectively manage the variability of some ocean energy sources at higher levels of deployment may require similar technical and institutional solutions as considered for wind and solar photovoltaic technologies, specifically, forecasting capability, increased system-wide flexibility, grid connection standards, demand flexibility and bulk energy storage. Other ocean energy technologies, on the other hand, have characteristics that may be similar to base-load or even partially dispatchable thermal generators, thereby not imposing concerns about operational integration, though new transmission infrastructure may still be required.

Social and environmental impacts: The social and environmental impacts of ocean energy projects are being evaluated as actual deployments multiply. Risk analysis and mitigation, using environmental impact assessments, will be essential components of early deployments. Competitive uses may preclude the availability of some high-quality sites, and environmental and ecological concerns are likely to impact deployment locations as well. A balanced approach to engaging coastal communities will be necessary, whilst maintaining a fair and responsible respect for existing coastal uses and ocean ecologies. That some forms of ocean energy have high levels of environmental reversibility may make them attractive for future development, but the early stage of ocean energy deployment creates uncertainty about the degree to which social and environmental concerns might eventually constrain development.

6.8.4 Conclusions regarding deployment

This preliminary presentation of scenarios that describe alternative levels of ocean energy deployment is among the first attempts to review the potential role of ocean energy in the medium- to long-term scenarios literature with the intention of establishing the potential contribution of ocean energy to future energy supplies and climate change mitigation. As shown by the limited number of existing scenarios, ocean energy has the potential to help mitigate long-term climate change by offsetting GHG emissions, with projected deployments resulting in energy delivery of up to 1,943 TWh/yr (~7 EJ/yr) by 2050. Other scenarios have been developed indicating deployment as low as 25 TWh/yr (0.9 EJ/yr) from ocean energy. The wide range in results is based in part on uncertainty about the degree to which climate change mitigation will drive energy sector transformation, but for ocean energy, is also based on inherent uncertainty as to when and if various ocean energy technologies will become commercially available at attractive costs. To better understand the possible role of ocean energy in climate change mitigation, not only will continued technical advances be necessary, but the scenarios modelling process will need to increasingly incorporate the range of potential ocean energy technology subtypes, with better data for resource potential, present and future investment costs, O&M costs and anticipated capacity factors. Improving the availability of the data at global and regional scales will be an important ingredient to improve coverage of ocean energy in the scenarios literature (see also Section 10.2.4).

References

AEA Technology and Poyry Energy Consulting (2006). *Additional Support for Marine Electricity Generation in Scotland*. Consultants report to Scottish Executive, Volume 1 – Summary Report ED02360. AEA Technology, Glengarnock, Scotland, 43 pp.

Allender, J., T. Audunson, S.F. Barstow, S. Bjerken, H.E. Krogstad, P. Steinbakke, L. Vartdal, L.E. Borgman, and C. Graham (1989). The WADIC Project: a comprehensive field evaluation of directional wave instrumentation. *Ocean Engineering*, **16**(5-6), pp. 505-536.

Andre, H. (1976). Operating experience with bulb units at the Rance tidal power plant and other French hydro-power sites. *Institute of Electrical and Electronics Engineers (IEEE) Transactions on Power Apparatus and Systems*. **PAS-95**, pp. 1038-1044.

Aqua-RET (2008). *Case study – European OWC pilot plant Pico/Azores* Aquatic Renewable Energy Technologies (Aqua-RET) Revision. Available at: www.wisions.net/files/tr_downloads/Aquaret_Pico_OWC.pdf.

AWATEA (2008). *Marine Energy Supply Chain: 2008 Directory*. Aotearoa Wave and Tidal Energy Association, Wellington, New Zealand.

Baker, A.C. (1991). *Tidal Power*. Peter Peregrinus Ltd. (pub.), London, UK.

Barber, N.F., and F. Ursell (1948). The generation and propagation of ocean waves and swell. I. Wave periods and velocities. *Philosophical transactions of the Royal Society of London. Series A, Mathematical and Physical Sciences*, **240**(824), pp. 527.

Battin, J. (2004). When good animals love bad habitats. *Conservation Biology*, **18**(6), pp. 1482-1491.

Bedard , R., M. Previsic, and A. Casavant (2006). *North America Tidal In-Stream Energy Conversion*. Electric Power Research Institute, Palo Alto, CA, USA.

Bernitsas, M.M., K. Raghavan, Y. Ben-Simon, and E.M.H. Garcia (2006). VIVACE (Vortex Induced Vibration Aquatic Clean Energy): A New Concept in Generation of Clean and Renewable Energy from Fluid Flow. In: *Proceedings of the 25th International Conference on Offshore Mechanics and Arctic Engineering (OMAE '06)*, Paper #92645, 4-9 June 2006, Hamburg, Germany. American Society of Mechanical Engineers, New York, NY, USA.

Binnie Black & Veatch (2001). *The commercial prospect of tidal stream power*. ETSU - T/06/00209/REP, UK Department of Trade and Industry New & Renewable Energy Programme in association with IT Power Ltd., London, UK.

Boehlert, G.W., and A.B. Gill (2010). Environmental and ecological effects of ocean renewable energy development: A current synthesis. *Oceanography*, **23**: 68-81.

Boehlert, G.W., G.R. McMurray, and C.E. Tortorici (2007). *Ecological Effects of Wave Energy Development in the Pacific Northwest. A Scientific Workshop*. Technical Memorandum NMFS-F/SPO-92, National Oceanic and Atmospheric Administration, Washington, DC, USA.

Bonnot-Courtois, C. (1993). Comparative study of dredging and flushing effects on sedimentation in the upper part of the Rance estuary. *La Houille Blanche*, **8**, pp. 539-550.

Bosc, J. (1997). Les groupes bulbes de la Rance après trente ans d'exploitation : Retour d'expérience; The Rance bulb turbines after 30 years' service: feedback of experience. In: *Congrès Colloque du 30e anniversaire de l'Usine Marémotrice de la Rance*, Société Hydrotechnique de France, Paris, France, pp. 18-24.

Buckley, W.H. (2005). *Extreme Waves for Ship and Offshore Platform Design: An Overview*. T&R Report, Society of Naval Architecture and Marine Engineering, Jersey City, NJ, USA, pp. 7-30.

BWEA (2005). *BWEA Briefing Sheet. Wind Turbine Technology*. British Wind Energy Association, London, UK.

California Energy Commission (2010). *California Renewable Energy Transmission Initiative*. California Energy Commission, Sacramento, CA, USA.

Callaghan, J. (2006). *Future Marine Energy - Results of the Marine Energy Challenge: Cost Competitiveness and Growth of Wave and Tidal Stream Energy*. CTC601, Carbon Trust, London, UK.

Carbon Trust (2004). *UK, Europe and Global Tidal Stream Energy Resource Assessment*. 107799/D/2100/05/1, Carbon Trust Marine Energy Challenge, Carbon Trust, London, UK.

Charlier, R.H. (2003). Sustainable co-generation from the tides: A review. *Renewable and Sustainable Energy Reviews*, **7**(3), pp. 187-213.

Charlier, R.H., and J.R. Justus (1993). *Ocean Energies: Environmental, Economic and Technological Aspects of Alternative Power Sources*. Elsevier Oceanography Series, Amsterdam, The Netherlands.

Chassignet, E.P., H.E. Hurlburt, O.M. Smedstad, G.R. Halliwell, P.J. Hogan, A.J. Wallcraft, R. Baraille, and R. Bleck (2007). The HYCOM (HYbrid Coordinate Ocean Model) data assimilative system. *Journal of Marine Systems*, **65**(1-4), pp. 60-83.

Cohen, R. (2009). An overview of ocean thermal energy technology, potential market applications, and technical challenges. In: *Proceedings 2009 Offshore Technology Conference*, Houston, TX, USA, 4-7 May 2009.

CEC (1996). *Wave Energy Project Results: The Exploitation of Tidal Marine Currents*. - DGXII - Report EUR16683EN, Commission of the European Communities, Brussels, Belgium.

CEC (1998). *Promotion of New Energy Sources in the Zhejiang Province, China, Final Report*. - DGXVII - Program SYNERGY Contract N° 4.1041/D/97-09, Commission of the European Communities, Brussels, Belgium.

Cooper, D.J., L.J. Meyer, and R.J. Varley (2009). *OTEC Commercialization Challenges*. In: *Proceedings 2009 Offshore Technology Conference*, Houston, TX, USA, 4-7 May 2009.

Cornett, A.M. (2006). *Inventory of Canada's Marine Renewable Energy Resources*. Canadian Hydraulics Centre, Ottawa, Ontario, Canada.

Cornett, A.M. (2008). A global wave energy resource assessment. In: *Proceedings of the Eighteenth (2008) International Society of Offshore and Polar Engineers*, Vancouver, BC, Canada, 6-11 July 2008, pp. 318-326. Available at: www.isope.org/publications/proceedings/ISOPE/ISOPE%202008/papers/I08TPC-579Corn.pdf.

De Laleu, V. (2009). La Rance Tidal Power Plant. 40-year operation feedback – Lessons learnt. In: *British Hydropower Association Annual Conference*, Liverpool, UK, 14-15 October 2009.

Denniss, T. (2005). *Energetech Wave Energy System*. Doc No. 05-0145 O'ahu Power Plant LOL T-12, Hawaii Public Utilities Commission, Honolulu, HI, USA.

Ecofys (2007). *Energie uit zout en zoet water met osmose: Een visualisatie bij de Afsluitdijk*. Ecofys Netherlands B.V., Utrecht, The Netherlands.

Engineering Business (2003). *Stingray Tidal Stream Energy Device – Phase 2*. Report prepared for the DTI New and Renewable Energy Programme T/06/00218/00/REP; URN 03/1433. The Crown Publishing Group, New York, NY, USA.

EOEA (2010). *Oceans of Energy: European Ocean Energy Roadmap 2010-2050*. European Ocean Energy Association, Brussels, Belgium.

ESF MB (2010). *Marine Renewable Energy*. Document 2, European Science Foundation Marine Board Vision, Ostend, Belgium.

Estefen, S.F., X. Castello, M.I. Lourenço, and R.M. Rossetto (2010). Design analysis applied to a Hyperbaric Wave Energy Converter. In: *Proceedings 11th International Symposium on Practical Design of Ships and Other Floating Structures*, Rio de Janeiro, Brazil, 19-24 September 2010.

ETSAP (2010a). *Marine Energy Technology Brief E13 - May, 2010*. Energy Technology Systems Analysis Program, Paris, France.

ETSAP (2010b). *Marine Energy Technology Brief E13 - November, 2010*. Energy Technology Systems Analysis Program, Paris, France.

Falcão, A. (2009). The development of wave energy utilization. In: *2008 Annual Report*, A. Brito-Melo and G. Bhuyan (eds.), International Energy Agency Implementing Agreement on Ocean Energy Systems, Lisboa, Portugal, pp. 30-37.

Falcão, A., C. Travassos, N. Marques, and R. Martino (2000). The shoreline OWC wave power plant at the Azores. In: *Proceedings of the 4th European Wave Power Conference*, Aalborg, Denmark, December 2000, Paper B1.

Francis, E.J. (1985). *Economics of OTEC*. Report prepared for the Solar Energy Research Institute, Golden, CO, USA, 126 pp.

Friends of the Earth (2004). *Severn Barrage or Tidal Lagoons? A Comparison*. Briefing, Friends of the Earth Cymru, Cardiff, Wales.

Garrett, C., and P. Cummins (2005). The Power potential of tidal currents in channels. Proceedings of the Royal Society A, 461, pp. 2563-2572.

Garrett, C., and P. Cummins (2008). Limits to tidal current power. *Renewable Energy*, 33, pp. 2485-2490.

Hagerman, G., B. Polagye, R. Bedard, and M. Previsic (2006). *Methodology for Estimating Tidal Current Energy Resources and Power Production by Tidal In-Stream Energy Conversion (TISEC) Devices*. Electric Power Research Institute, Palo Alto, CA, USA.

Harrison, G.P., and A.R. Wallace (2005). Sensitivity of wave energy to climate change. *IEEE Transactions on Energy Conversion*, 20(4), pp. 870-877.

Holmes, B., and K. Nielsen (2010). *Guidelines for the Development & Testing of Wave Energy Systems*. HMRC Hydraulics Maritime Research Centre report for OES-IA, Report T02-2.1, OES-IA Annex II: Task 2.1, Implementing Agreement for a Co-operative Programme on Ocean Energy Systems (OES-IA), Cork, Ireland. Available at: www.iea-oceans.org/_fich/6/Report_02-2.1(a).pdf.

House of Commons (2001). *Science and Technology – Seventh Report*. House of Commons Science and Technology Committee Publications, London, UK.

Huckerby, J.A., and P. McComb (2008). *Development of Marine Energy in New Zealand*. Published consultants' report for Energy Efficiency and Conservation Authority, Electricity Commission and Greater Wellington Regional Council, Wellington, New Zealand.

IEA (2009). *World Energy Outlook 2009*. International Energy Agency, Paris, France, 696 pp.

IEA (2010). *Energy Technology Perspectives 2010. Scenarios and Strategies to 2050*. International Energy Agency, Paris, France, 708 pp.

Johns, E., W.D. Wilson, and R.L. Molinari (1999). Direct observations of velocity and transport in the passages between the Intra-Americas Sea and the Atlantic Ocean, 1984-1996. *Journal of Geophysical Research*, 104(C11), pp. 25805-25820.

Karsten, R.H., J.M. McMillan, M.J. Lickley, and R.D. Haynes (2008). Assessment of tidal current energy in the Minas Passage, Bay of Fundy. *Proceedings of the Institution of Mechanical Engineers Part A: Journal of Power and Energy*, 222, pp. 493-507.

Kerr, D. (2007). Marine energy. *Philosophical Transactions of the Royal Society London, Series A (Mathematical, Physical and Engineering Sciences)*, 365(1853), pp. 971-992.

Khan, J., and G.S. Bhuyan (2009). *Ocean Energy: Global Technology Development Status*. Report prepared by Powertech Labs for the IEA-OES, Document T0104, Implementing Agreement for a Co-operative Programme on Ocean Energy Systems (OES-IA), Lisboa, Portugal. Available at: www.iea-oceans.org/_fich/6/ANNEX_1_Doc_T0104.pdf.

Khan, J., A. Moshref, and G. Bhuyan (2009). *A Generic Outline for Dynamic Modeling of Ocean Wave and Tidal Current Energy Conversion Systems*. Institute for Electrical and Electronic Engineering, Piscataway, NJ, USA, 6 pp.

Klein, J. (2009). *Comparative Costs of California Central Station Electricity Generation Technologies*. CEC-200-2009-017-SD, California Energy Commission. Sacramento, CA, USA.

Kobayashi, H., S. Jitsuhara, and H. Uehara (2004). The present status and features of OTEC and recent aspects of thermal energy conversion technologies. In: *24th Meeting of the UJNR Marine Facilities Panel*, Honolulu, HI, USA, 4-12 November 2001. Available at: www.nmri.go.jp/main/cooperation/ujnr/24ujnr_paper_jpn/Kobayashi.pdf.

Krewitt, W., K. Nienhaus, C. Kleßmann, C. Capone, E. Stricker, W. Graus, M. Hoogwijk, N. Supersberger, U. von Winterfeld, and S. Samadi (2009). *Role and Potential of Renewable Energy and Energy Efficiency for Global Energy Supply*. Climate Change 18/2009, ISSN 1862-4359, Federal Environment Agency, Dessau-Roßlau, Germany, 336 pp.

Krishna, I. (2009). *A SOPAC Desktop Study of Ocean-Based Renewable Energy Technologies*. SOPAC Miscellaneous Report 701, Secretariat of the Applied Geoscience Commission, Suva, Fiji Islands, 94 pp. Available at: dev.sopac.org.fj/VirLib/MR0701.pdf.

Lang, F. (2008). The Rance Tidal Power Plant: review of 40-years operation, environmental effects. In: *2nd International Conference on Ocean Energy*, Brest, France, 15-17 October 2008.

Leaman, K.D., R.L. Molinari, and P.S. Vertes (1987). Structure and variability of the Florida Current at 27°N. *Journal of Physical Oceanography*, 17(5), pp. 565-583.

Lennard, D.E. (ed.) (2004). Ocean Thermal Energy Conversion. In: *2004 Survey of Energy Resources*. World Energy Council, Elsevier, Amsterdam, The Netherlands, pp. 419-426.

Li, Y., and H.K. Florig (2006). *Modeling the Operation and Maintenance Costs of a Large Scale Tidal Current Turbine Farm*. Institute for Electrical and Electronic Engineering, Piscataway, NJ, USA, 6 pp.

Lighthill, J. (1978). *Waves in Fluids*. Cambridge University Press.

Liu, Z., H. Shi, and B.-S. Hyan (2009). Practical design and investigation of the breakwater facility in China. In: *Proceedings of the 8th European Wave and Tidal Energy Conference (EWTEC)*, Uppsala, Sweden, 7-10 September 2009, pp. 304-308.

Loeb, S., and R.S. Norman (1975). Osmotic power plants. *Science*, 189(4203), pp. 654-655.

Mackay, D.J.C. (2008). *Sustainable Energy – Without the Hot Air*. UIT Cambridge Ltd., Cambridge, UK.

MEG (2009). *Marine Energy Road Map*. Marine Energy Group, Forum for Renewable Energy Development in Scotland (FREDS), Edinburgh, Scotland. Available at: www.scotland.gov.uk/Resource/Doc/281865/0085187.pdf.

Marti, J.A. (2008). OTEC: Environmental and socio-economic aspects. Presentation at the *Convention of the College of Professional Engineers and Surveyors of Puerto Rico*, Fajardo, Puerto Rico, 7-8 August 2008. Available at: www.offinf.com/EnviroSocialCIAPR.pdf.

MCCIP (2008). *Marine Climate Change Impacts Annual Report Card 2007–2008*. Summary Report, Marine Climate Change Impacts Partnership (MCCIP), Lowestoft, UK.

Miller, A.K. (2010). Cold Water Pipe. In: *Technical Readiness of Ocean Thermal Energy Conversion (OTEC)*. Coastal Response Research Center, University of New Hampshire, Durham, NH, USA. Available at: coastalmanagement.noaa.gov/otec/docs/otectech1109.pdf.

MMS (2006). *Technology White Paper on Ocean Current Energy Potential on the U.S. Outer Continental Shelf*. Minerals Management Service Renewable Energy and Alternate Use Program, U.S Department of the Interior, Washington, DC, USA. Available at: ocsenergy.anl.gov/documents/docs/OCS_EIS_WhitePaper_Current.pdf.

Mørk, G., S. Barstow, M.T. Pontes, and A. Kabuth (2010). Assessing the global wave energy potential. In: *Proceedings of OMAE2010 (ASME), 29th International Conference on Ocean, Offshore Mechanics and Arctic Engineering*, Shanghai, China, 6-11 June 2010.

Mueller, M., and H. Jeffrey (2008). *UKERC Marine (Wave and Tidal Current) Renewable Energy Technology Roadmap: Summary Report*. UK Energy Research Centre, University of Edinburgh, Scotland.

NASA (2006). *TOPEX/Poseidon: Revealing Hidden Tidal Energy*. NASA Goddard Space Flight Center Scientific Visualization Studio, Greenbelt, Maryland. Available at: svs.gsfc.nasa.gov/stories/topex/.

Nicholls-Lee, R.F., S.R. Turnock, and S.W. Boyd (2008). Performance prediction of a free stream tidal turbine with composite bend-twist coupled blades. In: *Proceedings of the 2nd International Conference on Ocean Energy*, Brest, France, 15-17 October 2008.

Nihous, G.C. (2007). A preliminary assessment of ocean thermal energy conversion resources. *Journal of Energy Resources Technology*, **129**(March, 2007), pp. 10-17.

Nihous, G.C. (2009). Ocean Thermal Energy Conversion (OTEC) and derivative technologies: Status of development and prospects. In: *2008 Annual Report*. A. Brito-Melo and G. Bhuyan (eds.). International Energy Agency Implementing Agreement on Ocean Energy Systems, Lisboa, Portugal, pp. 47-51.

Nihous, G.C. (2010). Mapping available Ocean Thermal Energy Conversion resources around the main Hawaiian Islands with state-of-the-art tools. *Journal of Renewable and Sustainable Energy*, **2**, 043104.

Nihous, G.C., and L.A. Vega (1993). Design of a 100 MW OTEC-hydrogen plantship. *Marine Structures*, **6**(2-3), pp. 207-221.

Niiler, P.P., and W.S. Richardson (1973). Seasonal variability of the Florida Current. *Journal of Marine Research*, **31**, pp. 144-167.

Paik, D.H. (2008). Progress of the Sihwa tidal Power Project in South Korea. In: *Proceedings of the 2nd International Conference on Ocean Energy*, Brest, France, 15-17 October 2008.

Panchal, C.B. (2008). *OTEC Power Systems Developments*. Argonne National Laboratory, Argonne, IL, USA (internal publication).

Pelc, R., and R.M. Fujita (2002). Renewable energy from the ocean. *Marine Policy*, **26**(6), pp. 471-479.

Peyrard, C., C. Buvat, F. Lafon, and C. Abonnel (2006). Investigations of the wake effects in marine current farms through numerical modelling with the telemac system. In: *Proceedings of the 1st International Conference on Ocean Energy*, Bremerhaven, Germany, 23-24 October 2006.

Plocek, T.J., M. Laboy, and J.A. Marti (2009). Ocean Thermal Energy Conversion (OTEC): Technical viability, cost projections and development strategies. In: *Proceedings of the 2009 Offshore Technology Conference*, Houston, Texas, USA, 4-7 May 2009, OTC 19979.

Pontes, M.T., and M. Bruck (2008). Using remote sensed data for wave energy resource assessment. In: *Proceedings OMAE2008 (ASME), 27th International Conference on Offshore Mechanics and Arctic Engineering (OMAE2008)*, Estoril, Portugal, 15-20 June 2008.

Pontes, M.T., and A. Candelária (2009). *Wave Data Catalogue for Resource Assessment of IEA-OES Member Countries*. Report from INETI for the IEA-OES, Document T0103, Implementing Agreement for a Co-operative Programme on Ocean Energy Systems (OES-IA), Lisboa, Portugal. Available at: www.iea-oceans.org/_fich/6/Wave_Data_Catalogue-_9_April__2009_(1).pdf.

Previsic, M. (2004). *System Level Design, Performance, and Costs of California Pelamis Wave Power Plant*. Electric Power Research Institute, Palo Alto, CA, USA.

Pugh, D.T. (1987). *Tides, Surges and Mean-Sea Level: A Handbook for Engineers and Scientists*. Wiley, Chichester, UK.

Ravindran, M., and A. Raju (2002). The Indian 1 MW demonstration OTEC plant and the development activities. In: *OCEANS '02 MTS/IEEE*, Biloxi, Mississippi, 29-31 October 2002, **3**, pp. 1622-1628.

Ravindran, M., V. Jayashankar, P. Jalihal, and A.G. Pathak (1997). The Indian wave energy program - An overview. *TERI Information Digest on Energy*, **7**(3), pp. 173-188.

Raye, R. (2001). *Characterization Study of the Florida Current at 26.11° North Latitude, 79.50° West Longitude for Ocean Current Power Generation*. M.S. Thesis, College of Engineering, Florida Atlantic University, Boca Raton, FL, USA.

Relini, G., M. Relini, and M. Montanari (2000). An offshore buoy as a small artificial island and a fish-aggregating device (FAD) in the Mediterranean. *Hydrobiologia*, **440**(1), pp. 65-80.

REN21 (2010). *Renewables 2010: Global Status Report*. Renewable Energy Policy Network for the 21st Century Secretariat, Paris, France, 80 pp.

Renewable UK (2010). *Marine Renewable Energy. State of the Industry Report*. Renewable UK, London, UK.

Retiere, C., and R. Kirby (2006). Links between environmental consequences of La Rance and Severn Tidal Power Barrages. Presentation to the *South West Regional Assembly Meeting*, Taunton, UK, 20 Oct 2006. Available at: www.southwest-ra.gov.uk/media/SWRA/Assembly%20Papers/20th%20October%202006/10_LinksBetweenEnvironmentalConsequences2.pdf.

Rogner, H.-H., F. Barthel, M. Cabrera, A. Faaij, M. Giroux, D. Hall, V. Kagramanian, S. Kononov, T. Lefevre, R. Moreira, R. Nötstaller, P. Odell, and M. Taylor (2000). Energy resources. In: *World Energy Assessment. Energy and the Challenge of Sustainability*. United Nations Development Programme, United Nations Department of Economic and Social Affairs, World Energy Council, New York, USA, 508 pp.

Scråmestø, O.S., S.-E. Skilhagen, and W.K. Nielsen (2009). Power production based on osmotic pressure. In: *Waterpower XVI*, Spokane, WA, USA, 27-30 July 2009.

SEAI (2010). *Ocean Energy Roadmap*. Sustainable Energy Authority of Ireland, Dublin, Ireland.

SERI (1989). *Ocean Thermal Energy Conversion: An Overview*. SERI/SP-220-3024, Solar Energy Research Institute, Golden, CO, USA, 36 pp.

Sharmila, N., P. Jalihal, A.K. Swamy, and M. Ravindran (2004). Wave powered desalination system. *Energy*, **29**(11), pp. 1659-1672.

Shaw, L.T. (1997). Study of tidal power projects in the UK, with the exception of the Severn barrage. *La Houille Blanche*, **52**(3), pp. 57-65.

Sims, R.E.H., R.N. Schock, A. Adegbululgbe, J. Fenhann, I. Konstantinaviciute, W. Moomaw, H.B. Nimir, B. Schlamadinger, J. Torres-Martínez, C. Turner, Y. Uchiyama, S.J.V. Vuori, N. Wamukonya, and X. Zhang (2007). Energy supply. In: *Climate Change 2007: Mitigation of Climate Change. Contribution of Working Group III to the Fourth Assessment Report of the Intergovernmental Panel on Climate Change*. B. Metz, O.R. Davidson, P.R. Bosch, R. Dave and L.A. Meyer (eds.), Cambridge University Press, pp. 251-322.

Sinden, G.E. (2007). *Renewable Electricity Generation: Resource Characteristics and Implications of Wind, Wave and Tidal Stream Power in the UK*. Report prepared for the Renewables Advisory Board, UK Department of Energy and Climate Change, London, UK.

Statkraft (2009). *Crown Princess of Norway to open world's first osmotic power station*. Press release, 7 Oct 2009, Statkraft, Oslo, Norway. Available at: www.statkraft.com/presscentre/press-releases/2009/crown-princess-mette-marit-to-open-the-worlds-first-osmotic-power-plant.aspx.

Stewart, H.B. (1974). Current from the current. *Oceanus*, **18**, pp. 38-41.

Stommel, H. (1966). *The Gulf Stream*. University of California Press, Berkeley, CA, USA.

Stoutenburg, E.D., N. Jenkins, and M.Z. Jacobson (2010). Power output variations of co-located offshore wind turbines and wave energy converters in California. *Renewable Energy*, **35**(12), pp. 2781-2791.

Sustainable Development Commission (2007). *Turning the Tides: Tidal Power in the UK*. Sustainable Development Commission, London, UK.

Sutherland, G., M. Foreman, and C. Garrett (2008). Tidal current energy assessment for Johnstone Strait, Vancouver Island. *Proceedings of the Institution of Mechanical Engineers Part A: Journal of Power and Energy*, **221**, pp. 147-157.

TSB (2010). *Low-Cost, Low-Risk Tidal Power Technology*. Technology Strategy Board Innovation Results, Case Study 36, Technology Strategy Board, Swindon, UK. Available at: http://www.innovateuk.org/_assets/pdf/casestudies/tsbcase_study_low_cost_low_risk.pdf.

Teske, S., T. Pregger, S. Simon, T. Naegler, W. Graus, and C. Lins (2010). Energy [R]evolution 2010—a sustainable world energy outlook. *Energy Efficiency*, doi:10.1007/s12053-010-9098-y.

Thresher, R. (2010). *First Draft Roadmap: The United States Marine Hydrokinetic Renewable Energy Technology Roadmap*. National Renewable Energy Laboratory, Golden, Colorado, USA. Available at: http://www.oceanrenewable.com/wp-content/uploads/2010/05/1st-draft-roadmap-rwt-8april10.pdf.

Torre-Enciso, Y., I. Ortubia, L.I. Lopez de Aguileta, and J. Marques (2009). Mutriku Wave Power Plant: from the thinking out to the reality. In: *Proceedings of the 8th European Wave and Tidal Energy Conference*, Uppsala, Sweden, 7-10 September 2009.

UK Department of Trade and Industry (2004). *Atlas of UK Marine Renewable Energy Resources: Technical report*. UK Department of Trade and Industry, London, UK. Available at: www.renewables-atlas.info/.

UN (1984). *A Guide to OTEC Conversion for Developing Countries*. ST/ESA/134, UN Department of International Economic and Social Affairs, New York, NY, USA.

UNFCCC (2005). *Sihwa Tidal Power Plant CDM project*. CDM PDD Document, Version 2, Clean Development Mechanism, United Nations Framework Convention on Climate Change, Bonn, Germany.

US DOE (2010). *Energy Efficiency and Renewable Energy Marine and Hydrokinetic Database*. Energy Efficiency and Renewable Energy, US Department of Energy, Washington, DC, USA. Available at: www.eere.energy.gov/windandhydro/hydrokinetic/default.aspx.

Usachev, I.N. (2008). The outlook for world tidal power development. *International Journal on Hydropower and Dams*, **15**(5), pp. 100-105.

van den Ende, K., and F. Groeman (2007). *Blue Energy*. Leonardo Energy, KEMA Consulting, Arnhem, The Netherlands. Available at: www.leonardo-energy.org/webfm_send/161.

Van Walsum, E. (2003). Barriers to tidal power: Environmental effects. *International Water Power and Dam Construction*, **55**(10), pp. 38-43.

VanZwieten, J., F.R. Driscoll, A. Leonessa, and G. Deane (2005). Design of a prototype ocean current turbine–Part I: mathematical modeling and dynamics simulation. *Ocean Engineering*, **33**(11-12), pp. 1485-1521.

Vega, L.A. (1999). *Ocean Thermal Energy Conversion (OTEC): OTEC and the Environment*. Publisher unspecified. Available at: www.otecnews.org/otec-articles/ ocean-thermal-energy-conversion-otec-by-l-a-vega-ph-d/.

Vega, L.A. (2002). Ocean Thermal Energy Conversion Primer. *Marine Technology Society Journal*, **6**(4 Winter), pp. 25-35.

Venezia, W.A., and J. Holt (1995). Turbine under Gulf Stream: potential energy source. *Sea Technology*, **36**(9), pp. 10-14.

Wang, C., and W. Lu (2009). *Analysis Method and Reserves Estimation on Ocean Energy Resources*. Ocean Press, Beijing, China.

Willemse, R. (2007). *Case 27: Blue Energy (salinity power) in the Netherlands*. Create Acceptance, Petten, The Netherlands.

Wiser, R., and M. Bolinger (2009). *Wind Technologies Market Report*. US Department of Energy, Washington, DC, USA.

Wiser, R., and M. Bolinger (2010). *2009 Wind Technologies Market Report*. US Department of Energy, Washington, DC, USA.

World Energy Council (2000). *World Energy Assessment*. World Energy Council, London, UK.

Zhang, L. and K. Sun (2007). Tidal current energy developments in China, *Implementing Agreement for a Co-operative Programme on Ocean Energy Systems (OES-IA) Newsletter*, May 2007, p. 2.

7 Wind Energy

Coordinating Lead Authors:
Ryan Wiser (USA), Zhenbin Yang (China)

Lead Authors:
Maureen Hand (USA), Olav Hohmeyer (Germany), David Infield (United Kingdom),
Peter H. Jensen (Denmark), Vladimir Nikolaev (Russia), Mark O'Malley (Ireland),
Graham Sinden (United Kingdom/Australia), Arthouros Zervos (Greece)

Contributing Authors:
Naïm Darghouth (USA), Dennis Elliott (USA), Garvin Heath (USA), Ben Hoen (USA),
Hannele Holttinen (Finland), Jason Jonkman (USA), Andrew Mills (USA),
Patrick Moriarty (USA), Sara Pryor (USA), Scott Schreck (USA), Charles Smith (USA)

Review Editors:
Christian Kjaer (Belgium/Denmark) and Fatemeh Rahimzadeh (Iran)

This chapter should be cited as:
Wiser, R., Z. Yang, M. Hand, O. Hohmeyer, D. Infield, P. H. Jensen, V. Nikolaev, M. O'Malley, G. Sinden, A. Zervos, 2011: Wind Energy. In IPCC Special Report on Renewable Energy Sources and Climate Change Mitigation [O. Edenhofer, R. Pichs-Madruga, Y. Sokona, K. Seyboth, P. Matschoss, S. Kadner, T. Zwickel, P. Eickemeier, G. Hansen, S. Schlömer, C. von Stechow (eds)], Cambridge University Press, Cambridge, United Kingdom and New York, NY, USA.

Table of Contents

Executive Summary 539

7.1 Introduction 542

7.2 Resource potential 543

7.2.1 Global technical potential 544

7.2.2 Regional technical potential 546
7.2.2.1 Global assessment results by region 546
7.2.2.2 Regional assessment results 547

7.2.3 Possible impact of climate change on resource potential 548

7.3 Technology and applications 550

7.3.1 Technology development and status 550
7.3.1.1 Basic design principles 550
7.3.1.2 Onshore wind energy technology 551
7.3.1.3 Offshore wind energy technology 553

7.3.2 International wind energy technology standards 554

7.3.3 Power conversion and related grid connection issues 555

7.4 Global and regional status of market and industry development 556

7.4.1 Global status and trends 556

7.4.2 Regional and national status and trends 556

7.4.3 Industry development 558

7.4.4 Impact of policies 559

7.5 Near-term grid integration issues 560

7.5.1 Wind energy characteristics 560

7.5.2 Planning electric systems with wind energy 562
7.5.2.1 Electric system models 562
7.5.2.2 Wind power electrical characteristics and grid codes 562

| 7.5.2.3 | Transmission infrastructure | 563 |
| 7.5.2.4 | Generation adequacy | 563 |

7.5.3	**Operating electric systems with wind energy**	564
7.5.3.1	Integration, flexibility and variability	564
7.5.3.2	Practical experience with operating electric systems with wind energy	566

7.5.4	**Results from integration studies**	567
7.5.4.1	Methodological challenges	568
7.5.4.2	Increased balancing cost with wind energy	568
7.5.4.3	Relative cost of generation adequacy with wind energy	569
7.5.4.4	Cost of transmission for wind energy	569

7.6 Environmental and social impacts 570

7.6.1	**Environmental net benefits of wind energy**	570
7.6.1.1	Direct impacts	570
7.6.1.2	Indirect lifecycle impacts	571
7.6.1.3	Indirect variability impacts	571
7.6.1.4	Net environmental benefits	572

7.6.2	**Ecological impacts**	572
7.6.2.1	Bird and bat collision fatalities	572
7.6.2.2	Habitat and ecosystem modifications	573
7.6.2.3	Impact of wind power plants on the local climate	574

7.6.3	**Impacts on human activities and well-being**	574
7.6.3.1	Land and marine usage	574
7.6.3.2	Visual impacts	575
7.6.3.3	Noise, flicker, health and safety	575
7.6.3.4	Property values	576

| 7.6.4 | **Public attitudes and acceptance** | 576 |

| 7.6.5 | **Minimizing social and environmental concerns** | 576 |

7.7 Prospects for technology improvement and innovation 577

| 7.7.1 | **Research and development programmes** | 577 |

| 7.7.2 | **System-level design and optimization** | 578 |

7.7.3	**Component-level innovation opportunities**	578
7.7.3.1	Advanced tower concepts	578
7.7.3.2	Advanced rotors and blades	578

7.7.3.3	Reduced energy losses and improved availability	579
7.7.3.4	Advanced drive trains, generators, and power electronics	580
7.7.3.5	Manufacturing learning	580
7.7.3.6	Offshore research and development opportunities	580
7.7.4	**The importance of underpinning science**	582

7.8 Cost trends 583

7.8.1	**Factors that affect the cost of wind energy**	583
7.8.2	**Historical trends**	584
7.8.2.1	Investment costs	584
7.8.2.2	Operation and maintenance	584
7.8.2.3	Energy production	585
7.8.3	**Current conditions**	586
7.8.3.1	Investment costs	586
7.8.3.2	Operation and maintenance	587
7.8.3.3	Energy production	587
7.8.3.4	Levelized cost of energy estimates	588
7.8.4	**Potential for further reductions in the cost of wind energy**	589
7.8.4.1	Learning curve estimates	589
7.8.4.2	Engineering model estimates	590
7.8.4.3	Projected levelized cost of wind energy	590

7.9 Potential deployment 591

7.9.1	**Near-term forecasts**	591
7.9.2	**Long-term deployment in the context of carbon mitigation**	591
7.9.3	**Conclusions regarding deployment**	595

References 596

Executive Summary

Wind energy offers significant potential for near-term (2020) and long-term (2050) greenhouse gas (GHG) emissions reductions. A number of different wind energy technologies are available across a range of applications, but the primary use of wind energy of relevance to climate change mitigation is to generate electricity from larger, grid-connected wind turbines, deployed either on- or offshore. Focusing on these technologies, the wind power capacity installed by the end of 2009 was capable of meeting roughly 1.8% of worldwide electricity demand, and that contribution could grow to in excess of 20% by 2050 if ambitious efforts are made to reduce GHG emissions and to address the other impediments to increased wind energy deployment. Onshore wind energy is already being deployed at a rapid pace in many countries, and no insurmountable technical barriers exist that preclude increased levels of wind energy penetration into electricity supply systems. Moreover, though average wind speeds vary considerably by location, ample technical potential exists in most regions of the world to enable significant wind energy deployment. In some areas with good wind resources, the cost of wind energy is already competitive with current energy market prices, even without considering relative environmental impacts. Nonetheless, in most regions of the world, policy measures are still required to ensure rapid deployment. Continued advances in on- and offshore wind energy technology are expected, however, further reducing the cost of wind energy and improving wind energy's GHG emissions reduction potential.

The wind energy market has expanded rapidly. Modern wind turbines have evolved from small, simple machines to large, highly sophisticated devices, driven in part by more than three decades of basic and applied research and development (R&D). Typical wind turbine nameplate capacity ratings have increased dramatically since the 1980s, from roughly 75 kW to 1.5 MW and larger; wind turbine rotors now often exceed 80 m in diameter and are positioned on towers exceeding 80 m in height. The resulting cost reductions, along with government policies to expand renewable energy (RE) supply, have led to rapid market development. From a cumulative capacity of 14 GW by the end of 1999, global installed wind power capacity increased 12-fold in 10 years to reach almost 160 GW by the end of 2009. Most additions have been onshore, but 2.1 GW of offshore capacity was installed by the end of 2009, with European countries embarking on ambitious programmes of offshore wind energy deployment. From 2000 through 2009, roughly 11% of all global newly installed net electric capacity additions (in GW) came from new wind power plants; in 2009 alone, that figure was likely more than 20%. Total investment in wind power plant installations in 2009 equalled roughly USD$_{2005}$ 57 billion, while direct employment in the wind energy sector has been estimated at 500,000. Nonetheless, wind energy remains a relatively small fraction of worldwide electricity supply, and growth has been concentrated in Europe, Asia and North America. The top five countries in cumulative installed capacity by the end of 2009 were the USA, China, Germany, Spain and India. Policy frameworks continue to play a significant role in wind energy utilization.

The global technical potential for wind energy exceeds current global electricity production. Estimates of global technical potential range from a low of 70 EJ/yr (19,400 TWh/yr) (onshore only) to a high of 450 EJ/yr (125,000 TWh/yr) (onshore and near-shore) among those studies that consider relatively more development constraints. Estimates of the technical potential for offshore wind energy alone range from 15 EJ/yr to 130 EJ/yr (4,000-37,000 TWh/yr) when only considering relatively shallower and near-shore applications; greater technical potential is available if also considering deeper water applications that might rely on floating wind turbine designs. Economic constraints, institutional challenges associated with transmission access and operational integration, and concerns about social acceptance and environmental impacts are more likely to restrict growth than is the global technical potential. Ample technical potential also exists in most regions of the world to enable significant wind energy deployment relative to current levels. The wind resource is not evenly distributed across the globe nor uniformly located near population centres, however, and wind energy will therefore not contribute equally in meeting the needs of every country. Research into the effects of global climate change on the geographic distribution and variability of the wind resource is nascent, but research to date suggests that those effects are unlikely to be of a magnitude to greatly impact the global potential for wind energy deployment.

Analysis and operational experience demonstrate that successful integration of wind energy is achievable. Wind energy has characteristics that pose new challenges to electric system planners and operators, such as variable electrical output, limited (but improving) output predictability, and locational dependence. Acceptable wind electricity penetration limits and the operational costs of integration are system-specific, but wind energy has been successfully integrated into existing electric systems; in four countries (Denmark, Portugal, Spain, Ireland), wind energy in 2010 was already able to supply from 10 to roughly 20% of annual electricity demand. Detailed analyses and operating experience primarily from certain Organisation for Economic Co-operation and Development (OECD) countries suggest that, at low to medium levels of wind electricity penetration (up to 20% of total electricity demand), the integration of wind energy generally poses no insurmountable technical barriers and is economically manageable. Concerns about (and the costs of) wind energy integration will grow with wind energy deployment, however, and even at lower penetration levels, integration issues must be addressed. Active management through flexible power generation technologies, wind energy forecasting and output curtailment, and increased coordination and interconnection between electric systems are anticipated. Mass market demand response, bulk energy storage technologies, large-scale deployment of electric vehicles, diverting excess wind energy to fuel production or local heating and geographic diversification of wind power plant siting will also become increasingly beneficial as wind electricity penetration rises. Wind energy technology advances driven by electric system connection standards will increasingly enable wind power plants to become more active participants in maintaining the operability of the electric system. Finally, significant new transmission infrastructure, both on- and offshore, may be required to access areas with higher-quality wind resources. At low to medium levels of wind electricity penetration, the additional costs of managing variability and uncertainty, ensuring generation adequacy and adding new transmission to accommodate wind energy have been estimated to generally be in the range of US cents$_{2005}$ 0.7 to 3/kWh.

Environmental and social issues will affect wind energy deployment opportunities. The energy used and GHG emissions produced in the direct manufacture, transport, installation, operation and decommissioning of wind turbines are small compared to the energy generated and emissions avoided over the lifetime of wind power plants: the GHG emissions intensity of wind energy is estimated to range from 8 to 20 g CO_2/kWh in most instances, whereas energy payback times are between 3.4 and 8.5 months. In addition, managing the variability of wind power output has not been found to significantly degrade the GHG emissions benefits of wind energy. Alongside these benefits, however, wind energy also has the potential to produce some detrimental impacts on the environment and on human activities and well-being. The construction and operation of wind power plants impacts wildlife through bird and bat collisions and through habitat and ecosystem modifications, with the nature and magnitude of those impacts being site- and species-specific. For offshore wind energy, implications for benthic resources, fisheries and marine life must also be considered. Prominent social concerns include visibility/landscape impacts as well various nuisance effects and possible radar interference. Research is also underway on the potential impact of wind power plants on the local climate. As wind energy deployment increases and as larger wind power plants are considered, these existing concerns may become more acute and new concerns may arise. Though attempts to measure the relative impacts of various electricity supply technologies suggest that wind energy generally has a comparatively small environmental footprint, impacts do exist. Appropriate planning and siting procedures can reduce the impact of wind energy development on ecosystems and local communities, and techniques for assessing, minimizing and mitigating the remaining concerns could be further improved. Finally, though community and scientific concerns should be addressed, more proactive planning, siting and permitting procedures may be required to enable more rapid growth in wind energy utilization.

Technology innovation can further reduce the cost of wind energy. Current wind turbine technology has been developed largely for onshore applications, and has converged to three-bladed upwind rotors, with variable speed operation. Though onshore wind energy technology is already commercially manufactured and deployed on a large scale, continued incremental advances are expected to yield improved turbine design procedures, more efficient materials usage, increased reliability and energy capture, reduced operation and maintenance (O&M) costs and longer component lifetimes. In addition, as offshore wind energy gains more attention, new technology challenges arise and more radical technology innovations are possible (e.g., floating turbines). Wind turbine nameplate capacity ratings of 2 to 5 MW have been common for offshore wind power plants, but 10 MW and larger turbines are under consideration. Advances can also be made through more fundamental research to better understand the operating environment in which wind turbines must operate. For onshore wind power plants built in 2009, levelized generation costs in good to excellent wind resource regimes are estimated to average US cents$_{2005}$ 5 to 10/kWh, reaching US cents$_{2005}$ 15/kWh in lower resource areas. Offshore wind energy has typical levelized generation costs that are estimated to range from US cents$_{2005}$ 10/kWh to more than US cents$_{2005}$ 20/kWh for recently built or planned plants located in relatively shallow water. Reductions in the levelized cost of onshore wind energy of 10 to 30% by 2020 are often reported in the literature. Offshore wind energy is often found to have somewhat greater potential for cost reductions: 10 to 40% by 2020.

Wind energy offers significant potential for near- and long-term GHG emissions reductions. Given the commercial maturity and cost of onshore wind energy technology, wind energy offers the potential for significant near-term GHG emissions reductions: this potential is not conditioned on technology breakthroughs, and no insurmountable technical barriers exist that preclude increased levels of wind electricity penetration. As technology advances continue, greater contributions to GHG emissions reductions are possible in the longer term. Based on a review of the literature on the possible future contribution of RE supplies to meeting global energy needs under a range of GHG concentration stabilization scenarios, wind energy's contribution to global electricity supply could rise from 1.8% by the end of 2009 to 13 to 14% by 2050 in the median scenario for GHG concentration stabilization ranges of 440 to 600 and <440 ppm CO_2. At the 75th percentile of reviewed scenarios, and under similarly ambitious efforts to reduce GHG emissions, wind energy's contribution is shown to grow to 21 to 25% by 2050. Achieving the higher end of this range would be likely to require not only economic support policies of adequate size and predictability, but also an expansion of wind energy utilization regionally, increased reliance on offshore wind energy, technical and institutional solutions to transmission constraints and operational integration concerns, and proactive efforts to mitigate and manage social and environmental concerns. Additional R&D is expected to lead to incremental cost reductions for onshore wind energy, and enhanced R&D expenditures may be especially important for offshore wind energy technology. Finally, for those markets with good wind resources but that are new to wind energy deployment, both knowledge and technology transfer may help facilitate early wind power plant installations.

7.1 Introduction

This chapter addresses the potential role of wind energy in reducing GHG emissions. Wind energy (in many applications) is a mature renewable energy RE source that has been successfully deployed in many countries. It is technically and economically capable of significant continued expansion, and its further exploitation may be a crucial aspect of global GHG reduction strategies. Though average wind speeds vary considerably by location, the world's technical potential for wind energy exceeds global electricity production, and ample technical potential exists in most regions of the world to enable significant wind energy deployment.

Wind energy relies, indirectly, on the energy of the sun. A small proportion of the solar radiation received by the Earth is converted into kinetic energy (Hubbert, 1971), the main cause of which is the imbalance between the net outgoing radiation at high latitudes and the net incoming radiation at low latitudes. The Earth's rotation, geographic features and temperature gradients affect the location and nature of the resulting winds (Burton et al., 2001). The use of wind energy requires that the kinetic energy of moving air be converted to useful energy. As a result, the economics of using wind for electricity supply are highly sensitive to local wind conditions and the ability of wind turbines to reliably extract energy over a wide range of typical wind speeds.

Wind energy has been used for millennia (for historical overviews, see, e.g., Gipe, 1995; Ackermann and Soder, 2002; Pasqualetti et al., 2004; Musgrove, 2010). Sailing vessels relied on the wind from before 3,000 BC, with mechanical applications of wind energy in grinding grain, pumping water and powering factory machinery following, first with vertical axis devices and subsequently with horizontal axis turbines. By 200 BC, for example, simple windmills in China were pumping water, while vertical axis windmills were grinding grain in Persia and the Middle East. By the 11th century, windmills were used in food production in the Middle East; returning merchants and crusaders carried this idea back to Europe. The Dutch and others refined the windmill and adapted it further for industrial applications such as sawing wood, making paper and draining lakes and marshes. When settlers took this technology to the New World in the late 19th century, they began using windmills to pump water for farms and ranches. Industrialization and rural electrification, first in Europe and later in the USA, led to a gradual decline in the use of windmills for mechanical applications. The first successful experiments with the use of wind to generate electricity are often credited to James Blyth (1887), Charles Brush (1887), and Poul la Cour (1891). The use of wind electricity in rural areas and, experimentally, in larger-scale applications, continued throughout the mid-1900s. However, the use of wind to generate electricity at a commercial scale became viable only in the 1970s as a result of technical advances and government support, first in Denmark at a relatively small scale, then at a much larger scale in California (1980s), and then in Denmark, Germany and Spain (1990s).

The primary use of wind energy of relevance to climate change mitigation is to generate electricity from larger, grid-connected wind turbines, deployed either in a great number of smaller wind power plants or a smaller number of much larger plants. As of 2010, such turbines often stand on tubular towers exceeding 80 m in height, with three-bladed rotors that often exceed 80 m in diameter; commercial machines with rotor diameters and tower heights in excess of 125 m are operating, and even larger machines are under development. Wind power plants are commonly sited on land (termed 'onshore' in this chapter): by the end of 2009, wind power plants sited in sea- or freshwater were a relatively small proportion of global wind power installations. Nonetheless, as wind energy deployment expands and as the technology advances, offshore wind energy is expected to become a more significant source of overall wind energy supply.

Due to their potential importance to climate change mitigation, this chapter focuses on grid-connected on- and offshore wind turbines for electricity production. Notwithstanding this focus, wind energy has served and will continue to meet other energy service needs. In remote areas of the world that lack centrally provided electricity supplies, smaller wind turbines can be deployed alone or alongside other technologies to meet individual household or community electricity demands; small turbines of this nature also serve marine energy needs. Small island or remote electricity grids can also employ wind energy, along with other energy sources. Even in urban settings that already have ready access to electricity, smaller wind turbines can, with careful siting, be used to meet a portion of building energy needs. New concepts for higher-altitude wind energy machines are also under consideration. Moreover, in addition to electricity supply, wind energy can meet mechanical and propulsion needs in specific applications. Though not the focus of this chapter, some of these additional applications and technologies are briefly summarized in Box 7.1.

Drawing on available literature, this chapter begins by describing the global technical potential for wind energy, the regional distribution of that resource, and the possible impacts of climate change on the resource (Section 7.2). The chapter then reviews the status of and trends in modern onshore and offshore wind energy technology (Section 7.3). The chapter discusses the status of the wind energy market and industry developments, both globally and regionally, and the impact of policies on those developments (Section 7.4). Near-term issues associated with the integration of wind energy into electricity supply systems are addressed (Section 7.5), as is available evidence on the environmental and social impacts of wind energy (Section 7.6). The prospects for further technology improvement and innovation are summarized (Section 7.7), and historical, current and potential future cost trends are reviewed (Section 7.8). Based on the underpinnings offered in previous sections, the chapter concludes with an examination of the potential future deployment of wind energy, focusing on the GHG reduction and energy scenarios literature (Section 7.9).

Box 7.1 | Alternative wind energy applications and technologies

Beyond the use of large, modern wind turbines for electricity supply, a number of additional wind energy applications and technologies are currently employed or are under consideration, a subset of which are described here. Though these technologies and applications are at different phases of market development, and each holds a certain level of promise for scaled deployment, none are likely to compete with traditional large on- and offshore wind energy technology from the perspective of GHG emissions reductions, at least in the near to medium term.

Small wind turbines for electricity supply. Smaller-scale wind turbines are used in a wide range of applications. Though wind turbines from hundreds of watts to tens of kilowatts in size do not benefit from the economies of scale that have helped reduce the cost of larger wind turbines, they can be economically competitive with other supply alternatives in areas that do not have access to centrally provided electricity supply, providing electricity services to meet a wide variety of household or community energy needs (Byrne et al., 2007). For rural electrification or isolated areas, small wind turbines can be used on a stand-alone basis for battery charging or can be combined with other supply options (e.g., solar and/or diesel) in hybrid systems. As an example, China had 57 MW of cumulative small wind turbine (<100 kW) capacity installed by the end of 2008 (Li and Ma, 2009); 33 MW were reportedly installed in China in 2009. Small wind turbines are also employed in grid-connected applications for both residential and commercial electricity customers. The use of wind energy in these disparate applications can provide economic and social development benefits. In urban settings, however, where the wind resource is highly site-specific and can be poor, the GHG emissions savings associated with the displacement of grid electricity can be low or even zero once the manufacture and installation of the turbines are taken into account (Allen et al., 2008; Carbon Trust, 2008a). AWEA (2009) estimates annual global installations of <100 kW wind turbines from leading manufacturers at under 40 MW in 2008.

Wind energy to meet mechanical and propulsion needs. Among the first technologies to harness the energy from the wind were those that used the kinetic energy of the wind as a means of marine propulsion, grinding of grain and water pumping. Though these technologies were first developed long ago, opportunities remain for the expanded use of wind energy to meet a wide range of mechanical and propulsion needs. Using wind energy to pump water to serve domestic, agricultural and ranching needs remains important, for example, especially in certain remote areas (e.g., Purohit, 2007); the mechanical or electrical use of wind energy can also be applied for, among other things, water desalination and purification (e.g., Miranda and Infield, 2002). New concepts to harness the energy of the wind for propulsion are also under development, such as using large kites to complement diesel engines for marine transport. Demonstration projects and analytic studies have found that these systems may yield fuel savings of up to 50%, though this depends heavily on the technology and wind conditions (O'Rourke, 2006; Naaijen and Koster, 2007).

Higher-altitude wind electricity. Higher-altitude wind energy systems have recently received some attention as an alternative approach to generating electricity from the wind (Roberts et al., 2007; Archer and Caldeira, 2009; Argatov et al., 2009; Argatov and Silvennoinen, 2010; Kim and Park, 2010). A principal motivation for the development of this technology is the sizable wind resource present at higher altitudes. Two main approaches to higher-altitude wind energy have been proposed: (1) tethered wind turbines that transmit electricity to earth via cables, and (2) base stations that convert the kinetic energy from the wind collected via kites to electricity at ground level. A variety of concepts are under consideration, operating at altitudes of less than 500 m to more than 10,000 m. Though some research has been conducted on these technologies and on the size of the potential resource, the technology remains in its infancy, and scientific, economic and institutional challenges must be overcome before pilot projects are widely deployed and a realistic estimate of the GHG emissions reduction potential of higher-altitude wind energy can be developed.

7.2 Resource potential[1]

The theoretical potential for wind, as estimated by the global annual flux, has been estimated at 6,000 EJ/yr (Rogner et al., 2000). The global technical potential for wind energy, meanwhile, is not fixed, but is instead related to the status of the technology and assumptions made regarding other constraints to wind energy development. Nonetheless, a growing number of global wind resource assessments have demonstrated that the world's technical potential for wind energy exceeds current global electricity production, and that ample technical potential exists in most regions of the world to enable significant wind energy deployment relative to current levels. The wind resource is not evenly distributed across

---
1   See Annex I for definitions of the terms used to refer to various types of "resource potential."

the globe, however, and a variety of other regional factors are likely to restrict growth well before any absolute global technical resource limits are encountered. As a result, wind energy will not contribute equally in meeting the needs of every country.

This section summarizes available evidence on the size of the global technical potential of the wind energy resource (Section 7.2.1), the regional distribution of that resource (Section 7.2.2) and the possible impacts of climate change on wind energy resources (Section 7.2.3). It focuses on long-term average annual technical potential; for a discussion of interannual, seasonal and diurnal fluctuations and patterns in the wind resource, as well as shorter-term wind power output variability, see Section 7.5.

7.2.1 Global technical potential

A number of studies have evaluated the global technical potential for wind energy. In general, two methods can be used: first, available wind speed measurements can be interpolated to construct a surface wind distribution; and second, physics-based numerical weather prediction models can be applied. Studies of the global wind energy resource have used varying combinations of these two approaches.[2] Additionally, it is important to recognize that estimates of the technical potential for wind energy should not be viewed as fixed—the potential will change as wind energy technology develops (e.g., taller towers provide access to better wind, or foundation innovation allows offshore plants to be developed in greater water depths) and as more is learned about technical, environmental and social concerns that may influence development (e.g., land competition, distance from resource areas to electricity demand centres, etc.).

Synthesizing the available literature, the IPCC's Fourth Assessment Report identified 600 EJ/yr of onshore wind energy technical potential (IPCC, 2007). Using the direct equivalent method of deriving primary energy equivalence (where electricity supply, in TWh, is translated directly to primary energy, in EJ; see Annex II), the IPCC (2007) estimate of onshore wind energy technical potential is 180 EJ/yr (50,000 TWh/yr), more than two times greater than gross global electricity production in 2008 (73 EJ, or 20,200 TWh).[3] Of this 180 EJ/y, only 0.8 EJ (220 TWh, 0.4% of the estimated technical potential) was being used for wind energy supply in 2008 (IEA, 2010a).

More generally, a number of analyses have been undertaken to estimate the global technical potential for wind energy. The methods and results of these global assessments—some of which include offshore wind energy and some of which are restricted to onshore wind energy—are summarized in Table 7.1.

No standardized approach has been developed to estimate the global technical potential of wind energy: the diversity in data, methods, assumptions and even definitions for technical potential complicate comparisons. Consequently, the studies show a wide range of results. Specifically, estimates of global technical potential range from a low of 70 EJ/yr (19,400 TWh/yr) (onshore only) to a high of 450 EJ/yr (125,000 TWh/yr) (onshore and near-shore) among those studies that consider relatively more development constraints (identified as 'more constraints' in the table). This range equals from roughly one to six times global electricity production in 2008. If those studies that apply more limited development constraints are also included, the absolute range of technical potential is greater still, from 70 EJ/yr to 3,050 EJ/yr (19,400 to 840,000 TWh/yr). Results vary based in part on whether offshore wind energy is included (and under what assumptions), the wind speed data that are used, the areas assumed available for wind energy development, the rated output of wind turbines installed per unit of land area, and the assumed performance of wind power plants. The latter is, in part, related to hub height and turbine technology. These factors depend on technical assumptions as well as subjective judgements of development constraints, thus there is no single 'correct' estimate of technical potential.

Though research has generally found the technical potential for offshore wind energy to be smaller than for onshore wind energy, the technical potential is nonetheless sizable. Three of the studies included in Table 7.1 exclude the technical potential of offshore wind energy; even those studies that include offshore wind energy often do so only considering the wind energy technology likely to be deployed in the near to medium term in relatively shallower water and nearer to shore. In practice, the size of the offshore wind energy resource is, at least theoretically, enormous, and constraints are primarily economic rather than technical. In particular, water depth, accessibility and grid connection may limit development to relatively near-shore locations in the medium term, though technology improvements are expected, over time, to enable deeper water and more remote installations. Even when only considering relatively shallower and near-shore applications, however, study results span a range from 15 to 130 EJ/yr (4,000 to 37,000 TWh/yr),

---

[2] Wind power plant developers may rely upon global and regional wind resource estimates to obtain a general sense for the locations of potentially promising development prospects. However, on-site collection of actual wind speed data at or near turbine hub heights remains essential for most wind power plants of significant scale.

[3] The IPCC (2007) cites Johansson et al. (2004), which obtains its data from UNDP/UNEP/WEC (2000), which in turn references WEC (1994) and Grubb and Meyer (1993). To convert from TWh to EJ, the documents cited by IPCC (2007) use the standard conversion, and then divide by 0.3 (i.e.., a method of energy accounting in which RE supply is assumed to substitute for the primary energy of fossil fuel inputs into fossil power plants, accounting for plant conversion efficiencies). The direct equivalent method does not take this last step, and instead counts the electricity itself as primary energy (see Annex II), so this chapter reports the IPCC (2007) figure at 180 EJ/y, or roughly 50,000 TWh/y.

Table 7.1 | Global assessments of the technical potential for wind energy.

Study	Scope	Methods and Assumptions[1]	Results[2]
Krewitt et al. (2009)	Onshore and offshore	Updated Hoogwijk and Graus (2008), itself based on Hoogwijk et al. (2004), by revising offshore wind power plant spacing by 2050 to 16 MW/km^2	Technical (more constraints): 121,000 TWh/yr; 440 EJ/yr
Lu et al. (2009)	Onshore and offshore	>20% capacity factor (Class 1); 100 m hub height; 9 MW/km^2 spacing; based on coarse simulated model data set; exclusions for urban and developed areas, forests, inland water, permanent snow/ice; offshore assumes 100 m hub height, 6 MW/km^2, <92.6 km from shore, <200m depth, no other exclusions	Technical (limited constraints): 840,000 TWh/yr; 3,050 EJ/yr
Hoogwijk and Graus (2008)	Onshore and offshore	Updated Hoogwijk et al. (2004) by incorporating offshore wind energy, assuming 100 m hub height for onshore, and altering cost assumptions; for offshore, study updates and adds to earlier analysis by Fellows (2000); other assumptions as listed below under Hoogwijk et al. (2004); constrained technical potential defined here in economic terms separately for onshore and offshore	Technical/Economic (more constraints): 110,000 TWh/yr; 400 EJ/yr
Archer and Jacobson (2005)	Onshore and near-Shore	>Class 3; 80 m hub height; 9 MW/km^2 spacing; 48% average capacity factor; based on wind speeds from surface stations and balloon-launch monitoring stations; near-shore wind energy effectively included because resource data includes buoys (see study for details); constrained technical potential = 20% of total technical potential	Technical (limited constraints): 627,000 TWh/yr; 2,260 EJ/yr Technical (more constraints): 125,000 TWh/yr; 450 EJ/yr
WBGU (2004)	Onshore and offshore	Multi-MW turbines; based on interpolation of wind speeds from meteorological towers; exclusions for urban areas, forest areas, wetlands, nature reserves, glaciers, and sand dunes; local exclusions accounted for through corrections related to population density; offshore to 40 m depth, with sea ice and minimum distance to shore considered regionally; constrained technical potential (authors define as 'sustainable' potential) = 14% of total technical potential	Technical (limited constraints): 278,000 TWh/yr; 1,000 EJ/yr Technical (more constraints): 39,000 TWh/yr; 140 EJ/yr

Continued next Page →

while far greater technical potential is found when considering deeper water applications that might rely on floating wind turbine designs.[4]

---

4 Relatively few studies have investigated the global offshore technical wind energy resource potential, and neither Archer and Jacobson (2005) nor WBGU et al. (2004) report offshore potential separately from the total technical potential reported in Table 7.1. In one study of global technical potential considering development constraints, Leutz et al. (2001) estimate an offshore wind energy potential of 130 EJ/yr (37,000 TWh/yr) at depths less than 50 m. Building from Fellows (2000) and Hoogwijk and Graus (2008), Krewitt et al. (2009) estimate a global offshore wind energy technical potential of 57 EJ/yr by 2050 (16,000 TWh/yr). (Fellows (2000) provides an estimate of 15 EJ/yr, or more than 4,000 TWh/yr, whereas Hoogwijk and Graus (2008) estimate 23 EJ/yr, or 6,100 TWh/yr; see Table 7.1 for assumptions.) In another study, Siegfriedsen et al. (2003) calculate the technical potential of offshore wind energy outside of Europe as 17 EJ/yr (4,600 TWh/yr). Considering greater water depths and distances to shore, Lu et al. (2009) estimate an offshore wind energy resource potential of 540 EJ/yr (150,000 TWh/yr) at water depths less than 200 m and at distances less than 92.6 km from shore, of which 150 EJ/yr (42,000 TWh/yr) is available at depths of less than 20 m, though this study does not consider as many development constraints or exclusion zones as the other estimates listed here. Capps and Zender (2010) similarly do not consider many development constraints (except that the authors exclude all area within 30 km off shore), and find that the technical potential for offshore wind energy increases from 224 EJ/yr (62,000 TWh/yr) to 1,260 EJ/yr (350,000 TWh/yr) when maximum water depth increases from 45 m to 200 m. A number of regional studies have been completed as well, including (but not limited to) those that have estimated the size of the offshore wind energy resource in the EU (Matthies et al., 1995; Delft University et al., 2001; EEA, 2009), the USA (Kempton et al., 2007; Jiang et al., 2008; Schwartz et al., 2010) and China (CMA, 2006; Xiao et al., 2010).

There are two main reasons to believe that some these studies of on- and offshore wind energy may understate the global technical potential. First, several of the studies are dated, and considerable advances have occurred in both wind energy technology (e.g., hub height) and resource assessment methods. Partly as a result, the more recent studies listed in Table 7.1 often calculate larger technical potentials than the earlier studies. Second, even some of the more recent studies may understate the global technical potential for wind energy due to methodological limitations. The global assessments described in this section often use relatively simple analytical techniques with coarse spatial resolutions, rely on interpolations of wind speed data from a limited number (and quality) of surface stations, and apply limited validation from wind speed measurements in prime wind resource areas. Enabled in part by an increase in computing power, more sophisticated and finer geographic resolution atmospheric modelling approaches are beginning to be applied (and increasingly validated with higher-quality measurement data) on a country or regional basis, as described in more depth in Section 7.2.2. Experience shows that these techniques have often identified greater technical potential for wind energy than have earlier global assessments (see Section 7.2.2).

There are, however, at least two other issues that may suggest that the estimates of global technical potential have been overstated. First, global

Study	Scope	Methods and Assumptions[1]	Results[2]
Hoogwijk et al. (2004)	Onshore	>4 m/s at 10 m (some less than Class 2); 69 m hub height; 4 MW/km² spacing; assumptions for availability / array efficiency; based on interpolation of wind speeds from meteorological towers; exclusions for elevations >2000 m, urban areas, nature reserves, certain forests; reductions in use for many other land-uses; economic potential defined here as less than US cents$_{2005}$ 10/kWh	Technical (more constraints): 96,000 TWh/yr 350 EJ/yr Economic: (more constraints): 53,000 TWh/yr 190 EJ/yr
Fellows (2000)	Onshore and offshore	50 m hub height; 6 MW/km² spacing; based on upper-air model data set; exclusions for urban areas, forest areas, nature areas, water bodies and steep slopes; additional maximum density criterion; offshore assumes 60 m hub height, 8 MW/km² spacing, to 40 m depth, 5 to 40 km from shore, with 75% exclusion; constrained technical potential defined here in economic terms: less than US cents$_{2005}$ 23/kWh in 2020; focus on four regions, with extrapolations to others; some countries omitted altogether	Technical/Economic (more constraints): 46,000 TWh/yr 170 EJ/yr
WEC (1994)	Onshore	>Class 3; 8 MW/km² spacing; 23% average capacity factor; based on an early global wind resource map; constrained technical potential = 4% of total technical potential	Technical (limited constraints): 484,000 TWh/yr 1,740 EJ/yr Technical (more constraints): 19,400 TWh/yr 70 EJ/yr
Grubb and Meyer (1993)	Onshore	>Class 3; 50 m hub height; assumptions for conversion efficiency and turbine spacing; based on an early global wind resource map; exclusions for cities, forests and unreachable mountain areas, as well as for social, environmental and land use constraints, differentiated by region (results in constrained technical potential = ~10% of total technical potential, globally)	Technical (limited constraints): 498,000 TWh/yr 1,800 EJ/yr Technical (more constraints): 53,000 TWh/yr 190 EJ/yr

Notes: 1. Where used, wind resource classes refer to the following wind power densities at a 50 m hub height: Class 1 (<200 W/m²), Class 2 (200-300 W/m²), Class 3 (300-400 W/m²), Class 4 (400-500 W/m²), Class 5 (500-600 W/m²), Class 6 (600-800 W/m²) and Class 7 (>800 W/m²). 2. Reporting of resource potential and conversion between EJ and TWh are based on the direct equivalent method (see Annex II). Definitions for theoretical, technical, economic, sustainable and market potential are provided in Annex I, though individual authors cited in Table 7.1 often use different definitions of these terms. In particular, several of the studies included in the table report technical potential only below a maximum cost threshold. These are identified as 'economic potential' in the table though it is acknowledged that this definition differs from that provided in Annex I.

assessments may overstate the accessibility of the wind resource in remote areas that are far from population centres. Second, the assessments generally use point-source estimates of the wind resource, and assess the global technical potential for wind energy by summing local wind technical potentials. Large-scale atmospheric dynamics, thermodynamic limits, and array effects, however, may bound the aggregate amount of energy that can be extracted by wind power plants on a regional or global basis. Relatively little is known about the nature of these constraints, though early research suggests that the size of the effects are unlikely to be large enough to significantly constrain the use of wind energy in the electricity sector at a global scale (see Section 7.6.2.3).

Despite the limitations of the available literature, based on the above review, it can be concluded that the IPCC (2007) estimate of 180 EJ/yr (50,000 TWh/yr) likely understates the technical potential for wind energy. Moreover, regardless of the exact size of the technical potential, it is evident that the global wind resource is unlikely to be a limiting factor on global on- or offshore wind energy deployment. Instead, economic constraints associated with the cost of wind energy, institutional constraints and costs associated with transmission access and operational integration, and issues associated with social acceptance and environmental impacts are likely to restrict growth well before any absolute limit to the global technical potential for wind energy is encountered.

7.2.2 Regional technical potential

7.2.2.1 Global assessment results by region

The global assessments presented in Section 7.2.1 reach varying conclusions about the relative technical potential for onshore wind energy among different regions, with Table 7.2 summarizing results from a subset of these assessments. Differences in the regional results from these studies are due to differences in wind speed data and key input parameters, including the minimum wind speed assumed to be exploitable, land use constraints, density of wind energy development, and assumed wind power plant performance (Hoogwijk et al., 2004); differing regional categories also

Table 7.2 | Regional allocation of global technical potential for onshore wind energy.[1]

Grubb and Meyer (1993)		WEC (1994)		Krewitt et al. (2009)[2]		Lu et al. (2009)	
Region	%	Region	%	Region	%	Region	%
Western Europe	9	Western Europe	7	OECD Europe	5	OECD Europe	4
North America	26	North America	26	OECD North America	42	North America	22
Latin America	10	Latin America and Caribbean	11	Latin America	10	Latin America	9
Eastern Europe and Former Soviet Union	20	Eastern Europe and CIS	22	Transition Economies	17	Non-OECD Europe and Former Soviet Union	26
Africa	20	Sub-Saharan Africa	7	Africa and Middle East	9	Africa and Middle East	17
Australia	6	Middle East and North Africa	8	OECD Pacific	14	Oceania	13
Rest of Asia	9	Pacific	14	Rest of Asia	4	Rest of Asia	9
		Rest of Asia	4				

Notes: 1. Regions shown in the table are defined by each individual study. Some regions have been combined to improve comparability among the four studies. 2. Hoogwijk and Graus (2008) and Hoogwijk et al. (2004) show similar results.

complicate comparisons. Nonetheless, the technical potentials in OECD North America and Eastern Europe/Eurasia are found to be particularly sizable, whereas some areas of non-OECD Asia and OECD Europe appear to have more limited onshore technical potential. Visual inspection of Figure 7.1, a global wind resource map with a 5- by 5-km resolution, also demonstrates limited technical potential in certain areas of Latin America and Africa, though other portions of those continents have significant technical potential. Caution is required in interpreting these results, however, as other studies find significantly different regional allocations of global technical potential (e.g., Fellows, 2000), and more detailed country and regional assessments have reached differing conclusions about, for example, the wind energy resource in East Asia and other regions (Hoogwijk and Graus, 2008).

Hoogwijk et al. (2004) also compare onshore technical potential against regional electricity consumption in 1996. In most of the 17 regions evaluated, technical onshore wind energy potential exceeded electricity consumption in 1996. The multiple was over five in 10 regions: East Africa, Oceania, Canada, North Africa, South America, Former Soviet Union (FSU), Central America, West Africa, the USA and the Middle East. Areas in which onshore wind energy technical potential was estimated to be less than a two-fold multiple of 1996 electricity consumption were South Asia (1.9), Western Europe (1.6), East Asia (1.1), South Africa (1), Eastern Europe (1), South East Asia (0.1) and Japan (0.1), though again, caution is warranted in interpreting these results. More recent resource assessments and data on regional electricity consumption would alter these figures.

The estimates reported in Table 7.2 exclude offshore wind energy technical potential. Ignoring deeper water applications, Krewitt et al. (2009) estimate that of the 57 EJ/yr (16,000 TWh/yr) of technical offshore resource potential by 2050, the largest opportunities exist in OECD Europe (22% of global potential), the rest of Asia (21%), Latin America (18%) and the transition economies (16%), with lower but still significant technical potential in North America (12%), OECD Pacific (6%) and Africa and the Middle East (4%).

Overall, these studies find that ample technical potential exists in most regions of the world to enable significant wind energy deployment relative to current levels. The wind resource is not evenly distributed across the globe, however, and a variety of other regional factors (e.g., distance of resource from population centres, grid integration, social acceptance) are likely to restrict growth well before any absolute limit to the technical potential of wind energy is encountered. As a result, wind energy will not contribute equally in meeting the energy needs and GHG reduction demands of every region or country.

7.2.2.2 Regional assessment results

The global wind resource assessments described above have historically relied primarily on relatively coarse and imprecise estimates of the wind resource, sometimes relying heavily on measurement stations with relatively poor exposure to the wind (Elliott, 2002; Elliott et al., 2004).[5]

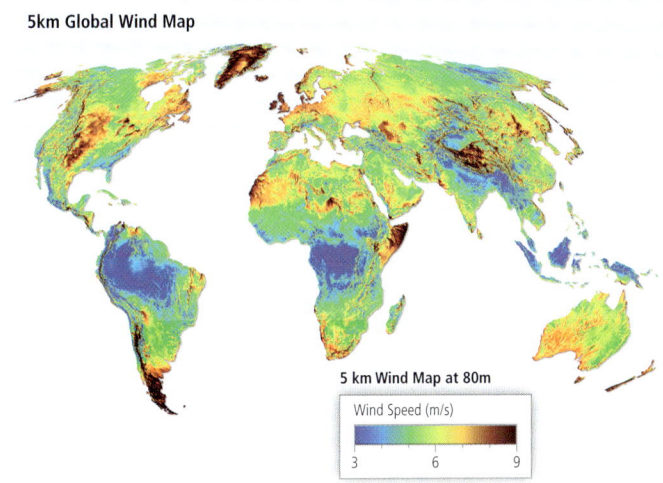

Figure 7.1 | Example global wind resource map with 5 km x 5 km resolution (3TIER, 2009).

---

[5] For more on the relative advantages and disadvantages of weather station measurement data and numerical weather prediction models, see Al-Yahyai et al. (2010).

The regional results from these global assessments, as presented in Section 7.2.2.1, should therefore be viewed with some caution, especially in areas where wind measurement data are of limited quantity and quality. In contrast, specific country and regional assessments have benefited from: wind speed data collected with wind resource estimation in mind; sophisticated numerical wind resource prediction techniques; improved model validation; and a dramatic growth in computing power. These advances have allowed the most recent country and regional resource assessments to capture smaller-scale terrain features and temporal variations in predicted wind speeds, and at a variety of possible turbine heights.

These techniques were initially applied in the EU[6] and the USA[7], but there are now publicly available high-resolution wind resource assessments covering a large number of regions and countries. The United Nations Environment Program's Solar and Wind Energy Resource Assessment, for example, provides wind resource information for a large number of its partner countries around the world;[8] the European Bank for Reconstruction and Development has developed RE assessments in its countries of operation (Black and Veatch, 2003); the World Bank's Asia Sustainable and Alternative Energy Program has prepared wind resource atlases for the Pacific Islands and Southeast Asia;[9] and wind resource assessments for portions of the Mediterranean region are available through Observatoire Méditerranéen de l'Energie.[10] A number of other publicly available country-level assessments have been produced by the US National Renewable Energy Laboratory,[11] Denmark's Risø DTU[12] and others. These assessments have sometimes proven especially helpful in catalyzing initial interest in wind energy. To illustrate the advances that have occurred outside of the EU and the USA, Box 7.2 presents details on the status of wind resource assessment in China (a country with significant wind energy deployment) and Russia (a country with significant wind energy technical potential).

These more detailed assessments have generally found the size of the wind resource to be greater than estimated in previous global or regional assessments. This is due primarily to improved data, spatial resolution and analytic techniques, but is also the result of wind turbine technology developments, for example, higher hub heights and improved machine efficiencies (see, e.g., Elliott, 2002; Elliot et al., 2004). Nevertheless, even greater spatial and temporal resolution and enhanced validation of model results with observational data are needed, as is an expanded geographic coverage of these assessments (see, e.g., Schreck et al., 2008; IEA, 2009). These developments will allow further refinement of estimates of the technical potential, and are likely to highlight regions with high-quality technical potential that have not previously been identified.

7.2.3 Possible impact of climate change on resource potential

Global climate change may alter the geographic distribution and/or the inter- and intra-annual variability of the wind resource, and/or the quality of the wind resource, and/or the prevalence of extreme weather events that may impact wind turbine design and operation. Research in this field is nascent, however, and global and regional climate models do not fully reproduce contemporary wind climates (Goyette et al., 2003) or historical trends (Pryor et al., 2009). Additional uncertainty in wind resource projections under global climate change scenarios derives, in part, from substantial variations in simulated circulation and flow regimes when using different climate models (Pryor et al., 2005, 2006; Bengtsson et al., 2009; Pryor and Schoof, 2010). Nevertheless, research to date suggests that it is unlikely that multi-year annual mean wind speeds will change by more than a maximum of ±25% over most of Europe and North America during the present century, while research covering northern Europe suggests that multi-year annual mean wind power densities will likely remain within ±50% of current values (Palutikof et al., 1987, 1992; Breslow and Sailor, 2002; Pryor et al., 2005, 2006; Walter et al., 2006; Bloom et al., 2008; Sailor et al., 2008; Pryor and Schoof, 2010). Fewer studies have been conducted for other regions of the world, though Brazil's wind resource was shown in one study to be relatively insensitive to (and perhaps to even increase as a result of) global climate change (de Lucena et al., 2009), and simulations for the west coast of South America showed increases in mean wind speeds of up to 15% (Garreaud and Falvey, 2009).

In addition to the possible impact of climate change on long-term average wind speeds, impacts on intra-annual, interannual and inter-decadal variability in wind speeds are also of interest. Wind climates in northern Europe, for example, exhibit seasonality, with the highest wind speeds during the winter (Rockel and Woth, 2007), and some analyses of the northeast Atlantic (1874 to 2007) have found notable differences in temporal trends in winter and summer (X. Wang et al., 2009). Internal climate modes have been found to be responsible for relatively high intra-annual, interannual and inter-decadal variability in wind climates in the mid-latitudes (e.g., Petersen et al., 1998; Pryor et al., 2009). The ability of climate models to accurately reproduce these conditions in current and possible future climates is the subject of intense research (Stoner et al., 2009). Equally, the degree to which historical variability and change in near-surface wind climates is attributable to global climate change or to other factors (Pryor et al., 2009; Pryor and Ledolter,

---

6   For the latest publicly available European wind resource map, see www.windatlas.dk/Europe/Index.htm. Publicly available assessments for individual EU countries are summarized in EWEA (2009); see also EEA (2009).

7   A large number of publicly available US wind resource maps have been produced at the national and state levels, many of which have subsequently been validated by the National Renewable Energy Laboratory (see www.windpoweringamerica.gov/wind_maps.asp).

8   See http://swera.unep.net/.

9   See go.worldbank.org/OTU2DVLIV0.

10 See www.omenergie.com/.

11 See www.nrel.gov/wind/international_wind_resources.html.

12 See www.windatlas.dk/World/About.html.

Box 7.2 | Advances in wind resource assessment in China and Russia

To illustrate the growing use of sophisticated wind resource assessment tools outside of the EU and the USA, historical and ongoing efforts in China and Russia to better characterize their wind resources are described here. In both cases, the wind energy resource has been found to be sizable compared to present electricity consumption, and recent analyses offer enhanced understanding of the size and location of those resources.

China's Meteorological Administration (CMA) completed its first wind resource assessment in the 1970s. In the 1980s, a second wind resource investigation was performed based on data from roughly 900 meteorological stations, and a spatial distribution of the resource was delineated. The CMA estimated the availability of 253 GW (510 TWh/yr at a 23% average capacity factor; 1.8 EJ/yr) of onshore technical potential (Xue et al., 2001). A third assessment was based on data from 2,384 meteorological stations, supplemented with data from other sources. Though still mainly based on measured wind speeds at 10 m, most data covered a period of over 50 years, and this assessment led to an estimate of 297 GW (600 TWh/yr at a 23% average capacity factor; 2.2 EJ/yr) of onshore technical potential (CMA, 2006). More recently, improved mesoscale atmospheric models and access to higher-elevation meteorological station data have facilitated higher-resolution assessments. Figure 7.2 (left panel) shows the results of these investigations, focused on onshore wind resources. Based on this research, the CMA has estimated 2,380 GW of onshore (4,800 TWh/yr at a 23% average capacity factor; 17 EJ/yr) and 200 GW of offshore (610 TWh/yr at a 35% average capacity factor; 2.2 EJ/yr) technical potential (Xiao et al., 2010). Other recent research has similarly estimated far greater technical potential than have past assessments (see, e.g., McElroy et al., 2009).

Considerable progress has also been made in understanding the magnitude and distribution of the wind energy resource in Russia (as well as the other Commonwealth of Independent States (CIS) countries and the Baltic countries), based in part on data from approximately 3,600 surface meteorological stations and 150 upper-air stations. An assessment by Nikolaev et al. (2010) uses these data and meteorological and statistical modelling to estimate the distribution of the wind resource in the region (Figure 7.2 (right panel)). Based on this work and after making assumptions about the characteristics and placement of wind turbines, Nikolaev et al. (2008) estimate that the technical potential for wind energy in Russia is more than 14,000 TWh/yr (50 EJ/yr). The more promising regions of Russia for wind energy development are in the western part of the country, the South Ural area, in western Siberia, and on the coasts of the seas of the Arctic and Pacific Oceans.

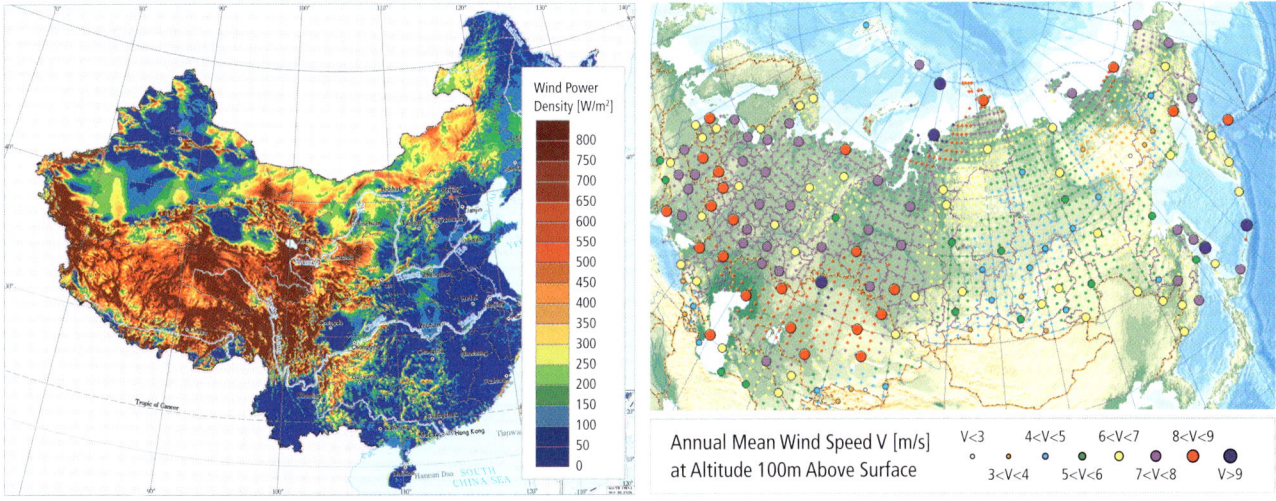

Figure 7.2 | Wind resource maps for (left panel) China (Xiao et al., 2010) and (right panel) Russia, CIS, and the Baltic (Nikolaev et al., 2010).

2010), and whether that variability will change as the global climate continues to evolve, is also being investigated.

Finally, the prevalence of extreme winds and the probability of icing have implications for wind turbine design and operation (X. Wang et al., 2009). Preliminary studies from northern and central Europe show some evidence of increased wind speed extremes (Pryor et al., 2005; Haugen and Iversen, 2008; Leckebusch et al., 2008), though changes in the occurrence of inherently rare events are difficult to quantify, and further research is warranted. Sea ice can impact turbine foundation loading for offshore plants, and changes in sea ice and/or permafrost conditions may also influence access for performing wind power plant O&M (Laakso et al., 2003). One study focusing on northern Europe found substantial declines in sea ice under reasonable climate change scenarios (Claussen et al., 2007). Other meteorological drivers of turbine loading may also be influenced by climate change but are likely to be secondary in comparison to changes in resource magnitude, weather extremes, and icing issues (Pryor and Barthelmie, 2010).

Additional research on the possible impact of climate change on the size, geographic distribution and variability of the wind resource is warranted, as is research on the possible impact of climate change on extreme weather events and therefore wind turbine operating environments. Overall, however, research to date suggests that these impacts are unlikely to be of a magnitude that will greatly impact the global potential of wind energy deployment.

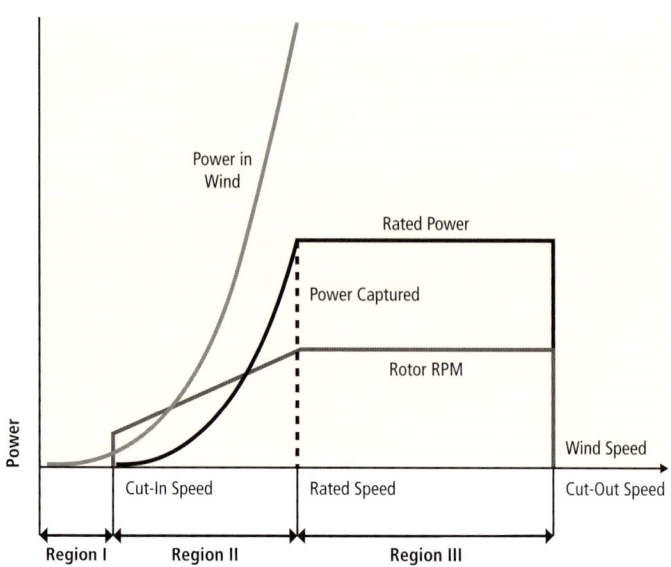

Figure 7.3 | Conceptual power curve for a modern variable-speed wind turbine (US DOE, 2008).

7.3 Technology and applications

Modern, commercial grid-connected wind turbines have evolved from small, simple machines to large, highly sophisticated devices. Scientific and engineering expertise and advances, as well as improved computational tools, design standards, manufacturing methods, and O&M procedures, have all supported these technology developments. As a result, typical wind turbine nameplate capacity ratings have increased dramatically since the 1980s (from roughly 75 kW to 1.5 MW and larger), while the cost of wind energy has substantially declined. Onshore wind energy technology is already being manufactured and deployed on a commercial basis. Nonetheless, additional R&D advances are anticipated, and are expected to further reduce the cost of wind energy while enhancing system and component performance and reliability. Offshore wind energy technology is still developing, with greater opportunities for additional advancement.

This section summarizes the historical development and current technology status of large grid-connected on- and offshore wind turbines (7.3.1), discusses international wind energy technology standards (7.3.2), and reviews power conversion and related grid connection issues (7.3.3); a later section (7.7) describes opportunities for further technical advances.

7.3.1 Technology development and status

7.3.1.1 Basic design principles

Generating electricity from the wind requires that the kinetic energy of moving air be converted to mechanical and then electrical energy, thus the engineering challenge for the wind energy industry is to design cost-effective wind turbines and power plants to perform this conversion. The amount of kinetic energy in the wind that is theoretically available for extraction increases with the cube of wind speed. However, a turbine only captures a portion of that available energy (see Figure 7.3).

Specifically, modern large wind turbines typically employ rotors that start extracting energy from the wind at speeds of roughly 3 to 4 m/s (cut-in speed). The Lanchester-Betz limit provides a theoretical upper limit (59.3%) on the amount of energy that can be extracted (Burton et al., 2001). A wind turbine increases power production with wind speed until it reaches its rated power level, often corresponding to a wind speed of 11 to 15 m/s. At still-higher wind speeds, control systems limit power output to prevent overloading the wind turbine, either through stall control, pitching the blades, or a combination of both (Burton et al., 2001). Most turbines then stop producing energy at wind speeds of approximately 20 to 25 m/s (cut-out speed) to limit loads on the rotor and prevent damage to the turbine's structural components.

Wind turbine design has centred on maximizing energy capture over the range of wind speeds experienced by wind turbines, while seeking to minimize the cost of wind energy. As described generally in Burton et al. (2001), increased generator capacity leads to greater energy capture when the turbine is operating at rated power (Region III). Larger rotor diameters for a given generator capacity, meanwhile, as well as aerodynamic design improvements, yield greater energy capture at lower wind speeds (Region II), reducing the wind speed at which rated power is achieved. Variable speed operation allows energy extraction at peak efficiency over a wider range of wind speeds (Region II). Finally, because the average wind speed at a given location varies with the height above ground level, taller towers typically lead to increased energy capture.

To minimize cost, wind turbine design is also motivated by a desire to reduce materials usage while continuing to increase turbine size, increase component and system reliability, and improve wind power plant operations. A system-level design and analysis approach is necessary to optimize wind turbine technology, power plant installation and O&M procedures for individual turbines and entire wind power plants. Moreover, optimizing turbine and power plant design for specific site conditions has become common as wind turbines, wind power plants and the wind energy market have all increased in size; site-specific conditions that can impact turbine and plant design include geographic and temporal variations in wind speed, site topography and access, interactions among individual wind turbines due to wake effects, and integration into the larger electricity system (Burton et al., 2001). Wind turbine and power plant design also impacts and is impacted by noise, visual, environmental and public acceptance issues (see Section 7.6).

7.3.1.2 Onshore wind energy technology

In the 1970s and 1980s, a variety of onshore wind turbine configurations were investigated, including both horizontal and vertical axis designs (see Figure 7.4). Gradually, the horizontal axis design came to dominate, although configurations varied, in particular the number of blades and whether those blades were oriented upwind or downwind of the tower (EWEA, 2009). After a period of further consolidation, turbine designs largely centred (with some notable exceptions) around the three-blade, upwind rotor; locating the turbine blades upwind of the tower prevents the tower from blocking wind flow onto the blades and producing extra aerodynamic noise and loading, while three-bladed machines typically have lower noise emissions than two-bladed machines. The three blades are attached to a hub and main shaft, from which power is transferred (sometimes through a gearbox, depending on design) to a generator. The main shaft and main bearings, gearbox, generator and control system are contained within a housing called the nacelle. Figure 7.5 shows the components in a modern wind turbine with a gearbox; in wind turbines without a gearbox, the rotor is mounted directly on the generator shaft.

In the 1980s, larger machines were rated at around 100 kW and primarily relied on aerodynamic blade stall to control power production from the fixed blades. These turbines generally operated at one or two rotational speeds. As turbine size increased over time, development went from stall control to full-span pitch control in which turbine output is controlled by pitching (i.e., rotating) the blades along their long axis (EWEA, 2009). In addition, a reduction in the cost of power electronics allowed variable speed wind turbine operation. Initially, variable speeds were used to smooth out the torque fluctuations in the drive train caused by wind turbulence and to allow more efficient operation in variable and gusty winds. More recently, almost all electric system operators require the continued operation of large wind power plants during electrical faults, together with being able to provide reactive power: these requirements have accelerated the adoption of variable-speed operation with power electronic conversion (see Section 7.3.3 for a summary of power conversion technologies, Section 7.5 for a fuller discussion of electric system integration issues, and Chapter 8 for a discussion of reactive power and broader issues with respect to the integration of RE into electricity systems). Modern wind turbines typically operate at variable speeds using full-span blade pitch control. Blades are commonly constructed with composite materials, and towers are usually tubular steel structures that taper from the base to the nacelle at the top (EWEA, 2009).

Over the past 30 years, average wind turbine size has grown significantly (Figure 7.6), with the largest fraction of onshore wind turbines installed globally in 2009 having a rated capacity of 1.5 to 2.5 MW; the average size of turbines installed in 2009 was 1.6 MW (BTM, 2010). As of 2010, wind turbines used onshore typically stand on 50- to 100-m towers, with rotors that are often 50 to 100 m in diameter; commercial

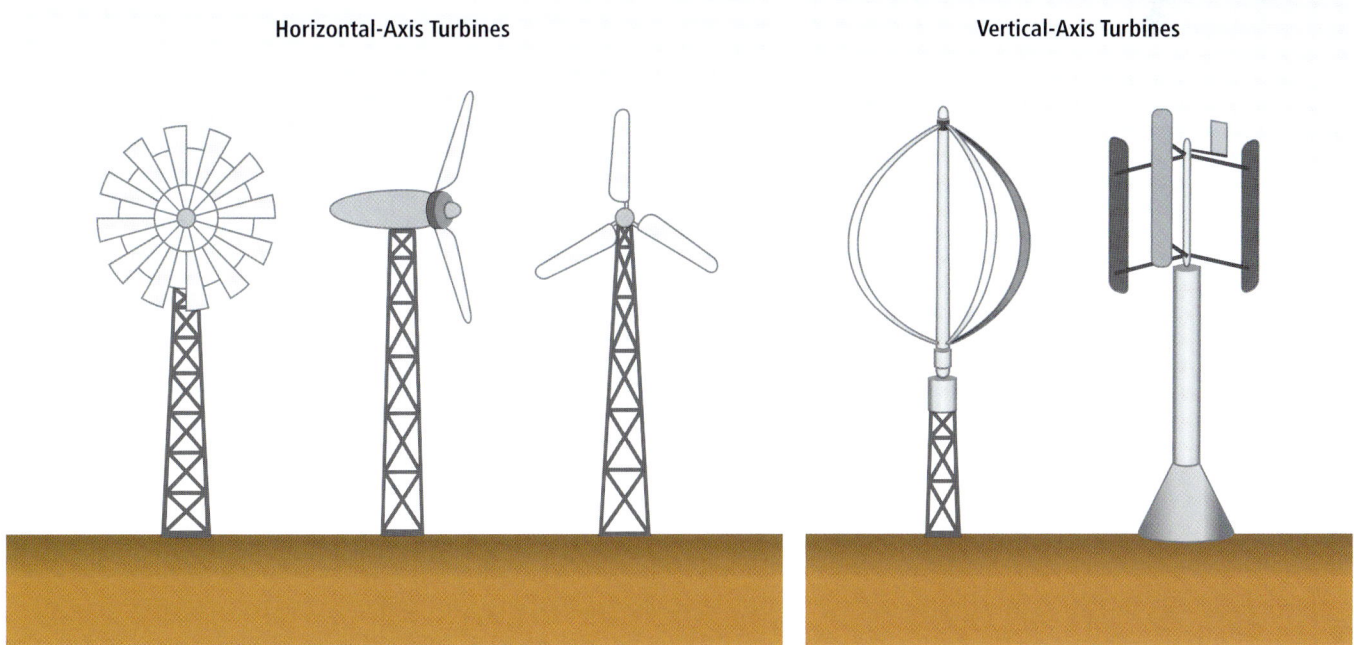

Figure 7.4 | Early wind turbine designs, including horizontal and vertical axis turbines (South et al., 1983).

Wind Energy Chapter 7

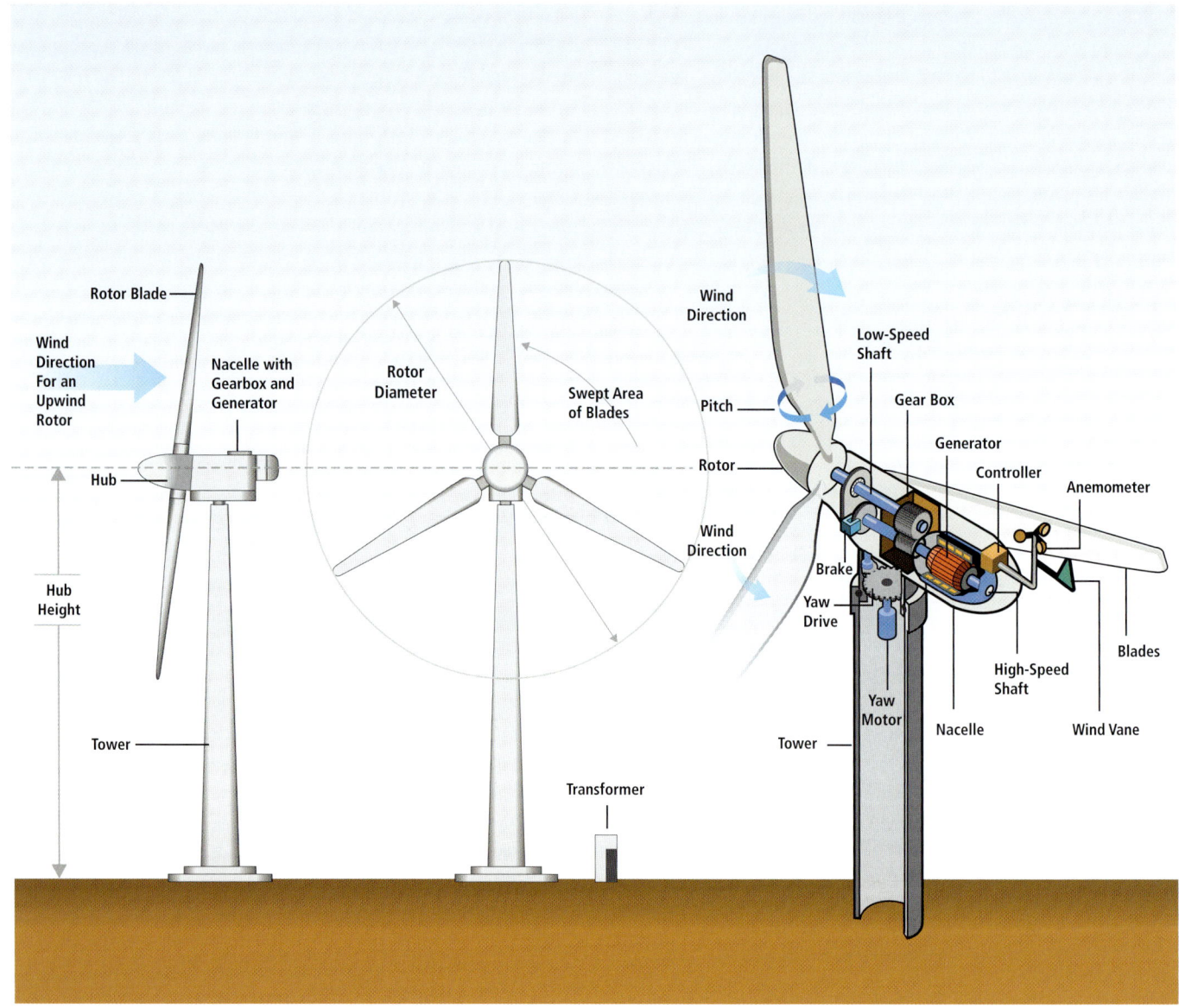

Figure 7.5 | Basic components of a modern, horizontal-axis wind turbine with a gearbox (Design by the National Renewable Energy Laboratory (NREL)).

machines with rotor diameters and tower heights in excess of 125 m are operating, and even larger machines are under development. Modern turbines operate with rotational speeds ranging from 12 to 20 revolutions per minute (RPM), which compares to the faster and potentially more visually disruptive speeds exceeding 60 RPM common of the smaller turbines installed during the 1980s.[13] Onshore wind turbines are typically grouped together into wind power plants, sometimes also called wind projects or wind farms. These wind power plants are often 5 to 300 MW in size, though smaller and larger plants do exist.

The main reason for the continual increase in turbine size to this point has been to minimize the levelized generation cost of wind energy by: increasing electricity production (taller towers provide access to a higher-quality wind resource, and larger rotors allow a greater exploitation of those winds as well as more cost-effective exploitation of lower-quality wind resource sites); reducing investment costs per unit of capacity (installation of a fewer number of larger turbines can, to a point, reduce overall investment costs); and reducing O&M costs (larger turbines can reduce maintenance costs per unit of capacity) (EWEA, 2009). For onshore turbines, however, additional growth in turbine size may ultimately be limited by not only engineering and materials usage constraints (discussed in Section 7.7), but also by the logistical constraints (or cost of resolving those constraints) of transporting the very large blades, tower, and nacelle components by road, as well as the cost of and difficulty in obtaining large cranes to lift the components into place. These same constraints are not as binding for offshore turbines, so future turbine scaling to the sizes shown in Figure 7.6 are more likely to be driven by offshore wind turbine design considerations.

---

13 Rotational speed decreases with larger rotor diameters. The acoustic noise resulting from tip speeds greater than 70 to 80 m/s is the primary design criterion that governs rotor speed.

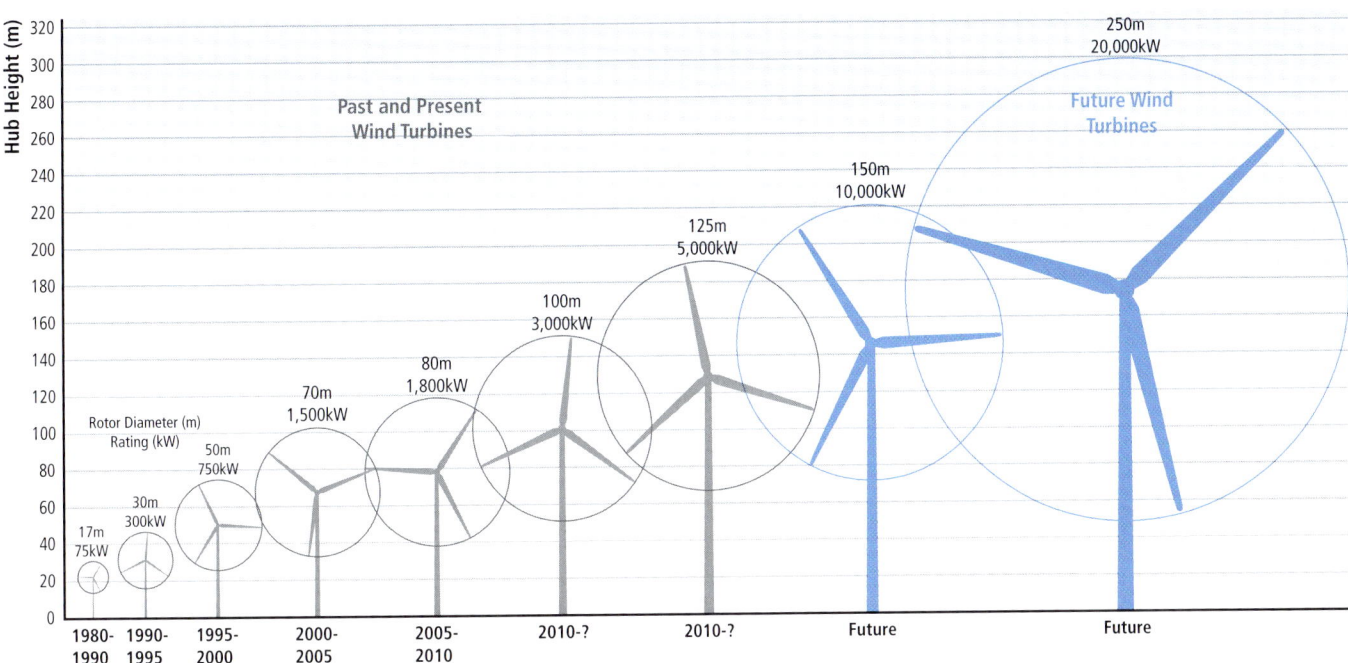

Figure 7.6 | Growth in size of typical commercial wind turbines (Design by NREL).

As a result of these and other developments, onshore wind energy technology is already being commercially manufactured and deployed on a large scale. Moreover, modern wind turbines have nearly reached the theoretical maximum of aerodynamic efficiency, with the coefficient of performance rising from 0.44 in the 1980s to about 0.50 by the mid 2000s.[14] The value of 0.50 is near the practical limit dictated by the drag of aerofoils and compares with the Lanchester-Betz theoretical limit of 0.593 (see Section 7.3.1.1). The design requirement for wind turbines is normally 20 years with 4,000 to 7,000 hours of operation (at and below rated power) each year depending on the characteristics of the local wind resource. Given the challenges of reliably meeting this design requirement, O&M teams work to maintain high plant availability despite component failure rates that have, in some instances, been higher than expected (Echavarria et al., 2008). Though wind turbines are reportedly under-performing in some contexts (Li, 2010), data collected through 2008 show that modern onshore wind turbines in mature markets can achieve an availability of 97% or more (Blanco, 2009; EWEA, 2009; IEA, 2009).

These results demonstrate that the technology has reached sufficient commercial maturity to allow large-scale manufacturing and deployment. Nonetheless, additional advances to improve reliability, increase electricity production and reduce costs are anticipated, and are discussed in Section 7.7. Additionally, most of the historical technology advances have occurred in developed countries. Increasingly, however, developing countries are investigating the use of wind energy, and opportunities for technology transfer in wind turbine design, component manufacturing and wind power plant siting exist. Extreme environmental conditions, such as icing or typhoons, may be more prominent in some of these markets, providing impetus for continuing research. Other aspects unique to less-developed countries, such as minimal transportation infrastructure, could also influence wind turbine designs if and as these markets grow.

7.3.1.3 Offshore wind energy technology

The first offshore wind power plant was built in 1991 in Denmark, consisting of eleven 450 kW wind turbines. Offshore wind energy technology is less mature than onshore, and has higher investment and O&M costs (see Section 7.8). By the end of 2009, just 1.3% of global installed wind power capacity was installed offshore, totalling 2,100 MW (GWEC, 2010a).

The primary motivation to develop offshore wind energy is to provide access to additional wind resources in areas where onshore wind energy development is constrained by limited technical potential and/or by planning and siting conflicts with other land uses. Other motivations for developing offshore wind energy include: the higher-quality wind resources located at sea (e.g., higher average wind speeds and lower shear near hub height; wind shear refers to the general increase in wind speed with height); the ability to use even larger wind turbines due to avoidance of certain land-based transportation constraints and the potential to thereby gain additional economies of scale; the ability to build larger power plants than onshore, gaining plant-level economies of scale; and a potential reduction in the need for new, long-distance, land-based transmission infrastructure

---

[14] Wind turbines achieve maximum aerodynamic efficiency when operating at wind speeds corresponding to power levels below the rated power level (see Region II in Figure 7.3). Aerodynamic efficiency is limited by the control system when operating at speeds above rated power (see Region III in Figure 7.3).

to access distant onshore wind energy[15] (Carbon Trust, 2008b; Snyder and Kaiser, 2009b; Twidell and Gaudiosi, 2009). These factors, combined with a significant offshore wind resource potential, have created considerable interest in offshore wind energy technology in the EU and, increasingly, in other regions, despite the typically higher costs relative to onshore wind energy.

Offshore wind turbines are typically larger than onshore, with nameplate capacity ratings of 2 to 5 MW being common for offshore wind power plants built from 2007 to 2009, and even larger turbines are under development. Offshore wind power plants installed from 2007 to 2009 were typically 20 to 120 MW in size, with a clear trend towards larger turbines and power plants over time. Water depths for most offshore wind turbines installed through 2005 were less than 10 m, but from 2006 to 2009, water depths from 10 to more than 20 m were common. Distance to shore has most often been below 20 km, but average distance has increased over time (EWEA, 2010a). As experience is gained, water depths are expected to increase further and more exposed locations with higher winds will be utilized. These trends will impact the wind resource characteristics faced by offshore wind power plants, as well as support structure design and the cost of offshore wind energy. A continued transition towards larger wind turbines (5 to 10 MW, or even larger) and wind power plants is also anticipated as a way of reducing the cost of offshore wind energy through turbine- and plant-level economies of scale.

To date, offshore turbine technology has been very similar to onshore designs, with some modifications and with special foundations (Musial, 2007; Carbon Trust, 2008b). The mono-pile foundation is the most common, though concrete gravity-based foundations have also been used with some frequency; a variety of other foundation designs (including floating designs) are being considered and in some instances used (Breton and Moe, 2009), especially as water depths increase, as discussed in Section 7.7. In addition to differences in foundations, modification to offshore turbines (relative to onshore) include structural upgrades to the tower to address wave loading; air conditioned and pressurized nacelles and other controls to prevent the effects of corrosive sea air from degrading turbine equipment; and personnel access platforms to facilitate maintenance. Additional design changes for marine navigational safety (e.g., warning lights, fog signals) and to minimize expensive servicing (e.g., more extensive condition monitoring, onboard service cranes) are common. Wind turbine tip speed could be chosen to be greater than for onshore turbines because concerns about noise are reduced for offshore power plants—higher tip speeds can sometimes lead to lower torque and lighter drive train components for the same power output. In addition, tower heights are sometimes lower than used for onshore wind power plants due to reduced wind shear offshore relative to onshore.

Lower power plant availabilities and higher O&M costs have been common for offshore wind energy relative to onshore wind both because of the comparatively less mature state of offshore wind energy technology and because of the inherently greater logistical challenges of maintaining and servicing offshore turbines (Carbon Trust, 2008b; UKERC, 2010). Wind energy technology specifically tailored for offshore applications will become more prevalent as the offshore market expands, and it is expected that larger turbines in the 5 to 10 MW range may come to dominate this market segment (EU, 2008). Future technical advancement possibilities for offshore wind energy are described in Section 7.7.

7.3.2 International wind energy technology standards

Wind turbines in the 1970s and 1980s were designed using simplified design models, which in some cases led to machine failures and in other cases resulted in design conservatism. The need to address both of these issues, combined with advances in computer processing power, motivated designers to improve their calculations during the 1990s (Quarton, 1998; Rasmussen et al., 2003). Improved design and testing methods have been codified in International Electrotechnical Commission (IEC) standards, and the rules and procedures for Conformity Testing and Certification of Wind Turbines (IEC, 2010) relies upon these standards. Certification agencies rely on accredited design and testing bodies to provide traceable documentation of the execution of rules and specifications outlined in the standards in order to certify turbines, components or entire wind power plants. The certification system assures that a wind turbine design or wind turbines installed in a given location meet common guidelines relating to safety, reliability, performance and testing. Figure 7.7(a) illustrates the design and testing procedures required to obtain a wind-turbine type certification. Plant certification, shown in Figure 7.7(b), requires a type certificate for the turbine and includes procedures for evaluating site conditions and turbine design parameters associated with that specific site, as well as other site-specific conditions including soil properties, installation and plant commissioning.

Insurance companies, financing institutions and power plant owners normally require some form of certification for plants to proceed, and the IEC standards therefore provide a common basis for certification to reduce uncertainty and increase the quality of wind turbine products available in the market (EWEA, 2009). In emerging markets, the lack of highly qualified testing laboratories and certification bodies limits the opportunities for manufacturers to obtain certification according to IEC standards and may lead to lower-quality products. As markets mature and design margins are compressed to reduce costs, reliance on internationally recognized standards is likely to become even more widespread to assure consistent performance, safety and reliability of wind turbines.

---

15 Of course, transmission infrastructure is needed to connect offshore wind power plants with electricity demand centres, and the per-kilometre cost of offshore transmission typically exceeds that for onshore lines. Whether offshore transmission needs are more or less extensive than those needed to access onshore wind energy varies by location.

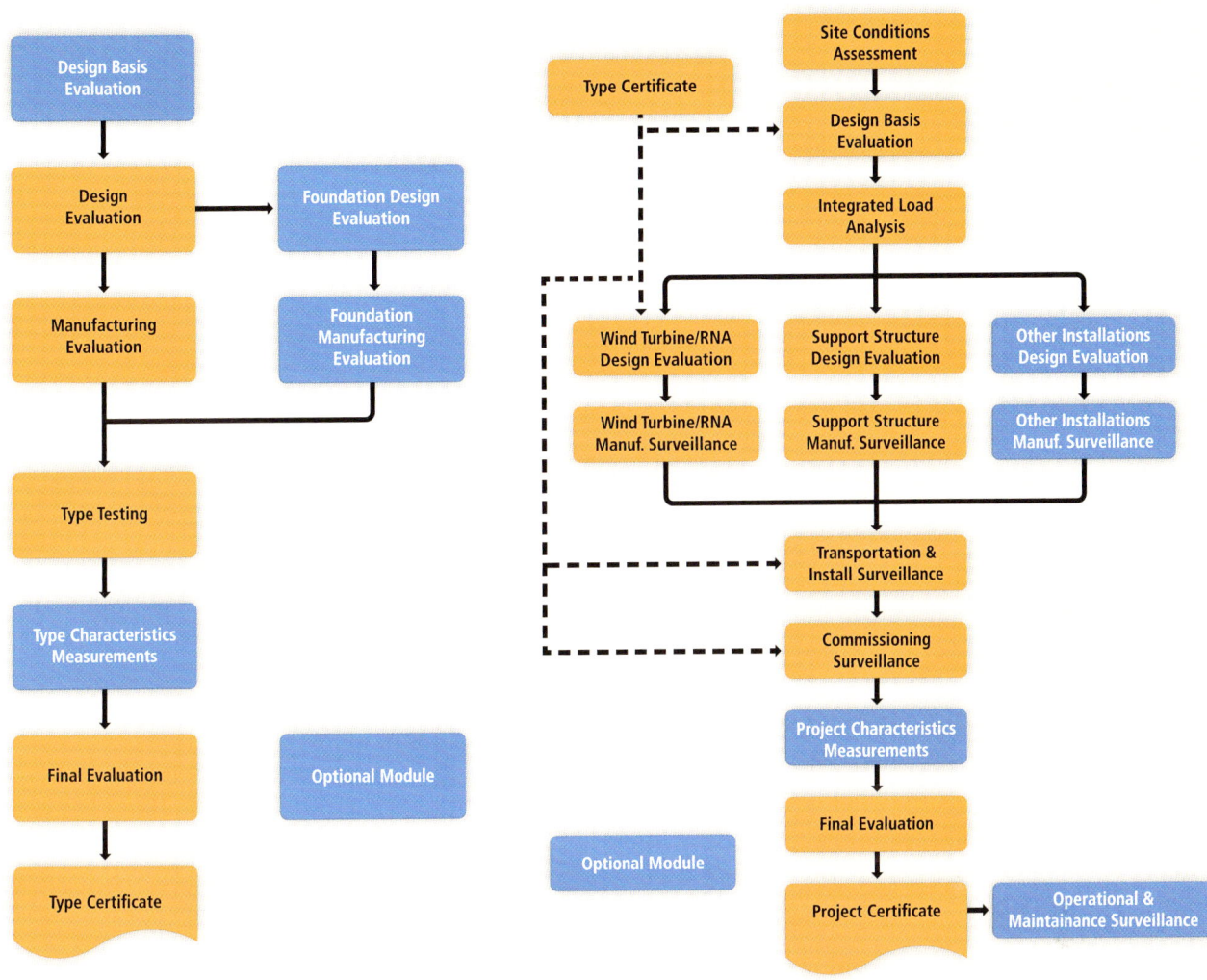

Figure 7.7 | Modules for (a) turbine type certification and (b) wind power plant certification (IEC, 2010).

Notes: RNA refers to Rotor Nacelle Assembly. The authors thank the IEC for permission to reproduce information from its International Standard IEC 61400-22 ed.1.0 (2010). All such extracts are copyright of IEC, Geneva, Switzerland. All rights reserved. Further information on the IEC is available from www.iec.ch. IEC has no responsibility for the placement and context in which the extracts and contents are reproduced by the authors, nor is IEC in any way responsible for the other content or accuracy therein. Copyright © 2010 IEC Geneva, Switzerland, www.iec.ch.

7.3.3 Power conversion and related grid connection issues

From an electric system reliability perspective, an important part of the wind turbine is the electrical conversion system. For large grid-connected turbines, electrical conversion systems come in three broad forms. Fixed-speed induction generators were popular in earlier years for both stall-regulated and pitch-controlled turbines; in these arrangements, wind turbines were net consumers of reactive power that had to be supplied by the electric system (see Ackermann, 2005). For modern turbines, these designs have now been largely replaced with variable-speed machines. Two arrangements are common, doubly-fed induction generators and synchronous generators with a full power electronic converter, both of which are almost always coupled with pitch-controlled rotors. These variable-speed designs essentially decouple the rotating masses of the turbine from the electric system, thereby offering a number of power quality advantages over earlier turbine designs (Ackermann, 2005; EWEA, 2009). For example, these turbines can provide real and reactive power as well as some fault ride-through capability, which are increasingly being required by electric system operators (these requirements and the institutional elements of wind energy integration are addressed in Section 7.5). These designs differ from the synchronous generators found in most large-scale fossil fuel-powered plants, however, in that they result in no intrinsic inertial response capability, that is, they do not increase (decrease) power output in synchronism with system power imbalances. This lack of inertial response is an important consideration for electric system planners because less overall inertia in the electric system makes the maintenance of stable system operation more challenging (Gautam et al., 2009). Wind turbine manufacturers have recognized this lack of intrinsic inertial response as a possible long-term impediment to wind energy and are actively pursuing a variety of solutions; for example, additional turbine controls can be added to provide inertial response (Mullane and O'Malley, 2005; Morren et al., 2006).

7.4 Global and regional status of market and industry development

The wind energy market expanded substantially in the 2000s, demonstrating the commercial and economic viability of the technology and industry, and the importance placed on wind energy development by a number of countries through policy support measures. Wind energy expansion has been concentrated in a limited number of regions, however, and wind energy remains a relatively small fraction of global electricity supply. Further expansion of wind energy, especially in regions of the world with little wind energy deployment to date and in offshore locations, is likely to require additional policy measures.

This section summarizes the global (Section 7.4.1) and regional (Section 7.4.2) status of wind energy deployment, discusses trends in the wind energy industry (Section 7.4.3) and highlights the importance of policy actions for the wind energy market (Section 7.4.4).

7.4.1 Global status and trends

Wind energy has quickly established itself as part of the mainstream electricity industry. From a cumulative capacity of 14 GW at the end of 1999, global installed wind power capacity increased 12-fold in 10 years to reach almost 160 GW by the end of 2009, an average annual increase in cumulative capacity of 28% (see Figure 7.8). Global annual wind power capacity additions equalled more than 38 GW in 2009, up from 26 GW in 2008 and 20 GW in 2007 (GWEC, 2010a).

The majority of the capacity has been installed onshore, with offshore installations constituting a small proportion of the total market. About 2.1 GW of offshore wind turbines were installed by the end of 2009; 0.6 GW were installed in 2009, including the first commercial offshore wind power plant outside of Europe, in China (GWEC, 2010a). Many of these offshore installations have taken place in the UK and Denmark. Significant offshore wind power plant development activity, however, also exists in, at a minimum, other EU countries, the USA, Canada and China (e.g., Mostafaeipour, 2010). Offshore wind energy is expected to develop in a more significant way in the years ahead as the technology advances and as onshore wind energy sites become constrained by local resource availability and/or siting challenges in some regions (BTM, 2010; GWEC, 2010a).

The total investment cost of new wind power plants installed in 2009 was USD$_{2005}$ 57 billion (GWEC, 2010a). Direct employment in the wind energy sector in 2009 has been estimated at roughly 190,000 in the EU and 85,000 in the USA. Worldwide, direct employment has been estimated at approximately 500,000 (GWEC, 2010a; REN21, 2010).

Despite these trends, wind energy remains a relatively small fraction of worldwide electricity supply. The total wind power capacity installed by the end of 2009 would, in an average year, meet roughly 1.8% of worldwide electricity demand, up from 1.5% by the end of 2008, 1.2% by the end of 2007, and 0.9% by the end of 2006 (Wiser and Bolinger, 2010).

7.4.2 Regional and national status and trends

The countries with the highest total installed wind power capacity by the end of 2009 were the USA (35 GW), China (26 GW), Germany (26 GW), Spain (19 GW) and India (11 GW). After its initial start in the USA in the 1980s, wind energy growth centred on countries in the EU and India during the 1990s and the early 2000s. In the late 2000s, however, the USA and then China became the locations for the greatest annual capacity additions (Figure 7.9).

Regionally, Europe continues to lead the market with 76 GW of cumulative installed wind power capacity by the end of 2009, representing 48% of the global total (Asia represented 25%, whereas North America

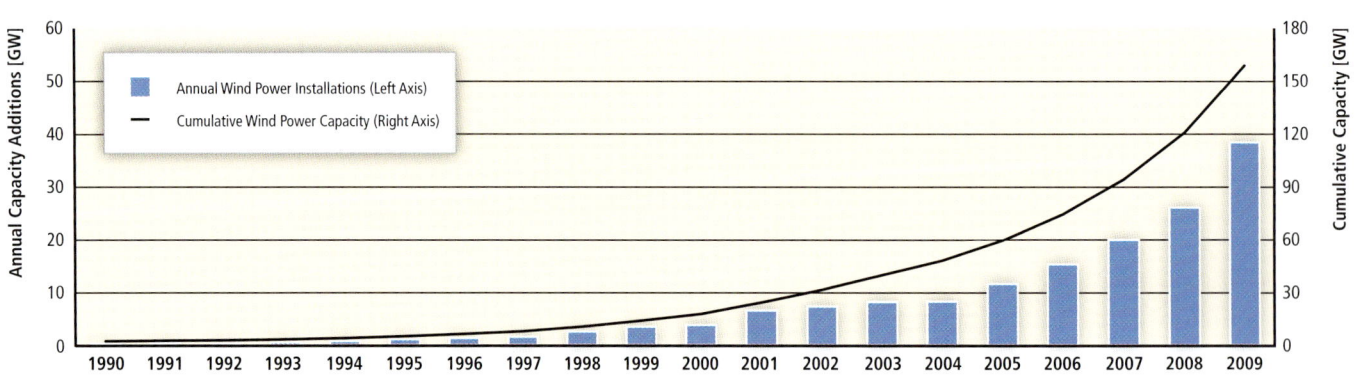

Figure 7.8 | Global annual wind power capacity additions and cumulative capacity (Data sources: GWEC, 2010a; Wiser and Bolinger, 2010).

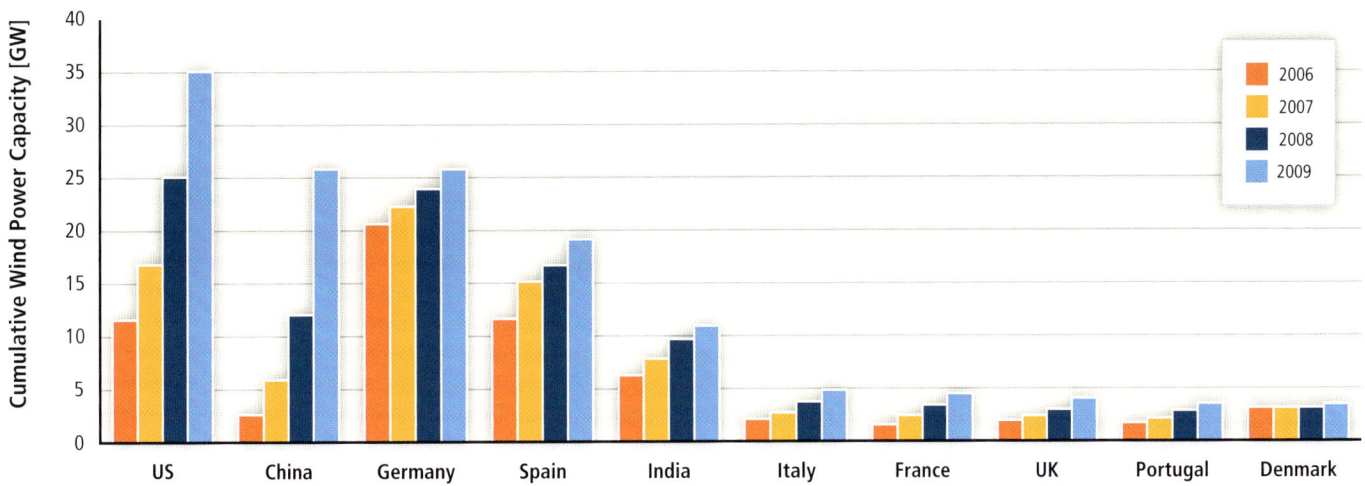

Figure 7.9 | Top-10 countries in cumulative wind power capacity (Date source: GWEC, 2010a).

represented 24%). Notwithstanding the continuing growth in Europe, the trend over time has been for the wind energy industry to become less reliant on a few key markets, and other regions of the world have increasingly become the dominant markets for wind energy growth. The annual growth in the European wind energy market in 2009, for example, accounted for just 28% of the total new wind power additions in that year, down from over 60% in the early 2000s (GWEC, 2010a). More than 70% of the annual wind power capacity additions in 2009 occurred outside of Europe, with particularly significant growth in Asia (40%) and North America (29%) (Figure 7.10). Even in Europe, though Germany and Spain have been the strongest markets during the 2000s, there is a trend towards less reliance on these two countries.

Despite the increased globalization of wind power capacity additions, the market remains concentrated regionally. As shown in Figure 7.10, Latin America, Africa and the Middle East, and the Pacific regions have installed relatively little wind power capacity despite significant technical potential in each region, as presented earlier in Section 7.2. And, even in the regions of significant growth, most of that growth has occurred in a limited number of countries. In 2009, for example, 90% of wind power capacity additions occurred in the 10 largest markets, and 62% was concentrated in just two countries: China (14 GW, 36%) and the USA (10 GW, 26%).

In both Europe and the USA, wind energy represents a major new source of electric capacity additions. From 2000 through 2009, wind energy was the second-largest new resource added in the USA (10% of all gross capacity additions) and EU (33% of all gross capacity additions) in terms of nameplate capacity, behind natural gas but ahead of coal. In 2009, 39% of all capacity additions in the USA and 39% of all additions in the EU came from wind energy (Figure 7.11). In China, 5% of the net capacity additions from 2000 to 2009 and 16% of the net additions in 2009 came from wind energy. On a global basis, from 2000 through 2009,

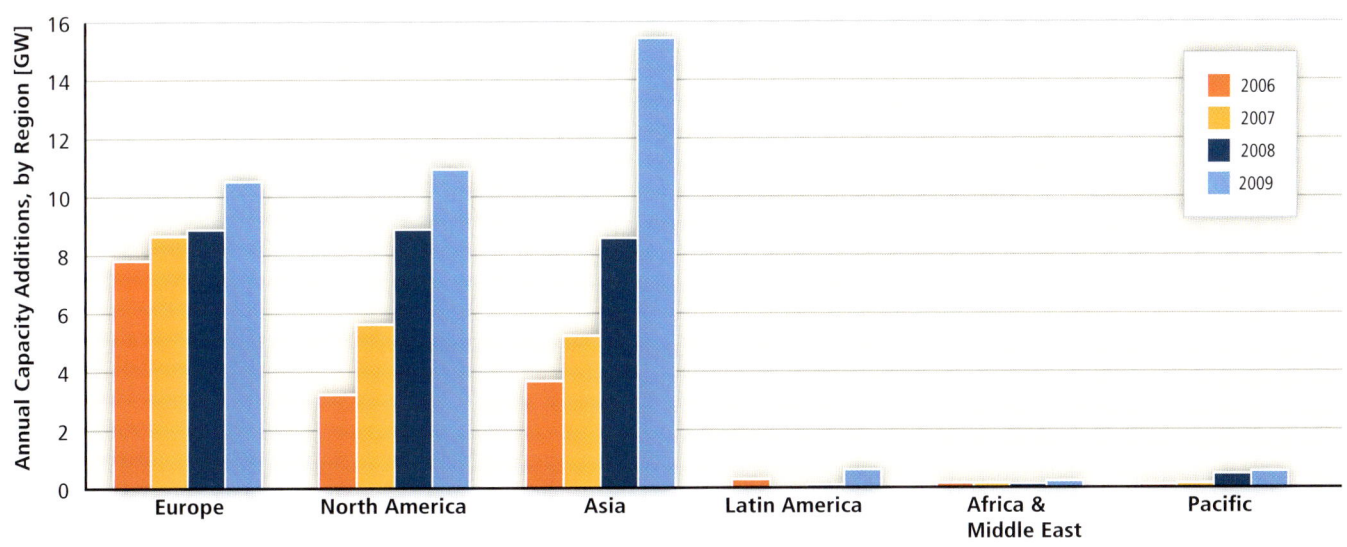

Figure 7.10 | Annual wind power capacity additions by region (Data source: GWEC, 2010a).

Note: Regions shown in the figure are defined by the study.

Wind Energy

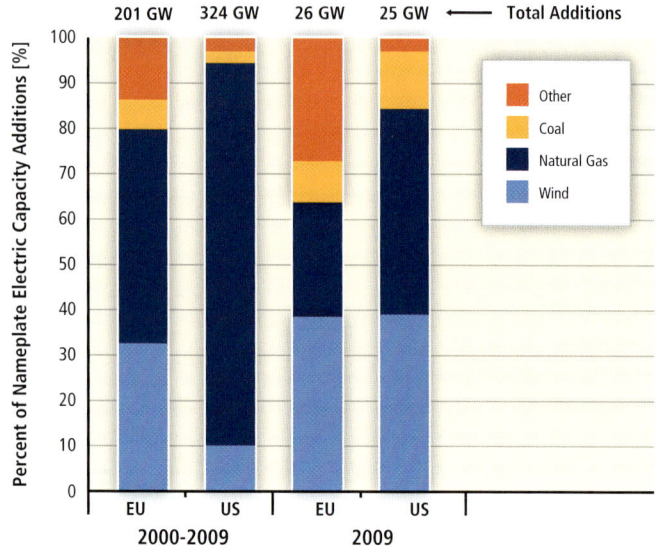

Figure 7.11 | Relative contribution of electricity supply types to gross capacity additions in the EU and the USA (Data sources: EWEA, 2010b; Wiser and Bolinger, 2010).

Note: The 'other' category includes other forms of renewable energy, nuclear energy, and fuel oil.

A number of countries are beginning to achieve relatively high levels of annual wind electricity penetration in their respective electric systems. Figure 7.12 presents data for the end of 2009 (and the end of 2006, 2007 and 2008) on installed wind power capacity, translated into projected annual electricity supply, and divided by electricity consumption. On this basis, and focusing only on the 20 countries with the greatest cumulative wind power capacity, at the end of 2009, wind power capacity was capable of supplying electricity equal to roughly 20% of Denmark's annual electricity demand, 14% of Portugal's, 14% of Spain's, 11% of Ireland's and 8% of Germany's (Wiser and Bolinger, 2010).[17]

7.4.3 Industry development

The growing maturity of the wind energy sector is illustrated not only by wind power capacity additions, but also by trends in the wind energy industry. In particular, major established companies from outside the traditional wind energy industry have become increasingly involved in the sector. For example, there has been a shift in the type of companies developing, owning and operating wind power plants, from relatively small independent power plant developers to large power generation

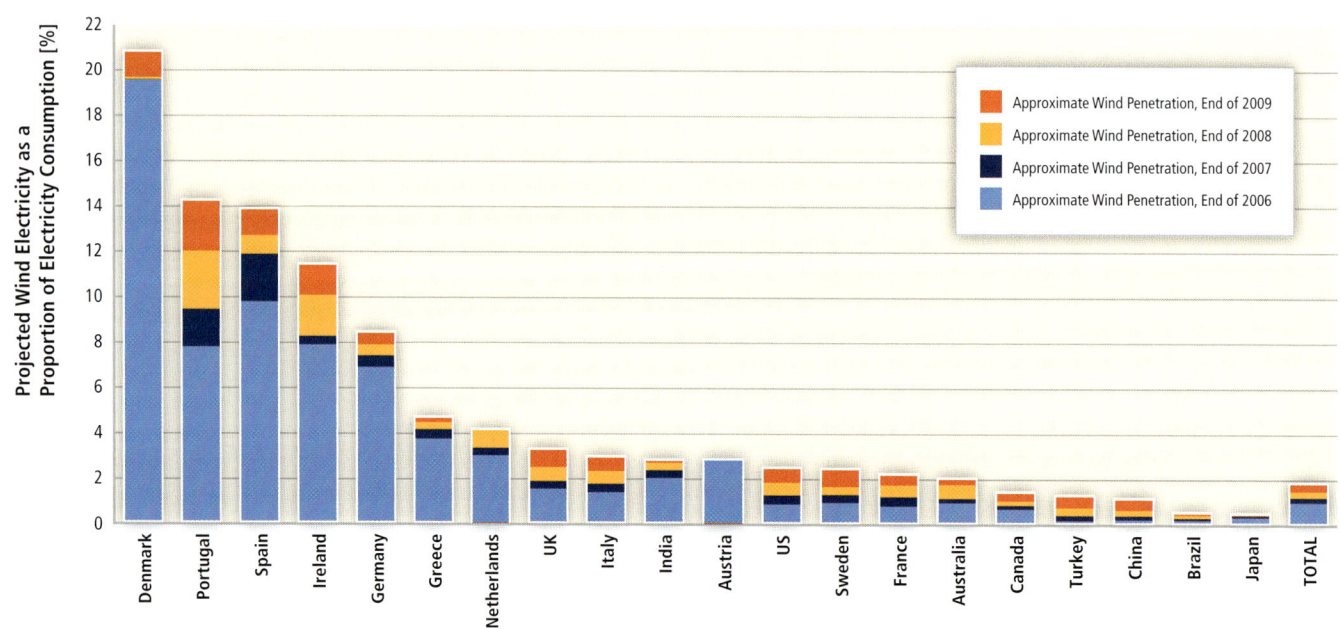

Figure 7.12 | Approximate annual average wind electricity penetration in the twenty countries with the greatest installed wind power capacity (Wiser and Bolinger, 2010).

roughly 11% of all newly installed net electric capacity additions came from new wind power plants; in 2009 alone, that figure was probably more than 20%.[16]

companies (including electric utilities) and large independent power plant developers. With respect to wind turbine and component manufacturing, the increase in the size and geographic spread of the wind energy market, along with manufacturing localization requirements in some countries, has brought in new players. The involvement of these new players has, in turn, encouraged a greater globalization of the industry. Manufacturer product strategies are shifting to address larger

---

16 Worldwide capacity additions from 2000 through 2007 come from historical data from the US Energy Information Administration. Capacity additions for 2008 and 2009 are estimated based on historical capacity growth from 2000 to 2007. The focus here is on capacity additions in GW terms, though it is recognized that electricity generation technologies often have widely divergent average capacity factors, and that the contribution of wind energy to new electricity demand (in GWh terms) may differ from what is presented here.

17 Because of interconnections among electricity grids, these percentages do not necessarily equate to the amount of wind electricity consumed within each country.

scale power plants, higher capacity and offshore turbines, and lower wind speeds. More generally, the significant contribution of wind energy to new electric capacity investment in several regions of the world has attracted a broad range of players across the industry supply chain, from local site-focused engineering firms to global vertically integrated utilities. The industry's supply chain has also become increasingly competitive as a multitude of firms seek the most profitable balance between vertical integration and specialization (BTM, 2010; GWEC, 2010a).

Despite these trends, the global wind turbine market remains somewhat regionally segmented, with just six countries hosting the majority of wind turbine manufacturing (China, Denmark, India, Germany, Spain and the USA). With markets developing differently, market share for turbine supply has been marked by the emergence of national industrial champions, the entry of highly focused technology innovators and the arrival of new start-ups licensing proven technology from other regions (Lewis and Wiser, 2007). Regardless, the industry continues to globalize: Europe's turbine and component manufacturers have penetrated the North American and Asian markets, and the growing presence of Asian manufacturers in Europe and North America is expected to become more pronounced in the years ahead. Chinese wind turbine manufacturers, in particular, are dominating their home market, and will increasingly seek export opportunities. Wind turbine sales and supply chain strategies are therefore expected to continue to take on a more international dimension as volumes increase.

Amidst the growth in the wind energy industry also come challenges. As discussed further in Section 7.8, from 2005 through 2008, supply chain difficulties caused by growing demand for wind energy strained the industry, and prices for wind turbines and turbine components increased to compensate for this imbalance. Commodity price increases, the availability of skilled labour and other factors also played a role in pushing wind turbine prices higher, while the underdeveloped supply chain for offshore wind power plants strained that portion of the industry. Overcoming supply chain difficulties is not simply a matter of ramping up the production of wind turbine components to meet the increased levels of demand. Large-scale investment decisions are more easily made based on a sound long-term outlook for the industry. In most markets, however, both the projections and actual demand for wind energy depend on a number of factors, some of which are outside of the control of the industry, such as political frameworks and policy measures.

7.4.4 Impact of policies[18]

The deployment of wind energy must overcome a number of challenges that vary in type and magnitude depending on the wind energy application and region.[19] The most significant challenges to wind energy deployment are summarized here. Perhaps most importantly, in many (though not all) regions of the world, wind energy is more expensive than current energy market prices, at least if environmental impacts are not internalized and monetized (NRC, 2010a). Wind energy also faces a number of other challenges, some of which are somewhat unique to wind energy or are at least particularly relevant to this sector. Some of the most critical challenges include: (1) concerns about the impact of wind energy's variability on electricity reliability; (2) challenges to building the new transmission infrastructure both on- and offshore (and within country and cross-border) needed to enable access to the most attractive wind resource areas; (3) cumbersome and slow planning, siting and permitting procedures that impede wind energy deployment; (4) the technical advancement needs and higher cost of offshore wind energy technology; and (5) lack of institutional and technical knowledge in regions that have not experienced substantial wind energy deployment to this point.

As a result of these challenges, growth in the wind energy sector is affected by and responsive to political frameworks and a wide range of government policies. During the past two decades, a significant number of developed countries and, more recently, a growing number of developing nations have laid out RE policy frameworks that have played a major role in the expansion of the wind energy market. These efforts have been motivated by the environmental, fuel diversity, and economic development impacts of wind energy deployment, as well as the potential for reducing the cost of wind energy over time. An early significant effort to deploy wind energy at a commercial scale occurred in California, with a feed-in tariff and aggressive tax incentives spurring growth in the 1980s (Bird et al., 2005). In the 1990s, wind energy deployment moved to Europe, with feed-in tariff policies initially established in Denmark and Germany, and later expanding to Spain and then a number of other countries (Meyer, 2007); renewable portfolio standards have been implemented in other European countries and, more recently, European renewable energy policies have been motivated in part by the EU's binding 20%-by-2020 target for renewable energy. In the 2000s, growth in the USA (Bird et al., 2005; Wiser and Bolinger, 2010), China (Li et al., 2007; Li, 2010; Liu and Kokko, 2010), and India (Goyal, 2010) was based on varied policy frameworks, including renewable portfolio standards, tax incentives, feed-in tariffs and government-overseen bidding. Still other policies have been used in a number of countries to directly encourage the localization of wind turbine and component manufacturing (Lewis and Wiser, 2007).

Though economic support policies differ, and a healthy debate exists over the relative merits of different approaches, a key finding is that both policy transparency and predictability are important (see Chapter 11). Moreover, though it is not uncommon to focus on economic policies for wind energy, as noted above and as discussed elsewhere in this chapter and in Chapter 11, experience shows that wind energy markets are also dependent on a variety of other factors (e.g., Valentine, 2010). These include local resource availability, site planning and approval procedures, operational integration into electric systems, transmission grid expansion, wind energy technology improvements, and the availability of institutional and technical knowledge in markets unfamiliar with

---

[18] Non-technology-specific policy issues are covered in Chapter 11 of this report.

[19] For a broader discussion of barriers and market failures associated with renewable energy, see Sections 1.4 and 11.1, respectively.

wind energy (e.g., IEA, 2009). For the wind energy industry, these issues have been critical in defining both the size of the market opportunity in each country and the rules for participation in those opportunities; many countries with sizable wind resources have not deployed significant amounts of wind energy as a result of these factors. Given the challenges to wind energy listed earlier, successful frameworks for wind energy deployment might consider the following elements: support systems that offer adequate profitability and that ensure investor confidence; appropriate administrative procedures for wind energy planning, siting and permitting; a degree of public acceptance of wind power plants to ease implementation; access to the existing transmission system and strategic transmission planning and new investment for wind energy; and proactive efforts to manage wind energy's inherent output variability and uncertainty. In addition, R&D by government and industry has been essential to enabling incremental improvements in onshore wind energy technology and to driving the improvements needed in offshore wind energy technology. Finally, for those markets that are new to wind energy deployment, both knowledge (e.g., wind resource mapping expertise) and technology transfer (e.g., to develop local wind turbine manufacturers and to ease grid integration) can help facilitate early installations.

7.5 Near-term grid integration issues[20]

As wind energy deployment has increased, so have concerns about the integration of that energy into electric systems (e.g., Fox et al., 2007). The nature and magnitude of the integration challenge will be system specific and will vary with the degree of wind electricity penetration. Moreover, as discussed in Chapter 8, integration challenges are not unique to wind energy: adding any type of generation technology to an electric system, particularly location-constrained variable generation, presents challenges. Nevertheless, analysis and operating experience primarily from certain OECD countries (where most of the wind energy deployment has occurred, until recently, see Section 7.4.2) suggest that, at low to medium levels of wind electricity penetration (defined here as up to 20% of total annual average electrical energy demand),[21] the integration of wind energy generally poses no insurmountable technical barriers and is economically manageable. In addition, increased operating experience with wind energy along with improved technology, altered operating and planning practices and additional research should facilitate the integration of even greater quantities of wind energy. Even at low to medium levels of wind electricity penetration, however, certain (and sometimes system-specific) technical and/or institutional challenges must be addressed.

---

20 Non-technology-specific issues related to integration of RE sources in current and future energy systems are covered in Chapter 8 of this report.

21 This level of penetration was chosen to loosely separate the integration needs for wind energy in the relatively near term from the broader, longer-term, and non-wind-specific discussion of electric system changes provided in Chapter 8. In addition, the majority of operational experience and literature on the integration of wind energy addresses penetration levels below 20%.

The integration issues covered in this section include how to address wind power variability and uncertainty, the possible need for additional transmission capacity to enable remotely located wind power plants to meet the needs of electricity demand centres, and the development of technical standards for connecting wind power plants with electric systems. The focus is on those issues faced at low to medium levels of wind electricity penetration (up to 20%). Even higher levels of penetration may depend on or benefit from the availability of additional flexibility options, such as: further increasing the flexibility of other electricity generation plants (fossil and otherwise); mass-market demand response; large-scale deployment of electric vehicles and their associated contributions to system flexibility through controlled battery charging; greater use of wind power curtailment and output control or diverting excess wind energy to fuel production or local heating; increased deployment of bulk energy storage technologies; and further improvements in the interconnections between electric systems. The deployment of a diversity of RE technologies may also help facilitate overall electric system integration. Many of these options relate to broader developments within the energy sector that are not specific to wind energy, however, and most are therefore addressed in Chapter 8.

This section begins by describing the specific characteristics of wind energy that present integration challenges (Section 7.5.1). The section then discusses how these characteristics impact issues associated with the planning (Section 7.5.2) and operation (Section 7.5.3) of electric systems to accommodate wind energy, including a selective discussion of actual operating experience. Finally, Section 7.5.4 summarizes the results of various studies that have quantified the technical issues and economic costs of integrating increased quantities of wind energy.

7.5.1 Wind energy characteristics

Several important characteristics of wind energy are different from those of many other generation sources. These characteristics must be considered in electric system planning and operation to ensure the reliable and economical operation of the electric power system.

The first characteristic to consider is that the quality of the wind resource and therefore the cost of wind energy is location dependent. As a result, regions with the highest-quality wind resources may not be situated near population centres that have high electricity demands (e.g., Hoppock and Patiño-Echeverri, 2010; Liu and Kokko, 2010). Additional transmission infrastructure is therefore sometimes economically justified (and is often needed) to bring wind energy from higher-quality wind resource areas to electricity demand centres as opposed to utilizing lower-quality wind resources that are located closer to demand centres and that may require less new transmission investment (see Sections 7.5.2.3 and 7.5.4.3).

The second important characteristic is that wind energy is weather dependent and therefore variable—the power output of a wind power plant varies from zero to its rated capacity depending on prevailing weather conditions. Variations can occur over multiple time scales, from shorter-term sub-hourly fluctuations to diurnal, seasonal, and even inter-annual fluctuations (e.g., Van der Hoven, 1957; Justus et al., 1979; Wan and Bucaneg, 2002; Apt, 2007; Rahimzadeh et al., 2011). The nature of these fluctuations and patterns is highly site- and region-specific. Figure 7.13 illustrates some elements of this variability by showing the scaled output of an individual wind turbine, a small collection of wind power plants, and a large collection of wind power plants in Germany over 10 consecutive days. An important aspect of wind power variability for electric system *operations* is the rate of change in wind power output over different relatively short time periods; Figure 7.13 demonstrates that the aggregate output of multiple wind power plants changes much more dramatically over relatively longer periods (multiple hours) than over very short periods (minutes). An important aspect of wind power variability for the purpose of electric sector *planning*, on the other hand, is the correlation of wind power output with the periods of time when electric system reliability is at greatest risk, typically periods of high electricity demand. In this case, the diurnal, seasonal, and even interannual patterns of wind power output (and the correlation of those patterns with electricity demand) can impact the capacity credit assigned by system planners to wind power plants, as discussed further in Section 7.5.3.4.

Third, in comparison with many other types of power plants, wind power output has lower levels of predictability. Forecasts of wind power output use various approaches and have multiple goals, and significant improvements in forecasting accuracy have been achieved in recent years (e.g., Costa et al., 2008). Despite those improvements, however, forecasts remain imperfect. In particular, forecasts are less accurate over longer forecast horizons (multiple hours to days) than over shorter periods (e.g., H. Madsen et al., 2005), which, depending on the characteristics of the electric system, can have implications for the ability of that system and related trading markets to manage wind power variability and uncertainty (Usaola, 2009; Weber, 2010).

The aggregate variability and uncertainty of wind power output depends, in part, on the degree of correlation between the outputs of different geographically dispersed wind power plants. This correlation between the outputs of wind power plants, in turn, depends on the geographic deployment of the plants and the regional characteristics of weather patterns, especially wind speeds. Generally, the output of wind power plants that are farther apart are less correlated with each other, and variability over shorter time periods (minutes) is less correlated than variability over longer time periods (multiple hours) (e.g., Wan et al., 2003; Sinden, 2007; Holttinen et al., 2009; Katzenstein et al., 2010). This lack of perfect correlation results in a smoothing effect associated with geographic diversity when the output of multiple wind turbines and power plants are combined, as illustrated in Figure 7.13: the aggregate scaled variability shown for groups of wind power

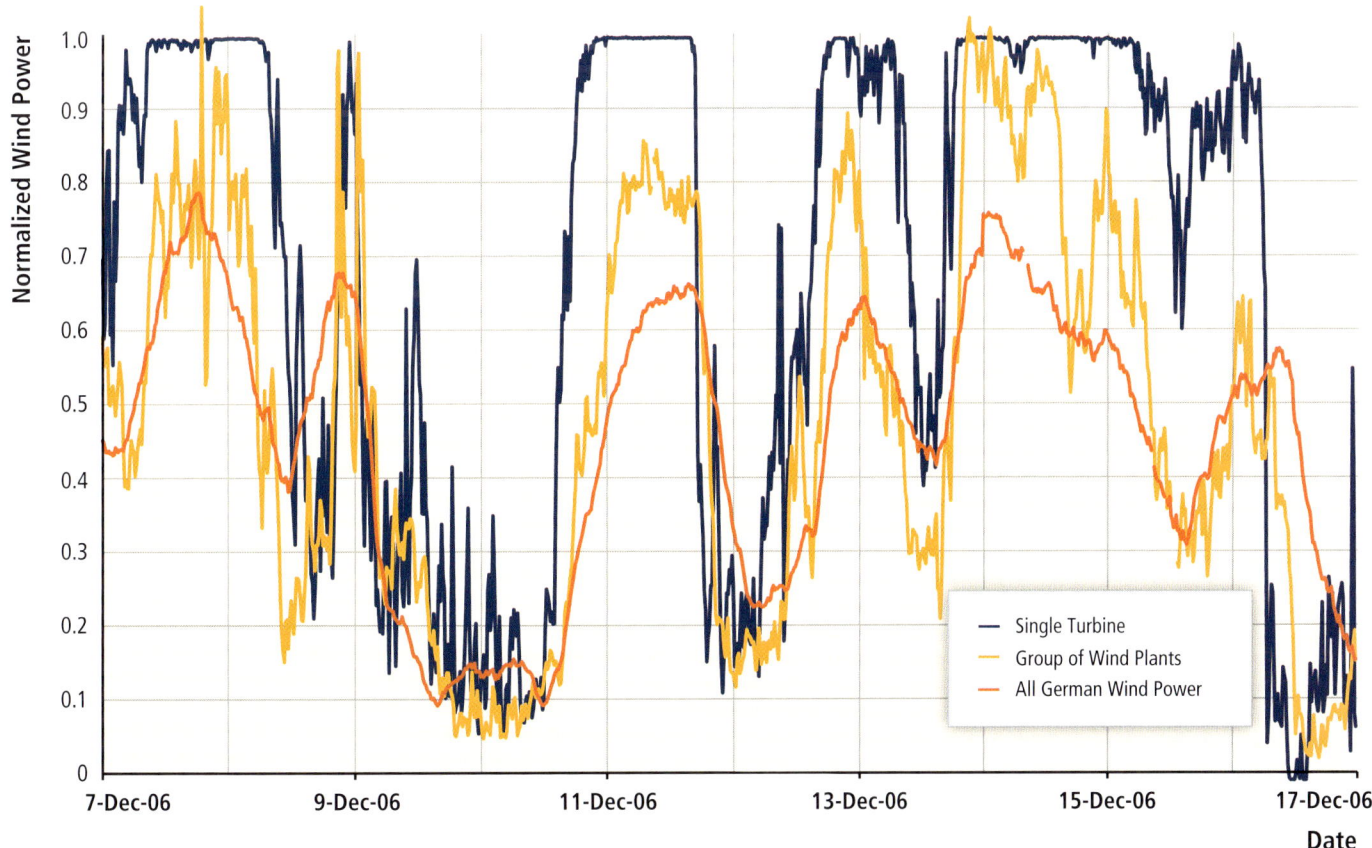

Figure 7.13 | Example time series of wind power output scaled to wind power capacity for a single wind turbine, a group of wind power plants, and all wind power plants in Germany over a 10-day period in 2006 (Durstewitz et al., 2008)

plants over a region is less than the scaled output of a single wind turbine. This apparent smoothing of aggregated output is due to the decreasing correlation of output between different wind power plants as distance between those plants increases. If, on the other hand, the output of multiple wind turbines and power plants was perfectly correlated, then the aggregate variability would be equivalent to the scaled variability of a single turbine. With sufficient transmission capacity between wind power plants, the observed geographic smoothing effect has implications for the variability of aggregate wind power output that electric systems must accommodate, and also influences forecast accuracy because accuracy improves with the number and diversity of wind power plants considered (e.g., Focken et al., 2002).

7.5.2 Planning electric systems with wind energy

Detailed system planning for new generation and transmission infrastructure is used to ensure that the electric system can be operated reliably and economically in the future. Advanced planning is required due, in part, to the long time horizons required to build new electricity infrastructure. More specifically, electric system planners[22] must evaluate the adequacy of transmission to deliver electricity to demand centres and the adequacy of generation to maintain a balance between supply and demand under a variety of operating conditions. Though not an exhaustive list, four technical planning issues are prominent when considering increased reliance on wind energy: the need for accurate electric system models of wind turbines and power plants; the development of technical standards for connecting wind power plants with electric systems (i.e., grid codes); the broader transmission infrastructure needs of electric systems with wind energy; and the maintenance of overall generation adequacy with increased wind electricity penetration.

7.5.2.1 Electric system models

Computer-based simulation models are used extensively to evaluate the ability of the electric system to accommodate new generation, changes in demand and changes in operational practices. An important role of electric system models is to demonstrate the ability of an electric system to recover from severe events or contingencies. Generic models of typical synchronous generators have been developed and validated over a period of multiple decades, and are used in industry standard software tools (e.g., power system simulators and analysis models) to study how the electric system and all its components will behave during system events or contingencies. Similar generic models of wind turbines and wind power plants are in the process of being developed and validated. Because wind turbines have electrical characteristics that differ from typical synchronous generators, this modelling exercise requires significant effort. As a result, though considerable progress has been made,

this progress is not complete, and increased deployment of wind energy will require improved and validated models to allow planners to better assess the capability of electric systems to accommodate wind energy (Coughlan et al., 2007; NERC, 2009).

7.5.2.2 Wind power electrical characteristics and grid codes

As wind power capacity has increased, so has the need for wind power plants to become more active participants in maintaining (rather than passively depending on) the operability and power quality of the electric system. Focusing here primarily on the technical aspects of grid connection, the electrical performance of wind turbines in interaction with the grid is often verified in accordance with international standards for the characteristics of wind turbines, in which methods to assess the impact of one or more wind turbines on power quality are specified (IEC, 2008). Additionally, an increasing number of electric system operators have implemented technical standards (sometimes called 'grid codes') that wind turbines and/or wind power plants (and other power plants) must meet when connecting to the grid to help prevent equipment or facilities from adversely affecting the electric system during normal operation and contingencies (see also Chapter 8). Electric system models and operating experience are used to develop these requirements, which can then typically be met through modifications to wind turbine design or through the addition of auxiliary equipment such as power conditioning devices. In some cases, the unique characteristics of specific generation types are addressed in grid codes, resulting in wind-specific grid codes (e.g., Singh and Singh, 2009).

Grid codes often require 'fault ride-through' capability, or the ability of a wind power plant to remain connected and operational during brief but severe changes in electric system voltage (Singh and Singh, 2009). The requirement for fault ride-through capability was in response to the increasing penetration of wind energy and the significant size of individual wind power plants. Electric systems can typically maintain reliable operation when small individual power plants shut down or disconnect from the system for protection purposes in response to fault conditions. When a large amount of wind power capacity disconnects in response to a fault, however, that disconnection can exacerbate the fault conditions. Electric system planners have therefore increasingly specified that wind power plants must meet minimum fault ride-through standards similar to those required of other large power plants. System-wide approaches have also been adopted: in Spain, for example, wind power output may be curtailed in order to avoid potential reliability issues in the event of a fault; the need to employ this curtailment, however, is expected to decrease as fault ride-through capability is added to new and existing wind power plants (Rivier Abbad, 2010). Reactive power control to help manage voltage is also often required by grid codes, enabling wind turbines to improve voltage stability margins particularly in weak parts of the electric system (Vittal et al., 2010). Requirements for wind turbine inertial response to improve system stability after disturbances are less common, but are under consideration (Hydro-Quebec TransEnergie,

---

22 Electric system planners (or organizations that plan electric systems) is used here as a generic term that refers to planners within any organization that regulates, operates components of, or builds infrastructure for the electric system.

2006; Doherty et al., 2010). Active power control (including limits on how quickly wind power plants can change their output) and frequency control are also sometimes required (Singh and Singh, 2009). Finally, controls can be added to wind power plants to enable beneficial dampening of inter-area oscillations during dynamic events (Miao et al., 2009).

7.5.2.3 Transmission infrastructure

As noted earlier, the highest-quality wind resources (whether on- or offshore) are often located at a distance from electricity demand centres. As a result, even at low to medium levels of wind electricity penetration, the addition of large quantities of wind energy in areas with the strongest wind resources may require significant new additions or upgrades to the transmission system (see also Chapter 8). Transmission adequacy evaluations must consider any tradeoffs between the costs of expanding the transmission system to access higher-quality wind resources and the costs of accessing lower-quality wind resources that require less transmission investment (e.g., Hoppock and Patiño-Echeverri, 2010). In addition, evaluations of new transmission capacity need to account for the relative smoothing benefits of aggregating wind power plants over large areas, the amount of transmission capacity devoted to managing the remaining variability of wind power output, and the broader non-wind-specific advantages and disadvantages of transmission expansion (Burke and O'Malley, 2010).

Irrespective of the costs and benefits of transmission expansion to accommodate increased wind energy deployment, one of the primary challenges is the long time it can take to plan, site, permit and construct new transmission infrastructure relative to the shorter time it often takes to add new wind power plants. Depending on the legal and regulatory framework in any particular region, the institutional challenges of transmission expansion, including cost allocation and siting, can be substantial (e.g., Benjamin, 2007; Vajjhala and Fischbeck, 2007; Swider et al., 2008). Enabling increased penetration of wind electricity may therefore require the creation of regulatory and legal frameworks for proactive rather than reactive transmission planning (Schumacher et al., 2009). Estimates of the cost of the new transmission required to achieve low to medium levels of wind electricity penetration in a variety of locations around the world are summarized in Section 7.5.4.

7.5.2.4 Generation adequacy

Though methods and objectives vary from region to region, generation adequacy evaluations are generally used to assess the capability of generation resources to reliably meet electricity demand. Planners often evaluate the long-term reliability of the electric system by estimating the probability that the system will be able to meet expected demand in the future, as measured by a statistical metric called the load-carrying capability of the system. Each electricity supply resource contributes some fraction of its nameplate capacity to the overall capability of the system, as indicated by the capacity credit assigned to the resource.[23] Although there is not a strict, uniform definition of capacity credit, the capacity credit of a generator is usually a 'system' characteristic in that it is determined not only by the generator's characteristics but also by the characteristics of the electric system to which that generator is connected, particularly the temporal profile of electricity demand (Amelin, 2009).

The contribution of wind energy to long-term reliability can be evaluated using standard approaches, and wind power plants are typically found to have a capacity credit of 5 to 40% of nameplate capacity (see Figure 7.14). The correlation between wind power output and electrical demand is an important determinant of the capacity credit of an individual wind power plant. In many cases, wind power output is uncorrelated or is weakly negatively correlated with periods of high electricity demand, reducing the capacity credit of wind power plants; this is not always the case, however, and wind power output in the UK, for example, has been found to be weakly positively correlated with periods of high demand (Sinden, 2007). These correlations are case specific as they depend on the diurnal, seasonal and yearly characteristics of both wind power output and electricity demand. A second important characteristic of the capacity credit for wind energy is that its value generally decreases as wind electricity penetration levels rise, because the capacity credit of a generator is greater when power output is well-correlated with periods of time when there is a higher risk of a supply shortage. As the level of wind electricity penetration increases, however, assuming that the outputs of wind power plants are positively correlated, the period of greatest risk will shift to times with low average levels of wind energy supply (Hasche et al., 2010). Aggregating wind power plants over larger areas may reduce the correlation between wind power outputs, as described earlier, and can slow the decline in capacity credit as wind electricity penetration increases, though adequate transmission capacity is required to aggregate the output of wind power plants in this way (Tradewind, 2009; EnerNex Corp, 2010).[24]

The relatively low average capacity credit of wind power plants (compared to fossil fuel-powered units, for example) suggests that systems with large amounts of wind energy will also tend to have significantly more total nameplate generation capacity (wind and non-wind) to meet the same peak electricity demand than will electric systems without large amounts of wind energy. Some of this generation capacity will operate infrequently, however, and the mix of other generation in an electric system with large amounts of wind energy will tend (on economic grounds) to increasingly shift towards more flexible 'peaking'

---

23 As an example, the addition of a very reliable 100 MW fossil unit in a system with numerous other reliable units will usually increase the load-carrying capability of the system by at least 90 MW, leading to a greater than 90% capacity credit for the fossil unit.

24 Generation resource adequacy evaluations are also beginning to include the capability of the system to provide adequate flexibility and operating reserves to accommodate more wind energy (NERC, 2009). The increased demand from wind energy for operating reserves and flexibility is addressed in Section 7.5.3.

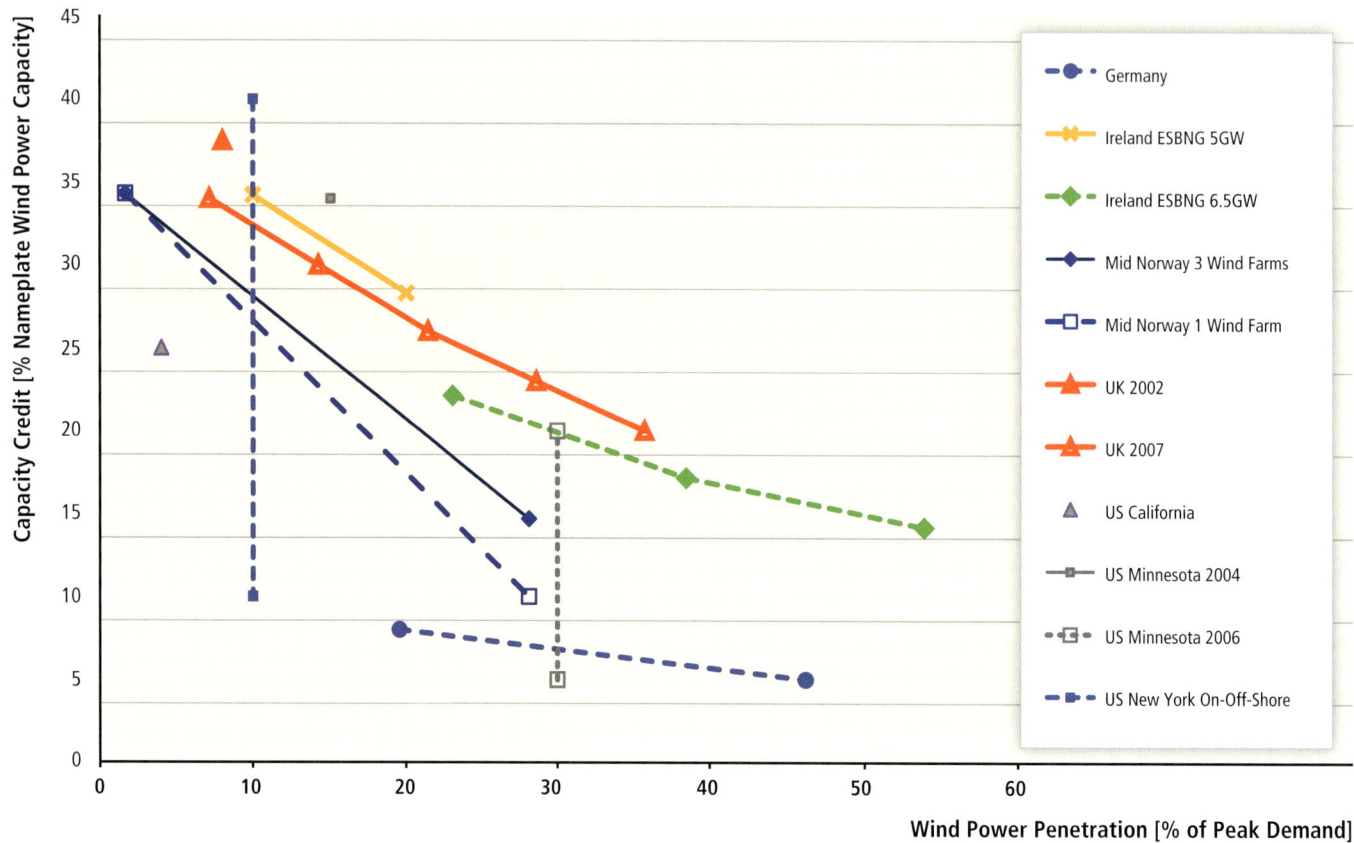

Figure 7.14 | Estimates of the capacity credit of wind power plants across several wind energy integration studies from Europe and the USA (Holttinen et al., 2009).

and 'intermediate' resources and away from 'base-load' resources (e.g., Lamont, 2008; Milborrow, 2009; Boccard, 2010).

7.5.3 Operating electric systems with wind energy

The unique characteristics of wind energy, and especially power output variability and uncertainty, also hold important implications for electric system operations. Here we summarize those implications in general (Section 7.5.3.1), and then briefly discuss three specific case studies of the integration of wind energy into real electricity systems (Section 7.5.3.2).

7.5.3.1 Integration, flexibility and variability

Because wind energy is generated with a very low marginal operating cost, it is typically used to meet demand when it is available, thereby displacing the use of generators that have higher marginal costs. This results in electric system operators and markets primarily dispatching other generators to meet demand minus any available wind energy (i.e., 'net demand').

As wind electricity penetration grows, the variability of wind energy results in an overall increase in the magnitude of *changes* in net demand, and also a decrease in the *minimum* net demand. For example, Figure 7.15 depicts demand and ramp duration curves for Ireland.[25] At relatively low levels of wind electricity penetration, the magnitude of changes in net demand, as shown in the 15-minute ramp duration curve, is similar to the magnitude of changes in total demand (Figure 7.15(c)). At higher levels of wind electricity penetration, however, changes in net demand are greater than changes in total demand (Figure 7.15(d)). Similar impacts on changes in net demand with increased wind energy have been reported in the USA (Milligan and Kirby, 2008). The figure also shows that, at high levels of wind electricity penetration, the magnitude of net demand across all hours of the year is lower than total demand, and that in some hours net demand is near or even below zero (Figure 7.15(b)).

As a result of these trends, wholesale electricity prices will tend to decline when wind power output is high (or is forecast to be high in the case of day-ahead markets) and transmission interconnection capacity to other energy markets is constrained, with a greater frequency of low or even negative prices (e.g., Jónsson et al., 2010; Morales et al., 2011). As with

---

25 Figure 7.15 presents demand and ramp duration curves for Ireland with (net demand) and without (demand) the addition of wind energy. A demand duration curve shows the percentage of the year that the demand exceeds a level on the vertical axis. Demand in Ireland exceeds 4,000 MW, for example, about 10% of the year. The ramp duration curves show the percentage of the year that changes in the demand exceed the level on the vertical axis. The 15-min change in demand in Ireland exceeds 100 MW/15minutes, for example, less than 10% of the year.

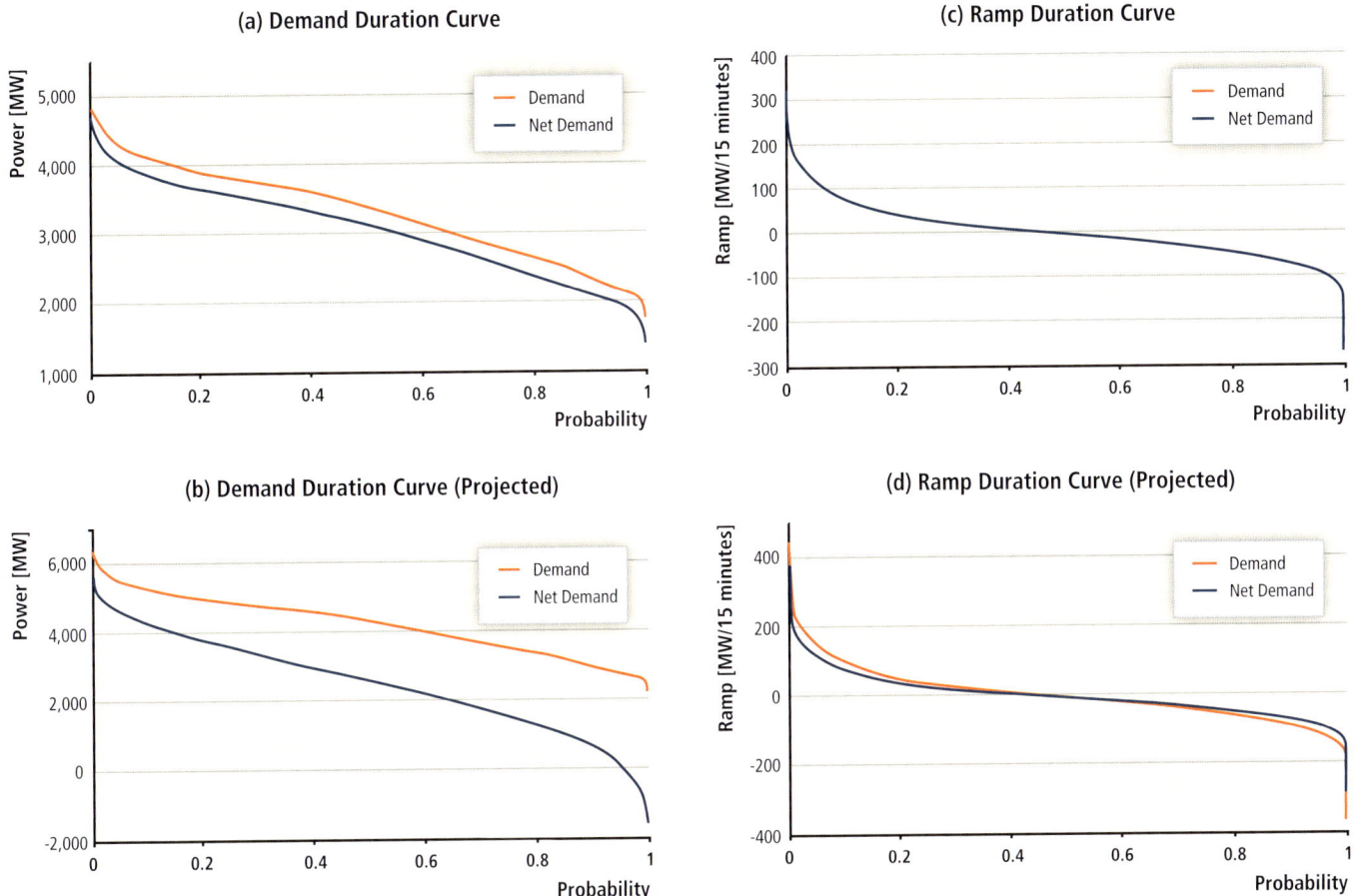

Figure 7.15 | Demand duration and 15-minute ramp duration curves for Ireland in (a, c) 2008 (wind energy represents 7.5% of total annual average electricity demand), and (b, d) projected for high wind electricity penetration levels (wind energy represents 40% of total annual average electricity demand).[1] Source: Data from www.eirgrid.com.

Note: 1. Projected demand and ramp duration curves are based on scaling 2008 data (demand is scaled by 1.27 and wind energy is scaled on average by 7). Ramp duration curves show the cumulative probability distributions of 15-minute changes in demand and net demand.

adding any low marginal cost resource to an electric system, increased wind electricity penetrations will therefore tend to reduce average wholesale prices in the short term (before changes are made to the mix of other generation sources) as wind energy displaces power sources with higher marginal costs. Price volatility will also tend to increase as the variability and uncertainty in wind power output ensures that wind energy will not always be available to displace higher marginal cost generators. In the long run, however, the average effect of wind energy on wholesale electricity prices is not as clear because the relationships between investment costs, O&M costs and wholesale price signals will begin to influence decisions about the expansion of transmission interconnections, generator retirement and the type of new generation that is built (Morthorst, 2003; Førsund et al., 2008; Lamont, 2008; Sáenz de Miera et al., 2008; Sensfuß et al., 2008; Söder and Holttinen, 2008; MacCormack et al., 2010).

These price impacts are a reflection of the fact that increased wind energy deployment will require some other generating units to operate in a more flexible manner than required without wind energy. At low to medium levels of wind electricity penetration, the increase in *minute-to-minute* variability will depend on the exact level of wind electricity penetration, the degree of geographic smoothing, and electric system size, but is generally expected to be relatively small and therefore inexpensive to manage in large electric systems (J. Smith et al., 2007). The more significant operational challenges relate to the variability and commensurate increased need for flexibility to manage changes in wind power output over *one to six hours* (Doherty and O'Malley, 2005; Holttinen et al., 2009). Incorporating state-of-the-art forecasting of wind energy over multiple time horizons into electric system operations can reduce the need for flexibility from other generators, and has been found to be especially important as wind electricity penetration levels increase (e.g., Doherty et al., 2004; Tuohy et al., 2009; GE Energy, 2010). Nonetheless, even with high-quality forecasts and geographically dispersed wind power plants, additional start-ups and shut-downs, part-load operation, and ramping will be required from fossil generation units to maintain the supply/demand balance (e.g., Göransson and Johnsson, 2009; Troy et al., 2010).

This additional flexibility is not free, as it increases the amount of time that fossil fuel-powered units are operated at less efficient part-load conditions (resulting in lower than expected reductions in production costs and emissions from fossil generators as described in Sections

7.5.4 and 7.6.1.3, respectively), increases wear and tear on boilers and other equipment, increases maintenance costs, and reduces power plant life (Denny and O'Malley, 2009). Various kinds of economic incentives can be used to ensure that the operational flexibility of other generators is made available to system operators. Some electricity systems, for example, have day-ahead, intra-day, and/or hour-ahead markets for electricity, as well as markets for reserves, balancing energy and other ancillary services. These markets can provide pricing signals for increased (or decreased) flexibility when needed as a result of rapid changes in or poorly predicted wind power output, and can therefore reduce the cost of integrating wind energy (J. Smith et al., 2007; Göransson and Johnsson, 2009). Markets with shorter scheduling periods have also been found to be more responsive to variability and uncertainty, thereby facilitating wind energy integration (Holttinen, 2005; Kirby and Milligan, 2008; Tradewind, 2009). In addition, coordinated electric system operations across larger areas has been shown to benefit wind energy integration, and increased levels of wind energy supply may therefore tend to motivate greater investments in and electricity trade across transmission interconnections (Milligan and Kirby, 2008; Denny et al., 2010). Where wholesale electricity markets do not exist, other planning methods or incentives would be needed to ensure that generating plants are flexible enough to accommodate increased deployment of wind energy.

Planning systems and incentives may also need to be adopted to ensure that new generating plants are sufficiently flexible to accommodate expected wind energy deployment. Moreover, in addition to flexible fossil fuel-powered units, hydropower stations, bulk energy storage, large-scale deployment of electric vehicles and their associated contributions to system flexibility through controlled battery charging, diverting excess wind energy to fuel production or local heating, and various forms of demand response can also be used to facilitate the integration of wind energy. The deployment of a diversity of RE technologies may also help facilitate overall electric system integration. The role of some of these technologies (as well as some of the operational and planning methods noted earlier) in electric systems is described in more detail in Chapter 8 because they are not all specific to wind energy and because some are more likely to be used at higher levels of wind electricity penetration than considered here (up to 20%). Wind power plants, meanwhile, can provide some flexibility by briefly curtailing output to provide downward regulation or, in extreme cases, curtailing output for extended periods to provide upward regulation. Modern controls on wind power plants can also use curtailment to limit or even (partially) control ramp rates (Fox et al., 2007). Though curtailing wind power output is a simple and often times readily available source of flexibility, there are sizable opportunity costs associated with curtailing plants that have low operating costs before reducing the output of other plants that have high fuel costs. These opportunity costs should be compared to the possible benefits of curtailment (e.g., reduced part-load efficiency penalties and wear and tear for fossil generators, and avoidance of certain transmission investments) when determining the prevalence of its use.

7.5.3.2 Practical experience with operating electric systems with wind energy

Actual operating experience in different parts of the world demonstrates that electric systems can operate reliably with increased contributions of wind energy (Söder et al., 2007). In four countries, as discussed earlier, wind energy in 2010 was already able to supply from 10 to roughly 20% of annual electricity demand. The three examples reported here demonstrate the challenges associated with this operational integration, and the methods used to manage the additional variability and uncertainty associated with wind energy. Naturally, these impacts and management methods vary across regions for reasons of geography, electric system design and regulatory structure, and additional examples of wind energy integration associated with operations, curtailment and transmission are described in Chapter 8. Moreover, as more wind energy is deployed in diverse regions and electric systems, additional knowledge about the impacts of wind power output on electric systems will be gained. To date, for example, there is little experience with severe contingencies (i.e., faults) during times with high instantaneous wind electricity penetration. Though existing experience demonstrates that electric systems can operate with wind energy, further analysis is required to determine whether electric systems are maintaining the same level of overall security, measured by the ability of the system to withstand major contingencies, with and without wind energy, and depending on various management options. Limited analysis (e.g., EirGrid and SONI, 2010; Eto et al., 2010) suggests that particular systems are able to survive such conditions but, if primary frequency control reserves are reduced as thermal generation is increasingly displaced by wind energy, additional management options may be needed to maintain adequate frequency response. The security of the electric system with high instantaneous wind electricity penetrations is described in more detail in Chapter 8.

Denmark has the highest wind electricity penetration of any country in the world, with wind energy supply equating to approximately 20% of total annual electricity demand. Total wind power capacity installed by the end of 2009 equalled 3.4 GW, while the peak demand in Denmark was 6.5 GW. Much of the wind power capacity (2.7 GW) is located in western Denmark, resulting in instantaneous wind power output exceeding total demand in western Denmark in some instances (see Figure 7.16). The Danish example demonstrates the benefits of having access to markets for flexible resources and having strong transmission interconnections to neighbouring countries. Denmark's electricity systems operate without serious reliability issues in part because the country is well interconnected to two different electric systems. In conjunction with wind power output forecasting, this allows wind energy to be exported to other markets and helps the Danish operators manage wind power variability. The interconnection with the Nordic system, in particular, provides access to flexible hydropower resources, and balancing the Danish system is much more difficult during periods when one of the interconnections is down. Even more flexibility is expected to be required, however, if Denmark markedly increases its penetration of wind electricity (Ea Energianalyse, 2007).

In contrast to the strong interconnections of the Danish system with other electric systems, the island of Ireland has a single synchronous system; its size is similar to the Danish system but interconnection capacity with other markets is limited to a single 500 MW high-voltage direct current link. The wind power capacity installed by the end of 2009 was capable of supplying roughly 11% of Ireland's annual electricity demand, and the Irish system operators have successfully managed that level of wind electricity penetration. The large daily variation in electricity demand in Ireland, combined with the isolated nature of the Irish system, has resulted in a relatively flexible electric system that is particularly well suited to integrating wind energy; flexible natural gas plants generated 65% of the electrical energy in the first half of 2010. As a result, despite the lack of significant interconnection capacity, the Irish system has successfully operated with instantaneous levels of wind electricity penetration of over 40% (see Figure 7.16). Nonetheless, it is recognized that as wind electricity penetration levels increase further, new challenges will arise. Of particular concern are: the possible lack of inertial response of wind turbines absent additional turbine controls, which could lead to increased frequency excursions during severe grid contingencies (Lalor et al., 2005); the need for even greater flexibility to maintain supply-demand balance; and the need to build additional high-voltage transmission (AIGS, 2008). Moreover, in common with the Danish experience, much of the wind energy is and will be connected to the distribution system, requiring attention to voltage control issues (Vittal et al., 2010). Figure 7.16 illustrates the high levels of instantaneous wind electricity penetration that exist in Ireland and West Denmark.

The Electric Reliability Council of Texas (ERCOT) operates a synchronous system with a peak demand of 63 GW and 8.5 GW of wind power capacity, and with a wind electricity penetration level of 6% of annual electricity demand by the end of 2009. ERCOT's experience demonstrates the importance of incorporating wind energy forecasts into system operations, and the need to schedule adequate reserves to accommodate system uncertainty. On 26 February 2008, a combination of factors, not all related to wind energy, led ERCOT to implement its emergency curtailment plan, which included the curtailment of 1,200 MW of demand that was voluntarily participating in ERCOT's 'Load Acting as a Resource' program. The factors involved in the event included wind energy scheduling errors, an incorrect day-ahead electricity demand forecast, and an unscheduled outage of a fossil fuel power plant. With regards to the role of wind energy, ERCOT experienced a decline in wind power output of 1,500 MW over a three-hour period on that day, roughly 30% of the 5 GW of installed wind power capacity in February 2008 (Ela and Kirby, 2008; ERCOT, 2008). The event was exacerbated by the fact that scheduling entities—which submit updated resource schedules to ERCOT one hour prior to the operating hour—consistently reported an expectation of more wind power output than actually occurred. A state-of-the-art forecast was available, but was not yet integrated into ERCOT system operations, and that forecast predicted the wind energy event much more accurately. As a result of this experience, ERCOT accelerated its schedule for incorporating the advanced wind energy forecasting system into its operations.

7.5.4 Results from integration studies

In addition to actual operating experience, a number of high-quality studies of the increased transmission and generation resources required to accommodate wind energy have been completed, primarily covering OECD countries. As summarized further below, these studies employ a wide variety of methodologies and have diverse objectives, but typically seek to evaluate the capability of the electric system to integrate increased penetrations of wind energy and to quantify the costs and benefits of operating the system with wind energy. The issues and costs often considered by these studies are reviewed in this section, and include: the increased operating reserves and balancing costs required

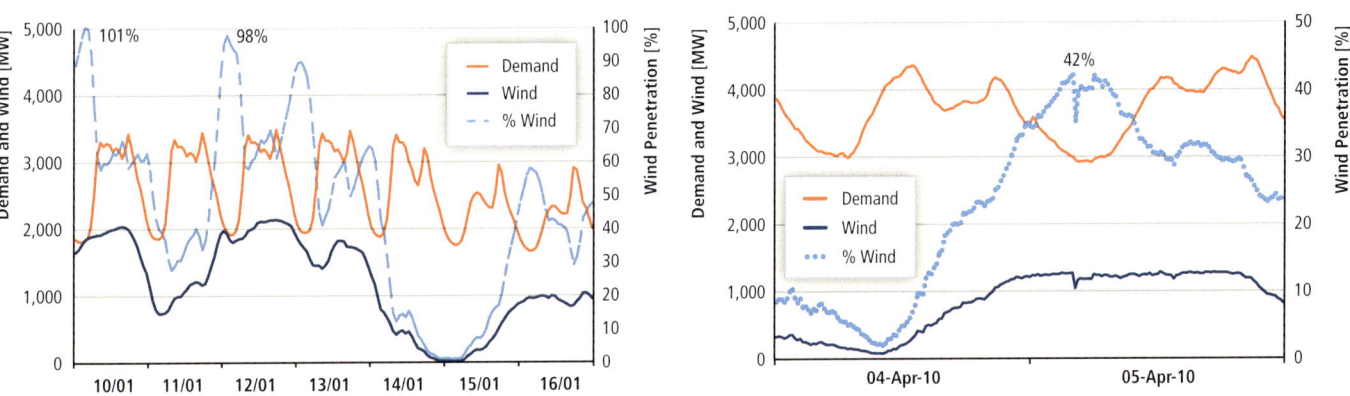

Figure 7.16 | Wind energy, electricity demand and instantaneous penetration levels in (left) West Denmark for a week in January 2005, and (right) the island of Ireland for two days in April 2010. Source: Data from (left) www.energinet.dk; (right) www.eirgrid.com and System Operator for Northern Ireland.

to accommodate the variability and uncertainty in net demand caused by wind energy; the requirement to maintain sufficient generation adequacy; and the possible need for additional transmission infrastructure. The studies also frequently analyze the benefits of adding wind energy, including avoided fossil fuel consumption and CO_2 emissions, though these benefits are not reviewed in this section. This section focuses on the general results of these studies as a whole; see Chapter 8 for brief descriptions of individual study results, including some studies that have investigated somewhat higher levels of wind electricity penetration than considered here.

7.5.4.1 Methodological challenges

Estimating the incremental impacts and costs of wind energy integration is difficult due to the complexity of electric systems and study data requirements. One of the most significant challenges in executing these studies is simulating wind power output data at high time resolutions for a chosen future wind electricity penetration level and for a sufficient duration for the results of the analysis to accurately depict worst-case conditions and correlations of wind and electricity demand. These data are then used in electric system simulations to mimic system planning and operations, thereby quantifying the impacts, costs and benefits of wind energy integration.

Addressing all integration impacts requires several different simulation models that operate over different time scales, and most individual studies therefore focus on a subset of the potential issues. The results of wind energy integration studies are also dependent on pre-existing differences in electric system designs and regulatory environments: important differences include generation capacity mix and the flexibility of that generation, the variability of demand and the strength and breadth of the transmission system. In addition, study results differ and are hard to compare because standard methodologies and even definitions have not been developed, though significant progress has been made in developing agreement on many high-level study design principles (Holttinen et al., 2009). The first-generation integration studies, for example, used models that were not designed to fully reflect the variability and uncertainty of wind energy, resulting in studies that addressed only parts of the larger system. More recent studies, on the other hand, have used models that can incorporate the uncertainty of wind power output from the day-ahead time scale to some hours ahead of delivery (e.g., Meibom et al., 2009; Tuohy et al., 2009). Integration studies are also increasingly simulating high wind electricity penetration scenarios over entire synchronized systems (not just individual, smaller balancing areas) (e.g., Tradewind, 2009; EnerNex Corp, 2010; GE Energy, 2010). Finally, only recently have studies begun to explore in more depth the capability of electric systems to maintain primary frequency control during system contingencies with high penetrations of wind energy (e.g., EirGrid and SONI, 2010; Eto et al., 2010).

Regardless of the challenges of executing and comparing such studies, the results, as described in more detail below, demonstrate that the cost of integrating up to 20% wind energy into electric systems is, in most cases, modest but not insignificant. Specifically, at low to medium levels of wind electricity penetration (up to 20% wind energy), the available literature (again, primarily from a subset of OECD countries) suggests that the additional costs of managing electric system variability and uncertainty, ensuring generation adequacy and adding new transmission to accommodate wind energy will be system specific but generally in the range of US cents$_{2005}$ 0.7 to 3/kWh.[26] Concerns about (and the costs of) wind energy integration will grow with wind energy deployment and, even at lower penetration levels, integration issues must be actively managed.

7.5.4.2 Increased balancing cost with wind energy

The additional variability and uncertainty in net demand caused by increased wind energy supply results in higher balancing costs, in part due to increases in the amount of short-term reserves procured by system operators. A number of significant integration studies from Europe and the USA have concluded that accommodating wind electricity penetrations of up to (and in a limited number of cases, exceeding) 20% is technically feasible, but not without challenges (R. Gross et al., 2007; J. Smith et al., 2007; Holttinen et al., 2009; Milligan et al., 2009). The estimated increase in short-term reserve requirements in eight studies summarized by Holttinen et al. (2009) has a range of 1 to 15% of installed wind power capacity at 10% wind electricity penetration, and 4 to 18% of installed wind power capacity at 20% wind electricity penetration. Those studies that predict a need for higher levels of reserves generally assume that day-ahead uncertainty and/or multi-hour variability of wind power output is handled with short-term reserves. In contrast, markets that are optimized for wind energy will generally be designed so that additional opportunities to balance supply and demand exist, reducing the reliance on more expensive short-term reserves (e.g., Weber, 2010). Notwithstanding the differences in results and methods, however, the studies reviewed by Holttinen et al. (2009) find that, in general, wind electricity penetrations of up to 20% can be accommodated with increased balancing costs of roughly US cents 0.14 to 0.56/kWh[27] of wind energy generated (Figure 7.17). State-of-the-art wind energy forecasts are often found to be a key factor in minimizing the impact of wind energy on market operations. Although definitions and methodologies for calculating increased balancing costs differ, and several open issues remain in estimating these costs, similar results are reported by R. Gross et al. (2007), J. Smith et al. (2007), and Milligan et al. (2009).

---

26 This cost range is based on the assumption that there may be electric systems where all three cost components (balancing costs, generation adequacy costs and transmission costs) are simultaneously at the low end of the range reported for each of these costs in the literature or conversely where all three cost components are simultaneously at the high end of the range. As reported below, the cost range for managing wind energy's variability and uncertainty (US cents$_{2005}$ 0.14 to 0.56/kWh), ensuring generation adequacy (US cents$_{2005}$ 0.58 to 0.96/kWh), and adding new transmission (US cents$_{2005}$ 0 to 1.5/kWh) sums to roughly US cents$_{2005}$ 0.7 to 3/kWh. Using a somewhat similar approach, IEA (2010b) developed estimates that are also broadly within this range.

27 Conversion to 2005 dollars is not possible given the range of study-specific assumptions.

7.5.4.3 Relative cost of generation adequacy with wind energy

The benefits of adding a wind power plant to an electric system are often compared to the benefits of a base-load, or fully utilized, plant that generates an equivalent amount of energy on an annual basis (a comparator plant). The comparator plant is typically assumed to have a high capacity credit, close to 100% of its nameplate capacity. Wind energy, on the other hand, was shown in Section 7.5.2.4 to have a capacity credit of 5 to 40% of its nameplate capacity. The resulting contribution of the wind plant to generation adequacy is therefore often lower than the contribution of an energy-equivalent comparator plant per unit of energy generated, and wind energy is typically less valuable than the comparator plant from the perspective of meeting generation adequacy targets. Using this framework, R. Gross et al. (2007) estimate that the difference between the contribution to generation adequacy of a wind power plant and an energy-equivalent base-load plant can result in a US cents$_{2005}$ 0.58 to 0.96/kWh generation adequacy cost for wind energy relative to a comparator plant at wind electricity penetration levels up to

to electricity demand, the geographic distribution of wind power plant siting and the level of wind electricity penetration will all impact the capacity credit estimated for wind energy, and therefore the relative cost of generation adequacy.

7.5.4.4 Cost of transmission for wind energy

Finally, a number of assessments of the need for and cost of upgrading or building large-scale transmission infrastructure between wind resource regions and demand centres have similarly found modest, but not insignificant, costs.[28] The transmission cost for achieving 20% wind electricity penetration in the USA, for example, was estimated to add about USD$_{2005}$ 150 to 290/kW to the investment cost of wind power plants (US DOE, 2008). The cost of this transmission expansion was found to be justified because of the higher quality of the wind resources accessed if the transmission were to be built relative to accessing only lower-quality wind resources with less transmission expansion. More

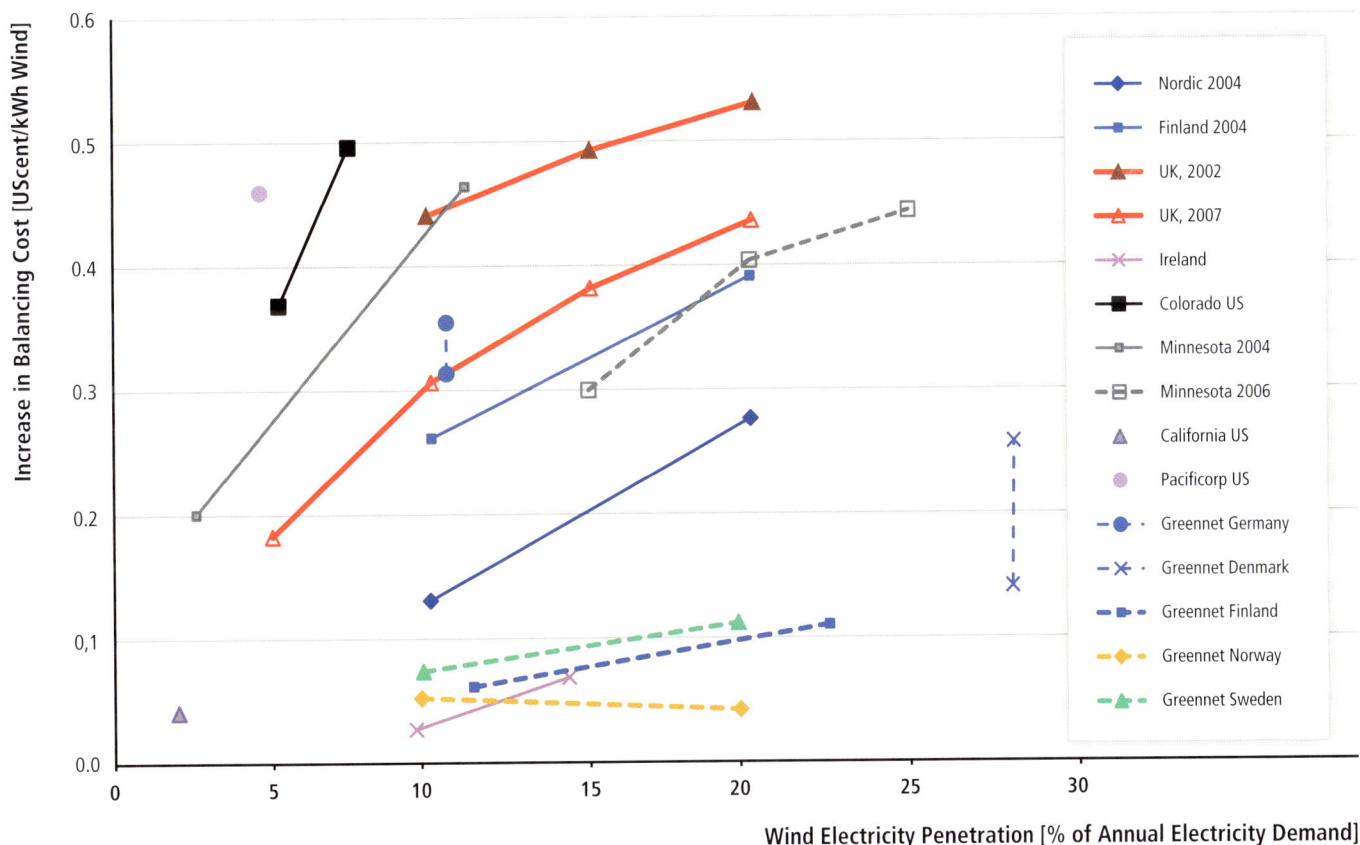

Figure 7.17 | Estimates of the increase in balancing costs due to wind energy from several wind energy integration studies in Europe and the USA (Holttinen et al., 2009).[1]

Note: 1. Conversion to 2005 dollars is not possible given the range of study-specific assumptions.

20%. Using a somewhat different approach, Boccard (2010) provides a comparable estimate of the generation adequacy cost of wind energy in several European countries. As discussed earlier, the methodology used to assess generation adequacy, the correlation of wind power output

[28] These costs are distinct from the costs to connect individual wind power plants to the transmission system; connection costs are often included in estimates of the investment costs of wind power plants (see Section 7.8).

detailed assessments of the transmission needed to accommodate increased wind energy deployment in the USA have found a wide range of results, with estimated costs ranging from very low to sometimes reaching (or even exceeding) USD$_{2005}$ 400/kW (JCSP, 2009; Mills et al., 2009a; EnerNex Corp, 2010). Large-scale transmission for cases with increased wind energy has also been considered in Europe (Czisch and Giebel, 2000) and China (Lew et al., 1998). Results from country-specific transmission assessments in Europe have resulted in varied estimates of the cost of new large-scale transmission; Auer et al. (2004) and EWEA (2005) identified transmission costs for a number of European studies, with cost estimates that are somewhat lower than those found in the USA. Holttinen et al. (2009) reviewed wind energy transmission costs from several European national case studies, and found costs ranging from USD$_{2005}$ 0/kW to as high as USD$_{2005}$ 310/kW.

Transmission expansion for wind energy can be justified by the reduction in congestion costs that would occur for the same level of wind energy deployment without transmission expansion. A European-wide study, for example, identified several transmission upgrades between nations and between high-quality offshore wind resource areas that would reduce transmission congestion and ease wind energy integration (Tradewind, 2009). The avoided congestion costs associated with transmission expansion were similarly found to justify transmission investments in two US-based detailed integration studies of high wind electricity penetrations (Milligan et al., 2009). At the same time, it is not always appropriate to fully assign the cost of transmission expansion to wind energy deployment. In some cases, these transmission expansion costs can be justified for reasons beyond wind energy, as new transmission can have wider benefits including increased electricity reliability, decreased pre-existing congestion and reduced market power (Budhraja et al., 2009). Moreover, wind energy is not unique in potentially requiring new transmission investment; other energy technologies may also require new transmission, and the costs summarized above do not all represent truly incremental costs.

Notwithstanding these important caveats, at the higher end of the range from the available literature (USD$_{2005}$ 400/kW), transmission expansion costs add roughly US cents$_{2005}$ 1.5/kWh to the levelized cost of wind energy. At the lower end, effectively no new transmission costs would need to be specifically assigned to the support of wind energy.

7.6 Environmental and social impacts[29]

Wind energy has significant potential to reduce (and already is reducing) GHG emissions, together with the emissions of other air pollutants, by displacing fossil fuel-based electricity generation. Because of the commercial readiness (Section 7.3) and cost (Section 7.8) of the technology, wind energy can be immediately deployed on a large scale (Section 7.9). As with other industrial activities, however, wind energy also has the potential to produce some detrimental impacts on the environment and on human activities and well-being, and many local and national governments have established planning, permitting and siting requirements to reduce those impacts. These potential concerns need to be taken into account to ensure a balanced view of the advantages and disadvantages of wind energy, especially if wind energy is to expand on a large scale.

This section summarizes the best available knowledge about the most relevant environmental net benefits of wind energy (Section 7.6.1), while also addressing ecological impacts (Section 7.6.2), impacts on human activities and well-being (Section 7.6.3), public attitudes and acceptance (Section 7.6.4) and processes for minimizing social and environmental concerns (Section 7.6.5).

7.6.1 Environmental net benefits of wind energy

The environmental benefits of wind energy come primarily from displacing the emissions from fossil fuel-based electricity generation. However, the manufacturing, transport, installation, operation and decommissioning of wind turbines induces some indirect negative effects, and the variability of wind power output also impacts the operations and emissions of fossil fuel-fired plants. Such effects need to be subtracted from the gross benefits of wind energy in order to estimate net benefits. As shown below, these latter effects are modest compared to the net GHG reduction benefits of wind energy.

7.6.1.1 Direct impacts

The major environmental benefits of wind energy (as well as other forms of RE) result from displacing electricity generation from fossil fuel-based power plants, as the operation of wind turbines does not directly emit GHGs or other air pollutants. Similarly, unlike some other generation sources, wind energy requires insignificant amounts of water, produces little waste and requires no mining or drilling to obtain its fuel supply (see Chapter 9).

Estimating the environmental benefits of wind energy is somewhat complicated by the operational characteristics of the electric system and the decisions that are made about investments in new power plants to economically meet electricity demand (Deutsche Energie-Agentur, 2005; NRC, 2007; Pehnt et al., 2008). In the short run, increased wind energy will typically displace the operations of existing fossil fuel-based plants that are otherwise on the margin. In the longer term, however, new generating plants may be needed, and the presence of wind energy can influence what types of power plants are built; specifically, increased wind energy will tend to favour on economic grounds flexible peaking/intermediate plants that operate less frequently over base-load plants (Kahn, 1979; Lamont, 2008). Because the impacts of these factors are both complicated and system specific, the benefits of wind energy will also be system specific and are difficult to forecast with precision.

---

[29] A comprehensive assessment of social and environmental impacts of all RE sources covered in this report can be found in Chapter 9.

Nonetheless, it is clear that the direct impact of wind energy is to reduce air pollutants and GHG emissions. Depending on the characteristics of the electric system into which wind energy is integrated and the amount of wind energy supply, the reduction of air pollution and GHG emissions may be substantial. Globally, it has been estimated that the roughly 160 GW of wind power capacity already installed by the end of 2009 could generate 340 TWh/yr (1.2 EJ/yr) of electricity and save more than 0.2 Gt CO_2/yr (GWEC, 2010b).[30]

7.6.1.2 Indirect lifecycle impacts

Some indirect environmental impacts of wind energy arise from the manufacturing, transport, installation and operation of wind turbines, and their subsequent decommissioning. Life-cycle assessment (LCA) procedures based on ISO 14040 and ISO 14044 standards (ISO, 2006) have been used to analyze these impacts. Though these studies may include a range of environmental impact categories, LCA studies for wind energy have often been used to determine the lifecycle GHG emissions per unit of wind electricity generated (allowing for full fuel-cycle comparisons with other forms of electricity production). The results of a comprehensive review of LCA studies published since 1980 are summarized in Figure 7.18.

Figure 7.18 shows that the majority of lifecycle GHG emission estimates cluster between about 8 and 20 g CO_2eq/kWh, with some estimates reaching 80 g CO_2eq/kWh.[31] Where studies have identified the significance of different stages of the lifecycle of a wind power plant, it is clear that emissions from the manufacturing stage dominate overall lifecycle GHG emissions (e.g., Jungbluth et al., 2005). Variability in estimates stems from differences in study context (e.g., wind resource, technological vintage), technological performance (e.g., capacity factor) and methods (e.g., LCA system boundaries).[32]

In addition to lifecycle GHG emissions, many of these studies also report on the energy payback time of wind power plants (i.e., the amount of time a wind power plant must operate in order produce an equivalent amount of energy that was required to build, operate and decommission it). Among 50 estimates from 20 studies passing screens for quality and relevance, the median reported energy payback time for wind power plants is 5.4 months, with a 25th to 75th percentile range of 3.4 months to 8.5 months (see also Chapter 9).

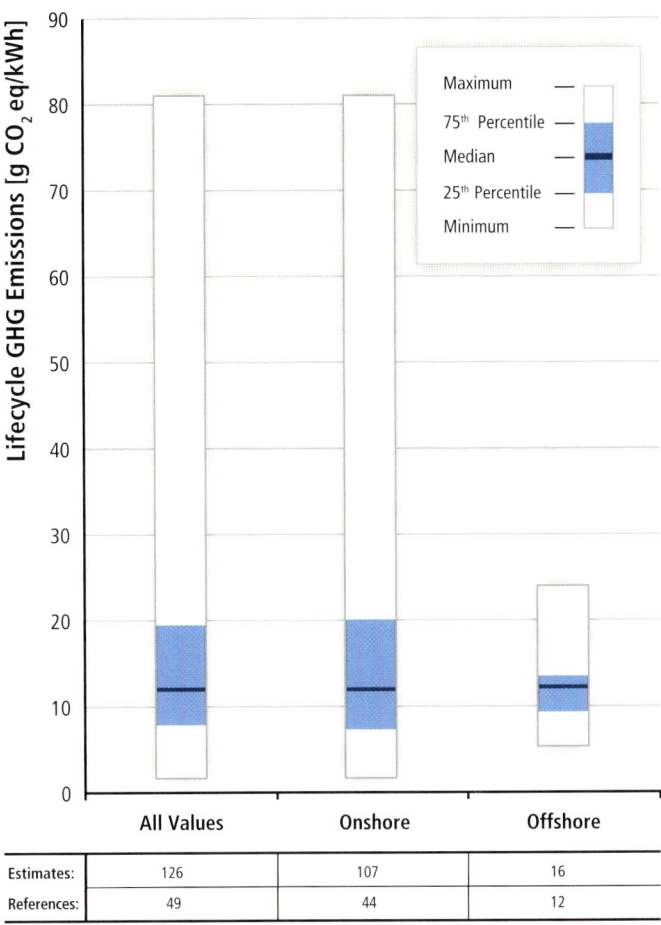

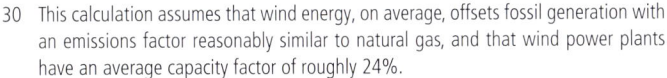

Figure 7.18 | Lifecycle GHG emissions of wind energy technologies (unmodified literature values, after quality screen). 'Offshore' represents relatively shallow offshore installations except for one floating offshore estimate. See Annex II.5.2 for details about the literature search and the literature citations contributing to the estimates displayed.

The lifecycle impacts of wind energy in comparison to other energy technologies are covered in Chapter 9, including not just GHG emissions and energy payback, but also local air pollutants, water consumption, land use and other impact categories.

7.6.1.3 Indirect variability impacts

Another concern that is sometimes raised is that the temporal variability and limited predictability of wind energy will limit the GHG emissions benefits of wind energy by increasing the short-term balancing reserves required for an electric system operator to maintain reliability (relative to the balancing reserve requirement without wind energy). Short-term reserves are generally provided by generating plants that are online and synchronized with the grid, and plants providing these reserves may be part-loaded to maintain the flexibility to respond to short-term fluctuations. Part-loading fossil fuel-based generators decreases the efficiency of the plants and therefore creates a fuel efficiency and GHG emissions penalty relative to a fully loaded plant. Analyses of the emissions benefits of wind energy do not always account for this effect.

---

30 This calculation assumes that wind energy, on average, offsets fossil generation with an emissions factor reasonably similar to natural gas, and that wind power plants have an average capacity factor of roughly 24%.

31 Note that the distributions shown in Figure 7.18 do not represent an assessment of likelihood; the figure simply reports the distribution of currently published literature estimates passing screens for quality and relevance. See Annex II.5.2 for a further description of the literature search methods.

32 Efforts to harmonize the methods and assumptions of these studies are recommended such that more robust estimates of central tendency and variability can be realized. Further LCA studies to increase the number of estimates for some technologies (e.g., floating offshore wind turbines) would also be beneficial.

R. Gross et al. (2007) performed an extensive literature review of the costs and impacts of variable electricity supply; over 200 reports and articles were reviewed. The review included a number of analyses of the fuel savings and GHG emissions benefits[33] of wind energy that accounted for the increase in necessary balancing reserves and the reduction in part-load efficiency of fossil fuel-powered plants. The efficiency penalty due to the variability of wind power output in four studies that explicitly addressed the issue ranged from near 0% to as much as 7%, for up to 20% wind electricity penetration (R. Gross et al., 2006). Pehnt et al. (2008) calculated an emission penalty of 3 to 8% for a wind electricity penetration of 12%, with the range reflecting varying types of other power plants built in future years.[34] In short, at low to medium levels of wind electricity penetration, "there is no evidence available to date to suggest that in aggregate efficiency reductions due to load following amount to more than a few percentage points" (Gross and Heptonstall, 2008).[35]

7.6.1.4 Net environmental benefits

The precise balance of positive and negative environmental and health effects of wind energy is system specific, but can in general be documented by the difference in estimated external costs for wind energy and other electricity supply options (see Chapter 10). Monetized figures for climate change damages, human health impacts, material damages and agricultural losses show significant benefits from wind energy (e.g., Krewitt and Schlomann, 2006). Krewitt and Schlomann (2006) also qualitatively assess the direction of possible impacts associated with other damage categories (ecosystem effects, large accidents, security of supply and geopolitical effects), finding that the net benefits of RE sources tend to be underestimated by not including these impacts in the monetized results. The environmental damages associated with other forms of electricity generation and benefits associated with wind energy have been summarized many times in the broader externalities literature (e.g., EC, 2003; Owen, 2004; Sundqvist, 2004; NRC, 2010a), and are highlighted in Chapters 9 and 10.

7.6.2 Ecological impacts

There are, nonetheless, ecological impacts that need to be taken into account when assessing wind energy. Potential ecological impacts of concern for onshore wind power plants include the population-level consequences of bird and bat collision fatalities and more indirect habitat and ecosystem modifications. For offshore wind energy, the aforementioned impacts as well as implications for benthic resources, fisheries and marine life more generally must be considered. Finally, the possible impacts of wind energy on the local climate have received attention. The focus here is on impacts associated with wind power plants themselves, but associated infrastructure also has impacts to consider (e.g., transmission lines, transportation to site etc.). In addition, though more systematic assessments are needed to evaluate the *relative* impacts of different forms of energy supply, especially within the context of the varying contributions of these energy sources towards global climate change, those comparisons are not provided here but are instead discussed in Chapter 9.

7.6.2.1 Bird and bat collision fatalities

Bird and bat fatalities through collisions with wind turbines are among the most publicized environmental concerns associated with wind power plants. Populations of many species of birds and bats are in decline, leading to concerns about the effects of wind energy on vulnerable species.

Though much remains unknown about the nature and population-level implications of these impacts, avian fatality rates are power plant- and species-specific, and can vary with region, site characteristics, season, weather, turbine size, height and design, and other factors. Focusing on all bird species combined, the US National Research Council (NRC) surveyed the available (limited) literature through early 2007 and found bird mortality estimates that range from 0.95 to 11.67/MW/yr (NRC, 2007); other results, including those from Europe, provide a reasonably similar range of estimates (e.g., De Lucas et al., 2004; Drewitt and Langston, 2006; Everaert and Stienen, 2007; Kuvlesky et al., 2007). Though most of the bird fatalities reported in the literature are of songbirds (Passeriformes), which are the most abundant bird group in terrestrial ecosystems (e.g., Erickson et al., 2005; NRC, 2007), raptor fatalities are considered to be of greater concern as their populations tend to be relatively small. Compared to songbird fatalities, raptor fatalities have been found to be relatively low; nonetheless, these impacts are site specific, and there are cases in which raptor fatalities (and the potential for population-level effects) have raised concerns (e.g., Barrios and Rodriguez, 2004; Kuvlesky et al., 2007; NRC, 2007; Smallwood and Thelander, 2008). As offshore wind energy has increased, concerns have also been raised about seabirds (e.g., Garthe and Hüppop, 2004). More research is needed and impacts will again be species specific (Desholm, 2009), but the limited research to date does not suggest that offshore plants pose a disproportionately large risk to birds relative to onshore wind energy (e.g., Dong Energy et al., 2006); Desholm and Kahlert (2005), for example, find that seabirds tend to detect and avoid large offshore wind power plants.

---

[33] Because GHG emissions are generally proportional to fuel consumption for a single fossil fuel-fired plant, the GHG emissions penalty is similar to the fuel efficiency penalty.

[34] Accounting for only the start-up and minimum load requirements of fossil generators (but not including the part-load efficiency penalty), Göransson and Johnsson (2009) estimate an emission penalty of 5%.

[35] Katzenstein and Apt (2009) conclude that the efficiency penalty could be as high as 20%, but inaccurately assume that every wind power plant requires spinning reserves equivalent to the nameplate capacity of the wind plant. Accounting for the smoothing benefits of geographic diversity (see Section 7.5) and the ability to commit and de-commit thermal plants lowers the estimated efficiency penalty substantially (Mills et al., 2009b).

Bat fatalities have not been researched as extensively as bird fatalities at wind power plants, and data allowing reliable assessments of bat fatalities are somewhat limited (Dürr and Bach, 2004; Kunz et al., 2007b; NRC, 2007; Cryan and Barclay, 2009). Several wind power plants have reported sizable numbers of bat fatalities, but other studies have shown low fatality rates. Surveying the available literature through early 2007, the NRC (2007) reported observed bat fatalities ranging from 0.8 to 41.1 bats per MW per year; a later review of 21 studies by Arnett et al. (2008) found fatality rates of 0.2 to 53.3 bats per MW per year. The specific role of different influences such as site characteristics, weather conditions, and turbine size, placement and operation remain somewhat uncertain due to the lack of extensive and comparable studies (e.g., Kunz et al., 2007b; Arnett et al., 2008). The impact of wind power plants on bat populations is of particular contemporary concern, because bats are long-lived and have low reproduction rates, because of the patterns of bat mortality at wind power plants (e.g., research has shown that bats may be attracted to wind turbine rotors), and because of uncertainty about the current size of bat populations (e.g., Barclay et al., 2007; Horn et al., 2008).

Significant uncertainty remains about the causal mechanisms underlying fatality rates and the effectiveness of mitigation measures, leading to limited ability to predict bird and bat fatality rates. Nonetheless, possible approaches to reducing fatalities that have been reported include siting power plants in areas with lower bird and bat population densities, placing turbines in areas with low prey density, and using different numbers, types and sizes of turbines. Recent research also suggests that limiting the operation of wind turbines during low wind situations may result in considerable reductions in bat fatalities (Baerwald et al., 2009; Arnett et al., 2011).

The magnitude and population-level consequences of bird and bat collision fatalities can also be viewed in the context of other fatalities caused by human activities. The number of bird fatalities at existing wind power plants appears to be orders of magnitude lower than other anthropogenic causes of bird deaths (e.g., vehicles, buildings and windows, transmission lines, communications towers, house cats, pollution and other contaminants) (Erickson et al., 2005; NRC, 2007). Moreover, it has been suggested that onshore wind power plants are not currently causing meaningful declines in bird population levels (NRC, 2007), and that other energy supply options also impact birds and bats through collisions, habitat modifications and contributions to global climate change (Lilley and Firestone, 2008; Sovacool, 2009; NABCI, 2010). These assessments are based on aggregate comparisons, however, and the cumulative population-level impacts of wind energy development on some species where biologically significant impacts are possible remain uncertain (especially vis-à-vis bats). Improved methods to assess these population-level impacts and their possible mitigation are needed (Kunz et al., 2007a), as are robust comparisons between the impacts of wind energy and other electricity supply options.

7.6.2.2 Habitat and ecosystem modifications

The habitat and ecosystem modification impacts of wind power plants on flora and fauna include, but are not limited to, avoidance of or displacement from an area, habitat destruction and reduced reproduction (e.g., Drewitt and Langston, 2006; NRC, 2007; Stewart et al., 2007). The relative biological significance of these impacts, compared to bird and bat collision fatalities, remains unclear. Moreover, the nature of these impacts will depend in part on the ecosystem into which wind power plants are integrated. Wind power plants are often installed in agricultural landscapes or on brown-field sites. In such cases, very different habitat and ecosystem impacts might be expected compared to wind power plants that are sited on previously undisturbed forested ridges or native grasslands. The development of wind power plants in largely undisturbed forests may, for example, lead to additional habitat destruction and fragmentation for intact forest-dependent species due to forest clearing for access roads, turbine foundations and power lines (e.g., Kuvlesky et al., 2007; NRC, 2007). Because habitat modification impacts are highly site and species specific (and affected by whether the wind power plant is located on- or offshore), they are ideally addressed (with mitigation measures) in the siting process; concerns for these impacts have also led to broader planning ordinances in some countries prohibiting the construction of wind power plants in ecologically sensitive areas.

The impacts of wind power plants on marine life have moved into focus as wind energy development starts to occur offshore and, as part of the licensing procedures for offshore wind power plants, a number of studies on the possible impacts of wind power plants on marine life and ecosystems have been conducted. As Michel et al. (2007) point out, there are "several excellent reviews...on the potential impacts of offshore wind parks on marine resources; most are based on environmental impact assessments and monitoring programs of existing offshore wind parks in Europe…". The localized impacts of offshore wind energy on marine life vary between the installation, operation and decommissioning phases, depend greatly on site-specific conditions, and may be negative or positive (e.g., Wahlberg and Westerberg, 2005; Dong Energy et al., 2006; Köller et al., 2006; P. Madsen et al., 2006; Michel et al., 2007; Wilhelmsson and Malm, 2008; Punt et al., 2009; Tougaard et al., 2009; Wilson and Elliott, 2009; Kikuchi, 2010). Potential negative impacts include underwater sounds and vibrations (especially during construction), electromagnetic fields, physical disruption and the establishment of invasive species. The physical structures may, however, create new breeding grounds or shelters and act as artificial reefs or fish aggregation devices (e.g., Wilhelmsson et al., 2006). Additional research is warranted on these impacts and their long-term and population-level consequences, especially in comparison to other sources of energy supply, but the impacts do not appear to be disproportionately large. In advance of conclusive findings, however, concerns about the impacts of offshore wind energy on marine life (and bird populations) have led to national zoning efforts in some countries that exclude the most sensitive areas from development.

7.6.2.3 Impact of wind power plants on the local climate

The possible impact of wind power plants on the local climate has also been the focus of some research. Wind power plants extract momentum from the air flow and thus reduce the wind speed behind the turbines, and also increase vertical mixing by introducing turbulence across a range of length scales (Petersen et al., 1998; Baidya Roy and Traiteur, 2010). These two processes are described by the term 'wind turbine wake (Barthelmie et al., 2004). Though intuitively turbine wakes must increase vertical mixing of the near-surface layer, and thus may increase the atmosphere-surface exchange of heat, water vapour and other parameters, the magnitude of the effect remains uncertain. One study using blade element momentum theory suggests that even very large-scale wind energy deployment, sufficient to supply global energy needs, would remove less than 1/10,000th of the total energy within the lowest 1 km of the atmosphere (Sta. Maria and Jacobson, 2009). Other studies have sought to quantify more local effects by treating large wind power plants as a block of enhanced surface roughness length or an elevated momentum sink in regional and global models. These studies have typically modelled scenarios of substantial wind energy deployment, and have found changes in local surface temperature of up to or even exceeding 1°C and in surface winds of several metres per second over (and even extending beyond) the areas of wind power plant installation (Keith et al., 2004; Kirk-Davidoff and Keith, 2008; C. Wang and Prinn, 2010); these local effects could also impact rainfall, radiation, clouds, wind direction and other climate variables. Though the global average impact of these local changes is much less pronounced, the local changes could have implications for ecosystems and human activities.

The assumptions and methods used by these studies may not, however, accurately represent the mechanisms by which wind turbines interact with the atmosphere. Studies often incorrectly assume that wind turbines act as invariant momentum sinks,[36] that turbine densities are above what is the norm, and that wind energy deployment occurs at a more substantial and geographically concentrated scale than is likely. Observed data from and models of large offshore wind power plants, for example, indicate that they may be of sufficient scale to perceptibly interact with the entire (relatively shallow) atmospheric boundary layer (Frandsen et al., 2006), but onsite measurements and remotely sensed near-surface wind speeds suggest that wake effects from large developments may no longer be discernible in near-surface wind speeds and turbulence intensity at approximately 20 km downwind (Christiansen and Hasager, 2005, 2006; Frandsen et al., 2009). As a result, the impact of wind energy on local climates remains uncertain. More generally, it should also be recognized that wind turbines are not the only structures to potentially impact local climate variables, and that any impacts caused by increased wind energy deployment should be placed in the context of other anthropogenic climate influences (Sta. Maria and Jacobson, 2009).

7.6.3 Impacts on human activities and well-being

In addition to ecological consequences, wind energy development impacts human activities and well-being in various ways. The primary impacts addressed here include: land and marine usage; visual impacts; proximal 'nuisance' impacts that might occur in close range to the turbines such as noise, flicker, health and safety; and property value impacts.

7.6.3.1 Land and marine usage

Wind turbines are sizable structures, and wind power plants can encompass a large area (5 to 10 MW per km2 is often assumed), thereby using space that might otherwise be used for other purposes.[37] The land footprint specifically disturbed by onshore wind turbines and their supporting roads and infrastructure, however, typically ranges from 2 to 5% of the total area encompassed by a wind power plant, allowing agriculture, ranching and certain other activities to continue within the area. Some forms of land use may be precluded from the area, such as housing developments, airport approaches and some radar installations. Nature reserves and historical and/or sacred sites are also often particularly sensitive. Somewhat similar issues apply to offshore wind power plants.

The possible impacts of wind power plants on aviation, shipping, fishing, communications and radar must also be considered, and depend on the placement of wind turbines and power plants. By avoiding airplane landing corridors and shipping routes, the interference of wind power plants with shipping and aviation can be kept to a minimum (Hohmeyer et al., 2005). Integrated marine spatial planning and integrated coastal zone management approaches are also starting to include offshore wind energy, thereby helping to assess the ecological impacts and economic and social benefits for coastal regions from alternative marine and coastal uses, and to minimize conflict among those uses (e.g., Murawski, 2007; Ehler and Douvere, 2009; Kannen and Burkhard, 2009).

Electromagnetic interference (EMI) associated with wind turbines can take various forms (e.g., Krug and Lewke, 2009). In general, wind turbines can interfere with detection of signals through reflection and blockage of electromagnetic waves and creation of large reflected radar returns, including Doppler produced by the rotation of turbine blades. Many EMI effects can be avoided by appropriate siting, for example, not locating wind turbines in close proximity to transmitters or receivers or relying on landscape terrain to mask the turbines (Summers, 2000; Hohmeyer et al., 2005). Moreover, there are no fundamental physical constraints preventing mitigation of EMI impacts (Brenner et al., 2008). In the case of military (or civilian) radar, reports have concluded that radar systems can sometimes be modified to ensure that aircraft safety and national defence are maintained (Butler and Johnson, 2003; Brenner et al., 2008). In particular, radar systems may have to be replaced or upgraded, or gap-filling and signal fusion systems installed, at some cost. In addition,

---

36 In these instances, the aerodynamic effect of wind turbines is treated via an increase in assumed surface roughness, in effect assuming that the turbines are operating all of the time to decrease wind speeds.

37 Chapter 9 addresses relative land use associated with multiple energy sources.

research is underway to investigate wind turbine design changes that may mitigate adverse impacts by making turbines less reflective to radar systems. EMI impacts can also extend to television, global positioning systems and communications systems, however, where they exist, these impacts can generally be managed by appropriate siting of wind power plants and through technical solutions.

7.6.3.2 Visual impacts

Visual impacts, and specifically how wind turbines and related infrastructures fit into the surrounding landscape, are often among the top concerns of communities considering wind power plants (Firestone and Kempton, 2007; NRC, 2007; Wolsink, 2007; Wustenhagen et al., 2007; Firestone et al., 2009; Jones and Eiser, 2009), of those living near existing wind power plants (Thayer and Hansen, 1988; Krohn and Damborg, 1999; Warren et al., 2005) and of institutions responsible for overseeing wind energy development (Nadaï and Labussière, 2009). Concerns have been expressed for on- and offshore wind energy (Ladenburg, 2009; Haggett, 2011). To capture the strongest and most consistent winds, wind turbines are often sited at high elevations and where there are few obstructions relative to the surrounding area. Moreover, wind turbines and power plants have grown in size, making the turbines and related transmission infrastructure more visible. Finally, as wind power plants increase in number and geographic spread, plants are being located in a wider diversity of landscapes (and, with offshore wind energy, unique seascapes as well), including areas that are more highly valued.

Though concerns about visibility cannot be fully mitigated, many jurisdictions require an assessment of visual impacts as part of the siting process, including defining the geographic scope of impact and preparing photo and video montages depicting the area before and after wind energy development. Other recommendations that have emerged to minimize visual intrusion include using turbines of similar size and shape, using light-coloured paints, choosing a smaller number of larger turbines over a larger number of smaller ones, burying connection cabling and ensuring that blades rotate in the same direction (e.g., Hohmeyer et al., 2005). More generally, a rethinking of traditional concepts of 'landscape' to include wind turbines has sometimes been recommended (Pasqualetti et al., 2002) including, for example, setting aside areas in advance where development can occur and others where it is precluded, especially when such planning allows for public involvement (Nadaï and Labussière, 2009).

7.6.3.3 Noise, flicker, health and safety

A variety of proximal 'nuisance' effects are also sometimes raised with respect to wind energy development, the most prominent of which is noise. Noise from wind turbines can be a problem, especially for those living within close range. Possible impacts can be characterized as both audible and sub-audible (i.e., infrasound). There are claims that sub-audible sound, that is, below the nominal audible frequency range, may cause health effects (Alves-Pereira and Branco, 2007), but a variety of studies (Jakobsen, 2005; Leventhall, 2006) and government reports (e.g., FANM, 2005; MDOH, 2009; CMOH, 2010; NHMRC, 2010) have not found sufficient evidence to support those claims to this point. Regarding audible noise from turbines, environmental noise guidelines (EPA, 1974, 1978; WHO, 1999, 2009) are generally believed to be sufficient to ensure that direct physiological health effects (e.g., hearing loss) are avoided (McCunney and Meyer, 2007). Some nearby residents, however, do experience annoyance from wind turbine sound (Pedersen and Waye, 2007, 2008; Pedersen et al., 2010), which can impact sleep patterns and well-being. This annoyance is correlated with acoustic factors (e.g., sound levels and characteristics) and also with non-acoustic factors (e.g., visibility of, or attitudes towards, the turbines) (Pedersen and Waye, 2007, 2008; Pedersen et al., 2010). Concerns about noise emissions may be especially great when hub-height wind speeds are high, but ground-level speeds are low (i.e., conditions of high wind shear). Under such conditions, the lack of wind-induced background noise at ground level coupled with higher sound levels from the turbines has been linked to increased audibility and in some cases annoyance (van den Berg, 2004, 2005, 2008; Prospathopoulos and Voutsinas, 2005).

Significant efforts have been made to reduce the sound levels emitted by wind turbines. As a result, mechanical sounds from modern turbines (e.g., gearboxes and generators) have been substantially reduced. Aeroacoustic noise is now the dominant concern (Wagner et al., 1996), and some of the specific aeroacoustic characteristics of wind turbines (e.g., van den Berg, 2005) have been found to be particularly detectable (Fastl and Zwicker, 2007) and annoying (Bradley, 1994; Bengtsson et al., 2009). Reducing aeroacoustic noise can be most easily accomplished by reducing blade speed, but different tip shapes and airfoil designs have also been explored (Migliore and Oerlemans, 2004; Lutz et al., 2007). In addition, the predictive models and environmental regulations used to manage these impacts have improved to some degree. Specifically, in some jurisdictions, both the wind shear and maximum sound power levels under all operating conditions are taken into account when establishing regulations (Bastasch et al., 2006). Absolute maximum sound levels during the day (e.g., 55 A-weighted decibels, dBA) and night (e.g., 45 dBA) can also be coupled with maximum levels that are set relative to pre-existing background sound levels (Bastasch et al., 2006). In other jurisdictions, simpler and cruder setbacks mandate a minimum distance between turbines and other structures (MOE, 2009). Despite these efforts, concerns about noise impacts remain a barrier to wind energy deployment in some areas.

In addition to sound impacts, rotating turbine blades can also cast moving shadows (i.e., shadow flicker), which may be annoying to residents living close to wind turbines. Turbines can be sited to minimize these concerns, or the operation of wind turbines can be stopped during acute periods (Hohmeyer et al., 2005). Finally, wind turbines can shed parts of or whole blades as a result of an accident or icing (or more broadly, blades can shed built-up ice, or turbines could collapse entirely). Wind energy technology certification standards are aimed at reducing such accidents (see Section 7.3.2), and setback requirements further reduce the remaining risks. In practice, fatalities and injuries have been rare (see Chapter 9 for a comparison of accident risks among energy generation technologies).

7.6.3.4 Property values

Concerns that the visibility of wind power plants may translate into negative impacts on residential property values at the local level have sometimes been expressed (Firestone et al., 2009; Graham et al., 2009; Jones and Eiser, 2009). Further, if various proximal nuisance effects are prominent, such as turbine noise or shadow flicker, additional impacts on local property values might occur. Although these concerns may be reasonable given effects found for other environmental disamenities (e.g., high-voltage transmission lines, fossil-fuelled power plants and landfills; see Simons, 2006), published research has not found strong evidence of any widespread effect for wind power plants (e.g., Sims and Dent, 2007; Sims et al., 2008; Hoen et al., 2011). This might be explained by the setbacks normally employed between homes and wind turbines; studies on the impacts of transmission lines on property values, for example, sometimes find that effects can fade at distances of 100 m (e.g., Des Rosiers, 2002). Alternatively, any effects may be too infrequent and/or small to distinguish statistically based on historical data. Finally, turbine noise and other effects might be difficult to assess when homes are sold, and therefore might not be fully priced into the market. More research is needed on the subject, but based on other disamenity research (e.g., Boyle and Kiel, 2001; T. Jackson, 2001; Simons and Saginor, 2006), it is likely that any effects that do exist are most pronounced within short distances from wind turbines and in the period immediately following a wind power plant announcement, when risks are most difficult to quantify (Wolsink, 2007).

7.6.4 Public attitudes and acceptance

Despite the possible impacts described above, surveys have consistently found wind energy to be widely accepted by the general public (e.g., Warren et al., 2005; Jones and Eiser, 2009; Klick and Smith, 2010; Swofford and Slattery, 2010). Translating this broad support into increased deployment (closing the 'social gap', see, e.g., Bell et al., 2005), however, often requires the support of local host communities and/or decision makers (Toke, 2006; Toke et al., 2008). To that end, a number of concerns exist that might temper the enthusiasm of these stakeholders about wind energy, such as land and marine use, and the visual, proximal and property value impacts discussed previously.

In general, research has found that public concern about wind energy development is greatest directly after the announcement of a wind power plant, but that acceptance increases after construction when actual impacts can be assessed (Wolsink, 1989; Warren et al., 2005; Eltham et al., 2008). Some studies have found that those most familiar with existing wind power plants, including those who live closest to them, are more accepting (or less concerned) than those less familiar and farther away (Krohn and Damborg, 1999; Warren et al., 2005), but other research has found the opposite to be true (van der Horst, 2007; Swofford and Slattery, 2010). Possible explanations for this apparent discrepancy include differences in attitudes towards proposed versus existing wind power plants (Swofford and Slattery, 2010), the pre-existing characteristics and values of the local community (van der Horst, 2007) and the degree of trust that the local community has concerning the development process and its outcome (Thayer and Freeman, 1987; Jones and Eiser, 2009). Research has also found that pre-construction attitudes can linger after the turbines are erected: for example, those opposed to a wind power plant's development have been found to consider the eventual plant to be noisier and more visually intrusive that those who favoured the same plant in the pre-construction time period (Krohn and Damborg, 1999; Jones and Eiser, 2009). Some research has found that concerns can be compounding. For instance, those who found turbines to be visually intrusive also found the noise from those turbines to be more annoying (Pedersen and Waye, 2004). Finally, in some contexts at least, there appears to be some preference for offshore over onshore wind energy development, though these preferences are dependent on the specific offshore power plant location (Ladenburg, 2009) and are far from universal (Haggett, 2011).

7.6.5 Minimizing social and environmental concerns

As wind energy deployment increases and as larger wind power plants are considered, existing concerns may become more acute and new concerns may arise. Regardless of the type and degree of social and environmental concerns, however, addressing them directly is an essential part of any successful wind power-planning and plant-siting process.[38] To that end, involving the local community in the planning and siting process has sometimes been shown to improve outcomes (Loring, 2007; Toke et al., 2008; Jones and Eiser, 2009; Nadaï and Labussière, 2009). This might include, for example, allowing the community to weigh in on alternative wind power plant and turbine locations, and improving education by hosting visits to existing wind power plants. Public attitudes have been found to improve when the development process is perceived as being transparent (Wolsink, 2000; C. Gross, 2007; Loring, 2007). Further, experience suggests that local ownership of wind power plants and other benefit-sharing mechanisms can improve public attitudes towards wind energy development (C. Gross, 2007; Wolsink, 2007; Jones and Eiser, 2009).

Proper planning for both on- and offshore wind energy developments can also help to minimize social and environmental impacts, and a number of siting guidelines have been developed (e.g., S. Nielsen, 1996; NRC, 2007; AWEA, 2008). Appropriate planning and siting will generally avoid placing wind turbines too close to dwellings, streets, railroad lines, airports, radar sites and shipping routes, and will avoid areas of

---

38 Chapter 11 provides a complementary summary of the extensive literature on planning and siting for RE.

heavy bird and bat activity; a variety of pre-construction studies are often conducted to define these impacts and their mitigation. Habitat fragmentation and ecological impacts both on- and offshore can often be minimized by careful placement of wind turbines and power plants and by proactive governmental planning for wind energy deployment. Examples of such planning can be found in many jurisdictions around the world. Planning and siting regulations vary dramatically by jurisdiction, however, with varying levels of stringency and degrees of centralization versus local control. These differences can impact the environmental and social outcomes of wind energy development, as well as the speed and ease of that development (e.g., Pettersson et al., 2010).

Although an all-encompassing numerical comparison of the full external costs and benefits of wind energy is impossible, as some impacts are very difficult to monetize, available evidence suggests that the positive environmental and social effects of wind energy generally outweigh the negative impacts that remain after careful planning and siting procedures are followed (see, e.g., Jacobson, 2009). In practice, however, complicated and time-consuming planning and siting processes are key obstacles to wind energy development in some countries and contexts (e.g., Bergek, 2010; Gibson and Howsam, 2010). In part, this is because even if the environmental and social impacts of wind energy are minimized through proper planning and siting procedures and community involvement, some impacts will remain. Efforts to better understand the nature and magnitude of these remaining impacts, together with efforts to minimize and mitigate those impacts, will therefore need to be pursued in concert with increasing wind energy deployment.

7.7 Prospects for technology improvement and innovation[39]

Over the past three decades, innovation in wind turbine design has led to significant cost reductions, while the capacity and physical size of individual turbines has grown markedly (EWEA, 2009). The 'square-cube law' is a mathematical relationship that states that as the diameter of a wind turbine increases, its theoretical energy output increases by the square of the rotor diameter, while the volume of material (and therefore its mass and cost) required to scale at the same rate increases as the cube of the rotor diameter, all else being equal (Burton et al., 2001). As a result, at some size, the cost of a larger turbine will grow faster than the resulting energy output and revenue, making further size increases uneconomic. To date, engineers have successfully worked around this relationship, preventing significant increases in the cost of wind energy as turbines have grown larger by optimizing designs with increasing turbine size, by reducing materials use and by using lighter, yet stronger, materials.

Significant opportunities remain for design optimization of on- and offshore wind turbines and power plants, and sizable cost reductions remain possible in the years ahead, though improvements are likely to be more incremental in nature than radical changes in fundamental design. Engineering around the 'square-cube law' remains a fundamental objective of research efforts aimed at further reducing the levelized cost of energy from wind, especially for offshore installations where significant additional up-scaling is anticipated. Breakthrough technologies from other fields may also find applications in wind energy, including new materials (e.g., superconducting generators) and sensors (providing active aerodynamic control along the entire span of a blade), which may yield even larger turbines in the future, up to or exceeding 10 MW.

This section describes R&D programs in wind energy (Section 7.7.1), system-level design and optimization approaches that may yield further reductions in the levelized generation cost of wind energy (Section 7.7.2), component-level opportunities for innovation in wind energy technology (Section 7.7.3) and the need to improve the scientific underpinnings of wind energy technology (Section 7.7.4).[40]

7.7.1 Research and development programmes

Public and private R&D programmes have played a major role in the technical advances seen in wind energy over the last decades (Klaassen et al., 2005; Lemming et al., 2009). Government support for R&D, in collaboration with industry, has led to system- and component-level technology advances, as well as improvements in resource assessment, technical standards, electric system integration, wind energy forecasting and other areas. From 1974 to 2006, government R&D budgets for wind energy in International Energy Agency (IEA) countries totalled USD$_{2005}$ 3.8 billion, representing an estimated 10% share of RE R&D budgets and 1% of total energy R&D expenditures (IEA, 2008; EWEA, 2009). In 2008, OECD research funding for wind energy totalled USD$_{2005}$ 180 million, or 1.5% of all energy R&D funding; additional funding was provided by non-OECD countries. Government-sponsored R&D programs have often emphasized longer-term innovation, while industry-funded R&D has focused on shorter-term production, operation and installation issues. Though data on industry R&D funding are scarce, EWEA (2009), Carbon Trust (2008b) and Wiesenthal et al. (2009) find that the ratio of turbine manufacturer R&D expenditures to net revenue typically ranges from 2 to 3%, while Wiesenthal et al. (2009) find that corporate wind energy R&D in the EU is three times as large as government R&D investments.

Wind energy research strategies have often been developed through government and industry collaborations, historically centred on Europe and the USA, though there has been growth in public and private R&D in other countries as well (e.g., Tan, 2010). In a study to explore the technical and economic feasibility of meeting 20% of electricity demand

---

[39] Section 10.5 offers a complementary perspective on drivers of and trends in technological progress across RE technologies.

[40] This section focuses on scientific and engineering challenges directly associated with reducing the cost of wind energy, but additional research areas of importance include: research on the integration of wind energy into electric systems and grid compatibility (e.g., forecasting, storage, power electronics); social science research on policy measures and social acceptance; and scientific research to understand the impacts of wind energy on the environment and on human activities and well-being. These issues are addressed only peripherally in this section.

in the USA with wind energy, the US Department of Energy (US DOE) found that key areas for further research included continued development of turbine technology, improved and expanded manufacturing processes, electric system integration of wind energy, and siting and environmental concerns (US DOE, 2008). The European Wind Energy Technology Platform (TPWind), meanwhile, has developed a roadmap through 2020 that is expected to form the basis for future European wind energy R&D strategies, with the following areas of focus: wind power systems (new turbines and components); offshore deployment and operation (offshore structures, installation and O&M protocols); wind energy integration (grid integration); and wind energy resources (wind resource assessment and design conditions) (EU, 2008; EC, 2009). In general, neither of these planning efforts requires a radical change in the fundamental design of wind turbines: instead, the path forward is seen as many evolutionary steps, executed through incremental technology advances, that may nonetheless result in significant improvements in the levelized cost of wind energy as well as larger turbines, up to or exceeding 10 MW.

7.7.2 System-level design and optimization

Wind power plants and turbines are sophisticated and complex systems that require integrated design approaches to optimize cost and performance. At the plant level, considerations include the selection of a wind turbine for a given wind resource regime, wind turbine siting, spacing, and installation procedures, O&M methodologies and electric system integration. Optimization of wind turbines and power plants therefore requires a whole-system perspective that evaluates not only the wind turbine as an individual aerodynamic device, mechanical structure and control system, but that also considers the interaction of the individual turbines at a plant level (EU, 2008).

Studies have identified a number of areas where technology advances could result in changes in the investment cost, annual energy production, reliability, O&M cost, and electric system integration of wind energy. Examples of studies that have explored the impacts of advanced concepts include those conducted by the US DOE under the Wind Partnership for Advanced Component Technologies (WindPACT) project (GEC, 2001; Griffin, 2001; Shafer et al., 2001; D. Smith, 2001; Malcolm and Hansen, 2006). One assessment of the possible impacts of technical advances on onshore wind energy production and turbine-level investment costs is summarized in Table 7.3 (US DOE, 2008). Though not all of these improvements may be achieved, there is sufficient potential to warrant continued R&D. The most likely scenario, as shown in Table 7.3, is a sizeable increase in energy production with a modest drop in investment cost (compared to 2002 levels, which is the baseline for the estimates in Table 7.3). Meanwhile, under the EU-funded UPWIND project, a system-level analysis of the potential challenges (e.g., manufacturing processes, installation processes and structural integrity) and design solutions for very large (up to 20 MW) onshore and offshore wind turbine systems is underway. This project similarly includes the development of a model to evaluate the impact of potential technical innovations on the system-level cost of wind energy (Sieros et al., 2011).

7.7.3 Component-level innovation opportunities

The potential areas of innovation outlined in Table 7.3 are further described in Sections 7.7.3.1 through 7.7.3.5. Though Table 7.3 is targeted towards wind turbines designed for onshore applications, the component-level innovations identified therein will impact both on- and offshore wind energy. In fact, some of these innovations will be more important for offshore wind energy technology due to the earlier state of and greater operational challenges facing that technology. Additional advances that are more specific to offshore wind energy are described in Section 7.7.3.6.

7.7.3.1 Advanced tower concepts

Taller towers allow the rotor to access higher wind speeds in a given location, increasing annual energy capture. The cost of large cranes and transportation, however, acts as a limit to tower height. As a result, research is being conducted into several novel tower designs that would eliminate the need for cranes for very high, heavy lifts. One concept is the telescoping or self-erecting tower, while other designs include lifting dollies or tower-climbing cranes that use tower-mounted tracks to lift the nacelle and rotor to the top of the tower. Still other developments aim to increase the height of the tower without unduly sacrificing material demands through the use of different materials, such as concrete and fibreglass, or different designs, such as space-frame construction or panel sections (see, e.g., GEC, 2001; Malcolm, 2004; Lanier, 2005).

7.7.3.2 Advanced rotors and blades

Due to technology advances, blade mass has been scaling at roughly an exponent of 2.4 to rotor diameter, compared to the expected exponent of 3.0 based on the 'square-cube' law (Griffin, 2001). The significance of this development is that wind turbine blades have become lighter for a given length over time. If advanced R&D can provide even better blade design methods, coupled with better materials (such as carbon fibre composites) and advanced manufacturing methods, then it will be possible to continue to innovate around the square-cube law in blade design. One approach to reducing cost involves developing new blade airfoil shapes that are much thicker where strength is most required, near the blade root, allowing inherently better structural properties and reducing overall mass (K. Jackson et al., 2005; Chao and van Dam, 2007). These airfoil shapes potentially offer equivalent aerodynamic performance, but have yet to be proven in the field. Another approach to increasing blade length while limiting increased material demand is to reduce the fatigue loading on the blade. Blade fatigue loads can be reduced by controlling the blade's aerodynamic response to turbulent wind by

Table 7.3 | Areas of potential technology improvement from a 2002 baseline onshore wind turbine (based on US DOE, 2008).[1]

Technical Area	Potential Advances	Increments from Baseline (Best/Expected/Least)	
		Annual Energy Production (%)	Turbine Investment Cost (%)
Advanced Tower Concepts	• Taller towers in difficult locations • New materials and/or processes • Advanced structures/foundations • Self-erecting, initial or for service	+11/+11/+11	+8/+12/+20
Advanced (Enlarged) Rotors	• Advanced materials • Improved structural-aero design • Active controls • Passive controls • Higher tip speed/lower acoustics	+35/+25/+10	-6/-3/+3
Reduced Energy Losses and Improved Availability	• Reduced blade soiling losses • Damage-tolerant sensors • Robust control systems • Prognostic maintenance	+7/+5/0	0/0/0
Advanced Drive Trains (Gearboxes and Generators and Power Electronics)	• Fewer gear stages or direct drive • Medium/low-speed generators • Distributed gearbox topologies • Permanent-magnet generators • Medium-voltage equipment • Advanced gear tooth profiles • New circuit topologies • New semiconductor devices • New materials	+8/+4/0	-11/-6/+1
Manufacturing Learning	• Sustained, incremental design and process improvements • Large-scale manufacturing • Reduced design loads	0/0/0	-27/-13/-3
Totals		+61/+45/+21	-36/-10/+21

Note: 1. The baseline for these estimates was a 2002 turbine system in the USA. There have already been sizeable improvements in capacity factor since 2002, from just over 30% to almost 35%, while investment costs have increased due to large increases in commodity costs in conjunction with a drop in the value of the US dollar. Therefore, working from a 2008 baseline, one might expect a more modest increase in capacity factor, but the 10% investment cost reduction is still quite possible (if not conservative), particularly from the higher 2008 starting point. Finally, the table does not consider any changes in the overall wind turbine design concept (e.g., two-bladed turbines).

using mechanisms that vary the angle of attack of the blade airfoil relative to the wind inflow. This is primarily accomplished with full-span blade pitch control. An elegant concept, however, is to build passive means of reducing loads directly into the blade structure (Ashwill, 2009). By carefully tailoring the structural properties of the blade using the unique attributes of composite materials, for example, blades can be built in a way that couples the bending deformation of the blade resulting from the wind with twisting deformation that passively mimics the motion of blade pitch control. Another approach is to build the blade in a curved shape so that the aerodynamic load fluctuations apply a twisting movement to the blade, which will vary the angle of attack (Ashwill, 2009). Because wind inflow displays a complex variation of speed and character across the rotor area, partial blade span actuation and sensing strategies to maximize load reduction are also promising (Buhl et al., 2005; Lackner and van Kuik, 2010). Devices such as trailing edge flaps and micro-tabs, for example, are being investigated, but new sensors may need to be developed for this purpose, with a goal of creating 'smart' blades with embedded sensors and actuators to control local aerodynamic effects (Andersen et al., 2006; Berg et al., 2009). To fully achieve these new designs, a better understanding of wind turbine aeroelastic, aerodynamic and aeroacoustic responses to complicated blade motion will be needed, as will control algorithms to incorporate new sensors and actuators in wind turbine operation.

7.7.3.3 Reduced energy losses and improved availability

Advanced turbine control and condition monitoring are expected to provide a primary means to improve turbine reliability and availability, reduce O&M costs and ultimately increase energy capture, for both individual turbines and wind power plants, on- and offshore. Advanced controllers are envisioned that can better control the turbine during turbulent winds and thereby reduce fatigue loading and extend blade life (Bossanyi, 2003; Stol and Balas, 2003; Wright, 2004), monitor and adapt to wind conditions to increase energy capture and reduce the impact of blade soiling or erosion (Johnson et al., 2004; Johnson and Fingersh, 2008; Frost et al., 2009) and anticipate and protect against damaging wind gusts by using new sensors to detect wind speeds immediately ahead of the blade (T. Larsen et al., 2004; Hand and Balas, 2007). Condition-monitoring systems of the future are expected to track and monitor ongoing conditions at critical locations in the turbine and report incipient failure possibilities and damage evolution, so that improved maintenance procedures can minimize outages and downtimes (Hameed et al., 2010). The full development of advanced control and monitoring systems of this nature will require considerable operational experience, and optimization algorithms will likely be turbine-specific; the general approach, however, should be transferable between turbine designs and configurations.

7.7.3.4 Advanced drive trains, generators, and power electronics

Several unique turbine designs are under development or in early commercial deployment to reduce drive train weight and cost while improving reliability (Poore and Lettenmaier, 2003; Bywaters et al., 2004; EWEA, 2009). One option, already in limited commercial use, is a direct-drive generator (removing the need for a gearbox); more than 10% of the additional wind power capacity installed in 2009 used first-generation direct drive turbines (BTM, 2010), but additional design advances are envisioned. The trade-off is that the slowly rotating generator must have a high pole count and be large in diameter, imposing a weight penalty. The availability and cost of rare-earth permanent magnets is expected to significantly affect the size and cost of future direct-drive generator designs, as permanent-magnet designs tend to be more compact and potentially lightweight, as well as reducing electrical losses in the windings.

Various additional drive train configurations are being explored and commercially deployed. A hybrid of the current geared and direct-drive approaches is the use of a single-stage drive using a low- or medium-speed generator. This allows the use of a generator that is significantly smaller and lighter than a comparable direct-drive design, and reduces (but does not eliminate) reliance on a gearbox. Another approach is the distributed drive train, where rotor torque is distributed to multiple smaller generators (rather than a single, larger one), reducing component size and (potentially) weight. Still other innovative drive train concepts are under development.

Power electronics that provide full power conversion from variable frequency alternating current (AC) electricity to constant frequency 50 or 60 Hz are also capable of providing ancillary grid services. The growth in turbine size is driving larger power electronic components as well as innovative higher-voltage circuit topologies. In the future, it is expected that wind turbines will use higher-voltage generators and converters than are used today (Erdman and Behnke, 2005), and therefore also make use of higher-voltage and higher-capacity circuits and transistors. New power conversion devices will need to be fully compliant with emerging grid codes to ensure that wind power plants do not degrade the reliability of the electric system.

7.7.3.5 Manufacturing learning

Manufacturing learning refers to the learning by doing achieved in serial production lines with repetitive manufacturing (see Section 7.8.4 for a broader discussion of learning in wind energy technology). Though turbine manufacturers already are beginning to operate at significant scale, as the industry expands further, additional cost savings can be expected. For example, especially as turbines increase in size, concepts such as manufacturing at wind power plant sites and segmented blades are being explored to reduce transportation challenges and costs. Further increases in manufacturing automation and optimized processes will also contribute to cost reductions in the manufacturing of wind turbines and components.

7.7.3.6 Offshore research and development opportunities

The cost of offshore wind energy exceeds that of onshore wind energy due, in part, to higher O&M costs as well as more expensive installation and support structures. The potential component-level technology advances described above will contribute to lower offshore wind energy costs, and some of those possible advances may even be largely driven by the unique needs of offshore wind energy applications. In addition, several areas of possible advancement are more specific to offshore wind energy, including O&M strategies, installation and assembly schemes, support structure design and the development of larger turbines, possibly including new turbine concepts.

Offshore wind turbines operate in harsh environments driven by both wind and wave conditions that can make access to turbines challenging or even impossible for extended periods (Breton and Moe, 2009). A variety of methods to provide greater access during a range of conditions are under consideration and development, including inflatable boats or helicopters (Van Bussel and Bierbooms, 2003). Sophisticated O&M approaches that include remote assessments of turbine operability and the scheduling of preventative maintenance to maximize access during favourable conditions are also being investigated, and employed (Wiggelinkhuizen et al., 2008). The development of more reliable turbine components, even if more expensive on a first-cost basis, is also expected to play a major role in reducing the overall levelized cost of offshore wind energy. Efforts are underway to more thoroughly analyze gearbox dynamics, for example, to contribute to more reliable designs (Peeters et al., 2006; Heege et al., 2007). A number of the component-level innovations described earlier, such as advanced direct-drive generators and passive blade controls, may also improve overall technology reliability.

Offshore wind turbine transportation and installation is not directly restricted by road or other land-based infrastructure limits. As a result, though offshore wind turbines are currently installed as individual components, concepts are being considered where fully assembled turbines are transported on special-purpose vessels and mounted on previously installed support structures. In addition to creating the vessels needed for such installation practices, ports and staging areas would need to be designed to efficiently perform the assembly processes.

Additional R&D is required to improve support structure design for offshore turbines. Foundation structure innovation offers the potential to access deeper waters, thereby increasing the technical potential of wind energy (Breton and Moe, 2009). Offshore turbines have historically been installed primarily in relatively shallow water, up to 30 m, on a mono-pile structure that is essentially an extension of the tower, but gravity-based structures have become more common. Other concepts that are more appropriate for deeper water depths include fixed-bottom space-frame structures, such as jackets and tripods, and floating platforms, such as spar- buoys, tension-leg platforms, semi-submersibles, or hybrids of these concepts. Offshore wind turbine support structures may undergo dynamic responses associated with wind and wave loads, requiring an integrated analysis of the rotor, tower and support structure supplemented with improved estimates of soil stiffness and scour conditions specific to offshore support structures (F. Nielsen et al., 2009). Floating wind turbines further increase the complexity of turbine design due to the additional motion of the base but, if cost effective, could: (1) offer access to significant additional wind resource areas; (2) encourage technology standardization whereby turbine and support structure design would be largely independent of water depths and seabed conditions; and (3) lead to simplified installation (e.g., full turbine assembly could occur in sheltered water) and decommissioning practices (EWEA, 2009). In 2009, the first full-scale floating wind turbine pilot plant was deployed off the coast of Norway at a 220 m depth. Figure 7.19 depicts some of the foundation concepts (left) in use or under consideration in the near term, while also (right) illustrating the concept of floating wind turbines, which are being considered for the longer term.

Future offshore wind turbines may be larger, lighter and more flexible. Offshore wind turbine size is not restricted in the same way as onshore wind energy technology, and the relatively higher cost of offshore foundations provides additional motivation for larger turbines (EWEA, 2009). As a result, turbines of 10 MW or larger are under consideration. Future offshore turbine designs can benefit from many of the possible component-level advances described previously. Nonetheless, the development of large turbines for offshore applications remains a significant research challenge, requiring continued advancement in component design and system-level analysis. Concepts that reduce the weight of the blades, tower and nacelle become more important as size increases, providing opportunities for greater advancement than may be incorporated in onshore wind energy technology. In addition to larger turbines, design criteria for offshore applications may be relaxed in cases where noise and visual impacts are of lesser concern. As a result, other advanced turbine concepts are under investigation, including two-bladed and downwind turbines. Downwind turbine designs may allow less-costly yaw mechanisms, and the use of softer, more flexible blades (Breton and Moe, 2009). Finally, innovative turbine concepts and significant up-scaling of existing designs will require improved turbine modelling to better capture the operating environment in which offshore turbines

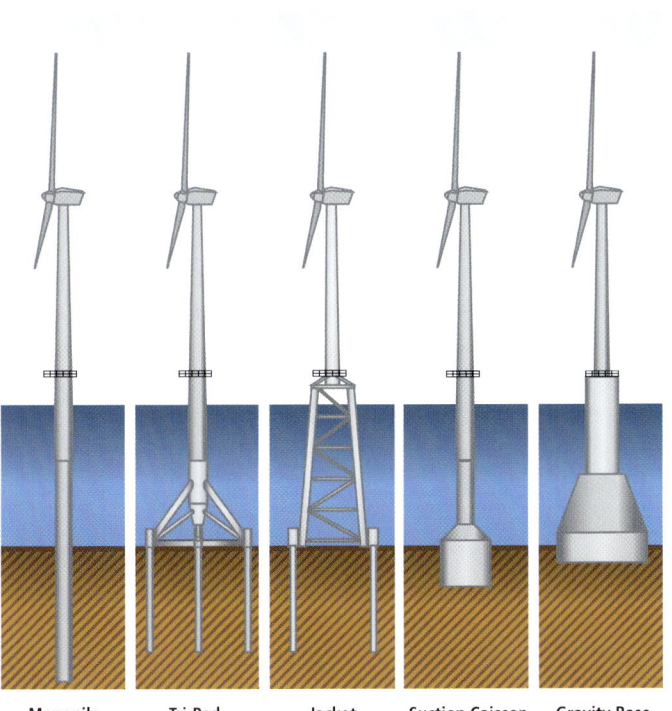
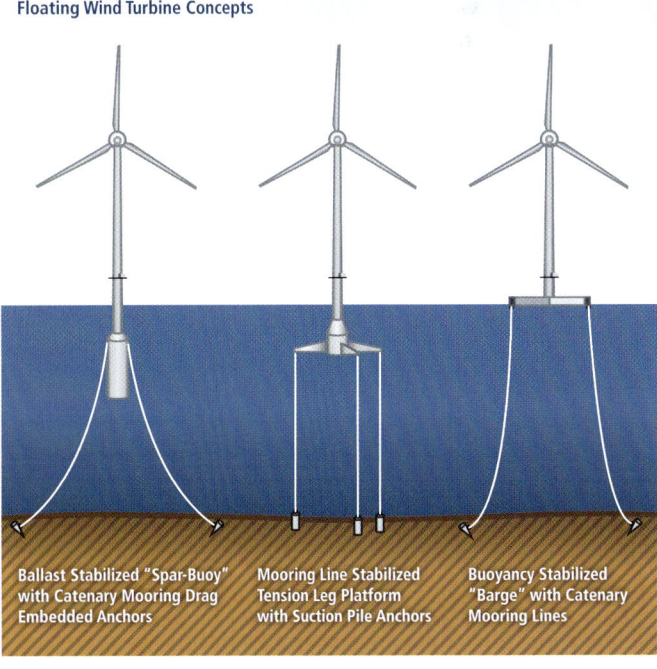

Figure 7.19 | Offshore wind turbine foundation designs: (left) near-term concepts and (right) floating offshore turbine concept. Sources: (left) UpWind (UpWind.eu) and (right) NREL.

are installed, including the dynamic response of turbines to wind and wave loading (see Section 7.7.4).

7.7.4 The importance of underpinning science

Although wind energy technology is being deployed at a rapid scale today, significant potential remains for continued innovation to further reduce cost and improve performance. International wind turbine design and safety standards dictate the level of analysis and testing required prior to commercializing new concepts. At the same time, technical innovation will push the design criteria and analysis tools to the limits of physical understanding. A significant effort is therefore needed to enhance fundamental understanding of the wind turbine and power plant operating environment in order to facilitate a new generation of reliable, safe, cost-effective wind turbines and to further optimize wind power plant siting and design.

Wind turbines operate in a challenging environment, and are designed to withstand a wide range of conditions with minimal attention. Wind turbines are complex, nonlinear, dynamic systems forced by gravity, centrifugal, inertial and gyroscopic loads as well as unsteady aerodynamic, hydrodynamic (for offshore) and corrosion impacts. Modern wind turbines also operate in a layer of the atmosphere (from 50 to 200 m) that is complex, and are impacted by phenomena that occur over scales ranging from microns to thousands of kilometres. Accurate, reliable wind measurements and computations across these scales are important. In addition, fundamental scientific research in a number of areas can improve physical understanding of this operating environment (including extreme weather events) and its impact on wind turbines and power plants. Research in the areas of aeroelastics, unsteady aerodynamics, aeroacoustics, advanced control systems and atmospheric sciences, for example, has yielded improved design capabilities in the past, and continued research in these areas is anticipated to continue to improve mathematical models and experimental data, which, in turn, will reduce the risk of unanticipated turbine failures, increase the reliability of the technology and encourage further design innovation.

Although the physics are strongly coupled, four primary spatio-temporal levels require additional research: (1) wind conditions that affect individual turbines; (2) wind power plant siting and array effects; (3) mesoscale atmospheric processes; and (4) global and local climate effects.

Wind conditions that affect individual turbines encompass detailed characterizations of wind flow fields and the interaction of those flows with wind turbines. Wind turbine aerodynamics are complicated by three-dimensional effects in rotating blade flow fields that are unsteady and create load oscillations linked to dynamic stall. Understanding these aerodynamic effects, however, is critical for making load predictions that are accurate enough for use in turbine design. To this point, these effects have been identified and quantified based on wind tunnel and field experiments (Schreck et al., 2000, 2001; Schreck and Robinson, 2003; H. Madsen et al., 2010), and empirical models of these effects have been developed (Bierbooms, 1992; Du and Selig, 1998; Snel, 2003; Leishman, 2006). Currently, these aerodynamic models rely on blade-element moment methods (Spera, 2009) augmented with analytically and empirically based models to calculate the aerodynamic forces along the span of the blade. The availability of effective computational fluid dynamics codes and their potential to deliver improved predictive accuracy, however, is prompting broader application (M.O. Hansen et al., 2006). Aeroelastic models, meanwhile, are used to translate aerodynamic forces into structural responses throughout the turbine system. As turbines grow in size and are optimized, the structural flexibility of the components will necessarily increase, causing more of the turbine's vibration frequencies to play a prominent role. To account for these effects, future aeroelastic tools will have to better model large variations in the wind inflow across the rotor, higher-order vibration modes, nonlinear blade deflection, and aeroelastic damping and instability (Quarton, 1998; Rasmussen et al., 2003; Riziotis et al., 2004; M.H. Hansen, 2007). The application of novel load-mitigation control technologies to blades (e.g., deformable trailing edges) (Buhl et al., 2005) will require analysis based on aeroelastic tools that are adapted for these architectures. Similarly, exploration of control systems that utilize wind speed measurements in advance of the blade, such as light detection and ranging (Harris et al., 2006) or pressure probe measurements (T. Larsen et al., 2004), will also require improved aeroelastic tools. Offshore wind energy will require that aeroelastic tools better model the coupled dynamic response of the wind turbine and the foundation/support platform, as subjected to combined wind and wave loads (Passon and Kühn, 2005; Jonkman, 2009). Finally, aeroacoustic noise (i.e., the noise of turbine blades) is an issue for wind turbines (Wagner et al., 1996), and increasingly sophisticated tools are under development to better understand and manage these effects (Wagner et al., 1996; Moriarty and Migliore, 2003; Zhu et al., 2005, 2007; Shen and Sörensen, 2007). As turbine aerodynamic, aeroelastic and aeroacoustic modelling advances, the crucial role (e.g., Simms et al., 2001) of research-grade turbine aerodynamics experiments (Hand et al., 2001; Snel et al., 2009) grows ever more evident, as does the need for future high-quality laboratory and field experiments. Even though wind turbines now extract energy from the wind at levels approaching the theoretical maximum, improved understanding of aerodynamic phenomena will allow more accurate calculation of loads and thus the development of lighter, less costly, more reliable and higher-performing turbines.

Wind power plant siting and array effects impact energy production and equipment reliability at the power plant level. As wind power plants grow in size and move offshore, such impacts become more important. Rotor wakes create aeroelastic effects on downwind turbines (G. Larsen et al., 2008). Improved models of wind turbine wakes (Thomsen and Sørensen, 1999; Frandsen et al., 2009; Barthelmie and Jensen, 2010) will therefore yield more reliable predictions of energy capture and better estimates of fatigue loading in large, multiple-row wind power plants, both on- and offshore. This improved understanding may then lead to wind turbine and power plant designs intended to minimize energy capture degradations and manage wake-based load impacts.

Planetary boundary layer research is important for accurately determining wind flow and turbulence in the presence of various atmospheric stability effects and complex land surface characteristics. Research in mesoscale atmospheric processes aims at improving the fundamental understanding of mesoscale and local wind flows (Banta et al., 2003; Kelley et al., 2004). In addition to its contribution towards understanding turbine-level aerodynamic and array wake effects, a better understanding of mesoscale atmospheric processes will yield improved wind energy resource assessments and forecasting methods. Physical and statistical modelling to resolve spatial scales in the 100- to 1,000-m range, a notable gap in current capabilities (Wyngaard, 2004), could occupy a central role of this research.

Finally, additional research is warranted on the interaction between global and local climate effects, and wind energy. Specifically, work is needed to identify and understand historical trends in wind resource variability in order to increase the reliability of future wind energy performance predictions. As discussed earlier in this chapter, further work is also warranted on the possible impacts of climate change on wind energy resource conditions, and on the impact of wind energy development on local, regional and global climates.

Significant progress in many of the above areas requires interdisciplinary research. Also crucial is the need to use experiments and observations in a coordinated fashion to support and validate computation and theory. Models developed in this way will help improve: (1) wind turbine design; (2) wind power plant performance estimates; (3) wind resource assessments; (4) short-term wind energy forecasting; and (5) estimates of the impact of large-scale wind energy deployment on the local climate, as well as the impact of potential climate change effects on wind resources.

7.8 Cost trends[41]

Though the cost of wind energy has declined significantly since the 1980s, policy measures are currently required to ensure rapid deployment in most regions of the world (e.g., NRC, 2010b). In some areas with good wind resources, however, the cost of wind energy is competitive with current energy market prices (e.g., Berry, 2009; IEA, 2009; IEA and OECD, 2010). Moreover, continued technology advances in on- and offshore wind energy are expected (Section 7.7), supporting further cost reductions. The degree to which wind energy is utilized globally and regionally will depend largely on the economic performance of wind energy compared to alternative power sources.

This section describes the factors that affect the cost of wind energy (Section 7.8.1), highlights historical trends in the cost and performance of wind power plants (Section 7.8.2), summarizes data and estimates the levelized generation cost of wind energy in 2009 (Section 7.8.3),

and summarizes forecasts of the potential for further cost reductions (Section 7.8.4). The economic competitiveness of wind energy in comparison to other energy sources, which necessarily must also include other factors such as subsidies and environmental externalities, is not covered in this section.[42] Moreover, the focus in this section is on wind energy generation costs; the costs of integration and transmission are generally not covered here, but are instead discussed in Section 7.5, though costs associated with grid connection are sometimes included in the investment cost figures presented in this section.

7.8.1 Factors that affect the cost of wind energy

The levelized cost of energy from on- and offshore wind power plants is affected by five primary factors: annual energy production, investment costs, O&M costs, financing costs and the assumed economic life of the plant.[43] Available support policies can also influence the cost (and price) of wind energy, as well as the cost of other electricity supply options, but these factors are not addressed here.

The nature of the wind resource, which varies geographically and temporally, largely determines the annual energy production from a prospective wind power plant, and is among the most important economic factors (Burton et al., 2001). Precise micro-siting of wind power plants and even individual turbines is critical for maximizing energy production. The trend towards turbines with larger rotor diameters and taller towers has led to increases in annual energy production per unit of installed capacity, and has also allowed wind power plants in lower-resource areas to become more economically competitive. Larger wind power plants, meanwhile, have led to consideration of array effects whereby the production of downwind turbines is affected by those turbines located upwind. Offshore power plants will, generally, be exposed to better wind resources than will onshore plants (EWEA, 2009).

Wind power plants are capital intensive and, over their lifetime, the initial investment cost ranges from 75 to 80% of total expenditure, with O&M costs contributing the balance (Blanco, 2009; EWEA, 2009). The investment cost includes the cost of the turbines (turbines, transportation to site, and installation), grid connection (cables, sub-station, connection), civil works (foundations, roads, buildings), and other costs (engineering, licensing, permitting, environmental assessments and monitoring equipment). Table 7.4 shows a rough breakdown of the investment cost components for modern wind power plants. Turbine costs comprise more than 70% of total investment costs for onshore wind power plants. The remaining investment costs are highly site-specific. Offshore wind power plants are dominated by these other costs, with the turbines often contributing less than 50% of the total. Site-dependent characteristics such as water depth and distance to shore significantly affect grid connection, civil works and

---

41 Discussion of costs in this section is largely limited to the perspective of private investors. Chapters 1 and 8 to 11 offer complementary perspectives on cost issues covering, for example, costs of integration, external costs and benefits, economy-wide costs and costs of policies.

42 The environmental impacts and costs of RE and non-RE sources are summarized in Chapters 9 and 10, respectively.

43 Decommissioning costs also exist, but are not expected to be sizable in most instances.

Table 7.4 | Investment cost distribution for on- and offshore wind power plants (Data sources: Blanco, 2009; EWEA, 2009).

Cost Component	Onshore (%)	Offshore (%)[1]
Turbine	71–76	37–49
Grid connection	10–12	21–23
Civil works	7–9	21–25
Other investment costs	5–8	9–15

Note: 1. Offshore cost categories consolidated from original study.

other costs. Offshore turbine foundations and internal electric grids are also considerably more costly than those for onshore power plants.

The O&M costs of wind power plants include fixed costs such as land leases, insurance, taxes, management, and forecasting services, as well as variable costs related to the maintenance and repair of turbines, including spare parts. O&M comprises approximately 20% of total wind power plant expenditure over a plant's lifetime (Blanco, 2009), with roughly 50% of total O&M costs associated directly with maintenance, repair and spare parts (EWEA, 2009). O&M costs for offshore wind energy are higher than for onshore due to the less mature state of technology as well as the challenges and costs of accessing offshore turbines, especially in harsh weather conditions (Blanco, 2009).

Financing arrangements, including the cost of debt and equity and the proportional use of each, can also influence the cost of wind energy, as can the expected operating life of the wind power plant. For example, ownership and financing structures have evolved in the USA that minimize the cost of capital while taking advantage of available incentives (Bolinger et al., 2009). Other research has found that the predictability of the policy measures supporting wind energy can have a sizable impact on financing costs, and therefore the ultimate cost of wind energy (Wiser and Pickle, 1998; Dinica, 2006; Dunlop, 2006; Agnolucci, 2007). Because offshore wind power plants are still relatively new, with greater performance risk, higher financing costs are experienced than for onshore plants (Dunlop, 2006; Blanco, 2009), and larger firms tend to dominate offshore wind energy development and ownership (Markard and Petersen, 2009).

7.8.2 Historical trends

7.8.2.1 Investment costs

From the beginnings of commercial wind energy deployment to roughly 2004, the average investment costs of onshore wind power plants dropped, while turbine size grew significantly.[44] With each generation of wind turbine technology during this period, design improvements and turbine scaling led to decreased investment costs. Historical investment cost data from Denmark and the USA demonstrate this trend (Figure 7.20). From 2004 to 2009, however, investment costs increased. Some of the reasons behind these increased costs are described in Section 7.8.3.

There is far less experience with offshore wind power plants, and the investment costs of offshore plants are highly site-specific. Nonetheless, the investment costs of offshore plants have historically been 50 to more than 100% higher than for onshore plants (BWEA and Garrad Hassan, 2009; EWEA, 2009). Moreover, offshore wind power plants built to date have generally been constructed in relatively shallow water and relatively close to shore (see Section 7.3); higher costs would be experienced for deeper water and more distant facilities. Figure 7.21 presents investment cost data for operating and announced offshore wind power plants. Offshore costs have been influenced by some of the same factors that caused rising onshore costs from 2004 through 2009 (as well as several unique factors), as described in Section 7.8.3, leading to a doubling of the average investment cost of offshore plants from 2004 through 2009 (BWEA and Garrad Hassan, 2009; UKERC, 2010).

7.8.2.2 Operation and maintenance

Modern turbines that meet IEC standards are designed for a 20-year life, and plant lifetimes may exceed 20 years if O&M costs remain at an acceptable level. Few wind power plants were constructed 20 or more years ago, however, and there is therefore limited experience in plant operations over this entire time period (Echavarria et al., 2008). Moreover, those plants that have reached or exceeded their 20-year lifetime tend to have turbines that are much smaller and less sophisticated than their modern counterparts. Early turbines were also designed using more conservative criteria, though they followed less stringent standards than today's designs. As a result, early plants only offer limited guidance for estimating O&M costs for more recent turbine designs.

In general, O&M costs during the first couple of years of a wind power plant's life are covered, in part, by manufacturer warranties that are included in the turbine purchase, resulting in lower ongoing costs than in subsequent years. Newer turbine models also tend to have lower initial O&M costs than older models, with maintenance costs increasing

---

44 Investment costs presented here and later in Section 7.8 (as well as all resulting levelized cost of energy estimates) generally include the cost of the turbines (turbines, transportation to site and installation), grid connection (cables, sub-station, connection, but not more general transmission expansion costs), civil works (foundations, roads, buildings), and other costs (engineering, licensing, permitting, environmental assessments, and monitoring equipment). Whether the cost of connecting to the grid is included varies by data source, and is sometimes unclear; costs associated with strengthening the 'backbone' transmission system are generally excluded.

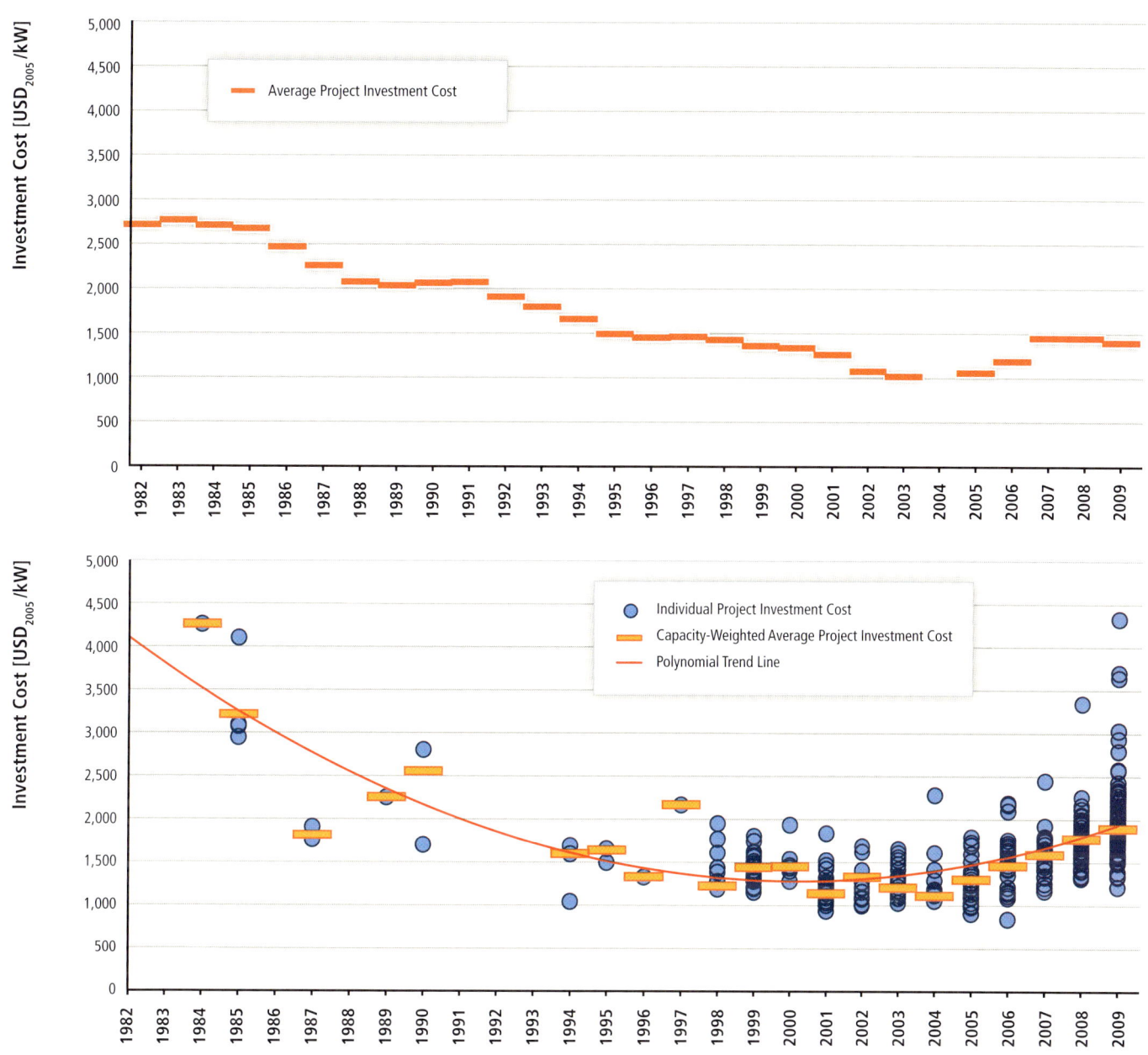

Figure 7.20. Investment cost of onshore wind power plants in (upper panel) Denmark (Data source: Nielson et al., 2010) and (lower panel) the USA (Wiser and Bolinger, 2010).

as turbines age (Blanco, 2009; EWEA, 2009; Wiser and Bolinger, 2010). Offshore wind power plants have historically incurred higher O&M costs than onshore plants (Junginger et al., 2004; EWEA, 2009; Lemming et al., 2009).

7.8.2.3 Energy production

The performance of wind power plants is highly site-specific, and is primarily governed by the characteristics of the local wind regime, which varies geographically and temporally. Wind power plant performance is also impacted by wind turbine design optimization, performance, and availability, however, and by the effectiveness of O&M procedures. Improved resource assessment and siting methodologies developed in the 1970s and 1980s played a major role in improved wind power plant productivity. Advances in wind energy technology, including taller towers and larger rotors, have also contributed to increased energy capture (EWEA, 2009).

Though plant-level capacity factors vary widely, data on average fleet-wide capacity factors[45] for a large sample of onshore wind power plants in the USA show a trend towards higher average capacity factors over time, as wind power plants built more recently have higher

---

[45] A wind power plant's capacity factor is only a partial indicator of performance (EWEA, 2009). Most turbine manufacturers supply variations on a given generator capacity with multiple rotor diameters and hub heights. In general, for a given generator capacity, increasing the hub height, the rotor diameter, or the average wind speed will result in an increased capacity factor. When comparing different wind turbines, however, it is possible to increase annual energy capture by using a larger generator, while at the same time decreasing the capacity factor.

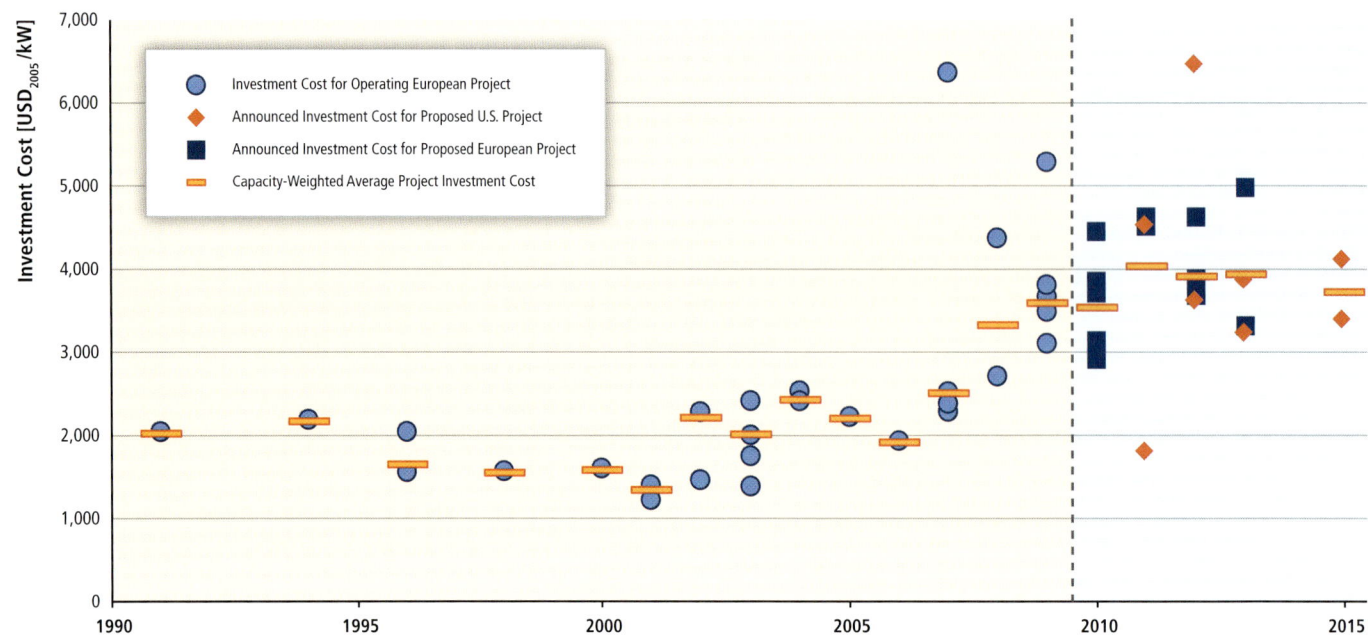

Figure 7.21 | Investment cost of operating and announced offshore wind power plants (Musial and Ram, 2010).

average capacity factors than those built earlier (Figure 7.22). Higher hub heights and larger rotor sizes are primarily responsible for these improvements, as the more recent wind power plants built in this time period and included in Figure 7.22 were, on average, sited in relatively lower-quality wind resource regimes.

Using a different metric for wind power plant performance, annual energy production per square meter of swept rotor area (kWh/m^2) for a given wind resource site, improvements of 2 to 3% per year over the last 15 years have been documented (IEA, 2008; EWEA, 2009).

7.8.3 Current conditions

7.8.3.1 Investment costs

The investment costs for onshore wind power plants installed worldwide in 2009 averaged approximately USD$_{2005}$ 1,750/kW, with many plants falling in the range of USD$_{2005}$ 1,400 to 2,100/kW (Milborrow, 2010); data in IEA Wind (2010) are reasonably consistent with this range. Wind power plants installed in the USA in 2009 averaged USD$_{2005}$ 1,900/kW (Wiser and Bolinger, 2010). Costs in some markets were lower: for

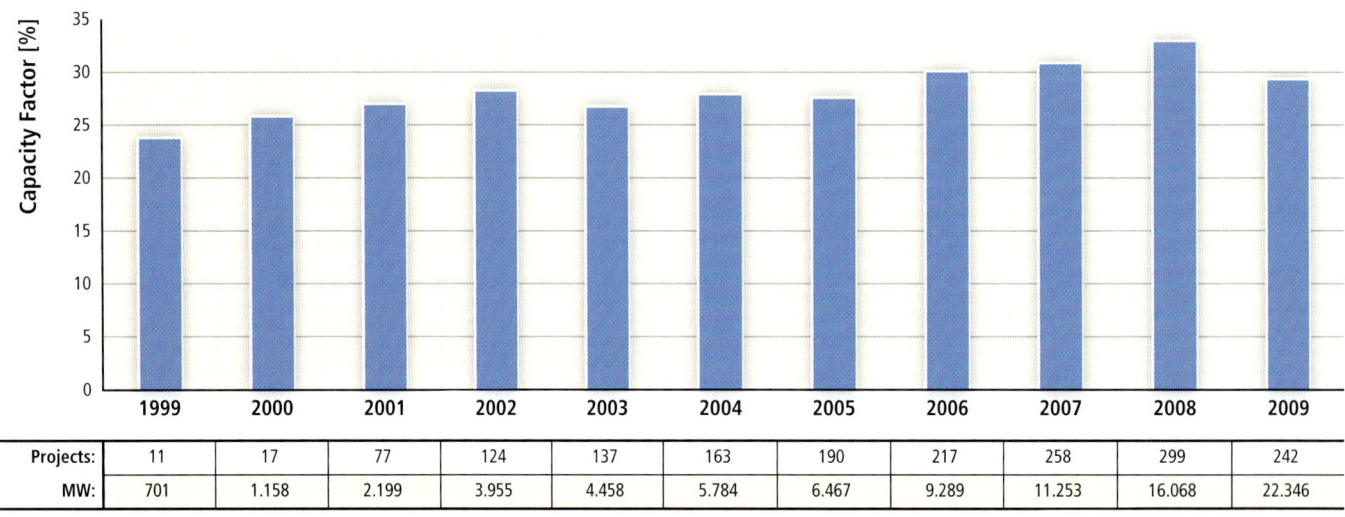

	1999	2000	2001	2002	2003	2004	2005	2006	2007	2008	2009
Projects:	11	17	77	124	137	163	190	217	258	299	242
MW:	701	1.158	2.199	3.955	4.458	5.784	6.467	9.289	11.253	16.068	22.346

Figure 7.22 | Fleet-wide average capacity factors for a large sample of wind power plants in the USA from 1999 to 2009 (Wiser and Bolinger, 2010).

example, average investment costs in China in 2008 and 2009 were around USD$_{2005}$ 1,000 to 1,350/kW, driven in part by the dominance of several Chinese turbine manufacturers serving the market with lower-cost wind turbines (China Renewable Energy Association, 2009; Li and Ma, 2009; Li, 2010).

Wind power plant investment costs rose from 2004 to 2009 (Figure 7.20), an increase primarily caused by the rising price of wind turbines (Wiser and Bolinger, 2010). Those price increases have been attributed to a number of factors. Increased rotor diameters and hub heights have enhanced the energy capture of modern wind turbines, for example, but those performance improvements have come with increased turbine costs, measured on a dollar per kW basis. The costs of raw materials, including steel, copper, cement, aluminium and carbon fibre, also rose sharply from 2004 through mid-2008 as a result of strong global economic growth. The strong demand for wind turbines over this period also put upward pressure on labour costs, and enabled turbine manufacturers and their component suppliers to boost profit margins. Strong demand, in excess of available supply, also placed particular pressure on critical components such as gearboxes and bearings (Blanco, 2009). Moreover, because many of the wind turbine manufacturers have historically been based in Europe, and many of the critical components have similarly been manufactured in Europe, the relative value of the Euro compared to other currencies also contributed to the wind turbine price increases in certain countries. Turbine manufacturers and component suppliers responded to the tight supply over this period by expanding or adding new manufacturing facilities. Coupled with reductions in materials costs that began in late 2008 as a result of the global financial crisis, these trends began to moderate wind turbine prices in 2009 (Wiser and Bolinger, 2010).

Due to the relatively small number of operating offshore wind power plants, investment cost data are sparse. Nonetheless, the average cost of offshore wind power plants is considerably higher than that for onshore plants, and the factors that have increased the cost of onshore plants have similarly affected the offshore sector. The limited availability of turbine manufacturers supplying the offshore market and of vessels to install such plants exacerbated cost increases since 2004, as has the installation of offshore plants in increasingly deeper waters and farther from shore, and the fierce competition among industry players for early-year (before 2005) demonstration plants (BWEA and Garrad Hassan, 2009; UKERC, 2010). As a result, offshore wind power plants over 50 MW in size, either built between 2006 and 2009 or planned for the early 2010s, had investment costs that ranged from approximately USD$_{2005}$ 2,000 to 5,000/kW (BWEA and Garrad Hassan, 2009; IEA, 2009; Snyder and Kaiser, 2009a; Musial and Ram, 2010). The most recently installed or announced plants cluster towards the higher end of this range, from USD$_{2005}$ 3,200 to 5,000/kW (Milborrow, 2010; Musial and Ram, 2010; UKERC, 2010). These investment costs are roughly 100% higher than costs seen from 2000 to 2004 (BWEA and Garrad Hassan, 2009; Musial and Ram, 2010; UKERC, 2010). Notwithstanding the increased water depth of offshore plants, the majority of the operating plants have been built in relatively shallow water. Offshore plants built in deeper waters, which are becoming increasingly common and are partly reflected in the costs for announced plants, will have relatively higher costs.

7.8.3.2   Operation and maintenance

Though fixed O&M costs such as insurance, land payments and routine maintenance are relatively easy to estimate, variable costs such as repairs and spare parts are more difficult to predict (Blanco, 2009). O&M costs can vary by wind power plant, turbine type and age, and the availability of a local servicing infrastructure, among other factors. Levelized O&M costs for onshore wind energy are often estimated to range from US cents$_{2005}$ 1.2 to 2.3/kWh (Blanco, 2009); these figures are reasonably consistent with costs reported in EWEA (2009), IEA (2010c), Milborrow (2010), and Wiser and Bolinger (2010).

Limited empirical data exist on O&M costs for offshore wind energy, due in large measure to the limited number of operating plants and the limited duration of those plants' operation. Reported or estimated O&M costs for offshore plants installed since 2002 range from US cents$_{2005}$ 2 to 4/kWh (EWEA, 2009; IEA, 2009, 2010c; Lemming et al., 2009; Milborrow, 2010; UKERC, 2010).

7.8.3.3   Energy production

Onshore wind power plant performance varies substantially, with capacity factors ranging from below 20 to more than 50% depending largely on local resource conditions. Among countries, variations in average performance also reflect differing wind resource conditions, as well as any difference in the wind turbine technology that is deployed: the average capacity factor for Germany's installed plants has been estimated at 20.5% (BTM, 2010); European country-level average capacity factors range from 20 to 30% (Boccard, 2009); average capacity factors in China are reported at roughly 23% (Li, 2010); average capacity factors in India are reported at around 20% (Goyal, 2010); and the average capacity factor for US wind power plants is above 30% (Wiser and Bolinger, 2010). Offshore wind power plants often experience a narrower range in capacity factors, with a typical range of 35 to 45% for the European plants installed to date (Lemming et al., 2009); some offshore plants in the UK, however, have experienced capacity factors of roughly 30%, in part due to relatively high component failures and access limitations (UKERC, 2010).

Because of these variations among countries and individual plants, which are primarily driven by local wind resource conditions but are also affected by turbine design and operations, estimates of the levelized cost of wind energy must include a range of energy production estimates. Moreover, because the attractiveness of offshore plants is enhanced by the potential for greater energy production than for onshore plants, performance variations among on- and offshore wind energy must also be considered.

7.8.3.4 Levelized cost of energy estimates

Using the methods summarized in Annex II, the levelized generation cost of wind energy is presented in Figure 7.23. For onshore wind energy, estimates are provided for plants built in 2009; for offshore wind energy, estimates are provided for plants built in 2008 and 2009 as well as those plants planned for completion in the early 2010s.[46] Estimated levelized costs are presented over a range of energy production estimates to represent the cost variation associated with inherent differences in the wind resource. The x-axis for these charts roughly correlates to annual average are used to produce levelized generation cost estimates.[48] Taxes, policy incentives, and the costs of electric system integration are not included in these calculations.[49]

The levelized cost of on- and offshore wind energy varies substantially, depending on assumed investment costs, energy production and discount rates. For onshore wind energy, levelized generation costs in good to excellent wind resource regimes are estimated to average US cents$_{2005}$ 5 to 10/kWh. Levelized generation costs can reach US cents$_{2005}$ 15/kWh in lower-resource areas. The costs of wind energy in China and

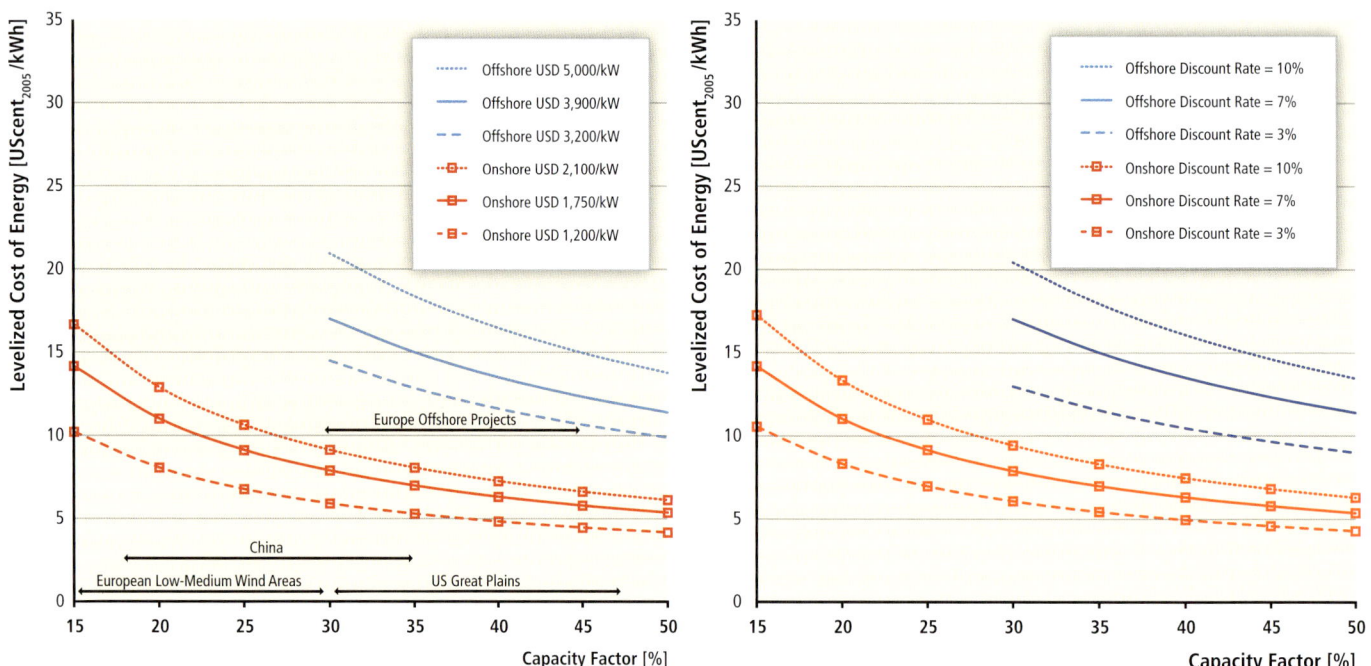

Figure 7.23 | Estimated levelized cost of on- and offshore wind energy, 2009: (left) as a function of capacity factor and investment cost* and (right) as a function of capacity factor and discount rate**.

Notes: * Discount rate assumed to equal 7%. ** Onshore investment cost assumed at USD$_{2005}$ 1,750/kW, and offshore at USD$_{2005}$ 3,900/kW.

wind speeds from 6 to 10 m/s. Onshore investment costs are assumed to range from USD$_{2005}$ 1,200 to 2,100/kW (with a mid-level cost of USD$_{2005}$ 1,750/kW); investment costs for offshore wind energy are assumed to range from USD$_{2005}$ 3,200 to 5,000/kW (mid-level cost of USD$_{2005}$ 3,900/kW).[47] Levelized O&M costs are assumed to average US cents$_{2005}$ 1.6/kWh and US cents$_{2005}$ 3/kWh over the life of the plant for onshore and offshore wind energy, respectively. A power plant design life of 20 years is assumed, and discount rates of 3 to 10% (mid-point estimate of 7%)

the USA tend towards the lower range of these estimates, due to lower average investment costs (China) and higher average capacity factors (USA); costs in much of Europe tend towards the higher end of the range due to relatively lower average capacity factors. Though the offshore cost estimates are more uncertain, offshore wind energy is generally more expensive than onshore, with typical levelized generation costs that are estimated to range from US cents$_{2005}$ 10/kWh to more than US cents$_{2005}$ 20/kWh for recently built or planned plants located in relatively

---

46 Because investment costs have risen in recent years, using the cost of recent and planned plants reasonably reflects the "current" cost of offshore wind energy.

47 Based on data presented earlier in this section, the mid-level investment cost for on- and offshore wind power plants does not represent the arithmetic mean between the low and high end of the range.

48 Though the same discount rate range and mid-point are used for on- and offshore wind energy, offshore wind power plants currently experience higher-cost financing than do onshore plants. As such, the levelized cost of energy from offshore plants may, in practice, tend towards the higher end of the range presented in the figure, at least in comparison to onshore plants.

49 Decommissioning costs are generally assumed to be low, and are excluded from these calculations.

shallow water. Where the exploitable onshore wind resource is limited, however, offshore plants can sometimes compete with onshore plants.

7.8.4 Potential for further reductions in the cost of wind energy

The wind energy industry has developed over a period of 30 years. Though the dramatic cost reductions seen in past decades will not continue indefinitely, the potential for further reductions remains given the many potential areas of technological advances described in Section 7.7. This potential spans both on- and offshore wind energy technologies; given the relatively less mature state of offshore wind energy, however, greater cost reductions can be expected in that segment. Two approaches are commonly used to forecast the future cost of wind energy, often in concert with some degree of expert judgement: (1) learning curve estimates that assume that future wind energy costs will follow a trajectory that is similar to an historical learning curve based on past costs; and (2) engineering-based estimates of the specific cost reduction possibilities associated with new or improved wind energy technologies or manufacturing capabilities (Mukora et al., 2009).

7.8.4.1 Learning curve estimates

Learning curves have been used extensively to understand past cost trends and to forecast future cost reductions for a variety of energy technologies (e.g., McDonald and Schrattenholzer, 2001; Kahouli-Brahmi, 2009; Junginger et al., 2010). Learning curves start with the premise that increases in the cumulative production of a given technology lead to a reduction in its costs. The principal parameter calculated by learning curve studies is the learning rate: for every doubling of cumulative production or installation, the learning rate specifies the associated percentage reduction in costs. Section 10.5 provides a more general discussion of learning curves as applied to renewable energy.

A number of published studies have evaluated historical learning rates for onshore wind energy (Table 7.5 provides a selective summary of the available literature).[50] The wide variation in results can be explained by differences in learning model specification (e.g., one-factor or multi-factor learning curves), variable selection and assumed system boundaries (e.g., whether investment cost, turbine cost, or levelized energy costs are explained, whether global or country-level cumulative installations are used, or whether country-level turbine production is used rather than

Table 7.5 | Summary of learning curve literature for onshore wind energy.

Authors	Learning By Doing Rate (%)	Global or National Independent Variable (cumulative capacity)	Dependent Variable	Data Years
Neij (1997)	4	Denmark[3]	Denmark (turbine cost)	1982–1995
Mackay and Probert (1998)	14	USA	USA (turbine cost)	1981–1996
Neij (1999)	8	Denmark[3]	Denmark (turbine cost)	1982–1997
Wene (2000)	32	USA[2]	USA (generation cost)	1985–1994
Wene (2000)	18	EU[2]	EU (generation cost)	1980–1995
Miketa and Schrattenholzer (2004)[1]	10	Global	Global (investment cost)	1971–1997
Junginger et al. (2005)	19	Global	UK (investment cost)	1992–2001
Junginger et al. (2005)	15	Global	Spain (investment cost)	1990–2001
Klaassen et al. (2005)[1]	5	Germany, Denmark, and UK	Germany, Denmark, and UK (investment cost)	1986–2000
Kobos et al. (2006)[1]	14	Global	Global (investment cost)	1981–1997
Jamasb (2007)[1]	13	Global	Global (investment cost)	1980–1998
Söderholm and Sundqvist (2007)	5	Germany, Denmark, and UK	Germany, Denmark, and UK (investment cost)	1986–2000
Söderholm and Sundqvist (2007)[1]	4	Germany, Denmark, and UK	Germany, Denmark, and UK (investment cost)	1986–2000
Neij (2008)	17	Denmark	Denmark (generation cost)	1981–2000
Kahouli-Brahmi (2009)	17	Global	Global (investment cost)	1979–1997
Nemet (2009)	11	Global	California (investment cost)	1981–2004
Ek and Söderholm (2010)[1]	17	Global	Germany, Denmark, Spain, Sweden, and UK (investment cost)	1986–2002
Wiser and Bolinger (2010)	9	Global	USA (investment cost)	1982–2009

Notes: 1. Two-factor learning curve that also includes R&D; others are one-factor learning curves. 2. Independent variable is cumulative production of electricity. 3. Cumulative turbine production used as independent variable; others use cumulative installations.

---

50   It is too early to develop a meaningful learning curve for offshore wind energy based on actual data from offshore plants. Studies have sometimes used learning rates to estimate future offshore costs, but those learning rates have typically been synthesized based on judgment and on learning rates for related industries and offshore subsystems (e.g., Junginger et al., 2004; Carbon Trust, 2008b).

installed wind power capacity), data quality, and the time period over which data are available. Because of these and other differences, the learning rates for wind energy presented in Table 7.5 range from 4 to 32%, but need special attention to be accurately interpreted and compared. Focusing *only* on the smaller set of studies completed in 2004 and later that have prepared estimates of learning curves based on total wind power plant *investment costs* and *global* cumulative installations, the range of learning rates narrows to 9 to 19%; the lowest figure within this range (9%) is the only one that includes data from 2004 to 2009, a period of increasing wind power plant investment costs.

There are also a number of limitations to the use of such models to forecast future costs (e.g., Junginger et al., 2010). First, learning curves typically (and simplistically) model how costs have decreased with increased installations in the past, but do not comprehensively explain the reasons behind the decrease (Mukora et al., 2009). In reality, costs may decline in part due to traditional learning and in part due to other factors, such as R&D expenditure and increases in turbine, power plant, and manufacturing facility size. Learning rate estimates that do not account for such factors may suffer from omitted variable bias, and may therefore be inaccurate. Second, if learning curves are used to forecast future cost trends, not only should the other factors that may influence costs be considered, but one must also assume that learning rates derived from historical data can be appropriately used to estimate future trends. As technologies mature, however, diminishing returns in cost reduction can be expected, and learning rates may fall (Arrow, 1962; Ferioli et al., 2009; Nemet, 2009). Third, the most appropriate cost measure for wind energy is arguably the levelized cost of energy, as wind energy generation costs are affected by investment costs, O&M costs and energy production (EWEA, 2009; Ferioli et al., 2009). Unfortunately, only two of the published studies calculate the learning rate for wind energy using a levelized cost of energy metric (Wene, 2000; Neij, 2008); most studies have used the more readily available metrics of investment cost or turbine cost. Fourth, a number of the published studies have sought to explain cost trends based on cumulative wind power capacity installations or production in individual countries or regions of the world; because the wind energy industry is global in scope, however, it is likely that much of the learning is now occurring based on cumulative global installations (e.g., Ek and Söderholm, 2010). Finally, from 2004 through 2009, wind turbine and power plant investment costs increased substantially, countering the effects of learning, in part due to materials and labour price increases and in part due to increased manufacturer profitability. Because production cost data are not generally publicly available, learning curve estimates typically rely upon price data that can be impacted by changes in materials costs and manufacturer profitability, resulting in the possibility of poorly estimated learning rates if dynamic price effects are not considered (Yu et al., 2011).

7.8.4.2   Engineering model estimates

Whereas learning curves examine aggregate historical data to forecast future trends, engineering-based models focus on the possible cost reductions associated with specific design changes and/or technical advances. Though limitations to engineering-based approaches also exist (Mukora et al., 2009), these models can lend support to learning curve predictions by defining the technology advances that can yield cost reductions and/or energy production increases.

These models have been used to estimate the impact of potential technology improvements on wind power plant investment costs and energy production, as highlighted in Section 7.7. Given the possible technology advances (in combination with manufacturing learning) discussed earlier, the US DOE (2008) estimates that onshore wind energy investment costs may decline by 10% by 2030, while energy production may increase by roughly 15%, relative to a 2008 starting point (see Table 7.3, and the note under that table).

There is arguably greater potential for technical advances in offshore than in onshore wind energy technology (see Section 7.7), particularly in foundation design, electrical system design and O&M costs. Larger offshore wind power plants are also expected to trigger more efficient installation procedures and dedicated vessels, enabling lower costs. Future levelized cost of energy reductions have sometimes been estimated by associating potential cost reductions with these technical improvements, sometimes relying on subsystem-level learning curve estimates from other industries (e.g., Junginger et al., 2004; Carbon Trust, 2008b).

7.8.4.3   Projected levelized cost of wind energy

A number of studies have developed forecasted cost trajectories for on- and offshore wind energy based on differing combinations of learning curve estimates, engineering models, and/or expert judgement. These estimates are sometimes—but not always—linked to certain levels of assumed wind energy deployment. Representative examples of this literature include Junginger et al. (2004), Carbon Trust (2008b), IEA (2008, 2010b, 2010c), US DOE (2008), EWEA (2009), Lemming et al. (2009), Teske et al. (2010), GWEC and GPI (2010) and UKERC (2010).

Recognizing that the starting year of the forecasts, the methodological approaches used, and the assumed deployment levels vary, these recent studies nonetheless support a range of levelized cost of energy reductions for onshore wind of 10 to 30% by 2020, and for offshore wind of 10 to 40% by 2020. Some studies focused on offshore wind energy technology even identify scenarios in which market factors lead to continued increases in the cost of offshore wind energy, at least in the near to medium term (BWEA and Garrad Hassan, 2009; UKERC, 2010). Longer-term projections are more reliant on assumed deployment levels and are subject to greater uncertainties, but for 2030, the same studies support reductions in the levelized cost of onshore wind energy of 15 to 35% and of offshore wind energy of 20 to 45%.

Using these estimates for the expected percentage cost reduction in levelized cost of energy, levelized cost trajectories for on- and offshore wind energy can be developed. Because longer-term cost projections are inherently more uncertain and depend, in part, on deployment levels and R&D expenditures that are also uncertain, the focus here is on relatively nearer-term cost projections to 2020. Specifically, Section 7.8.3.4 reported 2009 levelized cost of energy estimates for onshore wind energy of roughly US cents$_{2005}$ 5 to 15/kWh, whereas estimates for offshore wind energy were in the range of US cents$_{2005}$ 10 to 20/kWh. Conservatively, the *percentage* cost reductions reported above can be applied to these estimated 2009 levelized generation cost values to develop low and high projections for future levelized generation costs.[51]

Based on these assumptions, the levelized generation cost of onshore wind energy could range from roughly US cents$_{2005}$ 3.5 to 10.5/kWh by 2020 in a high cost-reduction case (30% by 2020), and from US cents$_{2005}$ 4.5 to 13.5/kWh in a low cost-reduction case (10% by 2020). Offshore wind energy is often anticipated to experience somewhat deeper cost reductions, with levelized generation costs that range from roughly US cents$_{2005}$ 6 to 12/kWh by 2020 in a high cost-reduction case (40% by 2020) to US cents$_{2005}$ 9 to 18/kWh in a low cost-reduction case (10% by 2020).[52]

Uncertainty exists over future wind energy costs, and the range of costs associated with varied wind resource strength introduces greater uncertainty. As installed wind power capacity increases, higher-quality resource sites will tend to be utilized first, leaving higher-cost sites for later development. As a result, the average levelized cost of wind energy will depend on the amount of deployment, not only due to learning effects, but also because of resource exhaustion. This 'supply-curve' effect is not captured in the estimates presented above. The estimates presented here therefore provide an indication of the technology advancement potential for on- and offshore wind energy, but should be used with some caution.

7.9 Potential deployment[53]

Wind energy offers significant potential for near- and long-term GHG emissions reductions. The wind power capacity installed by the end of 2009 was capable of meeting roughly 1.8% of worldwide electricity demand and, as presented in this section, that contribution could grow to in excess of 20% by 2050. On a global basis, the wind resource is unlikely to constrain further deployment (Section 7.2). Onshore wind energy technology is already being deployed at a rapid pace (Sections 7.3 and 7.4), therefore offering an immediate option for reducing GHG emissions in the electricity sector. In good to excellent wind resource regimes, the generation cost of onshore wind energy averages US cents$_{2005}$ 5 to 10/kWh (Section 7.8), and no insurmountable technical barriers exist that preclude increased levels of wind energy penetration into electricity supply systems (Section 7.5). Continued technology advances and cost reductions in on- and offshore wind energy are expected (Sections 7.7 and 7.8), further improving the GHG emissions reduction potential of wind energy over the long term.

This section begins by highlighting near-term forecasts for wind energy deployment (Section 7.9.1). It then discusses the prospects for and barriers to wind energy deployment in the longer term and the potential role of that deployment in reaching various GHG concentration stabilization levels (Section 7.9.2). Both subsections are largely based on energy market forecasts and GHG and energy scenarios literature published between 2007 and 2010. The section ends with brief conclusions (Section 7.9.3). Though the focus of this section is on larger on- and offshore wind turbines for electricity production, as discussed in Box 7.1, alternative technologies and applications for wind energy also exist.

7.9.1 Near-term forecasts

The rapid increase in global wind power capacity from 2000 to 2009 is expected by many studies to continue in the near to medium term (Table 7.6). From the roughly 160 GW of wind power capacity installed by the end of 2009, the IEA (2010b) 'New Policies' scenario and the EIA (2010) 'Reference case' scenario predict growth to 358 GW (forecasted electricity generation of 2.7 EJ/yr) and 277 GW (forecasted electricity generation of 2.5 EJ/yr) by 2015, respectively. Wind energy industry organizations predict even faster deployment rates, noting that past IEA and EIA forecasts have understated actual growth by a sizable margin (BTM, 2010; GWEC, 2010a). However, even these more aggressive forecasts estimate that wind energy will contribute less than 5% of global electricity supply by 2015. Asia, North America and Europe are projected to lead in wind power capacity additions over this period.

7.9.2 Long-term deployment in the context of carbon mitigation

A number of studies have tried to assess the longer-term potential of wind energy, often in the context of GHG concentration stabilization scenarios. As a variable, location-dependent resource with limited dispatchability, modelling the economics of wind energy expansion presents unique challenges (e.g., Neuhoff et al., 2008). The resulting differences among studies of the long-term deployment of wind energy may therefore reflect not just varying input assumptions and assumed policy and institutional contexts, but also differing modelling or scenario analysis approaches.

---

51 Because of the cost drivers discussed earlier in this section, wind energy costs in 2009 were higher than in some previous years. Applying the *percentage* cost reductions from the available literature to the 2009 starting point is, therefore, arguably a conservative approach to estimating future cost reduction possibilities; an alternative approach would be to use the *absolute* values of the cost estimates provided by the available literature. As a result, and also due to the underlying uncertainty associated with projections of this nature, future costs outside of the ranges presented here are possible.

52 As mentioned earlier, the 2009 starting point values for offshore wind energy are consistent with recently built or planned plants located in relatively shallow water.

53 Complementary perspectives on potential deployment based on a comprehensive assessment of numerous model-based scenarios of the energy system are presented in Sections 10.2 and 10.3 of this report.

Table 7.6 | Near-term global wind energy forecasts.

Study	Wind Energy Forecast			
	Installed Capacity (GW)	Generation (EJ/yr)	Percent of Global Electricity Supply (%)	Year
IEA (2010b)[1]	358	2.7	3.1	2015
EIA (2010)[2]	277	2.5	3.1	2015
GWEC (2010a)	409	N/A	N/A	2014
BTM (2010)	448	3.4	4.0	2014

Notes: 1. 'New Policies' scenario. 2. 'Reference case' scenario.

The IPCC's Fourth Assessment Report assumed that on- and offshore wind energy could contribute 7% of global electricity supply by 2030, or 8 EJ/yr (2,200 TWh/yr) (IPCC, 2007). Not surprisingly, this figure is higher than some commonly cited business-as-usual, reference-case forecasts (the IPCC estimate is not a business-as-usual case, but was instead developed within the context of efforts to mitigate global climate change). The IEA's World Energy Outlook 'Current Policies' scenario, for example, shows wind energy increasing to 6.0 EJ/yr (1,650 TWh/yr) by 2030, or 4.8% of global electricity supply (IEA, 2010b).[54] The US Energy Information Administration (EIA) forecasts 4.6 EJ/yr (1,200 TWh/yr) of wind energy in its 2030 reference case projection, or 3.9% of net electricity production from central producers (EIA, 2010).

A summary of the literature on the possible future contribution of RE supplies in meeting global energy needs under a range of GHG concentration stabilization scenarios is provided in Chapter 10. Focusing specifically on wind energy, Figures 7.24 and 7.25 present modelling results for the global supply of wind energy, in EJ/yr and as a percent of global electricity supply, respectively. About 150 different long-term scenarios underlie Figures 7.24 and 7.25. These scenario results derive from a diversity of modelling teams, and span a wide range of assumptions for—among other variables—electricity demand growth, the cost and availability of competing low-carbon technologies, and the cost and availability of RE technologies (including wind energy). Chapter 10 discusses how changes in some of these variables impact RE deployment outcomes, with Section 10.2.2 providing a description of the literature from which the scenarios have been taken. In Figures 7.24 and 7.25, the wind energy deployment results under these scenarios for 2020, 2030 and 2050 are presented for three GHG concentration stabilization ranges, based on the IPCC's Fourth Assessment Report: Baselines (>600 ppm CO_2), Categories III and IV (440 to 600 ppm) and Categories I and II (<440 ppm), all by 2100. Results are presented for the median scenario, the 25th to 75th percentile range among the scenarios, and the minimum and maximum scenario results.[55]

The baseline, or reference-case projections of wind energy's role in global energy supply span a broad range, but with a median among the reviewed scenarios of roughly 3 EJ/yr in 2020 (800 TWh/yr), 5 EJ/yr in 2030 (1,500 TWh/yr) and 16 EJ/yr in 2050 (4,400 TWh/yr) (Figure 7.24). Substantial growth of wind energy is therefore projected to occur even in the absence of climate change mitigation policies, with wind energy's median contribution to global electricity supply rising to nearly 9% by 2050 (Figure 7.25). Moreover, the contribution of wind energy grows as GHG reduction policies are assumed to become more stringent: by 2030, wind energy's median contribution among the reviewed scenarios equals roughly 11 EJ/yr (~9 to 10% of global electricity supply; 3,000 to 3,100 TWh/yr) in the 440 to 600 and <440 ppm CO_2 concentration stabilization ranges, increasing to 23 to 27 EJ/yr by 2050 (~13 to 14% of global electricity supply; 6,500 to 7,600 TWh/yr).[56]

The diversity of approaches and assumptions used to generate these scenarios is great, however, and results in a wide range of findings. Baseline case results for global wind energy supply in 2050 range from 2 to 58 EJ/yr (median of 16 EJ/yr), or 1 to 27% (median of 9%) of global electricity supply (500 to 16,200 TWh/yr). In the most stringent <440 ppm stabilization scenarios, wind energy supply in 2050 ranges from 7 to 113 EJ/yr (median of 27 EJ/yr), equivalent to 3 to 51% (median of 13%) of global electricity supply (2,000 to 31,500 TWh/yr).

Despite this wide range, the IPCC (2007) estimate for potential wind energy supply of roughly 8 EJ/yr (2,200 TWh/yr) by 2030 (which was largely based on literature available through 2005) appears somewhat conservative compared to the more recent scenarios literature presented here. Other recent forecasts of the possible role of wind energy in meeting global energy demands by RE organizations confirm this assessment, as the IPCC (2007) estimate is roughly one-third to one-half that shown in GWEC and GPI (2010) and Lemming et al. (2009). The IPCC (2007) estimate is more consistent with the IEA World Energy Outlook in its 'New Policies' scenario, but is 30% lower than that shown in the IEA's 450 ppm scenario (IEA, 2010b).

---

54 The IEA (2010b) 'Current Policies' scenario only reflects existing government policies, and is most similar to past IEA 'Reference case' forecasts. IEA (2010b) also presents a 'New Policies' scenario, in which stated government commitments are also considered, and in that instance wind energy grows to 8.2 EJ/yr (2,280 TWh/yr) by 2030, or 7% of global electricity supply.

55 In scenario ensemble analyses such as the review underlying the figures, there is a constant tension between the fact that the scenarios are not truly a random sample and the sense that the variation in the scenarios does still provide real and often clear insights into collective knowledge or lack of knowledge about the future (see Section 10.2.1.2 for a more detailed discussion).

56 In addition to the global scenarios literature, a growing body of work has sought to understand the technical and economic limits of wind energy deployment in regional electricity systems. These studies have sometimes evaluated higher levels of deployment than contemplated by the global scenarios, and have often used more sophisticated modelling tools. For a summary of a subset of these scenarios, see Martinot et al. (2007); examples of studies of this type include Deutsche Energie-Agentur (2005) (Germany); EC (2006); Nikolaev et al. (2008, 2010) (Russia); and US DOE (2008) (USA). In general, these studies confirm the basic findings from the global scenarios literature: wind energy deployment to 10% of global electricity supply and then to 20% or more is plausible, assuming that cost and policy factors are favourable.

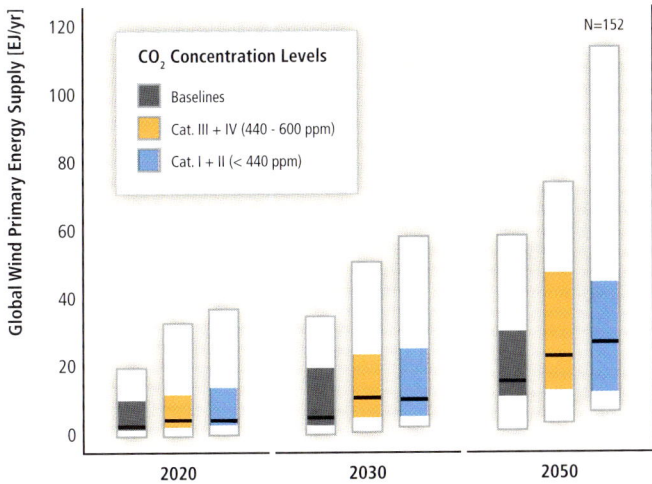

Figure 7.24 | Global primary energy supply of wind energy in long-term scenarios (median, 25th to 75th percentile range, and full range of scenario results; colour coding is based on categories of atmospheric CO_2 concentration level in 2100; the specific number of scenarios underlying the figure is indicated in the right upper corner) (adapted from Krey and Clarke, 2011; see also Chapter 10).

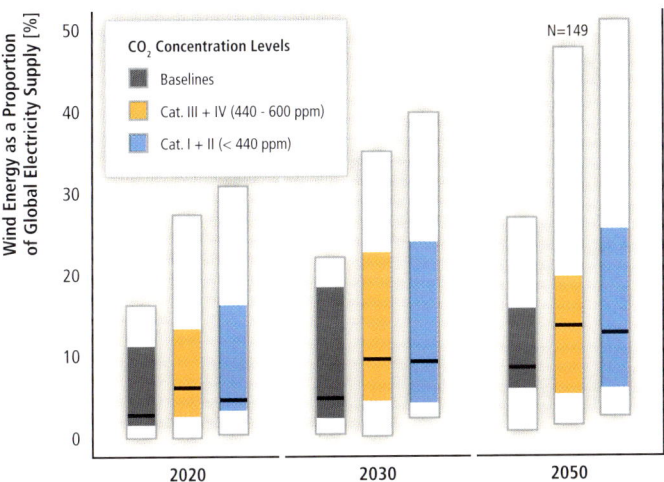

Figure 7.25 | Wind electricity share in total global electricity supply in long-term scenarios (median, 25th to 75th percentile range, and full range of scenario results; colour coding is based on categories of atmospheric CO_2 concentration level in 2100; the specific number of scenarios underlying the figure is indicated in the right upper corner) (adapted from Krey and Clarke, 2011; see also Chapter 10).

Though the literature summarized in Figures 7.24 and 7.25 shows an increase in wind energy with increasingly low GHG concentration stabilization levels, that impact is not as great as it is for biomass, geothermal and solar energy, where increasingly stringent GHG concentration stabilization ranges lead to more dramatic increases in technology deployment (see Chapter 10). One explanation for this result is that on shore wind energy is already comparatively economically competitive; as a result, continued deployment is predicted even in the absence of aggressive efforts to reduce GHG emissions.

The scenarios literature also shows that wind energy could play a significant long-term role in reducing global GHG emissions: by 2050, the median contribution of wind energy in the two GHG concentration stabilization scenarios is 23 to 27 EJ/yr (6,500 to 7,600 TWh/yr), increasing to 45 to 47 EJ/yr at the 75th percentile (12,400 to 12,900 TWh/yr), and to more than 100 EJ/yr in the highest scenario (31,500 TWh/yr). Achieving this contribution would require wind energy to deliver around 13 to 14% of global electricity supply by 2050 in the median scenario result, and 21 to 25% at the 75th percentile of the reviewed scenarios. By 2030, the corresponding wind electricity penetration levels are 9 to 10% in the median scenario result, increasing to 23 to 24% at the 75th percentile of the reviewed scenarios. Scenarios generated by wind energy and RE organizations are consistent with this median to 75th percentile range; Lemming et al. (2009), Teske et al. (2010), and GWEC and GPI (2010), for example, estimate the possibility of 31 to 39 EJ/yr (8,500 to 10,800 TWh/yr) of wind energy by 2050.

To achieve these levels of deployment, policies to reduce GHG emissions and/or increase RE supplies would likely be necessary, and those policies would need to be of adequate economic attractiveness *and* predictability to motivate substantial private investment (see Chapter 11). A variety of other possible challenges to aggressive wind energy growth also deserve discussion.

Resource Potential: Even the highest estimates for long-term wind energy supply in Figure 7.24 are below the global technical potential estimates for wind energy presented in Section 7.2, suggesting that—on a global basis, at least—technical potential is unlikely to be a limiting factor to wind energy deployment. Moreover, ample technical potential exists in most regions of the world to enable significant wind energy deployment relative to current levels. In certain countries or regions, however, higher deployment levels will begin to constrain the most economical resource supply, and wind energy will therefore not contribute equally in meeting the needs of every country.

Regional Deployment: Wind energy would need to expand beyond its historical base in Europe and, increasingly, the USA and China. The IEA WEO 'Current Policies' scenario projects the majority of wind energy deployment by 2035 to come from OECD Europe (36%), with lesser but still significant quantities from OECD North America (24%) and portions of non-OECD Asia (e.g., 18% in China and 4% in India) (IEA, 2010b). Under higher-penetration scenarios, however, a greater geographic distribution of wind energy deployment is likely to be needed. Scenarios from Teske et al. (2010), GWEC and GPI (2010) and IEA (2010c), for example, show non-OECD Asia (especially China), OECD North America, and OECD Europe to be the areas of greatest wind energy deployment, but also identify a number of other regions that are projected to be significant contributors to wind energy growth in high-penetration scenarios (Table 7.7).[57] Enabling this level of wind energy deployment in regions new to wind energy would be a challenge, and would benefit from institutional and technical knowledge transfer from those regions

---

57 Many of these other regions have lower expected electricity demands. As a result, some of the regions that are projected to make a small contribution to global wind electricity supply are still projected to obtain a sizable fraction of their own electricity supply from wind energy.

Table 7.7 | Regional distribution of global wind electricity supply (percentage of total worldwide wind electricity supply).

Region[1]	GWEC and GPI (2010) 2030 'Advanced' Scenario	Teske et al. (2010) 2050 'Energy Revolution' Scenario	(IEA, 2010c)[2] 2050 'BLUE Map' Scenario
Global Supply of Wind Energy	20 EJ/yr (5,400 TWh/yr)	31 EJ/yr (8,500 TWh/yr)	18 EJ/yr (4,900 TWh/yr)
OECD North America	27%	19%	13%
Latin America	4%	9%	8%
OECD Europe	22%	15%	21%
Eastern Europe / Eurasia	4%	8%	4%
OECD Pacific	5%	10%	7%
Developing Asia	35%	34%	39%
Africa	3%	2%	2%
Middle East	1%	3%	5%

Notes: 1. Regions are defined by each study, except that: GWEC and GPI (2010) estimates for 'Non-OECD Asia' are placed under 'Developing Asia'; IEA (2010c) estimates for 'U.S.' and 'Other OECD North America' are consolidated under 'OECD North America' while estimates for 'Eastern EU and Former Soviet Union' are placed under 'Eastern Europe / Eurasia'; and Teske et al. (2010) estimates for 'Transition Economies' are placed under 'Eastern Europe / Eurasia'. For all three studies, results for China and India are consolidated under 'Developing Asia'. (See also Annex II for definitions of regions and country groupings.) 2. For IEA (2010c), the percentage of worldwide wind power capacity investment through 2050 is presented.

that are already witnessing substantial wind energy activity (e.g., Lewis, 2007; IEA, 2009).

Supply Chain Issues: While *short-term* constraints will need to be addressed, no insurmountable *long-term* constraints to materials supply, labour availability, installation infrastructure or manufacturing capacity appear likely if policy frameworks for wind energy are sufficiently economically attractive *and* predictable (e.g., US DOE, 2008). The wind energy industry has scaled up rapidly over the last decades, resulting in greater globalization and competition throughout the supply chain (see Section 7.4). Supply-chain challenges have included the availability of skilled personnel and turbine component manufacturing, as well as turbine supply and installation infrastructure especially for offshore wind power plants (see Section 7.8). Nonetheless, annual additions and manufacturing volume reached 38 GW in 2009, and the significant further supply-chain scaling needed to meet the increased demands of higher-penetration scenarios (see also Section 10.3) appears challenging, but feasible in the long term.

Technology and Economics: Due to resource and siting constraints in some countries and regions, greater reliance on offshore wind energy, particularly in Europe, is likely to be required. Lemming et al. (2009) estimate that the proportion of total global wind energy supply likely to be delivered from offshore wind energy in 2050 is 18%, whereas the IEA's Energy Technology Perspectives BLUE Map Scenario forecasts a 32% share in capacity terms (IEA, 2010c). In another set of forecasts provided in the IEA's World Energy Outlook, offshore wind power capacity represents 15 to 24% of total wind power capacity by 2035, depending on the scenario (IEA, 2010b). Increases in offshore wind energy of this magnitude would require technological advances and cost reductions. Though R&D is expected to lead to incremental cost reductions for onshore wind energy technology, enhanced R&D expenditures by government and industry may be especially important for offshore wind energy technology given its less mature state compared to onshore wind energy (see Section 7.7).

Integration and Transmission: Proactive technical and institutional solutions to transmission constraints and operational integration concerns will need to be implemented. Analysis results and experience suggest that many electric systems can operate with up to roughly 20% wind energy with relatively modest integration costs (see Section 7.5 and Chapter 8). Additional studies have looked at wind electricity penetrations in excess of 20%, often using somewhat less-detailed analysis procedures than formal wind energy integration studies, and often involving the use of structural change in generation portfolios, electrical or thermal storage, plug-in hybrid vehicles and the electrification of transportation, demand response, and/or other technologies to manage the variability of wind power output (e.g., Grubb, 1991; Watson et al., 1994; Lund and Münster, 2003; Kempton and Tomic, 2005; Black and Strbac, 2006; DeCarolis and Keith, 2006; Denholm, 2006; Lund, 2006; Cavallo, 2007; Greenblatt et al., 2007; Hoogwijk et al., 2007; Benitez et al., 2008; Lamont, 2008; Leighty, 2008; Lund and Kempton, 2008; Kiviluoma and Meibom, 2010). These studies generally confirm that there are no insurmountable technical barriers to increased wind energy supply; instead, as deployment increases, transmission expansion and operational integration costs also increase, constraining growth on economic terms. These studies also find that new technical solutions that are not otherwise required at lower levels of wind energy deployment, such as expanded use of bulk energy storage and demand response, become increasingly valuable at higher levels of wind energy. Overall, the concerns about (and the costs of) operational integration and maintaining electric system reliability will grow with wind energy deployment, and efforts to ensure adequate system-wide flexibility, employ more restrictive grid connection standards, develop and use improved wind forecasting systems, and encourage demand flexibility and bulk energy storage are warranted.

Moreover, given the locational dependence of the wind resource, substantial new transmission infrastructure both on- and offshore would be required under even the more modest wind energy deployment scenarios presented earlier. Both cost and institutional barriers would need to be overcome to develop this needed transmission infrastructure (see Section 7.5 and Chapters 8 and 11).

Social and Environmental Concerns: Finally, given concerns about the social and environmental impacts of wind power plants summarized in Section 7.6, efforts to better understand the nature and magnitude of these impacts, together with efforts to minimize and mitigate those impacts, will need to be pursued in concert with increasing wind energy deployment. Prominent environmental concerns about wind energy include bird and bat collision fatalities and habitat and ecosystem modifications, while prominent social concerns include visibility and landscape impacts as well as various nuisance effects and possible radar interference. As wind energy deployment increases globally and regionally and as larger wind power plants are considered, existing concerns may become more acute and new concerns may arise. Though community and scientific concerns need to be addressed, more proactive planning, siting and permitting procedures for both on- and offshore wind energy may be required to enable the wind energy deployment envisioned under these scenarios (see also Chapter 11).

7.9.3 Conclusions regarding deployment

The literature presented in this section suggests that wind electricity penetration levels that approach or exceed 10% of global electricity supply by 2030 are feasible, assuming that cost and policy factors are favourable towards wind energy deployment. The scenarios further suggest that even more ambitious policies and/or technology improvements may allow wind energy to reach or exceed 20% of global electricity supply by 2050, and that these levels of supply may be economically attractive within the context of global climate change mitigation scenarios. However, a variety of challenges would need to be overcome if wind energy was to achieve these aggressive levels of penetration. In particular, the degree to which wind energy is utilized in the future will largely depend on: the economics of wind energy compared to alternative power sources; policies to directly or indirectly support wind energy deployment; local siting and permitting challenges; and real or perceived concerns about the ability to integrate wind energy into electric supply systems.

References

3TIER (2009). *The First Look Global Wind Dataset: Annual Mean Validation*. 3TIER, Seattle, WA, USA, 10 pp.

Ackermann, T. (ed.) (2005). *Wind Power in Power Systems*. John Wiley and Sons, London, UK.

Ackermann, T., and L. Soder (2002). An overview of wind energy-status 2002. *Renewable and Sustainable Energy Reviews*, **6**, pp. 67-127.

Agnolucci, P. (2007). The effect of financial constraints, technological progress and long-term contracts on tradable green certificates. *Energy Policy*, **35**, pp. 3347-3359.

AIGS (2008). *All Island Grid Study (AIGS). Workstream 4: Analysis of Impacts and Benefits*. Department of Communications, Energy and Natural Resources, Dublin, Ireland and UK Department of Enterprise, Trade and Investment, London, UK, 88 pp.

Allen, S.R., G.P. Hammond, and M.C. McManus (2008). Prospects for and barriers to domestic micro-generation: A United Kingdom perspective. *Applied Energy*, **85**, pp. 528-544.

Alves-Pereira, M., and C. Branco (2007). In-home wind turbine noise is conducive to vibroacoustic disease. Presented at the *Second International Meeting on Wind Turbine Noise INCE/Europe*, Lyon, France, 20-21 September, 2007.

Al-Yahyai, S., Y. Charabi, and A. Gastli (2010). Review of the use of Numerical Weather Prediction (NWP) Models for wind energy assessment. *Renewable and Sustainable Energy Reviews*, **14**, pp. 3192-3198.

Amelin, M. (2009). Comparison of capacity credit calculation methods for conventional power plants and wind power. *IEEE Transactions on Power Systems*, **24**, pp. 685-691.

Andersen, P.B., C.B. Mac Gaunaa, and T. Buhl (2006). Load alleviation on wind turbine blades using variable airfoil geometry. In: *2006 European Wind Energy Conference and Exhibition Proceedings*, Athens, Greece, 27 February – 2 March 2006.

Apt, J. (2007). The spectrum of power from wind turbines. *Journal of Power Sources*, **169**, pp. 369-374.

Archer, C.L., and M.Z. Jacobson (2005). Evaluation of global wind power. *Journal of Geophysical Research*, **110**, D12110.

Archer, C., and K. Caldeira (2009). Global assessment of high-altitude wind power. *Energies*, **2**, pp. 307-319.

Argatov, I., and R. Silvennoinen (2010). Energy conversion efficiency of the pumping kite wind generator. *Renewable Energy*, **35**, pp. 1052-1060.

Argatov, I., P. Rautakorpi, and R. Silvennoinen (2009). Estimation of the mechanical energy output of the kite wind generator. *Renewable Energy*, **34**, pp. 1525-1532.

Arnett, E.B., W.K. Brown, W.P. Erickson, J.K. Fiedler, B.L. Hamilton, T.H. Henry, A. Jain, G.D. Johnson, J. Kerns, and R.R. Koford (2008). Patterns of bat fatalities at wind energy facilities in North America. *Journal of Wildlife Management*, **72**, pp. 61-78.

Arnett, E.B., M.M.P. Huso, M.R. Schirmacher, and J.P. Hayes (2011). Altering turbine speed reduces bat mortality at wind-energy facilities. *Frontiers in Ecology and the Environment*, **9**(4), pp. 209-214.

Arrow, K.J. (1962). The economic implications of learning by doing. *The Review of Economic Studies*, **29**, pp. 155-173.

Ashwill, T.D. (2009). Materials and innovations for large blade structures: research opportunities in wind energy technology. In: *50th AIAA/ASME/ASCE/AHS/ASC Structures, Structural Dynamics, and Materials Conference*. Palm Springs, CA, USA, 4-7 May 2009, American Institute of Aeronautics and Astronautics, 20 pp.

Auer, H., M. Stadler, G. Resch, C. Huber, T. Schuster, H. Taus, L.H. Nielsen, J. Twidell, and D.J. Swider (2004). *Cost and Technical Constraints of RES-E Grid Integration: Working package 2*. European Commission, Brussels, Belgium, 82 pp.

AWEA (2008). *Wind Energy Siting Handbook*. American Wind Energy Association, Washington, DC, USA, 183 pp.

AWEA (2009). *AWEA Small Wind Turbine Global Market Study: Year Ending 2008*. American Wind Energy Association, Washington, DC, USA, 24 pp.

Baerwald, E.F., J. Edworthy, M. Holder, and R.M.R. Barclay (2009). A large-scale mitigation experiment to reduce bat fatalities at wind energy facilities. *Journal of Wildlife Management*, **73**, pp. 1077-1082.

Baidya, R.S., and J.J. Traiteur (2010). Impacts of wind farms on surface air temperatures. *Proceedings of the National Academy of Sciences*, **107**, pp. 17899-17904.

Banta, R.M., Y.L. Pichugina, and R.K. Newsom (2003). Relationship between low-level jet properties and turbulence kinetic energy in the nocturnal stable boundary layer. *Journal of the Atmospheric Sciences*, **60**, pp. 2549-2555.

Barclay, R.M.R., E.F. Baerwald, and J.C. Gruver (2007). Variation in bat and bird fatalities at wind energy facilities: assessing the effects of rotor size and tower height. *Canadian Journal of Zoology*, **85**, pp. 381-387.

Barrios, L., and A. Rodriguez (2004). Behavioral and environmental correlates of soaring bird mortality at onshore wind turbines. *Journal of Applied Ecology*, **41**, pp. 72-81.

Barthelmie, R.J., and L.E. Jensen (2010). Evaluation of wind farm efficiency and wind turbine wakes at the Nysted offshore wind farm. *Wind Energy*, **13**, pp. 573-586.

Barthelmie, R.J., G. Larsen, S.C. Pryor, H. Jörgensen, H. Bergström, W. Schlez, K. Rados, B. Lange, P. Völund, S. Neckelmann, S. Mogensen, G. Schepers, T. Hegberg, L. Folkerts, and M. Magnusson (2004). ENDOW (efficient development of offshore wind farms): Modelling wake and boundary layer interactions. *Wind Energy*, **7**, pp. 225-245.

Bastasch, M., J. van-Dam, B. Søndergaard, and A. Rogers (2006). Wind turbine noise – An overview. *Canadian Acoustics*, **34**, pp. 7-15.

Bell, D., T. Gray, and C. Haggett (2005). The 'social gap' in wind farm siting decisions: Explanations and policy responses. *Environmental Politics*, **14**, pp. 460-477.

Bengtsson, L., K.I. Hodges, and N. Keenlyside (2009). Will extratropical storms intensify in a warmer climate? *Journal of Climate*, **22**, pp. 2276-2301.

Benitez, L.E., P.C. Benitez, and G.C. van Kooten (2008). The economics of wind power with energy storage. *Energy Economics*, **30**, pp. 1973-1989.

Benjamin, R. (2007). Principles for interregional transmission expansion. *The Electricity Journal*, **20**, pp. 36-47.

Berg, D.E., D.G. Wilson, M.F. Barone, B.R. Resor, J.C. Berg, J.A. Paquette, J.R. Zayas, S. Kota, G. Ervin, and D. Maric (2009). The impact of active aerodynamic load control on fatigue and energy capture at low wind speed sites. In: *European Wind Energy Conference*, Marseille, France, 16-19 March 2009, pp. 16-19.

Bergek, A. (2010). Levelling the playing field? The influence of national wind power planning instruments on conflicts of interests in a Swedish county. *Energy Policy*, **38**, pp. 2357-2369.

Berry, D. (2009). Innovation and the price of wind energy in the US. *Energy Policy*, **37**, pp. 4493-4499.

Bierbooms, W.A.A.M. (1992). A comparison between unsteady aerodynamic models. *Journal of Wind Engineering and Industrial Aerodynamics*, **39**, pp. 23-33.

Bird, L., M. Bolinger, T. Gagliano, R. Wiser, M. Brown, and B. Parsons (2005). Policies and market factors driving wind power development in the United States. *Energy Policy*, **33**, pp. 1397-1407.

Black and Veatch (2003). *Strategic Assessment of the Potential for Renewable Energy: Stage 1*. European Bank for Reconstruction and Development, London, UK, 163 pp.

Black, M., and G. Strbac (2006). Value of storage in providing balancing services for electricity generation systems with high wind penetration. *Journal of Power Sources*, **162**, pp. 949-953.

Blanco, M.I. (2009). The economics of wind energy. *Renewable and Sustainable Energy Reviews*, **13**, pp. 1372-1382.

Bloom, A., V. Kotroni, and K. Lagouvardos (2008). Climate change impact of wind energy availability in the Eastern Mediterranean using the regional climate model PRECIS. *Natural Hazards and Earth System Sciences*, **8**, pp. 1249–1257.

Boccard, N. (2009). Capacity factor of wind power realized values vs. estimates. *Energy Policy*, **37**, pp. 2679-2688.

Boccard N. (2010). Economic properties of wind power: A European assessment. *Energy Policy*, **38**, pp. 3232-3244.

Bolinger, M., J. Harper, and M. Karcher (2009). A review of wind project financing structures in the USA. *Wind Energy*, **12**, pp. 295-309.

Bossanyi, EA. (2003). Individual blade pitch control for load reduction. *Wind Energy*, **6**, pp. 119-128.

Boyle, M.A., and K.A. Kiel (2001). A survey of house price hedonic studies of the impact of environmental externalities. *Journal of Real Estate Literature*, **9**, pp. 117-144.

Bradley, J.S. (1994). Annoyance caused by constant-amplitude and amplitude-modulated sounds containing rumble. *Noise Control Engineering Journal*, **42**, pp. 203-208.

Brenner, M., S. Cazares, M. Cornwall, F. Dyson, D. Eardley, P. Horowitz, D. Long, J. Sullivan, J. Vesecky, and P. Weinberger (2008). *Wind Farms and Radar*. US Department of Homeland Security, McLean, VA, USA, 21 pp.

Breslow, P.B., and D.J. Sailor (2002). Vulnerability of wind power resources to climate change in the continental United States. *Renewable Energy*, **27**, pp. 585-598.

Breton, S.P., and G. Moe (2009). Status, plans and technologies for offshore wind turbines in Europe and North America. *Renewable Energy*, **34**, pp. 646-654.

BTM (2010). *International Wind Energy Development. World Market Update 2009*. BTM Consult ApS, Ringkøbing, Denmark, 124 pp.

Budhraja, V.S., F. Mobasheri, J. Ballance, J. Dyer, A. Silverstein, and J. Eto (2009). Improving electricity resource-planning processes by considering the strategic benefits of transmission. *The Electricity Journal*, **22**, pp. 54-63.

Buhl, T., M. Gaunaa, and C. Bak (2005). Potential load reduction using airfoils with variable trailing edge geometry. *Journal of Solar Energy Engineering*, **127**, pp. 503-516.

Burke, D.J., and M.J. O'Malley (2010). Maximizing firm wind connection to security constrained transmission networks. *IEEE Transactions on Power Systems*, **25**, pp. 749-759.

Burton, T., D. Sharpe, N. Jenkins, and E. Bossanyi (2001). *Wind Energy: Handbook*. J. Wiley, Chichester, UK and New York, NY, USA, 642 pp.

Butler, M.M., and D.A. Johnson (2003). *Feasibility of Mitigating the Effects of Wind Farms on Primary Radar*. UK Department of Trade and Industry, London, UK, 210 pp.

BWEA and Garrad Hassan (2009). *UK Offshore Wind: Charting the Right Course*. British Wind Energy Association, London, UK, 42 pp.

Byrne, J., A. Zhou, B. Shen, and K. Hughes (2007). Evaluating the potential of small-scale renewable energy options to meet rural livelihoods needs: A GIS-and lifecycle cost-based assessment of Western China's options. *Energy Policy*, **35**, pp. 4391-4401.

Bywaters, G., V. John, J. Lynch, P. Mattila, G. Norton, J. Stowell, M. Salata, O. Labath, A. Chertok, and D. Hablanian (2004). *Northern Power Systems WindPACT Drive Train Alternative Design Study Report, Period of Performance: April 12, 2001 to January 31, 2005*. National Renewable Energy Laboratory, Golden, CO, USA, 404 pp.

Capps, S.B., and C.S. Zender (2010). Estimated global ocean wind power potential from QuikSCAT observations, accounting for turbine characteristics and siting. *Journal of Geophysical Research*, **115**, D09101.

Carbon Trust (2008a). *Small-scale Wind Energy: Policy Insights and Practical Guidance*. Carbon Trust, London, UK, 108 pp.

Carbon Trust (2008b). *Offshore Wind power: Big Challenge, Big Opportunity*. The Carbon Trust, London, UK, 40 pp.

Cavallo, A. (2007). Controllable and affordable utility-scale electricity from intermittent wind resources and compressed air energy storage (CAES). *Energy*, **32**, pp. 120-127.

Chao, D.D., and C.P. van Dam (2007). Computational aerodynamic analysis of a blunt trailing edge airfoil modification to the NREL Phase VI rotor. *Wind Energy*, **10**, pp. 529-550.

China Renewable Energy Association (2009). *Annual Report of New Energy and Renewable Energy in China, 2009*. China Renewable Energy Association, Beijing, China.

Christiansen, M.B., and C.B. Hasager (2005). Wake effects of large offshore wind farms identified from satellite SAR. *Remote Sensing of Environment*, **98**, pp. 251-268.

Christiansen, M.B., and C.B. Hasager (2006). Using airborne and satellite SAR for wake mapping offshore. *Wind Energy*, **9**, pp. 437-455.

Claussen, N.E., P. Lundsager, R.J. Barthelmie, H. Holttinen, T. Laakso, and S.C. Pryor (2007). Wind power. In: *Impacts of Climate Change on Renewable Energy Sources. Their Role in the Nordic Energy System*. J. Fenger (ed.), Nordic Council of Ministers, Copenhagen, Denmark, pp. 105-128.

CMA (2006). *The Report of Wind Energy Resource Assessment in China*. China Meteorological Administration, China Meteorological Press, Beijing, China.

CMOH (2010). *The Potential Health Impacts of Wind Turbines*. Ontario Chief Medical Officer of Health (CMOH), Ontario Ministry of Health and Long-Term Care, Toronto, Canada, 14 pp.

Costa, A., A. Crespo, J. Navarro, G. Lizcano, H. Madsen, and E. Feitosa (2008). A review on the young history of the wind power short-term prediction. *Renewable and Sustainable Energy Reviews*, **12**, pp. 1725-1744.

Coughlan, Y., P. Smith, A. Mullane, and M. O'Malley (2007). Wind turbine modelling for power system stability analysis—A system operator perspective. *IEEE Transactions on Power Systems*, **22**, pp. 929-936.

Cryan, P.M., and R.M.R. Barclay RMR (2009). Causes of bat fatalities at wind turbines: hypotheses and predictions. *Journal of Mammalogy*, **90**, pp. 1330-1340.

Czisch, G., and G. Giebel (2000). A comparison of intra-and extraeuropean options for an energy supply with wind power. In: *Proceedings of Wind Power for the 21st Century*, Kassel, Germany, 25-27 September 2000, pp. 69-73.

De Lucas, M., G.F.E. Janss, and M. Ferrer (2004). The effects of a wind farm on birds in a migration point: the Strait of Gibraltar. *Biodiversity and Conservation*, **13**, pp. 395-407.

de Lucena, P., A. Frossard, A.S. Szklo, R. Schaeffer, and R.M. Dutra (2009). The vulnerability of wind power to climate change in Brazil. *Renewable Energy*, **35**, pp. 904-912.

DeCarolis, J.F., and D.W. Keith (2006). The economics of large-scale wind power in a carbon constrained world. *Energy Policy*, **34**, pp. 395-410.

Delft University, Garrad Hassan & Partners, Tractebel Energy Engineering, Risø National Laboratory, Kvaerner Oil & Gas, Energi & Miljoe Undersoegelser (2001). *Concerted Action on Offshore Wind Energy in Europe*. Delft University, Delft, the Netherlands, 289 pp.

Denholm, P. (2006). Improving the technical, environmental and social performance of wind energy systems using biomass-based energy storage. *Renewable Energy*, **31**, pp. 1355-1370.

Denny, E., and M. O'Malley (2009). The impact of carbon prices on generation-cycling costs. *Energy Policy*, **37**, pp. 1204-1212.

Denny, E., A. Tuohy, P. Meibom, A. Keane, D. Flynn, A. Mullane, and M. O'Malley (2010). The impact of increased interconnection on electricity systems with large penetrations of wind generation: A case study of Ireland and Great Britain. *Energy Policy*, **38**, pp. 6946-6954.

Des Rosiers, F. (2002). Power lines, visual encumbrance and house values: a microspatial approach to impact measurement. *Journal of Real Estate Research*, **23**, pp. 275-301.

Desholm, M. (2009). Avian sensitivity to mortality: Prioritising migratory bird species for assessment at proposed wind farms. *Journal of Environmental Management*, **90**, pp. 2672-2679.

Desholm, M., and J. Kahlert (2005). Avian collision risk at an offshore wind farm. *Biology Letters*, **1**, pp. 296-298.

Deutsche Energie-Agentur GmbH (2005). *Energy Management Planning for the Integration of Wind Energy into the Grid in Germany, Onshore and Offshore by 2020*. Deutsche Energie-Agentur GmbH, Cologne, Germany.

Dinica, V. (2006). Support systems for the diffusion of renewable energy technologies – investor perspective. *Energy Policy*, **34**, pp. 461-480.

Doherty, R., and M. O'Malley (2005). A new approach to quantify reserve demand in systems with significant installed wind capacity. *IEEE Transactions on Power Systems*, **20**, pp. 587-595.

Doherty, R., L. Bryans, P. Gardner, and M. O'Malley (2004). Wind penetration studies on the island of Ireland. *Wind Engineering*, **28**, pp. 27-41.

Doherty, R., A. Mullane, G. Lalor, and D. Burke (2010). An assessment of the impact of wind generation on system frequency. *IEEE Transactions on Power Systems*, **25**, pp. 452-460.

Dong Energy, Vattenfall, Danish Energy Authority, and Danish Forest and Nature Agency (2006). *Danish Offshore Wind: Key Environmental Issues*. DONG Energy, Vattenfall, the Danish Energy Authority and the Danish Forest and Nature Agency, Copenhagen, Denmark, 142 pp.

Drewitt, A.L., and R.W. Langston (2006). Assessing the impacts of wind farms on birds. *Ibis*, **148**, pp. 29-42.

Du, Z., and M.S. Selig (1998). A 3-D stall-delay model for horizontal axis wind turbine performance prediction. In: *Proceedings of the 1998 ASME Wind Energy Symposium*. American Institute of Aeronautics and Astronautics (AIAA) and ASME International, Reno, NV, USA, 12-15 January 1998, pp. 9-19.

Dunlop, J. (2006). Wind power project returns – What should equity investors expect? *Journal of Structured Finance*, **12**, pp. 81-89.

Dürr, T., and L. Bach (2004). Bat deaths and wind turbines – a review of current knowledge, and of the information available in the database for Germany. *Bremer Beiträge für Naturkunde und Naturschutz*, **7**, pp. 253-264.

Durstewitz, M., B. Hahn, B. Lange, K. Rohrig, and A. Wessel (2008). *Windenergie Report Deutschland 2007*. Institut für Solare Energieversorgungstechnik (ISET), Kessel, Germany.

Ea Energianalyse (2007). *50% Wind Power in Denmark in 2025*. Ea Energieanalyse, Copenhagen, Denmark, 121 pp.

EC (2003). *External Costs: Research Results on Socio-Environmental Damages Due to Electricity and Transport*. European Commission (EC), Brussels, Belgium, 28 pp.

EC (2006). *European Energy and Transport: Scenarios on Energy Efficiency and Renewables*. European Commission (EC), Brussels, Belgium, 124 pp.

EC (2009). *Communication from the Commission on Investing in the Development of Low Carbon Technologies (SET-Plan). A Technology Roadmap*. Commission of the European Communities, Brussels, Belgium, 1295 pp.

Echavarria, E., B. Hahn, G.J.W. Van Bussel, and T. Tomiyama (2008). Reliability of wind turbine technology through time. *Journal of Solar Energy Engineering*, **130**, 031005.

EEA (2009). *Europe's Onshore and Offshore Wind Energy Potential: An Assessment of Environmental and Economic Constraints*. European Environmental Agency (EEA), Copenhagen, Denmark, 90 pp.

Ehler, C.N., and F. Douvere (2009). *Marine Spatial Planning: A Step-by-Step Approach Towards Ecosystem Based Management*. Intergovernmental Oceanographic Commission and Marine and Biosphere Programme, United Nations Educational, Scientific, and Cultural Organization, Paris, France, 99 pp.

EIA (2010). *International Energy Outlook, 2010*. Energy Information Agency, US Department of Energy, Washington, DC, USA, 231 pp.

EirGrid and SONI (2010). *All Island TSO Facilitation of Renewables Studies*. EirGrid and System Operator for Northern Ireland (SONI), Dublin, Ireland, 77 pp.

Ek, K., and P. Söderholm (2010). Technology learning in the presence of public R&D: The case of European wind power. *Ecological Economics*, **69**, pp. 2356-2362.

Ela, E., and B. Kirby (2008). *ERCOT Event on February 26, 2008: Lessons Learned*. USA: National Renewable Energy Laboratory, Golden, CO, USA, 12 pp.

Elliott, D. (2002). Assessing the world's wind resources. In: *IEEE Power Engineering Society Winter Meeting, 2002*, New York, NY, USA, 27-31 January 2002, pp. 346-347.

Elliott, D., M. Schwartz, and G. Scott (2004). Wind resource base. In: *Encyclopaedia of Energy*. Elsevier, Amsterdam, The Netherlands, 465-479.

Eltham, D.C., G.P. Harrison, and S. J. Allen (2008). Change in public attitudes towards a Cornish wind farm: Implications for planning. *Energy Policy*, **36**, pp. 23-33.

EnerNex Corp (2010). *Eastern Wind Integration and Transmission Study*. National Renewable Energy Laboratory, Golden, CO, USA, 242 pp.

EPA (1974). *Information on Levels of Environmental Noise Requisite to Protect Public Health and Welfare with an Adequate Margin of Safety*. Office of Noise Abatement and Control, US Environmental Protection Agency, Washington, DC, USA, 242 pp.

EPA (1978). *Protective Noise Levels*. Office of Noise Abatement and Control, US Environmental Protection Agency, Washington, DC, USA, 25 pp.

ERCOT (2008). *ERCOT Operations Report on the EECP Event of February 26, 2008*. The Electricity Reliability Council of Texas (ERCOT), Austin, TX, USA, 13 pp.

Erdman, W., and M. Behnke (2005). *Low Wind Speed Turbine Project Phase II: The Application of Medium-Voltage Electrical Apparatus to the Class of Variable Speed Multi-Megawatt Low Wind Speed Turbines*. National Renewable Energy Laboratory, Golden, CO, USA, 119 pp.

Erickson, W.P., G.D. Johnson, and D.P. Young Jr. (2005). *A Summary and Comparison of Bird Mortality from Anthropogenic Causes with an Emphasis on Collisions*. General Technical Report, United States Forest Service, Washington, DC, USA, 14 pp.

Eto, J., J. Undrill, P. Mackin, R. Daschmans, B. Williams, B. Haney, R. Hunt, J. Ellis, H. Illian, C. Martinez, M. O'Malley, K. Coughlin, and K. LaCommare (2010). *Use of Frequency Response Metrics to Assess the Planning and Operating Requirements for Reliable Integration of Variable Generation*. Lawrence Berkeley National Laboratory, Berkeley, California, USA, 90 pp.

EU (2008). *European Wind Energy Technology Platform for Wind Energy. Strategic Research Agenda - Market Development Strategy from 2008 to 2030. Synopsis*. European Union (EU), TPWind Secretariat, Brussels, Belgium, 15 pp.

Everaert, J., and E.W.M. Stienen (2007). Impact of wind turbines on birds in Zeebrugge (Belgium). *Biodiversity and Conservation*, **16**, pp. 3345-3359.

EWEA (2005). *Large Scale Integration of Wind Energy in the European Power Supply: Analysis, Issues and Recommendations*. European Wind Energy Association (EWEA), Brussels, Belgium, 172 pp.

EWEA (2009). *Wind Energy, the Facts*. European Wind Energy Association (EWEA), Brussels, Belgium, 488 pp.

EWEA (2010a). *The European Offshore Wind Industry – Key Trends and Statistics 2009*. European Wind Energy Association (EWEA), Brussels, Belgium, 9 pp.

EWEA (2010b). *Wind in Power: 2009 European Statistics*. European Wind Energy Association (EWEA), Brussels, Belgium, 14 pp.

FANM (2005). *Human Health Repercussions of Windmill Operation*. French Académie Nationale de Médecine (FANM), Paris, France, 17 pp.

Fastl, H., and E. Zwicker (2007). *Psychoacoustics: Facts and Models*. 3rd ed. Springer, Berlin and Heidelberg, Germany and New York, NY, USA, 428 pp.

Fellows, A. (2000). *The Potential of Wind Energy to Reduce Carbon Dioxide Emissions*. Garrad Hassan and Partners Ltd, Glasgow, Scotland, 146 pp.

Ferioli, F., K. Schoots, and B.C.C. Van der Zwaan (2009). Use and limitations of learning curves for energy technology policy: A component-learning hypothesis. *Energy Policy*, **37**, pp. 2525-2535.

Firestone, J., and W. Kempton (2007). Public opinion about large offshore wind power: Underlying factors. *Energy Policy*, **35**, pp. 1584-1598.

Firestone, J., W. Kempton, and A. Krueger (2009). Public acceptance of offshore wind power projects in the USA. *Wind Energy*, **12**, pp. 183-202.

Focken, U., M. Lange, K. Mönnich, H. Waldl, H.G. Beyer, and A. Luig (2002). Short-term prediction of the aggregated power output of wind farms – a statistical analysis of the reduction of the prediction error by spatial smoothing effects. *Journal of Wind Engineering and Industrial Aerodynamics*, **90**, pp. 231-246.

Førsund, F.R., B. Singh, T. Jensen, and C. Larsen (2008). Phasing in wind-power in Norway: Network congestion and crowding-out of hydropower. *Energy Policy*, **36**, pp. 3514-3520.

Fox, B., D. Flynn, L. Bryans, N. Jenkins, D. Milborrow, M. O'Malley, R. Watson, and O. Anaya-Lara (2007). *Wind power integration: connection and system operational aspects*. The Institution of Engineering and Technology, London, UK, 288 pp.

Frandsen, S.T., R.J. Barthelmie, S.C. Pryor, O. Rathmann, S. Larsen, J. Højstrup, and M. Thøgersen (2006). Analytical modelling of wind speed deficit in large offshore wind farms. *Wind Energy*, **9**, pp. 39-53.

Frandsen, S.T., H.E. Jørgensen, R.J. Barthelmie, O. Rathmann, J. Badger, K. Hansen, S. Ott, P. Rethore, S.E. Larsen, and L.E. Jensen (2009). The making of a second-generation wind farm efficiency model complex. *Wind Energy*, **12**, pp. 445-458.

Frost, S.A., M.J. Balas, and A.D. Wright (2009). Direct adaptive control of a utility-scale wind turbine for speed regulation. *International Journal of Robust and Nonlinear Control*, **19**, pp. 59-71.

Garreaud, R.D., and M. Falvey (2009). The coastal winds off western subtropical South America in future climate scenarios. *International Journal of Climatology*, **29**, pp. 543-554.

Garthe, S., and O. Hüppop (2004). Scaling possible adverse effects of marine wind farms on seabirds: developing and applying a vulnerability index. *Journal of Applied Ecology*, **41**, pp. 724-734.

Gautam, D., V. Vittal, and T. Harbour (2009). Impact of increased penetration of DFIG-based wind turbine generators on transient and small signal stability of power systems. *IEEE Transactions on Power Systems*, **24**, pp. 1426-1434.

GE Energy (2010). *Western Wind and Solar Integration Study*. General Electric (GE) Energy, National Renewable Energy Laboratory, Golden, CO, USA, 536 pp.

GEC (2001). *WindPACT Turbine Design Scaling Studies Technical Area 3–Self-Erecting Tower Structures*. Global Energy Concepts (GEC), National Renewable Energy Laboratory, Golden, CO, USA, 72 pp.

Gibson, E., and P. Howsam (2010). The legal framework for offshore wind farms: A critical analysis of the consents process. *Energy Policy*, **38**, pp. 4692-4702.

Gipe, P. (1995). *Wind Energy Comes of Age*. Wiley, New York, USA, 560 pp.

Göransson, L., and F. Johnsson (2009). Dispatch modeling of a regional power generation system - Integrating wind power. *Renewable Energy*, **34**, pp. 1040-1049.

Goyal, M. (2010). Repowering--Next big thing in India. *Renewable and Sustainable Energy Reviews*, **14**, pp. 1400-1409.

Goyette, S., O. Brasseur, and M. Beniston (2003). Application of a new wind gust parameterization: Multiscale case studies performed with the Canadian regional climate model. *Journal of Geophysical Research*, **108**, pp. 4374.

Graham, J.B., J.R. Stephenson, and I.J. Smith (2009). Public perceptions of wind energy developments: Case studies from New Zealand. *Energy Policy*, **37**, pp. 3348-3357.

Greenblatt, J.B., S. Succar, D.C. Denkenberger, R.H. Williams, and R.H. Socolow (2007). Baseload wind energy: modeling the competition between gas turbines and compressed air energy storage for supplemental generation. *Energy Policy*, **35**, pp. 1474-1492.

Griffin, D.A. (2001). *WindPACT Turbine Design Scaling Studies Technical Area 1 – Composite Blades for 80-to 120-Meter Rotor*. National Renewable Energy Laboratory, Golden, CO, USA, 44 pp.

Gross, C. (2007). Community perspectives of wind energy in Australia: The application of a justice and community fairness framework to increase social acceptance. *Energy Policy*, **35**, pp. 2727-2736.

Gross, R., and P. Heptonstall (2008). The costs and impacts of intermittency: An ongoing debate: "East is East, and West is West, and never the twain shall meet". *Energy Policy*, **36**, pp. 4005-4007.

Gross, R., P. Heptonstall, D. Anderson, T. Green, M. Leach, and J. Skea (2006). *The Costs and Impacts of Intermittency: An assessment of the evidence on the costs and impacts of intermittent generation on the British electricity network*. Imperial College London, London, UK, 112 pp.

Gross, R., P. Heptonstall, M. Leach, D. Anderson, T. Green, and J. Skea (2007). Renewables and the grid: understanding intermittency. *Energy*, **160**, pp. 31-41.

Grubb, M.J. (1991). Value of variable sources on power systems. *IEE Proceedings C Generation, Transmission and Distribution*, **138**, pp. 149-165.

Grubb, M.J., and N.I. Meyer (1993). Wind energy: Resources, systems and regional strategies. In: *Renewable Energy: Sources for Fuels and Electricity*. T.B. Johansson, H. Kelly, A.K. Reddy, and R.H. Williams (eds.), Island Press, Washington, DC, USA, pp. 157-212.

GWEC (2010a). *Global Wind 2009 Report*. Global Wind Energy Council (GWEC), Brussels, Belgium, 4 pp.

GWEC (2010b). *Global wind power boom continues despite economic woes*. Global Wind Energy Council (GWEC), Brussels, Belgium, 68 pp.

GWEC and GPI (2010). *Global Wind Energy Outlook 2010*. Global Wind Energy Council (GWEC) and Greenpeace International (GPI), Brussels, Belgium, 60 pp.

Haggett, C. (2011). Understanding public responses to offshore wind power. *Energy Policy*, **39**, pp. 503-510.

Hameed, Z., S.H. Ahn, and Y.M. Cho (2010). Practical aspects of a condition monitoring system for a wind turbine with emphasis on its design, system architecture, testing and installation. *Renewable Energy*, **35**, pp. 879-894.

Hand, M.M., and M.J. Balas (2007). Blade load mitigation control design for a wind turbine operating in the path of vortices. *Wind Energy*, **10**, pp. 339-355.

Hand, M.M., D.A. Simms, L.J. Fingersh, D.W. Jager, J.R. Cotrell, S. Schreck, S.M. Larwood (2001). *Unsteady aerodynamics experiment phase vi: Wind tunnel test configurations and available data campaigns*. National Renewable Energy Laboratory, Golden, CO, USA, 310 pp.

Hansen, M.H. (2007). Aeroelastic instability problems for wind turbines. *Wind Energy*, **10**, pp. 551-577.

Hansen, M.O., J.N. Sørensen, S. Voutsinas, N. Sørensen, and H.A. Madsen (2006). State of the art in wind turbine aerodynamics and aeroelasticity. *Progress in Aerospace Sciences*, **42**, pp. 285-330.

Harris, M., M. Hand, and A. Wright (2006). *Lidar for Turbine Control*. National Renewable Energy Laboratory, Golden, CO, USA.

Hasche, B., A. Keane, and M. O'Malley (2010). Capacity credit of wind power: calculation and data requirements; The Irish power system case. *IEEE Transactions on Power Systems*, **99**, pp. 1-11.

Haugen, J.E., and T. Iversen (2008). Response in extremes of daily precipitation and wind from a downscaled multi-model ensemble of anthropogenic global climate change scenarios. *Tellus A*, **60**, pp. 411-426.

Heege, A., J. Betran, and Y. Radovcic (2007). Fatigue load computation of wind turbine gearboxes by coupled finite element, multi-body system and aerodynamic analysis. *Wind Energy*, **10**, pp. 395-413.

Hoen, B., R. Wiser, P. Cappers, M. Thayer, and G. Sethi (2011). Wind energy facilities and residential properties: The effect of proximity and view on sales prices. *Journal of Real Estate Research*, accepted for publication.

Hohmeyer, O., D. Mora, and F. Wetzig (2005). *Wind Energy. The Facts. Volume 4*. European Wind Energy Association, Brussels, Belgium, 60 pp.

Holttinen, H. (2005). Optimal electricity market for wind power. *Energy Policy*, **33**, pp. 2052-2063.

Holttinen, H., P. Meibom, A. Orths, F. van Hulle, B. Lange, A. Tiedemann, M. O'Malley, J. Perik, B. Ummels, J.O. Tande, A. Estanqueiro, M. Matos, E. Gomez, L. Soder, G. Strbac, A. Shakoor, J.C. Smith, and M. Milligan (2009). *Design and Operation of Power Systems with Large Amounts of Wind Power: Phase One 2006-2008*. VTT Technical Research Centre of Finland, Espoo, Finland, 200 pp.

Hoogwijk, M., and W. Graus (2008). *Global Potential of Renewable Energy Sources: A Literature Assessment*. Ecofys, Utrecht, The Netherlands, 45 pp.

Hoogwijk, M., B. de Vries, and W. Turkenburg (2004). Assessment of the global and regional geographical, technical and economic potential of onshore wind energy. *Energy Economics*, **26**, pp. 889-919.

Hoogwijk, M., D. van Vuuren, B. de Vries, and W. Turkenburg (2007). Exploring the impact on cost and electricity production of high penetration levels of intermittent electricity in OECD Europe and the USA, results for wind energy. *Energy*, **32**, pp. 1381-1402.

Hoppock, D.C., and D. Patiño-Echeverri (2010). Cost of wind energy: comparing distant wind resources to local resources in the midwestern United States. *Environmental Science & Technology*, **44**, pp. 8758-8765.

Horn, J.W., E.B. Arnett, and T.H. Kunz (2008). Behavioral responses of bats to operating wind turbines. *Journal of Wildlife Management*, **72**, pp. 123-133.

Hubbert, M.K. (1971). The energy resources of the Earth. *Scientific American*, **225**, pp. 60-70.

Hydro-Quebec TransEnergie (2006). *Technical Requirements for the Connection of Generation Facilities to the Hydro-Quebec Transmission System*. Hydro-Quebec TransEnergie, Montreal, Canada, 17 pp.

IEA (2008). *Energy Technology Perspectives 2008. Scenarios and Strategies to 2050*. International Energy Agency, Paris, France, 646 pp.

IEA (2009). *Technology Roadmap – Wind energy*. International Energy Agency, Paris, France, 52 pp.

IEA (2010a). *Energy Balances of Non-OECD Countries*. International Energy Agency, Paris, France, 554 pp.

IEA (2010b). *World Energy Outlook 2010*. International Energy Agency, Paris, France, 736 pp.

IEA (2010c). *Energy Technology Perspectives 2010. Scenarios and Strategies to 2050*. International Energy Agency, Paris, France, 708 pp.

IEA and OECD (2010). *Projected Costs of Generating Electricity, 2010 Edition*. International Energy Agency and Organisation for Economic Co-operation and Development, Paris, France, 218 pp.

IEA Wind (2010). *IEA Wind Energy Annual Report 2009*. International Energy Agency Wind, Paris, France, 172 pp.

IEC (2008). *Wind Turbines – Part 21: Measurement and Assessment of Power Quality Characteristics of Grid Connected Wind Turbines; IEC 61400-21*. International Electrotechnical Commission, Delft, The Netherlands.

IEC (2010). *Wind Turbines - Part 22: Conformity Testing and Certification, IEC 61400-22*. International Electrotechnical Commission, Delft, The Netherlands.

IPCC (2007). *Climate Change 2007: Mitigation of Climate Change. Contribution of Working Group III to the Fourth Assessment Report of the Intergovernmental Panel on Climate Change*. B. Metz, O.R. Davidson, P.R. Bosch, R. Dave, and L.A. Meyer (eds.), Cambridge University Press, 851 pp.

ISO (2006). *Environmental Management - Life-Cycle Assessment - Requirements and Guidelines*. Geneva, Switzerland: International Organization for Standardization.

Jackson, K.J., M.D. Zuteck, C.P. van Dam, K.J. Standish, and D. Berry (2005). Innovative design approaches for large wind turbine blades. *Wind Energy*, **8**, pp. 141-171.

Jackson, T.O. (2001). The effects of environmental contamination on real estate: A literature review. *Journal of Real Estate Literature*, **9**, pp. 91-116.

Jacobson, M.Z. (2009). Review of solutions to global warming, air pollution, and energy security. *Energy & Environmental Science*, **2**, pp. 148-173.

Jakobsen, J. (2005). Infrasound emission from wind turbines. *Low Frequency Noise, Vibration and Active Control*, **24**, pp. 145-155.

Jamasb, T. (2007). Technical change theory and learning curves: patterns of progress in electricity generation technologies. *The Energy Journal*, **28**, pp. 51–71.

JCSP (2009). *Joint Coordinated System Plan 2008, Volume 1 - Economic Assessment*. Joint Coordinated System Plan (JCSP), Report of the Midwest Independent System Operator, PJM Interconnection, Southwest Power Pool Electric Energy Network, Tennessee Valley Authority and Mid-continent Area Power Pool, 113 pp.

Jiang, Q., J.D. Doyle, T. Haack, M.J. Dvorak, C.L. Archer, and M.Z. Jacobson (2008). Exploring wind energy potential off the California coast. *Geophysical Research Letters*, **35**, L20819.

Johansson, T.B., K. McCormick, L. Neij, and W. Turkenburg (2004). The Potentials of Renewable Energy. Thematic Background Paper for the *International Conference for Renewable Energies*, Bonn, Germany, 36 pp.

Johnson, K.E., and L.J. Fingersh (2008). Adaptive pitch control of variable-speed wind turbines. *Journal of Solar Energy Engineering*, **130**, pp. 031012-7.

Johnson, K.E., L.J. Fingersh, M.J. Balas, and L.Y. Pao (2004). Methods for increasing region 2 power capture on a variable-speed wind turbine. *Journal of Solar Energy Engineering*, **126**, pp. 1092-1100.

Jones, C.R., and J.R. Eiser (2009). Identifying predictors of attitudes towards local onshore wind development with reference to an English case study. *Energy Policy*, **37**, pp. 4604-4614.

Jonkman, J.M. (2009). Dynamics of offshore floating wind turbines-model development and verification. *Wind Energy*, **12**, pp. 459-492.

Jónsson. T., P. Pinson, and H. Madsen (2010). On the market impact of wind energy forecasts. *Energy Economics*, **32**, pp. 313-320.

Jungbluth, N., C. Bauer, R. Dones, and R. Frischknecht (2005). Life cycle assessment for emerging technologies: Case studies for photovoltaic and wind power. *The International Journal of Life Cycle Assessment*, **10**, pp. 24-34.

Junginger, M., A. Faaij, and W.C. Turkenburg (2004). Cost reduction prospects for offshore wind farms. *Wind Engineering*, **28**, pp. 97-118.

Junginger, M., A. Faaij, and W.C. Turkenburg (2005). Global experience curves for wind farms. *Energy Policy*, **33**, pp. 133-150.

Junginger, M., W.V. Sark, and A. Faaij (eds.) (2010). *Technological Learning in the Energy Sector: Lessons for Policy, Industry and Science*. Edward Elgar, Northampton, MA, USA.

Justus, C.G., K. Mani, and A.S. Mikhail (1979). Interannual and month-to-month variations of wind speed. *Journal of Applied Meteorology*, **18**, pp. 913-920.

Kahn, E. (1979). The compatibility of wind and solar technology with conventional energy systems. *Annual Review of Energy*, **4**, pp. 313-352.

Kahouli-Brahmi, S. (2009). Testing for the presence of some features of increasing returns to adoption factors in energy system dynamics: An analysis via the learning curve approach. *Ecological Economics*, **68**, pp. 1195-1212.

Kannen, A., and B. Burkhard (2009). Integrated assessment of coastal and marine changes using the example of offshore wind farms: The coastal futures approach. *GAIA - Ecological Perspectives for Science and Society*, **18**, pp. 229-238.

Katzenstein, W., and J. Apt (2009). Air emissions due to wind and solar power. *Environmental Science & Technology*, **43**, pp. 253-258.

Katzenstein, W., E. Fertig, and J. Apt (2010). The variability of interconnected wind plants. *Energy*, **38**, pp. 4400-4410.

Keith, D.W., J.F. DeCarolis, D.C. Denkenberger, D.H. Lenschow, S.L. Malyshev, S. Pacala, and P. J. Rasch (2004). The influence of large-scale wind power on global climate. *Proceedings of the National Academy of Sciences*, **101**, pp. 16115-16120.

Kelley, N., M. Shirazi, D. Jager, S. Wilde, J. Adams, M. Buhl, P. Sullivan, and E. Patton (2004). *Lamar Low-Level Jet Program Interim Report*. National Renewable Energy Laboratory, Golden, CO, USA, 216 pp.

Kempton, W., and J. Tomic (2005). Vehicle-to-grid power implementation: From stabilizing the grid to supporting large-scale renewable energy. *Journal of Power Sources*, **144**, pp. 280-294.

Kempton, W., C.L. Archer, A. Dhanju, R.W. Garvine, and M.Z. Jacobson (2007). Large CO_2 reductions via offshore wind power matched to inherent storage in energy end-uses. *Geophysical Research Letters*, **34**, L02817.

Kikuchi, R. (2010). Risk formulation for the sonic effects of offshore wind farms on fish in the EU region. *Marine Pollution Bulletin*, **60**, pp. 172-177.

Kim, J., and C. Park (2010). Wind power generation with a parawing on ships, a proposal. *Energy*, **35**, pp. 1425-1432.

Kirby, B., and M. Milligan (2008). An examination of capacity and ramping impacts of wind energy on power systems. *The Electricity Journal*, **21**, pp. 30-42.

Kirk-Davidoff, D.B., and D.W. Keith (2008). On the climate impact of surface roughness anomalies. *Journal of the Atmospheric Sciences*, **65**, pp. 2215-2234.

Kiviluoma, J., and P. Meibom (2010). Influence of wind power, plug-in electric vehicles, and heat storages on power system investments. *Energy*, **35**, pp. 1244-1255.

Klaassen, G., A. Miketa, K. Larsen, and T. Sundqvist (2005). The impact of R&D on innovation for wind energy in Denmark, Germany and the United Kingdom. *Ecological Economics*, **54**, pp. 227-240.

Klick, H., and E.R.A.N. Smith (2010). Public understanding of and support for wind power in the United States. *Renewable Energy*, **35**, pp. 1585-1591.

Kobos, P.H., J.D. Erickson, and T.E. Drennen (2006). Technological learning and renewable energy costs: implications for US renewable energy policy. *Energy Policy*, **34**, pp. 1645-1658.

Köller, J., J. Köppel, and W. Peters (eds.) (2006). *Offshore Wind Energy: Research on Environmental Impacts*. Springer, Berlin, Germany, 371 pp.

Krewitt, W., and B. Schlomann (2006). *Externe Kosten der Stromerzeugung aus erneuerbaren Energien im Vergleich zur Stromerzeugung aus fossilen Energieträgern*. Fraunhofer Institute for Systems and Innovation Research and DLR (German Center for Aeronautics and Astronautics), Karlsruhe and Stuttgart, Germany, 59 pp.

Krewitt, W., K. Nienhaus, C. Kleßmann, C. Capone, E. Stricker, W. Graus, M. Hoogwijk, N. Supersberger, U. von Winterfeld, and S. Samadi (2009). *Role and Potential of Renewable Energy and Energy Efficiency for Global Energy Supply*. Climate Change 18/2009, ISSN 1862-4359, Federal Environment Agency, Dessau-Roßlau, Germany, 336 pp.

Krey, V., and L. Clarke (2011). Role of renewable energy in climate change mitigation: a synthesis of recent scenarios. *Climate Policy*, **11**, pp. 1-28.

Krohn, S., and S. Damborg (1999). On public attitudes towards wind power. *Renewable Energy*, **16**, pp. 954-960.

Krug, F., and B. Lewke (2009). Electromagnetic interference on large wind turbines. *Energies*, **2**, pp. 1118-1129.

Kunz, T.H., E.B. Arnett, B.M. Cooper, W.P. Erickson, R.P. Larkin, T. Mabee, M.L. Morrison, M.D. Strickland, and J. M. Szewczak (2007a). Assessing impacts of wind-energy development on nocturnally active birds and bats: a guidance document. *Journal of Wildlife Management*, **71**, pp. 2449-2486.

Kunz, T.H., E.B. Arnett, W.P. Erickson, A.R. Hoar, G.D. Johnson, R.P. Larkin, M.D. Strickland, R.W. Thresher, and M.D. Tuttle (2007b). Ecological impacts of wind energy development on bats: questions, research needs, and hypotheses. *Frontiers in Ecology and the Environment*, **5**, pp. 315-324.

Kuvlesky, W.P., L.A. Brennan, M.L. Morrison, K.K. Boydston, B.M. Ballard, and F.C. Bryant (2007). Wind energy development and wildlife conservation: challenges and opportunities. *Journal of Wildlife Management*, **71**, pp. 2487-2498.

Laakso, T., H. Holttinen, G. Ronsten, L. Tallhaug, R. Horbaty, I. Baring-Gould, A. Lacroix, E. Peltola, and B. Tammelin (2003). *State-of-the-art of Wind Energy in Cold Climates*. VTT Technical Research Centre of Finland, Espoo, Finland, 50 pp.

Lackner, M., and G. van Kuik (2010). A comparison of smart rotor control approaches using trailing edge flaps and individual pitch control. *Wind Energy*, **13**, pp. 117-134.

Ladenburg, J. (2009). Stated public preferences for on land and offshore wind power generation—a review. *Wind Energy*, **12**, pp. 171-181.

Lalor, G., A. Mullane, and M. O'Malley (2005). Frequency control and wind turbine technologies. *IEEE Transactions on Power Systems*, **20**, pp. 1905-1913.

Lamont, A.D. (2008). Assessing the long-term system value of intermittent electric generation technologies. *Energy Economics*, **30**, pp. 1208-1231.

Lanier, M. (2005). *Low Wind Speed Technology Phase I Concept Study: Evaluation of Design and Construction Approaches for Economical Hybrid Steel/Concrete Wind Turbine Towers*. National Renewable Energy Laboratory, Golden, CO, USA, 698 pp.

Larsen, G.C., H.A. Madsen, K. Thomsen, and T.J. Larsen (2008). Wake meandering: a pragmatic approach. *Wind Energy*, **11**, pp. 377-395.

Larsen, T.J., H.A. Madsen, and K. Thomsen (2004). Active load reduction using individual pitch, based on local blade flow measurements. *Wind Energy*, **8**, pp. 67-80.

Leckebusch, G.C., A. Weimer, J.G. Pinto, M. Reyers, and P. Speth (2008). Extreme wind storms over Europe in present and future climate: a cluster analysis approach. *Meteorologische Zeitschrift*, **17**, pp. 67-82.

Leighty, W. (2008). Running the world on renewables: Hydrogen transmission pipelines and firming geologic storage. *International Journal of Energy Research*, **32**, pp. 408-426.

Leishman, J.G. (2006). *Principles of Helicopter Aerodynamics*. Cambridge University Press, 536 pp.

Lemming, J.K., P.E. Morthorst, N.E. Clausen, and P. Hjuler Jensen (2009). *Contribution to the Chapter on Wind Power in Energy Technology Perspectives 2008, IEA*. Risø National Laboratory, Roskilde, Denmark, 64 pp.

Leutz, R., T. Ackermann, A. Suzuki, A. Akisawa, and T. Kashiwagi (2001). Technical offshore wind energy potentials around the globe. In: *European Wind Energy Conference and Exhibition*. Copenhagen, Denmark, 2-6 July 2001, pp. 2-6.

Leventhall, G. (2006). Infrasound from wind turbines-fact, fiction or deception. *Canadian Acoustics*, **34**, pp. 29.

Lew, D.J., R.H. Williams, X. Shaoxiong, and Z. Shihui (1998). Large-scale baseload wind power in China. *Natural Resources Forum*, **22**(3), 165-184.

Lewis, J.I. (2007). Technology acquisition and innovation in the developing world: Wind turbine development in China and India. *Studies in Comparative International Development (SCID)*, **42**, pp. 208-232.

Lewis, J.I., and R.H. Wiser (2007). Fostering a renewable energy technology industry: An international comparison of wind industry policy support mechanisms. *Energy Policy*, **35**, pp. 1844-1857.

Li, J. (2010). Decarbonising power generation in China – Is the answer blowing in the wind? *Renewable and Sustainable Energy Reviews*, **14**, pp. 1154-1171.

Li, J., and L. Ma (2009). *Background Paper: Chinese Renewables Status Report*. Renewable Energy Policy Network for the 21st Century, Paris, France, 95 pp.

Li, J., H. Gao, J. Shi, L. Ma, H. Qin, and Y. Song (2007). *China Wind Power Report, 2007*. China Environmental Science Press, Beijing, China, 55 pp.

Lilley, M.B., and J. Firestone (2008). Wind power, wildlife, and the Migratory Bird Treaty Act: a way forward. *Environmental Law*, **38**, pp. 1167-1214.

Liu, Y., and A. Kokko (2010). Wind power in China: Policy and development challenges. *Energy Policy*, **38**, pp. 5520-5529.

Loring, J.M. (2007). Wind energy planning in England, Wales and Denmark: Factors influencing project success. *Energy Policy*, **35**, pp. 2648-2660.

Lu, X., M.B. McElroy, and J. Kiviluoma (2009). Global potential for wind-generated electricity. *Proceedings of the National Academy of Sciences*, **106**, pp. 10933-10939.

Lund, H. (2006). Large-scale integration of optimal combinations of PV, wind and wave power into the electricity supply. *Renewable Energy*, **31**, pp. 503-515.

Lund, H., and W. Kempton (2008). Integration of renewable energy into the transport and electricity sectors through V2G. *Energy Policy*, **36**, pp. 3578-3587.

Lund, H., and E. Münster (2003). Modelling of energy systems with a high percentage of CHP and wind power. *Renewable Energy*, **28**, pp. 2179-2193.

Lutz, T., A. Herrig, W. Wörz, M. Kamruzzaman, and E. Krämer (2007). Design and wind-tunnel verification of low-noise airfoils for wind turbines. *AIAA Journal*, **45**, pp. 779-785.

MacCormack, J., A. Hollis, H. Zareipour, and W. Rosehart (2010). The large-scale integration of wind generation: Impacts on price, reliability and dispatchable conventional suppliers. *Energy Policy*, **38**, pp. 3837-3846.

Mackay, R.M., and S.D. Probert (1998). Likely market-penetrations of renewable-energy technologies. *Applied Energy*, **59**, pp. 1-38.

Madsen, H., P. Pinson, G. Kariniotakis, H.A. Nielsen, and T. Nielsen (2005). Standardizing the performance evaluation of short term wind power prediction models. *Wind Engineering*, **29**, pp. 475-489.

Madsen, H.A., C. Bak, U.S. Paulsen, M. Gaunaa, N.N. Sørensen, P. Fuglsang, J. Romblad, N.A. Olsen, P. Enevoldsen, J. Laursen, and L. Jensen (2010). The DAN-AERO MW experiments. In: *48th AIAA Aerospace Sciences Meeting*. Orlando, FL, USA, 4-7 January 2010, pp. 2010-0645.

Madsen, P.T., M. Wahlberg, J. Tougaard, K. Lucke, and P. Tyack (2006). Wind turbine underwater noise and marine mammals: implications of current knowledge and data needs. *Marine Ecology-Progress Series*, **309**, pp. 279-295.

Malcolm, D.J. (2004). *WindPACT Rotor Design. Period of performance: June 29, 2000 - February 28, 2004*. National Renewable Energy Laboratory, Golden, CO, USA, 44 pp.

Malcolm, D.J., and A.C. Hansen (2006). *WindPACT turbine rotor design study: June 2000 - June 2002*. National Renewable Energy Laboratory, Golden, CO, USA, 84 pp.

Markard, J., and R. Petersen (2009). The offshore trend: Structural changes in the wind power sector. *Energy Policy*, **37**, pp. 3545-3556.

Martinot, E., C. Dienst, L. Weiliang, and C. Qimin (2007). Renewable energy futures: Targets, scenarios, and pathways. *Annual Review of Environment and Resources*, **32**, pp. 205-239.

Matthies, H.G., A.D. Garrad, M. Scherweit, C. Nath, M.A. Wastling, T. Siebers, T. Schellin, and D.C. Quarton (1995). *Study of Offshore Wind Energy in the European Community*. Germanischer Lloyd, Hamburg, Germany.

McCunney, R.J., and J. Meyer (2007). Occupational exposure to noise. In: *Environmental and Occupational Medicine*. 4th ed. W.N. Rom (ed.) Lippincott Williams and Wilkins, Baltimore, MD, pp. 1295-1238.

McDonald, A., and L. Schrattenholzer (2001). Learning rates for energy technologies. *Energy Policy*, **29**, pp. 255-261.

McElroy, M.B., X. Lu, C.P. Nielsen, and Y. Wang (2009). Potential for wind-generated electricity in China. *Science*, **325**, pp. 1378.

MDOH (2009). *Public Health Impacts of Wind Turbines*. Minnesota Department of Health (MDOH), Environmental Health Division, Minneapolis, MN, USA, 32 pp.

Meibom, P., C. Weber, R. Barth, and H. Brand (2009). Operational costs induced by fluctuating wind power production in Germany and Scandinavia. *IET Renewable Power Generation*, **3**, pp. 75-83.

Meyer, N. (2007). Learning from wind energy policy in the EU: lessons from Denmark, Sweden and Spain. *European Environment*, **17**, pp. 347-362.

Miao, Z., L. Fan, D. Osborn, and S. Yuvarajan (2009). Control of DFIG-based wind generation to improve interarea oscillation damping. *IEEE Transactions on Energy Conversion*, **24**, pp. 415-422.

Michel, J., H. Dunagan, C. Boring, E. Healy, W. Evans, J.M. Dean, A. McGillis, and J. Hain (2007). *Worldwide Synthesis and Analysis of Existing Information Regarding Environmental Effects of Alternative Energy Uses on the Outer Continental Shelf*. US Department of the Interior, Minerals Management Service, Herndon, VA, USA.

Migliore, P., and S. Oerlemans (2004). Wind tunnel aeroacoustic tests of six airfoils for use on small wind turbines. *Journal of Solar Energy Engineering*, **126**, pp. 974-985.

Miketa, A., and L. Schrattenholzer (2004). Experiments with a methodology to model the role of R&D expenditures in energy technology learning processes; first results. *Energy Policy*, **32**, pp. 1679-1692.

Milborrow, D. (2009). Quantifying the impacts of wind variability. *Proceedings of the Institution of Civil Engineers. Energy*, **162**, pp. 105-111.

Milborrow, D. (2010). Annual power costs comparison: What a difference a year can make. *Windpower Monthly*, **26**, pp. 41-47.

Milligan, M., and B. Kirby (2008). The impact of balancing area size and ramping requirements on wind integration. *Wind Engineering*, **32**, pp. 379-398.

Milligan, M., D.J. Lew, D. Corbus, P. Piwko, N. Miller, K. Clark, G. Jordan, L. Freeman, B. Zavadil, and M. Schuerger (2009). Large-Scale Wind Integration Studies in the United States: Preliminary Results. In: *8th International Workshop on Large Scale Integration of Wind Power and on Transmission Networks for Offshore Wind Farms*, Bremen, Germany, 14-15 October 2009, pp. 1-8.

Mills, A., R. Wiser, and K. Porter (2009a). *The Cost of Transmission for Wind Energy: A Review of Transmission Planning Studies*. Lawrence Berkeley National Laboratory, Berkeley, CA, USA, 66 pp.

Mills, A., R. Wiser, M. Milligan, and M. O'Malley (2009b). Comment on "Air Emissions Due to Wind and Solar Power". *Environmental Science & Technology*, **43**, pp. 6106-6107.

Miranda, M.S., and D. Infield (2002). A wind-powered seawater reverse-osmosis system without batteries. *Desalination*, **153**, pp. 9-16.

MOE (2009). *Development of Noise Setbacks for Wind Farms. Proposed Content for the Renewable Energy Approval Regulation - Section 47.3(1) - under the Environmental Protection Act*. Ontario Ministry of the Environment (MOE), Toronto, Ontario, Canada, 8 pp.

Morales, J.M., A.J. Conejo, and J. Perez-Ruiz (2011). Simulating the impact of wind production on locational marginal prices. *IEEE Transactions on Power Systems*, **26**(2), pp. 820-828.

Moriarty, P., and P. Migliore (2003). *Semi-Empirical Aeroacoustic Noise Prediction Code for Wind Turbines*. National Renewable Energy Laboratory, Golden, CO, USA, 39 pp.

Morren, J., S.W.H. de Haan, W.L. Kling, and J. Ferreira (2006). Wind turbines emulating inertia and supporting primary frequency control. *IEEE Transactions on Power Systems*, **21**, pp. 433-434.

Morthorst, P.E. (2003). Wind power and the conditions at a liberalized power market. *Wind Energy*, **6**, pp. 297-308.

Mostafaeipour, A. (2010). Feasibility study of offshore wind turbine installation in Iran compared with the world. *Renewable and Sustainable Energy Reviews*, **14**, pp. 1722-1743.

Mukora, A., M. Winskel, H.F. Jeffrey, and M. Mueller (2009). Learning curves for emerging energy technologies. *Proceedings of the Institution of Civil Engineers – Energy*, **162**, pp. 151-159.

Mullane, A., and M. O'Malley (2005). The inertial response of induction-machine-based wind turbines. *IEEE Transactions on Power Systems*, **20**, pp. 1496-1503.

Murawski, S.A. (2007). Ten myths concerning ecosystem approaches to marine resource management. *Marine Policy*, **31**, pp. 681-690.

Musgrove, P. (2010). *Wind Power*. Cambridge University Press, Cambridge, UK, 338 pp.

Musial, W. (2007). Offshore wind electricity: a viable energy option for the coastal United States. *Marine Technology Society Journal*, **41**, pp. 32-43.

Musial, W., and B. Ram (2010). *Large-Scale Offshore Wind Power in the United States: Assessment of Opportunities and Barriers*. National Renewable Energy Laboratory, Golden, CO, USA, 240 pp.

Naaijen, P., and V. Koster (2007). Performance of auxiliary wind propulsion for merchant ships using a kite. In: *2nd International Conference on Marine Research and Transportation*, Naples, Italy, 28-30 June 2007, pp. 45-53.

NABCI (2010). *The State of the Birds: 2010 Report on Climate Change*. North American Bird Conservation Initiative (NABCI), US Committee, US Department of the Interior, Washington, DC, USA.

Nadaï, A., and O. Labussière (2009). Wind power planning in France (Aveyron), from state regulation to local planning. *Land Use Policy*, **26**, pp. 744-754.

Neij, L. (1997). Use of experience curves to analyse the prospects for diffusion and adoption of renewable energy technology. *Energy Policy*, **25**, pp. 1099-1107.

Neij, L. (1999). Cost dynamics of wind power. *Energy*, **24**, pp. 375-389.

Neij, L. (2008). Cost development of future technologies for power generation – A study based on experience curves and complementary bottom-up assessments. *Energy Policy*, **36**, pp. 2200-2211.

Nemet, G.F. (2009). Interim monitoring of cost dynamics for publicly supported energy technologies. *Energy Policy*, **37**, pp. 825-835.

NERC (2009). *Accommodating High Levels of Variable Generation*. North American Electric Reliability Corporation (NERC), Princeton, NJ, USA, 104 pp.

Neuhoff, K., A. Ehrenmann, L. Butler, J. Cust, H. Hoexter, K. Keats, A. Kreczko, and G. Sinden (2008). Space and time: Wind in an investment planning model. *Energy Economics*, **30**, pp. 1990-2008.

NHMRC (2010). *Wind Turbines and Health: A Rapid Review of the Evidence*. National Health and Medical Research Council (NHMRC) of the Australian Government, Canberra, Australia, 11 pp.

Nielsen, F.G., K. Argyriadis, N. Fonseca, M. Le Boulluec, P. Liu, H. Suzuki, J. Sirkar, N.J. Tarp-Johansen, S.R. Turnock, J. Waegter, and Z. Zong (2009). Specialist Committee V.4: Ocean, wind and wave energy utilization. In: *17th International Ship and Offshore Structures Congress*, Seoul, Korea, 16-21 August 2009, pp. 201-257.

Nielsen, S.R. (1996). Wind energy planning in Denmark. *Renewable Energy*, **9**, pp. 776-771.

Nielson, P., J.K. Lemming, P.E. Morthorst, H. Lawetz, E.A. James-Smith, N.E. Clausen, S. Strøm, J. Larsen, N.C. Bang, and H.H. Lindboe (2010). *The Economics of Wind Turbines*. EMD International, Aalborg, Denmark, 86 pp.

Nikolaev, V.G., S.V. Ganaga, and K.I. Kudriashov (2008). *National Kadastr of Wind Resources of Russia and Methodological Grounds for their Determination*. Atmograph, Moscow, Russia, 590 pp.

Nikolaev, V.G., S.V. Ganaga, K.I. Kudriashov, R. Walter, P. Willems, and A. Sankovsky (2010). *Prospects of Development of Renewable Power Sources in Russian Federation.The results of TACIS project*. Europe Aid/116951/C/SV/RU. Atmograph, Moscow, Russia.

NRC (2007). *Environmental Impacts of Wind-Energy Projects*. National Research Council, The National Academy Press, Washington, DC, USA, 394 pp.

NRC (2010a). *Hidden Costs of Energy: Unpriced Consequences of Energy Production and Use*. National Research Council, The National Academy Press, Washington, DC, USA, 506 pp.

NRC (2010b). *Electricity from Renewable Resources: Status, Prospects, and Impediments*. National Research Council, The National Academy Press, Washington, DC, USA, 388 pp.

O'Rourke, R. (2006). *Navy Ship Propulsion Technologies: Options for Reducing Oil Use –Background for Congress*. Congressional Research Service, Washington, DC, USA, 41 pp.

Owen, A.D. (2004). Environmental externalities, market distortions and the economics of renewable energy technologies. *The Energy Journal*, **25**, pp. 127-156.

Palutikof, J.P., X. Guo, and J.A. Halliday (1992). Climate variability and the UK wind resource. *Journal of Wind Engineering and Industrial Aerodynamics*, **39**, pp. 243-249.

Palutikof, J.P., P.M. Kelly, T.D. Davies, and J.A. Halliday (1987). Impacts of spatial and temporal windspeed variability on wind energy output. *Journal of Climate and Applied Meteorology*, **26**, pp. 1124-1133.

Pasqualetti, M.J., P. Gipe, R.W. Righter (2002). *Wind Power in View: Energy Landscapes in a Crowded World*. Academic Press, San Diego, CA, USA, 248 pp.

Pasqualetti, M.J., R. Richter, and P. Gipe (2004). History of wind energy. In: *Encyclopaedia of Energy*. C.J. Cleveland (ed.), Elsevier, Amsterdam, the Netherlands, pp. 419-433.

Passon, P.,and M. Kühn (2005). State-of-the-art and development needs of simulation codes for offshore wind turbines. In: *Copenhagen Offshore Wind Conference*, Copenhagen, Denmark, 26-28 October, 2005, pp. 1-12.

Pedersen, E., and K. Waye (2004). Perception and annoyance due to wind turbine noise – a dose-response relationship. *The Journal of the Acoustical Society of America*, **116**, pp. 3460.

Pedersen, E., and K.P. Waye (2007). Wind turbine noise, annoyance and self-reported health and well-being in different living environments. *Occupational and Environmental Medicine*, **64**, pp. 480-486.

Pedersen, E., and K.P. Waye (2008). Wind turbines – low level noise sources interfering with restoration? *Environmental Research Letters*, **3**, pp. 1-5.

Pedersen, E., F. van den Berg, R. Bakker, and J. Bouma (2010). Can road traffic mask sound from wind turbines? Response to wind turbine sound at different levels of road traffic sound. *Energy Policy* **38**, pp. 2520-2527.

Peeters, J.L.M., D. Vandepitte, and P. Sas (2006). Analysis of internal drive train dynamics in a wind turbine. *Wind Energy*, **9**, pp. 141-161.

Pehnt, M., M. Oeser, and D.J. Swider (2008). Consequential environmental system analysis of expected offshore wind electricity production in Germany. *Energy*, **33**, pp. 747-759.

Petersen, E.L., N.G. Mortensen, L. Landberg, J. Højstrup, and H.P. Frank (1998). Wind power meteorology. Part I: Climate and turbulence. *Wind Energy*, **1**, pp. 2-22.

M. Pettersson, K. Ek, K. Söderholm, and P. Söderholm (2010). Wind power planning and permitting: Comparative perspectives from the Nordic countries. *Renewable and Sustainable Energy Reviews*, **14**, pp. 3116-3123.

Poore, R., and T. Lettenmaier (2003). *Alternative Design Study Report: WindPACT Advanced Wind Turbine Drive Train Designs Study*. National Renewable Energy Laboratory, Golden, CO, USA, 556 pp.

Prospathopoulos, J.M., and S.G. Voutsinas (2005). Noise propagation issues in wind energy applications. *Journal of Solar Energy Engineering*, **127**, pp. 234-241.

Pryor, S.C., and R.J. Barthelmie (2010). Climate change impacts on wind energy: A review. *Renewable and Sustainable Energy Reviews*, **14**, pp. 430-437.

Pryor, S.C., and J. Ledolter (2010). Addendum to: Wind speed trends over the contiguous USA. *Journal of Geophysical Research*, **115**, D10103.

Pryor, S.C., and J.T. Schoof (2010). Importance of the SRES in projections of climate change impacts on near-surface wind regimes. *Meteorologische Zeitschrift*, **19**, pp. 267-274.

Pryor, S.C., R.J. Barthelmie, and E. Kjellström (2005). Potential climate change impact on wind energy resources in northern Europe: analyses using a regional climate model. *Climate Dynamics*, **25**, pp. 815-835.

Pryor, S.C., J.T. Schoof, and R.J. Barthelmie (2006). Winds of change? Projections of near-surface winds under climate change scenarios. *Geophysical Research Letters*, **33**, L11702.

Pryor, S.C., R.J. Barthelmie, D.T. Young, E.S. Takle, R.W. Arritt, D. Flory, W. Gutowski Jr., A. Nunes, and J. Roads (2009). Wind speed trends over the contiguous United States. *Journal of Geophysical Research – Atmospheres*, **114**, D14105.

Punt, M.J., R.A. Groeneveld, E.C. van Ierland, and J.H. Stel (2009). Spatial planning of offshore wind farms: A windfall to marine environmental protection? *Ecological Economics*, **69**, pp. 93-103.

Purohit, P. (2007). Financial evaluation of renewable energy technologies for irrigation water pumping in India. *Energy Policy*, **35**, pp. 3134-3144.

Quarton, D.C. (1998). The evolution of wind turbine design analysis – a twenty year progress review. *Wind Energy*, **1**, pp. 5-24.

Rahimzadeh, F., A. Noorian, M. Pedram, and M. Kruk (2011). Wind speed variability over Iran and its impact on wind power potential: A case study of Esfehan Province. *Meteorological Applications*, doi:10.1002/met.229.

Rasmussen, F., M.H. Hansen, K. Thomsen, T.J. Larsen, F. Bertagnolio, J. Johansen, H.A. Madsen, C. Bak, and A.M. Hansen (2003). Present status of aeroelasticity of wind turbines. *Wind Energy*, **6**, pp. 213-228.

REN21 (2010). *Renewables 2010: Global Status Report*. Renewable Energy Policy Network for the 21st Century Secretariat, Paris, France, 80 pp.

Rivier Abbad, J. (2010). Electricity market participation of wind farms: the success story of the Spanish pragmatism. *Energy Policy*, **38**, pp. 3174-3179.

Riziotis, V.A., S.G. Voutsinas, E.S. Politis, and P.K. Chaviaropoulos (2004). Aeroelastic stability of wind turbines: the problem, the methods and the issues. *Wind Energy*, **7**, pp. 373-392.

Roberts, B.W., D.H. Shepard, K. Caldeira, M.E. Cannon, D.G. Eccles, A.J. Grenier, and J.F. Freidin (2007). Harnessing high-altitude wind power. *IEEE Transactions on Energy Conversion*, **22**, pp. 136-144.

B. Rockel, and K. Woth (2007). Extremes of near-surface wind speed over Europe and their future changes as estimated from an ensemble of RCM simulations. *Climatic Change*, **81**, pp. 267-280.

Rogner, H.-H., F. Barthel, M. Cabrera, A. Faaij, M. Giroux, D. Hall, V. Kagramanian, S. Kononov, T. Lefevre, R. Moreira, R. Nötstaller, P. Odell, M. Taylor (2000). Energy resources. In: *World Energy Assessment. Energy and the Challenge of Sustainability*. United Nations Development Programme, United Nations Department of Economic and Social Affairs, and World Energy Council, New York, NY, USA, pp. 135-171.

Sáenz de Miera, G., P. del Río González, and I. Vizcaíno (2008). Analysing the impact of renewable electricity support schemes on power prices: The case of wind electricity in Spain. *Energy Policy*, **36**, pp. 3345-3359.

Sailor, D.J., M. Smith, and M. Hart (2008). Climate change implications for wind power resources in the Northwest United States. *Renewable Energy*, **33**, pp. 2393-2406.

Schreck, S.J., and M.C. Robinson (2003). Boundary layer state and flow field structure underlying rotational augmentation of blade aerodynamic response. *Journal of Solar Energy Engineering*, **125**, pp. 448-457.

Schreck, S.J., M.C. Robinson, M.M. Hand, and D.A. Simms (2000). HAWT dynamic stall response asymmetries under yawed flow conditions. *Wind Energy*, **3**, pp. 215-232.

Schreck, S.J., M.C. Robinson, M.M. Hand, and D.A. Simms (2001). Blade dynamic stall vortex kinematics for a horizontal axis wind turbine in yawed conditions. *Journal of Solar Energy Engineering*, **123**, p. 272.

Schreck, S.J., J. Lundquist, and W. Shaw (2008). *U.S. Department of Energy Workshop Report: Research Needs for Wind Resource Characterization*. National Renewable Energy Laboratory, Golden, CO, USA, 116 pp.

Schumacher, A., S. Fink, and K. Porter (2009). Moving beyond paralysis: How states and regions are creating innovative transmission policies for renewable energy projects. *The Electricity Journal*, **22**, pp. 27-36.

Schwartz, M., D. Heimiller, S. Haymes, and W. Musial (2010). *Assessment of Offshore Wind Energy Resource for the United States*. National Renewable Energy Laboratory, Golden, CO, USA, 104 pp.

Sensfuß, F., M. Ragwitz, and M. Genoese (2008). The merit-order effect: A detailed analysis of the price effect of renewable electricity generation on spot market prices in Germany. *Energy Policy*, **36**, pp. 3086-3094.

Shafer, D.A., K.R. Strawmyer, R.M. Conley, J.H. Guidinger, D.C. Wilkie, T.F. Zellman, and D.W. Bernadett (2001). *WindPACT Turbine Design Scaling Studies: Technical Area 4 – Balance-of-Station Cost; 21 March 2000-15 March 2001*. National Renewable Energy Laboratory, Golden, CO, USA, 219 pp.

Shen, W.Z., and J.N. Sörensen (2007). Aero-acoustic modelling using large eddy simulation. *Journal of Physics: Conference Series*, **75**, pp. 012085.

Siegfriedsen, S., M. Lehnhoff, and A. Prehn (2003). Primary markets for offshore wind energy outside the European Union. *Wind Engineering*, **27**, pp. 419-429.

Sieros, G., P. Chaviaropoulos, J. Sørensen, B. Bulder, and P. Jamieson (2011). Upscaling wind turbines: theoretical and practical aspects and the impact on the cost of energy. *Wind Energy*, accepted for publication.

Simms, D.A., S. Schreck, M. Hand, and L.J. Fingersh (2001). *NREL Unsteady Aerodynamics Experiment in the NASA-Ames Wind Tunnel: A Comparison of Predictions to Measurements*. National Renewable Energy Laboratory, Golden, CO, USA, 51 pp.

Simons, R. (2006). Peer reviewed evidence on property value impacts by source of contamination. In: *When Bad Things Happen To Good Property*. Environmental Law Institute Press, Washington, DC, USA, pp. 63-112.

Simons, R.A., and J.D. Saginor (2006). A meta-analysis of the effect of environmental contamination and positive amenities on residential real estate values. *Journal of Real Estate Research*, **28**, pp. 71-104.

Sims, S., and P. Dent P (2007). Property stigma: wind farms are just the latest fashion. *Journal of Property Investment & Finance*, **25**, pp. 626-651.

Sims, S., P. Dent, and G.R. Oskrochi (2008). Modelling the impact of wind farms on house prices in the UK. *International Journal of Strategic Property Management*, **12**, pp. 251-269.

Sinden, G. (2007). Characteristics of the UK wind resource: long-term patterns and relationship to electricity demand. *Energy Policy*, **35**, pp. 112-127.

Singh, B., and S.N. Singh (2009). Wind power interconnection in power system: A review of grid code requirements. *The Electricity Journal*, **22**, pp. 54-63.

Smallwood, K.S., and C. Thelander (2008). Bird mortality in the Altamont Pass Wind Resource Area, California. *Journal of Wildlife Management*, **72**, pp. 215-223.

Smith, D.A. (2001). *WindPACT Turbine Design Scaling Studies Technical Area 2: Turbine, Rotor, and Blade Logistics; March 27, 2000 to December 31, 2000*. National Renewable Energy Laboratory, Golden, CO, USA, 224 pp.

Smith, J.C., M.R. Milligan, E.A. DeMeo, and B. Parsons (2007). Utility wind integration and operating impact state of the art. *IEEE Transactions on Power Systems*, **22**, pp. 900-908.

Snel, H. (2003). Review of aerodynamics for wind turbines. *Wind Energy*, **6**, pp. 203-211.

Snel, H., J.G. Schepers, and N. Siccama (2009). Mexico Project: the database and results of data processing and interpretation. In: *47th AIAA Aerospace Sciences Meeting Including the New Horizons Forum and Aerospace Exposition*. American Institute of Aeronautics and Astronautics (AIAA), Orlando, Florida, 5-8 January 2009, pp. 2009-1217.

Snyder, B., and M.J. Kaiser (2009a). A comparison of offshore wind power development in Europe and the US: Patterns and drivers of development. *Applied Energy*, **86**, pp. 1845-1856.

Snyder, B., and M.J. Kaiser (2009b). Ecological and economic cost-benefit analysis of offshore wind energy. *Renewable Energy*, **34**, pp. 1567-1578.

Söder, L., and H. Holttinen (2008). On methodology for modelling wind power impact on power systems. *International Journal of Global Energy Issues*, **29**, pp. 181-198.

Söder, L., L. Hofmann, A. Orths, H. Holttinen, Y. Wan, and A. Tuohy (2007). Experience from wind integration in some high penetration areas. *IEEE Transactions on Energy Conversion*, **22**, pp. 4-12.

Söderholm, P., and T. Sundqvist (2007). Empirical challenges in the use of learning curves for assessing the economic prospects of renewable energy technologies. *Renewable Energy*, **32**, pp. 2559-2578.

South, P., R. Mitchell, and E. Jacobs (1983). *Strategies for the Evaluation of Advanced Wind Energy Concepts*. Solar Energy Research Institute, Golden, CO, USA, 153 pp.

Sovacool, B.K. (2009). Contextualizing avian mortality: A preliminary appraisal of bird and bat fatalities from wind, fossil-fuel, and nuclear electricity. *Energy Policy*, **37**, pp. 2241-2248.

Spera, D. (ed.) (2009). *Wind Turbine Technology: Fundamental Concepts of Wind Turbine Engineering*. American Society of Mechanical Engineers Press, New York, NY, USA, 638 pp.

Sta. Maria, M.R.V., and M.Z. Jacobson (2009). Investigating the effect of large wind farms on energy in the atmosphere. *Energies*, **2**, pp. 816-838.

Stewart, G.B., A.S. Pullin, and C.F. Coles (2007). Poor evidence-base for assessment of windfarm impacts on birds. *Environmental Conservation*, **34**, pp. 1-11.

Stol, K.A., and M.J. Balas (2003). Periodic disturbance accommodating control for blade load mitigation in wind turbines. *Journal of Solar Energy Engineering*, **125**, pp. 379.

Stoner, A.M.K., K. Hayhoe, and D.J. Wuebbles (2009). Assessing general circulation model simulations of atmospheric teleconnection patterns. *Journal of Climate*, **22**, pp. 4348.

Summers, E. (2000). Operational effects of windfarm developments on air traffic control (ATC) radar procedures for Glasgow Prestwick International Airport. *Wind Engineering*, **24**, pp. 431-435.

Sundqvist, T. (2004). What causes the disparity of electricity externality estimates? *Energy Policy*, **32**, pp. 1753-1766.

Swider, D.J., L. Beurskens, S. Davidson, J. Twidell, J. Pyrko, W. Prüggler, H. Auer, K. Vertin, and R. Skema (2008). Conditions and costs for renewables electricity grid connection: Examples in Europe. *Renewable Energy*, **33**, pp. 1832-1842.

Swofford, J., and M. Slattery (2010). Public attitudes of wind energy in Texas: Local communities in close proximity to wind farms and their effect on decision-making. *Energy Policy*, **38**, pp. 2508-2519.

Tan, X. (2010). Clean technology R&D and innovation in emerging countries – Experience from China. *Energy Policy*, **38**, pp. 2916-2926.

Teske, S., T. Pregger, S. Simon, T. Naegler, W. Graus, and C. Lins (2010). Energy [R]evolution 2010—a sustainable world energy outlook. *Energy Efficiency*, **4**, pp. 409-433..

Thayer, R.L., and C.M. Freeman (1987). Altamont: Public perceptions of a wind energy landscape. *Landscape and Urban Planning*, **14**, pp. 379-398.

Thayer, R.L., and H. Hansen (1988). Wind on the land: Renewable energy and pastoral scenery vie for dominance in the siting of wind energy developments. *Landscape Architecture*, **78**, pp. 69-73.

Thomsen, K., and P. Sørensen (1999). Fatigue loads for wind turbines operating in wakes. *Journal of Wind Engineering and Industrial Aerodynamics*, **80**, pp. 121-136.

Toke, D. (2006). Explaining wind power planning outcomes: Some findings from a study in England and Wales. *Energy Policy*, **33**, pp. 1527-1539.

Toke, D., S. Breukers, and M. Wolsink (2008). Wind power deployment outcomes: How can we account for the differences? *Renewable and Sustainable Energy Reviews*, **12**, pp. 1129-1147.

Tougaard, J., J. Carstensen, J. Teilmann, H. Skov, and P. Rasmussen (2009). Pile driving zone of responsiveness extends beyond 20 km for harbor porpoises (*Phocoena phocoena* (L.)). *The Journal of the Acoustical Society of America*, **126**, pp. 11-14.

Tradewind (2009). *Integrating Wind. Developing Europe's Power Market for the Large Scale Integration of Wind*. Tradewind, Brussels, Belgium, 104 pp.

Troy, N., E. Denny, and M.J. O'Malley (2010). Base load cycling on a system with significant wind penetration. *IEEE Transactions on Power Systems*, **25**, pp. 1088-1097.

Tuohy, A., P. Meibom, E. Denny, and M. O'Malley (2009). Unit commitment for systems with significant wind penetration. *IEEE Transactions on Power Systems*, **24**, pp. 592-601.

Twidell, J., and G. Gaudiosi (eds.) (2009). *Offshore Wind Power*. Multi-Science Publishing, Brentwood, UK, 425 pp.

UKERC (2010). *Great Expectations: The Cost of Offshore Wind in UK Waters – Understanding the Past and Projecting the Future*. UK Energy Research Centre, London, UK, 112 pp.

US DOE (2008). *20% Wind Energy by 2030: Increasing Wind Energy's Contribution to U.S. Electricity Supply*. US Department of Energy, Washington, DC, USA, 248 pp.

Usaola, J. (2009). Probabilistic load flow in systems with wind generation. *IET Generation, Transmission & Distribution*, **3**, pp. 1031-1041.

Vajjhala, S.P., and P.S. Fischbeck (2007). Quantifying siting difficulty: A case study of US transmission line siting. *Energy Policy*, **35**, pp. 650-671.

Valentine, S.V. (2010). A STEP toward understanding wind power development policy barriers in advanced economies. *Renewable and Sustainable Energy Reviews*, **14**, pp. 2796-2807.

Van Bussel, G.J.W., and W. Bierbooms (2003). The DOWEC Offshore Reference Windfarm: analysis of transportation for operation and maintenance. *Wind Engineering*, **27**, pp. 381-391.

van den Berg, GP. (2004). Effects of the wind profile at night on wind turbine sound. *Journal of Sound and Vibration*, **277**, pp. 955-970.

van den Berg, G.P. (2005). The beat is getting stronger: the effect of atmospheric stability on low frequency modulated sound of wind turbines. *Noise Notes*, **4**, pp. 15-40.

van den Berg, G.P. (2008). Wind turbine power and sound in relation to atmospheric stability. *Wind Energy*, **11**, pp. 151-169.

van der Horst, D. (2007). NIMBY or not? Exploring the relevance of location and the politics of voiced opinions in renewable energy siting controversies. *Energy Policy*, **35**, pp. 2705-2714.

Van der Hoven, I. (1957). Power spectrum of horizontal wind speed in the frequency range from 0.0007 to 900 cycles per hour. *Journal of Atmospheric Sciences*, **14**, pp. 160-164.

Vittal, E., M.J. O'Malley, and A. Keane (2010). A steady-state voltage stability analysis of power systems with high penetrations of wind. *IEEE Transactions on Power Systems*, **25**, pp. 433-442.

Wagner, S., R. Bareiss, and G. Guidati (1996). *Wind Turbine Noise*. Springer-Verlag Telos, Emeryville, CA, USA, 204 pp.

Wahlberg, M., and H. Westerberg (2005). Hearing in fish and their reactions to sounds from offshore wind farms. *Marine Ecology Progress Series*, **288**, pp. 295-309.

Walter, A., K. Keuler, D. Jacob, R. Knoche, A. Block, S. Kotlarski, G. Muller-Westermeier, D. Rechid, and W. Ahrens (2006). A high resolution reference dataset of German wind velocity 1951-2001 and comparison with regional climate model results. *Meteorologische Zeitschrift*, **15**, pp. 585-596.

Wan, Y., and D. Bucaneg Jr. (2002). Short-term power fluctuations of large wind power plants. *Journal of Solar Energy Engineering*, **124**, p. 427.

Wan, Y., M. Milligan, and B. Parsons (2003). Output power correlation between adjacent wind power plants. *Journal of Solar Energy Engineering*, **125**, pp. 551-554.

Wang, C., and R.G. Prinn (2010). Potential climatic impacts and reliability of very large-scale wind farms. *Atmospheric Chemistry and Physics*, **10**, pp. 2053-2061.

Wang, X.L., F.W. Zwiers, V.R. Swail, and Y. Feng (2009). Trends and variability of storminess in the Northeast Atlantic region, 1874–2007. *Climate Dynamics*, **33**, pp. 1179-1195.

Warren, C., C. Lumsden, S. O'Dowd, and R. Birnie (2005). 'Green on Green': Public perceptions of wind power in Scotland and Ireland. *Journal of Environmental Planning and Management*, **48**, pp. 853-875.

Watson, S.J., L. Landberg, and J.A. Halliday (1994). Application of wind speed forecasting to the integration of wind energy into a large scale power system. *IEE Proceedings - Generation, Transmission and Distribution*, **141**, pp. 357-362.

WBGU, H. Graßl, J. Kokott, M. Kulessa, J. Luther, F. Nuscheler, R. Sauerborn, H.J. Schellnhuber, R. Schubert, E.D. Schulze (2004). *World in Transition: Towards Sustainable Energy Systems*. German Advisory Council on Global Change (WBGU), Earthscan, London, UK and Sterling, VA, USA, 352 pp.

Weber, C. (2010). Adequate intraday market design to enable the integration of wind energy into the European power systems. *Energy Policy*, **38**, pp. 3155-3163.

WEC (1994). *New Renewable Energy Resources: A Guide to the Future*. World Energy Council, Kogan Page, London, UK, 391 pp.

Wene, C.O. (2000). *Experience curves for energy technology policy*. International Energy Agency and Organisation for Economic Co-operation and Development, Paris, France, 127 pp.

WHO (1999). *Guidelines for Community Noise*. World Health Organization, Geneva, Switzerland, 141 pp.

WHO (2009). *Night Noise Guidelines for Europe*. World Health Organization, Geneva, Switzerland, 162 pp.

Wiesenthal, T., G. Leduc, H. Schwarz, and K. Haegeman (2009). *R&D Investment in the Priority Technologies of the European Strategic Energy Technology Plan*. European Commission, Joint Research Centre, Luxembourg, 84 pp.

Wiggelinkhuizen, E., T. Verbruggen, H. Braam, L. Rademakers, J. Xiang, and S. Watson (2008). Assessment of condition monitoring techniques for offshore wind farms. *Journal of Solar Energy Engineering*, **130**, 031004.

Wilhelmsson, D., and T. Malm (2008). Fouling assemblages on offshore wind power plants and adjacent substrata. *Estuarine, Coastal and Shelf Science*, **79**, pp. 459-466.

Wilhelmsson, D., T. Malm, and M.C. Ohman (2006). The influence of offshore windpower on demersal fish. *ICES Journal of Marine Science*, **63**, p. 775.

Wilson, J.C., and M. Elliott (2009). The habitat-creation potential of offshore wind farms. *Wind Energy*, **12**, pp. 203-212.

Wiser, R., and M. Bolinger (2010). *2009 Wind Technologies Market Report*. US Department of Energy, Washington, DC, USA, 88 pp.

Wiser, R.H., and S.J. Pickle (1998). Financing investments in renewable energy: the impacts of policy design. *Renewable and Sustainable Energy Reviews*, **2**, pp. 361-386.

Wolsink, M. (1989). Attitudes and expectancies about wind turbines and wind farms. *Wind Engineering*, **13**, pp. 196-206.

Wolsink, M. (2000). Wind power and the NIMBY-myth: institutional capacity and the limited significance of public support. *Renewable Energy*, **21**, pp. 49-64.

Wolsink, M. (2007). Planning of renewables schemes: Deliberative and fair decision-making on landscape issues instead of reproachful accusations of non-cooperation. *Energy Policy*, **35**, pp. 2692-2704.

Wright, A.D. (2004). *Modern Control Design for Flexible Wind Turbines*. National Renewable Energy Laboratory, Golden, CO, USA, 233 pp.

Wustenhagen, R., M. Wolsink, and M. Burer (2007). Social acceptance of renewable energy innovation: An introduction to the concept. *Energy Policy*, **35**, pp. 2683-2691.

Wyngaard, J.C. (2004). Toward numerical modeling in the "Terra Incognita". *Journal of the Atmospheric Sciences*, **61**, pp. 1816-1826.

Xiao, Z., Z. Rong, and L. Song (2010). *China Wind Energy Resource Assessment 2009*. China Meteorological Press, Beijing, China, 150 pp.

Xue, H., R.Z. Zhu, Z.B. Yang, and C.H. Yuan (2001). Assessment of wind energy reserves in China. *Acta Energiae Solaris Sinica*, **22**, pp. 168-170.

Yu, C., W. van Sark, and E. Alsema (2011). Unraveling the photovoltaic technology learning curve by incorporation of input price changes and scale effects. *Renewable and Sustainable Energy Reviews*, **15**, pp. 324-337.

Zhu, W.J., W.Z. Shen, and J.N. Sorensen (2007). Computational aero-acoustic using high-order finite-difference schemes. *Journal of Physics: Conference Series*, **75**, pp. 012084.

Zhu, W.J., N. Heilskov, W.Z. Shen, and J.N. Sørensen (2005). Modeling of aerodynamically generated noise from wind turbines. *Journal of Solar Energy Engineering*, **127**, p. 517.

8 Integration of Renewable Energy into Present and Future Energy Systems

Coordinating Lead Authors:
Ralph Sims (New Zealand), Pedro Mercado (Argentina), Wolfram Krewitt †(Germany)

Lead Authors:
Gouri Bhuyan (Canada), Damian Flynn (Ireland), Hannele Holttinen (Finland),
Gilberto Jannuzzi (Brazil), Smail Khennas (Senegal/Algeria), Yongqian Liu (China),
Lars J. Nilsson (Sweden), Joan Ogden (USA), Kazuhiko Ogimoto (Japan),
Mark O'Malley (Ireland), Hugh Outhred (Australia), Øystein Ulleberg (Norway),
Frans van Hulle (Belgium)

Contributing Authors:
Morgan Bazilian (Austria/USA), Milou Beerepoot (France), Trevor Demayo (USA/Canada),
Eleanor Denny (Ireland), David Infield (United Kingdom), Andrew Keane (Ireland),
Arthur Lee (USA), Michael Milligan (USA), Andrew Mills (USA), Michael Power (Ireland),
Paul Smith (Ireland), Lennart Söder (Sweden), Aidan Tuohy (USA),
Falko Ueckerdt (Germany), Jingjing Zhang (Sweden)

Review Editors:
Jim Skea (United Kingdom) and Kai Strunz (Germany)

This chapter should be cited as:
Sims, R., P. Mercado, W. Krewitt, G. Bhuyan, D. Flynn, H. Holttinen, G. Jannuzzi, S. Khennas, Y. Liu, M. O'Malley, L. J. Nilsson, J. Ogden, K. Ogimoto, H. Outhred, Ø. Ulleberg, F. van Hulle, 2011: Integration of Renewable Energy into Present and Future Energy Systems. In IPCC Special Report on Renewable Energy Sources and Climate Change Mitigation [O. Edenhofer, R. Pichs-Madruga, Y. Sokona, K. Seyboth, P. Matschoss, S. Kadner, T. Zwickel, P. Eickemeier, G. Hansen, S. Schlömer, C. von Stechow (eds)], Cambridge University Press, Cambridge, United Kingdom and New York, NY, USA.

Table of Contents

Executive Summary 612

8.1 Introduction 615

8.1.1 Objectives 618

8.1.2 Structure of the chapter 619

8.2 Integration of renewable energy into supply systems 619

8.2.1 Integration of renewable energy into electrical power systems 619
8.2.1.1 Features and structures of electrical power systems 620
8.2.1.2 Renewable energy generation characteristics 622
8.2.1.3 Integration of renewable energy into electrical power systems: experiences, studies and options 627

8.2.2 Integration of renewable energy into heating and cooling networks 640
8.2.2.1 Features and structure of district heating and cooling systems 640
8.2.2.2 Characteristics of renewable energy in district heating and cooling systems 641
8.2.2.3 Challenges associated with renewable energy integration into district heating and cooling networks 642
8.2.2.4 Options to facilitate renewable energy integration 643
8.2.2.5 Benefits and costs of large-scale penetration 645
8.2.2.6 Case studies 646

8.2.3 Integration of renewable energy into gas grids 647
8.2.3.1 Features and structure of existing gas grids 648
8.2.3.2 Characteristics of renewable energy with respect to integration 648
8.2.3.3 Challenges caused by renewable energy integration 649
8.2.3.4 Options to facilitate renewable energy integration 651

8.2.4 Integration of renewable energy into liquid fuel systems 654
8.2.4.1 Features and structure of liquid fuel supply systems 654
8.2.4.2 Characteristics with respect to renewable energy integration 654
8.2.4.3 Challenges of renewable energy integration 655
8.2.4.4 Options to facilitate renewable energy integration 656
8.2.4.5 Benefits and costs of large-scale renewable energy penetration 656
8.2.4.6 Case study: Brazil ethanol 657

8.2.5 Integration of renewable energy into autonomous energy systems 658
8.2.5.1 Characteristics with respect to renewable energy integration 658
8.2.5.2 Options to facilitate renewable energy integration and deployment 659
8.2.5.3 Benefits and costs of renewable energy integration and design 660
8.2.5.4 Constraints and opportunities for renewable energy deployment 660
8.2.5.5 Case studies 660

8.3 Strategic elements for transition pathways 661

8.3.1 Transport 662
8.3.1.1 Sector status and strategies 662
8.3.1.2 Renewable fuels and light-duty vehicle pathways 663
8.3.1.3 Transition pathways for renewable energy in light-duty transport 666
8.3.1.4 Comparisons of alternative fuel/vehicle pathways 668
8.3.1.5 Low-emission propulsion and renewable energy options in other transport sectors 670
8.3.1.6 Future trends for renewable energy in transport 672

8.3.2 Buildings and households 672
8.3.2.1 Sector status 673
8.3.2.2 Renewable energy and buildings in developed countries 674
8.3.2.3 Renewable energy and urban settlements in developing countries 677
8.3.2.4 Renewable energy and rural settlements in developing countries 678
8.3.2.5 Future trends for renewable energy in buildings 680

8.3.3 Industry 681
8.3.3.1 Sector status 681
8.3.3.2 Energy-intensive industries 682
8.3.3.3 Less energy-intensive industries and enterprises 684

8.3.4 Agriculture, forestry and fishing (primary production) 686
8.3.4.1 Sector status 686
8.3.4.2 Status and strategies 686
8.3.4.3 Pathways for renewable energy integration and adoption 687
8.3.4.4 Future trends for renewable energy in agriculture 687

References 690

Executive Summary

To achieve higher renewable energy (RE) shares than the low levels typically found in present energy supply systems will require additional integration efforts starting now and continuing over the longer term. These include improved understanding of the RE resource characteristics and availability, investments in enabling infrastructure and research, development and demonstrations (RD&D), modifications to institutional and governance frameworks, innovative thinking, attention to social aspects, markets and planning, and capacity building in anticipation of RE growth.

In many countries, sufficient RE resources are available for system integration to meet a major share of energy demands, either by direct input to end-use sectors or indirectly through present and future energy supply systems and energy carriers, whether for large or small communities in Organisation for Economic Co-operation and Development (OECD) or non-OECD countries. At the same time, the characteristics of many RE resources that distinguish them from fossil fuels and nuclear systems include their natural unpredictability and variability over time scales ranging from seconds to years. These can constrain the ease of integration and result in additional system costs, particularly when reaching higher RE shares of electricity, heat or gaseous and liquid fuels.

Existing energy infrastructure, markets and other institutional arrangements may need adapting, but there are few, if any, technical limits to the planned system integration of RE technologies across the very broad range of present energy supply systems worldwide, though other barriers (e.g., economic barriers) may exist. Improved overall system efficiency and higher RE shares can be achieved by the increased integration of a portfolio of RE resources and technologies. This can be enhanced by the flexible cogeneration of electricity, fuels, heating and cooling, as well as the utilization of storage and demand response options across different supply systems. Real-world case studies outlined throughout the chapter exemplify how different approaches to integration within a specific context have successfully achieved RE deployment by means of a combination of technologies, markets, and social and institutional mechanisms. Examples exist of islands, towns and communities achieving high shares of RE, with some approaching 100% RE electricity penetration and over a 50% share of liquid fuels for their light duty vehicle fleets.

Several mature RE technologies, including wind turbines, small and large hydropower generators, geothermal systems, bioenergy cogeneration plants, biomethane production, first generation liquid biofuels, and solar water heaters, have already been successfully integrated into the energy systems of some leading countries. Further integration could be encouraged by both national and local government initiatives. Over the longer term, integration of other less mature, pre-commercial technologies, including advanced biofuels, solar fuels, solar coolers, fuel cells, ocean energy technologies, distributed power generation, and electric vehicles, requires continuing investments in RD&D, infrastructure, capacity building and other supporting measures.

To reach the RE levels being projected in many scenarios over future decades will require integration of RE technologies at a higher rate of deployment than at present in each of the electricity generation, heating/cooling, gas and liquid fuel distribution, and autonomous energy supply systems.

RE can be integrated into all types of *electricity* supply systems, from large, interconnected, continental-scale grids to on-site generation and utilization in small, autonomous buildings. Technically and economically feasible levels of RE penetration depend on the unique characteristics of a system. These include the status of infrastructure development and interconnections, mix of generation technologies, control and communication capability, demand pattern and geographic location in relation to the RE resources available, market designs, and institutional rules.

The distribution, location, variability and predictability of the RE resources will also determine the scale of the integration challenge. Short time-variable wind, wave and solar resources can be more difficult to integrate than dispatchable reservoir hydro, bioenergy and geothermal resources, which tend to vary only over longer periods (years and decades). As variable RE penetration levels increase, maintaining system reliability becomes more challenging and costly. Depending on the specifics of a given electricity system, a portfolio of solutions to minimize the risks to the system and the costs of RE integration can include the development of complementary, flexible generation; strengthening and extending the network infrastructure;

interconnection; electricity demand that can respond in relation to supply availability; energy storage technologies (including hydro reservoirs); and modified institutional arrangements including regulatory and market mechanisms.

District heating (DH) and cooling (DC) systems offer flexibility with regard to the primary energy source, thereby enabling a gradual or rapid transition from the present use of fossil fuel sources to a greater share of RE. DH can use low temperature thermal RE inputs (such as solar or cascaded geothermal heat), or biomass with few competing uses (such as refuse-derived fuels or industrial wastes). DC systems are less common but also offer resource flexibility by being able to use a variety of natural waterways for the source of cold as well as ground source heat pumps. Thermal storage capability (hot or cold) can overcome the challenges of RE variability.

Injecting biomethane or, in the future, RE-derived hydrogen into *gas distribution grids* can be technically and economically achieved in order to meet a wide range of applications, including for transport, but successful integration requires that appropriate gas quality standards are met.

Liquid fuel systems can integrate biofuels either for cooking (such as ethanol gels and, in the future, dimethyl ether (DME)) or for transport applications when bioethanol or biodiesel esters are usually, but not always, blended with petroleum-based fuels to meet vehicle engine fuel specifications. Advanced biofuels developed in the future to tight specifications may be suitable for direct, unblended use in current and future engine designs used for road, aviation and marine applications.

Autonomous energy supply systems are typically small-scale and are often located in remote areas, small islands, or individual buildings where the provision of commercial energy is not readily available through grids and networks. The viability of autonomous RE systems depends upon the local RE resources available, the costs of RE technologies, future innovation, and the possible avoidance of construction costs for new or expanded infrastructure to service the location.

There are multiple pathways for increasing the share of RE through integration across the transport, building, industry and primary production end-use sectors, but the ease and additional costs of integration vary depending on the specific region, sector and technology.

Being contextual and complex, it is difficult to assess 'typical' system integration costs. These differ widely depending on the characteristics of the available RE resources; the geographic distance between the resource and the location of energy demand; the different integration approaches for large centralized systems versus decentralized, small-scale, local RE systems; the required balancing capacity; and the evolving status of the local and regional energy markets. The few comparative assessments in the literature, mainly for relatively low shares of RE (such as wind electricity in Europe and the USA and biomethane injection into European gas grids), show that the additional costs of integration are wide-ranging and site-specific.

To achieve higher RE shares across the end-use sectors requires planning, development and implementation of coherent frameworks and strategies. These will vary depending on the diverse range of existing energy supply systems in terms of scale, age and type. RE uptake can be achieved in all end-use sectors by either the direct use of RE (e.g., building-integrated solar water heating) or via energy carriers (e.g., blending of biofuels with gasoline or diesel at an oil refinery). Improved end-use energy efficiency and flexibility in the timing of energy use can further facilitate RE integration.

- The *transport sector* shows good potential for increasing RE shares over the next few decades, but from a low base. Currently the RE shares are mainly from liquid biofuels blended with petroleum products and some electric rail. To obtain higher shares in the future, the RE energy carriers of advanced biofuels, biomethane, hydrogen and electricity could all be produced either onsite or in centralized plants and used to displace fossil fuels. When, and to what extent, flex-fuel, plug-in hybrid, fuel cell or electric vehicles might gain a major share of the current light duty vehicle fleet partly depends on the availability of the energy carriers, the incremental costs of the commercial manufacturing of advanced drive trains, development of the supporting infrastructures, and the rate of technological developments of advanced biofuels, fuel cells and batteries. Integration of fuels and technologies for heavy duty vehicles, aviation and marine applications is more challenging. Advanced biofuels could become more fungible with petroleum fuels and distribution systems, but will need to become more cost competitive to gain greater market share. The cost and reliability of fuel cells and the limited range of electric vehicles are current constraints.

- The *building* sector currently uses RE to meet around 10% of its total consumer energy demand, excluding traditional biomass. In the future, RE can be integrated more easily into urban environments when combined with energy efficient 'green building' designs that facilitate time- and/or resource-flexible energy consumption. In rural areas in developing countries, many modest dwellings could benefit from the integration of RE technologies, often at the small scale, to provide basic energy services. RE technologies integrated into either new or existing building designs can enable the buildings to become net suppliers of electricity and heat. Individual heating systems using biomass (for cooking and space heating), geothermal (including hydrothermal and ground source heat pumps) and solar thermal (for water and space heating, and, to a lesser extent, for cooling) are already widespread at the domestic, community and district scales.

- For *industry*, integration of RE is site- and process-specific, whether for very large, energy-intensive 'heavy' industries or for 'light' small- and medium-sized processing enterprises. At the large industrial scale, RE integration can be combined with energy efficiency, materials recycling, and, perhaps in the future, carbon dioxide capture and storage (CCS). Some industries can also provide time-flexible, demand response services that can support enhanced RE integration into electricity supply systems. In the food and fibre processing industries, direct substitution of fossil fuels onsite can be feasible, for example by the use of biomass residues for heat and power. Many such industries (sugar, pulp and paper, rice processing) have the potential to become net suppliers of heat and electricity to adjacent grids. Electro-thermal processes, process hydrogen, and the use of other RE carriers provide good opportunities for increasing the shares of RE for industry in the future.

- *Agriculture*, ranging from large corporate-owned farms to subsistence peasant farmers, consumes relatively little energy as a sector. (Fertilizer and machinery manufacture is included in the industrial sector). Local RE sources such as wind, solar, crop residues and animal wastes are often abundant for the landowner or manager to utilize locally or to earn additional revenue by generating, then exporting, electricity, heat or biogas off-farm.

Parallel developments in transport (including electric vehicles), heating and cooling (including heat pumps), flexible demand response services (including the use of smart meters with real-time prices and net metering facilities) and more efficient thermal generation may lead to dramatic changes in future electrical power systems. Higher RE penetration levels and greater system flexibility could result (but also depend on nuclear power and CCS developments). Regardless of the present energy system, whether in energy-rich or energy-poor communities, higher shares of RE are technically feasible but require careful and consistent long-term planning and implementation of integration strategies and appropriate investments.

8.1 Introduction

This chapter examines the means by which larger shares of RE could be integrated into the wide range of energy supply systems and also directly into end-user sectors at national and local levels. It outlines how RE resources can be used through integration into energy supply networks that deliver energy to consumers using energy carriers with varying shares of RE embedded (Section 8.2) or directly by the transport, buildings, industry and agriculture end-use sectors (Section 8.3) (Figure 8.1).

Many energy systems exist globally, each with distinct technical, market, financial, and cultural differences. To enable RE to provide a greater share of electricity, heating, cooling and gaseous and liquid fuels than at present will require the adaptation of these existing energy supply and distribution systems so that they can accommodate greater supplies of RE. Integration solutions vary with location, scale and the current design of energy system and related institutions and regulations.

Established energy supply systems are relatively new in terms of human history, with only around 100 years elapsing since the original commercial deployment of internal combustion engines; approximately 90 years for national grid electricity; 80 years for the global oil industry; 50 years for the global gas industry; and only around 30 years for solid state electronic applications. Based upon the rate of development of these historical precedents, under enabling conditions and with societal acceptance, RE systems could conceivably become more prominent components of the global energy supply mix within the next few decades. Energy systems are continuously evolving, with the aims of improving conversion technology efficiencies, reducing losses, and lowering the cost of providing energy services to end users. As part of this evolution, it is technically feasible to continue to increase the shares of RE through integration with existing energy supply systems at national, regional and local scales as well as for individual buildings. To enable RE systems to provide a greater share of heating, cooling, transport fuels and electricity may require modification of current policies, markets and existing energy supply systems over time so that they can accommodate greater supplies of RE at higher rates of deployment than at present.

Regardless of the energy supply system presently in place, whether in energy-rich or energy-poor communities, over the long term and through measured system planning and integration, there are few, if any, technical limits to increasing the shares of RE, but other barriers would need to be overcome (Section 1.4). Specific technical barriers to increased deployment of individual RE technologies are discussed in chapters 2 through 7. This chapter outlines the more general barriers to integration (including social ones) that cut across all technologies and can therefore constrain achieving relatively high levels of RE integration. Where presented in the literature, solutions to overcoming these barriers are presented.

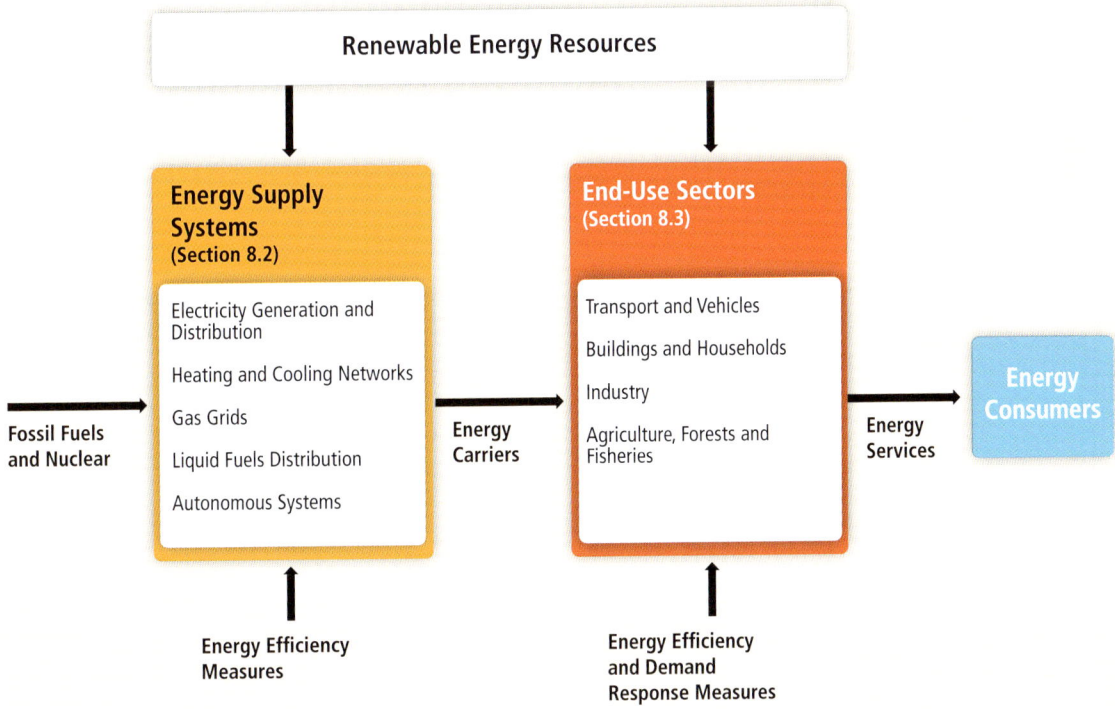

Figure 8.1 | Pathways for RE integration to provide energy services, either into energy supply systems or on-site for use by the end-use sectors.

Enhanced RE integration can provide a wide range of energy services for large and small communities in both developed and developing countries. The potential shares of RE depend on the scale and type of the existing energy supply system. Transition to low-carbon energy systems that accommodate high shares of RE integration can require considerable investments in new technologies and infrastructure, including more flexible electricity grids, expansion of district heating and cooling schemes, modifying existing distribution systems for incorporating RE-derived gases and liquid fuels, energy storage systems, novel methods of transport, and innovative distributed energy systems in buildings. The potential integration and rate of deployment of RE differs between geographic regions, depending on the current status of the markets and the varying political ambitions of all OECD and non-OECD countries.

All countries have access to some RE resources and in many parts of the world these are abundant. The characteristics of many of these resources distinguish them from fossil fuels and nuclear systems and have an impact on their integration. Some resources, such as solar, are widely distributed, whereas others, such as large hydro, are constrained by geographic location and hence integration options are more centralized. Some RE resources are variable and have limited predictability. Others have lower energy densities and different technical specifications from solid liquid and gaseous fossil fuels. Such RE resource characteristics can constrain their ease of integration and invoke additional system costs, particularly when reaching higher shares of RE.

Alongside RE, nuclear power and CCS linked with coal- or gas-fired power generation plants and industrial applications may well have a role to play in a low-carbon future (IPCC, 2007). However, for a country wishing to diversify its energy supply primarily by increasing domestic RE capacity to meet an increasing share of future energy demand, integrating a portfolio of local RE resources can be beneficial, and also make a positive contribution to improved energy supply security and system reliability (Awerbuch, 2006). Increasing RE integration can also offer a range of other opportunities and benefits (Sections 1.4.5 and 9.3) but carries its own risks, including natural variability (from seconds to years), physical threats to installed technologies from extreme weather events, locational dependence of some RE resources, additional infrastructure requirements, and other additional costs under certain conditions.

The future energy supply transition has been illustrated by many scenarios, the majority of which show increasing shares of RE over the next few decades (Section 10.2). The scenario used here as just one example (Figure 8.2) is based upon the International Energy Agency (IEA) World Energy Outlook 2010 '450 Policy Scenario' out to 2035. It illustrates that achieving high levels of RE penetration[1] will require a continuation of increasing market shares in all end-use sectors. The average annual RE growth increment required to meet this projection is almost 4 EJ/yr across all sectors; over three times the current RE growth rate.

---

[1] The terms 'shares' and 'penetration levels' of RE are used loosely throughout the text to indicate either the percentage of total installed capacity or total energy that comes from RE technologies.

In the 2010 World Energy Outlook (IEA, 2010b), the 22 EJ of final consumption RE (excluding traditional biomass) in 2008 is almost quadrupled in 2035 in the 450 Policy Scenario. This is due mainly to the power sector where the RE share in electricity supply rises from 19 to 32% over the same period. Government support for RE, projected to rise from USD 44 billion in 2008 to USD 205 billion in 2035, is a key driver along with projected lower RE investment costs and higher fossil fuel prices.

To achieve such increased shares of RE in total energy supply by 2035 and beyond will require overcoming the challenges of integration in each of the transport, building, industry and agriculture sectors. In order to gain greater RE deployment, strategic elements need to be better understood as do the social issues. Transition pathways for increasing the shares of each RE technology through integration should aim to facilitate a smoother integration with energy supply systems but depend on the specific sector, technology and region. Multiple benefits for energy consumers should be the ultimate aim.

Successful integration of high shares of RE with energy systems in recent years has been achieved in both OECD and non-OECD countries, including:

- Brazil, with over 50% of light duty transport fuels supplied from sugar cane ethanol (Zuurbier and Vooren, 2008) and 80% of electricity from hydro (BEN, 2010);

- China, where two-thirds of the world's solar water heaters have been installed (REN21, 2010);

- Denmark, with around 20% (7,180 GWh or 25.84 PJ) of total power supply in 2009 generated from wind turbines (Section 7.4) integrated with other forms of generation (mainly national coal- and gas-fired capacity, but also supported by interconnection to hydro-dominated systems) (DEA, 2009);

- Spain, where the 2000 Barcelona Solar Thermal Ordinance resulted in over 40% of all new and retrofitted buildings in the area having a solar water heating system installed (EC, 2006); and

- New Zealand and Iceland where the majority of electricity supply has been generated from hydro and geothermal power plants for several decades.

It is anticipated that increased urbanization will continue and that the 50% of the 6.4 billion world population living in cities and towns today will rise by 2030 to 60% of the then 8.2 billion people (UNDP, 2007). There is potential in many of these growing urban environments to capture local RE resources and thereby help meet an increasing share of future energy demands (MoP, 2006 Droege et al., 2010). The potential exists to integrate RE systems into the buildings and energy infrastructure as well as to convert municipal and industrial organic wastes to energy (Section 2.2.2). However, local government planning regulations

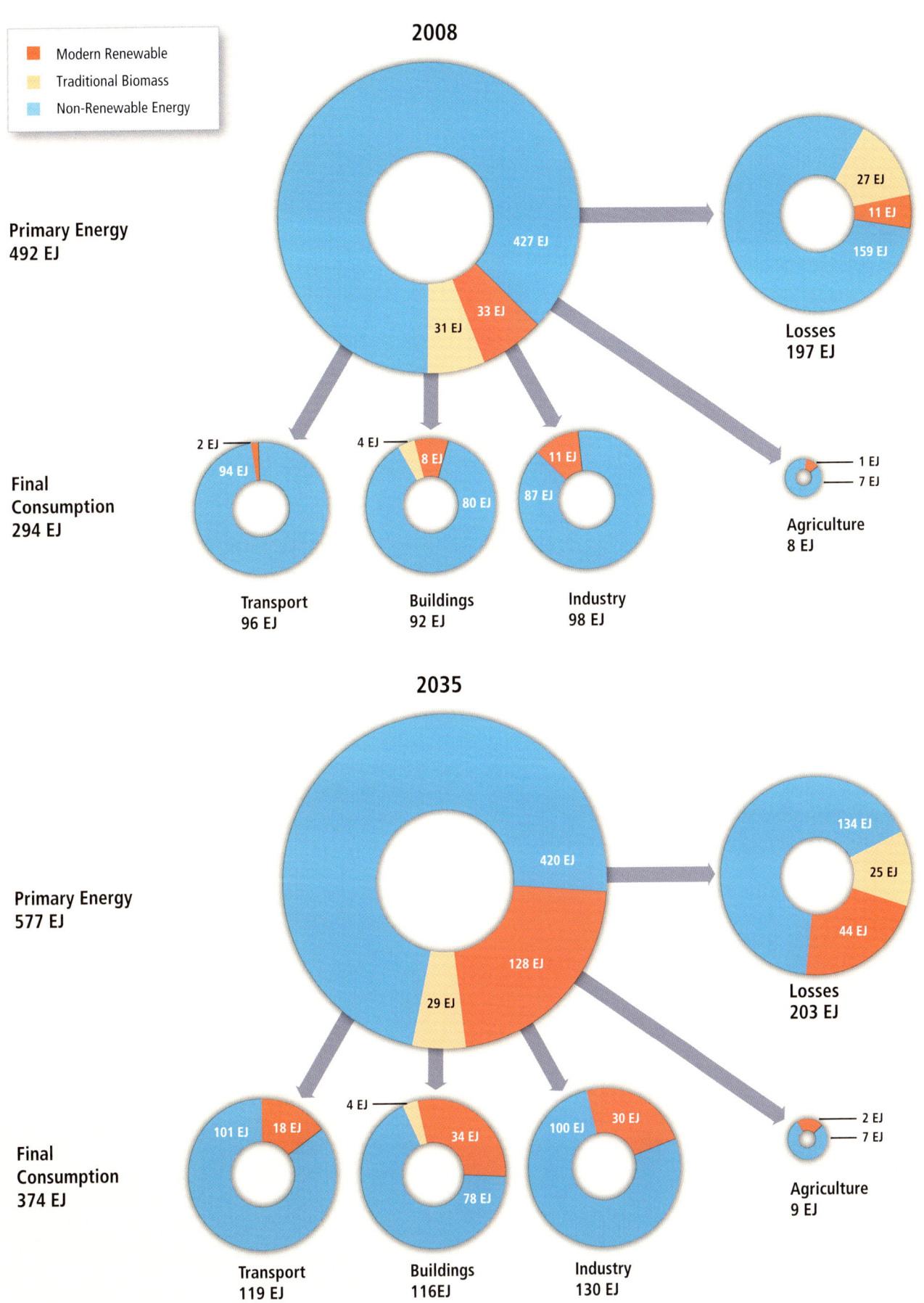

Figure 8.2 | (Preceding page) RE shares (red) of primary and final consumption energy in the transport, buildings (including traditional biomass), industry and agriculture sectors in 2008 and an indication of the projected increased RE shares needed by 2035 in order to be consistent with a 450 ppm CO_2eq stabilization target.

Notes: Areas of circles are approximately to scale. Energy system losses occur during the conversion, refining and distribution of primary energy sources to produce energy services for final consumption. 'Non-renewable' energy (blue) includes coal, oil, natural gas (with and without CCS by 2035) and nuclear power. This scenario example is based upon data taken from the IEA World Energy Outlook 2010 (IEA 2010d) but converted to direct equivalents (Annex II.4). Energy efficiency improvements above the baseline are included in the 2035 projection. RE in the buildings sector includes traditional solid biomass fuels (yellow) for cooking and heating for 2.7 billion people in developing countries (Section 2.2) along with some coal (UNDP and WHO, 2009). By 2035, some traditional biomass has been partly replaced by modern bioenergy conversion systems. Excluding traditional biomass, the overall RE system efficiency (when converting from primary to consumer energy) remains around 66% over the period.

may constrain the deployment of some RE technologies in the short term (IEA, 2009b).

Many energy scenarios have incorporated a wide range of energy efficiency initiatives (Sections 1.1.3 and 10.1). These reduce future energy demand baseline projections significantly across the building, industry, transport and energy supply sectors (IPCC, 2007). Lower energy demand reduces the required capacity, and hence cost, of an integrated RE system, which might facilitate having a greater share of RE in a growing energy market (Verbruggen, 2006; Pehnt et al., 2009a). For example, a building owner or developer could be encouraged to initially invest in energy saving measures and energy efficient building design before contemplating the installation of RE systems and hence reduce the installed capacity needed to meet the energy demand of the building occupiers (IEA, 2009b).

Integration of RE into the energy supply and infrastructure system of many OECD countries raises different challenges than those of non-OECD countries. For example, RE integration into dense urban regions that already have high shares of RE, or where cross-border energy supply options are possible, differs markedly from integration of RE into a small autonomous energy system in a remote rural region with limited energy infrastructure. In such districts, small-scale, distributed, RE systems may be able to avoid the high investment costs of constructing infrastructure presently deficient (ARE, 2009).

A technology that is successful in one region may not be so in another, even where RE resource conditions and supportive enabling environments are similar. Successful deployment can depend upon the local RE resources, current energy markets, population density, existing infrastructure, ability to increase supply capacity, financing options and credit availability. For any given location and energy market, issues relating to the integration of a RE project can be complex as there can be impacts on land and water use, adherence to national and local planning and consenting processes, variance due to the maturity of the technology (IEA, 2008b), co-benefits for stakeholders, and acceptance or rejection by the general public (as also would be the case for a fossil fuel, nuclear or CCS project).

8.1.1 Objectives

The objectives of this chapter are to

- assess the literature regarding the integration of RE into current and possible future energy systems;

- present the constraints that can exist when integrating RE into current electrical supply systems, heating and cooling networks, gas grids, liquid fuels and autonomous systems, particularly for RE shares that are significantly higher than at present; and

- determine whether increasing RE integration within present energy supply systems and facilitating the increased rate of deployment of RE technologies in the transport, building, industry and agricultural sectors are feasible propositions.

The chapter examines the complex cross-cutting issues that relate to RE integration across centralized, decentralized and autonomous energy supply systems and into the wide range of end-use technologies, buildings and appliances used to provide desirable energy services (heating, cooling, lighting, communication, entertainment, motor drives, mobility, comfort, etc.). These issues include energy distribution and transmission through energy carriers, system reliability and quality, energy supply/demand balances, system flexibility, storage systems, project ownership and financing, operation of the market, supply security and social acceptance. Regional differences between the integration of various RE systems are highlighted.

Due to the very specific nature of any individual energy supply system, it was not possible to provide general guidance on which policy intervention steps to follow logically in order to increase the share of RE through integration. The unique complexities of energy supply systems, due to their site-specificity, future cost uncertainties, and deficit of analysis in the literature, prohibited a detailed evaluation of the additional costs of RE system integration and deployment (other than for wind power; Section 7.5.4). The inability to determine 'typical' integration costs across the many differing systems and present them as 'representative' is a barrier to wider RE deployment and modelling scenarios. Further analysis would be useful.

8.1.2 Structure of the chapter

Section 8.2 discusses the integration of RE systems into existing and future centralized supply-side systems for both OECD and non-OECD regions. Where relevant, the benefits of system design and technology components to facilitate integration, operation and maintenance strategies, markets and costs are discussed.

Section 8.3 outlines the strategic elements, including non-technical issues, needed for transition pathways for each of the end-use sectors in order to gain greater RE deployment. The current status, possible pathways to enhance adoption of RE, related transition issues, and future trends are discussed for transport, buildings, industry and primary production.

Both sections endeavour to emphasize that though common solutions to RE integration exist there are sometimes differences between:

- RE integration into centralized, high voltage electricity systems, district heating schemes, and liquid fuel and gas pipelines, and

- RE integration into distributed, small-scale, energy systems such as low voltage electricity grids, heating and cooling of individual buildings, and liquid or gaseous fuel production for local transport use.

The case studies illustrate what has already been achieved, under a given set of circumstances.

8.2 Integration of renewable energy into supply systems

Energy supply systems have evolved over many decades to enable the efficient and cost-effective distribution of electricity, gas, heat and transport fuel energy carriers to provide useful energy services to end users. Increasing the deployment of RE systems requires their integration into these existing systems. This section outlines the issues and barriers involved as well as some possible solutions to overcome them in order to achieve increased RE penetration. The complexities of the various electricity supply systems and markets operating around the world result in marked differences in the approach to integration. Prerequisites for efficient and flexible energy conversion, mutual support between energy sectors, and an intelligent control strategy include coherent long-term planning and a holistic approach. Over time this could result in an inter-linked energy system to provide electricity, heating, cooling and mobility rather than having distinct sectors for each as at present. A significant increase in global electricity demand could result from a higher share being substituted for current fossil fuel demands in the heating and transport sectors.

8.2.1 Integration of renewable energy into electrical power systems

Modern electrical power systems (the grid) have been developing since the late 19th century and take different forms around the world. Some systems are very advanced and highly reliable but are at different scales, for example the Eastern Interconnection in the USA that serves 228 million consumers across 8.85 million square kilometres contrasts with smaller, more isolated systems such as Ireland serving a population of 6.2 million across 81,638 km^2 (NISRA, 2009; CSO, 2010). Other systems are not as well developed but are rapidly evolving. For example, China installed an average of 85 GW of plant per year from 2004 to 2008 and in the same period increased its electricity consumption by over 50% (J. Li, 2009). Other systems are not well developed either in terms of access or quality (e.g., many parts of sub-Saharan Africa). Autonomous and/or micro-scale systems also exist to serve small communities or single buildings or industrial plants (Section 8.2.5). Despite their variations, these systems have a common purpose: the provision of a reliable and cost-effective supply of electricity to loads by appropriate generation and use of network infrastructure.

The versatility of energy in electrical form, the ability to transport it across large distances (nearly) instantaneously, and its necessity for the deployment of modern technology and the advancement of economic and social development has resulted in a dramatic increase in the demand for electricity. This increase is projected to continue in a wide range of scenarios, including some of those that keep greenhouse gas (GHG) concentrations in the atmosphere below 450 ppm (e.g., IEA, 2010d; see also Section 10.2). The provision of modern energy services is recognized as a critical foundation for sustainable development (e.g., DFID, 2002; Modi et al., 2005; UNEA, 2009). This growth of electricity demand coupled with the geographically dispersed nature of many renewable sources makes electricity an attractive energy vector to harness RE where adequate network infrastructure is available. With the development of electric vehicles and heat pumps, electricity is also taking a growing share in the transport and heat markets (Kiviluoma and Meibom, 2010; Sections 8.3.1 and 8.3.2). Additionally, with the development of inexpensive and effective communications systems and technologies as well as smart meters, the electrical power system is experiencing dramatic change.[2] All these potential developments—RE, demand side participation, electric vehicles and any new thermal generation (i.e., fossil fuel or nuclear)—need to be integrated into electrical power systems. They collectively and individually pose common and unique challenges.

This section is comprised of three sub-sections that focus on the integration issues for renewable electricity and begins with a brief description of the basic principles of electrical power systems—how they are designed, planned and operated (Section 8.2.1.1). This is followed by a summary of the pertinent integration characteristics of renewable electricity sources and a high-level description of the integration

---

2 The term 'smart grid' is often used to refer to this mixture of new technologies but it is not used in this report.

challenges that result (Section 8.2.1.2). Finally, integration experiences, studies and options for existing and future electrical power systems are provided (Section 8.2.1.3).

8.2.1.1 Features and structures of electrical power systems

The first power plant used direct current (DC) that could transport electricity to consumers living close to the power station. However, a few years after the construction of this first power plant, alternating current (AC) electricity systems were developed (El-Sharkawi, 2009). Alternating currrent systems allow greater flexibility in the transmission of electricity across the various voltage levels in the electricity network and, as such, almost all electrical power systems across the world today use AC. However, DC is still used in the transmission of electricity over long distances, for interconnection of AC systems (sub-sea and over land), and in some very small domestic stand-alone systems. DC technology is developing rapidly and new application domains are being developed (Breuer et al., 2004; EASAC, 2009).

Integration of RE into any electrical power system poses a number of challenges (many shared with other technologies and developments) for the designers and operators of that system. In order to appropriately address these challenges, a basic understanding of the characteristics of electrical power systems is required and some salient elements of planning, design and operation are discussed here (Bergen and Vittal, 2000).

Electricity demand (including losses in the electrical power system) varies with the needs of the user; typically at a minimum at night and increasing to a peak during working hours. In addition, there are normally differences between working days and weekends/holidays and also between seasons; most systems also show an annual growth in consumption from year to year. Therefore, generators on a system must be scheduled (dispatched) to match these variations throughout the year and appropriate network infrastructure to transfer that power must be available. This balancing (of supply and demand) requires complex operational planning from the management of second-to-second changes in demand to the longer-term investment decisions in generation and transmission assets. The balancing is carried out by the system operator in balancing areas (or control areas), which often are parts of large interconnected AC systems.

In order to maintain an AC power system at its nominal frequency (e.g., 50 Hz in Europe and 60 Hz in North America), the instantaneous power supplied to the system must match the demand. Insufficient power results in a decreased frequency while excess power leads to an increased frequency. Either scenario is a threat to the security of the system, since the generators, interconnectors and loads that constitute the system are physically designed to operate within certain limits, and must be removed from the system once these limits are violated in order to ensure their integrity.

The electrical machines employed in the generation of electricity (and in the conversion of electricity to end-use energy) are an important component within electrical power systems. The traditional machine used for generation is the synchronous machine (El-Sharkawi, 2009). This machine is directly connected and synchronized to the frequency of the system. A synchronous electrical power system consists of (i) a network that connects (ii) synchronous generators to the (iii) demand. The network can further be divided into the transmission network, where large generators and consumers are connected and high voltages are used to transmit power over long distances; and the distribution network, which is used to transmit power to consumers at lower voltage levels and connect distributed generation. Synchronous machines maintain synchronism with one another through restoring forces that act whenever there are forces tending to accelerate or decelerate one or more generators with respect to other machines (Kundur, 2007). As a result of this, synchronous machines can detect and react to events on the system automatically; in particular inertial response to a frequency change. Generators also have governors that detect and react to frequency changes and this coupled with inertial response is of benefit to AC power systems as it allows for the support of frequency on an almost instantaneous basis.

Matching demand and supply (balancing) on a minute-to-minute basis is generally done by control of generation. This is known as regulation/load following and requires small to medium variations in the output of the power stations. It is usually controlled automatically or by a central electricity system operator, who is responsible for monitoring and operating equipment in the transmission system and in power generating stations. Dispatchable units are those that control their output between a minimum and maximum level. The output of some units such as wind generators cannot be fully controlled. Even here, however, some level of control is possible through a reduction of the output of the units, although such control strategies also lead to lost production. Units such as wind generators are therefore considered partially dispatchable as opposed to dispatchable.

Over slightly longer time periods (e.g., 30 minutes to 6 to 24 hours), decisions must be made regarding which power stations should turn on/turn off or ramp up/ramp down output to ensure the demand is met throughout the day (e.g., to meet low demand at night and high demand during the day). This is usually done using a method known as unit commitment (Wood and Wollenberg, 1996). Unit commitment involves complex optimizations that are conducted, typically one to two days ahead, to create an hourly or half-hourly schedule of generators required to reliably meet the forecasted demand at least cost. These schedules will usually instruct some units to run at their maximum capacity all day (these are known as base load units), some units to turn on in the morning and off at night (mid-merit units) and some units to just turn on during times of peak demand (peaking units). The running regime of a unit depends mainly on its operation cost (i.e., fuel used and efficiency), as well as other characteristics such as how long it takes to turn on or off, and the degree to which it can quickly change its output power.

Organized electricity markets have emerged in some countries/regions and they coordinate how the costs of the generators are included in the unit commitment methods. Trading of electricity between producers and consumers can be done in power exchanges (pools) or on a bilateral basis (Schweppe et al., 1988; Stoft, 2002). Sometimes these markets run on very short time horizons, for example, five minutes before the electricity is expected to be needed (Harris, 2006; AEMO, 2010), and in other cases the markets operate days, weeks or even months before the electricity is required. An important market parameter is the gate closure time, which is the time difference between bidding of generators into the market and the actual delivery of power. Properly functioning markets support the long-term financial investment in appropriate generation capacity and network infrastructure to ensure supply meets demand in a reliable manner and at least cost.

It should be noted that the principle of energy balance also applies to the smallest stand-alone autonomous systems. An autonomous electrical power system is one without interconnections to other systems and that cannot access the larger variety of balancing resources available to larger systems. In island systems, or developing economies, a common solution is often to use small autonomous systems in order to avoid the costs of transmission lines to areas with comparatively low consumption. Balancing in many such cases is provided by expensive battery energy storage and/or diesel generators and dump load resistors to absorb surplus energy that cannot be absorbed otherwise (Doolla and Bhatti, 2006). Autonomous systems can be as small as individual homes or groups of homes working on the low voltage distribution grid, sometimes referred to as microgrids (Tsikalakis and Hatziargyriou, 2008). Though the basic principles of electric power system operations do not differ between large interconnected networks and small autonomous systems, the practical implications of those principles can vary. Autonomous systems are addressed to some degree in this section, but are also covered in a more-dedicated fashion in Section 8.2.5.

Over an annual time frame, it is necessary to ensure that the electricity system always has enough generation capacity available to meet the forecasted demand. This means that maintenance schedules must be coordinated to ensure that all generating units and network infrastructure do not shut down for maintenance at the same time, while also considering the fact that units will break down unexpectedly. In addition, planning must also be done over much longer time horizons (5 to 20 years). The construction of generators and networks involves long lead times, high capital requirements, and long asset life and payback periods. Therefore, the electricity sector requires significant long-term planning to ensure that generation will continue to meet the demand in the decades ahead and network infrastructure is developed in a timely and economic manner.

A further important planning consideration is the geographic spread of generation. If a generator is located close to a demand centre then less transmission capacity will be required to deliver the electricity to the end user and less electricity will be lost in transmission

Electrical demand cannot always be met and there are many well known reliability metrics that can quantify this (Billinton and Allen, 1988), though the metrics themselves can vary widely among different electric power systems. For example, the value of lost load is different in a modern industrial economy than in a developing one. Electric systems that can accept lower levels of overall reliability may be able to manage the integration of RE into electrical power systems at lower costs than systems that demand higher levels of reliability, creating a trade-off that must be evaluated on a case-by-case basis.

A reliability metric known as the capacity credit[3] (also known as capacity value) (Keane et al., 2011a) gives an indication of the probability that a particular type of generation will reliably contribute to meeting demand, which generally means that it will be available to generate electricity during the peak demand hours. This is an important metric in the planning of future electricity systems. If a type of generation has a low capacity credit this indicates that its available output tends to be low during high demand periods. The total capacity credit for all generation on the system needs to be sufficient to cover peak demand with a certain level of reliability; usually systems also require an additional margin for reliability purposes (planning reserves). The capacity credit of generation depends on the generator availability (mechanical and fuel source), and the coincidence with electrical power system demand (in particular times of high demand).

To ensure system security and reliability, electrical power systems are designed and operated to withstand specific levels of contingencies. Generation contingencies result from the sudden loss of significant generation capacity; this could be the loss of a large generating unit or loss of a network connection. Reserves are carried by the system operator, usually in the form of other generators operating at reduced output, which rapidly replace the power that was lost during the contingency. Transmission systems are typically designed to withstand the loss of any single critical element, such as a transmission line, such that on the system (i.e., post fault), no other element on the network is overloaded and the system stays within prescribed limits. Faults on electrical power systems are detected and cleared by protection that continuously monitors the system for such events. Electrical power system protection is also critical to the maintenance of system integrity since generators and other critical equipment can be disconnected from the system if a fault on the system is not cleared quickly enough. Many of today's larger power systems use advanced energy management/network management systems to configure their systems in a secure manner, thus allowing them to withstand these contingencies, for example, fault ride through (FRT) capability of generators (and the associated capability

---

3   Note that capacity credit is different than capacity factor. The capacity factor of a power plant is the average output typically expressed as a percentage of its maximum (rated) output.

of providing frequency and voltage support during the fault). In order to ensure reliability and proper operability of the network, generators and large consumers connected to the network have to comply with the connection requirements published in the codes of the network operators. These include, for example, grid codes in Ireland (EirGrid, 2009) and Germany (Transmission Code, 2007) and connection standards in the USA (CAISO, 2010).

The power flows on the overhead lines and cables (feeders) of the system require careful management to ensure satisfactory voltage levels throughout the system and to respect the rating limits of individual feeders (El-Sharkawi, 2009). The power must be delivered to the loads via these feeders, and its efficient and reliable delivery is crucial. Key variables in this task are thermal ratings (heating caused by losses), voltage levels and stability limits. These requirements are managed at the planning stage when the network is designed and built and also on a shorter time frame as the network is reconfigured, generator output adjusted to influence the flows, or other control technologies employed to support system voltages (El-Sharkawi, 2009).

The AC nature of the electrical power system results in different voltages throughout the system, in the first instance determined by the demand and generation in the local area. In order to ensure an electricity supply of required quality and reliability, the voltages throughout the system must be maintained within defined limits. This is a challenge to the design and operation of electrical power systems across the world. The voltage levels can be affected by the size and characteristics of generators, transmission lines and consumers, and the design and location of these is one of the key parameters available when designing a reliable and economic electrical power system. Reactive power is a critical component of voltage control. It is distinct from the active power that supplies energy to loads and arises from the AC nature of modern electrical power systems (Taylor, 1994). The effective supply and demand of reactive power is a critical system support service in any AC electrical power system. Network users such as generators supply the different technical services, also called ancillary services, that are needed for proper operation of the network in normal operation (e.g., reactive power supply) and during network faults. Some of these services are delivered on a bilateral commercial basis, though ancillary service markets are emerging in many parts of the world (Cheung, 2008).

8.2.1.2 Renewable energy generation characteristics

Renewable electricity sources depend on energy flows in the natural environment, thus their power generation characteristics are very different in general from other generation based on stockpiles of fuel (with the exception of biomass-fuelled plants). In particular, they reflect the time-varying nature of the energy flows. Here, each of the RE generation technologies is dealt with in turn as it appears in Chapters 2 through 7. This section highlights supply characteristics of the technologies that are of direct relevance to integration into electrical power systems,

namely: (a) variability and predictability (uncertainty), which is relevant for scheduling and dispatch in the electrical power system; (b) location, which is a relevant indicator of the need for electrical networks; and (c) capacity factor, capacity credit and power plant characteristics, which are indicators relevant for comparison for example with thermal generation. These particular characteristics are outlined below, and a very brief summary for a selection of the technologies is given in Table 8.1. Further details are available in Chapters 2 through 7.

Bioenergy

Dedicated biopower plants are similar to fossil-fuel-powered plants in several respects; additionally, bioenergy can be blended with fossil fuels in fossil fuel plants that use co-firing. Biopower plants are powered by storable solid, gaseous or liquid fuel, and use similar types of technology and thermal cycles for the prime mover (e.g., steam turbine, diesel engine; Section 2.3.3). Temporal characteristics and output predictability are thus partly determined by operational decisions, and in part by the plant and biomass fuel availability, which can depend on how the fuel is prepared, stored and supplied to the plant and can exhibit daily, monthly, seasonal and annual variations.

The location of biopower plants is often determined by proximity to the fuel supply or fuel preparation plant. Biopower plant location is not as dependent on resource location as other renewable technologies as fuel can also be transported to the plant. A limitation to transporting fuel over long distances is the relatively low energy content of biomass fuels (in terms of kWh/m^3 or kWh/kg (kJ/m^3 or kJ/kg)). The high transport cost of biomass fuels means that it is generally more economical to locate the plant close to the fuel source (Section 2.3.2). Small biopower plants are very often connected at the distribution level. A single large plant, on the other hand, may be connected at the transmission level. The capacity credit of biopower plants is similar to combined heat and power (CHP) plants and thermal plants.

Biomass electricity production is often operated in CHP plants to achieve better fuel efficiency. As a result, there may be little flexibility in plant dispatch if the operation is heat-load driven. However, when heat storage is available, electricity can be produced in a flexible way (Lund and Münster, 2003; Kiviluoma and Meibom, 2010). Also, control characteristics (power, voltage) of biopower plant are similar to CHP and thermal plants. Plant sizes are mostly in the range from a few hundred kW to 100 MW and larger, particularly when co-fired with fossil fuels.

Direct solar energy

Direct electricity generation from solar takes two distinct forms: photovoltaic solar power (solar PV) in which sunlight is converted directly to electricity via the photovoltaic effect in a semiconductor; and concentrating solar power (CSP) in which a working fluid is heated to high temperature and used to drive a heat engine (e.g., a Rankine steam cycle or a Stirling cycle) that is connected to an electrical generator (Section 3.3). For both forms of generation the variability of the primary source, the available solar irradiation, is dependent on the level of aerosols in the atmosphere, the position of the sun in the sky, the potential

Table 8.1 | Summary of integration characteristics for a selection of renewable energy technologies.

Technology		Plant size range (MW)	Variability: Characteristic time scales for power system operation (Time scale)	Dispatchability (See legend)	Geographical diversity potential (See legend)	Predictability (See legend)	Capacity factor range (%)	Capacity credit range (%)	Active power, frequency control (See legend)	Voltage, reactive power control (See legend)
Bioenergy		0.1–100	Seasons (depending on biomass availability)	+++	+	++	50–90	Similar to thermal and CHP	++	++
Direct solar energy	PV	0.004–100 modular	Minutes to years	+	++	+	12–27	<25–75	+	+
	CSP with thermal storage*	50–250	Hours to years	++	++**	++	35–42	90	++	++
Geothermal Energy		2–100	Years	+++	N/A	++	60–90	Similar to thermal	++	++
Hydropower	Run of river	0.1–1,500	Hours to years	++	+	++	20–95	0–90	++	++
	Reservoir	1–20,000	Days to years	+++	+	++	30–60	Similar to thermal	++	++
Ocean energy	Tidal range	0.1–300	Hours to days	+	+	++	22.5–28.5	<10	++	++
	Tidal current	1–200	Hours to days	+	+	++	19–60	10–20	+	++
	Wave	1–200	Minutes to years	+	++	+	22–31	16	+	+
Wind energy		5–300	Minutes to years	+	++	+	20–40 onshore, 30–45 offshore	5–40	+	++

* Assuming CSP system with 6 hours of thermal storage in US Southwest. ** In areas with Direct Normal Irradiation (DNI) > 2,000 kWh/m²/yr (7,200 MJ/m²/yr).

Notes:

Plant size: range of typical rated plant capacity.

Characteristic time scales: time scales where variability significant for power system integration occurs.

Dispatchability: degree of plant dispatchability: + low partial dispatchability, ++ partial dispatchability, +++ dispatchable.

Geographical diversity potential: degree to which siting of the technology may mitigate variability and improve predictability, without substantial need for additional network: + moderate potential, ++ high diversity potential.

Predictability: Accuracy to which plant output power can be predicted at relevant time scales to assist power system operation: + moderate prediction accuracy (typical <10% Root Mean Squared (RMS) error of rated power day ahead), ++ high prediction accuracy.

Active power and frequency control: technology possibilities enabling plant to participate in active power control and frequency response during normal situations (steady state, dynamic) and during network fault situations (e.g., active power support during FRT): + good possibilities, ++ full control possibilities.

Voltage and reactive power control: technology possibilities enabling plant to participate in voltage and reactive power control during normal situations (steady state, dynamic) and during network fault situations (e.g., reactive power support during FRT): + good possibilities, ++ full control possibilities.

shadowing effect of obstacles (buildings, trees, etc.) and cloud cover. Depending on weather conditions, the latter two can be quite variable over time scales as short as seconds (Woyte et al., 2007). Because of their specific differences, the generation characteristics of solar PV and CSP are discussed separately.

Solar PV
The electrical output of PV panels changes nearly instantaneously as the solar radiation incident on the panels changes. The variability of a large solar PV plant will to some degree be smoothed due to the footprint of the plant, particularly over very short time scales (roughly less than about 10 minutes for plants of the order of about 100 MW) (Longhetto et al., 1989; Kawasaki et al., 2006; Curtright and Apt, 2008; Mills et al., 2009a; Marcos et al., 2011). The degree to which the variability and predictability of solar plants is smoothed will depend on the type of solar plants, the size of the individual plants, the geographic dispersion between sites, and prevailing weather patterns.

The aggregate variability of multiple solar plants will be smoothed by geographic diversity because clouds do not shade and un-shade dispersed plants simultaneously. This smoothing effect can substantially reduce the sub-hourly variability of the aggregate of several solar plants (Wiemken et al., 2001; Mills et al., 2009a; Murata et al., 2009; Hoff and Perez, 2010; Mills and Wiser, 2010). It can also lead to lower aggregate short-term forecast errors for multiple solar plants (Lorenz et al., 2009, 2010). This smoothing effect of geographic diversity was shown to lead to comparable variability for similarly sited wind and solar plants in one region of the USA (Mills and Wiser, 2010).

Solar electricity predictions have forecast errors in cloudy weather. There is no production during the night, and the morning and evening ramps as well as the overall diurnal variation are predictable. Locally, for distribution network control, prediction errors can be significant but decrease relatively in larger systems (Lorenz et al., 2009).

Although the solar resource varies from region to region, the sun does shine everywhere. This increases the versatility with which solar PV can be sited in contrast to many other more location-dependent renewable resources. With regard to the impact on network infrastructure, small and medium size solar PV is typically installed near to demand and connected at the distribution level. At low penetrations on distribution feeders (PV capacity < 100% peak load on feeder), PV may offset the need for distribution upgrades (where peak demand on the feeder occurs in daylight) and reduce losses. Large size PV plants, on the other hand, can be located far from the load centres, which typically requires additional network infrastructure.

Capacity factors of solar PV range between 12 and 27%. The lower capacity factors are for fixed tilt PV systems while the higher capacity factors typically utilize single axis tracking. Estimates of the capacity credit of PV range between 25 and 75% (Pelland and Abboud, 2008; Xcel Energy, 2009; GE Energy, 2010), though lower values are possible at high levels of solar penetration and in electricity systems where demand patterns and PV output are poorly correlated. Additional analysis indicates the potential for high capacity credit at low solar PV penetration when, as in many cases, there is a high degree of coincidence between solar PV production and demand (Perez et al., 2008). Network-connected PV systems use inverters for grid interfacing, enabling in principle control of electrical characteristics relevant for grid integration (McNutt et al., 2009). With additional controls it is possible for PV to even provide active power control through the plant inverters (Achilles et al., 2008), although this is always at a loss of PV production. Typical plant sizes range from a few kW to 100 MW but are increasing in size.

Concentrating solar power (CSP)
The smoothing effects due to geographic diversity for CSP are similar to those of solar PV. CSP, however, includes intrinsic thermal storage in its working fluid and thus can have substantial thermal inertia. Thermal inertia, to a degree, smooths the effects of short-term variations in solar radiation. This thermal inertia can be further enhanced through the storage of additional heated fluid. Adequate thermal storage coupled with an increased size in the solar collector field further smoothes plant output due to passing clouds and allows for extended plant operations into or through the night.

CSP plants can only use the direct-beam portion of solar irradiance. Sites with high direct normal irradiance, greater than approximately 2,000 kWh/m^2/yr (7,200 MJ/m^2/yr), are usually found in arid and semi-arid areas with reliably clear skies that typically lie at latitudes from 15° to 40° N or S and at higher altitudes (IEA, 2010c). The size of the plant in relation to local land availability determines the plant location, which is not necessarily close to load centres and therefore may often require new transmission infrastructure.

Capacity factors of CSP plants range from 22 to 26% without thermal storage and can reach as high as 74% with more than 10 hours of thermal storage (DOE and EPRI, 1997; Herrmann et al., 2004). In principle, without storage, the capacity credit of CSP can be similar to solar PV (Xcel Energy, 2009), whereas with storage, CSP's capacity credit could be 89 to 93%, or nearly as high as for thermal plants (GE Energy, 2010). Aside from the increased capacity factor and capacity credit, thermal storage allows CSP plants to provide improved dispatchability (i.e., from partially dispatchable to dispatchable). CSP plants with significant storage have similar electrical power plant characteristics to non-renewable thermal units and thereby enhance the overall grid flexibility to accommodate a larger share of variable energy sources. Plant sizes range from 50 MW to 250 MW and larger.

Geothermal energy
Geothermal resources can be utilized in a variety of sustainable power generating modes, including continuous low power rates, long-term (decades) cycles of high power rates separated by recovery periods, or uninterrupted high power rates sustained with effective fluid reinjection. Geothermal energy typically provides base load electrical generation,

but it has also been used for meeting peak demand. Geothermal plants represent major investment and have low variable costs and thus would tend to be operated at maximum, sustainable rated output. Operating in a flexible manner may be possible in some cases but it also may impact efficiency (D.W. Brown, 1996). As a result, while it may be possible to balance demand and/or variable generation with geothermal resources (Bromley et al., 2006), the overall economic effectiveness of this approach requires detailed evaluation at specific sites.

High-temperature hydrothermal-type geothermal reservoirs are geographically specific, and thus power generation will not always be near to population and load centres. Adding new geothermal resources often necessitates extending the transmission network and thus involves infrastructure investments (e.g., Mills et al., 2011). However, in the future enhanced geothermal systems will in principle have the potential of locating closer to demand (Tester et al., 2006). For new geothermal plants, capacity factors of 90% or higher are typically achieved (DiPippo, 2008), possibly declining over time with ageing. Geothermal plants use heat engines to drive electrical generators and as such they are in general dispatchable to the degree that dispatching the plant does not degrade the geothermal reservoir. In some cases it may be possible for geothermal plants to provide other network services such as frequency response and voltage control similar to thermal generation. The high availability of geothermal plants in California led to an estimated capacity credit of close to 100% (Shiu et al., 2006). Geothermal plant sizes can vary from small Stirling engine-based generators of a few kW up to steam plants of over 100 MW.

Hydropower (run-of-river, reservoir, pumped storage)
In addition to hydropower resources providing a source of RE, the generation characteristics of hydro resources further offer flexibility to the power system to manage the variable output of other renewable resources. Through integrated strategies, hydropower can buffer fluctuations in supply and demand, increasing the economic value of the power delivered (US DOE, 2004). Hydropower plants can be classified in three main categories according to operation and type of flow: run-of-river; reservoir based (storage hydropower); and pumped storage.

Run-of-river hydro facilities can exhibit substantial daily, monthly and seasonal variations depending on the precipitation and runoff in the catchment, and are built to operate with this variability. Some run-of-river plants may have limited balancing storage (e.g., diurnal) for meeting daily peak demand during periods of low water availability. Variability and predictability can also be influenced by hydrological restrictions, for example from mutual influences of plants operated in cascade along a given river. There can also be limits due to minimum flow in rivers or other similar hydrological factors. Variations in the water availability are in general well predicted at time scales relevant for system operation. In-stream technology using existing facilities like weirs, barrages, canals or falls generates power as per available water flow without any restriction and storage (Section 5.3.1).

For reservoir-based hydropower, when water is available, the electrical output of the plants is highly controllable and can offer significant flexibility for system operation. The reservoir capacity can vary from short term to seasonal to multi-seasonal. The energy storage in the reservoir allows hydro plants to operate in base load mode or as load following plants (Sections 5.3 and 5.5). Just like run-of-river hydro, the hydro plant flexibility can be limited by legally binding restrictions concerning minimum levels in the reservoirs, minimum river flows and other possible restrictions.

Pumped storage plants pump water from a lower reservoir into an upper storage basin using surplus electricity and reverse flow to generate electricity during the daily high demand period or other periods that require additional flexible generation (such as periods with high ramps). Pumped storage is a net consumer of energy due to pumping losses (not an energy source) (Section 5.3.1.3).

The geographic diversity potential of run-of-river hydropower is good; limiting factors are topography and precipitation conditions. The location of reservoir hydropower plants is very much geographically restricted and construction of large plants often requires substantial transmission network investments. Pumped hydro plants are similarly limited by economic constraints to areas that have suitable topography.

Capacity factors for run-of-river systems vary across a wide range (20 to 95%) depending on the geographic and climatological conditions, as well as technology and operational characteristics. For reservoir hydro, capacity factors are often in the range of 30 to 60% (Section 5.3.1.2). The capacity credit of run-of-river and reservoir hydro depends on the correlation of stream flows with periods of high demand and the size of the reservoir, as well as plant operational strategies. Hydro systems with large multi-seasonal reservoirs have capacity credits comparable to thermal plants (i.e., 97% in British Columbia, Canada; Wangdee et al. 2010). Such high capacity credit does depend on the size of the storage (Haldane and Blackstone, 1955; Billinton and Harrington, 1978) and the availability of other sources of energy during periods of regional drought (Barroso et al., 2003). A survey across a broad range of hydrologic and demand conditions for hydro lacking seasonal storage found capacity credits ranging between roughly 0 and 90% (Grimsrud et al., 1981). Some reservoir-based hydropower plants may be designed to operate as peaking power plants resulting in a low capacity factor but with a relatively high capacity credit (Section 5.5). The capacity factor and capacity credit for pumped storage are dependent on the energy storage capacity and the operational strategy, but the capacity credit would be expected to be high.

Electrical power plant characteristics of reservoir hydro plants using synchronous generators are similar to thermal generation; in fact, reservoir hydro can often provide rapid power control possibilities in excess of those possible with thermal units. Run-of-river plants use a variety of conversion systems, including variable speed systems with power electronic converters. As a consequence, electrical output characteristics of these run-of-river plants in terms of power and voltage control possibilities are comparable to wind power plants. The size of hydropower plants range from a few kW to over 20 GW.

Ocean energy (wave, tidal range, tidal and ocean currents, OTEC, salinity gradient)

Ocean energy comprises several different types of plant: wave energy; tidal range (due to the rise and fall of sea level, i.e., tidal barrages); tidal and ocean currents; Ocean Thermal Energy Conversion (OTEC); and salinity gradient. Virtually all ocean energy technologies are at best at the development or demonstration stage. Therefore, data are scarce in the scientific literature and much of what is available is heavily dependent on simulation studies with little operational field data.

The different forms of ocean energy are driven by very different natural energy flows and have different variability and predictability characteristics. Wave energy is a spatially integrated form of wind energy and daily variability may to some extent be less than wind energy. Seasonal variability has been reported to be similar to wind (Stoutenburg et al., 2010), however this is device dependent. Initial work on wave models and data shows that output can be forecasted and the models perform particularly well during high production situations (ECI, 2006). Forecasting performance for wave energy is reported to be comparable to wind and solar (Reikard, 2009).

Generation from both tidal range and tidal currents is variable in most configurations but production profiles are (almost) completely predictable. Phase differences in tidal currents between different locations within the same electrical power system could be exploited to realize significant power smoothing (Khan et al., 2009). Ocean currents have low variability at power system operational time scales. OTEC derives from thermal gradients that are reasonably well understood and near-continuous base load operations would be expected. Salinity gradient power generation is at an early stage of research and should the technology become commercial it is likely that plants would operate at constant output.

Although all ocean technology requires access to the ocean, the appropriateness of specific locations varies by the type of ocean technology. Wave energy can be collected on or reasonably near to the shore, and perhaps in the future further out into the oceans. Tidal plants and ocean current plants may locate in very specific locations, usually necessitating network infrastructure investments (University of Edinburgh, 2006). Large collections of ocean energy generators will also result in temporal smoothing of the power output (Salter et al., 2002), but are located some way from land and/or load centres.

There are a few studies with indicative values for capacity factors and capacity credit. Radtke et al. (2010) have shown that tidal range can have very low capacity credit (i.e., less than 10% for the example studied), while the capacity factor of tidal range is expected to be 22.5 to 28.5% (Section 6.3.3). Bryans et al. (2005) report capacity factors of 19 to 60% and capacity credit of 10 to 20% for tidal current. The higher end of the capacity factor and capacity credit range is achieved by downsizing the electrical generator and curtailing output during peak tidal currents. Stoutenburg et al. (2010) report capacity factors of 22 to 29% and capacity credit of 16% for wave energy off the coast of California. For Scottish wave energy, a capacity factor of 31% has been reported (University of Edinburgh, 2006).

Tidal range uses synchronous generators, and has electrical characteristics similar to thermal plants. Wave devices usually make use of power electronic converters for grid connection. Equally, tidal and ocean current turbines tend to be variable speed and thus converter connected. The electrical plant characteristics of wave, tidal current and ocean current may therefore be comparable to wind power plants. Plant sizes are 0.1 to 300 MW for tidal range, and will vary depending on the number of modules for other ocean energy technologies.

Wind energy

The electrical output of wind power plants varies with the fluctuating wind speed, with variations at all time scales relevant for power system planning, scheduling and operations (Holttinen et al., 2009). The variability of aggregated wind power output diminishes with geographical dispersion and area size, because of the decreasing correlation of wind speeds (Section 7.5.1). Prediction accuracy of wind power plant output decreases with the time span of prediction horizon, and improves with area size considered (Chapter 7). Control systems at the wind turbine, wind plant and area level (e.g., groups of distributed wind power plants) can be used to reduce the power output fluctuations when needed for secure power system operation (e.g., during extreme weather and low load situations), but at the cost of lost production.

In general, wind power plants are distributed over existing networks. However, access to areas of high wind resources, for example offshore or remote onshore, often requires extension of existing transmission networks.

Wind capacity factors depend on wind climate and technology used. Fleet-wide wind capacity factors are of the order of 20% to as high as 40% for onshore wind depending on the location, and even higher for offshore wind (Section 7.2). The capacity credit of aggregated wind power at low to medium penetrations is around 5 to 40%, depending on location, and diminishes with increasing penetration level (Section 7.5). Electrical power plant characteristics are determined by the type of conversion system and control characteristics of wind power plants. Although many existing wind plants have induction generators, as a general trend, modern wind power plants are connected to the power system via power electronic converters, and can be equipped to provide grid services such as active power, reactive power and voltage control, frequency response (inertial type response) FRT and power system support during network faults (Section 7.5.3). Recent onshore wind power plant sizes have typically ranged from 5 to 300 MW and offshore from 20 to 120 MW, though smaller and larger plant sizes do exist, including the recently commissioned 500 MW Greater Gabbard offshore plant in the UK.[4]

Challenges with integrating renewable resources into electrical power systems

Most RE resources are location specific. Therefore, renewable-generated electricity may need to be transported over considerable distances. For example, China's windy regions are often far from population and load centres. Scotland's tidal current resource is a long distance from a significant population. In the USA, the largest high quality wind resource regions and land with significant biomass production are located in the Midwest, a significant distance from the predominantly coastal population. In many of these cases, additional transmission infrastructure can be economically justified (and is often needed) to enable access to higher quality (and therefore lower cost) renewable resource regions by electricity load centres rather than utilizing lower quality renewable resources located closer to load centres. Many renewable sources can also be exploited as embedded generation in distribution networks, which may have benefits for the system when at moderate penetration levels, but also can pose challenges at higher penetration levels (e.g., voltage rise, see Masters (2002)).

Also, as discussed above, certain RE sources lack the flexibility needed to deal with certain aspects of power system operation, in particular balancing supply and demand. This is because they are subject to significant variability across a wide range of time scales important to electrical power systems and also experience more uncertainty in predicted output. Furthermore, renewable plants may displace non-renewable plants that have heretofore provided the required flexibility. Some renewable sources (hydropower with reservoirs and bioenergy) may help to manage this challenge by providing flexibility. However, overall balancing will become more difficult to achieve as partially dispatchable RE penetrations increase. Particular challenges to system balancing are situations where balancing resources are limited (e.g., low load situations with limited operational capacity).

Furthermore, increased penetration of RE production will require renewable generators to become more active participants in maintaining the stability of the grid during power system contingencies. Depending on local system penetration, network faults can trigger the loss of significant amounts of generation if the renewable generation resources are concentrated in a particular section of the power system and connection requirements have not properly accounted for this risk. A solution is to require renewable capacity to participate when possible in transient system voltage control thus supporting recovery from network faults (EirGrid, 2009, 2010b).

There are also challenges with regard to very short-term system balancing (i.e., frequency response). At high penetration levels the need for frequency response will increase unless supplementary controls are added (Pearmine et al., 2007). Many of the renewable technologies do not lend themselves easily to such service provision. In addition, RE interfaced through power electronics may displace synchronous generators, thereby reducing the overall system inertia and making frequency control more difficult. Research and development is in progress to deliver frequency response from time variable sources such as modern wind turbines,[5] and some equipment with frequency response and inertial response is already available (Section 7.7). This is a subject of ongoing research (Doherty et al., 2010) and development (Miller et al., 2010).

The output of the different renewable sources is not in general well correlated in time, so if power systems include a wide range of renewable sources, their aggregate output will be smoother thus easing the challenge of electrical power system balancing. Such a portfolio approach to generation should thus be assessed, but as noted above, many of the renewable resources are highly geographically specific so that beneficial combinations of renewable sources may not always be practicable.

Lastly there is the additional challenge of managing the transition from the predominant generation mixes of today to sustainable sources required for the low carbon power systems of the future. Major changes will be required to the generation plant mix, the electrical power systems infrastructure and operational procedures if such a transition is to be made. Specifically, major investments will be needed and will need to be undertaken in such a way, and far enough in advance, so as to not jeopardize the reliability and security of electricity supply.

8.2.1.3 Integration of renewable energy into electrical power systems: experiences, studies and options

As electrical power systems worldwide are different, there cannot be one recipe that fits all when examining the integration of RE. Dispatchable renewable sources (hydro, geothermal, bioenergy, CSP with storage[6]) may require network infrastructure but, in many cases, may offer extra flexibility for the system to integrate variable renewable sources (hydropower in particular). Partially dispatchable RE technologies (wind, solar PV, certain forms of ocean energy), on the other hand, will pose additional challenges to electrical power systems at higher penetration levels.

There is already significant experience in operating electrical power systems with large amounts of renewable sources (e.g., 2008 figures on an energy basis are: Iceland 100%; Norway 99%; Austria 69%; New Zealand 64%; and Canada 60% (IEA, 2010b)). High percentages of

---

4 www.sse.com/PressReleases2011/FirstElectricityGeneratedGreaterGabbardWalney/.

5 It is worth noting that older wind technologies provided this response inherently, although not as well as synchronous generation (see Mullane and O'Malley, 2005).

6 CSP without additional storage is partially dispatchable and with several hours of storage can be considered dispatchable.

renewable electricity generation generally involve dispatchable renewable sources, in particular hydropower and geothermal (e.g., 2008 figures on an energy basis are: Norway 99% hydro; Iceland 75% hydro and 25% geothermal (Nordel, 2008)). Large shares of bioenergy are not so common in electrical energy systems, but Finland produces 11% of its electrical energy from bioenergy (Statistics Finland, 2009). A number of other countries have managed operations with more than 10% of annual supply coming from wind energy. In addition, integration studies provide insight into possible options for future systems to cope with higher penetration of partially dispatchable renewable sources.

This subsection addresses the integration of RE in three ways. First, it discusses actual operational experience with RE integration. Second, it highlights RE integration studies that have evaluated the potential implications of even higher levels of RE supply. Finally, it discusses the technical and institutional solutions that can be used to help manage RE integration concerns. This section has a focus on the developed world as this is where most experience and studies exist to this point. Autonomous systems are covered here to a degree, while issues associated with such systems are covered in a more dedicated fashion in Section 8.2.5.

Integration experience

It is useful to distinguish between experience with RE generation plants that can be dispatched (hydro, bioenergy, geothermal, CSP with storage) and variable renewable sources that are only partially dispatchable (wind, solar PV, and certain types of ocean energy).

Dispatchable renewable sources (bioenergy, CSP with storage, geothermal, hydro)

Experience from biopower plants is similar to that from fossil fuel thermal power plants in power system operation. As the plants are, at least in principal, dispatchable they can also offer flexibility to the power system. Even with CHP plants there are ways to operate the plants so that the electricity production is not totally dependent on the heat load. In Finland, for example, the larger plants use back pressure turbines equipped with auxiliary condensing units making it possible to maintain efficient electricity production even when heat load is low (Alakangas and Flyktman, 2001). Experience from Denmark shows that when operating with thermal storage, small biopower CHP plants can provide electricity according to system needs (market prices) and thus help in providing flexibility (Holttinen et al., 2009).

A renewable integration cost report from California, analyzing real data from CSP plants from 2002 to 2004 shows consistently high generation during peak load periods given the natural tendency of solar generation to track demand that is largely driven by cooling loads. The auxiliary natural gas boilers on some of the CSP plants in the studied region augmented solar generation during the peak demand periods. The variability and ramping of the CSP plants was reported to be of the same (relative) magnitude as for wind power (Shiu et al., 2006).

Adding new geothermal resources has often meant extending the transmission network and thus infrastructure investments. For example, in New Zealand the construction of a 220 kV double circuit is planned to facilitate development of geothermal generation (up to 800 MW) in the North Island of New Zealand (TransPower, 2008; W. Brown, 2010). Geothermal resources typically produce power (and heat) on a stable basis and there is considerable experience with their use, mostly operating like base load units (Shiu et al., 2006). In California, the existing geothermal generation was assessed for integration impacts based on real output data from the years 2002 to 2004 and was found to impose a very small regulation cost. Because of the very low forced outage rates for geothermal units (0.66%) and low maintenance rates (2.61%) during the 2002 to 2004 period, geothermal plants were also able to provide more capacity credit to the system than the benchmark units (Shiu et al., 2006).

Adding new hydropower resources has meant extending the transmission network and thus required network investments. Examples include northern Sweden, northern Italy, the USA, and northern Quebec, Canada (Johansson et al., 1993) and more recently in China (X. Yang et al., 2010). The large seasonal and interannual variability of hydropower is usually tackled by building large reservoirs where possible. Aggregation of different regions can help in smoothing hydro resource variability, since the changes over weeks and years are not exactly the same in neighbouring areas. The experience from Nordic countries (Sweden, Norway, Finland, Denmark) shows that the large differences in inflow between a dry and a wet year (up to 86 TWh (309 PJ) when mean yearly hydro production is 200 TWh (720 PJ)) can be managed with strong interconnections to the large reservoir capacity of 120 TWh (432 PJ) in Norway and Sweden and thermal power availability in Finland and Denmark (Nordel, 1996, 2000). Interconnection to neighbouring systems has been shown to have a large impact on the way hydro is used, since it influences the plant mix and thus changes hydro scheduling (Gorenstin et al., 1992).

The operational cost of hydropower plants is very low; the challenge for scheduling is to use the limited amount of water as efficiently as possible (Sjelvgren et al., 1983). The flexibility of hydropower is often used as an effective balancing option in electrical power systems (Pérez-Díaz and Wilhelmi, 2010). Switzerland has a flexible hydro system with both reservoirs and pumping facilities, and that system is currently used for daily balancing in the whole interlinked system including Germany, France and Italy (Ochoa and van Ackere, 2009). The flexibility of hydropower can be observed by comparing the changes in the daily prices in different countries. In hydro-dominated systems the price differences are relatively small since water is easily moved from low price periods to high price periods until the price difference is small (Sandsmark and Tennbakk, 2010). Hydropower is a low cost balancing option for daily load following, as can be seen from the Nordic day-ahead market. Sandsmark and Tennbakk (2010) show that the normalized average hourly prices during working days, 2001 to 2003, varied much less in the

Nordic hydro-dominated system than in Germany where thermal power is used for balancing.

Partially dispatchable renewable sources (solar PV, ocean, wind)
Partially dispatchable renewable sources pose greater challenges to system operators. In essence these sources of generation cannot be fully controlled (dispatched) since they reflect the time-varying nature of the resource. The main way in which they can be controlled is through reduction of the output. This is in contrast to dispatchable generation that can be controlled by increasing or reducing fuel supply.

Solar PV penetration levels remain quite limited despite high growth rates of installed capacity in certain countries. For example, in Germany where active programmes of PV installation have been successful, about 10 GW of PV were installed by the end of 2009, producing 1.1% (6.6 TWh or 23.76 PJ) of German electrical energy in 2009 (BMU, 2010). Local penetration levels of PV are already higher in southern parts of Germany (Bavaria has the largest concentration of installations), however, and reinforcements have been needed in certain distribution networks, mainly in rural areas with weak grid feeders and high local penetration levels. In strong urban grids there has only been a marginal need for grid reinforcement. There is concern that severe grid disturbances with strong frequency deviations can be worsened by large amounts of PV systems (Strauss, 2009). Due to this, the German guideline for the connection to medium-voltage networks requires a defined frequency/power drop for frequencies above 50.2 Hz (BDEW, 2008). Protection systems in distribution grids also have to be adapted to ensure safety (Schäfer et al., 2010). In general, these adaptations and guidelines indicate that it is important that solar PV become a more active participant in electrical networks (Caamano-Martin et al., 2008). In Japan, several demonstration projects have provided experience with technologies related to over-voltage protection through reverse power flow control by generation curtailment and battery control, prevention of islanding (Ueda et al., 2008), and verification of grid stabilization with large-scale solar PV systems (Hara et al., 2009). In the USA, some infrastructure investments have been driven by solar energy. California has approved the Sunrise Powerlink, a 193 km, 500 kV line that will connect high-quality solar areas in the desert (for both PV and CSP plants), as well as geothermal resources, to the coastal demand centre of San Diego (U.S. Forest Service, 2010).

Some initial reports are also emerging that analyze the variability of groups of PV plants (Wiemken et al., 2001; Murata et al., 2009; Hoff and Perez, 2010; Mills et al., 2011). Local weather situations like clouds, fog and snow are factors that cause variability and challenge short-term forecasting. All of these studies, using data from different regions of the world, indicate that the variability of groups of PV plants is substantially smoothed relative to individual sites, particularly for sub-hourly variability. Day-ahead forecast errors using weather prediction models have been shown to provide forecasts with only slightly lower accuracy (still <5% forecast error normalized to installed power) (Lorenz et al., 2010).

Operational ocean energy capacity is effectively in the form of a few individual plants, typically of modest capacity, thus no extensive integration experience with larger installations or collections of plants exists.

The majority of the experience with partially dispatchable RE integration comes from the wind sector (Section 7.5.3.2). West Denmark has a 30% wind penetration and has hit instantaneous penetration levels of more than 100% of electricity demand coming from wind power (Söder et al., 2007). But West Denmark is a small control area that is synchronously well connected to the much larger Continental Europe system. Ireland has a small power system with very limited interconnection capacity to Great Britain. Ireland has an 11% wind energy penetration level (2009) and has coped with instantaneous power penetration levels of up to 50% (EirGrid, 2010b). Section 7.5.3.2 provides further information on the Danish and Irish systems. Spain and Portugal are medium size control areas with relatively weak synchronous connections to the rest of the Continental Europe system. They both have about 15% wind energy penetration and have coped, at times, with 54 and 71% instantaneous power penetration levels, respectively (Estanqueiro et al., 2010). There are also several wind-diesel systems where wind provides a large part of the energy for autonomous systems (e.g., in Alaska, USA, the Cape Verde islands, Chile and Australia (Lundsager and Baring-Gould, 2005)).

Many systems report the need for new grid infrastructure both inside the country/region as well as interconnection to neighbouring countries/regions. Grid planning includes grid reinforcements as well as new lines (or cables) for targeted wind power. Wind power is normally not the only driving force for the investments but it is a major factor (e.g., Ireland (EirGrid, 2008); Germany (Dena, 2010); Portugal, (REN, 2008); Europe (ENTSO-E, 2010); the USA (MTEP, 2008)). In the USA, a lack of transmission capacity to move the wind energy from the best wind resource areas, most of which are remote, to the distant load centres has been clearly identified. A challenge for transmission planning is to resolve the timing conflict of financing for the wind plants needing transmission access (i.e., wind plants can be permitted and constructed in 2 to 3 years while it may take 5 to 10 years to plan, permit and construct a transmission line). Another related issue is the need for cost recovery certainty (see Chapter 11). At the regional level in the USA, Texas has addressed these issues with the establishment of a Competitive Renewable Energy Zone (CREZ) process, which allows transmission to be built and paid for in advance of the construction of the wind plants. The completed CREZ transmission projects will eventually transmit approximately 18.5 GW of wind power. The estimated time of completion is the end of 2013 (CREZ, 2010). This model is being applied to other parts of the USA and is beginning to be explored in Europe. In Portugal, the investments reported for added transmission capacity to integrate wind production have been USD$_{2005}$ 185 million in the period 2004 to 2009 for increasing wind penetration from 3 to 13% (Smith et al., 2010a). The network investment plan for the period 2009 to 2019 is another USD$_{2005}$ 138 million dedicated to the connection of wind and other (comparatively small) independent producers (REN, 2008). China has rapidly become the world's largest market for wind power plant installations, and is therefore also beginning to confront the challenges of transmission and integration. Much of the wind power plant construction is occurring in northern and north-western China, in locations remote from major population centres, and is necessitating significant new transmission infrastructure (e.g., Liao et al., 2010; Liu and Kokko, 2010; Deng et al., 2011). The pace of wind power plant construction has also created a lag between the installation of wind power plants and the connection of those plants to the local grid (e.g., Liao et al., 2010; Deng et al., 2011).

In North Germany, a transitional solution allowing curtailments of wind power was made while waiting for the grid expansion in order to protect grid equipment such as overhead lines or transformers from overloads (Söder et al., 2007). Germany has also changed the standard transmission line rating calculation to increase the utilization of the existing grid. Dynamic line ratings, taking into account the cooling effect of the wind together with ambient temperature in determining the transmission constraints, can increase transmission capacity and/or delay the need for network expansion (Abdelkader et al., 2009; Hur et al., 2010). In the UK, some wind projects accept curtailments in order to lower the connection cost to the (distribution) grid that otherwise would need reinforcements (Jupe and Taylor, 2009; Jupe et al., 2010). Curtailment was particularly high in Texas in 2009 with 17% of all potential wind energy generation within the Electric Reliability Council of Texas curtailed (Wiser and Bolinger, 2010).

Many countries have already experienced high instantaneous wind penetration during low demand situations. Wind power is usually last to be curtailed. However, when all other units are already at minimum (and some shut down), system operators sometimes need to curtail wind power (Söder et al., 2007) to control frequency. Denmark has solved part of the curtailment issues by increasing flexible operation of CHP and by lowering the minimum production levels used in thermal plants (Holttinen et al., 2009). Experience from both Denmark and Spain shows that when reaching penetration levels of 5 to 10%, an increase in the use of reserves can be required, especially for reserves activated on a 10 to 15 minute time scale although, so far, no new reserve capacity has been built specifically for wind power (Söder et al., 2007; Gil et al., 2010). In Portugal and Spain, new pumped hydro is planned to be built to increase the flexibility of the power system, mainly to avoid curtailment of wind power (Estanqueiro et al., 2010). In small power systems such as those on islands, system balancing is more challenging due to a lack of load aggregation (Katsaprakakis et al., 2007). Power system operators have reported challenging situations for system balancing caused by high ramp rates for wind power production during storms when individual wind power plant production levels can drop from rated power to zero over a short time span, due to wind turbines cutting out. Due to aggregation effects, the impact on the power system/control area is often spread over 5 to 10 hours, however, and these events are rare (once in one to three years) (Holttinen et al., 2009).

In Ireland some curtailments have been due to concerns about low inertia (Dudurych, 2010b) and consequently susceptibility to instability in the system due to high instantaneous wind penetration and low system load. Currently, the issue of low inertia is unique to small systems like Ireland and possible solutions are being investigated (EirGrid, 2010b). In order to allow higher instantaneous penetration levels, the capability of wind power plants to provide (some) ancillary services must be improved. Equally, flexible balancing plants that can operate at low output levels and deliver stabilizing services would facilitate high instantaneous penetrations.

Low inertia has not, as yet, caused a problem for larger power systems but is being investigated (Vittal et al., 2009; Eto et al., 2010). Concerns about frequency regulation and stability have resulted in instantaneous penetration limits in the range of 30 to 40% for wind power on some Greek islands, including Crete (Caralis and Zervos, 2007a; Katsaprakakis et al., 2007; RAE, 2007). Frequency control and frequency response requirements associated with integration of Danish wind generation are reported to be virtually nonexistent (Eto et al., 2010) because the contribution of Danish wind generation is comparatively small in the large interconnected Continental Europe and Nordic systems (Denmark is connected to both). Experiences reported by the system operators in the Iberian Peninsula (Spain and Portugal) are consistent with those in Denmark in that no significant frequency impacts have been observed that are the result of wind power variation (Eto et al., 2010).

Formal forecasting methodologies are now implemented by system operators in many countries with high wind penetration (e.g., Denmark, Spain and Germany), with user acceptance/demonstration trials taking place in countries elsewhere (Ackermann et al., 2009; Grant et al., 2009). In Australia, the experience from a real-time, security-constrained, five-minute dispatch spot market, associated derivative and frequency control ancillary services markets, and a fully integrated wind energy forecasting system show that markets can in principle be designed to manage variable renewable sources (MacGill, 2010). Managing the variability and limited predictability of wind power output in China is made more complex by (1) China's reliance on coal-fired generation and the relatively low capacity of more flexible generation sources, especially in the regions where wind development is most rapid; (2) the still-developing structure of China's electricity and ancillary services markets; (3) the limited historical electricity trade among different regions of China; and (4) grid code requirements for wind plant installations that, historically, have been somewhat lenient (e.g., Yu et al. 2011). As a result of some of these factors, wind power plant curtailment has become common, especially in northern China. In Japan, the low flexibility of the power system has led to the development of certain options, such as requiring batteries in wind farms to reduce the night time variability (Morozumi et al., 2008).

There are short- and longer-term impacts of wind energy on wholesale electricity prices (Section 7.5.3.1). In Denmark, the Nordic electricity market is used for balancing wind power. The system operator balances the system net imbalance during the hour and passes this cost to all generators that have contributed to the imbalance, as balancing costs. Balancing costs for wind power are incurred when there are differences between the wind generation bid into the market (according to forecasts) and the actual production. The balancing cost of Danish wind power from the Nordic market has been approximately USD$_{2005}$ 1.37 to 2.98 per MWh (0.38 to 0.82 USD$_{2005}$/GJ) of wind energy (Holttinen et al., 2009). The Danish case also shows how interconnection benefits the balancing task: when Denmark is separated from the Nordic market area due to transmission constraints, the prices become very volatile with day-ahead market prices going to zero during windy low-load periods and with balancing prices also being affected (Ackermann and Morthorst, 2005; see also Section 7.5.3.2). There is already some initial experience in Germany and in the Denmark/Nordic market about how wind power impacts day-ahead electricity market prices—during hours with a lot of wind, the market prices are lowered (Munksgaard and Morthorst, 2008; Sensfuß et al., 2008). Other experience shows that wind power will increase the volatility in market prices when there is a high wind penetration in the market (Jónsson et al., 2010). Chapter 7 discusses the short- and longer-term impacts of wind energy on wholesale electricity prices (Section 7.5.3.1).

In Spain, the reliability impact of wind generation of greatest concern has been when network faults (for example short circuits) occur (Smith et al., 2010a). This concern is in part due to the older wind turbines deployed in Spain not being capable of FRT. Large amounts of wind power can therefore trip off the grid because of a short-lived transient disturbance of the grid (voltage drop). This problem has been addressed by new grid code requirements for wind power that have been adopted in many systems (Tsili and Papathanassiou, 2009) (Section 7.5.2.2). Germany has also changed the grid code to require FRT capability from wind turbines as simulated cases showed the possibility of losing more than 3,000 MW of wind power in a rather limited area in North Germany (Dena, 2005; Holttinen et al., 2009). The USA has also adopted a FRT requirement in FERC Order 661-A (FERC, 2005) as have a number of other jurisdictions (see Section 7.5.2.2 for more detail on grid codes for wind energy). The grid codes also require wind turbines to provide reactive power and in some regions also to take part in voltage and frequency control (Söder et al., 2007). Work in Spain has shown that wind power plants can contribute to voltage support in the network (Morales et al., 2008).

In Germany, wind and solar power have already created problematic flows through neighbouring systems (mainly the Netherlands and Poland; Ernst et al. (2010)).

Also of some concern is the possibility of low wind power production at times of high load. However, so far wind power has been built as additional generation and thus no problems with capacity adequacy were reported at least until 2007 (Söder et al., 2007).

Events in Germany in 2006 (UCTE, 2006) suggest that more and better information is needed in the control rooms of system operators, and also at the regional level (Section 7.5.3.2). Indeed, experiences from Denmark, Germany, Spain, Portugal and the USA show that system operators need to have on-line real-time variable renewable generation data together with forecasts of expected production (Holttinen et al., 2009). This can be challenging as variable renewable generation is sometimes from small units and is often connected to the distribution system. In Spain and Portugal, decentralized control centres have been established to collect on-line data and possibly to control smaller variable renewable power plants (Morales et al., 2008; J. Rodriguez et al., 2008). Experience from the USA shows that when most of the generation is connected to the transmission system, this is not as much of a problem, due to the requirement that wind plants provide supervisory control and data acquisition (SCADA) capability to transmit data and receive instructions from the transmission provider to protect system reliability (FERC, 2005).

Experience of a more institutional nature is the processing of large numbers of grid-connection applications that has led to group processing procedures in Ireland and Portugal (Holttinen et al., 2009; EirGrid, 2010a). Also the assessment of grid stability has required model development for wind turbines and wind power plants (Section 7.5.2.1). One high level experience that applies to integrating any form of generation into electrical power systems is the public opposition to overhead network infrastructure (Devine-Wright et al., 2010; Buijs et al., 2011). Evidence of this can be seen in Ireland and Denmark where needed transmission investment (not necessarily related to RE integration) is being opposed vigorously and burial options are being considered (Ecofys, 2008; Energinet.DK, 2008). Burying low voltage distribution networks is common practice, technically not challenging, but is more expensive. Burying high voltage transmission is rare, technically challenging and can be very costly (EASAC, 2009). The related issue of planning and permitting RE technologies is dealt with in detail in Section 11.6.4.

Results from integration studies for variable renewable sources
Numerous studies of RE integration have been undertaken over recent decades. It should be reiterated that integration issues are highly system specific and resource related and consequently there is a wide diversity of results and conclusions. To date most integration studies have focused on increasing levels of wind energy (typically above existing experience). Some recent large-scale studies look at both wind power and other renewable sources like solar and wave energy. There are very few dedicated and comprehensive solar or ocean integration studies, but there are some smaller-scale studies. Some of the results obtained from wind integration studies can also be applied to solar and wave integration.

The specific issues investigated in the wind integration studies vary and the methods applied have evolved over time, with studies building upon the experience gained in previous efforts (Section 7.5.4). Best practices are emerging and models are being improved (Smith et al., 2007; Söder and Holttinen, 2008; Holttinen et al., 2009). The main issues studied are the feasibility of integrating high levels of wind energy, the impact on the reliability and efficiency of the power system and the measures required to facilitate the increased levels of wind energy. Impacts typically considered include: effects on balancing at different time scales (e.g., any increase needed in reserves or ramping requirements); effects on the scheduling and efficiency of other power plants; impacts on grid reinforcement needs and stability; and impacts on generation adequacy and therefore long-term reliability. The large-scale studies briefly outlined below have been selected to illustrate key issues arising from wind integration into electrical power systems.[7] More detail on wind integration at low to medium penetration levels (i.e., <20%) can be found in Section 7.5.4.

A Danish analysis concluded that integration of a 50% penetration of wind power into the electricity system in Denmark by 2025 is technically possible without threatening security of supply (EA Energy Analyses, 2007). To do so would require new power system architectures that integrate local grids and consumers into system operation, coupling power generation, district heating (Section 8.2.2) and transport (Section 8.3.1), together with improved wind power forecasts and optimal reserve allocation. A strong transmission grid with connections to international markets will be needed, supported by a framework for improved international cooperation and harmonized operational procedures. In particular, the international electricity market must efficiently handle balancing and system reserve provisions across borders. Also, demand response would have to play a greater role as wind power penetrations increase (Energinet.DK, 2007; Eriksen and Orths, 2008).

The European Wind Integration Study (EWIS) and TradeWind are the first studies that examined wind integration at a European continental level. EWIS was led by a system operator consortium, and analyzed up to 185 GW of wind in 2015 (EWIS, 2010). TradeWind was led by a wind industry representative organization, the European Wind Energy Association, and analyzed up to 350 GW of wind in 2030 (TradeWind, 2009). Both studies identified the main interconnection upgrades needed (a total of 29 lines for 2015 by EWIS and a total of 42 lines for 2030 by TradeWind) and concluded that those interconnections would bring technical and economical benefits for the system in the short and long term. EWIS results pointed out that significant changes are needed in dispatch and interconnectors will be used more extensively. Additional measures needed to maintain system security include faster protection schemes, more reactive power compensation devices, faster ramping of other plants, and additional protection measures when using dynamic line rating for increasing network capacity. Future wind plants need to be equipped with state-of-the art FRT capability. The joint operation of the European network needs to be better coordinated, and dedicated control centres for renewable sources should be implemented similar to those in Spain (Morales et al., 2008; J. Rodriguez et al., 2008). Large-scale storage and demand side management were not found to bring significant benefits. The costs for upgrading the network for 185 GW wind by 2015 were found to be approximately 5.6 USD/MWh[8] (approximately 1.6 USD/GJ), while the additional deployment of reserves were estimated at 3.6 USD/MWh[8] (approximately 1.0 USD/GJ) (EWIS, 2010). TradeWind calculated the economic benefits of an offshore meshed transmission grid in the North Sea that could connect 100 GW wind power and improve electricity trade across the countries around the North Sea. Finally, the wind power capacity credit was found to be significantly higher when cross border transmission capacity in Europe was increased (TradeWind, 2009).

The U.S. Eastern Wind Integration and Transmission Study (EnerNex Corporation, 2010) examined three scenarios representing alternative build-outs of 20% wind energy, and a single build-out of 30% wind energy. The study found that new transmission would be required for all scenarios to avoid significant wind curtailment. In spite of the diverse locations of wind energy in the various scenarios, there is a common core of transmission that is required in each scenario. The study found that large regional control areas and significant changes in markets, tariffs and operations would be required. New transmission was found to enlarge the potential operating footprint, which decreases loss of load expectation and increases wind capacity credits. The wind capacity credit ranged from 16 to 23% in the lowest of three years, to 20 to 31% in the highest year. Adding new transmission increased the capacity credit of wind power by about 2 to 10 percentage points, depending on the year, scenario build-out and transmission additions.

The US Western Wind and Solar Integration Study (GE Energy, 2010) looks at a large regional electrical power system and finds that 30% wind and 5% "solar energy penetration is operationally feasible provided significant changes to current operating practice are made" (GE Energy, 2010). The changes include greater control area cooperation and sub-hourly generation and interchange scheduling. At penetration levels of 30% all available flexibility from coal and hydropower plants was found to be crucial for the operation of the power system. Up to a 20% penetration level relatively few new long distance interstate transmission additions were required assuming full utilization of existing transmission capacity. Wind was found to have a capacity credit of 10 to 15%, solar PV was 25 to 30% and CSP with six hours of thermal energy storage was 90 to 95%.

High system RE penetrations in the limited capacity and weakly interconnected Irish electricity system are anticipated to give rise to demanding integration challenges. Studies (AIGS, 2008; EirGrid, 2010b) have shown that 42% renewable sources including 34% wind is technically feasible at modest additional cost. Nonetheless, there will be a need for extensive transmission infrastructure development and a complementary flexible generation plant portfolio. Dynamic studies were also identified as a need, and the first stage of these was completed in 2010 (EirGrid, 2010b). It was confirmed that the technical performance of renewable and non-renewable generation to support high levels of renewable generation (mainly wind) is important. Operational limitations for non-synchronous generation, which may alter the fundamental characteristics of the electrical power system, may result in some curtailment of renewable generation but these operational restrictions will not prevent achievement of national targets for RE penetration (i.e., 40% electrical energy). However, these limitations will result in significant

---

7   Some of the studies also investigate other renewable sources but are dominated by wind.

8   Conversion to 2005 dollars is not possible given the range of study-specific assumptions.

curtailment if higher targets are set (assuming non-synchronous generation technology) and the economic barriers could be very significant. Similar operational limitations have also been reported for other island systems (Papathanassiou and Boulaxis, 2006).

The Hawaii Clean Energy Initiative (NREL, 2010) specifically identifies up to 400 MW of wind energy capacity offshore from Molokai and Lanai that could be brought by undersea cables (AC and/or DC) to Oahu as part of a diversified portfolio of RE technologies. The goal is 40% renewable electrical energy penetration. To accommodate the expected very high instantaneous penetration levels, the thermal generation minimum on-line level may need to be lowered and ramping capabilities increased. State-of-the-art wind and solar forecasting were also recommended.

There are also some studies on integration of wind in autonomous systems. On some islands, the maximum allowed wind power penetration has been restricted (Weisser and Garcia, 2005). Several studies have shown that this fixed limit does not guarantee system security and in some instances is not necessary. It has also been shown that it is possible to operate the power system of Crete with a high level of wind penetration while maintaining a high level of security when adequate and appropriate frequency and voltage control response from the other units are available (Karapidakis, 2007). Caralis and Zervos (2007b) investigate the use of storage in small autonomous Greek island systems where wind penetration is restricted for operational and dynamic reasons. They found that storage may reduce operational costs.

Many studies have specifically looked at the cost effectiveness of electricity storage to assist in integrating wind (Ummels et al., 2008; Denholm et al., 2010; Holttinen et al., 2011; Tuohy and O'Malley, 2011). Outside of autonomous energy systems, where storage may be more essential (Section 8.2.5), these studies have found that for wind penetration levels of as much as 50%, the cost effectiveness of building new electricity storage is still low when considering the need for wind integration alone due to the relatively higher cost of storage in comparison to other balancing options (excluding hydropower with large reservoirs and some pumped hydro). As and if storage costs decline, a greater role for storage in managing RE variability can be expected.

In general, the higher penetration studies have often been from island systems (Hawaii, Ireland). In such cases, the studies can be and need to be more detailed (AIGS, 2008; EirGrid, 2010b; NREL, 2010). Moreover, island systems (Hawaii, Ireland, Greek islands) are interesting as they can hit large penetrations faster, providing important early lessons for larger electric systems, and frequency control is more challenging. Another important trend, however, has been to study even larger areas in order to capture the impacts of variable renewable sources on a system wide basis, taking into account potentially valuable exchange possibilities (TradeWind, 2009; EnerNex Corporation, 2010; EWIS, 2010; GE Energy, 2010).

A useful attempt has been made to summarize the results of a number of recent wind integration studies (Holttinen et al., 2009). The studies cover different penetrations and systems and exhibit a wide range of results. Important conclusions include:

- Required increase in short-term reserve of 1 to 15% of installed wind power capacity at 10% penetration and 4 to 18% of installed wind power capacity at 20% penetration. The increased reserve requirement was calculated for the worst case (static, not dynamic) and does not necessarily require new investments for reserve capacity; rather generators that were formerly used to provide energy could now be used to provide reserves. The reserve requirements will be lower if shorter time scales are used in operation (gate closure time in markets).

- Increase in balancing costs at wind penetrations of up to 20% amounted to roughly 0.14 to 0.56 US cents/kWh[9] (roughly 0.4 to 1.6 USD/GJ) of wind power produced (see also Section 7.5.4.2). Balancing costs reflect increased use of reserves and less efficient scheduling of power plants. Though there is an increase in balancing costs and less efficient scheduling of power plants, the studies show a significant overall reduction of operational costs (fuel usage and costs) due to wind power even at higher penetration levels. Wind power is still found to lead to emission savings even with the increased integration effort (Denny and O'Malley, 2006; Mills et al., 2009b; Section 7.6.1.3).

- Capacity credit of wind is in the range of 5 to 40% of installed capacity depending on penetration, wind regime and correlation between wind and load (Keane et al., 2011a).

- The cost of grid reinforcements due to wind power is very dependent on where the wind power plants are located relative to load and grid infrastructure. Grid reinforcement costs roughly vary from 0 USD/kW to 378 USD/kW,[9] reflecting different systems, countries, grid infrastructure and calculation methodologies. The costs are not continuous; there can be single very high cost reinforcements. There can also be differences in how the costs are allocated to wind power.

While no large-scale and comprehensive studies have been conducted solely on the integration of solar there is a substantial body of work on the topic appearing in the literature. As PVs are installed predominantly locally, there is the possibility of reducing grid losses to the extent that the production coincides with demand (Wenger et al., 1994; Chowdhury and Sawab, 1996). At higher penetration, however, upgrades may be required to enable power to flow from the distribution feeder back to the transmission system without incurring large losses (Paatero and Lund, 2007; Liu and Bebic, 2008). In addition, voltage rise in distribution grids is an issue for PV integration (Widén et al., 2009). Thomson and Infield (2007), however, show that in a typical urban UK network with a very high PV penetration level (2,160 Wpeak on half of all houses), only small increases in average network voltages occur. Different studies propose solutions in order to avoid grid reinforcement such as

---

9 Conversion to 2005 dollars is not possible given the range of study-specific assumptions.

decentralized voltage control with reactive power (Braun et al., 2009). This could be performed by the PV inverters themselves (Stetz et al., 2010) or by other measures used for smart voltage control. Besides supporting frequency control and performing decentralized voltage control, other ancillary services could be provided by smart PV inverters. Such inverters can perform filtering/compensation of harmonics and support the fault behaviour of the power system with appropriate FRT capabilities (Notholt, 2008). In Japan, the target for PV is 28 GW in 2020 and 53 GW in 2030, which would supply around 3 and 6% of the total demand, respectively. Several demonstration projects in Japan addressed grid stabilization with large-scale PV systems by controlling PV generation and local demand (Kobayashi and Kurihara, 2009).

In some locations, adding solar PV to the system near demand centres may avoid the need to expand the transmission network. Kahn et al. (2008) illustrates a case in California where adding PV near coastal load centres would negate the need for significant transmission investments when compared with other renewable sources, in particular the transmission built to access solar PV, CSP, and geothermal in the desert described in the previous section. This benefit is likely to depend on local conditions and therefore vary greatly from region to region.

The capacity credit of solar varies in different parts of the world and by solar technology. In some electrical power systems due to high cooling demand at the peak load period, CSP with thermal energy storage can provide a capacity credit comparable to a thermal generator (GE Energy, 2010). The capacity credit for PV and CSP without thermal storage is much more dependent on the correlation of peak demand and the position of the sun (Pelland and Abboud, 2008; Perez et al., 2008; Xcel Energy, 2009; GE Energy 2010). The capacity credit of solar PV will drop as deployment increases (a similar characteristic to wind, see Section 7.5.2.4) due to the high degree of correlation between solar PV plants from the deterministic change of the position of the sun (Perez et al., 2008).

Managing the short-term variability of solar PV will be somewhat similar to that of wind power. The variability of solar PV systems can be considerable in partly cloudy weather and also with fog or snow (Lorenz et al., 2009; Mills et al., 2011). The ramping up and down during morning and evening of solar output, even if highly predictable and sometimes coinciding with load ramping, can also impose a large variation for electrical power systems with large amounts of solar PV energy (Denholm et al., 2009).

At increasingly high penetrations of solar PV and CSP without thermal storage (>10% annual energy production), the net demand (demand less solar production) will become increasingly low during the middle of the day when the sun is shining, while the night time net demand will not be reduced by these solar resources. Power systems with inflexible power plants may find it challenging to provide energy through the night, ramp down during daylight hours and then ramp back up at night. Inflexible electrical power systems are expected to therefore find integrating high levels of PV and CSP without thermal storage difficult without curtailing a significant amount of solar energy production (Denholm and Margolis, 2007).

Limited research exists in the published literature about ocean energy integration, but one review compared the integration of ocean energy with wind energy (Khan et al., 2009). Since there is little or no operational experience with ocean energy, the results are based only on simulations with little real data to validate the results. At an overall system level, however, the variability of ocean energy output is not expected to pose any greater challenges than the variability from wind power. However, short-term output fluctuations of wave energy plants could be greater than those from wind plants. Ocean wave resources are expected to have greater predictability than wind power because estimation of wave characteristics involves reduced uncertainties when compared to wind owing to its slower frequency of variation and direct dependence on wind conditions.

Bryans et al. (2005) explore methods of deployment and control of tidal current, including the down rating of the generator relative to turbine size and operational output reduction, to reduce the capital cost, increase capacity factor and reduce the impact on the grid system. The capacity credit (10 to 20%) and capacity factor (19 to 60%) of tidal current were also quantified. Denny (2009) used an electricity market model to determine the impact of tidal current generation on the operating schedules of the other units on the system and on the resulting cycling costs, emissions and fuel savings. It is found that for tidal current generation to produce positive net benefits for the case study, the capital costs would have to be less than USD$_{2005}$ 560/kW installed, which is currently an order of magnitude lower than the estimated capital cost of tidal current (Section 6.7).

Studies show that combining different variable renewable sources will be beneficial in smoothing the variability and decreasing overall uncertainty. A study undertaken in California, where the system load peak is driven by space cooling demand, shows that the average solar and wind plant profiles when considered in aggregate can be a good match to the load profile and hence improve the resulting composite capacity credit for variable generation (GE Energy Consulting, 2007). It should be noted that the negative correlation between wind and solar in California is not universal; there are many sites where positive correlation exists (e.g., Ireland, where the wind tends to peak in the late afternoon (Hasche et al., 2010). The combination of wind and hydro in British Columbia, Canada, was shown to lead to an improved capacity credit for hydro by using wind power to conserve water stored in the reservoir (Wangdee et al., 2010). Likewise, the independence of wind power and stream flows can reduce the risk of energy deficits in hydro-dominated systems (Denault et al., 2009). Additional analysis specifically on wind-hydro coordination is part of the ongoing IEA Wind Task 24.[10]

An analysis of high penetrations of RE in Denmark found that a mixture of wind, wave and solar power minimizes excess generation of RE. Wind

---
10 http://www.ieawind.org/Annex_XXIV.shtml

energy consistently contributed 50% of the RE mixture. The wave and solar share changed depending on the overall RE fraction (H. Lund, 2006). The potentials for reductions in variability when combining wave and wind energy have been reported for Scotland (University of Edinburgh, 2006), Ireland (Fusco et al., 2010) and California (Stoutenburg et al., 2010). How much of the reduction in variability is associated with the geographic diversity as opposed to the different resources remains an open question. Similarly, any benefits of technology diversity should be compared to the costs of diversifying the RE mix relative to the cost of a less diverse portfolio.

In summary, the results of integration studies for variable renewable sources vary depending on the system being analyzed, the level and type of renewable sources being considered and the methods and available data used in the analysis. However, some general messages can be drawn from the results. Studies show clearly that combining different variable renewable sources, and resources from larger geographical areas, will be beneficial in smoothing the variability and decreasing overall uncertainty for the power systems. The key issue is the importance of network infrastructure, both to deliver power from the generation plant to the consumer as well as to enable larger regions to be balanced; the options described below all need to be considered using a portfolio approach. There is a need for advanced techniques to optimize the infrastructure capacity required for variable renewable sources that have low capacity factors (Burke and O'Malley, 2010). The requirement to balance supply and demand over all time scales raises the need for access to flexible balancing resources (flexible generation, demand response, storage; NERC, 2010b) as well as the need to use advanced techniques for demand and supply forecasting and plant scheduling (NERC, 2010a). There is also a need for market or other mechanisms to ensure that all the complementary services necessary to balance supply and demand over all time scales are provided at a reasonable cost (Smith et al., 2010b; Vandezande et al., 2010).

Integration Options

The general form of the solutions required to accommodate a high penetration of renewable sources is largely known today. There is already considerable experience operating power systems with large amounts of renewable sources, and integration studies have also offered valuable insights into how high penetrations of renewable sources can be successfully achieved. This section examines in more detail the most important options identified to date. This should not be taken as a complete or definitive list since the future will no doubt open up new options and strategies. In addition, these options should not be viewed as competing in all circumstances, or that focussing on a single option will resolve all issues. Instead, for most electrical power systems, many, if not all of the options considered will be required, although the degree to which each is important may vary from one electrical power system to the next and over time (see Section 8.2.5 for a discussion of the autonomous systems and which of these options may be most appropriate in those circumstances).

Improving network infrastructure

Strengthening connections within an electrical power system, and introducing additional interconnections to other systems, can directly mitigate the impact of variable and uncertain RE sources. With strengthened connections, electrical energy can more easily be transmitted from where it is generated to where it can be consumed, without being constrained by bottlenecks or operational concerns. This argument also holds true for other generation and distributed loads, such that additional transmission may be viewed as of value to the entire system, rather than an integration cost associated with renewable generation. However, with much of this renewable generation being connected at the distribution level in some countries, greater cooperation and transparency will be required between distribution system operators and transmission system operators (Sebastian et al., 2008). Network expansion and refurbishment is an ongoing process to ensure security of supply and economic efficiency and to realize internal energy markets (ENTSO-E, 2010). Operating as part of a larger balancing area, or sharing balancing requirements across electrical power systems, reduces the integration cost associated with renewable generation and reduces the technical and operational challenges. The opportunity then also exists to exploit the geographical diversity of supply from RE sources to reduce net variability and uncertainty. This may also enable a wider range of renewable sources to be accessed, bringing further potential aggregation benefits due to the imperfect correlation between different renewable sources: for example, the concept of bringing together the solar-rich regions of northern Africa and the Middle East with the windy regions of mainland Europe (Pihl, 2009).

While power systems have traditionally employed AC connections to link dispersed generation to dispersed loads, there can be advantages to using DC connections instead (Meah and Ula, 2007). For example, for long point to point transmission lines (>500 km approximately) there will be a capital cost saving, while for underground or sub-sea connections, issues surrounding reactive power requirements are drastically reduced (Velasco et al., 2011). Consequently, DC connections are increasingly seen as attractive for capturing energy from offshore renewable sources, and for creating sub-sea interconnections between neighbouring countries/regions. However, issues surrounding meshed (rather than point-to-point) high voltage DC (HVDC) grids remain to be resolved (Henry et al., 2010). The investments required to put in place such infrastructure will be substantial and the value they add to the system needs to be carefully assessed (EASAC, 2009).

Employing communications technology to monitor and control larger electrical power system areas will enable more efficient use of the network infrastructure and reduce the likelihood of bottlenecks and other constraints. The cost of implementation of a secure and reliable communications and network infrastructure, however, could well be high, depending on previous investment in the networks and the geographical location of potential renewable generator sites relative to the existing network. The variability and uncertainty of some renewable sources may result in local network constraints, but such concerns may be solvable if the renewable (or embedded) generation can provide network support services such as reactive power (Keane et al., 2011b). This capability exists for modern wind generators, although incentives to exploit it are generally lacking (Martinez et al., 2008). Opportunities to realize the potential of flexible AC transmission system (FACTS) devices (which already exist, but have only been installed in small numbers) and other power flow control devices may also develop, as and when system stability issues arise (X.-P. Zhang et al., 2006; Hingorani, 2007; Tyll and Schettler, 2009).

Delivering new network infrastructure will face institutional challenges, in particular to provide incentives for the required transmission investments and to ensure social acceptance of new overhead lines or underground or sub-sea cables (see also Sections 11.6.4, 11.6.5 and 8.2.1.3). Investment in new transmission is, for example in Europe, the business of transmission system operators who recover their costs through transmission usage system charges. In some situations it is possible to divide the costs between different stakeholders. An effective framework should anticipate the need for transmission upgrades, so as not to inhibit investment in desirable new generation capacity (renewable or otherwise). Public opposition to new transmission lines can develop, traditionally linked with visual impacts (Devine-Wright et al., 2010), environmental concerns and the perceived impacts of electromagnetic fields on human health (Buijs et al., 2011). Underground cables are an available, but not necessarily preferable, option to alleviate such problems: cable reliability and maintenance concerns are potentially higher, and the investment cost will be much higher. With long underground connection distances (i.e., over 50 km approximately), DC will be the preferred technology (Schultz, 2007).

Increased generation flexibility
Thermal generation provides most of a power system's existing flexibility to cope with variability and uncertainty, through its collective ability to ramp up, turn down and cycle as needed (Troy et al., 2010). An increasing penetration of variable renewable sources implies a greater need to manage variability and uncertainty, and so greater flexibility is required from the generation mix. This can imply either investment in new flexible generation or improvements to existing power plants to enable them to operate in a more flexible manner. Retirement of existing inflexible generation may further accelerate this process, whereas the use of storage hydropower has been found to facilitate operational integration. Thermal power plants can be designed or retrofitted to ramp up and down faster and more frequently, but this will in general have a cost, both in capital and operational terms (Carraretto, 2006). A challenge is to achieve all of these aims in such a way that unit efficiency is not lowered so much that costs and emissions are significantly increased (Denny and O'Malley, 2006). Variable renewable generators can also be a focus for a degree of flexibility, for example limiting the rate at which they increase their output, and providing local voltage support for the network. Such capabilities are increasingly standard for wind generation (Z. Chen et al., 2009), but much less so for other variable renewable technologies. Increasing the flexibility of the generation fleet can occur progressively as power plants are modernized and investors see the need for more flexible operation to better respond to system or market needs. A significant future issue will be that as more variable generation comes online, dispatchable generation may be displaced thus reducing the amount of flexibility available. Ensuring that future power plants can maintain stable and profitable operation at output levels lower than at present will help to address this concern, but system operators will need to carefully monitor the dynamic stability of the power system to ensure safe and secure system operation.

In parallel with increasing targets for RE sources in electrical power systems across the world, it should also be noted that non-renewable options for low carbon generation, such as nuclear and fossil fuel with CCS are also in active development. With technology choices being made for economic, technical, social and political reasons, RE generation must recognize factors that may help, or in some cases hinder, future growth. For example, deployment of newer technologies such as integrated gasification combined cycle (IGCC) with carbon capture and sequestration and further deployment of nuclear technology (fission and also possibly fusion in the distant future) could have impacts on RE integration. These technologies may, for example, lack the required flexibility to help integrate variable renewable sources (Q. Chen et al., 2010), meaning that high penetrations of both RE and IGCC/CCS or nuclear may pose special integration challenges.

Synergies and connections also exist between the electricity sector and other energy sectors, so, for example, combining electricity and heat allows for greater flexibility in the electricity side as thermal storage options are already cost effective (Kiviluoma and Meibom, 2010). RE will also have impacts on the dispatch of gas deliveries in the systems where it is mainly gas power plants that react to increasing flexibility needs (Qadrdan et al., 2010).

Demand side measures
Flexible elements of demand, such as remotely switched night storage heating (Fox et al., 1998), have long been used, and often with good cost efficiency (Buckingham, 1965), to aid system operation. However, implementations tend to be proprietary in nature, installed over small geographical areas and with limited demand controllability actually offered. The development of advanced communications technology, with smart electricity meters linked to control centres, offers the potential to access much greater levels of flexibility from demand. One of the key opportunities is to make domestic demand flexible. Through pricing electricity differently at different times, and in particular higher prices during higher load periods, electricity users can be provided with incentives to modify and/or reduce their consumption. Such demand side management schemes, in which individual discretionary loads respond to price signals and/or external response 'request' signals, are seen as having a large potential (Brattle Group et al., 2009; Centolella, 2010). Thermal loads are ideal and include air conditioning, water heating, heat pumps and refrigeration, since the appliance can be temporarily switched on/off without significant impacts on service supply due to

intrinsic energy storage (Stadler, 2008). Water desalination, aluminium smelting, ice production, production line inventory, oil extraction from tar sands and shale deposits etc. can offer a similar flexibility (Kirby, 2007; Kirby and Milligan, 2010). Commercial entities may be particularly attractive, as installations will tend to be larger (load served), they are more likely to participate in schemes that deliver cost savings and they may be more willing to invest in necessary equipment. Electric vehicles represent an emerging load, but uncertainty exists about public uptake, battery performance and daily charging patterns. Vehicle battery charging, or even vehicle battery discharging, is potentially a further example of a discretionary load that can be controlled to assist in daily electrical power system operation (Kempton and Tomic, 2005).

All forms of demand side management require consumer engagement, in terms of changes in behavioural patterns, social acceptance and privacy/security issues. The implications of these various factors are not fully understood at present and more research is required. In addition, the amount of peaking plant that can be replaced by demand side measures is not fully understood (Earle et al., 2009; Cappers et al., 2010). Furthermore, a market or incentive system is required. Real-time electricity pricing (or some approximation) may be more widely adopted, whereby the electricity cost to the user more accurately reflects the cost of supply. However, demand side schemes are required that not only enable consumers to participate but actively encourage such behaviour, and correctly allocate charges and payments where required.

Although demand side measures have historically been implemented to reduce average demand or demand during peak load periods, demand side measures may potentially contribute to meeting electrical power system needs resulting from increased variable renewable generation. The low capacity credit of some types of variable generation, for instance, can be mitigated through demand side measures that reduce demand during peak load periods (Moura and de Almeida, 2010). Additionally, demand that can quickly be curtailed without notice during any time of the year can provide reserves (Huang et al., 2009), which have the potential to reduce electrical power system costs and emissions associated with short-term balancing of variable generation (Strbac, 2008; GE Energy, 2010). Demand that is flexible and can be met at anytime of the day can also participate in intra-day balancing, which mitigates day-ahead forecast errors for variable generation (Klobasa, 2010). Demand that responds to real-time electricity prices, on the other hand, may mitigate operational challenges for thermal plants that are expected to become increasingly difficult with variable generation, including minimum generation constraints and ramp rate limits (Sioshansi and Short, 2009). Challenges with managing electrical power systems during times with high wind generation and low demand, meanwhile, may be mitigated to a degree with demand resources that can provide frequency regulation (Kondoh, 2010). Off-peak electrical vehicle charging increases electrical demand and may reduce curtailment of variable renewable generation in high penetration scenarios (Lund and Kempton, 2008; Kiviluoma and Meibom, 2011).

The economic viability of any of these demand side measures should be evaluated relative to meeting the system needs with other resources, including renewable resources. Ultimately, however, accessing the flexibility of demand to mitigate variable renewable resources will depend on the integration of the demand side into system planning, markets and operations along with adequate communication infrastructure between power system operators and load aggregators/customers. It will also be necessary to engage, inform and provide incentives to users to participate in such schemes.

Demand side participation may have a particular role in small autonomous systems where there is limited access to other balancing resources.

Energy storage

At any given time, the amount of energy stored at plants in the form of fossil fuels or water reservoirs is large (Wilson et al., 2010). The amount of energy that can be converted into electricity and then converted back into stored energy, called electricity energy storage, is currently much more modest. The most common form of large-scale electrical energy storage is the mature technology of pumped hydro storage. Since the first pumped hydro storage plant was built in the late 1920s, over 300 plants with approximately 95 GW of pumped hydro capacity have been built in the world (Deane et al., 2010). Additionally, two large-scale commercial compressed air energy storage plants have been operating in Germany and the USA since 1978 and 1991, respectively, and a number of additional facilities are being planned or are under construction (H. Chen et al., 2009). Electrical energy storage is used in power systems to store energy at times when demand/price is low (i.e., off peak during the night /weekend) and generate when demand/price is high (i.e., at peak times during the afternoon). In addition, energy storage units can be very flexible resources for an electrical power system, and if correctly designed can respond quickly when needed (Mandle, 1988; Strunz and Louie, 2009). Technologies such as batteries or flywheels that store smaller amounts of energy (minutes to hours) can in theory be used to provide power in the intra-hour timeframe to regulate the balance between supply and demand in microgrids or in the internal network of the energy user (behind the electricity meter). Whether such technologies will be widely deployed will depend on capital costs, cycle efficiency and likely utilization (H. Chen et al., 2009; Ekman and Jensen, 2010). However, coupled with demonstration programs, market rules and tariffs are gradually being introduced to provide incentives for the participation of new technologies (Lazarewicz and Ryan, 2010; G. Rodriguez, 2010). Battery technology is an area of active research, with costs, efficiencies and other factors such as lifetime being improved continuously.

By storing electrical energy when renewable output is high and the demand low, and generating when renewable output is low and the demand high, the curtailment of RE will be reduced, and the base load units on the system will operate more efficiently (DeCarolis and Keith, 2006; Ummels et al., 2008; Lund and Salgi, 2009; Denholm et al., 2010; Loisel et al., 2010; Tuohy and O'Malley, 2011). Storage can also reduce transmission congestion and may reduce the need for, or delay, transmission upgrades (Denholm and Sioshansi, 2009). In autonomous systems, in particular, storage can play a particularly important role (Section 8.2.5).

When using storage to assist the integration of variable generation, storage should be viewed as a system asset to balance all forms of variability, including demand variations, as opposed to dedicating a storage unit to a single variable source. It is generally not cost effective to provide dedicated balancing capacity for variable generation in large power systems where the variability of all loads and generators is effectively reduced by aggregation, in the same way as it is not effective to have dedicated storage for outages of a certain thermal power plant, or to have specific plants following the variation of a certain load.

Market prices or system costs should determine how the storage asset is best used. The value of storage depends on the characteristics of the power system in question: its generation mix; its demand profile; connectivity to other systems; and the characteristics of the variable renewable generation plant (Tuohy and O'Malley, 2011). This is true for all power systems, including small autonomous systems (Caralis and Zervos, 2007a; Katsaprakakis et al., 2007). Storage must ultimately compete against increased interconnection to other electrical power systems, greater use of demand side measures, and the other options outlined here (Denny et al., 2010). The most effective choice is likely to be system specific and the economics will be affected by any specific electricity market incentives. Large-scale development of energy storage at the present time, however, remains questionable due to the generally high capital cost and inherent inefficiency in operation, unless these costs and inefficiencies can be justified through a reduction in curtailment, better use of other flexible resources or more efficient operation of the system more generally (DeCarolis and Keith, 2006; Ummels et al., 2008; GE Energy, 2010; Nyamdash et al., 2010; Tuohy and O'Malley, 2011). At the same time, storage technologies have attributes that have not, to this point, been fully valued in all electricity markets. For example, storage technologies that can provide ancillary services and very fast injections of energy for short periods of time may be able to provide virtual inertia particularly on isolated or weakly connected power systems (Wu et al., 2008; Delille et al., 2010). As these additional benefits are valued and as storage costs decline, the role of electrical storage in balancing supply and demand and assisting in RE integration is likely to increase.

Improved operational/market and planning methods
Existing operational, planning and electricity market procedures are largely based around dispatchable generation and predictable load patterns. The software tools that support these activities are largely deterministic in nature. In order to cope with increased penetrations of variable and uncertain generation, however, there is a greater need to identify sources of flexibility in operating the system, to develop probabilistic (rather than deterministic) operations and planning tools (Bayem et al., 2009; Papaefthymiou and Kurowicka, 2009) and to develop more advanced methods to maintain the electrical stability of the electrical systems. More fundamentally, real-time operations and long-term planning have traditionally been viewed as separate, decoupled activities. With high renewable penetrations, the two processes must come closer together such that a system is planned that can actually be operated in an economic and reliable manner (Swider and Weber, 2007).

To help cope with the variability and uncertainty associated with variable generation sources, forecasts of their output can be combined with stochastic unit commitment methods to determine both the required reserve to maintain the demand-generation balance, and also the expected optimal unit commitment (Meibom et al., 2011). This ensures less costly, more reliable operation of the system than conventional techniques. Wind (generation) forecasting systems have been developed that include ensemble probabilistic forecasting, and the technology is reaching maturity, with high forecast accuracies now achievable (NERC, 2010a Giebel et al., 2011). Forecasting systems for other variable RE sources (e.g., wave and solar) will need to be developed in parallel with commercial implementation of the devices. In addition, future forecasting systems, for all renewable sources, must include the ability to adequately predict extreme conditions, persistent high or low resource availability and exceptional power ramp rates (Greaves et al., 2009; Larsen and Mann, 2009).

Moving to larger balancing areas, or shared balancing between areas, is also desirable with large amounts of variable generation, due to the aggregation benefits of multiple, dispersed renewable sources (Milligan et al., 2009). Institutional changes may be required to enable such interaction with neighbouring systems and electricity markets (e.g., policies on transmission pricing), with the underlying assumption that adequate interconnection capacity is in place. The creation of the European Network of Transmission System Operators for Electricity as the first continental transmission system operators association with legal obligations to establish binding rules for cross-border network management and a pan-European grid plan follows this principle. Similarly, by making decisions closer to real time (i.e., shorter gate closure time in markets) and more frequently, a power system can use newer, more accurate information and thus dispatch generating units more economically (TradeWind, 2009; EWIS, 2010; Weber, 2010). Using a higher time resolution (intra-day, with resolutions of five minutes or less) provides a better representation of variability and the required balancing (Milligan et al., 2009), and so also enables more optimal decisions to be made closer to real time. In addition, institutional or electricity market structures must evolve such that they can quantify the flexibility requirements of the power system, and put measures in place to reward it (Arroyo and Galiana, 2005). In addition, reduced utilization of thermal generation may require an examination of market mechanisms to reduce investor risk (e.g., capacity payments, longer-term contracts) (Newbery, 2005, 2010).

Advanced planning methods are also required to optimally plan the upgrade and expansion of the electrical networks to ensure that variable generation can be connected in an efficient manner, especially considering the large geographical and remote areas that will sometimes be involved. Methods should ensure best usage of the existing transmission and distribution networks, as well as the best locations for upgrades or extensions (Keane and O'Malley, 2005). Planning methods should also move from 'snapshot' type studies, where the times of greatest system risk are well known, towards studies that consider the variable nature of renewable generation, recognizing correlations between different renewable sources and daily/seasonal patterns, and how this can cause risk at different times throughout the year (Burke and O'Malley, 2010). New metrics, similar to those already used in long-term resource planning, also need to be developed to ensure that sufficient short-term flexibility is planned for (NERC, 2009; Lannoye et al., 2010). This will require an understanding of the variability and uncertainty that variable renewable sources bring to different time scales, and how these increase the existing load variability and uncertainty in the short and long term (capacity adequacy). Detailed modelling of all sources of flexibility will be required, including generation and demand response, such that planning studies reflect the operational potential (NERC, 2010b,c).

On-line stability analysis tools must also be developed to ensure that the electrical power system is secure and robust against plausible eventualities (Dudurych, 2010a; P. Zhang et al., 2010), with optimal network configurations determined, and system recovery strategies identified in advance. Effective operation and management of the potentially large numbers of generation units will be very challenging and require a sophisticated information and communication infrastructure (J. Rodriguez et al., 2008). The emergence of more sophisticated network monitoring and control, coupled with demand side management and storage options, will ease the integration of RE sources into electrical power systems, but the control systems and decision-making systems required to monitor and manage the resulting complexity at both the distribution network level and transmission network level remain to be developed.

Summary and knowledge gaps:
RE can be integrated into all types of electrical power systems, from large interconnected continental-scale systems to small autonomous systems. System characteristics including the network infrastructure, demand pattern and its geographic location, generation mix, control and communication capability combined with the location, geographical footprint, and variability and predictability of the renewable resources determine the scale of the integration challenge. As the amounts of RE resources increase, additional electricity network infrastructure (transmission and/or distribution) will generally have to be constructed. Time variable renewable sources, such as wind, can be more difficult to integrate than non-variable renewable sources, such as bioenergy, and with increasing levels maintaining reliability becomes more challenging and costly. These challenges and costs can be minimized by deploying a portfolio of options including electrical network interconnection, the development of complementary flexible generation, larger balancing areas, sub-hourly markets, storage technologies and better forecasting and system operating and planning tools.

Parallel developments such as a move towards the use of electric vehicles, an increase in electric heating (including heat pumps), demand side control through the use of smart meters and thermal generation are providing complementary physical flexibility and together with the expansion of renewable power generation are driving dramatic changes in electrical power systems. These changes also include altered institutional arrangements including regulatory and market mechanisms (where markets exist), in particular those required to facilitate demand response and that reward the desired electrical power system portfolio. In addition, should variable RE penetration levels increase, deployment could increase in both developed and developing countries and the range of technologies could become more diverse (for example, if ocean energy technologies become competitive). These changes and developments lead to several gaps in our knowledge related to integration options that may become important in the future, including:

- Fundamental characteristics of future power systems due to wide spread deployment of non-synchronous generation, aspects of which were explored in EirGrid (2010b);

- Protection and interoperability of meshed HVDC networks, relevant for the connection of offshore wind and ocean energy (Henry et al., 2010);

- Changes to protective relaying to ensure system reliability and safety (Jenkins et al., 2010);

- New probabilistic methods for planning in the context of high proportions of variable stochastic generation (Bayem et al., 2009);

- Greater understanding of inter-area constraints and operational challenges (GE Energy, 2010);

- Changes in the non-renewable generation portfolio (e.g., impact of retirements, flexibility characteristics and the value of possible fleet additions or upgrades) (Doherty et al., 2006);

- Quantification of the potential for load participation or demand response (McDonough and Kraus, 2007) to provide the grid services needed to integrate RE (Sioshansi and Short, 2009; Klobasa, 2010);

- Impacts of the integration of the electricity sector with other energy sectors (Lund and Kempton, 2008);

- Integration needs in new and emerging markets that differ from those in which variable renewable sources have been integrated in the past (e.g., China);

- Benefits and costs of combining multiple RE resources in a complementary fashion (H. Lund, 2006); and

- Better market arrangements for variable renewable and flexible sources (Glanchant and Finon, 2010; Smith et al. 2010b).

8.2.2 Integration of renewable energy into heating and cooling networks

Heating, cooling and hot water account for a large share of energy use, particularly in the building and industry sectors. These energy services can be provided by using a range of fuels and technologies at the individual building level (Section 8.3.2) as can process heat and refrigeration for individual industries (Sections 8.3.3 and 8.3.4). District heating and cooling (DHC) is the alternative approach and this section deals with RE integration into such distribution networks.

8.2.2.1 Features and structure of district heating and cooling systems

DHC networks enable the carrying of energy from one or several production units, using multiple energy sources, to many energy users. The energy carrier, usually hot or cold water or steam, is typically pumped through underground insulated pipelines to the point of end use and then back to the production unit through return pipes. The temperatures in district heating (DH) outward pipes typically average 80 to 90°C, dropping to 45 to 60°C in return pipes after heat extraction. Heat exchangers are normally used to transfer the heat from the network to a hydronic heating system with radiators or to a hot water system (Werner, 2004).

Heat and CHP production have historically been dominated by oil and coal but, after the oil crises in the 1970s, oil was replaced by other fuels in most systems. In Western Europe, where DH systems commonly occur, the most popular fuels are natural gas and coal, although oil and biomass (Section 2.4; Figure 2.8) are also used. Coal still dominates in China and Eastern Europe. Waste heat from industrial processes, heat from waste incineration, geothermal heat and solar heat are feasible alternatives but less commonly used (Oliver-Solà et al., 2009).

Large DHC systems offer relatively high flexibility with respect to the energy source. Centralized heat production in DHC facilities can use low quality fuels often unsuitable for individual boilers and furnaces in buildings.[11] They also require pollution control equipment. Improved urban air quality and the possibility to cogenerate heat and electricity at low cost were, and still are, important motivations for DH (IEA, 2009c).

A good example of a central DHC plant is in Lillestrøm, Norway (Figure 8.3). It uses several energy sources, including a heat pump based on sewage effluent, to deliver heat and cold to commercial and domestic buildings. This system, and other DHC systems generally, includes an accumulator tank for hot water storage to even out fluctuations in demand over the day(s) to facilitate more stable production conditions (Section 8.2.2.4). The total investment is estimated to be around USD$_{2005}$ 25 million with completion planned in 2011.

Different production units dispatch heat in optimal ways to meet the varying demand (including the use of dedicated fast-response boilers and storage to meet peak demand). Higher overall system efficiencies can be obtained by combining the production of heat, cold and electricity and by using diurnal and seasonal storage of heat and cold. Using heat and cold sources in the same distribution network is possible and the selection of conversion technologies depends strongly on local conditions, including demand patterns. As a result, the energy supply mix varies widely between different countries and systems (Werner, 2006a).

DHC systems can be most economically viable in more densely populated urban areas where the concentration of heating and cooling demand is high. DHC schemes have typically been developed where strong planning powers exist and where a centralized planning body can build the necessary infrastructure, such as centrally planned economies, American university campuses, countries with utilities providing multiple services as in Scandinavia, and urban areas controlled by local municipalities. Urbanization creates opportunities for new or expanded DHC systems, as demonstrated on a large scale in China (Section 8.2.2.6). Development of DHC systems in less dense or rural areas has been restricted by the relatively high costs of distribution and higher heat distribution losses (Oliver-Solà et al., 2009).

Development and expansion of most DHC systems took place after 1950 in countries with cold winters, but earlier examples exist, such as New York in 1882 and Dresden in 1900. World annual district heat deliveries have been estimated at nearly 11 EJ (Werner, 2004) (around 10% of total world heat demand; IEA, 2010b) but the data are uncertain. Several high-latitude countries have a DH market penetration of 30 to 50%, and in Iceland, with abundant geothermal resources, the share has reached 96% (Figure 8.4).

District cooling (DC) is becoming increasingly popular through the distribution of chilled or naturally cold water through pipelines, possibly using the pipes of a DH network in higher latitudes to carry water to buildings where it is passed through a heat exchanger system. The supply source, normally around 6 to 7°C, is returned at 12 to 17°C (Werner, 2004). Alternatively, heat from a DH scheme can be used during summer to run heat-driven absorption chillers.

---

11 An example is a DHC in Kalundborg, Denmark (Section 2.4.3) that has several bioenergy components, including a pilot lignocellulosic ethanol plant.

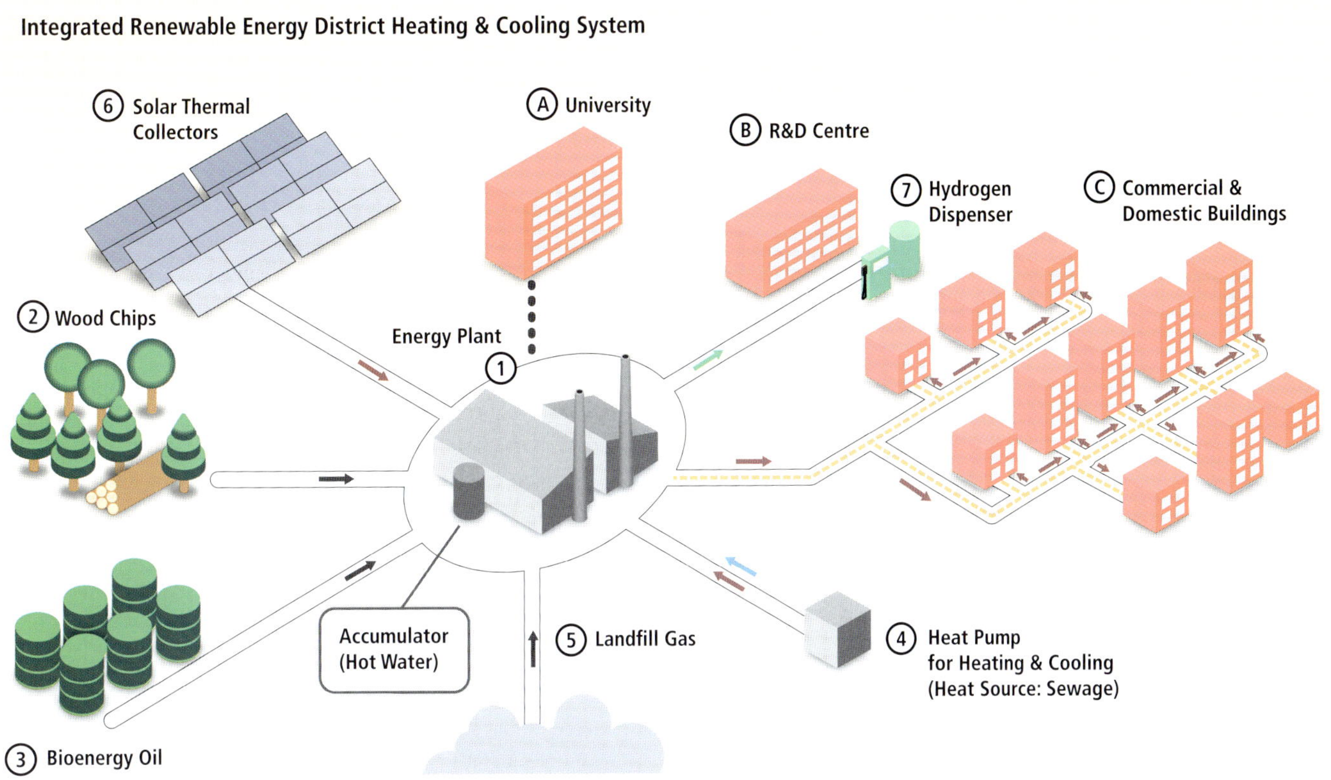

Integrated Renewable Energy District Heating & Cooling System

Figure 8.3 | An integrated RE-based energy plant in Lillestrøm, Norway, supplying the University, R&D Centre and a range of commercial and domestic buildings using a district heating and cooling system that incorporates a range of RE heat sources, thermal storage and a hydrogen production and distribution system (Akershus Energi, 2010).

Notes: (1) Central energy system with 1,200 m³ accumulator tank; (2) 20 MW$_{th}$ wood burner system (with flue gas heat recovery); (3) 40 MW$_{th}$ bio-oil burner; (4) 4.5 MW$_{th}$ heat pump; (5) 1.5 MW$_{th}$ landfill gas burner and a 5 km pipeline; (6) 10,000 m² solar thermal collector system (planned for completion in 2012); and (7) demonstration of RE-based hydrogen production (using water electrolysis and sorption-enhanced steam methane reforming of landfill gas) and fuel cell vehicle dispensing system planned for 2011.

Cooling demands in buildings are tending to grow because of increased internal heat loads from computers and other appliances, more stringent personal comfort levels and modern building designs having greater glazed areas that increase the incoming heat levels (IEA, 2007c). Recent warmer summers in many areas have also increased the global cooling demand, particularly to provide greater comfort for people living in many low-latitude, developing countries as their economies grow. Several modern DC systems, from 5 to 300 MW$_{th}$ capacity, have been operating successfully for many years including in Paris, Amsterdam, Lisbon, Stockholm and Barcelona (IEA, 2007d).

8.2.2.2 Characteristics of renewable energy in district heating and cooling systems

Over the past two decades, many DHC systems have been switched from fossil fuels to RE resources, initially in the 1980s to reduce oil dependence, but since then, to reduce carbon dioxide (CO_2) emissions. Centralized heat production can facilitate the use of low cost and/or low grade RE heat sources that are not suitable for use in individual heating systems. These include refuse-derived fuels, wood process residues and waste heat from CHP generation, industrial processes or biofuel production (Egeskog et al., 2009). In this regard, DHC systems can provide an enabling infrastructure for increased RE deployment.

The potential contribution and mix of RE in DHC systems depends strongly on local conditions, including the availability of RE resources. For biomass or geothermal systems it is not a technical problem to achieve high penetration levels as they can have high capacity factors. Hence many geothermal and biomass heating or CHP plants have been successfully integrated into DH systems operating under commercial conditions.

- Woody biomass, crop residues, pellets and solid organic wastes can be more efficiently used in a DH-integrated CHP plant than in individual small-scale burners (Table 2.6). Biomass fuels are important sources of district heat in several European countries where biomass is readily available, notably Sweden and Finland (Euroheat&Power, 2007). In Sweden, nearly half of the DH fuel share now comes from biomass (Box 11.11).

- Near-surface and low temperature geothermal resources are well suited to DH applications. Due to the often lower costs of competing fuels, however, the use of geothermal heat in DH schemes is

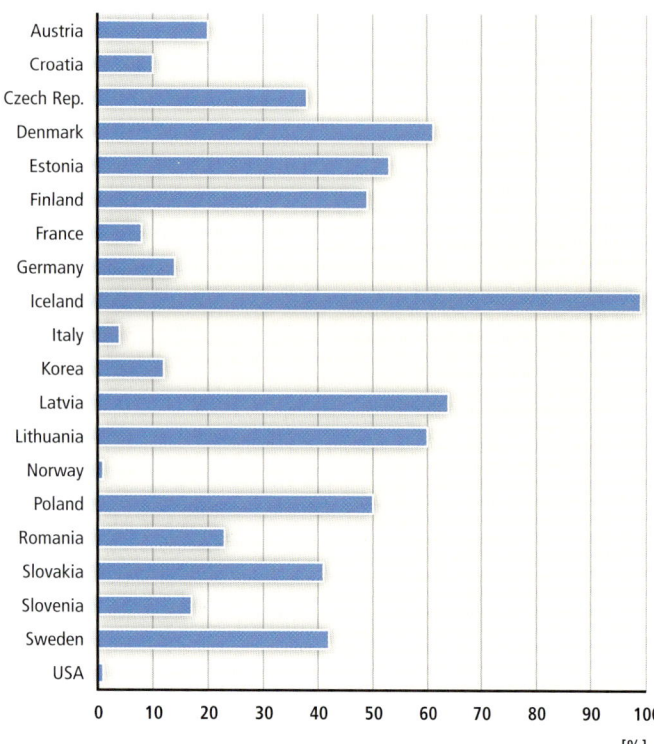

Figure 8.4 | Share of total heat demand in buildings supplied by district heating schemes for selected countries (Euroheat&Power, 2007).

low (with the exception of Iceland), even though the global technical potential of the resource is high (Section 4.2).

- The global installed capacity of solar thermal collectors in 2009 was 180 GW$_{th}$ (Section 3.4.1) but only a small fraction was used for DH (Weiss et al., 2009). Solar thermal DH plants are found mainly in Germany, Sweden, Austria and Denmark (Dalenbäck, 2010). In Denmark, several have large-scale collector areas of around 10,000 m² (Epp, 2009). At solar shares of up to 20%, the large number of customers connected to the DH system ensures a sufficiently large demand for hot water even in summer, so that high solar heat yields (~1,800 MJ/m²) can be achieved. Higher solar shares can be achieved by using seasonal thermal storage systems, for which integration into a DH system with a sufficiently high heat demand is an economic prerequisite. Pilot plants with a solar share of more than 50% equipped with seasonal heat storage have demonstrated the technical feasibility of such systems (Section 8.2.2.6).

Using RE through electricity sources in DH systems in situations with low or even negative electricity prices is possible through heat pumps and electric boilers, with thermal storage also an option (Lund et al., 2010). Through CHP plants, DH systems can also export electricity to the grid as well as provide demand response services that facilitate increased integration of RE into the local power system. Thermodynamically, using electricity to produce low grade heat may seem inefficient, but under some circumstances it can be a better economic option than spilling potential electricity from variable RE resources (Section 8.2.1).

DC systems that utilize natural aquifers, waterways, the sea or deep lakes as the source of cold can be classed as a RE resource. The potential for such cooling is difficult to estimate but many cities are located close to good water supplies that could easily provide a source of cold. Deep water cooling allows relatively high thermodynamic efficiency by utilizing water at a significantly lower heat rejection temperature than ambient temperature (Section 8.2.2.6). Often lake or sea water is sufficiently cold to cool buildings directly, which can, at times, enable the refrigeration portion of associated air-conditioning heat pump systems to be only operated to provide additional cooling when needed. All the excess building interior heat is transferred directly to the water heat sink.

To use RE cooling most efficiently in buildings from a quality perspective, a merit order of preferred cooling can be set up (as can also be done for heating) (IEA, 2007c). The order will differ due to specific local conditions and costs, but a typical example could be to supplement energy efficiency and passive cooling options by including active compression cooling and refrigeration powered by RE electricity; solar thermal, concentrating solar power, or shallow geothermal heat to drive active cooling systems (Section 3.7.2); and biomass-integrated systems to produce cold, possibly as tri-generation. The Swedish town of Växjö, for example, uses excess heat in summer from its biomass-fired CHP plant for absorption cooling in one district, and an additional 2 MW chiller is also planned (IEA, 2009b).

Ground source heat pumps can be used in summer for space cooling (air-to-ground) at virtually any location, as well as in winter for space heating (ground-to-air). They use the heat storage capacity of the ground as an earth-heat sink since the temperature at depths between 15 and 20 m remains fairly constant all year round, being around 12 to 14°C. They are commercially available at small to medium scales between 10 and 200 kW capacity.

8.2.2.3 Challenges associated with renewable energy integration into district heating and cooling networks

To meet growth in demand for heat or cold, and goals for integrating additional RE into energy systems, expansion of existing networks may be required. A DHC piping network involves up-front capital investment costs that are subject to large variations per kilometre depending on the local heat density and site conditions for constructing the underground, insulated pipes. Network capital investment costs and distribution losses per unit of delivered heat (or cold) are lower in areas with high annual demand (expressed as MJ/m²/yr, MW$_{peak}$/km² or GJ/m of pipe length/yr). Area heat densities can range up to 1,000 MJ/m² in dense urban, commercial and industrial areas down to below 70 MJ/m² in areas with dispersed, single family houses. Corresponding heat distribution losses can range from less than 5% in the former to more than 30% in the latter. The extent to which losses and network costs are considered an economic constraint depends on the cost and source of the heat. Under certain conditions, areas with either a heat density as low as 40 MJ/m²/yr, or a heat demand of 1.2 GJ/m of pipeline/yr, can be economically served by district heating (Zinco et al., 2008).

Energy efficiency measures in buildings and new building designs that meet high energy efficiency standards will reduce the demand for heating or cooling. As more buildings are built or retrofitted with low-energy and energy efficient designs, the total energy demand or density for existing DHC systems may decrease over time. Energy efficiency measures can also flatten the load demand profile by reducing peak heating or cooling demands. In these cases, the profitability of supplying district heat from either new DH plants or extending existing networks would be reduced (Thyholt and Hestnes, 2008). In Norway, Germany and Sweden the competition between low-energy building standards and DH development has received attention by policymakers working to design local or national energy policies (Thyholt and Hestnes, 2008). At the same time, while energy efficiency may be a challenge to the general economic viability of DH due to lower heating densities in the network, it may also facilitate higher shares of RE energy in individual heating systems (Verbruggen, 2006; IEA, 2009b).

The technical and economic challenges of heating and cooling using RE sources are not necessarily associated with the integration of the heat or cold into existing DHC networks that can be injected into a system for few additional costs. The challenges are instead primarily associated with assuring a consistent and reliable resource base from which the heat and cold can be produced.

- Combustion of wood residues or straw fuels can be challenging due to the varying composition of the fuel, the associated additional plant costs for storage and handling, fuel purchase costs and the need for a logistical supply chain to provide reliable supplies of biomass (Section 2.3.2).

- Extraction of geothermal heat is reliable but may entail local environmental impacts (Section 4.5).

- The variable nature of solar energy can be a challenge (Section 3.2) but is partly overcome by thermal storage. If used for DC, the need for diurnal and seasonal storage can be low because peak cooling demands often correlate relatively well with peak solar radiation levels.

In terms of cooling, the distance away from demand of the water to be used as the source of cold may also need costly infrastructure investment in order to integrate with DC systems. When using solar energy or biomass for absorption cooling, the challenges closely reflect those for heating.

In less densely populated areas, or those without a strong, centralized planning body, institutional barriers may pose challenges to developing or increasing the use of DHC, thereby posing indirect challenges to increasing the share of RE in the DHC networks. Constructing new capacity or expanding existing DHC networks usually requires planning consents and coordination of stakeholders and institutions.

8.2.2.4 Options to facilitate renewable energy integration

RE sources can be integrated into existing systems by replacing and retrofitting older production units or incorporating them into the designs of new DHC systems. DHC networks can be constructed or extended where a growing number of customers seek RE supply sources. These can be more cheaply integrated into existing systems at the slow natural rate of capital building stock turnover, or dedicated policies can speed up the grid connection process.

New technological options for heating

As new RE technologies are developed, additional technical options for increasing the shares of RE in DH systems are presented. Fuel switching and co-firing of biomass in existing fossil fuel-fired heat-only or CHP boilers present an option in the near term. The suitability of biomass fuels, their moisture contents, and whether they need to be pulverized or not, depend on the existing boiler design (whether grate, circulating or bubbling fluidized bed).

Heat from geothermal and solar thermal sources can be more readily integrated into existing DH systems. Enhanced geothermal systems (EGS) could be operated in CHP mode coupled with DH networks. The commercial exploitation of large heat flows is necessary to compensate for the high drilling costs of these deep geothermal systems (Thorsteinsson and Tester, 2010). Such a large heat demand is usually only available through DH networks or to supply major industries directly (Hotson, 1997).

Storage options

Heat storage systems can bridge the gap between variable and unsynchronized heat supply and demand. The capacity of a thermal storage system can range from a few MJ up to several TJ; the storage time from hours to months; and the temperature from 20°C up to 1,000°C. These wide ranges are made possible by choosing between solids, water, oil or salt as different thermal storage materials together with their corresponding storage mechanisms.

A hot water storage system design depends on the local geological and hydro-geological conditions, and the supply and demand characteristics of the DHC system. For short-term storage (hours and days) the thermal capacity of the distribution system itself can act as storage (Figure 8.5). Longer-term seasonal storage, usually between winter and summer, is less common. In this case, the main storage options include underground tanks, pits, boreholes and aquifers (Heidemann and Müller-Steinhagen, 2006). With geological storage, relatively small temperature differences are employed. In aquifers, heat may be injected during the

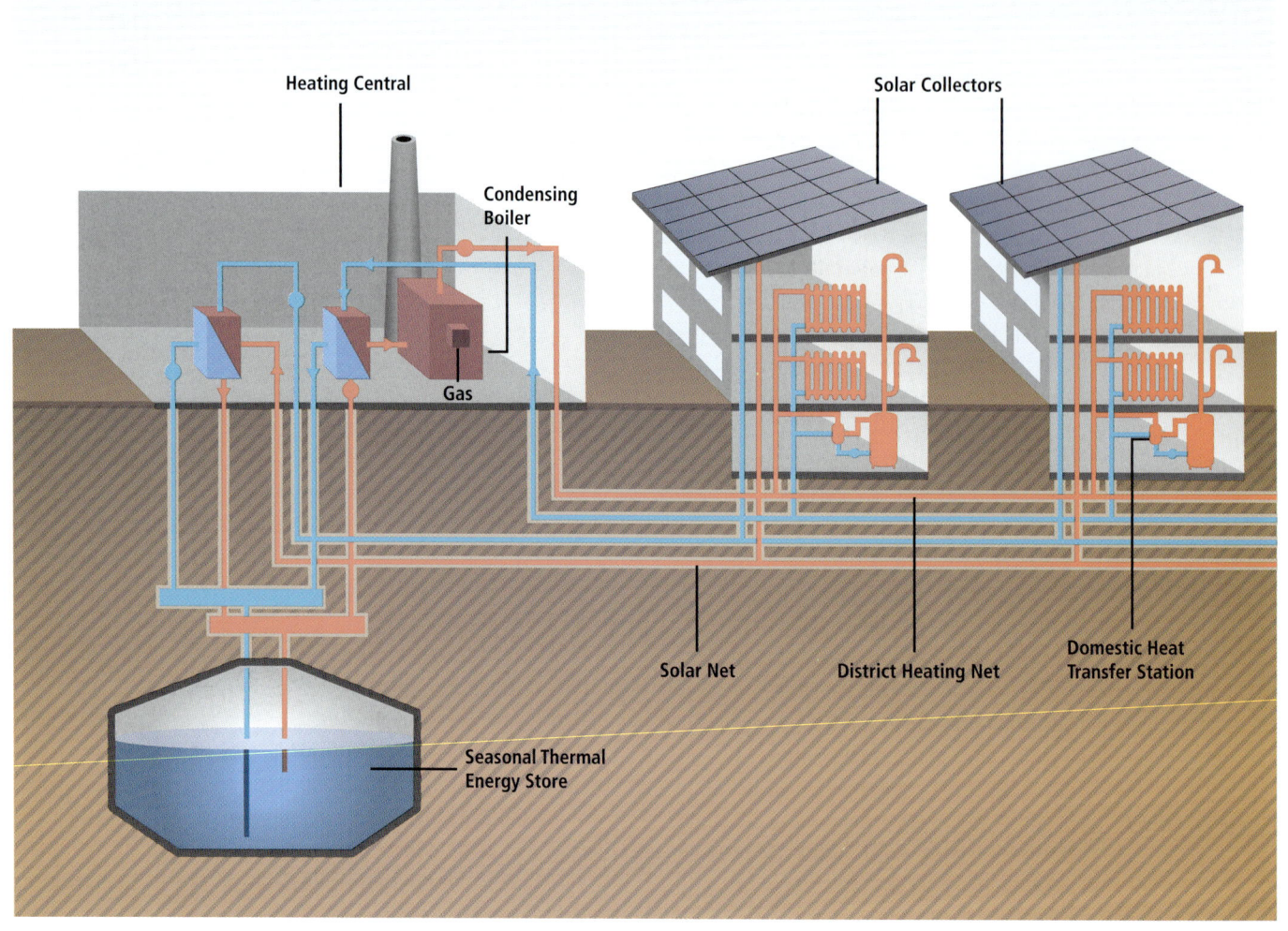

Figure 8.5 | Solar-supported centralized heating plant with seasonal tank storage connected to a district heating system (modified from Bodmann et al., 2005).

summer to increase the temperature and then be extracted during the winter. Seasonal storage is likely to become more important where high shares of solar thermal energy are used in DHC systems due to the seasonal and daily variability of the solar resource. In order to integrate large-scale solar systems into DH networks, the development of systems for seasonal heat storage (Figure 8.5) has made progress and several demonstration plants have been realized (Bauer et al., 2010).

Heat and cold storage systems using latent heat of fusion or evaporation, based on phase-change materials or the heat of sorption, offer relatively high thermal storage densities (Bajnóczy et al., 1999; Anant et al., 2008). Sorptive and thermo-chemical processes allow thermal storage for an almost unlimited period of time since heat supply or removal occurs only when the two physical or chemical reaction components are brought back into contact. However, both latent and sorptive heat storage technologies are in a relatively early phase of development.

Technological options for cooling

Cooling demands located remotely from a natural cold water source could be met using thermo-chemical sorption processes including chiller/heat pumps, absorption chillers or compression chillers (IEA, 2009b). Such active cooling systems can be used for centralized or decentralized conditioning and involve a range of technologies to produce cooling from a RE resource.

Solar-assisted cooling has been demonstrated in plants up to 3.6 MW$_{th}$ at Munich airport, but these technologies, being in their early stage of commercialization, tend to be relatively costly although the costs continue to decline with experience (IEA, 2007c). One main advantage of solar-assisted cooling technologies is that peak cooling demands often correlate with peak solar radiation levels and hence can offset peak electricity loads for conventional air conditioners.

Institutional and policy aspects

CHP as well as DHC developments do not always need financial incentives to compete in the marketplace, although government measures to address non-financial barriers, such as planning constraints, could aid greater deployment (IEA, 2008a) (Section 11.5.4). Some governments support investments in DH networks as well as provide incentives for using heat from deep geothermal and biomass CHP. In Germany, for example, if the share of RE is above 50%, a market incentive programme supports new DH schemes through investment grants in existing settlement areas, as well as for new development areas (BMU, 2009). In addition, the DH system operator receives a grant for each consumer connected to the new system.

In Sweden, high carbon taxes have provided strong incentives to switch to RE heating options (see case studies in Section 8.2.2.6 and Box 11.11). Targeted support under a climate investment programme has motivated investment in DH networks as well as new heating and CHP plants. Biomass CHP has also benefited from a quota obligation scheme (Section 11.5.3). DH, where available, is often competitive with alternative heating systems as a result of the carbon tax and other policy instruments (Figure 8.6). Similarly, under Danish conditions of high energy costs and carbon taxes, the integration of solar collectors into existing DH systems can be economically viable without additional targeted subsidies.

In the former centrally planned economies, DH prices were regulated because of a social policy to sell heat below the cost of supplying it. Today, in several countries with large DH schemes, an independent regulatory body ensures appropriate pricing where natural monopolies exist. In Denmark, for instance, the ownership of DH grids and the sale of heat as a monopoly are recognized, and hence the pricing and conditions of sale are regulated. The regulatory authority oversees the formation of prices and resolves disputes between consumers and utilities (Euroheat&Power, 2007).

In theory, third party access to DHC networks could lead to a more competitive market for heating services, stimulate independent producers of RE heating and cooling, and result in decreased heat prices for consumers. However, DHC plants operate and compete in markets that, by nature, are local, unlike national and regional electricity and natural gas markets. If a new competitor invests in a more efficient and less expensive production plant and is allowed to use the network of the existing DHC utility, then the incumbent utility may be unable to compete, the only choices then being to reduce the price and accept lost revenue. In this case, stranded asset costs could be higher than the customer benefits obtained from having a new third party producer, therefore resulting in the risk of a net overall loss. More pronounced competition could be obtained if several producers operate in the same network. However, most DHC systems are too small to host several producers. Thus, third party access into an existing DHC system must be evaluated on a case by case basis to ensure it is financially sustainable and beneficial for the customer.

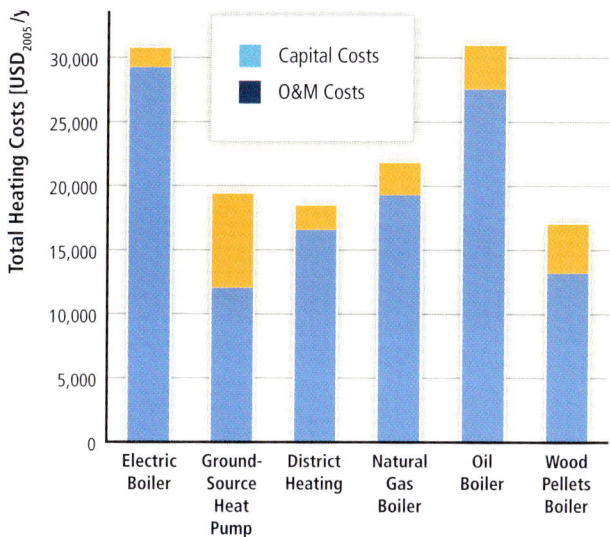

Figure 8.6 | Comparative average annual heating and unit costs (USD_{2005}), including climate, energy and carbon taxes, as seen by the end user in a typical Swedish 1,000 m^2 multi-family building with a heat demand around 700 GJ/yr.

Notes: Capital investment costs are for the end-user investment in the grid connection terminal, heat exchanger, boiler, heat pump etc. O&M costs are the end-user payments for electricity, district heat or fuel (including system capital costs, fuel, taxes, profit etc). For district heat, distribution cost is typically about 25% of total production and distribution costs and the distribution capital cost is about 35% of the total system capital cost. Data adapted from the Swedish Energy Markets Inspectorate (Ericsson, 2009). For the fuel mix of Swedish DH systems, see Figure 11.11.

8.2.2.5 Benefits and costs of large-scale penetration

The benefits and costs of increasing the contribution of RE sources in a DHC system depend on site-specific conditions such as the heating and cooling demand density, the availability of RE resources, and appropriate infrastructure. A Danish analysis of a future energy system, based upon achieving 100% RE by 2060, concluded that a gradual expansion of DH systems (using mainly heat pumps and biomass CHP), together with a switch to electric heat pumps for buildings that could not be connected to DH, would be the most efficient and least cost strategy for decarbonizing the heating of space and domestic water (Lund et al., 2010).

Large DHC systems offer benefits such as high overall system efficiencies (potentially obtained by combining the production of heat, cold and electricity and by using diurnal and seasonal storage of heat and cold) and relatively high flexibility with respect to energy source (as different production units can dispatch heat in an optimal way to meet varying demand). The incorporation of RE into DHC systems may provide additional benefits such as improved urban air quality, and the provision of heating and cooling at low or zero CO_2 emissions. (For a full discussion of these benefits of RE, see Chapter 9.)

DH networks represent a relatively mature technology. Expected reductions in heat network costs through improved design and reduced losses suggest that the expansion of DH will remain economically feasible in many locations, even in areas with relatively low heat densities (Bruus and Kristjansson, 2004). Improved designs include the co-insulation of paired, small diameter outward and return flow distribution pipes.

The total costs of a RE-based DHC system are highly contextual and site specific. The onsite heating of buildings using natural gas from grids in condensing boilers, small-scale heat pumps, biomass boilers, solar thermal systems or geothermal heat pumps can be strong competitors to DH in many locations. However, the ability of DHC systems to provide reliable supplies, avoid the need for maintenance of individual appliances, as well as integrate a broad spectrum of energy sources, facilitates competition among various heating/cooling sources, fuels and technologies (Gronheit and Mortensen, 2003). RE integration in itself does not lead to significant additional costs, except in the case where heat storage is necessary for high shares of solar thermal.

8.2.2.6 Case studies

Solar-assisted district heating system in Germany

As a demonstration project of proof-of-concept, a new residential area in Crailsheim with 260 houses, a school and sports hall has been designed for solar energy to displace about half the potential heat demand from a highly efficient, fossil fuel heating plant linked to the existing DH network. As a result, GHG emissions have been reduced by more than 1,000 t CO_2 per year (Wagner, 2009). Apartment blocks, new single houses and community buildings connected to the scheme are equipped with 3,800 m² of solar collectors with a further 3,500 m² installed on a noise protection wall that separates the residential and commercial areas. In 2010, a total annual heat demand of around 15 TJ is expected to be met by the solar collectors (Dalenbäck, 2010). Achieving such a high solar share was made possible by the use of a seasonal heat storage facility and a 100 m³ buffer tank used to directly meet instantaneous peak heat demands. Seasonal storage is provided by 75 55-m deep boreholes and a second 480 m³ buffer tank. The integration of a 350 kW heat pump allows the discharge of the borehole storage system down to a temperature of 20°C. This reduces the heat losses in the storage system and leads to a higher efficiency of the solar collectors due to lower return temperatures. The borehole storage system is designed to heat to 65°C by the end of summer and, at the end of the winter heating period, the lowest temperature is 20°C. Maximum temperatures during heat recharging will be above 90°C. In the second phase of the project, the heated residential area will be extended by 210 additional accommodation units, requiring an additional collector area of 2,200 m² and the seasonal storage system will need to be expanded to 160 boreholes (Mangold and Schmitt, 2006). Solar heat costs in this advanced proof-of-concept system are estimated to be around USD$_{2005}$ 67/GJ (Mangold et al., 2007). In less advanced systems without seasonal storage, the solar heat cost under northern European conditions is typically USD$_{2005}$ 14 to 28/GJ (Dalenbäck, 2010).

Biomass CHP district heating plant in Sweden

District heating in Sweden expanded rapidly between 1965 and 1985. Sweden used to be dependent on oil for the production of heat but after the 1979 oil crisis the fuel mix changed considerably. Since 2007, biomass has accounted for nearly half of total fuel supply in DH[12] (IEA, 2009c). The Enköping CHP plant is an illustrative case of this transition, driven by national CO_2 taxes, other policy instruments (Section 11.5.5, Box 11.8) and a local council decision to avoid fossil fuels (McKormick and Kåberger, 2005). The oil-fired DH system, constructed in the early 1970s, was converted after 1979 to use a mix of oil, solid biomass, coal, electric boilers and liquefied petroleum gas (LPG), until the construction in 1995 of a 45 MW$_{th}$, 24 MW$_e$ biomass-fired CHP plant enabled a transition to nearly 100% biomass by 1998.

The Enköping plant demonstrates an innovative approach to RE integration as a result of cooperation begun in 2000 among the local energy company, the nearby sewage treatment plant and a local landowner. The energy company wished to diversify fuel supply for the CHP plant fearing that there would not be enough forest residue biomass in the region to meet future heat and power demands. At the same time, the neighbouring municipal sewage plant was obligated to reduce its nitrogen discharges by 50%. The use of land treatment of the sewage effluent on to willow (*Salix*) was identified as a cost-effective solution. An 80 ha willow plantation acting as a 'nitrogen filter' was established on farmland adjacent to the sewage plant and close to the CHP plant. The farmer was remunerated for receiving the wastewater and sewage sludge on the land as well as by the market price for delivering biomass to the CHP plant. The success of this cooperation can be attributed to all parties being proactive and open to new solutions. Advisors worked as liaisons between parties, the regional and local authorities were positive and interested, and the risks were divided between the three main parties (Börjesson and Berndes, 2006). In 2008, the local area of willow plantations was increased to 860 ha and it is now the ambition of the energy company to further increase the biomass fuel share from the *Salix* to above the current 15%.

District heating in China

In China, the floor area of buildings served by DH has increased steadily from 277 million m² in 1991 to 3,489 million m² in 2008 (Figure 8.7), corresponding to an increase in heat deliveries from 0.4 EJ to nearly 2.6 EJ. About half of all Chinese cities, essentially those with colder winters, have DH systems (Kang and Zhang, 2008).

More than 95% of DH production in 2000 was based on coal. Nevertheless, the use of CHP results in lower emissions compared to the alternative of using individual boilers and coal-condensing power

---

12 The remaining heat production was based on 18% (35 PJ) from municipal solid waste, 10% industrial waste heat, 5% coal, 4% oil, 4% natural gas, 5% peat and 10% from heat pumps (Box 11.11).

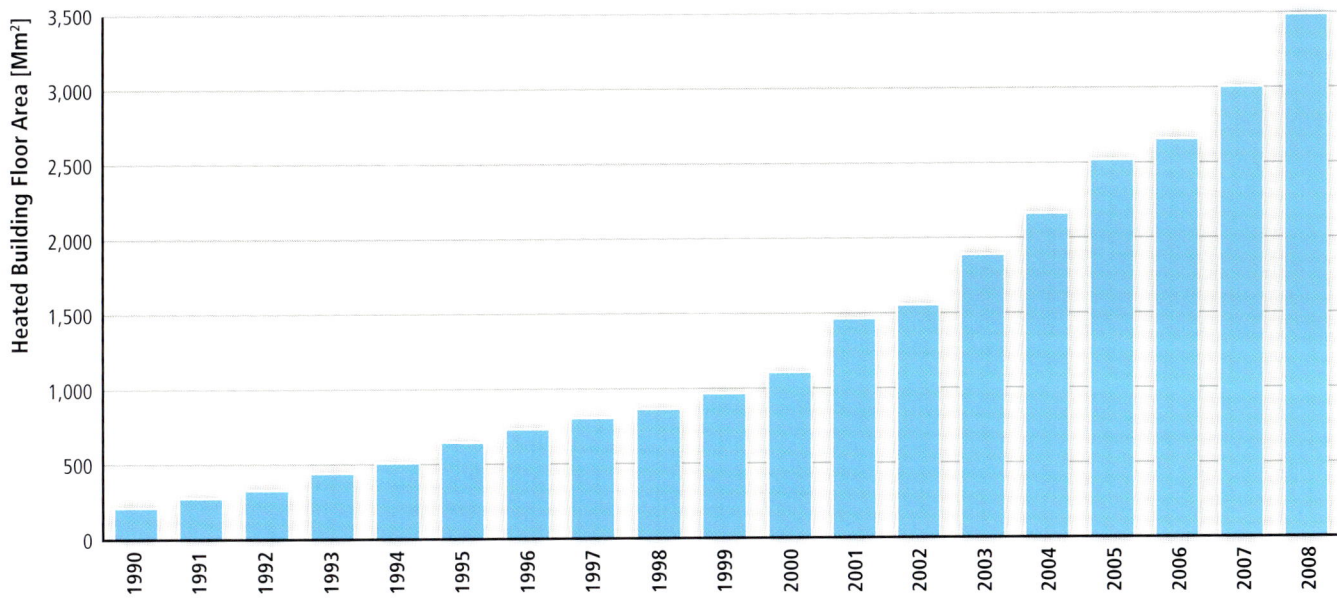

Figure 8.7 | Total area of building floor space served by district heating in China increased over twelve fold from 1990 through 2008 (adapted from Kang and Zhang (2008), updated with 2006-2008 data from the National Bureau of Statistics China, 2010).

generation plants. In the case of the city of Harbin, the result of installing a DH system was improved air quality in addition to 0.5 Mt/yr CO_2 emission reductions (WBCSD, 2008). Local air pollution concerns, as in Beijing and Tianjin, have motivated a shift from coal to natural gas in recent years, but interest in the integration of geothermal, biomass and solar thermal applications is now growing. For example, Shenyang, a leading city in the application of geothermal heat pumps, is meeting nearly one quarter of its 200 Mm^2 heated building floor area by taking water at 12 to 14°C from 80 to 160 m depths (Shenyang, 2006; Jiang and Hai, 2010). Eco-city developments, such as Caofeidian in Tangshan, are also fuelling a growing interest in RE for DHC systems.

District cooling in North America

Successful examples of DC installations include 51 MW of cooling at Cornell University, Ithaca, USA. Around 1,200 m^3/hr of 4°C water is pumped from the bottom of nearby Cayuga Lake through a heat exchanger before it is stored in a 20,000 m^3 stratified thermal storage tank (Zogg et al., 2008). A separate water loop runs back 2 km before passing through the air-conditioning systems of the 75 campus buildings and Ithaca High School. In this USD_{2005} 68 million scheme, the cooling water is discharged back to the lake at around 8 to 10°C and mixed with the surface water by 38 injection nozzles to maintain stable water temperatures. The 1.6 m diameter intake pipe has a screen at 76m depth and this, and the discharge nozzles, were carefully designed to minimize maintenance and environmental problems.

Compared with the original refrigeration-based cooling system, since the project started in 1999, GHG emissions have been reduced significantly due to both reducing the power demand for cooling by around 80 to 90% of the previous 25 GWh/yr (90 TJ/yr) and by avoiding the 12 to 13 t of chlorofluorocarbons (CFCs) that were used in the six chillers (Cornell, 2005).

The ecology, hydro-dynamics, temperature strata and geophysics of the lake have been closely monitored. There remain some concerns about bringing up phosphorus-rich sediments from the bed of the lake and discharging them near to the surface, which could possibly encourage algal growth.

In another example, Toronto, Canada, has pumped cold water drawn from nearby Lake Ontario to a 207 MW cooling plant since 2004. The DC system cools 3.2 million m^2 of office floor area in the financial district. The lake water intake pipe at 86 m depth runs 5 km out into the lake to ensure clean water is extracted, since this is also the supply for the city's domestic water system. No warm water return discharge impacts to the lake therefore result. Stockholm has a similar but smaller district cooling system based on extracting sea water from the harbour.

8.2.3 Integration of renewable energy into gas grids

The main objective of a gas grid is to transport gas from producers to consumers. The overall system consists of gas productions plants, main transmission and local distribution pipelines, storage tanks, and industrial or domestic gas consumers. The design of a gas system depends on the type and source of gas, location of the gas supply in relation to the consumers, and the supply volumes needed to meet peak demand. This section shows that replacing the combustion of some natural gas with biogas or biomass synthesis gas[13] to provide heat is relatively straightforward in the short term. Upgrading of biogas to biomethane (by removing carbon dioxide and hydrogen sulphide gases) and the cleaning of biomass synthesis gas (by removing tars) are necessary where the

---

13 Synthesis gas or syngas is a mixture of carbon monoxide, hydrogen, methane, higher hydrocarbon gases, and carbon dioxide also known as town gas or producer gas. It can be manufactured by gasification of coal or biomass (Section 2.2).

gas is to be injected into a natural gas grid or used to fuel an internal combustion engine in a vehicle or to power an electricity generator. In the longer term, demand for RE-produced hydrogen may expand but could need high capital investment in infrastructure.

8.2.3.1 Features and structure of existing gas grids

A gas distribution system is primarily designed to deliver adequate amounts of gas at a certain quality (heating value, pressure and purity) to downstream users. Existing gas processing, delivery and storage systems are quite complex. New designs depend on the existing energy system in the region where the gas grid is being considered. Consisting of different types of pipelines, some designs have been built just to supply local users with 'town gas'. Large grids, first developed in the 1960s, now traverse continents in order to distribute large volumes of gas from natural gas fields. For example, the natural gas grid in the USA is highly integrated, with more than 210 pipeline systems, 480,000 km of inter- and intra-state transmission pipelines and 394 underground storage facilities (EIA, 2007). Europe (EU27) has 1.8 million km of pipelines and 127 storage facilities to supply more than 110 million customers (Eurogas, 2008). To balance supply and demand, gas storage, usually in steel tanks, can be incorporated at various levels in the system. The capacity depends on how the gas is produced, how it can be integrated into the gas grid and the end-use applications. The volume of gas stored is normally minimized to reduce costs and safety hazards.

The gas flow rate depends on the scale and physical attributes of the gas (molecular weight, viscosity, specific heat) and the friction in the pipe (which depends on pipe layout, design and type). A pipeline designed with a large diameter and a high pressure drop can move more gas over a given distance than a smaller pipe at lower pressure (Mohitpour and Murray, 2000). There is an economic trade-off between increasing the diameter of the pipeline versus increasing the gas pressure.

The materials used in gas pipelines depend on the type of pipeline (transmission or distribution), location (sub-sea, overland, underground), operating conditions (pressure, temperature, moisture), and type and quality of gas or gases to be sent through the pipeline. Metallic materials are mainly used in larger transmission pipelines as they are tolerant to higher pressures and temperatures, but have the potential for internal and external corrosion problems (Castello et al., 2005). Plastics can be used in distribution gas grids operating at lower temperatures (<100°C) and pressures (<1,000 kPa).

Natural gas extraction points are normally connected to the pipeline head stations via trunk lines (at 7,000 to 10,000 kPa pressure). The gas is then pumped into long distance transmission pipelines (at 6,000 to 9,000 kPa) and sent to the takeoff stations from where it is transported to the control station of the regional distribution system (at 800 to 4,000 kPa), before it finally reaches industrial and household customers (at 5 to 10 kPa) (Castello et al., 2005). Distribution pipelines, including main feeders, station connections, valves and meters are contained on the property of the customer at the end-use point (EIGA, 2004).

Hydrogen pipelines are currently limited to a very few geographical areas that have large hydrogen consumers such as chemical and petrochemical industries (Castello et al., 2005). Blending of hydrogen (up to 20%) with natural gas on a large scale, and transporting this gas mix long distances in existing or new natural gas grids, could be a future option for the large-scale distribution of hydrogen (NATURALHY, 2009).

Once the energy feedstock for producing biogas or syngas has been established, the end-use application, whether for heating, combined heat and power (CHP), raw feedstock for the chemical industry, or transport fuels, needs to be determined. Local gas distribution systems have traditionally used gas-burning appliances to provide cooking, space and water heating. Using existing commercial internal combustion engine (ICE) and micro-turbine technologies, biomethane and syngas can also be used to fuel small- to large-scale CHP systems. The commercialization of highly efficient, small-scale fuel-cell-based CHP systems (with 80 to 90% overall efficiency) could contribute to a more energy efficient and cost effective use of existing and new gas grids in the longer term (DeValve and Olsommer, 2006; Zabalza et al., 2007).

8.2.3.2 Characteristics of renewable energy with respect to integration

Over the past decade there has been an increasing interest in 'greening' existing natural gas grids. In Europe the EU Directive 2003/55/EC of the European Parliament opened up the existing grid to carry alternative gases such as 'hythane' (a blend of hydrogen and natural gas), hydrogen and biogas (Persson et al., 2006; NATURALHY, 2009). Furthermore, an EU directive[14] included measures for increasing the share of biogas and enabling access to the gas grid. As a result, in Germany, for example, the target for 2020 is to substitute 20% (by volume; around 1.12 PJ/year) of compressed natural gas (CNG) used for transport with biomethane and a 2030 target is to substitute 10% of natural gas in all sectors with biomethane (382 PJ/year) (Müller-Langer et al., 2009). Similar proposals have been made for the natural gas grid running along the West Coast of North America, with a Bioenergy Action Plan having been introduced by the Governor of California (CEC, 2006).

Until recently, most of the raw biogas produced around the world (from landfills, urban sewage and industrial and agricultural wastes) has been used onsite or distributed in dedicated local gas systems, primarily for heating purposes. Biogas can be upgraded to biomethane of natural gas quality and suitable for blending with natural gas for transporting via gas grids. In a few cases biomethane has been transported via trucks to filling stations to supply gas-fuelled vehicles (Hagen et al., 2001; Persson et al., 2006). The biogas business is growing rapidly and is currently being commercialized by larger industrial players (Biogasmax, 2009).

---

14 Promotion of the use of energy from renewable sources. EU Directive 2009/28/EC.

These include major gas companies that are planning to upgrade and inject large quantities of biogas into national/regional transmission gas pipelines (NationalGrid, 2009) to offset some of the demand for natural gas in existing and future markets.

Synthesis gas can be produced via gasification (partial oxidation) of coal or biomass feedstocks (Section 2.2). The Lebon gasification process has been used since the beginning of the 19th century and the gas is already used for cooking, heating and power generation, especially in areas where natural gas is unavailable.

Hydrogen is today mainly produced from natural gas but it can also be produced from RE sources. The main current use is by industry (Section 8.3.3), but it can also be used as a transport fuel (Section 8.3.1). To establish a RE-based hydrogen economy, it will be necessary to develop more efficient small-scale distributed hydrogen production technologies such as water electrolyzers and steam methane reformers (Riis et al., 2006; NRC, 2008; Ogden and Yang, 2009). Small- to medium-scale hydrogen production based on wind (Section 8.3.4), solar or biomass has been evaluated favourably in some regions such as North Rhine-Westphalia, Germany (CEP, 2010) and the North American Great Plains (Leighty et al., 2006). Such RE technologies could conceivably provide the basis for large-scale hydrogen production in the future (IEA-HIA, 2006).

Several different options are available for hydrogen delivery, including road or rail transport of gaseous hydrogen compressed in cylinders of various sizes, trucking of cryogenic liquid hydrogen, and transmission by pipelines. The technical and economic competitiveness of each delivery option depends on the geographical area and gas volume demand. For small consumers, transport of liquefied or compressed hydrogen by trucks is the most viable option, while pipeline delivery can only be justified for a very high flow rate of hundreds of tonnes per day (Castello et al., 2005). The building of hydrogen production and distribution infrastructure over the next few decades could be a mix of centralized and decentralized systems (Bonhoff et al., 2009). Initially, hydrogen will mainly be distributed by trucks, while pipelines will only become important at a later stage as demand increases. For example, in North Rhine-Westphalia, a region with existing gas pipelines supplying industry, there are progressive plans to construct a hydrogen transport infrastructure based initially on the existing gas grid. Dedicated hydrogen pipelines could be needed after 2025 (Pastowski and Grube, 2009) and by 2050, about 80% of all hydrogen produced centrally in Germany could be transported by pipeline (Bonhoff et al., 2009).

Local gas distribution grids can complement heating and cooling networks (Section 8.2.2). At the national and regional scales, electricity and gas transmission grids can complement each other in the long-distance transport of energy carriers. The design of a future hydrogen infrastructure, for example, could depend strongly on its interaction with the electricity system (Sherif et al., 2005; C. Yang, 2008), which over time is expected to gain an increasing share of RE. Using surplus RE power to produce hydrogen by electrolysis is an example, possibly combining this with CO_2 (arising from biogas, fossil fuel combustion or extracted from the atmosphere) using the process of *methanation* to produce methane as an energy store and carrier (Sterner, 2009). Currently this process is not commercially viable.

8.2.3.3 Challenges caused by renewable energy integration

A few technical challenges exist related to gas source, composition and quality. The composition and specifications of fuel gases from different carbon-based sources vary widely (Table 8.2). Gas composition and heating values depend on the biomass source, gasification agent utilized in the process and reactor pressure. Such variations in quality may constitute a significant barrier to gas pipeline integration. Landfill gas or biogas from anaerobic digestion can be upgraded to reach a similar methane composition standard as natural gas (80 to 90% methane) by stripping out the CO_2 content before it is fed into a gas grid and/or used as a fuel in ICEs or high-temperature fuel cells. The composition of biomass-derived syngas depends on the type and moisture content of the organic feedstock and on the production method (e.g., using bubbling versus circulating fluidized bed gasifier designs).

Gas companies and/or authorities define the standard gas composition for injection into a gas grid on the basis of minimizing the risks associated with the infrastructure, the quality of combustion in industrial processes and domestic appliances, health, and emissions to the environment. In small-scale systems for stand-alone operations, the standards are mainly defined to minimize risks associated with the equipment and the processes themselves. Since only gases of a specific quality can be injected directly into a gas grid, meeting these standards can create market barriers for biogas and landfill gas producers (more than for syngas, which is relatively clean (Table 8.2) assuming tars can be avoided during gasification).

- CO_2 can be removed by several methods, but each have operational and cost issues (Persson et al., 2006).

  - Absorption (water scrubbing) requires large amounts of water. Blockage of the equipment by organic growth can also be a problem.
  - Absorption by organic solvents such as polyethylene glycols or alcohol amines requires large amounts of energy for regenerating the solvent.
  - Pressure swing adsorption requires dry gas.
  - Separation membranes, dry (gas-gas) or wet (gas-liquid) require handling of the methane in the permeate stream (which increases with high methane flow rates in the gas stream).
  - Cryogenic separation requires removal of water vapour and hydrogen sulphide (H_2S) prior to liquefaction of CO_2.

- Removal of corrosive H_2S from biogas is necessary to protect downstream metal pipelines, gas storage and end-use equipment. Micro-organisms can be used to reduce the concentration of H_2S by adding stoichiometric amounts of oxygen to the process (around 5% air to a digester or biofilter). Alternatively, simple vessels containing iron oxides can be used as they react with H_2S and can be easily regenerated once saturated by oxidation when placed in contact with air.

- Small volumes of siloxanes and organic silicon compounds (not shown in Table 8.2) can form extremely abrasive deposits on engine pistons, cylinder heads and turbine sections and cause damage to the internal components of an engine if not removed (Hagen et al., 2001; Persson et al., 2006).

- Other particulates and condensates may also need removal as there are normally low tolerances for impurities.

A community-scale biogas plant in Linköping, Sweden exemplifies an economically viable system for local use (IEA Bioenergy, 2010a). Multiple organic wastes are treated and processed in an anaerobic digester to produce biogas with similar properties to those shown in Table 8.2. The gas mixture is then upgraded to remove CO_2 and H_2S before the residual biomethane gas is distributed through a local grid to supply a slow overnight filling station for buses, 12 public refuelling stations for cars, taxis and fleet vehicles, and a refuelling system for a converted diesel train with 600 km range (IEA, 2010a). The system payback time is sensitive to the estimated long-term gas production and price, which in turn is affected by taxation and carbon values, the future end-use demands for the gas and the clean-up costs. The economic payback time to integrate 'scrubbed' biomethane into a gas grid depends on the location of injection. If injection is at the end of a pipeline as incremental capacity, then the cost can be relatively low. Local and regional differences in existing infrastructure make it difficult to make specific recommendations for planning and integration costs at a national and regional level.

Hydrogen transported via existing natural gas grids may first require some upgrading of the pipelines and components (Mohitpour and Murray, 2000; Huttenrauch and Muller-Syring, 2006). Since pure hydrogen has a lower volumetric density compared to natural gas, hydrogen pipelines will require either operation at higher pressures or around three times larger diameter pipes in order to carry the same amount of energy per unit time as a natural gas pipeline. In a dedicated hydrogen gas grid, depending on the hydrogen pathway but particularly if used with fuel cells rather than for direct combustion, the hydrogen needs to be purified and dried before it is stored and distributed. For

Table 8.2 | Examples of composition and parameters of gases from a range of carbon-based sources, using typical data for landfill gas, biogas from anaerobic digestion (AD), (Persson et al., 2006) and biomass-based syngas (Ciferno and Marano, 2002), and compared with natural gas.

Parameter	Unit	Landfill Gas	Biogas from AD[1]	Syngas from biomass[2]	North Sea natural gas
Lower heating value	MJ/Nm^3	16	23	4–18	40
Density	kg/Nm^3	1.3	1.2	—	0.84
Higher Wobbe index	MJ/Nm^3	18	27	—	55
Methane number		>130	>135	—	70
Methane, typical	vol-%	45	63	10	87
Methane, variation	vol-%	35–65	53–70	3–18	—
Higher hydrocarbons	vol-%	—	—	—	12
Hydrogen	vol-%	0–3	—	5–43	—
Carbon monoxide, typical	vol-%	—	—	30	—
Carbon monoxide, variation	vol-%	—	—	9–47	—
Carbon dioxide, typical	vol-%	40	47	25	1.2
Carbon dioxide, variation	vol-%	15–50	30–37	11–40	—
Nitrogen, typical	vol-%	15	0.2	35	0.3
Nitrogen variation	vol-%	5–40	—	13–56	—
Oxygen, typical	vol-%	1	—	<0.2	—
Oxygen, variation	vol-%	0–5	—	—	—
Hydrogen sulphide, typical	ppm	<100	<1,000	~0	1.5
Hydrogen sulphide, variation	ppm	0–100	0–10,000	—	1–2
Ammonia	ppm	5	<100	—	—
Total chlorine (as Cl$^-$)	mg/Nm^3	20–200	0–5	—	—
Tars	vol-%	—	—	<1	—

Notes: 1. Anaerobic digestion. 2. From gasification using bubbling or circulating fluidized beds with direct or indirect heating. Syngas followed by methanation can produce 83 to 97% methane and 1 to 8% hydrogen. Nm^3 is an uncompressed 'normal' cubic metre of gas at standard conditions of 0°C temperature and atmospheric pressure.

example, for fuel cell vehicles (Section 8.3.1), the hydrogen needs to be of a very high purity (>99.9995% H_2 and <1 ppm CO). Industrial hydrogen with lower purity can be transported in dedicated transmission and distribution pipelines so long as there is no risk of water vapour building up, or any other substances that could lead to internal corrosion. Regular checking for corrosion and material embrittlement in pipelines, seals and storage equipment is important when dealing with hydrogen (EIGA, 2004).

8.2.3.4 Options to facilitate renewable energy integration

Technical options

Pipeline compatibility and gas storage are the two main technical challenges when integrating RE-based gases into existing gas systems. For variable RE-based systems, a constant stream of gas may not be produced so some storage may be essential to balance supply with demand. Since RE-based gases can be produced regionally and locally, storage is likely to be located close to the demand of the end user. Hence, the size and shape of a storage facility will depend on the primary energy source and the end use. In small applications, pressure variations in the pipeline (Section 8.3.4) could act as storage depending on the varying rates of production and use (Gardiner et al., 2008). In cases where there are several complementary end users for the gas, infrastructure and storage costs can be shared.

Simpler system designs enable RE-derived gases with a lower volumetric energy density to be distributed locally in relatively cheap polymer pipelines. Such dedicated distribution gas pipelines can be operated at relatively low pressures but will then need a larger diameter to provide similar volume flow rates and energy delivery. After a RE gas has been upgraded, purified, dried, brought up to the prescribed gas quality, then safely injected into a distribution grid, the main operational challenge is to avoid leaks and regulate the pressure and flow rate so that it complies with the pipeline specifications. Continuously available compressors, safety pressure relief systems and gas buffer storage systems are used to maintain the optimum pressures and flow rates.

Small- to medium-sized gas buffer storage such as inflatable rubber or vinyl bags (normally with four or five days of gas demand capacity) can be used to collect and store biogas, biomethane or syngas produced from variable RE feedstocks to help balance supply and local demand. The options for large-scale storage of biomethane are similar to those of CNG or liquefied natural gas (LNG). In large landfill gas or industrialized biogas plants, upgraded biomethane gas can be stored at high pressures in steel storage cylinders (as used for CNG). These can be connected to a local dispenser, to a gas pipeline, or transported by truck to the place of demand. Liquefaction before transport is possible, as used for LNG, but this is likely to add significant cost and complexity to a system. Producing LNG requires a large amount of energy and is therefore mainly an option for gas transport by boat or truck over thousands of kilometres when it can compete with constructing new gas pipelines.

Small-scale storage of hydrogen can be achieved in steel cylinders at pressures around 20,000 to 45,000 kPa Commercial composite-based hydrogen gas cylinders can withstand pressures up to 70,000 kPa[15] and hydrogen stations with gas pipelines and tanks that can withstand pressures up to 100,000 kPa already exist (www.zeroregio.com). In integrated gas grids, it is suitable to use low pressure (1,200 to 1,600 kPa spherical containers that can store relatively large amounts (>30,000 m^3) of hydrogen above ground (Sherif et al., 2005). For safety reasons, such storage is normally situated in industrial areas away from densely populated and residential areas. Hydrogen can also be stored at low pressure in stationary metal hydrides, but these are relatively costly and can only be justified for small volumes of hydrogen or if compact storage is needed.

Large-scale hydrogen storage is normally as compressed gas, or cryogenically in liquid form. Liquefaction of hydrogen is more costly than liquefaction of biomethane due to its lower volumetric density and boiling temperature (-253°C). In practice, about 15 to 20% of the hydrogen energy content is required to compress it from atmospheric pressure to 20000 to 70000 kPa while around 30 to 40% is required to produce liquid cryogenic hydrogen (Riis et al., 2006). Natural underground storage options, such as caverns or aquifers, for large-scale seasonal storage can be found in various parts of the world, but their viability and safety must be evaluated on a case-by-case basis.

Institutional options

The main institutional challenges to integrating RE gases into existing gas systems are adequacy of supply, quality standards, pipeline security and safety issues (McCarthy et al., 2007).

- Adequacy of supply can be influenced by the variable and seasonal nature of some RE resources, while the capacity of the gas distribution system also needs to be able to meet demand.

- Meeting gas quality standards poses a barrier, but is not fundamentally technically challenging. For biomethane, this can often be achieved at relatively low additional costs. However, gas quality standards vary: Sweden and Germany, for example, have developed their own national standards for biomethane that differ widely (Persson et al., 2006) (Table 8.3). There is as yet no single international gas standard for pipeline quality RE-based gases.

- The security of a gas pipeline system involves assuring a primary supply and building robust networks that can withstand either natural or physical events. In order to enhance supply security, pipeline networks often include some degree of duplication and multiple pathways between suppliers and end users so that a disruption in a network cannot shut down the entire system. Assessing vulnerability to malicious attacks on an extensive pipeline system over thousands of kilometres is a daunting task, and may require technological solutions such as intelligent sensors that report back pipeline conditions

---

15 See www.dynetek.com.

Table 8.3 | National standards for biomethane to be met before allowing injection into Swedish and German natural gas grids (Persson et al., 2006).

Parameter	Unit	Demand in Standard
Sweden		
Lower Wobbe index	MJ/Nm³	43.9–47.3 (i.e., 95–99% methane)
MON (motor octane number)	—	>130 (calculated according to ISO 15403)
Water dew point	°C	$<T_{ambient} - 5$
$CO_2 + O_2 + N_2$	vol %	<5
O_2	vol %	<1
Total sulphur	mg/Nm³	<23
NH_3	mg/Nm³	20
Germany		
Higher Wobbe index	MJ/Nm³	46.1–56.5 (>97.5% HHV[1] methane)
	MJ/Nm³	37.8–46.8 (i.e., 87–98.5% LHV[2] methane)
Relative density	—	0.55–0.75
Dust	—	Technically zero
Water dew point	°C	<Ground temperature T_{ground}
CO_2	vol %	<6
O_2	vol %	<3 (in dry distribution grids)
S	mg/Nm³	<30

Notes: 1. HHV = higher heat value. 2. LHV = lower heat value.

via Global Positioning System (GPS) technology to allow rapid location of a problem and corrective action. Diverse local or regional RE resources used for gas production can offer more secure supply than a single source of imported gas.

- Safety procedures and regulations for hydrogen used in the chemical and petroleum refining industries are already in place. Industrial hydrogen pipeline standards and regulations for on-road transport of liquid and compressed hydrogen have also been established. However, there is a lack of safety information on components and systems, which poses a challenge to the commercialization of hydrogen energy technologies. Codes and standards are necessary to gain the confidence of local, regional and national officials involved in the planning of hydrogen and fuel cell projects, hence, several organizations are developing safety and operational standards.

Given relative costs, policy support for the integration of RE gases may be needed if higher rates of deployment are sought. For example, feed-in regulations could enable the injection of biomethane into natural gas grids, similarly to how RE power is fed into electricity grids (Section 11.5.2).

Benefits and costs of large-scale penetration of RE gases

Benefits and costs can be assessed using both economic (capital expenditure, operation and maintenance costs) and environmental (GHG emissions, local air pollution, energy input ratio, air pollution) indicators. The relevant parameters are significantly affected by the type of RE source, gas production technology, storage and distribution system, and end-use application being either transport (Section 8.3.1) or stationary (Sections 8.3.2 and 8.3.3).

The compatibility of biomethane for distribution in natural gas grids can facilitate the widespread production and use of biogas and landfill gas. The costs of distribution are similar to existing gas systems, which enables a straightforward transition path for integration. Biomethane is already well established for heating, cooking, power generation, CHP and transport fuels. The latter is mainly for vehicle fleets of only a few hundred associated with water treatment plants and some agricultural usage (Matic, 2006). By comparison, more than 9 million CNG and LNG vehicles are operating worldwide (Åhman, 2010).

The market for hydrogen-fuelled vehicles is presently limited to applications such as forklift trucks (that operate indoors and hence require zero emissions) and demonstration cars and buses. Several leading automobile manufacturers anticipate that hydrogen fuel cell vehicles will be commercially introduced from 2015 (Pastowski and Grube, 2009) (Section 8.3.1). Hydrogen distribution demonstration projects are currently being introduced. For example, in California, 7 new hydrogen stations are due for completion by 2011, resulting in 11 stations in two clusters around Los Angeles (Dunwoody, 2010). Germany plans to increase the number of hydrogen stations from around 10 in 2009 to more than 140 in 2015 (Bonhoff et al., 2009). Similar initiatives in Japan are described in the *Hydrogen and Fuel Cells Demonstration Project* (Uchida, 2010).

GHG emissions related to producing and upgrading a RE-based gas should be assessed before a system is implemented. Vehicles fuelled with landfill gas can reduce GHG emissions by around 75% compared to using CNG, or even more if using biogas produced from the anaerobic digestion of animal manure (NSCA, 2006). Methane leakage to the atmosphere during biogas upgrading, storage, distribution and vehicle filling processes, as well as GHG emissions from any heat and power consumed during the upgrading process, will affect the overall energy efficiency and total GHG emissions as assessed on a life cycle basis (Figure 8.8) (Pehnt et al., 2009b). For example, if biogas produced from animal manure is used to fuel a 500 kW$_e$ CHP system, assuming 20% utilization of the

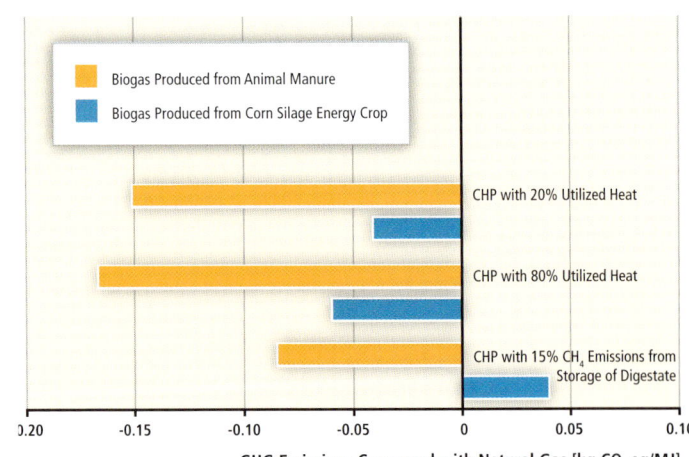

Figure 8.8 | For a 500 kW$_e$ CHP plant fuelled by biogas produced from either the anaerobic digestion of animal manure slurry or a corn silage energy crop, the potential reductions of GHG emissions can be compared with using natural gas to fuel the CHP plant. Methane leaks to atmosphere reduce the GHG reduction benefits (Pehnt et al., 2009b).

available useful heat, a 0.15 kg CO_2eq reduction in GHG emissions per MJ of energy output results, compared with a natural gas-fired CHP plant. If the biogas is produced from corn silage, only a 0.04 kg CO_2eq/MJ reduction would be achieved. If more heat can be utilized, the benefits increase, but should some methane leak to atmosphere, for example during storage of the process digestate, the GHG reduction benefits are considerably reduced (Figure 8.9).

To compete with other energy carriers, the cost of producing and upgrading biogas to the quality required for injection into an existing gas grid should be minimized. A comprehensive study of several biogas plants in Sweden showed that the electricity required to upgrade biogas is about 3 to 6% of the energy content of the cleaned gas, and the cost for upgrading is about USD_{2005} 0.005 to 0.02/MJ (Persson, 2003).

The cost per unit of energy delivered using a gas pipeline depends on the economies of scale and gas flow rate. The main cost is the pipe itself plus costs for installation, permits and rights of way. The cost of a local distribution pipeline is similar to that for district heating (Section 8.2.2) and depends mainly on the density of the urban demand. More dense systems yield a lower cost per unit of energy delivered. When designing a new gas grid, planning for anticipated future expansions is recommended because adding new pipes can be a costly option. Increasing the pressure to provide additional gas flow may be cheaper than adding larger diameter pipelines. The cost for distribution and dispensing of compressed biomethane at the medium scale is around USD_{2005} 15/GJ when transported by truck (Figure 8.9), which is substantially higher than by pipeline or as liquefied methane (Åhman, 2010).

In order to blend RE gases into a gas grid, the gas source needs to be located near the existing system to avoid high connection costs. More remotely located plants should ideally use the biomethane or hydrogen onsite to avoid the cost of gas distribution. Blending syngas or hydrogen into a natural gas system may require changes to the natural gas distribution and end-use equipment. Local networks in urban areas that currently carry fossil fuel-derived syngas (town gas) may also be suitable for biomass-derived syngas.

The limiting factors for hydrogen distribution are likely to be capital costs and the time involved to build a new infrastructure. In Germany, the cost for hydrogen production and distribution to supply some 7 million fuel cell light duty vehicles in 2030 is estimated to be around USD_{2005} 40 billion (Bonhoff et al., 2009). In the USA, for refuelling 200 million fuel cell vehicles, several hundred billion dollars would need to be invested over four decades (NRC, 2008). Incorporating variable RE sources would add to the cost due to the additional need for hydrogen storage.

In Europe, biomethane has been estimated to have the possibility of replacing 17.4 EJ of natural gas[16] in 2020 (Figure 8.10) (Müller-Langer

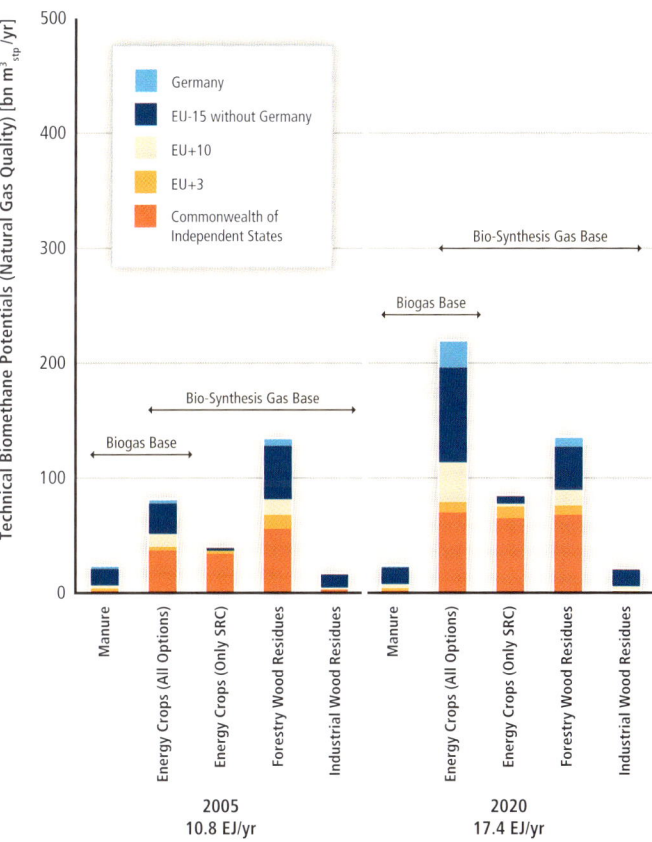

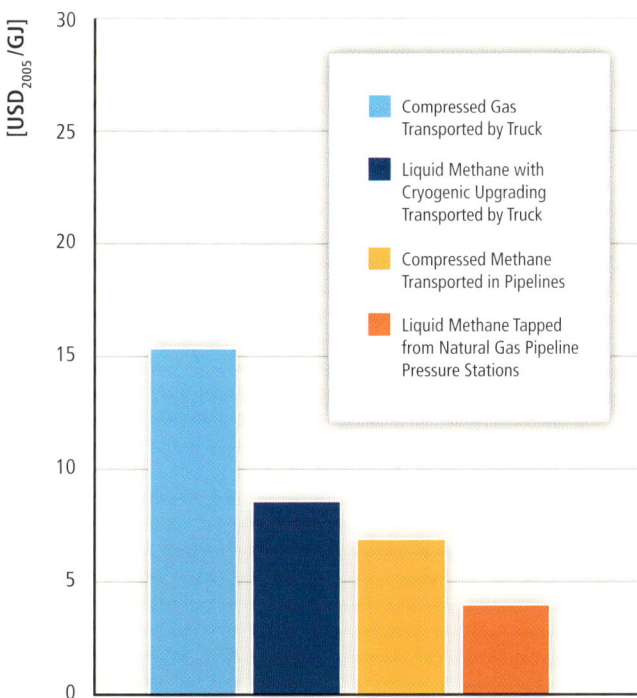

Figure 8.9 | Relative costs for distributing and dispensing biomethane (either compressed or liquefied) at the medium scale by truck or pipeline (Åhman, 2010).

Figure 8.10 | Technical annual potentials of biomethane at standard temperature and pressure (STP) as produced from a range of biomass feedstocks in the EU region in 2005 and 2020 (Müller-Langer et al., 2009).

Note: SRC = short rotation coppice; bn = billion = 10^9

16 Total natural gas consumption in OECD Europe (EU27) in 2008 was 19.1 EJ, 25% of total primary energy (IEA, 2010d).

et al., 2009), but this partly depends on the competition for the available biomass resources (Eurogas, 2008).

8.2.4 Integration of renewable energy into liquid fuel systems

8.2.4.1 Features and structure of liquid fuel supply systems

Renewable liquid fuels can take advantage of existing infrastructure components (storage, blending, transportation and dispensing) already used by petroleum-based fuels, with some adaptations. Integration issues may therefore be less problematic as compared to electricity or gas systems. The structure of a biomass-to-liquid fuel system for first generation biofuels is well understood (Figure 8.11) but sustainable production and land use remain controversial (Fritsche et al., 2010) (Sections 2.5.4 and 9.3.4.1).

The transport of bulky, low energy density biomass feedstocks to a biorefinery by road can be costly and normally produces GHGs. Rail transport can be a more efficient and cost effective delivery mode (Reynolds, 2000).

Biofuels can be blended with gasoline or diesel at oil refineries or blend centres during the distribution of petroleum fuels to vehicle refilling stations. Biofuels and blends can be stored at their production sites, alongside oil refineries or in underground storage tank facilities at service stations. As for petroleum products, similar care needs to be taken regarding safety and environmental protection. Due to the seasonality of agricultural crops grown specifically as biomass feedstocks, storage of the feedstock and/or the biofuel is crucial if the goal is to meet year-round demand (NAS, 2009), but this adds to the production costs. International trade can also play a role to provide a stable year-round supply (IEA, 2007a) (Section 2.4). Biodiesel is prone to variation in composition during storage due to the action of micro-organisms that can lead to rises in acidity and hence engine corrosion. Ethanol is more biologically stable.

8.2.4.2 Characteristics with respect to renewable energy integration

Currently most liquid biofuels are produced from sugar, carbohydrate and vegetable oil crops and integrated into existing fuel supplies by using blends, typically up to 25% (in volumetric terms) with gasoline and diesel (Sections 2.3.3 and 2.6.3). However, ethanol can be blended in any proportion with gasoline for use in flex-fuel vehicles (Section 8.3.1) and biodiesel can be used in compression ignition engines either neat (100% or B100) or blended with regular diesel. Modified diesel engines may also run on almost neat alcohol (E95) with an ignition improver (Scania, 2010). Several manufacturers produce trucks and agricultural machinery with engines certified for use with B20 and B100 fuels (NBB, 2010; New Holland Information Center, 2010; Power-Gen, 2009).

Solid lignocellulosic biomass sources can be converted to liquid fuels by means of biochemical processes such as enzymatic or acid hydrolysis, or by thermo-chemical processes to produce synthesis gas (mainly carbon monoxide (CO) and hydrogen (H_2)) followed by Fischer-Tropsch conversion

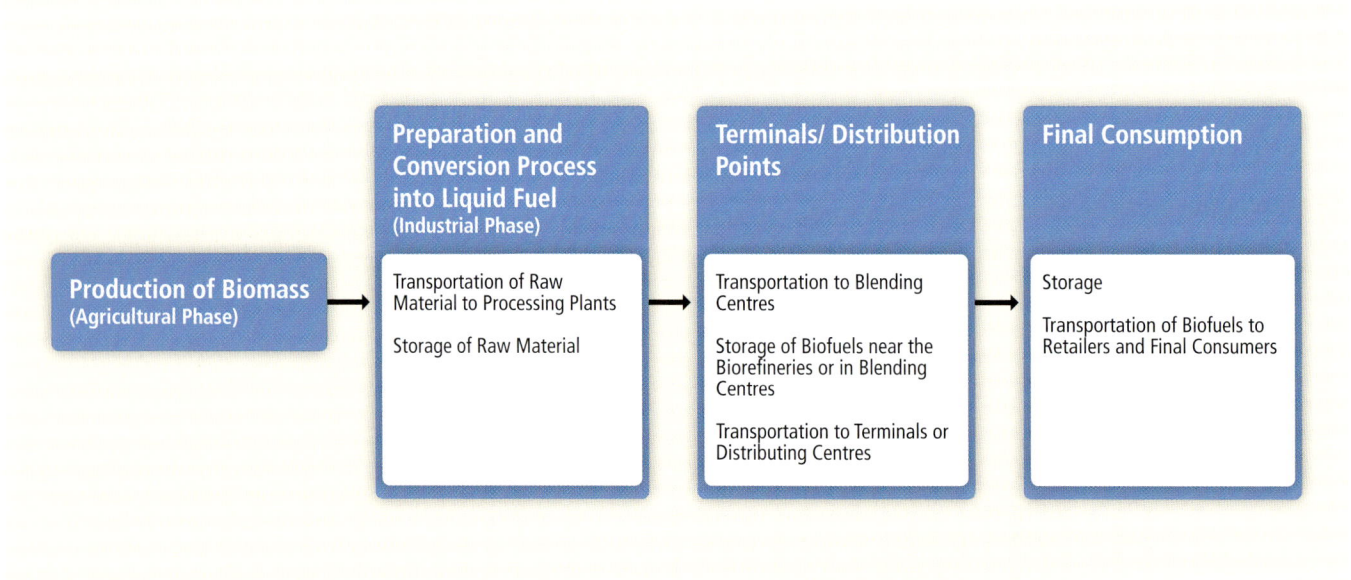

Figure 8.11 | The various phases in a typical biofuel production, blending and distribution system for transport fuel.

to produce a range of synthetic liquid fuels suitable for road transport, aviation, marine and other applications (Sims et al., 2008; Section 2.3.3). Fuel quality issues are important because they can affect the performance of vehicle engines and transport emissions (Section 8.3.1.2). Biomethane, if it meets appropriate specifications (Section 8.2.3), can also be combusted directly in spark-ignition ICEs as for compressed natural gas (CNG).

Most of the projected demand for liquid biofuels is for transport, though industrial demand for bio-lubricants, and chemicals such as methanol for use in chemical industries, could also increase (Section 2.6.3.5). Some biofuels are also used in stationary CHP engines as a substitute for petroleum fuels. The demand for large amounts of traditional solid biomass primarily in developing countries for cooking and heating could be replaced by more convenient gaseous fuels such as LPG, but also by liquid fuels produced from biomass such as ethanol gels (Utria, 2004; Rajvanshi et al., 2007) or dimethyl ether (DME) (Sims et al., 2008).

Liquid biofuels can be integrated into the existing oil product distribution infrastructure. Transition barriers are relatively low as biofuel blends could be introduced without costly modifications to existing petroleum storage and delivery systems, and could take advantage of existing infrastructure components (NAS, 2009). Some related costs could eventuate for blending and for additional technical modifications of fuel storage tanks and fuel pumps, or provision of new installations. The type of fuel storage and delivery system will vary depending on the properties of the biofuel and its compatibility with the existing petroleum-based fuel system. Most common biofuels have similar properties to gasoline and diesel so can be blended reasonably easily with these. Cold temperature conditions can represent difficulties, during transport and storage and in engines, especially for some biodiesels that may form polymer gels that restrict fuel flow. Overcoming these constraints is imperative if biofuels are to be used as aviation fuels.

Transport and delivery modes from refineries to terminals include trucks, barges, tankers and pipelines. From the terminals, trucks or distribution pipelines can supply the retail outlets depending on the distances and volume of biofuels involved. Storage and distribution costs would be similar to petroleum-based fuels.

Bio-refineries that produce biofuels and other co-products are generally much smaller in capacity than oil refineries and could be widely located in geographic regions where the resource exists. For example, numerous ethanol processing plants are situated throughout the mid-western and south-eastern corn belt of the USA, whereas a few oil refineries are concentrated along the coasts. Brazil already has many bio-refineries in operation producing sugar, ethanol, biodiesel, animal fodder, electricity, steam and heat.

Although the cost of fuel delivery is a small fraction of the overall production cost, the logistics and capital requirements for widespread expansion and integration could present hurdles if not well planned.

Technical issues regarding ethanol and gasoline blends (gasohol) during storage and transport can arise if water is absorbed by anhydrous ethanol in the pipelines (Section 8.2.4.3). However, in Brazil, ethanol produced from sugar cane has been successfully transported in pipelines also used for oil products for over 20 years, though the clean-up and maintenance procedures have increased. Since ethanol has around two-thirds the volumetric energy density of gasoline, larger storage systems, more rail cars or vessels, and larger capacity pipelines would be needed to store and transport a similar amount of energy and hence would increase the fuel storage and delivery costs compared to oil-based products.

The possibility exists to use by-products of biofuel production as raw materials for biogas production or electricity generation, for example from bagasse, the sugarcane residue. Integration with the existing electricity grid system has been successfully achieved in Brazil and elsewhere in cogeneration schemes after the energy demands of the processing plant have been met (Rodrigues et al., 2003; Pacca and Moreira, 2009). Anaerobic digestion of the by-products from bioethanol and biodiesel processing has the potential to be integrated with various existing bio-refinery models. The biogas can either be used for heat and electricity generation, as a vehicle fuel (Börjesson and Mattiasson, 2008), or injected into gas grids (Section 8.2.3).

8.2.4.3 Challenges of renewable energy integration

Although renewable liquid fuels can take advantage of the existing infrastructure components (storage, blending, transportation and dispensing) already used by the petroleum industry, some issues need to be addressed. Most biofuels have fairly similar properties to gasoline and diesel so can be blended reasonably easily with these fuels. Cold weather conditions can produce engine difficulties from higher viscosity and gel formation when using some biodiesels, and also produce difficulties for their storage and transport (NAS, 2009).

Sharing oil-product infrastructure with biofuels may lead to possible water contamination from hydrous ethanol, and the resulting corrosion may require using new materials to preserve the working life of the pipeline and equipment. Moisture resulting from condensation in oil-product pipelines can increase the water content of ethanol if being transported in the same lines. If it exceeds the technical specification for the bioethanol, additional distillation after delivery may then be required. Ethanol and biodiesel can also dissolve and carry any impurities present inside multi-product pipeline systems and these are potentially harmful to ICEs. Therefore a dedicated pipeline may be preferable if improved cleanup procedures between products being sent through multi-product pipelines are not successful. Moisture absorption and phase separation during pipeline shipment of ethanol can be avoided by shipping some hydrous ethanol first, which is then used directly by end users or distilled, followed by anhydrous ethanol that then remains suitable for direct blending with gasoline. An alternative strategy is to send a 'sacrificial buffer' of neat ethanol down a pipeline to absorb any moisture ahead of sending the primary batches of ethanol or blends. The buffer shot is discarded or re-distilled.

Ethanol in high concentrations can lead to accelerated stress corrosion cracking (SCC) in steel pipelines and storage tanks, especially at weld joints and bends (NAS, 2009). This can be avoided by adding tank liners, using selective post-weld heat treatments, and coating internal critical zones (at pipeline weld points, for example). However, these all increase system costs. Ethanol may degrade certain elastomers and polymers found in seals and valves in pipelines and terminals as well as some engines, so these may need replacement. New pipelines could be constructed with ethanol-compatible polymers in valves, gaskets and seals and be designed to minimize SCC (NAS, 2009).

8.2.4.4 Options to facilitate renewable energy integration

Technical options

Technologies will continue to evolve to produce biofuels that are more compatible with the existing petroleum infrastructure (Sims et al., 2008). Advanced biofuels in the future may need to meet stringent quality specifications in order to match the fuels with existing and new engine designs installed in heavy transport, marine and aviation applications. In some countries, the development of codes and standards for biofuels has been slow and delayed their integration into the supply system. Quality control procedures also need to be implemented to ensure that biofuels meet all applicable product specifications (Hoekman, 2009) and hence facilitate integration.

The facilitation of international trade in biofuels instigated a need for more homogeneous international standards to be developed. A comparison was made of existing biofuel standards (NCEP, 2007). The standards for biodiesel in Brazil and USA reflect its use only as a blending component in conventional mineral diesel fuel, whereas the European standard allows for its use either as a blend or as neat fuel. Variations also exist in current standards for regulating the quality of biodiesel reaching the market due to the different oil and fat feedstocks available. This translates to variations in the performance characteristics of each biofuel, less so for ethanol, which is a simple chemical compound compared with long-chain biodiesels. Bioethanol technical specifications differ with respect to the water content but do not constitute an impediment to international trade (NIST, 2007). Blending levels of biofuels need to account for regional differences in the predominant age and type of vehicle engines, ambient temperatures and local emission regulations.

Institutional aspects

Policy support has played an important role in creating a market for biofuels. For example, the mandatory blending of biodiesel and ethanol in diesel and gasoline respectively has been used in several countries (Section 11.5.5). Agencies in charge of regulating oil product markets could also include biofuels under their jurisdiction. These agencies are the most appropriate to deal with issues such as security of biofuel supplies, safety and technical specifications (or standards) and quality control at both the production and retail levels. This is currently the case for Brazil where the regulator for the oil sector also regulates biofuels (TN Petróleo, 2010).

Environmental agencies and related regulations (for example, low-carbon fuel standards and air quality controls), can facilitate greater penetration of biofuels and their integration into the existing energy system. National energy planning organizations can evaluate any impacts and additional costs associated with the large-scale integration of biofuel systems with existing and future energy production and delivery systems.

8.2.4.5 Benefits and costs of large-scale renewable energy penetration

Achieving a high share of biofuels should be relatively easy where unit production costs are similar to imported oil product costs since additional storage and transport costs are a relatively small portion of total costs. Existing infrastructure for oil distribution can be adapted and used for biofuels, especially at low blend levels. For large-scale penetration of biofuels, or where the use of E100 or B100 is envisaged, special provisions may need to be made.

Specialist equipment is needed at collection terminals at ports and oil refineries receiving biofuel shipments for blending of ethanol or biodiesel, and for loading the blended product on to barge, rail or road tankers (Reynolds, 2000). Existing transport, storage and dispensing equipment at vehicle refuelling stations need to be modified to handle biofuel blends, as has been successfully achieved in the USA, Brazil, Germany, Sweden and elsewhere. Underground storage tank systems, pumps and dispensers may need to be converted to be compatible with higher biofuel blends and to meet safety requirements. Issues relating to the retrofitting of existing facilities are similar to those associated with pipeline transport (Section 8.2.4.3) and include phase separation, SCC and the degradation of incompatible materials (NAS, 2009).

Ethanol terminals usually have one or more storage tanks ranging from 750 to 15,000 m^3 capacity. New ethanol storage tanks cost around USD_{2005} 180/m^3 capacity for small tanks up to USD_{2005} 60/m^3 for large ones (Reynolds, 2000). It may be possible to refurbish gasoline tanks to suit ethanol storage for lower costs than investing in new tanks.

In the USA, most ethanol is transported by rail, road tanker and barge (NCEP, 2007), but since 2008, batches in Florida have also been sent through gasoline pipelines (KinderMorgan, 2010). Capacities and costs vary for ethanol storage and delivery equipment (Table 8.4). As a point of reference, ethanol plants in the USA each produce 300 to 1,200 m^3/day, while the ethanol demand for 1 million cars using E10 would be about 400 to 800 m^3/day, and storage facilities typically hold between 4,000 and 12,000 m^3 for local and regional demands respectively.

Table 8.4 | Capacities and costs of a range of equipment suitable for ethanol storage and long-distance transport.

	Capacity	Cost (USD$_{2005}$)	Reference
Truck/trailer	25 m^3	103,000 141,000	EPA (2007) Reynolds (2000)
Rail car	90 m^3	85,000	EPA (2007)
River barge	Several linked units of 1,200 m^3/unit	5 million for one unit	EPA (2007)
Ocean ship	3,000–30,000 m^3	Not known	Reynolds (2000)
Pipeline (300 mm diameter)	12,000 m^3/day	0.34-0.85 million/km	
Terminal storage tank	3,000 m^3 6,000 m^3	360,000 540,000	Reynolds (2000) Reynolds (2000)
Retrofit a gasoline storage tank	1,200 m^3	18,800	EPA (2007)
Blending equipment		170,000-450,000	Reynolds (2000)
Total terminal refit	6,000 m^3	1.13 million	Reynolds (2000)

Rail shipment is generally the most cost effective delivery system for medium and longer distances (500 to 3,000 km) and to destinations without port facilities (Reynolds, 2000). Rail shipments require more handling at the terminals because of the greater number and smaller volumes of units compared to barges, as well as the more labour-intensive efforts for cargo loading, unloading and inspection, Trains containing up to 75 railcars have been proposed for ethanol as an alternative to pipeline development (Reynolds, 2000).

Barges are used for long distance transport when biofuel production plants have access to waterways. In the USA, for example, barges travel down the Mississippi River from ethanol plants in the Midwest to ports at the Gulf of Mexico where the ethanol is stored before being transferred to ships for transport to overseas or national coastal destination terminals for blending.

Estimates for the costs of transporting large ethanol volumes over long distances (Reynolds, 2000) (Section 2.6.2) range from USD$_{2005}$ 6 to 10/m^3 for ocean shipping, USD$_{2005}$ 20 to 90/m^3 for barge, USD$_{2005}$ 10 to 40/m^3 for rail and, over shorter haul distances, USD$_{2005}$ 10 to 20/m^3 for trucks (Section 2.6.2). In Brazil, depending on the origin of the biofuel, the costs of transporting ethanol from the producing regions to the export ports is around USD$_{2005}$ 35 to 64/m^3, which also includes storage costs at the terminal (Scandiffio and Leal, 2008). More pipelines are being planned to connect main rural ethanol producing centres to coastal export ports with the expected costs ranging from USD$_{2005}$ 20 to 29/m^3; 70% less than by road and 45% less than by rail (CGEE, 2009).

8.2.4.6 Case study: Brazil ethanol

After a relatively slow start, the ethanol distribution system, retailing of biofuel blends and manufacture of flex-fuel engines in Brazil have all proven successful in the past decade, so that in 2010, Brazil was the world's second largest producer of ethanol, after the USA (REN21, 2010). Integration of liquid biofuels with the oil distribution system began after the first global oil crisis when the government promoted sugarcane ethanol as a gasoline alternative (Box 11.9). The state oil company, Petrobras, was obliged to purchase all domestically produced ethanol, blend it with gasoline, and distribute it nationwide (Walter, 2006). In 1979, vehicles with engines designed to run on E100 were produced, so existing infrastructure was adapted for delivery of 100% ethanol nationwide, though production was regionally concentrated. Significant gains in sugarcane yields per hectare have since helped to increase ethanol output per unit of land area so that in 2008, ethanol production was 495 PJ, equivalent in energy terms to 85% of the gasoline consumed in Brazil that year (EPE, 2009).

About 60% of ethanol distilleries in Brazil are dual-purpose, producing sugar when world sugar prices are high, and converting it to ethanol at other times (Ministry for Agriculture Livestock and Supply, 2010). When world sugar prices rose in the 1990s, ethanol production declined and hence owners of dedicated E100 vehicles experienced fuel shortages. Vehicles with flexible fuel engines (Section 8.3.1.3) capable of using bioethanol blends ranging from E20 to E100 were therefore developed (de Moraes and Rodrigues, 2006) and have now largely replaced the dedicated E100 fleet. All present gasoline has a blended content of 20 to 25% anhydrous ethanol (by volume) and therefore, since their commercial introduction in 2003, the majority of new light duty vehicles sold today have 'flex-fuel' engines (Goldemberg, 2009).

Over the last 30 years, a country-wide ethanol storage and distribution system was implemented so that several biofuel blends up to E100 are available in practically all refuelling stations. All subsidies were removed in the 1990s (Box 11.9), but ethanol prices continued to decline steadily and remain competitive with gasoline when oil prices fluctuate around USD$_{2005}$ 70/barrel.

Since 1990, electricity and heat have been generated in sugar/ethanol plants by combusting the bagasse co-product in CHP systems (Cerri et al., 2007). Where the electricity grid is located nearby, any electricity that is surplus to onsite demand can be sold and fed into the national grid (Azevedo and Galiana, 2009). Technological improvements, better energy management and cogeneration schemes have enabled optimal use of the bagasse. Government programmes (PROINFA, 2010), regulatory changes and public auctions for electricity contracts were introduced to enable the electricity to be sold to local utilities or monitored and dispatched by the national transmission system operator (Section 8.2.1). Since the sugar cane harvesting period coincides with the dry season in Brazil, generation of electricity from bagasse complements the country's hydroelectric system. In 2009, the total installed capacity of bagasse-fuelled CHP was 5.6 GW$_e$ and generated around 4.75% of total electricity (BEN, 2010).

Brazil's experience suggests that the integration of high shares of biofuels can be successfully achieved by implementing blending mandates in combination with other policies to address economic, social and environmental barriers (Section 11.5).

8.2.5 Integration of renewable energy into autonomous energy systems

Not all buildings, communities or business enterprises are connected to electricity grids, district heating or cooling systems or gas grids, nor have easy access to liquid fuels. This section covers such autonomous energy supply systems, which are typically small scale and are often located in off-grid remote areas, on small islands or in individual buildings where the provision of commercial energy is not readily available through grids and networks. There is also growing interest by industry in the future potential for connecting decentralized energy supply systems[17] that could utilize advanced control systems and integrate numerous small heat and power generation technologies through smart meters and time-of-use and price-responsive appliances (Cheung and Wilshire, 2010). Overall system costs, benefits and constraints are uncertain, so RD&D, monitoring and evaluation have been undertaken by several governments in association with several leading electricity and information technology industries. Demonstration projects based on small, autonomous community micro-grids have been established in the USA, Japan, Denmark and elsewhere.

In principle, RE integration issues for autonomous systems are similar to large electrical power systems, for example for supply/demand balancing of electricity supply systems (Section 8.2.1), selection of heating and cooling options (Section 8.2.2), production of RE gases (Section 8.2.3) and liquid biofuel production for local use (Section 8.2.4). Autonomous systems also involve building-integrated RE technologies (Section 8.3.2).

Planning an autonomous system, often remotely located and with low energy demand, involves considering future fossil fuel supply options for the location, the local RE resources available, the costs of RE technologies, future technology innovation prospects and the possible avoidance of construction costs should new or expanded infrastructure be an option for the location (Nema et al., 2009).

---

17 Various terms are used in the literature to describe a possible future paradigm shift in energy supply such as 'distributed energy systems', 'digital energy', 'intelligent grids' and 'smart grids', but none are as yet clearly defined.

8.2.5.1 Characteristics with respect to renewable energy integration

Several types of autonomous systems exist and can make use of either single energy carriers (electricity, heat, liquid, gaseous or solid fuels) or a combination. A full range of energy services can usually be provided, including heating, lighting, drying, space cooling, refrigeration, desalination, water pumping (Bouzidi et al., 2009) and telecommunications.

Unlike large electrical power systems, smaller autonomous systems often have fewer RE supply options that are readily available at a local scale. Additionally, some of the technical and institutional options for managing integration within large electrical power systems, including sophisticated RE supply forecasting, stochastic unit commitment procedures, stringent fuel quality standards and benefiting from the smoothing effects of geographical and technical diversity, become more difficult or even implausible for smaller autonomous systems.

RE integration solutions typically become more restricted as supply systems become smaller, particularly where there are high shares of variable RE sources. Autonomous systems will naturally have a tendency to focus on storage, various types of demand response and highly flexible generation to help match supply and demand. RE supply options that better match local load profiles or that are dispatchable may be chosen over lower-cost RE supply options that do not have as strong a match with load patterns or that are variable. Managing RE integration within autonomous systems can, all else being equal, be more costly than in larger electrical power systems. One implication of this observation is that autonomous systems face harder tradeoffs between a desire for reliable/continuous supply and a desire to minimize overall supply costs than do larger networks. For those without ready access to electricity, cost comparisons with larger electrical power systems may be irrelevant and standards of reliability may vary.

For electricity generation in small to medium-sized autonomous systems, fossil fuels such as diesel, gasoline or LPG have been commonly used in stationary engines that drive generator sets (gensets). Due to the potential supply constraints and costs of delivering fossil fuel supplies to remote locations in developing countries, there is a growing trend towards using local RE resources where available. Supply/demand balancing problems associated with variable RE sources may emerge for autonomous electrical power systems, similarly as for larger centralized systems but perhaps more acutely. Discussion of the variability and predictability of different RE technologies and their effect on the reliability of electrical power system supply can be found in Section 8.2.1.2. In rural communities with small electric distribution networks, in small villages using simple, low voltage DC mini-grids, or in individual buildings, limited deployment of a single type of RE generation technology such as solar PV or micro-hydro is possible. However, in such cases, variable RE supplies will need to be coupled with other options such as demand side measures, energy storage and increasing generation flexibility to ensure reliability (Section 8.2.5.2).

Heating and cooling of off-grid autonomous buildings, often in rural locations, can use RE technologies, particularly where good solar, geothermal or biomass resources are available (IEA, 2007c). Variability again may be of some concern where solar thermal is used, but typically it can be addressed through the addition of thermal storage.

Domestic and commercial buildings in urban areas are normally connected to the network energy supply, though interest is growing in the possibility of more existing and new buildings becoming energy generators by installing integrated RE technologies (IEA, 2009b). Building-integrated solar PV (Bloem, 2008), off-grid system operation (Dalton et al., 2008), and distributed energy systems that include solar thermal, small bio-energy CHP plants, micro-hydro and small wind turbines (IEA, 2009a) have all been demonstrated with many successful technology examples surpassing the pre-commercial phase of development. Buildings can be designed to be energy efficient as well as using RE to generate as much energy as they consume. For example, the Net-Zero Energy Commercial Building Initiative of the US Department of Energy (USDOE, 2008a) aims to achieve marketable low-carbon building designs by 2025. Low-rise buildings can also become autonomous energy systems through the combination of energy efficiency (air-tight structure, high heat insulation, efficient ventilation, air conditioning, lighting, water heating etc.) and integration of RE technologies (Section 8.3.2).

8.2.5.2 Options to facilitate renewable energy integration and deployment

The integration of RE conversion technologies, balancing options and end-use technologies in an autonomous energy system depends on the site-specific availability of RE resources and the local energy demand, which can vary with local climate and lifestyles. The balance between cost and reliability is critical when designing and deploying autonomous power systems, particularly for rural areas of developing economies because, as noted earlier, the additional cost of providing continuous and reliable supply may be greater as autonomous systems grow smaller. The balancing options available to larger electrical power systems are also, in principle, available to autonomous electrical power systems, and are discussed extensively in Section 8.2.1.3. These include improving network infrastructure, increasing generation flexibility, demand side measures, electrical energy storage and improving operational/market and planning methods.

Prioritization among the available options for integrating variable RE into these systems will depend on a variety of factors including but not limited to the type of system, geographic location and expectations of reliability. As already discussed, however, as autonomous systems become smaller, several of the options for managing variability become impractical, and storage, flexible thermal generation and demand response often take precedence.

In terms of demand side measures, autonomous RE systems can be integrated with selected end-use technologies that use surplus electricity only when available. These include solar stills, humidifiers and dehumidifiers, membrane distillers, reverse osmosis or electro-dialysis water desalinators (Mathioulakis et al., 2007), water pumps using solar PV and an AC or DC motor (Delgado-Torres and Garcia-Rodriguez, 2007), solar adsorption refrigerators (Lemmini and Errougani, 2007) and oilseed presses (Mpagalile et al., 2005). Various other forms of load management may also be important for balancing autonomous systems that feature significant amounts of supply and demand variability.

Electrical energy storage technologies (Section 8.2.1.3) may often be the more attractive option for autonomous RE systems, despite their relatively high cost. Where, for example, pumped hydro is not an option, battery storage can be employed with installed capacity sufficient to meet two to three days of electricity demand. The cost of such storage options should be carefully evaluated against the level of reliability desired during the design and planning stages, alongside capital investment and operational costs of the system. Several simulation analyses, demonstration assessments and commercial operations of the application of energy storage technologies to autonomous systems have been reported. These include solar PV plus wind with hydrogen storage in Greece (Ipsakis et al., 2009), wind/hydrogen in Norway (Ulleberg et al., 2010; Section 8.2.5.5), pumped hydro systems plus wind integration on three Greek islands (Caralis and Zervos, 2007b, 2010) and a wind/solar/pumped hydro demonstration on the Spanish Canary Islands (Bueno and Carta, 2006; Section 8.2.5.5). Small PV systems coupled with battery storage are already in widespread use in many countries.

Alternatively, a portfolio of RE and non-RE technologies could be integrated to enhance system reliability. For example, small diesel- or gasoline-powered gensets and dump load (usually a resistance heater to use any excess electricity generated above the demand) could be cheaper to operate than having batteries for short periods when wind or solar resources are not available (Doolla and Bhatti, 2006).

Gaseous or liquid biofuels that are produced locally from biomass (Section 2.2.2) could be an option for heating or the fuelling of gensets or vehicles (Section 8.3.1). To maintain the desired supply reliability and flexibility of autonomous electricity system operation (Section 8.2.1), the present use of gasoline or diesel to fuel small gensets could, in future, be totally displaced by RE gases and biofuels. RE gases are easy to store under low or medium pressure in butyl containers or cylinders (Section 8.2.3) and liquid biofuels can be stored in steel or butyl rubber tanks (Section 8.2.4).

For many autonomous RE systems (with the possible exception of bioenergy CHP and certain run-of-river micro-hydro schemes), energy storage and low-energy utilization technologies are integral (Lone and Mufti, 2008). Autonomous micro-hydro schemes are popular in hilly regions, particularly in developing countries such as Nepal, to provide a resource-dependent continuous power supply (except possibly in dry seasons). For run-of-river hydro, a cost efficient solution for system balancing (Section 8.2.1) has been to use load control instead of controlling the power generation output (Paish, 2002).

Providing system reliability in a cost effective manner can prove difficult for autonomous systems, but possible future designs of autonomous heat and power supply systems based on the development of innovative system controls, smart meters and appliances that offer demand response services (Meenual, 2010) could provide solutions and enhance RE integration. Whether such technological solutions are appropriate for use in remote areas of less-developed countries, however, is unclear.

8.2.5.3 Benefits and costs of renewable energy integration and design

For remote rural areas, it is widely recognized that energy supplies can contribute to rural development through increased productivity per capita; enhanced social and business services such as education, establishment of markets and supply of water for drinking and irrigation; improved security due to street lighting; decreased poverty; and improved health and environmental conditions (Goldemberg, 2000; Johansson and Goldemberg, 2005; Takada and Charles, 2006; Takada and Fracchia, 2007). Issues of energy access are addressed in Section 9.3.2.

In developing countries, where suitable and sustainable biomass supplies are available, including organic wastes, their use can often be the least-cost option to provide basic services for cooking, water heating, small-scale power generation and lighting. In China, solar thermal water heating for isolated rural dwellings has brought environmental, social and economic benefits (Z. Li et al., 2007).

Electricity generated by an autonomous system is usually more expensive than using electricity where a grid connection is available. Therefore autonomous RE buildings have been uncommon in urban environments, though some interest in micro-grids and others concepts has been expressed. In remote areas, RE-based electricity autonomous systems may be the only or least-cost option, at least until a connection to external grids becomes available.

8.2.5.4 Constraints and opportunities for renewable energy deployment

Beyond those barriers already addressed, constraints to integration can arise from the wide-ranging RE technology specifications and the difficulties of their appropriate design, construction and maintenance. These can lead to capital investment and operational cost increases or inadequate maintenance. Should a technical failure occur, poor public perception of the technology could arise. Establishing standards, certifying products, integrating planning tools, training maintenance workers and developing a knowledge database could help avoid technology reliability problems (Kaldellis et al., 2009). Local capacity building, training, good planning and careful market establishment could result in lower operational and maintenance costs, an enhanced reputation, employment opportunities and other social benefits (Meah et al., 2008).

For each type of autonomous RE system, appropriate planning methods could assist developers to build projects (Giatrakos et al., 2009), though the variety of possible RE technologies that could be deployed and integrated makes development of broad planning guidelines difficult to achieve. To improve planning methodology, databases could be established from RD&D projects as well as from commercial experiences to enable comparisons between sustainability criteria (Igarashi et al., 2009), lifestyles (Amigun et al., 2008; Himri et al., 2008) and various combinations of technologies under specific site conditions.

The integration of RE into autonomous systems on a broad scale may also require policy measures to help cover the additional costs and to provide an enabling environment (Section 11.6). Even where an autonomous, RE-integrated system is assessed to be economically feasible over its lifetime, appropriate financial schemes to remove the barrier of high capital investment costs could be warranted.

8.2.5.5 Case studies

Seawater desalination in a rural area of Mexico
Baja California Sur is an arid, sparsely populated coastal state where underground aquifers are over-exploited due to population growth, agricultural irrigation demands and tourism. Around 70 seawater desalination plants are therefore operating using fossil fuel electricity and there are plans to construct more.

Small-scale desalination using solar PV is an alternative water supply option for the smaller, more remote communities in the state. Installed solar PV-powered seawater reverse osmosis plants can each produce 19 m^3 of fresh water per day with a total dissolved solids content of <250 ppm while consuming as little as 2.6 kWh/m^3 (~9.4 MJ/ m^3) of water (Contreras et al., 2007). The plants use an energy recovery device and integrate a battery bank to enable 24-hour operation. The balance between continuous, smooth operation and cost minimization depends on optimizing the integration and capacity of this battery bank. In the future, further integration of desalination plants and rural electrification could be beneficial to provide both clean water and sustainable energy supplies.

Wind/hydrogen demonstration, Utsira, Norway
An autonomous wind/hydrogen energy demonstration system located on the island of Utsira, Norway was officially launched by Norsk Hydro (now Statoil) and Enercon (a German wind turbine manufacturer) in July 2004. The main components of the system are a 600 kW$_e$ rated wind turbine, a water electrolyzer to produce 10 Nm^3/h of hydrogen, 2,400 Nm^3 of hydrogen storage (at 20,000 kPa), a hydrogen-powered internal combustion engine driving a 55 kW$_e$ generator, and a 10 kW$_e$ proton exchange membrane (PEM) fuel cell.[18] This innovative demonstration

---

18 Nm^3 is an uncompressed 'normal' cubic metre of gas at standard conditions of 0°C temperature and atmospheric pressure.

system supplies 10 households on the island providing two to three days of full energy autonomy (Ulleberg et al., 2010).

Operational experience and data collected from the plant over four to five years showed the electrical energy consumption for the hydrogen production system (electrolyzer, compressor, inverter, transformer, and auxiliary power system) under nominal operating conditions is about 6.5 kWh/Nm^3 (~23.4 MJ/Nm^3), equivalent to an efficiency of about 45% (based on lower heat value). The efficiency of the hydrogen engine/generator system is about 25%. Hence, the overall efficiency of the wind to AC-electricity to hydrogen to AC-electricity system, assuming no storage losses, is only about 10%. If the hydrogen engine was to be replaced by a 50 kW$_e$ PEM fuel cell, the overall efficiency would increase to 16 to 18%. Replacing the present electrolyzer with a more efficient unit (such as a PEM or a more advanced alkaline design), would increase the overall system efficiency to around 20% (Ulleberg et al., 2010).

The relatively low efficiency of the system illustrates the challenge for commercial hydrogen developments. More compact hydrogen storage systems and more robust and less costly fuel cells need to be developed before autonomous wind/hydrogen systems can become technically and economically viable (Gardiner et al., 2008). Nevertheless, this project has demonstrated that it is possible to supply remote area communities with wind power using hydrogen as the energy storage medium, but that further technical improvements and cost reductions need to be made to compete with a wind/diesel hybrid. The overall wind energy utilization of only 20% could be improved by installing more suitable and efficient load-following electrolyzers that allow for continuous and dynamic operation. Surplus wind power could also be used to meet local heating demands, both at the plant and in the households. In addition, the hydrogen could conceivably be utilized in other local applications, such as fuel for local vehicles and boats. The overall costs of the system are not known but are likely to be relatively high.

El Hierro – the Spanish Canary Islands
This, the smallest of the Canary Islands, used to meet the electricity demand of its 10,600 population (plus 60,000 tourists a year) with a 10 MWe diesel generating set, 100 kW$_e$ and 180 kW$_e$ wind turbines and a small, low voltage distribution grid going around the 276 km^2 island (IEA, 2009b). The annual imported diesel fuel costs were around USD$_{2005}$ 3 million per year. In 2005, the local government implemented a 100% RE electricity programme with a budget for wind of approximately USD$_{2005}$ 20 million, for hydro approximately USD$_{2005}$ 50 million, and for solar approximately USD$_{2005}$ 10 million. Energy saving is a key part of the project, which includes local government incentives to encourage solar water heating installations. The demonstration programme has a simple payback period of around 30 years, so is supported by a consortium of seven partners including the European Commission under its ALTENER programme.

The utility company Unelco-Endesa is developing the wind/hydro plant expected to be commissioned in 2011. The local government has a 70% stake in the project and the islanders can also purchase shares.

Five 2.2 MW turbines have been installed and surplus wind power will be used to pump water up a 3 km long, 0.5 m diameter pipe to the upper storage system, which is a lined volcanic crater giving 200,000 m^3 storage capacity and a 700 m head potential. This reservoir will be used to run a hydro plant to meet peak power demands and also act as balancing reserve during calm periods of up to seven days. Any surplus water could be used for irrigation purposes along with water from a desalination plant used to top up the system. The existing diesel generating plant remains for system balancing (and also as backup under extreme conditions) and is anticipated to initially meet around 20% of the annual total electricity demand.

Solar PV is used for street lighting and is also to be installed on 10 public buildings. Each 5 kW capacity system will feed any surplus power into the low voltage grid. Local, sustainably produced woody biomass is already used to meet a share of the heat demand. Biogas, produced from a range of feedstocks, is used to power a hybrid bus and could also be used for heat and power generation in the future (Insula, 2010). Electric vehicles are planned and the potential for ocean energy development is being assessed (Iglesias and Carballo, 2010). Successful initiation of the project resulted from major awareness campaigns undertaken for the islanders in 2005. Training sessions were also provided for locals so as to create a workforce of installers and maintenance personnel from within the population (de Angelis et al., 2010).

8.3 Strategic elements for transition pathways

For each of the transport, buildings, industry and primary production sectors, in order to increase the contribution of RE (Figure 8.2), possible strategies to overcome barriers and non-technical issues need to be better understood. Preparing transition pathways for each specific strategic element in a region can enable the effective integration of RE with existing energy supply systems to occur.

In the IPCC 4th Assessment Report (IPCC, 2007) the economic mitigation potentials for each of the sectors were analyzed at various carbon prices (Figure 8.12[19]). The substitution of fossil fuels by RE sources for heat and electricity generation was included in the energy supply sector (together with fuel switching, nuclear power and CCS). Integration of biofuels for transport, RE for heating/cooling of buildings, RE for process heat for industry and RE in food and fibre production were considered only to a limited degree.

The IPCC 4th Assessment Report was based mainly on information and data collected from 2004 or earlier as published in the latest literature at the time. Since then, RE technologies have continued to evolve and there has been increased deployment due to improved cost

---

19 In this chapter the 'Energy Supply' sector is covered in Section 8.2 and the 'Waste' sector discussion on biogas, landfill gas and municipal solid waste (MSW) incineration has been distributed between Sections 8.2.2, 8.2.3 and 8.3.4.

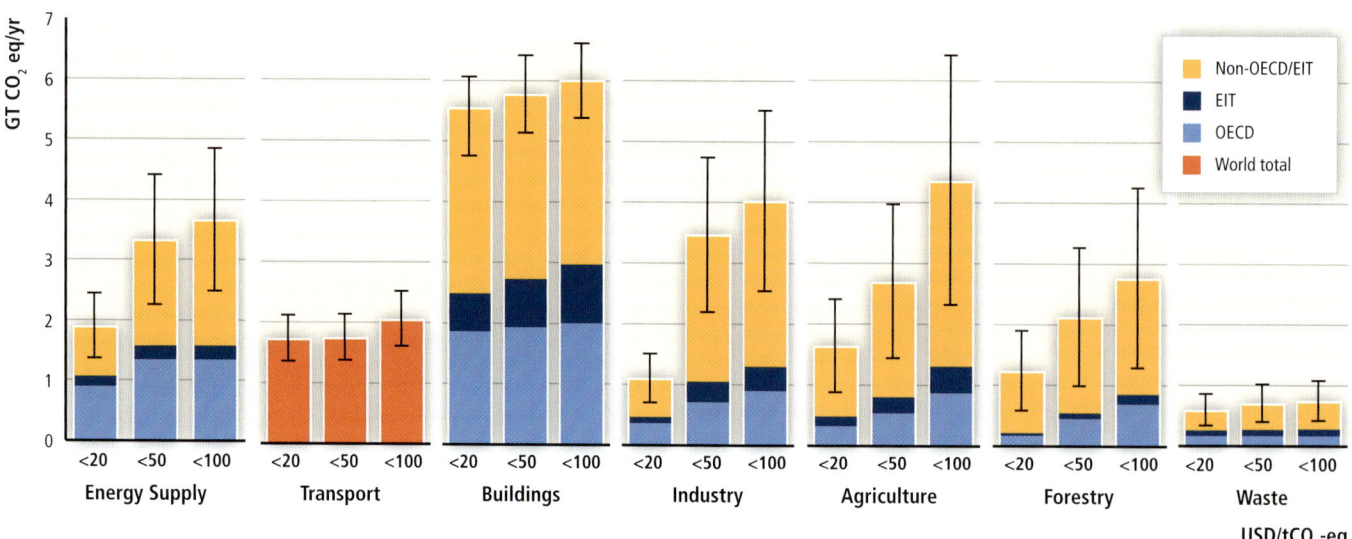

Figure 8.12 | Estimated economic potential ranges for GHG mitigation in the energy supply and end-use sectors, above the assumed baseline for different regions, as a function of the carbon price in 2030, and based on end-use allocations of emissions including from electricity generation (IPCC, 2007). Regional categories presented here include the entire world, member countries of the Organisation for Economic Co-operation and Development (OECD), Economy In Transition (EIT) countries, and Non-OECD/EIT countries.

competitiveness, more support policies and increased public concerns about the threats of an insecure energy supply and climate change. In the following sections, for each of the transport, residential and commercial building, industry and agriculture/forestry sectors, the current status of RE use, possible pathways to enhance increased adoption, the barriers to integration yet to be overcome, possible future trends and regional variations are discussed.

Ideally, the sectors need to be flexible enough to cope with future integration of the full range of RE systems as these continue to evolve. As market shares increase, competition between RE technologies, as well as with other low-carbon technologies, could result. For example, if domestic solar and geothermal heating systems for individual buildings become more cost competitive, an existing bioenergy district heating scheme supported by the local municipality could become a stranded asset as building owners disconnect. Similarly, at the larger scale, should a new large nuclear or coal-fired power plant with CCS be developed in a region to provide enough capacity to meet the future electricity demands of an energy-intensive industry, then this could compete with investment capital from the industry for developing a local geothermal, bioenergy or hydropower plant and potentially constrain such development for several decades, even where good RE resources exist. Failure to recognize future competition can result in an overestimation of the potential for integration of any single RE technology. Similarly, for road transport, it is uncertain how much investment in infrastructure for biofuel distribution, electric vehicle recharging or hydrogen production and storage will be required, or indeed how these technologies will compete.

8.3.1 Transport

Demand for mobility is growing rapidly, with the number of motorized vehicles projected to triple by 2050 (IEA, 2009c). Globally, about 94% of transport fuels come from petroleum sources, about 70% of which is traded (EIA, 2010). Decarbonizing and improving the efficiency of the transport sector will be critical for achieving long-term, deep cuts in carbon emissions. The potential exists for a transition by using larger quantities of RE as fuels (IEA, 2009c).

8.3.1.1 Sector status and strategies

In 2008, the direct combustion of oil products for transport accounted for around 18% of global primary energy use and produced approximately 22%[20] of energy-related GHG emissions (IEA, 2009d) and between 5 and 70% of air pollutant emissions, (varying with the pollutant and region). Of the total transport fuel consumption worldwide in 2008 (around 96 EJ, Figure 8.2), light duty vehicles (LDVs) consumed about half, with heavy duty vehicles (HDVs) accounting for 24%, aviation 11%, shipping 10% and rail 3% (IEA, 2009d). Future energy supply security is a serious concern for the sector.

To help meet future goals for both energy supply security and GHG reduction, oil use would need to be substantially reduced over a period of several decades. Many mitigation scenarios (Section 10.2) and other recent studies (C. Yang, 2008; IEA, 2008c; NRC, 2008) suggest that, other than diversifying the primary energy supply, a combination of

---

20 23% in 2005 on a well-to-wheel basis.

approaches will be needed to accomplish 50 to 80% reductions in transport-related GHG emissions by 2050 (compared to current values) whilst meeting the projected growth in demand (IEA, 2009c).[21]

- *Reduction of travel demand*. Less total vehicle kilometres travelled might be best achieved by encouraging greater use of car-pooling, cycling and walking, combining trips, or telecommuting. City and regional 'smart growth' practices could reduce GHG emissions as much as 25% by planning cities so that people do not have to travel as far to work, shop and socialize (Johnston, 2007; PCGCC, 2010).

- *Shift to more energy efficient modes* (in terms of *reduced MJ per kilometre*). For example, people could move from LDVs to mass transit (bus or rail)[22] or freight could be moved from trucks to rail or ships[23] (IEA, 2009c).

- *Improved energy efficiency of vehicles*. Reducing vehicle weight, aerodynamic streamlining, and improving the designs of engines, transmissions and drive-trains will continue. Examples include hybrid electric vehicles (HEVs), turbo-charging of internal combustion engines (ICEs) and down-sizing of installed vehicle engine power. Electric drive vehicles, employing either batteries or fuel cells, can be more efficient than their ICE counterparts, but the full well-to-wheel efficiency will depend on the source of the electricity or hydrogen (Kromer and Heywood, 2007; NRC, 2008; Section 8.3.1.3). Consumer acceptance of high efficiency drive-trains and lighter cars will depend on a host of factors including vehicle performance and purchase price, fuel price, and advancements in materials and safety. For light commercial trucks where high speeds are not needed, smaller, more efficient engines may be sufficient and could result in lower GHG emissions. In the HDV sub-sector for freight movement, and in aviation, there are also potentially significant energy efficiency improvements (Section 8.3.1.6).

- *Replacing petroleum-based fuels with low or near-zero carbon fuels*. These include biofuels, electricity or hydrogen produced from low-carbon sources such as RE, fossil energy with CCS or nuclear power. Other than biofuels, which provided around 2% of global road transport fuels in 2008 mostly as blends (Section 2.2), alternatives to petroleum-based fuels have had limited success thus far since the total number of internal combustion engine passenger vehicles (ICEVs) is currently more than 99% of the global on-road vehicle fleet (IEA, 2009c). Alternative fuels, including electricity for rail, presently represent about 5 to 6% of total transport energy use (IEA, 2009c). Exceptions include:

- Brazil, where sugar cane bioethanol and some biodiesel supply around 50% (by energy content) of total transport fuels used for LDVs (IEA, 2007b), representing about 15% of total energy use (EIA, 2010);

- Sweden, where imported ethanol is being encouraged through taxation policy; and

- The USA where ethanol, derived from corn or imported from Brazil, is currently blended with gasoline up to 10% by volume in some regions, although it still only accounts for about 3% of total US transport energy use (EIA, 2010).

Compressed natural gas (CNG) is widely used in LDV fleets mainly in Pakistan, Argentina, Iran, Brazil and India (IANGV, 2009). Liquefied petroleum gas (LPG) is also used in several countries while Sweden is encouraging the use of biomethane for vehicles (IEA Bioenergy, 2010b).[24] Electricity also makes a contribution to the transport sector in many countries, mostly limited to rail.[25] The context for alternative fuels is rapidly changing due to secure energy supply, oil price volatility and climate change concerns, and a host of policy initiatives in Europe, North America and Asia are driving towards lower carbon fuels and zero-emission vehicles.

8.3.1.2 Renewable fuels and light-duty vehicle pathways

A variety of more efficient vehicles and/or compatible alternative fuels have been proposed including gasoline and diesel plug-in hybrid electric vehicles (PHEVs), battery electric vehicles (EVs), hydrogen fuel cell electric vehicles (HFCVs) and liquid and gaseous biofuels. Possible fuel/vehicle pathways (Figure 8.13) begin with the primary energy source, its conversion to an energy carrier (or fuel), and then the end use in a vehicle drive train.

This section focuses on how the different RE pathways (including for liquid and gaseous biofuels; Sections 8.2.3, 8.2.4, and 2.6.3) can be integrated into the present transport system. Metrics include cost, GHG emissions from well-to-wheel (WTW),[26] energy use and air pollutant emissions (Section 9.3.4).

---

21 In IEA scenarios, vehicles become about twice as efficient by 2050. In the Energy Technology Perspectives 'Blue Map' scenario (50% GHG reduction by 2050), conventional gasoline and diesel-powered LDVs are largely replaced by a portfolio of vehicle drive trains (IEA 2010c). At least half of GHG emission reductions come from a mix of improved efficiency measures and alternative fuels (biofuels, electricity and hydrogen). These account for 25 to 50% of total transport fuel use in 2050, with liquid biofuels used more extensively in HDVs, aviation and marine applications.

22 Assuming that mass transit is operating at relatively high capacity. On a passenger-km basis, the transport modes with the lowest GHG intensity are rail, bus and two-wheel motor bikes, and the highest are LDVs and aviation.

23 For freight, shipping is the lowest GHG intensity mode on a tCO_2-km basis, followed by rail, and then HDVs and aviation by at least an order of magnitude higher.

24 In Sweden, 19% of biogas produced in 2006 was upgraded to biomethane and used in vehicles, but only represented about 1% of total domestic transport energy use.

25 For Germany as an example, in 2008, surface passenger transport amounted to 1,042 billion person-km of which roughly 8% was electric rail transport (DPG, 2010). Several regional rail networks purchase only RE electricity. This includes S-Bahn Hamburg that consequently avoids 60,000 t of CO_2/yr (www.s-bahn-hamburg.de/s_hamburg/view/aktuell/presse/2009_12_04.shtml).

26 Made up of 'well-to-tank' emissions upstream of the vehicle, plus 'tank-to-wheels' tail-pipe vehicle emissions.

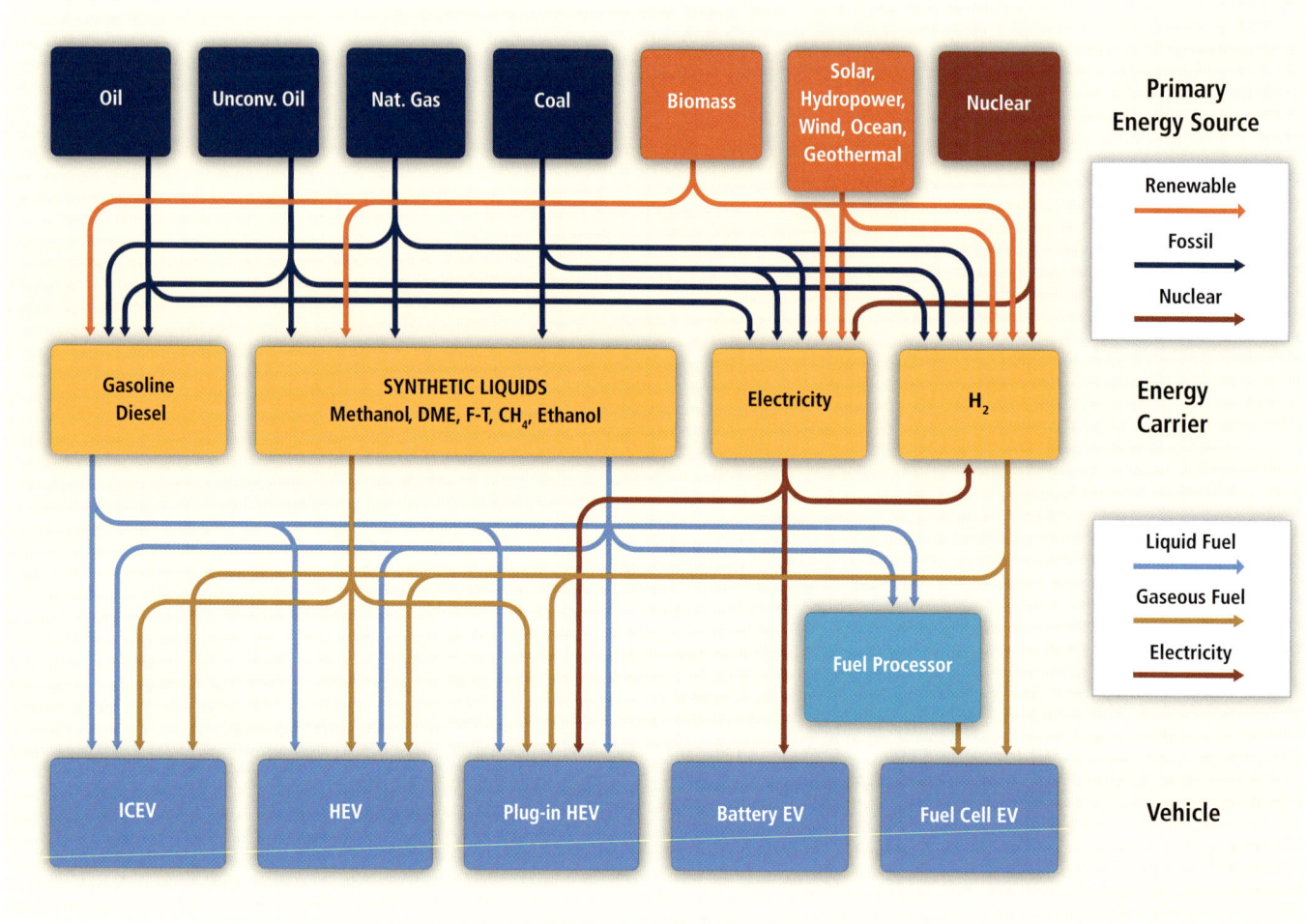

Figure 8.13 | Possible fuel/vehicle pathways from primary energy sources, through energy carriers (yellow, to vehicle end-use drive-train options (with RE resources highlighted in orange).

Notes: F-T= Fischer-Tropsch process. 'Unconventional oil' refers to oil sands, oil shale and other heavy crudes.

Each fuel/vehicle pathway has different environmental impacts, costs and benefits from a life-cycle perspective. WTW analyses (MacLean and Lave, 2003; CONCAWE, 2007; Bandivadekar et al., 2008; L. Wang, 2008) account for all emissions including those associated with primary resource extraction, processing, delivery, conversion to a useful fuel, distribution and dispensing, and vehicle use. Composite sustainable fuel indicators for future transport pathways include a variety of factors in addition to GHG emissions (Zah et al., 2007) such as air quality and a secure energy supply. Sustainability issues, such as land use and water (Section 2.5) may impose further constraints as well as the use of materials. Commercializing new vehicle drive technologies could require large amounts of scarce, hard to access mineral resources. For example, automotive fuel cells require platinum, electric motors require powerful lightweight magnets that may use neodymium and lanthanum (Delucchi and Jacobson, 2009; Margonelli, 2009; Mintzer, 2009), and the most likely next generation of advanced, lightweight, high-energy-density batteries will require lithium. Land use change impacts from biofuel feedstock production are sometimes but not always included (Fritsche et al., 2010; Section 2.5.3). Complementary discussions of these issues are provided in Chapters 2 and 9.

Status and prospects – vehicle technology

A variety of alternative vehicle drive-trains could use RE-based fuels, including advanced ICEVs using spark-ignition or compression-ignition engines, HEVs, PHEVs, EVs and HFCVs. Several recent studies have assessed the performance, technical status and cost of different vehicle types (CONCAWE, 2007; Kromer and Heywood, 2007; Bandivadekar et al., 2008; IEA, 2009c; Plotkin and Singh, 2009). Fuel economy and incremental costs of alternative-fuelled vehicles based upon these studies have been compared (Figures 8.14 and 8.15). Since each study employed different criteria and assumptions for vehicle design, technology status and development time frames (varying between 2010 and 2035), and since not all possible vehicle/fuel pathways were covered in all studies, the results have been normalized to those for an advanced gasoline ICEV, as defined in each study. The relative energy efficiency assumptions for

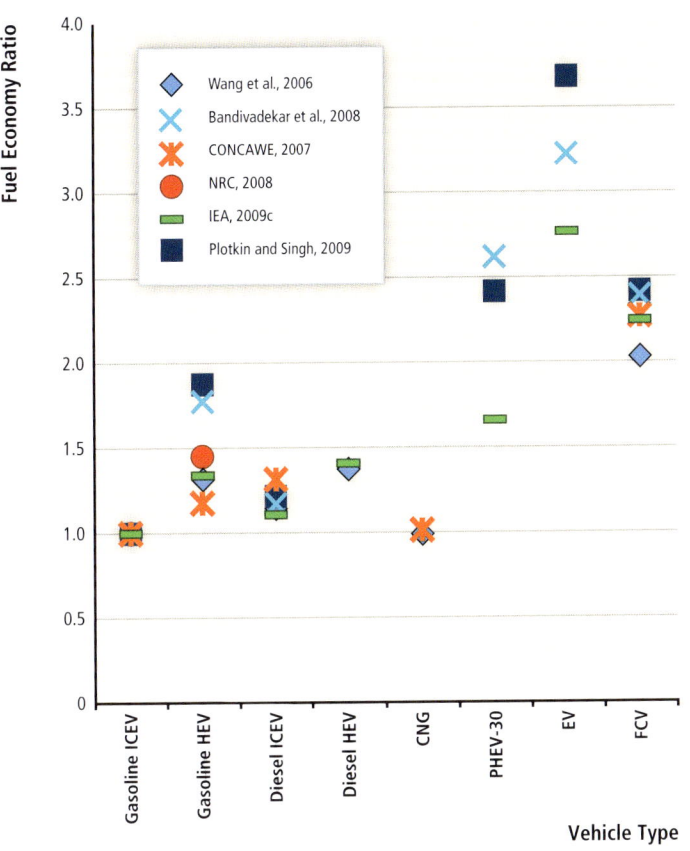

Figure 8.14 | Relative fuel economies of future alternative drive train light duty vehicles compared to advanced spark ignition, gasoline-fuelled, internal-combustion engine vehicles, based on several selected studies.

Notes: The comparative ratios only represent tank-to-wheel energy use. In a full analysis, well-to-tank energy use should also be considered (Section 8.3.1.2) with overall system losses typically 5 to 15% for gasoline and diesel extraction, refining and delivery; 20 to 80% for biofuels depending on the type and biomass feedstock; 40 to 80% for electricity; and 40 to 90% for hydrogen (M. Wang et al., 2006). Biofuels can be used in gasoline, diesel and hybrid drive train and biomethane in CNG engines. PHEV30 implies a 30 mile all-electric range (also termed PHEV 50km).

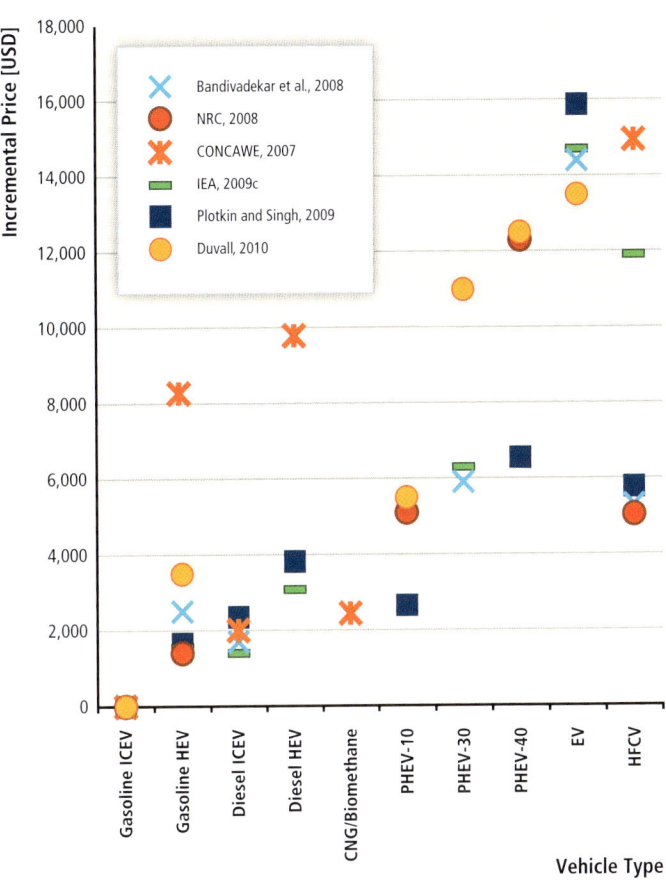

Figure 8.15 | Relative incremental retail price for future mid-sized alternative drive light duty vehicles compared to advanced gasoline, spark ignition, internal combustion engine vehicles as the reference price (= $0).

Notes: The reference gasoline ICEV had a price range of USD$_{2005}$ 21,000 to 24,000 as quoted in the various studies. Bandivadekar et al. (2008) gave projections for 2035. NRC (2008) assumed mature technologies with cost reductions due to experience learning and mass production post-2025. CONCAWE (2007) was for 2010 technologies; IEA (2009c) and Plotkin and Singh (2009) were for 2030 technology projections. The pure battery EVs in these studies had an assumed shorter range (typically 320 km) compared to the reference gasoline car because of imposed battery weight and cost limits. PHEV-10, -30 and -40 imply the range in miles on electricity only. Biofuels can be used in all gasoline and diesel vehicles.

different vehicle types also varied, especially for less mature technologies, although the overall findings of all studies were fairly consistent.

Several trends are apparent in fuel consumption of light-duty vehicles:

- There is significant potential to improve fuel economy by adopting new fuels and drivetrains and more advanced engines, improving aerodynamic design of the vehicle and employing lighter weight materials.

- HEVs increase efficiency and improve tank-to-wheel fuel economy of the vehicle by 15 to 70% over advanced conventional gasoline ICEVs.

- Although still under development and in the demonstration phase, HFCVs may be 2 to 2.5 times more efficient on a tank-to-wheel basis than non-hybrid gasoline ICEVs.

- EVs could operate around 3 to 4 times as efficiently as gasoline ICEVs on a tank-to-wheel basis, not including electric power generation or oil extraction and processing inefficiencies.

- On a total WTW fuel cycle basis, the relative efficiency improvements for HFCVs and EVs are considerably less when electricity generation and hydrogen production losses are included.

- Losses related to electricity generation, transmission and distribution range between 40 and 80%, depending on the source of power and transmission distance. A similar loss range occurs for hydrogen production, depending on the primary energy source, conversion technology and distribution infrastructure.

- In general, the higher the fuel economy, the higher the vehicle price (assuming similar size and performance).

- There is uncertainty in the fuel economy and cost projections, particularly for HFCVs and EVs that are not yet commercially produced at high volumes.

8.3.1.3 Transition pathways for renewable energy in light-duty transport

Historically, major changes in transport systems, such as building canals and railroads, paving highways and adopting gasoline cars, have taken many decades to complete for several reasons.

- Passenger vehicles have relatively long lifetimes: 15 years average in the USA (Davis et al., 2010), 10 to 13 years in EU countries (Christidis et al., 2003), 11 to 13 years in Japan (M. Wang et al., 2006) and 16 years in China (M. Wang et al., 2006). Even if a new technology rapidly moved to 100% of new vehicle sales, it would take many years for the present vehicle stock to 'turn over'. In practice, adoption of new vehicle technologies occurs slowly and can take 25 to 60 years for an innovation to be used in 35% of the on-road fleet (Kromer and Heywood, 2007). For example, research into gasoline HEVs in the 1970s and 1980s led to a decision to commercialize only in 1993 with the first vehicle becoming available for sale in 1997 in Japan. More than 13 years later, HEVs still represent only about 1% of new car sales[27] and less than 0.5% of the worldwide fleet (although low oil prices during part of this period were a possible factor). This slow turnover rate is also true for relatively modest technology changes such as the adoption of automatic transmissions, intermittent windscreen wipers and direct fuel injection (Kromer and Heywood, 2007; Bandivadekar et al., 2008). The time frame for new technologies relying on batteries, fuel cells or advanced biofuels could be even longer since they all need further RD&D investment and international standardization before they can become fully commercialized. Further cost reductions may also be needed to achieve wide customer acceptance.

- Changing fuel supply infrastructure, especially if switching on a major scale from liquids to gaseous fuels or electricity, will require a substantial amount of capital and take many decades to complete (IEA, 2009c; Plotkin and Singh, 2009). Developing new supply chains for RE and replacing existing fossil fuel systems will take time and require close coordination among fuel suppliers, vehicle manufacturers and policymakers.

Each fuel/vehicle pathway faces its own transition challenges that can vary by region. In terms of technology readiness of fuels and vehicles, challenges include infrastructure compatibility, consumer acceptance (costs, travel range, refuelling times, reliability and safety concerns), primary resource availability for fuel production, life-cycle GHG emissions, and environmental and sustainability issues including air pollutant emissions and competing demands for water, land and materials.

Millions of vehicles capable of running on liquid biofuels or biomethane are already commercially available in the global fleet. The cost, weight and life of present battery technologies are the main barriers to both EVs and PHEVs but the vehicles are undergoing rapid development, spurred by recent policy initiatives worldwide. Several companies have announced plans to commercialize them within the next few years, albeit in relatively small numbers initially (tens of thousands of vehicles per year) and at higher retail prices than comparable vehicles, even with proposed subsidies. Electric two-wheel motor-bikes and scooters are a large and fast-growing market in the developing world, especially in China with 20 million annual sales in 2007 (Kamakaté and Gordon, 2009). They have significant potential for fuel efficiency improvements and GHG reductions. HFCVs have been demonstrated, but are unlikely to be fully commercialized until at least 2015 to 2020 due to barriers of fuel cell durability and cost, on-board hydrogen storage and hydrogen infrastructure availability and cost. The timing for commercializing each technology is discussed in Section 8.3.1.4.

Liquid biofuel pathways
Biofuels are generally compatible with ICEV technologies and many vehicle owners already regularly choose liquid biofuels and blends, whereas only a small fraction of vehicles are adapted to run on gaseous fuels (CNG, LPG or hydrogen). Most of the existing gasoline and diesel ICEV and HEV fleet, however, can only operate on relatively low fraction biofuel blends. Blends above 10% by volume of ethanol or 5% of biodiesel risk possible adverse effects on some engine designs and, in some cases, higher air pollution levels. Over 22 million flexible fuel vehicles (FFVs), including motor bikes, have been designed to use either 100% gasoline or blends of ethanol between E20 and E100 in Brazil (Section 8.2.4.6), up to E85 in the USA and Canada, and up to E75 in Sweden under winter conditions. The incremental cost to produce an FFV is estimated to be only USD$_{2005}$ 50 to 100 per vehicle, so in many cases, manufacturers offer these vehicles at the same price as a comparable gasoline ICEV (EPA, 2010).

Biomass can be converted into liquid fuels using many different routes (Section 2.3.3). First generation processes are commercially available and second generation and more advanced processes, aiming to convert non-food, cellulosic materials and algae, are under development (Section 8.2.4). Advanced biofuels have potential for lower WTW GHG emissions than some first generation and petroleum-derived fuels, but these technologies are still several years from market (Sims et al., 2008) (Section 2.6.3).

---
27 In Japan adoption has been more rapid, with roughly 8% of the new car market in 2009 captured by HEVs.

An advantage of liquid biofuels is their relative compatibility with the existing liquid fuel infrastructure and ease of blending with petroleum-derived fuels (Section 8.2.4.1). In Brazil, for example, FFV users select their fuel blend based on price. Reduced vehicle range and fuel economy when using ethanol and, to a lesser extent, biodiesel, can also be a factor in consumer acceptance.

Primary biomass resource availability from sustainable production (Fritsche et al., 2010) can be a serious issue for biofuels. Recent studies (IEA, 2009c; Plotkin and Singh, 2009) have assessed the potential for biofuels to displace petroleum products. Environmental and land use concerns could limit production to 20 to 30% of total transport energy demand or about 35 to 50 EJ/yr of biofuel in 2050 (IEA, 2008e) though this remains under debate (Section 2.6.3). Given that certain transport sub-sectors such as aviation and marine require liquid fuels, it may be that biofuels will be used primarily for these applications (IEA, 2008c), whilst electric drive train vehicles (EVs, PHEVs or HFCVs), if successfully developed and cost effective, might eventually dominate the LDV sector.

Biomethane pathways

Biogas and landfill gas produced from organic wastes and green crops (Section 2.3.3) can be purified by stripping out the CO_2 (to give greater range per storage cylinder refill) and any H_2S (to reduce risk of engine corrosion) (Section 8.2.3.3) to provide biomethane. Various pathways include injection into existing natural gas distribution systems (Section 8.2.3) or direct use in ICEVs, mainly with spark-ignition engines designed or converted to run on biomethane using similar modifications as for CNG.

Hydrogen/fuel cell pathways

Hydrogen is a versatile energy carrier that can be produced in several ways (Section 8.2.3). WTW GHG emissions vary for different hydrogen fuel/vehicle pathways, but both RE and fossil hydrogen pathways can offer reductions compared to gasoline vehicles (Section 8.3.1.4).

Although hydrogen can be burned in a converted ICEV, more efficient HFCVs are attracting greater RD&D investment by engine manufacturers. Many of the world's major automakers have developed prototype HFCVs, and several hundred of these vehicles, including cars and buses, are being demonstrated worldwide. HFCVs are currently very costly, in part because they are not yet mass produced. Fuel cell lifetimes are also relatively short. It is projected that the costs of HFCVs will fall with further improvements resulting from R&D, economies of scale from mass production, and learning experience (NRC, 2008).

HFCVs could match current gasoline ICEVs in terms of vehicle performance and refuelling times. The maximum range of present HFCV designs of LDVs is acceptable at around 500 km[28] but hydrogen refilling availability and the high cost of both vehicle and fuel remain key barriers to consumer acceptance. Hydrogen is not yet widely distributed to consumers in the same way as gasoline, diesel and, depending on the market, electricity, natural gas and biofuels. Bringing hydrogen to large numbers of vehicle owners would require building a new refuelling infrastructure over several decades (Section 8.2.3.5). Hydrogen and fuel cells exhibit the 'chicken and egg' problem that vehicle makers will not introduce hydrogen cars until refuelling stations are in place, and fuel providers will not build refuelling stations until there are enough cars to use them. A solution is to introduce the first hydrogen vehicles and stations in a coordinated fashion in a series of demonstration projects (Gronich, 2006; CAFCP, 2009; Nicholas and Ogden, 2010).

Hydrogen can be produced regionally in industrial plants or locally onsite at vehicle refuelling stations or in buildings. The first steps to supply hydrogen to HFCV test fleets and demonstrate refuelling technologies in mini-networks have been constructed in Iceland, California, Germany and elsewhere.[29] System-level learning from these programmes is valuable and necessary, including development of safety codes and standards. In the longer term, in the USA for example, a mix of low-carbon resources including natural gas, coal (with CCS), biomass and wind power could supply ample hydrogen (NRC, 2008). The primary resources required to provide sufficient fuel for 100 million passenger vehicles in the USA using various gasoline and hydrogen pathways have been assessed (Ogden and Yang, 2009). Enough hydrogen could be produced from wind-powered electrolysis using about 13% of the technically available wind resource. However, the combined inefficiencies of producing the hydrogen via electrolysis from primary electricity sources, then converting it back into electricity on a vehicle via a fuel cell, loses more than 60% of the original RE inputs. Electricity would be used more efficiently in an EV or PHEV but hydrogen might be preferred in large vehicles requiring a longer range and faster refuelling times.

Hydrogen production and delivery pathways have a significant impact on the cost to the consumer. In addition, compared to industrial uses, fuel cell grade hydrogen needs to be >99.99% pure and generally compressed to 35 to 70 MPa before dispensing. Using optimistic assumptions in the near-term, hydrogen at the pump might cost around USD_{2005} 7 to 12/kg excluding taxes, potentially decreasing to USD_{2005} 3 to 4/kg[30] over time (NRC, 2008). However, estimates range from about USD 8 to 10/kg for dispensed hydrogen produced from natural gas reforming and about USD 10 to 13/kg for hydrogen from electrolysis using grid electricity (NREL, 2009). RE electricity may increase the electrolyzed hydrogen cost. Given the potential higher efficiency of fuel cell vehicles, the fuel cost per kilometre could eventually become competitive with ICEVs (Kromer and Heywood, 2007; NRC, 2008).

---

28 Some demonstration HFCVs have significantly higher ranges. The latest demonstration Toyota HFCV has 70 MPa compressed gas storage and achieves a range of 790 km under optimum conditions (www.cleanenergypartnership.de).

29 These include the GermanHy project (Bonhoff et al., 2009), Norway's Hynor project (www.hynor.no), the California Fuel Cell Partnership (www.fuelcellpartnership.org), Japan's Hydrogen and Fuel Cell Demonstration Project (www.nedo.go.jp), the European Clean Energy Partnership (www.cleanenergypartnership.de) and the EU Fuel Cells and Hydrogen - Joint Undertaking (ec.europa.eu/research/fch).

30 1 kg of hydrogen at 120.2 MJ (lower heat value) has a similar energy content to 1 US gallon (3.78 litres) of gasoline.

Several studies (Gielen and Simbolotti, 2005; Gronich, 2006; Greene et al., 2007; NRC, 2008) indicated that cost reductions were needed to bring down fuel cell vehicles to market clearing prices (through technological learning and mass production). In addition, to build the associated infrastructure over several decades could cost hundreds of billions of dollars (Section 8.2.3.5). The majority of this cost would be for the incremental costs of early hydrogen vehicles, with a lesser amount needed for early infrastructure. Even at high oil prices, government support policies may most likely be needed to subsidize these technologies in order to reach cost-competitive levels and gain customer acceptance.

Electric and hybrid vehicle pathways
EV drive trains are relatively efficient as is battery recharging as a way to store and use RE electricity. Combined EV efficiencies (motor/controller 90 to 95%; battery charge/discharge efficiencies ~90%) for electric plug-to-battery output-to-motor, are of the order of 81 to 86% (Kromer and Heywood, 2007), although electricity generation from primary energy sources including transmission and distribution losses is typically only 20 to 60% efficient (Graus et al., 2007; IEA, 2008c).

EV use is currently limited to neighbourhood and niche fleet vehicles, from golf carts to buses. There are also a limited number of operating passenger and light truck EVs that were sold by GM, Ford, Toyota, Honda and others during the 1990s and early 2000s. Limited commercialization of new designs of EVs and PHEVs is underway partly in response to policy measures (Kalhammer et al., 2007) with several automobile manufacturers making niche initial offerings. In Japan, Mitsubishi Motors and Fuji Heavy Industry launched EVs in 2009 and Nissan launched a model in 2010. GM has launched a PHEV in the USA and Toyota began road testing pre-commercial Prius PHEVs in 2010.

Today's lithium batteries cost USD 700 to 1,000/kWh (194 to 278 USD/MJ), three to five times the goal needed for an EV to compete with gasoline vehicles on a life cycle cost basis. The main transition issue is to bring down the cost and improve the performance of advanced batteries. Demonstrated lifetimes for advanced lithium battery technologies are presently only 3 to 5 years, whereas, ideally, a 10-year minimum life is required for automotive applications (Nelson et al., 2009).

For RE electricity to effectively serve growing EV markets, several innovations would need to occur, such as having flexibility in the charging schedule to reflect varying RE generation outputs (and possibly by encouraging off-peak charging at night) and optimizing peak-time charging loads. Additional power generation and distribution capacity would then not necessarily be needed and there may be an adequate temporal match with wind, solar or hydropower resources. Flexible grids, interconnections, energy storage etc. (Section 8.2.1) may also be ways to help control and balance vehicle recharging demands when using variable RE resources. In addition, upgrading the distribution grid to include smart meters and RE technologies could manage the added load (IEA, 2009b).

Public acceptance of EVs is yet to be demonstrated, but one attraction is that they can often be recharged at home, thereby avoiding trips to the refuelling station. However, home recharging would require new equipment that only 30 to 50% of households and apartments in the USA would be able to conveniently install (Kurani et al., 2009). Therefore a public recharging point infrastructure would need to be developed in some areas. Recharger technology costs vary with different levels.

- 'Level 1' home overnight charging, using a standard domestic plug socket at 110 V (e.g., in the USA) or 240 V (e.g., in Europe), could take several hours, compared with the quick refill time possible with liquid or gaseous fuels and the recharging system might cost USD$_{2005}$ 700 to 1,300 to install (USDOE, 2008b).

- 'Level 2' charging could take less time but would require a specialized higher power outlet and cost USD 800 to 1,900 to install.

- 'Level 3' fast-charge outlets at publicly accessible recharging stations might bring batteries to near full charge after only 10 to 15 minutes, faster than level 1 or 2 charging technologies, but taking more time than refilling an ICEV. They would costs tens of thousands of dollars for each recharging point.

An EV can have a range of 200 to 300 km under good conditions compared with a similar size ICEV of 500 to 900 km (Bandivadekar et al., 2008). While this range is adequate for 80% of car trips in urban/suburban areas, long distance EV travel would be less practical. This could be overcome by owners of small commuter EVs using rental or community cooperative car share HEVs or PHEVs[31] for longer journeys (IEA, 2009b).

The added vehicle cost for PHEVs, while still significant, is less than for a similar size EV and the range is comparable to a gasoline HEV. One strategy could be to introduce PHEVs initially while developing and scaling up battery technologies for EVs. This could help lead to more cost-competitive EVs. However, HEVs will always be cheaper to manufacture than PHEVs due to their smaller battery capacity. Any advances in battery technologies will apply to HEVs as well as to PHEVs and EVs.

8.3.1.4 Comparisons of alternative fuel/vehicle pathways

WTW GHG emissions differ, depending on the fuel/vehicle pathway. For petroleum fuels, most of the emissions are 'tank-to-wheels' and take place at the vehicles. The GHG emissions and environmental benefits of EVs depend on the marginal grid mix and the source of electricity used for vehicle charging. For PHEVs the source of electricity also impacts the life cycle GHG emissions (Figure 8.16) but to a lesser degree. With the current US grid being 45% dependent on coal, WTW emissions from

---
31 Community car sharing cooperatives exist in many cities in Europe, having started in Switzerland and Germany in 1987, and are now growing in North America (www.carsharing.net and www.cooperativeauto.net/).

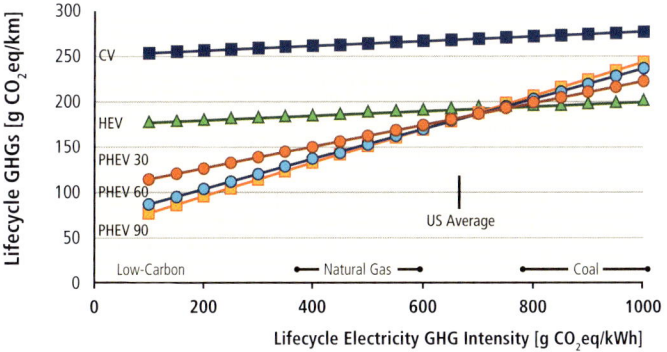

Figure 8.16 | Life cycle GHG emissions (excluding land use change) from a range of light duty vehicle types as a function of the GHG emission intensity of electricity generation systems using coal, natural gas or low-carbon technologies including nuclear and RE (Samaras and Meisterling, 2008).

Notes: The slight slopes of the conventional gasoline vehicle (CV) and HEV lines reflect the GHG emission intensity of the electricity used during production of the vehicles. Generation options correspond to various GHG intensities and provide insight into the impact of different generation mixes. For example, a 'low-carbon' portfolio could include nuclear, wind and coal with CCS. The vertical line at 670 g CO_2eq/kWh (186 CO_2eq/MJ) indicates the current US average life cycle GHG intensity. PHEV-30, -60, and -90 imply all-electric vehicle range in miles.

EVs would give only around 20 to 40% GHG emission reduction over efficient gasoline vehicles (Figure 8.17). By way of contrast, the French electric grid with a major share from nuclear power, or the Norwegian system dependent on hydropower, would give relatively low-carbon WTW emissions (Zgheib and Clodic, 2009).

For electricity and hydrogen, all emissions are 'well-to-tank' and the vehicle itself has zero GHG emissions except in the manufacturing process. For RE biofuel pathways, carbon emissions at the vehicle are partially offset by carbon uptake from the atmosphere by future biomass feedstocks. The degree of this offset is uncertain because of indirect land use issues (Searchinger et al., 2008; Fritsche et al., 2010; Section 2.5).

Various studies have developed scenarios for decarbonizing electricity grids over the next few decades (Sections 8.2.1 and 10.2), which would result in reduced WTW emissions for EVs, HEVs and PHEVs (EPRI, 2007; IEA, 2009c). Using larger fractions of RE or other low-carbon electricity, WTW emissions would, over time, become smaller than they are in many regions at present. EVs, having zero tailpipe emissions, can also reduce urban air pollution. However, if the electricity is produced

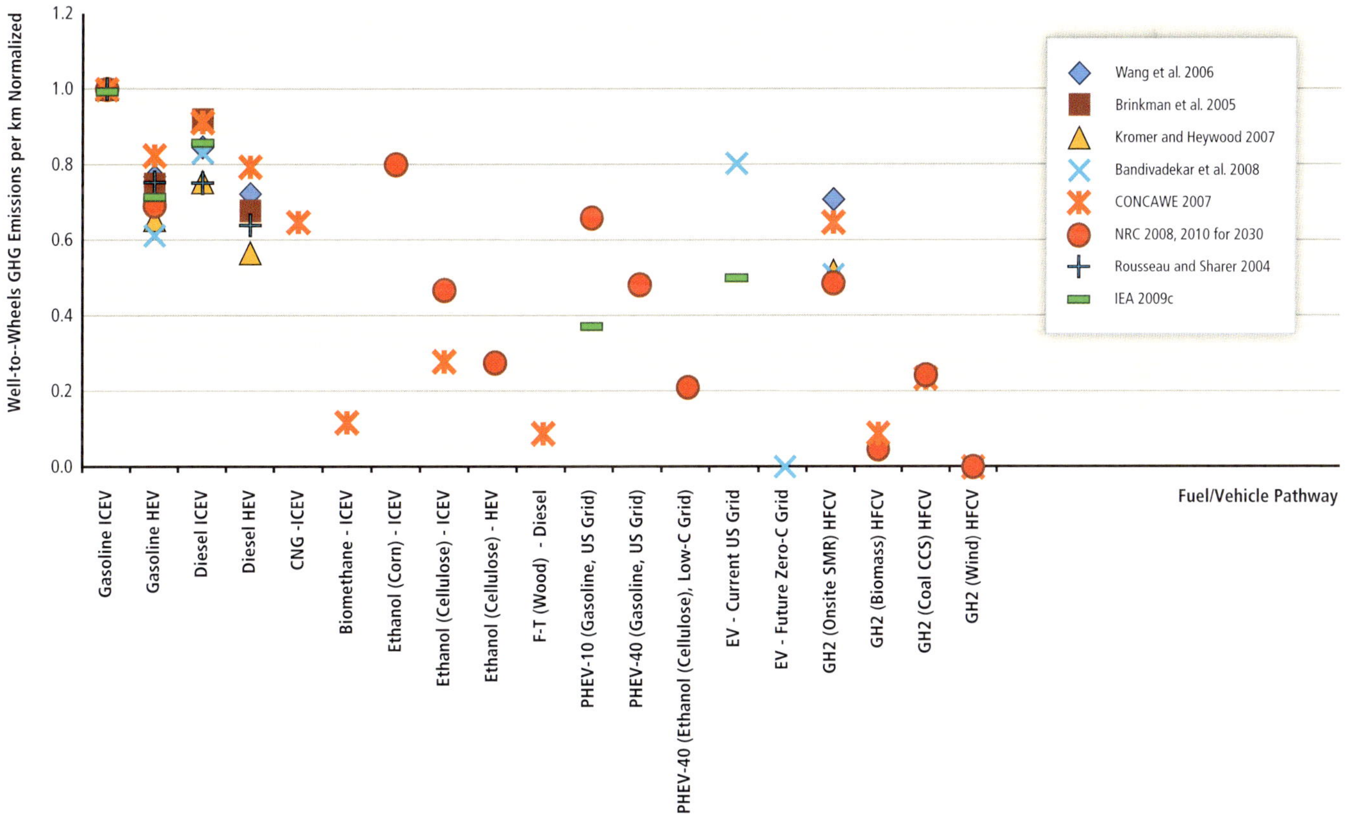

Figure 8.17 | Well-to-wheels GHG emissions per kilometre travelled from selected studies of alternative light duty fuel/vehicle pathways, normalized to the GHG emissions value of a gasoline internal combustion engine light duty vehicle (ICEV) but excluding land use change, vehicle manufacturing, and fuel supply equipment manufacturing impacts.

Notes: WTW GHG emissions per kilometre for the gasoline ICEV reference vehicle ('Gasoline ICEV' = 1 on the y-axis) were normalized to the average emissions taken from the gasoline ICEV in each study, which ranged from 170 to 394 gCO_2/km. For all hydrogen pathways, hydrogen is stored onboard the vehicle as a compressed gas (GH_2). SMR = steam methane reformer.

from an uncontrolled source (such as coal plants without particulate scrubbers), one source of pollution might simply be substituted for another but in a different location (Kromer and Heywood, 2007; Bandivadekar et al., 2008).

Making a transition to new fuels and types of vehicles is a complex process involving technology, cost, infrastructure, consumer acceptance and environmental and resource impacts. Transition issues vary for biofuels, hydrogen and electric vehicles. Biofuels have a clear start and could grow rapidly over the next decade (Section 2.8), but over the longer term, no one option is seen to be a clear 'winner' and all will take several decades to achieve large RE shares of the transport market.

8.3.1.5 Low-emission propulsion and renewable energy options in other transport sectors

Heavy duty vehicles
Globally, HDVs, consisting mainly of freight trucks and long-haul tractor-trailers, account for about 24% of transport-related energy use and a similar fraction of GHGs (IEA, 2009c). Other HDVs include buses and off-highway vehicles such as agriculture and construction equipment. As is the case for LDVs, several strategies can reduce fuel consumption and GHG emissions such as by:

- Partially switching to lower carbon fuels;

- Switching freight from trucks to more energy efficient modes such as rail and inland waterways;

- Streamlining operational logistics for handling freight and using GPS routing technology to avoid empty return trips; and

- Further increasing vehicle efficiency, perhaps by up to 30 to 40% by 2030 (IEA, 2009c), through more advanced engines, exhaust gas energy recovery (via advanced turbo-charging or turbo-compounding), hybrid vehicles (which may include either electric or hydraulic motors), weight reduction, lower rolling resistance tyres, use of aerodynamic technologies on the tractor and trailer, longer trains of more than one trailer, more efficient driving behaviour, optimized automatic gear shifting, speed reduction, and use of more efficient auxiliary power units (APUs) used when decoupled from the power train.

Presently, about 85% of freight-truck fuel is diesel, with the remainder being gasoline. Integrating biofuels into the fuel mix would be the most straightforward RE option. Second generation biofuels could become a more significant blend component in diesel fuel for trucks, possibly reaching as high as 20 to 30% by 2050 (IEA, 2008c). Due to the range and resulting fuel storage requirements for long-haul HDVs, the use of other lower-carbon fuel options such as CNG, LPG, compressed biomethane, hydrogen (for either HFCVs or ICEVs) or electricity would likely be limited to urban or short-haul HDVs, such as refuse trucks, delivery trucks and buses.[32] LNG might become an option for freight transport though it faces the key hurdles of limited driving range and lack of refuelling infrastructure. For example, an LNG truck could travel around 600 km between refuelling, only around half the range of some diesel trucks. The additional weight of onboard LNG tanks can pose constraints for vehicle payloads. For urban fleets where more stringent air pollution controls and a common refuelling site exist, LNG may be viable for applications such as refuse trucks (EIA, 2010). Another potential use of low-carbon hydrogen or electricity might be to power onboard fuel cell APUs or charge batteries, although neither of these options is yet cost effective. Trucks could also plug into an electrical energy source at a truck stop to run their accessories, but the GHG reduction benefit would depend on the carbon footprint of the local electricity source.

The reduction of fuel consumption and GHG emissions in HDVs may be more difficult than for LDVs due to more limited weight reduction potential, slower vehicle turnover, faster growth in vehicle kilometres travelled (VKT), less discretionary freight movement, and inherent economic drivers that continuously aim to minimize HDV operating costs. Many HDVs are purchased for fleet operations, so there could be an opportunity to integrate alternative fuels and vehicles by providing fleet-wide support for new fuelling infrastructure, technology maintenance and, if needed, driver training. According to the IEA's baseline scenario (IEA, 2008c), HDV energy use by 2050, even with improved energy efficiency of about 20%, is projected to increase by 50% due to double the current quantity of worldwide freight moved by trucks, mostly in non-OECD countries.

Aviation
Aviation energy demand accounts for about 11% of all transport energy and this could double or triple by 2050 (IEA, 2009c). Rapid growth of aviation emissions is due to the increase of air traffic volumes for both passenger and freight, with aviation usually having the highest energy and GHG intensity of all transport modes. About 90% of fuel use and GHG emissions occur in flight, mostly at cruising altitude (TRB, 2009). Efficiency improvements can play an important role in reducing aviation energy use by 30 to 50% in future aircraft designs compared with 2005 models (IEA, 2009c). These include improved aerodynamics, airframe weight reduction, higher engine efficiency, as well as improvements in operation and air traffic control management to give higher load factors, improved routing, and more efficient ground operations at airports (including gate electrification and use of low carbon-fuelled service vehicles) (TRB, 2009). A slow average fleet turnover of around 30 years (IEA, 2009c; TRB, 2009) will delay the penetration of advanced aircraft designs. Although reductions in energy use per passenger-km or per cargo tonne-km can be substantial, they are unlikely to be able to completely offset the expected increase in GHG emissions arising from higher demand for air freight and passenger transport.

---

32 An electric bus with a range of 200 km and recharged daily from 50 kWe of solar PV panels installed on the roof of the bus station, has been operating in Adelaide since 2009 (IEA, 2009b).

Aircraft will continue to rely mainly on liquid fuels due to the need for high energy density fuels in order to minimize fuel weight and volume. In addition, due to safety, the fuels need to meet more stringent requirements than for other transport modes, including thermal stability (to assure fuel integrity at high engine temperatures and to avoid freezing or gelling at low temperatures), specific viscosity, surface tension, ignition properties and compatibility with aircraft materials. Compared to other transport sectors, aviation has less potential for switching to lower carbon footprint fuels due to these special fuel requirements. In terms of RE, various aircraft have already flown test flights using various biofuel blends, but significantly more processing is needed than for road fuels to ensure that stringent aviation fuel specifications are met. Standards to allow greater biofuel blend fractions into conventional aviation fuel are currently under development. Industry and policy views on biofuels as a share of total aviation fuels by 2050 range from a few percent up to 30% (IEA, 2009c).

Liquid hydrogen is another long-term option, but faces significant hurdles due to its low volumetric energy density. Fundamental aircraft design changes would be needed to accommodate cryogenic storage, and airports would have to construct a hydrogen distribution and refuelling infrastructure. The most likely fuel alternatives to conventional jet fuel are therefore synthetic jet fuels (from natural gas, coal or biomass) since they have similar characteristics. Net carbon emissions will vary depending on the fuel source.

Maritime

Marine transport, the most efficient mode for moving freight, currently consumes about 9% of total transport fuel, 90% of which is used by international shipping (IEA, 2009c). Ships rely mainly on heavy fuel ('bunker') oil, but lighter marine diesel oil is also used. Heavy fuel oil accounts for nearly 80% of all marine fuels (IEA, 2009c). Its combustion releases sulphates that in turn create aerosols that may actually mitigate GHG impact by creating a cooling effect, though this will decline as ever more stringent air quality regulations aimed at reducing particulate emissions through cleaner fuels will require lower-sulphur marine fuels in the future. An expected doubling to tripling of shipping transport by 2050 will lead to greater GHG emissions from this sector.

Due to a fragmented industry where ship ownership and operation can occur in different countries, as well as a slow fleet turnover with typical ship replacement occurring about every 30 years (IEA, 2009c), energy efficiency across the shipping industry has not improved at the same rate as in the HDV and aviation sectors. Hence, significant opportunities exist to reduce fuel consumption through a range of technical and operational efficiency measures (IEA, 2009d; TRB, 2009) including improvements in:

- Vessel design (e.g., larger, lighter, more hydro-dynamic, lower drag hull coatings);
- Engine efficiency (e.g., diesel-electric drives, waste heat recovery, engine derating);
- Propulsion systems (e.g., optimized propeller design and operation, use of sails or kites);
- More efficient and lower GHG APUs; and
- Operation (e.g., speed reduction, routing optimization, better fleet utilization, reduced ballast).

These measures could potentially reduce energy intensity by as much as 50 to 70% for certain ship types (IEA, 2009c).

The key application of RE in marine transport could be through the use of biofuels. Existing ships could run on a range of fuels, including blends of biodiesel or lower quality fuels such as unrefined bio-crude oil produced from pyrolysis of biomass (Section 2.3.3). Engines would probably need to be modified in a manner similar to HDV road vehicles in order to operate reliably on high (80 to 100%) biofuel blends. Other RE and low-carbon options could include the use of on-deck hybrid solar PV and micro-wind systems to generate auxiliary power; solar thermal systems to provide hot water, space heating or cooling; wind kites for propulsion; and electric APU systems plugged into a RE grid source while at port. Although nuclear power has been used for decades by some navies, as well as ice breakers and a handful of other ships, widespread marine use would require large investments, demand for specialized crews and the need to deal with complex legal and security concerns. As a result, onboard nuclear marine power appears to be an unlikely and limited alternative for commercial ship propulsion (TRB, 2009).

Rail

Rail transport accounts for only about 3% of global transport energy use, but by 2050, rail freight volume is expected to increase by up to 50% with most of this growth occurring in non-OECD countries (IEA, 2009d). Rail moves more freight and uses an order of magnitude less energy per tonne-kilometre than road HDVs due to its much higher efficiency (IEA, 2009c). Rail transport is primarily powered by diesel fuel, especially for freight transport. However, electrification is increasing and accounted for 31% of global rail sector energy use, including both freight and passenger transport, in 2006 (IEA, 2009c). In certain economies including OECD Europe, the Former Soviet Union and Japan, over 50% of the rail sector is electric. Growth in high-speed electric rail technology continues rapidly in Europe, Japan, China and elsewhere. As with shipping, the use of high-sulphur fuels has helped to mitigate net GHG emissions due to the negative radiative forcing effect of sulphates, but this trend has other negative environmental consequences and will likely decline with stricter clean fuel regulations.

Rail sector efficiency increases of up to 20 to 25% are possible (IEA, 2009c; TRB, 2009) including:

- Upgrading locomotives to more efficient diesel engines, hybrids and APUs;
- Increasing load factors by reducing the empty weight of the rolling stock, lengthening trains and using double-stacked containers; and

- Operational improvements such as operator training, optimized logistics and reduced idling.

The two primary pathways for RE penetration in rail transport are through increased use of biodiesel, which may account for 2 to 20% of rail fuel use in 2050 (IEA, 2009d) and a further shift towards electrification. Compared to their diesel counterparts, all-electric locomotives can improve life cycle efficiency by up to 15%, (though less improvement if compared to a diesel hybrid-electric drive system that includes battery storage). GHG emissions can be further reduced as electricity generation switches to RE, nuclear power and fossil fuels with CCS. Although the use of hydrogen fuel cells may be limited due to range, energy storage and cost issues, the challenges for installing fuel cells on locomotives appear to be fewer than for passenger HFCVs. Compared with LDVs, a rail system provides more room for hydrogen storage, offers economies of scale for larger fuel cell systems and uses the electric traction motors already installed in diesel-electric locomotives.

8.3.1.6 Future trends for renewable energy in transport

The most important single trend facing the transport sector is the projected high growth of the road vehicle fleet worldwide, which is expected to triple from today's 700 million LDVs by 2050 (IEA, 2008c). Achieving a low-carbon, sustainable and secure transport sector will require substantial vehicle technology advancements and public acceptance of these new vehicles and alternative fuels, strong policy initiatives, monetary incentives, and possibly the willingness of customers to pay additional costs for fuels and vehicles. There is scope for RE transport fuel use to grow significantly over the next several decades, playing a major role in this transition.

In the future, a wider diversity of transport fuels and vehicle types is likely. These could vary by geographic region and transport sub-sector. For applications such as air and marine, liquid fuels are currently the only practical large-scale option. In the LDV sector, increased use of electric drive-train technologies has already begun, beginning with HEVs, and potentially progressing to PHEVs and EVs as well as possibly to HFCVs (IEA, 2008c). Historically, the electric and transport sectors have been developed separately, but, through grid-connected EVs, they are likely to interact in new ways by charging battery vehicles, or possibly 'vehicle-to-grid' electricity supply (Section 8.2.1; McCarthy et al., 2007).

Environmental and secure energy supply concerns are important motivations for new transport systems but sustainability issues may impose constraints on the use of alternative fuels or new vehicle drive trains. Understanding these issues will be necessary if a sustainable, low-carbon future transport system is to be achieved. Meeting future goals for GHG emissions and secure energy supplies will mean displacing today's ICEVs, planes, trains and ships with higher efficiency, lower GHG emission designs, switching to more efficient modes of transport and ultimately adopting new low- or zero- carbon fuels that can be produced cleanly and efficiently from diverse primary sources. There is considerable uncertainty in the various technology pathways, and further RD&D investment is needed for key technologies (including batteries, fuel cells and hydrogen storage) and for RE and low-carbon production methods for the energy carriers of biofuels, hydrogen and electricity.

Recent studies (IEA, 2008b, 2009d) see a major role for RE transport fuels in meeting future societal goals, assuming that strict carbon limits are put in place. Given uncertainties and the long timeline for change, however, it may be important to maintain a portfolio approach that includes behavioural changes (to reduce VKT), more efficient vehicles and a variety of low-carbon fuels. This approach may help recognize that people ultimately make vehicle purchase decisions, and that different technologies and fuel options will need to fit their various situations and preferences.

Present transport fuels and vehicle engine technologies represent sunk investments that, with experience and economies of scale, have progressed down their respective technological learning curves over the past century. Therefore, new alternative fuels and technologies are naturally disadvantaged (Section 11.11). Making the hydrogen, biofuel or electricity energy carriers more cost effective, efficient and reliable is one condition for providing RE for transport. Subsidies, tax exemptions and fuel standard exemptions for alternative fuel vehicles all have an impact on future market shares. To enable electricity or hydrogen from RE fuels to power transport vehicles, incentives such as low electricity prices relative to gasoline, carbon charges, subsidized low-carbon electricity and first-cost vehicle subsidies could be necessary to make EVs, PHEVs and HFCVs viable options (Avadikyan and Llerenaa, 2010). Policies could specifically provide incentives for infrastructure development that might enable biofuel production, trade and blending at high levels, public recharging of EVs, and hydrogen production and distribution. However, at this stage, it is not possible to determine which of these options will become dominant and should therefore receive the bulk of such incentives.

8.3.2 Buildings and households

Decarbonization of the building sector[33] can result from integration of RE in electric power systems (Section 8.2.1), heating and cooling networks (Section 8.2.2) and gas grids (Section 8.2.3) or by installing RE technologies onsite directly integrated into the building structure (Figure 8.1). RE deployment in a building can be combined with energy efficiency measures and encouraging energy conservation through education and behavioural change of the occupants (Pehnt et al., 2009a).

---

33 The 'building sector' is defined here as the combination of the 'residential' sector, the 'commercial and public services' sector and the 'non-specified' sector as segregated for IEA data.

8.3.2.1 Sector status

The building sector in 2008 accounted for about 92 EJ, or 32% of total global final energy consumption (IEA, 2010b; Figure 8.2). Around 4 EJ (±15%) of this total consumer energy was from combustion of around 31 EJ of traditional biomass for cooking and heating, assuming efficiency of combustion was around 15% (Section 2.1). Excluding this biomass, the residential sector consumed over half the total building energy demand followed by the commercial and public service buildings that slightly increased their share of the total since 1990 (IEA, 2010b). GHG emissions from the building sector, including through electricity use, were about 8.6 Gt CO_2 in 2004 with scope for significant reduction potential mainly from energy efficiency[34] (IPCC, 2007; IEA, 2009b).

Projections of energy demand for the building sector by region can vary considerably as a result of different assumptions of population growth rates, household numbers and service sector activity in each country. In OECD countries, decreasing energy use for heating buildings in OECD countries is expected as a result of energy efficiency and other policies (IEA, 2010b). For example, the EU Energy Performance of Buildings Directive, May 2010, demands that "member states shall ensure that by 2020, all new buildings are nearly zero-energy buildings" (EC, 2010). By contrast, non-OECD countries, as a result of significantly faster growing populations and increased average standard of building stock, will be faced with a potentially very large growth in energy demand, particularly for cooling. However, assuming stringent energy efficiency policies under the IEA 450 Policy Scenario, by 2035 the total sector demand could rise by only 25% above current levels to ~116 EJ (Figure 8.2).

A broad typology of the building sector includes

- Commercial buildings and high-rise apartment buildings in mega-cities;
- Small towns of mainly attached and detached dwellings;
- Historic quarters;
- New urban subdivision developments;
- Wealthy suburbs;
- Poor urban areas; and
- Small village settlements in developing countries that have limited access to energy services.

The composition of age class of the building stock of a country influences its future energy demand, especially for heating and cooling. Many buildings in developed countries have average life spans of 120 years and above, hence energy efficiency measures and the integration and deployment of RE technologies will need to result mainly from the retrofitting of existing buildings. Developing countries currently have stock turnover rates of 25 to 35 years on average with relatively high new building construction growth (IEA, 2010d), therefore offering good opportunities to integrate RE technologies through new building designs.

Energy service delivery systems for residential and commercial buildings vary depending on the energy carriers available, local characteristics of a region and its wealth. To support the basic human requirements, livelihoods and well-being of the people living and working in buildings in both developed and developing countries, the appliances used in these buildings provide a variety of basic energy services including for:

- Space heating, water heating, cooking;
- Cooling, refrigeration;
- Lighting, electronic and electrical appliances;
- Water pumping and waste treatment.

For both residential and commercial buildings, RE energy carriers and service delivery systems vary depending on the local characteristics of a region and its wealth (Section 9.3). In order to curb GHG emissions from the sector a combination of approaches are likely to be needed.

Reducing energy demand for heating and cooling

Whereas heating loads are generally large in OECD countries and economies in transition, in most developing countries, energy for cooling is often a higher demand. For both heating and cooling, the design of a building can contribute to lowering the energy demand. A UK regulation that began in the London Borough of Merton (IEA, 2009b) requires new building developers to integrate RE technologies to meet 10% of total energy demand. This has resulted in energy efficient building designs being constructed in order to minimize the additional costs of RE to meet the regulation and exemplifies the links between RE and efficiency. Where heat loads dominate, passive designs (that receive natural solar heat gain in winter and/or avoid excessive heating in summer), optimization of window surfaces, and insulation levels can contribute to reducing the demand for heating as well as facilitating natural lighting (see Chapter 3). In warm climates where cooling loads dominate, adapting bio-climatic principles of traditional designs to new building stock, such as extensive shading and natural ventilation, can contribute to decreasing energy demand.

Improving efficiency of appliances

Improved energy efficient designs of systems and appliances, such as gas condensing boilers, heat pumps, district heating from CHP plants (Section 8.2.2), electronic appliances when on standby, light-emitting diodes (LEDs) and compact fluorescent light bulbs (CFLs), can contribute to reduced energy demand. Since the life span of such technologies is relatively small compared with the building itself, policies to encourage uptake of energy efficient appliance designs can be key to achieving CO_2 reductions in the short term. In dwellings currently without access to electricity even for basic lighting (Lighting Africa, 2010), installing RE technologies such as small PV systems or micro-hydropower can be relatively expensive. So electricity demand should be minimized by use of energy efficient appliances such as LEDs and CFLs. Improved energy use and energy management systems in residential and commercial buildings continue to be found through R&D investment (Figure 8.18). For example, smart appliances that use less energy, and operate

---

34 Full details of the potential for energy efficiency and RE in the building sector were provided in Chapter 6 of the IPCC 4th Assessment Report – Mitigation (Levine et al., 2007).

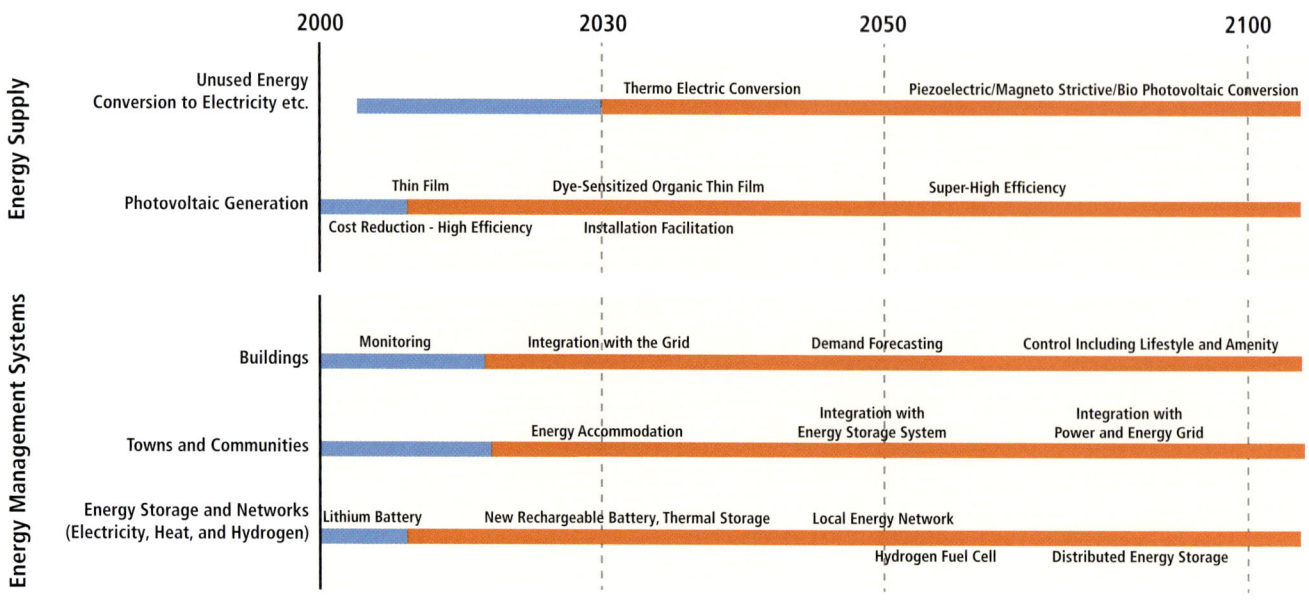

Figure 8.18 | Technology development pathways in Japan for future energy efficiency and RE technologies for use in residential and commercial buildings (METI, 2005).

automatically at off-peak times for use with future 'intelligent' electricity networks (Cheung and Wilshire, 2010), are beginning to reach the market.

Building management

An energy manager of a commercial or multi-unit apartment building is usually responsible for multiple objectives including the integration of RE as well as managing energy use, providing comfort for those living or working in the building, and reducing environmental impacts, all for minimal cost. Various building energy management systems and controls have been developed to balance these multiple objectives (Dounis and Caraiscos, 2009). Measuring and monitoring both energy use and the building environment are usually required (Figure 8.18) (Wei et al., 2009). Monitoring techniques can also be deployed in apartment buildings with home energy management standard technologies installed to control and actuate appliances as part of a distributed energy network.

RE technology deployment

Low or near-zero carbon fuels from modern biomass, geothermal and solar thermal currently supply around 3.5 to 4.5 EJ/yr, or about 6 to 8% of the total global heating demand for buildings (excluding traditional biomass) (IEA, 2007c). The share of RE for heating and cooling building space has the potential to be significantly increased in many regions using a range of new and improved RE technologies including cost-competitive and efficient enclosed pellet and other biomass stoves, heat pumps using low temperature heat available from ambient energy sources[35] (IEA, 2007c), solar thermal and PV systems, solar cooling systems and hybrid technologies such as combining solar thermal with biogas boilers, heat pumps or PV systems.

Policies to encourage the greater deployment of RE heating/cooling systems are not common, but several successful national and municipal approaches are in place (IEA, 2007c; Section 11.5.4). Electricity generated from RE sources is already widely utilized by the building sector. Increasing shares (Sections 8.2.1 and 10.3) could result in reduced sector GHG emissions (as could the use of electricity from the increased uptake of nuclear and CCS low-carbon supply side technologies). For air-tight, single-residential, multi-residential or commercial building designs, high energy demands for forced ventilation can be reduced through appropriate selection and hybridization of RE power generation, solar chimneys and wind cowls (Antvorskov, 2007). An innovative transition pathway to help decarbonize heat demand consists of using thermal storage systems that can also aid the balancing of variable electricity supplies (Hughes, 2010).

8.3.2.2 Renewable energy and buildings in developed countries

For any building class category in any given region, RE strategies and associated RE technical options can be developed based on the characteristics of the present or planned buildings, the building energy demand as a result of climatic conditions, and the RE resources available. This section examines the options to integrate RE into the built environment of developed countries. Following are options for urban (Section 8.3.2.3) and rural (Section 8.3.2.4) areas of developing countries. These contrasting situations face very different opportunities and challenges when endeavouring to accelerate RE uptake.

In the OECD and other major economies, most urban buildings are connected to electricity, water and sewage distribution schemes, and some to DHC schemes (Section 8.2.2). Many also use electricity, natural gas or LPG for heating and cooking, giving greater convenience to residents than using coal or oil products to provide these services. Woody biomass is also used for space and water heating, normally in efficient enclosed

---

[35] Ambient heat energy can be extracted from air, surface water or the ground (also referred to as shallow geothermal energy).

stoves more than open fires, but the fuel requires more handling and storage space than coal or oil with greater energy densities. Wood pellet stoves are therefore becoming popular, in part due to their operating convenience and the greater energy density of pellets compared to firewood logs (Section 2.3.2.1). Other RE conversion technologies such as solar water heaters and ground source heat pumps often have simple economic payback periods of five years or longer. Nevertheless, their integration in buildings is expanding in order to improve the quality of life of the residents whilst simultaneously realizing low carbon emission ambitions and security of future energy supplies (IEA, 2009b).

Challenges caused by RE integration

Greater integration of RE into the built environment is directly dependent on how urban planning, architectural design, engineering and a combination of technologies can be integrated. Tools and methods to assess and support strategic decisions for planning new building construction and retrofits are available (Doukas et al., 2008), including computer simulations to project the outcomes of a planning strategy (Dimoudi and Kostarela, 2008; Larsen et al., 2008). Therefore, to achieve more rapid RE deployment in the building sector of a city, town or municipality in an OECD country:

- A new vision for urban planning could be produced, based on the available RE energy resources;

- New buildings could be designed to accommodate the RE technologies for them to generate heat and electricity onsite rather than be consumers of imported energy as at present; and

- Assessments of the economic and non-economic barriers to RE technology deployment could be made and the need for supporting policies considered.

A transition from a fossil fuel-based, centralized energy supply system to a more distributed energy system with increased RE integration would need a comprehensive revision of how urban space has been traditionally planned and occupied. Changes in land and resource use, as well as modifying planning regulations to better accommodate RE technologies with the existing energy supply, are major strategic amendments that could be made to shape their integration.

The greater deployment of RE resources in an urban environment (IEA, 2009b) may require innovative use of roof and wall surfaces of the buildings to facilitate the uptake of RE technologies. This would affect the orientation and height of buildings in order to gain better access to solar radiation and wind resources without shading or sheltering neighbouring installations. Local seasonal storage of excess heat using ground source heat pumps may also contribute, along with more efficient bioenergy systems such as novel small-scale CHP systems that can run on natural gas or biogas (NZVCC, 2008; Aliabadi et al., 2010).

The technical challenges of integrating variable and distributed RE power and heat generation (Sections 8.2.1 and 8.2.2) can be partly resolved by the smart use of appliances in buildings. Technological advances can assist the integration of RE into the built environment, including energy storage technologies, real-time smart meters, demand side management and more efficient systems. Advanced electricity meters with bi-directional communication capability and the use of related information technologies interfacing with intelligent technology for appliances are expected to be widely deployed to gain the benefits of demand response and energy storage (possibly including electric vehicles in the future) in combination with distributed generation (NETL, 2008) (Section 8.2.5). If properly managed, appliances could contribute to maintaining the supply/demand balance of the energy system especially at higher penetration levels of variable RE sources. For some cities and towns, this could also require adaptation of the local electricity (Section 8.2.1) and/or heating/cooling distribution (Section 8.2.2) grids.

Without regulatory policies, efforts to improve energy efficiency and utilize RE sources are largely dependent on the motivation of building owners and occupiers. Institutional and financial measures such as energy auditing, appliance labelling, grants, regulations, incentives and automatic billing systems can lead to increased deployment (Section 11.5). Many buildings are leased to their occupiers, leading to the conundrum of owner/tenant benefits, also known as the 'split-incentive' (IEA, 2007d). Investing in energy efficiency or RE integration by the building owner usually benefits the tenants so that return on investment has to be recouped through higher rents.

Options to facilitate RE integration

New buildings in both hot and cold regions have demonstrated that 'importing' energy for heating or cooling can be minimized by innovative passive heating/cooling building designs, adequate insulation and thermal sinks. Building codes are steadily being improved to encourage the uptake of such technologies, so that new, well-designed buildings in future will require little, if any, heating or cooling using imported energy (EC, 2010). Many new building designs already demonstrate these passive solar concepts, but they remain a minority due to slow stock turnover.

Due to long life spans and low turnover rates, existing buildings can be retrofitted to significantly reduce their heating and cooling demand using energy efficient technologies such as triple glazing, cavity wall and ceiling insulation, shading and white painted roofs (Akbari et al., 2009; Oleson et al., 2010). The lower the energy consumption that the inhabitants of a building require to meet comfort standards and other energy services, then the more likely that RE can be employed to fully meet those demands (IEA, 2009b). RE tends to have a low energy density and often high capital investment costs, so reducing the energy demand by efficiency measures can help reduce the initial investment needed to meet the total energy demand of the building (Section 8.3.2.1).

Solar thermal and solar PV technologies can be integrated into building designs as components (such as roof tiles, wall facades, windows, balcony rails etc.). Innovative architects are beginning to incorporate such concepts into their designs. Integration of solar PV panels into roofs, window overhangs, and walls during construction can replace the function of traditional building materials and possibly improve building aesthetics relative to non-building-integrated solutions. Losses occurring during electricity distribution from centralized power stations can also be avoided.

In future, distributed energy systems could supply clusters of buildings on industrial estates or new residential developments using locally generated RE heat and power or RE-produced hydrogen for use in fuel cells at small to medium urban scales (Liu and Riffat, 2009). If sufficient heat and power is produced to meet local demands, any excess electricity or heat can be 'exported' off-site to gain revenue (IEA, 2009b). Bioenergy CHP combustion linked with steam engines, gas turbines and other conversion technologies is being undertaken at both medium (>50 MWe) and small (<5 kWe) scales, with ongoing research into fuel cells and other micro-CHP systems (Leilei et al., 2009).

Case study: RE house in Bruxelles, Belgium.
Among many buildings that have been retrofitted to enable high RE penetration levels for meeting their heating, cooling and electricity demands, the 'Renewable Energy House' in Bruxelles is a good example (EREC, 2008). Opened in 2006, it now houses the headquarters of the European Renewable Energy Council and fifteen RE industry associations. The aims of refurbishing the meeting facilities and offices of this historic, 120-year-old, 2,800-m² building were to reduce the annual energy consumption for heating, ventilation and air conditioning by 50% compared to a similar size reference building, and to meet the remaining energy demand for heating and cooling using solely RE sources (Figure 8.19). Key elements of the heating system are two biomass wood pellet boilers of 85 kW and 15 kW, 60 m² of solar thermal collectors (half being evacuated tubes and half flat plates), and four 115 m deep geothermal borehole loops in the courtyard connected to a 24 kW ground source heat pump (GSHP) also used in summer for cooling. Most cooling, however, comes from a 35 kW capacity (at 7 to 12°C) solar absorption cooler driven by relatively low-temperature solar heat (85°C) and a little electrical power for the controls and pumps.

In winter, the heating system mainly relies on the GSHP and the pellet boilers since the solar contribution is low. However, when available, any solar heat reduces the pellet fuel consumption since both are used to heat the same water storage tank. The GSHP operates on a separate circuit with borehole loops that absorb any excess low-grade summer heat and thus serve as a seasonal heat storage system. In summer, since high solar radiation levels usually coincide with cooling demands, the solar

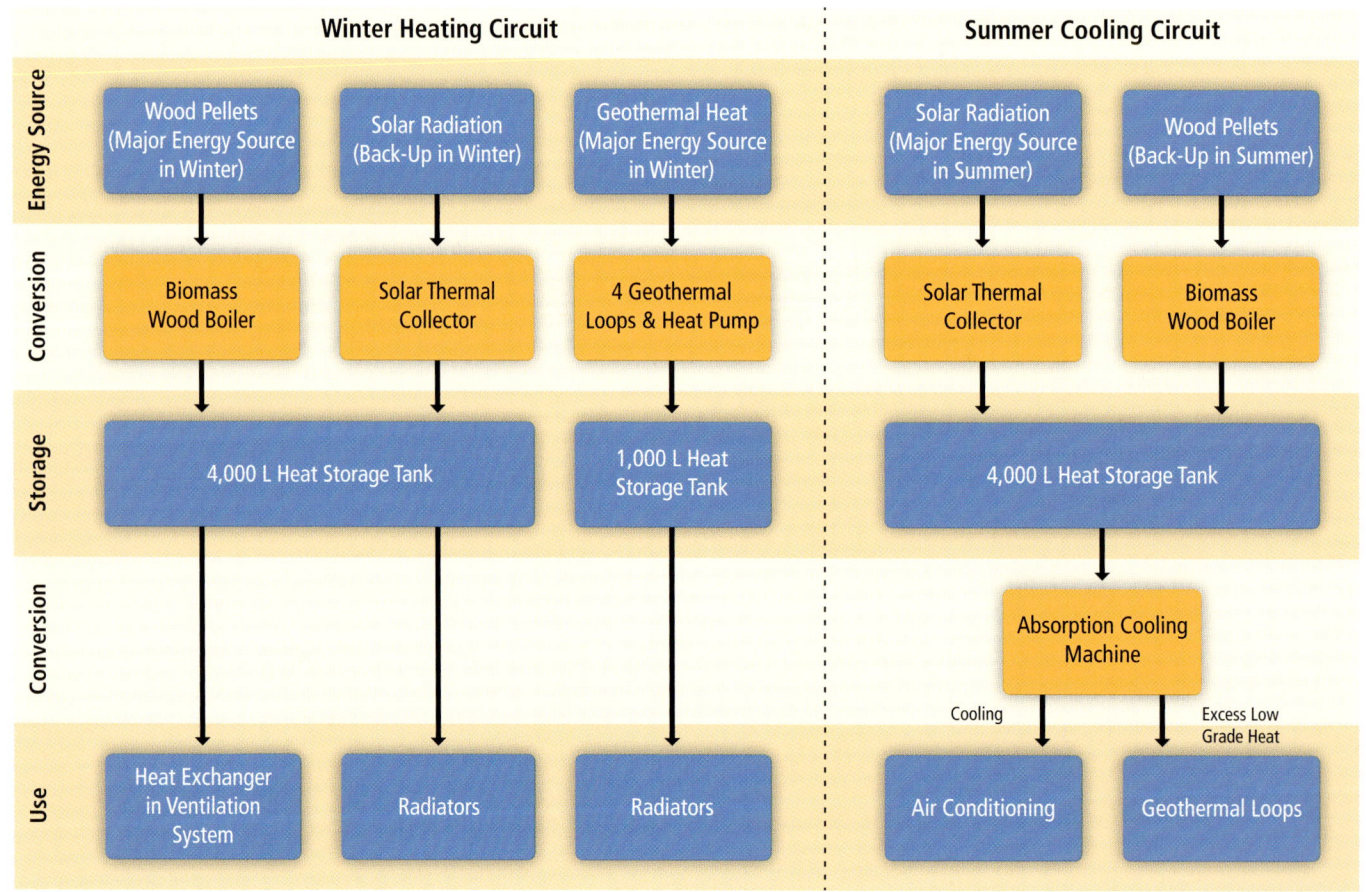

Figure 8.19 | RE integrated heating and cooling systems installed in a 120-year-old urban building in Bruxelles converted to commercial offices prior to the retrofit (EREC, 2008).

absorption cooler provides most of the cooling (backed up on cloudy days by heat from the biomass boiler). The thermally driven process of solar-assisted cooling is complex, being based on a thermo-chemical sorption process or a thermally driven open cooling cycle (IEA, 2009b). The technologies have not been widely applied and need more RD&D investment to gain reliability and sufficient cost reductions in order that they might compete with other cooling technologies such as heat pumps.

8.3.2.3 Renewable energy and urban settlements in developing countries

Urban energy consumption patterns of the more wealthy households in many developing countries resemble those of developed countries (Section 8.3.2.1). However, many poor urban households in low-income countries still rely mainly on collecting or purchasing traditional biomass that for many will probably remain their common fuel source for many years. In sub-Saharan Africa and elsewhere, many urban areas are experiencing a transition from burning fuelwood on open fires and small stoves to cleaner-burning charcoal for health reasons, though this trend can impact negatively on deforestation in the rural areas where charcoal is produced, given the growing demand and the very low energy conversion efficiency of traditional kilns used in the carbonization process (Section 2.3.2.2). Furthermore, the transport of charcoal from forest areas to urban demand centres often uses old and inefficient diesel trucks that contribute to GHG emissions. Modern RE technologies could provide an alternative option.

Challenges and options for RE integration
Biomass used by urban communities and households should be supplied from sustainably produced plantation forests. In a few places, community plantations have been grown to provide local biomass resources. To ensure the sustainability of such resources, a holistic approach to policy development would be useful that encompasses plantation biomass supplies, natural forest management as well as the demand side, such as fuel switching and the uptake of improved stoves and kilns (Figure 8.20). Such an approach may need fiscal policies (CILSS, 2008) in order to provide financial incentives to ensure the biomass is supplied from sustainable sources or to encourage the deployment of other RE technologies in the building sector such as small biogas digesters. In Nepal, for example, more than 200,000 domestic biogas plants had been installed as of December 2009 and 17 biogas appliance manufacturing businesses established as a result of recent supporting policies (Bajgain and Shakya, 2005).

In the majority of urban areas, grid electricity is available, although in some regions it can be unreliable, relatively expensive, and therefore often limited to providing basic needs. Along with small gasoline- or diesel-fuelled generating sets and coupled with energy storage, there is scope for increased penetration of independent, small-scale RE systems as backup support for when outages of the main grid electricity supply occur, but at additional costs.

Solar water heating (SWH) is considered to be a good RE option in grid-connected urban areas of many countries (as well as in off-grid rural areas without modern water heating services such as in China where over half the global SWH installations exist). Large-scale implementation can benefit both the customer and the utility. Where centralized switching (such as using ripple control communication over the power line) is used to manage electric water heater loads, the impact of solar water heaters is limited to energy savings. For utilities without this facility, the installation of a large number of solar water heaters may have the additional benefit of reducing peak electricity demand loads on the grid, especially in high sunshine regions where demand savings from using solar water heaters can correspond with high summer electrical demand for cooling. Hence there is a capacity benefit from load displacement of electric water heaters, particularly when used as a hybrid technology integrated with PV modules (Dubey and Tiwari, 2010). Markets for SWHs are apparent in the service sector such as hotels and lodges, in middle and high income households and for buildings not connected to the grid. Regulations and incentives could be necessary to reach a critical mass of installations in many urban areas (IEA, 2009b) and hence gain economies from greater dissemination.

Cooling demand in warmer climates has tended to rise where an increase in affluence occurs. Heat pump penetration rates in most developing countries are still low, but where coupled with high annual cooling degree days, could result in a future rapidly growing cooling demand as economies expand. This could cause peak power demand during periods of hot weather that, if exceeding the available supply capacity, could result in power outages. Offsetting cooling demand can be achieved by energy efficiency options such as reducing surface to volume ratios of new building designs, passive solar building designs and cooling towers (Chan et al., 2010). Active RE technologies for cooling include ground source heat pumps, district cooling using cold water sources (Section 8.2.3) and solar-assisted coolers (R. Wang et al., 2009). The latter technology offers the matching of peak cooling demand with peak solar radiation and hence with peak electricity demand for conventional air conditioners (air-to-air heat pumps). Another option is to use RE electricity to power conventional refrigeration appliances or air-to-air heat pumps (also known as 'air conditioners').

Case Study: Urban settlements in Brazil.
The rapid urbanization process in many developing countries has created peri-urban settlements near to central metropolitan areas. In Brazil, all major cities and a third of municipalities have a significant fraction of their population living in 'favelas'. Dwellings are usually precarious, fragile and temporary and frequently lack basic water, sanitation, gas and electricity distribution infrastructures (IBGE, 2008). Access to modern energy services is a challenge for many local governments and utilities. Energy planning is complex. Where an electricity distribution grid is available, it often does not comply with safety and regulatory standards of the utility. Furthermore, illegal connections with no meters are common practice. New integration of RE technologies could provide opportunities for improvements.

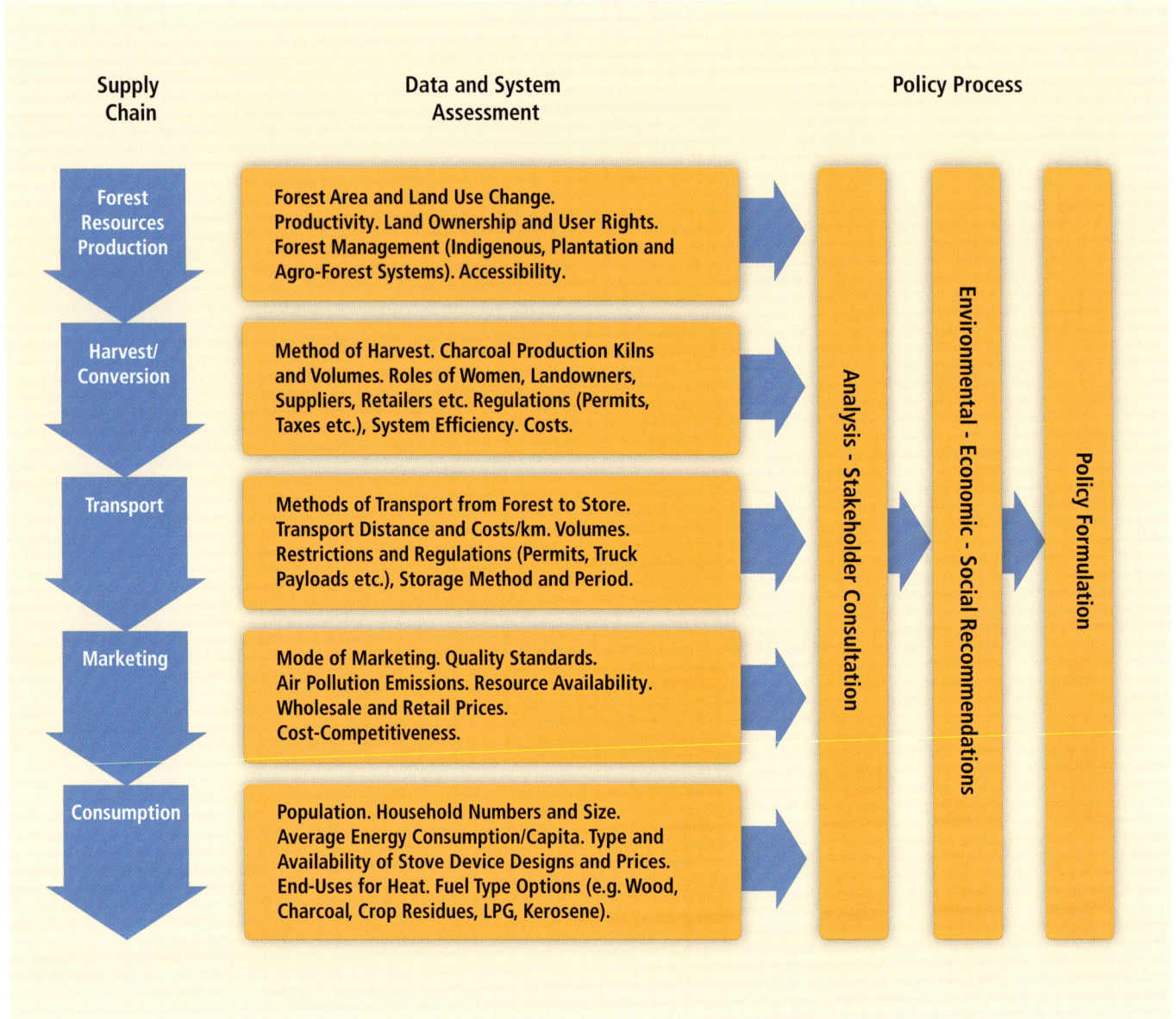

Figure 8.20 | A holistic approach using supply chain analysis for local or national policy development for the sustainable supply of biomass for domestic consumption in developing countries (Khennas et al., 2009). The formulation of policies can impact on the supply chain resulting in a continuous learning-by-doing feedback loop.

Under current regulations, Brazil's electricity utilities invest annually about USD$_{2005}$ 80M (half of their compulsory social investment) in energy efficiency programmes for low-income end users living in favelas. Complex issues still needing to be tackled include enforcing legal regulations, developing more creative and technical solutions to reduce theft of electricity and fraud, and improving the economic situation of the poor inhabitants. A pilot case study in one favela in São Paulo indicated that, as a result of promoting energy efficiency and solar water heating programmes, average household electricity consumption was reduced from 250 kWh/month to 151 kWh/month (~900 to 540 MJ/month) with a payback period of only 1.36 years (ICA, 2009). In addition there was opportunity for the uptake of state-of-the-art technologies including remote metering, real-time demand monitoring, more efficient transformers, new cabling systems and improved materials. The financial analysis identified a reduction in commercial and technical losses. Increased revenue resulted for the utility from a reduction in arrears and non-payments.

8.3.2.4 Renewable energy and rural settlements in developing countries

Rural households in developing countries relying on fuelwood, non-commercial crop residues and animal dung for their basic energy needs, and with zero or only limited access to modern energy services, are a constraint to eradicating poverty and improving health, education and social and economic development (Section 9.3.1). In several sub-Saharan Africa and other developing countries, traditional biomass accounts for more than 75% of total primary energy. The inefficiency of the whole supply chain, together with indoor air pollution problems, affect a large proportion of the population, particularly the many women who still rely on gathering fuelwood for their basic cooking and heating needs. Solutions to fuelwood scavenging include developing local forest plantations to be harvested sustainably, and improved natural forest management, though these are not always easy to accomplish due to land ownership, cost and social issues (CILSS, 2004).

Around one-quarter of the 2.7 billion people who rely on biomass (and another 0.3 billion on coal) now use improved cook stoves (UNDP and WHO, 2009). This amounts to 166 million households, around 70% being in China. Lighting demands met by relatively costly kerosene lamps, torches and candles, are being slowly replaced by RE electricity technologies that can deliver cost-effective high-quality lighting. For example, around 1 million solar lanterns (REN21, 2010) have been installed worldwide along with over 1.5 million solar PV home systems (also used for radio, television, refrigeration, communications and mobile phone charging). Solar PV-powered water pumps, micro-hydro schemes and mini-grids, small bioenergy gasifiers and biogas plants are all being widely deployed, but reliable statistics are not available to indicate rates of deployment with any accuracy (REN21, 2010).

Challenges and opportunities for RE integration

Although a variety of financial, regulatory and infrastructure barriers pose real challenges, they do not preclude RE having useful applications for reducing energy poverty in off-grid rural areas. RE applications, such as from solar PV systems, can provide income-generating activities to stimulate development of small and medium enterprises. To increase energy access as well as grid expansion, innovative and affordable delivery mechanisms could be developed, such as concessions coupled with subsidies and public/private partnerships (Section 11.5.6).

Some of the energy-poor may receive grid electricity during the next few decades as extension of the distribution network reaches more rural and peri-urban people (Section 9.4.2). Others in rural areas may benefit from local distributed energy supplies and mini-grids. Distributed energy supply technologies for buildings are under development (Section 8.2.5). The term 'digital energy' has been used to describe incorporation of the latest information technologies to effectively control domestic peak demand, energy storage equipment and RE generation systems in or around buildings (Cheung and Wilshire, 2010). Buildings that have been passive energy consumers could become energy producers and building managers could become operators of an energy network in collaboration with the local utilities (USDOE, 2008b). Whether such technologies are appropriate for use in rural areas in less developed countries has not yet been determined.

A combination of RE technologies suitable for rural communities or urban dwellings could be employed where suitable finance is available (Figure 8.21). Obtaining sufficient funding to purchase the electricity regardless of source could be challenging for new consumers, even for small amounts just to meet their basic needs. Innovative finance mechanisms (UNDP, 2009) can help ensure that the energy-poor better utilize local RE technologies as the least cost option.

Case Study: RE in the Democratic Republic of Congo (DR Congo).

A significant proportion of the rural population in the DR Congo, the largest and most populated country of the Congo Basin, has very limited access to modern energy services. Of its 70 million people, only around 5% have access to electricity compared with 12% in Angola, 18% in

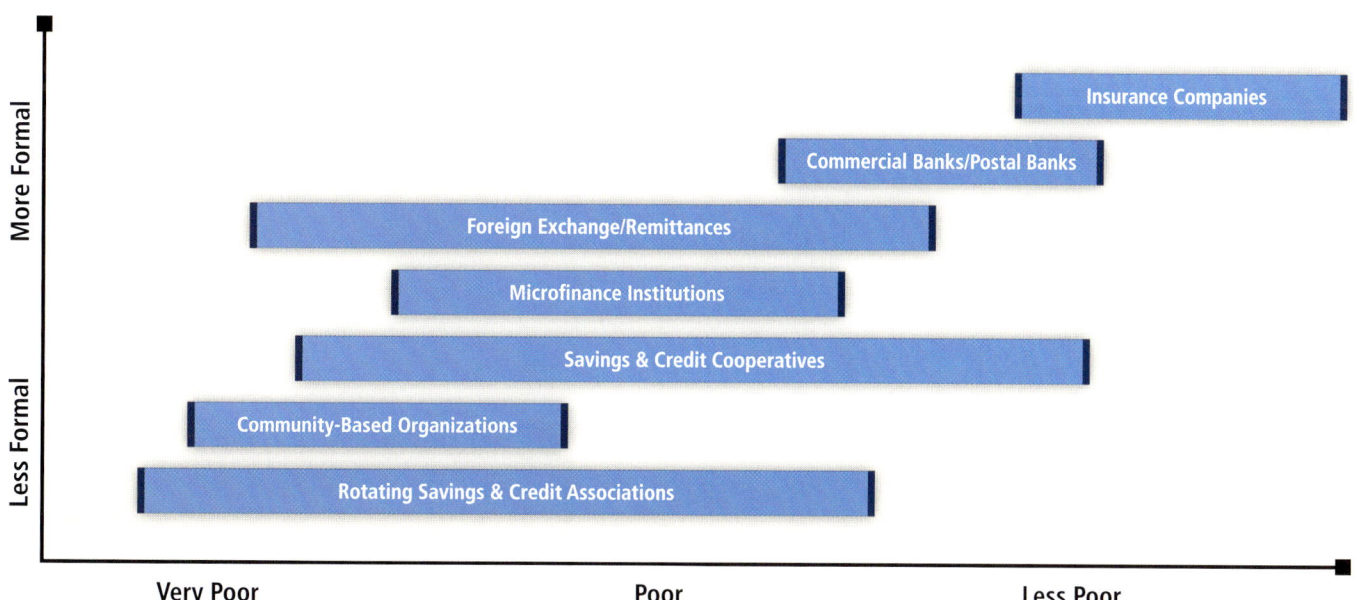

Figure 8.21 | Financing options to provide energy services for the poor, based on experiences in Burkina Faso, Kenya, Nepal and Tanzania (UNDP, 2009).

Congo, 46% in Cameroon and 47% in Gabon (IEA, 2006). Despite the high hydro potential in the region, the rural electrification growth rate is comparatively low at less than 1% of population per year. In addition to a good solar resource, some 325 potential hydro schemes have been identified and preliminary data gathered (Khennas et al., 2009). Developing this mini- and micro-hydro potential could dramatically increase the rural electrification rate and ultimately improve the livelihood of many poor rural households. The implementation of such a programme would dramatically increase the supply of RE for rural people to meet their needs for basic energy services. The Congo Basin, with the second largest tropical rainforest area in the world, is experiencing some deforestation (de Wasseige et al., 2009). Developing local RE resources could contribute to limiting deforestation around the villages by reducing the demand for traditional biomass.

8.3.2.5 Future trends for renewable energy in buildings

In many developed countries, heating and cooling and to a lesser extent lighting, have the highest potential to reduce energy demand in buildings and thereby offer increased opportunities for the cost-effective integration of RE by having to meet a lower demand (Section 8.3.2.1). A study, *Energy Efficiency in Buildings – Transforming the Market* (WBCSD, 2009), included several case studies:

- For office buildings in Japan, in parallel with energy efficiency initiatives for heating and cooling equipment and lighting, solar PV was the major RE source projected to be used onsite in 2050, but to a limited degree especially in high-rise building designs.

- Energy consumption of single-family houses in France is dominated by space heating (~60% of the total). Solar PV, along with solar water heaters, were projected to be integrated into improved energy efficient building designs by 2050 to meet a significant share of electricity demand.

Multi-family apartment blocks in China also have potential for numerous future energy efficiency improvements, especially for heating, ventilation and cooling. Only solar water heaters were projected to account for onsite RE potential in 2050. IEA scenario analysis (IEA, 2010c) forecast that there is potential for around 6 Gt CO_2 emission reductions below the baseline scenario coming from the building sector by 2050, with 10 to 25% of the total (depending on assumptions about rates of technological improvements and cost reductions) coming from RE and the remainder from energy efficiency measures including heat pumps, building design, lighting and appliances.

In developed countries, the trend is for new building developments, as well as building refurbishments, to continue towards achieving zero-energy buildings or even 'energy-positive' buildings where RE technologies will meet the energy demand of the inhabitants and generate more energy than the building consumes (Figure 8.22). Investment in both RE and energy efficiency in buildings can produce costs and CO_2 emissions reductions, but the comparative savings per unit of investment for either option will vary with the building type and location. In high-density urban areas, the energy demand per hectare of built land area usually greatly exceeds the local flows of RE, which are typically below a 10 kW/ha annual average. Therefore, RE integration to provide a high share of a building's total energy demand directly is more feasible in buildings located in rural and low-density urban areas. Therefore, compared with high-rise buildings, single-family homes could more easily become autonomous for their net energy needs (excluding transport) (Section 8.3.5). However, any savings in imported energy for such buildings located in rural or low-density urban areas could be partly offset by increased transport energy demands.

The market situation for RE integration during retrofitting of existing buildings is in the early development phase, as compared to integration into new buildings, but could strengthen in the near future as a result of policy attention shifting towards the existing building stock because of slow building stock turn-over.

In commercial buildings and urban and rural households in developing countries, the opportunities for integrating RE systems are considerable. To meet the future needs of the millions of people who currently rely on the inefficient combustion of traditional biomass (UN Energy, 2007), sustainable modern bioenergy systems, including small gasifiers, biogas engines, ethanol gels, pellet burners etc., coupled with efficient,

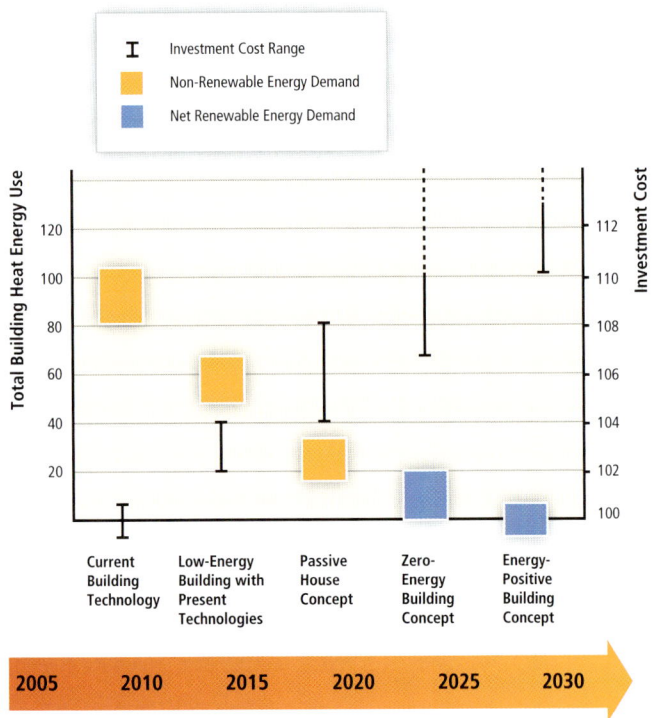

Figure 8.22 | Relative performance of current building technologies to meet heat energy demand compared with future designs of energy-efficient buildings with integrated RE systems and related investment costs (from base year 2005 = 100), based on a full-scale demonstration project in Finland (VTT, 2009).

affordable, well-designed and socially accepted appliances, particularly cooking stoves, could contribute significantly. The familiarity with the biomass resource could facilitate the transition. Poor access to modern energy services and the characteristics of energy demand in both urban and rural areas gives a comparative advantage to the integration of all forms of RE, which in the future could possibly be through decentralized RE supply options.

8.3.3 Industry

8.3.3.1 Sector status

Energy demand by manufacturing industries in 2008 was around 98 EJ of final energy (Figure 8.2), accounting for about one-third of total global consumer energy (IEA, 2010b), although the share differs markedly between countries. The industrial sector is highly diverse, ranging from 'heavy', very large, energy-intensive basic material manufacturers to small and medium sized enterprises (SMEs) with 'light' manufacturing. Energy-intensive iron and steel, non-ferrous metals, chemicals and fertilizers, petroleum refining, minerals, and pulp and paper industries account for approximately 85% of the sector's energy use (Bernstein et al., 2007). The production of these goods has grown strongly in the past 30 to 40 years and growth is projected to continue.

The sources of industrial CO_2 emissions are from use of fossil fuels in energy carriers (such as grid electricity) or used directly on site (such as coal for process heat) as well as from non-energy uses of fossil fuels in chemicals processing, and from non-fossil fuel sources arising from the process, mainly through the decarbonation of calcium carbonate ($CaCO_3$) in cement manufacturing. In most countries, CO_2 accounts for more than 90% of total industrial GHG emissions with the remainder coming from a range of gases including CFCs (IPCC, 2007). Direct and indirect CO_2 emissions from industry in 2006 were 7.2 and 3.4 Gt respectively, together equivalent to almost 40% of world energy and process CO_2 emissions (IEA, 2009d).

Carbon dioxide emissions from industry can be reduced by:

- Energy efficiency measures that reduce specific energy use, which therefore, for some bio-based industries, can make any excess RE heat, electricity and biogas available for sale off-site;

- Materials recovery and recycling that eliminate the energy-intensive primary extraction and conversion steps for many basic materials such as metals and paper pulp;

- RE integration and feedstock substitution to reduce the use of fossil fuels; and

- Carbon dioxide capture and storage (CCS) of emissions from both fossil and biomass fuels. Assuming that CCS becomes viable as a GHG mitigation element in future energy systems, this could also be an option for CO_2-producing industries and energy-intensive industries that consume biomass or fossil fuels for heating directly on-site.

Most of these measures are relevant also for integrating RE into present and future energy systems. The main opportunities for RE integration in industry, in no particular order, include:

- Direct use of process residues and biomass-derived fuels for on-site heat and CHP production and use as well as biogas and other bio-fuels, also used for transport applications (Sections 8.2.3, 8.3.1, and 2.3.3);

- Indirect use of RE through increased RE-based electricity demand, including for electro-thermal processes;

- Indirect use of RE through other purchased RE-based energy carriers including liquid fuels, biogas, heat and hydrogen (Section 8.2.3);

- Direct use of solar thermal energy for process heat and steam demands (Section 3.5.3); and

- Direct use of geothermal energy for process heat and steam demands (Section 4.3.5).

Other RE sources may also find industrial niche applications such as ocean energy for desalination (Section 6.3). There are no severe technical limits to increasing the direct and indirect use of RE in industry in the future. However, in the short term, integration may be limited by factors such as RE technology costs, capital turnover rates, space constraints or demands for high reliability and continuous operations.

The current direct use of RE by industry is dominated by biomass in the pulp and paper, sugar and ethanol industries where process by-products are important sources of cogenerated heat and electricity used mainly for the process but with potential to export off-site (Section 2.1). Thus, industry is not only a potential user of RE but also a potential supplier of RE as a co-product. Biomass is also an important fuel for many SMEs, such as the use of charcoal for brick making, notably in developing countries (Section 2.3.2). There is a growing interest in utilizing organic wastes and by-products for energy in, for example, the food industry through anaerobic digestion. Biogas production often replaces other forms of organic waste treatment due to waste and wastewater policies (Lantz et al., 2007). With the exception of biomass-based industries, the literature on RE in industry is relatively limited compared to other sectors.

Providing demand response services to enable electrical peak-load shifting as a form of load management is an important measure for industry. It is likely to achieve greater prominence in future electricity systems with increasing shares of variable RE generation (Section 8.2.1). It can also reduce the need for high marginal cost generation, offer low-cost system balancing and decrease grid reinforcement investment. The concept is already widely used to secure enough reserve- and peaking-capacity in many countries and is expected to become more important in the future. Existing programmes have mainly focused on industrial users that can shed relatively large loads through rescheduling, machinery interruption, and interruptible thermal energy storage, cool stores, electric boilers etc. Typically, industries are contracted to reduce or shut down load, sometimes remotely by the transmission system operator, according to pre-defined rules and with various means of financial compensation (Section 8.2.1.3). For industry, reduced production and risks of process equipment failure associated with demand response are important considerations. There are few published studies of the potential for demand response through industrial manageable power demand. In one example from Finland, the potential for demand response in the energy-intensive industries was estimated at 1,280 MW, equivalent to 9% of total system peak demand (Torriti et al., 2010).

8.3.3.2 Energy-intensive industries

The largest contributions of industry sector CO_2 emissions in 2006 came from iron and steel (29%), cement (25%) and chemicals and petrochemicals (17%) (IEA, 2009d). The pulp and paper industry accounted for only about 2% of industrial fossil fuel CO_2 emissions since it uses large amounts of biomass for process energy (bioenergy systems generally being low carbon emitting).

Overall, possible pathways for increased direct integration of RE vary between different industrial sub-sectors. The main options are to replace fossil fuels in boilers, produce biogas from wastewater with high organic content and switch from oil and gas to biomass for industrial processes, for example by using bark powder in lime kilns that produce calcium oxide for the preparation of pulping liquor. Biomass can be co-fired with, or completely replace, fossil fuels in boilers, kilns and furnaces and there are alternatives for replacing petrochemicals through switching to bio-based chemicals and materials.

Due to the scale of operations, access to sufficient volumes of biomass may be a constraint. Direct use of solar technologies can also be constrained by high energy demand and by the variability of the resource. Geothermal energy heat is suitable for use in industry due to its high capacity factors and energy densities but so far there are few applications in energy-intensive industries (Lund et al., 2010). Only around 500 MW of geothermal capacity, corresponding to 2.7% of worldwide direct applications of geothermal energy, is currently used for industrial process heat. Current utilization is about 12 PJ/yr with applications in dairies, laundries, leather tanning, beverages and pulp mills. The Kawerau, New Zealand geothermal plant provides steam to the Norske Skog Tasman pulp mill that accounts for around half the present global geothermal industrial heat use (White, 2009). Geothermal energy could meet more industrial process heat demands if heat pumps are used to elevate temperatures. The potential is large (Section 4.2) and high capacity factors relative to solar thermal energy make it an attractive alternative for industry. However cost and constrained resource locations have been barriers to date.

For many energy-intensive processes, an important future option is indirect RE integration through switching to electricity and hydrogen. Electricity is already the main energy input for producing aluminium using the electro-chemical Hall-Héroult process. The broad range of options for producing carbon-neutral electricity, and its versatility of use, implies that electro-thermal processes could become more important in the future for replacing fuels in drying, heating, curing and melting operations. Plasma technologies can deliver heat at several thousand degrees Celsius and replace fossil fuel combustion for high-temperature applications. Electro-thermal processes include heat pumps, electric boilers, electric ovens, resistive heating, electric arcs, plasma induction, radio frequency and microwaves, infrared and ultraviolet radiation, laser and electron beams (EPRI, 2009). These technologies are presently used where they offer distinct advantages (such as energy savings, higher productivity or product quality), or where there are no viable alternatives (such as for electric-arc furnaces). Deployment has been limited since direct combustion of fossil fuels is generally less expensive than electricity. However, relative prices may change considerably if climate policies place a value on carbon emissions. Electro-thermal processes must compete against a portfolio of other low-carbon process options even if electricity supply is RE-based or otherwise decarbonized.

Energy-intensive industries are generally capital intensive and the resulting long capital asset cycles constitute one of the main barriers to energy transition in this sector. Cyclical markets and periods of low profit margins are common where management focus is usually on cutting costs and extending asset life rather than on making investments and taking risks with new technologies. In existing plants, retrofit options may be constrained by space limitations, risk aversion and reliability requirements. Green-field investments are mainly taking place in developing countries, although enabling energy and climate policies are less common than in developed countries.

Energy-intensive industries are often given favourable treatment in developed countries that have ambitious climate policies since they are subject to international competition and hence carry risks of carbon leakage. Exemptions from energy and carbon taxes, or free allocation of emission permits in trading schemes, are prevalent. Bio-based industries, such as the pulp and paper industry, can benefit from, and respond to, RE policy (Ericsson et al., 2010). Sectoral approaches are considered in international climate policy in order to reduce carbon leakage risks and facilitate technology transfer and the financing of mitigation measures (Schmidt et al., 2008).

Iron and steel. Production of iron and steel involves ore preparation, coke making, and iron making in blast furnaces and basic oxygen furnaces by reducing the iron ore. Primary energy inputs are 13 to 14 GJ/t of iron, usually from coal. Natural gas for direct reduction of iron ore is also an established technology. Using electric-arc furnaces to recycle scrap steel, these energy-intensive steps can be bypassed and primary energy use reduced to around 4 to 6 GJ/t. However, the amount of scrap steel is limited and the increasing demand for primary steel is mainly met from iron ore. Various R&D efforts, some of which involve RE uptake, focus on reducing CO_2 emissions (Croezen and Korteland, 2010; Miwa and Okuda, 2010).

Charcoal was for a long time the main energy source for the iron and steel industry until coal and coke took over in the 1800s. During its traditional production, roughly only one-third of the total wood energy content is converted to charcoal, the rest being released as gases (Section 2.3.2). Higher efficiencies are attainable (Rossilo-Calle et al., 2000). Charcoal can provide the reducing agent in the production of iron in blast furnaces but coke has the advantage of higher heating value, purity and mechanical strength.

Present day steel mills mostly rely entirely on fossil fuels and electricity. Charcoal has not been able to compete, with the exception of use in a few blast furnaces in Brazil. Options for increasing the use of RE in the iron and steel industry in the near term include switching to RE electricity in electric-arc furnaces and substituting coal and coke with charcoal, subject to resource and sustainability constraints. Switching to biomethane is also an option. Research on electricity and hydrogen-based processes for reducing iron shows potential in the long term but CCS linked with coke combustion may be a less expensive option.

Cement. Production of cement involves extraction and grinding of limestone and heating to temperatures well above 950°C. Decomposition of calcium carbonate into calcium oxide takes place in a rotary kiln, driving off CO_2 in the process of producing the cement clinker. CO_2 emissions from this reaction account for slightly more than half of the total emissions with the remainder coming from the combustion of fossil fuels. Hence, even a complete switch to RE fuels would reduce emissions by less than half.

The cement process is not particularly sensitive to the type of fuel but sufficiently high flame temperatures are needed to heat the materials. Different types of waste, including used tyres, wood and plastics are already co-combusted in some cement kilns. A variety of biomass-derived fuels can be used to displace fossil fuels. Large reductions of CO_2 emissions from carbonate-based feedstock are not possible without CCS, but emissions could also be reduced by using non-carbonate-based feedstock (Phair, 2006).

Chemicals and petrochemicals. This sector is large and highly diverse. High-volume chemical manufacture of olefins and aromatics, methanol and ammonia account for more than 70% of total sector energy use (IEA, 2008c). The main feedstocks for providing the building blocks of chemical products are oil, natural gas and coal which are also consumed for energy (Ren and Patel, 2009). Chemicals such as ethanol and methanol may be considered both as fuels and as platform chemicals for a range of products.

Steam-cracking is a key process step in the production of olefins and aromatics. Combustion of various biomass fuels and wastes could be used for steam production. Methanol production is mostly based on natural gas but it can also be produced from biomass or by reacting CO_2 with hydrogen, possibly of renewable origin.

The potential for shifting to RE feedstocks in the chemicals sector is large (Hatti-Kaul et al., 2007). Many of the first man-made chemicals were derived from biomass through, for example, using ethanol as a platform chemical, before the shift was made to petroleum-based feedstocks. A shift back to bio-based chemicals would involve four principle approaches:

- Feedstocks converted using industrial biotechnology processes such as fermentation or enzymatic conversions (Section 2.3.3.3);
- Thermo-chemical conversion of biomass for the production of a range of chemicals, including methanol (Section 2.3.3.1);
- Naturally occurring polymers and other compounds extracted by various means; and
- Green biotechnology and plant breeding used to modify crops for non-food production.

In the fertilizer industry, ammonia production is an energy-intensive process that involves reacting hydrogen and nitrogen at high pressure. The energy embedded in fertilizer consumption by agriculture (Section 8.3.4) represents about 1% of global primary energy demand (Ramirez and Worrell, 2006). The nitrogen is obtained from the air and the source of hydrogen is typically natural gas, but also coal gasification, refinery gases and heavy oil products. Ammonia production gives a CO_2-rich stream and lends itself to CCS. Hydrogen from RE sources could also be used for the reaction and other nitrogen fixation processes are possible, including biological nitrogen fixation (Ahlgren et al., 2008).

Forest products. Forest harvesting operations and the transport of logs to saw mills, pulp and paper mills and wood processing industries involve handling large volumes of woody biomass. Residues and by-products all along the value chain can be used to provide energy for internal use as well as for export. For example, the bark component stripped from the logs can be combusted in separate boilers. Enough high-pressure steam can often be produced for CHP generation onsite to meet all the steam and electricity needs of a modern pulp mill. The onsite use of biomass as a by-product for heat and power generation means that the GHG intensity of the forest industry can be relatively low.

There are many different pulping processes but the two main routes are mechanical and chemical. For electricity-intensive mechanical pulping, after debarking and chipping, the wood chips are processed in large grinders and nearly all the fibre ends up in the pulp, which is used for producing paper such as newsprint. Heat is recovered from the mechanical pulping process and the steam produced is used for drying the paper and other processes. Chemical pulping is used to produce stronger high-quality fibres and involves dissolving the lignin in a chemical cooking process. About half of the wood, mainly lignin, ends up in the spent pulping liquor that is concentrated in evaporators. This 'black liquor' can be combusted in chemical recovery boilers. Changing from the traditional recovery boiler to black liquor gasification in chemical pulping would increase the efficiency of energy recovery and facilitate higher electricity-to-heat ratios in the CHP system with the syngas used for fuel production (see case study below).

Continuous incremental improvements in energy end-use efficiency, higher steam pressure in boilers and use of condensing steam turbines are reducing the need for importing purchased energy in the pulp and paper industry and can also free up a portion of the heat and electricity generated to be sold as co-products (Axegård et al., 2002).

Case study: Black liquor gasification for bio-DME production.
As an alternative to producing heat from black liquor in chemical recovery boilers, gasification is a technology that has been subject to R&D for more than 20 years and demonstrated in several pilot-scale plants (Kåberger and Kusar, 2008). The syngas produced (mainly CO and H_2) can be used with high efficiency in combined cycle CHP plants or for the production of biofuels via, for example, the Fischer-Tropsch process (Section 8.2.4). The first pilot plant demonstrating pressurized gasification for producing DME (dimethyl ether) was inaugurated in Piteå, Sweden, in September 2010 with a rated capacity of about 4t/day. Partner companies are Chemrec, Haldor Topsoe, Volvo, Preem, Total, Delphi and ETC with financial support from the Swedish Government and the European Commission. Compared to gasification of solid biomass, one advantage of black liquor is that it is easier to feed into a pressurized gasifier. Depending on the overall plant energy balance and layout there are often process integration advantages and potential for significant increases in energy efficiency. Energy that is tapped off for liquid or gaseous biofuel production (including DME) can be compensated for by using lower quality biomass to meet pulp and paper process energy demands. In addition to DME production, the project also involves four filling stations and 14 HDV trucks using DME for fuel to assess the viability of bio-DME.

8.3.3.3 Less energy-intensive industries and enterprises

Non-energy-intensive industries, although numerous, account for a smaller share of total energy use than energy-intensive industries but are more flexible and offer greater opportunities for the integration of RE. They include food processing, textiles, light manufacturing of appliances and electronics, automotive assembly plants, wood processing etc. Much of the energy demand in these 'light' industries is similar to energy use in commercial buildings such as lighting, space heating, cooling, ventilation and office equipment. Most industrial heating and cooling demands are for moderate temperature ranges that facilitate the application of solar thermal energy, geothermal energy and solar-powered cooling systems with absorption chillers (IEA, 2007c; Schnitzer et al., 2007). Almost 150 GW of solar thermal collector capacity was in operation worldwide in 2007 but less than 1% was used for industrial applications (IEA-SHC, 2010). Other than cost, part of the reason could be the variable nature of the solar resource providing insufficient reliability for an industrial process, although thermal storage, including for concentrating solar thermal systems (Section 3.2), could overcome the problem in some situations.

Typical process energy use is for low and medium temperature heating, cooking, cooling, washing, pumping and air-handling, coating, drying and dehydration, curing, grinding, preheating, product concentration, pasteurization and sterilization, and some chemical reactions. In addition, a range of mechanical operations use electric motors and compressed air to power tools and other equipment. Plants range in size from very small enterprises to larger-scale assembly plants and processing mills.

Many companies use hot water and steam for processes at temperatures between 50 and 120°C (Figure 8.23). When fossil fuels are used, installations that provide the heat are mostly run at temperatures

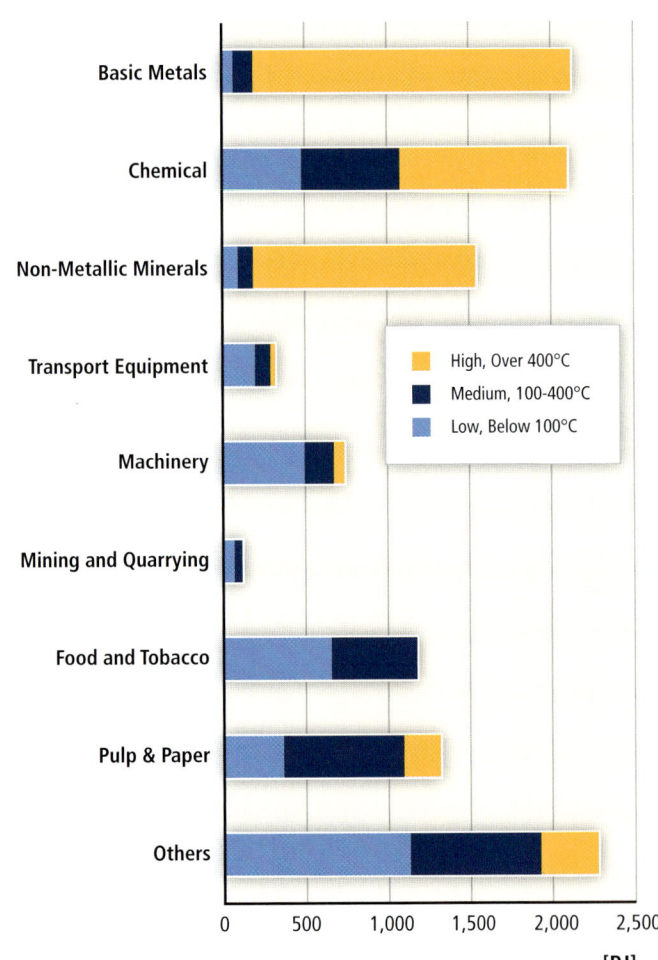

Figure 8.23 | Industrial heat demands by temperature quality and by manufacturing sector for 32 European countries (Werner, 2006b).

Note: Data created from German industry experiences and applied to the IEA energy database for the target region.

between 120 and 180°C since these higher process heat temperatures enable smaller areas of heat exchangers and heating networks to be utilized. Solar energy opportunities focus more on engineering designs for operating at lower temperatures in order to optimize the whole system. For temperatures <80°C, solar thermal collectors are on the market, but there is limited experience for applications that require temperatures up to 250°C (Schnitzer et al., 2007). Such higher temperatures are possible using heat pumps or, in appropriate areas, concentrating solar thermal systems.

Industrial electro-technologies can save primary heat energy from coal and gas by using electricity. Industrial CO_2 emissions can be reduced even if there are no primary energy savings, assuming electricity from RE resources replaces or saves fossil fuel-based thermal generation. Examples include freeze concentration instead of the thermal process of evaporation; dielectric heating (radio frequency and microwave heating) for drying; polymerization; and powder coatings using infra-red ovens for curing instead of solvent-based coatings and conventional convection ovens (Eurelectric, 2004). Other advantages include quick process start-up, improved process control and higher productivity (EPRI, 2009). The conventional wisdom that high quality (high exergy) electricity should not be used to provide low quality (low exergy) thermal applications may be challenged in the future once electricity systems become decarbonized.

Many SMEs in developing countries use substantial amounts of crop residues in the form of husks, straw and shells from nuts, coffee, coconuts, rice etc. for heat and power generation. These residues are low cost and often used, together with fuelwood and charcoal, as fuels to supply heat for other local industries. In some food- and fibre-processing industries, wastewater with high organic content can also be used for biogas production but the resource currently tends to be poorly utilized.

In developed countries, waste policies are an important factor driving the increased utilization of biomass residues for energy. Bioenergy is most common in the food- and fibre-processing industries where, as for forest products (Section 8.3.3.4), on-site biomass residues are widely used to meet internal energy needs or the energy is exported off-site for use elsewhere, which therefore avoids waste disposal problems. For example, sugar and ethanol plants in Brazil use the bagasse by-product to produce heat and power and sell any surplus to the grid (see case study below). Any waste heat can be used by other industries and in district heating systems (Section 8.2.2). Heated greenhouses and fish farming are potential users of low-grade heat.

Industrial ecology and symbiosis are relatively new concepts used to denote inter-firm exchanges of energy, water, by-products etc., although these are not new phenomena. An inventory of the Swedish forest industry found several examples of such inter-firm exchanges, but typically between different entities within the same company group (Wolf and Petersson, 2007). The potential for increasing the indirect use of RE in such innovative ways is difficult to estimate.

Dehydration of agricultural and other products is an important application of solar energy. In many developing countries, the traditional method of dehydration in open air can result in food contamination, nutritional deterioration and large product losses. Solar dryer technologies that improve product quality and reduce drying times have been demonstrated. Examples include a solar tunnel dryer for hot chilli peppers (Hossain and Bala, 2007) and a solar dryer with thermal storage and biomass backup heater for pineapple (Madhlopa and Ngwalo, 2007).

The potential for increasing the direct use of RE in both heavy and light industries in general is poorly understood due to the complexity and diversity of the sector, and the varying geographical and climatic conditions of various locations. Aggregate mitigation and typical RE integration cost estimates cannot be made for similar reasons.

Direct use of RE in industry has difficulty competing at present due to the relatively low fossil fuel prices and low- or zero-energy and carbon taxes for industry. Improved utilization of processing residues in biomass-based industries to substitute for fossil fuels offers near-term opportunities, particularly where biomass residue disposal costs can be avoided. Solar thermal technologies are promising but further development of collectors, thermal storage, balancing systems and process adaptation and integration is needed. Direct use of geothermal heat is already used where industrial heat demands are nearby. Increased use of energy carriers such as electricity and natural gas that are clean and convenient at the point of end use is a general trend in industry. Indirect RE integration using electricity generated from RE sources, and facilitated through electro-technologies, may therefore have a large impact in the near and long term. RE support policies in different countries tend to focus more on the energy, transport and building sectors than on industry. Consequently the RE potential for the industry sector is relatively uncharted.

Case studies

Sugar industry and CHP. Limited grid access and low prices offered by monopoly buyers of electricity and independent power producers have provided disincentives for industries to increase overall energy efficiency and electricity-to-heat ratios in CHP production. Process electricity consumption in sugar and sugar/ethanol mills, for example, is typically in the range of 20 to 30 kWh (72 to 108 MJ) per tonne of fresh cane. Most sugar mills have been designed to be self-sufficient in heat and electricity using mainly bagasse as a fuel in inefficient, low pressure boilers. With higher rates of residue recovery and the introduction of high pressure boilers and condensing extraction steam turbines, more than 100 kWh/t (360 MJ/t) can be produced for export. In Brazil, electricity generation is expected to increase from an average of about 9 kWh/t (32 MJ/t) of sugarcane in 2005 to 135 kWh/t (486 MJ/t) in 2020 (Macedo et al., 2008). However, sugar/ethanol mills provide opportunity for integrating a much higher level of biomass for energy in industry. The sugarcane tops and leaves are normally burned before harvest or left in the field after harvest. These could also be collected and brought to the mill to increase the potential export of electricity to more than 150 kWh/t (540 MJ/t). This could be further increased to more than 300 kWh/t (1,080 MJ/t) using gasification technology and combined-cycle power plants, or supercritical steam cycles (Larson et al., 2001). Integrating the utilization of biomass residues with feedstock logistics in sugar/ethanol mills offers cost and other advantages over separate handling and conversion of the residues.

Solar industrial process heat for industry. Solar thermal energy is well suited to many industrial processes. In 2003, the net industrial heat demand in Europe was estimated to be 8.7 EJ and the electricity demand was 4.4 EJ (Werner, 2006b). Heat demands were estimated in 2003 at low, medium and high temperature levels for several industries in the EU 25 countries, four accession countries and three European Free Trade Association countries (Figure 8.23). Industrial process heat accounted for around 28% of total primary energy consumption by the sector with more than half of this demand for temperatures below 400°C, which could be a suitable application for solar thermal energy (Vannoni et al., 2008).

Solar thermal energy technologies can be used to supply industrial heat, including concentrating solar thermal systems that can produce process steam directly in the collector. A pilot plant was inaugurated in 2010 in Ennepetal, Germany. This 'P3 project' demonstrated that direct steam generation from a small 100 m^2 area of parabolic trough collectors can be suitable for industrial applications (Hennecke et al., 2008; Krüger et al., 2009). Another solar thermal example is the installation of about 5,000 m^2 of solar collectors in 2008 by the Frito Lay food processing company at its plant in Modesto, California, to produce process steam and thereby reduce gas consumption and associated CO_2 emissions (Krüger et al., 2008).

8.3.4 Agriculture, forestry and fishing (primary production)

8.3.4.1 Sector status

In OECD countries, the energy demand of the primary production sector is typically around 5% of total consumer energy, while the overall global average is 3% (Figure 8.2). Excluding land use change, currently primary production accounts for around 15% of total GHG emissions including methane and nitrous oxide (IPCC, 2007). Integration of RE into primary production systems, either as energy suppliers or end users, has been successfully achieved in a myriad of examples at both medium scale (such as bioenergy CHP plants and mini-hydro projects) and small scale (such as biogas plants and wind-powered water pumps).

Complex relationships exist between energy inputs and crop yields, sustainable practices (including tillage and fertilizer practices), water use, land use change, biodiversity, landscape and recreation, and soil carbon balances. Large regional differences occur due to climate, soils and land management (IPCC, 2007).

Low input subsistence farming and fishing rely mainly on human energy and animal power, with traditional biomass also used for drying and heating applications (Section 2.4.2). Intensive, industrialized agriculture, forest and fish production depend on significant energy inputs, usually from fossil fuels. These are either combusted directly for heating, drying and powering boats, tractors and machinery, or used indirectly to manufacture fertilizers and agricultural chemicals (Section 8.3.3), produce and transport purchased animal feed, construct buildings and fences and generate electricity for water pumping, lighting, cooling and operating fixed equipment. Typically twice as much energy is used directly on-farm compared with the indirect energy inputs (Schnepf, 2006), though this varies with the enterprise type. Energy efficiency measures are being implemented and future opportunities also exist to reduce fertilizer and agricultural chemical inputs by using precision farming application methods (USDA, 2009) and less intensive, organic farming systems.

Energy input versus energy output ratios vary with product and system. For example, the total energy inputs for growing potatoes can exceed the food energy output value of the harvested crop (giving a negative ratio as a result) (Haj Seyed Hadi, 2006). Energy ratios depend upon the local farm management system, the boundaries used in the energy analysis, and other assumptions. Hence a positive energy ratio for potatoes has also been reported (Mohammadi et al., 2008).

Primary producers can have a dual role as energy users and as suppliers of RE (Table 8.5).[36] Landowners often have ready access to local RE resources including wind, biomass, solar radiation, the potential and kinetic energy in rivers and streams and biomass. Competition for land use to provide food, fibre, animal feed, energy crops for biofuels, recreation, biodiversity and conservation forests is growing and has come under close scrutiny (GMF, 2008; Fritsche et al., 2010).

Land investments have been made by some governments in countries other than their own in order to grow and export food such as wheat, rice and maize, but also energy crops for biofuels (Von Braun and Meizen-Dick, 2009). Possible exploitation of the existing rural communities has been a concern (WWICS, 2010), but benefits can accrue when the advantages of RE integration with land use are equitably shared, such as for sugar ethanol companies investing in Ghana (Sims, 2008). Developing a code of good conduct to share benefits, abide by national trade policies and respect customary rights of the family farm unit is being considered (UN Energy, 2007) as is the sustainable production of biomass (Section 2.5).

8.3.4.2 Status and strategies

The integration of RE with land use for primary production is well established. For example, wind turbines constructed on pasture and crop

---

36 Note that this section covers only on-farm and in-forest production and processing activities, including harvest and post-harvest operations up to the farm gate. Food and fibre processing operations are covered in Section 8.3.3.

lands can provide additional revenue to the landowner since only 2 to 5% of the total land area is taken out of agricultural production by the access roads, turbine foundations and control centre buildings (Section 7.6.3). Similar opportunities exist for small- and mini-hydropower projects. Many sites of old water-powered grain mills could be utilized for run-of-river micro-hydropower generation schemes (Section 5.3.1). Low-head turbines have been developed for operating in low-gradient water distribution channels to power irrigation pumps (EECA, 2008). Solar PV systems have been linked with water pumping and solar thermal systems have been commonly used for water heating and crop drying. Solar sorption technologies for air-conditioning, refrigeration, ice making and post-harvest chilling of fresh products remain at the development stage (Fan et al., 2007). Geothermal heat has been used for various applications including heating greenhouses, desiccation of fruit and vegetables, heating animal livestock houses, drying timber and heating water for fish and prawn farming (J. Lund, 2005).

Biomass resources produced in forests and on farms are commonly used to meet local agricultural and rural community heat energy demands but developing large-scale projects can be a challenge and possible removal of nutrients in the biomass a constraint for some soil types (IEA, 2007a). Returning some nutrients to the land as ash after combustion is feasible as is the production of biochar via pyrolysis, which can then be incorporated into the soil to improve the productivity as well as reduce atmospheric carbon concentrations if managed properly (Section 8.3.4.4).

Crop or forest residues are either collected and transported as a separate operation following the harvest of the primary product (grain or timber) or integrated as a harvesting operation of all co-products (Heikkilä et al., 2006). Privatization of the electricity industry in some countries has enabled sugar, rice and wood processing plant owners to invest in more efficient CHP plants that generate excess power for export (Section 8.3.3) and can also reduce local air pollution if the biomass is dry, combusted efficiently and displaces coal (Shanmukharadhya and Sudhakar, 2007).

Anaerobic digestion of animal manures, fish, food and fibre processing wastes, and green crops such as sorghum or maize is a well understood technology to produce biogas (Section 2.3.3). Gas storage is costly, so supply should be matched with demand where feasible (Section 8.2.3). The odourless, digested solid residues can be used for soil conditioning and nutrient replenishment. On-farm direct combustion of the biogas to supply heat is common practice, or after upgrading to biomethane (Section 8.2.3) it can be used in stationary gas engines for CHP or used as a transport fuel similar to compressed natural gas (Section 8.3.1).

8.3.4.3 Pathways for renewable energy integration and adoption

Much agricultural and forest land that produces food and fibre products could simultaneously be used for supplying RE, in many cases utilizing the heat and electricity on the property to displace the energy inputs purchased to run the enterprise (Table 8.5). Biofuels and biogas can also be produced on-farm, either for direct use on site (Section 8.2.3) or sold to the market. Market drivers for RE power generation on rural land and waterways include electrification of rural areas, a more secure energy supply and the avoidance of costly transmission line capacity upgrading in areas where demand loads are increasing (Section 8.2.1).

To meet the growing demands for primary products including biomass, increasing productivity of existing arable, pastoral and plantation forest lands by improving management and selecting higher yielding varieties is one option. (Changing diets to eat less animal products is another). Global average yields of staple crops have continued to increase over the past few decades (Figure 8.24). This trend could continue over the next few decades, with genetically modified crops possibly having a positive influence. Conversely, climate change trends including more frequent extreme weather events could offset some of the productivity gains expected from technological advances (Lobell and Field, 2007).

The primary production sector is making a slow transition to reducing its dependence on energy inputs as well as to better using its natural endowment of RE sources. Integration of land use for agriculture and energy purposes is growing but barriers to greater RE deployment in rural areas include high capital costs, lack of available financing, remoteness from energy demand (including access to electricity and gas grids), competition for land use, transport constraints, water supply limitations and lack of skills and knowledge in landowners and managers.

8.3.4.4 Future trends for renewable energy in agriculture

Distributed energy systems based on small-scale RE technologies (IEA, 2009b) have good potential in rural areas. The concept could also be applied to produce mini-power distribution grids (Section 8.2.1) in rural communities in developing countries where electricity services are not yet available.

A future opportunity for the agricultural sector is the concept of carbon sequestration in the soil as 'biochar' (Lehmann, 2007; Woolf et al.,

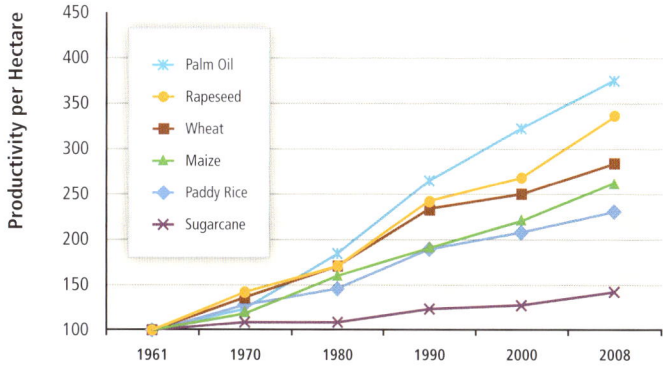

Figure 8.24 | Increased global productivity per hectare for a range of staple crops over the past few decades compared with average yields in base year 1961 shown at 100 (based on FAO (2009) data).

Table 8.5 | Primary production from industrial-scale enterprises requires direct energy inputs at various levels of energy use intensity (GJ input/unit of production or GJ/ha of land) that can either be purchased and brought in across the farm boundary or produced from on-farm RE systems with the potential to export any energy carriers based on excess RE sources as a revenue earner.

Type of enterprise	Direct energy inputs	Energy use intensity	Potential RE carriers	Energy export potential
Dairying	Electricity for milking facility, pumping of water and manure, refrigeration. Diesel for tractor. Diesel or electricity for irrigation.	High. Medium. High for pumped irrigation.	Manure for biogas. Heat from milk cooling. Solar for water heating. Solar PV electricity.	Biogas for CHP (combined heat and power). Solar heat or electricity is feasible on most land enterprises listed but not yet cost-effective.
Pastoral grazing animals (e.g., sheep, beef, deer, goat, llama).	Electricity for shearing. Diesel for farm machinery.	Very low, but higher if land is irrigated. Low (or medium where some pasture conserved).	Wind power if good hill sites. Hydropower from streams. Solar systems on buildings. Green crops for biogas.	Wind power. Hydropower. Biogas for CHP.
Intensive animal production.	Electricity for lighting, cooling, water pumping, cleaning. Diesel for tractor.	High if housed indoors. Medium to low if kept mainly outdoors. High for harvesting feed.	Manure for biogas. Combustion of poultry litter. Solar systems on buildings.	Biogas for CHP. Several multi-MW power plants operating in UK and USA.
Arable (e.g., wheat, maize, rapeseed, palm oil, cotton, sugarcane, rice, etc.).	Diesel fuel for tractors. Electricity for storage facilities. Conveyor motors, irrigation. Gas or LPG for drying.	Very high for machinery. Medium if rain-fed. High if crops irrigated. Low and seasonal.	Crop residues for heat, power and possibly biofuels. Energy crops. Wind and hydro if good sites.	High where energy crops are purpose-grown. Wind power possible but crops grown on land with few hills.
Vegetables – large scale (potatoes, onions, carrots, etc.).	Diesel fuel for tractors. Electricity for grading, conveying, irrigation, cooling.	High for machinery. High if land irrigated and for post-harvest chillers.	Dry residues for combustion. Wet residues for biogas.	Limited as would be mainly used onsite.
Market garden vegetables – small scale (wide range).	Diesel for machinery. Electricity for washing, grading.	Medium. Low for post-harvest. Medium for cool stores.	Residues and rejects for biogas (but too small and seasonal a resource for even onsite use).	Low.
Nursery cropping	Diesel for machinery. Heat for protected greenhouses.	Low. Medium.	Some residues and rejects for combustion.	Low.
Greenhouse crop production	Electricity for ventilation, lighting. Gas, oil, or biomass for heating.	High where heated. Medium if unheated.	Small volumes of residues and rejects for combustion.	Low.
Orchard (pip fruit, bananas, pineapple, olives, etc.).	Diesel for machinery. Electricity for grading, drip irrigation, cool stores, etc.	Medium. Medium if irrigated and post-harvest storage.	Combustion of pruning residue for heat. Reject fruit, bunches and residues for biogas or CHP.	Low.
Forest plantations (eucalyptus, spruce, pine, palm oil, etc.)	Diesel for planting, pruning and harvesting.	Low.	Forest residues. Short rotation forest crops. Spent oil palm bunches.	High—large volumes of biomass for CHP, or possibly for biofuels.
Fishing – large trawlers offshore.	Marine diesel/fuel oil. Electricity for refrigeration.	High. Medium.	Reject fish dumped at sea.	None.
Fish farm – near-shore or onshore.	Diesel for boats. Electricity for refrigeration.	Low. Medium if facilities offshore. Medium.	Fish wastes for biogas and oil. Ocean energy.	Low. Electricity from ocean energy possible in future.
Fishing – small boats near-shore.	Diesel/gasoline. Electricity for ice or refrigeration.	Low. Low.	Fish wastes for biogas and oil.	Low.

2010). When produced via gasification or pyrolysis using the controlled oxygen combustion of sustainably produced biomass, incorporation of the residual char into arable soils is claimed to enhance future plant growth and the carbon is removed from the atmosphere (Verheijen et al., 2010). Biochar properties vary with the biomass feedstock and various crops and soil types may respond in different ways in terms of their productivity. Further R&D is required to address the net energy and nutrient balances for the various types of biochar.

Case study: Distributed RE generation in a rural community.
The small community of Totara Valley, New Zealand, illustrates how local RE resources can be utilized to meet local demands for heat and power and provide revenue and social benefits. The hydropower generation potential, wind speeds and solar radiation levels in the vicinity were monitored and a method developed to show seasonal and daily variations and match these with electricity demand (Murray, 2005) (Figure 8.25). An electricity generation and/or a lines distribution company could have strong business interests in such a scheme by becoming a joint venture partner, not only to buy and sell the surplus electricity, but also to sell, hire or lease the RE equipment to the landowners (Jayamaha, 2003).

The Totara Valley small-scale demonstration project consists of solar PV, solar thermal panels and heat pumps on some of the houses, a biodiesel generating set, a 1 kW Pelton micro-hydro turbine, and, on a hill site selected for its average wind speeds and proximity to load, a 2.2 kW wind turbine. Due to the USD 13,000 cost estimate for installing 1.5 km of copper cable to connect the hill site to the community buildings, the wind turbine is instead used to power an adjacent electrolyzer (Sudol, 2009). The hydrogen produced is carried in an underground alkathene pipe to a fuel cell housed in the farm buildings. Storage and transfer losses in the pipe are only around 1% of total hydrogen production (Gardiner et al., 2008). The overall efficiency of the hydrogen system is low but is partly offset by it acting as an energy store for the community system. The demonstration has shown that integration of a portfolio of RE technologies with existing heat and power supply systems is feasible for an agricultural community, but economic assessment of the options is recommended on a site-by-site basis.

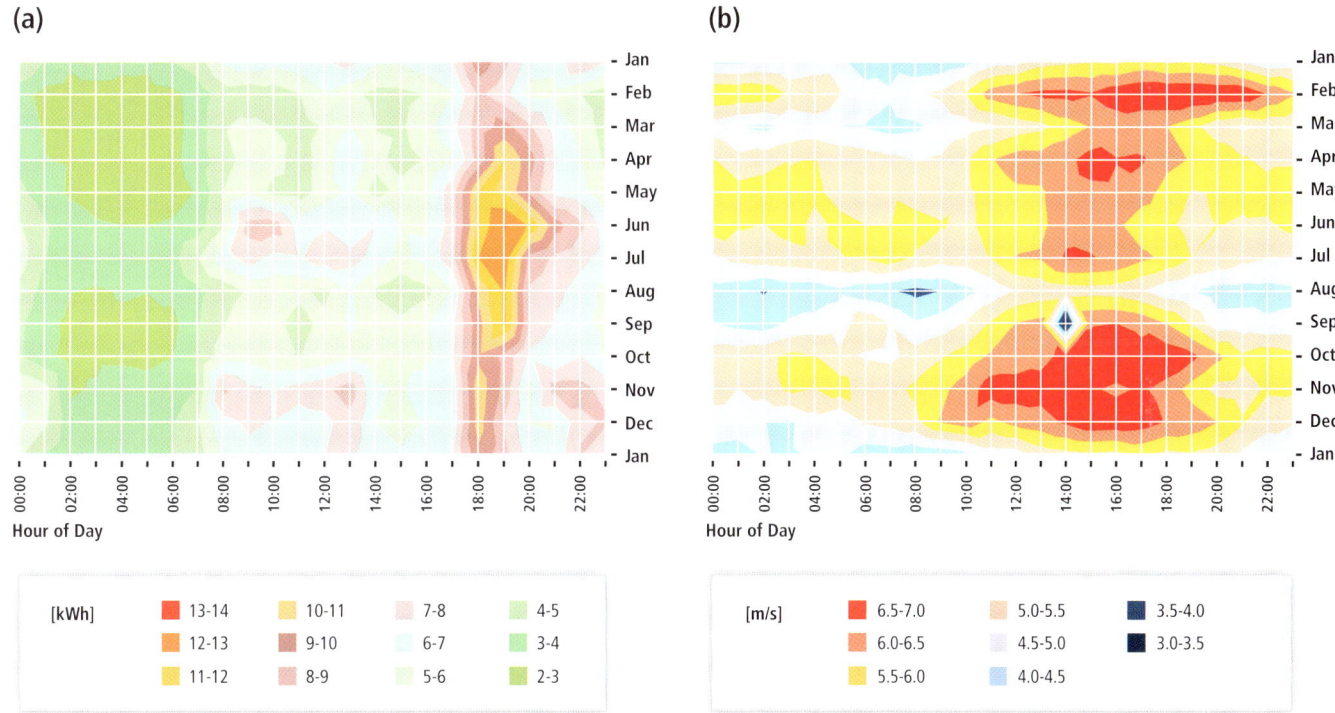

Figure 8.25 | (a) Average seasonal and daily electricity demand for the Totara Valley community (in kWh consumption per 30 minute periods), and (b) annual and daily wind data, showing some matching of wind power supply with evening and winter peak demands (Murray, 2005).

References

Abdelkader, S., S. Abbott, J. Fu, L. McClean, D. Flynn, B. Fox, and L. Bryans (2009). Dynamic monitoring of overhead line ratings in wind intensive areas. In: *European Wind Energy Conference (EWEC)*, Marseille, France, 16-19 March 2009.

Achilles, S., S. Schramm, and J. Bebic (2008). *Transmission System Performance Analysis for High-Penetration Photovoltaics, Subcontract Report*. National Renewable Energy Laboratory, Golden, CO, USA, 77 pp.

Ackermann, T., G. Ancell, L.D. Borup, P.B. Eriksen, B. Ernst, F. Groome, M. Lange, C. Mohrlen, A.G. Orths, J. O'Sullivan, and M. de la Torre (2009). Where the wind blows. *IEEE Power and Energy Magazine*, **7**(6), pp. 65-75.

Ackermann, T., and P. Morthorst (2005). Economic aspects of wind power in power systems. In: *Wind Power in Power Systems*. T. Ackermann (ed.), John Wiley & Sons, New York, NY, USA, pp. 383-410.

AEMO (2010). *An Introduction to Australia's National Electricity Market*. Australian Energy Market Operator, Melbourne, Australia, 28 pp.

Ahlgren, S., S. Bernesson, Å. Nordberg, and P.A. Hansson (2008). Nitrogen fertiliser production based on biogas – energy input, environmental impact and land use. *Bioresource Technology*, **101**(18), pp. 7181-7184.

Åhman, M. (2010). Biomethane in the transport sector – An appraisal of the forgotten option. *Energy Policy*, **38**(1), pp. 208-217.

AIGS (2008). *All Island Grid Study, Workstream 4: Analysis of Impacts and Benefits*. Department of Communications, Energy and Natural Resources, UK Department of Enterprise, Trade and Investment, Dublin, Ireland, 82 pp.

Akbari, H., S. Menon, and A. Rosenfeld (2009). Global cooling: increasing worldwide urban albedos to offset CO_2. *Climatic Change*, **94**(3-4), pp. 275-286.

Alakangas, E., and M. Flyktman (2001). *Biomass CHP technologies*. VTT Energy Reports 7/2001, VTT Energy, Jyvaskyla, Finland, 62 pp.

Aliabadi, A.A., J.M. Thomson, and J.S. Wallace (2010). Efficiency analysis of natural gas residential micro-cogeneration systems. *Energy Fuels*, **24**(3), pp. 1704-1710.

Amigun, B., R. Sigamoney, and H. Blottnitz (2008). Commercialization of biofuel industry in Africa: A review. *Renewable and Sustainable Energy Reviews*, **12**(3), pp. 690-711.

Anant, S., D. Buddhi, and R.L. Sawhney (2008). Thermal cycling test of selected inorganic and organic phase change materials. *Renewable Energy*, **33**(12), pp. 2606-2614.

Antvorskov, S. (2007). Introduction to integration of renewable energy in demand controlled hybrid ventilation systems for residential buildings. *Building and Environment*, **43**(8), pp. 1350-1353.

ARE (2009). *Hybrid Power Systems Based on Renewable Energies: A Suitable and Cost-Competitive Solution for Rural Electrification*. Position Paper, Alliance for Rural Electrification, Brussels, Belgium, 7 pp.

Akershus Energi (2010). Skedsmo Kommune, Akershus Energi Varme AS. Ecoheat4EU.www.ecoheat4.eu/en/Country-by-country-db/Norway/Local-Success-Stories/-print.

Arroyo, J.M., and F.D. Galiana (2005). Energy and reserve pricing in security and network-constrained electricity markets. *IEEE Transactions on Power Systems*, **20**(2), pp. 634-643.

Avadikyan, A., and P. Llerenaa (2010). A real options reasoning approach to hybrid vehicle investments. *Technological Forecasting and Social Change*, **77**(4), pp. 649-661.

Awerbuch, S. (2006). Portfolio-based electricity generation planning: policy implications for renewables and energy security. *Mitigation and Adaptation Strategies for Global Change*, **11**(3), pp. 693-710.

Axegård, P., B. Backlund, and B. Warnquist (2002). The Eco-Cyclic pulp mill: Focus on closure, energy efficiency and chemical recovery development. *Pulp & Paper Canada*, **103**(5), pp. 26-29.

Azevedo, J.M., and F.D. Galiana (2009). The sugarcane ethanol power industry in Brazil: obstacles, success and perspectives. In: *IEEE Electrical Power & Energy Conference*, Montreal, Canada, 22-23 October 2009.

Bajgain, S., and I.S. Shakya (2005). *The Nepal Biogas Support Program: A Successful Model of Public Private Partnership for Rural Household*. Netherlands Development Organisation and the Biogas Support Programme, The Hague, The Netherlands and Kathmandu, Nepal, 75 pp.

Bajnóczy, G., G. Palffy, E. Prépostffy, and A. Zöld (1999). Heat storage by two-grade phase change material. *Perodica Polytechnica Chemical Engineering*, **43**(2), pp. 137-147.

Bandivadekar, A., K. Bodek, L. Cheah, C. Evans, T. Groode, J. Heywood, E. Kasseris, M. Kromer, and M. Weiss (2008). *On the Road in 2035: Reducing Transportation's Petroleum Consumption and GHG Emissions*. Laboratory for Energy and the Environment, Massachusetts Institute of Technology, Cambridge, MA, USA.

Barroso, L.A., S. Granville, J. Trinkenreich, M.V. Pereira, and P. Lino. (2003). Managing hydrological risks in hydro-based portfolios. *IEEE Power Engineering Society General Meeting*, **2**(274).

Bauer, D., R. Marx, J. Nußbicker-Lux, F. Ochs, W. Heidemann, and H. Müller-Steinhagen (2010). German central solar heating plants with seasonal heat storage. *Solar Energy*, **84**(4), pp. 612-623.

Bayem, H., L. Capely, F. Dufourd, and M. Petit (2009). Probabilistic study of the maximum penetration rate of renewable energy in an island network. In: *IEEE PowerTech 2009*, Bucharest, Romania, 28 June – 2 July 2009, pp. 1-5.

BDEW (2008). *Technical Conditions for Connection to the Medium-Voltage Network*. German Association of Energy and Water Industries, Berlin, Germany.

BEN (2010). *Balanço Energetico Nacional (National Energy Balance Brazil)*. Ministry of Mines and Energy, Brazil. Available at: www.epe.gov.br/Estudos/Paginas/Balan%C3%A7o%20Energ%C3%A9tico%20Nacional%20%E2%80%93%20BEN/Estudos_13.aspx.

Bergen, A., and V. Vittal (2000). *Power Systems Analysis*. 2nd ed. Prentice-Hall Inc., NJ, USA, 619 pp.

Bernstein, L., J. Roy, K.C. Delhotal, J. Harnisch, R. Matsuhashi, L. Price, K. Tanaka, E. Worrell, F. Yamba, and Z. Fengqi (2007). Industry. In: *Climate Change 2007: Mitigation of Climate Change. Contribution of Working Group III to the Fourth Assessment Report of the Intergovernmental Panel on Climate Change*. B. Metz, O.R. Davidson, P.R. Bosch, R. Dave, and L.A. Meyer (eds.), Cambridge University Press, pp. 447-496.

Billinton, R. and R.N. Allan (1988). Concepts of power system reliability evaluation. *International Journal of Electrical Power & Energy Systems*, **10**(3), pp. 139-141.

Billinton, R., and P.G. Harrington (1978). Reliability evaluation in energy limited generating capacity studies. *IEEE Transactions on Power Apparatus and Systems*, **97**(6), pp. 2076-2085.

Biogasmax (2009). *European Biomethane Demonstration Project*. Available at: www.biogasmax.co.uk.

Bloem, J.J. (2008). Evaluation of a PV-integrated building application in a well-controlled outdoor test environment. *Building and Environment*, **43**(2), pp. 205-216.

BMU (2009). *Richtlinien zur Förderung von Maßnahmen zur Nutzung erneuerbarer*. Guidelines Energien im Wärmemarkt, German Federal Ministry for the Environment, Nature Conservation and Nuclear Safety (BMU), Berlin, Germany, 25 pp.

BMU (2010). *Renewable Energy Sources in Figures*. German Federal Ministry for the Environment, Nature Conservation and Nuclear Safety (BMU), Berlin, Germany, 76 pp.

Bodmann, M., D. Mangold, J. Nußbicker, S. Raab, A. Schenke, and T. Schmidt (2005). *Solar unterstützte Nahwärme und Langzeit-Wärmespeicher (Februar 2003 bis Mai 2005)*. Report 0329607F, Forschungsbericht zum BMWA/BMU-Vorhaben, Stuttgart, Germany, 159 pp.

Bonhoff, K., N. Parker, S. Joest, M. Fichtner, M. Wietschel, U. Bünger, C. Stiller, P. Schmidt, and F. Merten (2009). *Woher kommt der Wasserstoff in Deutschland bis 2050? [Where will the Hydrogen in Germany come from in 2050?]*. Report GermanHy Study, Nationale Organisation Wasserstoff- und Brennstoffzellentechnologie (NOW), Berlin, Germany, 62 pp.

Börjesson, P., and G. Berndes (2006). The prospects for willow plantations for wastewater treatment in Sweden. *Biomass & Bioenergy*, **30**, pp. 428-436.

Börjesson, P., and B. Mattiasson (2008). Biogas as a resource-efficient vehicle fuel. *Trends in Biotechnology*, **26**(1), pp. 7-13.

Bouzidi, B., M. Haddadi, and O. Belmokhtar (2009). Assessment of a photovoltaic pumping system in the areas of the Algerian Sahara. *Renewable and Sustainable Energy Reviews*, **13**(4), pp. 879-886.

Brattle Group, Freeman Sullivan & Co, and Global Energy Partners LLC (2009). *A National Assessment of Demand Response Potential*. Federal Energy Regulatory Commission, Washington, DC, USA, 254 pp.

Braun, T., T. Stetz, B. Reimann, G. Valov, and G. Arnold (2009). Optimal reactive power supply in distribution systems - Technological and economic assessment for PV systems. In: *24th European Photovoltaic Solar Energy Conference and Exhibition*, Hamburg, Germany, 21-25 September 2009.

Breuer, W., V. Hartmann, D. Povh, D. Retzmann, and E. Teltsch (2004). Application of HVDC for large power system interconnections. In: *International Council on Large Electric Systems, 2004 Session*, Paris, France, 29 August – 3 September 2004.

Brinkman, N., M. Wang, T Weber, and T. Darlington (2005). *Well-to-wheels analysis of advanced fuel/vehicle systems - a North American study of energy use*. Greenhouse Gas Emissions and Criteria Pollutant Emissions, General Motors Report, May. 173 pp http://www.transportation.anl.gov/pdfs/TA/339.pdf

Bromley, C.J., L. Rybach, M.A. Mongillo, and I. Matsunaga (2006). Geothermal resource utilization strategies to promote beneficial environmental effects and to optimize sustainability. In: *Proceedings RE 2006*, Chiba, Japan, 9-13 October 2006, pp. 1-4.

Brown, D.W. (1996). The geothermal analog of pumped storage for electrical demand load following. In: *Energy Conversion Engineering Conference (IECEC 96)*, Denver, Co, USA, 11-16 August 1996, pp. 1653-1656.

Brown, W. (2010). *The National Grid – Building for the Future*. Transpower New Zealand Ltd., Wellington, New Zealand, 48pp. Available at: www.transpower.co.nz/f1958,45849325/iet-wayne-11may2010.pdf.

Bruus, F., and H. Kristjansson (2004). Principal design of heat distribution. *Danish Board of District Heating Newsletter*, 2/2004. See:http://dbdh.dk/images/uploads/pdf-distribution/principal-design-of-heat-distribution.pdf.

Bryans, A.G., B. Fox, P.A. Crossley, and M. O'Malley (2005). Impact of tidal generation on power system operation in Ireland. *IEEE Transactions on Power Systems*, **20**(4), pp. 2034-2040.

Buckingham, G.S. (1965). Remote control of electricity supplies to the domestic consumer. *Electronics and Power*, **11**(3), p. 98.

Bueno, C., and J.A. Carta (2006). Wind powered pumped hydro storage systems, a means of increasing the penetration of renewable energy in the Canary Islands. *Renewable and Sustainable Energy Reviews*, **10**(4), pp. 312-340.

Buijs, P., D. Bekaert, S. Cole, D. Van Hertem, and R. Belmans (2011). Transmission investment problems in Europe: Going beyond standard solutions. *Energy Policy*, **39**(3), pp. 1794-1801

Burke, D., and M. O'Malley (2010). Maximizing firm wind connection to security constrained transmission networks. *IEEE Transactions on Power Systems*, **25**(2), pp. 749-759.

Caamano-Martin, E., H. Laukamp, M. Jantsch, T. Erge, J. Thornycroft, H. De Moor, S. Cobben, D. Suna, and B. Gaiddon (2008). Interaction between photovoltaic distributed generation and electricity networks. *Progress in Photovoltaics*, **16**(7), pp. 629-643.

CAFCP (2009). *Hydrogen Fuel Cell Vehicle and Station Deployment Plan: A Strategy for Meeting the Challenge Ahead. Action Plan*. California Fuel Cell Partnership (CAFCP), West Sacramento, CA, USA. Available at: www.fuelcellpartnership.org/sites/files/Action%20Plan%20FINAL.pdf.

CAISO (2010). *Fifth Replacement FERC Electric Tariff, Appendix V Large Generator Interconnection Agreement (LGIA)*. California Independent System Operator Corporation, Folsom, CA, USA, 68 pp.

Cappers, P., C. Goldman, and D. Kathan (2010). Demand response in US electricity markets: Empirical evidence. *Energy*, **35**(4), pp. 1526-1535.

Caralis, G., and A. Zervos, (2007a): Analysis of wind power penetration in autonomous Greek islands. *Wind Engineering*, **31**(6), pp. 487-502.

Caralis, G., and A. Zervos, (2007b): Analysis of the combined use of wind and pumped storage systems in autonomous Greek islands. *Renewable Power Generation*, **1**(1), pp. 49-60.

Caralis, G., and A. Zervos (2010). Value of wind energy on the reliability of autonomous power systems. *Renewable Power Generation*, **4**(2), pp. 186-197.

Carraretto, C. (2006). Power plant operation and management in a deregulated market. *Energy*, **31**(6-7), pp. 1000-1016.

Castello, P., E. Tzimas, P. Moretto, and S.D. Peteves (2005). *Techno-economic Assessment of Hydrogen Transmission and Distribution Systems in Europe in the Medium and Long Term*. Report EUR 21586 EN, Institute for Energy, Petten, The Netherlands, 137 pp. (ISBN: 92-894-9292-9).

CEC (2006). *Bioenergy Action Plan for California*. Action Plan CEC-600-2006-01, Bioenergy Interagency Working Group, California Energy Commission (CEC), Sacramento, CA, USA.

Centolella, P. (2010). The integration of price responsive demand into Regional Transmission Organization (RTO) wholesale power markets and system operations. *Energy*, **35**(4), pp. 1568-1574.

CEP (2010). *Clean Energy Partnership*. See: http://www.cleanenergypartnership.de/fileadmin/pdf/20110519_CEP_%20PM%20Beitritt_AL%20Honda_eng.pdf

Cerri, C.E.P., M. Easter, K. Paustian, K. Killian, K. Coleman, M. Bemoux, P. Falloon, D.S. Powlson, N.H. Batjes, E. Milne, and C.C. Cerri (2007). Predicted soil organic carbon stocks and changes in the Brazilian Amazon between 2000 and 2030. *Agriculture, Ecosystems and Environment*, **122**, pp. 58-72.

CGEE (2009). *Bioetanol como combustível: uma oportunidade para o Brasil*. Report Document Number 389b, Centro de Gestão e Estudos Estratégicos em Ciência (CGEE), Tecnologia e Inovação, Sala, Brazil, 536 pp.

Chan, H.-Y., B. Saffa, and J.Z. Riffat (2010). Review of passive solar heating and cooling technologies. *Renewable and Sustainable Energy Reviews*, **14**, pp. 781-789.

Chen, H.S., T.N. Cong, W. Yang, C.Q. Tan, Y.L. Li, and Y.L. Ding (2009). Progress in electrical energy storage system: A critical review. *Progress in Natural Science*, **19**(3), pp. 291-312.

Chen, Q., C. Kang, and Q. Xia (2010). Modeling flexible operation mechanism of CO_2 capture power plant and its effects on power-system operation. *IEEE Transactions on Energy Conversion*, **25**(3), pp. 853-861.

Chen, Z., J.M. Guerrero, and F. Blaabjerg (2009). A Review of the state of the art of power electronics for wind turbines. *IEEE Transactions on Power Electronics*, **24**(8), pp. 1859-1875.

Cheung, A., and M. Wilshire (2010). *White paper from Consortium on Digital Energy 2009-2010*. Bloomberg New Energy Finance, London, UK, 40 pp.

Cheung, K. (2008). Ancillary service market design and implementation in North America: From theory to practice. In: *Electric Utility Deregulation and Restructuring and Power Technologies, 2008 (DRPT 2008). Third International Conference*, Nanjuing, China, 6-9 April 2008, pp. 66-73

Chowdhury, B.H., and A.W. Sawab (1996). Evaluating the value of distributed photovoltaic generations in radial distribution systems. *IEEE Transactions on Energy Conversion*, **11**(3), pp. 595-600.

Christidis, P., I. Hidalgo, and A. Soria (2003). *Dynamics of the Introduction of New Passenger Car Technologies: The IPTS Transport Technologies Model*. Report EUR 20762 EN, European Commission, Brussels, Belgium.

Ciferno, J.P., and J.J. Marano (2002). *Benchmarking Biomass Gasification Technologies for Fuels, Chemicals and Hydrogen Production*. U.S. Department of Energy, National Energy Technology Laboratory, Pittsburgh, PA, USA, 58 pp.

CILSS (2004). *Capitalisation of the Sahelian experience in natural forest management for wood energy production. Regional workshop*. Comité Inter-Etate pour la Lutte contre la Sécheresse au Sahel (CILSS: Permanent Inter-State Committee for Drought Control in the Sahel), Niamey, Niger.

CILSS (2008). *Réglemenation et fiscalité sur le bois énergie: situation comparée, contexte, enjeux et défis. Regional workshop*. Comité Inter-Etate pour la Lutte contre la Sécheresse au Sahel (CILSS : Permanent Inter-State Committee for Drought Control in the Sahel), Niamey, Niger.

CONCAWE (2007). *Well-to-wheels Analysis of Future Automotive Fuels and Powertrains in the European Context*. Report Version 2c (March). Conservation of Clean Air and Water in Europe (CONCAWE), Brussels, Belgium. Available at: ies.jrc.ec.europa.eu/uploads/media/V3.1%20TTW%20Report%2007102008.pdf.

Contreras, A.B., M. Thomson, and D.G. Infield (2007). Renewable energy powered desalination in Baja California Sur, Mexico. *Desalination*, **220**(1-3), pp. 431-440.

Cornell (2005). *Lake Source Cooling*. Cornell University Facilities Services, Utilities and Energy Management, Ithaca, NY, USA. Available at: http://energyandsustainability.fs.cornell.edu/util/cooling/production/lsc/default.cfm.

CREZ (2010). *CREZ Progress Report No. 1*. Competitive Renewable Energy Zone (CREZ) Program Oversight, Public Utility Commission of Texas, Austin, TX, USA, 95 pp.

Croezen, H., and M. Korteland (2010). *A Long-Term View of CO_2 Efficient Manufacturing in the European Region*. Report 10 7207 47, CE Delft, Delft, The Netherlands, 87 pp.

CSO (2010). *Population and Migration Estimates*. Central Statistics Office, Cork, Ireland, 9 pp.

Curtright, A.E., and J. Apt (2008). The character of power output from utility-scale photovoltaic systems. *Progress in Photovoltaics*, **16**(3), pp. 241-247.

Dalenbäck, J.O. (2010). *Success factors in solar district heating* CIT Energy management AB, Project Report for Solar District Heating, (SDH) December. 62pp. http://www.solar-district-heating.eu/LinkClick.aspx?fileticket=c6-K2BVa2hM%3d&tabid=69

Dalton, G.J., D.A. Lockington, and T.E. Baldock (2008). Case study feasibility analysis of renewable energy supply options for small to medium-sized tourist accommodations. *Renewable Energy*, **34**(4), pp. 1134 -1144.

Davis, S., S. Diegel, and R. Boundy (2010). *Transportation Energy Data Book*. ORNL-6985 (Edition 29 of ORNL-5198), Oak Ridge National Laboratory, Center for Transportation Analysis, Energy and Transportation Science Division, Oak Ridge, TN, USA.

de Angelis, F., G. Magnante, N. Rossi, and A. Siniscalchi (2010). Local governance and sustainable communities. European benchmarking and EU convergence regions in Southern Italy. In: *MED3 Proceedings*, Tunis, Tunisia, 7-8 October 2010, pp. 1-17. Available at: med-eu.org/documents/MED3/DE%20ANGELIS-MAGNANTE-ROSSI-SINISCALCHI.pdf.

de Moraes, M.A.F.D., and L. Rodrigues (2006). *Brazil Alcohol National Program. Relatório de pesquisa*. No publisher specified, Piracicaba, Brazil, 54 pp. Available at: english.unica.com.br/multimedia/documentos/Default.asp?sqlPage=1.

de Wasseige, C., D. Devers, P. de Marcken, R. Eba'a Atyi, R. Nasi, and P. Mayaux (eds.) (2009). *Etat des Forêts 2008*. Congo Basin Forest Partnership, Publications Office of the European Union, 426 pp. (ISBN 978-92-79-132 11-7).

DEA (2009). *Danish Energy Agency*. Danish Energy Agency, Copenhagen, Denmark. See: www.ens.dk/en-US/supply/Renewable-energy/WindPower/Facts-about-Wind-Power/Key-figures-statistics/Sider/Forside.aspx

Deane, J., B.Ò. Gallachòir, and E.J. McGeogh (2010). Techno-economic review of existing and new pumped hydro energy storage plant. *Renewable and Sustainable Energy Reviews*, **14**(4), pp. 1293-1302.

DeCarolis, J., and D. Keith (2006). The economics of large-scale wind power in a carbon constrained world. *Energy Policy*, **34**(4), pp. 395-410.

Delgado-Torres, A.M., and L. García-Rodríguez (2007). Comparison of solar technologies for driving a desalination system by means of an organic Rankine cycle. *Desalination*, **216**(1-3), pp. 276-291.

Delille, G., B. François, and G. Malarange (2010). Dynamic frequency control support: A virtual inertia provided by distributed energy storage to isolated power systems. In: *Innovative Smart Grid Technologies Conference Europe (ISGT Europe), IEEE PES*, Gothenburg, Sweden, 11-13 October 2010.

Delucchi, M., and M. Jacobson (2009). A plan to power 100 percent of the planet with renewables. *Scientific American*, **November 2009**, pp. 58-65.

DENA (2005). *Energiewirtschaftliche Planung für die Netzintegration von Windenergie in Deutschland an Land und Offshore bis zum Jahr 2020*. Study by Energiewirtschaftliches Institut an der Universität zu Köln (EWI), DeutschesWindenergie-Institut (DEWI), E.ON Netz GmbH, RWE Transportnetz Strom GmbH, Vattenfall Europe Transmission GmbH. Report, Deutsche Energie-Agentur GmbH (DENA), Köln, Germany, 380 pp.

DENA (2010). *Dena-Netzstudie II – Integration erneuerbarer Energien in die deutsche Stromversorgung im Zeitraum 2015 – 2020 mit Ausblick 2025*. Deutsche Energie-Agentur GmbH (DENA), Berlin, Germany, 620 pp.

Denault, M., D. Dupuis, and S. Couture-Cardinal (2009). Complementarity of hydro and wind power: Improving the risk profile of energy inflows. *Energy Policy*, **37**(12), pp. 5376-5384.

Deng, Y., Z. Yu, and S. Liu (2011). A review on scale and siting of wind farms in China. *Wind Energy*, **14**(3), pp. 463-470.

Denholm, P., and R.M. Margolis (2007). Evaluating the limits of solar photovoltaics (PV) in electric power systems utilizing energy storage and other enabling technologies. *Energy Policy*, **35**(9), pp. 4424-4433.

Denholm, P., and R. Sioshansi (2009). The value of compressed air energy storage with wind in transmission-constrained electric power systems. *Energy Policy*, **37**(8), pp. 3149-3158.

Denholm, P., R.M. Margolis, and J.M. Milford (2009). Quantifying avoided fuel use and emissions from solar photovoltaic generation in the western United States. *Environmental Science & Technology*, **43**(1), pp. 226-232.

Denholm, P., E. Ela, and B. Kirby (2010). *The role of Energy Storage with Renewable Electricity Generation.* Technical report, National Renewable Energy Laboratory, Golden, CO, USA, 61 pp.

Denny, E. (2009). The economics of tidal energy. *Energy Policy*, **37**(5), pp. 1914-1924.

Denny, E., and M. O'Malley (2006). Wind generation, power system operation, and emissions reduction. *IEEE Transactions on Power Systems*, **21**(1), pp. 341-347.

Denny, E., A. Tuohy, P. Meibom, A. Keane, D. Flynn, A. Mullane, and M. O'Malley (2010). The impact of increased interconnection on electricity systems with large penetrations of wind generation: A case study of Ireland and Great Britain. *Energy Policy*, **38**(11), pp. 6946-6954.

DeValve, T., and B. Olsommer (2006). *Micro-CHP Systems for Residential Applications Final Report.* US DOE Report (Contract No. DE-FC26-04NT42217), United Technologies Research Center, East Hartford, CT, USA, 114 pp.

Devine-Wright, P., H. Devine-Wright, and F. Sherry-Brennan (2010). Visible technologies, invisible organisations: An empirical study of public beliefs about electricity supply networks. *Energy Policy*, **38**(8), pp. 4127-4134.

DFID (2002). *Energy for the Poor – Underpinning the Millennium Development Goals.* UK Department for International Development (DFID), London, England. Available at: http://www.ecn.nl/fileadmin/ecn/units/bs/JEPP/energyforthepoor.pdf

Dimoudi, A., and P. Kostarela (2008). Energy monitoring and conservation potential in school buildings in the C' climatic zone of Greece. *Renewable Energy*, **34**(1), pp. 289-296.

DiPippo, R. (2008). *Geothermal Power Plants: Principles, Applications, Case Studies and Environmental Impact.* 2nd ed. Elsevier Ltd., London, UK, 493 pp.

Doherty, R., H. Outhred, and M. O'Malley (2006). Establishing the role that wind generation may have in future generation portfolios. IEEE Transactions on Power Systems, 21(3), pp. 1415-1422.

Doherty, R., A. Mullane, G. Nolan, D.J. Burke, A. Bryson, and M. O'Malley (2010). An assessment of the impact of wind generation on system frequency control. *IEEE Transactions on Power Systems*, **25**(1), pp. 452-460.

Doolla, S., and T.S. Bhatti (2006). Load frequency control of an isolated small-hydro power plant with reduced dump load. *IEEE Transactions on Power Systems*, **21**(4), pp. 1912-1919.

Doukas, H., C. Nychtis, and J. Psarras (2008). Assessing energy-saving measures in buildings through an intelligent decision support model. *Building and Environment*, **44**(2), pp. 290-298.

Dounis, A.I., and C. Caraiscos (2009). Advanced control systems engineering for energy and comfort management in a building environment: A review. *Renewable and Sustainable Energy Reviews*, **13**(6-7), pp. 1246-1261.

DPG (2010). *Elektrizitat: Schlüssel zu einem nachhaltigen und klimaverträglichen Energiesystem.* Deutsch Physikalische Gesellschaft (DPG), Bad Honnef, Germany, 144 pp.

Droege, P., A. Radzi, N. Carlisle, and S. Lechtenböhmer (2010). *100% Renewable Energy and Beyond for Cities.* Report, HafenCity University Hamburg and World Future Council Foundation, Hamburg, Germany, 26 pp.

Dubey, S., and G.N. Tiwari (2010). Energy and exergy analysis of hybrid photovoltaic/thermal solar water heater considering with and without withdrawal from tank. *Journal of Renewable and Sustainable Energy*, **2**(4), 043106.

Dudurych, I.M. (2010a). On-line assessment of secure level of wind on the Irish power system. In: *Power and Energy Society General Meeting, IEEE*, 25-29 July, Minneapolis, pp. 1-29.

Dudurych, I.M. (2010b). Statistical analysis of frequency response of island power system under increasing wind penetration. In: *Power and Energy Society General Meeting, IEEE*, Minneapolis, MN, USA, 25-29 July 2010, pp. 1-29.

Dunwoody, C. (2010). Progress and Next Steps in California. In: *WHEC 2010 - 18th World Hydrogen Energy Conference.* D. Stolten (ed.), International Association for Hydrogen Energy, Essen, Germany, 16-21 May 2010.

Duvall, M. (2010). Benefits and impacts of plug-in hybrid and battery electric vehicles. In: *Electrification of the Transportation System: An MIT Energy Initiative Symposium*, Cambridge, MA, USA, 8 April 2010, 32 pp. Available at: web.mit.edu/mitei/docs/reports/electrification-transportation-system.pdf.

EA Energy Analyses (2007). *50% Wind Power in Denmark in 2025.* Energy Agency, Copenhagen, Denmark, 12 pp.

Earle, R., E.P. Kahn, and E. Macan (2009). Measuring the capacity impacts of demand response. *The Electricity Journal*, **22**(6), pp. 47-58.

EASAC (2009). *Transforming Europe's Electricity Supply – An Infrastructure Strategy for a Reliable, Renewable and Secure Power System.* Policy report 11, European Academies Science Advisory Council (EASAC), Halle, Germany, 41 pp.

EC (2006). *Spain's New Building Energy Standards Place the Country among the Leaders in Solar Energy in Europe.* Action Plan, European Commission Environmental Technologies, Brussels, Belgium, 2 pp.

EC (2010). *Energy Performance of Buildings Directive (Recast).* Report 2010/31/EU, European Commission, Brussels, Belgium.

ECI (2006). *Diversified Renewable Energy Resources: An Assessment of an Integrated Wind, Wave and Tidal Stream Electricity Generating System in the UK, and the Reliability of Wave Power Forecasting.* The Carbon Trust and the Environmental Change Institute (ECI), London and Oxford, UK, 42 pp.

Ecofys (2008). *Study on the Comparative Merits of Overhead Electricity Transmission Lines versus Underground Cables.* Report prepared for the Department of Communications, Energy and Natural Resources, Dublin, Ireland, 234 pp.

EECA (2008). *Hydro Energy on Your Farm, Technical Guide 5.0.* Energy Efficiency and Conservation Authority (EECA) of New Zealand, Wellington, New Zealand.

Egeskog, A., J. Hansson, G. Berndes, and S. Werner (2009). Co-generation of biofuels for transportation and heat for district heating systems – an assessment of the national possibilities in the EU. *Energy Policy*, **37**, pp. 5260-5272.

EIA (2007). *About U.S. Natural Gas Pipelines – Transporting Natural Gas.* Energy Information Administration, U.S. Department of Energy, Washington, DC, USA, 76 pp.

EIA (2010). *International Energy Outlook 2010.* Report DOE/EIA-0484(2010), Energy Information Administration, US Department of Energy, Washington, DC, USA, 338 pp. Available at: www.eia.doe.gov/oiaf/ieo/pdf/0484(2010).pdf.

EIGA (2004). *Hydrogen Transportation Pipelines.* Report IGC Doc 121/04, European Industrial Gases Association, Brussels, Belgium, 77 pp.

EirGrid (2008). *Grid 25: A Strategy for the Development of Ireland's Electricity Grid for a Sustainable and Competitive Future.* EirGrid, Dublin, Ireland, 47 pp.

EirGrid (ed.) (2009). *Grid Code Version 3.4*. EirGrid, Dublin, Ireland, 316 pp.

EirGrid (2010a). *Annual Renewable Report: Powering a Sustainable Future*. EirGrid, Dublin, Ireland, 32 pp.

EirGrid (2010b). *All Island TSO Facilitation of Renewable Studies*. EirGrid, Dublin, Ireland, 77 pp.

Ekman, C.K., and S.H. Jensen (2010). Prospects for large scale electricity storage in Denmark. *Energy Conversion and Management*, **51**(6), pp. 1140-1147.

El-Sharkawi, M.A. (2009). *Electric Energy – An Introduction*. CRC Press, Taylor & Francis LLC, Oxford, UK, 472 pp.

Energinet.DK (2007). *System Plan 2007*. Energinet.DK, Fredericia, Denmark, 64 pp.

Energinet.DK (2008). *Technical Report on the Future Expansion and Undergrounding of the Electricity Transmission Grid, Summary*. Energinet.DK, Fredericia, Denmark, 28 pp.

EnerNex Corporation (2010). *Eastern Wind Integration and Transmission study (EWITS)*. Subcontract Report National Renewable Energy Laboratory, Knoxville, TN, USA, 242 pp.

ENTSO-E (2010). *Ten-Year Network Development Plan 2010-2020*. European Network of Transmission System Operators for Electricity, Brussels, Belgium, 286 pp.

EPA (2007). *Regulatory Impact Analysis: Renewable Fuel Standard Program: Estimated Costs of Renewable Fuels, Gasoline and Diesel (Chapter 7)*. Report EPA420-R-07-004, Office of Transportation and Air Quality, U.S. Environmental Protection Agency, Assessment and Standards Division, Washington, DC, USA, 62 pp.

EPA (2010). *Renewable Fuel Standard Program (RFS-2) Regulatory Impact Analysis*. Report EPA-420-R-10-006, Assessment and Standards Division, Office of Transportation and Air Quality, U.S. Environmental Protection Agency, Washington, DC, USA, 62 pp.

EPE (2009). *Balanço Energético Nacional 2009: Ano base 2008*. Empresa de Pesquisa Energética, Rio de Janeiro, Brazil.

Epp, B. (2009). Dänemark: Solare Nahwärme ist wirtschaftlich. *Solarthemen*, **17**(24-25 September), pp. 312.

EPRI (2007). *Environmental Assessment of Plug-In Hybrid Electric Vehicles*. Report 1015325, Electric Power Research Institute, Palo Alto, California, USA. Available at: miastrada.com/yahoo_site_admin/assets/docs/epriVolume1R2.36180810.pdf.

EPRI (2009). *Program on Technology Innovation: Industrial Electrotechnology Development Opportunities*. Electric Power Research Institute, Palo Alto, CA, USA.

EREC (2008). *The Renewable Energy House*. European Renewable Energy Council, Brussels, Belgium. Available at: www.erec.org/reh.html.

Ericsson, K. (2009). *Introduction and Development of the Swedish District Heating Systems – Critical Factors and Lessons Learned (Policy Paper)*. Policy development for improving Renewable Energy Sources Heating and Cooling penetration in European Member States (Intelligent Energy Europe), CRES, Pikermi, Greece. Available at: www.res-h-policy.eu/downloads/Swedish_district_heating_case-study_(D5)_final.pdf.

Ericsson, K., L.J. Nilsson, and M. Nilsson (2010). New energy strategies in the Swedish pulp and paper industry – The role of national and EU climate and energy policies. *Energy Policy*, **39**, pp. 1439-1449.

Eriksen, P.B., and A.G. Orths (2008). The challenges and solutions of increasing from 20 to 50 percent of wind energy coverage in the Danish power system until 2025. In: *7th International Workshop on Large Scale Integration of Wind Power and on Transmission Networks for Offshore Wind Farms*. T. Akermann (ed.), Enernautics GmbH, Madrid, Spain, 26-27 May 2008, pp. 7.

Ernst, B., U. Schreier, F. Berster, J.H. Pease, C. Scholz, H.P. Erbring, S. Schlunke, and Y.V. Makarov (2010). *Large-scale Wind and Solar Integration in Germany*. Pacific Northwest National Laboratory, prepared for the U.S. Department of Energy, Oak Ridge, TN, USA, 52 pp.

Estanqueiro, A., C.B. Mateus, and R. Pestana (2010). Operational experience of extreme wind penetrations. In: *9th International Workshop on Large-Scale Integration of Wind Power into Power Systems as well as on Transmission Networks for Offshore Wind Power Plants*, Quebec City, Canada, 18-19 October 2010, pp. 34-39.

Eto, J., J. Undrill, P. Mackin, R. Daschmans, B. Williams, B. Haney, R. Hunt, J. Ellis, H. Illian, C. Martinez, M. O'Malley, K. Coughlin, and K.H. LaCommare (2010). *Use of Frequency Response Metrics to Assess the Planning and Operating Requirements for Reliable Integration of Variable Renewable Generation*. Lawrence Berkeley National Laboratory, Berkeley, CA, USA, 141 pp.

Eurelectric (2004). *Electricity for More Efficiency: Electric Technologies and Their Energy Savings Potential*. Union of the Electricity Industry, Brussels, Belgium.

Eurogas (2008). *Eurogas Annual Report 2007/2008*. Eurogas, Brussels, Belgium, 43 pp.

Euroheat & Power (2007). *District Heating and Cooling – Country by Country 2007 survey*. Report, Euroheat & Power, Brussels, Belgium.

EWIS (2010). *Towards a Successful Integration of Large Scale Wind Power into European Electricity Grids, Final Report*. European Wind Integration Study, Brussels, Belgium, 182 pp.

Fan, Y., L. Luo, and B. Souyri (2007). Review of solar sorption refrigeration technologies: development and applications. *Renewable and Sustainable Energy Reviews*, **11**(8), pp. 1758-1775.

FAO (2009). *FAO Statistical Databases*. UN Food and Agricultural Organization, Rome, Italy.

FERC (2005). *Order 661-A*. Federal Energy Regulatory Commission, Washington, DC, USA, 59 pp.

Fox, B., A.I. McCartney, and B.M. McCann (1998). Scheduling of radio-controlled heating load. *IEE Proceedings: Generation, Transmission and Distribution*, **145**, pp. 641-646.

Fritsche, U.R., R.E.H. Sims, and A. Monti (2010). Direct and indirect land use competition issues for energy crops and their sustainable production – an overview. *Biofuels, Bioproducts and Biorefining*, **4**, pp. 692-704.

Fusco, F., G. Nolan, and J.V. Ringwood (2010). Variability reduction through optimal combination of wind/wave resources – An Irish case study. *Energy*, **35**(1), pp. 314-325.

Gardiner, A.I., E.N. Pilbrow, S.R. Broome, and A.E. McPherson (2008). Hylink – a renewable distributed energy application for hydrogen. In: *3rd International Solar Energy Society Conference – Asia Pacific Region (ISES-AP-08)*, ISES, Sydney, Australia, 25-28 November 2008, pp. 11.

GE Energy (2010). *Western Wind and Solar Integration Study Report (WWSIS)*. Subcontract Report NREL/SR-550-47781, National Renewable Energy Laboratory, Golden, CO, USA, 536 pp.

GE Energy Consulting (2007). *Impact of Intermittent Generation on Operation of California Power Grid*. Subcontract Report, California Energy Commission Public Interest Energy Research Program, Sacramento, CA, USA, 242 pp.

Giatrakos, G.P., T.D. Tsoutsos, and N. Zografakis (2009). Sustainable power planning for the island of Crete. *Energy Policy*, **37**(4), pp. 1222-1238.

Giebel, G., R. Brownsword, G. Kariniotakis, M. Denhard, and C. Draxl (2011). *The State-Of-The-Art in Short-Term Prediction of Wind Power: A Literature Overview, 2nd Edition*. ANEMOS.plus Deliverable D-1.2, ARMINES, France.

Gielen, D., and G. Simbolotti (2005). *Prospects for Hydrogen and Fuel Cells*. International Energy Agency Publications, Paris, France, 253 pp.

Gil, A., M.d.l. Torre, and R. Rivas (2010). Influence of wind energy forecast in deterministic and probabilistic sizing of reserves. In: *9th International Workshop on Large-Scale Integration of Wind Power into Power Systems as well as on Transmission Networks for Offshore Wind Power Plants*, Quebec City, Canada, 18-19 October 2010.

Glanchant, J.M., and D. Finon (2010). Large-scale wind power in electricity markets. *Energy Policy*, 38(10), pp. 6384-6386.

GMF (2008). Feeding the world: are new global agricultural policies needed? German Marshall Fund report on *Wilton Park Conference 927*, Steyning, UK, 24-26 November 2008. Available at: cap2020.ieep.eu/2009/4/14/feeding-the-world-are-new-global-agricultural-policies-needed.

Goldemberg, J. (2000). Rural energy in developing countries. In: *World Energy Assessment: Energy and the Challenge of Sustainability*. United Nations Development Programme, United Nations Department of Economic and Social Affairs, and World Energy Council, New York, NY, USA, pp. 367-389. Available at: www.undp.org/energy/activities/wea/drafts-frame.html.

Goldemberg, J. (2009). The Brazilian experience with biofuels. *Innovations*, 4(4), pp. 91-107.

Gorenstin, B.G., N.M. Campodonico, J.P. Costa, and M.V.F. Pereira, 1992: Stochastic optimization of a hydrothermal system including network constraints. *IEEE Transactions on Power Systems*, 7(2), pp. 791-797.

Grant, W., D. Edelson, J. Dumas, J. Zack, M. Ahlstrom, J. Kehler, P. Storck, J. Lerner, K. Parks, and C. Finley (2009). Change in the air. *IEEE Power & Energy Magazine*, 7(6), pp. 47-58.

Graus, W., M. Voogt, and E. Worrell (2007). International comparison of energy efficiency of fossil power generation. *Energy Policy*, 35(7), pp. 3936-3951.

Greaves, B., J. Collins, J. Parkes, and A. Tindal (2009). Temporal forecast uncertainty for ramp events. *Wind Engineering*, 33(4), pp. 309-319.

Greene, D., P. Leiby, and D. Bowman (2007). *Integrated Analysis of Market Transformation Scenarios with HyTrans*. Report ORNL/TM-2007/094, Oak Ridge National Laboratory, Oak Ridge, TN, USA.

Grimsrud, G.P., L. Isaksen, M.L. Chan, and S.T.Y. Lee (1981). Marketability of low-head hydropower. *IEEE Transactions on Power Apparatus and Systems*, 100(1), pp. 184-189.

Gronheit, P.E., and B.O.G. Mortensen (2003). Competition in the market for space heating. District heating as the infrastructure for competition among fuels and technologies. *Energy Policy*, 31, pp. 817-826.

Gronich, S. (2006). Hydrogen & FCV implementation scenarios, 2010 – 2025. In: *USDOE Hydrogen Transition Analysis Workshop*. US Department of Energy, Washington, DC, USA, 26 January 2006. Available at: www1.eere.energy.gov/hydrogenandfuelcells/pdfs/transition_wkshp_summ.pdf.

Hagen, M., E. Polman, J.K. Jensen, A. Myken, O. Joensson, and A. Dahl (2001). *Adding Gas from Biomass to the Gas Grid*. Technical Report ISSN 1102-7371, Swedish Gas Center, Malmö, Sweden, 142 pp.

Haj Seyed Hadi, M.R. (2006). Energy efficiency and ecological sustainability in conventional and integrated potato production. In: *ATEF 2006 - Advanced Technology in the Environmental Field*, L. Ubertini (ed.), Acta Press, Lanzarote, Spain, 6-8 February 2006, 501-034.

Haldane, T.G.N., and P.L. Blackstone (1955). Problems of hydro-electric design in mixed thermal hydro-electric systems *Proceedings of the IEE - Part A: Power Engineering*, 102(3), pp. 311-322.

Hara, R., H. Kita, T. Tanabe, H. Sugihara, A. Kuwayama, and S. Miwa (2009). Testing the technologies: Demonstration grid-connected photovoltaic projects in Japan. *IEEE Power & Energy Magazine*, 7(3), pp. 77-85.

Harris, C. (2006). *Electricity Markets, Pricing, Structures and Economics*. John Wiley and Sons, Sussex, UK, 519 pp.

Hasche, B., A. Keane, and M. O'Malley (2010). Capacity credit of wind power, calculation and data requirements: The Irish power system case. *IEEE Transactions on Power Systems*, 26(1), pp. 1-11.

Hatti-Kaul, R., U. Törnvall, L. Gustafsson, and P. Börjesson (2007). Industrial biotechnology for the production of bio-based chemicals – a cradle-to-grave perspective. *Trends in Biotechnology*, 26, pp. 119-124.

Heidemann, W., and H. Müller-Steinhagen (2006). Central solar heating plants with seasonal heat storage. In: *International Conference on Renewable Energies and Water Technologies*, CIERTA, Almeria, Spain, 6-7 October 2006.

Heikkilä, J., J. Laitila, V. Tanttu, J. Lindblad, M. Sirén, and A. Asikainen (2006). Harvesting alternatives and cost factors of delimbed energy wood. *Forestry Studies (Metsanduslikud Uurimused)*, 45, pp. 49-56.

Hennecke, K., T. Hirsch, D. Krüger, A. Lokurlu, and M. Walder (2008). Pilot plant for solar process steam supply. In: *Eurosun 2008 – 1st International Congress on Heating, Cooling and Buildings*, ISES Europe, Lisbon, Spain, 7-10 October 2008.

Henry, S., A.M. Denis, and P. Panciatici (2010). Feasibility study of off-shore HVDC grids. In: *Power and Energy Society General Meeting, 2010 IEEE*, Minneapolis, MN, USA, 25-29 July 2010, pp. 1-5, 25-29.

Herrmann, U., B. Kelly, and H. Price (2004). Two-tank molten salt storage for parabolic trough solar power plants. *Energy*, 29(5-6), pp. 883-893.

Himri, Y., A.B. Stambouli, B. Draoui, and S. Himri (2008). Techno-economical study of hybrid power system for a remote village in Algeria. *Energy for Sustainable Development*, 33(7), pp. 1128-1136.

Hingorani, N.G. (2007). FACTS technology – State of the art, current challenges and the future prospects. In: *Power Engineering Society General Meeting, IEEE*, Tampa, FL, USA, 24-28 June 2007, pp. 1-4, 24-28.

Hoekman, S.K. (2009). Biofuels in the U.S. – Challenges and opportunities. *Renewable Energy*, 34(1), pp. 14-22.

Hoff, T., and R. Perez (2010). Quantifying PV power output variability. *Solar Energy*, 84(10), pp. 1782-1793.

Holttinen, H., P. Meibom, A. Orths, F. van Hulle, B. Lange, M. O'Malley, J. Pierik, B. Ummels, J.O. Tande, A. Estanqueiro, M. Matos, E. Gomez, L. Söder, G. Strbac, A. Shakoor, J. Ricardo, J.C. Smith, M. Milligan, and E. Ela (2009). *Design and Operation of Power Systems with Large Amounts of Wind Power*. Final Report IEA Wind Task 25 (2006-2008), VTT, Vuorimiehentie, Finland, 229 pp. (ISBN 978-951-38-7308-0).

Holttinen, H., P. Meibom, A. Orths, B. Lange, M. O'Malley, J.O. Tande, A. Estanqueiro, E. Gomez, L. Söder, G. Strbac, J.C. Smith, and F. van Hulle (2011). Impacts of large amounts of wind power on design and operation of power systems, results of IEA collaboration. *Wind Energy*, 14(2), pp. 179-192.

Hossain, M.A., and B.K. Bala (2007). Drying of hot chilli using a solar tunnel drier. *Solar Energy*, 81(1), pp. 85-92.

Hotson, G.W. (1997). Utilisation of geothermal energy in a pulp and paper mill. *Energy Sources*, 19(1), pp. 49-54.

Huang, S-H., Dumas, J., Gonzalez-Perez, C., Wei-Jen Lee (2009). Grid security through load reduction in the ERCOT market. *IEEE Transactions on Industry Applications*, **45**(2), pp. 555-559.

Hughes, L. (2010). Meeting residential space heating demand with wind-generated electricity. *Renewable Energy*, **35**(8), pp. 1765-1772.

Hur, K., M. Boddeti, N.D.R. Sarma, J. Dumas, J. Adams, and S.K. Chai (2010). High-wire act. *IEEE Power & Energy Magazine*, **8**(1), pp. 37-45.

Huttenrauch, J., and G. Muller-Syring (2006). *Assessment of repair And Rehabilitation Technologies relating to the Transport of hythane (hydrogen-methane-mixture)*. Naturalhy Project Report No. R0016-WP4-P-0, DBI Gas- und Umwelttechnik GmbH, Leipzig, Germany, 63 pp.

IANGV (2009). *International Association for Natural Gas Vehicles*. International Association for Natural Gas Vehicles, Auckland, New Zealand. See: http://www.iangv.org/tools-resources/statistics.html

IBGE (2008). *Perfil dos Municípios Brasileiros 2008*. Instituto Brasileiro de Geografia e Estatística, Rio de Janeiro, Brazil (ISBN 978-85-240-4061-0 (CD-ROM)). Available at: www.ibge.gov.br/home/estatistica/economia/perfilmunic/2008/munic2008.pdf.

ICA (2009). *A um passo da cidadania: Projeto Piloto para Eletrificação de Favelas e Redução de Perdas*. International Copper Association Ltd., São Paulo, Brazil, 94 pp. Available at: www.procobre.org/archivos/pdf/One_Step_Citizenship.pdf.

IEA (2006). Energy for cooking in developing countries. Chapter 15 in: *World Energy Outlook 2006*. International Energy Agency, Paris, France.

IEA (2007a). *Bioenergy Project Development and Biomass Supply – Good Practice Guidelines*. International Energy Agency, Paris, France, 66 pp.

IEA (2007b). *Energy Statistics for Brazil (2007)*. International Energy Agency, Paris, France. Available at: www.iea.org/stats.

IEA (2007c). *Mind the Gap - Quantifying Principal-Agent Problems in Energy Efficiency*. International Energy Agency, Paris, France, 160 pp. (ISBN 978-92-64-03884-4).

IEA (2007d). *Renewables for Heating And Cooling – Untapped Potential*. International Energy Agency, Paris, France, 209 pp.

IEA (2008a). *Combined Heat and Power: Evaluating the Benefits of Greater Global Investment*. International Energy Agency, Paris, France.

IEA (2008b). *Deploying Renewables: Principles for Effective Policies*. International Energy Agency, Paris, France, 250 pp. (ISBN: 978-92-64-04220-9).

IEA (2008c). *Energy Technology Perspectives 2008. Scenarios and Strategies to 2050*. International Energy Agency, Paris, France, 646 pp.

IEA (2008d). *World Energy Outlook 2008*. International Energy Agency, Paris, France, 578 pp.

IEA (2009a). *Cities, Towns and Renewable Energy – Yes In My Front Yard*. International Energy Agency, Paris, France, 194 pp. (ISBN 978-92-64-07687-7).

IEA (2009b). *Cogeneration and District Energy: Sustainable Energy Technologies for Today and Tomorrow*. International Energy Agency, Paris, France, 24 pp.

IEA (2009c). *Transport, Energy, and CO_2 – Moving Towards Sustainability*. International Energy Agency, Paris, France, 400 pp. (ISBN 978-92-64-07316-2).

IEA (2009d). *World Energy Outlook 2009*. International Energy Agency, Paris, France, 696 pp.

IEA (2010a). *Energy Technology Perspectives 2010. Scenarios and Strategies to 2050*. International Energy Agency, Paris, France, 708 pp.

IEA (2010b). *Key World Energy Statistics 2010*. International Energy Agency, Paris, France, 82 pp.

IEA (2010c). *Technology Roadmap: Concentrating Solar Power*. International Energy Agency, Paris, France, 52 pp.

IEA (2010d). *World Energy Outlook 2010*. International Energy Agency, Paris, France, 736 pp.

IEA Bioenergy (2010a). *100 % Biogas for Urban Transport in Linköping, Sweden* IEA Bioenergy Task 37, International Energy Agency, Svensk Biogas AB, Linköping, Sweden. Available at: www.iea-biogas.net/_download/linkoping_final.pdf.

IEA Bioenergy (2010b). *Country report Sweden* IEA Bioenergy Task 37, *Energy from Biogas*. IEA Bioenergy, International Energy Agency, Paris, France. See: http://www.iea-biogas.net/_download/publications/country-reports/april2011/Sweden_Country_Report.pdf.

IEA-HIA (2006). *Prospects for Hydrogen from Biomass*. IEA Hydrogen Implementing Agreement Task 16, Subtask B, Final Report, International Energy Agency, Paris, France, 69 pp. Available at: ieahia.org/pdfs/finalreports/Task16BFinal.pdf.

IEA-SHC (2010). *Solar Heating and Cooling Implementing Agreement*. International Energy Agency, Paris, France. See: http://www.iea-shc.org/publications/downloads/Solar_Heat_Worldwide-2010.pdf.

Igarashi, Y., K. Mochidzuki, M. Takayama, and A. Sakoda (2009). Development of local fuel system for sustainable regional biomass utilization: Climate change: Global risks, challenges and decisions. In: *IOP Conference Series: Earth and Environmental Science*, **6**(2009). IOP Publishing, Copenhagen, Denmark, doi:10.1088/1755-1315/6/1/001001.

Iglesias, G., and R. Carballo (2010). Wave resource in El Hierro – an island towards energy self-sufficiency. *Renewable Energy*, **36**(2), pp. 689-698.

Insula (2010). *El Hierro 100% RES*. International Scientific Council for Island Development, Paris, France. Available at: www.insula.org/index.php?option=com_wrapper&Itemid=22.

IPCC (2007). *Climate Change 2007: Mitigation of Climate Change. Contribution of Working Group III to the Fourth Assessment Report of the Intergovernmental Panel on Climate Change*. B. Metz, O.R. Davidson, P.R. Bosch, R. Dave, and L.A. Meyer (eds.), Cambridge University Press, 851 pp.

Ipsakis, D., S. Voutetakis, P. Seferlis, F. Stergiopoulos, and C. Elmasides (2009). Power management strategies for a stand-alone power system using renewable energy sources and hydrogen storage. *International Journal of Hydrogen Energy*, **34**(16), pp. 7081-7095.

Jayamaha, N.P. (2003). *Distributed Generation on Rural Electricity Networks – A Lines Company Perspective*. Massey University, Massey, New Zealand, 243 pp.

Jenkins, N., G. Strbac, and J. Ekanayake (2010). *Distributed Generation*. Institution of Engineering and Technology, Stevenage, UK, 272 pp.

Jiang, B., and Y. Hai (2010). China's heat pumps technology review (in Chinese). *China Construction*, **2010**(7), pp. 68-69.

Johansson, T.B., and J. Goldemberg (eds.) (2005). *World Energy Assessment Overview: 2004 Update*. United Nations Development Programme, United Nations Department of Economic and Social Affairs, and World Energy Council, New York, NY, USA, 85 pp. Available at: www.undp.org/energy/weaover2004.htm.

Johansson, T.B., H. Kelly, A. Reddy, R. Williams, and L. Burnham (1993). *Renewable Energy: Sources for Fuels and Electricity*. Island Press, Washington, DC, USA, 62 pp.

Johnston, R.A. (2007). *Review of U.S. and European Regional Modeling Studies of Policies Intended to Reduce Highway Congestion, Fuel Use, and Emissions*. Victoria Transport Policy Institute, Victoria, Canada, 9 pp.

Jónsson, T., P. Pinson, and H. Madsen (2010). On the market impact of wind energy forecasts. *Energy Economics*, **32**, pp. 313-320.

Jupe, S.C.E., and P.C. Taylor (2009). Distributed generation output control for network power flow management. *IET Renewable Power Generation*, **3**(4), pp. 371-386.

Jupe, S.C.E., P.C. Taylor, and A. Michiorri (2010). Coordinated output control of multiple distributed generation schemes. *IET Renewable Power Generation*, **4**(3), pp. 283-297.

Kåberger, T., and H. Kusar (2008). BioDME – beslut enskilt project, Volvo Powertrain Report, DNR 2008-000766, Project 31243-1 Swedish Energy Agency, Eskilstuna, Sweden. 10 pp.

Kaldellis, J.K., D. Zafirakis, and K. Kavadias (2009). Techno-economic comparison of energy storage systems for island autonomous electrical networks. *Renewable and Sustainable Energy Reviews*, **13**(2), pp. 378-392.

Kalhammer, F.A., B.M. Kopf, D.H. Swan, V.P. Roan, and M.P. Walsh (2007). *Status and Prospects for Zero Emission Vehicle Technology*. Report ARB Independent Expert Panel, State of California Air Resources Board, Sacramento, CA, USA, 207 pp.

Kamakaté, F. and D. Gordon (2009). *Managing Motorcycles: Opportunities to Reduce Pollution and Fuel Use from Two- and Three-Wheeled Vehicles*. International Council on Clean Transportation, Washington, DC, USA. Available at: www.theicct.org/pubs/managing_motorcycles.pdf.

Kang, Y., and J. Zhang (2008). Study on current status, barriers and recommendations for China's CHP/DHC market development. *Energy of China (In Chinese)*, **30**(10), pp. 8-13.

Karapidakis, E. (2007). Transient analysis of Crete's power system with increased wind power penetration. In: *International Conference on Power Engineering, Energy and Electrical Drives (POWERENG 2007)*, Setubal, Portugal, 12-14 April 2007, pp. 18-22.

Katsaprakakis, D.A., N. Papadakis, D.G. Christakis, and A. Zervos (2007). On the wind power rejection in the islands of Crete and Rhodes. *Wind Energy*, **10**(5), pp. 415-434.

Kawasaki, N., T. Oozeki, K. Otani, and K. Kurokawa (2006). An evaluation method of the fluctuation characteristics of photovoltaic systems by using frequency analysis. *Solar Energy Materials and Solar Cells*, **90**(18-19), pp. 3356-3363.

Keane, A., and M. O'Malley (2005). Optimal allocation of embedded generation on distribution networks. *IEEE Transactions on Power Systems*, **20**(3), pp. 1640-1646.

Keane, A., M. Milligan, C.J. Dent, B. Hasche, C. D'Annunzio, K. Dragoon, H. Holttinen, N. Samaan, L. Söder, and M. O'Malley (2011a). Capacity value of wind power. *IEEE Transactions on Power Systems*, **26**(2), pp. 564-572.

Keane, A., L.F. Ochoa, E. Vittal, C.J. Dent, and G.P. Harrison (2011b). Enhanced utilization of voltage control resources with distributed generation. *IEEE Transactions on Power Systems*, **26**(1), pp. 252-260.

Kempton, W., and J. Tomic (2005). Vehicle-to-grid power implementation: From stabilizing the grid to supporting large-scale renewable energy. *Journal of Power Sources*, **144**(1), pp. 280-294.

Khan, J., G. Bhuyan, and A. Moshref (2009). *Potential Opportunities and Differences Associated with Integration of Ocean Wave and Marine Current Energy Plants in Comparison to Wind Energy*. Report prepared by Powertech Labs for the IEA-OES Annex III, Powertech Labs Inc., Surrey, BC, Canada, 64 pp.

Khennas, S., C. Sepp, and S. Hunt (2009). *Review and Appraisal of Potential Transformative Rural Energy Interventions in the Congo Basin*. Final report by Practical Action Consulting and Eco Consulting, Department for International Development, London, UK.

KinderMorgan (2010). *Pipeline Products: Central Florida Pipeline Co.* Kinder Morgan, Dallas, Texas. Available at: www.kindermorgan.com/business/products_pipelines/central_florida.cfm

Kirby, B.J. (2007). Load response fundamentally matches power system reliability requirements. In: *Power Engineering Society General Meeting, IEEE*, Tampa, FL, USA, 24-28 June 2007, pp. 1-6.

Kirby, B., and M. Milligan (2010). Utilizing load response for wind and solar integration and power system reliability. In: *WindPower 2010*, Dallas, TX, 23-26 May 2010. Available at: www.nrel.gov/docs/fy10osti/48247.pdf.

Kiviluoma, J., and P. Meibom (2010). Influence of wind power, plug-in electric vehicles, and heat storages on power system investments. *Energy*, **35**(3), pp. 1244-1255.

Kiviluoma, J., and P. Meibom (2011). Methodology for modelling plug-in electric vehicles in the power system and cost estimates for a system with either smart or dumb electric vehicles. *Energy*, **36**(3), pp. 1758-1767.

Klobasa, M. (2010). Analysis of demand response and wind integration in Germany's electricity market. *IET Renewable Power Generation*, **4**(1), pp. 55-63.

Kobayashi, H., and I. Kurihara (2009). Research and development of grid integration of distributed generation in Japan. In: *IEEE Power and Energy Society General Meeting*, Calgary, AB, Canada, 26-30 July 2009.

Kondoh, J. (2010). Autonomous frequency regulation by controllable loads to increase acceptable wind power generation. *Wind Energy*, **13**(6), pp. 529-541.

Kromer, M.A., and J.B. Heywood (2007). *Electric Powertrains: Opportunities and Challenges in the U.S. Light-Duty Vehicle Fleet*. Report LFEE 2007-02 RP, Laboratory for Energy and the Environment, Massachusetts Institute of Technology, Cambridge, MA, USA, 153 pp. Available at: web.mit.edu/sloan-auto-lab/research/beforeh2/files/kromer_electric_powertrains.pdf.

Krüger, D., A. Anthrakidis, S. Fischer, A. Lokurlu, M. Walder, R. Croy, and V. Quaschning (2009). Experiences with solar steam supply for an industrial steam network in the P3 Project. In: *15th International SolarPACES Symposium*, Berlin, Germany, 18 September 2009, pp. 15.

Krüger, D.R., K. Hennecke, and S. Dathe (2008). *Parabolrinnen für prozesswärme, Projekte und Entwicklungen 11. Kölner Sonnenkolloqium*. German Aerospace Center (DLR) Solar Research, Köln, Germany. Available at: www.dlr.de/sf/en/Portaldata/73/Resources/dokumente/Soko/Soko2008/Poster/5_Parabolrinnen_fuer_Prozesswaerme.pdf.

Kundur, P., (2007). Power system stability. Chapter 7 in: *Power System Stability and Control*. L. Grigsby (ed.), CRC Press, Boca Raton, FL, USA.

Kurani, K.S., J. Axsen, N. Caperello, J. Davies, and T. Stillwater (2009). *Learning from Consumers: Plug-In Hybrid Electric Vehicle (PHEV) Demonstration and Consumer Education, Outreach, Market Research Program*. Research Report UCD-ITS-RR-09-21, Institute of Transportation Studies, University of California, Davis, CA, USA. Available at: pubs.its.ucdavis.edu/publication_detail.php?id=1310.

Lannoye, E., M. Milligan, J. Adams, A. Tuohy, H. Chandler, D. Flynn, and M. O'Malley (2010). Integration of variable generation: capacity value and evaluation of flexibility. In: *IEEE Power and Energy Society General Meeting*, Minneapolis, MN, 25-27 July 2010.

Lantz, M., M. Svensson, L. Björnsson, and P. Börjesson (2007). The prospects for an expansion of biogas systems in Sweden: incentives, barriers and potentials. *Energy Policy*, **35**(3), pp. 1830-1843.

Larsen, S.F., C. Filippin, A. Beascochea, and G. Lesino (2008). An experience on integrating monitoring and simulation tools in the design of energy-saving buildings. *Energy and Buildings*, **40**(6), pp. 987-997.

Larsen, X.G., and J. Mann (2009). Extreme winds from the NCEP/NCAR reanalysis data. *Wind Energy*, **12**(6), pp. 556-573.

Larson, E.D., R.H. Williams, M. Regis, and L.V. Leal (2001). A review of biomass integrated-gasifier/gas turbine combined cycle technology and its application in sugarcane industries, with an analysis for Cuba. *Energy for Sustainable Development*, **5**(1), pp. 54-76.

Lazarewicz, M.L., and T.M. Ryan (2010). Integration of flywheel-based energy storage for frequency regulation in deregulated markets. In: *IEEE Power and Energy Society General Meeting*, Minneapolis, MN, 25-27 July 2010, pp. 1-6, 25-29.

Lehmann, J. (2007). A handful of carbon. *Nature*, **447**, pp. 143-144.

Leighty, W.C., J. Holloway, R. Merer, B. Someday, C. Marchi, G. Keith, and D.E. White (2006). Compressorless hydrogen transmission pipelines deliver large-scale stranded renewable energy at competitive cost. In: *23rd World Gas Conference*, Amsterdam, The Netherlands, 5-9 June 2006. Available at: www.leightyfoundation.org/files/WGC-Amsterdam/WGC-Abstract310.pdf.

Leilei, D., H. Liu, and S. Riffat (2009). Development of small-scale and micro-scale biomass-fuelled CHP systems – a literature review. *Applied Thermal Engineering*, **29**(11-12), pp. 2119-2126.

Lemmini, F., and A. Errougani (2007). Experimentation of a solar adsorption refrigerator in Morocco. *Renewable Energy*, **32**(15), pp. 2629-2641.

Levine, M., D. Ürge-Vorsatz, K. Blok, L. Geng, D. Harvey, S. Lang, G. Levermore, A. Mongameli Mehlwana, S. Mirasgedis, A. Novikova, J. Rilling, and H. Yoshion (2007). Residential and commercial buildings. In: Climate Change 2007: Mitigation of Climate Change. Contribution of Working Group III to the Fourth Assessment Report of the Intergovernmental Panel on Climate Change, B. Metz, O.R. Davidson, P.R. Bosch, R. Dave and L.A. Meyer (eds.), Cambridge University Press, pp. 387-446.

Li, J. (2009). From strong to smart: the Chinese Smart Grid and its relation with the globe. *Asia Energy Platform News*, **September 2009**, pp. 10.

Li, Z.S., G.Q. Zhang, D.M. Li, J. Zhou, L.J. Li, and L.X. Li (2007). Application and development of solar energy in building industry and its prospects in China. *Energy Policy*, **35**(8), pp. 4121-4127.

Liao, C.P., E. Jochem, Y. Zhang, and N.R. Farid (2010). Wind power development and policies in China. *Renewable Energy*, **35**(9), pp. 1879-1886.

Lighting Africa (2010). *Lighting Africa – Catalyzing Markets for Modern Lighting*. International Finance Corporation and World Bank Programme. See: www.lightingafrica.org/.

Liu, D.H., and S. Riffat (2009). Development of small-scale and micro-scale biomass-fuelled CHP systems – A literature review. *Applied Thermal Engineering*, **29**(11-12), pp. 2119-2126.

Liu, E., and J. Bebic (2008). *Distribution System Voltage Performance Analysis for High-Penetration Photovoltaics*. Subcontract Report NREL/SR-581-42298, National Renewable Energy Laboratory, Golden, CO, USA.

Liu, Y.Q., and A. Kokko (2010). Wind power in China: Policy and development challenges. *Energy Policy*, **38**(10), pp. 5520-5529.

Lobell, D.B., and C.B. Field (2007). Global scale climate – crop yield relationships and the impacts of recent warming. *Environmental Research Letters*, **2**(1), pp. 1-7.

Loisel, R., A. Mercier, C. Gatzen, N. Elms, and H. Petric (2010). Valuation framework for large scale electricity storage in a case with wind curtailment. *Energy Policy*, **38**(11), pp. 7323-7337.

Lone, S.A., and M.D. Mufti (2008). Modelling and simulation of a stand-alone hybrid power generation system incorporating redox flow battery storage system. *International Journal of Modelling and Simulation*, **28**(3), pp. 337-346.

Longhetto, A., G. Elisei, and C. Giraud (1989). Effect of correlations in time and spatial extent on performance of very large solar conversion systems. *Solar Energy*, **43**(2), pp. 77-84.

Lorenz, E., J. Hurka, D. Heinemann, and H.G. Beyer (2009). Irradiance forecasting for the power prediction of grid-connected photovoltaic systems. *IEEE Journal of Selected Topics in Applied Earth Observations and Remote Sensing*, **2**(1), pp. 2-10.

Lorenz, E., T. Scheidsteger, J. Hurka, D. Heinemann, and C. Kurz (2010). Regional PV power prediction for improved grid integration. *Progress in Photovoltaics: Research and Applications*, doi:10.1002/pip.1033.

Lund, H. (2006). Large-scale integration of optimal combinations of PV, wind and wave power into the electricity supply. *Renewable Energy*, **31**(4), pp. 503-515.

Lund, H., and E. Munster (2003). Management of surplus electricity-production from a fluctuating renewable-energy source. *Applied Energy*, **76**(1-3), pp. 65-74.

Lund, H., and W. Kempton (2008). Integration of renewable energy into the transport and electricity sectors through V2G. *Energy Policy*, **36**(9), pp. 3578-3587.

Lund, H., and G. Salgi (2009). The role of compressed air energy storage (CAES) in future sustainable energy systems. *Energy Conversion and Management*, **50**(5), pp. 1172-1179.

Lund, J.W. (2005). Direct application of geothermal energy: 2005 worldwide review. *Geothermics*, **34**, pp. 691-727.

Lund, J.W., D.H. Freeston, and T.L. Boyd (2010). Direct utilisation of geothermal energy 2010 worldwide review. In: *Proceedings World Geothermal Congress 2010*, Bali, Indonesia, 25-30 April 2010. Available at: www.geothermal-energy.org/pdf/IGAstandard/WGC/2010/0007.pdf.

Lundsager, P., and I. Baring-Gould (2005). Isolated systems with wind power. In: *Wind Power in Power Systems*. T. Ackermann (ed.), John Wiley & Sons Ltd., London, UK, pp. 299-329.

Macedo, I.C., J.E.A. Seabra, and J.E.A.R. Silva (2008). Greenhouse gas emissions in the production and use of ethanol from sugarcane in Brazil: the 2005/2006 averages and a prediction for 2020. *Biomass & Bioenergy*, **32**, pp. 582-595.

MacGill, I. (2010). Electricity market design for facilitating the integration of wind energy: Experience and prospects with the Australian National Electricity Market. *Energy Policy*, **38**(7), pp. 3180-3191.

MacLean, H.L., and L.B. Lave (2003). Evaluating automobile fuel/propulsion technologies. *Progress in Energy and Combustion Science*, **29**, pp. 1-69.

Madhlopa, A., and G. Ngwalo (2007). Solar dryer with thermal storage and biomass-backup heater. *Solar Energy*, **81**(4), pp. 449-462.

Mandle, K.T. (1988). Dinorwig pumped-storage scheme. *Power Engineering Journal*, **2**(5), pp. 259-262.

Mangold, D., and T. Schmitt (2006). The new central solar heating plants with seasonal storage in Germany. In: *EuroSun 2006*, ISES (ed.), International Solar Energy Society, Glasgow, Scotland, 27-30 June 2006, pp. 6.

Mangold, D., M. Riegger, and T. Schmitt (2007). *Solare Nahwärme und Langzeit-Wärmespeicher*. Report 0329607L, Forschungsbericht zum BMU-Vorhaben, Stuttgart, Germany, 66 pp.

Marcos, J., L. Marroyo, E. Lorenzo, D. Alvira, and E. Izco (2011). From irradiance to output power fluctuations: the PV plant as a low pass filter. *Progress in Photovoltaics: Research and Applications*, doi:10.1002/pip.1063.

Margonelli, L. (2009). Clean energy's dirty little secret. *The Atlantic Magazine*, **May 2009**. Available at: www.theatlantic.com/magazine/archive/2009/05/clean-energy-apos-s-dirty-little-secret/7377.

Martinez, E., F. Sanza, J. Blanco, F. Daroca, and E. Jimenez (2008). Economic analysis of reactive power compensation in a wind farm: Influence of Spanish energy policy. *Renewable Energy*, **33**(8), pp. 1880-1891.

Masters, C.L. (2002). Voltage rise - the big issue when connecting embedded generation to long 11 kV overhead lines. *Power Engineering Journal*, **16**(1), pp. 5-12.

Mathioulakis, E., V. Belessiotis, and E. Delyannis (2007). Desalination by using alternative energy: Review and state-of-the-art. *Desalination*, **203**(1-3), pp. 346-365.

Matic, D. (2006). *Global Opportunities for Natural Gas as a Transportation Fuel for Today and Tomorrow*. Report from Working Committee 5 at the 23rd Word Gas Conference, Study Group 5.3 on Natural Gas for Vehicles (NGV), International Gas Union, Oslo, Norway, 144 pp.

McCarthy, R.W., J.M. Ogden, and D. Sperling (2007). Assessing reliability in energy supply systems. *Energy Policy*, **35**(4), pp. 2151-2162.

McDonough, C., and R. Kraus (2007). Does dynamic pricing make sense for mass market customers? *The Electricity Journal*, **20**(7), pp. 26-37.

McKormick, K., and T. Kåberger (2005). Exploring a pioneering bioenergy system: The case of Enköping in Sweden. *Journal of Cleaner Production*, **13**, pp. 1003-1014.

McNutt, P., J. Hambrick, M. Keesee, and D. Brown (2009). *Impact of SolarSmart Subdivisions on SMUD's Distribuion System*. Technical report NREL/TP-550-46093, National Renewable Energy Laboratory, Golden, CO, USA, 41 pp.

Meah, K., and S. Ula (2007). Comparative evaluation of HVDC and HVAC transmission systems. In: *IEEE Power Engineering Society General Meeting*, Tampa, FL, USA, 24-28 June 2007, pp. 1-5, 24-28.

Meah, K., S. Ula, and S. Barrett (2008). Solar photovoltaic water pumping – opportunities and challenges. *Renewable and Sustainable Energy Reviews*, **12**(4), pp. 1162 -1175.

Meenual, T. (2010). Roadmapping the Provincial Electricity Authority (PEA) smart grids. In: *2010 Proceedings of the International Conference on Energy and Sustainable Development (ESD): Issues and Strategies*, Chiang Mai, Thailand, 2-4 June 2010, 1-6pp. (ISBN: 978-1-4244-8563-5).

Meibom, P., R. Barth, B. Hasche, H. Brand, C. Weber, and M.J. O'Malley (2011). Stochastic optimisation model to study the operational impacts of high wind penetrations in Ireland. *IEEE Transactions on Power Systems*, doi:10.1109/TPWRS.2010.2070848.

METI (2005). *Energy Vision 2100, Strategic Technology Roadmap (Energy Sector)*. Ministry of Economy, Trade, and Industry, Tokyo, Japan, 42 pp. Available at: www.iae.or.jp/2100/main.pdf.

Miller, N., K. Clark, and M. Shao (2010). Impact of frequency responsive wind plant controls on grid performance. In: *Proceedings of the 9th International Workshop on Large Scale Integration of Wind Power and on Transmission Networks for Offshore Wind Farms*, Quebec City, Canada, 18-19 October 2010, pp. 371-382.

Milligan, M., B. Kirby, R. Gramlich, and M. Goggin (2009). *Impact of Electric Industry Structure on High Wind Penetration Potential*. Technical Report, National Renewable Energy Laboratory, Golden, CO, USA.

Mills, A., and R. Wiser (2010). *Implications of Wide-Area Geographic Diversity for Short-Term Variability of Solar Power*. LBNL-3884E, Environmental Energy Technologies Division, Lawrence Berkeley National Laboratory, Berkeley, CA, USA, 48 pp.

Mills, A., M. Ahlstrom, M. Brower, A. Ellis, R. George, T. Hoff, B. Kroposki, C. Lenox, N. Miller, J. Stein, and Y. Wan (2009a). *Understanding Variability and Uncertainty of Photovoltaics for Integration with the Electric Power System*. Lawrence Berkeley National Laboratory, Berkeley, CA, USA.

Mills, A., R. Wiser, M. Milligan, and M. O'Malley (2009b). Comment on "Air Emissions Due to Wind and Solar Power". *Environmental Science & Technology*, **43**(15), pp. 6106-6107.

Mills, A., A.Phadke, and R. Wiser (2011). Exploration of resource and transmission expansion decisions in the Western Renewable Energy Zone initiative. *Energy Policy*, **39**(3), pp. 1732-1745.

Ministry for Agriculture Livestock and Supply (2010). *Sugar Cane and Ethanol Distilleries*. Ministry for Agriculture Livestock and Supply, Brasilia, Brazil. (In Portuguese) Available at: http://www.agricultura.gov.br/arq_editor/file/1984_posicao_04_2010.pdf

Mintzer, I. (2009). Look before you leap: Exploring the implications of advanced vehicles for import dependence and passenger safety. In: *Plug-in Electric Vehicles: What Role for Washington?* D.B. Sandalow (ed.), Brookings Institution Press, Washington, DC, USA, pp 107-126. Available at: www.potomacenergyfund.com/files/Potomac%20Energy%20Fund%20-%20Look%20Before%20You%20Leap.pdf.

Miwa, T., and H. Okuda (2010). CO_2 ultimate reduction in steelmaking process by innovative technology for Cool Earth 50. *Journal of the Japan Institute of Energy*, **89**, pp. 28-35.

Modi, V., S. McDade, D. Lallement, and J. Saghir (2005). *Energy Services for the Millennium Development Goals*. Energy Sector Management Assistance Programme, United Nations Development Programme, UN Millennium Project, and World Bank, New York, NY, USA.

Mohammadi, A., A. Tabatabaeefar, S. Shahin, S. Rafiee, and A. Keyhani (2008). Energy use and economical analysis of potato production in Iran a case study: Ardabil province. *Energy Conversion and Management*, **49**(12), pp. 3566-3570.

Mohitpour, M., and A. Murray (2000). *Pipeline Design and Construction: A Practical Approach*. ASME Press, New York, NY, USA.

MoP (2006). Rural electrification policy, Resolution 44/26/05-RE. *Gazette of India, Ministry of Power*, **Vol. II** (23 August), pp. 17.

Morales, A., X. Robe, M. Sala, P. Prats, C. Aguerri, and E. Torres (2008). Advanced grid requirements for the integration of wind farms into the Spanish transmission system. *IET Renewable Power Generation*, **2**(1), pp. 47-59.

Morozumi, S., H. Nakama, and N. Inoue (2008). Demonstration projects for grid-connection issues in Japan. *e & i Elektrotechnik und Informationstechnik*, **125**(12), pp. 426-431.

Moura, P.S., and A.T. de Almeida (2010). The role of demand-side management in the grid integration of wind power. *Applied Energy*, **87**(8), pp. 2581-2588.

Mpagalile, J.J., M.A. Hanna, and R. Weber (2005). Design and testing of a solar photovoltaic operated multi-seeds oil press. *Renewable Energy*, **31**(12), pp. 1855-1866.

MTEP (2008). *The Midwest ISO Transmission Expansion Plan; Growing the grid across the Heartland*. Midwest ISO, Carmel, IN, USA, 402 pp.

Mullane, A., and M. O'Malley (2005). The inertial response of induction-machine-based wind turbines. *IEEE Transactions on Power Systems*, **20**(3), pp. 1496-1503.

Müller-Langer, F., F. Scholwin, and K Oehmichen (2009). *Biomethane for transport: a worldwide overview*. Presentation at IEA Bioenergy Task 39 Subtask Policy and Implementation Workshop: From today's to tomorrow's biofuels – From the Biofuels Directive to bio based transport systems in 2020; Dresden, Germany; June 3-5, 2009; adapted from: *Possible European biogas supply strategies*; Institute for Energy and Environment; Leipzig; 2007.

Munksgaard, J., and P.E. Morthorst (2008). Wind power in the Danish liberalised power market-Policy measures, price impact and investor incentives. *Energy Policy*, **36**(10), pp. 3940-3947.

Murata, A., H. Yamaguchi, and K. Otani (2009). A method of estimating the output fluctuation of many photovoltaic power generation systems dispersed in a wide area. *Electrical Engineering in Japan*, **166**(4), pp. 9-19.

Murray, P.E. (2005). *Designing Sustainable Distributed Generation Systems for Rural Communities*. Massey University, Palmerston North, New Zealand.

NAS (2009). *Liquid Transportation Fuels from Coal and Biomass: Technological Status, Costs, and Environmental Impacts*. National Academy of Sciences, National Academies Press, Washington, DC, USA, 388 pp. (ISBN-13: 978-0-309-13712-6).

National Bureau of Statistics China (2010). *City District Heating (1990-1999, 2004-2008)*. (In Chinese). Available at: www.stats.gov.cn.

NationalGrid (2009). *The Potential for Renewable Gas in the UK*. Paper, National Grid, Media Relations, UK.

NATURALHY (2009). *Strategic justification of the NATURALHY project*. Project Report, NATURALHY, Groningen, The Netherlands,19 pp. Available at: www.naturalhy.net/docs/Strategic_justification_NATURALHY.pdf.

NBB (2010). *Automakers' and Engine Manufacturers' Positions of Support for Biodiesel Blends*. National Biodiesel Board, Jefferson City, MO, USA. Available at: www.biodiesel.org/resources/oems/default.aspx.

NCEP (2007). *Task Force on Biofuel Infrastructure*. National Commission on Energy Policy, Washington, DC, USA. http://ourenergypolicy.org/docs/2/biofuels-task-force.pdf

Nelson, P.A., D.J. Santini, and J. Barnes (2009). Factors determining the manufacturing costs of lithium- ion batteries for PHEVs. In: *EVS-24 Conference*. Stavanger, Norway, 13-16 May 2009, pp. 12. Available at: www.cars21.com/files/papers/Nelson-Santini-Barnes-paper.pdf.

Nema, P., R.K. Nema, and S. Rangnekar (2009). A current and future state of art development of hybrid energy system using wind and PV-solar: A review. *Renewable and Sustainable Energy Reviews*, **13**(8), pp. 2096-2103.

NERC (2009). *Special Report: Accommodating High Levels of Variable Generation*. North American Reliability Corporation, Princeton, NJ, USA, 95 pp.

NERC (2010a). *NERC IVGTF Task 2.1 Report Variable Generation Power Forecasting for Operations*. North American Electric Reliability Corporation, Princeton, NJ, USA, 35 pp. Available at: www.nerc.com/files/Varialbe%20Generationn%20Power%20Forecasting%20for%20Operations.pdf.

NERC (2010b). *Special Report: Flexibility Requirements and Potential Metrics for Variable Generations: Implication for System Planning Studies*. North American Electric Reliability Corporation, Princeton, NJ, USA, 63 pp.

NERC (2010c) *Special Report: Potential Reliability Impacts of Emerging Flexible Resources*. North American Electric Reliability Corporation, Princeton, NJ, USA, 57 pp.

NETL (2008). *NETL Modern Grid Strategy – Powering Our 21st-Century Economy – Advanced Metering Infrastructure*. White Paper V1.0, National Energy Technology Laboratory, U.S. Department of Energy, Pittsburgh, PA, USA, 32 pp.

New Holland Information Center (2010). *Biodiesel Support*. New Holland Information Center, Racine, WI, USA. Available at: agriculture.newholland.com/us/en/information-center/Biodiesel-Support/Pages/default.aspx.

Newbery, D.M. (2005). Electricity liberalisation in Britain: the quest for a satisfactory wholesale market design. *Energy Journal, Special Issue on European Electricity Liberalisation*, **26**(Special I), pp. 43-70.

Newbery, D.M. (2010). Market design for a large share of wind power. *Energy Policy*, **38**(7), pp. 3131-3134.

Nicholas, M., and J. Ogden (2010). *An Analysis of Near-Term Hydrogen Vehicle Rollout Scenarios for Southern California*. Report UCD-ITS-RR-10-03, University of California, Institute of Transportation Studies, Davis, CA, USA.

NISRA (2009). *Northern Ireland Statistic & Research Agency. Population and Migration Estimates Northern Ireland (2009) – Statistical Report*. Northern Ireland Statistics & Research Agency, National Statistics, Belfast, Ireland, 19 pp.

NIST (2007). *White paper on Internationally Compatible Biofuel Standards (Tripartite task force: Brazil, European Union and United States of America)*. Report, National Institute of Standards and Technology, Gaithersburg, MD, USA, 93 pp.

Nordel (1996). *Annual Report*. Nordel, Helsinki, Finland, 64 pp.

Nordel (2000). *Annual Report*. Nordel, Helsinki, Finland, 70 pp.

Nordel (2008). *Annual Report*. Nordel, Helsinki, Finland, 23 pp.

Notholt, A. (2008). *Fault Ride Through Capabilities of Inverter-Based Distributed Generation Connected to Low and Medium Voltage Distribution Networks*. Kassel University, Kassel, Germany.

NRC (2008). *Transitions to Alternative Transportation Technologies: A Focus on Hydrogen*. The National Research Council, National Academies Press, Washington, DC, USA, 142 pp. (ISBN-13: 978-0-309-12100-2).

NRC (2010). *Transitions to alternative transportation technologies: plug-in hybrid vehicles*, The National Research Council, National Academies Press, Washington, DC, USA. 57pp. (ISBN-13: 978-0-309-14580-4).

NREL (2009). H_2 *Production Cost vs. Process Composite Data Product (CDP #15)*. National Renewable Energy Laboratory, Golden, CO, USA. Available at: www.nrel.gov/hydrogen/docs/cdp/cdp_15.jpg.

NREL (2010). *Oahu Wind Integration and Transmission Study: Summary Report*. National Renewable Energy Laboratory, Golden, CO, USA, 28 pp.

NSCA (2006). *Biogas as transport fuel*. Final Report, National Society for Clean Air and Environmental Protection, Brighton, UK, 46 pp. (ISBN 978 0 903 47461 1).

Nyamdash, B., E. Denny, and M. O'Malley (2010). The viability of balancing wind generation with large scale energy storage. *Energy Policy*, **38**(11), pp. 7200-7208.

NZVCC (2008). *University Research Commercialisation – Paying Dividends for New Zealand*. Report, New Zealand Vice-Chancellors' Committee, Wellington, New Zealand, 11 pp.

Ochoa, P., and A. van Ackere (2009). Policy changes and the dynamics of capacity expansion in the Swiss electricity market. *Energy Policy*, **37**(5), pp. 1983-1998.

Ogden, J.M., and C. Yang (2009). Build-up of a hydrogen infrastructure in the US. Chapter 15 in: *The Hydrogen Economy: Opportunities and Challenges*. M. Ball and M. Wietschel (eds.), Cambridge University Press, pp. 454-482.

Oleson, K.W., G.B. Bonan, and J. Feddema (2010). Effects of white roofs on urban temperature in a global climate model. *Geophysical Research Letters*, **37**, L03701, doi:10.1029/2009GL042194.

Oliver-Solà, J., X. Gabarrell, and J. Rieradevall (2009). Environmental impacts of the infrastructure for district heating in urban neighbourhoods. *Energy Policy*, **37**(11), pp. 4711-4719.

Paatero, J.V., and P.D. Lund (2007). Effects of large-scale photovoltaic power integration on electricity distribution networks. *Renewable Energy*, **32**(2), pp. 216-234.

Pacca, S., and J.R. Moreira (2009). Historical carbon budget of the Brazilian ethanol program. *Energy Policy*, **37**(11), pp. 4863-4873.

Paish, O. (2002). Small hydro power: technology and current status. *Renewable and Sustainable Energy Reviews*, **6**(6), pp. 537-556.

Papaefthymiou, G., and D. Kurowicka (2009). Using copulas for modeling stochastic dependence in power system uncertainty analysis. *IEEE Transactions on Power Systems*, **24**(1), pp. 40-49.

Papathanassiou, S.A., and N.G. Boulaxis (2006). Power limitations and energy yield evaluation for wind farms operating in island systems. *Renewable Energy*, **31**(4), pp. 457-479.

Pastowski, A., and T. Grube (2009). Scope and perspectives of industrial hydrogen production and infrastructure for fuel cell vehicles in North Rhine-Westphalia. *Energy Policy*, **38**(10), pp. 5382-5387.

PCGCC (2010). *State and Local Net Greenhouse Gas Emissions Reduction Programs*. Pew Center on Global Climate Change, Arlington, VA, USA. Available at: www.cleanairconstruction.org/content/legislative/AB1493%20-%20Greenhouse%20Gas%20Emissions%20Reduction.pdf.

Pearmine, R., Y.H. Song, and A. Chebbo (2007). Influence of wind turbine behaviour on the primary frequency control of the British transmission grid. *IET Renewable Power Generation*, **1**(2), pp. 142-150.

Pehnt, M., A. Paar, F. Merten, W. Irrek, and D. Schüwer (2009a). Intertwining renewable energy and energy efficiency: from distinctive policies to combined strategies. In: *ECEEE 2009 Summer Study*, ECEEE, La Colle sur Loup, Côte d'Azur, France, pp. 389-400.

Pehnt, M., A. Paar, P. Otter, F. Merten, T. Hanke, W. Irrek, D. Schüwer, N. Supersberger, and C. Zeiss (2009b). *Energiebalance - Optimale Systemlösungen für erneuerbare Energien und Energieeffizienz (Energy Balance – Optimum System Solutions for Renewable Energy and Energy Efficiency)*. Report FKZ 0327614, Institut für Energie-und Umweltforschung Heidelberg GmbH, Heidelberg, Germany, 440 pp.

Pelland, S., and I. Abboud (2008). Comparing photovoltaic capacity value metrics: A case study for the City of Toronto. *Progress in Photovoltaics*, **16**(8), pp. 715-724.

Perez, R., M. Taylor, T. Hoff, and J.P. Ross (2008). Reaching consensus in the definition of photovoltaics capacity credit in the USA: A practical application of satellite-derived solar resource data. *IEEE Journal of Selected Topics in Applied Earth Observations and Remote Sensing*, **1**(1), pp. 28-33.

Pérez-Díaz, J.I., and J.R. Wilhelmi (2010). Assessment of the economic impact of environmental constraints on short-term hydropower plant operation. *Energy Policy*, **38**(12), pp. 7960-7970.

Persson, M. (2003). *Utvärdering av uppgraderingstekniker för biogas (Study of Biogas Upgrading Techniques)*. Report SGC 142, ISSN 1102-7371, Swedish Gas Centre, Malmö, Sweden, 69 pp.

Persson, M., O. Jönsson, and A. Wellinger (2006). *Biogas Upgrading to Vehicle Fuel Standards and Grid Injection*. IEA Report, Swedish Gas Center (SGC), Malmö, Sweden, 34 pp.

Phair, J.W. (2006). Green chemistry for sustainable cement production and use. *Green Chemistry*, **8**, pp. 763-780.

Pihl, E. (2009). *Concentrating Solar Power*. Energy Committee of the Royal Swedish Academy of Sciences, Stockholm, Sweden, 32 pp.

Plotkin, S., and M. Singh (2009). *Multi-Path Transportation Futures Study: Vehicle Characterization and Scenario Analyses*. Report ANL/ESD/09-5, Argonne National Laboratory, Argonne, IL, USA, 310 pp. Available at: www.transportation.anl.gov/pdfs/TA/613.PDF.

Power-Gen (2009). *Cummins Approves B20 Biodiesel for 19- to 78-Litre High Horsepower Engines*. Power Engineering International, PennWell Publishing, Tulsa, OK, USA.

PROINFA (2010). *Programa de Incentivo às Fontes Alternativas de Energia Elétrica*, Minas e Energia, Brasília, Brazil. Available at: www.mme.gov.br/programas/proinfa.

Qadrdan, M., M. Chaudry, J.Z. Wu, N. Jenkins, and J. Ekanayake (2010). Impact of a large penetration of wind generation on the GB gas network. *Energy Policy*, **38**(10), pp. 5684-5695.

Radtke, J., C.J. Dent, and S.J. Couch (2010). Capacity value of large tidal barrages. *IEEE Transactions on Power Systems*, doi:10.1109/TPWRS.2010.2095433.

RAE (2007). Decision 85/2007, 25th April 2007: "Adoption of a methodology for determining the growth potential of RES plants in saturated networks in accordance with the provisions of Article 4 § 1 of the Licensing Rules for electricity Production from RES and CHP" (in Greek). Regulatory Authority for Energy, Greece, Official Gazette B 448/3.4.2007.

Rajvanshi, A.K., S.M. Patil, and B. Mendonca (2007). Low-concentration ethanol stove for rural areas in India. *Energy for Sustainable Development*, **11**(1), pp. 94-99.

Ramirez, C.A., and W. Worrell (2006). Feeding fossil fuels to the soil. An analysis of energy embedded and technological learning in the fertilizer industry. *Resources Conservation and Recycling*, **46**, pp. 75-93.

Reikard, G. (2009). Forecasting ocean wave energy: Tests of time-series models. *Ocean Engineering*, **36**(5), pp. 348-356.

REN (2008). *Plano de investimento e desenvolvimento da rede de transporte 2009-2014 (2019)*. Rede Eléctrica Nacional, Lisbon, Spain, 26 pp.

REN21 (2010). *Renewables 2010: Global Status Report*. Renewable Energy Policy Network for the 21st Century Secretariat, Paris, France, 80 pp.

Ren, T., and M.K. Patel (2009). Basic petrochemicals from natural gas, coal and biomass: Energy use and CO_2 emissions. *Resources, Conservation and Recycling*, **53**, pp. 513-528.

Reynolds, R. (2000). *The Current Fuel Ethanol Industry Transportation, Marketing, Distribution and Technical Considerations*. Report Oak Ridge National Laboratory Ethanol Project, Subcontract No. 4500010570, Downstream Alternatives Inc., Bremen, IN, USA, 263 pp.

Riis, T.R., G. Sandrock, E.F. Hagen, Ø. Ulleberg, and P.J.S. Vie (2006). *Hydrogen Production and Storage – R&D Priorities and Gaps*. Report (white paper), Hydrogen Implementing Agreement, International Energy Agency, Paris, France, 33 pp.

Rodrigues, M., A.P.C. Faaij, and A. Walter (2003). Techno-economic analysis of co-fired biomass integrated gasification/combined cycle systems with inclusion of economies of scale. *Energy*, **28**(12), pp. 1229-1258.

Rodriguez, G.D. (2010). A utility perspective of the role of energy storage in the smart grid. In: *IEEE Power and Energy Society General Meeting*, Minneapolis, MN, USA, 25-27 July 2010, pp. 1-2, 25-29.

Rodriguez, J.M., O. Alonso, M. Duvison, and T. Domingez (2008). The integration of renewable energy and the system operation: The Special Regime Control Centre (CECRE) in Spain. In: *IEEE Power and Energy Society General Meeting – Conversion and Delivery of Electrical Energy in the 21st Century*, Pittsburgh, PA, USA, 20-24 July 2008, pp. 1-6, 20-24.

Rossilo-Calle, F., S.V. Bajay, and H. Rothman (2000). *Industrial Uses of Biomass Energy: The Example of Brazil*. Taylor & Francis, London and New York.

Rousseau, A. And P. Sharer (2004). *Comparing Apples To Apples: Well-to-Wheel Analysis of Current ICE and Fuel Cell Vehicle Technologies*. Report No. 2004-01-1015, Argonne National Laboratory, Argonne, IL, USA.

Salter, S., J. Taylor, and N. Caldwell (2002). Power conversion mechanisms for wave energy. *Proceedings of the Institution of Mechanical Engineers, Part M: Journal of Engineering for the Maritime Environment*, **216**(1), pp. 1-27.

Samaras, C., and K. Meisterling (2008). Life cycle assessment of greenhouse gas emissions from plug-in hybrid vehicles: implications for policy. *Environmental Science & Technology*, **42**(9), pp. 3170-3176.

Sandsmark, M., and B. Tennbakk (2010). Ex post monitoring of market power in hydro dominated electricity markets. *Energy Policy*, **38**(3), pp. 1500-1509.

Scandiffio, M., and M.R.V. Leal (2008). Novo desenho logístico para exportação de etanol: uma visão de longo prazo. In: *7° Congresso Internacional sobre Geração Distribuída e Energia no Meio Rural*. L.B. Cortez (ed.), AGRENER, Fortaleza, Brazil, 23-26 September 2008. Available at: http://www.nipeunicamp.org.br/agrener/anais/2008/Artigos/72.pdf.

Scania (2010). *World's Largest Ethanol Bus Fleet Grows by 85 Scania Buses*. Press release, 21 June 2010, Scania, Södertälje, Sweden. Available at: www.scania.com/media/pressreleases/N10018EN.aspx.

Schäfer, N., T. Degner, J. Jäger, T.Teil, and A. Shustov (2010). Adaptive protection system for distribution networks with distributed energy resources. In: *10th International Conference on Developments in Power System Protection*, Manchester, UK, 29 March – 1 April 2010.

Schmidt, J., N. Helme, J. Lee, and M. Houdashelt (2008). Sector-based approach to the post-2012 climate change policy architecture. *Climate Policy*, **8**, pp. 494-515.

Schnepf, R. (2006). *Energy Use in Agriculture, Background And Issues - Updated*. Congressional Research Service report, Order Code RL32712, Library of Congress, Washington, DC, USA, 40 pp.

Schnitzer, H., C. Brunner, and G. Gwehenberger (2007). Minimizing greenhouse gas emissions through the application of solar thermal energy in industrial processes. *Journal of Cleaner Production*, **15**(13-14), pp. 1271-1286.

Schultz, R. (2007). HVDC options today [In my View]. *IEEE Power and Energy Magazine*, **5**(2), pp. 94-96.

Schweppe, F.C., M.C. Caramanis, R.D. Tabors, and R.E. Bohn (1988). *Spot Pricing of Electricity*. Kluwer Academic Publishers, Boston, MA, USA, 355 pp.

Searchinger, T., R., R.A. Heimlich, F. Houghton, A. Dong, J. Elobeid, S. Fabiosa, D. Tokgoz, and T.-H. Hayes (2008). Use of U.S. croplands for biofuels increases greenhouse gases through emissions from land use change. *Science*, **319**(5867), pp. 1238-1240.

Sebastian, M., J. Marti, and P. Lang (2008). Evolution of DSO control centre tool in order to maximize the value of aggregated distributed generation in smart grid. In: *SmartGrids for Distribution. IET-CIRED. CIRED Seminar*, Frankfurt, Germany, 23-24 June 2008, pp. 1-4, 23-24

Sensfuß, F., M. Ragwitz, and M. Genoese (2008). The merit-order effect: A detailed analysis of the price effect of renewable electricity generation on spot market prices in Germany. *Energy Policy*, **36**(8), pp. 3076-3084.

Shanmukharadhya, K.S., and K.G. Sudhakar (2007). Effect of fuel moisture on combustion in a bagasse-fired furnace. *Journal of Energy Resources Technology*, **129**(3), pp. 248-254.

Shenyang (2006). *Guidance on Promoting the Development and Application of Heat Pumps*. Document No.20 (in Chinese), Shenyang City Government, Shenyang, China.

Sherif, S.A., F. Barbir, and T.N. Veziroglu (2005). Towards a hydrogen economy. *The Electricity Journal*, **18**(6), pp. 62-76.

Shiu, H., M. Milligan, B. Kirby, and K. Jackson (2006). *California Renewables Portfolio Standard Renewable Generation Integration Cost Analysis: Multi-year analysis results and recommendations*. California Energy Commission, Sacramento, CA, USA, 134pp.

Sims, R. (2008). Reaching consensus on sustainable biofuels. *Renewable Energy World Magazine*, July 2008. Available at: www.renewableenergyworld.com/rea/news/article/2008/07/reaching-consensus-on-sustainable-biofuels-52692.

Sims, R., M. Taylor, J. Saddler, and W. Mabee (2008). *From 1st- to 2nd-Generation Biofuel Technologies - An Overview of Current Industry and RD&D Activities*. IEA, Paris, France, 120 pp.

Sioshansi, R., and W. Short (2009). Evaluating the impacts of real-time pricing on the usage of wind generation. *IEEE Transactions on Power Systems*, **24**(2), pp. 516-524.

Sjelvgren, D., S. Andersson, T. Andersson, U. Nyberg, and T.S. Dillon (1983). Optimal operations planning in a large hydrothermal power system. *IEEE Transactions on Power Apparatus and Systems*, **102**(11), pp. 3644-3651.

Smith, J.C., M.R. Milligan, E.A. DeMeo, and B. Parsons (2007). Utility wind integration and operating impact state of the art. *IEEE Transactions on Power Systems*, **22**(3), pp. 900-908.

Smith, J.C., S. Beuning, H. Durrwachter, E. Ela, D. Hawkins, B. Kirby, W. Lasher, J. Lowell, K. Porter, K. Schuyler, and P. Sotkiewicz (2010a). Impact of variable renewable energy on US electricity markets. In: *IEEE Power and Energy Society General Meeting*, Minneapolis, MN, USA, 25-27 July 2010, pp. 1-12, 25-29.

Smith, J.C., H. Holttinen, D. Osborn, R. Zavadil, W. Lasher, L. Gómez, T. Trötscher, J.O. Tande, M. Korpås, F. Van Hulle, A. Estanqueiro, and L. Dale (2010b). Transmission planning for wind energy: Status and prospects. In: *EWEC 2010 – European Wind Energy Conference*, Warsaw, Poland, 20-23 April 2010.

Söder, L., and H. Holttinen (2008). On methodology for modelling power system impact on power systems. *International Journal of Global Energy Issues*, **29**(1-2), pp. 181-198.

Söder, L., L. Hofmann, A. Orths, H. Holttinen, Y.H. Wan, and A. Tuohy (2007). Experience from wind integration in some high penetration areas. *IEEE Transactions on Energy Conversion*, **22**(1), pp. 4-12.

Stadler, I. (2008). Power grid balancing of energy systems with high renewable energy penetration by demand response. *Utilities Policy*, **16**(2), pp. 90-98.

Statistics Finland (2009). *Production of Electricity and Heat 2009*. Statistics Finland, Helsinki, Finland, 14 pp.

Sterner, M. (2009). *Bioenergy and Renewable Power Methane in Integrated 100% Renewable Energy Systems - Limiting Global Warming by Transforming Energy Systems*. PhD Thesis, Kassel University, Kassel, Germany (ISBN: 978-3-89958-798-2).

Stetz, T., W. Yan, and M. Braun (2010). Voltage control in distribution systems with high level PV-penetration – Improving absorption capacity for PV systems by reactive power supply. In: *25th European Photovoltaic Solar Energy Conference and Exhibition*, Valencia, Spain, 6-10 September 2010, pp. 5000-5006.

Stoft, S. (2002). *Power Systems Economics – Designing Markets for Electricity*. IEE Press Editorial Board, Piscataway, NJ, USA, 496 pp.

Stoutenburg, E.D., N. Jenkins, and M.Z. Jacobson (2010). Power output variations of co-located offshore wind turbines and wave energy converters in California. *Renewable Energy*, 35(12), pp. 2781-2791.

Strauss, P. (2009). *Einfluss des Frequenzverhaltens kleiner Generatoren und lasten auf Stromnetze unter besonderer Berücksichtigung großer Netzstörungen*. Dissertation, University of Kassel, Kassel, Germany, 128 pp.

Strbac, G. (2008). Demand side management: Benefits and challenges. *Energy Policy*, 36(12), pp. 4419-4426.

Strunz, K., and H. Louie (2009). Cache energy control for storage: Power system integration and education based on analogies derived from computer engineering. *IEEE Transactions on Power Systems*, 24(1), pp. 12-19.

Sudol, P. (2009). *Modelling and Analysis of Hydrogen Based Wind Energy Transmission and Storage Systems*. Massey University, Palmerston North, New Zealand.

Swider, D.J., and C. Weber (2007). The costs of wind's intermittency in Germany: application of a stochastic electricity market model. *European Transactions on Electrical Power*, 17(2), pp. 151-172.

Takada, M., and N.A. Charles (2006). *Energizing Poverty Reduction: A Review of the Energy-Poverty Nexus in Poverty Reduction Strategy Papers*. United Nations Development Programme, New York, NY, USA, 117 pp. Available at: www.undp.org/environment/sustainable-energy-library.shtml.

Takada, M., and S. Fracchia (2007). *A Review of Energy in National MDG Reports*. United Nations Development Programme, New York, NY, USA, 48 pp. Available at: www.undp.org/environment/sustainable-energy-library.shtml.

Taylor, C. (1994). *Power System Voltage Stability*. McGraw Hill, 273 pp.

Tester, J.W., B.J. Anderson, A.S. Batchelor, D.D. Blackwell, R. DiPippo, and E.M. Drake (eds.) (2006). *The Future of Geothermal Energy – Impact of Enhanced Geothermal Systems on the United States in the 21st Century*. Prepared by the Massachusetts Institute of Technology, under Idaho National Laboratory Subcontract No. 63 00019 for the U.S. Department of Energy, Assistant Secretary for Energy Efficiency and Renewable Energy, Office of Geothermal Technologies, Washington, DC, USA, 358 pp (ISBN-10: 0486477711, ISBN-13: 978-0486477718). Available at: geothermal.inel.gov/publications/future_of_geothermal_energy.pdf.

Thomson, M., and D.G. Infield (2007). Impact of widespread photovoltaics generation on distribution systems. *IET Renewable Power Generation*, 1(1), pp. 33-40.

Thorsteinsson, H.H., and J.W. Tester (2010). Barriers and enablers to geothermal district heating development in the United States. *Energy Policy*, 38, pp. 803-813.

Thyholt, M., and A.G. Hestnes (2008). Heat supply to low-energy buildings in district heating areas: Analyses of CO_2 emissions and electricity supply security. *Energy and Buildings*, 40(131-139)

TN Petróleo (2010). *ANP propõe regulação para biocombustível*. TN Petróleo, Brazil. Available at: www.tnpetroleo.com.br/clipping/4265/anp-propoe-regulacao-para-biocombustivel.

Torriti, J., M.G. Hassan, and M. Leach (2010). Demand response experience in Europe: Policies, programmes and implementation. *Energy*, 35(4), pp. 1575-1583.

TradeWind (2009). *Integrating Wind. Developing Europe's Power Market for the Large Scale Integration of Wind Power*. TradeWind, 104 pp.

Transmission Code (2007). *Transmission Code: Netz- und Systemregeln der deutschen Übertragungsnetzbetreiber*. Verband der Netzbetreiber, Berlin, Germany, 90pp.

TransPower (2008). *Grid upgrade plan 2008 Instalment 1, part III: Wairakei Ring Investment proposal*. Transpower New Zealand Ltd., Wellington, NZ, 43 pp.

TRB (2009). *Modal Primer on Greenhouse Gas and Energy Issues for the Transportation Industry*. Research Circular E-C143, Transportation Research Board, Washington, DC, USA (ISSN 0097-8515).

Troy, N., E. Denny, and M. O'Malley (2010). Base-load cycling on a system with significant wind penetration. *IEEE Transactions on Power Systems*, 25(2), pp. 1088-1097.

Tsikalakis, A.G., and N.D. Hatziargyriou (2008). Centralized control for optimizing microgrids operation. *IEEE Transactions on Energy Conversion*, 23(1), pp. 241-248.

Tsili, M., and S. Papathanassiou (2009). A review of grid code technical requirements for wind farms. *IET Renewable Power Generation*, 3(3), pp. 308-332.

Tuohy, A., and M. O'Malley (2011). Pumped storage in systems with very high wind penetration. *Energy Policy*, 39(4), pp. 1965-1974.

Tyll, H.K., and F. Schettler (2009). Power system problems solved by FACTS devices. In: *Power Systems Conference and Exposition*, Seattle, WA, USA, 16-18 March 2009, pp. 1-5, 15-18.

Uchida, H. (2010). Policy and action programs in Japan - Hydrogen energy as eco technology. In: *WHEC 2010 – 18th World Hydrogen Energy Conference*, International Association for Hydrogen Energy, Essen, Germany, 16-21 May 2010.

UCTE (2006). *Final Report: System Disturbance on 4 November 2006*. Union for the Co-Ordination of Transmission of Electricity, Brussels, Belgium, 85 pp.

Ueda, Y., K. Kurokawa, T. Tanabe, K. Kitamura, and H. Sugihara (2008). Analysis results of output power loss due to the grid voltage rise in grid-connected photovoltaic power generation systems. *IEEE Transactions on Industrial Electronics*, 55(7), pp. 2744-2751.

Ulleberg, Ø., T. Nakken, and A. Eté (2010). The wind/hydrogen demonstration system at Utsira in Norway: Evaluation of system performance using operational data and updated hydrogen energy system modeling tools. *International Journal of Hydrogen Energy*, 35(5), pp. 1841-1852.

Ummels, B.C., E. Pelgrum, and W.L. Kling (2008). Integration of large-scale wind power and use of energy storage in the Netherlands' electricity supply. *IET Renewable Power Generation*, 2(1), pp. 34-46.

UN Energy (2007). *Sustainable Bioenergy: A Framework for Decision Makers*. United Nations, New York, NY, USA, 64 pp. Available at: esa.un.org/un-energy/pdf/susdev.Biofuels.FAO.pdf.

UNEA (2009). *Energy for Sustainable Development: Policy Options for Africa*. United Nations Energy Africa, New York, NY, USA. Available at: www.uneca.org/eca_resources/publications/unea-publication-tocsd15.pdf.

UNDP (2007). *World Urbanization Prospects: The 2007 Revision*. United Nations Population Division, United Nations, New York, NY, USA, 230 pp. Available at: www.un.org/esa/population/publications/wup2007/2007WUP_Highlights_web.pdf.

UNDP (2009). *Small-Scale Finance for Modern Energy Services and the Role of Governments*. United Nations Development Programme, New York, NY, USA, 43 pp.

UNDP and WHO (2009). *The Energy Access Situation in Developing Countries - A Review Focusing on the Least Developed Countries and Sub-Saharan Africa.* United Nations Development Programme and World Health Organization, New York, NY, USA and Geneva, Switzerland. Available at: content.undp.org/go/newsroom/publications/environment-energy/www-ee-library/sustainable-energy/undp-who-report-on-energy-access-in-developing-countries-review-of-ldcs---ssas.en.

University of Edinburgh (2006). *Matching Renewable Electricity Generation with Demand: Academic study: Full report.* University of Edinburgh, Edinburgh, Scotland, 77 pp.

US DOE (2004). *Energy Information Administration Annual Energy Review 2003.* U.S. Department of Energy, Washington, DC, USA, 428 pp.

US DOE (2008a). *The Net-Zero Energy Commercial Building Initiative.* U.S. Department of Energy, Washington, DC, USA. Available at: www1.eere.energy.gov/buildings/initiative.html.

US DOE (2008b). *The Smart Grid: An Introduction.* U.S. Department of Energy, Washington, DC, USA. Available at: www.oe.energy.gov/DocumentsandMedia/DOE_SG_Book_Single_Pages(1).pdf.

US DOE and EPRI (1997). *Renewable Energy Technology Characterizations.* Topical Report, U.S. Department of Energy and EPRI, Washington, DC, USA and Palo Alto, CA, USA, 283 pp.

USDA (2009). *Precision, Geospatial and Sensor Technologies.* USDA Co-operative State Research Education and Extension Service, Washington, DC, USA. Available at: www.csrees.usda.gov/precisiongeospatialsensortechnologies.cfm.

U.S. Forest Service (2010). Record of Decision - San Diego Gas & Electric Special Use Authorization for the Sunrise Powerlink Transmission Line Project. San Diego County, California, USA. Available at: http://www.sdge.com/sunrisepowerlink/docs/ROD_SDGE_%20SpecialUse.pdf

Utria, B.E. (2004). Ethanol and gelfuel: clean renewable cooking fuels for poverty alleviation in Africa. *Energy for Sustainable Development*, **8**(3), pp. 107-114.

Vandezande, L., L. Meeus, R. Belmans, M. Saguan, and J.M. Glachant (2010). Well-functioning balancing markets: A prerequisite for wind power integration. *Energy Policy*, **38**(7), pp. 3146-3154.

Vannoni, C., R. Battisti, and S. Drigo (2008). *Potential for Solar Heat in Industrial Processes.* Booklet IEA SHC Task 33 and SolarPACES, CIEMAT, Madrid, Spain, 17 pp.

Velasco, D., C.L. Trujillo, and R.A. Pena (2011). Power transmission in direct current. Future expectations for Colombia. *Renewable & Sustainable Energy Reviews*, **15**(1), pp. 759-765.

Verbruggen, A. (2006). Electricity intensity backstop level to meet sustainable backstop supply technologies. *Energy Policy*, **34**, pp. 1310-1317.

Verheijen, F., I. Diafas, S. Jeffery, A.C. Bastos, and M. van der Velde (2010). *Biochar Application to Soils: A Critical Scientific Review of Effects on Soil Properties, Processes and Functions.* Science and Technical Report EUR 24099 EN – 2010, Joint Research Centre, Ispra, Italy, 166 pp.

Vittal, V., J. McCalley, V. Ajjarapu, and U. Shanbhag (2009). *Impact of Increased DFIG Wind Penetration on Power Systems and Markets: Final Project Report.* Power Systems Engineering Research Center, Tempe, AZ, USA, 221 pp.

Von Braun, J., and R. Meizen-Dick (2009). *Land Grabbing by Foreign Investors in Developing Countries – Risks and Opportunities.* IFPRI Policy Brief 13, International Food Policy Research Institute, Washington, DC, USA, 8 pp. Available at: www.ifpri.org/sites/default/files/publications/bp013all.pdf.

VTT (2009). *Energy Visions 2050* – Summary. VTT Technical Research Centre of Finland, Helsinki, Finland.

Wagner, J. (2009). Nahwärmekonzept Hirtenwiesen II. In: *Solarthermie 2009 - Heizen und Kühlen mit der Sonne*. VDI Verlag GmbH, Düsseldorf, Germany, pp. 121-132. Available at: www.fachbuch-erneuerbare-energien.de/solarthermie_2010_vdi_2074.htm.

Walter, A. (2006). Is Brazilian biofuels experience a model for other developing countries? *Entwicklung & Ländlicher Raum*, **40**, pp. 22-24.

Wang, L. (2008). Contemporary issues in thermal gasification of biomass and its application to electricity and fuel production. *Biomass and Bioenergy*, **32**(7), pp. 573-581.

Wang, M., H. Huo, L. Johnson, and D. He (2006). *Projection of Chinese Motor Vehicle Growth, Oil Demand, and CO_2 Emissions through 2050*. Report ANL/ESD/06-6, Argonne National Laboratory, Energy Systems Division, Argonne, IL, USA.

Wang, R.Z., T.S. Ge, C.J. Chen, Q. Ma, and Z.Q. Xiong (2009). Solar sorption cooling systems for residential applications: options and guidelines. *International Journal of Refrigeration*, **32**, pp. 638-660.

Wangdee, W., W. Li, and R. Billinton (2010). Coordinating wind and hydro generation to increase the effective load carrying capability. In: *IEEE 11th International Conference on Probabilistic Methods Applied to Power Systems (PMAPS)*, Singapore, 14-17 June 2010, pp. 337-342.

WBCSD (2008). *Efficient Heat and Power for One Million People (ABB Case Study)*. World Business Council for Sustainable Development, Geneva, Switzerland, 5 pp.

WBCSD (2009). *Transforming the Market - Energy Efficiency in Buildings*. World Business Council for Sustainable Development, Geneva, Switzerland, 72 pp. Available at: www.wbcsd.org/DocRoot/Ge2Laeua8uu2rkodeu7q/91719_EEBReport_WEB.pdf.

Weber, C. (2010). Adequate intraday market design to enable the integration of wind energy into the European power systems. *Energy Policy*, **38**(7), pp. 3155-3163.

Wei, N., W. Yong, S. Yan, and D. Zhongcheng (2009). Government management and implementation of national real-time energy monitoring system for China large-scale public building. *Energy Policy*, **37**(6), pp. 2087-2091.

Weiss, W., I. Bergmann, and R. Stelzer (2009). *Solar Heat World Wide: Markets and Contribution to the Energy Supply 2007*. IEA Solar Heating and Cooling Programme, International Energy Agency, Paris, France, 46 pp. Available at: www.energytech.at/pdf/SH_worldwide_2009.pdf.

Weisser, D., and R.S. Garcia (2005). Instantaneous wind energy penetration in isolated electricity grids: concepts and review. *Renewable Energy*, **30**(8), pp. 1299-1308.

Wenger, H.J., T.E. Hoff, and B.K. Farmer (1994). Measuring the value of distributed photovoltaic generation: final results of the Kerman grid-support project. In: *IEEE First World Conference on Photovoltaic Energy Conversion. Conference Record of the Twenty Fourth IEEE Photovoltaic Specialists Conference*, 5-9 December 1994, Waikoloa, HI, USA, pp. 792-796.

Werner, S. (2004). District heating and cooling. In: *Encyclopaedia of Energy, Vol. 1*. C.J. Cleveland (ed.), Elsevier, New York, NY, USA, pp. 841-848.

Werner, S. (2006a). *ECOHEATCOOL Work package 4 – The European Heat Market*. Final report prepared for the EU Intelligent Energy Europe Programme, Euroheat & Power, Brussels, Belgium, 73 pp.

Werner, S. (2006b). *ECOHEATCOOL Work package 1 - The European Heat Market Final Report*. Euroheat & Power, Brussels, Belgium, 18 pp.

White, B. (2009). *An Updated Assessment of Geothermal Direct Heat Use in New Zealand*. New Zealand Geothermal Association, Wellington, NZ, 36 pp.

Widén, J., E. Wäckelgård, and P.D. Lund (2009). Options for improving the load matching capability of distributed photovoltaics: Methodology and application to high-latitude data. *Solar Energy*, **83**(11), pp. 1953-1966.

Wiemken, E., H.G. Beyer, W. Heydenreich, and K. Kiefer (2001). Power characteristics of PV ensembles: experiences from the combined power production of 100 grid connected PV systems distributed over the area of Germany. *Solar Energy*, **70**(6), pp. 513-518.

Wilson, I.A.G., P.G. McGregor, and P.J. Hall (2010). Energy storage in the UK electrical network: Estimation of the scale and review of technology options. *Energy Policy*, **38**(8), pp. 4099-4106.

Wiser, R., and M. Bolinger (2010). *2009 Wind Technologies Market Report*. National Renewable Energy Laboratory, Golden, CO, USA, 88pp.

Wolf, A., and K. Petersson (2007). Industrial symbiosis in the Swedish forest industry. *Progress in Industrial Ecology*, **4**(5), pp. 348-362.

Wood, A.J., and B.F. Wollenberg (1996). *Power Generation Operation and Control*. 2nd ed. John Wiley and Sons, New York, NY, USA, 569 pp.

Woolf, D., J.E. Amonette, F.A. Street-Perrott, J. Lehmann, and S. Joseph (2010). Sustainable biochar to mitigate global climate change. *Nature Communications*, doi:10.1038/ncomms1053.

Woyte, A., R. Belmans, and J. Nijs (2007). Fluctuations in instantaneous clearness index: Analysis and statistics. *Solar Energy*, **81**(2), pp. 195-206.

Wu, C., W. Lee, C.L. Cheng, and H.-W. Lan (2008). Role and value of pumped storage units in an ancillary services market for isolated power systems – Simulation in the Taiwan power system. *IEEE Transactions on Industry Applications*, **44**(6), pp. 1924-1929.

WWICS (2010). *Land Grab: The Race for the World's Farmland (Event 5 May 2009)*. Woodrow Wilson International Centre for Scholars, Washington, DC, USA. Available at: www.wilsoncenter.org/index.cfm?fuseaction=events.event_summary&event_id=517903.

Xcel Energy (2009). *An Effective Load Carrying Capability Analysis for Estimating the Capacity Value of Solar Generation Resources on the Public Service Company of Colorado System*. Xcel Energy Services, Inc., Denver, CO, USA, 13 pp.

Yang, C. (2008). Hydrogen and electricity: Parallels, interactions, and convergence. *International Journal of Hydrogen Energy*, **33**, pp. 1977-1994.

Yang, X., Y. Song, G. Wang, and W. Wang (2010). A comprehensive review on the development of sustainable energy strategy and implementation in China. *IEEE Transactions on Sustainable Energy*, **1**(2), pp. 57-65.

Yu, D., J. Liang, X. Han, and J. Zhao (2011). Profiling the regional wind power fluctuation in China. *Energy Policy*, **39**(1), pp. 299-306.

Zabalza, I., A. Aranda, and M. Dolores de Gracia (2007). Feasibility analysis of fuel cells for combined heat and power systems in the tertiary sector. *International Journal of Hydrogen Energy*, **32**, pp. 1396-1403.

Zah, R., H. Böni, M. Gauch, R. Hischier, M. Lehmann, and P. Wäger (2007). *Life Cycle Assessment of Energy Products: Environmental Assessment of Biofuels*. Federal Office for the Environment (BFE), Bern, Switzerland, 16 pp. Available at: http://www.bfe.admin.ch/themen/00490/00496/index.html?lang=en&dossier_id=01273

Zgheib, E., and D. Clodic (2009). CO_2 emission and energy reduction evaluations of plug-in hybrids (Paper 2009-01-1234). In: *Advanced Hybrid Vehicle Powertrains*. SAE International, Warrendale, pp. 399.

Zhang, P., F. Li, and N. Bhatt (2010). Next-generation monitoring, analysis, and control for the future smart control center. *IEEE Transactions on Smart Grid*, **1**(2), pp. 186-192.

Zhang, X.-P., C. Rehtanz, and B. Pal (2006). *Flexible AC Transmission Systems: Modelling and Control.* Springer, Berlin, Germany, 383 pp.

Zinco, H., B. Bohm, H. Kristjansson, U. Ottosson, M. Rama, and K. Sipila (2008). *District Heating Distribution in Areas with Low Heat Demand Density*. IEA R&D Programme on District Heating and Cooling, including the integration of CHP, 117 pp. Available at: www.iea-dhc.org/reports/pdf/Energiteknik_IEA-Final-report-5.pdf.

Zogg, R., K. Roth, and J. Brodrick (2008). Lake-source district cooling systems. *ASHRAE Journal*, **February 2008**. Available at: findarticles.com/p/articles/mi_m5PRB/is_2_50/ai_n25376339/?tag=content;col1.

Zuurbier, P., and J.V.D. Vooren (eds.) (2008). *Sugarcane Ethanol:Contributions to Climate Change Mitigation and the Environment*. Wageningen Academic Publishers, Wageningen, The Netherlands.

9 Renewable Energy in the Context of Sustainable Development

Coordinating Lead Authors:
Jayant Sathaye (USA), Oswaldo Lucon (Brazil), Atiq Rahman (Bangladesh)

Lead Authors:
John Christensen (Denmark), Fatima Denton (Senegal/Gambia), Junichi Fujino (Japan), Garvin Heath (USA), Monirul Mirza (Canada/Bangladesh), Hugh Rudnick (Chile), August Schlaepfer (Germany/Australia), Andrey Shmakin (Russia)

Contributing Authors:
Gerhard Angerer (Germany), Christian Bauer (Switzerland/Austria), Morgan Bazilian (Austria/USA), Robert Brecha (Germany/USA), Peter Burgherr (Switzerland), Leon Clarke (USA), Felix Creutzig (Germany), James Edmonds (USA), Christian Hagelüken (Germany), Gerrit Hansen (Germany), Nathan Hultman (USA), Michael Jakob (Germany), Susanne Kadner (Germany), Manfred Lenzen (Australia/Germany), Jordan Macknick (USA), Eric Masanet (USA), Yu Nagai (Austria/Japan), Anne Olhoff (USA/Denmark), Karen Olsen (Denmark), Michael Pahle (Germany), Ari Rabl (France), Richard Richels (USA), Joyashree Roy (India), Tormod Schei (Norway), Christoph von Stechow (Germany), Jan Steckel (Germany), Ethan Warner (USA), Tom Wilbanks (USA), Yimin Zhang (USA)

Review Editors:
Volodymyr Demkine (Kenya/Ukraine), Ismail Elgizouli (Sudan), Jeffrey Logan (USA)

Special Advisor:
Susanne Kadner (Germany)

This chapter should be cited as:
Sathaye, J., O. Lucon, A. Rahman, J. Christensen, F. Denton, J. Fujino, G. Heath, S. Kadner, M. Mirza, H. Rudnick, A. Schlaepfer, A. Shmakin, 2011: Renewable Energy in the Context of Sustainable Development. In IPCC Special Report on Renewable Energy Sources and Climate Change Mitigation [O. Edenhofer, R. Pichs-Madruga, Y. Sokona, K. Seyboth, P. Matschoss, S. Kadner, T. Zwickel, P. Eickemeier, G. Hansen, S. Schlömer, C. von Stechow (eds)], Cambridge University Press, Cambridge, United Kingdom and New York, NY, USA.

Table of Contents

Executive Summary 710

9.1 Introduction 713

9.1.1 The concept of sustainable development 713

9.2 Interactions between sustainable development and renewable energies 713

9.2.1 Framework of Chapter 9 and linkages to other chapters of this report 714

9.2.2 Sustainable development goals for renewable energy and sustainable development indicators 715

9.3 Social, environmental and economic impacts: global and regional assessment 718

9.3.1 Social and economic development 718
9.3.1.1 Energy and economic growth 718
9.3.1.2 Human Development Index and energy 719
9.3.1.3 Employment creation 719
9.3.1.4 Financing renewable energy 720

9.3.2 Energy access 721

9.3.3 Energy security 724
9.3.3.1 Availability and distribution of resources 724
9.3.3.2 Variability and reliability of energy supply 726

9.3.4 Climate change mitigation and reduction of environmental and health impacts 728
9.3.4.1 Climate change 729
9.3.4.2 Local and regional air pollution 736
9.3.4.3 Health impacts 739
9.3.4.4 Water 741
9.3.4.5 Land use 743
9.3.4.6 Impacts on ecosystems and biodiversity 744
9.3.4.7 Accidents and risks 745

9.4 Implications of (sustainable) development pathways for renewable energy 747

9.4.1 Social and economic development 749
9.4.1.1 Social and economic development in scenarios of the future 749
9.4.1.2 Research gaps 751

9.4.2	**Energy access**	751
9.4.2.1	Energy access in scenarios of the future	751
9.4.2.2	Research gaps	752
9.4.3	**Energy security**	752
9.4.3.1	Energy security in scenarios of the future	753
9.4.3.2	Research gaps	755
9.4.4	**Climate change mitigation and reduction of environmental and health impacts**	756
9.4.4.1	Environmental and health impacts in scenarios of the future	756
9.4.4.2	Research gaps	756
9.5	**Barriers and opportunities for renewable energies in the context of sustainable development**	**757**
9.5.1	**Barriers**	757
9.5.1.1	Socio-cultural barriers	757
9.5.1.2	Information and awareness barriers	758
9.5.1.3	Market failures and economic barriers	759
9.5.2	**Opportunities**	760
9.5.2.1	International and national strategies for sustainable development	760
9.5.2.2	Local, private and nongovernmental sustainable development initiatives	763
9.6	**Synthesis**	**764**
9.6.1	Theoretical concepts and methodological tools for assessing renewable energy sources	764
9.6.2	Social and economic development	765
9.6.3	Energy access	765
9.6.4	Energy security	765
9.6.5	Climate change mitigation and reduction of environmental and health impacts	766
9.6.6	Conclusions	766
9.7	**Gaps in knowledge and future research needs**	**767**
References		**769**

Executive Summary

Historically, economic development has been strongly correlated with increasing energy use and growth of greenhouse gas (GHG) emissions. Renewable energy (RE) can help decouple that correlation, contributing to sustainable development (SD). In addition, RE offers the opportunity to improve access to modern energy services for the poorest members of society, which is crucial for the achievement of any single of the eight Millennium Development Goals.

Theoretical concepts of SD can provide useful frameworks to assess the interactions between SD and RE. SD addresses concerns about relationships between human society and nature. Traditionally, SD has been framed in the three-pillar model—Economy, Ecology, and Society—allowing a schematic categorization of development goals, with the three pillars being interdependent and mutually reinforcing. Within another conceptual framework, SD can be oriented along a continuum between the two paradigms of weak sustainability and strong sustainability. The two paradigms differ in assumptions about the substitutability of natural and human-made capital. RE can contribute to the development goals of the three-pillar model and can be assessed in terms of both weak and strong SD, since RE utilization is defined as sustaining natural capital as long as its resource use does not reduce the potential for future harvest.

The relationship between RE and SD can be viewed as a hierarchy of goals and constraints that involve both global and regional or local considerations. Though the exact contribution of RE to SD has to be evaluated in a country specific context, RE offers the opportunity to contribute to a number of important SD goals: (1) social and economic development; (2) energy access; (3) energy security; (4) climate change mitigation and the reduction of environmental and health impacts. The mitigation of dangerous anthropogenic climate change is seen as one strong driving force behind the increased use of RE worldwide. The chapter provides an overview of the scientific literature on the relationship between these four SD goals and RE and, at times, fossil and nuclear energy technologies. The assessments are based on different methodological tools, including bottom-up indicators derived from attributional lifecycle assessments (LCA) or energy statistics, dynamic integrated modelling approaches, and qualitative analyses.

Countries at different levels of development have different incentives and socioeconomic SD goals to advance RE. The creation of employment opportunities and actively promoting structural change in the economy are seen, especially in industrialized countries, as goals that support the promotion of RE. However, the associated costs are a major factor determining the desirability of RE to meet increasing energy demand and concerns have been voiced that increased energy prices might endanger industrializing countries' development prospects; this underlines the need for a concomitant discussion about the details of an international burden-sharing regime. Still, decentralized grids based on RE have expanded and already improved energy access in developing countries. Under favorable conditions, cost savings in comparison to non-RE use exist, in particular in remote areas and in poor rural areas lacking centralized energy access. In addition, non-electrical RE technologies offer opportunities for modernization of energy services, for example, using solar energy for water heating and crop drying, biofuels for transportation, biogas and modern biomass for heating, cooling, cooking and lighting, and wind for water pumping. RE deployment can contribute to energy security by diversifying energy sources and diminishing dependence on a limited number of suppliers, therefore reducing the economy's vulnerability to price volatility. Many developing countries specifically link energy access and security issues to include stability and reliability of local supply in their definition of energy security.

Supporting the SD goal to mitigate environmental impacts from energy systems, RE technologies can provide important benefits compared to fossil fuels, in particular regarding GHG emissions. Maximizing these benefits often depends on the specific technology, management, and site characteristics associated with each RE project, especially with respect to land use change (LUC) impacts. Lifecycle assessments for electricity generation indicate that GHG emissions from RE technologies are, in general, considerably lower than those associated with fossil fuel options, and in a range of conditions, less than fossil fuels employing carbon capture and storage (CCS). The maximum estimate for concentrating solar power (CSP), geothermal, hydropower, ocean and wind energy is less than or equal to 100 g CO_2eq/kWh, and median values for all RE range from 4 to 46 g CO_2eq/kWh. The GHG balances of bioenergy production, however, have considerable uncertainties, mostly related to land management and LUC. Excluding LUC, most bioenergy systems reduce GHG emissions compared to fossil-fuelled systems and can lead to avoided GHG emissions from residues and wastes in landfill disposals and co-products; the combination of bioenergy with CCS may provide for further reductions. For transport fuels, some first-generation biofuels result in relatively modest GHG mitigation potential, while most next-generation biofuels could provide greater climate benefits. To optimize benefits from bioenergy production, it is critical to reduce uncertainties and to consider ways to mitigate the risk of bioenergy-induced LUC.

RE technologies can also offer benefits with respect to air pollution and health. Non-combustion-based RE power generation technologies have the potential to significantly reduce local and regional air pollution and lower associated health impacts compared to fossil-based power generation. Impacts on water and biodiversity, however, depend on local conditions. In areas where water scarcity is already a concern, non-thermal RE technologies or thermal RE technologies using dry cooling can provide energy services without additional stress on water resources. Conventional water-cooled thermal power plants may be especially vulnerable to conditions of water scarcity and climate change. Hydropower and some bioenergy systems are dependent on water availability, and can either increase competition or mitigate water scarcity. RE specific impacts on biodiversity may be positive or negative; the degree of these impacts will be determined by site-specific conditions. Accident risks of RE technologies are not negligible, but the technologies' often decentralized structure strongly limits the potential for disastrous consequences in terms of fatalities. However, dams associated with some hydropower projects may create a specific risk depending on site-specific factors.

The scenario literature that describes global mitigation pathways for RE deployment can provide some insights into associated SD implications. Putting an upper limit on future GHG emissions results in welfare losses (usually measured as gross domestic product or consumption foregone), disregarding the costs of climate change impacts. These welfare losses are based on assumptions about the availability and costs of mitigation technologies and increase when the availability of technological alternatives for constraining GHGs, for example, RE technologies, is limited. Scenario analyses show that developing countries are likely to see most of the expansion of RE production. Increasing energy access is not necessarily beneficial for all aspects of SD, as a shift to modern energy away from, for example, traditional biomass could simply be a shift to fossil fuels. In general, available scenario analyses highlight the role of policies and finance for increased energy access, even though forced shifts to RE that would provide access to modern energy services could negatively affect household budgets. To the extent that RE deployment in mitigation scenarios contributes to diversifying the energy portfolio, it has the potential to enhance energy security by making the energy system less susceptible to (sudden) energy supply disruption. In scenarios, this role of RE will vary with the energy form. With appropriate carbon mitigation policies in place, electricity generation can be relatively easily decarbonized through RE sources that have the potential to replace concentrated and increasingly scarce fossil fuels in the building and industry sectors. By contrast, the demand for liquid fuels in the transport sector remains inelastic if no technological breakthrough can be achieved. Therefore oil and related energy security concerns are likely to continue to play a role in the future global energy system; as compared to today these will be seen more prominently in developing countries. In order to take account of environmental and health impacts from energy systems, several models have included explicit representation of these, such as sulphate pollution. Some scenario results show that climate policy can help drive improvements in local air pollution (i.e., particulate matter), but air pollution reduction policies alone do not necessarily drive reductions in GHG emissions. Another implication of some potential energy trajectories is the possible diversion of land to support biofuel production. Scenario results have pointed at the possibility that climate policy could drive widespread deforestation if not accompanied by other policy measures, with land use being shifted to bioenergy crops with possibly adverse SD implications, including GHG emissions.

The integration of RE policies and measures in SD strategies at various levels can help overcome existing barriers and create opportunities for RE deployment in line with meeting SD goals. In the context of SD, barriers continue to impede RE deployment. Besides market-related and economic barriers, those barriers intrinsically linked to societal and personal values and norms will fundamentally affect the perception and acceptance of RE technologies and related deployment impacts by individuals, groups and societies. Dedicated communication efforts are therefore a crucial component of any transformation strategy and local SD initiatives can play an important role in this context. At international and national levels, strategies should include: the removal of mechanisms that are perceived to work against SD; mechanisms for SD that internalize environmental and social externalities; and RE strategies that support low-carbon, green and sustainable development including leapfrogging.

The assessment has shown that RE can contribute to SD to varying degrees; more interdisciplinary research is needed to close existing knowledge gaps. While benefits with respect to reduced environmental and health impacts may appear more clear-cut, the exact contribution to, for example, social and economic development is more ambiguous. In order to improve the knowledge regarding the interrelations between SD and RE and to find answers to the question of an effective, economically efficient and socially acceptable transformation of the energy system, a much closer integration of insights from social, natural and economic sciences (e.g., through risk analysis approaches), reflecting the different (especially intertemporal, spatial and intra-generational) dimensions of sustainability, is required. So far, the knowledge base is often limited to very narrow views from specific branches of research, which do not fully account for the complexity of the issue.

9.1 Introduction

Sustainable development (SD) emerged in the political, public and academic arena in 1972 with the Founex report and again in 1987 with the publication of the World Commission on Environment and Development (WCED) report *Our Common Future*—also known as the 'Brundtland Report'. This *Special Report on Renewable Energy Sources and Climate Change Mitigation* follows the Brundtland definition that SD meets the needs of the present without compromising the ability of future generations to meet their own needs (WCED, 1987; Bojö et al., 1992). Due to the difficulty of putting such a concept into operation, many competing frameworks for SD have been put forward since then (Pezzey, 1992; Hopwood et al., 2005). In this chapter, some SD concepts will be introduced, links between SD and RE will be elucidated, and implications for decision making will be clarified.

SD was tightly coupled with climate change (and thence the IPCC) at the United Nations Conference on Environment and Development (UNCED) held in Rio de Janeiro, Brazil in 1992 that sought to stabilize atmospheric concentrations of greenhouse gases at levels considered to be safe. As a consequence, and building on the IPCC's First Assessment Report that focused on the technology and cost-effectiveness of mitigation activities, the Second Assessment Report included equity concerns in addition to social considerations (IPCC, 1996a). The Third Assessment Report addressed global sustainability comprehensively (IPCC, 2007b) and the Fourth Assessment (AR4) included chapters on SD in both Working Group (WG) II and III reports with a focus on a review of both climate-first and development-first literature (IPCC, 2007a,b).

9.1.1 The concept of sustainable development

Traditionally, sustainability has been framed in the three-pillar model: Economy, Ecology and Society are all considered to be interconnected and relevant for sustainability (BMU, 1998). The three-pillar model explicitly acknowledges the encompassing nature of the sustainability concept and allows a schematic categorization of sustainability issues. The United Nations General Assembly aims for action to promote the integration of the three components of SD—economic development, social development and environmental protection—as interdependent and mutually reinforcing pillars (UN, 2005a). This view subscribes to an understanding where a certain set of actions (e.g., substitution of fossil fuels with RE sources) can fulfil all three development goals simultaneously. The three-pillar model has been criticized for diluting a strong normative concept with vague categorization and replacing the need to protect natural capital with a methodological notion of trans-sectoral integration (Brand and Jochum, 2000).

Within another conceptual framework, SD can be oriented along a continuum between the two paradigms of weak sustainability and strong sustainability. The two paradigms differ in assumptions about the substitutability of natural and human-made capital (Hartwick, 1977; Pearce et al., 1996; Neumayer, 2003). Weak sustainability has been labelled the substitutability paradigm (Neumayer, 2003) and is based on the idea that only the aggregate stock of capital needs to be conserved—natural capital can be substituted with man-made capital without compromising future well-being. As such, it can be interpreted as an extension of neoclassical welfare economics (Solow, 1974; Hartwick, 1977). For example, one can argue that non-renewable resources, such as fossil fuels, can be substituted, for example, by renewable resources and technological progress as induced by market prices (Neumayer, 2003). Weak sustainability also implies that environmental degradation can be compensated for with man-made capital such as more machinery, transport infrastructure, education and information technology.

Whereas weak sustainability assumes that the economic system flexibly adapts to varying availability of forms of capital, strong sustainability starts from an ecological perspective with the intent of proposing guardrails for socioeconomic pathways. Strong sustainability can be viewed as the non-substitutability paradigm (Pearce et al., 1996; Neumayer, 2003), based on the belief that natural capital cannot be substituted, either for production purposes or for environmental provision of regulating, supporting and cultural services (Norgaard, 1994). As an example, limited sinks such as the atmosphere's capacity to absorb GHG emissions may be better captured by applying the constraints of the strong sustainability concept (Neumayer, 2003; IPCC, 2007b). In one important interpretation, the physical stock of specific non-substitutable resources (so-called 'critical natural capital') must be preserved (not allowing for substitution between different types of natural capital) (Ekins et al., 2003). Guardrails for remaining within the bounds of sustainability are often justified or motivated by nonlinearities, discontinuities, non-smoothness and non-convexities (Pearce et al., 1996). As a typical correlate, natural scientists warn of and describe specific tipping points, critical thresholds at which a tiny perturbation can qualitatively alter the state or development of Earth systems (Lenton et al., 2008). The precautionary principle argues for keeping a safe distance from guardrails, putting the burden of proof for the non-harmful character of natural capital reduction on those taking action (Ott, 2003).

RE can contribute to the development goals of the three-pillar model and can be assessed in terms of both weak and strong sustainability. Consumption of non-RE sources, such as fossil fuels and uranium, reduces natural capital directly. RE, in contrast, sustains natural capital as long as its resource use does not reduce the potential for future harvest.

9.2 Interactions between sustainable development and renewable energies

The relationship between RE and sustainability can be viewed as a hierarchy of goals and constraints that involve both global and regional or local considerations. In this chapter, and consistent with the conclusion of the AR4, a starting point is that mitigation of dangerous anthropogenic climate change will be one strong driving force behind increased use of RE technologies worldwide. To the extent that climate change stabilization levels (e.g., a maximum of 550 ppm CO_2eq atmospheric GHG concentration or a maximum of 2°C temperature increase with respect to the pre-industrial global average) are accepted, there is an implicit acknowledgement of a strong sustainability principle, as discussed in Section 9.1.

RE is projected to play a central role in most GHG mitigation strategies (Chapter 10), which must be technically feasible and economically efficient so that any cost burdens are minimized. Knowledge about technological capabilities and models for optimal mitigation pathways are therefore important. However, energy technologies, economic costs and benefits, and energy policies, as described in other chapters of this report, depend on the societies and natural environment within which they are embedded. Spatial and cultural variations are therefore another important factor in coherently addressing SD. Sustainability challenges and solutions crucially depend on geographic setting (e.g., solar radiation), socioeconomic conditions (e.g., inducing energy demand), inequalities within and across societies, fragmented institutions, and existing infrastructure (e.g., electric grids) (Holling, 1997; NRC, 2000), but also on a varying normative understanding of the connotation of sustainability (Lele and Norgaard, 1996). Analysts therefore call for a differentiation of analysis and solution strategies according to geographic locations and specific places (e.g., Wilbanks, 2002; Creutzig and Kammen, 2009) and a pluralism of epistemological and normative perspectives of sustainability (e.g., Sneddon et al., 2006).

These aspects underline the need to assess both the social and environmental impacts of RE technologies to ensure that RE deployment remains aligned with overall SD goals. Some of these important caveats are addressed in this chapter, like the extent to which RE technologies may have their own environmental impact and reduce natural capital, for example, by upstream GHG emissions, destroying forests, binding land that cannot be used otherwise and consuming water. Evaluating these impacts from the perspectives of the weak and strong sustainability paradigms elucidates potential tradeoffs between decarbonization and other sustainability goals.

Hence, efforts to ensure SD can impose additional constraints or selection criteria on some mitigation pathways, and may in fact compel policymakers and citizens to accept trade-offs. For each additional boundary condition placed on the energy system, some development pathways are eliminated as being unsustainable, and some technically feasible scenarios for climate mitigation may not be viable if SD matters. However, as also discussed in this chapter, the business-as-usual trajectories to which climate mitigation scenarios are compared are probably also insufficient to achieve SD.

9.2.1 Framework of Chapter 9 and linkages to other chapters of this report

This chapter provides an overview of the role that RE can play in advancing the overarching goal of SD. Chapter 1 in this report introduces RE and makes the link to climate change mitigation, and Chapters 2 through 7 assess the potential and impacts of specific RE technologies in isolation. Chapter 8 focuses on the integration of renewable sources into the current energy system, and Chapters 10 and 11 discuss the economic costs and benefits of RE and climate mitigation, and of RE policies, respectively. As an integrative chapter, this chapter assesses the role of RE from a SD perspective by comparing and reporting the SD impacts of different energy technologies, by drawing on still limited insights from the scenario literature with respect to SD goals, and by discussing barriers to and opportunities of RE deployment in relation to SD. Figure 9.1 illustrates the links of Chapter 9 to other chapters in this report.

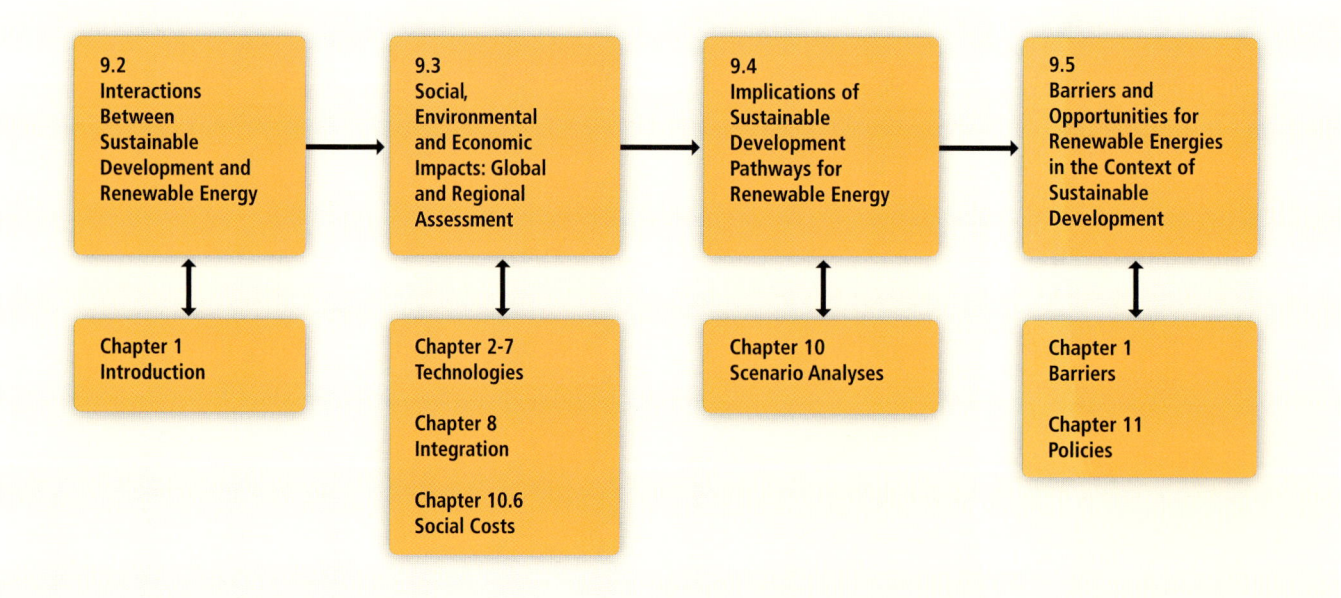

Figure 9.1 | Framework of Chapter 9 and linkages to other chapters.

For a conclusive and comprehensive assessment of sustainable RE deployment pathways, this chapter would need to integrate information on each specific energy technology, including associated economic costs and benefits and existing energy policies, as provided in the other chapters of this report. As a result, SD opportunities associated with RE deployment could be clearly outlined, informing policymakers about pathways and how to realize them while avoiding unintended side effects. However, given the diverse range of possible opportunities and the limitations of current modelling capacities, such comprehensive integrated assessments are not yet practicable. This chapter will focus its assessment on the clearly defined set of opportunities outlined in Section 1.4.1:

- Social and economic development,
- Energy access,
- Energy security, and
- Climate change mitigation and reduction of environmental and health impacts.

This set of opportunities can be viewed as goals that should be achieved for RE to contribute to SD. As will be discussed in the following section, the potential of RE to increase access to modern energy technologies can facilitate social and economic development. Energy access and social and economic development measures relate to current well-being and to some extent to intra-generational equity and sustainability, for example, through an emphasis on energy-related equity questions, including gender equity and empowerment. The potential contribution of RE to energy security, climate change mitigation and the reduction of environmental impacts addresses more explicitly the intertemporal and intergenerational well-being aspect inherent in sustainability. Energy access, social and economic development and energy security concerns are very often considered under the weak sustainability paradigm, because trade-offs are taken into account allowing for a balance between these goals. Environmental impacts, on the other hand, are usually evaluated under the strong sustainability paradigm because they are very often understood as constraints for transformation pathways. To enable responsible decision making, it is crucial to understand the implications and possible trade-offs of SD goals that result from alternative energy system choices.

This chapter provides an overview of the scientific literature on the relationship between these four SD goals and RE and, at times, fossil and nuclear energy technologies. SD aspects that need to be included in future and more comprehensive assessments of potential development pathways are outlined in a quantitative as well as in a qualitative and more narrative manner. Section 9.3 focuses on static bottom-up indicators based on currently available data (e.g., LCA) to assess the socioeconomic and environmental impacts of individual RE and other energy technologies. Section 9.4, on the other hand, aims to assess the interactions of future RE deployment and SD pathways in a more dynamic, top-down and integrated manner. Pathways are primarily understood as scenario results that attempt to address the complex interrelations among the different energy technologies at a global scale. Therefore the chapter mainly refers to global scenarios derived from large integrated models, which are also at the core of the analysis in Chapter 10. The analysis concludes with Section 9.5, which aims to analyze barriers and opportunities for RE in the context of SD.

To conclude, when evaluating RE with respect to the multi-dimensional challenge of SD, no single global answer is possible. Many solutions will depend strongly on local, regional and cultural conditions, and the approaches and emphases of developing and developed countries may also be different. Therefore, it is not possible for this chapter to provide a clear set of recommendations for a pathway towards SD using RE.

9.2.2 Sustainable development goals for renewable energy and sustainable development indicators

Energy indicators can assist countries in monitoring progress made in energy subsystems consistent with sustainability principles. Measurement and reporting of indicators not only gauges but also spurs the implementation of SD and can have a pervasive effect on decision making (Meadows, 1998; Bossel, 1999). However, measuring energy sustainability is surrounded by a wide range of conceptual and technical issues (Sathaye et al., 2007) and may require updated methodologies (Creutzig and Kammen, 2009).

Over the past two decades, progress has been made towards developing a uniform set of energy indicators for sustainable development which relate to the broad themes of economy, society and environment (Vera and Langlois, 2007). For RE technologies, quantitative indicators include price of generated electricity, GHG emissions during the full lifecycle of the technology, availability of renewable sources, efficiency of energy conversion, land requirements and water consumption (Evans et al., 2009). Other approaches develop a figure of merit to compare the different RE systems based upon their performance, net energy requirements, GHG emissions and other indicators (Varun et al., 2010).

Due to the need to expand the notion of economic development beyond the ubiquitously used gross domestic product (GDP), a variety of SD indicators have been suggested. Aggregate indicators of weak sustainability include green net national product, genuine savings (Hamilton, 1994; Hamilton and Clemens, 1999; Dasgupta, 2001), the index of sustainable economic welfare (ISEW) and the genuine progress indicator (GPI) (e.g., Daly, 2007), with the ISEW and GPI proposed as intermediate steps by proponents of strong sustainability. Notably, indicators that extend GDP, such as the latter two, tend to deviate qualitatively from the GDP since the 1970s or 1980s, stagnating (or in case of the UK decreasing) in many Organisation for Economic Co-operation and Development (OECD) countries (Lawn, 2003). Indicators more consistent with strong sustainability include carrying capacity, ecological footprint and resilience (Pearce et al., 1996), sustainable national income and sustainability gaps (Hueting, 1980; Ekins and Simon, 1999).

The use of aggregated indicators for economic development (e.g., the Human Development Index (HDI) or ISEW (Fleurbaey, 2009)), however, poses significant challenges. Resulting values are indexed with high uncertainty and are often challenged on methodological and epistemological grounds (Neumayer, 2003). Rigorous justification for specific choices for weighting the components of aggregate indicators is difficult to make and as many indicators are proxies, they may also convey a message of false quantitative accuracy. Also, it is often difficult to obtain reliable and internationally consistent data series across components of the composite indicator. Aggregate indicators of sustainability integrate many aspects of social and economic development, and hence, are ignorant of the specific sustainability impact of RE deployment. Sustainability assessment may instead require a well-identified dashboard of indicators (Stiglitz et al., 2009).

Section 9.3 evaluates RE in terms of static bottom-up measures while being cognizant of their limitations. The four SD goals, as defined in section 9.2.1, are used as guidelines to assess the contribution of RE to SD. Since sustainability is an open-boundary concept, and is confronted with tipping elements of unknown probability, doubts can be raised regarding the possibility of an ultimate coherent quantitative evaluation. Quantitative indicators, which might be adjusted as new challenges emerge and new data become available, reflect a suitable framework to assess the existing literature, but cannot close the considerable gaps in achieving a comprehensive and consistent measure of SD.

Social and economic development
The energy sector has generally been perceived as key to economic development with a strong correlation between economic growth and expansion of energy consumption. Indicators such as GDP or per capita GDP have been used as proxies for economic development for several decades (such as in integrated models, see Section 9.4.1) and the HDI has been shown to correlate well with per capita energy use (see Section 9.3.1). The HDI is used to assess comparative levels of development in countries and includes purchasing power parity-adjusted income, literacy and life expectancy as its three main matrices. The HDI is only one of many possible measures of the well-being of a society, but it can serve as a proxy indicator of development.

Due to the availability of data time series for these parameters (GDP, HDI), they will be used as indicators in this chapter (Sections 9.3.1.1 and 9.3.1.2). However, a key point is that aggregate macroeconomic parameters (GDP), or even extended versions of these economic indicators (HDI), are insufficient for obtaining a complete picture of the sustainability of social and economic development. A further indicator of technological development is decreasing energy intensity, that is, a decrease in the amount of energy needed to produce one dollar of GDP.

Beyond indicators that describe the efficiency characteristics of an economy, additional macroeconomic benefits are potentially associated with RE, for example, increased employment opportunities (see Section 9.3.1.3). Furthermore, under agreements such as that reached in Copenhagen in 2009, financial pledges have been made by wealthier nations to aid developing countries with climate change mitigation measures (see Section 9.3.1.4). Each of these latter points may have either positive or negative effects, depending on regional context and on the particular policies that are implemented.

Energy access
Access to modern energy services, whether from renewable or non-renewable sources, is closely correlated with measures of development, particularly for those countries at earlier development stages. Indeed, the link between adequate energy services and achievement of the Millennium Development Goals (MDGs) was defined explicitly in the Johannesburg Plan of Implementation that emerged from the World Summit on Sustainable Development in 2002 (IEA, 2010b). As emphasized by a number of studies, providing access to modern energy (such as electricity or natural gas) for the poorest members of society is crucial for the achievement of any single of the eight MDGs (Modi et al., 2006; GNESD, 2007a; Bazilian et al., 2010; IEA, 2010b).

Over the past few centuries, industrialized societies have transformed their quality of life by exploiting non-renewable fossil energy sources, nuclear energy and large-scale hydroelectric power. However, in 2010 almost 20% of the world population, mostly in rural areas, still lack access to electricity. Twice that percentage cook mainly with traditional biomass, mostly gathered in an unsustainable manner (IEA, 2010b). In the absence of a concerted effort to increase energy access, the absolute number of those without electricity and modern cooking possibilities is not expected to change substantially in the next few decades.

Concrete indicators to be discussed in more detail in Section 9.3.2 are per capita final energy consumption related to income, as well as breakdowns of electricity access (divided into rural and urban areas), and data for the number of those using coal or traditional biomass for cooking. Implicit in discussions of energy access is a need for models that can assess the sustainability of future energy system pathways with respect to decreasing the wide disparity between rural and urban areas (e.g., in terms of energy forms and quantities used or infrastructure reliability) within countries or regions (see Section 9.4.2).

Energy security
There is no commonly accepted definition of the term 'energy security' and its meaning is highly context-dependent (Kruyt et al., 2009). At a general level it can best be understood as robustness against (sudden) disruptions of energy supply (Grubb et al., 2006). Thinking broadly across energy systems, one can distinguish between different aspects of security that operate at varying temporal and geographical scales (Bazilian and Roques, 2008). Two broad themes can be identified that

are relevant to energy security, whether for current systems or for the planning of future RE systems: availability and distribution of resources, and variability and reliability of energy supply. Given the interdependence of economic growth and energy consumption, access to a stable energy supply is a major political concern and a technical and economic challenge facing both developed and developing economies, since prolonged disruptions would create serious economic and basic functionality problems for most societies (Larsen and Sønderberg Petersen, 2009).

In the long term, the potential for fossil fuel scarcity and decreasing quality of fossil reserves represents an important reason for a transition to a sustainable worldwide RE system. The issue of recoverable fossil fuel resource amounts is contentious, with optimists (Greene et al., 2006) countered by more pessimistic views (Campbell and Laherrère, 1998) and cautious projections of lacking investments falling between the two poles (IEA, 2009). However, increased use of RE permits countries to substitute away from the use of fossil fuels, such that existing reserves of fossil fuels are depleted less rapidly and the point at which these reserves will eventually be exhausted is shifted farther into the future (Kruyt et al., 2009).

Concerns about limited availability and distribution of resources are also a critical component of energy security in the short term. All else being equal, the more reliant an energy system is on a single energy source, the more susceptible the energy system is to serious disruptions. Examples include disruptions to oil supply, unexpectedly large and widespread periods of low wind or solar insolation (e.g., due to weather), or the emergence of unintended consequences of any supply source.

Dependence on energy imports, whether of fossil fuels or the technology needed for implementation of RE, represents a potential source of energy insecurity for both developing and industrialized countries. For example, the response of member states of the International Energy Agency (IEA; itself created in response to the first oil shock of the 1970s) to vulnerability to oil supply disruption has been to mandate that countries hold stocks of oil as reserves in the amount of 90 days of net imports. Compared to fossil fuels, RE resources are far more evenly distributed around the globe (WEC, 2007) and in general less traded on the world market; increasing their share in a country's energy portfolio can thus diminish the dependence on actual energy imports (Grubb et al., 2006). Hence, the extent to which RE sources contribute to the diversification of the portfolio of supply options and reduce an economy's vulnerability to price volatility (Awerbuch and Sauter, 2006) represent opportunities to enhance energy security at the global, the national as well as the local level (Awerbuch, 2006; Bazilian and Roques, 2008).

The introduction of renewable technologies that vary on different time scales, ranging from minutes to seasonal, adds a new concern to energy security. Not only will there be concerns about disruption of supplies by unfriendly agents, but also the vulnerability of energy supply to the vagaries of chance and nature (such as extreme events like drought). However, RE can also make a contribution to increasing the reliability of energy services, in particular in remote and rural areas that often suffer from insufficient grid access. Irrespective, a diverse portfolio of energy sources, together with good management and system design (for example, including geographical diversity of sources where appropriate) can help to enhance security.

Specific indicators for security are difficult to identify. Based on the two broad themes described above, the indicators used to provide information about the energy security criterion of SD are the magnitude of reserves, the reserves-to-production ratio, the share of imports in total primary energy consumption, the share of energy imports in total imports, as well as the share of variable and unpredictable RE sources.

Climate change mitigation and reduction of environmental and health impacts

As discussed in Chapter 1, reducing GHG emissions with the aim of mitigating climate change is one of the key driving forces behind a growing demand for RE technologies. However, to evaluate the overall burden from the energy system on the environment, and to identify potential trade-offs, other impacts and categories have to be taken into account as well. Mass emissions to water and air, and usage of water, energy and land per unit of energy generated must be evaluated across technologies. Whereas some parameters can be rigorously quantified, for others comprehensive data or useful indicators may be lacking. In addition, deriving generic impacts on human health or biodiversity is a challenging task, as they are mostly specific to given sites, exposure pathways and circumstances, and often difficult to attribute to single sources.

There are multiple methods to evaluate environmental impacts of projects, such as environmental impact statements/assessments and risk assessments. Most are site-specific, and often limited to direct environmental impacts associated with operation of the facility. To provide a clear framework for comparison, lifecycle assessment (LCA) has been chosen as a bottom-up measure in Section 9.3.4, complemented by a comparative assessment of accident risks to account for burdens resulting from outside normal operation. Most published LCAs of energy supply technologies only assemble lifecycle inventories; quantifying emissions to the environment (or use of resources) rather than reporting effects (or impacts) on environmental quality. A similar approach is followed in Section 9.3.4, as literature reporting lifecycle impacts or aggregate sustainability indicators is scarce. Partly, this is due to the incommensurability of different impact categories. Attempts to combine various types of indicators into one overall score (for example by joining their impact pathways into a common endpoint, or by monetization) have been made; however uncertainties associated with such scoring approaches are often so high that they preclude decision making (Hertwich et al., 1999; Rabl and Spadaro, 1999; Schleisner, 2000; Krewitt, 2002; Heijungs et al., 2003; Sundqvist, 2004; Lenzen et al.,

2006). Nevertheless, social costs are discussed in Chapter 10.6, and part of the analysis in Section 9.4.4 is based on monetization of impacts. The latter section analyzes the extent to which environmental impacts are represented in scenario analyses for RE deployment with a macro-perspective, with a focus on land use change and related GHG emissions, as well as local air pollution.

9.3 Social, environmental and economic impacts: global and regional assessment

Countries at different levels of development have different incentives to advance RE. For developing countries, the most likely reasons to adopt RE technologies are providing access to energy (see Section 9.3.2.), creating employment opportunities in the formal (i.e., legally regulated and taxable) economy, and reducing the costs of energy imports (or, in the case of fossil energy exporters, prolong the lifetime of their natural resource base). For industrialized countries, the primary reasons to encourage RE include reducing carbon emissions to mitigate climate change (see Chapter 1), enhancing energy security (see Section 9.3.3.), and actively promoting structural change in the economy, such that job losses in declining manufacturing sectors are softened by new employment opportunities related to RE. For a conceptual description of the four SD goals assessed in this chapter, see Section 9.2.2.

9.3.1 Social and economic development

This section assesses the potential contributions of RE to sustainable social and economic development. Due to the multi-dimensional nature of SD neither a comprehensive assessment of all mitigation options nor a full accounting of all relevant costs can be performed. Rather, the following section identifies key issues and provides a framework to discuss the relative benefits and disadvantages of RE and fossil fuels with respect to development.

9.3.1.1 Energy and economic growth

With the ability to control energy flows being a crucial factor for industrial production and socioeconomic development (Cleveland et al., 1984; Krausmann et al., 2008), industrial societies are frequently characterized as 'high-energy civilizations' (Smil, 2000). Globally, per capita incomes are positively correlated with per capita energy use and economic growth can be identified as the most relevant factor behind increasing energy consumption in the last decades. Nevertheless, there is no agreement on the direction of the causal relationship between energy use and increased macroeconomic output, as the results crucially depend on the empirical methodology employed as well as the region and time period under study (D. Stern, 1993; Asafu-Adjaye, 2000; S. Paul and Bhattacharya, 2004; Ang, 2007, 2008; Lee and Chang, 2008).

Industrialization brings about structural change in the economy and therefore affects energy demand. As economic activity expands and diversifies, demands for more sophisticated and flexible energy sources arise: while societies that highly depend on agriculture derive a large part of primary energy consumption from traditional biomass (Leach, 1992; Barnes and Floor, 1996), coal and liquid fuels—such as kerosene and liquid petroleum gas—gain in importance with rising income, and electricity, gas and oil dominate at high per capita incomes (Grübler, 2004; Marcotullio and Schulz, 2007; Burke, 2010; see Section 9.3.2 and Figure 9.5). From a sectoral perspective, countries at an early stage of development consume the largest part of total primary energy in the residential (and to a lesser extent agricultural) sector. In emerging economies the manufacturing sector dominates, while in fully industrialized countries services and transport account for steadily increasing shares (Schafer, 2005; see Figure 9.2). Furthermore, several authors (Jorgenson, 1984; Schurr, 1984) have pointed out that electricity—which offers higher quality and greater flexibility compared to other forms of energy—has been a driving force for the mechanization and automation of production in industrialized countries and a significant contributor to continued increases in productivity.

Despite the fact that as a group industrialized countries consume significantly higher amounts of energy per capita than developing ones, a considerable cross-sectional variation of energy use patterns across countries prevails: while some countries (such as, e.g., Japan) display high levels of per capita incomes at comparably low levels of energy use, others are relatively poor despite extensive energy consumption, especially countries abundantly endowed with fossil fuel resources, in which energy is often heavily subsidized (UNEP, 2008b). It is often asserted that developing and transition economies can 'leapfrog', that is, adopt modern, highly efficient energy technologies, to embark on less energy- and carbon-intensive growth patterns compared to the now fully industrialized economies during their phase of industrialization (Goldemberg, 1998). For instance, one study for 12 Eastern European EU member countries finds that between 1990 and 2000, convergence in per capita incomes (measured at purchasing power parity) between fully industrialized and transition economies has been accompanied by significant reductions of energy intensities in the latter (Markandya et al., 2006). For industrialized countries, one hypothesis suggests that economic growth can largely be decoupled from energy use by steady declines in energy intensity as structural change and efficiency improvements trigger the 'dematerialization' of economic activity (Herman et al., 1990). However, despite the decreasing energy intensities (i.e., energy consumption per unit of GDP) observed over time in almost all regions, declines in energy intensity historically often have been outpaced by economic growth and hence have proved insufficient to achieve actual reductions in energy use (Roy, 2000). In addition, it has been argued that decreases in energy intensity in industrialized countries can partially be explained by the fact that energy-intensive industries are increasingly moved to developing countries (G. Peters and Hertwich, 2008; Davis and Caldeira, 2010) and, as observed energy

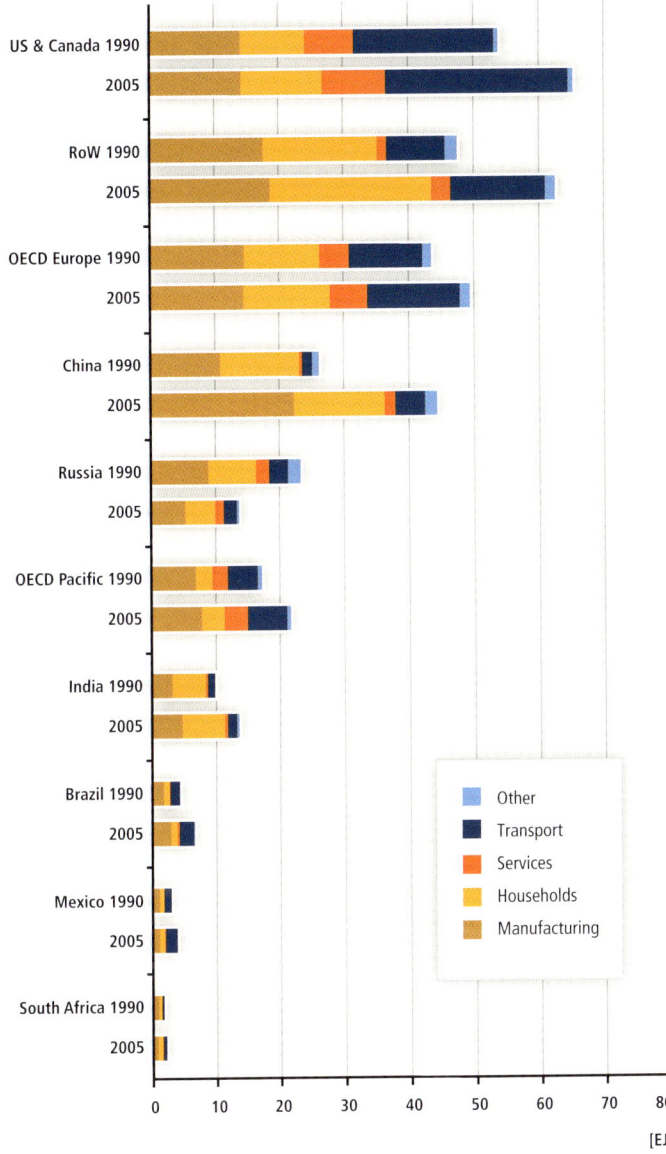

Figure 9.2 | Energy use (EJ) by economic sector. Note that the underlying data are calculated using the IEA physical content method, not the direct equivalent method[1] (IEA, 2008c). Note: RoW = Rest of World.

Note: 1. Historical energy data have only been available for energy use by economic sector. For a conversion of the data using the direct equivalent method, the different energy carriers used by each economic sector would need to be known.

efficiency improvements are largely driven by shifts to higher quality fuels, they cannot be expected to continue indeterminately (Cleveland et al., 2000; R.K. Kaufmann, 2004).

9.3.1.2 Human Development Index and energy

As already mentioned in Section 9.2.2, the industrialized societies' improvements in the quality of life have so far been mainly based on the exploitation of non-RE sources (while noting the important role of hydropower during the early stages of industrialization, as well as for many developing countries today). Apart from its significance for productive purposes, access to clean and reliable energy constitutes an important prerequisite for fundamental determinants of human development including health, education, gender equality and environmental safety (UNDP, 2007).

Figure 9.3 depicts the correlation between the HDI (see Section 9.2.2) and primary energy use per capita for 135 countries. The graph reveals a positive correlation between energy use and the HDI. In particular, countries with the highest levels of human development are also among the largest energy consumers. For countries with a relatively low energy demand (<84 GJ per capita), the picture is more diverse: while some are constrained to low HDI levels (<0.5), others display medium ones (between 0.5 and 0.8) at comparable energy consumption. With rising levels of energy consumption, saturation of the positive relationship between energy use and HDI sets in (Martinez and Ebenhack, 2008), which means that a certain minimum amount of energy is required to guarantee an acceptable standard of living. Goldemberg (2001) suggests 42 GJ per capita, after which raising energy consumption yields only marginal improvements in the quality of life.

9.3.1.3 Employment creation

According to a recent study prepared by UNEP (2008a), RE already accounts for about 2.3 million jobs worldwide and in many countries job creation is seen as one of the main benefits of investing in RE sources. A study by the German Environment Ministry finds that in 2006, about 236,000 people were employed in RE, up from roughly 161,000 two years earlier (BMU, 2009). Examples of the use of RE in India, Nepal and parts of Africa (Cherian, 2009) as well as Brazil (Goldemberg et al., 2008; Walter et al., 2011) indicate that in many parts of the developing world, RE can stimulate local economic and social development. Numerous governments have included substantial spending on clean energy technologies in their stimulus packages that were put into place in response to the financial and economic crisis (N. Bauer et al., 2009; Bowen et al., 2009). For the USA, one study (Houser et al., 2009) suggested that every USD_{2005} 1 billion spent on green fiscal measures had the potential to create about 33,000 jobs; another one, prepared by the Center for American Progress (Pollin et al., 2008), estimated that a green stimulus of USD_{2005} 90.7 billion could create roughly 2 million jobs. The Council of Economic Advisors to the US administration projects that the USD_{2005} 82 billion spending on clean energy included in the American Recovery and Reinvestment Act will create or safeguard 720,000 job-years through 2012. From a more long-term perspective, many national green growth strategies, for example, in China, Korea, Japan, the EU and the USA (UNEP, 2010), have stressed the deployment of RE as an important contribution to job creation and one study (Barbier, 2009) argues that a 'Global Green New Deal' could in the long run create more than 34 million jobs in low-carbon transportation and related activities alone.

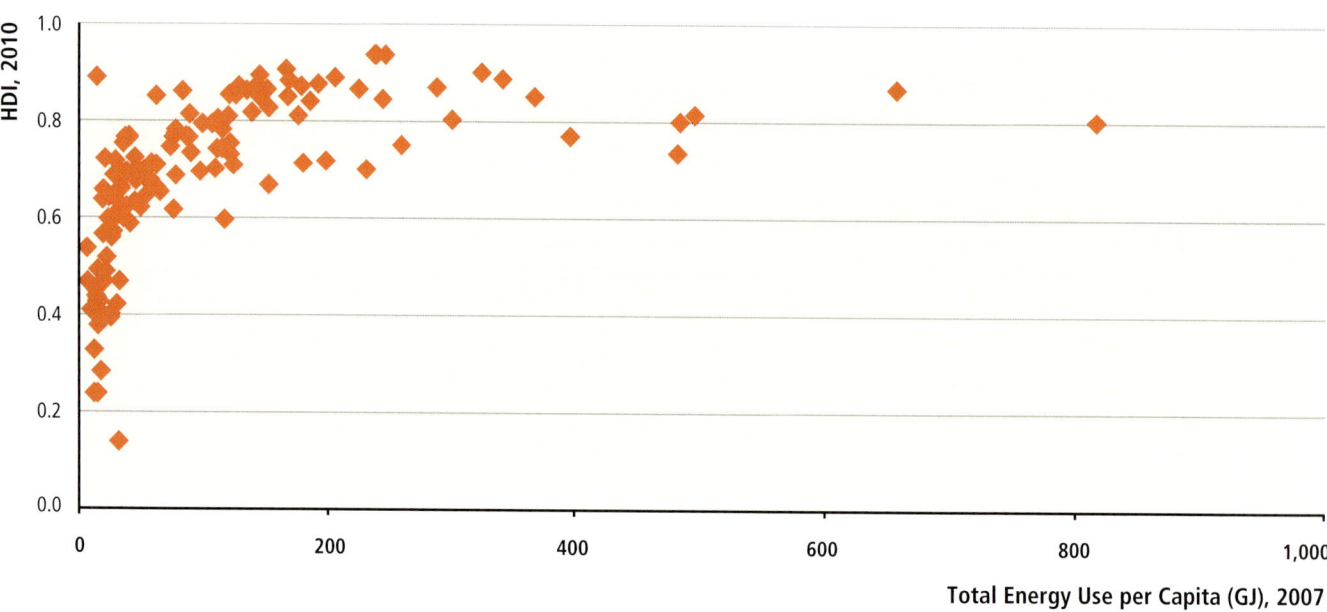

Figure 9.3 | Correlation between total energy use per capita (GJ) and the countries' Human Development Index (HDI). Note that the underlying data on energy use are calculated using the IEA physical content method, not the direct equivalent method.[1] Based on UNDP (2010) and World Bank (2010).

Note: 1. Historical energy data have only been available for energy use per capita by country. For a conversion of the data using the direct equivalent method, the different energy carriers used by each country would need to be known.

Other studies that also observe possible negative employment effects are more critical in this regard (Frondel et al., 2010) and the assertion of positive employment effects is further weakened by disagreements about the methodology used to calculate them (Sastresa et al., 2009). Evaluating the labour market effects of RE policies is in any case a challenging task that requires an assessment of how value chains and production patterns adjust in the mid-term and how structural adjustment and innovative activity respond in the long term (Fankhauser et al., 2008). RE should not be regarded as an instrument that can be employed to cure underlying inefficiencies in labour markets. For a comprehensive assessment, it would be necessary to factor in all social costs and benefits of a given technology (including interactions with labour market frictions) to be able to appropriately compare RE and fossil fuels on a level playing field. This includes the costs of support schemes for RE as well as subsidies for fossil fuels (see Section 9.5.2).

9.3.1.4 Financing renewable energy

An evaluation of the specific benefits of RE discussed in this section can only be undertaken in a country-specific context. Especially for developing countries, the associated costs are a major factor determining the desirability of RE to meet increasing energy demand, and concerns have been voiced that increased energy prices might endanger industrializing countries' development prospects (Mattoo et al., 2009). Yet, as will be discussed in more detail in Section 9.3.2, RE has been shown to bring about potential cost savings compared to fossil fuels (such as diesel generators) in poor rural areas without grid access (Casillas and Kammen, 2010). Nevertheless, in general the purely economic costs of RE exceed those of fossil fuel-based energy production in most instances (see Sections 2.7, 3.8, 4.7, 5.8, 6.7, 7.8 and 10.5) and further financial barriers to the adoption of RE are discussed in Section 11.4.3.

Overall, cost considerations cannot be discussed independently of the burden-sharing regime adopted, that is, without specifying who assumes the costs for the benefits brought about from reduced GHG emissions, which can be characterized as a global public good (N. Stern, 2007). For instance, the Copenhagen accord recognized that for the period 2010 to 2012 USD_{2005} 26 billion should be made available for climate measures in developing countries (including mitigation and adaptation), and that this sum should be scaled up to USD_{2005} 86 billion per year by 2020 (UNFCCC, 2009). Estimates of mid- to long-term financial flows to developing countries show considerable variation, depending to a high degree on the GHG stabilization level and burden-sharing scheme assumed to be in place. According to estimates assuming a 450 ppm atmospheric CO_2 stabilization scenario with an equal per capita distribution of emission permits, financial inflows related to climate finance could reach up to 10% of GDP for sub-Saharan Africa and up to 5% for India around 2020 (IMF, 2008). Obviously, such sizeable financial inflows can play an important role in supporting the transition towards RE-based energy systems. However, the appropriate governance of substantial financial inflows is also critically important, ensuring that these transfers result in actual SD benefits instead of undermining development by inducing rent-seeking behaviour and crowding out manufacturing activity (Strand, 2009). Insights from the governance of resource rents and aid flows can provide guidance on these issues, for example, by identifying best practices with regard to transparency and revenue management. Hence, this discussion emphasizes again that the decision to adopt RE cannot be based on a single criterion, but has to factor in a variety of aspects, including economic costs, ancillary benefits (such as energy access, energy security and reduced impacts on health and the environment), as well as additional funding possibilities by the means of climate finance.

9.3.2 Energy access

Significant parts of the global population today have no or limited access to modern and clean energy services. From a SD perspective, a sustainable energy expansion needs to increase the availability of energy services to groups that currently have no or limited access to them: the poor (measured by wealth, income or more integrative indicators), those in rural areas and those without connections to the grid. For households, the impacts from polluting and inefficient energy services on women have often been recognized (A. Reddy et al., 2000; Agbemabiese, 2009; Brew-Hammond, 2010).

Table 9.1 provides an estimate of the number of people without access to electricity, which totalled more than 1.4 billion in 2009. The regional distribution indicates that it is entirely a developing country issue, particularly in sub-Saharan Africa and South Asia.

A recent report from the UN Secretary General's advisory group on energy and climate change (AGECC, 2010) stresses the importance of universal access to modern energy sources by 2030 as a key part of enhancing SD. AGECC also suggests a new understanding of the term 'access', and identifies the specific contributions of RE to SD that go beyond the effects of increased energy access based on grid expansion or fossil technologies like diesel plants. This approach defines energy access as "access to clean, reliable and affordable energy services for cooking and heating, lighting, communications and productive uses" (AGECC, 2010) and illustrates the incremental process (Figure 9.4) involved in moving from servicing basic human needs to creating a self-sustaining process of SD.

Even a basic level of energy access, such as the provision of electricity for lighting, communication, healthcare and education, can result in substantial benefits for a community or household, including cost savings. However, AGECC argues for a broader definition of energy access and proposes that energy levels should provide not only for basic services but also for productive uses in order to improve livelihoods in the poorest countries and drive local economic development (see Figure 9.4). For a further discussion of energy access concepts, such as numerical minimum requirements for social and economic criteria, see Modi et al. (2005).

Access issues need to be understood in a local context[1] and in most countries there is a marked difference between electrification in urban and rural areas (Baumert et al., 2005; Bhattacharyya, 2005; World Bank, 2008b; UNDP and WHO, 2009; Brew-Hammond, 2010; IEA, 2010a). While this is especially true in the sub-Saharan African and South Asian regions, statistics show that rural access is still an issue of concern in developing regions with high overall national levels of electrification, illustrating that the rural-urban divide in modern energy services is still quite marked (see Table 9.1).

Decentralized grids based on RE are generally more competitive in rural areas with significant distances to the national grid (Baumert et al., 2005; Nouni et al., 2008; Deichmann et al., 2011) and the low levels of rural electrification offer significant opportunities for RE-based mini-grid systems. The role of RE in providing increased access to electricity in urban areas is less distinct. This relates either to the competitiveness

Table 9.1 | Millions of people without access to electricity in 2009 by region; projections to 2015 and 2030 under the IEA *World Energy Outlook 2010*, New Policies Scenario; and percentage of total populations with future access as a result of anticipated electrification rates (IEA, 2010b).

REGION	2009			2015	2030	2009	2015	2030
	Rural	Urban	Total	Total	Total	%	%	%
Africa	466	121	587	636	654	42	45	57
Sub-Saharan Africa	465	120	585	635	652	31	35	50
Developing Asia	716	82	799	725	545	78	81	88
China	8	0	8	5	0	99	100	100
India	380	23	404	389	293	66	70	80
Other Asia	328	59	387	331	252	65	72	82
Latin America	27	4	31	25	10	93	95	98
Developing Countries[1]	1,229	210	1,438	1,404	1213	73	75	81
World[2]	**1,232**	**210**	**1,441**	**1,406**	**1213**	**79**	**81**	**85**

Notes: 1. Includes Middle East countries. 2. Includes OECD and transition economies.

---

[1] See also the Earth trends database on electricity access: earthtrends.wri.org/searchable_db/index.php?theme=6.

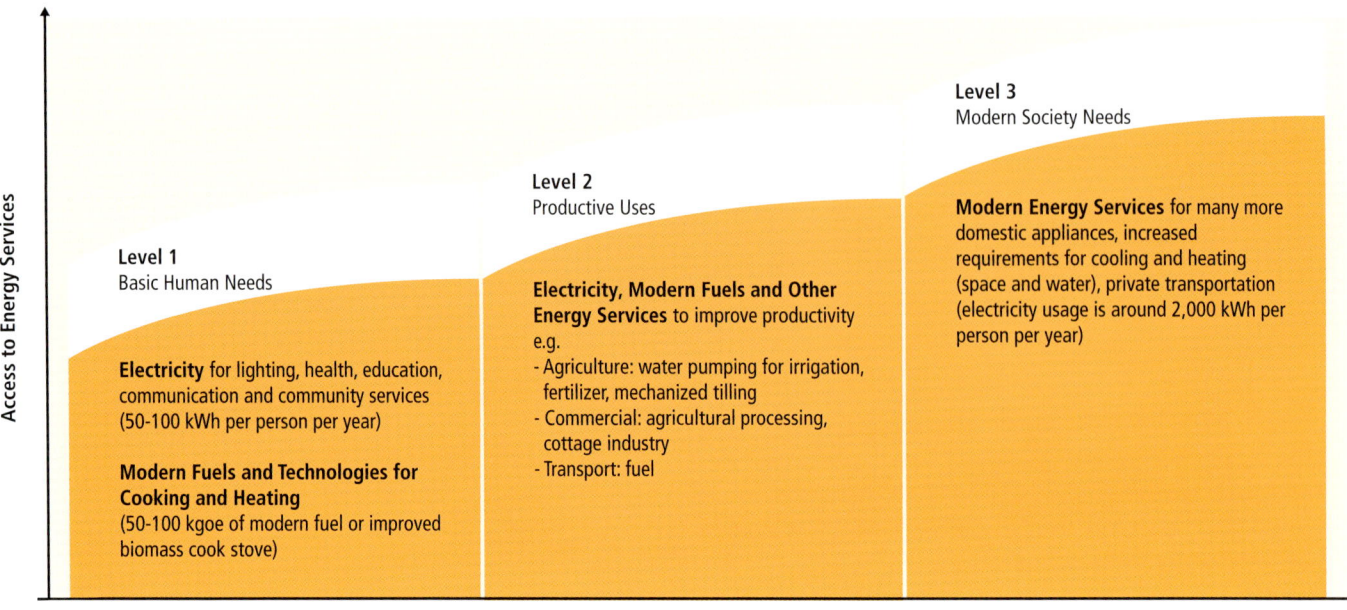

Figure 9.4 | Incremental level of access to energy services (AGECC, 2010; based on IEA data and analysis). Note: kgoe = kilogram(s) of oil equivalent.

with other grid supply options or to local social and economic issues at household or community levels; here, access is hampered by legal land issues or affordability.

Today, around 2.7 billion people rely on traditional biomass like wood, charcoal and dung for cooking energy and it is estimated that another half billion use coal (Table 9.2). Uncertainty in these estimates is high, but the span is limited across the different data sources (IEA, 2010a). In addition to the more than 1.4 billion with no access to electricity around another 1.3 billion people still use biomass, kerosene, coal or liquid propane gas (LPG) for energy-demanding services such as cooking despite having access to some form of electricity (Bravo et al., 2008; Karekezi et al., 2008; Dhingra et al., 2009, IEA, 2010b).

More detailed analysis of these statistics is generally hampered by very poor data about energy consumption among the poor in many developing countries. While an increasing number of national censuses include energy-related data, the coverage is still very limited for poor peri-urban and rural households with no official registration or land ownership (GNESD, 2008; Dhingra et al., 2009). The analytical constraints are compounded by the lack of well-defined and generally accepted indicators (IEA, 2010a).

The very dominant, and mainly indoor, use of traditional biomass fuels for cooking purposes has a number of documented negative effects. These include health impacts (Barnes et al., 2009; see Section 9.3.4.3), social effects, like the time spent gathering fuel or the high shares of income paid for small amounts of commercial biomass, and environmental aspects, like deforestation in areas where charcoal and market-based biomass are the dominant fuels.

A major challenge is to reverse the pattern of inefficient consumption of biomass by changing the present, often unsustainable, use to more sustainable and efficient alternatives. As illustrated by Figure 9.5 there is a strong correlation between low household income and use of low-quality fuels, illustrating that it is the poorest parts of the population who are at risk. The introduction of liquid or gaseous RE fuels, such as ethanol gels, to replace solid biomass for cooking could play a critical role whilst improving the health of millions of people (Lloyd and Visagle, 2007). While LPG has already displaced charcoal in some regions, it is a costly option for the majority of poor people and only a few countries have achieved significant penetration (Goldemberg et al., 2004). Replacing biomass or LPG with dimethyl ether produced from biomass shows some potential (Larson and Yang, 2004). The scale of liquid biofuel production required to meet cooking fuel demands is less than that for meeting transport fuel demand (Sections 8.2.4 and 8.3.1).

Table 9.2 | Number of people (millions) relying on traditional biomass for cooking in 2009 (IEA, 2010b).

REGION	Total
Africa	657
Sub-Saharan Africa	653
Developing Asia	1,937
China	423
India	855
Other Asia	659
Latin America	85
Developing Countries[1]	2,679
World[2]	2,679

Notes: 1. Includes Middle East countries. 2. Includes OECD and transition economies.

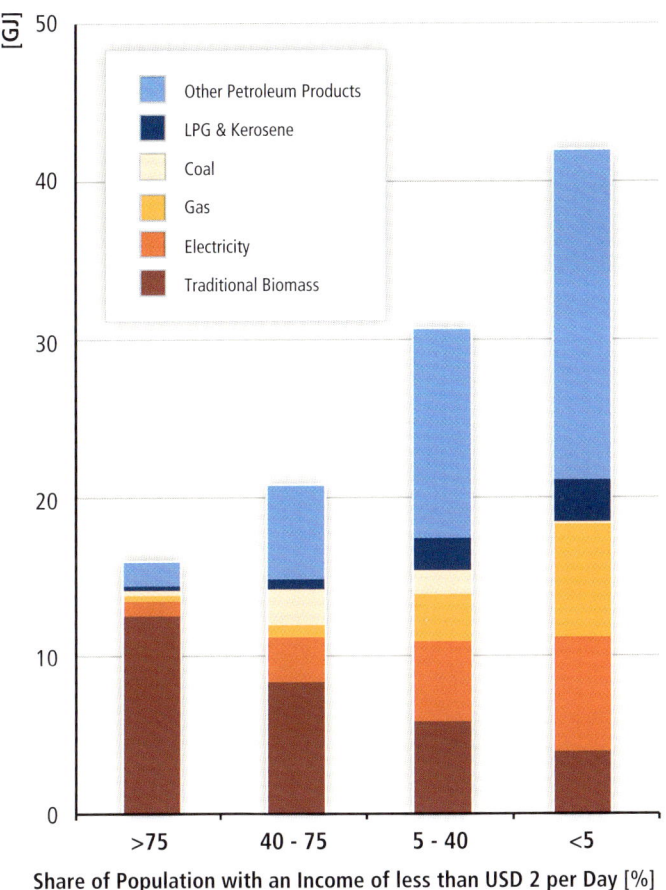

Figure 9.5 | The relationship between per capita final energy consumption and income in developing countries (IEA, 2010b). Data refer to the most recent year available during the period 2000 to 2008. Note: LPG = liquid petroleum gas.

Apart from the specific relevance of RE for electrification in remote areas, it is not well understood how contributions from RE sources can make a specific difference with regard to providing energy access in a more sustainable manner than other energy sources.

A study by the Global Network on Energy for Sustainable Development examined the options for RE technologies in making specific contributions to rural development (GNESD, 2007b). Several non-electrical technologies like using solar energy for water heating and crop drying, biofuels for transportation, biogas and modern biomass for heating, cooling, cooking and lighting, and wind for water pumping, etc. were found to serve priority household and productive energy needs (cooking, water heating, heating, water pumping) in areas with no access to electricity. This is also illustrated by the overview in Table 9.3, which outlines possible ways RE can provide basic energy services in rural off-grid areas. However, many of the options apply equally to the increasing number of slum communities in peri-urban areas where many households are not able to gain legal or economic access to even nearby electricity grids (Jain, 2010).

Energy access through some of these technologies allows local communities to widen their energy choices. As such, these technologies stimulate economies, provide incentives for local entrepreneurial efforts and meet basic needs and services related to lighting and cooking, thus providing ancillary health and education benefits. For example, the non-electrical technologies outlined above were found to exhibit a high potential for local job generation and increased economic activity through system manufacture and renewable resource extraction and processing (GNESD, 2007a).

Table 9.3 | Transition to renewable energy in rural (off-grid) areas (REN21, 2010).

Rural Energy Service	Existing Off-Grid Rural Energy Sources	Examples of New and Renewable Energy Sources
Lighting and other small electric needs (homes, schools, street lighting, telecom, hand tools, vaccine storage)	Candles, kerosene, batteries, central battery recharging by carting batteries to grid	• Hydropower (pico-scale, micro-scale, small-scale) • Biogas from household-scale digester • Small-scale biomass gasifier with gas engine • Village-scale mini-grids and solar/wind hybrid systems • Solar home systems
Communications (televisions, radios, cell phones)	Dry cell batteries, central battery recharging by carting batteries to grid	• Hydropower (pico-scale, micro-scale, small-scale) • Biogas from household-scale digester • Small-scale biomass gasifier with gas engine • Village-scale mini-grids and solar/wind hybrid systems • Solar home systems
Cooking (homes, commercial stoves and ovens)	Burning wood, dung, or straw in open fire at about 15% efficiency	• Improved cooking stoves (fuel wood, crop wastes) with efficiencies above 25% • Biogas from household-scale digester • Solar cookers
Heating and cooling (crop drying and other agricultural processing, hot water)	Mostly open fire from wood, dung, and straw	• Improved heating stoves • Biogas from small- and medium-scale digesters • Solar crop dryers • Solar water heaters • Ice making for food preservation • Fans from small grid renewable system
Process motive power (small industry)	Diesel engines and generators	• Small electricity grid systems from microhydro, gasifiers, direct combustion, and large biodigesters
Water pumping (agriculture and drinking water)	Diesel pumps and generators	• Mechanical wind pumps • Solar PV pumps • Small electricity grid systems from microhydro, gasifiers, direct combustion, and large biodigesters.

Implementation of RE-based energy access programs is expanding quite rapidly, but research on the sustainability-related aspects is still quite limited and there is hardly any literature on large-scale implementation. Instead, analysis has to rely on a few specific examples of actions where elements of energy access have been provided with a specific focus on the combination of social and productive services utilizing the potential for local job creation through small-scale business development (van der Vleuten et al., 2007; Nouni et al., 2008; Kaundinya et al., 2009; J. Peters et al., 2009; Urmee et al., 2009; Jonker Klunne and Michael, 2010). The assessment and case examples available, however, show that energy access is key for achievement of the MDGs and for economic development in general. RE technologies have the potential to make a significant contribution to improving the provision of clean and efficient energy services. But in order to ensure full achievement of the potential SD benefits from RE deployment, it is essential to put in place coherent, stable and supportive political and legal frameworks. The options for and barriers to such frameworks are further assessed in Chapter 11.

As a final caveat, it should also be noted that different RE facilities, that is, distributed versus central supply, face very different constraints, with the latter experiencing similar barriers as conventional energy systems, that is, high upfront investments, siting considerations, infrastructure and land requirements as well as network upgrade issues. Like for any other new technology, the introduction of RE will also face social and cultural barriers and implementation will need to be sensitive to social structures and local traditions like, for example, diets and cooking habits. There are many examples of improved stove programs failing due to lack of understanding of culture, staple food types and cooking habits (Slaski and Thurber, 2009).

9.3.3 Energy security

In addition to reducing energy consumption and improving energy efficiency, RE constitutes a further option that can enhance energy security. This section assesses the evidence for the potential contribution of RE technologies to energy security goals based on the two broad themes of energy security outlined in Section 9.2.2: availability and distribution of resources, and variability and reliability of energy sources.

The potential of RE to substitute for fossil energy—that is, theoretical and technical RE potentials—is summarized in Section 1.2 and discussed in detail in the respective technology chapters (Sections 2.2, 3.2, 4.2, 5.2, 6.2 and 7.2). Moreover, Section 11.3.3 discusses aspects of energy policies related to energy security.

9.3.3.1 Availability and distribution of resources

The ratio of proven reserves to current production (R/P), that is, for how many years production at current rates could be maintained before reserves are finally depleted, constitutes a popular measure to illustrate potential fossil fuel scarcities. According to this metric, recent estimates suggest that scarcity of coal (with a global R/P ratio of more than 100 years) is not a major issue at the moment, but at the current rate of production, global proven conventional reserves of oil and natural gas[2] would be exhausted in 41 to 45 and 54 to 62 years, respectively (BGR, 2009; BP, 2010; WEC, 2010).[3] While these figures only intend to give a sense of the magnitude of remaining fossil fuel reserves, they do not provide an assessment of when current reserves will actually be depleted. Proper interpretation of R/P ratios has to take many aspects into account, including the methodology of how reserves are classified and calculated, future changes in production and discovery of new reserves, as well as deterioration in the quality of reserves (Feygin and Satkin, 2004). A recent report that includes these factors in the analysis concludes with the projection of a likely peak of conventional oil before 2030 and a significant risk of a peak before 2020 (Sorrell et al., 2009).

As has been highlighted by the IEA (2008b) in its *World Energy Outlook 2008*, accelerated economic growth in many parts of the developing world is likely to raise global energy demand, which could further shorten the lifespan of remaining fossil fuel resources. Even though technological progress allows tapping reservoirs of oil from so-called non-conventional sources (such as, e.g., oil sands), usually large investments are required, which raise extraction costs and the price of oil and gas (Bentley, 2002). In addition, increasing amounts of energy are needed to produce a given quantity of usable energy from depleted conventional as well as from non-conventional reserves. Published estimates of the ratio of energy output-to-input (Energy Return on Energy Invested: EROEI, see Section 9.3.4) for conventional oil indicate that when the quality of reserves is taken into account there has been a substantial decline over time: while the EROEI reached its maximum of about 19 in 1972, it dropped to roughly 11 (i.e., about 42% lower) in 1997 (Cleveland, 2005). For non-conventional resources the EROEI is even lower (IEA, 2010b; Seljom et al., 2010). Thus, it is not surprising that the fossil fuel industry, particularly in the case of oil, has seen sharp increases in extraction costs over the past decade, although equipment, raw materials and labour demand have also played a role (EIA, 2009). Correlated with the increasing amounts of input energy to extract resources are the lifecycle carbon emissions from these resources.

As there is relatively little overlap between the location of fossil fuel reserves and the place of their consumption, fossil fuels are heavily traded and many countries with relatively scarce endowments rely to a large extent on imports of energy to meet desired levels of consumption.

---

2 Recent improvements in extraction technologies for shale gas and coal-bed methane are expected to result in notable production of natural gas from these non-conventional resources in the near future (IEA, 2008b).

3 Since 1990, proven conventional reserves of oil and natural gas have moderately grown due to revisions in official statistics, new discoveries and increased recovery factors. However, new discoveries have lagged behind consumption. Ultimately recoverable reserves (which include reserves that are yet to be discovered) are considerably larger than proven reserves; their actual size crucially depends on future oil prices and development costs (IEA, 2008b).

Due to the fact that a substantial share of global energy trade is channelled through a rather small number of critical geographical areas (so-called 'chokepoints'), it is highly vulnerable to accidents or terrorist attacks and importers face a considerable risk of supply disruption or price hikes (E. Gupta, 2008). Figure 9.6 shows that currently the European Union (EU-27), North America, and Asia and the Pacific region are net oil importers[4] supplying 85, 32, and 61% of their oil consumption from foreign producers, respectively. The EU-27 also relies on imports to meet more than half of its gas consumption, while for the Asia-Pacific region the import share is below 15% and North America almost fully meets demand for gas through domestic production. The Middle East, the Former Soviet Union (FSU), Africa and to some lesser extent Latin America are the most important exporters of oil and gas (for Africa, exports of both oil and gas exceed domestic consumption). Even though the EU-27 and the Middle East also rely on imports of coal,[5] energy security concerns are less salient: the former possesses reserves that exceed its annual consumption by a factor of more than 90, while for the latter coal only accounts for a marginal fraction of total energy use (BGR, 2009). This particular constellation of pronounced global imbalances in energy trade leads to a situation in which countries that heavily depend on energy imports frequently raise concerns that their energy consumption might be seriously affected by possible supply disruptions (Sen and Babali, 2007).

The spatial distribution of reserves, production and exports of fossil fuels is very uneven and highly concentrated in a few regions. Over 60% of coal reserves are located in just three regions (the USA, China and the FSU (BP, 2010)), and in 2009 China alone accounted for about half of global production of hard coal (IEA, 2010b). Over 75% of natural gas reserves are held by OPEC nations and states of the FSU, and 80% of the global gas market is supplied by the top 10 exporters (IEA, 2010b). This heavy concentration of energy resources, many of which are located in regions in which political events can have an adverse impact on the extraction or export of fossil fuel resources, creates a dependency for

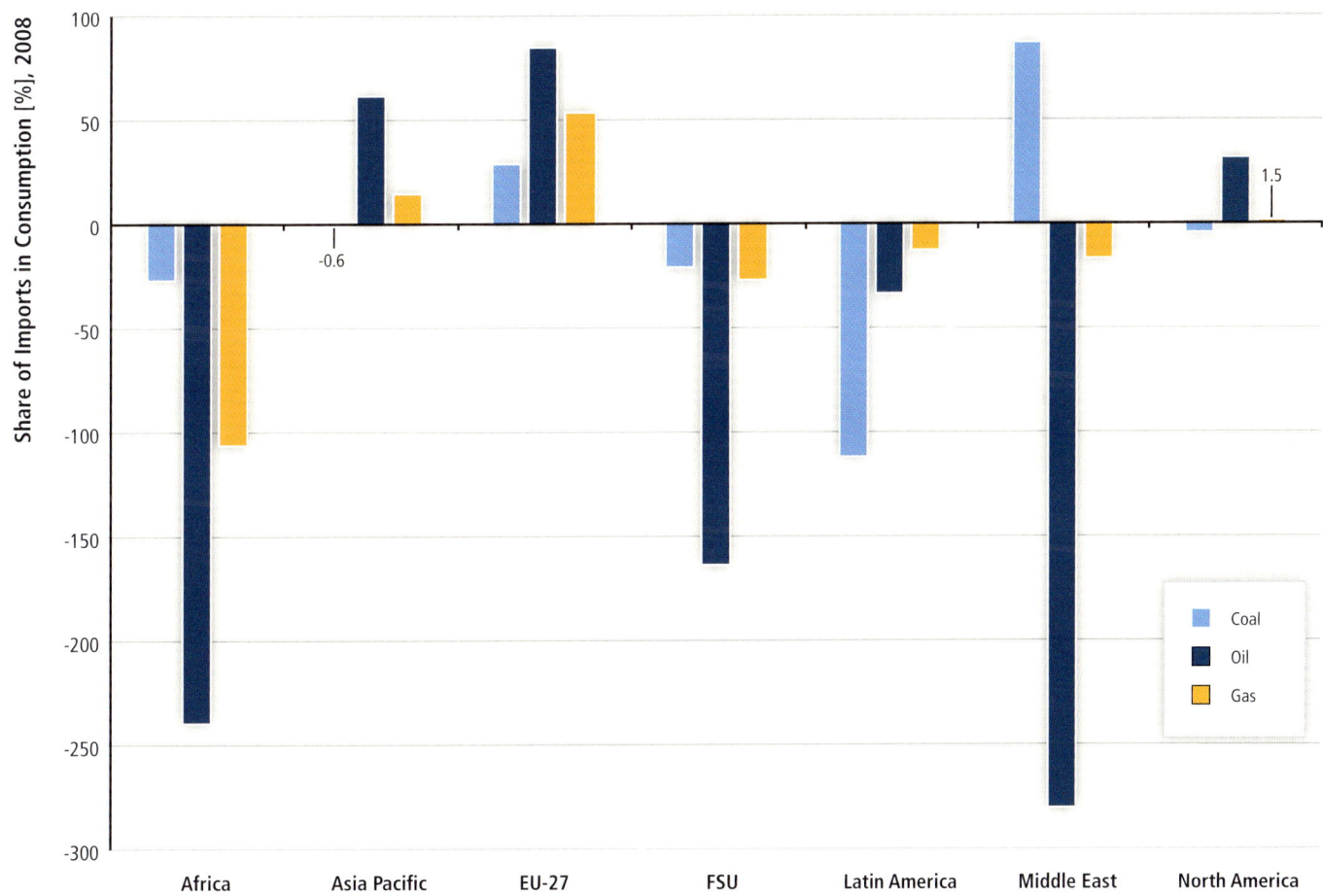

Figure 9.6 | Energy imports as the share of total primary energy consumption (%) for coal (hard coal and lignite), crude oil and natural gas for selected world regions in 2008. Negative values denote net exporters of energy carriers. Based on BGR (2009).

---

4   It should be noted that there is considerable heterogeneity within single regions (e.g., while the USA is a net oil importer, Canada is a net exporter).

5   Coal imports are hard coal; due to high transportation costs, lignite coal is in general not traded.

importers and raises the danger of energy supply disruptions (E. Gupta, 2008). That said, it should also be noted that exporting countries have a vested interest in maintaining income streams from the continued sale of fossil fuel supplies, so they are unlikely to limit exports for a prolonged period of time.

Further, for a number of countries (Moldova, Pakistan, Trinidad and Tobago, Madagascar, India, Ukraine, Tajikistan) the share of energy imports in total imports exceeded 25% for the period 2000 to 2005 and it was as high as 45% for Bahrain and 40% for Sierra Leone (World Bank, 2007b). A related indicator is the share that energy imports constitutes of export earnings and overall GDP. For example, Kenya and Senegal spend more than half of their export earnings for importing energy, while India spends over 45% (GNESD, 2010; Jain, 2010). Such dependence on energy imports exposes the affected economies to a potential risk of price fluctuations. The Energy Sector Management Program (ESMAP) of the World Bank has assessed the impacts of higher oil prices on low income countries and the poor (ESMAP, 2005).[6] Table 9.4, which summarizes these findings, illustrates that oil-importing developing countries are significantly affected by oil price increases and that a rise in oil prices of $USD_{1999-2001}$ 10 per barrel might result in GDP losses of almost 1.5% for the poorest countries (with per capita income less than $USD_{1999-2001}$ 300). The ESMAP national case studies also showed that the poorest households experienced the highest percentage changes in expenditures for commercial energy purchases of, for example, kerosene, LPG and diesel.

For these countries, increased uptake of RE technologies could further be an avenue to redirect foreign exchange flows away from energy imports towards imports of goods that cannot be produced locally, such as high-tech capital goods. For other developing countries that are net exporters of energy, promoting the domestic use of RE can extend the lifetime of their fossil resource base and prolong the time to diversify the scope of economic activities by decreasing the dependence on resource exports while strengthening their manufacturing and service sectors.

Governments frequently try to limit the impacts of international price increases in the short term by adjusting subsidies or providing targeted cash support to the poorest households, rationing supply or forcing supply companies to absorb some of the short-term effects (ESMAP, 2005, 2006, 2008). Since this may have significant effects both on state budgets and companies' abilities to maintain stable delivery (UNEP, 2008b), longer-term responses are focused more on efficiency measures and diversification. In this context, it needs to be noted that import dependencies do not only occur with respect to specific energy sources; the technologies needed for implementation of RE have their own specific risks for potential supply disruptions and price volatility (see Box 9.1).

9.3.3.2 Variability and reliability of energy supply

Besides the advantageous properties discussed above, renewable energy sources also possess some drawbacks. The variable long- or short-term availability of some RE due to seasonal, diurnal or weather changes can be addressed by storage and technical balancing to meet heat or power demand changes. In addition, institutional settings for energy markets can be optimized, such as regionally integrated electricity markets in which local fluctuations can be smoothed by means of geographic diversification (Roques et al., 2010), and a range of other solutions including grid flexibility may be implemented (see Section 8.2.1). The solutions to overcome variability constraints on an energy supply system can involve additional costs that should be taken into account when comparing the relative benefits of RE with conventional energy technology projects.

Analysis and operating experience primarily from certain OECD countries suggest that, at least for low to medium levels of wind electricity penetration (defined as up to 20% of total annual average electrical energy demand), the integration of wind energy generally poses no insurmountable technical barriers and is economically manageable. Nevertheless, concerns about (and the costs of) wind energy integration will grow with wind energy deployment and, even at lower penetration levels, integration issues must be actively managed. At low to medium levels of wind electricity penetration, the available literature suggests that the additional costs of managing electric system variability and uncertainty, ensuring generation adequacy and adding new transmission to accommodate wind energy will be system specific but generally in the range of US $cents_{2005}$ 0.7 to 3/kWh (Section 7.5).

Table 9.4 | Percentage change in GDP resulting from a $USD_{1999-2001}$ 10 per barrel rise in oil prices[1] (analytical results grouped by income levels) (ESMAP, 2005).

Net Oil Importers		Net Oil Exporters	
Income per capita ($USD_{1999-2001}$)	ΔGDP (%)	Income per capita ($USD_{1999-2001}$)	ΔGDP (%)
<300	-1.47	<300	+5.21
300–900	-0.76	900–9,000	+4.16
900–9,000	-0.56		
>9,000	-0.44		

Note: 1. As the grouping of countries in this table does not correspond to any regional grouping, it was not possible to convert monetary values to year 2005 USD due to a lack of appropriate conversion factors.

---

6   It should be noted that the data are based on a large number of country case studies and thus are not necessarily universally valid.

Box 9.1 | Access to raw materials for future renewable resources deployment.

While renewable resources can be a powerful instrument to mitigate fossil fuel depletion, scarcity of other raw materials may pose constraints to enhanced deployment of RE technologies. Securing access to required scarce inorganic mineral raw materials (IRM), above all precious rare earth and some specialty metals, at reasonable prices is an upcoming challenge for all industries. For the complex renewable energies sector no specific assessment of the structure and quantity of IRM demand is available. To identify potential areas of concern for future renewable resources deployment, a large set of technologies and possible technology pathways has to be considered; several reports are available as starting points for such analyses (Frondel et al., 2007; Reuscher et al., 2008; Angerer et al., 2009; Ziemann and Schebek, 2010; US DOE, 2010; EC, 2010; Kristof and Hennicke, 2010; Teipel, 2010).

The IRM supply chain has to be understood as a vulnerable system and is subject to various threats. Sources of potential market distortions are concentration processes and political instability of some major mining countries. Currently, 97% of rare earth elements, 60% of indium and 30% of gallium production are located in China, 56% of the global chromium supply is controlled by South Africa and Kazakhstan and 55% of cobalt is mined in politically instable regions in Africa (USGS, 2010).

With some notable exceptions (e.g., silver), future IRM constraints will be caused by imbalances of demand and supply rather than by depletion of geological resources (Angerer, 2010). Some metals are derived as by-products, mostly from ores of major or carrier metals in which they are present in low concentrations. Their production levels depend on the demand for the major metal as the main economic driver of extraction (Hagelüken and Meskers, 2010). Typical by-product metals are gallium, germanium, indium, tellurium and selenium. In some deposits, groups of metals may occur as 'coupled elements' without a real carrier metal. Notable examples include the platinum group metals and rare earth elements that generally have to be mined and processed together. In such cases, it may not be economically viable to increase production in response to rising demand for a certain element. As a result, complex price patterns and supply risks emerge. Market tensions also occur in response to unexpected changes in demand, for example, as a result of fast-rising prosperity in emerging and developing countries, or technology breakthroughs that cause a demand surge or drop.

In the future, demands for certain metals are projected to multiply significantly. Indicators that relate raw material demand by emerging technologies in 2030 to today's total world production show that as a result of expected technical innovations the demand for gallium and neodymium may be 6 and 3.8 times higher, respectively (Angerer et al., 2009; see Table 9.5). Demand drivers for gallium are thin-layer photovoltaics and high-speed integrated circuits, and for neodymium high-performance permanent magnets used in generators of wind turbines and energy efficient electric motors.

Table 9.5 | Estimated global demand for selected metals by emerging technologies in 2030 as a multiple of world production in 2006 (Angerer et al., 2009).

Element	Multiple
Gallium	6
Neodymium	3.8
Indium	3.3
Germanium	2.4
Scandium	2.3
Platinum	1.6
Tantalum	1
Silver, Tin	0.8
Cobalt	0.4
Palladium, titanium	0.3

The vulnerability of industrial sectors is especially large if there is no possibility for substitution. Current examples for such a lack of substitutes include chromium in stainless steels (e.g., for tidal power plants), cobalt in wear-resistant super alloys, scandium in lightweight alloys, indium in transparent indium-tin-oxide electrodes for photovoltaic panels and neodymium in strong permanent magnets. At the same time there are also competing uses of raw materials between industries. Cobalt, for instance, is needed for the varied and growing applications of lithium-ion rechargeable batteries, for catalysts in the Fischer-Tropsch process that may be used to produce future synthetic fuels from biomass, and is an essential component of extremely wear-resistant parts in automotive, mechanical and medical engineering. Table 9.6 gives an overview of critical raw materials in some essential components of renewable resources technologies.

An important future contribution to a secure IRM supply is the set-up of effective recycling systems. End-of-life products such as electronics, batteries or catalysts contain in total significant amounts of comparably enriched metals. For RE technologies it might become crucial to develop closed loop recycling concepts from the very beginning. Besides several environmental advantages, this could enhance the supply situation and long-term supply security of scarce raw materials and reduce dependency on (usually more energy intensive) primary supply while mitigating metal price volatility (Hagelüken and Meskers, 2010).

Table 9.6 | Critical raw materials content of renewable resources technologies.

Application	Component	Critical raw materials content
Wind and hydropower plants	Permanent magnets of synchronous generator	Neodymium, dysprosium, praseodymium, terbium
	Corrosion-resistant components	Chromium, nickel, molybdenum, manganese
Photovoltaics	Transparent electrode	Indium
	Thin film semiconductor	Indium, gallium, selenium, germanium, tellurium
	Dye-sensitized solar cell	Ruthenium, platinum, silver
	Electric contacts	Silver
Concentrating solar power (CSP)	Mirror	Silver
Fuel cell-driven electric vehicles	Hydrogen fuel cell	Platinum
	Electric motor	Neodymium, dysprosium, praseodymium, terbium, copper
Biomass to liquid (BtL)	Fischer-Tropsch synthesis	Cobalt, rhenium, platinum
Electricity storage	Redox flow rechargeable battery	Vanadium
	Lithium-ion rechargeable battery	Lithium, cobalt
Electricity grid	Low-loss high-temperature super-conductor cable	Bismuth, thallium, yttrium, barium, copper

A number of emerging regional power collaborations in East, West and Southern Africa, South and Central America and South East Asia aim to enhance the reliability of electricity grids and therefore local supply. ESMAP (2010) studied 12 sub-regional integration schemes and found that for most schemes energy security was one of the motivating factors. Larger integrated networks may also provide benefits in terms of cost efficiency, trade and more general economic development.

Many developing countries specifically include providing adequate and affordable access to all parts of the population as part of their definition of energy security and in this way link the access and security issues while broadening the concept to include stability and reliability of local supply. While regional interconnections may be an interesting way to ensure better supply security at the national level, it does not automatically 'trickle down' to the poorer segments of the population in terms of increased access or even stable and affordable supply for those who are connected. GNESD (2004) examined the effects of power sector reforms on access levels and found that only when there was strong political commitment to improve access to electricity for poor households did reforms deliver results. An explicit focus on poor households was found essential along with specific protection of funds for electrification.

While electricity connection is often used as a key indicator for access to modern energy services, it is important to underline that household connections have restrictions in terms of capacity, stability and outage problems, as illustrated by the data from the World Bank in Table 9.7.

Energy security at the micro level in developing countries may therefore have a number of social and economic effects that go beyond direct impacts of fuel price increases (Jain, 2010). Improving access to affordable and reliable energy supply will therefore not only provide improved energy services, but it may also broadly increase productivity and avoid parallel investments in infrastructure, from small-scale generation equipment to parallel lighting and cooking systems, where most households have at least two different options to hedge against unstable supply. However, decentralized RE is competitive mostly in remote and rural areas, while grid-connected supply generally dominates denser areas where the majority of households reside (Deichmann et al., 2011).

9.3.4 Climate change mitigation and reduction of environmental and health impacts

SD must ensure environmental quality and prevent undue environmental harm. No large-scale technology deployment comes without environmental trade-offs, and a large body of literature is available that assesses various environmental impacts of energy technologies from a bottom-up perspective.

The goal of this section is to review and compare available evidence about the environmental impacts associated with current and near-future energy technologies, including the full supply chain. This review is largely based on literature from lifecycle assessments (LCA). LCA does not attempt to determine a socially optimal energy supply portfolio; its aim is to aid technology comparisons in terms of environmental burden. While the development of sustainable strategies and portfolios needs to be viewed from a top-down, macro-economic and systemic perspective, bottom-up evidence from LCA provides valuable insights about the environmental performances of different technologies across categories. Similarly, the energy payback time (EPT, see Box 9.3) provides a measure for the lifecycle energy efficiency of individual technologies, which is helpful for identifying high-quality energy sources, but must additionally be viewed in the broader economic and

Table 9.7 | Indicators of the reliability of infrastructure services (World Bank, 2007a).

	Sub-Saharan Africa	Developing countries
Delay in obtaining electricity connection (days)	79.9	27.5
Electrical outages (days per year)	90.9	28.7
Value of lost output due to electrical outages (percent of turnover)	6.1	4.4
Firms maintaining own generation equipment (percent of total)	47.5	31.8

social context. As the following sections review the results of hundreds of LCA studies, the major characteristics and challenges of LCA in the context of energy technologies are introduced below (Box 9.2).

LCA allows a detailed investigation into the environmental consequences that are associated with manufacture, operation and decommissioning of a specific technology evaluated in the context of the current energy system. In doing so, LCAs complement economic assessments that focus on current costs, for example, the levelized cost of energy (LCOE; see Section 10.5.1). In the same way as future costs of RE technologies might decline (e.g., due to research and development (R&D) and learning by doing; see Section 10.5.2), the way future RE technologies are manufactured, operated and decommissioned might change as well. As a consequence, a comprehensive assessment of different RE expansion strategies should try to take these expected modifications into account. While marginal changes in the background energy system can be addressed by consequential LCA (see Box 9.2), non-marginal changes due to the ongoing evolution of the background systems can be accounted for in scenario analyses (see Sections 10.2 and 10.3). By extending scenario analyses to include lifecycle emissions and the energy requirements to construct, operate and decommission the different technologies explicitly, integrated models could provide useful information about the future mix of energy systems together with its associated lifecycle emissions and the total environmental burden.

It is not possible to cover all relevant environmental impacts[7] associated with energy supply technologies within the scope of this chapter. This section concentrates mostly on electricity generation and liquid transport fuels, as these areas are most frequently reported in the literature, including the technology chapters of this report. Heating and household energy are included in the assessments on air pollution and health, but omitted from most other sections due to a paucity of published work. Regarding the lifecycle impacts of heating fuels, the upstream impacts of fuel extraction and processing are in many cases similar to those of the corresponding transport or electricity generation chains. However, some renewable technologies such as heat pumps or passive solar may exhibit different properties. The discussion of transport fuels focuses on biofuels, as they are currently the only renewable fuels that can be considered mature and available for large-scale application. A discussion of renewable electricity generation for charging of electric battery vehicles, and other future pathways is provided in Section 8.3.1. A broader discussion of technology integration options is provided in Chapter 8.

Data available for different impact categories vary widely regarding the number and quality of sources. GHG emissions are generally well covered (Section 9.3.4.1). A significant number of studies report on air pollutant emissions (Section 9.3.4.2), related health impacts (Section 9.3.4.3) and operational water use (Section 9.3.4.4), but evidence is scarce for (lifecycle) emissions to water, land use (Section 9.3.4.5) and health impacts other than those linked to air pollution. Discussion of impacts on biodiversity and ecosystems is limited to qualitative summaries of potential areas of concern (Section 9.3.4.6), as no quantitative basis for comparison is available. To account for burdens associated with accidents as opposed to normal operation, Section 9.3.4.7 provides an overview about risks associated with energy technologies.

9.3.4.1 Climate change

This section reviews available estimates of lifecycle GHG emissions from renewable and non-renewable electricity generation technologies and liquid transportation fuels. Positive and negative emissions related to land use change (LUC) are omitted from both reviews, and discussed separately, albeit with a focus on biofuels.

LUC-related GHG emissions are potentially relevant to any technology, but are most significant for technologies that transform substantial amounts of land, and induce changes in carbon stocks of that land. For bioenergy systems, LUC impacts could reduce, negate or enhance potential GHG emission reduction benefits depending on the circumstance and assumptions. Methane emissions from submersed biomass or organic sediments may produce substantial emissions for certain hydropower reservoirs. However, the state of the science regarding actual net emissions from hydropower reservoirs is unresolved (see Section 5.6.3 for details). Research on LUC related to resource extraction for fossil fuels, for example, mountaintop-removal coal mining (Fox and Campbell, 2010) or oil production (Yeh et al., 2010), is nascent (Gorissen et al., 2010).

---
7   Within this subsection, the term impacts is not used in the strict sense of its definition within the field of LCA.

Box 9.2 | Lifecycle assessments of energy technologies.

LCA studies provide a well-established and comprehensive framework to compare RE with fossil-based and nuclear energy technologies. LCA methodologies have been evolving for a few decades and are now supported by international initiatives (UNEP and SETAC, 2010) and governed by standards (Cowie et al., 2006; ISO, 2006). Although LCA is increasingly applied to energy technologies, some methodological challenges persist (Udo de Haes and Heijungs, 2007).

The majority of the available literature on energy technologies is based on so-called attributional LCAs, which investigate the environmental impacts associated with the average product or technology lifecycle (Figure 9.7). A resulting key limitation is that changes in the energy system that might result from the decision to install additional renewable capacity are excluded. For instance, for wind power and solar PV, variability and limited predictability leads to an increased need for balancing reserves, and possibly efficiency penalties in the case of fossil power plants providing these reserves (R. Gross et al., 2007; Pehnt et al., 2008; see also Sections 3.5.4 and 7.6.1.3). In contrast, the recently developed approach of consequential LCA considers the marginal effects of implementing a technology, and displacing and changing the operation of other technologies, as reflected by market dynamic interactions between technologies and industries (Rebitzer et al., 2004; Brander et al., 2008; Finnveden et al., 2009). However, consequential LCAs form the minority of studies in the literature, and context dependency precludes the incorporation of the limited results available into the broader assessments presented here. Assumptions and changing characteristics of the background energy system (e.g., its carbon intensity) in turn particularly affect LCAs of most RE technologies, since their lifecycle impacts stem almost entirely from component manufacturing (see Lenzen and Wachsmann, 2004). Further challenges include the potential for double-counting when assessing large interconnected energy systems (Lenzen, 2009), and system boundary problems (Suh et al., 2003; Lenzen, 2008).

Substantial variability in published LCA results (as seen, for example, in Figure 9.8) is also due to technology characteristics (e.g., design, capacity factor, variability, service lifetime and vintage), geographic location, background energy system characteristics, data source type (empirical or theoretical), differences in LCA technique (e.g., process-based LCA or input-output LCA) and key methods and assumptions (e.g., co-product allocation, avoided emissions, study scope). Given these significant caveats, emphasis will be placed on the underlying reasons for uncertainties and variations when describing the results for selected energy technologies.

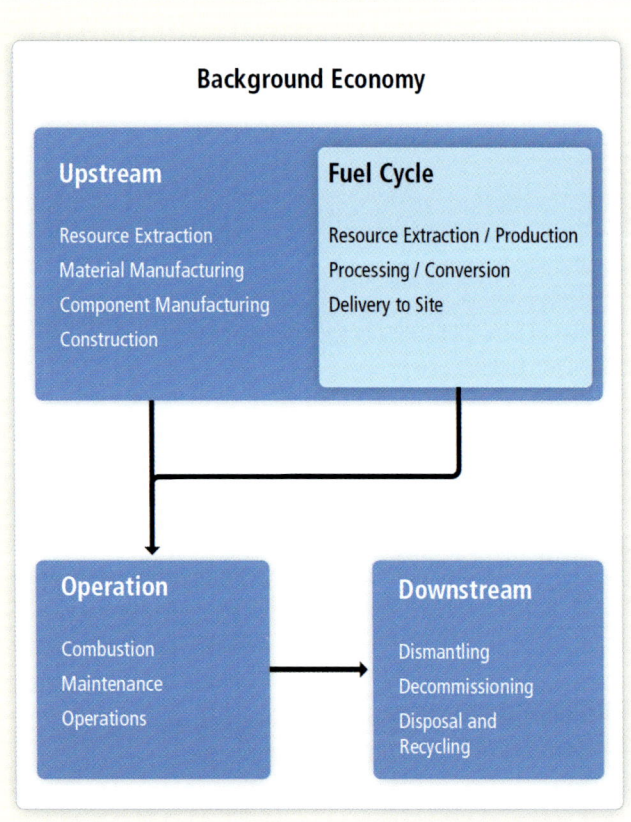

Figure 9.7 | Illustration of generalized lifecycle stages for an energy technology. Fuel cycle applies to fossil and nuclear chains and bioenergy.

LUC-related GHG emissions are excluded from the reviews for the following reasons:

1) significant gaps in available evidence for the full range of power technologies and fuels evaluated in this section preclude consistent comparisons; and

2) uncertainties in estimating GHG emissions from LUC are high relative to the understanding of GHG emissions more directly associated with the manufacture, operation and decommissioning of the technology itself.

Uncertainty in LUC estimates stems from many sources that are currently unresolved and inconsistent, including: modelling and estimation methods; data and modelling resolution (spatial, temporal, categorical); system boundary and vintage; allocation of impacts among primary products, co-products and residues; assumptions about the policy context and market size and characteristics; projections of technological

Box 9.3 | Energy payback of electricity generation.

The role of high-quality energy sources in the development of modern civilizations is widely recognized. The energy payback time (EPT) and similar concepts described below provide a measure for energetic efficiency of technologies or fuels. The following characterizes the balance between the energy expended for the manufacture, operation and decommissioning of electricity generating plants (the 'embodied' energy) and their energy output in terms of an EPT, that is, the operational time it would take the technology to recover its own embodied energy. For combustion technologies, this includes the energy requirements of fuel extraction and processing, but not the energy content of the fuel itself. The EPT is closely related to other common metrics such as the energy return on energy invested (EROEI) or the energy ratio. The latter quantities depend on assumptions about the expected lifetime of a plant, which is also shown below (see Annex II for definitions and further explanations). For some RE technologies, for example, wind and PV, EPTs have been declining rapidly over the last years due to technological advances and economies of scale. Fossil and nuclear power technologies are characterized by the continuous energy requirements for fuel extraction and processing. This might become increasingly important as qualities of conventional fuel supply decline and shares of unconventional fuels rise (Farrell and Brandt, 2006; Gagnon, 2008; Lenzen, 2008).

In addition to the common causes of variability in estimates of impacts from LCAs (Box 9.2), the ranges in Table 9.8 are mainly caused by variations in:

- Fuel characteristics (e.g., moisture content), cooling method, ambient and cooling water temperatures, and load fluctuations (coal and gas);
- Uranium ore grades and enrichment technology (nuclear);
- Crystalline or amorphous silicone materials (PV solar cells);
- Economies of scale in terms of power rating (wind); and
- Storage capacity and design (concentrating solar).

In addition, the location-specific capacity factor has a major bearing on the EPT, in particular that of variable RE technologies.

Table 9.8 | Energy payback times and energy ratios of electricity-generating technologies. Electricity from biomass is excluded, as the literature almost exclusively documents GHG instead of energy balances for this technology, and mostly covers the biofuel cycle only (Lenzen, 1999, 2008; Voorspools et al., 2000; Lenzen and Munksgaard, 2002; Lenzen et al., 2006; Gagnon, 2008; Kubiszewski et al., 2010).

Technology	Energy payback time (years)		Most commonly stated lifetime (years)	Energy ratio (kWh_e/kWh_{prim})	
	Low value	High value		Low value	High value
Brown coal, new subcritical	1.9	3.7	30	2.0	5.4
Black coal, new subcritical	0.5	3.6	30	2.5	20.0
Black coal, supercritical	1.0	2.6	30	2.9	10.1
Natural gas, open cycle	1.9	3.9	30	1.9	5.6
Natural gas, combined cycle	1.2	3.6	30	2.5	8.6
Heavy-water reactors	2.4	2.6	40	2.9	5.6
Light-water reactors	0.8	3.0	40	2.5	16.0
Photovoltaics	0.2	8.0	25	0.8	47.4
Concentrating solar	0.7	7.5	25	1.0	10.3
Geothermal	0.6	3.6	30	2.5	14.0
Wind turbines	0.1	1.5	25	5.0	40.0
Hydroelectricity	0.1	3.5	70	6.0	280.0

performance, background energy system and comparison reference case; and evaluation time horizon (Cherubini et al., 2009; Kline et al., 2009; Hertel et al., 2010).

Other uncertainties related to estimation of GHG emissions from bioenergy in particular include N_2O emissions from fertilization and soils (Crutzen et al., 2008; E. Davidson, 2009), how technologies perform

Renewable Energy in the Context of Sustainable Development

Chapter 9

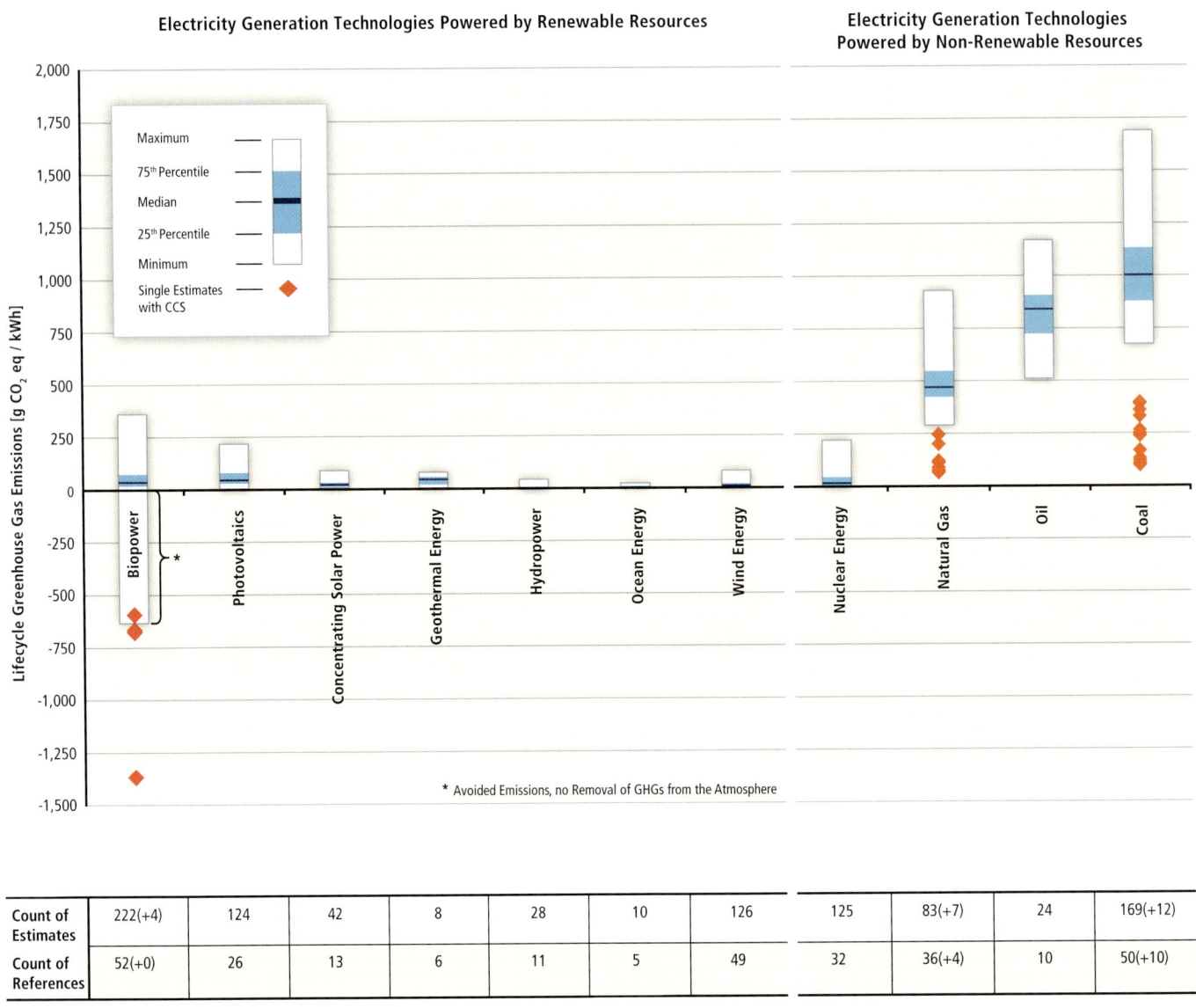

Figure 9.8 | Estimates of lifecycle GHG emissions (g CO_2eq/kWh) for broad categories of electricity generation technologies, plus some technologies integrated with CCS. Land-use related net changes in carbon stocks (mainly applicable to biopower and hydropower from reservoirs) and land management impacts are excluded; negative estimates[1] for biopower are based on assumptions about avoided emissions from residues and wastes in landfill disposals and co-products. References and methods for the review are reported in Annex II. The number of estimates is greater than the number of references because many studies considered multiple scenarios. Numbers reported in parentheses pertain to additional references and estimates that evaluated technologies with CCS. Distributional information relates to estimates currently available in LCA literature, not necessarily to underlying theoretical or practical extrema, or the true central tendency when considering all deployment conditions.

Note: 1. 'Negative estimates' within the terminology of lifecycle assessments presented in this report refer to avoided emissions. Unlike the case of bioenergy combined with CCS, avoided emissions do not remove GHGs from the atmosphere.

in practice compared to models and regulations now and in the future, lack of commercial-scale lignocellulosic feedstocks and fuels production, and other potentially significant indirect effects such as rebound effects in energy consumption due to changes in the price of energy after introduction of RE (Rajagopal et al., 2010). These uncertainties—along with the LCA-related caveats discussed in Box 9.2—should be kept in mind when considering the evidence presented in Section 9.3.4.1.

Lifecycle greenhouse gas emissions of electricity generation technologies

This section synthesizes evidence from a comprehensive review of published LCAs covering all regions of the world (literature collection, screening and analytical methods are described in Annex II). Without considering LUC, lifecycle GHG emissions normalized per unit of electrical output (g CO_2eq/kWh) from technologies powered by renewable resources are generally found to be considerably less than from those powered by fossil fuel-based resources (Figure 9.8). Nuclear power exhibits a similar inter-quartile range (IQR; 75th minus 25th percentile values) and median as do technologies powered by renewable resources. The maximum estimate for CSP, geothermal, hydropower, ocean and wind energy is less than or equal to 100 g CO_2eq/kWh and median values for all RE range from 4 to 46 g CO_2eq/kWh, although the number of references examining several of these technologies is small. The upper quartile of the distribution of estimates for photovoltaics and biopower extend 2 to 3 times above the maximum for other RE technologies, as it does for nuclear, mainly owing to differences in background energy system, assumed uranium ore grade (nuclear) and cases of suboptimal production processes (PV, biopower). Nevertheless, only the very highest estimates for biopower overlap with the range of a fossil-fuelled technology, and the central tendencies of all RE are between 400 and nearly 1,000 g CO_2eq/kWh lower than their fossil-fuelled counterparts (without CCS).

Cases of post-combustion carbon capture and storage (CCS) represent the emissions associated with the base technology plus CCS. As expected, their lifecycle GHG emissions are considerably lower than those of the base technology, and for fossil-fuelled technologies, can bring total lifecycle GHG emissions near the range of several RE technologies. Biopower with CCS can display significantly negative GHG emissions (without considering LUC). Because CCS is still not a mature technology, assumptions regarding the duration of sequestration and leakage rates contribute to the variability seen in Figure 9.8.

The proportion of GHG emissions from each lifecycle stage differs for technologies powered by renewable and non-renewable resources. For fossil-fuelled technologies, fuel combustion during operation of the facility emits the vast majority of GHGs. For nuclear and RE technologies, the majority of GHG emissions are upstream of operation. Most emissions for biopower are generated during feedstock production, where agricultural practices play an important role. For nuclear power, fuel processing stages are most important, and a significant share of GHG emissions is associated with construction and decommissioning. For other renewable technologies, most lifecycle GHG emissions stem from component manufacturing and, to a lesser extent, facility construction. The background energy system that, for instance, powers component manufacturing, will evolve over time, so estimates today may not reflect future conditions.

Variability in estimates of lifecycle GHG emissions from the evaluated technologies is caused both by factors related to methodological diversity in the underlying literature (see Box 9.2), and factors relating to diversity in the evaluated technologies. Expanding on the latter, for combustion technologies (fossil fuels and biopower), variability is most prominently caused by differences in capacity factor (which influences GHG emissions for many other technologies as well), combustion efficiency, carbon content of the fuel, and conditions under which the fuel is grown/extracted and transported. Biopower additionally is affected by assumptions regarding the reference use of the biomass feedstock; for instance, if landfilling of organic material can be avoided, the use of that biomass for power generation can be considered as avoiding methane emissions (seen in the non-CCS, negative emission estimates in Figure 9.8). Variability for PV stems from the rapidly evolving and multiple solar cell designs. For solar, geothermal,[8] ocean and wind technologies, the quality of the primary energy resource at the site significantly influences power output.

The state of knowledge on lifecycle GHG emissions from the electricity generation technologies was found to vary. The following synopses are based on an assessment of the number of references and estimates, the density of the distribution of estimates (IQR and range relative to the median), and an understanding of key drivers of lifecycle GHG emissions. Lifecycle GHG emissions from fossil-fuelled technologies and wind appear well understood.[9] Reasonably well known, but with some potentially important gaps in knowledge and a need for corroborative research, are those for biopower, hydropower, nuclear, some PV technologies and CSP. The current state of knowledge for geothermal and ocean energy is preliminary.

Lifecycle greenhouse gas emissions of selected petroleum fuels and biofuels

In this section, literature-derived estimates of lifecycle GHG emissions for first-generation biofuels (i.e., sugar- and starch-based ethanol, and oilseed-based biodiesel and renewable diesel (RD)), and selected next-generation biofuels derived from lignocellulosic biomass (i.e., ethanol and Fischer-Tropsch diesel (FTD)) are compared. Ranges of emissions for first-generation biofuels represent state-of-the-art technologies and projections of near-term technological improvements while those for next-generation ethanol and FTD from lignocellulosic biomass represent conceptual designs envisioned for commercial-scale biorefineries.

Emissions are reported on the basis of 1 MJ of fuel produced and used to propel a passenger vehicle. These results are nearly equivalent to a comparison per vehicle km travelled because the vehicle fuel efficiency (distance travelled per MJ) is virtually unchanged when considering the evaluated biofuels and the petroleum fuels they displace used in the same vehicle (Beer et al., 2002; Sheehan et al., 2004; CARB, 2009). Emissions from direct and indirect LUC are excluded for all fuels, and discussed in the following subsection (see also Sections 2.3.1 and 2.5.3). Readers should refer to Section 8.3.1 for a comparison of lifecycle GHG emissions of various fuels (including hydrogen and electricity) used in different vehicle configurations. Note that electric vehicles could have

---

8   Also, some existing formations may have high operational emissions of CO_2 due to configuration and high dissolved CO_2 concentrations in geothermal fluids, which are not reflected in LCA literature assessed. See Sections 4.5.1 and 4.5.2 for details.

9   In late 2010, some controversy emerged over potential revisions to the GHG profile of natural gas. Some observers believe that methane leakage associated with upstream production and transport of natural gas is higher than historically categorized. See EPA (2010a) and Lustgarten (2011) for views of this emerging controversy.

lower lifecycle GHG emissions compared to vehicles fuelled with existing biofuels if electricity from renewable sources is used, or higher emissions than petroleum-based fuels if carbon-intensive fossil-based power generation is used (Creutzig et al., 2009; van Vliet et al., 2011).

Results from the studies reviewed suggest that, without considering potential LUC-related GHG emissions, first- and next-generation biofuels have lower direct lifecycle GHG emissions compared to petroleum fuels from a variety of crude oil sources (Figure 9.9). By comparison, the range in estimates for biofuels is much wider than that for gasoline and diesel. This can be attributed to many factors, including the types of feedstocks utilized; variations in land productivity, crop management practices, conversion process, and process energy source; uncertainty in N_2O emissions from fertilization; and methodological choices in LCAs, for example, co-product allocation approaches and definition of system boundaries[10] (Williams et al., 2009; Hoefnagels et al., 2010; Cherubini and Strømman, 2011; see also Box 9.2).

Although there is significant overlap in the ranges of lifecycle GHG emissions for virtually all biofuels, not all biofuel systems are equally efficient in reducing GHG emissions compared to their petroleum counterparts. For example, ethanol from Brazilian sugarcane has lower GHG emissions than that produced from wheat and corn (von Blottnitz and Curran, 2007; S. Miller, 2010). Estimates are reasonably comparable for biodiesel derived from rapeseed and soybean (Hill et al., 2006; CONCAWE, 2008; Huo et al., 2009a; Hoefnagels et al., 2010). Without LUC, palm oil biodiesel could have similar lifecycle GHG emissions as rapeseed and soybean biodiesel when the palm plantation and palm oil mill effluent

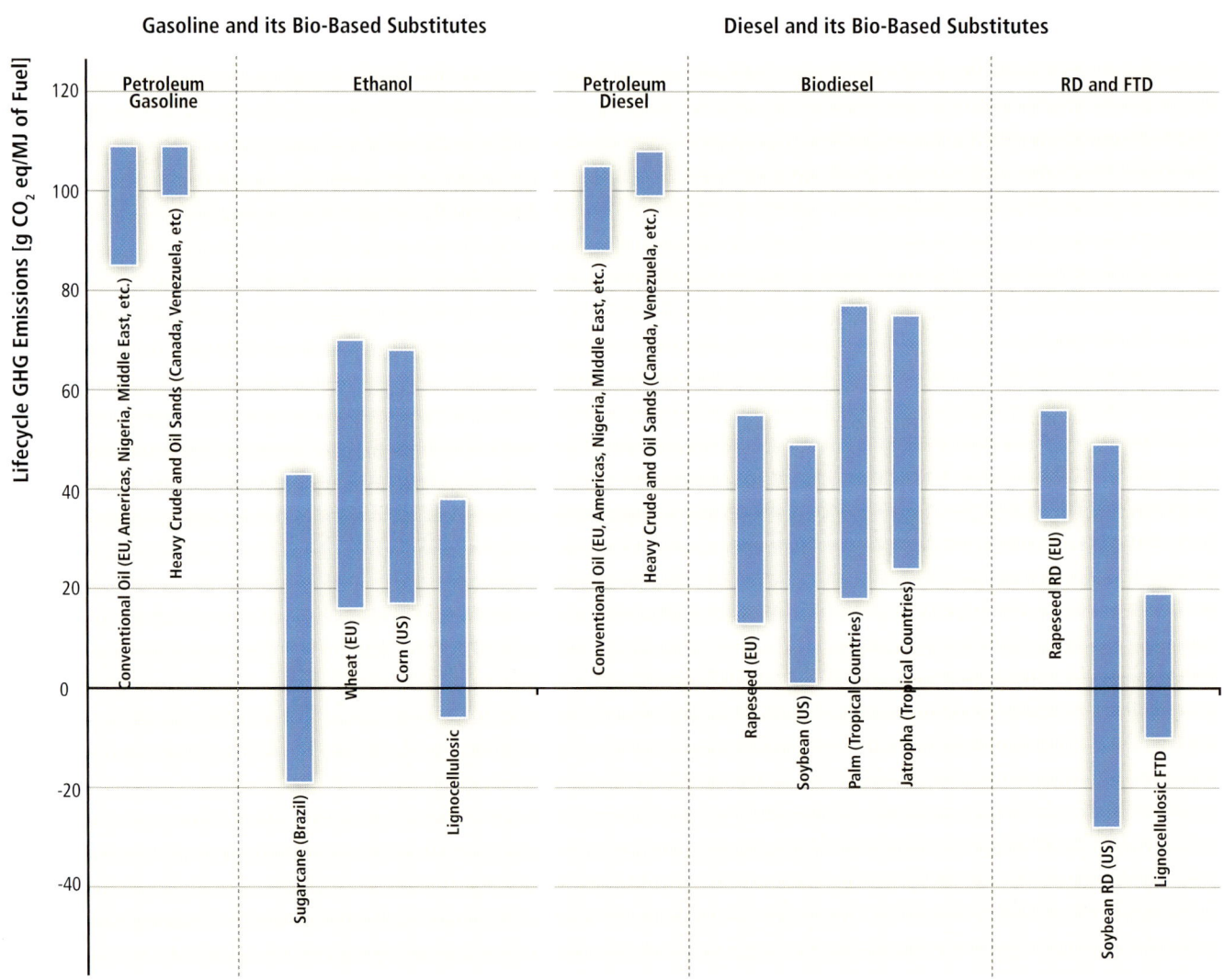

Figure 9.9 | Illustrative ranges in lifecycle GHG emissions of petroleum fuels, first-generation biofuels and selected next-generation lignocellulosic biofuels without considering land use change. (Sources for estimates plotted: Wu et al., 2005; Fleming et al., 2006; Hill et al., 2006, 2009; Beer et al., 2007; Wang et al., 2007; CONCAWE, 2008; Macedo and Seabra, 2008; NETL, 2008, 2009; CARB, 2009; Hoefnagels et al., 2010; Hsu et al., 2010; Kaliyan et al., 2010; Larson et al., 2010; Neely et al 2010). Note: FTD = Fischer-Tropsch diesel; RD = Renewable diesel (RD is different from biodiesel in processing and product properties). For common feedstock and fuel categories shown in both Figure 2.10 and above (e.g., sugarcane ethanol, FTD), the references cited and the ranges of GHG emission estimates are identical.

---

10 Sections 2.3 and 2.5 provide more detailed reviews of biofuel technologies and configurations, including lifecycle GHG emissions.

(POME) are properly managed, or higher emissions if methane release from POME is not captured (Beer et al., 2007; CONCAWE, 2008; Wicke et al., 2008; Achten et al., 2010; Hoefnagels et al., 2010). The range in GHG estimates for *Jatropha* biodiesel is comparable to that for palm oil biodiesel (Whitaker and Heath, 2010).

The lack of commercial-scale lignocellulosic feedstocks and fuels production leads to a high degree of uncertainty in estimates of lifecycle GHG emissions for these systems. Uncertainty analysis indicates that the GHG emissions of some projected lignocellulosic biofuel supply chains could be higher than shown in Figure 9.9 assuming a combination of worst-case conditions in different elements of the supply chain (e.g., poorly managed biomass production practices, and energy-intensive biomass pre-processing) (Soimakallio et al., 2009; Hsu et al., 2010). However, lignocellulosic biofuels under well-managed conditions can have lower GHG emissions than grain ethanol and oilseed biodiesel.

The total lifecycle GHG emissions of fuels critically depend on the sign and magnitude of direct and indirect LUC effects, which could potentially negate or exceed any GHG reduction benefit from the displacement of petroleum fuels by biofuels discussed in this section (Berndes et al., 2010).

Land use change-related greenhouse gas emissions and bioenergy

Conversion from one land cover type or use to another directly and indirectly affects terrestrial GHG stocks and flows, and historically has been a significant contributor to global GHG emissions (IPCC, 1996b; Le Quere et al., 2009). Agriculture and forestry systems are important drivers of these land use changes, with energy systems (especially bioenergy but also reservoir hydropower, mining and petroleum extraction) being an additional stressor (Schlamadinger, 1997). While GHG emissions from LUC are difficult to quantify, they are important to investigate and evaluate, since any potential GHG emission reduction benefits from increased use of bioenergy compared to fossil energy sources could be partially or wholly negated when LUC-related GHG emissions are considered.

Direct LUC (dLUC) occurs when bioenergy feedstock production modifies an existing land use, resulting in a change in above- and below-ground carbon stocks. dLUC-related GHG emissions are dependent on site-specific conditions such as the prior land use, soil type, local climate, crop management practices and the bioenergy crop to be grown. In the examples shown in Figure 9.10, the original land use is generally a more important factor in determining dLUC-related GHG emissions than the bioenergy feedstock type planted. The conversion of certain land types (e.g., rainforest and peatland) can lead to very large GHG emissions; conversely, the use of degraded land and sometimes former farmland (e.g., when using lignocellulosic feedstocks) can enhance carbon stocks. Any dLUC-related GHG emissions must be repaid over time before GHG emission reduction benefits for the use of bioenergy can accrue (Gibbs et al., 2008). Results reported in Figure 9.10 are totals averaged over a 30-year time horizon. Not considered in the analyses reviewed here is the time signature of these GHG emissions (an initial pulse followed by a long tail), which is an important determinant of GHG climate impacts.

Indirect LUC (iLUC) occurs when a change in the production level of an agricultural product (i.e., a reduction in food, feed or fibre production induced by agricultural land conversion to the production of bioenergy feedstocks) leads to a market-mediated shift in land management activities (i.e., dLUC) outside of where the primary driver occurs. iLUC is not directly observable, and is complex to model and attribute to a single cause. Important aspects of this complexity include model geographic resolution, interactions between bioenergy and other agricultural systems, how the systems respond to changes in market and policy, and assumptions about social and environmental responsibility for actions taken by multiple global actors. For example, estimates of iLUC-induced GHG emissions can depend on how land cover is modelled. Models using greater geographic resolution and number of land cover types have tended to produce lower estimates and tighter uncertainty ranges that those considering just, for example, pasture and forest, at lower resolution (Nassar et al., 2009; EPA, 2010b). Emission estimates also tend to increase if large future bioenergy markets and high growth rates are assumed. Despite similar evaluation methods, Al-Riffai et al. (2010) and Hiederer et al. (2010) report a LUC (direct and indirect) impact of 25 and 43 g CO_2eq/MJ, respectively, for a similar set of biofuels, partly because they evaluated different magnitudes of biofuels market growth (0.3 and 0.9 EJ, respectively).

Despite challenges in modelling iLUC attributable to bioenergy systems, improvements in methods and input biophysical data sets have been made. Some illustrative estimates of representative LUC-related (including d- and iLUC) GHG emissions are reported in Figure 9.11. See Section 2.5.3 for more published estimates and discussion of LUC.

The wide ranges of even the central tendency estimates reflect the uncertainty and variability remaining in the estimation of LUC-induced GHG emissions from bioenergy systems, but nonetheless point to a potentially significant impact of LUC relative to non-LUC lifecycle GHG emissions for many dedicated bioenergy systems. Thus, it is critical to continue research to improve LUC assessment methods and increase the availability and quality of information on current land use, bioenergy-derived products and other potential LUC drivers. It is also critical to consider ways to mitigate the risk of bioenergy-induced LUC, for instance Agro-Ecologic Zoning systems (EMBRAPA, 2009) coupled with adequate monitoring, enforcement and site-specific bioenergy carbon footprint evaluation; improvement of agricultural management and yields, for example, by intercropping and improved rotations systems; using lower LUC-risk lignocellulosic feedstocks or replacing dedicated biomass with residues or

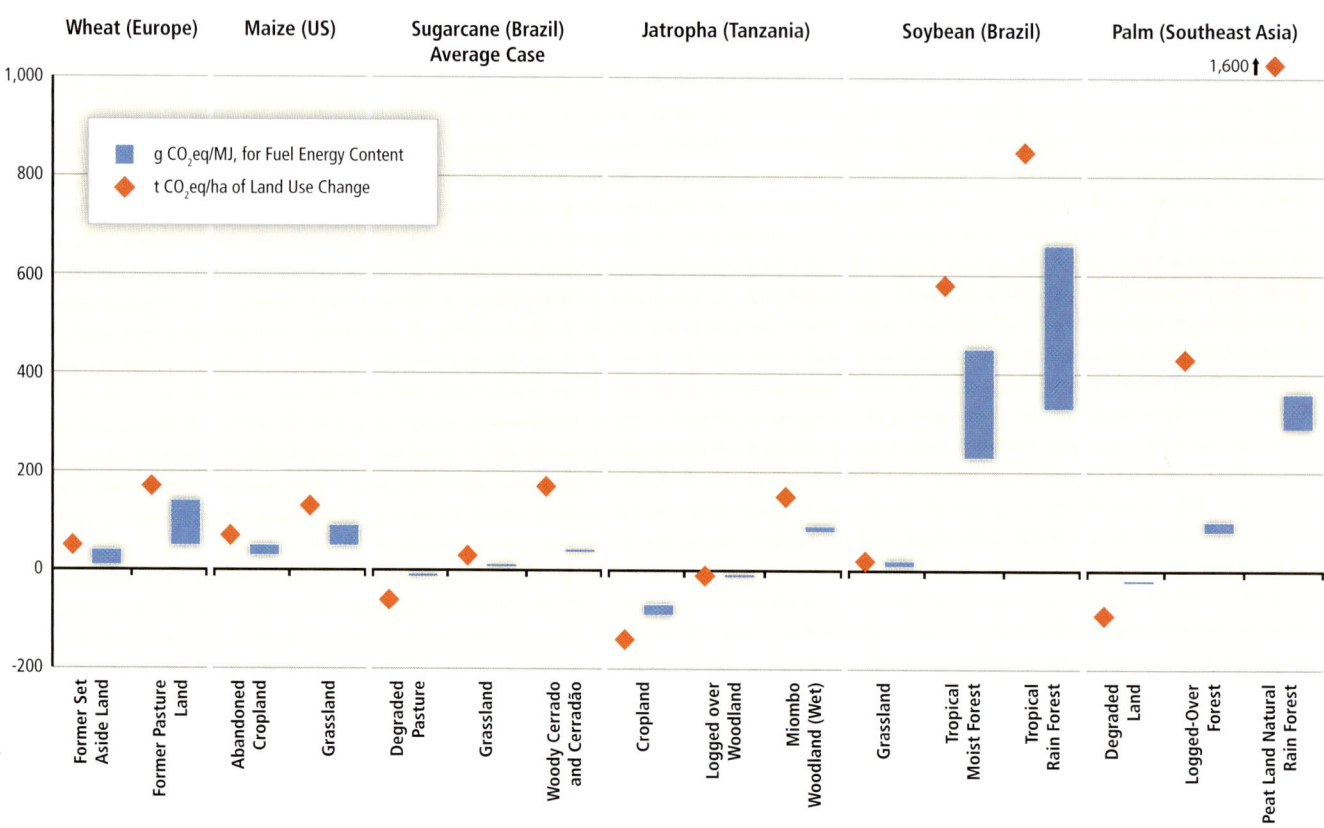

Figure 9.10 | Illustrative direct LUC-related GHG emission estimates from selected land use types and first-generation biofuel (ethanol and biodiesel) feedstocks. Results are taken from Hoefnagels et al. (2010) and Fargione et al. (2008) and, where necessary, converted (assuming a 30-year timeframe) to the functional units displayed using data from Hoefnagels et al. (2010) and EPA (2010b). Ranges are based on different co-product allocation methods (i.e., allocation by mass, energy and market value).

wastes; and promoting the use of degraded or marginal lands or sustainability certification systems (van Dam et al., 2009; Berndes et al., 2010; see Sections 2.2.4, 2.4.5, 2.5.2 and 2.8.4).

9.3.4.2 Local and regional air pollution

This section presents data on selected air pollutants that are emitted by energy technologies and that have the most important impacts on human health as indicated by the World Health Organization (WHO, 2006). These include particulate matter[11] (PM), nitrous oxides (NO_x), sulphur dioxide (SO_2) and non-methane volatile organic compounds (NMVOC). Their dispersion in the atmosphere entails significant impacts at the local and regional scale (up to a few thousand kilometres) (e.g., Hirschberg et al., 2004b). Black carbon, which constitutes a fraction of total PM emissions, and other aerosols can also have impacts on global and regional climate (see Box 9.4). The location-specific impacts from air pollutants depend on exposure, their concentrations in the atmosphere, as well as the concentrations of further pollutants acting as reactants, for example, for formation of secondary particulates (e.g., Kalberer et al., 2004; Andreani-Aksoyoglu et al., 2008; Hallquist et al., 2009). Air pollu-

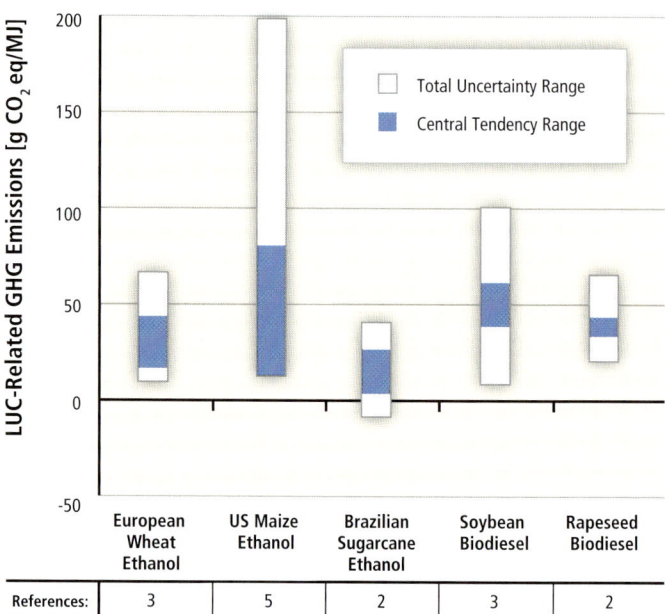

Figure 9.11 | Illustrative estimates of direct and indirect LUC-related GHG emissions induced by several first-generation biofuel pathways, reported here as ranges in central tendency and total reported uncertainty. Estimates reported here combine several different uncertainty calculation methods and central tendency measures and assume a 30-year time frame. Reported under the x-axis is the number of references with results falling within these ranges (Sources: Searchinger et al., 2008; Al-Riffai et al., 2010; EPA, 2010b; Fritsche et al., 2010; Hertel et al., 2010; Tyner et al., 2010).

---

11 PM emissions are specified as PM_d, where the subscript d indicates the largest diameter (in μm) of the particles that are included. Particles emitted by internal combustion engines are all very small and almost entirely included in the $PM_{2.5}$ measure.

Box 9.4 | Black carbon and aerosols: Climate effects of air pollutants.

Black carbon (BC) is a short-lived air pollutant formed by incomplete combustion of fossil or biomass fuels. Prime sources of BC are agricultural and forest fires, (diesel) combustion engines, in particular maritime vessels running on heavy oil, and residential use of heating and cooking fuels (Bond et al., 2004; Lack et al., 2008). BC emissions are particularly high in developing countries. BC has detrimental health effects (see Section 9.3.4.3), and can accelerate climate change both through its heat-absorbing properties in the atmosphere, and by reducing the albedo of cloud, snow and ice surfaces (Ramanathan and Carmichael, 2008; Flanner et al., 2009; Lau et al., 2010). BC is emitted together with organic carbon (OC), and other aerosols like sulphates, that have a negative effect on radiative forcing. Therefore, the net warming effect of aerosol emissions from combustion is source- and location-dependent, and still uncertain. Available literature suggests that contained combustion of fossil fuels and residential combustion of solid biomass results in net warming, while the net effects of open combustion (field fires) of biomass sources are negative, due to a higher ratio of reflective OC to absorptive BC aerosols (Bond et al., 2004; M. Jacobson, 2004; Hansen et al., 2005; Koch et al., 2007). Both processes play a prominent role in the formation of atmospheric brown clouds and other processes that exhibit strong regional climate impacts (Ramanathan et al., 2005, 2007), for example, alteration of the Indian Monsoon (Auffhammer et al., 2006) or larger warming in elevated regions of the tropics (Gautam et al., 2009).

BC abatement has been proposed as a significant means not only for climate change mitigation, but also for addressing additional sustainability concerns such as air pollution, inefficient energy services, and related health impacts on the poor (Grieshop et al., 2009). The provision of energy efficient and smoke-free cookers and soot-reducing technologies for coal combustion in small industries could have major benefits by reducing radiative forcing and combating indoor air pollution and respiratory diseases in urban centres (Ramanathan and Carmichael, 2008; see Sections 2.5.4 and 9.3.4.3). A switch from diesel to LPG in the public transport system in Delhi has resulted in net GHG savings and substantial reductions in BC loads (C. Reynolds and Kandlikar, 2008). However, it has been suggested that removing the 'masking' effect of *reflective* aerosols through air pollution control measures might accelerate the impacts from already-committed-to warming (Ramanathan and Feng, 2008; Carmichael et al., 2009).

---

tion also varies significantly between urban and rural areas. Therefore, cumulative lifecycle inventory results, that is, quantities of pollutants emitted per unit of energy delivered, must be interpreted with care regarding conclusions about potential impacts on human health and the environment (Torfs et al., 2007). The following results can only act as basic data for the estimation of specific impacts (see Section 9.3.4.3). Indoor air pollution caused by solid fuels in traditional cookstoves is discussed in Box 9.4 and Section 9.3.4.3.

Heat and electricity supply

For space heating and electricity production with fossil fuels and biomass (wood) combustion, the dominant contributor to lifecycle inventory results (per kWh of end-use energy) is the combustion stage, with typically a 70 to almost 100% share of the overall emissions (e.g., Jungbluth et al., 2005; C. Bauer, 2007; Dones et al., 2007) (see Figure 9.12). However, in the case of long distance transport of coal, natural gas, oil and wood fuel, the transport stage might become more important (e.g., C. Bauer, 2007, 2008). In general, natural gas causes the lowest emissions among fossil fuels. Contributions of different sections of the energy chains as well as total emissions vary within orders of magnitude with power plant technology, application of pollution control technologies (flue gas desulphurization, particulate filters, etc.) and characteristics of fuel feedstock applied, as indicated by minimum and maximum values in Figure 9.12.

In the case of space heating, for example, minimum and maximum figures represent the most and least efficient technology options among the datasets evaluated. Additionally, the type of fuel (e.g., wood logs, chips or pellets in case of biomass) affects the results. The figures for solar heating are valid for a certain location in central Europe, and variation in solar irradiation is not considered in the range shown. In the case of fossil electricity generation, the results include country-specific averages for current technology and fuel supply for all European and a few other countries, such as the USA and China. Minimum and maximum values therefore mainly represent the countries with the most and least efficient power plant and pollution control technology, respectively.

The results from this assessment show that non-combustion RE technologies and nuclear power cause comparatively minor emissions of air pollutants, only from upstream and downstream processes. Also, the variations in the results, depending on both technologies applied and site of power generation (in terms of, for example, solar irradiation (Jungbluth et al., 2009) and wind conditions (EWEA, 2004)), are in general much lower for RE and nuclear than for fossil power and heating systems. The potential increase in overall emissions from the power system due to a more flexible operation of fossil power plants in response to feed-in of variable renewable electricity is not taken into account. Although not shown in Figure 9.12, the type of electricity used for the operation of the geothermal heat pump has a significant impact on the performance of this technology (Heck, 2007).

LCA literature including results on air pollution in developing countries is scarce, and available case studies could not be integrated into the results displayed in a consistent way. However, emissions at the higher

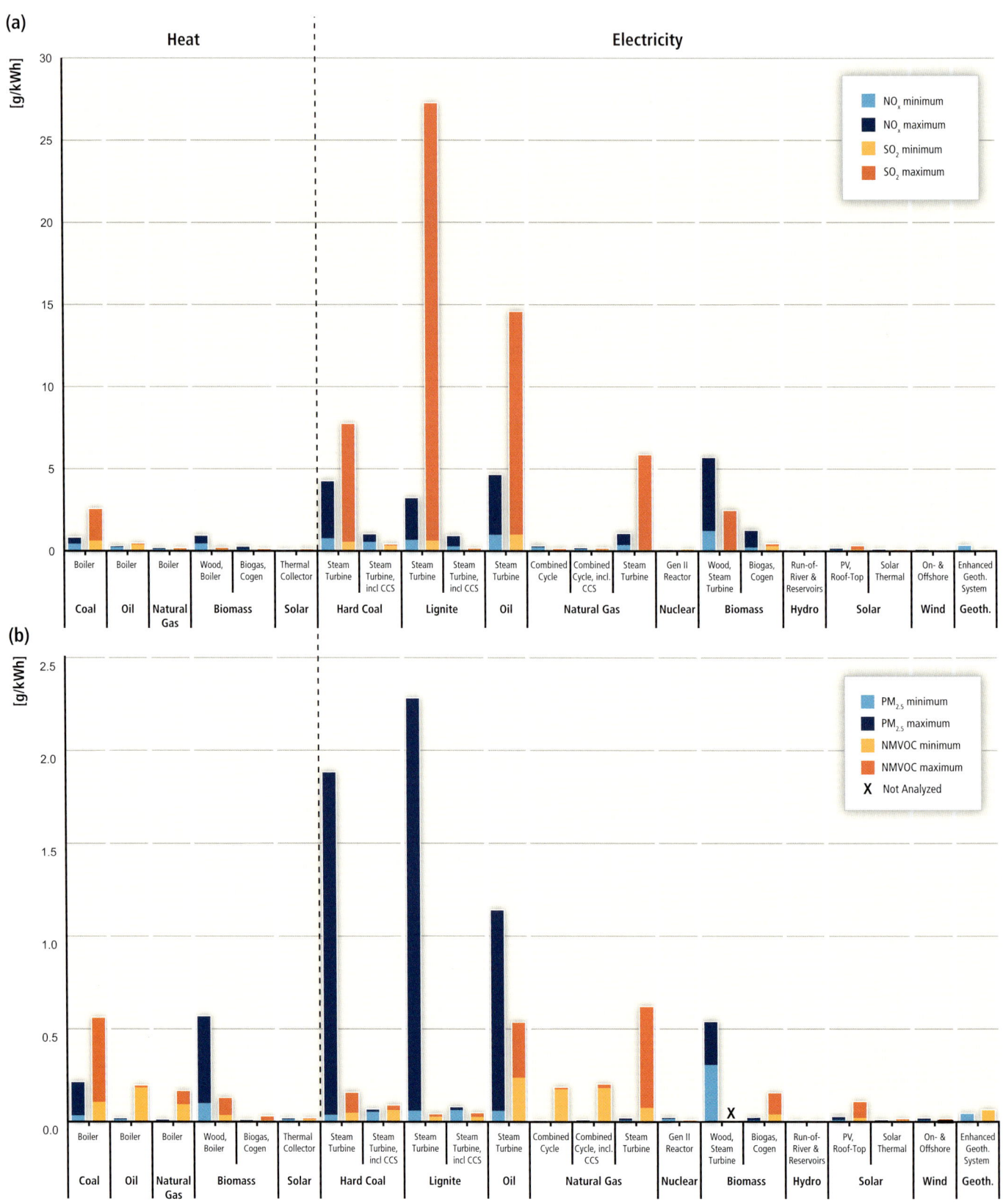

Figure 9.12 | Cumulative lifecycle emissions per unit of energy generated of (a) NOx and SO2 and (b) NMVOC and PM2.5 for current heat and electricity supply technologies (C. Bauer, 2008; Viebahn et al., 2008; Ecoinvent, 2009); traditional biomass use not considered. Figures for coal and gas power chains with CCS are valid for near-future forecasts (C. Bauer et al., 2009).

end of the ranges shown may typically apply to developing economies that use older technologies, have less pollution control measures in place and possibly consume lower-quality fuels. Also, lack of environmental regulation in developing countries results in comparatively higher emissions. Molina and Molina (2004) report outdoor urban air pollution in cities from industry, energy and transport that is a factor of 10 or higher than in developed nations; the location of the emission sources in combination with the prevailing meteorological conditions are important factors in this respect. Air pollution abatement has gained importance since the early 1990s, in particular in China, resulting in a slowdown of sulphur emissions in Asia (Carmichael et al., 2002). The substantial potential of RE to contribute to air pollution abatement has been studied in particular for emerging economies' electricity and transport sectors (Boudri et al., 2002; Aunan et al., 2004; Ramanathan and Carmichael, 2008; Creutzig and He, 2009; see Sections 9.4.4 and 10.6).

Transport fuels

Under a lifecycle approach, well-to-wheels air pollutant emissions of biomass fuel/vehicle systems differ significantly. These differences are caused by the feedstock used for fuel production, biomass yields, fuel production pathways and technologies, location of biomass growth and harvesting, as well as fuel characteristics and vehicle technologies (von Blottnitz and Curran, 2007; Cherubini and Strømman, 2011).

The use of gaseous fuels—both fossil and biomass origin—tends to reduce air pollution compared to liquid fuels (Zah et al., 2007). The effects of using biomass fuels and bioethanol and biodiesel blends on tailpipe emissions have been examined by numerous authors with varying results (Schifter et al., 2004, 2011; Niven, 2005; Coelho et al., 2006; Fernando et al., 2006; Goldemberg et al., 2008; Graham et al., 2008; Pang et al., 2008; Coronado et al., 2009; Costa and Sodré, 2009; Demirbas, 2009; Hilton and Duddy, 2009; Roayaei and Taheri, 2009; Yanowitz and McCormick, 2009; Yoon et al., 2009; Zhai et al., 2009; Park et al., 2010). Fuel blends, combustion and ambient temperatures as well as additives play a decisive role in air pollutant formation (Lucon et al., 2005; Coelho et al., 2006; Graham et al., 2008; Ginnebaugh et al., 2010). Overall, the studies tend to agree that carbon monoxide (CO) and hydrocarbon emissions are reduced by use of both ethanol and biodiesel blends compared to gasoline and diesel, respectively, while NO_x emissions seem to be higher. Increased NO_x and evaporative emissions from oxygenates of biofuel blends can lead to higher concentrations of tropospheric ozone (Schifter et al., 2004; Agarwal, 2007). Increased aldehyde emissions have been reported for bioethanol in Brazil, which are less toxic than the formaldehydes originating from fossil fuels (Goldemberg et al., 2008; Graham et al., 2008; Anderson, 2009). Second-generation and future biofuels are expected to improve performance, when the combustion system is specifically adapted (Pischinger et al., 2008; Ußner and Müller-Langer, 2009).

Notter et al. (2010) and Zackrisson et al. (2010) suggested that future electric or fuel cell vehicles (see Section 8.3.1) offer a substantial potential for reductions in air pollution (as well as other environmental burdens) if electricity or hydrogen from RE sources is used as the energy carrier.

Shifting emissions from urban to less-populated areas can result in less exposure and therefore reduced impacts on human health (see Section 9.3.4.3). Despite increases in total emissions, some bioethanol blends used in flex-fuel vehicles in Brazil contributed to reductions of up to 30% in urban emissions, as most emissions originated from farming equipment, fertilizer manufacture and ethanol plants located in rural areas (Huo et al., 2009b). Similarly, the formation of secondary pollutants as aerosols and ozone in towns might be reduced, depending on atmospheric conditions including background concentrations of pollutants.

9.3.4.3 Health impacts

The most important energy-related impacts on human health are those associated with air pollutant emissions by fossil fuel and biomass combustion (Ezzati et al., 2004; W. Paul et al., 2007). Air pollution, even at

Table 9.9 | Health impacts of important air pollutants (adapted from Bickel and Friedrich, 2005).

Primary Pollutants[1]	Secondary Pollutants[2]	Impacts
Particles (PM_{10}, $PM_{2.5}$, black carbon)		cardio-pulmonary morbidity (cerebrovascular and respiratory hospital admissions, heart failure, chronic bronchitis, upper and lower respiratory symptoms, aggravation of asthma), mortality
SO_2	sulphates	like particles[3]
NO_x	nitrates	morbidity, like particles[3]
NO_x+VOC	ozone	respiratory morbidity, mortality
CO		cardiovascular morbidity, mortality
Polyaromatic Hydrocarbon		cancers
Lead, Mercury		morbidity (neurotoxic and other)

Notes: 1. Emitted by pollution source. 2. created by chemical reactions in the atmosphere. 3. lack of specific evidence, as most available epidemiological studies are based on mass PM without distinction of components or characteristics.

current ambient levels, aggravates morbidity (especially respiratory and cardiovascular diseases) and leads to premature mortality (Table 9.9; Cohen et al., 2004; Curtis et al., 2006). Although the health effects of ambient air pollution result from a complex mixture of combustion products and are therefore difficult to attribute to a certain source or pollutant, negative effects have been most closely correlated with three species of pollutants in epidemiological studies: fine PM, SO_2, and tropospheric ozone (Ezzati et al., 2004; Curtis et al., 2006). Significant reductions in mass emissions of pollutants by deployment of RE should yield increased health benefits, and opportunities for policy measures combining climate change and (urban) air pollution mitigation are increasingly recognized (see Sections 9.4.4.1, 10.6 and 11.3.1).

Household environmental exposures, including indoor air pollution (IAP) from the combustion of solid heating and cooking fuels, generally decline with increased development, whereas community-level exposures have been found to increase initially, and then gradually decline, with important distinctions between rural and urban areas (Smith and Ezzati, 2005; HEI, 2010). Exposure to IAP from the combustion of coal and traditional biomass is recognized as one of the most important causes of morbidity and mortality in developing countries (Bruce et al., 2002; Ezzati et al., 2004; Smith and Ezzati, 2005; Zhang and Smith, 2007). For example, comparative quantifications of health risks showed that in 2000, more than 1.6 million deaths and over 38.5 million disability-adjusted life-years (DALYs) were attributable to indoor smoke from solid fuels (WHO, 2002; Smith and Mehta, 2003; Smith et al., 2004; Torres-Duque et al., 2008). Figure 9.13 illustrates the magnitude of the health problems associated with IAP, which is projected to exceed other major causes of premature deaths (e.g., HIV/AIDS, malaria and tuberculosis) by 2030 (IEA, 2010a).

Many health problems like chronic obstructive pulmonary disease, cataracts and pneumonia are most severe for women and children, which are most exposed to indoor emissions (Smith et al., 2000; Pokhrel et al., 2005; Barnes et al., 2009; Haines et al., 2009; UNDP and WHO, 2009), and generally affect the poorest segment of the population (see Section 9.3.2).

In traditional uses, biomass-based fuels yield worse results with respect to contaminant concentrations than charcoal or coal (Kim

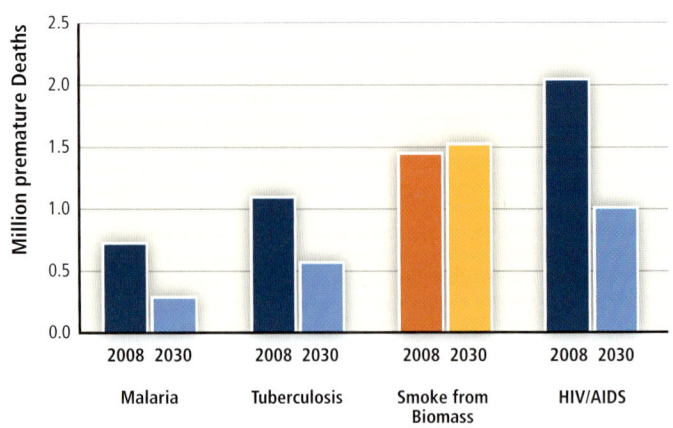

Figure 9.13 | Premature deaths from household air pollution and other diseases in 2008 and projected for 2030 (IEA, 2010a).

Oanh and Dung, 1999; Bailis and Cutler, 2004; Zhang and Smith, 2007). Mitigation options—besides the more costly switch to cleaner fuels (see Section 9.3.2)—for health impacts from IAP include improved cookstoves (ICS), ventilation and building design and behavioural changes (Smith et al., 2000; Bruce et al., 2004; Mehta and Shahpar, 2004; Palanivelraja and Manirathinem, 2010). Modern bioenergy technologies (ICS, biogas) can provide health benefits without fuel switching (Smith et al., 2007; Bailis et al., 2009), as well as additional environmental and social advantages (Haines et al., 2009) (see Section 2.5.7.2).

Non-combustion-related health impacts

Health impacts from energy technologies other than those described above can be regarded as relatively minor. Table 9.10 provides an overview of areas of concern for RE technologies as identified in this report.

For nuclear power, radiotoxicity of spent fuels and uranium tailings, including windblown radioactive dust dispersal, and radon gas from the mining stage are the most prominent health concerns (OECD/NEA, 2002; Abdelouas, 2006; Al-Zoughool and Krewski, 2009). Increased cancer risk for residents, particularly children, near nuclear power plants has been studied with contrasting results in different countries (Ghirga, 2010).

Table 9.10 | Overview of potential impacts on human health by RE technologies as reported in Sections 2.5, 4.6, 5.6 and 7.6. For solar and ocean technologies, no impacts were identified.

RE Technology	Potential Health Concerns
Bioenergy	Depending on feedstock and agricultural management, direct and indirect exposure to agrochemicals and derivatives like pesticides or nitrates, or smoke due to residue burning may cause local impacts Health impacts related to air pollutant emissions by combustion[1]
Geothermal Energy	For some operations, hydrogen sulphide emission may cause local impacts
Reservoir Hydropower	Standing water bodies can lead to spread of vector-borne diseases in tropical areas Concentrations of population and migrant workers during construction of large dams may cause public health concerns
Wind Energy	Nuisance from noise and flickering

Note: 1. See previous subsection for details.

9.3.4.4 Water

Water is a critical and highly localized resource with multiple and competing uses, including energy. The condition and amount of water resources in a given location will influence the selection, design and performance of an energy technology; impacts from energy technologies will also vary geographically and temporally. Hence, implications for the water-energy nexus must be considered within a SD context. Literature holistically evaluating the impacts of energy technologies on water resources is limited, especially from a lifecycle perspective. While some broad conclusions can be drawn from the evidence presented in the following sections, additional research is needed to confirm many of the results and fill existing knowledge gaps.

In 2006, the energy and industrial sectors accounted for 45% of freshwater withdrawals in Annex I countries and 10% of freshwater withdrawals in non-Annex I countries (Gleick, 2008). As lesser-developed countries industrialize and improve access to energy services, additional freshwater resources may be required to meet the water demands of increased energy production. However, various metrics indicate that many developing countries already experience water scarcity problems, and climate change may exacerbate water stress (Rijsberman, 2006; IPCC, 2008; Dai, 2011). Thermal power plants may be especially vulnerable to conditions of water scarcity and climate change due to their continuous water requirements. Also, hydropower and bioenergy are highly dependent on water availability, and exhibit potentials for both increased competition for and mitigation of water scarcity (see Sections 2.5.5.1 and 5.10).

Operational water use and water quality impacts of electricity generation
Electricity sector impacts involve both water withdrawal and consumption. Water withdrawal is the amount of water removed from the ground or diverted from a water source, while consumption is the amount of water that is lost through evaporation, transpiration, human consumption and incorporation into products (Kenny et al., 2009). Both metrics have an important impact on local water availability, and often with trade-offs such that using existing technology only one impact can be reduced at a time. Water consumption by industry and power plants, while accounting for less than 4% of global water consumption, is an important consideration for water-scarce regions; this is particularly relevant in the context of future resource development, with water being effectively removed from the system and not available for other uses, for example, agriculture or drinking water (Shiklomanov, 2000).

While water is used throughout the lifecycle of most technologies, operational cooling needs for thermal power plants result in the withdrawal and consumption of more water than any other lifecycle phase, with the exception of biomass feedstock production (Fthenakis and Kim, 2010). Figure 9.14 depicts the variability in operational water consumption rates associated with electricity generation units and cooling technologies. Water consumption varies widely both within cooling technology categories, but especially across categories. The choice of cooling system is often site-specific and based on water availability, local environmental regulations or quality impacts, parasitic energy loads, costs, or other considerations (J. Reynolds, 1980; Bloemkolk and van der Schaaf, 1996). Non-thermal technologies, with the exception of hydropower, are found to have the lowest operational and lifecycle withdrawal and consumptive water use values per unit electricity generated (Tsoutsos et al., 2005; Fthenakis and Kim, 2010). Substantial evaporation can occur from hydroelectric reservoirs, yet reservoirs often provide other beneficial services besides power production (e.g., flood control, freshwater supply, and recreation), and allocation schemes for determining water consumption from various reservoir uses can significantly influence reported water consumption values (Gleick, 1993; LeCornu, 1998; Torcellini et al., 2003). Research may be needed to determine the net effect of reservoir construction on evaporation in a specific watershed. Data shown in Figure 9.14 are from studies of US systems only, but represent a wide range of technology vintages and climatic conditions, both of which can affect water use rates (B. Miller et al., 1992), and thus their results are applicable and comparable to water use rates in other countries (EC, 2006).

Data for geothermal energy are not included in Figure 9.14 because in most situations, geothermal fluids are utilized for cooling before reinjection, and therefore no freshwater is consumed (Franco and Villani, 2009; see Section 4.5.3). Depending on technology, resource type and cooling system used, geothermal operational water consumption can range from near zero up to 15 m^3/MWh (Fthenakis and Kim, 2010).

Reduced water levels or higher temperatures in water bodies may require once-through cooled thermal power plants, which withdraw large volumes of water but consume comparatively little, to run at lower capacities or to shut down completely (Poumadère et al., 2005). Addressing this vulnerability by utilizing recirculating cooling technologies, which withdraw less water, could lead to increases in water consumption (Figure 9.14), reductions in plant-level thermal efficiencies and increases in operating and installed costs (Tawney et al., 2005). Ambient air temperature increases may lead to reduced plant-level thermal efficiency and cooling system performance, resulting in higher water use rates (B. Miller et al., 1992; Turchi et al., 2010). Thermal power plant vulnerability can be reduced by utilizing alternative water sources, such as municipal wastewater, or by utilizing a dry-cooling system, yet there are cost, performance and availability trade-offs and constraints (EPRI, 2003; Gadhamshetty et al., 2006). Reservoirs and river levels may also be affected by climate change, altering water availability and hydropower performance capabilities and output (Harrison and Whittington, 2002; IPCC, 2008).

Electricity generation units can affect water quality through thermal and chemical pollution. During normal operation, electricity generation units with once-through cooling systems can elevate the temperature of water bodies receiving the cooling water discharge, which can negatively affect aquatic ecosystems and reduce fish yields (Kelso and Milburn, 1979; Barnthouse, 2000; Poornima et al., 2005; Greenwood, 2008; Kesminas and Olechnoviciene, 2008; Shanthi and Gajendran, 2009). Deposition of air pollutant emissions from the combustion of fossil fuels to water bodies can also affect water quality (Larssen et

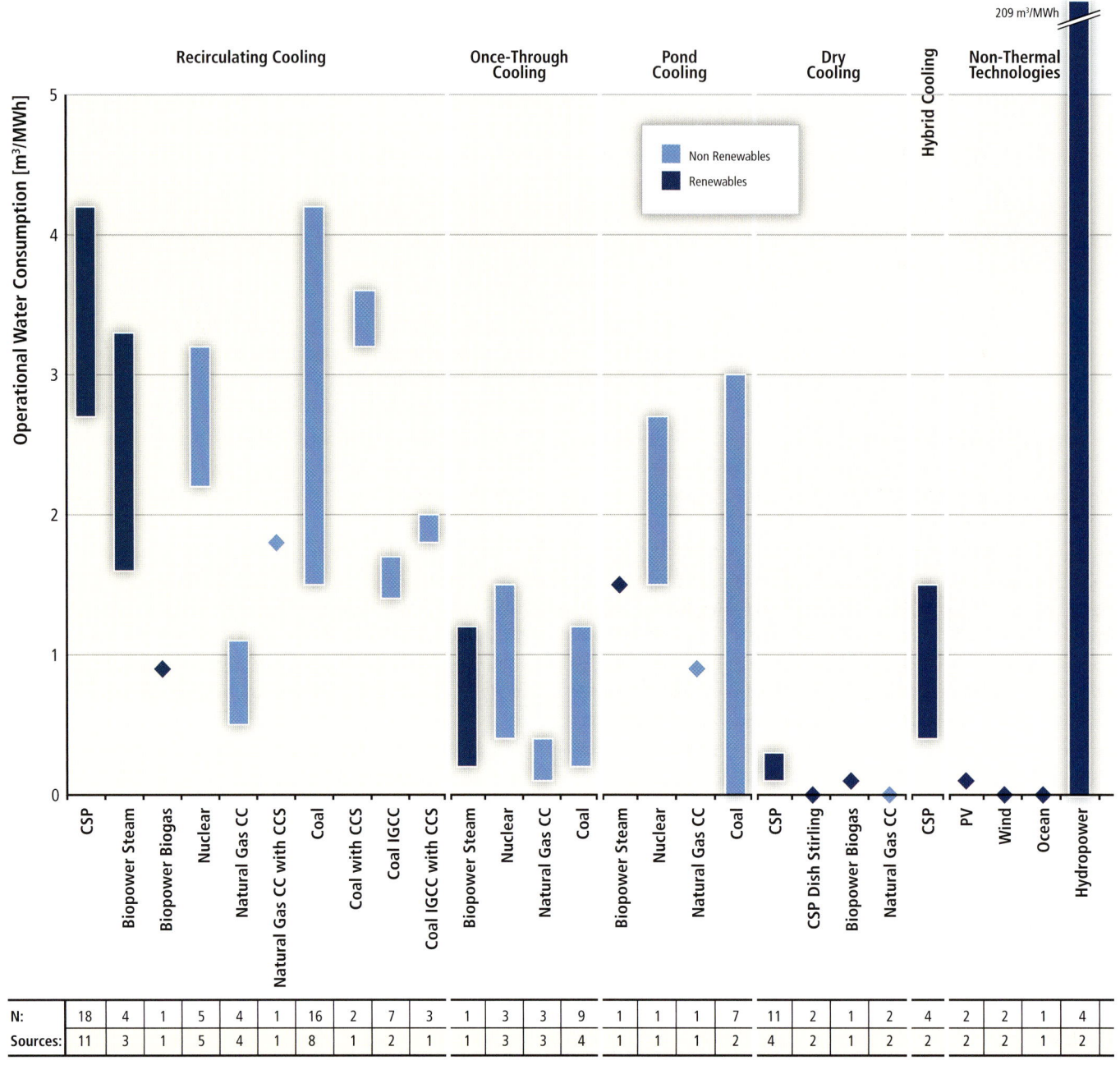

Figure 9.14 | Ranges of rates of operational water consumption by thermal and non-thermal electricity-generating technologies based on a review of available literature (m³/MWh). Bars represent absolute ranges from available literature, diamonds single estimates; n represents the number of estimates reported in the sources. Note that upper values for hydropower result from few studies measuring gross evaporation values, and may not be representative (see Box 5.2). Methods and references used in this literature review are reported in Annex II.

Notes: CSP: concentrating solar power; CCS: carbon capture and storage; IGCC: integrated gasification combined cycle; CC: combined cycle; PV: photovoltaic.

al., 2006). Hydroelectric facilities can impact both temperature and dissolved oxygen content of the released water while also altering the flow regime, disturbing ecosystems and disrupting the sediment distribution process (Cushman, 1985; Liu and Yu, 1992; Jager and Smith, 2008; see Section 5.6). Tidal energy facilities located at the mouths of estuaries could affect the hydrology and salinity of estuaries and ocean thermal energy conversion technologies can alter local water quality through the accidental release of toxic chemicals, such as ammonia and chlorine (Pelc and Fujita, 2002; Vega, 2002; see Section 6.5). Geothermal facilities can affect both surface and ground water quality through spillage of geothermal fluids at the surface during operation, leakage from surface storage impoundments, and through contamination of

Water use of upstream processes

Water use in upstream processes (see Figure 9.7) can be high for some energy technologies, particularly for fuel extraction and biomass feedstock production (Fthenakis and Kim, 2010). Specifically, unconventional fossil fuel (e.g., oil shale, shale gas) exploration and processing techniques can have significantly greater water use rates than conventional exploration techniques, and may require freshwater to be imported from other watersheds (GAO, 2010; Kargbo et al., 2010; Parfitt, 2010; Veil, 2010). Further research is necessary to determine water use as a function of output energy content of the extracted fuel in unconventional production to facilitate comparison to other conventionally produced fuels.

Biomass feedstock may be used for electricity generation or converted into liquid fuels. To account for both naturally variable precipitation and irrigation freshwater required in feedstock production, the water footprint metric is used (Gerbens-Leenes et al., 2009). The water footprint of feedstock production is highly dependent on feedstock type, geographic region and local climatic conditions, and crop management practices (Berndes, 2002, 2008; Gerbens-Leenes et al., 2009; Wu et al., 2009; Harto et al., 2010; Stone et al., 2010). These factors may change from year to year, and the water footprint for an individual case may differ substantially from the global average. Estimates of water footprints for biomass grown for multiple purposes can also vary significantly due to the choice of allocation method (S. Singh and Kumar, 2011).

The current water footprint of biomass feedstock production for electricity generation is approximately 70 to 400 times greater than operational water consumption requirements for thermal power plants (Gerbens-Leenes et al., 2009; S. Singh and Kumar, 2011). The current global average water footprint (weighted by production mass) of biofuel feedstock production ranges from about 60 to 600 litres per MJ fuel (Gerbens-Leenes et al., 2009). Biodiesel feedstock water footprints are nearly two to four times greater than the water footprint for ethanol crops, because oilseed crops are less water efficient (Gerbens-Leenes et al., 2009; S. Singh and Kumar, 2011). Refining and processing biofuels require around 0.1 to 0.5 litres of water per MJ fuel, which is far less than feedstock production requirements but still considerably higher than those of conventional petroleum products (Berndes, 2002; King and Webber, 2008; Wu et al., 2009; Harto et al., 2010; S. Singh and Kumar, 2011).

Without proper management, increased bioenergy production could therefore increase competition for water in critical areas (see Section 2.5.5.1; Dornburg et al., 2008; Berndes, 2010; Fingerman et al., 2010). However, the proportion of irrigation freshwater to total water consumed varies considerably, and the relationship between vegetation and hydrological processes at the landscape scale is complex. Certain feedstock production systems may drive land use towards systems with higher water productivity and decreased water competition, as, for example, woody crops grown in multi-year rotations. Some perennials can improve water retention functions on degraded lands, and considerable water efficiency gains are possible with improved agricultural management.

Quality impacts of upstream processes

Feedstock production, mining operations and fuel processing can also affect water quality (Larssen et al., 2006). Effluent from coal mining can degrade local water quality by lowering pH and increasing concentrations of solids and heavy metals; leachate water from overburden dumps can also have high metal concentrations (Tiwary, 2001). Effluent from uranium mining for nuclear fuel can increase concentrations of uranium, radium, selenium, molybdenum and nitrate in surrounding surface- and groundwater (R.F. Kaufmann et al., 1976; van Metre and Gray, 1992; Au et al., 1995; Voitsekhovitch et al., 2006; Carvalho et al., 2007). Radioactive water contamination can also occur from reprocessing of spent nuclear fuel, although releases can be greatly reduced through effective regulation (EC, 1999; Suzuki et al., 2008; Yamada and Zheng, 2008). Operational oil tanker discharges (i.e., dumping of oil during tanker cleaning operations) are a continuous source of water pollution (Jernelöv, 2010; Rogowska and Namiesnik, 2010). Most countries have established strict limits and safety standards to prevent water pollution, yet this does not always prevent accidents (see Section 9.3.4.7).

If conventional row-cropping production methods are used, bioenergy feedstock production can have water quality impacts from fertilizer and pesticide use similar to other row crops, yet second-generation feedstocks in many regions require lower chemical inputs for production than non-energy row crops (Paine, 1996; McLaughlin and Walsh, 1998; Lovett et al., 2009). Discharges of organic distillery wastes can pollute local water bodies, but can be reduced through existing anaerobic digestion technologies (Giampietro et al., 1997; Wilkie et al., 2000)

9.3.4.5 Land use

Most energy technologies have substantial land requirements when the whole supply chain is included. However, literature reporting lifecycle estimates for land use by energy technologies is scarce. The limited evidence available suggests that lifecycle land use by fossil energy chains can be comparable and higher than land use by RE sources (Hirschberg et al., 2006; Fthenakis and Kim, 2009).

A variety of metrics has been used in the literature to describe and compare land requirements by the dominating stage of different RE technologies, that is, the area occupied by the generating facility or cultivated for biomass feedstock. Examples are area occupied (m^2/kW) and percent effective land use (Trieb et al., 2009; Rovere et al., 2010) or land footprint (m^2 per capita) (Denholm and Margolis, 2008). Aspects that need to be considered for a proper interpretation and comparison of land requirements include:

- Properties and conditions of the land required (e.g., arable land or brown-fields, close or remote to centres of demand);

- Quality of land use (exclusive or allowing for multiple use); and

- Duration and reversibility of the land transformation (former land use/cover, reclamation times).

In particular, the assessment of environmental impacts of land transformation is very complex, with many methodological challenges yet to be solved (Dubreuil et al., 2007; Scholz, 2007). These include issues such as landscape fragmentation (Jordaan et al., 2009), impacts on life support functions and ecosystem services, impacts on naturalness of areas, like regeneration times after different types of use, and impacts on biodiversity (Lindeijer, 2000; Scholz, 2007; Schmidt, 2008) (see Section 9.3.4.6).

For fossil energy chains and nuclear power, land use is dominated by upstream and downstream processes (see Figure 9.7), depending on type of mining operations or extraction (e.g., onsite, leaching, surface or underground mining), quality of mineral deposits and fuel, and supply infrastructure (Hirschberg et al., 2006; Fthenakis and Kim, 2009; Jordaan et al., 2009). As a result of high ash content, waste disposal sites contribute significantly to land use of coal fired power stations (Mishra, 2004; NRC, 2010). Aboveground land transformation of nuclear power chains has lower ranges than do fossil fuel chains. However, the necessity of maintaining future disposal sites for high-level radioactive waste shielded from access for very long time spans (10,000 to 100,000 years) can increase the occupational land use of nuclear facilities substantially (Gagnon et al., 2002; Fthenakis and Kim, 2009).

For most RE sources, land use requirements are largest during the operational stage. An exception is the land intensity of bioenergy from dedicated feedstocks, which is significantly higher than for any other energy technology and shows substantial variations in energy yields per hectare for different feedstocks and climatic zones. If biomass from residues or organic wastes is used, additional land use is small (see Section 2.3.1).

To the extent that solar PV and solar thermal installations can be roof-mounted, operational land use is negligible, while for central PV plants and CSP design considerations can influence extent and exclusiveness of the land use (Tsoutsos et al., 2005; Denholm and Margolis, 2008; see Section 3.6.1). Geothermal generation has very low aboveground direct land use, but it increases considerably if the geothermal field is included for risk of land subsidence (Evans et al., 2009). The conservation of scenic landscapes and outstanding natural features, and related conflicts with tourism may arise as areas of concern (see Section 4.5.3.7). Similarly, the obstruction of landscape views both on- and offshore has emerged as an issue for wind energy (see Section 7.6.3.2).

Run-of-river hydropower has very low lifecycle land use, while the values for reservoir hydropower differ greatly depending on the physical conditions of the site (Gagnon et al., 2002). The impoundment and presence of a reservoir stands out as the most significant source of impacts (Egré and Milewski, 2002), with social issues such as involuntary population displacement or the destruction of cultural heritage adding a critical social dimension (see Sections 9.5.1 and 5.6.1.7). In the case of multipurpose reservoir use, inundation effects cannot be exclusively attributed to electricity generation (see Section 5.10). For wind, wave and ocean or tidal current energy, spacing between the facilities is needed for energy dissipation. Thus, the total land or ocean area transformed is quite large, but secondary uses such as farming, fishing and recreation activities are often feasible (Denholm et al., 2009; M. Jacobson, 2009), though constrained access for competing uses may be an issue for certain ocean technologies (see Section 6.5.2).

To conclude, it should be noted that land requirements for the establishment and upgrade of distribution and supply networks of future energy systems may be substantial, and may increase in the future with rising shares of variable renewable sources.

9.3.4.6 Impacts on ecosystems and biodiversity

Closely connected to land use are (site specific) impacts on ecosystems and biodiversity. Energy technologies impact ecosystems and biodiversity mainly through the following pathways:

- Direct physical destruction of habitats and ecosystems in the case of reservoir creation and alteration of rivers, surface mining, tidal barrages, waste deposits and land use changes from, for example, forest or grasslands to managed lands;

- Fragmentation of habitats, degradation of ecosystems and disturbance of certain species, for example, by infrastructure, harvesting operations or modifications in the built environment; and

- Deterioration of habitats due to air and water pollution.

While the latter is largely associated with fossil energy technologies and mining (M. Jacobson, 2009), thermal pollution, which is affecting aquatic life, constitutes a serious concern for all thermal technologies. Potential impacts of severe accidents in the extraction stage of fossil fuels can also be relevant (see Sections 9.3.4.4 and 9.3.4.7).

The assessment of impacts on biodiversity are not part of LCA methodologies, and even though efforts are made to establish and integrate indicators into the context of LCA (e.g., (Schmidt, 2008), no framework for the comparison of lifecycle impacts of different energy chains is currently available. An overview of potential concerns associated with RE technologies is provided in Table 9.11, followed by a short description of the status of knowledge. A broader discussion including potential benefits and mitigation measures is available in

Table 9.11 | Overview of potential negative impacts and concerns regarding ecosystems and biodiversity related to RE technologies as reported in Chapters 2 through 7 of this report; in depth discussion of technology-specific impacts and appropriate mitigation measures can be found in Sections 2.5.5, 3.6.1, 4.5.3, 5.6.1, 6.5.2, 7.6.2 and 7.6.5.

Bioenergy (dedicated feedstocks)	Loss of high quality natural habitats by conversion to managed lands, pressure on conservation areas, effects on agro-biodiversity and wildlife by agricultural intensification, soil degradation, eutrophication and pesticide emissions to aquatic habitats, introduction of invasive or genetically modified species
Bioenergy (residues)	Residue removal may lead to soil degradation, loss of woody debris habitats in forestry systems
Solar PV (field installations)	Disturbance through installation stage, plant community change due to shading effects
CSP	Disturbance of fragile desert ecosystems
Geothermal	Impacts of hazardous chemicals in brine fluids in case of surface disposal, modifications of habitats in conservation areas
Hydropower (general effects)	Alteration of littoral, riverine and lentic ecosystems, interference with fish migratory routes, reduced access to spawning grounds and rearing zones, change in sediment loads of the river
Hydropower (typical for reservoirs)	Habitat and special biotope loss through inundation (change of terrestrial to aquatic and riverine to lentic ecosystems), impacts of changes in chemical composition and water temperature (downstream), changes in seasonal flow and flooding regimes, extirpation of native species/introduction of non-native species, alteration of the hydrological cycle downstream
Ocean Tidal Barrage	Alteration of marine and coastal ecosystems, changes in water turbidity, salinity and sediment movements in estuary affecting vegetation, fish and bird breeding spaces
Ocean Salinity Gradient	Brackish waste water impacts on local marine and riverine environment
Ocean (Ocean Thermal Energy Conversion)	Up-welling effect of nutrient rich water to surface may impact aquatic life
Ocean (Wave energy, ocean and tidal current)	Rotating turbine blades, noise, vibration and electromagnetic fields may impact sensitive species (elasmobranchs, marine mammals), disturbance of pelagic habitats and benthic communities
Wind (Onshore)	Disturbance of air routes of migratory birds, collision fatalities of birds/raptors and bats, avoidance or displacement from an area, reduced reproduction
Wind (Offshore)	Sound waves during construction may negatively affect marine mammals, disturbance of benthic habitats

the technology chapters (see Sections 2.5.5, 3.6.1, 4.5.3, 5.6.1, 6.5.2, 7.6.2 and 7.6.5).

Scientific evidence regarding the impacts of RE technologies on biodiversity varies: for bioenergy, both local impacts of different feedstock production systems and consequences of large-scale deployment have been studied. There is evidence for both positive and negative local impacts of different feedstock production and management systems (including use of organic residues) on biodiversity (e.g., Semere and Slater, 2007; Firbank, 2008; Fitzherbert et al., 2008; Baum et al., 2009; Lovett et al., 2009; Schulz et al., 2009; Fletcher et al., 2011; Riffell et al., 2011). However, the exploitation of large bioenergy potentials is considered a reason for concern, with potential impacts on already fragmented and degraded areas that are rich in biodiversity and provide habitat for endangered and endemic species (e.g., Firbank, 2008; Sala et al., 2009; WBGU, 2009; Dauber et al., 2010; Beringer et al., 2011; see Sections 2.2.4., 2.5.5, 9.4.3.5, and 9.4.4). The overall impacts of bioenergy on biodiversity will also depend on the balance between the long-term positive effects of reduced future climate change, and the short-term negative effects of land use change (Dornburg et al., 2008).

For site-specific effects, ample evidence largely based on environmental impact assessments is available for hydropower (e.g., Rosenberg et al., 1997; Fearnside, 2001; IUCN, 2001; see Section 5.6), and to a certain extent for on- and offshore wind farms (see Section 7.6.2) and some solar technologies (e.g., Tsoutsos et al., 2005). Less evidence is available for geothermal energy, and the variety of marine and tidal devices—other than tidal barrages—are in a too early stage of development to assess their biodiversity effects. However, the long-term and population-level consequences of large-scale deployment need further research for all energy technologies.

9.3.4.7 Accidents and risks

The comparative assessment of accident risks associated with current and future energy systems is a pivotal aspect in a comprehensive evaluation of energy and sustainability. Accidental events can be triggered by natural hazards (e.g., Steinberg et al., 2008; Kaiser et al., 2009; Cozzani et al., 2010), technological failures (e.g., Hirschberg et al., 2004a; Burgherr et al., 2008), purposefully malicious action (e.g., Giroux, 2008), and human errors (e.g., Meshakti, 2007; Ale et al., 2008). This section compares risks from accidents of different energy technologies on the basis of objective information for the probability of an event and the consequences of that event, focusing on societal risk measures (e.g., Jonkman et al., 2003). Impacts from normal operation, intentional actions, and violations of ethical standards, as well as voluntary versus involuntary risks and aspects of risk internalization in occupational safety are not covered. Additional risks related to large-scale deployment of renewable technologies are also discussed.

The risks of energy technologies to society and the environment occur not only during the actual energy generation, but at all stages of the energy supply chain (Hirschberg et al., 1998; Burgherr and Hirschberg, 2008). It had already been recognized in the early 1990s that accidents in the energy sector form the second largest group of man-made accidents worldwide, however in terms of completeness and data quality their treatment was not considered satisfactory (Fritzsche, 1992). In response to this, the Energy-Related Severe Accident Database (ENSAD) was developed, established and is continuously updated by the Paul Scherrer Institute (e.g., Hirschberg et al., 1998, 2003; Burgherr and Hirschberg, 2008). The results presented here are focused on so-called severe accidents because they are most controversial in public perception and energy politics. A detailed description of the methodological approach is given in Annex II.

First, two complementary, fatality-based risk indicators are evaluated to provide a comprehensive overview. Fatalities were chosen because fatality data is typically most reliable, accurate and complete (Burgherr and Hirschberg, 2008); reducing risks to acceptable levels often includes fatalities since they are amenable to monetization (Viscusi, 2010); and actual or precursor events can provide an estimate for the maximum fatality potential of a technology (Vinnem, 2010). The fatality rate is based on the expected number of fatalities which occur in severe (≥5 fatalities) accidents, normalized to the electricity generation in GW-years. The maximum consequences are based on the maximum number of fatalities that are reasonably credible for a single accident of a specific energy technology.

Figure 9.15 shows risk assessment results for a broad range of currently operating technologies. For fossil energy chains and hydropower, OECD and EU 27 countries generally show lower fatality rates and maximum consequences than non-OECD countries. Among fossil chains, natural gas performs best with respect to both indicators. The fatality rate for coal in China (1994 to 1999) is distinctly higher than for the other non-OECD countries (Hirschberg et al., 2003; Burgherr and Hirschberg, 2007), however, data for 2000 to 2009 suggest that China is slowly approaching the non-OECD level (see Annex II). Among large centralized technologies, modern nuclear and OECD hydropower plants show the lowest fatality rates, but at the same time the consequences of extreme accidents can be very large. Experience with hydropower in OECD countries points to very low fatality rates, comparable to the representative Probabilistic Safety Assessment (PSA)-based results obtained for nuclear power plants, whereas in non-OECD countries, dam failures can claim large numbers of victims. Until 2010,[12] two core-melt events have occurred in nuclear power stations, one at Three Mile Island 2 (TMI-2, USA, 1979) and one at Chernobyl (Ukraine, 1986) (see Annex II). However, the Chernobyl accident is neither representative of operating plants in OECD countries using other and safer technologies, nor of today's situation in non-OECD countries (Hirschberg et al., 2004a; Burgherr and Hirschberg, 2008). New Generation III reactors are expected to have significantly lower fatality rates than currently operating power plants, but maximum consequences could increase due to the tendency towards larger plants (see Annex II). All other renewable technologies exhibit distinctly lower fatality rates than fossil chains, and are fully comparable to hydro and nuclear power in highly developed countries. Concerning maximum consequences, those renewable sources clearly outperform all other technologies because their decentralized nature strongly limits their catastrophic

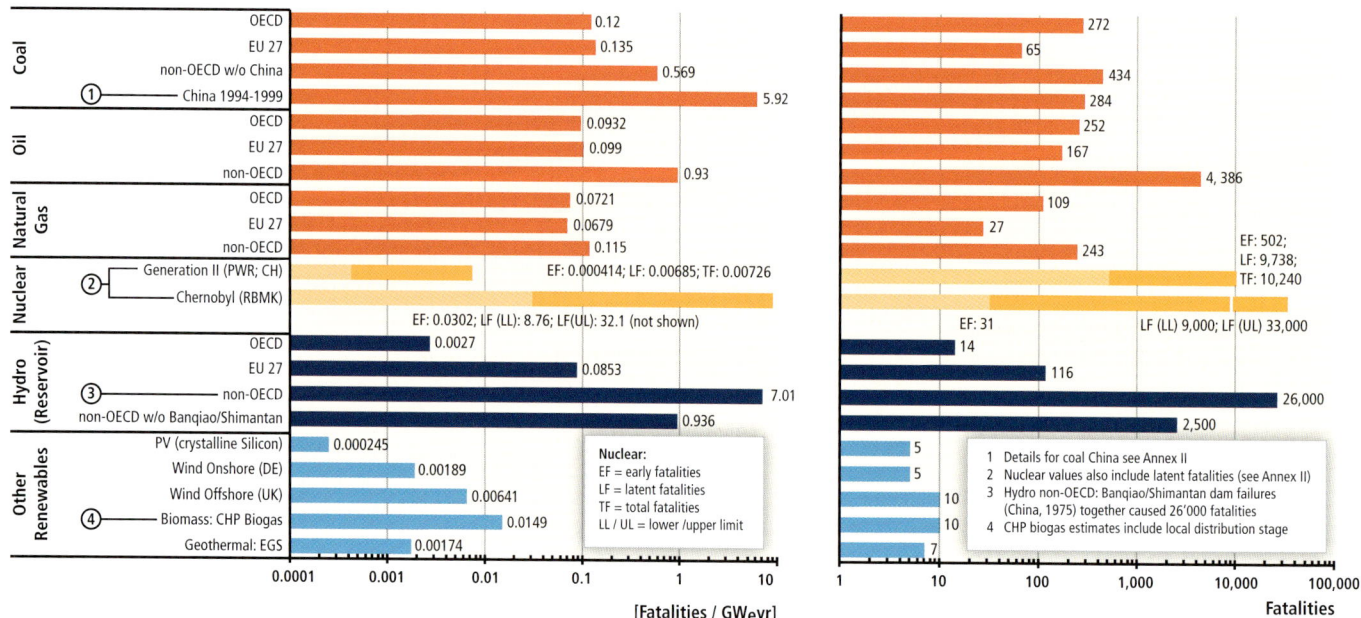

Figure 9.15 | Comparison of fatality rates and maximum consequences of currently operating large centralized and decentralized energy technologies. Fossil and hydropower is based on the ENSAD database (period 1970 to 2008); for nuclear PSA is applied; and for other renewable sources a combination of available data, literature survey and expert judgment is used. See Annex II for methodological details. Note: RBMK = reaktor bolshoy moshchnosty kanalny, a boiling water-cooled graphite moderated pressure tube type reactor; PWR = pressurized-water reactor; CHP = combined heat and power; EGS = Enhanced Geothermal Systems.

---

12 A third core-melt event that occurred in Fukushima, Japan, in March 2011 is not included in the current analysis.

Table 9.12 | Overview of selected additional risk aspects for various energy technologies.

Risk aspect	Affected technologies and references
Induced seismicity, subsidence	Oil and gas production, coal mining (Klose, 2007, 2010b; Suckale, 2009); hydropower reservoirs (H. Gupta, 2002; Kangi and Heidari, 2008; Klose, 2010a; Lei, 2010); geothermal (Bommer et al., 2006; Majer et al., 2007; Dannwolf and Ulmer, 2009); carbon capture and storage (IPCC, 2005; Benson, 2006; Holloway et al., 2007; Bachu, 2008; Ayash et al., 2009).
Resource competition	Bioenergy (Koh and Ghazoul, 2008; Ajanovic, 2011; Bartle and Abadi, 2010) reservoir hydro (Wolf, 1998; Sternberg, 2008; McNally et al., 2009).
Hazardous substances	Relevance for PV requires sector downscaling to allocate appropriate share of consequences (see Annex II) (Coburn and Cohen, 2004; Bernatik et al., 2008). In the case of geothermal, groundwater contamination may occur (Aksoy et al., 2009)
Long-term storage (public acceptance)	Disposal of nuclear waste (Adamantiades and Kessides, 2009; Sjöberg, 2009); carbon capture and storage (IPCC, 2005; Huijts et al., 2007; Ha-Duong et al., 2009; Wallquist et al., 2009).
Proliferation	Nuclear (Toth and Rogner, 2006; Yim, 2006; Adamantiades and Kessides, 2009).
Geopolitics, terrorist threat	Security and energy geopolitics of hydrocarbons and renewable sources (e.g., solar thermal) (Le Coq and Paltseva, 2009; Giroux, 2010; Toft et al., 2010; Lacher and Kumetat, 2010). Pirate attacks on oil/gas tankers (Hastings, 2009; Hong and Ng, 2010).

potential. However, it is important to assess additional risk factors of RE that are currently difficult to fully quantify, but could potentially impede their large-scale deployment (see Table 9.12).

Accidents can also result in the contamination of large land and water areas. Accidental land contamination due to the release of radioactive isotopes is only relevant for nuclear technologies (Burgherr et al., 2008). Regarding accidental releases of crude oil and its refined products into the maritime environment, substantial improvements have been achieved since the 1970s due to technical measures, but also to international conventions, national legislations and increased financial liabilities (Burgherr, 2007; Knapp and Franses, 2009; Kontovas et al., 2010). Still, accidental spills from the extraction and production of petroleum fuel are common and can affect both saline and freshwater resources (Kramer, 1982; Jernelöv, 2010; Rogowska and Namiesnik, 2010). Also, very disastrous events like the one of the drilling platform Deepwater Horizon (Gulf of Mexico, 2010; 670,000 t spill: Lubchenco et al., 2010) cannot be excluded in future. Furthermore, increased extraction of deep offshore resources (e.g., Gulf of Mexico, Brazil) as well as in extreme environments (e.g., the Arctic) provides an additional threat of accidents with potentially high environmental and economic impacts. Spills of chemicals can also occur via hydraulic fracturing during shale natural gas and geothermal operations, which can potentially result in local water contamination (Aksoy et al., 2009; Kargbo et al., 2010). Additional research is needed in this area as experience grows.

Table 9.12 and the following overview summarize a variety of risk aspects that are not amenable to full quantification yet because only limited data and experience are available or they cannot be fully covered by traditional risk indicators focusing mainly on consequences. The impact of induced seismicity from enhanced geothermal systems (EGS) has already been the cause of delays, and two major EGS projects in the USA and Switzerland were even permanently abandoned (Majer et al., 2007; Dannwolf and Ulmer, 2009). With the accelerating expansion of offshore wind parks, the risk analysis of ship collisions with offshore wind turbines and the subsequent implementation of risk-reducing measures becomes an import aspect; although the frequency of occurrence is low, the consequences could be large (Christensen et al., 2001; Biehl and Lehmann, 2006). With the installation of large renewable capacities in geopolitically less stable regions, threats to RE infrastructure (including the grid) and supply may become an important factor, including intentional supply cuts as well as physical or cyber attacks by non-state actors (e.g., sabotage, terrorism) (Lacher and Kumetat, 2010). Key issues for bioenergy include potential competition with food production and use of water resources (e.g., Koh and Ghazoul, 2008; see Sections 2.5.7.4 and 9.3.4.4). Despite numerous prototype installations and a few small commercial projects, tidal and wave power technologies are still at a relatively early stage of development, therefore their potential impacts and risks are yet rather poorly understood (Westwood, 2007; Güney and Kaygusuz, 2010; Langhamer et al., 2010; Shields et al., 2011).

In conclusion, accident risks of renewable technologies are not negligible, but their decentralized structure strongly limits the potential for disastrous consequences in terms of fatalities. However, various additional risks, complementing a purely fatality-based approach, should also be considered as outlined above because they may play an important role in public debate (e.g., risk aversion) and decision making (e.g., policies).

9.4 Implications of (sustainable) development pathways for renewable energy

In contrast to Section 9.3 that focused on the impacts of current and emerging renewable energy (RE) systems on the four sustainable development (SD) goals assessed in this chapter (for a conceptual description of these SD goals see Section 9.2), this section addresses SD pathways and future RE deployment. It will thus incorporate the intertemporal concerns of SD (see section 9.2.1).

However, only a few regional analyses address RE specifically in the context of SD pathways.[13] Even though these results indicate a positive relationship between SD pathways and RE deployment in general, they only offer limited insights with respect to the four goals that were discussed in Section 9.2. In addition, they are not explicit about the specific socioeconomic and biophysical constraints in terms of SD. Furthermore, they neglect complex global interrelations between different technologies for different energy services that significantly shape the future pathway of the global energy sector and its wider socioeconomic and environmental implications. Since the interaction of SD and RE deployment pathways[14] cannot be anticipated by relying on a partial analysis of individual energy technologies (see Section 9.3), the discussion in this section will be based on results from the scenario literature, which typically treats the portfolio of technological alternatives in the framework of a global or regional energy system.

The vast majority of the long-term scenarios reviewed in this section (and in Chapter 10) were constructed using computer-based modelling tools that capture, at a minimum, the interactions between different options for supplying, transforming and using energy. The models range from regional energy-economic models to integrated assessment models that couple models of global biogeophysical processes with models of key human systems including energy, the economy and land use. The value of these models in creating long-term scenarios, and their potential for understanding the linkages between SD and RE in particular, rests on their ability to explicitly consider interactions across a broad set of human activities (e.g., generating industrial emissions as well as leading to changes in land use and land cover), at global and regional scales, over annual to decadal to centennial time scales. Consistent with Chapter 10, these models are referred to as 'integrated models' for the remainder of the discussion in this section, since they do not look at individual technologies in isolation but rather explore the linkages between technologies, and between the energy system, the economy and other human and natural systems. Though integrated models are designed to be descriptive rather than policy prescriptive, they do offer policymakers insights into their actions that would otherwise be unavailable from focusing solely on traditional disciplinary research alone.

Integrated models have been used for many years to produce the sorts of detailed characterizations of the global energy system necessary to examine the role of RE in climate stabilization and its economic competition with other energy sources. These models also have a capability, to varying degrees, to examine issues related to the four SD goals laid out in Section 9.2. Models also vary in the degree to which they represent the biogeophysical processes that govern the fate of emissions in the atmosphere. Most models address some subset of human activities and interactions with ecosystems, but they do not in general capture feedbacks from other parts of the Earth system. In some cases, these feedbacks can be substantial.

While integrated models are powerful tools of analysis, and they will likely serve as the primary means to generate long-term scenarios in the near future, they are continually under development. Some of these developments will be relevant to the representation of sustainability concerns in future scenarios. Important areas of development include: improving their representation of resources and technology[15] to utilize them (including end-use technologies) to conserve energy resources; improving the representation of international and interregional trade; increasing both spatial and temporal resolution; allowing for a better representation of the distribution of wealth across the population; incorporating greater detail in human and physical Earth system characterization (e.g., water and the hydrological cycle), including climate feedbacks and impacts and adaptation to climate change; incorporating uncertainty and risk management; and exploring an increasingly diverse and complex policy environment.

Before turning to specific results, several caveats are in order. Although there has been some attempt at standardization among models, these are by no means 'controlled experiments'. For example, the models produce very different business-as-usual projections based upon non-standardized assumptions about a variety of critical factors, such as technology, population growth, economic growth, energy intensity and how the energy system will respond to changes in energy prices. These assumptions can have a profound effect on the energy system and welfare losses in mitigation scenarios. Even parameters that tend to be the focus of the analyses often differ across models, such as constraints on nuclear and CCS. Moreover, some but not all models use 'learning curves', that is, RE or other technology costs are assumed to decline as capacity grows. Additionally, some models allow for biomass plus CCS. As this technology option generates negative emissions, it can ease the transformation process and reduce the costs of mitigation (Wise et al., 2009; Edenhofer et al., 2010; Luckow et al., 2010; Tavoni and Tol, 2010; van Vuuren et al., 2010b). All of this leads to considerable variation among models. Importantly, however, the models basically agree on many fundamental insights (see Section 10.2).

This section will be structured along the lines of the four SD goals laid out in section 9.2: 1) social and economic development; 2) energy access; 3) energy security; and 4) climate change mitigation and reduction of environmental and health impacts. The section will give an overview of what can be learned from the literature on long-term scenarios with respect to the interrelation between SD pathways and RE. The aim of this section is twofold: first, to assess what long-term scenarios currently have to say with respect to SD pathways and the role of RE; and second, to evaluate

---

13 In a scenario analysis for India, for example, Shukla et al. (2008) found that the share of RE is higher for mitigation scenarios that include additional sustainability policies (47 versus 34% of primary energy). For Japan, several backcasting studies analyzing low-carbon society roadmaps emphasize the need for both supply-side and demand-side options including an increasing share of RE (Fujino et al., 2008; Suwa, 2009).

14 As already discussed in Section 9.2, pathways are thus primarily understood as scenario results that attempt to address the complex interrelations among SD on the one side and the different energy technologies on the other side at a global scale.

15 Unfortunately, until recently, such analyses have tended to pay insufficient attention to RE technologies and, indeed, to technology in general. The technological detail of the integrated models used to develop these scenarios is continually under development, and most of the models reviewed here and in Chapter 10 capture substantial improvements in the representations of technology with respect to the modelling capabilities available a decade ago.

how the modelling tools used to generate these scenarios can be improved to provide a better understanding of sustainability issues in the future.

9.4.1 Social and economic development

This section discusses the relationship between RE deployment and social and economic development in long-term scenarios. The integrated models used to generate these long-term scenarios generally take a strong macro-perspective and therefore ignore aspects like life expectancy or leisure time that would be relevant for alternative welfare indicators compared to GDP, such as the HDI (see Section 9.3.1). Therefore, this section will focus strongly on economic growth and related metrics. In general, growth of GDP by itself is an insufficient measure of sustainability (Fleurbaey, 2009). Most of the scenarios that are covered in Chapter 10 impose an upper limit on future cumulative GHG emissions. However, this report does not discuss to what extent the different carbon constraints are consistent with a policy avoiding dangerous climate change. Therefore, economic growth can only be used as an indicative welfare measure in the context of different stabilization pathways.

9.4.1.1 Social and economic development in scenarios of the future

There has been an enormous amount of analysis over the past two decades on the costs of reducing GHG emissions (see, e.g., IPCC, 1996a, 2001, 2007b). This work is typically based on cost-effectiveness analysis, in which the costs and means to meet a particular goal are explored, rather than cost-benefit analysis, in which the costs and benefits of mitigation and adaptation over centennial time scales are considered simultaneously, and a primary objective is to determine the optimal pattern of mitigation and adaptation over time. In cost-effectiveness studies, a long-term social goal is assumed, for example, limiting atmospheric GHG concentrations to no more than 450 ppm CO_2 equivalent. The limitation of emissions, concentrations, or more generally radiative forcing is used to study the most cost-effective pattern of emission reductions. These analyses are typically based on a variety of socioeconomic, technological and geopolitical assumptions extending over periods of decades to a century or more. When a constraint is imposed on GHG emissions, very often welfare losses are incurred. A variety of measures are used, ranging from direct estimates of social welfare loss to the more common aggregate measures such as GDP or consumption (a major component of GDP) foregone. Other concepts of welfare, as discussed in Section 9.3.1, for example, are usually not considered. Thus, at the heart of such calculations are assumptions about the availability and costs of, and GHG emissions generated by, those technologies used to satisfy energy demands—with and without a GHG constraint.

The scenario review in Chapter 10 gives an impression of possible welfare implications of RE. First note that, not surprisingly, GDP reductions are associated with a GHG constraint, independent from a particular technology portfolio. That is to say, mitigation in general decreases economic growth, at least in scenarios that do not consider the feedbacks from a changing climate, as is the case with the majority of the integrated scenarios that exist to date.

Second, by limiting the options available for constraining GHGs, GDP losses increase. It follows that economic development will be lower when the ability to deploy RE technologies is limited. A wide range of analyses over the last decade have explored the welfare implications of varying assumptions about the costs, performance and, more recently, the availability of RE (e.g., Kim Oanh and Dung, 1999; L. Clarke et al., 2008, 2009; Luderer et al., 2009; Edenhofer et al., 2010) for different levels of GHG stabilization. All of these studies have demonstrated that more pessimistic assessments of RE costs, performance and availability increase the costs of mitigation. Indeed, recent research indicates that very ambitious climate goals are not only more expensive, but may not be possible to achieve without a full portfolio of options, including RE. For example, several of the models in Edenhofer et al. (2010) could not find a feasible solution to reach a 400 ppm CO_2eq goal when constraining RE technologies to their baseline levels. The availability of bioenergy coupled with CCS is particularly important for meeting very aggressive climate goals (Azar et al., 2010; Edenhofer et al., 2010; van Vuuren et al., 2010b). More generally, scenarios do not find a clear indication that RE is more or less important in reducing costs than nuclear energy or fossil energy with CCS. For example, four of six models analyzed in Edenhofer et al. (2010) and Luderer et al. (2009) found that the economic costs of constraining RE were higher than those of constraining nuclear and fossil energy with CCS, however, of a comparable order of magnitude (see Figures 10.10 and 10.11 in Chapter 10). When other low-carbon energy technologies are constrained, not surprisingly, the share of primary energy provided by RE increases (see also the analysis provided in Chapter 10 and Figure 10.6). At the same time, higher mitigation costs result in decreasing overall energy consumption.

Looking at different sectors, a number of studies (Edmonds et al., 2006; L. Clarke et al., 2007, 2009; Fawcett et al., 2009; Luderer et al., 2009) have shown that the electricity sector can be more easily decarbonized than transportation due to the fact that many low-carbon options are available, including RE, nuclear energy and CCS. The result even proves to be robust when different low-carbon technologies are constrained as well as for developed and developing countries. The transportation sector proves to be more difficult to decarbonize and shows a significant share of fossil fuels in all models in the long term up to 2100. This can be explained by a lack of low-cost alternatives to oil (see also Section 9.4.3 on energy security), such as biofuels or the electrification of the transport sector (see, e.g., Turton and Moura, 2007 and Chapter 8). Many recent studies, for example, L. Clarke et al. (2009), include models that consider a wide range of passenger and commercial transport options such as electric vehicles and electric-hybrid vehicles. The development of a

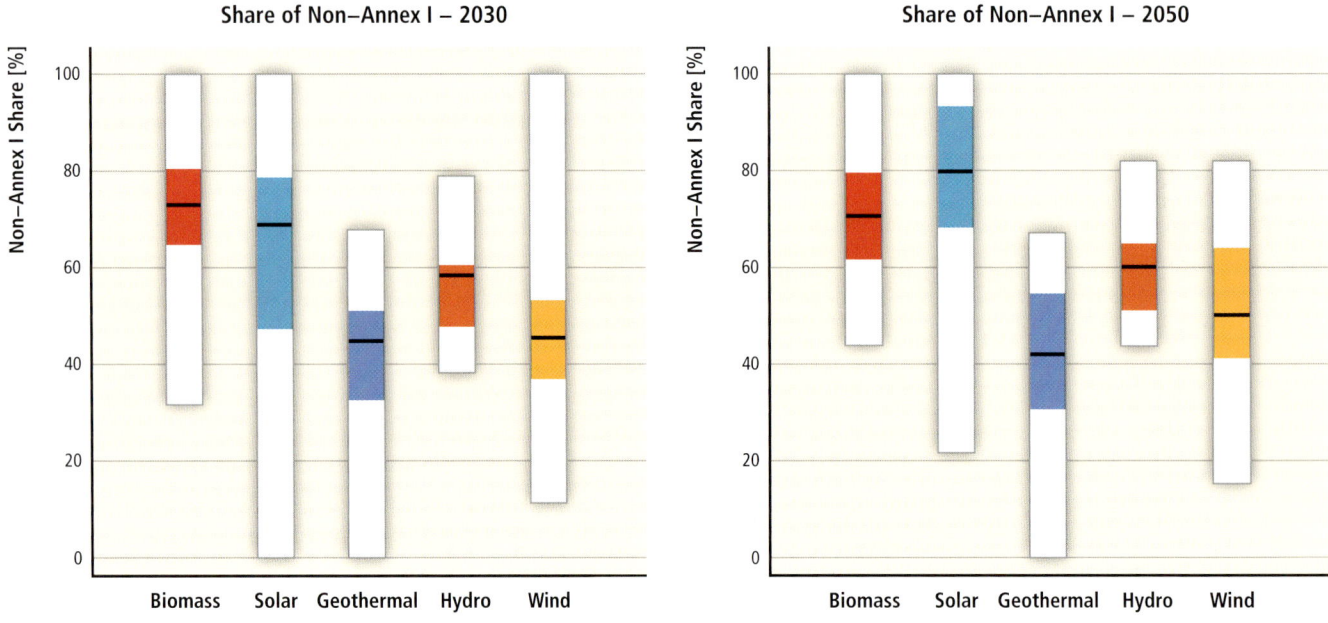

Figure 9.16 | Share of Non-Annex I countries in the global deployment of different RE sources in long-term scenarios by 2030 and 2050. The thick black line corresponds to the median, the coloured box corresponds to the inter-quartile range (25th to 75th percentile) and the white surrounding bars correspond to the total range across all reviewed scenarios (adapted from Krey and Clarke, 2011).

low-cost electric vehicle technology would make it easier and cheaper to reduce emissions in the transport sector (see, e.g., US DOT, 2010).

Although global average indicators of welfare are valuable for exploring the general relationships among RE, climate mitigation and economic growth, a great deal of interest centres not on global totals, but on the relative performance of developing and emerging economies. An important question is how mitigation in general and RE in particular influence economic growth.

Mitigation scenarios provide general insights into this issue. Overall, the same fundamental lessons about RE, mitigation and economic growth observed in global analyses are also found in analyses of developing countries. The economic growth effects are generally found to be larger in non-Annex I countries than in the Annex I countries. This is due to assumptions about more rapid economic growth and an increasingly large and dominant share of GHG mitigation over time in non-Annex I countries. Building upon the analysis in Chapter 10, Figure 9.16 shows the share of non-Annex I countries in global RE deployment for different RE sources, indicating that most future RE deployment is expected to take place in the developing world (Krey and Clarke, 2011). This is particularly important because developing countries have yet to go fully through their industrialization process. Even with huge advances in energy efficiency, their development process is likely to still involve substantial growth in energy consumption. The key challenge of deploying a carbon-free energy system in developing countries is to overcome the higher LCOEs of RE (and other low-carbon technologies) compared to current market prices (see Annex III). Successfully meeting this challenge could lead to leapfrogging the emission-intensive development paths that developed countries have taken so far.[16]

When all regions mitigate using the same economically efficient carbon price path, the resulting technology portfolio is independent of the allocation of emissions allowances (Coase, 1960). However, regional emissions mitigation will vary, depending on many factors such as technology availability, economic growth and population. When tradable allowances are allocated, each region's total cost is the sum of its mitigation costs plus (or minus) the value of permits that are purchased from (sold to) other regions. Total costs are thus reduced relative to domestic mitigation costs for permit sellers and increased for permit buyers, even though the global price of carbon is independent of the permit allocation.

If emissions mitigation obligations are distributed regionally and no trading is permitted, there is no reason to believe that marginal costs of emissions mitigation will be equal across regions and sectors, which in turn would impact the regional technology portfolio. In such circumstances, global total costs will be higher as compared to a situation where marginal costs are equal, for any given global emission mitigation level. However, the regional distribution of costs will depend on the particular assignment of mitigation obligations both initially and over time (Weyant, 1993; Edmonds et al., 1999; Scott et al., 2004; Luderer et al., 2009).

---

16 For a more detailed discussion of leap-frogging see also Section 9.5.2.

9.4.1.2 Research gaps

It should be stressed that the models used for the analyses mentioned above generally provide an incomplete measure of welfare losses because they focus on aggregate measures such as GDP or consumption losses. As noted in Section 9.2, GDP is considered by most economists as an inadequate measure of welfare. However, the use of other welfare indicators, such as, for example, life expectancy or leisure time, is difficult in the current set of integrated models. Also, losses are measured at the economy-wide level, which—although correlated with per capita GDP losses—can be misleading. Finally, the models do not give an indication of the distribution of wealth across the population. Is it concentrated among 'a few' or distributed more evenly across 'the many'?

Beyond the general insights presented in Section 9.4.1.1, particularly with respect to RE and other energy technologies, scenarios do not generally provide strong assessments of many of the forces that might make developing countries behave differently than developed countries; for example, differences in physical and institutional infrastructure and the efficiency and effectiveness of economic markets. The modelling structures used to generate long-term global scenarios generally assume perfectly functioning economic markets and institutional infrastructures across all regions of the globe, discounting the special circumstances that prevail in all countries, for example, in developing countries where these assumptions are particularly tenuous. These sorts of differences and the influence they might have on social and economic development among countries should be an area of active future research.

9.4.2 Energy access

9.4.2.1 Energy access in scenarios of the future

One of the fundamental goals of SD is the expansion of energy services, produced more cleanly, to those people who have only limited access to these services today (Goldemberg et al., 1985). While sustainable energy development comprises a number of elements (see Section 9.2; IPCC, 2000), this section focuses particularly on what different energy scenarios say about the future availability of energy services to different populations. Such services include basic household-level tasks (e.g., cooking, lighting, water heating, water collection, space heating, cooling, refrigeration); transportation (personal and freight); and energy for commerce, manufacturing and agriculture.

Integrated models have been used to evaluate and explore possible future energy systems for over three decades, but it is only in the last decade that analyses of energy access have been implemented in these models. Most, though not all, early versions of integrated models were based on the information and experiences of industrialized countries; energy systems of developing countries were often assumed to behave likewise, although some exceptions paid particular attention to differences between developed and developing regions (Shukla, 1995). In addition, for integrated modelling the data of industrialized countries were historically extrapolated to low-income countries, with no change in the underlying assumptions, to assess scenarios for developing countries. However, fundamental differences remain between the energy systems of developing countries and those of currently industrialized countries. As such, models grounded in developed country experience, and using developed country data, often fail to capture important and determinative dynamics in, for example, the choices to use traditional fuels, informal access to the electricity grid, informal economies, and structural changes in domestic economies, all of which exert a demonstrably large effect on access in many parts of the world (van Ruijven et al., 2008).

Although these factors are important for analyzing both the energy systems of developing countries and the dynamics of energy access, only a handful of integrated models explicitly account for them. A comparison study of 12 well-known integrated models by Urban et al. (2007) shows that there has been progress in addressing these issues for application in developing country contexts. All models covered electrification—though not all explicitly—and most models had implemented the use of traditional biomass and urban/rural dynamics. However, many of the models still lacked important factors such as potential supply shortages, informal economies, and investment decision making. Some of these issues are being implemented into revised models. For example, to understand how to avoid supply shortage during the peak hours, a higher temporal resolution and daily load curves to allow dynamic pricing of electricity were added to a MARKAL model of South Africa (Howells et al., 2005). Similarly, to reflect an aspect of the informal economy in fuel choices, a non-commercial 'inconvenience cost', related to using fuels, was added to MESSAGE (Ekholm et al., 2010). Several groups have attempted to increase the distributional resolution, and thereby to capture behavioural heterogeneity, by dividing populations into rural and urban categories, as well as diverse income groups (van Ruijven, 2008; Ekholm et al., 2010). Nevertheless, much more work remains ahead as models of energy access are typically limited to specific regions or countries due to lack of data or process resolution. Another obstacle is the relative difficulty of representing alternative pathways to receiving modern energy services, and specifically whether the models are really able to capture and analyze the range of distributed RE options: if models focus only on larger grid supply or cooking fuel, they only cover a part of the energy access issue.

While model resolution of energy access is improving, it remains imperfect for understanding rural dynamics. Nevertheless, it seems likely that rural populations in developing countries will continue to rely heavily on traditional fuel to satisfy their energy needs in the near future (see Table 9.1). Income growth is expected to alleviate some of the access issues, but linking this growth with fuel transitions carries much uncertainty. For example, a scenario analysis of India's energy system in 2050 showed more than a 10% difference in the future electrification rate depending on whether the Gini coefficients[17] approach the level of present day Italy or China (van Ruijven, 2008). To achieve a high penetration

---

17 The Gini coefficient is a numerical measure for the degree of inequality of income.

of modern energy, it is vital to put effective policies in place and to trigger major investments.

Electrification, whether by grid extension or off-grid distributed generation, is capital intensive and requires large investment. The IEA estimates that an investment of USD$_{2005}$ 558 billion from 2010 to 2030 is needed for universal modern energy access by 2030, of which USD$_{2005}$ 515 billion, or USD$_{2005}$ 24 billion per year on average, is needed to accomplish universal electricity access. If developing countries are not able to secure finance for electrification, the number of people without electricity is going to stay around the level of today (IEA, 2010b). During the build-up of new energy infrastructure, the combination of the availability of the low-cost traditional biomass and high initial investment cost for LPG will continue to make fuelwood and other forms of traditional biomass the main source of energy for cooking. Policies might induce higher penetration, but the structure of economic incentives must be calibrated to the local economic situation. A scenario analysis of cooking fuel in India by Ekholm et al. (2010) shows that without financing, a 50% subsidy for LPG is required for full penetration by 2020, but only a 20% subsidy is needed if improved financing for the purchase of appliances is also offered.

Having access to modern energy is not a guarantee to the path of SD. First, a shift to modern energy may be simply a shift to fossil fuels, which is not sustainable in the long run. Second, the distribution of energy use within a country with respect to income is an essential element of understanding access. For example, some countries have relatively equitable access to electricity (Norway, the USA), while others have highly unequal access depending on income (Kenya, Thailand) (A. Jacobson et al., 2005). Third, the use of RE can also have its own set of environmental or health impacts (see Section 9.3.4). However, to secure a sustainable use of energy, measures to alleviate the overall environmental burden while providing access to modern energy are essential. One aspect of such a shift would be an increasing fraction of energy supplied by RE technologies, both grid and decentralized. In addition, there is a social aspect of energy use, which relates to concerns that forced shifts to RE could affect household budgets and macroeconomic costs. In an analysis by Howells et al. (2005) on future rural household energy consumption in South Africa, a shift to electricity outside of lighting and entertainment services only occurred in the scenario which included health or other externalities from local combustion emissions.

9.4.2.2 Research gaps

Any sustainable energy expansion should increase availability of energy services to groups that currently tend to have less access to them: the poor (measured by wealth, income or more integrative indicators), those in rural areas, those without connections to the grid, and women (UNDP/UNDESA/WEC, 2000). From a development perspective, the distribution in the use and availability of energy technologies, and how they might change over time, is of fundamental importance in evaluating the potential for improvement in access (Baer, 2009). Since expanding access requires multiple changes in technology and the way services are delivered, understanding the starting distribution as well as the changes over time is necessary to evaluate the potential increase in access in one scenario relative to another. A second confounding factor in using model output to evaluate changes in access is the inability of many models to capture social phenomena and structural changes that underlie peoples' utilization of energy technologies.

These two aspects—lack of distributional resolution and structural rigidity—present particular challenges for integrated models. Models have historically focused much more on the technological and macroeconomic aspects of energy transitions, and in the process have produced largely aggregated measures of technological penetration or energy generated by particular sources of supply (Parson et al., 2007). Such measures can, of course, be useful for making broad comparisons, such as the relative share of low-carbon energy across countries. However, an explicit representation of the energy consequences for the poorest, women, specific ethnic groups within countries, or those in specific geographical areas, tends to be outside the range of current global model output.

Future modelling efforts could potentially address some of the problems highlighted in this section. Currently, access can be only estimated via proxies for aggregate statistics. However, the relationships between these aggregate statistics and access are clearly not consistent across countries and could change over time. Therefore, if access is a concern, then integrated models should incorporate the elements most likely to illuminate changes in energy access. Explicit representation of traditional fuels, modes of electrification, and income distribution could add some resolution to this process. More fundamentally, linking these to representation of alternate development pathways could provide a more comprehensive view of the possible range of options to provide access. For example, a dramatic expansion of distributed off-grid electricity generation coupled with efficient devices raises the possibility that large grid connectivity may not remain as fundamental a driver of access as it has been in the past. RE has historically been construed as relatively expensive in developing countries, but cost reductions and energy security concerns have in some cases recast it as a potentially useful source of supply in energy system studies (Goldemberg et al., 2000). RE, which is valuable in remote places due to the conversion of natural energy sources onsite, could play a major role in such scenarios (see Section 9.3.2).

9.4.3 Energy security

As noted in Sections 9.2 and 9.3.3, energy security, like SD, suffers from a lack of either a well-formed quantifiable or qualitative definition. In many countries, energy security is often taken to be inversely related to the level of oil imports. The focus on oil results from the fact that many countries are potentially vulnerable to supply disruptions, with many developed countries having experienced an oil supply disruption during the Organization of the Petroleum-Exporting Countries (OPEC) oil embargo of the mid-1970s. However, despite its importance, the real concern is not necessarily about oil, but about the vulnerability and resilience to sudden disruptions in energy supply and consequent price implications in general.

All other things being equal, the more reliant an energy system is on a single energy source, the more susceptible the energy system is to serious disruptions. This is true for energy security concerns with respect to both availability and distribution of resources, and the variability and reliability of energy sources, as discussed in Sections 9.2 and 9.3.3. At the same time, it is important to note that diversity of supply is only beneficial to the extent that the risks of disruptions are equal across sources. To the extent that risks are not equal, it is generally beneficial to rely more heavily on those sources with the lowest and most uncorrelated risks. The following discussion will address how RE influences energy security in scenarios of the future by focusing on diversity of supply and thereby energy suppliers' market power, particularly looking at the oil market; then the variability in energy supply associated with RE in the context of energy security will be assessed.

9.4.3.1 Energy security in scenarios of the future

Availability and distribution of resources: Diversity of supply and oil markets

RE deployment levels generally increase with climate change mitigation in long-term scenarios, leading to a more broadly diversified energy portfolio. To the extent that RE deployment in mitigation scenarios thus reduces the overall risk of disruption, this represents an energy security benefit. With fossil fuels continuing to dominate the energy system absent GHG mitigation (Grubb et al., 2006; L. Clarke et al., 2009), this would be particularly beneficial for regions with fossil fuel demand that can only be met by increasingly scarce or concentrated supplies.[18] Yet, market power in resource markets is typically not represented in large integrated models. This subsection thus focuses on the ability of RE to displace oil—the fossil fuel that is commonly perceived to cause the biggest energy security concerns, which are also triggered by the high price volatility (see Section 9.3.3).

The role of RE in reducing energy supply disruptions by diversifying energy supply will vary with the energy form. Hydropower, solar, wind, geothermal and ocean energy are often associated with electric power production, though some of these technologies also contribute to other end-use sectors. Reducing oil demand by increasing RE supplies in the electricity sector depends on the ability of electricity to supplant oil. This result is seen in mitigation scenarios for the buildings and industrial sectors and is caused by increasingly favourable relative electricity prices (as compared to fossil fuels). The demand for liquid fuels in the transport sector, however, is highly inelastic at present. Relatively little substitution of electricity for oil occurs without technology forcing or a technology breakthrough that makes electric power options competitive with liquid fuel transport options. This could only change if electric vehicle technology improves sufficiently in the future (see Sections 9.4.1 and 8.3.1).

Bioenergy, in contrast, is a versatile RE form that can be transformed into liquid fuels that can compete directly with liquid fossil fuels. In reference scenarios, liquids derived from biomass garner market share. The interaction between bioenergy and oil consumption is potentially sensitive to both policy and technology; the presence of a carbon price, for example, increases bioenergy's competitive advantage. However, the sector in which bioenergy is utilized depends strongly on whether or not CCS technology is available. Without CCS, bioenergy is used predominantly as a liquid fuel, whereas the availability of bioenergy with CCS shifts its use towards power generation—resulting in negative net carbon emissions for the system (Luckow et al., 2010; see Figure 9.17). Other studies show comparable results (van Vuuren et al., 2010b).

The emergence of bioenergy to supplant oil does not necessarily mean a reduction in the market power and volatility that surround markets for liquid fuels. While models generally assume that the emergence of bioenergy as a major energy form would take place in a market characterized by a large number of sellers with relatively little market power, this is by no means certain. If the bioenergy market were characterized by a small number of sellers, then buyers would be exposed to the same type of risk as is characteristic of the global oil market. However, this sort of risk-to-portfolio linkage is simply not explored by existing mitigation scenarios and a future bioenergy market might entail precisely the same volatility concerns as the current oil market.

The interaction between bioenergy production and food prices is another critical issue, since the linkage of food prices to potentially volatile energy markets has important implications for SD (see Section 2.5.7.4). A number of authors have critically assessed this relationship (Edmonds et al., 2003; Gurgel et al., 2007; Runge and Senauer, 2007; Gillingham et al., 2008; Wise et al., 2010) and some highlighted the importance of the policy environment and in particular the valuation of terrestrial carbon stocks (Calvin et al., 2009; Wise et al., 2009). Emissions mitigation policies that cause large bioenergy markets to form would clearly benefit the sellers of bioenergy and in general the owners of land, which would be more valuable. However, higher food prices clearly hurt the poor, even in scenarios with generally rising incomes. Burney et al. (2010) and Wise et al. (2009) also show the importance of traditional crop productivity in reducing GHG emissions due to the resulting higher biomass availability. Absent continued improvements in agricultural crop yields, bioenergy production never becomes a significant source of RE (Wise et al., 2010).

In the scenarios examined in Chapter 10, the consumption and price of oil do not change as significantly with more stringent mitigation as, for example, the consumption and price of coal. This more modest change in oil consumption is partly due to the fact that oil is primarily consumed in the transportation sector. Alternatives to oil, such as biofuels and

---

18 The concentration of energy supplies in the hands of a small number of sellers means that that a small group has the potential to control access. Diversification of the set of suppliers is one possible response to reduce the potential for energy supply disruptions.

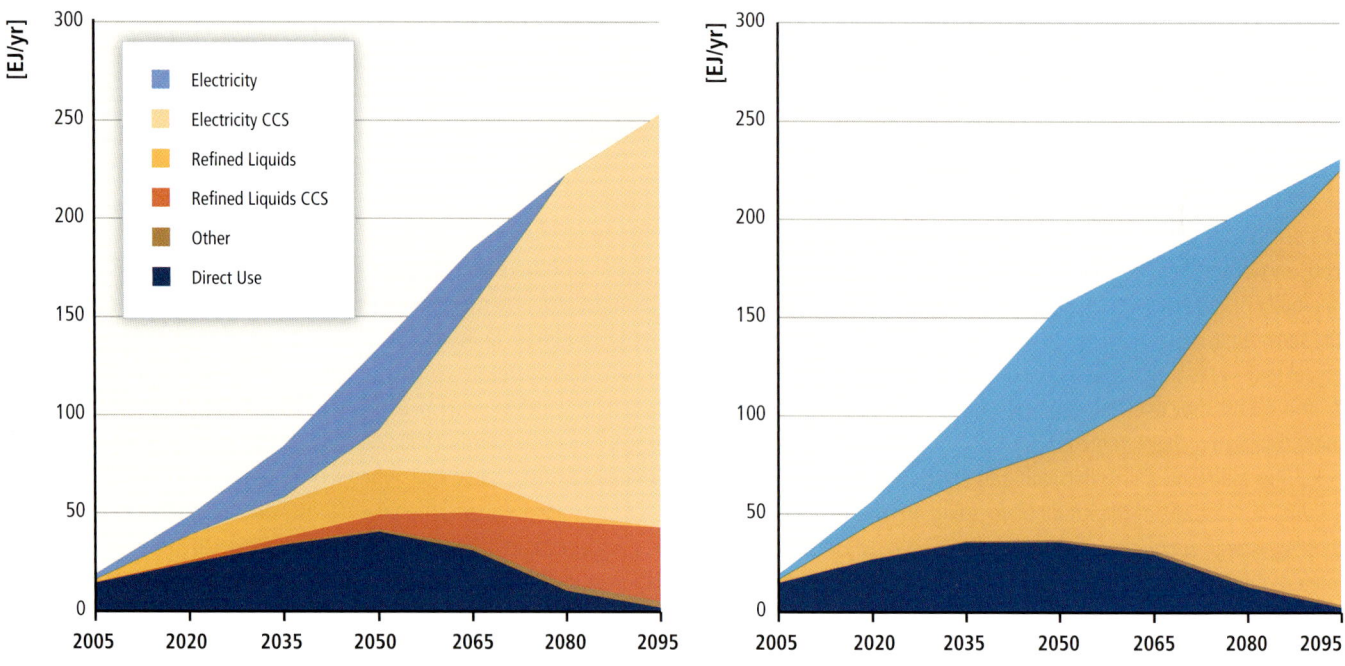

Figure 9.17 | Biomass consumption by use with (left) and without (right) CCS for a 450 ppm climate stabilization scenario using the GCAM model (Luckow et al., 2010).

electric vehicles, if included in the current generation of models, are still expensive and might have adverse impacts (e.g., first-generation biofuels, see Sections 9.4.1 and 2.5). These scenarios therefore do not see as dramatic differences between the baseline and policy scenarios with respect to cumulative oil consumption as they do for the consumption of coal. Compared to the baseline scenarios from Chapter 10, cumulative oil consumption decreases by 20% in the 440 to 600ppm CO_2 stabilization scenarios (Category III and IV, see Table 10.2) and by 40% in low stabilization scenarios (Category I and II, 400 to 440ppm CO_2) (see Figure 9.18, left).

To the extent that imports also decline, countries would be less vulnerable to oil supply disruptions than in a reference scenario. However, as discussed above, a move to bioenergy does not necessarily imply fewer liquid fuel supply disruptions in so far as bioenergy is a globally traded good. With oil still playing a major role in the mitigation scenarios of Chapter 10, energy security discussions concerning oil supply disruptions will thus remain relevant in the future. For developing countries, the issue will become even more important, as their share in global total oil consumption increases in nearly all scenarios, independent of the GHG concentration stabilization levels (Figure 9.18, right).

Furthermore, in scenarios that stabilize CO_2 concentrations, carbon prices generally rise to the point where unconventional oil supplies, such as oil shales, are more limited in supply compared to the baseline scenario (see, e.g., Figure 9.18, left). On the one hand, this effect would limit the environmental concerns (such as water pollution) that are generally associated with unconventional oil production. On the other hand, depending on a country's domestic resource base, this could increase (decrease) energy supply vulnerability for countries with (without) endowments of coal and unconventional liquids.

The effect of a GHG emissions constraint with respect to conventional oil is also notable in terms of consumption timing. Because conventional oil is relatively inexpensive to produce, the immediate suppression in demand, imports and the oil price to suppliers (consumer prices rise), is offset by an increase in oil use in later years. In other words, the effect of the cap in a CO_2 concentration stabilization scenario is to lower the peak in oil production and shift it further into the future. This has the effect of reducing near-term oil imports and increasing oil consumption in later years. As the allowable long-term CO_2 concentration declines, this effect is overwhelmed by declining cumulative allowable emissions (see, e.g., Bollen et al., 2010).

Energy security policies also have a noteworthy effect on RE and GHG emissions. A static general equilibrium model for the EU, which analyzed trade flows to and from the FSU, showed that policies to subsidize the domestic production of bioenergy simultaneously reduced fossil fuel CO_2 emissions and oil imports (Kuik, 2003). However, these policies were not seen as a cost-effective option for achieving climate goals in this study.

Variability and reliability of RE

Another source of energy supply vulnerability is exposure to unpredictable disruptive natural events. For example, wind power is vulnerable to periods of low wind. Other energy forms such as solar power or bioenergy are also susceptible to unusual weather episodes. Increased reliance on electricity generated from RE could have implications for grid stability and requires further research (see Section 8.2.1).

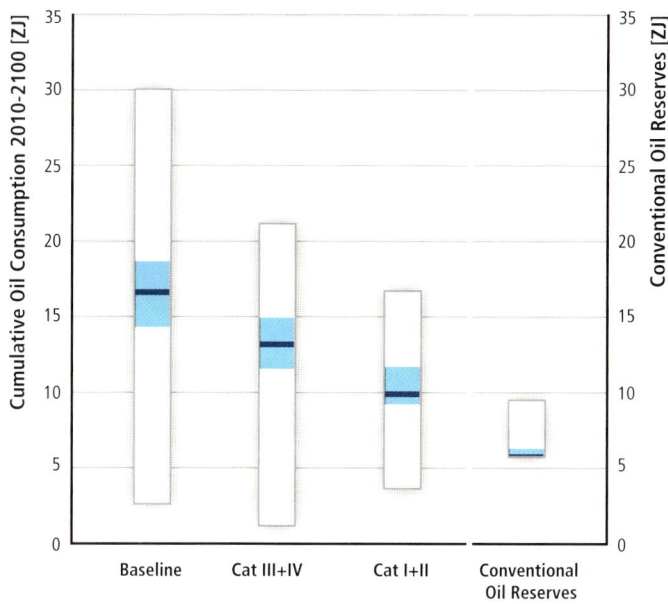
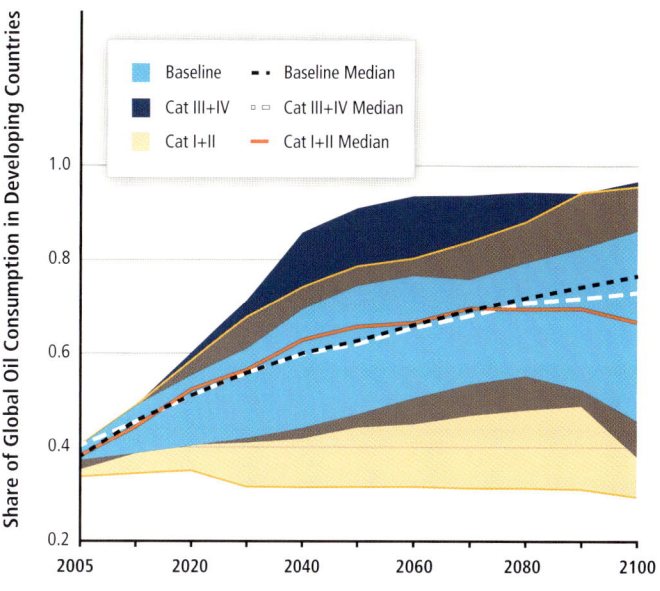

Figure 9.18 | Left: Conventional oil reserves compared to projected cumulative oil consumption (ZJ) from 2010 to 2100 in scenarios assessed in Chapter 10 for different scenario categories: baseline scenarios, category III and IV scenarios and low stabilization (category I+II) scenarios. The thick dark blue line corresponds to the median, the light blue bar corresponds to the inter-quartile range (25th to 75th percentile) and the white surrounding bar corresponds to the total range across all reviewed scenarios. The last column shows the range of proven recoverable conventional oil reserves (light blue bar) and estimated additional reserves (white surrounding bar) (Rogner, 1997).[1] Right: Share of global oil consumption in non-Annex I countries for different scenario categories over time, based on scenarios assessed in Chapter 10.

Note: 1. According to Rogner (1997), proved recoverable reserves are between 5.7 and 6.3 ZJ. In addition to that, estimated additional reserves range between 2.6 and 3.2 ZJ. This is in line with more recent estimates for proved recoverable reserves of conventional crude oil and natural gas liquids of 1,239 billion barrels (or 7.3 ZJ) (WEC, 2010). The total consumption of oil goes far beyond that in most scenarios reviewed in Chapter 10, which directly implies the use of unconventional reserves.

An important method for addressing energy supply stochasticity is holding stocks, which act to buffer the system (see Section 9.2.2). An increase in the role of bioenergy would likely lead to the creation of bioenergy stocks—either in the form of stocks of solid fuel or bioenergy liquids—as a hedge against uncertainty of supply.

RE forms such as wind, solar, geothermal and wave energy, which produce electricity, are generally not easily stored in their natural forms or as electricity. Energy supply variability can be reduced by increasing the geospatial diversity of supply. Additional efforts to increase system reliability will likely add costs and involve balancing needs (such as holding stocks of energy), the development of complementary flexible generation, strengthening network infrastructure and interconnections, energy storage technologies and modified institutional arrangements including regulatory and market mechanisms (see Sections 8.2.1 and 7.5).

9.4.3.2 Research gaps

The relationship between RE and energy security is characterized by numerous research gaps ranging from the lack of a clear quantifiable definition of energy security to the scarce scenario literature focusing on the relationship between RE and energy security. Consideration of energy security commonly focuses on the most prominent of energy security issues in recent memory, for example, disruptions to the global oil supply and security issues surrounding nuclear energy production. However, energy security issues go well beyond these aspects. For example, the supply of rare Earth metals and other critical inputs could constrain the production of some (renewable) energy technologies (see Box 9.1). These broader concerns as well as options for addressing them, e.g., recycling, are largely absent from future scenarios of mitigation and RE.

An important aspect of deploying RE sources at a large scale is their integration into the existing supply structure. Systems integration is most challenging for the variable and to a degree unpredictable electricity generation technologies such as wind power, solar PV and wave energy. A first-order proxy for the challenges related to systems integration is therefore the share of different variable and unpredictable RE sources at the global level (see also Figure 10.9). Again, those scenarios with high proportions of wind and solar PV electricity in the grid implicitly assume that any barriers to grid management in this context are largely overcome, for example, through electricity storage technologies, demand-side management options, and advances in grid management more generally (see Section 8.2.1). This is a strong assumption and managing storage, balancing generation, grid improvement and demand-side innovation will be essential to balancing variable RE generation and ensuring grid reliability. Improving the spatial and temporal resolution of integrated models to better reflect issues with respect to the integration of RE sources into the grid is an area of ongoing research (see also Section 9.4.4.2).

9.4.4 Climate change mitigation and reduction of environmental and health impacts

In addition to evaluating alternate scenarios with respect to the potential contribution to energy access and energy security, any assessment of energy futures under SD criteria must include a comparison of the environmental impacts of energy services. Fundamentally, reductions in environmental impacts can be derived from increases in the efficiency of providing services, changes in behaviour or shifting to lower-impact sources of supply.

9.4.4.1 Environmental and health impacts in scenarios of the future

As existing models include explicit representation of energy efficiency and energy supply mix, the scenarios they produce provide information on both of these dimensions of sustainability. In addition, several models have included explicit representation of factors that are linked to environmental or health impacts. For example, combustion of sulphur-containing coal without control technology can generate pollutants that are important at local and regional levels (e.g., sulphur oxides). This raises the possibility that a move away from sources of combustion would generate benefits not only via reductions in GHG emissions but also via reductions in local air pollution (see Section 9.3.4.2). Several models include sulphate pollution and therefore provide the basis for some estimation of the health or ecosystem consequences of this combustion by-product (van Ruijven et al. 2008). For example, van Vuuren et al. (2007) highlight the co-benefits in the form of reduced NO_x and SO_2 emissions when replacing fossil fuels with renewable sources and CCS. In standard scenarios, however, the link between regional pollutants and consequences is not explicit. Bollen et al. (2009) addressed this question by performing a cost-benefit analysis (using the MERGE model) that included both GHG and PM reductions. They found that climate policy can help drive improvements in local air pollution but that air pollution reduction policies do not necessarily drive reductions in GHG emissions. In addition, the external benefits were greatest when external costs of health effects due to particulate emissions and impacts of climate change were internalized (see Sections 9.3.4.3 and 10.6.4). Shrestha and Pradhan (2010) performed a broader co-benefits analysis within a specific country case, linking the MARKAL model to a model of Thailand's energy system. They found similarly that climate policy would lower the impacts from coal combustion.

Another implication of some potential energy trajectories is possible diversion of land to support biofuel production. While this has been a topic of intense discussion, many models have until recently not supported explicit links between energy supply options and land use. Early attempts to address the links were focused on trade-offs across energy supply and food production (Yamamoto et al., 2001) or used existing scenarios as a basis for estimating future bioenergy use (Hoogwijk and Faaij 2005). Subsequently, these approaches were combined by embedding bioenergy modules directly into integrated models (Gillingham et al., 2008). To date, substantial literature has, for example, become available related to emissions from indirect land use change (see Sections 9.3.4.1 and 2.5.3) (Yamamoto et al., 2001; Edmonds et al., 2003; McCarl and Schneider, 2003; Tilman et al., 2006; Searchinger et al., 2008; Calvin et al., 2009; Melillo et al., 2009; Wise et al., 2009). Wise et al. (2009) and Melillo et al. (2009) found that deforestation, land diversion and N_2O emissions were driven by biofuels expansion without proper policies in place. In both investigations, what might ostensibly have been seen as a 'sustainable' energy scenario (i.e., the increasing use of biofuels) was shown to have potential consequences that contravened the principles of SD.

Model scenarios can be useful in demonstrating scenarios of potentially unanticipated (or at least unquantified) environmental benefits as well as scenarios of unanticipated or unquantified environmental costs. However, a variety of approaches in addition to modelling are underway (e.g., Croezen et al., 2010), and other aggregate measures that could be amenable to analysis under current scenarios include, for example, water use intensity of energy (m^3/MWh) and land use (ha/MWh). These could be linked to other dimensions of sustainability, such as loss of biodiversity or changes in food security, though the appropriate treatment of this link is not defined.

9.4.4.2 Research gaps

Unfortunately, aside from the linkages discussed above (land use (change), SO_2 and PM emissions), the existing scenario literature does not explicitly treat the many non-emissions-related environmental elements of sustainable energy development such as water use, (where only very broad and non-technology-specific studies are available from the literature; see, e.g., Hanasaki et al., 2008; Shen et al., 2008) and the impacts of energy choices on household-level services or indoor air quality. These environmental aspects of sustainability depend to a much greater degree on the distribution of energy use and how each energy technology is used in practice. Analyzing this with the existing models might be difficult since models have been designed to look at fairly large world regions without looking at income or geographic distribution (see Section 9.4.2.2). Existing scenarios, rather, enable users to compare the outcomes of different possible 'futures' (L. Clarke et al., 2007; O'Neill and Nakicenovic, 2008) by allowing easy comparisons of aggregate measurements of sustainability—for example, national or sectoral GHG emissions. Although some models have also begun to allow for comparison across smaller geographic scales of impact, such as for regional air pollution and land use change, some environmental impacts remain opaque in the scenarios produced to date: the distribution of the use of traditional fuels, for example, can matter significantly for the health of billions of people (Bailis et al., 2005). In addition, most models face challenges in modelling local ecosystem impacts because of the small scales involved in many ecosystem processes. There is currently extensive discussion about the feasibility of and mechanisms for achieving finer resolution in space and time in future scenarios, not only for physical and ecosystem changes but also for social, demographic and economic factors (Moss et al., 2010). Some integrated assessment models have addressed issues of smaller scale through downscaling. However, these downscaling methods have been applied primarily to variables like emissions and demographics (Bengtsson et al., 2006; Grübler et al., 2007; van Vuuren et al., 2007, 2010a). Because the downscaling was focused on informing other questions, it does not meaningfully resolve questions about local sustainability. Finally, many models do not explicitly allow for an assessment of lifecycle impacts of the technologies used in different scenarios. What these impacts are, whether and how to compare them across categories, and whether they might be incorporated into future scenarios would constitute useful areas for future research.

9.5 Barriers and opportunities for renewable energies in the context of sustainable development

Pursuing a RE deployment strategy in the context of SD implies that all environmental, social and economic effects are taken explicitly into account. Integrated planning, policy and implementation processes can support this by anticipating and overcoming potential barriers to and exploiting opportunities of RE deployment. Barriers that are particularly pertinent in a SD context and that may either impede RE deployment or result in trade-offs with SD criteria are discussed in Section 9.5.1.[19] Section 9.5.2 focuses on how the integration of RE policies and measures in strategies for SD at various levels can help overcome such barriers and create opportunities for RE deployment that more fully meet SD goals.

9.5.1 Barriers

Integration of RE policymaking and deployment activities in SD strategy frameworks implies the explicit consideration of inter-linkages (synergies and trade-offs) with the three pillars of SD and related SD goals (see Section 9.2.1). In this way, RE policies as well as project planning, construction and operation are rooted in the specific social, economic and environmental context and support the strategic development objectives of a given society or project location. They should also remain aligned with multilateral environmental agreements. This section looks at some of the main socio-cultural, information and awareness, and economic barriers to RE deployment in a SD context addressed in the literature. For each category of barriers, links are provided to potential environmental, social or economic concerns that should be taken into account during RE policy development and deployment.

9.5.1.1 Socio-cultural barriers

Most communities have traditionally viewed RE applications as environmentally friendly and a high level of general public support for RE is documented in available studies and opinion polls (Devine-Wright, 2005; McGowan and Sauter, 2005; Wolsink, 2007b; BERR, 2008). However, public support of RE at the generic level does not necessarily translate into active support and acceptance of RE at the local implementation level, where RE deployment is often associated with direct impacts for individuals and groups (Painuly, 2001; Bell et al., 2005; Wustenhagen et al., 2007).[20] Increased public resistance to large, new installations has, for example, been experienced in many countries, often beyond the narrow 'not in my backyard' type of opposition (Wolsink, 2007b; Devine-Wright, 2009).

Socio-cultural barriers or concerns with respect to the deployment of RE and its potential SD trade-offs have different origins and are intrinsically linked to societal and personal values and norms (Sovacool and Hirsh, 2009). Such values and norms affect the perception and acceptance of RE technologies and the potential impacts of their deployment by individuals, groups and societies (GNESD, 2007b; Sovacool, 2009; West et al., 2010). From a SD perspective, barriers may arise from inadequate attention to such socio-cultural concerns, which include barriers related to behaviour; natural habitats and natural and human heritage sites, including impacts on biodiversity and ecosystems (see Sections 2.5.5.2 and 9.3.4.6); landscape aesthetics; and water/land use and water/land use rights (see Section 9.3.4.4 and 9.3.4.5) as well as their availability for competing uses. These barriers are briefly discussed below.

Deployment of RE technologies may be associated with *behavioural* implications that challenge social and cultural values, norms and perceptions (Painuly, 2001; S. Reddy and Painuly, 2004; GNESD, 2007b; Chaurey and Kandpal, 2010). In India, for example, multi-criteria analysis of domestic cooking devices (Pohekar and Ramachandran, 2006) reveals that behavioural concerns[21] are second most important in determining consumer preferences for cooking devices, only surpassed by technical criteria. Behavioural concerns limit uptake not only of the relatively new and technically advanced solar cookers. They also offer an important explanation for the non-use of installed improved fuelwood cookstoves in India, where only 6 million out of a total of 23 million installed improved fuelwood stoves were found to be functional (Neudoerffer et al., 2001; Pohekar and Ramachandran, 2006). Similar findings regarding the significance of behavioural barriers for dissemination and use

---

[19] Barriers are addressed in many chapters of the report. Chapter 1 provides a general overview of barriers to RE development and implementation, categorizing the barriers as socio-cultural, information and awareness, economic, and institutional. The technical chapters (2 to 7) cover the technology-specific barriers, with Chapter 8 addressing energy system lock-in and RE integration. Barriers to policymaking and financing are covered in Chapter 11.

[20] Local opposition to renewable energy projects may also depend on methods used to gather public opinion (van der Horst, 2007).

[21] Related to ease of operation; types of dishes cooked; cleanliness of utensils; need for additional cookstove; motivation to buy; taste of food; and aesthetics.

of improved cookstoves are found for other developing countries (Ben Hagan, 2003; Zuk et al., 2007; Bailis et al., 2009). Behavioural barriers to new RE technologies and systems may be relatively small as long as the transition seeks to emulate existing practices and properties of current technologies. However, they tend to increase with the extent of changes in behaviour or consumption levels (Kumar et al., 2009; Petersen and Andersen, 2009).

Although applicable, the precautionary principle is not always utilized to minimize impacts on *natural habitats and natural and human heritage sites* (Rylands and Brandon, 2005; Hreinsson, 2007; Nandy et al., 2007; S. Clarke, 2009; Hennenberg et al., 2010; Wolsink, 2010). This has led to public resistance to various types of RE development projects. Public perception of impacts related to *aesthetics* of altered landscapes associated with wind power developments in OECD countries is a barrier that is extensively analyzed in the literature (Wolsink, 2000, 2007b, 2010; Upreti, 2004; Jobert et al., 2007; Wustenhagen et al., 2007). Attitudes towards offshore wind farms visible from shore depend on, for example, the type and frequency of beach use with regular visitors perceiving coastal landscapes as more pristine resources and thus less suited for industrial usage (Ladenburg, 2010). See also Section 8.2.1.3 on public opposition with regard to new network infrastructure.

Displacement and resettlement of communities in project developments that involve large quantities of *land*, such as large-scale hydropower, may be significant (Richter et al., 2010). The World Commission on Dams (2000) estimates that worldwide, 40 to 80 million people have been displaced by large dams. This figure increases significantly when the associated impacts of alterations in river flows and freshwater *ecosystems* on downstream populations are included (Richter et al., 2010). Although more recent figures on the number of people affected by hydropower developments are available at the individual project and country level,[22] aggregate statistics seem to be limited to the 2000 report by the World Commission on Dams. Large-scale hydropower projects are in addition often associated with trade-offs related to competing uses of water, for example, for water supply for domestic and industrial purposes, flood control and irrigation (Moore et al., 2010). Resettlement of populations affected by large-scale hydropower developments is intrinsically linked to the issue of *land use rights* of indigenous people (Bao, 2010; Moore et al., 2010; Ölz and Beerepoot, 2010) and associated with complex resettlement and compensation issues (Chen, 2009; Mirza et al., 2009). For example, insufficient economic compensation may be offered to affected populations or to those affected by externalities such as losses in cultural heritage (Cernea, 1997; World Commission on Dams, 2000; Bao, 2010; Brown and Xu, 2010). Land use issues arising from commercial-scale energy crops are another area of increasing attention (IIED, 2009). Occupational concerns regarding human and labour rights, such as working conditions in field crop projects, are important to consider in this context (ILO, 2010). Finally, food security is another important social concern (see Section 2.5.7.4) to which certification schemes are paying increased attention (see Section 2.4.5). Public awareness and acceptance is, as indicated above, an important element in the need to rapidly and significantly scale-up RE deployment to help meet climate change mitigation goals. Large scale implementation can only be undertaken successfully with the understanding and support of the public (Zoellner et al., 2008). This may require dedicated communication efforts related to the achievements and the opportunities associated with wider-scale applications (Barry et al., 2008). At the same time, however, public participation in planning decisions as well as fairness and equity considerations in the distribution of the benefits and costs of RE deployment play an equally important role and cannot be side-stepped (see below and Section 9.5.2.2; Wolsink, 2007b; Malesios and Arabatzis, 2010).

9.5.1.2 Information and awareness barriers

A common argument to promote RE projects is their contribution to poverty reduction, with local communities benefiting from employment opportunities, skills development, investment opportunities and technology transfer (see Sections 9.3.1.3 and 11.3; UN, 2002; GNESD, 2004, 2007a,b, 2008; Goldemberg and Teixeira Coelho, 2004; Modi et al., 2006; Goldemberg et al., 2008; UNEP, 2008a; Barbier, 2009). Many RE pilot projects in developing countries give anecdotal evidence of the role that renewable sources can play in energy-poor communities (Karekezi and Kithyoma, 2003; Mondal et al., 2010). However, if the local community does not perceive these benefits, or their distribution is considered inequitable, project acceptance may be problematic (Upreti, 2004; Gunawardena, 2010; see Section 11.6.4). In developing countries, limited technical and business skills and absence of technical support systems are particularly apparent in the energy sector, where awareness of and information dissemination regarding available and appropriate RE options among potential consumers is a key determinant of uptake and market creation (Painuly, 2001; Ölz and Beerepoot, 2010). This gap in awareness is often perceived as the single most important factor affecting the deployment of RE and development of small and medium enterprises that contribute to economic growth. Ignoring the informational and perception concerns associated with decentralized units can often result in abandoned or dysfunctional systems (Werner and Schaefer, 2007).

In cases where the proprietary ownership of RE technology is in the hands of private sector companies and the diffusion of technologies also typically occurs through markets in which companies are key actors (Wilkins, 2002), there is a need to focus on the capacity of these actors to develop, implement and deploy RE technologies. Therefore, the importance of increasing technical and business capability as a part of capacity building (Section 11.6.6)—at the micro or firm level—needs to be addressed (Lall, 2002; Figueiredo, 2003).

Attitudes towards RE are shaped by more than knowledge and facts. Norms and values are important to consider, as illustrated in Section

---

22 See, for example, factsanddetails.com/china.php?itemid=323&catid=13&subcatid=85#01 for information on dams and hydropower in China and www.gms-eoc.org/CEP/Comp1/docs/Vietnam/Hydropower/SocialImpact.pdf for Vietnam.

9.5.1.1, and may affect public and personal perceptions of the implications of RE for consumption as well as for deeply held values regarding trust, control and freedom (Sovacool, 2009; Walker et al., 2010). This implies that attitudes towards RE in addition to rationality are driven by emotions and psychological issues (Bang et al., 2000; Devine-Wright, 2009). To be successful, RE deployment and information and awareness efforts and strategies need to take this explicitly into account (Jager, 2006; Nannen and van den Bergh, 2010; Litvine and Wüstenhagen, 2011), particularly as barriers to information and awareness may have implications for RE uptake, markets, uncertainty and hence capital costs (Painuly, 2001; Ölz and Beerepoot, 2010).

9.5.1.3 Market failures and economic barriers

The economics of RE are discussed in nearly all chapters of this report (Chapters 2 through 7 in cost sections, Chapter 10 on externalities, Chapter 11 on policy case studies). To assess the economics of RE in the context of SD, social costs and benefits need to be explicitly considered. RE should be assessed against quantifiable criteria targeted at cost effectiveness, regional appropriateness, and environmental and distributional consequences (C. Gross, 2007; Creutzig and He, 2009). From a social perspective, a level economic playing field is required to support rational RE investment decisions. This implies that market distortions, such as taxes and subsidies and their structure, as well as market imperfections and failures must be considered carefully with respect to their implications for the deployment of RE and the internalization of social costs, such as damages from GHG emissions, health, and environmental costs (Rao and Kishore, 2010; see Sections 9.5.2 and 10.6).

Grid size and technologies are key determinants of the *economic viability* of RE and of the competitiveness of RE compared to non-RE. Appropriate RE technologies that are economically viable are often found to be available for expanding rural off-grid energy access (Bishop and Amaratunga, 2008; Ravindranath and Balachandra, 2009; Thompson and Duggirala, 2009; Deichmann et al., 2011; see Section 9.3.2). For smaller off-grid applications, there is some evidence that several RE technologies, including wind, mini-hydro and biomass-electric, can deliver the lowest levelized generation costs of electrification, that is, including the levelized costs of transmission and distribution (ESMAP, 2007). Several RE technologies, including biomass (particularly biogas digesters and biomass gasifiers), geothermal, wind and hydro, are also potentially the least-cost mini-grid generation technology (ESMAP, 2007).[23] However, non-renewable power generation technologies remain more economically viable than RE in many contexts (van Alphen et al., 2007; Cowan et al., 2009). This is particularly the case for most large grid-connected applications, even with increases in oil price forecasts (ESMAP, 2007) and when likely RE technology cost reductions over the next 20 years are considered (Deichmann et al., 2011).

Assessments of the economic viability of RE are based on and subject to assumptions regarding the *availability and cost of the renewable resource*. The lack of adequate resource potential data directly affects uncertainty regarding resource availability, which may translate into higher risk premiums by investors and project developers, as appears to be the case with geothermal electricity development in Indonesia (Ölz and Beerepoot, 2010). An emerging area of attention relates to the potential impacts of climate variability and climate change on energy services and resources, where the timing and availability of RE resources are immediately impacted (World Bank, 2011). Impacts of climate variability and extreme events (e.g., hurricanes and typhoons, heat waves, floods, and droughts) on energy services and resources are already being experienced. In Eastern Africa, for example, where power supply is heavily reliant on hydropower, recent droughts were associated with estimated annual costs of the order of 1 to 3.3% of annual GDP (Eberhard et al., 2008; Karekezi et al., 2009). For issues related to the higher costs of RE due to their variable availability, see Section 8.2.

In cases where deployment of RE is viable from an economic perspective, other economic and financial barriers may affect the deployment of RE. High upfront costs of investments, including high installation and grid connection costs, are examples of frequently identified barriers to RE deployment (Painuly, 2001; Limmeechokchai and Chawana, 2007; Kassenga, 2008; Mathews, 2008; Monroy and Hernandez, 2008; Rao and Kishore, 2010; Green and Vasilakos, 2011). Particularly in low-income countries, high upfront costs of RE technologies may inhibit uptake by consumers. Consumers may prefer to keep the initial cost low rather than minimizing the operating costs, which run over a longer period of time, or they may have no choice if they lack access to cash and/or credit (S. Reddy and Painuly, 2004). Hence, the successful uptake of RE technologies depends to some degree on the choice and set-up of the dissemination model, such as donations, cash sales, consumer credits or fee-for-service schemes (Nieuwenhout et al., 2000).

Policy and entrepreneurial support systems are needed along with RE deployment to stimulate economic growth and SD and catalyze rural and peri-urban cash economies (O. Davidson et al., 2003). Investments are, for example, required to ensure availability of the technical capacity required to operate and maintain the systems, which is a significant barrier for harnessing available RE sources in developing countries (Ölz and Beerepoot, 2010). A new set of thinking is also gradually emerging, treating RE as an integral component of a market-based energy economy and more strongly involving the private sector (GNESD, 2007b, 2008).

High upfront costs may also reflect high-risk perceptions of investors and a general lack of financing instruments as well as fragmented or underdeveloped financial sectors (Brunnschweiler, 2010). In this way, anecdotal evidence from South East Asia suggests that a lack of experience with and understanding of RE systems among financial institutions and investors leads to low participation by national financiers, which may increase the cost of capital for RE projects through higher risk

---

23 Mini-grid applications are village- and district-level isolated networks with loads between 5 and 500 kW.

premiums (see Section 11.4.3). In Indonesia, biomass-based power projects are viewed as facing additional hurdles linked to a general lack of experience in bioenergy project development and related feedstock supply issues among banks and national investors (Ölz and Beerepoot, 2010).

The effects of the timing of the stream of costs and benefits from RE investments lead to a trade-off with respect to sustainability, for example in cases where decision makers in developing countries have to choose between investments in non-RE with shorter payback time, but higher external costs, and RE investments with longer payback time, but higher positive externalities for example, for job creation, health, GHG emission reduction, etc. Barriers to RE financing are also addressed in Sections 9.3.1.4 and 11.4.3.

Externalities result from market distortions and are central when RE deployment is addressed in the context of SD. The structure of subsidies and/or taxes may, for example, favour non-RE with adverse implications for the competitiveness of RE (see Section 9.5.2.1). Similarly, existing grid networks and engineering capacities will advantage some forms of energy over others, with implications for the path dependency of energy deployment (see Section 11.6.1). Path dependencies may lock in societies into energy or infrastructure options that may be inferior in terms of cost efficiency or accumulated social costs in the long term (Unruh, 2000). In many cases, internalization of environmental externalities has considerable effects for the levelized costs of RE technologies (Cowan et al., 2009; Harmon and Cowan, 2009; Fahlen and Ahlgren, 2010) and subsequently their non-inclusion presents a barrier for RE deployment. Internalization of damage costs resulting from combustion of fossil fuels into the price of the resulting output of electricity could, for example, lead to a number of renewable technologies being financially competitive with generation from coal plants (Owen, 2006; see Section 10.6). Similar conclusions were reached for PV mini-grids for three remote rural regions in Senegal, where levelized electricity costs from PV technologies were found to be lower than the cost of energy from grid extension when environmental externalities are taken into account (Thiam, 2010).

A number of recent studies include several social and environmental sustainability indicators in assessing and ranking energy options. In addition to GHG emissions, these sustainability indicators include land requirements, water consumption, social impacts and availability of renewable sources, providing additional insight into potential barriers for RE deployment in a sustainability context (Afgan et al., 2007; Becerra-Lopez and Golding, 2008; Brent and Kruger, 2009; Evans et al., 2009; Brent and Rogers, 2010; Browne et al., 2010; Carrera and Mack, 2010; see Section 9.5.2.1).

9.5.2   Opportunities

Strategies for SD at international, national and local levels as well as in private and nongovernmental spheres of society can help overcome barriers and create opportunities for RE deployment by integrating RE and SD policies and practices. At international and national levels strategies include: removal of mechanisms that are perceived as to work against SD; mechanisms for SD that internalize environmental and social externalities; and integration of RE and SD strategies. At the local level, SD initiatives by cities, local governments, and private and non-governmental organizations can be drivers of change and contribute to overcome local resistance to RE installations.

9.5.2.1   International and national strategies for sustainable development

The need for cross-sectoral SD strategies has been articulated at the multilateral level since the 1972 Stockholm Conference on the Human Environment (Founex Committee, 1971; Engfeldt, 2009). The concerns were reinforced in the goals of Agenda 21 (UNCED, 1992), aiming at the adoption of strategies to harmonize these different sectoral processes (Steurer and Martinuzzi, 2007). In the Johannesburg Plan of Implementation adopted at the World Summit on Sustainable Development in 2002, governments were called upon with a sense of urgency to substantially increase the global share of RE and to take immediate steps towards national strategies for SD by 2005 (UN, 2002). In the formulation of such National Sustainable Development Strategies (NSDS), countries have usually prioritized strategic policy areas and concrete objectives for which national circumstances and international commitments required swift action, such as limiting climate change and increasing the use of RE (OECD, 2002; UNDESA, 2008). Such prioritization may contribute to productivity, income growth, health and education, gender equality, reduced social impacts associated with energy extraction, human development, and macroeconomic stability and governance (World Bank, 2001). RE technologies, in particular, can add other benefits (see Section 9.3). In addition, integrating RE policy into NSDS provides a framework for countries to select specific policy instruments, to incorporate concerns of other countries into their own, and to align with international policy measures (OECD, 2002).

Removal of mechanisms that work against sustainable development

The removal of fossil fuel subsidies has the potential to open up opportunities for more extensive use or even market entry of RE. It decreases the artificially widened competitive advantage of fossil fuels and may free spending on fossil fuel subsidies to be redirected to R&D and deployment of RE technologies. With the 2009 G-20 Summit having agreed to phase out 'inefficient fossil fuel subsidies' over the medium term (G-20, 2009), this may offer some co-benefits for RE technologies. A report by the IEA, OECD and World Bank (2010), prepared for the subsequent G-20 Summit, finds that government support of fossil fuels is geographically concentrated. In 2009, 37 economies, mainly non-OECD, accounted for more than 95% of fossil fuel subsidies worldwide representing a total value of USD$_{2005}$ 268 billion.[24] Government support of fossil fuels is predominant in economies where supported energy carriers are abundant, for example, Iran and Saudi Arabia.[25] Supported fuels are mainly oil (USD$_{2005}$ 108 billion) and natural gas (USD$_{2005}$ 73 billion), and may also implicitly cover electricity (USD$_{2005}$ 82 billion), if largely generated by these fuels. In contrast, global coal subsidies are comparatively small at only USD$_{2005}$ 5 billion.

A general concern when reforming these subsidies is how they affect the poor; they need to be carefully designed as low-income households are likely to be disproportionally affected (IEA, 2010b). However, subsidies are often regressive and there is a substantial benefit leakage to higher-income groups (Del Granado et al., 2010). For example, in Iran the richest 30% percent consume 70% of all government support (Nikou, 2010), and in Indonesia the bottom 40% of low-income families reap only 15% of all energy subsidies (IEA, 2008a). By and large this includes most supported fuels, for instance, electricity in several African countries (Angel-Urdinola and Wodon, 2007), LPG in India (Gangopadhyay et al., 2005) and petroleum products worldwide (Coady et al., 2010). In the case of kerosene, however, the picture is less clear and subsidies are relatively better targeted (Coady et al., 2004).

Accordingly, reforming subsidies towards the use of RE technologies should necessarily go along with addressing the specific needs of the poor. In order to do so, two general directions appear suitable. The first direction is expanding rural electrification, as poor households tend to live in areas without electricity service (Angel-Urdinola and Wodon, 2007). Successful programs have been initiated in Ethiopia and Vietnam (IEA/OECD/World Bank, 2010), and the phase-out of concurrent fossil fuel subsidies may create further incentives for business activities (Barnes and Halpern, 2001). Increasing electrification could be complemented with additional support for RE technologies in centralized power supplies, which would then also become available to the poor. Second, if electrification is not viable or better low-cost options exist, RE off-grid technologies are an alternative. In Nepal, for example, financial aids have significantly increased the awareness levels in adopting RE off-grid technologies and the willingness to pay for electricity (Mainali and Silveira, 2011). Moreover, for domestic lighting in India, solar photovoltaics and modern bioenergy systems are better options in rural areas compared to traditional kerosene-based lighting (Mahapatra et al., 2009).

It is likely that many more such opportunities exist, but to identify potential gains for RE and evaluate efficiency further case-specific analysis is needed. Without such analysis it is neither clear that RE technologies directly benefit from a phase-out of fossil fuel subsidies, nor whether the phase-out as such is potentially harmful.

The importance of *eliminating barriers to trade in RE supplies and associated technologies* as part of a broader strategy to reduce dependence on more-polluting and less secure energy sources has been stressed in several studies and events. This is the case for, among others, PV, wind turbines and biofuels (Steenblik, 2005; Lucon and Rei, 2006; OECD, 2006). As outlined in Section 2.4.6.2, barriers to the market penetration and international trade of bioenergy include tariff barriers, technical standards, inappropriately restrictive sustainability criteria and certification systems for biomass and biofuels, logistical barriers, and sanitary requirements. More generally, the elimination or reduction of barriers to trade can facilitate access to RE and other environmental goods that can contribute to climate change mitigation by fostering a better dissemination of technologies at lower costs. Elimination of both tariffs and non-tariff barriers to clean technologies could potentially result in a 14% increase in trade in these products (WTO, 2010).

As parties to the Kyoto Protocol of the UN Framework Convention on Climate Change develop and implement policies and measures to achieve GHG concentration stabilization, compatibility with World Trade Organization (WTO) rules could become a recurrent issue. More generally, the nexus of investment rules inside and outside the WTO with the climate regime needs further attention (Brewer, 2004). Interactions that are the most problematic include the potential use of border measures to offset cross-national differences in the energy costs of goods, Clean Development Mechanism (CDM) and Joint Implementation projects in relation to the WTO subsidies agreement, efficiency standards in relationship to the WTO technical barriers agreement and carbon sequestration in relationship to the WTO agriculture agreement (Tamiotti et al., 2009).

Mechanisms for sustainable development that internalize environmental and social externalities

There is a constant need for mechanisms for SD that internalize environmental or social externalities. Diffusion of RE technologies is driven by policies and incentives that help overcome high upfront costs and lack of a level playing field (Rao and Kishore, 2010). However, when external costs (see Section 10.6) are included, the relative advantage of renewable energies is highlighted—especially regarding GHG emissions (Onat and Bayar, 2010; Varun et al., 2010). Incorporating external costs requires good indicators. A methodological limitation found in studies of different energy production systems is their use of an insufficient number of comparable sustainability indicators, which may lead to biases and flaws in the ranking of energy sources and technologies against sustainability (Brent and Kruger, 2009; Eason et al., 2009; Kowalski et al., 2009). Although multi-criteria decision analysis and approaches contribute significantly, it is recognized that appraising the contribution of RE options to SD is a complex task, considering the different aspects of SD, the imprecision and uncertainty of the related information as well as the qualitative aspects embodied that cannot be represented solely by numerical values (Cavallaro, 2009; Michalena et al., 2009; Donat Castello et al., 2010; Doukas et al., 2010).

---

24 Even though the underlying price gap approach has some limitations, it may serve as a first estimate.

25 For more information on subsidy rates see www.iea.org/subsidy/index.html.

The CDM established under the Kyoto Protocol is a practical example of a mechanism for SD.[26] RE to substitute for fossil fuels constitutes 61% of projects and 35% of expected Certified Emission Reductions by 2012 under the CDM (UNEP Risø Pipeline, 2011). The CDM is widely acknowledged as one of the most innovative features of the Kyoto Protocol with the involvement of 69 developing countries in the creation of a global carbon market worth billions of US dollars. It is, however, also widely known that its contribution to sustainable and low-carbon development paths in host countries is questionable (Figueres and Streck, 2009). CDM projects are submitted for sustainability screening and approval at the national level by the Designated National Authority (DNA; see also Sections 11.5.3.3, 11.6,11.6.6.1). There is, however, no international standard for sustainability assessment to counter weaknesses in the existing system of sustainability approval (Olsen and Fenhann, 2008b). Thus, DNAs have an important role in meeting national SD priorities—as well as in attracting investment (Winkler et al., 2005). Literature reviews of the CDM (Paulsson, 2009) and its contribution to SD (Olsen, 2007) find that one of the main weaknesses of the market mechanism is that of cheap emission reduction projects being preferred over more expensive projects that often are associated with higher SD benefits (Sutter and Parreño, 2007). Voluntary standards exist, such as the Gold Standard and the Climate, Community and Biodiversity Standards, that aim to attract investors who are willing to pay a premium for emission reductions with guaranteed co-benefits (Nussbaumer, 2009). The Gold Standard applies to RE and energy efficiency projects, where the most common RE projects are wind, biogas, biomass energy, hydro, landfill and solar. These labelled projects, however, make up a small share of the total volume of CDM projects and as voluntary standards, they are successful in rewarding high-quality projects rather than improving low- or unsustainable projects (Wood, 2011). As input to the negotiations for a post-2012 climate regime, much literature has addressed how to reform the CDM to better achieve new and improved mechanisms for SD (Hepburn and Stern, 2008; Olsen and Fenhann, 2008a; Wara, 2008; Figueres and Streck, 2009; Schneider, 2009). Ideas include an up-scaling of mitigation actions through sector no-lose targets (Ward, 2008), introduction of new sectoral approaches (Marcu, 2009), differentiation of developing country eligibility for CDM crediting (Murphy et al., 2008) and structural changes for the CDM to contribute to long-term benefits for a low-carbon economy (Americano, 2008).

Mechanisms for SD may also be addressed from a wider perspective than sustainability assessments. The idea that developing countries might be able to follow more sustainable, low-carbon development pathways than industrialized countries have is particularly attractive. Such decisions are both political and societal, but depend intrinsically on the understanding of the concept of leapfrogging (see Box 9.5).

Integrating renewable energy and sustainable development strategies
Opportunities for RE to play a role in national strategies for SD can be approached in two ways: 1) by integrating SD and RE goals into development policies and plans such as budgeting processes and Poverty Reduction Strategy Plans; and 2) by development of sectoral strategies for RE contributing to goals for green growth, low-carbon and sustainable development.

Though the idea of *National Sustainable Development Strategies* (NSDS) was born at the international level, the actual implementation of strategies takes place at the national level. By 2009, 106 countries corresponding to 55% of Member States to the United Nations had reported to the Commission on Sustainable Development that they were implementing an NSDS. The overall idea of NSDS is to integrate principles for SD such as the three pillars of sustainability, participation, ownership, comprehensive and coordinated policymaking, as well as targeting, resourcing and monitoring (i.e., the measurement and monitoring of development outcomes) into a country's existing development process (George and Kirkpatrick, 2006). NSDS should not be a new, separate strategy but are meant to integrate SD concerns into a country's existing governance and decision-making framework. As countries differ in their institutional, developmental and geographical conditions no blueprint exists for NSDS, but generally they are structured into three levels: 1) major goals and policy areas such as dealing with climate change and energy security; 2) concrete objectives and issues such as transport, energy efficiency and RE; and 3) aims and actions such as implementing a RE strategy, liberalizing energy markets or using the CDM to support small RE power projects (UNDESA, 2008). When it comes to implementation of NSDS, however, the record of progress has been limited (George and Kirkpatrick, 2006). Volkery et al. (2006) found that many countries are still at early stages of learning and a key challenge is coordination of NSDS with other strategy processes such as the national budget, sectoral and sub-national strategy processes. In most countries, the NSDS provides a summary of existing strategies and as such it works as a post-rationalization rather than an overarching framework guiding and stimulating new action (George and Kirkpatrick, 2006; Volkery et al., 2006). Compared to the rich institutional landscape for economic cooperation and development, the institutional landscape for SD is still relatively small but may be improved through better ownership of SD strategies central to government.

RE strategies for low-carbon, green and sustainable development are increasingly important as a means to achieve goals such as GHG concentration stabilization, energy security, energy access for the poor and the creation of green jobs (IEA, 2010b; SARI, 2010; Lund et al., 2011; see Section 9.3). Policy targets for RE can be helpful to mobilize people and resources and to monitor progress. By 2010, more than 85 countries worldwide had adopted policy targets for the share of RE; typically 5 to 30% for electricity production. Examples of targets for final energy are 15% by 2020 in China, 20% by 2020 in the EU and 100% by 2013 in the small island states of Fiji and Tonga (REN21, 2010). The policy targets are specific to RE but represent important elements in overall strategies for low-carbon, green and sustainable development (UN, 2005b; SARI, 2010; Offer et al., 2011).

Essentially, RE strategies describe the challenges and possible solutions of phasing out unsustainable fossil fuels and technologies while

---

26 The CDM has the twin objectives of promoting SD in developing countries and assisting developed countries to achieve their emission reduction targets cost-effectively.

> **Box 9.5 | Leapfrogging.**
>
> 'Leapfrogging' relates to the opportunity for developing countries to avoid going through the same pollution intensive stages of industrial development as industrialized countries have experienced in the past (see Annex I for definition). Three different types of 'environmental leapfrogging' are distinguished: leapfrogging within overall development pathways, leapfrogging within industrial development, and leapfrogging in the adoption and use of technologies. A sufficient level of absorptive capacity is at the core of successful leapfrogging; it includes the existence of technological capabilities to instigate and manage change and the support of appropriate national and international institutions (Sauter and Watson, 2008).
>
> Any leapfrogging strategy involves risks, but latecomer countries can benefit if initial risks of developing new products and establishing markets have been borne in 'frontrunner' countries. Once a market is established, developing countries can catch up through rapid adoption of new technologies and/or the development of manufacturing capacity. More radical innovation—due to a shift in technological paradigms—can provide additional 'windows of opportunity' for developing countries. Different factors have been identified for the success of this process and since there is no standard model of development, trial-and-error learning needs to be accepted as part of leapfrogging strategies (Hobday, 2003; Sauter and Watson, 2008). Technological leapfrogging in RE has been reported by several studies (L. Clarke et al., 2007; Moreno et al., 2007; R. Singh, 2007; Tarik-ul-Islam and Ferdousi, 2007; Karakosta et al., 2010; Reiche, 2010; Saygin and Cetin, 2010), although current energy technologies may prevent the energy sector from being as conducive to leapfrogging as other sectors like information technology (World Bank, 2008a). Overall, experience has shown that the embarkment on a fundamentally cleaner development pathway needs to be accompanied by ongoing and targeted policy support and guidance, improved institutional capabilities and far-reaching political will in both developing and developed countries (Perkins, 2003; Gallagher, 2006).

phasing in RE systems (Lund, 2007; Verbruggen and Lauber, 2009). To harness the full potential of RE sources, major technological changes are needed along with policies and regulation to ensure a sustainable, effective and efficient use of energy sources and technologies. To ensure the sustainable use of RE sources and technologies, detailed scientific differentiation and qualification of renewable electricity sources and technologies is required to assess the huge diversity in the field (Verbruggen and Lauber, 2009). Further methodological development of sustainability criteria for, indicators for, and assessments of RE sources and technologies based on their attributes (such as types, density, variability, accessibility, scale, maturity, costs etc.), would allow improved fine-tuned regulation for sustainable RE solutions (Verbruggen and Lauber, 2009). In Norway, environmental concerns have led to a more sustainable use of hydropower (see Box 9.6).

9.5.2.2 Local, private and nongovernmental sustainable development initiatives

At the local level, cities and local governments in alliance with business and citizen interests can be drivers of change for RE deployment (REN21, 2009). In response to enabling framework conditions at international and national levels, cities and local governments can independently use their legislative and purchasing power to implement RE initiatives in their own operations and the wider community (see Section 11.6). Typically, local policy initiatives are motivated by sustainability goals such as low GHG concentration stabilization, the share of renewable electricity production or total energy consumption (Ostergaard and Lund, 2010). Other types of local RE policies and SD initiatives are urban planning that incorporates RE, inclusion of RE in building codes or permitting, regulatory measures such as blending of biofuels, RE in municipal infrastructure and operations and voluntary actions to support RE and serve as a role model for business and citizens (REN21, 2009). To share experiences and inspire local actions a range of networks and initiatives have emerged such as the World Mayors and Local Governments Climate Protection Agreement, the Local Government Climate Roadmap, Solar Cities, 100% renewable energy regions, ICLEI's Local Renewables Initiative, the European Green Cities Network, Green Capital Awards and many others. Common to these initiatives is a broad recognition of the local SD benefits RE may bring (del Rio and Burguillo, 2008, 2009), such as a local supply of energy, saving energy and money, creating local jobs and involving the private sector in playing a role in providing RE services (Hvelplund, 2006).

Involvement of community-based organizations can mitigate local opposition to RE installations by facilitating local ownership and sharing of benefits (Rogers et al., 2008; Zografakis et al., 2009). The creation of local energy markets can provide opportunities for local private investors (Hvelplund, 2006) and thereby ensure public acceptance of integrating an increasing number of local RE installations (windmills, solar panels, biogas plants etc.) into the energy system. Positive impacts on the local economy further improve public attitudes towards RE developments (Jobert et al., 2007; Maruyama et al., 2007; Aitken, 2010; Warren and McFadyen, 2010). Case studies evaluating the success of wind energy projects in France and Germany found that the familiarity of the developer with local circumstances and concerns

> **Box 9.6 | Sustainable hydropower in Norway.**
>
> For about a century, hydropower, 'the white coal of Norway', has been a strong driving force in the industrialization of the country (Skjold, 2009). By early 2010, installed capacity was about 29 GW and the average annual generation was about 122 TWh, meeting 98 to 115% of Norway's annual electricity demand, depending on rainfall (NVE, 2009). After intense exploitation during the 1970s and 1980s, newly heightened environmental awareness led to a period of relative standstill in the development of hydropower plants in general, and in 1973 the Norwegian government adopted its initial national protection plan (today there are four in total). As a result, approximately 400 rivers are now protected. In 1986, the first version of a master plan for hydropower was passed; it categorizes potential projects according to economic and technical viability, but also strongly emphasizes potential environmental and social conflicts (Thaulow et al., 2010). Of the estimated feasible potential of 205 TWh of hydropower from Norway's rivers, 122 TWh are utilized, 46 TWh are protected, and about 37 TWh are sorted into acceptable/not acceptable projects in the National Master Plan for hydropower (Thaulow et al., 2010). The last 30 years have seen improved environmental and social impact assessment procedures, guidelines and criteria, increased involvement of stakeholders, and better licensing procedures; all efforts to make hydropower more sustainable for the long term.

(Jobert et al., 2007) as well as transparency, provision of information and participation of the local population in the planning process from the early stages on (Wolsink, 2007a) are crucial factors for public acceptance. In the context of developing countries, this also includes the empowerment of rural women in order to seek the best solutions for community energy needs (Omer, 2003; Oikonomou et al., 2009; A. Singh, 2009).

9.6 Synthesis

The renewable energy (RE) technologies discussed in this report will play an increasingly important role in the world energy system over the next several decades. Mitigation of climate change caused by the combustion of fossil fuels provides one key motivation for a drastic transformation of the world energy system. Additional factors pointing towards the desirability of increasing reliance on RE include concerns about uneven distribution and future supply scarcity of fossil fuel resources, the affordable provision of modern energy services and reductions of burdens on the environment and human health. Given the heavy reliance of modern societies on fossil fuels, any proposed transformation pathway must be carefully analyzed for feasibility and its implications for SD.

In order to be seen as advancing SD, any energy technology has to contribute to a number of SD goals. In the context of this report, these have been identified as social and economic development, energy access, energy security, and the reduction of adverse impacts on health and the environment. To date, RE has often been claimed to advance these four goals and the assessment of this chapter has focused on validating these assumptions. In the following sections, the theoretical concepts and methodological tools used in the analyses are briefly presented. Building on that, results from the bottom-up and integrated assessments of Sections 9.3 and 9.4 are combined to provide clear insights into where the contribution of RE to SD may remain limited and where it shows significant potential.

9.6.1 Theoretical concepts and methodological tools for assessing renewable energy sources

SD has predominantly been framed in the context of the three-pillar model, that is, the contribution to economic and social development and environmental protection. SD is also oriented along a continuum between the weak and strong sustainability paradigms, which differ in assumptions about the substitutability of natural and human-made capital. RE technologies can be evaluated within both concepts: the contribution of RE to the development targets of the three-pillar model and the prioritization of goals according to the weak and strong sustainability framework. As such, SD concepts provide useful frameworks for policymakers to assess the contribution of RE to SD and to formulate appropriate economic, social and environmental measures.

The assessments carried out in this chapter are based on different methodological tools, including bottom-up indicators derived from attributional lifecycle assessments (LCA) or energy statistics, dynamic integrated modelling approaches, and qualitative analyses. Naturally, each of these assessment techniques comes with its own set of limitations. For example, general conclusions from results of individual LCAs are thwarted by potential system boundary problems, differences in technology and background energy system characteristics, geographic location, data source type and other central methods and assumptions. Yet LCA provides a standardized framework for comparison, and bottom-up evidence allows valuable insights about environmental performances of different technologies across categories. In a complementary approach, scenario results of global integrated models were analyzed to derive conclusions about the contribution of RE deployment to the named SD goals within a macro-economic and systemic perspective. However, any interpretation of these results needs to be accompanied by the recognition that integrated models in existence today were generated around a relatively specific set of tasks. These relate to understanding the effects of policy or economics on the energy portfolios of fairly large world regions and the emissions trajectories implied by changes in those energy portfolios over time. While expanding the models beyond these tasks can be challenging, there is room for improving treatment of sustainability in the future. For example, questions relating to the ability of integrated models to accurately represent cultural dimensions of energy use and the impact of non-price policies on behaviour and investment are not resolved.

One of the key points that emerged from the literature assessment is that the evaluation of energy system impacts (beyond GHG emissions), climate mitigation scenarios and SD goals has for the most part proceeded in parallel without much interaction. Effective, economically efficient and socially acceptable transformations of the energy system will require a much closer integration of results from all three of these research areas. While the assessment carried out within the context of this report generated a number of important insights, it also disclosed some of these shortcomings. For example, it highlights the need for the inclusion of additional boundaries (e.g., environmental) and more complex energy system models within an integrated model framework to improve the representation of specific local conditions, variability or biophysical constraints. However, it is also evident that for the multi-dimensional challenge of integrating RE and SD, no single global answer is possible. Many solutions will depend strongly on local and regional cultural conditions, and the approaches and emphases of developing and developed countries may also be different.

9.6.2 Social and economic development

The energy sector has generally been perceived as key to economic development with a strong correlation between economic growth and expansion of energy consumption. Historically, increased energy use has also strongly correlated with growth in GHG emissions. While considerable cross-sectional variation of energy use patterns across countries prevails, the correlation is confirmed by both analyses of single measures such GDP as well as composite indicators such as the Human Development Index. Developing and transition economies may have the opportunity to 'leapfrog' to less energy- and carbon-intensive growth patterns. This requires strong policy and institutional frameworks, as experiences show that rapid economic growth can outpace any declines in energy or carbon intensity.

The contribution of RE to social and economic development may differ between developed and developing countries. To the extent that developing countries can avoid expensive energy imports by deploying economically more efficient RE technologies, they can redirect foreign exchange flows towards imports of other goods that cannot be produced locally. However, generation costs of RE today are generally higher than current energy market prices, although further cost reductions are expected. In poor rural areas lacking grid access, RE can already lead to substantial cost savings today. Creating employment opportunities and actively promoting structural change in the economy are seen, especially in industrialized countries, as goals that support the promotion of RE.

Results from the scenario literature highlight the role of RE for cost-efficient mitigation efforts in the long run—particularly for low-GHG stabilization levels. In developing countries, for which large-scale integrated models suggest a higher share of global RE deployment over time, RE may help accelerate the deployment of low-carbon energy systems. Climate finance is expected to play a crucial role in providing the funding required for large-scale adoption of RE.

9.6.3 Energy access

Enhancing access to clean, reliable and affordable energy sources is a key part of SD and RE has potential to contribute significantly to this goal. Currently, around 1.4 billion people have no access to electricity and about 2.7 billion rely on traditional biomass for cooking (Section 9.3.2). Access to modern energy services is an important precondition for many fundamental determinants of human development, including health, education, gender equality and environmental safety. Even at basic levels, substantial benefits can be provided to a community or household, for example, by improved lighting, communication or healthcare opportunities. In developing countries, decentralized grids based on RE have expanded and improved energy access in rural areas with significant distances to the national grid. In addition, non-electrical RE technologies offer opportunities for direct modernization of energy services, for example, using solar energy for water heating and crop drying, biofuels for transportation, biogas and modern biomass for heating, cooling, cooking and lighting, and wind for water pumping (see Table 9.3). Model analyses confirm that income growth tends to lead to increased energy access, but this is also dependent on the level of income distribution within a society. If developing countries are able to secure dedicated financing for enhanced energy access and apply tailored policies, the number of people with access to modern energy services can expand more rapidly.

9.6.4 Energy security

The role of RE in shaping economies' energy security is complex and depends on the development level of a given country. For example, for developing and transition economies, RE can make a contribution to economizing foreign exchange reserves and help to increase the reliability of energy services. For many developing countries, the definition of energy security specifically includes the provision of adequate and affordable access to all parts of the population and thus exhibits strong links to energy access aspects. Hence, the definition of energy security, that is, the risk of supply disruptions, is broadened from resource availability and distribution of resources, and variability of supplies, to include the reliability of local energy supply.

Scenario analysis confirms that RE can help to diversify energy supply and thus enhance energy security. Local RE options can substitute for increasingly scarce or concentrated fossil fuel supplies, diversifying energy supply and diminishing dependence on a small number of suppliers. As long as RE markets (e.g., bioenergy) are not characterized by concentrated supply, this may help reduce economic vulnerability to price volatility. However, due to the variable output profiles of some RE technologies, technical and institutional measures appropriate to local conditions are often necessary to minimize new insecurities. Also, supply constraints of certain inorganic raw materials may affect enhanced deployment of RE.

The degree to which RE can substitute for liquid fossil fuels used in transport will depend on technology, market and institutional developments. Even with these advances, oil and related energy security concerns will likely continue to play a dominant role in the global energy system of the future.

9.6.5 Climate change mitigation and reduction of environmental and health impacts

RE technologies can provide important environmental benefits compared to fossil fuels, including reduced GHG emissions. Maximizing these benefits often depends on the specific technology, management and site characteristics associated with each RE project. While all energy technologies deployed at scale will create environmental impacts—determined in large measure by local implementation decisions—most RE options can offer advantages across categories, in particular regarding impacts on climate, water resources and air quality. The environmental advantages of RE over other options are not always clear-cut. Significant differences exist between technologies, and some might potentially result in difficult SD trade-offs.

In particular, bioenergy has a special role. It is the only RE based on combustion, leading to associated burdens such as air pollution and cooling water needs. Other impacts from bioenergy production may be positive or negative and relate to land and water use, as well as water and soil quality. These require special attention due to bioenergy's inherent connection to agriculture, forestry and rural development. The net effects of bioenergy production, in particular in terms of lifecycle GHG emissions, are strongly influenced by land and biomass resource management practices, and the prior condition of the land converted for feedstock production. While most models do not yet include land use and terrestrial carbon stocks, those scenarios that have focused on direct and indirect land use change highlight the possible negative consequences for SD. These result from high expansion rates without proper policies in place and large future bioenergy markets, and can lead to deforestation, land diversion and increased GHG emissions. Proper governance of land use, zoning and choice of biomass production systems are key to achieving desired outcomes.

RE has the potential to significantly reduce local and regional air pollution from power generation and associated health impacts. Scenarios that explicitly address regional air pollutants, for example, PM and sulphur emissions, found that climate policy can lead to important co-benefits in that area. Indoor air pollution caused by the use of solid fuels in traditional systems is a major health problem at a global scale, and improved technologies and fuels could also address other SD concerns. Careful decisions based on local resources are needed to ensure that water scarcity does not become a barrier to SD, and that increasing access to energy services does not exacerbate local water problems. Non-thermal RE technologies (e.g., wind and PV) can provide clean electricity without putting additional stress on water resources, whereas operational water needs make thermal power plants and hydropower vulnerable to changes in water availability. While accident risks of RE technologies are not negligible, their often decentralized structure strongly limits the potential for disastrous consequences in terms of fatalities. However, dams associated with some hydropower projects may create a specific risk depending on site-specific factors.

Insights from the modelling approaches show that integrated assessment models might be well suited to include some important environmental indicators in addition to GHG emissions (e.g., air pollutant emission, water use), but may be challenged by addressing localized impacts, for example, related to energy choices at the household level. Resulting scenarios could be useful to demonstrate unanticipated or unquantified environmental benefits or costs.

9.6.6 Conclusions

The previous sections have shown that RE can contribute to SD and the four goals assessed in this chapter to varying degrees. While benefits with respect to reduced environmental and health impacts may appear more clear-cut, the exact contribution to, for example, social and economic development is more ambiguous. Also, countries may prioritize the four SD goals according to their level of development. To some extent, however, these SD goals are also strongly interlinked. Climate change mitigation constitutes in itself a necessary prerequisite for successful social and economic development in many developing countries.

Following this logic, climate change mitigation can be assessed under the strong SD paradigm, if mitigation goals are imposed as constraints on future development pathways. If climate change mitigation is balanced against economic growth or other socioeconomic criteria, the problem is framed within the paradigm of weak SD, allowing for trade-offs between these goals and using cost-benefit type analyses to provide guidance in their prioritization.

However, the existence of uncertainty and ignorance as inherent components of any development pathway, as well as the existence of associated and possibly 'unacceptably high' opportunity costs (Neumayer, 2003), will make continued adjustments crucial. In the future, integrated models may be in a favourable position to better link the weak and strong SD paradigms for decision-making processes. Within well-defined guardrails, integrated models could explore scenarios for different mitigation pathways, taking account of the remaining SD goals by including important and relevant bottom-up indicators. According to model type, these alternative development pathways might be optimized for socially beneficial outcome. Equally, however, the incorporation of GHG emission-related LCA data will be crucial for a clear definition of appropriate GHG concentration stabilization levels in the first place.

Despite the potential existence of several technically, economically and environmentally feasible development pathways, it is the human component that will ultimately define the success of any such strategy. Important barriers, especially in the SD context, are those relating to socio-cultural and information and awareness aspects. In particular, barriers intrinsically linked to societal and personal values and norms will fundamentally affect the perception and acceptance of RE technologies and related deployment impacts by individuals, groups and societies. Dedicated communication efforts, addressing these subjective and psychological aspects in the same manner as the more objective opportunities associated with wider-scale RE applications are therefore a crucial component of any transformation strategy. Local SD initiatives by cities, local governments, and private and nongovernmental organizations can act as important drivers of change in this context.

Local initiatives, however, also need to be embedded in coherent SD strategies at the national level. The clear integration of SD and RE goals into development policies and the development of sectoral strategies for RE can provide an opportunity for contributing to goals for green growth, low-carbon and sustainable development, including leapfrogging.

9.7 Gaps in knowledge and future research needs

This chapter has described part of the interactions between SD and RE and focused on SD goals such as social and economic development, energy access, energy security, climate change mitigation and the reduction of environmental and health impacts. An assessment of indicators related to these goals has revealed several gaps in knowledge.

Beginning with the more conceptual discussion of SD, there is a tremendous gap between intertemporal measures of human well-being (sustainability) and measurable sub-indicators that needs to be narrowed. In addition, possibilities for relating the two opposite paradigms of sustainability, weak and strong sustainability, need to be explored. One possibility would be to allow for nonlinearities, tipping points, and uncertainty about nonlinearities in intertemporal measures, or to provide formal guidelines for consideration of the precautionary principle. In the context of this report, this also means that specific indicators of weak sustainability like genuine savings, ISEW or GPI, but also those of strong sustainability (e.g., land use boundaries) need to be statistically and logically related to RE indicators.

Apart from the definitions and indicators, data that are necessary to assess sustainability and RE are insufficiently available. There is a clear need for better information and data on energy supply and consumption for non-electrified households and also low-end electricity consumers. Furthermore, there is a need for analysis of RE-based mini-grid experiences for improving access and for the energy security implications of regional power integration. The electrification of the transport sector and its implications for energy security, environmental impacts and GHG emissions also deserves attention.

Many aspects of the assessment of environmental impacts of energy technologies require additional research to resolve key scientific questions, or provide confirmatory research for less contentious but also less-studied aspects. Two key issues regarding GHG emissions caused by energy technologies are direct and indirect land use change. For RE technologies, these issues mainly concern the production of biomass for bioenergy systems and hydropower impoundments, but land use change associated with some non-RE technologies deserve investigation as well (e.g., carbon emission from soils exposed by mountaintop removal coal mining). Several energy technologies are lacking substantial or any studies of lifecycle GHG emissions: geothermal, ocean energy and some types of PV cells. Water use has not been consistently or robustly evaluated for any energy technology across its lifecycle. The state of knowledge about land use, especially when considered on a lifecycle basis, is in a condition similar to water. For both, metrics to quantify water and land use need consensus as well as substantial additional study using those metrics. More is known about air pollutants, at least for the operation of combustion systems, but this knowledge has not been well augmented on a lifecycle basis, and the interpretation of air pollutant emissions on a lifecycle basis needs to be enhanced since the important effects of pollutants should not be summarized by summing masses over time and space. For LCAs as a whole, heterogeneity of methods and assumptions thwarts fair comparison and pooling of estimates from different studies. Ex post facto harmonization of the methods of previous research (and meta-analysis) and perhaps stronger standards guiding the conduct of new LCAs is critical to clarifying results and producing robust estimates.

Assessments of the scenario literature have provided some useful insights on how SD pathways will interact with RE and vice versa. However, in the past, models have focused on the technological and macro-economic aspects of energy transitions and the evaluation of SD pathways therefore mostly needs to rely on proxies that are not always informative. One major difficulty is the models' macro perspective, while some issues for SD are relevant at a micro and regional level. Thus, when focusing more specifically on different SD criteria, major drawbacks can be found for all of them:

- With respect to sustainable social and economic development, the scenario literature has a strong focus on consumption and GDP. Even though models address multiple criteria for welfare, they are generally not sufficiently specific to inform about distributional issues. Differentiations between income groups, urban and rural populations and so on are difficult to make.

- The distribution and availability of energy services, and how they change over time, are aspects that are not broadly included in most energy-economy models so far, which makes the evaluation of energy access challenging.

- Regarding energy security, the current representation of the grid structure in most of the models does not allow for a thorough analysis of possible difficulties related to large-scale integration of RE. Possible barriers are mostly assumed to be overcome without difficulties, particularly when thinking of storage and variability issues that might occur. Possible co-benefits of renewable sources, such as growing diversity of supply and possibilities to electrify rural areas, are also poorly covered in the literature as, for example, fuel supply risks are usually not taken into account in the models.

- The existing scenario literature does not give an explicit treatment to many non-emissions- related aspects of sustainable energy development, for example, water use, biodiversity impacts, or the impacts of energy choices on household-level services or indoor air quality. In addition to that, regarding Section 9.3.4 of this chapter, emissions are generally not treated over the lifecycles of technology choices, which might be an interesting aspect of future research.

In conclusion, knowledge regarding the interrelations between SD and RE in particular is still very limited. Finding answers to the question of how to achieve effective, economically efficient and socially acceptable transformations of the energy system will require a much closer integration of insights from social, natural and economic sciences (e.g., through risk analysis approaches) in order to reflect the different dimensions of sustainability. So far, the knowledge base is often limited to very narrow views from specific branches of research, which do not fully account for the complexity of the issue.

References

Abdelouas, A. (2006). Uranium mill tailings: Geochemistry, mineralogy, and environmental impact. *Elements*, **2**(6), pp. 335-341.

Achten, W.M.J., L.R. Lene R Nielsen, R. Aerts, A.G. Lengkeek, E.D. Erik D Kjær, A. Trabucco, J.K. Hansen, W.H. Maes, L. Lars Graudal, F.K. Festus, K. Akinnifesi, and B. Muys (2010). Towards domestication of *Jatropha curcas*. *Biofuels*, **1**(1), pp. 91-107.

Adamantiades, A., and I. Kessides (2009). Nuclear power for sustainable development: Current status and future prospects. *Energy Policy*, **37**, pp. 5149-5166.

Afgan, N.H., F. Begic, and A. Kazagic (2007). Multi-criteria sustainability assessment – A tool for evaluation of new energy system. *Thermal Science*, **11**(3), pp. 43-53.

Agarwal, A.K. (2007). Biofuels (alcohols and biodiesel) applications as fuels for internal combustion engines. *Progress in Energy and Combustion Science*, **33**(3), pp. 233-271.

Agbemabiese, L. (2009). A framework for sustainable energy development beyond the grid: Meeting the needs of rural and remote populations. *Bulletin of Science, Technology & Society*, **29**(2), pp. 151-158.

AGECC (2010). *Energy for a Sustainable Future*. United Nations Secretary General's Advisory Group on Energy and Climate (AGECC), New York, NY, USA.

Aitken, M. (2010). Wind power and community benefits: Challenges and opportunities. *Energy Policy*, **38**(10), pp. 6066-6075.

Ajanovic, A. (2011). Biofuels versus food production: Does biofuels production increase food prices? *Energy*, **36**(4), pp. 2070-2076.

Aksoy, N., C. Şimşek, and O. Gunduz (2009). Groundwater contamination mechanism in a geothermal field: A case study of Balcova, Turkey. *Journal of Contaminant Hydrology*, **103**, pp. 13-28.

Al-Riffai, P., B. Dimaranan, and L. Laborde (2010). *Global Trade and Environmental Impact Study of the EU Biofuels Mandate*. International Food Policy Institute (IFPRI) for the Directorate General for Trade of the European Commission, Brussels, Belgium, 123 pp. Available at: www.ifpri.org/sites/default/files/publications/biofuelsreportec.pdf.

Al-Zoughool, M., and D. Krewski (2009). Health effects of radon: a review of the literature. *International Journal of Radiation Biology*, **85**(1), pp. 57-69.

Ale, B.J.M., H. Baksteen, L.J. Bellamy, A. Bloemhof, L. Goossens, A. Hale, M.L. Mude, J.I.H. Oh, I.A. Papazoglou, J. Post, and J.Y. Whiston (2008). Quantifying occupational risk: The development of an occupational risk model. *Safety Science*, **46**, pp. 176-185.

Americano, B. (2008). CDM in Brazil: Towards structural change for sustainable development in some sectors. In: *A Reformed CDM*. K.H. Olsen and J. Fenhann (eds.), UNEP Risø Centre, Roskilde, Denmark, pp. 23-46.

Anderson, L. (2009). Ethanol fuel use in Brazil: air quality impacts. *Energy & Environmental Science*, **2**, pp. 1015-1037.

Andreani-Aksoyoglu, S., J. Keller, A.S.H. Prévôt, U. Baltensperger, and J. Flemming (2008). Secondary aerosols in Switzerland and northern Italy: Modeling and sensitivity studies for summer 2003. *Journal of Geophysical Research*, **113**, pp. 131-143.

Ang, J.B. (2007). CO_2 emissions, energy consumption, and output in France. *Energy Policy*, **35**(10), pp. 4772-4778.

Ang, J.B. (2008). Economic development, pollutant emissions and energy consumption in Malaysia. *Journal of Policy Modeling*, **30**(2), pp. 271-278.

Angel-Urdinola, D., and Q. Wodon (2007). Do utility subsidies reach the poor? Framework and evidence for Cape Verde, Sao Tome, and Rwanda. *Economics Bulletin*, **9**(4), pp. 1-7.

Angerer, G. (2010). Der Einfluss des technischen Fortschritts und der Weltwirtschaft auf die Rohstoffnachfrage. In: *Rohstoffeffizienz und Rohstoffinnovationen*. U. Teipel (ed.), Fraunhofer Verlag, Stuttgart, Germany, pp. 61-68.

Angerer, G., E. Erdmann, F. Marscheider-Weidemann, M. Scharp, A. Lüllmann, V. Handke, and M. Marwede (2009). *Rohstoffe für Zukunftstechnologien*. Fraunhofer Verlag, Stuttgart, Germany.

Asafu-Adjaye, J. (2000). The relationship between energy consumption, energy prices and economic growth: time series evidence from Asian developing countries. *Energy Economics*, **22**(6), pp. 615-625.

Au, W., R. Lane, M. Legator, E. Whorton, G. Wilkinson, and G. Gabehart (1995). Biomarker monitoring of a population residing near uranium mining activities. *Environmental Health Perspectives*, **103**(5), pp. 466-470.

Auffhammer, M., V. Ramanathan, and J.R. Vincent (2006). Integrated model shows that atmospheric brown clouds and greenhouse gases have reduced rice harvests in India. *Proceedings of the National Academy of Sciences*, **103**(52), pp. 19668-19672.

Aunan, K., J. Fang, H. Vennemo, K. Oye, and H.M. Seip (2004). Co-benefits of climate policy – lessons learned from a study in Shanxi, China. *Energy Policy*, **32**(4), pp. 567-581.

Awerbuch, S. (2006). Portfolio-based electricity generation planning: Policy implications for renewables and energy security. *Mitigation and Adaptation Strategies for Global Change*, **11**(3), pp. 693-710.

Awerbuch, S., and R. Sauter (2006). Exploiting the oil-GDP effect to support renewables deployment. *Energy Policy*, **34**(17), pp. 2805-2819.

Ayash, S.C., A.A. Dobroskok, J.A. Sorensen, S.L. Wolfe, E.N. Steadman, and J.A. Harju (2009). Probabilistic approach to evaluating seismicity in CO_2 storage risk assessment. *Energy Procedia*, **1**, pp. 2487-2494.

Azar, C., K. Lindgren, M. Obersteiner, K. Riahi, D.P. van Vuuren, K.M. den Elzen, K. Möllersten, and E.D. Larson (2010). The feasibility of low CO_2 concentration targets and the role of bio-energy with carbon capture and storage (BECCS). *Climatic Change*, **100**(1), pp. 195-202.

Bachu, S. (2008). CO_2 storage in geological media: Role, means, status and barriers to deployment. *Progress in Energy and Combustion Science*, **34**, pp. 254-273.

Baer, P. (2009). Equity in climate–economy scenarios: the importance of subnational income distribution. *Environmental Research Letters*, **4**(1), 015007.

Bailis, R., and J.C. Cutler (2004). Wood in household energy use. In: *Encyclopedia of Energy*. Elsevier, New York, NY, USA, pp. 509-526.

Bailis, R., M. Ezzati, and D.M. Kammen (2005). Mortality and greenhouse gas impacts of biomass and petroleum energy futures in Africa. *Science* **308**(5718), pp. 98-103.

Bailis, R., A. Cowan, V. Berrueta, and O. Masera (2009). Arresting the killer in the kitchen: The promises and pitfalls of commercializing improved cookstoves. *World Development*, **37**(10), pp. 1694-1705.

Bang, H.K., A.E. Ellinger, J. Hadjimarcou, and P.A. Traichal (2000). Consumer concern, knowledge, belief, and attitude toward renewable energy: An application of the reasoned action theory. *Psychology and Marketing*, **17**(6), pp. 449-468.

Bao, G.J. (2010). Study on the relevance of cultural system and hydropower resettlement project in Nujiang River. In: *Advance in Resources & Environmental Economics Research*. Scientific Research Publishing, California, USA, pp. 360-364.

Barbier, E.B. (2009). *Rethinking the Economic Recovery: A Global Green New Deal*. Report prepared for the Green Economy Initiative and the Division of Technology, Industry and Economics of the UN Environment Programme, Available at: http://www.unep.org/greeneconomy/portals/30/docs/GGND-Report-April2009.pdf.

Barnes, B., A. Mathee, E. Thomas, and N. Bruce (2009). Household energy, indoor air pollution and child respiratory health in South Africa. *Journal of Energy in South Africa*, 20(1), pp. 4-13.

Barnes, D.F., and W.M. Floor (1996). Rural energy in developing countries: A challenge for economic development. *Annual Review of Energy and the Environment*, 21, pp. 497-530.

Barnes, D., and J. Halpern (2001). Reaching the poor: Designing energy subsidies to benefit those that need it. *Refocus*, 2(6), pp. 32-34.

Barnthouse, L. (2000). Impacts of power-plant cooling systems on estuarine fish populations: the Hudson River after 25 years. *Environmental Science & Policy*, 3, pp. 341-348.

Barry, J., G. Ellis, and C. Robinson (2008). Cool rationalities and hot air: A rhetorical approach to understanding debates on renewable energy. *Global Environmental Politics*, 8(2), pp. 67-98.

Bartle, J.R., and A. Abadi (2010). Toward sustainable production of second generation bioenergy feedstocks. *Energy Fuels*, 24, pp. 2-9.

Bauer, C. (2007). *Holzenergie*. Paul Scherrer Institut and Swiss Centre for Life Cycle Inventories, Villigen and Duebendorf, Switzerland.

Bauer, C. (2008). *Life Cycle Assessment of Fossil and Biomass Power Generation Chains*. Paul Scherrer Institut, Villigen, Switzerland.

Bauer, C., T. Heck, R. Dones, O. Mayer-Spohn, and M. Blesl (2009). *Final Report on Technical Data, Costs, and Life Cycle Inventories of Advanced Fossil Power Generation Systems*. Deliverable n° 7.2 - RS 1a, European Commission, Brussels, Belgium

Bauer, N., A. Bowen, S. Brunner, O. Edenhofer, C. Flachsland, M. Jakob, and N. Stern (2009). *Towards a Global Green Recovery. Recommendations for Immediate G20 Action*. Report prepared on behalf of the German Foreign Office, Potsdam Institute for Climate Impact Research (PIK), The Grantham Research Institute on Climate Change and Environment (GRI LSE), 49 pp.

Baum, S., M. Weih, G. Busch, F. Kroiher, and A. Bolte (2009). The impact of Short Rotation Coppice plantations on phytodiversity. *Landbauforschung vTI Agriculture and Forestry Research*, 3, pp. 163-170.

Baumert, K., T. Herzog, and J. Pershing (2005). *Navigating the Numbers. Greenhouse Gas Data and International Climate Policy*. World Resource Institute, Washington, DC, USA.

Bazilian, M., and F. Roques (eds.) (2008). *Analytical Methods for Energy Diversity and Security. Portfolio Optimization in the Energy Sector: A Tribute to the work of Dr. Shimon Awerbuch*. Elsevier Science, Oxford, UK and Amsterdam, The Netherlands.

Bazilian, M., P. Nussbaumer, E. Haites, M. Levi, M. Howells, and K. Yumkella (2010). Understanding the scale of investments for universal energy access. *Geopolitics of Energy*, 32(10-11), pp. 21-42.

Becerra-Lopez, H.R., and P. Golding (2008). Multi-objective optimization for capacity expansion of regional power-generation systems: Case study of far west Texas. *Energy Conversion and Management*, 49(6), pp. 1433-1445.

Beer, T., T. Grant, D. Williams, and H. Watson (2002). Fuel-cycle greenhouse gas emissions from alternative fuels in Australian heavy vehicles. *Atmospheric Environment*, 36(4), pp. 753-763.

Beer, T., T. Grant, and P.K. Campbell (2007). *The Greenhouse and Air Quality Emissions of Biodiesel Blends in Australia*. Report Number KS54C/1/F2.29, Report for Caltex Australia Limited, Commonwealth Scientific and Industrial Research Organisation (CSIRO), Clayton South, Australia, 126 pp.

Bell, D., T. Gray, and C. Haggett (2005). The 'social gap' in wind farm siting decisions: Explanations and policy responses. *Environmental Politics*, 14(4), pp. 460-477.

Ben Hagan, E. (2003). Woodfuels consumption in Ghana – Environmental issues and challenges. In: *Proceedings of 1st International Conference on Energy and Environment*, Changsha, China, 11-14 October 2003, pp. 495-500.

Bengtsson, M., Y. Shen, and T. Oki (2006). A SRES-based gridded global population dataset for 1990-2100. *Population and Environment*, 28, pp. 113-131.

Benson, S.M. (2006). *Carbon Dioxide Capture and Storage: Assessment of Risks from Storage of Carbon Dioxide in Deep Underground Geological Formations*. Lawrence Berkeley National Laboratory, Earth Sciences Division, Berkeley, CA, USA.

Bentley, R.W. (2002). Global oil and gas depletion: an overview. *Energy Policy*, 30(3), pp. 189-205.

Beringer, T.I.M., W. Lucht, and S. Schaphoff (2011). Bioenergy production potential of global biomass plantations under environmental and agricultural constraints. *Global Change Biology - Bioenergy*, doi:10.1111/j.1757-1707.2010.01088.x.

Bernatik, A., W. Zimmerman, M. Pitt, M. Strizik, V. Nevrly, and Z. Zelinger (2008). Modelling accidental releases of dangerous gases into the lower troposphere from mobile sources. *Process Safety and Environmental Protection*, 86(3), pp. 198-207.

Berndes, G. (2002). Bioenergy and water – the implications of large-scale bioenergy production for water use and supply. *Global Environmental Change*, 12, pp. 253-271.

Berndes, G. (2008). Future biomass energy supply: The consumptive water use perspective. *International Journal of Water Resources Development*, 24, pp. 235-245.

Berndes, G. (2010). Bioenergy and water: risks and opportunities. *Biofuels, Bioproducts and Biorefining*, 4(5), pp. 473-474.

Berndes, G., N. Bird, and A. Cowie (2010). *Bioenergy, Land Use Change and Climate Change Mitigation*. IEA Bioenergy: ExCo:2010:03, International Energy Agency, Whakarewarewa, Rotorua, New Zealand, 20 pp. Available at: www.ieabioenergy.com/LibItem.aspx?id=6770.

BERR (2008). *Renewable Energy Awareness And Attitudes Research. Management Summary*. URN 08/657, Department for Business Enterprise and Regulatory Reform (BERR), London, UK.

BGR (2009). *Reserven, Ressourcen und Verfügbarkeit von Energierohstoffen*. Federal Institute for Geosciences and Natural Resources (BGR), Hannover, Germany.

Bhattacharyya, S.C. (2005). Energy access problem of the poor in India: Is rural electrification a remedy? *Energy Policy*, 34(18), pp. 3383-3397.

Bickel, P., and R. Friedrich (2005). *Externalities of Energy Methodology 2005 Update*. EUR 21951, Institut für Energiewirtschaft und Rationelle Energieanwendung, Universität Stuttgart, Stuttgart, Germany.

Biehl, F., and E. Lehmann (2006). Collisions of ships with offshore wind turbines: Calculation and risk evaluation. In: *Offshore Wind Energy: Research on Environmental Impacts*. J. Köller, J. Köppel, and W. Peters (eds.), Springer-Verlag, Berlin and Heidelberg, Germany, pp. 281-304.

Bishop, J.D.K., and G.A.J. Amaratunga (2008). Evaluation of small wind turbines in distributed arrangement as sustainable wind energy option for Barbados. *Energy Conversion and Management*, **49**(6), pp. 1652-1661.

Bloemkolk, J., and R. van der Schaaf (1996). Design alternatives for the use of cooling water in the process industry: minimization of the environmental impact from cooling systems. *Journal of Cleaner Production*, **4**(1), pp. 21-27.

BMU (1998). *Nachhaltige Entwicklung in Deutschland, Entwurf eines umweltpolitischen Schwerpunktprogramms*. Bundesministeriums für Umwelt, Naturschutz und Reaktorsicherheit (BMU), Bonn, Germany.

BMU (2009). *Umweltwirtschaftsbericht 2009*. Bundesministeriums für Umwelt, Naturschutz und Reaktorsicherheit (BMU), Bonn, Germany.

Bojö, J., K.-G. Maler, and L. Unemo (1992). *Environment and Development: An Economic Approach*. Kluwer Academic Publishers, Dordrecht, The Netherlands and Boston, MA, USA.

Bollen, J., B. van der Zwaan, C. Brinka, and H. Eerensa (2009). Local air pollution and global climate change: A combined cost-benefit analysis. *Resource and Energy Economics*, **31**(3), pp. 161-181.

Bollen, J., S. Hers, and B. van der Zwaan (2010). An integrated assessment of climate change, air pollution, and energy security policy. *Energy Policy*, **38**, pp. 4021-4030.

Bommer, J.J., S. Oates, J.M. Cepeda, Lindholm, Conrad, J. Bird, R. Torres, G. Marroquin, and J. Rivas (2006). Control of hazard due to seismicity induced by a hot fractured rock geothermal project. *Engineering Geology*, **86**, pp. 287-306.

Bond, T.C., D.G. Streets, K.F. Yarber, S.M. Nelson, J.H. Woo, and Z. Klimont (2004). A technology-based global inventory of black and organic carbon emissions from combustion. *Journal of Geophysical Research - Atmospheres*, **109**, D14203, doi:10.1029/2003JD003697.

Bossel, H., 1999: *Indicators for Sustainable Development: Theory, Methods, Applications*. International Institute for Sustainable Development, Winnipeg, MB, Canada, 138 pp.

Boudri, J.C., L. Hordijk, C. Kroeze, M. Amann, J. Cofala, I. Bertok, L. Junfeng, D. Lin, Z. Shuang, H. Runquing, T.S. Panwar, S. Gupta, D. Singh, A. Kumar, M.C. Vipradas, P. Dadhich, N.S. Prasad, and L. Srivastava (2002). The potential contribution of renewable energy in air pollution abatement in China and India. *Energy Policy*, **30**(5), pp. 409-424.

Bowen, A., S. Fankhauser, N. Stern, and D. Zenghelis (2009). *An Outline of the Case for 'Green' Stimulus*. The Grantham Research Institute on Climate Change and the Environment, The Centre for Climate Change Economics and Policy, London, UK, 11 pp.

BP (2010). *Statistical Review of World Energy 2010*. BP, London, UK. Available at: www.bp.com/productlanding.do?categoryId=6929&contentId=7044622.

Brand, K.-W., and G. Jochum (2000). *Die Struktur des deutschen Diskurs zu nachhaltiger Entwicklung*. MPS-Texte 1/2000, Münchner Projektgruppe für Sozialforschung e.V., München, Germany.

Brander, M., R. Tipper, C. Hutchison, and G. Davis (2009). *Consequential and Attributional Approaches to LCA: a Guide to Policy Makers with Specific Reference to Greenhouse Gas LCA of Biofuels*. Technical Paper TP-090403-A, Ecometrica Press, Edinburgh, Scotland, 14 pp. Available at: d3u3pjcknor73l.cloudfront.net/assets/media/pdf/approachesto_LCA3_technical.pdf.

Bravo, G., R. Kozluj, and R. Landaveri (2008). Energy access in urban and peri-urban areas of Buenos Aires. *Energy for Sustainable Development*, **12**(4), pp. 56-72.

Brent, A.C., and W.J.L. Kruger (2009). Systems analyses and the sustainable transfer of renewable energy technologies: A focus on remote areas of Africa. *Renewable Energy*, **34**(7), pp. 1774-1781.

Brent, A.C., and D.E. Rogers (2010). Renewable rural electrification: Sustainability assessment of mini-hybrid off-grid technological systems in the African context. *Renewable Energy*, **35**(1), pp. 257-265.

Brew-Hammond, A. (2010). Energy access in Africa: Challenges ahead. *Energy Policy*, **38**(5), pp. 2291-2301.

Brewer, T.L. (2004). The WTO and the Kyoto Protocol: Interaction issues. *Climate Policy*, **4**, pp. 3-12.

Brophy, P. (1997). Environmental advantages to the utilization of geothermal energy. *Renewable Energy*, **10**, pp. 367-377.

Brown, P.H., and K. Xu (2010). Hydropower development and resettlement policy on China's Nu River. *Journal of Contemporary China*, **19**(66), pp. 777-797.

Browne, D., B. O'Regan, and R. Moles (2010). Use of multi-criteria decision analysis to explore alternative domestic energy and electricity policy scenarios in an Irish city-region. *Energy*, **35**(2), pp. 518-528.

Bruce, N., R. Albalak, and P. Perez-Padilla (2002). *The Health Effects of Indoor Air Pollution Exposure in Developing Countries*. WHO/SDE/OEH/02.05, World Health Organization, Geneva, Switzerland.

Bruce, N., J. McCracken, R. Albalak, M.A. Schei, K.R. Smith, V. Lopez, and C. West (2004). Impact of improved stoves, house construction and child location on levels of indoor air pollution exposure in young Guatemalan children. *Journal of Exposure Analysis and Environmental Epidemiology*, **14**(S1), pp. S26-S33.

Brunnschweiler, C.N. (2010). Finance for renewable energy: an empirical analysis of developing and transition economies. *Environment and Development Economics* **15**(3), pp. 241-274.

Burgherr, P. (2007). In-depth analysis of accidental oil spills from tankers in the context of global spill trends from all sources. *Journal of Hazardous Materials*, **140**(1-2), pp. 245-256.

Burgherr, P., and S. Hirschberg (2007). Assessment of severe accident risks in the Chinese coal chain. *International Journal of Risk Assessment and Management*, **7**(8), pp. 1157-1175.

Burgherr, P., and S. Hirschberg (2008). A comparative analysis of accident risks in fossil, hydro, and nuclear energy chains. *Human and Ecological Risk Assessment*, **14**(5), pp. 947-973.

Burgherr, P., S. Hirschberg, and E. Cazzoli (2008). *Final Report on Quantification of Risk Indicators for Sustainability Assessment of Future Electricity Supply Options*. NEEDS Deliverable No. D7.1 - Research Stream 2b. New Energy Externalities Developments for Sustainability, Brussels, Belgium.

Burke, P.J. (2010). Income, resources, and electricity mix. *Energy Economics*, **32**(3), pp. 616-626.

Burney, J.A., S.J. Davis, and D.B. Lobell (2010). Greenhouse gas mitigation by agricultural intensification. *Proceedings of the National Academy of Sciences*, **107**(26), pp. 12052-12057.

Calvin, K., J. Edmonds, B. Bond-Lamberty, L. Clarke, S.H. Kim, P. Kyle, S.J. Smith, A. Thomson, and M. Wise (2009). 2.6: Limiting climate change to 450 ppm CO_2 equivalent in the 21st century. *Energy Economics*, **31**(Supplement 2), pp. S107-S120.

Campbell, C.J., and J.H. Laherrère (1998). The end of cheap oil. *Scientific American*, **March 1998**, pp. 80-85.

CARB (2009). *Low Carbon Fuel Standard Program. Fuel Pathways Documents*. California Air Resources Board, Sacramento, CA, USA.

Carmichael, G.R., D.G. Streets, G. Calori, M. Amann, M.Z. Jacobson, J. Hansen, and H. Ueda (2002). Changing trends in sulfur emissions in Asia: Implications for acid deposition, air pollution, and climate. *Environmental Science & Technology*, 36(22), pp. 4707-4713.

Carmichael, G.R., B. Adhikary, S. Kulkarni, A.D. Allura, Y. Tang, D. Streets, Q. Zhang, T.C. Bond, V. Ramanathan, A. Jamroensan, and P. Marrapu (2009). Asian aerosols: Current and year 2030 distributions and implications to human health and regional climate change. *Environmental Science & Technology*, 43(15), pp. 5811-5817.

Carrera, D.G., and A. Mack (2010). Sustainability assessment of energy technologies via social indicators: Results of a survey among European energy experts. *Energy Policy*, 38(2), pp. 1030-1039.

Carvalho, F., J. Oliveira, I. Lopes, and A. Batista (2007). Radionuclides from past uranium mining in rivers of Portugal. *Journal of Environmental Radioactivity*, 98(3), pp. 298-314.

Casillas, C.E., and D.M. Kammen (2010). Environment and development. The energy-poverty-climate nexus. *Science*, 330(6008), pp. 1181-1182.

Cavallaro, F. (2009). Multi-criteria decision aid to assess concentrated solar thermal technologies. *Renewable Energy*, 34(7), pp. 1678-1685.

Cernea, M. (1997). The risks and reconstruction model for resettling displaced populations. *World Development*, 25(10), pp. 1569-1587.

Chaurey, A., and T.C. Kandpal (2010). Assessment and evaluation of PV based decentralized rural electrification: An overview. *Renewable & Sustainable Energy Reviews*, 14(8), pp. 2266-2278.

Chen, S.J. (2009). Discussion of resettlement cost externalization of water resources and hydropower projects. *Advances in Water Resources and Hydraulic Engineering, Vols 1-6*, pp. 1427-1432.

Cherian, A. (2009). *Bridging the Divide Between Poverty Reduction and Climate Change through Sustainable and Innovative Energy Technologies*. Environment and Energy Group, United Nations Development Programme, New York, NY, USA, 55 pp.

Cherubini, F., and A.H. Strømman (2011). Life cycle assessment of bioenergy systems: state of the art and future challenges. *Bioresource Technology*, 102(2), pp. 437-451.

Cherubini, F., N.D. Bird, A. Cowie, G. Jungmeier, B. Schlamadinger, and S. Woess-Gallasch (2009). Energy- and greenhouse gas-based LCA of biofuel and bioenergy systems: Key issues, ranges and recommendations. *Resources, Conservation and Recycling*, 53(8), pp. 434-447.

Christensen, C.F., L.W. Andersen, and P.H. Pedersen (2001). Ship collision risk for an offshore wind farm. In: *Structural Safety and Reliability: Proceedings of the Eighth International Conference, ICOSSAR '01*, Newport Beach, CA, USA, 17-22 June 2001. Swets & Zeitlinger B.V., Lisse, The Netherlands.

Clarke, L., J. Edmonds, H. Jacoby, H. Pitcher, J. Reilly, and R. Richels (2007). *Scenarios of Greenhouse Gas Emissions and Atmospheric Concentrations*. Sub-report 2.1, Department of Energy, Office of Biological and Environmental Research, Washington, DC, USA, 154 pp.

Clarke, L., P. Kyle, M. Wise, K. Calvin, J. Edmonds, S. Kim, M. Placet, and S. Smith (2008). CO_2 *Emissions Mitigation and Technological Advance: An Updated Analysis of Advanced Technology Scenarios*. Technical Report PNNL-18075, Pacific Northwest National Laboratory, Richland, WA, USA.

Clarke, L., J. Edmonds, V. Krey, R. Richels, S. Rose, and M. Tavoni (2009). International climate policy architectures: Overview of the EMF 22 International Scenarios. *Energy Economics*, 31(Supplement 2), pp. S64-S81.

Clarke, S. (2009). Balancing environmental and cultural impact against the strategic need for wind power. *International Journal of Heritage Studies*, 15(2-3), pp. 175-191.

Cleveland, C.J. (2005). Net energy from the extraction of oil and gas in the United States. *Energy*, 30(5), pp. 769-782.

Cleveland, C.J., R. Costanza, C.A.S. Hall, and R. Kaufmann (1984). Energy and the U.S. economy: a biophysical perspective. *Science*, 225(4665), pp. 890-897.

Cleveland, C.J., R.K. Kaufman, and D.I. Stern (2000). Aggregation and the role of energy in the economy. *Ecological Economics*, 32(2), pp. 301-317.

Coady, D., M. Grosh, and J. Hoddinott (2004). *Targeting of Transfers in Developing Countries: Review of Lessons and Experience*. The World Bank, Washington, DC, USA.

Coady, D., R. Gillingham, R. Ossowski, J. Piotrowski, T. Shamsuddin, and J. Tyson (2010). *Petroleum Product Subsidies: Costly, Inequitable, and Rising*. IMF Staff Position Note SPN/10/05, International Monetary Fund, Washington, DC, USA.

Coase, R.H. (1960). The problem of social cost. *Journal of Law and Economics*, 3, pp. 1-44.

Coburn, A., and A. Cohen (2004). *Catastrophe, Injury, and Insurance. The Impact of Catastrophes on Workers Compensation, Life, and Health Insurance*. Risk Management Solutions, Newark, CA, USA, 80pp.

Coelho, S.T., J. Goldemberg, O. Lucon, and P. Guardabassi (2006). Brazilian sugarcane ethanol: lessons learned. *Energy for Sustainable Development*, 10(2), pp. 26-39.

Cohen, A.J., H.R. Anderson, B. Ostro, K.D. Pandey, M. Krzyzanowski, N. Künzli, K. Gutschmidt, C.A. Pope III, I. Romieu, J.M. Samet, and K. Smith (2004). Urban air pollution. In: *Global and Regional Burden of Disease Attributable to Selected Major Risk Factors: Comparative Quantification of Health Risks*. World Health Organization, Geneva, Switzerland, pp. 1353-1434.

CONCAWE (2008). *Well-to-Wheels Analysis of Future Automotive Fuels and Powertrains in the European Context*. European Council for Automotive R&D (EUCAR), European Association for Environment, Health and Safety in Oil Refining and Distribution (CONCAWE), and the Institute for Environment and Sustainability of the EU Commission's Joint Research Centre (JRC/IES). Brussels, Belgium and Ispra, Italy.

Coronado, C.R., J.A. de Carvalho Jr., J.T. Yoshioka, and J.L. Silveira (2009). Determination of ecological efficiency in internal combustion engines: The use of biodiesel. *Applied Thermal Engineering*, 29(10), pp. 1887-1892.

Costa, R.C., and J.R. Sodré (2009). Hydrous ethanol vs. gasoline-ethanol blend: Engine performance and emissions. *Fuel*, 89(2), pp. 287-293.

Cowan, K.R., T. Daim, W. Wakeland, H. Fallah, G. Sheble, L. Lutzenhiser, A. Ingle, R. Hammond, and M. Nguyen (2009). Forecasting the adoption of emerging energy technologies: Managing climate change and evolving social values. In: *Portland International Conference on Management of Engineering and Technology, PICMET 2009*, Portland, OR, USA, 2-6 Aug 2009, pp. 2964-2974.

Cowie, A., P. Smith, and D. Johnson (2006). Does soil carbon loss in biomass production systems negate the greenhouse benefits of bioenergy? *Mitigation and Adaptation Strategies for Global Change*, 11(5), pp. 979-1002.

Cozzani, V., M. Campedela, E. Renni, and E. Krausmann (2010). Industrial accidents triggered by flood events: Analysis of past accidents. *Journal of Hazardous Materials*, **175**, pp. 501-509.

Creutzig, F., and D. He (2009). Climate change mitigation and co-benefits of feasible transport demand policies in Beijing. *Transportation Research Part D: Transport and Environment*, **14**(2), pp. 120-131.

Creutzig, F., and D. Kammen (2009). The Post-Copenhagen roadmap towards sustainability: differentiated geographic approaches, integrated over goals. *Innovations*, **4**(4), pp. 301-321.

Creutzig, F., A. Papson, L. Schipper, and D.M. Kammen (2009). Economic and environmental evaluation of compressed-air cars. *Environmental Research Letters*, **4**, 044011.

Croezen, H.J., G.C. Bergsma, M.B.J. Otten, and M.P.J. van Valkengoed (2010). *Biofuels: Indirect Land Use Change and Climate Impact*. 10 8169 49, CE Delft, Delft, The Netherlands, 62 pp.

Crutzen, P.J., A.R. Mosier, K.A. Smith, and W. Winiwarter (2008). N_2O release from agro-biofuel production negates global warming reduction by replacing fossil fuels. *Atmospheric Chemistry and Physics*, **8**, pp. 389-395.

Curtis, L., W. Rea, P. Smith-Willis, E. Fenyves, and Y. Pan (2006). Adverse health effects of outdoor air pollutants. *Environment International*, **32**(6), pp. 815-830.

Cushman, R.M. (1985). Review of ecological effects of rapidly varying flows downstream from hydroelectric facilities. *North American Journal of Fisheries Management*, **5**, pp. 330-339.

Dai, A. (2011). Drought under global warming: a review. *Wiley Interdisciplinary Reviews: Climate Change*, **2**(1), pp. 45-65.

Daly, H. (2007). *Ecological Economics and Sustainable Development, Selected Essays of Herman Daly*. Edward Elgar Publishing, Cheltenham, UK.

Dannwolf, U.S., and F. Ulmer (2009). *AP 6000 Report – Technology risk comparison of the geothermal DHM project in Basel, Switzerland – Risk appraisal including social aspects*. RC006, RiskCom, Pforzheim, Germany.

Dasgupta, P. (2001). *Human Well-Being and the Natural Environment*. Oxford University Press, Oxford, UK.

Dauber, J., M.B. Jones, and J.C. Stout (2010). The impact of biomass crop cultivation on temperate biodiversity. *Global Change Biology Bioenergy*, **2**(6), pp. 289-309.

Davidson, E.A. (2009). The contribution of manure and fertilizer nitrogen to atmospheric nitrous oxide since 1860. *Nature Geoscience*, **2**(9), pp. 659-662.

Davidson, O., K. Halsnæs, S. Huq, M. Kok, B. Metz, Y. Sokona, and J. Verhagen (2003). The development and climate nexus: the case of sub-Saharan Africa. *Climate Policy*, **3**(Supplement 1), pp. S97-S113.

Davis, S.J., and K. Caldeira (2010). Consumption-based accounting of CO_2 emissions. *Proceedings of the National Academy of Sciences*, **107**(12), pp. 5687-5692.

Deichmann, U., C. Meisner, S. Murray, and D. Wheeler (2011). The economics of renewable energy expansion in rural sub-Saharan Africa. *Energy Policy*, **39**(1), pp. 215-227.

Del Granado, J.A., D. Coady, and R. Gillingham (2010). *The Unequal Benefits of Fuel Subsidies: A Review of Evidence for Developing Countries*. IMF Working Paper WP/10/202, International Monetary Fund, Washington, DC, USA.

del Rio, P., and M. Burguillo (2008). Assessing the impact of renewable energy deployment on local sustainability: Towards a theoretical framework. *Renewable and Sustainable Energy Reviews*, **12**(5), pp. 1325-1344.

del Rio, P., and M. Burguillo (2009). An empirical analysis of the impact of renewable energy deployment on local sustainability. *Renewable and Sustainable Energy Reviews*, **13**(6-7), pp. 1314-1325.

Demirbas, A. (2009). Emission characteristics of gasohol and diesohol. *Energy Sources, Part A: Recovery, Utilization and Environmental Effects*, **31**(13), pp. 1099-1104.

Denholm, P., and R.M. Margolis (2008). Land-use requirements and the per-capita solar footprint for photovoltaic generation in the United States. *Energy Policy*, **36**(9), pp. 3531-3543.

Denholm, P., M. Hand, M. Jackson, and S. Ong (2009). *Land Use Requirements of Modern Wind Power Plants in the United States*. Technical report NREL/TP-6A2-45834, National Renewable Energy Laboratory, Golden, CO, USA.

Devine-Wright, P. (2005). Beyond NIMBYism: towards an integrated framework for understanding public perceptions of wind energy. *Wind Energy*, **8**(2), pp. 125-139.

Devine-Wright, P. (2009). Rethinking NIMBYism: The role of place attachment and place identity in explaining place-protective action. *Journal of Community & Applied Social Psychology*, **19**(6), pp. 426-441.

Dhingra, C., S. Gandhi, A. Chaurey, and P.K. Agarwal (2009). Access to clean energy services for the urban and peri-urban poor: a case-study of Delhi, India. *Energy for Sustainable Development*, **12**(4), pp. 49-55.

Dogdu, M.S., and C.S. Bayari (2004). Environmental impact of geothermal fluids on surface water, groundwater and streambed sediments in the Akarcay Basin, Turkey. *Environmental Geology*, **47**, pp. 325-340.

Donat Castello, L., D. Gil-Gonzalez, C. Alvarez-Dardet Diaz, and I. Hernandez-Aguado (2010). The Environmental Millennium Development Goal: progress and barriers to its achievement. *Environmental Science and Policy*, **13**(2), pp. 154-163.

Dones, R., C. Bauer, and A. Röder (2007). *Kohle*. Paul Scherrer Institut and Swiss Centre for Life Cycle Inventories, Villigen and Dübendorf, Switzerland.

Dornburg, V., A. Faaij, P. Verweij, H. Langeveld, G. van de Ven, F. Wester, H. van Keulen, K. van Diepen, M. Meeusen, M. Banse, J. Ros, D.P. van Vuuren, G.J. van den Born, M. van Oorschot, F. Smout, J. van Vliet, H. Aiking, M. Londo, H. Mozaffarian, and K. Smekens (2008). *Assessment of Global Biomass Potentials and Their Links to Food, Water, Biodiversity, Energy Demand and Economy*. The Netherlands Environmental Assessment Agency, Wageningen, The Netherlands.

Doukas, H., C. Karakosta, and J. Psarras (2010). Computing with words to assess the sustainability of renewable energy options. *Expert Systems with Applications*, **37**(7), pp. 5491-5497.

Dubreuil, A., G. Gaillard, and R. Müller-Wenk (2007). Key elements in a framework for land use impact assessment within LCA. *The International Journal of Life Cycle Assessment*, **12**(1), pp. 5-15.

Eason, T.N., Y.A. Owusu, and H. Chapman (2009). A systematic approach to assessing the sustainability of the Renewable Energy Standard (RES) under the proposed American Renewable Energy Act (H.R. 890). *International Journal of Global Energy Issues*, **32**(1-2), pp. 139-159.

Eberhard, A., V. Foster, C. Briceño-Garmendia, F. Ouedraogo, D. Camos, and M. Shkaratan (2008). *Underpowered: The State of the Power Sector in Sub-Saharan Africa*. World Bank, Washington, DC, USA.

EC (1999). *Pilot Study for the Update of the MARINA Project on the Radiological Exposure of the European Community from Radioactivity in North European Marine Waters*. European Commission, Brussels, Belgium, 77 pp.

EC (2006). *Reference Document on Best Available Techniques for Large Combustion Plants*. sic/tm/32, European Commission, Joint Research Centre, Institute for Prospective Technological Studies, Seville, Spain.

EC (2010). *Critical Raw Materials for the EU*. European Commission, Enterprise and Industry, Brussels, Belgium.

Ecoinvent (2009). *The Ecoinvent LCI Database, Data v2.2*. Swiss Centre for Life Cycle Inventories, Duebendorf, Switzerland.

Edenhofer, O., B. Knopf, M. Leimbach, and N. Bauer (2010). The economics of low stabilization. *The Energy Journal*, **31**(Special Issue 1), pp. 11-48.

Edmonds, J.A., M.J. Scott, J.M. Roop, and C.N. MacCracken (1999). *International Emission Trading and the Cost of Greenhouse Gas Emissions Mitigation*. Pew Center for Global Climate Change, Washington, DC, USA.

Edmonds, J., T. Wilson, M. Wise, and J. Weyant (2006). Electrification of the economy and CO_2 emissions mitigation. *Environmental Economics and Policy Studies*, **7**(3), pp. 175-203.

Edmonds, J.A., J. Clarke, J. Dooley, S.H. Kim, R. Izaurralde, N. Rosenberg, and G. Stokes (2003). The potential role of biotechnology in addressing the long-term problem of climate change in the context of global energy and ecosystems. In: *Proceedings of the Sixth International Conference on Greenhouse Gas Control Technologies*, Kyoto, Japan, 1-4 October 2002, **2**, pp. 1427-1432.

Egré, D., and J.C. Milewski (2002). The diversity of hydropower projects. *Energy Policy*, **30**(14), pp. 1225-1230.

EIA (2009). *International Energy Outlook 2009*. U.S. Department of Energy, Energy Information Administration (EIA), Washington, DC, USA.

Ekholm, T., V. Krey, S. Pachauri, and K. Riahi (2010). Determinants of household energy consumption in India. *Energy Policy*, **38**(10), pp. 5696-5707.

Ekins, P., and S. Simon (1999). The sustainability gap: a practical indicator of sustainability in the framework of the national accounts. *International Journal of Sustainable Development*, **2**(1), pp. 32-58.

Ekins, P., S. Simon, L. Deutsch, C. Folke, and R. De Groot (2003). A framework for the practical application of the concepts of critical natural capital and strong sustainability. *Ecological Economics*, **44**(2-3), pp. 165-185.

EMBRAPA (2009). *Zoneamento Agroecológico da Cana de Açúcar. Expandir a produção, preservar a vida, garantir o futuro*. 110, Empresa Brasileira de Pesquisa Agropecuária Setembro, Centro Nacional de Pesquisa de Solos, Ministério da Agricultura, Pecuária e Abastecimento, Rio de Janeiro, Brazil.

Engfeldt, L.-G. (2009). *From Stockholm to Johannesburg and Beyond*. Press, Information and Communication Department, Ministry for Foreign Affairs, Stockholm, Sweden.

EPA (2010a). *Greenhouse Gas Emissions Reporting from the Petroleum and Natural Gas Industry*. U.S. Environmental Protection Agency, Climate Change Division, Washington, DC, USA.

EPA (2010b). *Renewable Fuel Standard Program (RFS2) Regulatory Impact Analysis*. U.S. Environmental Protection Agency, Washington, DC, USA.

EPRI (2003). *Use of Degraded Water Sources as Cooling Water in Power Plants*. Electric Power Research Institute (EPRI), Palo Alto, CA, USA.

ESMAP (2005). *The Impacts of Higher Oil Prices on Low Income Countries and the Poor: Impacts and Policies*. Energy Sector Management Assistance Program, World Bank, Washington, DC, USA.

ESMAP (2006). *Coping with Higher Oil Prices*. Energy Sector Management Assistance Program, World Bank, Washington, DC, USA.

ESMAP (2007). *Technical and Economic Assessment of Off-grid, Mini-grid and Grid Electrification Technologies*. Energy Sector Management Assistance Program, World Bank, Washington, DC, USA.

ESMAP (2008). *Coping with Oil Price Volatility*. Energy Sector Management Assistance Program, World Bank, Washington, DC, USA.

ESMAP (2010). *Regional Power Sector Integration Lessons from Global Case Studies and a Literature Review*. Energy Sector Management Assistance Program, World Bank, Washington, DC, USA.

Evans, A., V. Strezov, and T.J. Evans (2009). Assessment of sustainability indicators for renewable energy technologies. *Renewable and Sustainable Energy Reviews*, **13**(5), pp. 1082-1088.

EWEA (2004). *Wind Energy – The Facts. An Analysis of Wind Energy in the EU-25*. European Wind Energy Association (EWEA), Brussels, Belgium.

Ezzati, M., R. Bailis, D.M. Kammen, T. Holloway, L. Price, L.A. Cifuentes, B. Barnes, A. Chaurey, and K.N. Dhanapala (2004). Energy management and global health. *Annual Review of Environment and Resources*, **29**(1), pp. 383-419.

Fahlen, E., and E.O. Ahlgren (2010). Accounting for external costs in a study of a Swedish district-heating system – An assessment of environmental policies. *Energy Policy*, **38**(9), pp. 4909-4920.

Fankhauser, S., F. Sehlleier, and N. Stern (2008). Climate change, innovation and jobs. *Climate Policy*, **8**(4), pp. 421-429.

Fargione, J., J. Hill, D. Tilman, S. Polasky, and P. Hawthorne (2008). Land clearing and the biofuel carbon debt. *Science*, **319**(5867), pp. 1235-1238.

Farrell, A.E., and A.R. Brandt (2006). Risks of the oil transition. *Environmental Research Letters*, **1**, 014004.

Fawcett, A.A., K.V. Calvin, F.C. de la Chesnaye, J.M. Reilly, and J.P. Weyant (2009). Overview of EMF 22 U.S. transition scenarios. *Energy Economics* **31**, pp. 198-211.

Fearnside, P.M. (2001). Environmental impacts of Brazil's Tucuruí dam: Unlearned lessons for hydroelectric development in Amazonia. *Environmental Management*, **27**(3), pp. 377-396.

Fernando, S., C. Hall, and S. Jha (2006). NO_x reduction from biodiesel fuels. *Energy and Fuels*, **20**(1), pp. 376-382.

Feygin, M., and R. Satkin (2004). The oil reserves-to-production ratio and its proper interpretation. *Natural Resources Research*, **13**(1), pp. 57-60.

Figueiredo, P.N. (2003). Learning, capability accumulation and firms differences: evidence from latecomer steel. *Industrial and Corporate Change*, **12**(3), pp. 607-643.

Figueres, C., and C. Streck (2009). The evolution of the CDM in a post-2012 climate agreement. *Journal of Environment & Development*, **18**, pp. 227-246.

Fingerman, K., D. Kammen, S. Torn, and M. O'Hare (2010). Accounting for the water impacts of ethanol production. *Environmental Research Letters*, **5**(1), 014020.

Finnveden, G., M.Z. Hauschild, T. Ekvall, J. Guinee, R. Heijungs, S. Hellweg, A. Koehler, D. Pennington, and S. Suh (2009). Recent developments in Life Cycle Assessment. *Journal of Environmental Management*, **91**, pp. 1-21.

Firbank, L. (2008). Assessing the ecological impacts of bioenergy projects. *BioEnergy Research*, **1**(1), pp. 12-19.

Fitzherbert, E.B., M.J. Struebig, A. Morel, F. Danielsen, C.A. Brühl, P.F. Donald, and B. Phalan (2008). How will oil palm expansion affect biodiversity? *Trends in Ecology & Evolution*, **23**(10), pp. 538-545.

Flanner, M.G., C.S. Zender, P.G. Hess, N.M. Mahowald, T.H. Painter, V. Ramanathan, and P.J. Rasch (2009). Springtime warming and reduced snow cover from carbonaceous particles. *Atmospheric Chemistry and Physics*, **9**(7), pp. 2481-2497.

Fleming, J.S., S. Habibi, and H.L. MacLean (2006). Investigating the sustainability of lignocellulose-derived fuels for light-duty vehicles. *Transportation Research Part D-Transport and Environment*, **11**(2), pp. 146-159.

Fletcher, R.J., B.A. Robertson, J. Evans, P.J. Doran, J.R.R. Alavalapati, and D.W. Schemske (2011). Biodiversity conservation in the era of biofuels: risks and opportunities. *Frontiers in Ecology and the Environment*, **9**(3) pp. 161-168.

Fleurbaey, M. (2009). Beyond GDP: The quest for a measure of social welfare. *Journal of Economic Literature*, **47**(4), pp. 1029-1075.

Founex Committee (1971). *The Founex Report on Development and Environment*. Founex Committee, Founex, Switzerland.

Fox, J., and J.E. Campbell (2010). Terrestrial carbon disturbance from mountaintop mining increases lifecycle emissions for clean coal. *Environmental Science and Technology*, **44**(6), pp. 2144-2149.

Franco, A., and M. Villani (2009). Optimal design of binary cycle power plants for water-dominated, medium-temperature geothermal fields. *Geothermics*, **38**(4), pp. 379-391.

Fritsche, U., K. Hennenberg, and K. Hünecke (2010). *The "iLUC Factor" as a Means to Hedge Risks of GHG Emissions from Indirect Land Use Change*. Oeko Institute, Darmstadt, Germany, 64 pp.

Fritzsche, A.F. (1992). Editorial – Severe accidents: can they occur only in the nuclear production of electricity? *Risk Analysis*, **12**, pp. 327-329.

Frondel, M., P. Grösche, D. Huchtemann, A. Oberheitmann, J. Peters, G. Angerer, C. Sartorius, P. Buchholz, S. Röhling, and M. Wagner (2007). *Trends der Angebots- und Nachfragesituation bei mineralischen Rohstoffen*. Rheinisch-Westfälisches Institut für Wirtschaftsforschung (RWI), Fraunhofer-Institut für System- und Innovationsforschung (ISI), and Bundesanstalt für Geowissenschaften und Rohstoffe (BGR), Essen, Karlsruhe, and Hannover, Germany.

Frondel, M., N. Ritter, C.M. Schmidt, and C. Vance (2010). Economic impacts from the promotion of renewable energy technologies: The German experience. *Energy Policy*, **38**, pp. 4048-4056.

Fthenakis, V., and H.C. Kim (2009). Land use and electricity generation: A life-cycle analysis. *Renewable and Sustainable Energy Reviews*, **13**(6-7), pp. 1465-1474.

Fthenakis, V., and H.C. Kim (2010). Life-cycle uses of water in U.S. electricity generation. *Renewable and Sustainable Energy Reviews*, **14**(7), pp. 2039-2048.

Fujino, J., G. Hibino, T. Ehara, Y. Matsuoka, T. Masui, and M. Kainuma (2008). Back-casting analysis for 70% emission reduction in Japan by 2050. *Climate Policy*, **8**, pp. S108-S124.

G-20 (2009). *Leaders' Statement: The Pittsburgh Summit*. The Pittsburgh G-20 Summit. U.S. Department of State, Pittsburgh, PA, USA.

Gadhamshetty, V., N. Nirmalakhandan, M. Myint, and C. Ricketts (2006). Improving air-cooled condenser performance in combined cycle power plants. *Journal of Energy Engineering*, **132**, pp. 81-88.

Gagnon, L. (2008). Civilisation and energy payback. *Energy Policy*, **36**(9), pp. 3317-3322.

Gagnon, L., C. Bélanger, and Y. Uchiyama (2002). Life-cycle assessment of electricity generation options: The status of research in year 2001. *Energy Policy*, **30**(14), pp. 1267-1278.

Gallagher, K.S. (2006). Limits to leapfrogging in energy technologies? Evidence from the Chinese automobile industry. *Energy Policy*, **34**(4), pp. 383-394.

Gangopadhyay, S., B. Ramaswami, and W. Wadhwa (2005). Reducing subsidies on household fuels in India: how will it affect the poor? *Energy Policy*, **33**(18), pp. 2326-2336.

GAO (2010). *Energy-Water Nexus: A Better and Coordinated Understanding of Water Resources could Help Mitigate the Impacts of Potential Oil Shale Development*. GAO-11-35, U.S. Government Accountability Office (GAO), Washington, DC, USA.

Gautam, R., N.C. Hsu, K.-M. Lau, S.-C. Tsay, and M. Kafatos (2009). Enhanced pre-monsoon warming over the Himalayan-Gangetic region from 1979 to 2007. *Geophysical Research Letters*, **36**(7), L07704.

George, C., and C. Kirkpatrick (2006). Assessing national sustainable development strategies: Strengthening the links to operational policy. *Natural Resources Forum*, **30**(2), pp. 146-156.

Gerbens-Leenes, W., A.Y. Hoekstra, and T.H. van der Meer (2009). The water footprint of bioenergy. *Proceedings of the National Academy of Sciences*, **106**, pp. 10219-10223.

Ghirga, G. (2010). Cancer in children residing near nuclear power plants: an open question. *Italian Journal of Pediatrics*, **36**(1), pp. 60.

Giampietro, M., S. Ulgiati, and D. Pimentel (1997). Feasibility of large-scale biofuel production. *BioScience*, **47**, pp. 587-600.

Gibbs, H.K., M. Johnston, J.A. Foley, T. Holloway, C. Monfreda, N. Ramankutty, and D. Zaks (2008). Carbon payback times for crop-based biofuel expansion in the tropics: the effects of changing yield and technology. *Environmental Research Letters*, **3**, 034001.

Gillingham, K., S. Smith, and R. Sands (2008). Impact of bioenergy crops in a carbon dioxide constrained world: an application of the MiniCAM energy-agriculture and land use model. *Mitigation and Adaptation Strategies for Global Change*, **13**(7), pp. 675-701.

Ginnebaugh, D.L., J. Liang, and M.Z. Jacobson (2010). Examining the temperature dependence of ethanol (E85) versus gasoline emissions on air pollution with a largely-explicit chemical mechanism. *Atmospheric Environment*, **44**(9), pp. 1192-1199.

Giroux, J. (2008). Turmoil in the Delta: trends and implications. *Perspectives on Terrorism*, **2**(8), pp. 11-22.

Giroux, J. (2010). A portrait of complexity: new actors and contemporary challenges in the global energy system and the role of energy infrastructure security. *Risk, Hazards & Crisis in Public Policy*, **1**(1), pp. 34-56.

Gleick, P. (1993). *Water in Crisis: A Guide to the World's Fresh Water Resources*. Oxford University Press, New York, NY, USA, 504 pp.

Gleick, P. (2008). *The World's Water 2008-2009*. Island Press, Washington, DC, USA.

GNESD (2004). *Energy Access – Making Power Sector Reform Work for the Poor*. Global Network on Energy for Sustainable Development (GNESD), Roskilde, Denmark.

GNESD (2007a). *Reaching the Millennium Development Goals and Beyond: Access to Modern Forms of Energy as a Prerequisite*. Global Network on Energy for Sustainable Development (GNESD), Roskilde, Denmark.

GNESD (2007b). *Renewable Energy Technologies and Poverty Alleviation: Overcoming Barriers and Unlocking Potentials*. Global Network on Energy for Sustainable Development (GNESD), Roskilde, Denmark.

GNESD (2008). *Clean Energy for the Urban Poor: An Urgent Issue*. Global Network on Energy for Sustainable Development (GNESD), Roskilde, Denmark.

GNESD (2010). *Energy Security*. Global Network on Energy for Sustainable Development (GNESD), Roskilde, Denmark.

Goldemberg, J. (1998). Leapfrog energy technologies. *Energy Policy*, **26**(10), pp. 729-741.

Goldemberg, J. (2001). *Energy and Human Well Being*. Human Development Occasional Paper HDOCPA-2001-02, United Nations Development Program, New York, NY, USA.

Goldemberg, J., and S. Teixeira Coelho (2004). Renewable energy – Traditional biomass vs. modern biomass. *Energy Policy*, **32**(6), pp. 711-714.

Goldemberg, J., T.B. Johansson, A.K.N. Reddy, and R.H. Williams (1985). An end-use oriented global energy strategy. *Annual Review of Energy*, **10**(1), pp. 613-688.

Goldemberg, J., A.K.N. Reddy, K.R. Smith, and R.H. Williams (2000). Rural energy in developing countries. In: *World Energy Assessment: Energy and the Challenge of Sustainability*. United Nations Development Program, New York, NY, USA.

Goldemberg, J., T.B. Johansson, A.K.N. Reddy, and R.H. Williams (2004). A global clean cooking fuel initiative. *Energy for Sustainable Development*, **8**(3), pp. 1-12.

Goldemberg, J., S.T. Coelho, and P. Guardabassi (2008). The sustainability of ethanol production from sugarcane. *Energy Policy*, **36**(6), pp. 2086-2097.

Gorissen, L., V. Buytaert, D. Cuypers, T. Dauwe, and L. Pelkmans (2010). Why the debate about land use change should not only focus on biofuels. *Environmental Science & Technology*, **44**(11), pp. 4046-4049.

Graham, L.A., S.L. Belisle, and C.-L. Baas (2008). Emissions from light duty gasoline vehicles operating on low blend ethanol gasoline and E85. *Atmospheric Environment*, **42**(19), pp. 4498-4516.

Green, R., and N. Vasilakos (2011). The economics of offshore wind. *Energy Policy*, **39**(2), pp. 496-502.

Greene, D., J. Hopson, and J. Li (2006). Have we run out of oil yet? Oil peaking analysis from an optimist's perspective. *Energy Policy*, **34**(5), pp. 515-531.

Greenwood, M. (2008). Fish mortality by impingement on the cooling-water intake screens of Britain's largest direct-cooled power station. *Marine Pollution Bulletin*, **56**(4), pp. 723-739.

Grieshop, A.P., C.C.O. Reynolds, M. Kandlikar, and H. Dowlatabadi (2009). A black-carbon mitigation wedge. *Nature Geoscience*, **2**(8), pp. 533-534.

Gross, C. (2007). Community perspectives of wind energy in Australia: The application of a justice and community fairness framework to increase social acceptance. *Energy Policy*, **35**(5), pp. 2727-2736.

Gross, R., P. Heptonstall, D. Anderson, T. Green, M. Leach, and J. Skea (2006). *The Costs and Impacts of Intermittency: An Assessment of the Evidence on the Costs and Impacts of Intermittent Generation on the British Electricity Network*. UK Energy Research Centre, London, UK.

Gross, R., P. Heptonstall, M. Leach, D. Anderson, T. Green, and J. Skea (2007). Renewables and the grid: understanding intermittency. *Energy*, **160**, pp. 31-41.

Grubb, M., L. Butler, and P. Twomey (2006). Diversity and security in UK electricity generation: The influence of low-carbon objectives. *Energy Policy*, **34**(18), pp. 4050-4062.

Grübler, A. (2004). Transitions in energy use. In: *Encyclopedia of Energy*, **6**, pp. 163-177. Available at: www.eoearth.org/article/Energy_transitions.

Grübler, A., B. O'Neill, K. Riahi, V. Chirkov, A. Goujon, P. Kolp, I. Prommer, S. Scherbov, and E. Slentoe (2007). Regional, national, and spatially explicit scenarios of demographic and economic change based on SRES. *Technological Forecasting and Social Change*, **74**(7), pp. 980-1029.

Gunawardena, U.A.D.P. (2010). Inequalities and externalities of power sector: A case of Broadlands hydropower project in Sri Lanka. *Energy Policy*, **38**(2), pp. 726-734.

Güney, M.S., and K. Kaygusuz (2010). Hydrokinetic energy conversion systems: A technology status review. *Renewable and Sustainable Energy Reviews*, **14**, pp. 2996-3004.

Gupta, E. (2008). Oil vulnerability index of oil-importing countries. *Energy Policy*, **36**(3), pp. 1195-1211.

Gupta, H.K. (2002). A review of recent studies of triggered earthquakes by artificial water reservoirs with special emphasis on earthquakes in Koyna, India. *Earth-Science Reviews*, **58**(3-4), pp. 279-310.

Gurgel, A., J. Reilly, and S. Paltsev (2007). Potential land use implications of a global biofuels industry. *Journal of Agricultural & Food Industrial Organization*, **5**(2), Article 9.

Ha-Duong, M., A. Nadai, and A.S. Campos (2009). A survey on the public perception of CCS in France. *International Journal of Greenhouse Gas Control*, **3**, pp. 633-640.

Hagelüken, C., and C.E.M. Meskers (2010). Complex lifecycles of precious and special metals. In: *Linkages of Sustainability*. T.E. Graedel and E. van der Voet (eds.), MIT Press, Cambridge, MA, USA, pp. 163-198.

Haines, A., K. Smith, D. Anderson, P. Epstein, A. McMichael, I. Roberts, P. Wilkinson, J. Woodcock, and J. Woods (2009). Policies for accelerating access to clean energy, improving health, advancing development, and mitigating climate change. *The Lancet*, **370**(9594), pp. 1264-1281.

Hallquist, M., J.C. Wenger, U. Baltensperger, Y. Rudich, D. Simpson, M. Claeys, J. Dommen, N.M. Donahue, C. George, A.H. Goldstein, J.F. Hamilton, H. Herrmann, T. Hoffmann, Y. Iinuma, M. Jang, M.E. Jenkin, J.L. Jimenez, A. Kiendler-Scharr, W. Maenhaut, G. McFiggans, T.F. Mentel, A. Monod, A.S.H. Prévôt, J.H. Seinfeld, J.D. Surratt, R. Szmigielski, and J. Wildt (2009). The formation, properties and impact of secondary organic aerosol: current and emerging issues. *Atmospheric Chemistry and Physics*, **9**, pp. 5155-5236.

Hamilton, K. (1994). Green adjustments to GDP. *Resources Policy*, **20**(3), pp. 155-168.

Hamilton, K., and M. Clemens (1999). Genuine savings rates in developing countries. *The World Bank Economic Review*, **13**(2), pp. 333-356.

Hanasaki, N., S. Kanae, T. Oki, K. Masuda, K. Motoya, N. Shirakawa, Y. Shen, and K. Tanaka (2008). An integrated model for the assessment of global water resources. Part 2: Applications and assessments. *Hydrology and Earth System Sciences*, **12**(4), pp. 1027-1037.

Hansen, J., M. Sato, R. Ruedy, L. Nazarenko, A. Lacis, G.A. Schmidt, G. Russell, I. Aleinov, M. Bauer, S. Bauer, N. Bell, B. Cairns, V. Canuto, M. Chandler, Y. Cheng, A. Del Genio, G. Faluvegi, E. Fleming, A. Friend, T. Hall, C. Jackman, M. Kelley, N. Kiang, D. Koch, J. Lean, J. Lerner, K. Lo, S. Menon, R. Miller, P. Minnis, T. Novakov, V. Oinas, J. Perlwitz, J. Perlwitz, D. Rind, A. Romanou, D. Shindell, P. Stone, S. Sun, N. Tausnev, D. Thresher, B. Wielicki, T. Wong, M. Yao, and S. Zhang (2005). Efficacy of climate forcings. *Journal of Geophysical Research*, **110**(D18), D18104.

Harmon, R.R., and K.R. Cowan (2009). A multiple perspectives view of the market case for green energy. *Technological Forecasting and Social Change*, **76**(1), pp. 204-213.

Harrison, G.P., and H.W. Whittington (2002). Vulnerability of hydropower projects to climate change. *IEE Proceedings - Generation, Transmission and Distribution*, **149**(3), pp. 249.

Harto, C., R. Meyers, and E. Williams (2010). Life cycle water use of low-carbon transport fuels. *Energy Policy*, **38**(9), pp. 4933-4944.

Hartwick, J. (1977). Intergenerational equity and the investing of rents from exhaustible resources. *The American Economic Review*, **67**(5), pp. 972-974.

Hastings, J.V. (2009). Geographies of state failure and sophistication in maritime piracy hijackings. *Political Geography*, **28**, pp. 213-223.

Heck, T. (2007). *Wärmepumpen*. Ecoinvent report No. 6-X, Paul Scherrer Institut and Swiss Centre for Life Cycle Inventories, Villigen and Duebendorf, Switzerland.

HEI (2010). *Outdoor Air Pollution and Health in the Developing Countries of Asia: A Comprehensive Review*. Special Report 18, Health Effects Institute, Boston, MA, USA.

Heijungs, R., M.J. Goedkoop, J. Struijs, S. Effting, M. Sevenster, and G. Huppes (2003). *Towards a Life Cycle Impact Assessment Method which Comprises Category Indicators at the Midpoint and the Endpoint Level*. PRé Consultants, Amersfoort, The Netherlands.

Hennenberg, K.J., C. Dragisic, S. Haye, J. Hewson, B. Semroc, C. Savy, K. Wiegmann, H. Fehrenbach, and U.R. Fritsche (2010). The power of bioenergy-related standards to protect biodiversity. *Conservation Biology*, **24**(2), pp. 412-423.

Hepburn, C., and N. Stern (2008). A new global deal on climate change. *Oxford Review of Economic Policy*, **24**(2), pp. 259-279.

Herman, R., S.A. Ardekani, and J.H. Ausubel (1990). Dematerialization. *Technological Forecasting and Social Change*, **38**(4), pp. 333-347.

Hertel, T.W., A.A. Golub, A.D. Jones, M. O'Hare, R.J. Plevin, and D.M. Kammen (2010). Effects of US maize ethanol on global land use and greenhouse gas emissions: Estimating market-mediated responses. *BioScience*, **60**(3), pp. 223-231.

Hertwich, E.G., T.E. McKone, and W.S. Pease (1999). Parameter uncertainty and variability in evaluative fate and exposure models. *Risk Analysis*, **19**(6), pp. 1193-1204.

Hiederer, R., F. Ramos, C. Capitani, R. Koeble, V. Blujdea, O. Gomez, D. Mulligan, and L. Marelli (2010). *Biofuels: A New Methodology to Estimate GHG Emissions from Global Land Use Change, A methodology involving spatial allocation of agricultural land demand and estimation of CO_2 and N_2O emissions*. EUR 24483 EN - 2010, Joint Research Center, European Commission, Brussels, Belgium.

Hill, J., E. Nelson, D. Tilman, S. Polasky, and D. Tiffany (2006). Environmental, economic, and energetic costs and benefits of biodiesel and ethanol biofuels. *Proceedings of the National Academy of Sciences*, **103**(30), pp. 11206-11210.

Hill, J., S. Polasky, E. Nelson, D. Tilman, H. Huo, L. Ludwig, J. Neumann, H.C. Zheng, and D. Bonta (2009). Climate change and health costs of air emissions from biofuels and gasoline. *Proceedings of the National Academy of Sciences*, **106**(6), pp. 2077-2082.

Hilton, B., and B. Duddy (2009). The effect of E20 ethanol fuel on vehicle emissions. *Proceedings of the Institution of Mechanical Engineers, Part D: Journal of Automobile Engineering*, **223**(12), pp. 1577-1586.

Hirschberg, S., G. Spiekerman, and R. Dones (1998). *Severe Accidents in the Energy Sector – First Edition*. PSI Report No. 98-16, Paul Scherrer Institut, Villigen, Switzerland.

Hirschberg, S., P. Burgherr, G. Spiekerman, E. Cazzoli, J. Vitazek, and L. Cheng (2003). Assessment of severe accident risks. In: *Integrated Assessment of Sustainable Energy Systems in China. The China Energy Technology Program – A framework for decision support in the electric sector of Shandong province*. Alliance for Global Sustainability Series Vol. 4. Kluwer Academic Publishers, Amsterdam, The Netherlands, pp. 587-660.

Hirschberg, S., P. Burgherr, G. Spiekerman, and R. Dones (2004a). Severe accidents in the energy sector: comparative perspective. *Journal of Hazardous Materials*, **111**(1-3), pp. 57-65.

Hirschberg, S., T. Heck, U. Gantner, Y. Lu, J.V. Spadaro, A. Trunkenmüller, and Y. Zhao (2004b). Health and environmental impacts of China's current and future electricity supply, with associated external costs. *International Journal of Global Energy Issues*, **22**(2/3/4), pp. 155-179.

Hirschberg, S., R. Dones, T. Heck, P. Burgherr, W. Schenler, and C. Bauer (2006). Strengths and weakness of current energy chains in a sustainable development perspective. *atw - International Journal for Nuclear Power*, **51**(7), pp. 447-457.

Hobday, M. (2003). Innovation in Asian industrialization: A Gerschenkronian perspective. *Oxford Development Studies*, **31**(3), pp. 293-314.

Hoefnagels, R., E. Smeets, and A. Faaij (2010). Greenhouse gas footprints of different biofuel production systems. *Renewable and Sustainable Energy Reviews*, **14**(7), pp. 1661-1694.

Holling, C.S. (1997). Regional responses to global change. *Conservation Ecology*, **1**(2), Article 3.

Holloway, S., J.M. Pearce, V.L. Hards, T. Ohsumi, and J. Gale (2007). Natural emissions of CO_2 from the geosphere and their bearing on the geological storage of carbon dioxide. *Energy*, **32**, pp. 1194-1201.

Hong, N., and A.K.Y. Ng (2010). The international legal instruments in addressing piracy and maritime terrorism: A critical review. *Research in Transportation Economics*, **27**, pp. 51-60.

Hoogwijk, M., and A. Faaij (2005). Potential of biomass energy out to 2100, for four IPCC SRES land-use scenarios. *Biomass and Bioenergy*, **29**(4), pp. 225-257.

Hopwood, B., M. Mellor, and G. O'Brien (2005). Sustainable development: mapping different approaches. *Sustainable Development*, **13**(1), pp. 38-52.

Houser, T., S. Mohan, and R. Heilmayr (2009). *A Green Global Recovery? Assessing US Economic Stimulus and the Prospects for International Coordination*. Petersen Institute for International Economics, World Resources Institute, Washington, DC, USA.

Howells, M.I., T. Alfstad, D.G. Victor, G. Goldstein, and U. Remme (2005). A model of household energy services in a low-income rural African village. *Energy Policy*, **33**(14), pp. 1833-1851.

Hreinsson, E.B. (2007). Environmental, technical, economics and policy issues of the master plan for the renewable hydro and geothermal energy resources in Iceland. In: *42nd Universities Power Engineering Conference, Vols 1-3*, Brighton, UK, 4-6 September 2007, pp. 726-731.

Hsu, D.D., D. Inman, G.A. Heath, E.J. Wolfrum, M.K. Mann, and A. Aden (2010). Life cycle environmental impacts of selected US ethanol production and use pathways in 2022. *Environmental Science & Technology*, **44**(13), pp. 5289-5297.

Hueting, R. (1980). *New Scarcity and Economic Growth: More Welfare through Less Production?* North-Holland Publishing Company, Amsterdam, The Netherlands and New York, NY, USA.

Huijts, N.M.A., C.J.H. Midden, and A.L. Meijnders (2007). Social acceptance of carbon dioxide storage. *Energy Policy*, **35**, pp. 2780-2789.

Huo, H., M. Wang, C. Bloyd, and V. Putsche (2009a). Life-cycle assessment of energy use and greenhouse gas emissions of soybean-derived biodiesel and renewable fuels. *Environmental Science & Technology*, **43**(3), pp. 750-756.

Huo, H., Y. Wu, and M. Wang (2009b). Total versus urban: Well-to-wheels assessment of criteria pollutant emissions from various vehicle/fuel systems. *Atmospheric Environment*, **43**(10), pp. 1796-1904.

Hvelplund, F. (2006). Renewable energy and the need for local energy markets. *Energy*, **31**(13), pp. 2293-2302.

IEA (2008a). *Energy Policy Review of Indonesia*. International Energy Agency, Paris, France.

IEA (2008b). *World Energy Outlook 2008*. International Energy Agency, Paris, France, 578 pp.

IEA (2008c). *Worldwide Trends in Energy Use and Efficiency. Key Insights from IEA Indicator Analysis*. International Energy Agency, Paris, France.

IEA (2009). *World Energy Outlook 2009*. International Energy Agency, Paris, France, 696 pp.

IEA (2010a). *Energy Poverty – How to make modern energy access universal. Special excerpt from WEO 2010 with UNIDO and UNDP*. International Energy Agency, Paris, France.

IEA (2010b). *World Energy Outlook 2010*. International Energy Agency, Paris, France, 736 pp.

IEA/ OECD/ World Bank (2010). *The Scope of Fossil-Fuel Subsidies in 2009 and a Roadmap for Phasing Out Fossil Fuel Subsidies*. International Energy Agency (IEA), Organisation for Economic Co-Operation and Development (OECD), and The World Bank,. Paris, France and Washington, DC, USA. Available at: www.iea.org/weo/docs/second_joint_report.pdf.

IIED (2009). *'Land grabs' in Africa: Can the Deals Work for Development? September Briefing*. International Institute for Environment and Development (IIED), London, UK.

ILO (2010). *Occupational Hazard Datasheets – Field Crop Worker*. International Labour Organization (ILO), International Occupational Safety and Health Information Centre (CIS), Geneva, Switzerland.

IMF (2008). *Fiscal Implications of Climate Change*. Fiscal Affairs Department, International Monetary Fund, Washington, DC, USA.

IPCC (1996a). *Climate Change 1995: Economic and Social Dimensions of Climate Change. Contribution of Working Group III to the Second Assessment Report of the Intergovernmental Panel on Climate Change*. J. Bruce, H. Lee, and E.F. Haites (eds.), Cambridge University Press, 448 pp.

IPCC (1996b). *Climate Change 1995: Impacts, Adaptation, and Mitigation of Climate Change - Scientific-Technical Analysis. Contribution of Working Group II to the Second Assessment Report of the Intergovernmental Panel on Climate Change*. R.T. Watson, M.C. Zinyowera, and R.H. Moss (eds.), Cambridge University Press, 879 pp.

IPCC (2000). *Special Report on Emissions Scenarios*. N. Nakicenovic and R. Swart (eds.), Cambridge University Press, 570 pp.

IPCC (2001). *Climate Change 2001: Mitigation. Contribution of Working Group III to the Third Assessment Report of the Intergovernmental Panel on Climate Change*. B. Metz, O. Davidson, R. Swart, and J. Pan (eds.), Cambridge University Press, 700 pp.

IPCC (2005). *Special Report on Carbon Dioxide Capture and Storage*. B. Metz, O. Davidson, H. de Coninck, M. Loos, and L. Meyer (eds.), Cambridge University Press, 431 pp.

IPCC (2007a). *Climate Change 2007: Impacts, Adaptation and Vulnerability. Contribution of Working Group II to the Fourth Assessment Report of the Intergovernmental Panel on Climate Change, 2007*. M.L. Parry, O.F. Canziani, J.P. Palutikof, P.J. van der Linden, and C.E. Hanson (eds.), Cambridge University Press, 979 pp.

IPCC (2007b). *Climate Change 2007: Mitigation. Contribution of Working Group III to the Fourth Assessment Report of the Intergovernmental Panel on Climate Change, 2007*. B. Metz, O.R. Davidson, P.R. Bosch, R. Dave, and L.A. Meyer (eds.), Cambridge University Press, 851 pp.

IPCC (2008). *Climate Change and Water. Technical Paper of the Intergovernmental Panel on Climate Change*. B.C. Bates, Z.W. Kundzewicz, S. Wu, and J.P. Palutikof (eds.), Cambridge University Press.

ISO (2006). *ISO 14040:2006, Environmental management - Life cycle assessment - Principles and framework*. Internet site, International Organization for Standardization (ISO), Geneva, Switzerland. Available at: http://www.iso.org/iso/catalogue_detail?csnumber=37456.

IUCN (2001). *Biodiversity Impacts of Large Dams*. World Conservation Union (IUCN), Gland, Switzerland.

Jacobson, A., A.D. Milman, and D.M. Kammen (2005). Letting the (Energy) Gini out of the bottle: Lorentz curves of cumulative electricity consumption and Gini coefficients as metrics of energy distribution and equity. *Energy Policy*, **33**(14), pp. 1825-1832.

Jacobson, M.Z. (2004). Climate response of fossil fuel and biofuel soot, accounting for soot's feedback to snow and sea ice albedo and emissivity. *Journal of Geophysical Research*, **109**(D21), D21201.

Jacobson, M.Z. (2009). Review of solutions to global warming, air pollution, and energy security. *Energy & Environmental Science*, **2**(2), pp. 148-173.

Jager, H.I., and B.T. Smith (2008). Sustainable reservoir operation: can we generate hydropower and preserve ecosystem values? *River Research and Applications*, **24**(3), pp. 340-352.

Jager, W. (2006). Stimulating the diffusion of photovoltaic systems: A behavioural perspective. *Energy Policy*, **34**(14), pp. 1935-1943.

Jain, G. (2010). Energy security issues at household level in India. *Energy Policy*, **38**(6), pp. 2835-2845.

Jernelöv, A. (2010). The threats from oil spills: Now, then, and in the future. *Ambio*, **39**(5-6), pp. 353-366.

Jobert, A., P. Laborgne, and S. Mimler (2007). Local acceptance of wind energy: Factors of success identified in French and German case studies. *Energy Policy*, **35**(5), pp. 2751-2760.

Jonker Klunne, W., and E.G. Michael (2010). Increasing sustainability of rural community electricity schemes – case study of small hydropower in Tanzania. *International Journal of Low-Carbon Technologies*, **5**(3), pp. 144-147.

Jonkman, S.N., P.H.A.J.M. van Gelder, and J.K. Vrijling (2003). An overview of quantitative risk measures for loss of life and economic damage. *Journal of Hazardous Materials*, **A99**, pp. 1-30.

Jordaan, S.M., D.W. Keith, and B. Stelfox (2009). Quantifying land use of oil sands production: a life cycle perspective. *Environmental Research Letters*, **4**(2), 024004.

Jorgenson, D.W. (1984). The role of energy in productivity growth. *American Economic Review*, **74**(2), pp. 26-30.

Jungbluth, N., C. Bauer, R. Dones, and R. Frischknecht (2005). Life cycle assessment for emerging technologies: Case studies for photovoltaic and wind power. *International Journal of Life Cycle Assessment*, **10**(1), pp. 24-34.

Jungbluth, N., M. Stucki, and R. Frischknecht (2009). *Photovoltaics*. Swiss Centre for Life Cycle Inventories, Duebendorf, Switzerland.

Kaiser, M.J., Y. Yu, and C.J. Jablonowski (2009). Modeling lost production from destroyed platforms in the 2004–2005 Gulf of Mexico hurricane seasons. *Energy*, **34**(9), pp. 1156-1171.

Kalberer, M., D. Paulsen, M. Sax, M. Steinbacher, J. Dommen, A.S.H. Prevot, R. Fisseha, E. Weingartner, V. Frankevich, R. Zenobi, and U. Baltensperger (2004). Identification of polymers as major components of atmospheric organic aerosols. *Science*, **303**(5664), pp. 1659-1662.

Kaliyan, N., R.V. Morey, and D.G. Tiffany (2010). Reducing life cycle greenhouse gas emissions of corn ethanol. In: *American Society of Agricultural and Biological Engineers (ASABE) Annual International Meeting*. Pittsburgh, PA, USA, 20-23 June 2010.

Kangi, A., and N. Heidari (2008). Reservoir-induced seismicity in Karun III dam (Southwestern Iran). *Journal of Seismology*, **12**(4), pp. 519-527.

Karakosta, C., H. Doukas, and J. Psarras (2010). Technology transfer through climate change: Setting a sustainable energy pattern. *Renewable and Sustainable Energy Reviews*, **14**(6), pp. 1546-1557.

Karekezi, S., and W. Kithyoma (2003). Renewable Energy in Africa: Prospects and Limits. In: *Republic of Senegal and United Nations Workshop for African Energy Experts on Operationalizing the NEPAD Energy Initiative*, Dakar, Senegal, 2-4 June 2003, 30 pp. Available at: www.un.org/esa/sustdev/sdissues/energy/op/nepadkarekezi.

Karekezi, S., J. Kimani, and O. Onguru (2008). Energy access among the urban poor in Kenya. *Energy for Sustainable Development*, **12**(4), pp. 38-48.

Karekezi, S., J. Kimani, O. Onguru, and W. Kithyoma (2009). *Large Scale Hydropower, Renewable Energy and Adaptation to Climate Change. Climate Change and Energy Security in East and Horn of Africa.* Energy, Environment and Development Network for Africa, Nairobi, Kenya.

Kargbo, D.M., R.G. Wilhelm, and D.J. Campbell (2010). Natural gas plays in the Marcellus Shale: challenges and potential opportunities. *Environmental Science & Technology*, **44**, pp. 5679-5684.

Kassenga, G.R. (2008). The status and constraints of solar photovoltaic energy development in Tanzania. *Energy Sources Part B-Economics Planning and Policy*, **3**(4), pp. 420-432.

Kaufmann, R.F., G.G. Eadie, and C.R. Russell (1976). Effects of uranium mining and milling on ground water in the Grants mineral belt, New Mexico. *Ground Water*, **14**, pp. 296-308.

Kaufmann, R.K. (2004). The mechanisms for autonomous energy efficiency increases: A cointegration analysis of the US energy/GDP ratio. *Energy Journal*, **25**(1), pp. 63-86.

Kaundinya, D.P., P. Balachandra, and N.H. Ravindranath (2009). Grid-connected versus stand alone energy systems for decentralized power – a review of literature. *Renewable and Sustainable Energy Reviews*, **13**(8), pp. 2041-2050.

Kelso, J.R.M., and G.S. Milburn (1979). Entrainment and impingement of fish by power plants in the Great Lakes which use the once-through cooling process. *Journal of Great Lakes Research*, **5**, pp. 182-194.

Kenny, J.F., N.L. Barber, S.S. Hutson, K.S. Linsey, J.K. Lovelace, and M.A. Maupin (2009). *Estimated Use of Water in the United States in 2005*. Circular 1344, U.S. Geological Survey, Reston, VA, USA, 52 pp.

Kesminas, V., and J. Olechnoviciene (2008). Fish community changes in the cooler of the Ignalina nuclear power plant. *Ekologija*, **54**(2), pp. 124-131.

Kim Oanh, N.T., and N.T. Dung (1999). Emission of polycyclic aromatic hydrocarbons and particulate matter from domestic combustion of selected fuels. *Environmental Science & Technology*, **33**(16), pp. 2703-2709.

King, C.W., and M.E. Webber (2008). Water intensity of transportation. *Environmental Science & Technology*, **42**, pp. 7866-7872.

Kline, K.L., V.H. Dale, R. Efroymson, A. Goss Eng, and Z. Haq (2009). *Land-Use Change and Bioenergy: Report from the 2009 Workshop*. ORNL/CBES-001, U.S. Department of Energy, Office of Energy Efficiency and Renewable Energy and Oak Ridge National Laboratory, Center for Bioenergy Sustainability, Vonore, TN, USA.

Klose, C. (2007). Mine water discharge and flooding: A cause of severe earthquakes. *Mine Water and the Environment*, **26**(3), pp. 172-180.

Klose, C.D. (2010a). Evidence for surface loading as trigger mechanism of the 2008 Wenchuan earthquake. *arXiv*:1007.2155v2 [physics.geo-ph].

Klose, C.D. (2010b). Human-triggered earthquakes and their impacts on human security. In: *Achieving Environmental Security: Ecosystem Services and Human Welfare*. NATO Science for Peace and Security Series - E: Human and Societal Dynamics, Vol. 69. P.H. Liotta, W.G. Kepner, J.M. Lancaster, and D.A. Mouat (eds.), IOS Press, Amsterdam, The Netherlands, pp. 13-19.

Knapp, S., and P.H. Franses (2009). Does ratification matter and do major conventions improve safety and decrease pollution in shipping? *Marine Policy*, **33**, pp. 826-846.

Koch, D., T.C. Bond, D. Streets, N. Unger, and G.R. van der Werf (2007). Global impacts of aerosols from particular source regions and sectors. *Journal of Geophysical Research*, **112**(D2), D02205.

Koh, L.P., and J. Ghazoul (2008). Biofuels, biodiversity, and people: Understanding the conflicts and finding opportunities. *Biological Conservation*, **141**, pp. 2450-2460.

Kontovas, C.A., H.N. Psaraftis, and N.P. Ventikos (2010). An empirical analysis of IOPCF oil spill cost data. *Marine Pollution Bulletin*, **60**, pp. 1455-1466.

Kowalski, K., S. Stagl, R. Madlener, and I. Omann (2009). Sustainable energy futures: Methodological challenges in combining scenarios and participatory multi-criteria analysis. *European Journal of Operational Research*, **197**(3), pp. 1063-1074.

Kramer, W.H. (1982). Ground-water pollution from gasoline. *Ground Water Monitoring & Remediation*, **2**, pp. 18-22.

Krausmann, F., H. Schandl, and R.P. Sieferle (2008). Socio-ecological regime transitions in Austria and the United Kingdom. *Ecological Economics*, **65**(1), pp. 187-201.

Krewitt, W. (2002). External cost of energy – do the answers match the questions? Looking back at 10 years of ExternE. *Energy Policy*, **30**, pp. 839-848.

Krey, V., and L. Clarke (2011). The role of renewable energy in climate change mitigation: a synthesis of recent scenarios. *Climate Policy*, in press.

Kristof, K., and P. Hennicke (2010). *Materialeffizienz und Ressourcenschonung*. Wuppertal Institute for Climate, Environment and Energy, Wuppertal, Germany.

Kruyt, B., D.P. van Vuuren, H.J.M. de Vries, and H. Groenenberg (2009). Indicators for energy security. *Energy Policy*, **37**(6), pp. 2166-2181.

Kubiszewski, I., C.J. Cleveland, and P.K. Endres (2010). Meta-analysis of net energy return for wind power systems. *Renewable Energy*, **35**(1), pp. 218-225.

Kuik, O.J. (2003). Climate change policies, energy security and carbon dependency. Trade-offs for the European Union in the longer term. *International Environmental Agreements: Politics, Law and Economics*, **3**, pp. 221-242.

Kumar, P., R. Britter, and N. Gupta (2009). Hydrogen fuel: Opportunities and barriers. *Journal of Fuel Cell Science and Technology*, **6**(2), 021009, doi:10.1115/1.3005384.

Lacher, W., and D. Kumetat (2010). The security of energy infrastructure and supply in North Africa: Hydrocarbons and renewable energies in comparative perspective. *Energy Policy*, doi:10.1016/j.enpol.2010.10.026.

Lack, D., B. Lerner, C. Granier, T. Baynard, E. Lovejoy, P. Massoli, A.R. Ravishankara, and E. Williams (2008). Light absorbing carbon emissions from commercial shipping. *Geophysical Research Letters*, **35**(13), L13815.

Ladenburg, J. (2010). Attitudes towards offshore wind farms – The role of beach visits on attitude and demographic and attitude relations. *Energy Policy*, **38**(3), pp. 1297-1304.

Lall, S. (2002). Linking FDI and technology development for capacity building and strategic competitiveness. *Transnational Corporations*, **11**(3), pp. 39-88.

Langhamer, O., K. Haikonen, and J. Sundberg (2010). Wave power – Sustainable energy or environmentally costly? A review with special emphasis on linear wave energy converters. *Renewable and Sustainable Energy Reviews*, **14**, pp. 1329-1335.

Larsen, H., and L. Sønderberg Petersen (2009). *Risø Energy Report 8. The intelligent energy system infrastructure for the future*. Risø-R-1695(EN), Risø National Laboratory for Sustainable Energy, Technical University of Denmark, Roskilde, Denmark.

Larson, E.D., and H. Yang (2004). Dimethyl ether (DME) as a household cooking fuel in China. *Energy for Sustainable Development*, **8**(3), pp. 115-126.

Larson, E.D., G. Fiorese, G.J. Liu, R.H. Williams, T.G. Kreutz, and S. Consonni (2010). Co-production of decarbonized synfuels and electricity from coal plus biomass with CO_2 capture and storage: an Illinois case study. *Energy & Environmental Science*, **3**(1), pp. 28-42.

Larssen, T., E. Lydersen, D. Tang, Y. He, J. Gao, H. Liu, L. Duan, H.M. Seip, R.D. Vogt, J. Mulder, M. Shao, Y. Wang, H. Shang, X. Zhang, S. Solberg, W. Aas, T. Okland, O. Eilertsen, V. Angell, Q. Li, D. Zhao, R. Xiang, J. Xiao, and J. Luo (2006). Acid rain in China. *Environmental Science & Technology*, **40**, pp. 418-425.

Lau, W.K.M., M.-K. Kim, K.-M. Kim, and W.-S. Lee (2010). Enhanced surface warming and accelerated snow melt in the Himalayas and Tibetan Plateau induced by absorbing aerosols. *Environmental Research Letters*, **5**(2), pp. 025204.

Lawn, P.A. (2003). A theoretical foundation to support the Index of Sustainable Economic Welfare (ISEW), Genuine Progress Indicator (GPI), and other related indexes. *Ecological Economics*, **44**(1), pp. 105-118.

Le Coq, C., and E. Paltseva (2009). Measuring the security of external energy supply in the European Union. *Energy Policy*, **37**, pp. 4474-4481.

Le Quere, C., M.R. Raupach, J.G. Canadell, G. Marland, L. Bopp, P. Ciais, T.J. Conway, S.C. Doney, R.A. Feely, P. Foster, P. Friedlingstein, K. Gurney, R.A. Houghton, J.I. House, C. Huntingford, P.E. Levy, M.R. Lomas, J. Majkut, N. Metzl, J.P. Ometto, G.P. Peters, I.C. Prentice, J.T. Randerson, S.W. Running, J.L. Sarmiento, U. Schuster, S. Sitch, T. Takahashi, N. Viovy, G.R. van der Werf, and F.I. Woodward (2009). Trends in the sources and sinks of carbon dioxide. *Nature Geoscience*, **2**(12), pp. 831-836.

Leach, G. (1992). The energy transition. *Energy Policy*, **20**(2), pp. 116-123.

LeCornu, J. (ed.) (1998). *Dams and Water Managment*. Report of the Secretary General, International Commission on Large Dams to the *Conference Internationale Eau et Developpement Durable*, Paris, France, 19-21 March 1998.

Lee, C.C., and C.P. Chang (2008). Energy consumption and economic growth in Asian economies: A more comprehensive analysis using panel data. *Resource and Energy Economics*, **30**(1), pp. 50-65.

Lei, X. (2010). Possible roles of the Zipingpu Reservoir in triggering the 2008 Wenchuan earthquake. *Journal of Asian Earth Sciences*, **40**(4), pp. 844-854.

Lele, S., and R.B. Norgaard (1996). Sustainability and the scientist's burden. *Conservation Biology*, **10**(2), pp. 354-365.

Lenton, T.M., H. Held, E. Kriegler, J.W. Hall, W. Lucht, S. Rahmstorf, and H.J. Schellnhuber (2008). Tipping elements in the Earth's climate system. *Proceedings of the National Academy of Sciences*, **105**(6), pp. 1786-1793.

Lenzen, M. (1999). Greenhouse gas analysis of solar-thermal electricity generation. *Solar Energy*, **65**(6), pp. 353-368.

Lenzen, M. (2008). Life cycle energy and greenhouse gas emissions of nuclear energy: A review. *Energy Conversion and Management*, **49**(8), pp. 2178-2199.

Lenzen, M. (2009). Double-counting in life-cycle calculations. *Journal of Industrial Ecology*, **12**(4), pp. 583-599.

Lenzen, M., and J. Munksgaard (2002). Energy and CO_2 analyses of wind turbines – review and applications. *Renewable Energy*, **26**(3), pp. 339-362.

Lenzen, M., and U. Wachsmann (2004). Wind energy converters in Brazil and Germany: an example for geographical variability in LCA. *Applied Energy*, **77**, pp. 119-130.

Lenzen, M., C. Dey, C. Hardy, and M. Bilek (2006). *Life-Cycle Energy Balance and Greenhouse Gas Emissions of Nuclear Energy in Australia*. ISA, University of Sydney, Sydney, Australia.

Limmeechokchai, B., and S. Chawana (2007). Sustainable energy development strategies in the rural Thailand: The case of the improved cooking stove and the small biogas digester. *Renewable and Sustainable Energy Reviews*, **11**(5), pp. 818-837.

Lindeijer, E. (2000). Review of land use impact methodologies. *Journal of Cleaner Production*, **8**(4), pp. 273-281.

Litvine, D., and R. Wüstenhagen (2011). Helping "light green" consumers walk the talk: Results of a behavioural intervention survey in the Swiss electricity market. *Ecological Economics*, **70**(3), pp. 462-474.

Liu, J.K., and Z.T. Yu (1992). Water quality changes and effects on fish populations in the Hanjiang River, China, following hydroelectric dam construction. *Regulated Rivers: Research & Management*, **7**, pp. 359-368.

Lloyd, P.J.D., and E.M. Visagle (2007). A comparison of gel fuels with alternate cooking fuels. *Journal of Energy in Southern Africa*, **18**(3), pp. 26-31.

Lovett, A.A., G.M. Sünnenberg, G.M. Richter, A.G. Dailey, A.B. Riche, and A. Karp (2009). Land use implications of increased biomass production identified by GIS-based suitability and yield mapping for Miscanthus in England. *BioEnergy Research*, **2**, pp. 17-28.

Lubchenco, J., M. McNutt, B. Lehr, M. Sogge, M. Miller, S. Hammond, and W. Conner (2010). *BP Deepwater Horizon Oil Budget: What Happened to the Oil?* No publisher specified. Available at: www.usgs.gov/foia/budget/08-03-2010...Oil%20Budget%20description%20FINAL.pdf.

Luckow, P., M.A. Wise, J.J. Dooley, and S.H. Kim (2010). Large-scale utilization of biomass energy and carbon dioxide capture and storage in the transport and electricity sectors under stringent CO_2 concentration limit scenarios. *International Journal of Greenhouse Gas Control*, **4**(5), pp. 865-877.

Lucon, O., and F. Rei (2006). *Identifying Complementary Measures to Ensure the Maximum Realisation of Benefits from the Liberalisation of EG&S. Case study: Brazil*. Working Paper No. 2004-04, COM/ENV/TD(2003)116/FINAL, Organisation for Economic Co-operation and Development, Trade and Environment, Paris, France, 35 pp. Available at: www.oecd.org/dataoecd/18/53/37325499.pdf.

Lucon, O., S.T. Coelho, and J.O. Alvares (2005). Bioethanol: the way forward. In: *International Symposium on Alcohol Fuels, ISAF XV*, University of California at Riverside, CA, USA, 26-28 September 2005.

Luderer, G., V. Bosetti, J. Steckel, H. Waisman, N. Bauer, E. Decian, M. Leimbach, O. Sassi, and M. Tavoni (2009). *The Economics of Decarbonization – Results from the RECIPE Model Intercomparison*. Potsdam Institute for Climate Impact Research, Potsdam, Germany.

Lund, H. (2007). Renewable energy strategies for sustainable development. *Energy*, **32**(6), pp. 912-919.

Lund, H., P.A. Ostergaard, and I. Stadler (2011). Towards 100% renewable energy systems. *Applied Energy*, **88**(2), pp. 419-421.

Lustgarten, A. (2011). Climate benefits of natural gas may be overstated. *Scientific American*, **26 January 2011**, Available at: www.scientificamerican.com/article.cfm?id=climate-benefits-natural-gas-overstated.

Macedo, I.C., and J.E.A. Seabra (2008). Mitigation of GHG emissions using sugarcane bioethanol. In: *Sugarcane Ethanol: Contributions to Climate Change Mitigation and the Environment*. P. Zuurbier and J. van de Vooren (eds.), Wageningen Academic Publishers, Wageningen, The Netherlands, pp. 95-110 (ISBN: 978-90-8686-090-6).

Mahapatra, S., H.N. Chanakya, and S. Dasappa (2009). Evaluation of various energy devices for domestic lighting in India: Technology, economics and CO_2 emissions. *Energy for Sustainable Development*, **13**(4), pp. 271-279.

Mainali, B., and S. Silveira (2011). Financing off-grid rural electrification: Country case Nepal. *Energy*, **36**(4), pp. 2194-2201.

Majer, E.L., R. Baria, M. Stark, S. Oates, J. Bommer, B. Smith, and H. Asanumag (2007). Induced seismicity associated with Enhanced Geothermal Systems. *Geothermics*, **36**, pp. 185-222.

Malesios, C., and G. Arabatzis (2010). Small hydropower stations in Greece: The local people's attitudes in a mountainous prefecture. *Renewable and Sustainable Energy Reviews*, **14**(9), pp. 2492-2510.

Marcotullio, P.J., and N.B. Schulz (2007). Comparison of energy transitions in the United States and developing and industrializing economies. *World Development*, **35**(10), pp. 1650-1683.

Marcu, A. (2009). Sectoral approaches in greenhouse gas markets: A viable proposition? In: *NAMAs and the Carbon Market. Nationally Appropriate Mitigation Actions of Developing Countries*. UNEP Risø Centre, Roskilde, Denmark, pp. 97-111.

Markandya, A., S. Pedroso-Galinato, and D. Streimikiene (2006). Energy intensity in transition economies: Is there convergence towards the EU average? *Energy Economics*, **28**(1), pp. 121-145.

Martinez, D.M., and B.W. Ebenhack (2008). Understanding the role of energy consumption in human development through the use of saturation phenomena. *Energy Policy*, **36**(4), pp. 1430-1435.

Maruyama, Y., M. Nishikido, and T. Iida (2007). The rise of community wind power in Japan: Enhanced acceptance through social innovation. *Energy Policy*, **35**(5), pp. 2761-2769.

Mathews, J.A. (2008). How carbon credits could drive the emergence of renewable energies. *Energy Policy*, **36**(10), pp. 3633-3639.

Mattoo, A., A. Subramanian, D. van der Mensbrugghe, and J. He (2009). *Can Global De-Carbonization Inhibit Developing Country Industrialization?* Center for Global Development, Washington, DC, USA, 39 pp. Available at: www.cgdev.org/content/publications/detail/1423203.

McCarl, B.A., and U.A. Schneider (2003). Greenhouse gas mitigation in U.S. agriculture and forestry. *Science*, **294**(5551), pp. 2481-2482.

McGowan, F., and R. Sauter (2005). *Public Opinion on Energy Research: A Desk Study for the Research Councils*. Sussex Energy Group, Science and Technology Policy Research, University of Sussex, Brighton, UK.

McLaughlin, S., and M.E. Walsh (1998). Evaluating environmental consequences of producing herbaceous crops for bioenergy. *Biomass and Bioenergy*, **14**, pp. 317-324.

McNally, A., D. Magee, and A.T. Wolf (2009). Hydropower and sustainability: Resilience and vulnerability in China's powersheds. *Journal of Environmental Management*, **90**, pp. 286-293.

Meadows, D.H. (1998). *Indicators and Information Systems for Sustainable Development*. A Report to the Balaton Group, The Sustainability Institute, Hartland, VT, USA.

Mehta, S., and C. Shahpar (2004). The health benefits of interventions to reduce indoor air pollution from solid fuel use: A cost-effectiveness analysis. *Energy for Sustainable Development*, **8**(3), pp. 53-59.

Melillo, J., J.M. Reilly, D.W. Kicklighter, A.C. Gurgel, T.W. Cronin, S. Paltsev, B.S. Felzer, X. Wang, A.P. Sokolov, and C.A. Schlosser (2009). Indirect emissions from biofuels: How important? *Science*, **326**(5958), pp. 1397-1399.

Meshakti, N. (2007). The safety and reliability of complex energy processing systems. *Energy Sources Part B - Economics Planning and Policy*, **2**(2), pp. 141-154.

Michalena, E., J. Hills, and J.-P. Amat (2009). Developing sustainable tourism, using a multicriteria analysis on renewable energy in Mediterranean Islands. *Energy for Sustainable Development*, **13**(2), pp. 129-136.

Miller, B.A., V. Alavian, M.D. Bender, D.J. Benton, J.P. Ostrowski, J.A. Parsly, and M.C. Shiao (1992). Integrated assessment of temperature change impacts on the TVA reservoir and power supply systems. In: *Hydraulic Engineering: Saving a Threatened Resource – In Search of Solutions: Proceedings of the Hydraulic Engineering sessions at Water Forum '92*, Baltimore, MD, USA, 2-6 August 1992, pp. 563-568.

Miller, S.A. (2010). Minimizing land use and nitrogen intensity of bioenergy. *Environmental Science & Technology*, **44**(10), pp. 3932-3939.

Mirza, U.K., N. Ahmad, K. Harijan, and T. Majeed (2009). Identifying and addressing barriers to renewable energy development in Pakistan. *Renewable and Sustainable Energy Reviews*, **13**(4), pp. 927-931.

Mishra, U.C. (2004). Environmental impact of coal industry and thermal power plants in India. *Journal of Environmental Radioactivity*, **72**(1-2), pp. 35-40.

Modi, V., S. McDade, D. Lallement, and J. Saghir (2005). *Energy Services for the Millennium Development Goals*. Energy Sector Management Assistance Programme, United Nations Development Programme, UN Millennium Project and World Bank, New York, NY, USA.

Molina, M.J., and L.T. Molina (2004). Megacities and atmospheric pollution. *Journal of the Air & Waste Management Association*, **54**(6), pp. 644.

Mondal, M.A.H., L.M. Kamp, and N.I. Pachova (2010). Drivers, barriers, and strategies for implementation of renewable energy technologies in rural areas in Bangladesh – An innovation system analysis. *Energy Policy*, **38**(8), pp. 4626-4634.

Monroy, C.R., and A.S.S. Hernandez (2008). Strengthening financial innovation energy supply projects for rural communities in developing countries. *International Journal of Sustainable Development and World Ecology*, **15**(5), pp. 471-483.

Moore, D., J. Dore, and D. Gyawali (2010). The World Commission on Dams + 10: Revisiting the large dam controversy. *Water Alternatives*, **3**(2), pp. 3-13.

Moreno, A., F. Fontana, and S. Grande (2007). ENEA e-learn platform for development and sustainability with international renewable energies network. *Data Science Journal*, **6**(Supplement 9), pp. S92-S98.

Moss, R.H., J.A. Edmonds, K.A. Hibbard, M.R. Manning, and S.K. Rose (2010). The next generation of scenarios for climate change research and assessment. *Nature*, **463**, pp. 747-756.

Murphy, D., A. Cosbey, and J. Drexhage (2008). Market mechanisms for sustainable development in a post-2012 climate regime: Implications for the Development Dividend. In: *A Reformed CDM – Including New Mechanisms for Sustainable Development*. K.H. Olsen and J. Fenhann (eds.), UNEP Risø Centre, Roskilde, Denmark, pp 9-23. Available at: www.cd4cdm.org/Publications/Perspectives/ReformedCDM.pdf .

Nandy, S., S.P.S. Kushwaha, and S. Mukhopadhyay (2007). Monitoring the Chilla-Motichur wildlife corridor using geospatial tools. *Journal for Nature Conservation*, **15**(4), pp. 237-244.

Nannen, V., and J.C.J.M. van den Bergh (2010). Policy instruments for evolution of bounded rationality: Application to climate-energy problems. *Technological Forecasting and Social Change*, **77**(1), pp. 76-93.

Nassar, A., L. Harfuch, M.M.R. Moreira, L.C. Bachion, L.B. Antoniazzi, and G. Sparovek (2009). *Impacts on Land Use and GHG Emissions from a Shock on Brazilian Sugarcane Ethanol Exports to the United States using the Brazilian Land Use Model (BLUM)*. The Brazilian Institute for International Negotiations, Sao Paulo, Brazil.

Neely, J.G., A.E. Magit, J.T. Rich, C.C.J. Voelker, E.W. Wang, R.C. Paniello, B. Nussenbaum, and J.P. Bradley (2010). A practical guide to understanding systematic reviews and meta-analyses. *Otolaryngology-Head and Neck Surgery*, **142**, pp. 6-14.

NETL (2008). *Development of Baseline Data and Analysis of Life Cycle Greenhouse Gas Emissions of Petroleum-Based Fuels*. DOE/NETL-2009/1362, National Energy Technology Laboratory (NETL), Pittsburgh, PA, USA.

NETL (2009). *An Evaluation of the Extraction, Transport and Refining of Imported Crude Oils and the Impact on Life Cycle Greenhouse Gas Emissions*. DOE/NETL-2009/1362, National Energy Technology Laboratory (NETL), Pittsburgh, PA, USA.

Neudoerffer, R.C., P. Malhotra, and P.V. Ramana (2001). Participatory rural energy planning in India – a policy context. *Energy Policy*, **29**(5), pp. 371-381.

Neumayer, E. (2003). *Weak versus Strong Sustainability: Exploring the Limits of Two Opposing Paradigms*. 2nd ed. Edward Elgar, Northampton MA.

Nieuwenhout, F.D.J., A. van Dijk, V.A.P. van Dijk, D. Hirsch, P.E. Lasschuit, G. van Roekel, H. Arriaza, M. Hankins, B.D. Sharma, and H. Wade (2000). *Monitoring and Evaluation of Solar Home Systems. Experiences with Applications of Solar PV for Households in Developing Countries*. Report ECN-C--00-089, Netherlands Energy Research Foundation, Department of Science, Technology and Society of Utrecht University, Utrecht, The Netherlands.

Nikou, S.N. (2010). *Iran's Subsidies Conundrum*. USIP PeaceBrief 49. United States Institute of Peace, Washington, DC, USA. Available at: www.usip.org/files/resources/pb49_0.pdf.

Niven, R.K. (2005). Ethanol in gasoline: Environmental impacts and sustainability review article. *Renewable and Sustainable Energy Reviews*, **9**(6), pp. 535-555.

Norgaard, R. (1994). *Development Betrayed: The End of Progress and a Co-evolutionary Revisioning of the Future*. Routledge, London, UK.

Notter, D., M. Gauch, R. Widmer, P. Wager, A. Stamp, R. Zah, and H.J. Althaus (2010). Contribution of Li-ion batteries to the environmental impact of electric vehicles. *Environmental Science & Technology*, **44**(17), pp. 6550-6556.

Nouni, M.R., S.C. Mullick, and T.C. Kandpai (2008). Providing electricity access to remote areas in India: Niche areas for decentralized electricity supply. *Renewable Energy*, **34**(2), pp. 430-434.

NRC (2000). *Our Common Journey: A Transition toward Sustainability*. National Research Council (NRC), National Academies Press, Washington, DC, USA.

NRC (2010). *Hidden Costs of Energy. Unpriced Consequences of Energy Production and Use*. National Research Council (NRC), National Academies Press, Washington, DC, USA.

Nussbaumer, P. (2009). On the contribution of labelled Certified Emission Reductions to sustainable development: A multi-criteria evaluation of CDM projects. *Energy Policy*, **37**(1), pp. 91-101.

NVE (2009). *Energy in Norway*. Norwegian Water Resource and Energy Directorate (NVE), Oslo, Norway. Available at: www.nve.no/en/Energy/Energy-in-Norway---a-brief-annual-presentation/.

O'Neill, B., and N. Nakicenovic (2008). Learning from global emissions scenarios. *Environmental Research Letters*, **3**(045014), pp. 1-9.

OECD (2002). *Governance for Sustainable Development – Five OECD Case Studies*. Organisation for Economic Co-operation and Development (OECD), Paris, France.

OECD (2006). *Environmental and Energy Products: The Benefits of Liberalising Trade*. Organisation for Economic Co-operation and Development (OECD), Paris, France (ISBN-92-64-02481-6).

OECD/NEA (2002). *Accelerator-driven Systems (ADS) and Fast Reactors (FR) in Advanced Nuclear Fuel Cycles – A Comparative Study*. Organisation for Economic Co-operation and Development (OECD) and Nuclear Energy Agency (NEA), Paris, France.

Offer, G., N. Meah, and A. Coke (2011). *Enabling a Transition to Low Carbon Economies in Developing Countries. Case Study: Bangladesh*. Imperial College, London, UK, 21 pp.

Oikonomou, E.K., V. Kilias, A. Goumas, A. Rigopoulos, E. Karakatsani, M. Damasiotis, D. Papastefanakis, and N. Marini (2009). Renewable energy sources (RES) projects and their barriers on a regional scale: The case study of wind parks in the Dodecanese islands, Greece. *Energy Policy*, **37**(11), pp. 4874-4883.

Olsen, K.H. (2007). The Clean Development Mechanism's contribution to sustainable development: a review of the literature. *Climatic Change*, **84**(1), pp. 59-73.

Olsen, K.H., and J. Fenhann (2008a). *A Reformed CDM Including New Mechanisms for Sustainable Development*. UNEP Risø Centre, Roskilde, Denmark, 184 pp.

Olsen, K.H., and J. Fenhann (2008b). Sustainable development benefits of clean development mechanism projects: A new methodology for sustainability assessment based on text analysis of the project design documents submitted for validation. *Energy Policy*, **36**(8), pp. 2819-2830.

Ölz, S., and M. Beerepoot (2010). *Deploying Renewables in Southeast Asia. Trends and potentials*. Organisation for Economic Co-operation and Development and International Energy Agency, Paris, France.

Omer, A.M. (2003). Implications of renewable energy for women in Sudan: Challenges and opportunities. *International Journal of Sustainable Development*, **6**(2), pp. 246-259.

Onat, N., and H. Bayar (2010). The sustainability indicators of power production systems. *Renewable and Sustainable Energy Reviews*, **14**(9), pp. 3108-3115.

Ostergaard, P.A., and H. Lund (2010). A renewable energy system in Frederikshavn using low-temperature geothermal energy for district heating. *Applied Energy*, **88**(2), pp. 479-487.

Ott, K. (2003). The case for strong sustainability. In: *Greifswald's Environmental Ethics*. Steinbecker Verlag Ulrich Rose, Greifswald, Germany, pp. 59-64.

Owen, A.D. (2006). Renewable energy: Externality costs as market barriers. *Energy Policy*, **34**(5), pp. 632-642.

Paine, L. (1996). Some ecological and socio-economic considerations for biomass energy crop production. *Biomass and Bioenergy*, **10**, pp. 231-242.

Painuly, J.P. (2001). Barriers to renewable energy penetration; a framework for analysis. *Renewable Energy*, **24**(1), pp. 73-89.

Palanivelraja, S., and K.I. Manirathinem (2010). Studies on indoor air quality in a rural sustainable home. *International Journal of Engineering and Applied Sciences*, **6**(2), pp. 70-74.

Pang, X., Y. Mu, J. Yuan, and H. He (2008). Carbonyls emission from ethanol-blended gasoline and biodiesel-ethanol-diesel used in engines. *Atmospheric Environment*, **42**(5), pp. 1349-1358.

Parfitt, B. (2010). *Fracture Lines: Will Canada's Water be Protected in the Rush to Develop Shale Gas*. Munk School of Global Affairs, University of Toronto, Toronto, Canada.

Park, C., Y. Choi, C. Kim, S. Oh, G. Lim, and Y. Moriyoshi (2010). Performance and exhaust emission characteristics of a spark ignition engine using ethanol and ethanol-reformed gas. *Fuel*, **89**(8), pp. 2118-2125.

Parson, E., V. Burkett, K. Fisher-Vanden, D. Keith, L. Mearns, H. Pitcher, C. Rosenzweig, and M. Webster (2007). *Global Change Scenarios: Their Development and Use*. Department of Energy, Office of Biological and Environmental Research, Washington, DC, USA, 106 pp.

Paul, S., and R.N. Bhattacharya (2004). Causality between energy consumption and economic growth in India: a note on conflicting results. *Energy Economics*, **26**(6), pp. 977-983.

Paul, W., R.S. Kirk, J. Michael, and H. Andrew (2007). A global perspective on energy: health effects and injustices. *Lancet*, **370**(9591), pp. 965-978.

Paulsson, E. (2009). A review of the CDM literature: from fine-tuning to critical scrutiny? *International Environmental Agreements-Politics Law and Economics*, **9**(1), pp. 63-80.

Pearce, D., K. Hamilton, and G. Atkinson (1996). Measuring sustainable development: progress on indicators. *Environment and Development Economics*, **1**, pp. 85-101.

Pehnt, M., M. Oeser, and D.J. Swider (2008). Consequential environmental system analysis of expected offshore wind electricity production in Germany. *Energy*, **33**(5), pp. 747-759.

Pelc, R., and R. Fujita (2002). Renewable energy from the ocean. *Marine Policy*, **26**, pp. 471-479.

Perkins, R. (2003). Environmental leapfrogging in developing countries: A critical assessment and reconstruction. *Natural Resources Forum*, **27**(3), pp. 177-188.

Peters, G.P., and E.G. Hertwich (2008). CO_2 embodied in international trade with implications for global climate policy. *Environmental Science & Technology*, **42**(5), pp. 1401-1407.

Peters, J., M. Harsdorff, and F. Ziegler (2009). Rural electrification, Accelerating impacts with complementary services. *Energy for Sustainable Development*, **13**(1), pp. 38-42.

Petersen, L.K., and A.H. Andersen (2009). *Socio-cultural Barriers to the Development of a Sustainable Energy System – The Case of Hydrogen*. 248, National Environmental Research Institute, Aarhus, Denmark.

Pezzey, J. (1992). Sustainability – An interdisciplinary guide. *Environmental Values*, **1**, pp. 321-362.

Pischinger, S., M. Müther, F. Fricke, and A. Kolbeck (2008). From fuel to wheel: How modern fuels behave in combustion engines. *Erdoel Erdgas Kohle*, **124**(2), pp. 58-63.

Pohekar, S.D., and M. Ramachandran (2006). Multi-criteria evaluation of cooking devices with special reference to utility of parabolic solar cooker (PSC) in India. *Energy*, **31**(8-9), pp. 1215-1227.

Pokhrel, A.K., K.R. Smith, A. Khalakdina, A. Deuja, and M.N. Bates (2005). Case control study of indoor cooking smoke exposure and cataract in Nepal and India. *International Journal of Epidemiology*, **34**(3), pp. 702-708.

Pollin, R., H. Garrett-Peltier, J. Heintz, and H. Scharber (2008). *Green Recovery – A Program to Create Good Jobs and Start Building a Low-Carbon Economy*. Centre for American Progress and Political Economy Research Institute (PERI), University of Massachusetts, Washington, DC and Amherst, MA, USA.

Poornima, E., M. Rajadurai, T. Rao, B. Anupkumar, R. Rajamohan, S. Narasimhan, V. Rao, and V. Venugopalan (2005). Impact of thermal discharge from a tropical coastal power plant on phytoplankton. *Journal of Thermal Biology*, **30**, pp. 307-316.

Poumadère, M., C. Mays, S. Le Mer, and R. Blong (2005). The 2003 heat wave in France: dangerous climate change here and now. *Risk Analysis*, **25**(6), pp. 1483-94.

Rabl, A., and J.V. Spadaro (1999). Damages and costs of air pollution: an analysis of uncertainties. *Environment International*, **25**(1), pp. 29-46.

Rajagopal, D., G. Hochman, and D. Zilberman (2010). Indirect fuel use change (IFUC) and the lifecycle environmental impact of biofuel policies. *Energy Policy*, **39**(1), pp. 228-233

Ramanathan, V., and G. Carmichael (2008). Global and regional climate changes due to black carbon. *Nature Geoscience*, **1**, pp. 221-228.

Ramanathan, V., and Y. Feng (2008). On avoiding dangerous anthropogenic interference with the climate system: Formidable challenges ahead. *Proceedings of the National Academy of Sciences*, **105**(38), pp. 14245-14250.

Ramanathan, V., C. Chung, D. Kim, T. Bettge, L. Buja, J.T. Kiehl, W.M. Washington, Q. Fu, D.R. Sikka, and M. Wild (2005). Atmospheric brown clouds: Impacts on South Asian climate and hydrological cycle. *Proceedings of the National Academy of Sciences*, **102**(15), pp. 5326-5333.

Ramanathan, V., M.V. Ramana, G. Roberts, D. Kim, C. Corrigan, C. Chung, and D. Winker (2007). Warming trends in Asia amplified by brown cloud solar absorption. *Nature*, **448**(7153), pp. 575-578.

Rao, K.U., and V.V.N. Kishore (2010). A review of technology diffusion models with special reference to renewable energy technologies. *Renewable and Sustainable Energy Reviews*, **14**(3), pp. 1070-1078.

Ravindranath, N.H., and R. Balachandra (2009). Sustainable bioenergy for India: Technical, economic and policy analysis. *Energy*, **34**(8), pp. 1003-1013.

Rebitzer, G., T. Ekvall, R. Frischknecht, D. Hunkeler, G. Norris, T. Rydberg, W.P. Schmidt, S. Suh, B.P. Weidema, and D.W. Pennington (2004). Review: Life cycle assessment. Part 1: Framework, goal and scope definition, inventory analysis and applications. *Environment International*, **30**, pp. 701-720.

Reddy, A.K.N., W. Annecke, K. Blok, D. Bloom, B. Boardman, A. Eberhard, J. Ramakrishna, Q. Wodon, and A.K.M. Zaidi (2000). Energy and social issues. In: *World Energy Assessment: Energy and the Challenge of Sustainability*. United Nations Development Programme, UN Department of Economic and Social Affairs and the World Energy Council, New York, NY, USA and London, UK, pp 40-60. Available at: www.undp.org/energy/activities/wea/drafts-frame.html.

Reddy, S., and J.P. Painuly (2004). Diffusion of renewable energy technologies – barriers and stakeholders' perspectives. *Renewable Energy*, **29**(9), pp. 1431-1447.

Reiche, D. (2010). Renewable energy policies in the Gulf countries: A case study of the carbon-neutral "Masdar City" in Abu Dhabi. *Energy Policy*, **38**(1), pp. 378-382.

REN21 (2009). *Renewables Global Status Report: 2009 Update*. Renewable Energy Policy Network for the 21st Century Secretariat, Paris, France, 42 pp.

REN21 (2010). *Renewables 2010: Global Status Report*. Renewable Energy Policy Network for the 21st Century Secretariat, Paris, France, 80 pp.

Reuscher, G., C. Ploetz, V. Grimm, and A. Zweck (2008). *Innovationen gegen Rohstoffknappheit. Zukünftige Technologien*. Zukünftige Technologien Consulting der VDI Technologiezentrum GmbH, Düsseldorf, Germany.

Reynolds, C.C.O., and M. Kandlikar (2008). Climate impacts of air quality policy: Switching to a natural gas-fueled public transportation system in New Delhi. *Environmental Science & Technology*, **42**(16), pp. 5860-5865.

Reynolds, J.Z. (1980). Power plant cooling systems: policy alternatives. *Science*, **207**, pp. 367-372.

Richter, B.D., S. Postel, C. Revenga, T. Scudder, B. Lehner, A. Churchill, and M. Chow (2010). Lost in development's shadow: The downstream human consequences of dams. *Water Alternatives*, **3**(2), pp. 14-42.

Riffell, S., J. Verschuyl, D. Miller, and T.B. Wigley (2011). Biofuel harvests, coarse woody debris, and biodiversity – A meta-analysis. *Forest Ecology and Management*, **261**(4), pp. 878-887.

Rijsberman, F. (2006). Water scarcity: Fact or fiction? *Agricultural Water Management*, **80**(1-3), pp. 5-22.

Roayaei, E., and K. Taheri (2009). Test run evaluation of a blend of fuel-grade ethanol and regular commercial gasoline: Its effect on engine efficiency and exhaust gas composition. *Clean Technologies and Environmental Policy*, **11**(4), pp. 385-389.

Rogers, J.C., E.A. Simmons, I. Convery, and A. Weatherall (2008). Public perceptions of opportunities for community-based renewable energy projects. *Energy Policy*, **36**(11), pp. 4217-4226.

Rogner, H.H. (1997). An assessment of world hydrocarbon resources. *Annual Review of Energy and the Environment*, **22**, pp. 217-262.

Rogowska, J., and J. Namiesnik (2010). Environmental implications of oil spills from shipping accidents. *Reviews of Environmental Contamination and Toxicology*, **206**, pp. 95-114.

Roques, F., C. Hiroux, and M. Saguan (2010). Optimal wind power deployment in Europe – A portfolio approach. *Energy Policy*, **38**(7), pp. 3245-3256.

Rosenberg, D.M., F. Berkes, R.A. Bodaly, R.E. Hecky, C.A. Kelly, and J.W.M. Rudd (1997). Large scale impacts of hydroelectric development. *Environmental Reviews*, **5**, pp. 27-54.

Rovere, E.L.L., J.B. Soares, L.B. Oliveira, and T. Lauria (2010). Sustainable expansion of electricity sector: Sustainability indicators as an instrument to support decision making. *Renewable and Sustainable Energy Reviews*, **14**, pp. 422-429.

Roy, J. (2000). The rebound effect: some empirical evidence from India. *Energy Policy*, **28**(6-7), pp. 433-438.

Runge, C.F., and B. Senauer (2007). Biofuel: corn isn't the king of this growing domain. *Nature*, **449**(7163), pp. 637.

Rylands, A.B., and K. Brandon (2005). Brazilian protected area. *Conservation Biology*, **19**(3), pp. 612-618.

Sala, O.E., D. Sax, and H. Leslie (2009). Biodiversity consequences of biofuel production. In: *Biofuels: Environmental Consequences and Interactions with Changing Land Use. Proceedings of the Scientific Committee on Problems of the Environment (SCOPE) International Biofuels Project Rapid Assessment*. R.W. Howarth and S. Bringezu (eds.), Gummersbach, Germany, 22-25 September 2008, pp. 127-137.

SARI (2010). *Unlocking South Africa's Green Growth Potential. The South African Renewables Initiative*. South African Renewables Initiative (SARI), Department of Trade and Industry, Department for Public Enterprises, Pretoria, South Africa.

Sastresa, E.L., A.A. Usón, A.Z. Bribián, and S. Scarpellin (2009). Local impact of renewables on employment: assessment methodology and case study. *Renewable and Sustainable Energy Reviews*, **14**(2), pp. 679-690.

Sathaye, J., A. Najam, C. Cocklin, T. Heller, F. Lecocq, J. Llanes-Regueiro, J. Pan, G. Petschel-Held, S. Rayner, J. Robinson, R. Schaeffer, Y. Sokona, R. Swart, H. Winkler, S. Burch, J. Corfee Morlot, R. Dave, L. Pinter, and A. Wyatt (2007). Sustainable development and mitigation. In: *Climate Change 2007: Mitigation. Contribution of Working Group III to the Fourth Assessment Report of the Intergovernmental Panel on Climate Change, 2007*. B. Metz, O.R. Davidson, P.R. Bosch, R. Dave, and L.A. Meyer (eds.), Cambridge University Press, pp. 691-743.

Sauter, R., and J. Watson (2008). *Technology Leapfrogging: A Review of the Evidence*. A report for DFID, Sussex Energy Group, Science and Technology Policy Research, University of Sussex, Brighton, UK.

Saygin, H., and F. Cetin (2010). New energy paradigm and renewable energy: Turkey's vision. *Insight Turkey*, **12**(3), pp. 107-128.

Schafer, A. (2005). Structural change in energy use. *Energy Policy*, **33**(4), pp. 429-437.

Schifter, I., L. Díaz, M. Vera, E. Guzmán, and E. López-Salinas (2004). Fuel formulation and vehicle exhaust emissions in Mexico. *Fuel*, **83**(14-15Spec), pp. 2065-2074.

Schifter, I., L. Díaz, R. Rodríguez, and L. Salazar (2011). Assessment of Mexico's program to use ethanol as transportation fuel: impact of 6% ethanol-blended fuel on emissions of light-duty gasoline vehicles. *Environmental Monitoring and Assessment*, **173**(1-4), pp. 343-360.

Schlamadinger, B. (1997). Forests for carbon sequestration or fossil fuel substitution? A sensitivity analysis. *Biomass and Bioenergy*, **13**(6), pp. 389-397.

Schleisner, L. (2000). Comparison of methodologies for externality assessment. *Energy Policy*, **28**, pp. 1127-1136.

Schmidt, J.H. (2008). Development of LCIA characterisation factors for land use impacts on biodiversity. *Journal of Cleaner Production*, **16**(18), pp. 1929-1942.

Schneider, L. (2009). A Clean Development Mechanism with global atmospheric benefits for a post-2012 climate regime. *International Environmental Agreements-Politics Law and Economics*, **9**(2), pp. 95-111.

Scholz, R. (2007). Assessment of land use impacts on the natural environment. Part 1: An analytical framework for pure land occupation and land use change. *The International Journal of Life Cycle Assessment*, **12**(1), pp. 16-23.

Schulz, U., O. Brauner, and H. Gruss (2009). Animal diversity on short-rotation coppices – a review. *Landbauforschung vTI Agriculture and Forestry Research*, **3**, pp. 171-182.

Schurr, S.H. (1984). Energy use, technological change, and productive efficiency – an economic-historical interpretation. *Annual Review of Energy*, **9**, pp. 409-425.

Scott, M.J., J.A. Edmonds, N. Mahasenan, J. Roop, A.L. Brunello, and E.F. Haites (2004). International emission trading and the cost of greenhouse gas emissions mitigation and sequestration. *Climatic Change*, **64**(3), pp. 257-287.

Searchinger, T., R. Heimlich, R.A. Houghton, F. Dong, A. Elobeid, J. Fabiosa, S. Tokgoz, D. Hayes, and T.H. Yu (2008). Use of U.S. croplands for biofuels increases greenhouse gases through emissions from land-use change. *Science*, **319**(5867), pp. 1238-1240.

Seljom, P., G. Simbolotti, and G. Tosato (2010). *Unconventional Oil and Gas Production*. IEA Energy Technology Systems Analysis Program, Paris, France.

Semere, T., and F.M. Slater (2007). Ground flora, small mammal and bird species diversity in miscanthus (Miscanthus×giganteus) and reed canary-grass (*Phalaris arundinacea*) fields. *Biomass and Bioenergy*, **31**(1), pp. 20-29.

Sen, S., and T. Babali (2007). Security concerns in the Middle East for oil supply: Problems and solutions. *Energy Policy*, **35**(3), pp. 1517-1524.

Shanthi, V., and N. Gajendran (2009). The impact of water pollution on the socio-economic status of the stakeholders of Ennore Creek, Bay of Bengal (India): Part I. *Indian Journal of Science and Technology*, **2**(3), pp. 66-79.

Sheehan, J., A. Aden, K. Paustian, K. Killian, J. Brenner, M. Walsh, and R. Nelson (2004). Energy and environmental aspects of using corn stover for fuel ethanol. *Journal of Industrial Ecology*, **7**(3-4), pp. 117-146.

Shen, Y., T. Oki, N. Utsumi, S. Kanae, and N. Hanasaki (2008). Projection of future world water resources under SRES scenarios: water withdrawal. *Hydrological Sciences Journal*, **53**(1), pp. 11-33.

Shields, M.A., D.K. Woolf, E.P.M. Grist, S.A. Kerr, A.C. Jackson, R.E. Harris, M.C. Bell, R. Beharie, A. Want, E. Osalusi, S.W. Gibb, and J. Side (2011). Marine renewable energy: The ecological implications of altering the hydrodynamics of the marine environment. *Ocean & Coastal Management*, **54**, pp. 2-9.

Shiklomanov, I.A. (2000). Appraisal and assessment of world water resources. *Water International*, **25**(1), pp. 11-32.

Shrestha, R.M., and S. Pradhan (2010). Co-benefits of CO_2 emission reduction in a developing country. *Energy Policy* **38**, pp. 2586-2597.

Shukla, P.R. (1995). Greenhouse gas models and abatement costs for developing nations : A critical assessment. *Energy Policy*, **23**(8), pp. 677-687.

Shukla, P.R., S. Dhar, and D. Mahapatra (2008). Low-carbon society scenarios for India. *Climate Policy*, **8**, pp. S156-S176.

Singh, A. (2009). The sustainable development of Fiji's energy infrastructure: A status report. *Pacific Economic Bulletin*, **24**(2), pp. 141-154.

Singh, R. (2007). Advancing a "carrot and stick" framework for effective CARICOM environmental cooperation and governance. *Penn State Environmental Law Review*, **16**(1), pp. 199-256.

Singh, S., and A. Kumar (2011). Development of water requirement factors for biomass conversion pathway. *Bioresource Technology*, **102**(2), pp. 1316-1328.

Sjöberg, L. (2009). Precautionary attitudes and the acceptance of a local nuclear waste repository. *Safety Science*, **47**, pp. 542-546.

Skjold, D.O. (2009). *Power for Generations: The Development of Statkraft and the Role of the State in Norwegian Electrification 1890–2009*. Universitetsforlaget, Oslo, Norway, 284 pp.

Slaski, X., and M. Thurber (2009). *Cookstoves and Obstacles to Technology Adoption by the Poor*. Program on Energy and Sustainable Development, Freeman Spogli Institute for International Studies, Stanford University, Stanford, CA, USA.

Smil, V. (2000). Energy in the twentieth century: Resources, conversions, costs, uses, and consequences. *Annual Review of Energy and the Environment*, **25**, pp. 21-51.

Smith, K.R., and M. Ezzati (2005). How environmental health risks change with development: The epidemiologic and environmental risk transitions revisited. *Annual Review of Environment and Resources*, **30**(1), pp. 291-333.

Smith, K.R., and S. Mehta (2003). The burden of disease from indoor air pollution in developing countries: comparison of estimates. *International Journal of Hygiene and Environmental Health*, **206**, pp. 279-289.

Smith, K.R., J.M. Samet, I. Romieu, and N. Bruce (2000). Indoor air pollution in developing countries and acute lower respiratory infections in children. *Thorax*, **55**(6), pp. 518-532.

Smith, K.R., S. Mehta, and M. Maeusezahl-Feuz (2004). Indoor air pollution from household use of solid fuels: comparative quantification of health risks. In: *Global And Regional Burden of Disease Attributable to Selected Major Risk Factors*. WHO, Geneva, Switzerland, pp. 1435-1493.

Smith, K.R., K. Dutta, C. Chengappa, P.P.S. Gusain, O.M.a.V. Berrueta, R. Edwards, R. Bailis, and K.N. Shields (2007). Monitoring and evaluation of improved biomass cookstove programs for indoor air quality and stove performance: conclusions from the Household Energy and Health Project. *Energy for Sustainable Development*, **11**(2), pp. 5-18.

Sneddon, C., R.B. Howarth, and R.B. Norgaard (2006). Sustainable development in a post-Brundtland world. *Ecological Economics*, **57**(2), pp. 253-268.

Soimakallio, S., T. Mäkinen, T. Ekholm, K. Pahkala, H. Mikkola, and T. Paappanen (2009). Greenhouse gas balances of transportation biofuels, electricity and heat generation in Finland – Dealing with the uncertainties. *Energy Policy*, **37**(1), pp. 90-90.

Solow, R.M. (1974). Intergenerational equity and exhaustible resources. *The Review of Economic Studies*, **41**, pp. 29-45.

Sorrell, S., J. Speirs, R. Bentley, A. Brandt, and R. Miller (2009). *Global Oil Depletion – An Assessment of the Evidence for a Near-Term Peak in Global Oil Production*. UK Energy Research Centre, London, UK.

Sovacool, B.K. (2009). The cultural barriers to renewable energy and energy efficiency in the United States. *Technology in Society*, **31**(4), pp. 365-373.

Sovacool, B.K., and R.F. Hirsh (2009). Beyond batteries: An examination of the benefits and barriers to plug-in hybrid electric vehicles (PHEVs) and a vehicle-to-grid (V2G) transition. *Energy Policy*, **37**(3), pp. 1095-1103.

Steenblik, R. (2005). *Liberalisation of Trade in Renewable-Energy Products and Associated Goods: Charcoal, Solar Photovoltaic Systems and Wind Pumps and Turbines*. COM/ENV/TD(2005)23/FINAL, Organisation for Economic Co-operation and Development, Paris, France.

Steinberg, L.J., H. Sengul, and A.M. Cruz (2008). Natech risk and management: an assessment of the state of the art. *Natural Hazards*, **46**, pp. 143-152.

Stern, D.I., 1993: Energy and economic-growth in the USA – a multivariate approach. *Energy Economics*, **15**(2), pp. 137-150.

Stern, N. (2007). *The Economics of Climate Change*. Cambridge University Press, 712 pp. Available at: webarchive.nationalarchives.gov.uk/+/http://www.hm-treasury.gov.uk/sternreview_index.htm.

Sternberg, R. (2008). Hydropower: Dimensions of social and environmental coexistence. *Renewable and Sustainable Energy Reviews*, **12**, pp. 1588-1621.

Steurer, R., and A. Martinuzzi (2007). From environmental plans to sustainable development strategies. *European Environment*, **17**(3), pp. 147-151.

Stiglitz, J. E., Sen, A. and J.-P. Fitoussi (2009). *Report by the Commission on the Measurement of Economic Performance and Social Progress*. No publisher specified. Available at: www.stiglitz-sen-fitoussi.fr.

Stone, K.C., P.G. Hunt, K.B. Cantrell, and K.S. Ro (2010). The potential impacts of biomass feedstock production on water resource availability. *Bioresource Technology*, **101**, pp. 2014-2025.

Strand, J. (2009). *Revenue Management Effects Related to Financial Flows Generated by Climate Policy*. Policy Research Working Paper, World Bank, Washington, DC, USA, 37 pp.

Suckale, J. (2009). Induced seismicity in hydrocarbon fields. *Advances in Geophysics*, **51**, pp. 55-106.

Suh, S., M. Lenzen, G.J. Treloar, H. Hondo, A. Horvath, G. Huppes, O. Jolliet, U. Klann, W. Krewitt, Y. Moriguchi, J. Munksgaard, and G. Norris (2003). System boundary selection in life-cycle inventories using hybrid approaches. *Environmental Science & Technology*, **38**(3), pp. 657-664.

Sundqvist, T. (2004). What causes the disparity of electricity externality estimates? *Energy Policy*, **32**, pp. 1753-1766.

Sutter, C., and J.C. Parreño (2007). Does the current Clean Development Mechanism (CDM) deliver its sustainable development claim? An analysis of oficially registered CDM projects. *Climatic Change*, **84**(1), pp. 75-90.

Suwa, A. (2009). Soft energy paths in Japan: a backcasting approach to energy planning. *Climate Policy*, **9**, pp. 185-206.

Suzuki, T., S. Kabuto, and O. Togawa (2008). Measurement of iodine-129 in seawater samples collected from the Japan Sea area using accelerator mass spectrometry: Contribution of nuclear fuel reprocessing plants. *Quaternary Geochronology*, **3**(3), pp. 268-275.

Tamiotti, L., A. Olhoff, R. Teh, B. Simmons, V. Kulaçoğlu, and H. Abaza (2009). *Trade and Climate Change*. A Report by the United Nations Environment Programme and the World Trade Organization, World Trade Organization, Geneva, Switzerland.

Tarik-ul-Islam, M.D., and S. Ferdousi (2007). Renewable energy development – Challenges for Bangladesh. *Energy and Environment*, **18**(3-4), pp. 421-430.

Tavoni, M., and R.S.J. Tol (2010). Counting only the hits? The risk of underestimating the costs of stringent climate policy. *Climatic Change*, 100(3-4), pp. 769-778.

Tawney, R., Z. Khan, and J. Zachary (2005). Economic and performance evaluation of heat sink options in combined cycle applications. *Journal of Engineering for Gas Turbines and Power*, **127**(2), pp. 397-403.

Teipel, U. (2010). *Rohstoffeffizienz und Rohstoffinnovationen*. Fraunhofer Verlag, Stuttgart, Germany.

Thaulow, H., A. Tvede, T.S. Pedersen, and K. Seelos (2010). Managing catchments for hydropower generation. In: *Handbook of Catchment Management*. R. Ferrier and A. Jenkins (eds.), Wiley-Blackwell, Oxford, UK, pp. 253-287.

Thiam, D.R. (2010). Renewable decentralized in developing countries: Appraisal from microgrids project in Senegal. *Renewable Energy*, **35**(8), pp. 1615-1623.

Thompson, S., and B. Duggirala (2009). The feasibility of renewable energies at an off-grid community in Canada. *Renewable and Sustainable Energy Reviews*, **13**(9), pp. 2740-2745.

Tilman, D., J. Hill, and C. Lehman (2006). Carbon-negative biofuels from low-input high-diversity grassland biomass. *Science*, **314**(5805), pp. 1598-1600

Tiwary, R.K. (2001). Environmental impact of coal mining on water regime and its management. *Water, Air & Soil Pollution*, **132**, pp. 185-199-199.

Toft, P., A. Duero, and A. Bieliauskas (2010). Terrorist targeting and energy security. *Energy Policy*, **38**, pp. 4411-4421.

Torcellini, P., N. Long, and R. Judkoff (2003). *Consumptive Water Use for U.S. Power Production*. National Renewable Energy Laboratory, Golden, CO, USA.

Torfs, R., F. Hurley, B. Miller, and A. Rabl (2007). *A Set of Concentration-Response Functions*. European Commission, Brussels, Belgium.

Torres-Duque, C., D. Maldonado, R. Perez-Padilla, M. Ezzati, and G. Viegi (2008). Biomass fuels and respiratory diseases: A review of the evidence. *Proceedings of the American Thoracic Society*, **5**(5), pp. 577-590.

Toth, F.L., and H.-H. Rogner (2006). Oil and nuclear power: Past, present, and future. *Energy Economics*, **28**, pp. 1-25.

Trieb, F.S., C. Schillings, M. O'Sullivan, T. Pregger, and C. Hoyer-Klick (2009). Global Potential of Concentrating Solar Power. In: *SolarPACES Conference*, 15-18 September 2009, Berlin, Germany.Available at: www.dlr.de/tt/Portaldata/41/Resources/dokumente/institut/system/publications/Solar_Paces_Paper_Trieb_Final_Colour_corrected.pdf.

Tsoutsos, T., N. Frantzeskaki, and V. Gekas (2005). Environmental impacts from the solar energy technologies. *Energy Policy*, **33**, pp. 289-296.

Turchi, C., M. Wagner, and C. Kutscher (2010). *Water Use in Parabolic Trough Power Plants: Summary Results from Worley Parsons' Analyses*. NREL/TP-5500-49468, National Renewable Energy Laboratory, Golden, CO, USA.

Turton, H., and F. Moura (2007). Vehicle-to-grid systems for sustainable development: An integrated energy analysis. *Technological Forecasting and Social Change*, **75**(8), pp. 1091-1108

Tyner, W., F. Taheripour, Q. Zhuang, D. Birur, and U. Baldos (2010). *Land Use Changes and Consequent CO_2 Emissions due to U.S. Corn Ethanol Production: A Comprehensive Analysis*. GTAP Resource 3288, Department of Agricultural Economics, Purdue University, West Lafayette, IN, USA.

Udo de Haes, H.A., and R. Heijungs (2007). Life-cycle assessment for energy analysis and management. *Applied Energy*, **84**, pp. 817-827.

UN (2002). *Report of the World Summit on Sustainable Development*. A/CONF.199/20*, United Nations, Johannesburg, South Africa and New York, NY, USA, 173 pp.

UN (2005a). *2005 World Summit Outcome. Resolution Adopted by the General Assembly*. A/RES/60/1, United Nations, New York, NY, USA.

UN (2005b). *Beijing Declaration on Renewable Energy for Sustainable Development*. United Nations, New York, NY, USA. Available at: www.un.org/esa/sustdev/whats_new/beijingDecl_RenewableEnergy.pdf.

UNCED (1992). *Agenda 21*. UN Conference on Environment and Development (UNCED), UN Department of Economic and Social Affairs, New York, NY, USA.

UNDESA (2008). *Addressing Climate Change in National Sustainable Development Strategies – Common Practices*. Background Paper No. 12, DESA/DSD/2008/12, Comission on Sustainable Development, UN Department of Economic and Social Affairs (UNDESA), New York, NY, USA, 62 pp.

UNDP (2007). *Human Development Report 2007/2008*. United Nations Development Programme (UNDP), New York, NY, USA (ISBN 978-0-230-54704-9).

UNDP (2010). *Human Development Report 2010*. United Nations Development Programme (UNDP), New York, NY, USA.

UNDP and WHO (2009). *The Energy Access Situation in Developing Countries, A Review Focusing on the Least Developed Countries and sub-Saharan Africa*. United Nations Development Programme (UNDP) and the World Health Organization (WHO), New York, NY, USA.

UNDP/UNDESA/WEC (2000). *World Energy Assessment: Energy and the Challenge of Sustainability*. United Nations Development Programme, United Nations Department of Economic and Social Affairs, and World Energy Council, New York, NY, USA.

UNEP (2008a). *Green Jobs: Towards Decent Work in a Sustainable, Low-Carbon World*. United Nations Environment Programme (UNEP), Nairobi, Kenya (ISBN: 978-92-807-5).

UNEP (2008b). *Reforming Energy Subsidies. Opportunities to Contribute to the Climate Change Agenda*. Division of Technology, Industry and Economics, United Nations Environment Programme (UNEP), Paris, France.

UNEP (2010). *Global Trends in Sustainable Energy Investment 2010. Analysis of Trends and Issues in the Financing of Renewable Energy and Energy Efficiency*. United Nations Environment Programme (UNEP) and Bloomberg New Energy Finance, Nairobi, Kenya.

UNEP and SETAC (2010). *The Life Cycle Initiative*. Division of Technology, Industry and Economics, United Nations Environment Programme (UNEP), Society for Environmental Toxicology and Chemistry (SETAC), Paris, France.

UNEP Risø Pipeline (2011). *UNEP Risø CDM/JI Pipeline*. UNEP Risø Centre, Roskilde, Denmark. Available at: www.cdmpipeline.org.

UNFCCC (2009). *The Copenhagen Accord 2/CP.15*. FCCC/CP/2009/11/Add.1, United Nations Framework Convention on Climate Change (UNFCCC), Bonn, Germany. Available at: unfccc.int/resource/docs/2009/cop15/eng/11a01.pdf.

Unruh (2000). Understanding carbon lock-in. *Energy Policy*, **28**, pp. 817-830.

Upreti, B.R. (2004). Conflict over biomass energy development in the United Kingdom: some observations and lessons from England and Wales. *Energy Policy*, **32**(6), pp. 785-800.

Urban, F., R.M.J. Benders, and H.C. Moll (2007). Modelling energy systems for developing countries. *Energy Policy*, **35**(6), pp. 3473-3482.

Urmee, T., D. Harrie, and A. Schlapfer (2009). Issues related to rural electrification using renewable energy in developing countries of Asia and Pacific. *Renewable Energy*, **34**(2), pp. 354-357.

US DOE (2010). *Critical Materials Strategy*. U.S. Department of Energy (DOE), Washington, DC, USA, 166 pp.

US DOT (2010). *Transportation's Role in Reducing U.S. Greenhouse Gas Emissions*. U.S. Department of Transportation, Washington, DC, USA.

USGS (2010). *Mineral Commodity Summaries 2010*. United States Department of the Interior, United States Geological Survey (USGS), Washington, DC, USA, 193 pp.

Ußner, M., and F. Müller-Langer (2009). Biofuels today and tomorrow: effects of fuel composition on exhaust gas emissions. *Accreditation and Quality Assurance: Journal for Quality, Comparability and Reliability in Chemical Measurement*, **14**(12), pp. 685-691.

van Alphen, K., W.G.J.H.M. van Sark, and M.P. Hekkert (2007). Renewable energy technologies in the Maldives – determining the potential. *Renewable and Sustainable Energy Reviews*, **11**(8), pp. 1650-1674.

van Dam, J., A.P.C. Faaij, J. Hilbert, H. Petruzzi, and W.C. Turkenburg (2009). Large-scale bioenergy production from soybeans and switchgrass in Argentina. Part B: Environmental and socio-economic impacts on a regional level. *Renewable and Sustainable Energy Reviews*, **13**(8), pp. 1679-1709.

van der Horst, D. (2007). NIMBY or not? Exploring the relevance of location and the politics of voiced opinions in renewable energy siting controversies. *Energy Policy*, **35**(5), pp. 2705-2714.

van der Vleuten, F., N. Stam, and R. van der Plas (2007). Putting solar home systems programmes into perspective: What lessons are relevant. *Energy Policy*, **34**(3), pp. 1439-1451.

van Metre, P.C., and J. Gray (1992). Effects of uranium-mining releases on groundwater quality in the Puerco River Basin, Arizona and New Mexico. *Hydrological Sciences*, **37**, pp. 463-480.

van Ruijven, B. (2008). *Energy and Development – A Modelling Approach*. PhD Thesis, Department of Science, Technology and Society, Utrecht University, Utrecht, The Netherlands.

van Ruijven, B., F. Urban, R.M.J. Benders, H.C. Moll, J.P. van der Sluijs, B. de Vries, and D.P. van Vuuren (2008). Modeling energy and development: An evaluation of models and concepts. *World Development*, **36**(12), pp. 2801-2821.

van Vliet, O., A.S. Brouwer, T. Kuramochi, M. van den Broek, and A. Faaij (2011). Energy use, costs and CO_2 emissions of electric cars. *Journal of Power Sources*, **196**(4), pp. 2298-2310.

van Vuuren, D.P., P.L. Lucas, and H. Hilderink (2007). Downscaling drivers of global environmental change: Enabling use of global SRES scenarios at the national and grid levels. *Global Environmental Change*, **17**(1), pp. 114-130.

van Vuuren, D.P., S.J. Smith, and K. Riahi (2010a). Downscaling socioeconomic and emissions scenarios for global environmental change research: a review. *Interdisciplinary Reviews: Climate Change*, **1**(3), pp. 393-404.

van Vuuren, D.P., E. Bellevrat, A. Kitous, and M. Isaac (2010b). Bio-energy use and low stabilization scenarios. *Energy Journal* **31**(Special Issue 1), pp. 192-222.

Varun, R. Prakash, and I.K. Bhat (2010). A figure of merit for evaluating sustainability of renewable energy systems. *Renewable and Sustainable Energy Reviews*, **14**(6), pp. 1640-1643.

Vega, L.A. (2002). Ocean thermal energy conversion primer. *Marine Technology Society Journal*, **36**, pp. 25-35.

Veil, J. (2010). *Water Management Technologies Used by Marcellus Shale Gas Producers*. ANL/EVS/R-10/3, Argonne National Laboratory, Argonne, IL, USA.

Vera, I., and L. Langlois (2007). Energy indicators for sustainable development. *Energy*, **32**(6), pp. 875-882.

Verbruggen, A., and V. Lauber (2009). Basic concepts for designing renewable electricity support aiming at a full-scale transition by 2050. *Energy Policy*, **37**(12), pp. 5732-5743.

Viebahn, P., S. Kronshage, F. Trieb, and Y. Lechon (2008). *Final Report on Technical Data, Costs, and Life Cycle Inventories of Solar Thermal Power Plants*. European Commission, Brussels, Belgium.

Vinnem, J.E. (2010). Risk indicators for major hazards on offshore installations. *Safety Science*, **48**, pp. 770-787.

Viscusi, K.W. (2010). The heterogeneity of the value of statistical life: Introduction and overview. *Journal of Risk and Uncertainty*, **40**, pp. 1-13.

Voitsekhovitch, O., Y. Soroka, and T. Lavrova (2006). Uranium mining and ore processing in Ukraine – radioecological effects on the Dnipro River water ecosystem and human health. *Radioactivity in the Environment*, **8**, pp. 206-214.

Volkery, A., D. Swanson, K. Jacob, F. Bregha, and L. Pintér (2006). Coordination, challenges, and innovations in 19 national sustainable development strategies. *World Development*, **34**(12), pp. 2047-2063.

von Blottnitz, H., and M.A. Curran (2007). A review of assessments conducted on bio-ethanol as a transportation fuel from a net energy, greenhouse gas, and environmental life cycle perspective. *Journal of Cleaner Production*, **15**(7), pp. 607-619.

Voorspools, K.R., E.A. Brouwers, and W.D. D'haeseleer (2000). Energy content and indirect greenhouse gas emissions embedded in 'emission-free' plants: results from the Low Countries. *Applied Energy*, **67**, pp. 307-330.

Walker, G., P. Devine-Wright, S. Hunter, H. High, and B. Evans (2010). Trust and community: Exploring the meanings, contexts and dynamics of community renewable energy. *Energy Policy*, **38**(6), pp. 2655-2663.

Wallquist, L., V.H.M. Visschers, and M. Siegrist (2009). Lay concepts on CCS deployment in Switzerland based on qualitative interviews. *International Journal of Greenhouse Gas Control*, **3**, pp. 652-657.

Walter, A., P. Dolzan, O. Quilodrán, J.G. de Oliveira, C. da Silva, F. Piacente, and A. Segerstedt (2011). Sustainability assessment of bio-ethanol production in Brazil considering land use change, GHG emissions and socio-economic aspects. *Energy Policy*, doi:10.1016/j.enpol.2010.07.043.

Wang, M., M. Wu, and H. Hong (2007). Life-cycle energy and greenhouse gas emission impacts of different corn ethanol plant types. *Environmental Research Letters*, **2**(2), 024001.

Wara, M. (2008). Measuring the Clean Development Mechanism's performance and potential. *UCLA Law Review*, **55**(6), pp. 1759-1803.

Ward, M. (2008). Sector no-lose targets: A new scaling up mechanism for developing countries. In: *A Reformed CDM Including New Mechanisms for Sustainable Development*. K.H. Olsen and J. Fenhann (eds.), UNEP Risø Centre, Roskilde, Denmark, pp. 147-163.

Warren, C.R., and M. McFadyen (2010). Does community ownership affect public attitudes to wind energy? A case study from south-west Scotland. *Land Use Policy*, **27**(2), pp. 204-213.

WBGU (2009). *World in Transition – Future Bioenergy and Sustainable Land Use*. German Advisory Council on Global Change (WBGU), Earthscan, London, UK.

WCED (1987). *Our Common Future*. World Commission on Environment and Development (WCED), Oxford University Press, Oxford, UK and New York, NY, USA.

WEC (2007). *Survey of Energy Resources*. World Energy Council (WEC), London, UK. Available at: www.worldenergy.org/documents/ser2007_final_online_version_1.pdf.

WEC (2010). *Survey of Energy Resources 2010*. World Energy Council (WEC), London, UK. Available at: www.worldenergy.org/documents/ser_2010_report_1.pdf

Werner, M., and A.I. Schaefer (2007). Social aspects of a solar-powered desalination unit for remote Australian communities. *Desalination*, **203**(1-3), pp. 375-393.

West, J., I. Bailey, and M. Winter (2010). Renewable energy policy and public perceptions of renewable energy: A cultural theory approach. *Energy Policy*, **38**(10), pp. 5739-5748.

Westwood, A. (2007). Wave and tidal – project review. *Renewable Energy Focus* **8**(4), pp. 30-33.

Weyant, J.P. (1993). Costs of reducing global carbon emissions. *The Journal of Economic Perspectives*, **7**(4), pp. 27-46.

Whitaker, M., and G. Heath (2010). *Life Cycle Assessment Comparing the Use of Jatropha Biodiesel in the Indian Road and Rail Sectors*. NREL/TP-6A2-47462, National Renewable Energy Laboratory, Golden, CO, USA.

WHO (2002). *The World Health Report. Reducing Risks, Promoting Healthy Life*. World Health Organization (WHO), Geneva, Switzerland.

WHO (2006). *Air Quality Guidelines. Global Update 2005. Particulate Matter, Ozone, Nitrogen Dioxide and Sulfur Dioxide*. World Health Organization (WHO) Regional Office for Europe, Copenhagen, Denmark.

Wicke, B., V. Dornburg, M. Junginger, and A. Faaij (2008). Different palm oil production systems for energy purposes and their greenhouse gas implications. *Biomass and Bioenergy*, **32**(12), pp. 1322-1337.

Wilbanks, T.J. (2002). Geographic scaling issues in integrated assessments of climate change. *Integrated Assessment*, **3**(2-3), pp. 100-114.

Wilkie, A.C., K.J. Riedesel, and J.M. Owens (2000). Stillage characterization and anaerobic treatment of ethanol stillage from conventional and cellulosic feedstocks. *Biomass and Bioenergy*, **19**, pp. 39.

Wilkins, G. (2002). *Technology Transfer for Renewable Energy. Overcoming Barriers in Developing Countries*. The Royal Institute of International Affairs, Earthscan Publications, London, UK.

Williams, P.R.D., D. Inman, A. Aden, and G.A. Heath (2009). Environmental and sustainability factors associated with next-generation biofuels in the US: What do we really know? *Environmental Science & Technology*, **43**(13), pp. 4763-4775.

Winkler, H., O. Davidson, and S. Mwakasonda (2005). Developing institutions for the clean development mechanism (CDM): African perspectives. *Climate Policy*, **5**, pp. 209-220.

Wise, M., K. Calvin, A. Thomson, L. Clarke, B. Bond-Lamberty, R. Sands, S.J. Smith, A. Janetos, and J. Edmonds (2009). Implications of limiting CO_2 concentrations for land use and energy. *Science*, **324**(5931), pp. 1183-1186.

Wise, M., G. Kyle, J. Dooley, and S. Kim (2010). The impact of electric passenger transport technology under an economy-wide climate policy in the United States: Carbon dioxide emissions, coal use, and carbon dioxide capture and storage. *International Journal of Greenhouse Gas Control*, **4**(2), pp. 301-308.

Wolf, A.T. (1998). Conflict and cooperation along international waterways. *Water Policy*, **1**(2), pp. 251-265.

Wolsink, M. (2000). Wind power and the NIMBY-myth: institutional capacity and the limited significance of public support. *Renewable Energy*, **21**(1), pp. 49-64.

Wolsink, M. (2007a). Planning of renewables schemes: Deliberative and fair decision-making on landscape issues instead of reproachful accusations of non-cooperation. *Energy Policy*, **35**(5), pp. 2692-2704.

Wolsink, M. (2007b). Wind power implementation: The nature of public attitudes: Equity and fairness instead of 'backyard motives'. *Renewable and Sustainable Energy Reviews*, **11**(6), pp. 1188-1207.

Wolsink, M. (2010). Near-shore wind power – Protected seascapes, environmentalists' attitudes, and the technocratic planning perspective. *Land Use Policy*, **27**(2), pp. 195-203.

Wood, R.G. (2011). *Carbon Finance and Pro-Poor Co-benefits: The Gold Standard and Climate, Community and Biodiversity Standards*. International Institute for Environment and Development, London, UK, 24pp.

World Bank (2001). *The World Bank Group's Energy Program – Poverty Reduction, Sustainability and Selectivity*. Energy and Mining Sector Board, The World Bank, Washington, DC, USA.

World Bank (2007a). *African Development Indicators 2007*. World Bank, Washington, DC, USA, 176 pp.

World Bank (2007b). *World Development Indicators 2007*. World Bank, Washington, DC, USA.

World Bank (2008a). *Global Economic Prospects*. The International Bank for Reconstruction and Development, The World Bank, Washington, DC, USA, 224 pp.

World Bank (2008b). *The Welfare Impact of Rural Electrification: A Reassessment of the Costs and Benefits*. The World Bank, Washington, DC, USA, 178 pp.

World Bank (2010). *World Development Indicators 2010*. World Bank, Washington, DC, USA, 489 pp.

World Bank (2011). *Climate Impacts on Energy Systems. Key Issues for Energy Sector Adaptation. A World Bank Study*. The World Bank, Washington, DC, USA (ISBN: 978-0-8213-8697-2).

World Commission on Dams (2000). *Dams and Development – A New Framework for Decision-Making*. Earthscan Publications Ltd, London, UK and Sterling, VA, USA.

WTO (2010). *Background Note: Trade and Environment in the WTO*. World Trade Organization, Geneva, Switzerland.

Wu, M., Y. Wu, and M. Wang (2005). *Mobility Chains Analysis of Technologies for Passenger Cars and Light-Duty Vehicles Fueled with Biofuels: Application of the GREET Model to the Role of Biomass in America's Energy Future (RBAEF) Project*. ANL/ESD/07-11, Argonne National Laboratory, Argonne, IL, USA.

Wu, M., M. Mintz, M. Wang, and S. Arora (2009). Water consumption in the production of ethanol and petroleum gasoline. *Environmental Management*, **44**, pp. 981-97.

Wustenhagen, R., M. Wolsink, and M.J. Burer (2007). Social acceptance of renewable energy innovation: An introduction to the concept. *Energy Policy*, **35**(5), pp. 2683-2691.

Yamada, M., and J. Zheng (2008). Determination of ^{240}Pu/^{239}Pu atom ratio in coastal surface seawaters from the western North Pacific Ocean and Japan Sea. *Applied Radiation and Isotopes*, **66**(1), pp. 103-107.

Yamamoto, H., J. Fujino, and K. Yamaji (2001). Evaluation of bioenergy potential with a multi-regional global-land-use-and-energy model. *Biomass and Energy*, **21**, pp. 185-203.

Yanowitz, J., and R.L. McCormick (2009). Effect of E85 on tailpipe emissions from light-duty vehicles. *Journal of the Air and Waste Management Association*, **59**(2), pp. 172-182.

Yeh, S., S. Jordaan, A. Brandt, M. Turetsky, S. Spatari, and D. Keith (2010). Land use greenhouse gas emissions from conventional oil production and oil sands. *Environmental Science and Technology*, **44**, pp. 8766-8772.

Yim, M.-S. (2006). Nuclear nonproliferation and the future expansion of nuclear power. *Progress in Nuclear Energy*, **48**, pp. 504-524.

Yoon, S.H., S.Y. Ha, H.G. Roh, and C.S. Lee (2009). Effect of bioethanol as an alternative fuel on the emissions reduction characteristics and combustion stability in a spark ignition engine. *Proceedings of the Institution of Mechanical Engineers, Part D: Journal of Automobile Engineering*, **223**(7), pp. 941-951.

Zackrisson, M., L. Avellan, and J. Orlenius (2010). Life cycle assessment of lithium-ion batteries for plug-in hybrid electric vehicles – Critical issues. *Journal of Cleaner Production*, **18**(15), pp. 1519-1529.

Zah, R., H. Böni, M. Gauch, R. Hischier, M. Lehmann, and P. Wäger (2007). *Ökobilanz von Energieprodukten: Ökologische Bewertung von Biotreibstoffen*. Bundesamtes für Energie, Bundesamt für Umwelt, Bundesamt für Landwirtschaft, Bern, Switzerland.

Zhai, H., H.C. Frey, N.M. Rouphail, G.A. Gonçalves, and T.L. Farias (2009). Comparison of flexible fuel vehicle and life-cycle fuel consumption and emissions of selected pollutants and greenhouse gases for ethanol 85 versus gasoline. *Journal of the Air and Waste Management Association*, **59**(8), pp. 912-924.

Zhang, J., and K.R. Smith (2007). Household air pollution from coal and biomass fuels in China: Measurements, health impacts, and interventions. *Environmental Health Perspectives*, **115**(6), pp. 848-855.

Ziemann, S., and L. Schebek (2010). Substitution knapper Metalle – Ein Ausweg aus der Rohstoffknappheit? *Chemie, Ingenieur, Technik*, 82(11), pp. 1965-1975.

Zoellner, J., P. Schweizer-Ries, and C. Wemheuer (2008). Public acceptance of renewable energies: Results from case studies in Germany. *Energy Policy*, **36**(11), pp. 4136-4141.

Zografakis, N., E. Sifaki, M. Pagalou, G. Nikitaki, V. Psarakis, and K.P. Tsagarakis (2009). Assessment of public acceptance and willingness to pay for renewable energy sources in Crete. *Renewable & Sustainable Energy Reviews*, **14**(3), pp. 1088-1095.

Zuk, M., L. Rojas, S. Blanco, P. Serrano, J. Cruz, F. Angeles, G. Tzintzun, C. Armendariz, R.D. Edwards, M. Johnson, H. Riojas-Rodriguez, and O. Masera (2007). The impact of improved wood-burning stoves on fine particulate matter concentrations in rural Mexican homes. *Journal of Exposure Science and Environmental Epidemiology*, **17**(3), pp. 224-232.

10 Mitigation Potential and Costs

Coordinating Lead Authors:
Manfred Fischedick (Germany) and Roberto Schaeffer (Brazil)

Lead Authors:
Akintayo Adedoyin (Botswana), Makoto Akai (Japan), Thomas Bruckner (Germany), Leon Clarke (USA), Volker Krey (Austria/Germany), Ilkka Savolainen (Finland), Sven Teske (Germany), Diana Ürge-Vorsatz (Hungary), Raymond Wright † (Jamaica)

Contributing Authors:
Gunnar Luderer (Germany)

Review Editors:
Erin Baker (USA) and Keywan Riahi (Austria)

This chapter should be cited as:
Fischedick, M., R. Schaeffer, A. Adedoyin, M. Akai, T. Bruckner, L. Clarke, V. Krey, I. Savolainen, S. Teske, D. Ürge-Vorsatz, R. Wright, 2011: Mitigation Potential and Costs. In IPCC Special Report on Renewable Energy Sources and Climate Change Mitigation [O. Edenhofer, R. Pichs-Madruga, Y. Sokona, K. Seyboth, P. Matschoss, S. Kadner, T. Zwickel, P. Eickemeier, G. Hansen, S. Schlömer, C. von Stechow (eds)], Cambridge University Press, Cambridge, United Kingdom and New York, NY, USA.

Table of Contents

Executive Summary 794

10.1 Introduction 798

10.2 Synthesis of mitigation scenarios for different renewable energy strategies 799

10.2.1 State of scenario analysis 799
10.2.1.1 Types of scenario methods 799
10.2.1.2 Strengths and weaknesses of quantitative scenarios 800

10.2.2 The role of renewable energy sources in scenarios 800
10.2.2.1 Overview of the scenarios reviewed in this section 800
10.2.2.2 Overview of the role of renewable energy in the scenarios 801
10.2.2.3 Setting the scale of renewable energy deployment: Energy system growth and long-term climate goals 803
10.2.2.4 Competition between renewable energy sources and other forms of low-carbon energy 805
10.2.2.5 Renewable energy deployment by technology, over time and by region 806
10.2.2.6 Renewable energy and the costs of mitigation 808

10.2.3 The deployment of renewable energy sources in scenarios from the technology perspective 812

10.2.4 Knowledge gaps 812

10.3 Assessment of representative mitigation scenarios for different renewable energy strategies 813

10.3.1 Sectoral breakdown of renewable energy sources 813
10.3.1.1 Renewable energy deployment in the electricity sector 816
10.3.1.2 Renewable energy deployment in the heating and cooling sector 818
10.3.1.3 Renewable energy deployment in the transport sector 820
10.3.1.4 Global renewable energy primary energy contribution 820

10.3.2 Regional breakdown – technical potential versus market deployment 820
10.3.2.1 Regional renewable energy supply curves 820
10.3.2.2 Primary energy by region, technology and sector 825

10.3.3 Greenhouse gas mitigation potential of renewable energy in aggregate and as individual options 826

10.3.4 Comparison of the results of the in-depth scenario analysis and knowledge gaps 830

10.4 Regional cost curves for mitigation with renewable energies 832

10.4.1 Introduction 832

10.4.2	Cost curves: concept, strengths and limitations	832
10.4.2.1	The concept	832
10.4.2.2	Limitations of the supply curve method	833
10.4.3	Review of regional energy and abatement cost curves from the literature	834
10.4.3.1	Introduction	834
10.4.3.2	Regional and global renewable energy supply curves	834
10.4.3.3	Regional and global carbon abatement cost curves	834
10.4.4	Review of selected technology resource cost curves	836
10.4.5	Gaps in knowledge	840

10.5 Costs of commercialization and deployment ... 841

10.5.1	Introduction: Review of present technology costs	841
10.5.2	Prospects for cost decreases	846
10.5.3	Deployment cost curves and learning investments	849
10.5.4	Time-dependent expenditures	849
10.5.5	Market support and research, development, demonstration and deployment	851
10.5.6	Knowledge gaps	851

10.6 Social and environmental costs and benefits ... 851

10.6.1	Background and objective	851
10.6.2	Review of studies on external costs and benefits	853
10.6.2.1	Climate change	853
10.6.2.2	Health impacts due to air pollution	854
10.6.2.3	Other impacts	854
10.6.3	Social and environmental costs and benefits by energy sources and regional considerations	854
10.6.4	Synergistic strategies for limiting damages and external costs	857
10.6.5	Knowledge gaps	857

References ... 858

Executive Summary

Renewable energy (RE) has the potential to play an important and increasing role in achieving ambitious climate mitigation targets. Many RE technologies are increasingly becoming market competitive, although some innovative RE technologies are not yet mature, economic alternatives to non-RE technologies. However, assessing the future role of RE requires not only consideration of the cost and performance of RE technologies, but also an integrative perspective that takes into account the interactions between various forces and the overall systems behaviours.

An increasing number of integrated scenario analyses are available in the published literature. They are able to provide relevant insights into the potential contribution of RE to future energy supplies and climate change mitigation. A review of 164 scenarios from 16 different large-scale integrated models was conducted through an open call. Although a collection of scenarios from the literature does not represent a truly random sample suitable for rigorous statistical analysis, a scenario overview can provide some critical and strategic insights about the role of RE in climate mitigation, in spite of the uncertainties involved.

Although it is not possible to precisely link long-term climate goals and global RE deployment levels, RE deployment significantly increases in the scenarios with ambitious greenhouse gas (GHG) concentration stabilization levels. Ambitious GHG concentration stabilization levels lead on average to higher RE deployment compared to the baseline. However, for any given long-term GHG concentration goal, the scenarios exhibit a wide range of RE deployment levels. In scenarios that stabilize the atmospheric carbon dioxide (CO_2) concentration at a level of less than 440 ppm, the median RE deployment levels are 139 EJ/yr in 2030 and 248 EJ/yr in 2050, with the highest levels reaching 252 EJ/yr in 2030 and up to 428 EJ/yr in 2050. This range is a result of differences in assumptions about factors such as: developments in RE technologies and their associated resource bases and costs; comparative attractiveness of competing mitigation options (i.e., end-use energy efficiency, nuclear energy and fossil energy with carbon capture and storage (CCS)); fundamental drivers of energy services demand (including population, economic growth); the ability to integrate variable RE sources into power grids; fossil fuel resources; specific policy approaches to mitigation; and emissions pathways towards long-term goals (e.g., overshoot versus stabilization). However, despite the observed variation, the scenarios indicate that, all else being equal, more ambitious mitigation generally leads to greater deployment of RE.

The majority of the 164 recent scenarios indicate a substantial increase in the deployment of RE by 2030, 2050 and beyond. In 2008, total RE production stood at roughly 64 EJ/yr (12.9% of total primary energy supply) with more than 30 EJ/yr of this being traditional biomass. More than 50% of the scenarios project levels of RE deployment in 2050 of more than 173 EJ/yr reaching up to over 400 EJ/yr in some cases. Given that traditional biomass demand decreases in most scenarios, an increase in the production level of RE (excluding traditional biomass) anywhere from roughly three-fold to more than ten-fold is projected. The global primary energy supply share of RE differs substantially among the scenarios. More than half of the scenarios show a contribution from RE in excess of a 17% share of primary energy supply in 2030, rising to more than 27% in 2050. The scenarios with the highest RE shares reach approximately 43% in 2030 and 77% in 2050. In other words, it is likely that RE will have a significantly larger role (in absolute and relative numbers) in the global energy system in the future than today.

Even without efforts to address climate change RE can be expected to expand. Most baseline scenarios with no assumed climate mitigation policy show RE deployments significantly above the 2008 level of 64 EJ/yr—up to 120 EJ/yr by 2030. By 2050 many baseline scenarios reach RE deployment levels of more than 100 EJ/yr, in some cases up to about 250 EJ/yr. These substantial deployment levels result from a range of assumptions, including, for example, the assumption that energy service demand will continue to grow substantially throughout the century and assumptions about the ability of RE to contribute to increased energy access and the limited long-term availability of fossil resources. Other assumptions (e.g., improved costs and performance of RE technologies) render RE technologies increasingly economically competitive in many applications even in the absence of climate policy.

RE deployment significantly increases in scenarios with low GHG stabilization concentrations. Low GHG stabilization scenarios lead on average to higher RE deployment compared to the baseline. However, for any given long-term GHG concentration goal, the scenarios exhibit a wide range of RE deployment levels (Figure 10.2). In scenarios that stabilize atmospheric CO_2 concentrations at a level of less than 440 ppm, the median RE deployment level in 2050 is 248 EJ/yr (139 EJ/yr in 2030), with the highest levels reaching 428 EJ/yr by 2050.

Many combinations of low-carbon energy supply options and energy efficiency improvements can contribute to given low GHG concentration levels, with RE becoming the dominant low-carbon energy supply option by 2050 in the majority of scenarios. Ambitious GHG concentration stabilization levels lead, on average, to higher RE deployment compared to the baseline, with above 400 EJ/yr by 2050 as the upper limit of RE deployment. Many scenarios were constructed as sensitivities with explicit limits on the deployment of nuclear energy and CCS, and RE played an increasingly important role in these scenarios. Yet even in scenarios with no explicit limits on these competing low-carbon options, RE often represents well over 50% of the global primary energy supply.

Scenarios generally indicate that growth in RE will be widespread around the world. Although the precise distribution of RE deployment across regions substantially varies across scenarios, they are largely consistent in indicating widespread growth in RE deployment around the globe. In addition, scenarios suggest that RE deployment levels will be higher over the long term in the group of non-Annex I countries than in the group of Annex I countries, in part a reflection of the fact that non-Annex I countries are expected to represent an increasing share of total global energy demand over the coming decades.

Scenarios do not indicate an obvious single dominant RE technology at a global level. Besides the aspect that all RE obtains a more important role in the scenarios over time, a general trend is that bioenergy (predominantly modern biomass), wind energy and solar energy are commonly characterized by the largest contributions to the energy system among RE technologies by 2050.

Individual studies indicate that if RE deployment is limited, mitigation costs increase and low GHG stabilization concentrations may not be achieved. A number of studies have pursued scenario sensitivities that assume constraints on the deployment of individual mitigation options, including RE as well as nuclear and fossil energy with CCS. These studies indicate that mitigation costs are higher when options, including RE, are not available, but there is little agreement on the precise magnitude of the increase in costs. They also indicate that more ambitious GHG concentration goals may not be achievable when RE options are not available.

An in-depth analysis of four selected illustrative scenarios from the larger set of 164 scenarios allowed a more detailed look at the possible contribution of specific RE technologies in different regions and sectors. Even within this smaller set, the role of RE varies substantially, in part because the scenarios are aimed at different long-term climate goals, and because they are based on different assumptions about technology costs and also on distinct scenario methodologies.

In the four representative scenarios, the RE-based electricity generation develops most quickly, at least in the medium term, followed by RE for heating/cooling and transport. For RE-based electricity generation, the highest market shares are expected in the analyzed time span. In contrast, currently the heating sector in many regions of the world is one of the most dominant demand sectors. Its RE share is high, especially in non-Annex I countries, but it is mainly based on traditional bioenergy. The total share of RE-based electricity production for the four illustrative scenarios varies for the year 2050 (2030) from 24% (20%) up to 95% (61%) (cf. 19% RE-based electricity share in 2008). The corresponding range for the contribution of RE to the heating sector for these four scenarios lies for the year 2050 (2030) between 21% (20%) and 91% (49%). In most of the scenarios the heating and, particularly, the transport sector are less highlighted, showing that more importance should be given to thermal and transport RE applications in future studies.

Scenarios indicate that overall global technical potentials will not constrain the future contribution of RE. Although deployment of the different RE technologies significantly increases over time, the resulting contribution of RE in the scenarios for most technologies is much lower than their corresponding technical potentials. In the four illustrative scenarios, for instance, despite significant technological and regional differences less than 2.5% of the global available technical RE potential is used. In this sense, scenario results confirm that technical potentials will not be the limiting factors for the expansion of RE on a global scale.

Increasing sectoral shares of RE can substantially contribute to GHG mitigation. The four in-depth analyzed illustrative scenarios span a range of global cumulative CO_2 savings, from about 220 to 560 Gt CO_2 between 2010 and 2050 compared to about 1,530 Gt CO_2 cumulative fossil and industrial CO_2 emissions in the IEA World Energy Outlook 2009 Reference Scenario during the same period. The precise attribution of mitigation potentials to RE not only depends on the role scenarios attribute to specific mitigation technologies, but also on complex systems behaviours and, in particular, on the energy sources that RE displaces. Therefore, attribution of precise mitigation potentials to RE should be viewed with appropriate caution.

Scenarios often do not directly associate mitigation potentials with different technological options. Instead, abatement cost curves are often used to discuss and to compare different mitigation strategies. Abatement cost curves and energy supply curves are an approach that is very often used for discussing mitigation strategies and prioritizing abatement options. One of the most important strengths of this method is that the results can be understood easily and that the outcomes of these methods give, at first glance, a clear orientation as they rank available options in order of cost-effectiveness. On the other hand, abatement cost curves have important limitations. In contrast to scenario analysis, they are not able to reflect the complex system behaviour and corresponding interdependencies. Thus they have to rely on simplified assumptions about the substituted non-RE supply and corresponding emission factors. In general, it is very difficult to compare data and findings from RE abatement cost and supply curves, as there have been very few studies using a comprehensive and consistent approach and detailing their methodologies, and most studies use different assumptions. Many of the regional and country studies provide less than 10% abatement of the baseline CO_2 emissions over the medium term at abatement costs under around USD_{2005} 100/t CO_2. The resulting low-cost abatement potentials are quite low compared to the reported mitigation potentials of many of the scenarios reviewed here.

Some RE technologies are broadly competitive with current market energy prices. Many of the other RE technologies can provide competitive energy services in certain circumstances, for example, in regions with favourable resource conditions or that lack the infrastructure for other low-cost energy supplies. In most regions of the world, however, policy measures are still required to ensure rapid deployment of many RE sources.

In the field of RE, significant opportunities exist to further improve the energy efficiencies, and/or to decrease the costs of producing and installing the respective technologies. Together, these effects are expected to decrease the levelized cost of energy of many innovative RE-sourcing technologies in the future. Over time, energy generation costs of many RE technologies have shown significant declines. In general, historical cost decreases can be described by experience curves with global learning rates (the relationship between the reduction in cost and a doubling of production).

To realize the learning effects and to allow an increase in the competitiveness of RE technologies, upfront investments in deployment, as well as research and development, will be needed, which will result in new market opportunities for RE suppliers. The four illustrative scenarios analyzed in detail in this Special Report estimate global cumulative RE investments (in the power generation sector only) ranging from USD_{2005} 1,360 to 5,100 billion for the decade 2011 to 2020, and from USD_{2005} 1,490 to 7,180 billion for the decade 2021 to 2030. The lower values refer to the IEA World Energy Outlook 2009 Reference Scenario and the higher ones to a scenario that seeks to stabilize atmospheric CO_2 (only) concentration at 450 ppm. The annual averages of these investment needs are all smaller than 1% of the world's gross domestic product (GDP). The average annual investments in the reference scenario are slightly lower than the respective investments reported for 2009. Between 2011 and 2020, the higher end of the range of the annual averages of the RE electricity sector investments approximately correspond to a three-fold increase in the current global investments in this field. For the next decade (2021 to 2030), a five-fold increase is projected.

Increasing the installed capacity of RE power plants will reduce the amount of fossil and nuclear fuels that otherwise would be needed in order to meet a given electricity demand. In addition to investment, operation and maintenance (O&M) and (where applicable) feedstock costs related to RE power plants, any assessment of the overall economic burden that is associated with their application will therefore have to consider avoided fuel and substituted investment costs as well.

Assessments of the costs of future paths of RE deployment and mitigation have to consider the whole range of costs, including external costs and co-benefits. Literature on long-term scenarios does not normally take into consideration external costs (dominated typically by climate change and health impacts due to air pollution) of different energy technologies. Although the uncertainty is relatively high, in most cases RE sources have rather low external costs assessed on a lifecycle basis when compared to fossil fuel-based technologies. Particularly, the external costs of RE-based power generation technologies have most frequently been reported as being lower than those of fossil supply options.

In summary, scenarios strongly indicate that RE will become increasingly important over time, even without but particularly with GHG emissions constraints. However, the resulting contribution of RE in the various studies available in the literature is much lower than their corresponding technical potentials. Moreover, even if substantial growth rates are combined with future RE deployment paths, they are, in general, lower than what has been achieved by the RE industry during the past 10 years.

10.1 Introduction

The evolution of future GHG emissions is highly dependent on various future factors, including, among other things, economic growth, population growth, the associated demand for energy, energy resources and the future costs and performance of energy supply and end use technologies (IPCC, 2007; Chapter 1). Not only must all these different forces be considered when exploring the role of RE in climate mitigation, but also it is not possible to know today with any certainty how these different key forces might evolve decades into the future. Against that background, this chapter discusses the mitigation potentials and costs of RE technologies with a particular focus on a systems perspective and on an explicit consideration of the wide range of ways in which these various forces may evolve and shape the future.

Section 10.2 provides context for understanding the role of RE in climate mitigation through the review of 164 medium- to long-term scenarios from large-scale, integrated models. The review explores the range of global RE deployment levels emerging in recent scenarios and identifies some of the key forces that drive the variation among them. It does so at the scale of RE as a whole, but also in the context of individual RE technologies. The review highlights the importance of interactions and competition with other mitigation technologies as well as the evolution of energy demand more generally. Section 10.2 also considers the linkage between RE and mitigation costs in scenarios, and ends with a discussion, gleaned from Chapters 2 through 7, of the factors that might influence the ability to meet the deployment levels achieved in scenarios (e.g., technology and economic aspects).

Section 10.3 complements the large-scale review with a more detailed review using 4 of the 164 scenarios as illustrative examples. The four scenarios span a range from a more baseline-oriented future development of RE to optimistic expectations about RE's future, and cover different GHG stabilization levels and underlying modelling methodologies. This section provides a next level of detail for exploring the role of RE in climate change mitigation. Section 10.3 provides the details of particular futures, giving more minute treatment to the regional and sectoral (e.g., power generation, heating, cooling, transport) character of RE deployment. Within this more detailed context, it considers such issues as required generation capacity, annual growth rates and estimates of the corresponding mitigation potentials of RE deployment. Additionally, and as another perspective on scenario results, Section 10.3 uses the methodology of supply cost curves to give a sense of how RE technologies are deployed in the four scenarios as a function of costs.

In this context, particularly for comparing RE with non-RE technologies or even biomass with other RE technologies, it is important to note that the direct equivalent method is used to calculate primary energy in this chapter and throughout this report. In comparison to other conventions, this approach tends to indicate lower primary energy shares for RE than other primary equivalent approaches (see Box 1.1 in Chapter 1 for further details).

Section 10.4 provides a more general discussion about cost curves. It starts with an assessment of the strengths and shortcomings of supply curves for RE and GHG mitigation, and then reviews the existing literature on regional RE supply curves, as well as abatement cost curves, as they pertain to mitigation using RE sources. The second part of the section includes a summary of technology-specific supply and cost curves, including consideration of uncertainty.

Section 10.5 addresses the costs of RE commercialization and deployment. It reviews current RE technology costs, as well as expectations about how these costs might evolve into the future. Learning by research (triggered by research and development (R&D) expenditures) and learning by doing (fostered by capacity expansion programs) might result in a considerable long-term decline in RE technology costs. The section, therefore, presents historic data on R&D funding as well as on observed learning rates. In order to allow an assessment of future market volumes and investment needs, investments in RE are discussed in particular with respect to what is required if ambitious climate protection goals are to be achieved, and compared with investment needs in RE following more or less a baseline pathway. To provide a consistent thread throughout the chapter, the discussion of investment needs is based on the four illustrative scenarios that are explored in Section 10.3.

Finally, Section 10.6 expands the consideration of cost beyond standard measures of technology and mitigation costs. It synthesizes and discusses social and environmental costs and benefits from increased deployment of RE in relation to climate change mitigation and sustainable development; costs that are often not considered in scenarios, but are important for an overall assessment of different future paths. It builds on the discussions in Chapter 9, but it is more focused on economic aspects.

Gaps in knowledge and uncertainties associated with RE technical potentials and costs are discussed at the end of each of the sections of the chapter.

The following guiding questions were used to structure the development of insights and themes:

- What roles are RE sources likely to play in the future and particularly in contributing to GHG-mitigation pathways?

- What factors influence the possible deployment of RE sources in meeting GHG mitigation pathways (e.g., energy demand, cost and performance, competing mitigation options, barriers, social factors, co-benefits, policies)?

- What is the resulting role of RE regarding specific RE technologies, demand sectors and regions?

- How do possible RE deployment paths from the literature mesh with the technical potentials at global and regional levels?

- What are the costs of RE commercialization and deployment and what are the resulting investment needs for RE deployment?

- To what extent are the non-market costs and benefits relevant for social and environmental factors?

- How uncertain are the possible answers to all these questions, and what are the robust findings despite all uncertainties involved?

10.2 Synthesis of mitigation scenarios for different renewable energy strategies

This section reviews 164 recent medium- to long-term scenarios from 16 global energy-economic and integrated assessment models. These scenarios are among the most sophisticated explorations of how the future might evolve to address climate change; as such, they provide a window into current understanding of the role of RE technologies in climate mitigation.

The discussion in this section is motivated primarily by three strategic questions. First, what RE deployment levels are consistent with different CO_2 concentration goals; or, put another way, what is the linkage between CO_2 concentration goals and RE deployments? Second, over what time frames and where will RE deployments occur and how might that differ by RE technology? Third, how do the costs of mitigation relate to RE deployments and the availability, cost and performance of RE?

(Note that Sections 10.2.1 and 10.2.2 rely heavily on, and largely follow, Krey and Clarke (2011), in terms of both analysis and discussion. Krey and Clarke's (2011) publication was produced in parallel with this report. It provides a more thorough and extensive review and discussion of the methodology and results of an analysis of 162 of the 164 scenarios reviewed in this section.)

10.2.1 State of scenario analysis

10.2.1.1 Types of scenario methods

The climate change mitigation scenario literature largely consists of two distinct approaches to scenario development: quantitative modelling and qualitative narratives (see Morita et al., 2001; Fisher et al., 2007 for a more extensive review). Several attempts have also been made to integrate narratives and quantitative modelling approaches (IPCC, 2000; Morita et al., 2001; Carpenter et al., 2005). The review in this section exclusively relies on scenarios developed through quantitative modelling. These scenarios provide estimates of RE deployments and other important parameters for understanding the role of RE in climate mitigation, and they do so based on models that follow a systems approach and thus explicitly and formally represent the interactions between RE technologies, other mitigation technologies and the various other factors that influence the characteristics of mitigation.

Although all of the scenarios in this review were developed using quantitative modelling, it is important to observe that there is enormous variation in the detail and structure of the models used to construct the scenarios. Many authors have, in the past, attempted to categorize models as either bottom-up or top-down. For several reasons (see Box 10.1), this review will not rely on the top-down/bottom-up taxonomy. Instead, the models are referred to generically as large-scale, integrated models. The important methodological characteristics of the scenarios reviewed in this section, and the models used to generate them, are: (1) they take an integrated view of the energy system so that they can

Box 10.1 | Moving beyond top-down versus bottom-up?

In previous IPCC reports (e.g., Herzog et al., 2005; Barker et al., 2007), quantitative scenario modelling approaches were broadly separated into two groups: top-down and bottom-up. Although this classification may have made sense in the past, recent developments make it decreasingly appropriate. Most importantly, (i) the transition between the two categories is continuous, and (ii) many models, although rooted in one of the two traditions (e.g., macro-economic or energy-engineering models), incorporate important aspects of the other approach and thus belong to the class of so-called hybrid models (Hourcade et al., 2006; van Vuuren et al., 2009).

In addition, the terms top-down and bottom-up can be misleading, because they are context dependent and used differently in different scientific communities. For example, in previous IPCC assessments, all integrated modelling approaches were classified as top-down models regardless of whether they included significant technology information (van Vuuren et al., 2009). In the energy-economic modelling community, macro-economic approaches are traditionally classified as top-down models and energy-engineering models as bottom-up. However, in engineering sciences, even the more detailed energy-engineering models that represent individual technologies such as power plants, but essentially treat them as 'black boxes', are characterized as top-down models because they do not assume a component-based view, which would be considered bottom-up. For these reasons, the modelling tools used to generate scenarios in this review are simply referred to as large-scale, integrated models.

capture the interactions, at least at an aggregate scale, between competing energy technologies; (2) they have a basis in economics in the sense that decision making is largely based on economic criteria; (3) they are long-term and global in scale, but with some regional detail; (4) they include the policy levers necessary to meet emissions outcomes; and (5) they have sufficient technology detail to explore RE deployment levels at both regional and global scales. Many also have integrated views beyond the energy system, for example, fully coupled models of agriculture and land use.

10.2.1.2 Strengths and weaknesses of quantitative scenarios

Scenarios are a tool for understanding, but not predicting, the future. They provide a plausible description of how the future may develop based on a coherent and internally consistent set of assumptions about key driving forces (e.g., rate of technological change, prices) and relationships (IPCC, 2007). In the context of this report, scenarios are thus a means to explore the potential contribution of RE to future energy supplies and to identify the drivers of renewable deployment.

The benefit of scenarios generated using large-scale, integrated models, such as those reviewed in this section, is that they capture many of the key interactions with other technologies (including competing mitigation technologies such as fossil energy with CCS, nuclear energy and demand reduction options), other parts of the energy system, other relevant human systems (e.g., agriculture, the economy as a whole) and important physical processes associated with climate change (e.g., the carbon cycle), that serve as the environment in which RE technologies will be deployed. This integration provides an important degree of internal consistency. In addition, they explore these interactions over at least several decades to a full century into the future and at a global scale. This degree of spatial and temporal coverage is crucial for establishing the strategic context for RE.

The design, assumptions and focus of the scenarios covered in this assessment vary greatly: some are based on a more detailed representation of individual renewable and other energy technologies and aspects of systems integration of RE, while others focus on the implications of RE deployment for the economy as a whole. This variation in methods, assumptions and focus provides a window into the deep uncertainties associated with future dynamics of the energy system and the role of RE sources in climate change mitigation.

As discussed in Krey and Clarke (2011), two important caveats must be kept in mind when interpreting the scenarios in this section. First, maintaining a global, long-term, integrated perspective involves tradeoffs in terms of detail. For example, the models do not represent all the forces that govern decision making at the national or even the company or individual scale, in particular in the short term. Further, these are not power system models or engineering models, and they therefore employ stylized representations of many details that influence the performance and deployment of RE, for example, the challenges of incorporating variable electricity generation into the electric grid. The level of sophistication in representing these details varies substantially across models. An outcome of these simplifications is that integrated global and regional scenarios are most useful for the medium- to long-term outlook, say 2020 onwards. For shorter time horizons, tools such as market outlooks or short-term national analyses that explicitly address all existing policies and regulations are more suitable sources of information.

Second, the scenarios do not represent a random sample of possible scenarios that could be used for formal uncertainty analysis. They were developed for different purposes and are not a set of 'best guesses'. Many of the scenarios represent sensitivities, particularly along the dimensions of future technology availability and the timing of international action on climate change, and are therefore related to one another. Some modelling groups provided substantially more scenarios than others. In scenario ensemble analyses based on collecting scenarios from different studies, such as the review here, there is a constant tension between the fact that the scenarios are not truly a random sample and the sense that the variation in the scenarios does still provide real and often clear insights into our collective lack of knowledge about the future.

10.2.2 The role of renewable energy sources in scenarios

10.2.2.1 Overview of the scenarios reviewed in this section

The 164 scenarios reviewed in this section were collected through an open call to modellers for RE data from recently published scenarios. All scenarios that were submitted were included in the review. The bulk of the scenarios in this assessment (see Table 10.1) come from three coordinated, multi-model studies: the Energy Modeling Forum (EMF) 22 international scenarios (Clarke et al., 2009), the Adaptation and Mitigation Strategies (ADAM) project (Knopf et al., 2009; Edenhofer et al., 2010) and the Report on Energy and Climate Policy in Europe (RECIPE) comparison (Edenhofer et al., 2009; Luderer et al., 2009). These three exercises harmonize some scenario dimensions, such as baseline assumptions or climate policies, across the participating models. The remaining scenarios come from individual publications. Although the 164 scenarios are clearly not exhaustive of recent literature, nor do they represent a truly random sample, the set is large and extensive enough to provide robust insights into current understanding of the role of RE in climate change mitigation.

The full set of scenarios covers a large range of CO_2 concentrations (350 to 1,050 ppm atmospheric CO_2 concentration by 2100, see Table 10.2), representing both mitigation and no-policy, or baseline, scenarios. The full set of scenarios also covers the time horizon 2050 to 2100, and all of the scenarios are global in scope.

There are several characteristics of the scenarios included in this review that make them particularly valuable for this discussion. First, they come from the most recent work of the integrated modelling community;

all of the scenarios in this study were published during or after 2006. The scenarios therefore reflect the most recent understanding of key underlying parameters and the most up-to-date representations of the dynamics of the underlying human and Earth systems. The scenarios are also valuable in that they include a relatively large number of scenarios that represent less optimistic views on international action to deal with climate change (second-best policy) or address consequences of limited technology portfolios (constrained technology). The assumptions regarding second-best policy vary considerably across the scenarios, but are mostly taken from the EMF 22 study (Clarke et al., 2009) and the RECIPE project (Edenhofer et al., 2009; Luderer et al., 2009) and capture delayed action by developing countries. Technology availability is not defined homogenously across all scenarios in the analyzed set, but the limited technology portfolio studies that are highlighted here are those with limitations on the deployment of fossil energy with CCS, nuclear energy and RE. Finally, data regarding RE deployment were collected at a level of detail beyond that found in most published papers or existing scenario databases, for example, those compiled for IPCC reports (Morita et al., 2001; Hanaoka et al., 2006; Nakicenovic et al., 2006). Whereas RE deployment information was often collected in the past in terms simply of bioenergy and non-biomass renewable sources, the data reviewed here explicitly include information on the deployment of wind energy, solar energy, bioenergy, geothermal energy, hydroelectric power and ocean energy.

10.2.2.2 Overview of the role of renewable energy in the scenarios

A fundamental question relating to the role of RE in climate mitigation is how closely correlated are RE deployment levels and long-term climate concentration or related climate goals. As background to understanding the relationship of RE deployments to climate goals, it is important to first observe that, consistent with past scenario literature (Fisher et al., 2007), there is a strong correlation between fossil and industrial CO_2 emissions pathways and long-term CO_2 concentration goals across the

Table 10.1 | Energy-economic and integrated assessment models considered in this analysis. The total number of scenarios per model varies significantly. Scenarios are further classified by the inclusion of delayed participation in mitigation (second-best policy) and constraints on and/or variations in the deployment of fossil energy with CCS, nuclear energy and RE (constrained technology). Adapted from Krey and Clarke (2011), modified to include IEA (2009) and Teske et al. (2010).

Model	Number of scenarios	Baseline scenarios	Policy Scenarios				Comparison project	Citation
			First best	Constrained technology[1]	Second-best policy	Constrained technology & second-best policy		
AIM/CGE	3	1	1	0	1	0	—	Masui et al. (2010)
DNE21	7	1	3	3	0	0	—	Akimoto et al. (2008)
GRAPE	2	1	1	0	0	0	—	Kurosawa (2006)
GTEM	7	1	4	0	2	0	EMF 22	Gurney et al. (2009)
IEA-ETP	3	1	2	0	0	0	—	IEA (2008b)
IEA-WEM	1	1	0	0	0	0	—	IEA (2009); extension to 2050, Teske et al. (2010)
IMACLIM	8	1	2	4	1	0	RECIPE	Luderer et al. (2009)
IMAGE	17	3	5	6	0	3	EMF 22 / ADAM	van Vuuren et al. (2007, 2010); van Vliet et al. (2009)
MERGE-ETL	19	4	3	12	0	0	ADAM	Magne et al. (2010)
MESAP/PlaNet	2	0	0	2	0	0	—	Krewitt et al. (2009); Teske et al. (2010)
MESSAGE	15	2	4	7	2	0	EMF 22	Riahi et al. (2007); Krey and Riahi (2009)
MiniCAM	15	1	5	4	3	2	EMF 22	Calvin et al. (2009)
POLES	15	4	3	8	0	0	ADAM	Kitous et al. (2010)
ReMIND	28	4	6	14	4	0	ADAM / RECIPE	Luderer et al. (2009); Leimbach et al. (2010)
TIAM	10	1	5	0	4	0	EMF 22	Loulou et al. (2009)
WITCH	12	1	4	4	3	0	EMF 22 / RECIPE	Bosetti et al. (2009); Luderer et al. (2009)
TOTAL	164	27	48	64	20	5		

Note: 1. While in the vast majority of constrained technology scenarios, the deployment of individual technologies or technology clusters has actually been constrained, in a few cases included under this category, the potential for bioenergy was expanded compared to the model's default assumption.

Table 10.2 | Categorization of the 164 scenarios reviewed in this section based on CO_2 concentration levels in 2100, the inclusion of delayed participation in mitigation (second-best policy), and constraints on and/or variations in the deployment of fossil energy with CCS, nuclear energy and RE. The CO_2 concentration categories are defined consistently with those in the IPCC Fourth Assessment Report (AR4), WGIII (Fisher et al., 2007). Note that Categories V and above are not included here and Category IV is extended to 600 ppm from 570 ppm, because all stabilization scenarios lie below 600 ppm CO_2 in 2100 and because the lowest baseline scenarios reach concentration levels of slightly more than 600 ppm by 2100.[1] Data adapted from Krey and Clarke (2011) modified to include two additional scenarios.

	CO_2 concentration by 2100 (ppm)	Number of scenarios	Policy Scenarios			
			First-best	Constrained technology	Second-best policy	Constrained technology & second-best policy
Baselines	>600	27	—	—	—	—
Category IV	485–600	32	11	13	6	2
Category III	440–485	63	20	29	11	3
Category II	400–440	14	7	6	1	0
Category I	<400	28	10	16	2	0

Note: 1. This definition of CO_2 concentration stabilization categories is consistent with that used in the AR4. Section 3.3.5 in Fisher et al. (2007) explains that most scenarios assessed in the AR4 stabilize concentrations between 2100 and 2150 while the definition used here is based on CO_2 concentrations in 2100. Stabilization after 2100 is typically relevant for scenarios with high CO_2 concentration targets, that is, Categories V and higher, which have not been assessed here and for very low stabilization scenarios in Category I that show a temporary overshoot in concentrations before reaching the final target. The latter does not influence the assignment to categories, since Category I is not bounded from below. In addition, it should be noted that CO_2 concentrations are affected by assumptions about the carbon cycle that may result in differences across models.

scenarios (Figure 10.1, as depicted by close grouping of the coloured categories). An important reason for this correlation is similarity across scenarios in assumptions regarding the key physical processes underlying the global carbon cycle. Any variation in emissions pathways reflects remaining differences in assumptions about the carbon cycle as well as assumptions regarding factors that determine the allocation of emissions over time in mitigation scenarios. This includes the rate of technological improvements, underlying drivers of emissions in general such as economic growth, and methodological approaches for allocating emissions over time, including discount rates and the choice of overshoot and not-to-exceed pathways.

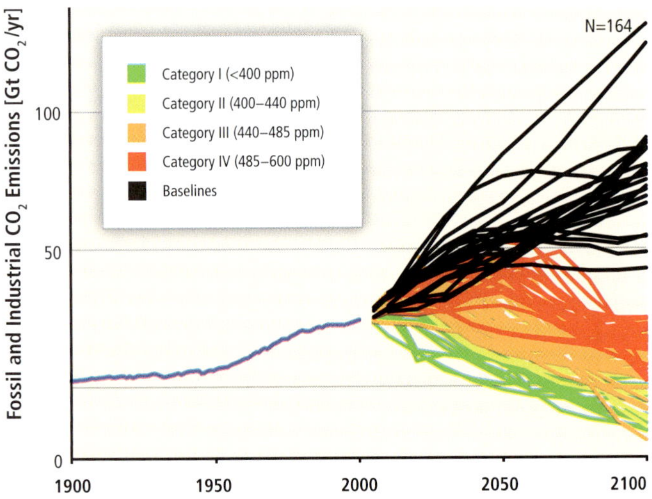

The relationship between RE deployment and CO_2 concentration goals is far less robust (Figure 10.2). On the one hand, RE deployment is generally increasing with the stringency of the CO_2 concentration goal, particularly several decades into the future and beyond. In other words, all other things being equal, more stringent CO_2 concentration goals will generally lead to larger RE deployment. At the same time, there is enormous variation among RE deployment levels for any CO_2 concentration goal. This variation is a reflection of uncertainty regarding the precise role that RE might play in climate mitigation, illustrating a lack of consensus among scenario developers as to what degree of RE deployment would be associated with any particular climate goal.

At the same time, it is also important to note that despite the variation, the absolute magnitudes of RE deployment are dramatically higher than those of today in the vast majority of the scenarios. In 2008, global renewable primary energy supply in direct equivalent stood at 63.6 EJ/yr (IEA, 2010d),[1] with more than 30 EJ/yr of this being traditional biomass. In contrast, by 2030 many scenarios indicate a doubling of RE deployment or more compared to today, and this is accompanied in most scenarios by a reduction in traditional biomass, implying substantial growth in modern sources. By 2050, RE deployment levels in most scenarios are higher than 100 EJ/yr (median at 173 EJ/yr), reach 200 EJ/yr in many of the scenarios and more than 400 EJ/yr in some cases. Given that traditional biomass use decreases in most scenarios, the scenarios represent an increase in RE production (excluding traditional biomass) of anywhere from roughly three- to more than ten-fold. Similarly, the global primary energy supply share of RE differs substantially among

Figure 10.1 | Historic global fossil and industrial CO_2 emissions and projections from 164 long-term scenarios. Colour coding is based on categories of atmospheric CO_2 concentration in 2100 as defined in the IPCC AR4, WGIII (Fisher et al., 2007), with historic emission data from Nakicenovic et al. (2006). Figure and data adapted from Krey and Clarke (2011), modified to include two additional scenarios.

1 Note that there is a small difference from the value of 65.6 EJ published by the IEA (and shown in Figure 8.2) due to the different primary energy accounting methods used. See Box 1.1 in Chapter 1, Section 1.2.1 and Appendix A.II.4 for additional background on this topic.

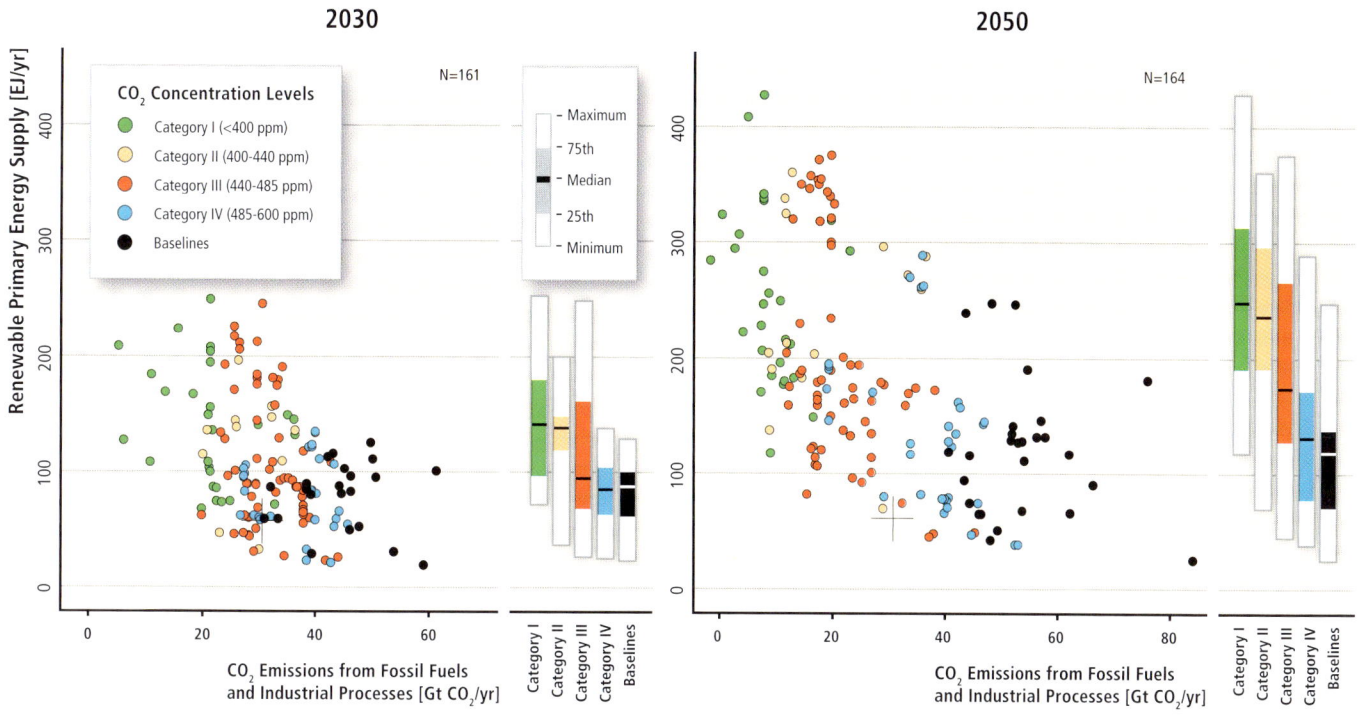

Figure 10.2 | Global RE primary energy supply (direct equivalent) from 164 long-term scenarios versus fossil and industrial CO_2 emissions in 2030 and 2050. Colour coding is based on categories of atmospheric CO_2 concentration level in 2100 (Fisher et al., 2007). The panels to the right of the scatterplots show the deployment levels of RE in each of the atmospheric CO_2 concentration categories. The thick black line corresponds to the median, the coloured box corresponds to the inter-quartile range (25th to 75th percentile) and the ends of the white surrounding bars correspond to the total range across all reviewed scenarios. The crossed-lines show the relationship in 2007. Pearson's correlation coefficients for the two data sets are -0.40 (2030) and -0.55 (2050). For data reporting reasons only, 161 scenarios are included in the 2030 results shown here, as opposed to the full set of 164 scenarios. RE deployment levels below those of today are a result both of model output as well as differences in the reporting of traditional biomass. Figure and data adapted from Krey and Clarke (2011), modified to include two additional scenarios.

the scenarios. More than half of the scenarios show a contribution of RE in excess of a 17% share of primary energy supply in 2030, rising to more than 27% in 2050. The scenarios with the highest RE shares reach approximately 43% in 2030 and 77% in 2050. RE deployment levels in 2100 are substantially larger than these, reflecting continued growth throughout the century.

Indeed, RE deployment is quite large in many of the baseline scenarios; that is, scenarios without any explicit climate policy. By 2030, RE deployment levels of up to about 120 EJ/yr are projected, with many baseline scenarios reaching more than 100 EJ/yr in 2050 and in some cases up to 250 EJ/yr. These large RE baseline deployments result directly from the assumption that energy consumption will continue to grow substantially throughout the century and assumptions that render RE technologies economically competitive in many applications absent climate policy.

10.2.2.3 Setting the scale of renewable energy deployment: Energy system growth and long-term climate goals

Section 10.2.2.2 demonstrated the large variation in RE deployment levels across scenarios for a given CO_2 concentration goal. This section explores the variation primarily through the lens of energy system growth. Section 10.2.2.4 then explores the competition with other low-carbon energy supply sources.

A first step in unpacking the variation in RE deployment levels is to note that there is only a weak correlation between primary energy consumption and long-term climate goals across the 164 scenarios (Figure 10.2). For example, in scenarios that stabilize atmospheric CO_2 concentrations at a level of less than 440 ppm (Categories I and II), the median RE deployment levels are 139 EJ/yr in 2030 and 248 EJ/yr in 2050, with the highest levels reaching 252 EJ/yr in 2030 and up to 428 EJ/yr in 2050. These levels are considerably higher than the corresponding RE deployment levels in baseline scenarios, while it has to be acknowledged that the range of RE deployment in each of the CO_2 stabilization categories is wide. Although, all other things being equal, CO_2 mitigation puts downward pressure on total global energy consumption,[2] the magnitude of this effect is highly varied across scenarios, and often small enough so that there is far less correlation in the scenarios between total primary energy consumption and long-term climate goals (Figure 10.3) than there is for CO_2 emissions and long-term climate goals (Figure 10.1). In other words, the effect of mitigation on primary energy consumption is variable across models and scenarios. In addition, variation in primary energy consumption under mitigation is heavily influenced by variation in assumptions about the fundamental drivers of energy consumption, such as economic growth and associated demand for energy services, that drive baseline primary energy consumption. The variation results from

---

2 Note that this is not always true. Scenarios exist in which primary energy increases because of large-scale electrification in response to climate policy (see, e.g., Loulou et al., 2009).

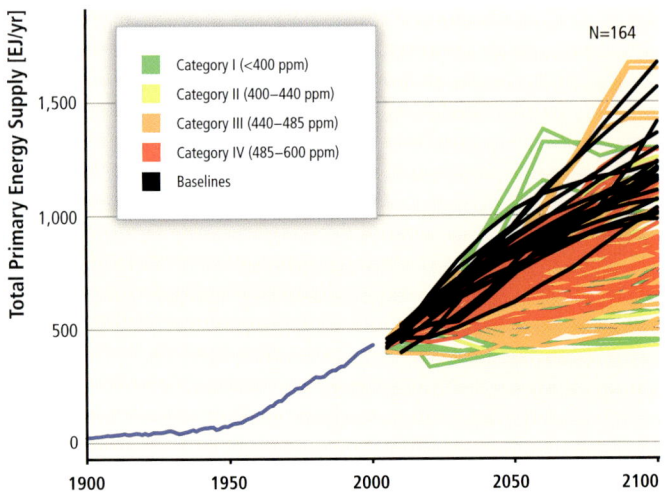

Figure 10.3 | Historic global total primary energy supply (direct equivalent) and projections from 164 long-term scenarios. Colour coding is based on categories of atmospheric CO_2 concentration level in 2100 (Fisher et al., 2007), with historic data from Grubler (2008). Figure and data adapted from Krey and Clarke (2011), modified to include two additional scenarios.

notably the global carbon cycle, put bounds on the levels of CO_2 emissions that are associated with meeting any particular long-term goal; this, in turn, bounds the amount of energy that can be produced from freely-emitting fossil energy sources. Factors leading to remaining variation in freely-emitting fossil energy associated with a given level of CO_2 emissions include the ability to switch between fossil sources with different carbon contents (e.g., natural gas has a lower carbon content than coal per unit of energy) and the potential to achieve negative emissions by utilizing bioenergy with CCS (see Section 2.6.3.3) or forest sink enhancements. The relationship between CO_2 emissions and long-term goals is influenced by differences in the time path of emissions reductions over time as a result of differing underlying model structures, assumptions about technology and emissions drivers, and representations of physical systems such as the carbon cycle.

RE is only one of three major low-carbon supply options. The other two options are nuclear energy and fossil energy with CCS. The demand for low-carbon energy (the total of all three) is, in the context of the discussion here, simply the difference between total primary energy demand and the production of freely-emitting fossil energy (see Figure 10.5). That is to say, whatever energy cannot be supplied from freely-emitting fossil energy because of climate constraints must be supplied either by low-carbon energy or by measures that reduce energy consumption. Given, as discussed above, that the demand response from mitigation is swamped by variability in demand more generally across a scenario set such as the one explored here, the result is that although there is a strong correlation between the CO_2 concentration goal and low-carbon energy (see also Clarke et al., 2009; O'Neill et al., 2010), there is still

the lack of consensus about these fundamental drivers; these are forces that simply cannot be understood with any degree of certainty today.

In contrast to the variation in total primary energy, the production of freely-emitting fossil energy (fossil sources without CCS) is tightly constrained by CO_2 emissions at any point in time (Figure 10.4). Meeting long-term climate goals requires a reduction in the CO_2 emissions from energy and other anthropogenic sources. Important Earth systems, most

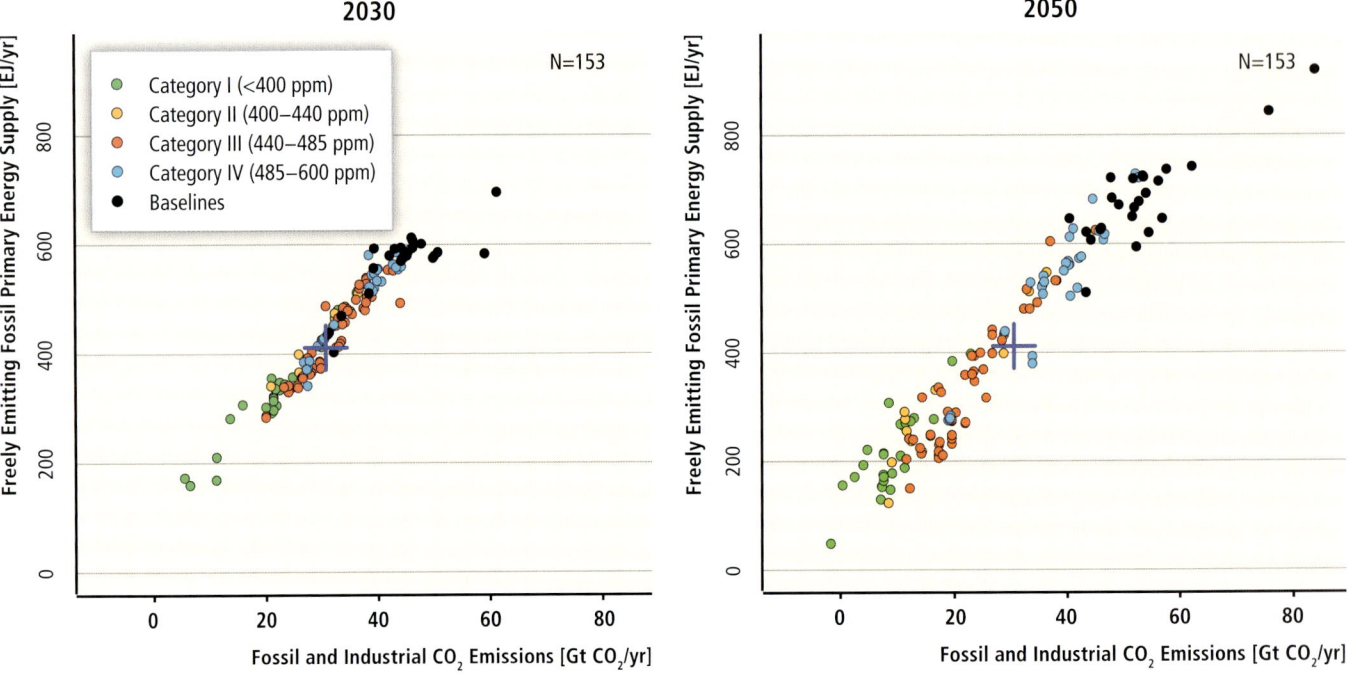

Figure 10.4 | Global freely-emitting fossil primary energy supply (direct equivalent) from 164 long-term scenarios by 2030 and 2050 as a function of fossil and industrial CO_2 emissions. Colour coding is based on categories of atmospheric CO_2 concentration level in 2100 (Fisher et al., 2007). The blue crossed lines show the relationship in 2007. Pearson's correlation coefficients for the two data sets are 0.96 (2030) and 0.97 (2050). For data reporting reasons only 153 scenarios are included in the 2030 and 2050 results shown here, as opposed to the full set of 164 scenarios. Figure and data adapted from Krey and Clarke (2011), modified to include two additional scenarios.

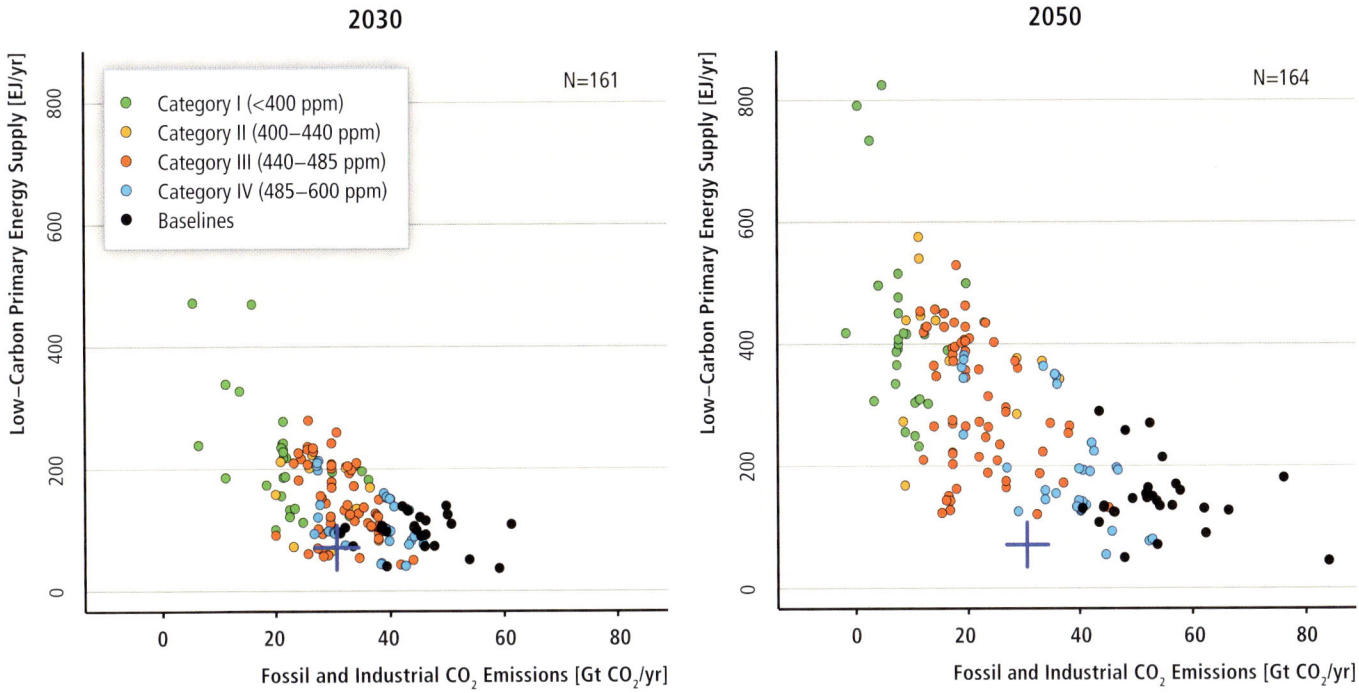

Figure 10.5 | Global low-carbon primary energy supply (direct equivalent) in 164 long-term scenarios by 2030 and 2050 as a function of fossil and industrial CO_2 emissions. Low-carbon energy refers to energy from RE, fossil energy with CCS, and nuclear energy. Colour coding is based on categories of atmospheric CO_2 concentration level in 2100 (Fisher et al., 2007). The blue crossed lines show the relationship in 2007. Pearson's correlation coefficients for the two data sets are -0.60 (2030) and -0.68 (2050). For data reporting reasons, only 161 scenarios are included in the 2030 results shown here, as opposed to the full set of 164 scenarios. Figure and data adapted from Krey and Clarke (2011), modified to include two additional scenarios.

substantial variability in low-carbon energy for any given CO_2 concentration goal. The competition between RE, nuclear energy and fossil energy with CCS then adds another layer of variability in the relationship between RE deployment and CO_2 concentration goal (Figure 10.2).

10.2.2.4 Competition between renewable energy sources and other forms of low-carbon energy

This section addresses the competition between RE and the two other low-carbon supply options: nuclear energy and fossil energy with CCS. Many of the 164 scenarios are characterized by explicit limits on the deployment of one or both of these two options. The constrained CCS scenarios simply excluded the option to install CCS either on new or existing power plants or other energy conversion facilities with fossil or bioenergy as an input (e.g., refining). The constrained nuclear energy scenarios take on three forms. Two approaches maintain nuclear deployments at or below today's levels, allowing existing power plants to retire over time and not allowing any new installations, or maintain the total deployment of nuclear at current levels, which might reflect either lifetime extensions or just enough new installations to counteract retirements. A third option applied in a number of scenarios is to maintain nuclear deployment over time in mitigation scenarios at baseline levels. The difficulty in interpreting this third category of scenarios is that nuclear energy expands to substantially different degrees across baseline scenarios, limiting comparability (see caption of Figure 10.6 for details).

All other things being equal, when competing options are not available or are otherwise constrained, RE deployments are higher (Figure 10.6). Two effects simultaneously contribute to the increase in the renewable primary energy share. First, with fewer competing options, RE will constitute a larger share of low-carbon energy. Second, higher mitigation costs resulting from the lack of options put downward pressure on total energy consumption, because end-use options become increasingly economically attractive. The relative influence of these two forces varies across models.

At the same time, it is important to reemphasize that technology competition is only one factor influencing RE deployment levels; it cannot by itself explain the variation in RE deployments associated with different mitigation levels. The discussion to this point should make clear that for any mitigation level, the fundamental drivers of energy demand—economic growth, population growth, energy intensity of economic growth and energy end-use improvements—along with the technology characteristics of RE technologies themselves are equally critical drivers of RE deployments. Nonetheless, if environmental, social or national security barriers largely inhibit *both* fossil energy with CCS and nuclear energy, then it is appropriate to assume that RE will be required to provide the bulk of low-carbon energy (Figure 10.7). Independent of the availability of these non-renewable low-carbon energy supply options, the majority of scenarios relies to a greater extent on RE sources than on nuclear energy and fossil energy with CCS to provide low-carbon energy by 2050 (see upper left triangle of Figure 10.7). If only one of these options is limited, then the RE deployment

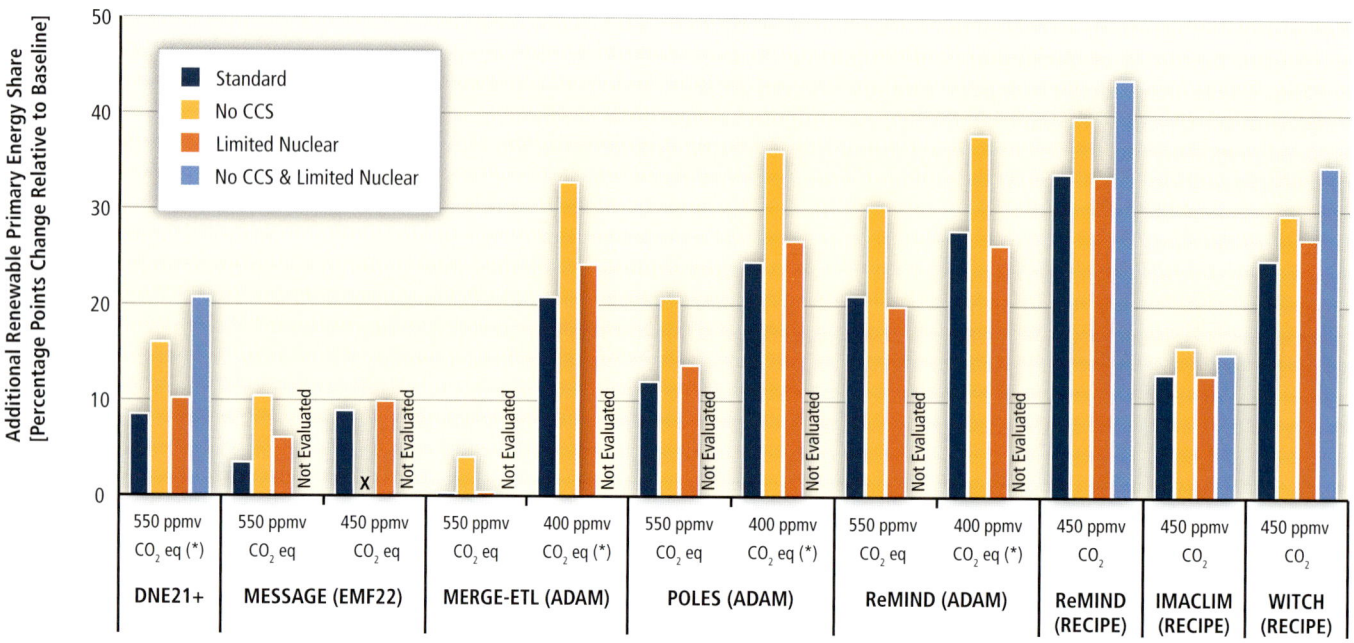

Figure 10.6 | Increase in global renewable primary energy share (direct equivalent) in 2050 in selected constrained technology scenarios compared to the respective baseline scenarios. The 'X' indicates that the respective concentration level for the scenario was not achieved. The definition of 'Limited Nuclear' and 'No CCS' cases varies across models. The DNE21+, MERGE-ETL and POLES scenarios represent nuclear phase-outs at different speeds; the MESSAGE scenarios limit the deployment to 2010; and the ReMIND, IMACLIM and WITCH scenarios limit nuclear energy to the contribution in the respective baseline scenarios, which can still imply a significant expansion compared to current deployment levels. The REMIND (ADAM) 400 ppm no CCS scenario refers to a scenario in which cumulative CO_2 storage is constrained to 120 Gt CO_2. The MERGE-ETL 400 ppm no CCS case allows cumulative CO_2 storage of about 720 Gt CO_2. The POLES 400 ppm CO_2eq no CCS scenario was infeasible and therefore the respective concentration level of the scenario shown here was relaxed by approximately 50 ppm CO_2. The DNE21+ scenario is approximated at 550 ppm CO_2eq based on emissions pathways through 2050. Figure adapted from Krey and Clarke (2011).

proportions of low-carbon energy are generally higher than they would otherwise be, but the degree of this effect is dependent on the ability of the other of these options to take up the slack in lieu of RE. In many modelling paradigms, fossil energy with CCS and nuclear energy are assumed to be close substitutes for the production of baseload electricity production. When one is not available, the majority of the generation it would have provided is provided instead by the other rather than by RE sources, because solar, wave and wind energy are variable. At the same time, it is important to note that reservoir hydropower, bioenergy and geothermal energy can be dispatchable base load (Section 8.2.1).

A fundamental question raised by limited technology scenarios is whether one or more energy supply options are 'necessary' this century to meet low stabilization goals; that is, could the goal still be met if these technologies were not available. One way to explore this issue is to identify scenarios that were attempted with limited technology, but that could not be produced by the associated models. These attempts give a sense of the difficulty of meeting stabilization goals with limited technology options, although, in most cases, they cannot truly be considered as indications of physical feasibility (Clarke et al., 2009). These attempted scenarios tell a mixed story. In some cases, models could not achieve stabilization without nuclear and CCS; however, in others, models were able to produce these scenarios (Figure 10.6). Several studies found that limits on RE deployments kept models from achieving stabilization goals (see, e.g., Figure 10.11). Other studies have indicated that it is the combination of RE, in the form of bioenergy, with CCS that makes low stabilization goals substantially easier through negative emissions (Azar et al., 2006; van Vuuren et al., 2007; Clarke et al., 2009; Edenhofer et al., 2010; Tavoni and Tol, 2010).

10.2.2.5 Renewable energy deployment by technology, over time and by region

There is great variation in the deployment characteristics of individual technologies (Figures 10.8 and 10.9). Several dimensions of this variation bear mention. First, the absolute scales of deployments vary considerably among technologies. Bioenergy, wind and solar energy generally show higher incremental deployment levels than hydropower and geothermal energy, although the variation is large enough that there are clearly scenarios with minimal penetration of wind and solar relative to hydropower and geothermal energy. Ocean energy is currently only represented in very few scenarios and will therefore not be discussed here (see also Section 10.2.4). Further, deployment magnitudes are characterized by greater variation for some technologies relative to others. For example, variation in hydroelectric deployment is far less than in geothermal deployment. The high deployment scenarios for geothermal energy probably assume competitive electricity from enhanced geothermal systems and/or wide application of geothermal heat pumps (see Sections 4.2 and 4.8). It is important to use some caution in interpreting the bioenergy numbers in Figures 10.8 and 10.9 relative to those associated with the other renewable energy technologies. This analysis is being conducted using the direct equivalent

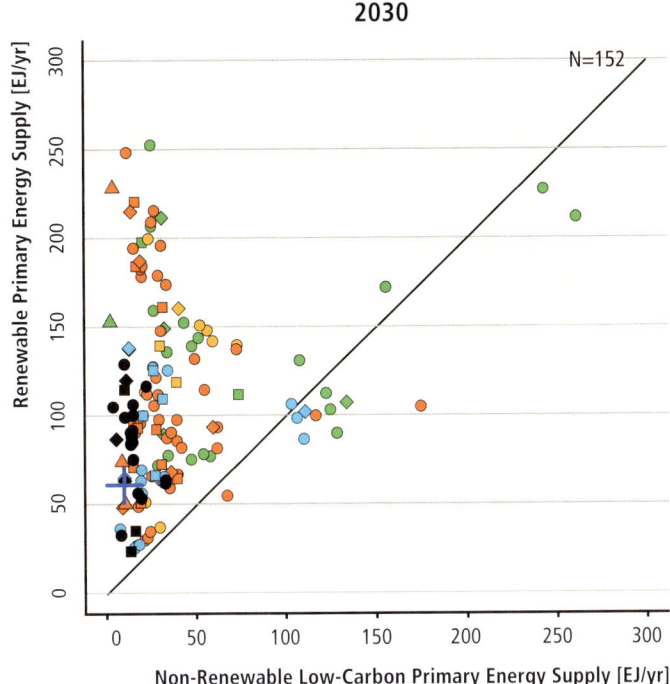

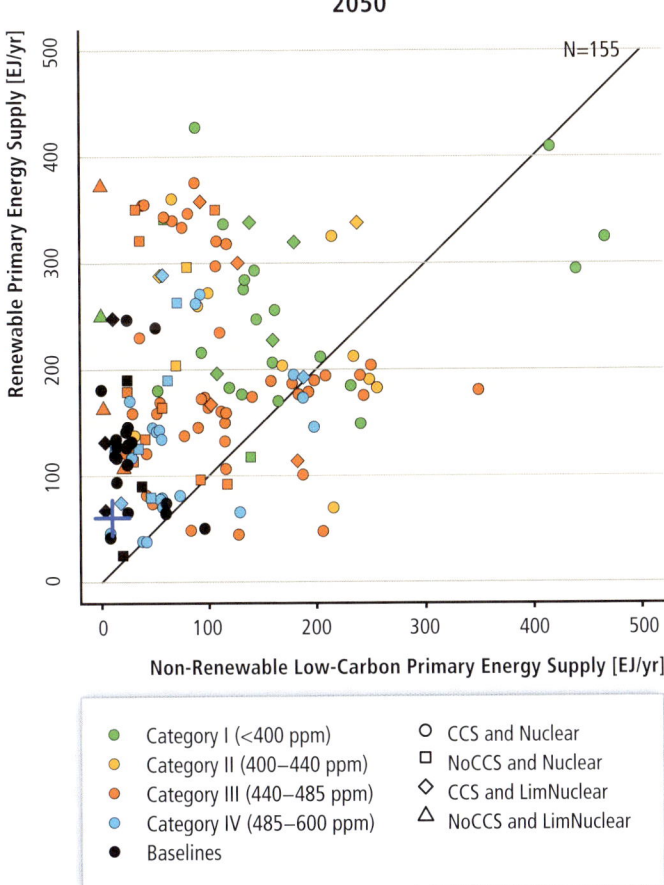

Figure 10.7 | Global RE primary energy supply (direct equivalent) plotted against non-renewable low-carbon energy primary energy supply (direct equivalent) in 2030 and 2050. Colour coding is based on categories of atmospheric CO_2 concentration level in 2100 (Fisher et al., 2007). The shapes identify constraints on the availability of the competing low-carbon energy supply options, fossil with CCS and nuclear. Note that limited nuclear scenarios include nuclear phase-outs, constraints on the production of new nuclear energy and scenarios in which nuclear production is constrained to baseline levels. The blue crossed lines show the relationship in 2007. For data reporting reasons, only 152 and 155 scenarios are included in the 2030 and 2050 results shown here, as opposed to the full set of 164 scenarios. Figure and data adapted from Krey and Clarke (2011), modified to include two additional scenarios.

accounting method. Bioenergy is accounted for prior to conversion to fuels such as ethanol or electricity when it is used in those applications. In contrast, the other technologies generally produce electricity, and they are accounted for as electricity produced in these cases. If they were to be converted to primary energy by using the substitution method, then they might be roughly three times larger, based on average fossil electricity efficiencies.

Second, the time scale of deployment varies across different RE technologies (Figures 10.8 and 10.9), in large part representing differences in deployment levels today and (often) associated assumptions about relative technological maturity. For example, hydroelectric power experiences only modest growth across scenarios (a 1.7-fold increase in the median case and a 3-fold increase in the highest scenario by 2050 compared to today); wind energy grows more rapidly, beginning from lower deployment levels today; and solar energy grows most rapidly, beginning from only minimal deployment today, as well as in 2020 in most scenarios. Indeed, much of the growth in solar energy occurs after 2030, indicating a general consistency among scenarios that solar energy at a large scale is a longer-term option than several other options. Global bioenergy production includes both traditional uses of biomass (more than 30 EJ/yr or roughly two-thirds of all bioenergy consumption in 2008, see Chapter 2) as well as more advanced methods, including cellulosic approaches. Traditional biomass use is typically assumed to decline as economic development progresses, implying that the growth in bioenergy is largely in modern applications. It is also useful to note that some technologies appear to be more clearly influenced by the climate policy than others. For example, solar energy deployment levels are noticeably higher in the most ambitious climate scenarios than in the other scenarios. All of the technologies experience this effect but to varying degrees.

Finally, scenarios generally indicate that RE deployment is larger in non-Annex I countries over time than in the Annex I countries (Figure 10.8 and Krey and Clarke, 2011). Virtually all scenarios include the assumption that economic and energy demand growth will be larger in the non-Annex I countries than in the Annex I countries (Clarke et al., 2007, 2009). The result is that the non-Annex I countries account for an increasingly large proportion of CO_2 emissions in baseline, or no-policy, cases and must therefore make larger emissions reductions over time. All other things being equal, larger reductions imply larger deployment of low-carbon supply options, including RE. Hence, it is not surprising that scenarios generally indicate larger RE deployment levels in non-Annex I regions.

At the same time, it is important to note that the actual deployment levels, particularly in the nearer term, will depend not only on the long-term

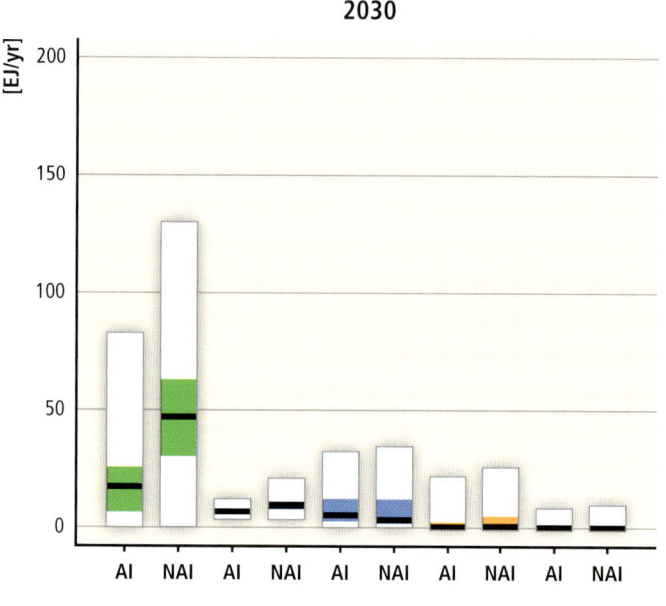

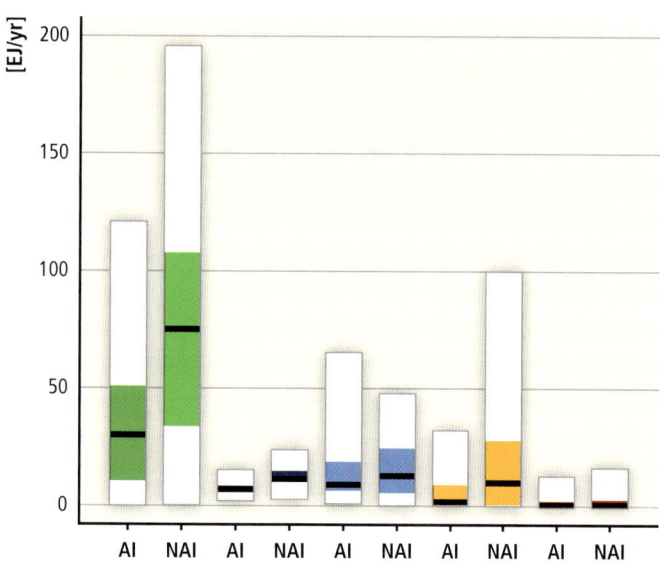

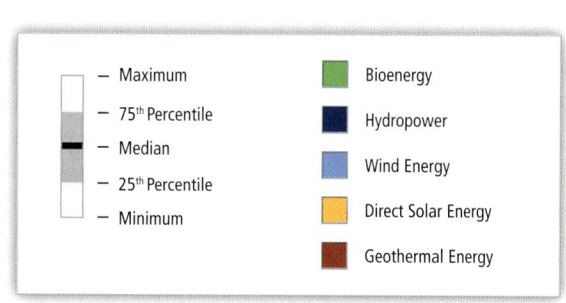

Figure 10.8 | Global RE primary energy supply (direct equivalent) by source in Annex I (AI) and Non-Annex I (NAI) countries in 164 long-term scenarios by 2030 and 2050. The thick black line corresponds to the median, the coloured box corresponds to the inter-quartile range (25th to 75th percentile) and the ends of the white surrounding bars correspond to the total range across all reviewed scenarios. Depending on the source, the number of scenarios underlying these figures varies between 122 and 164. Note that ocean energy is represented in very few scenarios, insufficient to generate a similar graph. Although instructive for interpreting the information, it is important to note that the 164 scenarios are not explicitly a random sample meant for formal statistical analysis. (One reason that bioenergy supply appears larger than supplies from other sources is that the direct equivalent method is used to represent primary energy in this figure. Bioenergy is accounted for prior to conversion to fuels such as ethanol or electricity. The other technologies produce primarily (but not entirely) electricity and they are accounted for based on the electricity produced. If primary equivalents were used, based on the substitution method, rather than direct equivalents, then energy production from non-biomass renewable sources would be of the order of three times larger than shown here.) Figure and data adapted from Krey and Clarke (2011), modified to include two additional scenarios.

goal, but also on the degree to which countries take action towards the long-term goal. For example, in scenarios in which some countries delay participation in global emissions reductions, RE deployment is necessarily lower than it is in scenarios with full global participation (Clarke et al., 2009; Krey and Clarke, 2011). Nonetheless, because stabilization of CO_2 concentrations means bringing CO_2 emissions to near zero, all countries must eventually bring their emissions to this point, and those with larger energy consumption will require more low-carbon energy than others, regardless of which countries may have initiated action on climate the soonest. It is also important to note that countries may take different approaches to mitigation, some focusing on price-based policies where others use regulatory policies that could include mandates for RE, and this could influence the spatial character of RE deployments. The scenarios described here mostly rely exclusively on price-based mitigation and therefore do not capture this sort of variation.

10.2.2.6 Renewable energy and the costs of mitigation

RE's role in climate mitigation might be observed not only through the lens of RE deployment levels, but also by an exploration of the manner in which RE availability and deployment influences the economic consequences, or costs, of mitigation. One way that researchers have attempted to link particular technologies to mitigation costs is to build mitigation cost curves; that is, relationships that indicate how much mitigation might be achieved by particular technologies at a given carbon price. In the context of RE, these curves attempt to answer the question: how much CO_2 abatement and at what cost can be provided by RE technologies? Such mitigation cost curves are not provided here for reasons discussed more thoroughly in Section 10.4. It is noted here only that assigning mitigation to particular technologies is not a primary output of integrated models. Integrated models provide information on prices, emissions and deployments, but in general they do not assign emissions to the presence or absence of specific technologies. Such assignments are the result of post-processing, offline accounting calculations that rely on analyst judgment about key assumptions. Applying these post-processing assumptions to the scenarios would constitute new analysis rather than synthesis, and it would blur the signal from the scenarios themselves. A sense of the variation of CO_2 emission mitigation due to the use of different methods is given in Section 10.3 on the basis of 4 selected scenarios from the whole set of 164 analyzed in this section. In addition, these analyses do not account for the benefits of climate mitigation (e.g., less severe climate change impacts in the long term, reduced need for adaptation), secure energy supply and air pollution

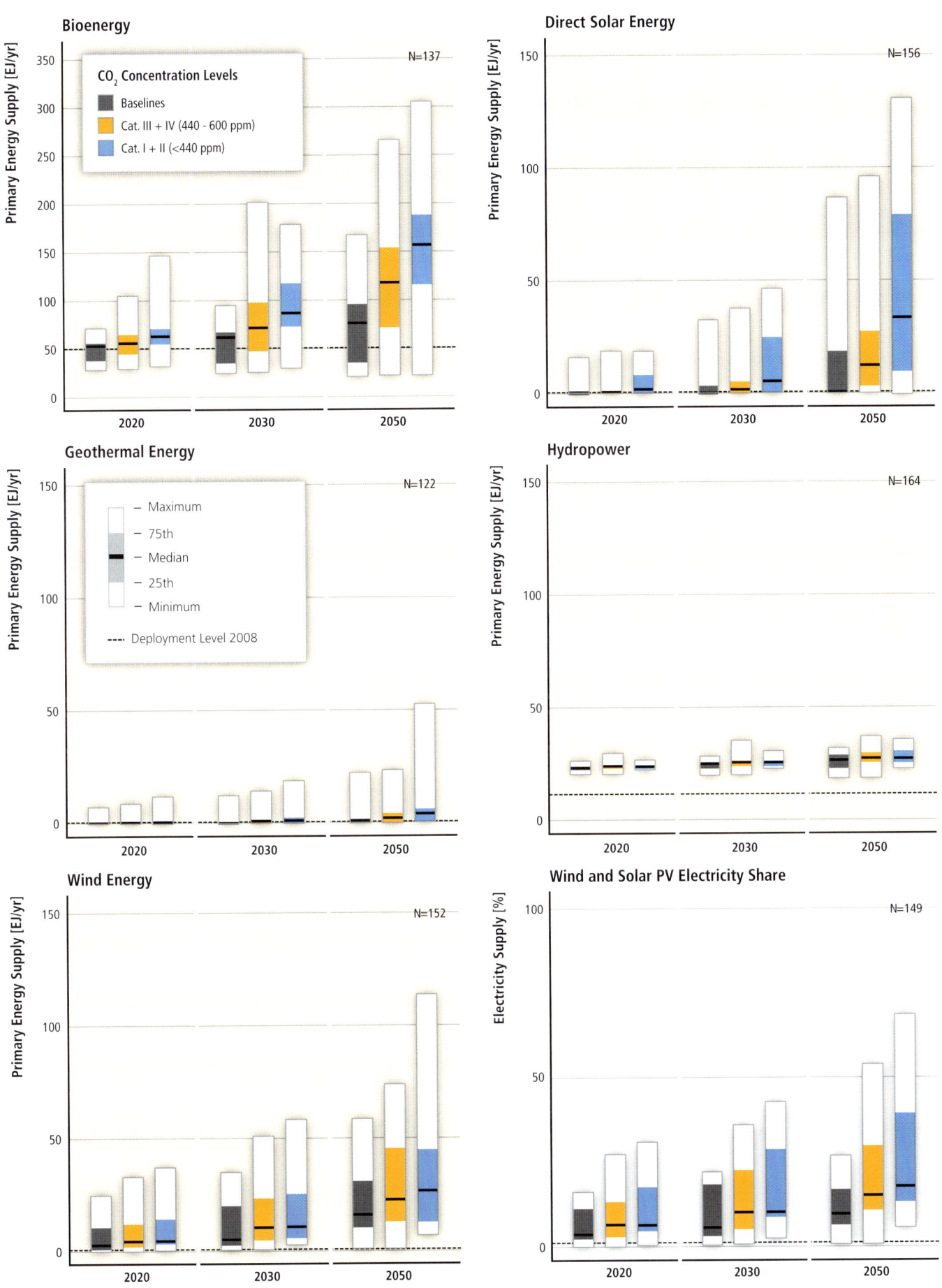

Figure 10.9 | (Preceding page) Global primary energy supply (direct equivalent) of biomass, wind, solar, hydro, and geothermal energy and share of variable RE (wind and solar photovoltaic) in global electricity generation in 164 long-term scenarios in 2020, 2030 and 2050, and grouped by different categories of atmospheric CO_2 concentration level in 2100 (Fisher et al., 2007). Following the direct equivalent methodology, biomass primary energy supply is accounted for prior to conversion whereas the other RE options are accounted for based on secondary energy produced. The thick black line corresponds to the median, the coloured box corresponds to the inter-quartile range (25th to 75th percentile) and the ends of the white surrounding bars correspond to the total range across all reviewed scenarios. Although instructive for interpreting the information, it is important to note that the 164 scenarios are not explicitly a random sample meant for formal statistical analysis. For data reporting reasons, the number of scenarios included in each of the panels shown here varies considerably. The number of scenarios (N) underlying the individual panels, as opposed to the full set of 164 scenarios, is indicated in the right upper corner of each panel. Figure and data adapted from Krey and Clarke (2011), modified to include two additional scenarios.

(e.g., reduced health expenditures) due to the deployment of RE technologies (see e.g., Nemet et al., 2010). A more detailed discussion of co-benefits can be found in Section 10.6.

Another possible view into the relationship between RE and mitigation costs is afforded by considering the relationship between RE deployment levels and carbon prices across scenarios. This approach attempts to answer the question: how much RE will be deployed at a given carbon price? The 164 scenarios demonstrate no meaningful correlation between RE deployment and carbon prices (see Figure 10.10). All the forces that blur the relationship between RE deployment levels and long-term concentration goals, as discussed in Sections 10.2.2.2, 10.2.2.3 and 10.2.2.4, influence the relationship between RE deployment and carbon prices. In addition, integrated energy models are characterized by a wide range of carbon prices based both on parameter assumptions and model structure (Clarke et al., 2007, 2009). The result is little ability to link RE deployment levels to carbon prices when looking across a wide range of models.

CO_2 prices are only a limited metric for cost because they represent the marginal costs of abatement and not the total cost. A range of other cost measures have been used in the literature to capture the economic consequences of mitigation. These include changes in gross domestic product (GDP) or consumption, or total mitigation costs, that is, the additional cost to deploy and operate an energy system with lower GHG emissions, which can provide a broader sense of the cost implications of RE. In general, mitigation tends to reduce GDP (Fisher et al., 2007).[3] However, these measures do not necessarily lead to a stronger correlation with RE deployment than carbon prices. For example, the overall variation of GDP in the baseline scenarios reviewed in this section (a factor of 1.8 in 2050 between the lowest and the highest GDP) is much larger than the changes in GDP as the result of climate mitigation (up to a few percent of baseline GDP by 2050), which can be derived by comparing the GDP in mitigation scenarios to their respective baseline for those models that include feedbacks to GDP. The dominance of, and variation in, baseline GDP would further obscure any relationship between total GDP and RE deployment.

A different reflection of the relationship between the economic consequences of mitigation and RE deployments can be ascertained by exploring how mitigation costs would change under differing

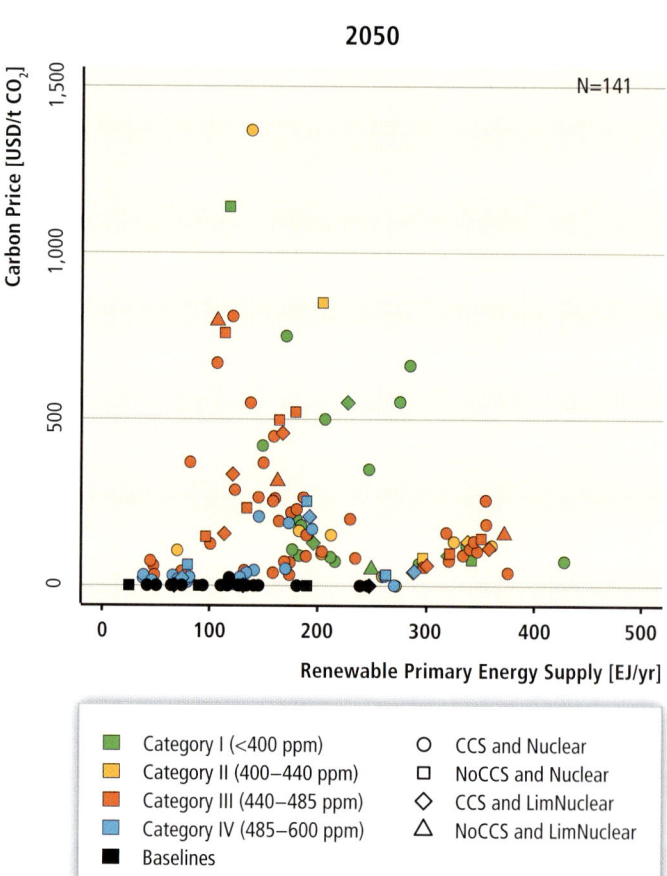

Figure 10.10 | Carbon prices (in USD_{2005}) as a function of global RE primary energy supply (direct equivalent) in 2050. Colour coding is based on categories of atmospheric CO_2 concentration level in 2100 (Fisher et al., 2007). Different symbols in the graph denote the availability of CCS and nuclear energy. Note that limited nuclear scenarios include nuclear phase-outs, constraints on the production of new nuclear and scenarios in which nuclear production is constrained to baseline levels. For data reporting reasons, only 141 scenarios are included in the 2050 results shown here, as opposed to the full set of 164 scenarios. Figure and data adapted from Krey and Clarke (2011), modified to include two additional scenarios.

assumptions about the availability or cost and performance of RE as well as competing mitigation options. A number of researchers have explored this issue (see, e.g., Clarke et al., 2008; Luderer et al., 2009; Edenhofer et al., 2010; Tavoni and Tol, 2010). Consistent with intuition, these studies demonstrate that the presence of RE technologies or improvements in the cost and performance of RE technologies reduces mitigation costs. This is not surprising: more or better options should not increase costs. More important is the relative magnitude of the change in mitigation costs resulting from increases in the availability, cost or performance of RE technologies relative to the change in mitigation costs resulting from

---

3 Note that a minority of researchers have argued that climate mitigation could lead to increased economic output (e.g., Barker et al., 2006). The basic argument is that under specific assumptions, induced technological change due to a carbon price increase leads to additional investments that trigger higher economic growth.

increases in the availability of fossil energy with CCS and/or nuclear energy. For example, in both the ADAM (Edenhofer et al., 2010) and RECIPE projects (Luderer et al., 2009), each involving three models, the cost increase that results from the absence of the option to expand RE deployment is not of a distinctly different order of magnitude than the cost increase from the absence of the option to implement fossil energy with CCS or expand production of nuclear energy beyond today's levels or beyond baseline levels (see Figures 10.11 and 10.12). Indeed, in several scenarios, constraining RE results in larger cost increases than constraining nuclear power or fossil energy with CCS. The value of RE availability, cost and performance may also vary with the degree of ambition. For example, the availability of bioenergy with CCS has been identified as a particularly valuable technology combination for meeting tight stabilization constraints (Azar et al., 2006; van Vuuren et al., 2007; Clarke et al., 2009; Edenhofer et al., 2010; Tavoni and Tol, 2010). To summarize, while there is an agreement in the literature that mitigation costs will increase if the deployment of RE technologies is constrained and that more ambitious stabilization levels may not be reachable, there is little agreement on the precise magnitude of the cost increase.

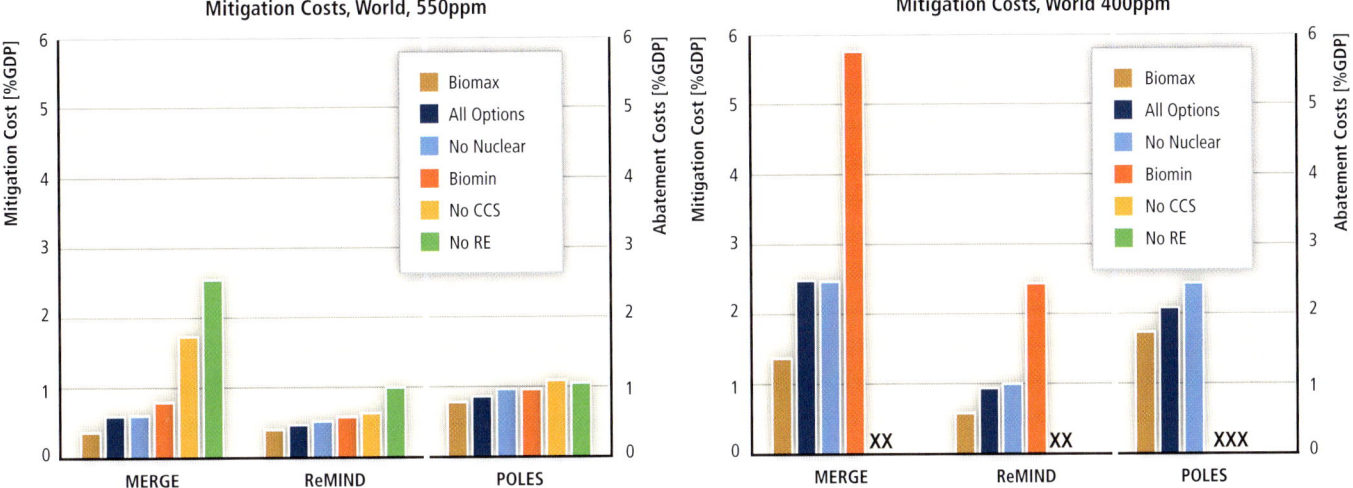

Figure 10.11 | Global mitigation costs from the ADAM project under varying assumptions regarding technology availability for long-term stabilization levels of 550 and 400 ppm CO_2eq (Edenhofer et al., 2010). Mitigation costs are given as aggregated GDP losses (MERGE, REMIND) or increase of abatement costs (POLES) up to 2100 relative to baseline in % of GDP. 'All Options' refers to the standard technology portfolio assumptions in the different models, while 'Biomax' and 'Biomin' assume double and half the standard technical potential of biomass of 200 EJ, respectively. 'No CCS' excludes CCS from the mitigation portfolio and 'No Nuclear' and 'No RE' constrain the deployment levels of nuclear and RE to the baseline level, which still potentially means a considerable expansion compared to today. The '**x**' in the right panel indicates non-attainability of the 400 ppm CO_2eq level in the case of limited technology options.

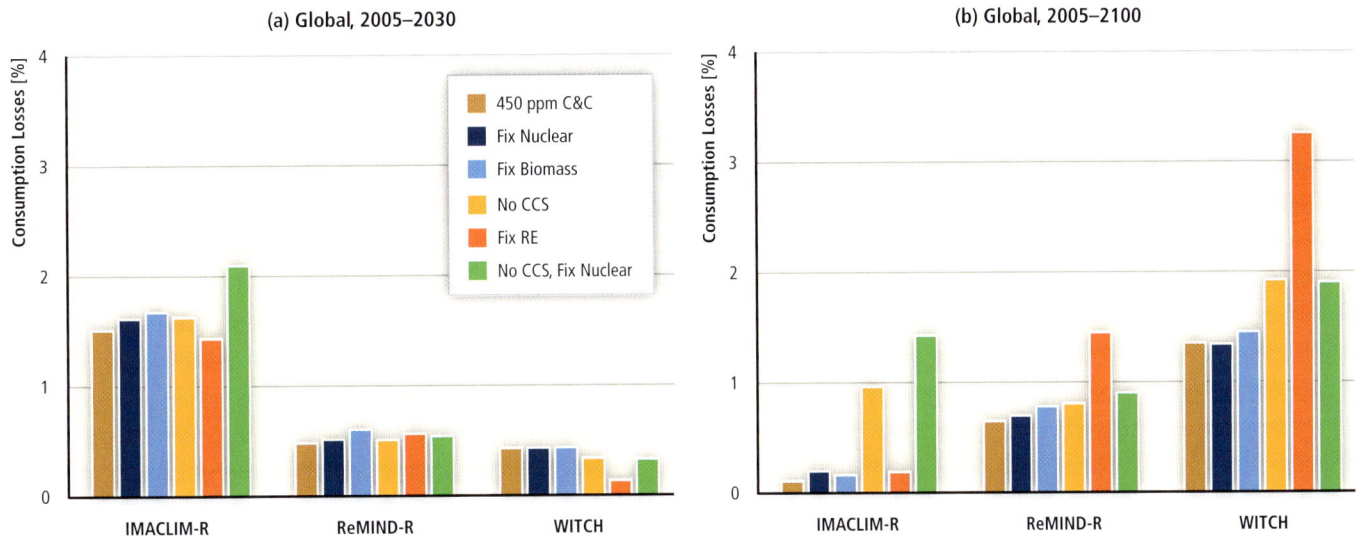

Figure 10.12 | Mitigation costs from the RECIPE project under varying assumptions regarding technology availability for a long-term stabilization level of 450 ppm CO_2 (Luderer et al., 2009). Option values of technologies in terms of consumption losses for scenarios in which the option indicated is foregone (CCS) or limited to baseline levels (all other technologies) for the periods (a) 2005 to 2030 and (b) 2005 to 2100. Option values are calculated as differences in consumption losses for a scenario in which the use of certain technologies is limited with respect to the baseline scenario. Note that for WITCH, the generic backstop technology was assumed to be unavailable in the 'Fix RE' scenario.

10.2.3 The deployment of renewable energy sources in scenarios from the technology perspective

The scenarios in this section were produced using global, integrated models. These models have several advantages, but they also have the weakness that they pay only limited attention to many critical factors that ultimately will influence the deployment of RE. As a means to better understand the role of these forces, the scenarios from this section are briefly explored in the 'long-term deployment in the context of carbon mitigation' sections of Chapters 2 through 7. The aim of these individual technology explorations is to identify potential barriers that an expansion of RE may face and enabling factors to achieve the higher RE deployments levels as found in the scenario literature. This section briefly summarizes the key elements of those sections.

Resource Potential: In general, even the highest deployment levels were not considered to be constrained by the available technical potentials at the global level for all of the RE categories. However, because RE resources are regionally heterogeneous, some of the higher deployment levels may begin to constrain the economically most attractive sites, for example, for hydro and wind energy. For most RE sources, availability is geographically constrained, for example, for certain forms of ocean, geothermal, biomass and solar energy, as well as hydropower and wind energy. In the case of bioenergy, the supply levels in the scenarios with low GHG stabilization levels of up to about 300 EJ/yr by 2050 almost exactly coincide with the upper range of possible deployment levels as discussed in Chapter 2 (see Section 2.8.4 and Figure 2.8.3).

Regional Deployment: Economic development and technology maturity are primary determinants of regional deployment levels. Regional policy frameworks for RE need to be economically attractive and predictable. For mature technologies such as large hydropower, a large fraction of available technical potential in Organisation for Economic Cooperation and Development (OECD) countries has been exhausted and the largest future expansion is expected in the non-OECD countries of Asia and Latin America. For wind energy, which has seen high expansion rates, mostly in Europe and North America over the past decade as well as in China and India more recently, a greater geographical distribution of deployment than currently observed is likely to be needed to achieve the higher deployments indicated by the scenario literature. The other, less mature technologies are likely to initially focus on expansion in affluent regions (Europe, North America, Australia and parts of Asia) where financing conditions and infrastructure integration are favourable.

Supply Chain Issues: In general, no insurmountable medium- to long-term constraints of materials, labour and manufacturing capacity were identified that would prevent higher deployment levels in the scenarios. For example, the wind industry has witnessed rapid expansion over the past that led to globalization of the production chain, but further scaling up of the industry will be needed to reach the capacity addition rates seen in the more stringent scenarios. It is also important to recognize that markets and supply chains for some technologies are global (e.g., wind, solar photovoltaic (PV)) while others (e.g., passive solar and low-temperature solar thermal) to date are largely local. As markets expand, they are likely to become more global in scope. Past rates of growth suggest that, assuming that policy and market signals are clear, no absolute long-term constraints exist.

Technology and Economics: Because the maturity of the renewable technologies is highly variable, so is the need for cost and technological advances. On the one end of the spectrum, hydropower is competitive with thermal power plants, while on the other end of the spectrum, commercial-scale demonstration plants for most ocean energy technologies do not yet exist. For offshore wind energy, more remote offshore locations will require further technology advances; further, cost reductions will impact deployment outcomes. Similarly, concentrating solar power (CSP), solar PV, geothermal heat pumps, and enhanced geothermal systems (EGS) will require technological improvements, but in particular further reductions in electricity generation costs. Technical progress is similarly required for advanced biofuels and bio-refineries with potential for commercialization around 2020 given R&D investment and near-market support.

Systems Integration and Infrastructure: Systems integration is challenging for the variable electricity generation technologies wind, solar PV and wave energy (Section 8.2.1). Technical (e.g., balancing generation capacity, inter-connection and storage) and institutional (e.g., market design and operations, market access and tariff structure) solutions will need to be implemented to address operational integration concerns. Additionally, substantial new transmission infrastructure may be required under even modest expansion scenarios to connect remote resources, for example, off- but also onshore wind power, central station CSP and PV, hydrothermal geothermal power and hydropower. A greater reliance on offshore wind power is likely for regions such as Europe, which will require the development of offshore transmission infrastructure; certain forms of ocean energy face similar integration challenges and synergies may therefore exist in the deployment of these technologies. To gain greater penetration into the energy supply systems, other RE-based energy carriers such as heat, biogas, liquid bio-fuels, solid biomass and hydrogen all need appropriate integration into existing system infrastructure as outlined in Section 8.2.

10.2.4 Knowledge gaps

The primary knowledge gap with respect to the assessment of RE in large-scale, integrated models is the representation of RE technologies themselves within these models. The coverage of different RE sources in the scenario literature varies significantly. Mature technologies such hydropower were included by all models reviewed in this assessment, while less mature technologies or those not deployed today at large scale—for example, ocean energy, offshore wind, concentrating solar power and geothermal energy—were addressed by smaller sets of scenarios. One reason is that there is less demand to specifically address less mature technologies or those that are a priori assumed to have lower contributions. A second reason is that there is a lack of high-quality global resource data (preferably gridded) for some renewable resources (e.g., geothermal energy, the various ocean energy forms), which is a precondition for constructing resource supply curves that are inputs to energy-economic and integrated assessment models. More broadly, beyond representations of the technologies themselves, many issues related the implementation of RE technologies require further research and inclusion in large-scale integrated models. Important areas in this regard include the integration of RE into the electricity grid and the relationship between bioenergy production, crop production and deforestation.

However, it is important to note that improved representations of RE technologies and associated systems will not entirely eliminate the uncertainty regarding the role of RE in climate mitigation. As was discussed throughout this section, a range of other uncertainties, unrelated to RE technologies, such as economic and population growth, the availability, cost and performance of competing technologies, and the nature of mitigation approaches and ambitions will influence the role of RE in climate mitigation. Uncertainty derived simply from the design of different modelling platforms can also influence results. Therefore, an important research priority for the future is to improve the understanding of why model results vary with respect to RE and to attribute these differences in model outcomes to differences in assumptions and methodologies.

10.3 Assessment of representative mitigation scenarios for different renewable energy strategies

Section 10.2, coming from a more statistical perspective, gave a comprehensive overview of the possible role RE technologies could play in different mitigation pathways. In contrast, this section goes beyond the more aggregated data level and focuses on regional and sectoral perspectives. For this in-depth analysis, four scenarios from the previous section's full set of the scenario assessments have been chosen to represent different illustrative energy and emission pathways (see Table 10.3). The scenarios differ in assumptions, mitigation goals and in the types of underlying models used. For a description of the scenarios and models, see Box 10.2. Primary data for this analysis go beyond what has been published to date, and were provided at special request by the scenario authors and institutions.[4]

---

4 The International Energy Agency (IEA) and the German Aerospace Center for IEA-WEO2009 Baseline; the Potsdam Institute for Climate Impact Research for ReMIND-RECIPE; the Pacific Northwest National Laboratory for MiniCAM-EMF22; and the German Aerospace Center for ER-2010.

10.3.1 Sectoral breakdown of renewable energy sources

The amount of RE deployed in the scenarios depends on a large number of variables, assumptions and input data (see also Section 10.2.1, especially Section 10.2.1.1). Often most influential are the cost and performance assumptions for the different RE technologies. They help determine the comparative attractiveness of competing low-carbon supply options (i.e., nuclear energy and fossil energy with CCS), but also of end-use energy efficiency measures. Underestimation of costs leads to overestimation of RE deployment and vice versa. The share of RE calculated is furthermore determined by the general availability of competing options. Constraints on alternative mitigation options mean that more RE deployment will occur for a given level of GHG mitigation. Assumptions about infrastructure restrictions and system integration options are further important determinants. In this context, a significant factor relates to assumptions about how the power grid would adapt to significant amounts of variable renewable resources. In contrast, the overall technical potential for RE—that is, the total amount of energy that can be produced taking into account the primary resources, the socio-geographical constraints and the technical losses in the conversion process (see definition in Annex I)—is not considered to be a limiting factor at the global level as the technical potential supersedes the current and projected future demand by orders of magnitude (see Section 1.2.2). Thus, to fully exploit the entire technical RE potential is neither needed nor necessary.

In practice, deployment of RE resources should respect sustainability criteria in order to achieve an environmentally friendly future energy supply (see Chapters 1 and 9). Public acceptance is crucial to the expansion of RE sources as well. Some RE applications, such as rooftop PV and solar thermal as well as bioenergy cogeneration plants and onshore wind, are often decentralized energy production facilities and may be located near or even at demand centres. Other RE applications are more likely to involve industrial-scale energy production facilities located at some distance from demand centres and requiring large-scale transmission, for example, large onshore wind parks, offshore wind energy, concentrated solar power in deserts, hydrothermal geothermal plants, and hydropower. In both cases, public acceptance concerns can constrain development if not carefully managed. The use of biomass has been especially controversial recently, as issues have arisen over competition with other land uses, food production and ecosystem preservation, as well as possible direct or indirect GHG emissions due to land use change (see Sections 2.5, 9.3.4 and 10.6). On the other hand, RE deployment is positively driven by sustainability criteria since it has the potential to provide energy access in remote areas without some of the environmental and health impacts usually associated with fossil fuels (see Sections 9.3.2, 9.3.4 and 10.6). Therefore, non-economic criteria have a significant influence on the resulting RE deployment and corresponding assumptions are crucial for scenario results.

Table 10.3 | Overview of key parameters of the illustrative scenarios based on assumptions that are exogenous to the models' respective endogenous model results. Dark grey marks exogenous input; dark yellow marks endogenous model results. Note that the concentration categories are defined in terms of CO_2 (only) concentrations, while other metrics, predominantly CO_2-equivalent concentrations—of Kyoto gases or of all forcing agents—are used in the literature. (Sources: IEA-WEO2009-Baseline (IEA, 2009; Teske et al., 2010), ReMIND-RECIPE (Luderer et al., 2009), MiniCAM-EMF22 (Calvin et al., 2009), ER-2010 (Teske et al., 2010)).

Category	Units	Status Quo	Baseline		Category III+IV (440 - 600ppm)		Category I+ II (< 440ppm)		Category I+ II (< 440ppm)	
Scenario name			IEA-WEO2009-Baseline		ReMind-RECIPE		MiniCAM-EMF22		ER-2010	
Model					ReMind		MiniCAM		MESAP/PlaNet	
	yr	2007	2030	2050[1]	2030	2050	2030	2050	2030	2050
Technology pathway[2]										
Renewables			all[3]	all	solar: PV and CSP not differentiated		solar: PV and CSP not differentiated, ocean energy not included		all	all
CCS			+	+	+	+	+	+	-	-
Nuclear			+	+	+	+	+	+	+	-
Population	billion	6.67	8.31	9.15	8.32	9.19	8.07	8.82	8.31	9.15
GDP/capita[4]	thousand USD_{2005}/capita	10.9	17.4	24.3	12.4	18.2	9.7	13.9	17.4	24.3
Energy Demand (direct equivalent)	EJ/y	469	645	749	590	674	608	690	474	407
Energy Intensity	MJ/USD_{2005}	6.5	4.5	3.4	5.7	4.0	7.8	5.6	3.3	1.8
Renewable Energy	%	13	14	15	32	48	24	31	39	77
Fossil & Industrial CO_2 Emissions	Gt CO_2/y	27.4	38.5	44.3	26.6	15.8	29.9	12.4	18.4	3.7
Carbon Intensity	kg CO_2/GJ	58.4	57.1	56.6	45.0	23.5	49.2	18.0	36.7	7.1

Notes: 1. IEA (2009) does not cover the years 2031 until 2050. As the IEA's projection only covers a time horizon up to 2030 for this scenario exercise, an extrapolation of the scenario has been used that was provided by the German Aerospace Agency (DLR) by extrapolating the key macroeconomic and energy indicators of WEO 2009 forward to 2050 (Teske et al., 2010). 2. (-): Technology not included; (+): Technology included. 3. This includes: Solar photovoltaics, CSP, solar water heating, wind (on- and offshore), geothermal power, heating and cogeneration, bioenergy power, heating and cogeneration, hydropower, ocean energy. 4. The data are either input for the model or endogenous model results.

Last but not least, climate and energy policy frameworks are highly relevant to RE deployment in scenario analysis. Market forces and constraints are relevant for the deployment of RE and determine the market potential. As market potential also includes opportunities, it may in theory be larger than the economic potential due to support programs, but usually the market potential is lower because of a variety of constraining market failures for RE and other new technologies (Sections 1.4.2 and 11.4). Market potential analyses have to take into account the behaviour of private economic agents under their specific conditions, which are partly shaped by public authorities (see Sections 11.5 and 11.6). In this context, the energy policy framework has a profound impact on the expansion of RE sources respective to corresponding assumptions for the scenario results.

RE deployment is driven and hindered by a variety of factors and very much depends on how the different determinants and their impacts are being assessed; uncertainties about future development are generally high and determined by specific assumptions. In this context, energy scenarios bundling a consistent set of specific assumptions are an approximation of what can be expected for the future under specific conditions. As a comparison of different scenarios spans a range of possible futures, it can show overarching commonalities and trends and can make differences and uncertainties visible and more transparent.

Selection of four illustrative scenarios for an in-depth analysis
Scenario results are determined not only by parameter assumptions, but also by the underlying modelling architecture and model-specific restrictions (e.g., upper deployment bound for specific RE technologies). The four scenarios were selected to present a wide range of different modelling architectures, demand projections and technology portfolios for the supply side (see Box 10.2). The IEA-WEO2009-Baseline Reference Scenario (IEA, 2009; extension to 2050: Teske et al., 2010) (henceforth IEA-WEO2009-Baseline) is the only baseline scenario in this set, that is, it does not incorporate any climate policy targets beyond those implemented by 2009. It is characterized by a comparatively high demand projection with low RE deployment. In two of the three mitigation scenarios, ReMIND RECIPE 450 ppm Stabilization Scenario (Luderer et al., 2009) (henceforth ReMIND-RECIPE) and MiniCAM EMF 22 first-best 2.6 W/m² Overshoot Scenario (Calvin et al., 2009) (henceforth MiniCAM-EMF22), high demand expectation and a significant increase in RE is combined with the possibility of employing CCS and nuclear power plants. Low demand (e.g., due to a significant increase in energy efficiency) is combined with high

Box 10.2 | Overview of the four illustrative scenarios and their underlying models.

IEA-WEO2009-Baseline: This scenario uses a typical baseline scenario approach. As such, it calculates the possible energy pathway without any substantial change in government policy (IEA, 2009, p. 44) and under the assumption of a minimal to moderate fossil fuel cost increase. The scenario does not include specific GHG emissions constraints. As the IEA (2009) projection only covers a time horizon up to 2030 for this scenario exercise, an extrapolation of the scenario has been used that was provided by the German Aerospace Center (DLR) that uses the key macroeconomic and energy indicators of IEA (2009) and brings them forward to 2050 (Teske et al., 2010). Regarding fossil and industrial CO_2 emissions, the baseline scenario expects an increase from 27.4 Gt CO_2/yr in 2007 to 44.3 Gt CO_2/yr by 2050. (Scenario 'IEA WEO 2009 Reference Scenario' from IEA (2009) extended beyond 2030 by Teske et al. (2010).)

ReMIND-RECIPE: This scenario describes a mitigation path aiming to stabilize atmospheric CO_2 (only) concentration at 450 ppm (corresponding to fossil and industrial CO_2 emissions of 15.8 Gt CO_2/yr by 2050). It was generated with the energy-economy-climate model ReMIND-R, which computes welfare-optimized transformation trajectories under full 'where-flexibility' (emission reductions are performed where it is cheapest), 'when-flexibility' (emission reductions are performed when they are cheapest) and 'what-flexibility' (emission reductions are performed by choosing the least expensive combination of technologies) conditions. Another crucial assumption is perfect foresight: investment decisions are made knowing in advance the future changes in prices and technology developments. The model is characterized by a high level of integration: the macro-economy and the energy system are treated within an integrated optimization framework, thus fully accounting for the macro-economic feedbacks of the climate mitigation effort. The complex integrated formulation requires compromises in terms of the sectoral and technological resolution of the energy system. ReMIND-RECIPE accounts for a variety of RE sources (wind, solar, biomass, hydro and geothermal) and conversion technologies. Wind power and solar PV are parameterized as learning technologies. RE technologies can be deployed at the industrial scale at optimal sites and be transported within world regions (up to continental scale) to demand centres, whereby the model implicitly assumes that bottlenecks (e.g., with respect to grid infrastructure) are avoided by early and anticipatory planning. (Scenario '450 ppm stabilization scenarios' from Luderer et al. (2009).)

MiniCAM-EMF22: The MiniCAM-EMF22 scenario was developed as part of the Energy Modelling Forum study 22 (EMF 22), which looks at possible approaches to long-term climate goals. The scenario was generated using the MiniCam integrated assessment model, the precursor to the Global Change Assessment Model (GCAM) integrated assessment model. The scenario is an overshoot scenario that reaches 450 ppm CO_2eq (Kyoto gases)[1] by 2100, after peaking at 525 ppm CO_2eq in 2050, and assumes full international participation in emissions reductions. The specific concentration levels correspond with fossil and industrial CO_2 emissions of 12.4 Gt CO_2/yr by 2050. The underlying characteristics of the scenario include global population growth that peaks at approximately 9.0 billion people in 2070 and then declines to 8.7 billion in 2100. The scenario considers the availability of a wide range of energy supply options, including major RE options, nuclear power and both fossil energy and bioenergy equipped with CCS technology. The presence of bioenergy with CCS is particularly important in the scenario because it allows for the option to create negative emissions, primarily in electricity production (Calvin et al., 2009; Clarke et al., 2009). (Scenario 'First-best 2.6 W/m² Overshoot Scenario' from Calvin et al. (2009).)

ER-2010: The ER-2010 scenario (Teske et al., 2010) is based on the socioeconomic assumptions of the IEA-WEO2009-Baseline scenario, but assumes an increase in fossil fuel costs and a price for carbon from 2010 onwards. The scenario has a key constraint that limits worldwide CO_2 emissions to a level of 3.7 Gt CO_2 per year by 2050. To achieve this, the scenario is characterized by significant efforts to fully exploit the large potential for energy efficiency, using currently available best practice technology, and to foster the use of RE. In all sectors, the latest market development projections and the resulting cost reductions for the RE industry have been taken into account, and a stable development of the RE sector is pursued. To accelerate the market penetration of RE, various additional measures have been assumed, such as a speedier introduction of electric vehicles combined with the implementation of effective communications systems and technologies, smart meters and faster expansion of super grids to allow a higher share of variable RE power generation (PV and wind) to be employed. The methodological background of the scenario is the simulation model PlaNet of the energy and environmental planning package MESAP (see Krewitt et al. (2009), which was created for long-term strategic planning on a national, regional or local level. The model is characterized by a very detailed technology breakdown for each sector. Following the simulation approach, activities and drivers of demand (e.g., mobility demand), as well as relevant market shares of technologies, amongst other factors, are specified exogenously by the user. (Scenario 'Advanced Energy [R]evolution 2010' from Teske et al. (2010).)

Note: 1. Note that atmospheric CO_2 (only) concentrations reach about 385 ppm by 2100, that is, the scenario falls into concentration category 1 (<400 ppm); see also Table 10.2.

RE deployment, no employment of CCS and a global nuclear phase-out by 2045 in the third mitigation scenario, Advanced Energy [R]evolution 2010 (Teske et al., 2010) (henceforth ER-2010).

Table 10.3 shows key parameters for the four illustrative scenarios. Depending on the model, some of the assumptions may be exogenously applied or be determined endogenously. All scenarios project a significant increase in global population and assume or calculate a significant increase in GDP. The IEA-WEO2009-Baseline GDP projections are based on forecasts by the International Monetary Fund (IMF, 2009) and the OECD. Those GDP projections have been used as input parameters for the ER-2010 model as well. In contrast, GDP projections from MiniCAM-EMF22 and ReMIND-RECIPE are endogenously determined. Both population and GDP changes are major driving forces for future energy demand (which is endogenously calculated in all models) and therefore at least indirectly determine the resulting shares of RE.

For the set of the four illustrative scenarios, the following sections give an overview of the available data for each of the different sectors. Global energy scenarios often provide detailed information on RE electricity generation. Information about the current and future RE power market is often publicly accessible, while suitable data sets about the RE heating sector and RE application in the transport sector are often not available or less detailed than for the power sector. These sectors deserve more attention, particularly because RE heating shows a significant technical potential and is in many cases already cost-effective (Aitken, 2003; Seyboth et al., 2007).

10.3.1.1 Renewable energy deployment in the electricity sector

The RE electricity sector scenarios analyzed here show more dynamic development and larger RE shares over the midterm compared to either the heating or transport sector scenarios.

Factors for market development in the RE electricity sector

Technology cost and performance assumptions are among the most influential variables affecting energy deployment in the scenarios. The largest variations in the cost assumptions can be found for solar PV, CSP, and ocean energy. As an illustrative example: for 2020, the highest cost projections for solar PV in the analyzed scenarios was USD$_{2005}$ 5,406/kW and the lowest projection was less than half of that at USD$_{2005}$ 2,177/kW. The upper limit is in the range of current market prices (see Section 3.8.3), although all scenarios assume cost reductions in the future. This demonstrates a typical problem in scenario analysis covering a new technology market where numbers in scenarios are often superseded by recent developments. The different cost assumptions lead to very different market development pathways in the scenarios, spanning a range for solar PV-based electricity generation, even in the mitigation-oriented scenarios, from 115 TWh (414 PJ) up to 594 TWh (2,138 PJ) in 2020 (see Table 10.4), corresponding to annual market growth rates of between 18% and 42%, respectively.

However, cost projections for installed PV systems in 2050 had a significant lower level of variability, ranging from USD$_{2005}$ 753/kW in the low case to USD$_{2005}$ 1,125/kW in the high case. Nevertheless, the expected deployment rates in the scenarios are quite different. With regard to the PV-based electricity generation in 2050, there is a 25-fold difference between two of the mitigation oriented scenarios: 20,790 TWh/yr (74,844 PJ/yr) in the ReMIND-RECIPE scenario versus 822 TWh/yr (2,959 PJ/yr) in MinCam-EMF22. This example illustrates the complexity of the analysis, as the resulting deployment path for PV depends not only on cost assumptions, but also on many other factors (e.g., availability and characteristics of alternative mitigation technologies like CCS and nuclear power in the case of MinCam-EMF 22).

Among all RE technologies for electricity generation, onshore wind energy saw the least variation in cost projections among the models, ranging around ±10% over the entire time frame. Cost-optimization energy models use cost assumptions for each technology as one of the main determinants of market expansion or reduction, and the input cost assumptions will therefore play a major role in determining the scenario energy mix.

Annual market potential for the RE electricity sector

Based on the energy parameters of the analyzed scenarios, the required annual production capacity (representing the annual market volume) has been either calculated ex-post (IEA-WEO2009-Baseline, ReMIND-RECIPE, MinCam-EMF 22) or has been provided by the scenario authors (ER-2010). These calculated manufacturing capacities (Table 10.4) do not include the additional needs for re-powering (i.e., replacement of old wind turbines with new ones). Annual market growth rates in the analyzed scenarios are very different, as are the expectations about how the current dynamic of the market might change. In some cases, drastic reductions in the current average market growth rates have been outlined, even in those scenarios aiming for an ambitious GHG stabilization level. The global PV industry had an average annual growth rate of 35% between 1998 and 2008 (EPIA, 2008). The wind industry experienced a 30% annual growth rate over the same time period (Sawyer, 2009). While the advanced technology roadmaps from the PV, CSP and wind industry indicate these annual growth rates can be maintained over the next decade (Sawyer, 2009; EPIA, 2010) and will decline later, most of the analyzed integrated energy scenarios expect much slower annual growth for all RE electricity supply technologies. The MiniCAM-EMF22 scenario, in particular did not project a stabilization of the growth rates at the current level, but instead found alternative non-RE mitigation technologies or other RE options (like biomass technologies) to be more cost-competitive than solar PV. Furthermore, as MiniCAM-EMF22 is representing an overshoot scenario in the medium term, the pressure to further deploy RE is much lower than in scenarios with more ambitious GHG stabilization levels for 2030 (e.g., ER-2010). Additionally, while MiniCAM-EMF22 and ReMIND-RECIPE are predominantly cost driven, in the ER-2010 scenario the market development is simulated and based on exogenous settings. With these settings, ER-2010 seeks to avoid large fluctuations in annual RE markets in order to achieve stable development and employment in the RE sector.

Chapter 10 — Mitigation Potential and Costs

Table 10.4 | Overview of scenario results for four illustrative scenarios: renewable electricity generation, resulting RE market shares, annual market growth rates and required annual manufacturing capacity. Both the IEA-WEO2009-Baseline and ER-2010 have a separate category for bioenergy and geothermal combined heat and power (CHP) and power-generation-only power plants—heat generation is excluded and listed in Table 10.5. "N/A": data not available, "NSM": not specifically modelled. Sources: IEA-WEO2009-Baseline (IEA, 2009; Teske et al., 2010), ReMIND-RECIPE (Luderer et al., 2009), MiniCAM-EMF22 (Calvin et al., 2009), ER-2010 (Teske et al., 2010).

	Energy Parameter									Market Development							
	Generation [EJ/y]				Percent of global demand based on the demand projection of the analysed scenario [%]					Annual Market growth [%/y]					Annual Market Volume [GW/y]			
	IEA-WEO2009-Baseline	ReMIND-RECIPE	MiniCAM-EMF22	ER-2010	IEA-WEO 2009-Baseline	ReMIND-RECIPE	MiniCAM-EMF22	ER-2010	>600 ppm IEA WEO 2008	IEA-WEO 2009-Baseline	ReMIND-RECIPE	MiniCAM-EMF22	ER-2010	IEA-WEO 2009-Baseline	ReMIND-RECIPE	MiniCAM-EMF22	ER-2010
Total projected energy demand by scenario:																	
2020	98.1	117.9	103.4	92.9													
2030	123.5	146.3	124.8	111.2													
2050	167.6	228.2	222.4	158.1													
Solar																	
PV 2020	0.4	0.8	0.4	2.1	0.4	0.7	0.4	2.3					42	5	12	6	36
PV 2030	1.0	9.3	1.0	7.0	0.8	6.4	0.8	6.3		17	27	18	14	18	163	17	120
PV 2050	2.3	74.8	3.0	24.6	1.4	32.8	1.3	15.6		11	32	10	7	40	651	25	211
CSP2020	0.1		0.7	2.5	0.1		0.7	2.7		4	12	6	62	1		3	12
CSP2030	0.4	N/A	2.0	9.8	0.4	N/A	1.5	8.8		17	N/A	40	17	2	N/A	9	45
CSP2050	0.9		5.6	32.4	0.5		2.5	20.5		14		13	6	4		11	66
Wind										4		6					
on+offshore2020	3.6	16.7	8.6	10.3	3.7	14.2	8.4	11.0			33	23	26	26	175	83	101
on+offshore2030	5.5	35.2	15.8	21.1	4.5	24.0	11.9	19.0		12	9	7	8	60	381	171	229
on+offshore2050	9.1	51.4	28.3	39.0	5.4	22.6	12.5	24.7		5	2	3	3	93	262	146	202
Geothermal										3							
for power generation																	
2020	0.4			1.3	0.4			1.4		6	33	NSM	20	1	NSM	NSM	4
2030	0.6	NSM	NSM	4.6	0.5	NSM	NSM	4.1		4	NSM		15	2	NSM	NSM	18
2050	1.0			10.7	0.6			6.8		2			5	4			21
heat & power																	
2020	0.0			0.2	0.0			0.3		13	NSM	NSM	47	0	NSM	NSM	1
2030	0.0	NSM	NSM	0.9	0.0	NSM	NSM	0.8		5			16	0			5
2050	0.1			4.5	0.0			2.9		4			9	0			11

In addition to the specific RE cost projections and assumptions for other supply side mitigation technologies (e.g., CCS, nuclear power), the future of electricity demand may help determine the future role of RE sources in terms of absolute market share. In all scenarios, high energy demand does not necessarily coincide with high deployment of RE. ReMIND-RECIPE and MiniCAM-EMF22 both project a large increase in electricity demand, but whereas MiniCAM-EMF22 predicts a low RE market share, ReMIND-RECIPE expects a high one. The ER-2010 has the lowest demand projection of all analyzed scenarios and the highest RE share. However, the RE market projections of the ER-2010 (in absolute numbers) for solar and wind are amongst the scenarios in the medium and high range, respectively, but in the lower range for hydro and biomass. High electricity demand in some of the scenarios arises from relatively low expectations about the role that energy (electricity) efficiency is expected to play in the future.

The underlying assumptions for future RE deployment growth in the scenarios do not always correspond with current manufacturing capacity and thus are not able to reflect the market behaviour (interactions) in practice. The IEA-WEO2009-Baseline scenario, for example, expects lower global deployment of wind power in 2020 than currently available manufacturing capacity,[5] which could lead to overcapacity and lower market prices for wind turbines. Lower prices for wind would, all else being equal, lead to greater deployment. This shows once more the problem of dealing with a very dynamic (and in this case policy-driven) sector using scenario analysis. On the other hand, the high scenario for wind in ReMIND-RECIPE requires an annual production capacity of 175 GW by 2020, which would represent a four-fold increase in production capacity at a global level. Both the ER-2010 and MiniCAM-EMF22 scenarios require this production capacity about a decade later (by 2030), leading to a global wind power share of 12 to 19% under the demand projections of these scenarios. The highest global wind share occurs in the ReMIND-RECIPE scenario, with a 24% portion by 2030, a share that is reached in the ER-2010 scenario only by 2050. One reason the ReMIND-RECIPE scenario projects such a high share of RE penetration is because it allows for RE learning and therefore endogenously considers technological progress as well as cost reduction effects. Moreover, the underlying model assumes perfect foresight and assumes potential bottlenecks with regard to RE integration to be resolved by anticipatory planning of grid infrastructure and storage (see Box 10.2). The deployment of wind in 2030 is lower in ER-2010 as the scenario limits the expansion of wind due to long-term integration costs and the limited possibility to reallocate the labour force between the renewable energy sector and the rest of the economy.

Figure 10.13 summarizes the resulting range of electricity generation by RE sources in the different scenario projections for 2050. Solar PV, CSP and wind power have the largest expected market potential beyond 2020. Hydropower remains at a relatively high and stable level in almost all scenarios (10 to 15% by 2030), indicating a high correlation among projections. The total renewable electricity generation market potential in the lowest case (IEA-WEO2009-Baseline) is 9% above the 2008 level with a 24% share by 2050. The highest RE electricity shares are 95% (ER-2010) and 72% (ReMIND-RECIPE) by 2050, while the MiniCAM-EMF22 scenario achieves a global renewable electricity share of 35%.

Hence, all scenarios project a significant increase in RE electricity generation. The required increase in manufacturing capacities for RE electricity generation technologies has not been identified as a fundamental barrier to growth, but certainly could represent a challenge to the growth envisioned by some of the scenarios. The availability of different mitigation technologies besides RE (e.g., fossil CCS and nuclear) and corresponding policy pathways lead to significantly different—in most cases lower—renewable energy deployment.

10.3.1.2 Renewable energy deployment in the heating and cooling sector

The heating sector is one of the largest demand sectors and the RE share—mainly traditional bioenergy—is currently high, especially in non-Annex I countries. RE for heating could also be used for cooling, which offers new and additional market opportunities for countries with Mediterranean, subtropical, or tropical climates. RE for cooling—in combination with solar architecture—can be applied for instance for air-conditioning and would in that context reduce electricity demand for electric air-conditioning significantly. RE heating and cooling technologies represent a variety of different technology pathways and require different infrastructure. Electricity-based geothermal heat pumps, small- and large-scale solar collectors and district heating with a network of bioenergy cogeneration plants are to some extent competing technologies. Low-energy buildings, for example, are a limiting factor for cogeneration networks and could make electrical heating systems such as heat pumps the preferred choice (see Section 8.2.2).

Factors for market development in the RE heating and cooling sector

Besides cost aspects, policy choices in favour of specific RE technologies and associated infrastructure (e.g., district heating networks) as well as oil and gas price projections have a significant impact on the projected deployment for each RE heating technology. Only the ER-2010 scenario indicates a significant increase in the global RE share, from 24%[6] in 2007 (IEA-WEO2009-Baseline) up to 90% by 2050, while the other of the four illustrative scenarios expect only a slight increase of RE heat to a maximum of 30% (MiniCAM-EMF22) by 2020 and a decrease again to 2007 levels by 2050. All studies indicate that electricity demand increases in the heating sector at the expense of fuel consumption.

---

5 Global annual installation of wind turbines in 2009 was 38.3 GW according to the Global Wind Report 2009 of the Global Wind Energy Council (GWEC).

6 Excluding traditional biomass for cooking and heating, RE provides around 5 to 6% of total global heating demand and very little cooling (Seyboth et al., 2007).

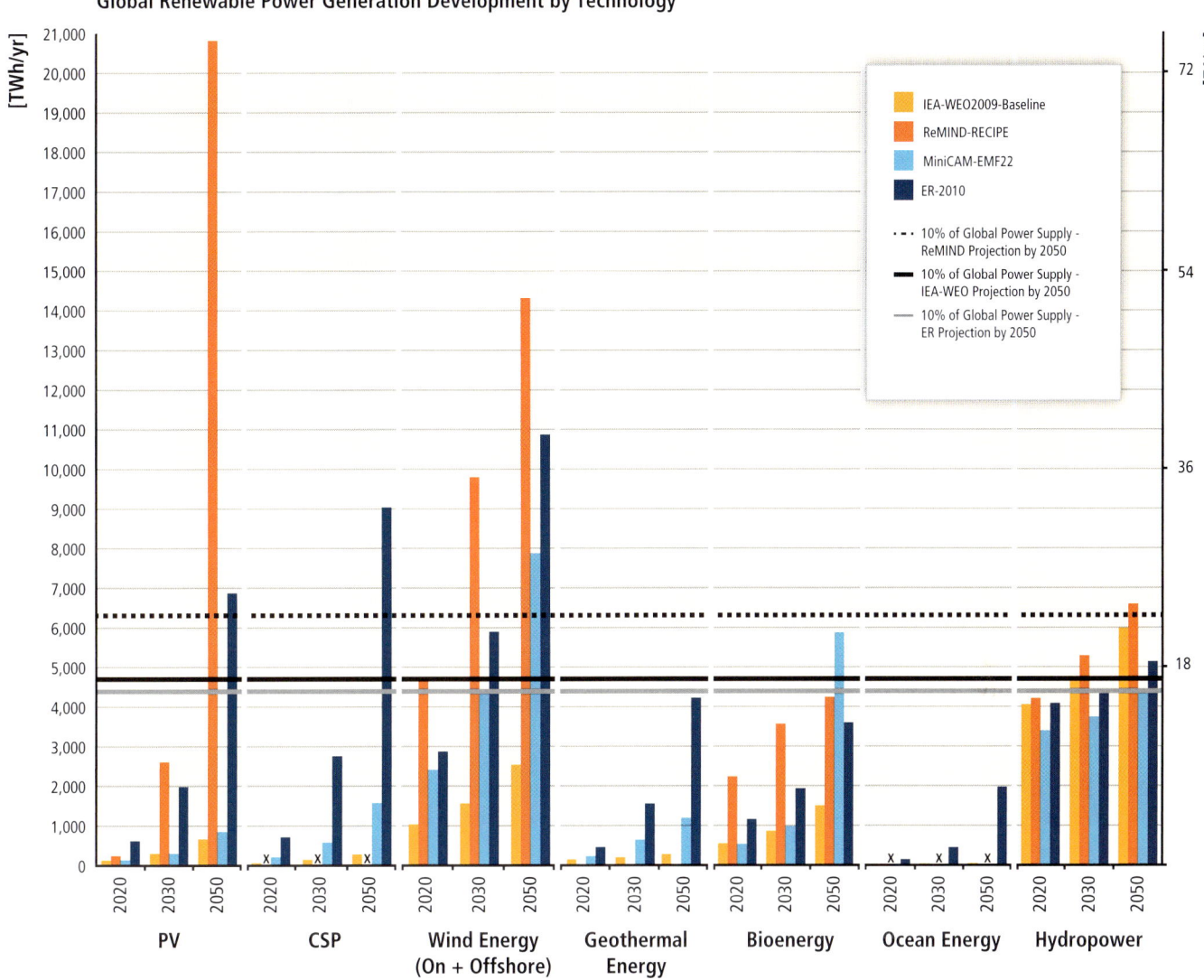

Figure 10.13 | Global RE electricity generation (development projections by technology and shares of global power generation for the four illustrative scenarios for comparison). The total renewable power generation by 2050 is 11,159 TWh/yr (IEA-WEO2009-Baseline), 63,384 TWh/yr (ReMIND-RECIPE), 21,660 TWh/yr (MiniCAM-EMF22) and 41,500 TWh/yr (ER-2010) respectively. Sources: IEA-WEO2009-Baseline (IEA, 2009; Teske et al., 2010), ReMIND-RECIPE (Luderer et al., 2009), MiniCAM-EMF22 (Calvin et al., 2009), ER-2010 (Teske et al., 2010).

Annual market potential for RE heating and cooling

The RE heating sector shows for the various technologies much lower growth rate projections than outlined for the electricity sector. The highest growth rates are expected for solar heating—especially solar collectors for water heating and space heating—followed by geothermal heating. Geothermal heating includes heat pumps, while geothermal cogeneration plants are presented in Section 10.3.2.1 under RE electricity generation.

In the ER-2010 scenario, solar heating systems show a significant increase with market growth rates of above 35% until 2020 and a minimum of 10% afterwards up to the end of the projection in the year 2050 (see Section 3.4).

A shift from the traditional and sometimes unsustainable use of bioenergy for heating towards modern and more sustainable uses of bioenergy heating such as wood pellet ovens or biogas burners are assumed in all scenarios. The more efficient use of biomass would increase the share of biomass heating without the necessity to increase the overall demand for biomass. However, only one of the analyzed scenarios provides information about the specific breakdown of traditional versus modern biomass use. Therefore, it is not possible to estimate the real annual market development of the different bioenergy heating systems.

The market potential at both domestic and industrial scales for RE heating technologies such as solar collectors, geothermal heat pumps or pellet heating systems overlaps with the market potential analysis of the RE power sector. While the solar collector market is independent from the electricity sector, biomass cogeneration provides electricity as well as heat. Geothermal heat pumps use electricity for their operation and therefore increase the demand for electricity. RE heating and cooling is more dispersed than RE electricity generation, which, together with lack of metering, is why statistical data are poor and further research is needed. Based on the energy parameters of the four scenarios analyzed, the required annual market volume has been calculated in order to identify the needed manufacturing capacities and how they relate to current capacities. Table 10.5 provides an overview of the projected annual market volumes.

Manufacturing capacities for all RE heating and cooling technologies must be expanded significantly in order to realize the projected RE heat production in all scenarios. The annual market volume for solar collectors is projected to triple from less than 35 PJ/yr in 2020 to 100 PJ/yr in 2030 in the IEA-WEO2009-Baseline case and up to 1,162 PJ/yr in the ER-2010 case. Due to the diverse technology options for bio- and geothermal energy heating systems and the low level of information in all analyzed scenarios, it is not possible to provide here a full set of specific market size data by technology.

The total share of RE heating systems in all scenarios by 2050 significantly varies, from a market share of around 23% (IEA-WEO2009-Baseline, ReMIND-RECIPE and MiniCAM-EMF22) to 91% (ER-2010). The resulting shares for RE technologies for heating and cooling are significantly driven by the scenario assumptions (including assumptions about infrastructure changes such as the expansion of district heating networks, as well as improvements in building efficiency and industrial processes). The large share of RE heating systems in ER-2010 depends, for instance, on the assumption that district heating systems for the distribution of solar-, geothermal- and bioenergy-generated heat would be available and competitive after 2020 (see Table 10.5).

10.3.1.3 Renewable energy deployment in the transport sector

The use of RE in the transport sector in all analyzed studies was limited to liquid biofuels, biomethane from biogas and RE-based electric vehicles for private use or public transport. Most of the scenario literature does not take into account new technologies such as second-generation sails for ships. Additionally, different reporting and categorization within the underlying scenario models do not support a stringent comparison of scenario results. However, even this comparison shows the substantial influence of driving forces (e.g., GHG stabilization levels) on the resulting RE share, which differs between scenarios by up to an order of magnitude (see Table 10.6).

10.3.1.4 Global renewable energy primary energy contribution

Figure 10.14 provides an overview of the projected primary energy production (using the direct equivalent methodology, see Section 1.1.8) by source for the four selected scenarios for 2020, 2030 and 2050, and compares the numbers with different projected global primary energy demands. Bioenergy has the highest market share, on average, across all of the scenarios, followed by solar energy, though scenario-specific results vary. This is largely driven by the fact that bioenergy (see Chapter 2) can be used across all sectors (electricity, heating and cooling as well as transport) in combination with the selected primary energy accounting methodology. As the available land for bioenergy is limited and competition with nature conservation issues as well as food and materials production is crucial, the sectoral use for the available bioenergy significantly depends on scenario assumptions and underlying priorities (see Sections 2.2, 2.5 and 9.3.4). Solar energy can be used in direct form for heating and cooling and electricity generation (and indirectly via electricity for transport purposes), but solar technology starts from a relatively low level. The relatively lower average primary energy share for wind and hydropower may in part be due to their exclusive use in the electricity sector, though some scenarios show substantial contributions from wind in particular.

The total RE share in the primary energy mix by 2050 has a substantial variation across all four scenarios. With 15% by 2050—compared to 12.9% in 2008—the IEA-WEO2009-Baseline projects the lowest primary RE share, while ER-2010 reaches 77%, the MiniCAM-EMF22 achieves 31% and ReMIND-RECIPE 48% of the worlds primary energy demand with RE. While it is not surprising that without constraining GHG concentration levels, RE deployment rates are rather low (IEA-WEO2009-Baseline), it is worth mentioning that there is even a significant difference (more than a factor of two with regard to the relative RE shares) between the mitigation-oriented scenarios. Once more, this is a result of many aspects; that is, technology-specific assumptions (e.g., costs) and model characteristics (e.g., inclusion of endogenous learning), assumptions about the availability of other mitigation technologies and the expected energy demand. The overall total global RE deployment by 2050 in all analyzed scenarios represents less than 2% of the available technical RE potential (see Section 10.3.2.2). The wide range of RE shares is a function of different assumptions about policy, technology costs, chosen mitigation technologies (e.g., availability of CCS) and future energy demand projections.

10.3.2 Regional breakdown – technical potential versus market deployment

This section focuses on the regional perspective and provides an overview of the regional market penetration paths given in the four scenarios. A comparison with the technical potential per region for each technology indicates to what level the regional technical potentials will be exploited. Additionally, an in-depth cost curve analysis of three regions (China, India and Europe) provides deeper insights into the assumed cost development of renewable electricity generation.

10.3.2.1 Regional renewable energy supply curves

Regional energy supply cost curves can serve as 'snapshots' of the selected scenarios and are thus an alternative perspective on scenario

Table 10.5 | Projected RE heat production, possible market shares, annual growth rates and annual market volumes for the four illustrative scenarios—excluding additional needs for re-powering. "N/A": data not available. Sources: IEA-WEO2009-Baseline (IEA, 2009; Teske et al., 2010), ReMIND-RECIPE (Luderer et al., 2009), MiniCAM-EMF22 (Calvin et al., 2009), ER-2010 (Teske et al., 2010).

Energy Parameter	Generation [EJ/yr]				Percent of global demand based on demand projections of the scenarios (incl. CHP) [%]				Market Development							
									Annual Market growth [%/yr]				Annual Market Volume [PJ/yr]			
	IEA-WEO 2009-Baseline	ReMIND-RECIPE	MiniCAM-EMF22	ER-2010	IEA-WEO 2009-Baseline	ReMIND-RECIPE	MiniCAM-EMF22	ER-2010	IEA-WEO 2009-Baseline	ReMIND-RECIPE	MiniCAM-EMF22	ER-2010	IEA-WEO 2009-Baseline	ReMIND-RECIPE	MiniCAM-EMF22	ER-2010
Total projected heat demand by scenario:																
2020	158	190	135	152												
2030	174	198	145	156												
2050	205	160	151	152												
Solar																
Solar Thermal 2020	0.8	N/A	N/A	6.5	0.5	N/A	N/A	4.5	10	N/A	N/A	39	32	N/A	N/A	409
Solar Thermal 2030	1.6	N/A	N/A	15.8	0.9	N/A	N/A	12.2	8	N/A	N/A	12	100	N/A	N/A	1162
Solar Thermal 2050	3.1	N/A	N/A	38.7	1.5	N/A	N/A	33.7	3	N/A	N/A	5	187	N/A	N/A	1568
Geothermal heating																
2020	0.6	0.1	N/A	4.4	0.4	0.1	N/A	3.0	14	-6	N/A	41	6	-1	N/A	63
2030	0.9	0.2	N/A	9.3	0.5	0.1	N/A	7.0	4	7	N/A	10	12	3	N/A	149
2050	1.6	4.6	N/A	26.5	0.8	2.8	N/A	26.4	3	18	N/A	7	22	66	N/A	283
Bioenergy heating																
2020	36.2	40.8	40.4	41.7	23.0	21.5	30.0	27.6	not specifically modelled				28	112	104	130
2030	38.2	39.8	39.0	45.4	22.0	20.1	27.0	29.7					678	698	686	811
2050	43.6	32.4	31.7	48.1	21.3	20.2	21.0	31.7					270	191	186	295
Total renewables heating																
2020	37.7	40.9	40.4	52.6	23.9	21.6	20.0	35.0	1	N/A	N/A	5	66	N/A	N/A	601
2030	40.7	40.0	39.0	70.5	23.4	20.2	27.0	48.7	1	N/A	N/A	4	791	N/A	N/A	2122
2050	48.4	37.0	31.7	113.3	23.6	23.1	21.0	90.8	1	N/A	N/A	3	479	N/A	N/A	2146

Mitigation Potential and Costs

Table 10.6 | Projected RE shares in the transportation sector for the four illustrative scenarios. (Note: The electricity share includes RE- and non-RE-based electricity as well as hydrogen produced with electricity. For the IEA-WEO2009-Baseline, MiniCAM-EMF22 and ER-2010 the RE share in the electricity sector has been used to identify the RE share of the electricity used for the transport sector. Therefore the total RE share within the transport sector is lower than the sum of the percentages.) Sources: IEA-WEO2009-Baseline (IEA, 2009; Teske et al., 2010), ReMIND-RECIPE (Luderer et al., 2009), MiniCAM-EMF22 (Calvin et al., 2009), ER-2010 (Teske et al., 2010).

RE share in Transport Sector		IEA-WEO2009-baseline (%)	ReMIND-RECIPE (%)	MiniCAM-EMF22 (%)	ER-2010 (%)
Biofuels	2020	4.3	2.2	6.8	5.4
	2030	4.6	12.9	9.5	9.3
	2050	5.0	26.8	10,2	14.0
Electricity (including conventional generation+ hydrogen)	2020	1.4	0.1	2.5	4.4
	2030	1,5	1.0	4.1	14.7
	2050	1,6	6.7	11.2	57.4
Total RE share	2020	4.6	2.3	7.5	7.3
	2030	4.9	13.9	10.8	19.1
	2050	5.4	33.6	15.6	68.9

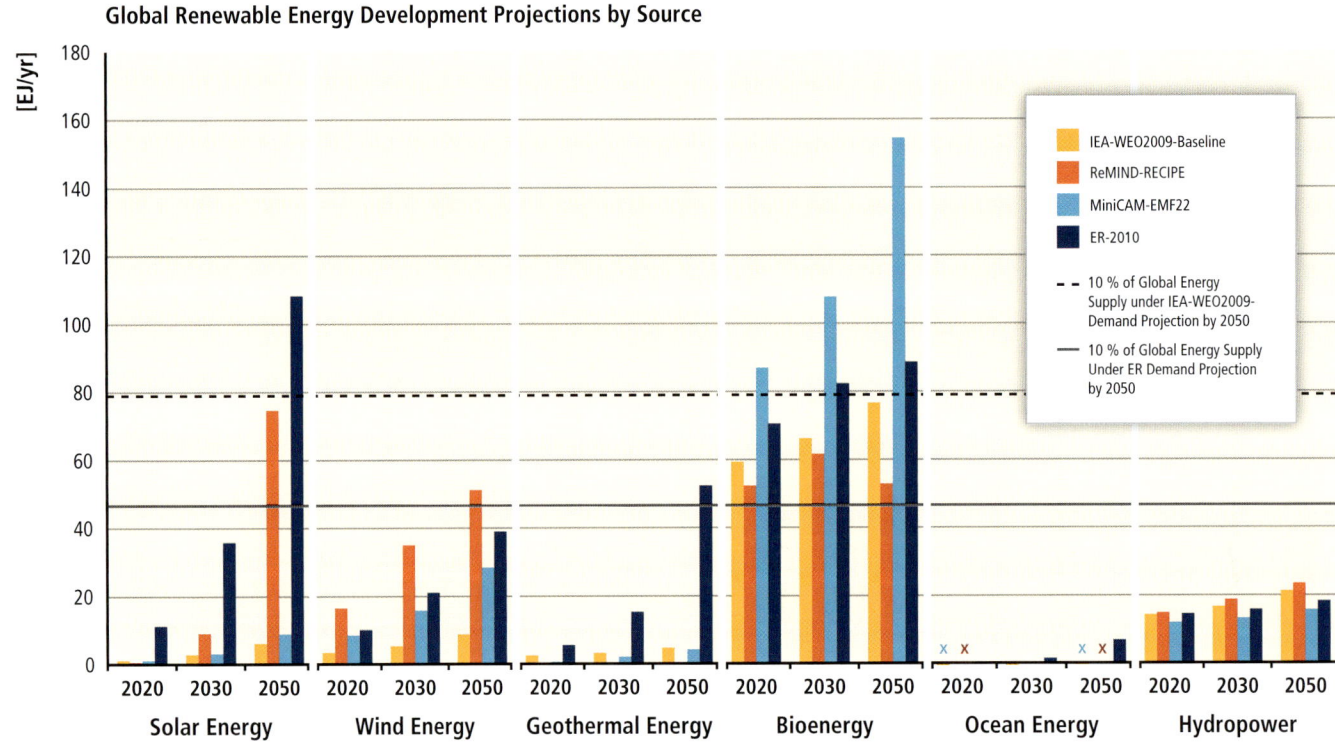

Figure 10.14 | Global RE development projections by source and global renewable primary energy shares (direct equivalent) by source for a set of four illustrative scenarios. The total renewable energy deployment projected for 2050 is 117 EJ/yr (IEA-WEO2009-Baseline), 214 EJ /yr (ReMIND-RECIPE), 323 EJ/yr (MiniCAM-EMF22) and 314 EJ/yr (ER-2010) respectively. Sources: IEA-WEO2009-Baseline (IEA, 2009; Teske et al., 2010), ReMIND-RECIPE (Luderer et al., 2009), MiniCAM-EMF22 (Calvin et al., 2009), ER-2010 (Teske et al., 2010).

results. The following curves (see Figures 10.15, 10.16 and 10.17) are illustrative examples and represent a cross-section of three of the four scenarios (specific data for MiniCAM-EMF22 are not available for this exercise).[7] The regional energy supply cost curves focus on a specific target year and relate the deployment of certain RE electricity technologies in the different regions (as a result of the specific scenarios) to their cost levels in discrete steps. Thus, the curves report scenario results (potential deployment) and are not a reflection of RE technical potentials.

This presentation alleviates two major shortcomings of the cost curve method (which are discussed in a more general and comprehensive way in Section 10.4). First, recognizing the crucial determinant role of carbon emission factors, energy pricing and fossil fuel policies in the ultimate shape of abatement cost curves, only RE supply cost curves are created (and not mitigation cost curves). Second, in order to capture the uncertainties in cost projections, several scenarios were reviewed. Using dynamic scenarios that span a longer time horizon to create the curves as done here also prevents the problem of following a static perspective.

Beyond the general issues about cost curves detailed in Section 10.4, it is important to note a few points for proper interpretation of the curves.

---

7 Unlike other parts of this section, IEA-WEO2008-Baseline and not IEA-WEO2009-Baseline is used to represent a baseline scenario here due to data constraints.

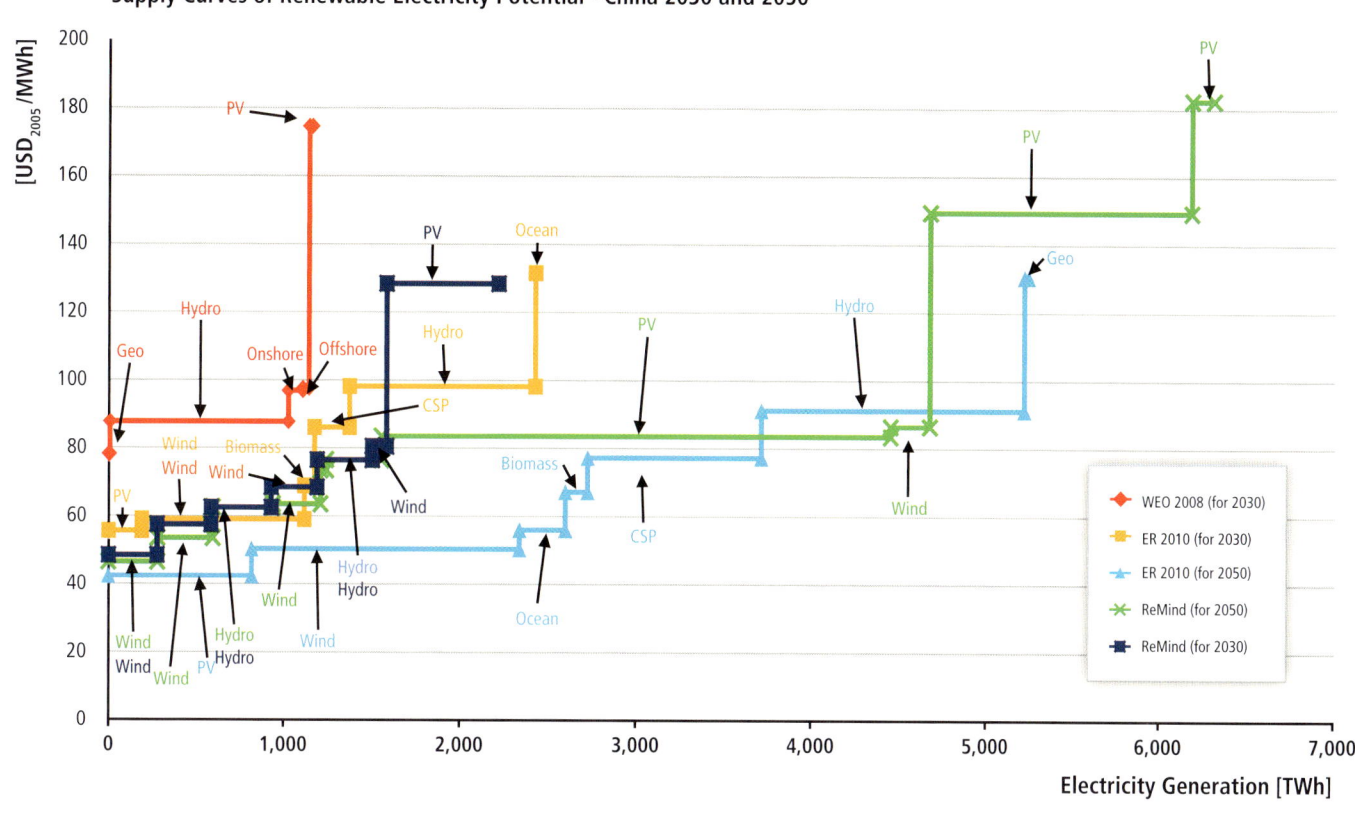

Figure 10.15 | Illustrative RE electricity supply curves for China for the years 2030 and 2050. The curves report scenario results (level of deployment) and are not a reflection of RE technical potential.

First, the ER-2010 and the IEA-WEO2008-Baseline scenario data were not as detailed in cost data as was the ReMIND-RECIPE scenario. For the former two scenarios, each technology in a region is represented by a single average cost. Second, average costs for a technology for a whole region can mask the more cost-effective sub-categorization of technologies and sites into an average. Thus, with this approach it is not possible to highlight the cheaper (or more expensive) sites and sub-technologies.

It was not possible to deduct existing capacity from the RE deployment by cost level. Thus, values include all capacity that can be installed in the target year allowed by the different constraints assumed. Due to space and data constraints, only curves for the three regions and the electricity sector are shown.

The figures illustrate several important trends. Perhaps the most important message is the importance of a long-term vision for RE. RE deployment is consistently and significantly larger for 2050 than for 2030 in all regions and scenarios (caused by cost degression effects), often doubling at medium cost levels, except for OECD Europe. Even in this region, there is a large increase in RE deployment between these two time periods, although the ER-2010 scenario does not envision a larger than approximately 50% increase in RE deployment at most cost levels. On the other hand, a more than doubling of the potential deployment in both China and India in both scenarios during this period can be seen.

When comparing the three models, the IEA-WEO2008-Baseline projects the highest costs and lowest RE deployment in all three regions, while typically the ReMIND-RECIPE scenario envisions the lowest cost levels and highest RE deployment.[8] While in some regions the curves from different models are close to each other and project similar deployment levels at similar cost levels, the technologies they consider the most promising are often different. For instance, the ReMIND-RECIPE scenarios see the largest promise in PV and in 2050 the lion's share of its cost-effective RE deployment comes from this technology in all three regions. Projected RE deployment in the ER-2010 scenario consists of a balance of wind (on- and offshore), PV, concentrating solar power (CSP), hydropower and geothermal energy. The IEA-WEO2008-Baseline projects mainly wind and hydropower through 2030, and considers PV as too expensive in all regions. This is the technology for which the scenarios differ the most both in terms of costs and deployment level. For instance, the ReMIND-RECIPE's highest PV cost band for 2050 in OECD Europe is approximately one-fourth of the average PV cost projected by the IEA-WEO2008-Baseline by 2030,

---

8   ReMIND-RECIPE assumes that RE technologies will be deployed at the industrial scale at optimal sites and transported over large distances (up to continental scale) to demand centres. It implicitly assumes that bottlenecks, for example, with respect to grid infrastructure, are avoided by early and anticipatory planning. This results in high capacity factors in ReMIND-RECIPE compared to other scenarios, which in turn has a strong effect on electricity generation costs and deployment levels.

Mitigation Potential and Costs Chapter 10

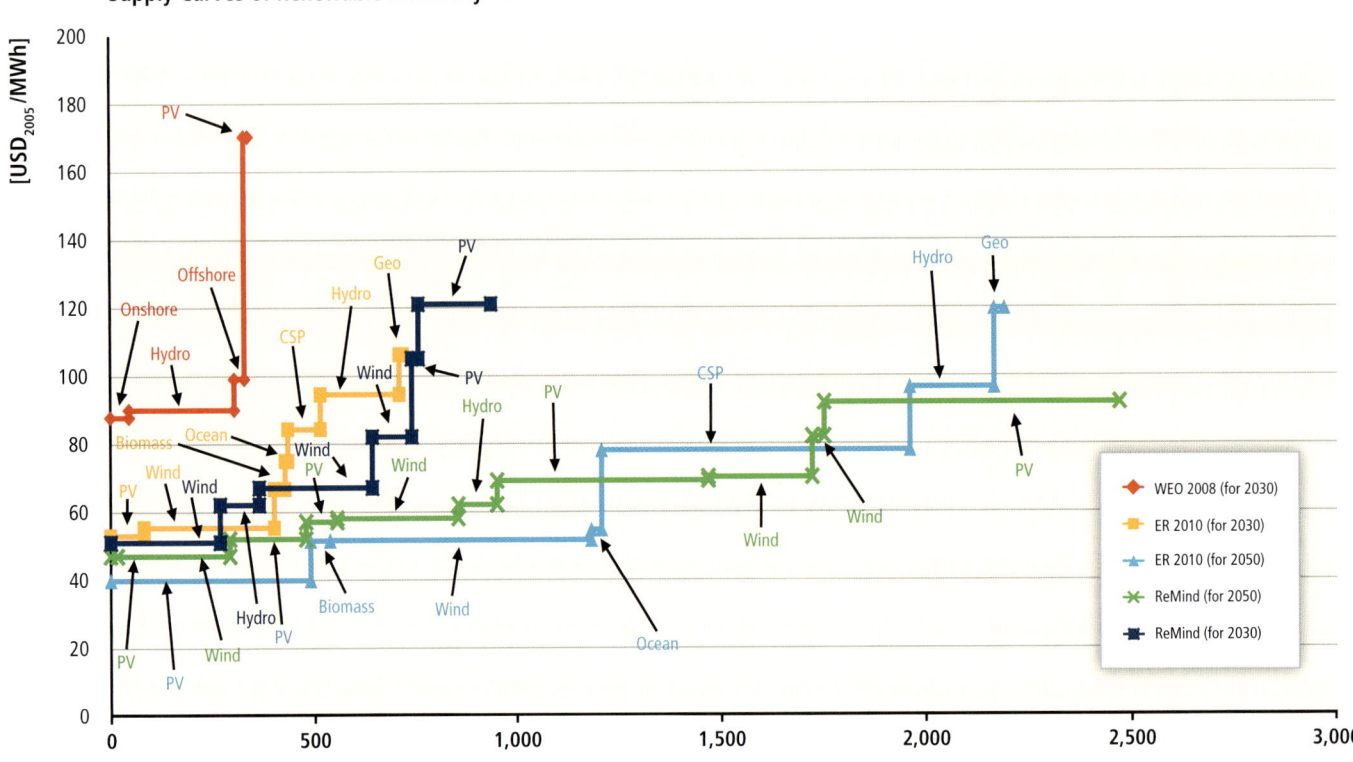

Figure 10.16 | Illustrative RE electricity supply curves for India for the years 2030 and 2050. The curves report scenario results (level of deployment) and are not a reflection of RE technical potential.

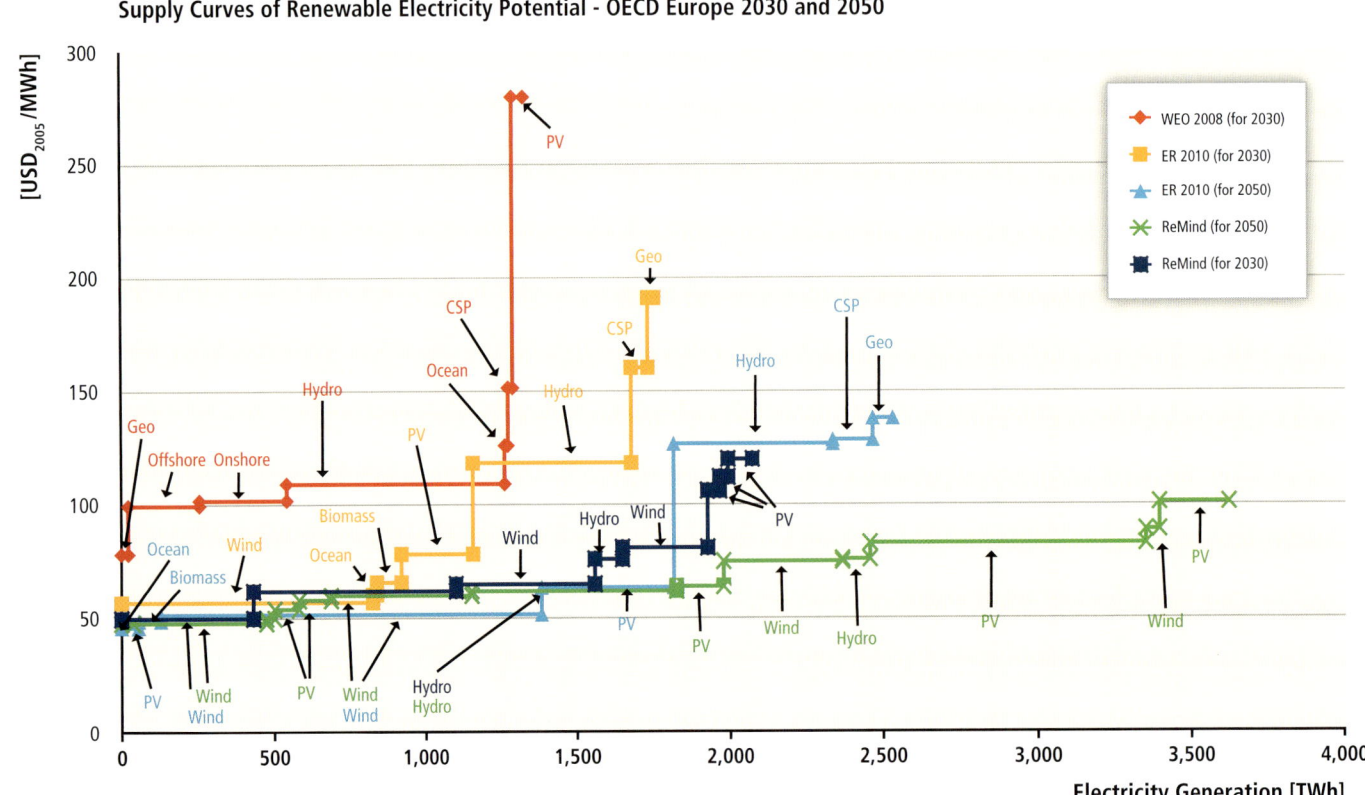

Figure 10.17 | Illustrative RE electricity supply curves for OECD Europe for the years 2030 and 2050. The curves report scenario results (level of deployment) and are not a reflection of RE technical potential.

and even the highest cost band in 2030 is half the average PV cost projected by that same study.

The different scenarios see different roles and costs for CSP. ReMIND-RECIPE considers a generic solar technology parameterized based on PV, and thus this technology was not specifically modelled in this scenario. The ER-2010 scenarios see a larger role for CSP than for PV in both China and India in the longer term, albeit at a higher cost. Neither of the models attributes a major deployment of geothermal energy, but they see its costs very differently. The costs of this electricity generation source in the IEA-WEO2008-Baseline is approximately half of that in the ER-2010 scenarios for the same target year (2030), and even in 2050 the ER-2010 cost projections are significantly higher for this technology than in the IEA-WEO2008-Baseline scenario in 2030—although the deployment levels at this cost are several times higher than projected by the other scenarios, making a noticeable contribution to the total deployment in 2050 in India and OECD Europe from among the examined regions. The ReMIND-RECIPE scenarios do not consider geothermal power.

With regard to the quality of electricity supply, it is also important to keep in mind that the presented supply curves do not distinguish between highly variable, and sometimes unpredictable, energy sources and dispatchable energy sources. In this context, a cost premium due to a higher reliability level that might be needed is also not considered as additional backup costs for highly variable RE sources.

10.3.2.2 Primary energy by region, technology and sector

This section provides an overview of the potential deployment paths given in the four scenarios versus the technical potential per region. For each technology, deployment shares indicate to what level the regional technical potential has been exploited. Figure 10.19 compares the resulting primary energy contribution of RE in relation to the technical potential by region and technology for the four scenarios, while Figure 10.18 gives an overview for all scenarios, but for RE as a whole by region, compared to the demand projections by 2050 and the current regional primary energy demand.

The maximum deployment share out of the overall technical potential for RE in 2050 was found for India with a total of 22.1% (ER-2010), followed by China with a total of 17.7% (ER-2010) and OECD Europe 15.3% (ER-2010). Two regions had deployment rates of about 5 to 7% of the regional available technical RE potential by 2050: 6.9% in developing Asia (MiniCAM-EMF22) and 5.5% for OECD North America (ER-2010). The remaining five regions used less than 4.5% of the

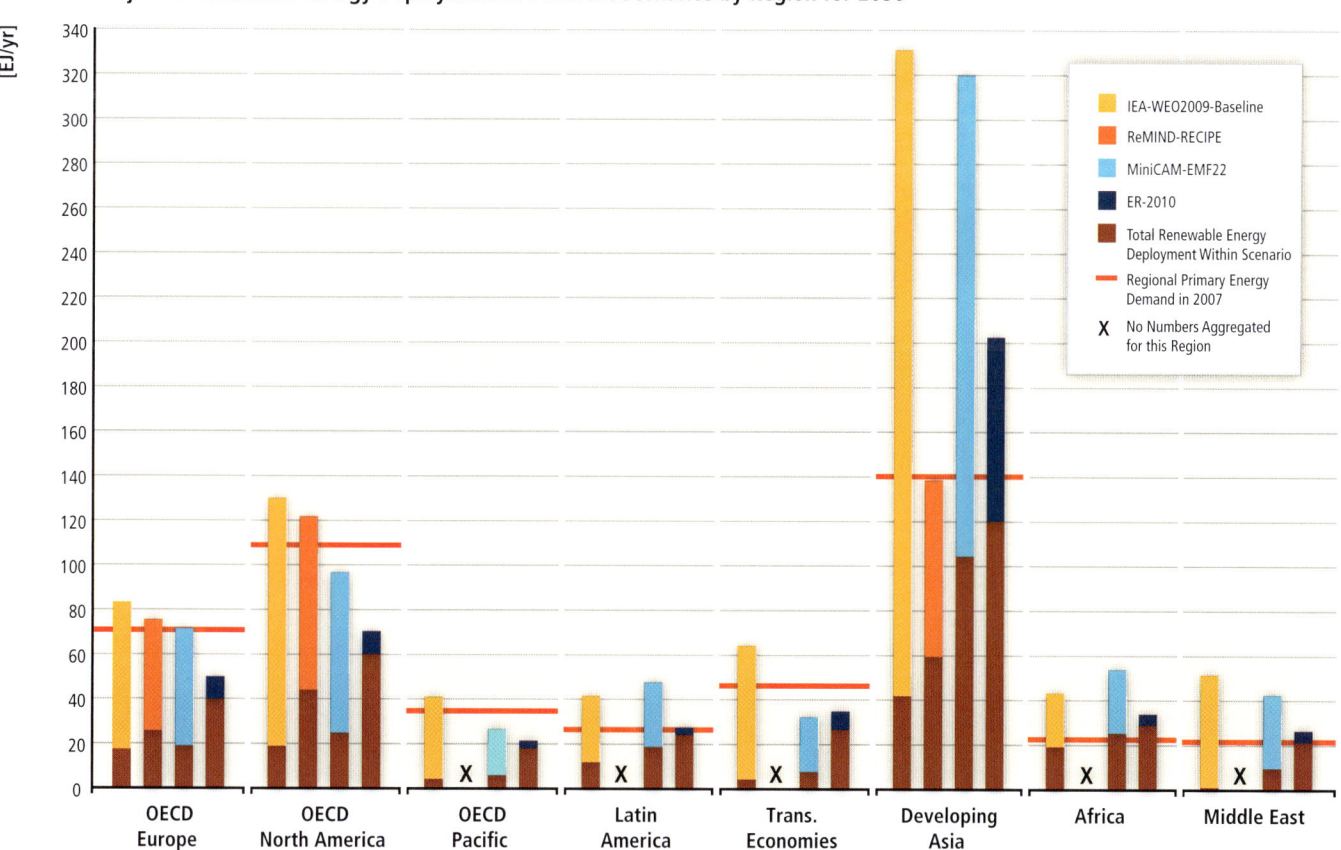

Figure 10.18 | Regional breakdown of possible energy demand and RE potential deployment for the selected set of four scenarios in 2050 (direct equivalent). Sources: IEA-WEO2009-Baseline (IEA, 2009; Teske et al., 2010), ReMIND-RECIPE (Luderer et al., 2009), MiniCAM-EMF22 (Calvin et al., 2009), ER-2010 (Teske et al., 2010). For comparison, total primary energy demand in 2007 is given (IEA, 2009).

available technical potential for RE. Wind energy has been exploited to a much larger extent in all regions than solar energy. Geothermal energy does not reach the technical potential limit in any of the scenarios analyzed, with the deployment rate remaining below 5% at both the regional and global level. Apart from some specific regions (e.g., China, India and Europe), the same is the case for ocean energy as a very young technology form. The established hydropower potential deployment at a global level covers roughly one-third of the technical potential, while in some specific regions the estimated capacity for 2050 is already very close to the maximum possible capacity.

While the overall technical potential for RE exceeds current global primary energy by an order of magnitude (see Chapter 1), even the two most ambitious scenarios in terms of RE deployment with comparable high growth rates for RE did not exceed 2.5% (ER 2010: 2.3%; MiniCAM-EMF22: 1.8%) of the given technical RE potential for 2050 at a global level.

10.3.3 Greenhouse gas mitigation potential of renewable energy in aggregate and as individual options

This section focuses on the question of how much RE can contribute to climate change mitigation, both in aggregate and as individual technologies. The numbers given in this section are derived from the results of the four illustrative scenarios (e.g., the underlying deployment paths of different RE technologies). As the amount of GHGs mitigated by renewable technologies greatly depends on the GHG intensity of the energy mix and on whether it is assumed that RE substitutes for fossil fuels only or also possibly other energy generation technologies (e.g., nuclear, other REs), the GHG mitigation potentials are provided over a range in this section to reflect the given uncertainties. Note that besides the fact that numbers are shown only for a limited number of scenarios, the following calculation is necessarily based on simplified assumptions and can only be seen as indicative.

For the power sector, the range is defined by the following three cases:

- Upper case: Substitution of the specific average CO_2 emissions of the fossil generation mix under the baseline scenario.

- Medium case: Substitution of the specific average CO_2 emissions of the overall generation mix under the baseline scenario.

- Lower case: Substitution of the specific average CO_2 emissions of the generation mix of the particular analyzed scenario.

For the electricity sector, Table 10.7 shows the underlying assumptions for the calculation of the CO_2 mitigation potential. The specific carbon

Table 10.7 | Assumptions for the CO_2 mitigation potential calculation: average specific CO_2 emissions from electricity generation or heat supply being substituted in the different scenarios. Sources for the underlying RE deployment: IEA-WEO2009-Baseline (IEA, 2009; Teske et al., 2010), ReMIND-RECIPE (Luderer et al., 2009), MiniCAM-EMF22 (Calvin et al., 2009), ER-2010 (Teske et al., 2010).

Average specific CO_2 Emissions		IEA-WEO2009-Baseline	ReMIND-RECIPE	MiniCAM-EMF22	ER-2010
Power Sector					
Upper Case	2020 [g CO_2/kWh] 2030 [g CO_2/kWh] 2050 [g CO_2/kWh]		812 768 716		
Medium Case	2020 [g CO_2/kWh] 2030 [g CO_2/kWh] 2050 [g CO_2/kWh]		625 580 531		
Lower case	2020 [g CO_2/kWh] 2030 [g CO_2/kWh] 2050 [g CO_2/kWh]	599 564 500	543 370 190	487 374 147	544 345 123
Heating + Cooling Sector					
Upper Case (Medium + 10%)	2020 [kt CO_2/PJ] 2030 [kt CO_2/PJ] 2050 [kt CO_2/PJ]		78.1[1] 78.1[1] 78.1[1]		
Medium Case	2020 [kt CO_2/PJ] 2030 [kt CO_2/PJ] 2050 [kt CO_2/PJ]		72[2] 72[2] 72[2]		
Lower Case (Medium -10%)	2020 [kt CO_2/PJ] 2030 [kt CO_2/PJ] 2050 [kt CO_2/PJ]		63.9[3] 63.9[3] 63.9[3]		

Notes: The medium case for the power sector was defined by taking the average of the baseline scenarios of the studies IEA-WEO2009, ReMIND-RECIPE and MiniCAM-EMF22 (ER-2010, being based on IEA-WEO2009, has no baseline of its own). The upper case is defined by only taking the fossil fuel component of the above baseline scenarios. The lower case assumes the substitution of the specific average CO_2 emissions of the generation mix of the particular analyzed scenario. As a pragmatic assumption for direct heat bioenergy 50% of the emission factor for heating and cooling have been applied to consider that relevant GHG emission occur in the process chain. (1) 39 kt CO_2/PJ (2) 36 kt CO_2/PJ (3) 32 kt CO_2/PJ.

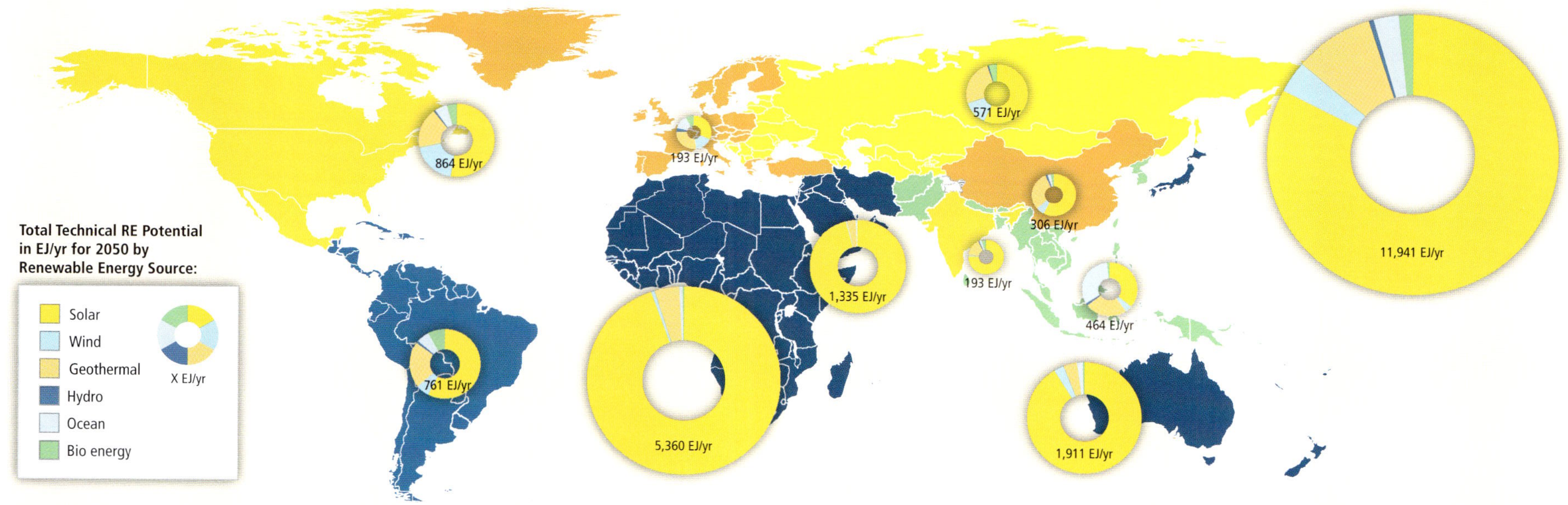

Chapter 10 — Mitigation Potential and Costs

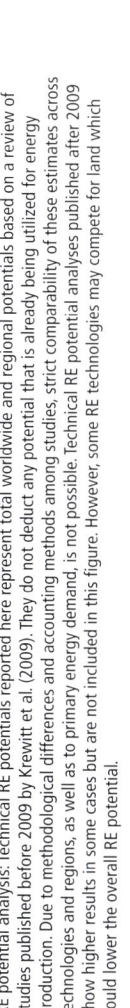
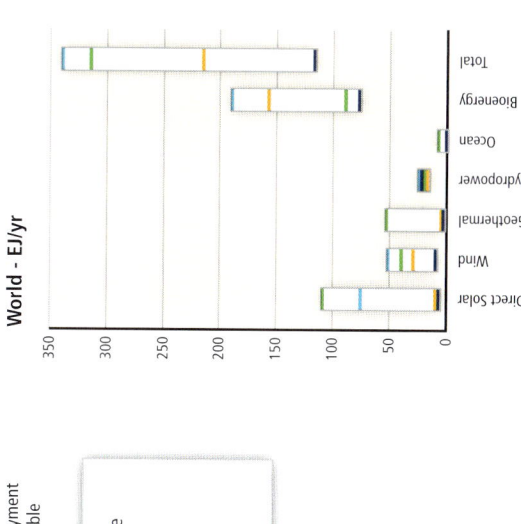

Total Technical RE Potential in EJ/yr for 2050 by Renewable Energy Source:
- Solar
- Wind
- Geothermal
- Hydro
- Ocean
- Bio energy

Technical RE Potential Can Supply the 2007 Primary Energy Demand by a Factor of:
0–2.5 | 2.6–5.0 | 5.1–7.5 | 7.6–10 | 10–12.5 | 12.6–15 | 15.1–17.5 | 17.6–20 | 20.1–22.5 | 22.6–25 | 25–50 | Over 50

Range graphs: Level of RE Deployment in 2050 by Scenario and Renewable Energy, in EJ/yr:
- IEA-WEO2009-Baseline
- ReMIND-RECIPE
- MiniCAM-EMF22
- ER-2010
- Range

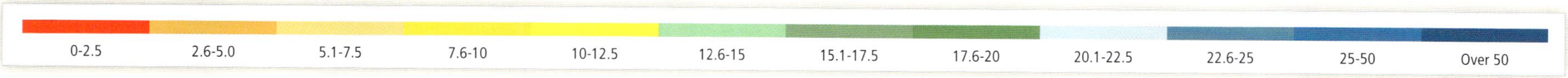

RE potential analysis: Technical RE potentials reported here represent total worldwide and regional potentials based on a review of studies published before 2009 by Krewitt et al. (2009). They do not deduct any potential that is already being utilized for energy production. Due to methodological differences and accounting methods among studies, strict comparability of these estimates across technologies and regions, as well as to primary energy demand, is not possible. Technical RE potential analyses published after 2009 show higher results in some cases but are not included in this figure. However, some RE technologies may compete for land which could lower the overall RE potential.

Scenario data: IEA WEO 2009 Reference Scenario (International Energy Agency (IEA), 2009; Teske et al., 2010), ReMIND-RECIPE 450ppm Stabilization Scenario (Luderer et al., 2009), MiniCAM EMF22 1st-best 2.6 W/2 Overshoot Scenario (Calvin et al., 2009), Advanced Energy [R]evolution 2010 (Teske et al., 2010)

Mitigation Potential and Costs

Chapter 10

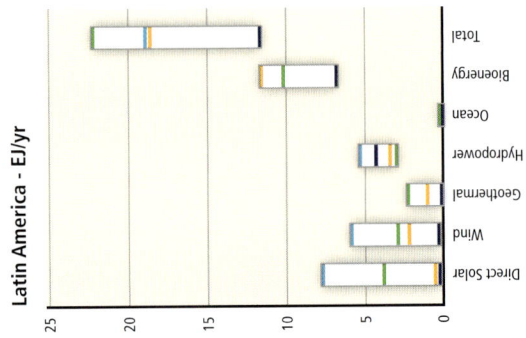
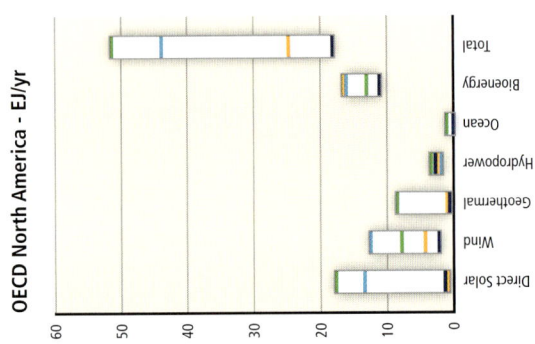
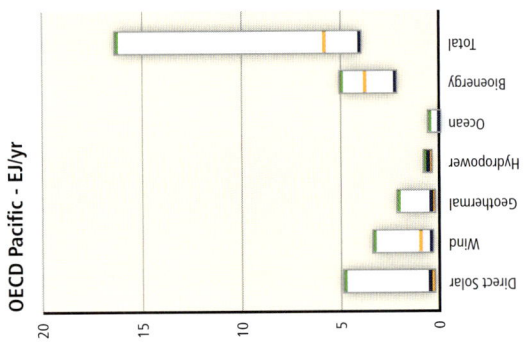
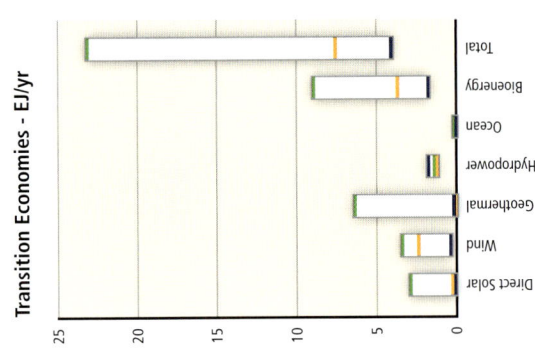
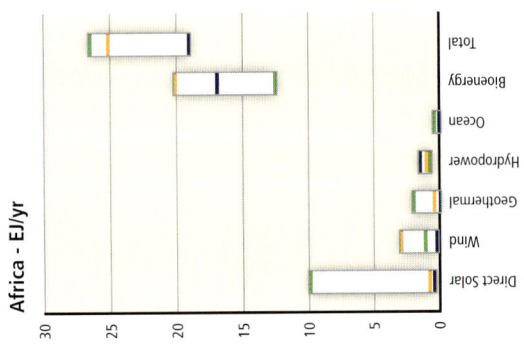
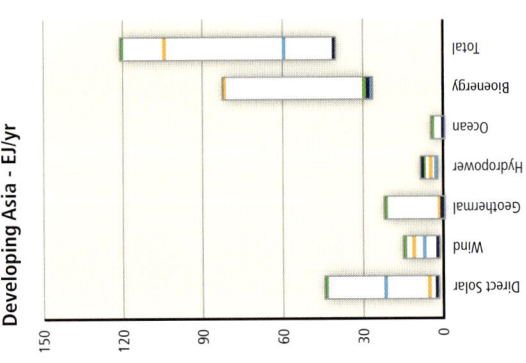
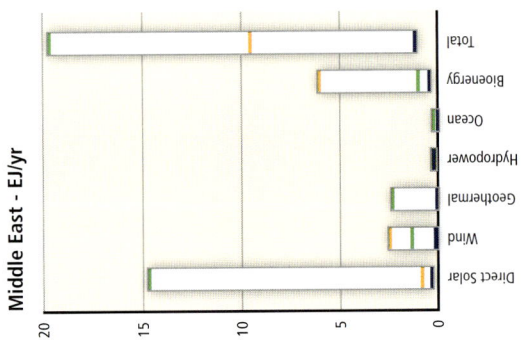
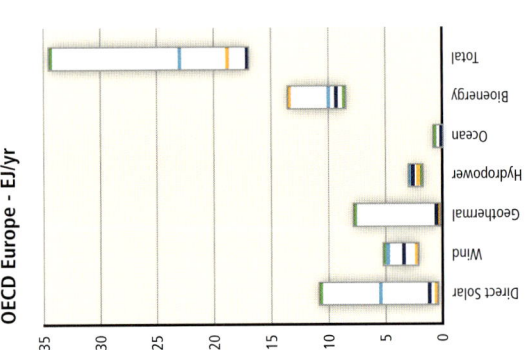

Figure 10.19 | (Preceding pages) Overview of the relation between the primary energy contribution of RE (direct equivalent) and the corresponding technical potential for different technologies and regions for 2050 for the selected set of four scenarios. Due to differences in regional aggregation not all models provide data for all regions.

Note: Data for technical potential presented in Chapters 2 through 7 may disagree with the figures in Krewitt et al. (2009) due to differences in assessed studies and the underlying methodologies (see also Chapter 1, in which Krewitt et al. (2009) worldwide RE technical potential estimates are compared to a range of values in the literature presented in Chapters 2 through 7).

emissions factor for the year 2050 ranges from 716 g CO_2/kWh (199 g CO_2/kJ) (upper case) to between 123 and 190 g CO_2/kWh (34 to 53 g CO_2/kJ) (lower case) for the selected mitigation scenarios. As noted in the table, a range of emission factors was also assumed for RE used in heating and cooling applications. In contrast to electricity generation, no specific information for these applications was available from the different scenarios. Against that background for the calculation, a pragmatic approach was selected for the underlying emission factors starting with a substitution of oil for the medium case and considering an uncertainty range.

Biofuels and other RE options for transport are excluded in the calculation due to limited data availability. To reflect the embedded GHG emissions saved due to bioenergy used for direct heating, only half of the theoretical CO_2 savings have been considered in the calculation. Given the high uncertainties and variability of embedded GHG emissions (see Chapter 2 for the discussion of indirect GHG emissions from the whole biomass process chain and Chapter 9 for a more general discussion on lifecycle assessment of different RE sources) this is necessarily once more a simplifying assumption.

Figure 10.20 shows the resulting annual CO_2 reduction potential by RE source for all scenarios for 2030 and 2050. The black line at 2.9 Gt CO_2/yr identifies 10% of the global energy-related CO_2 emissions; the red line here indicates 33% of total energy-related CO_2 emissions (base year for both lines is 2008).

The three mitigation scenarios of the illustrative scenarios show a wide range of possible RE contribution. While in all three, hydropower and wind energy play leading roles in 2030, in two of the scenarios (ReMIND-RECIPE, ER-2010) solar energy supersedes the other technologies by 2050. In contrast, as discussed earlier, due to the specific primary energy accounting approach the primary energy share ranking is led by bioenergy (see Section 10.3.1.4). This shows that the contributions (and effectiveness) of RE technologies vary by what perspective is taken (GHG mitigation or primary energy perspective). Further, the dependence of the resulting impacts on underlying assumptions is of great importance.

The resulting GHG reduction potential of all RE technologies heavily depends on the complex system behaviour determining the substituted

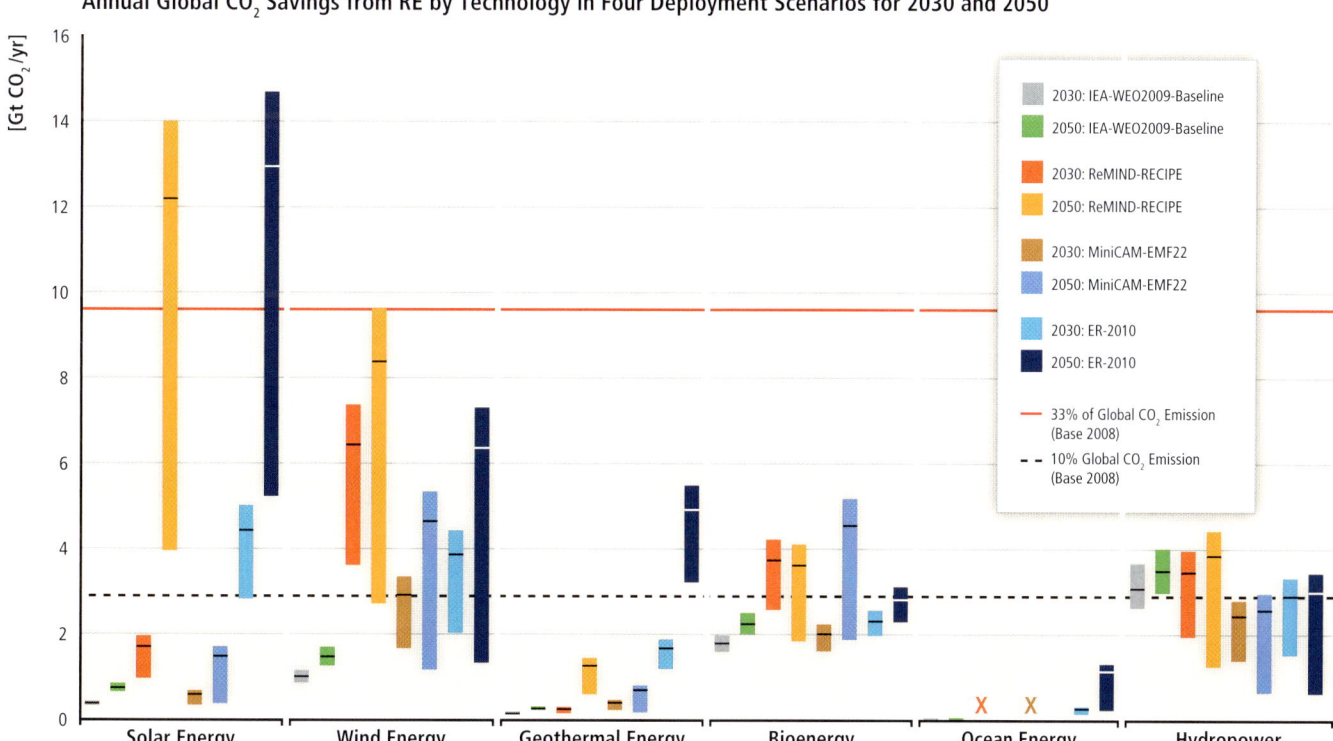

Figure 10.20 | Expected range of annual global CO_2 savings from RE for the four illustrative scenarios for 2030 and 2050. Biofuels for transport are excluded, and biomass used for direct heating only accounts for half the CO_2 savings due to imbedded GHG emissions from bioenergy. The presented range marks the high uncertainties regarding the substituted energy source: While the upper limit assumes full substitution of high-carbon fossil fuels, the lower limit considers specific CO_2 emissions of the analyzed scenario itself. Sources: IEA-WEO2009-Baseline (IEA, 2009; Teske et al., 2010), ReMIND-RECIPE (Luderer et al., 2009), MiniCAM-EMF22 (Calvin et al., 2009), ER-2010 (Teske et al., 2010). For comparison, global CO_2 emissions in 2008 are given (IEA 2010d).

energy sources. Considering the limitations of the rough approximations applied here, in the four scenarios the corresponding annual CO_2 reduction potential in 2050 reaches from 4.2 Gt CO_2/yr (MiniCAM-EMF22 lower case) to 35.3 Gt CO_2/yr (ER-2010 upper case) (Figure 10.21). At the upper level, this is equal to approximately 80% of the energy-related CO_2 emissions of the analyzed baseline scenario (IEA-WEO2009-Baseline) in the year 2050.

Cumulative CO_2 reduction potentials from RE sources up to 2020, 2030 and 2050 (Figure 10.22) have been calculated on the basis of the annual average CO_2 savings shown in Figure 10.21.[9] Based on this, the analyzed scenarios would have a cumulative reduction potential (2010 to 2050) in the medium case approach of between 244 Gt CO_2 (IEA-WEO2009-Baseline) under the baseline conditions, 297 Gt CO_2 (MiniCAM-EMF22), 482 Gt CO_2 (ER-2010) and 490 Gt CO_2 (ReMIND-RECIPE scenario). The full range across all calculated cases and scenarios for the cumulative CO_2 savings is between 218 Gt CO_2 (IEA-WEO2009-Baseline) and 561 Gt CO_2 (ReMIND-RECIPE), compared to about 1,530 Gt CO_2 cumulative fossil and industrial CO_2 emissions in the IEA-WEO2009-Baseline scenario during the same period.

Again, these numbers exclude CO_2 savings from RE use in the transport sector (including biofuels and electric vehicles). The overall CO_2 mitigation potential can therefore be higher.

10.3.4 Comparison of the results of the in-depth scenario analysis and knowledge gaps

All in-depth scenarios analyzed here show an increase in RE sources across all sectors. However, the electricity sector is in the forefront of all sectors and here the most dynamic increase in RE capacity is projected. Hydropower is expected to play the dominant role in the RE electricity sector in the near term and on a global basis, but based largely on already-existing installed generation capacity. Wind is expected in all three mitigation scenarios to overtake hydropower in terms of global electricity supply by 2030. The results for all other technologies are far more diverse. Two scenarios see solar PV as an important player in the electricity sector after 2030, with a share of more than 10% by 2050, while the baseline scenario projects PV remaining at marginal levels. In all but the ER-2010 scenario, the foreseen role for geothermal energy remains low at levels

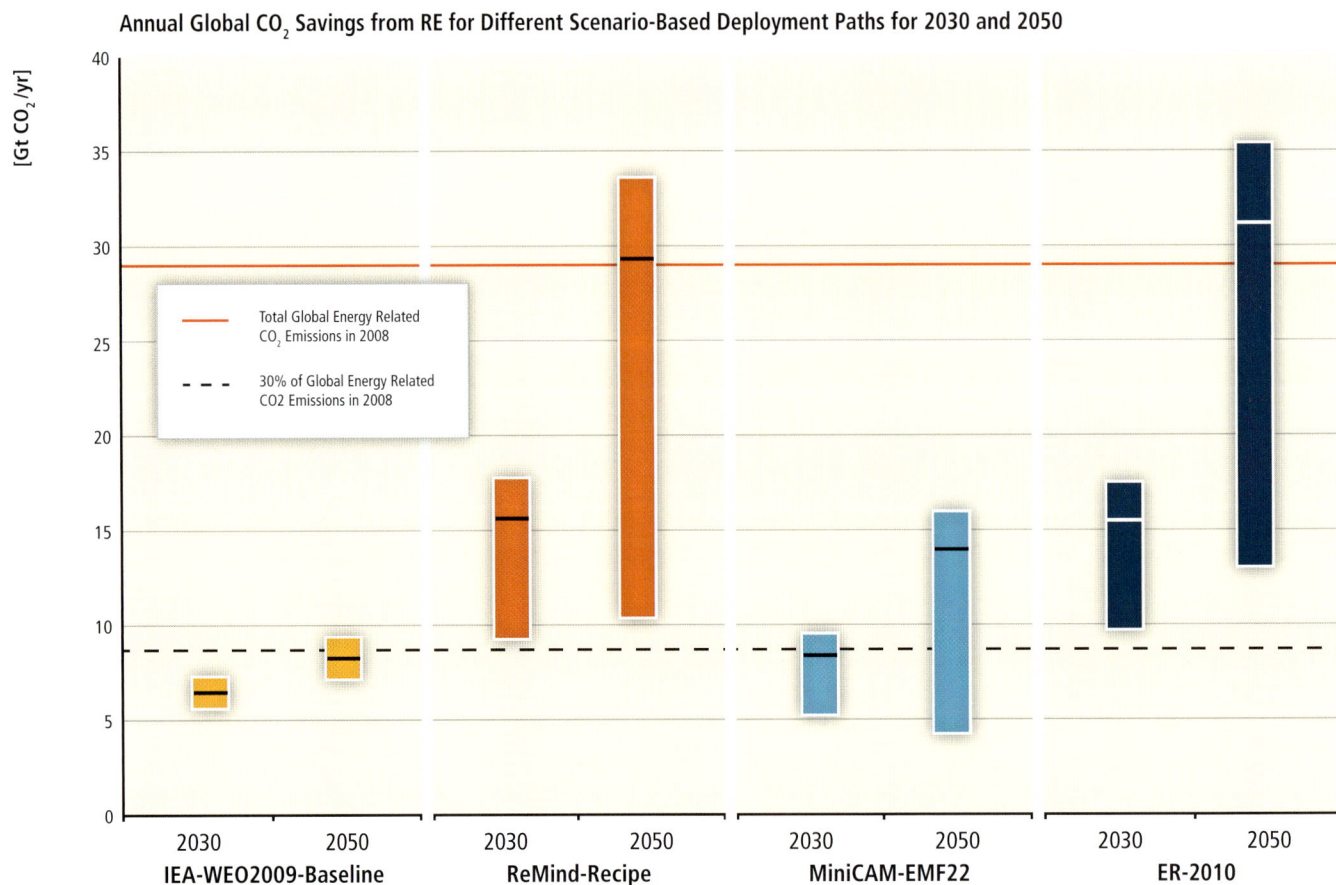

Figure 10.21 | Range of annual global CO_2 savings from RE in total for a set of four illustrative scenarios for 2030 and 2050 (Note: biofuels for transport are excluded, and biomass used for direct heating only accounts for half the CO_2 savings due to embedded GHG emissions from bioenergy). (The presented range marks the high uncertainties regarding the substituted energy source: while the upper limit assumes a full substitution of high-carbon fossil fuels, the lower limit considers specific CO_2 emissions of the analyzed scenario itself.) Sources: IEA-WEO2009-Baseline (IEA, 2009; Teske et al., 2010), ReMIND-RECIPE (Luderer et al., 2009), MiniCAM-EMF22 (Calvin et al., 2009), ER-2010 (Teske et al., 2010).

---

[9] For the integration, the time periods 2020 to 2030 and 2030 to 2050 were linearly interpolated.

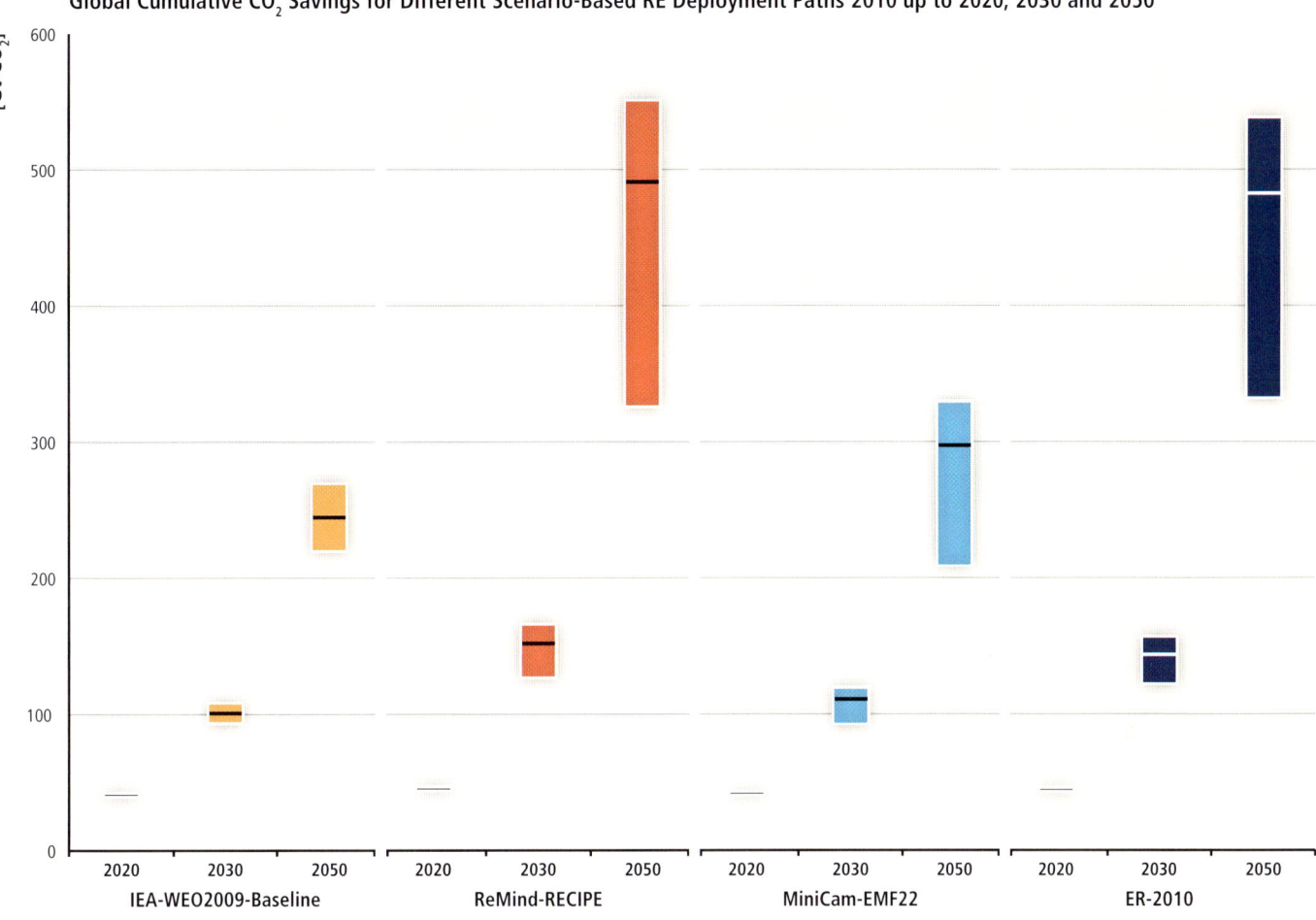

Figure 10.22 | Expected range of global cumulative CO_2 savings up to 2020, 2030 and 2050. The presented range marks the high uncertainties regarding the substituted conventional energy source: while the upper limit assumes a full substitution of high-carbon fossil fuels, the lower limit considers specific CO_2 emissions of the analyzed scenario itself. Sources: IEA-WEO2009-Baseline (IEA, 2009; Teske et al., 2010), ReMIND-RECIPE (Luderer et al., 2009), MiniCAM-EMF22 (Calvin et al., 2009), ER-2010 (Teske et al., 2010).

well below 5% of the global electricity supply. The scenario results for the heating and cooling sector include significant uncertainties as the models use different accounting methods, for example, for geothermal heat pumps. In terms of primary energy share, bioenergy has the greatest share—especially in the heating sector. Wind and solar energy are projected to become important players by and after 2030.

As already stressed in the comprehensive scenario survey (see Section 10.2), there are many reasons why the investigated scenarios reach different results. Each of the in-depth scenarios follows a different strategy. Significant differences in the demand projections and whether or not a shift towards more electricity within the transport and/or heating sector are projected to have a significant impact on the selected technologies and their deployment rates. Moreover, other mitigation technologies, such as CCS and/or nuclear, have a significant impact on the resulting role of RE sources in a future energy mix. In practice, a high RE deployment can only be achieved if system-relevant policy decisions are made many years ahead of the intended market penetration. The assumptions of expanded energy infrastructure such as transmission grids (see Chapter 8) or district heating networks can change the RE deployment of the scenario entirely. Even if the analyzed models do not include grid modelling via system integration aspects, these issues are at least covered implicitly in the scenarios (integration restrictions).

Due to comparably long lifetime expectations, the energy system is relatively inflexible and investment decisions have long-lasting impacts. A high share of relatively inflexible 'base load' power plants—such as coal, lignite and nuclear power plants, for instance—will reduce the technical and economic 'space' of variable renewable electricity generation like solar photovoltaic and wind. Technology choices and preferences predetermine the future RE deployment as well as the assumed RE cost developments and corresponding fossil fuel price projections. The overall share of RE in primary energy demand within the three in-depth mitigation scenario ranges from 24% (MiniCAM-EMF22) to 39% (ER-2010) by 2030 and 31% (MiniCAM-EMF22) to 77% (ER-2010) by 2050. Lower RE shares are due to the availability of competing low-carbon technologies such as CCS and nuclear, while scenarios not allowing access to these technologies expect higher RE shares, but not necessarily higher absolute numbers.

In addition to the comprehensive scenario survey in the previous section (see Section 10.2), the in-depth analyses of the four illustrative scenarios could deliver further specific insights into the specific RE technology deployment and the corresponding driving forces. However, often data availability limits detailed investigations. Against that background the following knowledge gaps can be identified:

- Lack of consistent RE technical potential estimates across the globe, and especially in developing countries (consistent economic potential estimates are an important input basis for the models).

- Modelling of the heating and transport sectors in most of the existing models is less detailed than modelling of the electricity sector, although both sectors are substantially contribute to GHG emissions. More generally, there is a severe lack of data for the heating and transport sector especially on a sectoral or regional basis.

- New RE technologies, such as ocean energy, are not represented in most of the current energy scenarios.

- The reporting system, for example, for geothermal heat pumps, is very different in all scenarios and sometimes not transparent, which makes it difficult to compare the results.

- The interaction of the technology pathways with the effects on deployment costs (learning effects) are treated differently in the scenarios and underlying assumptions or implemented calculation rules are sometimes not very well reported.

- Simplified calculations of the resulting CO_2 mitigation potential of RE deployment can give an orientation, but are associated with severe shortcomings. Comparative model runs (with and without RE) are necessary to consider the energy system behaviour in an appropriate way.

10.4 Regional cost curves for mitigation with renewable energies

10.4.1 Introduction

Governments and decision makers face limited financial and institutional resources and capacities for mitigation, and therefore tools that assist them in strategizing how these limited resources are prioritized have become very popular. Among these tools are abatement cost curves—a tool that relates the mitigation potential of a mitigation option to its marginal cost. Recent years have seen a major interest among decision- and policymakers in abatement cost curves, witnessed by the proliferation in the number of such studies and institutions/companies engaged in preparing such reports (e.g., Next Energy, 2004; Creyts et al., 2007; Dornburg et al., 2007; McKinsey&Company, 2007, 2008a, 2009b,c; IEA, 2008b). However,

while abatement curves are very practical and can provide important strategic overviews, it is pertinent to understand that their use for decision making has many limitations.

The aims of this section are to: (a) review the concept of abatement cost curves briefly and appraise their strengths and shortcomings (Section 10.4.2); (b) review the existing literature on regional abatement cost curves as they pertain to mitigation using RE (Section 10.4.3); and (c) review the literature on (regional) RE technology resource supply cost curves (Section 10.4.4). The section thus covers supply curves of RE on the one hand, which evaluate the unit costs of energy generation and the possibilities of utilizing the technical potential depending on the technology deployed, and on the other hand carbon abatement cost curves, which describe the mitigation potentials and marginal costs of emission mitigation (usually per tonne of CO_2eq.) through the deployment of renewable energy sources.

10.4.2 Cost curves: concept, strengths and limitations

10.4.2.1 The concept

The concepts of supply curves of carbon abatement, energy, or conserved energy all rest on the same foundation. They are curves consisting typically of discrete steps, each step relating the marginal cost of the abatement measure/energy generation technology or measure to conserve energy to its potential; these steps are ranked according to their cost. Graphically, the steps start at the lowest cost on the left with the next highest cost added to the right and so on, making an upward sloping left-to-right marginal cost curve. As a result, a curve is obtained that can be interpreted similarly to the concept of supply curves in traditional economics.

Supply curves of conserved energy were first introduced by Arthur Rosenfeld (see Meier et al., 1983) and became a popular concept in the 1980s (Stoft, 1995). The methodology has since been revised and upgraded, and the field of its application extended to energy generation supply curves including RE cost curves; as well as carbon abatement from the 1990s (Rufo, 2003). One of the benefits of the method was that it provided a framework for comparing otherwise different options, such as the cost-effectiveness of different energy supply options compared to energy conservation options, and therefore was a practical tool for some decision-making approaches, such as integrated resource planning. Although Stoft (1995) explains why the supply curves used in the studies by Meier et al. (1983) cannot be regarded as 'true' supply curves, including the fact that markets associated with the different types of options depicted in them, such as energy efficiency and energy supply markets, differ in many aspects, he maintains that they are useful for their purpose.

Despite the widespread use of supply curves and their advantages discussed above, there are some inherent limitations to the method that have attracted criticism from various authors that are important to review before reviewing the literature on them or presenting the regional cost curves.

10.4.2.2   Limitations of the supply curve method

The concept of abatement, energy and conservation supply curves has common and specific limitations. Much of the criticism in the early and some later literature focuses on the notion of options with negative costs. For instance, IEA (2008b) raises an objection based on the perfect market theory from neoclassical economics, arguing that it is not possible to have negative cost options as under perfect market conditions someone must have realized those options complying with rational economic behaviour. The existence of untapped 'profitable' (i.e., negative cost) opportunities represents a realm of debates ongoing for decades between different schools of thought (e.g., see Carlsmith et al., 1990; Sutherland, 1991; Koomey et al. 1998; Gumerman et al., 2001). Those accepting negative cost opportunities argue, among other things, that certain barriers prevent those investments from taking place on a purely market basis, but policy interventions can remove these barriers and unlock these profitable opportunities. Therefore the barriers prevailing in RE markets, detailed in other sections of this report, such as insufficient information, limited access to capital, uncertainty about future fuel prices (e.g., in the case of fossil fuels or biomass) or misplaced incentives (e.g., fossil fuel subsidies for social or other reasons) hinder a higher rate of investments into RE technologies, potentially resulting in negative cost options (Novikova, 2009).

A further concern about supply curves is raised by Gordon et al. (2008), who argue that the methodology simplifies reality. In their view, the curves do not reflect the real choices of actors, who accordingly do not always implement the available options in the order suggested by the curve. Both Gordon et al. (2008) and IEA (2008b) agree that there is the problem of high uncertainty in the use of supply curves for the future. This uncertainty is related to both economic and technological perspectives. Additional uncertainty arising from the methodology is the sensitivity of mitigation curves relative to the baseline assumption of the analysis (Kuik et al., 2009). Baker et al. (2008) have demonstrated that aggregation may also trigger significant uncertainty in abatement cost curves. For any given hour with given load and fuel prices, the expected monotonically rising (although not necessarily convex) relationship between price and abatement can be observed. However, when hours are aggregated into days, weeks, months and years, the constancy of the relationship will be completely lost. Perhaps one of the key shortcomings of the cost curves are that they consider and compare mitigation options individually (whereas typically a package of measures are applied together), therefore potentially missing synergistic and integrational opportunities, or potential overlaps. Optimized, strategic packages of measures may have lower average costs than the average of the individual measures applied using a piecemeal approach. Conversely, some measures may be more expensive or even become unviable when other measures are implemented. Any measures that compete against each other are substitutable, in some part or entirely (Sweeney and Weyant, 2008).

For GHG abatement cost curves, a key input that largely influences the results is the carbon intensity, or emission factor, of the country or area to which it is applied, and the uncertainty in projecting this into the future. This may lead to a situation where the option in one locality is shown to be a much more attractive mitigation measure as compared to an alternative than in another locality simply as a result of the differences in emission factors (Fleiter et al., 2009). As a result, a carbon abatement curve for a future date may say more about expected policies for fossil fuels than about the actual measures analyzed by the curves, and the ranking of the individual measures is also very sensitive to the developments in carbon intensity of energy supply.

Some concerns are emerging in relation to abatement cost curves that are not yet fully documented in the peer-reviewed literature (see Box 10.3). For instance, the costs of a RE technology in a future year largely depend on the deployment pathway of the technology in the years preceding—that is, the policy environment in the previous decades. The abatement cost of a RE option heavily depends also on the prices of fossil fuels, which are also very uncertain to predict. Furthermore, for variable (and sometimes to a degree unpredictable) RE generation technologies, the additional costs associated are not just a function of the amount of technology deployed. They are also a function of the fraction of the load met by the technology (higher fractions require more ancillary services, e.g., operating reserves), the flexibility of the existing generation portfolio, the location of the technology deployed relative to loads and existing transmission lines, etc.

Economic data, such as technological costs or retail rates, are derived from past and current economic trends that may obviously not be valid for the future, as sudden technological leaps, policy interventions or unforeseeable economic changes may occur—as has often been observed in the field of RE technology proliferation. These uncertainties can be mostly alleviated through the use of scenarios, which may result in multiple curves, such as for example in van Dam et al., (2007), and as presented in Sections 10.2 and 10.3. Some of the key uncertainty factors are the discount rates used and energy price developments assumed. The uncertainty about discount rates stems both from the fact that it is difficult to project them for the future, and because it is difficult to decide what discount rate to use, that is, social versus market discount rates (e.g., see Dasgupta et al., 2000). A number of studies (see e.g., Nichols, 1994) have discussed that in the case of investments in energy efficiency or RE, individual companies or consumers often use higher discount rates than would be otherwise expected for other types of, for example, financial investments. On the other hand, as Fleiter et al. (2009) note, society faces a lower risk in the case of such investments, therefore a lower discount rate could be considered appropriate from that perspective. Kuik et al. (2009) demonstrated that depending on the method used to construct them, abatement cost

> **Box 10.3 | Overview of selected key limitations of the cost/supply curve method:**
>
> - Controversy among scientists about opportunities at negative costs;
>
> - Strong focus on costs as selection criteria, while in reality actors base their decisions also on other criteria than those reflected in the curves;
>
> - Economic and technological uncertainty inherent to predicting the future, including energy price developments and discount rates;
>
> - Further uncertainty due to strong level of aggregation of the databases used (e.g., site- and technology-specific differences);
>
> - High sensitivity relative to baseline assumptions and the whole future generation and transmission portfolio;
>
> - Consideration of individual measures separately, ignoring interdependencies between measures applied together or in different order (including path dependency issues and treatment of transmission and integration aspects); and
>
> - For carbon abatement curves, high sensitivity to (uncertain) emission factor assumptions.

curves are affected by policies abroad. Essentially, policies abroad create a shift in the baseline for a country through changes in prices in energy markets as well as in price developments in RE technologies.

While several of these shortcomings can be addressed or mitigated to some extent in a carefully designed study, including those related to cost uncertainty, others cannot, and thus when cost curves are used for decision making, these limitations need to be kept in mind while discussing regional cost curves reviewed from the literature in the following section as well as regarding the regional cost curves out of the scenario results in Section 10.3.

10.4.3 Review of regional energy and abatement cost curves from the literature

10.4.3.1 Introduction

This section reviews key studies that have produced national or regional cost curves for RE and its application for mitigation. First, the section reviews work that looks at RE supply curves, followed by a review of the role of RE in overall abatement cost curves—since designated cost curves for RE alone are rare.

10.4.3.2 Regional and global renewable energy supply curves

In an attempt to review the existing literature on regional and global RE supply curves, a number of studies were identified, as summarized in Table 10.8. As discussed in the previous section, the assumptions used in these studies have a major influence on the shape of the curve, ranking of options and the opportunities identified by the curves. Therefore, the table also reviews the most important characteristics and assumptions of the models/calculations as well as their key findings.

In general, it is very difficult to compare data and findings from different RE supply curves, as there have been very few studies using a comprehensive and consistent approach and detailing their methodology, and most studies use different assumptions (technologies reviewed, base resource data, target year, discount rate, energy prices, deployment dynamics, technology learning etc.). Therefore, country or regional findings in Table 10.8 need to be compared with caution, and for the same reasons findings for the same country can be very different in different studies.

One of the weaknesses of many regional or technology studies is that they usually do not account for the competition for land and other resources among the various energy sources (except for probably the various plant species in the case of biomass). In studies that do take this into account (such as de Vries et al., 2007), technical potentials substantially decline in case of exclusive land use.

10.4.3.3 Regional and global carbon abatement cost curves

Table 10.9 summarizes the findings and characterizes the assumptions in the studies reviewed that construct regional/national/global carbon abatement cost curves with the perspective of the role of RE technology deployment. They have a different focus, goal and approach as compared to RE supply curve studies, and are broader in scope, examining RE within a wider portfolio of mitigation options.

Table 10.8 | Summary of RE supply curves for world, regions and countries, with the data grouped into cost categories. Baseline refers to the expected projection of the energy type, the details of which are described in the notes by the target year; most typically the projected total primary energy supply for the particular country, unless otherwise noted in the notes. Currency values are given as in the respective sources as base years are often not specified and conversion to USD$_{2005}$ is not possible.

Country/region		Cost (USD/MWh)	Total RE (TWh/yr) [EJ/yr]	Percent of baseline (%)	Discount rate (%)	Notes	Source
Global		<100	200,000–300,000 [720–1,080]	>100	10	• Combined data for onshore wind, solar PV and biomass given land use constraints and technology scenarios • Sources of uncertainty considered	de Vries et al. (2007), baseline: WEC (2004b) and Hoogwijk et al. (2004). Target year: 2050
Global (Biomass)		<100	97,200 [350]	N/A	10	• Study claims biomass production under this price can exceed present electricity consumption multiple times	Hoogwijk et al. (2003). Target year not specified
Global	Wind	<40 <60 <80 <100	2,000 [7.2] 23,000 [83] 39,000 [140] 42,000 [151]	6 72 123 133	10	• Liquid transport fuel and electricity from biomass, onshore wind, PV • Capacity calculated for the whole world; grid connections, supply-demand relationships etc. not incorporated • Global technical potential for electricity generation • High technology development scenario (IPCC SRES (IPCC, 2000) A1 scenario) with stabilizing world population and fast and widespread yield improvements.	RE data: de Vries et al. (2007) Target year: 2050 Baseline data: IEA (2003)
	Biomass	<60	59,000 [212]	187			
	PV	<80 <100	400,000 [1,440] 1,850,000 [6,660]	1,268 5,868			
Global		<70 <100	21,000 [76] 53,000 [191]	600–700 -	10	• Technical potential for onshore wind based on wind strength and land use issues; grid availability, network operation and energy storage issues are ignored • Baseline refers to 2001 world electricity consumption	Hoogwijk et al. (2004) Based on 2001 state of technology, no target year specified.
	Former USSR	<70 <100	2,000 [7.2] 7,000 [25]	160 550			
	USA	<70 <100	3,000 [11] 13,000 [47]	80 350			
	East Asia	<70 <100	0 [0] 50 [.2]	0 3			
	Western Europe	<70 <100	1,000 [3.6] 2,000 [7.2]	40 80			
Global		<50	121,805 [438]	N/A	10	• Biomass energy from short-rotation crops on abandoned cropland and unused rest land • Four IPCC SRES (2000) land use scenarios for the year 2050 • Land productivity improvement over time, cost reductions due to learning and capital-labour substitution • Present world electricity consumption (20 PWh/yr) may be generated at costs below USD 45/MWh (IPCC SRES (IPCC, 2000) A1 B1 scenarios) and USD 50/MWh (IPCC SRES (IPCC, 2000) A2 B2 scenarios) in 2050	Hoogwijk et al. (2009). Target year: 2050
	Former USSR		23,538 [85]				
	USA		9,444 [34]				
	East Asia		17,666 [64]				
	OECD Europe		3,194 [12]				
Central and Eastern Europe		<100	3,233 [12]	74	N/A	• Biomass only, best scenario with willow being the selected energy crop (highest yield) • Countries: Bulgaria, Czech Republic, Estonia, Hungary, Latvia, Lithuania, Poland, Romania, Slovakia • Baseline data includes Slovenia, however, its share is rather low, therefore resulting distortion is not so high.	RE data: van Dam et al. (2007) Target year: 2030 Baseline data: Solinski (2005)
Czech Republic		<100	101 [.4]	20	4	• Only biomass production • Best-case scenario where future yields equal the level of the Netherlands	RE data: Lewandowski et al (2006) Target year: 2030 Baseline data: IEA (2005)

Continued next Page →

Country/region	Cost (USD/MWh)	Total RE (TWh/yr) [EJ/yr]	Percent of baseline (%)	Discount rate (%)	Notes	Source
India	<100	56 [.2]	3.4	10	• Small hydro • Grid availability not expected to be a serious concern • Baseline refers to 2005 electricity consumption	Pillai and Banerjee (2009) Target year: 2030
	<200	90 [.3]	5.6		• Wind • Grid availability not expected to be a serious concern • Baseline refers to 2005 electricity consumption	
Netherlands	<100	22 [.08]	2.1	N/A	• Included: onshore and offshore wind, PV, biomass and hydro • Discount rate is not available, however, this option is a scenario where sustainable production is calculated. Therefore they use 5% internal rate of return (IRR) assuming that there are governmental support • Baseline is total primary energy supply forecast for 2020 by IEA	RE data: Junginger et al., 2004 Target year: 2020 Baseline data: IEA (2006)
	<200	23 [.08]	2.2			
	<300	24 [.09]	2.3			
UK	<100	81 [.3]	22	7.9	• Included: 'Low-cost technologies' (landfill gas, onshore wind, sewage gas, hydro) • Costs: capital, operating and financing elements • Baseline is all electricity generated in the UK forecasted for 2015	RE data: Enviros Consulting Ltd. (2005) Target year: 2015 Baseline data: UK SSEFRA (2006)
	<200	119 [.4]	33			
USA	<100	3,421 12]	15	N/A	• Wind energy only	RE data: Milligan (2007) Target year: 2030 Baseline data: EIA (2009)
USA (WGA)	<100	177 [.6]	0.77	N/A	• Only the WGA region • CSP, biomass, and geothermal • Geothermal reaches maximum capacity under USD 100/MWh • CSP has a large technical potential, but full range is between USD 100 and 200/MWh	RE data:(Mehos and Kearney, 2007; Overend and Milbrandt, 2007; Vorum and Tester, 2007) Target year: 2030 Baseline data: EIA (2009)
	<200	1,959 [7]	8.5			
	<300	1,971 [7]	8.6			
USA (Arizona 2025)	<100	0.28 [.001]	N/A	Biomass and PV: 7.5 Rest: 8	• State of Arizona, USA • RE: wind, biomass, solar, hydro, geothermal • Discount rates vary between energy sources	RE data: Black & Veatch Corporation (2007) Target year: 2025
	<200	10.5 [.04]	N/A			
	<300	20 [.07]	N/A			

One general trend can be observed based on this illustrative sample of a limited number of selected studies. Abatement cost curve studies tend to find lower potentials for mitigation through RE than those focusing on RE for energy supply. Even for the same country these two approaches may find very different mitigation potentials.

One factor contributing to this general trend is that RE supply studies typically examine a broader portfolio of RE source technologies, while the carbon mitigation studies reviewed focus on selected resources/technologies to keep models and calculations within reasonable complexity levels.

The highest figure in carbon mitigation potential share by the deployment of RE, as shown in Table 10.9, is for Australia: 13.4% under USD 100/t CO_2eq by 2030. This has to be seen in contrast with the much higher shares as a percentage of national total primary energy supply (TPES) reported in the previous section (data from McKinsey&Company, 2008a). Besides Australia, countries with the most promising abatement potentials through RE sources identified in the sample of studies are China and Poland—both having high emission factors.

10.4.4 Review of selected technology resource cost curves

The energy and abatement cost curves discussed above provide a more aggregated picture (see Sections 10.4.2 and 10.4.3). For selected technologies, this section ends with the discussion of illustrative examples of resource cost curves. In this context, some studies are highlighted that were already part of the general overview in Section 10.4.3.

Table 10.9 | Summary of carbon abatement cost curves for world, regions and countries (cells including grey literature are coloured in grey).

Country/region	Year	Cost (USD/tCO$_2$eq)	Mitigation potential (Mt CO$_2$)	Percent of baseline (%)	Discount rate (%)	Notes	Source
Global	2050	<200	46,195	85	N/A	• Key sensitivities: lower technical potential for wind, hydro or CCS, lower uranium resources raise abatement costs by 2 to 5%	Syri et al. (2008) Baseline model: global ETSAP/TIAM Baseline Scenario: IEA (2009)
Global	2030	<100	6,390	9.1	4	• Scenario A (maximum growth of RE and nuclear) • Scenario B (50% growth of RE and nuclear)	McKinsey&Company (2009b)
		<100	4,070	5.8			
Annex I	2020	<100	2,818	20	N/A	• Different abatement allocations analyzed depending (equal marginal cost, per capita emission right convergence, equal percentage reduction) • CO$_2$ equivalent emissions six Kyoto GHGs, but exclude LULUCF • Costs in 2005 USD	den Elzen et al. (2009) Baseline Scenario: IEA WEO (IEA, 2009)
Australia	2020	<100	74	9.5	N/A		(McKinsey&Company, 2008a)
Australia	2030	<100	105	13			
Australia (NSW Region)	2014	<100	8.1	1.0	N/A	• New South Wales region • Includes governmental support for RES	Abatement data: Next Energy (2004) Baseline data: McKinsey&Company (2008a)
		<300	8.5	1.1			
China	2030	<100	1,560	11	4		(McKinsey&Company, 2009a)
China	2030	<50	3,484	27	N/A	• Storylines do not describe all possible development (e.g., disaster scenarios, explicit new climate policies) • Main abatement (half of total) is efficiency, the rest is renewable and fuel switch from coal	van Vuuren et al. (2003) Baseline Scenario: ERI 2009
China	2030	<100	2,323	18	N/A	• Main factor influencing abatement cost is constraints on the rollout of nuclear power • Baseline seems to be underestimated as 2010 power consumption is 40% below fact.	Chen, 2005 Baseline Scenario: ERI (2009)
Czech Republic	2030	<100	9.3	6.2	N/A	• Scenario with maximum use of RE sources	McKinsey&Company (2008b)
		<200	11.9	8.0			
		<300	16.6	11			
Germany	2020	<100	20	1.9	7	• Societal costs (governmental compensation not included)	McKinsey&Company (2007)
		<200	31	3.0			
		<300	34	3.2			
Poland	2015	<100	50	11	6	• Only biomass • Best case scenario	Abatement data: Dornburg et al. (2007) Baseline data: EEA (2007)
		<200	55.9	12			

Continued next Page →

Country/region	Year	Cost (USD/tCO$_2$eq)	Mitigation potential (Mt CO$_2$)	Percent of baseline (%)	Discount rate (%)	Notes	Source
Switzerland	2030	<100	0.9	1.6	2.5	• Base case scenario	McKinsey&Company (2007)
South Africa	2050	<100	83	5.2	10	• Renewable electricity to 50% scenario	Hughes et al. (2007)
Sweden	2020	<100	1.26	1.9	N/A		McKinsey&Company (2008c)
USA	2030	<100	380	3.7	7		Creyts et al. (2007)
UK	2020	<100	4.38	0.46	N/A		Confederation of British Industry (CBI, 2007)
		<200	8.76	0.93			
UK	2020	<100	7	4.0	3.5		Committee on Climate Change (2008)
		<200	33	18.8			

Resource cost curves have to be seen in context with the discussion of the energy and cost aspects in the various technology chapters (Chapters 2 through 7).

Summary of biomass resource cost curves.[10] The analyses of biomass resource cost curves in the literature use typically different land use scenarios (de Vries et al., 2007; Hoogwijk et al., 2009). They take into account geographical specificities (crop productivity and land availability) as well as capital and labour input. Hoogwijk et al. (2009) find that biomass can supply about 40 to 70% of the present primary energy consumption (130 to 270 EJ/yr) by 2050 at costs below USD 2/GJ/yr, which is the present lower limit of the cost of coal (see Figure 10.23).

Regions of low production cost and relatively high technical potential are the former USSR, Oceania, eastern and western Africa and East Asia. Cost reductions are due to land productivity improvements over time, learning and capital-labour substitution. Biomass-derived electricity costs are at present slightly higher than electricity base-load costs. The present world electricity consumption of around 20 PWh/yr (72 EJ/yr) may be generated in 2050 at costs below USD 12.5/GJ in two scenarios,

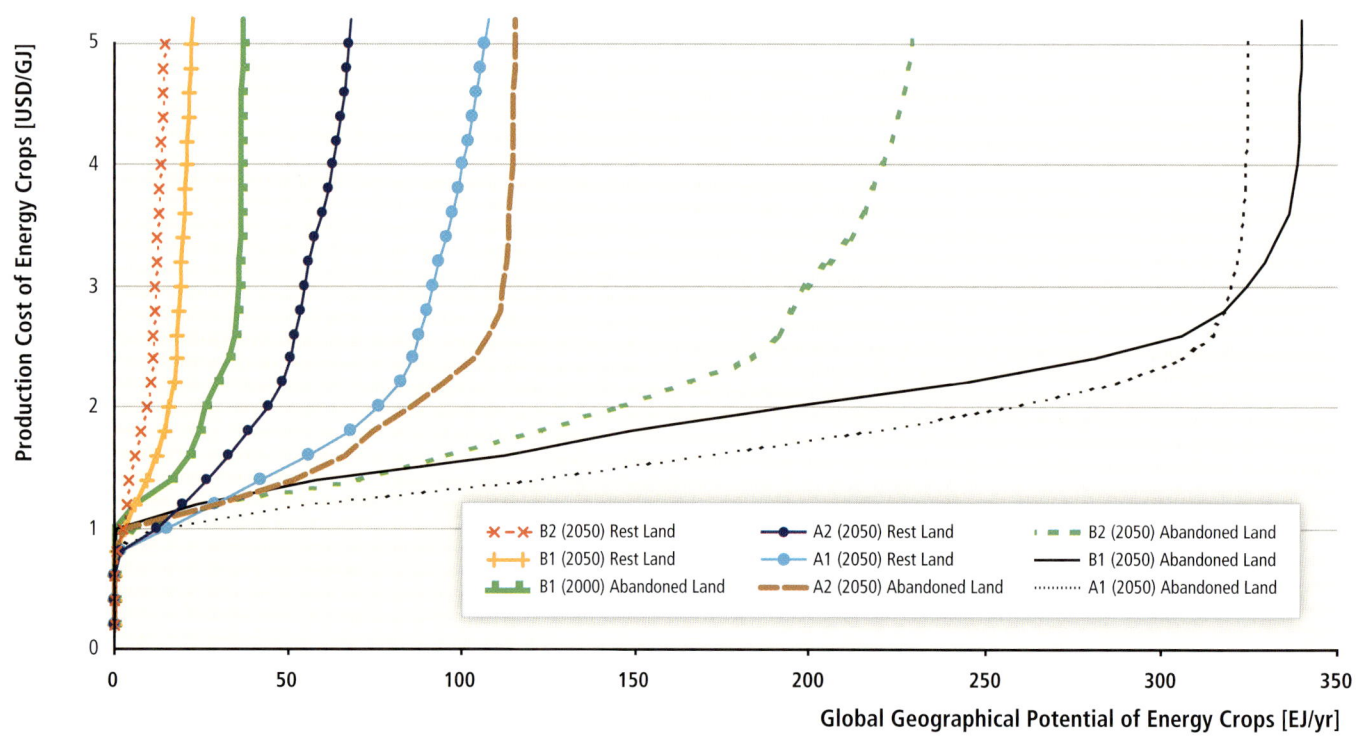

Figure 10.23 | Global average cost-supply curve for the production of bioenergy plants on the two land categories 'abandoned land' (agricultural land not required for food) and 'rest land' in 2050. The curves are generated based on IMAGE 2.2 modelling of four SRES scenarios. The cost supply curve for abandoned agricultural land in 2000 (SRES B1 scenario) is also shown. Source: Hoogwijk et al. (2009). The scenarios A1, A2, B1 and B2 correspond to the storylines developed for the IPCC Special Report on Emission Scenarios (IPCC, 2000).

10 For further details, see Section 2.2.

while below USD 15.3/GJ in two others. At costs of USD 16.7/GJ, about 18 to 53 PWh/yr (65 to 191 EJ/yr) of electricity can be produced in 2050. The global curve that sums all regional curves is found to be relatively flat up to 300 EJ/yr; land rental costs and the substitution of capital for labour represent the highest sensitivity.

In the study of de Vries et al. (2007), another trade-off is addressed: food versus energy. The authors assess four land use scenarios, each corresponding to different levels of food trade, technology development and population. A low technical potential estimate in the A2 scenario is a direct consequence of more people, hence higher food demand and lower yield (improvement), hence more land demand for food production (see Figure 10.24).

For a cost range of electricity from biomass of USD 13.9 to 27.8/GJ, there were 7 PWh (25 EJ) of technical potential in the year 2000, while for a projected cost range between USD 8.3 and 27.8/GJ, there is an estimated technical potential of 59 PWh (212 EJ) by 2050 (with a sensitivity of 30 to 85 PWh/yr (108 to 310 EJ/yr), depending upon discount rates, land use patterns, technology assumptions and land use implementation fractions).

Summary of PV resource cost curves. De Vries et al. (2007) estimate PV electricity generation technical potential at 4,105 PWh/yr (4,778 EJ/yr) in 2050 at the cost of USD 16.7 to 69.4/GJ. Since the technical potential for the year 2050 depends primarily on cost-reducing innovations, for a cut-off cost level of USD 27.8/GJ, a non-zero technical potential emerges only under specific scenario conditions (e.g., high economic growth vs. low population growth, or medium economic and population growth), as in the IPCC (2000) A1 and B1 scenarios (see Figure 10.25).

In this particular study, solar PV economic potential is sensitive to competition for land. If the technological breakthroughs do not take place, a large part of the major technical potential is unlikely to become economic. Its capital-intensive nature also makes it sensitive to changes in discount rates. High or low exclusion factors also affect the solar PV technical potential. For the technical potential, land is not a constraint as even with a high exclusion factor, the technical potential is over 20 times the 2000 world electricity demand (de Vries et al., 2007).

Summary of onshore wind cost curves. Papers assessing wind technical potentials usually base their data on climatic models of wind speeds or interpolation of wind speed measurements (Hoogwijk et al., 2004; de Vries et al., 2007; Changliang and Zhanfeng, 2009). Hoogwijk et al. (2009) have made explicit assumptions about the average turbine availability, wind farm array efficiency and spacing, and, related to this, power density; this has not differentiated across grid cells, that is, one global parameter has been used. The estimated global technical potential that can be realized at relatively low cost is largely confined to three regions (Figure 10.26). These are the USA, the Former USSR and Oceania (Hoogwijk et al., 2004; McElroy et al., 2009). Wind power might even be generated at costs below USD 11.1/GJ in scenarios assuming either high

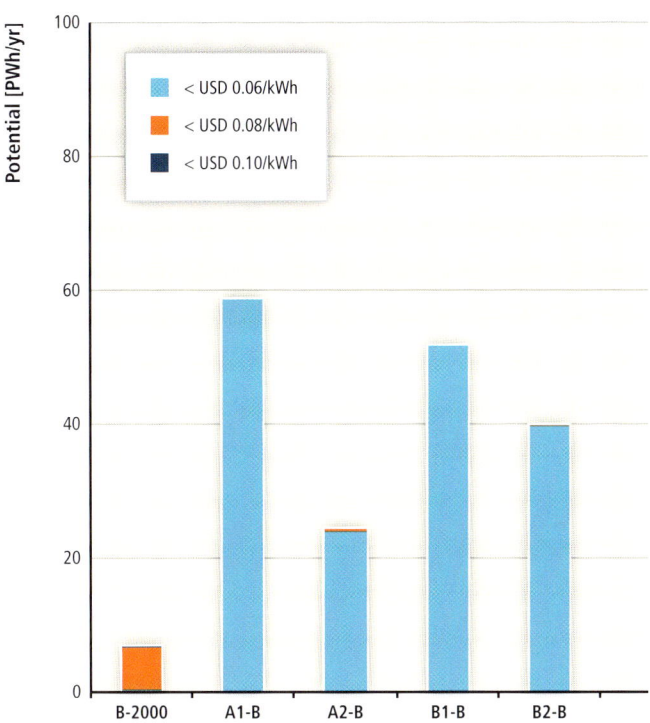

Figure 10.24 | The global technical potential for electricity from biomass in 2000 and in four IPCC SRES (IPCC, 2000) scenarios for 2050 for four production categories (de Vries et al., 2007).

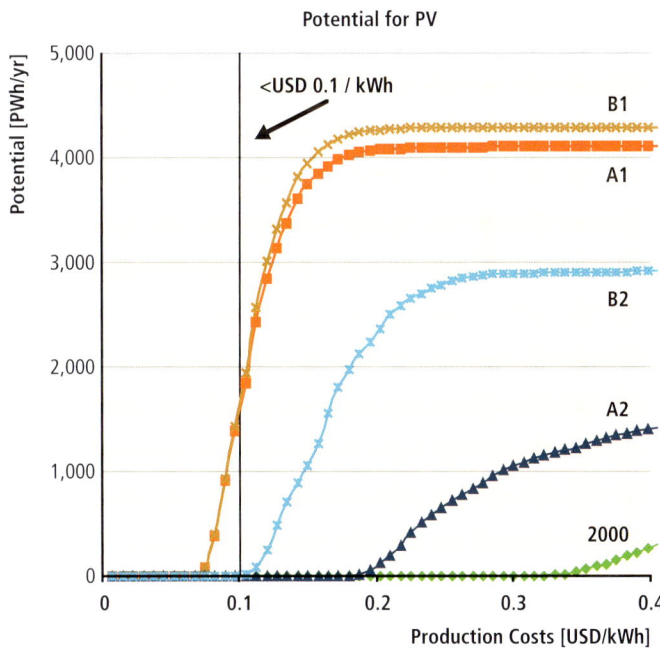

Figure 10.25 | Resource supply cost curve for PV for four IPCC SRES (IPCC, 2000) scenarios in 2050. The figure also shows the USD 0.1/kWh (USD 0.03/MJ) line used in the paper as the cut-off cost in determining the economic potential (de Vries et al., 2007).

economic growth and low population growth or medium economic and population growth (IPCC SRES (IPCC, 2000) A1 and B1 scenarios), which is significantly lower than the current cost level (see Chapter 7).

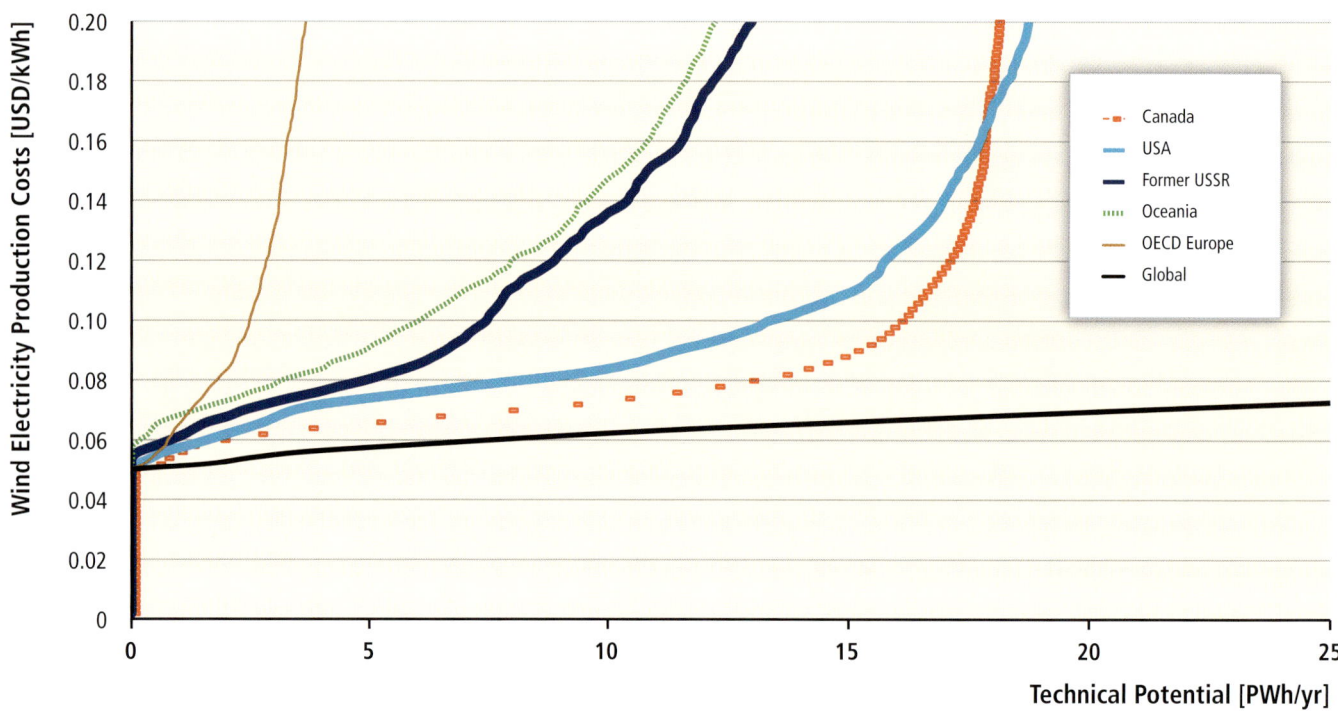

Figure 10.26 | Global, regional and country cost-supply curves for wind energy (USD/kWh versus PWh/yr) (Hoogwijk et al., 2004).

Finally, none of the studies reviewed here fully consider transmission and integration issues (see Chapter 8). In one study that did seek to account for these factors, wind remains an important contributor to the worldwide economic potential at less than USD 27.8/GJ, with an economic potential between 8 and 43 PWh/yr (29 and 155 EJ/yr)—or 50 to 300% of the 2000 world electricity demand (de Vries et al., 2007).

Summary of offshore wind cost curves. For offshore wind, the technical potential and costs are strongly determined by the distance of the installation from the shore and the water depth. In a recent study by EEA (2007), the lower limit of wind speed at hub height has been set to 5.0 m/s to consider the wind power plant economically viable. At an average production cost of USD$_{2005}$ 0.024/MJ (6.9 Eurocents/kWh) in 2030, 5,800 GW of offshore wind power could be developed in Europe (Figure 10.27).

Various studies have assessed the technical potential for offshore wind. Nevertheless, only Fellows (2000) presents the assessments at a global level (except Norway and Canada), including cost estimates for the time frame to 2020. Hoogwijk and Graus (2008) have added values for Canada and updated the data for the technological development for 2020 to 2050. High technical potentials are found in OECD Europe and Latin America, the latter having high shares of unexplored low-cost technical potentials. An economic potential of 1.2 PWh/yr (4.3 EJ/yr) for OECD Europe and Latin America is found at costs lower than USD 27.8/GJ. At costs above USD 13.9/GJ, 0.3 PWh/yr (1 EJ/yr) is available in OECD Europe, and 0.55 PWh/yr (1.98 EJ/yr) in Latin America. The lowest technical potentials are found in the Middle East, where even at less than USD 27.8/GJ only 0.18 PWh/yr (0.65 EJ/yr) capacity is available (Hoogwijk and Graus, 2008).

Summary of technology resource cost curves. This section has reviewed selected resource cost curves for selected RE technologies for which such curves were found. It is important to emphasize that such studies are comparable only to a limited extent due to the use of different methodologies and potentially conflicting assumptions (such as related to land use), thus they should not be directly used for potential summation or comparison purposes. These results also significantly differ from the integrated technology cost curves produced based on scenarios presented in Section 10.3.2.1, since these present potential deployment levels taking into account many more constraints than the technical potential/cost studies in Section 10.3.

10.4.5 Gaps in knowledge

There is a major gap in knowledge for RE heat and transport fuel technical potentials on a regional basis, especially as a function of cost. Additionally, the real benefit of the cost curve method (to identify the really cost-effective opportunities) in practice cannot be fully utilized with the given data sets. Average costs for a technology for a whole region mask the really cost-effective technical potentials and sites into an average, compromised by the inclusion of less attractive sites or sub-technologies. Therefore, significant, globally coordinated further research is needed for refining these curves into sub-steps by sites and sub-technologies in order to identify the most attractive opportunities

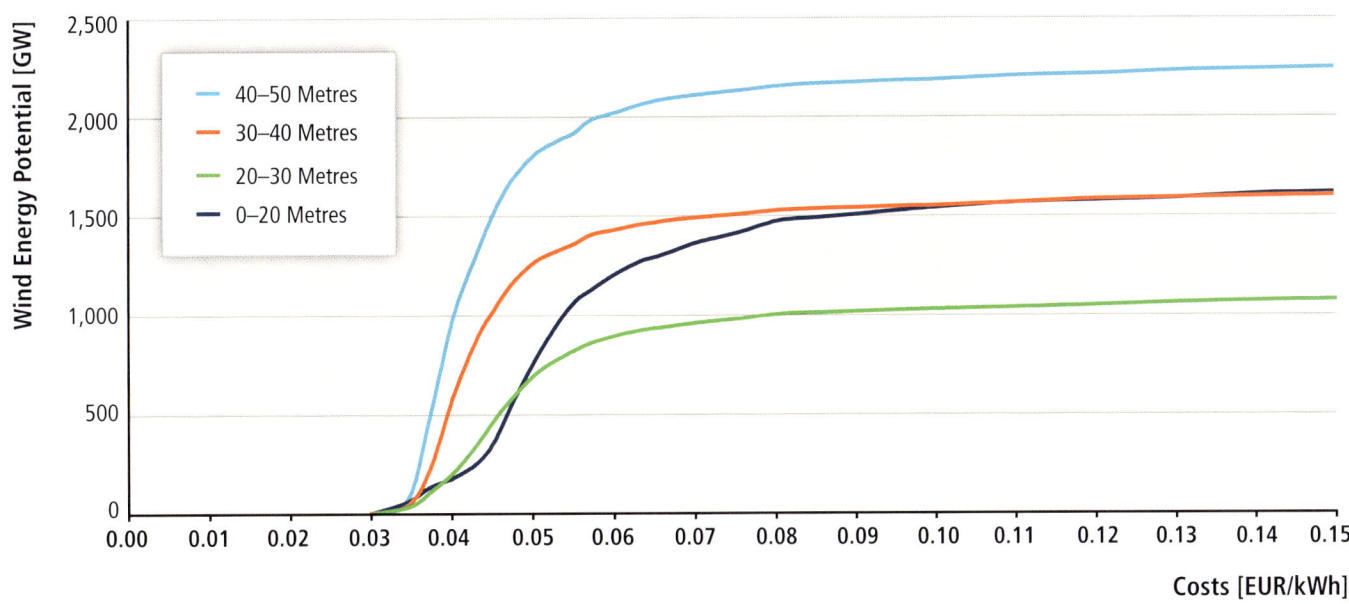

Figure 10.27 | Technical potential for offshore wind energy generation at different water depths in 2030 for Europe (EEA, 2009).

broken out of otherwise less economic technologies (such as more attractive wind sites, higher productivity biomass technologies/plants/sites, etc.). Finally, global data sets on deployment rates as a function of energy production costs as well as the cost of additional system balancing and transmission are a key requisite for integrated assessment modelling studies. The lack of such comprehensive data sets (with the laudable exception of Hoogwijk and Graus data) is striking, and is an important knowledge gap.

10.5 Costs of commercialization and deployment

Some RE technologies are broadly competitive with current market energy prices. Many of the other RE technologies can provide competitive energy services in certain circumstances, for example, in regions with favourable resource conditions or that lack the infrastructure for other low-cost energy supplies. In most regions of the world, however, policy measures are still required to ensure rapid deployment of many RE sources.

The aforementioned statement, which is consistent with recent publications of the IEA (IEA, 2007a, 2010a,d), is based on a consideration of the resource base, the energy services requested as well as technology-specific assessments of current costs of investment, financing, operation and maintenance as presented in the cost sections of the various technology chapters (see Sections 2.7, 3.8, 4.7, 5.8, 6.7 and 7.8).

Under favourable conditions, inter alia, modern combustible biomass to produce heat (IEA, 2007a), solar thermal energy (e.g., solar water heaters in China (IEA, 2010d)), selected off-grid PV applications (IEA, 2010c), large-scale hydropower (IEA, 2008a), larger geothermal projects (>30 MWe (IEA, 2007b)) and (if the cost of carbon is reflected in the markets) wind onshore power plants (IEA, 2010a) are already competitive. Provided that sufficient policy support is available, grid parity of PV (i.e., competitiveness with grid retail prices) is envisioned in many countries by 2020 (IEA, 2010c). Other technologies, such as CSP and offshore wind power, will require further support in order to compete with wholesale prices in the long term.

Currently and in the mid-term, the application of RE technologies can result in additional private costs compared to energy supply from other sources.[11] Starting with a review of present technology costs (i.e., current costs observed and published in the last few years), the remainder of this section will focus on expectations about how these costs might decline in the future, for instance, due to extended R&D efforts, technological learning associated with increased deployment, or spill-over effects (see IPCC, 2007). In addition, historic R&D expenditures and future investment needs will be discussed. It must be emphasized that Section 10.5 focuses on technology costs only. Integration aspects are discussed in Chapter 8; externalities and the associated social costs in Chapter 9 and Section 10.6.

10.5.1 Introduction: Review of present technology costs

In the field of RE, energy supply costs are mainly determined by investment costs. Nevertheless, operation and maintenance costs (O&M costs), and—if applicable—fuel costs (in the case of biomass), may play an important role as well. The respective cost components are discussed

---

11 Within this section, the external costs of other technologies are not considered. Although the term 'private' will be omitted in the remainder of this section, the reader should be aware that all costs discussed here are private costs in the sense of Section 10.6. Externalities therefore are not taken into account.

in detail in the technology chapters (Sections 2.7, 3.8, 4.7, 5.8, 6.7 and 7.8) and recent values are summarized in Annex III (Tables 1 through 3), where, inter alia, technology-specific values for typical device sizes (in MW), recent specific investment costs (in USD/kW), annual O&M costs (in USD/kW or US cents/kWh), capacity factors (in %) and economic lifetimes (in years) can be found. At a global scale, the respective values are highly uncertain for the various RE technologies. As recent years have shown, investment costs, for instance, might be considerably influenced by changes in material (e.g., steel) and engineering costs as well as by technological learning and mass market effects (IEA, 2010a,b).

Levelized costs of energy (LCOE, also called levelized unit costs or levelized generation costs; see Annex II for more information and illustrative calculations) are defined as 'the ratio of total lifetime expenses versus total expected outputs, expressed in terms of the present value equivalent' (IEA, 2005, p.174). LCOE therefore capture the full costs (i.e., investment costs, O&M costs, fuel costs and decommissioning costs) of an energy conversion installation and allocate these costs over the energy output during its lifetime. In general, LCOE do not take into account subsidies, policy incentives or integration costs.

The LCOE that can be derived from the values given in Annex III (Tables 1 to 3) are shown in Figures 10.28 through 10.31. Though these represent LCOE estimates for recent renewable energy plants, LCOE are different at different locations as discount rates, investment cost, O&M costs, capacity factors (especially due to the local RE resource availability) and fuel prices are site dependent (Heptonstall, 2007; IEA, 2010b).

The cost ranges in the background of Figure 10.28 display the global ranges of indicative values for the cost of energy supply options using fossil fuels. For electricity, the range is based on a recent assessment of LCOE for new coal and gas-fired power plants (IEA, 2010b). The values refer to centralized power plants. In contrast to IEA (2010b), a carbon price mark-up has not been included.

Following IEA (2007a), the (levelized) cost of oil and gas based heat supply options are estimated by taking into account retail fuel prices and conversion losses only. The investment costs for conventional boilers were neglected, because their contribution to overall LCOH is small (and because conventional heating facilities are often needed as a back-up for RE conversion technologies). Retail prices are used as most RE heating technologies have to compete at the final consumer level. For conversion efficiencies the values proposed by IEA (2007a) are applied. The indicative cost range depicted in Figure 10.28 is based on differing national retail prices (including taxes) for light fuel oil and natural gas as reported in the recently published IEA Key World Energy Statistics (IEA, 2010f). The lower bound of the range refers to natural gas-fired industrial heating applications; the higher bound to light fuel oil use in households.

According to the IEA (2010d), the cost of conventional transport fuels is strongly correlated with the underlying (historical) Brent crude oil spot price. In order to facilitate an investigation of the competitiveness of biofuels in times of highly fluctuating crude oil prices, the indicative transport fossil fuel cost range depicted in Figure 10.28 refers to a variation in the underlying crude oil spot price between USD 40 and 130/barrel.

As RE technologies are often characterized by high shares of investment costs relative to O&M costs and fuel costs, the applied discount rate has a prominent influence on the LCOE (see Figures 10.29, 10.30 and 10.31). The discount rate itself refers to a risk-free rate of return (assessed to be broadly of the order of 3%/yr) adjusted by a project-dependent risk premium (IEA, 2005, Appendix 6). According to IEA (2010b) (see Chapter 8 in this report), a discount rate of 5% is typically adopted by US investors facing a low risk in a fairly stable environment. Prominent examples are a public monopolist acting in a regulated market or a private investor investing in a low-risk technology in a favourable market environment. In the case where the investor is facing substantially greater financial, technological and price risks, a real discount rate of 10% can be justified (IEA, 2010b, p.154). As discussed in Appendix II, this report uses three values of real discount rates (3, 7 and 10%) in order to allow for an easy comparison between different projects and/or technologies. Note that in liberalized markets, private investors might ask for a higher real rate of returns than those characterized by a discount rate of 10% (IEA, 2005).

The LCOE ranges depicted in Figures 10.28 through 10.31 can be traced back to variations in the underlying parameters, which, in turn, can be grouped into:

a) The considered range of the performance parameter (characterized by the capacity factor) that heavily depends on the local resource base (e.g., wind velocities or solar radiation).

b) The global spread of the technology-dependent parameters (i.e., lifetime as well as investment and O&M costs) that are influenced by local technology maturity, market conditions and wages.

c) The range of the different real discount rate selected for this study (3 to 10%).

The lowest LCOE values depicted in Figures 10.28 through 10.31 correspond to best-case conditions (highest achievable capacity factor and highest lifetime, lowest investment and O&M costs, and lowest bound on the discount rate). The upper range of the LCOE is characterized by high, but still reasonable values for costs; low, but still realistic values for the lifetime; low, but still observed capacity factors; and a discount rate of 10% (if not indicated otherwise). Less favourable conditions can yield substantially higher costs compared to those shown in the figures.

The results presented in Figures 10.28 through 10.31 warrant some discussion in comparison to the cost data presented in other chapters. Most of the technology chapters show the levelized cost as a function of a) the capacity factor, b) the investment costs and c) the discount rate (Sections 2.7, 3.8, 4.6, 5.8, 6.7 and 7.8). In order to facilitate a comparison between different technologies, Figures 10.28 through 10.31 do not repeat showing the respective sensitivities in an explicit way. As discussed above, the

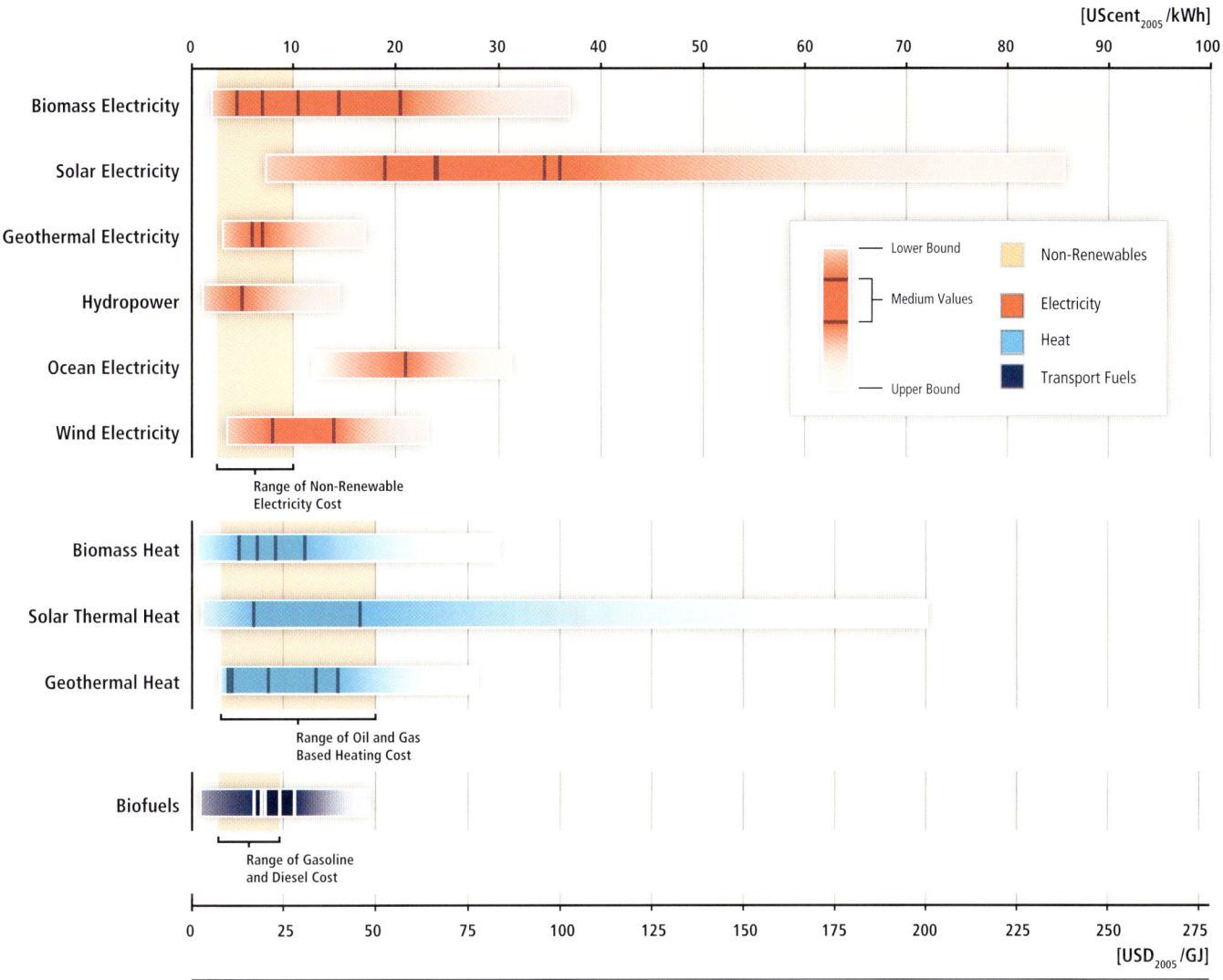

Figure 10.28 | Range in recent levelized cost of energy for selected commercially available RE technologies in comparison to recent non-renewable energy costs. Technology subcategories and discount rates were aggregated for this figure. For related figures with less or no such aggregation, see Annex III. Additional information concerning the cost of non-renewable energy supply options is given below.

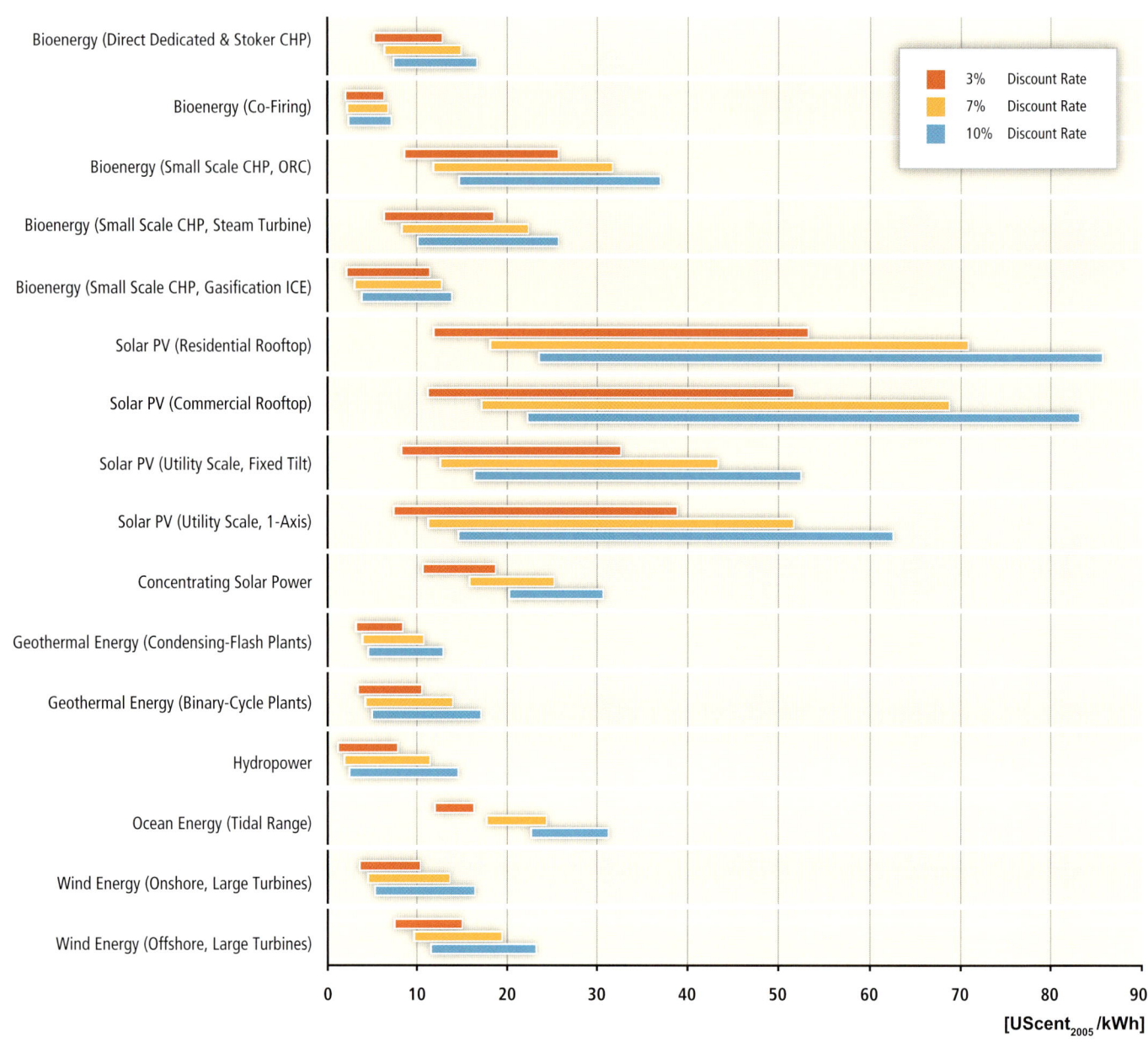

Figure 10.29 | Levelized cost of electricity for commercially available RE technologies at 3, 7 and 10% discount rates. The levelized cost estimates for all technologies are based on input data summarized in Annex III and the methodology outlined in Annex II. The lower bound of the levelized cost range is based on the low ends of the ranges of investment, operations and maintenance (O&M), and (if applicable) feedstock cost and the high ends of the ranges of capacity factors and lifetimes as well as (if applicable) the high ends of the ranges of conversion efficiencies and by-product revenue. The higher bound of the levelized cost range is accordingly based on the high end of the ranges of investment, O&M and (if applicable) feedstock costs and the low end of the ranges of capacity factors and lifetimes as well as (if applicable) the low ends of the ranges of conversion efficiencies and by-product revenue. Note that conversion efficiencies, by-product revenue and lifetimes were in some cases set to standard or average values. For data and supplementary information see Annex III. (CHP: combined heat and power; ORC: organic Rankine cycle, ICE: internal combustion engine).

figures nevertheless show the range of LCOE that originates from varying the capacity factors and investment costs within reasonable bounds.

In contrast to the aforementioned LCOE sensitivity diagrams that are contained in the technology chapters, the supply cost curves presented in Section 10.4.4 (Figures 10.23, 10.25, 10.26 and 10.27) provide additional information about the available resource base. Instead of showing the sensitivity with respect to the capacity factor, they allow an insight into the amount of RE that can be harnessed up to a prescribed level of the LCOE. This additional information comes from studies that made their own assumptions about other factors (beyond site-dependent capacity factors) that have an influence on the LCOE (e.g., discount rates, investment and O&M costs, and lifetimes). As a result, these results might not be fully compatible with the LCOE calculations summarized in Annex III.

The supply cost curves discussed in Section 10.3.2.1 (Figures 10.15 through 10.17) exhibit the amount of RE that is harnessed (once again as a function of the associated LCOE) in different regions once specific

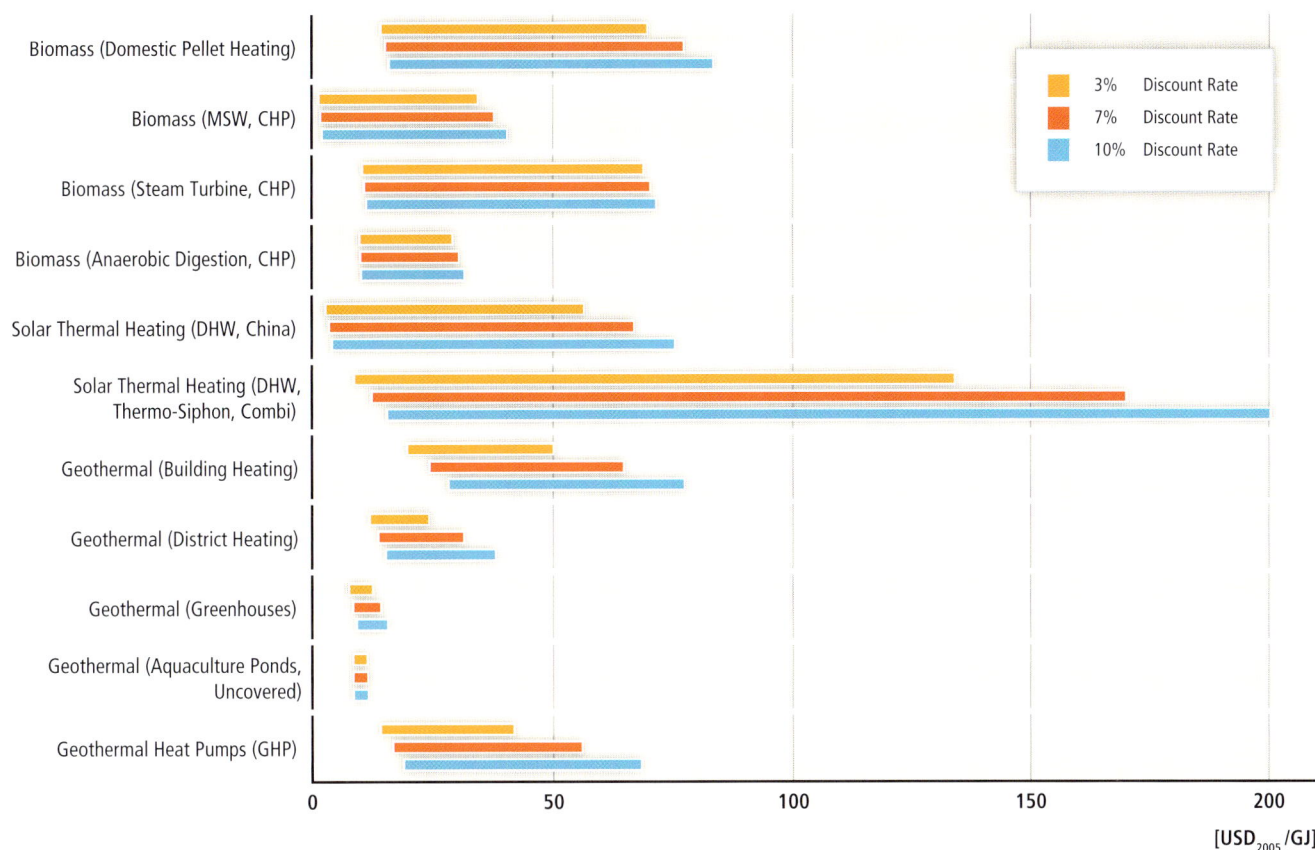

Figure 10.30 | Levelized cost of heat (LCOH) for commercially available RE technologies at 3, 7 and 10% discount rates. The LCOH estimates for all technologies are based on input data summarized in Annex III and the methodology outlined in Annex II. The lower bound of the levelized cost range is based on the low ends of the ranges of investment, operations and maintenance (O&M), and (if applicable) feedstock cost and the high ends of the ranges of capacity factors and lifetimes as well as (if applicable) the high ends of the ranges of conversion efficiencies and by-product revenue. The higher bound of the levelized cost range is accordingly based on the high end of the ranges of investment, O&M and (if applicable) feedstock costs and the low end of the ranges of capacity factors and lifetimes as well as (if applicable) the low ends of the ranges of conversion efficiencies and by-product revenue. Note that capacity factors and lifetimes were in some cases set to standard or average values. For data and supplementary information see Annex III. (MSW: municipal solid waste; DHW: domestic hot water).

trajectories for the expansion of RE are followed. As the results clearly show, the respective numbers are heavily dependent on the peculiarities (e.g., applied assumptions) of the underlying models.

In addition, it must be emphasized that most of the supply cost curves refer to future points in time (e.g., 2030 or 2050), whereas the levelized costs given in the cost sections of the technology chapters as well as those shown in Figures 10.28 through 10.31 (and in Annex III) refer to current costs.

The LCOE presented in Figures 10.28 through 10.31 are based on literature reviews and represent the most current cost data available. The corresponding data are summarized in Tables 1 to 3 of Annex III. The LCOE ranges are rather broad as the values vary across the globe depending on the RE resource base and the local costs of investment, financing, operation and maintenance. Comparison between different technologies therefore should not be based on the cost data provided here; instead, site-, project- and investor-specific conditions should be taken into account. The technology chapters (Sections 2.7, 3.8, 4.6, 5.8, 6.7 and 7.8) provide useful sensitivities in this respect.

Similar to LCOE, wholesale and retail prices of electricity that might be used in order to assess the competitiveness of centralized and decentralized RE power plants are country specific as well. The same holds true for the cost of fuels used for heating and transport purposes. A comparison of RE LCOE with those of other technologies or market prices should therefore be project-based as well.

The LCOE of a technology is not the sole determinant of its value or economic competitiveness. In addition to integration and transmission costs, relative environmental impacts must be considered, as well as the contribution of a technology to meeting specific energy services, for example, peak electricity demands.

Nevertheless, and despite the existing uncertainties, summarizing the information contained in Figures 10.28 through 10.31, Sections 2.7, 3.8, 4.6, 5.8, 6.7 and 7.8 as well as in recent benchmark studies (IEA, 2010a,b,c,d), the following conclusions can be drawn:

A comparison of LCOE of RE technologies with those of other technologies (nuclear, gas and coal power plants) shows that—at least as long as

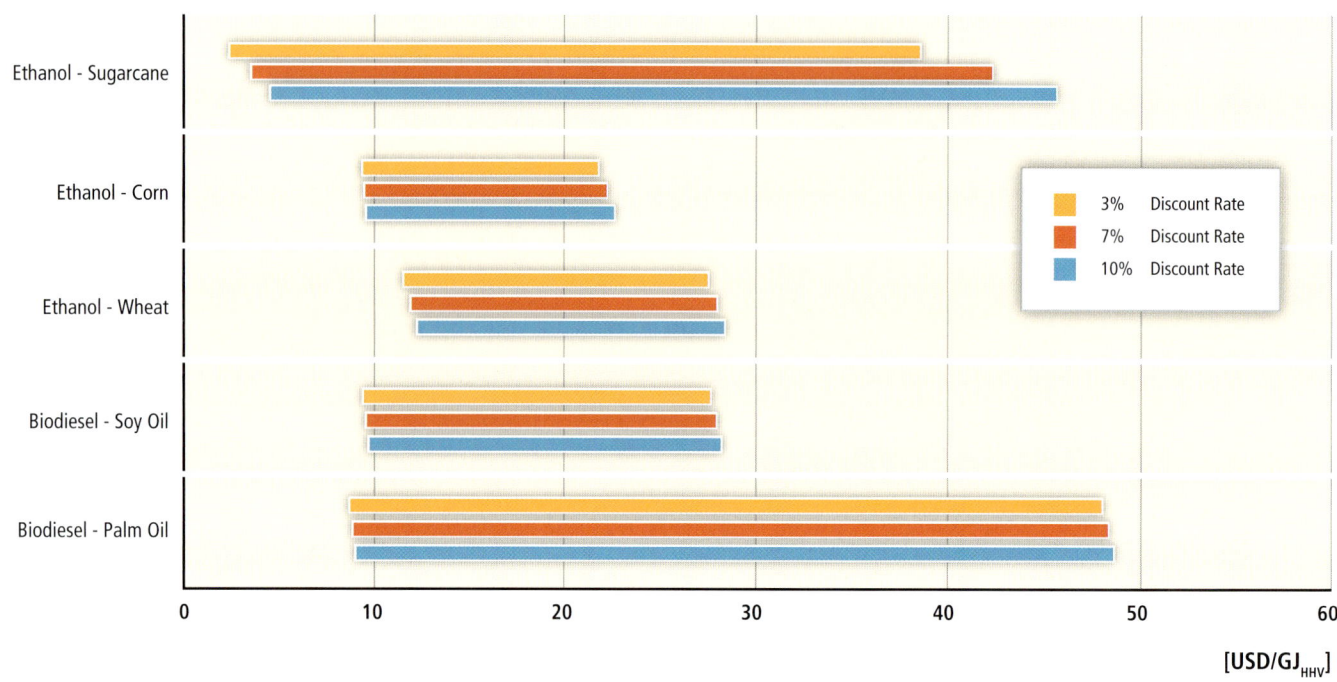

Figure 10.31 | Levelized cost of fuels (LCOF) for commercially available biomass conversion technologies at 3, 7 and 10% discount rates. LCOF estimates for all technologies are based on input data summarized in Annex III and the methodology outlined in Annex II. The lower bound of the levelized cost range is based on the low ends of the ranges of investment, operations and maintenance (O&M) and feedstock costs. The higher bound of the levelized cost range is accordingly based on the high end of the ranges of investment, O&M and feedstock costs. Note that conversion efficiencies, by-product revenue, capacity factors and lifetimes were set to average values. For data and supplementary information see Annex III. (HHV: higher heating value).

externalities are not taken into account—RE sources are often not yet competitive with other sources, especially if they both feed into the electricity grid. If the respective technologies are used in a decentralized mode, private investors would compare their production cost with the retail consumer power price, which is much higher. In this case, niche markets might exist that facilitate the market introduction of new technologies. The same holds true for applications in remote areas, where often no grid-based electricity is available (IEA, 2010c). Similar trends exist outside of the power sector for the use of RE in heating and transportation applications (IEA, 2007a).

Given suitable conditions, the lower end of the LCOE ranges indicate (see Figure 10.28) that some RE technologies already can compete with traditional forms at current energy market prices in many regions of the world. That said, the graphs provide no indication of the resource potential that can be utilized at low cost. Sections 10.3 and 10.4 provide more information in this regard.

10.5.2 Prospects for cost decreases

In the field of RE, significant opportunities exist to further improve the energy efficiencies and/or to decrease the costs of producing and installing the respective technologies (see Sections 2.7, 3.8, 4.7, 6.7 and 7.8). Together, these effects are expected to decrease the LCOE of many innovative RE sourcing technologies in the future (IEA, 2008b, 2010a). According to Junginger et al. (2006), the list of the most important mechanisms causing cost reductions comprises:

- *Learning by searching*, that is, improvements due to research, development and demonstration (RD&D)—especially, but not exclusively in the stage of invention;

- *Learning by doing* (in the strict sense), that is, improvements in the production process (e.g., increased labour efficiency, work specialization);

- *Learning by using*, that is, improvements triggered by user experience feedbacks occur once the technology enters (niche) markets;

- *Learning by interacting (or 'spill-overs')* (IPCC, 2007; Clarke et al., 2008), that is, the reinforcement of the above-mentioned mechanism due to an increased interaction among various actors in the diffusion phase;

- *Upsizing of technologies* (e.g., up-scaling of wind turbines); and

- *Economies of scale* (i.e., mass production) once the stage of large-scale production is reached.

The various mechanisms may occur simultaneously at various stages of the innovation chain. In addition, they may reinforce each other. As a consequence of the aforementioned mechanisms, many technologies applied in the field of RE sources showed a significant cost decrease in the past (IEA, 2000, 2008a). This empirical observation is highlighted by *experience (or 'learning') curves*, which describe how costs have declined with accumulated experience and corresponding cumulative production or installed capacity. An illustrative experience curve (referring to wind energy) is shown in Figure 10.32. Further examples concerning bioenergy use and photovoltaic modules can be found in Section 2.7.2 (Figure 2.21) and in Section 3.8.3 (Figure 3.17), respectively.

For a doubling of the (cumulatively) installed capacity, many technologies showed a more or less constant percentage decrease in the specific investment costs (or in the levelized costs or unit price, depending on the selected cost indicator). The corresponding *learning rate (LR)* is defined as the percentage cost reduction for each doubling of the cumulative capacity. A summary of observed learning rates is provided in Table 10.10. Occasionally, the *progress ratio (PR)* is used as a substitute for the learning rate. It is defined as PR = 1 - LR (e.g., a learning rate of 20% would imply a progress ratio of 80%). Frequently, energy supply costs (e.g., electricity generation costs) and the cumulative energy supplied

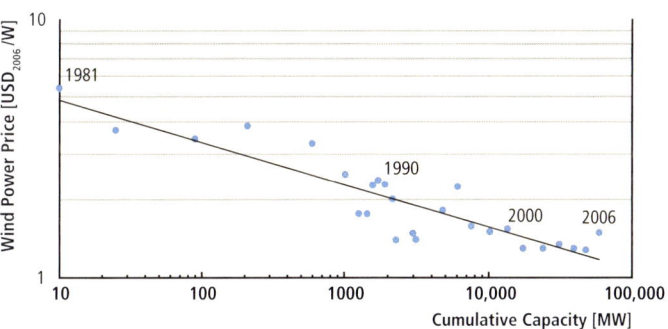

Figure 10.32 | Illustrative experience curve for wind turbines. Source: Nemet (2009).

by the respective technology (e.g., the cumulative electricity production) are used as substitutes for investment costs and the cumulative installed capacity, respectively. If the learning rate is time-independent, the empirical experience curve can be fitted by a power law. In this case, representing costs against cumulative installed capacity in a graph with double logarithmic scales shows the experience curve as a straight line (Junginger et al., 2010) (see Figure 10.32).

As there is no natural law that costs *have* to follow a power law (Junginger et al., 2010), care must be taken if historic experience curves are extrapolated in order to predict future costs (Nemet, 2009). Obviously, the cost reduction cannot go *ad infinitum* and there might be some unexpected steps in the curve in practice (e.g., caused by technology breakthroughs). As technologies mature, learning rates may fall (Ferioli et al., 2009; Nemet, 2009). In order to avoid implausible results, projections that extrapolate experience cost curves in order to assess future costs should therefore constrain the cost reduction by appropriate *floor costs* (see Edenhofer et al., 2006).

Concerning levelized costs or turnkey investment costs, a significant share of these floor costs might arise from balance of system and installation costs, which, in turn, are often dominated by labour costs. Although installers might gain experience, the future decrease in this cost component is limited (Yang, 2010). Unfortunately, *cost* data are not easily obtained in a competitive market environment. Indicators that are intended to serve as a substitute, for example, product *prices*, do not necessarily reveal the actual improvement achieved (Yu et al., 2011). Instead, they might be heavily influenced by an imbalance of supply and demand. This refers to both the final product itself (e.g., if financial support stipulates a high demand) and the cost of production factors, which might be temporarily scarce (e.g., steel prices due to supply bottlenecks). A deviation from price-based experience curves, as especially observed for PV modules in the years between 2004 and 2008 (see Section 3.8.3, Figure 3.17), therefore does not necessarily imply that a fundamental cost limit has been reached (Nemet, 2009). Instead, it might simply indicate that producers were able to make extra profits while the cost reduction takes place in the background. After a subsequent 'shakeout' phase, the short-term deviation from the long-term experience curve might be largely removed (Junginger et al., 2005b). In the field of solar PV, for instance, the recent development is characterized by overcapacities and a resulting increased competition between PV companies (see Chapter 3). As a result, PV system prices fell by 40% between 2008 and 2009 (IEA, 2010c; and see Section 3.8.3, Figure 3.17).

A summary of observed learning rates is provided in Table 10.10. Learning rates referring to investment costs (or turnkey investment costs) are often lower than those derived from electricity generation costs. Although the cost reduction in the specific investment costs of wind power plants, for instance, might be small, the scale-up results in higher hub heights and an associated significant increase in the capacity factor (and consequently in the amount of energy delivered). The ultimate goal of technological progress in the field of RE is a reduction of the energy production costs per kWh (in other words, the LCOE), not of the investment costs per se (see Section 7.8.4.1; EWEA, 2009; Ferioli et al., 2009).

Any efforts to assess future costs by extrapolating historic experience curves must take into account the uncertainty of learning rates as well as the caveats and knowledge gaps discussed in Sections 10.5.6 and 7.8.4.1. As a supplementary approach, expert elicitations could be used to gather additional information about future cost reduction potentials (Curtright et al., 2008), which might be contrasted with the assessments gained by using learning rates. Furthermore, engineering model analyses to identify technology improvement potentials could also provide additional information for developing cost projections (see Sections 2.6, 3.7, 4.6, 6.6 and 7.7)

Important potential technological advances and associated cost reductions, for instance, are expected in (but are not limited to) the following application fields: next-generation biofuels and bio-refineries (see Section 2.6); advanced PV and CSP technologies and manufacturing processes (see Section 3.7); enhanced geothermal systems (see Section

Mitigation Potential and Costs — Chapter 10

Table 10.10 | Observed learning rates for various electricity supply technologies. Source: IEA, 2008b, p. 205, extended and updated with a select list of additional literature (this report). (Note that values cited by older publications are less reliable as these refer to shorter time periods. In addition, only values for single-factor learning curves are shown. As a consequence there is some, albeit restricted, overlap with the learning rate information offered by Chapters 2 through 7.)

Technology	Source	Country / region	Period	Learning rate (%)	Performance measure
Onshore wind					
	Neij, 1997	Denmark	1982-1995	4	Price of wind turbine (USD/kW)
	Mackay and Probert, 1998	USA	1981-1996	14	Price of wind turbine (USD/kW)
	Neij, 1999	Denmark	1982-1997	8	Price of wind turbine (USD/kW)
	Durstewitz, 1999	Germany	1990-1998	8	Price of wind turbine (USD/kW)
	IEA, 2000	USA	1985-1994	32	Electricity production cost (USD/kWh)
	IEA, 2000	EU	1980-1995	18	Electricity production cost (USD/kWh)
	Kouvaritakis et al., 2000	OECD	1981-1995	17	Price of wind turbine (USD/kW)
	Neij, 2003	Denmark	1982-1997	8	Price of wind turbine (USD/kW)
	Junginger et al., 2005a	Spain	1990-2001	15	Turnkey investment costs (EUR/kW)
	Junginger et al., 2005a	UK	1992-2001	19	Turnkey investment costs (EUR/kW)
	Söderholm and Sundqvist, 2007	Germany, UK, Denmark	1986-2000	5	Turnkey investment costs (EUR/kW)
	Neij, 2008	Denmark	1981-2000	17	Electricity production cost (USD/kWh)
	Kahouli-Brahmi, 2009	Global	1979-1997	17	Investment costs (USD/kW)
	Nemet, 2009	Global	1981-2004	11	Investment costs (USD/kW)
	Wiser and Bolinger, 2010	Global	1982-2009	9	Investment costs (USD/kW)
Offshore wind					
	Isles, 2006	8 EU countries	1991-2006	3	Investment cost of wind farms (USD/kW)
Photovoltaics (PV)					
	Harmon, 2000	Global	1968-1998	20	Price PV module (USD/Wpeak)
	IEA, 2000	EU	1976-1996	21	Price PV module (USD/Wpeak)
	Williams, 2002	Global	1976-2002	20	Price PV module (USD/Wpeak)
	ECN, 2004	EU	1976-2001	20-23	Price PV module (USD/Wpeak)
	ECN, 2004	Germany	1992-2001	22	Price of balance of system costs
	van Sark et al., 2007	Global	1976-2006	21	Price PV module (USD/Wpeak)
	Kruck and Eltrop, 2007	Germany	1977-2005	13	Price PV module (EUR/Wpeak)
	Kruck and Eltrop, 2007	Germany	1999-2005	26	Price of balance of system costs
	Nemet, 2009	Global	1976-2006	15-21	Price PV module (USD/Wpeak)
Concentrating Solar Power (CSP)					
	Enermodal, 1999	USA	1984-1998	8-15	Plant investment cost (USD/kW)
Biomass					
	IEA, 2000	EU	1980-1995	15	Electricity production cost (USD/kWh)
	Goldemberg et al., 2004	Brazil	1985-2002	29	Prices for ethanol fuel (USD/m^3)
	Junginger et al., 2005b	Sweden, Finland	1975-2003	15	Forest wood chip prices (EUR/GJ)
	Junginger et al., 2006	Denmark	1984-1991	15	Biogas production costs (EUR/Nm^3)
	Junginger et al., 2006	Sweden	1990-2002	8-9	Biomass CHP power (EUR/kWh)
	Junginger et al., 2006	Denmark	1984-2001	0-15	Biogas production costs (EUR/Nm^3)
	Junginger et al., 2006	Denmark	1984-1998	12	Biogas plants (€/m^3 biogas/day)
	Van den Wall Bake et al., 2009	Brazil	1975-2003	19	Ethanol from sugarcane (USD/m^3)
	Goldemberg et al., 2004	Brazil	1980-1985	7	Ethanol from sugarcane (USD/m^3)
	Goldemberg et al., 2004	Brazil	1985-2002	29	Ethanol from sugarcane (USD/m^3)
	Van den Wall Bake et al., 2009	Brazil	1975-2003	20	Ethanol from sugarcane (USD/m^3)
	Hettinga et al., 2009	USA	1983-2005	18	Ethanol from corn (USD/m^3)
	Hettinga et al., 2009	USA	1975-2005	45	Corn production costs (USD/t corn)
	Van den Wall Bake et al., 2009	Brazil	1975-2003	32	Sugarcane production costs (USD/t)

4.7); multiple emerging ocean technologies (see Section 6.6); and foundation and turbine designs for offshore wind energy (see Section 7.7). Further cost reductions for hydropower are likely to be less significant than some of the other RE technologies, but R&D opportunities exist to make hydropower projects technically feasible in a wider range of natural conditions and improve the technical performance of new and existing projects (see Sections 5.3, 5.7 and 5.8).

10.5.3 Deployment cost curves and learning investments

According to the definition used by the IEA (2008b, p. 208), "*deployment costs* represent the *total* costs of cumulative production needed for a new technology to become competitive with the current, incumbent technology." As the innovative technologies replace O&M costs, investment needs and fuel costs of other technologies, the *learning investments* are considerably lower. The *learning investments* are defined as the *additional* investment needs of the new technology. They are therefore equal to the deployment costs minus (replaced) cumulative costs of the incumbent technology.

Although not directly discussed in IEA (2008b)—to give the full picture—the cost difference could be extended to take into account variable costs as well (Figure 10.33). Because of fuel costs, the latter is evident for fossil fuel and biomass technologies. Once variable costs are taken into account, avoided carbon costs contribute to a further reduction of the *additional* investment needs (IEA, 2008b). Figure 10.33 shows a schematic presentation of experience curves, deployment costs and learning investments. The deployment costs are equal to the integral below the experience curve, calculated up to the break-even point.

In the beginning of the deployment phase, additional costs are expected to be positive ('expenditures'). Due to technological learning (in the broadest sense) and the possibility of increasing fossil fuel prices, additional costs could become negative after some decades (IEA, 2008b, 2010a). A least-cost approach towards a decarbonized economy therefore should not focus solely on the additional costs that are incurred until the break-even point with other technologies has been achieved (learning investments). After the break-even point, the innovative technologies considered are able to supply energy with costs lower than the traditional supply. As these costs savings occur then (after the break-even point) and indefinitely thereafter, their present value might be able to compensate the upfront investments (additional investment needs). Whether this is the case depends on various factors: the discount rate, the stringency of the selected climate stabilization goal and—most important—the future cost development of all its potentially competitive alternatives (see Section 10.2; Edenhofer et al., 2006; Clarke et al., 2009).

An answer to the question of whether or not upfront investments in a specific innovative technology are justified therefore cannot be given as long as this technology is treated in isolation (Kverndokk and

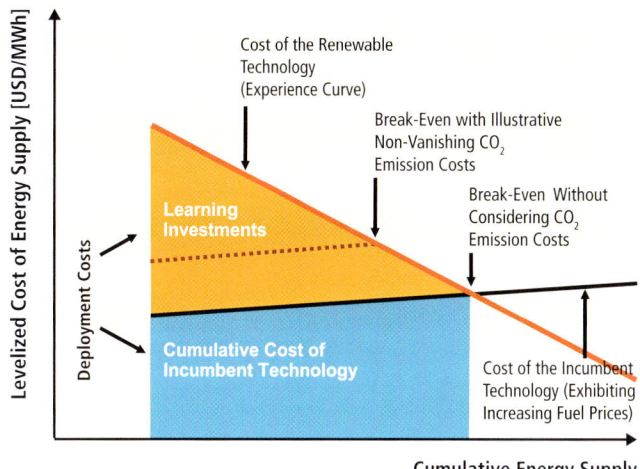

Figure 10.33 | Schematic representation of experience curves, deployment costs and learning investments (modified version of the diagram depicted in IEA, 2008b, p. 204).

Rosendahl, 2007). In a first attempt to clarify this issue and, especially, to investigate the mutual competition of prospective climate protection technologies, integrated assessment modellers have started to model technological learning in an endogenous way (Edenhofer et al., 2006, 2009, 2010; Clarke et al., 2009; Knopf et al., 2009). The results obtained from these modelling comparison exercises indicate that—in the context of stringent climate goals—upfront investments in learning technologies can be justified in many cases. However, as the different scenarios considered in Figure 10.34 and other studies clearly show, considerable uncertainty surrounds the exact volume and timing of these investments.

In reality, incentives for private investments in climate-friendly technologies are often low. In fact, private sector innovation market failures distort private sector investments in technological progress. The main problem is that private investors developing new technologies might not be able to benefit from the cost savings that are related to the application of these technologies in a couple of decades. Furthermore, as long as external environmental effects are not completely internalized, the use of fossil fuels appears to be cheaper than justified (Jaffe et al., 2005; Montgomery and Smith, 2007; van Benthem et al., 2008).

10.5.4 Time-dependent expenditures

A comprehensive survey of past investments in renewable energies is given in Section 11.2.2. This section therefore will constrain itself to a discussion of future investment estimates.

In Figure 10.34, future investments in different RE technologies are shown for the four illustrative scenarios discussed in detail in Section 10.3 (see Box 10.2). The resulting cumulative global investment estimates (in the power generation sector only) range from USD_{2005} 1,360 to 5,100 billion for the decade 2011 to 2020, and from USD_{2005} 1,490

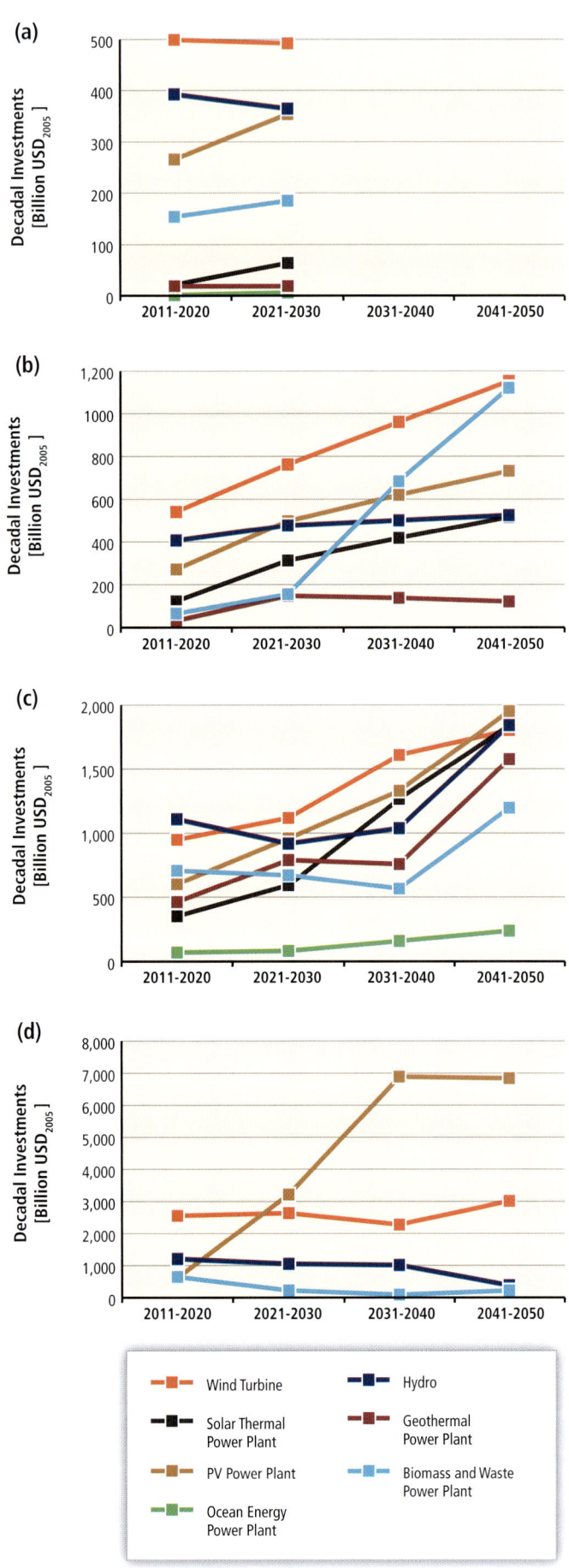

Figure 10.34 | Illustrative global *decadal* investments (in billion USD$_{2005}$) needed in order to achieve ambitious climate protection goals: (b) MiniCAM-EMF22 (first-best 2.6 W/m² overshoot scenario, nuclear and carbon capture technologies are permitted); (c) ER-2010 (450 ppm CO_2eq, nuclear and carbon capture technologies are not permitted); and (d) ReMIND-RECIPE (450 ppm CO_2, nuclear plants and carbon capture technologies are permitted). Compared to the other scenarios, the PV share is high in (d) as concentrating solar power has not been considered. For comparison, (a) shows the IEA-WEO2009-Baseline (baseline scenario without climate protection). Sources: (a) IEA (2009); (b) Calvin et al. (2009); (c) Teske et al. (2010); and (d) Luderer et al. (2009).

to 7,180 billion for the decade 2021 to 2030. The lower values refer to the IEA World Energy Outlook 2009 Reference Scenario and the higher ones to a scenario that seeks to stabilize atmospheric CO_2 (only) concentration at 450 ppm. The average annual investments in the reference scenario are slightly lower than the respective investments reported for 2009 (see Section 11.2.2). Between 2011 and 2020, the higher values of the annual averages of the RE power sector investment approximately correspond to a three-fold increase in the current global investments in this field. For the next decade (2021 to 2030), a five-fold increase is projected. Even the upper level is smaller than 1% of the world GDP (IEA, 2009). Additionally, increasing the installed capacity of RE power plants will reduce the amount of fossil and nuclear fuels (and the related fuel costs) that otherwise would be needed in order to meet a given electricity demand. These numbers indicate how much money will be spent in the sector of RE sources if these scenarios materialize. The given numbers are useful to inform investors who are interested in the expected market volume. Data on energy delivered by the corresponding scenarios can be found in Sections 10.3 and 10.4.

Specific investment costs of RE sources are still often higher than those of other energy supply technologies. In order to assess the additional costs arising from using RE sources, two effects must be taken into account. Due to their capacity credit, investments in RE sources reduce investment needs for other technologies (see Sections 7.5.2.4 and 8.2.1.1). In addition, fossil fuel costs (and O&M costs) will be reduced as well. As a consequence, investment needs do not indicate the overall mitigation costs societies face if these scenarios materialize. In calculating the net total cost, replaced other investments and avoided variable costs must be considered as well (see IEA, 2008b, 2010a). As the latter are dependent on the development of fossil fuel prices, the overall net cost balance could be positive from a mid- or long-term perspective (for a national study, see Winkler et al., 2009).

Many integrated assessment models used to derive the scenarios considered in Section 10.2 consider avoided costs and take them into account during the respective calculation runs. However, the results for total avoided investments in other plants, and the overall avoided fuel costs are seldom published. In addition, there is a lack of global scenario exercises that attribute avoided costs to distinguished technologies—although this information would be extremely useful in order to carry out a fair assessment of learning investments or (net) deployment costs.

In the absence of technology-specific results, aggregated avoided costs will be discussed for an illustrative climate protection scenario (the BLUE

Map scenario) that has been designed by the International Energy Agency (IEA, 2010a). In order to deliver a 50% cut in CO_2 emissions by 2050 (compared to 2005), different technologies are applied. Their respective shares in delivering the requested emission reduction are: end-use fuel and electricity efficiency 38%, end-use fuel switching 15%, power generation efficiency and fuel switching 5%, CCS 19%, nuclear 6% wand RE 17%. Between 2010 and 2050, the additional investment of the BLUE Map scenario (compared to the Baseline scenario) is USD_{2005} 41.72 trillion. In the same time period, the undiscounted fuel cost savings of this scenario are estimated to be USD_{2005} 101.59 trillion. Taken together, the total undiscounted net savings approach USD_{2005} 59.87 trillion. Even at a 10% discount rate, the fuel savings outweigh the additional incremental investment needs of the BLUE Map scenario.

Note that the results do not only take into account investments into RE sources. Other low-carbon technologies (energy efficiency improvements, nuclear energy, carbon capture and storage) are considered as well. Nevertheless, the results highlight the importance of comparing investment needs on the one hand and associated avoided (investment, O&M and fuel) costs of the substituted technologies on the other.

10.5.5 Market support and research, development, demonstration and deployment

Whereas the list in Section 10.5.2 summarizes different *causes* for technological progress and associated cost reductions, an alternative nomenclature focuses on how these effects can be *triggered*. Following this kind of reasoning, (Jamasb, 2007) distinguishes:

- *Learning by research* triggered by R&D expenditures that intend to achieve a *supply push* and

- *Learning by doing* (in the broader sense) resulting from capacity expansion promotion programs that intend to establish a *demand pull*.

Figure 10.35 depicts the historic RD&D support for RE research in relation to other technologies. Note that for fossil and nuclear technologies, the large-scale government support in the early stages of their respective innovation chain (i.e., well before the 1970s) is not shown.

As the IEA emphasizes, the role of governments is most effective if it combines 'supply push' and 'demand pull' programs depending on the position of the considered technology in the innovation chain (IEA, 2008b, 2010a). RD&D funding is particularly appropriate for infant technologies. Market entry support and demand pull programs (e.g., via norms, feed-in tariffs, renewable quota schemes, tax credits, bonus and malus systems) focus on the deployment and commercialization phase (Foxon et al., 2005; González, 2008), but can also help to trigger private investment in RD&D. A detailed description of corresponding policy options can be found in Chapter 11.

10.5.6 Knowledge gaps

At present, experience curves are often an integral part of integrated assessment models that seek to treat technological learning in an endogenous way. Unfortunately, small variations in the assumed learning rates can have a significant influence on the results of models that use experience curves. Empirical studies therefore should strive to provide error bars for the derived learning rates (van Sark et al., 2007; Mukora et al., 2009). In addition, a better understanding of the processes that result in cost reductions would be extremely valuable (Sagar and van der Zwaan, 2005; van den Wall-Bake et al., 2009). Furthermore, there is a severe lack of information that is necessary to decide whether short-term deviations from the experience curve can be attributed to supply bottlenecks, or whether they already indicate that the cost limit (in the sense of floor costs) is reached (Nemet, 2009). In addition, there is a need for studies that quantitatively investigate the extent to which spillovers to other firms are able to endanger the opportunity of innovating firms to harvest the innovation benefits (see Kverndokk and Rosendahl, 2007). If available at all, cost discussions in the literature mostly focus on investment needs. Unfortunately, many global studies neither display total cost balances (including estimates about operational costs and cost savings) nor externalities like social, political and environmental costs (e.g., side benefits like employment effects or the role of RE sources in reducing the risks associated with fossil fuel price volatility (Awerbuch, 2006; Gross and Heptonstall, 2008). Another crucial issue is that of optimal timing of RD&D versus demand pull programs as well as investigations into how a premature lock-in in sub-optimal technologies can be avoided (Sagar and van der Zwaan, 2005).

Although some assessments of externalities have taken place at a national level (see Chapter 9 and Section 10.6), a comprehensive global investigation and an associated cost-benefit analysis is highly recommended.

In addition, as Section 8.1 shows, there is a further need for comprehensive assessments of the additional costs arising from integrating RE sources into existing and future energy systems (Gross and Heptonstall, 2008).

10.6 Social and environmental costs and benefits

10.6.1 Background and objective

Energy production typically causes direct and indirect costs and benefits for the energy producer and for society. Energy producers, for instance, incur private costs, such as plant investment and operating costs, and receive private benefits, such as income from the energy market. Private costs and benefits are defined as costs or benefits accounted for by the

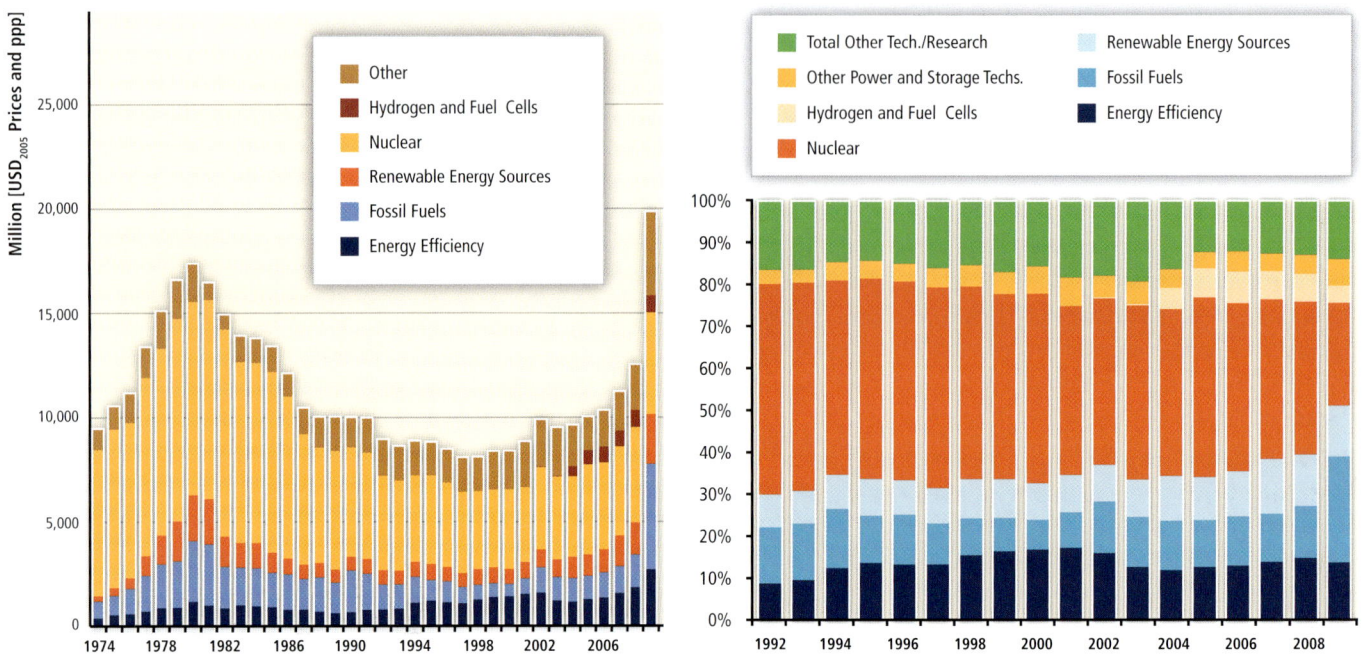

Figure 10.35 | Government budgets on energy RD&D of IEA countries (left panel) and technology shares of government energy RD&D expenditures in IEA countries (right panel) (IEA, 2008b, pp. 172-173, updated with data from IEA, 2010g).

agents responsible for the activity. The operations of energy producers often cause external impacts, which may be beneficial or detrimental but which are not covered by the energy producers or the price mechanisms. The costs and benefits due to external impacts are called external costs or external benefits, correspondingly (for the definition, see Annex I). External costs are usually indirect and they arise, for example, from pollutant emissions. The reduction of detrimental impacts caused by pollutant emissions can be seen as an external benefit from the system point of view when RE replaces some more harmful energy sources. Additionally, external benefits might occur if energy production and consumption result in positive effects for the society. Social costs are assumed to include here both private costs and external costs (Ricci, 2009a,b), although other definitions have also been used in the past (e.g., Hohmeyer, 1992).

In non-RE production, private costs are usually lower than the private benefits, which means that the energy production is normally profitable. On the other hand, the external costs can be high, on occasions exceeding the total (social) benefits. Alternatively, energy derived from RE technologies can often be unprofitable for the energy producer if not supported by incentive schemes. If the external costs (including environmental costs) are taken into account, the production of RE can, however, as a whole be more profitable from a social point of view than other energy production (Owen, 2006).

Typical factors causing external costs include atmospheric emissions from fossil fuel-based energy production, especially from combustion but also from other parts of the fuel chain. As shown in Chapter 9, the emissions can, among other things, consist of GHGs, acidifying emissions and particulate matter. These types of emissions can often but not always be lowered if RE is used to replace fossil fuels (Weisser, 2007).[12] Increasing the share of RE often contributes positively to access to energy,[13] energy security and the trade balance and it limits the negative effects from fluctuating prices of fossil-based energy (Section 9.3; Berry and Jaccard, 2001; Bolinger et al., 2006; Chen et al., 2007). However, various types of RE have their own private and external costs and benefits, depending on the energy source and the technology utilized. Chapter 9 addresses these issues comprehensively, based on the available literature.

Costs and benefits can be addressed in cost-benefit analyses to support decision making. However, the value of RE is not strictly intrinsic to renewable technologies themselves, but rather to the character of the energy system in which they are applied (Kennedy, 2005). The benefits of an increased use of RE are to a large part attributable to the reduced use of non-RE in the energy system.

The coverage and monetary valuation of the external impacts in general are difficult. The assessment of external costs is often tentative, may be inaccurate and might be seen impossible in many cases. As a result,

---

12 Note that in particular biomass applications can also cause particulate emissions.

13 About 1.4 billion people are still without access to electricity (Table 9.3.2); the RE sources due to their distributed character can at least to some extent help to alleviate this problem.

the cost-benefit analysis of some measure or policy, where the benefit arises from decreases in an environmental or external impact, is often contentious. In contrast, the difference between benefits and costs can be made clear even though the concrete numbers of the cost and benefit terms are uncertain. The long time spans associated with climate change and its impacts are not easy to consider in cost-benefit analyses. Discounting of impacts over long time horizons is at least to some extent problematic (Weitzman, 2007; Dietz and Stern, 2008). Further, many environmental impacts are not well understood or highly complex and their consideration and monetary valuation is difficult. Moreover, there are usually no compensation mechanisms that could balance costs and benefits among different stakeholders (Soderholm and Sundqvist, 2003). These aspects might limit the use of cost-benefit analysis and require other approaches, such as public consultation and direct setting of environmental targets and cost-benefit or cost-effectiveness analyses under these targets (Krewitt, 2002; Soderholm and Sundqvist, 2003; Grubb and Newbery, 2008).

Against this background, the objective of this section is to synthesize and discuss external costs and benefits of increased RE use in relation to climate change mitigation. The results are presented by technology at global and regional levels. Therefore, the section defines the cost categories considered and identifies quantitative estimates or qualitative assessments for costs by category type, by RE type, and as far as possible also by geographical area.

This section has links to the other chapters of this report, such as Chapters 1 and 9. Parts of this section consider the same topics, but from the viewpoints of external costs and benefits. The external costs and benefits considered in this section complement the cost considerations in the other parts of the chapter, forming a more holistic picture of costs from the social viewpoint.

10.6.2 Review of studies on external costs and benefits

Energy extraction, conversion and use cause significant environmental impacts and social costs. Many environmental impacts can be lowered by reducing emissions with advanced emission control technologies (Amann, 2008).

Although replacing fossil fuel-based energy with RE can reduce GHG emissions and also to some extent other environmental impacts and social costs caused by them, RE can also have environmental impacts and external costs, depending on the energy source and technology (da Costa et al., 2007). These impacts and costs should be lowered and of course should be considered if a comprehensive cost assessment is required.

This section considers studies in a cost and benefit category and presents a summary regarding energy sources as well. Some of the studies are global in nature, and to some extent regional studies, mostly for Europe and North America, will also be quoted. The number of studies for other regions is still limited. Many studies consider only one energy source or technology, but some studies cover a wider list of energy sources and technologies.

In the case of energy production technologies based on combustion, the impacts and external costs, in particular the environmental costs, mainly arise from emissions to air, especially if the greenhouse impact and health impact are considered. The lifecycle approach, including impacts via all stages of the energy production chain, is, however, necessary in order to recognize and account for total impact (Section 9.3.4). This holds true also in the case of non-combustible energy sources (WEC, 2004a; Kirkinen et al., 2008; Ricci, 2009a,b).

10.6.2.1 Climate change

The damage due to changing climate is often described by linking CO_2 emissions with the social costs of their impacts. This relation is called social costs of carbon (SCC), which is expressed as social costs per tonne of carbon or CO_2 released. A number of studies have been published on this subject and on the use of SCC in decision making (e.g., Anthoff, 2007; Grubb and Newbery, 2008; Watkiss and Downing, 2008).

The monetary evaluation of the impacts of the changing climate is difficult, however. To a large extent, the impacts manifest themselves slowly over a long period of time. In addition, the impacts can arise very far from a polluter in ecosystems and societies that are very different from the ecosystems and the society found at the polluter's location. It is for this reason that, for example, the methods used by the Stern (2007) review for damage cost accounting on a global scale are criticized, but they can also be seen as a choice for producing reasonable qualitative estimates. Apart from the question about discount rate, which is quite relevant considering the long term impacts of GHG emissions, considerable uncertainty exists in areas such as climate sensitivity, damages due to climate change, valuation of damages and equity weighting (Watkiss and Downing, 2008).

A German study (Krewitt and Schlomann, 2006) addressing external costs uses the values of USD 17/t CO_2, USD 90/t CO_2 and USD 350/t CO_2 (€ 14, 70 and 280/t CO_2) for the lower limit, best guess and upper limit for SCC, respectively, referring to Downing et al. (2005) and Watkiss and Downing (2008). The study assesses that the range of the estimated SCC values covers three orders of magnitude, which can be explained by the many different choices possible in modelling and approaches to quantifying the damages. As a benchmark lower limit for global decision making, they give a value of about USD$_{2005}$ 17/t CO$_2$ (£35/t CO$_2$). They do not give any best guess or upper limit benchmark value, but recommend that further studies should be done on the basis of long-term climate change mitigation stabilization levels.

The price of carbon can also be considered from other standpoints, for example, what price level of CO$_2$ emissions is needed in order to limit the atmospheric concentration to a given stabilization level. Emission trading gives also a price for carbon that is linked to the total allotted amount of emissions. Another way is to see the SCC as insurance for reducing the risks of climate change (Grubb and Newbery, 2008).

RE sources have usually quite low GHG emissions per each energy unit produced (see Chapter 9.3; WEC, 2004a; IPCC, 2007; Krewitt, 2007), so the impacts through climate change and the external costs they cause are usually low. There can also be exceptions, for example, in some cases of fuels requiring long refining chains like transportation biofuels produced under unfavourable conditions (Hill et al., 2006; Soimakallio et al., 2009) or land clearing for increasing biofuel production (Edwards et al., 2008; Searchinger et al., 2008).

Increasing the use of RE sources often displaces fossil energy sources that have relatively high GHG emissions and external costs (Koljonen et al., 2008). The net impact of an increase in RE supply is therefore positive external benefits if the whole system is considered. The magnitude of these positive impacts will depend in large part on the properties of the original energy system (Kennedy, 2005).

10.6.2.2 Health impacts due to air pollution

Combustion of both renewable fuels and fossil fuels often causes emissions of particulates and gases that have health impacts (Section 9.3.4; Krewitt, 2002; Torfs et al., 2007; Amann, 2008; Smith et al., 2009; Committee on Health, 2010). Exposure to smoke aerosols can be exceptionally large in primitive traditional burning of solid fuels, for example, in cooking of food in developing countries (see Section 9.3; Bailis et al., 2005). Also, emissions to the environment from stacks can reach people living far from the emission sources. The exposure and the number of health impacts depend on the physical and chemical character of the particulates, their concentrations in the air and population density (Krewitt, 2007). The exposure statistically leads to increased morbidity and mortality. The relationships between exposure and health impacts are estimated on the basis of epidemiological studies (e.g., Torfs et al., 2007). The external costs of increased mortality can be assessed using, for example, the concepts of value of life years lost (Preiss, 2009; Ricci, 2010) or value of statistical life (Committee on Health, 2010).

The results depend on many assumptions in the modelling, calculations and epidemiological studies. Krewitt (2002) describes how the estimated external costs of fossil-based electricity production have changed by a factor of ten during the ExternE project period between the years 1992 and 2002. ExternE is a major research programme launched by the European Commission at the beginning of the 1990s to provide a scientific basis for the quantification of energy-related externalities. The cost estimates have been increased by extension of the considered area (more people affected) and by inclusion of the chronic mortality. Furthermore, the cost estimates have been lowered by changing the indicator for costs arising from deaths and by using new exposure-impact models. It can be argued that the results include considerable uncertainty (Torfs et al., 2007).

The typical specific external costs through various impact chains per tonne of emissions have been assessed, for example, in Krewitt and Schlomann (2006), Preiss (2009) and Committee on Health (2010), to be for sulphur dioxide (SO$_2$) about USD 4,000 to 10,000/t, for nitrous oxides (NO$_x$) about USD$_{2005}$ 2,000 to 10,000/t, and for particulates PM2.5 about USD 10,000 to 30,000/t. The wide ranges of values give a picture of variability and uncertainty.

When RE is used to replace fossil energy, the total social costs of the total energy system due to health impacts usually decrease (Kennedy, 2005; Bollen et al., 2009), which can be interpreted to lead to social benefits linked to the increase of RE. However, this is not always the case, as discussed in this section, but requires a more detailed analysis.

10.6.2.3 Other impacts

RE can have impacts on waters, land use, soil, ecosystems and biodiversity (Section 9.3.4). It can also have a positive influence on energy security and trade balance and rural employment or have impacts on other socioeconomic aspects. Some of these impacts are not in a strict sense external as they are covered by price mechanisms, although they can be of importance from the viewpoints of the society. Most of these impacts have been considered in the technology Chapters 2 to 7 or in Chapter 9 in detail. The external costs due to these impacts are usually lower than the external costs due to GHG emissions or due to health effects caused by pollutant emissions (Krewitt and Schlomann, 2006; Preiss, 2009; Committee on Health, 2010; Ricci, 2010). However, in some cases specific impacts may cause considerable external costs that should be evaluated on the project by project basis. Some information on the magnitudes of the impacts can be found in Section 10.6.3.

10.6.3 Social and environmental costs and benefits by energy sources and regional considerations

Most of the studies covered in this section consider North America (Gallagher et al., 2003; Roth and Ambs, 2004; Kennedy, 2005; Chen et al., 2007; Committee on Health, 2010; Kusiima and Powers, 2010) and Europe (Groscurth et al., 2000; Bergmann et al., 2006; Krewitt and Schlomann, 2006; Ricci, 2009b), whilst some are more general without a specific geographical area.

Some studies consider developing countries. Da Costa et al. (2007) discuss social features of energy production and use in Brazil. Fearnside (1999, 2005) and Oliveira and Rosa (2003) studied large hydropower projects and the technical potential of wastes in Brazil, respectively. Sparovek et al. (2009) investigated the impacts of the extension of sugarcane production in Brazil. Bailis et al. (2005) considered biomass- and petroleum-based domestic energy scenarios in Africa and their impacts on mortality on the basis of particulate emissions. Spalding-Fecher and Matibe (2003) studied total external costs of coal-fired power generation in South Africa. Amann (2008) studied cost-effective reduction of emissions of air pollutants and GHGs in China.

Studies concerning different areas of the globe are still sparse. More investigations, articles and reports are needed to provide information on external costs and their possible variation in the ecosystems and societies of different geographical areas.

To calculate the net impact in terms of social costs of an extension of RE sources, two things have to be done. First, (a) the external costs and benefits can be assessed on the basis of the lifecycle approach for each technology in the conditions typical for that technology so that only the direct impacts of that technology are taken into account (Pingoud et al., 1999; Roth and Ambs, 2004; Krewitt and Schlomann, 2006; Ricci, 2009b). The other thing (b) is to consider the RE technologies as parts of the total energy system and society, when the impacts of a possible increase in the use of the RE technologies can be assessed as causing decreases in the use and external costs of other energy sources. These decreases in external costs can be seen as external benefits of the RE technologies for society (Kennedy, 2005; Loulou et al., 2005; Koljonen et al., 2009).

An assessment of external costs in Central European conditions is presented in Table 10.11 (Krewitt and Schlomann, 2006). It can be seen that the social costs due to climate change and health impacts dominate the results in Table 10.11. The other impacts make a lesser contribution to the final results, keeping in mind that not all impacts are quantifiable. Even if the low-end SCC value of USD 17/t CO_2 assumed in the reference is used in Table 10.11 instead of USD 90/t CO_2, the climate impact still dominates in the total social costs of fossil-based technologies, but for renewable technologies the health impacts would be dominant.

Figure 10.36 shows the large uncertainty ranges of two dominant external cost components, namely climate- and health-related external costs. As one example, a recent extensive study made for the conditions in the USA (Committee on Health, 2010) arrived at almost similar

Table 10.11 | External costs (US cents/kWh (3,600 kJ)) due to electricity production based on RE sources and fossil energy in Central European conditions. Valuation of climate change is based on an SCC value of 90 USD/t CO_2 (Krewitt and Schlomann, 2006). Uncertainty ranges are not reported in the table. For uncertainty estimates, see Figure 10.36.

	PV (2000)	PV (2030)	Hydro 300 kW	Wind 1.5 MW Onshore	Wind 2.5 MW Offshore	Geo-thermal	Solar Thermal	Lignite η=40%	Lignite Comb.C η=48%	Coal η=43%	Coal Comp.C η=46%	Natural Gas η=58%
Climate change	0.86	0.48	0.11	0.09	0.08	0.33	0.11	9.3	8.0	7.4	6.9	3.4
Health	0.43	0.25	0.075	0.09	0.04	0.15	0.11	0.63	0.35	0.46	0.33	0.21
Ecosystems	●	●	●	●	●	●	●	○	○	○	○	○
Material damages	0.011	0.008	0.001	0.001	0.001	0.004	0.002	0.019	0.010	0.016	0.01	0.006
Agricultural losses	0.006	0.004	0.001	0.002	0.0005	0.002	0.001	0.013	0.005	0.011	0.006	0.005
Large accidents	●	●	●	●	●	●	●	●	●	●	●	●
Proliferation	●	●	●	●	●	●	●	●	●	●	●	●
Energy security	●	●	●	●	●	●	●	●	●	●	●	○
Geo-political effects	●	●	●	●	●	●	●	●	●	●	●	○
Sum	~1.3	~0.74	~0.19	~0.18	~0.12	~0.49	~0.22	>9.9	>8.4	>7.9	>7.2	>3.6

Notes: ● 'green light': no significant impacts or external costs worth mentioning (Krewitt and Schlomann, 2006). ○ 'yellow': impacts will arise that cannot be neglected and that will cause external costs. Comb.C: combined gas turbine and steam cycles; η: efficiency factor.

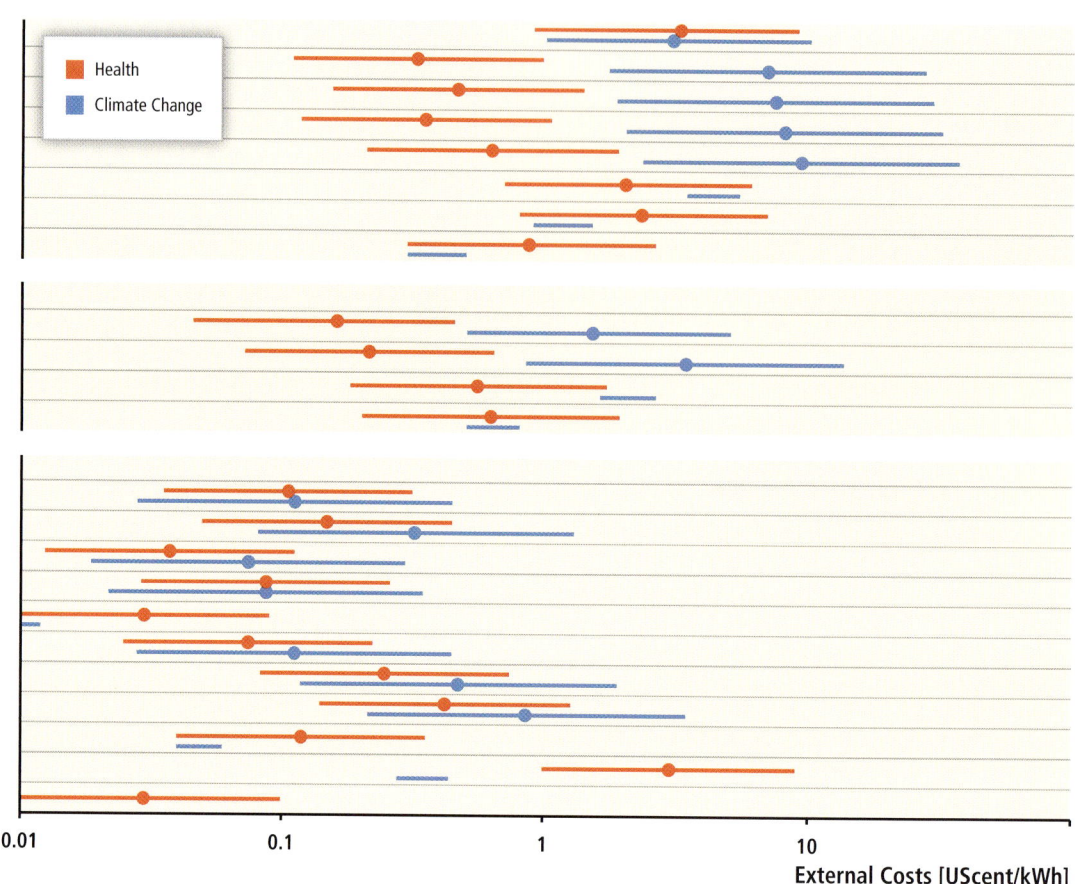

Figure 10.36 | Illustration of external costs due to the life-cycle of electricity production based on RE and fossil energy. The blue lines indicate the range of the external cost due to climate change and the red lines indicate the range of the external costs due to air pollutant health effects. External costs due to climate change mainly dominate in fossil energy if not equipped with carbon capture and storage (CCS). Comb.C: Combined Cycle; Postcom: Post-Combustion; η: efficiency factor. The results are based on four studies having different assumptions: (A) Committee on Health (2010): Existing power plants in the USA, SCC central estimate USD 30/t CO_2, range from USD 10 to 100/t CO_2, assumed value of statistical life USD 6 million; (B) Krewitt and Schlomann (2006): Central European conditions, SCC central estimate USD 90/t CO_2 and range from USD 17 to 350/t CO_2; (C) Results from the NEEDS project (Preiss, 2009; Ricci, 2010): Central European conditions in 2025, value of life year about USD 50,000, SCC range for the considered case is from USD 40 to 65/t CO_2; (D) As biomass case of (C) but particulate emissions reduced by electrostatic precipitators (ESP) (estimated on the basis of Sippula et al. (2009)) and the external costs presented per fuel energy. The uncertainty for the external costs of health impacts is assumed to be a factor of three (based on Preiss (2009); Krewitt and Schlomann (2006); and Krewitt (2002)).

results to those of Krewitt and Schlomann (2006) and Preiss (2009) for natural gas-based electricity production but clearly higher external cost levels for coal-based production due to higher non-climate impacts.

As shown in Figure 10.36, within the portfolio of RE technologies, offshore wind energy seems to cause the smallest external costs. In contrast, small-scale biomass-fired CHP plants cause relatively high external costs due to health effects via particulate emissions (Figure 10.36) based on the specific technology considered in the New Energy Externalities Development for Sustainability (NEEDS) study (Gärtner, 2008; Preiss, 2009). It should be noted that inexpensive technical solutions like electrostatic precipitators or fabric filters can lower particulate emissions considerably in plants of moderate size classes as measured and reported, for example, by Sippula et al. (2009).

External cost estimates for nuclear power are not reported here because the character of external costs and risk from release of radionuclides due to low probability accidents or due to leakages from waste repositories in a distant future are very different, for example, from climate change and air pollution, which are practically unavoidable. Those external impacts related to nuclear power can be, however, considered by discussion and judgment in the society. Also not included here is a quantitative assessment of accident risks, though Chapter 9 covers this issue in some depth, and accident risks in terms of fatalities due to various energy production chains (e.g., coal, oil, gas and hydropower) seem be to clearly higher in non-OECD countries than in OECD countries (Burgherr and Hirschberg, 2008) (see Chapter 9).

Following the results of Figure 10.36, in most cases the environmental damages and related external costs decrease when fossil fuels are replaced by RE. Also the social benefits from the supply of RE usually increase. In some cases, however, there can be trade-offs between RE expansion and some aspects of sustainable development. Therefore, it is important to conduct environmental impact assessments for specific RE projects under consideration in order to be sure that essential requirements for the implementation of the projects are realized. Chapter 9 discusses this topic in more detail.

Figure 10.36 can only summarize a part of the available literature. Some additional studies have, for example, considered the external costs from alternative transportation biofuels and other energy sources for automobiles (Hill et al., 2006, 2009; Committee on Health, 2010). The results suggest that lower external costs per vehicle kilometre than from fossil fuels can be achieved in many cases by using biofuels, but not always. Case-specific studies are needed to assess the impacts of considered feedstock cultivation and harvest, as well as fuel processing and use.

10.6.4 Synergistic strategies for limiting damages and external costs

Many environmental impacts and external costs follow from the use of energy sources and energy technologies that cause GHG emissions, particulate emissions and acidifying emissions—fossil fuel combustion being a prime example. Therefore, it might be beneficial to consider the reduction of emission-related impacts using integrated strategies (Amann, 2008; Bollen et al., 2009).

Bollen et al. (2009) have made global cost-benefit studies using the MERGE model (Manne and Richels, 2005). In their studies, the external costs of health effects due to particulate emissions and impacts of climate change were internalized. According to the study (Figure 10.37), the external benefits were greatest when both external cost types were internalized, although the mitigation costs were high as they work in a shorter time frame. The discounted benefits from the control of particulate emissions are clearly larger than the discounted benefits from the mitigation of climate change. The difference is, according to a sensitivity study, mostly greater by at least a factor of two, but of course depends on the specific assumptions. The countries would therefore benefit from combined strategies quite rapidly due to decreased external costs stemming from the reduced air pollution health impacts.

Amann (2008) reached quite similar conclusions in a case study for China. According to the study, the reduction of GHG emissions in China caused considerable benefits when there is a desire to reduce local air pollution. Also a study (Syri et al., 2002) considering the impacts of the reduction of GHG emissions in Finland stated that particulate emissions are also likely to decrease.

A study by Spalding-Fecher and Matibe (2003) is one of the few for developing countries. They found that, in South Africa, the total external costs of coal-fired power generation are 40 and 20% of industrial and residential charges for electricity. They concluded also that a reduction in GHG emissions lessens air-borne particulates that led to respiratory disorders and other diseases.

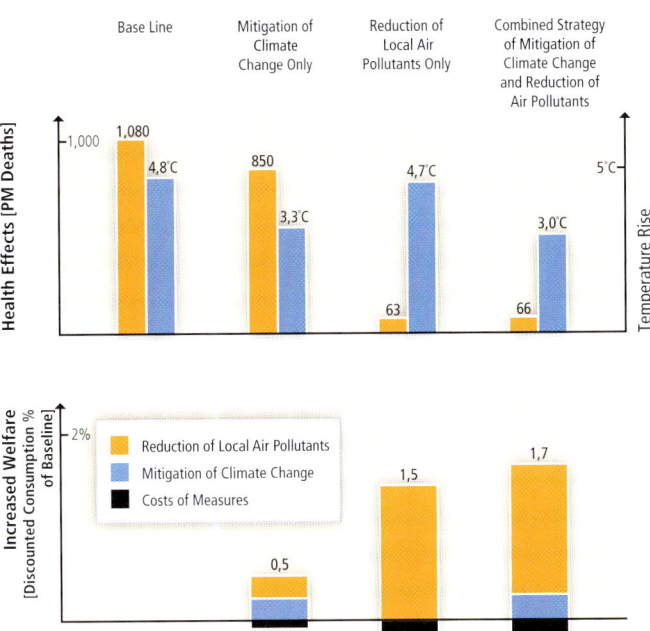

Figure 10.37 | Illustration of changes in costs, benefits and global welfare for three scenarios ('reduction of local air pollutants', 'mitigation of climate change', and 'combined strategy of mitigation of climate change and reduction of local air pollutants'), expressed as percentage consumption change (welfare increase) in comparison to the baseline (lower panel). The global temperature rise (degrees Celsius compared to the pre-industrial level) and number of deaths due to air pollution (millions) are given in the upper panel for each scenario. In the scenario 'mitigation of climate change only', the external costs of climate change have been internalized; in the scenario 'reduction of local air pollutants only', the external costs of local air pollutants have been internalized; and in the scenario of 'combined strategy', both external cost components have been internalized. The 'combined strategy' is most beneficial for society according to the results. In the baseline, the number of particulate matter (PM) deaths due to air pollutants would be around 1,000 million and the temperature rise 4.8°C (Bollen et al., 2009).

10.6.5 Knowledge gaps

Considerable uncertainties exist in the assessment and valuation of external impacts of energy sources. The assessment of physical, biological and health damages includes considerable uncertainty and the estimates are based typically on purely quantitative models, the results of which are often difficult to validate. The damages or changes seldom have market values that could be used in cost estimation, thus indirect information or other approaches must be used for damage valuation. Further, many of the damages will take place far in the future or in societies very different from those benefiting from the use of the considered energy production, which complicates the considerations. These factors contribute to the uncertainty about external costs.

However, the knowledge about external costs and benefits due to alternative energy sources can give some guidance for society to select best alternatives and to steer the energy system towards overall efficiency and high welfare gains.

References

Aitken, D. (2003). *Transitioning to a Renewable Energy Future.* White Paper, International Solar Energy Society, Freiburg, Germany, 55 pp. Available at: www.ises.org/shortcut.nsf/to/wp.

Akimoto, K., F. Sano, J. Oda, T. Homma, U.K. Rout, and T. Tomoda (2008). Global emission reductions through a sectoral intensity target scheme. *Climate Policy*, **8**(Supplement), pp. S46-S59.

Amann, M. (2008). *GAINS ASIA Scenarios for Cost-Effective Control of Air Pollution and Greenhouse Gases in China.* International Institute for Applied Systems Analysis, Laxenburg, Austria.

Anthoff, D. (2007). *Marginal external damage costs inventory of greenhouse gas emissions.* Delivery No. 5.4 - RS 1b, New Energy Externalities Developments for Sustainability, Rome, Italy.

Awerbuch, S. (2006). Portfolio-based electricity generation planning: Policy implications for renewables and energy security. *Mitigation and Adaptation Strategies for Global Change*, **11**(3), pp. 693-710.

Azar, C., K. Lindgren, E. Larson, and K. Möllersten (2006). Carbon capture and storage from fossil fuels and biomass – Costs and potential role in stabilizing the atmosphere. *Climatic Change*, **74**(1), pp. 47-79.

Bailis, R., M. Ezzati, and D.M. Kammen (2005). Mortality and greenhouse gas impacts of biomass and petroleum energy futures in Africa. *Science*, **308**(5718), pp. 98-103.

Baker, E., L. Clarke, and E. Shittu (2008). Technical change and the marginal cost of abatement. *Energy Economics*, **30**(6), pp. 2799-2816.

Barker, T., I. Bashmakov, A. Alharthi, M. Amann, L. Cifuentes, J. Drexhage, M. Duan, O. Edenhofer, B. Flannery, M. Grubb, M. Hoogwijk, F.I. Ibitoye, C.J. Jepma, W.A. Pizer, and K. Yamaji (2007). Mitigation from a cross-sectoral perspective. In: *Climate Change 2007: Mitigation of Climate Change. Contribution of Working Group III to the Fourth Assessment Report of the Intergovernmental Panel on Climate Change.* B. Metz, O.R. Davidson, P.R. Bosch, R. Dave, and L.A. Meyer (eds.), Cambridge University Press, pp. 619-690.

Barker, T., H. Pun, J. Köhler, R. Warren, and S. Winne (2006). Decarbonizing the global economy with induced technological change: Scenarios to 2100 using E3MG. *Energy Journal*, **27**(Special Issue 1), pp. 241-258.

Bergmann, A., M. Hanley, and R. Wright (2006). Valuing the attributes of renewable energy investments. *Energy Policy*, **34**(9), pp. 1004-1014.

Berry, T., and M. Jaccard (2001). The renewable portfolio standard: design considerations and an implementation survey. *Energy Policy*, **29**(4), pp. 263-277.

Black & Veatch Corporation (2007). *Arizona Renewable Energy Assessment. Final Report.* Black & Veatch Corporation, Lamar, Kansas.

Bolinger, M., R. Wiser, and W. Golove (2006). Accounting for fuel price risk when comparing renewable to gas-fired generation: the role of forward natural gas prices. *Energy Policy*, **34**(6), pp. 706-720.

Bollen, J., B. van der Zwaan, C. Brink, and H. Erens (2009). Local air pollution and global climate change: A combined cost-benefit analysis. *Resource and Energy Economics*, **31**, pp. 161-181.

Bosetti, V., C. Carraro, and M. Tavoni (2009). Climate change mitigation strategies in fast-growing countries: The benefits of early action. *Energy Economics*, **31**(Supplement 2), pp. S144-S151.

Burgherr, P., and S. Hirschberg (2008). A comparative analysis of accident risks in fossil, hydro and nuclear energy chains. *Human and Ecological Risk Assessment*, **14**, pp. 947-973.

Calvin, K., J. Edmonds, B. Bond-Lamberty, L. Clarke, S.H. Kim, P. Kyle, S.J. Smith, A. Thomson, and M. Wise (2009). 2.6: Limiting climate change to 450 ppm CO_2 equivalent in the 21st century. *Energy Economics*, **31**(Supplement 2), pp. S107-S120.

Carlsmith, R.S., W.U. Chandler, J.E. McMahon, and D.J. Santini (1990). *Energy Efficiency: How Far can We Go?* ORNL/TM-11441, Oak Ridge National Laboratory, Oak Ridge, TN, USA, 83 pp.

Carpenter, S.R., P.L. Pingali, E.M. Bennet, and M.B. Zurek (eds.) (2005). *Ecosystems and Human Wellbeing: Scenarios.* Island Press, Chicago, IL, USA, 561 pp.

CBI (2007). *Climate Change: Everyone's Business.* Confederation of British Industry Climate Change Task Force, London, UK.

Changliang, X., and S. Zhanfeng (2009). Wind energy in China: Current scenario and future perspectives. *Renewable and Sustainable Energy Reviews*, **13**(8), pp. 1966-1974.

Chen, W. (2005). The Costs of Mitigating Carbon Emissions in China: Findings from China MARKALMACRO modeling. *Energy Policy*, **33**, pp. 885-896.

Chen, C., R. Wiser, and M. Bolinger (2007). *Weighing the Costs and Benefits of State Renewables Portfolio Standards: A Comparative Analysis of State-Level Policy Impact Projections.* LBNL-61580, Ernest Orlando Lawrence Berkeley National Laboratory, Berkeley, CA, USA, 71 pp.

Clarke, L., J. Edmonds, H. Jacoby, H. Pitcher, J. Reilly, and R. Richels (2007). *Scenarios of Greenhouse Gas Emissions and Atmospheric Concentrations. Sub-report 2.1A of Synthesis and Assessment Product 2.1.* U.S. Department of Energy, Office of Biological and Environmental Research, Washington, DC, USA, 154 pp.

Clarke, L., J.P. Weyant, and J. Edmonds (2008). On the sources of technological change: what do the models assume. *Energy Economics*, **30**(2), pp. 409-424.

Clarke, L., J. Edmonds, V. Krey, R. Richels, S. Rose, and M. Tavoni (2009). International climate policy architectures: Overview of the EMF 22 International Scenarios. *Energy Economics*, **31**(Supplement 2), pp. 64-81.

Committee on Climate Change (2008). *Building a Low-Carbon Economy – The UK's Contribution to Tackling Climate Change.* The Stationary Office, Norwich, UK. Available at: www.theccc.org.uk/reports/building-a-low-carbon-economy.

Committee on Health (2010). *Hidden Costs of Energy: Unpriced Consequences of Energy Production and Use.* Committee on Health, Environmental, and Other External Costs and Benefits of Energy Production and Consumption, National Research Council, National Academies Press, Atlanta, GA, USA, 506 pp.

Creyts, J., A. Derkach, S. Nyquist, K. Ostrowski, and J. Stephenson (2007). *Reducing U.S. Greenhouse Gas Emissions: How Much at What Cost?* McKinsey & Company, USA, 83 pp.

Curtright, A.E., M.G. Morgan, and D.W. Keith (2008). Expert assessments of future photovoltaic technologies. *Environmental Science & Technology*, **42**(24), pp. 9031-9038.

da Costa, M.M., C. Cohen, and R. Schaeffer (2007). Social features of energy production and use in Brazil: Goals for a sustainable energy future. *Natural Resources Forum*, **31**(1), pp. 11-20.

Dasgupta, P., K.-G. Mäler, and S. Barrett (2000). *Intergenerational Equity, Social Discount Rates and Global Warming.* University of Cambridge, Cambridge, UK.

de Vries, B.J.M., D.P. van Vuuren, and M.M. Hoogwijk (2007). Renewable energy sources: Their global potential for the first-half of the 21st century at a global level: An integrated approach. *Energy Policy*, **35**, pp. 2590-2610.

den Elzen, M., N. Höhne, and J. van Vliet (2009). Analysing comparable greenhouse gas mitigation efforts for Annex I countries. *Energy Policy*, **37**, pp. 4114-4131.

Dietz, S., and N. Stern (2008). Why economic analysis supports strong action on climate change: A response to the Stern review's critics. *Review of Environmental Economics and Policy*, **2**(1), pp. 94-113.

Dornburg, V., J. van Dam, and A. Faaij (2007). Estimating GHG emission mitigation supply curves of large-scale biomass use on a country level. *Biomass and Bioenergy*, **31**(1), pp. 46-65.

Downing, T., D. Anthoff, R. Butterfield, M. Ceronsky, M. Grubb, J. Guo, C. Hepburn, C. Hope, A. Hunt, A. Li, A. Markandya, S. Moss, A. Nyong, R. Tol, and P. Watkiss (2005). *Social Cost of Carbon: A Closer Look at Uncertainty*. Stockholm Environment Institute, Oxford, UK.

Durstewitz, M., and M. Hoppe-Klipper (1999). Wind energy experience curve for the German "250 MW Wind Program". In: *IEA International Workshop on Experience Curves for Policy Making - The Case of Energy Technologies*. Stuttgart, Germany, 10-11 May 1999. Available at: www.iset.uni-kassel.de/abt/FB-I/publication/99-05-10_exp_curves_iea.pdf.

ECN (2004). *Learning from the Sun: Final Report of the Photes Project*. Energy Research Center of the Netherlands (ECN), Petten, The Netherlands.

Edenhofer, O., C. Carraro, J. Köhler, and M. Grubb (guest eds.) (2006). Endogenous technological change and the economics of atmospheric stabilization. *Energy Journal*, **27**(Special Issue 1), 284 pp.

Edenhofer, O., C. Carraro, J.-C. Hourcade, K. Neuhoff, G. Luderer, C. Flachsland, M. Jakob, A. Popp, J. Steckel, J. Strohschein, N. Bauer, S. Brunner, M. Leimbach, H. Lotze-Campen, V. Bosetti, E.d. Cian, M. Tavoni, O. Sassi, H. Waisman, R. Crassous-Doerfler, S. Monjon, S. Dröge, H.v. Essen, P.d. Río, and A. Türk (2009). *The Economics of Decarbonization – Report of the RECIPE Project*. Potsdam Institute for Climate Impact Research, Potsdam, Germany.

Edenhofer, O., B. Knopf, T. Barker, L. Baumstark, E. Bellevrat, B. Chateau, P. Criqui, A. Isaac, A. Kitous, S. Kypreos, M. Leimbach, K. Lessmann, B. Magne, Å. Scrieciu, H. Turton, and D.P. Van Vuuren (2010). The economics of low stabilization: Model comparison of mitigation strategies and costs. *Energy Journal*, **31**(Special Issue), pp. 11-48.

Edwards, R., S. Szekeres, F. Neuwahl, and V. Mahieu (2008). *Biofuels in the European Context: Facts, Uncertainties and Recommendations*. JRC Institute for Energy, Petten, The Netherlands, 30 pp.

EEA (2007). *Greenhouse gas emission trends and projections in Europe 2007 – Country profile: Poland*. European Energy Agency, Brussels, Belgium.

EEA (2009). *Europe's onshore and offshore wind energy potential: An assessment of environmental and economic constraints*. European Environment Agency, Copenhagen, Denmark.

EIA (2009). *Annual Energy Outlook 2009 with Projections to 2030*. Energy Information Administration, Washington, DC, USA.

Enermodal (1999). *Cost Reduction Study for Solar Thermal Power Plants: Final Report*. Enermodal Engeneering Limited, Kitchener, ON, Canada.

Enviros Consulting Ltd. (2005). *The Costs of Supplying Renewable Energy*. Enviros Consulting Ltd., London, UK.

EPIA (2008). *Solar Generation*. European Photovoltaic Industry Association (EPIA), Brussels, Belgium.

EPIA (2010). *Global Market Outlook for Photovoltaics until 2014*. European Photovoltaic Industry Association (EPIA), Brussels, Belgium, 8 pp.

ERI (2009). *2050 China Energy and CO_2 Emissions Report (CEACER)*. Energy Research Institute, Beijing: Science Press. In Chinese. As cited in: *China's Energy and Carbon Emissions Outlook to 2050*, N. Zhou and D. Fridley (2011). Lawrence Berkeley National Laboratory.

EWEA (2009). *Wind Energy, the Facts*. European Wind Energy Association (EWEA), Brussels, Belgium.

Fearnside, P.M. (1999). Social impacts of Brazil's Tucurui Dam. *Environmental Management*, **24**(4), pp. 483-495.

Fearnside, P.M. (2005). Brazil's Samuel Dam: Lessons for hydroelectric development policy and the environment in Amazonia. *Environmental Management*, **35**(1), pp. 1-19.

Fellows, A. (2000). *The Potential of Wind Energy to Reduce Carbon Dioxide Emissions*. Garrad Hassan, Glasgow, Scotland.

Ferioli, F., K. Schoots, and B.C.C. Van der Zwaan (2009). Use and limitations of learning curves for energy technology policy: A component-learning hypothesis. *Energy Policy*, **37**, pp. 2525-2535.

Fisher, B.S., N. Nakicenovic, K. Alfsen, J. Corfee Morlot, F. de la Chesnaye, J.-C. Hourcade, K. Jiang, M. Kainuma, E. La Rovere, A. Matysek, A. Rana, K. Riahi, R. Richels, S. Rose, D. van Vuuren, and R. Warren (2007). Issues related to mitigation in the long term context. In: *Climate Change 2007: Mitigation of Climate Change. Contribution of Working Group III to the Fourth Assessment Report of the Intergovernmental Panel on Climate Change*. B. Metz, O.R. Davidson, P.R. Bosch, R. Dave, and L.A. Meyer (eds.), Cambridge University Press, pp.169-250.

Fleiter, T., M. Hagemann, S. Hirzel, W. Eichhammer, and M. Wietschel (2009). Costs and potentials of energy savings in European industry – a critical assessment of the concept of conservation supply curves. In: *Proceedings of ECEEE 2009 Summer Study*, European Council for an Energy Efficient Economy (ed.), La Colle sur Loup, France, 1-6 June 2009, Panel 5, pp. 1261-1272. Available at: www.eceee.org/conference_proceedings/eceee/2009/Panel_5/5.376/.

Foxon, T.J., R. Gross, A. Chase, J. Howes, A. Arnall, and D. Anderson (2005). UK innovation systems for new and renewable energy technologies: drivers, barriers and system failures. *Energy Policy*, **33**, pp. 2123-2137.

Gallagher, P.W., M. Dikeman, J. Fritz, E. Wailes, W. Gauthier, and H. Shapouri (2003). Supply and social cost estimates for biomass from crop residues in the United States. *Environmental & Resource Economics*, **24**(4), pp. 335-358.

Gärtner, S. (2008). *Final Report on Technical Data, Costs and Life Cycle Inventories of Biomass CHP Plants*. Deliverable No. 13.2 – RS 1a, New Energy Externalities Developments for Sustainability, Rome, Italy.

Goldemberg, J., S. Coelhob, P.M. Nastaric, and O. Lucon (2004). Ethanol learning curve: The Brazilian experience. *Biomass & Bioenergy*, **26**(2004), pp. 301-304.

González, P.d.R. (2008). Policy implications of potential conflicts between short-term and long-term efficiency in CO_2 emissions abatement. *Ecological Economics*, **65**, pp. 292-303.

Gordon, F., L. Garth, T. Eckman, and C. Grist (2008). *Beyond Energy Supply Curves*. University of California, Davis, CA, USA.

Groscurth, H.M., A. de Almeida, A. Bauen, F.B. Costa, S.O. Ericson, J. Giegrich, N. von Grabczewski, D.O. Hall, O. Hohmeyer, K. Jorgensen, C. Kern, I. Kuhn, R. Lofstedt, J.D. Mariano, P.M.G. Mariano, N.I. Meyer, P.S. Nielsen, C. Nunes, A. Patyk, G.A. Reinhardt, F. Rosillo-Calle, I. Scrase, and B. Widmann (2000). Total costs and benefits of biomass in selected regions of the European Union. *Energy*, **25**(11), pp. 1081-1095.

Gross, R., and P. Heptonstall (2008). The costs and the impacts of intermittency: An ongoing debate. *Energy Policy*, **36**(2008), pp. 4005-4007.

Grubb, M., and D. Newbery (2008). Pricing carbon for electricity generation: national and international dimensions. In: *Delivering a Low Carbon Electricity System: Technologies, Economics and Policy*. M. Grubb, T. Jamasb, and M.G. Pollitt (eds.), Cambridge University Press.

Grubler, A. (2008). Energy transitions. In: *Encyclopedia of Earth*. C.J. Cleveland (ed.), Environmental Information Coalition, National Council for Science and the Environment, Washington, DC, USA.

Gumerman, E., J.G. Koomey, and M.A. Brown (2001). Strategies for cost-effective carbon reductions: a sensitivity analysis of alternative scenarios. *Energy Policy*, **29**, pp. 1313-1323.

Gurney, A., H. Ahammad, and M. Ford (2009). The economics of greenhouse gas mitigation: Insights from illustrative global abatement scenarios modelling. *Energy Economics*, **31**(Supplement 2), pp. S174-S186.

Hanaoka, T., R. Kawase, M. Kainuma, Y. Matsuoka, H. Ishii, and K. Oka (2006). *Greenhouse Gas Emissions Scenarios Database and Regional Mitigation Analysis*. National Institute for Environmental Studies (NIES), Tsukuba, Japan, 106 pp.

Harmon, C. (2000). *Experience Curves for Photovoltaic Technology*. IR-00-014, International Institute for Applied Systems Analysis, Laxenburg, Austria.

Heptonstall, P. (2007). *A Review of Electricity Unit Cost Estimates, Working Paper*. UK Energy Research Centre, London, UK.

Herzog, H., K. Smekens, P. Dadhich, J. Dooley, Y. Fujii, O. Hohmeyer, and K. Riahi (2005). Cost and economic potential. In: *IPCC Special Report on Carbon Dioxide Capture and Storage*. B. Metz, O. Davidson, H. de Coninck, M. Loos, and L. Meyer (eds.), Cambridge University Press, pp. 339-362.

Hettinga, W.G., H.M. Junginger, S.C. Dekker, M. Hoogwijk, A.J. McAloon, and K.B. Hicks (2009). Understanding the reductions in US corn ethanol production costs: An experience curve approach. *Energy Policy*, **37**(1), pp. 190-203.

Hill, J., E. Nelson, D. Tilman, S. Polasky, and D. Tiffany (2006). Environmental, economic, and energetic costs and benefits of biodiesel and ethanol biofuels. *Proceedings of the National Academy of Sciences*, **103**(30), pp. 11206-11210.

Hill, J., S. Polasky, E. Nelson, D. Tilman, H. Huo, L. Ludwig, J. Neumann, H. Zheng, and D. Bonta (2009). Climate change health costs of air emissions from biofuels and gasoline. *Proceedings of the National Academy of Sciences*, **106**(6), pp. 2077-2082.

Hohmeyer, O. (1992). Renewables and the full costs of energy. *Energy Policy*, **20**(4), pp. 365-375.

Hoogwijk, M., A. Faaij, R. van den Broek, G. Berndes, D. Gielen, and W. Turkenburg (2003). Exploration of the ranges of the global potential of biomass for energy. *Biomass and Bioenergy*, **25**, pp. 119-133.

Hoogwijk, M., B. de Vries, and W. Turkenburg (2004). Assessment of the global and regional geographical, technical and economic potential of onshore wind energy. *Energy Economics*, **26**, pp. 889-919.

Hoogwijk, M., and W. Graus (2008). *Global Potential of Renewable Energy Sources: A Literature Assessment*. Ecofys, Utrecht, The Netherlands, 45 pp.

Hoogwijk, M., A. Faaij, B. de Vries, and W. Turkenburg (2009). Exploration of regional and global cost-supply curves of biomass energy from short-rotation crops at abandoned croplands and rest land under four IPCC SRES land-use scenarios. *Biomass and Bioenergy*, **33**, pp. 26-43.

Hourcade, J.-C., M. Jaccard, C. Bataille, and F. Ghersi (2006). Hybrid modeling: New answers to old challenges – Introduction to the Special Issue of The Energy Journal. *Energy Journal*, **27**(Special Issue October), pp. 1-12.

Hughes, A., M. Haw, H. Winkler, A. Marquard, and B. Merven (2007). *Energy Modeling: A Modelling Input into the Long Term Mitigation Scenarios Process. LTMS Input Report*. Energy Research Centre, Cape Town, South Africa.

IEA (2000). *Experience Curves for Energy Technology Policy*. International Energy Agency (IEA), Paris, France, 133 pp.

IEA (2003). *Energy to 2050. Scenarios for a Sustainable Future*. International Energy Agency (IEA), Paris, France.

IEA (2005). *Projected Costs of Generating Electricity*. OECD/IEA, Paris, France.

IEA (2006). *Standard Review of the Netherlands*. International Energy Agency (IEA), Paris, France.

IEA (2007a). *Renewables for Heating and Cooling, Untapped Potential*. International Energy Agency (IEA), Paris, France.

IEA (2007b). *Renewables in Global Energy Supply: An IEA Fact Sheet*. International Energy Agency (IEA), Paris, France.

IEA (2008a). *Deploying Renewable Energies: Principles for Effective Policies*. International Energy Agency (IEA), Paris, France.

IEA (2008b). *Energy Technology Perspectives 2008. Scenarios and Strategies to 2050*. International Energy Agency (IEA), Paris, France, 646 pp.

IEA (2009). *World Energy Outlook 2009*. International Energy Agency (IEA), Paris, France, 696 pp.

IEA (2010a). *Energy Technology Perspectives 2010. Scenarios and Strategies to 2050*. International Energy Agency (IEA), Paris, France, 708 pp.

IEA (2010b). *Projected Costs of Generating Electricity - 2010 Edition*. International Energy Agency (IEA), Paris, France.

IEA (2010c). *Technology Roadmap Solar Photovoltaic Energy*. International Energy Agency (IEA), Paris, France.

IEA (2010d). *World Energy Outlook 2010*. International Energy Agency (IEA), Paris, France, 736 pp.

IEA (2010e). *Energy Balances of Non-OECD Countries*. International Energy Agency (IEA), Paris, France.

IEA (2010f). *Key Energy Statistics 2010*. International Energy Agency (IEA), Paris, France.

IEA (2010g). RD&D Budget. *IEA Energy Technology R&D Statistics* (database), doi: 10.1787/data-00488-en. Available at: www.oecd-ilibrary.org/energy/data/iea-energy-technology-r-d-statistics/rd-d-budget_data-00488-en.

IMF (2009). *World Economic Outlook Update: Contractionary Forces Receding but Weak Recovery Ahead*, International Monetary Fund (IMF), Washington D.C., USA

IPCC (2000). *Special Report on Emissions Scenarios*. N. Nakicenovic and R. Swart (eds.), Cambridge University Press, 570 pp.

IPCC (2007). *Climate Change 2007: Mitigation of Climate Change. Contribution of Working Group III to the Fourth Assessment Report of the Intergovernmental Panel on Climate Change*. B. Metz, O.R. Davidson, P.R. Bosch, R. Dave, and L.A. Meyer (eds.), Cambridge University Press, 851 pp.

Isles, L. (2006). *Offshore Wind Farm Development: Cost Reduction Potential*. PhD Thesis, Lund University, Lund, Sweden.

Jaffe, A.B., R.G. Newell, and R.N. Stavins (2005). A tale of two market failures: Technology and environmental policy. *Ecological Economics*, **54**(2-3), pp. 164-174.

Jamasb, T. (2007). Technical change theory and learning curves: Patterns of progress in electricity generation technologies. *Energy Journal*, **28**(3), pp. 133-150.

Junginger, M., S. Agterbosch, A. Faaij and W.C. Turkenburg (2004). Renewable electricity in the Netherlands. *Energy Policy*, **32**, pp. 1053-1073.

Junginger, M., A. Faaij, R. Björheden, and W.C. Turkenburg (2005a). Technological learning and cost reductions in wood fuel supply chains in Sweden. *Biomass and Bioenergy*, **29**, pp. 399-418.

Junginger, M., A. Faaij, and W.C. Turkenburg (2005b). Global experience curves for wind farms. *Energy Policy*, **33**, pp. 133-150.

Junginger, M., E. Visser, K. Hjort-Gregersen, J. Koornneef, R. Raven, A. Faaij, and W. Turkenburg (2006). Technological learning in bioenergy systems. *Energy Policy*, **34**, pp. 4024-4041.

Junginger, M., W.V. Sark, and A. Faaij (eds.) (2010). *Technological Learning in the Energy Sector: Lessons for Policy, Industry and Science.* Edward Elgar, Northampton, MA, USA.

Kahouli-Brahmi (2009). Testing for the presence of some features of increasing returns to adoption factors in energy system dynamics: An analysis via the learning curve approach. *Ecological Economics*, **68**, pp. 1195-1212.

Kennedy, S. (2005). Wind power planning: assessing long-term costs and benefits. *Energy Policy*, **33**(13), pp. 1661-1675.

Kirkinen, J., T. Palosuo, K. Holmgren, and I. Savolainen (2008). Greenhouse impact due to the use of combustible fuels: Life cycle viewpoint and relative radiative forcing commitment. *Environmental Management*, **42**(3), pp. 458-469.

Kitous, A., P. Criqui, E. Bellevrat, and B. Chateau (2010). Transformation patterns of the worldwide energy system – Scenarios for the century with the POLES model. *Energy Journal*, **31**(Special Issue), pp. 49-82.

Knopf, B., O. Edenhofer, T. Barker, N. Bauer, L. Baumstark, B. Chateau, P. Criqui, A. Held, M. Isaac, M. Jakob, E. Jochem, A. Kitous, S. Kypreos, M. Leimbach, B. Magne, S. Mima, W. Schade, S. Scrieciu, H. Turton, and D.v. Vuuren (2009). The economics of low stabilisation: implications for technological change and policy. In: *Making Climate Change Work for Us: European Perspectives on Adaptation and Mitigation Strategies.* M. Hulme and H Neufeldt (eds.), Cambridge University Press.

Koljonen, T., H. Ronde, A. Lehtilä, T. Ekholm, I. Savolainen, and S. Syri (2008). Greenhouse gas emission mitigation and energy security, a scenario results and practical programmes in some Asian countries. In: *2nd IAEE Asian Conference*, International Association for Energy Economics, Perth, Australia, 5-7 November 2008.

Koljonen, T., M. Flyktman, A. Lehtilä, K. Pahkala, E. Peltola, and I. Savolainen (2009). The role of CCS and renewables in tackling climate change. *Energy Procedia*, **1**(1), pp. 4323-4330.

Koomey, J.G., Richey, R.C., Laitner, J.A., Markel, R.J., and Marnay, C. (1998). *Technology and Greenhouse Gas Emissions: An Integrated Analysis using the LBNL–NEMS Model.* Lawrence Berkeley National Laboratory, Berkeley, CA, USA.

Kouvaritakis, N., A. Soria, and S. Isoard (2000). Modelling energy technology dynamics: methodology for adaptive expectations models with learning by doing and learning by searching. *International Journal of Global Issues*, **14**, pp. 1-4.

Krewitt, W. (2002). External costs of energy – do the answers match the questions? Looking back at 10 years of ExternE. *Energy Policy*, **30**(10), pp. 839-848.

Krewitt, W. (2007). Die externen Kosten der Stromerzeugung aus erneuerbaren Energien im Vergleich zur fossilen Stromerzeugung. *Umweltwissenschaften und Schadstoff-Forschung*, **19**(3), pp. 144-151.

Krewitt, W., and B. Schlomann (2006). *Externe Kosten der Stromerzeugung aus erneuerbaren Energien im Vergleich zur Stromerzeugung aus fossilen Energieträgern.* DLR, Institut für Technische Thermodynamik, Fraunhofer Institut für System-und Innovationsforschung, Gutachten im Auftrag des ZSW im Rahmen von Beratungsleistungen für das BMU, 59 pp.

Krewitt, W., S. Teske, S. Simon, T. Pregger, W. Graus, E. Blomen, S. Schmid, and O. Schäfer (2009). Energy [R]evolution 2008 – a sustainable world energy perspective. *Energy Policy*, **37**(12), pp. 5764-5775.

Krey, V., and K. Riahi (2009). Implications of delayed participation and technology failure for the feasibility, costs, and likelihood of staying below temperature targets – Greenhouse gas mitigation scenarios for the 21st century. *Energy Economics*, **31**(Supplement 2), pp. S94-S106.

Krey, V., and L. Clarke (2011). Role of renewable energy in climate mitigation: a synthesis of recent scenarios. *Climate Policy*, in press.

Kruck, C., and L. Eltrop (2007). *Perspektiven der Stromerzeugung aus Solar- und Windenergienutzung: Endbericht.* FKZ A204/04, IER (Institut für Energiewirtschaft und Rationelle Energieanwendung), Universität Stuttgart, Stuttgart, Germany.

Kuik, O., L. Brander, and R.S.J. Tol (2009). Marginal abatement costs of greenhouse gas emissions: A meta-analysis. *Energy Policy*, **37**, pp. 1395-1403.

Kurosawa, A. (2006). Multigas mitigation: An economic analysis using GRAPE model. *Energy Journal*, **27**(Special Issue November), pp. 275-288.

Kusiima, J.M., and S.E. Powers (2010). Monetary value of the environmental and health externalities associated with production of ethanol from biomass feedstocks. *Energy Policy*, **38**(6), pp. 2785-2796.

Kverndokk, S., and K.E. Rosendahl (2007). Climate policies and learning by doing: Impacts and timing of technology subsidies. *Resource and Energy Economics*, **29**(6), pp. 2785-2796.

Leimbach, M., N. Bauer, L. Baumstark, M. Lüken, and O. Edenhofer (2010). Technological change and international trade – Insights from REMIND-R. *Energy Journal*, **31**(Special Issue), pp. 109-136.

Lewandowski, I., J. Weger, A. van Hooijdonk, K. Havlickova, J. van Dam, and A. Faaij (2006). The potential biomass for energy production in the Czech Republic. *Biomass and Bioenergy*, **30**, pp. 405-421.

Loulou, R., U. Remme, A. Kanudia, A. Lehtila, and G. Goldstein (2005). *Documentation for the TIMES Model.* IEA Energy Technology Systems Analysis Programme, Paris, France.

Loulou, R., M. Labriet, and A. Kanudia (2009). Deterministic and stochastic analysis of alternative climate targets under differentiated cooperation regimes. *Energy Economics*, **31**(Supplement 2), pp. S131-S143.

Luderer, G., V. Bosetti, J. Steckel, H. Waisman, N. Bauer, E. Decian, M. Leimbach, O. Sassi, and M. Tavoni (2009). *The Economics of Decarbonization – Results from the RECIPE model Intercomparison.* Potsdam Institute for Climate Impact Research, Potsdam, Germany (Peer-reviewed version accepted for publication: Luderer, G., V Bosetti, M Jakob, M Leimbach, J Steckel, H Waisman, O Edenhofer (2011). *The Economics of Decarbonizing the Energy System – Results and Insights from the RECIPE Model Intercomparison.* Climatic Change, doi: 10.1007/s10584-011-0105-x).

Mackay, R., and S. Probert (1998). Likely market-penetrations of renewable-energy technologies. *Applied Energy*, **59**, pp. 1-38.

Magne, B., S. Kypreos, and H. Turton (2010). Technology options for low stabilization pathways with MERGE. *Energy Journal*, **31**(Special Issue), pp. 83-108.

Manne, A., and R. Richels (2005). *Merge: An Integrated Assessment Model for Global Climate Change.* In: *Energy and Environment.* R. Loulou, J.-P. Waaub, and G. Zaccour (eds.), Springer, pp. 175-189.

Masui, T., S. Ashina, and J. Fujino (2010). *Analysis of 4.5 W/m2 Stabilization Scenarios with Renewable Energies and Advanced Technologies using AIM/CGE[Global] model.* AIM Team, National Institute for Environmental Studies, Tsukuba, Japan.

McElroy, M.B., X. Lu, C.P. Nielsen, and Y. Wang (2009). Potential for wind-generated electricity in China. *Science*, **325**, pp. 1380.

McKinsey&Company (2007). *Costs and Potentials of Greenhouse Gas Abatement in Germany.* McKinsey&Company.

McKinsey&Company (2008a). *An Australian Cost Curve for Greenhouse Gas Reduction.* McKinsey&Company.

McKinsey&Company (2008b). *Costs and Potentials for Greenhouse Gas Abatement in the Czech Republic.* McKinsey&Company.

McKinsey&Company (2009a). *China's Green Revolution.* McKinsey&Company.

McKinsey&Company (2009b). *Pathway to a Low-Carbon Economy.* McKinsey&Company.

McKinsey&Company (2009c). *Swiss Greenhouse Gas Abatement Cost Curve.* McKinsey&Company.

Mehos, M.S., and D.W. Kearney (2007). Potential carbon emissions reductions from concentrating solar power by 2030. In: *Tackling Climate Change in the U.S.: Potential Carbon Emissions Reductions from Energy Efficiency and Renewable Energy by 2030.* C.F. Kutscher (ed.), American Solar Energy Society, Boulder, CO, USA, pp. 79-90. Available at: http://ases.org/images/stories/file/ASES/climate_change.pdf.

Meier, A., J. Wright, and A.H. Rosenfeld (1983). *Supplying Energy through Greater Efficiency: The Potential for Conservation in California's Residential Sector.* University of California Press, Berkeley, CA, USA.

Milligan, M. (2007). *Potential carbon emissions reductions from wind by 2030.* In: *Tackling Climate Change in the U.S.: Potential Carbon Emissions Reductions from Energy Efficiency and Renewable Energy by 2030.* C.F. Kutscher (ed.), American Solar Energy Society, Boulder, CO, USA, pp. 101-112. Available at: http://ases.org/images/stories/file/ASES/climate_change.pdf.

Montgomery, W.D., and A.E. Smith (2007). Price, quantity, and technology strategies for climate change policy. In: *Human-induced Climate Change: An Interdisciplinary Assessment.* M.E. Schlesinger, H.S. Kheshgi, J. Smith, F.C. de la Chesnaye, J.M. Reilly, T. Wilson, and C. Kolstad (eds.), Cambridge University Press, pp. 328-342.

Morita, T., J. Robinson, A. Adegbulugbe, J. Alcamo, D. Herbert, E. Lebre la Rovere, N. Nakicenivic, H. Pitcher, P. Raskin, K. Riahi, A. Sankovski, V. Solkolov, B.d. Vries, and D. Zhou (2001). Greenhouse gas emission mitigation scenarios and implications. In: *Climate Change 2001: Mitigation of Climate Change; Contribution of Working Group III to the Third Assessment Report of the IPCC.* Cambridge University Press, pp. 115-166.

Mukora, A., M. Winskel, H.F. Jeffrey, and M. Müller (2009). Learning curves for emerging energy technologies. *Proceedings of the Institution of Civil Engineers - Energy*, **162**, pp. 151-159.

Nakicenovic, N., P. Kolp, K. Riahi, M. Kainuma, and T. Hanaoka (2006). Assessment of emissions scenarios revisited. *Environmental Economics and Policy Studies*, **7**(3), pp. 137-173.

Neij, L. (1997). Use of experience curves to analyse the prospects for diffusion and adoption of renewable energy technology. *Energy Policy*, **25**, pp. 1099-1107.

Neij, L. (1999). Cost dynamics of wind power. *Energy Policy*, **24**, pp. 375-389.

Neij, L. (2003). *Final Report of EXTOOL - Experience Curves: A Tool for Energy Policiy Programme Assessment.* KFS AB, Lund, Sweden. Available at: www.iset.uni-kassel.de/extool/Extool_final_report.pdf.

Neij, L. (2008). Cost developments of future technologies for power generation – A study based on experience curves and complementary bottom-up assessments. *Energy Policy* **36**, pp. 2200-2211.

Nemet, G.F. (2009). Interim monitoring of cost dynamics for publicly supported energy technologies. *Energy Policy*, **37**, pp. 825-835.

Nemet, G.F., T. Holloway, and P. Meier (2010). Implications of incorporating air-quality co-benefits into climate change policymaking. *Environmental Research Letters*, **5**(1), 014007.

Next Energy (2004). *Cost Curve for NSW Greenhouse Gas Abatement.* Next Energy Pty Ltd., Sydney, Australia. Available at: www.environment.nsw.gov.au/resources/climatechange/costcurve.pdf.

Nichols, L.A. (1994). Demand-side management. *Energy Policy*, **22**, pp. 840-847.

Novikova, A. (2009). *Sustainable Energy and Climate Mitigation Solutions and Policies: 3. Renewable Energy.* Department of Environmental Sciences, Central European University, Budapest, Hungary.

O'Neill, B.C., K. Riahi, and I. Keppo (2010). Mitigation implications of midcentury targets that preserve long-term climate policy options. *Proceedings of the National Academy of Sciences*, **107**(3), pp. 1011-1016.

Oliveira, L.B., and L.P. Rosa (2003). Brazilian waste potential: energy, environmental, social and economic benefits. *Energy Policy*, **31**(14), pp. 1481-1491.

Overend, R.P., and A. Milbrandt (2007). Potential carbon emissions reductions from biomass by 2030. In: *Tackling Climate Change in the U.S.: Potential Carbon Emissions Reductions from Energy Efficiency and Renewable Energy by 2030.* C.F. Kutscher (ed.), American Solar Energy Society, Boulder, CO, USA, pp. 113-130. Available at: http://ases.org/images/stories/file/ASES/climate_change.pdf.

Owen, A.D. (2006). Renewable energy: Externality costs as market barriers. *Energy Policy*, **34**(5), pp. 632-642.

Pillai, R., and R. Banerjee (2009). Renewable energy in India: Status and potential. *Energy*, **34**, pp. 970-980.

Pingoud, K., H. Mälkki, M. Wihersaari, M. Hongisto, S. Siitonen, A. Lehtilä, M. Johansson, P. Pirilä, and T. Otterström (1999). *ExternE National Implementation in Finland.* VTT, Espoo, Finland, 131 pp.

Preiss, P. (2009). *Report on the Application of the Tools for Innovative Energy Technologies.* Deliverable No. 7.2 – RS 1b, New Energy Externalities Developments for Sustainability (NEEDS), Rome, Italy.

Riahi, K., A. Grübler, and N. Nakicenovic (2007). Scenarios of long-term socio-economic and environmental development under climate stabilization. *Technological Forecasting and Social Change*, **74**(7), pp. 887-935.

Ricci, A. (2009a). *NEEDS : A Summary Account of the Final Debate.* New Energy Externalities Development for Sustainability (NEEDS), Rome, Italy.

Ricci, A. (2009b). *NEEDS : Policy Use of NEEDS Results.* New Energy Externalities Development for Sustainability (NEEDS), Rome, Italy.

Ricci, A. (2010). *Policy Use of the NEEDS report.* Final integrated report, Deliverable No. 5.3 - RS In, New Energy Externalities Developments for Sustainability (NEEDS), Rome, Italy.

Roth, I.F., and L.L. Ambs (2004). Incorporating externalities into a full cost approach to electric power generation life-cycle costing. *Energy*, **29**(12-15), pp. 2125-2144.

Rufo, M. (2003). *Developing Greenhouse Gas Mitigation Supply Curves for In-State Resources*. P500-03-025FAV, California Energy Commission, Sacramento, CA, USA.

Sagar, A.D., and B. van der Zwaan (2005). Technological innovation in the energy sector: R&D, deployment, and learning-by-doing. *Energy Policy*, **34**, pp. 2601-2608.

Sawyer, S. (2009). The Global status of wind power. In: *Global Wind Report 2009*. Global Wind Energy Council, Brussels, Belgium, pp. 8-13. Available at: www.gwec.net/index.php?id=167.

Searchinger, T., R. Heimlich, R.A. Houghton, F.X. Dong, A. Elobeid, J. Fabiosa, S. Tokgoz, D. Hayes, and T.H. Yu (2008). Use of US croplands for biofuels increases greenhouse gases through emissions from land-use change. *Science*, **319**(5867), pp. 1238-1240.

Seyboth, K., L. Beurskens, O. Langniss, and R.E.H. Sims (2007). Recognising the potential for renewable energy heating and cooling. *Energy Policy*, **36**(7), pp. 2460-2463.

Sippula, O., J. Hokkinen, H. Puustinen, P. Yli-Pirilä, and J. Jokiniemi (2009). Comparison of particle emissions from small heavy fuel oil and wood-fired boilers. *Atmospheric Environment*, **43**, pp. 4855-4864.

Smith, K.R., M. Jerrett, H.R. Anderson, R.T. Burnett, V. Stone, R. Derwent, R.W. Atkinson, A. Cohen, S.B. Shonkoff, D. Krewski, C.A. Pope, M.J. Thun, and G. Thurston (2009). Public health benefits of strategies to reduce greenhouse-gas emissions: health implications of short-lived greenhouse pollutants. *Lancet*, **374**(9707), pp. 2091-2103.

Soderholm, P., and T. Sundqvist (2003). Pricing environmental externalities in the power sector: ethical limits and implications for social choice. *Ecological Economics*, **46**(3), pp. 333-350.

Soimakallio, S., T. Makinen, T. Ekholm, K. Pahkala, H. Mikkola, and T. Paapanen (2009). Greenhouse gas balances of transportation biofuels, electricity and heat generation in Finland – Dealing with the uncertainties. *Energy Policy*, **37**(1), pp. 80-90.

Solinski, J. (2005). Primary energy balances of the CEE region and the countries dependence on energy import. In: *International Conference on "Policy and strategy of sustainable energy development for Central and Eastern European Countries until 2030"*, Warsaw, Poland, 22-23 November 2005.

Soderholm, P., and T. Sundqvist (2007). Empirical challenges in the use of learning curves for assessing the economic prospects of renewable energy technologies. *Renewable Energy*, **32**, pp. 2559-2578.

Spalding-Fecher, R., and D.K. Matibe (2003). Electricity and externalities in South Africa. *Energy Policy*, **31**(8), pp. 721-734.

Sparovek, G., A. Barretto, G. Berndes, S. Martins, and R. Maule (2009). Environmental, land-use and economic implications of Brazilian sugarcane expansion 1996 - 2006. *Mitigation and Adaptation Strategies for Global Change*, **14**(3), pp. 285-298.

Stern, N. (2007). *The Economics of Climate Change*. Cambridge University Press, 712 pp. Available at: webarchive.nationalarchives.gov.uk/+/http://www.hm-treasury.gov.uk/sternreview_index.htm.

Stoft, S. (1995). *The Economics of Conserved-Energy "Supply" Curves*. University of California Energy Institute, Berkeley, CA, USA.

Sutherland, R.J. (1991). Market barriers to energy efficiency investments. *Energy Journal*, **3**(12), pp. 15-35.

Sweeney, J., and J. Weyant (2008). *Analysis of Measures to Meet the Requirements of California's Assembly Bill 32. Discussion draft, September 2008.* Stanford University Precourt Institute for Energy Efficiency, Stanford, CA, USA.

Syri, S., N. Karvosenoja, A. Lehtila, T. Laurila, V. Lindfors, and J.P. Tuovinen (2002). Modeling the impacts of the Finnish Climate Strategy on air pollution. *Atmospheric Environment*, **36**(19), pp. 3059-3069.

Syri, S., A. Lehtilä, T. Ekholm, and I. Savolainen (2008). Global energy and Emissions Scenarios for effective climate change mitigation – Deterministic and stochastic scenarios with the TIAM model. *International Journal of Greenhouse Gas Control*, **2**, pp. 274-285.

Tavoni, M., and R. Tol (2010). Counting only the hits? The risk of underestimating the costs of stringent climate policy. *Climatic Change*, **100**(3), pp. 769-778.

Teske, S., T. Pregger, S. Simon, T. Naegler, W. Graus, and C. Lins (2010). Energy [R]evolution 2010—a sustainable world energy outlook. *Energy Efficiency*, doi:10.1007/s12053-010-9098-y.

Torfs, R., L. Hurley, B. Miller, and A. Rabl (2007). *A Set of Concentration-Response Functions*. Deliverable 3.7 – RS1b/WP3, New Energy Externalities Development for Sustainability, Rome, Italy.

UK SSEFRA (2006). *Climate Change. The UK Programme 2006.* Secretary of State for the Environment Food and Rural Affairs (UK SSEFRA), The Stationary Office, Norwich, UK.

van Benthem, A., K. Gillingham, and J. Sweeney (2008). Learning-by-doing and the optimal solar policy in California. *Energy Journal*, **29**(3), pp. 131-151.

van Dam, J., A. Faaij, I. Lewandowski, and G. Fischer (2007). Biomass production potentials in Central and Eastern Europe under different scenarios. *Biomass and Bioenergy*, **31**(6), pp. 345-366.

van den Wall-Bake, J.D., M. Junginger, A. Faij, T. Poot, and A.d.S. Walter (2009). Explaining the experience curve: Cost reductions of Brazilian ethanol from sugarcane. *Biomass & Bioenergy*, **33**(4), pp. 644-658.

van Sark, W., A.E. Alsema, H. Junginger, H. de Moor, and G.J. Schaeffer (2007). Accuracy of progress ratios determined from experience curves: the case of crystalline silicon photovoltaic module technology development. *Progress in Photovoltaics: Research and Applications*, **16**, pp. 441-453.

van Vliet, J., M.G.J. den Elzen, and D.P. van Vuuren (2009). Meeting radiative forcing targets under delayed participation. *Energy Economics*, **31**(Supplement 2), pp. S152-S162.

van Vuuren, D., Z. Fengqi, B. De Vries, J. Kejun, C. Graveland, and L. Yun (2003). Energy and emission scenarios for China in the 21st century – Exploration of baseline development and mitigation options. *Energy Policy*, **31**, pp. 369-387.

van Vuuren, D., M. den Elzen, P. Lucas, B. Eickhout, B. Strengers, B. van Ruijven, S. Wonink, and R. van Houdt (2007). Stabilizing greenhouse gas concentrations at low levels: an assessment of reduction strategies and costs. *Climatic Change*, **81**(2), pp. 119-159.

van Vuuren, D.P., M. Hoogwijk, T. Barker, K. Riahi, S. Boeters, J. Chateau, S. Scrieciu, J. van Vliet, T. Masui, K. Blok, E. Blomen, and T. Kram (2009). Comparison of top-down and bottom-up estimates of sectoral and regional greenhouse gas emission reduction potentials. *Energy Policy*, **37**(12), pp. 5125-5139.

van Vuuren, D.P., M. Isaac, M.G.J. Den Elzen, E. Stehfest, and J. Van Vliet (2010). Low stabilization scenarios and implications for major world regions from an integrated assessment perspective. *Energy Journal*, **31**(Special Issue), pp. 165-192.

Vorum, M., and J. Tester (2007). Potential carbon emissions reductions from geothermal power by 2030. In: *Tackling Climate Change in the U.S.: Potential Carbon Emissions Reductions from Energy Efficiency and Renewable Energy by 2030.* C.F. Kutscher (ed.), American Solar Energy Society, Boulder, CO, USA, pp. 145-162. Available at: http://ases.org/images/stories/file/ASES/climate_change.pdf.

Watkiss, P., and T. Downing (2008). The social cost of carbon: Valuation estimates and their use in UK policy. *Integrated Assessment*, **8**(1), pp. 85-105.

WEC (2004a). *Comparison of energy systems using life cycle assessment. A special report.* World Energy Council, London, UK.

WEC (2004b). *Energy end-use technologies for the 21st century.* World Energy Council, London, UK

Weisser, D. (2007). A guide to life-cycle greenhouse gas (GHG) emissions from electric supply technologies. *Energy*, **32**(9), pp. 1543-1559.

Weitzman, M.L. (2007). Review: A Review of "The Stern Review on the Economics of Climate Change". *Journal of Economic Literature*, **45**(3), pp. 703-724.

Williams, R.H. (2002). Facilitating widespread deployment of wind and photovoltaic technologies. In: *2001 Annual Report.* The Energy Foundation, San Francisco, CA, USA, pp. 19-30.

Winkler, H., H. A., and M. Mary Hawb (2009). Technology learning for renewable energy: Implications for South Africa's long-term mitigation scenarios. *Energy Policy*, **37**, pp. 4987-4996.

Wiser, R., and M. Bolinger (2010). *2009 Wind Technologies Market Report.* US Department of Energy, Washington, DC, USA.

Yang, C. (2010). Reconsidering solar grid parity. *Energy Policy*, **38**, pp. 3270-3273.

Yu, C., W. van Sark, and E. Alsema (2011). Unraveling the photovoltaic technology learning curve by incorporation of input price and scale effects. *Renewable and Sustainable Energy Reviews*, **15**, pp. 324-337.

11 Policy, Financing and Implementation

Coordinating Lead Authors:
Catherine Mitchell (United Kingdom), Janet L. Sawin (USA), Govind R. Pokharel (Nepal), Daniel Kammen (USA), Zhongying Wang (China)

Lead Authors:
Solomone Fifita (Fiji/Tonga), Mark Jaccard (Canada), Ole Langniss (Germany), Hugo Lucas (United Arab Emirates/Spain), Alain Nadai (France), Ramiro Trujillo Blanco (Bolivia), Eric Usher (Sweden/Canada), Aviel Verbruggen (Belgium), Rolf Wüstenhagen (Switzerland/Germany), Kaoru Yamaguchi (Japan)

Contributing Authors:
Douglas Arent (USA), Greg Arrowsmith (Belguim/United Kingdom), Morgan Bazilian (Austria/USA), Lori Bird (USA), Thomas Boermans (Germany), Alex Bowen (United Kingdom), Sylvia Breukers (The Netherlands), Thomas Bruckner (Germany), Sebastian Busch (Austria/Germany), Elisabeth Clemens (Norway), Peter Connor (United Kingdom), Felix Creutzig (Germany), Peter Droege (Liechtenstein/Germany), Karin Ericsson (Sweden), Chris Greacen (USA), Renata Grisoli (Brazil), Erik Haites (Canada), Kirsty Hamilton (United Kingdom), Jochen Harnisch (Germany), Cameron Hepburn (United Kingdom), Suzanne Hunt (USA), Matthias Kalkuhl (Germany), Heleen de Koninck (Netherlands), Patrick Lamers (Germany), Birger Madsen (Denmark), Gregory Nemet (USA), Lars J. Nilsson (Sweden), Supachai Panitchpakdi (Switzerland/Thailand), David Popp (USA), Anis Radzi (Liechtenstein/Australia), Gustav Resch (Austria), Sven Schimschar (Germany), Kristin Seyboth (Germany/USA), Sergio Trindade (USA/Brazil and USA), Bernhard Truffer (Switzerland), Sarah Truitt (USA), Dan van der Horst (United Kingdom/The Netherlands), Saskia Vermeylen (United Kingdom), Charles Wilson (United Kingdom), Ryan Wiser (USA)

Review Editors:
David de Jager (The Netherlands), Antonina Ivanova Boncheva (Mexico/Bulgaria)

This chapter should be cited as:
Mitchell, C., J. L. Sawin, G. R. Pokharel, D. Kammen, Z. Wang, S. Fifita, M. Jaccard, O. Langniss, H. Lucas, A. Nadai, R. Trujillo Blanco, E. Usher, A. Verbruggen, R. Wüstenhagen, K. Yamaguchi, 2011: Policy, Financing and Implementation. In IPCC Special Report on Renewable Energy Sources and Climate Change Mitigation [O. Edenhofer, R. Pichs-Madruga, Y. Sokona, K. Seyboth, P. Matschoss, S. Kadner, T. Zwickel, P. Eickemeier, G. Hansen, S. Schlömer, C. von Stechow (eds)], Cambridge University Press, Cambridge, United Kingdom and New York, NY, USA.

Table of Contents

Executive Summary 869

11.1 Introduction 871

11.1.1 The rationale of renewable energy policies 871

11.1.2 Policy timing and strength 873

11.1.3 Roadmap for the chapter 873

11.2 Current trends: Policies, financing and investment 874

11.2.1 Trends in renewable energy policies 874

11.2.2 Trends in renewable energy finance 876
11.2.2.1 Trends along the financing continuum 877
11.2.2.2 Financing technology research and development 877
11.2.2.3 Financing technology commercialization 877
11.2.2.4 Financing manufacturing and sales 877
11.2.2.5 Financing construction 878
11.2.2.6 Refinancing and sale of companies 878

11.2.3 Global investment transition 878

11.3 Key drivers, opportunities and benefits 878

11.3.1 Climate change mitigation and reduction of environmental and health impacts 879

11.3.2 Energy access 879

11.3.3 Energy security 880

11.3.4 Social and economic development 880

11.4 Barriers to renewable energy policymaking, implementation and financing 880

11.4.1 Barriers to renewable energy policymaking 880

11.4.2 Barriers to implementation of renewable energy policies 881

11.4.3 Barriers to renewable energy financing 882

11.5	**Experience with and assessment of policy options**	882
11.5.1	Criteria for policy evaluation	883
11.5.2	Research, development and deployment policies for renewable energy	884
11.5.2.1	Why and when public research and development is needed	884
11.5.2.2	Public research and development measures	885
11.5.2.3	Lessons learned	886
11.5.2.4	Positive feedbacks from combining research and development policies with deployment policies	888
11.5.3	Policy options for renewable energy deployment	889
11.5.3.1	Fiscal incentives	889
11.5.3.2	Public finance	892
11.5.3.3	Regulations	894
11.5.4	Policies for deployment – electricity	895
11.5.4.1	Fiscal incentives	895
11.5.4.2	Public finance	895
11.5.4.3	Regulations	895
11.5.5	Policies for deployment – heating and cooling	907
11.5.5.1	Fiscal incentives	907
11.5.5.2	Public finance	908
11.5.5.3	Regulations	908
11.5.5.4	Policy for renewable energy sources of cooling	910
11.5.6	Policies for deployment – transportation	911
11.5.6.1	Fiscal incentives	911
11.5.6.2	Public finance	912
11.5.6.3	Regulations	912
11.5.7	Synthesis	913
11.5.7.1	Assessment of RE policies	913
11.5.7.2	Macroeconomic impacts and cost-benefit analysis	916
11.5.7.3	Interactions and potential unintended consequences of renewable energy and climate policies	916
11.6	**Enabling environment and regional issues**	917
11.6.1	Innovation in the energy system	919
11.6.2	Complementing renewable energy policies and non-renewable energy policies	920
11.6.3	Reducing financial and investment risk	920

11.6.4	**Planning and permitting at the local level**	921
11.6.4.1	Aligning stakeholder expectations and interests	921
11.6.4.2	Learning about the importance of context for RE deployment	921
11.6.4.3	Adopting benefit-sharing mechanisms	921
11.6.4.4	Timing: pro-active national and local government	923
11.6.4.5	Building collaborative networks	923
11.6.4.6	Mechanisms for articulating conflict and negotiation	923
11.6.5	**Providing infrastructures, networks and markets for renewable energy**	924
11.6.5.1	Infrastructure building and connection to networks	924
11.6.5.2	Access to and injection of renewable energy into the network	925
11.6.5.3	Network standards	925
11.6.5.4	Increasing resilience of the system	925
11.6.6	**Technology transfer and capacity building**	925
11.6.6.1	Technology transfer and intellectual property rights	926
11.6.6.2	Technology transfer and international institutions	926
11.6.6.3	Technology transfer and energy access	927
11.6.7	**Institutional learning**	927
11.6.8	**A role for cities and communities**	927
11.6.8.1	Community and individual links	927
11.6.8.2	A role for individuals as part of civil society	927

11.7 A structural shift ... 929

11.7.1	The link between scenarios and policies	930
11.7.2	Structural shifts result from a combination of technology and behaviour change	930
11.7.3	Addressing the challenges of governing long-term energy transitions	930
11.7.4	Co-evolution of 'bricolage' versus 'breakthrough'	931
11.7.5	Specific policy options for an accelerated transition to a high renewable energy world	931

References ... 933

Executive Summary

Renewable energy can provide a host of benefits to society. In addition to the reduction of carbon dioxide (CO_2) emissions, governments have enacted renewable energy (RE) policies to meet a number of objectives including the creation of local environmental and health benefits; facilitation of energy access, particularly for rural areas; advancement of energy security goals by diversifying the portfolio of energy technologies and resources; and improving social and economic development through potential employment opportunities. Energy access and social and economic development have been the primary drivers in developing countries whereas ensuring a secure energy supply and environmental concerns have been most important in developed countries.

An increasing number and variety of RE policies—motivated by a variety of factors—have driven substantial growth of RE technologies in recent years. Government policies have played a crucial role in accelerating the deployment of RE technologies. At the same time, not all RE policies have proven effective and efficient in rapidly or substantially increasing RE deployment. The focus of policies is broadening from a concentration almost entirely on RE electricity to include RE heating and cooling and transportation.

RE policies have promoted an increase in RE capacity installations by helping to overcome various barriers. Barriers specific to RE policymaking (e.g., a lack of information and awareness), to implementation (e.g., a lack of an educated and trained workforce to match developing RE technologies) and to financing (e.g., market failures) may further impede deployment of RE. A broad application of RE would require policies to address these barriers, and to help overcome challenges such as the lack of infrastructure necessary for integrating RE into the existing system.

Policy mechanisms enacted specifically to promote RE are varied and can apply to all energy sectors. They include fiscal incentives such as tax credits and rebates; public financing policies such as low-interest loans; regulations such as quantity-driven policies like quotas and price-driven policies including feed-in tariffs for electricity, heat obligations, and biofuels blending requirements. Policies can be sector specific and can be implemented at the local, state/provincial, national and in some cases regional level and can be complemented by bilateral, regional and international cooperation.

Public research and development (R&D) investments are most effective when complemented by other policy instruments, particularly RE deployment policies that simultaneously enhance demand for new RE technologies. Together, R&D and deployment policies create a positive feedback cycle, inducing private sector investment in R&D. Enacting deployment policies early in the development of a given technology can accelerate learning by inducing private R&D, which in turn further reduces costs and provides additional incentives for using the technology.

Some policy elements have been shown to be more effective and efficient in rapidly increasing RE deployment, but there is no one-size-fits-all policy, and the mix of policies and their design and implementation are also important. Key policy elements for ensuring effectiveness and efficiency can include adequate value to cover costs and account for social benefits, guaranteed access to networks and markets, long-term contracts to reduce risk, inclusiveness and ease of administration.

- Several studies have concluded that some feed-in tariffs have been effective and efficient at promoting RE electricity, mainly due to the combination of long-term fixed price or premium payments, network connections, and guaranteed purchase of all RE electricity generated. Quota policies can be effective and efficient if designed to reduce risk; for example, with long-term contracts.

- An increasing number of governments are adopting fiscal incentives for RE heating and cooling. Obligations to use RE heat are gaining attention for their potential to encourage growth independent of public financial support.

- In the transportation sector, RE fuel mandates or blending requirements are key drivers in the development of most modern biofuel industries. Other policies include direct government payments or tax reductions. Policies have influenced the development of an international biofuel trade.

The flexibility to adjust as technologies, markets and other factors evolve is important. The details of design and implementation are critical in determining the effectiveness and efficiency of a policy. Policy frameworks that are transparent and sustained can reduce investment risks and facilitate deployment of RE and the evolution of low-cost applications.

A mix of policies is generally needed to address the various barriers to RE. Further, experience shows that different policies or combinations of policies can be more effective and efficient depending on factors such as the level of technological maturity, availability of affordable capital and the local and national RE resource base.

If the goal is to transform the energy sector over the next several decades to one based on low-carbon fuels and technologies, it is important to minimize costs over this entire period, not only in the near term. It is also important to include all costs and benefits to society in that calculation. Conducting an integrated analysis of costs and benefits associated with RE is extremely demanding because so many elements are involved in determining net impacts; thus, such efforts face substantial limitations and uncertainties. Few studies have examined such impacts on national or regional economies; however, those that have been carried out have generally found net positive economic impacts.

Two separate market failures create the rationale for the additional support of innovative RE technologies that have high potential for technological development, even if an emission market (or GHG pricing policy in general) exists. The first market failure refers to the external cost of GHG emissions. The second market failure is in the field of innovation: if firms underestimate the future benefits of investments into learning RE technologies or if they cannot appropriate these benefits, they will invest less than is optimal from a macroeconomic perspective. In addition to GHG pricing policies, RE-specific policies may be appropriate from an economic point of view if the related opportunities for technological development are to be addressed (or if other goals beyond climate mitigation are pursued). Potentially adverse consequences such as lock-in, carbon leakage and rebound effects must be taken into account in the design of a portfolio of policies.

RE technologies can play a greater role in climate change mitigation if they are implemented in conjunction with 'enabling' policies. A favourable, or enabling, environment for RE can be created by encouraging innovation in the energy system; addressing the possible interactions of a given policy with other RE policies as well as with other energy and non-energy policies (e.g., those targeting agriculture, transportation, water management and urban planning); by understanding the ability of RE developers to obtain finance and planning permission to build and site a project; by removing barriers for access to networks and markets for RE installations and output; by enabling technology transfer; and by increasing education and awareness. In turn, existence of an 'enabling' environment can increase the efficiency and effectiveness of policies to promote RE.

The literature indicates that long-term objectives for RE and flexibility to learn from experience would be critical to achieve cost-effective and high penetrations of RE. The energy scenarios analyzed in Chapter 10 show RE penetrations of up to 77% of primary energy by 2050, depending on the rate of installation. To achieve GHG concentration stabilization levels with high shares of RE, a structural shift in today's energy systems will be required over the next few decades. Such a transition to low-carbon energy differs from previous ones (e.g., from wood to coal, or coal to oil) because the available time span is restricted to a few decades, and because RE must develop and integrate into a system constructed in the context of an existing energy structure that is very different from what might be required under higher-penetration RE futures.

A structural shift would require systematic development of policy frameworks that reduce risks and enable attractive returns that provide stability over a timeframe relevant to RE and related infrastructure investments. An appropriate and reliable mix of instruments is even more important where energy infrastructure is still developing and energy demand is expected to increase in the future.

11.1 Introduction

The potential for RE to play a role in the mitigation of climate change is significant, as discussed in previous chapters. RE capacity is increasing rapidly around the world, and government interest in renewable technologies is driven by a range of factors including climate mitigation, access to energy, secure energy supply, job creation and others. But a number of barriers continue to hold back further RE advances.

The scenarios in Chapter 10 show that the role RE can play in mitigating climate change can range from relatively minor to very significant depending on the rate of RE deployment. This rate, in turn, will depend on choices of societies and governments regarding how best to address climate change, as one among several energy related challenges that also include energy access or security. If RE is to contribute substantially to the mitigation of climate change, and to do so quickly, various forms of economic support policies as well as policies to create an enabling environment are likely to be required.

RE policies can be sector specific and can be implemented at all levels of government—from local to state/provincial to national and international—and can be complemented by bilateral, regional and international cooperation. International agencies such as the International Energy Agency (IEA) are able to advise members about energy sources and policies; some, like the European Commission, can enact Directives while others mainly enhance understanding and awareness and distribute information (e.g., the Renewable Energy Policy Network for the 21st Century (REN21) and the International Renewable Energy Agency (IRENA)). National governments can enact laws, assign different policies, and adapt or create regulations and other enabling environment dimensions. State, provincial or regional, and municipal or local initiatives may provide important support for local policies. In some countries, regulatory agencies and public utilities may be given responsibility for, or on their own initiative, design and implement support mechanisms for RE. The extent to which governments of all levels can 'learn' (Thelen, 1999; Breukers and Wolsink, 2007a)—whether from other governments, institutions, companies, communities and/or individuals—and are flexible or reflexive to be able to evaluate past policies, to experiment and look for best practice (Smith et al., 2005) is also helpful. This chapter examines the roles of all of these actors, but focuses primarily on national governments and policymakers.

RE policies range from basic R&D for technology development through to support for deployment of RE systems or the electricity, heat or fuels they produce. Deployment policies include fiscal incentives (tax policies, rebates, grants etc.), public finance mechanisms (loans, guarantees etc.) and regulations (e.g., feed-in tariffs, quotas, building mandates and biofuels blending mandates).

RE projects and production covered by policies can be qualified by RE source (type, location, flow or stock character, variability, density), by technology (type, vintage, maturity, scale of the projects), by ownership (households, cooperatives, independent companies, electric utilities) and other attributes that are in some way measurable (Jacobsson and Lauber, 2006; Mendonça, 2007; Verbruggen and Lauber, 2009). RE may be measured by additional qualifiers such as time and reliability of delivery (availability) and other metrics related to RE's integration into networks (Klessmann et al., 2008; Langniß et al., 2009). There is also much that governments and other actors can do to create an environment conducive for RE deployment. This chapter examines the options available for policymakers and the role of policies in advancing RE. Policies can advance technologies and stimulate markets, but complementary non-RE policies provide comfort for investors, thereby further enabling deployment. Thus, this chapter addresses the role of policies and an enabling environment in making financing available and affordable. It assesses policies based on a number of criteria, including effectiveness, efficiency, equity and institutional feasibility. It provides policymakers with a range of options for achieving the desired level of RE deployment and penetration, and aims to answer the following questions in each of the identified sections:

- Why, and under what conditions, is RE-specific policy support needed (Section 11.1)?
- What are the current trends globally in RE policies, finance and investment (Section 11.2)?
- What are the factors, in addition to climate change mitigation, driving policymakers to enact policies to advance RE? How do these drivers differ between developing and developed countries (Section 11.3)?
- What are the barriers to RE policy making, implementation and finance (Section 11.4), and how can policies help to overcome the various barriers to RE (Sections 11.5, 11.6 and 11.7)?
- What policy options are available to advance RE in different end-use sectors (Section 11.5)?
- What have been the experiences with these policy options to date, and which are most successful and under what conditions (Sections 11.5 and 11.6)?
- How do RE policies interact with climate policies (Section 11.5) and other types of policies (Section 11.6)?
- What combinations of policy packages can overcome the barriers necessary to achieve varying levels of RE penetration desired for mitigating climate change (Section 11.7)?

The remainder of this section begins to address some of the above questions, starting with a summary of the literature on the conditions that may make RE-specific policies necessary alongside climate policies (carbon pricing) in order to mitigate climate change.

11.1.1 The rationale of renewable energy policies

Renewable energies can provide a host of benefits to society. In addition to carbon dioxide emissions reduction, RE technologies are associated with local environmental and health benefits (Sections 11.3.1 and 9.3.4); can facilitate energy access particularly in rural areas (Sections 11.3.2 and 9.3.2); can increase energy security by increasing the portfolio of energy technologies and resources (Sections 11.3.3 and 9.3.3); and improve social and economic development (Sections 11.3.4 and 9.3.1) by creating employment opportunities and economic growth.

Some RE technologies are broadly competitive with current market energy prices. Of the other RE technologies that are not yet broadly competitive, many can provide competitive energy services in certain circumstances, for example, in regions with favourable resource conditions or that lack infrastructure for other low-cost energy supplies. In most regions of the world, however, policy measures are still required to facilitate an increasing deployment of RE (Section 10.5).

From a macro-economic perspective, government intervention can be justified where market distortions exist. There are two market failures particularly pertinent to RE:[1]

1. Imperfect appropriability of benefits from innovation: Specifically, research and development (R&D), innovation, diffusion and adoption of new low-carbon technologies often create wider benefits to society than those captured by the innovator (Jaffe, 1986; Griliches, 1992; Jaffe et al., 2003, 2005; Edenhofer et al., 2005; Popp, 2006b). If firms underestimate the (future) benefits of investments into learning technologies or if they cannot appropriate these benefits, they will invest less than is optimal from a macro-economic perspective. Hence, *specific RE policies* (e.g., feed-in tariffs or quota systems) can be justified in order to address the market failures associated with technological learning and spill-over effects.

2. External costs of burning fossil fuels: Damages from global warming and local pollution are not usually considered by firms unless the associated external costs are purposefully internalized (Pigou, 1920; Cropper and Oates, 1992). As a consequence, there is an under-investment in energy efficiency improvements as well as in low-carbon technologies including RE. Where implemented, *carbon pricing* (via carbon taxes, emission trading schemes, or implicitly through regulation) is expected to yield a cost-efficient mix of mitigation measures—provided that no additional market failures introduce further distortions (Stern, 2007).

Where two market failures exist, two types of policies may be required to obtain a socially optimal outcome. With regard to the two market failures that are relevant to RE, carbon pricing and support for research, development and diffusion of new technologies would be required. Otherwise, the two objectives (internalizing the cost of greenhouse gas (GHG) emissions and encouraging innovation and deployment of low-carbon technologies) would have to be traded off against one another—possibly sacrificing one of the objectives to some extent. For instance, carbon pricing on its own is likely to under-deliver investment in R&D for new low-carbon technologies (Rosendahl, 2004; Rivers and Jaccard, 2006; Stern, 2007, Ch. 16; Fischer, 2008; Fischer and Newell, 2008; Otto et al., 2008).

There are further barriers that impede RE technologies, including oligopoly and imperfect competition, existing subsidies, network economies, information failures, labour market failures and non-internalized environmental and health effects beyond the impact of climate change (Sorell and Sijm, 2003; Sjögren, 2009; see also Sections 1.4.2, 9.5.1, and 9.5.2.1) Energy utilities whose incumbent technologies may have benefited from economies of scale might resist the entry of low-carbon competitors. Past investments into carbon-intensive infrastructure and engineering knowledge based upon that infrastructure may have created a lock-in into related technologies, impeding innovation and integration of RE (Unruh 2000; Acemoglu et al., 2009).

Transforming the energy system would require substantial investment, potentially binding capital for multiple decades. Hence, for such a target, investors would need clear and stable framing regulatory conditions as well as well-developed capital, insurance and futures markets to diversify investment risks. Information asymmetries (regarding, e.g., the innovation, learning and potential deployment of technologies) on capital markets increase the perceived risks and thus also the cost of investments. This is particularly relevant for some RE technologies, which as capital-intensive technologies suffer from high capital costs (Section 11.4.3).

Since, in practice, governments have not yet implemented 'ideal' carbon pricing or 'ideal' support for low-carbon R&D, there may be a role for additional 'second-best' government intervention, including stronger RE deployment policies to tackle more effectively the climate externality. Carbon prices are often nonexistent or lower than estimated associated social costs (Stern, 2007; Tol, 2009), and have not provided a sufficiently credible basis for a large-scale shift towards low-carbon investment (see, for example, Committee on Climate Change 2010 (CCC, 2010) for the UK). Further, because governments are unable to pre-commit for the long term, there is a general lack of belief in government policies on long-term carbon pricing (Ulph and Ulph, 2009). Uncertainty over future regulation and, thus, over the future role of RE in the energy mix, discourages capital-intensive long-term investments. That is a salutary reminder that policymakers in the real world are subject to lobbying and rent-seeking as well as uncertainty about the costs and benefits of policies, including the costs of public administration of those policies.

The uncertainty of costs and the complex linkage of RE-specific market failures and barriers make it difficult to determine the optimal level of RE deployment for each of the drivers and co-benefits of RE. The remainder of this chapter presumes that decision makers aim to increase RE deployment as a means to achieve any number of social objectives—mitigating climate change is considered as one objective among many. Nonetheless, the complex interplay of RE policies with climate policies is revisited later in the chapter (see Section 11.5.7.3) as an important component for consideration, as the two policies might influence each other and lead to unintended consequences.

---

[1] Both market failures must be taken into account simultaneously for those RE technologies that are prone to cost reductions via R&D and technological learning.

11.1.2 Policy timing and strength

The timing, strength and level of coordination of R&D versus deployment policies have implications for the efficiency and effectiveness of the policies, and for the total cost to society, in three main ways:

1. Whether a country promotes RE immediately or waits until costs have declined further. Although many RE technologies currently are not yet competitive with the energy market prices, the levelized cost of energy generated by RE has declined substantially in the past. As many of these technologies are still in early phases of their respective development chains, further cost reductions are expected in the future, especially if these technologies are appropriately supported by research, development, demonstration and deployment programs (RDD&D) (IEA, 2008b, 2010a). Chapter 10 concludes that in order to achieve full competitiveness with fossil fuel technologies, significant up-front investments will be required until the break-even point is achieved. When those investments should be made depends on the goal. If the international community aims to stabilize the average global temperature increase at 2°C, then investments in low-carbon technologies must start almost immediately. If a less stringent level were chosen there would be more time;

2. Once a country has decided to support RE, the timing, strength and coordination of when R&D policies give way to deployment policies (Nemet, 2006; Junginger et al., 2010), discussed in Section 11.5.2; and

3. The critical debate of the cost and benefit of accelerated versus slower 'market demand' policy implementation. This debate concerns the dual objectives of rapid deployment of clean energy technologies to 'jump start' market growth, generally at higher up-front costs but with significant ability to evolve technologies down the cost curve (Langniß and Neij, 2004) to reduce GHG emissions, versus slower deployment that may not have as rapid a climate benefit, but which comes at a lower up-front capital and political cost.

11.1.3 Roadmap for the chapter

An increasing number of governments around the world are investing in RE and enacting RE policies to address climate change and for a variety of other reasons. As described in the introduction, the chapter aims to answer a number of questions about policy needs and experiences to date. The next section (11.2) begins by highlighting recent trends in RE policies to promote deployment, and then discusses trends in financing and research and development funding. Section 11.3 examines various drivers of RE policies, and Section 11.4 briefly reviews the barriers that impede RE policymaking and implementation, and barriers to financing.

Section 11.5 presents the various RE-specific policy options available to advance RE technology development and deployment. Tables 11.1 and 11.2, found near the beginning of the section, list and define a range of policies currently used specifically to promote RE, and Table 11.2 notes which policies have been applied to which end-use sectors (electricity, heating and cooling, transportation). The section provides some assessment of how various policy options stand up to a range of different criteria, primarily effectiveness and efficiency, and provides a discussion of key elements to consider when selecting and designing RE policies.

In Section 11.6, an enabling environment is defined and explained. An environment that is enabling includes a skilled workforce, capacity for technology transfer, access to affordable financing, access to networks and markets, transparency in the process of obtaining permitting, etc. While it is not a critical prerequisite to have all elements of an enabling environment in place for the successful deployment of RE, the ease with which RE projects interact with these dimensions will match the ease with which RE is deployed.

This chapter concludes with Section 11.7, which focuses on broader considerations and requirements for a structural shift to a sustainable, low-carbon energy economy, particularly one based on RE and energy efficiency.

A number of case studies appear in text boxes in Sections 11.5 and 11.6. These aim to highlight key messages of the chapter and to provide insights into specific policy experiences that offer lessons for other regions or countries.

The issue of finance and RE can be examined in several ways, including: an assessment of the current trends in RE finance (Section 11.2.2); a review of existing barriers to financing of RE (Section 11.4.3); a review of public finance instruments as a policy option available to governments (Section 11.5.3); and a discussion of the relationship between RE project financing and broader financial market conditions that may contribute to the success of a project (Section 11.6.3). Because of the cross-cutting nature of finance, relevant aspects for RE are addressed in most sections of the chapter.

Available RE resources vary from place to place, and maturity levels vary among the different RE technologies; further, political, economic, social, financial, ecological and cultural needs and conditions differ from one city, state, region or country to another, thereby leading to different options and constraints. Thus there is no one-size-fits-all policy package, and the optimal mix of RE policies will differ from one place to the next. Clearly, it is not possible to cover everything in a single chapter. However, there are valuable and transferable lessons to be learned from experiences to date, and this chapter aims to elucidate them.

In general, this chapter does not include technology-specific policy needs and related experiences.

11.2 Current trends: Policies, financing and investment

The number of RE-specific policies enacted and implemented by governments, and the number of countries with RE policies, is increasing rapidly around the globe (Figure 11.1). The focus of RE policies is shifting from a concentration almost entirely on electricity to include the heating/cooling and transportation sectors. These trends are matched by increasing success in the development of a range of RE technologies and their manufacture and implementation (see Chapters 2 through 7), as well as by a rapid increase in annual investment in RE and a diversification of financing institutions. This section describes recent trends in RE policies and in public and private finance and investment, from research and development (R&D) through to refinancing and the sale of RE companies.

11.2.1 Trends in renewable energy policies

While several factors are driving rapid growth in RE markets, government policies have played a crucial role in accelerating the deployment of RE technologies to date (Sawin, 2001, 2004; Meyer, 2003; Renewables 2004, 2004; Rickerson et al., 2007; REN21, 2009b; IEA, 2010d).

Until the early 1990s, few countries had enacted policies to promote RE. Since then, and particularly since the early- to mid-2000s, policies have begun to emerge in a growing number of countries at the municipal, state/provincial, national and international levels (REN21, 2005, 2009b). Initially, most policies adopted were in developed countries, but an increasing number of developing countries have enacted policy frameworks at various levels of government to promote RE since the late 1990s and early 2000s (Wiser and Pickle, 2000; Martinot et al., 2002; REN21, 2010).

According to the Renewable Energy Policy Network for the 21st Century (REN21), which is believed to be the only source that tracks RE policies annually on a global and comprehensive basis,[2] the number of countries with some kind of RE target and/or deployment policy related to RE almost doubled from an estimated 55 in early 2005 to more than 100 in early 2010 (REN21, 2010). By early 2010, at least 85 countries, including all 27 EU member states, had adopted RE targets at the national level—for specific shares of electricity, or shares of primary or final energy from RE; sub-national targets exist in a number of additional countries (REN21, 2010). This is up from 43 countries with national targets in mid-2005 (plus 2 countries with state/provincial level targets) (REN21, 2006). An estimated 83 countries were known to have RE policies in place by early 2010.

There is much overlap between these two categories (countries with policies and those with targets); some countries have adopted policies specifically to deliver their targets, while others have enacted policies but do not have official targets at the national level. Further, a significant number of developing countries have adopted targets but have not yet enacted national RE policies. Most countries with RE policies have more than one type of policy in place, and many existing policies and targets have been strengthened over time (REN21, 2010).

Existing RE policies are directed to all end-use sectors—electricity, heating and transportation. (See Section 11.5 and Tables 11.1 and 11.2 for full discussion of RE policy options.) By the date of publication, however, most RE deployment policies focused on the electricity sector. At least 83 countries had adopted some sort of policy to promote RE power generation by early 2010 (IEA, 2010c; REN21, 2010), up from an estimated 48 countries in mid-2005 (REN21, 2006). These policies included fiscal incentives such as investment subsidies and tax credits; government financing such as low-interest loans; and regulations such as feed-in tariffs (FITs), quotas and net metering. Of those countries with RE electricity policies, approximately half were developing countries from every region of the world (REN21, 2010).

Although governments use a variety of policies to promote RE electricity, the most common ones in use as of publication were FITs and quotas or Renewable Portfolio Standards (RPS). By early 2010, at least 45 countries had FITs at the national level (including much of Europe), with a further 4 countries using them at the state/provincial/territorial and/or municipal levels (Mendonça, 2007; Rickerson et al., 2007, 2008; REN21, 2010). RPS or quotas are also widely used and, by early 2010, were in force in an estimated 10 countries at the national level, and at least 4 additional countries at the state, provincial or regional level, including 29 US states, at least 12 Indian states, and some provinces and regions in Canada and Belgium (REN21, 2010).

An increasing number of governments are adopting incentives and mandates to advance renewable transport fuels and renewable heating technologies (IEA, 2007b; Rickerson et al., 2009). For example, in the 12 countries analyzed for the International Energy Agency, the number of policies introduced to support renewable heating either directly or indirectly increased from 5 in 1990 to more than 55 by May 2007 (IEA, 2007b; REN21, 2009b).

By early 2010, at least 41 states/provinces and 24 countries at the national level had adopted mandates for blending biofuels with gasoline or diesel fuel, while others had set production or use targets (REN21, 2009b). Most mandates require blending relatively small (e.g., up to 10%) percentages of ethanol or biodiesel with petroleum-based fuels for transportation. Brazil has been an exception, with ethanol blending shares required in the 20 to 25% range, although many vehicles in Brazil operate on 100% ethanol, which is also readily available (Goldemberg, 2009). Production subsidies and tax exemptions for biofuels have also increased in use in developed and developing countries (REN21, 2010).

---

2 Note that the International Energy Agency database focuses on the Organisation for Economic Cooperation and Development (OECD), BRICS (Brazil, Russia, India, China and South Africa) and other countries that supply information, but is not as comprehensive as REN21 (which relies on the IEA database and other sources).

2005

Early 2011

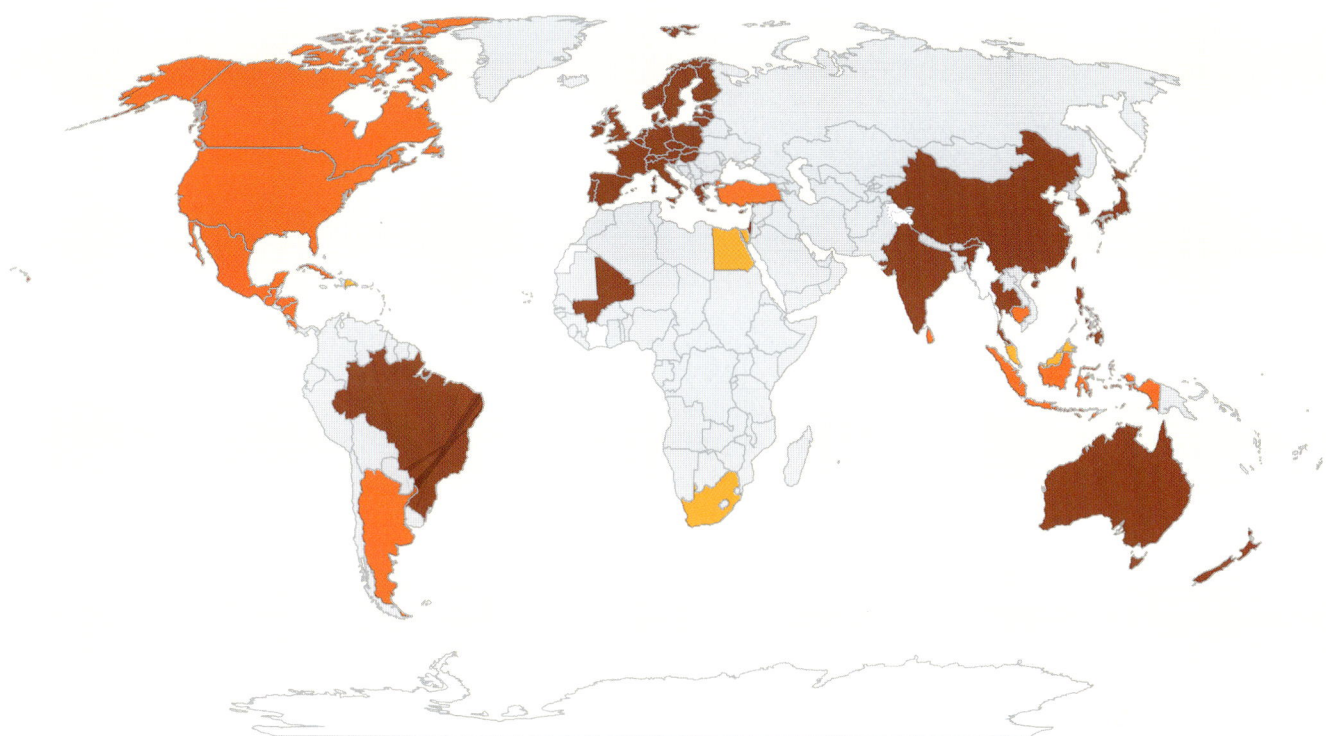

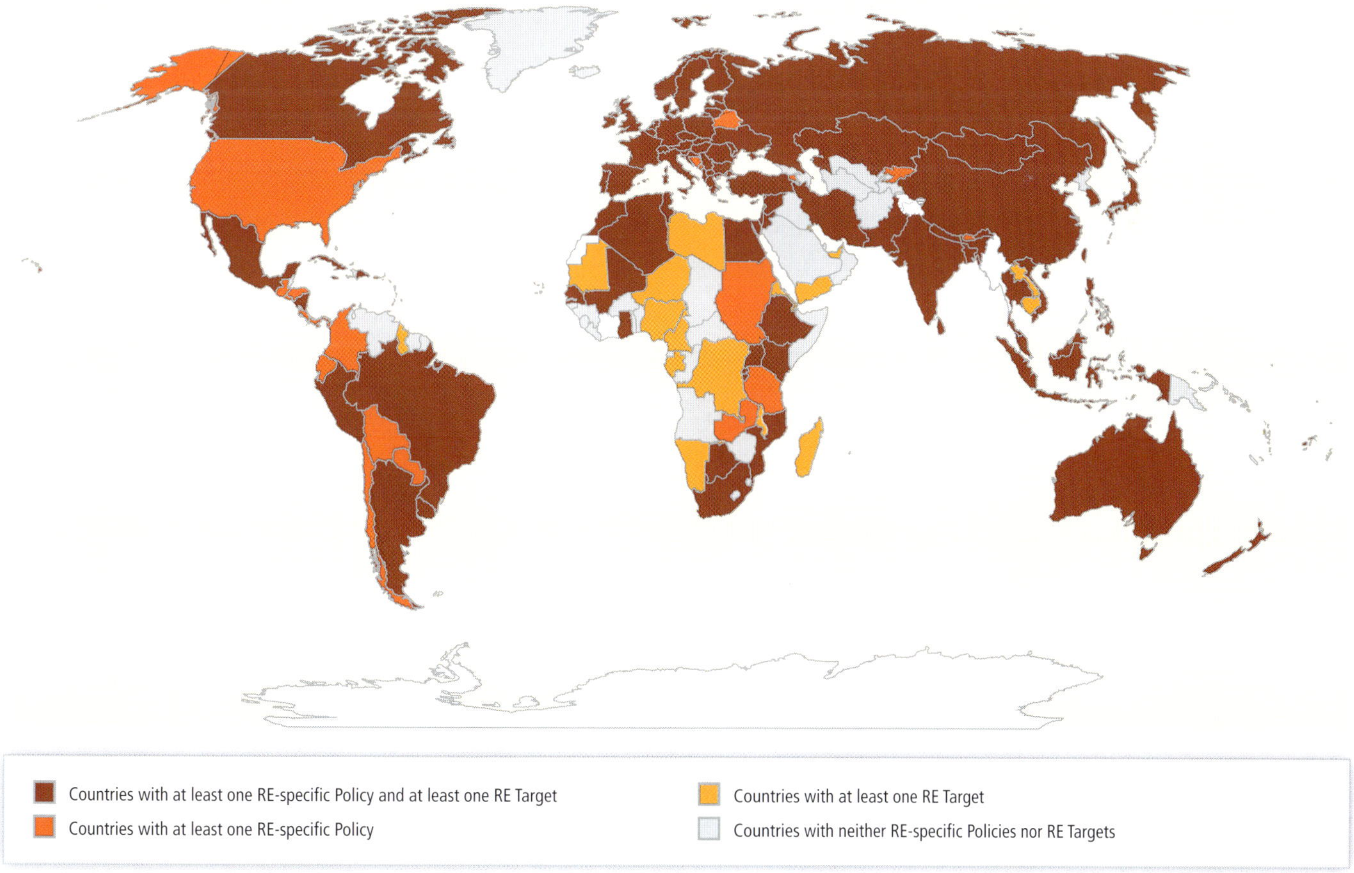

Figure 11.1 | Countries with at least one RE-specific deployment target and/or at least one RE-specific deployment policy in mid-2005 and in early 2011. This figure includes only national-level targets and policies (not municipal or state/provincial) and is not necessarily all-inclusive (RECIPES, 2005; REN21, 2005, 2010, 2011; CIPORE, 2011; Austrian Energy Agency, 2011; IEA, 2011; REEGLE, 2011; DSIRE, 2011).

Another policy trend seen particularly with bioenergy, and biofuels especially, is the adoption of environmental and other sustainability standards, including regulations on associated lifecycle CO_2 emissions, such as the U.S. Renewable Fuel Standard and mandatory sustainability standards under the EU Renewable Energy Directive (European Commission, 2009b; USEPA, 2010b). (For more on sustainability standards, see Section 2.4.5.2.)

Beyond national policies, the number of international policies and partnerships is increasing. The EU Renewables Directive entered into force in June 2009, setting a binding target to source 20% of EU final energy consumption from RE by 2020; all member states have been assigned targets for 2020 that are driving RE policies at the national level (European Commission, 2009a; REN21, 2009b). Another example is the Mediterranean Solar Plan, an agreement among countries in the region for research and deployment of 20 GW of RE by 2020 (Resources and Logistics, 2010).

Several hundred city and local governments around the world have also established goals or enacted renewable deployment policies and other mechanisms to spur local RE development (Droege, 2009; REN21, 2009b). Innovative policies such as Property-Assessed Clean Energy (PACE) have begun to emerge on this level (Fuller et al., 2009a) (see Box 11.3). Indeed, some of the most rapid transformations from fossil fuels to RE-based systems have taken place at the local level, with entire communities and cities—including Samsø in Denmark and Güssing in Austria (see Box 11.14)—devising innovative means to finance RE and making the transition towards 100% RE systems (Droege, 2009; Sawin and Moomaw, 2009).

The IEA (IEA et al., 2010) estimates that in 2009, governmental RE deployment support—including subsidies, renewable portfolio standards/quotas, FITs, green certificates and several fiscal incentives (but excluding R&D support)—totalled USD_{2005} 49 billion (USD_{2009} 57 billion). This compares with USD_{2005} 38 billion (USD_{2008} 44 billion) in government support during 2008 and USD_{2005} 35 billion (USD_{2007} 41 billion) in 2007.

The vast majority of capacity or generation for most RE technologies is still in a relatively small number of countries. However, as RE policies are enacted by an increasing number of governments, new countries and regions are emerging as important manufacturers and installers of RE (GWEC, 2008, 2010; REN21, 2010).

11.2.2 Trends in renewable energy finance

In response to the increasingly supportive policy environment, the overall RE sector globally has seen a significant rise in the level of investment since 2004–2005. According to UNEP and Bloomberg New Energy Finance (BNEF), USD_{2005} 101.1 billion were newly invested in RE electricity (not including hydropower plants) and biofuels technologies in 2009. This was up from USD_{2005} 16.9 billion in 2004 (UNEP and BNEF, 2010), although down from USD_{2005} 110.7 billion in 2008 due to the financial downturn (Figure 11.2). Using a different methodology,[3] REN21 (2010) identified a total investment figure for 2009 that was significantly higher than the findings of UNEP and BNEF (2010).

Meanwhile, global investment in hydropower facilities increased from approximately USD_{2005} 6.2 billion in 2004 to USD_{2005} 58.5 billion in 2009 (IJHD, 2009) (Figure 11.3).

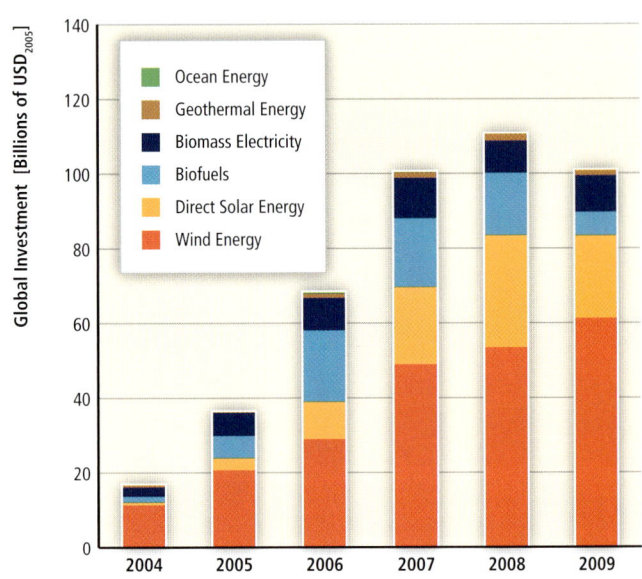

Figure 11.2 | Global investment in RE electricity (excluding hydropower) and biofuels, by technology, 2004 to 2009 (UNEP and NEF, 2009).

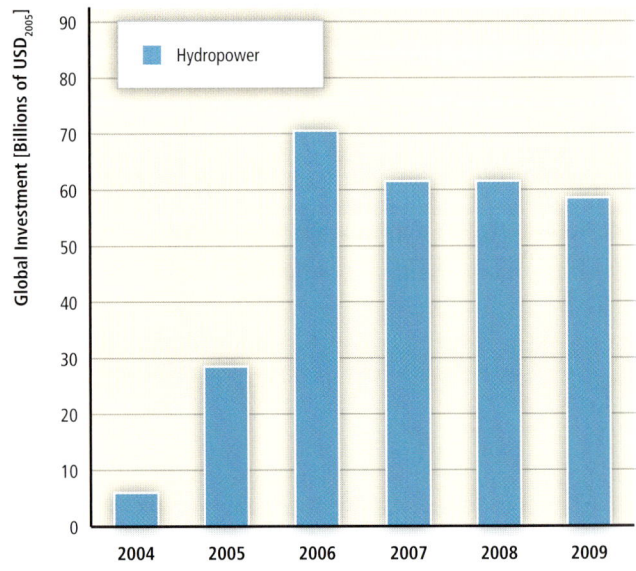

Figure 11.3 | Global investment in hydropower plants, 2004 to 2009 (IJHD, 2009).

3   The REN21 estimates were higher than BNEF/UNEP estimates for two reasons: REN21 data for small-scale projects included (1) global investment in solar hot water (estimated at USD_{2005} 12 billion); and (2) balance-of-plant costs for distributed grid-connected solar photovoltaics (PV) (<200 kW), while BNEF/UNEP included only PV module costs (REN21, 2010).

11.2.2.1 Trends along the financing continuum

Financing occurs over what is known as the 'continuum' or stages of technology development. The five segments of the continuum are: i) R&D; ii) technology development and commercialization; iii) equipment manufacture and sales; iv) project construction; and v) the refinancing and sale of companies, largely through mergers and acquisitions. Literature is available that examines financing along this continuum for biofuels and all RE electricity technologies except hydropower. For these technologies, financing has been increasing all along the continuum. These trends represent successive steps in the innovation process and provide indicators of the RE sector's current and expected growth, as follows:

- Trends in (i) R&D funding and (ii) technology development and commercialization (Sections 11.2.2.2 and 11.2.2.3) are indicators of the long- to mid-term expectations for the sector—investments are being made that will usually only begin to pay off in several years' time, once the technology is fully commercialized.

- Trends in (iii) manufacturing and sales investment (Section 11.2.2.4) are an indicator of near-term expectations for the sector—essentially, that the growth in market demand will continue.

- Trends in (iv) construction investment (Section 11.2.2.5) are an indicator of current sector activity, including the extent to which internalizing costs associated with GHGs can result in new financial flows to RE projects.

- Trends in (v) industry mergers and acquisitions (Section 11.2.2.6) can reflect the overall maturity of the sector, and increasing refinancing activity over time indicates that larger, more conventional investors are entering the sector, buying up successful early investments from first movers.

Each of these trends is discussed in the following sub-sections. The sum of the funds invested in each segment, in biofuels and RE electricity (except hydropower) equals the amount shown for the technologies included in Figure 11.2. In some segments of the continuum, public finance (funds from governments) and regulatory support mechanisms, which provide certainty of revenue, also play an important complementary role, as discussed in Section 11.5.

Although the concept of a continuum infers a smooth transition among the different types of financing involved, the reality is that financiers each have their own risk and return expectations and have different external drivers that make the various segments of the continuum less or more attractive for commercial investment.

11.2.2.2 Financing technology research and development

Governments fund most of the basic research aimed at increasing the understanding of fundamental principles, often with no direct or immediate commercial benefits. Large corporations fund most of the applied research and development aimed at a specific commercial or client-driven purpose. Worldwide public investment in RE R&D grew most rapidly from 1974 to 1980, when it peaked; it then declined throughout the 1980s and remained low in the 1990s. Since 2000, it has steadily risen to close to USD$_{2005}$ 1.81 billion (USD$_{2008}$ 2 billion) as of 2008 (IEA, 2010b), although that level is below investment in the 1978 to 1982 period. Private sector investment has followed a similar path (Nemet and Kammen, 2007). Another source reports higher levels of government sponsored non-hydro RE R&D, increasing from USD$_{2005}$ 0.9 billion in 2004 to USD$_{2005}$ 2.3 billion in 2009, a compound annual growth rate of 19% (UNEP and BNEF, 2010). (See also Section 10.5.5.)

11.2.2.3 Financing technology commercialization

Venture capital is a type of private equity capital typically provided for high-potential technology companies in the early market deployment phase in the interest of generating a return on investment through a trade sale of the company or an eventual listing on a public stock exchange. Venture capitalists begin to play a role once technologies are ready to move from the lab bench to the early market deployment phase, often working with and through government public-private demonstration and commercialization programmes.

According to Moore and Wüstenhagen, venture capitalists were initially slow to pick up on the emerging opportunities in the energy technology sector (Moore and Wüstenhagen, 2004), with RE accounting for only 1 to 3% of venture capital investment in most countries in the early 2000s. However, between 2002 and 2009, venture capital investment in RE technology firms increased markedly. Venture capital into RE electricity (excluding hydro) and biofuels companies grew from USD$_{2005}$ 392 million in 2004 to USD$_{2005}$ 1.41 billion in 2009 (UNEP and BNEF, 2010), representing a compound annual growth rate of 29%. This growth trend in technology investment now appears to be a leading indicator that the finance community expects continued significant growth in the RE sector. Downturns such as that experienced in 2008/2009 may slow or reverse the trend in the short term (as seen in Figure 11.2), but in the longer term, increased engagement of financial investors is foreseen in RE technology development (UNEP and NEF, 2009).

11.2.2.4 Financing manufacturing and sales

Once a technology has passed the demonstration phase, the capital needed to set up manufacturing and sales facilities usually comes initially from private equity investors (i.e., investors in unlisted companies) and subsequently from public equity investors who buy shares of companies listed on the public stock markets. Private equity investment is capital provided by investors and funds directly into private companies, often for setting up a manufacturing operation or other business activity, whereas public equity investment is capital provided by investors into publicly listed companies. These forms of capital are also used to finance some of the working capital requirements of companies, with the rest coming from bank loans.

Private and public equity investment in RE electricity (excluding hydro) and biofuels grew from USD$_{2005}$ 691 million in 2004 to USD$_{2005}$ 13.5 billion in 2009, representing a compound annual growth rate of 81% (UNEP and NEF, 2009). Even with this very fast growth in manufacturing investments, several technologies had supply bottlenecks through early 2008 that delayed sector growth and pushed up prices. In 2008, stock markets in general dropped sharply, but RE shares fared worse due to the energy price collapse and the fact that investors shunned stocks with any sort of technology or execution risk, particularly those with high capital requirements (UNEP and NEF, 2009). Financing for manufacturing facilities has also been negatively affected by some policy-induced boom and bust cycles that have made long-term production planning difficult (see for instance Box 11.5).

11.2.2.5 Financing construction

Financing RE generating facilities involves a mix of equity investment from project owners and loans from banks ('private debt') or capital markets ('public debt') raised through bond offerings). Both types of finance are combined into the term 'asset finance', which represents all forms of financing secured for RE projects (whether from internal funds, debt finance or equity finance). Regulatory RE policies (see Section 11.5), which create a quota for RE or ensure a certain price, may be important and complementary factors.

Asset financing of RE electricity (excluding hydro) and biofuels grew from USD$_{2005}$ 15.3 billion in 2004 to USD$_{2005}$ 88.7 billion in 2009, representing a compound annual growth rate of 42% (UNEP and NEF, 2009).

By 2007, the capital flows available to RE projects had become more mainstream and had broadened, meaning that the industry had gained access to a far wider range of financial sources and products than it had around 2004/2005 (UNEP and NEF, 2008). For instance, the largest component of total RE capital flows by 2009 was through project finance investment (DBCCA, 2010), an approach that mobilizes large flows of private sector investment in infrastructure.

Consumer loans, micro-finance and leasing are some of the instruments that banks offer to households and other end users to finance the purchase of small-scale technologies. However most investment in such systems comes from the end user themselves, usually through purchases made on a cash basis. Total global investment in residential RE projects was USD$_{2005}$ 16.43 billion in 2009 (UNEP and NEF, 2008), about 14% of total investment in RE projects. REN21, however, reported a much larger figure of USD$_{2005}$ 46 billion in 2009 using a broader methodology that includes balance of systems costs for photovoltaics (PV) and small-scale solar water heating systems (REN21, 2010).

Multilateral and development bank support has increased significantly in recent years, with USD$_{2005}$ 19.2 billion loaned to RE in 2009, up from USD$_{2005}$ 6.1 billion in 2007. According to de Jager et al. (2010), from mid-2008 onwards the multilateral banks aimed to fill the void in the project finance market caused by the financial crisis.

11.2.2.6 Refinancing and sale of companies

In 2009, USD$_{2005}$ 53.1 billion worth of mergers and acquisitions (M&A) took place involving the refinancing and sale of RE companies and projects (excluding hydro larger than 50 MW), up from USD$_{2005}$ 9.3 billion in 2004, or 42% compound annual growth (UNEP and NEF, 2009). M&A transactions usually involve the sale of generating assets or project pipelines, or sale of companies that develop or manufacture technologies and services. Increasing M&A activity in the short term is a sign of industry consolidation, as larger companies buy smaller, less well-capitalized competitors. In the longer term, increasing M&A activity provides an indication of the increasing mainstreaming of the sector, as larger entrants prefer to buy their way into the industry rather than developing RE businesses from the ground up.

11.2.3 Global investment transition

The recent trends in RE policies and finance have been generally positive for the RE sector. Even despite the financial downturn, total investment in 2009 in new RE capacity was greater than investment in new fossil fuel capacity in the electricity sector, for the second year running (UNEP and BNEF, 2010). This trend was driven in large part by that fact that more than half the world's countries had some type of policy target or promotion policy in place for RE (REN21, 2010). These inter-linked trends underline that RE was not a by-product of the ill-fated credit boom, but part of a global investment transition that is likely to strengthen over time (UNEP and BNEF, 2010). The next section examines the drivers, opportunities and benefits associated with this transition.

11.3 Key drivers, opportunities and benefits

A number of environmental, economic, social and security opportunities provided by RE are discussed in Chapters 1 and 9. In the policy context, they are considered as drivers, or factors that drive governments to adopt policies for RE development and deployment.

The motivations of policymakers to promote RE are described with specific examples from selected countries for illustrative reasons. The relative importance of the drivers for RE differ from country to country, and may vary over time. Without ranking them, key drivers for policies to advance RE are outlined below.

In general, economic opportunities drive policies in most developing countries, where RE is sometimes the only affordable means for providing energy access (e.g., Bolivia (REN21, 2009b), Bangladesh (Urmee et al., 2009), Brazil (Lucena et al., 2009), China (Standing Committee of the National People's Congress, 2005), India (Hiremath et al., 2009), Pakistan (Government of Pakistan, 2006), Tonga (Government of the Kingdom of Tonga, 2010), South Africa (Department of Minerals and Energy, 2003) and Zambia (Haanyika, 2008)) (Domac et al., 2005). So in terms of the share of global population concerned, this driver has been most important. In most developed countries, the desire to reduce environmental impacts of energy supplies, including climate change mitigation, and to decrease dependence on energy imports have been the primary drivers for RE promotion (for instance Australia, California in the USA, the EU, Quebec in Canada (Domac et al., 2005)). Thus, in terms of RE capacity added globally between 1990 and 2010, these drivers have been most important. In addition, in some countries the possibility of developing a new industry with related jobs is considered an opportunity; such motivations are of increasing importance in many emerging and developing economies as well.

11.3.1 Climate change mitigation and reduction of environmental and health impacts

RE can be a major tool for climate change mitigation (Section 9.3.4), although the degree to which RE mitigates climate change depends on many factors (Sections 10.2 and 10.3).

RE is an integral aspect of government strategies for reducing CO_2 (and other) emissions in many countries, including all member states of the EU (e.g., BMU, 2006; European Parliament and of the Council, 2009); and several US states, including California (CEC and CPUC, 2008) and Washington (CTED, 2009). Developing countries are also enacting RE policies in order to address climate change, among other goals. India's National Action Plan on Climate Change, launched in 2008, specifically mentions RE, and the country's National Solar Mission aims to constitute a major contribution by India to the global effort to meet the challenges of climate change (JNNSM, 2009). The 2009 meeting of Leaders of Pacific Island Countries observed that in addition to RE offering the promise of cost-effective, reliable energy services to rural households it will also provide a contribution to global GHG mitigation efforts (PIFS, 2009a).

In numerous cities, from Chicago (Parzen, 2009) and Miami (City of Miami, 2008) in the USA to Rizhao in China and Waitakere in New Zealand (IEA, 2009a), RE is playing an important role in climate mitigation strategies. By March 2010, more than 1,300 European municipalities had joined the Covenant of Mayors, committing to reduce CO_2 emissions beyond the EU objective of 20% by 2020 with the help of RE deployment, among other tools (European Commission, 2010).

The benefits of RE to the broader environment and human health (Section 9.3.4) are also driving governments to enact RE policies. At the same time, manufacture, construction and disposal of RE systems can have direct non-climate change impacts on the natural environment, including land use and aesthetics, and problems associated with chemicals required for manufacture and others. Policymakers can implement processes to minimize these negative outcomes while benefiting from the opportunities and benefits. Chapter 9 explores these issues in detail, while Chapters 2 through 7 review technology-specific impacts.

In China, for example, a major driver for the promotion of clean energy technologies, including RE, has been the goal of reducing or avoiding negative local and regional environmental impacts associated with energy (Standing Committee of the National People's Congress, 2005; Gan and Yu, 2008). The government of Pakistan intends to develop RE in order to avoid local environmental and health impacts of unsustainable and inefficient traditional biomass fuels and fossil fuel-powered electricity generation (Government of Pakistan, 2006). The South African government recognizes that millions of people are routinely exposed to noxious gases and particulates from the burning of fossil fuels due to inadequate living conditions and a lack of infrastructure in much of the country; the need to improve air quality has been a motivating factor in government plans to deploy RE technologies (Department of Minerals and Energy, 2003). In light of increasing concerns about water scarcity, many governments are turning to RE to reduce water consumption associated with energy production (Inhaber, 2004).

Growing awareness of the potential for RE to avoid some of the harmful impacts of fuel extraction on biodiversity of plant and animal species (IPCC, 2002) has led some governments to establish targets, or to adopt other policies, to increase RE deployment. For example, the Commonwealth of the Bahamas pays special attention to RE technology as a means to sustain vulnerable ecosystem services (National Energy Policy Committee, 2008). In Nepalese villages, modern RE systems have been deployed to mitigate negative impacts on biodiversity and deforestation resulting from the unsustainable use of biomass (Zahnd and Kimber, 2009).

11.3.2 Energy access

RE can enhance access to reliable, affordable and clean modern energy services (DBCCA, 2009), it is particularly well-suited for remote rural populations, and in many instances can provide the lowest cost option for energy access (Lucena et al., 2009; Mahapatra et al., 2009; Section 9.3.2). Many developing countries—including Bolivia (REN21, 2009b), Bangladesh (Urmee et al., 2009), Brazil (Lucena et al., 2009), China (Standing Committee of the National People's Congress, 2005), India (Hiremath et al., 2009), Pakistan (Government of Pakistan, 2006), Tonga (Government of the Kingdom of Tonga, 2010), South Africa (Department of Minerals and Energy, 2003) and Zambia (Haanyika, 2008)—have adopted RE policies, such as connection targets and subsidies, in order to provide access to energy services in rural areas.

11.3.3 Energy security

RE can improve security of energy supply in a variety of ways, including reducing dependence on imported fuels, helping to diversify supply, enhancing the national balance of trade and reducing vulnerability to price fluctuations (Section 9.3.3). These various benefits are driving a number of governments around the world to adopt policies to promote RE.

Since the early 1970s, Brazil has promoted ethanol from sugarcane as an alternative to fossil transport fuels in order to decrease dependency on imported fuels (Pousa et al., 2007; see Box 11.10). China established its 2005 Renewable Energy Law in part to diversify energy supplies and safeguard a secure energy supply (Standing Committee of the National People's Congress, 2005; see Box 11.11), and the Jamaican government aims to diversify its energy portfolio by incorporating RE into the mix, reducing reliance on imported oil (Government of Jamaica, 2006). A number of municipalities and communities from across Canada (St. Denis and Parker, 2009) to Güssing in Austria (see Box 11.14) and elsewhere are adopting RE plans to become more energy self-sufficient. Many governments have regarded RE (particularly biofuels) as a means to enhance their national balance of trade by substituting domestic RE fuels for imported fuels (National Greenhouse Strategy, 1998; Department of Minerals and Energy, 2003; DTI, 2007; Smitherman, 2009).

The relationship between public RE R&D funding and movements in the price of oil illustrate the significant role that the security of supply consideration has on government decisions to fund research into alternative sources of energy such as RE. Figures collected by the IEA (2008c) show that spending on RE peaked in 1981, and as oil prices dropped in the 1980s, RE R&D spending declined by more than two thirds, hitting a low in 1989. RE R&D funding has gradually increased since then, but not to earlier levels, as discussed in Section 11.2.2.2. The IEA (2008a) has argued that governments choose to focus their attention on technologies that can tap into their most abundant domestic natural resources. Non-IEA countries also justify focusing on a particular energy resource by pointing to its relative local abundance, like solar energy in India (JNNSM, 2009) and Singapore (SERIS, 2009). But there are important exceptions. Germany, for instance, spends more on PV R&D than any other country in Europe (European Commission, 2009a), but with a view to growing a competitive export industry (IEA, 2008c).

11.3.4 Social and economic development

Policymakers in many countries are enacting RE policies with the purpose of advancing economic development and/or creating jobs. (See Section 9.3.1 for a full discussion of RE in relation to social and economic development.) For example, the EU has highlighted the potential of RE to create new jobs, especially in rural and isolated areas (European Parliament and of the Council, 2009). Creating employment opportunities was an important driver in creation of the German Renewable Energy Act in 2000 (Jacobsson and Lauber, 2006), and Germany's fast-growing RE industries have motivated policymakers there to maintain strong promotion policies. A main target of the Greek government's RE promotion policies is to strengthen employment (Tsoutsos et al., 2008).

The development of domestic markets for RE is also seen as a means to attract new industries that may in turn supply international markets, thereby gaining competitive advantages (Lewis, 2007; Lund, 2008). One example is the case of Japan (see Box 11.2) and its PV industry. However, if combined with policies that promote domestic/local content and provide subsidies to protect domestic industries, conflicts can arise over international trade rules (International Center for Trade and Sustainable Development, 2010).

Rural development is often tied to the deployment of RE, whether in developed or developing countries. The biogas program operated by the Nepalese Alternative Energy Promotion Center together with the Netherlands Development Organization (SNV) has linked the deployment of RE with its socioeconomic development program (Mendis and van Nes, 1999). Bangladesh has been exploring the potential for RE to aid in rural development, with public and nongovernmental organizations working together to develop rural RE projects (Mondal et al., 2010). Rural development is also a key driver for RE policies in India, such as the country's support for biofuels (Bansal, 2009).

11.4 Barriers to renewable energy policymaking, implementation and financing

While there are a number of drivers, opportunities and benefits associated with RE, there are also a number of barriers to the development and deployment of RE. If RE is to play a significant role in mitigating climate change, it is important to address these barriers. Chapter 1 of this report offers an overview of barriers to RE development and deployment, while Chapters 2 through 7 cover technology-specific challenges, Chapter 8 addresses barriers to integration of RE at high shares, and Chapter 9 discusses barriers to RE in the context of sustainable development. This section summarizes some of the numerous barriers to successful policymaking, implementation and financing, which can also hamper the development and deployment of RE.

11.4.1 Barriers to renewable energy policymaking

Barriers to making and enacting policy include a lack of information and awareness about RE resources, technologies and policy options; lack of understanding about 'best' policy design or how to undertake energy transitions; difficulties associated with quantifying and internalizing external costs and benefits; and lock-in to existing technologies and policies.

A lack of information and awareness can affect policymaking in the design and enactment stages. Many policymakers lack the required knowledge and experience of RE policies: for example, the available policy options; how they work and should be implemented; how much they cost; what their benefits and difficulties are; and experiences to date in other countries. Best practices for successful RE policy, such as setting clear goals for sustainable technology innovation and communications with stakeholders, may not be effectively conveyed among policymakers, from the local to the international level (IEA, 2006; van den Bergh and Bruinsma, 2008). Further, lack of information about the effectiveness of policies, once implemented, can impede the redesign and improvement of existing policies or design of potential new policies. The failure of past policies can also create resistance to new policies to promote RE (Sawin, 2001).

Added to this, RE technological development is uncertain, dynamic, systemic and cumulative (Grubler, 1998; Fri, 2003; Foxon and Pearson, 2008). RE sources are local and circumstantial, and doing an inventory of resource potential and possibilities for development requires multi-disciplinary expertise (Twiddell and Weir, 2006). This means that even if policymakers have a general understanding of RE, time and effort are required to understand local conditions and develop connections to practitioner and scientific communities.

Further, there are a number of technology options available to policymakers wishing to pursue low-carbon energy futures—including RE, energy efficiency improvements, fast-track development of carbon capture and storage, or nuclear power—and assessments of the various portfolio options based on transparent sets of criteria are generally lacking (IEA, 2006, 2008a). Even once a portfolio of options has been selected, many policymakers lack the knowledge and expertise required to design policies that can proactively and effectively integrate RE supplies with other low-carbon options, with other policy goals (such as poverty alleviation, spatial planning), and across different but interconnected sectors (e.g., agriculture, housing, education, health, water and transportation) (Section 11.6.2). There are still differences of opinion about the linkages and interactions between climate policies (i.e., carbon pricing through tax or cap and trade) and RE policies (Section 11.5.7.3).

Although there is some understanding of how energy transitions occurred in the centuries past (R. Fouquet, 2008), there is no clear roadmap to a transition. Nevertheless, there is increasing analysis of how to undertake transitions to RE (e.g., van den Bergh and Bruinsma, 2008). This new generation of governance approaches aims at inducing and navigating the complex processes of socio-technical change by means of deliberation, probing and learning. Some argue that policy design should be longer term and be flexible, adaptive and reflexive (Voß et al., 2009). Others argue that a transformation to a low-carbon energy system can emerge only from interactions among multiple interest groups as well as wider institutional and social constituencies (Smith et al., 2005; Verbong and Geels, 2007).

Any or all of these factors can make policy design difficult; they can also make it difficult to reach a consensus and to enact specific policies (C. Mitchell, 2010). In addition, regulatory authorities and policymakers face an asymmetry of information between established and newer technologies, and they may also be captured by incumbent technology interest groups, leading to decisions on energy policy that do not optimize social welfare (Laffont and Tirole, 1998; Helm, 2010).

There are also economic barriers related to RE costs and externalities associated with energy production and use. Policymakers may not recognize the value of RE due to the higher costs of many RE technologies relative to current energy market prices. Further, although there is growing acceptance that the social costs and risks of energy use should be incorporated into the price of energy (Stern, 2007), it is difficult to quantify and internalize these costs (Stirling, 1994). If societies could reach a policy consensus on how much RE is socially desirable, in terms of how much extra society is prepared to pay, and/or in terms of a specific share of energy to be derived from RE sources, public policies could be implemented to reflect this social consensus. However, it is difficult for societies to make a rational choice about technology without full information.

Further, the existing energy system exerts a strong momentum for its own continuation (Hughes, 1987), which locks existing technologies and policies (mostly fossil fuel-based (IEA, 2009d)) in place and locks out new technologies and ways of doing things (Unruh, 2000). This dampens the drive for new policies while also making it harder for them to be put into practice because implementation occurs within the existing energy system. In addition, incumbents of the existing energy system enjoy greater organizational strength, more influential networks and increased lobbying power over newer RE technologies (Hughes, 1986), and thus have greater potential to influence policy design and enactment.

11.4.2 Barriers to implementation of renewable energy policies

Once policies have been enacted, challenges can arise related to implementation. These include conflicts with existing regulations; lack of skilled workers; and/or lack of institutional capacity to implement RE policies.

Regulation of markets and networks, including existing standards and licensing practices that were established to aid and maintain the existing energy system, can erect barriers to RE (Beck and Martinot, 2004; P. Baker et al., 2009; M. Baker, 2010). Existing administrative procedures often make it a lengthy and difficult process to change the scope or applicability of economic regulation to accommodate RE technologies (P. Baker et al., 2009; C. Mitchell, 2010).

In addition, workforce education and training generally reinforce incumbent technologies and lag behind the emergence of new technologies, constraining the rate of RE installation and maintenance. Even when programmes are in place, ramping up skills takes time. This lack of educational and skills base in turn constrains the knowledge about emerging options, and it aggravates a low awareness and acceptance by authorities, companies and the public (IEEE PES, 2009; Bird and Institute for Public Policy Research, 2009; Energy Skills Queensland, 2009; MERC Partners, 2009; European Centre for Development of Vocational Training, 2010).

Institutional barriers also hold back RE policymaking and implementation at all levels of government. Planning frameworks and institutional coordination for RE policy are often rudimentary or may not yet exist (ECLAC, 2009). Further, lack of coordination among overlapping national and local authorities, regarding such aspects as spatial planning for accommodation of RE installations, may lead to a long process for obtaining necessary permits (Ragwitz et al., 2007). In addition, in some municipalities, states/provinces or countries, the institutions needed to administer RE policies might not yet be in place (de Jager and Rathmann, 2008).

11.4.3 Barriers to renewable energy financing

As discussed in Section 11.2.2, financing is critical in every stage of technology development. Yet there are also many barriers that affect the availability of financing.

First, and most importantly, many RE technologies are not economically competitive with current energy market prices, making them financially unprofitable for investors absent various forms of policy support, and thereby restricting investment capital.

Second is a lack of information. To operate effectively, markets rely on timely, appropriate and truthful information. But energy markets are far from perfect; this is particularly true of markets in technological and structural transition, such as the RE market. As a result of insufficient information, underlying project risk tends to be overrated and transaction costs can increase as compared to conventional fossil fuel technologies (Sonntag-O'Brien and Usher, 2004).

Compounding this lack of information is the issue of financial structure. RE projects typically have higher investment costs and lower operating costs than fossil fuel technologies do. Their financial structure therefore requires a higher level of financing that must be amortized over the life of the project. This makes an RE investment's risk exposure a longer-term challenge than that faced with fossil fuel generating plants, which often have lower investment costs (Sonntag-O'Brien and Usher, 2004).

In addition to higher investment costs, financiers face other issues of concern that are related to RE projects. Besides having more assets at risk and over a longer time period, other aspects of risk also come into play. According to de Jager et al. (2010), private investors lack experience on the technology side (upstream) with new types of sponsors, business models, the markets and/or technologies involved. On the project side (downstream), their concerns often relate to the performance of the installation, the experience and reliability of the developer or owner, and difficulties in obtaining operating licenses, the purchase power agreement (PPA) and other administrative hurdles (de Jager et al., 2010).

The issue of project scale can also act as a barrier to RE financing. Since RE projects are typically smaller than traditional fossil or nuclear projects, the transaction costs are disproportionately higher. Any investment requires initial feasibility and due diligence work, and the costs for this work do not vary significantly with project size. As a result, pre-investment costs, including legal and engineering fees, consultants and permitting costs have a proportionately higher impact on the transaction costs of RE projects. Furthermore, the generally smaller nature of RE projects results in lower gross returns, even though the rate of return may be well within market standards of what is considered an attractive investment (Sonntag-O'Brien and Usher, 2004).

Developers of RE projects are often under-financed and have limited track records. Financiers therefore perceive them as being high risk and are reluctant to provide non-recourse project finance where the financier cannot recover the loan beyond the value of that specific project's assets and revenues. Lenders wish to see experienced construction contractors, suppliers with proven equipment and experienced operators. Additional development costs imposed by financiers on under-capitalized developers during due diligence can significantly jeopardize a project (Sonntag-O'Brien and Usher, 2004).

Further, institutional weakness including imperfect capital markets and insufficient access to affordable financing can inhibit private sector engagement in RE project finance. In many countries, the financial sectors are not developed sufficiently to provide the form of long-term debt that RE and related infrastructure projects require (UNEP, 2008). This is a particular problem in many developing countries. A lack of appropriate financing mechanisms available to end users in developing countries is another significant barrier to RE uptake (Derrick, 1998). Stronger intervention may be necessary to unlock private sector investment in new technologies (UNEP Finance Initiative, 2009), particularly for off-grid and rural markets.

11.5 Experience with and assessment of policy options

This section explains the policies currently available and in use around the world to support RE technologies—from their infant stages, to demonstration and pre-commercialization, and through to maturity and wide-scale deployment—in order to address existing barriers outlined in Section 1.4 and many of the barriers in Section 11.4, and to enable RE to play a significant role in mitigating climate change. These include government R&D policies (supply-push) for advancing RE technologies, and deployment policies (demand-pull), which aim to create a market for RE technologies. This section focuses on policies directly supporting RE, based on the assumption that policymakers are aiming to increase

RE levels based on drivers of their choosing. For those policymakers targeting climate change mitigation goals, the interplay between RE and climate policies is discussed in Section 11.5.7.3.

Policies could be categorized in a variety of ways and there exists no globally agreed list of RE policy options or groupings. For the purpose of simplification, this chapter organizes R&D and deployment policies within the following categories:

- **Fiscal incentive:** actors (individuals, households, companies) are allowed a reduction of their contribution to the public treasury via income or other taxes or are provided payments from the public treasury in the form of rebates or grants.

- **Public finance:** public support for which a financial return is expected (loans, equity) or financial liability is incurred (guarantee); and

- **Regulation:** rule to guide or control conduct of those to whom it applies.

RE policies are often linked to national or regional targets, such as the EU RE Directive, which calls for RE to provide 20% of energy used in the EU by 2020. Literature is lacking that provides evidence of whether targets, absent obligatory mandates or implementing policies, make RE policies more efficient or effective within the energy system. Although targets are a central component of policies, policies in place may not need specific targets to be successful. Further, targets without policies to deliver them are unlikely to be met, as seen in the Pacific Island States where RE targets and financing without appropriate RE policies have been insufficient to achieve significant progress with RE (See Box 11.1).

After a discussion on policy evaluation criteria (Section 11.5.1), this section first summarizes the policy options for R&D and the important interactions of R&D policies with deployment policies (Section 11.5.2). Most of the section then focuses on policies for RE deployment, with a general overview of policy options (Section 11.5.3) and then sector-specific (electricity, Section 11.5.4; heating and cooling, Section 11.5.5; transportation, Section 11.5.6) assessments and lessons learned based on experiences to date. The section concludes with some general findings, a discussion of the macroeconomic impacts of RE policies, and a review of the possible positive or negative interactions between RE and carbon policies. Only those policies specifically targeting RE advancement are covered in this section; a full discussion of policies required to create an enabling environment for RE is provided in Section 11.6.

11.5.1 Criteria for policy evaluation

The success of policy instruments is determined by how well they are able to achieve various objectives or criteria. To the extent that literature is available, this section assesses policies based on a variety of criteria that have been used for evaluating policy instruments (Bohm and Russell, 1985; Hanley et al., 1997; Aldy et al., 2003; Hanley et al., 2004; Huber et al., 2004; Sawin, 2004; Gupta et al., 2007; Bergek and Jacobsson, 2010; European Commission, 2010; Verbruggen, 2010; among others). These criteria include the following:

- **Effectiveness:** the extent to which intended objectives are met, for instance the actual increase in the amount of RE electricity generated or share of RE in total energy supply within a specified time period. Beyond quantitative targets, factors may include achieved degrees of technological diversity (promotion of different RE technologies), which is considered a crucial factor for dynamic effectiveness (long-term sustained growth that enables innovation and the development of a manufacturing base), or of spatial diversity (geographical distribution of RE supplies).

- **Efficiency:** the ratio of outcomes to inputs, or RE targets realized on economic resources spent, mostly measured at one point in time (*static efficiency*); also called cost-effectiveness. *Dynamic efficiency* adds a future time dimension by including how much technology development and innovation is triggered by the policy instrument. Reducing the risks to investors is crucial for minimizing costs of financing, which in turn reduces project costs.

- **Equity:** the incidence and distributional consequences of a policy, including dimensions such as fairness, justice and respect for the rights of indigenous peoples. Equity can be assessed, in part, by looking at the *distribution* of costs and benefits of a policy (e.g., a policy that follows the polluter pays principle is generally considered to be fair (Heyward, 2007)), and/or by evaluating the extent to which it allows the *participation* of a wide range of different stakeholders (e.g., equal rights to independent power producers and to incumbent utilities). *Excess profits*, created by suboptimal policy designs, transfer money from rate- or taxpayers to mostly incumbent power producers, undermining equity (Verbruggen, 2009; Bergek and Jacobsson, 2010).

- **Institutional feasibility:** the extent to which a policy instrument is likely to be viewed as legitimate, gain acceptance, and be adopted and implemented. Institutional feasibility is high when policies are well adapted to existing institutional constraints. Economists traditionally evaluate instruments for environmental policy under ideal theoretical conditions; however, those conditions are rarely met in practice, and instrument design and implementation must take political realities into account. In reality, policy choices must be both acceptable to a wide range of stakeholders and supported by institutions. In market economies, instruments need to be compatible with markets. An important dimension of institutional feasibility addresses the ability to implement policies once they have been designed and adopted.

Other criteria are also examined in the literature, including subcategories of the four set out above. But most literature focuses on effectiveness and efficiency of policies, which are therefore the main criteria that

> **Box 11.1 | Lessons from the Pacific Island States: Renewable energy target setting.**
>
> The Pacific Islands, home to more than 1.5 million people, are among the most vulnerable places in the world to the impacts of climate change. Although their contribution to global GHG emissions is negligible, the islands are blessed with significant RE resources and are receiving significant donor assistance that is specific to RE: the Global Environment Facility (GEF) contributed approximately USD 30 million during 2000 through 2009 (SIS, 2009), and development partners have allocated a further estimated USD 300 million in funding for 2010 to 2015 (SPC, 2010).[1] RE is increasingly viewed as a means for achieving energy security—supporting accessibility, affordability, productivity and clean energy (SPC, 2010).
>
> In response to these factors, the Pacific Island countries have adopted national RE targets and made commitments to pursue a RE development path. For example, Fiji targets at least 90% of its energy needs to be met with RE by 2011, Nauru targets 50% of its energy to be derived from RE by 2015 and Vanuatu's power utility will generate 25% of its electricity from RE by 2012 (PEMM, 2009). Both Tonga and Tuvalu have incorporated RE targets into their national energy strategies (PIFS, 2009a). Tonga originally set itself a 50% RE target in three years, but has since redirected its approach by adopting a Tonga Energy Roadmap (TERM) with the objective of finding a least-cost implementation plan that involves energy efficiency improvements and a shift from fossil-based electricity generation to RE (Government of the Kingdom of Tonga, 2010). At their annual meeting in 2010, the Pacific Island Leaders adopted a regional framework for Energy Security in the Pacific which is based on the premise of 'Many Partners One Team One Plan' (PEMM, 2009; PIFS, 2010).
>
> However, the RE target commitments made are ambitious and require a full understanding of RE resource potential, RE investment costs, and their technical and economic viabilities. Thus far the general progress towards the RE target is slow. Experiences imply that setting RE targets and having significant amounts of financing available are both important factors in advancing RE, but they are not sufficient—they need to be backed by appropriate policies and they must be realistic and practical (PIFS, 2009b, 2010).
>
> ---
> Note: 1. Conversion to 2005 dollars is not possible given the range of study-specific assumptions.

serve as the basis of some of the discussion in Section 11.5. Ultimately, however, criteria for judging how well policies work will depend on the policy goals of the jurisdiction that enacts and implements those policies.

11.5.2 Research, development and deployment policies for renewable energy

11.5.2.1 Why and when public research and development is needed

While private sector engagement in the R&D process is essential, and ultimately comprises the majority of investment, governments play a crucial role in funding RE R&D for several reasons. First, it is difficult for private companies to fully appropriate investments in some R&D activities, especially early stage ones (Nelson, 1959), which reduces incentives to invest (Jaffe et al., 2005). Second, firms may be reluctant to take on the risk associated with investing in a new technology that may not ultimately succeed (Siddiqui et al., 2007; Popp, 2010). Third, the time involved with bringing a technology from the R&D phase to adoption in the marketplace sufficient to pay back investments may be beyond that required by private investors (Meijer et al., 2007a,b; Kenney, 2010). And fourth, expected future payoffs may not stimulate private sector R&D because future markets for RE technology may be considered too uncertain, especially because RE markets are typically heavily influenced by policy decisions, which can change and thus make markets volatile and risky (Yang et al., 2008; Blyth et al., 2009; Nemet, 2010b). It is for these reasons that the R&D and innovation market failure was described earlier as a key factor motivating the need for policy intervention beyond carbon pricing to most efficiently address climate mitigation.

Not all countries can afford to support R&D with public funds, but in the majority of countries where some level of support is possible, public R&D for RE enhances the performance of nascent technologies so that they can meet the demands of initial adopters and it improves existing technologies that already function in commercial environments. Investments falling under the rubric of R&D span a wide variety of activities along the technology development lifecycle, from RE resource mapping to improvements in commercial RE technologies. The magnitudes of investments required in each stage vary substantially; importantly, the costs of progressing from one stage to the next generally increase (NSB, 2010). Several studies claim that current levels of public (and private) investment in RE R&D are too low to address energy-related concerns including climate change (Schock et al., 1999; Holdren and Baldwin, 2001; Davis and Owens, 2003; Nemet and Kammen, 2007; Weiss and Bonvillian, 2009).

As with any new technology, RE technologies at some point are likely to traverse the point just before a technology has proven itself and is ready for widespread deployment. The so-called 'valley of death' is a particular problem associated with the integration of R&D and demand side (or deployment) policies (Murphy and Edwards, 2003; Weyant, 2010). This stage of development is characterized by a troublesome combination of a substantial increase in the scale of investment required, unproven technical reliability, uncertain market receptiveness and outcomes that are likely to be highly beneficial to companies other than those making an investment. One way of putting it is that social returns to investment at this stage far exceed private returns; a lack of investment by both the public and private sector has been a typical result.

This stage of the technology innovation process is particularly amenable to cost sharing between governments and private firms, and industrial consortia, as with PV in Japan (Watanabe et al., 2004). In the USA and Europe, public-private partnerships for demonstration (where industry-led projects demonstrate new technologies with government co-funding) are increasingly viewed as one appropriate vehicle to vault this 'valley' (Strategic Energy Technology Plan, 2007; House of Commons, 2008; US DOE, 2009).

The need for R&D continues even after technologies reach commercial deployment. Scale economies and learning by doing may dominate innovation at the deployment stage, but codification of experience-derived changes, improvement of manufacturing processes, increasing reliability and the development of supporting innovations may all benefit from sustaining R&D during deployment. Continuing R&D support offers many opportunities to accelerate cost reductions and performance improvements (Neuhoff, 2005). Examples of important post-deployment R&D programs include wind power in Germany and Denmark (Langniß and Neij, 2004) (see Boxes 11.6 and 11.12), concentrating solar thermal electric generation in California in the 1980s (Lotker, 1991; Cohen et al., 1999) and the US PV manufacturing program in the 1990s and 2000s (R. Mitchell et al., 2002; Jayanthi et al., 2009).

While RE R&D investment is typically associated with the accumulation of new knowledge, technical know-how developed through R&D can lose its value over time. Knowledge depreciates when employees turn over and tacit knowledge in researchers' heads is lost, and existing knowledge becomes obsolete once it is no longer suitable for application to updated processes and techniques (Argote et al., 1990). Depreciation of R&D assets may be especially problematic in RE where funding levels are volatile and technological change is rapid, for example in PV (Watanabe et al., 2000). Stable funding levels, retention of personnel, as well as codification of techniques and experimental outcomes, can avoid the waste associated with preventable depreciation of R&D investments.

An essential element of R&D projects is the stochastic nature of the results: the outcomes of R&D investments are inherently unknowable in advance. Moreover, analysis of past energy R&D investments shows that benefits attributable to a small number of successful projects more than make up for the investments in projects that did not result in commercial applications (NRC, 2001). Further, an important determinant of the social value of RE investments is how quickly they become adopted by the market (Moore et al., 2007). One implication of unknowable ex ante technical and market outcomes is that evaluation of RE R&D is best suited to considering investments as 'insurance' (Schock et al., 1999), a 'hedge' (E. Baker et al., 2003), and as having 'option value' (Davis and Owens, 2003; Siddiqui et al., 2007). Prospectively, an important way to address inherently uncertain returns on R&D is to make use of an aggregation of expert opinions on expected future technology outcomes (NRC, 2007). Finally, these features of RE R&D investments make them particularly amenable to consideration of them as portfolios of investments (Frenken et al., 2004; Richels and Blanford, 2008; Blanford, 2009). Key considerations in portfolio design are: level of tolerance for risk; when to support diversity and when to eliminate options; whether investments are characterized by critical minimal scale or diminishing returns; and how to populate the probabilities of successful outcomes (Nemet, 2009; Sovacool, 2009b).

Critics of public investment in R&D for RE cite the possibility that public spending crowds out private investment (Goolsbee, 1998; David et al., 2000), the mixed record of success in past investments (Cohen and Noll, 1991), and the tendency to isolate scientific understanding from technical knowledge (Stokes, 1997). However, recent work on RE finds limited evidence of crowding out (Popp and Newell, 2009).

11.5.2.2 Public research and development measures

Table 11.1 presents a list of RE policies for R&D and their definitions. One general trend is that policy measures in the RD&D sphere are becoming more collaborative and innovative as governments seek new means of tapping into potential financiers, investors and innovators. Collaboration encourages 'buy-in' from partners as early as possible in the technology development spectrum, and intends to use public money as efficiently and effectively as possible.

Fiscal incentives available to policy makers include the following, and more, as outlined in Table 11.1:

Contingent grants can serve to cover some of the costs during the highest-risk development stages and in some cases increase investor confidence, thereby leveraging highly needed risk capital.

Technology incubators can assist developers in covering operating costs, provide advice on business development and raising capital, help to create and mentor management teams, and provide energy-related market research. An example is the UK Carbon Trust Incubator Programme, which furnishes an important stepping stone to commercialization for new sustainable energy and 'low carbon' technologies (UNEP, 2005).

Public Research Centres can provide a means for 'open innovation', a way for companies to acquire intellectual property by jointly contracting with one or more public R&D centres, while endorsing both the costs and benefits associated with the innovation. It is currently developed for silicon PV cells in Belgium and the Indian government wants to explore a similar scheme (IMEC, 2009a,b; JNNSM, 2009).

Table 11.1 | Definitions of existing R&D policy mechanisms.

Policy	Definition
PUBLIC R&D POLICIES	
FISCAL INCENTIVES	
Academic R&D funding	Investment monies provided to academics for undertaking creative work to increase stock of knowledge in a particular field and use it to devise new applications.
Grant	Funding for R&D and demonstration with no repayment requirements. Challenge grants are provided alongside industry commitments, often targeting product innovations or early manufacturing facilities. Contingent grants are loans that do not require repayment unless and until technologies and intellectual property have been successfully exploited.
Incubation support	Assistance to entrepreneurs including business development and raising financing.
National/International Public Research Centre	Research facility funded by local, national or international government bodies or publicly funded organizations.
Public-private partnership	Arrangement typified by collaboration between the public and private sectors. Can cover delivery of policies, services, technologies and infrastructure.
Prize	Awarded to winning competitors to help finance costs of private R&D; generally used in innovation stage.
Tax credit	Allows investments in RE R&D to be fully or partially deducted from tax obligations or income.
Voucher scheme	Provides companies access to R&D centres for the purpose of doing research.
PUBLIC FINANCE	
Venture capital	Financing aimed at turning promising research into new products and services; invested independently or with matching private investors.
Soft/convertible loan	Financing instrument available at pre-commercial stage to promote and commercialize RE technologies; often loans are repayable only once technology reaches commercialization.

Public-private partnerships in research can include co-funded research, which has the benefit of creating direct research networking among different sectors (academy, industry), disciplines or locations. It may enable partners to take bigger risks, move off the beaten track, and to build a supply chain and ultimately realize a product, process or business model. Research networks can draft joint action plans in order to meet short-, medium- and long-term goals for technology performance and cost (IEA, 2008a); governments can then scrutinize and adopt these plans. Road mapping is one example of collaborative R&D that has been outlined in Japan for PV technology (see Box 11.2), and in the European region (Strategic Energy Technology Plan, 2007; NEDO, 2009).

Prizes are sometimes used to foster technology development. While the R&D risk lands on the shoulders of the competitors, they have freedom in the way they approach innovation and the competition process is sometimes easier than applying for public grants (contracting, reporting, control) (Peretz and Acs, 2011).

Besides R&D support, public funding is also needed to help move technology innovations through the product development stages towards commercialization. To convince investors, developers must prove that their technology will be able to perform in real market conditions and be commercially viable (UNEP, 2005). In addition, governments are starting to implement new financing mechanisms that are capitalized by public sources, such as convertible loans and publicly backed venture capital, in order to push technology innovation towards the market and to engage commercial investment in the RE sector (UNEP, 2005).

Various government agencies in the USA, Australia and the UK have been experimenting with *venture capital mechanisms* as part of their overall industrial and economic development policy aimed at turning promising research into new products and services (SEF Alliance, 2008). More than one mechanism can be used at a time—for example, the US state of Connecticut combines grant support for demonstration projects with a soft loan that is repayable if the technology reaches commercialization.

11.5.2.3 Lessons learned

Successful subsidies lead technology innovators towards commercialization and help attract early and later risk capital investment that otherwise would not be available because investors see high risk and protracted investment horizons. Further, experience has shown that it is important that subsidies for R&D (and beyond) are designed to have an 'exit strategy' whereby the subsidies are progressively phased out as the technology commercializes, leaving a functioning and sustainable sector in place (ICCEPT, 2003). Subsidy policies can be designed to avoid dependence (i.e., a tendency to keep technologies at the R&D and first demonstration stages rather than moving them on to deployment) and instead to grow a new technology area while minimizing market distortions. Grant-support models that are linked to performance, for example, can allow developers to build a track record, which is not possible if only traditional up-front grants are used.

Successful outcomes from R&D programmes are not solely related to the total amount of funding allocated, but are also related to the consistency of funding from year to year. On-off operations in R&D are detrimental to technical learning, and learning and cost reductions depend on continuity, commitment and organization of effort, and where and how funds are directed, as much as they rely on the scale of effort (Grubler, 1998; Sawin, 2001). Karnoe (1990) compared the early US and Danish wind

Box 11.2 | Lessons from Japan: Coupling supply-push with demand-pull for PV.

Japan turned to RE in search of energy security and stable supply after the first oil shock seriously weakened the nation's economy (Sugiyama, 2008). Starting in 1974, MITI (Japan's Ministry of International Trade and Industry) launched the 'Sunshine Project', which aimed to achieve technological progress with new energy technologies, and significant funds were directed to PV R&D. The principal long-term target has been the development of highly efficient low-cost solar cells (Takahashi, 1989).

MITI worked to link its PV project to Japan's industrial development. Although the primary goal was development of solar energy technologies, MITI expected that technological advances could provide benefits that went well beyond the energy field. It was hoped that the national investment in PV R&D would lead not only to provision of electric power on a large scale and realization of a domestic supply of energy, but also to new international markets for solar calculators and other appliances (Watanabe et al., 2000).

The investment paid off with the global increase in demand for electronic appliances and the expansion of a semiconductor market for computer 'chips'. By 1990, when MITI established an R&D consortium for PV development (Photovoltaic Power Generation Technology Research Association), electronic machinery companies like Sanyo and Sharp were the major players. The result was a dramatic decrease in solar cell prices between 1974 and 1994, from 26,120 yen/W (38,580 yen_{2005}/W (USD_{2005} 350)) to 650 yen/W (USD_{2005} 5.4) (Watanabe et al., 2000). Based on this achievement, in 1992 Japan's electric utility companies voluntarily started to purchase surplus PV power, helping to expand the market for grid-connected PV systems and to demonstrate PV's potential to meet domestic power needs.

In 1993, the purpose of RE advancement expanded to encompass sustainable development objectives, including CO_2 reductions, and Japan made the transition to the 'New Sunshine Project'. Parallel to its R&D efforts, Japan established targets for PV deployment and initiated a gradually-declining subsidy for residential rooftop PV systems, in exchange for operational data, with the goal of driving down PV costs through economies of scale and commercial competition among manufacturers. To create market awareness, the government began promoting PV through a variety of avenues, including television and newspapers (IEA, 2003a).

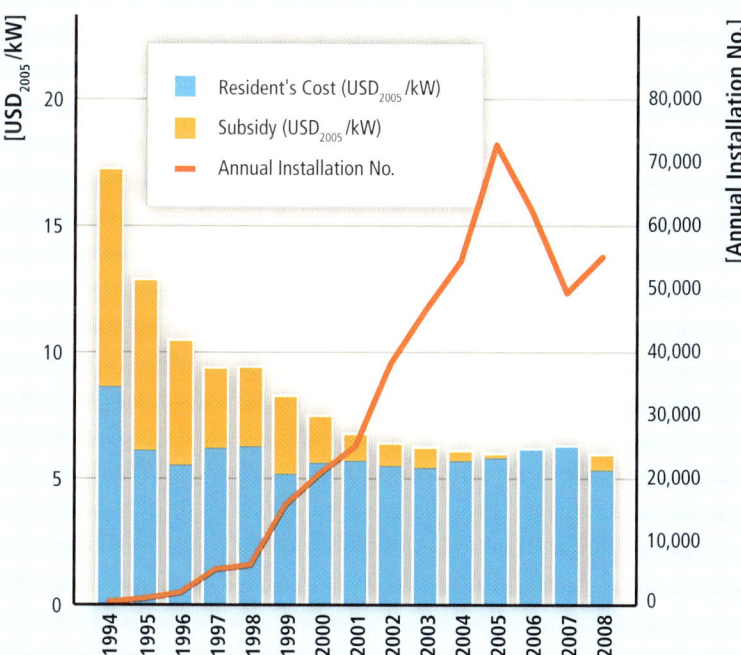

Figure 11.4 | Japan's rooftop PV: annual costs, subsidies and number of systems installed annually, 1994 through 2008 (Ito, 2003; Kobayashi, 2003; NEPC, 2009).

The result was a dramatic increase in installed capacity and accompanying reduction in PV costs. Japan rose from a minor player to become the world's largest PV producer in less than a decade. Over the 1994 to 2004 period, system costs declined by two-thirds, from USD_{2005} 18 (USD_{2005} 1.2/kWh[1] based on 2000 yen/W) in 1994 to USD_{2005} 6 (USD_{2005} 0.4 /kWh; based on 660 yen/W) in 2004 (NEDO, 2009), and annual installations increased more than 1,000-fold over this period, from 1.8 MW in 1994 to 2,002 MW in 2004 (Ito, 2003; Kobayashi, 2003; NEPC, 2009). Despite these advances, market growth slowed after the subsidy program ended in 2005 (see Figure 11.4), and Japan's role in global PV manufacturing has subsequently declined in part as a result of the rising dominance of China's solar manufacturing sector.

In 2009, for the purpose of job creation and increased competitiveness in the international marketplace, the government established a buy-back system for residential rooftop PV (residential producers can sell excess power to the utility company at about twice the retail rate). The purpose was to further accelerate the introduction of PV and provide an incentive for customers to minimize their own

Note: 1. Levelized cost estimates based on the following assumptions: duration 20 years; capacity factor 12%; and discount/interest 4%. Assumptions are based on the practice of the Government of Japan.

electricity use in order to sell as much as possible to their utility (METI, 2009). In April 2010, a revised subsidy system started again, further boosting the domestic PV market.

For most of the past three decades, Japan has enacted effective and consistent policies to promote PV and has retained them even through major budget crises. Its experience suggests the importance of long-term targets and planning, the potential to link RE development to other applications and industries, as well as the positive feedback of declining costs, technology advances and increasing deployment that result from coupling supply push (R&D) with policies to create a market.

energy R&D programmes and found that, while the USA had invested a great deal more in funding, they were less successful in turbine development due to their focus on scale and other factors rather than reliability (Karnoe, 1990; Sawin, 2001). Garud and Karnøe (2003) argue that 'bricolage not breakthrough'—or progress via research aiming at incremental improvements versus radical technological advances—is the more successful approach to R&D policy. Nemet (2009) also analyzed the value of incremental versus non-incremental approaches (see Section 11.7.4 for a longer discussion). Successful technology development occurring via the incremental approach is supported by detailed studies of RE technology development in Europe (Jacobsson and Johnson, 2000) as well as experiences in Japan and Thailand (see Boxes 11.2 and 11.7). However, others argue that both approaches are required simultaneously (O'Reilly and Tushman, 2004; Hockerts and Wüstenhagen, 2010).

Additionally, several key considerations exist for improving the effectiveness of future RE R&D investments. Improved measurement and documentation of R&D investment outcomes continues to be needed and can inform future decisions. Promising approaches to optimize public R&D investments include those informed by option value, portfolio analysis, and aggregation of expert opinion (NRC, 2007). Evaluation of programs based on the results of the overall portfolio, rather than individual investments, may lead to different incentives than exist today. The results of past investments have the potential to substantially improve the management of and budget allocations for government RE R&D programs. Still, several types of decisions remain crucial, for example: how much diversification is optimal, given that there may be increasing returns to the scale of R&D investment; consideration of whether public managers have incentives to take on more early stage technical risk than the private sector is willing to accept; when to patiently continue support and when to terminate programs with low likelihoods of success; and when to switch from emphasizing R&D to emphasizing demand-side support (Nemet, 2010a).

11.5.2.4 Positive feedbacks from combining research and development policies with deployment policies

The timing of R&D policies, and their balance with deployment policies, is also important (Langniß and Neij, 2004; Neij, 2008). One of the most robust findings, from both the theoretical literature and technology case studies, is that R&D investments are most effective when complemented by other policy instruments—particularly, but not limited to, policies that simultaneously enhance demand for new RE technologies. Relatively early deployment policies in a technology's development accelerate learning, whether learning through R&D or learning through utilization (as a result of manufacture) (Neij, 2008), as seen in Japan and Denmark, for example (see Boxes 11.2 and 11.12). Disentangling the contributions of public R&D spending and economies of scale to cost reduction is difficult, especially since the commercialization of the technology stimulates private sector investment in R&D (Schaeffer et al., 2004). Nonetheless, existing literature suggests that R&D and deployment policies used simultaneously can best induce innovation (Mowery and Rosenberg, 1979; Johnstone et al., 2010). Successful innovations show the ability to connect, or 'couple' a technical opportunity with a market opportunity (Freeman, 1974; Grubb, 2004), while studies of the effectiveness of technology policy for RE support this general consensus that both are needed (Grubler et al., 1999b; Norberg-Bohm, 1999; Requate, 2005; Horbach, 2007).

It is not simply that both factors contribute; they also interact because a positive feedback exists between R&D and deployment (Watanabe et al., 2000) (see Figure 11.5). This cycle of positive feedback, and its resulting benefits, can also cause a positive feedback to the policy cycles (from agenda and target setting to policy implementation and evaluation), increasing acceptance for (more) ambitious policies. This dynamic mechanism in countries like China and Germany (see Boxes 11.11 and 11.6) has encouraged policymakers to introduce stricter RE targets (Jacobsson and Lauber, 2006; Jänicke, 2010). Real-world deployment experience can also reveal new challenges that require investments in R&D to overcome them; it can facilitate the incorporation of market feedback about what customers actually want into subsequent R&D decisions; and commercialization generally increases the ability of firms to profit from their inventions, heightening the incentives for private sector investment in R&D (Nordhaus, 2010). An important result to consider in allocating between the two is that R&D typically dominates investment in the early stages of the innovation process, while deployment mechanisms are more important in the later stages (Dosi, 1988; Freeman and Perez, 1988). Moreover, not only are both types of policies needed, many different parties are likely to be needed in the commercialization of R&D programs (Mowery et al., 2010).

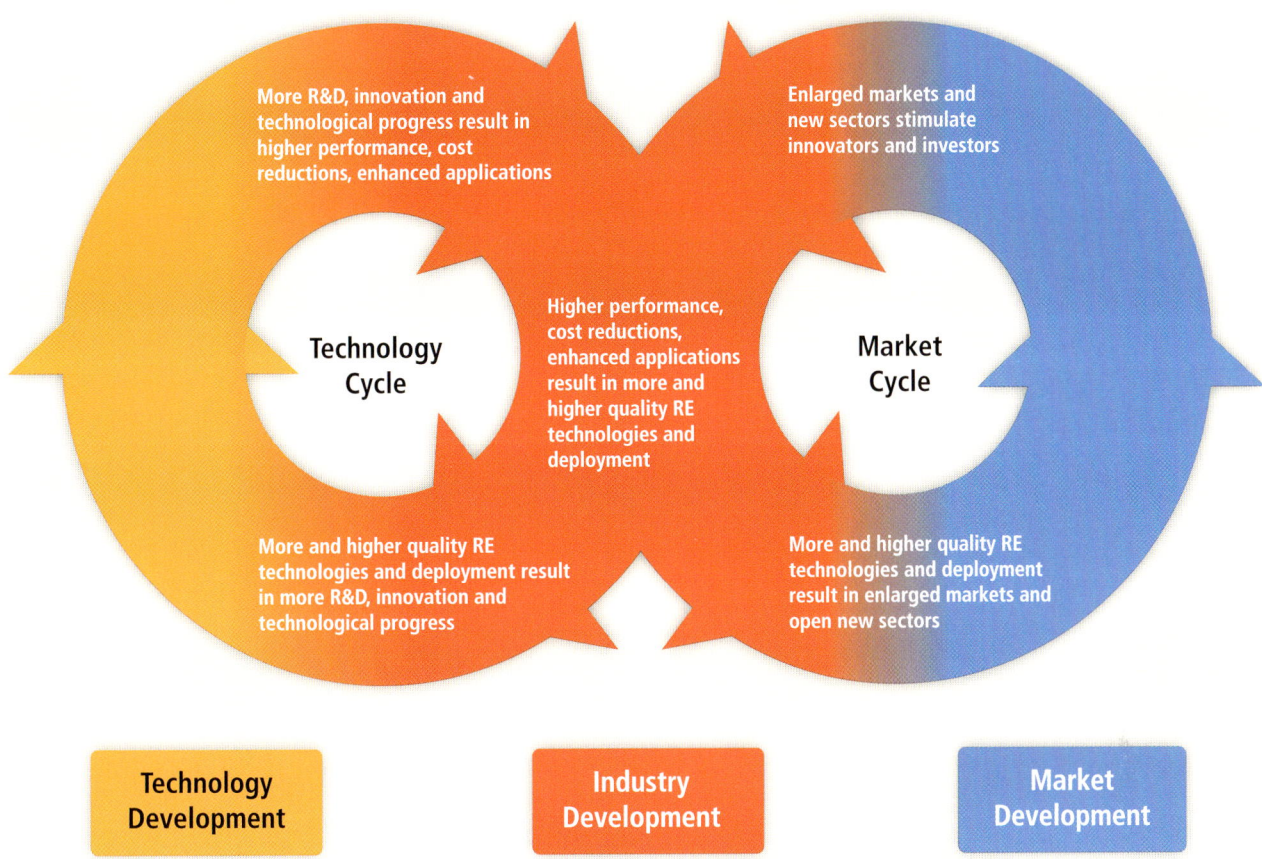

Figure 11.5 | The mutually reinforcing cycles of technology development and market deployment drive down technology costs (Based on IEA, 2003b).

11.5.3 Policy options for renewable energy deployment

This section provides an introduction to the RE-specific policy options for RE deployment—or demand-pull policies—that create demand for RE technologies in the marketplace, as set out in Table 11.2. This section focuses primarily on fiscal incentives and public financing tools, which apply generally to all end-use sectors, though a brief discussion is also provided on regulatory policies. To the extent possible, analysis of these policy options is provided relative to the assessment criteria set out early in Section 11.5, with a focus primarily on effectiveness and efficiency. Most discussion surrounding regulatory policies along with sector-specific experiences and analysis can be found in the end-use sector-specific Sections 11.5.4 (electricity), 11.5.5 (heat) and 11.5.6 (transportation).

11.5.3.1 Fiscal incentives

Financial incentives of various forms—based on investment or production, and including tax credits, reductions and exemptions; accelerated or variable depreciation of investment expenditure; and rebates and grants (all set out in Table 11.2)—can reduce the costs and risks of investing in RE by lowering the upfront investment costs associated with installation, reducing the cost of production or increasing the payment received for energy generated with renewable sources. Fiscal incentives compensate for the various market failures that leave RE at a competitive disadvantage compared to energy market prices (Section 1.4.2), and help to reduce the financial burden of investing in RE. Setting the correct level of incentive requires care to ensure expansion without an excessive public burden (IEA, 2007a).

Grants and rebates

Some countries, like Japan and several US states, have promoted RE deployment by subsidizing investment through grants or rebates (Sawin, 2004). Grants consist of money provided up front to help finance an investment, whereas rebates are refunds provided after an investment has been made.

Capital grants and rebates assist directly with reducing the upfront investment cost of a plant, with a government typically providing a certain level of financial support, for example a refund per megawatt of installed capacity or a percentage of total investment, up to a specified limit. They can apply from the small scale, for example, a domestic solar thermal or PV system, through to large-scale generating stations such as biomass combined heat and power (CHP) plants.

Policy, Financing and Implementation — Chapter 11

Table 11.2 | Definitions of existing RE-specific deployment policies and their use by sector.

Electricity (E), Heating (H) and Transport (T)

Policy	Definition	E	H	T
DEPLOYMENT POLICIES				
FISCAL INCENTIVES				
Grant	Monetary assistance that does not have to be repaid and that is bestowed by a government for specified purposes to an eligible recipient. Usually conditional upon certain qualifications as to the use, maintenance of specified standards, or a proportional contribution by the grantee or other grantor(s). Grants (and rebates) help reduce system investment costs associated with preparation, purchase or construction of RE equipment or related infrastructure. In some cases grants are used to create concessional financing instruments (e.g., allowing banks to offer low-interest loans for RE systems).	X	X	X
Energy production payment	Direct payment from the government per unit of renewable energy produced.	X	X	X
Rebate	One-time direct payment from the government to a private party to cover a percentage or specified amount of the investment cost of a RE system or service. Typically offered automatically to eligible projects after completion, not requiring detailed application procedures.	X	X	X
Tax credit (production or investment)	Provides the investor or owner of qualifying property with an annual income tax credit based on the amount of money invested in that facility or the amount of energy that it generates during the relevant year. Allows investments in RE to be fully or partially deducted from tax obligations or income.	X	X	X
Tax reduction/exemption	Reduction in tax—including but not limited to sales, value-added, energy or carbon tax—applicable to the purchase (or production) of renewable energy or RE technologies.	X	X	X
Variable or accelerated depreciation	Allows for reduction in income tax burden in first years of operation of RE equipment. Generally applies to commercial entities.	X	X	X
PUBLIC FINANCE				
Investment	Financing provided in return for an equity ownership interest in a RE company or project. Usually delivered as a government-managed fund that directly invests equity in projects and companies, or as a funder of privately managed funds (*fund of funds*).	X	X	X
Guarantee	Risk-sharing mechanism aimed at mobilizing domestic lending from commercial banks for RE companies and projects that have high perceived credit (i.e., repayment) risk. Typically a guarantee is partial, that is, it covers a portion of the outstanding loan principal with 50 to 80% being common.	X	X	X
Loan	Financing provided to a RE company or project in return for a debt (i.e., repayment) obligation. Provided by government, development bank or investment authority usually on concessional terms (e.g., lower interest rates or with lower security requirements).	X	X	X
Public procurement	Public entities preferentially purchase RE services (such as electricity) and/or RE equipment.	X	X	X
REGULATIONS				
Quantity-Driven				
Renewable Portfolio Standard/ Quota obligation or mandate	Obligates designated parties (generators, suppliers, consumers) to meet minimum (often gradually increasing) RE targets, generally expressed as percentages of total supplies or as an amount of RE capacity, with costs borne by consumers. Building codes or obligations requiring installation of RE heat or power technologies, often combined with efficiency investments. RE heating purchase mandates. Mandates for blending biofuels into total transportation fuel in percent or specific quantity.	X	X	X
Tendering/ Bidding	Public authorities organize tenders for given quota of RE supplies or supply capacities, and remunerate winning bids at prices mostly above standard market levels.	X		
Price-Driven				
Fixed payment feed-in tariff (FIT)	Guarantees RE supplies with priority access and dispatch, and sets a fixed price varying by technology per unit delivered during a specified number of years.	X	X	
Premium payment FIT	Guarantees RE supplies an additional payment on top of their energy market price or end-use value.	X	X	
Quality-Driven				
Green energy purchasing	Regulates the supply of voluntary RE purchases by consumers, beyond existing RE obligations.	X	X	X
Green labelling	Government-sponsored labelling (there are also some private sector labels) that guarantees that energy products meet certain sustainability criteria to facilitate voluntary green energy purchasing. Some governments require labelling on consumer bills, with full disclosure of the energy mix (or share of RE).	X	X	X
Access				
Net metering (also net billing)	Allows a two-way flow of electricity between the electricity distribution grid and customers with their own generation. The meter flows backwards when power is fed into the grid, with power compensated at the retail rate during the 'netting' cycle regardless of whether instantaneous customer generation exceeds customer demand.	X		

Continued next Page →

Electricity (E), Heating (H) and Transport (T)				
Policy	Definition	E	H	T
Priority or guaranteed access to network	Provides RE supplies with unhindered access to established energy networks.	X	X	
Priority dispatch	Mandates that RE supplies are integrated into energy systems before supplies from other sources.	X	X	

Notes: Assumes that transport is only liquid fuel-based and heat is only non-electric; electric-based transport or heat are covered under the electricity category.

Grants and rebates do not require a long-term policy and financial commitment to each specific project, and they can play a significant role in increasing deployment of small, customer-sited projects particularly for emerging renewable technologies (Wiser and Pickle, 1997). However, they have often failed to provide the stable conditions required to promote market growth and thus may not be effective at driving broad adoption of RE (Lantz and Doris, 2009). This is in part because they can be vulnerable to fluctuations in budgets to the detriment of stable demand growth, as with the German Market Incentive Program (Nast et al., 2007) and the UK's Low Carbon Building Programme (BERR, 2008).

Rebate programs function well when the rebate amount is tailored to existing market and policy conditions, when they are matched with a clear set of goals, and when used to advance technologies from the prototype stage to mass production (Lantz and Doris, 2009). Automatic rebates for eligible projects may be especially valuable for smaller-scale RE facilities that face investment cost barriers and where competitively awarded grants or other policy approaches may be less suitable due to the transaction costs of incentive administration.

Capital grants have both advantages and disadvantages (DEFRA/BERR, 2007; Connor et al., 2009). From the point of view of the recipient, they are very low risk, in the sense that payment is not subject to the vicissitudes of future policy. From the point of view of the payer, the value of the grant is known and does not create, at least in principle, any future liabilities. But while a grant may help get a facility built, without post-installation follow up it does not ensure that a project will operate or operate efficiently. Grants generally require oversight to ensure that certain preconditions are met, that the quality of new generating capacity meets at least a minimum standard, and that effective operation of installed systems is achieved. This implies additional administrative costs (DEFRA/BERR, 2007; Connor et al., 2009).

If the project fails, either under construction or subsequently, the grantor generally has little recourse. Grants are therefore most attractive for facilities that have significant investment costs, but relatively low operating costs. There is an argument that they are best suited to less mature technologies. Grants provide a straightforward way to stimulate investment and, potentially, to draw new investors. Grantors can increase the efficiency of grants through competitive awards, though this can increase administration costs and may be more effective for larger-scale developments due to the relevance of experience in preparing bids (van Dijk et al., 2003; Bürger et al., 2008; Connor et al., 2009).

The volume of funding and the continuous availability of grants or rebates can significantly influence their effectiveness in driving RE deployment. For example, there is some evidence that if funding runs out early in a program, consumers might delay an investment that they would have made without the grant, thus potentially reducing investment and the economic efficiency of applying public funds. Early exhaustion of funds also tends to indicate that the grant or rebate levels may have been set too high, since it implies that some projects not at the margin have received funding (van Dijk et al., 2003; Bürger et al., 2008; Connor et al., 2009).

Tax Policies

Tax credits, reductions or exemptions amount to tax-deductible sums that involve foregone government revenue and that are calculated as predefined fixed amounts or a percentage of total investment in an installation or on the basis of energy delivered. In theory at least, tax incentives are flexible tools that can be gradually increased or decreased as technologies and supply chains develop and as markets evolve. They can be targeted to specific technologies and/or selected markets, or applied more broadly (de Jager et al., 2010).

Tax policies can influence supply and demand sides. For example, production tax credits encourage an increase in production, whereas tax credits or exemptions for the use of RE electricity, heat or fuels affect the demand side. Investment tax credits focus on initial investment costs, whereas production tax credits address operating production costs. Tax reductions and exemptions may also cover property, sales, energy, carbon and value-added tax and act directly on the total payable tax, thereby reducing its magnitude and thus the total cost associated with development (Connor et al., 2009).

A study for the IEA Renewable Energy Technology Deployment implementing agreement determined that the effectiveness of fiscal incentives such as tax reductions or exemptions (e.g., from energy, carbon or other taxes) depends on the applicable tax rate (de Jager and Rathmann, 2008). In the Nordic countries, which apply relatively high energy tax rates, such tax exemptions can be sufficient to stimulate the use of renewable electricity; however, in countries with relatively low energy tax rates, they must be combined with other measures (European Commission, 2005). The current US federal investment and production tax credits (which provide a credit against income tax for each kWh or MJ of electricity produced) have created strong growth in the nation's wind and solar markets, but only when the credits have been in place for multiple years, allowing enough time from project planning through to completion (Sawin, 2004; Wiser et al., 2007).

Accelerated or variable depreciation that can be used as a means of reducing taxable income in the early years of an investment and therefore improving the economics of that investment, has been successful in encouraging small-scale wind development in Sweden and Denmark, in particular. In Denmark, this policy contributed to a significant increase in farmer-owned wind turbines during the mid-1990s (Buen, 2005; Barry and Chapman, 2009). Accelerated depreciation has also been extensively used in the USA for most RE technologies and in India for wind energy. Policies such as the Netherlands Willekeurige Afschrijving Milieu-Investeringen (VAMIL) programme, Canada's Accelerated Capital Cost Allowance and the UK's Enhanced Capital Allowance Scheme are examples of programmes that have been successful in the RE heating sector (Worrell and Graus, 2005; IEA, 2007b).

Assessment of fiscal incentives

The impacts of production and investment support instruments like investment grants, rebates and tax policies are difficult to measure as they are generally used as supplementary policy tools (European Commission, 2005; Klein et al., 2008a). In the EU, for example, only Finland and Malta used tax incentives and investment grants as their main support schemes as of 2008 (Klein et al., 2008a). Fiscal incentives have also been used as the primary means of support at the national level in the USA, although most US states have additional RE incentives or mandates in place (DSIRE, 2011).

Despite the difficulties in measuring their impact, some studies have found that financial incentives tend to be most effective when combined with other policy mechanisms (IEA, 2008a). Japan's solar roofs program of the 1990s and early 2000s combined rebates that declined over time with net metering, low interest loans and public education. This expanded capacity, which helped to drive down system costs, made Japan the world's leading manufacturer of solar PV, at least temporarily (Watanabe et al., 2000) (see Box 11.2).

In general, those countries that have relied heavily on tax-based incentives have often struggled with unstable or insufficient markets for wind power or biogas, for example (Lewis and Wiser, 2005). In the USA, this is due in part to the frequent expiration of the available tax credits, as seen in Box 11.5. It could also result from the fact that only a small number of players have enough tax liability to take direct advantage of the tax credits, meaning the value of the credit varies according to legal standing, income level or tax rate (Metcalf, 2008). This challenge can be addressed by making tax policies more inclusive or finding other policies that encourage broader participation (Mendonça et al., 2009). Generally, tax credits work best in countries where there are numerous profitable, tax-paying private sector firms that are in a position to take advantage of them.

Experience with wind energy policies suggests that cash payments may be preferable to tax credits because the benefits of payments and rebates are equal for people of all income levels and thus promote broader investment and use. Also, because they are generally provided at or near the time of purchase or production, they result in more even growth over time (rather than the tendency to invest in most capacity toward the end of a tax period) (Sawin, 2001). According to a 2009 UN Environment Programme report, the global economic slow-down of 2008-2009 made clear that markets driven by tax credits are generally not effective in a downturn (UNEP and NEF, 2009). Responding to the inability of investors to take advantage of federal tax credits during the economic crisis, the US government temporarily offered cash grants in their stead (Wiser and Bolinger, 2010) (see Box 11.5).

Incentives that subsidize production are generally preferable to investment subsidies because they promote the desired outcome—energy generation (Sawin, 2001); they encourage market deployment while also promoting increases in efficiency (Neuhoff, 2004). However, policies must be tailored to particular technologies and stages of maturation, and investment subsidies can be helpful when a technology is still relatively expensive or when the technology is applied on a small scale (e.g., small rooftop solar systems), particularly if they are paired with technology standards and certification to ensure a minimum quality of systems and installation (Sawin, 2001). Many have argued, for example, that wind power never would have taken off in California in the 1980s without investment credits because the risks and investment costs were high. Alternatively, production incentives can be paired with other policies that help to reduce the cost of financing (Sawin, 2001).

11.5.3.2 Public finance

The provision of public finance can also be of great importance for supporting RE uptake. RE projects generally operate with the same financing structures that apply to conventional fossil-fuelled energy projects. The main forms of capital involved include equity investment from the owners of the project, loans from banks, insurance to cover some of the risks, and possibly other forms of financing, depending on the specific project needs.

For many RE projects the availability of commercial financing is still limited, particularly in developing countries, where the elevated risks and weaker institutional capacities frequently inhibit private sector engagement. Often the gaps can be filled only with financial products created through the help of public finance mechanisms, which help commercial financiers act within a national policy framework, filling gaps and sharing risks where the private sector is initially unwilling or unable to act on its own (UNEP, 2008).

Public finance mechanisms have a twofold objective: to directly mobilize or leverage commercial investment into RE projects, and to indirectly create scaled up and commercially sustainable markets for these technologies. It is important to design policies such that their direct short-term benefits do not create market distortions that indirectly hinder the growth of sustainable, long-term markets (UNEP and BNEF, 2010).

Investments

Public finance mechanisms can take the form of government funds set up to invest equity in private transactions, termed private equity. A public institution's role in the operation of private equity funds can be either as the fund manager, directly investing in projects or companies, or as a fund of funds, whereby they pool their monies alongside other investors in a private sector managed fund. Either way, the funds can be structured to provide a range of financial products, from venture capital for new technology developments, to early stage equity for project development activities, to late stage equity for projects that are already fully permitted and ready for construction (UNEP, 2008).

Guarantees

Guarantees can mobilize domestic lending by sharing credit risk, thereby reducing what local banks might perceive as a high credit risk (i.e., repayment risk) associated with some RE projects. Guarantees help banks to gain experience managing portfolios of RE loans, putting them in a better position to evaluate true project risks and thus addressing perceptions of elevated risk associated with RE projects (UNEP, 2008), as discussed in Section 11.4.3.

Loans

Loans (debt financing) account for the bulk of the financing needed for RE projects (London School of Economics, 2009). The challenges for mobilizing this debt relate to access and risk. As mentioned in Section 11.4.3, the financial sectors in many countries are not developed sufficiently to provide long-term debt required for RE and related infrastructure projects. Public finance mechanisms can be used to provide financing directly to projects or as credit lines that deliver financing through locally based commercial financial institutions.

Credit lines are generally preferable because they help build local capacity for RE financing (UNEP, 2008). For example, credit lines from the World Bank, Kreditanstalt für Wiederaufbau (KfW, Reconstruction Credit Institute) and the Asian Development Bank (ADB) helped the Indian Renewable Energy Development Agency (IREDA) become an important lender to India's RE sector, and key to its success. Incorporated in 1987, IREDA invests mainly as a senior lender and provides debt financing that covers up to 80% of project investment costs for terms up to 10 years. About one-third of its capital is now raised domestically, through bank borrowing and the issuance of tax-free bonds. IREDA is now working with state governments in India to replicate its capability through state energy conservation funds (UNEP, 2008).

Public loans are usually offered at concessional rates, or 'softened', and are relatively easy to administer (IEA, 2007b). Soft loans have long been a feature of German efforts in support of RE technologies; Norway and Spain also have loan programs relating to RE heat, and Japan and Sweden have employed soft loans for RE in the past (IEA, 2007b). Alternatively, approaches such as subordinated loans, which take a higher risk position in the financial structure (i.e., they get paid out only after the senior lenders are paid), can leverage higher levels of commercial financing (London School of Economics, 2009).

Public funds can also be used to buy down the interest rate, while a commercial finance institution provides the bulk of the financing. This reduces the interest rate seen by borrowers, effectively reducing the cost of financing. This approach has been applied successfully in India for domestic solar thermal and solar PV systems, in Tunisia for solar thermal and in Germany for a range of RE technologies (UNEP, 2008).

Other innovative lending mechanisms are arising at various levels of government, including the municipal level. For example, Property Assessed Clean Energy (PACE), which first emerged in the USA, has the potential to provide access to affordable financing while also helping to overcome the market failure of split incentives (see Section 1.4.2 and Box 11.3) With such mechanisms even small investors, such as home owners, are able to repay loans over the lifetime of their systems, with repayment essentially matched by energy savings (Fuller et al., 2009a).

Public procurement

Public procurement of RE technologies and energy supplies is a frequently cited but not often utilized mechanism to stimulate the market for RE. Governments can support RE development by making commitments to purchase RE for their own facilities or encouraging clean energy options for consumers. The potential of this mechanism is significant: in many nations, state and national energy purchases are the largest components of public expenditures, and also in many nations the state is the largest consumer of energy (IEA, 2009c).

Assessment of public finance

Public finance is most commonly employed today in developing countries where the commercial financial sector is usually less mature and therefore unable to provide RE companies and projects with the many types of financing they require (UNEP, 2008). In the developing world, development agencies and financing institutions partner with governments and the private sector to develop frameworks conducive for RE investments; they demonstrate innovative technologies, provide soft loans for sector investment plans and pave the road for market introduction. And they promote technology deployment by means of international carbon finance, in part by stimulating the use of the Kyoto Protocol's Clean Development Mechanism (CDM). Their work builds institutional capacities and is important for reducing financial and investment risk.

Development agencies and financing institutions include multilateral development banks, such as the World Bank and international development banks, and bilateral development banks that are supervised by individual developed countries. These two groups have been major drivers of RE deployment in some developing countries (SEI, 2009). International development finance institutions frequently work closely with national development banks in developing countries. Government development agencies and international environment programmes have also played an important role in disseminating best practices, supporting strategy and policy development, setting up training programmes for decision makers and strengthening institutions like Designated National Authorities under the CDM (UNEP, 2008).

> **Box 11.3 | Innovative financing: Berkeley Sustainable Energy Financing District.**
>
> In 2007, the US city of Berkeley, California, established a Sustainable Energy Financing District (or Property Assessed Clean Energy, PACE) for which it issued bonds and used the proceeds to provide loans to property owners for energy efficiency improvements and/or the installation of solar PV systems. The loans to property owners typically have 20-year terms, allowing repayment to be matched with energy savings; thus, costs are not front-loaded but paid for during the period of use, and purchase decisions do not depend on the need for a quick payback. In existing and proposed programmes, the structure has allowed for locally appropriate and cost-effective technology choices (Fuller et al., 2009a). The city bears the credit risk of the loans but collects loan payments on the property tax bill. The tax assessment belongs to the property, rather than the individual end user, even when the property is sold, protecting the purchaser of the RE system from loss if they sell their home before their investment has been paid back in the form of energy savings.
>
> Several other U S cities and counties have implemented PACE districts and more than 20 US states have enacted enabling legislation to launch PACE programmes; efforts are also underway in Germany, Italy and Portugal (Fuller et al., 2009b).
>
> By late 2010, PACE programs across the USA were on hold, however, due to the severe US recession, which produced a record number of property foreclosures. As a result, the Treasury Department ruled that any policy increasing the debt burden was to be avoided, at least temporarily, and it was required that all PACE loans be paid off in full before the sale or refinancing of properties. Aside from the current US situation, PACE programmes are considered a positive force when economies are stable or growing (Kammen, 2009).

Coordination of public finance mechanisms is increasingly important as the number of funding initiatives increase and because there is a multitude of decentralized activities. The Paris Declaration and Accra Agenda for Action (OECD, 2008) have both formalized and helped to implement principles to improve the effectiveness of international development cooperation, leading to better coordination of international development cooperation in the climate change field, among others. However, financing RE projects and developing national frameworks through international donor coordination and alignment remain challenges. Decentralized and centralized models (e.g., Reed et al., 2009; Müller, 2010) are thus under discussion at the level of international climate negotiations in order to make best use of the capacity and experience of existing development and financing institutions in full alignment with newly created institutions.

A subject of growing research interest is the leveraging of private international investment flows by means of public funding that is delivered via development finance institutions (UNEP, 2008). Results and leverage factors are specific to the technologies, country conditions and the instruments applied (UNEP, 2008).

11.5.3.3 Regulations

As set out in Table 11.2, regulatory policies include quantity- and price-driven policies including quotas and feed-in tariffs, quality aspects and incentives, and access instruments such as net metering. Below are short descriptions of each policy type. Details are provided here only for quality incentives, which are not discussed in Sections 11.5.4 through 11.5.6.

Quantity- and price-driven policies

Quantity-driven policies set the quantity to be achieved and allow the market to determine the price, whereas price-driven policies set the price and allow the market to determine quantity. Quantity-driven policies can be used in all three end-use sectors in the form of obligations or mandates. The best examples of price-driven policies to date are feed-in tariffs (FITs). Sections 11.5.4, 11.5.5 and 11.5.6 discuss these options in detail.

Quality incentives

Quality incentives include green energy purchasing and green labelling programs (occasionally mandated by governments, but not always), which provide information to consumers about the quality of energy products to enable consumers to make voluntary decisions and drive demand for RE.

In the USA, some states have required utilities to provide consumers with green energy options (in many places such options are also voluntary on the part of utilities), which enable consumers and institutions to procure RE for a portion or all of their energy needs. To date, most such programs have been in the electricity sector. Green energy can typically be purchased from utilities, retail suppliers in markets with retail competition, or in the form of RE certificates (RECs) that are sold separate from electricity (or heat/fuels). Retail premiums for green power products vary, but have generally declined in recent years (Bird and Sumner, 2010).

While voluntary commitments to purchase RE can help provide support for and awareness of the importance of RE, they may not be as effective as direct financial incentives or regulatory policies in driving new RE development because they rely on voluntary, often short-term commitments by purchasing entities (Gillenwater, 2008). However, voluntary markets may provide additional revenue streams and alternative markets for output that reduce risks for developers (Bird and Lokey, 2007). The impact on new development also depends on whether or not purchases are additional to regulatory requirements, such as quota obligations.

Green labelling of products is another example of quality incentives or regulations. For instance, the EU Guarantee of Origin (GO) is an electronic document with the sole function of providing proof to a final customer that a given share or quantity of energy was produced from renewable sources. GOs are used for green electricity products and quality labels, as these are systems based on voluntary participation. However, because these labels and products are based on demand for RE over and above that already being generated, they are likely to require implementation of a fully consistent and transparent system that can be audited to demonstrate additionality (Vrolijk et al., 2004).

Access policies

RE projects need to connect to networks in order to sell their electricity, heat, or fuels for heating, cooking and transportation. The ease and cost of doing this is also central to the ability of project developers to raise finance. Once connected, the generation has to be sold or 'taken' by the network. Connection and then sale of generation are two different requirements and it is important to overcome barriers to both. Access to markets—both physical connection and sale of energy or fuels produced—is provided via different policy mechanisms in each of the end-use sectors (i.e., access rules for electricity (Section 11.5.4), third party access (TPA) for heating (Section 11.5.5), blending mandates for biofuels (Section 11.5.6)).

11.5.4 Policies for deployment – electricity

To date, far more policies have been enacted to promote RE for electricity generation than for heating and cooling or for transport, and this is reflected in the vast literature available regarding RE electricity policies. It is important to note, however, that much of the literature describing and comparing these instruments, including their costs, is European, and grey, stimulated largely by the need of EU countries to fulfil their RE Directive requirements by 2020 (e.g., Haas et al., 2011).

After a short discussion of fiscal incentives and public finance, this section describes quantity-driven regulatory instruments, including quota obligations and tendering/bidding regulations, as well as price-driven regulatory policies. It then assesses these regulatory options relative to the criteria set out at the beginning of Section 11.5, particularly effectiveness and efficiency. The section concludes with a brief discussion of access policies.

11.5.4.1 Fiscal incentives

The range of fiscal incentives set out in Table 11.2 has been used to promote RE in the electricity sector. Assessment of policy options and impacts is found in Section 11.5.3.1.

11.5.4.2 Public finance

Loans and other public finance policies have been used to advance deployment of RE electricity technologies, for PV in Spain, for example (see Box 11.8), and innovative financing in many municipalities, as described in Section 11.5.3. Concessional loans, guarantees and even equity investments have been used frequently in other contexts as well, including in developing countries. Government procurement is also an option that is of increasing significance in some countries, including the USA. For example, the US Energy Policy Act of 2005 requires federal agencies to obtain 7.5% of their electricity needs from renewable sources by 2013 and thereafter (US DOE, 2008b). In addition, many US state and local governments have made voluntary commitments to purchase renewable electricity for government facilities (USEPA, 2010a).

11.5.4.3 Regulations

Quantity-based policy

Quota obligations. Quota obligations are also known as Renewable Portfolio Standards (RPS) (among others) in the USA, Renewable Electricity Standards (RES) in India, Renewables Obligations (RO) in the UK, and Renewable Energy Targets in Australia (Lewis and Wiser, 2005). By early 2010, quotas were in place in 56 states, provinces or countries, including more than half of the US states (REN21, 2009b).

Under quota systems, governments typically mandate a minimum amount or share of capacity, generation or sales to come from renewable sources. Quotas tend to be placed on a purchasing authority, with any additional costs of RE generally borne by electricity consumers. There are significant variations of design from one scheme to the next (e.g., Verbruggen, 2009; Bergek and Jacobsson, 2010), even among various state-level policies in the USA (Wiser et al., 2007) and India (MNRE, 2010).

Quotas can be linked to certificate trading, for example 'tradable green certificates' (TGCs) in Europe, or 'renewable energy credits/certificates' (RECs) in the USA (Sawin, 2004; C. Mitchell et al., 2006; Ford et al., 2007; Fouquet and Johansson, 2008). Generally, certificates are awarded to producers for the renewable electricity they generate, and add flexibility by enabling actors with quota obligations to trade, sell or buy credits to meet their obligations—provided there is sufficient liquidity in the marketplace (Sawin, 2004). Electricity suppliers, or other agents in the power sector, 'prove' they have met their obligations by showing the regulator (or other executive body) the number of certificates equal to their obligation. Most quotas have built-in penalties for actors who do not comply with the quota (C. Mitchell, 2008).

One of the intrinsic effects of uniform RE quotas, for example in Sweden, is that only lowest-cost RE options achieve notable levels of deployment. This is because such policies fail "to trigger immediate deployment, enhancements and cost reduction of (RE) technologies which are currently still more expensive" (Resch et al., 2009). To overcome this drawback, technology-specific support can be introduced either via a banding approach (e.g., UK and Italy) or via 'carve-outs', which are sub-quotas reserved for specific technologies (popular in many U.S. states).

Quota schemes with banding enable less mature/more expensive RE technologies to receive a greater number of certificates per MWh generation (i.e., two ROCs/MWh in the UK rather than one ROC/MWh received for wind generation), which increases the value of the RE to the generator (ASIF, 2009). In a quota with carve-outs, a prescribed part of the overall target can be met by only a particular type, or types, of RE. In practice, this leads to a market separation and narrows the tradable volume within each sub-quota.

Experiences in Sweden (see Box 11.4), the USA (see Box 11.5) and Australia demonstrate that the effectiveness of quota schemes can be high and compliance levels achieved if RE certificates are delivered under well-designed policies with long-term contracts that mute (if not eliminate) price volatility and reduce risk (Lauber, 2004; van der Linden et al., 2005; Agnolucci, 2007; Rickerson et al., 2007; Toke, 2007; Wiser et al., 2007). More than 50% of total US wind power capacity additions between 2001 and 2006 were driven at least in part by state RPS laws (Wiser et al., 2007). As discussed in Box 11.5, the US experience has also shown the benefits of longer-term certainty provided by RPS laws in combination with stable and consistent fiscal incentives to address various barriers to RE deployment.

In some instances—including some US states (Wiser et al., 2007) and the UK—targets under quota schemes have not been achieved. For example, under the UK Renewables Obligation, eligible sources rose from 4.0% of electricity generation in 2005 to 5.4% in 2008, rather than the obligated increase from 5.5 to 9.1%. Between 2005 and 2008, only 59 to 73% of each annual obligation was met, with an annual average of 65% (DUKES, 2009). In the USA, experiences in meeting set-asides (or carve-outs) have also been mixed, with only three of nine states with solar or distributed generation set-aside obligations in 2008 achieving their targets. One reason is caps set on the costs that utilities may bear, which have sometimes been set below the amount required to achieve existing targets. Despite such challenges, state RPS programs resulted in more than 250 MW of new solar capacity through the end of 2009 (Wiser et al., 2010).

Electricity policy in the Canadian province of British Columbia provides evidence that it is possible for a quota system to achieve a very high rate of RE investment if the quota is high enough and backed by credible policy and legal requirements (Jaccard et al., 2011). In 2007, the province implemented a 93% clean energy requirement that is now backed by legislation (GBC, 2010). This step resulted in the cancellation of two proposed coal-fired plants (BC Hydro, 2006-2008) and accelerated RE deployment. As of late 2010, all new electricity investment (2,260 MW) had been in RE capacity (BC Hydro, 2007-2010), acquired at the lowest possible cost because of the confidential, closed-envelope bidding system and the freedom of BC Hydro to pick the lowest bids (Jaccard et al., 2011).

RE tendering or bidding. An alternative to the quota or price-driven mechanisms are bidding schemes, for example, the Non Fossil Fuel Obligation (NFFO) that was in place in the UK from 1990 to 1998 (C. Mitchell, 1995, 2000). Under the NFFO, a generator put in a bid to produce a specific amount of electricity from a particular technology at a certain price. The government accepted the cheapest bids up to a maximum, predetermined level. Generators had five years to install approved projects before forfeiting their contract. An NFFO contract provided generators with a fixed price for a certain number of years and a guaranteed a purchase contract for all generation (rather like a FIT), which could be used as the basis of financing. Problems with the NFFO included intense competition resulting from limited available funds (unlike a FIT), and a lack of penalties for failing to implement a contract, which led to bids at unattainably low prices. As a result, the NFFO did not deliver much deployment (C. Mitchell, 2000).

Bidding procedures for large onshore wind power plants and, later, wind turbines and offshore wind power plants, have also been common in China as one of two key policies driving growth in wind power plant installations since 2003 (the other being regionally differentiated FIT prices; see Yu et al. (2009); Liu and Kokko (2010); and Box 11.11). As in the UK, wind power plant bidding for both on- and offshore wind energy has led to concerns about price competition and the resulting low profitability of plant ownership (Han et al., 2009; Yu et al., 2009; Liao et al., 2010). A large number of wind power plants have come online as a result of the program, however, and bidding has also led to some level of price transparency that has been used in establishing FIT prices (Yu et al., 2009; Wang et al., 2010). More recently, somewhat similar bidding procedures have been extended to solar plants in China, for both PV and concentrating solar power (CSP).

Lessons learned. The most effective and efficient quantity-based mechanisms have included most if not all of the following elements, particularly those that minimize risk (Sawin, 2004; van der Linden et al., 2005; Wiser et al., 2005):

- Application to a large segment of the market (quota only);
- Clearly defined eligibility rules including eligible resources and actors (applies to quotas and tendering/bidding);
- Well-balanced supply-demand conditions with a clear focus on new capacities—quotas should exceed existing supply but be achievable at reasonable cost (quota only);

- Long-term contracts/specific purchase obligations and end dates, and no time gaps between one quota and the next (quota only);
- Adequate penalties for non-compliance, and adequate enforcement (applies to quotas and tendering/bidding);
- Long-term targets, of at least 10 years (quota only);
- Technology-specific bands or carve-outs to provide differentiated support (applies to quotas and tendering/bidding); and
- Minimum payments to enable adequate return and financing (applies to quotas and tendering/bidding).

Box 11.4 | Lessons from Sweden: Success with tradable renewable electricity certificates and bio-energy.

The Swedish quota obligation scheme with tradable renewable electricity certificates (TRECs) went into force in May 2003. Its aim was to increase RE electricity generation 10 TWh (36 PJ) above 2002 levels by 2010. The scheme has subsequently been revised and extended several times, with the growth target raised in 2009 to 25 TWh (90 PJ) above 2002 levels by 2020. Electricity production eligible for TRECs includes all RE except hydropower greater than 1.5 MW and, since 2004, peat used in CHP production. Plants that were commissioned before introduction of the policy are entitled to certificates through 2012, while others can receive TRECs for 15 years, or until the end of 2035, whichever is earlier.

RE electricity is sold at the market electricity price. However, in addition to income from the sale of electricity, RE producers receive income from the sale of TRECs, which are traded separately. Electricity suppliers are obliged to purchase TRECs corresponding to a certain proportion (legislated quota) of the electricity they sell. Only electricity used in manufacturing processes in electricity-intensive industries is excluded from the required quota. Suppliers annually submit the required amount of TRECs to the Swedish Energy Agency, one of the two authorities responsible for the scheme. The other authority, Svenska Kraftnät, is the state-owned company that administers and runs the national electrical grid. In case of non-compliance, a supplier must pay a penalty fee of 150% of the average annual price of TRECs.

The TREC scheme more than doubled eligible RE electricity production over a seven-year period, from 6.5 TWh (24.3 PJ) in 2002 to 14.7 TWh (52.9 PJ) in 2009—or 15.6 TWh (56.2 PJ) in 2009 including peat (Swedish Energy Agency, 2010a). Biomass-based electricity production in CHP plants has experienced steady growth under the scheme, accounting for 63% of the TRECs in 2009. About half of the biomass CHP electricity is produced in district heating systems (see Box 11.9) and the other half in the pulp and paper industry.

Investments in wind power were initially restricted by the short time frame of the scheme, but conditions improved in 2006 after the scheme was extended and a 15-year support period was established. Wind power investments took off after that but have been slowed down by permitting and planning procedures. The permitting procedure for wind power was simplified in 2009, when two parallel processes were replaced by one. At the same time, however, local governments were given the legal right to veto wind power investments in their municipality, something that has become an important obstacle to wind power investments. In 2009, wind power producers received 16% of the TRECs (Swedish Energy Agency, 2010a).

The annual average price of TRECs has varied between USD_{2005} 22 and 41/MWh (approximately USD_{2005} 6.1-11.4/GJ). In 2009, the scheme generated USD_{2005} 573 million in income for RE electricity producers, while it increased the average cost of electricity to consumers by USD_{2005} 6.6/MWh (approximately USD_{2005} 1.83/GJ) (Swedish Energy Agency, 2010a).

Since 2006, the TREC scheme has fulfilled RE electricity targets by providing stable investment conditions. However, the scheme has been criticized for overcompensating biomass CHP, a fairly mature technology, and not driving technology development, requiring additional support for nascent technologies (Bergek and Jacobsson, 2010). So far the price of TRECs has been too low to generate investments in more expensive RE technologies; for example, solar electricity has received a negligible amount of TRECs.

Sweden's experiences with the TREC scheme show that this instrument, if appropriately designed (i.e., long time frame), can provide stable investment conditions and fulfil RE electricity targets. The scheme stimulates investments in the least expensive RE technology, and thus does not drive technology development unless specifically designed to do so. The experience with wind power shows that additional policies addressing non-economic barriers, such as the adoption of clearer permitting procedures, are also important for the diffusion of RE technologies.

Box 11.5 | Lessons from the USA: Mix of stable and consistent policies for wind power development.

In the USA, installed wind energy capacity grew from 2.6 GW in 2000 to more than 40 GW in 2010 (Wiser and Bolinger, 2010; AWEA, 2011). Federal tax incentives, state RPS, other RE incentives and the improving economics of wind drove this development, most of which occurred towards the end of the decade (Menz and Vachon, 2006; Wiser et al., 2007; Adelaja et al., 2010).

From 1999 to 2004, failure to consistently renew the federal production tax credit (PTC), which provides approximately two cents per kilowatt-hour for the production from wind facilities for the first 10 years of operation, created a boom and bust cycle for wind development (Bird et al., 2005). Figure 11.6 shows the impact of allowing the PTC to expire at the end of 1999, 2001, and 2003, as installations peaked before the expiration and fell in subsequent years.

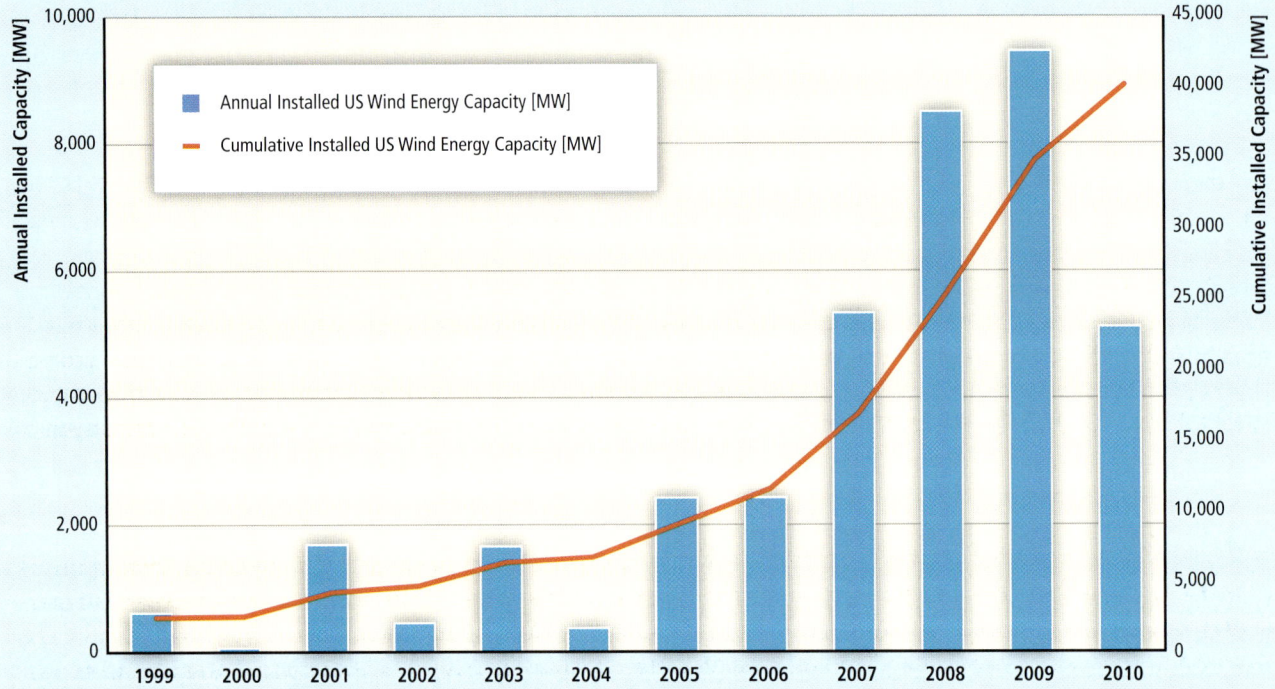

Figure 11.6 | US wind power annual installations and cumulative capacity, 1999 to 2010 (Wiser and Bolinger, 2010; AWEA, 2011).

However, between 2005 and 2009, the rate of annual installations climbed steadily, as federal tax credits were re-authorized before expiring, more states adopted RPS laws and many states strengthened pre-existing RPS targets. As of June 2010, 29 states had adopted an RPS and another 7 had established nonbinding renewable energy goals. Many states require electricity providers to obtain 20% or more of the power needed to serve their loads from RE sources by 2020. Collectively, these state RPS policies call for more than 65 GW of new RE by 2020 (Wiser and Barbose, 2008).

Some states have seen rapid RE growth through these policies, and Texas achieved its 2015 RPS target of 5 GW of installed renewable capacity six years early (ERCOT, 2010). However, the socio-political context and siting barriers have impeded development in other states (Fischlein et al., 2010), demonstrating the need to address barriers, such as siting and transmission, in addition to establishing targets and financial incentives.

Collectively, the combination of policies establishing binding, long-term state RE mandates and federal and state financial incentives, and efforts to address siting and financing barriers, have created greater market certainty and reduced regulatory risk, which in turn have contributed to investments in manufacturing capacity. Companies have also sought local manufacturing to reduce transportation costs and currency risks (Wiser and Bolinger, 2009, 2010). Between 2004 and 2009, US domestic manufacturing of wind turbines and their components increased 12-fold and, as of 2009, 16 turbine manufacturers had opened or announced plans for factories in the USA, up from only 1 turbine manufacturer in 2004 (AWEA, 2010).

Starting in 2008, the federal government provided RE support as part of its effort to help fuel economic recovery. In response to the inability of investors to utilize tax incentives during the recession, the government provided project developers with the short-term option to receive cash grants in lieu of the federal tax credits and extended the tax credits for wind through 2012. This policy, which provided an important response to financial barriers to wind development, contributed to a record number of new wind power installations in 2009 (Wiser and Bolinger, 2010). However, installations slowed considerably in 2010 (AWEA, 2011). The slowdown resulted from a drop in wholesale power market prices driven by lower natural gas prices, and by reduced demand for RE because of a slowing in electricity consumption and the large amount of wind that came online the previous year, putting some states temporarily ahead of their RPS targets (Wiser and Bolinger, 2010).

Overall, the US wind industry experience over the last decade indicates the importance of a mix of stable, consistent and responsive long-term policies that address economic and other barriers to create investor and developer confidence and lead to a robust market and steady growth in manufacturing for renewable energy. State RPS requirements have provided greater market certainty and have influenced the location of development, while federal tax incentives have helped improve the cost-effectiveness of wind and other renewable technologies.

Price-driven policies

Price-driven policies set a price for RE electricity and let the market determine the quantity supplied (except for those systems with capacity caps, such as Spain with PV). They have been called feed-in tariffs (FITs), premium payments, standard offer contracts, minimum price payments, renewable energy payments, and advanced renewable tariffs (Couture and Gagnon, 2009; Couture et al., 2010). Price-driven instruments generally guarantee connection and access to the network, but not always. They have different impacts on investor certainty and payment, ratepayer payments, the speed of deployment, and transparency and complexity of the system, depending on details of their design (Couture, 2009).

The most important distinction is between FITs that set a fixed price that is independent of electricity market prices (e.g., used in Germany (see Box 11.6) and Greece), and those with premium payments (e.g., Denmark, the Netherlands and Thailand (see Box 11.7)), which provide fixed premiums on top of market prices for electricity. The four main approaches used to set FIT payments are levelized costs of RE generation, value of RE generation, simple fixed-price incentives based on neither generation costs nor notion of value, and auction-based mechanisms (Couture et al., 2010).

The fixed-price FIT typically also ensures connection to the network at a pre-agreed price and guarantees the purchase of all generation, sometimes with limited exceptions. These three factors (a set price independent of the electricity price, network connection, and guaranteed purchase) lead to an almost risk-free contract from the point of view of generators (Couture et al., 2010). European FIT policies generally extend eligibility to anyone who is able to invest (Couture et al., 2010). Rules concerning the costs of connection differ amongst different FIT schemes (for example, in Denmark, Germany and Spain these costs are capped) as does whether the generation has guaranteed purchase.

Premium payment systems have gained some ground in recent years. In some countries they are the primary form of support, whereas in others (e.g., Spain and the Czech Republic) they operate in parallel with fixed-price FITs. Premiums can be linked to electricity price developments (e.g., limited by a floor price or cap), or set adders; the former provides higher certainty and less risk of overcompensation. These systems provide a secure additional return for producers but, compared to fixed-price FITs, they provide less certainty for investors because producers are exposed to electricity price risk. This, in turn, implies higher risk premiums and a higher cost of capital. The advantage of premiums is that they encourage producers to adjust generation in response to market price signals (de Jager et al., 2010).

FITs can be very simple and available for one technology only, such as wind power, or they can be quite complex. For example, fixed payments can vary by technology according to state of development and/or generating costs. FITs are suited to incremental adjustments and payments can be increased or decreased as necessary to meet policy goals or to account for technology advances or changes in the marketplace. The costs of FITs can be covered by energy taxes, supplementary means such as auction of carbon allowances or, more frequently, by an additional per-kilowatt hour charge spread across electricity consumers, sometimes with exemptions, for example major electricity users in Germany (BMU, 2010).

To limit FIT-related expenditures and/or provide support where the benefit is greatest, tariffs can be 'stepped' so that payment levels are linked to available resource, location or time of day generated (Mendonça, 2007; Couture and Gagnon, 2009; BMU, 2010; Couture et al., 2010). Most price-driven policies include a regularly scheduled tariff degression (i.e., reduction in the tariff as applied to new eligible RE plants).

It is important to set the right price to avoid overpayment and overstimulation of the market, as well as high costs that might result from supporting significant installation of more expensive RE technologies. To this end, some countries (e.g., Spain) have established caps on annual payments or set limits on capacity that can qualify for payment. The

Box 11.6 | Lessons from Germany: From a single policy to a comprehensive approach.

Germany has devoted significant resources to RE technology development and market deployment since the 1970s, driven by the oil crises and the anti-nuclear movement (Jacobsson and Lauber, 2006). As a result of public R&D efforts, by the mid-1980s many technologies were ready for deployment even though they were not yet cost-competitive (IEA, 2004a). But in the 1980s and beyond, RE faced a largely hostile political-economic structure in Germany. Declining oil prices and surplus electric capacity in the late 1980s made it difficult for RE to compete in the market, while the electricity supply system was dominated by large utilities that opposed all small and decentralized forms of generation as uneconomic and foreign to the system (Jacobsson and Lauber, 2006).

In 1989, the government established a subsidy (€0.031/kWh, USD_{2005} 0.053/kWh or approximately €8.6/GJ, USD_{2005} 14.7/GJ) for the first 100 MW of wind power installed in Germany. Beneficiaries were obliged to report on performance so that a common knowledge base was established. In 1990, Germany's first FIT law was enacted, requiring utilities to connect RE power plants to the grid, purchase the generated power and buy the electricity at a specified percentage of the retail rate: for wind and solar energy, this amounted to 90% of the average tariff for final customers (Lauber and Mez, 2004).

The FIT was revised and broadened into the Renewable Energy Sources Act (Erneuerbare Energien Gesetz – EEG) in 2000, adding geothermal and large biomass power plants and introducing cost-based tariffs that are guaranteed to all RE generators for at least 20 years (Lipp, 2007). The remuneration decreases for new plants at a predetermined annual rate (Langniß et al., 2009). It obligates grid operators and electricity suppliers to purchase RE electricity (Langniß et al., 2009).

The EEG sets a target for 30% of Germany's power to come from RE by 2020 (Büsgen and Dürrschmidt, 2009). It has been amended twice, reflecting progress in technology development and stringent requirements for RE integration (Büsgen and Dürrschmidt, 2009).

As installations increase, particularly for more expensive PV, the extra burden to consumers of financing the EEG has been discussed more widely. The total additional cost from PV support alone, granted through the EEG during 2000 through 2008, was an estimated €$_{2007}$ 35 billion (USD_{2005} 41.6 billion) (Frondel et al., 2010); in 2007, the additional annual cost amounted to €4.3 billion (USD_{2005} 5.12 billion) (Büsgen and Dürrschmidt, 2009). Benefits include avoided CO_2 emissions, saved fossil fuels, employment (Lehr et al., 2008) and merit-order effects (Sensfuß et al., 2008).

Several other policies have been used to promote deployment of RE electricity, to support further R&D and to level the playing field (Laird and Stefes, 2009). Federal banks offered low-interest loans with favourable payment conditions, easing access to capital. Changes to German building codes granted RE the same legal status as other power generation technologies, and municipalities were required to allocate potential sites to wind power facilities in their land development plans (IEA, 2004b).

As a result, Germany has seen rapid growth of electricity generation from RE. Germany's share of electricity from RE rose from 3.1% in 1991 to 7.8% in 2002, and more than doubled again by the end of 2009 to 16.9% (Wüstenhagen and Bilharz, 2006; BMU, 2009). Wind energy has experienced the greatest increase, but bioenergy and solar PV have grown substantially under this policy as well. (Note that wind-generated electricity declined towards the end of this period due to below average annual winds, but installed capacity continued to increase (BWE, 2011).) (See Figure 11.7.)

Since 2000, the focus of Germany's RE promotion policies has broadened to include heat and transport fuel markets. A comprehensive 'market acceleration programme' introduced to award investment grants and soft loans for RE heat systems was supplemented in 2009 with a mandate requiring a minimum share of RE heating/cooling in new buildings. Initially promoted through tax exemptions (Bomb et al., 2007), RE transport fuels are now mandated through a blending quota for fuel suppliers.

The government's overarching frame for RE development has been creation of ambitious targets for the use of RE in individual sectors and for the economy as a whole. The share of RE in total primary energy supply increased steadily from 1.3% in 1990 to 8.9% in 2009[1] (BMU, 2010; BWE, 2011).

Note: 1. Note that the BMU reports data based on statistics that rely on the physical content method for primary energy conversion, whereas this report uses the direct equivalent method.

The German example shows how rapidly RE can advance when supported by ambitious policies that convey clear and consistent signals and that adapt to technical and market changes. RE deployment policies can start with simple incentives, evolving towards stable and predictable policies and frameworks to address the long-term nature of developing and integrating RE into existing energy systems. However, integration of RE remains a constant challenge as indicated by recent limitations of the German electricity network to absorb rising shares of RE, and the cost implications of Germany's program have also begun to attract concern.

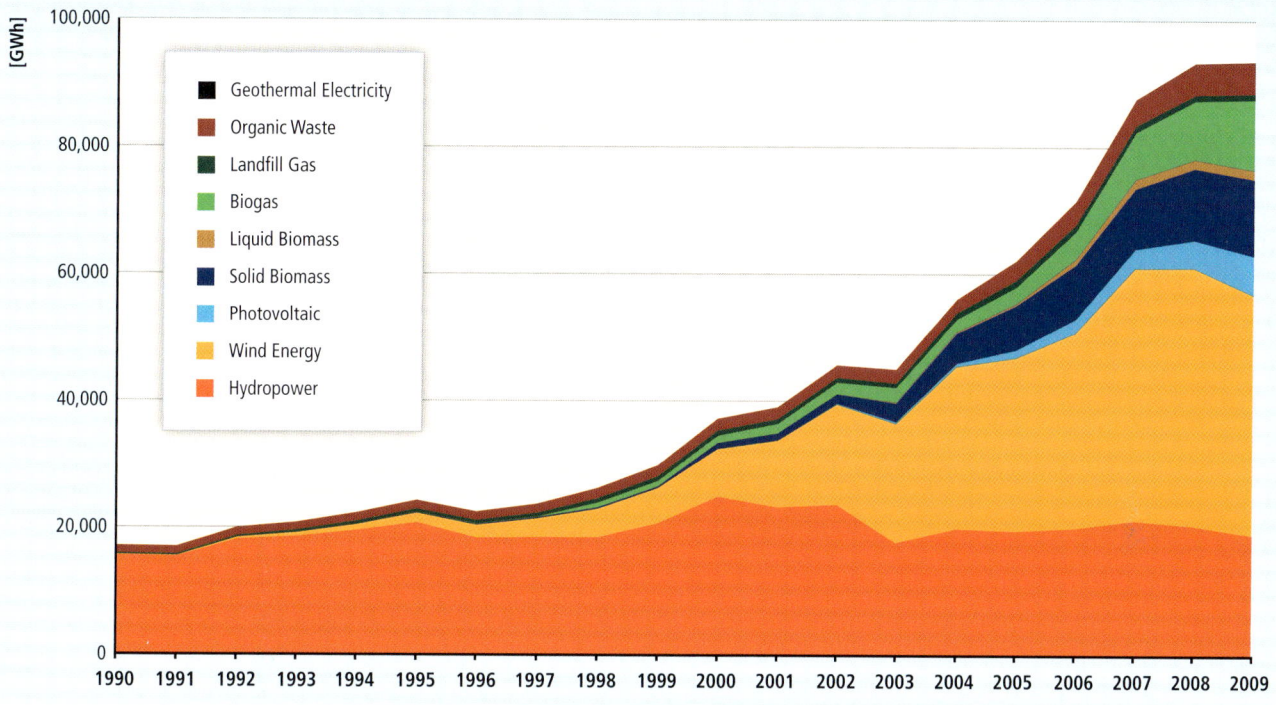

Figure 11.7 | Germany's electricity generation from RE, 1990 to 2009 (BMU, 2010).

downside of caps is that they reduce investment stability and cause frequent stop-and-go in the market. Thus, some countries (e.g., Germany for PV) have established 'growth corridors' with continuous automatic adjustments of tariffs (BMU, 2010). Market growth above the corridor results in a stepped-up tariff degression; if growth is lower than desired, the rate of tariff degression is decreased. The higher the frequency of adjustments (e.g., quarterly instead of annually) and the higher the degression rate in case of overshoot, the greater the control of support cost but the lower the stability for investors. Although this option preserves investment stability to a higher degree than a cap does, it may be less effective in limiting the increase in support expenditures (de Jager et al., 2010).

An advantage of the FIT with a fixed price is the long-term certainty of receiving a fixed payment, which lowers investment risk. Guaranteed network connection and priority access further reduce investor risk because investors are assured a market for the electricity they produce. An advantage of the premium payment is that RE generators participate to a greater degree in the electricity market and, if they have fuel costs, they can be given incentives to produce electricity when the market needs it most.

Although they have not succeeded in every country that has enacted them, price-driven policies have resulted in rapid renewable electric capacity growth and strong domestic industries in several countries—most notably Germany (See Box 11.6) and Spain (See Box 11.8) but more recently in China and other countries as well—and have spread rapidly across Europe and around the world (REN21, 2006, 2009b; Mendonça, 2007; Rickerson et al., 2007; Girardet and Mendonca, 2009). (See Boxes 11.7, 11.11 and 11.12.)

The success of FIT policies depends on the details. The most effective and efficient policies have included most or all of the following elements (Sawin, 2004; Mendonça, 2007; Klein et al., 2008a; Couture, 2009):

- Utility purchase obligation;
- Priority access and dispatch;
- Tariffs based on cost of generation and differentiated by technology type and project size, with carefully calculated starting values;
- Regular long-term design evaluations and short-term payment level adjustments, with incremental adjustments built into law in order to reflect changes in technologies and the marketplace, to encourage innovation and technological change, and to control costs;

Box 11.7 | Lessons from Thailand: Gradual expansion of RE policies.

Decentralized, grid-connected RE has made a substantial and rapidly increasing contribution to Thailand's electricity supply. As of March 2010, 1,364 MW of private sector RE was online and an additional 4,104 MW with signed power purchase agreements (PPAs) were in the pipeline. Biomass makes up the bulk of this capacity with 1,292 MW (online) and 2,119 MW (PPA only). Solar electricity is second but rapidly catching up, with 78 MW online and signed PPAs for an additional 1,759 MW (EPPO, 2010b,c). Strong market growth has been due to plentiful agricultural residues and a comprehensive set of policies including streamlined grid interconnection access, a FIT based on premium payments, tax breaks and low-cost financing (Amranand, 2009; Fox, 2010).

Policies to accommodate grid interconnection of customer-owned RE started in 1992 with the Small Power Producer (SPP) program, which included standardized interconnection and PPAs for generators up to 90 MW (Greacen and Greacen, 2004). By 2007 the program had saturated at 53 RE generators (mostly bagasse cogeneration) with combined nameplate capacity of 967 MW (EPPO, 2007b).

In 2002, Thailand adopted Very Small Power Producer (VSPP) regulations, modelled on US net metering legislation, further streamlining utility interconnection requirements for generators up to 1 MW (Greacen et al., 2003). This and other policies helped to foster the development of integrated biorefineries for sugarcane and rice, enabling simultaneous production of food, ethanol, heat and electric power, and the recovery of some of the fertilizer value. By 2008, for electricity production sold to the grid, there were 42 biomass-based VSPP projects using a variety of biomass residues and 31 biomass-based SPPs, for example, from bagasse and rice husks. The generating capacity of these projects totalled 1,689 MW; about half of this produced power for the grid (Amranand, 2009; Jenvanitpanjakul and Bhandhubanyong, 2009).

In 2006, the Thai government enacted a FIT premium payment that provides an adder paid on top of utility avoided costs, differentiated by technology type and generator size, and guaranteed for 7 to 10 years. Additional per-kilowatt hour subsidies are provided for projects that offset diesel use in remote areas (on mini-grid systems), and utilities are provided further incentives to accommodate VSPPs. Incremental costs are passed through to consumers; however, electricity is subsidized for small consumers (<150 kWh/month or <540 MJ/month) such that they pay less than marginal cost and are not negatively affected by the FIT (Amranand, 2008). In 2010, the additional burden associated with the FIT was USD$_{2005}$[1] 0.001/kWh or approximately USD 2.78/GJ) (ERC, 2010); the Thai government expects that by 2022 the FIT adder will be about double that amount. In response to the FIT adder, RE online capacity increased sharply, from 992 MW in February 2007 to 1,364 MW by March 2010 (EPPO, 2007a, 2010c).

The government's decision to adopt a FIT premium payment was driven by concerns about increasing reliance on imported fossil fuels; difficulty siting new coal and natural gas plants; interest in reducing GHG emissions; encouragement from the Thai RE industry; and a national target of 8% RE by 2011 (Prommin Lertsuriyadej, 2003; Thai Ministry of Energy, 2003; Amranand, 2008). Other important incentives for RE include an eight-year corporate tax holiday; reduction or exemption of import duties; technical assistance; and low-interest loans and government equity financing (Yoohoon, 2009).

Further, the government has worked to address challenges as they have arisen. For example, in response to companies that applied for PPAs only to sell them to developers, the government requires a reimbursable bid bond for projects over 100 kW, and projects must produce power within one year of the scheduled date of commissioning to receive subsidies (Tongsopit, 2010). The variability of RE and small size of individual generators has been difficult to accommodate using traditional planning methods (Greacen, 2007). This was acknowledged and partially addressed in the 2010 revision of the Power Development Plan (EPPO, 2010a).

Thailand's experience demonstrates that well-designed and effectively implemented policies can lead to substantial deployment of RE in developing countries. The FIT adder has been instrumental in increasing RE capacity and encouraging a diversity of RE sources. Explicit financial incentives for Thai utilities to purchase VSPP power helps overcome their reluctance to accommodate interconnection, grid operations and billing challenges that can accompany distributed generation. The sequence of regulation, starting with interconnection policies and later adoption of FITs, has allowed utilities to 'learn by doing' as they ramp up programs to accommodate distributed RE.

Note: 1. The 2010 monetary figure has been deflated to USD$_{2005}$ for the years 2009 to 2005, as the 2010 data was not yet available. Thus, the given number is only an approximation.

- Tariffs for all potential generators, including utilities;
- Tariffs guaranteed for a long enough time period to ensure adequate rate of return;
- Integration of costs into the rate base and shared equally across country or region;
- Clear connection standards and procedures to allocate costs for transmission and distribution;
- Streamlined administrative and application processes; and
- Attention to preferred exempted groups, for example, major users on competitiveness grounds or low-income and other vulnerable customers.

Assessment of quantity- and price-based policies

This section reviews the literature assessing quantity- and price-based policies, with a focus on quotas and FITs. More than 100 countries, states, and provinces, and even some municipalities around the world have had experience with one or both of these mechanisms (REN21, 2010). For several years, particularly in Europe and to a lesser extent in the USA, there has been debate regarding the efficiency and effectiveness of FITs versus quota systems (Rickerson et al., 2007; Commission of the European Communities, 2008; Cory et al., 2009). As a result, there is a wealth of literature assessing these policy options, with most analysis focused on effectiveness and efficiency.

Effectiveness

As defined above, effectiveness is the extent to which intended policy objectives are met, and can include the amount or share of RE generation and/or degrees of technological and/or geographical diversity of installed capacity.

Many US states have successfully achieved their targets with RPS, although others have not due to overly aggressive targets, insufficient enforcement and/or lack of long-term contracting (van der Linden et al., 2005; Wiser et al., 2007). Ragwitz et al. (2009) and Resch et al. (2009), in reviews of European policies, found that countries with FITs were typically more effective at generally moderate support levels, with the exception of France, where rapid wind development was found to be prevented by administrative barriers.

The IEA argues that the key for countries like Germany, Spain and Denmark has been high investment security coupled with low administrative and regulatory barriers (IEA, 2008c). The IPCC's Fourth Assessment Report, in comparing quantity-based mechanisms and FITs, noted that: "In theory, this difference should not exist as bidding prices that are set at the same level as feed-in tariffs should logically give rise to comparable capacities being installed. The discrepancy can be explained by the higher certainty of current feed-in tariff schemes and the stronger incentive effect of guaranteed prices." (Sims et al., 2007). Likewise, Stern (2007) concluded that "feed-in mechanisms achieve larger [RE] deployment at lower cost. Central to this is the assurance of long-term price guarantees [that come with FITs].... Uncertainty discourages investment and increases the cost of capital as the risks associated with the uncertain rewards require greater rewards.". Bürer and Wüstenhagen (2009) found that, because FITs effectively reduce risk, venture capital and private equity investors perceive FITs to be the most effective policy to stimulate investment in RE technologies (Bürer and Wüstenhagen, 2009).

With regard to technological diversity, quantity-based systems have been found to benefit the most mature, least-cost technologies (Espey, 2001; Sawin, 2004; Jacobsson et al., 2009), although quantity-based mechanisms can address this if they distinguish among RE options or are paired with other incentives (de Jager et al., 2010). In Sweden (as seen in Box 11.4), the UK and Flanders, TGC systems have advanced mainly biomass generation and some wind power, but have done little to advance other RE (Jacobsson et al., 2009). In the USA, between 1998 and 2007, 93% of non-hydropower additions under state RPS laws came from wind power, 4% from biomass, with only 2% from solar and 1% from geothermal (Wiser and Barbose, 2008). As a result, a large number of states have created set-asides of various forms to encourage diversity (DSIRE, 2011. FITs have encouraged both technological (Huber et al., 2004) and geographic diversity (Sawin, 2004), and have been found to be more suitable for promoting projects of varying sizes (Mitchell and Connor, 2004; van Alphen et al., 2008).

Efficiency

As noted early in Section 11.5, static efficiency can be measured as cost-effectiveness or a comparison of total support received relative to generation costs, and dynamic efficiency accounts for future technology development that is triggered by a policy.

A number of studies have concluded that FITs have consistently delivered new supply, from a variety of technologies, more effectively and at lower cost than alternative mechanisms, including quotas, although they have not succeeded in every country that has enacted them (Ragwitz et al., 2005; Stern, 2007; de Jager and Rathmann, 2008).

Recent studies (Resch et al., 2009; de Jager et al., 2010) of quota systems in Europe found that Italy, the UK, Poland and Belgium had experienced high producer profits resulting from high investment risks and low growth rates. Other studies have reached similar conclusions (D. Fouquet et al., 2005; New Energy Finance Limited, 2007; Jacobsson et al., 2009; Verbruggen and Lauber, 2009). Such profits primarily benefit incumbent actors and relatively mature, low-cost technologies, and can be costly for consumers (Jacobsson et al., 2009). The exception among European countries using a quota obligation is Sweden, which has experienced a high rate of RE growth coupled with relatively low producer profits. This was because quota systems tend to favour least-cost RE and Sweden has an abundance of biomass (see Box 11.4).

The higher risk under quota systems includes price risk (fluctuating power and certificate prices), volume risk (no purchase guarantee), and balancing risk; all three risks increase the cost of capital (C. Mitchell et al., 2006). While quota and tendering systems theoretically make optimum use of market forces, government tendering systems in particular have often had a stop-and-go nature that has not been conducive to

Box 11.8 | Lessons from Spain: Policy issues for PV deployment.

To provide a predictable and transparent framework to attract private investments, the Spanish government enacted a FIT in 1998 and published indicative 2010 targets for installed capacity in the Plan to Promote Renewable Energies 2000-2010 (MIyE, 1998; IDAE, 2009).

Due to the immaturity of the market, initially the FIT was not enough to develop the PV sector despite Spain's significant solar resource and, in 2001, a combination of investment subsidies and low-interest loans was established. They remained in place until 2005, and total direct subsidies to PVs during the period amounted to USD_{2005} 64.6 million (IDAE, 2009).

The FIT was revised in April 2004 (Ministerio de Economía, 2004) and again in May 2007 (MITyC, 2007). In addition to raising the tariff for PV, both acts increased the maximum capacity of projects that could receive the high tariff (from a maximum of 100 kW to 10 MW starting in May 2007), and made projects of up to 50 MW eligible to receive 25-year fixed price contracts. Cost benefits associated with the economies of scale of larger projects combined with the 2007 policy changes to encourage development of several new ground-mounted projects of 10 MW. Newly installed capacity increased from 21 MW in 2005 to 107 MW in 2006 and 555 MW in 2007 (IDAE, 2008).

In September 2007, 85% of Spain's RE target had been achieved, setting off a one-year deadline for the government to publish new targets and tariffs, and for developers to complete projects under the existing scheme. This period was fine for most RE projects already under development, with relatively long lead times; but PV projects can be developed quite quickly. The one-year notice set off a mad rush to install PV systems before the existing system expired. As a result, 2,575 MW of PV were added in 2008, breaking all past records and making Spain the world leader for PV installations that year (IDAE, 2009; MITyC, 2009).

Because the country's 2010 targets had been exceeded, in September 2008 the government established a new economic regime for future installations (MITyC, 2008). For the first time, a differentiated tariff was adopted for building-integrated PV (BIPV). In addition, annual caps were set for new capacity, with separate caps for ground-mounted (up to 10 MW) and rooftop (under 20 kW; and 20 kW to 2 MW) PV projects. The caps adjust automatically depending on the previous year's installations, while the tariff for ground-mounted projects continues to decrease over time. The new scheme aimed to: provide long-term predictability; better control the cost of the FIT; guarantee profits more appropriate for a regulated market; encourage declining investment costs; increase competitiveness; and encourage distributed generation through BIPV. The policy change resulted in a significant increase in distributed rooftop projects (IDAE, 2010).

At the same time, uncertainty about the design of the new framework scheduled for adoption in late 2008, the reduction in market size due to the cap on ground-mounted systems, and lack of experience with the new administrative procedures led to a significant reduction in new capacity installations (MITyC, 2008) (see Figure 11.8).

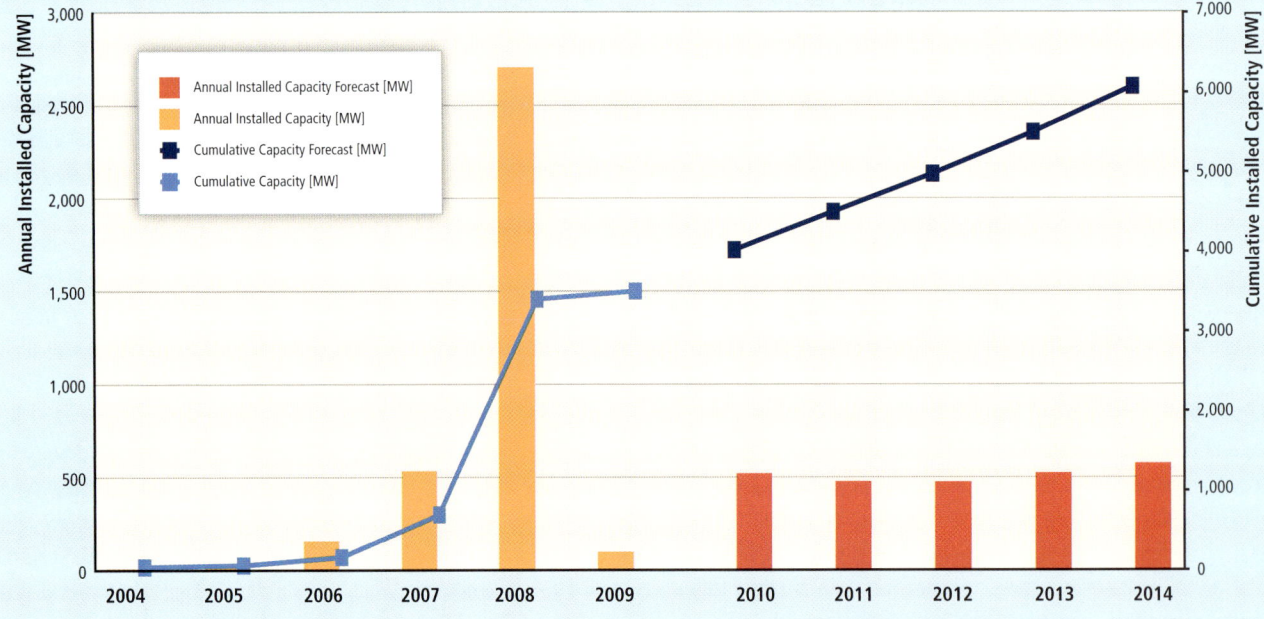

Figure 11.8 | Spanish PV annual installations and cumulative capacity, actual (2004 to 2009) and projected (2010 to 2014) (IDAE, 2010).

Spain's story highlights the importance of learning from experience and of building forward-looking flexibility into policy to avoid the need for frequent regulatory changes. Overall, lessons from Spain's experience include: a combination of support schemes can be important for advancing RE technologies, particularly when the market is immature; ambitious long-term targets are critical as are stable, predictable policies; and transitional incentives that decrease over time internalizing technology development and therefore keeping constant a reasonable internal rate of return for each new project, can foster technological innovation and control total costs.

stable investment conditions. In addition to private investment-related risks, there is also the risk that low-bid projects may not be implemented (European Commission, 2005).

However, experience in the USA demonstrates that the effectiveness and efficiency of quota schemes can be high and compliance levels achieved at reasonable cost and with lower producer profits if RE certificates are delivered under well-designed policies with long-term contracts that mute (if not eliminate) price volatility and reduce risk (Lauber, 2004; van der Linden et al., 2005; Agnolucci, 2007; Rickerson et al., 2007; Toke, 2007; Wiser et al., 2007). Others have concluded that more challenging targets and better enforcement could improve the results of TGC systems (Mitchell and Connor, 2004; C. Mitchell et al., 2006; Fouquet and Johansson, 2008), and that quota systems in many states and countries are still quite new and thus in a transitional phase (Wiser et al., 2007; Commission of the European Communities, 2008).

While Spain has been very successful in terms of deployment, recent experiences there demonstrate that even FITs can bring uncertainty and temporarily high per unit costs with frequent and unpredictable policy adjustments (see Box 11.8) that have increased political risk for all FITs (CITI, 2010) while having a significant short-term impact on the solar industry.

In the USA, there is little evidence of a sizable impact on electricity costs associated with quotas, but cost impacts have varied from state to state and significant REC price fluctuations are possible, impeding development (Wiser et al., 2007). Toke (2007) notes that success of the US RPS in states like Texas, and their ability to achieve targets cost-effectively, is greatly due to the federal production tax credit (Toke, 2007) (see Box 11.5).

With respect to competitiveness, another element of efficiency, a 2008 analysis found that market competition (number of players) was stronger among wind turbine producers and constructors under the German FIT than under the quota scheme used in the UK (Butler and Neuhoff, 2008). Except in the case of Spain, where the premium option attracts mostly incumbent power generators, FITs have been more successful at bringing new players into the market (Verbruggen and Lauber, 2009). FITs encourage competition among manufacturers rather than investors (Held et al., 2007). FITs have been found to encourage development of domestic manufacturing industries, which leads to a large number of companies and thereby creates competition (Sawin, 2004). FITs shift competition from electricity price to equipment price, which some analysts have argued is more appropriate competition for capital-intensive RE technologies (Wagner, 1999; Hvelplund, 2001).

Verbruggen and Lauber (2009) demonstrate that well-designed FITs provide dynamic incentives to reduce long-run marginal costs of a variety of RE technologies because investment money is assigned to investors accordingly; more efficient producers obtain greater rents by lowering costs, and the FIT payment rates are regularly adjusted to avoid excessive rents.

Equity

Concerns about distributional impacts of RE policies on poorer consumers (see Section 11.5.7.2) arise most frequently in countries where FITs have led to significant increases in RE capacity, particularly for relatively high-cost technologies such as PVs, because of resulting increases in total electricity costs. This becomes a greater problem as the total costs of the RE policy increase (Frondel et al., 2010). There are ways to address such impacts, as seen in Thailand where small electricity consumers receive subsidized electricity and are unaffected by the national Premium Payment FIT (see Box 11.7).

Concerns have been raised about electric rate impacts of quota systems as well, especially among sensitive industrial customers in US states with RPS requirements, despite the fact that RPS requirements are typically predicted to have a modest impact on average retail electricity rates. As a result, several state RPS programs have specifically exempted certain industrial loads from the RPS, or have established low caps on the extra costs that may be imposed on these customers (van der Linden et al., 2005). Such exemptions in the USA and Sweden, for example, might also be cause for equity concerns, but have generally been required to gain acceptance of quota regulations (van der Linden et al., 2005).

Another equity-related concern is related to participation. In the USA, for example, publicly owned utilities are sometimes exempt from RPS requirements, leading to equity concerns among other providers (van der Linden et al., 2005). At the same time, detailed analysis of which companies gain from quota systems suggest that it is primarily incumbent actors that continue to benefit from the new market (Girardet and Mendonca, 2009; Jacobsson et al., 2009; Verbruggen and Lauber, 2009). The transaction and administrative costs of a TGC system are higher than with FIT, making participation of small-scale new entrants cumbersome, and therefore limited (C. Mitchell et al., 2006).

In contrast, FITs tend to favour ease of entry, local ownership and control of RE systems (Sawin, 2004; Lipp, 2007; Farrell, 2009), and thus can result in wider public support for RE (Damborg and Krohn, 1998; Sawin, 2001, 2004; Hvelplund, 2006; Mendonça et al., 2009). Such ease of entry has also proved a powerful means for unleashing capital towards the deployment of RE projects (Couture et al., 2010). Mendonça et al. (2009) found that steady, sustainable growth of RE would require policies that ensure diverse ownership structures and broad support for RE, and they propose that local acceptance will become increasingly important as RE technologies continue to grow in both size and number (Mendonça et al., 2009). This is supported by studies in New Zealand and elsewhere (Barry and Chapman, 2009).

Institutional feasibility
FITs generally have lower administrative costs than quota policies (Haas et al., 2011) and are considered easier to implement (van der Linden et al., 2005), though tariff setting can be challenging, particularly if there are very dynamic cost developments (as with PV in recent years). Quotas, particularly those operating with tradable certificates, appear to be more complex because of the need to set both penalty prices and quantities. Transaction costs are also generally higher for such quota systems. Complexities also arise from the need for trading platforms under quotas with tradable certificates, and tendering schemes require administrative capacity to deal with the bidding process (Sawin, 2004; de Jager et al., 2010).

With regard to market compatibility, the policies are quite different. Under a FIT with fixed payment or tariff, a single buyer sells all generated electricity into the power market; with all other systems (including premium payments under FITs), generators must sell into the power markets. Because electricity market prices do not influence the remuneration of generators in fixed-payment FIT systems, there is generally no incentive to produce power according to market demand and/or to react to price signals (de Jager et al., 2010).

In summary, a number of historical studies, including those carried out for the European Commission, have concluded that well-designed and well-implemented FITs are the most efficient (defined as comparison of total support received and generation cost) and effective (ability to deliver increase in the share of RE electricity consumed) support policies for promoting RE electricity (Ragwitz et al, 2005; de Jager et al, 2010; Sawin, 2004; European Commission, 2005; Stern, 2007; Mendonça, 2007; Ernst & Young, 2008; Klein et al., 2008b; Couture and Gagnon, 2009; Held et al, 2010; Ragwitz et al., 2011). It is important to note that there are FITs that have been very effective and efficient and FITs that have not; quotas that have been effective and efficient, and some that have not (Sawin, 2004). Policy design and implementation play an important role in determining how well these policy options measure up against the various criteria, and governments are continuing to adjust details and to learn how these policy options might meet changing needs.

Access instruments
Net Metering. Net metering, or net billing, enables small producers to 'sell' into the grid, at the retail rate, any renewable electricity that they generate in excess of their total demand in real time as long as that excess generation is compensated for by excess customer load at other times during the designated netting period. It is essentially a means for customers to use their own generation to offset consumption (through inter-temporal shifting) over a netting period by allowing their electric meter to spin backwards at times when generation exceeds demand. In general, customers have either two unidirectional meters spinning in opposite directions, or one bi-directional meter that can spin in both directions so that net metering customers pay only for their net electricity draw from the grid over the entire netting period (Klein et al., 2008a). Any net export over a specified period (typically a month or a year) is typically compensated at below the retail rate, if at all (DSIRE, 2011).

Net metering is most commonly used as a policy in the USA, where it has been enacted in most states (DSIRE, 2011), but the mechanism is also used in some countries in Europe and elsewhere around the world (Klein et al., 2008b; REN21, 2010).

Net metering is considered an easily administered tool for motivating customers to invest in small-scale, distributed power and to feed it into the grid, while also benefiting providers by improving load factor if RE electricity is produced during peak demand periods (US DOE, 2008a). It has been introduced in some countries (e.g., Italy) with the aim to decrease the grid load and to limit support expenditures (Ragwitz et al., 2010). According to Rose et al. (2008), the best results are achieved when net metering laws do not limit system size or overall capacity, allow credit for excess electricity (meaning that if generation is greater than use in any particular month, the excess generation is credited to the next month), allow customers to keep their RE credits, permit all renewable technologies and customer classes to participate, and protect customers from unnecessary red tape (Rose et al., 2008). In addition, it is important that net metering policies evolve as markets expand and change (IREC, 2010).

However, Klein et al. (2010) found that, at least in the USA, the remuneration is generally insufficient to stimulate substantial growth of less competitive technologies like PV, since generation costs are significantly higher than retail prices (Klein et al., 2010). Instead, distributed PV has been encouraged in the USA by a combination of federal tax policy, state rebates and performance incentives, state RPS programs and net metering (Sherwood, 2010). Based on impacts seen on small wind systems in the USA, Forsyth et al. (2002) concluded that net metering alone provides only minimal incentives for consumers to invest in RE systems, particularly where people must deal with cumbersome zoning and interconnection issues. However, when combined with public education and/or other financial incentives, net metering might encourage greater participation (Forsyth et al., 2002).

Priority access to network and priority dispatch. In the EU, the Directive 2001/77/EC on the promotion of electricity produced from renewable energy sources states that EU member states must ensure that transmission and distribution system operators 'guarantee grid access for electricity generated by RE' (European Parliament and of the Council, 2009). This is for both connection to the network and off-take (i.e., injection into the grid). As a result of the EU Directive, some European countries, particularly those that have FITs, have implemented connection regulations that guarantee access to the network. 'Priority' grid access in these countries means that electricity generated by RE projects is given priority access to the network and all is taken into the grid.

However, from a power integration point of view, priority access is different from dispatch. Generation may have access to the network, but it does not necessarily mean that it is dispatched; and whether the RE generator receives remuneration for the dispatched or non-dispatched generation will depend on the policy, network or market rules in place. The Spanish FIT does provide for priority dispatch in the event of a constraint, providing security and quality of the supply is guaranteed. Priority access and dispatch are considered in more detail in Section 11.6.5 (see also Section 8.2.1).

11.5.5 Policies for deployment – heating and cooling

In 2008, traditional biomass, modern biomass, solar thermal and geothermal together met 27% of the total global demand for heat (the majority from traditional biomass) (IEA, 2010d), while RE cooling technologies provided a much smaller share of global cooling demand. For modern RE to meet a growing share of total demand, political support will be needed to overcome barriers (e.g., the initial capital barrier to system purchase) to RE heating and cooling (RE H/C).

Support for RE H/C presents policymakers with a unique challenge due to the often distributed nature of heating and cooling technologies. Heating and cooling services can be provided via small- to medium-scale installations that service a single dwelling, or can be used in large-scale applications to provide district heating[4] (DH)/cooling (IEA RETD, 2010). Policy instruments for both RE heating (RE-H) and cooling (RE-C) need to specifically address the more heterogeneous characteristics of resources, including their wide range in scale, varying ability to deliver different levels of temperature, widely distributed demand, relationship to heat load, variability of use and the absence of a central delivery or trading mechanism (IEA, 2007b; Seyboth et al., 2008; Connor et al., 2009).

---

4 District heating is the distribution of heat generated at one or a few centralized production units through a network of pipelines to residential and commercial buildings that use the heat for space heating and water heating (see Section 8.2.2). DH networks vary in scale from single multi-occupier buildings to city-wide installations.

Similar to RE electricity and RE transport, RE H/C policies will be better suited to particular circumstances/locations if, in their design, the state of maturity of the particular technology, of the existing markets and of the existing supply chains are taken into consideration (Haas et al., 2004). RE-H/C technologies vary in maturity (see Table 1.2), and the maturity of the markets and infrastructure for a given technology may vary by region (e.g., some solar water heating systems are closer to being competitive in China or Israel than in Europe (Xiao et al., 2004)) and in terms of supply chains (manufacturing, integration, infrastructure, maintenance). Though in some regions the infrastructure to support development and installation of RE H/C technologies may not yet exist at all, in others it is well developed. Examples of well-developed RE-H infrastructure include solar water heating in China and geothermal energy in Iceland, where geothermal energy for space heating on a commercial scale began in 1930, and in 2005 supplied 89% of space heat (Lund and Freeston, 2001; IEA, 2007b).

The number of policies to support RE sources of heating and cooling has increased in recent years, resulting in increasing generation of RE H/C (IEA, 2007b). However, a majority of support mechanisms have been focused on RE-H. Policies in place to promote RE-H include fiscal incentives such as rebates and grants, tax reductions and tax credits (Section 11.5.5.1); public finance policies like loans (Section 11.5.5.2); regulations such as use obligations (Section 11.5.5.3); and educational efforts (Section 11.6). To date, fiscal incentives have been the prevalent policy in use (DEFRA/BERR, 2007; Bürger et al., 2008; Seyboth et al., 2008; Connor et al., 2009), though there is increasing interest in regulatory mechanisms.

This section describes the aforementioned policies strictly as they relate to RE H/C. A more general description of the mechanisms themselves can be found in Section 11.5.3. The section concludes with a brief discussion of issues relevant only to RE-C.

11.5.5.1 Fiscal incentives

Grants, rebates, and production incentives

Rebates and grants are the most commonly applied policy for RE-H (and RE-C to a lesser extent), with various applications in multiple countries and regions including Austria, Canada, Greece, Germany, Ireland, Japan, the Netherlands, Poland and the UK (IEA, 2007b; Bürger et al., 2008; Connor et al., 2009). Production-based incentives could also be used to support the production of RE H/C. For H/C, however, production-based incentives are often complicated by the distributed nature of the heat supply where there are few cost-effective metering or monitoring procedures (IEA, 2007b). Production incentives may therefore be most effective for larger H/C systems, such as district heating grids.

Cash incentives, however designed, will have implications for the public budget, which must be carefully considered. Fluctuations, or stop-and-go funding, have been shown to have a direct impact on the resulting deployment of RE H/C technologies (IEA, 2007b; IEA RETD, 2010). For example, the German Market Incentive Program (MAP), while successful in increasing the deployment of solar thermal technologies in Germany, experienced complications when demand for the incentive exceeded availability, and as funding fluctuated annually.

Tax policies

Tax incentives have often been implemented in support of RE-H alongside support for RE electricity technologies (IEA, 2007b). Indirect support for RE H/C, such as exemptions from eco-taxes, carbon and energy charges levied on fossil fuels used for heating, has also been successful in the promotion of RE-H, for example, in Sweden (see Box 11.9).

For RE-H/C, both investment and production tax credits are possible. As production tax credits provide incentive for the amount of RE H/C actually produced, they may be advantageous in assuring the generation of RE H/C as well as the increased quality of installation (IEA, 2007b). Similar to cash incentives, however, the application of production tax credits for distributed heat generation is complicated due to the lack of cost-effective metering or monitoring procedures.

Tax credits available after the installation of a RE-H system (i.e., ex-post) may be logistically advantageous compared with grants, for example, which require pre-approval before installation. For instance, in France, the 2005 Finance Law included a tax rebate system that allowed owners to recover costs via an income tax declaration, suggesting an easy-to-administer, simple and straightforward promotion system (IEA, 2007b; Roulleau and Lloyd, 2008; Walker, 2008; Gillingham, 2009). This law effectively shifted the French system—previously largely based on direct investment incentives (e.g., grants)—to a tax rebate system. After this shift, substantial growth occurred in the solar thermal market, likely the result of simplified procedures (IEA, 2007b).

11.5.5.2 Public finance

Public finance policies such as guarantees, loans and public procurement to promote RE-H are much less common than the aforementioned fiscal incentives, though have in some cases been implemented. For example, the Crediting System in Favour of Energy Management (FOGIME) in France began a guarantee of up to 70% of the total investment on bank loans requested for RE (including RE-H) and energy efficiency projects (IEA, 2007b). Various types of public finance programs have also been used in less developed countries to support the use of modern biomass, residential solar heating and other modern RE technologies.

11.5.5.3 Regulations

Though most support policies for RE H/C technologies to date have been fiscal incentives, regulatory policies like use obligations and quotas have attracted increased interest for their potential to encourage growth of RE H/C independent of public budgets (Bürger et al., 2008; Seyboth et al., 2008).

Use obligation

A use obligation, or building regulation, requires the installation of RE systems in new construction or buildings undergoing substantial renovation. Use obligations are advantageous in that they support the installation of RE heating technologies and related infrastructure at the time of construction, when installation is most cost-effective. They also address the market failure of split incentives (Section 1.4.2), which might otherwise discourage builders or owners from RE-H investments if they won't be paying to heat a building (CCC, 2009).

Initially adopted in various municipalities in Spain, Germany (Nast, 2010), Italy, Ireland, Portugal and the UK, use obligations are now employed at the national level in Spain and Germany. Variations exist regarding eligible technologies and whether the energy has to be onsite or can be located elsewhere (Bürger et al., 2008; Puig, 2008). Use obligations can be applied at different levels of governance and for DH as well as household systems.

However, there are a number of problems associated with this policy. For example, a gradual increase in the obligation level implies that a building stock compliant with the early use obligation may need to be retrofitted later to meet a more stringent future use obligation. It also imposes costs unequally across society because early obligated parties pay relatively higher costs, while later obligated parties may benefit from cost reductions resulting from volume demand and greater skill capacity. There is also the potential for the policy to motivate a delay in replacement of inefficient technologies as building owners wait for the obligation to come into effect and the requirements to become more clear (Connor et al., 2009), or to delay substantial retrofits to avoid the extra cost of compliance.

Ideally, compulsory refurbishment would also include protection for the economically vulnerable (Bürger et al., 2008; Connor et al., 2009). One simple and less onerous application is to mandate the inclusion of basic connection technologies in new buildings to allow for later integration of RE H/C. Integration of the technology for later connection to district heating or cooling is one potential application that might have a good fit with later investment (Connor et al., 2009).

The application of a system of standards to ensure a minimum quality of hardware, installation and design planning when implementing use obligations for RE-H is likely to be essential to ensuring proper compliance; a monitoring system including periodic examinations of installations and/or minimum quality standards is advisable, though this will increase administrative costs (Connor et al., 2009). A high level of compliance is fundamental to the success of the use obligation (Bürger et al., 2008).

Box 11.9 | Further lessons from Sweden: Biomass district heat and value of infrastructure

Sweden's experience with DH illustrates how fiscal incentives for RE-H and the existence of an enabling infrastructure can support a shift to RE sources for heating. Between 1980 and 2007, the biomass share in DH production increased from zero to 44% (90 PJ) (IEA, 2009b).

Sweden's shift to a large share of biomass-based heat was facilitated by the existence of two infrastructure systems (IEA, 2007b). First is Sweden's rich biomass resource (about 52% of the total land area is productive forest) and its forestry industry, which has a long history and a well-established infrastructure (IEA, 2007b). Second is the country's DH system, which as of 2008 accounted for 56% of heating in the residential and service sectors (Swedish Energy Agency, 2009a).

The main expansion of the system occurred during the period 1965 to 1985, when municipal administrations and companies built, owned and operated Sweden's DH system. The shift was driven in the 1980s by high oil prices and taxes on oil products; opportunities for combined heat and power (CHP) production, fuel flexibility, economic efficiency, and better pollution control compared to individual boilers also motivated development of DH infrastructure. Expansion was also facilitated by strong local planning powers and high acceptance for solutions driven by the public sector (Ericsson and Svenningsson, 2009).

In 1991, the Swedish government implemented a carbon tax at USD_{2005} 41 per tonne of CO_2 (this tax gradually increased and reached USD_{2005} 130 per tonne in 2007). Biomass was exempt from the tax, making it the least expensive fuel for DH systems. As a result, the use of biomass expanded rapidly as seen in Figure 11.9, from 14 PJ in 1990 to 60 PJ in 1996 (Ericsson and Svenningsson, 2009). Sweden's carbon tax also accelerated the phase-out of oil for heating of individual buildings, to the benefit of DH, ground-source heat pumps and wood pellets (Ericsson and Svenningsson, 2009).

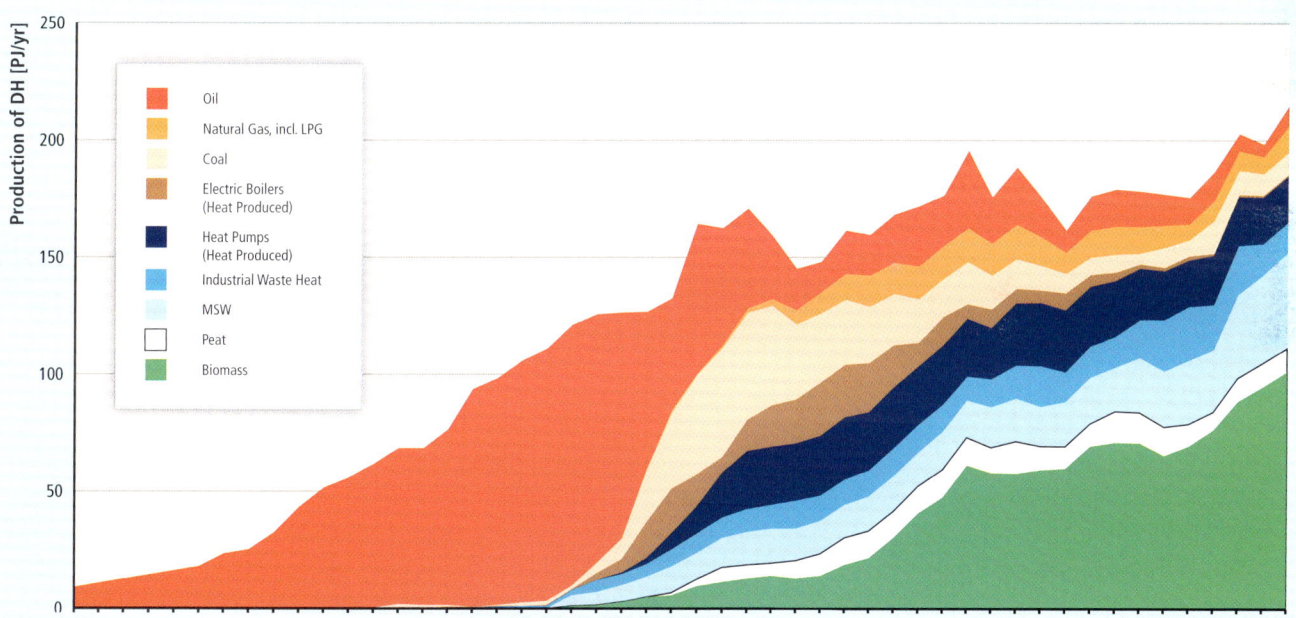

Figure 11.9 | Sweden's district heat production, by fuels and energy sources, 1960 to 2009.

Note: Curves are not corrected for outdoor temperature variations (Swedish District Heating Association, 2001; Ericsson and Svenningsson, 2009; Swedish Energy Agency, 2009b, 2010b).

In addition to the tax exemptions for biomass, investment subsidies were made available for biomass-based CHP from 1991 to 2002, further helping to fuel growth. In 2003, largely driven by the desire to replace nuclear power, the government introduced an electricity quota obligation combined with a green certificates scheme. This led to a further significant increase in heat (and electricity) generation from biomass-based CHP. In response to these policies, district heat from CHP increased from 22 PJ in 1990 to 71 PJ in 2007 (SCB, 2009), and electricity from CHP increased from about 2 TWh (7.2 PJ) in 1990 to 7.5 TWh (27 PJ) in 2007; of this, 41% was from biomass (IEA, 2009b; Bergek and Jacobsson, 2010) (see Box 11.4).

Bonus mechanisms
The bonus mechanism legislates a fixed payment for each unit of heat generated, with potential for setting different levels of payment according to technology (Bürger et al., 2008). Payments can be given as a result of either metered output or some form of estimation of output. They can be capped for a fixed period or for a fixed output, and they can be designed to vary with technology and/or building size to complement energy conservation efforts. Degression can be applied annually to reduce impacts on government budgets.

Bonus mechanisms are similar to price-driven instruments for electricity such as FITs (see Section 11.5.4.3), and differ primarily in two ways: potential scope (many more RE heat than electricity generators might be expected to result), and the likelihood that heat will be used where it is generated. These factors have the potential to make a bonus programme relatively complex and costly, due to the scale of metering and administration required. Consolidation offers a potential solution; for example, a third party organization could aggregate and distribute the benefits of the bonus payments to a large number of its members, reducing the burden of utility or government administration. Further, bonus funds could be paid on a limited number of occasions, perhaps two to three over the lifetime of an installed technology (Bürger et al., 2008), thereby minimizing administrative costs.

There has been little experience with bonus mechanisms to date. However, because of the limited impact on the public budget if payments are made by utilities suppliers (rather than government), it has received increased interest. For example, the UK adopted legislation for a RE-H bonus mechanism with a projected April 2011 adoption, selected largely on the grounds that it would have lower impact on the public budget than other policy options (BERR/NERA, 2008; DECC, 2009).

Quota obligations
Quota obligations, also known as RPS, have largely been deployed in support of RE electricity (see Section 11.5.4.3). In some such cases (e.g., in Australia's Mandatory RE Target (Buckman and Diesendorf, 2010), in Japan's Law on Special Measure for the Utilization of New Energy (IEA, 2007b) and in some US states (DSIRE, 2011), the eligibility of RE technologies has included RE-H technologies such as solar hot water heaters.

Although they have been discussed in Germany and the UK, for example, there is very little experience with quota obligations specifically targeting RE-H (IEA, 2007b). Quota obligations for electricity often include a system of tradable certificates, awarded to producers for the renewable energy they generate. Because of the distributed nature of heat generation and use (except in the case of DH/C systems), such certificate systems for RE-H introduce additional challenges, though in theory RE-H users, their designated agents, or companies in the RE heat supply chain would be eligible to receive tradable certificates if they produced evidence of RE heat use. Market participants could sell certificates to suppliers to earn revenues to offset their costs (Radov et al., 2008).

Network access for district heating
Third party access (TPA) to DH systems can allow greater levels of competition to drive down costs, and provide increased access to a market (Section 8.2.2). There is little experience with TPA for DH systems to date, but some countries (e.g., Sweden (Ericsson and Svenningsson, 2009)) have considered their implementation. However there is some concern that widening TPA might increase costs for DH providers as a result of both increased administration costs and increased price uncertainty and volatility (SOU, 2005; Wårell and Sundqvist, 2009).

Wårell and Sundqvist (2009) identify three possible forms of TPA in DH: 1) regulated TPA generally means new companies can access the grid if they meet certain conditions, a stipulation that is typical in the electricity sector; 2) negotiated TPA comprises ex-post agreement between the network owner and heat provider; and 3) the single buyer model, under which a single consolidator negotiates with all suppliers and sells to all consumers on a regulated basis; rates account for system costs and a certain permitted rate of return.

Variable local conditions will determine the most appropriate form of TPA regulation; these include:

- Scale of heating networks and their potential for expansion. Lithuania, for example, regulates systems that supply above 10 GWh (36 TJ) per year (Gatautis et al., 2009);

- Availability of different heat sources;

- Potential administrative costs; and

- Political and/or public perspectives regarding the opening of markets.

11.5.5.4 Policy for renewable energy sources of cooling

RE-C can include passive cooling measures, solar-assisted, CSP or shallow geothermal technologies driving active cooling systems (e.g., via absorption cooling), biomass adsorption or absorption cooling (though still at early stages of development), or active compression cooling and refrigeration powered by RE electricity (DG TREN, 2007; IEA, 2007b).

Though there are some examples of policies supporting RE-C technologies, in general policy aiming to drive deployment of RE-C solely is considerably less well-developed than that for RE-H. Many of the mechanisms described in the sections above could also be applied to RE-C, generally with similar advantages and disadvantages. Most policy support for RE-C to date has been integrated into programs supporting other RE technologies, including RE-H (IEA, 2007b). Such examples have almost exclusively been fiscal incentives. Spain offered grants directly for solar cooling installations as part of its Renewable Energy Plan for 2005-2010 (IDAE, 2006). Similarly, in Germany, the Solarthermie 2000 Plus program provides grants for solar air-conditioning installations as well as for solar thermal and solar-assisted DH installations (IEA, 2007b).

The lack of experience with deployment policies for RE-C is likely linked to the early levels of technological development of many RE-C technologies. R&D support as well as policy support to develop the early market and supply chains may be of particular importance for increasing the deployment of RE-C technologies in the near future.

11.5.6 Policies for deployment – transportation

A range of policies has been implemented to support the deployment of RE for transport around the world. Because the vast majority of these policies have related to biofuels, this section focuses primarily on biofuel policies. Even for biofuel policies, many of which have been put in place only over the last three to four years, the literature has gaps in assessing effectiveness, efficiency, equity and institutional feasibility.

An increasing number of countries have implemented national biofuel strategies in recent years—for example, Argentina, EU member countries, India, Indonesia, Mexico, Thailand and the USA (Altenburg et al., 2008; Felix-Saul, 2008). Many countries, in particular across South America, with favourable climatic conditions for sugar cane—including Peru (USDA/FAS, 2009b) and Guatemala (USDA/FAS, 2009a)—aim to follow what is seen as Brazil's successful experience with fuel ethanol (see Box 11.10).

Biofuel support policies aim to promote domestic consumption via fiscal incentives (e.g., tax exemptions for biofuel at the pump) or regulations (e.g., blending mandates), or to promote domestic production via public finance (e.g., loans) for production facilities, via feedstock support or via tax incentives (e.g., excise tax exemptions). In addition, trade related measures can be applied to either shield local production through protective measures (e.g., import tariffs, standards) or prevent exports by installing export tariffs (Junginger et al., 2011; Lamers et al., 2011). (See 2.4.4 and 2.4.6 for more information on trade issues.)

11.5.6.1 Fiscal incentives

Tax policies

Tax incentives are commonly used to support biofuels and act to change the cost-competitiveness of biofuels relative to fossil fuels. They can be instituted along the whole biofuel value chain, but are most commonly provided to either biofuel producers (e.g., excise tax exemptions/credits) and/or to end consumers (e.g., tax reductions for biofuels at the pump).

For example, in the USA, Volumetric Excise Tax Credits for the blending of fuel ethanol and biodiesel have been provided to biofuel producers under the American Jobs Creation Act (US Congress, 2004) since 2004. In the EU, the Energy Taxation Directive permits exemptions or reductions from energy taxation for biofuels (Directive 2003/96/EC). Currently, all but two EU member states (Finland and the Netherlands) provide some sort of tax exemption or deduction; the majority are aimed at final consumption (see e.g., European Commission (2011)). Partial or total tax exemptions for biofuels have proven to be critical for the promotion of biofuels across the EU in the past (Wiesenthal et al., 2009). Because the tax exemption given to biofuels must not exceed the level of the fossil fuel tax, the instrument has proven most successful in those EU member states with fossil fuel tax levels high enough to compensate for the additional production costs of biofuels as compared to their fossil fuel alternative (Wiesenthal et al., 2009).

Experiences in Germany and the UK demonstrate that excise duty exemptions can stimulate investments in biofuels, particularly in the early stages of a biofuel market development (Bomb et al., 2007). However, removal of tax breaks can have unintended consequences, as seen in Germany. Prior to August 2006, biodiesel (including pure vegetable oil) was exempt from excise taxes in Germany and the industry flourished, selling 520,000 tonnes of biodiesel in 2005 (Hogan, 2007). By 2006, Germany was the single largest global producer and consumer of biodiesel (REN21, 2007; Eurostat, 2010). However, that year the German government began to gradually phase out tax exemptions for biodiesel and introduced a biofuel mandate as of 2007. This led to a sharp decline in biodiesel consumption (in particular pure vegetable oil). By late 2009, German biodiesel sales had dropped to an estimated 200,000 tonnes (Hogan, 2009). It is estimated that this policy shift reduced biofuels' share of total national fuel consumption from 7.2% in 2007 to 5.9% in 2009 (BMU, 2009).

Several other European and G8+5 countries have begun gradually shifting from the use of tax breaks for biofuels to blending mandates (FAO/GBEP, 2007). This shift has been driven by the potential advantages of mandates as well as disadvantages associated with the use of tax policy (see Section 11.5.3.1).

Fiscal incentives and public finance (see below) have also helped to trigger private sector investments in biofuel production facilities. At the same time, fiscal incentives that are designed cautiously and adapted on a regular basis regarding fossil fuel and biofuel production cost developments are more apt to create market stimuli while avoiding over-compensation.

It is important to note that the introduction of absolute mandates in combination with existing tax credits—as has occurred, for example, in the USA—could have detrimental effects, such as an increased consumption of petrol at the expense of ethanol, Under a mandate, the blenders' ethanol input prices and the ethanol production level will most likely not decline; however, blenders could increase profits by lowering the retail price of fuel and gaining market share, thus reducing the implicit price paid by consumers for the blended fuel (de Gorter and Just, 2010). This could lead to an increase of total fuel consumption while ethanol consumption remains constant under an

> **Box 11.10 | Lessons from Brazil: Gradual expansion of policies to deliver a competitive RE fuel source.**
>
> Brazil was hit hard by the first world oil crisis in the mid-1970s. In 1975, taking advantage of its position as a leading sugar producer, the government established the Brazilian Alcohol Program (PROALCOOL) to promote sugarcane ethanol as a gasoline alternative through production targets and producer subsidies (Goldemberg, 2009).
>
> As part of this policy, Brazil's government mandated that ethanol be blended with gasoline in proportions from 20-25%. Production was supported by subsidies, low-interest loans and guaranteed purchase by the state-owned petroleum company (Petrobras), with parallel research to develop engines that could run on pure ethanol (Dias de Moraes and Rodrigues, 2006).
>
> Responding to government pressure due to concerns about fluctuating ethanol supply and prices that began in the mid-1980s, auto manufacturers introduced flex-fuel motors in 2003 (Goldemberg, 2009). Other early challenges included the need for a network for production and use, which was initially addressed through government activities and eventually turned over to the private sector (Goldemberg, 2006; Walter, 2006). (See Section 8.2.4.6 for more on integration.)
>
> To address social and environmental sustainability concerns that have arisen with an increase in ethanol production, several measures have been enacted at the federal and state levels. These include ecological (AgroEcological Zoning for sugarcane or seed oil plants; see Section 2.2.3) and economic zoning laws that dictate where sugarcane and ethanol production can occur and regulations governing water usage (Goldemberg et al., 2008).
>
> Bagasse (fibrous residue from sugarcane) is used for heat and power generation in the sugarcane refining process to ethanol and sugar, lowering associated carbon emissions, and improving the economics of production (Cerri et al., 2007). The mills meet their own energy needs and sell excess electricity to the grid, which provides another source of income (Section 2.2.3). Early production was stimulated through incentives; today, mill owners sell directly into the grid through contracts or auctions, although lack of access to grid connections is still a barrier for some (Azevedo and Galiana, 2009).
>
> Although ethanol production was initiated as a highly subsidized program, improvements in sugarcane and ethanol production technologies and economies of scale drove down production costs (Section 2.7.2). Ethanol subsidies were removed in the 1990s, and by 2004 ethanol in Brazil was economically competitive with gasoline without subsidies (Goldemberg et al., 2004). The only related incentives by 2010 were reduced taxes for flex-fuel cars. Studies have found that the economic costs of Brazil's ethanol policies over the years were more than outweighed by avoided expenditures associated with imported oil (Moreira and Goldemberg, 1999; Goldemberg et al., 2004). By 2010, Brazil was the world's second largest producer of ethanol, after the USA (Section 2.4.4; REN21, 2010; UNICA, 2010).
>
> Brazil's experience suggests the importance of blending mandates for biofuels in combination with other policies to address economic and other barriers.

absolute mandate (de Gorter and Just, 2010). Partial solutions could be tax structures that self-adjust depending on market developments in the price of oil and in biofuel production. So-called price collars establish lower and upper limits on the price of an RE fuel to address the impacts of market price volatility of competing petroleum fuels and give some assurance to both suppliers and consumers.

11.5.6.2 Public finance

A number of countries, including China (IISD, 2008) and Indonesia (Dillon et al., 2008), provide direct support for biofuels via public finance. Direct financial supports have the advantage of providing easily quantified results, but their outcomes tend to be limited to individual projects. These supports are generally paid for directly out of government budgets (FAO/GBEP, 2007).

As in the electricity sector, public procurement is an option for driving market growth. The government of Thailand, for example, requires all of its fleets to be fuelled with gasohol (gasoline blended with up to 20% ethanol) (Milbrandt and Overend, 2008).

11.5.6.3 Regulations

Renewable fuel mandates and targets
Renewable fuel mandates are key drivers in the development and growth of most modern biofuels industries. Such mandates have been enacted in

at least 41 states/provinces and 24 countries at the national level (REN21, 2010); Russia is the only G8+5 country that has not created a transport biofuel target (FAO/GBEP, 2007; REN21, 2010). Brazil first mandated ethanol blending with gasoline starting in the 1970s, but most countries started blending renewable fuel with voluntary targets. However, mandatory blending mandates, enforceable via legal mechanisms, are becoming increasingly utilized and with greater effect, notably in the EU and in the USA (Canadian Food Grains Bank, 2008).

The distinction between voluntary and mandatory is critical because voluntary targets can be influential but do not have the impact of legally binding mandates. The original EU biofuel strategy (in Directive 2003/30) posed indicative, not mandatory, targets for all member states. The voluntary targets were not influential for most EU countries—only three members (Germany, Austria and Sweden) met the 2005 target (FAO/GBEP, 2007). Under the current EU Fuel Quality Directive (FQD), all member states are required to ensure a 10% share of RE in final energy demand in the transport sector by 2020 (European Commission, 2009a). Each member state of the European Union has its own blending mandates for ethanol and biodiesel, and most mandates allow for flexibility in how to meet the mandate (Flach et al., 2009). Generally, blending mandates are able to provide the desired market signals without the need for government funding.

As the recent biofuel policy development in the EU shows, those countries with the highest shares of biofuels in transport fuel consumption have had hybrid systems that combine mandates (including penalties) with fiscal incentives (foremost tax exemptions). However, it is difficult to assess the level of support under biofuel mandates because prices implied by these obligations are generally not public (in contrast to the electricity sector, for example) (Held et al., 2010).

While mandates have proven to be an effective instrument for the promotion of biofuels in general, they are found to be less appropriate for the promotion of specific biofuel types because fuel suppliers tend to blend low-cost biofuels (Wiesenthal et al. 2009). In the European context, this has led to the abolishment of small-scale, distributed regional biofuel production facilities for large-scale production centres in harbours or along strategic inland waterways, which enjoy a greater access to (cheaper) international (feedstock) imports (Lamers et al., 2011). Further, mandates have been criticized for inducing global food insecurity (Pimentel et al., 2009), indirect land use effects such as market-induced deforestation and associated ineffectiveness in reducing GHG emissions (Searchinger et al., 2008; Creutzig and Kammen, 2009; Hertel et al., 2010; Lapola et al., 2010), and negative impacts on water quality (Vitousek et al., 1997) (Section 2.5.3).

Such impacts can be reduced or avoided if additional criteria are mandated. For example, the US Renewable Fuels Standard 2 indicates maximum GHG emission thresholds for different biofuels (USEPA, 2010b). The EU FQD and RED set minimum requirements for GHG savings for biofuels and outline sustainability standards (Section 2.5.7.1). All policies also define specific lifecycle accounting methodologies, assumptions and default values because, as discussed in Chapter 2, GHG emission estimates for biofuels are hugely varying, especially if indirect land use change is taken into account (Plevin et al., 2010).

Biofuel production and/or blending mandates (of energy or volume content) have proven to be effective in rapidly increasing domestic biofuel production and consumption (Wiesenthal et al., 2009; European Commission, 2011). They are the most important policy option evaluated in terms of effectiveness and institutional feasibility. By nature, however, they need to be carefully designed and accompanied by further requirements in order to reach a broader level of distributional equity. This is particularly the case for biofuels in terms of sustainability criteria such as GHG emission reductions (Section 2.5.4) or land use (Sections 2.5.3 and 2.5.7).

As in the electricity and heating/cooling sectors, governments generally enact a combination of policy options. As noted above, Brazil is a case in point, with a mandate as well as subsidies that were in place for many years, and the USA has had mandates alongside tax credits and other policies. Another example is Thailand, where the government has provided incentives for various ethanol blends through excise tax waivers and fuel price incentives, is building a distribution infrastructure, provides soft loans to farmers growing palm crops and supports R&D of new crops like jatropha (Johansson et al., 2004; Milbrandt and Overend, 2008; Nilkuha, 2009).

11.5.7 Synthesis

11.5.7.1 Assessment of RE policies

Policy mechanisms enacted specifically to promote RE are varied and can apply to all energy sectors. They include fiscal incentives such as tax credits, grants and rebates; government finance policies such as guarantees and loans; and regulations such as quantity-driven policies like quotas and price-driven policies like FITs for electricity, mandates for heating and biofuels blending requirements. Policies can be enacted by local, state/provincial, national and international authorities.

RE R&D and deployment policies have promoted an increase in RE shares by helping to overcome various barriers that impede technology development and deployment of RE. Table 11.3 lists some possible policy options for addressing the various barriers to RE set out in Chapter 1.

Experience shows that public R&D investments are most effective when complemented by other policy instruments, particularly RE deployment policies that simultaneously enhance demand for new RE technologies and create a steadily increasing market. Together, R&D and deployment policies create a positive feedback cycle, inducing private sector investment in R&D. Enacting deployment policies early in the development of a given technology can accelerate learning by inducing private R&D, which in turn further reduces costs and provides additional incentives for using the technology, as seen in Japan with PV and Denmark with wind power.

Table 11.3 | Barriers to RE deployment and policies to address them.

Type of barrier	Potential policy instruments include
Market failures and economic barriers (Section 1.4.2.1) • Cost barriers • Financial risk • Allocation of government financial support • Trade barriers	Public support for RE R&D; deployment policies that support private investment, including fiscal incentives, public finance, and regulatory mechanisms (e.g., FITs, quotas, use standards)
Information and awareness barriers (Section 1.4.2.2) • Deficient data about natural resources • Skilled human resources (capacity) • Public and institutional awareness	Resource assessments; energy standards; green labelling; public procurement; information campaigns; education, training and capacity building
Institutional and policy barriers (Section 1.4.2.3) • Existing infrastructure and energy market regulation • Intellectual property • Industry structure	Enabling environment for innovation; economic regulation to enable access to networks and markets and investment in infrastructure; revised technical regulations; international support for technology transfer (e.g., under UNFCCC); microfinance; technical training
Issues relevant to policy (Section 1.4.3) • Social acceptance	Information campaigns; community projects; public procurement; governmental (national and local) policy cooperation; improved processes for land use planning

Some policy elements have been shown to be more effective and efficient than others for rapidly increasing RE deployment and enabling government/society to achieve specific targets. Institutional feasibility and equity are also important, but these criteria have not been analyzed as fully. Synthesizing the previous sections, key elements of policies that make them most likely to meet these criteria include:

- Adequate value derived from subsidies, FITs etc. to cover costs such that investors are able to recover their investment at a rate of return that matches their risk;

- Guaranteed access to networks and markets or at a minimum clearly defined exceptions to that guaranteed access; and

- Long-term contracts to reduce risk and thereby reduce financing costs.

Note: the three preceding bullets are all important for reducing key risks and encouraging greater levels of private investment. Reducing risk helps to improve access to and lower the cost of financing (because profitability expected is lower (Haas et al., 2011)), which can reduce project costs as well as end costs of delivered energy paid by consumers.

- Provisions that account for diversity of technologies and applications. RE technologies are at varying levels of maturity and with different characteristics, often facing very different barriers. Multiple RE sources and technologies may be needed to mitigate climate change, and some that are currently less mature and/or more costly than others could play a significant role in the future in meeting energy needs and reducing GHG emissions.

- Incentives that decline predictably over time as technologies and/or markets advance, such as the declining grant for wind in Denmark (see Box 11.12), or degressive tariffs in Germany (see Box 11.6).

- Policy that is transparent and easily accessible so that actors can understand the policy and how it works, as well as what is required to enter the market and/or to be in compliance. Also includes longer-term transparency of policy goals, such as medium- and long-term policy targets.

- Inclusive, meaning that potential for participation is as broad as possible on both 1) the supply side (traditional producers, distributors of technologies or energy supplies, whether electricity, heat or fuel), and 2) the demand side (businesses, households etc.), which can 'self-generate' with distributed RE, enabling broader participation that unleashes more capital for investment, helps to build broader public support for RE (as in Denmark and Germany) and creates greater competition.

- Attention to preferred exempted groups, for example, major users on competitiveness grounds or low-income and vulnerable customers on equity and distributional grounds.

It is also important to recognize that there is no one-size-fits-all policy, and policymakers can benefit from the ability to learn from experience and adjust programs as necessary. Policies need to respond to local political, economic, social, ecological, cultural and financial needs and conditions, as well as factors such as the level of technological maturity, availability of affordable capital, and the local and national RE resource base. In addition, a mix of policies is generally needed to address the various barriers to RE, as highlighted by China's experience (see Box 11.11). As seen in the case studies in this and the following sections, more than one policy has been utilized to advance RE—for example, FITs and low-interest loans, grants, or tax credits in combination with quota obligations.

Finally, transparent, sustained, consistent signals—from predictability of a specific policy, to pricing of carbon and other externalities, to

Box 11.11 | Lessons from China: Mixed policy approach to energy access and large-scale RE.

China has relied increasingly on RE to help meet its rising energy demand, improve its energy structure, reduce environmental pollution, stimulate economic growth and create jobs (Zhang et al., 2009). China installed more wind power capacity during 2009 than any other country and, by the end of the year, ranked first globally for total RE electricity generation capacity and third for non-hydro RE capacity (REN21, 2010). China is, by far, the leading global market for solar hot water systems and, in 2009, was the third largest producer of ethanol (REN21, 2010). In addition, a strong domestic manufacturing industry for wind power, PV and solar thermal collectors has emerged, triggered in part by policies that have encouraged industry development along with technology deployment (Han et al., 2010; Liu et al., 2010; Q. Wang, 2010).

The Chinese government has devoted significant attention to RE development in recent decades, both for rural energy access and large-scale grid-connected projects. China began developing wind power in the early 1970s for the primary purpose of supplying power to remote areas (Changliang and Zhanfeng, 2009). Grid-connected wind power started in the 1980s with small-scale demonstration projects and evolved to a main source of power supply by 2003, when the Wind Farm Concession Program was established through which bidding procedures were used to develop larger wind power plants (Q. Wang, 2010). Solar water heaters have been applied since the 1970s (Han et al., 2010), and biogas digesters have been promoted since the 1980s (Peidong et al., 2009).

Under the Township Electrification Programme, more than 1,000 townships in nine western provinces were electrified in just 20 months, bringing power to almost one million rural Chinese (NREL, 2004). Important to the success of China's rural electrification efforts have been education of local and national decision makers, training and capacity building, technical and implementation standards and community access to revolving credit (Wallace et al., 1998; NREL, 2004; Ku et al., 2005).

For grid-connected RE, China's national Renewable Energy Law took effect in 2006, creating a national framework to support RE and to institutionalize several support policies, including mandatory grid connection standards, RE planning, and promotion funding (Zhang et al., 2009). The law has been followed by a large number of specific regulations and measures to support the development of wind, solar and biomass sources. For example, the Medium and Long-term Renewable Energy Development Plan, released in 2007, set a national target for RE to meet 10% of total energy consumption by 2010 and 15% by 2020 (the latter 15% target was later revised to refer to all non-fossil energy sources) (Q. Wang, 2010), while also establishing RE technology-specific targets. The 30 GW wind power target for 2020, as specified by The 11th Five Year Plan for Renewable Energy in 2008, was achieved a decade ahead of schedule (B. Wang, 2010).

Under the Renewable Energy Law and its implementing regulations, a wide variety of promotional policies have been employed to support the continued growth of renewable electricity (e.g., Yu et al., 2009; Liao et al., 2010; Wang et al., 2010; Zhao et al., 2011). Feed-in tariffs have been established for wind and biomass power plants, while bidding procedures have been used for offshore wind power plants, for wind turbine purchases to serve China's seven planned large-scale wind bases, and increasingly for solar power plants. Grid-connected (and off-grid) PV systems have also benefited from grants. Funding for many of these programs has come from a national electricity surcharge and resulting RE fund, while the Kyoto Protocol's Clean Development Mechanism (CDM) has also played a role in improving project profitability (Lewis, 2010).

In addition to these policies and the national RE targets, the country's largest generating companies have been called upon to expand their renewable power capacity to 3% of their total capacity by 2010, and at least 8% by 2020. China provides a clear example of a country that has relied upon a diversity of mechanisms to achieve policy goals.

China continues to address challenges as they arise by developing and revising RE policies and measures, including enhancing technical skills; establishing institutions to support R&D development and a national RE research institute; extending electricity transmission to ensure that new RE capacity can be effectively brought online; creating a domestic market to stimulate demand and avoid over-reliance on overseas markets; and establishing a national RE industry association to coordinate development and formally bridge the industry and policymaking processes (Martinot and Junfeng, 2007; REN21, 2009a). By addressing the wide variety of RE technologies and applications in a coherent long-term manner and with a sizable mix of policies, China has been able to establish RE as a significant bulk energy carrier. This creates good prospects for further growth in deployment and manufacturing of RE technologies.

long-term targets for RE—have been found to be crucial for reducing the risk of investment sufficiently to enable appropriate rates of deployment and the evolution of low-cost applications.

11.5.7.2 Macroeconomic impacts and cost-benefit analysis

Payment for supply-push, or R&D, type support tends to come from public budgets (multinational, national, local) and therefore taxpayers, whereas the cost of demand-pull, or deployment, policies often lands on the end users of energy. For example, if a fiscal incentive is added to electricity, the additional cost of this incentive is borne by consumers, although exemptions or re-allocations can reduce costs for industrial or vulnerable customers where necessary, or for equity or other reasons (Jacobsson et al., 2009).

If the goal is to transform the energy sector over the next several decades, then it is important to minimize costs over this entire period, not only in the near term; it is also important to include all costs and benefits to society in that calculation. Moreover, as mentioned above, the timing, strength and level of coordination of R&D versus deployment policies will affect this calculation.

Conducting an integrated analysis of costs and benefits associated with RE is extremely demanding because so many elements are involved in determining net impacts. Concepts that try, at least partly, to balance costs and benefits (as the concept of external costs tries to do in terms of environmental aspects) face substantial limitations and are confronted with significant uncertainties (see Section 10.6). Breitschopf et al. (2010, in German only with translation from the German Environmental Ministry (BMU (2010)) conclude that effects fall under three categories, including direct and indirect costs of the system as well as benefits of RE expansion; distributional effects (which economic actors or groups enjoy benefits of, or suffer burdens from, RE support); and macroeconomic aspects such as impacts on the gross domestic product or employment. For example, potential economic growth and job creation are key drivers for RE policies (see Section 11.3.4), but measuring net effects is complex and uncertain because the additional costs of RE support create distribution and budget effects on the economy.

Because of this complexity, there are few studies that examine the economic impacts in this way on a country's or region's economy. Ragwitz et al. (2009) analyzed these effects for the EU, accounting for positive and negative impacts for two possible scenarios: business-as-usual, leading to a 14% RE share in final energy consumption by 2020; and an 'accelerated deployment policies' scenario, achieving the EU 20% target by 2020. They found that RE support policies have a slight positive impact on gross domestic product (GDP) and employment, and that benefits are greater for the higher RE share. Houser et al. (2010) analyzed the potential impacts of Proposed American Power Act on the USA from the perspective of energy security, environmental impact and employment effects, all of which were net positive while the macroeconomic perspective of GDP was broadly neutral. It is important to note that these studies focus on specific geographical areas and that findings could differ for other regions and varying conditions. Most such studies focus on analysing the net effects of RE policy on one economic sector. For example, Lehr et al. (2008) focused on Germany and net employment, and also found positive economic impacts.

These macroeconomic studies are important for gaining an understanding of the distributional impacts across society. While the costs of subsidies are often spread broadly through an economy, the economic benefits tend to be more concentrated (IPCC, 2007). As such, support mechanisms can shift economic wealth from some groups in society to others. Such impacts may simultaneously meet effectiveness, efficiency and equity concerns, or they may cause conflicts among these concerns. Providing energy access, for example, is generally expected to increase equity (Casillas and Kammen, 2010). (See Section 11.5.1 for more on effectiveness, efficiency and equity.)

Distributional impacts are less clear if the cost of a RE policy is assessed relative to an alternative use by government of the same funds or in foregone spending by individuals (Frondel et al., 2010), or in relation to the effects of that policy on different segments of society (Bergek and Jacobsson, 2010). If the costs of a policy are spread across all consumers, poorer people pay a relatively larger share of their income to support RE than do others, unless there are policies in place to mitigate such impacts (Boardman, 2009).

11.5.7.3 Interactions and potential unintended consequences of renewable energy and climate policies

If each externality and each market failure of RE deployment were addressed by the 'ideal' first-best instrument—for example, a carbon price for the climate externality, R&D and deployment subsidies for innovation spillovers, and financial instruments to reduce inappropriate investment risks— the result would be an economically optimal deployment of low-carbon technologies. In reality, however, due to overlapping drivers and rationales for RE deployment (Section 11.3) and overlapping jurisdictions (local versus national versus international level) there may be substantial interplay among policies at times and with unintended consequences. Due to the barriers to policy development discussed in part in Section 11.4 (e.g., informational and political constraints (Bennear and Stavins, 2007)), policymakers often do not implement policies that address market failures in an 'ideal' way. A clear understanding of the interplay among policies and the cumulative effects of multiple policies is crucial in order to address counterintuitive or unintended consequences. This section addresses the interplay between climate change policies, such as carbon pricing, and RE policies. A discussion of the interplay between RE policies and non-RE policies that goes beyond climate change policies (e.g., agricultural policies) can be found in Section 11.6.2.

Firstly, in order to be effective and efficient, both carbon pricing and RE- specific policies must apply over long time periods. Therefore a careful consideration of dynamic incentive effects is required—in particular with respect to the supply of fossil fuel resources. If not applied globally and comprehensively, both carbon pricing and RE policies create risks of 'carbon leakage': RE policies in one jurisdiction or sector reduce the demand for fossil fuel energy in that jurisdiction or sector, which *ceteris paribus* reduces fossil fuel prices globally and hence increases demand for fossil energy in other jurisdictions or sectors. Similarly, climate change policies in one jurisdiction increase the relative cost of emitting in that jurisdiction, providing firms with an incentive to shift production from plants facing carbon prices or regulation to plants in countries with weaker climate change policy (Ritz, 2009). Hence, the impact of carbon pricing and RE policies on emission reduction could potentially become small or even zero. The scope of offset provisions within a carbon cap-and-trade system (the Kyoto Protocol's Clean Development Mechanism or Joint Implementation, for example) can also affect the RE objective by giving firms an alternative to domestic emissions reductions, thereby reducing the incentive to deploy RE technologies in the country to which the policy applies (del Río González et al., 2005).

Even if implemented globally, suboptimal carbon prices and RE policies could potentially lead to higher carbon emissions (Sinn, 2008; Gerlagh, 2010; Grafton et al., 2010; Van der Ploeg and Withagen, 2010). For example, there is a potential danger that as soon as RE policies start to allow RE to compete with fossil fuel technologies in the market place, fossil fuel prices could fall, discouraging further RE deployment and thereby restoring the competitiveness of fossil fuels. If fossil fuel resource owners fear more supportive RE deployment policies in the long term, they could increase resource extraction as long as RE support is moderate. Similarly, the prospect of future carbon price increases may encourage owners of oil and gas wells to extract resources more rapidly, while carbon taxes are lower, undermining policymakers' objectives for both the climate and the spread of RE technology. The conditions of such a 'green paradox' are rather specific: carbon pricing would have to begin at low levels and increase quickly (Sinn, 2008; Hoel, 2010; Edenhofer and Kalkuhl, 2011). Simultaneously, subsidized RE would have to remain more expensive than fossil fuel-based technologies (Van der Ploeg and Withagen, 2010). If carbon prices and RE subsidies begin at high levels from the beginning, such green paradoxes become unlikely. Moreover, quantity instruments like emissions trading schemes and green quotas (if globally applied) eliminate the risk of green paradoxes altogether.

Secondly, carbon pricing and RE policies administered at the same time create complex changes in the incentives for the deployment of energy technologies (de Miera et al., 2008; de Jonghe et al., 2009; Fischer and Preonas, 2010). The cumulative effect of combining policies that set fixed carbon prices, like carbon taxes, with RE subsidies is largely additive: in other words, extending a carbon tax with RE subsidies decreases emissions and increases the deployment of RE.

However, the effect on the energy system of combining endogenous price policies, like emissions trading and/or RE quota obligations, is usually not as straightforward. This is because several feedback mechanisms have an effect on the resulting price signals for fossil and low-carbon technologies. Adding RE policies on top of an emissions trading scheme usually reduces carbon prices (Amundsen and Mortensen, 2001; Fankhauser et al., 2010), which, in turn, makes carbon-intensive (e.g., coal-based) technologies more attractive compared to other non-RE abatement options such as natural gas, nuclear energy and/or energy efficiency improvements (Blyth et al., 2009; Böhringer and Rosendahl, 2010; Fischer and Preonas, 2010). In such cases, although overall emissions remain fixed by the cap, RE policies reduce the costs of compliance and/or improve social welfare only if RE technologies experience specific externalities and market barriers to a greater extent than other energy technologies. If that is not the case, the RE support cannot be economically justified on climate policy grounds alone.

However, if an emissions cap were chosen in anticipation of the contribution from well-designed RE deployment policies—whether FITs, fiscal incentives or other policies—that were targeted at RE-specific market failures, RE support can play a role a role in removing those market failures (Fischer and Preonas, 2010). Further, a quantity-based instrument like a quota obligation could become non-binding (implying zero prices) if other instruments are very stringent. For example, CO_2 allowance prices within an emissions trading scheme could fall to zero if a strong RE policy (in terms of high RE quotas or subsidies) is in place. Equally, the price of tradable RE certificates could fall to zero if carbon prices are very high due to ambitious emissions caps or high carbon taxes (Unger and Ahlgren, 2005; de Jonghe et al., 2009).

Finally, RE policies alone (i.e., without carbon pricing) are not necessarily an efficient instrument to reduce carbon emissions because they do not provide enough incentives to use all available least-cost mitigation options including non-RE low-carbon technologies and energy efficiency improvements (Fischer and Newell, 2008). The implementation of an appropriate carbon pricing scheme remains crucial if the goal of policymakers is to efficiently reduce carbon emissions (Stern 2007, p. xviii, Ch. 14; IPCC 2007, p. 19).

In conclusion, the combination of carbon pricing and RE policies is most efficient in reaching climate change mitigation goals if RE policies address RE-specific market failures and carbon pricing policies address the climate externality. Carbon pricing is expected by many to be the most important policy to reduce carbon emissions. Poorly designed RE policies, in particular in cases without carbon pricing policies, may increase mitigation costs or can, in extreme cases, even increase carbon emissions. At the same time, if carefully designed, RE policies can be a useful supplement to carbon pricing, removing associated market failures and decreasing mitigation costs.

11.6 Enabling environment and regional issues

An environment that is 'enabling' of RE-specific policies is made up of cross-cutting domains as presented in Table 11.4. An enabling environment encompasses different factors such as institutions,

Table 11.4 | Factors and participants contributing to a successful RE governance regime.

Dimensions of an Enabling Environment >> Factors and actors contributing to the success of RE policy	11.6.2 Integrating policies (national/ supranational policies)	11.6.3 Reducing financial and investment risk	11.6.4 Planning and permitting at the local level	11.6.5 Providing infrastructures, networks and markets for RE technology	11.6.6 Technology transfer and capacity building	11.6.7-8 Learning from actors beyond government
Institutions	Integrating RE policies with other policies at the design level reduces potential for conflict among government policies	Development of financing institutions and agencies can aid cooperation between countries, provide soft loans or international carbon finance (CDM). Long-term commitment can reduce the perception of risk	Planning and permitting processes enable RE policy to be integrated with non-RE policies at the local level	Policymakers and regulators can enact incentives and rules for networks and markets, such as security standards and access rules	Reliability of RE technologies can be ensured through certification. Institutional agreements enable technology transfer	Openness to learning from other actors can complement design of policies and enhance their effectiveness by working within existing social conditions
Civil society (individuals, households, nongovernment organizations, unions etc.)	Municipalities or cities can play a decisive role in integrating state policies at the local level	Community investment can share and reduce investment risk. Public-private partnerships in investment and project development can contribute to reducing risks associated with policy instruments. Appropriate international institutions can enable an equitable distribution of funds	Participation of civil society in local planning and permitting processes might allow for selection of the most socially relevant RE projects	Civil society can become part of supply networks through co-production of energy and new decentralized models.	Local actors and NGOs can be involved in technology transfer through new business models bringing together multi-national companies / NGOs / small and medium enterprises (SMEs)	Civil society participation in open policy processes can generate new knowledge and induce institutional change. Municipalities or cities may develop solutions to make RE technology development possible at the local level. People (individually or collectively) have a potential for advancing energy-related behaviours when policy signals and contextual constraints are coherent
Finance and business communities		Public private partnerships in investment and project development can contribute to reducing risks associated with policy instruments	RE project developers can offer know-how and professional networks in : i) aligning project development with planning and permitting requirements ; and ii) adapting planning and permitting processes to local needs and conditions. Businesses can be active in lobbying for coherent and integrated policies	Clarity of network and market rules improves investor confidence	Financing institutions and agencies can partner with national governments, provide soft loans or international carbon finance (CDM).	Multi-national companies can involve local NGOs or SMEs as partners in new technology development (new business models). Development of corporations and international institutions reduces risk of investment
Infrastructures	Policy integration with network and market rules can enable development of infrastructure suitable for a low-carbon economy	Clarity of network and market rules reduces risk of investment and improves investor confidence		Clear and transparent network and market rules are more likely to lead to infrastructures complementary to a low-carbon future		City and community level frameworks for the development of long-term infrastructure and networks can sustain the involvement of local actors in policy development

Continued next Page →

Dimensions of an Enabling Environment >> Factors and actors contributing to the success of RE policy v v	11.6.2 Integrating policies (national/supranational policies)	11.6.3 Reducing financial and investment risk	11.6.4 Planning and permitting at the local level	11.6.5 Providing infrastructures, networks and markets for RE technology	11.6.6 Technology transfer and capacity building	11.6.7-8 Learning from actors beyond government
Politics (international agreements / cooperation, climate change strategy, technology transfer etc.)	Supra-national guidelines (e.g., EU on 'streamlining', ocean planning, impact study) may contribute to integrating RE policy with other policies	Long-term political commitment to RE policy reduces investors risk in RE projects	Supra-national guidelines may contribute to evolving planning and permitting processes	Development cooperation helps sustain infrastructure development and allows easier access to low-carbon technologies	CDM, Intellectual property rights and patent agreements can contribute to technology transfer	Appropriate input from non-government institutions stimulates more agreements that are socially connected

UNFCCC process mechanisms such as Expert Group on Technology Transfer, the Global Environment Facility, and the Clean Development Mechanism and Joint Implementation may provide guidelines to facilitate the involvement of non-state actors in RE policy development |

infrastructures (e.g., networks) and political outcomes (e.g., international agreements/cooperation, climate change strategy) and different actors or participants (e.g., the finance community, business community, civil society, government), each of which influences the success of RE-specific policies while interacting in different configurations. For example, these factors can influence how change may occur within a country; how risky investment in RE may be; how economic regulation encourages (or not) RE deployment; and how communities react to RE. These various configurations present different challenges to RE deployment, depending on the countries and their states of development, and local needs and conditions. This section highlights the potential contribution of these individual factors and participants to a governance of RE that can strengthen, and goes beyond, government action.

11.6.1 Innovation in the energy system

If RE is to play a major role in climate change mitigation, then an overarching and parallel step is to implement policies that enable change to occur in the energy system. A number of studies have reconstructed the historical emergence and formation of socio-technical systems that are taken for granted today (e.g., transition from horses to the internal combustion engine (Geels, 2005); transition from cesspool to sewer systems in urban hygiene (Geels, 2004)). A widely accepted conclusion is that established socio-technical systems tend to narrow the diversity of innovations because the prevailing technologies develop a fitting institutional environment (David, 1985). This environment supports these technologies by making it easier and cheaper to develop and deploy them, or to develop technologies that do not require a profound transformation of the energy system (Grubler et al., 1999a; Unruh, 2000). Actors, institutions and even the very structure of the economy evolve to depend, to some degree, on the existing socio-technical systems. This may give rise to strong path dependencies and exclude (or lock out) rivalling and potentially better-performing alternatives (Nelson and Winter, 1982).

For these reasons, socio-technical system change takes time, and it involves change that is systemic rather than linear. Recent studies have focused on ongoing innovation processes in order to understand the preconditions under which radical transformations of socio-technical systems could occur (Carlsson et al., 2002; Jacobsson and Bergek, 2004; Hekkert et al., 2007; Markard and Truffer, 2008). These studies emphasized that the interplay between existing institutional contexts and technology development was important for explaining the effectiveness (or failure) of specific promotional policies, such as RE policies.

RE technologies are being integrated into an energy system that, in much of the world, was constructed to benefit the existing energy supply mix. As a result, infrastructure favours the currently dominant fuels, and there are existing lobbies and interests that all need to be taken into account (e.g., Verbong and Geels, 2007). In light of this situation, RE deployment policies can be more efficient and effective if the environment around them becomes more conducive to change.

Due to the intricacies of technological change, it is important that all levels of government (from local through to international) encourage RE development through policies, and that nongovernmental actors also be involved in policy formulation and implementation. In recent years, public-private partnerships, civil society and business actors have played increasingly influential roles in the formulation and implementation of policies (Rotmans et al., 2001; van den Bergh and Bruinsma, 2008). In response, the focus of political science literature is shifting from "government" to "governance" related research (Rosenau and Czempiel, 1992; Rhodes, 1996; Newig and Fritsch, 2009), focusing increasingly on understanding the interplay between governments and other societal actors and the implications for the success of policy implementation. Some argue that policy action is more effective and efficient when it includes non-state actors, networks and coalitions in building guiding visions, and formulating and implementing public policy (Rotmans et al., 2001; van den Bergh and Bruinsma, 2008).

11.6.2 Complementing renewable energy policies and non-renewable energy policies

Government policies are more likely to be effective and efficient if they complement one another (Peters, 1998). Further, the design of individual RE policies will also affect their coordination with other policies (both other RE-specific policies and policies targeting other sectors). Although such coordination has been described as a lynchpin for implementation or realization of sustainable development (Jordan and Lenschow, 2000; Lenschow, 2002), it remains a rather elusive principle that is open to divergent interpretations (Jordan and Lenschow, 2000; Persson, 2004). There is a clear need for strong central coordination to eliminate contradictions and conflicts among sectoral policies and to simultaneously coordinate action at more than one level of governance (Jordan and Lenschow, 2000). However, there are few 'best practices' for coordination that can be shared easily at the international level (Jordan and Lenschow, 2000).

Attempting to actively promote the complementarity of policies (for example, agricultural and energy policies) while also considering the independent objectives of each, is not an easy task and may create win-lose and/or win-win situations, with possible tradeoffs (e.g., economic versus environmental, long- versus short-term) (Lenschow, 2002; Resch et al., 2009), as seen in relation to RE transportation, to take one example.

A number of policies that are not directly aimed at promoting RE in the transport sector can have an influence on the effectiveness and efficiency of RE-specific policies. On the 'negative' side, because nearly all liquid biofuels for transport are currently produced from conventional agricultural crops, the removal of agricultural crop subsidies may have a direct impact on the development of liquid biofuels for transportation (see Sections 11.5.5, 2.4.5, 2.5.7 and 2.8.4). In contrast, urban transport policies that aim to regulate transport demand through price signals (e.g., parking fees and congestion charges) can also induce a shift to alternative fuel vehicles through fee exemptions and thereby facilitate deployment of RE transportation (Prud'homme and Bocajero, 2005; Creutzig and He, 2009). Further, carbon-intensity fuel standards—such as the California Low Carbon Fuels Standard—and the EU Emissions Trading Scheme can provide incentives for low-carbon RE transport fuels by helping to level the playing field (Sperling and Yeh, 2009; Creutzig et al., 2010).

RE policies and demand-side measures can complement each other by taking advantage of synergies between RE and energy efficiency, as discussed in Sections 1.2.5 and 11.7. For example, the use of smart meters, time-differentiated pricing and responsive demand can enable a shift in demand load that can both benefit system operation and match demand to RE supply (Sioshansi and Short, 2010; Sections 11.6.5 and 8.2.1).

11.6.3 Reducing financial and investment risk

A broader enabling environment includes a financial sector that can offer access to financing on terms that reflect the specific risk/reward profile of a RE technology or project. The cost of financing and access to it depends on the broader financial market conditions prevalent at the time of investment, and on the specific risks of a project, technology and actors involved. Beyond RE-specific policies, broader conditions can include political and currency risks, and energy-related issues such as competition for investment from other parts of the energy sector, and the state of energy sector regulations or reform (ADB, 2007)The fundamental principle of modern global capital markets is that private capital will flow to those countries, or markets, where regulatory frameworks and policies governing investment are transparent, well-considered and consistent, providing confidence to investors over a time period that is appropriate to the life cycle of their investment (ADB, 2007).

Improving access to finance is necessary but not always sufficient to promote RE project deployment, particularly in developing countries. Successful public finance mechanisms typically combine access to finance with technical assistance programmes that are designed to help prepare projects for investment and to build the capacity of the various actors involved. There are numerous examples of finance facilities that were created but that never disbursed funds because they failed to find and generate sufficient demand for the financing (UNEP, 2008). As seen in the Pacific Islands, access to financing and even targets are not necessarily enough; it is also necessary to have specific policies in support of RE (see Box 11.1).

Government RE policies can play an important role in creating an environment conducive to investment. Long-term commitment contributes to the effectiveness and efficiency of RE policy because it reduces uncertainty about expected returns from investing in RE projects, as described in Section 11.5. However, linking RE policies to permitting policies for RE projects (Section 11.6.4), to the economic regulation of networks and markets (Section 11.6.5), to policies to encourage and enable technology transfer (Section 11.6.6) and to attitudes towards RE beyond government (Section 11.6.7) reduces investor attitudes to risk, thereby freeing up more investment. One specific example can be seen on the ground in Nepal, where it has been shown that development of local capacity can play a major role in attracting private financing in developing countries (UNDP and AEPC, 2010; see Box 11.13).

11.6.4 Planning and permitting at the local level

Deployment of RE technologies has the potential to interfere with existing and traditional resource uses, conservation values or commercial interests. Rules are needed to integrate RE policy with other (e.g., environmental, landscape, agriculture) policies, to resolve potential conflicts at the local level, and to ensure sustainable deployment of RE technologies (see Chapter 9 for a full discussion). This section addresses the challenges of balancing planning regulation that supports RE deployment while also ensuring public oversight and environmental protection, and it provides some general lessons from experiences to date. Technology-specific planning issues are covered in the relevant technology chapters.

Spatial planning (land/sea space, landscape) processes are social processes (Ellis et al., 2009). It is often in the process of preparing, designing, planning, deciding and implementing a specific project, whether RE or otherwise, that differences in perspectives, expectations and interests become manifest. The system of spatial planning provides for a framework—a set of legal, formal rules and procedures—to address and mediate conflicting interests and values (Owens and Driffill, 2008; Ellis et al., 2009). An appropriate planning framework can reduce hurdles at the project level, making it easier for RE developers, communities or households to access the RE resource and succeed with their projects. It can also provide protection against developments that may not be beneficial to the local community or local environment.

This framework needs to be in line with the national or local political culture and reflects historically evolved 'ways of doing'—for example, traditions of administrative coordination between levels of government, with more or less autonomy for local governments in making decisions on local land use (e.g., Kahn, 2003; Söderholm et al., 2007; Bergek and Jacobsson, 2010).

Whether conflict related to project siting is likely to occur depends greatly on the specific context and on the type of project under consideration. For instance, potential wind energy projects might face significant barriers in locations where landscape amenity is a cultural-historical value (Cowell, 2010; Nadaï and Labussière, 2010), but have less trouble gaining acceptance where this is not the case (Toke et al., 2008).

The successful deployment of RE technologies to date has depended on a combination of favourable procedures at both national and local levels. Universal procedural fixes, such as 'streamlining' of permitting applications, are unlikely to resolve conflicts among stakeholders at the level of project deployment because they would ignore place- and scale-specific conditions (Breukers and Wolsink, 2007b; Agterbosch et al., 2009; Ellis et al., 2009). Recent evidence in the siting and planning of RE points to the need for systems that are pro-active, positive and place- and scale-sensitive. Following are elements that such planning systems might include.

11.6.4.1 Aligning stakeholder expectations and interests

Several case studies in RE planning processes have shown the importance of aligning interests among various stakeholders (Devine-Wright, 2005; Warren and McFadyen, 2010). This can be done in a variety ways, including adopting procedures for project development that are judged fair by the different parties (Gross, 2007), or identifying (creating, negotiating) during the 'pre-application process' multiple benefits that a RE project may bring for different stakeholders (Heiskanen et al., 2008a; Ellis et al., 2009).

11.6.4.2 Learning about the importance of context for RE deployment

Those who object to projects are often very knowledgeable (Ellis et al., 2007) and cannot be dismissed as simply ignorant or misinformed. Understanding the local societal context of RE could help RE planning processes overcome the hurdles they face (Breukers and Wolsink, 2007a; Raven et al., 2008).

11.6.4.3 Adopting benefit-sharing mechanisms

Benefits associated with RE projects (for example, social, environmental, or financial/economic (Madlener, 2007; J. Rogers et al., 2008; Walker, 2008)) accrue mostly to the project developer and to broader society (beyond the area directly affected by a specific project) (e.g., D. Bell et al., 2005).

An acknowledgement that benefits, costs and risks are unequally distributed, followed by efforts to arrive at a more equitable benefit sharing, is helpful. Participation of local communities in the benefits generated by development of a specific project, may include co-ownership (Deepchand, 2002; Meyer, 2007; Walker, 2008; Warren and McFadyen, 2010), as seen in Denmark (see Box 11.12); local employment by making use of/setting up local contractors and services (Faulin et al., 2006; Agterbosch and Breukers, 2008; Heiskanen et al., 2008a); direct reinvestment by developers into infrastructures of the local community (Upreti and Van Der Horst, 2004; Aitken, 2010); transfer of benefits through lump sum or business tax to local communities (Faulin et al., 2006; Nadaï, 2007); energy price reduction (Deepchand, 2002); or environmental compensation (Cowell, 2007). Some studies have shown that local economic involvement favoured a better acceptance of RE projects (Jobert et al., 2007; Maruyama et al., 2007).

Box 11.12 | Lessons from Denmark: The value of a comprehensive approach and individual and community ownership.

Since the 1970s, wind power has developed into a mainstream technology in the Danish energy system, generating 20% of Denmark's electricity by 2009. In 2009, the Danish wind industry was the country's largest manufacturing industry, employing some 24,000 people (Danish Wind Industry Association, 2010) and accounting for 20% of the global market (BTM Consult ApS, 2010).

The first oil crisis brought concern about energy security, and energy efficiency and RE became top political priorities. In the 1980s and beyond, energy security, creation of domestic jobs and export markets were the major drivers for transformation of the Danish energy sector (Danish Ministry of Energy, 1981).

A combination of policy mechanisms, guided by national energy plans with long-term targets, has facilitated RE development. A publicly funded R&D programme began in 1976 with the goal of designing and testing megawatt-scale turbines. A small turbine test station was established at Risø National Laboratory; interaction between the test station and small enterprises in the industry helped feed experience back into the field to improve basic knowledge about turbine design (Sawin, 2001; Madsen, 2009).

In 1979, the government introduced its first and most important policy to stimulate the market, based on a 30% investment grant to purchasers of 'system-approved' wind turbines. This 10-year programme saw regular reductions in the grant level as technology improvements and economies of scale reduced costs. The investment grants to end users (private investors) created a small but strong industry by the early 1980s (Madsen, 2009). In 1985, the government enacted a per-kilowatt hour subsidy for all wind power fed into the grid, funded in part through a tax on CO_2. A voluntary feed-in tariff (equivalent to 85% of the retail rate) paid by utilities to wind producers was fixed by law in 1992 (Sawin, 2001; Madsen, 2009).

Private investors, often organized in small cooperatives, owned more than 80% of total installed capacity through the 1990s. This was largely due to a number of government policies, from special tax breaks to ownership limitations, to encourage local individual and cooperative ownership (Madsen, 2009). During the pioneering period, incentives for individuals and cooperatives encouraged municipalities to set aside specific areas for turbines. In 1992, the Danish Planning Agency launched guidelines that accelerated the permitting process and established capacity targets for all Danish counties, thereby eliminating uncertainty about siting while giving communities control over where projects were located (Danish Ministry of the Environment, 1993; Sawin, 2001).

Also important were Ministry of Energy 'contract policies', which required utilities to participate in wind power development. Under the first such contract, initiated in 1985, utilities were required to construct 100 MW of wind capacity over five years. The utility mandate was extended twice, and the first requirement for offshore capacity was issued in 1990 (Sawin, 2001).

Nearly three decades of consistent policy were interrupted in the early 2000s when leadership changed, the per-kilowatt hour subsidy was significantly reduced, and deregulation of the electricity sector created uncertainty (see Figure 11.10). Little new capacity was added until 2008 because most projects were not economically feasible, and changes in planning structure delayed siting and installation of larger turbines (Madsen, 2009).

The government has since changed its position, announcing a political target of a '100% fossil-free' energy system by 2050. As of 2009, Denmark aimed to get nearly 20% of total energy from RE sources by 2012 and 30% by 2020, with wind power playing a major role (European Union, 2009). As a result, development has picked up again.

Consistent support for public R&D in Denmark played a critical role in the advancement of wind power technology, education of technical experts and development of a manufacturing base. Market stimulation in the form of direct grants and later fixed feed-in tariffs, which reduced risk to investors, was essential for increasing deployment, reducing costs and creating broad-based support and a strong domestic industry, but significant policy changes and uncertainty stalled development for several years. Finally, Denmark's experience demonstrates that local ownership of wind power plants can facilitate market development.

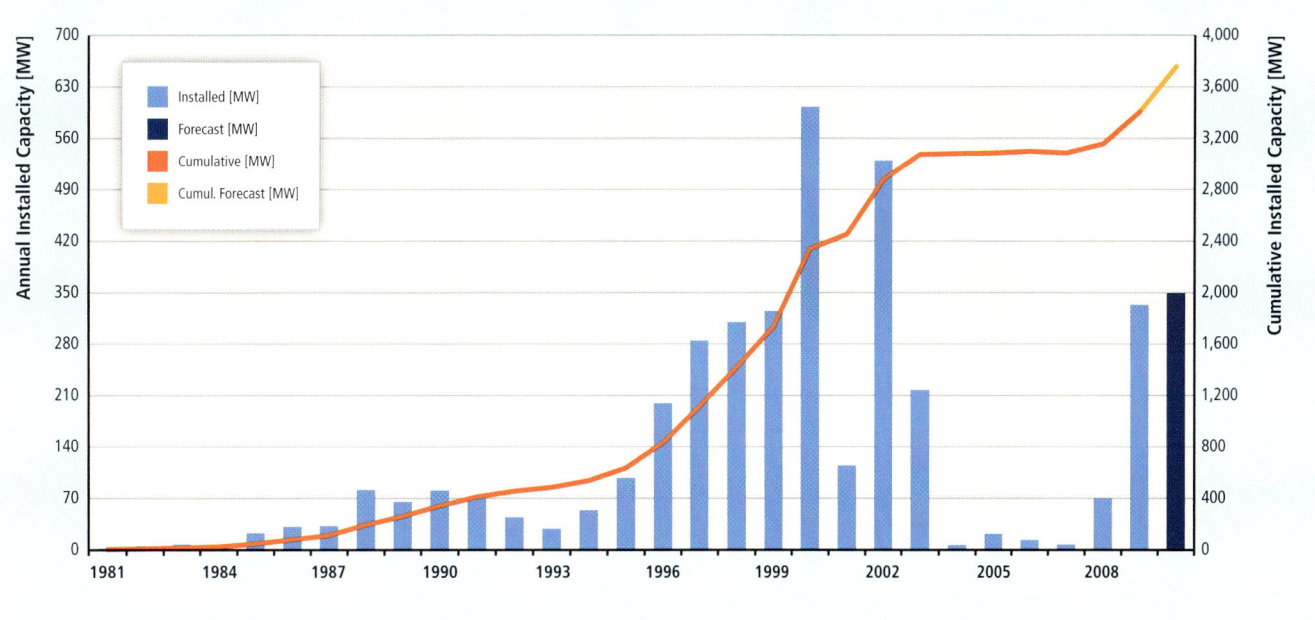

Figure 11.10 | Denmark's annual and cumulative installed wind capacity, 1995 to 2010 (BTM Consult ApS, 2010).

11.6.4.4 Timing: pro-active national and local government

Clear procedural rules (e.g., requirements for permitting, ground for court appeal, allocation of responsibilities and timing of the process) are important to reduce risks for the developer and to ensure legal security for other stakeholders.

National planning policies sometimes lag behind initiatives of those deploying innovative technologies, and therefore may hamper these innovations. Legislative changes or case-by-case approaches that account for technology- and scale-specific challenges might be required. For example, ocean energy projects at an early commercial stage occasionally face a 'catch-22' situation in which the existing permitting regime requires project impact data that could be produced only if they were granted temporary authorization (IEA, 2009a). In such cases, project license leases, pilot development zones, or specific site agreements have been used as tailored solutions.

Local governments are also often caught by surprise when a project developer presents a RE project proposal (Breukers and Wolsink, 2007a; Nadaï and Labussière, 2010). Organizing local participation in the development of comprehensive plans and identifying main siting areas before any projects are planned makes it easier to create an open and non-polarized discussion, as seen in Denmark (Sussman, 2008).

Finally, explicit political support for RE at the national level can reduce local polarization by encouraging the perception of RE and associated impacts as public rather than private issues (Bergek and Jacobsson, 2010).

11.6.4.5 Building collaborative networks

If relevant stakeholders are brought into the RE project process and become part of the agreement for RE deployment, their long-term acceptance and lasting commitment toward a project are more likely to come about than if this does not occur. Further, networks that result can be important 'vehicles' for exchanging experience and knowledge; this in turn supports learning processes that stimulate change, such as policies or institutions that further help RE development (Breukers and Wolsink, 2007b; Mallett, 2007; Negro et al., 2007; Dinica, 2008; Heiskanen et al., 2008b; Suurs and Hekkert, 2009). Or, collaboration could bring about radical innovation in 'ways of doing', such as finding innovative ways to renew landscape values or protect birds in relation to wind power (e.g., Ellis et al., 2007; Nadaï and Labussière, 2009, 2010).

11.6.4.6 Mechanisms for articulating conflict and negotiation

The deployment of a RE project will rarely serve the interests of all stakeholders. Yet, existing formal avenues to voice opposition usually offer only the opportunity to object to ready-made project proposals (Wolsink, 2000). This can lead to polarization and be counterproductive (Healey, 1997). It is useful to enable the articulation of differing perspectives to allow parties to reach subsequent solutions or compromises through constructive deliberation (Cuppen et al., 2010). For example, following enactment of the Energy Policy Act of 2005, the US Departments of Energy and the Interior identified 24 tracts of land for large-scale solar energy development in six Western states, and then encouraged public participation in the studies of those areas through public scoping meetings, public comment on the draft programmatic environmental impact statement, and via a comprehensive project website (US Department of the Interior, 2008; ANL, 2010).

11.6.5 Providing infrastructures, networks and markets for renewable energy

After a RE project receives planning permission, investment to build it is only forthcoming once its economic connection to a network is agreed; when it has a contract for the 'off-take' of its production into the network; and when its sale of energy, usually via a market, is assured. The ability, ease and cost of fulfilling these requirements is central to the feasibility of a RE project. Moreover, the methods by which RE is integrated into the energy system will have an effect on the total system cost of RE integration (see Chapter 8) and the cost of different scenario pathways (see Chapter 10). This section discusses integration as it relates to enabling policies and available solutions. It is heavily weighted towards electricity because most experience has been in this sector; electricity is also relevant to both RE electric water and space heating and cooling, and to RE electric transportation. (See Section 8.2.1 for details related to technical integration.)

The economic regulation overseeing these areas is often technology- and fuel- 'blind', meaning that there is no differentiation made between technologies or fuels. Even so, however, it is possible for policies to be implemented to facilitate RE connection to networks and access to markets and to ensure that infrastructure requirements specific to RE are made in a timely and cost-effective fashion.

11.6.5.1 Infrastructure building and connection to networks

Planning and investment in network infrastructure present challenges due to the large economies of scale in network investments (or the 'lumpiness' of transmission) and the broad impacts and beneficiaries of network expansion (Keller and Wild, 2004). These issues are particularly challenging in countries and regions where vertical separation exists between the generation, transmission and distribution of electricity to electrical customers. Significant debate and diverse policies regarding network investment exist throughout North America and Europe, for example; both regions where generation is largely vertically separated from transmission (see Joskow, 2005; Buijs et al., 2010).

One of the key policy debates regarding network infrastructure investments is that of cost allocation. Most policies generally fall between the two extremes of 1) socialized cost allocation, in which all network users share the burden of covering the cost of any network expansion, and 2) 'beneficiaries pay', where only those network users that benefit from specific network upgrades are responsible for paying the network investment costs (Krapels, 2010).

The connection of RE to networks and the expansion of the network to accommodate increased power flow between RE generation and demand will occur within this broader framework and may, due to the unique characteristics of RE, exacerbate some of the challenges. RE resources, for example, are often concentrated in areas where existing electricity networks have limited extra capacity for transporting additional electricity. These areas also may be a long way from centres of energy demand (see Section 8.2.1.2). With regard to RE, proponents of a 'beneficiary pays' type of mechanism argue that socialized network expansion costs may lead to inefficient siting of RE projects if individual projects do not bear any of the costs of network expansion. RE projects may locate in areas with the highest quality resources but, due to the additional network costs, these areas may not always be as economically efficient as RE resources in lower-quality regions that are closer to demand centres or existing network capacity (e.g., Hoppcock and Patiño-Echeverri, 2010).

Proponents of socialized cost-type mechanisms point out that network investments are long-term infrastructure investments and that they benefit a broad range of network users that may change as the system evolves. Furthermore, the large economies of scale involved with network expansion and the large size of RE resources relative to individual RE projects often leads to the most cost-effective network expansion, far exceeding the size required by an individual RE project. Policies that require individual RE projects to finance network expansion may therefore stifle efficient development of properly sized transmission investment (Puga and Lesser, 2009). Moreover, if the individual RE project must bear all of the costs of the larger, more efficiently sized network expansion, a project that otherwise may be economically efficient may become economically infeasible (Access Reform Options Development Group, 2006).

A further challenge is that the time it takes to plan, site and build transmission infrastructure sometimes well exceeds the time it takes to plan, site and build certain RE facilities. This difficulty can be exacerbated because most economic regulation of networks is based on the principle of 'ex-ante' cost regulation (Baldwin and Black, 2010). This means that network operators often must have regulatory approval in advance of undertaking the strengthening of the network. Before approving individual network reinforcements, however, regulators may require a clear financial commitment from generators or customers of their intention to connect to the network and utilize network assets. However, potential RE generators are unlikely to be able to commit financially to network reinforcement without planning consent; and they may be loathe to spend money on achieving planning consent without knowing the costs of connection. This presents a 'catch-22' situation, which is often further complicated by the disparity between RE project and network reinforcement commissioning time scales (Locke Lord Bissell & Liddell, 2007).

In order to ensure the timely expansion and reinforcement of infrastructure and connection of RE projects, economic regulators may need to allow 'anticipatory' or 'proactive' network investment and/or allow projects to connect in advance of full infrastructure reinforcement (Araneda et al., 2010) (see Section 8.2.1.3 for examples of these policies being applied in practice). Traditionally within economic regulation, allowing anticipatory investment is thought to increase the risk of stranded assets. Policies that provide incentives could be allowed to the network operators to account for the extra risk of such investment decisions, for example by allowing enhanced rates of return on investments (Ofgem, 2008), or otherwise end-use customers could be asked to front the cost of the necessary transmission upgrades.

11.6.5.2 Access to and injection of renewable energy into the network

The rules and costs of how energy is injected into the network, whether a system operator has the right to refuse the RE, and whether the RE project is paid if it is refused access to the network all have major implications for the economics of electricity power plants and their ability to obtain investment (Strbac, 2007).

RE-specific policies can sometimes bypass these complex negotiations. In the EU, the Directive 2001/77/EC on the promotion of electricity produced from RE sources states that EU member states must ensure that transmission and distribution system operators guarantee network access for electricity generated by RE (European Commission, 2009a). This is both connection and off-take (i.e., injection into the grid). In general, but not always, a fundamental design feature of a FIT is a project's connection to the network, and the off-take of the electricity, according to a defined process and remuneration. As a result of the EU Directive, some European countries, particularly those which have FITs, have implemented interconnecting regulations that guarantee access to the network.

In other regions, access may be granted to new RE generation, but electricity generated by RE can be curtailed for economic or reliability reasons. Recent experience with curtailment of wind demonstrates that there are many different policies in place that restrict the injection of wind into networks under constrained conditions and many different policies to compensate wind generation during times where curtailment occurs (Fink et al., 2009).

11.6.5.3 Network standards

Historically, network design standards identify the reinforcement requirements triggered by an energy plant connecting to them to reach a particular level of network security. Alteration of network standards, ahead of time, that take account of RE technical characteristics and that maintain system security can avoid connection and system operation concerns. The UK, for example, has had a series of Work Groups since 2001 whose role is to highlight and recommend how to overcome potential concerns ahead of time (see DTI/Ofgem Embedded Generation Working Group, 2001; National Grid, 2008). In addition to standards for network reinforcements, network operators may also impose minimum performance or equipment requirements on generators in order to allow the plant to be connected. These requirements are often called 'grid codes' or 'interconnection standards' (see Sections 7.5.2.2 and 8.2.1.1).

11.6.5.4 Increasing resilience of the system

One of the significant challenges for integrating RE into the electricity sector in particular is dealing with the variability and uncertainty of some RE resources. As the percentage of RE increases there is an increasing requirement for resilience within the energy system (P. Baker et al., 2009), which is determined by a system's capacity to integrate variable energy output while matching energy demand. Policies can be put in place to facilitate such integration.

Policies might first recognize the variability smoothing effects of diversity for RE production (i.e., aggregation reduces forecasting and integration challenges (IEA, 2008a)). Similarly, policies might ensure the incorporation of aggregate RE production data (actual and forecasted) into electricity market operations by creating new mechanisms or altering rules. Spain, for example, has chosen to encourage RE by requiring the mandatory aggregation of all wind power plant data in Delegated Control Centres, which involves online communication with the National Renewable Energy Control Centre (Morales et al., 2008; Rodriguez, et al., 2008).

Similarly, since variable output RE such as wind cannot be forecast as accurately as far in advance as other energy resources, RE can be accommodated by 'balancing' the electricity as near to real time as possible, such as an hour ahead rather than three hours ahead or a day ahead. Flexible electricity trading rules can reduce the impact of forecast errors on electricity market operations (IEA, 2008a). There are also several changes to the power system that can increase the ability of the system to manage variable and uncertain RE generation. These changes will often require revisions to existing policies. In addition to the already-mentioned examples, increasing interconnection capacity within systems, adopting demand-side management measures that include real-time pricing (e.g., Sioshansi and Short, 2010), increasing storage capacity, using more flexible thermal generation, and improving planning methods are all examples of the measures that would also help to integrate variable RE (Alonso et al., 2008) (see Section 8.2.1.3 for further details).

11.6.6 Technology transfer and capacity building

Barriers to technology transfer in RE and other low-carbon technologies have been identified as being institutional, economic, informational, technological and social (UNFCCC, 1998; IPCC, 2000; Wilkins, 2002; Kline et al., 2004). It has been argued that many developing nations are unlikely to 'leapfrog' pollution-intensive stages of industrial development without access to clean technologies that have been developed in more advanced economies (Gallagher, 2006; Sauter and Watson, 2008). The reality is that most low-carbon technologies, including RE technologies, are developed and concentrated in a few countries. A recent study (UNEP et al., 2010) of patenting in selected RE technologies finds that six countries—Japan, the USA, Germany, the Republic of Korea, the UK and France—account for almost 80% of all patent applications. Accessing, adapting and diffusing these technologies to developing (and other developed) countries could greatly facilitate their ability to contribute to the mitigation of climate change.

Technology transfer is not the exclusive domain of any one actor, and technologies can be transferred from developed countries to other developed or even developing countries, not just from the developed to developing world. Also important is that clean technologies typically do not flow across borders unless environmental policies in the recipient country provide incentives for their adoption (e.g., Jha, 2009; Lovely and Popp, 2011).

An important insight in the evolution of technology and innovation (Mytelka, 2007; Roffe and Tesfachew, undated) in the past thirty years is the recognition that technology transfer is not just an end in itself, but a means to achieving a greater strategy of technological capacity building. Technology transfer is a process, not a one-off transaction. It occurs primarily between firms via the market, through the consumption of products or services that incorporate a specific technology; through licensing the capability to produce such products, either by an indigenous firm or through a joint venture arrangement or foreign direct investment (Kim, 1991, 1997; UNCTAD, 2010c).

Nor should technology transfer be considered only the transfer of hardware from one country to another (Dosi, 1982). Technology transfer can take place within countries (e.g., from urban to rural areas), between industries, academia and nongovernmental organizations. And in most cases it also includes transfer of skills and know-how, as well as knowledge and expertise embedded in the technology (M. Bell, 1990, 2007; IPCC, 2000; Ockwell et al., 2010)—in other words, a combination of 'hardware, software and orgware' (Fodella, 1989). Figure 11.11 illustrates the different types of technological content of technology transfer between countries.

11.6.6.1 Technology transfer and intellectual property rights

The role of intellectual property rights (IPRs) in the technology transfer process has been the source of much debate and controversy in the context of international climate change negotiations. Some empirical studies (Ockwell et al., 2010) suggest that intellectual property protection is a necessary but insufficient condition for the success of low-carbon technology transfer. The most recent empirical study (UNEP et al., 2010), carried out by UNEP, the International Centre for Trade and Sustainable Development (ICTSD) and the European Patent Office, finds that firms attach slightly more importance to scientific infrastructure, human capital, favourable market conditions and investment climates than IPR in their licensing decisions. The same study also revealed that 70% of the respondents were prepared to offer flexible licensing agreements to poor developing countries. However, there is evidence that technology transfer is inhibited in countries with high tariffs and lax intellectual property rights.

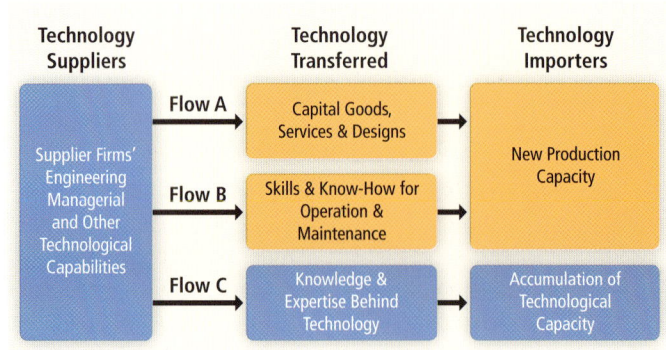

Figure 11.11 | The different types of technological content in technology transfer between countries (Ockwell et al., 2010; based on M. Bell, 1990).

11.6.6.2 Technology transfer and international institutions

Development cooperation plays a major role in driving the adoption of RE in developing countries, many of which are undergoing considerable economic and infrastructure development that could result in lock-in to fossil fuel technologies without easy access to low-carbon technologies (IPCC, 2007). Mechanisms established within the UN Framework Convention on Climate Change (UNFCCC) process to facilitate development and transfer of clean technologies include an Expert Group on Technology Transfer (EGTT), the Global Environment Facility (GEF), the Clean Development Mechanism (CDM) and Joint Implementation (JI) (UNFCCC, 2007b). Development agencies and financing institutions demonstrate innovative technologies, provide soft loans for sector investment plans and pave the road for market introduction or promote technology deployment by means of international carbon finance, all of which is conducive for investment.

Incentives for technology transfer are currently included in mechanisms under the Kyoto Protocol, including the CDM. The CDM allows developed countries to meet their Kyoto Protocol commitments by financing emission reduction projects in developing countries. Even though the first projects were not registered until 2004, an analysis of international transfer of wind power technologies, covering 100 countries during the period 1988 to 2007, found that the CDM had a significant impact (Haščič and Johnstone, 2009).

Several studies have analyzed technology transfer associated with CDM projects (Haites et al., 2006; de Coninck et al., 2007; UNFCCC, 2007a, 2008, 2010; Dechezleprêtre et al., 2008; Schneider et al., 2008; Seres et al., 2009), and determined that roughly 40% of projects, which accounted for about 60% of the emission reductions up to 2009, involved technology transfer. The decline in the rate of technology transfer for CDM projects over time suggests that mitigation technologies are being developed in, or transferred to, host countries through conventional channels such as trade, foreign direct investment and licensing (Hoekman et al., 2004; UNFCCC, 2010).

11.6.6.3 Technology transfer and energy access

Looking at the sub-national level, the rural poor in developing countries who lack access to modern energy services are increasingly left out of the technology transfer debate. The type of innovative capabilities required tend to involve the adoption and adaptation of technologies to suit local conditions and needs, or supply chain management, rather than innovating at the technological frontier as technology producers. In order to have the capacity to adapt, install, maintain, repair and improve on RE technologies in remote and rural communities, investment in technology transfer must be complemented by investment in community-based extension services that provide expertise, advice and training regarding installation, technology adaptation, repair and maintenance (Ockwell et al., 2009; UNCTAD, 2010a) (see Box 11.13).

The United Nations Commission on Science and Technology (CSTD) (UNCTAD, 2010b) suggests that new, international collaborative approaches to low-carbon technology research and development are needed to facilitate North-South and South-South technology transfer. It calls on the UN Conference on Trade and Development (UNCTAD) as well as other UN entities to explore the structure of internationally collaborative R&D mechanisms that might be effective in facilitating low-carbon technology transfer and learning with and from actors beyond national governments.

11.6.7 Institutional learning

In addition to technology transfer, institutional learning plays an important role in advancing deployment of RE. Institutional learning is conducive to institutional change, which provides space for institutions to improve the choice and design of RE policies. It also encourages a stronger institutional capacity at the deeper, often more local, level where numerous decisions on siting and investments in RE projects need to be made (Thelen, 1999; Breukers and Wolsink, 2007a). Private actors and civil society (e.g., regional energy distributors, small wind power entrepreneurs, local mayors, researchers) develop new social skills such as management styles and informal contacts through collaboration. They also rely on existing social conditions (e.g., trust or social coherence) in order to move through the prevailing institutional structure—including electricity regulation, nature conservation norms and planning procedures— in order to get RE projects developed (Agterbosch et al., 2009). Their insights can inform and influence policies to improve RE deployment. Institutional learning can occur if policymakers are able draw on these nongovernmental actors for collaborative approaches in policymaking. Others emphasize the gain in being flexible and reflexive because policymakers can learn from what happens, experiment, look for best practice, re-evaluate and so on (Smith et al., 2005; Stirling, 2009)

11.6.8 A role for cities and communities

Cities, towns, local authorities and communities, which often incorporate RE into their policies, have the potential to play an important role in climate change mitigation (Droege, 2009; IEA, 2009a) (see Box 11.14). Droege (2009) argues that whether and how cities and communities are able to implement climate change and RE policies both depend on their spatial, environmental, social and economic capacities to implement RE. Nearly 20% of city and local governments surveyed for a REN21 study have some sort of building code or permitting policy that incorporates RE. Mandates for solar water heating in new construction are in place in many countries, states and cities worldwide. Other mandates include designing buildings to include features that ease future installations of renewable energy technologies (REN21, 2010).

Both Droege (2009) and the IEA (2009a) conclude that local initiatives occur in places where there are people who understand the technical aspects of RE (i.e., technically literate) and that positive local experiences reinforce other local experiences. Local policymakers have support groups (for example, Local Governments for Sustainability (ICLEI), an association of 1,200 local government members).

11.6.8.1 Community and individual links

Communities provide the social experiences that individuals encounter beyond their own households. A growing body of research has found that social norms influence energy-related behaviour and that 'social visibility' of energy underlies social norms (Nolan et al., 2008; Wilson, 2008). Social visibility describes the extent to which people's attitudes and behaviour towards RE is communicated through social networks (Schultz, 2002). This type of social communication is central to the diffusion process for innovations, including many examples of distributed RE (Archer et al., 1987; E. Rogers, 2003; Jager,W, 2006). The physical visibility of residential wind or solar may help RE become a day-to-day talking point, and so enhance its 'social visibility' (Hanson et al., 2006) and the converse is true of poorly visible technologies such as micro-CHP or energy efficiency. Demonstration projects help promote 'social visibility' and allow potential adopters to observe, learn and communicate about, and test RE technologies vicariously. With solar PV for example, demonstration projects helped breed familiarity and reduce perceived risks for Dutch homeowners and US utility managers alike (Kaplan, 1999; Jager, 2006).

11.6.8.2 A role for individuals as part of civil society

The influence of supportive social norms may also be limited. In a household context, RE technologies have been described as limited by ritual and lifestyle (Sovacool, 2009a). Past experiences and habits are a key

Box 11.13 | Lessons from Nepal: Importance of upfront public investments in capacity building.

The National Micro-Hydropower Programme in Nepal aims to enhance rural livelihoods and human development by accelerating the achievement of the Millennium Development Goals, primarily through the delivery of community-managed micro-hydropower systems (MHS). The Programme is coordinated by the Alternative Energy Promotion Centre (AEPC), a centre established under the Ministry of Environment to serve as a national agency for coordinating and monitoring alternative energy development programmes in Nepal (UNDP and AEPC, 2010).

Field experiences from the programme between 1996 and 2006 revealed that capacity development is central to successfully scaling up decentralized energy access programmes and attracting private financing. Capacity development efforts went far beyond training and management to include: planning, oversight, and monitoring; situational analysis; facilitation of stakeholder dialogue, communications and community mobilization; training; setting up and/or strengthening institutions, implementation capacities and management support; and the provision of policy advice (UNDP and AEPC, 2010).

Given the considerable planning, situational analysis and institution set-up efforts, especially at the national level, more than 90% of the early programme costs went to capacity development. As such, the upfront, publicly-funded investment (from government and donors) was essential to developing the functional capacities needed to scale up the rural energy programme (UNDP and AEPC, 2010).

However, when capacity development is created by systematic interventions, programme successes and maturation over time, it can enable market transformation to occur. Indeed, the study found that the share of public financing for the micro-hydro programme gradually declined to about 50%, attracting substantial private sector funding in later stages of the programme. This indicates the important role of public investment in capacity development for attracting private financing sources, particularly decentralized sources among a project's many users/beneficiaries. Communities provided cash, acquired bank loans and supplied in-kind labour contributions—by digging channels for the MHS, for example—making up a significant portion of the overall financing needs (UNDP and AEPC, 2010).

Productive uses of the resulting energy services fuelled rural economies and increased the possibility for attracting further private investments, including through micro-finance. Fostering ownership also proved to be a necessary sustainability component, providing an incentive for users to use and maintain the technology properly (UNDP and AEPC, 2010).

Local action on the ground, which resulted from training and community mobilization, informed local and district institutions, which were created as a result of capacity development in the form of institutional set-up and strengthening. That, in turn, informed institutions at the national level, which used the knowledge gained to provide the functional capacity of 'policy development and advice'. Although this functional capacity makes up only a small proportion of the total capacity development cost, policy development and advice plays a major role in informing policy and regulation development, supporting overall programme success and sustainability, such as the enactment of a Rural Energy Policy in 2006 (UNDP and AEPC, 2010).

Knowledge gained through the positive experiences of alternative energy development programmes was used to develop Nepal's Rural Energy Policy, which aims to motivate and mobilize local institutions, rural energy users groups, nongovernmental organizations, cooperatives and private sector organizations for the development and expansion of rural energy resources for the purposes of providing energy access and furthering rural economic development and job creation (Government of Nepal, 2006).

In summary, the Nepal programme found that capacity building, broadly defined, was critical for successful scale-up. Further, involving stakeholders in the local community and promoting a sense of ownership was important for sustainability of the projects. It concluded that considerable upfront public investment is needed to develop local and national capacities through systematic interventions and to inform policy development to scale-up rural energy service delivery; however, once these upfront investments are made, they can attract substantial financing from private sources at later stages of the programme, and subsequently reap significant economic, social and environmental benefits (UNDP and AEPC, 2010).

> **Box 11.14 | Lessons from Güssing, Austria: Potential for rapid transition in a community's energy production and use.**
>
> Güssing in Austria was the first town in the EU to reduce its carbon emissions by 90% (below 1992 levels) and today is a model for environmentally friendly energy production based on energy saving, self-sufficiency and environmental protection. Thirty RE plants—solid biomass, biodiesel, biogas and PV facilities—operate within 10 km of Güssing and meet the town's fuel demands for transportation, residential heating and electricity. Electricity produced locally and sold into the grid has increased local revenue, with profits reinvested into the community and its RE projects. By 2009, Güssing's renewable profile had attracted 60 companies wanting to run on clean energy, creating at least 1,000 new jobs (Droege, 2009).
>
> The town's transformation began in the late 1980s when a massive fuel debt prompted the local mayor to enforce energy-saving measures and begin phasing out fossil fuel use in all sectors, replacing it with locally supplied RE (Droege, 2009). The municipal government initiated and supported financially the construction of local RE plants, which were locally managed and provided the town and greater region with energy services (BMVIT, 2007). It also implemented policies to manage and sustain local farms and forests to produce raw material for generating bioenergy (Droege, 2009). Several local and regional public and private research institutions provided technological assistance, while grants from regional authorities, the Austrian government and the European Commission helped with construction of new infrastructure, such as the district heating system (Droege, 2009).
>
> A municipal marketing program promoted RE through the internet, brochures, exhibitions and conferences as a means to attract companies to the area. But the municipality is also working to export its model, and Güssing's specialized centre on RE has helped to raise public awareness about clean energy and climate protection goals (Droege, 2009).
>
> Within two years of embarking on this path, Güssing's energy expenditures were reduced drastically. By 2001, Güssing was 100% self-sufficient and meeting all power and heat needs with RE (Droege, 2009).

element of an individual's behavioural context (Brennan, 2007). RE has to be seen as part of a social and technical system of energy provision and use, characterized by deeply embedded routines, social practices, patterns of time use, lifestyles and so on (Shove, 2003). These contextual factors point to the importance of collective action as a more effective, albeit more complex medium for change than individual action. This supports coordinated, systemic policies that go beyond narrow 'attitude-behaviour-change' policies if a policymaker wishes to involve individuals in the RE transition.

Information and education are often emphasized as key policy tools for influencing energy-related behaviours. They are relatively low-cost, uncontroversial and potentially empowering instruments of autonomous choice, favoured over coercion from an individual standpoint (Attari et al., 2009). However, impacts on behaviour are diffuse, long-term and hard to measure because values concerning the environment do not have a strong correlation with behaviour (Gatersleben et al., 2002; Poortinga et al., 2004). This cautions against an over-reliance on information- and education-based policies alone.

Individuals as part of civil society can play an important part in moving to a low-carbon economy, as seen in the Austrian town of Güssing (Box 11.14), as well as in many of the scenarios reviewed in Chapter 10. There is no universal model or understanding of what motivates such behaviours. Rather, a host of factors and constraints influences energy-related behaviours, but these factors do not necessarily exert influence directly. Some sources of influence are intentional. These include information policy, public education or policy signals (such as energy prices, financial incentives). Other influences are part of an individual's everyday environment. These include household routines and relationships, social practices and the inter-personal networks through which individuals communicate (Poortinga, 2004).

11.7 A structural shift

There is now substantial evidence that RE policies have had an impact on technology development and RE deployment in many countries, and that some policies or specific elements of policies have been more effective and efficient in advancing RE. However, RE's share of energy production is still limited in most countries. On a global basis, RE accounted for an estimated 12.9 % of primary energy supply in 2008 (Section 1.3; IEA, 2010d). And although some countries can now look back on two to three decades of national experience with and lessons from RE policy, a shorter time series of data is available in most countries. Therefore, trying to assess what is needed for achieving a high share of RE is subject to substantial uncertainties. Further research is also needed to fully understand the effectiveness and efficiency of combinations of policy instruments designed to achieve a very high share of RE in the long term.

11.7.1 The link between scenarios and policies

The scenarios presented in Chapter 10 demonstrate that a wide range of energy futures is possible, differing in their shares of RE up to 77% of primary energy by 2050. Conceptually, the scenarios can be distinguished into the four quadrants of potential energy futures, as seen in Table 11.5.

Table 11.5 | Conceptual placement of Chapter 10 scenarios against RE and energy efficiency levels.

(3) High energy efficiency; low shares of RE	(4) High energy efficiency; high shares of RE
(1) Low energy efficiency; low shares of RE	(2) Low energy efficiency; high shares of RE

When comparing these quadrants, a few policy differences become apparent. First, those scenarios that fall into quadrant (Q) 2 seem to bear a higher risk of overshooting global carbon targets than do scenarios in Q4. Second, given the need to create energy systems at a larger scale in a world characterized by high energy demand, scenarios in Q2 are more capital intensive on the supply side, although the necessary investment for RE and attendant infrastructure depends on the absolute contribution of RE. Thirdly, there are different societal risks involved in the two kinds of high RE scenarios (Q2 and Q4). Those scenarios that combine high RE with high energy efficiency rely on either active energy efficiency policies (which may create barriers to political acceptance) or they assume significant fuel (oil and gas) price shocks and an appropriate reaction from the consumer side and policy (for instance supporting structures or quick substitutions of fossil fuel and/or nuclear power technologies with low energy efficiency). On the other hand, the high RE, low energy efficiency scenarios rely on greater levels of deployment of RE supply infrastructure, which in turn could become an issue of social acceptance in many countries.

11.7.2 Structural shifts result from a combination of technology and behaviour change

An important, yet often implicit dimension of energy scenarios is whether the scenarios assume changes to be mainly driven by technological development, or whether they assume changes in behaviour as a driver for future development of energy systems. Scenarios that assume changes through technological development can also be differentiated between futures characterized by incremental technological changes and those based on disruptive technological change (Christensen, 1997). Similarly, the scenarios that assume changes in behaviour can also be differentiated between those that are incremental and those that are disruptive. 'Disruptive' refers to a new, low-cost, often simpler technology that displaces an existing technology and, in doing so, radically transforms or destroys existing markets in order to make way for new technologies or systems (Christensen, 1997). It implies the opposite of gradual or incremental changes.

Most 'business-as-usual' scenarios are based on the assumption that both technological development and behaviour change remain incremental. As a consequence, a high share of RE is relatively unlikely in these scenarios. However, in the disruptive, technology-optimistic world (Friedrichs, 2010), the scenarios reflect a leap in the competitiveness of RE, leading to higher market penetration (for similar arguments related to other examples of low-carbon technologies, see Von Weizsäcker et al. (1998); Lovins et al. (2004)). However, behaviours and lifestyles in these scenarios resemble the business-as-usual world, and hence levels of energy consumption remain high.

In order to achieve a future energy mix based on a high share of RE and high energy efficiency, or to be in Q4, it is likely that disruptive changes will need to occur in both technology and behaviour.

11.7.3 Addressing the challenges of governing long-term energy transitions

Given that many RE technologies still have to reap considerable learning economies, there is the potential that short term-oriented policy assessment will undervalue the longer-term benefits that could accrue from supporting technology development today. If we are to achieve a structural shift towards high shares of RE, however, what sort of policy framework might that require?

Long-term policymaking was popular between the mid-1940s and into the 1970s. At that time, it was mostly implemented in the form of government-centred, hierarchical planning processes (Hiller and Healey, 2008). The demise of this approach was due to its low ability to predict major societal transformations (e.g., the oil crisis) and its incapability to provide solutions for the ever-increasing societal and environmental problems.

However, this concept of policymaking has experienced a revival in political sciences (Voß et al., 2009; for an example, see Box 11.15). In an effort to overcome the limitations of the earlier approach, today it is framed as 'long-term policy design', an interactive process of constructing and shaping socioeconomic transformation processes (Schneider and Ingram, 1997) that look two to three decades into the future, extending well beyond the attention spans that are generally prevalent in political processes (electoral cycles, standard government programs, hiring spans of civil servants etc.). In order to support long-term structural shifts, policies have to interact with many transformative changes as they unfold. Long-term policy design thus needs to be flexible, adaptive and reflexive (Voß et al., 2009).

This new generation of approaches to governance aims at navigating and spurring the complex processes of socio-technical change by means of deliberation, probing and learning. Emphasis is put on the

Box 11.15 | The Dutch technology and innovation frameworks.

A notable example of recent innovation and technology policy frameworks aiming at a substantial increase of RE technologies is the Dutch Transition Management framework (Kemp and Rotmans, 2009). Since 2001, the Dutch ministry of economic affairs has been committed to a long-term sustainability program under the label of 'Transition Management'. It encompasses the elaboration of long-term transformation goals and associated policy mixes in sectors like energy, transport, food or housing (Loorbach, 2007). The particular approach to policy design in transition management comprises five main components: (1) Establishing a transition arena (i.e., a broad constituency of representatives from industry, politics and society that accompany the ongoing planning and implementation process); (2) developing a vision of a future sustainable sector structure; (3) identifying pathways towards these future states by means of backcasting methods; (4) setting up experiments for particularly interesting development options; and (5) monitoring, evaluation and revisions (Loorbach, 2007).

These experiences have gained considerable attention from researchers and policymakers alike. Still, many important conceptual and implementation problems remain unresolved (Kern and Howlett, 2009), and it is fair to say that the current state of Transition Management theory and praxis does not represent a readily available recipe that other countries could easily copy. Nevertheless, the issue of long-term policy design deserves considerable attention in future policy research and implementation, if policymakers decide to pursue ambitious goals of high RE shares (Meadowcroft, 2007).

---

interaction among different segments in society (government, civil society, industry etc.). Explorative scenarios, experimentation and learning therefore constitute important elements in specific policy mixes.

11.7.4 Co-evolution of 'bricolage' versus 'breakthrough'

As noted earlier, disruptive change for both technologies and behaviour is likely to be required to reach the high RE-high energy efficiency scenarios of quadrant 4 (Table 11.5). When developing a long-term policy framework for how to achieve such change, policymakers can choose amongst policies that attempt a technological 'bricolage' (aimed at change through resourcefulness and improvization on the part of involved actors, and more incremental) and/or policies that attempt technological 'breakthrough' (which is taken to evoke an image of actors attempting to generate dramatic and more disruptive outcomes (Garud and Karnøe, 2003)). Counter-intuitively, achieving disruptive technological or behavioural change is more likely to occur if bricolage and breakthrough policies are pursued together. O'Reilly and Tushman (2004) refer to ambidextrous organizations as those that master the art of simultaneously pursuing incremental and disruptive innovation (O'Reilly and Tushman, 2004). Similarly, if achieving the sustainable transformation of an industry requires a fine-tuned mix of disruptive and incremental innovation, then this implies a balanced development of emerging technologies and greening existing technologies rather than single-mindedly focusing on only one of these paths (Hockerts and Wüstenhagen, 2010).

11.7.5 Specific policy options for an accelerated transition to a high renewable energy world

Facilitating disruptive change that enables a structural shift to a low-carbon energy future, particularly one that relies heavily on RE, will require more active policy approaches for the following reasons:

- Substantial new investment is needed. In the absence of stable and predictable policy frameworks and clearly communicated long-term targets (SRU, 2010; Teske et al., 2010), investors will shy away from such investment due to perceived policy risk (IEA, 2007a; Bürer and Wüstenhagen, 2009).

- The necessary infrastructure investment may require some level of public funding or public-private partnerships (for example grid connection for offshore wind power, intercontinental trading of concentrating solar power, new storage facilities) (IEA, 2010a).

- While low levels of RE penetration can be achieved with a relatively limited number of technologies, a high-RE world is likely to rely on a broader portfolio of RE sources with differing levels of maturity. Sustained efforts of research, development and deployment at significantly higher levels than today will be required to bring these different technologies to market over time (Sanden and Azar, 2005; Neuhoff et al., 2009; IEA, 2010a).

- Technology R&D alone is not likely sufficient to ensure commercialization of new energy technologies, and there is a general consensus that both R&D and RE deployment policies are needed (Grubler et al., 1999b; Norberg-Bohm, 1999; Requate, 2005; Horbach, 2007). RE R&D investments are most effective at advancing technology and reducing costs when complemented by policies that simultaneously enhance demand for new RE technologies, thereby stimulating private sector investment in R&D.

- Strategic frameworks and long-term commitments and planning, along with flexibility to learn from experience will be critical for bringing about a structural shift. Countries like the Netherlands have implemented specific deployment policies to create protected spaces for experimentation with new energy technologies, and subsequent scale-up of promising concepts (Sanden and Azar, 2005; Voß et al., 2009).

- Two of the currently fastest growing renewable technologies, wind and solar, differ in their generation profile from current power generation technologies. A further sustained growth of these variable resources will require adaptation of electricity market rules if inefficiencies are to be avoided (Teske et al., 2010).

- Most high-RE scenarios simultaneously assume a substantial increase in energy efficiency. While some scenarios assume high shares of renewable sources at relatively high levels of energy consumption, and technical potential is high for many renewable sources, a high RE and high energy consumption scenario (quadrant 2) tends to face tighter constraints when it comes to capital requirements and social acceptance issues than does a high RE scenario that simultaneously increases energy efficiency (see Section 11.7.2). Such energy efficiency increases may be driven by market forces (e.g., fuel price shocks) or by active policies (e.g., carbon pricing, energy taxes, efficiency standards, labelling) (Teske et al., 2010).

- Both the level of energy consumption and the share of fossil and/or nuclear energy in the mix depend on strategic choices made today that are heavily interconnected to other policy areas, notably urban planning and transportation policies (Dowall, 1980; Hankey and Marshall, 2010). Achieving a high-RE world will depend on early policy integration.

- The magnitude of changes needed will require public consent to a variety of policies, which in turn implies increased efforts to raise public awareness of renewable energy (IEA, 2010a; SRU, 2010; West et al., 2010).

Synthesis

Significant investments will be required to make the transition to a low carbon future, whatever technologies are pursued (Section 10.5). Such a shift will require additional policies to attract large increases in private investment into technologies and infrastructure. From an investor's perspective, further deployment of RE technologies will result in new market opportunities.

The literature indicates that long-term objectives for RE and flexibility to learn from experience would be critical to achieve cost-effective and high penetrations of RE. To achieve GHG concentration stabilization levels with high shares of RE, a structural shift in today's energy systems will be required over the next few decades. This would require systematic development of policy frameworks that reduce risks and enable attractive returns that provide stability over a timeframe relevant to the RE and related infrastructure investments (Sections 11.6 and 11.7). The appropriate and reliable mix of instruments is even more important where energy infrastructure is still developing and energy demand is expected to increase in the future.

References

Access Reform Options Development Group (2006). *A Framework for Considering Reforms to how Generators gain Access to the GB Electricity Transmission System*. A report by the Access Reform Options Development Group, UK.

Acemoglu, D., P. Aghion, L. Bursztyn, and D. Hemous (2009). *The Environment and Directed Technical Change*. NBER Working Paper 15451, National Bureau of Economic Research. Cambridge, MA, USA.

Adelaja, A., Y.Jailu, C. Mckeown, and A. Tekle (2010). Effects of renewable energy policies on wind industry development in the U.S. *Journal of Natural Resources Policy Research*, **2**(3), pp. 245-262.

Agnolucci, P. (2007). The effect of financial constraints, technological progress and long-term contracts on tradable green certificates. *Energy Policy*, **35**(6), pp. 3347-3359.

Agterbosch, S., and S. Breukers (2008). Socio-political embedding of onshore wind power in the Netherlands and North Rhine-Westphalia. *Technology Analysis & Strategic Management*, **20**(5), pp. 633-648.

Agterbosch, S., R.M. Meertens, and W.J.V. Vermeulen (2009). The relative importance of social and institutional conditions in the planning of wind power projects. *Renewable and Sustainable Energy Reviews*, **13**(2), pp. 393-405.

Aitken, M. (2010). Wind power and community benefits: Challenges and opportunities. *Energy Policy*, **38**(10), pp. 6066-6075.

Aldy, J.E., S. Barrett, and R.N. Stavins (2003). 13+1: A Comparison of Global Climate Change Policy Architectures. *Climate Policy*, **3**(2003), pp. 373-397.

Alonso, O., J. Revuelta, M. de la Torre, and L. Coronado (2008). Spanish experience in wind energy integration. In: *Power-Gen Conference and Exhibition*, Milan, Italy, 3-5 June 2008.

Altenburg, T., H. Schmitz, and A. Stamm (2008). Breakthrough? China's and India's transition from production to innovation. *World Development*, **36**(2), pp. 325-344.

Amranand, P. (2008). *Alternative Energy, Cogeneration and Distributed Generation: Crucial Strategy for Sustainability of Thailand's Energy Sector*. Energy Policy and Planning Office (EPPO), Ministry of Energy, Bangkok, Thailand.

Amranand, P. (2009). Keynote Address: The role of renewable energy, cogeneration and distributed generation in sustainable energy development in Thailand. In: *World Renewable Energy Congress 2009 Asia, BITEC*, Bangkok, Thailand, 19-22 May 2009.

Amundsen, E.S. and J.B. Mortensen. (2001). The Danish Green Certificate Scheme: Some simple analytical results. *Energy Economics*, **23**(5), pp. 489–509.

ANL (2010). *Solar Energy Development Programmatic Environmental Impact Statement Web site*. Argonne National Laboratory (ANL), U.S. Department of Energy, Argonne, IL, USA. Available at: solareis.anl.gov/.

Araneda, J.C., S. Mocarquer, R. Moreno, and H. Rudnick (2010). Challenges on integrating renewables into the Chilean grid. In: *International Conference on Power System Technology, 2010*, Hangzhou, China, 24-28 October 2010.

Archer, D., T. Pettigrew, M. Constanzo, B. Iritani, I. Walker, and L. White (1987). Energy conservation and public policy: The mediation of individual behavior. In: *Energy Efficiency: Perspectives on Individual Behavior*. W. Kempton and M. Neiman (eds.). American Council for an Energy Efficient Economy, Washington, DC, USA, pp. 69-92.

Argote, L., S.L. Beckman, and D. Epple (1990). The Persistence and Transfer of Learning in Industrial Settings. *Management Science*, **36**(2), pp. 140-154.

ADB (2007). *Investing in Clean Energy and Low Carbon Alternatives in Asia*. Asian Development Bank (ADB), Manila, the Philippines. Available at: aequero.com/docs/Publications_Presentations/Investing_in_Clean_Energy_and_Low_Carbon_Alternatives_in_Asia_-_November_2007.pdf.

ASIF (2009). Informe anual 2009. *Hacia la consolidación de la energía Fotovoltaica en España*. Asociación de la industria Fotovoltaica (ASIF), Madrid, Spain.

Attari, S., M. Schoen, C. Davidson, M. DeKay, W. Bruine de Bruin, R. Dawes, and M. Small (2009). Preferences for change: Do individuals prefer voluntary actions, soft regulations, or hard regulations to decrease fossil fuel consumption? *Ecological Economics*, **68**(6), pp. 1701-1710.

Austrian Energy Agency (2011). *Energy in Central & Eastern Europe*. Online database, Vienna, Austria. Available at: http://www.enercee.net/.

AWEA (2010). *Windpower Outlook 2010*. American Wind Energy Association (AWEA), Washington, DC, USA.

AWEA (2011). *Fourth Quarter 2010 Market Report*. American Wind Energy Association (AWEA), Washington, DC, USA.

Azevedo, J. M. and F. D. Galiana (2009). The sugarcane ethanol power industry in Brazil: Obstacles, success and perspectives. In: *IEEE Electrical Power & Energy Conference*, Montreal, Canada, 22-23 October 2009.

Baker, E., L. Clarke, and J. Weyant (2003). *R&D as a Hedge against Climate Damages*. University of Massachusetts, Amherst, MA, USA.

Baker, P. (2010). Electricity market design for a sustainable future. UKERC Working Paper Draft, UK Energy Research Centre, London, UK. Available at: geography.exeter.ac.uk/PhilBaker/marketdesignproject2.pdf.

Baker, P., C. Mitchell, and B. Woodman (2009). *The Extent to which Economic Regulation Enables the Transition to a Sustainable Electricity System*. UKERC/WP/ESM/2009/013, UK Energy Research Centre, London, UK.

Baldwin, R., and J. Black (2010). Really responsive risk-based regulation. *Law and Policy*, **32**(2), pp. 181-213.

Bansal, R. (2009). *Identifying 'Key Value Drivers' and 'Determinants for Backward Integration' in Biofuel Industry*. Imperial College London, Business School, London, UK.

Barry, M., and R. Chapman (2009). Distributed small-scale wind in New Zealand: Advantages, barriers and policy support instruments. *Energy Policy*, **37**(9), pp. 3358-3369.

BC Hydro (2006-2008). *Annual Report*. BC Hydro, Vancouver, Canada.

BC Hydro (2007-2010). *Annual Report*. BC Hydro Regeneration,Vancouver, Canada.

Beck, F., and E. Martinot (2004). Renewable energy barriers and policies. In: *Encyclopedia of Energy*. C. Cleveland (ed.), Academic Press/Elsevier Science, San Diego, CA, USA, pp. 365-383.

Bell, D., T. Gray, and C. Haggett (2005). The 'Social Gap' in wind farm citing decisions: explanations and policy responses. *Environmental Politics*, **14**, pp. 460-477.

Bell, M. (1990). *Continuing Industrialisation, Climate Change and International Technology Transfer*. SPRU, University of Sussex, Brighton, UK.

Bell, M. (2007). *Technological Learning and the Development of Production and Innovative Capacities in the Industry and Infrastructure Sectors of the Least Developed Countries: What Roles for ODA?* University of Sussex: SPRU - Science and Technology Policy Research, Brighton, UK.

Bennear, L.S. and R.N. Stavins. (2007). Second-best theory and the use of multiple policy instruments. *Environmental and Resource Economics*, **37**(1), pp. 111–129.

Bergek, A., and S. Jacobsson (2010). Are tradable green certificates a cost-efficient policy driving technical change or a rent-generating machine? Lessons from Sweden 2003–2008. *Energy Policy*, **38**(3), pp. 1227-1606.

BERR (2008). *Low Carbon Buildings Programme*. Department for Business, Enterprise and Regulatory Reform (BERR), Department of Energy and Climate Change (DECC). London, UK. Available at: www.lowcarbonbuildings.org.uk/.

BERR/NERA (2008). *Qualitative Evaluation of Financial Instruments for Renewable Heat*. Department for Business, Enterprise and Regulatory Reform (BERR) and National Economic Research Associates (NERA), London, UK.

Bird, J., and Institute for Public Policy Research (2009). *A fair wind for a Green Economy*. Institute for Public Policy Research, London, UK.

Bird, L., M. Bolinger, T. Gagliano, R. Wiser, M. Brown, and B. Parsons (2005). Policies and market factors driving wind power development in the United States. *Energy Policy*, **33**, pp. 1397-1407.

Bird, L., and E. Lokey (2007). *Interaction of Compliance and Voluntary Renewable Energy Markets*. NREL/TP-670-42096, National Renewable Energy Laboratory, Golden, CO, USA.

Bird, L., and J. Sumner (2010). *Green Power Marketing in the United States: A Status Report (2009 Data)*. NREL/TP-6A20-49403, National Renewable Energy Laboratory, Golden, CO, USA.

Blanford, G.J. (2009). R&D investment strategy for climate change. *Energy Economics*, **31**(Supplement 1), pp. S27-S36.

Blyth, W., D. Bunn, J. Kettunen, and T. Wilson (2009). Policy interactions, risk and price formation in carbon markets. *Energy Policy*, **37**(12), pp. 5192-5207.

BMU (2006). *Renewable Energy Sources Act (EEG), Development of electricity generation from renewable energies up to 2020 and financial impacts*. Bundesministerium fur Umwelt, Naturschultz und Reaktorsicherheit (BMU), Berlin, Germany.

BMU (2009). *Renewable Energy Sources in Figures: States, National and International Development*, German Federal Ministry for the Environment, Nature Conservation and Nuclear Safety (BMU), Berlin, Germany, 80 pp.

BMU (2010). *Zeutreihen zur Entwicklung der erneuerbaren Energien in Deutschland [Development of Renewable Energy Sources in Germany]*. Bundesministerium fur Umwelt, Naturschultz und Reaktorsicherheit (BMU), Berlin, Germany.

BMVIT (2007). *Model Region Güssing: Self-sufficient energy supply based on regionally available renewable resources and sustainable regional development*. Bundesministerium für Verkehr, Innovation und Technologie (BMVIT: Austrian Federal Ministry for Transport, Innovation and Technology), Vienna, Austria.

Boardman, B. (2009). *Fixing Fuel Poverty: Challenges and Solutions*. Earthscan Ltd., Oxford, UK, 270 pp.

Bohm, P., and C.S. Russell (1985). Comparative analysis of alternative policy instruments. In: *Handbook of Natural Resource and Energy Economics*. A.V. Kneese and J. Sweeney (eds.), Elsevier Science Publishers, pp. 395-460.

Bohringer, C. and K. E. Rosendahl (2010). Green promotes the dirtiest: On the interaction between black and green quotas in energy markets. *Journal of Regulatory Economics*, **37**(3), pp. 316-325.

Bomb, C., K. McCormick, E. Deurwaarder, and T. Kaberger (2007). Biofuels for transport in Europe: lessons from Germany and the UK. *Energy Policy*, **35**(2007), pp. 2256-2267.

Bouille, A. (2010). *Financing Renewable Energy in the European Energy Market*. Project No. PECPNL084659, A study led by Ecofys funded by the European Commission, DG Energy, Brussels, Belgium. Available at: ec.europa.eu/energy/renewables/studies/doc/renewables/2011_financing_renewable.pdf.

Breitschopf, B., M. Klobasa, F. Sensfuß, J. Steinbach, M. Ragwitz, U. Lehr, J. Horst, U. Leprich, J. Diekmann, F. Braun, and M. Horn (2010). *Einzel- und gesamtwirtschaftliche Analyse von Kosten- und Nutzenwirkungen des Ausbaus Erneuerbarer Energien im deutschen Strom- und Wärmemarkt (in German)*. Interim report of a study led by Fraunhofer ISI done on behalf of the German Federal Environment Ministry, Bonn, Germany.

Brennan, T.J. (2007). Consumer preference not to choose: methodological and policy implications. *Energy Policy*, **35**(3), pp. 1616-1627.

Breukers, S., and M. Wolsink (2007a). Wind energy policies in the Netherlands: Institutional capacity-building for ecological modernisation. *Environmental Politics*, **16**(1), pp. 92-112.

Breukers, S., and M. Wolsink (2007). Wind power implementation in changing institutional landscapes: An international comparison. *Energy Policy*, **35**, pp. 2737-2750.

BTM Consult ApS (2010). *World Market Update 2009*. BTM Consult ApS, Ringkøbing, Denmark.

Buckman, G., and M. Diesendorf (2010). Design limitations in Australian renewable electricity policies. *Energy Policy*, **38**(7), pp. 3365-3376.

Buen, J. (2005). Danish and Norwegian wind industry: the relationship between policy instruments, innovation and diffusion. *Energy Policy*, **34**(18), pp. 3887-3897.

Buijs, P., D. Bekaert, and R. Belmans (2010). Seams issues in European transmission investments. *The Electricity Journal* **23**(10), pp. 18-26.

Bürer, M.J., and R. Wüstenhagen (2009). Which renewable energy policy is a venture capitalist's best friend? Empirical evidence from a survey of international cleantech investors. *Energy Policy*, **37**(12), pp. 4997-5006.

Bürger, V., S. Klinski, U. Lehr, U. Leprich, M. Nast, and M. Ragwitz (2008). Policies to support renewable energies in the heat market. *Energy Policy*, **36**(8), pp. 3150-3159.

Büsgen, U., and W. Dürrschmidt (2009). The expansion of electricity generation from renewable energies in Germany: A review based on the Renewable Energy Sources Act Progress Report 2007 and the new German feed-in legislation. *Energy Policy*, **1**(2009), pp. 2536-2545.

Butler, L., and K. Neuhoff (2008). Comparison of feed-in tariff, quota and auction mechanisms to support wind power development. *Renewable Energy*, **33**(8), pp. 1854-1867.

BWE (2011). *Windjahr in Prozent zum langähringen Mittel*. Statistics, Bundesverband Windenergie e.V. (BWE), Berlin, Germany. Available at: www.wind-energie.de/de/statistiken/.

CEC and CPUC (2008). *Final Opinion and Recommendations on Greenhouse Gas Regulatory Strategies*. California Energy Commission (CEC) and California Public Utilities Commission (CPUC), Sacramento, CA, USA, 297 pp.

Canadian Food Grains Bank (2008). *Can A Hungry World Afford Biofuels?* Canadian Food Grains Bank, Winnipeg, Canada.

Carlsson, B., S. Jacobsson, M. Holmén, and A. Rickne (2002). Innovation systems: analytical and methodological issues. *Research Policy*, **31**(2), pp. 233-245.

Casillas, C., and D.M. Kammen (2010). The energy-poverty-climate nexus. *Science*, **330**(6008), pp. 1181-1182

CCC (2009). Reducing Emissions in Buildings and Industry. Chapter 5 in: *Meeting Carbon Budgets – The Need for a Step Change. Progress Report to Parliament, London, October 2009*. Committee on Climate Change (CCC), London, UK, pp. 151-188. Available at: downloads.theccc.org.uk/21667%20CCC%20Report%20AW%20WEB.pdf.

CCC (2010). *Meeting carbon budgets: Ensuring a Low-Carbon Recovery*. Committee on Climate Change (CCC), London, UK.

Cerri, C.E.P., M. Easter, K. Paustian, K. Killian, K. Coleman, M. Bemoux, P. Falloon, D.S. Powlson, N.H. Batjes, E.Milne, and C.C. Cerri (2007). Predicted soil organic carbon stocks and changes in the Brazilian Amazon between 2000 and 2030. *Agriculture, Ecosystems and Environment*, **122**, pp. 58-72.

Changliang, X., and S. Zhanfeng (2009). Wind energy in China: Current scenario and future perspectives. *Renewable and Sustainable Energy Reviews*, **13**, pp. 1966-1974.

Christensen, C. (1997). *The Innovator's Dilemma: When New Technologies Cause Great Firms to Fail*. Harvard Business School Press, Cambridge, MA, USA.

CIPORE (2011). Caribbean Information Platform on Renewable Energy (CIPORE). Online database, Caribbean Information Platform on Renewable Energy (CIPORE), Kingston, Jamaica. Available at: www.cipore.org/.

CITI (2010). *Europe Utilities (Citigroup Global Markets)*, September 2010. Citi Investment Research & Analysis, London, UK.

City of Miami (2008). *MiPlan: City of Miami Climate Action Plan*. City of Miami, Miami, FL, USA, 48 pp.

Cohen, G.E., D.W. Kearney, and G.J. Kolb (1999). *Final Report on the Operation and Maintenance Improvement Program for Concentrating Solar Power Plants*. Sandia National Laboratories, Oak Ridge, TN, USA.

Cohen, L.R., and R.G. Noll (1991). *The Technology Pork Barrel*. Brookings, Washington, DC, USA.

Commission of the European Communities (2008). *Commission Staff Working Document: The Support of Electricity from Renewable Energy Sources*. Commission of the European Communities, Brussels, Belgium, 38 pp.

Connor, P., V. Bürger, L. Beurskens, K. Ericsson, and C. Egger (2009). *Overview of RES-H/RES-C Support Options*. D4 of WP2 from the RES-H Policy project. University of Exeter, Exeter, UK. Available at: www.res-h-policy.eu/downloads/RES-H_Policy-Options_(D4)_final.pdf.

Cory, K., T. Couture, and C. Kreycik (2009). *Feed-in Tariff Policy: Design, Implementation and RPS Policy Interactions*. NREL/TP-6A2-45549, National Renewable Energy Laboratory (NREL), Golden, CO, USA, 17 pp.

Couture, T. (2009). *State Clean Energy Policy Analysis: Renewable Energy Feed-in Tariffs. SCEPA Webinar*. National Renewable Energy Laboratory, Golden, CO, USA, 23 pp.

Couture, T., and Y. Gagnon (2009). *An Analysis of Feed-in Tariff Policy Design Options for Renewable Energy Sources*. Universite de Moncton, Moncton, Canada.

Couture, T.D., K. Cory, C. Kreycik, and E. Williams (2010). *A Policymaker's Guide to Feed In Tariff Policy Design*. NREL/TP-6A2-44849, National Renewable Energy Laboratory (NREL) Golden, CO, USA.

Cowell, R. (2007). Wind power and 'the planning problem': the experience of Wales. *European Environment*, **17**(5), pp. 291-306.

Cowell, R. (2010). Wind power, landscape and strategic, spatial planning – The construction of 'acceptable locations' in Wales. *Land Use Policy*, **27**(2), pp. 222-232.

Creutzig, F., and D. He (2009). Climate change mitigation and co-benefits of feasible transport demand policies in Beijing. *Transportation Research D*, **14**, pp. 120-131.

Cruetzig, F.S. and D.M. Kammen (2010). Getting the carbon out of transportation fuels. In: *Global Sustainability: A Nobel Cause*. H.-J. Schellenhuber, M. Molina, N. Stern, V. Huber and S. Kadner (eds), Cambridge University Press, pp. 307-318.

Creutzig, F., E. McGlynn, J. Minx, and O. Edenhofer (2010). *Climate Policies for Road Transport Revisited: Evaluation of the Current Framework, Working Papers 1*. Department of Climate Change Economics, TU Berlin, Berlin, Germany.

Cropper, M.L. and W.E. Oates (1992). Environmental economics: A survey. *Journal of Economic Literature*, **30**, pp. 675-740.

CTED (2009). *Washington State's Green Economy: A Strategic Framework. Discussion Draft*. Washington State Department of Community, Trade and Economic Development (CTED), Olympia, WA, USA, 100 pp.

Cuppen, E., S. Breukersb, M. Hisschemöllera, and E. Bergsmaa (2010). Q methodology to select participants for a stakeholder dialogue on energy options from biomass in the Netherlands. *Ecological Economics*, **69**(3), pp. 579-591.

Damborg, S., and S. Krohn (1998). *Public Attitudes towards Wind Power*. Danish Wind Turbine Manufacturers Association, Copenhagen, Denmark.

Danish Ministry of Energy (1981). *Energiplan 81*. Energiministeriet, Copenhagen, Denmark.

Danish Ministry of the Environment (1993). *Cirkulære om primærkommuners planlægning for vindmøller (til alle kommunalbestyrelser)*. Miljøiministeriet, Copenhagen, Denmark.

Danish Wind Industry Association (2010). *Danish Wind Industry Maintains High Export Figures In 2009 Despite Financial Crisis*. Danish Wind Industry Association (Vindmølleindustrien), Frederiksberg, Denmark, Available at: www.windpower.org/en/news/news.html.

David, P.A. (1985). Clio and the Economics of QWERTY. *American Economic Review*, **75**, pp. 332-337.

David, P.A., B.H. Hall, and A.A. Toole (2000). Is public R&D a complement or substitute for private R&D? A review of the econometric evidence. *Research Policy*, **29**(4-5), pp. 497-529.

Davis, G.A., and B. Owens (2003). Optimizing the level of renewable electric R&D Expenditures using real options analysis. *Energy Policy*, **31**, pp. 1589-1608.

DBCCA (2009). *Infrastructure Investments in Renewable Energy*. Deutsche Bank Climate Change Advisors (DBCCA), Deutsche Bank, New York, NY, USA.

DBCCA (2010). *Investing in Climate Change 2010: A Strategic Asset Allocation Perspective*. Deutsche Bank Climate Change Advisors (DBCCA), Deutsche Bank, New York, NY, USA.

de Coninck, H.C., F. Haake, and N. van der Linden (2007). Technology transfer in the Clean Development Mechanism. *Climate Policy*, **7**(5), pp. 444-456.

de Gorter, H., and D.R. Just (2010). The Social costs and benefits of biofuels: The intersection of environmental, energy and agricultural policy. *Applied Economic Perspectives and Policy*, **32**(1), pp. 4-32.

de Jager, D., and M. Rathmann (2008). *Policy Instrument Design to Reduce Financing Costs in Renewable Energy Technology Projects*. Ecofys, Utrecht, The Netherlands, 142 pp.

de Jager, D., C. Klessman, E. Stricker, T. Winkel, E. de Visser, M. Koper, M. Ragwitz, A. Held, G. Resch, S. Busch, C. Panzer, A. Gazzo, T. Roulleau, P. Gousseland, M. Henriet, and A. Bouille (2010). *Financing Renewable Energy in the European Energy Market*. Project No. PECPNL084659, A study led by Ecofys funded by the European Commission, DG Energy, Brussels, Belgium.

de Jonghe, C., E. Delarue, R. Belmans, and W. D'haeseleer (2009). Interactions between measures for the support of electricity from renewable energy sources and CO_2 mitigation. *Energy Policy*, **37**(11), pp. 4743-4752.

de Miera, G.S., P. del Río González, and I. Vizcaíno (2008). Analysing the impact of renewable electricity support schemes on power prices: the case of wind electricity in Spain. *Energy Policy*, **36**(9), pp. 3345-3359.

de Saravia, C.F., and A. Diego Rosell (2011). Coup de Grâce: A New Royal Decree Slashes Tariffs and Opens the Door to Retroactive Changes for Spanish PV. *Photon International,* pp. 66-68.

DECC (2009). *The UK Renewable Energy Strategy*. Department of Energy and Climate Change (DECC). HM Government, London, UK.

Dechezleprêtre, A., M. Glachant, and Y. Ménière (2008). The Clean Development Mechanism and the international diffusion of technologies: An empirical study. *Energy Policy*, **36**, pp.1273-1283.

Deepchand, K. (2002). Promoting equity in large-scale renewable energy development: the case of Mauritius. *Energy Policy*, **30**(11-12), pp. 1129-1142.

DEFRA/BERR (2007). *Renewable Heat Support Mechanisms*. Department for Environment, Food and Rural Affairs (DEFRA) and Department for Business, Enterprise and Regulatory Reform (BERR), London, UK.

del Río González, P., F. Hernández, and M. Gual (2005). The implications of the Kyoto project mechanisms for the deployment of renewable electricity in Europe. *Energy Policy*, **33**(15), pp. 2010-2022.

Department of Minerals and Energy (2003). *White Paper on Renewable Energy*. Department of Minerals and Energy, Pretoria, Republic of South Africa.

Derrick, A. (1998). Financing mechanism for renewable energy. *Renewable Energy*, **15**(1998), pp. 211-214.

Devine-Wright, P. (2005). Beyond Nimbyism: Towards an integrated framework for understanding public perceptions of wind energy. *Wind Energy*, **8**(2), pp. 125-139.

DG TREN (2007). *Heating and Cooling from Renewable Energies: Costs of National Policies and Administrative Barriers*. Contract TREN/D1/2006-7/S07.67170, MVV Consulting. Brussels, Belgium.

Dias de Moraes, M. A. F., and L. Rodrigues (2006). *Brazil Alcohol National Program*. Relatório de pesquisa. Piracicaba, Brazil: 54.

Dillon, H. S., T. Laan, and H.S. Dillon (2008). *Biofuels – At What cost? Government Support For Ethanol and Biodiesel in Indonesia*. Global Subsidies Initiative, Geneva, Switzerland. Available at: http://www.globalsubsidies.org/files/assets/Indonesia_biofuels.pdf.

Dinica, V. (2008). Initiating a sustained diffusion of wind power: The role of public-private partnerships in Spain. *Energy Policy*, **36**(9), pp. 3562-3571.

Domac, J., K. Richards, and S. Risovic (2005). Socio-economic drivers in implementing bioenergy projects. *Biomass and Bioenergy*, **28**(2), pp. 97-106.

Dosi, G. (1982). Technological paradigms and technological trajectories: A suggested interpretation of the determinants and directions of technical change. *Research Policy*, **11**, pp. 147-162.

Dosi, G. (1988). Sources, procedures, and microeconomic effects of innovation. *Journal of Economic Literature*, **26**(3), pp. 1120-1171.

Dowall, D.E. (1980). US land use and energy policy – assessing potential conflicts. *Energy Policy*, **8**(1), pp. 50-60.

Droege, P. (2009). *100% Renewable: Energy Autonomy in Action*. Earthscan, London, UK.

DSIRE (2011). US Database of State Incentives for Renewables and Efficiency. Online database, US Database of State Incentives for Renewables and Efficiency (DSIRE), North Carolina State University, Raleigh, NC, USA. Available at: www.dsireusa.org/.

DTI (2007). *Meeting the Energy Challenge: A White Paper on Energy*. Department of Trade and Industry (DTI), The Stationery Office, London, UK.

DTI/Ofgem Embedded Generation Working Group (2001). *Report into Network Access Issues*, Department of Trade and Industry (DTI) and Ofgem, London, UK.

DUKES (2009). *Digest of United Kingdom Energy Statistics (DUKES)*. Department of Energy and Climate Change, London, UK. Available at: www.decc.gov.uk/en/content/cms/statistics/publications/dukes/dukes.aspx.

ECLAC (2009). *Contribution of Energy Services to the Millennium Development Goals and to Poverty Alleviation in Latin America and the Caribbean*. LC/W.281-P/I, Economic Commission for Latin America and the Caribbean (ECLAC), Club de Madrid, GTZ and UNDP, United Nations, Santiago, Chile. Available at: www.eclac.org/publicaciones/xml/0/38790/lcw281i.pdf.

Edenhofer, O. and M. Kalkuhl (2011). When do increasing carbon taxes accelerate global warming? A note on the green paradox. *Energy Policy*, **39**(4), pp. 2208-2212.

Edenhofer, O., N. Bauer, and E. Kriegler (2005). The impact of technological change on climate protection and welfare: insights from the model MIND. *Ecological Economics*, **54**(2-3), pp. 277-292.

EGWG (2001). *The Embedded Generation Working Group Report into Network Access Issues*. Embedded Generation Working Group (EGWG), UK. Available at: www.dti.gov.uk/energy/domestic_markets/network_access_elec/egwp_report/.

Ellis, G., J. Barry, and C. Robinson (2007). Many ways to say 'no', different ways to say 'yes': applying Q-methodology to understanding public acceptance of wind farm proposals. *Journal of Environmental Planning and Management*, **50**(4), pp. 517-555.

Ellis, G., R. Cowell, C. Warren, P. Strachan, and J. Szarka (2009). Wind Power and the 'Planning Problem'. *Journal of Planning Theory and Practice*, **10**(4), pp. 521-547.

Energy Skills Queensland (2009). *Research report: Sustainable Energy Skills Formation Strategy*. Energy Skills Queensland. Rocklea, Queensland, Australia.

EPPO (2007a). *VSPP (As of April 2007)*. Energy Policy and Planning Office (EPPO), Ministry of Energy, Thailand. Bangkok, Thailand.

EPPO (2007b). สรุปการรับซื้อไฟฟ้าจากผู้ผลิตไฟฟ้ารายเล็ก (สถานภาพ ณ เดือนกุมภาพันธ์ 2549. Summary of electricity purchased from SPPs (as of February 2007). Energy Policy and Planning Office (EPPO), Ministry of Energy, Thailand. Bangkok, Thailand.

EPPO (2010a). *Thailand Power Development Plan*. Energy Policy and Planning Office (EPPO), Ministry of Energy, Thailand. Bangkok, Thailand.

EPPO (2010b). สถานภาพการรับซื้อไฟฟ้าจาก SPP จำแนกตามประเภทเชื้อเพลิง (ณ วันที่ 24 มีนาคม 2553. (Electricity purchased from SPPs by fuel type - as of 24 March 2010). Energy Policy and Planning Office (EPPO), Ministry of Energy, Thailand. Bangkok, Thailand.

EPPO (2010c). สถานภาพการรับซื้อไฟฟ้าจาก VSPP จำแนกตามประเภทเชื้อเพลิง (ณ วันที่ 24 มีนาคม 2553. (Electricity purchased from VSPP by fuel type as of 24 March, 2010). Energy Policy and Planning Office (EPPO), Ministry of Energy, Thailand. Bangkok, Thailand.

ERCOT (2010). *Texas Posts Record Increase in Voluntary Renewable Energy Credits: State Exceeds Legislature's 2025 Goal 15 Years Early*. Press release, Electric Reliability Council of Texas (ERCOT), Austin, TX, USA. Available at: www.ercot.com/news/press_releases/2010/nr-05-14-10.

Ericsson, K., and P. Svenningsson (2009). *Introduction and Development of the Swedish District Heating Systems: Critical Factors and Lessons Learned*. Lund University, Lund, Sweden.

Ernst & Young (2008). *Renewable Energy Country Attractiveness Indices: Global Highlights*. Ernst & Young, London, UK, 24 pp.

Espey, S. (2001). Renewables portfolio standard: a means for trade with electricity from renewable energy sources? *Energy Policy*, **29**(7), pp. 557-566.

European Centre for Development of Vocational Training (2010). *Skills for Green Jobs: European Synthesis Report*. European Centre for Development of Vocational Training, Publications Office of the European Union, Luxembourg. Available at: www.cedefop.europa.eu/EN/Files/3057_en.pdf.

European Commission (2005). *The Support of Renewable Energy Sources*. European Commission, Brussels, Belgium.

European Commission (2009a). Directive 2009/28/EC of the European Parliament and of the Council of 23 April 2009 on the promotion of the use of energy from renewable sources and amending and subsequently repealing Directives 2001/77/EC and 2003/30/EC (2009). *Official Journal of the European Union*, pp. 16-61.

European Commission (2009b). *Investing in the Development of Low Carbon Technologies: A Technology Roadmap*. Commission to the European Parliament, the Council, the European Economic and Social Committee and the Committee of the Regions, Brussels, Belgium.

European Commission (2010). *Covenant of Mayors: Cities take the Lead to Tackle Climate Change*. News release, EP President Press Service, Available at: www.eumayors.eu/IMG/pdf/EP_PR.pdf.

European Commission (2011). *Commission staff working document: Recent progress in developing renewable energy sources and technical evaluation of the use of biofuels and other renewable fuels in transport in accordance with Article 3 of Directive 2001/77/EC and Article 4(2) of Directive 2003/30/EC*. Accompanying document to the Communication from the Commission to the European Parliament and the Council: Renewable Energy: Progressing towards the 2020 target; COM(2011) 31 final. European Commission, Brussels, Belgium.

European Parliament and of the Council (2009). Directive 2009/28/EC of the European Parliament and of the Council of 23 April 2009 on the promotion of the use of energy from renewable sources and amending and subsequently repealing Directives 2001/77/EC and 2003/30/EC. *Official Journal of the European Union*.

European Union (2009). *Annex I of Directive 2009/28/EC of the European Parliament and of the Council of 23 April 2009 on the Promotion and Use of Energy from Renewable Sources and Amending and Subsequently Repealing Directives 2001/77/EC and 2003/30/EC*. European Union, Brussels, Belgium.

Eurostat (2010). *Supply, transformation, consumption – renewables (biofuels) – annual data (nrg_1073a)*. European Commission, Brussels, Belgium. Available at: appsso.eurostat.ec.europa.eu/nui/show.do?dataset=nrg_1073a&lang=en.

Fankhauser, S., C. Hepburn, and J. Park (2010). Combining multiple climate policy instruments: How not to do it. *Climate Change Economics*, **1**, pp. 209-225

FAO/GBEP (2007). *A Review of the Current State of Biofuel Development in G8+5 Countries*. Food and Agriculture Organization (FAO) and Global Bioenergy Partnership (GBEP), Rome, Italy.

Farrell, J. (2009). *Feed-in tariffs in America: Driving the Economy with Renewable Energy Policy that Works*. The New Rules Project, Minneapolis, MN, USA, 30 pp.

Faulin, J., F. Lera, J.M. Pintor, and J. Garcia (2006). The outlook for renewable energy in Navarre: An economic profile. *Energy Policy*, **34**, pp. 2201-2216.

Felix-Saul, R. (2008). Assessing the impact of Mexico's Biofuels Law. *Baker & McKenzie Biomass Magazine*. Available at: biomassmagazine.com/articles/1678/assessing-the-impact-of-mexico's-biofuels-law/.

Fink, S., C. Mudd, K. Porter, and B. Morgenstern (2009). *Wind Energy Curtailment Case Studies*. National Renewable Energy Laboratory, Golden, CO, USA.

Fischer, C. (2008). Emissions pricing, spillovers, and public investment in environmentally friendly technologies. *Energy Economics*, **30**(2), pp. 487-502.

Fischer, C. and R.G. Newell (2008). Environmental and technology policies for climate mitigation. *Journal of Environmental Economics and Management*, **55**, pp. 142-162.

Fischer, C., and L. Preonas (2010). Combining policies for renewable energy: Is the whole less than the sum of its parts? *International Review of Environmental and Resource Economics*, **4**(1), pp. 51-92.

Fischlein, M., J. Larson, D. Hall, R. Chaudhry, T.R. Peterson, J. Stephens, and E. Wilson (2010). Policy stakeholders and deployment of wind power in the sub-national context: A comparison of four U.S. states. *Energy Policy*, **38**(8), pp. 4429-4439.

Flach, B., S. Lieberz, K. Bendz, B. Dahlbacka and D. Achilles (2009). *EU-27 Biofuels Annual Report 2009*. United States Department of Agriculture Foreign Agricultural Service, Washington, DC, USA.

Fodella, G. (1989) Orgware: the key of Japanese success. *Rivista Internazionale di Scienze Economiche Ecommerciali*, **36**(12), pp. 1057-1062.

Ford, A., K. Vogstad, and H. Flynn (2007). Simulating price patterns for tradable green certificates to promote electricity generation from wind. *Energy Policy*, **35**(1), pp. 91-111.

Forsyth, T.L., M. Pedden, and T. Gagliano (2002). *The Effects of Net Metering on the Use of Small-Scale Wind Systems in the United States*. NREL/TP-500-32471, NREL, Golden, CO, USA, 20 pp.

Fouquet, D., and T.B. Johansson (2008). European renewable energy policy at crossroads – Focus on electricity support mechanisms. *Energy Policy*, **36**(11), pp. 4079-4092.

Fouquet, D., C. Grotz, J.L. Sawin, and N. Vassilakos (2005). *Reflections on a Possible Unified EU-Financial Support Scheme for Renewable Energy Systems (RES): A Comparison of Minimum-Price and Quota Systems and an Analysis of Market Conditions*. European Renewable Energies Federation and Worldwatch Institute, Brussels, Belgium and Washington, DC, USA.

Fouquet, R. (2008). *Heat, Power And Light: Revolutions in Energy Services*. Edward Elgar Publishing Ltd., Cheltenham, UK.

Fox, J. (2010). *Renewable Energy in Thailand: Green Policies Take Off*. Thailand Law Forum. Available at: www.thailawforum.com/green-policies-take-off.html.

Foxon, T., and P. Pearson (2008). Overcoming barriers to innovation and diffusion of cleaner technologies: some features of a sustainable innovation policy regime. *Journal of Cleaner Production*, **16**(S1), pp. S148-S161.

Freeman, C. (1974). *The Economics of Industrial Innovation*. The MIT Press, Cambridge, MA USA.

Freeman, C., and C. Perez (1988). Structural crises of adjustment, business cycles, and investment behavior. In: *Technical Change and Economic Theory*. G. Dosi, C. Freeman, R. Nelson, G. Silverberg and L. Soete (eds.), Pinter, London, UK and New York, NY, USA, pp. 38-66.

Frenken, K., M. Hekkert, and P. Godfroij (2004). R&D portfolios in environmentally friendly automotive propulsion: Variety, competition and policy implications. *Technological Forecasting and Social Change*, **71**(5), pp. 485-507.

Fri, R.W. (2003). The role of knowledge: Technological innovation in the energy system. *The Energy Journal*, **24**(4), pp. 51-74.

Friedrichs, J. (2010). Global energy crunch: how different parts of the world would react to a peak oil scenario. *Energy Policy*, **38**(8), pp. 4562-4569.

Frondel, M., N. Ritter, C.M. Schmidt, and C. Vance (2010). Economic impacts from the promotion of renewable energy technologies: The German experience. *Energy Policy*, **38**(2010), pp. 4048-4056.

Fuller, M.C., C. Kunkel, and D.M. Kammen (2009a). *Guide to Energy Efficiency and Renewable Energy Financing Districts for Local Governments, prepared for the city of Berkeley, California*. Renewable and Appropriate Energy Laboratory, University of California, Berkeley, CA, USA.

Fuller, M.C., S. Portis, and D.M. Kammen (2009b). Towards a low-carbon economy: municipal financing for energy efficiency and solar power. *Environment Magazine*, **51**(2), pp. 22-32.

Gallagher, K.S. (2006). Limits to leapfrogging in energy technologies? Evidence from the Chinese automobile industry. *Energy Policy*, **34**, pp. 383-394.

Gan, Z. and Yu, L. (2008). Bioenergy transition in rural China: Policy options and co-benefits. *Energy Policy*, **36**(2), pp: 531-540.

Garud, R., and P. Karnøe (2003). Bricolage versus breakthrough: distributed and embedded agency in technology entrepreneurship. *Research Policy*, **32**, pp. 277-300.

Gatautis, R., I. Konstantinaviciute, D. Tarvydas, and V. Bobinaite (2009). *Current State of Heating and Cooling Markets in Lithuania Kaunas*. Lithuanian Energy Institute, Kaunas, Lithuania.

Gatersleben, B., L. Steg, and C. Vlek (2002). Measurement and determinants of environmentally significant consumer behavior. *Environment and Behavior*, **34**(3), pp. 335-362.

GBC (2010). *Clean Energy Act, Statutes of British Columbia, 2010. Bill 17*. Government of British Columbia (GBC), Victoria, Canada. Available at: www.leg.bc.ca/39th2nd/1st_read/gov17-1.htm.

Geels, F.W. (2004). From sectoral systems of innovation to sociotechnical systems: insights about dynamics and change from sociology and institutional theory. *Research Policy*, **33** (6-7), pp. 897-920.

Geels, F.W. (2005). The dynamics of transitions in socio-technical systems: A multi-level analysis of the transition pathway from horse-drawn carriages to automobiles (1860-1930). *Technology Analysis & Strategic Management*, **17**(4), pp. 445-476.

Gerlagh, R. (2010). *Too much oil*. CESifo Economic Studies.

Gillenwater, M. (2008). Redefining RECs (Part 1): Untangling attributes and offsets. *Energy Policy*, **36**(6), pp. 2109-2119.

Gillingham, K. (2009). Economic efficiency of solar hot water policy in New Zealand. *Energy Policy*, **37**(9), pp. 3336-3347.

Girardet, H., and M. Mendonca (2009). *A Renewable World: Energy, Ecology, Equality*. Green Books, Devon, UK.

Goldemberg, J. (2006). The ethanol program in Brazil. *Environmental Research Letters*, **1**, 014008.

Goldemberg, J. (2009). The Brazilian experience with Biofuels. *Innovations*, **4**(4), pp. 91-107.

Goldemberg, J., S.T. Coelho, P.M. Nastari, and O. Lucon (2004). Ethanol learning curve – the Brazilian experience. *Biomass and Bioenergy*, **26**(3), pp. 301-304.

Goldemberg, J., S.T. Coelho, and P.M. Guardabassi (2008). The sustainability of ethanol production from sugarcane. *Energy Policy*, **36**(6), pp. 2086-2097.

Goolsbee, A. (1998). Does government R&D policy mainly benefit scientists and engineers? *American Economic Review*, **88**(2), pp. 298-302.

Government of Jamaica (2006). *Green Paper: The Jamaica Energy Policy 2006-2020*. Government of Jamaica, Kingston, Jamaica.

Government of Nepal (2006). *Rural Energy Policy*. Government of Nepal, Ministry of Environment, Kathmandu, Nepal.

Government of Pakistan (2006). *Policy for Development of Renewable Energy for Power Generation: Employing Small Hydro, Wind, and Solar Technologies*. Government of Pakistan, Islamabad, Pakistan.

Government of the Kingdom of Tonga (2010). *Tonga Energy Roadmap (TERM) 2010-2020: A Ten Year Road Map to Reduce Tonga's Vulnerability to Oil Price Shocks & Achieve an Increase in Quality Access to Modern Energy Services in an Environmentally Sustainable Manner*. HM Government of the Kingdom of Tonga, Nuku'aLofa, Tonga.

Grafton, R.Q., T. Kompas, and N. Van Long (2010). *Biofuels Subsidies and the Green Paradox*. CESifo Working Paper Series 2960, CESifo Group, Munich, Germany.

Greacen, C. (2007). An emerging light: Thailand gives the go-ahead to distributed energy. *Cogeneration & On-Site Power Production Magazine*, pp. 65-73.

Greacen, C., and C. Greacen (2004). Thailand's electricity reforms: privatization of benefits and socialization of costs and risks. *Pacific Affairs*, **77**(4), 517-541.

Greacen, C., C. Greacen, and R. Plevin (2003). Thai power: Net metering comes to Thailand. *ReFocus*, **4**(6), pp. 34-37. Available at: netmeter.org/en/docs/NetMeteringRefocusNov2003.pdf.

Griliches, Z. (1992). The search for R&D spillovers. *Scandinavian Journal of Economics*, **94**, pp. S29-S47.

Grimaud, A., and G. Lafforgue (2008). *Climate change mitigation policies: are R&D subsidies preferable to a carbon tax?* University of Toulouse, Toulouse, France.

Gross, C. (2007). Community perspectives of wind energy in Australia: The application of a justice and community fairness framework to increase social acceptance. *Energy Policy*, **35**, pp. 2727-2736.

Grubb, M. (2004). Technology innovation and climate change policy: An overview of issues and options. *Keio Economic Studies*, **41**(2), pp. 103-132.

Grubler, A. (1998). *Technology and Global Change*. Cambridge University Press.

Grubler, A., N. Nakicenovic, and D.G. Victor (1999a). Dynamics of energy technologies and global change. *Energy Policy*, **27**(5), pp. 247-280.

Grubler, A., N. Nakicenovic, and D.G. Victor (1999b). Modeling technological change: Implications for the global environment. *Annual Review of Energy and the Environment*, **24**, pp. 545-569.

Gupta, S., D.A. Tirpak, N. Burger, J. Gupta, N. Höhne, A.I. Boncheva, G.M. Kanoan, C. Kolstad, J.A. Kruger, A. Michaelowa, S. Murase, J. Pershing, T. Saijo, and A. Sari (2007). *Policies, Instruments and Co-operative Arrangements*. Cambridge University Press.

GWEC (2008). *Global Wind 2007 Report*. Global Wind Energy Council (GWEC), Brussels, Belgium.

GWEC (2010). *Global Wind 2009 Report*. Global Wind Energy Council (GWEC), Brussels, Belgium.

Haanyika, C.M. (2008). Rural electrification in Zambia: A policy and institutional analysis. *Energy Policy*, **36**(3), pp. 1044-1058.

Haas, R., W. Eichhammer, C. Huber, O. Langniss, A. Lorenzoni, R. Madlener, P. Menanteau, P.-E. Morthorst, A. Martins, A. Oniiszk, J. Schleich, A. Smith, Z. Vass, and A. Verbruggen (2004). How to promote renewable energy systems successfully and effectively. *Energy Policy*, **32**(6), pp. 833-839.

Haas, R., C. Panzer, G. Resch, M. Ragwitz, G. Reece, and A. Held (2011). A historical review of promotion strategies for electricity from renewable energy sources in EU countries. *Renewable and Sustainable Energy Reviews*, **15**, pp. 1003-1034.

Haites, E., M. Duan, and S. Seres (2006). Technology transfer by CDM projects. *Climate Policy*, **6**(3), pp. 327-344.

Han, J., A. Mol, Y. Lu, and L. Zhang (2009). Onshore wind power development in China: Challenges behind a successful story. *Energy Policy*, **37**(8), pp. 2941-2951.

Han, J., A.P.J. Mol, and Y. Lu (2010). Solar water heaters in China: A new day dawning. *Energy Policy*, **38**, pp. 383-391.

Hankey, S., and J.D. Marshall (2010). Impacts of urban form on future US passenger-vehicle greenhouse gas emissions. *Energy Policy*, **38**(9), pp. 4880-4887.

Hanley, N., J.F. Shogren, and B. White (1997). *Environmental Economics. In Theory and Practice*. MacMillan, 464 pp.

Hanley, N., A. Bergmann, and R. Wright (2004). *Valuing the Environmental and Employment Impacts of Renewable Energy Investments in Scotland*. Scottish Economic Policy Network.

Hanson, M., M. Bernstein, and R. Hammon (2006). The role of energy efficiency in homebuying decisions: Results of initial focus group discussions. In: *ACEEE Summer Study on Energy Efficiency in Buildings*, Asilomar, CA, 13-18 August 2006, American Council for an Energy Efficient Economy.

Haščič, I., and N. Johnstone (2009). *The Clean Development Mechanism and International Technology Transfer: Empirical Evidence on Wind Power using Patent Data*. Organisation for Economic Co-operation and Development, Paris, France.

Healey, P. (1997). *Collaborative Planning. Shaping Places in Fragmented Societies*. MacMillan Press, London, UK.

Heiskanen, E., M. Hodson, R.M. Mourik, R.P. J.M. C.F.J. Feenstra, T. Alcantud Torrent, B. Brohmann, A. Daniels, B. Difiore, B. Farkas, U.R. Fritsche, J. Fucskó, E. Jolivet, M.H. Maack, K. Matschoss, A. Oniszk-Poplawska, B.M. Poti, G. Prasad, R. Willemse, B. Schaefer, and K. Hünecke (2008a). *Factors Influencing the Societal Acceptance of New Energy Technologies: Meta-Analysis of Recent European Projects*. D3-D4 Create Acceptance. ECN-E-07-058 Project co-funded by the European Commission within the Sixth Framework Programme (2002-2006), Energy Research Centre of the Netherlands, Petten, The Netherlands.

Heiskanen, E., K. Jarvela, A. Pulliainen, M. Saastamoinen, and P. Timonen (2008b). Qualitative research and consumer policy: Focus group discussions as a form of consumer participation. *The Qualitative Report*, **13**(2), pp. 152-172.

Hekkert, M., R.A.A. Suurs, S. Negro, S. Kuhlmann, and R. Smits (2007). Functions of innovation systems: A new approach for analysing technological change. *Technological Forecasting and Social Change*, **74**(4), pp. 413-432.

Held, A., M. Ragwitz, C. Huber, G. Resch, T. Faber, and K. Vertin (2007). *Feed-in Systems in Germany, Spain and Slovenia: A Comparison*. Fraunhofer Institute Systems and Innovation Research, Karlsruhe, Germany.

Held, A., M. Ragwitz, E. Merkel, M. Rathmann, and C. Klessmann (2010). *Indicators Assessing the Performance of Renewable Energy Support Policies in 27 Member States*. RE-Shaping project report to the European Commission under the Intelligent Energy for Europe Program (Grant agreement no. EIE/08/517/SI2.529243). Fraunhofer ISI and Ecofys, Karlsruhe, Germany and Utrecht, The Netherlands.

Helm, D. (2010). Government failure, rent-seeking, and capture: the design of climate change policy. *Oxford Review of Economic Policy*, **26**(2), pp. 182-196.

Hertel, T.W., A.A. Golub, A.D. Jones, M. O'Hare, R.J. Plevin, and D.M. Kammen (2010). Effects of US maize ethanol on global land use and greenhouse gas emissions: Estimating market-mediated responses. *BioScience*, **60**, pp. 223–231.

Heyward, M. (2007). Equity and international climate change negotiations: a matter of perspective. *Climate Policy*, **7**, pp. 518-534.

Hiller, J., and P. Healey (2008). *Contemporary Movements in Planning Theory. Critical Essays in Planning Theory*. Ashgate, Aldershot, UK.

Hiremath, R.B., B. Kumar, P. Balachandra, N.H. Ravindranath, and B.N. Raghunandan (2009). Decentralised renewable energy: Scope, relevance and applications in the Indian context. *Energy for Sustainable Development*, **13**(1), pp. 4-10.

Hockerts, K., and R. Wüstenhagen (2010). Greening Goliaths versus emerging Davids –Theorizing about the role of incumbents and new entrants in sustainable entrepreneurship. *Journal of Business Venturing*, **25**, pp. 481-492.

Hoekman, B.M., K.E. Maskus, and K. Saggi (2004). *Transfer of Technology to Developing Countries: Unilateral and Multilateral Policy Options*. World Bank, Washington, DC, USA.

Hoel, M. (2010). *Is there a Green Paradox?* CESifo Working Paper Series 3168. CESifo Group, Munich, Germany.

Hogan, M. (2007). *German tax hits Europe's biggest biodiesel market*. Reuters, 2 Feb 2007. Available at: uk.reuters.com/article/2007/02/02/biofuels-germany-idUKL0231973020070202.

Hogan, M. (2009). *German biodiesel firms say U.S. imports escape duty*. Reuters, 30 Nov 2009. Available at: www.reuters.com/article/2009/11/30/us-germany-biodiesel-us-idUSTRE5AT3QG20091130.

Holdren, J.P., and S.F. Baldwin (2001). The PCAST energy studies: Toward a national consensus on energy research, development, demonstration, and deployment policy. *Annual Review of Energy and Environment*, **26**, pp. 391-434.

Hoppock, D.C., and D. Patiño-Echeverri (2010). Cost of wind energy: Comparing distant wind resources to local resources in the midwestern United States. *Environmental Science & Technology*, **44**(22), pp. 8758-8765.

Horbach, J. (2007). Determinants of environmental innovation – New evidence from German panel data sources. *Research Policy*, **37**(1), pp. 163-173.

House of Commons (2008). *Renewable Electricity – Generation Technologies*. Innovation, Universities, Science and Skills Committee, The Stationery Office Limited, London, UK.

Houser, T., S. Mohan, and I. Hoffman (2010). *Assessing the American Power Act: The Economic, Employment, Energy Security, and Environmental Impact of Senator Kerry and Senator Lieberman's Discussion Draft*. No PB10-12, Policy Briefs from Peterson Institute for International Economics, Washington, DC, USA.

Huber, C., T. Faber, R. Haas, G. Resch, J. Green, S. Olz, S. White, H. Cleijne, W. Ruijgrok, P.E. Morthorst, K. Skytte, M. Gual, P. Del Rio, F. Hernandez, A. Tacsir, M. Ragwitz, J. Schleich, W. Orasch, M. Bokermann, and C. Lins (2004). *Green-X: Deriving Optimal Promotion Strategies for Increasing the Share of RES-E in a Dynamic European Electricity Market*. TU Wien, Energy Economics Group, Technical University of Denmark (DTU), Riso National Laboratory for Sustainable Energy, Consejo Superior de Investigaciones Cientificas (CSIC), Madrid, Fraunhofer-Institut fur Systemtechnik und Innovationsforschung (ISI), Karlsruhe, Vienna, Austria.

Hughes, T. (1986). The Seamless Web: Technology, Science, Etcetera, Etcetera. *Social Studies of Science*, **16**(2), pp. 281-292.

Hughes, T.P. (1987). The Evolution of large technological systems. In: *The Social Construction of Technological Systems*. W.E. Bijker, T.P. Hughes, and T. Pinch (eds.), MIT Press, Cambridge, MA, USA, pp. 51-82.

Hvelplund, F. (2001). Political prices or political quantities? A comparison of renewable support systems. *New Energy*, **5**, pp. 18-23.

Hvelplund, F. (2006). Renewable energy and the need for local energy markets. *Energy*, **31**(13), pp. 2293-2302.

ICCEPT (2003). *Innovation in Long Term Renewables Options in the UK – Overcoming Barriers and 'Systems Failures'*. ICCEPT report for the DTI Renewable Innovation Review, Centre for Energy Policy and Technology (ICCEPT), Imperial College, London, UK. Available at: webarchive.nationalarchives.gov.uk/+/http://www.berr.gov.uk/files/file22072.pdf.

IDAE (2006). *Plan de energias renovables 2005-2010, description de medidas, actuaciones y requisitos*. Institute for the Saving and Diversification of Energy (IDEA), Madrid, Spain.

IDAE (2008). *Seguimiento del Plan de Energías Renovables en España (PER) 2005-2010*. Memoria 2008. Institute for the Saving and Diversification of Energy (IDAE), Ministerio de Industria, Turismo y Comercio Madrid, Spain.

IDAE (2009). *La biomasa en el marco de los Planes de las Energías Renovables de España. Aspectos económicos y sociales de la agroenergética*. Institute for the Saving and Diversification of Energy (IDAE), Ministerio de Industria, Turismo y Comercio, Madrid, Spain.

IDAE (2010). *La industria fotovoltaica española en el contexto europeo*. Institute for the Saving and Diversification of Energy (IDAE), Ministerio de Industria, Turismo y Comercio, Madrid, Spain.

IEA (2003a). *National Survey Report of PV Power Applications in Japan 2002, prepared by Kiyoshi Shino*. International Energy Agency, Paris, France, 34 pp.

IEA (2003b). *Renewables for Power Generation: Status & Prospects*. International Energy Agency, Paris, France.

IEA (2004a). *Renewable Energy – Market and Policy Trends in IEA Countries*. International Energy Agency and Organisation for Economic Co-operation and Development, Paris, France.

IEA (2004b). *World Energy Outlook 2004*. International Energy Agency, Paris, France.

IEA (2006). *Barriers, Challenges and Opportunities*. International Energy Agency, Paris, France.

IEA (2007a). *Climate Policy Uncertainty and Investment Risk. In Support of the G8 Plan of Action*. International Energy Agency, Paris, France.

IEA (2007b). *Renewables for Heating and Cooling: Untapped Potential*. Renewable Energy Technology Deployment, International Energy Agency and Organisation for Economic Co-operation and Development, Paris, France.

IEA (2008a). *Deploying Renewables: Principles for Effective Policies*. International Energy Agency, Paris, France, 200 pp.

IEA (2008b). *Energy Technology Perspectives 2008. Scenarios and Strategies to 2050*. International Energy Agency, Paris, France, 646 pp.

IEA (2008c). *World Energy Outlook 2008*. International Energy Agency, Paris, France, 578 pp.

IEA (2009a). *Cities, Towns and Renewable Energy – Yes In My Front Yard*. International Energy Agency, Paris, France

IEA (2009b). *Cogeneration and District Energy: Sustainable Energy Technologies for Today and Tomorrow*. International Energy Agency and Organisation for Economic Co-operation and Development, Paris, France, 24 pp.

IEA (2009c). *Statistics & Balances*. International Energy Agency, Paris, France, 2009.

IEA (2009d). *World Energy Outlook 2009*. International Energy Agency, Paris, France, 696 pp.

IEA (2010a). *Energy Technology Perspectives 2010. Scenarios and Strategies to 2050*. International Energy Agency, Paris, France, 708 pp.

IEA (2010b). *Energy Technology R&D Statistics*. International Energy Agency, Paris, France.

IEA (2010c). *Global Renewable Energy Policies and Measures Database*. International Energy Agency, Paris, France

IEA (2010d). *World Energy Outlook 2010*. International Energy Agency, Paris, France, 736 pp.

IEA (2011). *Global Renewable Energy Policies and Measures Database, March 2011*. International Energy Agency, Paris, France.

IEA, OECD, and World Bank (2010). *The Scope of Fossil Fuel Subsidies in 2009 and a Roadmap for Phasing Out Fossil-Fuel Subsidies*. International Energy Agency, Organisation for Economic Co-operation and Development, and the World Bank, Paris, France and New York, NY, USA.

IEA RETD (2010). *Best Practices for the Deployment of RE for Heating and Cooling in the Residential Sector*. International Energy Agency Renewable Energy Technology Development (IEA RETD), Paris, France.

IEEE PES (2009). *Preparing the US Foundation for Future Electric Energy Systems: A Strong Power and Energy Engineering Workforce*. IEEE US Power and Energy Engineering Workforce Collaborative, Piscataway, NJ, USA. Available at: www.ieee-pes.org/images/pdf/US_Power_&_Energy_Collaborative_Action_Plan_April_2009_Adobe72.pdf.

IISD (2008). *Biofuels – At what Cost? Government Support for Ethanol and Biodiesel in China*, Global Subsidies Initiative, International Institute for Sustainable Development (IISD), Geneva, Switzerland. Available at: www.iisd.org/pdf/2008/biofuels_subsidies_aus.pdf.

IJHD (2009). *Annual Directory, Years 2004 to 2009*. International Journal of Hydropower and Dams, Surrey, UK.

IMEC (2009a). *SCHOTT Solar joins IMEC research program on silicon photovoltaics*. Available at: www2.imec.be/be_en/press/imec-news/archive-2009/schott-solar-joins-imec-research-program-on-silicon-photovoltaics.html.

IMEC (2009b). *Total, GDF SUEZ, and Photovoltech join IMEC's silicon solar cell research program*. Press release, IMEC, Leuven, Belgium. Available at: www2.imec.be/be_en/press/imec-news/archive-2009/total-gdf-suez-and-photovoltech-join-imec-08217-s-silicon-solar-cell-research.html.

Inhaber, H. (2004). Water use in renewable and conventional electricity production. *Energy Sources*, **26**, pp. 309-322.

International Center for Trade and Sustainable Development (2010). *US, EU Join Japan in Row over Canadian Green Energy Incentives*. International Center for Trade and Sustainable Development, Geneva, Switzerland. Available at: ictsd.org/i/news/bridgesweekly/86146/.

IPCC (2000). *Methodological and Technological Issues in Technology Transfer*. B. Metz, O.R. Davidson, J.-W. Martens, S.N.M. van Rooijen, and L. Van Wie McGrory (eds.). Cambridge University Press, 432 pp.

IPCC (2002). *Climate Change and Biodiversity*. H. Gitay, A. Suárez, RT. Watson, and D.J. Dokken (eds.), IPCC Technical Paper V, IPCC, Geneva, Switzerland, 85 pp.

IPCC (2007). *Climate Change 2007: Mitigation of Climate Change. Contribution of Working Group III to the Fourth Assessment Report of the Intergovernmental Panel on Climate Change*. B. Metz, O.R. Davidson, P.R. Bosch, R. Dave, and L.A. Meyer (eds.), Cambridge University Press, 851 pp.

IREC (2010). *2010 Updates and Trends*. Interstate Renewable Energy Council (IREC), Latham, NY, USA, 38 pp.

Ito, H. (2003). *Japan's New and Renewable Energy Policies*. Ministry of Economy, Trade and Industry, Tokyo, Japan. Available at: www.asiapacificpartnership.org/pdf/REDGTF/1st_meeting/Japan1-country_report.pdf.

Jaccard, M., N. Melton, and J. Nyboer (2011). Institutions and processes for scaling up renewables: Run-of-river hydropower in British Columbia. *Energy Policy*, doi:10.1016/j.enpol.2011.02.035.

Jacobsson, S., and A. Johnson (2000). The diffusion of renewable energy technology: an analytical framework and key issues for research. *Energy Policy*, **28**(9), pp. 625-640.

Jacobsson, S., and A. Bergek (2004). Transforming the energy sector: the evolution of technological systems in renewable energy technology. *Industrial and Corporate Change*, **13**(5), pp. 815-849.

Jacobsson, S., and V. Lauber (2006). The politics and policy of energy system transformation - explaining the German diffusion of renewable energy technology. *Energy Policy*, **34**(3), pp. 256-276.

Jacobsson, S., A. Bergek, D. Finon, V. Lauber, C. Mitchell, D. Toke, and A. Verbruggen (2009). EU renewable energy support policy: Faith or facts? *Energy Policy*, **37**(6), pp. 2143-2146.

Jaffe, A.B. (1986). Technological opportunity and spillover of R&D: evidence from firms' patents, profits, and market value. *American Economic Review*, **76**, pp. 984-1001.

Jaffe, A.B., R.G. Newell, and R.N. Stavins (2003). Technological change and the environment. In: *Handbook of Environmental Economics*. K.-G. Mäler and J. Vincent (eds.), Elsevier Science, Amsterdam, The Netherlands, pp. 461- 516.

Jaffe, A.B., R.G. Newell, and R.N. Stavins (2005). A tale of two market failures: Technology and environmental policy. *Ecological Economics*, **54**(2-3), pp. 164-174.

Jager, W. (2006). Stimulating the diffusion of photovoltaic systems: A behavioural perspective." Energy Policy 34(14): 1935-1943.

Jänicke, M. (2010). Das Innovationstempo in der Klimapolitik forcieren! In: *Jahrbuch Ökologie 2011*, S. Hirzel, Stuttgart, Germany, pp. 138-147. Available at: www.jahrbuch-oekologie.de/Jaenicke2011.pdf.

Jayanthi, S., E.C. Witt, and V. Singh (2009). Evaluation of potential of innovations: A DEA-based application to US photovoltaic industry. *IEEE Transactions on Engineering Management*, **56**(3), pp. 478-493.

Jenvanitpanjakul, P., and P. Bhandhubanyong (2009). *Rice and sugar energy complex model*. In: *6th Biomass-Asia Workshop*, Hiroshima, Japan, 18-20 November 2009. Available at: http://www.biomass-asia-workshop.jp/biomassws/06workshop/presentation/18_Peesamai.pdf.

Jha, V. (2009). *Trade Flows, Barriers and Market Drivers in Renewable Energy Supply Goods: The Need to Level the Playing Field*. International Centre for Trade and Sustainable Development, Geneva, Switzerland.

JNNSM (2009). *Towards Building SOLAR INDIA*. Jawaharlal Nehru National Solar Mission (JNNSM), Ministry of New and Renewable Energy, Government of India, New Delhi, India. Available at: mnre.gov.in/pdf/mission-document-JNNSM.pdf.

Jobert, A., P. Laborgne, and S. Mimler (2007). Local acceptance of wind energy: Factors of success identified in French and German case studies. *Energy Policy*, **35**, pp. 2751-2760.

Johansson, T.B., U.R. Fritsche, C. Flavin, J. Sawin, D. Aßmann, and T.C. Herberg (2004). Policy recommendations for renewable energies. Prepared for the *International Conference for Renewable Energies under the guidance of the Conveners of the Conference*. Bonn, Germany, 1-4 June 2004. Available at: www.oei.es/salactsi/recommendations_final.pdf.

Johnstone, N., I. Haščič, and D. Popp (2010). Renewable energy policies and technological innovation: Evidence based on patent counts. *Environmental and Resource Economics*, **45**(1), pp. 133-155.

Jordan, A.J., and A. Lenschow (2000). Greening' the European Union: What can be learned from the leaders of EU environmental policy? *European Environment*, **10**(3), pp. 109-120.

Joskow, P.L. (2005). Transmission policy in the United States. *Utilities Policy* **13**(2), pp. 95-115.

Junginger, M., W. van Sark, and A. Faaij (eds.) (2010). *Technological Learning in the Energy Sector: Lessons from Policy, Industry and Science*. Edward Elgar, Cheltenham, UK, 352 pp.

Junginger, M., J. van Dam, S. Zarrilli, F. Ali Mohamed, D. Marchal, and A. Faaij (2011). Opportunities and barriers for global bioenergy trade. *Energy Policy*, **39**(4), pp. 2028-2042.

Kahn, J. (2003). Wind Power planning in three Swedish municipalities. *Journal of Environmental Planning and Management*, **46**(4), pp. 563-581.

Kammen, D.M. (2009). Financing energy efficiency with taxes. *Scientific American*, **21**(Earth 3.0), March 2009.

Kaplan, A.W. (1999). From passive to active about solar electricity: innovation decision process and photovoltaic interest generation. *Technovation*, **19**(8). pp. 467-481.

Karnoe, P. (1990). Technological innovation and industrial organization in the Danish wind industry. *Entrepreneurship & Regional Development*, **2**(2), pp. 105-124.

Keller, K., and J. Wild (2004). Long-term investment in electricity: a trade-off between co-ordination and competition? *Utilities Policy* **12**(4), pp. 243-251.

Kemp, R., and J. Rotmans (2009). Transitioning policy: Co-production of a new strategic framework for energy innovation policy in the Netherlands. *Policy Sciences*, **42**(4), pp. 303-322.

Kenney, M. (2010). Venture capital investment in the greentech industries: a provocative essay. In: *Handbook of Research on Energy Entrepreneurship*. R. Wustenhagen and R. Wuebker (eds.), Edward Elgar, Cheltenham, UK.

Kern, F., and F. Howlett (2009). Implementing transition management as policy reforms: A case study of the Dutch energy sector. *Policy Sciences*, **42**(4), pp. 391-408.

Kim, L. (1991). Pros and cons of international technology transfer: A developing country view. In: *Technology Transfer in International Business*. T. Agmon and M.A. Von Glinow (eds.), Oxford University Press.

Kim, L. (1997). *Imitation to Innovation: The Dynamics of Korea's Technological Learning*. Harvard Business School Press, Cambridge, MA, USA.

Klein, A., A. Held, M. Ragwitz, G. Resch, and T. Faber (2008a). *Evaluation of Different Feed-in Tariff Design Options – Best Practice Paper for the International Feed-in Cooperation*. Fraunhofer Institute Systems and Innovation Research and Energy Economics Group, Karlsruhe, Germany and Vienna, Austria.

Klein, A., B. Pfluger, A. Held, M. Ragwitz, G. Resch and T. Faber (2008b). *Evaluation of Different Feed-in Tariff Design Options – Best Practice Paper for the International Feed-In Cooperation, 2nd edition*. Energy Economics Group and Fraunhofer Institute Systems and Innovation Research, Vienna, Austria and Karlsruhe, Germany.

Klein, A., E. Merkel, B. Pfluger, A. Held, M. Ragwitz, G. Resch, and S. Busch (2010). *Evaluation of Different Feed-in Tariff Design Options – Best Practice Paper for the International Feed-In Cooperation, 3rd edition*. Energy Economics Group and Fraunhofer Institute Systems and Innovation Research, Vienna, Austria and Karlsruhe, Germany.

Klessmann, C., C. Nabe, and K. Burges (2008). Pros and cons of exposing renewables to electricity market risks – A comparison of the market integration approaches in Germany, Spain, and the UK. *Energy Policy*, **36**, pp. 3646-3661.

Kline, D.M., L. Vimmerstedt, and R. Benioff (2004). Clean energy technology transfer: A review of programs under the UNFCCC. *Mitigation and Adaptation Strategies for Global Change*, **9**, pp. 1-35.

Kobayashi, T. (2003). Vision of the future of the photovoltaic industry in Japan. In: Proceedings of 3rd World Conference on Photovoltaic Energy Conversion, Osaka, Japan, 18-18 May 2003, pp. 2538–2543.

Krapels, E.N. (2010). The terrible trio impeding transmission development: Siting, cost allocation, and interconnection animus. *The Electricity Journal*, **23**(1), pp. 34-38.

Ku, J., E.I. Baring-Gould, and K. Stroup (2005). *Renewable Energy Applications for Rural Development in China*. NREL/CP-710-37605, National Renewable Energy Laboratory, Golden, CO, USA, 5 pp.

Laffont, J.J. and Tirole, J. (1988). The dynamics of incentive contracts. *Econometrica*, 56, pp. 1153-1176.

Laird, F.M., and C. Stefes (2009). The diverging paths of German and United States policies for renewable energy: Sources of difference. *Energy Policy*, **37**(2009), pp. 2619-2629.

Lamers, P., C. Hamelinck, M. Junginger, and A. Faaij (2011). International bioenergy trade – a review of past developments in the liquid biofuels market. *Renewable & Sustainable Energy Reviews*, **15**(6), pp. 2655-2676.

Langniß, O., and L. Neij (2004). National and international learning with wind power. *Energy & Environment*, **15**(2), pp. 175-185.

Langniß, O., J. Diekmann, and U. Lehr (2009). Advanced mechanisms for the promotion of renewable energy. Models for the future evolution of the German Renewable Energy Act. *Energy Policy*, **37**(2009), pp. 1289-1297.

Lantz, E., and E. Doris (2009). *State Clean Energy Practices: Renewable Energy Rebates*. NREL/TP-6A2-45039, National Renewable Energy Laboratory, Golden, CO, USA, 38 pp.

Lapola, D.M. R. Schaldach, J. Alcamo, A. Bondeau, J. Koch, C. Koelking, and J.A. Priess (2010). Indirect land use changes can overcome carbon savings from biofuels in Brazil. *Proceedings of the National Academy of Sciences*, **107**(8), pp. 3388-3393.

Lauber, V. (2004). REFIT and RPS: options for a harmonised Community framework. *Energy Policy*, **32**(12), pp. 1405-1414.

Lauber, V., and L. Mez (2004). Three decades of renewable electricity policies in Germany. *Energy and Environment*, **15**(4), pp. 599-623.

Lehr, U., J. Nitsch, M. Kratzat, C. Lutz, and D. Edler (2008). Renewable energy and employment in Germany. *Energy Policy*, **36**(1), pp. 108-117.

Lenschow, A. (2002). *Environmental Policy Integration: Greening Sectoral Policies in Europe*. Earthscan, London, UK.

Lewis, J.I. (2007). Technology acquisition and innovation in the developing world: Wind turbine development in China and India. *Studies in Comparative International Development*, **42**, pp. 208-232.

Lewis, J. (2010). The evolving role of carbon finance in promoting renewable energy development in China. *Energy Policy*, **38**(6). pp. 2875-2886.

Lewis, J., and R. Wiser (2005). *Fostering a Renewable Energy Technology Industry: An International Comparison of Wind Industry Policy Support Mechanisms*. LBNL-59116, Ernest Orlando Lawrence Berkeley National Laboratory, Berkeley, CA, USA, 30 pp.

Liao, C., E. Jochem, Y. Zhang, and N.R. Farid (2010). Wind power development and policies in China. *Renewable Energy*, **35**, pp. 1879-1886.

Lipp, J. (2007). Lessons for effective renewable electricity policy from Denmark, Germany and the United Kingdom. *Energy Policy*, **35**(11), pp. 5481-5495.

Liu, Y. and A. Kokko (2010). Wind power in China: Policy and development challenges. *Energy Policy*, **38**, pp. 5520-5529.

Liu, L.-q., Z.-x. Wanga, H.-q. Zhang, and Y.-c. Xue (2010). Solar energy development in China—A review. *Renewable and Sustainable Energy Reviews*, **14**, pp. 301-311.

Locke Lord Bissell & Liddell (2007). *Transmission Access Challenges for Wind Generation*. Locke Lord Bissell & Liddell, Chicago, IL, USA.

London School of Economics (2009). *Meeting the Climate Challenge: Using Public Funds to Leverage Private Investment in Developing Countries*. London School of Economics, London, UK.

Loorbach, D. (2007). *Transition Management. New Mode of Governance for Sustainable Development*. International Books, Utrecht, The Netherlands.

Lotker, M. (1991). *Barriers to Commercialization of Large-Scale Solar Electricity: Lessons Learned from the LUZ Experience*. Contractor Report, Sandia National Laboratories, Oak Ridge, TN, USA.

Lovely, M., and D. Popp (2011). Trade, technology and the environment: Does access to technology promote environmental regulation? *Journal of Environmental Economics and Management*, **61**(1), pp. 16-35.

Lovins, A.B., K. Datta, O.-E. Bustnes, J. Koomey, and N. Glasgow (2004). *Winning the Oil Endgame: Innovation for Profits, Jobs and Security*. Rocky Mountain Institute, Snowmass, CO, USA.

Lucena, A.F.P., A.S. Szklo, R. Schaeffer, R.R. Souza, B.S.M.C. Borba, and I.V.L. Costa (2009). The vulnerability of renewable energy to climate change in Brazil. *Energy Policy*, **37**, pp. 879-889.

Lund, J.W., and D.H. Freeston (2001). World-wide direct uses of geothermal energy 2000. *Geothermics*, **30**(1), pp. 29-68.

Lund, P.D. (2008). Effects of energy policies on industry expansion in renewable energy. *Renewable Energy*, **34**(1), pp. 53-64.

Madlener, R. (2007). Innovation diffusion, public policy and local initiative: the case of wood-fuelled district heating systems in Austria. *Energy Policy*, **35**, pp. 1992-2008.

Madsen, B.T. (2009). Public initiatives and industrial development after 1979. In: *The Danish Way: From Poul La Cour to Modern Wind Turbines*. Poul La Cour Foundation, Askov, Denmark.

Mahapatra, S., H.N. Chanakya, and S. Dasappa (2009). Evaluation of various energy devices for domestic lighting in India: Technology, economics and CO_2 emissions. *Energy for Sustainable Development*, **13**(4), pp. 271-279.

Mallett, A. (2007). Social acceptance of renewable energy innovations: The role of technology cooperation in urban Mexico. *Energy Policy*, **35**(5), pp. 2790-2798.

Markard, J., and B. Truffer (2008). Technological innovation systems and the multi-level perspective: towards an integrated framework. *Research Policy*, **37**(4), pp. 596-615.

Martinot, E., and L. Junfeng (2007). *Powering China's Development – The Role of Renewable Energy*. Worldwatch Institute, Washington, DC, USA.

Martinot, E., A. Chaurey, D. Lew, J.R. Moreira, and N. Wamukonya (2002). Renewable energy markets in developing countries. *Annual Review of Energy and Environment*, **27**, pp. 309-48.

Maruyama, Y., M. Nishikido, and T. Iido (2007). The rise of community wind power in Japan: Enhanced acceptance through social innovation. *Energy Policy*, **35**, pp. 2761-2769.

Meadowcroft, J. (2007). National sustainable development strategies: Features, challenges, and reflexivity. *European Environment*, **17**(3), pp. 152-163.

Meijer, I.S.M., M.P. Hekkert, and J.F.M. Koppenjan (2007a). How perceived uncertainties influence transitions; the case of micro-CHP in the Netherlands. *Technological Forecasting and Social Change*, **74**(4), pp. 519-537.

Meijer, I.S.M., M.P. Hekkert, and J.F.M. Koppenjan (2007b). The influence of perceived uncertainty on entrepreneurial action in emerging renewable energy technology; biomass gasification projects in the Netherlands. *Energy Policy*, **35**(11), pp. 5836-5854.

Mendis, M.S., and W.J. van Nes (1999). *The Nepal Biogas Support Programme, Elements for Success in Rural Household Energy Supply, Policy and Best Practice Document 4*. Ministry of Foreign Affairs, The Netherlands.

Mendonça, M. (2007). *Feed-In Tariffs: Accelerating the Deployment of Renewable Energy*. Earthscan, London, UK.

Mendonça, M., S. Lacey, and F. Hvelplund (2009). Stability, participation and transparency in renewable energy policy: Lessons from Denmark and the United States. *Policy and Society*, **27**, pp. 379-398.

Menz, F.C., and S. Vachon (2006). The effectiveness of different policy regimes for promoting wind power: experiences from the States. *Energy Policy*, **34**(14), pp. 1786-1796.

MERC Partners (2009). *Staffing the Energy Industry: A survey on current and future skills needs*. MERC Partners, Dublin, Ireland

Metcalf, G.E. (2008). *Tax Policy for Financing Alternative Energy Equipment*. Tufts University Economics Department Working Paper series, Tufts University, Medford, MA, USA.

METI (2009). *About New Buyback System of PV. The 35th meeting (2009). Reference Material 1*. New Energy Committee, Ministry of Economy, Trade and Industry (METI), Japan.

Meyer, N.I. (2003). European schemes for promoting renewables in liberalised markets. *Energy Policy*, **31**(7), pp. 665-676.

Meyer, N.I. (2007). Learning from wind energy policy in the EU: lessons from Denmark, Sweden and Spain. *European Environment*, **17**(5), pp. 347-362.

Milbrandt, A. and R.P. Overend (2008). *Survey of Biomass Resource Assessments and Assessment Capabilities in APEC Economies*. NREL/TP-6A2-43710; APEC#208-RE-01.9, Asia-Pacific Economic Cooperation and National Renewable Energy Laboratory, Singapore and Golden, CO, USA, 155 pp. Available at: www.nrel.gov/docs/fy09osti/43710.pdf.

Ministerio de Economía (2004). Real Decreto 436/2004, de 12 de marzo, por el que se establece la metodología para la sistematización y actualización del régimen jurídico y económico de la actividad de producción de energía eléctrica en régimen especial. *Boletín Oficial del Estado*, Madrid, Spain.

Mitchell, C. (1995). The Renewable NFFO – A review. *Energy Policy*, **23**(12), pp. 1077-1091.

Mitchell, C. (2000). The Non-Fossil Fuel Obligation and its future. *Annual Review of Energy and Environment*, **25**, pp. 285-312.

Mitchell, C. (2008). *The Political Economy of Sustainable Energy*. Palgrave MacMillan, Hampshire, UK.

Mitchell, C. (2010). Forging European responses to the challenge of climate change and energy resource supply. In: *International Science and Technology Cooperation in a Globalised World: the External Dimension of the European Research Area*. H. Prange-Gstohl (ed.), Edward Elgar Publishers, Cheltenham, UK.

Mitchell, C., and P. Connor (2004). Renewable energy policy in the UK 1990-2003. *Energy Policy*, **32**(17), pp. 1935-1947.

Mitchell, C., D. Bauknecht, and P.M. Connor (2006). Effectiveness through risk reduction: a comparison of the renewable obligation in England and Wales and the feed-in system in Germany. *Energy Policy*, **34**(3), pp. 297-305.

Mitchell, R.L., C.E. Witt, R. King, and D. Ruby (2002). PVMaT advances in the photovoltaic industry and the focus of future PV manufacturing R&D. In: *Conference Record of the 29th IEEE Photovoltaic Specialists Conference*, New Orleans, LA, USA, 19-24 May 2002, pp. 1444-1447.

MITyC (2007). Real Decreto 661/2007, de 25 de mayo, por el que se regula la actividad de producción de energía eléctrica en régimen especial. *Boletín Oficial del Estado*, Madrid, Spain.

MITyC (2008). Real Decreto 1578/2008, de 26 de septiembre, de retribución de la actividad de producción de energía eléctrica mediante tecnología solar fotovoltaica para instalaciones posteriores a la fecha límite de mantenimiento de la retribución del Real Decreto 661/2007, de 25 de mayo, para dicha tecnología. *Boletín Oficial del Estado*, Madrid, Spain (in Spanish). Available at: www.boe.es/aeboe/consultas/bases_datos/doc.php?id=BOE-A-2008-15595.

MITyC (2009). *La Energía en España 2008;* Ministerio de Industria Turismo y Comercio, Madrid, Spain. Available at: www.mityc.es/energia/balances/Balances/LibrosEnergia/ENERGIA_2008.pdf.

MIyE (1998). *Real Decreto 2818/1998, de 23 de diciembre, sobre producción de energía eléctrica por instalaciones de abastecidas por recursos o fuentes de energías renovables, residuos y cogeneración, Boletín Oficial del Estado núm. 312, de 30 de diciembre de 1998*. Ministerio de industria y Energía, Madrid, Spain.

MNRE (2010). *Annual Report 2009-10*. Ministry of New and Renewable Energy, Government of India, New Delhi, India.

Mondal, M.A.H., L.M. Kamp, and N.I. Pachova (2010). Drivers, barriers, and strategies for implementation of renewable energy technologies in rural areas in Bangladesh – An innovation system analysis. *Energy Policy*, **38**(8), pp. 4626-4634.

Moore, B., and R. Wüstenhagen (2004). Innovative and sustainable energy technologies: The role of venture capital. *Business Strategy and the Environment*, **13**, pp. 235-245.

Moore, M.C., D.J. Arent, and D. Norland (2007). R&D advancement, technology diffusion, and impact on evaluation of public R&D. *Energy Policy*, **35**(3), pp. 1464-1473.

Morales, A., X. Robe, M. Sala, P. Prats, C. Aguerri, and E. Torres (2008). Advanced grid requirements for the integration of wind farms into the Spanish transmission system. *Iet Renewable Power Generation*, **2**(1), pp. 47-59.

Moreira, J.R., and J. Goldemberg (1999). The alcohol program. *Energy Policy*, **27**(4), pp. 229-245.

Mowery, D., and N. Rosenberg (1979). The influence of market demand upon innovation: a critical review of some recent empirical studies. *Research Policy*, **8**(2), pp. 102-153.

Mowery, D.C., R.R. Nelson, and B.R. Martin (2010). Technology policy and global warming: Why new policy models are needed (or why putting new wine in old bottles won't work). *Research Policy*, **39**(8), pp. 1011-1023.

Müller, B. (2010). *The Reformed Financial Mechanism of the UNFCCC Part II: The Question of Oversight Post Copenhagen Synthesis*. Report EV52, Oxford Institute for Energy Studies, Oxford, UK.

Murphy, L.M., and P.L. Edwards (2003). *Bridging the Valley of Death: Transitioning from Public to Private Sector Financing*. National Renewable Energy Laboratory, Golden, CO, USA.

Mytelka, L. (2007). *Technology Transfer Issues in Environmental Goods and Services: An Illustrative Analysis of Sectors Relevant to Air-pollution and Renewable Energy*. International Centre for Trade and Sustainable Development, Geneva, Switzerland

Nadaï, A. (2007). "Planning", "siting" and the local acceptance of wind power: some lessons from the French case. *Energy Policy*, **35**(5), pp. 2715-2726.

Nadaï, A., and O. Labussière (2009). Wind power planning in France (Aveyron), from state regulation to local planning. *Land Use Policy*, **26**(3), pp. 744-754.

Nadaï, A., and O. Labussière (2010). Birds, wind, and the making of wind power landscapes in Aude, Southern France. *Landscape Research*, **35**(2), pp. 209-233.

Nast, M. (2010). Renewable energies heat act and government grants in Germany. *Renewable Energy*, **35**(8), pp. 1852-1856.

Nast, M., O. Langniss, and U. Leprich (2007). Instruments to promote renewable energy in the German heat market – Renewable Heat Sources Act. *Renewable Energy*, **32**, pp. 1127-1135.

National Energy Policy Committee (2008). *The Bahamas National Energy Policy*. National Energy Policy Committee, Jamaica. Available at: cipore.org/bahamas-national-energy-policy/.

National Greenhouse Strategy (1998). *Strategic Framework for Advancing Australia's Greenhouse Response*. Australian Government, Canberra, Australia.

National Grid (2008). *GB SQSS Consultation Document – Review of Onshore Intermittent Generation*. National Grid, UK. Available at: www.nationalgrid.com/uk/Electricity/Codes/gbsqsscode/reviews/.

NEDO (2009). *The Roadmap PV 2030+*. New Energy and Industrial Technology Development Organization (NEDO), Kawasaki, Japan.

Negro, S.O., M.P. Hekkert, and R.E.H.M. Smits (2007). Explaining the failure of the Dutch innovation system for biomass digestion – A functional analysis. *Energy Policy*, **35**(2), pp. 925-938.

Neij, L. (2008). Cost development of future technologies for power generation – A study based on experience curves and complementary bottom-up assessments. *Energy Policy*, **36**(6), 2200-2211.

Nelson, R.R. (1959). The simple economics of basic scientific research. *Journal of Political Economy*, **67**(3), pp. 297-306.

Nelson, R.R., and S.G. Winter (1982). *An Evolutionary Theory of Economic Change*. Belknap Press, Cambridge, MA, USA and London, UK.

Nemet, G. (2006). Beyond the learning curve: factors influencing cost reductions in photovoltaics. *Energy Policy*, **34**(2006), pp. 3218-3232.

Nemet, G.F. (2009). Demand-pull, technology-push, and government-led incentives for non-incremental technical change. *Research Policy*, **38**(5), pp. 700-709.

Nemet, G. (2010a). Benefit cost analysis of R&D as a solution to climate change. In: *Smart Solutions to Climate Change: Comparing Costs and Benefits*. B. Lomborg (ed.), Cambridge University Press, pp. 349-359.

Nemet, G.F. (2010b). Robust incentives and the design of a climate change governance regime. *Energy Policy*, **38**(11), pp. 7216-7225.

Nemet, G.F., and D.M. Kammen (2007). U.S. energy research and development: Declining investment, increasing need, and the feasibility of expansion. *Energy Policy*, **35**(1), pp. 746-755.

NEPC (2009). *Status of introduction of PV System by fiscal year and by prefecture (Nendobetu todofukenbetu jutakuyou taiyouko-hatsuden-sisutemu donyu-jokyou)*. New Energy Promotion Council (NEPC), Tokyo, Japan (in Japanese). Available at: nepc.or.jp/topics/pdf/090605_1.pdf.

Neuhoff, K. (2004). *Large Scale Deployment of Renewables for Electricity Generation*. Organisation for Economic Co-operation and Development, Paris, France, 40 pp.

Neuhoff, K. (2005). Large-scale deployment of renewables for electricity generation. *Oxford Review of Economic Policy*, **21**(1), pp. 88-110.

Neuhoff, K., S. Dröge, O. Edenhofer, C. Flachsland, H. Held, M. Ragwitz, J. Strohschein, A. Türk, and A. Michaelowa (2009). *Translating Model Results to Economic Policies*. RECIPE Background Paper, Potsdam Institute for Climate Impact Research (PIK), Potsdam, Germany.

New Energy Finance Limited (2007). *RECs, ROCs, Feed-in Tariffs: What is the Best Incentive Scheme for Wind Power Investors?* Bloomberg New Energy Finance.

Newig, J., and O. Fritsch (2009). Environmental governance: participatory, multi-level – and effective? *Environmental Policy and Governance*, **19**, pp. 197-214.

Nilkuha, K. (2009). National Biofuels Policy, Deployment and Plans – Thailand. In: *Bangkok Biofuels 2009. Sustainable development of biofuels*, 7-8 September 2009, Bangkok, Thailand

Nolan, J.M., P.W. Schultz, R.B. Cialdini, V. Griskevicius, and N.J. Goldstein (2008). Normative social influence is underdetected. *Personality & Social Psychology Bulletin*, **34**(7), pp. 913-923.

Norberg-Bohm, V. (1999). Stimulating green technological innovation: An analysis of alternative policy mechanisms. *Policy Sciences*, **32**(1), pp. 13-38.

Nordhaus, W. (2010). Designing a friendly space for technological change to slow global warming. *Energy Economics*, doi:10.1016/j.eneco.2010.08.005. Abstract available at: nordhaus.econ.yale.edu/documents/sm_052610.pdf.

NRC (2001). *Energy Research at DOE: Was It Worth It? Energy Efficiency and Fossil Energy Research 1978 to 2000*. National Research Council (NRC), National Academies Press, Washington, DC, USA.

NRC (2007). *Prospective Evaluation of Applied Energy Research and Development at DOE (Phase Two)*. National Research Council (NRC), The National Academies Press, Washington, DC, USA.

NREL (2004). *Renewable Energy in China: Township Electrification Program*. NREL/FS-710-35788, National Renewable Energy Laboratory (NREL), Golden, CO, USA.

NSB (2010). *Science and Engineering Indicators 2010*. National Science Board (NSB), National Science Foundation, Arlington, VA, USA.

O'Reilly, C.A.I., and M.L. Tushman (2004). The ambidextrous organization. *Harvard Business Review*, **April**, pp. 74-81.

Ockwell, D., A. Ely, A. Mallett, O. Johnson, and J. Watson (2009). *Low Carbon Development: The Role of Local Innovative Capabilities: Paper 31*. STEPS Centre and Sussex Energy Group, SPRU, University of Sussex, Brighton, UK. Available at: www.sussex.ac.uk/sussexenergygroup/documents/ockwell-et-al-paper-31.pdf.

Ockwell, D.G., R. Hauma, A. Mallett, and J. Watson (2010). Intellectual property rights and low carbon technology transfer: Conflicting discourses of diffusion and development. *Global Environmental Change*, **20**, pp. 729-738.

OECD (2008). *The Paris Declaration on Aid Effectiveness and the Accra Agenda for Action*. Organisation for Economic Co-operation and Development (OECD), Paris, France. Available at: www.oecd.org/dataoecd/11/41/34428351.pdf.

Ofgem (2008). *Transmission Access Review - Initial Consultation on Enhanced Investment Incentives*. Ofgem, London, UK. Available at: www.ofgem.gov.uk/Networks/Trans/ElecTransPolicy/tar/Documents1/081219_TOincentives_consultation_FINAL.pdf.

Otto, V.M., A. Löschel, and J. Reilly (2008). Directed technical change and differentiation of climate policy. *Energy Economics*, **30**(6), pp. 2855–2878.

Owens, S., and L. Driffill (2008). How to change attitudes and behaviours in the context of energy. *Energy Policy*, **36**(12), pp. 4412-4418.

Parzen, J. (2009). *Lessons Learned: Creating the Chicago Climate Action Plan*. Prepared for the City of Chicago Department of the Environment, Chicago, IL, USA, 38 pp.

Peidong, Z., Y. Yanli, T. Yongsheng, Y. Xutong, Z. Yongkai, Z. Yonghong, and W. Lisheng (2009). Bioenergy industries development in China: Dilemma and solution. *Renewable and Sustainable Energy Reviews*, **13**, pp. 2571-2579.

PEMM (2009). *Proceedings of the Pacific Energy Ministers' Meeting (PEMM): A CROP Energy Working Group Report*, 20-24 April 2009, Nuku'alofa, Tonga. Available at: dev.sopac.org.fj/VirLib/JC0200.pdf.

Peretz, N., and Z. Acs (2011). Driving energy innovation through ex ante incentive prizes. In: *Handbook on Research in Energy Entrepreneurship*. R. Wüstenhagen and R. Wuebker (eds.), Edward Elgar Publishing, Cheltenham, UK.

Persson, A. (2004). *Environmental Policy Integration: an Introduction, PINTS Background Paper*. Stockholm Environment Institute, Stockholm, Sweden.

Peters, G.B. (1998). *Managing Horizontal Government*. Research Paper 21. Canadian Centre for Management Development, Ottawa, Canada.

PIFS (2009a). *Final Communique of 40th Pacific Islands Forum*, Fortieth Pacific Islands Forum, Cairns, Australia, 5-6 August 2009.

PIFS (2009b). Summary of Decisions. Eighteenth Smaller Island States Leaders' Meeting. In: *Pacific Islands Forum Secretariat*, Cairns, Australia, 4 August 2009, pp. 7.

PIFS (2010). Pacific Islands Forum Secretariat. In: *Forum Communiqué, Forty-first Pacific Islands Forum*, Port Vila, Vanuatu, 4-5 August 2010, pp. 15.

Pigou, A.C. (1920). *The Economics of Welfare*. Macmillan, London, UK.

Pimentel, D., A. Marklein, M.A. Toth, M.N. Karpoff, G.S. Paul, R. McCormack, J. Kyriazis and T. Krueger (2009). Food versus Biofuels: Environmental and Economic Costs. *Human Ecology*, **37**, pp. 1-12.

Plevin, R.J., M. O'Hare, A.D. Jones, M.S. Torn, and H.K. Gibbs (2010). Greenhouse gas emissions from biofuels' indirect land use change are uncertain but may be much greater than previously estimated. *Environmental Science & Technology*, **44**(21), pp. 8015-8021.

Poortinga, W., L. Steg, and C. Vlek (2004). Values, environmental concern, and environmental behavior: a study into household energy use. *Environment and Behavior*, **36**(1), pp. 70-93.

Popp, D. (2006a). Comparison of climate policies in the ENTICE-BR model. *Energy Journal* (Special Issue), pp. 163-174.

Popp, D. (2006b). ENTICE-BR: The effects of backstop technology R&D on climate policy models. *Energy Economics*, **28**, pp. 188-222.

Popp, D. (2010). Innovation and climate policy. *Annual Review of Resource Economics*, **2**(1), pp. 275-298.

Popp, D., and R.G. Newell (2009). *Where Does Energy R&D Come From? Examining Crowding Out from Environmentally-Friendly R&D*. NBER Working Paper No. 15423, National Bureau of Economic Research, Cambridge, MA, USA.

Pousa, G.P.A.G., A.L.F. Santos, and A.Z. Suarez (2007). History and policy of biodiesel in Brazil. *Energy Policy*, **35**(11), pp. 5393-5398.

Prommin Lertsuriyadej (2003). Energy Strategy for Competitiveness Workshop. In: *Energy Strategy for Competitiveness Workshop. Chaired by Prime Minister Thaksin Shinawatra*, Bangkok, Thailand, 28 August 2003.

Prud'homme, R., and J.P. Bocajero (2005). The London congestion charge: a tentative economic appraisal. *Transport Policy*, **12**(3), pp. 279-287.

Puga, J.N., and J.A. Lesser (2009). Public policy and private interests: Why transmission planning and cost-allocation methods continue to stifle renewable energy policy goals. *The Electricity Journal*, **22**(10), pp. 7-19.

Puig, J. (2008). Barcelona and the power of solar ordinances: Political will, capacity building and people's participation. In: *Urban Energy Transition: From Fossil Fuels to Renewable Power*. P. Droege (ed.), Elsevier, London, UK, pp. 433-450.

Radov, D., P. Klevnas, and S. Carter (2008). *Qualitative Evaluation of Financial Instruments for Renewable Heat*. National Economic Research Associates, London, UK.

Ragwitz, M., A. Held, G. Resch, T. Faber, C. Huber, and R. Haas (2005). *Final Report: Monitoring and Evaluation of Policy Instruments to Support Renewable Electricity in EU Member States*. Fraunhofer Institute Systems and Innovation Research and Energy Economics Group, Karlsruhe, Germany and Vienna, Austria.

Ragwitz, M. A. Held, F. Sensfuss, C. Huber, G. Resch, T. Faber, R. Haas, R. Coenraads, A. Morotz, S.G. Jensen, P.E. Morthorst, I. Konstantinaviciute, and B. Heyder (2006). *OPTRES – Assessment and Optimisation of Renewable Support Schemes in the European Electricity Market*. Intelligent Energy Europe. Available at: www.optres.fhg.de/OPTRES_FINAL_REPORT.pdf.

Ragwitz, M., W. Schade, B. Breitschopf, R. Walz, N. Helfrich, M. Rathmann, G. Resch, C. Panzer, T. Faber, R. Haas, C. Nathani, M. Holzhey, I. Konstantinaviciute, P. Zagamé, A. Fougeyrollas, and B.L. Le Hir (2009). *The Impact of Renewable Energy Policy on Economic Growth and Employment in the European Union*. EMployRES Final report, Contract No. TREN/D1/474/2006, 27 April, European Commission DG Energy and Transport, Brussels, Belgium. Available at: ec.europa.eu/energy/renewables/studies/doc/renewables/2009_employ_res_report.pdf.

Ragwitz, M., A. Held, E. Stricker, A. Krechting, G. Resch, and C. Panzer (2010). *Recent Experiences with Feed-in Tariff Systems in the EU – A research paper for the International Feed-In Cooperation*. A report commissioned by the Ministry for the Environment, Nature Conservation and Nuclear Safety (BMU), Bonn, Germany.

Ragwitz, M., A. Held, B. Breitschopf, M. Rathman, C. Klessmann, G. Reche, C. Panzer, S. Busch, K. Neuhoff, M. Junginger, R. Hoefinagels, N. Cusumano, A. Lorenzoni, J. Burgers, M. Boots, I. Konstantinaviciute, and B. Weores (2011). *D8 Report Review on Support Schemes for Renewable Electricity and heating in Europe*. A Report compiled for RE-Shaping, No. EIE/08/517/512.529243, Intelligent Energy Europe. Available at: www.reshaping-res-policy.eu/downloads/D8%20Review%20Report_final%20(RE-Shaping).pdf.

Raven, R., E. Heiskanen, R. Lovio, M. Hodson, and B. Brohmann (2008). The contribution of local experiments and negotiation processes to field-level learning in emerging (niche) technologies: Meta-analysis of 27 new energy projects in Europe. *Bulletin of Science, Technology and Society*, **28**, pp. 464-477.

RECIPES (2005). *Country Reports, 2005-2006*. Developing Renewables, 'Renewable Energy in developing countries: Current situation, market Potential and recommendations for a win-win-win for EU industry, the Environment and local Socio-economic development' (RECIPES).

Reed, D., A. Kutter, A. Ballesteros, E. Fendley, M. del Socorro Flores Liera, J. Harnisch, S. Huq, and H.-O. Ibrekk (2009). *The Institutional Architecture for Financing a Global Climate Deal: An Options Paper*. Technical Working Group on the Institutional Architecture for Climate Finance. Available at: www.usclimatenetwork.org/resource-database/Options%20Paper%20Final%20May%2028.pdf.

REEGLE (2011). *Clean Energy Information Portal (REEGLE)*. Online database. Available at: www.reegle.info/.

REN21 (2005). *Renewables 2005 Global Status Report: Notes and References Companion Document*. Worldwatch Institute and Renewable Energy Policy Network for the 21st Century (REN21) Secretariat, Washington, DC, USA and Paris, France.

REN21 (2006). *Changing Climates: The Role of Renewable Energy in a Carbon-Constrained World*. Renewable Energy Policy Network for the 21st Century (REN21) Secretariat, Paris, France.

REN21 (2007). *Renewables 2007: Global Status Report*. Renewable Energy Policy Network for the 21st Century (REN21), Paris, France.

REN21 (2009a). *Recommendations for Improving Effectiveness of Renewable Energy Policies in China*. Renewable Energy Policy Network for the 21st Century (REN21), Paris, France.

REN21 (2009b). *Renewables Global Status Report: 2009 Update*. Renewable Energy Policy Network for the 21st Century Secretariat, Paris, France, 42 pp.

REN21 (2010). *Renewables 2010: Global Status Report*. Renewable Energy Policy Network for the 21st Century Secretariat, Paris, France, 80 pp.

REN21 (2011). *Renewables Interactive Map*. Renewable Energy Policy Network for the 21st Century (REN21). Paris, France. Available at: www.ren21.net/REN21Activities/InteractiveTools/RenewablesMap/tabid/5444/Default.aspx.

Renewables (2004). Conference Report: Outcomes & Documentation – Political Declaration / International Action Programme/Policy Recommendations for Renewable Energies. In: *Renewables 2004 - International Conference for Renewable Energies*, Bonn, Germany, 1- 4 June 2004.

Requate, T. (2005). Dynamic incentives by environmental policy instruments – a survey. *Ecological Economics*, **54**(2-3), pp. 175-195.

Resch, G., C. Panzer, M. Ragwitz, T. Faber, C. Huber, M. Rathmann, G. Reece, A. Held, R. Haas, P.E. Morthorst, S. Grenna, L. Jawowski, I. Konstantinaviciute, R. Pasinetti, and K. Vertin (2009). *futures-e – deriving a future European policy for renewable electricity*. Final Report of the Research Project futures-e, with support from the European Commission. Contract No. EIE/06/143/S12.444285, DG TREN, EACI under the Intelligent Energy for Europe Programme, Vienna, Austria.

Resources and Logistics (2010). *Identification Mission for the Mediterranean Solar Plan, Final Report*. Funded by the European Union, Resources and Logistics. available at: ec.europa.eu/energy/international/international_cooperation/doc/2010_01_solar_plan_report.pdf.

Rhodes, R.A.W. (1996). The new governance: Governing without government. *Political Studies*, **44**(4), pp. 652-667.

Richels, R.G., and G.J. Blanford (2008). The value of technological advance in decarbonizing the U.S. economy. *Energy Economics*, **30**(6), pp. 2930-2946.

Rickerson, W.H., J.L. Sawin, and R.C. Grace (2007). If the shoe FITs: Using feed-in tariffs to meet U.S. renewable electricity targets. *The Electricity Journal*, **20**(4), pp. 73-86.

Rickerson, W.H., F. Bennhold, and J. Bradbury (2008). *Feed-in Tariffs and Renewable Energy in the USA – a Policy Update*. North Carolina Solar Center, Heinrich Boll Foundation, and World Future Council, Raleigh, NC, USA, Berlin, Germany, and Washington, DC, USA.

Rickerson, W.H., T. Halfpenny, and S. Cohan (2009). The emergence of renewable heating and cooling policy in the United States. *Policy and Society*, **27**(4), pp. 365-377.

Ritz, R. (2009). *Carbon Leakage under Incomplete Environmental Regulation: An Industry Level Approach*. Discussion Paper 461, Oxford University Department of Economics, Oxford, UK.

Rivers, N. and M. Jaccard (2006). Choice of environmental policy in the presence of learning by doing. *Energy Economics*, **28**(2), pp. 223-242.

Rodriguez, J.M., O. Alonso, M. Duvison, and T. Domingez (2008). The integration of renewable energy and the system operation: The Special Regime Control Centre (CECRE) in Spain. In: *Power and Energy Society General Meeting – Conversion and Delivery of Electrical Energy in the 21st Century, IEEE*, Pittsburgh, PA, USA, 20-24 July 2008, pp. 1-6, 20-24

Roffe, P., and T. Tesfachew (2002). Revisiting the technology transfer debate: lessons for the new WTO Working Group. *Bridges*, **6**(2), February 2002, International Centre for Trade and Sustainable Development.

Rogers, E.M. (2003). *Diffusion of Innovations*. Free Press, New York, NY, USA.

Rogers, J., E. Simmons, I. Convery, and A. Weatherall (2008). Public perceptions of opportunities for community-based renewable energy projects. *Energy Policy*, pp. 4217-4226.

Rose, J., E. Webber, A. Browning, S. Chapman, G. Rose, C. Eyzaguirre, J. Keyes, K. Fox, R. Haynes, K. McAllister, M. Quinlan, and C. Murchie (2008). *Freeing the Grid: Best and Worst Practices in State Net Metering Policies and Interconnection Standards*. Network for New Energy Choices, New York, NY, USA.

Rosenau, J.N., and E.O. Czempiel (eds.) (1992). *Governance without Government: Order and Change in World Politics*. Cambridge University Press.

Rosendahl, K.E. (2004). Cost-effective environmental policy: implications of induced technological change. *Journal of Environmental Economics and Management*, **48**(3), pp. 1099-1121.

Rotmans, J., R. Kemp, and M. Van Asselt (2001). More evolution than revolution: Transition management in public policy. *Foresight*, **3**(1), pp.15–31.

Roulleau, T., and C.R. Lloyd (2008). International policy issues regarding solar water heating, with a focus on New Zealand. *Energy Policy*, **36**(6), pp. 1843-1857.

Sanden, B.A., and C. Azar (2005). Near-term technology policies for long-term climate targets--economy wide versus technology specific approaches. *Energy Policy*, **33**(12), pp. 1557-1576.

Sauter, R., and J. Watson (2008). *Technology Leapfrogging: A Review of the Evidence: A Report for DFID*. Sussex Energy Group – Science & Technology Policy Research (SRPU), University of Sussex, Brighton, UK.

Sawin, J.L. (2001). *The Role of Government in the Development and Diffusion of Renewable Energy Technologies: Wind Power in the United States, California, Denmark and Germany, 1970-2000*. PhD Thesis, Fletcher School of Law and Diplomacy, Tufts University, Medford, MA, USA, 672 pp.

Sawin, J.L. (2004). *National Policy Instruments: Policy Lessons for the Advancement and Diffusion of Renewable Energy Technologies Around the World – Thematic Background Paper. International Conference for Renewable Energies*. Secretariat of the International Conference for Renewable Energies. Bonn, Germany.

Sawin, J.L., and W.R. Moomaw (2009). An enduring energy future. In: *State of the World 2009: Into a Warming World*. W.W. Norton & Company, Inc., Washington, DC, USA.

SCB (2009). *Electricity Supply, District Heating and Supply of Natural and Gasworks Gas 2008*. EN 11 SM 1002 (and older reports in that publishing series), Statistics Sweden (SCB), Örebro, Sweden.

Schaeffer, G.J., E. Alsema, A. Seebregt, L. Beurskens, H. de Moor, W. van Sark, M. Durstewitz, M. Perrin, P. Boulanger, H. Laukamp, and C. Zuccaro (2004). *Learning from the Sun: Analysis of the Use of Experience Curves for Energy Policy Purposes: The Case of Photovoltaic Power*. ECN-C--04-035, Energy Research Centre of the Netherlands, Petten, The Netherlands.

Schmidt, R.C., and R. Marschinski (2009). A model of technological breakthrough in the renewable energy sector. *Ecological Economics*, **69**, 2: 435-444

Schneider, A.L., and H. Ingram (1997). *Policy design for democracy*. University of Kansas Press, Lawrence, KS, USA.

Schneider, M., A. Holzer, and Volker H. Hoffmann (2008). Understanding the CDM's contribution to technology transfer. *Energy Policy*, **36**, pp. 2920-2928.

Schock, R.N., W. Fulkerson, M.L. Brown, R.L.S. Martin, D.L. Greene, and J. Edmonds (1999). How much is energy research and development worth as insurance? *Annual Review of Energy and Environment*, **24**, pp. 487-512.

Schultz, P.W. (2002). Knowledge, information, and household recycling: Examining the knowledge-deficit model of behavior change. In: *New Tools for Environmental Protection: Education, Information, and Voluntary Measures*. T. Dietz and P.C. Stern (eds.), National Academy Press, Washington, DC, pp. 67-82.

Searchinger, T., R. Heimlich, and R.A. Houghton (2008). Use of U.S. croplands for biofuels increases greenhouse gases through emissions from land use change. *Science*, **319**, pp. 1238-1240.

SEF Alliance (2008). *Public Venture Capital Study*. Bloomberg New Energy Finance. Available at: www.sefalliance.org/fileadmin/media/sefalliance/docs/specialised_research/NEF_Public_VC_Study_Final.pdf.

SEI (2009). *Bilateral Finance Institutions and Climate Change – A Mapping of Climate Portfolios, Working Paper*. Stockholm Environment Institute, Stockholm, Sweden.

Sensfuß, F., M. Ragwitz, and M. Genoese (2008). The merit-order effect: A detailed analysis of the price effect of renewable electricity generation on spot market prices in Germany. *Energy Policy*, **36**, pp. 3086-3094.

Seres, S., E. Haites, and K. Murphy (2009). Analysis of technology transfer in CDM projects: An update. *Energy Policy*, **37**, pp. 4919-4926.

SERIS (2009). *Annual Report*. National University of Singapore and Singapore Economic Development Board, Singapore.

Seyboth, K., L. Beurskens, O. Langniss, and R.E.H. Sims (2008). Recognising the potential for renewable energy heating and cooling. *Energy Policy*, **36**(7), pp. 2460-2463.

Sherwood, L. (2010). *U.S. Solar Market Trends 2009*. Interstate Renewable Energy Council, Latham, NY, USA, 24 pp. Available at: irecusa.org/wp-content/uploads/2010/07/IREC-Solar-Market-Trends-Report-2010_7-27-10_web1.pdf.

Shove, E. (2003). *Comfort, Cleanliness, and Convenience: The Social Organisation of Normality*. Berg Publishers, Oxford, UK.

Siddiqui, A.S., C. Marnay, and R.H. Wiser (2007). Real options valuation of US federal renewable energy research, development, demonstration, and deployment. *Energy Policy*, **35**(1), pp. 265-279.

Sims, R.E.H., R.N. Schock, A. Adegbululgbe, J. Fenhann, I. Konstantinaviciute, W. Moomaw, H.B. Nimir, B. Schlamadinger, J. Torres-Martinez, C. Turner, U. Y., S.J.V. Vuori, N. Wamukonya, and X. Zhang (2007). Energy Supply. In: *Climate Change 2007: Mitigation of Climate Change. Contribution of Working Group III to the Fourth Assessment Report of the Intergovernmental Panel on Climate Change*. B. Metz, O.R. Davidson, P.R. Bosch, R. Dave, L.A. Meyer (eds.), Cambridge University Press, pp. 251-322.

Sinn, H.-W. (2008). Public policies against global warming: a supply-side approach. *International Tax and Public Finance*, **15**, pp. 360-394.

Sioshansi, R and W. Short (2009). Evaluating the impacts of real-time pricing on the usage of wind generation. *IEEE Transactions on Power Systems*, **24**(2), pp. 516-524.

SIS (2009). Renewable energy and energy efficiency in the SIS. In: *SIS Leaders' Summit*, Cairns, Australia, 4 August 2009, Pacific Islands Forum Smaller Island States (SIS).

Sjögren, T. (2009). *Optimal Taxation and Environmental Policy in a Decentralized Economic Federation with Environmental and Labor Market Externalities*. Department of Economics, Umeå University, Umeå, Sweden.

Smith, A., A. Stirling, and F. Berkhout (2005). The governance of sustainable socio-technical transitions. *Research Policy*, **34**(10), pp. 1491-1510.

Smitherman, G. (2009). *An Act to enact the Green Energy Act, 2009 and to build a green economy, to repeal the Energy Conservation Leadership Act, 2006 and the Energy Efficiency Act and to amend other statutes*. Ministry of Energy and Infrastructure, Legislative Assembly of Ontario, Canada.

Söderholm, P., K. Ek, and M. Pettersson (2007). Wind power development in Sweden: Global policies and local obstacles. *Renewable and Sustainable Energy Reviews*, **11**(3), pp. 365-400.

Sonntag-O'Brien, V., and E. Usher (2004). *Mobilising Finance for Renewable Energies*. Thematic Background Paper for the International Conference for Renewable Energies. Secretariat of the International Conference for Renewable Energies, Bonn, Germany.

Sorrell, S. and J. Sijm (2003). Carbon trading in the policy mix. *Oxford Review of Economic Policy*, **19**(3), pp. 420-437.

Sovacool, B.K. (2009a). The importance of comprehensiveness in renewable electricity and energy-efficiency policy. *Energy Policy*, **37**(4), pp. 1529-1541.

Sovacool, B.K. (2009b). Resolving the impasse in American energy policy: The case for a transformational R&D strategy at the U.S. Department of Energy. *Renewable and Sustainable Energy Reviews*, **13**(2), pp. 346-361.

SOU (2005). *Fjärrvärme och kraftvärme i framtiden*. Statens offentliga utredningar (SOU), Betänkande av Fjärrvärmeutredningen, Stockholm, Sweden.

SPC (2010). *Towards an Energy Secure Pacific: A Framework for Action on Energy Security in the Pacific*. Secretariat of the Pacific Community (SPC), Noumea, New Caledonia. Available at: www.sprep.org/att/irc/ecopies/pacific_region/686.pdf.

Sperling, D., and S. Yeh (2009). Low carbon fuel standards. *Issues in Science and Technology*, **2**, pp. 57-66.

SRU (2010). *Climate-Friendly, Reliable, Affordable: 100% Renewable Electricity Supply by 2050*. German Advisory Council on the Environment (SRU), Berlin, Germany.

St. Denis, G., and P. Parker (2009). Community energy planning in Canada: The role of renewable energy. *Renewable and Sustainable Energy Reviews*, **13**(8), pp. 2088-2095.

Standing Committee of the National People's Congress (2005). *The Renewable Energy Law of the People's Republic of China*. Standing Committee of the National People's Congress (NPC) of the People's Republic of China in the 14th Session.

Stern, N. (2007). *The Economics of Climate Change*. Cambridge University Press, 712 pp. Available at: webarchive.nationalarchives.gov.uk/+/http://www.hm-treasury.gov.uk/sternreview_index.htm.

Stirling, A. (1994). Diversity and ignorance in electricity supply investment: Addressing the solution rather than the problem. *Energy Policy*, **22**(3), pp. 195-216.

Scrase, I., A. Stirling, F.W. Geels, A. Smith, and P. Van Zwanenberg (2009). *Transformative Innovation*. A report to the Department for Environment, Food and Rural Affairs, SPRU – Science and Technology Policy Research, University of Sussex, Brighton, UK.

Stokes, D.E. (1997). *Pasteur's Quadrant: Basic Science and Technological Innovation*. Brookings Institution Press, Washington, DC, USA, 180 pp.

Strategic Energy Technology Plan (2007). *A European Strategic Energy Technology Plan – Towards a Low Carbon Future*. European Commission, Brussels, Belgium.

Strbac, P.C. (2007). *Transmission Investment, Access and Pricing in Systems with Wind Generation*. DTI Centre for Distributed Generation and Sustainable Electrical Energy Draft Summary Report. Centre for Sustainable Electricity & Distributed Generation, UK.

Sugiyama, K. (2008). Recall of the first oil shock (Daiitiji oil-shock wo kaiko suru). *Quarterly Review of International Trade and Investment (Kikan Kokusai-boeki to toshi)*, No. 71. Institute for International Trade and Investment (in Japanese). Available at: www.iti.or.jp/kikan71/71echo.pdf.

Sussman, E. (2008). Reshaping municipal and county laws to foster green building, energy efficiency, and renewable energy, N.Y.U. *Environmental Law Journal*, **15**(1) pp. 1-44.

Suurs, R.A.A., and M.P. Hekkert (2009). Cumulative causation in the formation of a technological innovation system: The case of biofuels in the Netherlands. *Technological Forecasting & Social Change*, **76**(8), pp. 1003-1020.

Swedish District Heating Association (2001). *Fjärrvärme 2000; fakta och statistik*. Swedish District Heating Association, Stockholm, Sweden.

Swedish Energy Agency (2009a). *Energy statistics for dwellings and non-residential premises 2008*. ES 2009:10, Swedish Energy Agency, Eskilstuna, Sweden.

Swedish Energy Agency (2009b). *Facts and figures - Energy in Sweden 2009*. ET2009:29, Swedish Energy Agency, Eskilstuna, Sweden.

Swedish Energy Agency (2010a). *The Electricity Certificate System 2010 (in Swedish)*. ET2010:25, Swedish Energy Agency, Eskilstuna, Sweden.

Swedish Energy Agency (2010b). *Facts and figures - Energy in Sweden 2010*. ET2010:46, Swedish Energy Agency, Eskilstuna, Sweden.

Takahashi, K. (1989). Sunshine Project in Japan – Solar photovoltaic program. *Solar Cells*, **26**, pp. 87-96.

Teske, S., T. Pregger, S. Simon, T. Naegler, W. Graus, and C. Lins (2010). Energy [R]evolution 2010—a sustainable world energy outlook. *Energy Efficiency*, doi:10.1007/s12053-010-9098-y.

Thai Ministry of Energy (2003). *Energy Strategy for Competitiveness*. Thai Ministry of Energy. Bangkok, Thailand.

Thelen, K. (1999). Historical institutionalism in comparative politics. *Annual Review of Political Science*, **2**, pp. 369-404.

Toke, D. (2007). Renewable financial support systems and cost-effectiveness. *Journal of Cleaner Production*, **15**(3), pp. 280-287.

Toke, D., S. Breukers, and M. Wolsink (2008). Wind power deployment outcomes: How can we count for the differences? *Renewable and Sustainable Energy Reviews*, **12**(4), pp. 1129-1147.

Tol, R.S.J. (2009). The economic effects of climate change. *Journal of Economic Perspectives*, **23**(2), pp. 29-51.

Tongsopit, J. (2010). *Thailand's VSPP Program. Technical Visit of the Delegation from the United Republic of Tanzania to Thailand regarding Thailand's Very Small Power Producer (VSPP) program*. Bankgok, Thailand.

Tsoutsos, T., E. Papadopoulou, A. Katsiri, and A.M. Papadopoulos (2008). Supporting schemes for renewable energy sources and their impact on reducing the emissions of greenhouse gases in Greece. *Renewable and Sustainable Energy Reviews*, **12**, pp. 1767-1788.

Twiddell, J., and T. Weir (2006). *Renewable Energy Resources*. 2nd ed. Taylor & Francis, Oxford, UK and New York, NY, USA.

Ulph, A., and D. Ulph (2009). *Optimal Climate Change Policies When Governments Cannot Commit*. Discussion Paper Series, Department of Economics, University of St. Andrews, Fife, Scotland. Available at: www.st-andrews.ac.uk/academic/economics/papers/dp0909.pdf.

UNCTAD (2010a). *Report of the Expert Meeting on Green and Renewable Technologies as Energy Solutions for Rural Development*. TD/B/C.I/EM.3/3. United Nations Conference on Trade and Development, Geneva, Switzerland.

UNCTAD (2010b). *Report of the Secretary general on New and emerging technologies: renewable energy for development*. E/CN.16/2010/4. United Nations Conference on Trade and Development, Geneva, Switzerland.

UNCTAD (2010c). *World Investment Report*. United Nations Conference on Trade and Development, Geneva, Switzerland.

UNDP and AEPC (2010). *Capacity development in Scaling up Decentralized Energy Access Programmes. Lessons from Nepal on its role, costs and financing*. Alternative Energy Promotion Centre (AEPC), United Nations Development Programme (UNDP), New York, NY, USA.

UNEP (2005). *Public Finance Mechanisms to Catalyse Sustainable Energy Sector Growth*. United Nations Environment Programme (UNEP), Paris, France. Available at: www.uneptie.org/energy/activities/sefi/pdf/SEFI_PublicFinanceReport.pdf.

UNEP (2008). *Public Finance Mechanisms to Mobilise Investment in Climate Change Mitigation: An overview of mechanisms being used today to help scale up the climate mitigation markets, with a particular focus on the clean energy sector*. United Nations Environment Programme (UNEP), Paris, France. Available at: www.unep.fr/energy/finance/documents/pdf/UNEP_PFM%20_Advance_Draft.pdf.

UNEP and BNEF (2010) *Global Trends in Sustainable Energy Investment 2010: Analysis of Trends and Issues in the Financing of Renewable Energy and Energy Efficiency*. United Nations Environment Programme (UNEP), Paris, France and Bloomberg New Energy Finance (BNEF). Available at: bnef.com/WhitePapers/download/30.

UNEP, EPO, and ICTSD (2010). *Patent and Clean Energy, Bridging the Gap between Evidence and Policy*. United Nations Environment Programme (UNEP), Paris, France, European Patent Office and International Centre for Trade and Sustainable Development, Geneva, Switzerland.

UNEP and NEF (2008). *Global Trends in Sustainable Energy Investment 2008: Analysis of Trends and Issues in the Financing of Renewable Energy and Energy Efficiency*. Division of Technology, Industry and Economics, United Nations Environment Programme (UNEP) Sustainable Energy Finance Initiative and New Energy Finance (NEF) Limited, Paris, France.

UNEP and NEF (2009). *Global Trends in Sustainable Energy Investment 2009: Analysis of Trends and Issues in the Financing of Renewable Energy and Energy Efficiency*. Division of Technology, Industry and Economics, United Nations Environment Programme (UNEP) Sustainable Energy Finance Initiative and New Energy Finance (NEF) Limited, Paris, France.

UNEP Finance Initiative (2009). *Financing a Global Deal on Climate Change: A Green Paper produced by the UNEP Finance Initiative Climate Change Working Group*. United Nations Environment Programme (UNEP), Geneva, Switzerland.

UNFCCC (1998). *Technical Paper on Terms of Transfer of Technology and know-how*. FCCC/TP/1998/1. United Nations Framework Convention on Climate Change (UNFCCC), Bonn, Germany. Available at: unfccc.int/resource/docs/tp/tp0198.pdf.

UNFCCC (2007a). *Analysis of technology transfer in CDM projects*. United Nations Framework Convention on Climate Change (UNFCCC), Bonn, Germany.

UNFCCC (2007b). *Investment and financial flows relevant to the development of an effective and appropriate international response to Climate Change*. United Nations Framework Convention on Climate Change (UNFCCC), Bonn, Germany.

UNFCCC (2008). *A summary of submissions to the Ad hoc Working Group on Long-term Cooperative Action*. FCCC/AWGLCA/2008/16, United Nations Framework Convention on Climate Change (UNFCCC), Bonn, Germany.

UNFCCC (2010). *Analysis of the contribution of the clean development mechanism to technology*. United Nations Framework Convention on Climate Change (UNFCCC), Bonn, Germany.

Unger, T., and E.O. Ahlgren (2005). Impacts of a common green certificate market on electricity and CO_2-emission markets in the Nordic countries. *Energy Policy*, **33**(16), pp. 2152-2163.

UNICA (2010). *Quotes and Stats. Ethanol production*. Brazilian Sugarcane Industry Association (UNICA), Sao Paulo, Brazil. Available at: english.unica.com.br/dadosCotacao/estatistica/.

Unruh, G.C. (2000). Understanding carbon lock-in. *Energy Policy*, **28**(12), pp. 817-830.

Upreti, B.R., and D. Van Der Horst (2004). National renewable energy policy and local opposition in the UK: The failed development of a biomass electricity plant. *Biomass and Bioenergy*, **26**(1), pp. 61-69.

Urmee, T., D. Harries, and A. Schlapfer (2009). Issues related to rural electrification using renewable energy in developing countries of Asia and Pacific. *Renewable Energy*, **34**(2), pp. 354-357.

US Congress (2004). *American Jobs Creation Act of 2004*. Congress, (2004). In: *GovTrack.us (database of federal legislation)*. Available at: www.govtrack.us/congress/bill.xpd?bill=h108-4520.

US Department of the Interior (2008). *Energy Corridors Designated in Eleven Western States*. Press Release, U.S. Department of the Interior. Washington, DC, USA.

US DOE (2008a). *Green Power Markets: Net Metering Policies*. Office of Energy Efficiency and Renewable Energy, U.S. Department of Energy, Washington, DC, USA.

US DOE (2008b). *Renewable Energy Requirement Guidance for EPACT 2005 and Executive Order 13423*. Office of Energy Efficiency and Renewable Energy Federal Energy Management Program, U.S. Department of Energy, Washington, DC, USA.

US DOE (2009). *Technology Pathway Partnerships*. U.S. Department of Energy, Washington, DC, USA. Available at: www1.eere.energy.gov/solar/technology_pathway_partnerships.html.

USDA/FAS (2009a). *Guatemala Biofuels Annual*. U.S. Department of Agriculture Foreign Agricultural Service, Washington, DC, USA. Available at: gain.fas.usda.gov/Recent%20GAIN%20Publications/General%20Report_Guatemala_Guatemala_5-26-2009.pdf.

USDA/FAS (2009b). *Peru Biofuels Annual*. Lima, Peru. U.S. Department of Agriculture Foreign Agricultural Service, Washington, DC, USA. Available at: gain.fas.usda.gov/Recent%20GAIN%20Publications/Biofuels%20Annual_Lima_Peru_8-31-2010.pdf.

USEPA (2010a). *Green Power Partnership*. U.S. Environmental Protection Agency. Washington, DC

USEPA (2010b). *Regulatory Announcement – EPA Finalizes Regulations for the National Renewable Fuel Standard Program for 2010 and Beyond*. EPA-420-F-10-007, U.S. Environmental Protection Agency, Washington, DC, USA. Available at: www.epa.gov/oms/renewablefuels/420f10007.pdf.

van Alphen, K., H.S. Kunz, and M.P. Hekkert (2008). Policy measures to promote the widespread utilization of renewable energy technologies for electricity generation in the Maldives. *Renewable and Sustainable Energy Reviews*, **12**(7), pp. 1959-1973.

van den Bergh, J.C.J.M., and F.R. Bruinsma (2008). *Managing the Transition to Renewable Energy: Theory and Practice from Local, Regional and Macro Perspectives*. Edward Elgar Publishing Limited, Cheltenham, UK.

van der Linden, N.H., M.A. Uyterlinde, C. Vrolijk, L.J. Nilsson, J. Khan, K. Astrand, K. Ericsson, and R. Wiser (2005). *Review of International Experience with Renewable Energy Obligation Support Mechanisms*. ECN-C-05-025, Energy Research Centre of the Netherlands. Petten, The Netherlands.

van der Ploeg, F. and C.A.Withagen. (2010). Is There Really a Green Paradox? *CESifo Working Paper 2963*

van Dijk, A.L., L.W.M. Beurskens, M.G. Boots, M.B.T. Kaal, T.J. de Lange, E.J.W. van Sambeek and M.A. Uyterlinde (2003). *Renewable Energy Policies and Market Developments*. ECN, Amsterdam, the Netherlands.

Verbong, G., and F. Geels (2007). The ongoing energy transition: Lessons from a socio-technical, multi-level analysis of the Dutch electricity system (1960–2004). *Energy Policy*, **35 (2)**, 1025-1037

Verbruggen, A. (2009). Performance evaluation of renewable energy support policies, applied on Flanders' tradable certificates system. *Energy Policy*, **37**(4), pp. 1385-1394.

Verbruggen, A. (2010). Preparing the design of robust policy architectures. *International Environmental Agreements: Politics, Law and Economics*, doi:10.1007/s10784-010-9130-x.

Verbruggen, A., and V. Lauber (2009). Basic concepts for designing renewable electricity support aiming at a full-scale transition by 2050. *Energy Policy*, **37**(12), pp. 5732-5743.

Vitousek, P.M. H.A. Mooney, J. Lubchenco, and J.M. Melillo (1997). Human domination of Earth's ecosystems. *Science*, **277**(5325), pp. 494-499.

Von Weizsäcker, E., A.B. Lovins, and L.H. Lovins (1998). *Factor Four: Doubling Wealth, Halving Resource Use*. Earthscan, London, UK.

Voß, J.-P., A. Smith, and J. Grin (2009). Designing long-term policy: rethinking transition management. *Policy Sciences*, **42**, pp. 275-302.

Vrolijk, C., J. Green, V. Bürger, C. Timpe, N. van der Linden, J. Jansen, M. Uyterlinde, C. García Barquero, P. Yerro, and F. Rivero (2004). *Renewable Energy Guarantees of Origin: implementation, interaction and utilisation: Summary of the RE-GO project*. IT Power, Madrid, Spain.

Wagner, A. (1999). Wind power on "Liberalised Markets": Maximum Market Penetration with Minimum Regulation. In: *European Wind Energy Conference*, Nice, France, 1-5 March 1999. Available at: www.oregon.gov/ENERGY/RENEW/Wind/OWWG/docs/FGWWagnerWindpoweronLiberalisedMarkets1999.pdf?ga=t.

Walker, G. (2008). What are the barriers and incentives for community-owned means of energy production and use? *Energy Policy*, **36**(12), pp. 4401-4405.

Wallace, W.L., L. Jingming, and G. Shangbin (1998). *The Use of Photovoltaics for Rural Electrification in Northwestern China*. NREL/CP-520-23920, National Renewable Energy Laboratory and Chinese Ministry of Agriculture, Golden, CO, USA and Beijing, China.

Walter, A. (2006). Is Brazilian Biofuels Experience a Model for Other Developing Countries? *Entwicklung & Ländlicher Raum*, **40**, pp. 22-24.

Wang, B. (2010). Can CDM bring technology transfer to China? An empirical study of technology transfer in China's CDM projects. *Energy Policy*, **38**, pp. 2572-2585.

Wang, F. and S. HaitaoYin (2010). China's renewable energy policy: Commitments and challenges. *Energy Policy*, **38**, pp. 1872-1878.

Wang, Q. (2010). Effective policies for renewable energy – the example of China's wind power-lessons for China's photovoltaic power. *Renewable and Sustainable Energy Reviews*, **14**, pp. 702-712.

Wårell, L. and T. Sundqvist (2009). *Market Opening in Local District Heating Networks*. Lund University, Lund, Sweden.

Warren, C.R., and M. McFadyen (2010). Does community ownership affect public attitudes to wind energy? A case study from south-west Scotland. *Land Use Policy*, **27**(2), pp. 204-213.

Watanabe, C., K. Wakabayashi, and T. Miyazawa (2000). Industrial dynamism and the creation of a 'Virtuous cycle' between R&D, market growth and price reduction: The case of photovoltaic power generation (PV) development in Japan. *Technovation*, **20**(6), pp. 299-312.

Watanabe, C., M. Kishioka, and A. Nagamatsu (2004). Effect and limit of the government role in spurring technology spillover – a case of R&D consortia by the Japanese government. *Technovation*, **24**(5), pp. 403-420.

Weiss, C., and W.B. Bonvillian (2009). Stimulating a revolution in sustainable energy technology. *Environment*, **51**(4), pp. 10-20.

West, J., I. Bailey, and M. Winter (2010). Renewable energy policy and public perceptions of renewable energy: A cultural theory approach. *Energy Policy*, **38**(10), pp. 5739-5748.

Weyant, J.P. (2010). Accelerating the development and diffusion of new energy technologies: Beyond the "valley of death". *Energy Economics*, doi:10.1016/j.eneco.2010.08.008.

Wiesenthal, T., G. Leduc, P. Christidis, B. Schade, L. Pelkmans, L. Govaerts, and P. Georgopoulos (2009). Biofuel support policies in Europe: Lessons learnt for the long way ahead. *Renewable and Sustainable Energy Reviews*, **13**(4), pp. 789-800.

Wilkins, G. (2002). *Technology Transfer for Renewable Energy – Overcoming Barriers in Developing Countries*. Royal Institute of International Affairs/Chatham House, London, UK.

Wilson, C. (2008). Social norms and policies to promote energy efficiency in the home. *Environmental Law Reporter*, **38**, pp. 10882-10888.

Wiser, R., and G. Barbose (2008). *Renewables Portfolio Standards in the United States: A Status Report with Data Through 2007*. LBNL-154E, Lawrence Berkeley National Laboratory, Berkeley, CA, USA.

Wiser, R., and M. Bolinger (2009). *2008 Wind Technologies Market Report*. U.S. Department of Energy, Washington, DC, USA. Available at: eetd.lbl.gov/ea/emp/reports/2008-wind-technologies.pdf.

Wiser, R. and M. Bolinger (2010). *2009 Wind Technologies Market Report*. LBNL-3716E. U.S. Department of Energy, Washington, DC, USA. Available at: eetd.lbl.gov/ea/emp/reports/lbnl-3716e.pdf.

Wiser, R., and S. Pickle (1997). *Financing Investments in Renewable Energy: The Role of Policy Design and Restructuring*. LBNL-39826, Lawrence Berkeley National Laboratory, Berkeley, CA, USA.

Wiser, R., and S. Pickle (2000). *Renewable Energy Policy Options for China: Feed In Laws and Renewable Portfolio Standards Compared*. Center for Resource Solutions, San Francisco, CA, USA.

Wiser, R., K. Porter, and R. Grace (2005). Evaluating experience with renewables portfolio standards in the United States. *Mitigation and Adaptation Strategies for Global Change*, **10**, pp. 237-263.

Wiser, R., M. Bolinger and G. Barbose (2007). Using the Federal Production Tax Credit to build a durable market for wind power in the United States, *The Electricity Journal*, **20**(9), pp. 77-88.

Wiser, R., G. Barbose, and E. Holt (2010). *Supporting Solar Power in Renewables Portfolio Standards: Experience from the United States*. Lawrence Berkeley National Laboratory, Berkeley, CA, USA.

Wolsink, M. (2000). Wind power and the NIMBY-myth: institutional capacity and the limited significance of public support. *Renewable Energy*, **21**(1), pp. 49-64.

Worrell, E., and W. Graus (2005). *Tax and Fiscal Policies for Promotion of Industrial Energy Efficiency: A Survey of International Experience*. Ernest Orlando Lawrence Berkeley National Laboratory, Berkeley, CA, USA.

Wüstenhagen, R., and M. Bilharz (2006). Green energy market development in Germany: effective public policy and emerging customer demand. *Energy Policy*, **34**(13), pp. 1681-1696.

Xiao, C., H. Luo, R. Tang, and H. Zhong (2004). Solar thermal utilization in China. *Renewable Energy*, **29**(9), pp. 1549-1556.

Yang, M., W. Blyth, R. Bradley, D. Bunn, C. Clarke, and T. Wilson (2008). Evaluating the power investment options with uncertainty in climate policy. *Energy Economics*, **30**(4), pp. 1933-1950.

Yoohoon, A. (2010). *Low Carbon Development Path for Asia and the Pacific: Challenges and Opportunities to the Energy Sector*. ESCAP Energy Resources Development Series ST/ESCAP/2589, Beijing, China.

Yu, J., F. Ji, L. Zhang, and Y. Chen (2009). An over painted oriental arts: Evaluation of the development of Chinese renewable energy market using wind power market as model. *Energy Policy*, **37**, pp. 5221-5225.

Zahnd, A., and H.M. Kimber (2009). Benefits from a renewable energy village electrification system. *Renewable Energy*, **34**(2), pp. 362-368.

Zhang, X., W. Ruoshui, H. Molin, and E. Martinot (2010). A study of the role played by renewable energies in China's sustainable energy supply. *Energy & Environment*, **35**(11), pp. 4392-4399.

Zhao, Z., J. Zuo, L.-L. Fan, and G. Zillante (2011). Impacts of renewable energy regulations on the structure of power generation in China – A critical analysis. *Renewable Energy*, **36**, pp. 24-30.

IV
Annexes I to VI

ANNEX I

Glossary, Acronyms, Chemical Symbols and Prefixes

Editors:
Aviel Verbruggen (Belgium), William Moomaw (USA), John Nyboer (Canada)

This annex should be cited as:

Verbruggen, A., W. Moomaw, J. Nyboer, 2011: Annex I: Glossary, Acronyms, Chemical Symbols and Prefixes. In IPCC Special Report on Renewable Energy Sources and Climate Change Mitigation [O. Edenhofer, R. Pichs-Madruga, Y. Sokona, K. Seyboth, P. Matschoss, S. Kadner, T. Zwickel, P. Eickemeier, G. Hansen, S. Schlömer, C. von Stechow (eds)], Cambridge University Press, Cambridge, United Kingdom and New York, NY, USA.

Glossary, Acronyms, Chemical Symbols and Prefixes

Glossary entries (highlighted in **bold**) are by preference subjects; a main entry can contain **subentries**, in bold italic, for example, ***Final Energy*** is defined under the entry **Energy**. The Glossary is followed by a list of acronyms/abbreviations, a list of chemical names and symbols, and a list of prefixes (international standard units). Some definitions are adapted from C.J. Cleveland and C. Morris, 2006: *Dictionary of Energy*, Elsevier, Amsterdam. Definitions of regions and country groupings are given in Section A.II.6 of Annex II of this report.

Glossary

Adaptation: Initiatives and measures to reduce the vulnerability or increase the resilience of natural and human systems to actual or expected climate change impacts. Various types of adaptation exist, for example, anticipatory and reactive, private and public, and autonomous and planned. Examples are raising river or coastal dikes, retreating from coastal areas subject to flooding from sea level rise or introducing alternative temperature-appropriate or drought-adapted crops for conventional ones.

Aerosols: A collection of airborne solid or liquid particles, typically between 0.01 and 10 μm in size and residing in the atmosphere for at least several hours. Aerosols may be of natural or anthropogenic origin. See also black carbon.

Afforestation: Direct human-induced conversion of land that has not been forested historically to forested land through planting, seeding and/or the human-induced promotion of natural seed sources.[1] See also deforestation, reforestation, land use.

Annex I countries: The group of countries included in Annex I (as amended since Malta was added after that date) to the UNFCCC, including developed countries and some countries with economies in transition. Under Articles 4.2 (a) and 4.2 (b) of the Convention, Annex I countries were encouraged to return individually or jointly to their 1990 levels of greenhouse gas emissions by 2000. The group is largely similar to the Annex B countries to the Kyoto Protocol. By default, the other countries are referred to as **Non-Annex I countries**. See also UNFCCC, Kyoto Protocol.

Annex B countries: This is the subset of Annex I countries that have specified greenhouse gas reduction commitments under the Kyoto Protocol. The group is largely similar to the Annex I countries to the UNFCCC. By default, the other countries are referred to as Non-Annex I countries. See also UNFCCC, Kyoto Protocol.

Anthropogenic: Related to or resulting from the influence of human beings on nature.

Anthropogenic emissions of greenhouse gases, greenhouse gas precursors and aerosols result from burning fossil fuels, deforestation, land use changes, livestock, fertilization, industrial, commercial and other activities that result in a net increase in emissions.

Availability (of a production plant): The percentage of time a plant is ready to produce, measured as uptime to total time (total time = uptime + downtime due to maintenance and outages).

Balancing power/reserves: Due to instantaneous and short-term fluctuations in electric loads and uncertain availability of power plants there is a constant need for spinning and quick-start generators that balance demand and supply at the imposed quality levels for frequency and voltage.

Barrier: Any obstacle to developing and deploying a renewable energy (RE) potential that can be overcome or attenuated by a policy, programme or measure. Barriers to RE deployment are unintentional or intentionally constructed impediments made by man (e.g., badly oriented buildings or power grid access criteria that discriminate against independent RE generators). Distinct from barriers are issues like intrinsically natural properties impeding the application of some RE sources at some place or time (e.g., flat land impedes hydropower and night the collection of direct solar energy).

Barrier removal includes correcting market failures directly or reducing the transactions costs in the public and private sectors by, for example, improving institutional capacity, reducing risk and uncertainty, facilitating market transactions and enforcing regulatory policies.

Baseline: The reference scenario for measurable quantities from which an alternative outcome can be measured, for example, a non-intervention scenario is used as a reference in the analysis of intervention scenarios. A baseline may be an extrapolation of recent trends, or it may assume frozen technology or costs. See also business as usual, models, scenario.

---

1 For a discussion of the term *forest* and related terms such as *afforestation*, *reforestation* and *deforestation*, see IPCC 2000: *Land Use, Land-Use Change, and Forestry*, A Special Report of the IPCC [R.T. Watson, I.A. Noble, B. Bolin, N.H. Ravindranath, D.J. Verardo, D.J. Dokken (eds.)], Cambridge University Press, Cambridge, United Kingdom and New York, NY, USA.

Benchmark: A measurable variable used as a baseline or reference in evaluating the performance of a technology, a system or an organization. Benchmarks may be drawn from internal experience, from external correspondences or from legal requirements and are often used to gauge changes in performance over time.

Biodiversity: The variability among living organisms from all sources including, inter alia, terrestrial, marine and other aquatic ecosystems and the ecological complexes of which they are part; this includes diversity within species, among species and of ecosystems.

Bioenergy: Energy derived from any form of biomass.

Biofuel: Any liquid, gaseous or solid fuel produced from biomass, for example, soybean oil, alcohol from fermented sugar, black liquor from the paper manufacturing process, wood as fuel, etc. Traditional biofuels include wood, dung, grass and agricultural residues.

First-generation manufactured biofuel is derived from grains, oilseeds, animal fats and waste vegetable oils with mature conversion technologies.

Second-generation biofuel uses non-traditional biochemical and thermochemical conversion processes and feedstock mostly derived from the lignocellulosic fractions of, for example, agricultural and forestry residues, municipal solid waste, etc.

Third-generation biofuel would be derived from feedstocks like algae and energy crops by advanced processes still under development. These second- and third-generation biofuels produced through new processes are also referred to as next-generation or advanced biofuels or advanced biofuel technologies.

Biomass: Material of biological origin (plants or animal matter), excluding material embedded in geological formations and transformed to fossil fuels or peat. The International Energy Agency (*World Energy Outlook 2010*) defines **traditional biomass** as biomass consumption in the residential sector in developing countries that refers to the often unsustainable use of wood, charcoal, agricultural residues and animal dung for cooking and heating. All other biomass use is defined as **modern biomass**, differentiated further by this report into two groups.

Modern bioenergy encompasses electricity generation and combined heat and power (CHP) from biomass and municipal solid waste (MSW), biogas, residential space and hot water in buildings and commercial applications from biomass, MSW, and biogas, and liquid transport fuels.

Industrial bioenergy applications include heating through steam generation and self generation of electricity and CHP in the pulp and paper industry, forest products, food and related industries.

Black carbon: Operationally defined aerosol species based on measurement of light absorption and chemical reactivity and/or thermal stability; consists of soot, charcoal and/or light-absorbing refractory organic matter.

Business as usual (BAU): The future is projected or predicted on the assumption that operating conditions and applied policies remain what they are at present. See also baseline, models, scenario.

Capacity: In general, the facility to produce, perform, deploy or contain.

Generation capacity of a renewable energy installation is the maximum power, that is, the maximum quantity of energy delivered per unit of time.

Capacity credit is the share of the capacity of a renewable energy unit counted as guaranteed available during particular time periods and accepted as a 'firm' contribution to total system generation capacity.

Capacity factor is the ratio of the actual output of a generating unit over a period of time (typically a year) to the theoretical output that would be produced if the unit were operating uninterruptedly at its **nameplate capacity** during the same period of time. Also known as rated capacity or nominal capacity, **nameplate capacity** is the facility's intended output level for a sustained period under normal circumstances.

Capacity building: In the context of climate change policies, the development of technical skills and institutional capability (the art of doing) and capacity (sufficient means) of countries to enable their participation in all aspects of adaptation to, mitigation of and research on climate change. See also mitigation capacity.

Carbon cycle: Describes the flow of carbon (in various forms, e.g., carbon dioxide, methane, etc) through the atmosphere, oceans, terrestrial biosphere and lithosphere.

Carbon dioxide (CO_2): CO_2 is a naturally occurring gas and a by-product of burning fossil fuels or biomass, of land use changes and of industrial processes. It is the principal anthropogenic greenhouse gas that affects Earth's radiative balance. It is the reference gas against which other greenhouse gases are measured and therefore it has a global warming potential of 1.

Carbon dioxide capture and storage (CCS): CO_2 from industrial and energy-related sources is separated, compressed and transported to a storage location for long-term isolation from the atmosphere.

Cellulose: The principal chemical constituent of the cell walls of plants and the source of fibrous materials for the manufacturing of various goods like paper, rayon, cellophane, etc. It is the main input for manufacturing second-generation biofuels.

Clean Development Mechanism (CDM): A mechanism under the Kyoto Protocol through which developed (Annex B) countries may finance greenhouse gas emission reduction or removal projects in developing (Non-Annex B) countries, and receive credits for doing so which they may apply for meeting mandatory limits on their own emissions.

Climate Change: Climate change refers to a change in the state of the climate that can be identified (e.g. using statistical tests) by changes in the mean and/or the variability of these properties and that persists for an extended period, typically decades or longer. Climate change may be due to natural internal processes or external forcings, or to persistent anthropogenic changes in the composition of the atmosphere or in land use. Note that Article 1 of the UNFCCC defines 'climate change' as "a change of climate which is attributed directly or indirectly to human activity that alters the composition of the global atmosphere and which is in addition to natural climate variability observed over comparable time periods". The UNFCCC thus makes a distinction between 'climate change' attributable to human activities altering atmospheric composition, and 'climate variability' attributable to natural causes.

CO_2-equivalent emission (CO_2eq): The amount of CO_2 emission that would cause the same radiative forcing as an emitted amount of a greenhouse gas or of a mixture of greenhouse gases, all multiplied by their respective global warming potentials, which take into account the differing times they remain in the atmosphere. See also global warming potential.

Co-benefits: The ancillary benefits of targeted policies that accrue to non-targeted, valuable objectives, for example, a wider use of renewable energy may also reduce air pollutants while lowering CO_2 emissions. Different definitions exist in the literature with co-benefits either being addressed intentionally (character of an opportunity) or gained unintentionally (character of a windfall profit). The term co-impact is more generic in covering both benefits and costs. See also drivers and opportunities.

Cogeneration: At thermal electricity generation plants otherwise wasted heat is utilized. The heat from steam turbines or hot flue gases exhausted from gas turbines may be used for industrial purposes, heating water or buildings or for district heating. Also referred to as combined heat and power (CHP).

Combined-cycle gas turbine (CCGT): A power plant that combines two processes for generating electricity. First, gas or light fuel oil feeds a gas turbine that exhausts hot flue gases (> 600°C). Second, heat recovered from these gases, with additional firing, is the source for producing steam that drives a steam turbine. The turbines rotate separate alternators. It becomes an **integrated CCGT** when the fuel is syngas from a coal or biomass gasification reactor with exchange of energy flows between the gasification and CCGT plants.

Compliance: Compliance is whether and to what extent countries adhere to the provisions of an accord or individuals or firms adhere to regulations. Compliance depends on implementing policies ordered, and on whether measures follow up the policies.

Conversion: Energy shows itself in numerous ways, with transformations from one type to another called energy conversions. For example, kinetic energy in wind flows is captured as rotating shaft work further converted to electricity; solar light is converted into electricity by photovoltaic cells. Also, electric currents of given characteristics (e.g., direct/alternating, voltage level) are converted to currents with other characteristics. A **converter** is the equipment used to realize the conversion.

Cost: The consumption of resources such as labour time, capital, materials, fuels, etc. as the consequence of an action. In economics, all resources are valued at their **opportunity cost**, which is the value of the most valuable alternative use of the resources. Costs are defined in a variety of ways and under a variety of assumptions that affect their value. The negative of costs are benefits and often both are considered together, for example, net cost is the difference between gross costs and benefits.

Private costs are carried by individuals, companies or other entities that undertake the action.

Social costs include additionally the external costs for the environment and for society as a whole, for example, **damage costs** of impacts on ecosystems, economies and people due to climate change.

Total cost includes all costs due to a specific activity; **average (unit, specific) cost** is total costs divided by the number of units generated; **marginal or incremental cost** is the cost of the last additional unit.

Project costs of a renewable energy project include **investment cost** (costs, discounted to the starting year of the project, of making the renewable energy device ready to commence production); **operation and maintenance (O&M) costs** (which occur during operation of the renewable energy facility); and **decommissioning costs** (which occur once the device has ceased production to restore the state of the site of production).

Lifecycle costs include all of the above discounted to the starting year of a project.

Levelized cost of energy (see Annex II) is the unique cost price of the outputs (US cent/kWh or USD/GJ) of a project that makes the present value of the revenues (benefits) equal to the present value of the costs over the lifetime of the project. See also discounting and present value.

There are many more categories of costs labelled with names that are often unclear and confusing, for example, installation costs may refer to the hardware equipment installed, or to the activities to put the equipment in place.

Cost–benefit analysis: Monetary measurement of all negative and positive impacts associated with a given action. Costs and benefits are compared in terms of their difference and/or ratio as an indicator of how a given investment or other policy effort pays off seen from the society's point of view.

Cost-effectiveness analysis: A reduction of cost–benefit analysis in which all the costs of a portfolio of projects are assessed in relation to a fixed policy goal. The policy goal in this case represents the benefits of the projects and all the other impacts are measured as costs or as negative costs (benefits). The policy goal can be, for example, realizing particular renewable energy potentials.

Deforestation: The natural or anthropogenic process that converts forest land to non-forest. See also afforestation, reforestation and land use.

Demand-side management: Policies and programmes for influencing the demand for goods and/or services. In the energy sector, demand-side management aims at reducing the demand for electricity and other forms of energy required to deliver energy services.

Density: Quantity or mass per unit volume, unit area or unit length.

Energy density is the amount of energy per unit volume or mass (for example, the heating value of a litre of oil).

Power density is typically understood as the capacity deliverable of solar, wind, biomass, hydropower or ocean power per unit area (watts/m^2). For batteries the capacity per unit weight (watts/kg) is used.

Direct solar energy - See solar energy

Discounting: A mathematical operation making monetary (or other) amounts received or expended at different points in time (years) comparable across time (see Annex II). The operator uses a fixed or possibly time-varying discount rate (>0) from year to year that makes future value worth less today. A **descriptive discounting approach** accepts the discount rates that people (savers and investors) actually apply in their day-to-day decisions (**private discount rate**). In a **prescriptive (ethical** or **normative) discounting approach**, the discount rate is fixed from a social perspective, for example, based on an ethical judgement about the interests of future generations (**social discount rate**).

In this report, potentials of renewable energy supplies are assessed using discount rates of 3, 7 and 10%.

Dispatch (power dispatching / dispatchable): Electrical power systems that consist of many power supply units and grids are governed by system operators. They allow generators to supply power to the system for balancing demand and supply in a reliable and economical way. Generation units are fully dispatchable when they can be loaded from zero to their nameplate capacity without significant delay. Not fully dispatchable are variable renewable sources that depend on natural currents, but also large-scale thermal plants with shallow ramping rates in changing their output. See also balancing, capacity, grid.

District heating (DH): Hot water (steam in old systems) is distributed from central stations to buildings and industries in a densely occupied area (a district, a city or an industrialized area). The insulated two-pipe network functions like a water-based central heating system in a building. The central heat sources can be waste heat recovery from industrial processes, waste incineration plants, geothermal sources, cogeneration power plants or stand-alone boilers burning fossil fuels or biomass. More and more DH systems also provide cooling via cold water or slurries (**district heating and cooling - DHC**).

Drivers: In a policy context, drivers provide an impetus and direction for initiating and supporting policy actions. The deployment of renewable energy is, for example, driven by concerns about climate change or energy security. In a more general sense, a driver is the leverage to bring about a reaction, for example, emissions are caused by fossil fuel consumption and/or economic growth. See also opportunities.

Economies of scale (scale economies): The unit cost of an activity declines when the activity is extended, for example, more units are produced.

Ecosystem: An open system of living organisms, interacting with each other and with their abiotic environment, that is capable of self-regulation to a certain degree. Depending on the focus of interest or study the extent of an ecosystem may range from very small spatial scales to the entire planet.

Electricity: The flow of passing charge through a conductor, driven by a difference in voltage between the ends of the conductor. Electrical power is generated by work from heat in a gas or steam turbine or from wind, oceans or falling water, or produced directly from sunlight using a photovoltaic device or chemically in a fuel cell. Being a current, electricity cannot be stored and requires wires and cables for its transmission (see grid). Because electric current flows immediately, the demand for electricity must be matched by production in real time.

Emissions: Direct emissions are released and attributed at points in a specific renewable energy chain, whether a sector, a technology or an activity. For example, methane emissions from decomposing submerged organic materials in hydropower reservoirs, or the release of CO_2 dissolved in hot water from geothermal plants, or CO_2 from biomass combustion. **Indirect emissions** are due to activities outside the considered renewable energy chain but which are required to realize the renewable energy deployment. For example, emissions from increased production of fertilizers used in the cultivation of biofuel crops or emissions from displaced crop production or deforestation as the result of biofuel crops. **Avoided emissions** are emission reductions arising from mitigation measures like renewable energy deployment.

Emission factor: An emission factor is the rate of emission per unit of activity, output or input.

Emissions trading: A market-based instrument to reduce greenhouse gas or other emissions. The environmental objective or sum of total allowed emissions is expressed as an emissions cap. The cap is divided in tradable emission permits that are allocated—either by auctioning or handing out for free (grandfathering)—to entities within the jurisdiction of the trading scheme. Entities need to surrender emission permits equal to the amount of their emissions (e.g., tonnes of CO_2). An entity may sell excess permits. Trading schemes may occur at the intra-company, domestic or international level and may apply to CO_2, other greenhouse gases or other substances. Emissions trading is also one of the mechanisms under the Kyoto Protocol.

Energy: The amount of work or heat delivered. Energy is classified in a variety of types and becomes available to human ends when it flows from one place to another or is converted from one type into another. Daily, the sun supplies large flows of radiation energy. Part of that energy is used directly, while part undergoes several conversions creating water evaporation, winds, etc. Some share is stored in biomass or rivers that can be harvested. Some share is directly usable such as daylight, ventilation or ambient heat.

> *Primary energy* (also referred to as energy sources) is the energy embodied in natural resources (e.g., coal, crude oil, natural gas, uranium, and renewable sources). It is defined in several alternative ways. The International Energy Agency utilizes the physical energy content method, which defines primary energy as energy that has not undergone any anthropogenic conversion. The method used in this report is the direct equivalent method (see Annex II), which counts one unit of secondary energy provided from non-combustible sources as one unit of primary energy, but treats combustion energy as the energy potential contained in fuels prior to treatment or combustion. Primary energy is transformed into **secondary energy** by cleaning (natural gas), refining (crude oil to oil products) or by conversion into electricity or heat. When the secondary energy is delivered at the end-use facilities it is called **final energy** (e.g., electricity at the wall outlet), where it becomes **usable energy** in supplying services (e.g., light).

Embodied energy is the energy used to produce a material substance (such as processed metals or building materials), taking into account energy used at the manufacturing facility (zero order), energy used in producing the materials that are used in the manufacturing facility (first order), and so on.

Renewable energy (RE) is any form of energy from solar, geophysical or biological sources that is replenished by natural processes at a rate that equals or exceeds its rate of use. Renewable energy is obtained from the continuing or repetitive flows of energy occurring in the natural environment and includes low-carbon technologies such as solar energy, hydropower, wind, tide and waves and ocean thermal energy, as well as renewable fuels such as biomass. For a more detailed description see specific renewable energy types in this glossary, for example, biomass, solar, hydropower, ocean, geothermal and wind.

Energy access: People are provided the ability to benefit from affordable, clean and reliable energy services for basic human needs (cooking and heating, lighting, communication, mobility) and productive uses.

Energy carrier: A substance for delivering mechanical work or transfer of heat. Examples of energy carriers include: solid, liquid or gaseous fuels (e.g., biomass, coal, oil, natural gas, hydrogen); pressurized/heated/cooled fluids (air, water, steam); and electric current.

Energy efficiency: The ratio of useful energy or other useful physical outputs obtained from a system, conversion process, transmission or storage activity to the input of energy (measured as kWh/kWh, tonnes/kWh or any other physical measure of useful output like tonne-km transported, etc.). Energy efficiency is a component of energy intensity.

Energy intensity: The ratio of energy inputs (in Joules) to the economic output (in dollars) that absorbed the energy input. Energy intensity is the reciprocal of energy productivity. At the national level, energy intensity is the ratio of total domestic primary (or final) energy use to gross domestic product (GDP). The energy intensity of an economy is the weighted sum of the energy intensities of particular activities with the activities' shares in GDP as weights. Energy intensities are obtained from available statistics (International Energy Agency, International Monetary Fund) and published annually for most countries in the world. Energy intensity is also used as a name for the ratio of energy inputs to output or performance in physical terms (e.g., tonnes of steel output, tonne-km transported, etc.) and in such cases, is the reciprocal of energy efficiency.

Energy productivity: The reciprocal of energy intensity.

Energy savings: Decreasing energy intensity by changing the activities that demand energy inputs. Energy savings can be realized by technical, organizational, institutional and structural actions and by changed behaviour.

Energy security: The goal of a given country, or the global community as a whole, to maintain an adequate energy supply. Measures encompass safeguarding access to energy resources; enabling development and deployment of technologies; building sufficient infrastructure to generate, store and transmit energy supplies; ensuring enforceable contracts of delivery; and access to energy at affordable prices for a specific society or groups in society.

Energy services: Energy services are the tasks to be performed using energy. A specific energy service such as lighting may be supplied by a number of different means from daylighting to oil lamps to incandescent, fluorescent or light-emitting diode devices. The amount of energy used to provide a service may vary over a factor of 10 or more, and the corresponding greenhouse gas emissions may vary from zero to a very high value depending on the source of energy and the type of end-use device.

Energy transfer: Energy is transferred as work, light or heat. **Heat transfer** spontaneously occurs from objects at higher temperature to objects at lower temperature and is classified as conduction (when the objects have contact), convection (when a fluid like air or water takes the heat from the warmer object and is moved to the colder object to deliver the heat) and radiation (when heat travels through space in the form of electromagnetic waves).

Externality / external cost / external benefit: Externalities arise from a human activity, when agents responsible for the activity do not take full account of the activity's impact on others' production and consumption possibilities, and no compensation exists for such impacts. When the impact is negative, they are external costs. When positive they are referred to as external benefits.

Feed-in tariff: The price per unit of electricity that a utility or power supplier has to pay for distributed or renewable electricity fed into the grid by non-utility generators. A public authority regulates the tariff. There may also be a tariff for supporting renewable heat supplies.

Financing: Raising or providing money or capital by individuals, businesses, banks, venture funds, public instances, etc. for realizing a project or continuing an activity. Depending on the financier the money is raised and is provided differently. For example, businesses may raise money from internal company profits, debt or equity (shares).

>*Project financing* of renewable energy may be provided by financiers to distinct, single-purpose companies, whose renewable energy sales are usually guaranteed by power purchase agreements.

>*Non-recourse financing* is known as off-balance sheet since the financiers rely on the certainty of project cash flows to pay back the loan, not on the creditworthiness of the project developer.

>*Public equity financing* is capital provided for publicly listed companies.

>*Private equity financing* is capital provided directly to private companies.

>*Corporate financing* by banks via debt obligations uses 'on-balance sheet' assets as collateral and is therefore limited by the debt ratio of companies that must rationalize each additional loan with other capital needs.

Fiscal incentive: Actors (individuals, households, companies) are granted a reduction of their contribution to the public treasury via income or other taxes.

Fuel cell: A fuel cell generates electricity in a direct and continuous way from the controlled electrochemical reaction of hydrogen or another fuel and oxygen. With hydrogen as fuel it emits only water and heat (no CO_2) and the heat can be utilized (see cogeneration).

General equilibrium models: General equilibrium models consider simultaneously all the markets and feedback effects among them in an economy leading to market clearance.

Generation control: Generation of electricity at a renewable energy plant may be subject to various controls.

>*Active control* is a deliberate intervention in the functioning of a system (for example, wind turbine **pitch control**: changing the orientation of the blades for varying a wind turbine's output).

>*Passive control* is when natural forces adjust the functioning of a system (for example, wind turbine **stall control**: the design of the blade shape such that at a desired speed the blade spills the wind in order to automatically control the wind turbine's output).

Geothermal energy: Accessible thermal energy stored in the Earth's interior, in both rock and trapped steam or liquid water (hydrothermal resources), which may be used to generate electric energy in a thermal power plant, or to supply heat to any process requiring it. The main sources of geothermal energy are the residual energy available from planet formation and the energy continuously generated from radionuclide decay.

Geothermal gradient: Rate at which the Earth's temperature increases with depth, indicating heat flowing from the Earth's warm interior to its colder parts.

Global warming potential (GWP): GWP is an index, based upon radiative properties of well-mixed greenhouse gases, measuring the radiative forcing of a unit mass of a given well-mixed greenhouse gas in today's atmosphere integrated over a chosen time horizon, relative to that of CO_2. The GWP represents the combined effect of the differing lengths of time that these gases remain in the atmosphere and their relative effectiveness in absorbing outgoing infrared radiation. The Kyoto Protocol ranks greenhouse gases on the basis of GWPs from single pulse emissions over subsequent 100-year time frames. See also climate change and CO_2-equivalent emission.

Governance: Governance is a comprehensive and inclusive concept of the full range of means for deciding, managing and implementing policies and measures. Whereas govern*ment* is defined strictly in terms of the nation-state, the more inclusive concept of gover*nance*, recognizes the contributions of various levels of government (global, international, regional, local) and the contributing roles of the private sector, of nongovernmental actors and of civil society to addressing the many types of issues facing the global community.

Greenhouse gases (GHGs): Greenhouse gases are those gaseous constituents of the atmosphere, both natural and anthropogenic, that absorb and emit radiation at specific wavelengths within the spectrum of thermal infrared radiation emitted by the Earth's surface, the atmosphere and clouds. This property causes the greenhouse effect. Water vapour (H_2O), carbon dioxide (CO_2), nitrous oxide (N_2O), methane (CH_4) and ozone (O_3) are the primary greenhouse gases in the Earth's atmosphere. Moreover, there are a number of entirely human-made greenhouse gases in the atmosphere, such as the halocarbons and other chlorine- and bromine-containing substances, dealt with under the Montreal Protocol. Besides CO_2, N_2O and CH_4, the Kyoto Protocol deals with the greenhouse gases sulphur hexafluoride (SF_6), hydrofluorocarbons (HFCs) and perfluorocarbons (PFCs).

Grid (electric grid, electricity grid, power grid): A network consisting of wires, switches and transformers to transmit electricity from power sources to power users. A large network is layered from low-voltage (110-240 V) distribution, over intermediate voltage (1-50 kV) to high-voltage (above 50 kV to MV) transport subsystems. Interconnected grids cover large areas up to continents. The grid is a power exchange platform enhancing supply reliability and economies of scale.

Grid connection for a power producer is mostly crucial for economical operation.

Grid codes are technical conditions for equipment and operation that a power producer must obey for getting supply access to the grid; also consumer connections must respect technical rules.

Grid access refers to the acceptance of power producers to deliver to the grid.

Grid integration accommodates power production from a portfolio of diverse and some variable generation sources in a balanced power system. See also transmission and distribution.

Gross Domestic Product (GDP): The sum of gross value added, at purchasers' prices, by all resident and non-resident producers in the economy, plus any taxes and minus any subsidies not included in the value of the products in a country or a geographic region for a given period, normally one year. It is calculated without deducting for depreciation of fabricated assets or depletion and degradation of natural resources.

Heat exchanger: Devices for efficient **heat transfer** from one medium to another without mixing the hot and cold flows, for example, radiators, boilers, steam generators, condensers.

Heat pump: Installation that transfers heat from a colder to a hotter place, opposite to the natural direction of heat flows (see energy transfer). Technically similar to a refrigerator, heat pumps are used to extract heat from ambient environments like the ground (geothermal or ground source), water or air. Heat pumps can be inverted to provide cooling in summer.

Human Development Index (HDI): The HDI allows the assessment of countries' progress regarding social and economic development as a composite index of three indicators: 1) health measured by life expectancy at birth; 2) knowledge as measured by a combination of the adult literacy rate and the combined primary, secondary and tertiary school enrolment ratio; and 3) standard of living as gross domestic product per capita (in purchasing power parity). The HDI only acts as a broad proxy for some of the key issues of human development; for instance, it does not reflect issues such as political participation or gender inequalities.

Hybrid vehicle: Any vehicle that employs two sources of propulsion, most commonly a vehicle that combines an internal combustion engine with an electric motor and storage batteries.

Hydropower: The energy of water moving from higher to lower elevations that is converted into mechanical energy through a turbine or other device that is either used directly for mechanical work or more commonly to operate a generator that produces electricity. The term is also used to describe the kinetic energy of stream flow that may also be converted into mechanical energy of a generator through an in-stream turbine to produce electricity.

Informal sector/economy: The informal sector/economy is broadly characterized as comprising production units that operate at a small scale and at a low level of organization, with little or no division between labour and capital as factors of production, and with the primary objective of generating income and employment for the persons concerned. The economic activity of the informal sector is not accounted for in determining sectoral or national economic activity.

Institution: A structure, a mechanism of social order or cooperation, which governs the behaviour of a group of individuals within a human community. Institutions are intended to be functionally relevant for an extended period, able to help transcend individual interests and help govern cooperative human behaviour. The term can be extended to also cover regulations, technology standards, certification and the like.

Integrated assessment: A method of analysis that combines results and models from the physical, biological, economic and social sciences, and the interactions between these components in a consistent framework to evaluate the status and the consequences of environmental change and the policy responses to it. See also models.

Kyoto Protocol: The Kyoto Protocol to the UNFCCC was adopted at the Third Session of the Conference of the Parties in 1997 in Kyoto. It contains legally binding commitments, in addition to those included in the UNFCCC. Annex B countries agreed to reduce their anthropogenic greenhouse gas emissions (CO_2, methane, nitrous oxide, hydrofluorocarbons, perfluorocarbons and sulphur hexafluoride) by at least 5% below 1990 levels in the commitment period 2008 to 2012. The Kyoto Protocol came into force on 16 February 2005. See also UNFCCC.

Land use (change; direct and indirect): The total of arrangements, activities and inputs undertaken in a certain land cover type. The social and economic purposes for which land is managed (e.g., grazing, timber extraction and conservation).

Land use change occurs whenever land is transformed from one use to another, for example, from forest to agricultural land or to urban areas. Since different land types have different carbon storage potential (e.g., higher for forests than for agricultural or urban areas), land use changes may lead to net emissions or to carbon uptake.

Indirect land use change refers to market-mediated or policy-driven shifts in land use that cannot be directly attributed to land use management decisions of individuals or groups. For example, if agricultural land is diverted to fuel production, forest clearance may occur elsewhere to replace the former agricultural production. See also afforestation, deforestation and reforestation.

Landfill: A solid waste disposal site where waste is deposited below, at or above ground level. Limited to engineered sites with cover materials, controlled placement of waste and management of liquids and gases. It excludes uncontrolled waste disposal. Landfills often release methane, CO_2 and other gases as organic materials decay.

Leapfrogging: The ability of developing countries to bypass intermediate technologies and jump straight to advanced clean technologies. Leapfrogging can enable developing countries to move to a low-emissions development trajectory.

Learning curve / rate: Decreasing cost-prices of renewable energy supplies shown as a function of increasing (total or yearly) supplies. Learning improves technologies and processes over time due to experience, as production increases and/or with increasing research and development. The **learning rate** is the percent decrease of the cost-price for every doubling of the cumulative supplies (also called **progress ratio**).

Levelized cost of energy – See Cost.

Lifecycle analysis (LCA): LCA aims to compare the full range of environmental damages of any given product, technology, or service (see Annex II). LCA usually includes raw material input, energy requirements, and waste and emissions production. This includes operation of the technology/facility/product as well as all upstream processes (i.e., those occurring prior to when the technology/facility/product commences operation) and downstream processes (i.e., those occurring after the useful lifetime of the technology/facility/product), as in the 'cradle to grave' approach.

Load (electrical): The demand for electricity by (thousands to millions) power users at the same moment aggregated and raised by the losses in transport and delivery, and to be supplied by the integrated power supply system.

Load levelling reduces the amplitude of the load fluctuations over time.

Load shedding occurs when available generation or transmission capacity is insufficient to meet the aggregated loads.

Peak load is the maximum load observed over a given period of time (day, week, year) and of short duration.

Base load is power continuously demanded over the period.

Loans: Loans are money that public or private lenders provide to borrowers mandated to pay back the nominal sum increased with interest payments.

Soft loans (also called soft financing or concessional funding) offer flexible or lenient terms for repayment, usually at lower than market interest rates or no interest. Soft loans are provided customarily by government agencies and not by financial institutions.

Convertible loans entitle the lender to convert the loan to common or preferred stock (ordinary or preference shares) at a specified conversion rate and within a specified time frame.

Lock-in: Technologies that cover large market shares continue to be used due to factors such as sunk investment costs, related infrastructure development, use of complementary technologies and associated social and institutional habits and structures.

Carbon lock-in means that the established technologies and practices are carbon intensive.

Low-carbon technology: A technology that over its lifecycle causes very low to zero CO_2eq emissions. See emissions.

Market failure: When private decisions are based on market prices that do not reflect the real scarcity of goods and services, they do not generate an efficient allocation of resources but cause welfare losses. Factors causing market prices to deviate from real economic scarcity are environmental externalities, public goods and monopoly power.

Measures: In climate policy, measures are technologies, processes or practices that reduce greenhouse gas emissions or impacts below anticipated future levels, for example renewable energy technologies, waste minimization processes, public transport commuting practices, etc. See also policies.

Merit order (of power plants): Ranking of all available power generating units in an integrated power system, being the sequence of their short-run marginal cost per kWh starting with the cheapest for delivering electricity to the grid.

Millennium Development Goals (MDG): A set of eight time-bound and measurable goals for combating poverty, hunger, disease, illiteracy, discrimination against women and environmental degradation. These were agreed to at the UN Millennium Summit in 2000 together with an action plan to reach these goals.

Mitigation: Technological change and changes in activities that reduce resource inputs and emissions per unit of output. Although several social, economic and technological policies would produce an emission reduction, with respect to climate change, mitigation means implementing policies to reduce greenhouse gas emissions and enhance sinks. Renewable energy deployment is a mitigation option when avoided greenhouse gas emissions exceed the sum of direct and indirect emissions (see emissions).

Mitigation capacity is a country's ability to reduce anthropogenic greenhouse gas emissions or to enhance natural sinks, where ability refers to skills, competencies, fitness and proficiencies that a country has attained and depends on technology, institutions, wealth, equity, infrastructure and information. Mitigation capacity is rooted in a country's sustainable development path.

Models: Models are structured imitations of a system's attributes and mechanisms to mimic appearance or functioning of systems, for example, the climate, the economy of a country, or a crop. Mathematical models assemble (many) variables and relations (often in a computer code) to simulate system functioning and performance for variations in parameters and inputs.

Bottom-up models aggregate technological, engineering and cost details of specific activities and processes.

Top-down models apply macroeconomic theory, econometric and optimization techniques to aggregate economic variables, like total consumption, prices, incomes and factor costs.

Hybrid models integrate bottom-up and top-down models to some degree.

Non-Annex I countries – See Annex I countries.

Non-Annex B countries – See Annex B countries.

Ocean energy: Energy obtained from the ocean via waves, tidal ranges, tidal and ocean currents, and thermal and saline gradients (note: submarine geothermal energy is covered under geothermal energy and marine biomass is covered under biomass energy).

Offset (in climate policy): A unit of CO_2-equivalent (CO_2eq) that is reduced, avoided or sequestered to compensate for emissions occurring elsewhere.

Opportunities: In general: conditions that allow for advancement, progress or profit. In the policy context, circumstances for action with the attribute of a chance character. For example, the anticipation of additional benefits that may go along with the deployment of renewable energy (enhanced energy access and energy security, reduced local air pollution) but are not intentionally targeted. See also co-benefits and drivers.

Path dependence: Outcomes of a process are conditioned by previous decisions, events and outcomes, rather than only by current actions. Choices based on transitory conditions can exert a persistent impact long after those conditions have changed.

Payback: Mostly used in investment appraisal as **financial payback**, which is the time needed to repay the initial investment by the returns of a project. A **payback gap** exists when, for example, private investors and micro-financing schemes require higher profitability rates from renewable energy projects than from fossil-fired ones. Imposing an x-times higher financial return on renewable energy investments is equivalent to imposing an x-times higher technical performance hurdle on delivery by novel renewable solutions compared to incumbent energy expansion. **Energy payback** is the time an energy project needs to deliver as much energy as had been used for setting the project online. **Carbon payback** is the time a renewable energy project needs to deliver as much net greenhouse gas savings (with respect to the fossil reference energy system) as its realization has caused greenhouse gas emissions from a perspective of lifecycle analysis (including land use changes and loss of terrestrial carbon stocks).

Photosynthesis: The production of carbohydrates in plants, algae and some bacteria using the energy of light. CO_2 is used as the carbon source.

Photovoltaics (PV): The technology of converting light energy directly into electricity by mobilizing electrons in solid state devices. The specially prepared thin sheet semiconductors are called PV cells. See solar energy.

Policies: Policies are taken and/or mandated by a government—often in conjunction with business and industry within a single country, or collectively with other countries—to accelerate mitigation and adaptation measures. Examples of policies are support mechanisms for renewable energy supplies, carbon or energy taxes, fuel efficiency standards for automobiles, etc.

Common and co-ordinated or *harmonized policies* refer to those adopted jointly by parties. See also measures.

Policy criteria: General: a standard on which a judgment or decision may be based. In the context of policies and policy instruments to support renewable energy, four inclusive criteria are common:

Effectiveness (efficacy) is the extent to which intended objectives are met, for instance the actual increase in the output of renewable electricity generated or shares of renewable energy in total energy supplies within a specified time period. Beyond *quantitative* targets, this may include factors such as achieved degrees of *technological diversity* (promotion of different renewable energy technologies) or of *spatial diversity* (geographical distribution of renewable energy supplies).

Efficiency is the ratio of outcomes to inputs, for example, renewable energy targets realized for economic resources spent, mostly measured at one point of time (*static efficiency*), also called cost-effectiveness. *Dynamic efficiency* adds a future time dimension by including how much innovation is triggered to improve the ratio of outcomes to inputs.

Equity covers the incidence and distributional consequences of a policy, including fairness, justice and respect for the rights of indigenous peoples. The equity criterion looks at the *distribution* of costs and benefits of a policy and at the *inclusion* and *participation* of wide ranges of different stakeholders (e.g., local populations, independent power producers).

Institutional feasibility is the extent to which a policy or policy instrument is seen as legitimate, able to gain acceptance, and able to be adopted and implemented. It covers **administrative feasibility** when compatible with the available information base and administrative capacity, legal structure and economic realities. **Political feasibility** needs acceptance and support by stakeholders, organizations and constituencies, and compatibility with prevailing cultures and traditions.

Polluter pays principle: In 1972 the OECD agreed that polluters should pay the costs of abating the own environmental pollution, for example by installation of filters, sanitation plants and other add-on techniques. This is the narrow definition. The extended definition is when polluters would additionally pay for the damage caused by their residual pollution (eventually also historical pollution). Another extension is the precautionary polluter pays principle where potential polluters are mandated to take insurance or preventive measures for pollution that may occur in the future. The acronym PPP has also other meanings, such as Preventing Pollution Pays-off, Public Private Partnership, or Purchasing Power Parity.

Portfolio analysis: Examination of a collection of assets or policies that are characterized by different risks and payoffs. The objective function is built up around the variability of returns and their risks, leading up to the decision rule to choose the portfolio with highest expected return.

Potential: Several levels of renewable energy supply potentials can be identified, although every level may span a broad range. In this report, **resource potential** encompasses all levels for a specific renewable energy resource.

Market potential is the amount of renewable energy output expected to occur under forecast market conditions, shaped by private economic agents and regulated by public authorities. Private economic agents realize private objectives within given, perceived and expected conditions. Market potentials are based on expected private revenues and expenditures, calculated at private prices (incorporating subsidies, levies and rents) and with private discount rates. The private context is partly shaped by public authority policies.

Economic potential is the amount of renewable energy output projected when all social costs and benefits related to that output are included, there is full transparency of information, and assuming exchanges in the economy install a general equilibrium characterized by spatial and temporal efficiency. Negative externalities and co-benefits of all energy uses and of other economic activities are priced. Social discount rates balance the interests of consecutive human generations.

Sustainable development potential is the amount of renewable energy output that would be obtained in an *ideal setting* of perfect economic markets, optimal social (institutional and governance) systems and achievement of the sustainable flow of environmental goods and services. This is distinct from economic potential because it explicitly addresses inter- and intra-generational equity (distribution) and governance issues.

Technical potential is the amount of renewable energy output obtainable by full implementation of demonstrated technologies or practices. No explicit reference to costs, barriers or policies is made.

Technical potentials reported in the literature being assessed in this report, however, may have taken into account practical constraints and when explicitly stated there, they are generally indicated in the underlying report.

Theoretical potential is derived from natural and climatic (physical) parameters (e.g., total solar irradiation on a continent's surface). The theoretical potential can be quantified with reasonable accuracy, but the information is of limited practical relevance. It represents the upper limit of what can be produced from an energy resource based on physical principles and current scientific knowledge. It does not take into account energy losses during the conversion process necessary to make use of the resource, nor any kind of barriers.

Power: Power is the rate in which energy is transferred or converted per unit of time or the rate at which work is done. It is expressed in watts (joules/second).

Present value: The value of a money amount differs when the amount is available at different moments in time (years). To make amounts at differing times comparable and additive, a date is fixed as the 'present.' Amounts available at different dates in the future are discounted back to a present value, and summed to get the present value of a series of future cash flows. **Net present value** is the difference between the present value of the revenues (benefits) and the present value of the costs. See also discounting.

Project cost – see Cost.

Progress ratio – see Learning curve / rate.

Public finance: Public support for which a financial return is expected (loans, equity) or financial liability is incurred (guarantee).

Public good: Public goods are simultaneously used by several parties (opposite to private goods). Some public goods are fully free from rivalry in use; for others the use by some subtract from the availability for others, creating congestion. Access to public goods may be restricted dependent on whether public goods are commons, state-owned or res nullius (no one's case). The atmosphere and climate are the ultimate public goods of mankind. Many renewable energy sources are also public goods.

Public-private partnerships: Arrangements typified by joint working between the public and private sector. In the broadest sense, they cover all types of collaboration across the interface between the public and private sectors to deliver services or infrastructure.

Quota (on renewable electricity/energy): Established quotas obligate designated parties (generators or suppliers) to meet minimum (often gradually increasing) renewable energy targets, generally expressed as percentages of total supplies or as an amount of renewable energy capacity, with costs borne by consumers. Various countries use different names for quotas, for example, Renewable Portfolio Standards, Renewable Obligations. See also tradable certificates

Reactive power: The part of instantaneous power that does no real work. Its function is to establish and sustain the electric and magnetic fields required to let active power perform useful work.

Rebound effect: After implementation of efficient technologies and practices, part of the expected energy savings is not realized because the accompanying savings in energy bills may be used to acquire more energy services. For example, improvements in car engine efficiency lower the cost per kilometre driven, encouraging consumers to drive more often or longer distances, or to spend the saved money on other energy-consuming activities. Successful energy efficiency policies may lead to lower economy-wide energy demand and if so to lower energy prices with the possibility of the financial savings stimulating rebound effects. The rebound effect is the ratio of non-realized energy and resource savings compared to the potential savings in case consumption would have remained constant as before the efficiency measures were implemented. For climate change, the main concern about rebound effects is their impact on CO_2 emissions (carbon rebound).

Reforestation: Direct human-induced conversion of non-forested land to forested land through planting, seeding and/or the human-induced promotion of natural seed sources, on land that was previously forested but converted to non-forested land. See also afforestation, deforestation and land use.

Regulation: A rule or order issued by governmental executive authorities or regulatory agencies and having the force of law. Regulations implement policies and are mostly specific for particular groups of people, legal entities or targeted activities. Regulation is also the act of designing and imposing rules or orders. Informational, transactional, administrative and political constraints in practice limit the regulator's capability for implementing preferred policies.

Reliability: In general: reliability is the degree of performance according to imposed standards or expectations.

Electrical reliability is the absence of unplanned interruptions of the current by, for example, shortage of supply capacity or by failures in parts of the grid. Reliability differs from security and from fluctuations in power quality due to impulses or harmonics.

Renewable energy – see Energy

Scenario: A plausible description of how the future may develop based on a coherent and internally consistent set of assumptions about key relationships and driving forces (e.g., rate of technological change, prices) on social and economic development, energy use, etc. Note that scenarios are neither predictions nor forecasts, but are useful to provide a view of the implications of alternative developments and actions. See also baseline, business as usual, models.

Seismicity: The distribution and frequency of earthquakes in time, magnitude and space, for example, the yearly number of earthquakes of magnitude between 5 and 6 per 100 km^2 or in some region.

Sink: Any process, activity or mechanism that removes a greenhouse gas or aerosol, or a precursor of a greenhouse gas or aerosol, from the atmosphere.

Solar collector: A device for converting solar energy to thermal energy (heat) of a flowing fluid.

Solar energy: Energy from the Sun that is captured either as heat, as light that is converted into chemical energy by natural or artificial photosynthesis, or by photovoltaic panels and converted directly into electricity.

Concentrating solar power (CSP) systems use either lenses or mirrors to capture large amounts of solar energy and focus it down to a smaller region of space. The higher temperatures produced can operate a thermal steam turbine or be used in high-temperature industrial processes.

Direct solar energy refers to the use of solar energy as it arrives at the Earth's surface before it is stored in water or soils.

Solar thermal is the use of direct solar energy for heat end-uses, excluding CSP.

Active solar needs equipment like panels, pumps and fans to collect and distribute the energy.

Passive solar is based on structural design and construction techniques that enable buildings to utilize solar energy for heating, cooling and lighting by non-mechanical means.

Solar irradiance: The rate of solar power incidence on a surface (W/m^2). Irradiance depends on the orientation of the surface, with as special orientations: (a) surfaces perpendicular to the beam solar radiation; (b) surfaces horizontal with or on the ground. **Full sun** is solar irradiance that is approximately 1,000 W/m^2.

Solar radiation: The sun radiates light and heat energy in wavelengths from ultraviolet to infrared. Radiation arriving at surfaces may be absorbed, reflected or transmitted.

Global solar radiation consists of **beam** (arriving on Earth in a straight line) and **diffuse radiation** (arriving on Earth after being scattered by the atmosphere and by clouds).

Standards: Set of rules or codes mandating or defining product performance (e.g., grades, dimensions, characteristics, test methods and rules for use).

Product, *technology* or *performance standards* establish minimum requirements for affected products or technologies.

Subsidy: Direct payment from the government or a tax reduction to a private party for implementing a practice the government wishes to encourage. The reduction of greenhouse gas emissions is stimulated by lowering existing subsidies that have the effect of raising emissions (such as subsidies for fossil fuel use) or by providing subsidies for practices that reduce emissions or enhance sinks (e.g., renewable energy projects, insulation of buildings or planting trees).

Sustainable development (SD): The concept of sustainable development was introduced in the World Conservation Strategy of the International Union for Conservation of Nature in 1980 and had its roots in the concept of a sustainable society and in the management of renewable resources. Adopted by the World Council for Environment and Development in 1987 and by the Rio Conference in 1992 as a process of change in which the exploitation of resources, the direction of investments, the orientation of technological development and institutional change are all in harmony and enhance both current and future potential to meet human needs and aspirations. SD integrates the political, social, economic and environmental dimensions, and respects resource and sink constraints.

Tax: A **carbon tax** is a levy on the carbon content of fossil fuels. Because virtually all of the carbon in fossil fuels is ultimately emitted as CO_2, a carbon tax is equivalent to an **emission tax** on CO_2 emissions. An **energy tax**—a levy on the energy content of fuels—reduces demand for energy and so reduces CO_2 emissions from fossil fuel use. An **eco-tax** is a carbon, emissions or energy tax designed to influence human behaviour (specifically economic behaviour) to follow an ecologically benign path. A **tax credit** is a reduction of tax in order to stimulate purchasing of or investment in a certain product, like greenhouse gas emission-reducing technologies. A **levy** or **charge** is used as synonymous for tax.

Technological change: Mostly considered as technological *improvement*, that is, more or better goods and services can be provided from a given amount of resources (production factors). Economic models distinguish autonomous (exogenous), endogenous and induced technological change.

Autonomous (exogenous) technological change is imposed from outside the model (i.e., as a parameter), usually in the form of a time trend affecting factor or/and energy productivity and therefore energy demand or output growth.

Endogenous technological change is the outcome of economic activity *within* the model (i.e., as a variable) so that factor productivity or the choice of technologies is included within the model and affects energy demand and/or economic growth.

Induced technological change implies endogenous technological change but adds further changes *induced* by policies and measures, such as carbon taxes triggering research and development efforts.

Technology: The practical application of knowledge to achieve particular tasks that employs both technical artefacts (hardware, equipment) and (social) information ('software', know-how for production and use of artefacts).

Supply push aims at developing specific technologies through support for research, development and demonstration.

Demand pull is the practice of creating market and other incentives to induce the introduction of particular sets of technologies (e.g., low-carbon technologies through carbon pricing) or single technologies (e.g., through technology-specific feed-in tariffs).

Technology transfer: The exchange of knowledge, hardware and associated software, money and goods among stakeholders, which leads to the spread of technology for adaptation or mitigation. The term encompasses both diffusion of technologies and technological cooperation across and within countries.

Tradable certificates (tradable green certificates): Parties subject to a renewable energy quota meet the annual obligation by delivering the appropriate amount of tradable certificates to a regulatory office. The certificates are created by the office and assigned to the renewable energy producers to sell or for their own use in fulfilling their quota. See quota.

Transmission and distribution (electricity): The network that transmits electricity through wires from where it is generated to where it is used. The distribution system refers to the lower-voltage system that actually delivers the electricity to the end user. See also grid.

Turbine: Equipment that converts the kinetic energy of a flow of air, water, hot gas or steam into rotary mechanical power, used for direct drive or electricity generation (see wind, hydro, gas or steam turbines). **Condensing steam turbines** exhaust depleted steam in a heat exchanger (called condenser) using ambient cooling from water (river, lake, sea) or air sources (cooling towers). A **backpressure steam turbine** has no condenser at ambient temperature conditions, but exhausts all steam at higher temperatures for use in particular heat end-uses.

United Nations Framework Convention on Climate Change (UNFCCC): The Convention was adopted on 9 May 1992 in New York and signed at the 1992 Earth Summit in Rio de Janeiro by more than 150 countries and the European Economic Community. Its ultimate objective is the "stabilization of greenhouse gas concentrations in the atmosphere at a level that would prevent dangerous anthropogenic interference with the climate system". It contains commitments for all parties. Under the Convention, parties included in Annex I aimed to return greenhouse gas emissions not controlled by the Montreal Protocol to 1990 levels by the year 2000. The convention came into force in March 1994. In 1997, the UNFCCC adopted the Kyoto Protocol. See also Annex I countries, Annex B countries and Kyoto Protocol.

Valley of death: Expression for a phase in the development of some technology when it is generating a large and negative cash flow because development costs increase but the risks associated with the technology are not reduced enough to entice private investors to take on the financing burden.

Value added: The net output of a sector or activity after adding up all outputs and subtracting intermediate inputs.

Values: Worth, desirability or utility based on individual preferences. Most social science disciplines use several definitions of value. Related to nature and environment, there is a distinction between intrinsic and instrumental values, the latter assigned by humans. Within instrumental values, there is an unsettled catalogue of different values, such as (direct and indirect) use, option, conservation, serendipity, bequest, existence, etc.

Mainstream economics define the total value of any resource as the sum of the values of the different individuals involved in the use of the resource. The economic values, which are the foundation of the estimation of costs, are measured in terms of the willingness to pay by individuals to receive the resource or by the willingness of individuals to accept payment to part with the resource.

Vent (geothermal/hydrothermal/submarine): An opening at the surface of the Earth (terrestrial or submarine) through which materials and energy flow.

Venture capital: A type of private equity capital typically provided for early-stage, high-potential technology companies in the interest of generating a return on investment through a trade sale of the company or an eventual listing on a public stock exchange.

Well-to-tank (WTT): WTT includes activities from resource extraction through fuel production to delivery of the fuel to vehicle. Compared to WTW, WTT does not take into consideration fuel use in vehicle operations.

Well-to-wheel (WTW): WTW analysis refers to specific lifecycle analysis applied to transportation fuels and their use in vehicles. The WTW stage includes resource extraction, fuel production, delivery of the fuel to vehicle, and end use of fuel in vehicle operations. Although feedstocks for alternative fuels do not necessarily come from a well, the WTW terminology is adopted for transportation fuel analysis.

Wind energy: Kinetic energy from air currents arising from uneven heating of the Earth's surface. A **wind turbine** is a rotating machine including its support structure for converting the kinetic energy to mechanical shaft energy to generate electricity. A **windmill** has oblique vanes or sails and the mechanical power obtained is mostly used directly, for example, for water pumping. A **wind farm**, **wind project** or **wind power plant** is a group of wind turbines interconnected to a common utility system through a system of transformers, distribution lines, and (usually) one substation.

Acronyms

AA-CAES	Advanced adiabatic compressed air energy storage	DDG	Distillers dried grains
AC	Alternating current	DDGS	Distillers dried grains plus solubles
AEM	Anion exchange membrane	DH	District heating
AEPC	Alternative Energy Promotion Centre	DHC	District heating or cooling
AFEX	Ammonia fibre expansion	DHW	Domestic hot water
APU	Auxiliary power unit	DLR	Deutsches Zentrum für Luft- und Raumfahrt (German Aerospace Centre)
AR4	4th assessment report (of the IPCC)		
AR5	5th assessment report (of the IPCC)	DLUC	Direct land use change
BC	Black carbon	DME	Dimethyl ether
BCCS	Biological carbon sequestration	DNI	Direct-normal irradiance
Bio-CCS	Biomass with carbon capture and storage	DPH	Domestic pellet heating
BIPV	Building-integrated photovoltaic	DSSC	Dye-sensitized solar cell
BMU	Bundesministerium für Umwelt, Naturschutz und Reaktorsicherheit (German Federal Ministry for the Environment, Nature Conservation and Nuclear Safety)	EGS	Enhanced geothermal systems
		EGTT	Expert Group on Technology Transfer
		EIA	Energy Information Administration (USA)
		EIT	Economy In Transition
BNEF	Bloomberg New Energy Finance	EMEC	European Marine Energy Centre
BOS	Balance of systems	EMF	Energy Modelling Form
BSI	Better Sugarcane Initiative	EMI	Electromagnetic interference
CAES	Compressed air energy storage	ENSAD	Energy-Related Severe Accident Database
CBP	Consolidated bioprocessing	EPRI	Electric Power Research Institute (USA)
CC	Combined cycle	EPT	Energy payback time
CCIY	China Coal Industry Yearbook	E[R]	Energy [R]evolution
CCS	Carbon dioxide capture and storage	ER	Energy ratio
CDM	Clean Development Mechanism	ERCOT	Electric Reliability Council of Texas
CEM	Cation exchange membrane	EREC	European Renewable Energy Council
CER	Certified Emissions Reduction	EROEI	Energy return on energy investment
CF	Capacity factor	ESMAP	Energy Sector Management Program (World Bank)
CFB	Circulating fluid bed	ETBE	Ethyl tert-butyl ether
CFD	Computational fluid dynamics	ETP	Energy Technology Perspectives
CFL	Compact fluorescent lightbulb	EU	European Union
CHP	Combined heat and power	EV	Electric vehicle
CIGSS	Copper indium/gallium disulfide/(di)selenide	FACTS	Flexible AC transmission system
CIS	Commonwealth of Independent States	FASOM	Forest and Agricultural Sector Optimization Model
CMA	China's Meteorological Administration	FAO	Food and Agriculture Organization (of the UN)
CNG	Compressed natural gas	FFV	Flexible fuel vehicle
CoC	Chain of custody	FQD	Fuel quality directive
COP	Coefficient of performance	FIT	Feed-in tariff
CPP	Captive power plant	FOGIME	Crediting System in Favour of Energy Management
CPV	Concentrating photovoltaics	FRT	Fault ride through
CREZ	Competitive renewable energy zone	FSU	Former Soviet Union
CRF	Capital recovery factor	FTD	Fischer-Tropsch diesel
CSIRO	Commonwealth Scientific and Industrial Research Organisation	GBD	Global burden of disease
		GBEP	Global Bioenergy Partnership
CSP	Concentrating solar power	GCAM	Global Change Assessment Model
CPV	Concentrating photovoltaics	GCM	Global climate model; General circulation model
CSTD	Commission on Science and Technology (UN)	GDP	Gross domestic product
DALY	Disability-adjusted life year	GEF	Global Environment Facility
dBA	A-weighted decibels	GHG	Greenhouse gas
DC	Direct current or district cooling	GHP	Geothermal heat pump

GIS	Geographic information system	LDV	Light duty vehicle
GM	Genetically modified	LED	Light-emitting diode
GMO	Genetically modified organism	LHV	Lower heating value
GO	Guarantee of origin	LNG	Liquefied natural gas
GPI	Genuine progress indicator	LPG	Liquefied petroleum gas
GPS	Global positioning system	LR	Learning rate
GSHP	Ground source heat pump	LUC	Land use change
HANPP	Human appropriation of terrestrial NPP	M&A	Mergers and acquisitions
HCE	Heat collection element	MDG	Millennium Development Goals
HDI	Human Development Index	MEH	Multiple-effect humidification
HDR	Hot dry rock	MHS	Micro-hydropower systems
HDV	Heavy duty vehicle	MITI	Ministry of International Trade and Industry (Japan)
HFCV	Hydrogen fuel cell electric vehicle	MSW	Municipal solid waste
HFR	Hot fractured rock	NASA	National Aeronautics and Space Administration (USA)
HHV	Higher heating value	NDRC	National Development and Reform Commission (China)
HPP	Hydropower plant		
HRV	Heat recovery ventilator	NFFO	Non Fossil Fuel Obligation
HEV	Hybrid electric vehicle	NG	Natural gas
HVAC	Heating, ventilation and air-conditioning	NGO	Nongovernmental organization
HVDC	High voltage direct current	Nm^3	Normal cubic metre (of gas) at standard temperature and pressure
HWR	Hot wet rock		
IA	Impact assessment	NMVOC	Non-methane volatile organic compounds
IAP	Indoor air pollution	NPP	Net primary production
IBC	interdigitated back-contact	NPV	Net present value
ICE	Internal combustion engine	NRC	National Research Council (USA)
ICEV	Internal combustion engine vehicle	NREL	National Renewable Energy Laboratory (USA)
ICLEI	Local Governments for Sustainability	NSDS	National Sustainable Development Strategies
ICOLD	International Commission on Large Dams	O&M	Operation and maintenance
ICS	Improved cookstove or Integral collector storage (Ch 3)	OB	Oscillating-body
		OC	Organic carbon
ICTSD	International Centre for Trade and Sustainable Development	OECD	Organisation for Economic Co-operation and Development
IEA	International Energy Agency		
IEC	International Electrotechnical Commission	OM	Organic matter
IEEE	Institute of Electrical and Electronics Engineers	OPV	Organic photovoltaic
IHA	International Hydropower Association	ORC	Organic Rankine Cycle
ILUC	Indirect land use change	OTEC	Ocean thermal energy conversion
IGCC	Integrated gasification combined cycle	OWC	Oscillating water column
IPCC	Intergovernmental Panel on Climate Change	PACE	Property Assessed Clean Energy
IPR	Intellectual property rights	PBR	Photobioreactor
IQR	Inter-quartile range	PCM	Phase-change material
IREDA	Indian Renewable Energy Development Agency	PDI	Power density index
IRENA	International Renewable Energy Agency	PEC	Photoelectrochemical
IRM	Inorganic mineral raw materials	PHEV	Plug-in hybrid electric vehicle
ISCC	Integrated solar combined-cycle	PM	Particulate matter
ISES	International Solar Energy Society	POME	Palm oil mill effluent
ISEW	Index of sustainable economic welfare	PPA	Purchase power agreement
ISO	International Organization for Standardization	PRO	Pressure-retarded osmosis
J	Joule	PROALCOOL	Brazilian Alcohol Program
JI	Joint implementation	PSA	Probabilistic safety assessment
LCA	Lifecycle assessment	PSI	Paul Scherrer Institute
LCOE	Levelized cost of energy (or of electricity)	PSP	Pumped storage plants
LCOF	Levelized cost of fuel	PTC	Production tax credit
LCOH	Levelized cost of heat	PV	Photovoltaic

PV/T	Photovoltaic/thermal	SSCF	Simultaneous saccharification and co-fermentation
PWR	Pressurized water reactor	SSF	Simultaneous saccharification and fermentation
R&D	Research and development	SSP	Space-based solar power
RBMK	Reaktor bolshoy moshchnosty kanalny	STP	Standard temperature and pressure
RCM	Regional climate model	SWH	Solar water heating
RD&D	Research, development and demonstration	TBM	Tunnel-boring machines
R/P	Reserves to current production (ratio)	TERM	Tonga Energy Roadmap
RD	Renewable diesel	TGC	Tradable green certificate
RE	Renewable energy	TPA	Third-party access
RE-C	Renewable energy cooling	TPES	Total primary energy supply
RE-H	Renewable energy heating	TPWind	European Wind Energy Technology Platform
RE-H/C	Renewable energy heating/cooling	TS	Technical Summary or thermosyphon
REC	Renewable energy certificate	US	United States of America (adjective)
RED	Reversed electro dialysis	USA	United States of America (noun)
REN21	Renewable Energy Policy Network for the 21st Century	UN	United Nations
		UNCED	United Nations Conference on Environment and Development
RES	Renewable electricity standard	UNCTAD	United Nations Conference on Trade and Development
RM&U	Renovation, modernization and upgrading		
RMS	Root mean square	UNDP	United Nations Development Programme
RNA	Rotor nacelle assembly	UNEP	United Nations Environment Programme
RO	Renewables obligation	UNFCCC	United Nations Framework Convention on Climate Change
RoR	Run of river		
RPS	Renewable portfolio standard	USD	US dollar
RSB	Roundtable for Sustainable Biofuels	USDOE	US Department of Energy
SCADA	Supervisory control and data acquisition	V	Volt
SCC	Stress corrosion cracking	VKT	Vehicle kilometres travelled
SD	Sustainable development	VRB	Vanadium redox battery
SEGS	Solar Electric Generating Station (California)	W	Watt
SHC	Solar heating and cooling	W_e	Watt of electricity
SHP	Small-scale hydropower plant	W_p	Watt peak of PV installation
SI	Suitability index	WBG	World Bank Group
SME	Small and medium sized enterprises	WCD	World Commission on Dams
SNG	Synthesis gas	WCED	World Commission on Environment and Development
SNV	Netherlands Development Organization	WEA	World Energy Assessment
SPF	Seasonal performance factor	WEO	World Energy Outlook
SPM	Summary for Policymakers	WindPACT	Wind Partnership for Advanced Component Technologies
SPP	Small power producer		
SPS	Sanitary and phytosanitary	WTO	World Trade Organization
SR	Short rotation	WTW	Well to wheel
SRES	Special Report on Emission Scenarios (of the IPCC)		
SRREN	Special Report on Renewable Energy Sources and Climate Change Mitigation (of the IPCC)		

Chemical Symbols

a-Si	Amorphous silicon	H_2S	Hydrogen sulphide
C	Carbon	HFC	Hydrofluorocarbons
CdS	Cadmium sulphide	K	Potassium
CdTe	Cadmium telluride	Mg	Magnesium
CH_4	Methane	N	Nitrogen
CH_3CH_2OH	Ethanol	N_2	Nitrogen gas
CH_3OCH_3	Dimethyl ether (DME)	N_2O	Nitrous oxide
CH_3OH	Methanol	Na	Sodium
CIGS(S)	Copper indium gallium diselenide (disulfide)	NaS	Sodium-sulfur
Cl	Chlorine	NH_3	Ammonia
CO	Carbon monoxide	Ni	Nickel
CO_2	Carbon dioxide	NiCd	Nickel-cadmium
CO_2eq	Carbon dioxide equivalent	NO_x	Nitrous oxides
c-Si	Crystalline silicon	O_3	Ozone
Cu	Copper	P	Phosphorus
$CuInSe_2$	Copper indium diselenide	PFC	Perfluorocarbon
DME	Dimethyl ether	SF_6	Sulfur hexafluoride
Fe	Iron	Si	Silicon
GaAs	Gallium arsenide	SiC	Silicon carbide
H_2	Hydrogen gas	SO_2	Sulfur dioxide
H_2O	Water	ZnO	Zinc oxide

Prefixes (International Standard Units)

Symbol	Multiplier	Prefix	Symbol	Multiplier	Prefix
Z	10^{21}	zetta	d	10^{-1}	deci
E	10^{18}	exa	c	10^{-2}	centi
P	10^{15}	peta	m	10^{-3}	milli
T	10^{12}	tera	µ	10^{-6}	micro
G	10^{9}	giga	n	10^{-9}	nano
M	10^{6}	mega	p	10^{-12}	pico
k	10^{3}	kilo	f	10^{-15}	femto
h	10^{2}	hecto	a	10^{-18}	atto
da	10	deca			

Annex II

Methodology

Lead Authors:
William Moomaw (USA), Peter Burgherr (Switzerland), Garvin Heath (USA),
Manfred Lenzen (Australia, Germany), John Nyboer (Canada), Aviel Verbruggen (Belgium)

This annex should be cited as:
Moomaw, W., P. Burgherr, G. Heath, M. Lenzen, J. Nyboer, A. Verbruggen, 2011: Annex II: Methodology. In IPCC Special Report on Renewable Energy Sources and Climate Change Mitigation [O. Edenhofer, R. Pichs-Madruga, Y. Sokona, K. Seyboth, P. Matschoss, S. Kadner, T. Zwickel, P. Eickemeier, G. Hansen, S. Schlömer, C. von Stechow (eds)], Cambridge University Press, Cambridge, United Kingdom and New York, NY, USA.

Table of Contents

A.II.1   Introduction 975

A.II.2   Metrics for analysis in this report 975

A.II.3   Financial assessment of technologies over project lifetime 975
    A.II.3.1   Constant (real) values 975
    A.II.3.2   Discounting and net present value 975
    A.II.3.3   Levelized cost 976
    A.II.3.4   Annuity factor or capital cost recovery factor 976

A.II.4   Primary energy accounting 976

A.II.5   Lifecycle assessment and risk analysis 978
    A.II.5.1   Energy payback time and energy ratio 979
    A.II.5.2   Review of lifecycle assessments of electricity generation technologies 979
    A.II.5.2.1   Review methodology 980
    A.II.5.2.2   List of references 981
    A.II.5.3   Review of operational water use of electricity generation technologies 991
    A.II.5.3.1   Review methodology 991
    A.II.5.3.2   List of references 992
    A.II.5.4   Risk analysis 993

A.II.6   Regional definitions and country groupings 994

A.II.7   General conversion factors for energy 997

References 998

A.II.1 Introduction

Parties need to agree upon common data, standards, supporting theories and methodologies. This annex summarizes a set of agreed upon conventions and methodologies. These include the establishment of metrics, determination of a base year, definitions of methodologies and consistency of protocols that permit a legitimate comparison between alternative types of energy in the context of climate change phenomena. This section defines or describes these fundamental definitions and concepts as used throughout this report, recognizing that the literature often uses inconsistent definitions and assumptions.

This report communicates uncertainty where relevant, for example, by showing the results of sensitivity analyses and by quantitatively presenting ranges in cost numbers as well as ranges in the scenario results. This report does not apply formal IPCC uncertainty terminology because at the time of approval of this report, IPCC uncertainty guidance was in the process of being revised.

A.II.2 Metrics for analysis in this report

A number of metrics can simply be stated or are relatively easy to define. Annex II provides the set of agreed upon metrics. Those which require further description are found below. The units used and basic parameters pertinent to the analysis of each RE type in this report include:

- International System of Units (SI) for standards and units
- Metric tonnes (t) CO_2, CO_2eq
- Primary energy values in exajoules (EJ)
- IEA energy conversion factors between physical and energy units
- Capacity: GW thermal (GW_t), GW electricity (GW_e)
- Capacity factor
- Technical and economic lifetime
- Transparent energy accounting (e.g., transformations of nuclear or hydro energy to electricity)
- Investment cost in USD/kW (peak capacity)
- Energy cost in USD_{2005}/kWh or USD_{2005}/EJ
- Currency values in USD_{2005} (at market exchange rate where applicable, no purchasing power parity is used)
- Discount rates applied = 3, 7 and 10%
- World Energy Outlook (WEO) 2008 fossil fuel price assumptions
- Baseline year = 2005 for all components (population, capacity, production, costs). Note that more recent data may also be included (e.g., 2009 energy consumption)
- Target years: 2020, 2030 and 2050.

A.II.3 Financial assessment of technologies over project lifetime

The metrics defined here provides the basis from which one renewable resource type (or project) can be compared to another. To make projects or resources comparable, at least in terms of costs, costs that may occur at various moments in time (e.g., in various years) are represented as a single number anchored at one particular year, the reference year (2005). Textbooks on investment appraisal provide background on the concepts of constant values, discounting, net present value calculations, and levelized costs, for example (Jelen and Black, 1983).

A.II.3.1 Constant (real) values

The analyses of costs are in constant or real[1] dollars (i.e., excluding the impacts of inflation) based in a particular year, the base year 2005, in USD. Specific studies on which the report depends may use market exchange rates as a default option or use purchasing power parities, but where these are part of the analysis, they will be stated clearly and, where possible, converted to USD_{2005}.

When the monetary series in the analyses are in real dollars, consistency requires that the discount rate should also be real (free of inflationary components). This consistency is often not obeyed; studies refer to 'observed market interest rates' or 'observed discount rates', which include inflation or expectations about inflation. 'Real/constant' interest rates are never directly observed, but derived from the ex-post identity:

$$(1+m) = (1+i) \times (1+f) \qquad (1)$$

where

m = nominal rate (%)
i = real or constant rate (%)
f = inflation rate (%)

The reference year for discounting and the base year for anchoring constant prices may differ in studies used in the various chapters; where possible, an attempt was made to harmonize the data to reflect discount rates applied here.

A.II.3.2 Discounting and net present value

Private agents assign less value to things further in the future than to things in the present because of a 'time preference for consumption' or to reflect a 'return on investment'. Discounting reduces future cash flows by a value less than 1. Applying this rule on a series of net cash flows in real USD, the net present value (NPV) of the project can be ascertained and, thus, compared to other projects using:

$$NPV = \sum_{j=0}^{n} \frac{Net\ cash\ flows\ (j)}{(1+i)^j} \qquad (2)$$

where

n = lifetime of the project
i = discount rate

---

[1] The economists' term 'real' may be confusing because what they call real does not correspond to observed financial flows ('nominal', includes inflation); 'real' reflects the actual purchasing power of the flows in constant dollars.

Methodology

This report's analysts have used three values of discount rates ($i = 3$, 7 and 10%) for the cost evaluations. The discount rates may reflect typical rates used, with the higher ones including a risk premium. The discount rate is open to much discussion and no clear parameter or guideline can be suggested as an appropriate risk premium. This discussion is not addressed here; the goal is to provide an appropriate means of comparison between projects, renewable energy types and new versus current components of the energy system.

A.II.3.3 Levelized cost

Levelized costs are used in the appraisal of power generation investments, where the outputs are quantifiable (MWh generated during the lifetime of the investment). The levelized cost is the unique break-even cost price where discounted revenues (price x quantities)[2] are equal to the discounted net expenses:

$$C_{Lev} = \frac{\sum_{j=0}^{n} \frac{Expenses_j}{(1+i)^j}}{\sum_{j=0}^{n} \frac{Quantities_j}{(1+i)^j}} \quad (3)$$

where

C_{Lev} = levelized cost
n = lifetime of the project
i = discount rate

A.II.3.4 Annuity factor or capital cost recovery factor

A very common practice is the conversion of a given sum of money at moment 0 into a number n of constant annual amounts over the coming n future years:

Let A = annual constant amount in payments over n years
Let B = cash amount to pay for the project in year 0

A is obtained from B using a slightly modified equation 2: the lender wants to receive B back at the discount rate i. The NPV of the n times A receipts in the future therefore must exactly equal B:

$$\sum_{j=1}^{n} \frac{A}{(1+i)^j} = B, \text{ or}: A\sum_{j=1}^{n} \frac{1}{(1+i)^j} = B \quad (4)$$

We can bring A before the summation because it is a constant (not dependent on j).

The sum of the discount factors (a finite geometrical series) is deductible as a particular number. When this number is calculated, A is found by dividing B by this number. This is known as the *Capital Recovery Factor*

---
[2] This is also referred to as Levelized Price. Note that, in this case, MWh would be discounted.

(CRF) but may be known as the *Annuity Factor* 'δ'. Like NPV, the annuity factor δ depends on the two parameters i and n:

$$\delta = \frac{i \times (1+i)^n}{(1+i)^n - 1}$$

The CRF (or δ) can be used to quickly calculate levelized costs for very simple projects where investment costs during one given year are the only expenditures and where production remains constant over the lifetime (n):

$$C_{Lev} \times Q = B \times \delta, \text{ or}: C_{Lev} = (B \times \delta)/Q \quad (5)$$

or where one can assume that operation and maintenance (O&M) costs do not change from year to year:

$$C_{Lev} = \frac{B \times \delta + O\&M}{Q} \quad (6)$$

where

C_{Lev} = levelized cost
B = investment cost
Q = production
O&M = annual operating and maintenance costs
n = life time of the project
i = discount rate

A.II.4 Primary energy accounting

This section introduces the primary energy accounting method used throughout this report. Different energy analyses use different accounting methods that lead to different quantitative outcomes for reporting both current primary energy use and energy use in scenarios that explore future energy transitions. Multiple definitions, methodologies and metrics are applied. Energy accounting systems are utilized in the literature often without a clear statement as to which system is being used as noted by Lightfoot, 2007 and Martinot et al., 2007. An overview of differences in primary energy accounting from different statistics has been described (Macknick, 2009) and the implications of applying different accounting systems in long-term scenario analysis were illustrated by Nakicenovic et al., (1998).

Three alternative methods are predominantly used to report primary energy. While the accounting of combustible sources, including all fossil energy forms and biomass, is unambiguous and identical across the different methods, they feature different conventions on how to calculate primary energy supplied by non-combustible energy sources, i.e., nuclear energy and all renewable energy sources except biomass.

These methods are:

- *The physical energy content method* adopted, for example, by the Organisation for Economic Cooperation and Development (OECD), the International Energy Agency (IEA) and Eurostat (IEA/OECD/Eurostat, 2005),

- *The substitution method*, which is used in slightly different variants by BP (2009) and the US Energy Information Administration (EIA online glossary), each of which publish international energy statistics, and

- *The direct equivalent method* that is used by UN Statistics (2010) and in multiple IPCC reports that deal with long-term energy and emission scenarios (Nakicenovic and Swart, 2000; Morita et al., 2001; Fisher et al., 2007).

For non-combustible energy sources, the *physical energy content method* adopts the principle that the primary energy form should be the first energy form used downstream in the production process for which multiple energy uses are practical (IEA/OECD/Eurostat, 2005). This leads to the choice of the following *primary* energy forms:

- Heat for nuclear, geothermal and solar thermal energy; and
- Electricity for hydro, wind, tide/wave/ocean and solar photovoltaic (PV) energy.

Using this method, the primary energy equivalent of hydropower and solar PV, for example, assumes a 100% conversion efficiency to 'primary electricity', so that the gross energy input for the source is 3.6 MJ of primary energy = 1 kWh electricity. Nuclear energy is calculated from the gross generation by assuming a 33% thermal conversion efficiency,[3] that is, 1 kWh = (3.6 ÷ 0.33) = 10.9 MJ. For geothermal energy, if no country-specific information is available, the primary energy equivalent is calculated using 10% conversion efficiency for geothermal electricity (so 1 kWh = (3.6 ÷ 0.1) = 36 MJ), and 50% for geothermal heat.

The *substitution method* reports primary energy from non-combustible sources as if they had been substituted for combustible energy. Note, however, that different variants of the substitution method use somewhat different conversion factors. For example, BP applies a 38% conversion efficiency to electricity generated from nuclear and hydropower, whereas the World Energy Council used 38.6% for nuclear and non-combustible renewable sources (WEC, 1993) and the EIA uses still different values. Macknick (2009) provides a more complete overview. For useful heat generated from non-combustible energy sources, other conversion efficiencies are used.

The *direct equivalent method* counts one unit of secondary energy provided from non-combustible sources as one unit of primary energy, that is, 1 kWh of electricity or heat is accounted for as 1 kWh = 3.6 MJ of primary energy. This method is mostly used in the long-term scenarios literature, including multiple IPCC reports (IPCC, 1995; Nakicenovic and Swart, 2000; Morita et al., 2001; Fisher et al., 2007), because it deals with fundamental transitions of energy systems that rely to a large extent on low-carbon, non-combustible energy sources.

In this report, IEA data are utilized, but energy supply is reported using the *direct equivalent method*. The major difference between this and the *physical energy content method* will appear in the amount of primary energy reported for electricity production by geothermal heat, concentrating solar thermal, ocean temperature gradients or nuclear energy. Table A.II.1 compares the amounts of global primary energy by source and percentages using the *physical energy content*, the *direct equivalent* and a variant of the *substitution method* for the year 2008 based on IEA data (IEA, 2010a). In current statistical energy data, the main differences in absolute terms appear when comparing nuclear and hydropower. Since they both produced a comparable amount of electricity globally in 2008, under both *direct equivalent* and *substitution methods*, their share of meeting total final consumption is similar, whereas under the *physical energy content method*, nuclear is reported at about three times the primary energy of hydropower.

The alternative methods outlined above emphasize different aspects of primary energy supply. Therefore, depending on the application, one method may be more appropriate than another. However, none of them is superior to the others in all facets. In addition, it is important to realize that total primary energy supply does not fully describe an energy system, but is merely one indicator amongst many. Energy balances as published by the IEA (2010a) offer a much wider set of indicators, which allows tracing the flow of energy from the resource to final energy use. For instance, complementing total primary energy consumption with other indicators, such as total final energy consumption and secondary energy production (e.g., electricity, heat), using different sources helps link the conversion processes with the final use of energy. See Figure 1.16 and the associated discussion for a summary of this approach.

For the purpose of this report, the *direct equivalent method* is chosen for the following reasons.

- It emphasizes the secondary energy perspective for non-combustible sources, which is the main focus of the analyses in the technology chapters (Chapters 2 through 7).

- All non-combustible sources are treated in an identical way by using the amount of secondary energy they provide. This allows the comparison of all non-CO_2-emitting renewable and nuclear energy sources on a common basis. Primary energy of fossil fuels and biomass combines both the secondary energy and the thermal energy losses from the conversion process. When fossil fuels or biofuels are replaced by nuclear systems or other renewable technologies than biomass, the total of reported primary energy decreases substantially (Jacobson, 2009).

- Energy and CO_2 emissions scenario literature that deals with fundamental transitions of the energy system to avoid dangerous anthropogenic interference with the climate system over the long term (50 to 100 years) has used the direct equivalent method most frequently (Nakicenovic and Swart, 2000; Fisher et al., 2007).

---

3   As the amount of heat produced in nuclear reactors is not always known, the IEA estimates the primary energy equivalent from the electricity generation by assuming an efficiency of 33%, which is the average for nuclear power plants in Europe (IEA, 2010b).

Table A.II.1 | Comparison of global total primary energy supply in 2008 using different primary energy accounting methods (data from IEA, 2010a).

	Physical content method		Direct equivalent method		Substitution method[1]	
	EJ	%	EJ	%	EJ	%
Fossil fuels	418.15	81.41	418.15	85.06	418.15	79.14
Nuclear	29.82	5.81	9.85	2.00	25.90	4.90
Renewable:	65.61	12.78	63.58	12.93	84.27	15.95
Bioenergy[2]	50.33	9.80	50.33	10.24	50.33	9.53
Solar	0.51	0.10	0.50	0.10	0.66	0.12
Geothermal	2.44	0.48	0.41	0.08	0.82	0.16
Hydro	11.55	2.25	11.55	2.35	30.40	5.75
Ocean	0.00	0.00	0.00	0.00	0.01	0.00
Wind	0.79	0.15	0.79	0.16	2.07	0.39
Other	0.03	0.01	0.03	0.01	0.03	0.01
Total	513.61	100.00	491.61	100.00	528.35	100.00

Notes:

1. For the substitution method, conversion efficiencies of 38% for electricity and 85% for heat from non-combustible sources were used. BP uses the conversion value of 38% for electricity generated from hydro and nuclear sources. BP does not report solar, wind and geothermal in its statistics; here, 38% for electricity and 85% for heat is used.

2. Note that IEA reports first-generation biofuels in secondary energy terms (the primary biomass used to produce the biofuel would be higher due to conversion losses, see Sections 2.3 and 2.4).

Table A.II.2 shows the differences in the primary energy accounting for the three methods for a scenario that would produce a 550 ppm CO_2eq stabilization by 2100.

While the differences between applying the three accounting methods to current energy consumption are modest, differences grow significantly when generating long-term lower CO_2 emissions energy scenarios where non-combustion technologies take on a larger relative role (Table A.II.2). The accounting gap between the different methods becomes bigger over time (Figure A.II.1). There are significant differences in individual non-combustible sources in 2050 and even the share of total renewable primary energy supply varies between 24 and 37% across the three methods (Table A.II.2). The biggest absolute gap (and relative difference) for a single source is for geothermal energy, with about 200 EJ difference between the direct equivalent and the physical energy content method, and the gap between hydro and nuclear primary energy remains considerable. The scenario presented here is fairly representative and by no means extreme. The chosen 550 ppm stabilization target is not particularly stringent nor is the share of non-combustible energy very high.

A.II.5 Lifecycle assessment and risk analysis

This section describes methods and underlying literature and assumptions of analyses of energy payback times and energy ratios (A.II.5.1),

Table A.II.2 | Comparison of global total primary energy supply in 2050 using different primary energy accounting methods based on a 550 ppm CO_2eq stabilization scenario (Loulou et al., 2009).

	Physical content method		Direct equivalent method		Substitution method	
	EJ	%	EJ	%	EJ	%
Fossil fuels	581.6	55.2	581.56	72.47	581.6	61.7
Nuclear	81.1	7.7	26.76	3.34	70.4	7.8
Renewable:	390.1	37.1	194.15	24.19	290.4	30.8
Bioenergy	120.0	11.4	120.0	15.0	120.0	12.7
Solar	23.5	2.2	22.0	2.8	35.3	3.8
Geothermal	217.3	20.6	22.9	2.9	58.1	6.2
Hydro	23.8	2.3	23.8	3.0	62.6	6.6
Ocean	0.0	0.0	0.0	0.0	0.0	0.0
Wind	5.5	0.5	5.5	0.7	14.3	1.5
Total	1,052.8	100	802.5	100	942.4	100

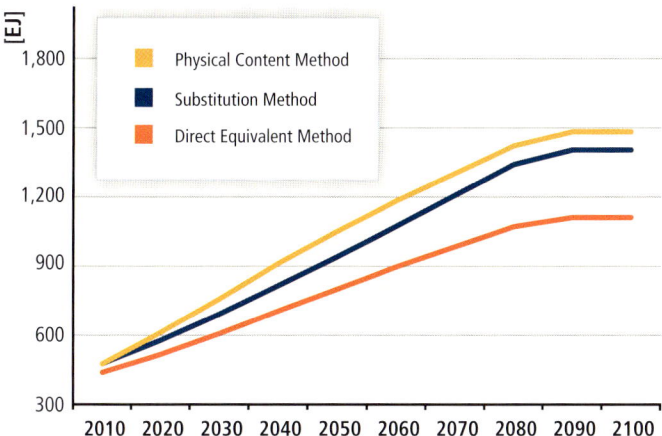

Figure A.II.1 | Comparison of global total primary energy supply between 2010 and 2100 using different primary energy accounting methods based on a 550 ppm CO₂eq stabilization scenario (Loulou et al., 2009).

lifecycle GHG emissions (A.II.5.2), operational water use (A.II.5.3) and hazards and risks (A.II.5.4) of energy technologies as presented in Chapter 9. Results of the analysis carried out for lifecycle GHG emissions are also included in Sections 2.5, 3.6, 4.5, 5.6, 6.5 and 7.6. Please note that the literature bases for the reviews in A.II.5.2 and A.II.5.3 are included as lists within the respective sections.

A.II.5.1 Energy payback time and energy ratio

The Energy Ratio, ER (also referred to as the energy payback ratio, or the Energy Return on Energy Investment, $EROEI$; see Gagnon, 2008), of an energy supply system of power rating P and load factor λ, is defined as the ratio

$$ER = \frac{E_{life}}{E} = \frac{P \times 8760\, hy^{-1} \times \lambda \times T}{E}$$

of the lifetime electricity output E_{life} of the plant over its lifetime T, and the total (gross) energy requirement E for construction, operation and decommissioning (Gagnon, 2008). In calculating E, it is a convention to a) exclude the energy from human labour, energy in the ground (fossil and minerals), energy in the sun, and hydrostatic potential, and b) not to discount future against present energy requirements (Perry et al., 1977; Herendeen, 1988). Further, in computing the total energy requirement E, all its constituents must be of the same energy quality (for example only electricity, or only thermal energy, see the 'valuation problem' discussed in Leach (1975), Huettner (1976), Herendeen (1988), and especially Rotty et al. (1975, pp. 5-9 for the case of nuclear energy)). Whilst E may include derived and primary energy forms (for example electricity and thermal energy), it is usually expressed in terms of primary energy, with the electricity component converted to primary energy equivalents using the thermal efficiency $R_{conv} \approx 0.3$ of a typical subcritical black-coal-fired power station as the conversion factor. This report follows these conventions. E is sometimes reported in units of kWh_e/MJ_{prim}, and sometimes in units of kWh_e/kWh_{prim}. Whilst the first option chooses the most common units for either energy form, the second option allows the reader to readily understand the percentage or multiple connecting embodied energy and energy output. Moreover, it has been argued (see Voorspools et al., (2000, p. 326)) that in the absence of alternative technologies, electricity would have to be generated by conventional means. We therefore use kWh_e/kWh_{prim} in this report.

Applying the lifecycle energy metric to an energy supply system allows defining an *energy payback time*. This is the time t_{PB} that it takes the system to supply an amount of energy that is equal to its own energy requirement E. Once again, this energy is best measured in terms of the primary energy equivalent $\frac{E_{PB}}{R_{conv}}$ of the system's electricity output E_{PB} over the payback time. Voorspools et al. (2000, p. 326) note that were the system to pay back its embodied primary energy in equal amounts of electricity, energy payback times would be more than three times as long.

Mathematically, the above condition reads

$$E = \frac{E_{PB}}{R_{conv}} = \frac{P \times 8760\, hy^{-1} \times \lambda \times t_{PB}}{R_{conv}} \quad , \text{and leads to}$$

$$t_{PB} = \frac{E}{\dfrac{P \times 8760\, hy^{-1} \times \lambda}{R_{conv}}} = \frac{E}{\dfrac{E_{out\,annual}}{R_{conv}}}$$

(which, for example, coincides with the standard German VDI 4600 definition). Here, $\dfrac{E_{out\,annual}}{R_{conv}}$ is the system's annual net energy output expressed in primary energy equivalents. It can be shown that the Energy Ratio ER (or $EROEI$) and the energy payback time t_{PB} can be converted into each other according to

$$t_{PB} = \frac{E\,T}{\dfrac{E_{out\,annual}}{R_{conv}} T} = \frac{E\,T}{\dfrac{E_{life}}{R_{conv}}} = \frac{R_{conv}}{ER} T \quad .$$

Note that the energy payback time is not dependent on the lifetime T, because

$$t_{PB} = \frac{E\,R_{conv}}{P \times 8760\, hy^{-1} \times \lambda} \quad .$$

Energy payback times have been partly converted from energy ratios found in the literature (Lenzen, 1999, 2008; Lenzen and Munksgaard, 2002; Lenzen et al., 2006; Gagnon, 2008; Kubiszewski et al., 2010) based on the assumed average lifetimes given in Table 9.8 (Chapter 9). Note that energy payback as defined in the glossary (Annex I) and used in some technology chapters refers to what is defined here as energy payback time.

A.II.5.2 Review of lifecycle assessments of electricity generation technologies

The National Renewable Energy Laboratory (NREL) carried out a comprehensive review of published lifecycle assessments (LCAs) of electricity generation technologies. Of 2,165 references collected, 296 passed screens, described below, for quality and relevance and were entered into a database. This database forms the basis for the assessment of lifecycle greenhouse gas (GHG) emissions from electricity generation technologies in this report. Based on estimates compiled in the database, plots of published estimates of lifecycle GHG emissions appear in each technology chapter of this report (Chapters 2 through 7) and in Chapters 1 and 9, where lifecycle GHG emissions from RE technologies are compared to those from fossil and nuclear electricity generation technologies. The following subchapters describe the methods applied in this review (A.II.5.2.1), and list all references that are shown in the final results, sorted by technology (A.II.5.2.2).

A.II.5.2.1 Review methodology

Broadly, the review followed guidelines for *systematic reviews* as commonly performed, for instance, in the medical sciences (Neely et al., 2010). The methods of reviews in the medical sciences differ somewhat from those in the physical sciences, in that there is an emphasis on multiple, independent reviews of each candidate reference using predefined screening criteria; the formation of a review team composed of, in this case, LCA experts, technology experts and literature search experts that meets regularly to ensure consistent application of the screening criteria; and an exhaustive search of published literature to ensure no bias by, for instance, publication type (journal, report, etc.).

It is critical to note at the outset that this review did not alter (except for unit conversion) or audit for accuracy the estimates of lifecycle GHG emissions published in studies that pass the screening criteria. Additionally, no attempt was made to identify or screen for outliers, or pass judgment on the validity of input parameter assumptions. Because estimates are plotted as published, considerable methodological inconsistency is inherent, which limits comparability of the estimates both within particular power generation technology categories and across the technology categories. This limitation is partially counteracted by the comprehensiveness of the literature search and the breadth and depth of literature revealed. Few attempts have been made to broadly review the LCA literature on electricity generation technologies. Those that do exist tend to focus on individual technologies and are more limited in comprehensiveness compared to the present review (e.g., Lenzen and Munksgaard, 2002; Fthenakis and Kim, 2007; Lenzen, 2008; Sovacool, 2008b; Beerten et al., 2009; Kubiszewski et al., 2010).

The review procedure included the following steps: literature collection, screening and analysis.

Literature collection

Starting in May of 2009, potentially relevant literature was identified through multiple mechanisms, including searches in major bibliographic databases (e.g., Web of Science, WorldCat) using a variety of search algorithms and combinations of key words, review of reference lists of relevant literature, and specialized searches on websites of known studies series (e.g., European Union's ExternE and its descendants) and known LCA literature databases (e.g., the library contained within the SimaPro LCA software package). All collected literature was first categorized by content (with key information from every collected reference recorded in a database) and added to a bibliographic database.

The literature collection methods described here apply to all classes of electricity generation technologies reviewed in this report except for oil and hydropower. LCA data for hydropower and oil were added at a later stage to the NREL database and have therefore undergone a less comprehensive literature collection process.

Literature screening

Collected references were independently subjected to three rounds of screening by multiple experts to select references that met criteria for quality and relevance. References often reported multiple GHG emission estimates based on alternative scenarios. Where relevant, the screening criteria were applied at the level of the scenario estimate, occasionally resulting in only a subset of scenarios analyzed in a given reference passing the screens.

References having passed the first quality screen included peer-reviewed journal articles, scientifically detailed conference proceedings, PhD theses, and reports (authored by government agencies, academic institutions, non-governmental organizations, international institutions, or corporations) published after 1980 and in English. Attempts were made to obtain English versions of non-English publications and a few exceptions were translated. The first screen also ensured that the accepted references were LCAs, defined as analyzing two or more lifecycle phases (with exceptions for PV and wind energy given that the literature demonstrates that the vast majority of lifecycle GHG emissions occur in the manufacturing phase (Frankl et al., 2005; Jungbluth et al., 2005)).

All references passing the first screen were then directly judged based on more stringent quality and relevance criteria:

- Employed a currently accepted attributional LCA and GHG accounting method (consequential LCAs were not included because their results are fundamentally not comparable to results based on attributional LCA methods; see Section 9.3.4 for further description of attributional and consequential LCAs);

- Reported inputs, scenario/technology characteristics, important assumptions and results in enough detail to trace and trust the results; and

- Evaluated a technology of modern or future relevance.

For the published results to be analyzed, estimates had to pass a final set of criteria:

- To ensure accuracy in transcription, only GHG emission estimates that were reported numerically (i.e., not only graphically) were included.

- Estimates duplicating prior published work were not included.

- Results had to have been easily convertible to the functional unit chosen for this study: grams of CO_2eq per kWh generated.

Table A.II.3 reports the counts of references at each stage in the screening process for the broad classes of electricity generation technologies considered in this report.

Analysis of estimates

Estimates of lifecycle GHG emissions from studies passing both screens were then analyzed and plotted. First, estimates were categorized by technology within the broad classes considered in this report, listed in Table A.II.3. Second, estimates were converted to the common functional unit of g CO_2eq per kWh generated. This conversion was performed using no exogenous assumptions; if any were required, that estimate was not included. Third, estimates of total lifecycle GHG emissions that included contributions from either land use change (LUC) or heat production (in cases of cogeneration) were removed. This step required that studies that considered LUC- or heat-related GHG emissions had to report those contributions separately such that estimates included here pertain to the generation of electricity alone. Finally, distributional information required for display in box and whisker plots were calculated: minimum, 25th percentile value, 50th percentile value, 75th percentile value and maximum. Technologies with data sets composed of less than five estimates (e.g., geothermal) have been plotted as discrete points rather than superimposing synthetic distributional information.

The resulting values underlying Figure 9.8 are shown in Table A.II.4. Figures displayed in technology chapters are based on the same data set, yet displayed with a higher level of resolution regarding technology sub-categories (e.g., on- and offshore wind energy).

A.II.5.2.2 List of references

Below, all references for the review of lifecycle assessments of greenhouse gas emissions from electricity generation that are shown in the final results in this report are listed, sorted by technology and in alphabetical order.

Biomass-based power generation (52)

Beals, D., and D. Hutchinson (1993). *Environmental Impacts of Alternative Electricity Generation Technologies: Final Report.* Beals and Associates, Guelph, Ontario, Canada, 151 pp.

Beeharry, R.P. (2001). Carbon balance of sugarcane bioenergy systems. *Biomass & Bioenergy*, **20**(5), pp. 361-370.

Corti, A., and L. Lombardi (2004). Biomass integrated gasification combined cycle with reduced CO_2 emissions: Performance analysis and life cycle assessment (LCA). *Energy*, **29**(12-15), pp. 2109-2124.

Cottrell, A., J. Nunn, A. Urfer, and L. Wibberley (2003). *Systems Assessment of Electricity Generation Using Biomass and Coal in CFBC.* Cooperative Research Centre for Coal in Sustainable Development, Pullenvale, Qld., Australia, 21 pp.

Cowie, A.L. (2004). *Greenhouse Gas Balance of Bioenergy Systems Based on Integrated Plantation Forestry in North East New South Wales, Australia: International Energy Agency (IEA)Bioenergy Task 38 on GHG Balances of Biomass and Bioenergy Systems.* IEA, Paris, France. 6 pp. Available at: www.ieabioenergy-task38.org/projects/task38casestudies/aus-brochure.pdf.

Table A.II.3 | Counts of LCAs of electricity generation technologies ('references') at each stage in the literature collection and screening process and numbers of scenarios ('estimates') of lifecycle GHG emissions evaluated herein.

Technology category	References reviewed	References passing the first screen	References passing the second screen	References providing lifecycle GHG emissions estimates	Estimates of lifecycle GHG emissions passing screens
Biopower	369	162	84	52	226
Coal	273	192	110	52	181
Concentrating solar power	125	45	19	13	42
Geothermal Energy	46	24	9	6	8
Hydropower	89	45	11	11	28
Natural gas	251	157	77	40	90
Nuclear Energy	249	196	64	32	125
Ocean energy	64	30	6	5	10
Oil	68	45	19	10	24
Photovoltaics	400	239	75	26	124
Wind Energy	231	174	72	49	126
TOTALS	**2165**	**1309**	**546**	**296**	**984**
% of total reviewed		60%	25%	14%	
% of those passing first screen			42%	23%	
% of those passing second screen				54%	

Note: Some double counting is inherent in the totals given that some references investigated more than one technology.

Table A.II.4 | Aggregated results of literature review of LCAs of GHG emissions from electricity generation technologies as displayed in Figure 9.8 (g CO_2eq/kWh).

Values	Bio-power	Solar		Geothermal Energy	Hydropower	Ocean Energy	Wind Energy	Nuclear Energy	Natural Gas	Oil	Coal
		PV	CSP								
Minimum	-633	5	7	6	0	2	2	1	290	510	675
25th percentile	360	29	14	20	3	6	8	8	422	722	877
50th percentile	18	46	22	45	4	8	12	16	469	840	1001
75th percentile	37	80	32	57	7	9	20	45	548	907	1130
Maximum	75	217	89	79	43	23	81	220	930	1170	1689
CCS min	-1368								65		98
CCS max	-594								245		396

Note: CCS = Carbon capture and storage, PV = Photovoltaic, CSP = Concentrating solar power.

Cuperus, M.A.T. (2003). *Biomass Systems: Final Report.* Environmental and Ecological Life Cycle Inventories for Present and Future Power Systems in Europe (ECLIPSE): N.V. tot Keuring van Electrotechnische Materialen (KEMA) Nederland B.V., Arnhem, The Netherlands, 83 pp.

Damen, K., and A.P.C. Faaij (2003). *A Life Cycle Inventory of Existing Biomass Import Chains for "Green" Electricity Production.* NW&S-E-2003-1, Universiteit Utrecht Copernicus Institute, Department of Science, Technology and Society, Utrecht, The Netherlands, 76 pp.

Daugherty, E.C. (2001). *Biomass Energy Systems Efficiency: Analyzed Through a Life Cycle Assessment.* M.S. Thesis, Lund University, Lund, Sweden, 39 pp.

Dones, R., C. Bauer, R. Bolliger, B. Burger, T. Heck, A. Roder, M.F. Emenegger, R. Frischknecht, N. Jungbluth, and M. Tuchschmid (2007). *Life Cycle Inventories of Energy Systems: Results for Current Systems in Switzerland and Other UCTE Countries.* Ecoinvent Report No. 5, Paul Scherrer Institute, Swiss Centre for Life Cycle Inventories, Villigen, Switzerland, 185 pp. Available at: www.ecolo.org/documents/documents_in_english/Life-cycle-analysis-PSI-05.pdf.

Dowaki, K., H. Ishitani, R. Matsuhashi, and N. Sam (2002). A comprehensive life cycle analysis of a biomass energy system. *Technology,* **8**(4-6), pp. 193-204.

Dowaki, K., S. Mori, H. Abe, P.F. Grierson, M.A. Adams, N. Sam, P. Nimiago, J. Gale, and Y. Kaya (2003). A life cycle analysis of biomass energy system tanking [sic] sustainable forest management into consideration. In: *Greenhouse Gas Control Technologies – 6th International Conference,* Kyoto, Japan, 1-4 October 2002. Pergamon, Oxford, pp. 1383-1388.

Dubuisson, X., and I. Sintzoff (1998). Energy and CO_2 balances in different power generation routes using wood fuel from short rotation coppice. *Biomass & Bioenergy,* **15**(4-5), pp. 379-390.

Elsayed, M.A., R. Matthews, and N.D. Mortimer (2003). *Carbon and Energy Balances for a Range of Biofuel Options.* Resources Research Institute, Sheffield Hallam University, Sheffield, UK, 341 pp.

European Commission (1999). National Implementation. *ExternE: Externalities of Energy.* European Commission, Directorate-General XII, Luxembourg, **20**, 534 pp.

Faaij, A., B. Meuleman, W. Turkenburg, A. van Wijk, B. Ausilio, F. Rosillo-Calle, and D. Hall (1998). Externalities of biomass based electricity production compared with power generation from coal in the Netherlands. *Biomass and Bioenergy,* **14**(2), pp. 125-147.

Faix, A., J. Schweinle, S. Scholl, G. Becker, and D. Meier (2010). (GTI-tcbiomass) life-cycle assessment of the BTO-Process (biomass-to-oil) with combined heat and power generation. *Environmental Progress and Sustainable Energy,* **29**(2), pp. 193-202.

Forsberg, G. (2000). Biomass energy transport – Analysis of bioenergy transport chains using life cycle inventory method. *Biomass & Bioenergy,* **19**(1), pp. 17-30.

Froese, R.E., D.R. Shonnard, C.A. Miller, K.P. Koers, and D.M. Johnson (2010). An evaluation of greenhouse gas mitigation options for coal-fired power plants in the US Great Lakes states. *Biomass and Bioenergy,* **34**(3), pp. 251-262.

Gaunt, J.L., and J. Lehmann (2008). Energy balance and emissions associated with biochar sequestration and pyrolysis bioenergy production. *Environmental Science & Technology,* **42**(11), pp. 4152-4158.

Gmünder, S.M., R. Zah, S. Bhatacharjee, M. Classen, P. Mukherjee, and R. Widmer (2010). Life cycle assessment of village electrification based on straight Jatropha oil in Chhattisgarh, India. *Biomass and Bioenergy,* **34**(3):347-355.

Hanaoka, T., and S.-Y. Yokoyama (2003). CO_2 mitigation by biomass-fired power generation in Japan. *International Energy Journal,* **4**(2), pp. 99-103.

Hartmann, D., and M. Kaltschmitt (1999). Electricity generation from solid biomass via co-combustion with coal - Energy and emission balances from a German case study. *Biomass & Bioenergy,* **16**(6), pp. 397-406.

Heller, M.C., G.A. Keoleian, M.K. Mann, and T.A. Volk (2004). Life cycle energy and environmental benefits of generating electricity from willow biomass. *Renewable Energy,* **29**(7), pp. 1023-1042.

Herrera, I., C. Lago, Y. Lechon, R. Saez, M. Munarriz, and J. Gil (2008). Life cycle assessment of two biomass power generation plants. In: *16th European Biomass Conference & Exhibition,* Valencia, Spain, 2-6 June 2008, pp. 2606-2613.

Hong, S.W. (2007). *The Usability of Switchgrass, Rice Straw, and Logging Residue as Feedstocks for Power Generation in East Texas.* M.S. Thesis, Texas A&M University, College Station, TX, USA, 83 pp.

IEA (2002). *Environmental and Health Impacts of Electricity Generation. A Comparison of the Environmental Impacts of Hydropower with those of Other Generation Technologies.* International Energy Agency (IEA), Paris, France, 239 pp.

Jungmeier, G., and J. Spitzer (2001). Greenhouse gas emissions of bioenergy from agriculture compared to fossil energy for heat and electricity supply. *Nutrient Cycling in Agroecosystems,* **60**(1-3), pp. 267-273.

Jungmeier, G., J. Spitzer, and G. Resch (1998). Environmental burdens over the entire life cycle of a biomass CHP plant. *Biomass and Bioenergy*, **15**(4-5), pp. 311-323.

Lettens, S., B. Muys, R. Ceulemans, E. Moons, J. Garcia, and P. Coppin (2003). Energy budget and greenhouse gas balance evaluation of sustainable coppice systems for electricity production. *Biomass and Bioenergy*, **24**(3), pp. 179-197.

Ma, X., F. Li, Z. Zhao, C. Wu, and Y. Chen (2003). Life cycle assessment on biomass gasification combined cycle and coal fired power plant. In: *Energy and the Environment – Proceedings of the International Conference on Energy and the Environment*, Shanghai, China, 22-24 May, 2003. Shanghai Scientific and Technical Publishers, Shanghai, China, **1**, pp. 209-214.

Malkki, H., and Y. Virtanen (2003). Selected emissions and efficiencies of energy systems based on logging and sawmill residues. *Biomass and Bioenergy*, **24**, pp. 321-327.

Mann, M.K., and P.L. Spath (1997). *Life Cycle Assessment of a Biomass Gasification Combined-Cycle System*. NREL/TP-430-23076, National Renewable Energy Laboratory, Golden, CO, USA, 157 pp.

Mann, M.K., and P.L. Spath (2001). A life-cycle assessment of biomass cofiring in a coal-fired power plant. *Clean Products and Processes*, **3**, pp. 81-91.

Mohan, T. (2005). *An Integrated Approach for Techno-economic and Environmental Analysis of Energy from Biomass and Fossil Fuels*. M.S. Thesis, Texas A&M University, College Station, TX, USA, 200 pp.

Pehnt, M. (2006). Dynamic life cycle assessment (LCA) of renewable energy technologies. *Renewable Energy*, **31**(1), pp. 55-71.

Rafaschieri, A., M. Rapaccini, and G. Manfrida (1999). Life cycle assessment of electricity production from poplar energy crops compared with conventional fossil fuels. *Energy Conversion and Management*, **40**(14), pp. 1477-1493.

Ramjeawon, T. (2008). Life cycle assessment of electricity generation from bagasse in Mauritius. *Journal of Cleaner Production*, **16**(16), pp. 1727-1734.

Renouf, M.A. (2002). *Preliminary LCA of Electricity Generation from Sugarcane Bagasse*. Environmental Energy Center, University of Queensland, Queensland, Australia, 10 pp. Available at: www.docstoc.com/docs/39528266/PRELIMINARY-LCA-OF-ELECTRICITY-GENERATION-FROM-SUGARCANE-BAGASSE.

Robertson, K. (2003). *Greenhouse Gas Benefits of a Combined Heat and Power Bioenergy System in New Zealand*. FORCE Consulting, Kirkland, WA, USA, 16 pp. Available at: www.ieabioenergy-task38.org/projects/task38casestudies/nz_full-report.pdf.

Saskatchewan Energy Conservation and Development Authority (1994). *Levelized Cost and Full Fuel Cycle Environmental Impacts of Saskatchewan's Electric Supply Options*. SECDA Publication No. T800-94-004, Saskatoon, SK, Canada, 205 pp.

Schaffner, B., K. Persson, U. Nilsson, and J. Peterson (2002). *Environmental and Health Impacts of Electricity Generation. A Comparison of the Environmental Impacts of Hydropower with Those of Other Generation Technologies*. International Energy Agency (IEA), Paris, France, 221 pp. Available at: www.ieahydro.org/reports/ST3-020613b.pdf.

Searcy, E., and P. Flynn (2008). Processing of straw/corn stover: Comparison of life cycle emissions. *International Journal of Green Energy*, **5**(6), pp. 423-437.

Setterwall, C., M. Munter, P. Sarkozi, and B. Bodlund (2003). *Bio-fuelled Combined Heat and Power Systems. Environmental and Ecological Life Cycle Inventories for Present and Future Power Systems in Europe* (ECLIPSE). N.V. tot Keuring van Electrotechnische Materialen (KEMA) Nederland B.V., Arnhem, The Netherlands.

Sikkema, R., M. Junginger, W. Pichler, S. Hayes, and A.P.C. Faaij (2010). The international logistics of wood pellets for heating and power production in Europe: Costs, energy-input and greenhouse gas balances of pellet consumption in Italy, Sweden and the Netherlands. *Biofuels, Bioproducts and Biorefining*, **4**(2), pp. 132-153.

Spath, P.L., and M.K. Mann (2004). *Biomass Power and Conventional Fossil Systems with and without CO_2 Sequestration – Comparing the Energy Balance, Greenhouse Gas Emissions and Economics*. NREL/TP-510-32575. National Renewable Energy Laboratory, Golden, CO, USA, 28 pp.

Styles, D., and M.B. Jones (2007). Energy crops in Ireland: Quantifying the potential life-cycle greenhouse gas reductions of energy-crop electricity. *Biomass & Bioenergy*, **31**(11-12), pp. 759-772.

Tiwary, A., and J. Colls (2010). Mitigating secondary aerosol generation potentials from biofuel use in the energy sector. *Science of the Total Environment*, **408**(3), pp. 607-616.

Wibberley, L. (2001). *Coal in a Sustainable Society*. Australian Coal Association Research Program, Brisbane, Queensland, Australia.

Wibberley, L., J. Nunn, A. Cottrell, M. Searles, A. Urfer, and P. Scaife (2000). *Life Cycle Analysis for Steel and Electricity Production in Australia*. Australian Coal Association Research Program, Brisbane, Queensland, Australia, 36 pp.

Wicke, B., V. Dornburg, M. Junginger, and A. Faaij (2008). Different palm oil production systems for energy purposes and their greenhouse gas implications. *Biomass and Bioenergy*, **32**(12), pp. 1322-1337.

Yoshioka, T., K. Aruga, T. Nitami, H. Kobayashi, and H. Sakai (2005). Energy and carbon dioxide (CO_2) balance of logging residues as alternative energy resources: System analysis based on the method of a life cycle inventory (LCI) analysis. *Journal of Forest Research*, **10**(2), pp. 125-134.

Zhang, Y.M., S. Habibi, and H.L. MacLean (2007). Environmental and economic evaluation of bioenergy in Ontario, Canada. *Journal of the Air and Waste Management Association*, **57**(8), pp. 919-933.

Coal-fired power generation (52)

Akai, M., N. Nomura, H. Waku, and M. Inoue (1997). Life-cycle analysis of a fossil-fuel power plant with CO_2 recovery and a sequestering system. *Energy*, **22**(2-3), pp. 249-256.

Bates, J.L. (1995). *Full Fuel Cycle Atmospheric Emissions and Global Warming Impacts from UK Electricity Generation*. Report Number: ETSU-R-88, Energy Technical Support Unit (ETSU), London, UK, 51 pp. (ISBN 011 515 4027).

Corrado, A., P. Fiorini, and E. Sciubba (2006). Environmental assessment and extended exergy analysis of a "Zero CO_2 Emission," high-efficiency steam power plant. *Energy*, **31**(15), pp. 3186-3198.

Cottrell, A., J. Nunn, A. Urfer, and L. Wibberley (2003). *Systems Assessment of Electricity Generation Using Biomass and Coal in CFBC*. Cooperative Research Centre for Coal in Sustainable Development, Pullenvale, Qld., Australia, 21 pp.

Damen, K., and A.P.C. Faaij (2003). *A Life Cycle Inventory of Existing Biomass Import Chains for "Green" Electricity Production*. NW&S-E-2003-1, Universiteit Utrecht Copernicus Institute, Department of Science, Technology and Society, Utrecht, The Netherlands, 76 pp.

Dolan, S.L. (2007). *Life Cycle Assessment and Emergy Synthesis of a Theoretical Offshore Wind Farm for Jacksonville, Florida*. M.S. Thesis, University of Florida, 125 pp.

Dones, R., U. Ganter, and S. Hirschberg (1999). Environmental inventories for future electricity supply systems for Switzerland. *International Journal of Global Energy Issues*, **12**(1-6), pp. 271-282.

Dones, R., X. Zhou, and C. Tian (2004). Life cycle assessment (LCA) of Chinese energy chains for Shandong electricity scenarios. *International Journal of Global Energy Issues*, **22**(2/3/4), pp. 199-224.

Dones, R., C. Bauer, R. Bolliger, B. Burger, T. Heck, A. Roder, M.F. Emenegger, R. Frischknecht, N. Jungbluth, and M. Tuchschmid (2007). *Life Cycle Inventories of Energy Systems: Results for Current Systems in Switzerland and Other UCTE Countries*. Ecoinvent Report No. 5, Paul Scherrer Institute, Swiss Centre for Life Cycle Inventories, Villigen, Switzerland, 185 pp. Available at: www.ecolo.org/documents/documents_in_english/Life-cycle-analysis-PSI-05.pdf.

Dones, R., C. Bauer, T. Heck, O. Mayer-Spohn, and M. Blesl (2008). Life cycle assessment of future fossil technologies with and without carbon capture and storage. *Life-Cycle Analysis for New Energy Conversion and Storage Systems*, **1041**, pp. 147-158.

European Commission (1995). Coal & Lignite. *ExternE: Externalities of Energy*. Luxembourg, European Commission, Directorate-General XII. **3**, 573 pp.

European Commission (1999). National Implementation. *ExternE: Externalities of Energy*. Luxembourg, European Commission, Directorate-General XII. **20**, 534 pp.

Fiaschi, D., and L. Lombardi (2002). Integrated gasifier combined cycle plant with integrated CO_2 - H_2S removal: Performance analysis, life cycle assessment and exergetic life cycle assessment. *International Journal of Applied Thermodynamics*, **5**(1), pp. 13-24.

Friedrich, R., and T. Marheineke (1996). Life cycle analysis of electric systems: Methods and results. In: *IAEA Advisory Group Meeting on Analysis of Net Energy Balance and Full-energy-chain Greenhouse Gas Emissions for Nuclear and Other Energy Systems*, Beijing, China, 7 October 1994, International Atomic Energy Agency, pp. 67-75. Available at: www.iaea.org/inis/collection/NCLCollectionStore/_Public/28/013/28013414.pdf.

Froese, R.E., D.R. Shonnard, C.A. Miller, K.P. Koers, and D.M. Johnson (2010). An evaluation of greenhouse gas mitigation options for coal-fired power plants in the US Great Lakes States. *Biomass and Bioenergy*, **34**(3), pp. 251-262.

Gorokhov, V., L. Manfredo, M. Ramezan, J. Ratafia-Brown (2000). *Life Cycle Assessment of IGCC*. Systems Phase II Report, Science Applications International Corporation (SAIC), McLean, VA, USA, 162 pp.

Hartmann, D., and M. Kaltschmitt (1999). Electricity generation from solid biomass via co-combustion with coal - Energy and emission balances from a German case study. *Biomass & Bioenergy*, **16**(6), pp. 397-406.

Heller, M.C., G.A. Keoleian, M.K. Mann, and T.A. Volk (2004). Life cycle energy and environmental benefits of generating electricity from willow biomass. *Renewable Energy*, **29**(7), pp. 1023-1042.

Herrick, C.N., A. Sikri, L. Greene and J. Finnell (1995). *Assessment of the Environmental Benefits of Renewables Deployment: A Total Fuel Cycle Analysis of the Greenhouse Gas Impacts of Renewable Generation Technologies in Regional Utility Systems*. DynCorp EENSP, Inc, Alexandria, VA, USA.

Hondo, H. (2005). Life cycle GHG emission analysis of power generation systems: Japanese case. *Energy*, **30**(11-12), pp. 2042-2056.

Jaramillo, P., W.M. Griffin, and H.S. Matthews (2006). *Comparative Life Cycle Carbon Emissions of LNG Versus Coal and Gas for Electricity Generation*, no publisher given, 16 pp. Available at: www.ce.cmu.edu/~gdrg/readings/2005/10/12/Jaramillo_LifeCycleCarbonEmissionsFromLNG.pdf.

Koornneef, J., T. van Keulen, A. Faaij, and W. Turkenburg (2008). Life cycle assessment of a pulverized coal power plant with post-combustion capture, transport and storage of CO_2. *International Journal of Greenhouse Gas Control*, **2**(4), pp. 448-467.

Kreith, F., P. Norton, and D. Brown (1990). CO_2 *Emissions from Coal-fired and Solar Electric Power Plants*. Solar Energy Research Institute (SERI), Golden, CO, USA, 44 pp.

Krewitt, W., P. Mayerhofer, R. Friedrich, A. Truckenmüller, T. Heck, A. Gressmann, F. Raptis, F. Kaspar, J. Sachau, K. Rennings, J. Diekmann, and B. Praetorius (1997). *ExternE National Implementation in Germany*. University of Stuttgart, Stuttgart, Germany, 189 pp.

Lee, K.-M., S.-Y. Lee, and T. Hur (2004). Life cycle inventory analysis for electricity in Korea. *Energy*, **29**(1), pp. 87-101.

Lee, R. (1994). Estimating externalities of coal fuel cycles. In: *External Costs and Benefits of Fuel Cycles, Vol. 3*. Oak Ridge National Laboratory, Oak Ridge, TN, USA, 719 pp.

Lenzen, M. (2008). Life cycle energy and greenhouse gas emissions of nuclear energy: A review. *Energy Conversion and Management*, **49**, pp. 2178-2199. Available at: www.isa.org.usyd.edu.au/publications/documents/ISA_Nuclear_Report.pdf.

Markewitz, P., A. Schreiber, S. Vögele, and P. Zapp (2009). Environmental impacts of a German CCS strategy. *Energy Procedia*, **1**(1), pp. 3763-3770.

Martin, J.A. (1997). A total fuel cycle approach to reducing greenhouse gas emissions: Solar generation technologies as greenhouse gas offsets in U.S. utility systems. In: *Solar Energy (Selected Proceeding of ISES 1995: Solar World Congress. Part IV)*, **59**(4-6), pp. 195-203.

May, J.R. and D.J. Brennan (2003). Life cycle assessment of Australian fossil energy options. *Process Safety and Environmental Protection: Transactions of the Institution of Chemical Engineers, Part B*, **81**(5), pp. 317-330.

Meier, P.J., P.P.H. Wilson, G.L. Kulcinski, and P.L. Denholm (2005). US electric industry response to carbon constraint: A life-cycle assessment of supply side alternatives. *Energy Policy*, **33**(9), pp. 1099-1108.

Meridian Corporation (1989). *Energy System Emissions and Materiel Requirements*. Meridian Corporation, Alexandria, VA, USA, 34 pp.

Odeh, N.A. and T.T. Cockerill (2008). Life cycle analysis of UK coal fired power plants. *Energy Conversion and Management*, **49**(2), pp. 212-220.

Odeh, N.A. and T.T. Cockerill (2008). Life cycle GHG assessment of fossil fuel power plants with carbon capture and storage. *Energy Policy*, **36**(1), pp. 367-380.

Pacca, S.A. (2003). *Global Warming Effect Applied to Electricity Generation Technologies*. PhD Thesis, University of California, Berkeley, CA, USA, 191 pp.

Peiu, N. (2007). Life cycle inventory study of the electrical energy production in Romania. *International Journal of Life Cycle Assessment*, **12**(4), pp. 225-229.

Ruether, J.A., M. Ramezan, and P.C. Balash (2004). Greenhouse gas emissions from coal gasification power generation systems. *Journal of Infrastructure Systems*, **10**(3), pp. 111-119.

San Martin, R.L. (1989). *Environmental Emissions from Energy Technology Systems: The Total Fuel Cycle*. U.S. Department of Energy, Washington, DC, USA, 21 pp.

Saskatchewan Energy Conservation and Development Authority (1994). *Levelized Cost and Full Fuel Cycle Environmental Impacts of Saskatchewan's Electric Supply Options*. SECDA Publication No. T800-94-004, Saskatoon, SK, Canada, 205 pp.

Schreiber, A., P. Zapp, and W. Kuckshinrichs (2009). Environmental assessment of German electricity generation from coal-fired power plants with amine-based carbon capture. *International Journal of Life Cycle Assessment*, **14**(6), pp. 547-559.

SENES Consultants Limited (2005). *Methods to Assess the Impacts on the Natural Environment of Generation Options.* Prepared by SENES Consultants for the Ontario Power Authority, Richmond Hill, ON, Canada, 166 pp.

Shukla, P.R. and D. Mahapatra (2007). Full Fuel Cycle for India. In: *CASES: Cost Assessment of Sustainable Energy Systems.* Document No. 7.1, Indian Institute of Management Ahmedabad (IIMA), Vestrapur, Ahemdabad, India, 10 pp.

Spath, P.L., and M.K. Mann (2004). *Biomass Power and Conventional Fossil Systems with and without CO_2 Sequestration – Comparing the Energy Balance, Greenhouse Gas Emissions and Economics.* NREL/TP-510-32575. National Renewable Energy Laboratory, Golden, CO, USA, 28 pp.

Spath, P.L., M.K. Mann, and D.R. Kerr (1999). *Life Cycle Assessment of Coal Fired Power Production.* National Renewable Energy Laboratory, Golden, CO, USA, 172 pp.

Styles, D., and M.B. Jones (2007). Energy crops in Ireland: Quantifying the potential life-cycle greenhouse gas reductions of energy-crop electricity. *Biomass & Bioenergy*, **31**(11-12), pp. 759-772.

Uchiyama, Y. (1996). Validity of FENCH-GHG study: Methodologies and databases. comparison of energy sources in terms of their full-energy-chain emission factors of greenhouse gases. In: *IAEA Advisory Group Meeting on Analysis of Net Energy Balance and Full-energy-chain Greenhouse Gas Emissions for Nuclear and Other Energy Systems*, Beijing, China, 4-7 Oct 1994, International Atomic Energy Agency (IAEA), pp. 85-94. Available at: www.iaea.org/inis/collection/NCLCollectionStore/_Public/28/013/28013414.pdf.

White, S.W. (1998). *Net Energy Payback and CO_2 Emissions from Helium-3 Fusion and Wind Electrical Power Plants.* PhD Thesis, University of Wisconsin, Madison, WI, USA, 166 pp.

Wibberley, L. (2001). *Coal in a Sustainable Society.* Australian Coal Association Research Program, Brisbane, Queensland, Australia.

Wibberley, L., J. Nunn, A. Cottrell, M. Searles, A. Urfer, and P. Scaife (2000). *Life Cycle Analysis for Steel and Electricity Production in Australia.* Australian Coal Association Research Program, Brisbane, Queensland, Australia, 36 pp.

Zerlia, T. (2003). Greenhouse gases in the life cycle of fossil fuels: Critical points in the assessment of pre-combustion emissions and repercussions on the complete life cycle. *La Rivista dei Combustibili*, **57**(6), pp. 281-293.

Zhang, Y.M., S. Habibi, and H.L. MacLean (2007). Environmental and economic evaluation of bioenergy in Ontario, Canada. *Journal of the Air and Waste Management Association*, **57**(8), pp. 919-933.

Zhang, Y.M., J. McKechnie, D. Cormier, R. Lyng, W. Mabee, A. Ogino, and H.L. MacLean (2010). Life cycle emissions and cost of producing electricity from coal, natural gas, and wood pellets in Ontario, Canada. *Environmental Science & Technology*, **44**(1), pp. 538-544.

Concentrating solar power (13)

Burkhardt, J., G. Heath, and C. Turchi (2010). Life cycle assessment of a model parabolic trough concentrating solar power plant with thermal energy storage. In: *ASME 4th International Conference on Energy Sustainability*, American Society of Mechanical Engineers (ASME), Phoenix, AZ, USA, 17-22 May 2010.

Cavallaro, F., and L. Ciraolo (2006). Life Cycle Assessment (LCA) of Paraboloidal-dish Solar Thermal Power Generation System. In: *1st International Symposium on Environment Identities and Mediterranean Area, ISEIM*, IEEE, Corte-Ajaccio, France, 10-13 July 2006, pp. 260-265.

German Aerospace Center (DLR) (2006). *Trans-Mediterranean Interconnection for Concentrating Solar Power. Final Report.* Institute of Technical Thermodynamics, and Section Systems Analysis and Technology Assessment, German Aerospace Center (DLR), Stuttgart, Germany, 190 pp.

Jacobson, M.Z. (2009). Review of solutions to global warming, air pollution, and energy security. *Energy & Environmental Science*, **2**, pp. 148-173.

Kreith, F., P. Norton, and D. Brown (1990). CO_2 *Emissions from Coal-fired and Solar Electric Power Plants.* SERI/TP-260-3772, Solar Energy Research Institute (SERI), Golden, CO, USA, 44 pp.

Lenzen, M. (1999). Greenhouse gas analysis of solar-thermal electricity generation. *Solar Energy*, **65**(6), pp. 353-368.

Ordóñez, I., N. Jiménez, and M.A. Silva (2009). Life cycle environmental impacts of electricity production by dish/Stirling systems in Spain. In: *SolarPACES 2009*, Berlin, Germany, 15-18 September 2009, 8 pp.

Pehnt, M. (2006). Dynamic life cycle assessment (LCA) of renewable energy technologies. *Renewable Energy*, **31**(1), pp. 55-71.

Piemonte, V., M.D. Falco, P. Tarquini, and A. Giaconia (2010). Life cycle assessment of a high temperature molten salt concentrated solar power plant. In: *20th European Symposium on Computer Aided Process Engineering – ESCAPE20*, Pierucci, S., and G.B. Ferraris (eds.), Elsevier, Naples, Italy, 6-9 June 2010, 6 pp.

Vant-Hull, L. (1992). Solar thermal electricity: An environmentally benign and viable alternative. *Perspectives in Energy*, **2**, pp. 157-166.

Viebahn, P., S. Kronshage, and F. Trieb (2008). *Final Report on Technical Data, Costs, and Life Cycle Inventories of Solar Thermal Power Plants.* Project no: 502687. New Energy Externalities Developments for Sustainability (NEEDS), Rome, Italy, 95 pp. Available at: www.needs-project.org/docs/results/RS1a/RS1a%20D12.2%20Final%20report%20concentrating%20solar%20thermal%20power%20plants.pdf.

Weinrebe, G., M. Bohnke, and F. Trieb (1998). Life cycle assessment of an 80 MW SEGS plant and a 30 MW PHOEBUS power tower. In: *International Solar Energy Conference. Solar Engineering.* ASME, Albuquerque, NM, USA, 14-17 June 1998, pp. 417-424.

Wibberley, L. (2001). *Coal in a Sustainable Society.* Australian Coal Association Research Program, Brisbane, Queensland, Australia.

Geothermal power generation (6)

Frick, S., M. Kaltschmitt, and G. Schroder (2010). Life cycle assessment of geothermal binary power plants using enhanced low-temperature reservoirs. *Energy*, **35**(5), pp. 2281-2294.

Hondo, H. (2005). Life cycle GHG emission analysis of power generation systems: Japanese case. *Energy*, **30**(11-12), pp. 2042-2056.

Karlsdottir, M.R., O.P. Palsson, and H. Palsson (2010). Factors for Primary Energy Efficiency and CO2 Emission of Geothermal Power Production. In: *World Geothermal Congress 2010*, International Geothermal Association, Bali, Indonesia, 25-29 April 2010, 7 pp.

Rogge, S., and M. Kaltschmitt (2003). Electricity and heat production from geothermal energy – An ecologic comparison. *Erdoel Erdgas Kohle/EKEP*, **119**(1), pp. 35-40.

Rule, B.M., Z.J. Worth, and C.A. Boyle (2009). Comparison of life cycle carbon dioxide emissions and embodied energy in four renewable electricity generation technologies in New Zealand. *Environmental Science & Technology*, **43**(16), pp. 6406-6413.

Uchiyama, Y. (1997). Environmental life cycle analysis of geothermal power generating technology; Chinetsu hatsuden gijutsu no kankyo life cycle bunseki. *Denki Gakkaishi (Journal of the Institute of Electrical Engineers in Japan)*, **117**(11), pp. 752-755.

Hydropower (11)

Barnthouse, L.W., G.F. Cada, M.-D. Cheng, C.E. Easterly, R.L. Kroodsma, R. Lee, D.S. Shriner, V.R. Tolbert, and R.S. Turner (1994). *Estimating Externalities of the Hydro Fuel Cycles. Report 6.* Oak Ridge National Laboratory, Oak Ridge, TN, USA, 205 pp.

Denholm, P., and G.L. Kulcinski (2004). Life cycle energy requirements and greenhouse gas emissions from large scale energy storage systems. *Energy Conversion and Management*, **45**(13-14), pp. 2153-2172.

Dones, R., T. Heck, C. Bauer, S. Hirschberg, P. Bickel, P. Preiss, L.I. Panis, and I. De Vlieger (2005). *Externalities of Energy: Extension of Accounting Framework and Policy Applications: New Energy Technologies.* ENG1-CT-2002-00609, Paul Scherrer Institute (PSI), Villigen, Switzerland, 76 pp.

Dones, R., C. Bauer, R. Bolliger, B. Burger, T. Heck, A. Roder, M.F. Emenegger, R. Frischknecht, N. Jungbluth, and M. Tuchschmid (2007). *Life Cycle Inventories of Energy Systems: Results for Current Systems in Switzerland and Other UCTE Countries.* Ecoinvent Report No. 5, Paul Scherrer Institute, Swiss Centre for Life Cycle Inventories, Villigen, Switzerland, 185 pp. Available at: www.ecolo.org/documents/documents_in_english/Life-cycle-analysis-PSI-05.pdf.

Horvath, A. (2005). *Decision-making in Electricity Generation Based on Global Warming Potential and Life-cycle Assessment for Climate Change.* University of California Energy Institute, Berkeley, CA, USA, 16 pp. Available at: repositories.cdlib.org/ucei/devtech/EDT-006.

IEA (1998). *Benign Energy? The Environmental Implications of Renewables.* International Energy Agency, Paris, France, 128 pp.

Pacca, S. (2007). Impacts from decommissioning of hydroelectric dams: A life cycle perspective. *Climatic Change*, **84**(3-4), pp. 281-294.

Rhodes, S., J. Wazlaw, C. Chaffee, F. Kommonen, S. Apfelbaum, and L. Brown (2000). *A Study of the Lake Chelan Hydroelectric Project Based on Life-cycle Stressor-effects Assessment. Final Report.* Scientific Certification Systems, Oakland, CA, USA, 193 pp.

Ribeiro, F.d.M., and G.A. da Silva (2009). Life-cycle inventory for hydroelectric generation: a Brazilian case study. *Journal of Cleaner Production*, **18**(1), pp. 44-54.

Vattenfall (2008). *Vattenfall AB Generation Nordic Certified Environmental Product Declaration EPD® of Electricity from Vattenfall's Nordic Hydropower.* Report No. S-P-00088, Vattenfall, Stockholm, Sweden, 50 pp.

Zhang, Q., B. Karney, H.L. MacLean, and J. Feng (2007). Life-Cycle Inventory of Energy Use and Greenhouse Gas Emissions for Two Hydropower Projects in China. *Journal of Infrastructure Systems*, **13**(4), pp. 271-279.

Natural gas-fired power generation (40)

Audus, H., and L. Saroff (1995). Full Fuel Cycle Evaluation of CO2 Mitigation Options for Fossil Fuel Fired Power Plant. *Energy Conversion and Management*, **36**(6-9), pp. 831-834.

Badea, A.A., I. Voda, and C.F. Dinca (2010). Comparative Analysis of Coal, Natural Gas and Nuclear Fuel Life Cycles by Chains of Electrical Energy Production. *UPB Scientific Bulletin, Series C: Electrical Engineering*, **72**(2), pp. 221-238.

Bergerson, J., and L. Lave (2007). The Long-term Life Cycle Private and External Costs of High Coal Usage in the US. *Energy Policy*, **35**(12), pp. 6225-6234.

Bernier, E., F. Maréchal, and R. Samson (2010). Multi-Objective Design Optimization of a Natural Gas-combined Cycle with Carbon Dioxide Capture in a Life Cycle Perspective. *Energy*, **35**(2), pp. 1121-1128.

Berry, J.E., M.R. Holland, P.R. Watkiss, R. Boyd, and W. Stephenson (1998). *Power Generation and the Environment: a UK Perspective.* AEA Technology, Oxfordshire, UK, 275 pp.

Dolan, S.L. (2007). *Life Cycle Assessment and Emergy Synthesis of a Theoretical Offshore Wind Farm for Jacksonville, Florida.* M.S. Thesis, University of Florida, 125 pp. Available at: http://etd.fcla.edu/UF/UFE0021032/dolan_s.pdf.

Dones, R., S. Hirschberg, and I. Knoepfel (1996). Greenhouse gas emission inventory based on full energy chain analysis. In: *IAEA Advisory Group Meeting on Analysis of Net Energy Balance and Full-energy-chain Greenhouse Gas Emissions for Nuclear and Other Energy Systems.* Beijing, China, 4-7 October 1994, pp. 95-114. Available at: www.iaea.org/inis/collection/NCLCollectionStore/_Public/28/013/28013414.pdf.

Dones, R., U. Ganter, and S. Hirschberg (1999). Environmental inventories for future electricity supply systems for Switzerland. *International Journal of Global Energy Issues*, **12**(1-6), pp. 271-282.

Dones, R., T. Heck, and S. Hirschberg (2004). Greenhouse gas emissions from energy systems, comparison and overview. *Encyclopedia of Energy*, **3**, pp. 77-95, doi:10.1016/B0-12-176480-X/00397-1.

Dones, R., X. Zhou, and C. Tian (2004). Life cycle assessment (LCA) of Chinese energy chains for Shandong electricity scenarios. *International Journal of Global Energy Issues*, **22**(2/3/4), pp. 199-224.

Dones, R., T. Heck, C. Bauer, S. Hirschberg, P. Bickel, P. Preiss, L.I. Panis, and I. De Vlieger (2005). *Externalities of Energy: Extension of Accounting Framework and Policy Applications: New Energy Technologies.* ENG1-CT-2002-00609, Paul Scherrer Institute (PSI), Villigen, Switzerland, 76 pp.

Dones, R., C. Bauer, R. Bolliger, B. Burger, T. Heck, A. Roder, M.F. Emenegger, R. Frischknecht, N. Jungbluth, and M. Tuchschmid (2007). *Life Cycle Inventories of Energy Systems: Results for Current Systems in Switzerland and Other UCTE Countries.* Ecoinvent Report No. 5, Paul Scherrer Institute, Swiss Centre for Life Cycle Inventories, Villigen, Switzerland, 185 pp. Available at: www.ecolo.org/documents/documents_in_english/Life-cycle-analysis-PSI-05.pdf.

European Commission (1995). *Oil & Gas. ExternE: Externalities of Energy.* European Commission, Directorate-General XII, Luxembourg, **4**, 470 pp.

Frischknecht, R. (1998). *Life Cycle Inventory Analysis for Decision-Making: Scope-Dependent Inventory System Models and Context-Specific Joint Product Allocation.* Dissertation, Swiss Federal Institute of Technology Zurich, Zurich, Switzerland, 256 pp.

Gantner, U., M. Jakob, and S. Hirschberg (2001). Total greenhouse gas emissions and costs of alternative Swiss energy supply strategies. In: *Fifth International Conference on Greenhouse Gas Control Technologies (GHGT-5).* CSIRO Publishing, Cairns, Australia, 13-16 August 2000, pp. 991-996.

Herrick, C.N., A. Sikri, L. Greene, and J. Finnell (1995). *Assessment of the Environmental Benefits of Renewables Deployment: A Total Fuel Cycle Analysis of the Greenhouse Gas Impacts of Renewable Generation Technologies in Regional Utility Systems.* DynCorp EENSP, Inc., Alexandria, VA, USA.

Hondo, H. (2005). Life cycle GHG emission analysis of power generation systems: Japanese case. *Energy,* 30(11-12), pp. 2042-2056.

IEA (2002). *Environmental and Health Impacts of Electricity Generation. A Comparison of the Environmental Impacts of Hydropower with those of Other Generation Technologies.* International Energy Agency (IEA), Paris, France, 239 pp. Available at: www.ieahydro.org/reports/ST3-020613b.pdf.

Kannan, R., K.C. Leong, R. Osman, and H.K. Ho (2007). Life cycle energy, emissions and cost inventory of power generation technologies in Singapore. *Renewable and Sustainable Energy Reviews,* 11, pp. 702-715.

Kato, S., and A. Widiyanto (1999). A life cycle assessment scheme for environmental load estimation of power generation systems with NETS evaluation method. In: *International Joint Power Generation Conference.* S.R.H. Penfield and R. McMullen (eds.). American Society of Mechanical Engineers (ASME), Burlingame, CA, USA, 25-28 July 1999, 2, pp. 139-146.

Krewitt, W., P. Mayerhofer, R. Friedrich, A. Truckenmüller, T. Heck, A. Gressmann, F. Raptis, F. Kaspar, J. Sachau, K. Rennings, J. Diekmann, and B. Praetorius (1997). *ExternE National Implementation in Germany.* University of Stuttgart, Stuttgart, Germany, 189 pp.

Lee, R. (1998). *Estimating Externalities of Natural Gas Fuel Cycles. External Costs and Benefits of Fuel Cycles: A Study by the U.S. Department of Energy and the Commission of the European Communities.* Report No. 4, Oak Ridge National Laboratory and Resources for the Future, Oak Ridge, TN, USA, 440 pp.

Lenzen, M. (1999). Greenhouse gas analysis of solar-thermal electricity generation. *Solar Energy,* 65(6), pp. 353-368.

Lombardi, L. (2003). Life cycle assessment comparison of technical solutions for CO_2 emissions reduction in power generation. *Energy Conversion and Management,* 44(1), pp. 93-108.

Martin, J.A. (1997). A total fuel cycle approach to reducing greenhouse gas emissions: Solar generation technologies as greenhouse gas offsets in U.S. utility systems. *Solar Energy (Selected Proceeding of ISES 1995: Solar World Congress. Part IV),* 59(4-6), pp. 195-203.

Meier, P.J. (2002). *Life-Cycle Assessment of Electricity Generation Systems and Applications for Climate Change Policy Analysis.* PhD Thesis, University of Wisconsin, Madison, WI, USA, 147 pp.

Meier, P.J., and G.L. Kulcinski (2001). The Potential for fusion power to mitigate US greenhouse gas emissions. *Fusion Technology,* 39(2), pp. 507-512.

Meier, P.J., P.P.H. Wilson, G.L. Kulcinski, and P.L. Denholm (2005). US electric industry response to carbon constraint: A life-cycle assessment of supply side alternatives. *Energy Policy,* 33(9), pp. 1099-1108.

Norton, B., P.C. Eames, and S.N.G. Lo (1998). Full-energy-chain analysis of greenhouse gas emissions for solar thermal electric power generation systems. *Renewable Energy,* 15(1-4), pp. 131-136.

Odeh, N.A., and T.T. Cockerill (2008). Life cycle GHG assessment of fossil fuel power plants with carbon capture and storage. *Energy Policy,* 36(1), pp. 367-380.

Pacca, S.A. (2003). *Global Warming Effect Applied to Electricity Generation Technologies.* PhD Thesis, University of California, Berkeley, CA, USA, 191 pp.

Phumpradab, K., S.H. Gheewala, and M. Sagisaka (2009). Life cycle assessment of natural gas power plants in Thailand. *International Journal of Life Cycle Assessment,* 14(4), pp. 354-363.

Raugei, M., S. Bargigli, and S. Ulgiati (2005). A multi-criteria life cycle assessment of molten carbonate fuel cells (MCFC) – A comparison to natural gas turbines. *International Journal of Hydrogen Energy,* 30(2), pp. 123-130.

Riva, A., S. D'Angelosante, and C. Trebeschi (2006). Natural gas and the environmental results of life cycle assessment. *Energy,* 31(1), pp. 138-148.

Saskatchewan Energy Conservation and Development Authority (1994). *Levelized Cost and Full Fuel Cycle Environmental Impacts of Saskatchewan's Electric Supply Options.* SECDA Publication No. T800-94-004, Saskatoon, SK, Canada, 205 pp.

SENES Consultants Limited (2005). *Methods to Assess the Impacts on the Natural Environment of Generation Options.* Prepared by SENES Consultants for the Ontario Power Authority, Richmond Hill, Ontario, Canada, 166 pp.

Spath, P.L., and M.K. Mann (2000). *Life Cycle Assessment of a Natural Gas Combined-Cycle Power Generation System.* NREL/TP-570-27715, National Renewable Energy Laboratory, Golden, CO, USA, 54 pp.

Spath, P.L., and M.K. Mann (2004). *Biomass Power and Conventional Fossil Systems with and without CO_2 Sequestration – Comparing the Energy Balance, Greenhouse Gas Emissions and Economics.* NREL/TP-510-32575. National Renewable Energy Laboratory, Golden, CO, USA, 28 pp.

Uchiyama, Y. (1996). Validity of FENCH-GHG study: Methodologies and databases. comparison of energy sources in terms of their full-energy-chain emission factors of greenhouse gases. In: *IAEA Advisory Group Meeting on Analysis of Net Energy Balance and Full-energy-chain Greenhouse Gas Emissions for Nuclear and Other Energy Systems,* Beijing, China, 4-7 Oct 1994, International Atomic Energy Agency (IAEA), pp. 85-94. Available at: www.iaea.org/inis/collection/NCLCollectionStore/_Public/28/013/28013414.pdf.

World Energy Council (2004). *Comparison of Energy Systems Using Life Cycle Assessment.* World Energy Council, London, UK, 67 pp.

Nuclear power (32)

AEA Technologies (2005). *Environmental Product Declaration of Electricity from Torness Nuclear Power Station.* British Energy, London, UK, 52 pp.

AEA Technologies (2006). *Carbon Footprint of the Nuclear Fuel Cycle.* British Energy, London, UK, 26 pp.

Andseta, S., M.J. Thompson, J.P. Jarrell, and D.R. Pendergast (1998). Candu reactors and greenhouse gas emissions. In: *Canadian Nuclear Society 19th Annual Conference.* D.B. Buss and D.A. Jenkins (eds.), Canadian Nuclear Association, Toronto, Ontario, Canada, 18-21 October 1998.

AXPO Nuclear Energy (2008). *Beznau Nuclear Power Plant.* Axpo AG, Baden, Germany, 21 pp.

Badea, A.A., I. Voda, and C.F. Dinca (2010). Comparative analysis of coal, natural gas and nuclear fuel life cycles by chains of electrical energy production. *UPB Scientific Bulletin, Series C: Electrical Engineering,* 72(2), pp. 221-238.

Beerten, J., E. Laes, G. Meskens, and W. D'haeseleer (2009). Greenhouse gas emissions in the nuclear life cycle: A balanced appraisal. *Energy Policy,* 37(12), pp. 5056-5058.

Dones, R., S. Hirschberg, and I. Knoepfel (1996). Greenhouse gas emission inventory based on full energy chain analysis. In: *IAEA Advisory Group Meeting on Analysis of Net Energy Balance and Full-energy-chain Greenhouse Gas Emissions for Nuclear and Other Energy Systems.* Beijing, China, 4-7 October 1994, pp. 95-114. Available at: www.iaea.org/inis/collection/NCLCollectionStore/_Public/28/013/28013414.pdf.

Dones, R., X. Zhou, and C. Tian (2004). Life cycle assessment (LCA) of Chinese energy chains for Shandong electricity scenarios. *International Journal of Global Energy Issues*, **22**(2/3/4), pp. 199-224.

Dones, R., T. Heck, C. Bauer, S. Hirschberg, P. Bickel, P. Preiss, L.I. Panis, and I. De Vlieger (2005). *Externalities of Energy: Extension of Accounting Framework and Policy Applications: New Energy Technologies.* ENG1-CT-2002-00609, Paul Scherrer Institute (PSI), Villigen, Switzerland, 76 pp.

Dones, R., C. Bauer, R. Bolliger, B. Burger, T. Heck, A. Roder, M.F. Emenegger, R. Frischknecht, N. Jungbluth, and M. Tuchschmid (2007). *Life Cycle Inventories of Energy Systems: Results for Current Systems in Switzerland and Other UCTE Countries.* Ecoinvent Report No. 5, Paul Scherrer Institute, Swiss Centre for Life Cycle Inventories, Villigen, Switzerland, 185 pp. Available at: www.ecolo.org/documents/documents_in_english/Life-cycle-analysis-PSI-05.pdf.

Dones, R., C. Bauer, and T. Heck (2007). *LCA of Current Coal, Gas and Nuclear Electricity Systems and Electricity Mix in the USA.* Paul Scherrer Institute, Villigen, Switzerland, 4 pp.

Frischknecht, R. (1998). *Life Cycle Inventory Analysis for Decision-Making: Scope-Dependent Inventory System Models and Context-Specific Joint Product Allocation.* Dissertation, Swiss Federal Institute of Technology Zurich, Zurich, Switzerland, 256 pp.

Fthenakis, V.M., and H.C. Kim (2007). Greenhouse-gas emissions from solar electric- and nuclear power: A life-cycle study. *Energy Policy*, **35**(4), pp. 2549-2557.

Hondo, H. (2005). Life cycle GHG emission analysis of power generation systems: Japanese case. *Energy*, **30**(11-12), pp. 2042-2056.

Kivisto, A. (1995). Energy payback period and carbon dioxide emissions in different power generation methods in Finland. In: *IAEE International Conference.* International Association for Energy Economics, Washington, D.C., 5-8 July 1995, pp. 191-198.

Krewitt, W., P. Mayerhofer, R. Friedrich, A. Truckenmüller, T. Heck, A. Gressmann, F. Raptis, F. Kaspar, J. Sachau, K. Rennings, J. Diekmann, and B. Praetorius (1997). *ExternE National Implementation in Germany.* University of Stuttgart, Stuttgart, Germany, 189 pp.

Lecointe, C., D. Lecarpentier, V. Maupu, D. Le Boulch, and R. Richard (2007). *Final Report on Technical Data, Costs and Life Cycle Inventories of Nuclear Power Plants.* D14.2 – RS 1a, New Energy Externalities Developments for Sustainability (NEEDS), Rome, Italy, 62 pp. Available at: www.needs-project.org/RS1a/RS1a%20D14.2%20Final%20report%20on%20nuclear.pdf.

Lenzen, M., C. Dey, C. Hardy, and M. Bilek (2006). *Life-cycle Energy Balance and Greenhouse Gas Emissions of Nuclear Energy in Australia.* ISA, University of Sydney, Sydney, Australia, 180 pp.

Meridian Corporation (1989). *Energy System Emissions and Materiel Requirements.* Meridian Corporation, Alexandria, VA, USA, 34 pp.

Rashad, S.M., and F.H. Hammad (2000). Nuclear power and the environment: Comparative assessment of environmental and health impacts of electricity-generating systems. *Applied Energy*, **65**(1-4), pp. 211-229.

San Martin, R.L. (1989). *Environmental Emissions from Energy Technology Systems: The Total Fuel Cycle.* U.S. Department of Energy, Washington, DC, USA, 21 pp.

Saskatchewan Energy Conservation and Development Authority (1994). *Levelized Cost and Full Fuel Cycle Environmental Impacts of Saskatchewan's Electric Supply Options.* SECDA Publication No. T800-94-004, Saskatoon, SK, Canada, 205 pp.

Tokimatsu, K., T. Asami, Y. Kaya, T. Kosugi, and E. Williams (2006). Evaluation of lifecycle CO_2 emissions from the Japanese electric power sector in the 21st century under various nuclear scenarios. *Energy Policy*, **34**(7), pp. 833-852.

Uchiyama, Y. (1996). Validity of FENCH-GHG study: Methodologies and databases. comparison of energy sources in terms of their full-energy-chain emission factors of greenhouse gases. In: *IAEA Advisory Group Meeting on Analysis of Net Energy Balance and Full-energy-chain Greenhouse Gas Emissions for Nuclear and Other Energy Systems*, Beijing, China, 4-7 Oct 1994, International Atomic Energy Agency (IAEA), pp. 85-94. Available at: www.iaea.org/inis/collection/NCLCollectionStore/_Public/28/013/28013414.pdf.

Uchiyama, Y. (1996). Life cycle analysis of electricity generation and supply systems: Net energy analysis and greenhouse gas emissions. In: *Electricity, Health and the Environment: Comparative Assessment in Support of Decision Making*, International Atomic Energy Agency (IAEA), Vienna, Austria, 16-19 October 1995, pp. 279-291.

Vattenfall (2007). *Summary of Vattenfall AB Generation Nordic Certified Environmental Product Declaration, EPD® of Electricity from Ringhals Nuclear Power Plant.* S-P-00026 2007-11-01, Vattenfall, Stockholm, Sweden, 4 pp.

Vattenfall (2007). *Vattenfall AB Generation Nordic Certified Environmental Product Declaration, EPD, of Electricity from Forsmark Nuclear Power Plant.* Report No. S-P-00088, Vattenfall, Stockholm, Sweden, 59 pp.

Voorspools, K.R., E.A. Brouwers, and W.D. D'Haeseleer (2000). Energy content and indirect greenhouse gas emissions embedded in 'emission-free' power plants: Results for the low countries. *Applied Energy*, **67**(3), pp. 307-330.

White, S.W., and G.L. Kulcinski (1999). *'Birth to Death' Analysis of the Energy Payback Ratio and CO_2 Gas Emission Rates from Coal, Fission, Wind, and DT Fusion Power Plants.* University of Wisconsin, Madison, WI, USA, 17 pp.

Wibberley, L. (2001). *Coal in a Sustainable Society.* Australian Coal Association Research Program, Brisbane, Queensland, Australia.

Yasukawa, S., Y. Tadokoro, and T. Kajiyama (1992). Life cycle CO_2 emission from nuclear power reactor and fuel cycle system. In: *Expert Workshop on Life-cycle Analysis of Energy Systems, Methods and Experience.* Paris, France, 21-22 May 1992, pp. 151-160.

Yasukawa, S., Y. Tadokoro, O. Sato, and M. Yamaguchi (1996). Integration of indirect CO_2 emissions from the full energy chain. In: *IAEA Advisory Group Meeting on Analysis of Net Energy Balance and Full-energy-chain Greenhouse Gas Emissions for Nuclear and Other Energy Systems.* Beijing, China, pp. 139-150. Available at: www.iaea.org/inis/collection/NCLCollectionStore/_Public/28/013/28013414.pdf.

Ocean energy (5)

Parker, R.P.M., G.P. Harrison, and J.P. Chick (2008). Energy and carbon audit of an offshore wave energy converter. *Proceedings of the Institution of Mechanical Engineers, Part A: Journal of Power and Energy*, **221**(8), pp. 1119-1130.

Rule, B.M., Z.J. Worth, and C.A. Boyle (2009). Comparison of life cycle carbon dioxide emissions and embodied energy in four renewable electricity generation technologies in New Zealand. *Environmental Science & Technology*, **43**(16), pp. 6406-6413.

Sorensen, H.C., and S. Naef (2008). *Report on Technical Specification of Reference Technologies (Wave and Tidal Power Plant)*. New Energy Externalities Developments for Sustainability (NEEDS), Rome, Italy and SPOK Consult, Kopenhagen, Denmark, 59 pp.

Wibberley, L. (2001). *Coal in a Sustainable Society*. Australian Coal Association Research Program, Brisbane, Queensland, Australia.

Woollcombe-Adams, C., M. Watson, and T. Shaw (2009). Severn Barrage tidal power project: Implications for carbon emissions. *Water and Environment Journal*, **23**(1), pp. 63-68.

Oil-fired power generation (10)

Bates, J.L. (1995). *Full Fuel Cycle Atmospheric Emissions and Global Warming Impacts from UK Electricity Generation*. ETSU, London, UK, 51 pp.

Berry, J.E., M.R. Holland, P.R. Watkiss, R. Boyd, and W. Stephenson (1998). *Power Generation and the Environment: a UK Perspective*. AEA Technology, Oxfordshire, UK, 275 pp.

Dones, R., S. Hirschberg, and I. Knoepfel (1996). Greenhouse gas emission inventory based on full energy chain analysis. In: *IAEA Advisory Group Meeting on Analysis of Net Energy Balance and Full-energy-chain Greenhouse Gas Emissions for Nuclear and Other Energy Systems*. Beijing, China, 4-7 October 1994, pp. 95-114. Available at: www.iaea.org/inis/collection/NCLCollectionStore/_Public/28/013/28013414.pdf.

Dones, R., U. Ganter, and S. Hirschberg (1999). Environmental inventories for future electricity supply systems for Switzerland. *International Journal of Global Energy Issues*, **12**(1-6), pp. 271-282.

Dones, R., T. Heck, C. Bauer, S. Hirschberg, P. Bickel, P. Preiss, L.I. Panis, and I. De Vlieger (2005). *Externalities of Energy: Extension of Accounting Framework and Policy Applications: New Energy Technologies*. ENG1-CT-2002-00609, Paul Scherrer Institute (PSI), Villigen, Switzerland, 76 pp.

Dones, R., C. Bauer, R. Bolliger, B. Burger, T. Heck, A. Roder, M.F. Emenegger, R. Frischknecht, N. Jungbluth, and M. Tuchschmid (2007). *Life Cycle Inventories of Energy Systems: Results for Current Systems in Switzerland and Other UCTE Countries*. Ecoinvent Report No. 5, Paul Scherrer Institute, Swiss Centre for Life Cycle Inventories, Villigen, Switzerland, 185 pp. Available at: www.ecolo.org/documents/documents_in_english/Life-cycle-analysis-PSI-05.pdf.

European Commission (1995). Oil & Gas. *ExternE: Externalities of Energy*. European Commission, Directorate-General XII, Luxembourg, **4**, 470 pp.

Gagnon, L., C. Belanger, and Y. Uchiyama (2002). Life-cycle assessment of electricity generation options: The status of research in year 2001. *Energy Policy*, **30**, pp. 1267-1279.

Hondo, H. (2005). Life cycle GHG emission analysis of power generation systems: Japanese case. *Energy*, **30**(11-12), pp. 2042-2056.

Kannan, R., C.P. Tso, R. Osman, and H.K. Ho (2004). LCA-LCCA of oil fired steam turbine power plant in Singapore. *Energy Conversion and Management*, **45**, pp. 3091-3107.

Solar photovoltaic (26)

Alsema, E.A. (2000). Energy pay-back time and CO_2 emissions of PV systems. *Progress in Photovoltaics*, **8**(1), pp. 17-25.

Alsema, E.A., and M.J. de Wild-Scholten (2006). Environmental Impacts of Crystalline Silicon Photovoltaic Module Production. In: *13th CIRP International Conference on Life Cycle Engineering*, Leuven, Belgium, 31 May - 2 Jun, 2006. Available at: www.mech.kuleuven.be/lce2006/Registration_papers.htm.

Dones, R., T. Heck, and S. Hirschberg (2004). Greenhouse gas emissions from energy systems, comparison and overview. *Encyclopedia of Energy*, **3**, pp. 77-95.

Frankl, P., E. Menichetti, M. Raugei, S. Lombardelli, and G. Prennushi (2005). *Final Report on Technical Data, Costs and Life Cycle Inventories of PV Applications*. Ambiente Italia, Milan, Italy, 81 pp.

Fthenakis, V.M., and E. Alsema (2006). Photovoltaics energy payback times, greenhouse gas emissions and external costs: 2004 - early 2005 status. *Progress in Photovoltaics: Research and Applications*, **14**(3), pp. 275-280.

Fthenakis, V., and H.C. Kim (2006). Energy use and greenhouse gas emissions in the life cycle of CdTe photovoltaics. In: *Life-Cycle Analysis Tools for "Green" Materials and Process Selection, Materials Research Society Symposium 2006*. S. Papasavva and V.M.P.O. Fthenakis (eds.), Materials Research Society, Boston, MA, 28-30 November 2005, **895**, pp. 83-88.

Fthenakis, V.M., and H.C. Kim (2007). Greenhouse-gas emissions from solar electric- and nuclear power: A life-cycle study. *Energy Policy*, **35**(4), pp. 2549-2557.

Garcia-Valverde, R., C. Miguel, R. Martinez-Bejar, and A. Urbina (2009). Life cycle assessment study of a 4.2 kW(p) stand-alone photovoltaic system. *Solar Energy*, **83**(9), pp. 1434-1445.

Graebig, M., S. Bringezu, and R. Fenner (2010). Comparative analysis of environmental impacts of maize-biogas and photovoltaics on a land use basis. *Solar Energy*, **84**(7), pp. 1255-1263.

Greijer, H., L. Karlson, S.E. Lindquist, and A. Hagfeldt (2001). Environmental aspects of electricity generation from a nanocrystalline dye sensitized solar cell system. *Renewable Energy*, **23**(1), pp. 27-39.

Hayami, H., M. Nakamura, and K. Yoshioka (2005). The life cycle CO_2 emission performance of the DOE/NASA solar power satellite system: a comparison of alternative power generation systems in Japan. *IEEE Transactions on Systems, Man, and Cybernetics, Part C: Applications and Reviews*, **35**(3), pp. 391-400.

Hondo, H. (2005). Life cycle GHG emission analysis of power generation systems: Japanese case. *Energy*, **30**(11-12), pp. 2042-2056.

Ito, M., K. Kato, K. Komoto, T. Kichimi, H. Sugihara, and K. Kurokawa (2003). An analysis of variation of very large-scale PV (VLS-PV) systems in the world deserts. In: *3rd World Conference on Photovoltaic Energy Conversion (WCPEC)*. WCPEC, Osaka, Japan, 11-18 May 2003, **C**, pp. 2809-2814.

Kannan, R., K.C. Leong, R. Osman, H.K. Ho, and C.P. Tso (2006). Life cycle assessment study of solar PV systems: An example of a 2.7 kWp distributed solar PV System in Singapore. *Solar Energy*, **80**(5), pp. 555-563.

Lenzen, M., C. Dey, C. Hardy, and M. Bilek (2006). *Life-cycle Energy Balance and Greenhouse Gas Emissions of Nuclear Energy in Australia*. ISA, University of Sydney, Sydney, Australia, 180 pp.

Muneer, T., S. Younes, P. Clarke, and J. Kubie (2006). *Napier University's School of Engineering Life Cycle Assessment of a Medium Sized PV Facility in Edinburgh. EuroSun*. ES06-T10-0171, The Solar Energy Society, Glasgow, 157 pp.

Pacca, S.A. (2003). *Global Warming Effect Applied to Electricity Generation Technologies.* PhD Thesis, University of California, Berkeley, CA, USA, 191 pp.

Pehnt, M. (2006). Dynamic life cycle assessment (LCA) of renewable energy technologies. *Renewable Energy*, **31**(1), pp. 55-71.

Pehnt, M., A. Bubenzer, and A. Rauber (2002). Life cycle assessment of photovoltaic systems–Trying to fight deep-seated prejudices. In: *Photovoltaics Guidebook for Decision Makers.* A. Bubenzer and J. Luther (eds.), Springer, Berlin, Germany, pp. 179-213.

Reich-Weiser, C. (2010). *Decision-Making to Reduce Manufacturing Greenhouse Gas Emissions.* PhD Thesis, University of California, Berkeley, CA, USA, 101 pp.

Reich-Weiser, C., T. Fletcher, D.A. Dornfeld, and S. Horne (2008). Development of the Supply Chain Optimization and Planning for the Environment (SCOPE) tool - Applied to solar energy. In: *2008 IEEE International Symposium on Electronics and the Environment.* IEEE, San Francisco, CA, 19-21 May 2008, 6 pp.

Sengul, H. (2009). *Life Cycle Analysis of Quantum Dot Semiconductor Materials.* PhD Thesis, University of Illinois, Chicago, IL, USA, 255 pp.

Stoppato, A. (2008). Life cycle assessment of photovoltaic electricity generation. *Energy*, **33**(2), pp. 224-232.

Tripanagnostopoulos, Y., M. Souliotis, R. Battisti, and A. Corrado (2006). Performance, cost and life-cycle assessment study of hybrid PVT/AIR solar systems. *Progress in Photovoltaics: Research and Applications*, **14**(1), pp. 65-76.

Uchiyama, Y. (1997). Life cycle analysis of photovoltaic cell and wind power plants. In: *IAEA Advisory Group Meeting on the Assessment of Greenhouse Gas Emissions from the Full Energy Chain of Solar and Wind Power*, International Atomic Energy Agency, Vienna, Austria, 21-24 October 1996, pp. 111-122.

Voorspools, K.R., E.A. Brouwers, and W.D. D'Haeseleer (2000). Energy content and indirect greenhouse gas emissions embedded in 'emission-free' power plants: Results for the low countries. *Applied Energy*, **67**(3), pp. 307-330.

Wind energy (49)

Ardente, F., M. Beccali, M. Cellura, and V. Lo Brano (2008). Energy performances and life cycle assessment of an Italian wind farm. *Renewable & Sustainable Energy Reviews*, **12**(1), pp. 200-217.

Berry, J.E., M.R. Holland, P.R. Watkiss, R. Boyd, and W. Stephenson (1998). *Power Generation and the Environment: a UK Perspective.* AEA Technology, Oxfordshire, UK, 275 pp.

Chataignere, A., and D. Le Boulch (2003). *Wind Turbine (WT) Systems: Final Report.* Energy de France (EDF R&D), Paris, France, 110 pp.

Crawford, R.H. (2009). Life cycle energy and greenhouse emissions analysis of wind turbines and the effect of size on energy yield. *Renewable and Sustainable Energy Reviews*, **13**(9), pp. 2653-2660.

Dolan, S.L. (2007). *Life Cycle Assessment and Emergy Synthesis of a Theoretical Offshore Wind Farm for Jacksonville, Florida.* M.S. Thesis, University of Florida, 125 pp. Available at: http://etd.fcla.edu/UF/UFE0021032/dolan_s.pdf.

Dones, R., T. Heck, C. Bauer, S. Hirschberg, P. Bickel, P. Preiss, L.I. Panis, and I. De Vlieger (2005). *Externalities of Energy: Extension of Accounting Framework and Policy Applications: New Energy Technologies.* ENG1-CT-2002-00609, Paul Scherrer Institute (PSI), Villigen, Switzerland, 76 pp.

Dones, R., C. Bauer, R. Bolliger, B. Burger, T. Heck, A. Roder, M.F. Emenegger, R. Frischknecht, N. Jungbluth, and M. Tuchschmid (2007). *Life Cycle Inventories of Energy Systems: Results for Current Systems in Switzerland and Other UCTE Countries.* Ecoinvent Report No. 5, Paul Scherrer Institute, Swiss Centre for Life Cycle Inventories, Villigen, Switzerland, 185 pp. Available at: www.ecolo.org/documents/documents_in_english/Life-cycle-analysis-PSI-05.pdf.

DONG Energy (2008). *Life Cycle Approaches to Assess Emerging Energy Technologies: Final Report on Offshore Wind Technology.* DONG Energy, Fredericia, Denmark, 60 pp.

Enel SpA (2004). *Certified Environmental Product Declaration of Electricity from Enel's Wind Plant in Sclafani Bagni (Palermo, Italy).* Enel SpA, Rome, Italy, 25 pp.

European Commission (1995). Wind & Hydro. *ExternE: Externalities of Energy.* European Commission, Directorate-General XII, Luxembourg, **6**, 295 pp.

Frischknecht, R. (1998). *Life Cycle Inventory Analysis for Decision-Making: Scope-Dependent Inventory System Models and Context-Specific Joint Product Allocation.* Dissertation, Swiss Federal Institute of Technology Zurich, Zurich, Switzerland, 256 pp.

Hartmann, D. (1997). FENCH-analysis of electricity generation greenhouse gas emissions from solar and wind power in Germany. In: *IAEA Advisory Group Meeting on Assessment of Greenhouse Gas Emissions from the Full Energy Chain of Solar and Wind Power.* IAEA, Vienna, Austria, 21-24 October 1996, pp. 77-87.

Hondo, H. (2005). Life cycle GHG emission analysis of power generation systems: Japanese case. *Energy*, **30**(11-12), pp. 2042-2056.

Jacobson, M.Z. (2009). Review of solutions to global warming, air pollution, and energy security. *Energy & Environmental Science*, **2**, pp. 148-173.

Jungbluth, N., C. Bauer, R. Dones, and R. Frischknecht (2005). Life cycle assessment for emerging technologies: Case studies for photovoltaic and wind power. *International Journal of Life Cycle Assessment*, **10**(1), pp. 24-34.

Khan, F.I., K. Hawboldt, and M.T. Iqbal (2005). Life cycle analysis of wind-fuel cell integrated system. *Renewable Energy*, **30**(2), pp. 157-177.

Krewitt, W., P. Mayerhofer, R. Friedrich, A. Truckenmüller, T. Heck, A. Gressmann, F. Raptis, F. Kaspar, J. Sachau, K. Rennings, J. Diekmann, and B. Praetorius (1997). *ExternE National Implementation in Germany.* University of Stuttgart, Stuttgart, Germany, 189 pp.

Kuemmel, B., and B. Sørensen (1997). *Life-cycle Analysis of the Total Danish Energy System.* IMFUFA, Roskilde Universitetscenter, Roskilde, Denmark, 219 pp.

Lee, Y.-M., and Y.-E. Tzeng (2008). Development and life-cycle inventory analysis of wind energy in Taiwan. *Journal of Energy Engineering*, **134**(2), pp. 53-57.

Lenzen, M., and U. Wachsmann (2004). Wind turbines in Brazil and Germany: An example of geographical variability in life-cycle assessment. *Applied Energy*, **77**(2), pp. 119-130.

Liberman, E.J. (2003). *A Life Cycle Assessment and Economic Analysis of Wind Turbines Using Monte Carlo Simulation.* M.S. Thesis, Air Force Institute of Technology, Wright-Patterson Air Force Base, OH, USA, 162 pp.

Martínez, E., F. Sanz, S. Pellegrini, E. Jiménez, and J. Blanco (2009). Life-cycle assessment of a 2-MW rated power wind turbine: CML method. *The International Journal of Life Cycle Assessment*, **14**(1), pp. 52-63.

McCulloch, M., M. Raynolds, and M. Laurie (2000). *Life-Cycle Value Assessment of a Wind Turbine.* The Pembina Institute, Drayton Valley, Alberta, Canada, 14 pp.

Nadal, G. (1995). Life cycle direct and indirect pollution associated with PV and wind energy systems. In: *ISES 1995: Solar World Congress.* Fundacion Bariloche, Harare, Zimbabwe, 11-15 September 1995, pp. 39

Pacca, S.A. (2003). *Global Warming Effect Applied to Electricity Generation Technologies.* PhD Thesis, University of California, Berkeley, CA, USA, 191 pp.

Pacca, S.A., and A. Horvath (2002). Greenhouse gas emissions from building and operating electric power plants in the upper Colorado River Basin. *Environmental Science & Technology*, **36**(14), pp. 3194-3200.

Pehnt, M. (2006). Dynamic life cycle assessment (LCA) of renewable energy technologies. *Renewable Energy*, **31**(1), pp. 55-71.

Pehnt, M., M. Oeser, and D.J. Swider (2008). Consequential environmental system analysis of expected offshore wind electricity production in Germany. *Energy*, **33**(5), pp. 747-759.

Proops, J.L.R., P.W. Gay, S. Speck, and T. Schröder (1996). The lifetime pollution implications of various types of electricity generation. An input-output analysis. *Energy Policy*, **24**(3), pp. 229-237.

Rule, B.M., Z.J. Worth, and C.A. Boyle (2009). Comparison of life cycle carbon dioxide emissions and embodied energy in four renewable electricity generation technologies in New Zealand. *Environmental Science & Technology*, **43**(16), pp. 6406-6413.

Rydh, J., M. Jonsson, and P. Lindahl (2004). *Replacement of Old Wind Turbines Assessed from Energy, Environmental and Economic Perspectives.* University of Kalmar, Department of Technology, Kalmar, Sweden, 33 pp.

Saskatchewan Energy Conservation and Development Authority (1994). *Levelized Cost and Full Fuel Cycle Environmental Impacts of Saskatchewan's Electric Supply Options.* SECDA Publication No. T800-94-004, Saskatoon, SK, Canada, 205 pp.

Schleisner, L. (2000). Life cycle assessment of a wind farm and related externalities. *Renewable Energy*, **20**(3), pp. 279-288.

Spitzley, D.V., and G.A. Keoleian (2005). *Life Cycle Environmental and Economic Assessment of Willow Biomass Electricity: A Comparison with Other Renewable and Non-renewable Sources.* Report No. CSS04-05R, University of Michigan, Center for Sustainable Systems, Ann Arbor, MI, USA, 69 pp.

Tremeac, B., and F. Meunier (2009). Life cycle analysis of 4.5 MW and 250 W wind turbines. *Renewable and Sustainable Energy Reviews*, **13**(8), pp. 2104-2110.

Uchiyama, Y. (1997). Life cycle analysis of photovoltaic cell and wind power plants. In: *IAEA Advisory Group Meeting on the Assessment of Greenhouse Gas Emissions from the Full Energy Chain of Solar and Wind Power*, International Atomic Energy Agency, Vienna, Austria, 21-24 October 1996, pp. 111-122.

van de Vate, J.F. (1996). Comparison of the greenhouse gas emissions from the full energy chains of solar and wind power generation. In: *IAEA Advisory Group Meeting organized by the IAEA Headquarters.* IAEA, Vienna, Austria, 21-24 October 1996, pp. 13.

Vattenfall AB (2003). *Certified Environmental Product Declaration of Electricity from Vattenfall AB's Swedish Windpower Plants.* Vattenfall, Stockholm, Sweden, 31 pp.

Vattenfall AB (2010). *Vattenfall Wind Power Certified Environmental Product Declaration EPD of Electricity from Vattenfall's Wind Farms.* Vattenfall Wind Power, Stockholm, Sweden, 51 pp.

Vestas Wind Systems A/S (2006). *Life Cycle Assessment of Electricity Produced from Onshore Sited Wind Power Plants Based on Vestas V82-1.65 MW turbines.* Vestas, Randers, Denmark, 77 pp.

Vestas Wind Systems A/S (2006). *Life Cycle Assessment of Offshore and Onshore Sited Wind Power Plants Based on Vestas V90-3.0 MW Turbines.* Vestas, Randers, Denmark, 60 pp.

Voorspools, K.R., E.A. Brouwers, and W.D. D'Haeseleer (2000). Energy content and indirect greenhouse gas emissions embedded in 'emission-free' power plants: Results for the low countries. *Applied Energy*, **67**(3), pp. 307-330.

Waters, T.M., R. Forrest, and D.C. McConnell (1997). Life-cycle assessment of wind energy: A case study based on Baix Ebre Windfarm, Spain. In: *Wind Energy Conversion 1997: Proceedings of the Nineteenth BWEA Wind Energy Conference*, R. Hunter (ed.), Mechanical Engineering Publications Limited, Heriot-Watt University, Edinburgh, UK, 16-18 July 1997, pp. 231-238.

Weinzettel, J., M. Reenaas, C. Solli, and E.G. Hertwich (2009). Life cycle assessment of a floating offshore wind turbine. *Renewable Energy*, **34**(3), pp. 742-747.

White, S. (2006). Net energy payback and CO_2 emissions from three Midwestern wind farms: An update. *Natural Resources Research*, **15**(4), pp. 271-281.

White, S.W., and G.L. Kulcinski (1998). *Net Energy Payback and CO_2 Emissions from Wind-Generated Electricity in the Midwest.* UWFDM-1092, University of Wisconsin, Madison, WI, USA, 72 pp.

White, S.W., and G.L. Kulcinski (1999). *'Birth to Death' Analysis of the Energy Payback Ratio and CO_2 Gas Emission Rates from Coal, Fission, Wind, and DT Fusion Power Plants.* University of Wisconsin, Madison, WI, USA, 17 pp.

Wibberley, L. (2001). *Coal in a Sustainable Society.* Australian Coal Association Research Program, Brisbane, Queensland, Australia.

World Energy Council (2004). *Comparison of Energy Systems Using Life Cycle Assessment.* World Energy Council, London, UK, 67 pp.

A.II.5.3 Review of operational water use of electricity generation technologies

This overview describes the methods of a comprehensive review of published estimates of operational water withdrawal and consumption intensity of electricity generation technologies. Results are discussed in Section 9.3.4.4 and shown in Figure 9.14.

A.II.5.3.1 Review methodology

Lifecycle water consumption and withdrawal literature for electricity generating technologies was reviewed, but due to lack of quality and breadth of data, the review focused exclusively on operational water use. Lifecycle literature considered here are studies that passed the screening process used in this report's review of lifecycle GHG emissions from electricity generation technologies (see A II.5.2). Upstream water use for biofuel energy crops is not subject of this section.

This review did not alter (except for unit conversion) or audit for accuracy the estimates of water use published in studies that passed the screening criteria. Also, because estimates are used as published, considerable methodological inconsistency is inherent, which limits comparability. A few attempts have been made to review the operational water use literature for electricity generation technologies, though all of these were limited in their comprehensiveness of either technologies or of primary literature considered (Gleick, 1993; Inhaber, 2004; NETL, 2007a,b; WRA, 2008; Fthenakis and Kim, 2010). The present review therefore informs the discourse of this report in a unique way.

Literature collection

The identification of relevant literature started with a core library of references held previously by the researchers, followed by searching in major bibliographic databases using a variety of search algorithms and combinations of key words, and then reviewing reference lists of every collected reference. All collected literature was added to a bibliographic database. The literature collection methods described here apply to all classes of electricity generation technologies reviewed in this report.

Literature screening

Collected references were independently subjected to screening to select references that met criteria for quality and relevance. Operational water use studies must have been written in English, addressed operational water use for facilities located in North America, provided sufficient information to calculate a water use intensity factor (in cubic metres per megawatt-hour generated), made estimates of water consumption that did not duplicate others previously published, and have been in one of the following formats: journal article, conference proceedings, or report (authored by government agencies, nongovernmental organizations, international institutions, or corporations). Estimates of national average water use intensity for particular technologies, estimates of existing plant operational water use, and estimates derived from laboratory experiments were considered equally. Given the paucity of available estimates of water consumption for electricity generation technologies and that the estimates that have been published are being used in the policy context already, no additional screens based on quality or completeness of reporting were applied.

Analysis of estimates

Estimates were categorized by fuel technology and cooling systems. Certain aggregations of fuel technology types and cooling system types were made to facilitate analysis. Concentrating solar power includes both parabolic trough and power tower systems. Nuclear includes pressurized water reactors and boiling water reactors. Coal includes subcritical and supercritical technologies. For recirculating cooling technologies, no distinction is made between natural draft and mechanical draft cooling tower systems. Similarly, all pond-cooled systems are treated identically. Estimates were converted to the common functional unit of cubic meters per MWh generated. This conversion was performed using no exogenous assumptions; if any were required, that estimate was not analyzed.

A.II.5.3.2 List of references

CEC (2008). *2007 Environmental Performance Report of California's Electrical Generation System.* California Energy Commission (CEC) Final Staff Report, CA, USA.

Cohen, G., D.W. Kearney, C. Drive, D. Mar, and G.J. Kolb (1999). *Final Report on the Operation and Maintenance Improvement Program for Concentrating Solar Plants.* Sandia National Laboratories Technical Report-SAND99-1290, doi:10.2172/8378, Albuquerque, NM, USA.

Dziegielewski, B., and T. Bik (2006). *Water Use Benchmarks for Thermoelectric Power Generation.* Research Report of the Department of Geography and Environmental Resources, Southern Illinois University, Carbondale, IL, USA.

EPRI (2002). *Water and sustainability (Volume 2): an assessment of water demand, supply, and quality in the U.S.-the next half century.* Technical Report 1006785, Electric Power Research Institute (EPRI). Palo Alto, CA, USA.

EPRI and US DOE (1997). *Renewable Energy Technology Characterizations.* EPRI Topical Report-109496, Electric Power Research Institute (EPRI) and U.S. Department of Energy (US DOE), Palo Alto, CA and Washington, DC, USA.

Feeley, T.J., L. Green, J.T. Murphy, J. Hoffmann, and B.A. Carney (2005). *Department of Energy / Office of Fossil Energy's Power Plant Water Management R & D Program.* National Energy Technology Laboratory, Pittsburgh, PA, USA, 18 pp. Available at: www.netl.doe.gov/technologies/coalpower/ewr/pubs/IEP_Power_Plant_Water_R%26D_Final_1.pdf.

Feeley, T.J., T.J. Skone, G.J. Stiegel, A. Mcnemar, M. Nemeth, B. Schimmoller, J.T. Murphy, and L. Manfredo (2008). Water: A critical resource in the thermoelectric power industry. *Energy*, 33, pp. 1-11.

Fthenakis, V., and H.C. Kim (2010). Life-cycle uses of water in U.S. electricity generation. *Renewable and Sustainable Energy Reviews*, 14, pp. 2039-2048.

Gleick, P. (1992). Environmental consequences of hydroelectric development: The role of facility size and type. *Energy*, 17(8), pp. 735-747.

Gleick, P. (1993). *Water in Crisis: A Guide to the World's Fresh Water Resources.* Oxford University Press, New York, NY, USA.

Hoffmann, J., S. Forbes, and T. Feeley (2004). *Estimating Freshwater Needs to Meet 2025 Electricity Generating Capacity Forecasts.* National Energy Technology Laboratory Pittsburgh, PA, USA, 12 pp. Available at: www.netl.doe.gov/technologies /coalpower/ewr/pubs/Estimating%20Freshwater%20Needs%20to%20 2025.pdf.

Inhaber, H. (2004). Water use in renewable and conventional electricity production. *Energy Sources, Part A: Recovery, Utilization, and Environmental Effects*, 26, pp. 309-322, doi:10.1080/00908310490266698.

Kelly, B. (2006). *Nexant Parabolic Trough Solar Power Plant Systems Analysis-Task 2: Comparison of Wet and Dry Rankine Cycle Heat Rejection.* Subcontractor Report-NREL/SR-550-40163, National Renewable Energy Laboratory (NREL), Golden, CO, USA. Available at: www.nrel.gov/csp/troughnet/pdfs/40163.pdf.

Leitner, A. (2002). *Fuel from the Sky: Solar Power's Potential for Western Energy Supply.* Subcontractor Report-NREL/SR 550-32160, National Renewable Energy Laboratory (NREL), Golden, CO, USA. Available at: www.nrel.gov/csp/pdfs/32160.pdf.

Mann, M., and P. Spath (1997). *Life Cycle Assessment of a Biomass Gasification Combined-Cycle System.* Technical Report-TP-430-23076, National Renewable Energy Laboratory (NREL), Golden, CO, USA. Available at: www.nrel.gov/docs/legosti/fy98/23076.pdf.

Meridian (1989). *Energy System Emissions and Material Requirements.* Meridian Corporation Report to U.S. Department of Energy (DOE), Washington, DC, USA.

NETL (2007). *Cost and Performance Baseline for Fossil Energy Plants-Volume 1: Bituminous Coal and Natural Gas to Electricity Final Report.* DOE/NETL-2007/1281, National Energy Technology Laboratory (NETL), Pittsburgh, PA, USA. Available at www.netl.doe.gov/energy-analyses/pubs/BitBase_FinRep_2007.pdf.

NETL (2007). *Power Plant Water Usage and Loss Study. 2007 Update.* National Energy Technology Laboratory (NETL), Pittsburgh, PA, USA. Available at: www.netl.doe.gov/technologies/coalpower/gasification/pubs/pdf/WaterReport_Revised%20 May2007.pdf.

NETL (2009). *Estimating Freshwater Needs to Meet Future Thermoelectric Generation Requirements.* DOE/NETL-400/2009/1339, National Energy Technology Laboratory (NETL), Pittsburgh, PA, USA. Available at: www.netl.doe.gov/energy-analyses/pubs/2009%20Water%20Needs%20Analysis%20-%20Final%20%289-30-2009%29.pdf.

NETL (2009). *Existing Plants, Emissions and Capture – Setting Water-Energy R&D Program Goals.* DOE/NETL-2009/1372, National Energy Technology Laboratory (NETL), Pittsburgh, PA, USA. Available at: www.netl.doe.gov/technologies/coalpower/ewr/water/pdfs/EPEC%20water-energy%20R%26D%20goal%20update%20v.1%20may09.pdf.

Sargent & Lundy (2003). *Assessment of Parabolic Trough and Power Tower Solar Technology Cost and Performance Forecasts.* NREL/SR-550-34440, National Renewable Energy Laboratory (NREL), Golden, CO, USA. Available at: www.nrel.gov/docs/fy04osti/34440.pdf.

Stoddard, L., J. Abiecunas, and R.O. Connell (2006). *Economic, Energy, and Environmental Benefits of Concentrating Solar Power in California.* NREL/SR-550-39291, National Renewable Energy Laboratory (NREL), Golden, CO, USA. Available at: www.nrel.gov/docs/fy06osti/39291.pdf.

Torcellini, P., N. Long, and R. Judkoff (2003). *Consumptive Water Use for U.S. Power Production.* Technical Report-TP-550-33905, National Renewable Energy Laboratory (NREL), Golden, CO, USA. Available at: www.nrel.gov/docs/fy04osti/33905.pdf.

Turchi, C., M. Wagner, and C. Kutscher (2010). *Water Use in Parabolic Trough Power Plants: Summary Results from WorleyParsons' Analyses.* NREL/TP-5500-49468, National Renewable Energy Laboratory (NREL), Golden, CO, USA. Available at: www.nrel.gov/docs/fy11osti/49468.pdf.

US DOE (2009). *Concentrating Solar Power Commercial Application Study: Reducing Water Consumption of Concentrating Solar Power Electricity Generation.* Report to Congress. U.S. Department of Energy (DOE), Washington, DC, USA.

Viebahn, P., S. Kronshage, F. Trieb, and Y. Lechon (2008). *Final Report on Technical Data, Costs, and Life Cycle Inventories of Solar Thermal Power Plants.* Project 502687, New Energy Externalities Developments for Sustainability (NEEDS), Brussels, Belgium, 95 pp. Available at: www.needs-project.org/RS1a/RS1a%20D12.2%20Final%20report%20concentrating%20solar%20thermal%20power%20plants.pdf.

WorleyParsons (2009). *Analysis of Wet and Dry Condensing 125 MW Parabolic Trough Power Plants.* WorleyParsons Report No. NREL-2-ME-REP-0002-R0, WorleyParsons Group, North Sydney, Australia.

WorleyParsons (2009). *Beacon Solar Energy Project Dry Cooling Evaluation.* WorleyParsons Report No. FPLS-0-LI-450-0001, WorleyParsons Group, North Sydney, Australia.

WorleyParsons (2010). *Material Input for Life Cycle Assessment Task 5 Subtask 2: O&M Schedules.* WorleyParsons Report No. NREL-0-LS-019-0005, WorleyParsons Group, North Sydney, Australia.

WorleyParsons (2010). *Parabolic Trough Reference Plant for Cost Modeling with the Solar Advisor Model.* WorleyParsons Report, WorleyParsons Group, North Sydney, Australia.

WRA (2008). *A Sustainable Path: Meeting Nevada's Water and Energy Demands.* Western Resource Advocates (WRA), Boulder, CO, USA, 43 pp. Available at: www.westernresourceadvocates.org/water/NVenergy-waterreport.pdf.

Yang, X., and B. Dziegielewski (2007). Water use by thermoelectric power plants in the United States. *Journal of the American Water Resources Association*, **43**, pp. 160-169.

A.II.5.4 Risk analysis

This section introduces the methods applied for the assessment of hazards and risks of energy technologies presented in Section 9.3.4.7, and provides references and central assumptions (Table A.II.5).

A large variety of definitions of the term risk exists, depending on the field of application and the object under study (Haimes, 2009). In engineering and natural sciences, risk is frequently defined in a quantitative way: risk (R) = probability (p) × consequence (C). This definition does not include subjective factors of risk perception and aversion, which can also influence the decision-making process, that is, stakeholders may make trade-offs between quantitative and qualitative risk factors (Gregory and Lichtenstein, 1994; Stirling, 1999). Risk assessment and evaluation is further complicated when certain risks significantly transcend everyday levels; their handling posing a challenge for society (WBGU, 2000). For example, Renn et al. (2001) assigned risks into three categories or areas, namely (1) the normal area manageable by routine operations and existing laws and regulations, (2) the intermediate area, and (3) the intolerable area (area of permission). Kristensen et al. (2006) proposed a modified classification scheme to further improve the characterization of risk. Recently, additional aspects such as critical infrastructure protection, complex interrelated systems and 'unknown unknowns' have become a major focus (Samson et al., 2009; Aven and Zio, 2011; Elahi, 2011).

The evaluation of the 'hazards and risks' of various energy technologies as presented in Section 9.3.4.7 builds upon the approach of comparative risk assessment as it has been established at the Paul Scherrer Institut (PSI) since the 1990s;[4] at the core of which is the Energy-Related Severe Accident Database (ENSAD) (Hirschberg et al., 1998, 2003a; Burgherr et al., 2004, 2008; Burgherr and Hirschberg, 2005). The consideration of full energy chains is essential because an accident can happen in any chain stage from exploration, extraction, processing and storage, long distance transport, regional and local distribution, power and/or heat generation, waste treatment, and disposal. However, not all these stages are applicable to every energy chain. For fossil energy chains (coal, oil, natural gas) and hydropower, extensive historical experience is contained in ENSAD for the period 1970 to 2008. In the case of nuclear power, Probabilistic Safety Assessment (PSA) is employed to address hypothetical accidents (Hirschberg et al., 2004a). In contrast, consideration of renewable energy technologies other than hydropower is based on available accident statistics, literature review and expert judgment because of limited or lacking historical experience. It should be noted that available analyses have limited scope and do not include

---

4 In a recent study, Felder (2009) compared the ENSAD database with another energy accident compilation (Sovacool, 2008a). Despite numerous and partially substantial differences between the two data sets, several interesting findings with regard to methodological and policy aspects were addressed. However, the study was based on the first official release of ENSAD (Hirschberg et al., 1998), and thus disregarded all subsequent updates and extensions. Another study by Colli et al. (2009) took a slightly different approach using a rather broad set of so-called Risk Characterization Indicators, however the actual testing with illustrative examples was based on ENSAD data.

probabilistic modelling of hypothetical accidents. This may have bearing particularly on results for solar PV.

No consensus definition of the term 'severe accident' exists in the literature. Within the framework of PSI's database ENSAD, an accident is considered to be severe if it is characterized by one or several of the following consequences:

- At least 5 fatalities or
- At least 10 injured or
- At least 200 evacuees or
- An extensive ban on consumption of food or
- Releases of hydrocarbons exceeding 10,000 metric tons or
- Enforced clean-up of land and water over an area of at least 25 km^2 or
- Economic loss of at least 5 million USD$_{2000}$

For large centralized energy technologies, results are given for three major country aggregates, namely for OECD and non-OECD countries as well as EU 27. Such a distinction is meaningful because of the substantial differences in management, regulatory frameworks and general safety culture between highly developed countries (i.e., OECD and EU 27) and the mostly less-developed non-OECD countries (Burgherr and Hirschberg, 2008). In the case of China, coal chain data were only analyzed for the years 1994 to 1999 when data on individual accidents from the China Coal Industry Yearbook (CCIY) were available, indicating that previous years were subject to substantial underreporting (Hirschberg et al., 2003a,b). For the period 2000 to 2009, only annual totals of coal chain fatalities from CCIY were available, which is why they were not combined with the data from the previous period. For renewable energy technologies except hydropower, estimates can be considered representative for developed countries (e.g., OECD and EU 27).

Comparisons of the various energy chains were based on data normalized to the unit of electricity production. For fossil energy chains the thermal energy was converted to an equivalent electrical output using a generic efficiency factor of 0.35. For nuclear, hydropower and new renewable technologies the normalization is straightforward since the generated product is electrical energy. The Gigawatt-electric-year (GW$_e$yr) was chosen because large individual plants have capacities in the neighbourhood of 1 GW of electrical output (GW$_e$). This makes the GW$_e$yr a natural unit to use when presenting normalized indicators generated within technology assessments.

A.II.6 Regional definitions and country groupings

The IPCC SRREN uses the following regional definitions and country groupings, largely based on the definitions of the *World Energy Outlook 2009* (IEA, 2009). Grouping names and definitions vary in the published literature, and in the SRREN in some instances there may be slight deviations from the standard below. Alternative grouping names that are used in the SRREN are given in parenthesis.

Africa

Algeria, Angola, Benin, Botswana, Burkina Faso, Burundi, Cameroon, Cape Verde, Central African Republic, Chad, Comoros, Congo, Democratic Republic of Congo, Côte d'Ivoire, Djibouti, Egypt, Equatorial Guinea, Eritrea, Ethiopia, Gabon, Gambia, Ghana, Guinea, Guinea-Bissau, Kenya, Lesotho, Liberia, Libya, Madagascar, Malawi, Mali, Mauritania, Mauritius, Morocco, Mozambique, Namibia, Niger, Nigeria, Reunion, Rwanda, Sao Tome and Principe, Senegal, Seychelles, Sierra Leone, Somalia, South Africa, Sudan, Swaziland, United Republic of Tanzania, Togo, Tunisia, Uganda, Zambia and Zimbabwe.

Annex I Parties to the United Nations Framework Convention on Climate Change

Australia, Austria, Belarus, Belgium, Bulgaria, Canada, Croatia, Czech Republic, Denmark, Estonia, Finland, France, Germany, Greece, Hungary, Iceland, Ireland, Italy, Japan, Latvia, Liechtenstein, Lithuania, Luxembourg, Monaco, Netherlands, New Zealand, Norway, Poland, Portugal, Romania, Russian Federation, Slovak Republic, Slovenia, Spain, Sweden, Switzerland, Turkey, Ukraine, United Kingdom and United States.

Eastern Europe/Eurasia (also sometimes referred to as 'Transition Economies')

Albania, Armenia, Azerbaijan, Belarus, Bosnia and Herzegovina, Bulgaria, Croatia, Estonia, Georgia, Kazakhstan, Kyrgyzstan, Latvia, Lithuania, the former Yugoslav Republic of Macedonia, the Republic of Moldova, Romania, Russian Federation, Serbia, Slovenia, Tajikistan, Turkmenistan, Ukraine, and Uzbekistan. For statistical reasons, this region also includes Cyprus, Gibraltar and Malta.

European Union

Austria, Belgium, Bulgaria, Cyprus, Czech Republic, Denmark, Estonia, Finland, France, Germany, Greece, Hungary, Ireland, Italy, Latvia, Lithuania, Luxembourg, Malta, Netherlands, Poland, Portugal, Romania, Slovak Republic, Slovenia, Spain, Sweden and United Kingdom.

G8

Canada, France, Germany, Italy, Japan, Russian Federation, United Kingdom and United States.

Latin America

Antigua and Barbuda, Aruba, Argentina, Bahamas, Barbados, Belize, Bermuda, Bolivia, Brazil, the British Virgin Islands, the Cayman Islands,

Table A.II.5 | Overview of data sources and assumptions for the calculation of fatality rates and maximum consequences.

Coal

- ENSAD database at PSI; severe (≥5 fatalities) accidents.[1]
- OECD: 1970-2008; 86 accidents; 2,239 fatalities. EU 27: 1970-2008; 45 accidents; 989 fatalities. Non-OECD without China: 1970-2008; 163 accidents; 5.808 fatalities (Burgherr et al., 2011).
  Previous studies: Hirschberg et al. (1998); Burgherr et al. (2004, 2008).
- China (1994-1999): 818 accidents; 11,302 fatalities (Hirschberg et al., 2003a; Burgherr and Hirschberg, 2007).
- China (2000-2009): for comparison, the fatality rate in the period 2000 to 2009 was calculated based on data reported by the State Administration of Work Safety (SATW) of China.[2] Annual values given by SATW correspond to total fatalities (i.e., severe and minor accidents). Thus for the fatality rate calculation it was assumed that fatalities from severe accidents comprise 30% of total fatalities, as has been found in the China Energy Technology Program (Hirschberg et al., 2003a; Burgherr and Hirschberg, 2007). Chinese fatality rate (2000-2009) = 3.14 fatalities/GW$_e$yr.

Oil

- ENSAD database at PSI; severe (≥5 fatalities) accidents.[1]
- OECD: 1970-2008; 179 accidents; 3,383 fatalities. EU 27: 1970-2008; 64 accidents; 1,236 fatalities. Non-OECD: 1970-2008; 351 accidents; 19,376 fatalities (Burgherr et al., 2011).
  Previous studies: Hirschberg et al. (1998); Burgherr et al. (2004, 2008).

Natural Gas

- ENSAD database at PSI; severe (≥5 fatalities) accidents.[1]
- OECD: 1970-2008; 109 accidents; 1,257 fatalities. EU 27: 1970-2008; 37 accidents; 366 fatalities. Non-OECD: 1970-2008; 77 accidents; 1,549 fatalities (Burgherr et al., 2011).
  Previous studies: Hirschberg et al. (1998); Burgherr et al. (2004, 2008); Burgherr and Hirschberg (2005).

Nuclear

- Generation II (Gen. II) - Pressurized Water Reactor, Switzerland; simplified Probabilistic Safety Assessment (PSA) (Roth et al., 2009).
- Generation III (Gen. III) - European Pressurized Reactor (EPR) 2030, Switzerland; simplified PSA (Roth et al., 2009).
  Available results for the above described EPR point towards significantly lower fatality rates (early fatalities (EF): 3.83E-07 fatalities/GW$_e$yr; latent fatalities (LF): 1.03E-05 fatalities/GW$_e$yr; total fatalities (TF): 1.07E-05 fatalities/GW$_e$yr) due to a range of advanced features, especially with respect to Severe Accident Management (SAM) active and passive systems. However, maximum consequences of hypothetical accidents may increase (ca. 48,800 fatalities) due to the larger plant size (1,600 MW) and the larger associated radioactive inventory.
- In the case of a severe accident in the nuclear chain, immediate or early (acute) fatalities are of minor importance and denote those fatalities that occur in a short time period after exposure, whereas latent (chronic) fatalities due to cancer dominate total fatalities (Hirschberg et al., 1998). Therefore, the above estimates for Gen. II and III include immediate and latent fatalities.
- Three Mile Island 2, TMI-2: The TMI-2 accident occurred as a result of equipment failures combined with human errors. Due to the small amount of radioactivity released, the estimated collective effective dose to the public was about 40 person-sievert (Sv). The individual doses to members of the public were extremely low: <1 mSv in the worst case. On the basis of the collective dose one extra cancer fatality was estimated. However, 144,000 people were evacuated from the area around the plant. For more information, see Hirschberg et al. (1998).
- Chernobyl: 31 immediate fatalities; PSA-based estimate of 9,000 to 33,000 latent fatalities (Hirschberg et al., 1998).
- PSI's Chernobyl estimates for latent fatalities range from about 9,000 for Ukraine, Russia and Belarus to about 33,000 for the entire northern hemisphere in the next 70 years (Hirschberg et al., 1998). According to a recent study by numerous United Nations organizations, up to 4,000 persons could die due to radiation exposure in the most contaminated areas (Chernobyl Forum, 2005). This estimate is substantially lower than the upper limit of the PSI interval, which, however, was not restricted to the most contaminated areas.

Hydro

- ENSAD Database at PSI; severe (≥5 fatalities) accidents.[1]
- OECD: 1970-2008; 1 accident; 14 fatalities (Teton dam failure, USA, 1976). EU 27: 1970-2008; 1 accident; 116 fatalities (Belci dam failure, Romania, 1991) (Burgherr et al., 2011).
- Based on a theoretical model, maximum consequences for the total failure of a large Swiss dam range between 7,125 and 11,050 fatalities without pre-warning, but can be reduced to 2 to 27 fatalities with 2 hours pre-warning time (Burgherr and Hirschberg, 2005, and references therein).
- Non-OECD: 1970-2008; 12 accidents; 30,007 fatalities. Non-OECD without Banqiao/Shimantan 1970-2008; 11 accidents; 4,007 fatalities; largest accident in China (Banqiao/Shimantan dam failure, China, 1975) excluded (Burgherr et al., 2011).
- Previous studies: Hirschberg et al. (1998); Burgherr et al. (2004, 2008).

Photovoltaic (PV)

- Current estimates include only silicon (Si) technologies, weighted by their 2008 market shares, i.e., 86% for c-Si and 5.1% for a-Si/u-Si.
- The analysis covers risks of selected hazardous substances (chlorine, hydrochloric acid, silane and trichlorosilane) relevant in the Si PV life cycle.
- Accident data were collected for the USA (for which a good coverage exists), and for the years 2000 to 2008 to ensure that estimates are representative of currently operating technologies.
- Database sources: Emergency Response Notification System, Risk Management Plan, Major Hazard Incident Data Service, Major Accidents Reporting System, Analysis Research and Information on Accidents, Occupational Safety and Health Update.
- Since collected accidents were not only from the PV sector, the actual PV fatality share was estimated, based on the above substance amounts in the PV sector as a share of the total USA production, as well as data from the ecoinvent database.
- Cumulated fatalities for the four above substances were then normalized to the unit of energy production using a generic load factor of 10% (Burgherr et al., 2008).
- Assumption that 1 out of 100 accidents is severe.[3]
- Current estimate for fatality rate: Burgherr et al. (2011).
- Maximum consequences represent an expert judgment due to limited historical experience (Burgherr et al., 2008).
- Previous studies: Hirschberg et al. (2004b); Burgherr et al. (2008); Roth et al. (2009).
- Other studies: Ungers et al. (1982); Fthenakis et al. (2006); Fthenakis and Kim (2010).

Continued next Page →

Wind Onshore

- Data sources: Windpower Death Database (Gipe, 2010) and Wind Turbine Accident Compilation (Caithness Windfarm Information Forum, 2010).
- Fatal accidents in Germany in the period 1975-2010; 10 accidents; 10 fatalities. 3 car accidents, where driver distraction from wind farm is given as reason, were excluded from the analysis.
- Assumption that 1 out of 100 accidents is severe.[3]
- Current estimate for fatality rate: Burgherr et al. (2011).
- Maximum consequences represent an expert judgment due to limited historical experience (Roth et al., 2009).
- Previous study: Hirschberg et al. (2004b).

Wind Offshore

- Data sources: see onshore above.
- Up to now there were 2 fatal accidents during construction in the UK (2009 and 2010) with 2 fatalities, and 2 fatal accidents during research activities in the USA (2008) with 2 fatalities.
- For the current estimate, only UK accidents were used, assuming a generic load factor of 0.43 (Roth et al., 2009) for the currently installed capacity of 1,340 MW (Renewable UK, 2010).
- Assumption that 1 out of 100 accidents is severe.[3]
- Current estimate for fatality rate: Burgherr et al. (2011).
- Maximum consequences: see onshore above.

Biomass: Combined Heat and Power (CHP) Biogas

- ENSAD Database at PSI; severe (≥5 fatalities) accidents.[1] Due to limited historical experience, the CHP Biogas fatality rate was approximated using natural gas accident data from the local distribution chain stage.
- OECD: 1970-2008; 24 accidents; 260 fatalities (Burgherr et al., 2011).
- Maximum consequences represent an expert judgment due to limited historical experience (Burgherr et al., 2011).
- Previous studies: Roth et al. (2009).

Enhanced Geothermal System (EGS)

- For the fatality rate calculations, only well drilling accidents were considered. Due to limited historical experience, exploration accidents in the oil chain were used as a rough approximation because of similar drilling equipment.
- ENSAD Database at PSI; severe (≥5 fatalities) accidents.[1]
- OECD: 1970-2008; oil exploration, 7 accidents; 63 fatalities (Burgherr, et al. 2011).
- For maximum consequences an induced seismic event was considered to be potentially most severe. Due to limited historical experience, the upper fatality boundary from the seismic risk assessment of the EGS project in Basel (Switzerland) was taken as an approximation (Dannwolf and Ulmer, 2009).
- Previous studies: Roth et al. (2009).

Notes: 1. Fatality rates are normalized to the unit of energy production in the corresponding country aggregate. Maximum consequences correspond to the most deadly accident that occurred in the observation period. 2. Data from SATW for the years 2000 to 2005 were reported in the China Labour News Flash No. 60 (2006-01-06) available at www.china-labour.org.hk/en/node/19312 (accessed December 2010). SATW data for the years 2006 to 2009 were published by Reuters, available at www.reuters.com/article/idUSPEK206148 (2006), uk.reuters.com/article/idUKPEK32921920080112 (2007), uk.reuters.com/article/idUKTOE61D00V20100214 (2008 and 2009), (all accessed December 2010). 3. For example, the rate for natural gas in Germany is about 1 out of 10 (Burgherr and Hirschberg, 2005), and for coal in China about 1 out of 3 (Hirschberg et al., 2003b).

Chile, Colombia, Costa Rica, Cuba, Dominica, the Dominican Republic, Ecuador, El Salvador, the Falkland Islands, French Guyana, Grenada, Guadeloupe, Guatemala, Guyana, Haiti, Honduras, Jamaica, Martinique, Montserrat, Netherlands Antilles, Nicaragua, Panama, Paraguay, Peru, St. Kitts and Nevis, Saint Lucia, Saint Pierre et Miquelon, St. Vincent and the Grenadines, Suriname, Trinidad and Tobago, the Turks and Caicos Islands, Uruguay and Venezuela.

Middle East

Bahrain, the Islamic Republic of Iran, Iraq, Israel, Jordan, Kuwait, Lebanon, Oman, Qatar, Saudi Arabia, Syrian Arab Republic, the United Arab Emirates and Yemen. It includes the neutral zone between Saudi Arabia and Iraq.

Non-OECD Asia (also sometimes referred to as 'developing Asia')

Afghanistan, Bangladesh, Bhutan, Brunei Darussalam, Cambodia, China, Chinese Taipei, the Cook Islands, East Timor, Fiji, French Polynesia, India, Indonesia, Kiribati, the Democratic People's Republic of Korea, Laos, Macau, Malaysia, Maldives, Mongolia, Myanmar, Nepal, New Caledonia, Pakistan, Papua New Guinea, the Philippines, Samoa, Singapore, Solomon Islands, Sri Lanka, Thailand, Tonga, Vietnam and Vanuatu.

North Africa

Algeria, Egypt, Libyan Arab Jamahiriya, Morocco and Tunisia.

OECD – Organisation for Economic Cooperation and Development

OECD Europe, OECD North America and OECD Pacific as listed below. Countries that joined the OECD in 2010 (Chile, Estonia, Israel and Slovenia) are not yet included in the statistics used in this report.

OECD Europe

Austria, Belgium, the Czech Republic, Denmark, Finland, France, Germany, Greece, Hungary, Iceland, Ireland, Italy, Luxembourg, the Netherlands, Norway, Poland, Portugal, the Slovak Republic, Spain, Sweden, Switzerland, Turkey and the United Kingdom.

OECD North America

Canada, Mexico and the United States.

OECD Pacific

Australia, Japan, Korea and New Zealand.

OPEC (Organization of Petroleum Exporting Countries)

Algeria, Angola, Ecuador, Islamic Republic of Iran, Iraq, Kuwait, Libya, Nigeria, Qatar, Saudi Arabia, United Arab Emirates and Venezuela.

Sub-Saharan Africa

Africa regional grouping excluding the North African regional grouping and South Africa.

A.II.7 General conversion factors for energy

Table A.II.6 provides conversion factors for a variety of energy-related units.

Table A.II.6 | Conversion factors for energy units (IEA, 2010b).

To: From:	TJ	Gcal	Mtoe	MBtu	GWh
	multiply by:				
TJ	1	238.8	2.388×10^{-5}	947.8	0.2778
Gcal	4.1868×10^{-3}	1	10^{-7}	3.968	1.163×10^{-3}
Mtoe	4.1868×10^{4}	10^{7}	1	3.968×10^{7}	11,630
MBtu	1.0551×10^{-3}	0.252	2.52×10^{-8}	1	2.931×10^{-4}
GWh	3.6	860	8.6×10^{-5}	3,412	1

Notes: MBtu: million British thermal unit; GWh: gigawatt hour; Gcal: gigacalorie; TJ: terajoule; Mtoe: megatonne of oil equivalent.

References

Aven, T., and E. Zio (2011). Some considerations on the treatment of uncertainties in risk assessment for practical decision making. *Reliability Engineering and System Safety*, **96**, pp. 64-74.

Beerten, J., E. Laes, G. Meskens, and W. D'haeseleer (2009). Greenhouse gas emissions in the nuclear life cycle: A balanced appraisal. *Energy Policy*, **37**(12), pp. 5056-5058.

BP (2009). *BP Statistical Review of World Energy.* BP, London, UK.

Burgherr, P., and S. Hirschberg (2005). *Comparative assessment of natural gas accident risks.* PSI Report No. 05-01, Paul Scherrer Institut, Villigen, Switzerland.

Burgherr, P., and S. Hirschberg (2007). Assessment of severe accident risks in the Chinese coal chain. *International Journal of Risk Assessment and Management*, **7**(8), pp. 1157-1175.

Burgherr, P., and S. Hirschberg (2008). A comparative analysis of accident risks in fossil, hydro and nuclear energy chains. *Human and Ecological Risk Assessment*, **14**(5), pp. 947 - 973.

Burgherr, P., S. Hirschberg, and E. Cazzoli (2008). *Final report on quantification of risk indicators for sustainability assessment of future electricity supply options. NEEDS Deliverable no D7.1 - Research Stream 2b. NEEDS project.* New Energy Externalities Developments for Sustainability, Brussels, Belgium.

Burgherr, P., S. Hirschberg, A. Hunt, and R.A. Ortiz (2004). *Severe accidents in the energy sector. Final Report to the European Commission of the EU 5th Framework Programme "New Elements for the Assessment of External Costs from Energy Technologies" (NewExt).* DG Research, Technological Development and Demonstration (RTD), Brussels, Belgium.

Burgherr, P., P. Eckle, S. Hirschberg, and E. Cazzoli (2011). *Final Report on Severe Accident Risks including Key Indicators.* SECURE Deliverable No. D5.7.2a. Security of Energy Considering its Uncertainty, Risk and Economic implications (SECURE), Brussels, Belgium. Available at: gabe.web.psi.ch/pdfs/secure/SECURE%20-%20Deliverable_D5-7-2%20-%20Severe%20Accident%20Risks.pdf.

Caithness Windfarm Information Forum (2010). *Summary of Wind Turbine Accident data to 30th September 2010.* Caithness Windfarm Information Forum, UK. Available at: www.caithnesswindfarms.co.uk/fullaccidents.pdf.

Chernobyl Forum (2005). *Chernobyl's legacy: health, environmental and socio-economic impacts and recommendations to the governments of Belarus, the Russian Federation and Ukraine. The Chernobyl Forum: 2003–2005.* Second revised version. International Atomic Energy Agency (IAEA), Vienna, Austria.

Colli, A., D. Serbanescu, and B.J.M. Ale (2009). Indicators to compare risk expressions, grouping, and relative ranking of risk for energy systems: Application with some accidental events from fossil fuels. *Safety Science*, **47**(5), pp. 591-607.

Dannwolf, U.S., and F. Ulmer (2009). *AP6000 Report - Technology risk comparison of the geothermal DHM project in Basel, Switzerland - Risk appraisal including social aspects.* SERIANEX Group - Trinational Seismis Risk Analysis Expert Group, RiskCom, Pforzheim, Germany.

Elahi, S. (2011). Here be dragons…exploring the 'unknown unknowns'. *Futures*, **43**(2), pp. 196-201.

Felder, F.A. (2009). A critical assessment of energy accident studies. *Energy Policy*, **37**(12), pp. 5744-5751.

Fisher, B.S., N. Nakicenovic, K. Alfsen, J. Corfee Morlot, F. de la Chesnaye, J.-C. Hourcade, K. Jiang, M. Kainuma, E. La Rovere, A. Matysek, A. Rana, K. Riahi, R. Richels, S. Rose, D. van Vuuren, and R. Warren (2007). Issues related to mitigation in the long term context. In: *Climate Change 2007: Mitigation. Contribution of Working Group III to the Fourth Assessment Report of the Intergovernmental Panel on Climate Change.* B. Metz, O.R. Davidson, P.R. Bosch, R. Dave, and L.A. Meyer (eds.), Cambridge University Press, pp. 169-250.

Frankl, P., E. Menichetti and M. Raugei (2005). *Final Report on Technical Data, Costs and Life Cycle Inventories of PV Applications.* NEEDS: New Energy Externalities Developments for Sustainability. Ambiente Italia, Milan, Italy, 81 pp.

Fthenakis, V.M., and H.C. Kim (2007). Greenhouse-gas emissions from solar electric- and nuclear power: A life-cycle study. *Energy Policy*, **35**(4), pp. 2549-2557.

Fthenakis, V.M., and H.C. Kim (2010). Life-cycle uses of water in U.S. electricity generation. *Renewable and Sustainable Energy Reviews*, **14**(7), pp. 2039-2048.

Fthenakis, V.M., H.C. Kim, A. Colli, and C. Kirchsteiger (2006). Evaluation of risks in the life cycle of photovoltaics in a comparative context. In: *21st European Photovoltaic Solar Energy Conference*, Dresden, Germany, 4-8 September 2006.

Gagnon, L. (2008). Civilisation and energy payback. *Energy Policy*, **36**, pp. 3317-3322.

Gipe, P. (2010). *Wind Energy Deaths Database - Summary of Deaths in Wind Energy.* No publisher specified. Available at: www.wind-works.org/articles/BreathLife.html.

Gleick, P. (1993). *Water in Crisis: A Guide to the World's Fresh Water Resources.* Oxford University Press, New York, NY, USA.

Gregory, R., and S. Lichtenstein (1994). A hint of risk: tradeoffs between quantitative and qualitative risk factors. *Risk Analysis*, **14**(2), pp. 199-206.

Haimes, Y.Y. (2009). On the complex definition of risk: A systems-based approach. *Risk Analysis*, **29**(12), pp. 1647-1654.

Herendeen, R.A. (1988). Net energy considerations. In: *Economic Analysis of Solar Thermal Energy Systems.* R.E. West and F. Kreith (eds.), The MIT Press, Cambridge, MA, USA, pp. 255-273.

Hirschberg, S., G. Spiekerman, and R. Dones (1998). *Severe Accidents in the Energy Sector - First Edition.* PSI Report No. 98-16. Paul Scherrer Institut, Villigen PSI, Switzerland.

Hirschberg, S., P. Burgherr, G. Spiekerman, and R. Dones (2004a). Severe accidents in the energy sector: Comparative perspective. *Journal of Hazardous Materials*, **111**(1-3), pp. 57-65.

Hirschberg, S., P. Burgherr, G. Spiekerman, E. Cazzoli, J. Vitazek, and L. Cheng (2003a). Assessment of severe accident risks. In: *Integrated Assessment of Sustainable Energy Systems in China. The China Energy Technology Program - A framework for decision support in the electric sector of Shandong province. Alliance for Global Sustainability Series Vol. 4.* Kluwer Academic Publishers, Amsterdam, The Netherlands, pp. 587-660.

Hirschberg, S., P. Burgherr, G. Spiekerman, E. Cazzoli, J. Vitazek, and L. Cheng (2003b). *Comparative Assessment of Severe Accidents in the Chinese Energy Sector.* PSI Report No. 03-04. Paul Scherrer Institut, Villigen PSI, Switzerland.

Hirschberg, S., R. Dones, T. Heck, P. Burgherr, W. Schenler, and C. Bauer (2004b). *Sustainability of Electricity Supply Technologies under German Conditions: A Comparative Evaluation.* PSI-Report No. 04-15. Paul Scherrer Institut, Villigen, Switzerland.

Huettner, D.A. (1976). Net energy analysis: an economic assessment. *Science*, **192**(4235), pp. 101-104.

IEA (2009). *World Energy Outlook 2009.* International Energy Agency, Paris, France, pp. 670-673.

IEA (2010a). *Energy Balances of Non-OECD Countries; 2010 Edition.* International Energy Agency, Paris, France.

IEA (2010b). *Key World Energy Statistics.* International Energy Agency, Paris France.

IEA/OECD/Eurostat (2005). *Energy Statistics Manual.* Organisation for Economic Co-operation and Development and International Energy Agency, Paris, France.

Inhaber, H. (2004). Water use in renewable and conventional electricity production. *Energy Sources*, **26**(3), pp. 309-322.

IPCC (1996). *Climate Change 1995: Impacts, Adaptation, and Mitigation of Climate Change - Scientific-Technical Analysis. Contribution of Working Group II to the Second Assessment Report of the Intergovernmental Panel on Climate Change.* R.T. Watson, M.C. Zinyowera, and R.H. Moss (eds.), Cambridge University Press, 879 pp.

IPCC (2000). *Special Report on Emissions Scenarios.* N. Nakicenovic and R. Swart (eds.), Cambridge University Press, 570 pp.

Jacobson, M.Z. (2009). Review of solutions to global warming, air pollution, and energy security. *Energy and Environmental Science*, **2**(2), pp. 148-173.

Jelen, F.C., and J.H. Black (1983). *Cost and Optimization Engineering.* McGraw-Hill, New York, NY, USA, 538 pp.

Jungbluth, N., C. Bauer, R. Dones and R. Frischknecht (2005). Life cycle assessment for emerging technologies: Case studies for photovoltaic and wind power. *International Journal of Life Cycle Assessment*, **10**(1), pp. 24-34.

Kristensen, V., T. Aven, and D. Ford (2006). A new perspective on Renn and Klinke's approach to risk evaluation and management. *Reliability Engineering and System Safety*, **91**, pp. 421-432.

Kubiszewski, I., C.J. Cleveland, and P.K. Endres (2010). Meta-analysis of net energy return for wind power systems. *Renewable Energy*, **35**(1), pp. 218-225.

Leach, G. (1975). Net energy analysis - is it any use? *Energy Policy*, **3**(4), pp. 332-344.

Lenzen, M. (1999). Greenhouse gas analysis of solar-thermal electricity generation. *Solar Energy*, **65**(6), pp. 353-368.

Lenzen, M. (2008). Life cycle energy and greenhouse gas emissions of nuclear energy: A review. *Energy Conversion and Management*, **49**(8), pp. 2178-2199.

Lenzen, M., and J. Munksgaard (2002). Energy and CO_2 analyses of wind turbines – review and applications. *Renewable Energy*, **26**(3), pp. 339-362.

Lenzen, M., C. Dey, C. Hardy, and M. Bilek (2006). *Life-Cycle Energy Balance and Greenhouse Gas Emissions of Nuclear Energy in Australia.* Report to the Prime Minister's Uranium Mining, Processing and Nuclear Energy Review (UMPNER), ISA, University of Sydney, Sydney, Australia. Available at: http://www.isa.org.usyd.edu.au/publications/documents/ISA_Nuclear_Report.pdf.

Lightfoot, H.D. (2007). Understand the three different scales for measuring primary energy and avoid errors. *Energy*, **32**(8), pp. 1478-1483.

Loulou, R., M. Labriet, and A. Kanudia (2009). Deterministic and stochastic analysis of alternative climate targets under differentiated cooperation regimes. *Energy Economics*, **31**(Supplement 2), pp. S131-S143.

Macknick, J. (2009). *Energy and Carbon Dioxide Emission Data Uncertainties.* International Institute for Applied Systems Analysis (IIASA) Interim Report, IR-09-032, IIASA, Laxenburg, Austria.

Martinot, E., C. Dienst, L. Weiliang, and C. Qimin (2007). Renewable energy futures: Targets, scenarios, and pathways. *Annual Review of Environment and Resources*, **32**(1), pp. 205-239.

Morita, T., J. Robinson, A. Adegbulugbe, J. Alcamo, D. Herbert, E. Lebre la Rovere, N. Nakicenivic, H. Pitcher, P. Raskin, K. Riahi, A. Sankovski, V. Solkolov, B.d. Vries, and D. Zhou (2001). Greenhouse gas emission mitigation scenarios and implications. In: *Climate Change 2001: Mitigation; Contribution of Working Group III to the Third Assessment Report of the IPCC.* Metz, B., Davidson, O., Swart, R., and Pan, J. (eds.), Cambridge University Press, pp. 115-166.

Nakicenovic, N., A. Grubler, and A. McDonald (eds.) (1998). *Global Energy Perspectives.* Cambridge University Press.

Neely, J.G., A.E. Magit, J.T. Rich, C.C.J. Voelker, E.W. Wang, R.C. Paniello, B. Nussenbaum, and J.P. Bradley (2010). A practical guide to understanding systematic reviews and meta-analyses. *Otolaryngology-Head and Neck Surgery*, **142**, pp. 6-14.

NETL (2007a). *Cost and Performance Baseline for Fossil Energy Plants-Volume 1: Bituminous Coal and Natural Gas to Electricity Final Report.* DOE/NETL-2007/1281, National Energy Technology Laboratory, Pittsburgh, PA, USA.

NETL (2007b). *Power Plant Water Usage and Loss Study. 2007 Update.* National Energy Technology Laboratory, Pittsburgh, PA, USA. Available at: www.netl.doe.gov/technologies/coalpower/gasification/pubs/pdf/WaterReport_Revised%20May2007.pdf.

Perry, A.M., W.D. Devine, and D.B. Reister (1977). *The Energy Cost of Energy - Guidelines for Net Energy Analysis of Energy Supply Systems.* ORAU/IEA(R)-77-14, Institute for Energy Analysis, Oak Ridge Associated Universities, Oak Ridge, TN, USA, 106 pp.

Renewable UK (2010). *Offshore Windfarms Operational.* Renewable UK. Available at: www.renewable-manifesto.com/ukwed/offshore.asp.

Renn, O., A. Klinke, G. Busch, F. Beese, and G. Lammel (2001). A new tool for characterizing and managing risks. In: *Global Biogeochemical Cycles in the Climate System.* E.D. Schulze, M. Heimann, S. Harrison, E. Holland, J. Lloyd, I. Prentice, and D. Schimel (eds.), Academic Press, San Diego, CA, USA, pp. 303-316.

Roth, S., S. Hirschberg, C. Bauer, P. Burgherr, R. Dones, T. Heck, and W. Schenler (2009). Sustainability of electricity supply technology portfolio. *Annals of Nuclear Energy*, **36**, pp. 409–416.

Rotty, R.M., A.M. Perry, and D.B. Reister (1975). *Net Energy from Nuclear Power.* IEA Report, Institute for Energy Analysis, Oak Ridge Associated Universities, Oak Ridge, TN, USA.

Samson, S., J. Reneke, and M.M. Wiecek (2009). A review of different perspectives on uncertainty and risk and analternative modeling paradigm. *Reliability Engineering and System Safety*, **94**, pp. 558-567.

Sovacool, B.K. (2008a). The cost of failure: a preliminary assessment of major energy accidents, 1907–2007. *Energy Policy*, **36**, pp. 1802-1820.

Sovacool, B.K. (2008b). Valuing the greenhouse gas emissions from nuclear power: A critical survey. *Energy Policy*, **36**(8), pp. 2950-2963.

Stirling, A. (1999). Risk at a turning point? *Journal of Environmental Medicine*, **1**, pp. 119-126.

UN Statistics (2010). *Energy Balances and Electricity Profiles – Concepts and definitions.* UN Statistics, New York, NY, USA. Available at: unstats.un.org/unsd/energy/balance/concepts.htm.

Ungers, L.J., P.D. Moskowitz, T.W. Owens, A.D. Harmon, and T.M. Briggs (1982). Methodology for an occupational risk assessment: an evaluation of four processes for the fabrication of photovoltaic cells. *American Industrial Hygiene Association Journal*, **43**(2), pp. 73-79.

Voorspools, K.R., E.A. Brouwers, and W.D. D'haeseleer (2000). Energy content and indirect greenhouse gas emissions embedded in 'emission-free' plants: results from the Low Countries. *Applied Energy*, **67**, pp. 307-330.

WBGU (2000). *World in Transition: Strategies for Managing Global Environmental Risks. Flagship Report 1998.* German Advisory Council on Global Change (WBGU). Springer, Berlin, Germany.

WEC (1993). *Energy for Tomorrow's World. WEC Commission global report.* World Energy Council, London, UK.

WRA (2008). *A Sustainable Path: Meeting Nevada's Water and Energy Demands.* Western Resource Advocates (WRA), Boulder, CO, USA, 43 pp. Available at: www.westernresourceadvocates.org/water/NVenergy-waterreport.pdf.

Recent Renewable Energy Cost and Performance Parameters

ANNEX III

Lead Authors:
Thomas Bruckner (Germany), Helena Chum (USA/Brazil),
Arnulf Jäger-Waldau (Italy/Germany), Ånund Killingtveit (Norway),
Luis Gutiérrez-Negrín (Mexico), John Nyboer (Canada), Walter Musial (USA),
Aviel Verbruggen (Belgium), Ryan Wiser (USA)

Contributing Authors:
Dan Arvizu (USA), Richard Bain (USA), Jean-Michel Devernay (France), Don Gwinner (USA),
Gerardo Hiriart (Mexico), John Huckerby (New Zealand), Arun Kumar (India),
José Moreira (Brazil), Steffen Schlömer (Germany)

This annex should be cited as:
Bruckner, T., H. Chum, A. Jäger-Waldau, Å. Killingtveit, L. Gutiérrez-Negrín, J. Nyboer, W. Musial, A. Verbruggen, R. Wiser, 2011: Annex III: Cost Table. In IPCC Special Report on Renewable Energy Sources and Climate Change Mitigation [O. Edenhofer, R. Pichs-Madruga, Y. Sokona, K. Seyboth, P. Matschoss, S. Kadner, T. Zwickel, P. Eickemeier, G. Hansen, S. Schlömer, C. von Stechow (eds)], Cambridge University Press, Cambridge, United Kingdom and New York, NY, USA.

Annex III Recent Renewable Energy Cost and Performance Parameters

Annex III is intended to become a 'living document', which will be updated in the light of new information in order to serve as an input to the IPCC Fifth Assessment Report (AR5). Scientists that are interested in supporting this process are invited to contact the IPCC WG III Technical Support Unit (TSU) (using srren_cost@ipcc-wg3.de) in order to get further information concerning the submission process.[1] Comments and new data input will be considered for inclusion in Volume 3 of the IPCC AR5 according to the procedures of the IPCC review system.

This Annex contains recent cost and performance parameter information for currently commercially available renewable power generation technologies (Table A.III.1), heating technologies (Table A.III.2) and biofuel production processes (Table A.III.3). It summarizes information that determines the levelized cost of energy or energy carriers supplied by the respective technologies.

The input ranges are based on assessments of various studies by authors of the respective technology chapters (Chapters 2 through 7). If not stated otherwise, the data ranges provided here are worldwide aggregates. Data are generally for 2008, but can be as recent as 2009. They represent roughly the mid-80% of values found in the literature, hence, excluding outliers. The availability and quality of different sources of data varies significantly across individual technologies for a variety of reasons.[2] Some expert judgment is therefore required to determine data ranges that are representative of particular classes of technologies and specific periods of time and valid globally.

The references to specific information are quoted in the footnotes. If the full dataset is based on one particular reference, it is included in the reference column of the green part of the table. Further information on the data reported in the table is provided in the footnotes and in Chapters 2 through 7 (see in particular Sections 2.7, 3.8, 4.7, 5.8, 6.7 and 7.8).

The levelized cost of electricity (LCOE), heat (LCOH) and transport fuels (LCOF)[3] are calculated based on the data compiled here and the methodology described in Annex II, using three different real discount rates (3, 7 and 10%). They represent the full range of possible levelized cost values resulting from the lower and upper bounds of input data in this table. More precisely, the lower bound of the levelized cost ranges is based on the low ends of the ranges of investment, operation and maintenance (O&M) and (if applicable) feedstock cost and the high ends of the ranges of capacity factors and lifetimes as well as (if applicable) the high ends of the ranges of conversion efficiencies and by-product revenue stated in this table. The higher bound of the levelized cost ranges is accordingly based on the high end of the ranges of investment, O&M and (if applicable) feedstock costs and the low end of the ranges of capacity factors and lifetimes as well as (if applicable) the low ends of the ranges of conversion efficiencies and by-product revenue.[4]

These levelized cost figures (violet parts of the tables) are discussed in Sections 1.3.2 and 10.5.1 of the main report. Most technology chapters (Chapters 2 through 7) provide more detail on the sensitivity of the levelized costs to particular input parameters beyond discount rates (see in particular Sections 2.7, 3.8, 4.7, 5.8, 6.7 and 7.8). These sensitivity analyses provide additional insights into the relative weight of the large number of parameters that determine the levelized costs under more specific conditions.

In addition to the technology-specific sensitivity analysis in the respective chapters (Chapters 2 through 7) and the discussions in Sections 1.3.2 and 10.5.1, Figures A.III.2 through A.III.4 (a, b) show the sensitivity of the levelized cost in a complementary way using so-called tornado graphs (Figures A.III.2 through A.III.4 a) as well as their 'negatives' (Figures A.III.2 through A.III.4 b).

Figures A.III.1a and A.III.1b show schematic versions of the tornado graphs and their 'negatives', respectively, explaining how to read them correctly.

---

1 No individual responses can be guaranteed, but all emails as well as relevant material attached to those emails will be archived and made available in appropriate form to the authors involved in the AR5 process.

2 No standardized uncertainty language has been used in this report. Nonetheless, the authors of this Annex have carefully assessed available data and highlighted data limitations and uncertainties in the footnotes. A fair impression of the breadth of the reference base can be deduced from the list of references in this Annex.

3 The levelized cost represents the cost of an energy generating system over its lifetime. It is calculated as the per unit price at which energy must be generated from a specific source over its lifetime to break even. The levelized costs usually include all private costs that accrue upstream in the value chain, but they do not include the downstream cost of delivery to the final customer, the cost of integration, or external environmental or other costs. Subsidies for RE generation and tax credits are not included. However, indirect taxes and subsidies on inputs or commodities affecting the prices of inputs and, hence, private cost, cannot be fully excluded.

4 This approach assumes that input parameters to the LCOE/LCOH/LCOF calculation are independent from each other. This is a simplifying assumption that implies that the lower ranges of LCOE/LCOH/LCOF (as a combination of best-case input values) may in some cases be lower than is most often the case, while the upper range of LCOE/LCOH/LCOFs (as a combination of worst-case input values) may in some cases be higher than what is generally considered economically attractive from a private investors' perspective. The extent to which this approach introduces a structural bias in the LCOE/LCOH/LCOF ranges, however, is reduced by taking a rather conservative approach to the range of input values (partly involving expert judgement), that is, by restricting input values roughly to the medium 80% range where possible.

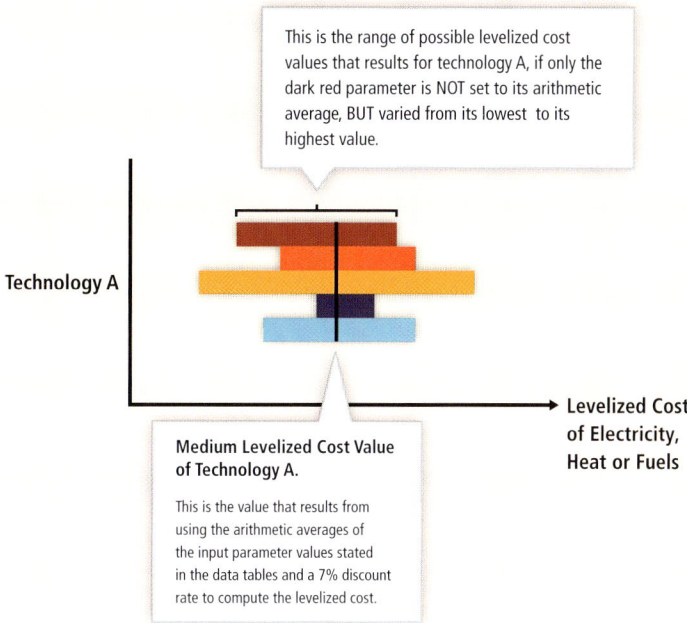

Figure A.III.1a | Tornado graph. Starting from the medium levelized cost value at a 7% interest rate, a broader range of levelized cost values becomes possible if individual parameters are varied over the full of range of values that these parameters may take on under different conditions. If the LCOE/LCOH/LCOF of a technology is very sensitive to variation of a particular parameter, then the corresponding bar will be broad. This means that a variation of that particular parameter may lead to LCOE/LCOH/LCOF values that can deviate strongly from the medium LCOE/LCOH/LCOF value. If the LCOE/LCOH/LCOF of a technology is robust for variations of the respective parameter, the bars will be narrow and only slight deviations from the medium LCOE/LCOH/LCOF value may result from variation of that parameter. Note, however, that no or narrow bars may also be the result of no or limited variation of the input parameters.

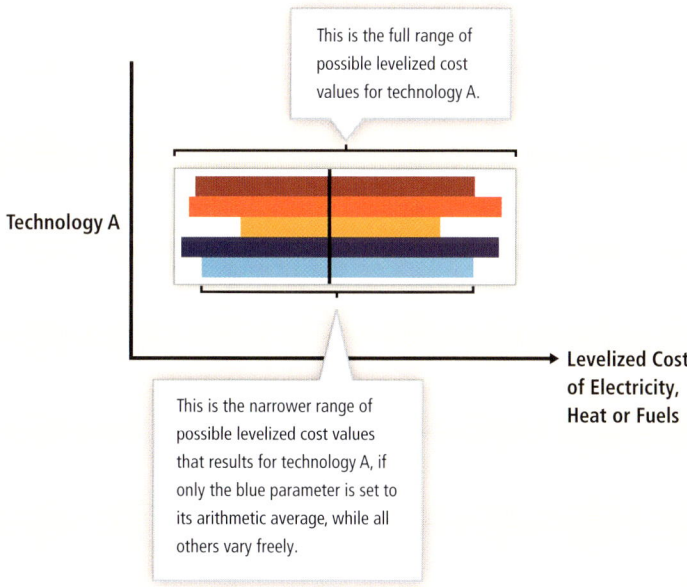

Figure A.III.1b | 'Negative' of tornado graph. Starting from the low and high bounds of the full range of levelized cost values at a 3% and 10% interest rate, respectively, a narrower range of levelized cost values remains possible if individual parameters are fixed at their respective medium values. If the LCOE/LCOH/LCOF of a technology is very sensitive to variations of a particular parameter, then the corresponding bar that remains will be narrowed to a large degree. Such parameters are of particular importance in determining the LCOE/LCOH/LCOF under more specific conditions. If the LCOE/LCOH/LCOF of a technology is robust for variations of the respective parameter, the remaining range will remain close to the full range of possible LCOE/LCOH/LCOF values. Such parameters are of less importance in determining the LCOE/LCOH/LCOF more precisely. Note, however, that no or small deviations from the full range may also be the result of no or limited variation of the input parameters.

Table A.III.1 | Cost-performance parameters for RE power generation technologies.[i]

Resource	Technology	Typical size of the device (MW)[ii]	Investment cost (USD/kW)	O&M cost, fixed annual (USD/kW) and/or (non-feed) variable (US¢/kWh)	By-product revenue (US¢/kWh)[iii]	Feedstock cost (USD/GJ$_{feed, HHV}$[iv])	Feedstock conversion efficiency$_{el}$ (%)	Capacity factor (%)	Economic design lifetime (years)	References	LCOE[v] (US¢/kWh) Discount rate 3%	7%	10%
Bioenergy	Dedicated Biopower CFB[vi]	25–100	2,700–4,100[vii]	87 USD/kW and 0.40 US¢/kWh	N/A[viii]	1.25–5.0[ix]	28	70–80	20	McGowin (2008)	6.1–13	6.9–15	7.9–16
	Dedicated Biopower Stoker[x]	See above	2,600–4,000[vii]	84 USD/kW and 0.34 US¢/kWh	N/A[viii]	See above	27	See above	See above		5.6–13	6.7–15	7.7–16
	Dedicated Biopower (Stoker CHP[xi])	See above	2,800–4,200[vii]	86 USD/kW and 0.35 US¢/kWh	1.0[xii]	See above	24	See above	See above		5.1–13	6.3–15	7.3–17
	Co-firing: Co-feed	20–100	430–500[xiii]	12 USD/kW and 0.18 US¢/kWh	N/A[viii]	See above	36	See above	See above	McGowin (2008)	2.0–5.9	2.2–6.2	2.3–6.4
	Co-firing: Separate Feed	See above	760–900[xiii]	18 USD/kW	N/A[viii]	See above	36	See above	See above	Bain (2011)	2.3–6.3	2.6–6.7	2.9–7.1
	CHP (ORC[xiv])	0.65–1.6	6,500–9,800	59–80 USD/kW and 4.3–5.1 US¢/kWh	7.7[xv, xvi]	See above	14	55–68	See above	Obernberger et al. (2008)	8.6–26	12–32	15–37
	CHP (Steam Turbine)	2.5–10	4,100–6,200[xvii]	54 USD/kW and 3.5 US¢/kWh	5.4[xv, xviii]	See above	18	See above	See above		6.2–18	8.3–22	10–26
	CHP (Gasification ICE)[xix]	2.2–13	1,800–2,100	65–71 USD/kW and 1.1–1.9 US¢/kWh	1.0–4.5[xv, xx]	See above	28–30	See above	See above		2.1–11	3.0–13	3.8–14
Direct Solar Energy	PV (Residential Rooftop)	0.004–0.01	3,700–6,800[xxi]	19–110 USD/kW[xxii]	N/A[viii]	N/A[viii]	N/A[viii]	12–20[xxiii]	20–30	see Section 3.8 and footnotes	12–53	18–71	23–86
	PV (Commercial Rooftop)	0.02–0.5	3,500–6,600[xxi]	18–100 USD/kW[xxii]	N/A[viii]	N/A[viii]	N/A[viii]	See above	See above		11–52	17–69	22–83
	PV (Utility Scale, Fixed Tilt)	0.5–100[xxiv]	2,700–5,200[xxi]	14–69 USD/kW[xxii]	N/A[viii]	N/A[viii]	N/A[viii]	15–21[xxiii]	See above		8.4–33	13–43	16–52
	PV (Utility Scale, One-Axis)	0.5–100[xxiv]	3,100–6,200[xxi]	16–75 USD/kW[xxii]	N/A[viii]	N/A[viii]	N/A[viii]	15–27[xxiii]	See above		7.4–39	11–52	15–62
	CSP	50–250[xxv]	6,000–7,300[xxvi]	60–82 USD/kW[xxvii]	N/A[viii]	N/A[viii]	N/A[viii]	35–42[xxviii]	See above		11–19	16–25	20–31
Geothermal Energy	Geothermal Energy (Condensing-Flash Plants)	10–100	1,800–3,600[xxix]	150–190 USD/kW[xxx]	N/A[viii]	N/A[viii]	N/A[viii]	60–90[xxxi]	25–30[xxxii]	see Section 4.7 and footnotes	3.1–8.4	3.8–11	4.5–13
	Geothermal Energy (Binary-Cycle Plants)	2–20	2,100–5,200[xxix]	See above	N/A[viii]	N/A[viii]	N/A[viii]	See above	See above		3.3–11	4.1–14	4.9–17
Hydropower	All	<0.1 – >20,000[xxxiii]	1,000–3,000[xxxiv]	25–75 USD/kW[xxxv]	N/A[viii]	N/A[viii]	N/A[viii]	30–60[xxxvi]	40–80[xxxvii]	see Chapter 5 and footnotes	1.1–7.8	1.8–11	2.4–15
Ocean Energy	Tidal Range[xxxviii]	<1 – >250[xxxix]	4,500–5,000[viii]	100 USD/kW[xxxviii]	N/A[viii]	N/A[viii]	N/A[viii]	22.5–28.5[xl]	40[xli, xxxviii]	see Section 6.7 and footnotes	12–16	18–24	23–32

Continued next page →

Annex III — Recent Renewable Energy Cost and Performance Parameters

Resource	Technology	Input data							References	Output data
		Typical size of the device (MW)[ii]	Investment cost (USD/kW)	O&M cost, fixed annual (USD/kW) and/or (non-feed) variable (US¢/kWh)	By-product revenue (US¢/kWh)[iii]	Feedstock cost (USD/GJ$_{feed, HHV}$[iv])	Feedstock conversion efficiency$_{el}$ (%)	Capacity factor (%)	Economic design lifetime (years)		LCOE[v] (US¢/kWh) Discount rate: 3% / 7% / 10%
Wind Energy	Wind Energy (Onshore, Large Turbines)	5–300[xiii]	1,200–2,100[xiii]	1.2–2.3 US¢/kWh	N/A[viii]	N/A[viii]	N/A[viii]	20–40[xiv]	20[xiv]	see Chapter 7	3.5–10 / 4.4–14 / 5.2–17
	Wind Energy (Off-Shore, Large Turbines)	20–120[xiii]	3,200–5,000[xiv]	2.0–4.0 US¢/kWh	N/A[viii]	N/A[viii]	N/A[viii]	35–45[xiii]	See above		7.5–15 / 9.7–19 / 12–23

General remarks/notes:

i All data are rounded to 2 significant digits. Most technology chapters (Chapters 2 through 7) provide additional and/or more detailed cost and performance information in the respective chapters' sections on cost trends. Direct comparison between levelized cost estimates taken directly from the literature should take the underlying assumptions into due consideration.

ii Device sizes are intended to be representative of current/recent sizes. If future sizes are expected to differ from these values, this is included in the footnotes to the relevant technologies.

iii For combined heat and power (CHP) plants, heat production is considered as a by-product in the calculation of the levelized cost of electricity providing full capital cost information as a stand-alone plant.

iv HHV: Higher heating value. LHV: Lower heating value.

v LCOE: Levelized cost of electricity. The levelized cost usually includes all private costs that accrue upstream in the value chain of electricity production, but they do not include the cost of transmission and distribution to the final customer. Output subsidies for RE generation and tax credits are not included. However, indirect taxes and subsidies on inputs or commodities affecting the prices of inputs and, hence, private cost, cannot be fully excluded. Depending on the context of discussion, LCOE may also stand for levelized cost of energy.

Bioenergy:

vi A circulating fluid bed (CFB) is a turbulent (high gas flow) fluid bed where solid particles are captured and returned to the bed. A fluid bed itself is a collection of small solid particles suspended and kept in motion by an upward flow of fluid, typically a gas.

vii The reference data are for a 50 MW plant. Investment costs for larger and smaller plants have been rescaled according to the power law: Specific investment cost$_{size\,2}$ = Investment cost$_{size\,1}$ x (Size 2/Size 1)$^{n-1}$, where the scaling factor n = 0.7. Capital cost estimates include facilities for fuel handling and preparation, boiler and air quality control, steam turbine and auxiliaries, balance of plant, general facilities and engineering fee, project and process contingency, allowance for funds used during construction, owner costs, and taxes and fees.

viii The abbreviation 'N/A' means here 'not applicable'.

ix Feedstock is wood with HHV = 20.0 GJ/t, LHV = 18.6 GJ/t.

x A mechanical stoker is a machine or device that feeds fuel to a boiler.

xi CHP: Combined heat and power.

xii The calculation of the by-product revenue for the large-scale CHP plant assumes: heat output used for industrial applications is 5.38 GJ of heat per MWh electricity; steam is valued at USD$_{2005}$ 4.85/GJ (75% of US pulp and paper purchased steam price) (EIA, 2009, Table 7.2); and 75% of heat output is sold.

xiii The reference data are for a 50 MW plant. Investment costs for larger and smaller plants have been rescaled according to the power law: Specific investment cost$_{size\,2}$ = Investment cost$_{size\,1}$ x (Size 2/Size 1)$^{n-1}$, where the scaling factor n = 0.9 (Peters et al., 2003). The cofiring investment costs estimates were developed for retrofits of existing coal-fired power plants in the USA and include facilities for fuel handling and preparation, additional expenditures for boiler modifications, balance of plant, general facilities and engineering, project and process contingency, allowance for funds used during construction, owner costs, and taxes and fees. Cofiring cost estimate protocols in the USA do not include prorated boiler costs.

xiv ORC: Organic Rankine Cycle.

xv For the calculation of the by-product revenue for small-scale CHP plants, hot water is valued at USD$_{2005}$ 12.51/GJ (average of Rauch (2010) and Skjoldborg (2010)), 33% of gross value is taken into account, because the operator can only recover a portion of the value and because use of hot water is seasonal.

xvi Heat output used for hot water is 18.51 GJ of heat per MWh electricity.

xvii The reference data are for a 5 MW CHP plant. Investment costs for larger and smaller plants have been rescaled according to the power law: Specific investment cost$_{size\,2}$ = Investment cost$_{size\,1}$ x (Size 2/Size 1)$^{n-1}$, where the scaling factor n = 0.7 (Peters et al., 2003).

Continued next page →

xviii   Heat output used for hot water is 12.95 GJ of heat per MWh electricity.

xix   ICE: Internal combustion engine.

xx   Heat output used for hot water is in the range of 2.373 to 10.86 GJ/MWh.

Direct solar energy – photovoltaic (PV) systems:

xxi   In 2009, wholesale factory PV module prices decreased by more than 50%. As a result, the market prices for installed PV systems in Germany, the most competitive market, decreased by over 30% in 2009 compared to about 10% in 2008 (see Section 3.8.3). 2009 market price data from Germany is used as the lower bound for investment costs of residential rooftop systems (Bundesverband Solarwirtschaft e.V., 2010) and for utility-scale fixed tilt systems (Bloomberg, 2010). Based on US market data for 2008 and 2009, larger, commercial rooftop systems are assumed to have a 5% lower investment cost than the smaller, residential rooftop systems (NREL, 2011b; see also section 3.8.3). Tracking systems are assumed to have a 15-20% higher investment cost than the one-axis, non-tracking systems considered here (NREL, 2011a; see also Section 3.8.3). Capacity-weighted averages of investment costs in the USA in 2009 (NREL, 2011b) are used as upper bound to capture the investment cost ranges typical of roughly 80% of global installations in 2009 (see Section 3.4.1 and Section 3.8.3).

xxii   O&M costs of PV systems are low and are given in a range between 0.5 and 1.5% annually of the initial investment costs (Breyer et al., 2009; IEA, 2010c).

xxiii   The main parameter that influences the capacity factor of a PV system is the actual annual solar irradiation in $kWh/m^2/yr$ at a given location and the type of system. Capacity factors of some recently installed systems are provided in Sharma (2011).

xxiv   The upper limit of utility-scale PV systems represents current status. Much larger systems (up to 1 GW) are in the proposal and development phase and might be realized within the next decade.

Direct solar energy – concentrating solar power (CSP):

xxv   Project sizes of CSP plants can minimally match the size of a single power generating system (e.g., a 25 kW dish/engine system). However, the range provided is typical for projects being built or proposed today. 'Power Parks' consisting of multiple CSP plants in a single location are also being proposed at sizes of up to or exceeding 1 GW (4 x 250 MW).

xxvi   Cost ranges are for parabolic trough plants with six hours of thermal energy storage in 2009. Investment cost includes direct plus indirect costs where indirect costs include engineering, procurement and construction mark-up, owner costs, land, and taxes. Investment costs are lower for plants without storage and higher for plants with larger storage capacity. The IEA (2010a) estimates investment costs as low as USD_{2005} 3,800/kW for plants without storage and as high as USD_{2005} 7,600/kW for plants with large storage (assumed currency base year: 2009). Capacity factors vary as well, if thermal storage is installed (see note xxviii).

xxvii   The IEA (2010a) states O&M costs relative to energy output as US¢ 1.2 to 2.7/kWh (assumed currency base year: 2009). Depending on actual energy output this may result in lower or higher annual O&M cost compared to the range stated here.

xxviii   Capacity factor for a parabolic trough plant with six hours of thermal energy storage for solar resource classes typical of the southwest USA. Depending on the size of the thermal storage capacity, capacity factors as well as investment costs vary substantially. Apart from the Solar Electric Generating Station plants in California, new CSP plants only became operational from 2007 onwards, thus few actual performance data are available and most of the literature just gives estimated or predicted capacity factors. Sharma (2011) reports multi-year (1998-2002) average capacity factors of 12.4 to 27.7% for plants without thermal storage, but with natural gas backup. The IEA (2010a) states that plants in Spain with 15 hours of storage may produce up to 6,600 hours per year. This is equivalent to a 75% capacity factor, if production occurs at full capacity during the 6,600 hours. Larger storage also increases investment costs (see note xxvi).

Geothermal energy:

xxix   Investment cost includes: exploration and resource confirmation; drilling of production and injection wells; surface facilities and infrastructure; and the power plant. For expansion projects (i.e., new plants in the same geothermal field) investment costs can be 10 to 15% lower (see Section 4.7.1). Investment cost ranges are based on Bromley et al. (2010) (see also Figure 4.7).

xxx   O&M costs are based on Hance (2005). In New Zealand, O&M costs range from US¢ 1 to 1.4/kWh for 20 to 50 MW_e plant capacity (Barnett and Quinlivan, 2009), which are equivalent to USD 83 to 117/kW/yr, i.e. considerably lower than those given by Hance (2005). For further information see Section 4.7.2.

xxxi   The current (data for 2008-2009) worldwide capacity factor (CF) for condensing (flash) and binary-cycle plants in operation is 74.5%. Excluding some outliers, the lower and upper bounds can be estimated as 60 and 90%. Typical CFs for new geothermal power plants are over 90% (Hance, 2005; DiPippo, 2008; Bertani, 2010). The worldwide average CF for 2020 is projected to be 80%, and could be 85% in 2030 and as high as 90% in 2050 (see Sections 4.7.3 and 4.7.5).

xxxii   25 to 30 years is the common lifetime of geothermal power plants worldwide. This payback period allows for refurbishment or replacement of the aging surface plant at the end of its lifetime, but is not equivalent to the economic resource lifetime of the geothermal reservoir, which is typically much longer (e.g., Larderello, Wairakei, The Geysers: Section 4.7.3). In some reservoirs, however, the possibility of resource degradation over time is one of several factors that affect the economics of continuing plant operation.

Hydropower:

xxxiii   The mid-80% of project sizes is not well documented for hydropower. The range stated here is indicative of the full range of project sizes. Hydropower projects are always site-specific as they are designed to use the flow and head at each site. Therefore, projects can be very small, down to a few kW in a small stream, and up to several thousand MW, for example 18,000 MW for the Three Gorges project in China (which will be 22,400 MW when completed) (see Section 5.1.2). 90% of the installed hydropower capacity and 94% of hydropower energy production today is in hydropower plants >10 MW in size (IJHD, 2010).

xxxiv   The investment cost for hydropower projects can be as low as USD 400 to 500/kW but most realistic projects today lie in the range of USD 1,000 to 3,000/kW (Section 5.8.1).

xxxv   O&M costs are usually given as a percentage of investment cost for hydropower projects. Typical values range from 1 to 4%, while the table relies on an average value of 2.5% applied to the range of investment costs. This will usually be sufficient to cover refurbishment of mechanical and electrical equipment like turbine overhaul, generator rewinding and reinvestments in communication and control systems (Section 5.8.1).

Continued next page →

xxxvi Capacity factors (CF) will be determined by hydrological conditions, installed capacity and plant design, and the way the plant is operated (i.e., the degree of plant output regulation). For power plant designs intended for maximum energy production (base-load) and with some regulation, CFs will often be from 30 to 60%. Figure 5.20 shows average CFs for different world regions. For peaking-type power plants the CF will be much lower, down to 20%, as these stations are designed with much higher capacity in order to meet peaking needs. CFs for run-of-river systems vary across a wide range (20 to 95%) depending on the geographical and climatological conditions, technology and operational characteristics (see Section 5.8.3).

xxxvii Hydropower plants in general have very long physical lifetimes. There are many examples of hydropower plants that have been in operation for more than 100 years, with regular upgrading of electrical and mechanical systems but no major upgrades of the most expensive civil structures (dams, tunnels, etc.). The IEA (2010d) reports that many plants built 50 to 100 years ago are still operating today. For large hydropower plants, the lifetime can, hence, safely be set to at least 40 years, and an 80-year lifetime is used as upper bound. For small-scale hydropower plants the typical lifetime can be set to 40 years, in some cases even less. The economic design lifetime may differ from actual physical plant lifetimes, and will depend strongly on how hydropower plants are owned and financed (see Section 5.8.1).

Ocean Energy:

xxxviii The data supplied for tidal range power plants are based on a very small number of installations (see subsequent footnotes). Therefore, all data should be considered with appropriate caution.

xxxix The only utility-scale tidal power station in the world is the 240 MW La Rance power station, which has been in successful operation since 1966. Other smaller projects have been commissioned since then in China, Canada and Russia with 3.9 MW, 20 MW and 0.4 MW, respectively. The 254 MW Sihwa barrage is expected to be commissioned in 2011 and will then become the largest tidal power station in the world. Numerous projects have been identified, some of them with very large capacities, including in the UK (Severn Estuary, 9.3 GW), India (1.8 GW), Korea (740 MW) and Russia (the White Sea and Sea of Okhotsk, 28 GW). None have been considered to be economic yet and many of them face environmental objections (Kerr, 2007). The projects at the Severn Estuary have been evaluated by the UK government and recently been deferred.

xl An earlier assessment suggests capacity factors in the range of 25 to 35% (Charlier, 2003).

xli Tidal barrages resemble hydropower plants, which in general have very long design lives. Many hydropower plants have been in operation for more than 100 years, with regular upgrading of electro-mechanical systems but no major upgrades of the most expensive civil structures (dams, tunnels etc). Tidal barrages are therefore assumed to have a similar economic design lifetime as large hydropower plants, which can safely be set to at least 40 years (see Chapter 5).

Wind energy:

xlii Typical size of the device is taken as the power plant (not turbine) size. For onshore wind energy, 5 to 300 MW plants were common from 2007 to 2009, though both smaller and larger plants are prevalent. For offshore wind energy, 20 to 120 MW plants were common from 2007 to 2009, though much larger plant sizes are expected in the future. As a modular technology, a wide range of plant sizes is common, driven by market and geographic conditions.

xliii The lowest cost onshore wind power plants have been installed in China, with higher costs experienced in the USA and Europe. The range reflects the majority of onshore wind power plants installed worldwide in 2009 (the most recent year for which solid data exist as of writing), but plants installed in China have average costs that can be even below this range (USD 1,000 to 1,350/kW is common in China). In most cases, the investment cost includes the cost of the turbines (turbines, transportation to site, and installation), grid connection (cables, sub-station, interconnection, but not more general transmission expansion costs), civil works (foundations, roads, buildings), and other costs (engineering, licensing, permitting, environmental assessments, and monitoring equipment).

xliv Capacity factors depend in part on the strength of the underlying wind resource, which varies by region and site, as well as by turbine design.

xlv Modern wind turbines that meet International Electrotechnical Commission standards are designed for a 20-year life, and turbine lifetimes may even exceed 20 years if O&M costs remain at an acceptable level. Wind power plants are typically financed over a 20-year time period.

xlvi For offshore wind power plants, the range in investment costs includes the majority of offshore wind power plants installed in the most recent years (through 2009) as well as those plants planned for completion in the early 2010s. Because costs have risen in recent years, using the cost of recent and planned projects reasonably reflects the 'current' cost of offshore wind power plants. In most cases, the investment cost includes the cost of the turbines (turbines, transportation to site, and installation), grid connection (cables, sub-station, interconnection, but not more general transmission expansion costs), civil works (foundations, roads, buildings), and other costs (engineering, licensing, permitting, environmental assessments, and monitoring equipment).

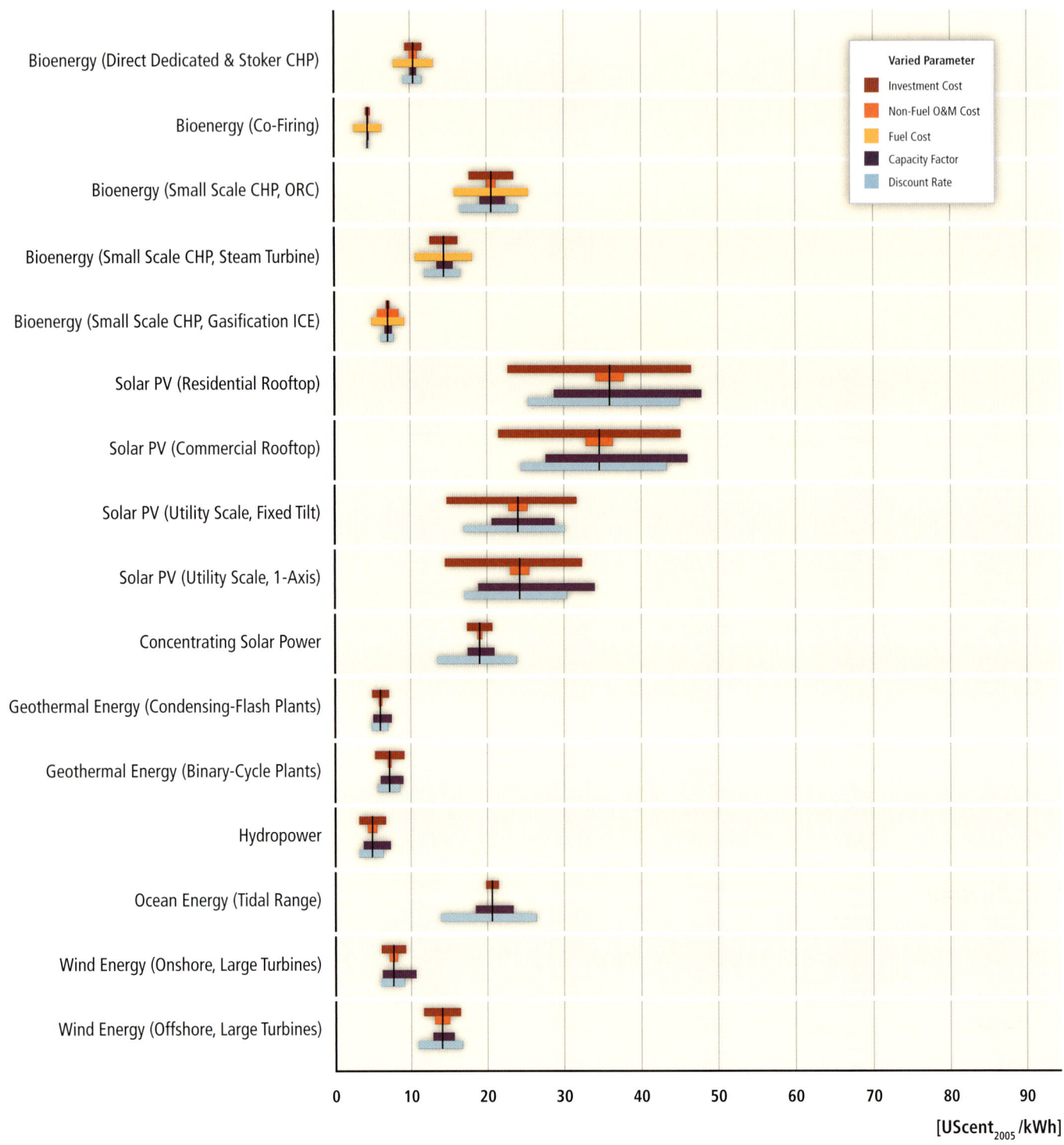

Figure A.III.2a | Tornado graph for renewable power technologies. For further explanation see Figure A.III.1a.

Annex III — Recent Renewable Energy Cost and Performance Parameters

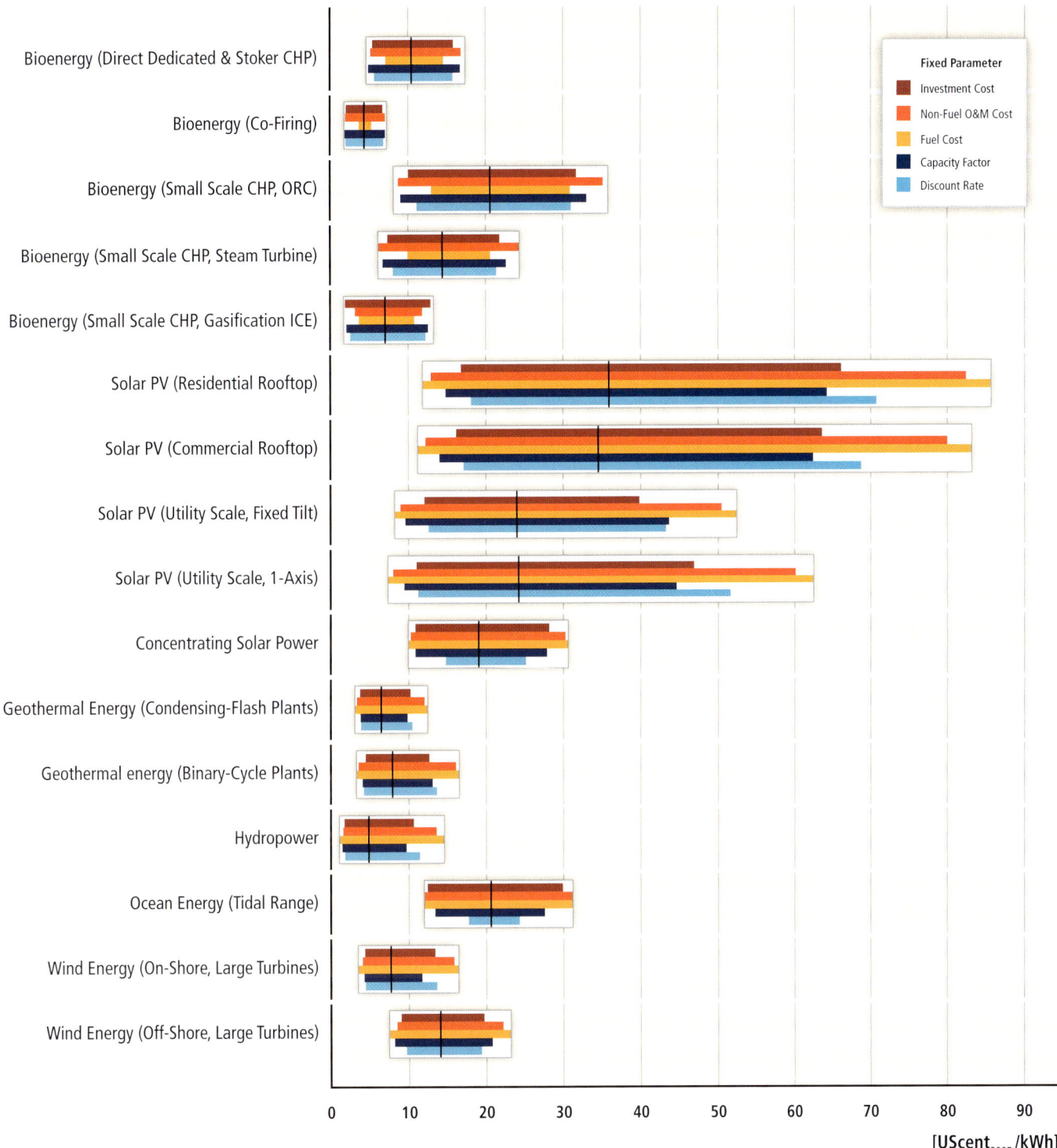

Figure A.III.2b | 'Negative' of tornado graph for renewable power technologies. For further explanation see Figure A.III.1b.

Note: The upper bounds of both geothermal energy technologies are calculated based on an assumed construction time of 4 years. In the simplified approach used for the sensitivity analysis shown here, this assumption was not taken into account, resulting in upper bounds that were below those based on the more accurate methodology. The ranges were rescaled, however, to yield the same results as the more accurate approach.

Recent Renewable Energy Cost and Performance Parameters — Annex III

Table A.III.2 | Cost-performance parameters for RE heating technologies.[i]

Resource	Technology	Typical size of the device (MW$_{th}$)	Investment cost (USD/kW$_{th}$)	O&M cost, fixed annual (USD/kW) and/or variable (USD/GJ)	By-product revenue (USD/GJ$_{feed}$)[ii]	Feedstock cost (USD/GJ$_{feed}$)	(Feedstock) conversion efficiency (%)	Capacity factor (%)	Economic design lifetime (years)	References	LCOH[iii] (USD/GJ) 3%	LCOH 7%	LCOH 10%
Bioenergy	Biomass (DPH)[iv]	0.005–0.1[v]	310–1,200[vi]	13–43 USD/kW[vii]	N/A[viii]	10–20	86–95	13–29	10–20	IEA (2007b)	14–70	15–77	16–84
	Biomass (MSW[ix], CHP)	1–10[xi]	370–3,000[xii, xiii]	15–130 USD/kW[vii]	N/A[viii]	0–3	20–40[xiv]	80–91	10–20		1.4–34	1.8–38	2.1–41
	Biomass (Steam Turbine, CHP)[xv]	12–14	370–1,000[xii]	1.2–2.5 USD/GJ[vii]	N/A[viii]	3.7–6.2	10–40	63–74	10–20		10–69	11–70	11–72
	Biomass (Anaerobic Digestion, CHP)	0.5–5[xi]	170–1,000[xii, xvi]	37–140 USD/kW[vii]	N/A[viii]	2.5–3.7[xvii]	20–30[xviii]	68–91	15–25		10–29	10–30	10–32
Solar Energy	Solar Thermal Heating (DHW[ix], China)	0.0017–0.01[xx]	120–540[xxi]	1.5–10 USD/GJ[vii]	N/A[viii]	N/A[viii]	20–80[xxiii]	4.1–13[xxiv]	10–15[xxv]	see Section 3.8.2 and footnotes	2.8–56	3.6–67	4.2–75
	Solar Thermal Heating (DHW, Thermo-siphon, Combi-systems)	0.0017–0.07[xx]	530–1,800	5.6–22 USD/kW[xxii]	N/A[viii]	N/A[viii]	20–80[xxiii]	4.1–13[xxiv]	15–25	IEA (2007b)	8.8–134	12–170	16–200
Geothermal Energy	Geothermal (Building Heating)	0.1–1	1,600–3,900[xxvi]	8.3–11 USD/GJ[xxvi]	N/A[viii]	N/A[viii]	N/A[viii]	25–30	20	see Section 4.7.6	20–50	24–65	28–77
	Geothermal (District Heating)	3.8–35	600–1,600[xxvi]	8.3–11 USD/GJ[xxvi]	N/A[viii]	N/A[viii]	N/A[viii]	25–30	25		12–24	14–31	15–38
	Geothermal (Greenhouses)	2–5.5	500–1,000[xxvi]	5.6–8.3 USD/GJ[xxvi]	N/A[viii]	N/A[viii]	N/A[viii]	50	20		7.7–13	8.6–14	9.3–16
	Geothermal (Aquaculture Ponds, Uncovered)	5–14	50–100[xxvi]	8.3–11 USD/GJ[xxvi]	N/A[viii]	N/A[viii]	N/A[viii]	60	20		8.5–11	8.6–12	8.6–12
	Geothermal Heat Pumps (GHP)	0.01–0.35	900–3,800[xxvi]	7.8–8.9 USD/GJ[xxvi]	N/A[viii]	N/A[viii]	N/A[viii]	25–30	20		14–42	17–56	19–68

Continued next page →

General remarks/notes:

i All data are rounded to 2 significant digits. Most technology chapters (Chapters 2 through 4) provide additional and/or more detailed cost and performance information in the respective chapters' sections on cost trends. The assumptions underlying some of the production cost estimates quoted directly from the literature may, however, not be as transparent as the data sets in this Annex and should therefore be considered with caution.

ii CHP plants produce both, heat and electricity. Calculating the levelized cost of one product only, that is, either heat or electricity, can be done in different ways. One way is to assign a (discounted) market value to the 'by-product' and subtract this additional income from the remaining expenses. This has been done in the calculation of the LCOE of bioenergy CHP plants. The calculation of LCOH has been done in a different way according to the methodology used in IEA (2007) which served as main reference for the input data: Instead of considering electricity as a 'by-product' and subtracting its value from the remaining expenses for the supply of heat, the total expenses over the lifetime of the investment project were split according to the average heat/electricity output ratio and only the heat shares of investment and O&M costs were taken into account. For this reason no by-product revenue is stated in the heat table. Both methodologies come with different advantages/disadvantages.

iii LCOH: Levelized cost of heat supply. The levelized cost does not include the cost of transmission and distribution in the case of district heating systems. Output subsidies for RE generation and tax credits are also excluded. However, indirect taxes and subsidies on inputs or commodities affecting the prices of inputs and, hence, private cost, cannot be fully excluded.

Bioenergy:

iv   DPH: Domestic pellet heating.

v   This range is typical of a low-energy single family dwelling (5 kW) or an apartment building (100 kW).

vi   Investment costs of a biomass pellet heating system for the combustion plant only (including controls) range from USD_{2005} 100 to 640/kW. The higher range stated above includes civil works and fuel and heat storage (IEA, 2007).

vii   Fixed annual O&M costs include costs of auxiliary energy. Auxiliary energy needs are 10 to 20 kWh/kW_{th}/yr. Electricity prices are assumed to be USD_{2005} 0.1 to 0.3/kWh. O&M costs for CHP options include heat share only.

viii   The abbreviation 'N/A' means here 'not applicable'.

ix   MSW: Municipal solid waste.

x   CHP: Combined heat and power.

xi   Typical size based on expert judgment and cost data from IEA (2007).

xii   Investment costs for CHP options include heat share only. The electricity data in Table A.III.1 provides examples of total investment cost (see Section 2.4.4).

xiii   Investment costs of MSW installations are mainly determined by the cost of flue gas cleaning, which can be allocated to waste treatment rather than to heat production (IEA, 2007).

xiv   Heat-only MSW incinerators (as used in Denmark and Sweden) could have a thermal efficiency of 70 to 80%, but are not considered (IEA, 2007).

xv   The ranges provided in this category are mainly based on two plants in Denmark and Austria and have been taken from IEA (2007).

xvi   Investment costs for anaerobic digestion are based on literature values provided relative to electric capacity. For conversion to thermal capacity an electric efficiency of 37% and a thermal efficiency of 55% were used (IEA, 2007).

xvii   For anaerobic digestion, fuel prices are based on a mix of green crop maize and manure feedstock. Other biogas feedstocks include source-separated wastes and landfill gas, but are not considered here (IEA, 2007).

xviii   Conversion efficiencies include auxiliary heat input (8 to 20% for process heat) as well as use of any co-substrate that might increase process efficiency. For source-separated wastes, the efficiency would be lower (IEA, 2007).

Solar Energy:

xix   DHW: Domestic hot water.

xx   1 m² of collector area is converted into 0.7 kW_{th} of installed capacity (see Section 3.4.1).

xxi   70% of the 13.5 million m² sales volume in 2004 was sold below Yuan 1,500/m² (USD_{2005} ~190/kW) (Zhang et al., 2010). The lower bound is based on data collected during standardized interviews in the Zhejiang Province, China, in 2008 (Han et al., 2010). The higher bound is based on Chang et al. (2011).

xxii   Fixed annual operating cost is assumed to be 1 to 3% of investment cost (IEA, 2007) plus annual cost of auxiliary energy. Annual auxiliary energy needs are 2 to 10 kWh/m². Electricity prices are assumed to be USD_{2005} 0.1 to 0.3/kWh.

xxiii   The conversion efficiency of a solar thermal system tends to be larger in regions with lower solar irradiance. This partly offsets the negative effect of lower solar irradiance on cost as energy yields per m² of collector area will be similar (Harvey, 2006, p. 461). Conversion efficiencies, which affect the resulting capacity factor, have not been used in LCOH calculations directly.

xxiv   Capacity factors are based on an assumed annual energy yield of 250 to 800 kWh/m² (IEA, 2007).

xxv   Expected design lifetimes for Chinese solar water heaters are in the range of 10 to 15 years (Han et al., 2010).

Geothermal energy:

xxvi   For geothermal heat pumps (GHP) the bounds of investment costs include residential and commercial or institutional installations. For commercial and institutional installations, costs are assumed to include drilling costs, but for residential installations drilling costs are not included.

xxvii   Average O&M costs expressed in USD_{2005}/kWh_{th} are: 0.03 to 0.04 for building and district heating and for aquaculture uncovered ponds, 0.02 to 0.03 for greenhouses, and 0.028 to 0.032 for GHP.

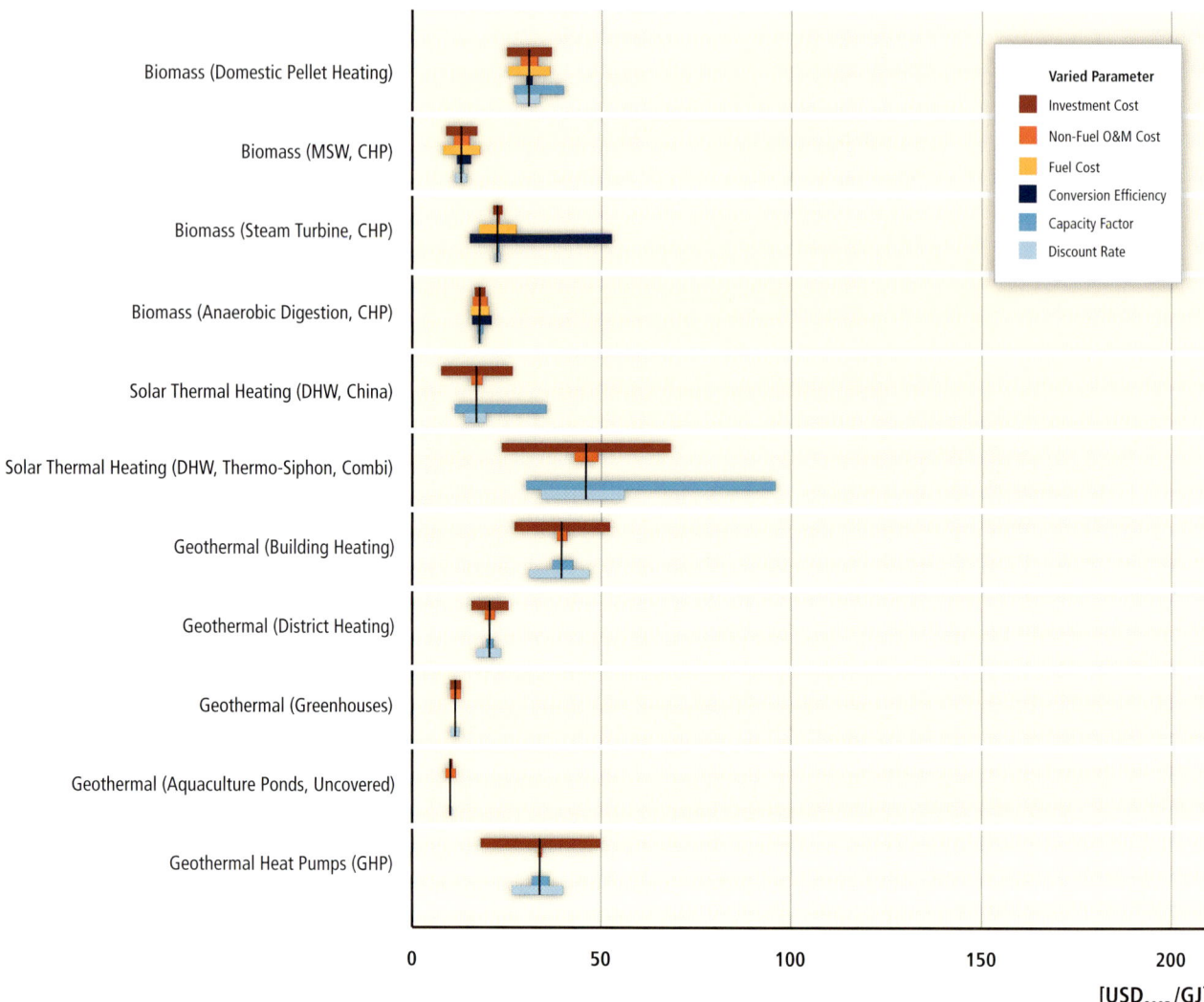

Figure A.III.3a | Tornado graph for renewable heat technologies. For further explanation see Figure A.III.1a.

Note: It may be somewhat misleading that solar thermal and geothermal heat applications do not show any sensitivity to variations in conversion efficiencies. This is due to the fact that the energy input for solar and geothermal has zero cost and that the effect of higher conversion efficiencies of the energy input on LCOH works solely via an increase in annual output. Variations in annual output, in turn, are fully captured by varying the capacity factor.

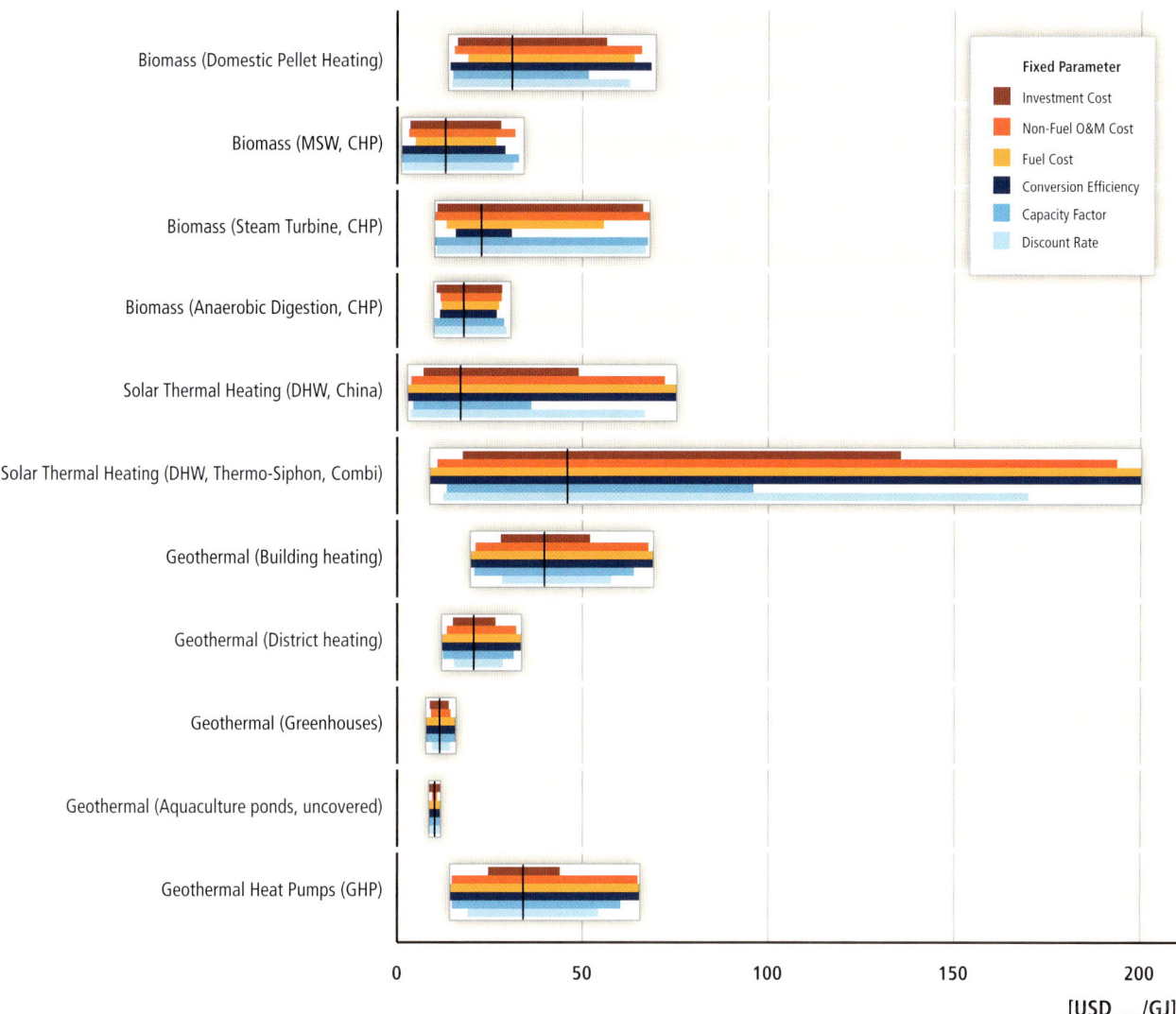

Figure A.III.3b | 'Negative' of tornado graph for renewable heat technologies. For further explanation see Figure A.III.1b.

Table A.III.3 | Cost-performance parameters for biofuels.[i]

Feedstock	Fuel, Region	Input data							Output data
		Typical size of the device (MW_{th})	Investment cost (USD/kW_{th})[ii]	O&M cost, fixed annual (USD/kW_{th}) and non-feed variable (USD/GJ_{feed})	By-product Revenue (USD/GJ_{feed})	Feedstock cost (USD/GJ_{feed})	Feedstock conversion efficiency[iii] (%) Product only (product + by-product)	Capacity factor (%)	Economic design lifetime (years)	References	LCOF[iv] USD/GJ_{HHV}[v] Discount rate
											3%	7%	10%
Ethanol													
Sugarcane	Overall	170–1,000	83–360	16–35 USD/kW$_{th}$ and 0.87 USD/GJ$_{feed}$	Co-product: sugar[vi] 4.3	2.1–7.1	17 (39)	50%	20	Alfstad (2008), Bain (2007), Kline et al. (2007)	2.4–39	3.5–42	4.5–46
	Brazil, Case A[vii]	See above	100–330	20–32 USD/kW$_{th}$ and 0.87 USD/GJ$_{feed}$	See above	2.1–6.5[viii]	See above	See above	See above	Bohlmann and Cesar (2006), Oliverio (2006), van den Wall Bake et al. (2009)	2.4–38	3.5–41	4.5–44
	Argentina	See above	110–340	21–34 USD/kW$_{th}$ and 0.87 USD/GJ$_{feed}$	See above	6.5[ix]	See above	See above	See above	Oliverio and Riberio (2006), see also row 'Overall' above	28–39	30–42	31–46
	Caribbean Basin[x, xi]	See above	110–360	22–35 USD/kW$_{th}$ and 0.87 USD/GJ$_{feed}$	See above	2.6–6.2	See above	See above	See above	Rosillo-Calle et al. (2000) see also row 'Overall' above	6.4–38	7.7–42	8.8–46
	Colombia	See above	100–320	20–31 USD/kW$_{th}$ and 0.87 USD/GJ$_{feed}$	See above	5.6	See above	See above	See above	McDonald and Schrattenholzer (2001), Goldemberg (1996), see also row 'Overall' above	23–32	24–36	25–39
	India	See above	110–340	21–33 USD/kW$_{th}$ and 0.87 USD/GJ$_{feed}$	See above	2.6–6.2	See above	See above	See above	see row 'Overall' above	5.9–37	7.1–41	8.2–44
	Mexico	See above	83–260	16–25 USD/kW$_{th}$ and 0.87 USD/GJ$_{feed}$	See above	5.2–7.1	See above	See above	See above	see row 'Overall' above	19–37	19–40	20–42
	USA	See above	100–320	20–31 USD/kW$_{th}$ and 0.87 USD/GJ$_{feed}$	See above	6.2	See above	See above	See above	see row 'Overall' above	27–36	28–40	29–43

Continued next page →

Annex III — Recent Renewable Energy Cost and Performance Parameters

Feedstock	Fuel, Region	Input data								Output data		
		Typical size of the device (MW_{th})	Investment cost (USD/kW_{th})[ii]	O&M cost, fixed annual (USD/kW_{th}) and non-feed variable (USD/GJ_{feed})	By-product Revenue (USD/GJ_{feed})	Feedstock cost (USD/GJ_{feed})	Feedstock conversion efficiency[iii] (%) Product only (product + by-product)	Capacity factor (%)	Economic design lifetime (years)	References	LCOF[iv] USD/GJ_{HHV}[v] Discount rate		
											3%	7%	10%
Ethanol													
	Overall	N/A	160–310	9–27 USD/kW_{th} and 1.98 USD/GJ_{feed}	By-product: DDGS[xii] 1.56	4.2–10[xiii]	54 (91)	95%	20	Alfstad (2008), Bain (2007), Kline et al. (2007)	9.3–22	9.5–22	10–23
Corn	USA	140–550[xiv]	160–240	9–18 USD/kW_{th} and 1.98 USD/GJ_{feed}	See above	4.2–10[xv]	See above	See above	See above	Delta-T Corporation (1997), Ibsen et al. (2005), Jechura (2005), see also row 'Overall' above			
	Argentina	See above	170–260	9–17 USD/kW_{th} and 1.98 USD/GJ_{feed}	See above	7.5	See above	See above	See above	McAloon et al. (2000). RFA (2011), University of Illinois (2011), see also row 'Overall' above	16–17	16–17	17–18
	Canada	See above	200–310	13–27 USD/kW_{th} and 1.98 USD/GJ_{feed}	See above	4.8–5.7	See above	See above	See above	see row 'Overall' above	11–15	12–15	12–16
Ethanol													
	Overall	150–610	140–280[xvi]	8–25 USD/kW_{th} and 1.41 USD/GJ_{feed}	By-product: DDGS[xii] 1.74	5.1–13	49 (91)	95%	20	Alfstad (2008), Bain (2007), Kline et al. (2007)	12–28	12–28	12–28
Wheat	USA	See above	140–220	8–17 USD/kW_{th} and 1.41 USD/GJ_{feed}	See above	6.3–13	See above	See above	See above	OECD (2002), Shapouri and Salassi (2006), USDA (2007), see also 'Overall'	13–28	14–28	14–28
	Argentina	See above	150–230	8–16 USD/kW_{th} and 1.41 USD/GJ_{feed}	See above	6.5–7	See above	See above	See above	see row 'Overall' above	14–16	14–16	14–17
	Canada	See above	190–280	12–25 USD/kW_{th} and 1.41 USD/GJ_{feed}	See above	5.1–6.9	See above	See above	See above	see row 'Overall' above	12–16	12–17	12–17

Continued next page →

Recent Renewable Energy Cost and Performance Parameters

Feedstock	Fuel, Region	Typical size of the device (MW$_{th}$)	Investment cost (USD/kW$_{th}$)[ii]	O&M cost, fixed annual (USD/kW$_{th}$) and non-feed variable (USD/GJ$_{feed}$)	By-product Revenue (USD/GJ$_{feed}$)	Feedstock cost (USD/GJ$_{feed}$)	Feedstock conversion efficiency[iii] (%) Product only (product + by-product)	Capacity factor (%)	Economic design lifetime (years)	References	LCOF[iv] USD/GJ$_{HHV}$[v] Discount rate 3%	7%	10%
	Biodiesel[xvii]				By-product: Glycerin[xviii]								
	Overall	44–440	160–320	9–46 USD/kW$_{th}$ and 2.58 USD/GJ$_{feed}$	0.58	7.0–24	103 (107)19	95%	20	Alfstad (2008), Bain (2007), Kline et al. (2007), Haas et al. (2006), Sheehan et al. (2006)	9.4–28	10–28	10–28
Soy Oil	Argentina	See above	170–320	12–42 USD/kW$_{th}$ and 2.58 USD/GJ$_{feed}$	See above	14–16[ix]	See above	See above	See above	Chicago Board of Trade (2006), see also row 'Overall' above	16–19	16–19	17–20
	Brazil	See above	160–310	9–27 USD/kW$_{th}$ and 2.58 USD/GJ$_{feed}$	See above	7.0–18[ix]	See above	See above	See above	Chicago Board of Trade (2006), see also row 'Overall' above	9.4–21	10–21	10–21
	USA	See above	160–300	12–46 USD/kW$_{th}$ and 2.58 USD/GJ$_{feed}$	See above	9.7–24	See above	See above	See above	USDA (2006), see also row 'Overall' above	12–28	12–28	12–28
	Biodiesel				By-product: Glycerin[xviii]								
	Overall	44–440	160–340	10–46 USD/kWth and 2.58 USD/GJfeed	0.58	6.1–45	103 (107)	95%	20	Alfstad (2008), Bain (2007), Kline et al. (2007), Haas et al. (2006), Sheehan et al. (1998)	8.7–48	8.9–48	9.0–49
Palm Oil	Colombia	See above	160–300	10–34 USD/kW$_{th}$ and 2.58 USD/GJ$_{feed}$	See above	6.1–45	See above	See above	See above	see row 'Overall' above	8.7–48	8.8–48	9.0–49
	Caribbean Basin[x]	See above	180–340	13–46 USD/kW$_{th}$ and 2.58 USD/GJ$_{feed}$	See above	11–45	See above	See above	See above	see row 'Overall' above	14–48	14–48	14–48
	Pyrolytic Fuel Oil				By-product: Electricity[xxi]								
Wood, Bagasse, other	Overall	110–440	160–240	12–44 USD/kWth and 0.42 USD/GJfeed	0.07	0.44–5.5[xxii]	67 (69)	95%	20	Ringer et al. (2006)	2.3–12	2.6–12	2.8–12
	USA	See above	160–230	19–44 USD/kW$_{th}$ and 0.42 USD/GJ$_{feed}$	See above	1.4–5.5	See above	See above	See above	see row 'Overall' above	4.0–12	4.3–12	4.5–12
	Brazil	See above	160–240	12–24 USD/kW$_{th}$ and 0.42 USD/GJ$_{feed}$	See above	0.44–5.5	See above	See above	See above	see row 'Overall' above	2.3–11	2.5–11	2.8–11

Continued next page →

Annex III — Recent Renewable Energy Cost and Performance Parameters

General remarks/notes:

i   All data are rounded to two significant digits. Chapter 2 provides additional cost and performance information in the section on cost trends. The assumptions underlying some of the production cost estimates quoted directly from the literature may, however, not be as transparent as the data sets in this Annex and should therefore be considered with caution.

ii Investment cost is based on plant capacity factor and not at 100% stream factor, which is the normal convention.

iii The feedstock conversion efficiency measured in energy units of input relative to energy units of output is stated for biomass only. Conversion factors for a mixture of biomass and fossil inputs are generally lower.

iv LCOF: Levelized Cost of Transport Fuels. The levelized costs of transport fuels include all private costs that accrue upstream in the bioenergy system, but do not include the cost of transportation and distribution to the final customers. Output subsidies for RE generation and tax credits are also excluded. However, indirect taxes and subsidies on inputs or commodities affecting the prices of inputs and, hence, private cost, cannot be fully excluded.

v   HHV: Higher heating value. LHV: Lower heating value.

vi Price of / revenue from sugar assumed to be USD_{2005} 22/GJ_{sugar} based on average 2005 to 2008 world refined sugar price.

vii A cane sucrose content of 14% is used in the calculations of case A with the additional assumption that 50% of the total sucrose is used for sugar production (97% extraction efficiency) and the other 50% of the total sucrose is used for ethanol production (90% conversion efficiency). The bagasse content of cane used is 16%. The HHVs used are bagasse: 18.6 GJ/t; sucrose: 17.0 GJ/t; and as received cane: 5.3 GJ/t.

viii Brazilian feedstock costs have declined by 60% in the time period of 1975 to 2005 (Hettinga et al, 2009). For a more detailed discussion of historical and future cost trends see also Sections 2.7.2, 2.7.3 and 2.7.4.

ix 55.2% of feed used is bagasse. More detailed information on feedstock characteristics can, for instance, be found in Section 2.3.1.

x   Caribbean Basin Initiative Countries: Guatemala, Honduras, Nicaragua, Dominican Republic, Costa Rica, El Salvador, Guyana, and others.

xi Mixed ethanol/sugar mill: 50/50. More detailed information on sugar mills can be found in Section 2.3.4.

xii DDGS: Distillers dried grains plus solubles.

xiii For international feed range, supply curves from Kline et al. (2007) were used. For more information on feedstock supply curves and other economic considerations in biomass resource assessments see Chapter section 2.2.3.

xiv Plant size range (140-550 MW is the equivalent of 25-100 million gallons per year (mmgpy) of anhydrous ethanol) is representative of the US corn ethanol industry (RFA, 2011).

xv Corn prices in the USA have declined by 63% in the period from 1975 to 2005 (Hettinga et al., 2009). For a more detailed discussion of historical and future cost trends see also Sections 2.7.2, 2.7.3 and 2.7.4.

xvi Based on corn mill costs, corrected for HHV, and distillers dried grain (DDG) yields for wheat. More detailed information on milling can be found in Section 2.3.4.

xvii Installation basis is soy oil, not soybeans. Crush spread is used to convert from soybean prices to soy oil price. HHV soy oil = 39.6 GJ/t.

xviii Glycerine is also referred to as glycerol and is a simple polyol compound (1,2,3-propanetriol), and is central to all lipids known as triglycerides. Glycerine is a by-product of biodiesel production.

xix The yield is higher than 100% because methanol (or other alcohol) is incorporated into the product.

xx Soy oil prices are estimated from soybean prices (Kline et al., 2007) and crush spread (Chicago Board of Trade, 2006).

xxi Process-derived gas and residual solids (char) are used for process heat and power. Excess electricity is exported as a by-product.

xxii Feedstock cost range is based on bagasse residue and wood residue prices (Kline et al. 2007). High range is for wood-based pyrolysis, low range is typical of pyrolysis of bagasse. For more information on pyrolysis see Section 2.3.3.2. For a discussion of historical and future cost trends see also Sections 2.7.2, 2.7.3 and 2.7.4.

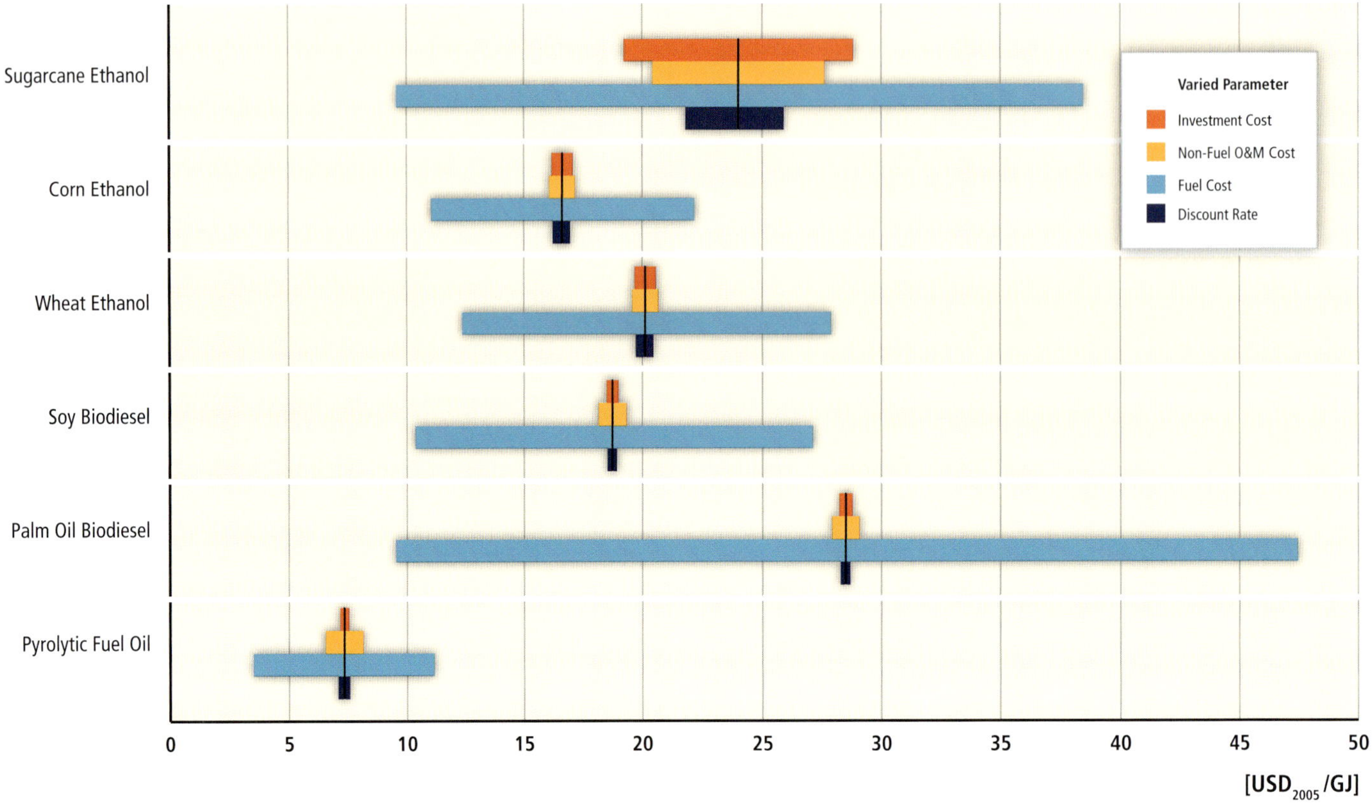

Recent Renewable Energy Cost and Performance Parameters   Annex III

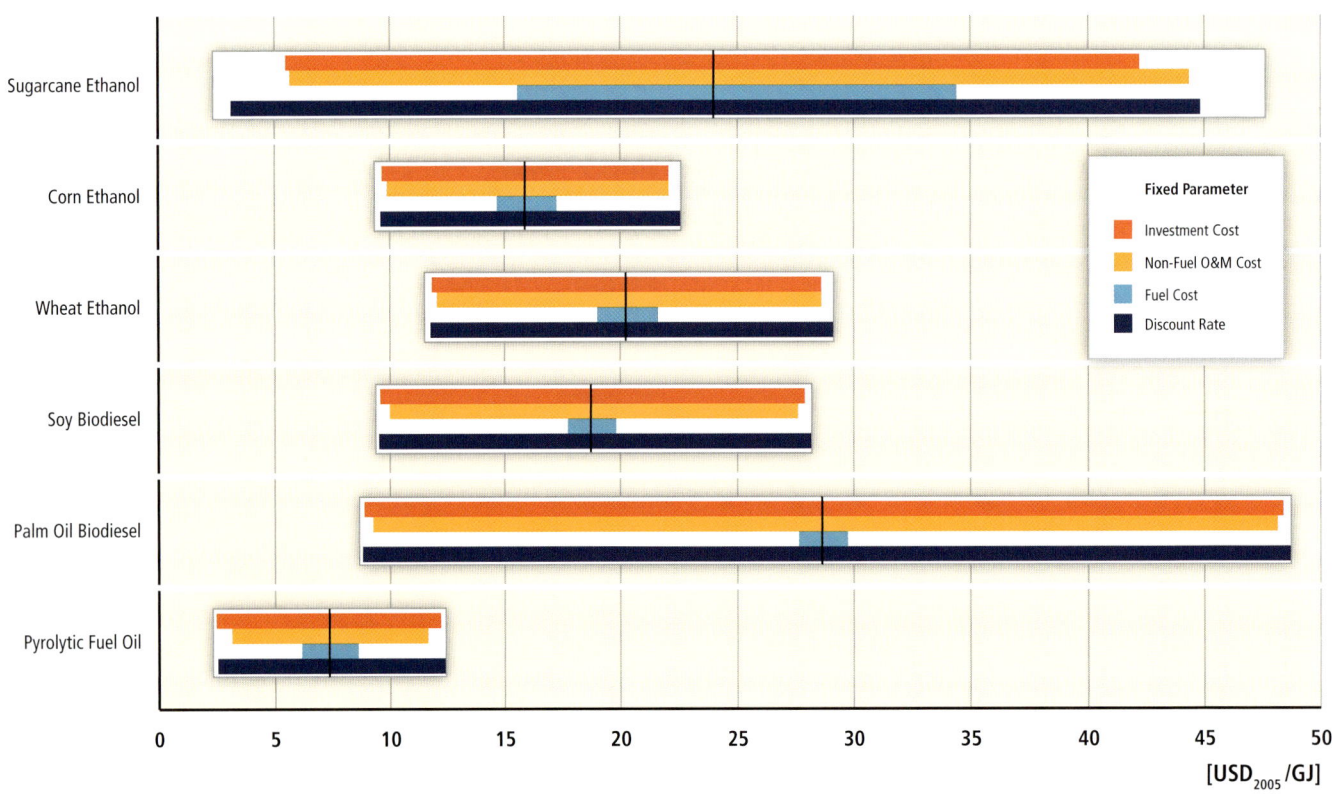

Figure A.III.4a | Tornado graph for biofuels. For further explanation see Figure A.III.1a.

Figure A.III.4b | 'Negative' of tornado graph for biofuels. For further explanation see Figure A.III.1b.

Note: Aggregation of input data over various regions and subsequent LCOF calculations leads to slightly larger LCOF ranges than those obtained if region-specific LCOF values are calculated first and these regional LCOF values are subsequently aggregated. In order to allow for a broad sensitivity analysis the first approach was followed here. The broader ranges were, however, rescaled to yield the same results as the latter approach, which is more accurate and is used in the remainder of the report.

References

The references in this list have been used in the assessment of the cost and performance data of the individual technologies summarized in the tables. Only some of them are quoted in the text of this Annex to support specific information included in the explanatory text. All references are sorted by energy type/carrier and by technology.

Electricity

Bioenergy

Remark 1: Further references on cost have been assessed in the body of Chapter 2. These have served to cross-check the reliability of the results from the meta-analysis based on the data sources listed here.

Bain, R.L. (2007). *World Biofuels Assessment, Worldwide Biomass Potential: Technology Characterizations.* NREL/MP-510-42467, National Renewable Energy Laboratory, Golden, CO, USA, 140 pp.

Bain, R.L. (2011). *Biopower Technologies in Renewable Electricity Alternative Futures.* National Renewable Energy Laboratory, Golden, CO, USA, in press.

Bain, R.L., W.P. Amos, M. Downing, and R.L. Perlack (2003). *Biopower Technical Assessment: State of the Industry and the Technology.* TP-510-33123, National Renewable Energy Laboratory, Golden, CO, USA, 277 pp.

DeMeo, E.A., and J.F. Galdo (1997). *Renewable Energy Technology Characterizations.* TR-109496, U.S. Department of Energy and Electric Power Research Institute, Washington, DC, USA, 283 pp.

EIA (2009). *2006 Energy Consumption by Manufacturers—Data Tables.* Table 7.2. Energy Information Administration, US Department of Energy, Washington, DC, USA. Available at: eia.doe.gov/emeu/mecs/mecs2006/2006tables.html.

McGowin, C. (2008). *Renewable Energy Technical Assessment Guide.* TAG-RE: 2007, Electric Power Research Institute (EPRI), Palo Alto, CA, USA.

Neij, L. (2008). Cost development of future technologies for power generation – A study based on experience curves and complementary bottom-up assessments. *Energy Policy*, **36**(6), pp. 2200-2211.

OANDA (2011). *Historical Exchange Rates.*

Obernberger, I., and G. Thek (2004). *Techno-economic evaluation of selected decentralised CHP applications based on biomass combustion in IEA partner countries.* BIOS Bioenergiesysteme GmbH, Graz, Austria, 87 pp.

Obernberger, I., G. Thek, and D. Reiter (2008). *Economic Evaluation of Decentralised CHP Applications Based on Biomass Combustion and Biomass Gasification.* BIOS Bioenergiesysteme GmbH, Graz, Austria, 19 pp.

Peters, M, K. Timmerhaus, and R. West (2003). *Plant Design and Economics for Chemical engineers, Fifth Edition*, McGraw –Hill Companies, NY, USA, 242 pp. (ISBN 0-07-239266-5).

Rauch, R. (2010). Indirect Gasification. In: *IEA Joint Task 32 &33 Workshop*, Copenhagen, Denmark, 7 October 2010. Available at: www.ieabcc.nl/meetings/task32_Copenhagen /09%20TU%20Vienna.pdf.

Skjoldborg, B. (2010). Optimization of I/S Skive District Heating Plant. In: *IEA Joint Task 32 & 33 Workshop*, Copenhagen, Denmark, 7 October 2010. Available at: www.ieabcc.nl/meetings/task32_Copenhagen/11%20Skive.pdf.

Direct Solar Energy

Bloomberg (2010). *Bloomberg New Energy Finance—Renewable Energy Data.* Available at: bnef.com/.

Breyer, C., A. Gerlach, J. Mueller, H. Behacker, and A. Milner (2009). Grid-parity analysis for EU and US regions and market segments - Dynamics of grid-parity and dependence on solar irradiance, local electricity prices and PV progress ratio. In: *Proceedings of the 24th European Photovoltaic Solar Energy Conference*, 21-25 September 2009, Hamburg, Germany, pp. 4492-4500.

Bundesverband Solarwirtschaft e.V. (2010). *Statistische Zahlen der deutschen Solarstrombranche (photovoltaik).* Bundesverband Solarwirtschaft e.V. (BSW Solar), Berlin, Germany, 4 pp.

IEA (2010a). *Energy Technology Perspectives: Scenarios & Strategies to 2050.* International Energy Agency, Paris, France, 710 pp.

IEA (2010b). *Technology Roadmap, Concentrating Solar Power.* International Energy Agency, Paris, France, 48 pp.

IEA (2010c). *Technology Roadmap, Solar Photovoltaic Energy.* International Energy Agency, Paris, France, 48 pp.

NEEDS (2009). *New Energy Externalities Development for Sustainability (NEEDS). Final Report and Database.* New Energy Externalities Development for Sustainability, Rome, Italy.

NREL (2011a). Solar PV Manufacturing Cost Model Group: Installed Solar PV System Prices. Presentation to *SEGIS_ADEPT Power Electronic in Photovoltaic Systems Workshop*, Arlington, VA, USA, 8 February 2011. NREL/PR-6A20-50955.

NREL (2011b). *The Open PV Project.* Online database. Available at: openpv.nrel.org.

Sharma, A. (2011). A comprehensive study of solar power in India and world. *Renewable and Sustainable Energy Reviews*, **15**(4), pp. 1767-1776.

Trieb, F., C. Schillings, M. O'Sullivan, T. Pregger, and C. Hoyer-Klick (2009). Global potential of concentrating solar power. In: *SolarPACES Conference*, Berlin, Germany, 15-18 September 2009.

Viebahn, P., Y. Lechon, and F. Trieb (2010). The potential role of concentrated solar power (CSP) in Africa and Europe: A dynamic assessment of technology development, cost development and life cycle inventories until 2050. *Energy Policy*, doi: 10.1016/j.enpol.2010.09.026.

Geothermal Energy

Barnett, P., and P. Quinlivan (2009). *Assessment of Current Costs of Geothermal Power Generation in New Zealand (2007 basis).* Report by SKM for New Zealand Geothermal Association, Wellington, NZ. Available at: www.nzgeothermal.org.nz\industry_papers.html.

Bertani, R. (2010). Geothermal electric power generation in the world: 2005-2010 update report. In: *Proceedings of the World Geothermal Congress 2010*, Bali, Indonesia, 25-30 April 2010. Available at: www.geothermal-energy.org/pdf/IGAstandard/WGC/2010/0008.pdf.

Bromley, C.J., M.A. Mongillo, B. Goldstein, G. Hiriart, R. Bertani, E. Huenges, H. Muraoka, A. Ragnarsson, J. Tester, and V. Zui (2010). Contribution of geothermal energy to climate change mitigation: the IPCC renewable energy report. In: *Proceedings of the World Geothermal Congress 2010*, Bali, Indonesia, 25-30 April 2010. Available at: www.geothermal-energy.org/pdf/IGAstandard/WGC/2010/0225.pdf.

Cross, J., and J. Freeman (2009). *2008 Geothermal Technologies Market Report*. Geothermal Technologies Program of the US Department of Energy, Washington, DC, USA, 46 pp. Available at: www1.eere.energy.gov/geothermal/pdfs/2008_market_report.pdf.

Darma, S., S. Harsoprayitno, B. Setiawan, Hadyanto, R. Sukhyar, A.W. Soedibjo, N. Ganefianto, and J. Stimac (2010). Geothermal energy update: Geothermal energy development and utilization in Indonesia. In: *Proceedings World Geothermal Congress 2010*, Bali, Indonesia, 25-29 April, 2010. Available at: www.geothermal-energy.org/pdf/IGAstandard/WGC/2010/0128.pdf.

DiPippo, R. (2008). *Geothermal Power Plants: Principles, Applications, Case Studies and Environmental Impact*. Elsevier, London, UK, 493 pp.

GTP (2008). *Geothermal Tomorrow 2008*. DOE-GO-102008-2633, Geothermal Technologies Program of the US Department of Energy, Washington, DC, USA, 36 pp.

Gutiérrez-Negrín, L.C.A., R. Maya-González, and J.L. Quijano-León (2010). Current status of geothermics in Mexico. In: *Proceedings World Geothermal Congress 2010*, Bali, Indonesia, 25-29 April 2010. Available at: www.geothermal-energy.org/pdf/IGAstandard/WGC/2010/0101.pdf.

Hance, C.N. (2005). *Factors Affecting Costs of Geothermal Power Development*. Geothermal Energy Association, Washington, DC, USA, 64 pp. Available at: www.geo-energy.org/reports/Factors%20Affecting%20Cost%20of%20Geothermal%20Power%20Development%20-%20August%202005.pdf.

Hjastarson, A., and J.G. Einarsson (2010). *Geothermal resources and properties of HS Orka, Reyjanes Peninsula, Iceland*. Independent Technical Report prepared by Mannvit Engineering for Magma Energy Corporation, 151 pp. Available upon request at: www.mannvit.com.

Kutscher, C. (2000). *The Status and Future of Geothermal Electric Power*. Publication NREL/CP-550-28204, National Renewable Energy Laboratory, US Department of Energy, Washington, DC, USA, 9 pp. Available at: www.nrel.gov/docs/fy00osti/28204.pdf.

Lovekin, J. (2000). The economics of sustainable geothermal development. In: *Proceedings World Geothermal Congress 2000*, Kyushu-Tohoku, Japan, 28 May – 10 June 2000 (ISBN: 0473068117). Available at: www.geothermal-energy.org/pdf/IGAstandard/WGC/2000/R0123.PDF.

Lund, J.W., K. Gawell, T.L. Boyd, and D. Jennejohn (2010). The United States of America country update 2010. In: *Proceedings World Geothermal Congress 2010*, Bali, Indonesia, 25-30 April 2010. Available at: www.geothermal-energy.org/pdf/IGAstandard/WGC/2010/0102.pdf.

Owens, B. (2002). *An Economic Valuation of a Geothermal Production Tax Credit*. Publication NREL/TP-620-31969, National Renewable Energy Laboratory, US Department of Energy, Washington, DC, USA, 24 pp. Available at: www.nrel.gov/docs/fy02osti/31969.pdf.

Stefansson, V. (2002). Investment cost for geothermal power plants. *Geothermics*, **31**, pp. 263-272.

Hydropower

Avarado-Anchieta, and C. Adolfo (2009). Estimating E&M powerhouse costs. *International Water Power and Dam Construction*, **61**(2), pp. 21-25.

BMU (2008). *Further development of the 'Strategy to increase the use of renewable energies' within the context of the current climate protection goals of Germany and Europe*. German Federal Ministry for the Environment, Nature Conservation and Nuclear Safety (BMU), Bonn, Germany, 118 pp.

Hall, D.G., G.R. Carroll, S.J. Cherry, R.D. Lee, and G.L. Sommers (2003). *Low Head/Low Power Hydropower Resource Assessment of the North Atlantic and Middle Atlantic Hydrologic Regions*. DOE/ID-11077, U.S. Department of Energy Idaho Operations Office, Idaho Falls, ID, USA.

IEA (2008a). *World Energy Outlook 2008*. International Energy Agency, Paris, France, 578 pp.

IEA (2008b). *Energy Technology Perspectives 2008. Scenarios and Strategies to 2050*. International Energy Agency, Paris, France, 646 pp.

IEA (2010d). *Renewable Energy Essentials: Hydropower*. International Energy Agency, Paris, France. 4 pp.

IEA (2010e). *Projected Costs of Generating Electricity*. International Energy Agency, Paris, France, 218 pp.

IJHD (2010). *World Atlas & Industry Guide*. International Journal on Hydropower and Dams (IJHD), Wallington, Surrey, UK, 405 pp.

Krewitt, W., K. Nienhaus, C. Klebmann, C. Capone, E. Stricker, W. Grauss, M. Hoogwijk, N. Supersberger, U.V. Winterfeld, and S. Samadi (2009). *Role and Potential of Renewable Energy and Energy Efficiency for Global Energy Supply*. Climate Change 18/2009, ISSN 1862-4359, Federal Environment Agency, Dessau-Roßlau, Germany, 336 pp.

Lako, P., H. Eder, M. de Noord, and H. Reisinger (2003). *Hydropower Development with a Focus on Asia and Western Europe: Overview in the Framework of VLEEM 2*. Verbundplan ECN-C-03-027. Energy Research Centre of the Netherlands, Petten, The Netherlands.

REN21 (2010). *Renewables 2010 Global Status Report*. Renewable Energy Policy Network for the 21st Century (REN21), Paris, France, 80 pp.

Teske, S., T. Pregger, S. Simon, T. Naegler, W. Graus, and C. Lins (2010). Energy [R]evolution 2010—a sustainable world energy outlook. *Energy Efficiency*, doi:10.1007/s12053-010-9098-y.

UNDP/UNDESA/WEC (2004). *World Energy Assessment: Overview 2004 Update*. Bureau for Development Policy, UN Development Programme, New York, New York, USA, 85 pp.

Ocean Energy

Charlier, R.H. (2003). Sustainable co-generation from the tides: A review. *Renewable and Sustainable Energy Reviews*, **7**(3), pp. 187-213.

ETSAP (2010b). *Marine Energy Technology Brief E13 - November, 2010*. Energy Technology Systems Analysis Programme, International Energy Agency, Paris, France. Available at: www.etsap.org/E-techDS/PDF/E08-Ocean%20Energy_GSgct_Ana_LCPL_rev30Nov2010.pdf.

Kerr, D. (2007). Marine energy. *Philosophical Transactions of the Royal Society London, Series A (Mathematical, Physical and Engineering Sciences)*, **365**(1853), pp. 971-92.

Wind Energy

Blanco, M.I. (2009). The economics of wind energy. *Renewable and Sustainable Energy Reviews*, **13**, pp. 1372-1382.

Boccard, N. (2009). Capacity factor of wind power realized values vs. estimates. *Energy Policy*, **37**, pp. 2679-2688.

BTM Consult ApS (2010). *International Wind Energy Development. World Market Update 2009.* BTM Consult ApS, Ringkøbing, Denmark, 124 pp.

BWEA and Garrad Hassan (2009). *UK Offshore Wind: Charting the Right Course.* British Wind Energy Association, London, UK, 42 pp.

China Renewable Energy Association (2009). *Annual Report of New Energy and Renewable Energy in China, 2009.* China Renewable Energy Association, Beijing, China.

EWEA (2009). *Wind Energy, the Facts.* European Wind Energy Association, Brussels, Belgium, 488 pp.

Goyal, M. (2010). Repowering – Next big thing in India. *Renewable and Sustainable Energy Reviews*, **14**, pp. 1400-1409.

IEA (2009). *Technology Roadmap – Wind Energy.* International Energy Agency, Paris, France, 52 pp.

IEA (2010a). *Energy Technology Perspectives: Scenarios & Strategies to 2050.* International Energy Agency, Paris, France, 710pp.

IEA Wind (2010). *IEA Wind Energy Annual Report 2009.* International Energy Agency Wind, International Energy Agency, Paris, France, 172 pp.

Lemming, J.K., P.E. Morthorst, N.E. Clausen, and J.P. Hjuler, (2009). *Contribution to the Chapter on Wind Power in Energy Technology Perspectives 2008, IEA.* Risø National Laboratory, Roskilde, Denmark, 64 pp.

Li, J. (2010). Decarbonising power generation in China – Is the answer blowing in the wind? *Renewable and Sustainable Energy Reviews*, **14**, pp. 1154-1171.

Li, J., and L. Ma (2009). *Background Paper: Chinese Renewables Status Report.* Renewable Energy Policy Network for the 21st Century, Paris, France, 95 pp.

Milborrow, D. (2010). Annual power costs comparison: What a difference a year can make. *Windpower Monthly*, **26**, pp. 41-47.

Musial, W., and B. Ram (2010). *Large-Scale Offshore Wind Power in the United States: Assessment of Opportunities and Barriers.* National Renewable Energy Laboratory, Golden, CO, USA, 240 pp.

Nielson, P., J.K. Lemming, P.E. Morthorst, H. Lawetz, E.A. James-Smith, N.E. Clausen, S. Strøm, J. Larsen, N.C. Bang, and H.H. Lindboe (2010). *The Economics of Wind Turbines.* EMD International, Aalborg, Denmark, 86 pp.

Snyder, B., and M.J. Kaiser (2009). A comparison of offshore wind power development in Europe and the US: Patterns and drivers of development. *Applied Energy*, **86**, pp. 1845-1856.

UKERC (2010). *Great Expectations: The Cost of Offshore Wind in UK Waters – Understanding the Past and Projecting the Future.* United Kingdom Energy Research Centre, London, England, 112 pp.

Wiser, R., and M. Bolinger (2010). *2009 Wind Technologies Market Report.* US Department of Energy, Washington, DC, USA, 88 pp.

Heat

Bioenergy

Remark: Further references on cost have been assessed in the body of Chapter 2. These have served to cross-check the reliability of the results from the meta-analysis based on the data sources listed here.

Obernberger, I., and G. Thek (2004). Techno-economic evaluation of selected decentralised CHP applications based on biomass combustion in IEA partner countries. BIOS Bioenergiesysteme GmbH, Graz, Austria, 87 pp.

IEA (2007). *Renewables for Heating and Cooling – Untapped Potential.* International Energy Agency, Paris, France, 209 pp.

Direct Solar Energy

Chang, K.-C., W.-M. Lin, T.-S. Lee, and K.-M. Chung (2011). Subsidy programs on diffusion of solar water heaters: Taiwan's experience. *Energy Policy*, **39**, pp. 563-567.

Han, J., A.P.J. Mol, and Y. Lu (2010). Solar water heaters in China: A new day dawning. *Energy Policy*, **38**(1), pp. 383-391.

Harvey, L.D.D. (2006). *A Handbook on Low-Energy Buildings and District-Energy Systems: Fundamentals, Techniques and Examples.* Earthscan, Sterling, Virginia, USA, 701 pp.

IEA (2007). *Renewables for Heating and Cooling – Untapped Potential*, International Energy Agency, Paris, France, 209 pp.

Zhang, X., W. Ruoshui, H. Molin, and E. Martinot (2010). A study of the role played by renewable energies in China's sustainable energy supply. *Energy*, **35**(11), pp. 4392-4399.

Geothermal Energy

Balcer, M. (2000). Infrastruktura techniczna zakladu geotermalnego w Mszczonowie (in Polish). In: *Symposium on the Role of Geothermal Energy in the Sustainable Development of the Mazovian and Lodz Regions (Rola energii geotermalnej w zrównowazonym rozwoju regionów Mazowieckiego i Lodzkiego)*, Mineral and Energy Economy Research Institute, Polish Academy of Sciences, Cracow, Poland, 4-6 October 2000, pp. 107-114 (ISBN 83-87854-62-X).

Lund, J.W. (1995). Onion dehydration. *Transactions of the Geothermal Resources Council*, **19**, pp. 69-74.

Lund, J.W., and T.L. Boyd (2009). Geothermal utilization on the Oregon Institute of Technology campus, Klamath Falls, Oregon. *Proceedings of the 34th Workshop on Geothermal Reservoir Engineering*, Stanford University, CA, USA (ISBN: 9781615673186).

Radeckas, B., and V. Lukosevicius (2000). Klaipeda Geothermal demonstration project. In: *Proceedings World Geothermal Congress 2000*, Kyushu-Tohoku, Japan, 28 May – 10 June 2000, pp. 3547-3550 (ISBN: 0473068117). Available at: www.geothermal-energy.org/pdf/IGAstandard/WGC/2000/R0237.PDF.

Reif, T. (2008). Profitability analysis and risk management of geothermal projects. *Geo-Heat Center Quarterly Bulletin*, **28**(4), pp. 1-4. Available at: geoheat.oit.edu/bulletin/bull28-4/bull28-4-all.pdf.

Biofuels

Remark: Further references on cost have been assessed in the body of Chapter 2. These have served to cross-check the reliability of the results from the meta-analysis based on the data sources listed here.

General References

Alfstad, T. (2008). *World Biofuels Study: Scenario Analysis of Global Biofuels Markets.* BNL-80238-2008, Brookhaven National Laboratory, New York, NY, USA, 67 pp.

Bain, R.L. (2007). *World Biofuels Assessment, Worldwide Biomass Potential: Technology Characterizations.* NREL/MP-510-42467, National Renewable Energy Laboratory, Golden, CO, USA, 140 pp.

Goldemberg, J. (1996). The evolution of ethanol costs in Brazil. *Energy Policy*, **24**(12), pp. 1127-1128.

Hettinga, W.G., H.M. Junginger, S.C. Dekker, M. Hoogwijk, A.J. McAloon, and K.B. Hicks (2009). Understanding the reductions in US corn ethanol production costs: An experience curve approach. *Energy Policy*, **37**(1), pp. 190-203.

Kline, K.L., G. Oladosu, A. Wolfe, R.D. Perlack, and M. McMahon (2007). *Biofuel Feedstock Assessment for Selected Countries.* ORNL/TM-2007/224, Oak Ridge National Laboratory, Oak Ridge, TN, USA, 243 pp.

Corn Ethanol

Delta-T Corporation (1997). *Proprietary information.* Williamsburg, VA, USA.

Ibsen, K., R. Wallace, S. Jones, and T. Werpy (2005). *Evaluating Progressive Technology Scenarios in the Development of the Advanced Dry Mill Biorefinery.* FY05-630, National Renewable Energy Laboratory, Golden, CO, USA.

Jechura, J. (2005). *Dry Mill Cost-By-Area: ASPEN Case Summary.* National Renewable Energy Laboratory, Golden, CO, USA, 2 pp.

McAloon, A., F. Taylor, W. Lee, K. Ibsen, and R. Wooley (2000). *Determining the Cost of Producing Ethanol from Corn Starch and Lignocellulosic Feedstocks.* NREL/TP-580-28893, National Renewable Energy Laboratory, Golden, CO, USA, 43 pp.

RFA (2011). *Biorefinery Plant Locations.* Renewable Fuels Association (RFA), Washington, DC, USA. Available at: www.ethanolrfa.org/bio-refinery-locations/.

University of Illinois (2011). *farmdoc: Historical Corn Prices.* University of Illinois, Urbana, IL, USA. Available at: www.farmdoc.illinois.edu/manage/pricehistory/price_history.html.

Wheat Ethanol

Kline, K., G. Oladosu, A. Wolfe, R. Perlack, V. Dale and M. McMahon (2007). *Biofuel feedstock assessment for selected countries*, ORNL/TM-2007/224, Oak Ridge National Laboratory, Oak Ridge, TN, USA, 243 pp.

Shapouri, H., and M. Salassi (2006). *The Economic Feasibility of Ethanol Production in the United States.* US Department of Agriculture, Washington, DC, USA, 69 pp.

USDA (2007). *Wheat Data: Yearbook Tables.* Economic Research Service, US Department of Agriculture (USDA), Washington, DC, USA.

Sugarcane

Bohlmann, G.M., and M.A. Cesar (2006). The Brazilian opportunity for biorefineries. *Industrial Biotechnology*, **2**(2), pp. 127-132.

Oliverio, J.L. (2006). Technological evolution of the Brazilian sugar and alcohol sector: Dedini's contribution. *International Sugar Journal*, **108**(1287), pp. 120-129.

Oliverio, J.L., and J.E. Riberio (2006). Cogeneration in Brazilian sugar and bioethanol mills: Past, present and challenges. *International Sugar Journal*, **108**(191), pp. 391-401.

Rosillo-Calle, F., S.V. Bajay, and H. Rothman (2000). *Industrial Uses of Biomass Energy: The Example of Brazil.* Taylor & Francis, London, UK.

van den Wall Bake (2006). *Cane as Key in Brazilian Ethanol Industry.* Master's Thesis, NWS-1-2006-14, University of Utrecht, Utrecht, The Netherlands.

van den Wall Bake, J.D., M. Junginger, A. Faaij, T. Poot, and A. Walter (2009). Explaining the experience curve: Cost reductions of Brazilian ethanol from sugarcane. *Biomass and Bioenergy*, **33**(4), pp. 644-658.

Biodiesel

Chicago Board of Trade (2006). *CBOT® Soybean Crush Reference Guide.* Board of Trade of the City of Chicago, Chicago, IL, USA.

Haas, M.J., A.J. McAloon, W.C. Yee, and T.A. Foglia (2006). A process model to estimate biodiesel production costs. *Bioresource Technology*, **97**(4), pp. 671-678.

Sheehan, J., V. Camobreco, J. Duffield, M. Graboski, and H. Shapouri (1998). *Life Cycle Inventory of Biodiesel and Petroleum Diesel for Use in an Urban Bus.* NREL/SR-580-24089. National Renewable Energy Laboratory, Golden, CO, USA.

Pyrolysis Oil

Ringer, M., V. Putsche, and J. Scahill (2006). *Large-Scale Pyrolysis Oil Production: A Technology Assessment and Economic Analysis.* TP-510-37779, National Renewable Energy Laboratory, Golden, CO, USA, 93 pp.

Contributors to the IPCC Special Report

Contributors to the IPCC Special Report

Coordinating lead authors, lead authors, contributing authors and review editors are listed alphabetically by surname; when citizenship is different from country of residence it is mentioned second.

ABDULLA, Amjad
Ministry of the Environment
Republic of Maldives

ABERLE, Armin
National University of Singapore
Singapore / Germany

ADEDOYIN, Akintayo
University of Botswana
Botswana

AHENKORAH, Alfred K. Ofosu
Energy Commission
Ghana

AKAI, Makoto
National Institute of Advanced Industrial Science and Technology (AIST)
Japan

ANGERER, Gerhard
Fraunhofer Institute for Systems and Innovation Research (ISI)
Germany

ARENT, Douglas J.
National Renewable Energy Laboratory (NREL)
United States of America

ARROWSMITH, Greg
European Renewable Energy Centres (EUREC) Agency
Belgium / United Kingdom

ARVIZU, Dan E.
National Renewable Energy Laboratory (NREL)
United States of America

ATHIENITIS, Andreas
Concordia University
Canada

BAIN, Richard
National Renewable Energy Laboratory (NREL)
United States of America

BAKER, Erin
University of Massachusetts
United States of America

BALAYA, Palani
National University of Singapore
Singapore / India

BAUER, Christian
Paul Scherrer Institute (PSI)
Switzerland / Austria

BAZILIAN, Morgan
United Nations Industrial Development Organization (UNIDO)
Austria / United States of America

BEEREPOOT, Milou
International Energy Agency
France

BERNDES, Göran
Chalmers University of Technology
Sweden

BERTANI, Ruggero
Enel Green Power S.p.A.
Italy

BHARATHAN, Desikan
National Renewable Energy Laboratory (NREL)
United States of America

BHUYAN, Gouri S.
Powertech Labs Inc.
Canada

BIRD, Lori
National Renewable Energy Laboratory (NREL)
United States of America

BLACKWELL, David P.
Southern Methodist University
United States of America

BOERMANS, Thomas
Ecofys
Germany

BOWEN, Alex
Grantham Research Institute on Climate Change and the Environment
United Kingdom

BRANCHE, Emmanuel
Electricité de France (EDF)
France

BRECHA, Robert
Potsdam Institute for Climate Impact Research (PIK)
Germany / United States of America

BREUKERS, Sylvia
Energy Research Centre of the Netherlands (ECN)
The Netherlands

BROMLEY, Christopher J.
GNS Science
New Zealand

BRUCKNER, Thomas
University of Leipzig
Germany

BURGHERR, Peter
Paul Scherrer Institute (PSI)
Switzerland

Annex IV

Contributors to the IPCC Special Report

BURKHARDT, John
National Renewable Energy Laboratory (NREL)
United States of America

BUSCH, Sebastian
Vienna University of Technology
Austria / Germany

CABEZA, Luisa F.
University of Lleida
Spain

CACERES RODRIGUEZ, J. Rodolfo
Comisión Ejecutiva Hidroeléctrica del Río Lempa
El Salvador

CALVO, Eduardo
Universidad Nacional Mayor de San Marcos
Peru

CHIANG, Ranyee
U.S. Department of Energy
United States of America

CHRISTENSEN, John M.
Risø National Laboratory for Sustainable Energy
Denmark

CHUM, Helena L.
National Renewable Energy Laboratory (NREL)
United States of America / Brazil, USA

CLARKE, Leon
Joint Global Change Research Institute
United States of America

CLEMENS, Elisabeth
United Nations Development Program (UNDP)
Norway

CONNOR, Peter
University of Exeter
United Kingdom

COWLIN, Shannon
National Renewable Energy Laboratory (NREL)
United States of America

CREUTZIG, Felix
Technische Universität Berlin
Germany

DARGHOUTH, Naïm R.
University of California
United States of America

DAWE, David
Food and Agriculture Organization of the United Nations (FAO)
Thailand / United States of America

DE JAGER, David
Ecofys
The Netherlands

DEMAYO, Trevor N.
Chevron Energy Technology Co.
United States of America / Canada

DEMKINE, Volodymyr
United Nations Environment Programme (UNEP)
Kenya / Ukraine

DENNY, Eleanor
Trinity College Dublin
Ireland

DENTON, Fatima
International Development Research Centre (IDRC)
Senegal / Gambia

DEVERNAY, Jean-Michel E.
Electricité de France (EDF)
France

DHAMIJA, Parveen
Ministry of New & Renewable Energy
India

DIAZ MOREJON, Cristobal
Ministry of Science, Technology and the Environment
Cuba

DONG, Hongmin
Chinese Academy of Agricultural Sciences
People's Republic of China

DROEGE, Peter
University of Liechtenstein
Liechtenstein, Germany

EDENHOFER, Ottmar
Potsdam Institute for Climate Impact Research (PIK)
Germany

EDMONDS, James A.
Joint Global Change Research Institute
United States of America

ELGIZOULI, Ismail A. R.
Higher Council for Environment & Natural Resources
Sudan

ELLIOTT, Dennis
National Renewable Energy Laboratory (NREL)
United States of America

ERICSSON, Karin
Lund University
Sweden

ESTEFEN, Segen F.
Universidade Federal do Rio de Janeiro (UFRJ)
Brazil

FAAIJ, André P. C.
Utrecht University
The Netherlands

FIFITA, Solomone
Secretariat of the Pacific Community
Republic of Fiji / Tonga

FISCHEDICK, Manfred
Wuppertal Institute for Climate, Environment and Energy
Germany

FLYNN, Damian
University College Dublin
Ireland

FREITAS, Marcos A. V.
Universidade Federal do Rio de Janeiro (UFRJ)
Brazil

FUJINO, Junichi
National Institute for Environmental Studies
Japan

GABRIELLE, Benoît X.
AgroParisTech
France

GIRARDIN, L. Osvaldo
Bariloche Foundation
Argentina

GOLDSTEIN, Barry A.
Government of South Australia
Australia

GOSS ENG, Alison M.
U.S. Department of Energy
United States of America

GREACEN, Chris
Consultant
United States of America

GRISOLI, Renata
University of São Paulo
Brazil

GUTIRREZ-NEGRIN, Luis
Asociación Geotérmica Mexicana
Mexico

GWINNER, Don
National Renewable Energy Laboratory (NREL)
United States of America

HAGELÜKEN, Christian
Umicore Precious Metals Refining
Germany

HALL, Douglas G.
Idaho National Laboratory
United States of America

HAMILTON, Kirsty S.
Chatham House
United Kingdom

HAND, M. Maureen
National Renewable Energy Laboratory (NREL)
United States of America

HANSEN, Gerrit
IPCC WGIII TSU
Germany

HANSON, Howard
Florida Atlantic University
United States of America

HARNISCH, Jochen
KfW Entwicklungsbank
Germany

HEATH, Garvin
National Renewable Energy Laboratory (NREL)
United States of America

HEPBURN, Cameron
New College
United Kingdom

HIRIART, Gerardo L.
Energías Alternas, Estudios y Proyectos
Mexico

HOEN, Ben
Lawrence Berkeley National Laboratory
United States of America

HOHMEYER, Olav
University of Flensburg
Germany

HOLLANDS, Terry K. G.
University of Waterloo
Canada

HOLTTINEN, Hannele K.
VTT Technical Research Centre of Finland
Finland

HUCKERBY, John
Power Projects Ltd.
New Zealand

HUENGES, Ernst
Deutsches GeoForschungsZentrum
Germany

HULD, Thomas
European Commission Joint Research Center (JRC)
Italy / Denmark

HULTMAN, Nathan
University of Maryland
United States of America

HUNT, Suzanne
HuntGreen LLC
United States of America

INFIELD, David
University of Strathclyde
United Kingdom

IVANOVA BONCHEVA, Antonina
Universidad Autónoma de Baja California Sur (UABCS)
Mexico / Bulgaria, Mexico

JACCARD, Marc
Simon Fraser University
Canada

JÄGER-WALDAU, Arnulf A.
European Commission Joint Research Centre (JRC)
Italy / Germany

JAKOB, Michael
Potsdam Institute for Climate Impact Research (PIK)
Germany

JAMES, Ted
National Renewable Energy Laboratory (NREL)
United States of America

JANNUZZI, Gilberto de Martino
Universidade Estadual de Campinas (UNICAMP)
Brazil

JENSEN, Peter Hjuler
Risø National Laboratory for Sustainable Energy
Denmark

JONKMAN, Jason
National Renewable Energy Laboratory (NREL)
United States of America

JUNGINGER, Martin
Utrecht University
The Netherlands

KADNER, Susanne
IPCC WGIII TSU
Germany

KAHN RIBEIRO, Suzana
Universidade Federal do Rio de Janeiro (UFRJ)
Brazil

KALKUHL, Matthias
Potsdam Institute for Climate Impact Research (PIK)
Germany

KAMIMOTO, Masayuki
National Institute of Advanced Industrial Science and Technology (AIST)
Japan

KAMMEN, Daniel
University of California
United States of America

KAZMERSKI, Lawrence
National Renewable Energy Laboratory (NREL)
United States of America

KEANE, Andrew
University College Dublin
Ireland

KHENNAS, Smail
United Nations Development Programme (UNDP)
Senegal / Algeria

KILLINGTVEIT, Ånund
Norwegian University of Science and Technology (NTNU)
Norway

KJAER, Christian
European Wind Energy Association (EWEA)
Belgium / Denmark

KONDO, Michio
National Institute of Advanced Industrial Science and Technology (AIST)
Japan

KONSEIBO, Charles D.
Centre Ecologique Albert Schweitzer du Burkina Faso
Burkina Faso

KREWITT, Wolfram †
Deutsches Zentrum für Luft- und Raumfahrt
Germany

KREY, Volker
International Institute for Applied Systems Analysis (IIASA)
Austria / Germany

KRUG, Thelma
Instituto Nacional de Pesquisas Espaciais (INPE)
Brazil

KUMAR, Arun
Indian Institute of Technology Roorkee (IITR)
India

LAMERS, Patrick
Ecofys
Germany

LANGNISS, Ole
Fichtner GmbH & Co. KG
Germany

LEE, Arthur
Chevron Corporation
United States of America

LEE, Kwang Soo
Korea Ocean Research and Development Institute (KORDI)
Republic of Korea

LENZEN, Manfred
University of Sydney
Australia / Germany

LEWIS, Anthony
University College Cork
Ireland

LIU, Yongqian
North China Electric Power University
People's Republic of China

LIU, Zhiyu
Ministry of Water Resources
People's Republic of China

LOGAN, Jeffrey
National Renewable Energy Laboratory (NREL)
United States of America

LOUIS, Frederic
Electricité de France (EDF)
France

LUCAS, Hugo
International Renewable Energy Agency (IRENA)
United Arab Emirates / Spain

LUCHT, Wolfgang
Potsdam Institute for Climate Impact Research (PIK)
Germany

LUCON, Oswaldo
São Paulo State Environment Secretariat
Brazil

LUDERER, Gunnar
Potsdam Institute for Climate Impact Research (PIK)
Germany

LUND, John W.
Oregon Institute of Technology
United States of America

MACKNICK, Jordan
National Renewable Energy Laboratory (NREL)
United States of America

MADSEN, Birger
BTM Consult ApS
Denmark

MANN, Margaret
National Renewable Energy Laboratory (NREL)
United States of America

MAPAKO, Maxwell
Council for Scientific and Industrial Research (CSIR)
South Africa / Zimbabwe

MASANET, Eric
Lawrence Berkeley National Laboratory
United States of America

MASERA CERUTTI, Omar
Universidad Nacional Autónoma de México (UNAM)
Mexico

MATSCHOSS, Patrick
IPCC WGIII TSU
Germany

MATSUBARA, Koji
Osaka University
Japan

MAURICE, Lourdes Q.
Federal Aviation Administration
United States of America

MCINTYRE, Terry C.
Environment Canada
Canada

MEIER, Anton
Paul Scherrer Institute (PSI)
Switzerland

MELESHKO, Valentin P.
Voeikov Main Geophysical Observatory
Russia

MERCADO, Pedro E.
Universidad Nacional de San Juan
Argentina

MILLIGAN, Michael
National Renewable Energy Laboratory (NREL)
United States of America

MILLS, Andrew
Lawrence Berkeley National Laboratory
United States of America

MINOWA, Tomoaki
National Institute of Advanced Industrial Science and Technology (AIST)
Japan

MIRZA, Monirul
Environment Canada
Canada / Bangladesh

MITCHELL, Catherine H. C.
University of Exeter
United Kingdom

MONGILLO, Michael A.
International Energy Agency Geothermal Implementing Agreement
New Zealand

MOOMAW, William R.
The Fletcher School
United States of America

MOREIRA, José R.
Brazilian Reference Center on Biomass, Institute of Electrotechnology and Energy, Universidade de Sao Paulo
Brazil

MORENO, José M.
University of Castilla-La Mancha
Spain

MORIARTY, Patrick
National Renewable Energy Laboratory (NREL)
United States of America

MUJUMDAR, Arun
National University of Singapore
Singapore

MURAOKA, Hirofumi
Hirosaki University
Japan

MUSIAL, Walter
National Renewable Energy Laboratory (NREL)
United States of America

NADAI, Alain
Centre Cired
France

NAGAI, Yu
International Institute for Applied Systems Analysis (IIASA)
Austria / Japan

NEMET, Gregory
Nelson Institute for Environmental Studies
United States of America

NEWELL, David
Chevron Geothermal and Power
Indonesia / United States of America

NIKOLAEV, Vladimir G.
Research & Information Center "ATMOGRAPH"
Russia

NILSSON, Lars J.
Lund University
Sweden

NYBOER, John G.
Simon Fraser University
Canada

O´MALLEY, Mark J.
University College Dublin
Ireland

OGDEN, Joan
University of California
United States of America

OGIMOTO, Kazuhiko
Institute of Industrial Science
Japan

OLHOFF, Anne
Risø National Laboratory for Sustainable Energy
United States of America / Denmark

OLSEN, Karen Holm
Risø National Laboratory for Sustainable Energy
Denmark

OOZEKI, Takashi
Tokyo University of Agriculture and Technology
Japan

OUTHRED, Hugh
University of New South Wales
Australia

PAHLE, Michael
Hertie School of Governance
Germany

PAN, Jiahua
Institute of Urban Environment
People's Republic of China

PANITCHPAKDI, Supachai
United Nations Conference on Trade and Development (UNCTAD)
Switzerland / Thailand

PARÉ, David
Laurentian Forestry Centre
Canada

PATEL, Martin
Utrecht University
The Netherlands

PINGOUD, Kim
VTT Technical Research Centre of Finland
Finland

POKHAREL, Govind R.
SNV Nepal Country Office
Nepal

PONTES, Teresa
The National Institute of Engineering, Technology and Innovation
Portugal

POPP, David
Syracuse University
United States of America

POWER, Michael
University College Dublin
Ireland

PRYOR, Sara
Indiana University
United States of America

RABL, Ari
ARMINES/Ecole de Mines
France

PICHS-MADRUGA, Ramón
Centro de Investigaciones de la Economía Mundial (CIEM)
Cuba

RADZI, Anis
University of Liechtenstein
Liechtenstein / Australia

RAGNARSSON, Arni
ISOR Iceland GeoSurvey
Iceland

RAHIMZADEH, Fatemeh
Atmospheric Science and Meteorological Research Center (ASMERC)
Islamic Republic of Iran

RAHMAN, Atiq
Bangladesh Centre for Advanced Studies
Bangladesh

RESCH, Gustav
Vienna University of Technology
Austria

RIAHI, Keywan
International Institute for Applied Systems Analysis (IIASA)
Austria

RICHELS, Richard
Electric Power Research Institute (EPRI)
United States of America

ROMANI, Mattia
Grantham Research Institute on Climate Change and the Environment
United Kingdom / Italy

ROY, Joyashree
Jadavpur University
India

RUDNICK, Hugh
Pontificia Universidad Católica de Chile
Chile

SANOGO, Oumar
Institut de Recherche en Sciences Appliquées et Technologies (IRSAT)
Burkina Faso

SANTAMOURIS, Matheos
National and Kapodistrian University of Athens
Greece

SANYAL, Subir K.
GeothermEx, Inc.
United States of America

SATHAYE, Jayant
Lawrence Berkeley National Laboratory
United States of America

SAVOLAINEN, Ilkka
VTT Technical Research Centre of Finland
Finland

SAWIN, Janet L.
Worldwatch Institute
United States of America

SCHAEFFER, Roberto
Universidade Federal do Rio de Janeiro (UFRJ)
Brazil

SCHEI, Tormod A.
Statkraft AS
Norway

SCHIMSCHAR, Sven
Ecofys
Germany

SCHLAEPFER, August
University of Flensburg
Germany / Australia

SCHLÖMER, Steffen
IPCC WGIII TSU
Germany

SCHMID, Jürgen
Fraunhofer Institute for Windenergy and Energy System Technology (IWES)
Germany

SCHRECK, Scott
National Renewable Energy Laboratory (NREL)
USA

SCRÅMESTØ, Sandvik Ø.
Statkraft AS
Norway

SEELOS, Karin
Statkraft AS
Norway

SEYBOTH, Kristin
IPCC WGIII TSU
Germany / United States of America

SHMAKIN, Andrey B.
Institute of Geography
Russia

SIMS, Ralph E. H.
Massey University
New Zealand

SINDEN, Graham
The Carbon Trust Ltd
United Kingdom / Australia

SKEA, Jim
UK Energy Research Centre (UKERC)
United Kingdom

SMITH, Charles
Utility Wind Integration Group (UWIG)
United States of America

SMITH, Paul
University College Dublin
Ireland

SÖDER, Lennart
KTH Royal Institute of Technology
Sweden

SOKONA, Youba
United Nations Economic Commission for Africa (UNECA)
Ethiopia / Mali

STECKEL, Jan
Potsdam Institute for Climate Impact Research (PIK)
Germany

STEIN, Wesley H.
The Commonwealth Scientific and Industrial Research Organisation (CSIRO)
Australia

STERNER, Michael
Fraunhofer Institute for Windenergy and Energy System Technology (IWES)
Germany

STRATTON, Russell
Massachusetts Institute of Technology (MIT)
United States of America

STRUNZ, Kai
Technische Universität Berlin
Germany

Annex IV — **Contributors to the IPCC Special Report**

TAMAURA, Yutaka
Tokyo Institute of Technology
Japan

TEKLEMARIAM ZEMEDKUN, Meseret
Consultant
Ethiopia

TESKE, Sven
Greenpeace
Germany

TESTER, Jefferson W.
Massachusetts Institute of Technology (MIT)
United States of America

TORRES-MARTINEZ, Julio
Cubasolar
Cuba

TRINDADE, Sergio C.
SE2T International, Ltd.
United States of America / Brazil, USA

TRUFFER, Bernhard
Swiss Federal Institute of Aquatic Science and Technology (Eawag)
Switzerland

TRUITT, Sarah
National Renewable Energy Laboratory (NREL)
United States of America

TRUJILLO BLANCO, Ramiro J.
United Nations Development Programme (UNDP)
Bolivia

TUOHY, Aidan
Electric Power Research Institute (EPRI)
United States of America

UECKERDT, Falko
Potsdam Institute for Climate Impact Research (PIK)
Germany

ULLEBERG, Øystein
Institute for Energy Technology
Norway

URAMA, Kevin
African Technology Policy Studies (ATPS) Network
Kenya / Nigeria

ÜRGE-VORSATZ, Diana
Central European University
Hungary

USHER, Eric
United Nations Environment Programme (UNEP)
Sweden

VAN DER HORST, Dan
University of Birmingham
United Kingdom / The Netherlands

VAN HULLE, Frans J. L.
XP Wind
Belgium

VAN YPERSELE, Jean-Pascal
Université catholique de Louvain
Belgium

VERBRUGGEN, Aviel
University of Antwerp
Belgium

VERMEYLEN, Saskia
Lancaster University
United Kingdom / Belgium

VON STECHOW, Christoph
IPCC WGIII TSU
Germany

WANG, Zhongying
Energy Research Institute
People's Republic of China

WARNER, Ethan
National Renewable Energy Laboratory (NREL)
United States of America

WEIR, Tony A. D.
University of the South Pacific
Republic of Fiji / Australia

WEYERS, Paul
Energy Research Centre of the Netherlands (ECN)
The Netherlands

WILBANKS, Thomas
Oak Ridge National Laboratory
United States of America

WILLIAMSON, Kenneth H.
Consultant
United States of America

WILSON, Charlie
London School of Economics (LSE)
United Kingdom

WISER, Ryan
Lawrence Berkeley National Laboratory
United States of America

WRATT, David
National Institute of Water & Atmospheric Research (NIWA)
New Zealand

WRIGHT, Raymond M. †
Petroleum Corporation of Jamaica (PCJ)
Jamaica

WÜSTENHAGEN, Rolf
University of St. Gallen
Switzerland / Germany

WYBORN, Doone
Geodynamics Limited
Australia

XU, Honghua
Institute of Electrical Engineering
People's Republic of China

YAMAGUCHI, Kaoru
Institute of Energy Economics
Japan

YAMBA, Francis
Centre for Energy Environment and Engineering
Zambia

YANG, Joyce
U.S. Department of Energy
United States of America

YANG, Zhenbin
Chinese Academy of Meteorological Sciences
People's Republic of China

YOU, Yage
Guangzhou Institute of Energy Conversion
People's Republic of China

ZERVOS, Arthouros
National Technical University of Athens
Greece

ZHANG, Jingjing
Lund University
Sweden

ZHANG, Yimin
National Renewable Energy Laboratory (NREL)
United States of America

ZILLES, Roberto
University of São Paulo
Brazil

ZUI, Vladimir I.
Republican Unitary Enterprise "Belarussian Research Geological Exploration Institute"
Republic of Belarus

ZWICKEL, Timm
IPCC WGIII TSU
Germany

ANNEX V

Reviewers of the IPCC Special Report

Reviewers of the IPCC Special Report

ALGERIA

SENOUCI, Mohamed
IHFR

ARGENTINA

BLANCO, Gabriel
Universidad Nacional del Centro de la Provincia de Buenos Aires

BOUILLE, Daniel
Fundación Bariloche

CARACCIA, Maria Eugenia
Secretary of Energy

CASTILLO MARIN, Nazareno
Secretaria de Ambiente y Desarrollo Sustentable

GIRARDIN, Leonidas
Fundación Bariloche

NADAL, Gustavo
Fundacion Bariloche

PARACCA, Juan Ignacio
Secretary of Energy

PEDACE, Alberto Roque
Buenos Aires University & Maestria Polititica y Gestioniencia y Tenologia

QUILES, Ernesto
Ministerio de Agricultura, Ganaderia y Pesca

SERVANT, Monica
Secretary of Energy

AUSTRALIA

BUDD, Anthony
Geoscience Australia

CLARKE, Drew
Department of Resources, Energy and Tourism

COLDREY, Olivia
Australian Solar Institute

GLEESON, Trish
Australian Bureau of Agricultural and Resource Economics

HITCHENS, Michael
Australian Industry Greenhouse Network

HOUSTON, Anne
Australian Academy of Technical Sciences and Engineering

JENNINGS, Philip
Murdoch University

OUTHRED, Hugh
University of New South Wales

PAGE, Brad
Electricity Supply Association of Australia

SMITHAM, Jim
Commonwealth Scientific and Industrial Research Organisation

STEIN, Wes
The Commonwealth Scientific and Industrial Research Organisation

AUSTRIA

BAZILIAN, Morgan
United Nations Industrial Development Organisation

HABERL, Helmut
University of Klagenfurt

KREY, Volker
International Institute for Applied Systems Analysis (IIASA)

LAUBER, Volkmar
University of Salzburg

ORGIS, Manfred
Ministry of Environment

RADUNSKY, Klaus
Umweltbundesamt

ROGNER, Hans-Holger
International Atomic Energy Agency

BAHRAIN

ABDEL-GELIL, Ibrahim
Arabian Gulf University

BANGLADESH

GORISSEN, Leen
Flemish Institute for Technological Research

GUISSON, Ruben
Flemish Institute for Technological Research

ISLAM, Sirajul
North South University

BELARUS

ZUI, Vladimir
Republican Unitary Enterprise

BELGIUM

DAUWE, Tom
Flemish Institute for Technological Research

DE PAEPE, Michel
Ghent University

DRIESEN, Johan
K.U. Leuven

EGGERMONT, Gilbert

GAINO, Bruna
Université Catholique de Louvain

MARBAIX, Philippe
Université Catholique de Louvain

RODRIGUES, Glória
European Wind Energy Association (EWEA)

SAWYER, Steve
Global Wind Energy Council

SCOWCROFT, John
EURELECTRIC

STRUYF, Igor
Public Planning Service Science Policy

SUL, Jung-ui
Sidley Austin, LLP

TURBELIN, Elise
Université Catholique de Louvain

VAN HULLE, Frans
European Wind Energy Association (EWEA)

VANDERSTRAETEN, Martine
Public Planning Service Science Policy

VERBRUGGEN, Aviel
University of Antwerp

VERHOEST, Chrystelle
LABORELEC

WOYTE, Achim
3E s.a.

BRAZIL

AMARAL, Luiz Fernando
UNICA Brazilian Sugarcane Industry Association

CABRAL, Marco Tulio
MRE

DE CAMPOS, Christiano Pires
Petrobras

DE CAMPOS BARBOSA, Paulo Cesar
Petrobras

FONTES LIMA, Francisco
Petrobras

GONZALEZ MIGUEZ, José Domingos
Ministry of Science and Technology

GUTIERRES, Ricardo
Petrobras

HORTA NOGUEIRA, Luiz A.
Instituto de Recursos Naturais

JANNUZZI, Gilberto
University of Campinas

LEITE DRACHMANN, Marcia
Petrobras

MARQUES, Fabio
The Plantar Group

MOREIRA, Jose Roberto
Brazilian Reference Center on Biomass

MOUTINHO DOS SANTOS, Edmilson
Universidade de Sao Paulo

PACCA, Sergio
University of Sao Paulo

PINHO, João
Institute of Technology

RODRIGUES CUNHA, Paulo Cesar
Petrobras

SANTANA MUSSE, Ana Paula
Petrobras

SCHMALL, Vicente
Petrobras

SOLIANO PEREIRA, Osvaldo
Universidade Salvador

TEXEIRA COELHO, Suani
Institute of Electrotechnics and Energy

TOLMASQUIM, Mauricio
Empresa de Pesquisa Energética

BURKINA FASO

COULIBALY, Yezouma
International Institute for Water and Environmental Engineering

PHILIPPE, Girard
International Institute for Water and Environmental Engineering

CANADA

AMANDEEP, Garcha
Natural Resources Canada

ANGEN, Meara
Environment Canada

AYOUB, Josef
Natural Resources Canada

BERNIER, Pierre
Natural Resources Canada

BHUYAN, Gouri
Powertech Labs

BLAIS, Caroline
Environment Canada

BLAIS, Darcy
Natural Resources Canada

BLANDFORD, Laurence
Environment Canada

Reviewers of the IPCC Special Report Annex V

BRANDON, Robert
Natural Resources Canada

BURKE, David
Environment Canada

BUSH, Elizabeth
Environment Canada

CAMPBELL, Chris
Ocean Renewable Energy Group

CHARBONNEAU, Maxime
Environment Canada

COATES, Laura
Environment Canada

CUMMINS, Patrick
Fisheries and Oceans Canada

CUNNINGHAM, Don
Natural Resources Canada

DALLAIRE, Lynne
Indian and Northern Affairs Canada

DAVISON, Matt
University of Western Ontario

DODDS, Karen
Environment Canada

DUNCAN, Tracy
Natural Resources Canada

ESSAJEE, Samina
Environment Canada

FERGUSON, Grant
St. Francis Xavier University

GAGNON, Luc
Hydro-Quebec

GILLETT, Nathan
Environment Canada

GILSENAN, Rory
Natural Resources Canada

GOUR, Christian
Foreign Affairs and International Trade Canada

GRAY, Brian
Environment Canada

JENSEN, Jack
Natural Resources Canada

JUTZI, Dan
Environment Canada

KAPOOR, Anoop
Natural Resources Canada

KOSTELZ, Tony
Environment Canada

KRAMER, Amanda
Environment Canada

LABIB, Herzel
Natural Resources Canada

LACROIX, Antoine
Natural Resources Canada

LEI, Cecilia
Environment Canada

LEMMEN, Don
Natural Resources Canada

LITTLE, Brad
Environment Canada

LOW, Heather
Foreign Affairs and International Trade Canada

LUNDY, Katie
Environment Canada

MARTIN, Laura
Natural Resources Canada

MCKENNEY, Dan
Great Lakes Forestry Centre

NADEAU, Melanie
Natural Resources Canada

ODHIAMBO, Joseph
Environment Canada

PAUNESCU, Michael
Natural Resources Canada

RADOVAN, Rock
Environment Canada

ROYER, Jimmy
Natural Resources Canada

SAMSON, Rachel
Environment Canada

SCHUBERT, Philip
Canadian International Development Agency

SMITH, Donald L.
McGill University

STRAUSS, Jessica
Environment Canada

TITUS, Brian
Natural Resources Canada

TUDVIER, Simon
Environment Canada

TUTHILL, Jennifer
Natural Resources Canada

VANDELIGT, Kelly
Environment Canada

VELJKOVIC, Maja
National Research Council of Canada

WALSH, Elizabeth
Natural Resources Canada

WELSH, Leslie
Environment Canada

YU, Wei
Environment Canada

ZWIERS, Francis
Pacific Climate Impacts Consortium

CHILE

FARÍAS, Fernando
CONAMA

GALETIVIC, Alexander
Chilean Ministry of Economy

GARCIA, Javier
Renewable Energy Center

JADRIJEVIC, Maritza
Chilean National Environmental Commission

JARA TIRAPEGUI, Wilfredo
Endesa Eco S.A.

WOISCHNIK, Alwine
Chilean National Environmental Commission

PEOPLE'S REPUBLIC OF CHINA

CAI, Fengbo
Chinese Wind Energy Association

CHAI, Qimin
Tsinghua University

CHEN, Dayong
Ministry of Water Resources

CHEN, Guohai
Hydrochina Hudong Engineering Corporation

CHEN, Mozi
China Electric Power Research Institute

CHEN, Zhenghong
Hubei Service Center of Meteorological Science & Technology

DING, Yi
CNOOC New Energy Investment Co. Ltd.

DING, Yongyao
First Institute of Oceangraphy

DONG, Hongmin
Chinese Academy of Agricultural Sciences

GAO, Hu
Energy Research Institute

GAO, Lin
Chinese Academy of Sciences

GAO, Yun
China Meteorological Administration

GUO, Shiyi
Tsinghua University

HAO, Aibin
China Geological Survey

HE, Dexin
Chinese Wind Energy Association

HONG, Hao
Peking University

HU, Xiulian
National Development and Reform Commission

JIA, Jinsheng
ICOLD

JIANG, Jianchun
Chinese Academy of Forestry

JIN, Hongguang
Chinese Academy of Sciences

LI, Jingmin
Ministry of Agriculture

LI, Junfeng
Energy Research Institute

LIAO, Wengen
China Institute of Water Resources and Hydropower Research

LIU, Fuyou
National Ocean Technology Center

LIU, Mingliang
China General Certification Center

LIU, Yongqian
North China Electric Power University

MA, Weibin
Chinese Academy of Sciences

PAN, Jiahua
Chinese Academy of Social Sciences

PANG, Zhonghe
Institute of Geology and Geophysics

REN, Dongming
Energy Research Institute

REN, Xiangkun
Beijing Research Institute

RUAN, Honghua
Nanjing Forestry University

SHEN, Yanbo
China Meteorological Administration

SHI, Jingli
Energy Research Institute

SHI, Pengfei
China Hydropower Engineering Consulting Group Co.

SHI, Zuomin
Chinese Academy of Forestry

SICHENG, Wang
Beijing Jike Energy New Tech. Development Co.

SUN, Pengsen
Ministry of Environment and Protection

TANG, Feiwen
COFCO

TENG, Fei
Tsinghua University

TIAN, Zhongxing
Ministry of Water Resources

WANG, Bingzhen
National Ocean Technology Center

WANG, Yi
Institute of Policy and Management, Chinese Academy of Sciences

WANG, Weisheng
China Electric Power Research Institute

WANG, Zhifeng
Chinese Academy of Sciences

WANG, Zhongying
Energy Research Institute

WEI, Dongyuan
Chinese Academy of Science and Technology for Development

WU, Shurirong
Chinese Academy of Forestry

XIA, Jianxin
Minzu University of China

XING, Yuanyue
Ministry of Water Resources

XU, Ruina
Tsinghua University

XU, Honghua
Institute of Electrical Engineering

YANG, Xiaosheng
China Longyuan Power Group Coproration Ltd.

YOU, Yage
Guangzhou Institute of Energy Conversion

YU, Zhouwen
National Marine Environment Forecast Center

ZHANG, Chengyi
China Meteorological Administration

ZHANG, Guobin
Academy of Forest Inventory & Planning

ZHANG, Liang
Harbin Engineering University

ZHANG, Xiliang
Tsinghua University

ZHANG, Xuejin
IN-SHP

ZHANG, Yanru
National Bio Energy Co., Ltd.

ZHAO, Lixin
Chinese Academy of Agricultural Engineering

ZHAO, Ping
Institute of Geology and Geophysics

ZHAO, Zongci
China Meteorological Administration

ZHENBIN, Yang
China Meteorological Administration

ZHENG, Guoguang
China Meteorological Administration

ZHONG, Deyu
Tsinghua University

ZHOU, Dadi
National Development and Reform Commission

ZHU, Rong
China Meteorological Administration

ZHU, Xiaoqing
Novozymes (China) Investment Co.

ZHUANG, Huiyong
National Bio Energy Co. Ltd.

COSTA RICA

BALLESTERO, Johnny Montenegro
National Meteorological Institute

HEINRICH, Kristel
Instituto Meteorologico Nacional

CUBA

ACOSTA MORENO, Roberto
CITMA

ALFREDO, Curbelo
Cubaenergia

GUTIÉREZ-PÉREZ, Tomás
Instituto de Meteorología

FERNANDEZ DIAZ-SILVEIRA, Modesto
Ministry of Science, Technology and Environment

HERNÁNDEZ, Gladys
Centro de Investigaciones de la Economía Mundial

LIMIA, Miriam Ester
Instituto de Meteorologica de la Republica de Cuba

LLANES-REQUEIRO, Juan F.
University of Havana

PICHS-MADRUGA, Ramón
Centre de Investigaciones de la Economía Mundial (CIEM)

RODRÍGUEZ, Carlos
National Institute for Physical Planning

ROJAS, Nazareth
National Meteorological Institute

SOMOZA, José
University of Havana

DENMARK

ANDERSEN, Katrine Krogh
Danish Meteorological Institute

BARBU, Anca-Diana
European Environment Agency

CLUBB, David
European Environment Agency

ERIKSEN, Peter Børre
Energinet.dk

JORGENSEN, Anne Mette K.
Danish Meteorological Institute

JUUL-KRISTENSEN, Bjarne
Danish Energy Agency

KARLSSON, Kenneth
University of Denmark

PETERSEN, Leif Sønderberg
University of Denmark

SCHOU, Annette
Danish Energy Agency

STIESDAL, Henrik
Siemens Wind Power

ECUADOR

HERVAS JATIVA, Istvan
National Electricity Council of Ecuador

MOGOLLÓN ZAPATA, Galo Fernando
Ministry of Environment

PALACIOS CABRERA, Teresa Alejandra
Ministry of Environment

EGYPT

ABD EL-WAHAB, Mohamed Kadry
Zagazig University

ABED, Kamal
National Research Centre

EL-HINNAWI, Essam
National Research Centre

EL SALVADOR

SAUERBREY, Mauricio
Energie Renovable

ETHIOPIA

SOKONA, Youba
African Climate Policy Centre, United Nations Economic Commission for Africa (UNECA)

TEKLEMARIAM ZEMEDKUN, Meseret
Geological Survey of Ethiopia

FIJI

AHMED, Rafiuddin
The University of the South Pacific

GONELEVU, Arieta
International Union for Conservation of Nature

JOHNSTON, Peter
Environmental & Energy Consultants Ltd.

RATURI, Atul
The University of the South Pacific

WEIR, Tony
University of the South Pacific

FINLAND

ANTIKAINEN, Riina
Finnish Environment Institute

ASIKAINEN, Antti
Finnish Forest Research Institute

HAKALA, Kaija
MTT Agrifood Research Finland

HALME, Janne
Aalto University

HANNINEN, Seppo
VTT Technical Research Center of Finland

HEIKINHEIMO, Pirkko
Prime Minister's Office

HELYNEN, Satu
VTT Technical Research Center of Finland

HOLTTINEN, Hannele
VTT Technical Research Center of Finland

JÄRVENPÄÄ, Markku
MTT Agrifood Research Finland

Reviewers of the IPCC Special Report

KAHILUOTO, Helena
MTT Agrifood Research Finland

KANGAS, Markku
Finish Meteorological Institute

KATI, Koponen
VTT Technical Research Centre of Finland

KIRKINEN, Johanna
Sitra, the Finnish Innovation Fund

KIVILUOMA, Juha
VTT Technical Research Centre of Finland

KOPONEN, Kati
VTT Technical Research Centre of Finland

LEHTINEN, Kari
Finnish Meteorological Institute

LUND, Peter
Helsinki University of Technology

OJALA, Jaakko
Ministry of the Environment

PERÄLÄ, Hanna
Prime Minister's office

PERRELS, Adriaan
Finnish Meteorological Institute &
Government Institute for Economic Research

PINGOUD, Kim
VTT Technical Research Centre

PIRILÄ, Pekka
Aalto University

RIINA, Antikainen
Finnish Environment Institute

SAMPO, Soimakallio
VTT Technical Research Centre of Finland

SAVOLAINEN, Ilkka
VTT Technical Research Centre of Finland

SEPPÄLÄ, Jyri
Finnish Environment Institute

SOIMAKALLIO, Sampo
VTT Technical Research Centre of Finland

TAALAS, Petteri
Finnish Meteorological Institute

TUOMAS, Helin
VTT Technical Research Centre of Finland

UUSIVUORI, Jussi
Finnish Forest Research Institute Metla

VAPAAVUORI, Elina
Finnish Forest Research Institute Metla

FRANCE

AELBRECHT, Denis
Électricité de France

AGBEMABIESE, Lawrence
United Nations Environment Programme (UNEP)

ALLAL, Houda
Observatoire Méditerranéen de l'Energie (OME)

ARGIRI, Maria
International Energy Agency (IEA)

BERIOT, Nicolas
Ministry of Ecology, Energy, Sustainable Development and the Sea (MEEDDM)

BONDUELLE, Antoine
EE Consultant

BRANCHE, Emmanuel
Electricité de France (EDF)

CANEILL, Jean-Yves
Electricité de France

COZZI, Laura
International Energy Agency (IEA)

DARRAS, Marc
GDF SUEZ

DEVERNAY, Jean-Michel
Electricité de France

GABRIELLE, Benoît
AgroParisTech

LOUIS, Frederic
EDF Hydro Engineering Centre

MARCHAL, Julien
MEEDDM

MENICHETTI, Emanuela
Observatoire Méditerranéen de l'Energie

NADAI, Alain
CIRED

PETIT, Michel
CGIET

PHILIBERT, Cédric
International Energy Agency

POUFFARY, Stephanie
Energies 2050

SONNTAG-O'BRIEN, Virginia
REN21

GAMBIA

MANNEH, Pa Abdoulie
Ministry of Finance and Economic Affairs

GERMANY

AVENHAUS, Wibke
Potsdam Institute for Climate Impact Research (PIK)

Annex V — Reviewers of the IPCC Special Report

BAUER, Nico
Potsdam Institute for Climate Impact Research (PIK)

BEHRENDT, Frank
Institute for Energy Engineering

BONHOFF, Klaus
NOW GmbH National Organization Hydrogen and Fuel Cell Technology

BRECHA, Robert
Potsdam Institute for Climate Impact Research (PIK)

BRUCKNER, Thomas
University of Leipzig

BRUNNER, Steffen
Potsdam Institute for Climate Impact Research (PIK)

CREUTZIG, Felix
TU Berlin

DEUTSCH, Matthias
Prognos

EDENHOFER, Ottmar
Potsdam Institute for Climate Impact Research (PIK)

EICKEMEIER, Patrick
IPCC WGIII TSU

FISCHEDICK, Manfred
Wuppertal Institute for Climate, Environment, Energy

GIFFORD, Mary Louise
Potsdam Institute for Climate Impact Research (PIK)

GOERNER, Marlen
IPCC WGIII TSU

GRASSL, Hartmut
Max Planck Institute for Meteorology

GUENTHER, Edeltraud
TU Dresden

HALLER, Markus
Postdam Institute for Climate Impact Research (PIK)

HANSEN, Gerrit
IPCC WGIII TSU

HAUM, Rüdiger
German Advisory Council on Global Change

HERBENER, Reinhard
German Federal Environment Agency

VON HIRSCHHAUSEN, Christian
TU Berlin

HOHMEYER, Olav
University of Flensburg

HÜBLER, Michael
Centre for European Economic Research (ZEW)

JAKOB, Michael
Potsdam Institute for Climate Impact Research (PIK)

KADNER, Susanne
IPCC WGIII TSU

KAUP, Felix
Potsdam Institute for Climate Impact Research (PIK)

KLASEN, Stephan
Ibero-American Institute

KLEIDON, Axel
Max Planck Institute of Biogeochemistry

KLEIN, David
Potsdam Institute for Climate Impact Research (PIK)

KLESSMAN, Corinna
Ecofys Germany

KNOPF, Brigitte
Potsdam Institute for Climate Impact Research (PIK)

LAMERS, Patrick
Ecofys Germany

LEHMANN, Harry
German Federal Environment Agency

LIPSIUS, Kai
German Federal Environment Agency

LOHSE, Christiane
German Federal Environment Agency

LOTZE-CAMPEN, Hermann
Potsdam Institute for Climate Impact Research (PIK)

LUDERER, Gunnar
Potsdam Institute for Climate Impact Research (PIK)

LUDIG, Sylvie
Potsdam Institute for Climate Impact Research (PIK)

MASTIAUX, Frank
EON Climate & Renewables

MATSCHOSS, Patrick
IPCC WGIII TSU

MEINSHAUSEN, Malte
Potsdam Institute for Climate Impact Research (PIK)

MUELLER, Richard
Climate Monitoring Satellite Application Facility, DWD

NEUHOFF, Karsten
German Institute for Economic Research (DIW Berlin)

OETZEL, Nicolas
Federal Ministry for the Environment, Nature Conservation and Nuclear Safety

PEHNT, Martin
Institute for Energy and Environmental Research

PIETZCKER, Robert
Potsdam Institute of Climate Impact Research (PIK)

POPP, Alexander
Potsdam Institute of Climate Impact Research (PIK)

PRAESSLER, Thomas
Potsdam Institute of Climate Impact Research (PIK)

RAUCH, Ernst
Munich Reinsurance Company (Munich Re)

RECH, Bernd
Helmholtz-Zentrum Berlin für Materialien und Energie GmbH

RIECKE, Wolfgang
Deutscher Wetterdienst

RUFIN, Julia
Federal Ministry for the Environment, Nature Conservation and Nuclear Safety

SCHEFFRAN, Jürgen
University of Hamburg

SCHLÖMER, Steffen
IPCC WGIII TSU

SCHULZ, Astrid
German Advisory Council on Global Change (WBGU)

SCHWEIZERHOF, Henriette
Federal Ministry for the Environment, Nature Conservation and Nuclear Safety

SEVEN, Jan
German Federal Environment Agency

SEYBOTH, Kristin
IPCC WGIII TSU

VON STECHOW, Christoph
IPCC WGIII TSU

STECKEL, Jan
Potsdam Institute for Climate Impact Research (PIK)

STENGLER, Ella
CEWEP

TEXTOR, Christiane
German Aerospace Center

THRÄN, Daniela
DBFZ / UFZ

TREBER, Manfred
Germanwatch e.V.

UECKERDT, Falko
Potsdam Institute for Climate Impact Research (PIK)

VAHRENHOLT, Fritz
RWE Innogy GmbH

VENGHAUS, Sandra
Potsdam Institute for Climate Impact Research (PIK)

WALZ, Rainer
Fraunhofer Systems and Innovation Research

WEIMANN, Joachim
Otto von Guericke University

WEISSBACH, Sven
German Federal Environment Agency

WEINHOLD, Michael
Siemens AG

WILKE, Nicole
Federal Ministry for the Environment, Nature Conservation and Nuclear Safety

ZWICKEL, Timm
IPCC WG III TSU

GREECE

BALARAS, Constantinos
National Observatory of Athens

CHAVIAROPOULOS, Panagiotis
Centre for Renewable Energy Sources and Saving

DIAKOULAKI, Danae
National Technical University of Athens

GEORGOPOLOU, Elena
National Observatory of Athens

GLINOU, Georgia
Regulatory Authority for Energy

KANELOPOULOS, Dimitrios
Public Power Corporation-Renewables S.A.

LINGOS, Elias
Public Power Corporation-Renewables S.A.

MIRASGEDIS, Sebastian
National Observatory of Athens

SARAFIDIS, Yiannis
National Observatory of Athens

TSILINGIRIDIS, George
Aristotle University of Thessaloniki

HUNGARY

PÁLVÖLGYI, Tamás
Budapest University of Technology and Economics

SOMOGYI, Zoltán
Hungarian Forest Research Institute

ICELAND

FRIDLEIFSSON, Ingvar
United Nations University Geothermal Training Programme

INDIA

HEDGE, Ishwar
Suzlon Energy Ltd.

MURTY, Maddipati Narasimha
Institute of Economic Growth

PATWARDHAN, Anand
Indian Institute of Technology-Bombay

ISLAMIC REPUBLIC OF IRAN

SEHAT KASHAMI, Saviz
Atmospheric Sciences and Meteorological Research Center

RAHIMI, Mohammad
IRIMO

IRELAND

DODD, David
Environmental Protection Agency of Ireland

LEAHY, Paul
University College Cork

O'MALLEY, Mark
University College Cork

O'SULLIVAN, Dara
University College Cork

POWER, Michael
University College Cork

SMITH, Paul
University College Cork

ITALY

CASTELLARI, Sergio
Instituto Nazionale di Geofisica e Vulcanologia

CONTALDI, Mario
Institute for Environmental Protection and Research (ISPRA)

GAUDIOSO, Domenico
Institute for Environmental Protection and Research (ISPRA)

GRACCEVA, Francesco
Italian National Agency for New Technologies, Energy and Sustainable Economic Development (ENEA)

JÄGER-WALDAU, Arnulf
European Commission

TAVONI, Massimo
FEEM/CMCC

JAPAN

AKIMOTO, Keigo
Research Institute of Innovative Technology for the Earth (RITE)

FUKUI, Kiroyuki
Toyota

HONGO, Takashi
Japan Bank for International Cooperation

KAMIMOTO, Masayuki
National Institute of Advanced Industrial Science and Technology (AIST)

KANO, Takehiro
The Japanese Ministry of Foreign Affairs

KAWASATO, Taro
Ministry of the Environment, Japan

KIMURA, Osamu
Central Research Institute of Electric Power Industry

KOBAYASHI, Shigeki
Toyota R&D Labs Inc.

MAEDA, Ichiro
The Federation of Electric Power Companies of Japan

NAKAO, Shinsuke
National Institute of Advanced Industrial Science and Technology (AIST)

OGIMOTO, Kazuhiko
The University of Tokyo

SUGIYAMA, Taishi
Central Research Institute of Electric Power Industry (CRIEPI)

TAGASHIRA, Naoto
Central Research Institute of Electric Power Industry (CRIEPI)

TAKEUCHI, Hiromi
Advanced Industrial Science and Technology

TANI, Saeko
Ministry of Economy, Trade and Industry

KENYA

DE OLIVEIRA, Thierry
United Nations Environmental Programme (UNEP)

Reviewers of the IPCC Special Report

REPUBLIC OF KOREA

BOO, Kyung-Jin
Korea Energy Economics Institute

KIM, Hyun-Kyung
Korea Meteorological Administration (KMA)

SHIM, Sung-Hee
Korea Energy Economics Institute

MEXICO

DE LA VEGA NAVARRO, Angel
National Autonomous University of Mexico (UNAM)

GARCIA-GUTIERREZ, Alfonso
Instituto de Investigaciones Electricas

GUTIERREZ-NEGRIN, Luis C. A.
Mexican Geothermal Association

NEPAL

POKHAREL, Govind
SNV Netherlands Development Organisation, Nepal

THE NETHERLANDS

BEURSKENS, Jos
ECN Wind Energy

HAAK, Hein
Royal Dutch Meteorological Institute (KNMI)

LONDO, Marc
Energy Research Centre of the Netherlands (ECN)

PAGNIER, Henk
TNO

SINKE, Wim
Energy research Centre of the Netherlands (ECN)

NEW ZEALAND

HAWKE, Richard
Ministry of Economic Development

HUCKERBY, John
Power Projects Ltd.

JACK, Michael
Scion - New Zealand Forest Research Institute Ltd.

KRIEBLE, Todd
Ministry of the Environment

SIMS, Ralph
MUCER

NORWAY

ALFSEN, Knut
Cicero

ANKER-NIELSEN, Per
Confederation of Norwegian Enterprise (NHO)

ASPHEJELL, Torgrim
Climate and Pollution Agency

BARSTAD, Idar
Uni Research AS

BERRE, Inga
University of Bergen

BEVANGER, Kjetil
Norwegian Institute for Nature Research (NINA)

BRYHN JACOBSEN, Linn
Climate and Pollution Agency

CHRISTOPHERSEN, Øyvind
Climate and Pollution Agency

GLOMNES RUDI, Anne
NORAD

GÖSSLING, Stefan
Western Norway Research Institute

GRIMSRUD, Ole
Scatec AS

GRØNSTAD, Christoffer
Climate and Pollution Agency

HAUG, Trond Espen
Statkraft

HAUGLAND, Hege
Climate and Pollution Agency

HAUGLAND, Svein
Agder Energi AS

HARBY, Atle
SINTEF Energy Research

HERTWICH, Edgar
Norwegian University of Science and Technology (NTNU)

HESTNES, Anne Grete
Norwegian University of Science and Technology (NTNU)

JENSEN, Trond Arnljot
Statnett

JOHANSEN, Øivind Jan
Ministry of Petroleum and Energy

KOLSTAD, Anne-Grethe
Climate and Pollution Agency

LEFFERTSTRA, Harold
Climate and Pollution Agency

MOE, Geir
Norwegian University of Science and Technology (NTNU)

MOSTAD, Helle
Statoil

NIELSEN, Finn Gunnar
Statoil

OFSTAD, Elizabeth Baumann
Statoil ASA

PETTERSEN, Marit Victoria
Ministry of Foreign Affairs

RANDERS, Jorgen
Norwegian School of Management (BI)

SCHEI, Tormod
Statkraft AS

TORVANGER, Asbjørn
Cicero

TVETEN, Åsa
Climate and Pollution Agency

ULLEBERG, Øystein
Institute for Energy Technology

ULSETH, Oluf
Statkraft AS

VESTRENG, Vigdis
Climate and Pollution Agency

WITTGENS, Bernd
SINTEF Materials and Chemistry

PAKISTAN

CHAUDHRY, Qamar-uz-Zaman

IQBAL, Muhammad Mohsin
Global Change Impact Studies Centre (GCISC)

AGUIAR, Ricardo
National Laboratory for Energy and Geology (LNEG)

POLAND

FILIPIAK, Janusz
Institute of Meteorology and Water Management

KAMINSKI, Jacek
Ekoprognoza

LOBOCKI, Lech
Warsaw University of Technology

STRUZEWSKA, Joanna
Warsaw University of Technology

ROMANIA

BADESCU, Viorel
Polytechnic University of Bucharest

BOJARIU, Roxana
Meteo Romania

GLUCK, Peter
Info Kappa

RUSSIA

GOGOLEV, George
Geography of the Russian Academy of Sciences

REUTOV, Boris
Federal Agency for Science and Innovation

SENEGAL

SARR, Babacar
ENERTEC-SARL

SOUTH AFRICA

KRUGER, Andries
South African Weather Service

WINKLER, Harald
University of Cape Town

SPAIN

AAGESEN-MUÑOZ, Sara
Spanish Bureau for Climate Change. Ministry for Environment, Rural and Marine Affairs

BONNET FERNÁNDEZ-TRUJIL, Jorge
Agencia Canaria de Desarrollo Sostenible y Cambio Climático

CHANES-VICENTE, Rodrigo
Ministry of Industry, Tourism and Trade

DÍAZ-RUIZ, Prado
Ministry for Industry, Tourism and Trade

FERNÁNDEZ-LOPEZ, Carlos
Institute for Diversification and Energy Saving.

GONZALEZ-FERNANDEZ, Eduardo
Spanish Bureau for Climate Change. Ministry for Environment, Rural and Marine Affairs

LOPEZ –MONLLOR, Carlos
Spanish Bureau for Climate Change. Ministry for Environment, Rural and Marine Affairs

MARBÁN, Gregorio
Instituto Nacional del Carbón (CSIC)

MARTÍNEZ CHAMORRO, Jorge
Agencia Canaria de Desarrollo Sostenible y Cambio Climático

MARTÍNEZ-LOPE, Concepcion
Ministry for Environment, Rural and Marine Affairs

Reviewers of the IPCC Special Report

PIERNAVIEJA, Gonzalo
Instituto Tecnológico de Canarias (ITC)

RÜBBELKE, Dirk
Basque Centre for Climate Change/ IKERBASQUE

RUBIERA, Fernando
Instituto Nacional del Carbon (CSIC)

RUIZ-CASTELLO, Pablo
Ministry for Environment, Rural and Marine Affairs

SABIDO-MARTIN, Alberto
Spanish Bureau for Climate Change. Ministry for Environment, Rural and Marine Affairs

SANCHEZ, Juan Jose
Ministry of the Environment, and Rural and Marine Affairs

VELASCO, Teresa M.
Ministry of Industry, Tourism and Trade

SWEDEN

ÅHMAN, Max
Swedish Environmental Protection Agency

BJÖRCK, Anders
Elforsk AB

DI LUCA, Lorenzo
Lund University

ERIKSSON, Karin
Lund University

GULDBRAND, Lars
Swedish Energy Agency

LILLIESKÖLD, Marianne
Swedish Environmental Protection Agency

MÖLLERSTEN, Kenneth
Swedish Energy Agency

NILSSON, Lars J.
Lund University

NILSSON, Måns
Stockholm Environment Institute

OLSSON, Larsolov
Swedish Environmental Protection Agency

RANTIL, Michael
Swedish Energy Agency

SANDÉN, Björn
Chalmers University of Technology

SÖDER, Lennart
Royal Institute of Technology

SÖDERHOLM, Patrik
Luleå University of Technology

SWITZERLAND

ALLEN, Simon
IPCC WGI TSU

BAUER, Christian
Paul Scherrer Institute

DE HAAN, Peter
Ernst Basler/Partner AG

FISCHLIN, Andreas
ETH Zurich

IVAR, Baste
United Nations Environmental Programme (UNEP)

KRYSIAK, Frank
University of Basel

McCORMICK, Nadine
International Union for Conservation of Nature (IUCN)

MICHAELOWA, Axel
University of Zurich

NAUELS, Alex
IPCC WGI TSU

PANITCHPAKDI, Supachai
United Nations Conference on Trade and Development

PITTEL, Karen
ETH Zurich

PLATTNER, Gian-Kasper
IPCC WGI TSU

ROMERO, Jose
Swiss Federal Office for the Environment

RYBACH, Ladislaus
Geowatt AG Zurich

TRUFFER, Bernhard
Eawag

THAILAND

LIMMEECHOKCHAI, Bundit
Thammasat University

WADE, Herbert

UNITED KINGDOM

CAREY, Liz
Cambridge Centre for Climate Change Mitigation Research

CHARLES, Amanda
BIS

CONBOY, Alison
Department of Energy and Climate Change

CRAWFORD-BROWN, Doug
Cambridge Centre for Climate Change Mitigation Research

DAVEY, James
Department of Energy and Climate Change

EVANS, Geraint
NNFCC

FELGENHAUER, Tyler
Cambridge Centre for Climate Change
Mitigation Research

GAMBHIR, Ajay
Department of Energy and Climate Change

GRIFITHS, Rhodri
Sustainable Energy and Industry Wales

HAMILTON, Kirsty
Chatham House

HASLETT, Andrew
Manchester University

HAYES, Lucy
Department of Energy and Climate Change

HOSKYNS, John
Department of Energy and Climate Change

INFIELD, David
University of Strathclyde

JONES, Leanne
Department for International Development

JOSLIN, Tim
Cambridge Centre for Climate Change
Mitigation Research

KESSELS, John
International Energy Agency (IEA)

KHENNAS, Smail

KNIGHT, Oliver
Department for International Development

KNOX, Catriona
Department of Energy and Climate Change

KYTE, William
E.ON AG

LAIL, Davinder
Defra

LA PORTA, Filomena
Technology Strategy Board

MILBORROW, David
Consultant

OFFER, Greg
Department of Energy and Climate Change

RAI, Kavita
Global Village Energy Partnership
International

RATCLIFFE, Simon
Policy and Research Team

SCHARLEMANN, Jörn
United Nations Environment Programme
World Conservation Monitoring Centre
(UNEP-WCMC)

SEWELL, Martin
Cambridge Centre for Climate Change
Mitigation Research

SINDEN, Graham
Carbon Trust

SIVETER, Robert
IPIECA

SKEA, Jim
UK Energy Research Council

STAUNTON, Garry
Carbon Trust

TANG, Lily
Department of Energy and Climate Change

TAYLOR, Richard
International Hydropower Association (IHA)

THORNLEY, Patricia
School of Mechanical, Aerospace and Civil
Engineering

TWIDELL, John
AMSET Centre

UPHAM, Paul
Manchester Business School

WARRILOW, David
Department of Energy and Climate Change

WICKINS, Chris
Department of Energy and Climate Change

WYATT, Stephen
Carbon Trust

UNITED STATES OF AMERICA

ADAMANTIADES, Misha
White House Council on Environmental
Quality

ADEN, Andy
National Renewable Energy Laboratory
(NREL)

ARENT, Doug
National Renewable Energy Laboratory
(NREL)

ASHWILL, Thomas
Sandia National Laboratories

BALDWIN, Sam
U.S. Department of Energy

BEDARD, Roger
Electric Power Research Institute

Reviewers of the IPCC Special Report

BENIOFF, Ron
National Renewable Energy Laboratory (NREL)

BHATT, Vatsal
Brookhaven National Laboratory

BILELLO, Dan
National Renewable Energy Laboratory (NREL)

BIRD, Lori
National Renewable Energy Laboratory (NREL)

BLANKENSHIP, Doug
Sandia National Laboratories, New Mexico

BODNER, Paul
U.S. Department of State

BOTTERUD, Audun
Argonne National Laboratory

BRAITSCH, Jay
U.S. Department of Energy

BRANDT, Adam
Stanford University

BROWN, Austin
U.S. Department of Energy

BROWN, Nathan
U.S. Department of Energy

CALLOWAY, Thomas
Savannah River National Laboratory

CAMERON, Christopher
Sandia National Laboratories

CHIPMAN, Peter
U.S. Department of Transportation

CHUM, Helena
National Renewable Energy Laboratory (NREL)

CLARK, Charlton
U.S. Department of Energy

CLOUSE, Matt
U.S. Environmental Protection Agency

CONKLIN, Russell
U.S. Department of Energy

CONZELMANN, Guenter
Argonne National Laboratory

COOPER, Craig
Idaho National Laboratory

COSTA, Stephen
U.S. Department of Transportation

CRITCHFIELD, James
U.S. Environmental Protection Agency

DALE, Bruce
Michigan State University

DARIN, Thomas
U.S. Department of Energy

DARMSTADTER, Joel
Resources for the Future

DEMAYO, Trevor
Chevron Energy Technology Co.

DENHOLM, Paul
National Renewable Energy Laboratory (NREL)

DHAM, Rajesh
U.S. Department of Energy

DIAMOND, David
U.S. Department of Energy

DIEHL, Timothy
U.S. Geological Survey

DRURY, Easan
National Renewable Energy Laboratory (NREL)

DUNN, Seth
GE Energy

EBI, Kristie
IPCC WGII TSU

ELLIOTT, Dennis
National Renewable Energy Laboaratory (NREL)

EMERY, Keith
National Renewable Energy Laboaratory (NREL)

FLANK, Shalom
Pareto Energy

FORSGREN, Christopher
Idaho National Laboratory

FTHENAKIS, Vasilis
Centre for Life Cycle Analysis

FULTON, Mark
Deutsche Bank

GARDLAND, Rebecca
U.S. Department of Energy

GILMAN, Patrick
U.S. Department of Energy

GROL, Eric
National Energy Technology Laboratory

GULLIVER, John
University of Minnesota

GUTTROMSON, Ross
Pacific Northwest National Laboratory

HAMILTON, Bruce
National Science Foundation

HAND, Maureen
National Renewable Energy Laboratory (NREL)

Annex V

Reviewers of the IPCC Special Report

HAQ, Zia
U.S. Department of Energy

HEAL, Geoffrey
Columbia University

HORNE, Roland
Stanford University

JARAMILLO, Paulina
Carnegie Mellon University

JOHANSSON, Bob
U.S. Department of Agriculture

JOHNSON, Sarah
Office of Science and Technology Policy

KAMMEN, Daniel

KEMPTON, Willett
University of Delaware

KENNEDY, Mack B.
Lawrence Berkeley National Laboratory

KEOLEIAN, Gregory
Center for Sustainable Systems

KHESHGI, Haroon
ExxonMobil Research and Engineering Company

KING, Carey
University of Texas

KING, Eric
Bonneville Power Administration

KLEMICK, Heather
U.S. Environmental Protection Agency

KORITAROV, Vladimir
Argonne National Laboratory

KOSKE, Burton
Idaho National Laboratory

KOZLOFF, Keith
U.S. Department of the Treasury

KUTSCHER, Charles
National Renewable Energy Laboartory (NREL)

LEE, Arthur
Chevron Corporation

LEE, Audrey
U.S. Department of the Treasury

LESTER, Lave
Carnegie Mellon University

LOGAN, Jeffrey
National Renewable Energy Laboratory (NREL)

MAINZER, Elliott
Bonneville Power Administration

MALTZER, Eric
U.S. Department of State

MARGOLIS, Robert
National Renewable Energy Laboratory (NREL)

MARLAY, Robert
U.S. Department of Energy

MARRIOTT, Joe
Booz Allen Hamilton and University of Pittsburg

MAURICE, Lourdes
Federal Aviation Administration

MEYER, David
U.S. Department of Energy

MILLIGAN, Michael
National Renewable Energy Laboratory (NREL)

MILLS, Andrew
Lawrence Berkeley National Laboratory

MINES, Greg
Idaho National Laboratory

MUÑOZ, Miquel
Boston University

MUSIAL, Walt
National Renewable Energy Laboratory (NREL)

NAGELHOUT, Peter
U.S. Environmental Protection Agency

NATHWANI, Jay
U.S. Department of Energy

NEWMARK, Robin
National Renewable Energy Laboratory (NREL)

NIHOUS, Gerard
University of Hawaii at Manoa

PARADES, Juan Roberto
Inter-American Development Bank

PETRI, Mark
Argonne National Laboratory

PHELAN, Patrick
Arizona State University

PIWKO, Richard
General Electric Company

PLEVIN, Richard
UC Berkeley

PRENTICE, Geoffrey
National Science Foundation

PUGH, Graham
White House Council on Environmental Quality

RABL, Veronika
Vision & Results

RAM, Bonnie
Energetics Inc.

REED, Michael
U.S. Department of Energy

REGALBUTO, John
University of Illinois at Chicago

RENNE, Dave
National Renewable Energy Laboratory (NREL)

RENNER, Joel
Idaho National Laboratory (retired)

ROBINSON, Michael
National Renewable Energy Laboratory (NREL)

ROEGIERS, Jean-Claude
University of Oklahoma

ROSINSKI, Stan
Electric Power Research Institute

RYPINSKI, Arthur
U.S. Department of Transportation

SALE, Mike
Oak Ridge National Laboratory (retired)

SARGENT, Keith
U.S. Environmental Protection Agency

SAWIN, Janet
Sunna Research/Worldwatch Institute

SCHWABE, Paul
National Renewable Energy Laboratory (NREL)

SEDJO, Roger
Resources for the Future

SHORT, Walter
National Renewable Energy Laboratory (NREL)

SILVERMAN, Linda
U.S. Department of Energy

SMITH, Steven
PNNL

SMITH, Charlie
Utility Wind Interest Group

SMITH, Kirk
University of California

STENHOUSE, Jeb
U.S. Environmental Protection Agency

STRASSER, Alan
U.S. Department of Transportation

SURLES, Terrence
University of Hawaii at Manoa

TALLEY, Trigg
U.S. Department of State

TAYLOR, Cody
U.S. Department of Energy

TAYLOR, Roger
National Renewable Energy Laboratory (NREL)

THEIS, Joel
U.S. Department of Energy

THOMSON, Allison
Pacific Northwest National Laboratory

THOMPSON, Griffin
U.S. Department of State

THRESHER, Robert
National Renewable Energy Laboratory (NREL)

VEERS, Paul
Sandia National Laboratories

VERDUZCO, Laura
Chevron Corporation

VISCONTI, Gloria
Inter-American Development Bank

VISSER, Charlie
National Renewable Energy Laboratory (NREL)

WANG, Jianhui
Argonne National Laboratory

WANG, Michael
Argonne National Laboratory

WASHBURN, Morning
U.S. Agency for International Development

WILBANKS, Thomas
ORNL

WILLAMSON, Kenneth
Chevron Corporation

WISER, Ryan
Lawrence Berkeley National Laboratory

WOLVERTON, Ann
U.S. Environmental Protection Agency

WRIGHT, Alan
National Renewable Energy Laboratory (NREL)

ZAMUDA, Craig
U.S. Department of Energy

VIETNAM

TRAN, Thuc
Vietnam Institute of Meteorology, Hydrology and Environment

ANNEX VI

Permissions to Publish

Permissions to Publish

Permissions to publish have been granted by the following copyright holders:

Fig. 2.2: From Bauen, A. and Co-authors, 2009. *Bioenergy; A Sustainable and Reliable Energy Source: A Review of Status and Prospects.* IEA Bioenergy: ExCo:2009:06 108pp. Reprinted with permission from IEA Bioenergy Implementing Agreement.

Fig. 2.3: From Dornburg, V. and Co-authors, 2010. Bioenergy Revisited: Key Factors in Global Potentials of Bioenergy. *Energy & Environmental Science*, **3**, pp. 258-267. Reprinted with permission from the Royal Society of Chemistry.

Fig. 2.4: From Fischer, G., E. Hizsnyik, S. Prieler, M. Shah, and H. van Velthuizen, 2009. *Biofuels and Food Security.* The OPEC Fund for International Development (OFID) and International Institute of Applied Systems Analysis (IIASA), Vienna, Austria, 228 pp. Reprinted with permission from International Institute for Applied Systems Analysis.

Fig. 2.5(b): From de Wit, M., and A. Faaij, 2010. European biomass resource potential and costs. *Biomass and Bioenergy*, **34**(2), pp. 188-202. Reprinted with permission from Elsevier Ltd.

Fig. 2.6: From Bauen, A. and Co-authors, 2009. *Bioenergy; A Sustainable and Reliable Energy Source: A Review of Status and Prospects.* IEA Bioenergy: ExCo:2009:06 108 pp. Reprinted with permission from IEA Bioenergy Implementing Agreement.

Fig. 2.8: From Sikkema, R., and Co-authors, 2011: The European wood pellet markets: current status and prospects for 2020. *Biofuels, Bioproducts and Biorefining*, **5**(3), pp. 250-278, DOI: 10.1002/bbb.277. Reprinted with permission of John Wiley and Sons.

Fig. 2.12 (a,b): From Gibbs, H.K., and Co-authors, 2008. Carbon payback times for crop-based biofuel expansion in the tropics: the effects of changing yield and technology. *Environmental Research Letters*, **3**(3), 034001 (10 pp). Reprinted with permission from IOP Publishing Ltd.

Fig. 2.14: From Bailis, R. and Co-Authors, 2009. Arresting the Killer in the Kitchen: The Promises and Pitfalls of Commercializing Improved Cookstoves. *World Development*, **37**(10), pp. 1694-1705. Reprinted with permission from Elsevier Ltd.

Fig. 2.16: From Hamelinck, C.N., and A.P.C. Faaij, 2006. Outlook for advanced biofuels. *Energy Policy*, **34**(17), pp. 3268-3283. Reprinted with permission from Elsevier Ltd.

Fig. 2.17: From Hoogwijk, M. and Co-authors, 2009. Exploration of regional and global cost-supply curves of biomass energy from short-rotation crops at abandoned cropland and rest land under four IPCC SRES land-use scenarios. *Biomass and Bioenergy*, **33**(1), pp. 26-43. Reprinted with permission from Elsevier Ltd.

Fig. 2.21: From van den Wall Bake, J.D. and Co-authors, 2009. Explaining the experience curve: Cost reductions of Brazilian ethanol from sugarcane. *Biomass and Bioenergy*, **33**(4), pp. 644-658. Reprinted with permission from Elsevier Ltd.

Fig. 2.23: From Krey, V., and L. Clarke, 2011. Role of renewable energy in climate mitigation: A synthesis of recent scenarios. *Climate Policy*, **in press**. Reprinted with permission from the Taylor & Francis Group.

Fig. 2.24(a): From *World Energy Outlook 2010*. Reprinted with permission from the International Energy Agency.

Fig. 2.25: From Dornburg, V. and Co-authors, 2010. Bioenergy Revisited: Key Factors in Global Potentials of Bioenergy. *Energy & Environmental Science*, **3**, pp. 258-267. Reprinted with permission of the Royal Society of Chemistry.

Fig. 2.26: From Hoogwijk, M. and Co-authors, 2005. Potential of biomass energy out to 2100, for four IPCC SRES land-use scenarios. *Biomass Bioenergy*, **29**(4), pp. 225-257. Reprinted with permission from Elsevier Ltd.

Table 2.3: From Fischer, G. and Co-authors, 2009. *Biofuels and Food Security.* The OPEC Fund for International Development (OFID) and International Institute of Applied Systems Analysis (IIASA), Vienna, Austria, 228 pp. Reprinted with permission from International Institute for Applied Systems Analysis.

Table 2.11: From GBEP, 2008. *A Review of the Current State of Bioenergy Development in G8+5 Countries.* Food and Agriculture Organization of the United Nations, Rome, Italy, 278 pp. Reprinted with permission from the Food and Agriculture Organization of the United Nations.

Table 2.16: From Hoogwijk, M. and Co-authors, 2009. Exploration of regional and global cost-supply curves of biomass energy from short-rotation crops at abandoned cropland and rest land under four IPCC SRES land-use scenarios. *Biomass and Bioenergy*, **33**(1), pp. 26-43. Reprinted with permission from Elsevier Ltd.

Fig. 3.6: From IEA, 2009c. *Trends in Photovoltaic Applications: Survey Report of Selected IEA Countries between 1992 and 2008.* IEA Photovoltaic Power Systems Program (PVPS), International Energy Agency, Paris, France, 44 pp. Reprinted with permission from Net Nowak Energy & Technology Ltd.

Fig. 3.7: From Richter, C., S. Teske, and R. Short, 2009. *Concentrating Solar Power: Global Outlook 2009 – Why Renewable Energy is Hot.* Greenpeace International, SolarPACES and ESTELA, 88 pp. Reprinted with permission from Greenpeace International.

Fig. 3.8: From Steinfeld, A., and A. Meier, 2004. Solar Fuels and Materials. In: *Encyclopedia of Energy.* Vol. 5. Elsevier, Amsterdam, The Netherlands, pp. 623-637. Reprinted with permission from Elsevier Ltd. And from Steinfeld, A., 2005. Solar thermochemical production of hydrogen - a review. *Solar Energy*, **78**(5), pp. 603-615. Reprinted with permission from Elsevier Ltd.

Fig. 3.20: From A.T. Kearney, 2010. Solar Thermal Electricity 2025--Clean Electricity On Demand: Attractive STE Cost Stabilize Energy Production. A.T. Kearney GmbH, Duesseldorf, Germany, 52 pp. Reprinted with permission from A.T. Kearney GmbH.

Fig. 3.22: From Krey, V. and L. Clarke, 2011. Role of renewable energy in climate change mitigation: a synthesis of recent scenarios. *Climate Policy*, **in press**. Adapted and printed with permission from the Taylor & Francis Group.

Table 3.3: From NEEDS, 2009. *New Energy Externalities Development for Sustainability (NEEDS). Final Report and Database.* New Energy Externalities Development for Sustainability, Rome, Italy. Reprinted with permission from the Instituto di Studi per l'Integrazione dei Sistemi.

Table 3.4: From NEEDS, 2009. *New Energy Externalities Development for Sustainability (NEEDS). Final Report and Database.* New Energy Externalities Development for Sustainability, Rome, Italy. Reprinted with permission from the Instituto di Studi per l'Integrazione dei Sistemi.

Table 3.6: From Graf, D. and Co-authors, 2008. Economic comparison of solar hydrogen generation by means of thermochemical cycles and electrolysis. *International Journal of Hydrogen Energy*, **33**(17), pp. 4511-4519. Reprinted with permission from International Journal of Hydrogen Energy.

Fig. 4.5: From Hamza, V.M. and Co-Authors, 2008. Spherical harmonic analysis of Earth's conductive heat flow. *International Journal of Earth Sciences*, **97**(2), pp. 205-226. Reprinted with permission from Springer GmbH.

Fig. 4.9(a): From Krey, V., and L. Clarke, 2011. Role of renewable energy in climate change mitigation: a synthesis of recent scenarios. *Climate Policy*, **in press**. Reprinted with permission from the Taylor & Francis Group.

Fig. 5.9: From Vinogg, L., and I. Elstad, 2003. *Mechanical Equipment.* Norwegian University of Science and Technology, Trondheim, Norway, 130 pp. Reprinted with permission from the Norwegian University of Science and Technology.

Fig. 5.10: From Zare, S., and A. Bruland, 2007. Progress of drill and blast tunnelling efficiency with relation to excavation time and cost. In: *33rd ITA World Tunnel Congress*, Prague, Czech Republic, 5-10 May 2007, pp. 805-809. Reprinted with permission from CRC Press.

Fig. 5.13: From Vennemann, P., L. Thiel, and H.C. Funke, 2010. Pumped storage plants in the future power supply system. *Journal VGB Power Tech*, **90**(1/2), pp. 44-49. Reprinted with permission from the authors.

Fig. 5.16: From Guerin, F., 2006. *Emissions de Gaz a Effet de Serre (CO_2 CH_4) par une Retenue de Barrage Hydroelectrique en Zone Tropicale (Petit-Saut, Guyane Francaise): Experimentation et Modelization*. Thèse de doctorat de l'Université Paul Sabatier (Toulouse III). Reprinted with permission from the author.

Fig. 5.17: From Alvarado-Ancieta, C.A., 2009. Estimating E&M powerhouse costs. *International Water Power and Dam Construction*, 17 February 2009, pp 21-25. Reprinted with permission from the International Journal for Water power and Dam Construction.

Fig. 5.21: From Krey, V., and L. Clarke, 2011. Role of renewable energy in climate change mitigation: a synthesis of recent scenarios. *Climate Policy*, **in press**. Reprinted with permission from the Taylor & Francis Group.

Fig. 5.22: From Krey, V., and L. Clarke, 2011. Role of renewable energy in climate change mitigation: a synthesis of recent scenarios. *Climate Policy*, **in press**. Reprinted with permission from the Taylor & Francis Group.

Fig. 6.1: From Cornett, A.M., 2008. A global wave energy resource assessment. In: *Proceedings of the Eighteenth (2008) International Society of Offshore and Polar Engineers*, Vancouver, BC, Canada, 6-11 July 2008, pp. 318-326. Reprinted with permission from International Society of Offshore and Polar Engineers.

Fig. 6.4: From Nihous, G.C., 2010. Mapping available Ocean Thermal Energy Conversion resources around the main Hawaiian Islands with state-of-the-art tools. *Journal of Renewable and Sustainable Energy*, **2**, 043104. Reprinted with permission from Journal of Renewable Sustainable Energy.

Fig. 6.5: From Falcão, 2009. The Development of Wave Energy Utilization. In: *2008 Annual Report*. International Energy Agency Implementing Agreement on Ocean Energy Systems (IEA-OES). Reprinted with permission from the author.

Table 6.1: From Mørk, G., S. Barstow, M.T. Pontes, and A. Kabuth, 2010. Assessing the global wave energy potential. In: *OMAE, Shanghai, China*. Reprinted with permission from ASME.

Table 6.2: From Huckerby, J.A., and P. McComb, 2008. *Development of marine energy in New Zealand*. Published consultants' report for Energy Efficiency and Conservation Authority, Electricity Commission and Greater Wellington Regional Council, Wellington. Reprinted with permission from Power Projects Ltd.

Fig. 7.1: From 3TIER, 2009. The First Look Global Wind Dataset: Annual Mean Validation. 3TIER, Seattle, WA, USA, 10 pp. Reprinted with permission from 3TIER, Inc.

Fig. 7.2(a): From Xiao, Z. and Co-authors, 2010. *China Wind Energy Resource Assessment 2009*. China Meteorological Press, Beijing, China, 150 pp. Reprinted with permission from China Meteorological Press.

Fig. 7.2(b): From Nikolaev VG, Ganaga SV, Kudriashov KI, Walter R, Willems P, Sankovsky A., 2010. *Prospects of Development of Renewable Power Sources in Russian Federation*. *The results of TACIS project*. Europe Aid/116951/C/SV/RU. Moscow, Russia: ATMOGTRAPH. Reprinted with permission from ATMOGRAPH.

Fig. 7.7: From IEC, 2010. *Wind Turbines - Part 22: Conformity Testing and Certification, IEC 61400-22*. International Electrotechnical Commission, Delft, The Netherlands. Reprinted with permission from International Electrotechnical Commission.

Fig. 7.13: From Durstewitz, M. and Co-authors, 2008. *Windenergie Report Deutschland 2007*. Institut für Solare Energieversorgungstechnik (ISET), Kassel, Germany. Reprinted with permission from the Institut für Solare Energieversorgungstechnik.

Fig. 7.14: From Holttinen, H. and Co-authors, 2009. *Design and Operation of Power Systems with Large Amounts of Wind Power: Phase One 2006-2008*. VTT Technical Research Centre of Finland, Espoo, Finland, 200 pp. Reprinted with permission from the authors.

Fig. 7.17: From Holttinen, H. and Co-authors, 2009. *Design and Operation of Power Systems with Large Amounts of Wind Power: Phase One 2006-2008*. VTT Technical Research Centre of Finland, Espoo, Finland, 200 pp. Reprinted with permission from the authors.

Fig. 7.24: From Krey, V., and L. Clarke, 2011. Role of renewable energy in climate change mitigation: a synthesis of recent scenarios. *Climate Policy*, **in press**. Reprinted with permission from the Taylor & Francis Group.

Fig. 7.25: From Krey, V., and L. Clarke, 2011. Role of renewable energy in climate change mitigation: a synthesis of recent scenarios. *Climate Policy*, **in press**. Reprinted with permission from the Taylor & Francis Group.

Fig. 8.3: From Akershus Energi, 2010. Akershus Energi Varme AS. Lillestrøm, Norway. Reprinted with permission from Akershus Energi Varme AS.

Fig. 8.4: From Euroheat & Power, 2007. *District Heating and Cooling – Country by Country 2007 survey*. Report, Euroheat & Power, Brussels, Belgium. Reprinted with permission from Euroheat and Power.

Fig. 8.5: From Bodmann, M. and Co-authors, 2005. *Solar unterstützte Nahwärme und Langzeit-Wärmespeicher (Februar 2003 bis Mai 2005)*. Report 0329607F, Forschungsbericht zum BMWA/BMU-Vorhaben, Stuttgart, Germany, 159 pp. Reprinted with permission from Solar- und Wärmetechnik Stuttgart.

Fig. 8.8: From Pehnt, M., A. Paar, F. Merten, W. Irrek, and D. Schüwer, 2009. Intertwining renewable energy and energy efficiency: from distinctive policies to combined strategies. In: *ECEEE 2009 Summer Study*, ECEEE, La Colle sur Loup, Côte d'Azur, France, pp. 389-400. Reprinted with permission from the Institut für Energie- und Umweltforschung.

Fig. 8.9: From Åhman, M., 2010. Biomethane in the transport sector - An appraisal of the forgotten option. *Energy Policy*, **38**(1), pp. 208-217. Reprinted with permission from Elsevier Ltd.

Fig. 8.10: Müller-Langer, F., 2007. BIOMETHANE FOR TRANSPORT - A worldwide overview; Presentation at IEA Bioenergy Task 39 Subtask Policy and Implementation Workshop: From today's to tomorrow's biofuels – *From the Biofuels Directive to bio based transport systems in 2020*; June 3-5, 2009; Dresden, Germany; adapted from the Study: Possible European biogas supply strategies; Institute for Energy and Environment; Leipzig; 2007. Reprinted with permission from the German Biomass Research Centre.

Fig. 8.16: From Samaras, C. and K. Meisterling, 2008. Life cycle assessment of greenhouse gas emissions from plug-in hybrid vehicles: implications for policy. *Environmental Science & Technology*, **42**(9), pp. 3170-3176. Reprinted with permission from American Chemical Society.

Fig. 8.18: From METI, 2005. *Energy vision 2100, Strategic technology roadmap (energy sector)*. Ministry of Economy, Trade, and Industry, Tokyo, 42 pp. Reprinted with permission from the Ministry of Economy, Trade, and Industry, Japan.

Fig. 8.19: From EREC, 2008. *The Renewable Energy House*. European Renewable Energy Council, Brussels, Belgium. Reprinted with permission from EREC.

Fig. 8.22: From VTT, 2009. Energy Visions 2050 – *summary*. VTT Technical Research Centre of Finland, Helsinki. Reprinted with permission from VTT Technical Research Center of Finland.

Fig. 8.23: From Werner, S., 2006. *ECOHEATCOOL Work package 4 – The European Heat Market*. Final report prepared for the EU Intelligent Energy Europe Programme, Euroheat & Power, Brussels, Belgium, 73 pp. Reprinted with permission from Euroheat and Power.

Table 8.3: From Persson, M., O. Jönsson, and A. Wellinger, 2006. *Biogas Upgrading to Vehicle Fuel Standards and Grid Injection*. IEA Report, Swedish Gas Center (SGC), Malmö, Sweden, 34pp. Reprinted with permission from the IEA Bioenergy Implementing Agreement.

Fig. 9.2: From *2008 Worldwide Trends in Energy Use and Efficiency*. Reprinted with permission from the International Energy Agency.

Fig. 9.5: From *World Energy Outlook 2010*. Reprinted with permission from the International Energy Agency.

Fig. 9.13: From *World Energy Outlook 2010*. Reprinted with permission from the International Energy Agency.

Fig. 9.16: From Krey, V., and L. Clarke, 2011. Role of renewable energy in climate change mitigation: a synthesis of recent scenarios. *Climate Policy*, **in press**. Reprinted with permission from the Taylor & Francis Group.

Fig. 9.17: From Luckow, P. and Co-authors, 2010. Large-scale utilization of biomass energy and carbon dioxide capture and storage in the transport and electricity sectors under stringent CO_2 concentration limit scenarios. *International Journal of Greenhouse Gas Control*, **4**(5), pp. 865-877. Reprinted with permission from Elsevier Ltd.

Table 9.1: From *World Energy Outlook 2010*. Reprinted with permission from the International Energy Agency.

Table 9.2: From *World Energy Outlook 2010*. Reprinted with permission from the International Energy Agency.

Table 9.3: From REN21, 2010. *Renewables 2010 Global Status Report.* REN21 Renewable Energy Policy Network for the 21st Century, Paris, France. Reprinted with permission from REN21.

Table 9.4: From ESMAP, 2005. *The Impacts of higher oil prices on low income countries and the poor: Impacts and policies.* Energy Sector Management Assistance Program, World Bank, Washington, DC, US. Reprinted with permission from the World Bank.

Table 9.5: From Angerer, G. and Co-authors, 2009. *Rohstoffe für Zukunftstechnologien.* Fraunhofer Verlag, Stuttgart, Germany. Reprinted with permission from the Frauenhofer Institute.

Fig. 10.1: From Krey, V., and L. Clarke, 2011. Role of renewable energy in climate change mitigation: a synthesis of recent scenarios. *Climate Policy*, **in press**. Reprinted with permission from the Taylor & Francis Group.

Fig. 10.2: From Krey, V., and L. Clarke, 2011. Role of renewable energy in climate change mitigation: a synthesis of recent scenarios. *Climate Policy*, **in press**. Reprinted with permission from the Taylor and Francis Group.

Fig. 10.3: From Krey, V., and L. Clarke, 2011. Role of renewable energy in climate change mitigation: a synthesis of recent scenarios. *Climate Policy*, **in press**. Reprinted with permission from the Taylor & Francis Group.

Fig. 10.4: From Krey, V., and L. Clarke, 2011. Role of renewable energy in climate change mitigation: a synthesis of recent scenarios. *Climate Policy*, **in press**. Reprinted with permission from the Taylor & Francis Group.

Fig. 10.5: From Krey, V., and L. Clarke, 2011. Role of renewable energy in climate change mitigation: a synthesis of recent scenarios. *Climate Policy*, **in press**. Reprinted with permission from the Taylor & Francis Group.

Fig. 10.6: From Krey, V., and L. Clarke, 2011. Role of renewable energy in climate change mitigation: a synthesis of recent scenarios. *Climate Policy*, **in press**. Reprinted with permission from the Taylor & Francis Group.

Fig. 10.7: From Krey, V., and L. Clarke, 2011. Role of renewable energy in climate change mitigation: a synthesis of recent scenarios. *Climate Policy*, **in press**. Reprinted with permission from the Taylor & Francis Group.

Fig. 10.8: From Krey, V., and L. Clarke, 2011. Role of renewable energy in climate change mitigation: a synthesis of recent scenarios. *Climate Policy*, **in press**. Reprinted with permission from the Taylor & Francis Group.

Fig. 10.9: From Krey, V., and L. Clarke, 2011. Role of renewable energy in climate change mitigation: a synthesis of recent scenarios. *Climate Policy*, **in press**. Reprinted with permission from the Taylor & Francis Group.

Fig. 10.10: From Krey, V., and L. Clarke, 2011. Role of renewable energy in climate change mitigation: a synthesis of recent scenarios. *Climate Policy*, **in press**. Reprinted with permission from the Taylor & Francis Group.

Fig. 10.11: From Edenhofer, O. and Co-authors, 2010. The economics of low stabilization: Model comparison of mitigation strategies and costs. *Energy Journal*, 31(Special Issue), pp. 11-48. Reprinted with permission from the International Association for Energy Economics.

Fig. 10.12: From Luderer, G. and Co-authors, 2009. *The Economics of Decarbonization – Results from the RECIPE model Intercomparison*. Potsdam Institute for Climate Impact Research, Potsdam, Germany. Reprinted with permission from the authors.

Fig. 10.23: From Hoogwijk, M. and Co-authors, 2009. Exploration of regional and global cost-supply curves of biomass energy from short-rotation crops at abandoned croplands and rest land under four IPCC SRES land-use scenarios. *Biomass and Bioenergy*, **33**(1), pp. 26-43. Reprinted with permission from Elsevier Ltd.

Fig. 10.24: From de Vries, B.J.M. and Co-authors, 2007. Renewable energy sources: Their global potential for the first-half of the 21st century at a global level: An integrated approach. *Energy Policy*, **35**(4), pp. 2590-2610. Reprinted with permission from Elsevier Ltd.

Fig. 10.25: From de Vries, B.J.M. and Co-authors, 2007. Renewable energy sources: Their global potential for the first-half of the 21st century at a global level: An integrated approach. *Energy Policy*, **35**(4), pp. 2590-2610. Reprinted with permission from Elsevier Ltd.

Fig. 10.26: From Hoogwijk, M. and Co-authors, 2004. Assessment of the global and regional geographical, technical and economic potential of onshore wind energy. *Energy Economics*, **26**(5), pp. 889-919. Reprinted with permission from Elsevier Ltd.

Fig. 10.32: From Nemet, G.F., 2009. Interim monitoring of cost dynamics for publicly supported energy technologies. *Energy Policy*, **37**(3), pp. 825-835. Reprinted with permission from Elsevier Ltd.

Fig. 10.33: From *Energy Technology Perspectives 2008*. Reprinted with permission from the International Energy Agency.

Fig. 10.34(d): From Luderer, G. and Co-authors, 2009. *The Economics of Decarbonization – Results from the RECIPE model Intercomparison*. Potsdam Institute for Climate Impact Research, Potsdam, Germany. Reprinted with permission from the authors.

Fig. 10.35: From *Energy Technology Perspectives 2008*. Reprinted with permission from the International Energy Agency.

Table 10.1: From Krey, V., and L. Clarke, 2011. Role of renewable energy in climate change mitigation: a synthesis of recent scenarios. *Climate Policy*, **in press**. Reprinted with permission from the Taylor & Francis Group.

Table 10.10: From *Energy Technology Perspectives 2008*. Reprinted with permission from the International Energy Agency.

Fig. 11.5: From IEA, 2003b. *Renewables for Power Generation: Status & Prospects.* Reprinted with permission from NET Nowak Energy & Technology Ltd.

Fig. 11.10: From BTM Consult ApS, 2010. *World Market Update 2009*. BTM Consult ApS, Ringkøbing, Denmark. Reprinted with permission from BTM Consult ApS – A part of Navigant Consulting.

Fig. 11.11: From Ockwell, D.G. and Co-authors, 2010. Intellectual property rights and low carbon technology transfer: Conflicting discourses of diffusion and development. *Global Environmental Change*, **20**(4), pp. 729-738. Reprinted with permission from Elsevier Ltd.

Index

Indexer:
Marilyn Anderson (United States)

Index

Note: Glossary terms are indicated by an asterisk (*). Bold page numbers indicate page spans for entire chapters. Italicized page numbers denote tables, figures and boxed material.

A

Abatement costs, 808-811, *810*, *811*, 834-836, *835-838*
Acceptance of RE. *See* Public attitudes and acceptance
Access to energy. *See* Energy access
Accidents and risks, 745-747, *746*, *855*
Adaptation[*], 40, 44, 48, 192
Aerosols[*], 174-175, 263, 736, *737*, *739*, 854
Afforestation[*], 227
Agriculture, 119, 235-236, *662*, 686-689
    integration of RE, 17, 119, 614, *617-618*, 687-689
    solar drying (of crops), 346, 351
Air-conditioning, solar-powered, 349
Air pollution, 20, 124-125, 271, 373, 736-739
    climate effects, *737*
    components of, 736, *737*, *739*
    costs of, 854
    health impacts, 739-740, *740*, 756-757, 854
    RE benefits for, 711
Air quality:
    bioenergy and, 268, 271-272
    solar energy and, 373
Aircraft. *See* Aviation
Albedo, increasing, 175
Alcohol. *See* Ethanol; Methanol
Algae, 54, *218*, 277-278, *282*, 285, 303-304
Anaerobic digestion, 46, 50, 53, *55-56*, 111, *218*, *235*, 240, 285, 687
Annex I countries[*]:
    abatement cost curves, *837*
    RE deployment scenarios, 132, 807, *808*
Anthropogenic emissions[*], 7, 33, 84, 119, 164, 167, 168
Aquatic biomass, 277-278
Aquatic feedstocks, 48, 53, *56*, *218*, 277-278
Aqueous phase reforming, 285
Aquifers:
    geothermal energy from, 406, *408*
    geothermal energy impacts on, 419-420
    thermal energy storage, 349-350
Assessment:
    integrated[*], 800-801, *801*
    of RE strategies, 133-135, 813-832
    See also Lifecycle analysis; Scenarios
Austria, energy transition in, *929*
Autonomous energy supply systems, 113, 613, 658-661
Average cost[*], *823*, *833*, *838*, 840

Aviation, 670-671
Avoided costs, 797, 849-851
Avoided emissions[*], 99, 122-123, *124*, *177*, 261
Awareness barriers, 44, 129, 194-195, 758-759, 881, *914*

B

Backpressure steam turbines[*], 460
Balancing power/reserves[*], 620, 621
Barrier removal[*], 24-25, 712, 761
    RE policies and, 869, *914*
Barriers[*], 44, 192-196, *194*, 871, 872, *914*
    categorization of, 193, *194*
    implementation of RE policies, 148-150, 881-882
    RE financing, 150, 882
    RE policymaking, 148-150, 880-881
    sustainable development, 129-130, 712, 757-760
    See also specific technologies
Baseline[*], 164, 794-796, 798, *801-803*, 803, 814-816, *814*, *815*, *817*
    See also Scenarios
Batteries, 62, 107, 109, *114*, 115, 185, 621, 637-638, 659, *664*, 674
    electric vehicle, 45, 99, 114-115, 186, 199, 637, 663-664, *665*
    energy density of, 358
    solar radiation/energy and, 62, 659, 660
    wind energy and, 630
Beam (solar) radiation[*], 60, 62, 69, 341, 342, 355, 367
Belgium, heating and cooling example, 676-677, *676*
Benchmark[*], 188, 845, 854
Benefit-sharing, 921-923
Bidding/tenders, *890*, 896
Bio-based products, 286-287
Biodiesel, *218*, *235*, 243, *245*, 285
    costs, 53, *846*
    global trade in, 252, *252*
    integration of, 613
Biodiversity[*], 20, 48, 52, 744-745, *744*, 879
    bioenergy and, 229, 231, 269, 305, *745*
    hydropower and, 465-466, *745*
Bioenergy[*], *8*, 46-60, *182*, **209-311**
    barriers, 250-251, 255-257
    biodiversity and, 229, 231, 269, 305, *745*
    carbon payback, *264*
    climate change and, 232, 259-267
    climate change indicators, 261-263, *262*

    conversion, 217, *218*, *235*, 238-240, *239*
    conversion, improvements in, 53, 280-287, *281-283*
    cost scenarios, 293-295, *294*, *296*
    costs, 53-55, 215, 227-228, *242-243*, 243-244, *244-245*, 288-296, *289*, *856*, *1004*
    current use and trends, 216-219, *216*, 246-248, *247*, *248*
    deployment, 48-50, 55-60, 214, 296-307, *298*, *308*
    environmental and social impacts, 50-52, 215, 219, 257-276, *258*, *275*, 304-306, *745*
    feedstocks, 48, *49*, 233-236, *234*, 270, 276-278, *277*, *1014-1016*
    global trade in, 251-253, *252*, 297
    health impacts, 739-740, *740*
    heating, *821*, *1010*
    impact assessments (IAs), 258, 305
    improvements in, 215
    incentives and barriers, 249-251
    integration, 53, 622, *623*, 628
    key messages about, 57-60, 306-307, *308*
    land use change and, 50-51, 215, 219, 263-267, *266*, 275, 304-305, 735-736, *736*
    learning curves, 292-293, *293*, *848*
    levelized costs, 288-292, *290-292*, *843-845*
    lifecycle assessments, 258-261, *260*
    logistics and supply chains, 278-280, *279*, 302-304
    marginal lands, use of, 231
    market and industry development, 48-50, 246-257
    market potential, 227-228, *228*
    mitigation potential, 214, 220
    modern[*], *216*, 217, 263-267
    opportunities, 218, 255
    policies for, *251*, 253-255, 257
    positive and negative aspects, 218
    pretreatment technologies, 279-280
    previous IPCC assessments, 219-220
    projections for, *22-23*, 753-754, *754*
    scenarios, 806, *808*, *822*, 827-829
    share of global primary energy, *174*
    sustainability and, 52, 215, 254-255, 271
    systems and chains, 240-244, *242-243*
    technical potential, 47-48, 214, 220, *221*, 223-227, 300-302, *301*
    technologies and applications, 48, 53, 233-246, 276-287
    theoretical potential, 46-47, *183*, 220, 222
    water issues, 227, 233, 257-258, 268-269, 305
    See also Biofuel; Biomass; Feedstocks

1060

Index

Biofuels[*], *218*, 241-243
    advanced, 613
    blending requirements, 24, 25, 869, 874, 895, 911
    certification, 256
    costs, 243-244, *244-245*, *281-283*, *1014-1016*
    efficiency of, 186
    environmental impacts, 257-258, *258*
    experience curves, *56*
    first-generation manufactured[*], 50, 612, 733
    global trade in, 251-253, *252*, 297
    historical use of, *11*, *176*
    integration of, 17, 112-113, *112*, *116*, 612-614, 654-658, *654*
    land use change and, 735-736, *736*
    levelized cost of energy, *14*, *843*, *845*
    levelized cost of fuel (LCOF), *14*, *42-43*, *55*, *144*, 1002, *1003*, *1014-1018*
    lifecycle GHG emissions from, 733-735, *734*
    lignocellulosic, 54, 215, *218*, 294-295, *296*, 303, 735
    liquid biofuels pathways, 666-667
    mandates, 874
    mandates and targets, 912-913
    policies, 869, 874-875, 911-913
    production issues, 270-271, 275-276
    production subsidies, 874-876
    production technologies, 281-287, *281-283*
    scenarios, *822*
    second-generation[*], 50, 238, 247, 250, 257
    sustainability criteria, 256
    tariffs, 255-256
    transition issues, *116*
    See also Bioenergy; Fuels
Biogas, 250, 280, 648-649, *650*
    costs, 652-653
    experience curves, *56*
Biomass[*], 214
    aquatic, 277-278
    biomass combined with CCS, 215, 286, 304
    biomass combined with solar thermal, 367
    climate change effects, 268
    combustion, 238
    conversion paths, 217, *218*, 238-240, *239*
    conversion to liquid fuels, 666-667
    cook stoves, 249, 268, 271, 722
    cost curves, 838-839, *838*, *839*
    current use and trends, 216-219, *216*
    electricity production, 622, *819*, 839
    energy demand, 219
    external costs, *856*
    feedstocks, 50, 53, 218-219, *218*, 233-236, *234*
    forests and, 687
    future projections for, *754*
    gasification, 239-240, *282*
    global trade in, 251-253, *252*, 297
    health effects, 44, 192, 740
    historical use of, *11*, 17, *176*
    incentives and barriers, 249-250
    inefficient consumption of, 722
    learning rates, *848*
    levelized cost of energy, *14*, *843*, *845*
    logistics and supply chains, 236-238, 278-280, *279*, 302-304
    mitigation potential, 214, 220
    modern[*], 46, *216*, 217, 250-251
    number of people relying on, 722, *722*
    plantations, 224-226, *225*, 230
    policies for, *251*, 253-255
    share of primary energy supply, 165, *176*, 214, *217*
    technical potential, *12*, *184*, *206*, 219-220, 222-233, *224*, *228*, 300-302, *301*
    technology improvements, 248-250
    traditional[*], 46, 216-217, *216*
    See also Bioenergy
Biomethane, 17, *218*, 612, 613, 667
    conversion routes, *235*
    costs, 652, 653-654
Birds, wind energy and, 572-573
Black carbon[*], 174-175, 218, 263, 736, *737*
Black liquor gasification, 684
Blending requirements for biofuels, 24, 25, 869, 874, 895, 911
Bonus mechanisms, 82, 910
Bottom-up indicators, 119, 122-125, 130, 710, 728-747
    See also Lifecycle analysis
Bottom-up models[*], 799, *799*
Brazil:
    ethanol use/integration, *15*, 657-658, 663, 880, *912*
    policies for RE fuel, *912*
    sugar industry in, 616, *912*
    urban settlements in, 677-678
Bricolage, 931
Building regulations, 24, 908
Building sector:
    autonomous energy systems, 658
    building managements, 674
    case studies, 676-678, *676*, 679-680
    efficiency in, 186
    integration of RE, 17, 117, 614, *617-618*, *662*, 672-681
    low-energy/green buildings, 346, 614
    mitigation potential, *662*, 672-681
    net-zero-energy solar buildings, 346, 356
    passive solar use, 344-346
    Passivhaus design, 186, 349
    status of, 673-674
    technology development pathways, 674, *674*

C

California, innovative financing example, *894*
Canary Islands, wind-hydro plant, 661
Cancun Agreements, 34, 164, 169
Capacity[*], 35-36
    generation capacity[*], 35, 77, 99
    installed, *See also specific technologies*
    installed capacity, 797, 869
    nameplate capacity[*], 99, 539, 541, 550, 554, 557, *558*, 563
    peaking, 442, 459-460
Capacity building[*], *918-919*, 925-927, *928*
Capacity credit[*], 621, *623*, 850
Capacity factor[*], *623*
    geothermal energy, 404, 409, 425-426, *425-426*, *623*, 625
    ocean energy, *623*, 626
    wind energy, *623*, 626-627
Carbon, social costs of (SCC), 853-854
Carbon budget, 37, 175
Carbon capture and storage. *See* Carbon dioxide capture and storage (CCS)
Carbon cycle[*], 84, 263, 267
Carbon debt, 264
Carbon dioxide (CO_2)[*]:
    climate change and, 168
    CO_2-equivalent emission, 74
    drivers of emissions, 169-172
    emissions and stocks, *172*, *173*
    in freshwater reservoirs, 472, *473*
    global emissions, historical, *171*, *802*
    hydropower and, 485-486, *486*
    increase in, 168
    industry sector and, 681-684
    lifecycle assessment, 192, *193*
    RE deployment and, 802, *803*
    scenarios, 800-801, *802*, *803*, 826-830, *829*
    top emitting countries, 172, *173*
    See also Greenhouse gases; Mitigation
Carbon dioxide capture and storage (CCS)[*], 616, *814*
    combined with biomass conversion, 215, 286, 304
    combined with fossil energy, *133*, 804-880
    GHG emissions from, *124*, 733
Carbon intensity:
    decrease in, 169, *169*, 170
    GDP and, 170
    in scenarios, *814*

1061

Index

Carbon leakage, 40, 186, 198
Carbon lock-in[*]**,** 147, 148, 870, 872, 881, 926
Carbon monoxide, 218, *650*
Carbon payback[*]**,** bioenergy, 264-265, *264*
Carbon pricing, 198, 853-854, 872
    hydropower and, 457
    policies and, 916-917
    RE deployment and, 810, *810*
Carbon sinks, 174, 229, 264
Carbon stocks, terrestrial, 20, 264
Carbon tax[*]**,** 24, 872, *890*, *909*, 917
Cellulose[*]. *See* Lignocellulosic biofuels/crops
Cement industry, 682, 683
Certificate trading, 895-896
Charcoal, 40, 46, *47*, 237
Charge. *See* Taxes
Chemical and petrochemical industry, 682, 683
China:
    bioenergy in, 217
    CO_2 emissions, 172, *173*
    district heating in, 646-647
    RE development in, 190, 198, 616, 879
    RE electricity supply curves, 823, *823*
    RE policies, *915*
Chokepoints, 725
Cities, role for, 927-929
Clean Development Mechanism (CDM)[*]**,** 129-130, 761-762
Climate change[*]**,** 168-169
    air pollution and, *737*
    bioenergy and, 232, 259-267
    external costs, *855*
    geothermal energy and, 410
    greenhouse gas emissions and, 164, 168, *168*
    hydropower and, 447-449, *448*
    policy interactions, 154-155, 916-917
    renewable energy and, 7, 33-45, 132, **161-207**
    solar energy and, 343, 387
    wind energy and, 548-550
    See also Greenhouse gases (GHGs); Mitigation
Climate change adaptation. *See* Adaptation
Climate change mitigation. *See* Mitigation
Climate protection goals, 37, 175, 794, 803-805
Co-benefits[*]**,** 33, 51, 71, 164, 756, 760, 766, 768
CO_2-equivalent emission[*]**,** 74
    See also Direct equivalent method
Coal, 34, 44, 192, *856*
    GHG emissions from, 33, *52*, 53, 169, *172*
    percentage of primary energy supply, *10*, *35*
Cogeneration[*]**,** 74, 118
Combined-cycle power plants, 53, 65
Commercialization, 137-144, 841-851, 877
    costs, 137-144, 841-851

Communities, role for, 927-929
Community ownership, 922-923
Compliance[*]**,** 152-153, 154, 896, 905, 908
Compressed natural gas (CNG), 663
Concentrating photovoltaics (CPV), 361, 375-376
Concentrating solar power (CSP)[*]**,** 62-63, 355-358, *356*, 365, 369
    costs, 69-71, 369, *369*, 382-385, *385*
    electricity generation, 67, 355-358, 377, 388
    environmental effects, 370-372, *372*, *745*
    integration, *623*, 624
    learning rates, 71, *848*
    scenarios, *817*
    thermal storage for, 357-358
Condensing steam turbines[*]**,** 412
Conflict, resolving, 921, 923-924
Conversion[*]**,** 178, *180*, *181*, *193*
    biomass/bioenergy, 217, *218*, 235, 238-240, *239*
    solar energy, 337, 340, 377
    thermal conversion, 178, 337, 377
Conversion factors for energy, *997*
Convertible loans[*]**,** 886, *886*
Cook stoves, biomass, 249, 268, 271, 722
Cooling systems. *See* Heating and cooling
Corn. *See* Maize
Cost[*]**,** 798, **1001-1022**
    annuity factor, 976
    assessment of future costs, 797
    average cost[*], 823, 833, *838*, 840
    carbon. *See* Carbon pricing
    competitiveness and, 13, 40, 165, 796
    damage costs[*], 369
    decommissioning cost[*], 138, 481
    deployment costs, 849, *849*
    environmental costs, 851-857, *855-857*
    external costs, 144-146, 797, 851-857, *855-857*
    learning curves[*] and, 796-797, 846-849, *847*, *848*
    levelized cost of energy[*], 13, *14*, 40, 165, 187-190, *188-189*, 796, 842-846, *843-846*, 976
    lifecycle costs[*], 82, 417
    marginal (incremental) cost[*], 135, 810, 832
    mitigation costs, 24, 130-146, 795, 808-811, *810*, *811*
    operation and maintenance (O&M), 480-481, 584-585, 587
    opportunity cost[*], 130
    policies and, 870
    private costs[*], 852
    project costs[*], 85, 883, 914, 975-976
    RE-based electricity production, *855*
    RE commercialization and deployment, 137-144, 798, 841-851
    RE integration, 15-16

    RE technologies, *14*, 165, 187-190, 796, 841-846, *843-846*
    recent RE cost parameters, 1001-1022
    reduction of RE costs, 13, 40, 796, 846-849
    social and environmental, 144-146, 851-857, *855-857*
    social costs of carbon (SCC), 853-854
    total cost[*], 40, 53, 86, 147, 873, 891, 905
    uncertainties of, 872
    up-front investment, 194, 366, 796-797
    See also Financing; Fiscal incentives; Investment cost; Levelized cost of energy; *specific technologies*
Cost-benefit analysis[*]**,** 852
    policies and, 870, 916
Cost curves, 135-137, 798, 832-841
    abatement, 834-836, *835-838*
    biomass resource, 838-839, *838*, *839*
    concept, 832-833
    deployment curves, 849, *849*
    limitations of, 833-834, *834*
    regional, 135-137
    technology resource, 836-840
Cost-effectiveness analysis[*]**,** 870, 903
Cracking, solar/thermal, 366
Credit lines, 893
Criteria, policy evaluation, 883-884
CSP. *See* Concentrating solar power

D

Damage costs[*]**,** 369
Daylighting, 338, 344-346, 373-374, 378-379
Decommissioning, *85*, 92, 99, 100, 481
Decommissioning cost[*]**,** 138, 481
Deforestation[*]**,** 44, 51, 128, 192, 813
Demand pull[*]**,** 150, 154, 851, *887-888*
Democratic Republic of Congo, 679-680
Denmark, 198, 616
    policy approach, *922-923*
Density[*]**:**
    energy density[*], 44, 53, 87, 113, 215, 236-237
    power density[*], 44, 82
Deployment, 9, 34-36, 612, 803-808
    barriers to, *914*
    bioenergy, 214, 296-307, *308*
    carbon prices and, 810, *810*
    combining R&D with, 888, *889*
    costs, 798, 841-851, *849*
    deployment curves, 849, *849*
    energy security and, 18
    external costs and, 797
    financing, 889-895, *890-891*
    geothermal energy, 78-79, 428-432
    hydropower, 484-488, *485*

Index

large networks (centralized), 7, 165, 181
mitigation and, 802-803
mitigation costs and, 795
ocean energy, 502, 526-529, *527*
opportunities from, 760-764
point of use (decentralized), 7, 165, 181
policies, 13, 25, 34, 198, 869-870, 871, 874-876, *875*, 876
policy options, 151-154, 889-895, *890-891*
policies: electricity sector, 152-153, 895-907
policies: heating and cooling sector, 907-911
policies: transportation sector, 911-913
projected increase in, 20-21, *21*, 794
regional, 812
scenarios, 131-132, 794-795, 803-805, *804*, 812, *830*, 841-851
solar energy, 71, *71*, 339, 386-390, *387*, 388
support and fiscal incentives, 876
technology perspective, 812
time scale, 807, *808-810*
wind energy, 103, 539-540, 591-595
Depreciation, variable or accelerated, RE deployment, *890*, 892
Desalination, in Mexico, 660
Developed countries:
 CO_2 emissions, 172, *173*
 RE and buildings, 674-677
Developing countries, 41
 CO_2 emissions, 172, *173*
 future RE scenarios, 711
 per capita income, *122*
 RE integration in, 677-680
Development:
 energy use and, 18, 191, 710, 716, 718, *719*
 See also Economic development; Sustainable development
Diesel:
 biodiesel, 243, *245*, 252, *252*, 285, 613, *846*
 lifecycle GHG emissions, 733-735, *734*
 renewable, *235*
 solar-generated, 358, 359
 substitution, 285
Diffuse (solar) radiation[*], 60, 341, 342
Dimethyl ether (DME), *218*, 358, 613, 684
Direct emissions[*], 74
Direct equivalent method, *10*, 178-181, *180*, 798, 802, *803-810*, 820, 977, *977-978*, *978-979*
Direct solar energy. *See* Solar energy
Discounting[*], 40, 802, 833, *835-837*, 975-976
 hydropower and, 482
 of impacts, 853

levelized cost of energy with, 138, 842, *844-846*
social discount rate[*], 833
Dispatch (power dispatching)[*], 44, *108*, *623*, 628-631, *891*, 907
 dispatchable RE sources, 39, 63, 73, 107, *108*, *623*, 624-625, 628-629
 dispatchable units, 620
 non-dispatchable RE sources, *108*, *623*
 partially-dispatchable RE sources, 39, 107, *108*, 109, 620, *623*, 624, 627-628, 629-631
 priority dispatch, *891*, 907
 RE technologies, 628-631
Displacement factor, 261, *262*
District cooling, 16, 185, 613, 640-641, 647
District heating (DH)[*], 16, 185, 613
 case studies, 646-647
 geothermal energy and, 412
 integration, 110, 640-647, *641*
 solar energy and, 367
 See also Heating and cooling
Diversity (technological), 883, 885, *902*, 903, 914, 925
 See also Biodiversity
DME. *See* Dimethyl ether
Drivers[*], 148, 878-880
 of emissions, 169-172
Dutch technology and innovation, *931*
Duty exemptions, 911
Dye-sensitized solar cells (DSSCs), 352-353

E

Eco-tax[*], 908
Economic barriers, 759-760, 881
 See also Market failures
Economic costs, *912*, *914*, 916
Economic development, 41, 120-121, 126-127
 energy use and, 18, *121*, 191, 710, 716, 718, *719*
 policies and, 880
 RE and, 18, 191
 sustainable development and, 120-127, *570*, 718-721, 749-751, 765
Economic opportunities, 879
Economic regulation, 881, *914*, 919, 920, 924
Economies of scale (scale economies)[*], 65, 67, 97, 123, 357, 367, 377, 382, *384*
Ecosystems[*], 33, 48, 65-66, 100, 744-745, *744*
 external costs, *855*
 See also Environmental impacts
Effectiveness[*], 150, 883, 903
Efficiency[*], 150, 185-186, 883
 of appliances, 673-674
 of policies, 883, 903-905

El Hierro, Spanish Canary Islands, 661
Electric vehicles, 17, 666, 668, *822*
 batteries for, 45, 99, 114-115, 186, 199, 637, 663-664, *665*, 668
 hybrid, 45, 114, *114*
Electricity[*]:
 air pollution from, 737-739, *738*
 alternating current (AC), 620, 622
 autonomous systems and, 658
 bioenergy, 622, *819*, 839
 capacity, 621-622
 capacity credit, 621
 concentrating solar power, 63-64, 67, 355-358, *356*, 365, 377, *388*, *819*
 costs. *See specific technologies*
 direct current (DC), 620
 electrical power systems, features of, 620-622
 energy payback[*], *731*
 external costs, *855-856*
 generation contingencies, 621-622
 generation control[*], 105
 geothermal energy, 408-409, *409*, 415-416, *415*, *416*, *819*
 global investment in, 876, *876*
 hydropower, 442, 455-456, *456*, 458-460, *486*, *487*, *819*
 integration of RE, 15-16, 17, 45, 107-109, 184-185, 612-613, 619-640
 levelized cost of (LCOE), *14*, *42-43*, *55*, 68, *70*, 93, *142*, 1002, *1003*
 levelized cost of renewable energies for, *14*, 188, *843-844*
 lifecycle analysis[*] for, 18, *19*, 122-123, 173-174, *177*, 732-733, *732*, 979-981
 load[*], 62, 65, 73-74, 82, 113
 ocean energy, *819*
 photovoltaics (PV), 64, 337, 338, 351-355, 363-364, 375-376, *388*, *819*, 839
 policies and, 152-153, 869, 874, *890-891*, 895-907, *897*, *898-899*
 primary energy sources for, *175*
 RE characteristics and, 622-627
 scenarios for, 795, 816-818, *817*, *819*, 823-825, *823*, *824*, 826-829, *826*
 solar energy, 63-64, 355-358, 377, *388*, 622-624
 storage (batteries), 45, 62, 99, 114-115, 621, *664*, *674*
 storage (thermal), 344, *345*, 349-350, *674*
 supply curves, 823-825, *823*, *824*
 transmission and distribution[*], 563
 water resources/quality and, 741-743, *742*, 991-993
 wind energy. *See* Wind power

1063

Index

Electricity demand, 620, 621
   balancing, 620, 621
Electricity grid[*], *See also* Grid
Electricity grid[*]. *See* Grid
Electrolysis, 358, 362, 377
Emission factor[*], 134, 137, *172*
Emission tax[*]. *See* Taxes
Emissions[*]:
   anthropogenic[*], 7, 33, 84, 119, 164, 167, 168
   avoided[*], 99, 122-123, *124*, *177*, 261
   direct emissions[*], 74
   emissions cap, 917
   zero- or near-zero, 37, *37*, 74, 114, 115, *115*, *116*, 175
   See also Greenhouse gases; *specific gases*
Emissions trading[*], 895-896, *897*
Employment creation, 719-720
Enabling environment, 25, 45, 155-158, 199, 870, 873, *909*, 917-929, *918-919*
Energy[*]:
   conversion of, 178, *180*, *181*
   efficiency, 185-186
   final energy[*], *56*, 117, 120, *122*, *617-618*
   increasing consumption of, 165
   kinetic, 542, 550
   paths from source to service, *181*
   primary. *See* Primary energy
   renewable. *See* Renewable energy
Energy access[*], 18, 41, 121, 191, 716, 721, 869, 879, 925
   future scenarios, 127, 751-752
   net metering, *890*, 906-907
   number of people without access, 721, *721*
   policies and, 879-880, *890-891*, 895, 906-907
   priority access to network, 24, *891*, 907
   priority dispatch, *891*, 907
   regulations, *890*
   solar energy and, 372-373
   sustainable development and, 121, 127, 716, 721-724, *721-723*, 751-752, 765
   technology transfer and, 157, 927
   third party access (TPA), 895, 910
Energy carrier[*], 37, 38, *38*, 46-48, 53, *56*
Energy conversion. *See* Conversion
Energy demand, 7, *814*
   reducing, 186, 618, 673
Energy density[*], 44, 53, 87, 113, 215, 236-237
Energy efficiency[*], 37, *37*, 170, 185-186, 796
   of appliances, 673-674
   RE and, 37, *37*, 40, 175-178, *177*, 185-186, 796
   of vehicles, 664-665
Energy flows, global, 187, *187*

Energy imports, 717, *725*
   See also Energy security
Energy indicators, 715-716
Energy intensity[*], 164, 185
   GDP and, 168, *168*
   projected decrease in, 169, *169*
   in scenarios, *814*
Energy payback[*], 123, *470*, 728-729, 979
   electricity generation, *731*
   hydropower, *470*
   solar energy, 338, 370
   See also Carbon payback
Energy payback ratio, 979
Energy ratio, 979
Energy Revolution-2010 scenario, 814-816, *814*, *815*, *817*, 830-831
Energy savings[*], 40, 185
Energy security[*], 18, 41-43, 122, 127-128, 165, 191, 716-717, 724, 869
   availability and distribution of resources, 724-725, 753-754
   external costs, *855*
   future scenarios, 752-756
   policies and, 880
   raw materials and, *728*, *729*
   sustainable development and, 122, 127-128, 716-717, 724-728, 752-756, 765-766
Energy services[*], 40, *104*, 164, 178, 185, 187-190
Energy storage. *See* Storage of energy
Energy supply curves, 820-825, *823*, *824*
Energy system:
   innovation in, 919-920
   structural shift in, 45, 158, 199, 870, 929-932
Energy tax[*]. *See* Taxes
Enhanced geothermal systems (EGS) 13, 412, 416, 422-423, 432
Environmental costs, 144-146, 851-857, *855-857*
Environmental impacts:
   bioenergy, 50-52, 257-271
   biofuels, 257-258, *258*
   as driver for policies, 879
   evaluation of, 717-718
   geothermal energy, 74-76, 404-405, 418-421, *418*, 432
   hydropower, 83-84, 442, 461-474, 487
   ocean energy, 92-93, 502, 517-520, *518*, 528
   solar energy, 65-66, 369-372, 390
   sustainable development and, 128-129, 710-711, 717-718, 728-739, 766
   wind energy, 99-100, 540, 570-575, 595
Equity[*], 150, 883, 905-906
Equity investments, 877-878, 892, 895, 903

Ethanol, 9, *15*, 173, *218*, *235*, 241-243, 246, 880
   costs, *1014-1015*
   experience curves, *56*
   global trade in, 252, *252*
   integration case study, 657-658, 663
   integration issues, 613, 656
   levelized cost of, *846*
   lifecycle GHG emissions, 733-735, *734*
   policies and, *912*
   production technologies and costs, 281-285, *281*
Europe:
   EU Renewables Directive, 876
   RE electricity costs, 855, *855*
   RE electricity supply curves, 823-825, *824*
   RE energy scenarios, *825*
Experience curves. *See* Learning curves
External cost/benefit[*], 144-146, 851-857, *855-857*, 872
Externality[*], 881

F

Fatality rates, for RE technologies, 20, 746, *746*
Feasibility[*], 150, 883, 906
Feed-in tariffs (FITs)[*], 24, 25, 45, 869
   electricity sector, 874, 899-907
   RE deployment, *890*
   success, factors for, 901-903
Feedstocks, 48, *49*, 233-236, *234*, 270, 276-278, *277*, *1014-1016*
   aquatic, 48, 53, *56*, *218*, 277-278
   biomass[*], 50, 53, 218-219, 233-236, *234*
   carbon payback, *264*
   climate change and, 214
   conversion routes, *235*
   costs, 53-55, *55*, *56*, 215, *234*, *245*
   deployment projections, 56-57
   environmental issues, 270
   experience curves, *56*
   improvements in, 215, 276-278
   integration of, 112-113, *112*, *116*, 117, 119
   land use change and, 50-51, 123-125
   lignocellulosic, 50-51, 53, *56*, 215, *218*, 219, *226*, 233, *234*, *235*, 266-267, *277*
   maize (corn), *56*, 215, *225*, *234*, *264*, 265, *266*, *277*
   modern bioenergy, 217
   oil crops, 53, *218*, 233, *234*, *235*, 243, *245*, *264*
   soy, *56*, *218*, *234*, *235*, 243, *245*, *264*, *266*, *277*
   sugarcane, *56*, 215, *218*, 233, *234*, *235*, *245*, *264*, *266*, *277*
   switchgrass, 217, *225*, 230, *234*, 269, 270, *277*, *281*
   wood pellets, *56*
Fertilizer industry, 683

Index

Final energy[*], *56*, 117, 120, *617-618*
   income and, *122*
Financial payback[*]. *See* Payback
Financial risk, 194, *918-919*, 920-921
Financing[*], 24-26, 146-158, 720-721, **865-950**
   allocation of, 196
   barriers, 150, 882
   continuum, 877
   non-recourse[*], 882
   policies and, 873, *918-919*
   private equity financing[*], 877, 893, 903
   project financing[*], 873, 877-878, 882
   public equity[*], 877-878
   public finance[*], 45, 150, 197, 883-886, *886*, *890*
   RE deployment policy options, 889-895, *890-891*
   research and development (R&D), 885-886, *886*
   trends in, 148, 876-878, *876*
   See also Policies
First-generation biofuels[*], 50, 612, 733
Fiscal incentives[*], 45, 150, 197, 869, 883
   assessment of, 892
   electricity sector, 895
   grants and rebates, 889-891
   heating and cooling sector, 907-908
   RE deployment, 889-892, *890*
   research and development, 885-886, *886*
   tax policies, 891-892
   transportation sector, 911-912
Fiscal policies. *See* Financing; Policies
Fish migration, 466-467, 475
Fishing, 119, 686-689, *688*
Food security, risks from bioenergy, 273
Forest products, 119, 216, 219, 229-230, 233, *234*, 235-236, *277*, 683-684
   experience curves, *56*
   See also Bioenergy; Lignocellulosic crops/feedstocks
Forest protection, 267
Fossil fuels, 192, 744, 805
   combined with CCS, 804-805
   compared to RE technologies, 710-711
   decarbonization of, 358
   energy system momentum, 881
   external costs, *856*
   future scenarios, 717, 724, 753-754, *755*
   global emissions from, *171*, *802*, *803*
   health impacts, 854
   imports of, 717
   increase in demand/use, 7
   projected emissions from, *21*, *22*, 169, *172*, *804*, *806*
   replacement of, 663
   scenario parameters, *814*
   subsidies, removal of, 760-761
Fourth Assessment Report (AR4), 167
Fuel cells[*], 114, 186, 668, *674*
   microbial, 286
Fuels:
   air pollution from, 739, 854
   aviation, 670-671
   levelized cost of, *189*, *846*
   levelized cost of transport fuel (LCOF), *14*, *42-43*, *55*, *144*, 1002, *1003*, *1014-1018*
   lifecycle GHG emissions, 123, 733-735, *734*
   mandates and targets, 912-913
   mitigation potential, 662-672, *662*
   policies and, 869, 912-913
   RE integration issues, 112-113, 612-614, 654-658
   solar production costs, 385, *386*
   solar production of, 63-64, 358-359, *359*, 361-362, 365-366, 377-378
   See also Biofuels; Fossil fuels

G

Gas distribution grids, 17, 185, 647-654
   features of, 648
   integration issues, 111-112, 613, 647-654
Gases:
   composition and parameters of, *650*
   quality standards, 651, *652*
   See also Greenhouse gases (GHGs)
Gasification, *218*, *235*, 239-240, *282-283*, 284, 304
Gasoline, lifecycle GHG emissions, 733-735, *734*
Generation capacity[*], 35, 77, 99
Generation control[*], 105
Genetically engineered crops, 230, 270, 275
Geographical chokepoints, 725
Geothermal energy[*], *8*, 71-80, *182*, **401-436**
   barriers, 405, 417
   climate change and, 410
   cogeneration plants, 412
   conductive systems, 406, *407*, *408*
   convective (hydrothermal) systems, 406, *406*, *408*
   costs, 77-78, 405, 423-428, *856*, *1004*, *1010*
   costs, future trends in, 426-427, *427*
   deep aquifers, 406, *408*
   depletion of reservoirs, 408
   deployment, 78-79, 428-432, *430*
   direct uses, 410, 412-414, *414*, 416-417, *416*, 432
   electricity generation, 408-409, *409*, 415-416, *415*, *416*, *819*
   electricity generation costs, 405, 424-426, *424*, *426*
   enhanced geothermal systems (EGS), 13, 412, 416, 422-423, 432
   environmental and social impacts, 74-76, 404-405, 418-421, 432, *745*
   exploration and drilling, 411
   geothermal heat pumps (GHPs), 406, 413-414
   health impacts, *740*
   historical use of, *11*, *176*, 404, 406, 410, 414
   installed capacity, 415-416, *415*, *416*, *425*
   integration, 431-432, *623*, 624-625, 628
   investment costs, 405, 423, 424, *424*, *427*
   land use and, 76, 420, *420*
   levelized costs, *14*, 405, 423, 425-426, *426*, *843-845*
   market and industry development, 74, 414-417
   maturity of technologies, 404, 431, 612
   mitigation potential, 406, 429-432, *430*
   policies and, 417
   power plants, 412, *413*
   production wells, 410-411, *413*
   projections for, *22-23*
   regional deployment, 428, *429*, 431
   regional technical potential, 410, *411*
   reservoir engineering, 411-412
   scenarios, 806, *808*, *817*, *821*, *822*, *827-829*
   seismicity and, 76, 420
   submarine vents, 410
   sustainability and, 404
   technical potential, *12*, 72-73, *184*, *206*, 404, 408-410, *409*
   technology and applications, 73-74, 410-414
   technology improvement and innovation, 77, 405, 421-423
   technology-specific challenges, 13
   temperatures and depths, 406, *408*
   theoretical potential, *183*, 408, 409
Geothermal heat pumps, 406, 413-414
Geothermal vent[*], 410
Germany:
   RE development approach, *900-901*
   solar-assisted district heating, 646
Global primary energy scenarios, 20-24, *21-23*, *803*, *804*, *806*
Global solar radiation[*], 351, *351*
Global warming potential (GWP)[*], 164, 168, *170*
Governance[*]:
   enabling environment/policies, 917-929, *918-919*
   energy transitions and, 881, 930-931
   RE integration and, 612
Grants, 198, *886*
   capital, 891
   contingent, 885
   heating and cooling sector, 907-908
   for RE deployment, 889-891, *890*

Index

Green energy purchasing, *890*
Green labelling, *890*, 895
Greenhouse gases (GHGs)[*]:
  bioenergy and, 259-267, *259*, *260*
  climate change and, 33, 164, 168, *168*
  displacement factor, 261, *262*
  economic development and, 710
  electricity generation, 122-123, 732-733, *732*
  energy services and, 7, 33, 164
  fuels: petroleum and biofuels, 7, 733-735, *734*
  future, factors influencing, 798
  geothermal energy and, 418-419, *418*
  hydropower and, 442, 470-474, *471*, 483
  land use change-related, 18-19, 123-124, 729-732
  lifecycle analysis, 18, *19*, *177*, 732-735, *732*, *734*
  lifecycle analysis, literature review, 981-993
  ocean energy and, 517-518, *518*
  projections for, *170*
  RE deployment scenarios and, 794
  RE sources, 122-124, *177*, 854
  reduction with RE, 7, 18, 21, 37, 164
  savings indicators, 261-262, *262*
  scenarios, 164, 798, 826-830, *826*, *829*, *830-831*
  stabilization level scenarios, 794-795, 870
  structural shift, 870
  sustainable development and, 717-718
  top emitting countries, 172, *173*
  zero- or near-zero emissions, 37, *37*, 74, 114, 115, *115*, *116*, 175
  See also Emissions; Lifecycle analysis; Mitigation
Grid[*]:
  assumptions in future scenarios, 813
  decentralized, 710
  gas distribution, 17, 623, 647-654
  infrastructure, 629-630
Grid codes[*], 562-563
Grid connection[*], 62, 354, *354*, 376
Grid integration[*], 98-100, 560-570, 624
Gross Domestic Product (GDP)[*], 715
  climate change and, 168, *168*, 170
  projections, *168*
  in scenarios, *814*
Guarantees, 24, *890*, 893

H

Habitat loss/modification, 269, 573
Harmonized policies[*], 50, 66
Hazardous substances. *See* Toxic substances
Health:
  air pollution and, 739-740, *740*, 756-757, 854
  bioenergy and, 44, 192
  as driver for policies, 879
  energy-related impacts, 128-129, 739-740, *740*, 766
  external costs, *855*
  future health impacts, 756-757
  hydropower and, 467-468
  RE benefits for, 20, 165, 192, 711
  solar energy and, 372, 373
Heat pump[*], 406, 413-414
Heating and cooling:
  air pollution from, 737-739, *738*
  autonomous systems and, 658
  Belgium, case study, 676-677, *676*
  bioenergy and, *821*
  costs, *14*, *189*, 405, 843, *845*, *1010*
  district heating and cooling, 16, 110, 185, 367, 613, 640-647, *641*
  geothermal energy, 405, 410, 412-414, *414*, *821*
  integration of RE technologies, 110-111, 613, 640-647
  levelized cost of heat (LCOH), *14*, *42-43*, *55*, *69*, *79*, *143*, *1002*, *1003*, *1010-1013*
  market development and potential, 818-820
  new options for, 643, 644-645
  policies, 153, 869, *890-891*, 907-911
  RE integration, 16-17, 45, 185
  scenarios for, 795, *817*, 818-820, *821*, *826*
  solar energy and, 337, 340-351, *347*, 374-375, *821*
  storage options (for energy), 110-111, 643-644, *644*
Housing. *See* Building sector
Human Development Index (HDI)[*], 41, 716, 719, *720*
Hybrid models[*], *799*
Hybrid vehicles[*], 45, 114, *114*, 668
  See also Electric vehicles
Hydrogen (H$_2$), *218*, *279*, 285-286, 377-378, *650*
  concentrating solar power and, 357
  costs, 362, 366, 385, 652
  delivery options, 649, 650-651
  energy density of, 358
  fuel synthesis pathways, 358-359, *359*, 366, 377-378, 385, 667
  integration into gas grids, 613, 649
  natural gas and, 366, 385
  safety and, 652
  solar energy and, 337, 358-359, *359*, 362, 366, 377-378, *650*
  storage, 114
  transition issues, *116*
Hydrogen sulfide, *650*
Hydrogenation (of oils), *235*, 240, *282*, 285

Hydropower[*], *8*, 80-87, *182*, **437-496**
  adaptation and, 192
  classification of projects, 450-452, *450-452*, 458
  climate change impact on, 447-449, *448*
  costs, 84-86, 441, 442, 474-484, *478*, *479-480*, *856*, *1004*
  decommissioning cost, 481
  deployment, 86-87, 484-488, *485*
  efficiency, 452-453, *453*
  electricity generation, 442, 455-456, *456*, 458-460, *819*
  energy payback, *470*
  environmental and social impacts, 83-84, 442, 461-474, 487, *745*
  fish migration and, 466-467, 475
  global investment in, 876, *876*
  guidelines and regulations for impacts, 468-470, *469*
  health impacts, *740*
  historical use of, *11*, *176*
  history of, 443-444, *444*
  hydrokinetic turbines, 475-476
  in-stream, 452, *452*
  installed capacity, 441, *446*, 449, *461*
  integration into energy systems, 82-83, 458-461, 487, *623*, *625*, *628*
  integration into water management systems, 87, 488-490
  investment costs, 441, 477-480, *479-481*
  levelized cost of energy, *14*, 481-483, *483*, *843-844*
  lifecycle assessments, 470-474, *471*
  market and industry development, 82, 455-458
  matrix technology, 475
  maturity of technology, 452-454, 612
  mitigation potential, 441, 442, 462-468, 485-487, *486*
  modernizing/upgrading, 454-455
  multiplier effects, *462*
  operation and maintenance costs, 480-481
  policies and, 457
  power generation support by, 460
  power system services from, 459-460
  in primary energy supply, *176*
  projections for, *22-23*
  pumped storage, 451-452, *452*, 460, *462*, 625
  regional deployment, 476-477, *487*
  regional technical potential, *81*, 444-447, *445*, *446*
  reservoirs and, 451, *451*, *462*, 464, 625
  river flow/runoff and, 447, *448*, 463-464
  run-of-river, 451, *451*, *462*, 625
  rural electrification, 458-459
  scenarios, *22-23*, 806, *808*, *822*, 827-829

1066

Index

sedimentation and, 447, 454, 465
size of projects, 82, 443, 450, 458
social/economic development and, 441
source of power, 443
storage of energy, 80-81, 82-83, 451-452, *451*
storage (reservoir) hydropower, 451, *451*, *462*
sustainable hydropower in Norway, *764*
technical potential, *12*, 80, *184*, *206*, 441, 444-449
technology and applications, 80-82, 449-455, *462*
technology improvement and innovation, 84, 474-476
technology-specific challenges, 13
theoretical potential, *183*, 444
tunnelling capacity, 453-454, *454*, 476
variable-speed technology, 475
water management services and, 441, *462*, 488-490
water quality and, 464-465
Hydrothermal energy, 406, *406*
Hydrothermal liquefaction, 285
Hydrothermal vent[*]**,** 410

I

Iceland, 616
Implementation, 24-26, 146-158, **865-950**
    barriers to RE policy implementation, 881-882
    See also Environmental impacts; Financing; Policies; Social impacts
Imports of energy. *See* Energy imports
Incentives, 874
India:
    CO_2 emissions, *173*
    emissions, *173*
    RE electricity supply curves, 823-825, *824*
Indicators:
    aggregate, 715-716
    climate change, 261-263, *262*
    energy, 715-716
    energy access, 728
    infrastructure, *729*
    sustainable development, 715-716, 719, 728
Indirect land use change (iLUC). *See* Land use change
Individual ownership, *922-923*
Individuals. *See* Stakeholders
Industry capacity. *See* Market and industry development
Industry sector:
    energy-intensive industries, 682-684
    less energy-intensive industries, 684-686
    RE integration in, 117-119, 614, *617-618*, *662*, 681-686
    status of, 681-682

Informational barriers, 44, 194-195, 758-759, 881, *914*
Infrastructure, 812, *918-919*, 924-925
    assumptions about, 813
    building, 924-925
    enabling, 17, 199, *909*
    fuel supply/vehicles, 666
    grid, 629-630
    indicators of, *729*
    integration and, 17, 612, 618, 635-636
    investments in, 24
    strengthening, 16
    value of, *909*
Innovation. *See* Technology improvement and innovation
Institutional barriers, 196, *914*
Institutional feasibility[*]**,** 150, 883, 906
Institutional learning, 157, 927
Institutions[*]**,** *918-919*
Integrated assessment[*]**,** 800-801, *801*
Integrated assessment models, 120, 125-126, 748, 798-799, *799*, 800, 800-801, *801*
Integrated solar combined-cycle (ISCC) systems, 369
Integration, 15-18, *16*, 39-40, 103-119, 184-185, *609-705*
    agriculture (primary production) sector, 119, 614, *622*, 686-689
    autonomous energy supply systems, 113, 613, 658-661
    bioenergy, 53, 622, *623*, 628
    buildings sector, 117, 614, *662*, 672-681
    case studies/experiences, 616, 627-631, 646-647, 657-658, 660-661, 676-677, 677-678, 679-680, 689
    challenges of, 627, 642-643, 649-671, 655-656, 675, 677
    characteristics of RE and, 15, 622-627, *623*, 641-642, 648-649
    complexities of, 618-619
    costs and challenges of, 15-16, 17, 612
    current RE integration, 15
    dispatchable RE sources, 628-629
    electrical power systems, 612-613, 619-640
    end-use sectors, 113-119, 616, 661-689, *662*
    energy storage and, 637-638
    energy supply systems, 107-109, 619-661
    facilitating, 675-676
    gas grids, 111-112, 647-654
    geothermal energy, 431-432, *623*, 624-625, 628
    heating and cooling networks, 110-111, 640-647
    holistic approach/supply chain analysis, 105, *678*

    hydropower, 82-83, 458-461, *623*, 625, 628
    industry sector, 117-119, 614, *662*, 681-686
    infrastructure and, 17, 612, 618, 629-630, 635-636
    knowledge gaps, 639-640
    liquid fuel systems, 112-113, 654-658
    ocean energy, *623*, 626, 634-635
    options, 635-639
    partially dispatchable RE sources, 629-631
    pathways for, *16*, 17, *615*
    planning and, 638-639
    policies and, 166, 645, 656
    rate of, 18
    resource characteristics and, 616
    solar energy, 367-369, 612, 622-624, *623*, *624*, 629, 633-634
    status of, *617-618*
    studies on, 631-635
    transition pathways, 113-119, 661-689
    transportation sector, 113-117, 613-614
    wind energy, 98-99, 540, 560-570, 612, *623*, 626-627, 629-632
Intellectual property rights, 196, 926
Intensity. *See* Energy intensity
International Energy Agency (IEA). *See* World Energy Outlook scenarios
Invasive species, 100, 270
Investment, 25, 872, 877-878
    capital-intensive technologies, 872
    costs, recent, *1004*
    financing, *890*, 893
    in infrastructure and RE technologies, 24
    policies and, 25
    transition, 878
    trends in, 148, 876-878, *876*
    See also Financing
Investment cost[*]**:**
    additional, 849
    hydropower, 424, 441, 477-480, *479-481*
    learning investments, 849, *849*
    scenarios, 849-851, *850*
    time-dependent expenditures, 849-851, *850*
    up-front, 194, 366, 796-797
Investment risks, 872, *918-919*, 920-921
IPCC:
    Fourth Assessment Report (AR4), 167
    Special Report on Renewable Energy Sources and Climate Change Mitigation (SRREN), 33, 167-168, *167*
Iron and steel industries, 682-683
Irradiance. *See* Solar irradiance; Solar radiation
Issues, 44, 196-197

Index

J
Japan, energy investment strategies, *887-888*
Johannesburg Plan of Implementation, 716

K
Kaya identity, 34, *35*, 170, *173*
Knowledge:
    advancement, 26, 37, 178, *179*
    barriers, 44, 194-195
Knowledge gaps, 37
    integration, 639-640
    scenarios, 798-799, 812-813, 832, 840-841, 851, 857, 897-899
    sustainable development, 130, 712, 767-768
Kyoto Protocol[*], 761

L
Land use[*], 125, 743-745
    agriculture and, 686
    geothermal energy and, 76, 420, *420*
    management practices, 19-20
    wind energy and, 574-575
Land use change (LUC)[*], 18-19
    bioenergy and, 50-51, 215, 219, 263-267, *266*, 275, 304-305
    direct, 735
    fossil fuel production and, 192
    GHG emissions from, 123-124, 729-732, 735-736, *736*
    indirect (iLUC)[*], 263, 267, 305, 735, 913
    solar energy and, 370-371
Landfills, 122, *239*
    GHG emissions from, *124*, *260*
    landfill gas, *46*, *110*, 111, 217, 220, *242*, 285
    toxic substances and, 271
Leapfrogging[*], 120, *763*
Learning curves[*], 796-797, 846-849, *847*, *848*
    bioenergy, *56*, 292-293, *293*, *848*
    deployment costs and investments and, 849, *849*
    institutional learning, 157, 927
    solar energy, 338-339, *848*
Learning investments, 849, *849*
Learning rate[*], *847*, *848*
Levelized cost of energy (LCOE)[*], 13, *14*, 40, 137-141, *142-144*, 165, 842-846, *843-846*, 1002, *1003-1018*
    calculation of, 976
    competitiveness of technology and, 13
    of electricity (LCOE), *14*, *42-43*, *55*, *68*, *70*, *93*, *142*, 1002, *1003*
    of heat (LCOH), *14*, *42-43*, *55*, *69*, *79*, *143*, 1002, *1003*, *1010-1013*
    projected decrease in, 796, 846-849

    tornado graphs, *1003*, *1008-1009*, *1012-1013*, *1018*
    of transport fuel (LCOF), *14*, *42-43*, *55*, *144*, 1002, *1003*, *1014-1018*
    See also Annex II; Annex III; specific technologies
Levy. *See* Taxes
Lifecycle analysis (LCA)[*], 18, *19*, 122-124, 173-174, *177*, 192, *193*, 710, 978-993
    bioenergy, 258-261, *260*
    electricity generation, 18, *19*, 122-123, 173-174, 732-733, 979-981
    hydropower, 470-474, *471*
    ocean energy, 517-518, *518*
    sustainable development and, 122-124, 710, 717, 728-729
    wind energy, 571, *571*
Lifecycle costs[*], 82, 417
    See also Lifecycle analysis
Lignocellulosic biofuels, 54, 215, *218*, 294-295, *296*, 303, 735
    conversion routes, *235*
    costs, *53*, *56*, 215, *234*
Lignocellulosic crops/feedstocks, 50-51, 53, 219, *234*, 266-267
    technical potential, *226*
    See also Forest products
Liquid fuel systems, 17, 112-113, 654-658
    See also Biofuels; Fossil fuels; Fuels
Load (electrical)[*], 62, 65, 73-74, 82, 113
    base load[*], 73, 77, 80, 88
    excess load, 153
Loans[*], 24, *890*, 893
    convertible loans[*], 886, *886*
    public loans, 893
    soft loans[*], *886*, 893
Lock-in[*], 147, 148, 870, 872, 881, 926

M
Macroeconomics, 870, 872, 916
Maize (corn), 215, *234*, 265, *266*, 277
    carbon payback, *264*
    costs of fuel production, *53*
Mandates, 874, 911-912
    absolute, 911-912
    biofuel, 911
    renewable fuel, 912-913
Marginal (incremental) cost[*], 135, 810, 832
Maritime transport, 671
Market and industry development:
    bioenergy, 48-50, 246-257
    domestic markets for RE, 880
    geothermal energy, 74, 414-417
    hydropower, 82, 455-458

    ocean energy, 90-92, 513-517
    scenarios, 816-826, *817*, *821*
    solar energy, 63-65, 359-367
    wind energy, 97, 556-560
Market failures[*], 25-26, 44, 193-194, 759-760, 872
    geothermal energy and, 417
    policies and, 872, *914*
Market potential[*], 220, 796-797
    RE electricity, 816-818
    RE heating and cooling, 819-820
Maturity of technologies, *182-183*, 612
    bioenergy, 238-240, 261
    geothermal energy, 404, 431, 612
    hydropower, 452-454, 612
    ocean energy, 501, 522, 525
    solar energy, 337, 340, 350, 389-390
    wind energy, 541, 553, 558-560
Measures[*], 872, 885-886, 891, 911, 920, 925
Mergers and acquisitions, 878
Methane, 50, 285, *650*
    hydropower reservoirs and, 472, *473*, 729
    reduction with bioenergy, 218
    solar, 359
Methanol, *218*, 358, 359
Methanol-dimethyl ether (DME), 285-286
Methodology, 973-1000
Metrics, 975
Mexico, seawater desalination, 660
Microbial fuel cells, 286
Microwave power beaming, 68, 378
Millennium Development Goals (MDG)[*], 18, 716
MiniCAM-EMF22 scenario, 814-816, *814*, *815*, *817*, *830-831*
Mitigation[*], 20-24, 43, 130-146, 174-177, 192, *791-864*
    assessment of RE strategies for, 813-822
    biomass and, 210, 214
    building sector, *662*, 672-681
    as driver for policies, 879
    energy supply, 661, *662*
    geothermal energy and, 406, 429-432, *430*
    hydropower and, 441, 442, 462-468, 485-487, *486*
    ocean energy and, 502, 527-528
    options for, 174-177
    RE technologies and, 7, 18-24, *22-23*, 43-44, 164, 165, 172-177, *177*, 192, 714, 794
    solar energy and, 337, 345-346, 386-390
    sustainable development and, 122-125, 128-129, 714, 717-718, 729-736, 756-757, 766
    transport sector, 662-672, *662*
    wind energy and, 539, 541, 591-593
    See also Mitigation scenarios

Mitigation costs, 24, 130-146, 795, 832-857
    abatement cost curves, 834-836, *835-838*
    cost curves, 798, 832-841
    direct and indirect costs, 851-852
    external costs, 797
    RE and, 808-811, *810*, *811*
    social and environmental costs, 851-857
Mitigation scenarios, 795-797
    assessment of RE strategies, 133-135, 813-832
    gaps in knowledge and uncertainties, 798-799, 812-813, 832, 840-841, 851, 857, 897-899
    by sectors and RE sources, 826-830, *826-830*
    synthesis for RE strategies, 131-133, 799-813
    See also Scenarios
Models[*], 120, 131, 794, 798-799
    bottom-up models[*], *799*
    hybrid models[*], *799*
    integrated models, 120, 125-126, 748, 798-799, 800
    quantitative modelling, 799-800
    top-down models[*], *799*
    See also Scenarios
Modern bioenergy[*], *216*, 217, 263-267
Modern biomass[*], 46, *216*, 217, 250-251
Multipurpose use, 441, 457, *462*, *470*, 471, 476, 488-489

N

Nairobi Programme of Action, 177
Nameplate capacity[*], 99, 539, 541, 550, 554, 557, *558*
National Sustainable Development Strategies (NSDS), 762
Natural gas, 34, 37, 43, *52*, 66, *111*, *126*, *147*, 366, 385, 663
    blending with hydrogen, 112
    compressed, 114, *115*
    extent of supplies, 122
    GHG emissions, *124*
    grids, 105, 111
    imports of, *123*
    synthetic, 53, *56*
Negotiation, 923-924
Nepal, public investments in, *928*
Net metering (net billing), *890*, 906-907
Netherlands, *931*
Networks, *918-919*
    access to, *891*, 907, 925
    collaborative, 923
    heating and cooling, 110-111, 640-647, 910
    infrastructure and, 923-924
    large (centralized), 7, 165, 181
    third party access (TPA), 895, 910

New Zealand, 616, 689, *689*
Nitrogen, *650*
Nitrous oxide (N$_2$O), 50, 261, 275, 736, *739*, 854
Non-Annex I countries, *750*
    RE deployment scenarios, 132, 807, *808*
    See also Annex I countries
Non-methane volatile organic compounds (NMVOC), *124*, 736
Non-recourse financing[*], *882*
North America:
    district cooling, 647
    RE energy scenarios, *825*
Norway:
    autonomous wind/hydrogen demonstration, 660-661
    sustainable hydropower in, *764*
Nuclear energy, *43*, 44, 616, 744, 804, 805
    GHG emissions from, *124*
    health impacts, 740
    percentage of primary energy supply, *10*, *35*, *134*, *174*
    RE integration of, *104*, *114*
    risks, *746*, 856
    scenarios, *814*

O

Obligations, 869, 908
    quota obligations (renewable portfolio standards), 65, 152, 874, *890*, 895-897, 910
Ocean energy[*], *9*, 87-95, *182*, 497-533
    costs, 93-94, 502, 522-526, *523*, *526*, *1004*
    deployment, 94-95, 502, 526-529, *527*
    environmental and social impacts, 92-93, 502, 517-520, 528, *745*
    global distribution of resources, *89*
    historical use of, *11*, *176*
    implementation/integration, 93, 501, 520-522, 528, *623*, 626, 634-635
    levelized cost of energy, *14*, 843-844
    market and industry development, 90-92, 513-517
    maturity of technologies, 501, 522, 525
    mitigation potential, 502, 527-528
    ocean currents, 87, 503, 506, *506*, 510-511, 515-516, 521, *523*
    ocean thermal, *523*
    ocean thermal energy conservation (OTEC), 87-88, 503, 507, *507*, 511-512, 516, 521, *523*, 525
    policies and, 501, 516-517, *517*
    in primary energy supply, *176*
    regional deployment, 528
    regional theoretical potential, *504*
    salinity gradients, 88, 503, 507, 512-513, *512*, *513*, 516, 521, *523*, 526

    scenarios, *822*, *827-829*
    sources of, 503
    supply chain, 528
    technical potential, *12*, 87-88, *184*, *206*, 503-507, 527-528
    technology and applications, 89-90, 507-513
    technology improvement and innovation, 93, 520-522
    technology-specific challenges, 13
    theoretical potential, *183*, 501, 503, *504*
    tidal currents, 87, 503, 506, 510-511, *511*, 515-516, 521, 522-524, *523*
    tidal range, 87, 503, 505, *505*, 510, 515, 521, *523*, 524-525
    wave energy, 87, 503-505, *504*, 507-510, *508*, *509*, 514-515, 520-521, 522-524, *523*
Ocean fertilization, 175
Off-grid systems, *882*, *915*
    autonomous energy systems, 613, 658-661
    PV, 62, 353-354, *354*, 376
Offset (in climate policy)[*], *917*
Oil:
    future scenarios, 753-754, *755*
    lifecycle GHG emissions from, 733-735, *734*
    percentage of primary energy supply, *10*
Oil crops, *218*, 233, *277*
    carbon payback, *264*
    conversion routes, *235*
    costs, *53*, *245*
    hydrogenation of, *235*, 240, 282, 285
Operation and maintenance (O&M) costs[*], 480-481, 584-585, 587
Opportunities[*], 40-44, 148, 165-166, 191-192, 878-880
    bioenergy, 218
    sustainable development and, 712, 760-764
Opportunity cost[*], 130
Organic PV (OPV) cells, 353
Osmotic power. *See* Salinity gradients
Oxygen gas, 362, *650*

P

Pacific Island States, RE target setting in, *884*
Palm oil, *234*, *264*, *277*
    costs of fuel production, *53*, *234*
Participation, 883, 892, 895, 905-906, 914, *918*, 921, 923
Particulates, 736, *737*, *739*, 854
Payback[*], 621, 650, 728-729
    carbon payback[*], 264-265, *264*
    energy payback[*], *123*, 338, 370, *470*, 728-729, *731*, 979
Peaking, 442, 445, 459-460, 466, 481

Index

Performance parameters, recent, 1001-1022
Performance standards[*], 17, 97
Permitting, 921-924
Photosynthesis[*], 226, 232, 359
Photosynthetic algae, 54-55, *218*, 285, 303-304
Photovoltaics (PV)[*], *15*, 62, 337, 338, 351-355, *351*
    applications, 353-355, *354*
    concentrating photovoltaics (CPV), 361, 375-376
    cost curves, 839, *839*
    costs, 68-69, *369*, 370, 380-382, *381-384*, *856*
    electricity generation, 351-355, 363-364, *363*, 375-376, *388*
    emerging technologies, 66, 352-353
    environmental and social effects, 370, *371*, 372-373
    existing technologies, 351-352
    generation and smoothing effect, 368-369
    installed capacity, 63, 360-361, *361*
    integration, *623*, 624
    learning rates, *848*
    lifecycle GHG emissions, 370, *371*, 376
    novel technologies, 353
    photovoltaic systems, 353
    scenarios, *817*
    toxic chemicals and, 352, 370
Physical energy content method, *180*, 977, *978-979*
Planning, *918-919*, 921-924
Policies[*], 7, 24-26, 146-158, 166, 177-178, 197-198, 865-950
    assessment of, 150-155, 869-870, 882-917, 913-916
    barriers, 148-150, 196, 880-882, *914*
    barriers, overcoming, 869, *914*
    combining R&D with deployment policies, 888, *889*
    complementing, 869, 920
    cost-benefit analysis and, 870
    current trends, 148, 874-876, *875*
    demand pull[*], 150, 851, *887-888*
    drivers, opportunities and benefits, 148, 878-880
    effectiveness[*], 150, 883, 903
    efficiency[*], 150, 883, 903-905
    electricity sector, 152-153, 874, *890-891*, 895-907
    enabling, 25, 45, 155-158, 198, 199, 870, 873, 917-929, *918-919*
    energy sectors and, 869
    equity[*], 150, 883, 905-906
    evaluation criteria[*], 883-884
    experience with, 882-917
    finance and RE, 873
    fiscal incentives[*], 150, 883, *886*, *890*
    harmonized[*], 50, 66
    heating and cooling sector, 153, 907-911
    institutional feasibility[*], 150, 883, 906
    interactions and unintended consequences, 154-155, 916-917
    international, 177-178
    key policy instruments, *251*
    lessons learned, 886-888, 896-897
    link with scenarios, 930, *930*
    mix of policies, 869-870
    options, 150-155, 882-917
    overcoming barriers with, 24-25
    public finance, 45, 150, 197, *866*, 883, 883-886, *886*, *890*
    rationale of RE policies, 146-147, 871-872
    RE and climate policy interactions, 154-155, 916-917
    RE deployment and, 7, 13, 20, 24, 814, 871
    RE deployment policies, 151-154, 869-870, 871, 874-876, *875*, 889-895, *890-891*
    RE growth and, 869
    RE integration and, 166
    RE promotion through, 869
    regulations, 150, 883, *890*
    research and development (R&D), 150-151, 869, 871, 884-888, 913-914
    role of, 44-45, 197-198
    scenario analysis, 814
    sector-specific, 871
    structural shift, 158, 870, 929-932
    supply push[*], 150, 851, *887-888*, 916
    sustainable development and, 712, 760-763
    timing and strength, 147-148, 873, 923
    tracking, 874, *875*
    transparency and flexibility in, 870
    transportation sector, 153-154, 911-913
    See also Policy case studies; *specific policy options, energy sectors and technologies*
Policy case studies, 873
    Berkeley, California: innovative financing, *894*
    Brazil: expansion of policies, *912*
    China: mixed policy approach, *915*
    Denmark: comprehensive approach and individual/community ownership, *922-923*
    Dutch technology and innovation, *931*
    Germany: comprehensive approach, *900-901*
    Güssing, Austria: rapid transition in, *929*
    Japan: coupling supply-push with demand-pull, *887-888*
    Nepal: upfront public investments, *928*
    Pacific Island States: RE target setting, *884*
    Spain: PV deployment, *904-905*
    Sweden: biomass district heating and infrastructure, *909*
    Thailand: expansion of RE policies, *902*
Policy criteria[*], 883-884
Population, 34, *35*, 48, 130
    climate change and, 170
    in Kaya identity, 34, *35*
    in scenarios, *58*, 814
Potential[*]. See Deployment; Market potential; Technical potential
Power density[*], 44, 82
Power dispatching[*]. See Dispatch
Power grid[*]. See Grid
Pressure-retarded osmosis (PRO), 93, 513, *513*
Primary energy[*], *174-176*, 617-618
    accounting of, *10*, 178-181, *180*, 798, 976-978
    direct equivalent method, *10*, 178-181, *180*, 798, 802, *803-810*, 820, 977-978, *978-979*
    historical development in, *11*, *176*, *804*
    physical content method, *180*, 977, *978-979*
    projections/scenarios for, 20-24, *21-23*, *169*, *803*, *804*, *806*, 820, 825-826, *825*, *827-829*
    RE percentage of, 9, *10*, 34, 165, *174*, 820
    substitution method, *180*, 807, *808*, 977, *978-979*
Primary production, 119, 614, *622*, 686-689, *688*
    See also Agriculture
Priority access, 24, *891*, 907
Priority dispatch, *891*, 907
Private costs[*], 852
Private equity financing[*], 877, 893, 903
Prizes, 886, *886*
Process intensification, 284-285
Production payments/incentives, *890*, 907-908
Progress ratio[*], 847
Project costs[*], 85, 883, 914, 975-976
Project financing[*], 873, 877-878, 882
    See also Financing
Property Assessed Clean Energy (PACE), 893, *894*
Public attitudes and acceptance, 100, 129, 576, 813
Public equity financing[*], 877-878
Public finance[*], 45, 150, 197, 883
    assessment of, 893-894
    electricity sector, 895
    heating and cooling sector, 908
    Nepal: upfront public investments, 927
    RE deployment, 890, 892-894
    research and development, 884-886, *886*
    transportation sector, 912
Public good[*], 121, 720
Public-private partnerships[*], 886, *886*
Public procurement, 893
Public Research Centres, 885, *886*
Pumped storage, 451-452, *452*, 460, *462*, 625
Pyrolysis, *218*, *235*, 238-239, 279-280, *282*, 285

Index

Q

Quotas[*], 24, 25, 65, 152, 366, 869, *890*
  electricity sector, 874, 895-897
  quota obligations (renewable portfolio standards), 65, 152, 874, *890*, 895-897, 910
  risk under, 903-905

R

Rail transport, 671-672
Raw materials, *728*, *729*
RE. *See* Renewable energy (RE)
Rebates:
  heating and cooling sector, 907-908
  RE deployment, 889-891, *890*
Rebound effect[*], 40, 147, 158, 185-186
Refinancing, 874, 877, 878
Reforestation[*], 219, *224*
Regional definitions and country groupings, 993-997
Regulations[*], 45, 150, 197, 883
  access, *890*
  electricity sector, 895-907
  heating and cooling sector, 908-910
  price-driven, *890*, 894, 899-907
  quality-driven, *890*, 894-895
  quantity-driven, *890*, 894, 895-897
  RE deployment, *890*, 894-895
  transportation sector, 912-913
Reliability[*], 39, 41-42, 77, 84, 97, 99, 100, 101
ReMIND-RECIPE scenario, 814-816, *814*, *815*, *817*, *830-831*
Renewable energy (RE)[*]:
  accidents and risks, 745-747, *746*
  assessment of strategies, 813-832
  benefits of, 7, 18-20, 191-192, *193*, 710-711, 869
  climate change and, 7, 33-45, **161-207**
  comparison of sources, *10*, 178-181, *180*
  current cost of, 187-190, *188-189*
  current energy flows, 187, *187*
  as dominant low-carbon energy supply by 2050, 795
  economic competition and, 13, 40, 164-165, 177
  energy efficiency, 185-186
  integration of, 15-18, *16*, 39-40, 184-185, *609-705*
  lifecycle GHG emissions, 122-124, *177*, 854
  mitigation potential, 7, 20-24, 43-44, 164, 165, 172-177, *177*, 192, 714, 813-832
  opportunities, barriers and issues, 40-44, 190-197, 757-760
  percentage of primary energy supply, 9, *10*, 34, 165, *174*
  portfolio of technologies, 16, *177*, 616, 795, 869

  projected growth in, 22, *22-23*
  raw materials for, *728*, *729*
  regional aspects, 190
  scenarios[*], **791-864**
  sources of, *8-9*, 38-40, 178-186, *182-183*
  sustainable, 174
  sustainable development[*] and, 18-20, 119-130, **707-789**
  technical potential[*], 10-11, *12*, 39, 181-184, *184*, 206
  theoretical potential[*], 39, 165, 181, *183*
  transition to, 616, 661-689, 881, *890-891*
  trends in, 874-876, *875*
  variability/reliability of sources, 8, 165, *623*, 726-728, 754-755
Renewable energy technologies, 7-15, *8-9*
  See also Deployment; Technology; *specific technologies*
Renewable energy tenders, *890*
Renewable Fuels for Europe project, 294-295
Renewable portfolio standard (RPS, quota obligation), 65, 152, 874, *890*, 895-897
Research and development (R&D), 45, 197-198, 851, 880, 932
  combining with deployment, 888, *889*
  financing, 877, 885-886, *886*
  government budgets for, *852*
  investments in, 796-797
  learning by research, 851
  lessons learned, 886-888, *887-888*, 902
  need for, 884-885
  policies and, 150-151, 869, 871, 884-888, *886*, *889*, 913-914
  RE integration and, 612
  wind energy, 577-578
Reservoirs (hydropower), 451, *451*, *462*, 464
Resource potential[*]. *See* Technical potential; Theoretical potential
Reversed electro dialysis (RED), *92*, 512, *512*
Rio de Janeiro, Brazil, UNCED conference, 713
Risks, 745-747, *746*
  fatalities, 20, 746, *746*
  financial, 194, *918-919*
  financial, reduction of, 920-921
  of policy decisions, 903-905
Rural areas:
  bioenergy and, 274
  decentralized grids and, 710
  development and RE, 880
  RE integration in, 614, 618, 678-680, 689
  transitions to RE in, 723, *723*

S

Salinity gradients, 88, 503, 507, 512-513, *512*, *513*, 516, 521
  costs, *523*, 526
  environmental impacts, 520, *745*
Sanitary and phytosanitary (SPS) measures, 256
Scale economies. *See* Economies of scale
Scenarios[*], 20-24, 131-146, 164, **791-864**
  assessment of RE strategies, 133-135, 813-832
  bioenergy use, 214
  comparison of results, 830-832
  cost curves, 135-137, 832-841
  costs of commercialization and deployment, 137-144, 841-851
  emissions, 164, *168-169*, *171*
  emissions mitigation, 711
  ER-2010, 814-816, *814*, *815*, *817*, *830-831*
  four illustrative scenarios, 795, 814-816, *814*, *815*, *817*, 830-831, *830-831*
  GHG emissions, 164, 798, 800-801, *802*, *803*
  GHG mitigation, 795-813
  global, 796, *827-828*
  global primary energy, 20-24, *21-23*, *803*, *804*, *806*
  integrated models, 748, 794, 800-801, *801*
  key parameters of, *814*
  knowledge gaps, 812-813, 832, 840-841, 851, 857, 897-899
  long-term, 748
  MiniCAM-EMF22, 814-816, *814*, *815*, *817*, *830-831*
  mitigation costs and RE, 808-811, *810*, *811*
  overview of, 800-803, *801*, *802*
  policies and, 814, 930, *930*
  quantitative modelling, 799-800
  RE competition with other low-carbon energies, 805-806, *806*
  RE deployment, 20-22, *21*, 131, 794, 803-805, *804*, 806-808, *808-810*, *830-831*
  RE deployment, technology perspective, 812
  RE dominance in future, 795
  RE sources, role of, 800-811
  regional, 806-808, *808-809*, 820-826, *823-825*, *827-828*, *835-838*
  ReMIND-RECIPE, 814-816, *814*, *815*, *817*, *830-831*
  sectorial, 795-796, 813-820, 826-830, *826*, *829*
  state of scenario analysis, 799-800
  sustainable development, 125-129, 711, 747-757
  synthesis for RE strategies, 131-133, 799-813
  technical potential vs. market deployment, 820-826
  World Energy Outlook, 616, 814-816, *814*, *815*, *817*, *830-831*
  See also specific technologies and sectors

1071

Index

Second-generation biofuels[*], 50, 238, 247, 250, 257
Security. *See* Energy security
Sedimentation, 447, 454, 465
Seismicity[*], 76, 420
Shipping (maritime transport), 671
Silicon photovoltaics, 351-352, 364-365
Sink[*], 174, 229, 264
Skilled human resources, 195
Smart meters, 920
Social costs[*], 144-146, 851-857, *855-857*
    of carbon (SCC), 853-854
Social development, 41, 120-121, 126-127
    energy use and, *121*, 716
    policies and, 880
    RE and, 18, 41, 191, 718-721
    sustainable development, 120-127, *570*, 718-721, 749-751, 765
Social discount rate[*], 833
Social equity, 150, 883, 905-906
Social impacts:
    bioenergy, 52, 271-274, 306
    geothermal energy, 404-405, 420-421, 432
    hydropower, 83-84, 442, 461, *462*, 463, 467-468, 487
    ocean energy, 92-93, 502, 518, 528
    solar energy, 66, 372-373, 390
    sustainable development, 717
    wind energy, 100, 540, 574-577, 595
    See also Health
Socio-cultural barriers, 44, 129, 195, 757-758
Soft loans[*], *886*, 893
Soil resources, bioenergy and, 229, 269-270
Solar cells, photovoltaic, 351-353
Solar collectors[*], 346-349, *347*, *348*, 360, 377
Solar cracking, 366
Solar drying, 346, 351
Solar energy[*], *8*, 60-71, **333-400**
    abundance and potential of, 337, 340, *341*
    active solar[*], 346-351, 360, 374-375, 379
    adaptation and, 192
    climate change and, 343, 387
    concentrating solar power (CSP)[*], 62-63, 355-358, *356*, 365, 369
    conversion technologies, 337, 340, 377
    costs, 68-71, 338-339, 372, 378-385, *380*, 390, *856*, *1004*
    daylighting, 338, 344-346, 373-374
    deployment, 71, *71*, 339, 386-390, *388*
    direct solar energy[*], *8*, *12*, *22-23*, 60-71, *182*, *184*, **333-400**
    energy payback[*], 338, 370

    environmental and social impacts, 65-66, 369-373, 390, *745*
    historical use of, *11*, *176*
    history of technologies, 340
    industry capacity and supply chain, 64-65, 362-366
    installed capacity and generated energy, 63, 359-362
    integration, 65, 338, 362, 367-369, 390, 612, 622-624, *623*, 629, 633-634
    learning curves, 71, 338-339, 385, *848*
    levelized cost of energy, *14*, 843-844
    market and industry development, 63-65, 359-367
    maturity of technologies, 337, 340, 350, 389-390
    Mediterranean Solar Plan, 876
    mitigation potential, 337, 345-346, 369-390
    net-zero-energy solar buildings, 346, 356
    opportunities, 338
    passive solar[*], 338, 344-346, 360, 373-374, 378-379
    photovoltaics (PV)[*], 62, 68-69, 338, 351-355, *351*
    policies and, 65, 338, 366-367, 390
    in primary energy supply, *176*
    regional deployment, 389
    regional technical potential, 342, *342*
    scenarios, 806, *808*, *817*, *822*, *827-829*
    solar cooling, 349, 350-351
    solar fuel production, 63-64, 358-359, *359*, 361-362, 365-366, 377-378
    solar heating, 346-349, *347*, 350-351
    solar thermal[*], 60-62, *61*, 63, 66, 68, 360, 686, *821*, *1010*
    space-based solar power (SSP), 378
    supply chain issues, 389
    technical potential, *12*, 60, *184*, *206*, 341-343, 389
    technology and applications, 60-63, 343-359
    technology improvement and innovation, 66-68, 373-378
    technology-specific challenges, 13
    theoretical potential, *183*, 341
    thermal storage, 344, *345*, 349-350, 357, 374
    variability of, 340, *623*
    See also Concentrating solar power (CSP); Photovoltaics (PV)
Solar irradiance[*], 60, 341, *341*, 342-343, *343*
Solar radiation[*]:
    beam[*], 60, 341, 342, 355, 367
    diffuse[*], 60, 341, 342
    global[*], 351, *351*
Soy, *218*, *234*, 243, *266*, *277*
    carbon payback, *264*
    conversion routes, *235*
    costs of fuel production, *53*, *234*, *245*

Spain, 616
    policies for PV deployment, *904-905*, 905
Stakeholders, 53, 157, 883, *918*, 921, 923
    individual and community ownership, *922*
    individuals, 45, *156*, 157-158, 927-929
    resolving conflicts among, 921, 923-924
Standards[*], 874, 876, 881, 882, 892, 895, 903, 908, *914*
    network, 925
    performance[*], 17, 97
    sustainability, 913
    technology, 554, *555*
Starch crops, *234*, *245*, *277*
Steam-cracking, 683
Storage of energy, 637-638, 659, *664*, *674*
    autonomous systems and, 659
    batteries, 62, 107, 109, 114-115, *114*, 185, 358, 621, 630, 637, 659, 664, *664*, 668, *674*
    concentrating solar power systems, 63, 65
    costs and, 69
    heating and cooling, 110-111, 643-644, *644*
    hydropower, 80-83, 451-452
    integration and, 637-638
    thermal storage, 80, 344, *345*, 349-350, *674*
    See also Thermal storage
Structural shift (in energy systems), 45, 158, 199, 870, 929-932
Subsidies[*], 872, 874-876, 880, 886, *912*, 913, 914, 916-917
    biofuel production subsidies, 874-876
    costs of, 916
    fossil fuel, removal of, 760-761
    investment, 874, 892, *904*
Substitution method, *180*, 807, *808*, 977, *978-979*
Sugar/sugarcane, *15*, *218*, 233, *234*, 241-243, *266*, *277*, 616, 685-686, *912*, *1014*
    carbon payback, *264*
    conversion routes, *235*
    costs of fuel production, *53*, 215, *245*
    experience curves, *56*
Sulphur dioxide (SO_2), 124, 736, *739*, 854
Supply chains, 812
    bioenergy, 278-280, *279*, 302-304
    See also Market and industry development
Supply curves, 820-825, *823*, *824*, 832-834, *835-836*
    limitations of, *834*
Supply push[*], 150, 154, 851, *887-888*, 916
Sustainability:
    bioenergy, 52, 215, 254-255, 271
    biofuels, 256
    geothermal energy, 404
    transport systems and, 672
    of watersheds, 489-490

Index

Sustainable development (SD)[*], 18-20, 119-130, 707-789
    accidents and risks, 745-747, *746*
    air pollution, 124-125, 736-739
    barriers to, 129-130, 712, 757-760
    Clean Development Mechanism (CDM)[*], 129-130, 761-762
    climate change mitigation and, 122-125, 128-129, 714, 717-718, 729-736, 756-757, 766
    concept and definition, 713
    electricity generation, GHG emissions from, 122-123, 732-733, *732*
    employment creation, 719-720
    energy access and, 121, 127, 716, 721-724, *721-723*, 751-752, 765
    energy security and, 122, 127-128, 716-717, 724-728, *728*, *729*, 752-756, 765-766
    energy use and, 710, 716
    environmental benefits, 711
    environmental impacts, 128-129, 710-711, 728-745, *745*, 766
    environmental impacts, evaluation of, 717-718
    fuels, GHG emissions from, 733-735, *734*
    future scenarios, 125-129, 711, 747-757
    goals, 710, 715-718
    health benefits, 711
    health impacts, 128-129, 739-740, *740*, 756-757, 766
    indicators of, 715-716, 719, 728
    integrating RE and SD strategies, 119-120, 129-130, 762-763
    interactions of RE and SD, 18-19, 710, 713-718, 762-763
    internalizing environmental/social externalities, 761-762
    international and national strategies for, 129, 760-763
    knowledge gaps and research needs, 130, 712, 767-768
    land use/land use change, 123-124, 735-736, *736*, 743-745
    leapfrogging[*], *763*
    lifecycle assessments and, 122-124, 710, 717, 728-729
    linkages to other chapters/topics, 714-715, *714*
    local, private, and nongovernmental initiatives, 763-764, *764*
    National Sustainable Development Strategies (NSDS), 762
    opportunities, 712, 760-764
    policies and, 712
    removal of mechanisms working against, 760-761
    resource assessment, 764-765
    social and economic development, 120-125, 126-127, *570*, 710, 716, 718-721, 749-751, 765
    theoretical concepts and tools, 764-765
    variability/reliability of RE, 726-728, 754-755
    water resources and, 125, 741-743, *742*
    weak vs. strong sustainability, 713, 715-716
    World Summit on (2002), 716
Sustainable energy, 174
Sustainable Energy Financing District, *894*
Sweden:
    biomass CHP heating, 646
    biomass district heating and infrastructure, *909*
    tradable RE certificates, *897*
Switchgrass, 217, *225*, 230, *234*, 269, 270, *277*, *281*
Syngas, 280, 358, *359*, 378, 649, *650*
Systems integration, 812

T

Targets, 874-876, *875*, 879-880, 883, *884*, 896-897
    voluntary vs. mandatory, 913
Tariffs, 196
    biofuels, 255-256
    feed-in tariffs[*], 24, 25, *890*, 899-907
Tax credit[*], *886*
    heating and cooling sector, 908
    for RE deployment, *890*, 891-892
Tax policies, *890*, 891-892
    accelerated/variable depreciation, *890*, 892
    heating and cooling sector, 908
    reduction/exemption, *890*, 891-892
    transportation sector, 911-912
Taxes[*]:
    carbon tax[*], 24, 872, *890*, *909*, 917
    eco-tax[*], 908
    tax credit[*], *886*, *890*, 891-892
    See also Fiscal incentives
Technical potential[*], 10-11, *12*, 39, 181-184, *184*, *206*
    climate change impact on, 12
    RE contribution not constrained by, 23, 796
    scenarios, 796, 820-826, *827-829*
    See also specific technologies
Technological change[*], 930
    bricolage vs. breakthrough, 931
Technology[*]:
    challenges in, 13
    commercialization of, 137-144, 841-851, 877
    competitiveness of, 13, 40, 164-165, 796, 805-806
    cost curves, 836-840
    costs, present, 841-846, *843-846*
    demand pull[*], 150, 851, *887-888*
    economics and, 812
    maturity of, *182-183*, 612
    policies and, 869, 931-932
    portfolio of RE technologies, 16, *177*, 795, 869
    RE integration and, 17-18
    RE technologies, 7-15, *8-9*, 38-40, *182-183*
    supply push[*], 150, 851, *887-888*, 916
    trends in, 874-876, *875*
    types of content, *926*
    zero- or low-carbon technologies, 37, *37*, 74, 114, 115, *115*, *116*, 175
Technology cycle, 888, *889*
Technology improvement and innovation, 919-920
    bioenergy, 53, 276-287
    geothermal energy, 77, 405, 421-423
    hydropower, 84, 474-476
    ocean energy, 520-522
    policies and, 885
    solar energy, 66-68, 373-378
    wind energy, 540-541
Technology incubators, 885
Technology standards[*], 554, *555*
Technology transfer[*], *918-919*, 925-927
Temperature, global, 33, 164, 169
    agreements to limit, 34, 164, 169
Tendering/bidding, *890*, 896
Thailand, RE policies in, *902*
Theoretical potential[*], 39, 165, 181, *183*
    bioenergy, 46-47, *183*, 220, 222
    geothermal energy, *183*, 408, 409
    hydropower, *183*, 444
    ocean energy, *183*, 501, 503, *504*
    solar energy, *183*, 341
    wind energy, *183*, 543-544
    See also Technical potential
Thermal conversion, 178, 337, 377
Thermal energy:
    geothermal[*], **401-436**, *821*
    ocean thermal, 87-88, 503, 507, *507*, 511-512, 516, 521, *523*, 525
    solar thermal[*], 60-62, *61*, 63, 66, 68, 360, 686, *821*
Thermal storage, 80, 344, *345*, 349-350, *674*
    aquifer thermal energy storage, 349-350
    concentrating solar power systems, 69, 357
    latent heat storage, 349
    of solar energy, 344, *345*, 349-350, 357, 374
    sorption heat storage, 349
    thermochemical heat storage, 349
    underground thermal heat storage, 349-350
Thermolysis, 358
Third party access (TPA), 895, 910
Tidal energy. See Ocean energy

1073

Index

Time:
    policy timing, 147-148, 873, 923
    time-dependent expenditures, 849-851, *850*
    time scale of deployment, 807, *808-810*
Top-down models[*], 799, *799*
Tornado graphs, *1003*
    biofuels, *1018*
    heat technologies, *1012-1013*
    RE power technologies, *1008-1009*
Torrefied wood, 279
Total cost[*], 40, 53, 86, 147, 873, 891, 905
Toxic substances:
    geothermal energy and, 419
    nuclear power and, 740
    photovoltaics (PV) and, 352, 370
Tradable certificates[*], 895-896, *897*
Transesterification, 240
Transition economies, *73, 79*, 84, 120, *140-141*
Transitions, 24, 37, 41, 103-107, 158
    end-use sector integration, 113-119, 616, 661-689
    global investment, 878
    policies and, 881, *890-891*
    RE in rural areas, 723, *723*
    structural shift and, 45, 158, 199, 870, 929-932
    See also Integration
Transmission and distribution (electricity)[*], 563
Transportation sector:
    air pollution and, 124-125, 739
    aviation, 670-671
    biofuels and, *299*, 300, 612-614, *822*
    efficiency in, 186
    evolution of fuel consumption, *299*, 300
    fuel mandates and, 25
    heavy-duty vehicles, 670
    integration of RE, 17, 40, 45, 113-117, 613-614, *617-618*
    levelized cost of RE, *14*
    levelized cost of transport fuel (LCOF), *14, 42-43, 55, 144*, 1002, *1003, 1014-1018*
    light-duty vehicles, 665-670
    maritime transport, 671
    mitigation potential, 662-672, *662*
    policies, 153-154, 911-913
    rail transport, 671-672
    scenarios for, 795, 820, *822*
    solar fuels and, 366
    vehicle technology, 664-666
    See also Biofuels
Trees. *See* Bioenergy; Forest products; Forest protection; Lignocellulosic crops/feedstocks
Turbines[*]:
    backpressure steam turbines[*], 460
    condensing steam turbines[*], 412
    fish-friendly, 466, 475
    geothermal energy, 410-412
    hydrokinetic, 451, 475-476
    hydropower, 443, 450-455, *453*, 474-476
    ocean energy, 506, 508-512, *509, 511*, 516
    solar energy, 355-357, 369, 377
    wind turbines[*], 550-556, *551-553*, 580-582

U

UN Conference on Environment and Development, 177
UN Conference on New and Renewable Sources of Energy, 177
Uncertainties, 169, 177, *221*
UNFCCC. *See* United Nations Framework Convention on Climate Change
United Nations Conference on Environment and Development (UNCED), 713
United Nations Framework Convention on Climate Change (UNFCCC)[*], 761
United States. *See* USA
Upstream processes, 125, 743, 744
Urbanization, 616-618
    Brazil case study, 677-678
USA:
    California, innovative financing example, *894*
    quota use in, 905
    RE policies, *898-899*, 905
Use obligation, 908

V

Valley of death[*], 53, 884-885
Value added[*], *890*, 891
Values[*], 129, 712, 757-759, 767
Variability of RE sources, 8, 165, *623*, 726-728, 754-755
Vent (geothermal/hydrothermal/submarine)[*], 410
Venture capital[*], 886, *886*
Vouchers, *886*

W

Water:
    electrolysis of, 358, 362, 377
    purification, with solar energy, 350-351
    RE technology and, 20, 192, 741-743, *742*
    storage facilities (reservoirs), 87, 451, *451*
    sustainable development and, 741-743, *742*
    upstream processes and, 743, 744
    See also Hydropower
Water management services:
    climate-driven, 488
    hydropower and, 87, 441, *462*, 488-490
    multipurpose use of resources, 488-489
    regional cooperation in, 489-490
    sustainable watersheds, 489-490
Water resources, impacts on, 44, 125, 192, 854
    bioenergy, 227, 233, 257-258, 268-269, 305
    concentrating solar power (CSP), 338
    electricity generation, 741-743, *742*, 991-993
    hydropower, 464-465
Waves, 503
    See also Ocean energy
Well-to-tank (WTT)[*], 663, *665*, 669
Well-to-wheel (WTW)[*], *115*, 123, *282*
Wheat, 234, *266, 277*
    costs of fuel production, *53*, 234
Wind energy[*], *9*, 95-103, *183*, **535-608**
    alternative applications and technologies, *543*
    autonomous energy examples, 660-661
    balancing cost with, 568-569, *569*
    bird and bat collision fatalities, 572-573
    characteristics of, 560-563
    climate change and, 548-550
    cost curves, 839-840, *840, 841*
    costs, 101-103, 568-570, 580, 583-591, *584, 856, 1005*
    costs, reduction of, 541, 589-591
    in Denmark, *922-923*
    deployment, 103, 539-540, 591-595
    electricity from. *See* Wind power (electricity)
    energy source, 542
    environmental and social impacts, 99-100, 540, 570-577, 595, *745*
    health impacts, *740*
    historical use of, *11, 176*, 542
    industry development, 558-559
    integration, 98-99, 540, 560-570, 612, *623*, 626-627, 629-632
    learning curves, 589-590, *589, 847, 848*
    levelized cost of energy, *14*, 588-589, *588*, 590-591, *843-844*
    lifecycle GHG emissions, 571, *571*
    market and industry development, 97, 539, 556-560
    mechanical and propulsion applications, *543*
    mitigation potential, 539, 541, 591-593
    offshore, 553-554, 580-582, *581, 584*
    onshore, 551-553, *584*
    policies and, 198, 559-560, *898-899*
    in primary energy supply, *176*
    projections for, *22-23*
    public attitudes and acceptance, 100, 576
    regional and national deployment, 556-558, *557*, 593-594, *594*
    regional technical potential, 546-548, *547*

Index

resource assessment in China and Russia, *549*
scenarios, *22-23*, 806, *808*, *817*, *822*, *827-829*
scientific research and, 582-583
small wind turbines, *543*
supply chain issues, 594
technical potential, *12*, 95-96, *184*, *206*, 539, 544-548, *545-546*
technology and applications, 96-97, 550-556
technology improvement and innovation, 100-102, 540-541, 577-583
technology-specific challenges, 13
technology standards, 554, *555*
theoretical potential, *183*, 543-544
variability in, 564-566, *565*, 571-572, 612, *623*
See also Wind power; Wind turbines

Wind farms[*], 552, *564*

Wind power (electricity), 539, 542, 543-550, *543*, 555-556
capacity, *556-558*, *564*
conversion and grid connection issues, 555-556
costs, 585-586, 587
electrical characteristics and grid codes, 562-563
flexibility and variability, 564-566, *565*
higher-altitude, *543*
integration issues, 560-570, 594-595
planning for, 562-564
practical experience with, 566-567
regional technical potential, 546-548, *547*
scenarios, *819*
top ten countries in, *557*
transmission issues, 563, 594-595

Wind power plants[*], 578-582
certification procedure, 554, *555*
investment costs, 584, *585*, 586-587
operation and maintenance costs, 584-585, 587

Wind project[*], 552, *555*

Wind turbines[*], 550-556, *551-553*
certification procedure, 554, *555*
foundation designs, *101*
offshore technology, 553-554, 580-582, *581*
onshore technology, 551-553
pitch control[*], 551, 555, 579
power curve for, *550*
size of, *96*
small, *543*
stall control[*], 550, 551, 555
technology advances in, 578-580, *579*

Windmills[*], 542

Wood. *See* Forest products

Wood pellets, 237-238, 253
experience curves, *56*

Wood, torrefied, 279

World Energy Outlook scenarios, 616, 724, 797, 814-816, *814*, *815*, *817*, *830-831*
See also Scenarios

Z

Zero- or low-carbon emission technologies, 37, *37*, 74, 114, 115, *115*, *116*, 175